MediaLink Tool Periodic Table

10	11 IB	12 IIB	13 III IIIA	14 IV IVA	15 V VA	16 VI VIA	17 VII VIIA	18 VIII VIIIA
								2 He 4.00 $1s^2$
			5 B 10.81 $2s^2 2p^1$	6 C 12.01 $2s^2 2p^2$	7 N 14.01 $2s^2 2p^3$	8 O 16.00 $2s^2 2p^4$	9 F 19.00 $2s^2 2p^5$	10 Ne 20.18 $2s^2 2p^6$
			13 Al 26.98 $3s^2 3p^1$	14 Si 28.09 $3s^2 3p^2$	15 P 30.97 $3s^2 3p^3$	16 S 32.07 $3s^2 3p^4$	17 Cl 35.45 $3s^2 3p^5$	18 Ar 39.95 $3s^2 3p^6$
28 Ni 58.69 $3d^8 4s^2$	29 Cu 63.55 $3d^{10} 4s^1$	30 Zn 65.39 $3d^{10} 4s^2$	31 Ga 69.72 $3d^{10} 4s^2 4p^1$	32 Ge 72.61 $3d^{10} 4s^2 4p^2$	33 As 74.92 $3d^{10} 4s^2 4p^3$	34 Se 78.96 $3d^{10} 4s^2 4p^4$	35 Br 79.90 $3d^{10} 4s^2 4p^5$	36 Kr 83.80 $3d^{10} 4s^2 4p^6$
46 Pd 106.42 $4d^{10}$	47 Ag 107.87 $4d^{10} 5s^1$	48 Cd 112.41 $4d^{10} 5s^2$	49 In 114.82 $4d^{10} 5s^2 5p^1$	50 Sn 118.71 $4d^{10} 5s^2 5p^2$	51 Sb 121.76 $4d^{10} 5s^2 5p^3$	52 Te 127.60 $4d^{10} 5s^2 5p^4$	53 I 126.90 $4d^{10} 5s^2 5p^5$	54 Xe 131.29 $4d^{10} 5s^2 5p^6$
78 Pt 195.08 $4f^{14} 5d^9 6s^1$	79 Au 196.97 $4f^{14} 5d^{10} 6s^1$	80 Hg 200.59 $4f^{14} 5d^{10} 6s^2$	81 Tl 204.38 $4f^{14} 5d^{10} 6s^2 6p^1$	82 Pb 207.2 $4f^{14} 5d^{10} 6s^2 6p^2$	83 Bi 208.98 $4f^{14} 5d^{10} 6s^2 6p^3$	84 Po (209) $4f^{14} 5d^{10} 6s^2 6p^4$	85 At (210) $4f^{14} 5d^{10} 6s^2 6p^5$	86 Rn (222) $4f^{14} 5d^{10} 6s^2 6p^6$
110 Uun	111 Uuu	112 Uub	113 Uut	114 Uuq		116 Uuh		118 Uuo

Metals ⟵ ⟶ Nonmetals

Metalloids

62 Sm 150.36 $4f^6 6s^2$	63 Eu 151.96 $4f^7 6s^2$	64 Gd 157.25 $4f^7 5d^1 6s^2$	65 Tb 158.93 $4f^9 6s^2$	66 Dy 162.50 $4f^{10} 6s^2$	67 Ho 164.93 $4f^{11} 6s^2$	68 Er 167.26 $4f^{12} 6s^2$	69 Tm 168.93 $4f^{13} 6s^2$	70 Yb 173.04 $4f^{14} 6s^2$
94 Pu (244) $5f^6 7s^2$	95 Am (243) $5f^7 7s^2$	96 Cm (247) $5f^7 6d^1 7s^2$	97 Bk (247) $5f^9 7s^2$	98 Cf (251) $5f^{10} 7s^2$	99 Es (252) $5f^{11} 7s^2$	100 Fm (257) $5f^{12} 7s^2$	101 Md (258) $5f^{13} 7s^2$	102 No (259) $5f^{14} 7s^2$

FREQUENTLY USED TABLES AND FIGURES

CHEMICAL PRINCIPLES

The Quest for Insight

About the cover: The cover image is a part of a computer-generated molecular model of an enzyme regulator (aspartate carbamoyltransferase). Models such as these help scientists understand how biomolecules such as enzymes control the processes of life—in this case, the biosynthesis of pyrimidine. Molecular representations and graphics in general have transformed the way that chemists do research and are an essential part of learning chemistry and developing insight into the nature and properties of matter. (Francis Leroy/Biocosmos/SPL Photo Researchers.)

CHEMICAL PRINCIPLES

The Quest for Insight

SECOND EDITION

PETER ATKINS
Oxford University

LORETTA JONES
University of Northern Colorado

W. H. FREEMAN AND COMPANY
NEW YORK

Acquisitions Editor:	Jessica Fiorillo
Publisher:	Michelle Russel Julet
Development Editor:	Marjorie Singer Anderson
Assistant Editor:	Robert Jordan
New Media and Supplements Editor:	Charles Van Wagner
Marketing Director:	John A. Britch
Senior Project Editor:	Georgia Lee Hadler
Copy Editor:	Jodi Simpson
Cover and Text Designer:	Cambraia Fonseca Fernandes
Illustration Coordinator:	Cecilia Varas
Illustrations:	Peter Atkins with Network Graphics
Photo Editor:	Meg Kuhta
Production Coordinator:	Paul W. Rohloff
Composition:	Black Dot Graphics
Manufacturing:	RR Donnelley & Sons Company

Library of Congress Cataloging-in Publication Data

Atkins, P.W. (Peter William)
Chemical principles: the quest for insight/Peter Atkins, Loretta Jones.—2d ed.
p. cm.
Includes index.
ISBN 0-7167-3923-2
1. Chemistry. I. Jones, Loretta. II. Title.
QD33.2.A74 2001
540—dc21

2001023890

Printed in the United States of America

Second printing, 2002

Contents in Brief

CONTENTS

CHAPTER 1 ATOMS: THE QUANTUM WORLD

CHAPTER 2 CHEMICAL BONDS

p. 64

p. 112

Chapter 4 The Properties of Gases

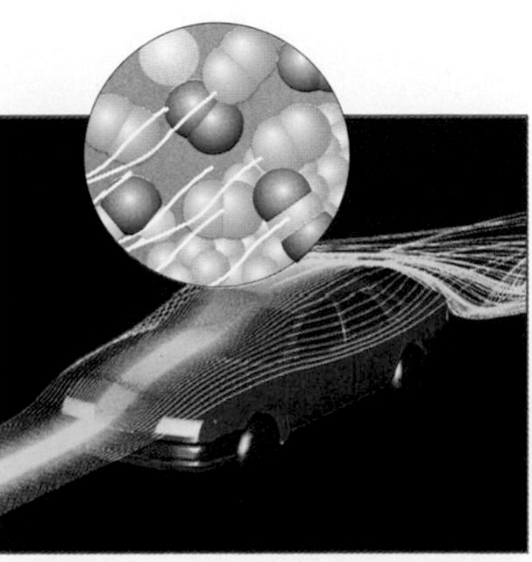

p. 170

p. 220

Chapter 5 Liquids and Solids

Chapter 6 Thermodynamics: The First Law

CHAPTER 7 THERMODYNAMICS: THE SECOND AND THIRD LAWS

p. 340

p. 390

CHAPTER 8 PHYSICAL EQUILIBRIA

Chapter 9 Chemical Equilibria

Chapter 10 Acids and Bases

p. 512

p. 578

Chapter 11 Aqueous Equilibria

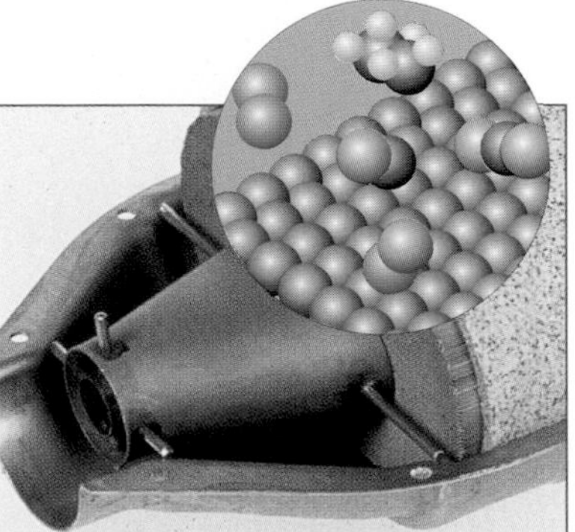

p. 684

p. 748

CHAPTER 14 THE ELEMENTS: THE FIRST FOUR MAIN GROUPS

CHAPTER 15 THE ELEMENTS: THE LAST FOUR MAIN GROUPS

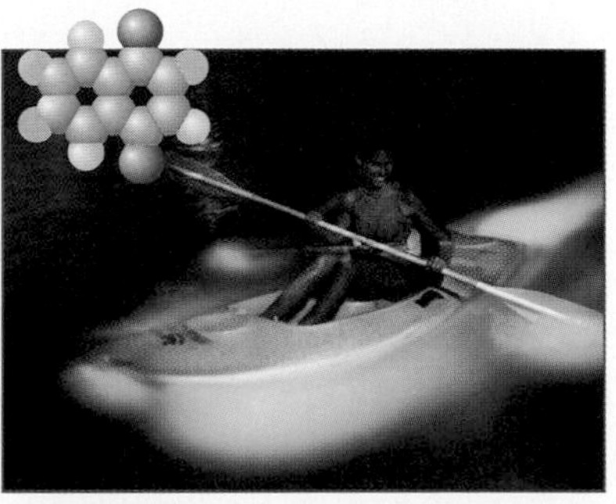

p. 808

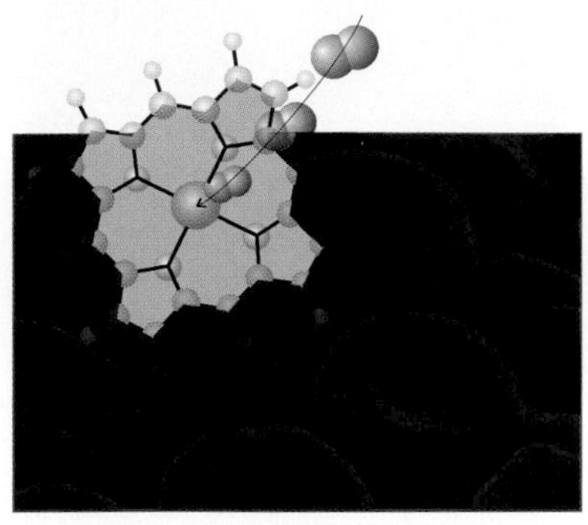

p. 856

Chapter 16 The *d* Block: Metal in Transition

Chapter 17 Nuclear Chemistry

Chapter 18 Organic Chemistry I: The Hydrocarbons

Chapter 19 Organic Chemistry II: Functional Groups

p. 952

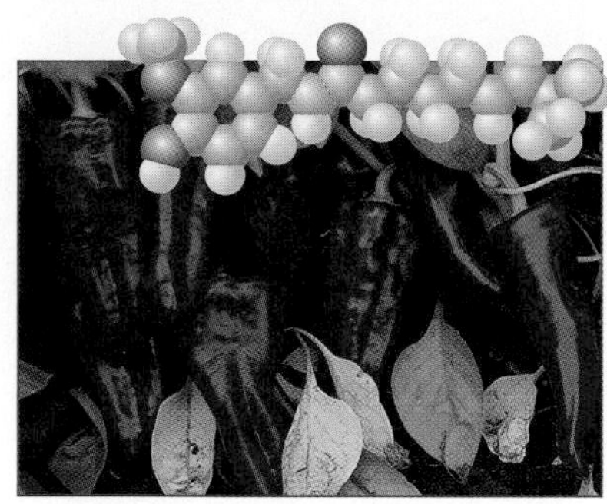

p. 986

Letter from the Authors

Dear Colleague,

We are pleased to present this new edition of our text *Chemical Principles*, which is designed for a rigorous course in introductory chemistry. We were gratified by the reception of the first edition and have listened carefully to your comments as we prepared the second edition. A popular feature of the first edition was the "atoms-first" approach; most instructors found this topic order to be a logical sequence that works well in the classroom. We have retained that approach in this edition but have made numerous changes in light of your comments.

Another feature that was well received was the *Fundamentals* section at the start of the book. Instructors found that they could use this section in two ways. Some found that it was a useful, succinct review of basic material that their students already knew and referred to it when necessary. Others found that they could start their courses with a quick survey of this material before starting on the main text of the book. We have retained the *Fundamentals* section in this edition but have streamlined and focused the presentation by moving some of the topics that are not required for the first few chapters forward into the main text. For example, mole fraction is now taught with gaseous mixtures in Chapter 4.

A principal change in this edition is the enhancement of the book's Web connections. Although we believe that a paper textbook has a great deal of merit on its own, we have been eager to make use of the enriching possibilities of the Internet. In this edition we feel we have made important advances in the use of the World Wide Web. Throughout the book you will see that many of the figures and exercises are accompanied by *Media Link* annotations that direct students to Web-based resources. Some figures have thumbnail images of animations that show dynamically the concepts presented in static two dimensions in the figure. Many of the graphs are "living graphs" in the sense that the Web site allows students to modify the parameters of the graph and explore the consequences. Additional tools, such as molecular models, an interactive periodic table, and graphing tools, can be used to build conceptual understanding. Media exercises help students learn how to make use of these resources.

Despite these fundamental changes in the presentation of the material, the central theme of the text remains unchanged. We want to encourage your students to question and we want to help them build insight. We start each chapter by addressing the typical student question: *Why do we need to know this material?* We continue by presenting the required background for the chapter in answer to another typical student question: *What do we need to know already?* In the chapter text we encounter a task and ask *How do we do that?* and show that the answers to many questions can be formulated by drawing on the power of mathematics, including calculus. Scattered through the text are other questions we want students to learn to ask, such as *How do we know that . . . ?* which introduces the experiments used to collect the data and motivate the concepts in the text. Boxes that ask *What has this to do with . . . ?* show how the material is relevant to staying alive (human health) and to the environment. *Frontiers of Chemistry* boxes introduce students to exciting and relevant areas of research that portray the dynamic nature of chemistry and ask *How might you contribute?* Short worked examples that ask *How can we use this information . . . ?* show how to use an equation or a concept immediately after it is introduced.

This edition is designed to guide students on the path to a scientific, engineering, or medical career. To that end, it emphasizes the nature of chemistry and how chemical knowledge is obtained. To help students build problem-solving skills and understanding, this edition includes enhanced problem-solving guidance throughout and an enhanced art program that helps students to see the connection between the macroscopic world and the underlying chemical concepts. We have also expanded the treatment of the molecular basis for the gas laws and thermodyamics and have strengthened the emphasis on building scientific models. A range of new end-of-chapter exercises has been added by Ken Whitmire, including more challenging problems.

We are confident that this new edition captures the pedagogical and scientific spirit of the first edition, but has become far more dynamic and interactive through its use of Web-based material. We have written it with the modern student and the modern instructor in mind, and we very much hope you like what you see.

Yours sincerely,

Peter Atkins

Loretta Jones

PREFACE

ATOMS-FIRST ORGANIZATION

The unique atoms-first sequence of *Chemical Principles* develops student understanding of the concepts of chemistry by starting with the behavior of atoms and molecules and building to more complex properties and interactions:

- **Atoms and molecules** come first (including discussions of quantum mechanics and molecular orbitals), providing the foundation for understanding bulk properties and models of gases, liquids, and solids.
- Next comes an exploration of **thermodynamics and equilibrium** based on conceptual understanding of free energy and entropy. This integrated presentation lays a common foundation for these concepts and provides a basis for the form of the equilibrium constant.
- **Kinetics** then shows the dynamic nature of chemistry and the crucial role of insight and model building in identifying reaction mechanisms.
- A section on **fundamentals,** streamlined and updated in this new edition, precedes Chapter 1 for a flexible overview of the basics.

VISUALIZATION

Conceived and executed by the authors, the art in *Chemical Principles* is designed to help students make the connections between what chemists imagine and what they perceive, between chemical models and laboratory observations. More than 15 percent of the art in the second edition is new.

- **Connections between the microscopic and macroscopic.** Many illustrations are paired with photographs or drawings of chemical systems, starting with the molecular model superimposed on each chapter's opening photograph. The combination of representations, from macroscopic to microscopic, abstract to concrete, and graphical to pictorial, helps students bridge the gap between the imagined and the perceived.
- **Realistic images show molecular dynamics.** Multistep graphics show how molecules and ions move and interact, highlighting the sense of motion so critical to chemical understanding. Most of these sequences are further illustrated by actual animations available on the Web site.

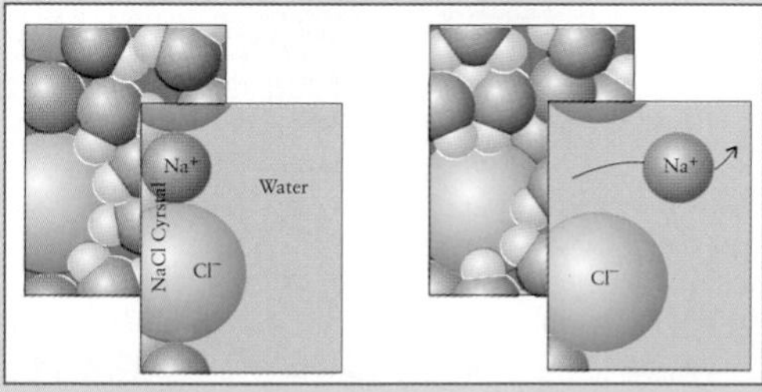

CONTEMPORARY CHEMISTRY

- **Emphasis on thinking like a chemist.** Throughout the book, *Chemical Principles* shows students how chemists propose models based on insight, develop those models with mathematics, and refine them through experimentation. This approach is used to help students

MAJOR TECHNIQUE 2 ULTRAVIOLET AND VISIBLE SPECTROSCOPY

ll the colors of vegetation around us arise from the :lective absorption and reflection of visible light. If ur eyes were sensitive to ultraviolet radiation—the ectromagnetic radiation at shorter wavelengths iigher frequencies) than violet light—we would see an ven richer tapestry of color, because substances also bsorb and reflect selectively in the ultraviolet region. Ioneybees do have this ability, and they can "see" idiation in the ultraviolet as well as in the visible inge. This selective absorption and transmittance of isible light and ultraviolet radiation is the basis of a :chnique for identifying compounds and for etermining their concentration in samples.

The Technique

7hen electromagnetic radiation falls on a molecule, the ectrons in the molecule can be excited to a higher iergy state. If the frequency of the radiation is ν (nu), can raise the molecule to a state that differs in energy om the first state by an amount ΔE, where

$$\Delta E = h\nu \quad (1)$$

his is the *Bohr frequency condition*, which was icountered in Section 1.6; h is the Planck constant. or ultraviolet wavelengths of 300 nm and less orresponding to a freq ency of abo t 10^{15} Hz or

hence be absorbed. Therefore, the study of visible and ultraviolet absorption gives us information about the electronic energy levels of molecules.

Visible and ultraviolet (UV) absorption spectra are measured in an absorption spectrometer. The source gives out intense visible light or ultraviolet radiation. The wavelengths can be selected with a glass prism for visible light and with a quartz prism or a diffraction grating for ultraviolet radiation (which is absorbed by glass). A typical absorption spectrum, that of chlorophyll, is shown in Fig. 1. Note that chlorophyll absorbs red and blue light, leaving the green light present in white light to be reflected. That is why most vegetation looks green. The spectrum can help us to assess the absorption quantitatively and to make a precise analysis of the energy-capturing power of the molecule.

2 Chromophores

The presence of certain absorption bands in visible and ultraviolet spectra can often be traced to the presence of characteristic groups of atoms in the molecules. These groups of atoms are called *chromophores*, from the Greek words for "color bringer."

One important chromophore is a carbon-carbon double bond. In terms of the language of molecular orbital theory, the electronic transition that occurs when energy is absorbed is the excitation of an electron

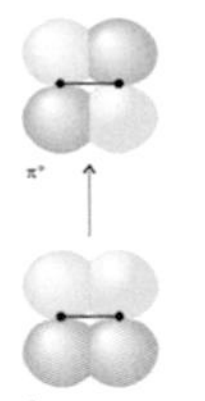

FIGURE 2 In a π-to-π^* transition, an electron in a bonding π-orbital is excited into an empty antibonding π^*-orbital.

The transition can be brought into the visible region—by decreasing the separation between bonding and antibonding orbitals—if the molecule has a chain of alternating single and double bonds. Such "conjugated" double bonding occurs in the compound carotene

for the color of carrots, mangoes, and persimmons. Related compounds account for the colors of shrimps and flamingoes. This transition is also responsible for the primary act of vision, as explained in Section 3.8.

Another important chromophore is the carbonyl group, >C=O, which absorbs at about 280 nm. The transition responsible for the absorption is the excitation of a lone-pair electron on the oxygen atom (a "nonbonding" electron, denoted n) into the empty antibonding π^*-orbital of the C=O double bond (Fig. 3). This transition is therefore called an *n-to-π^* (n to pi star) transition.*

A *d*-metal ion may also be responsible for color, as is apparent from the varied colors of many *d*-metal complexes (see Chapter 16). Two types of transitions may be involved. In one, which is called a *d-to-d transition*, an electron is excited from a *d*-orbital of one energy to a *d*-orbital of higher energy. Because the energy differences between *d*-orbitals are quite small, visible light brings enough energy to cause this excitation, so colors are absorbed from white light and the sample takes on colors complementary to those observed (Fig. 4). A second type of transition is called a *charge-transfer transition*, in which electrons migrate from the atoms attached to the central metal atom into the latter's *d*-orbitals, or vice versa. This transfer of charge can result in very intense absorption; it is

THE IMPACT ON MATERIALS

Metals are the mainstays of technology, and it is hardly surprising that so much effort has gone into the development and understanding of their properties. We dealt with their electrical properties in Section 3.14. Here we explore some of their other physical properties.

5.14 The Properties of Metals

In a metallic solid, cations lie in a regular array and are surrounded by a sea of electrons. This structure results in unique properties. The characteristic luster of metals is due to the mobility of their electrons (Section 3.14). An incident light wave is an oscillating electromagnetic field. When it strikes the surface, the electric field of the radiation pushes the mobile electrons backward and forward. These oscillating electrons radiate light, and we see it as a luster—essentially a reflection of the incident light

understand key concepts and also to expand students' ideas of how to apply these concepts to modern research.

- **Major Techniques interchapters** introduce students to important experimental methods, connecting the classroom, the laboratory, and the world.
- **New and updated Impact on Materials and Impact on Biology** sections explore the burgeoning fields of materials chemistry and biological chemistry while reinforcing chemical principles. These sections introduce students to new fields and innovations.

BOX 3.1 FRONTIERS OF CHEMISTRY: DRUGS BY DESIGN AND DISCOVERY

The search for new drugs relies on the skills of synthetic organic chemists. Because there are so many millions of compounds, it would take too long to start with the elements, combine them in different ways, then test them. Instead, chemists usually start either by *drug discovery*, the identification and possibly the modification of promising medicines that already exist, or by *rational drug design*, the identification of characteristics of a target bacterium or parasite and the design of new compounds to react with it.

In drug discovery, a chemist usually begins by investigating compounds that have already shown medicinal value. A fruitful path is to find a *natural product*, a substance found in nature, that has been shown to have healing characteristics. Nature is the best of all synthetic chemists, with billions of chemicals that fulfill as many different needs. The challenge is to find compounds that have curative powers. These substances are found in different ways: random or "blind" collection of samples that are then tested or collection of specific samples identified by native healers as medically effective.

Observation of the properties of plants and animals can help to guide a random search. For example, if certain types of fruits remain fresh while others rot or become moldy, we might expect them to contain antifungal agents. One example of this type of collection is the gathering of tunicates and sponges in the Caribbean. The chemists harvest the samples by diving from research vessels (see Fig. F.1). The samples are tested for antiviral and antitumor activity in a chemical laboratory on the ship. The antiviral drug didemnin-C and

A field biologist examines a plant in a South American rainforest. The plant produces chemicals that will be investigated for their medicinal value.

the anticancer drug bryostatin 1 were discovered in marine organisms.

The guided route involves the testing of fewer samples because the chemist works with a native healer, the ancient lore guiding the modern chemistry. Often an ethnobotanist, a specialist in plants used for native healing, joins the team. This approach saves time for the scientists and can provide an economic benefit for the healers and their nations as well. Drugs that have been discovered in this way include a variety of anticancer and antimalarial drugs, blood-clotting agents, antibiotics, and medicines for the heart and digestive system.

- **Frontiers of Chemistry** boxed essays explore modern chemical research and communicate the excitement of working at the cutting edge of chemistry. Each Frontiers of Chemistry box asks students, **"How Might You Contribute?"**—challenging them to lead the way in solving global problems by using their understanding of chemistry.

PROBLEM SOLVING

Students at all levels benefit from the support of strong pedagogy. These proven strategies that build student confidence in their mastery of the concepts have been improved and enhanced.

- **"Bottom line" statements** summarize the key ideas in each section.

Atoms of all metallic elements are held in the solid or liquid state by "metallic bonds," which are discussed in Section 5.9.

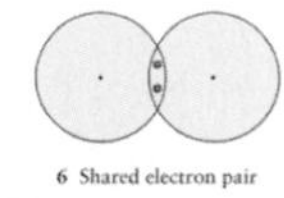

6 Shared electron pair

and the polyatomic species P_4 and S_8. Elements near the border between metals and nonmetals can form solids with an extended network of atoms, such as crystalline silicon and the graphite or diamond forms of carbon. Covalent bonding is also important in the formation of compounds. There are also countless examples of diatomic, polyatomic, and extended network compounds between different nonmetallic elements, including the millions of organic compounds.

Lewis thought of the shared pair of electrons as lying between the two atoms and interacting with both their nuclei (**6**). The two atoms are bound together as a result of the coulombic attraction between the two electrons and the nuclei. Neither atom has to release an electron totally, so neither atom has to be supplied with its full ionization energy.

Nonmetallic elements form covalent bonds to one another by sharing pairs of electrons.

PROBLEM SOLVING

EXAMPLE 4.1 Using Boyle's law

Assume that when you press in the piston of a bicycle pump, the volume inside the pump is decreased from about 100 cm^3 to 20 cm^3 before the compressed air flows into the tire. Suppose that the compression is isothermal; estimate the final pressure of the compressed air in the pump, given an initial pressure of 1.00 atm.

STRATEGY In calculations of this kind, it is a good idea to summarize in a table the data supplied and the information required. We expect the final pressure, P_2, to be higher than the initial pressure, P_1, because the volume occupied by the air has been decreased. Because the product of pressure and volume is a constant, we can use Eq. 4 to write

$$P_2V_2 = P_1V_1 \quad \text{or} \quad \frac{P_2}{P_1} = \frac{V_1}{V_2}$$

where V_1 is the initial volume and V_2 is the final volume. We then rearrange this expression into one for P_2 in terms of P_1 and the ratio of volumes, and substitute the data.

SOLUTION The information given and required is summarized in the following table:

	Pressure, P	Volume, V	Temperature, T	Amount, n
1. Initial	1.00 atm	100 cm^3	same	same
2. Final	?	20 cm^3	same	same

- **Worked Examples** include **Strategy** sections that show students how to organize information and solve problems.

- The new edition also introduces the **"How can we use this information . . . ?"** feature. These brief demonstrations of mathematical formulae and concepts immediately follow and reinforce new material.

How can we use this information . . . to determine the temperature of the surface of a star?

Wien's law can be used to determine the surface temperature of stars, which can be treated as hot black bodies. For example, the maximum intensity of solar radiation occurs at 490 nm, so the surface temperature of the Sun must be about

$$T = \frac{c_2}{5\lambda_{max}} = \frac{1.44 \times 10^{-2}\ \text{K·m}}{5 \times 4.90 \times 10^{-7}\ \text{m}} = 5.88 \times 10^3\ \text{K}$$

or about 6000 K.

Self-Test 1.2A In 1965, it was discovered that the universe is pervaded by electromagnetic radiation with a maximum at 1.05 mm (in the microwave region). What is the temperature of "empty" space?

[*Answer:* 2.74 K]

Self-Test 1.2B A red giant is a dying star. The average wavelength maximum at 700 nm shows that a red giant cools as it dies. What is the surface temperature of a red giant?

- **Toolboxes** show students how to tackle major types of calculations. They are designed as learning aids and handy summaries of key material. In the new edition, each Toolbox now begins with explicit statements of that Toolbox's **conceptual basis** and **procedure.** Each Toolbox is immediately followed by a related example or sample calculation.

78 CHAPTER 2 CHEMICAL BONDS

TOOLBOX 2.1 How to write the Lewis structure of a polyatomic species

Conceptual Basis

We look for ways of using all the valence electrons to complete the octets (or duplets).

Procedure

We take the following steps, which are illustrated for HCN in Fig. 2.8:

Step 1 Count the number of valence electrons on each atom. Divide the total number of valence electrons in the molecule by 2 to obtain the number of electron pairs.

Step 2 Predict the most likely arrangements of atoms by using common patterns and the clues indicated in the text, then write the chemical symbols of the atoms to show their layout in the molecule.

Step 3 Place one electron pair between each pair of bonded atoms.

Step 4 Complete the octet (or duplet, in the case of H) of each atom by placing any remaining electron pairs around the atoms. If there are not enough electron pairs, form multiple bonds.

Step 5 Represent each bonded electron pair by a line.

To check on the validity of a Lewis structure, verify that each atom has an octet or a duplet.

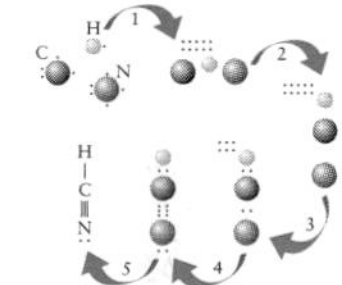

FIGURE 2.8 The five steps used to write the Lewis structure of HCN, as described in this Toolbox.

Self-Test 1.15A Arrange each pair of ions, (a) Mg^{2+} and Al^{3+}, and (b) O^{2-} and S^{2-}, in order of increasing ionic radius.

[*Answer:* (a) $r(Al^{3+}) < r(Mg^{2+})$; (b) $r(O^{2-}) < r(S^{2-})$]

Self-Test 1.15B Arrange each pair of ions, (a) Ca^{2+} and K^+, and (b) S^{2-} and Cl^-, in order of increasing ionic radius.

EXAMPLE 2.3 Writing the Lewis structure of a molecule or an ion

Write the Lewis structures of (a) water, H_2O; (b) methanal, H_2CO; and (c) the chlorite ion, ClO_2^-.

STRATEGY We can write the Lewis structures of molecules and ions by following the steps in Toolbox 2.1 and remembering to adjust the number of electrons for ions. Therefore, we add one electron for the negative charge of ClO_2^-. Find the central atom by inspecting formulas. In methanal, the central atom must be C, because it has a central position and the H atoms can each form only one bond.

SOLUTION

	(a) H_2O	(b) H_2CO	(c) ClO_2^-
1 Number of valence electrons	8	12	20
Number of electron pairs	4	6	10
2 Layout of structure	H O H	H C O H	O Cl O

- **Self-Tests.** All worked examples and "How can we use this information . . . ?" features end with paired self-tests, one answered immediately, the other with a mini-solution in the back of the book.

- **Exercises.** Kenton Whitmire of Rice University has enriched the exercises with additional questions at all levels of difficulty, including questions that require students to use the media on the Web site for this book.

2.101 Quinone, $C_6H_4O_2$, is an organic molecule that can be reduced to give a dianion: $[C_6H_4O_2]^{2-}$. (a) Draw the Lewis structure of the reduced product. (b) On the basis of formal charges derived from the Lewis structure, predict which atoms in the molecule are most negatively charged. (c) If two protons are added to the reduced product, where are they most likely to bond?

Quinone

2.102 In air, the NO radical can react with the NO_2 radical. What is the most likely product of the reaction? Answer this question by drawing Lewis structures of the reactants and products.

different types of hydrogen atoms will be found in the following molecules: (a) C_2H_2; (b) C_2H_4; (c) C_2H_5Cl; (d) *cis*-$C_2H_2Cl_2$; (e) *trans*-$C_2H_2Cl_2$. (f) In the molecule C_2H_5Cl, the hydrogen atoms could all be different; yet by a variety of experimental techniques, only two types of hydrogen atoms are found. Propose an explanation.

2.106 Predict which bond will absorb light of shorter wavelength and explain why: C—H or C—Cl′. Refer to Major Technique 1 on infrared spectroscopy, which follows these exercises.

2.107 Infrared spectra show absorption due to C—H bond stretching at 3.38 μm for a methyl (—CH_3) group and at 3.1 μm for an alkyne (—C≡C—H) group. Which C—H bond is stiffer (has the larger force constant k), assuming that the vibrating atoms have the same effective mass? Refer to Major Technique 1 on infrared spectroscopy, which follows these exercises.

2.108 Vibrational spectra are often so complicated that assignment of a particular absorption to a given bond is difficult. One way to confirm that an assignment is

PROBLEM SOLVING

NEW! INTEGRATED WITH MEDIA LINKS TO . . .

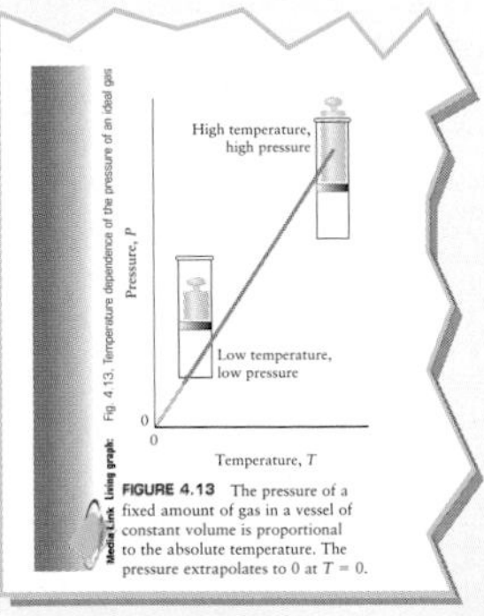

FIGURE 4.13 The pressure of a fixed amount of gas in a vessel of constant volume is proportional to the absolute temperature. The pressure extrapolates to 0 at $T = 0$.

- **Living Graphs.** Selected graphs in the text are available on the Web site. Students can manipulate parameters and explore the consequences.

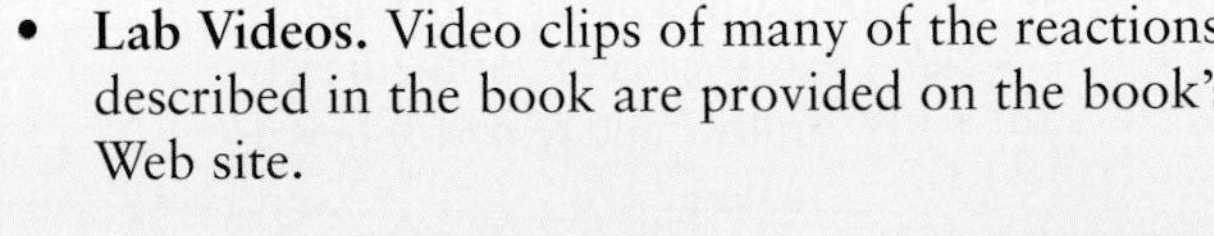

- **Lab Videos.** Video clips of many of the reactions described in the book are provided on the book's Web site.

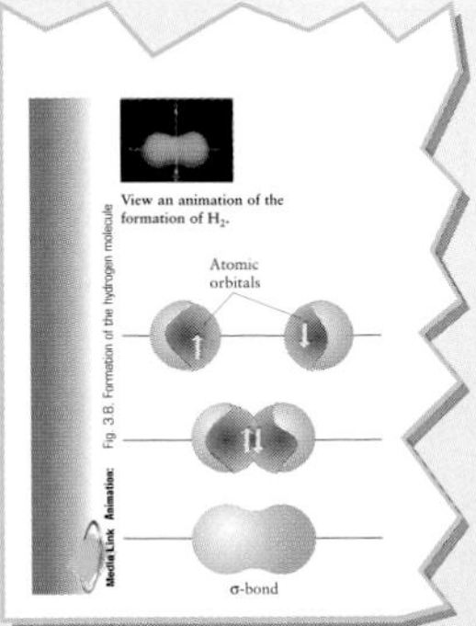

- **Animations.** Selected art in the text is supported by media. Students can view motion, three dimensional images, atomic and molecular interactions and learn to visualize like chemists—at a microscopic level.

- **Tools.** Tools on the book's Web site allow the study of chemical calculations, graphing, and the exploration of periodic properties from different points of view.

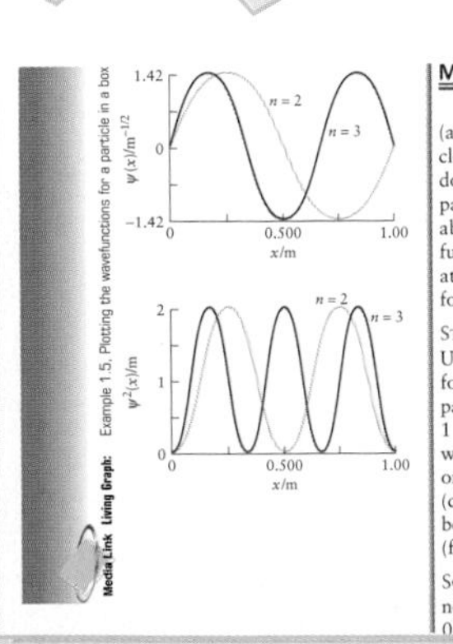

MEDIA EXAMPLE 1.5 Plotting the wavefunctions fo in a box

(a) Using the Living Graphs on the Web site for this boo cle-in-a-box wavefunction for $n = 2$ and $L = 1$ m. (b) does the wavefunction have? Where do these nodes parts (a) and (b) for $n = 3$. (d) What general conclusi about the relation between n and the number of nodes function? (e) Convert the $n = 2$ plot to a probability de at what values of x is it most likely to find the particle? for $n = 3$.

STRATEGY The wavefunctions for a particle in a box ar Use the Living Graph "Particle-in-a-Box Wavefunctions for this book to plot them. (a) To graph the wavefu parameters $L = 1$ m and $n = 2$, set the range for the ent 1 m), check the $f(x)$ box, and hit New Plot. (b) The where the curve passes through 0. To find the value of on the plot at that point. (c) Create a new plot as befo (d) Look for the patterns in node number and n. (e) Rep box checked and look for the maxima in the pr (f) Repeat part (e) for $n = 3$.

SOLUTION The plots are shown on the left. (b) The plot node at $x = 0.50$ m. (c) The plot for $n = 3$ has two node

- **In-Chapter Examples.** Designed to encourage students to use the media, these examples show students how to use the tools and art provided on the book's Web site.

4.73 Calculate the root mean square speeds of (a) methane; (b) ethane; and (c) propane molecules, all at −20°C.

4.74 Calculate the root mean square speed of (a) fluorine gas; (b) chlorine gas; and (c) bromine gas molecules, all at 350°C.

4.75 Consider the Maxwell distribution of speeds found in Fig. 4.28. (a) From the graph, find the location that represents the most probable speed of the molecules. (b) What happens to the percentage of molecules having the most probable speed as the temperature increases?

4.76 (a) Using the Living Graphs program on the Web site for this text, create and print a plot of the Maxwell distribution of molecular speeds for the noble gases He, Ne, Ar, Xe, and Kr from $s = 0$ to $2000\ m\cdot s^{-1}$. (b) Describe the trends observed in the graph in part (a). (c) Again using the Living Graphs program, create graphs for He and for Kr at 100 K and at 300 K. (d) Describe the trends observed in part (c).

(b) Under these conditions, which term has the larger effect on the real pressure of CO_2, the intermolecular attractions or the molar volume? (c) If we consider to be ideal those gases for which the observed pressure differs by no more than 5% from the ideal value, at what pressure does CO_2 become a "real" gas?

4.82 Calculate the pressure exerted by 1.00 mol $C_2H_6(g)$ behaving as (a) an ideal gas; (b) a van der Waals gas when it is confined under the following conditions: (1) at 273.15 K in 22.414 L; (2) at 1.00×10^3 K in 0.100 L.

4.83 Plot pressure against volume for 1 mol of (a) an ideal gas; (b) ammonia; (c) oxygen gas, for the range $V = 0.05$ L to 1.0 L at 298 K. Use a spreadsheet or the graphical plotter on the Web site for this book. Compare the plots and explain the origin of any differences you find.

4.84 Plot pressure against volume for 1 mol of (a) an ideal gas; (b) carbon dioxide, (c) ammonia; (d) benzene for the range $V = 0.1$ L to 1.0 L at 298 K. Use the van der Waals

- **End-of-Chapter Exercises.** Selected exercises direct students to use media to solve problems.

MEDIA LINKS

SUPPLEMENTS FOR *CHEMICAL PRINCIPLES*, SECOND EDITION

Media

Chemical Principles Web site at www.whfreeman.com/chemicalprinciples

Web-based media support for *Chemical Principles* offers students a range of tools for problem solving and chemical exploration.

- Animations that allow students to visualize reactions on a molecular level
- "Living Graphs," which allow the user to control the parameters
- Graphing tools
- Question and Answer—self-scoring quizzes for every chapter and for the Fundamentals provide instant feedback and help students review their knowledge of chemistry.
- Web links that keep students up to date in the world of chemistry
- A calculator adapted for solving equilibrium calculations
- An interactive periodic table of the elements

In addition, all the illustrations and tables from the textbook are available for the instructor in a downloadable format.

Bridging to the Lab (BTL)

The Bridging to the Lab modules are computer-based laboratory simulations with engaging activities that emphasize experimental design and visualization of structures and processes at the molecular level. Every module has a built-in accountability feature that records section completion for use in setting grades.

Used either as prelaboratory preparation for related laboratory activities or to expose students to additional laboratory activities not available in their program, these modules motivate students to learn by proposing real-life problems in a virtual environment. Students make decisions on experimental design, observe reactions, record data, interpret these data, perform calculations, and draw conclusions from their results. Following a summary of the experiment, students test their understanding by applying what they have learned to new situations or by analyzing the effect of experimental errors.

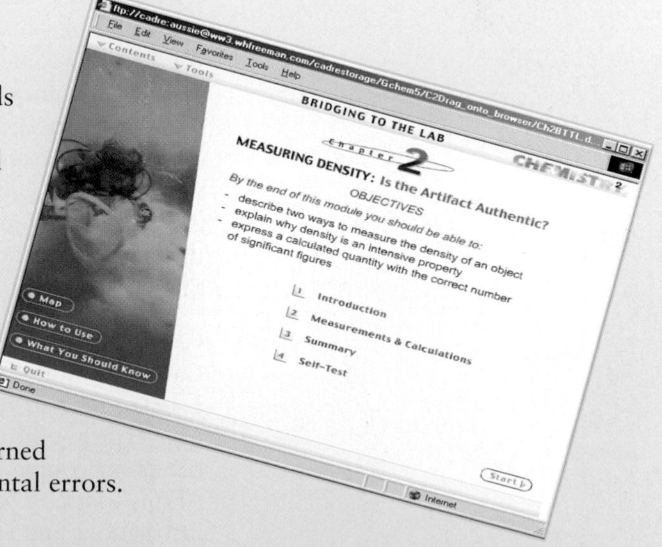

Assessment

Online Quizzing, powered by Question Mark

Access via the *Chemical Principles* Web site

An excellent on-line quizzing bank from which instructors can easily quiz students using prewritten, multiple-choice questions for each text chapter (not from the Test Bank). Students receive instant feedback and can retake quizzes multiple times. Instructors can go into a protected Web site to view results by quiz, student, or question, or can get weekly results via e-mail. Excellent for practice testing or homework or both.

Test Bank CD-ROM (Windows and Mac versions on one disk)
Robert Balahura of the University of Calgary, 0-7167-4434-1;
printed: 0-7167-4433-3
The Test Bank offers more than 1400 multiple-choice, fill-in-the-blank, and essay questions. The easy-to-use CD version includes Windows and Mac versions on a single disk, in a format that lets instructors add, edit, and resequence questions to suit their needs.

Diploma Online Testing, from the Brownstone Research Group
With Diploma, instructors can easily create and administer secure exams over a network and over the Internet, with questions that incorporate multimedia and interactive exercises. The program allows them to restrict tests to specific computers or time blocks, and includes an impressive suite of gradebook and result-analysis features. For more information on Diploma, please visit their Web site: www.brownstone.net

Printed

CUSTOMIZED LAB

Create a Customized Lab Manual to accompany Atkins/Jones, *Chemical Principles,* Second Edition

For the Laboratory
W. H. Freeman and Company is pleased to offer you the ability to create a unique lab manual for your course in just minutes. Drawing from our diverse library of experiments, you can pick and choose the labs you want to use, and even add your own materials. Organize the material as you like, customize the cover, view your newly created book online in PDF format, and then order it on-line with the click of a button. W. H. Freeman's state-of-the-art custom publishing system is easy to use and an excellent alternative to traditional lab manuals. Try it today at http://custompub.whfreeman.com

Working with Chemistry Lab Manual, Donald J. Wink and
Sharon Fetzer Gislason, University of Chicago, Illinois, and
Julie Ellefson Kuehn, William Rainey Harper College, 0-7167-3549-0
With this inquiry-based program, students build skills by using important chemical concepts and techniques to design a solution to a scenario drawn from a professional environment. The scenarios are drawn from the lives of people who work with chemistry every day, ranging from field ecologists to chemical engineers, to health professionals. Instructors: you can download and class-test sample labs from *Working with Chemistry* at the Freeman Web site, www.whfreeman.com/chemistry/wwc

***Chemistry in the Laboratory,* Fifth Edition,** James M. Postma,
California State University, Chico, and Julian L. Roberts, Jr., and
J. Leland Hollenberg, University of Redlands, 0-7167-3547-4
This clearly written, class-tested manual has long given students essential hands-on training with key experiments. Exceptionally compatible with Atkins/Jones, *Chemical Principles,* the manual is known for its clear instructions and illustrations. All experiments are available as lab separates.

HSG Molecular Structure Model Set, Maruzen Company, Ltd.
Molecular modeling helps students understand physical and chemical properties by providing a way to visualize the three-dimensional arrangement of atoms. This model set uses polyhedra to represent atoms, and plastic connectors to represent bonds (scaled to correct bond lengths). Plastic plates representing orbital lobes are included for indicating lone pairs of electrons, radicals, and multiple bonds—a feature unique to this set.

Chemistry Laboratory Student Notebook, Second Edition, 0-7167-3900-3
114 pages of graph paper in a convenient $8\frac{1}{2} \times 11$, three-hole punched format, carbon included. The new edition adds tables and graphs that make the Notebook a handy reference as well.

For the Instructor

Instructor's Resource Manual, Kenton H. Whitmire, Rice University, and Charles Trapp, University of Louisville, 0-7167-4432-5
This useful manual contains worked-out solutions and answers to the even-numbered exercises in the textbook.

For Students

Student's Study Guide, John Krenos and Joseph Potenza, Rutgers University, 0-7167-4436-8
Designed to help students avoid common mistakes and understand the material better, this *Study Guide* is an authoritative and thorough introduction to chemistry for students anticipating careers in science or engineering disciplines. The parallel between topical coverage, the symbols, concepts, and style of the *Study Guide* and the textbook enable the student to move easily back and forth between the two.

Individual chapters of the *Student's Study Guide* and the *Student's Solutions Manual* can be purchased and downloaded instantly, one chapter at a time, whenever they are needed through the Freeman On-line Store at whfreeman.com/chemicalprinciples.

Student's Solutions Manual, Kenton H. Whitmire, Rice University, and Charles Trapp, University of Louisville, 0-7167-4435-X
This manual contains worked-out solutions and answers to the odd-numbered exercises in the textbook. It also explains important problem-solving techniques.

Acknowledgments

The preparation of the second edition of this textbook was an enjoyable yet exacting task that was helped by the generous advice of many who had used the first edition and by the support of many of our colleagues. It is not possible to list everyone who influenced and inspired us, but we wish to recognize those who have made a major impact on our ideas and those who have contributed their expertise to this edition. We would like above all to thank those who so conscientiously reported on the first edition and who read drafts of this edition. Their insights are now part of the text:

Thomas Albrecht-Schmidt, *Auburn University*
David Baker, *Delta College*
Mario Baur, *University of California, Los Angeles*
Paul Braterman, *University of North Texas*
Paul Charlesworth, *Michigan Tech University*
William Cleaver, *University of Vermont*
James Davis, *Harvard University*
Ronald Drucker, *City College of San Francisco*
Robert Eierman, *University of Wisconsin*
Lawrence Fong, *City College of San Francisco*
Regina Frey, *Washington University*
Dennis Gallo, *Augustana College*
David Harris, *University of California, Santa Barbara*
Michael Henchman, *Brandeis University*
Robert Kolodny, *Armstrong Atlantic State University*
Mariusz Kozik, *Canisius College*
Julie Ellefson Kuehn, *William Rainey Harper College*

Donald Parkhurst, *The Walker School*
George Schatz, *Northwestern University*
Michael Sommer, *Harvard University*
Tim Su, *City College of San Francisco*
Larry Thompson, *University of Minnesota*
James Whitten, *University of Massachusetts, Lowell*
Martin Zysmilich, *Massachusetts Institute of Technology*

The contributions of the reviewers of the first edition remain embedded in the text, so we also wish to renew our thanks to:

Matthew P. Augustine, *University of California, Davis*
Ken Brooks, *New Mexico State University*
Robert K. Bohn, *University of Connecticut*
Julia R. Burdge, *University of Akron*
William R. Brennan, *University of Pennsylvania*
David Dalton, *Temple University*
J. M. D'Auria, *Simon Fraser University*
James E. Davis, *Harvard University*
Jimmie Doll, *Brown University*
Jetty Duffy-Matzner, *State University of New York, Cortland*
Donald D. Fitts, *University of Pennsylvania*
David Harris, *University of California, Santa Barbara*
Geoffrey Herring, *University of British Columbia*
Paul Hunter, *Michigan State University*
Petra van Koppen, *University of California, Santa Barbara*
Alison McCurdy, *Harvey Mudd College*
Noel Owen, *Brigham Young University*
Montgomery Pettitt, *University of Houston*
Philip J. Reid, *University of Washington*
Barbara Sawrey, *University of California, San Diego*
Carl Seliskar, *University of Cincinnati*
Herb Silber, *San Jose State University*
Lee G. Sobotka, *Washington University*
Dino Tinti, *University of California, Davis*
Sidney Toby, *Rutgers University*
Ken Whitmire, *Rice University*

The contributions of Ken Whitmire, Rice University, to this book have been substantial. He had full responsibility for the end-of-chapter exercises, generating new exercises, rewriting many of the existing ones, and writing the *Student's Solutions Manual* and *Instructor's Resource Manual*. In the process he gave more than was required and used his expertise in inorganic chemistry to critique the text as well. Some others contributed in substantial ways. We would particularly like to thank Regina Frey, Washington University, and Jamie Nossal, Rice University, for painstakingly checking the accuracy of the worked examples and Self-Tests and providing the Self-Test mini-solutions. We would also like to thank Jetty Duffy-Matzner, Augustana College, and the late Ken Brooks, New Mexico State University, for reading and thoroughly checking final sections of the text. Exercise solutions were double-checked by Maximilienne Bishop and David Price, Rice University. The supplement authors, particularly John Krenos, Joseph Potenza, and Ken Whitmire, have offered us much useful and sage advice. As before, our students have been our most important critics; they helped us to see the world of chemistry through their eyes and made many important suggestions.

We are grateful to our publishers, who understood our vision and helped bring it to fruition. The enthusiasm and skill with which they applied their considerable talents to this project have earned our gratitude. In particular, we would like to acknowledge Jessica Fiorillo, chemistry editor, who kept our spirits high and was perpetually enthusiastic and helpful; Michelle Russel Julet, publisher, who lent her insight and perspective to the project; Marjorie Singer Anderson, our perceptive development editor, who replaced murk with clarity and won our admiration; Jodi Simpson, our copy editor, who pinpointed and corrected flaws with accuracy and precision; Georgia Lee Hadler, senior project editor, who molded raw manuscript into final form; Meg Kuhta, who has an eye for exactly the right photographs; and Charlie van Wagner, who oversaw the development and production of the substantial array of supplements and media.

FUNDAMENTALS

Welcome to chemistry! You are about to embark on an extraordinary voyage, during which you will acquire new insights into the world around you. You will stand in the center of science. Looking in one direction, toward physics, you will see how the principles of chemistry are based on the behavior of atoms and molecules. Looking in another direction, toward biology, you will see how chemists contribute to an understanding of that most awesome property of matter, life. You will be able to look at an everyday object, see in your mind's eye its composition in terms of atoms, and understand its properties. Chemistry gives us the tools to become magicians with matter and to make forms of matter that have never existed before and perhaps do not exist anywhere else in the universe.

INTRODUCTION AND ORIENTATION

Chemistry is the science of matter and the changes it can undergo. The world of chemistry therefore embraces everything material around us—the stones we stand on, the food we eat, the flesh we are made of, and the silicon we build into computers. There is nothing material beyond the reach of chemistry, be it living or dead, vegetable or mineral, on Earth or in a distant star.

Chemistry and Society

We have only to look around us to appreciate the impact of chemistry on technology and society. In the earliest days of civilization, when the Stone Age gave way to the Bronze Age and then to the Iron Age, people did not realize that they were doing chemistry. But they were. They were changing the material they found as stones—we would now call them *minerals*—into metals (Fig. 1). The possession of metals gave them a new

FIGURE 1 Copper is easily extracted from its ores and was one of the first metals worked. The Bronze Age began following the discovery that adding some tin to copper made the metal harder and stronger. These four bronze swords date from 1250–850 BCE, the Late Bronze Age, and are from a collection in the Naturhistorisches Museum, Vienna, Austria. From bottom to top, they are a short sword, an antenna-type sword, a tongue-shaped sword, and a Liptau-type sword.

FIGURE 2 Cold weather triggers a reduction of the amount of the green chlorophyll in leaves, allowing the colors of various other pigments to show.

power over their environment, and treacherous nature became less brutal. The trappings of civilization emerged as skills in transforming materials grew: glass, jewels, coins, ceramics, and, inevitably, weapons became more varied and effective. Art, agriculture, and warfare became more sophisticated. None of this would have happened without chemistry.

The development of steel accelerated the profound impact of chemistry on society. Better steel led to the Industrial Revolution, when muscles gave way to steam and giant enterprises could be contemplated. With improved transport and greater output from factories came more extensive trade, and the world became simultaneously a smaller but busier place. None of this would have happened without chemistry.

With the twentieth century came enormous progress in the development of the chemical industry. Chemistry transformed agriculture. Synthetic fertilizers provided the means of feeding the enormous, growing population of the world. Chemistry transformed communication and transportation. It provided advanced materials, such as polymers for fabrics, ultrapure silicon for computers, and glass for optical fibers; it developed more efficient and renewable fuels. It gave us the tough, light alloys that are needed for modern aircraft and space travel. Chemistry transformed medicine, substantially extended life expectancy, and provided the foundations of genetic engineering. None of this would have happened without chemistry.

We have to admit, though, that the price of all these benefits has been high. The rapid growth of industry and agriculture, for instance, has stressed the Earth and damaged our inheritance. Now, at the dawn of the twenty-first century, there is a rising concern about the preservation of our extraordinary planet. It will be up to you and your contemporaries to draw on chemistry—in whatever career you choose—to build on what has already been achieved. Perhaps you will help to start a new phase of civilization based on new materials, just as semiconductors transformed society during the twentieth century. Perhaps you will help to do so in a way that reduces the harshness of the impact of progress on our environment. To do that, you will need chemistry.

Media Link Lab video: Fig. 3, Reaction of magnesium and oxygen

FIGURE 3 When magnesium burns in air, it gives off a lot of heat and light. The gray-white powdery product looks like smoke.

Chemistry: A Science at Two Levels

Chemistry operates on two levels. At one level, chemistry is about matter and its transformations. At this level, we can actually see the changes, as when a fuel burns, a leaf changes color in the fall (Fig. 2), or magnesium burns brightly in air (Fig. 3). This level is the **macroscopic level,** the level dealing with the properties of large, visible objects. However, there is an underworld of change, a world that we cannot see directly. At this deeper **microscopic level,** chemistry interprets these phenomena in terms of the rearrangements of atoms (Fig. 4).

The **symbolic language** of chemistry, the expression of chemical phenomena in terms of chemical symbols and mathematical equations, ties the two levels together. Chemical equations describe observations at the macroscopic level in a way that can be interpreted in terms of events at the microscopic level. It has been said that a chemist thinks at the microscopic level, conducts experiments at the macroscopic level, and represents both symbolically. Success in chemistry depends on our ability to move

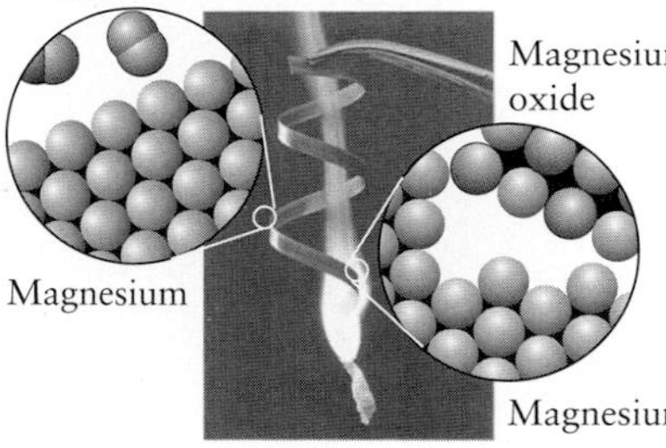

FIGURE 4 When a chemical reaction takes place, atoms exchange partners. As a result, two forms of matter (left inset) are changed into another form of matter (right inset). Atoms are neither created nor destroyed in chemical reactions.

Oxygen
Magnesium oxide
Magnesium
Magnesium

between these levels and relate them symbolically. To foster communication between scientists, international standards have been developed. For example, measurements from around the world are reported in *Système International* units (SI units); these units are described in Appendix 1B.

How Science Is Done

Science pursues ideas in an ill-defined but effective way called the **scientific method.** The scientific method takes many forms. There is no strict rule of procedure that leads you from a good idea to a Nobel prize, or even a publishable discovery. Some scientists are meticulously careful, others are highly creative. The best scientists are probably both creative and careful. Although there are various scientific methods in use, a typical approach consists of a series of steps (Fig. 5). The first step is to collect **data** by making observations and measurements. These measurements are usually made on small **samples** of matter, representative pieces of the material we want to study.

Scientists are always on the lookout for patterns. When a pattern is observed in the data, it can be stated as a scientific **law,** a succinct summary of a wide range of observations. For example, one of the earliest laws of chemistry was the "law of constant composition," which states that *a compound has the same composition regardless of the source of the sample.* Thus, a sample of water has twice as many hydrogen atoms as oxygen atoms, regardless of where the sample is collected.

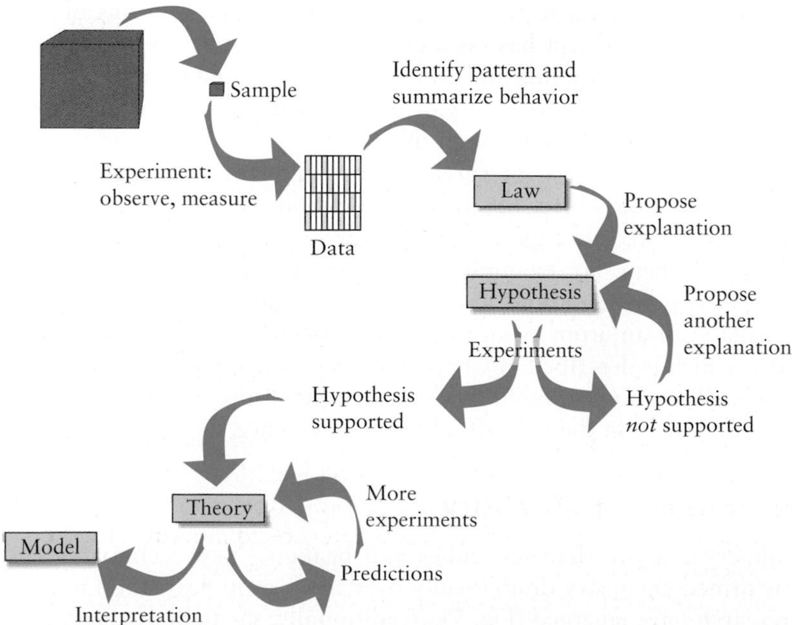

FIGURE 5 A summary of the principal activities that constitute the scientific method. At each stage, the crucial activity is experiment and its comparison with prevailing views.

Not all observations are summarized by laws. There are many properties of matter (such as superconductivity, the ability of a few cold solids to conduct electricity without any resistance) that are currently at the forefront of research but are not described by grand "laws" that embrace hundreds of different compounds. Formulating a law is just one way, not the only way, of summarizing data.

Once they have detected patterns, scientists develop **hypotheses,** possible explanations of the laws—or the observations—in terms of more fundamental concepts. Observation requires careful attention to detail, but the development of a hypothesis requires insight, imagination, and creativity. In 1807, John Dalton interpreted experimental results to propose the hypothesis that matter consists of atoms. Although Dalton could not see individual atoms, he was able to imagine them and formulate his *atomic hypothesis.* Dalton's hypothesis was a monumental insight that helped others understand the world in a new way. The process of scientific discovery never stops. In 1985, chemists found molecules composed of 60 carbon atoms in soot produced when graphite was struck by a laser beam; their hypothesis that these molecules were shaped like soccer balls generated great excitement (Fig. 6). The hypothesis—since confirmed—that buckminsterfullerene has a certain structure is not on the grand scale of Dalton's discovery, but it might nevertheless have a profound impact on the future of chemistry. With luck and application, you may acquire that kind of insight as you read through this text, and one day you may make your own extraordinary hypotheses.

The postulates of Dalton's atomic hypothesis are described in Section B.

In 1996, the Nobel prize in chemistry was awarded to R. F. Curl, R. E. Smalley, and H. W. Kroto for their discovery of C_{60} molecules.

After formulating a hypothesis, scientists design further **experiments,** carefully controlled tests, to verify their hypothesis. Experiments often require ingenuity and sometimes good luck. If the results of repeated experiments support the hypothesis, scientists may go on to formulate a **theory,** a formal explanation of a law. Quite often, and particularly in the physical sciences, of which chemistry is a part, the theory is expressed mathematically. An enormously powerful procedure in science is the conversion of a **qualitative** concept (a concept expressed in words or pictures) into a **quantitative** form (the same concept expressed in terms of mathematics). Once a concept has been expressed quantitatively, it can be subjected to rigorous experimental confirmation and used to make numerical predictions. You will have plenty of practice with this important aspect of chemistry while working through this text.

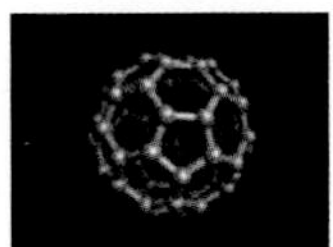

View an animated structure of a buckminsterfullerene molecule.

Scientists commonly interpret a theory in terms of a **model,** a simplified version of the object of study. Like hypotheses, theories and models must be subjected to experiment and revised if experimental results do not support them. For example, our current model of the atom has gone through many formulations and progressive revisions, starting from Dalton's vision of an atom as an uncuttable sphere to our current elaborate model, which is described in Chapter 1. One of the main goals of this text is to show you how to build models, turn them into a testable form, and then refine them in the light of additional evidence.

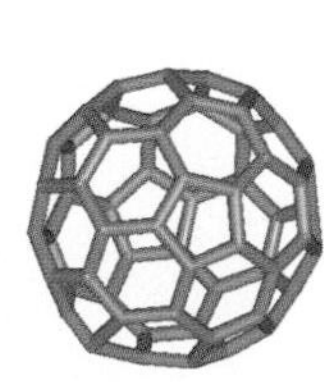

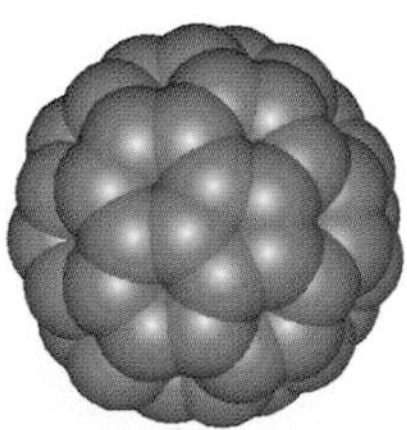

Media Link Animation: Fig. 6, Bucky ball

FIGURE 6 Buckminsterfullerene, C_{60}, consists of 60 carbon atoms arranged in five- and six-membered rings. The structure on the left is a line structure that shows only the bonds between the atoms. That on the right shows how the atoms occupy space.

The Branches of Chemistry

Chemistry is more than test tubes and beakers. New technologies have transformed chemistry dramatically over the last 50 years, and new areas of research have emerged (Fig. 7). Traditionally, the field of chemistry has been organized into three main branches:

organic chemistry, the study of compounds of carbon;

inorganic chemistry, the study of all the other elements and their compounds; and

physical chemistry, the study of the principles of chemistry.

New regions of study have developed as information has been acquired in specialized areas or as a result of the use of particular techniques. We now speak of many more branches of chemistry, including

biochemistry, the study of biologically important substances, processes, and reactions;

analytical chemistry, the study of techniques for identifying substances and measuring their amounts;

theoretical chemistry, the study of molecular structure and properties in terms of mathematical models; and

chemical engineering, the study of industrial chemical processes.

Various branches of knowledge emerging from chemistry have adopted names that do not seem related to their origin; nevertheless, they are rooted in chemistry. They include

molecular biology, the study of the functions of living organisms in terms of their molecular composition; and

materials science, the study of materials in terms of their chemical structure and composition.

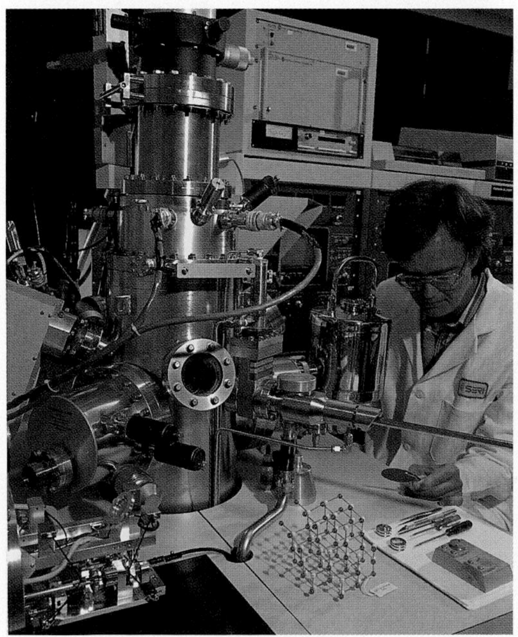

FIGURE 7 Scientific research today often requires sophisticated equipment and computers. This chemist is using an Auger electron spectrometer to probe the surface of a crystal. The data collected will allow the chemist to determine which elements are present in the surface.

In fact, all sciences, medicine, and many fields of commercial activity draw on chemistry. You can be confident that whatever career you choose in a scientific or technical field, it will make use of the concepts discussed in this text. Chemistry is truly central to science.

Mastering Chemistry

The following lettered sections summarize the basic information you need to begin your chemistry course. You might already have a strong background in chemistry and some of its basic concepts. These introductory pages with a blue border will provide you with a focused summary of the fundamental principles of chemistry. You can use them to refresh your memory of concepts or to see how the principles are formulated in a systematic way. Your instructor will advise you on how to use these sections to prepare yourself for the chapters in the text itself.

On the other hand, you might have done little chemistry before. These pages are for you, too. They contain a brief but systematic summary of the basic concepts and calculations of chemistry that represent the knowledge you need for the chapters in the text. You can return to them whenever you feel you need more background or need to refine your skills.

What can you expect to find in the following pages? We want you to enjoy your study; and at the end of your course, we want you to have a sharpened insight into matter. You will see how models are devised, tested, and refined. You will see how qualitative ideas are turned into testable theories. You will come to have insights into matter that allow you to understand how the world of observation springs from the underworld of atoms.

A MATTER AND ENERGY

Just to whet your appetite: matter consists of fermions bound together by bosons.

Whenever we touch, pour, or weigh something, we are working with matter. The whole of chemistry is concerned with the properties of matter, and particularly the conversion of one form of matter into another kind. But what is matter? Matter is in fact quite difficult to define precisely without drawing on advanced ideas from elementary particle physics, but a straightforward working definition is that **matter** is anything that has mass and takes up space. Thus, gold, water, and flesh are forms of matter; electromagnetic radiation (which includes light) and justice are not.

One characteristic of science is that it gives precise meanings to common words from our everyday language. In everyday language, a "substance" is just another name for matter. However, in chemistry, a **substance** is a *single, pure form of matter*. Thus, gold and water are distinct substances. Flesh is a mixture of many different substances, and, in the technical sense used in chemistry, it is not a "substance." Air is matter; but, being a mixture of several gases, it is not a single substance.

Substances, and matter in general, come in different forms, called **states of matter**. The three most common states of matter are solid, liquid, and gas:

A **solid** is a rigid form of matter.

A **liquid** is a fluid form of matter that has a well-defined surface and takes the shape of the part of the container it occupies.

A **gas** is a fluid form of matter that fills any vessel containing it.

The term **vapor** denotes the gaseous form of a substance that is normally a solid or liquid. Thus, we speak of ice (the solid form of water), liquid water, and water vapor.

A.1 Physical Properties

Chemistry is concerned with the "properties" of matter, the distinguishing characteristics of matter; and a large part of this text is devoted to the investigation of properties. However, even this everyday term has been refined and sharpened. Thus, chemists distinguish between "physical" properties and "chemical" properties. A **physical property** of a substance is a characteristic that we can observe or measure without changing the identity of the substance. For example, a physical property of water is its mass; another is its temperature. Physical properties include characteristics such as melting point (the temperature at which a solid turns into a liquid), hardness, color, state of matter (solid, liquid, or gas), and density (see below). A **chemical property** refers to the ability of a substance to change into another substance. A chemical property of the gas hydrogen, for instance, is that it reacts with oxygen (it burns in oxygen) to produce water.

Properties are also classified according to their dependence on the mass of a sample. An **intensive property** is a property that is independent of the mass of the sample. For example, temperature is an intensive property, because we could take a sample of any size from a uniform bath of water and measure the same temperature (Fig. A.1). An **extensive property** is a property that does depend on the mass ("extent") of the sample. Volume is an extensive property: 2 kg of water occupies twice the volume of 1 kg of water.

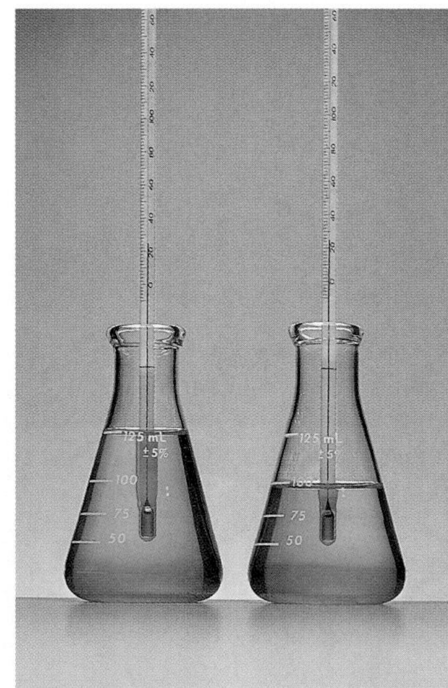

FIGURE A.1 Mass is an extensive property, but temperature is intensive. Both these samples of iron(II) sulfate solution were taken from the same well-mixed supply: they have different masses but the same temperature.

FUNDAMENTALS

Some intensive properties are ratios of two extensive properties. For example, the property we call density is a ratio: the **density**, d, of a sample is its mass, m, divided by its volume, V:

$$\text{Density} = \frac{\text{mass}}{\text{volume}} \quad \text{or} \quad d = \frac{m}{V} \tag{1}$$

The density of a substance is independent of the size of the sample because doubling the volume also doubles the mass, so the ratio of mass to volume remains the same. Density is therefore an intensive property. When we say, in casual conversation, that iron is heavier than water, what we really mean is that the density of iron is greater than that of water: the first statement is not necessarily true, because a cubic meter of water is in fact heavier than a cubic millimeter of iron; the second statement is always true, and independent of the size of the sample.

How can we use this information . . . to calculate the volume of a sample?

We can find the volume of a sample of given mass if we know the density of the substance. For example, the density of silver is 10.50 $g{\cdot}cm^{-3}$, so the volume of a 5.0-g sample of silver is

$$V = \frac{m}{d} = \frac{5.0\ \text{g}}{10.50\ \text{g}{\cdot}\text{cm}^{-3}} = 0.48\ \text{cm}^3$$

Self-Test A.1A The density of selenium is 4.79 $g{\cdot}cm^{-3}$. What is the mass of a 6.5-cm^3 sample of selenium?

[*Answer:* 31 g]

Self-Test A.1B The density of helium gas at 273 K and 1.00 atm is 0.176 85 $g{\cdot}L^{-1}$. What is the volume of a balloon containing 10.0 g of helium under the same conditions?

Units for physical quantities and temperature scales are discussed in Appendix 1B.

We distinguish different substances by their intensive properties. Thus, we might be able to recognize a sample as water by noting its color, density (1.00 $g{\cdot}cm^{-3}$), melting point (0°C), boiling point (100°C), and the fact that it is a liquid.

Physical properties are those that do not involve changing the identity of a substance. Chemical properties are those that involve changing the identity of a substance. Extensive properties depend on the mass of the sample; intensive properties do not.

A.2 Energy

Some chemical changes give off a lot of energy (Fig. A.2); others absorb energy. We shall see that energy is the key to understanding chemical phenomena and the structures of atoms and molecules. But just what is energy? Energy occurs so commonly in everyday language that most people have a general sense of what energy means; however, to get a technical

FIGURE A.2 When bromine is poured on red phosphorus, a chemical change occurs in which a lot of energy is released as heat and light.

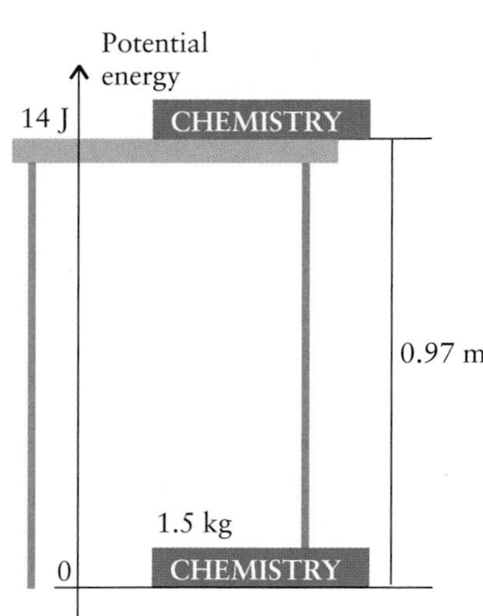

FIGURE A.3 The energy required to raise this book from floor to tabletop is approximately 14 J. The same energy would be released if the book fell from table top to floor.

Media Link Lab video: Fig A.2, Reaction of phosphorus and bromine

answer to this question, we would have to delve into the theory of relativity, which is far beyond the scope of this book. In chemistry, we use a practical definition of **energy**; we treat it as a measure of the capacity to do work. Thus, energy is needed to raise a weight or to force an electric current through a circuit. The greater the energy of an object, the greater its capacity to do work.

To whet your appetite again: energy is curved spacetime.

Three forms of energy are important in chemistry: kinetic energy, potential energy, and electromagnetic energy. The SI unit for energy is the **joule** (J). As explained in Appendix 1B,

$$1\ \text{J} = 1\ \text{kg·m}^2\text{·s}^{-2}$$

The joule is named for James Joule, a nineteenth-century English scientist who made many contributions to the study of heat.

As explained in Appendix 1B, kg stands for kilogram, m for meter, and s for second.

Each beat of the human heart uses about 1 J of energy, and to raise this book (of mass close to 1.5 kg) from the floor to a tabletop about 0.97 m above the floor requires about 14 J (Fig. A.3). Because energy changes in chemical reactions tend to be of the order of thousands of joules for the amounts usually studied, it is more common in chemistry to use the kilojoule, kJ, which is defined as $1\ \text{kJ} = 10^3\ \text{J}$.

Kinetic energy, E_K, is the energy a body possesses because of its motion. For a body of mass m traveling at a speed v, the kinetic energy is

$$E_K = \tfrac{1}{2}mv^2 \qquad (2)$$

A heavy body traveling rapidly has a high kinetic energy. A body at rest (stationary, $v = 0$) has zero kinetic energy.

How can we use this information . . . to calculate kinetic energy?

To calculate the energy it takes to accelerate a person and a bicycle of total mass 75 kg to 20 mph (8.9 m·s^{-1}), starting from rest and ignoring friction and wind resistance, we substitute the data into Eq. 2:

$$E_K = \tfrac{1}{2} \times (75\ \text{kg}) \times (8.9\ \text{m·s}^{-1})^2 = 3.0 \times 10^3\ \text{kg·m}^2\text{·s}^{-2} = 3.0\ \text{kJ}$$

You would have to expend more energy to maintain that speed against friction and wind resistance.

Self-Test A.2A Calculate the kinetic energy of a ball of mass 0.050 kg traveling at 25 m·s^{-1}.

[*Answer:* 16 J]

Self-Test A.2B Calculate the kinetic energy of this book (mass 1.5 kg) just before it lands on your foot after falling off a table, when it is traveling at 3.0 m·s^{-1}.

The concept of force is explained in Section A.3. A field is a region where a force acts.

The second kind of energy we shall consider is **potential energy**, E_P, the energy a body possesses on account of its position in a field of force. There is no single formula given for the potential energy of an object because the potential energy depends on the nature of the forces being experienced by the object. However, there are two simple cases, gravitational potential energy (for a particle in a gravitational field) and Coulomb potential energy (for a charged particle in an electrostatic field).

A body of mass m at a height h above the surface of the Earth has a gravitational potential energy

$$E_P = mgh \qquad (3)$$

relative to its potential energy on the surface itself (Fig. A.4), where g is the constant known as the *acceleration of free fall* (and commonly, the "acceleration of gravity"); the value of g is 9.81 m·s^{-2} (see Appendix 1B for a discussion of units). This formula shows that the greater the altitude of an object, the greater is its gravitational potential energy.

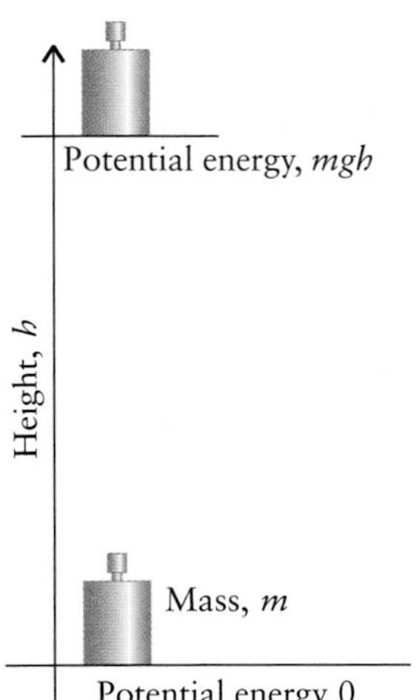

FIGURE A.4 The potential energy of a mass m in a gravitational field is proportional to its height h above a point (the "floor") that is taken to correspond to zero potential energy.

How can we use this information . . . to calculate the gravitational potential energy?

Suppose a person of mass 65 kg walks up a flight of stairs between two floors of a building that are separated by 3.0 m. The potential energy of the person on the upper floor relative to that on the lower floor is

$$\begin{aligned} E_P &= (65\ \text{kg}) \times (9.81\ \text{m·s}^{-2}) \times (3.0\ \text{m}) \\ &= 1.9 \times 10^3\ \text{kg·m}^2\text{·s}^{-2} = 1.9\ \text{kJ} \end{aligned}$$

This energy has to be provided by a chemical process: the digestion of food.

Self-Test A.3A What is the gravitational potential energy of this book (mass 1.5 kg) when it is on a table of height 0.82 m relative to its potential energy when it is on the floor?

[*Answer:* 12 J]

Self-Test A.3B How much energy has to be expended to carry a can of soda (mass 0.350 kg) to the top of the Sears Tower in Chicago (height 443 m)?

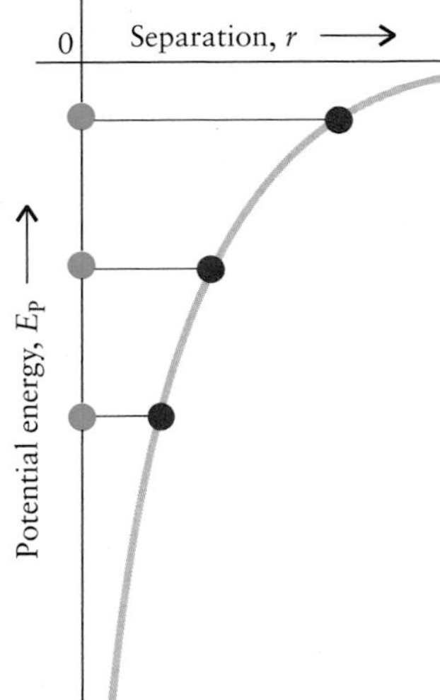

FIGURE A.5 The variation of the Coulomb potential energy of two opposite charges (one represented by the red sphere, the other by the green sphere) with their separation.

The Coulomb potential energy is especially important in chemistry and arises from the attractions and repulsions of electric charges. The **Coulomb potential energy** of a particle of charge q_1 at a distance r from another particle of charge q_2 is proportional to the two charges and inversely proportional to the distance between them:

$$E_P \propto \frac{q_1 q_2}{r} \qquad (4)$$

The Coulomb potential energy approaches 0 as the distance between two particles approaches infinity. If the particles have the same charge—if they are two electrons, for instance—then the numerator q_1q_2 is positive and the potential energy *rises* (becomes more strongly positive) as the distance between them is decreased. If they have opposite charges—an electron and an atomic nucleus, for instance—then the numerator is negative and the potential energy *decreases* (becomes more strongly negative) as the separation is decreased (Fig. A.5).

Subatomic particles, such as electrons and nuclei, are introduced in Section B.

The third form of energy important in chemistry, "electromagnetic energy," is the energy of the **electromagnetic field**, such as the energy carried by radio waves, light waves, and x-rays. An electromagnetic field is generated by the acceleration of charged particles and consists of an oscillating **electric field** and an oscillating **magnetic field** (Fig. A.6). An electric field is an influence that affects charged particles. A magnetic field is an influence that affects moving charged particles.

The **total energy**, E, of a particle is the sum of its kinetic and potential energies:

$$\text{Total energy} = \text{kinetic energy} + \text{potential energy},$$
$$\text{or} \qquad E = E_K + E_P \qquad (5)$$

A very important feature of the total energy is that, provided there are no outside influences, it is constant. Kinetic energy and potential energy can change into each other, but their sum is constant. We summarize this observation by saying that *energy is conserved.* For instance, a ball thrown up into the air initially has high kinetic energy and zero potential energy. At the top of its flight, it has zero kinetic energy and a high potential energy. However, as it returns to Earth, its kinetic energy rises and its potential energy approaches 0 again. At each stage, its *total* energy is the same as it was initially (Fig. A.7). Once it strikes the Earth, the ball is no longer isolated, and its energy is dissipated as **thermal motion**, the chaotic, random motion of atoms and molecules. If we counted up all their

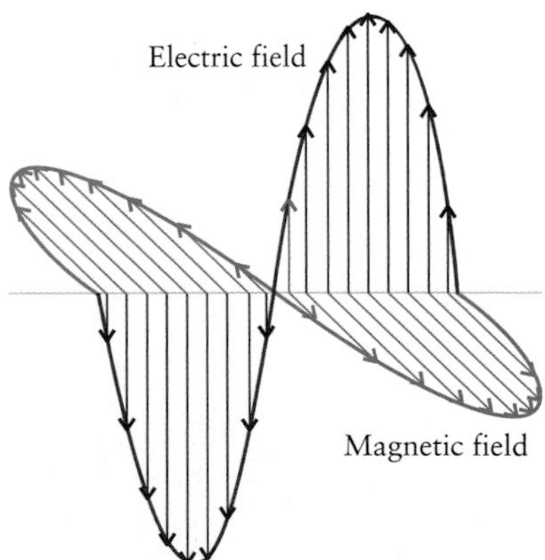

FIGURE A.6 An electromagnetic field oscillates in time and space. The magnetic field is perpendicular to the electric field. The length of an arrow at any point represents the strength of the field at that point.

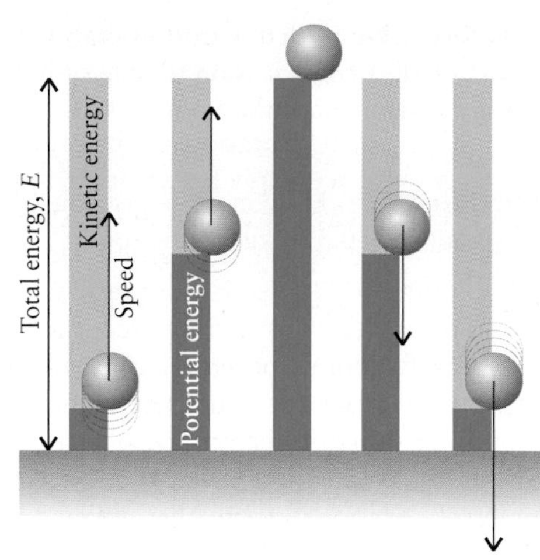

FIGURE A.7 Kinetic energy (represented by the height of the light green bar) and potential energy (the dark green bar) are interconvertible, but their sum (the total height of the bar) is a constant in the absence of external influences, such as air resistance. A ball thrown up from the ground loses kinetic energy as it slows, but gains potential energy. The reverse happens as it falls back to Earth.

kinetic and potential energies, we would find that their total energy had increased by exactly the same amount as the ball had lost. No one has ever observed any exception to the **law of conservation of energy**, the observation that energy can be neither created nor destroyed. One region of the universe—an individual atom, for instance—can lose energy, but another region must gain that energy.

Energy and mass are equivalent and can be interconverted; Einstein deduced that $E = mc^2$.

A change in energy of a body—such as the energy of a single atom—is written ΔE. The Greek letter Δ (delta) usually denotes a change in the property being measured and always indicates the final value minus the initial value:

$$\Delta E = E_{\text{final}} - E_{\text{initial}} \tag{6}$$

When the energy increases, ΔE is positive. When the energy decreases, ΔE is negative.

Kinetic energy results from motion, potential energy from position. An electromagnetic field carries energy through space.

A.3 Force

A **force**, F, is an influence that changes the state of motion of an object. For instance, we exert a force to open a door—to start the door swinging open—and we exert a force on a ball when we hit it with a bat. According to Newton's second law of motion, when an object experiences a force, it is accelerated. The **acceleration**, a, of the object, the rate of change of its velocity, is proportional to the force it experiences. Explicitly,

$$\text{Mass} \times \text{acceleration} = \text{force}, \quad \text{or} \quad ma = F \tag{7}$$

where m is the mass of the object. A stronger force is required to accelerate a heavy object to a given velocity than to accelerate a light object to the same velocity.

Velocity, the rate of change of position, has both magnitude and direction, so when a force acts, it can change the magnitude alone, the direction alone, or both simultaneously (Fig. A.8). The magnitude of the velocity of an object is called its **speed**. When we accelerate a car in a straight line, we change its speed, but not its direction, by applying a force through the rotation of the wheels and their contact with the road. To stop a car, we apply a force that opposes the motion. However, a force can also act without changing the speed: if a body is forced to travel in a different direction at the same speed, it undergoes acceleration because velocity involves direction as well as magnitude. For example, when a ball bounces on the floor, the force exerted by the floor reverses the ball's direction of travel without affecting its speed very much. We shall see a chemical application of this concept when we consider gases, for the molecules of gases hit the walls of the containers they occupy and exert a force on the walls that we experience as the "pressure" of the gas (Chapter 4). When we squeeze a gas-filled balloon or a bicycle tire, the force we exert on the outside of the walls is opposed by the force being exerted by the gas molecules on the inside of the walls.

A force can also be used to move an object between locations at which it has different potential energies. For example, a force is needed to raise this book from the floor against the pull of gravity (another force). A force is also needed to stretch a spring and to pull one atom away from another in a molecule.

Acceleration, the rate of change of velocity, is proportional to applied force: ma = F.

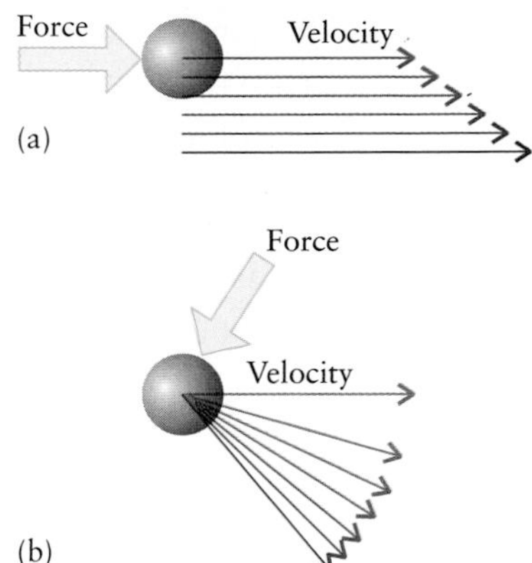

FIGURE A.8 (a) When a force acts along the direction of travel, the speed (the magnitude of the velocity) changes, but not the direction of motion. (b) The direction of travel can be changed without affecting the speed if the force is applied in an appropriate direction. Both changes in velocity correspond to acceleration.

FUNDAMENTALS

SKILLS YOU SHOULD HAVE MASTERED

1 Identify properties as chemical or physical, intensive or extensive.

2 Use the density of a substance in calculations.

3 Calculate the kinetic energy of an object.

4 Calculate the gravitational potential energy of an object.

5 Distinguish among the forms of energy described in this section.

EXERCISES

A.1 Classify the following as chemical or physical properties: (a) objects made of silver become tarnished; (b) the red color of rubies is due to the presence of Cr^{3+} ions; (c) the boiling point of ethanol is 78°C.

A.2 A chemist investigates the boiling point, melting point, and flammability of acetone, a component of fingernail polish remover. Which of these properties are physical properties and which are chemical properties?

A.3 Identify all the physical properties and changes in the statement "The temperature of the land is an important factor for the ripening of oranges, because it affects the evaporation of water and the humidity of the surrounding air."

A.4 Identify all the chemical properties and changes in the statement "Copper is a red-brown element obtained from copper sulfide ores by heating in air, which forms copper oxide. Heating the copper oxide with carbon produces impure copper, which is purified by electrolysis."

A.5 State whether the following are extensive or intensive properties: (a) the temperature at which ice melts; (b) the color of nickel chloride; (c) the amount of energy produced upon burning gasoline; (d) the cost of gasoline.

A.6 State which of the following properties are extensive and which are intensive: (a) the price of platinum; (b) the humidity of the atmosphere; (c) the air pressure in a tire; (d) the hardness of concrete.

A.7 When a piece of metal of mass 112.32 g is dropped into a graduated cylinder containing 23.45 mL of water, the water level rises to 29.27 mL. What is the density of the metal (in grams per cubic centimeter)?

A.8 When a piece of metal of mass 10.00 g is dropped into a graduated cylinder containing 8.75 mL of water, the water level rises to 9.87 mL. What is the density of the metal (in grams per cubic centimeter)?

A.9 The density of diamond is 3.51 $g{\cdot}cm^{-3}$. The international (but non-SI) unit for weighing diamonds is the "carat" (1 carat = 200 mg). What is the volume of a 0.750-carat diamond?

A.10 What volume (in cm^3) of lead (of density 11.3 $g{\cdot}cm^{-3}$) has the same mass as 375 cm^3 of a piece of redwood (of density 0.38 $g{\cdot}cm^{-3}$)?

A.11 Calculate the average density of a single uranium atom by assuming that it is a uniform sphere of radius 138 pm and that the mass of a uranium atom is 3.95×10^{-22} g. The volume of a sphere is $\frac{4}{3}\pi r^3$, where r is its radius. Express the density of uranium in grams per cubic centimeter. The density of metallic uranium is 18.95 $g{\cdot}cm^{-3}$. What does your answer suggest about the way the atoms are packed together in the metal?

A.12 Assume that the entire mass of an atom is concentrated in its nucleus, a sphere of radius 1.5×10^{-5} pm. (a) If the mass of a carbon atom is 2.0×10^{-23} g, what is the density of a carbon nucleus? The volume of a sphere is $\frac{4}{3}\pi r^3$, where r is its radius. (b) What would the radius of the Earth be if its matter were compressed to the same density as a carbon nucleus? (The Earth's average radius is 6.4×10^3 km and its average density is 5.5 $g{\cdot}cm^{-3}$.)

A.13 The maximum ground speed of a chicken is 14 $km{\cdot}h^{-1}$. Calculate the kinetic energy of a chicken of mass 4.2 kg crossing a road at its maximum speed.

A.14 The maximum ground speed of a three-toed sloth is 0.25 $km{\cdot}h^{-1}$. Calculate the maximum kinetic energy of a three-toed sloth of mass 21 kg as it traverses the forest floor.

A.15 Mars orbits the Sun at 25 $km{\cdot}s^{-1}$. A space ship attempting to land on Mars must match its orbital speed. If the mass of the spaceship is 2.5×10^5 kg, what is its kinetic energy when its speed has matched that of Mars?

A.16 (a) An automobile with a mass of 3.0×10^3 kg is traveling on a highway at 70 mph (113 $km{\cdot}h^{-1}$). What is its kinetic energy? (b) If that same automobile is speeding at 100 mph (161 $km{\cdot}h^{-1}$), what is its kinetic energy?

A.17 What is the minimum energy that a football player must expend to kick a football of mass 0.51 kg over a goalpost of height 3.0 m?

A.18 It has been said in jest that the only exercise some people get is raising a fork to their lips. How much energy do we expend to raise a loaded fork of mass 40.0 g a height of 0.50 m 30 times in the course of a meal?

A.19 A skier makes a straight and virtually frictionless run down a black-diamond slope called the Cyclone. If the mass of the skier with equipment is 99.5 kg and the elevation of the ski run is 244 m, what is the kinetic energy of the skier at the bottom of the slope? Ignore wind resistance.

A.20 A ball of mass 0.95 kg is dropped onto a floor from a height of 13.9 m. (a) What is its kinetic energy after it has fallen halfway? (b) What is its kinetic energy when it hits the floor?

A.21 An electron in a cathode-ray tube experienced a constant acceleration of 1.2×10^6 $m{\cdot}s^{-2}$ over a distance of 0.85 cm. What force in newtons had to be exerted on the electron? The mass of an electron is 9.1×10^{-31} kg.

A.22 Pressure is defined as force divided by the area over which the force is exerted. Derive the SI units of pressure (refer to Appendix 1B).

Science is a quest for simplicity. Although the complexity of the world appears boundless, this complexity springs from an underlying simplicity that science attempts to describe. Chemistry's contribution to this quest is to show how everything around us—mountains, trees, people, computers, brains, concrete, oceans—is in fact made up of a handful of simple entities.

The ancient Greeks had much the same idea. They thought there were four elements—earth, air, fire, and water—that could produce all other substances when combined in the right proportions. Their concept of an element is similar to our own; but on the basis of experiments, we now know that there are actually more than one hundred elements, which—in various combinations—make up all the matter on Earth (Fig. B.1).

Appendix 2D lists the names and chemical symbols of the elements and gives the origins of their names.

B.1 Atoms

The Greeks asked what would happen if they continued to cut matter into ever smaller pieces. Is there a point at which they would have to stop because the pieces no longer had the same properties as the whole, or could they go on cutting forever? We now know that there is a point at which we have to stop. That is, matter consists of almost unimaginably tiny particles. The smallest particle of an element that can exist is called an **atom.** The story of the development of the modern model of the atom is an excellent illustration of how scientific models are constructed and revised.

The name *atom* comes from the Greek word for "not cuttable."

The first convincing argument for atoms was made in 1807 by the English schoolteacher and chemist John Dalton (Fig. B.2). He made many measurements of the ratios of the masses of elements that combine together to form the substances we call "compounds" (see Section C) and was able to detect consistent mass ratios. For example, he found that, in every sample of water, 8 g of oxygen were present for every 1 g of hydrogen. These data led Dalton to develop his *atomic hypothesis:*

FIGURE B.1 Samples of common elements. Clockwise from the red-brown liquid bromine are the silvery liquid mercury and the solids iodine, cadmium, red phosphorus, and copper.

FIGURE B.2 John Dalton (1766–1844), the English schoolteacher who used experimental measurements to argue that matter consists of atoms.

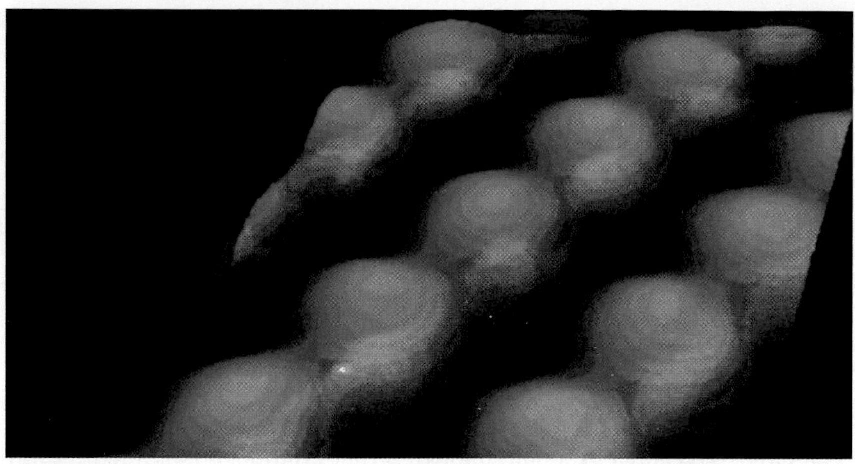

FIGURE B.3 Individual atoms can be seen as bumps on the surface of a solid by the technique called scanning tunneling microscopy (STM). This is an image of the surface of gallium arsenide. The gallium atoms are shown as blue and the arsenic atoms as red (these are not their actual colors: false color helps to distinguish the two types of atoms).

1 All the atoms of a given element are identical.

2 The atoms of different elements have different masses.

3 A compound is a specific combination of atoms of more than one element.

4 In a chemical reaction, atoms are neither created nor destroyed but exchange partners to produce new substances.

Dalton based his hypothesis on a large number of laborious measurements. Today, modern instrumentation provides much more direct evidence of atoms (Fig. B.3). There is no longer any doubt that atoms exist and that they are the units that make up the elements. Indeed, we are now so confident that atoms exist that chemists use their existence as the definition of an element: an **element** is a substance composed of only one kind of atom.

The atoms of an element are not all *exactly* the same, because they can differ slightly in mass. We examine this point in Section B.3.

When scientists make a new element in one of their giant accelerators, they recognize it as new by checking—in ways we shall examine later—whether its atoms are different from those of all known elements. By 2000, 115 elements had been discovered or created, but in some cases in only very small amounts. For instance, when element 111 was made, only two atoms were produced, and even they lasted for only a tiny fraction of a second before disintegrating.

The technique of mass spectrometry, described later in Section B and also in Major Technique 5 following Chapter 19, is used to measure the masses of atoms.

All matter is made up of various combinations of the simple forms of matter called the chemical elements. An element is a substance that consists of only one kind of atom.

B.2 The Nuclear Model

Two hundred years ago, Dalton pictured atoms as featureless spheres like billiard balls. Today we know that atoms have an internal structure: they are built from even smaller **subatomic particles.** By investigating the internal structure of atoms, we can come to see how one element differs from another. The differences depend on the number of each type of subatomic particle in the atoms.

FIGURE B.4 Joseph John Thomson (1856–1940), with the apparatus he used to discover the electron.

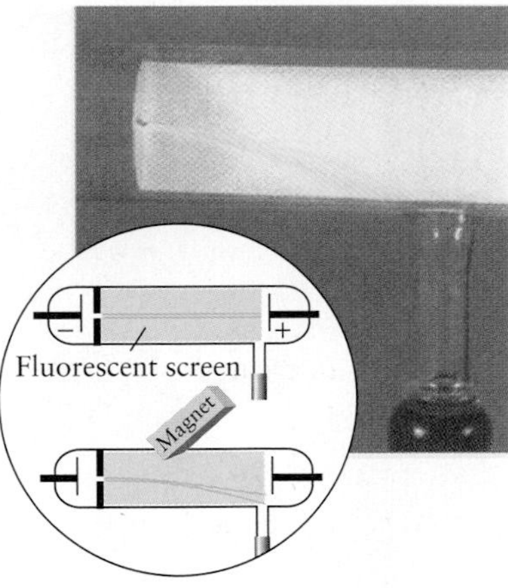

FIGURE B.5 A close-up of the glowing path formed by a stream of electrons on a fluorescent screen near a cathode in a simple cathode-ray tube, an apparatus like that used by Thomson. Note the cathode ray and the deflection of the ray by a magnetic field.

The earliest experimental evidence for the internal structure of atoms was the discovery in 1897 of the first subatomic particle, the electron. The British physicist J. J. Thomson (Fig. B.4) was investigating "cathode rays," the rays that are emitted when a high potential difference (a high voltage) is applied between two electrodes (metal contacts) in an evacuated glass tube (Fig. B.5). Thomson showed that cathode rays are streams of negatively charged particles. They came from inside the atoms that made up the negatively charged electrode, which is called the *cathode*. Thomson found that the charged particles were the same regardless of the metal he used for the cathode. He concluded that they are part of the makeup of *all* atoms. These particles were named **electrons** (denoted e^-).

Thomson was able to measure the value of *e/m*, the ratio of the size of the electron's charge *e* to its mass *m*. Later workers, most notably the American physicist Robert Millikan, carried out experiments that enabled them to determine the charge of the electron. Millikan designed an ingenious apparatus in which he could measure the mass of tiny oil droplets by observing the rate at which they fell through a chamber containing ionized gas particles that stuck to the oil droplets (Fig. B.6). Then he set up an electric field that attracted the droplets upward until they could be suspended. From the strength of the field required, he determined the values of the charges on the particles. Because each oil droplet contained more than one additional electron, he took the charge on one electron to be the smallest increment of charge between droplets, 1.602×10^{-19} C, where C is the abbreviation for the SI unit of charge, the coulomb. This value is considered to be "one unit" of negative charge. The mass of the electron was calculated by combining this charge with the *e/m* ratio measured by Thomson. It was found to be only 9.1×10^{-31} kg.

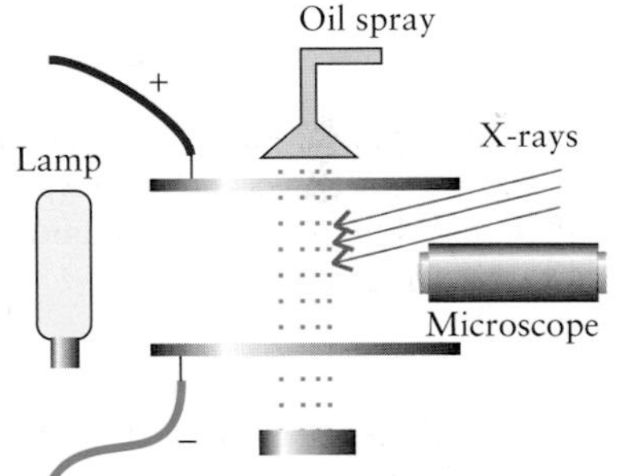

FIGURE B.6 A schematic diagram of Millikan's oil-drop experiment. Oil was sprayed as a fine mist (represented by the small green spheres) into a chamber containing a charged gas, and the rate of fall between two plates was measured with a microscope. Oil was used because it does not evaporate readily. Ions (charged atoms) from the gas stuck to the oil droplets, and their rate of fall was measured first without an electrical charge across the plates, then with a charge. The charge opposed the rate of fall and allowed the exact charge on individual droplets to be determined.

FIGURE B.7 Ernest Rutherford (1871–1937), who was responsible for many discoveries about the structure of the atom and its nucleus.

In later experiments, gold foil was used, with similar results.

Although electrons have a negative charge, atoms overall have zero charge. Therefore, scientists at the beginning of the twentieth century knew that each atom must contain enough positive charge to cancel the negative charge. But where was the positive charge? Thomson suggested a model of an atom as a blob of a positively charged, jellylike material, with the electrons suspended in it like raisins in pudding. However, this model was overthrown in 1908 by a simple experiment. Ernest Rutherford (Fig. B.7) was training some of his students to use a newly constructed apparatus. Rutherford knew that some elements, including radon, emit streams of positively charged particles, which he called **α particles** (alpha particles). He asked two students, Hans Geiger and Ernest Marsden, to shoot α particles toward a piece of platinum foil only a few atoms thick (Fig. B.8). If atoms were indeed like blobs of positively charged jelly, then all the α particles would easily pass through the diffuse positive charge of the foil, with only occasional slight deflections in their paths.

Geiger and Marsden's observations astonished everyone. Although almost all the α particles did pass through and were deflected only very slightly, about 1 in 20 000 was deflected through more than 90°, and a few α particles bounced straight back in the direction from which they had come. "It was almost as incredible," said Rutherford, "as if you had fired a 15-inch shell at a piece of tissue paper and it had come back and hit you."

The explanation had to be that an atom is not a blob of positively charged jelly with electrons suspended in it like raisins. Instead, the results of the experiment suggested a model of an atom in which there is a dense

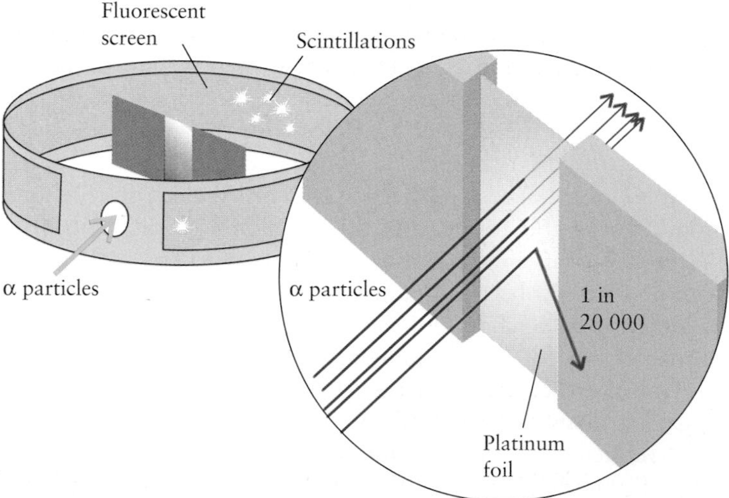

FIGURE B.8 Part of the experimental arrangement used by Geiger and Marsden. The α particles came from a sample of the radioactive gas radon. They were directed through a slit in the side of a cylinder with a zinc sulfide coating on the inside. The α particles struck the platinum foil mounted inside the cylinder, and their deflections were measured by observing flashes of light (scintillations) where they struck the screen. About 1 in 20 000 α particles was deflected through very large angles; most went through the foil with almost no deflection.

pointlike center of positive charge surrounded by a large volume of mostly empty space. Rutherford called the pointlike positively charged region the **atomic nucleus.** He reasoned that a positively charged α particle that scored a direct hit on one of the minute but heavy platinum nuclei was strongly repelled by the positive charge of the nucleus and deflected through a large angle, like a tennis ball bouncing off a stationary cannonball (Fig. B.9).

According to the current **nuclear model** of the atom, electrons are thinly distributed throughout the space around the nucleus. Compared with the size of the nucleus (about 10^{-14} m in diameter), the space occupied by the electrons is enormous (about 10^{-9} m in diameter; a hundred thousand times greater). If the nucleus of an atom were the size of a fly at the center of a baseball field, then the space occupied by the surrounding electrons would be about the size of the entire baseball stadium (Fig. B.10).

The positive charge of the nucleus exactly cancels the negative charge of the surrounding electrons. So, for each electron outside the nucleus, there must be a matching positively charged particle inside the nucleus. These positively charged particles are called **protons** (denoted p); their properties are given in Table B.1. A proton is nearly 2000 times heavier than an electron; each proton has one unit of positive charge.

The number of protons in an element's atomic nucleus is called the **atomic number,** Z, of that element. For example, hydrogen has $Z = 1$, so we know that the nucleus of a hydrogen atom has one proton; helium has $Z = 2$, so its nucleus contains two protons. Henry Moseley, a young British scientist, was the first to determine atomic numbers unambiguously, shortly before he was killed in action in World War I. Moseley knew that when elements are bombarded with rapidly moving electrons (the cathode rays of Thomson's experiment), they emit x-rays. He found that the properties of the x-rays emitted by an element depend on its atomic number; and, by studying the x-rays of many elements, he was able to determine the values of Z for them. Scientists have since determined the atomic numbers of all the known elements (see the list of elements inside the back cover).

In the nuclear model of the atom, all the positive charge and almost all the mass is concentrated in the tiny nucleus, and the negatively charged electrons surround the nucleus. The atomic number is the number of protons in the nucleus; there is an equal number of electrons outside the nucleus.

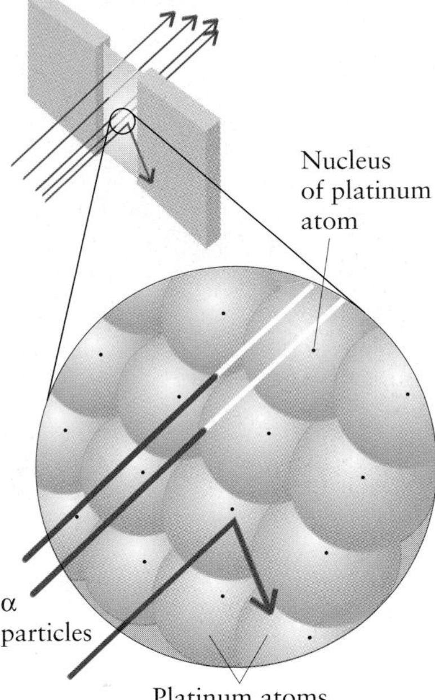

FIGURE B.9 Rutherford's model of the atom explains why most α particles pass almost straight through the platinum foil, whereas a very few—those scoring a direct hit on the nucleus—undergo very large deflections. Most of the atom is nearly empty space thinly populated by the atom's electrons. The nuclei are much smaller relative to the atom than shown here.

FIGURE B.10 Think of a fly at the center of this stadium: that is the relative size of the nucleus of an atom when the atom is magnified to the size of the stadium.

TABLE B.1 Properties of Subatomic Particles

Particle	Symbol	Charge*	Mass, kg
electron	e^-	−1	9.109×10^{-31}
proton	p	+1	1.673×10^{-27}
neutron	n	0	1.675×10^{-27}

* Charges are given as multiples of the charge on a proton, which in SI units is 1.602×10^{-19} C (see Appendix 1B).

FUNDAMENTALS

B.3 Neutrons

Technological advances in electronics early in the twentieth century led to the invention of the **mass spectrometer,** a device for determining the mass of an atom (Fig. B.11). Mass spectrometers have been improved greatly since then, and they have become a very valuable part of the equipment of chemical laboratories. They are described more fully in Major Technique 5 following Chapter 18.

Let's imagine that we are using one of the early mass spectrometers. First, we would pump all the air out of the instrument. Then we would let a gaseous element (neon, for instance) or the vapor of a liquid or a solid element (such as mercury or zinc) enter the ionization chamber. In this chamber, the atoms of the gas or vapor are exposed to a beam of rapidly moving electrons. When one of the accelerated electrons collides with an atom, it knocks another electron out of the atom, thereby leaving the atom with a positive charge. In other words, the collision creates a **positive ion,** an atom with a positive electrical charge. The positive ions are accelerated out of the chamber by a strong electric field applied between two metal grids. The speeds reached by the ions depend on their masses: lighter ions reach higher speeds than heavy ones.

Each rapidly moving ion passes between the poles of an electromagnet. The magnetic field bends the paths of the ions to an extent that depends on their mass, their speed, and the strength of the field. When an ion arrives at the detector, it produces a signal. The instrument combines the signals into a **mass spectrum,** a plot of the detector signal against the magnetic field. The positions of the peaks give the masses of the ions, and the heights of the peaks indicate the relative numbers of ions of various masses.

Mass spectrometry has been used to determine the masses of the atoms of all the elements. We now know, for example, that the mass of a hydrogen atom is 1.67×10^{-27} kg and that of a carbon atom is 1.99×10^{-26} kg. Even the heaviest atoms have masses of only about

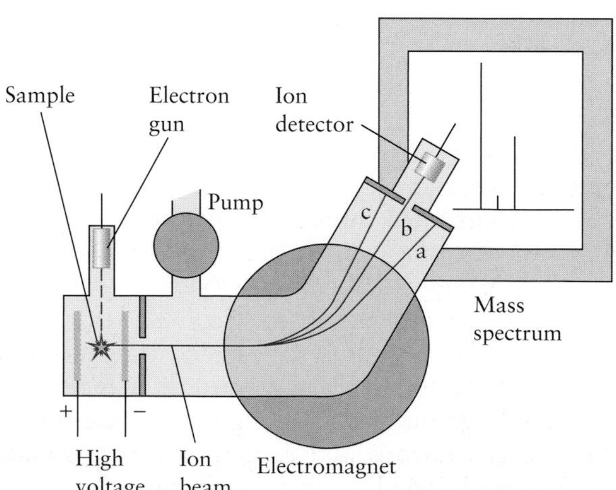

FIGURE B.11 A mass spectrometer is used to measure the masses of atoms. As the strength of the magnetic field is changed, the path of the accelerated ions moves from a to c. When the path is at b, the ion detector sends a signal to the recorder. The mass of the ion is proportional to the strength of the magnetic field needed to move the beam into position.

5×10^{-25} kg. Once we know the mass of an individual atom, we can determine the number of those atoms in a given mass of element simply by dividing the mass of the sample by the mass of one atom.

How can we use this information . . . to determine the number of atoms in a sample?

To calculate the number of atoms in a sample of carbon of mass 10.0 g, we divide the mass of the sample by the mass of one carbon atom:

$$\text{Number of C atoms} = \frac{\text{mass of sample}}{\text{mass of one atom}} = \frac{1.00 \times 10^{-2}\ \text{kg}}{1.99 \times 10^{-26}\ \text{kg}} = 5.03 \times 10^{23}$$

Note how we have converted the mass of the sample to the same units as the mass of the atom. Note, too, how even a small sample of a substance contains a vast number of atoms.

Self-Test B.1A The mass of an iron atom is 9.29×10^{-26} kg. How many iron atoms are present in an iron magnet of mass 25.0 g?

[*Answer:* 2.69×10^{23}]

Self-Test B.1B A miner panning for gold in an Alaskan creek collects 12.3 g of the fine pieces of gold known as "gold dust." The mass of one gold atom is 3.27×10^{-25} kg. How many gold atoms has the miner collected?

As happens so often in science, a new and more precise technique of measurement led to a major discovery. When scientists first used mass spectrometers in much the same way as we have described, they found that the mass of an atom is not proportional to the number of protons: as the number of protons increases, the mass of an atom of an element increases more rapidly. They also found—much to their surprise—that not all the atoms of a single element have the same mass. In a sample of perfectly pure neon, for example, most of the atoms have mass 3.32×10^{-26} kg, which is about 20 times greater than the mass of a hydrogen atom. Some neon atoms, however, are found to be about 22 times heavier than hydrogen. Others are about 21 times heavier (Fig. B.12). All three types of atoms have the same atomic number, so they are definitely atoms of neon.

The observation of differences in mass among atoms of a single element helped scientists refine the nuclear model still further. They realized that an atomic nucleus must contain subatomic particles other than protons and proposed that it also contains electrically neutral particles called **neutrons** (denoted n). Because neutrons have no electric charge, their presence does not affect the nuclear charge or the number of electrons in the atom. However, they do add substantially to the mass of the nucleus, so different numbers of neutrons in a nucleus give rise to atoms of different masses, even though the atoms belong to the same element. As we can see from Table B.1, neutrons and protons are very similar apart from their charge; they are jointly known as **nucleons**.

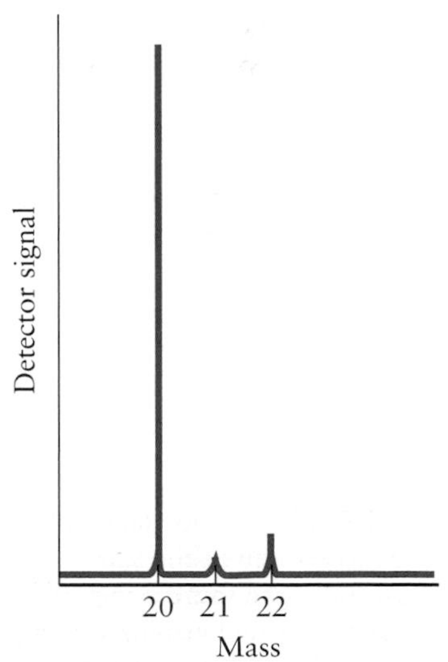

FIGURE B.12 The mass spectrum of neon. The locations of the peaks tell us the relative masses of the atoms, and the intensities tell us the relative numbers of atoms with each mass.

We can now summarize the nuclear model of the atom:

1 Atoms are made up of subatomic particles called electrons, protons, and neutrons.
2 The protons and neutrons form a tiny, dense, central body called the nucleus of the atom.
3 The electrons are distributed in space around the nucleus.

Almost all the mass of an atom is located in the very tiny nucleus, which has a diameter only about one hundred-thousandth that of the atom. In fact, the nucleus of an atom is so dense that, if a particle the size of a grain of sand were as dense, it would have a mass of 50 million tons! Such densities are encountered in *neutron stars,* in which all the electrons of the atoms have combined with the protons, so the star consists only of neutrons and resembles a single huge blob of nuclear matter.

Neutrons are uncharged subatomic particles with nearly the same mass as protons. They are located in the nuclei of atoms.

B.4 Isotopes

Another, better, name for the mass number is *nucleon number.*

The total number of protons and neutrons in a nucleus is called the **mass number,** *A*, of the atom. A nucleus of mass number *A* is about *A* times heavier than a hydrogen atom, which has a nucleus that consists of a single proton. Conversely, if we know that an atom is a certain number of times heavier than a hydrogen atom, then we can infer the mass number of the atom. For example, because mass spectrometry shows that the three varieties of neon atoms are 20, 21, and 22 times heavier than a hydrogen atom, we know that the mass numbers of the three types of neon atoms are 20, 21, and 22. Because for each of them $Z = 10$, these neon atoms must contain 10, 11, and 12 neutrons, respectively (Fig. B.13).

The name *isotope* comes from the Greek words for "equal place."

Atoms with the same atomic number (belonging to the same element) but with different mass numbers are called **isotopes** of the element. All isotopes of an element have exactly the same atomic number; hence they have the same number of protons and electrons. An isotope is named by writing its mass number after the name of the element, as in neon-20, neon-21, and neon-22. Its symbol is obtained by writing the mass number as a superscript to the left of the chemical symbol of the element, as in ^{20}Ne, ^{21}Ne, and ^{22}Ne. You will occasionally see the atomic number included as a subscript on the lower left, as in the symbol $^{22}_{10}Ne$ used in Fig. B.13.

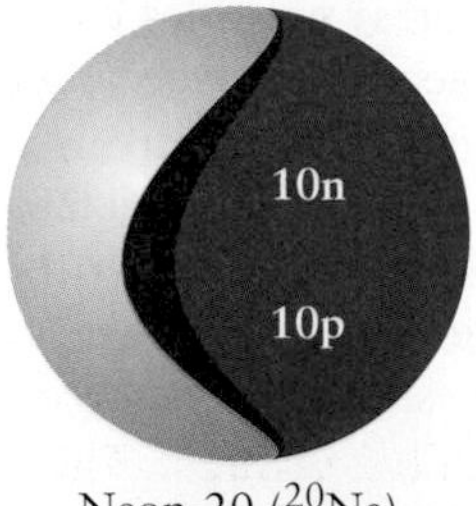

Neon-20 ($^{20}_{10}Ne$)

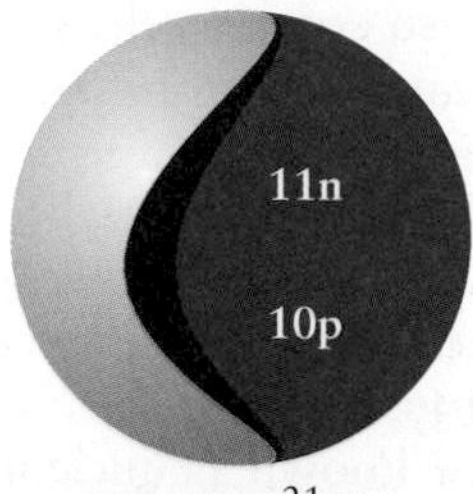

Neon-21 ($^{21}_{10}Ne$)

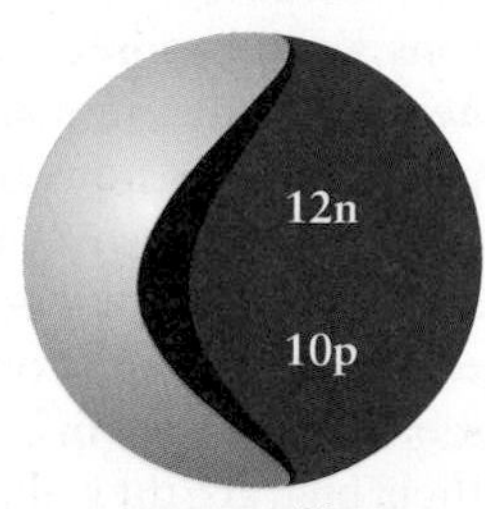

Neon-22 ($^{22}_{10}Ne$)

FIGURE B.13 The nuclei of different isotopes of the same element have the same numbers of protons but different numbers of neutrons. These three diagrams show the composition of the nuclei of the three isotopes of neon. On this scale, the atom itself would be about 1 km in diameter. These diagrams make no attempt to show how the protons and neutrons are arranged inside the nucleus.

TABLE B.2 Some Isotopes of Common Elements

Element	Symbol	Atomic number, Z	Mass number, A	Abundance, %
hydrogen	^{1}H	1	1	99.985
deuterium	^{2}H or D	1	2	0.015
tritium	^{3}H or T	1	3	—*
carbon-12	^{12}C	6	12	98.90
carbon-13	^{13}C	6	13	1.10
oxygen-16	^{16}O	8	16	99.76

* Radioactive, short-lived.

Because isotopes of the same element have the same number of protons and the same number of electrons, they have essentially the same chemical and physical properties. However, the mass differences between isotopes of hydrogen are relatively large, leading to noticeable differences in some physical properties and slight variations in some of their chemical properties. Hydrogen has three isotopes (Table B.2). The most common (^{1}H) has no neutrons, so its nucleus is a lone proton. The other two isotopes are less common but nevertheless so important in chemistry and nuclear physics that they are given special names and symbols. One isotope (^{2}H) is called *deuterium* (D), and the other (^{3}H) is called *tritium* (T).

Isotopes of an element have the same atomic number but different mass numbers. Their nuclei have the same number of protons but different numbers of neutrons.

Self-Test B.2A How many protons, neutrons, and electrons are present in (a) an atom of nitrogen-15; (b) an atom of iron-56?

[*Answer:* (a) 7, 8, 7; (b) 26, 30, 26]

Self-Test B.2B How many protons, neutrons, and electrons are present in (a) an atom of oxygen-16; (b) an atom of uranium-236?

B.5 The Organization of the Elements

There are over 100 elements, and at first sight the prospect of learning their properties might seem overwhelming. The task is made much easier—and more interesting—by one of the most important discoveries in the history of chemistry. Chemists have found that, when the elements are listed in order of their atomic number and arranged in rows of certain lengths, they form families that show regular trends in properties. The arrangement of elements that shows their family relationships is called the **periodic table** (it is printed inside the front cover of this book and repeated schematically in Fig. B.14).

The story of the discovery of periodic relationships by Dmitri Mendeleev can be found in Box 1.2.

The vertical columns of the periodic table are called **groups.** These groups identify the principal families of elements. The taller columns

In some versions of the periodic table, you will see a different notation for groups, with the noble gases belonging to Group VIII, VIIIA, or VIIIB. These alternatives are given in the table printed inside the front cover. Here we show the Roman numeral as well as the arabic number, as in 18/VIII, for Groups 13–18.

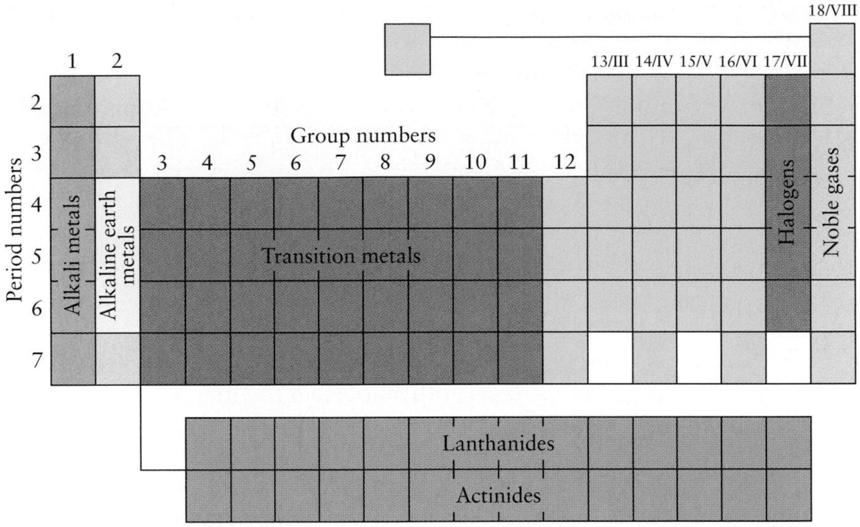

Media Link Tool: Periodic table

FIGURE B.14 The structure of the periodic table, showing the names of the regions and groups. The groups are the vertical columns, numbered 1 through 18. The periods are the horizontal rows, numbered 1 through 7. The main-group elements are those in Groups 1, 2, and 13–18, together with hydrogen. Some versions of the table use different notations for groups. We use both notations for Groups 13–18.

(Groups 1, 2, and 13/III through 18/VIII) are called the **main groups** of the table. The horizontal rows are called **periods** and are numbered from the top down. The four rectangular regions of the periodic table are called **blocks** and, for reasons related to atomic structure (Section 1.13), are labeled *s, p, d,* and *f.*

The members of the *d* block, with the exception of the elements in Group 12 (the zinc group) are called **transition metals.** As we shall see, these elements are transitional in character between the vigorously reactive metals in the *s* block and the less reactive metals on the left of the *p* block. The members of the *f* block, which is shown below the main table (to save space), are the **inner transition metals.** The upper row of this block, following lanthanum (element 57) in Period 6, consists of the **lanthanides;** and the lower row, following actinium (element 89) in Period 7, consists of the **actinides.**

Some of the main groups have special names. The elements in Group 1 are called the **alkali metals.** All of them are soft, silvery metals that melt at low temperatures. They all produce hydrogen when they come in contact with water (Fig. B.15)—lithium gently, but with increasing violence down the group. The elements calcium (Ca), strontium (Sr), and barium (Ba) in Group 2 are called the **alkaline earth metals,** but the name is often extended to all the members of the group. The Group 2 metals share many properties with the Group 1 metals, but their reactions are less vigorous.

FIGURE B.15 The alkali metals react with water, producing gaseous hydrogen and heat. Potassium reacts vigorously, producing so much heat that the hydrogen produced from the water is ignited.

Media Link Lab video: Fig B.15, Reaction of sodium and water

On the far right of the table, in Group 18/VIII, are the elements known as the **noble gases.** They are so called because they combine with very few elements—they are chemically aloof. In fact, until the 1960s, they were called the "inert gases" because it was thought that they did not combine with any elements at all. All the Group 18/VIII elements are colorless, odorless gases. Next to the noble gases are the **halogens** of Group 17/VII. Many of the properties of the halogens show regular variations from fluorine (F) through chlorine (Cl) and bromine (Br) to iodine (I). Fluorine, for instance, is a very pale yellow, almost colorless, gas, chlorine a yellow-green gas, bromine a red-brown liquid, and iodine a purple-black solid (Fig. B.16).

At the head of the periodic table, standing alone, is hydrogen. Some tables place hydrogen in Group 1, others place it in Group 17/VII, and yet others place it in both groups. We treat it as a very special element and place it in none of the groups.

The elements are classified as metals, nonmetals, and metalloids:

A **metal** conducts electricity, has a luster, and is malleable and ductile.

A **nonmetal** does not conduct electricity and is neither malleable nor ductile.

A **metalloid** has the appearance and some properties of a metal but behaves chemically like a nonmetal.

Some chemists refer to metalloids as "semimetals."

A "malleable" substance (from the Latin word for "hammer") is one that can be hammered into thin sheets (Fig. B.17). A "ductile" substance (from

FIGURE B.16 The halogens are colored elements. From left to right, chlorine is a yellow-green gas, bromine is a red-brown liquid (its vapor fills the flask), and iodine is a dark purple-black solid (note the small crystals).

FIGURE B.17 All metals can be deformed by hammering. Gold can be hammered into a sheet so thin that light can pass through it. Here it is possible to see the light of a flame through the sheet of gold.

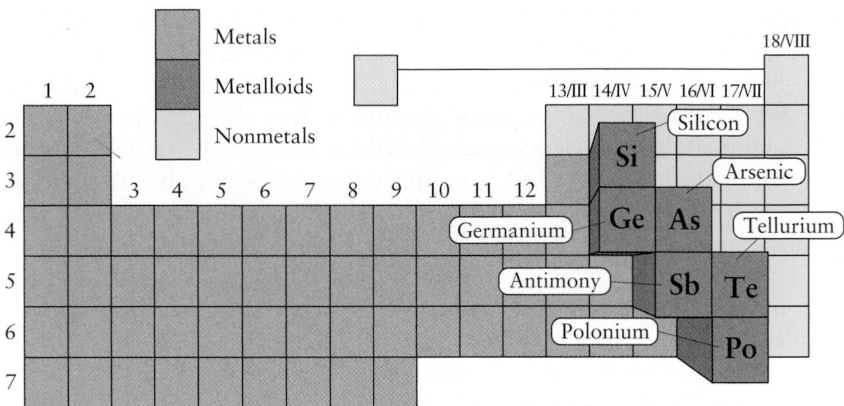

FIGURE B.18 The location of the seven elements commonly regarded as metalloids: these elements have characteristics of both metals and nonmetals. Other elements, notably beryllium and bismuth, are sometimes included in the classification. Boron, although not resembling a metal in appearance, is included because it resembles silicon chemically.

the Latin word for "drawing out") is one that can be drawn out into wires. Copper, for example, is a metal. It conducts electricity, has a luster when polished, and is malleable. It is so ductile that it is readily drawn out to form electrical wires. Sulfur, on the other hand, is a nonmetal. This brittle yellow solid does not conduct electricity, cannot be hammered into thin sheets, and cannot be drawn out into wires. The distinctions between metals and metalloids and between metalloids and nonmetals are not very precise (and not always made), but the metalloids are often taken to be the seven elements shown in Fig. B.18 on a diagonal band between the metals on the left and the nonmetals on the right.

The periodic table is a very useful summary of the properties of the elements. Even if we have never heard of osmium, we can glance at the periodic table and see that it is a metal because it lies to the left of the metalloids (all the *d*-block elements are metals). Similarly, even though chemists might not have investigated the reactions of radon—it is a dangerously radioactive gas, and even chemists draw the line somewhere—we can anticipate that it will resemble the other noble gases, particularly its immediate neighbor in Group 18/VIII, xenon, and form very few compounds. As you work through this text and encounter a new element, it is a good idea to refer to the periodic table, identify the neighbors of the element, and try to predict its properties.

The periodic table is an arrangement of the elements that reflects their family relationships; members of the same group typically show a smooth trend in properties.

SKILLS YOU SHOULD HAVE MASTERED

1 Describe the structure of an atom.

2 Find the number of atoms in a given mass of an element.

3 Find the numbers of neutrons, protons, and electrons in an isotope, Self-Test B.2.

4 Describe the organization of the periodic table and the characteristics of elements in different regions of the table.

EXERCISES

B.1 The mass of an atom of beryllium is 1.50×10^{-26} kg. How many beryllium atoms are present in a 0.210-g beryllium film used as a window on an x-ray tube?

B.2 (a) Which part of Dalton's atomic hypothesis has been disproved by experiments? (b) Summarize the evidence disproving this hypothesis.

B.3 At the time J. J. Thomson conducted his experiments on cathode rays, the nature of the electron was in doubt. Some considered it to be a form of radiation, like light; others believed the electron to be a particle. Some of the observations made on cathode rays were used to advance one view or the other. Show how each of the following properties of cathode rays supports either the radiation or particle model of the electron. (a) They pass through metal foils. (b) They travel at speeds slower than that of light. (c) If an object is placed in their path, they cast a shadow. (d) Their path is deflected when they are passed between electrically charged plates.

B.4 Millikan measured the charge of the electron in *electrostatic units*, esu. The data he collected included the following series of charges found on oil drops: 9.60×10^{-10} esu, 1.92×10^{-9} esu, 2.40×10^{-9} esu, 2.88×10^{-9} esu, and 4.80×10^{-9} esu. (a) Find from this series the likely charge on the electron in esu. (b) Predict the number of electrons on an oil drop with the charge 6.72×10^{-9} esu.

B.5 Give the number of protons, neutrons, and electrons in an atom of (a) deuterium, ^{2}H; (b) ^{127}I; (c) nitrogen-15; (d) ^{209}Bi.

B.6 Give the number of protons, neutrons, and electrons in an atom of (a) boron-11; (b) ^{10}B; (c) phosphorus-31; (d) ^{238}U.

B.7 Identify the isotope that has atoms with (a) 104 neutrons, 71 protons, and 71 electrons; (b) 68 neutrons, 50 protons, and 50 electrons; (c) 3 neutrons, 3 protons, and 3 electrons.

B.8 Identify the isotope that has atoms with (a) 117 neutrons, 77 protons, and 77 electrons; (b) 12 neutrons, 10 protons, and 10 electrons; (c) 28 neutrons, 23 protons, and 23 electrons.

B.9 (a) What characteristics do atoms of carbon-12, carbon-13, and carbon-14 have in common? (b) In what ways are they different? (Consider the numbers and types of subatomic particles.)

B.10 (a) What characteristics do atoms of argon-40, potassium-40, and calcium-40 have in common? (b) In what ways are they different? (Consider the numbers and types of subatomic particles.)

B.11 (a) What characteristics do atoms of manganese-55, iron-56, and nickel-58 have in common? (b) In what ways are they different? (Consider the numbers and types of subatomic particles.)

B.12 The bromine molecule, Br_2, is known to contain two atoms of bromine. A mass spectrum of the molecular ions formed showed three Br_2^+ peaks, with the mass numbers 158, 160, and 162. Use this information to determine which isotopes of bromine occur in nature.

B.13 (a) Determine the total number of protons, neutrons, and electrons in one ammonia molecule, NH_3. (b) What is the total mass of protons, of neutrons, and of electrons in one ammonia molecule? (Calculate three masses.)

B.14 Determine the fraction of the total mass of an ^{56}Fe atom that is due to (a) neutrons; (b) protons; (c) electrons. (d) What is the mass of protons in a 1000-kg automobile? Assume that the total mass of the vehicle is due to ^{56}Fe.

B.15 Name the elements (a) Si; (b) Sr; (c) Sn; (d) Sb. List their group numbers in the periodic table. Identify each as a metal, a nonmetal, or a metalloid.

B.16 Name the elements (a) Rb; (b) Ra; (c) Ru; (d) Rn. List their group numbers in the periodic table, and identify each as a metal, a nonmetal, or a metalloid.

B.17 Write the symbol of (a) silver; (b) germanium; (c) krypton. Classify each as a metal, a nonmetal, or a metalloid.

B.18 Write the symbol of (a) osmium; (b) thallium; (c) astatine. Classify each as a metal, a nonmetal, or a metalloid.

B.19 List the names, symbols, and atomic numbers of the alkali metals. Characterize their reactions with water and describe their trend in melting points.

B.20 List the names, symbols, and atomic numbers of the halogens. Identify the normal physical state of each.

B.21 State three physical properties that are typical of (a) metals; (b) nonmetals.

B.22 Identify the blocks to which the following elements belong: (a) zirconium; (b) Sr; (c) Kr; (d) gold; (e) lead; (f) Fe.

C COMPOUNDS

The relatively small number of elements that make up our world combine to produce matter in a seemingly limitless variety of forms. We only have to look at the vegetation, flesh, landscapes, fabrics, building materials, and other things around us to appreciate the wonderful variety of the material world. A part of chemistry is **analysis:** the discovery of which elements have combined together to form a substance. Another aspect of chemistry is **synthesis:** the production of new combinations of atoms, new materials, to improve on nature or to produce a greater abundance of substances that are in short supply. If the elements are the alphabet of chemistry, then the compounds are its plays, its poems, and its novels.

C.1 What Are Compounds?

FIGURE C.1 Elemental sulfur burns in air with a blue flame and produces the dense gas sulfur dioxide, a compound of sulfur and oxygen.

A **compound** is an electrically neutral substance that consists of two or more different elements with their atoms present in a definite ratio. A **binary compound** consists of only two elements. Water, for instance, is a binary compound of hydrogen and oxygen, with two hydrogen atoms for each oxygen atom. Whatever the source of the water, it has exactly the same composition; indeed, a substance with a different ratio of atoms would not be water! Chemists took a big step forward when they first noticed this invariance of composition, for it suggested an underlying order in nature. They summarized the observation as the **law of constant composition.** The law was important in the historical quest for an understanding of matter, because it suggested to chemists that compounds consisted of specific combinations of atoms.

Compounds are classified as either organic or inorganic. **Organic compounds** contain the element carbon and usually hydrogen, too. They include fuels such as methane and propane, sugars such as glucose and sucrose, and most medicines. Millions of other substances are also organic compounds, and thousands of new ones are synthesized, identified, and reported each year. These compounds are called organic because it was once believed, incorrectly, that they could be formed only by living organisms. **Inorganic compounds** are all the other compounds; they include water, calcium sulfate, ammonia, silica, hydrochloric acid, and many, many more. In addition, some very simple carbon compounds, particularly carbon dioxide and the carbonates, which include chalk (calcium carbonate), are treated as inorganic compounds.

Bonding is discussed in more detail in Chapters 2 and 3.

The elements in a compound are not just mixed together. Their atoms are actually joined, or *bonded,* to one another in a specific way. The result is a substance with chemical and physical properties different from those of the elements that form it. For example, when sulfur is ignited in air, it combines with oxygen from the air to form the compound sulfur dioxide. Solid yellow sulfur and odorless oxygen gas produce a colorless, pungent, and poisonous gas (Fig. C.1).

Chemists have found that atoms bond together to form molecules, or atoms can be present in compounds as ions:

The name *ion* comes from the Greek word for "go," because charged particles go either toward or away from a charged electrode.

A **molecule** is a discrete group of atoms bonded together in a specific arrangement.

An **ion** is a positively or negatively charged atom or molecule.

A positively charged ion is called a **cation,** and a negatively charged ion is called an **anion.** For instance, a positively charged sodium atom is a cation and is denoted Na^+; a negatively charged chlorine atom is an anion and is denoted Cl^-. An example of a "polyatomic" (many-atom) cation is the ammonium ion, NH_4^+, and an example of a polyatomic anion is the carbonate ion, CO_3^{2-}; note that the latter has two negative charges. A **molecular compound** consists of molecules; an **ionic compound** consists of ions. In general, binary compounds of two nonmetals are molecular, whereas binary compounds formed by a metal and a nonmetal are ionic. For example, water (H_2O) is an example of a binary molecular compound and sodium chloride (NaCl) an example of a binary ionic compound. As we shall see, these two types of compounds have characteristic properties, and knowing the type of compound we are studying gives us insight into its properties.

The prefixes *cat-* and *an-* come from the Greek words for "down" and "up." Oppositely charged ions travel in opposite directions in an electric field.

Compounds are combinations of elements in which the atoms of the different elements are present in a characteristic, constant ratio. A compound is classified as molecular if it consists of molecules and as ionic if it consists of ions.

C.2 Molecules and Molecular Compounds

The **chemical formula** of a compound represents its composition in terms of chemical symbols. Subscripts show the numbers of atoms of each element present in the smallest unit that is representative of the compound. For molecular compounds, it is common to give the **molecular formula,** a chemical formula that shows how many atoms of each type of element are present in a single molecule of the compound. For instance, the molecular formula for water is H_2O: each molecule contains one O atom and two H atoms. The molecular formula for estrone, a female sex hormone, is $C_{18}H_{22}O_2$, showing that a single molecule of estrone consists of 18 C atoms, 22 H atoms, and 2 O atoms. A molecule of a male sex hormone, testosterone, differs by only a few atoms: its molecular formula is $C_{19}H_{28}O_2$.

Some elements also exist in molecular form. Except for the noble gases, all the elements that are gases at ordinary temperatures are found as **diatomic molecules,** molecules that consist of two atoms. For example, molecules of hydrogen gas contain two hydrogen atoms bonded together, and are represented as H_2. Oxygen exists as O_2 and chlorine as Cl_2. Solid sulfur exists as S_8 molecules and phosphorus as P_4 molecules.

A **structural formula** indicates how the atoms are linked together. For instance, the molecular formula of ethanol is C_2H_6O, and its structural formula is shown in (**1**): each line represents a chemical bond and each symbol an atom. Structural formulas are very clear, but they are cumbersome. So, chemists condense them and write, for instance, CH_3CH_2OH to represent the structure of ethanol. This "condensed" structural formula indicates the groupings of the atoms and summarizes the full structural formula. In most cases, symbols with subscripts represent atoms connected to the previous element in the formula. A group of atoms attached to another atom in the molecule is set off with parentheses. For example, methylpropane (**2**) has a *methyl* group ($—CH_3$) attached

```
   H   H   H
   |   |   |
H— C — C — O
   |   |
   H   H
```

1 Ethanol, CH_3CH_2OH

```
          H
          |
   H   H— C —H   H
   |      |      |
H— C ———— C ———— C —H
   |      |      |
   H      H      H
```

2 Methylpropane, $CH_3CH(CH_3)CH_3$

to the central carbon atom, and the condensed version of its structural formula is written $CH_3CH(CH_3)CH_3$. It is always possible to inflate a condensed formula into a full structural formula by inserting the lines representing the bonds.

The richness of organic chemistry arises in part from the fact that carbon atoms nearly always form four bonds and can form chains and rings of almost limitless variety. Another source of its richness is that the carbon atoms can be linked together by different kinds of chemical bonds. In Chapter 2, we shall see that atoms can be joined by single bonds, indicated by a single line (C—C); double bonds, represented by a double line (C=C); and triple bonds, represented by a triple line (C≡C). A carbon atom can form four single bonds, two double bonds, or any combination that results in four bonds, such as one single and one triple bond.

(a)

(b)

3 2-Chlorobutane, $CH_3CHClCH_2CH_3$

Organic chemists have found a way to draw these potentially very complex molecular structures in a very simple way, by not showing the C and H atoms explicitly. A **line structure** represents a chain of carbon atoms by a zigzag line, where each short line indicates a bond and the end of each line represents a carbon atom. Atoms other than C and H are shown by their symbols. Sometimes carbon atoms form double bonds between themselves or to other atoms. These bonds are represented by a double line. Because carbon almost always forms four bonds in organic compounds, there is no need to show the C—H bonds explicitly. We just fill in the correct number of hydrogen atoms mentally: compare the line structure of 2-chlorobutane, $CH_3CHClCH_2CH_3$ (**3a**) with its structural form (**3b**). Line structures are particularly useful for complex molecules, like those of testosterone (**4**). They are used occasionally in the early chapters of this text and are explained in more detail in Chapter 18.

4 Testosterone, $C_{19}H_{28}O_2$

The next most important aspect of a molecular compound is its shape. The pictorial representation of molecules that most accurately shows their shapes comes from the use of computer graphics that show computed structures. An example is the **space-filling model** of an ethanol molecule shown in (**5**). The atoms are represented by colored spheres (they are not the actual colors of the atoms!) that fit into one another. Another representation of the same molecule, called a **ball-and-stick model,** is shown in (**6**). Each ball represents the location of an atom, and the sticks represent the bonds. Although this kind of model does not represent the actual molecular shape as well as a space-filling model does, it shows bond lengths and angles more clearly. It is also easier to draw and interpret.

5 Ethanol, CH_3CH_2OH

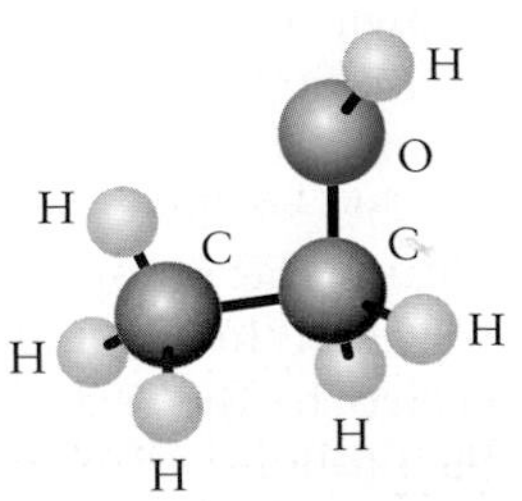

6 Ethanol, CH_3CH_2OH

There are several other kinds of images that chemists use to depict molecular structure. A **tube structure** is like a ball-and-stick structure without the balls: different atoms are depicted by different colors at the ends of the tube (7). More sophisticated than any of these representations are images that show the distribution of electrons in the molecule. The cloud in (8), for instance, indicates how the electrons are spread throughout an ethanol molecule: it is an example of a **density isosurface** and gives a very real sense of the shape of the molecule (but it is hard to interpret in terms of the location of the atoms). Structure (9) shows the same diagram with the density isosurface made transparent and a tube structure inside, which helps to clarify the significance of the isosurface. Finally, chemists also need to know the distribution of electric charge in a molecule, because that distribution affects its physical and chemical properties. To do so, they sometimes use an **electrostatic potential surface** (an "elpot" surface), in which the net charge is calculated at each point of the density isosurface and depicted by different colors (10). A blue tint indicates that the positive charge due to the nuclei outweighs the negative charge of the electrons at that point; a red tint indicates the opposite.

7 Tube structure, C_2H_5OH

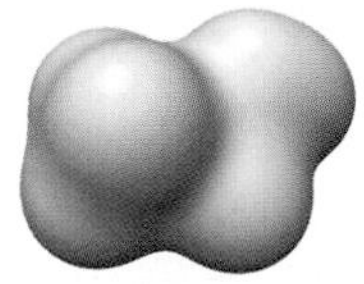

8 Density isosurface, C_2H_5OH

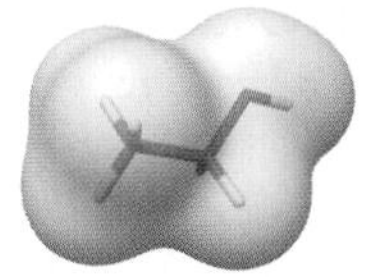

9 Density isosurface, C_2H_5OH

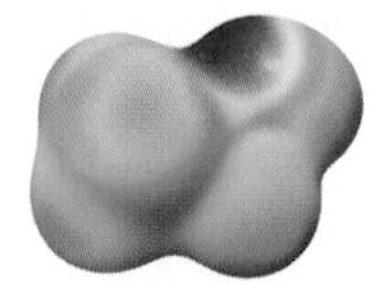

10 Elpot surface, C_2H_5OH

A molecular formula shows the composition of a molecule in terms of the numbers of atoms of each element present.

FUNDAMENTALS

EXAMPLE C.1 Drawing structural formulas from condensed structural formulas and stick formulas

Draw the structural formulas of (a) *N*-methylglucamine,

$$CH_2(OH)CH(OH)CH(OH)CH(OH)CH(OH)CH_2(NHCH_3)$$

which is used to make dyes and pharmaceuticals, and (b) cyclohexene (**11**), which is found in coal tar.

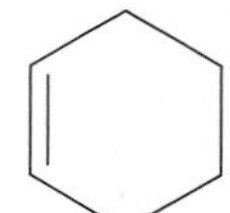

11 Cyclohexene

STRATEGY (a) Insert single bonds between all atoms, unless double bonds are indicated. A group of atoms enclosed in parentheses is attached to the preceding atom through the first atom in the group, unless the parenthetical group is at the beginning of the chain, in which case it is attached to the following atom. (b) Draw a C atom at the end of each line segment. Add enough H atoms to give each C atom four bonds (a double bond counts as two for this purpose).

SOLUTION (a) Draw a chain of six C atoms, with each H atom and the groups in parentheses attached to the preceding C atom (**12**). (b) Draw six C atoms in a hexagon, with a double line between two C atoms and single lines between the others. Add two H atoms to the four C atoms with only single bonds and one H atom to each of the two C atoms connected with a double bond (**13**).

```
    H   H   H   H   H   H   H   H
    |   |   |   |   |   |   |   |
H — C — C — C — C — C — C — N — C — H
    |   |   |   |   |   |       |
    OH  OH  OH  OH  OH  H       H
```

12 N-Methylglucamine

```
          H H
           \/
   H       C        H
    \    /   \     /
      C        C — H
      ||       |
      C        C — H
    /   \    /   \
   H       C       H
           /\
          H H
```

13 Cyclohexene, C_6H_{10}

OH

14 2-Propanol

Self-Test C.1A Draw the structural formulas of (a) 2,2-dimethylbutane, $(CH_3)_3CCH_2CH_3$, which is obtained from petroleum, and (b) 2-propanol (**14**), which is commonly sold as isopropyl alcohol.

[*Answer:* (a) (**15**); (b) (**16**)]

```
             H
             |
    H    H — C — H   H   H
    |        |       |   |
H — C ————— C ————— C — C — H
    |        |       |   |
    H    H — C — H   H   H
             |
             H
```

15 2,2-Dimethylbutane

```
            H
           /
    H   O   H
    |   |   |
H — C — C — C — H
    |   |   |
    H   H   H
```

16 2-Propanol

O

17 Diethyl ether

Self-Test C.1B Draw the structural formulas of (a) 2-methyl-1-butanethiol, $CH_3CH_2CH(CH_3)CH_2SH$, which is used to synthesize organic compounds; and (b) diethyl ether (**17**), which has been used as a surgical anesthetic.

C.3 Ions and Ionic Compounds

If we want to imagine what ionic compounds look like, we have to visualize a huge number of cations and anions stacked together in a regular three-dimensional formation. The ions bond together by the attraction between their opposite charges. Each crystal of sodium chloride, for example, consists of an orderly array of a vast number of alternating Na^+ and Cl^- ions (Fig. C.2). When you take a pinch of salt, each crystal you are picking up consists of more ions than there are stars in the visible universe.

The nuclear model of the atom readily explains the existence of **monatomic ions** (single-atom ions). When an electron is removed from a neutral atom, the charge of the remaining electrons no longer fully cancels the positive charge of the nucleus (Fig. C.3). Because an electron has one unit of negative charge, each electron removed from a neutral atom leaves behind a cation with one additional unit of positive charge. For example, a sodium cation, Na^+, is a sodium atom that has lost one electron. When a calcium atom loses two electrons, it becomes the doubly positively charged calcium ion, Ca^{2+}.

Each electron gained by an atom increases the negative charge by one unit (Fig. C.4). So, when a fluorine atom gains an electron, it becomes the singly negatively charged fluoride ion, F^-. When an oxygen atom gains two electrons, it becomes the doubly charged oxide ion, O^{2-}. When a nitrogen atom gains three electrons, it becomes the triply charged nitride ion, N^{3-}.

The periodic table can help us decide what type of ion an element forms and what charge to expect the ion to have. Fuller details will be given in Chapters 1 and 2, but we can begin to see the patterns. One major pattern is that metallic elements—those toward the left of the periodic table—typically form cations by electron loss. Nonmetallic elements—those toward the right of the table—typically form anions by gaining electrons. Thus, the alkali metals form cations and the halogens form anions.

Figure C.5 shows another pattern in the charges of ions. For elements in Groups 1 and 2, for instance, the charge of the ion is equal to the group number. Thus, cesium in Group 1 forms Cs^+ ions; barium in Group 2 forms Ba^{2+} ions. Why this is so will become clear in Chapter 2. Figure C.5 also shows that atoms of the *d*-block elements and some of the heavier metals of Groups 13/III and 14/IV can form cations with different

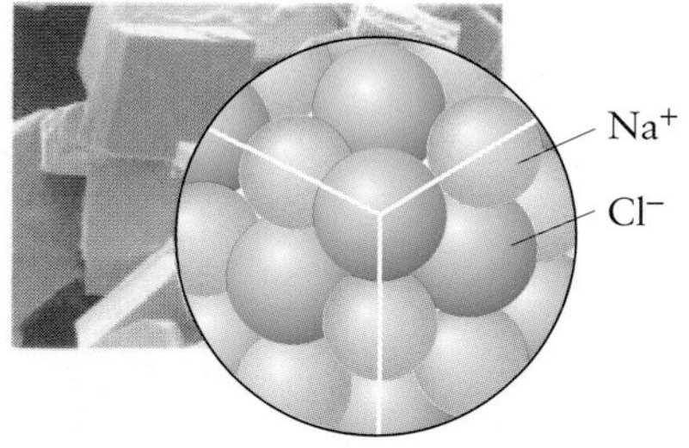

FIGURE C.2 An ionic solid consists of an array of cations and anions stacked together. This illustration shows the arrangement of sodium cations (Na^+) and chlorine anions (chloride ions, Cl^-) in a crystal of sodium chloride (common table salt). The faces of the crystal are where the stacks of ions come to an end.

FUNDAMENTALS

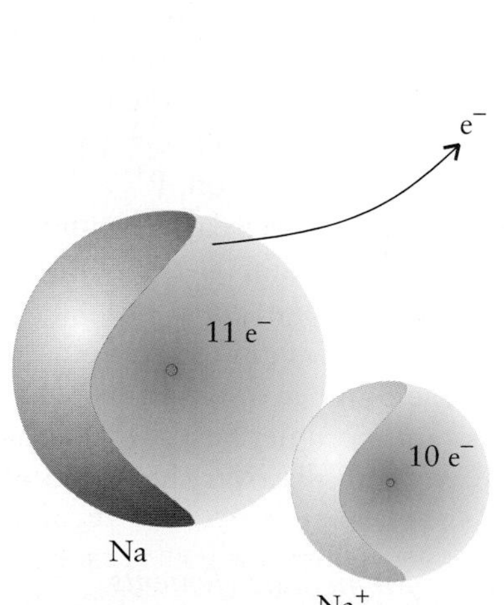

FIGURE C.3 A neutral sodium atom (left) consists of a nucleus that contains 11 protons and is surrounded by 11 electrons. When one electron is lost, the remaining 10 electrons cancel only 10 of the proton charges, and the resulting ion (right) has one overall positive charge.

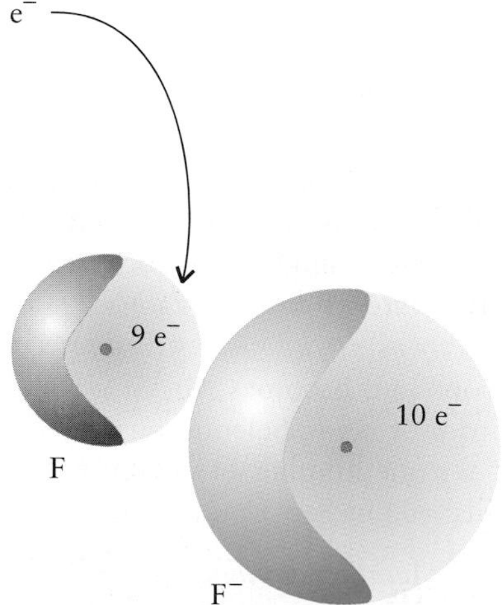

FIGURE C.4 A neutral fluorine atom (left) consists of a nucleus that contains 9 protons and is surrounded by 9 electrons. When the atom gains 1 electron, the 9 proton charges cancel all but one of the 10 electron charges, and the resulting ion (right) has one overall negative charge.

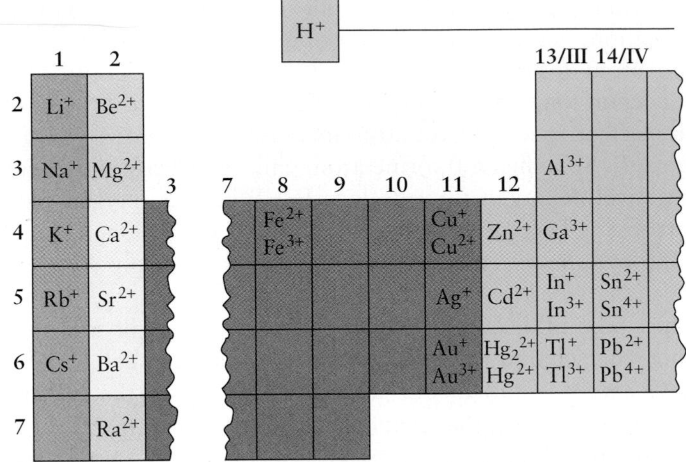

FIGURE C.5 The typical cations formed by a selection of elements in the periodic table. The transition metals form a wide variety of cations; we have shown only a few.

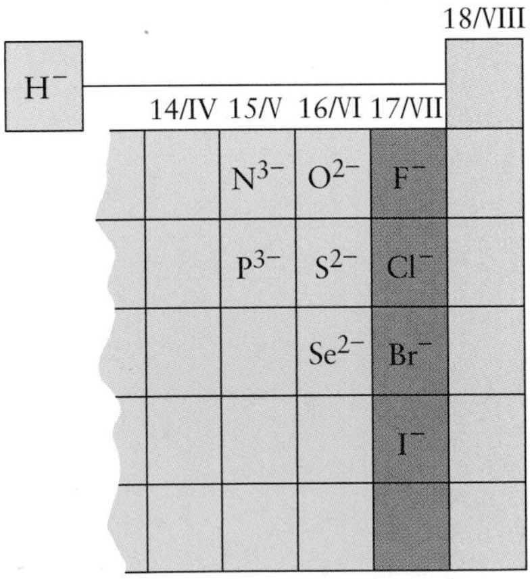

FIGURE C.6 The typical monatomic anions formed by a selection of elements in the periodic table. Only the nonmetals are shown, for only they form monatomic anions.

charges. An iron atom, for instance, can lose two electrons to become Fe^{2+} or three electrons to become Fe^{3+}. Copper can lose either one electron to form Cu^{+} or two electrons to become Cu^{2+}. Later in the text, we shall see why some elements form cations with various charges.

Some of the anions formed by the elements are listed in Fig. C.6. We see another pattern emerging: an element on the right of the table forms an anion with a negative charge equal to its distance from the noble gas group. Specifically, if the group number of the element is N (in the 1 to 18 system), the charge of the anions it forms is $N - 18$. Thus, oxygen is two groups away from the noble gases and forms the oxide ion, O^{2-}; phosphorus, which is three groups away, forms the phosphide ion, P^{3-}.

The pattern of ion formation by main-group elements can be summarized by a single rule: for atoms toward the left or right of the periodic table, *atoms lose or gain electrons until they have the same number of electrons as the nearest noble gas atom.* Thus, magnesium loses two electrons and becomes Mg^{2+}, which has the same number of electrons as an atom of neon. Selenium gains two electrons and becomes Se^{2-}, which has the same number of electrons as krypton. We shall discover the origin of this rule in Section 2.3.

Metallic elements typically form cations, nonmetallic elements typically form anions; the charges of monatomic ions are related to the group to which they belong in the periodic table.

How can we use this information . . . to predict the likely charge of an ion?

(a) Nitrogen (N) is a nonmetal, so it can be expected to form an anion. It belongs to Group 15/V. Because $15 - 18 = -3$, it can be expected to form the anion N^{3-}. (b) Calcium (Ca) is a metal, so it tends to form cations. It belongs to Group 2 and forms the cation Ca^{2+}, the calcium ion.

Self-Test C.2A What ions are (a) iodine and (b) aluminum likely to form?
[*Answer:* (a) Iodide ion, I^-; (b) aluminum ion, Al^{3+}]

Self-Test C.2B What ions are sulfur and potassium likely to form?

Many ions are **diatomic,** meaning that they consist of two atoms bonded together, or **polyatomic,** meaning that they consist of three or more atoms bonded together. In each case, they have an overall positive or negative charge. For example, the cyanide ion, CN^-, is diatomic and the ammonium ion, NH_4^+, is polyatomic. The most common polyatomic anions are the **oxoanions,** polyatomic anions that contain oxygen. They include the carbonate, CO_3^{2-}, nitrate NO_3^-, phosphate, PO_4^{3-}, and sulfate, SO_4^{2-}, anions. A carbonate ion (**18**), for example, has three O atoms bonded to a C atom, and an additional two electrons. Similarly, a phosphate ion (**19**) consists of four O atoms bonded to a P atom, with three additional electrons.

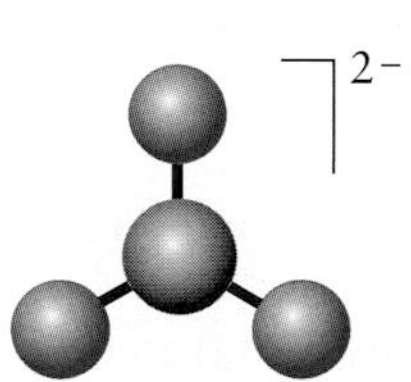

18 Carbonate ion, CO_3^{2-}

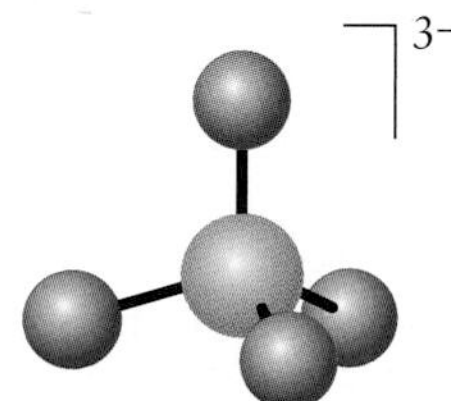

19 Phosphate ion, PO_4^{3-}

The formulas of ionic compounds and molecular compounds have a different significance. Each crystal of sodium chloride has a different total number of cations and anions. We cannot simply specify the numbers of ions present as the formula of this ionic compound, for each crystal would have a different formula and the subscripts would be enormous numbers. However, the *ratio* of the number of cations to the number of anions is the same in all the crystals, and the chemical formula shows this ratio. In sodium chloride, there is one Na^+ ion for each Cl^- ion, so its formula is NaCl. Sodium chloride is an example of a **binary ionic compound,** a compound formed from the ions of two elements.

The formulas of compounds containing polyatomic ions follow similar rules. In sodium carbonate, there are two Na^+ (sodium) ions per CO_3^{2-} (carbonate) ion, so its formula is Na_2CO_3. When a subscript has to be added to a polyatomic ion, the ion is written within parentheses, as in $(NH_4)_2SO_4$, where $(NH_4)_2$ means that there are two NH_4^+ (ammonium) ions for each SO_4^{2-} (sulfate) ion in ammonium sulfate. In each case, the ions combine in such a way that the positive and negative charges cancel: *all compounds are electrically neutral overall.*

We do not usually speak of a "molecule" of an ionic compound. However, it is useful to be able to refer to a representative group of ions with the number of atoms given by the formula. This group of ions is called a **formula unit.** For example, the formula unit of sodium chloride, NaCl, consists of one Na^+ ion and one Cl^- ion; and the formula unit of ammonium sulfate, $(NH_4)_2SO_4$, consists of two NH_4^+ ions and one SO_4^{2-} ion.

Clusters of ions, such as Na^+Cl^-, do exist in the gas phase, where they are commonly termed "ion pairs" rather than molecules.

Occasionally, you will find formula units referred to as "molecules." Take your instructor's advice on this point.

The chemical formula of an ionic compound shows the ratio of the numbers of atoms of each element present in the compound in terms of the smallest whole numbers of ions. A formula unit of an ionic compound is a group of ions with the same number of each element as appears in its formula.

How can we use this information . . . to write the likely formula of a binary ionic compound?

We can predict the formula of a binary ionic compound (one made from two elements), such as magnesium and phosphorus, if we know the charges on the ions. We obtain the charges from the locations of the elements in the periodic table. Magnesium, for example, is in Group 2, so it forms ions with charge +2. Phosphorus is in Group 15/V and forms ions with charge $15 - 18 = -3$. Because a compound is electrically neutral, the ions must combine in such a way that their charges cancel. Thus, we need three Mg^{2+} ions to produce a charge of +6 and two P^{3-} ions to produce a charge of −6. The formula of the compound is therefore Mg_3P_2.

Self-Test C.3A Write the formula of the binary ionic compound formed by (a) calcium and chlorine; (b) aluminum and oxygen.

[***Answer:*** (a) $CaCl_2$; (b) Al_2O_3]

Self-Test C.3B Write the formula of the binary ionic compound formed by (a) lithium and nitrogen; (b) strontium and bromine.

SKILLS YOU SHOULD HAVE MASTERED

1 Distinguish among molecules, ions, and atoms.

2 Identify compounds as organic or inorganic and as molecular or ionic.

3 Use various means of representing molecules and write formulas from molecular structures, Example C.1.

4 Predict the anion or cation a main-group element is likely to form.

5 Interpret chemical formulas in terms of the number of each type of atom present.

6 Predict the formulas of binary ionic compounds.

EXERCISES

C.1 What is meant by (a) an ionic compound and (b) a molecular compound? What are the typical properties of the two classes of compounds?

C.2 Can an element be a molecular substance? Explain.

C.3 Draw the structural formulas of (a) 2,2,4-trimethylpentane, $(CH_3)_3CCH_2CH(CH_3)_2$, which is also called isooctane and is used to raise the octane rating of gasoline; (b) 2-methyl-1-propanol, which is commonly sold as isobutyl alcohol and is used to make fruit flavorings.

OH

2-Methyl-1-propanol

C.4 Draw the structural formulas of (a) 2-ethyl-1,3-hexanediol,

$$CH_3CH_2CH_2CH(OH)CH(CH_2CH_3)CH_2OH,$$

which is used as an insect repellent; and (b) 3-methyl-1-butanol, also known as isoamyl alcohol and used to determine the fat content of milk.

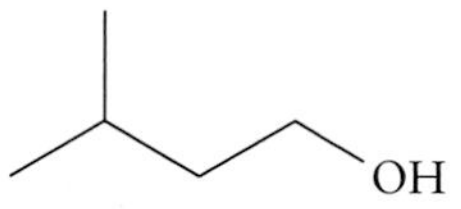

OH

3-Methyl-1-butanol

C.5 State whether or not the following elements are more likely to form a cation or an anion and write the formula for the ion most likely to be formed: (a) cesium; (b) iodine; (c) selenium; (d) calcium.

C.6 State whether or not the following elements are more likely to form a cation or an anion and write the formula for the ion most likely to be formed: (a) sulfur; (b) lithium; (c) phosphorus; (d) oxygen.

C.7 How many protons, neutrons, and electrons are present in (a) $^{4}He^{2+}$; (b) $^{15}N^{3-}$; (c) $^{127}I^{-}$; (d) $^{80}Se^{2-}$?

C.8 How many protons, neutrons, and electrons are present in (a) $^{107}Ag^{+}$; (b) $^{140}Ce^{4+}$; (c) $^{17}O^{2-}$; (d) $^{40}Ca^{2+}$?

C.9 Write the symbol of the ion that has (a) 9 protons, 10 neutrons, and 10 electrons; (b) 12 protons, 12 neutrons, and 10 electrons; (c) 52 protons, 76 neutrons, and 54 electrons; (d) 37 protons, 49 neutrons, and 36 electrons.

C.10 Write the symbol of the ion that has (a) 11 protons, 13 neutrons, and 10 electrons; (b) 13 protons, 14 neutrons, and 10 electrons; (c) 34 protons, 45 neutrons, and 36 electrons; (d) 24 protons, 28 neutrons, and 22 electrons.

C.11 Write the formula of a compound formed by combining (a) Al and Te; (b) Mg and O; (c) Na and S; (d) Rb and I.

C.12 Write the formula of a compound formed by combining (a) Na and P; (b) Al and Cl; (c) Sr and F; (d) Li and Se.

D THE NOMENCLATURE OF COMPOUNDS

Every field of science has a language that enables specialists to communicate concisely but precisely. The business of learning chemistry includes mastering the language that chemists use, particularly the rules for naming compounds.

Many compounds were given informal, **common names** before their compositions were known. Common names include water, salt, sugar, ammonia, and quartz. A **systematic name,** on the other hand, reveals which elements are present and, in some cases, the arrangement of atoms. The systemic naming of compounds, which is called **chemical nomenclature,** follows a set of rules, so that the name of each compound need not be memorized, only the rules.

D.1 Names of Cations

The name of a monatomic cation is the same as the name of the element forming it, with the addition of the word *ion*, as in sodium ion for Na^+. When an element can form more than one kind of cation, such as Cu^+ and Cu^{2+} from copper, we use the **oxidation number,** the charge of the cation, written as a Roman numeral in parentheses. Thus, Cu^+ is a copper(I) ion and Cu^{2+} is a copper(II) ion. Similarly, Fe^{2+} is an iron(II) ion and Fe^{3+} is an iron(III) ion. Figure C.5 showed a selection of elements that form cations with more than one charge. Most transition metals form more than one kind of ion, so it is usually necessary to include the oxidation number in the names of their compounds.

In chemical nomenclature, the oxidation number is sometimes called the Stock number for the German chemist Alfred Stock, who devised this numbering system. Oxidation numbers are discussed in more detail in Sections K and 2.9.

Some older systems of nomenclature are still in use. For example, some cations were once denoted by the endings *-ous* and *-ic* for the ions with lower and higher charges, respectively. To make matters worse, these endings were in some cases added to the Latin form of the element's name. Thus, iron(II) ions were called ferrous ions and iron(III) ions were called ferric ions (see Appendix 3C). We do not use this system in this text, but you will sometimes come across it and should be aware of it.

The name of a monatomic cation is the name of the element plus the word ion, except that, for elements that can form more than one type of cation, the oxidation number, a Roman numeral indicating the charge, is included.

D.2 Names of Anions

Monatomic anions, like the Cl^- ions that occur in sodium chloride and the O^{2-} ions that occur in *quicklime* (CaO), are named by adding the suffix *-ide* and the word *ion* to the first part of the name of the element (the "stem" of its name), as shown in the list of anions in Table D.1; thus, S^{2-} is a sulfide ion and O^{2-} is an oxide ion. There is usually no need to give the charge, because most elements that form monatomic anions form only one kind of ion. The ions formed by the halogens are collectively called *halide ions* and include fluoride (F^-), chloride (Cl^-), bromide (Br^-), and iodide (I^-) ions.

We first met polyatomic ions in Section C. Many of the most important and common polyatomic anions are the oxoanions, which are ions that contain oxygen (Table D.1). If only one oxoanion of an element exists, its name is formed by adding the suffix *-ate* to the stem of the

TABLE D.1 Common Anions and Their Parent Acids

Anion	Parent acid
fluoride ion, F^-	hydrofluoric acid,* HF (hydrogen fluoride)
chloride ion, Cl^-	hydrochloric acid,* HCl (hydrogen chloride)
bromide ion, Br^-	hydrobromic acid,* HBr (hydrogen bromide)
iodide ion, I^-	hydroiodic acid,* HI (hydrogen iodide)
oxide ion, O^{2-}	water, H_2O
hydroxide ion, OH^-	
sulfide ion, S^{2-}	hydrosulfuric acid,* H_2S (hydrogen sulfide)
hydrogen sulfide ion, HS^-	
cyanide ion, CN^-	hydrocyanic acid,* HCN (hydrogen cyanide)
acetate ion, $CH_3CO_2^-$	acetic acid, CH_3COOH
carbonate ion, CO_3^{2-}	carbonic acid, H_2CO_3
hydrogen carbonate (bicarbonate) ion, HCO_3^-	
nitrite ion, NO_2^-	nitrous acid, HNO_2
nitrate ion, NO_3^-	nitric acid, HNO_3
phosphate ion, PO_4^{3-}	phosphoric acid, H_3PO_4
hydrogen phosphate ion, HPO_4^{2-}	
dihydrogen phosphate ion, $H_2PO_4^-$	
sulfite ion, SO_3^{2-}	sulfurous acid, H_2SO_3
hydrogen sulfite ion, HSO_3^-	
sulfate ion, SO_4^{2-}	sulfuric acid, H_2SO_4
hydrogen sulfate ion, HSO_4^-	
hypochlorite ion, ClO^-	hypochlorous acid, HClO
chlorite ion, ClO_2^-	chlorous acid, $HClO_2$
chlorate ion, ClO_3^-	chloric acid, $HClO_3$
perchlorate ion, ClO_4^-	perchloric acid, $HClO_4$

* The name of the aqueous solution of the compound. The name of the compound itself is in parentheses.

name of the element, as in the carbonate ion, CO_3^{2-}. Some elements can form two types of oxoanions, with different numbers of oxygen atoms, so we need names that distinguish them. Nitrogen, for example, forms both NO_2^- and NO_3^-. In such cases, the ion with the larger number of oxygen atoms is given the suffix *-ate,* and that with the smaller number of oxygen atoms is given the suffix *-ite*. Thus, NO_3^- is nitrate and NO_2^- is nitrite. The nitrate ion is an important source of nitrogen for plants and is included in some fertilizers (such as ammonium nitrate, NH_4NO_3).

Some elements—particularly the halogens—form more than two kinds of oxoanions. The name of the oxoanion with the smallest number of oxygen atoms is formed by adding the prefix *hypo-* to the *-ite* form of the name, as in the *hypo*chlor*ite* ion, ClO^-. The oxoanion with more oxygen atoms than the *-ate* oxoanion is named with the prefix *per-* added to the *-ate* form of the name. An example is the *per*chlor*ate* ion, ClO_4^-. As can be seen from Table D.1, chlorine forms oxoanions that span the range. The rules for naming polyatomic ions are summarized in Appendix 3A.

Hypo **comes from the Greek word for "under."**

Per **is the Latin word for "all over," suggesting that the element's ability to combine with oxygen is finally satisfied.**

Hydrogen is present in some anions. Two examples are HS^- and HCO_3^-. The names of these anions begin with "hydrogen." Thus, HS^- is the hydrogen sulfide ion and HCO_3^- is the hydrogen carbonate ion. You might also see the name of the ion written as a single word, as in hydrogencarbonate ion. In an older system of nomenclature, which is still quite widely used, an anion containing hydrogen is named with the prefix *bi-*, as in bicarbonate ion for HCO_3^-.

An **oxoacid** is an acidic molecular compound that contains oxygen. Oxoacids are the parents of the oxoanions in the sense that an oxoanion is formed by removing one or more hydrogen ions from an oxoacid molecule (see Table D.1). Conversely, the formulas of oxoacids are derived from those of the corresponding oxoanions by allowing enough hydrogen ions to combine with the oxoanion to balance the charges. In general, *-ic* oxoacids are the parents of *-ate* oxoanions and *-ous* oxoacids are the parents of *-ite* oxoanions. For example, the sulfate ion, SO_4^{2-}, needs two H^+ ions to cancel its negative charge, so sulfuric acid is the molecular compound H_2SO_4. Similarly, the sulfite ion, SO_3^{2-}, also needs two H^+ ions, so its parent acid is the molecular compound H_2SO_3, sulfurous acid.

Names of monatomic anions end in -ide. Oxoanions are anions that contain oxygen. Oxoacids are molecular acids that contain oxygen. Within a series of oxoacids, the suffixes -ate and -ic acid indicate a greater number of oxygen atoms than the suffixes -ite and -ous acid.

Self-Test D.1A Write (a) the name of HIO and (b) the formula of carbonic acid.

[*Answer:* (a) Hypoiodous acid; (b) H_2CO_3]

Self-Test D.1B Write (a) the name of H_3PO_4 and (b) the formula of chloric acid.

D.3 Names of Ionic Compounds

An ionic compound is named with the cation name first, followed by the name of the anion; the word *ion* is omitted in each case. The oxidation number of the cation is given if more than one charge is possible. However, if the cation comes from an element that exists in only one charge state (as listed in Fig. C.5), then the oxidation number is omitted. Typical names include potassium chloride (KCl), a compound containing K^+ and Cl^- ions; and ammonium nitrate (NH_4NO_3), which contains NH_4^+ and NO_3^- ions. The copper chloride that contains Cu^+ ions (CuCl) is called copper(I) chloride; and the chloride that contains Cu^{2+} ions ($CuCl_2$) is called copper(II) chloride.

Some ionic compounds form crystals that incorporate a definite proportion of molecules of water as well as the ions of the compound itself. These compounds are called **hydrates.** For example, copper(II) sulfate normally occurs as blue crystals of composition $CuSO_4 \cdot 5H_2O$. The raised dot in this formula is used to separate the *water of hydration* from the rest of the formula. This formula indicates that there are five H_2O molecules for each $CuSO_4$ formula unit. Hydrates are named by first giving the name of the compound, then adding the word *hydrate* with a Greek prefix indicating how many molecules of water are found in each formula unit (Table D.2). For example, the name of $CuSO_4 \cdot 5H_2O$, the common blue form of this compound, is copper(II) sulfate pentahydrate. When this compound is heated, the water of hydration is lost and the blue crystals crumble to a white powder (Fig. D.1). The white powder is $CuSO_4$ itself. When we wish to emphasize that the compound has lost its water of hydration, we call it **anhydrous.** Thus, $CuSO_4$ is anhydrous copper(II) sulfate. The distinction can be commercially important: if you buy sodium sulfate decahydrate, $Na_2SO_4 \cdot 10H_2O$, instead of its anhydrous form, Na_2SO_4, you are buying a lot of water!

Ionic compounds are named by giving the name of the cation (with its oxidation number if more than one charge is possible), followed by the name of the anion; hydrates are named by adding the word hydrate, preceded by a Greek prefix indicating the number of water molecules in the formula unit.

TABLE D.2 Prefixes Used for Naming Compounds

Prefix	Meaning
mono-	1
di-	2
tri-	3
tetra-	4
penta-	5
hexa-	6
hepta-	7
octa-	8
nona-	9
deca-	10
undeca-	11
dodeca-	12

FUNDAMENTALS

The Greek prefixes in the table are used throughout chemistry, so it is a good idea to learn them now.

EXAMPLE D.1 Naming ionic compounds

Give the systematic names of (a) $CrCl_3 \cdot 6H_2O$ and (b) $Ba(ClO_4)_2$.

STRATEGY First identify the cation and the anion, by referring to Table D.1 or Appendix 3A, if necessary. If the cation is a transition metal or belongs to Groups 12 through 15/V, give its oxidation number. To determine the oxidation number, note the charge of the anion and the number of anions and cations in the formula; then decide what cation charge is required to cancel the total negative charge. If water molecules are included in the formula, then the compound is a hydrate, so add the word *hydrate* and a Greek prefix corresponding to the number in front of H_2O in the formula.

SOLUTION (a) The cation is a chromium ion and the anion is a chloride ion, Cl^-. For the compound to be electrically neutral, the cation must be Cr^{3+}; so chromium (a transition metal) is present as chromium(III). The $6H_2O$ indicates that the compound is a hexahydrate; so the compound is chromium(III) chloride hexahydrate. (b) The cation is a barium ion, Ba^{2+} (the only cation that barium forms). The anion is a perchlorate ion, ClO_4^-. Hence, the compound is barium perchlorate.

Self-Test D.2A Name the compounds (a) $FeCl_2 \cdot 2H_2O$; (b) $AlBr_3$; (c) $Cr(ClO_3)_2$.

[*Answer:* (a) Iron(II) chloride dihydrate; (b) aluminum bromide; (c) chromium(II) chlorate]

Self-Test D.2B Name the compounds (a) $AuCl_3$; (b) CaS; (c) Mn_2O_3.

Media Link **Lab video:** Fig D.1, Dehydration of copper(II) sulfate pentahydrate

FIGURE D.1 Blue crystals of copper(II) sulfate pentahydrate ($CuSO_4 \cdot 5H_2O$) lose water above 150°C and form a white anhydrous powder ($CuSO_4$). The color is restored when water is added; and, in fact, anhydrous copper sulfate has such a strong attraction for water that it is usually colored a very pale blue from reaction with the water in air.

D.4 Names of Inorganic Molecular Compounds

Many simple inorganic molecular compounds are named by using the Greek prefixes in Table D.2 to indicate the number of each type of atom present. Usually, we do not use a prefix if only one atom of an element is present; for example, NO_2 is nitrogen dioxide. An important exception to this rule is carbon monoxide, CO. Most of the common binary molecular

TABLE D.3 Common Names for Some Simple Molecular Compounds

Formula*	Common name
NH_3	ammonia
N_2H_4	hydrazine
NH_2OH	hydroxylamine
PH_3	phosphine
NO	nitric oxide
N_2O	nitrous oxide
C_2H_4	ethylene
C_2H_2	acetylene

* For historical reasons, the molecular formulas of binary hydrogen compounds of Group 15/V elements are written with the Group 15/V element first.

compounds—molecular compounds built from two elements—have at least one element from Group 16/VI or 17/VII. These elements are named second, with their endings changed to *-ide:*

phosphorus trichloride PCl_3 dinitrogen oxide N_2O
sulfur hexafluoride SF_6 dinitrogen pentoxide N_2O_5

Some exceptions to these rules are the phosphorus oxides and compounds that are generally known by their common names. The phosphorus oxides are distinguished by the oxidation number of phosphorus, which is calculated as though phosphorus were a metal and the oxygen present as O^{2-}. Thus, P_4O_6 is thought of as $(P^{3+})_4(O^{2-})_6$ and named phosphorus(III) oxide, and P_4O_{10} is thought of as $(P^{5+})_4(O^{2-})_{10}$ and named phosphorus(V) oxide. These compounds, though, are molecular. Certain binary molecular compounds, such as NH_3 and H_2O, have widely used common names (Table D.3).

Both the names and the molecular formulas of compounds formed between hydrogen and nonmetals in Groups 16/VI or 17/VII are written with the H atom first: for example, the formula for hydrogen chloride is HCl and that for hydrogen sulfide is H_2S. All the oxoacids are written in this way, as illustrated in Table D.1. Note that, when these compounds are dissolved in water, many act as acids and are named as acids. Binary acids are named by adding the prefix *hydro-* and changing the ending of the element name to *-ic acid,* as in hydrochloric acid for HCl in water and hydrosulfuric acid for H_2S in water. An **aqueous solution**, a solution in water, is indicated by (aq) following the formula.

Binary molecular compounds are named by using Greek prefixes to indicate the number of atoms of each element present; the element named second has its ending changed to -ide.

Self-Test D.3A Name the compounds (a) N_2O; (b) BCl_3; (c) IF_5.
[*Answer:* (a) Dinitrogen oxide; (b) boron trichloride; (c) iodine pentafluoride]

Self-Test D.3B Name the compounds (a) PCl_3; (b) SO_3; (c) HBr(aq).

EXAMPLE D.2 Writing the formula of a binary compound from its name

Write the formulas of (a) cobalt(II) chloride hexahydrate; (b) diboron trisulfide; (c) silicon tetrachloride.

STRATEGY First check to see whether the compounds are ionic or molecular. Many compounds that contain a metal are ionic. The symbol of the metal is written first, followed by the symbol of the nonmetal. The charge on the cation (the metal ion) is governed by its position in the periodic table if it is a Group 1 or 2 metal, or aluminum, or gallium; the charge is indicated by the oxidation number written as a Roman numeral in parentheses, if the metal ion can have more than one charge. The charge on the anion is governed by its group in the periodic table. Subscripts are chosen to balance charges. Compounds of two nonmetals are normally molecular.

Their formulas are written by listing the symbols of the elements in the same order as in the name and giving them subscripts corresponding to the Greek prefixes used.

SOLUTION (a) The combination of metal and nonmetal tells us that the compound is ionic. The II in cobalt(II) signifies that the metal cobalt has a charge of +2 in the compound. Chlorine is in Group 17/VII, so the chloride ion has a charge of 17 − 18 = −1. We can balance the charge of the +2 cation with two −1 anions, so the formula is $CoCl_2$. This compound is a hydrate, so we then add the six water molecules, as indicated by the Greek prefix *hexa-*; the complete formula is therefore $CoCl_2{\cdot}6H_2O$. The subscripts for the individual atoms in a molecular compound are taken from the Greek prefixes, so we write (b) B_2S_3 and (c) $SiCl_4$.

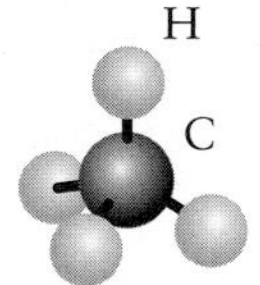

1 Methane, CH_4

Self-Test D.4A Write the formulas for (a) vanadium(V) oxide; (b) beryllium carbide; (c) sulfur tetrafluoride; (d) dinitrogen trioxide.

[*Answer:* (a) V_2O_5; (b) Be_2C; (c) SF_4; (d) N_2O_3]

Self-Test D.4B Write the formulas for (a) cesium sulfide tetrahydrate; (b) manganese(VII) oxide; (c) hydrogen cyanide (a poisonous gas); (d) disulfur dichloride.

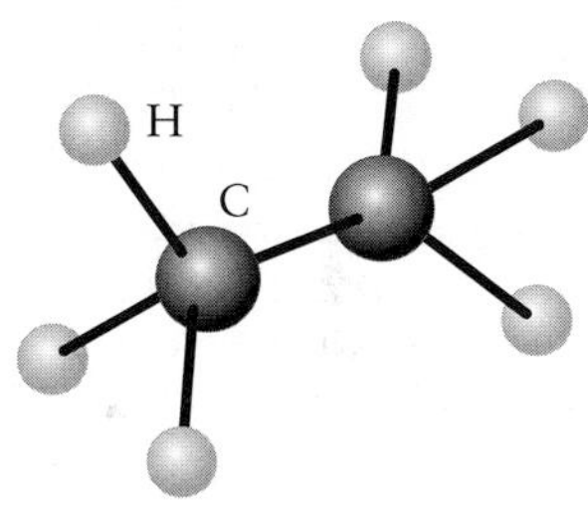

2 Ethane, C_2H_6

D.5 Names of Organic Compounds

Organic compounds are named by using a different set of rules. There are millions of organic compounds; many consist of highly intricate molecules, so their names can be very complicated. You could, for example, find yourself asking for α-D-glucopyranosyl(1→2)-β-D-fructofuranose when all you wanted was sugar (sucrose). However, for most of this text, you will need to know only a few simple organic compounds. Chapters 18 and 19 present a more complete introduction to the nomenclature of organic compounds.

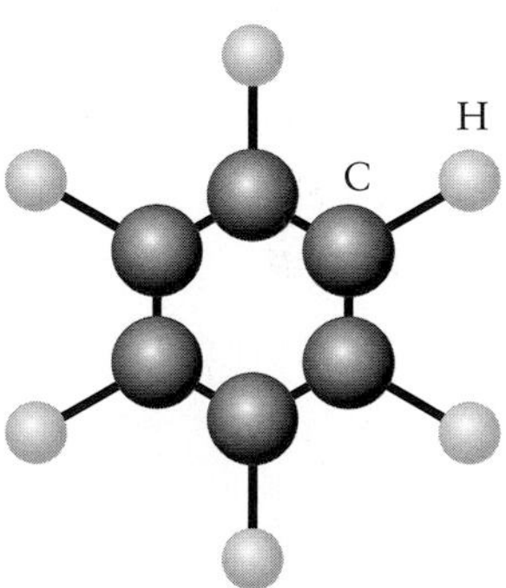

3 Benzene, C_6H_6

Compounds of hydrogen and carbon are called **hydrocarbons.** They include methane, CH_4 (**1**); ethane, C_2H_6 (**2**); and benzene, C_6H_6 (**3**). Hydrocarbons that have no carbon-carbon multiple bonds are called **alkanes.** Thus, methane and ethane are both alkanes. The unbranched alkanes with up to 12 carbon atoms are listed in Table D.4. Notice that Greek prefixes are used to name all the alkanes with five or more carbon atoms. Hydrocarbons with double bonds are called **alkenes.** Ethene, $CH_2{=}CH_2$, is the simplest example of an alkene. It used to be (and still widely is) called ethylene; it is used in the manufacture of polyethylene. Benzene is a hydrocarbon with double bonds, and it has such distinct properties that it is regarded as the parent hydrocarbon of a whole new class of compounds called—for historical reasons—**aromatic compounds.** The benzene ring is exceptionally stable and can be found in many important compounds.

The hydrocarbons are the basic framework for all organic compounds, with one or more of the hydrogen atoms replaced by other atoms or groups of atoms. For example, ethanol, CH_3CH_2OH (**4**), the "alcohol" of beer and wine, is an ethane molecule in which one H atom has been replaced by an —OH group. In organic chemistry, the name of one

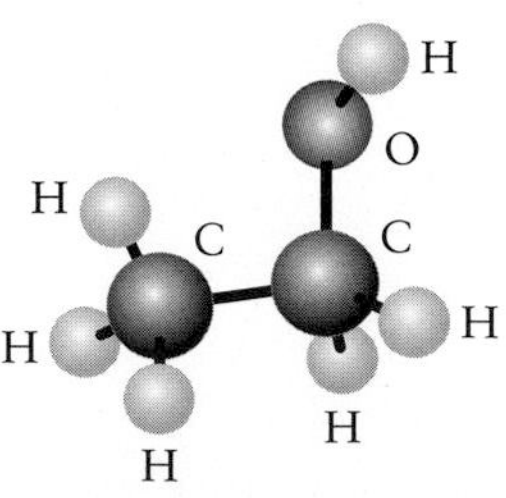

4 Ethanol, CH_3CH_2OH

FUNDAMENTALS

TABLE D.4 Alkane Nomenclature

Number of carbon atoms	Formula	Name of alkane	Name of alkyl group
1	CH_4	methane	methyl
2	CH_3CH_3	ethane	ethyl
3	$CH_3CH_2CH_3$	propane	propyl
4	$CH_3(CH_2)_2CH_3$	butane	butyl
5	$CH_3(CH_2)_3CH_3$	pentane	pentyl
6	$CH_3(CH_2)_4CH_3$	hexane	hexyl
7	$CH_3(CH_2)_5CH_3$	heptane	heptyl
8	$CH_3(CH_2)_6CH_3$	octane	octyl
9	$CH_3(CH_2)_7CH_3$	nonane	nonyl
10	$CH_3(CH_2)_8CH_3$	decane	decyl
11	$CH_3(CH_2)_9CH_3$	undecane	undecyl
12	$CH_3(CH_2)_{10}CH_3$	dodecane	dodecyl

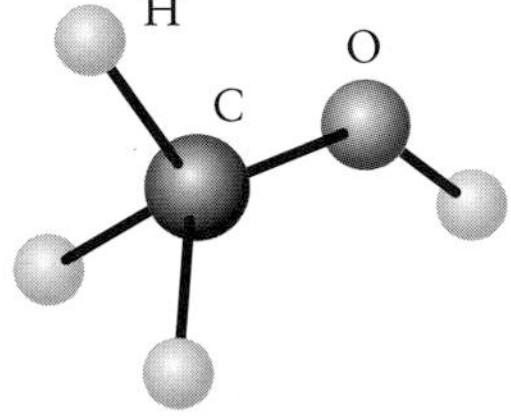

5 Methanol, CH_3OH

compound often inspires the name for a whole class of related compounds. Thus, an **alcohol** is an organic compound that contains an —OH group; and CH_3OH (**5**) is methanol, or wood alcohol. The sugars glucose and sucrose (cane sugar) have molecules in which several —OH groups are present, so they can be regarded as a special type of alcohol.

Note how the names of alcohols are obtained by replacing the final *-e* of the name of the parent hydrocarbon by the suffix *-ol*. Thus, we get the name ethanol from ethane and methanol from methane. Alcohols can also be named by regarding the hydrocarbon chain as a group attached to the rest of the molecule. To name an alkane as a group, replace the ending by *-yl*, as in $-CH_3$, methyl, and $-CH_2CH_3$, ethyl. Thus, CH_3CH_2OH is also called ethyl alcohol.

Another important class of organic compounds that we shall meet frequently, even in the early chapters of the text, are the **carboxylic acids.** These compounds are characterized by the **carboxyl group,** —COOH (**6**). As their names suggest, these compounds are acids. The most common example is acetic acid, CH_3COOH (7), the acid that gives vinegar its sharp taste. Another simple carboxylic acid is formic acid, HCOOH (**8**), the acid of ant venom. These names are the common names we shall use throughout the text. The systematic names of carboxylic acids are presented in Chapter 19.

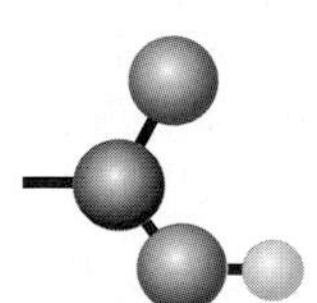
6 Carboxyl group, —COOH

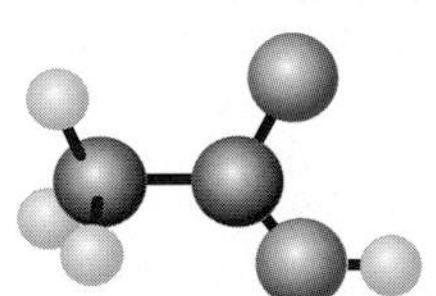
7 Acetic acid, CH_3COOH

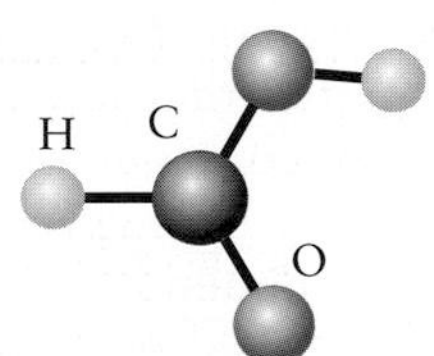

8 Formic acid, HCOOH

Another type of organic compound can be formed by replacing hydrogen atoms in hydrocarbons by halogen atoms. The resulting "haloalkanes" include chloromethane, CH_3Cl (**9**), and trichloromethane, $CHCl_3$ (**10**). The latter compound has the common name chloroform and was an early anesthetic. Note how one of the Greek prefixes in Table D.2 is used here to show the number of halogen atoms present in the molecule.

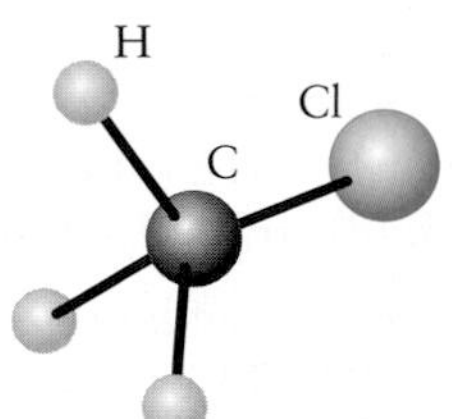

9 Chloromethane, CH_3Cl

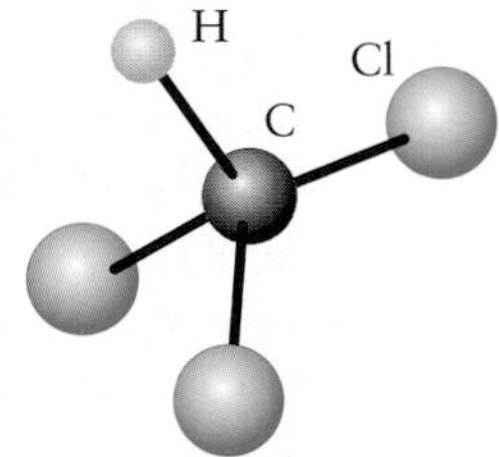

10 Trichloromethane, $CHCl_3$

The names of organic compounds are based on the names of the parent hydrocarbons; alcohols contain —OH groups, carboxylic acids contain —COOH groups, and haloalkanes contain halogen atoms.

Self-Test D.5A (a) Name CH_2BrCl. (b) What type of compound is $CH_3CH(OH)CH_3$?

[*Answer:* (a) Bromochloromethane; (b) alcohol]

Self-Test D.5B (a) Name $CH_3CH_2CH_2CH_2CH_3$. (b) What type of compound is CH_3CH_2COOH?

SKILLS YOU SHOULD HAVE MASTERED

1 Name ions, binary inorganic compounds, oxoacids, compounds with common polyatomic ions, and hydrates, and write their formulas, Self-Tests D.1 and D.3, and Examples D.1 and D.2.

2 Name simple hydrocarbons, Section D.5.

3 Identify alcohols and carboxylic acids from their formulas, Self-Test D.5.

EXERCISES

D.1 Write the formula of (a) aluminum oxide (alumina); (b) strontium phosphate; (c) aluminum carbonate; (d) lithium nitride.

D.2 Write the formula of (a) barium hydroxide; (b) aluminum phosphate; (c) zinc chloride; (d) ferric chloride.

D.3 Name the following ionic compounds. Write both the old and modern names wherever appropriate. (a) $Ca_3(PO_4)_2$, the major inorganic component of bones; (b) SnF_2; (c) V_2O_5; (d) Cu_2O.

D.4 The following ionic compounds are commonly found in laboratories. Write the modern name of (a) $NaHCO_3$ (baking soda); (b) Hg_2Cl_2 (calomel): (c) NaOH (lye); (d) ZnO (calamine).

D.5 Write the formula of (a) titanium dioxide; (b) silicon tetrachloride; (c) carbon disulfide; (d) sulfur tetrafluoride; (e) lithium sulfide; (f) antimony pentafluoride; (g) dinitrogen pentoxide; (h) iodine heptafluoride.

D.6 Write the formula of (a) dinitrogen tetroxide: (b) hydrogen sulfide; (c) dichlorine heptoxide; (d) nitrogen triiodide; (e) sulfur dioxide; (f) hydrogen fluoride; (g) diiodine hexachloride.

D.7 Name each binary molecular compound: (a) SF_6; (b) N_2O_5; (c) NI_3; (d) XeF_4; (e) $AsBr_3$; (f) ClO_2.

D.8 The following compounds are often found in chemical laboratories. Name each compound: (a) SiO_2 (silica); (b) SiC (carborundum); (c) N_2O (a general anesthetic); (d) P_4O_{10} (a drying agent for organic solvents); (e) CS_2 (a solvent); (f) SO_2 (a bleaching agent); (g) NH_3 (a common reagent).

D.9 The following aqueous solutions are common laboratory acids. What are their names? (a) HCl(aq); (b) H_2SO_4(aq); (c) HNO_3(aq); (d) CH_3COOH(aq); (e) H_2SO_3(aq); (f) H_3PO_4(aq).

D.10 The following acids are also used in chemical laboratories, although they are less common than those in the preceding exercise. Write the formula of (a) perchloric acid; (b) hypochlorous acid; (c) hypoiodous acid; (d) hydrofluoric acid; (c) phosphorous acid; (f) periodic acid.

D.11 Write the formula for the ionic compound formed from (a) sodium and oxide ions; (b) potassium and sulfate ions; (c) silver and fluoride ions; (d) zinc and nitrate ions; (e) aluminum and sulfide ions.

D.12 Write the formula for the ionic compound formed from (a) calcium and chloride ions; (b) iron(III) and sulfate ions; (c) ammonium and iodide ions; (d) lithium and sulfide ions; (e) calcium and phosphide ions.

D.13 Name each compound: (a) Na_2SO_3; (b) Fe_2O_3; (c) FeO; (d) $Mg(OH)_2$; (e) $NiSO_4{\cdot}6H_2O$; (f) PCl_5; (g) $Cr(H_2PO_4)_3$; (h) As_2O_3; (i) $RuCl_2$.

D.14 Name each compound: (a) $CrCl_3{\cdot}6H_2O$; (b) $Co(NO_3)_2{\cdot}6H_2O$; (c) TlCl; (d) BrCl; (e) MnO_2; (f) $Hg(NO_3)_2$; (g) $Ni(ClO_4)_2$; (h) $Ca(OCl)_2$; (i) Nb_2O_5.

D.15 Name each organic compound: (a) $CH_3CH_2CH_2CH_2CH_2CH_2CH_3$; (b) $CH_3CH_2CH_3$; (c) $CH_3CH_2CH_2CH_2OH$; (d) $CH_3CH_2CH_2Cl$.

D.16 Name each organic compound: (a) CH_4; (b) $CH_3CH_2CH_2CH_3$; (c) CH_3CH_2Br; (d) $CH_3CH_2CH_2OH$.

D.17 Name each compound, using analogous compounds with phosphorus and sulfur as a guide: (a) H_2TeO_4; (b) Na_3AsO_4; (c) $CaSeO_3$; (d) $Ba_3(SbO_4)_2$; (e) H_3AsO_4; (f) $Co_2(TeO_4)_3$.

D.18 Name each compound, using Appendix 3A: (a) UO_2Cl_2; (b) NaN_3; (c) CaC_2O_4; (d) VOS; (e) $K_2Cr_2O_7$; (f) LiSCN.

E MOLES AND MOLAR MASSES

In most of their calculations, chemists are more interested in the numbers of atoms, ions, or molecules present than in the mass of the sample or the volume of the solution. Knowing the *types* of atoms is fundamental to *qualitative* chemistry—understanding the properties of substances, for instance. Knowing the *numbers* of atoms is fundamental to *quantitative* chemistry—the calculation of the values of these properties.

It is inconvenient to refer to large numbers like 2.0×10^{25} atoms, just as wholesalers find it inconvenient to report individual items instead of dozens (12) or gross (144). To keep track of the enormous numbers of atoms, ions, or molecules in a sample, we need an efficient way of determining and reporting these numbers.

E.1 The Mole

Chemists report numbers of atoms, ions, and molecules in terms of a unit called a "mole." A mole is the analogue of the wholesaler's "dozen." A "dozen" could be defined as the number of soda cans in a "twelve-pack" carton supplied by a wholesaler. Even if you could not open the carton to count the number of cans inside, you could determine the meaning of "1 doz cans" by weighing the carton and dividing the mass of the carton by the mass of one can. A similar approach is used to define a **mole** (abbreviated mol):

The name *mole* comes from the Latin word for "massive heap." Coincidentally, the animal of the same name makes massive heaps of soil on lawns.

1 mole is the number of atoms in exactly 12 g of carbon-12.

So, to tell someone what we mean by 1 mol, we could give them 12 g of carbon-12 and invite them to count the atoms (Fig. E.1). However, counting atoms directly is impractical, so we use the indirect route based on the mass of one atom. The mass of a carbon-12 atom has been found—by mass spectrometry—to be $1.992\,65 \times 10^{-23}$ g. It follows that the number of atoms in exactly 12 g of carbon-12 is

$$\text{Number of carbon-12 atoms} = \frac{12 \text{ g}}{1.9926 \times 10^{-23} \text{ g}} = 6.0221 \times 10^{23}$$

Because the mole tells us the number of atoms in a sample, it follows that 1 mol of atoms of any element is 6.0221×10^{23} atoms of the element (Fig. E.2). The same is true of 1 mol of any objects, including atoms, ions, and molecules:

1 mol of objects means 6.0221×10^{23} of those objects.

Just as 1 g and 1 m are units for physical properties, so too is the unit 1 mol. The mole is the unit for the physical property formally called the **amount of substance**, n. However, that name has found little favor among chemists, who colloquially refer to n as "the number of moles." A compromise, which is gaining acceptance, is to refer to n as the "chemical amount" or simply "amount" of entities present in a sample. Thus, 1.0000 mol of hydrogen atoms (6.0221×10^{23} hydrogen atoms), which is written 1.0000 mol H, is the chemical amount of hydrogen atoms in the sample. Take the advice of your instructor on whether or not to use the official term. Like any SI unit, the mole can be used with prefixes. For

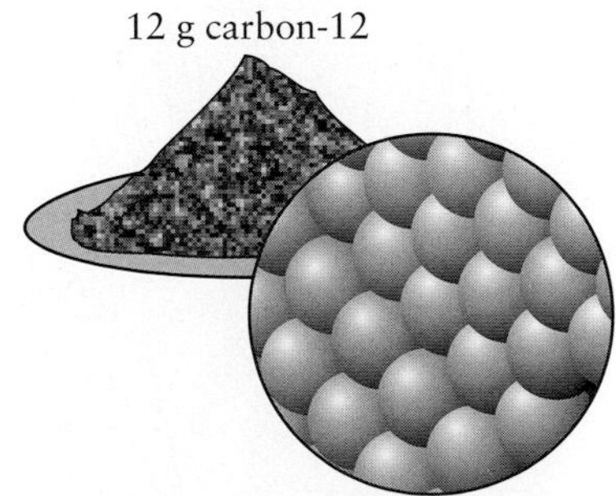

FIGURE E.1 The definition of the mole: if we measure out exactly 12 g of carbon-12, then we have exactly 1 mol of carbon-12 atoms. There will be exactly an Avogadro's number of atoms in the pile.

FIGURE E.2 Each sample consists of 1 mol of atoms of the element. Clockwise from the upper right are 32 g of sulfur, 201 g of mercury, 207 g of lead, 64 g of copper, and 12 g of carbon.

example, 1 mmol = 10^{-3} mol and 1 μmol = 10^{-6} mol. Chemists encounter small quantities like these when dealing with rare or expensive natural products.

Note that the Avogadro constant is a constant with units, not a pure number. You will often hear people referring to *Avogadro's number:* they mean the pure number 6.0221×10^{23}.

The number of objects per mole, $6.0221 \times 10^{23}\ \text{mol}^{-1}$, is called the **Avogadro constant,** N_A, in honor of the nineteenth-century Italian scientist Amedeo Avogadro (Fig. E.3), who helped to establish the existence of atoms. The Avogadro constant is used to convert between the chemical amount (the number of moles) and the number of atoms, ions, or molecules in that amount:

$$\begin{aligned}\text{Number of objects} &= \text{number of moles} \times \text{number of objects per mole}\\ &= \text{number of moles} \times N_A\end{aligned}$$

If we denote the number of objects by N and the amount of substance (in moles) by n, this relation is written

$$N = nN_A \qquad (1)$$

There is an important point to remember when working with moles, just as there is when working with dozens. To be unambiguous when using moles, it is important to be specific about which species (that is, which atoms, molecules, formula units, or ions) we mean. For example, hydrogen occurs naturally as a gas, with each molecule built from two atoms, which is why it is denoted H_2. We write 1 mol H if we mean hydrogen atoms or 1 mol H_2 if we mean hydrogen molecules.

The amounts of atoms, ions, or molecules in a sample are expressed in moles, and the Avogadro constant, N_A, is used to convert between numbers of these particles and the numbers of moles.

FIGURE E.3 Lorenzo Romano Amedeo Carlo Avogadro, Count of Quaregna and Cerreto (1776–1856).

EXAMPLE E.1 Converting number of atoms to moles

Suppose a sample of vitamin C is known to contain 1.29×10^{24} hydrogen atoms (as well as other kinds of atoms). What is the chemical amount (in moles) of hydrogen atoms in the sample?

STRATEGY A good strategy in chemistry is to estimate an approximate answer before going through the calculation because we can then detect major errors quickly. In this case, because the number of atoms in the sample is greater than 6×10^{23}, we anticipate that more than 1 mol of atoms is present. For the calculation, we rearrange Eq. 1 to give us the quantity we want (the number of moles) in terms of the information given (the number of hydrogen atoms) and then substitute the data:

$$n = \frac{N}{N_A}, \quad \text{or} \quad \text{Amount of H atoms (mol)} = \frac{\text{number of H atoms}}{N_A}$$

SOLUTION Substituting the data (the number of H atoms) and the value of the Avogadro constant gives

$$\text{Amount of H atoms (mol)} = \frac{1.29 \times 10^{24}\ \text{H}}{6.0221 \times 10^{23}\ \text{mol}^{-1}} = 2.14\ \text{mol H}$$

Self-Test E.1A A sample of a drug extracted from a fruit used by the Achuar Jivaro people of Peru to treat fungal infections is found to contain 2.58×10^{24} oxygen atoms. What is the chemical amount (in moles) of O atoms in the sample?

[*Answer:* 4.28 mol O]

Self-Test E.1B A small cup of coffee contains 3.14 mol of H_2O molecules. What is the number of H atoms present?

E.2 Molar Mass

How can we determine the amount of substance present if we can't count the atoms directly? We can find the amount if we know the mass of the sample and the **molar mass**, *M*, the mass per mole of particles. All we have to do to find the number of moles, *n*, is to divide the total mass, *m*, of the sample by the mass per mole:

$$\text{Amount (mol)} = \frac{\text{mass of sample}}{\text{mass per mole}}, \quad \text{or} \quad n = \frac{m}{M} \qquad (2)$$

How can we use this information . . . to determine the chemical amount of atoms in a sample?
To find the amount (in moles) of F atoms in 22.5 g of fluorine, we need to know the molar mass of fluorine (that is, the mass per mole of F atoms), which is 19.00 g·mol^{-1}. Then

$$\text{Amount of F atoms (mol)} = \frac{22.5\ \text{g}}{19.00\ \text{g·mol}^{-1}} = 1.18\ \text{mol}$$

To calculate the actual number of atoms in the sample, we multiply the amount by the Avogadro constant:

$$N = (1.18 \text{ mol F}) \times (6.022 \times 10^{23} \text{ atoms}\cdot\text{mol}^{-1})$$
$$= 7.11 \times 10^{23} \text{ F atoms}$$

Self-Test E.2A The mass of a copper coin is 3.20 g. Suppose it were pure copper. (a) How many moles of Cu atoms would the coin contain, given a molar mass of Cu of 63.55 g·mol^{-1}? (b) How many Cu atoms are present?

[*Answer:* 0.0504 mol Cu; 3.03×10^{22} Cu atoms]

Self-Test E.2B In one day, 5.4 kg of aluminum was collected from a recycling bin. (a) How many moles of Al atoms did the bin contain, given that the molar mass of Al is 26.98 g·mol^{-1}? (b) How many Al atoms are present?

The *molar mass of an element* is the mass per mole of its atoms; the *molar mass of a molecular compound* is the mass per mole of its molecules; the *molar mass of an ionic compound* is the mass per mole of its formula units. The units of molar mass in each case are grams per mole (g·mol^{-1}).

The molar masses of elements are determined by using mass spectrometry to measure the masses of the individual isotopes and their abundances. The mass per mole of atoms is the mass of an individual atom multiplied by the Avogadro constant (the number of atoms per mole). However, there is a complication. Most elements occur in nature as a mixture of isotopes; we saw in Section B, for instance, that neon occurs as three isotopes, each with a different mass. In chemistry, we almost always deal with natural samples of elements, which have the natural abundance of isotopes. So, we need the *average* molar mass, the molar mass calculated by taking into account the masses of the isotopes and their relative abundances in typical samples. *All molar masses quoted in this text refer to these average values.* Their values are given in Appendix 2D. They are also included in the periodic table inside the front cover and in the alphabetical list of elements inside the back cover.

EXAMPLE E.2 Evaluating an average molar mass

There are two naturally occurring isotopes of chlorine, chlorine-35 and chlorine-37. The mass of an atom of chlorine-35 is 5.807×10^{-23} g and that of an atom of chlorine-37 is 6.139×10^{-23} g. In a typical natural sample of chlorine, 75.77% of the sample is chlorine-35 and 24.23% is chlorine-37. What is the molar mass of a typical sample of chlorine?

STRATEGY First, calculate the average atomic mass of the isotopes by adding together the individual masses, each multiplied by the fraction that represents its abundance. Then obtain the molar mass, the mass per mole of atoms, by multiplying the average atomic mass by the Avogadro constant.

SOLUTION The average mass of an atom of chlorine in a natural sample is

$$\begin{aligned}\text{Average mass of a Cl atom}\\ &= 0.7577 \times (5.807 \times 10^{-23}\ \text{g}) + 0.2423 \times (6.139 \times 10^{-23}\ \text{g})\\ &= (4.400 \times 10^{-23}\ \text{g}) + (1.487 \times 10^{-23}\ \text{g})\\ &= 5.887 \times 10^{-23}\ \text{g}\end{aligned}$$

It follows that the molar mass of a typical sample of chlorine atoms is

$$\begin{aligned}\text{Molar mass of chlorine}\\ &= \text{average mass of a Cl atom} \times \text{number of Cl atoms per mole}\\ &= (5.887 \times 10^{-23}\ \text{g}) \times (6.022 \times 10^{23}\ \text{mol}^{-1}) = 35.45\ \text{g·mol}^{-1}\end{aligned}$$

Self-Test E.3A In a typical sample of magnesium, 78.99% is magnesium-24 (with atomic mass 3.983×10^{-23} g), 10.00% magnesium-25 (4.149×10^{-23} g), and 11.01% magnesium-26 (4.315×10^{-23} g). Calculate the molar mass of a typical sample of magnesium, given the atomic masses (in parentheses).

[*Answer:* 24.31 g·mol^{-1}]

Self-Test E.3B Calculate the molar mass of copper, given that a natural sample typically consists of 69.17% copper-63, which has a molar mass of 62.94 g·mol^{-1}, and 30.83% copper-65, which has a molar mass of 64.93 g·mol^{-1}.

To calculate the molar masses of molecular and ionic compounds, we use the molar masses of the elements present: *the molar mass of a compound is the sum of the molar masses of the elements that make up the molecule or the formula unit.* We need only note how many times each atom or ion appears in the molecular formula or the formula unit of the ionic compound. For example, the molar mass of the ionic compound Na_2SO_4 is

$$\begin{aligned}\text{Molar mass of } Na_2SO_4\\ &= 2(\text{molar mass of Na}) + (\text{molar mass of S}) + 4(\text{molar mass of O})\\ &= 2(22.99\ \text{g·mol}^{-1}) + (32.07\ \text{g·mol}^{-1}) + 4(16.00\ \text{g·mol}^{-1})\\ &= 142.05\ \text{g·mol}^{-1}\end{aligned}$$

Self-Test E.4A Calculate the molar mass of (a) ethanol, C_2H_5OH; (b) copper(II) sulfate pentahydrate.

[*Answer:* (a) 46.07 g·mol^{-1}; (b) 249.70 g·mol^{-1}]

Self-Test E.4B Calculate the molar mass of (a) phenol, C_6H_5OH; (b) sodium carbonate decahydrate.

Two terms still widely used throughout the chemical literature are *atomic weight* and *molecular weight*. The **atomic weight** of an element is the numerical value of its molar mass. For instance, the atomic weight of hydrogen (molar mass 1.0079 g·mol^{-1}) is 1.0079, and the atomic weight of chlorine (molar mass 35.45 g·mol^{-1}) is 35.45. Similarly, the **molecular weight** of a molecular compound or the **formula weight** of an ionic compound is the numerical value of its molar mass. The molecular weight of water (molar mass 18.02 $g \cdot mol^{-1}$) is 18.02, and the formula weight of sodium chloride (molar mass 58.44 $g \cdot mol^{-1}$) is 58.44. These two terms are traditional and deeply ingrained in chemical conversations, even though the numbers are not "weights." The mass of an object is a measure of the *quantity of matter* it contains, whereas the weight of an object is a measure of the *gravitational pull* it experiences. Mass and weight are proportional to each other, but they are not identical. An astronaut would have the same mass (contain the same quantity of matter) but different weights on Earth and on Mars.

Once the molar mass of a compound has been calculated, we can use the same technique used for elements to determine how many moles of molecules or formula units are in a sample of a given mass.

Self-Test E.5A Calculate the amount of urea, $OC(NH_2)_2$, molecules in 2.3×10^5 kg of urea, which is used in facial creams and, on a bigger scale, as an agricultural fertilizer.

[*Answer:* 3.8×10^6 mol]

Self-Test E.5B Calculate the amount of $Ca(OH)_2$ formula units in 1.00 kg of slaked lime (calcium hydroxide), which is used to adjust the acidity of soils.

Molar mass is important when we need to know the number of atoms in a sample. It would be impossible to count out 6×10^{23} atoms of an element, but it is very easy to measure out a mass equal to the molar mass of the element in grams. For instance, because the molar mass of iron is 55.85 $g \cdot mol^{-1}$, the mass of 1.000 mol Fe is

$$\text{Mass of iron} = (1.000 \text{ mol}) \times (55.85 \text{ g} \cdot \text{mol}^{-1}) = 55.85 \text{ g}$$

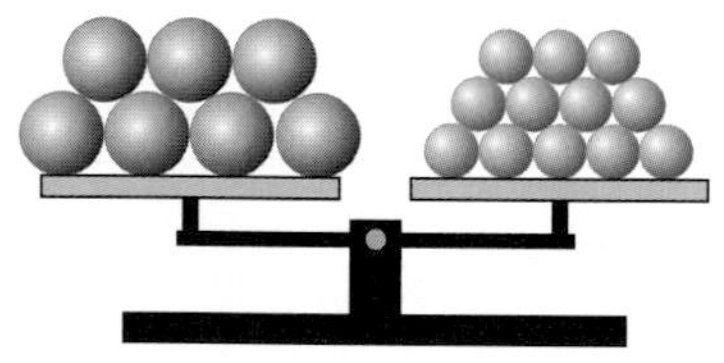

(a) Equal masses

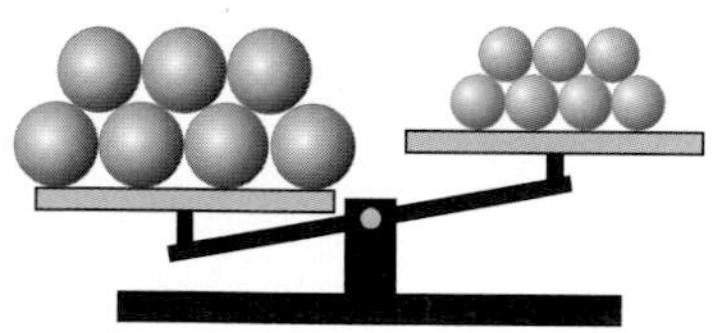

(b) Equal amounts

FIGURE E.4 (a) The two samples have the same mass, but because the atoms on the right are lighter, the sample on the right consists of a greater number of atoms. (b) The two samples contain the same number of atoms, but because the atoms on the right are lighter than those on the left, the mass of the sample on the right is the smaller of the two. Equal amounts (equal numbers of moles) of atoms do not necessarily have the same mass.

Therefore, to obtain 1.000 mol Fe, we weigh out 55.85 g of iron. Similarly, to obtain 1.000 mol Al, we weigh out 26.98 g of aluminum, and so on. Each of the samples shown in Fig. E.2 was obtained in this way: each sample contains the same number of atoms of the element (6.022×10^{23}), but the masses vary because the masses of the atoms are different (Fig. E.4). The same rule applies to compounds. Hence, if we measure out 58.44 g of sodium chloride, we obtain a sample that contains 1.000 mol NaCl formula units (Fig. E.5).

If we want an amount other than 1 mol, we use the molar mass to convert from the stated number of moles to the mass required:

$$\text{Mass} = \text{amount (in moles)} \times \text{mass per mole, or } m = nM \qquad (3)$$

In practice, chemists rarely try to measure out an exact mass. Instead, they estimate the mass required and spoon out that mass approximately. Then they measure the mass of the sample precisely and convert it to moles to find the precise amount they have obtained.

The molar mass of a compound, the mass per mole of its molecules or formula units, is used to convert between the mass of a sample and the amount of molecules or formula units it contains.

FIGURE E.5 Each sample contains 1 mol of formula units of an ionic compound. From left to right are 58 g of sodium chloride (NaCl), 100 g of calcium carbonate ($CaCO_3$), 278 g of iron(II) sulfate heptahydrate ($FeSO_4{\cdot}7H_2O$), and 78 g of sodium peroxide (Na_2O_2).

How can we use this information . . . to calculate the mass of sample corresponding to a given amount in moles?

Suppose we are preparing a solution of potassium permanganate, $KMnO_4$, and need about 0.10 mol of the compound. Because the molar mass of potassium permanganate is 158.04 $g{\cdot}mol^{-1}$, this amount corresponds to

$$\text{Mass of } KMnO_4 = (0.10 \text{ mol}) \times (158.04 \text{ g}{\cdot}\text{mol}^{-1}) = 16 \text{ g}$$

So, we measure out about 16 g of $KMnO_4$. If, when we weigh the sample, we find that its actual mass is 14.87 g, then we can conclude that we actually weighed out

$$\text{Moles of } KMnO_4 = \frac{14.87 \text{ g}}{158.04 \text{ g}{\cdot}\text{mol}^{-1}} = 0.094\ 09 \text{ mol}$$

Self-Test E.6A What mass of anhydrous sodium hydrogen sulfate should you weigh out to obtain about 0.20 mol $NaHSO_4$?

[*Answer:* About 24 g]

Self-Test E.6B What mass of acetic acid should you weigh out to obtain 1.5 mol CH_3COOH?

SKILLS YOU SHOULD HAVE MASTERED

1 Use the Avogadro constant to convert between amount in moles and the number of atoms, molecules, or ions in a sample, Example E.1.

2 Calculate the molar mass of an element, given its isotopic composition, Example E.2.

3 Calculate the molar mass of a compound, given its chemical formula.

4 Convert between mass and amount in moles by using the molar mass.

EXERCISES

E.1 (a) The approximate population of Earth is 6.0 billion people. How many moles of people inhabit Earth? (b) If all people were pea pickers and pea counters, then how long would it take for the entire population of the Earth to count out 1 mol of peas at the rate of one pea per second, working 24 hours per day, 365 days per year?

E.2 (a) About 1000 metric tons of sand contain about a trillion (10^{12}) grains of sand. How many metric tons of sand are needed to provide 1 mol of grains of sand? (b) Assuming the volume of a grain of sand is 1 mm^3 and the land area of the continental United States is 3.6×10^6 mi^2, how deep would the sand pile over the United States be if this area were evenly covered with 1 mol of grains of sand?

E.3 The nuclear power industry extracts ^6Li but not ^7Li from natural samples of lithium. As a result, the molar mass of commercial samples of lithium is increasing. The current abundances of the two isotopes are 7.42% and 92.58%, respectively, and the masses of their atoms are 9.988×10^{-24} g and 1.165×10^{-23} g, respectively. (a) What is the current molar mass of a natural sample of lithium? (b) What will the molar mass be when the abundance of ^6Li is reduced to 5.67%?

E.4 Calculate the molar mass of sulfur in a natural sample, which consists of 95.0% ^{32}S (molar mass 31.97 $g{\cdot}mol^{-1}$), 0.8% ^{33}S (molar mass 32.97 $g{\cdot}mol^{-1}$), and 4.2% ^{34}S (molar mass 33.97 $g{\cdot}mol^{-1}$).

E.5 The molar mass of boron atoms in a natural sample is 10.81 $g{\cdot}mol^{-1}$. The sample is known to consist of ^{10}B (molar mass 10.013 $g{\cdot}mol^{-1}$) and ^{11}B (molar mass 11.093 $g{\cdot}mol^{-1}$). What are the percentage abundances of the two isotopes?

E.6 Calculate the molar mass of the noble gas krypton in a natural sample, which consists of 0.3% ^{78}Kr (molar mass 77.92 $g{\cdot}mol^{-1}$), 2.3% ^{80}Kr (molar mass 79.91 $g{\cdot}mol^{-1}$), 11.6% ^{82}Kr (molar mass 81.91 $g{\cdot}mol^{-1}$), 11.5% ^{83}Kr (molar mass 82.92 $g{\cdot}mol^{-1}$), 56.9% ^{84}Kr (molar mass 83.91 $g{\cdot}mol^{-1}$), and 17.4% ^{86}Kr (molar mass 85.91 $g{\cdot}mol^{-1}$).

E.7 Which sample in each of the following pairs contains the greater number of moles of atoms? (a) 75 g of indium or 80 g of tellurium; (b) 15.0 g of P or 15.0 g of S; (c) 7.36×10^{27} atoms of Ru or 7.36×10^{27} atoms of Fe.

E.8 Calculate the mass, in micrograms, of (a) 5.68×10^{15} Hg atoms; (b) 7.924×10^{-9} mol Hf; (c) 3.49 μmol Gd; (d) 6.29×10^{24} Sb atoms.

E.9 What mass of rhodium contains as many atoms as there are (a) nitrogen atoms in 57 g of nitrogen; (b) zirconium atoms in 57 g of zirconium?

E.10 Determine the mass of copper that has the same number of atoms as there are in (a) 7.35 mg of potassium; (b) 7.35 mg of gold.

E.11 Calculate the amount (in moles) and the number of molecules and formula units (or atoms, if indicated) in (a) 10.0 g of alumina, Al_2O_3; (b) 25.92 mg of hydrogen fluoride, HF; (c) 1.55 mg of hydrogen peroxide, H_2O_2; (d) 1.25 kg of glucose, $C_6H_{12}O_6$; (e) 4.37 g of nitrogen as N atoms and as N_2 molecules.

E.12 Convert each of the following masses to amount (in moles) and to number of molecules (and atoms, if indicated). (a) 1.80 kg of H_2O; (b) 49 kg of benzene (C_6H_6); (c) 100.0 g of phosphorus, as P atoms and as P_4 molecules; (d) 5.0 g of CO_2; (e) 5.0 g of NO_2.

E.13 Calculate the amount (in moles) of (a) Ag^+ ions in 2.00 g of AgCl; (b) UO_3 in 6.00×10^2 g of UO_3; (c) Cl^- ions in 4.19 mg of $FeCl_3$; (d) H_2O in 1.00 g of $AuCl_3{\cdot}2H_2O$.

E.14 Calculate the amount (in moles) of (a) CN^- in 1.00 g of KCN; (b) H atoms in 2.00×10^2 mg of H_2O; (c) $CaCO_3$ in 5.00 kg of $CaCO_3$; (d) H_2O in 5.00 g of $La_2(SO_4)_3{\cdot}9H_2O$.

E.15 (a) Determine the number of formula units in 0.750 mol KNO_3. (b) What is the mass (in milligrams) of 2.39×10^{20} formula units of Ag_2SO_4? (c) Estimate the number of formula units in 3.429 g of $NaHCO_2$, sodium formate, which is used in dyeing and printing fabrics.

E.16 (a) How many NaH formula units are present in 5.294 g of CaH_2? (b) Determine the mass of 6.25×10^{24} formula units of $NaBF_4$, sodium tetrafluoroborate. (c) Calculate the amount (in moles) of 9.54×10^{21} formula units of CeI_3, cerium(III) iodide, a bright yellow, water-soluble solid.

E.17 (a) Calculate the mass, in grams, of one water molecule. (b) Determine the number of H_2O molecules in 1.00 kg of H_2O.

E.18 Octane, C_8H_{18}, is typical of the molecules found in gasoline. (a) Calculate the mass of one octane molecule. (b) Determine the number of C_8H_{18} molecules in 1.00 mL of C_8H_{18}, the mass of which is 0.82 g.

E.19 A chemist measured out 7.35 g of copper(II) bromide tetrahydrate, $CuBr_2{\cdot}4H_2O$. (a) How many moles of $CuBr_2{\cdot}4H_2O$ were measured out? (b) How many moles of Br^- are present in the sample? (c) How many water molecules are present in the sample? (d) What fraction of the total mass of the sample was due to oxygen?

E.20 A chemist wants to extract the gold from 35.25 g of gold(III) chloride dihydrate, $AuCl_3{\cdot}2H_2O$, by electrolysis of an aqueous solution (this technique is described in Chapter 12). What mass of gold could be obtained from the sample?

E.21 Metallic copper reacts with elemental sulfur at elevated temperatures to form CuS(s). (a) How many S atoms are required for complete reaction with 43.4 g of Cu? (b) How many sulfur molecules, S_8, are necessary? (c) What mass of sulfur is needed for the reaction?

E.22 Copper metal can be extracted from a copper(II) sulfate solution by electrolysis (as described in Chapter 12). If 29.50 g of copper(II) sulfate pentahydrate, $CuSO_4{\cdot}5H_2O$, is dissolved in 100 mL of water and all the copper is electroplated out, what mass of copper would be obtained?

E.23 Epsom salts consists of magnesium sulfate heptahydrate. Write its formula. (a) How many atoms of oxygen are in 5.15 g of Epsom salts? (b) How many formula units of the compound are present in 5.15 g? (c) How many moles of water molecules are in 5.15 g of Epsom salts?

F DETERMINATION OF CHEMICAL FORMULAS

Many new drugs are discovered by studying the properties of compounds extracted from plants or animals (Fig. F.1). Once a chemist has extracted an active compound from a natural product, the first step in identifying its molecular structure is to determine the empirical and molecular formulas of the compound. The **empirical formula** shows the relative numbers of atoms of each element present in the compound. For example, the empirical formula of glucose, which is CH_2O, tells us that carbon, hydrogen, and oxygen atoms are present in the ratio 1:2:1. The elements are present in these proportions regardless of the size of the sample. After the empirical formula has been determined, the next step is to determine the molecular formula (Section C): a *molecular formula* tells us the *actual* numbers of atoms of each element in a molecule. The molecular formula for glucose, which is $C_6H_{12}O_6$, tells us that each glucose molecule consists of 6 carbon atoms, 12 hydrogen atoms, and 6 oxygen atoms (**1**, which shows one common configuration of the glucose molecule). Note that different compounds with different molecular formulas can have the same empirical formula. For example, formaldehyde, CH_2O (the preservative in formalin solution), acetic acid, $C_2H_4O_2$ (the acid in vinegar), and lactic acid, $C_3H_6O_3$ (the acid in sour milk), all have the same empirical formula (CH_2O) as glucose.

FIGURE F.1 The research vessel *Alpha Helix* is used by chemists at the University of Illinois at Urbana-Champaign to search for marine organisms that contain compounds of medicinal value.

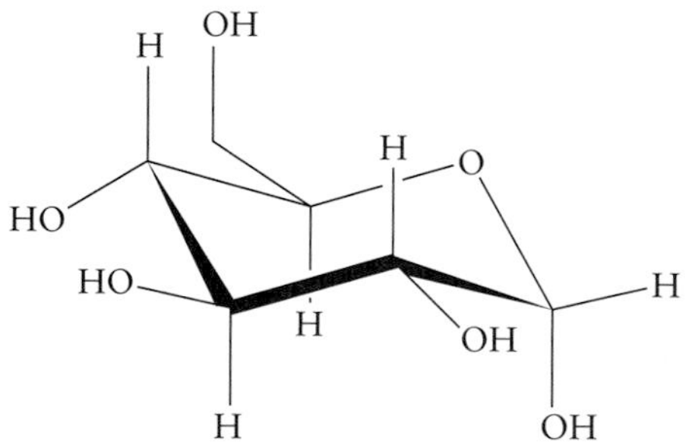

1 α-D-Glucose

F.1 Mass Percentage Composition

To determine the empirical formula of a compound, we begin by measuring the mass of each element present in a sample. The result is usually reported as the **mass percentage composition**, that is, the mass of each element expressed as a percentage of the total mass:

$$\text{Mass percentage of element} = \frac{\text{mass of element in the sample}}{\text{total mass of sample}} \times 100\% \qquad (1)$$

Because the mass percentage composition is independent of the size of the sample—in the language of Section A, it is an intensive property—it is representative of the composition of any sample of the substance. One of the principal techniques for determining the mass percentage composition of an unknown organic compound is combustion analysis. Chemists commonly use a laboratory or agency to carry out combustion analysis on their samples and receive the results as mass percentage composition.

See Section M for a description of combustion analysis.

Self-Test F.1A For centuries, the Australian aborigines have used the leaves of the eucalyptus tree to alleviate sore throats and other pains. The primary active ingredient has been identified and named eucalyptol. The analysis of a sample of eucalyptol of total mass 3.16 g gave its composition as 2.46 g of carbon, 0.373 g of hydrogen, and 0.329 g of oxygen. Determine the mass percentages of carbon, hydrogen, and oxygen in eucalyptol.

[*Answer:* 77.8% C, 11.8% H, 10.4% O]

Self-Test F.1B The compound α-pinene, a natural antiseptic found in the resin of the piñon tree, has been used since ancient times by Zuni healers. A 7.50-g sample of α-pinene contains 6.61 g of carbon and 0.89 g of hydrogen. What are the mass percentages of carbon and hydrogen in α-pinene?

If the formula of a compound is already known, its mass percentage composition can be obtained from the formula.

Mass percentage composition is found by calculating the fraction of the total mass contributed by each element present in a compound and expressing the fraction as a percentage.

How can we use this information . . . to calculate the mass percentage of an element in a compound?

Suppose we are generating hydrogen from water to use as a fuel: we might need to find the mass percentage of hydrogen in water in order to know how much hydrogen a given mass of water could provide. We could simply find the mass of H atoms in 1 mol H_2O molecules, noting that there are 2 mol H in 1 mol H_2O, and divide that mass by the mass of 1 mol H_2O:

$$\begin{aligned}\text{Mass percentage of H} &= \frac{\text{total mass of H atoms}}{\text{mass of H}_2\text{O molecules}} \times 100\% \\ &= \frac{(2\text{ mol}) \times (1.0079\text{ g}\cdot\text{mol}^{-1})}{(1\text{ mol}) \times (18.02\text{ g}\cdot\text{mol}^{-1})} \times 100\% \\ &= 11.19\%\end{aligned}$$

Self-Test F.2A Calculate the mass percentage of Cl in NaCl.

[*Answer:* 60.66%]

Self-Test F.2B Calculate the mass percentage of Ag in $AgNO_3$.

F.2 Determining Empirical Formulas

Suppose we receive the results of a combustion analysis from the laboratory and want to convert the mass percentage composition to an empirical formula. For this step, we need to determine the relative amounts of each type of atom. The simplest procedure is to imagine that we have a sample of mass 100 g exactly. That way, the mass percentage composition tells us the mass in grams of each element. Then we can use the molar mass of each element to convert these masses into amount in moles and go on to find the relative numbers of moles of each type of atom. Let's do that for vitamin C, which was once identified in this way, and suppose that the laboratory has reported that the sample we supplied is 40.9% carbon, 4.58% hydrogen, and 54.5% oxygen.

How can we use this information . . . to calculate the amount in moles for a given mass percentage?

The mass of carbon in a sample of vitamin C of mass 100 g is 40.9 g. Because the molar mass carbon is 12.01 $g{\cdot}mol^{-1}$, we know that

$$\text{Amount of C atoms (mol)} = \frac{40.9\ g}{12.01\ g{\cdot}mol^{-1}} = 3.41\ mol$$

Self-Test F.3A Find the amounts of H and O atoms in the same sample of vitamin C.

[*Answer:* 4.54 mol H, 3.41 mol O]

Self-Test F.3B Use the information in Self-Test F.1A to find the amounts of C, H, and O atoms present in 100 g of eucalyptol.

Because the number of moles is proportional to the number of atoms, we know from the preceding calculation and Self-Test F.3A that atoms of each element are present in vitamin C in the ratio (3.41 C atoms):(4.54 H atoms):(3.41 O atoms), or 3.41:4.54:3.41. To get the empirical formula, we must express the ratios of numbers of atoms as the simplest whole numbers. First, we divide each number by the smallest value (3.41), which gives a ratio of 1.00:1.33:1.00. Molecules contain only whole numbers of atoms and one of these numbers is still not a whole number. Hence we must multiply each number by the correct factor so that all numbers can be rounded off to whole numbers. Because 1.33 is $\frac{4}{3}$ (within experimental error), we multiply through by 3 to obtain 3.00:3.99:3.00, or approximately 3:4:3. Now we know that the empirical formula of vitamin C is $C_3H_4O_3$.

The calculated values are not exact because they are derived from experimental data, which are subject to experimental error.

The empirical formula of a compound is determined from the mass percentage composition and the molar masses of the elements present.

EXAMPLE F.1 Determining the empirical formula from mass percentage composition

The mass percentage composition of a compound that assists in the coagulation of blood is 76.71% C, 7.02% H, and 16.27% N. Determine the empirical formula of the compound.

STRATEGY Mass percentages tell us the mass of each element in 100 g of the compound. Convert each mass percentage to the number of moles found in 100 g of the vitamin by dividing by the molar mass of the element. Divide the number of moles of each element by the smallest number. If fractional numbers result, then multiply by a factor that will give the smallest whole numbers of moles.

SOLUTION The masses of each element in 100 g of the compound are

Carbon	76.71 g
Hydrogen	7.02 g
Oxygen	16.27 g

Convert each mass to number of moles by using the molar mass of the element:

$$\text{Amount of C atoms} = \frac{76.71\ \text{g C}}{12.01\ \text{g·mol}^{-1}} = 6.387\ \text{mol C}$$

$$\text{Amount of H atoms} = \frac{7.02\ \text{g H}}{1.0079\ \text{g·mol}^{-1}} = 6.96\ \text{mol H}$$

$$\text{Amount of N atoms} = \frac{16.27\ \text{g N}}{14.01\ \text{g·mol}^{-1}} = 1.161\ \text{mol N}$$

Divide each amount by the smallest number of moles (1.161 mol):

$$\text{Carbon} \quad \frac{6.387\ \text{mol}}{1.161\ \text{mol}} = 5.501$$

$$\text{Hydrogen} \quad \frac{6.96\ \text{mol}}{1.161\ \text{mol}} = 5.99$$

$$\text{Oxygen} \quad \frac{1.161\ \text{mol}}{1.161\ \text{mol}} = 1.000$$

Noting that 5.501 is approximately $\frac{11}{2}$, we multiply all the numbers by 2 to get the ratios 11.002:12.0:2.000. The empirical formula is therefore $C_{11}H_{12}N_2$.

Self-Test F.4A Use the molar composition of eucalyptol calculated in Self-Test F.3B to determine its empirical formula.

[*Answer:* $C_{10}H_{18}O$]

Self-Test F.4B The mass percentage composition of the compound thionyl difluoride is 18.59% O, 37.25% S, and 44.16% F. Calculate its empirical formula.

F.3 Determining Molecular Formulas

So far, we know that the empirical formula of vitamin C is $C_3H_4O_3$. However, all this formula tells us is that the C, H, and O atoms are present in the sample in the ratio 3:4:3. We do not yet know the numbers of atoms in individual molecules. The same empirical formula would be obtained for $C_3H_4O_3$, $C_6H_8O_6$, $C_9H_{12}O_9$, or any other whole-number multiple of the empirical formula.

Once we have found the empirical formula, we need one more piece of information to find the molecular formula of a compound—its molar mass. Then all we have to do is to calculate how many formula units are needed to account for the molar mass. One of the best ways of determining the molar mass of an organic compound is by mass spectrometry. We saw this technique applied to atoms in Section B; it can be applied to molecules, too; and although there are important changes of detail, the technique is essentially the same. Notice that in going from mass percentage composition to molecular formula, we have used molar masses twice:

More details about how mass spectrometry is used to determine the molar masses of compounds can be found in Major Technique 5, following Chapter 18.

$$\text{Mass percentage} \xrightarrow{\text{molar masses of elements}} \text{empirical formula} \xrightarrow{\text{molar mass of compound}} \text{molecular formula}$$

The molecular formula of a compound is found by determining how many empirical formula units are needed to account for the measured molar mass of the compound.

How can we use this information . . . to determine the molecular formula, given the empirical formula?

Mass spectrometry has been used to show that the molar mass of vitamin C is 176.12 g·mol^{-1}. The molar mass of a $C_3H_4O_3$ formula unit is

$$\begin{aligned}\text{Molar mass of } C_3H_4O_3 &\\ &= 3(12.01\ \text{g·mol}^{-1}) + 4(1.008\ \text{g·mol}^{-1}) + 3(16.00\ \text{g·mol}^{-1})\\ &= 88.06\ \text{g·mol}^{-1}\end{aligned}$$

To find the number of $C_3H_4O_3$ formula units needed to account for the observed molar mass of vitamin C, we divide the molar mass of the molecule by the molar mass of the empirical formula unit:

$$\frac{176.14\ \text{g·mol}^{-1}}{88.06\ \text{g·mol}^{-1}} = 2.000$$

We conclude that the molecular formula of vitamin C is $2 \times (C_3H_4O_3)$, or $C_6H_8O_6$.

Self-Test F.5A The molar mass of styrene, which is used in the manufacture of the plastic polystyrene, is 104 g·mol^{-1}, and its empirical formula is CH. Deduce its molecular formula.

[*Answer:* C_8H_8]

Self-Test F.5B The molar mass of oxalic acid, a toxic substance found in rhubarb leaves, is 90.0 g·mol^{-1}, and its empirical formula is CHO_2. What is its molecular formula?

SKILLS YOU SHOULD HAVE MASTERED

1 Calculate the mass percentage of an element in a compound from a formula.

2 Calculate the empirical formula of a compound from its mass percentage composition, Example F.1.

3 Determine the molecular formula of a compound from its empirical formula and its molar mass.

EXERCISES

F.1 What is the mass percentage composition of testosterone, $C_{19}H_{28}O_2$, a male sex hormone?

F.2 What is the mass percentage composition of aspartame, $C_{14}H_{18}N_2O_5$, an artificial sweetener sold as NutraSweet?

F.3 Determine the chemical formulas from the following analyses. (a) The mass percentage composition of cryolite, a compound used in the production of aluminum, is 32.79% Na, 13.02% Al, and 54.19% F. (b) A compound used to generate O_2 gas in the laboratory has mass percentage composition 31.91% K and 28.93% Cl, the remainder being oxygen. (c) A fertilizer is found to have the following mass percentage composition: 12.2% N, 5.26% H, 26.9% P, and 55.6% O.

F.4 Determine the empirical formula of each compound from the following data. (a) Talc (used in talcum powder) has mass composition 19.2% Mg, 29.6% Si, 42.2% O, and 9.0% H. (b) Saccharin, a sweetening agent, has mass composition 45.89% C, 2.75% H, 7.65% N, 26.20% O, and 17.50% S. (c) Salicylic acid, used in the synthesis of aspirin, has mass composition 60.87% C, 4.38% H, and 34.75% O.

F.5 In an experiment, 4.14 g of the element phosphorus combined with chlorine to produce 27.8 g of a white solid compound. What is the empirical formula of the compound?

F.6 A chemist found that 4.69 g of sulfur combined with fluorine to produce 15.81 g of a gas. What is the empirical formula of the gas?

F.7 L-Dopa, a drug used for the treatment of Parkinson's disease, is 54.82% C, 5.62% H, 7.10% N, and 32.46% O. What is the empirical formula of the compound?

F.8 Iron forms a compound called ferrocene. It has the composition 64.56% C, 5.42% H, and 30.02% Fe. What is the empirical formula of ferrocene?

F.9 Osmium forms a number of molecular compounds with carbon monoxide. One light yellow compound was analyzed to give the following elemental composition: 15.89% C, 21.18% O, and 62.93% Os. (a) What is the empirical formula of this compound? (b) From the mass spectrum of the compound, it was determined that the molecule has a molar mass of 907 $g\cdot mol^{-1}$. What is its molecular formula?

F.10 Paclitaxel, which is extracted from the Pacific yew tree *Taxus brevifolia*, has antitumor activity for ovarian and breast cancer. It is sold under the tradename Taxol. Upon analysis, it gives the following percentage composition: 66.11% C, 6.02% H, and 1.64% N, with the balance being oxygen. What is the empirical formula of paclitaxel?

F.11 Caffeine, a stimulant in coffee and tea, has molar mass 194.19 $g\cdot mol^{-1}$ and mass percentage composition 49.48% C, 5.19% H, 28.85% N, and 16.48% O. What is the molecular formula of caffeine?

F.12 Cacodyl, which has an intolerable garlicky odor and is used in the manufacture of cacodylic acid, a cotton herbicide, has mass percentage composition 22.88% C, 5.76% H, and 71.36% As and molar mass 209.96 $g\cdot mol^{-1}$. What is the molecular formula of cacodyl?

F.13 In determining the composition of an unknown compound, elemental analyses are often considered to have an error of ±0.5%. Using elemental analyses alone, would it be possible to determine whether a sample was glucose ($C_6H_{12}O_6$) or sucrose ($C_{12}H_{22}O_{11}$)?

F.14 In 1978, scientists extracted a compound with antitumor and antiviral properties from marine animals in the Caribbean Sea. A 2.52-mg sample of the compound, didemnin-C, was analyzed and found to have the following composition: 1.55 mg C, 0.204 mg H, 0.209 mg N, and 0.557 mg O. The molar mass of didemnin-C was found to be 1014 $g\cdot mol^{-1}$. What is its molecular formula?

G MIXTURES AND SOLUTIONS

Most materials are neither pure elements nor pure compounds, so they are not pure substances; they are **mixtures** of these simpler substances, with one substance mingled with another. For example, air, blood, and seawater are mixtures. Many alloys, which are mixtures of metals, are formulated for maximum strength and resistance to corrosion. A medicine, such as a cough syrup, is often a mixture of various ingredients that has been formulated to achieve an overall biological effect. Much the same can be said of a perfume.

Chemists need to be able to specify the composition of mixtures *quantitatively*. For example, a chemist may need to monitor a pollutant, administer a dosage, or transfer a known amount of a solute. "Molar concentration," which is commonly referred to informally as the "molarity," is the most common measure of concentration, and we introduce it in this section.

G.1 Classifying Mixtures

A compound has a fixed composition, whereas a mixture may have any desired composition. There are always two H atoms for each O atom in a sample of the compound water, but sugar and sand, for instance, can be mixed in any proportions. Because the components of a mixture are merely mingled with one another, they retain their own chemical properties in the mixture. In contrast, a compound has specific properties that differ from those of its components. The differences between mixtures and compounds are summarized in Table G.1.

We can identify the different components of some mixtures with a microscope, or even the unaided eye (Fig. G.1). Such a patchwork of different substances is called a **heterogeneous mixture**. Many of the rocks that form the landscape are heterogeneous mixtures. Milk, which looks like a pure substance, is in fact a heterogeneous mixture; through a microscope, we can see the individual globules of butterfat floating in a liquid that is largely water. Human bodies are heterogeneous mixtures of thousands of compounds.

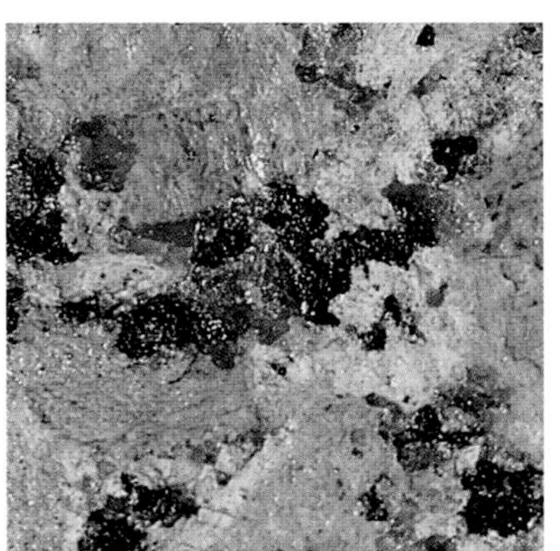

FIGURE G.1 This piece of granite is a heterogeneous mixture of several substances.

In some mixtures, we cannot make out distinct components even with a very powerful microscope because the molecules or ions of the components are so well mixed that the composition is the same throughout, no matter how small the sample. Such a mixture is called a **homogeneous mixture** (Fig. G.2). For example, syrup is a homogeneous mixture of sugar and water. The molecules of sugar are dispersed and mixed so thoroughly with the water that we cannot see separate regions or particles.

TABLE G.1 Differences Between Mixtures and Compounds

Mixture	Compound
Components can be separated by using physical techniques	Components cannot be separated by using physical techniques
Composition is variable	Composition is fixed
Properties are related to those of its components	Properties are unlike those of its components

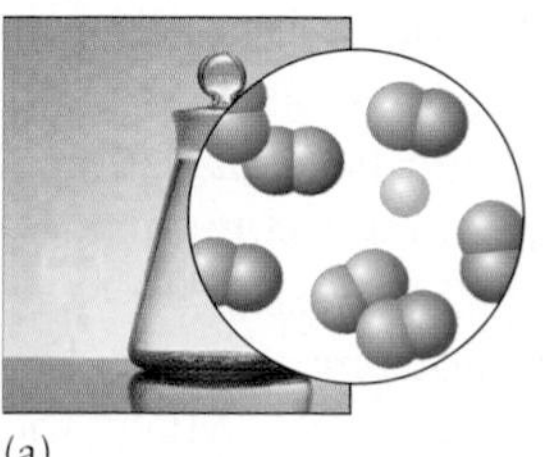

(a)

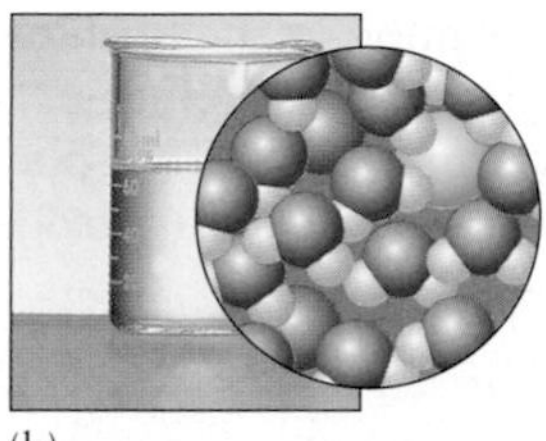

(b)

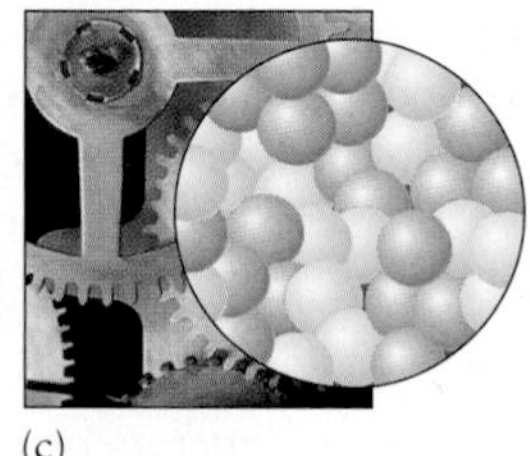

(c)

FIGURE G.2 Three examples of homogeneous mixtures. (a) Air is a homogeneous mixture of many gases, including the nitrogen, oxygen, and argon depicted here. (b) Table salt dissolved in water consists of sodium ions and chloride ions among water molecules. (c) Many alloys are solid homogeneous mixtures of two or more metals.

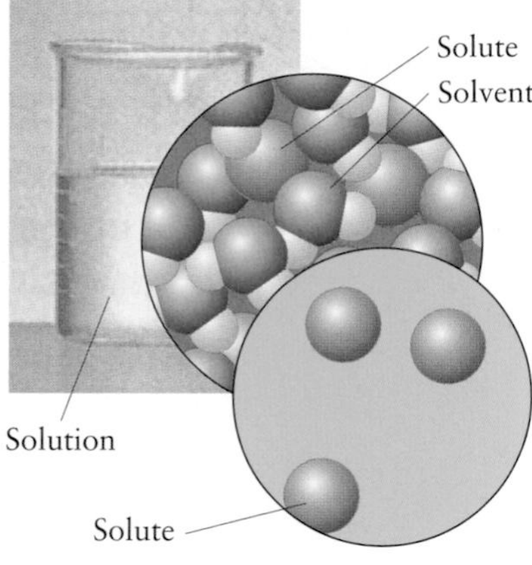

FIGURE G.3 Solutions are homogeneous mixtures in which one substance, the solvent (here water), is usually in large excess. A dissolved substance is called a solute.

FUNDAMENTALS

Even using a microscope, we cannot distinguish between a pure substance and a homogeneous mixture.

Homogeneous mixtures are also called **solutions**. Many of the materials around us are solutions. Root beer is a solution consisting mostly of water, along with sugar, plant extracts, and various additives. Filtered seawater is a solution of common salt (sodium chloride) and many other substances in water. When we use the everyday term *dissolving*, we mean the process of producing a solution. Usually the component of the solution present in the larger amount (water in these examples) is called the **solvent**, and the dissolved substances are the **solutes** (Fig. G.3). However, if we want to emphasize the role of one of the substances as the one the others are "dissolved in," we may identify that substance as the solvent. For example, in a heavy syrup, the sugar may be present in a much greater amount than the water, but water is considered the solvent. Normally the solvent determines the physical state of the solution (whether it is a solid, a liquid, or a gas).

Crystallization occurs when the solute slowly comes out of solution as crystals, perhaps as the solvent evaporates. For example, salt crystals that form when water evaporates line the shores of the Great Salt Lake in Utah. In **precipitation**, a solute comes out of solution so rapidly that a single crystal does not have time to form. Instead, the solute forms a finely divided powder called a precipitate. Precipitation is often almost instantaneous (Fig. G.4).

Beverages and seawater are examples of **aqueous solutions**, solutions in which the solvent is water. Aqueous solutions are very common in everyday life and in chemical laboratories; for that reason, most of the solutions mentioned in this text are aqueous. **Nonaqueous solutions** are solutions in which the solvent is not water. Although they are less common than aqueous solutions, they have important uses. In "dry cleaning," the grease and dirt on fabrics are dissolved in a nonaqueous liquid solvent such as tetrachloroethene, C_2Cl_4. There are also **solid solutions**, in which

FIGURE G.4 Precipitation occurs when an insoluble substance is formed. Here lead(II) iodide, PbI_2, which is an insoluble yellow solid, precipitates when we mix solutions of lead(II) nitrate, $Pb(NO_3)_2$, and potassium iodide, KI.

Media Link **Lab video:** Fig G.4, Formation of lead iodide the solvent is a solid. An example is the form of brass that is a solution of copper in zinc. Although gaseous solutions exist, they are normally referred to as gaseous mixtures. The atmosphere could be regarded as a vast gaseous solution of various gases in nitrogen, its major component.

Mixtures display the properties of their constituents; they differ from compounds, as summarized in Table G.1. Mixtures are classified as homogeneous or heterogeneous, and solutions are homogeneous mixtures.

G.2 Separation Techniques

Natural products, such as enzymes and vitamins, are almost invariably extracted from mixtures. To analyze the composition of any sample that we suspect is a mixture, we first separate its components by physical means and then identify each individual substance present (Fig. G.5). Common physical separation techniques are decanting, filtration, chromatography, and distillation.

One of the simplest separation techniques, **decanting**, makes use of differences in density. One liquid floats on another liquid or lies above a solid and is poured off. **Filtration** makes use of differences in solubility (the ability to dissolve in a particular solvent). The sample is stirred with a liquid and then poured through a fine mesh, the filter. Soluble components of the mixture dissolve in the liquid and pass through the filter, whereas the insoluble solid components do not dissolve and are captured by the filter. The technique can be used to separate sugar from sand because sugar is soluble in water but sand is not. Filtration is a common first stage in the treatment of domestic water supplies. A related technique, one of the most sensitive techniques available for separating a mixture, is **chromatography**, which relies on the different abilities of substances to **adsorb**, or stick, to surfaces (Fig. G.6). Chromatography is discussed in detail in Major Technique 4, following Chapter 8.

The colored patterns that appear gave rise to the name *chromatography,* which comes from the Greek words for "color writing."

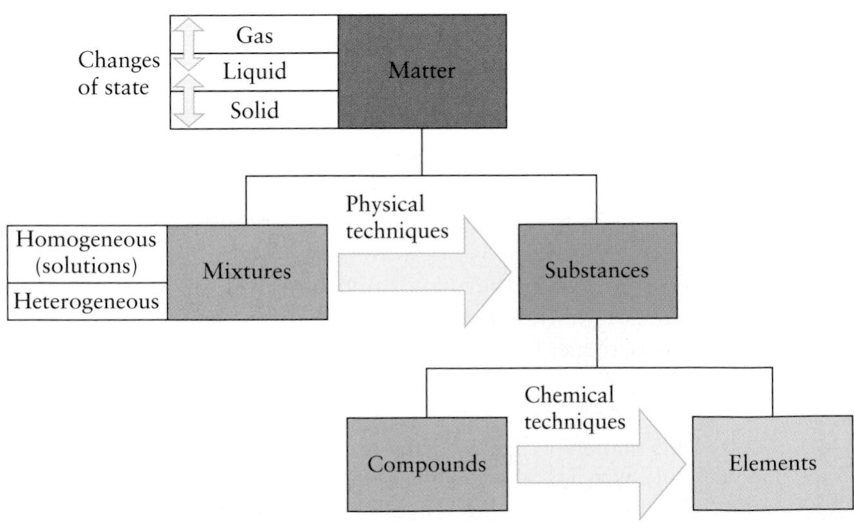

FIGURE G.5 The hierarchy of materials: matter, mixtures and substances, compounds, and elements. Physical techniques of separation are used to separate mixtures into pure substances.

FIGURE G.6 In paper chromatography, the components of a mixture are separated by washing them along a paper— the support— with a solvent. A primitive form of the technique is shown here. On the left is a dry filter paper to which a drop of food coloring was applied. Solvent was then poured on to the center of the filter paper. The filter paper on the right was allowed to dry after the solvent had spread out to the edges of the paper, carrying two dyes to different distances as it spread. The dried support showing the separated components of a mixture is called a chromatogram.

Distillation makes use of differences in boiling points to separate mixtures. When a solution is distilled, the components of a mixture boil away at different temperatures and condense in a cooled tube (Fig. G.7). Distillation can be used to remove water (which boils at 100°C in its pure state) from table salt (sodium chloride), which boils at 1465°C; the salt is left behind when the water evaporates.

The first known distillation apparatus, assembled in the first century AD, is attributed to Mary the Jewess, an alchemist who lived in the Middle East.

Mixtures are separated by making use of the differences in physical properties of the components; techniques based on physical differences include decanting, filtration, chromatography, and distillation.

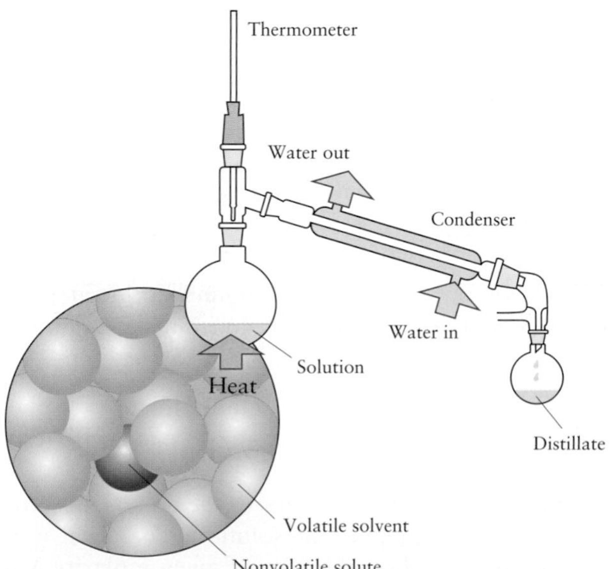

FIGURE G.7 The technique of distillation, which is used to separate a low boiling liquid from a dissolved solid or liquid with a much higher boiling point. When the solution is heated, the low boiling liquid boils off, condenses in the water-jacketed tube (the condenser), and is collected as the distillate.

G.3 Molarity

The **molar concentration,** c, of a solute in a solution, which is widely called the **molarity** of the solute, is the amount of solute molecules or formula units (in moles) divided by the volume of the solution (in liters):

$$\text{Molarity} = \frac{\text{amount of solute (moles)}}{\text{volume of solution (liters)}} \qquad \text{or} \qquad c = \frac{n}{V} \tag{1}$$

The units of molarity are moles per liter ($\text{mol}\cdot\text{L}^{-1}$), often denoted M:

$$1\ \text{M} = 1\ \text{mol}\cdot\text{L}^{-1}$$

The symbol M is often read "molar." Note that 1 $\text{mol}\cdot\text{L}^{-1}$ is the same as 1 $\text{mmol}\cdot\text{mL}^{-1}$. Chemists working with very low concentrations of solutes also report molar concentrations as millimoles per liter ($\text{mmol}\cdot\text{L}^{-1}$) and micromoles per liter ($\mu\text{mol}\cdot\text{L}^{-1}$).

How can we use this information . . . to calculate the molarity of a solution?

Suppose we dissolved 10.0 g of cane sugar in enough water to make 200 mL of solution, which we might do (with less precision) if we were making a glass of lemonade. Cane sugar is sucrose ($C_{12}H_{22}O_{11}$); and from its molar mass ($342\ \text{g}\cdot\text{mol}^{-1}$), we can calculate the amount of sucrose molecules added. It follows from Eq. 1 that the molarity of the solute in the solution is

$$c = \frac{\overbrace{(10.0\ \text{g})/(342\ \text{g}\cdot\text{mol}^{-1})}^{n\,=\,m/M}}{\underbrace{0.200\ \text{L}}_{V}}$$

$$= \frac{10.0\ \text{g}}{(342\ \text{g}\cdot\text{mol}^{-1}) \times (0.200\ \text{L})} = 0.146\ \text{mol}\cdot\text{L}^{-1}$$

We report this concentration as 0.146 M $C_{12}H_{22}O_{11}(aq)$. The (aq) indicates an aqueous solution. If instead of 10.0 g, we were to dissolve 20.0 g of cane sugar in the same volume of solution, the sugar would be twice as concentrated: its molarity would be 0.292 M $C_{12}H_{22}O_{11}(aq)$.

Self-Test G.1A What is the molarity of sodium chloride in a solution prepared by dissolving 12.0 g in enough water to make 250 mL of solution?

[*Answer:* 0.821 M NaCl(aq)]

Self-Test G.1B What is the molarity of sodium sulfate in a solution prepared by dissolving 15.5 g in enough water to make 350 mL of solution?

Note that molarity is defined in terms of the volume of solution, not the volume of solvent used to prepare the solution. The usual way to prepare an aqueous solution of a solid substance of given molarity is to transfer a known mass of the solid into a volumetric flask, dissolve it in a little water, fill the flask up to the mark with water, and then mix the solution thoroughly by tipping the flask end over end (Fig. G.8).

To use the molarity of a solution to calculate the amount of solute in a given volume of solution, we use Eq. 1 rearranged into

$$n\ (\text{mol}) = c\ (\text{mol·L}^{-1}) \times V\ (\text{L}) \qquad (2)$$

where c is the molarity, V is the volume, and n is the amount. Because we can convert moles to mass by using the molar mass of the solute, we can also use this formula to calculate the mass of solute present in a given volume of solution, and hence estimate the mass of solute we need to make up a solution of given concentration.

How can we use this information . . . to calculate the amount of solute present in a solution?

Suppose we want to know the amount (in moles) of sucrose molecules in 15 mL (0.015 L) of 0.10 M $C_{12}H_{22}O_{11}$(aq) solution. Substitution of the data into Eq. 2 gives

$$\begin{aligned}\text{Amount of } C_{12}H_{22}O_{11} &= (0.10\ \text{mol·L}^{-1}) \times (0.015\ \text{L}) \\ &= 1.5 \times 10^{-3}\ \text{mol} = 1.5\ \text{mmol}\end{aligned}$$

Self-Test G.2A What amount (in moles) of NaCl formula units is present in 25.00 mL of 1.85 M NaCl(aq)?

[*Answer:* 4.62×10^{-2} mol NaCl]

Self-Test G.2B What amount (in moles) of urea molecules, $(NH_2)_2CO$, is present in 100.0 mL of 1.45×10^{-2} M $(NH_2)_2CO$(aq)?

FUNDAMENTALS

FIGURE G.8 The steps involved in making up a solution of known molarity. A known mass of the compound is dispensed into a volumetric flask (left). Some water is added to dissolve it (center). Finally, water is added up to the mark on the stem of the flask (right). The bottom of the solution's meniscus (the curved top surface of the liquid) should be level with the mark.

How can we use this information . . . to calculate the mass of solute required for a given concentration?

Suppose we were asked to prepare 250 mL of a solution that was approximately 0.0380 M $CuSO_4(aq)$ and we had available only copper(II) sulfate pentahydrate, $CuSO_4 \cdot 5H_2O$, from the stockroom. To calculate the mass of the solid that we need to use, first we need to know the amount of $CuSO_4$ in 250 mL (0.250 L) of solution:

$$\begin{aligned}\text{Amount of } CuSO_4 \text{ (mol)} &= (0.0380 \text{ mol}\cdot\text{L}^{-1}) \times (0.250 \text{ L}) \\ &= 0.0380 \times 0.250 \text{ mol}\end{aligned}$$

Because 1 mol $CuSO_4 \cdot 5H_2O$ contains 1 mol $CuSO_4$ and the molar mass of copper(II) sulfate pentahydrate is 249.6 $g \cdot mol^{-1}$, this amount corresponds to the following mass of the pentahydrate:

$$\begin{aligned}\text{Mass of } CuSO_4 \cdot 5H_2O &= (0.0380 \times 0.250 \text{ mol}) \times (249.6 \text{ g}\cdot\text{mol}^{-1}) \\ &= 2.37 \text{ g}\end{aligned}$$

We conclude that we should measure out about 2.37 g of copper(II) sulfate pentahydrate. In practice, we might find that we had spooned out 2.403 g, in which case the molarity would be 0.0385 M $CuSO_4(aq)$.

Self-Test G.3A Calculate the mass of glucose needed to prepare 150 mL of 0.442 M $C_6H_{12}O_6(aq)$.

[*Answer:* 11.9 g]

Self-Test G.3B Calculate the mass of oxalic acid needed to prepare 50.00 mL of 0.125 M $C_2H_2O_4(aq)$.

The molarity is also used to calculate the volume of solution, V, that contains a given amount of solute. For this type of calculation, we rearrange Eq. 1 into

$$V \text{ (L)} = \frac{n \text{ (mol)}}{c \text{ (mol}\cdot\text{L}^{-1})} \quad (3)$$

and then substitute the data.

The molarity of a solute in a solution is the amount of solute in moles divided by the volume of the solution in liters.

How can we use this information . . . to calculate the volume of solution that contains a given amount of solute?

Suppose we want to measure out 0.760 mmol of CH_3COOH, acetic acid, an acid found in vinegar and often used in the laboratory, and we have available 0.0380 M $CH_3COOH(aq)$. To find the correct volume of solution to use, we substitute the data for n and c into Eq. 3:

$$\text{Volume of solution} = \frac{0.760 \times 10^{-3} \text{ mol}}{0.0380 \text{ mol}\cdot\text{L}^{-1}} = 2.00 \times 10^{-2} \text{ L}$$

We should therefore transfer 20.0 mL of the acetic acid solution by using a buret or a pipet. The flask will then contain 0.760 mmol CH_3COOH.

Self-Test G.4A What volume of 1.25×10^{-3} M $C_6H_{12}O_6(aq)$ should you transfer to obtain a solution that contains 1.44×10^{-6} mol of glucose molecules?

[*Answer:* 1.15 mL]

Self-Test G.4B What volume of 0.358 M HCl(aq) should you transfer to have a sample that contains 2.55 mmol HCl?

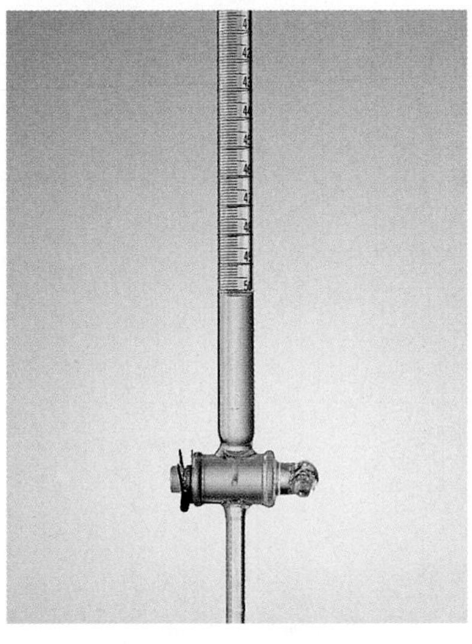

FIGURE G.9 A buret is calibrated so that the volume of liquid delivered can be measured.

G.4 Dilution

A common space-saving practice in chemistry is to store a solution in a concentrated form called a **stock solution** and then to **dilute** it, or reduce its concentration, to whatever concentration is needed. Anyone working in a chemical laboratory, therefore, needs to be able to dilute solutions accurately. Moreover, the use of solutions and of techniques such as dilution give chemists very precise control over the amounts of the substances they are handling, even in very small quantities. For example, pipetting 25.0 mL of a 1.50×10^{-3} M $KMnO_4(aq)$ solution corresponds to transferring as little as 37.5 μmol $KMnO_4$, or only 5.93 μg of the compound. A mass this small would be difficult to weigh out accurately.

To dilute a solution to a desired concentration, we first use a pipet to transfer the appropriate volume of solution to a **volumetric flask**, a flask calibrated to contain a specified volume. **Pipets** and **burets** are narrow tubes used to transfer liquids accurately (Fig. G.9). Then we add enough solvent to increase the volume of the solution to its final value. Toolbox G.1 shows how to calculate the correct initial volume of stock solution.

When a solution is diluted to a larger volume, the total number of moles of solute in the solution does not change, but the concentration of solute is reduced.

FUNDAMENTALS

TOOLBOX G.1 How to calculate the volume of stock solution required for a given dilution

Conceptual Basis

This procedure is based on a simple idea: although we may add more solvent to a given volume of solution, we do not change the amount of solute (Fig. G.10). After dilution, the same amount of solute simply occupies a larger volume of solution.

Procedure

The procedure has two steps:

Step 1 Calculate the amount of formula units in the final solution. The result of this calculation is the amount of solute to transfer into the volumetric flask. To calculate the amount of solute in the dilute solution, we use the expression

$$n = c_{\text{final}} V_{\text{final}}$$

where c_{final} is the molarity of the final solution and V_{final} is its volume.

Step 2 Calculate the volume of the initial stock solution that contains this amount of solute. For this step, we use

$$V_{\text{initial}} = \frac{n}{c_{\text{initial}}}$$

Because the amount of solute, n, is the same in these two expressions, they can be combined into

$$V_{\text{initial}} = \frac{c_{\text{final}} V_{\text{final}}}{c_{\text{initial}}} \quad (4a)$$

This expression is easy to remember when it is rearranged into the form

$$c_{\text{initial}} V_{\text{initial}} = c_{\text{final}} V_{\text{final}} \tag{4b}$$

That is, the amount of solute in the final solution (the product on the right) is the same as the amount of solute in the initial volume of solution (the product on the left).

Before dilution After dilution

FIGURE G.10 When a solution is diluted, the same number of solute molecules occupy a larger volume. Therefore, the same volume (as depicted by the square) will contain fewer molecules than in the concentrated solution.

How can we use this information . . . to calculate the volume of stock solution to dilute?

Suppose we want to know the volume of 0.0380 M $KMnO_4(aq)$ stock solution that we should use to prepare 250 mL of 1.50×10^{-3} M $KMnO_4(aq)$. First, we rearrange Eq. 4b to give the quantities we know on the right and the quantity we want to know on the left:

$$V_{\text{initial}} = \frac{c_{\text{final}} V_{\text{final}}}{c_{\text{initial}}}$$

Then substitute the data:

$$V_{\text{initial}} = \frac{(1.50 \times 10^{-3}\ \text{mol·L}^{-1}) \times (0.250\ \text{L})}{0.0380\ \text{mol·L}^{-1}} = 9.87 \times 10^{-3}\ \text{L}$$

We can conclude that 9.87 mL of the stock solution should be measured into a 250-mL volumetric flask (using a buret) and water added up to the mark (Fig. G.11).

Self-Test G.5A Calculate the volume of 0.0155 M HCl(aq) that we should use to prepare 100 mL of 5.23×10^{-4} M HCl(aq).

[*Answer:* 3.37 mL]

Self-Test G.5B Calculate the volume of 0.152 M $C_6H_{12}O_6(aq)$ that should be used to prepare 25.00 mL of 1.59×10^{-5} M $C_6H_{12}O_6(aq)$.

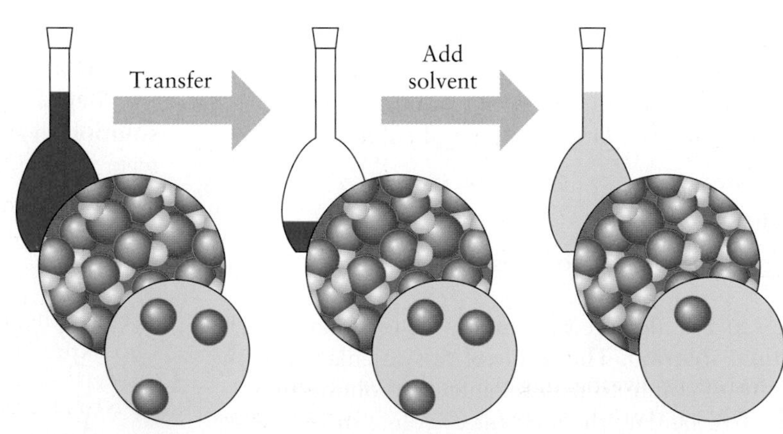

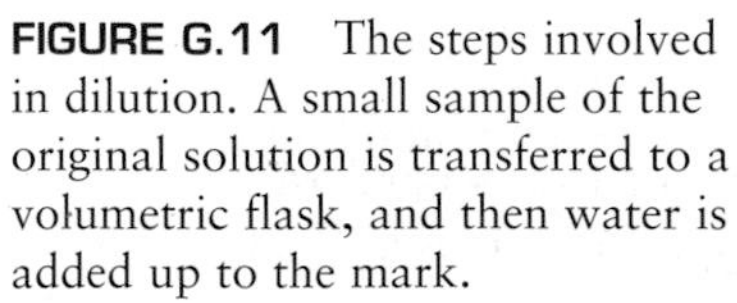

FIGURE G.11 The steps involved in dilution. A small sample of the original solution is transferred to a volumetric flask, and then water is added up to the mark.

SKILLS YOU SHOULD HAVE MASTERED

1 Distinguish heterogeneous and homogeneous mixtures and describe methods of separation.

2 Calculate the molarity of a solute in a solution, volume of solution, and mass of solute, given the other two quantities.

3 Determine the volume of stock solution needed to prepare a dilute solution of a given molarity, Toolbox G.1.

EXERCISES

G.1 What physical properties are used for separation of the components of a mixture by (a) filtration; (b) chromatography; (c) distillation?

G.2 Water purification, in which impure water is made drinkable, can be achieved by several methods. On the basis of the information in this chapter, suggest a physical process for converting seawater to pure water.

G.3 Identify the following as homogeneous or heterogeneous mixtures and suggest a technique for separating their components: (a) alcohol and water; (b) chalk and table salt; (c) salt water.

G.4 Identify the following as homogeneous or heterogeneous mixtures and suggest a technique for separating their components: (a) gasoline and motor oil; (b) carbonated water; (c) charcoal and sugar.

G.5 A chemist studying the properties of photographic emulsions needed to prepare 500.0 mL of 0.179 M $AgNO_3$(aq). What mass of silver nitrate must be placed into a 500.0-mL volumetric flask, dissolved, and diluted to the mark with water?

G.6 (a) A chemist prepared a solution by dissolving 2.345 g of $NaNO_3$ in enough water to make 200.0 mL of solution. What molar concentration of sodium nitrate should appear on the label? (b) If the chemist had mistakenly used a 250.0-mL volumetric flask instead of the 200.0-mL flask in (a), what molar concentration of potassium chloride has the chemist actually prepared?

G.7 A student prepared a solution of sodium carbonate by adding 2.111 g of the solid to a 250.0-mL volumetric flask and adding water to the mark. Some of this solution was transferred to a buret. What volume of solution should the student run into a flask to transfer (a) 2.15 mmol Na^+; (b) 4.98 mmol CO_3^{2-}; (c) 50.0 mg Na_2CO_3?

G.8 A student investigating the properties of solutions containing carbonate ions prepared a solution containing 7.112 g of Na_2CO_3 in a 250.0-mL volumetric flask. Some of the solution was transferred to a buret. What volume of solution should be dispensed from the buret to provide (a) 5.112×10^{-3} mol Na_2CO_3; (b) 3.451×10^{-3} mol CO_3^{2-}?

G.9 Explain how you would prepare an aqueous solution of 0.010 M $KMnO_4$(aq) by starting with (a) solid $KMnO_4$; (b) 0.050 M $KMnO_4$(aq).

G.10 (a) A 12.56-mL sample of 1.345 M K_2SO_4(aq) is diluted to 250.0 mL. What is the molar concentration of K_2SO_4 in the diluted solution? (b) A 25.00-mL sample of 0.366 M HCl(aq) is drawn from a reagent bottle with a pipet. The sample is transferred to a 125.00-mL volumetric flask and diluted to the mark with water. What is the molar concentration of the dilute hydrochloric acid solution?

G.11 (a) What volume of a 0.778 M Na_2CO_3(aq) solution should be diluted to 150.0 mL with water to reduce its concentration to 0.0234 M Na_2CO_3(aq)? (b) An experiment requires the use of 60.0 mL of 0.50 M NaOH(aq). The stockroom assistant can only find a reagent bottle of 2.5 M NaOH(aq). How is the 0.50 M NaOH(aq) solution to be prepared?

G.12 A chemist dissolves 0.094 g of $CuSO_4{\cdot}5H_2O$ in water and dilutes the solution to the mark in a 500.0-mL volumetric flask. A 2.00-mL sample of this solution is then transferred to a second 500.0-mL volumetric flask and diluted. (a) What is the molarity of $CuSO_4$ in the final solution? (b) To prepare the solution directly, what mass of $CuSO_4{\cdot}5H_2O$ would need to be weighed out?

G.13 (a) Determine the mass of anhydrous copper(II) sulfate that must be used to prepare 250 mL of a 0.20 M $CuSO_4$(aq). (b) Determine the mass of $CuSO_4{\cdot}5H_2O$ that must be used to prepare 250 mL of a 0.20 M $CuSO_4$(aq).

G.14 The ammonia solution that is purchased for a stockroom has a molarity of 15.0 $mol{\cdot}L^{-1}$; all ammonia solutions for experiments are prepared by dilution of this stock solution. (a) Determine the volume of 15.0 M NH_3(aq) that must be diluted to 500 mL to prepare a 1.25 M NH_3(aq). (b) An experiment requires a 0.32 M NH_3(aq). The stockroom manager estimates that 15.0 L of the base is needed. What volume of 15.0 M NH_3(aq) must be used for the preparation?

G.15 A student prepares a solution by placing 0.376 g of $NiSO_4 \cdot 6H_2O$ and 0.129 g of $NiCl_2 \cdot 6H_2O$ in a 250.0-mL volumetric flask. A small amount of distilled water is then added, and the compounds are swirled in the flask until they dissolve. Once they are dissolved, the solution is diluted to the mark. (a) What is the molar concentration of chloride ions in the final solution? (b) What is the molar concentration of Ni^{2+} ions in the final solution?

G.16 To prepare a very dilute solution, it is advisable to perform successive dilutions of a single prepared reagent solution, rather than to weigh a very small mass or to measure a very small volume of stock chemical. A solution was prepared by transferring 0.661 g of $K_2Cr_2O_7$ to a 250.0-mL volumetric flask and adding water to the mark. A 1.000-mL sample of this solution was transferred to a 500-mL volumetric flask and diluted to the mark with water. Then 10.0 mL of the diluted solution was transferred to a 250-mL flask and diluted to the mark with water. (a) What is the final concentration of $K_2Cr_2O_7$ in solution? (b) What mass of $K_2Cr_2O_7$ is in this final solution? (The answer to the last question gives the amount that would have had to have been weighed out if the solution had been prepared directly.)

G.17 What mass (in grams) of anhydrous solute is needed to prepare each of the following solutions? (a) 1.00 L of 0.125 M K_2SO_4(aq); (b) 375 mL of 0.015 M NaF(aq); (c) 500 mL of 0.35 M $C_{12}H_{22}O_{11}$(aq).

G.18 Describe the preparation of each solution, starting with the anhydrous solute and water and using the indicated size of volumetric flask: (a) 75.0 mL of 5.0 M NaOH(aq); (b) 5.0 L of 0.21 M $BaCl_2$(aq); (c) 300 mL of 0.0340 M $AgNO_3$(aq).

H CHEMICAL EQUATIONS

A **chemical reaction** is the *process* of chemical change, the conversion of one or more substances into other substances. The starting materials are called the **reactants.** The substances formed are called the **products**. The chemicals available in a laboratory are called **reagents.** Only when a reagent is being used in a particular reaction is it called a reactant. In this section, we see how to describe chemical reactions in a simple, symbolic form.

H.1 Symbolizing Chemical Reactions

Chemists have devised a helpful way of summarizing chemical reactions. First a chemical reaction is symbolized by an arrow:

$$\text{Reactants} \longrightarrow \text{products}$$

For example, sodium is a soft, shiny metal that reacts vigorously with water. When we drop a small lump of sodium metal into a container of water, hydrogen gas forms rapidly and sodium hydroxide is left in solution (Fig. H.1). To summarize this reaction, we could write

$$\text{Sodium} + \text{water} \longrightarrow \text{sodium hydroxide} + \text{hydrogen}$$

However, using the chemical formulas of the species provides a way to present important quantitative information about the reaction concisely and unambiguously. We can summarize the chemical changes in the reaction as

$$Na + H_2O \longrightarrow NaOH + H_2$$

This expression is called a **skeletal equation** because it shows the bare bones of the reaction (the identities of the reactants and products) in terms of chemical formulas. A skeletal equation is a *qualitative* summary of a chemical reaction.

To summarize reactions *quantitatively*, we need to note that atoms are neither created nor destroyed in a chemical reaction: they simply change their partners (see Fig. H.1). The principal evidence for this conclusion is that there is no overall change in mass when a reaction takes place in a sealed container. The observation that the total mass is constant during a chemical reaction is called the **law of conservation of mass.** Because atoms are neither created nor destroyed, chemists multiply formulas by factors to show the same numbers of atoms of each element on both sides of the arrow; the resulting expression is said to be **balanced** and is called a **chemical equation.** For example, there are two H atoms on the left of the skeletal equation above, but three H atoms on the right. So, we rewrite the expression as

$$2\,Na + 2\,H_2O \longrightarrow 2\,NaOH + H_2$$

Now there are four H atoms, two Na atoms, and two O atoms on each side, and the equation conforms to the law of conservation of mass. The number multiplying an *entire* chemical formula in a chemical equation (for example, the 2 multiplying H_2O) is called the **stoichiometric coefficient** of the substance. A coefficient of 1 (as for H_2) is not written explicitly.

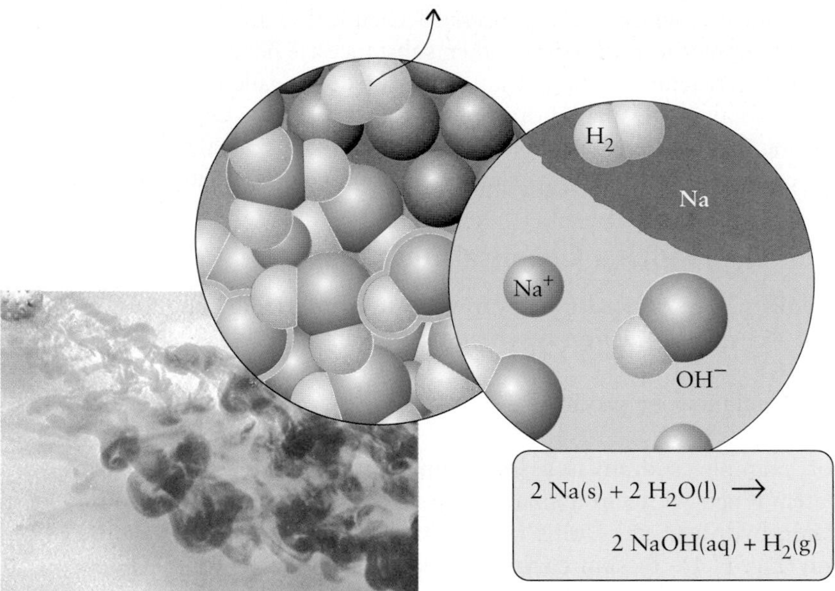

FIGURE H.1 When a small piece of sodium is dropped into water, a vigorous reaction takes place. Hydrogen gas and sodium hydroxide are formed, and the heat released is enough to melt the sodium into a sphere. The pink color is due to the presence of a dye that changes color in the presence of sodium hydroxide. Note that two sodium atoms give rise to two sodium ions and that two water molecules give rise to one hydrogen molecule (which escapes as a gas) and two hydroxide ions. There is a rearrangement of partners, not a creation or annihilation of atoms. The unreacted water molecules are not shown in the inset on the right.

A chemical equation typically also shows the physical state of each reactant and product by using a **state symbol:**

(s): solid (l): liquid (g): gas (aq): aqueous solution

For the reaction between solid sodium and water, the complete, balanced chemical equation is therefore

$$2\,\text{Na(s)} + 2\,\text{H}_2\text{O(l)} \longrightarrow 2\,\text{NaOH(aq)} + \text{H}_2\text{(g)}$$

When we want to show that a reaction requires a high temperature, we write the Greek letter Δ (delta) over the arrow. For example, the conversion of limestone to quicklime takes place at about 800°C, and we write

$$\text{CaCO}_3\text{(s)} \xrightarrow{\Delta} \text{CaO(s)} + \text{CO}_2\text{(g)}$$

Sometimes a **catalyst,** a substance that increases the rate of a reaction but is not itself consumed in the reaction, is added. For example, vanadium pentoxide, V_2O_5, is a catalyst in one step of the industrial process for the production of sulfuric acid. The presence of a catalyst is indicated by writing the formula of the catalyst above the reaction arrow:

$$2\,\text{SO}_2\text{(g)} + \text{O}_2\text{(g)} \xrightarrow{\text{V}_2\text{O}_5} 2\,\text{SO}_3\text{(g)}$$

Now we come to an important interpretation of a chemical equation. First, we note that the equation for the reaction of sodium with water tells us that

- When any 2 *atoms* of sodium react with 2 *molecules* of water, they produce 2 *formula units* of NaOH and one *molecule* of hydrogen.

We can multiply through by 6.022×10^{23} and conclude that

- When 2 *moles* of Na atoms react with 2 *moles* of H_2O molecules, they produce 2 *moles* of NaOH formula units and 1 *mole* of H_2 molecules.

In other words, the stoichiometric coefficients multiplying the chemical formulas in any balanced chemical equation tell us the relative number of moles of each substance that reacts or is produced in the reaction.

A balanced chemical equation symbolizes both the qualitative and quantitative changes that occur in a chemical reaction. The stoichiometric coefficients tell us the relative numbers of moles of reactants and products taking part in the reaction.

H.2 Balancing Chemical Equations

In some cases, the stoichiometric coefficients we need to balance an equation are obvious and can be written down without much thought. For example, let's consider the reaction in which hydrogen and oxygen gases combine to form water. We start by summarizing the qualitative information as a skeletal equation:

$$H_2 + O_2 \longrightarrow H_2O \quad \triangle$$

We use the international *Hazard!* road sign ⚠ to warn that a skeletal equation is not balanced. Then we see that hydrogen and oxygen are both balanced if we write

$$2\,H_2 + O_2 \longrightarrow 2\,H_2O$$

There are four H atoms and two O atoms on each side of the arrow. At this stage, we insert the state symbols:

$$2\,H_2(g) + O_2(g) \longrightarrow 2\,H_2O(l)$$

Figure H.2 is a pictorial representation of this reaction.

An equation must never be balanced by changing the subscripts in the chemical formulas. That change would imply that different substances were taking part in the reaction. For example, changing H_2O to H_2O_2 in the skeletal equation and writing

$$H_2 + O_2 \longrightarrow H_2O_2$$

certainly results in a balanced equation. However, it is a summary of a different reaction—the formation of hydrogen peroxide (H_2O_2) from its elements. Nor should we write

$$2\,H + O \longrightarrow H_2O$$

Although this equation is balanced, it summarizes the reaction between hydrogen and oxygen atoms, not the molecules that are the actual starting materials.

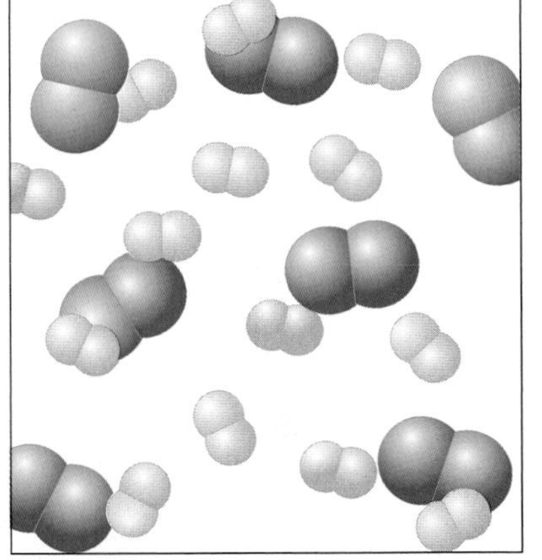

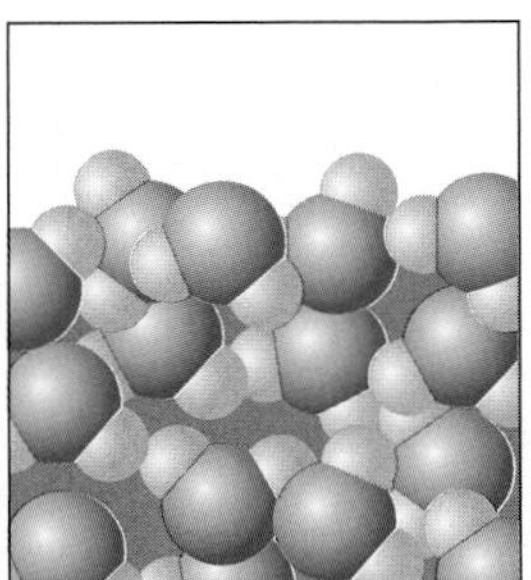

FIGURE H.2 A representation of the reaction between hydrogen and oxygen, with the production of water. No atoms are created or destroyed; they simply change partners. For every two hydrogen molecules that react, one oxygen molecule is consumed and two water molecules are formed.

A chemical equation expresses a chemical reaction in terms of chemical formulas; the stoichiometric coefficients are chosen to show that atoms are neither created nor destroyed in the reaction.

EXAMPLE H.1 Balancing a chemical equation

Methane, CH_4, is the main ingredient of natural gas (Fig. H.3). It burns in oxygen to form carbon dioxide and water, both formed initially as gases. Write and balance the chemical equation for this reaction.

STRATEGY Construct the skeletal equation for the reaction by writing the formulas of the reactants on the left of the reaction arrow and the formulas of the products on the right. Balance first the element that occurs in the fewest formulas and last the element that occurs in the most formulas. Verify that the coefficients are the smallest whole numbers and add the state of matter after each formula.

SOLUTION To write the balanced equation for the reaction, first write the skeletal equation:

$$CH_4 + O_2 \longrightarrow CO_2 + H_2O \quad ⚠$$

Because C and H occur in two formulas and O occurs in three, we begin with C and H. The C atoms are already balanced. We balance the H atoms by using a stoichiometric coefficient of 2 for H_2O to give four H atoms on each side:

$$CH_4 + O_2 \longrightarrow CO_2 + 2\ H_2O \quad ⚠$$

Now only O remains to be balanced. Because there are four O atoms on the right but only two on the left, the O_2 needs a stoichiometric coefficient of 2. The result is

$$CH_4 + 2\ O_2 \longrightarrow CO_2 + 2\ H_2O$$

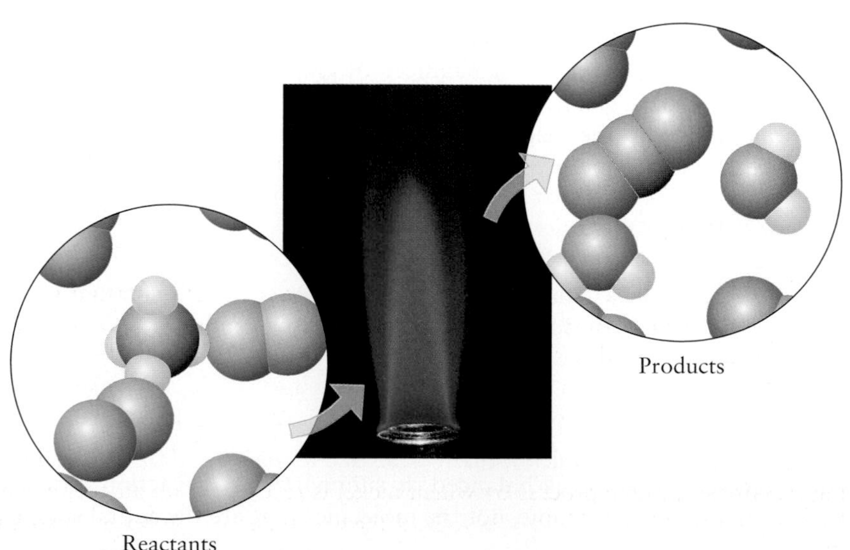

FIGURE H.3 When methane burns, it forms carbon dioxide and water. The blue color is due to the presence of C_2 molecules in the flame. If the oxygen supply is inadequate, these carbon molecules can stick together and form soot, thereby producing a smoky flame. Note that one carbon dioxide molecule and two water molecules are produced for each methane molecule that is consumed. The hydrogen atoms in the water molecules do not necessarily all come from the same methane molecule: the illustration depicts the overall outcome, not the specific outcome of the reaction of one molecule. Note that the excess oxygen remains unreacted.

We verify that the equation is balanced by counting the number of atoms of each element on each side of the arrow. At this stage, we specify the states. Because water is produced as a vapor, we write

$$CH_4(g) + 2\ O_2(g) \longrightarrow CO_2(g) + 2\ H_2O(g)$$

Self-Test H.1A When aluminum is melted and heated with solid barium oxide, a vigorous reaction takes place, and elemental molten barium and solid aluminum oxide are formed. Write the chemical equation for the reaction.

[*Answer:* $2\ Al(l) + 3\ BaO(s) \xrightarrow{\Delta} Al_2O_3(s) + 3\ Ba(l)$]

Self-Test H.1B Write the balanced chemical equation for the combustion of propane gas, C_3H_8, to carbon dioxide gas and liquid water.

SKILLS YOU SHOULD HAVE MASTERED

1 Explain the role of stoichiometric coefficients.

2 Write, balance, and label a chemical equation on the basis of information given in a sentence, Example H.1.

EXERCISES

H.1 Balance the following skeletal chemical equations:
(a) $BCl_3(g) + H_2O(l) \rightarrow B(OH)_3(aq) + HCl(aq)$
(b) $NaNO_3(s) \rightarrow NaNO_2(s) + O_2(g)$
(c) $Ca_3(PO_4)_2(s) + SiO_2(s) + C(s) \xrightarrow{\Delta} CaSiO_3(s) + CO(g) + P_4(s)$
(d) $Fe_2P(s) + S(s) \rightarrow P_4S_{10}(s) + FeS(s)$

H.2 Balance the following skeletal chemical equations:
(a) $AgNO_3(s) \rightarrow Ag(s) + NO_2(g) + O_2(g)$
(b) $P_2S_5(s) + PCl_5(s) \rightarrow PSCl_3(g)$
(c) $BF_3(g) + NaH \rightarrow B_2H_6(g) + NaF(s)$
(d) $LnC_2(s) + H_2O(l) \rightarrow Ln(OH)_3(s) + C_2H_2(g) + H_2(g)$

H.3 Write balanced chemical equations for the following reactions: (a) Sodium metal reacts with water to produce hydrogen gas and sodium hydroxide. (b) The reaction of sodium oxide, Na_2O, and water produces sodium hydroxide. (c) Hot lithium metal reacts in a nitrogen atmosphere to produce lithium nitride, Li_3N. (d) The reaction of calcium metal with water leads to the evolution of hydrogen gas and the formation of calcium hydroxide, $Ca(OH)_2$.

H.4 Write balanced chemical equations for the following reactions: (a) One process by which nickel is recovered from nickel(II) sulfide ores is to heat the ore in air. During this "roasting" process, molecular oxygen reacts with the nickel(II) sulfide to produce solid nickel(II) oxide and sulfur dioxide gas. (b) The diamondlike abrasive silicon carbide, SiC, is made by reacting silicon dioxide with elemental carbon at 2000°C to produce silicon carbide and carbon monoxide. (c) The reaction of elemental hydrogen and nitrogen is used for the commercial production of ammonia in the Haber process. (d) Magnesium metal reacts with solid boron oxide, B_2O_3, to form elemental boron and magnesium oxide.

H.5 In one stage in the commercial production of iron metal in a blast furnace, the iron(III) oxide, Fe_2O_3, reacts with carbon monoxide to form Fe_3O_4 and carbon dioxide. In a second stage, the Fe_3O_4 reacts further with carbon monoxide to produce elemental iron and carbon dioxide. Write the balanced equation for each stage in the process.

H.6 An important role of stratospheric ozone, O_3, is to remove damaging ultraviolet radiation from sunlight. One result is the eventual dissociation of ozone into molecular oxygen. Write the balanced equation for the reaction.

H.7 When nitrogen and oxygen react in the cylinder of an automobile engine, nitric oxide, NO, is formed. After it escapes into the atmosphere with the other exhaust gases, the nitric oxide reacts with oxygen to produce nitrogen dioxide, one of the precursors of acid rain. Write the two balanced equations for the reactions leading to the formation of nitrogen dioxide.

H.8 The reaction of boron trifluoride, $BF_3(g)$, with sodium borohydride, $NaBH_4(s)$, leads to the formation of sodium tetrafluoroborate, $NaBF_4(s)$, and diborane gas, $B_2H_6(g)$. The diborane reacts with the oxygen in air, forming boron oxide, $B_2O_3(s)$, and water. Write the two balanced equations leading to the formation of boron oxide.

H.9 Hydrofluoric acid is used to etch grooves in glass because it reacts with the silica, $SiO_2(s)$, in glass. The products of the reaction are silicon tetrafluoride and water. Write the balanced equation for the reaction.

H.10 The compound $Sb_4O_5Cl_2(s)$, which has been investigated because of its interesting electrical properties, can be prepared by warming a mixture of antimony(III) oxide and antimony(III) chloride, both of which are solids. Write the balanced equation for the reaction.

H.11 Write a balanced equation for the complete combustion (reaction with oxygen) of octane, C_8H_{18}, a component typical of the hydrocarbons in gasoline, to carbon dioxide and water vapor.

H.12 Aspartame, $C_{14}H_{18}N_2O_5$, is used as an artificial sweetener. Write the balanced equation for its combustion to carbon dioxide, liquid water, and nitrogen.

H.13 The psychoactive drug sold as methamphetamine ("speed"), $C_{10}H_{15}N$, undergoes a series of reactions in the body; the net result of these reactions is the oxidation of solid methamphetamine by oxygen to produce carbon dioxide, liquid water, and nitrogen. Write the balanced equation for this net reaction.

H.14 Aspirin is the analgesic acetylsalicylic acid, $C_9H_8O_4$. Write the balanced equation for the combustion of acetylsalicylic acid to carbon dioxide and liquid water.

H.15 Sodium thiosulfate, which as the pentahydrate, $Na_2S_2O_3{\cdot}5H_2O$, forms the large white crystals used as "photographer's hypo," can be prepared by bubbling oxygen through a solution of sodium polysulfide, Na_2S_5, in alcohol and adding water. Sulfur dioxide gas is also formed as a by-product. Sodium polysulfide is made by the action of hydrogen sulfide gas on a solution of sodium sulfide, Na_2S, in alcohol, which, in turn, is made by the reaction of hydrogen sulfide, H_2S, with sodium hydroxide. Write the three chemical equations that show how hypo is prepared from hydrogen sulfide and sodium hydroxide. Use (alc) to label the state of species dissolved in alcohol.

H.16 The first stage in the production of nitric acid by the Ostwald process is the reaction of ammonia with oxygen, producing nitric oxide, NO, and water. The nitric oxide further reacts with oxygen to produce nitrogen dioxide, which, when dissolved in water, produces nitric acid and nitric oxide. Write the three balanced equations that lead to the production of nitric acid.

I AQUEOUS SOLUTIONS AND PRECIPITATION

This and the next two sections introduce three of the main types of chemical reactions: precipitation reactions, acid-base reactions, and redox reactions, all of which are discussed in more depth in later chapters. (The fourth type of reactions discussed in this text, Lewis acid-base reactions, are introduced in Section 2.13.) Many chemical reactions take place in solution, particularly in water; hence we need to know what is present in an aqueous solution and what kinds of reactions can take place between the substances present.

I.1 Electrolytes

A **soluble substance** is one that dissolves to a significant extent in a specified solvent. When we refer simply to solubility without mentioning the solvent, we mean "soluble in water." An **insoluble substance** is one that does not dissolve significantly in a specified solvent; substances are often regarded as "insoluble" if they do not dissolve to more than about 0.1 $mol{\cdot}L^{-1}$. Unless otherwise specified in this text, we use the term *insoluble* to mean "insoluble in water." For instance, calcium carbonate, $CaCO_3$, which makes up limestone and chalk, dissolves to form a solution that contains only 0.01 $g{\cdot}L^{-1}$ (corresponding to 1×10^{-4} $mol{\cdot}L^{-1}$), and we regard $CaCO_3$ as insoluble. This insolubility is important for landscapes: chalk hills and limestone buildings do not wash away in natural rainwater.

A solute may be present as ions or as molecules. We can identify the form of the solute by noting whether the solution conducts an electric current. Because a current is a flow of electric charge, only solutions that contain ions conduct electricity. There is such a tiny concentration of ions in pure water (about 10^{-7} $mol{\cdot}L^{-1}$) that water alone does not conduct electricity. A substance that dissolves to give a solution that conducts electricity is called an **electrolyte**. Electrolyte solutions (solutions of electrolytes), which conduct electricity because they contain ions, include aqueous solutions of ionic compounds, such as sodium chloride and potassium nitrate. The ions are not *formed* when an ionic solid dissolves; they exist as separate ions in the solid but become free to move apart in the presence of water (Fig. I.1). Acids are electrolytes: although they are

There are a few ions present because H_2O forms some H_3O^+ and OH^- ions. However, these ions are present in *very* low concentrations (see Section 10.4).

FUNDAMENTALS

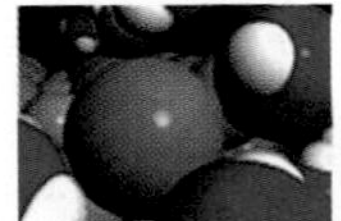

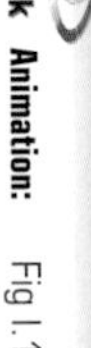

Media Link Animation: Fig I.1, NaCl dissolving

View an animation of NaCl dissolving.

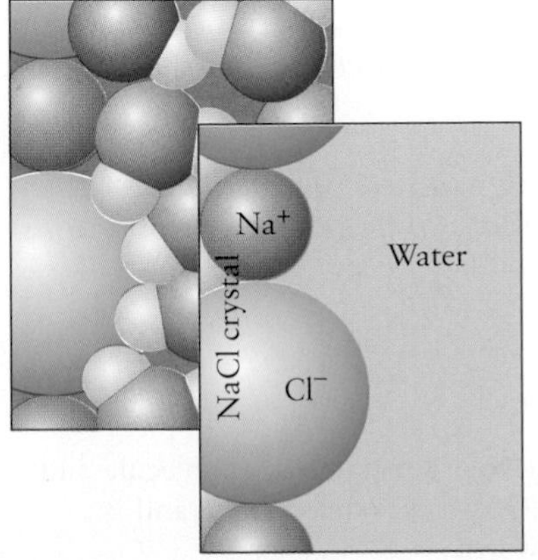

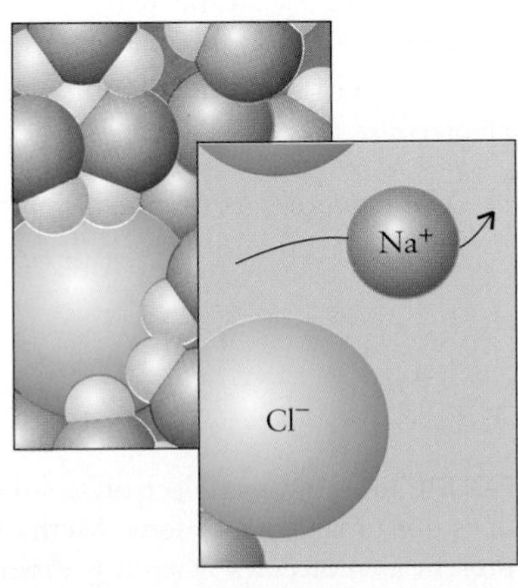

FIGURE I.1 Sodium chloride consists of sodium ions and chloride ions. When it comes in contact with water (left), the ions are able to separate, and they spread through the solvent (right). The solution consists of water, sodium ions, and chloride ions. There are no NaCl molecules present at any stage.

molecular compounds, acids form ions when they dissolve. One example is the gas hydrogen chloride, which exists as HCl molecules. In solution, however, HCl is present as hydrogen ions (attached to water molecules) and chloride ions, and the solution is called hydrochloric acid.

A **nonelectrolyte** is a substance that dissolves to form a solution that does not conduct electricity. Nonelectrolyte solutions (solutions of nonelectrolytes) do not contain ions. Aqueous solutions of acetone (**1**) and glucose (**2**) are nonelectrolyte solutions.

Sweetened water contains intact sugar molecules moving among the water molecules. If we could see the individual molecules in a nonelectrolyte solution, we would see the intact solute molecules dispersed among the solvent molecules (Fig. I.2).

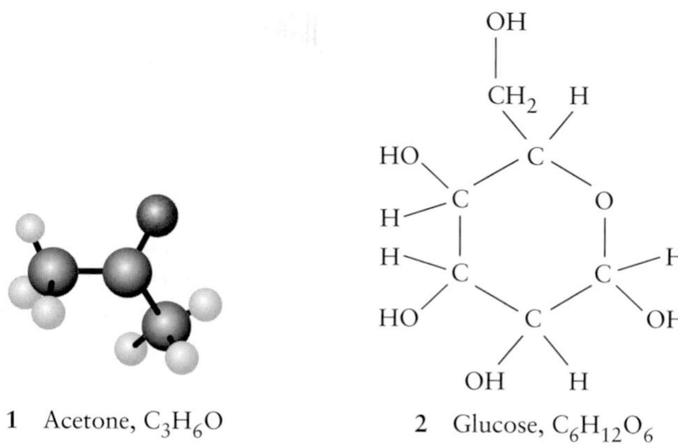

1 Acetone, C_3H_6O

2 Glucose, $C_6H_{12}O_6$

A **strong electrolyte** forms a solution in which the solute is present almost entirely as ions. Hydrogen chloride is a strong electrolyte; so is sodium chloride. Most soluble ionic compounds are strong electrolytes. There are very few molecular compounds that are strong electrolytes: the acids HCl, HBr, and HI are three examples. A **weak electrolyte** forms a solution in which the solute is incompletely ionized in solution; in other words, most of the molecules remain intact. Acetic acid is a weak electrolyte: in aqueous solution at normal concentrations, only a small fraction of CH_3COOH molecules separate into hydrogen ions and acetate ions, $CH_3CO_2^-$. One way to distinguish strong and weak electrolytes is to measure their abilities to conduct electricity: for the same molar concentration of solute, a strong electrolyte is a better conductor than a weak electrolyte (Fig. I.3).

An aqueous strong electrolyte solution consists of ions that are free to move through the solvent. The solutes in nonelectrolyte solutions

Methanol molecule

FIGURE I.2 In a nonelectrolyte solution, the solute remains as a molecule and does not break up into ions. Methanol, CH_3OH, is a nonelectrolyte and is present as molecules when it is dissolved in water.

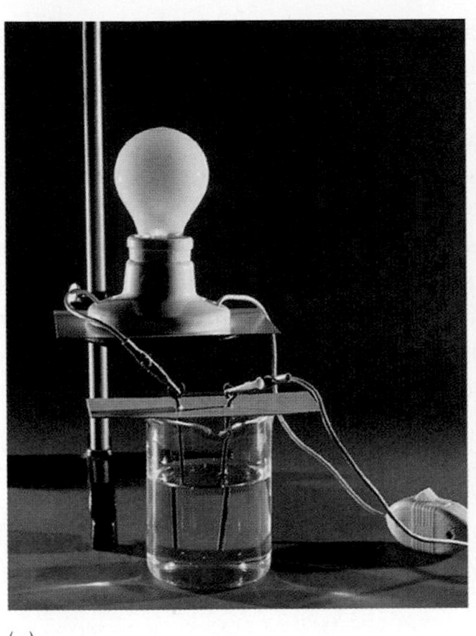
(a)

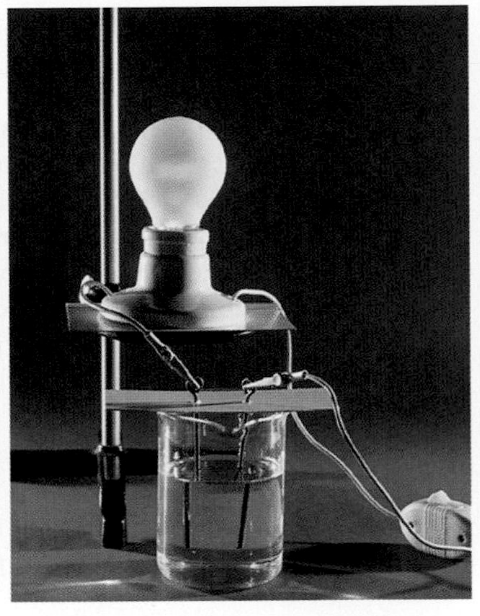
(b)

(c)

FIGURE I.3 Pure water is a poor conductor of electricity, as shown by the very dim glow of the bulb in the circuit on the left (a). However, when ions are present, as in an electrolyte solution, the solution does conduct. The ability of the solution to conduct is low when the solute is a weak electrolyte (b), but significant when the solute is a strong electrolyte (c), even when the solute concentration is the same in each instance.

Media Link **Lab video:** Fig I.3, The conductance of electrolytes

are present as molecules. Only a small fraction of the solute molecules in weak electrolyte solutions are present as ions.

Self-Test I.1A Identify each of the following substances as an electrolyte or a nonelectrolyte and predict which will conduct electricity when dissolved in water: (a) NaOH; (b) Br_2.

[*Answer:* (a) Ionic compound, so a strong electrolyte, conducts electricity; (b) molecular compound, so a nonelectrolyte, does not conduct electricity]

Self-Test I.1B Identify each of the following substances as an electrolyte or a nonelectrolyte and predict which will conduct electricity when dissolved in water: (a) ethanol, $CH_3CH_2OH(aq)$; (b) $Pb(NO_3)_2(aq)$.

I.2 Precipitation Reactions

Consider what happens when we pour a solution of sodium chloride (a strong electrolyte) into a solution of silver nitrate (another strong electrolyte). A solution of sodium chloride contains Na^+ cations and Cl^- anions. Similarly, a solution of silver nitrate, $AgNO_3$, contains Ag^+ cations and NO_3^- anions. When we mix these two aqueous solutions, we

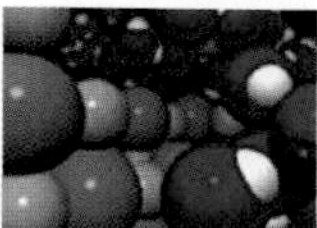

View animation of AgCl precipitating.

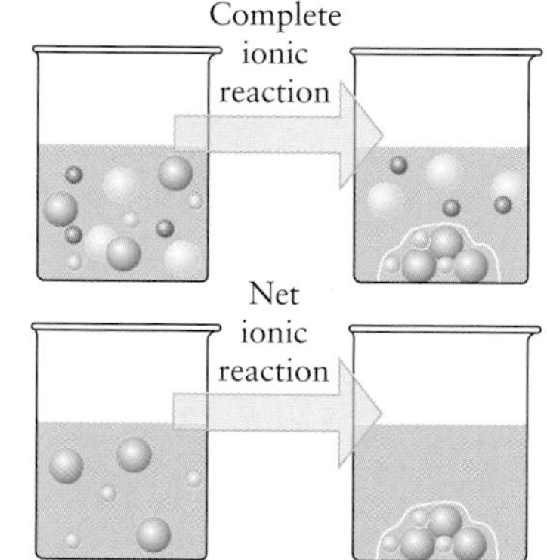

Media Link Animation: Fig I.4, Silver chlorate precipitation

FIGURE I.4 Silver chloride precipitates immediately when silver nitrate solution is added to a solution of sodium chloride. If we imagine the removal of the spectator ions from the complete ionic reaction (middle), we can focus on the essential process, the net ionic reaction (bottom).

immediately get a white **precipitate**, a cloudy, finely divided solid deposit. Analysis shows that the precipitate is silver chloride, AgCl, an insoluble white solid (Fig. I.4). The colorless solution remaining above the precipitate in our example contains dissolved Na^+ cations and NO_3^- anions. These ions remain in solution because sodium nitrate, $NaNO_3$, is soluble in water.

In a **precipitation reaction**, an insoluble solid product forms when we mix two strong electrolyte solutions. When an insoluble substance is formed in water, it immediately precipitates. In the chemical equation for a precipitation reaction, we use (aq) to indicate substances that are dissolved in water and (s) to indicate the solid that has precipitated:

$$AgNO_3(aq) + NaCl(aq) \longrightarrow AgCl(s) + NaNO_3(aq)$$

A precipitation reaction occurs when solutions of two strong electrolytes are mixed and react to form an insoluble solid.

I.3 Ionic and Net Ionic Equations

A complete ionic equation for a precipitation reaction shows all the species of ions in solution explicitly. For example, the complete ionic equation for the silver chloride precipitation reaction shown in Fig. I.4 is

$$Ag^+(aq) + NO_3^-(aq) + Na^+(aq) + Cl^-(aq) \longrightarrow AgCl(s) + Na^+(aq) + NO_3^-(aq)$$

Because the Na^+ and NO_3^- ions appear as both reactants and products, they play no direct role in the reaction. They are **spectator ions,** ions that are present while the reaction takes place but remain unchanged, like spectators at a sports event. Because spectator ions remain unchanged, we can cancel them on each side of the arrow in the ionic equation:

$$Ag^+(aq) + \cancel{NO_3^-(aq)} + \cancel{Na^+(aq)} + Cl^-(aq) \longrightarrow AgCl(s) + \cancel{Na^+(aq)} + \cancel{NO_3^-(aq)}$$

This cancellation leaves the **net ionic equation** for the reaction, the chemical equation that displays the net change occurring in the reaction:

$$Ag^+(aq) + Cl^-(aq) \longrightarrow AgCl(s)$$

In other words, a net ionic equation shows only the net chemical change that occurs (see Fig. I.4). It shows that the Ag^+ ions supplied by one solution combine with the Cl^- ions supplied by the other solution and that these ions precipitate as solid silver chloride, AgCl. A net ionic equation focuses our attention on the actual process taking place and emphasizes the change that results from the chemical reaction.

A complete ionic equation expresses a reaction in terms of the ions that are present in solution; a net ionic equation is the chemical equation that remains after the cancellation of the spectator ions.

EXAMPLE I.1 Writing a net ionic equation

The chemical equation for the precipitation reaction that occurs when concentrated aqueous solutions of barium nitrate, $Ba(NO_3)_2$, and ammonium iodate, NH_4IO_3, are mixed to form insoluble barium iodate, $Ba(IO_3)_2$, is

$$Ba(NO_3)_2(aq) + 2\ NH_4IO_3(aq) \longrightarrow Ba(IO_3)_2(s) + 2\ NH_4NO_3(aq)$$

Write the net ionic equation for the reaction.

STRATEGY First, write and balance the complete ionic equation, showing all the dissolved ions as they actually exist in solution, as separate, charged ions. Insoluble solids are shown as complete compounds. Next, cancel the spectator ions, the ions in solution on both sides of the arrow.

SOLUTION The complete ionic equation, with all the dissolved ions written as they exist in the solutions before and after mixing, is

$$Ba^{2+}(aq) + \cancel{2\ NO_3^-(aq)} + \cancel{2\ NH_4^+(aq)} + 2\ IO_3^-(aq) \longrightarrow Ba(IO_3)_2\ (s) + \cancel{2\ NH_4^+(aq)} + \cancel{2\ NO_3^-(aq)}$$

The spectator ions are NH_4^+ and NO_3^-. When they are canceled, we obtain the net ionic equation:

$$Ba^{2+}(aq) + 2\ IO_3^-(aq) \longrightarrow Ba(IO_3)_2\ (s)$$

Notice how much easier it is to see which species actually react by looking at the net ionic equation rather than at either the complete ionic equation or the chemical equation itself.

Self-Test I.2A Write the net ionic equation for the reaction in which aqueous solutions of colorless silver nitrate and yellow sodium chromate react to give a precipitate of red silver chromate.

[*Answer:* $2\ Ag^+(aq) + CrO_4^{2-}\ (aq) \longrightarrow Ag_2CrO_4(s)$]

Self-Test I.2B The mercury(I) ion, Hg_2^{2+}, consists of two Hg^+ ions joined together. Write the net ionic equation for the reaction in which colorless aqueous solutions of mercury(I) nitrate, $Hg_2(NO_3)_2$, and potassium phosphate, K_3PO_4, react to give a white precipitate of mercury(I) phosphate.

I.4 Putting Precipitation to Work

Precipitation reactions have two main applications. One is to make compounds. The strategy is to choose starting solutions that form a precipitate of the desired insoluble compound when they are mixed. Then we can separate the insoluble compound from the reaction mixture by filtration. A second application is in chemical analysis. In **qualitative analysis**—the determination of the substances present in a sample—the formation of a precipitate is used to confirm the identity of certain ions. In **quantitative analysis**, the aim is to determine the amount of each sub-

Section 11.16 describes the use of precipitates in qualitative analysis in more detail.

stance or element present. In particular, in **gravimetric analysis**, the amount of substance present is determined by measurements of mass. In this application, an insoluble compound is precipitated, the precipitate is filtered off and weighed, and from its mass the amount of one of the original substances in one of the solutions is calculated (Fig. I.5). Gravimetric analysis is used in environmental monitoring to find out how much of a heavy metal ion, such as lead or mercury, is in a sample of water.

For this step, we use the techniques described in Section L.

Table I.1 summarizes the observed solubility patterns of common ionic compounds in water. Notice that all nitrates and all common compounds of the Group 1 metals are soluble, so they make useful starting solutions for precipitation reactions. Any spectator ions can be used, provided they remain in solution and do not otherwise react. For example, the table shows that mercury(I) iodide, Hg_2I_2, is insoluble. It is formed as a precipitate whenever solutions containing Hg_2^{2+} ions and I^- ions are mixed:

$$Hg_2^{2+}(aq) + 2\,I^-(aq) \longrightarrow Hg_2I_2(s)$$

The net ionic equation will be the same when any soluble mercury(I) compound is mixed with any soluble iodide.

The solubility rules in Table I.1 are used to predict the outcome of precipitation reactions.

FIGURE I.5 A step in a gravimetric analysis. An ion has come out of solution as part of a precipitate and is being filtered. The filter paper, which has a known mass, will then be dried and weighed, thereby allowing the mass of the precipitate to be determined.

TABLE I.1 Solubility Rules for Inorganic Compounds

Soluble Compounds	Insoluble Compounds
compounds of Group 1 elements ammonium (NH_4^+) compounds	carbonates (CO_3^{2-}), chromates (CrO_4^{2-}), oxalates ($C_2O_4^{2-}$), and phosphates (PO_4^{3-}), **except** those of the Group 1 elements and NH_4^+
chlorides (Cl^-), bromides (Br^-), and iodides (I^-), **except** those of Ag^+, Hg_2^{2+}, and Pb^{2+}*	
nitrates (NO_3^-), acetates ($CH_3CO_2^-$), chlorates (ClO_3^-), and perchlorates (ClO_4^-)	sulfides (S^{2-}), **except** those of the Group 1 and 2 elements and NH_4^+
sulfates (SO_4^{2-}), **except** those of Ca^{2+}, Sr^{2+}, Ba^{2+}, Pb^{2+}, Hg_2^{2+}, and Ag^+‡	hydroxides (OH^-) and oxides (O^{2-}), **except** those of the Group 1 and 2 elements†

* $PbCl_2$ is slightly soluble.

† $Ca(OH)_2$ and $Sr(OH)_2$ are sparingly (slightly) soluble; $Mg(OH)_2$ is only very slightly soluble.

‡ Ag_2SO_4 is slightly soluble.

EXAMPLE I.2 Predicting the outcome of a precipitation reaction

Predict the precipitate likely to be formed when aqueous solutions of sodium phosphate and lead(II) nitrate are mixed. Write the net ionic equation for the reaction.

STRATEGY Decide which ions are present in the mixed solutions and consider all possible combinations. Use the solubility rules in Table I.1 to decide which combination corresponds to an insoluble compound and write the net ionic equation to match.

SOLUTION The mixed solution will contain Na^+, PO_4^{3-}, Pb^{2+}, and NO_3^- ions. All nitrates and all compounds of Group 1 metals are soluble, but phosphates of other elements are generally insoluble. Hence, we can predict that Pb^{2+} and PO_4^{3-} ions will form an insoluble compound and that lead(II) phosphate, $Pb_3(PO_4)_2$, will precipitate.

$$3\ Pb^{2+}(aq) + 2\ PO_4^{3-}(aq) \longrightarrow Pb_3(PO_4)_2(s)$$

The Na^+ and NO_3^- ions are spectators so they are omitted from the net ionic equation.

Self-Test I.3A Predict the identity of the precipitate that forms, if any, when aqueous solutions of ammonium sulfide and copper(II) sulfate are mixed, and write the net ionic equation for the reaction.
[*Answer:* Copper(II) sulfide; $Cu^{2+}(aq) + S^{2-}(aq) \longrightarrow CuS(s)$]

Self-Test I.3B Suggest two solutions that can be mixed to prepare strontium sulfate, and write the net ionic equation for the reaction.

SKILLS YOU SHOULD HAVE MASTERED

1 Identify substances as electrolytes or nonelectrolytes on the basis of the formulas of the solutes, Self-Test I.1.

2 Construct balanced complete ionic and net ionic equations for reactions involving ions, Example I.1.

3 Use the solubility rules to select appropriate solutions that, when mixed, will produce a desired precipitate.

4 Identify any precipitate that may form on mixing two given solutions, Example I.2.

EXERCISES

I.1 Classify the following substances as strong electrolytes, weak electrolytes, or nonelectrolytes: (a) CH_3OH; (b) $CaBr_2$; (c) KI.

I.2 Classify the following substances as strong electrolytes, weak electrolytes, or nonelectrolytes: (a) HCl; (b) KOH; (c) CH_3COOH.

I.3 Use the information in Table I.1 to classify the following ionic compounds as soluble or insoluble in water: (a) potassium phosphate K_3PO_4; (b) lead(II) chloride, $PbCl_2$; (c) cadmium sulfide, CdS; (d) barium sulfate, $BaSO_4$.

I.4 Use the information in Table I.1 to classify the following ionic compounds as soluble or insoluble in water: (a) zinc acetate, $Zn(CH_3CO_2)_2$; (b) iron(III) hydroxide, $Fe(OH)_3$; (c) silver iodide, AgI; (d) copper(II) acetate, $Cu(CH_3CO_2)_2$.

I.5 What are the principal solute species present in aqueous solutions of (a) NaI; (b) Ag_2CO_3; (c) $(NH_4)_3PO_4$; (d) $FeSO_4$?

I.6 What are the principal solute species present in aqueous solutions of (a) $PbSO_4$; (b) K_2CO_3; (c) K_2CrO_4; (d) Hg_2Cl_2?

I.7 (a) When aqueous solutions of iron(III) sulfate, $Fe_2(SO_4)_3$, and sodium hydroxide were mixed, a precipitate formed. Write the formula of the precipitate. (b) Does a precipitate form when aqueous solutions of silver nitrate, $AgNO_3$, and potassium carbonate are mixed? If so, write the formula of the precipitate. (c) Aqueous solutions of lead(II) nitrate, $Pb(NO_3)_2$, and sodium acetate are mixed. Does a precipitate form? If so, write the formula of the precipitate.

I.8 (a) Solid ammonium nitrate, NH_4NO_3, and solid calcium chloride were placed in water and the mixture stirred. Is the formation of a precipitate expected? If so, write the formula of the precipitate. (b) Solid magnesium carbonate, $MgCO_3$, and solid sodium nitrate were mixed, placed in water, and stirred. What is observed? If a precipitate is present, write its formula. (c) Aqueous solutions of sodium sulfate, Na_2SO_4, and barium chloride are mixed. What is observed? If a precipitate is present, write its formula.

I.9 When the solution in Beaker 1 is mixed with the solution in Beaker 2, a precipitate forms. Using the table below, write the net ionic equation describing the formation of the precipitate, and then identify the spectator ions.

Beaker 1	Beaker 2
(a) $FeCl_2(aq)$	$Na_2S(aq)$
(b) $Pb(NO_3)_2(aq)$	$KI(aq)$
(c) $Ca(NO_3)_2(aq)$	$K_2SO_4(aq)$
(d) $Na_2CrO_4(aq)$	$Pb(NO_3)_2(aq)$
(e) $Hg_2(NO_3)_2(aq)$	$K_2SO_4(aq)$

I.10 The contents of Beaker 1 are mixed with those of Beaker 2. If a reaction occurs, write the net ionic equation and indicate the spectator ions.

Beaker 1	Beaker 2
(a) $K_2SO_4(aq)$	$NaNO_3(aq)$
(b) $H_3PO_4(aq)$	$CaCl_2(aq)$
(c) $K_2S(aq)$	$NH_4NO_3(aq)$
(d) $NiSO_4(aq)$	$(NH_4)_2CO_3(aq)$
(e) $HNO_3(aq)$	$Ba(OH)_2(aq)$

I.11 The five procedures below each result in the formation of a precipitate. For each reaction, write the chemical equations describing the formation of the precipitate: the overall equation, the complete ionic equation, and the net ionic equation. Identify the spectator ions.
(a) $(NH_4)_2CrO_4(aq)$ is mixed with $BaCl_2(aq)$.
(b) $CuSO_4(aq)$ is mixed with $Na_2S(aq)$.
(c) $FeCl_2(aq)$ is mixed with $(NH_4)_3PO_4(aq)$.
(d) Potassium oxalate, $K_2C_2O_4(aq)$, is mixed with $Ca(NO_3)_2(aq)$.
(e) $NiSO_4(aq)$ is mixed with $Ba(NO_3)_2(aq)$.

I.12 The five procedures below each result in the formation of a precipitate. For each reaction, write the chemical equations describing the formation of the precipitate: the overall equation, the complete ionic equation, and the net ionic equation. Identify the spectator ions.
(a) $AgNO_3(aq)$ is mixed with $Na_2CO_3(aq)$.
(b) $Pb(NO_3)_2(aq)$ is mixed with $KI(aq)$.
(c) $Ba(OH)_2(aq)$ is mixed with $H_2SO_4(aq)$.
(d) $(NH_4)_2S(aq)$ is mixed with $Cd(NO_3)_2(aq)$.
(e) $KOH(aq)$ is mixed with $CuCl_2(aq)$.

I.13 Write the balanced, complete ionic, and net ionic equations corresponding to each of the following reactions:
(a) $Ba(CH_3CO_2)_2(aq) + Li_2CO_3(aq) \rightarrow BaCO_3(s) + LiCH_3CO_2(aq)$
(b) $NH_4Cl(aq) + Hg_2(NO_3)_2(aq) \rightarrow NH_4NO_3(aq) + Hg_2Cl_2(s)$
(c) $Cu(NO_3)_2(aq) + Ba(OH)_2(aq) \rightarrow Cu(OH)_2(s) + Ba(NO_3)_2(aq)$

I.14 Write the balanced, complete ionic, and net ionic equations corresponding to each of the following reactions:
(a) $Pb(NO_3)_2(aq) + K_3PO_4(aq) \rightarrow Pb_3(PO_4)_2(s) + KNO_3(aq)$
(b) $K_2S(aq) + Bi(NO_3)_3(aq) \rightarrow Bi_2S_3(s) + KNO_3(aq)$

(c) $Na_2C_2O_4(aq) + Mn(CH_3CO_2)_2(aq) \rightarrow MnC_2O_4(s) + NaCH_3CO_2(aq)$

I.15 Suggest two soluble ionic compounds that, when mixed together in water, result in the following net ionic equations:
(a) $2\ Ag^+(aq) + CrO_4^{2-}(aq) \rightarrow Ag_2CrO_4(s)$
(b) $Ca^{2+}(aq) + CO_3^{2-}(aq) \rightarrow CaCO_3(s)$, the reaction responsible for the deposition of chalk hills and sea urchin spines
(c) $Cd^{2+}(aq) + S^{2-}(aq) \rightarrow CdS(s)$, a yellow substance used to color glass

I.16 Suggest two soluble ionic compounds that, when mixed together in water, result in the following net ionic equations:
(a) $2\ Ag^+(aq) + SO_4^{2-}(aq) \rightarrow Ag_2SO_4(s)$
(b) $Mg^{2+}(aq) + 2\ OH^-(aq) \rightarrow Mg(OH)_2(s)$, the suspension present in milk of magnesia
(c) $3\ Ca^{2+}(aq) + 2\ PO_4^{3-}(aq) \rightarrow Ca_3(PO_4)_2(s)$, the major component of phosphate rock

I.17 Write the net ionic equation for the formation of each insoluble compound in aqueous solution: (a) silver sulfate, Ag_2SO_4; (b) mercury(II) sulfide, HgS, used as an electrolyte in some primary batteries; (c) calcium phosphate, $Ca_3(PO_4)_2$, a component of bones and teeth. (d) Select two soluble ionic compounds that, when mixed in solution, form each of the above insoluble compounds. Identify the spectator ions.

I.18 Write the net ionic equation for the formation of each of the following insoluble compounds in aqueous solution: (a) lead(II) chromate, $PbCrO_4$, the yellow pigment that has been used for centuries in oil paints; (b) aluminum phosphate, $AlPO_4$, used in cements and as an antacid (c) iron(II) hydroxide, $Fe(OH)_2$; (d) Select two soluble ionic compounds that, when mixed in solution, form each of the above insoluble compounds. Identify the spectator ions.

J ACIDS AND BASES

Early chemists applied the term *acid* to substances that had a sharp or sour taste. Vinegar, for instance, contains acetic acid, CH_3COOH. Aqueous solutions of substances they called *bases* were recognized by their soapy feel. Fortunately, there are less hazardous ways of recognizing acids and bases. For instance, acids and bases change the color of certain dyes known as indicators (Fig. J.1). One of the most famous indicators is litmus, a vegetable dye obtained from a lichen. Aqueous solutions of acids turn litmus red; aqueous solutions of bases, which are also known as *alkalis*, turn it blue. Later (in Chapter 10) we shall see that the electronic instrument known as a "pH meter" provides a rapid way of identifying a solution as acidic or basic: a pH reading *lower* than 7 denotes an acidic solution and a reading *higher* than 7 denotes a basic solution. The advantage of an electronic instrument is that it can easily be connected to monitoring equipment.

The pH scale is described in more detail in Section 10.5.

J.1 Acids and Bases in Aqueous Solution

Chemists debated the concepts of acid and base for many years before precise definitions emerged. Among the first useful definitions was the one proposed by the Swedish chemist Svante Arrhenius in about 1884:

> An **acid** is a compound that contains hydrogen and reacts with water to form hydrogen ions.
>
> A **base** is a compound that produces hydroxide ions in water.

We shall call such substances **Arrhenius acid and bases.** For instance, HCl is an Arrhenius acid, because it releases a hydrogen ion, H^+ (a proton), when it dissolves in water; CH_4 is not an Arrhenius acid, because it does not release hydrogen ions in water. Sodium hydroxide is an Arrhenius base because OH^- ions go into solution when it dissolves; ammonia is also an Arrhenius base, because it produces OH^- ions by reacting with water:

$$NH_3(aq) + H_2O(l) \longrightarrow NH_4^+(aq) + OH^-(aq) \qquad (1)$$

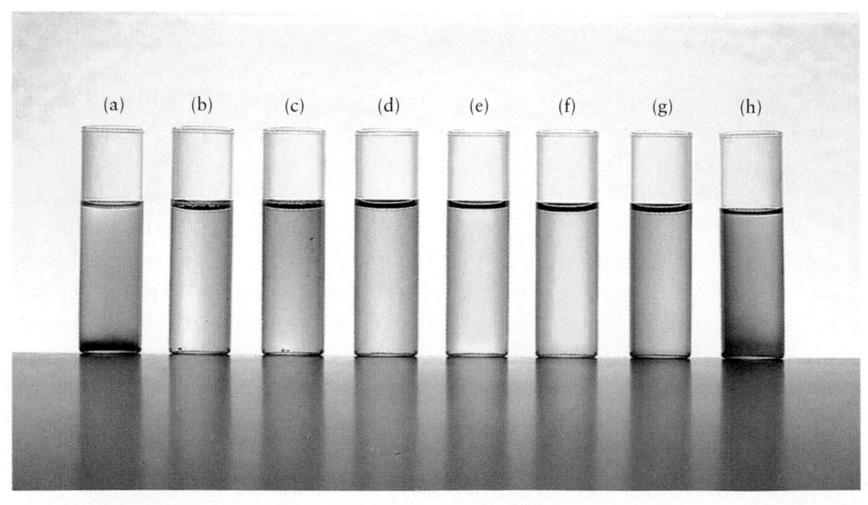

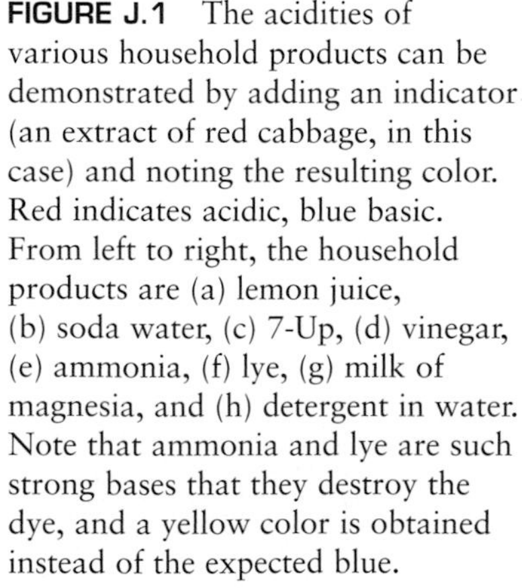
FIGURE J.1 The acidities of various household products can be demonstrated by adding an indicator (an extract of red cabbage, in this case) and noting the resulting color. Red indicates acidic, blue basic. From left to right, the household products are (a) lemon juice, (b) soda water, (c) 7-Up, (d) vinegar, (e) ammonia, (f) lye, (g) milk of magnesia, and (h) detergent in water. Note that ammonia and lye are such strong bases that they destroy the dye, and a yellow color is obtained instead of the expected blue.

Sodium metal, although it produces OH^- ions when it reacts with water, does not count as an Arrhenius base because it is an element, not a compound, as the definition requires.

The problem with the Arrhenius definitions is that they are specific to one particular solvent, water. When chemists studied nonaqueous solvents, such as liquid ammonia, they found that a number of substances showed the same pattern of acid-base behavior, but plainly the Arrhenius definitions could not be used. A major advance in our understanding of what it means to be an acid or a base came in 1923 when two chemists working independently, Thomas Lowry in England and Johannes Brønsted in Denmark, came up with the same idea. Their insight was to realize that the key process responsible for the properties of acids and bases was the transfer of a proton (a hydrogen ion) from one substance to another. The **Brønsted-Lowry definition** of acids and bases is as follows:

Acids and bases are described in more detail in Chapter 10.

An **acid** is a proton donor.

A **base** is a proton acceptor.

We call such substances "Brønsted acids and bases" or just plain "acids and bases" because the Brønsted-Lowry definition is the one commonly accepted today and the one we shall use throughout this text.

The definitions of acid and base are expanded even further in Section 2.13, which introduces Lewis acids and bases.

We consider acids first. When a molecule of an acid dissolves in water, it donates a hydrogen ion, H^+, to one of the water molecules and forms a *hydronium ion*, H_3O^+ (**1**). For example, when hydrogen chloride, HCl, dissolves in water, it releases a hydrogen ion, and the resulting solution consists of hydronium ions and chloride ions:

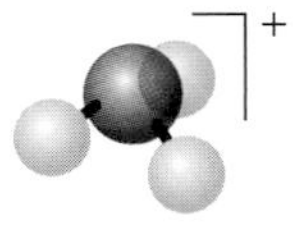

1 Hydronium ion, H_3O^+

$$HCl(aq) + H_2O(l) \longrightarrow H_3O^+(aq) + Cl^-(aq)$$

Hydrogen chloride, HCl, and nitric acid, HNO_3, are Brønsted acids. The molecules of both compounds contain hydrogen atoms that they can donate as protons to other substances. Both compounds form hydronium ions in water. Methane, CH_4, is not a Brønsted acid. Although it contains hydrogen, it does not release hydrogen ions to other substances readily. Acetic acid, CH_3COOH, releases *one* hydrogen ion (from the hydrogen atom of the carboxyl group, –COOH (**2**)) to water and other substances. Like HCl and HNO_3, acetic acid is a **monoprotic acid**, an acid that can donate only one proton from each molecule. Sulfuric acid, H_2SO_4, can release both its hydrogens as ions—one much more readily than the other—so it is an example of a **polyprotic acid**, an acid that can donate more than one proton from each molecule.

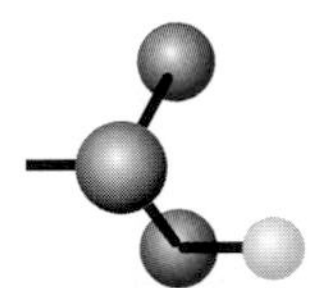

2 Carboxyl group, –COOH

The **acidic hydrogen atom** in a compound is the hydrogen atom that can be released as a proton. It is often written as the first element in a molecular formula, as in HCl and HNO_3. However, for organic acids, such as acetic acid, it is more informative to write the formulas to show the carboxyl group, —COOH, explicitly, thereby making it easier to remember that the H atom in this group of atoms is the acidic one. Thus, we can immediately recognize from their formulas that HCl, H_2CO_3 (carbonic acid), H_2SO_4 (sulfuric acid), and C_6H_5COOH (benzoic acid) are acids, but that CH_4, NH_3 (ammonia), and $CH_3CO_2^-$ (the acetate ion) are not. The common oxoacids, acids containing oxygen, were introduced in Section D and are listed in Table D.1.

Now consider bases, species that accept protons. First, we can see that hydroxide ions are bases, because they accept protons from acids to form molecules of water:

$$OH^-(aq) + CH_3COOH(aq) \longrightarrow H_2O(l) + CH_3CO_2^-(aq)$$

It follows that sodium hydroxide, because it supplies OH^- ions, which accept protons, is a Brønsted base. Ammonia is a base because, as we see from Eq. 1, it accepts protons from water and forms NH_4^+ ions.

Acids are molecules or ions that are proton donors. Bases are molecules or ions that are proton acceptors.

Self-Test J.1A Which of the following are (Brønsted) acids or bases in water? (a) HNO_3; (b) C_6H_6; (c) KOH; (d) C_3H_5COOH.

[*Answer:* (a) and (d) are acids; (b) is neither an acid nor a base; (c) is a base]

Self-Test J.1B Which of the following are (Brønsted) acids or bases in water? (a) KCl; (b) HClO; (c) HF; (d) $Ca(OH)_2$.

J.2 Strong and Weak Acids and Bases

Acids and bases are electrolytes and are classified in similar terms. We saw in Section I.1 that electrolytes are classified as strong or weak according to the extent to which they are present as ions in solution. Strong electrolytes are present in solution virtually entirely as ions, whereas weak electrolytes are present in solution as both ions and molecules. When these concepts are extended to the formation of ions by proton transfer, we use the term **deprotonation** to mean loss of a proton and **protonation** to mean the gain of a proton. We then classify acids and bases as follows:

The terms *ionized* and *dissociated* are also commonly used instead of "deprotonated."

A **strong acid** is completely deprotonated in solution.

A **weak acid** is incompletely deprotonated in solution.

A **strong base** is completely protonated in solution.

A **weak base** is incompletely protonated in solution.

The term *ionized* is also widely used in this context to mean "protonated."

By "completely deprotonated," we mean that each acid molecule or ion has lost the proton of its acidic hydrogen atom by transferring it as a hydrogen ion to a solvent molecule. "Completely protonated" means that each base species has acquired a proton. By "incompletely deprotonated" or "incompletely protonated," we mean that only a fraction (and usually a very tiny fraction) of the acid molecules or ions has lost acidic hydrogen atoms as protons, or only a tiny fraction of the base species has acquired protons.

Once again, we examine acids first. Hydrogen chloride is a strong acid in water. A solution of hydrogen chloride and water, which we call hydrochloric acid, contains hydronium ions, chloride ions, and virtually no HCl molecules.

Why do we say "virtually"? *Complete* deprotonation is an ideal model; in practice, an insignificantly small fraction of acid molecules may keep their protons.

$$HCl(g) + H_2O(l) \longrightarrow H_3O^+(aq) + Cl^-(aq)$$

Acetic acid, on the other hand, is a weak acid in water. Only a small fraction of its molecules undergo deprotonation, according to the equation

$$CH_3COOH(aq) + H_2O(l) \longrightarrow H_3O^+(aq) + CH_3CO_2^-(aq)$$

and the solution consists principally of CH_3COOH molecules (Fig. J.2). Later in the text (Chapter 10), we shall see how to show that 0.1 M $CH_3COOH(aq)$ contains only about one $CH_3CO_2^-$ ion for every hundred acetic acid molecules that were added.

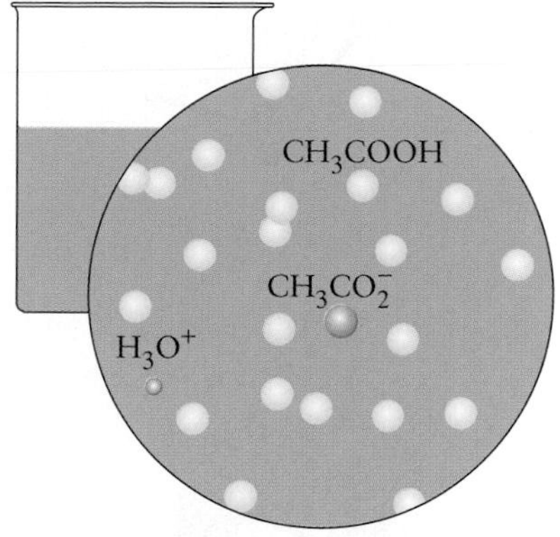

FIGURE J.2 Acetic acid, like all carboxylic acids, is a weak acid in water. This classification means that most of it remains as acetic acid molecules, CH_3COOH; however, a small proportion of these molecules donate a hydrogen ion to a water molecule to form hydronium ions, H_3O^+, and acetate ions, $CH_3CO_2^-$.

Table J.1 lists all the common strong acids in water. They include three acids that are often found as reagents in laboratories—hydrochloric acid, nitric acid, and sulfuric acid (with respect to the loss of one proton from each H_2SO_4 molecule). Most acids are weak in water. All carboxylic acids are weak in water.

Now consider strong and weak bases. The common strong bases are easy to remember: they are the alkali metal and alkaline earth metal hydroxides and oxides, such as calcium oxide (Table J.1). When the oxide dissolves in water, the oxide ions, O^{2-}, accept protons to form hydroxide ions:

$$O^{2-}(aq) + H_2O(l) \longrightarrow 2\ OH^-(aq)$$

Hydroxides, such as sodium hydroxide and calcium hydroxide, are also strong bases in water. There is an interesting point here: if an OH^- ion accepts a proton from a water molecule, then the water molecule donating the proton becomes a hydroxide ion, so there is no net change:

$$H_2O(l) + OH^-(aq) \longrightarrow OH^-(aq) + H_2O(l)$$

Even though a hydroxide ion is a strong base and is protonated in water, it effectively survives! All other common bases are weak bases in water. Ammonia is a weak base in water. In its aqueous solutions, it exists almost entirely as NH_3 molecules, with just a small proportion—usually fewer than one in a hundred molecules at normal concentrations—of NH_4^+ cations and OH^- anions. Other common weak bases are the

TABLE J.1 The Strong Acids and Bases in Water

Strong acids	Strong bases
hydrobromic acid, $HBr(aq)$	Group 1 hydroxides
hydrochloric acid, $HCl(aq)$	alkaline earth metal hydroxides*
hydroiodic acid, $HI(aq)$	Group 1 and 2 oxides
nitric acid, HNO_3	
perchloric acid, $HClO_4$	
chloric acid, $HClO_3$	
sulfuric acid, H_2SO_4 (to HSO_4^-)	

* $Ca(OH)_2$, $Sr(OH)_2$, $Ba(OH)_2$.

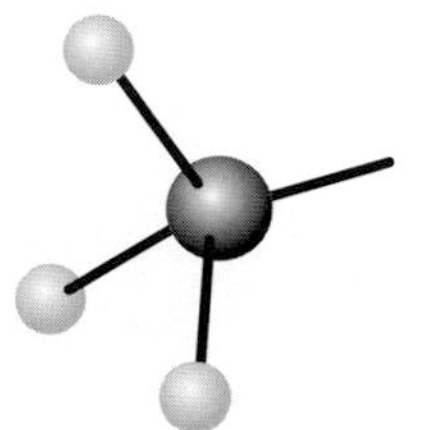

3 Methyl, $-CH_3$

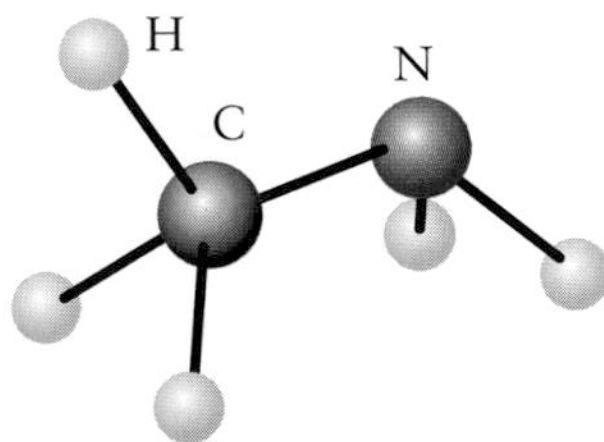

4 Methylamine, CH_3NH_2

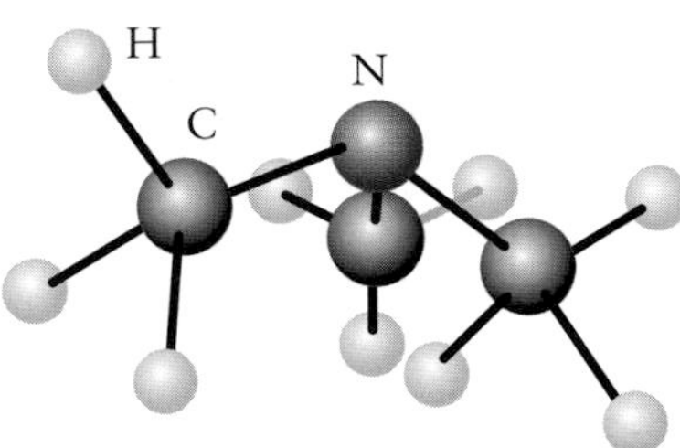

5 Trimethylamine, $(CH_3)_3N$

amines, smelly compounds that are derived from ammonia by replacement of one or more of its hydrogen atoms by an organic group. For example, the replacement of one hydrogen atom in NH_3 by a methyl group, $-CH_3$ (**3**), results in methylamine, CH_3NH_2 (**4**). The replacement of all three hydrogen atoms in NH_3 by methyl groups results in trimethylamine, $(CH_3)_3N$ (**5**), a substance found in decomposing fish and on unwashed dogs.

Strong acids (the acids listed in Table J.1) are completely deprotonated in solution; weak acids (all other acids) are not. Strong bases (the metal oxides and hydroxides listed in Table J.1) are completely protonated in solution. Weak bases (ammonia and its organic derivatives, the amines) are only partially protonated in solution.

J.3 Neutralization

The reaction between an acid and a base is called a **neutralization reaction**, and the ionic compound produced in the reaction is called a **salt.** The general form of a neutralization reaction of a strong acid and strong base in water is

$$\text{Acid} + \text{base} \longrightarrow \text{salt} + \text{water}$$

The name *salt* is taken from ordinary table salt, sodium chloride, the ionic product of the reaction between hydrochloric acid and sodium hydroxide:

$$HCl(aq) + NaOH(aq) \longrightarrow NaCl(aq) + H_2O(l)$$

In any neutralization reaction, the cation of the salt is provided by the base and the anion is provided by the acid. Another example is the reaction between nitric acid and barium hydroxide:

$$2\,HNO_3(aq) + Ba(OH)_2(aq) \longrightarrow Ba(NO_3)_2(aq) + 2\,H_2O(l)$$

The barium nitrate remains in solution as Ba^{2+} and NO_3^- ions.

We saw in Section I that the net chemical change in a precipitation reaction is clarified by writing its net ionic equation. The same is true of neutralization reactions. First, we write the complete ionic equation for the neutralization reaction between nitric acid and barium hydroxide in water:

$$2\,H^+(aq) + 2\,NO_3^-(aq) + Ba^{2+}(aq) + 2\,OH^-(aq) \longrightarrow Ba^{2+}(aq) + 2\,NO_3^-(aq) + 2\,H_2O(l)$$

The ions common to both sides now cancel, and the net ionic equation of this reaction is therefore

$$2\,H^+(aq) + 2\,OH^-(aq) \longrightarrow 2\,H_2O(l)$$

which simplifies to

$$H^+(aq) + OH^-(aq) \longrightarrow H_2O(l)$$

The net outcome of *any* neutralization reaction between a strong acid and a strong base in water is the formation of water from hydrogen ions and hydroxide ions.

When we write the net ionic equation for the neutralization of a weak acid or a weak base, we use the molecular form of the weak acid or base, because molecules are the dominant species in solution. For example, we would write the net ionic equation for the reaction of the weak acid HCN with the strong base NaOH in water as

$$HCN(aq) + OH^-(aq) \longrightarrow H_2O(l) + CN^-(aq)$$

Similarly, the net ionic equation for the reaction of the weak base ammonia with the strong acid HCl in water is

$$NH_3(aq) + H^+(aq) \rightarrow NH_4{}^+(aq)$$

In a neutralization reaction in water, an acid reacts with a base to produce a salt (and water if the base is strong); the net outcome of the reaction between solutions of a strong acid and a strong base is the formation of water from hydrogen ions and hydroxide ions.

Self-Test J.2A What acid and base solutions could you use to prepare rubidium nitrate? Write the chemical equation for the neutralization.
[***Answer:*** $HNO_3(aq) + RbOH(aq) \longrightarrow RbNO_3(aq) + H_2O(l)$]

Self-Test J.2B Write the chemical equation for a neutralization reaction in which calcium phosphate is produced.

FUNDAMENTALS

SKILLS YOU SHOULD HAVE MASTERED

1 Describe the chemical properties of acids and bases.

2 Classify substances as acids or bases, Self-Test J.1.

3 Identify common strong acids and bases, Table J.1.

4 Predict the outcome of neutralization reactions and write their chemical equations, Self-Test J.2.

EXERCISES

J.1 Identify each compound as either an acid or a base: (a) NH_3; (b) HCl; (c) NaOH; (d) H_2SO_4; (e) $Ba(OH)_2$.

J.2 Classify each compound as either an acid or a base: (a) HNO_3; (b) CH_3NH_2, a derivative of ammonia; (c) CH_3COOH; (d) KOH; (e) $HClO_4$.

J.3 Complete and write the overall equation, the complete ionic equation, and the net ionic equation for the following acid-base reactions. If the substance is a weak acid or base, leave it in its molecular form in writing the equations.
(a) $HF(aq) + NaOH(aq) \rightarrow$
(b) $(CH_3)_3N(aq) + HNO_3(aq) \rightarrow$
(c) $LiOH(aq) + HI(aq) \rightarrow$

J.4 Complete and write the overall equation, the complete ionic equation, and the net ionic equation for the following acid-base reactions. If the substance is a weak acid or base, leave it in its molecular form in writing the equations.
(a) $H_3PO_4(aq) + KOH(aq) \rightarrow$
(Phosphoric acid H_3PO_4 is a triprotic acid. Write the equation for complete reaction with KOH.)

(b) $Ba(OH)_2(aq) + CH_3COOH(aq) \rightarrow$
(c) $Mg(OH)_2(aq) + HClO_3(aq) \rightarrow$

J.5 Select an acid and a base for a neutralization reaction that results in the formation of (a) potassium bromide; (b) zinc nitrite; (c) calcium cyanide, $Ca(CN)_2$; (d) potassium phosphate. Write the balanced equations for each reaction.

J.6 Identify the salt that is produced from the acid-base neutralization reaction between (a) potassium hydroxide and acetic acid, CH_3COOH; (b) ammonia and hydroiodic acid; (c) barium hydroxide and sulfuric acid (both H atoms react); (d) sodium hydroxide and hydrocyanic acid, HCN. Write the full ionic equation for each reaction.

J.7 Identify the acid and the base in the following reactions:
(a) $CH_3NH_2(aq) + H_3O^+(aq) \rightarrow CH_3NH_3^+(aq) + H_2O(l)$
(b) $C_2H_5NH_2(aq) + HCl(aq) \rightarrow C_2H_5NH_3^+(aq) + Cl^-(aq)$
(c) $CaO(s) + 2\ HI(aq) \rightarrow CaI_2(aq) + H_2O(l)$

J.8 Identify the acid and the base in the following reactions:
(a) $CH_3COOH(aq) + NH_3(aq) \rightarrow NH_4^+(aq) + CH_3CO_2^-(aq)$
(b) $(CH_3)_3N(aq) + HCl(aq) \rightarrow (CH_3)_3NH^+(aq) + Cl^-(aq)$
(c) $O^{2-}(aq) + H_2O(l) \rightarrow 2\ OH^-(aq)$

The third of the three major classes of chemical reactions are the redox reactions. The variety of these reactions is remarkable. Many common reactions, like combustion, corrosion, photosynthesis, the metabolism of food, and the extraction of metals from their ores appear to be completely different. However, to a chemist's eye, they are all examples of a single type of process. Once again, we see how chemistry gives us the ability to see the extraordinary complexity of the world as the outcome of much simpler events going on below the surface.

K.1 Oxidation and Reduction

Let's examine some of these redox reactions and look for their common characteristic. First, consider the reaction between magnesium and oxygen, which produces magnesium oxide (Fig. K.1); this reaction is used in fireworks to produce white sparks. It is also used, less agreeably, in tracer ammunition and incendiary devices. The reaction of magnesium and oxygen is a classic example of an *oxidation reaction*, in the original sense of the term—"reaction with oxygen." In the course of the reaction, the Mg atoms in the solid magnesium lose electrons to form Mg^{2+} ions and the O atoms in the molecular oxygen gain electrons to form O^{2-} ions:

$$2\,Mg(s) + O_2(g) \longrightarrow 2\,Mg^{2+}(s) + 2\,O^{2-}(s) \quad (\text{as } 2\,MgO(s))$$

A similar reaction takes place when magnesium reacts with chlorine to produce magnesium chloride:

$$Mg(s) + Cl_2(g) \longrightarrow Mg^{2+}(s) + 2\,Cl^{-}(s) \quad (\text{as } MgCl_2(s))$$

Because the *pattern* of reaction is the same, it makes sense to regard the second reaction as an "oxidation" of magnesium even though no oxygen is involved. In each case, the common feature is the loss of electrons from magnesium and their transfer to another reactant. The loss of electrons from one species to another is now recognized as the essential event in an oxidation, and chemists now define **oxidation** as the loss of electrons, regardless of the species to which the electrons migrate.

We can often recognize loss of electrons by noting the increase in the charge of a species. This rule also applies to anions, as in the oxidation of bromide ions (charge −1) to bromine (charge 0) in a reaction like the one used commercially to make bromine (Fig. K.2):

$$2\,NaBr(s) + Cl_2(g) \longrightarrow 2\,NaCl(s) + Br_2(l)$$

Here, sodium bromide (more specifically, the bromide ion) is oxidized to bromine by the chlorine gas.

The name *reduction* originally referred to extraction of a metal from its oxide, often by reaction with hydrogen, carbon, or carbon monoxide. One example is the reduction of iron(III) oxide by carbon monoxide in the manufacture of steel:

$$Fe_2O_3(s) + 3\,CO(g) \xrightarrow{\Delta} 2\,Fe(l) + 3\,CO_2(g)$$

Media Link **Lab video:** Fig. K.1, Reactions of magnesium and oxygen

FIGURE K.1 An example of an oxidation reaction: magnesium burning brightly in air. Magnesium also burns brightly in water and carbon dioxide; consequently, magnesium fires are very difficult to extinguish.

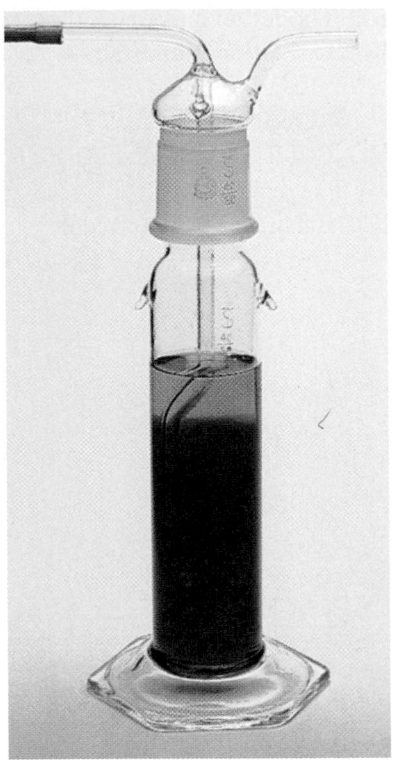

FIGURE K.2 When chlorine is bubbled through a solution of bromide ions, it oxidizes the ions to bromine, which colors the solution reddish brown.

In this reaction, an oxide of an element is converted to the free element, so we can suspect that oxidation and reduction are closely related. Is there a common underlying pattern? Specifically, can reduction also be discussed in terms of the transfer of electrons?

It is easy to identify the pattern. In the reduction of iron(III) oxide, Fe^{3+} ions present in Fe_2O_3 are converted into uncharged Fe atoms. For this conversion to take place, the Fe^{3+} ions must gain electrons to neutralize their positive charges. That turns out to be the pattern common to all reductions: in a **reduction**, an atom *gains* electrons from another species. Whenever the charge on a species is decreased (as from Fe^{3+} to Fe), we say that reduction has taken place. The same rule applies if the charge is negative. For example, when chlorine is converted to chloride ions in the reaction

$$2\ NaBr(s) + Cl_2(g) \longrightarrow 2\ NaCl(s) + Br_2(l)$$

the charge decreases from 0 (in Cl_2) to -1 (in Cl^-), and we conclude that in this reaction chlorine has been reduced.

Self-Test K.1A Identify the species that have been oxidized or reduced in the reaction $3\ Ag^+(aq) + Al(s) \longrightarrow 3\ Ag(s) + Al^{3+}(aq)$.

[*Answer:* Al(s) is oxidized, $Ag^+(aq)$ is reduced]

Self-Test K.1B Identify the species that have been oxidized or reduced in the reaction $2\ Cu^+(aq) + I_2(s) \longrightarrow 2\ Cu^{2+}(aq) + 2\ I^-(aq)$.

We have seen that oxidation is electron loss and reduction is electron gain. Electrons are real particles and cannot just be "lost." Therefore, whenever a reaction involves the oxidation of a species, it must be accompanied by the reduction of another species. Oxidation or reduction taken separately is like one hand clapping: one transfer must occur in conjunction with the other for reaction to take place. For instance, in the reaction between chlorine and sodium bromide, the bromide ions are oxidized and the chlorine molecules are reduced. Because oxidation is always accompanied by reduction, chemists speak of **redox reactions**, oxidation-reduction reactions, rather than simply oxidation reactions or reduction reactions.

Oxidation is electron loss; reduction is electron gain. A redox reaction is a combination of oxidation and reduction.

K.2 Oxidation Numbers: Keeping Track of Electrons

How do we recognize redox reactions? We need to decide whether electrons have migrated from one species to another. The loss or gain of electrons is easy to identify for monatomic ions, because we can monitor the charges of the species. Thus, when Br^- ions are converted to bromine atoms (which go on to form Br_2 molecules), we know that each Br^- ion must have lost an electron and hence that it has been oxidized. When O_2 forms oxide ions, O^{2-}, we know that each oxygen atom must have gained two electrons and therefore that it has been reduced. The difficulty arises when the transfer of electrons is accompanied by the transfer of atoms. For example, is chlorine gas oxidized or reduced when it is converted to hypochlorite ions, ClO^-? The presence of oxygen in the ion suggests oxidation, but the presence of the negative charge suggests reduction!

Chemists have found a way to keep track of electrons by assigning an "oxidation number" to each element. The **oxidation number** is defined so that

The concept of oxidation number was first introduced in Section D.

Oxidation corresponds to an *increase* in oxidation number.

Reduction corresponds to a *decrease* in oxidation number.

A redox reaction, therefore, is any reaction in which there are changes in oxidation numbers.

The oxidation number of an element in a monatomic ion is the same as its charge. So, the oxidation number of magnesium is +2 when it is present as Mg^{2+} ions, and the oxidation number of chlorine is −1 when it is present as Cl^- ions. The oxidation number of the elemental form of an element is 0, so magnesium metal has oxidation number 0 and chlorine in the form of Cl_2 molecules also has oxidation number 0. When magnesium combines with chlorine, the oxidation numbers change as follows:

$$Mg(s) + Cl_2(g) \longrightarrow MgCl_2(s)$$

Mg(0) → Mg(+2), Cl(0) → Cl(−1)

We see that magnesium has been oxidized and chlorine has been reduced. Similarly, in the reaction between sodium bromide and chlorine,

$$2\ NaBr(s) + Cl_2(g) \longrightarrow 2\ NaCl(s) + Br_2(l)$$

Na(+1) → Na(+1), Br(−1) → Br(0), Cl(0) → Cl(−1)

In this reaction, bromine has been oxidized and chlorine has been reduced, but the sodium ions remain unchanged as Na^+.

When an element is part of a compound or a polyatomic ion, we assign its oxidation number by using the procedure in Toolbox K.1. Note that you will hear chemists speaking of both "oxidation number" and "oxidation state." An oxidation *number* is the number assigned according to the rules in the Toolbox. An **oxidation state** is the actual condition of a species with a specified oxidation number. Thus, an element *has* a certain oxidation number and *is in* the corresponding oxidation state. For example, Mg^{2+} is the +2 oxidation state of magnesium; and in that state, magnesium has oxidation number +2.

Oxidation increases the oxidation number of an element; reduction decreases the oxidation number.

FUNDAMENTALS

TOOLBOX K.1 How to assign oxidation numbers

Conceptual Basis

Oxidation numbers are assigned according to a set of rules based on an ionic model of bonding. To assign an oxidation number, we imagine that each atom in a molecule, formula unit, or polyatomic ion is present in ionic form (which, of course, it may not be). The oxidation number is then taken to be the charge on each "ion." The "anion" is usually oxygen as O^{2-} or the element furthest to the right in the periodic table (actually, the most *electronegative* element, see Section 2.14). We then assign charges to the other atoms so as to balance the charge on the "anions."

Procedure

To assign an oxidation number to an element, we start with two simple rules:

1. The oxidation number of an element uncombined with other elements is 0.
2. The sum of the oxidation numbers of all the atoms in a species is equal to its total charge.

The oxidation numbers of elements in the compounds we are likely to meet at this stage in the text are assigned by using these two rules in conjunction with the following specific values:

- The oxidation number of hydrogen is +1 in combination with nonmetals and −1 in combination with metals.
- The oxidation number of elements in Groups 1 and 2 is equal to their group number.
- The oxidation number of all the halogens is −1 unless the halogen is in combination with oxygen or another halogen higher in the group. The oxidation number of fluorine is −1 in all its compounds.
- The oxidation number of oxygen is −2 in most of its compounds. Exceptions are its compounds with fluorine (in which case, the previous statement takes precedence) and its occurrence as peroxides (O_2^{2-}), superoxides (O_2^-), and ozonides (O_3^-).

How can we use this information . . . to determine the oxidation number of an element in a compound?

Here we assign the oxidation numbers of sulfur in (a) SO_2 and (b) SO_4^{2-}. First, we represent the oxidation number of sulfur by x. The oxidation number of oxygen is −2 in both compounds. (a) By rule 2, the sum of oxidation numbers of the atoms in the compound must be 0:

$$\text{Oxidation number of S} + [2 \times (\text{oxidation number of O})] = 0$$

That is, $x + [2 \times (-2)] = 0$. Therefore, the oxidation number of sulfur is $x = +4$. (b) By rule 2, the sum of oxidation numbers of the atoms in the ion is −2, so $x + [4 \times (-2)] = -2$. Therefore, $x = +6$. We can conclude that sulfur is more highly oxidized in the sulfate ion than in sulfur dioxide despite the negative charge of the ion.

Self-Test K.2A Find the oxidation numbers of sulfur, phosphorus, and nitrogen in (a) H_2S; (b) P_4O_6; (c) NO_3^-, respectively.

[*Answer:* (a) −2; (b) +3; (c) +5]

Self-Test K.2B Find the oxidation numbers of sulfur, nitrogen, and chlorine in (a) SO_3^{2-}; (b) NO_2^-; (c) $HClO_3$, respectively.

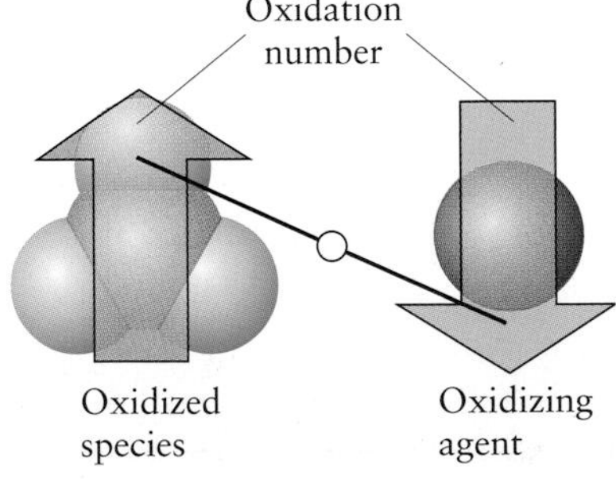

FIGURE K.3 The oxidizing agent is the species containing the element that undergoes a decrease in oxidation number. Here we see how the oxidation number of the species on the left is driven upward as the species on the right (the oxidizing agent) gains electrons.

K.3 Oxidizing and Reducing Agents

The recognition of a whole class of redox reactions introduces some new words into the language of chemistry. The species that *causes* oxidation in a redox reaction is called the **oxidizing agent** (or "oxidant"). The oxidizing agent in the reaction between sodium bromide and chlorine is the chlorine; in a combustion reaction, the oxidizing agent is oxygen. When an oxidizing agent acts, it accepts the electrons released by the species being oxidized (the chlorine becomes chloride ions, for instance). In other words, the oxidizing agent contains an element that undergoes a *decrease* in oxidation number (Fig. K.3). That is,

- The oxidizing agent in a redox reaction is the species that is reduced.

For instance, oxygen removes electrons from magnesium. Because it accepts those electrons, its oxidation number decreases from 0 to −2 (a reduction). Oxygen is therefore the oxidizing agent in this reaction. Oxidizing agents can be elements, ions, or compounds, and we will come across many examples later in the text.

The species that brings about reduction is called the **reducing agent** (or "reductant"). Because the reducing agent supplies electrons to the species being reduced, the reducing agent loses electrons. That is, the reducing agent contains an element that undergoes an increase in oxidation number (Fig. K.4). In other words,

- The reducing agent in a redox reaction is the species that is oxidized.

For example, magnesium metal supplies electrons to oxygen, bringing about the reduction of oxygen. In doing so, the oxidation number of magnesium increases from 0 to +2 (an oxidation). It is the reducing agent in the reaction of magnesium and oxygen.

It follows from this discussion that, to identify the oxidizing and reducing agents in a redox reaction, we have to examine the oxidation numbers of the elements in the species involved. The reactant that contains an element that is reduced in the reaction is the oxidizing agent, and the reactant that contains an element that is oxidized is the reducing agent. For example, when a piece of zinc is placed in a copper(II) solution (Fig. K.5), the reaction is

$$Zn(s) + Cu^{2+}(aq) \rightarrow Zn^{2+}(aq) + Cu(s)$$

We see that the oxidation number of zinc changes from 0 to +2 (an oxidation) whereas that of copper decreases from +2 to 0 (a reduction). Therefore, because zinc is oxidized, zinc metal is the reducing agent in this reaction. Conversely, because copper is reduced, we can regard the copper(II) ions as the oxidizing agent.

Oxidation is brought about by an oxidizing agent, a species that contains an element that undergoes reduction. Reduction is brought about by a reducing agent, a species that contains an element that undergoes oxidation.

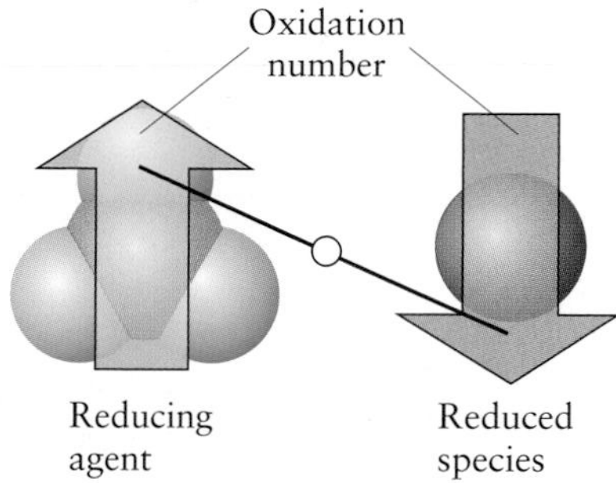

FIGURE K.4 The reducing agent is the species containing the element that undergoes an increase in oxidation number. Here we see how the oxidation number of the species on the right is driven down as the species on the left (the reducing agent) loses electrons.

Media Link **Lab video:** Fig. K.5, The oxidation of zinc metal by copper ions

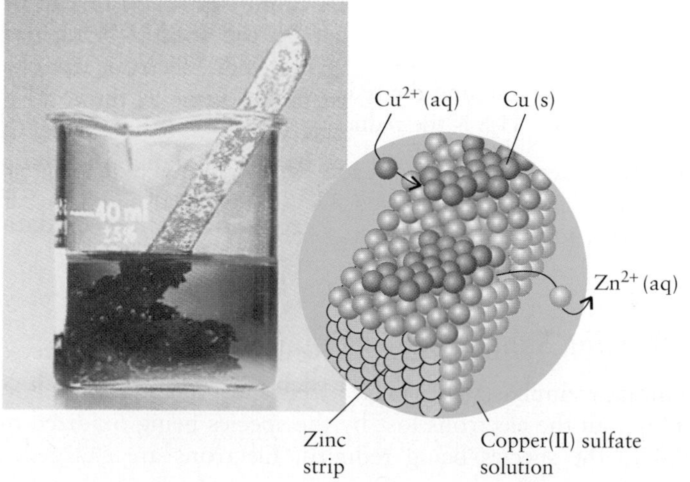

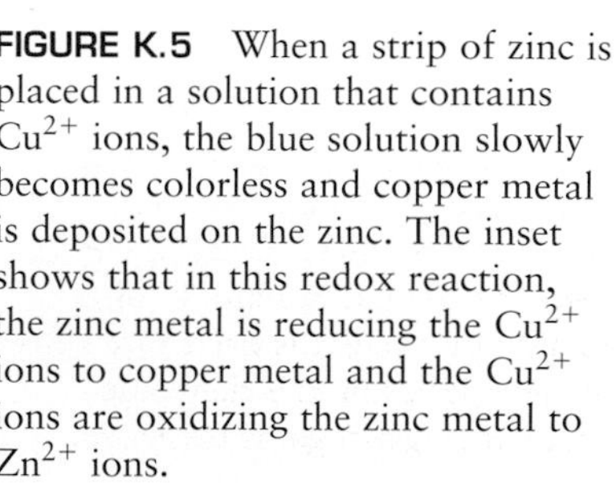

FIGURE K.5 When a strip of zinc is placed in a solution that contains Cu^{2+} ions, the blue solution slowly becomes colorless and copper metal is deposited on the zinc. The inset shows that in this redox reaction, the zinc metal is reducing the Cu^{2+} ions to copper metal and the Cu^{2+} ions are oxidizing the zinc metal to Zn^{2+} ions.

EXAMPLE K.1 **Identifying oxidizing agents and reducing agents**

When an acidic solution of potassium dichromate, $K_2Cr_2O_7$, is mixed with an iron(II) chloride solution, iron(III) ions and chromium(III) ions are formed. Identify the oxidizing agent and the reducing agent.

STRATEGY First, write the formulas for the species present and determine the oxidation numbers of the elements they contain. Oxidation is signaled by an increase in oxidation number; reduction is signaled by a decrease in oxidation number. The oxidizing agent is the species that, in the course of carrying out the oxidation, is itself reduced. Similarly, the reducing agent is the species that is itself oxidized.

SOLUTION Consider the fate of each reactant separately: $K_2Cr_2O_7$ produces Cr^{3+} ions; $FeCl_2$ produces Fe^{3+} ions. No changes to the potassium or chloride ions are mentioned, so we assume they are spectator ions. We need the oxidation number of chromium in the $Cr_2O_7^{2-}$ ion. Symbolizing that number by x, we calculate it from

$$\underset{Cr_2O_7^{2-}}{2x} + \underset{Cr_2O_7^{2-}}{[7 \times (-2)]} = \underset{Cr_2O_7^{2-}}{-2}$$

where $2x$ represents the total oxidation number of the two Cr atoms. Because $x = +6$, the oxidation number of chromium decreases from $+6$ to $+3$ in the reaction. The dichromate ion therefore undergoes reduction, so it is the oxidizing agent. At the same time, the oxidation number of iron increases from $+2$ to $+3$, so it undergoes oxidation. Fe^{2+} is the reducing agent.

Oxidizing agent:

$$K_2Cr_2O_7 + \ldots \longrightarrow 2\,Cr^{3+} + \ldots$$

Cr(+6) → Cr(+3), reduction

Reducing agent:

$$FeCl_2 + \ldots \longrightarrow Fe^{3+} + \ldots$$

Fe(+2) → Fe(+3), oxidation

Self-Test K.3A In the Claus process for the recovery of sulfur from natural gas and petroleum, hydrogen sulfide reacts with sulfur dioxide to form elemental sulfur and water: $2\,H_2S(g) + SO_2(g) \longrightarrow 3\,S(s) + 2\,H_2O(l)$. Identify the oxidizing agent and the reducing agent.

[*Answer:* H_2S is the reducing agent, SO_2 is the oxidizing agent]

Self-Test K.3B When sulfuric acid reacts with sodium iodide, sodium iodate and sulfur dioxide are produced. Identify the oxidizing and reducing agents in this reaction.

K.4 Balancing Simple Redox Equations

We have already emphasized that electrons can be neither lost nor created. Therefore, all the electrons lost by the species being oxidized must be transferred to the species being reduced. Electrons are charged, so the total charge of the reactants must be the same as the total charge of the products. This requirement means that, when balancing the chemical equation for a redox reaction, we have to balance the charges as well as the atoms.

For example, consider the net ionic equation for the oxidation of copper metal to copper(II) ions by silver ions (Fig. K.6):

$$Cu(s) + Ag^{+}(aq) \longrightarrow Cu^{2+}(aq) + Ag(s) \quad ⚠$$

At first glance, the equation appears to be balanced, because we see the same number of each kind of atom on each side. However, each copper atom has lost two electrons, whereas each silver atom has gained only one. To balance the electrons, we have to balance the charge. Therefore, we need to use a stoichiometric coefficient and write

$$Cu(s) + 2\,Ag^{+}(aq) \longrightarrow Cu^{2+}(aq) + 2\,Ag(s)$$

Now all the electrons lost by copper have been gained by silver, and everything is in balance.

Self-Test K.4A When tin metal is placed in contact with a solution of Fe^{3+} ions, it reduces the iron to iron(II) and is itself oxidized to tin(II) ions. Write the net ionic equation for the reaction.

[*Answer:* $Sn(s) + 2\,Fe^{3+}(aq) \longrightarrow Sn^{2+}(aq) + 2\,Fe^{2+}(aq)$]

Self-Test K.4B In aqueous solution, cerium(IV) ions oxidize iodide ions to solid diatomic iodine and are themselves reduced to cerium(III) ions. Write the net ionic equation for the reaction.

FUNDAMENTALS

Media Link **Animation:** Fig. K.6, Reaction of silver ions with copper metal

View an animation of silver ions reacting with copper.

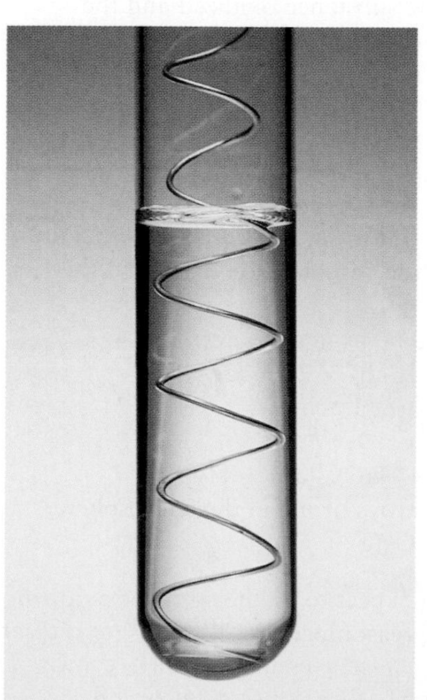

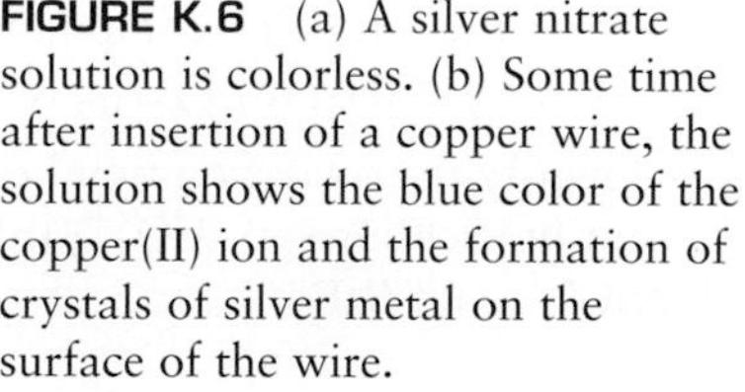

FIGURE K.6 (a) A silver nitrate solution is colorless. (b) Some time after insertion of a copper wire, the solution shows the blue color of the copper(II) ion and the formation of crystals of silver metal on the surface of the wire.

Some redox reactions, particularly those involving oxoanions, have complex chemical equations that require special balancing procedures. We meet examples, and see how to balance them, in Chapter 12.

When balancing the chemical equation for a redox reaction involving ions, the total charge on each side must be balanced.

SKILLS YOU SHOULD HAVE MASTERED

1 Determine the oxidation number of an element in an ion and in a compound, Toolbox K.1.

2 Identify the oxidizing and reducing agents in a reaction, Example K.1.

3 Write and balance chemical equations for simple redox reactions, Self-Test K.4.

EXERCISES

K.1 Write the balanced equations for the following skeletal redox reactions:
(a) $NO_2(g) + O_3(g) \rightarrow N_2O_5(g) + O_2(g)$
(b) $S_8(s) + Na(s) \rightarrow Na_2S(s)$
(c) $Cr^{2+}(aq) + Sn^{4+}(aq) \rightarrow Cr^{3+}(aq) + Sn^{2+}(aq)$
(d) $As(s) + Cl_2(g) \rightarrow AsCl_3(l)$

K.2 Write the balanced equations for the following skeletal redox equations:
(a) $Hg^{2+}(aq) + Fe(s) \rightarrow Hg_2^{2+}(aq) + Fe^{3+}(aq)$
(b) $Pt^{4+}(aq) + H_2(g) \rightarrow Pt^{2+}(aq) + H^+(aq)$
(c) $Al(s) + Fe_2O_3(s) \rightarrow Fe(s) + Al_2O_3(s)$
(d) $La(s) + Br_2(l) \rightarrow LaBr_3(s)$

K.3 Write the balanced equations for the following redox reactions:
(a) Displacement of copper(II) ion from solution by magnesium metal:
$Mg(s) + Cu^{2+}(aq) \rightarrow Mg^{2+}(aq) + Cu(s)$
(b) Formation of iron(III) ion in the following reaction:
$Fe^{2+}(aq) + Ce^{4+}(aq) \rightarrow Fe^{3+}(aq) + Ce^{3+}(aq)$
(c) Synthesis of hydrogen chloride from its elements:
$H_2(g) + Cl_2(g) \rightarrow HCl(g)$
(d) Formation of rust (a simplified equation):
$Fe(s) + O_2(g) \rightarrow Fe_2O_3(s)$

K.4 Determine the oxidation number of the italicized element in the following compounds: (a) $\mathit{S}OCl_2$; (b)$\mathit{Se}O_3$; (c) N_2O_5; (d) $\mathit{N}O_2$; (e) $H\mathit{Br}O_2$; (f) $\mathit{Xe}F_2$.

K.5 Determine the oxidation number of the italicized element in the following compounds: (a) $H_2\mathit{C}O_3$; (b) $\mathit{Ge}O_2$; (c) N_2H_4; (d) P_4O_{10}; (e) S_2Cl_2; (f) P_4.

K.6 Identify the oxidation number of the italicized element in each ion: (a) $\mathit{Al}O_2^-$; (b) $\mathit{N}O_2^-$; (c) $\mathit{S}O_3^{2-}$; (d) $\mathit{Ni}Cl_4^{2-}$; (e) $\mathit{Br}F_3^+$.

K.7 Identify the oxidation number of the italicized element in each ion: (a) $\mathit{Zn}(OH)_4^{2-}$; (b) $\mathit{Pd}Cl_4^{2-}$; (c) $\mathit{U}O_2^{2+}$; (d) $\mathit{Si}F_6^{2-}$; (e) $\mathit{I}O^-$.

K.8 Some compounds of hydrogen and oxygen are exceptions to the common observation that H has an oxidation number of +1 and O has an oxidation number of −2. Assuming that each metal has the oxidation number of its most common ion, find the oxidation numbers of H and O in the following compounds: (a) KO_2; (b) $LiAlH_4$; (c) Na_2O_2; (d) NaH; (e) KO_3.

K.9 For the following reactions that are redox reactions, identify the substance oxidized and the substance reduced by the change in oxidation numbers.
(a) $CH_3OH(aq) + O_2(g) \rightarrow HCOOH(aq) + H_2O(l)$
(b) $2\ MoCl_5(s) + 5\ Na_2S(s) \rightarrow 2\ MoS_2(s) + 10\ NaCl(s) + S(s)$
(c) $3\ Tl^+(aq) \rightarrow 2\ Tl(s) + Tl^{3+}(aq)$

K.10 In each of the following reactions, use oxidation numbers to identify the substance oxidized and the substance reduced.
(a) Production of iodine from seawater:
$Cl_2(g) + 2\ I^-(aq) \rightarrow I_2(aq) + 2\ Cl^-(aq)$
(b) Reaction to prepare bleach:
$Cl_2(g) + 2\ NaOH(aq) \rightarrow NaCl(aq) + NaOCl(aq) + H_2O(l)$
(c) Reaction that destroys ozone in the stratosphere:
$NO(g) + O_3(g) \rightarrow NO_2(g) + O_2(g)$

K.11 Which do you expect to be the stronger oxidizing agent? Explain your reasoning. (a) Cl_2 or Cl^-; (b) N_2O_5 or N_2O.

K.12 Which do you expect to be the stronger oxidizing agent? Explain your reasoning. (a) KBrO or $KBrO_3$; (b) MnO_4^- or Mn^{2+}.

K.13 Identify the oxidizing agent and the reducing agent in each of the following reactions:
(a) $Zn(s) + 2\ HCl(aq) \rightarrow ZnCl_2(aq) + H_2(g)$, a simple means of preparing H_2 gas in the laboratory
(b) $2\ H_2S(g) + SO_2(g) \rightarrow 3\ S(s) + 2\ H_2O(l)$, a reaction used to produce sulfur from hydrogen sulfide, the "sour gas" in natural gas
(c) $B_2O_3(s) + 3\ Mg(s) \rightarrow 2\ B(s) + 3\ MgO(s)$, a preparation of elemental boron

K.14 Identify the oxidizing agent and the reducing agent in each of the following reactions:
(a) $2\ Al(l) + Cr_2O_3(s) \xrightarrow{\Delta} Al_2O_3(s) + 2\ Cr(l)$, an example of a thermite reaction used to obtain some metals from their ores
(b) $6\ Li(s) + N_2(g) \rightarrow 2\ Li_3N(s)$, a reaction that shows the similarity of lithium and magnesium
(c) $2\ Ca_3(PO_4)_2(s) + 6\ SiO_2(s) + 10\ C(s) \rightarrow P_4(g) + 6\ CaSiO_3(s) + 10\ CO(g)$,
a reaction for the preparation of elemental phosphorus

K.15 Would you choose an oxidizing agent or a reducing agent to make the following conversions?
(a) $ClO_3^-(aq) \rightarrow ClO_2(g)$
(b) $SO_4^{2-}(aq) \rightarrow S^{2-}(aq)$
(c) $Mn^{2+}(aq) \rightarrow MnO_2(s)$
(d) HCHO (formaldehyde) → HCOOH (formic acid)

K.16 The industrial production of sodium metal and chlorine gas makes use of the Downs process, in which molten sodium chloride is electrolyzed (Chapter 12). Write the balanced equation for the production of the two elements from molten sodium chloride. Which element is produced by oxidation and which by reduction?

K.17 Identify the oxidizing agent and the reducing agent for each of the following reactions:

(a) Production of tungsten metal from its oxide by the reaction:
$WO_3(s) + 3\ H_2(g) \rightarrow W(s) + 3\ H_2O(l)$
(b) Generation of hydrogen gas in the laboratory:
$Zn(s) + 2\ HCl(aq) \rightarrow H_2(g) + ZnCl_2(aq)$
(c) Production of metallic tin from tin(IV) oxide, the mineral cassiterite:

$SnO_2(s) + 2\ C(s) \xrightarrow{\Delta} Sn(l) + 2\ CO(g)$
(d) Hydrazine has a low molar mass and releases a significant amount of energy in its reaction with dinitrogen tetroxide. The combination is used as a rocket propellant:
$2\ N_2H_4(g) + N_2O_4(g) \rightarrow 3\ N_2(g) + 4\ H_2O(g)$

K.18 Classify the following reactions as precipitation, acid-base neutralization, or redox. If a precipitation reaction, write a net ionic equation; if a neutralization reaction, identify the acid and the base; if a redox reaction, identify the oxidizing agent and the reducing agent.
(a) Reaction used to measure the concentration of carbon monoxide in a gas stream:
$5\ CO(g) + I_2O_5(s) \rightarrow I_2(s) + 5\ CO_2(g)$
(b) Reaction often used to determine the amount of iodine in a sample:
$I_2(aq) + 2\ S_2O_3^{2-}(aq) \rightarrow 2\ I^-(aq) + S_4O_6^{2-}(aq)$
(c) Test for bromide ions in solution:
$AgNO_3(aq) + Br^-(aq) \rightarrow AgBr(s) + NO_3^-(aq)$
(d) Heating of uranium tetrafluoride with magnesium, one stage in the purification of uranium metal:
$UF_4(g) + 2\ Mg(s) \rightarrow U(s) + 2\ MgF_2(s)$

L REACTION STOICHIOMETRY

The quantitative interpretation of chemical reactions is the part of chemistry called **reaction stoichiometry**. Recall from Section H that a stoichiometric coefficient in a chemical equation tells us the relative number of moles of a substance that reacts or is produced. Thus, the stoichiometric coefficients in

$$N_2(g) + 3\,H_2(g) \longrightarrow 2\,NH_3(g)$$

tell us that if 1 mol N_2 reacts, then 3 mol H_2 will be consumed and 2 mol NH_3 will be produced. We summarize this information by writing

$$1\text{ mol } N_2 \simeq 3\text{ mol } H_2 \qquad 1\text{ mol } N_2 \simeq 2\text{ mol } NH_3$$

The sign $\simeq$ is read "is chemically equivalent to," and these expressions are called **stoichiometric relations.**

The term "chemically equivalent" refers only to relative amounts in a specific chemical reaction. Different reactions entail different stoichiometric relations.

Self-Test L.1A Write the stoichiometric relation between P_4 and Cl_2 in (a) the formation of phosphorus trichloride; (b) the formation of phosphorus pentachloride.

[*Answer:* (a) 1 mol $P_4 \simeq$ 6 mol Cl_2; (b) 1 mol $P_4 \simeq$ 10 mol Cl_2]

Self-Test L.1B Write the stoichiometric relation between N_2 and O_2 in (a) the formation of dinitrogen trioxide; (b) the formation of dinitrogen oxide.

L.1 Mole-to-Mole Predictions

Stoichiometry has many practical applications, such as predicting how much product can be formed in a reaction. For example, in the space shuttle fuel cell, oxygen reacts with hydrogen to produce water (Fig. L.1). Space shuttle engineers know how much oxygen the fuel cell can consume in a given period of time and also need to be able to calculate what amount of water can be produced in that time. Let's look at the calculation they would have to do to find out how much water is formed when 0.25 mol O_2 reacts with hydrogen.

First, we write the chemical equation for the reaction:

$$2\,H_2\,(g) + O_2\,(g) \longrightarrow 2\,H_2\,O(l)$$

FIGURE L.1 One of the three hydrogen-oxygen fuel cells used on the space shuttle. Although only one cell is needed to provide life-support electricity and drinking water, shuttle flight rules require that all three must be functional. In April 1987, a mission of space shuttle *Columbia* was cut short when one of the fuel cells malfunctioned.

Then we summarize the information that 1 mol O_2 reacts to form 2 mol H_2O by writing the stoichiometric relation between oxygen (the given substance) and water (the required substance):

$$1\text{ mol } O_2 \simeq 2\text{ mol } H_2O$$

Next, we use the stoichiometric relation to set up a conversion factor relating the given substance to the required substance:

$$\frac{\text{Substance required}}{\text{Substance given}} = \frac{2\text{ mol } H_2O}{1\text{ mol } O_2}$$

This factor, which is commonly called the **mole ratio** for the reaction, allows us to relate the amount of O_2 molecules consumed to the amount of H_2O molecules produced. We use it in the same way we use a conversion factor when we are converting units, as illustrated in Appendix 1B:

$$\text{Amount of } H_2O \text{ produced (mol)} = (0.25 \text{ mol } O_2) \times \left(\frac{2 \text{ mol } H_2O}{1 \text{ mol } O_2}\right) = 0.50 \text{ mol } H_2O$$

Note that the unit *mol* and the species (in this case, O_2 molecules) cancel. The general strategy for this type of calculation is summarized in (**1**).

Stoichiometric coefficients are exact numbers, so they do not limit the significant figures of stoichiometric calculations (see Appendix 1C).

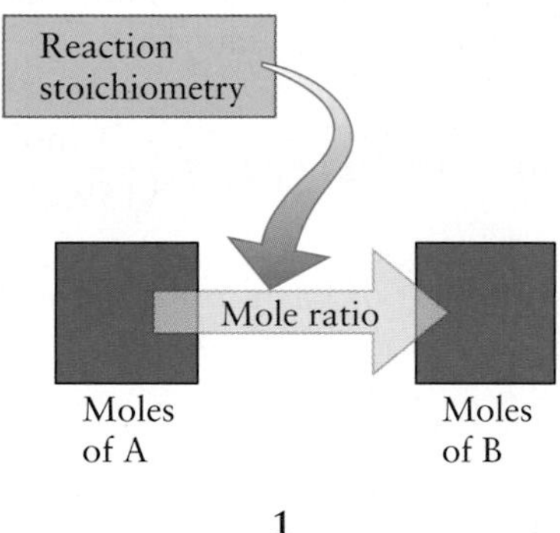

1

The balanced chemical equation for a reaction is used to set up the mole ratio, a factor that is used to convert the amount of one substance to the amount of another.

Self-Test L.2A What amount of NH_3 can be produced from 2.0 mol H_2 in the reaction $N_2(g) + 3\ H_2(g) \rightarrow 2\ NH_3(g)$?

[*Answer:* 1.3 mol NH_3]

Self-Test L.2B What amount of Fe atoms can be extracted from 25 mol Fe_2O_3?

L.2 Mass-to-Mass Predictions

To find out the mass of a product that can be formed from a known mass of a reactant, we use the mole ratio from the balanced equation, but convert between moles and masses by using molar masses. Essentially, we go through the following steps:

$$\text{Mass of reactant} \longrightarrow \text{moles of reactant} \longrightarrow \text{moles of product} \longrightarrow \text{mass of product}$$

For example, suppose we want to know the mass of iron that can be obtained from 10.0 g of iron(III) oxide, Fe_2O_3, present in iron ore by reduction with CO in a blast furnace:

$$Fe_2O_3(s) + 3\ CO(g) \longrightarrow 2\ Fe(s) + 3\ CO_2(g)$$

FUNDAMENTALS

To convert moles of Fe_2O_3 to moles of Fe, we use the stoichiometric relation between them,

$$1 \text{ mol } Fe_2O_3 \simeq 2 \text{ mol Fe}$$

in the form of a mole ratio. The molar mass of iron(III) oxide is 159.69 g·mol^{-1}, so we convert from grams to moles of Fe_2O_3:

$$\text{Amount of } Fe_2O_3 \text{ (mol)} = 10.0 \text{ g } Fe_2O_3 \times \left(\frac{1 \text{ mol}}{159.69 \text{ g}}\right)$$

Then we apply the mole ratio as a conversion factor to calculate the amount of Fe atoms produced in the reaction:

$$\text{Amount of Fe (mol)} = 10.0 \text{ g } Fe_2O_3 \times \left(\frac{1 \text{ mol}}{159.69 \text{ g}}\right) \times \left(\frac{2 \text{ mol Fe}}{1 \text{ mol } Fe_2O_3}\right)$$

Finally, we convert the amount of Fe atoms into the mass of iron by using the molar mass of iron, which is 55.85 g·mol^{-1}:

$$\begin{aligned}\text{Mass of Fe} \\ &= 10.0 \text{ g } Fe_2O_3 \times \left(\frac{1 \text{ mol}}{159.69 \text{ g}}\right) \times \left(\frac{2 \text{ mol Fe}}{1 \text{ mol } Fe_2O_3}\right) \times \left(\frac{55.85 \text{ g}}{1 \text{ mol}}\right) \\ &= 6.99 \text{ g Fe}\end{aligned}$$

In practice, because $n = m/M$ and $m = nM$ (Section E.2), most chemists find it easier to simply divide or multiply by the molar mass rather than setting it out as a conversion factor:

$$\begin{aligned}\text{Mass of Fe} &= \left(\frac{10.0 \text{ g } Fe_2O_3}{159.69 \text{ g·mol}^{-1}}\right) \times \left(\frac{2 \text{ mol Fe}}{1 \text{ mol } Fe_2O_3}\right) \times (55.85 \text{ g·mol}^{-1}) \\ &= 6.99 \text{ g Fe}\end{aligned}$$

This is the method used in the remainder of this textbook. Whichever method you use, to avoid rounding errors, carry out the conversions in a single step, as shown here.

In a mass-to-mass calculation, convert the given mass to amount in moles, apply the mole-to-mole conversion factor to obtain the amount required, and finally convert amount in moles to mass.

TOOLBOX L.1 How to carry out mass-to-mass calculations

Conceptual Basis

A chemical equation tells us the relations between the amounts (in moles) of each reactant and product. By using the molar masses as conversion factors, we can express these relations in terms of masses.

Procedure

The general procedure for mass-to-mass calculations, summarized in diagram (2), is

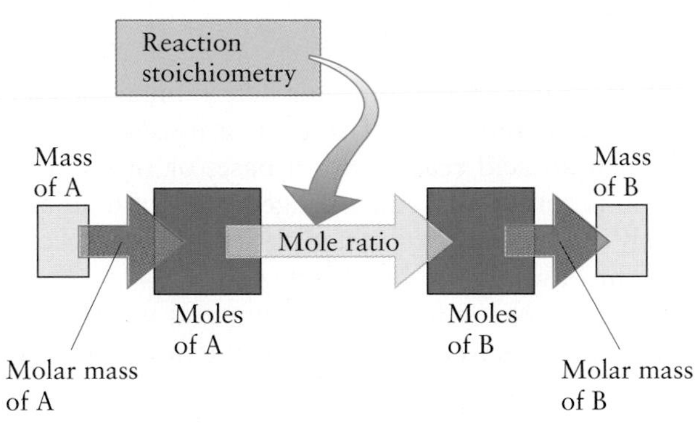

2

Step 1 Convert the given mass of one substance (in grams) to amount in moles by using its molar mass.

Step 2 Write the balanced chemical equation and use the mole ratio derived from the stoichiometric coefficients to convert from the given amount of one substance to the amount in moles of the other substance.

Step 3 Convert from amount in moles of the second substance to mass (in grams) by using its molar mass.

EXAMPLE L.1 Calculating the mass of reactant required to react with another reactant

What mass of aluminum is needed to reduce 10.0 kg of chromium(III) oxide to produce chromium metal? The chemical equation for the reaction is

$$2\ \text{Al(l)} + \text{Cr}_2\text{O}_3\text{(s)} \xrightarrow{\Delta} \text{Al}_2\text{O}_3\text{(s)} + 2\ \text{Cr(l)}$$

STRATEGY We use the three-step procedure in Toolbox L.1 but write the actual calculation as a single step. Convert any mass given in another mass unit or a multiple of grams (such as kilograms) to a mass in grams.

SOLUTION The conversion from kilograms to grams uses 1 kg = 10^3 g. The molar mass of Cr_2O_3 is 152.00 g·mol^{-1}. The stoichiometric relation between the two reactants is

$$1\ \text{mol Cr}_2\text{O}_3 \simeq 2\ \text{mol Al}$$

The conversion from amount of Al to mass makes use of the molar mass of aluminum, which is 26.98 g·mol^{-1}. Therefore,

$$\begin{aligned}\text{Mass of aluminum (g)} &= \left(\frac{10.0\ \text{kg Cr}_2\text{O}_3}{152.00\ \text{g}\cdot\text{mol}^{-1}}\right) \times \left(\frac{10^3\ \text{g}}{1\ \text{kg}}\right) \\ &\quad \times \left(\frac{2\ \text{mol Al}}{1\ \text{mol Cr}_2\text{O}_3}\right) \times (26.98\ \text{g}\cdot\text{mol}^{-1}) \\ &= 3.55 \times 10^3\ \text{g Al}\end{aligned}$$

We see that we need to use 3.55 kg of aluminum.

Self-Test L.3A Calculate the mass of potassium needed to react with 0.450 g of hydrogen gas to produce potassium hydride, KH.

[*Answer:* 17.5 g]

Self-Test L.3B Carbon dioxide can be removed from power plant exhaust gases by combining it with an aqueous slurry of calcium silicate: $2\ CO_2(g) + H_2O(l) + CaSiO_3(s) \rightarrow SiO_2(s) + Ca(HCO_3)_2(aq)$. What mass of $CaSiO_3$ (of molar mass 116.17 g·mol^{-1}) is needed to react completely with 0.300 kg of carbon dioxide?

FUNDAMENTALS

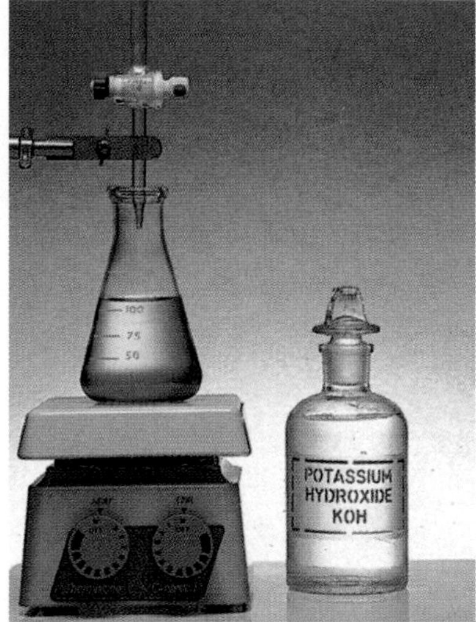

FIGURE L.2 The apparatus typically used for a titration: magnetic stirrer; flask containing the analyte; clamp; buret containing the titrant, in this case, potassium hydroxide.

The stoichiometric point is also called the "equivalence point."

L.3 Volumetric Analysis

One of the most common laboratory techniques for determining the concentration of a solute is **titration** (Fig. L.2). Titrations are usually either **acid-base titrations**, in which an acid reacts with a base, or **redox titrations**, in which the reaction is between a reducing agent and an oxidizing agent. Titrations are widely used to monitor water purity and blood composition and for quality control in the food industry.

In a titration, the volume of one solution is known and we measure the volume of the other solution required for complete reaction. The solution being analyzed is called the **analyte**, and a known volume is transferred into a flask, usually with a pipet. Then a solution containing a known concentration of reactant is measured into the flask from a narrow calibrated cylinder called a buret until all the analyte has reacted. The solution in the buret is called the **titrant**, and the difference between the initial and final volume readings of the buret tells us the volume of titrant that has drained into the flask. The determination of concentration or amount by measuring volume is called **volumetric analysis**.

In a typical acid-base titration, the analyte is a solution of a base and the titrant is a solution of an acid, or vice versa. An indicator, a water-soluble dye (Section J), helps us detect the **stoichiometric point**, the stage at which the volume of titrant added is exactly that required by the stoichiometric relation between titrant and analyte. For example, if we titrate hydrochloric acid containing a few drops of the indicator phenolphthalein, the solution is initially colorless. Because 1 mol NaOH reacts with each mole of HCl, at the stoichiometric point, the amount of NaOH added is exactly equal to the amount of HCl in the analyte. After the stoichiometric point, when excess base is present, the solution in the flask is basic and the indicator is pink. The indicator color change is sudden, so it is easy to detect the stoichiometric point. We then note the precise volume of titrant needed to cause the indicator to change color (Fig. L.3). Toolbox L.2 shows how to interpret a titration; the procedure is summarized in diagram (3), where solution A is the titrant and solution B is the analyte.

FIGURE L.3 An acid-base titration at the stoichiometric point. The indicator is phenolphthalein.

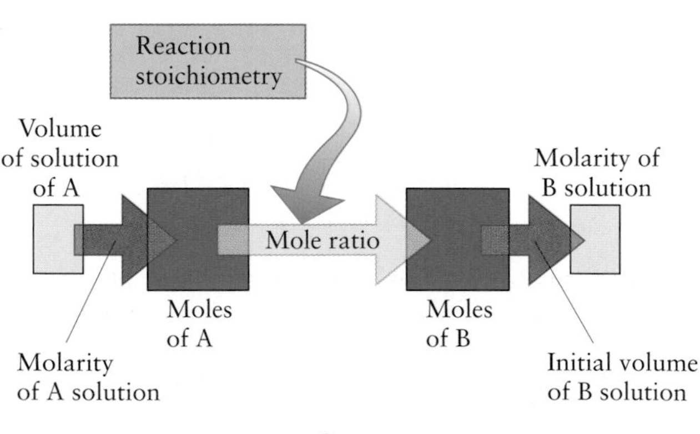

The stoichiometric relation between analyte and titrant species, together with the molarity of the titrant, is used in titrations to determine the molarity of the analyte.

TOOLBOX L.2 How to interpret a titration

Conceptual Basis

To interpret a titration, we need the stoichiometric relation from the chemical equation for the reaction. This relation is used to write the mole ratio in the usual way. The only new steps are to use the molarity of the titrant to convert from the volume of titrant to the amount of reactant in the titrant and the volume of analyte to convert to the concentration of reactant or the amount of reactant in the analyte.

Procedure

The data in a typical titration are the volume of analyte, the volume of titrant needed to reach the stoichiometric point, and the molarity of the titrant. Our goal is to determine the molarity of the analyte or the mass of reactant in the analyte. The strategy for reaching this goal has three steps:

Step 1 Use the volume of titrant and its molarity to calculate the amount in moles of titrant species added from the buret (see Toolbox G.1):

$$\text{Amount of titrant (moles)} = \text{volume of titrant used (liters)} \times \text{molarity of titrant (moles per liter)}$$

Because we need to keep track of the substances, we must be sure to indicate the exact species and concentration units, by writing, for example, 1.0 (mol HCl)·L^{-1} or 1.0 M HCl(aq).

Step 2 Determine the stoichiometric relation between the titrant and analyte from the chemical equation for the reaction and use it to convert the amount of titrant to amount of analyte. This step gives us a mole ratio of the form

$$\frac{\text{Substance required}}{\text{Substance given}}, \quad \text{which in this case is} \quad \frac{\text{analyte}}{\text{titrant}}$$

We then multiply the amount of titrant by this ratio:

$$\text{Amount of analyte present (moles)} = \text{amount of titrant used (moles)} \times \text{mole ratio}$$

Step 3 Calculate the initial molarity of the analyte by dividing the amount of solute by the initial volume of the solution:

$$\text{Molarity of analyte} = \frac{\text{amount of solute (moles)}}{\text{volume of solution (liters)}}$$

This procedure is illustrated in Example L.2.

If the mass of reactant in the analyte is required, instead of step 3, the molar mass of the reactant is used to convert moles to grams. This procedure is demonstrated in Example L.3. As usual in calculations with multiple steps, we minimize rounding errors by doing the numerical calculation in one step at the end.

FUNDAMENTALS

EXAMPLE L.2 Determining the molarity of an acid by titration

Oxalic acid is a poisonous compound but an excellent reducing agent. Suppose that 25.00 mL of a solution of oxalic acid, $H_2C_2O_4$ (**4**), which has two acidic protons, is titrated with 0.100 M NaOH(aq) and that the stoichiometric point is reached when 38.0 mL of the solution of base is added. Find the molarity of the oxalic acid solution.

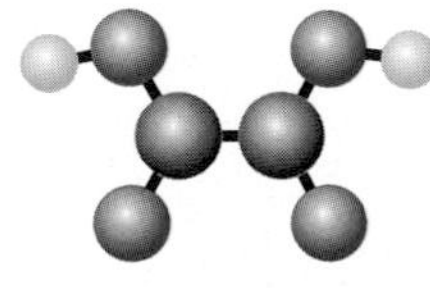

4 Oxalic acid, $(COOH)_2$

STRATEGY Because each acid molecule provides two protons, if the concentrations of acid and base were the same, the titration would require a volume of base equal to *twice* the volume of acid. The volume of base added to reach the stoichiometric point is less than twice the volume of acid, so we can anticipate that the acid is less concentrated than the base. Proceed as in Toolbox L.2.

SOLUTION *Step 1* The amount of NaOH added is

$$\begin{aligned}\text{Amount of NaOH} &= (38.0 \times 10^{-3}\ \text{L}) \times (0.100\ \text{mol NaOH}\cdot\text{L}^{-1}) \\ &= 38.0 \times 10^{-3} \times 0.100\ \text{mol NaOH}\end{aligned}$$

Step 2 The neutralization reaction is

$$H_2C_2O_4(aq) + 2\ NaOH(aq) \longrightarrow Na_2C_2O_4(aq) + 2\ H_2O(l)$$

It follows that the stoichiometric relation we require is

$$2\ \text{mol NaOH} \mathrel{\hat{=}} 1\ \text{mol}\ H_2C_2O_4$$

Therefore, the amount of $H_2C_2O_4$ in the original analyte is

$$\begin{aligned}&\text{Amount of } H_2C_2O_4\\ &\quad = (38.0 \times 10^{-3} \times 0.100\ \text{mol NaOH}) \times \frac{1\ \text{mol}\ H_2C_2O_4}{2\ \text{mol NaOH}}\\ &\quad = \tfrac{1}{2} \times (38.0 \times 10^{-3} \times 0.100)\ \text{mol}\ H_2C_2O_4\end{aligned}$$

Step 3 The molarity of the acid is therefore

$$\begin{aligned}\text{Molarity of } H_2C_2O_4(aq) &= \frac{\tfrac{1}{2} \times (38.0 \times 10^{-3} \times 0.100)\ \text{mol}\ H_2C_2O_4}{25.00 \times 10^{-3}\ \text{L}}\\ &= 0.0760\ (\text{mol}\ H_2C_2O_4)\cdot\text{L}^{-1}\end{aligned}$$

That is, the solution is 0.0760 M $H_2C_2O_4(aq)$.

Self-Test L.4A A student used a sample of hydrochloric acid that was known to contain 0.72 g of hydrogen chloride in 500.0 mL of solution to titrate 25.0 mL of a solution of calcium hydroxide. The stoichiometric point was reached when 15.1 mL of acid had been added. What was the molarity of the calcium hydroxide solution?

[*Answer:* 0.012 M $Ca(OH)_2(aq)$]

Self-Test L.4B Many abandoned mines have exposed nearby communities to the problem of acid mine drainage. Certain minerals, such as pyrite (FeS_2), decompose when exposed to air, forming solutions of sulfuric acid. The acidic mine water then drains into lakes and creeks, killing fish and other animals. At a mine in Colorado, a 16.45-mL sample of mine water was neutralized with 25.00 mL of 0.255 M KOH(aq). What is the molar concentration of H_2SO_4 in the water?

EXAMPLE L.3 Interpreting a redox titration

The iron content of ores can be determined by titrating a sample with a solution of potassium permanganate, $KMnO_4$. The ore is dissolved in hydrochloric acid, forming iron(II) ions, and the latter react with MnO_4^-:

$$5\ Fe^{2+}(aq) + MnO_4^-(aq) + 8\ H^+(aq) \longrightarrow 5\ Fe^{3+}(aq) + Mn^{2+}(aq) + 4\ H_2O(l)$$

The stoichiometric point is reached when the purple color of the permanganate ion persists. A sample of ore of mass 0.202 g was dissolved in hydrochloric acid and the resulting solution needed 16.7 mL of 0.0108 M $KMnO_4(aq)$ to reach the stoichiometric point. (a) What mass of iron(II) ions is present? (b) What is the mass percentage of iron in the ore sample?

STRATEGY (a) To obtain the amount of iron(II) in the analyte, we use the volume and concentration of the titrant. We follow the first two steps of the procedure in Toolbox L.2. Then we convert moles of Fe^{2+} ions to mass by using the molar mass of Fe^{2+}: because the mass of electrons is so small, use the molar mass of elemental iron for the molar mass of iron(II) ions. (b) Divide the mass of iron by the mass of the ore sample and multiply by 100%.

SOLUTION (a) From the chemical equation, the stoichiometric relation between the iron and permanganate ions is

$$5 \text{ mol } Fe^{2+} \simeq 1 \text{ mol } MnO_4^-$$

We can set up the mass calculation in one step:

$$\begin{aligned}\text{Mass of iron(II)} &= \overbrace{(1.67 \times 10^{-2}\text{ L}) \times (0.0108\text{ (mol } MnO_4^-)\cdot L^{-1})}^{\text{Amount of } MnO_4^-} \\ &\qquad \times \frac{5 \text{ mol } Fe^{2+}}{1 \text{ mol } MnO_4^-} \times 55.85 \text{ g}\cdot\text{mol}^{-1} \\ &= 0.0504 \text{ g } Fe^{2+}\end{aligned}$$

The sample contains 0.0504 g of iron.
(b) The mass percentage of iron in the ore is

$$\text{Mass percentage iron} = \frac{\overbrace{0.0504\text{ g}}^{\text{Mass of Fe}}}{\underbrace{0.202\text{ g}}_{\text{Mass of sample}}} \times 100\% = 25.0\%$$

Self-Test L.5A A 20.750-g sample of clay for use in making ceramics was analyzed to determine its iron content. The clay was washed with hydrochloric acid and the iron converted to iron(II) ions. The resulting solution was titrated with cerium(IV) sulfate solution:

$$Fe^{2+}(aq) + Ce^{4+}(aq) \longrightarrow Fe^{3+}(aq) + Ce^{3+}(aq)$$

In the titration, 13.45 mL of 1.340 M Ce_2SO_4(aq) was needed to reach the stoichiometric point. What is the mass percentage of iron in the clay?

[*Answer:* 9.70%]

Self-Test L.5B The amount of arsenic(III) oxide in a mineral can be determined by dissolving the mineral and titrating it with potassium permanganate:

$$24\, H^+(aq) + 5\, As_4O_6(s) + 8\, MnO_4^-(aq) + 18\, H_2O(l) \longrightarrow 8\, Mn^{2+}(aq) + 20\, H_3AsO_4(aq)$$

A sample of industrial waste was analyzed for the presence of arsenic(III) oxide by titration with 0.0100 M $KMnO_4$. It took 28.15 mL of the titrant to reach the stoichiometric point. How many grams of arsenic(III) oxide did the sample contain?

SKILLS YOU SHOULD HAVE MASTERED

1 Carry out mole-to-mole, mass-to-mole, and mass-to-mass calculations for any two species involved in a chemical reaction, Toolbox L.1 and Example L.1.

2 Calculate the molar concentration (molarity) of a solute from titration data, Toolbox L.2 and Example L.2.

3 Calculate the mass of a solute from titration data, Example L.3.

EXERCISES

L.1 Sodium thiosulfate, photographer's hypo, reacts with unexposed silver bromide in the film emulsion to form sodium bromide and a soluble compound of formula $Na_3[Ag(S_2O_3)_2]$:

$$2\ Na_2S_2O_3(aq) + AgBr(s) \longrightarrow NaBr(aq) + Na_3[Ag(S_2O_3)_2](aq)$$

(a) How many moles of $Na_2S_2O_3$ are needed to dissolve 1.0 mg of AgBr? (b) Calculate the mass of silver bromide that will produce 0.033 mol $Na_3[Ag(S_2O_3)_2]$.

L.2 Impure phosphoric acid for use in the manufacture of fertilizers is produced by the reaction of sulfuric acid on phosphate rock, of which a principal component is $Ca_3(PO_4)_2$. The reaction is

$$Ca_3(PO_4)_2(s) + 3\ H_2SO_4(aq) \longrightarrow 3\ CaSO_4(s) + 2\ H_3PO_4(aq)$$

(a) How many moles of H_3PO_4 can be produced from the reaction of 200 kg of H_2SO_4? (b) Determine the mass of calcium sulfate that is produced as a by-product of the reaction of 200 mol $Ca_3(PO_4)_2$.

L.3 The solid fuel in the booster stage of the space shuttle is a mixture of ammonium perchlorate and aluminum powder. Upon ignition, the reaction that occurs is

$$6\ NH_4ClO_4(s) + 10\ Al(s) \longrightarrow 5\ Al_2O_3(s) + 3\ N_2(g) + 6\ HCl(g) + 9\ H_2O(g)$$

(a) What mass of aluminum should be mixed with 1.325 kg of NH_4ClO_4 for this reaction? (b) Determine the mass of Al_2O_3 (alumina, a finely divided white powder that is produced as billows of white smoke) formed in the reaction of 3.500×10^3 kg of aluminum.

L.4 The compound diborane, B_2H_6, was at one time considered for use as a rocket fuel. Its combustion reaction is

$$B_2H_6(g) + 3\ O_2(l) \longrightarrow 2\ HBO_2(g) + 2\ H_2O(l)$$

The fact that HBO_2, a reactive compound, was produced rather than the relatively inert B_2O_3 was a factor in the discontinuation of the investigation of diborane as a fuel. (a) What mass of liquid oxygen (LOX) would be needed to burn 257 g of B_2H_6? (b) Determine the mass of HBO_2 produced from the combustion of 106 g of B_2H_6.

L.5 The camel stores the fat tristearin, $C_{57}H_{110}O_6$, in its hump. As well as being a source of energy, the fat is also a source of water because, when it is used, the reaction

$$2\ C_{57}H_{110}O_6(s) + 163\ O_2(g) \longrightarrow 114\ CO_2(g) + 110\ H_2O(l)$$

takes place. (a) What mass of water is available from 1.00 pound (454 g) of this fat? (b) What mass of oxygen is needed to oxidize this amount of tristearin?

L.6 Potassium superoxide, KO_2, is utilized in a closed-system breathing apparatus to remove carbon dioxide and water from exhaled air. The removal of water generates oxygen for breathing by the reaction

$$4\ KO_2(s) + 2\ H_2O(l) \longrightarrow 3\ O_2(g) + 4\ KOH(s)$$

The potassium hydroxide removes carbon dioxide from the apparatus by the reaction

$$KOH(s) + CO_2(g) \longrightarrow KHCO_3(s)$$

(a) What mass of potassium superoxide generates 115 g of O_2? (b) What mass of CO_2 can be removed from the apparatus by 75.0 g of KO_2?

L.7 When a hydrocarbon burns, water is produced as well as carbon dioxide. (For this reason, clouds of condensed water droplets are often seen coming from automobile exhausts, especially on a cold day.) The density of gasoline is 0.79 $g{\cdot}mL^{-1}$. Assume gasoline to be represented by octane, C_8H_{18}, for which the combustion reaction is

$$2\ C_8H_{18}(l) + 25\ O_2(g) \longrightarrow 16\ CO_2(g) + 18\ H_2O(l)$$

Calculate the mass of water produced from the combustion of 1.0 gallon (3.8 L) of gasoline.

L.8 The density of oak wood is 0.72 $g{\cdot}cm^{-3}$. Assuming oak wood to have the empirical formula CH_2O, calculate the mass of water produced when a log of dimensions 12 cm × 14 cm × 25 cm is burned.

L.9 A 15.00-mL sample of sodium hydroxide was titrated to the stoichiometric point with 17.40 mL of 0.234 M HCl(aq). (a) What is the initial molarity of NaOH in the solution? (b) Calculate the mass of NaOH in the solution.

L.10 A 35.25-mL sample of oxalic acid, $H_2C_2O_4$ (with two acidic protons), was titrated to the stoichiometric point with 25.67 mL of 0.327 M NaOH(aq). (a) What is the molarity of the oxalic acid? (b) Determine the mass of oxalic acid in the solution.

L.11 A 9.670-g sample of barium hydroxide is dissolved and diluted to the mark in a 250-mL volumetric flask. It was found that 11.56 mL of this solution was needed to reach the stoichiometric point in a titration of 25.0 mL of a nitric acid solution. (a) Calculate the molarity of the HNO_3 solution. (b) What mass of HNO_3 is in the initial solution?

L.12 A 10.0-mL volume of 3.0 M KOH(aq) is transferred to a 250-mL volumetric flask and diluted to the mark. It was found that 38.5 mL of this diluted solution was needed to reach the stoichiometric point in a titration of 10.0 mL of a phosphoric acid solution according to the reaction

$$3\ KOH(aq) + H_3PO_4(aq) \longrightarrow K_3PO_4(aq) + 3\ H_2O(l)$$

(a) Calculate the molarity of H_3PO_4 in the solution. (b) What mass of H_3PO_4 is in the initial solution?

L.13 In a titration, a 3.25-g sample of an acid, HX, requires 68.8 mL of a 0.750 M NaOH(aq) solution for complete reaction. What is the molar mass of the acid?

L.14 In a titration, 14.56 mL of 0.115 M NaOH(aq) was required to titrate 0.2037 g of an unknown acid, HX. What is the molar mass of the acid?

L.15 A solution of hydrochloric acid was prepared by measuring 10.00 mL of the concentrated acid into a 1.000-L volumetric flask and adding water up to the mark. Another solution was prepared by adding 0.832 g of anhydrous sodium carbonate to a 100.0-mL volumetric flask and adding water up to the mark. Then, 25.00 mL of the latter solution was pipetted into a flask and titrated with the diluted acid. The stoichiometric point was reached after 31.25 mL of acid had been added. (a) Write the balanced equation for the reaction of HCl(aq) with Na_2CO_3(aq). (b) What is the molarity of the original hydrochloric acid?

L.16 A vitamin C tablet was analyzed to determine whether it did in fact contain, as the manufacturer claimed, 1.0 g of the vitamin. A tablet was dissolved in water to form a 100.00-mL solution, and a 10.0-mL sample was titrated with iodine (as potassium triiodide). It required 10.1 mL of 0.0521 M I_3^-(aq) to reach the stoichiometric point in the titration. Given that 1 mol I_3^- reacts with 1 mol vitamin C in the reaction, is the manufacturer's claim correct? The molar mass of vitamin C is 176 $g{\cdot}mol^{-1}$.

M LIMITING REACTANTS

Stoichiometric calculations of the amount or mass of product formed in a reaction are based on an ideal view of the world. They suppose, for instance, that all the reactants react exactly as described in the chemical equation. In practice, that might not be so. Some of the starting materials may be consumed in a **competing reaction**, a reaction occurring at the same time as the one we are interested in and using some of the same reactants. Another possibility is that the reaction might not be complete at the time we make our measurements. A third possibility—one that is of major importance in chemistry and that occupies several chapters of this text—is that many reactions do not go to completion. They appear to stop once a certain proportion of the reactants has been consumed.

M.1 Reaction Yield

The **theoretical yield** of a reaction is the *maximum* quantity (moles, mass, or volume) of product that can be obtained from a given quantity of reactant. The quantities of products calculated from a given mass of reactant in Section L were all theoretical yields. The **percentage yield** is the fraction of the theoretical yield actually produced, expressed as a percentage:

$$\text{Percentage yield} = \frac{\text{actual yield}}{\text{theoretical yield}} \times 100\% \qquad (1)$$

EXAMPLE M.1 Calculating the percentage yield of a product

In a test of an automobile engine to monitor the combustion of 1.00 L of octane (702 g) under certain conditions, 1.84 kg of carbon dioxide is produced. What is the percentage yield of carbon dioxide?

STRATEGY Begin by writing the chemical equation for the complete oxidation of octane to carbon dioxide and water. Then calculate the theoretical yield (in grams) of CO_2 by using the procedure in Toolbox L.1. To avoid rounding errors, do all the numerical work at the end of the calculation. To obtain the percentage yield, divide the actual mass produced by the theoretical mass of product and multiply by 100%.

SOLUTION The molar mass of C_8H_{18} is 114.2 $g\cdot mol^{-1}$ and that of CO_2 is 44.01 $g\cdot mol^{-1}$. The chemical equation for the combustion of octane is

$$2\ C_8H_{18}(l) + 25\ O_2(g) \longrightarrow 16\ CO_2(g) + 18\ H_2O(l)$$

We now calculate that when 702 g of octane burns in a plentiful supply of oxygen, the mass of carbon dioxide produced is

$$\begin{aligned}\text{Mass of } CO_2 &= (702\ g\ C_8H_{18}) \times \frac{1\ mol\ C_8H_{18}}{114.2\ g\ C_8H_{18}} \\ &\quad \times \frac{16\ mol\ CO_2}{2\ mol\ C_8H_{18}} \times \frac{44.01\ g\ CO_2}{1\ mol\ CO_2} \\ &= 2.16 \times 10^3\ g\ CO_2\end{aligned}$$

The theoretical yield of carbon dioxide is therefore 2.16 kg; this is the maximum mass that can be produced by the combustion of 1.00 L octane. Because only 1.84 kg was produced, the percentage yield of carbon dioxide is

This mass corresponds to about 8 kg of CO_2 per gallon of fuel. The average car emits about 0.4 kg of CO_2 per mile of travel.

$$\text{Percentage yield of } CO_2 = \frac{1.84 \text{ kg}}{2.16 \text{ kg}} \times 100\% = 85.2\%$$

Self-Test M.1A When 24.0 g of potassium nitrate was heated with lead, 13.8 g of potassium nitrite was formed in the reaction $Pb(s) + KNO_3(s) \xrightarrow{\Delta} PbO(s) + KNO_2(s)$. Calculate the percentage yield of potassium nitrite.

[*Answer:* 68.3%]

Self-Test M.1B Reduction of 15 kg of iron(III) oxide in a blast furnace produced 8.8 kg of iron. What is the percentage yield of iron?

The theoretical yield of a product is the maximum quantity that can be expected on the basis of the stoichiometry of a chemical equation. The percentage yield is the percentage of the theoretical yield actually achieved.

M.2 The Limits of Reaction

The **limiting reactant** in a reaction is the reactant that governs the maximum yield of product. A limiting reactant is like a part in a motorcycle factory: if there are eight wheels and seven motorcycle frames, then the maximum number of motorcycles is limited by the number of wheels. Because each frame requires two wheels, there are enough wheels for only four motorcycles, so the wheels play the role of the limiting reactant. When all the wheels have been used, three frames remain unused. The frames were present in excess.

In some cases, we must determine which is the limiting reactant. For example, from the equation

$$N_2(g) + 3\,H_2(g) \xrightarrow{\text{Fe}} 2\,NH_3(g)$$

we see that, in the synthesis of ammonia, 1 mol $N_2 \simeq$ 3 mol H_2. To decide which reactant is the limiting one, we compare the number of moles of each reactant supplied with the stoichiometric coefficients. For example, suppose we had available 1 mol N_2 but only 2 mol H_2. Because this amount of hydrogen is less than is required by the stoichiometric relation, hydrogen would be the limiting reactant. Once we have identified the limiting reactant, we can calculate the amount of product that can be formed. We can also calculate the amount of excess reactant that remains at the end of the reaction.

The limiting reactant in a reaction is the reactant supplied in an amount smaller than that required by the stoichiometric relation between the reactants.

TOOLBOX M.1 How to identify the limiting reactant

Conceptual basis

The limiting reactant is the reactant that will be completely used up. All other reactants are in excess. Because the limiting reactant is the one that limits the amounts of products that can be formed, the amount of product predicted is calculated from the amount of the limiting reactant.

Procedure

There are two ways of determining which reactant is the limiting one.

Method 1:

In this approach, we use the mole ratio from the chemical equation to determine whether there is enough of one reactant to react with another.

Step 1 Calculate the amount of each reactant in moles, converting any masses to amounts by using the molar mass.

Step 2 Choose one of the substances and use the stoichiometric relation to calculate the theoretical amount of the second reactant needed for complete reaction with the first.

Step 3 Compare the actual amount of the second reactant to that calculated in step 2. If the actual amount is greater than the amount needed, then the second reactant is present in excess and not all of it will react; in this case, the first substance is the limiting reactant. If the actual amount of the second reactant is less than that calculated, then all of it will react, so it is the limiting reactant and the first reactant is in excess.

Method 2:

An alternative approach is to calculate the theoretical molar yield of one of the products for each reactant separately, by using the procedure in Toolbox L.1. This method is a good one to use when there are more than two reactants. The reactant that would produce the smallest amount of product is the limiting reactant.

Step 1 Calculate the amount of each reactant in moles, if necessary by converting masses to amounts by using the molar masses of the substances.

Step 2 Select one of the products. For each reactant, calculate how many moles of the product it can form.

Step 3 Compare the amount of product each reactant can form. The reactant that can produce the least product is the limiting reactant.

EXAMPLE M.2 Identifying the limiting reactant

Calcium carbide, CaC_2, reacts with water to form calcium hydroxide and the flammable gas ethyne (acetylene). This reaction was once used for lamps on bicycles, because the reactants are easily transported. (a) Which is the limiting reactant when 1.00×10^2 g of water reacts with 1.00×10^2 g of calcium carbide? (b) What mass of ethyne can be produced? (c) What mass of excess reactant remains after reaction is complete? The chemical equation is

$$CaC_2(s) + 2\,H_2O(l) \longrightarrow Ca(OH)_2(aq) + C_2H_2(g)$$

STRATEGY For this calculation, we suppose that the calcium carbide is pure and that all the ethyne is collected. We then follow the procedure in Method 1 of Toolbox M.1.

SOLUTION (a) The molar mass of calcium carbide is 64.10 $g{\cdot}mol^{-1}$ and that of water is 18.02 $g{\cdot}mol^{-1}$. Therefore, the amounts of reactants are

$$\text{Amount of } CaC_2\text{ (mol)} = \frac{100\text{ g}}{64.10\text{ g}{\cdot}\text{mol}^{-1}} = \frac{100}{64.10}\text{ mol} = 1.56\text{ mol}$$

$$\text{Amount of } H_2O \text{ (mol)} = \frac{100 \text{ g}}{18.02 \text{ g}\cdot\text{mol}^{-1}} = \frac{100}{18.02} \text{ mol} = 5.55 \text{ mol}$$

Next, we use the stoichiometric relation 1 mol CaC_2 ≏ 2 mol H_2O to work out the amount of H_2O that is needed to react with 1.56 mol CaC_2:

$$\begin{aligned}\text{Amount of } H_2O \text{ (mol)} &= \left(\frac{100}{64.10} \text{ mol } CaC_2\right) \times \left(\frac{2 \text{ mol } H_2O}{1 \text{ mol } CaC_2}\right) \\ &= 3.12 \text{ mol } H_2O\end{aligned}$$

Because 3.12 mol H_2O is required and 5.55 mol H_2O is supplied, all the calcium carbide can react; so the calcium carbide is the limiting reactant and water is present in excess. (b) The reaction stoichiometry implies that 1 mol CaC_2 ≏ 1 mol C_2H_2. It follows that the mass of ethyne (of molar mass 26.04 g·mol^{-1}) that can be produced is

$$\begin{aligned}\text{Mass of } C_2H_2 &= \left(\frac{100 \text{ g } CaC_2}{64.10 \text{ g}\cdot\text{mol}^{-1}}\right) \\ &\quad \times \left(\frac{1 \text{ mol } C_2H_2}{1 \text{ mol } CaC_2}\right) \times (26.04 \text{ g}\cdot\text{mol}^{-1}) \\ &= 40.6 \text{ g } C_2H_2\end{aligned}$$

(c) The reactant remaining unreacted will be the reactant initially in excess, namely, water. Because 5.55 mol H_2O was supplied and 3.12 mol H_2O has been consumed, the amount of H_2O remaining is 5.55 − 3.12 mol = 2.43 mol. Therefore, the mass of the excess reactant remaining at the end of the reaction is

$$\text{Mass of } H_2O \text{ remaining} = (2.43 \text{ mol}) \times (18.02 \text{ g}\cdot\text{mol}^{-1}) = 43.8 \text{ g}$$

Self-Test M.2A (a) Identify the limiting reactant in the reaction 6 Na(l) + Al_2O_3(s) → 2 Al(l) + 3 Na_2O(s) when 5.52 g of sodium is heated with 5.10 g of Al_2O_3. (b) What mass of aluminum can be produced? (c) What mass of excess reactant remains at the end of the reaction?

[*Answer:* (a) Sodium; (b) 2.16 g Al; (c) 1.02 g Al_2O_3]

Self-Test M.2B (a) What is the limiting reactant for the preparation of urea from ammonia in the reaction 2 NH_3(g) + CO_2(g) → $OC(NH_2)_2$(s) + H_2O(l) when 14.5 kg of ammonia is available to react with 22.1 kg of carbon dioxide? (b) What mass of urea can be produced? (c) What mass of excess reactant remains at the end of the reaction?

M.3 Combustion Analysis

In Section F, we saw that one technique used in modern chemical laboratories is "combustion analysis." This technique has been used since the early days of chemistry; it is still used to determine the empirical formulas of organic compounds and, in combination with mass spectrometry, their molecular formulas. It is used both for newly synthesized compounds and to identify active compounds in natural products. We are now in a position to understand the basis of the technique, for it makes use of the concept of limiting reactant.

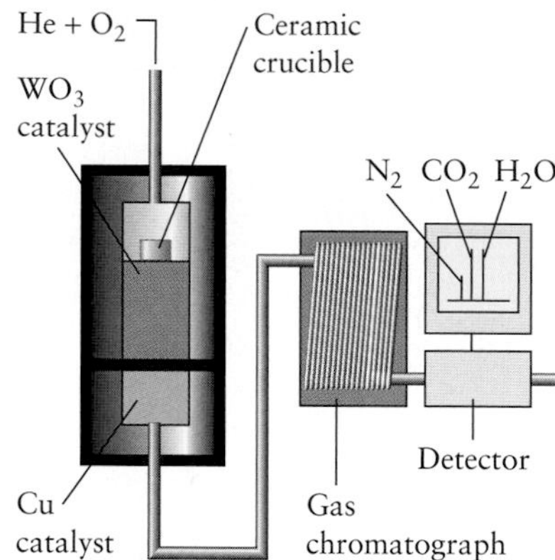

FIGURE M.1 The apparatus used for a combustion analysis. An oxygen-helium mixture is passed over the sample, which is oxidized. The masses of carbon dioxide and water produced are obtained by separating the product gases and measuring their thermal conductivities. A catalyst ensures that any CO produced is oxidized to CO_2.

The sample is ignited and allowed to burn in a tube while a plentiful supply of oxygen passes through (Fig. M.1). The use of excess oxygen ensures that the sample is the limiting reactant. All the hydrogen in the compound is converted to water and all the carbon is converted to carbon dioxide. In the modern version of the technique, the product gases are separated chromatographically and their relative amounts determined by measuring the thermal conductivity (the ability to conduct heat) of the gas streaming from the apparatus.

Let's see how to analyze the data, which are the initial mass of the compound and the masses of the water and carbon dioxide produced in the combustion of the compound. In the presence of excess oxygen, each carbon atom in the compound ends up in one molecule of carbon dioxide (Fig. M.2). Therefore, we can write

$$1 \text{ mol C in sample} \simeq 1 \text{ mol } CO_2 \text{ as product}$$

or, more briefly, 1 mol C $\simeq$ 1 mol CO_2. This relation is true for any organic compound, even if we do not yet know its formula. Hence, by measuring the mass of carbon dioxide produced and converting that mass to amount of C atoms, we can find the number of moles of C atoms in the original sample.

Similarly, in the presence of excess oxygen, each hydrogen atom in a compound contributes to a water molecule when the compound burns. We can infer that

$$2 \text{ mol H in sample} \simeq 1 \text{ mol } H_2O \text{ as product}$$

or, more briefly, 2 mol H $\simeq$ 1 mol H_2O. This relation is true for the combustion of any compound containing hydrogen. Therefore, if we measure the mass of water produced when the compound burns in plenty of oxygen, we can find the amount of H atoms (in moles) in the sample.

Many organic compounds also contain oxygen. Provided the compound contains only carbon, hydrogen, and oxygen, we can calculate the mass of oxygen originally present by subtracting the masses of carbon and hydrogen in the sample from the original mass of the sample. That mass of oxygen can be converted to amount of O atoms (in moles) by using the molar mass of oxygen atoms (16.00 $g\cdot mol^{-1}$). We can then calculate the empirical formula.

In a combustion analysis, the amounts of C, H, and O atoms in a sample of a compound, and thus its empirical formula, are determined from the masses of carbon dioxide and water produced when the compound burns in excess oxygen.

EXAMPLE M.3 Determining an empirical formula by combustion analysis

A combustion analysis was carried out on 1.621 g of a newly synthesized compound, which was known to contain only C, H, and O. The masses of water and carbon dioxide produced were 1.902 g and 3.095 g, respectively. What is the empirical formula of the compound?

STRATEGY We need to use the stoichiometric relations given earlier to find the amounts of carbon and hydrogen atoms in the sample and then convert those moles to masses. The mass of oxygen in the sample is obtained by subtraction of the total mass of carbon and hydrogen from the mass of the original sample. It is then converted to the number of moles of O atoms. Finally, the relative numbers of atoms are expressed as an empirical formula.

SOLUTION To convert the mass of carbon dioxide to amount of C atoms (in moles), we use the molar mass of carbon dioxide (44.01 g·mol^{-1}) and the stoichiometric relation 1 mol C $\simeq$ 1 mol CO_2:

$$\text{Amount of C (mol)} = \left(\frac{3.095 \text{ g } CO_2}{44.01 \text{ g·mol}^{-1}}\right) \times \left(\frac{1 \text{ mol C}}{1 \text{ mol } CO_2}\right)$$
$$= \frac{3.095}{44.01} \text{ mol C} = 0.070\ 32 \text{ mol C}$$

The mass of carbon in the sample is therefore

$$\text{Mass of C} = \left(\frac{3.095}{44.01} \text{ mol C}\right) \times (12.01 \text{ g·mol}^{-1}) = 0.8446 \text{ g C}$$

To find the amount of H atoms in the sample, we use the molar mass of water (18.02 g·mol^{-1}) and the stoichiometric relation 1 mol H_2O $\simeq$ 2 mol H to obtain

$$\text{Amount of H (mol)} = \left(\frac{1.902 \text{ g } H_2O}{18.02 \text{ g·mol}^{-1}}\right) \times \left(\frac{2 \text{ mol H}}{1 \text{ mol } H_2O}\right)$$
$$= 0.2111 \text{ mol H}$$

To convert this amount to mass, we use the molar mass of hydrogen atoms (1.008 g·mol^{-1}):

$$\text{Mass of hydrogen} = \left(\frac{1.902 \times 2}{18.02} \text{ mol H}\right) \times (1.008 \text{ g·mol}^{-1})$$
$$= 0.2128 \text{ g H}$$

The total mass of carbon and hydrogen is 0.8446 g + 0.2128 g = 1.0574 g, so the mass of oxygen in the sample is

$$\text{Mass of oxygen} = 1.621 \text{ g} - 1.0574 \text{ g} = 0.564 \text{ g}$$

Therefore,

$$\text{Amount of O (mol)} = \frac{0.564 \text{ g}}{16.00 \text{ g·mol}^{-1}} = 0.0352 \text{ mol}$$

We now know that the amount of each element in the sample, and therefore the relative numbers of atoms of each element in the compound, are

$$\text{C:H:O} = 0.070\ 32 : 0.2111 : 0.0352$$

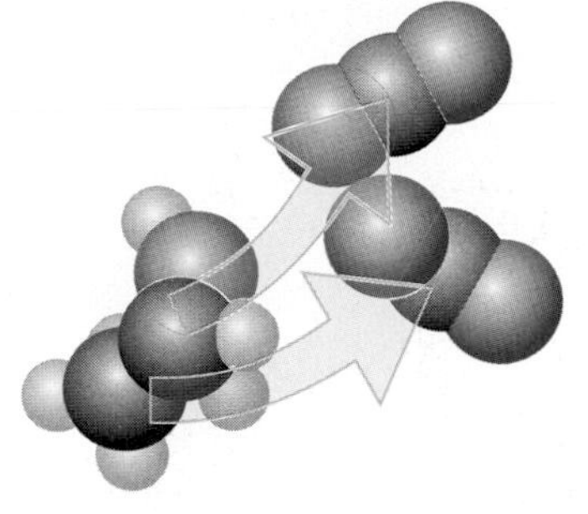

(a)

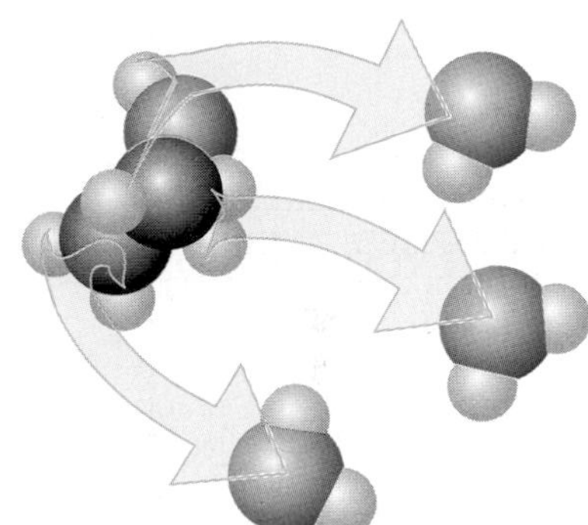

(b)

FIGURE M.2 When an organic compound burns in a plentiful supply of oxygen, (a) each C atom produces one CO_2 molecule and (b) each pair of H atoms ends up as one H_2O molecule (not necessarily in the same molecule).

Division by the smallest number, 0.0352, gives

$$C{:}H{:}O = 2.00 : 6.00 : 1.00$$

We conclude that the empirical formula of the new compound is C_2H_6O.

Self-Test M.3A When 0.528 g of sucrose (a compound of carbon, hydrogen, and oxygen) is burned, 0.306 g of water and 0.815 g of carbon dioxide are formed. Deduce the empirical formula of sucrose.

[*Answer:* $C_{12}H_{22}O_{11}$]

Self-Test M.3B When 0.236 g of aspirin is burned in oxygen, 0.519 g of carbon dioxide and 0.0945 g of water are formed. Deduce the empirical formula of aspirin.

SKILLS YOU SHOULD HAVE MASTERED

1 Calculate the theoretical and percentage yields of the products of a reaction, given the mass of starting material, Example M.1.

2 Identify the limiting reactant of a reaction and calculate the amount of excess reactant remaining after reaction is complete, given the initial mass of each reactant, Toolbox M.1 and Example M.2.

3 Use the limiting reactant to calculate the yield of a product, Example M.2.

4 Determine the empirical formula of an organic compound by combustion analysis, Example M.3.

EXERCISES

M.1 When limestone rock, which is principally $CaCO_3$, is heated, carbon dioxide and quicklime, CaO, are produced by the reaction $CaCO_3(s) \xrightarrow{\Delta} CaO(s) + CO_2(g)$. If 17.5 g of CO_2 was produced from the thermal decomposition of 42.73 g of $CaCO_3$, what is the percentage yield of the reaction?

M.2 Phosphorus trichloride, PCl_3, is produced from the reaction of white phosphorus, P_4, and chlorine: $P_4(s) + 6\ Cl_2(g) \rightarrow 4\ PCl_3(g)$. A 300.5-g sample of PCl_3 was collected from the reaction of 77.25 g of P_4 with excess chlorine. What is the percentage yield of the reaction?

M.3 The souring of wine occurs when ethanol, C_2H_5OH, is converted by oxidation into acetic acid: $C_2H_5OH(aq) + O_2(g) \rightarrow CH_3COOH(aq) + H_2O(l)$. A 1.00-L bottle of wine is labeled as being 8.5% by volume ethanol. Upon inspection, it was found that the cork in the bottle had not sealed properly. Analysis of the wine indicated that 1.00 mL, contained 0.0275 g of acetic acid. The density of ethanol is $0.816\ g{\cdot}mL^{-1}$ and that of water is $1.00\ g{\cdot}ml^{-1}$. (a) What mass of oxygen must have leaked into the bottle in order to cause the conversion of ethanol to acetic acid? (b) What is the percentage yield for the conversion of ethanol to acetic acid?

M.4 Slaked lime, $Ca(OH)_2$, is formed from quicklime, CaO, by the addition of water: $CaO(s) + H_2O(l) \rightarrow Ca(OH)_2(s)$. What mass of slaked lime can be produced from a mixture of 30.0 g of CaO and 10.0 g of H_2O?

M.5 A reaction vessel contains 5.77 g of white phosphorus and 5.77 g of oxygen. The first reaction to occur is the formation of phosphorus(III) oxide, P_4O_6: $P_4(s) + 3\ O_2(g) \rightarrow P_4O_6(s)$. If enough oxygen is present, it can further react with this oxide to produce phosphorus(V) oxide, P_4O_{10}: $P_4O_6(s) + 2\ O_2(g) \rightarrow P_4O_{10}(s)$. (a) What is the limiting reactant for the formation of P_4O_{10}? (b) How much P_4O_{10} is produced? (c) How many grams of the excess reactant remain in the reaction vessel?

M.6 A mixture of 10.325 g of iron(II) oxide and 5.734 g Al as aluminum metal is placed in a crucible and heated in a high-temperature oven, where a reduction of the oxide occurs: $3\ FeO(s) + 2\ Al(l) \rightarrow$

3 Fe(l) + Al_2O_3(s). (a) What is the limiting reactant? (b) Determine the maximum amount of iron (in moles of Fe) that can be produced. (c) Calculate the mass of excess reactant remaining in the crucible.

M.7 Vitamin B_{12}, also known as cobalamin, has the molecular formula $C_{63}H_{88}CoN_{14}O_{14}P$. What mass of CO_2 and H_2O would be produced during a combustion analysis of 0.1674 g of cobalamin?

M.8 Ephedrine, which is used as a bronchodilator, has the chemical formula $C_{10}H_{15}NO$. What mass of CO_2 and of H_2O would be produced during a combustion analysis of 0.057 32 g of ephedrine?

M.9 The stimulant in coffee and tea is caffeine, a substance of molar mass 194 $g\cdot mol^{-1}$. When 0.376 g of caffeine was burned, 0.682 g of carbon dioxide, 0.174 g of water, and 0.110 g of nitrogen were formed. Determine the empirical and molecular formulas of caffeine, and write the equation for its combustion.

M.10 Nicotine, the stimulant in tobacco, causes a very complex set of physiological effects in the body. It is known to have a molar mass of 162 $g\cdot mol^{-1}$. When a sample of mass 0.385 g was burned, 1.072 g of carbon dioxide, 0.307 g of water, and 0.068 g of nitrogen were produced. What are the empirical and molecular formulas of nicotine? Write the equation for its combustion.

M.11 When aqueous solutions of calcium nitrate and phosphoric acid are mixed, a white solid precipitates. (a) What is the formula of the solid? (b) How many grams of the solid can be formed from 206 g calcium nitrate and 150 g phosphoric acid?

M.12 Small amounts of chlorine gas can be generated in the laboratory from the reaction of manganese (IV) oxide with hydrochloric acid: 4 HCl(aq) + MnO_2(s) → 2 H_2O(l) + $MnCl_2$(s) + Cl_2(g). (a) What mass of Cl_2 can be produced from 42.7 g of MnO_2 with an excess of HCl(aq)? (b) What volume of chlorine gas (of density 3.17 $g\cdot L^{-1}$) will be produced from the reaction of 300 mL of 0.100 M HCl(aq) with an excess of MnO_2? (c) Suppose only 150 mL of chlorine was produced in the reaction in (b). What is the percentage yield of the reaction?

M.13 In addition to determining elemental *composition* of *pure* unknown compounds, combustion analysis can be used to determine the *purity* of known compounds. A sample of 2-naphthol $C_{10}H_7OH$, which is used to prepare antioxidants to incorporate into synthetic rubber, was found to be contaminated with a small amount of LiBr. The combustion analysis of this sample gave the following results: 77.48% C and 5.20% H. Assuming that the only species present are 2-naphthol and LiBr, calculate the percentage purity by mass of the sample.

M.14 An organic compound with formula $C_{14}H_{20}O_2N$ was recrystallized from 1,1,2,2-tetrachloroethane, $C_2H_2Cl_4$. A combustion analysis of the compound gave the following data: 68.50% C, 8.18% H. Because the data were considerably different from that expected for pure $C_{14}H_{20}O_2N$, the sample was examined and found to contain a significant amount of 1,1,2,3-tetrachloroethane. Assuming that only these two compounds are present, what is the percentage purity by mass of the $C_{14}H_{20}O_2N$?

CHAPTER 1

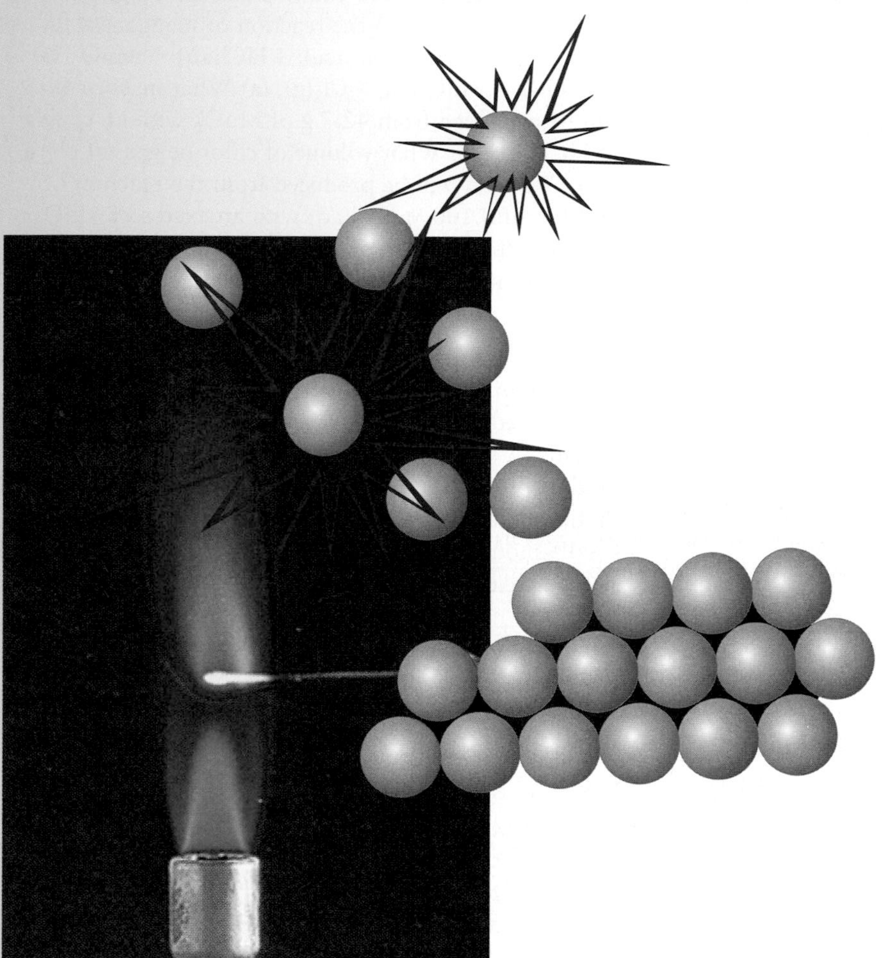

The color of this flame is due to atoms of lithium that are vaporized from the solid lithium compound on the wire and then excited energetically. As the excited atoms return to states of lower energy, they discard the excess energy as electromagnetic radiation. To understand the behavior of atoms and molecules, we have to understand how the electrons in them are arranged, which in turn means knowing about quantum mechanics. The photographs at the beginning of each chapter throughout this text will illustrate how objects look through chemists' eyes.

Why Do We Need to Know This Material?

Atoms are the fundamental building blocks of matter. Almost all chemical properties can be explained in terms of the properties of atoms, so the information in this chapter is central to developing an understanding of chemistry—including why elements have such strikingly different physical and chemical characteristics. The topics in this chapter also account for the structure of the periodic table, the scheme that helps to organize the whole of chemistry, and provide a basis for understanding how elements combine to form compounds. This chapter is also important because it introduces the theory of matter known as quantum mechanics, which is essential for understanding how electrons behave.

What Do We Need to Know Already?

We need to be familiar with the nuclear model of the atom and the general layout of the periodic table (Section B). We also need the concepts of kinetic and potential energy (Section A).

ATOMS: THE QUANTUM WORLD

You need insight to think like a chemist. When you look at an everyday object or a sample of a chemical, you need to be able to imagine the atoms that make it up. Not only that, you need to have insight into the *interiors* of atoms and be able to imagine their internal structures. So we need to plunge, in our mind's eye, deep into the center of matter and discover the internal structure of atoms. In particular, because chemical properties are related to the arrangements of electrons, we need to understand the **electronic structure** of an atom, the description of how electrons are arranged around its nucleus.

As soon as we start this journey, we encounter an extraordinary feature of our world. When Rutherford proposed the nuclear model of the atom (Section B), he expected to be able to use **classical mechanics**, the laws of motion proposed by Newton in the seventeenth century, to describe its electronic structure. After all, classical mechanics had been tremendously successful for describing the motion of visible objects such as balls and planets. However, it soon became clear that classical mechanics fails when applied to electrons in atoms. New laws, which came to be known as **quantum mechanics,** were developed in the early part of the twentieth century.

This chapter follows the development of the modern nuclear model of the atom. First, we see how experiments have led to our modern concepts of the nature of matter and radiation. In the nineteenth century, scientists thought that matter consisted of particles and that radiation consisted of waves. Our modern model is quite different and has been forced on us by experiments showing that the earlier models are inadequate. Next, we see how additional experiments led to a profound refinement of Dalton's model of an atom as an uncuttable sphere and have clarified the structure of Rutherford's nuclear model of the atom (see Section B).

OBSERVING ATOMS

How do chemists know anything about the internal structure of objects as small as atoms? They study atoms by observing the properties of electromagnetic radiation that atoms emit. They then propose a model of the structure of the atom that accounts for these properties. The analysis of the electromagnetic radiation emitted or absorbed by substances is a branch of chemistry called **spectroscopy**. We shall see how to use atomic spectroscopy—spectroscopy applied to atoms—to determine their internal structure.

1.1 The Characteristics of Electromagnetic Radiation

A ray of **electromagnetic radiation** consists of **oscillating** (time-varying) electric and magnetic fields that travel through empty space at $3.00 \times 10^8\ \text{m}\cdot\text{s}^{-1}$, or at just over 670 million miles per hour. This speed is denoted c and called the "speed of light." Visible light is a form of electromagnetic

As we saw in Section A, we can think of a *field* as a region of influence, like the gravitational field of the Earth.

You will find values of the fundamental constants inside the back cover of the book.

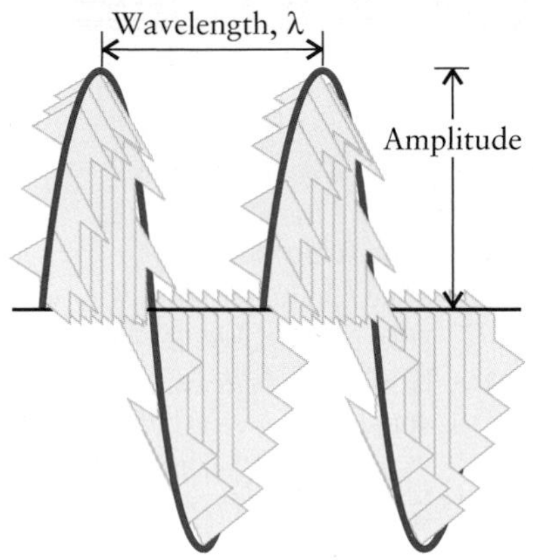

FIGURE 1.1 The electric field of electromagnetic radiation oscillates in space and time. This diagram represents a "snapshot" of an electromagnetic wave at a given instant. The length of an arrow at any point represents the strength of the force that the field exerts on a charged particle at that point. The distance between the peaks is the wavelength of the radiation, and the height of the wave above the center line is the amplitude. The magnetic field is perpendicular to the electric field; both fields are perpendicular to the direction of propagation.

radiation, as are radio waves, microwaves, and x-rays. All these forms of radiation transfer energy from one region of space to another. The warmth we feel from the Sun is a tiny proportion of the total energy it emits, and it is brought to us through space as electromagnetic radiation.

An electric field pushes on charged particles such as electrons. As a light ray passes an electron, its electric field pushes the electron first in one direction, then in the opposite direction, over and over again. That is, the field oscillates in both direction and strength (Fig. 1.1). The number of **cycles** (complete reversals of direction away from and back to the initial strength and direction) per second is called the **frequency,** ν (the Greek letter nu), of the radiation. The unit of frequency, 1 hertz (1 Hz), is defined as 1 cycle per second:

$$1\ \text{Hz} = 1\ \text{s}^{-1}$$

Electromagnetic radiation of frequency 1 Hz would push a charge in one direction, then the opposite direction, and return to the original direction once per second. The frequency of electromagnetic radiation we see as visible light is close to 10^{15} Hz, so its electric field changes direction at about a thousand trillion (10^{15}) times a second as it travels past a given point.

An instantaneous snapshot of a wave of electromagnetic radiation spread through space would look like Fig. 1.1. The wave is characterized by its amplitude and wavelength. The **amplitude** is the height of the wave above the center line. The square of the amplitude determines the **intensity,** or brightness, of the radiation. The **wavelength,** λ (the Greek letter lambda), is the peak-to-peak distance. The wavelengths of visible light are

TABLE 1.1 Color, Frequency, and Wavelength of Electromagnetic Radiation

Radiation type	Frequency, 10^{14} Hz	Wavelength, nm (2 sf)*	Energy per photon, 10^{-19} J
x-rays and γ-rays	$\geq 10^3$	≤ 3	$\geq 10^3$
ultraviolet	8.6	350	5.7
visible light			
violet	7.1	420	4.7
blue	6.4	470	4.2
green	5.7	530	3.8
yellow	5.2	580	3.4
orange	4.8	620	3.2
red	4.3	700	2.8
infrared	3.0	1000	2.0
microwaves and radio waves	$\leq 10^{-3}$	$\geq 3 \times 10^6$	$\leq 10^{-3}$

* sf denotes the number of significant figures in the data.

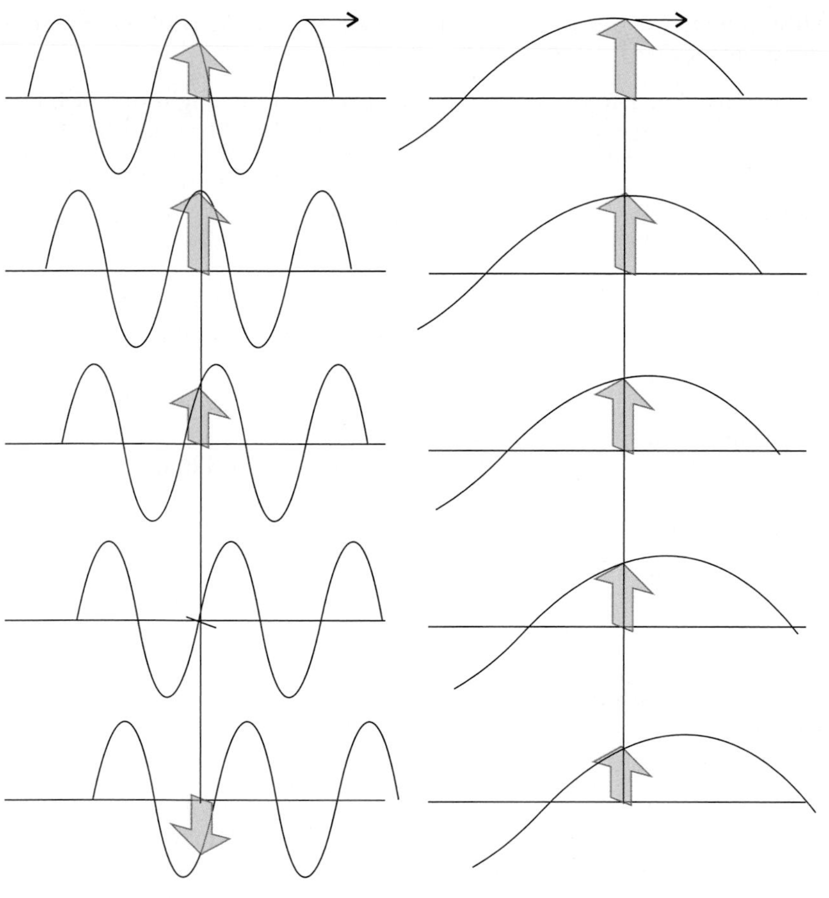

FIGURE 1.2 (a) Short-wavelength radiation: the vertical arrow shows how the electric field changes markedly at five successive instants. (b) For the same five instants, the electric field of the long-wavelength radiation changes much less. The horizontal arrows show that in each case the wave has traveled the same distance. Short-wavelength radiation has a high frequency, whereas long-wavelength radiation has a low frequency.

close to 500 nm. Although 500 nm is only half of one-thousandth of a millimeter (so you might *just* be able to imagine it), it is much longer than the diameters of atoms, which are typically close to 0.2 nm. Different wavelengths of electromagnetic radiation correspond to different regions of the spectrum (see Table 1.1). Our eyes detect electromagnetic radiation with wavelengths in the range 700 nm (red light) to 400 nm (violet light). Radiation in this range is called **visible light**, and the frequency of visible light determines its color. White light, which includes sunlight, is a mixture of all wavelengths of visible light.

Now imagine the wave in Fig. 1.1 as zooming along at its actual speed, the speed of light, c. If the wavelength of the light is very short, very many complete oscillations pass a given point in a second (Fig. 1.2a). If the wavelength is long, the light still travels at the speed c, but fewer complete oscillations pass the point in a second (Fig. 1.2b). A short wavelength therefore corresponds to high-frequency radiation and a long wavelength corresponds to low-frequency radiation. The precise relation is

$$\text{Wavelength} \times \text{frequency} = \text{speed of light, or } \lambda\nu = c \qquad (1)$$

How can we use this information . . . to calculate the wavelength from the frequency?

We can use Eq. 1 to calculate the wavelength of blue light once we know that the frequency is 6.4×10^{14} Hz. Noting that 1 Hz = s^{-1}, we write

$$\lambda = \frac{c}{\nu} = \frac{3.00 \times 10^{8}\ \text{m}\cdot\text{s}^{-1}}{6.4 \times 10^{14}\ \text{s}^{-1}} = 4.7 \times 10^{-7}\ \text{m}$$

or about 470 nm. Blue light, which has a relatively high frequency, has a shorter wavelength than low-frequency red light (700 nm).

Self-Test 1.1A Calculate the wavelengths of the light from traffic signals as they change (use the frequencies given in Table 1.1)

[*Answer:* Green, 530 nm; yellow, 580 nm; red, 700 nm]

Self-Test 1.1B What is the wavelength of a radio station transmitting at 98.4 MHz?

Modern theories suggest that our concept of space breaks down on a scale of 10^{-34} m, so that may constitute a lower bound to wavelength.

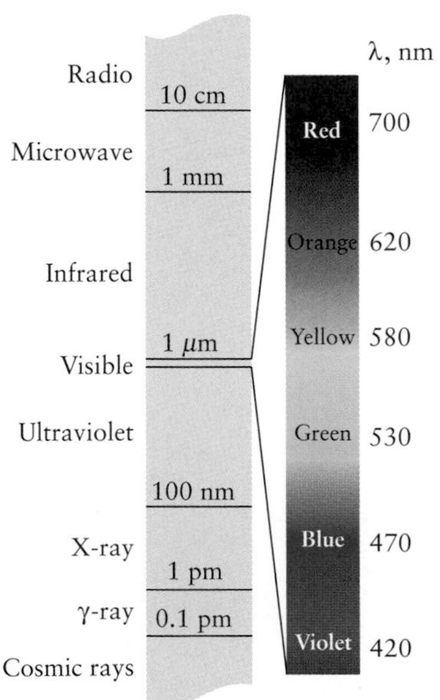

FIGURE 1.3 The electromagnetic spectrum and the names of its regions. The region we call "visible light" occupies a very narrow range of wavelengths. The regions are not drawn to scale.

As far as we know, there is neither an upper limit nor a lower limit to the wavelength of electromagnetic radiation (Fig. 1.3). **Ultraviolet radiation** is radiation at higher frequency than violet light; its wavelength is less than about 400 nm. This damaging component of radiation from the Sun is responsible for sunburn and tanning, but it is largely prevented from reaching the surface of the Earth by the ozone layer. **Infrared radiation,** the radiation we experience as heat, has a lower frequency and longer wavelength than red light; its wavelength is greater than about 800 nm. Microwaves, which are used in radar and microwave ovens, have wavelengths in the millimeter to centimeter range.

The color of light depends on its frequency or wavelength; long-wavelength radiation has a lower frequency than short-wavelength radiation.

1.2 Radiation, Quanta, and Photons

As an object is heated to high temperatures, it glows more brightly and the color of light it gives off changes from red through orange and yellow toward white. Those are *qualitative* observations. To study the effect *quantitatively*, scientists had to measure the intensity of radiation at each wavelength and repeat the measurements at a variety of different temperatures. These experiments caused one of the greatest revolutions that has ever occurred in science. Figure 1.4 shows some of the experimental results. The "hot object" is known as a **black body** (even though it might be glowing white hot!). The name signifies that the object does not favor one wavelength over another. The curves in Fig. 1.4 show the intensity of **black-body radiation**, the radiation emitted at different wavelengths by a heated black body, for a series of temperatures.

In 1879, Josef Stefan investigated the increasing brightness of a black body as it is heated and discovered that the total intensity of radiation

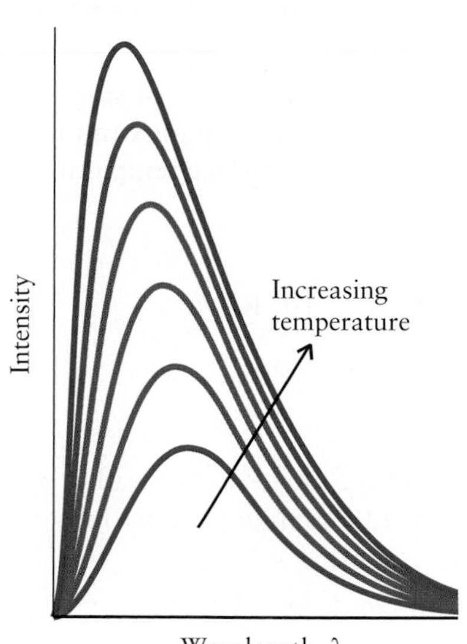

FIGURE 1.4 The intensity of radiation emitted by a heated black body as a function of wavelength. As the temperature increases, the total energy emitted (the area under the curve) increases sharply, and the maximum intensity of emission moves to shorter wavelengths.

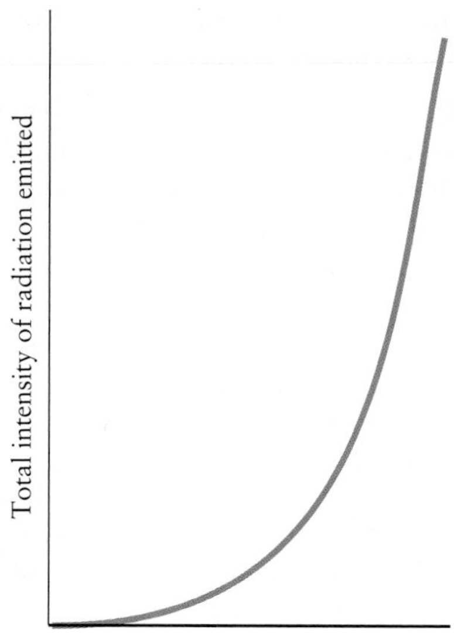

FIGURE 1.5 The total intensity of radiation emitted by a heated black body increases as the fourth power of the temperature, so a body at 1000 K emits over 120 times as much energy as the same body at 300 K.

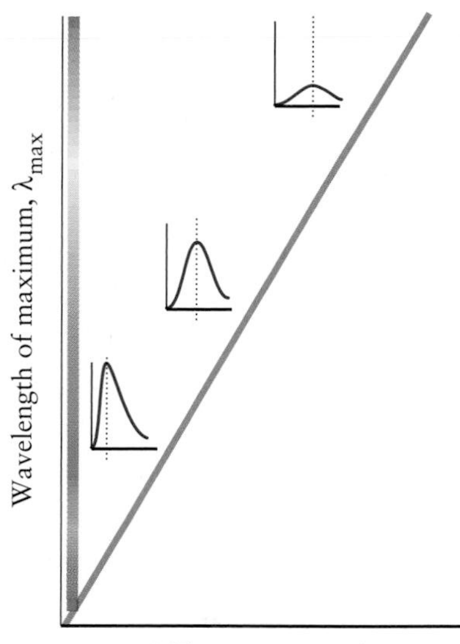

FIGURE 1.6 As the temperature is raised ($1/T$ decreases), the wavelength of maximum emission shifts to smaller values. The insets show the full black-body radiation curves and the shift in their maxima.

emitted over all wavelengths increases as the fourth power of the temperature (Fig. 1.5). This quantitative relation is now called the **Stefan-Boltzmann law** and is usually written

The name recognizes Ludwig Boltzmann's theoretical contribution.

$$\frac{\text{Power emitted (watts)}}{\text{Surface area (meter}^2\text{)}} = \text{constant} \times T^4 \qquad (2)$$

where T signifies an absolute temperature, one reported on the Kelvin scale (Appendix 1B). The experimental value of the constant is $5.67 \times 10^{-8}\ \text{W·m}^{-2}\text{·K}^{-4}$, where W denotes watts ($1\ \text{W} = 1\ \text{J·s}^{-1}$). A few years later, in 1893, Wilhelm Wien examined the shift in color of black-body radiation as the temperature increases and discovered that the wavelength corresponding to the maximum in the intensity, λ_{max}, is inversely proportional to the temperature, $\lambda_{max} \propto 1/T$, and therefore that $\lambda_{max} \times T$ is a constant (Fig. 1.6). This quantitative result is now called **Wien's law** and is normally written

$$T\lambda_{max} = \tfrac{1}{5}c_2 \qquad (3)$$

where c_2 is called the **second radiation constant.** The empirical (experimental) value of c_2 is 1.44×10^{-2} K·m.

How can we use this information . . . to determine the temperature of the surface of a star?

Wien's law can be used to determine the surface temperature of stars, which can be treated as hot black bodies. For example, the maximum intensity of solar radiation occurs at 490 nm, so the surface temperature of the Sun must be about

$$T = \frac{c_2}{5\lambda_{\text{max}}} = \frac{1.44 \times 10^{-2}\ \text{K·m}}{5 \times 4.90 \times 10^{-7}\ \text{m}} = 5.88 \times 10^3\ \text{K}$$

or about 6000 K.

Self-Test 1.2A In 1965, it was discovered that the universe is pervaded by electromagnetic radiation with a maximum at 1.05 mm (in the microwave region). What is the temperature of "empty" space?

[*Answer:* 2.74 K]

Self-Test 1.2B A red giant is a dying star. The average wavelength maximum at 700 nm shows that a red giant cools as it dies. What is the surface temperature of a red giant?

For nineteenth-century scientists, the obvious way to account for the laws of black-body radiation was to set up a model of electromagnetic radiation in terms of waves and to use classical physics to derive its characteristics. However, much to their surprise and dismay, they found that the characteristics they deduced did not match any of their observations. Worst of all was the **ultraviolet catastrophe:** classical physics predicted that any black body at any nonzero temperature should emit intense ultraviolet radiation and even x-rays and γ-rays! According to classical physics, any warm object would devastate the countryside with high-frequency radiation. Even a human body at 37°C would glow in the dark. There would, in fact, be no darkness.

The word *quantum* comes from the Latin for amount—literally, "How much?"

The suggestion that resolved the problem came in 1900 from the German physicist Max Planck. He proposed that the exchange of energy between matter and radiation occurs in **quanta**, or packets of energy. Planck focused his attention on the hot, rapidly oscillating atoms of the black body. His central idea was that an atom oscillating at a frequency ν can exchange energy with its surroundings only in packets of magnitude

$$E = h\nu \qquad (4)$$

The constant h, now called the **Planck constant**, has the value 6.626×10^{-34} J·s. If the oscillating atom releases an energy E into the surroundings, radiation of frequency $\nu = E/h$ will be detected. It is important to note that the intensity of radiation is an indication of the *number* of energy packets generated, whereas E is a measure of the energy of each individual packet.

Planck's hypothesis implies that radiation of frequency ν can be generated only if enough energy is available. The atoms of a cool body do not have enough energy to generate high-frequency, ultraviolet radiation, so the intensity curves in Fig. 1.4 die away at high frequencies (short wavelengths) and the ultraviolet catastrophe is avoided. The hypothesis is *quantitatively* successful, too, for Planck was able to use his proposal to derive the Stefan-Boltzmann and Wien laws. He was also able to calculate the variation of intensity with wavelength and obtained a curve that matched the experimental curve almost exactly.

To achieve this success, Planck had to discard classical physics, which puts no restriction on how small an energy may be transferred from one object to another. Planck proposed instead that energy is transferred in discrete packets. To justify such a dramatic revolution, more evidence was needed. That evidence came from the **photoelectric effect**, the ejection of electrons from a metal when its surface is exposed to ultraviolet radiation (Fig. 1.7). The experimental observations were as follows:

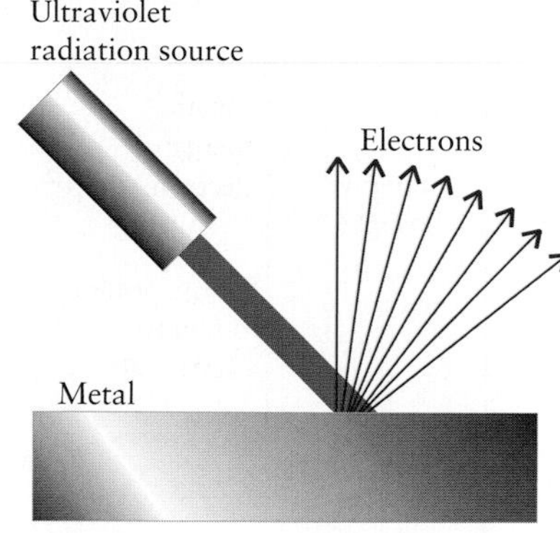

FIGURE 1.7 When a metal is illuminated with ultraviolet radiation, electrons are ejected provided the frequency is above a threshold frequency that is characteristic of the metal.

1 No electrons are ejected unless the radiation has a frequency above a certain threshold value characteristic of the metal.

2 Electrons are ejected immediately, however low the intensity of the radiation.

3 The kinetic energy of the ejected electrons increases linearly with the frequency of the incident radiation.

We say that a property y varies linearly with x if the relation between y and x can be written $y = b + mx$, where b and m are constants. See Appendix 1E.

Einstein found an explanation of these observations and, in the process, profoundly changed our conception of the electromagnetic field. He proposed that electromagnetic radiation consists of particles, which were later called **photons**. Each photon can be regarded as a packet of energy, and the energy of a single photon is related to the frequency of the radiation by Eq. 4 ($E = h\nu$). Equation 4 implies that photons of blue light are more energetic than photons of red light and that ultraviolet photons are more energetic than photons of visible light. According to this photon model of electromagnetic radiation, we can visualize a beam of red light as a stream of photons of one particular energy, yellow light as a stream of photons of higher energy, and green light as a stream of photons of higher energy still.

It is because ultraviolet photons in sunlight are so energetic that they can damage skin.

How can we use this information . . . to calculate the energy of a photon?

According to Eq. 4, each photon of blue light of frequency 6.4×10^{14} Hz has an energy

$$E = (6.626 \times 10^{-34}\ \text{J·s}) \times (6.4 \times 10^{14}\ \text{Hz}) = 4.2 \times 10^{-19}\ \text{J}$$

To derive this value, we have used $1\ \text{Hz} = 1\ \text{s}^{-1}$, so

$$(\text{J·s}) \times \text{Hz} = (\text{J·s}) \times \text{s}^{-1} = \text{J}$$

Self-Test 1.3A What is the energy of a photon of yellow light of frequency 5.2×10^{14} Hz?

[*Answer:* 3.4×10^{-19} J]

Self-Test 1.3B What is the energy of a photon of orange light of frequency 4.8×10^{14} Hz?

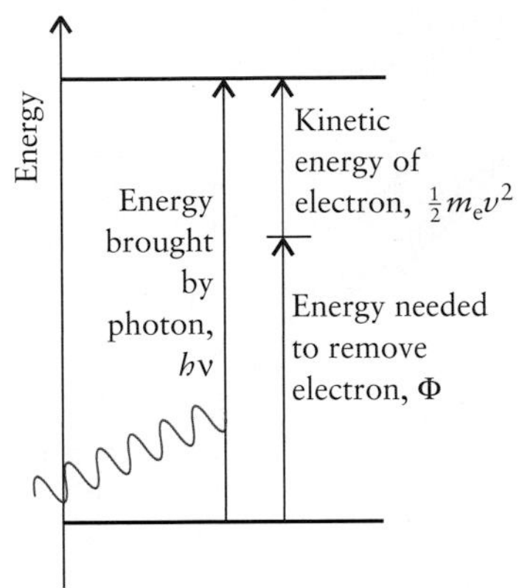

FIGURE 1.8 In the photoelectric effect, a photon with energy $h\nu$ strikes the surface of a metal and its energy is absorbed by an electron. If the energy of the photon is greater than the work function, Φ, of the metal, the electron absorbs enough energy to break away from the metal. The kinetic energy of the ejected electron is the difference between the energy of the photon and the work function: $\frac{1}{2}m_ev^2 = h\nu - \Phi$.

The characteristics of the photoelectric effect are easy to explain in terms of photons. If the incident radiation has frequency ν, it consists of a stream of photons of energy $h\nu$. These particles collide with the electrons in the metal. If the energy of a photon is less than the energy required to remove an electron from the metal, then an electron will not be ejected, regardless of the intensity of the radiation. The energy required to remove an electron from the surface of a metal is called the **work function** of the metal and denoted Φ (uppercase phi). However, if the energy of the photon, $h\nu$, is greater than Φ, then an electron is ejected with a kinetic energy, $E_K = \frac{1}{2}m_ev^2$, equal to the difference between the energy of the incoming photon and the work function: $E_K = h\nu - \Phi$ (Fig. 1.8). It follows that

$$\overbrace{\tfrac{1}{2}m_ev^2}^{\text{Kinetic energy of ejected electron}} = \overbrace{h\nu}^{\text{Energy supplied by photon}} - \overbrace{\Phi}^{\text{Energy required to eject electron}} \qquad (5)$$

We can now interpret the experimental observations of the photoelectric effect in light of Einstein's theory:

1 An electron can be driven out of the metal only if it receives at least a certain minimum energy, Φ, from the photon during the collision. Therefore, the frequency of the radiation must have a certain minimum value if electrons are to be ejected. This minimum frequency depends on the work function and hence on the identity of the metal (Fig. 1.9).

2 Provided a photon has enough energy, a collision results in the immediate ejection of an electron.

3 The kinetic energy of the electron ejected from the metal increases linearly with the frequency of the incident radiation according to Eq. 5.

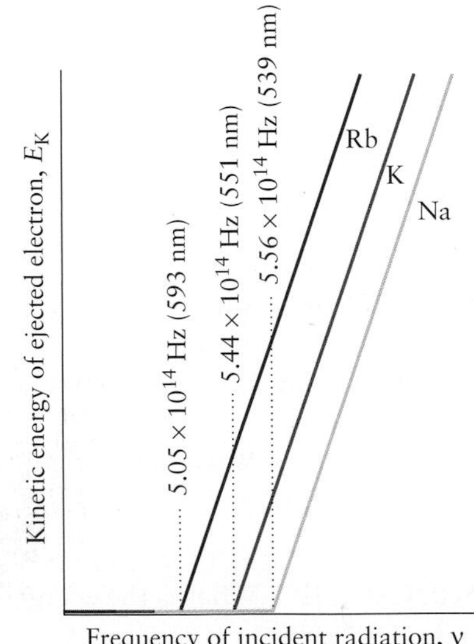

FIGURE 1.9 When photons strike a metal, no electrons are ejected unless the incident radiation has a frequency above a value characteristic of the metal. The kinetic energy of the ejected electrons varies linearly with the frequency of the incident radiation.

The photoelectric effect strongly supports the view that electromagnetic radiation consists of photons that behave like particles. However, there is plenty of evidence to show that electromagnetic radiation behaves like waves! The most compelling evidence is the observation of **diffraction**, the pattern of high and low intensities generated by an object in the path of a ray of light (Fig. 1.10). A diffraction pattern results when the peaks and troughs of waves traveling along one path interfere with the peaks and troughs of waves traveling along another path. If the peaks coincide, the amplitude of the wave (its height) is enhanced; we call this enhancement **constructive interference** (Fig. 1.11a). If the peaks of one wave coincide with the troughs of another wave, the amplitude of the wave is diminished by **destructive interference** (Fig. 1.11b). This effect is the basis of a useful technique for studying matter. For example, x-ray diffraction is one of the most important tools for studying the structures of molecules (see Major Technique 3, following Chapter 5).

Let's review where we are. Some experiments (the photoelectric effect) compel us to the view that electromagnetic radiation is particlelike. Other experiments (diffraction) compel us equally firmly to the view that electromagnetic radiation is wavelike. Thus we are brought to the heart of modern physics. Experiments oblige us to accept the **wave-particle duality** of electromagnetic radiation, in which the concepts of waves and particles blend together. In the wave model, the intensity of the radiation is proportional to the square of the amplitude of the wave. In the particle model, intensity is proportional to the number of particles present at each instant.

Studies of black-body radiation led to Planck's hypothesis of the quantization of electromagnetic radiation. The photoelectric effect provides evidence of the particulate nature of electromagnetic radiation; diffraction provides evidence of its wave nature.

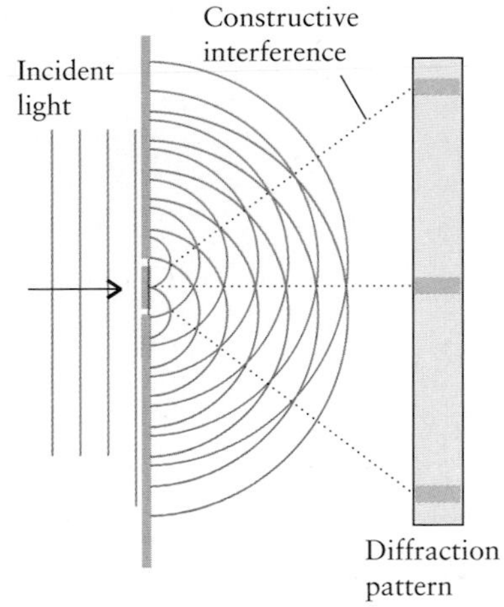

FIGURE 1.10 In this illustration, the peaks of the waves of electromagnetic radiation are represented by orange lines. When radiation incident from the left (the vertical lines) passes through a pair of closely spaced slits, circular waves are generated at each slit. These waves interfere with each other. Where they interfere constructively (as indicated by the positions of the dotted lines), a bright line is seen on the screen behind the slits; where the interference is destructive, the screen is dark.

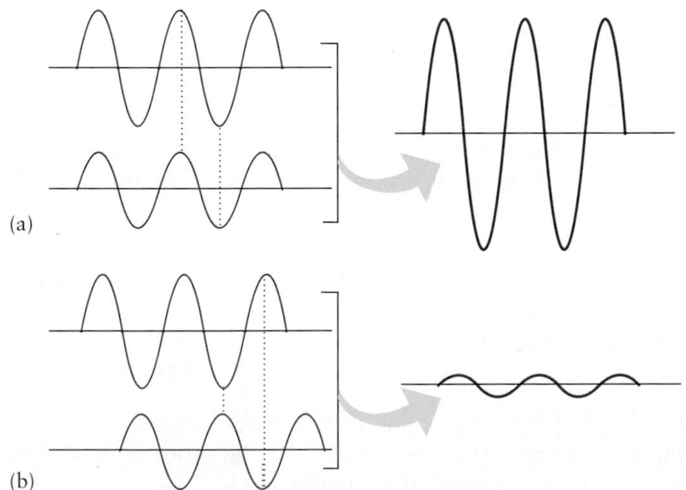

FIGURE 1.11 (a) Constructive interference. The two component waves (left) are "in phase" in the sense that their peaks and troughs coincide. The resultant (right) has an amplitude that is the sum of the amplitudes of the components. The wavelength of the radiation is not changed by interference, only the amplitude. (b) Destructive interference. The two component waves are "out of phase" in the sense that the troughs of one coincide with the peaks of the other. The resultant has a much lower amplitude.

EXAMPLE 1.1 Interpreting a light ray in terms of photons

A discharge lamp rated at 25 W (or 25 $J{\cdot}s^{-1}$) emits yellow light of wavelength 580 nm. How many photons of yellow light does the lamp generate in 1.0 s?

STRATEGY The number of photons that accounts for a given energy is that energy divided by the energy of a single photon. We expect a large number of photons because the energy of a single photon is very small. We calculate the energy of a photon from its wavelength by using Eq. 1 to convert the given wavelength to a frequency and Eq. 4 to convert frequency to energy. To reduce rounding errors, it is good practice to work through a calculation symbolically as far as possible and to introduce numerical data at the last possible stage.

SOLUTION It follows from Eq. 1 that the frequency of the light is

$$\nu = \frac{c}{\lambda}$$

Therefore, the energy of each photon of wavelength λ is

$$E_{\text{photon}} = h\nu = \frac{hc}{\lambda}$$

The number of photons that accounts for a total energy E_{total} is therefore

$$\text{Number of photons} = \frac{E_{\text{total}}}{E_{\text{photon}}} = \frac{E_{\text{total}}}{(hc/\lambda)} = \frac{E_{\text{total}}\lambda}{hc}$$

where

$$\begin{aligned} E_{\text{total}} &= \text{power} \times \text{time lamp is on} \\ &= (25\ \text{J}{\cdot}\text{s}^{-1}) \times (1.0\ \text{s}) = 25\ \text{J} \end{aligned}$$

Now we express the wavelength in meters and insert the data:

$$\begin{aligned} \text{Number of photons} &= \frac{(25\ \text{J}) \times (5.80 \times 10^{-7}\ \text{m})}{(6.626 \times 10^{-34}\ \text{J}{\cdot}\text{s}) \times (3.00 \times 10^{8}\ \text{m}{\cdot}\text{s}^{-1})} \\ &= 7.3 \times 10^{19} \end{aligned}$$

This result means that when you turn on the lamp, it generates about 10^{20} photons of yellow light each second. As predicted, this is a very large number.

Self-Test 1.4A Another discharge lamp produces 5.0 J of energy per second in the blue region of the spectrum. How many photons of blue (470 nm) light would the lamp generate if it were left on for 8.5 s?

[*Answer:* 1.0×10^{20}]

Self-Test 1.4B In 1.0 s, a lamp that produces 25 J of energy per second in a certain region of the spectrum emits 5.5×10^{19} photons of light in that region. What is the wavelength of the emitted light?

1.3 The Wave-Particle Duality of Matter

If electromagnetic radiation, which for a long time had been regarded as wavelike, has a dual character, could it be that matter, which since Dalton's day had been regarded as consisting of particles, also has wavelike properties? On the basis of theoretical considerations, the French scientist Louis de Broglie suggested that all particles should be regarded as having wavelike properties. He went on to suggest that the wavelength associated with the "particle wave" is inversely proportional to its mass, m, and speed, v, and proposed that

Be careful to distinguish the symbol for speed, v (for velocity), from the symbol for frequency, ν (the Greek letter nu).

$$\lambda = \frac{h}{mv} \tag{6a}$$

The product of mass and speed is called the **linear momentum,** p, of a particle, so this expression is more simply written as the **de Broglie relation:**

$$\lambda = \frac{h}{p} \tag{6b}$$

It is easy to see why the wavelike properties of particles had not been noticed. According to the de Broglie relation, a particle of mass 1 g traveling at 1 $m{\cdot}s^{-1}$ has a wavelength of

$$\begin{aligned}\lambda &= \frac{h}{p} = \frac{h}{mv} = \frac{6.626 \times 10^{-34}\ \text{J}{\cdot}\text{s}}{(1 \times 10^{-3}\ \text{kg}) \times (1\ \text{m}{\cdot}\text{s}^{-1})}\\ &= 7 \times 10^{-31}\ \frac{\text{kg}{\cdot}\text{m}^2{\cdot}\text{s}^{-1}}{\text{kg}{\cdot}\text{m}{\cdot}\text{s}^{-1}} = 7 \times 10^{-31}\ \text{m}\end{aligned}$$

(In the second numerical step, we have used $1\ \text{J} = 1\ \text{kg}{\cdot}\text{m}^2{\cdot}\text{s}^{-2}$.) This wavelength is undetectably small; the same is true for any macroscopic (visible) object traveling at normal speeds.

The wavelike character of electrons was detected by showing that they could be diffracted. The experiment was first performed in 1925 by two American scientists, Clinton Davisson and Lester Germer, who directed a beam of fast electrons at a single crystal of nickel. The regular array of atoms in the crystal, with centers separated by 250 pm, acts as a grid that diffracts waves; and a diffraction pattern was observed (Fig. 1.12). Electron diffraction is now an important technique for determining the structures of molecules and exploring the structures of solid surfaces.

Electrons have both wavelike and particlelike properties.

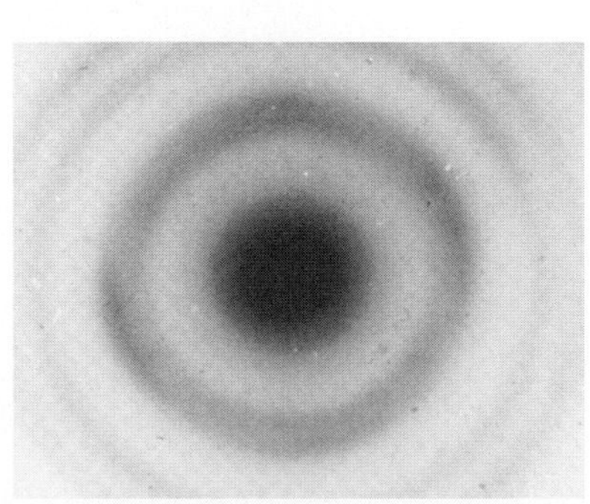

FIGURE 1.12 Davisson and Germer showed that electrons produced a diffraction pattern when reflected from a crystal. G. P. Thomson, working in Aberdeen, Scotland, showed that they also produce a diffraction pattern when they pass through a very thin gold foil. The latter is shown here. G. P. Thomson was the son of J. J. Thomson, who identified the electron (Section B). Both received Nobel prizes: J. J. for showing that the electron is a particle and G. P. for showing that it is a wave.

EXAMPLE 1.2 **Estimating the wavelength of a particle**

Estimate the wavelength of (a) a proton moving at 1/100 the speed of light; (b) a marble of mass 5.00 g traveling at 1.00 $m{\cdot}s^{-1}$.

STRATEGY Use the de Broglie relation, Eq. 6a. The mass of a proton is given in Table B.1, and the speed of light is given inside the back cover. Expect the macroscopic particle, the marble, to have a very short wavelength. Remember to express all quantities in kilograms, meters, and seconds, and use 1 J = 1 $kg{\cdot}m^2{\cdot}s^{-2}$.

SOLUTION (a) The mass of a proton is 1.673×10^{-27} kg and the speed of light is $2.998 \times 10^8\ m{\cdot}s^{-1}$; therefore,

$$\lambda = \frac{h}{mv} = \frac{6.626 \times 10^{-34}\ \mathrm{J{\cdot}s}}{(1.673 \times 10^{-27}\ \mathrm{kg}) \times (2.998 \times 10^{8}\ \mathrm{m{\cdot}s^{-1}})/100}$$
$$= 1.32 \times 10^{-13}\,\frac{(\mathrm{kg{\cdot}m^2{\cdot}s^{-2}}) \times \mathrm{s}}{\mathrm{kg} \times (\mathrm{m{\cdot}s^{-1}})} = 1.32 \times 10^{-13}\ \mathrm{m}$$

This wavelength corresponds to 0.132 pm. (b) The mass of the marble is 5.00×10^{-3} kg, so

$$\lambda = \frac{6.626 \times 10^{-34}\ \mathrm{J{\cdot}s}}{(5.00 \times 10^{-3}\ \mathrm{kg}) \times (1.00\ \mathrm{m{\cdot}s^{-1}})} = 1.33 \times 10^{-31}\ \mathrm{m}$$

As expected, the wavelength of the marble is undetectably small. However, the wavelength of the rapidly moving proton is not very much smaller than the diameter of an atom (about 200 pm).

Self-Test 1.5A Calculate the wavelength of an electron traveling at 1/1000 the speed of light.

[*Answer:* 2.43 nm]

Self-Test 1.5B Calculate the wavelength of a rifle bullet of mass 5.0 g traveling at twice the speed of sound (the speed of sound is 331 $m{\cdot}s^{-1}$).

1.4 The Uncertainty Principle

The wave-particle duality not only changes our understanding of electromagnetic radiation and matter, but it also sweeps away the foundations of classical physics. In classical mechanics, a particle has a definite **trajectory**, or path on which location and linear momentum are specified at each instant. However, we cannot specify the precise location of a particle if it behaves like a wave: think of a wave in a guitar string, which is spread out all along the string, not localized at a precise point. A particle with a precise linear momentum has a precise wavelength; but because it is meaningless to speak of the location of a wave, it follows that we cannot specify the location of a particle that has a precise linear momentum.

The difficulty will not go away. Wave-particle duality denies the possibility of specifying the location if the linear momentum is known, so we cannot specify the trajectory of particles. If we know that a particle is *here* at one instant, we can say *nothing* about where it will be an instant later! The impossibility of knowing the position with arbitrarily great precision if the linear momentum is known precisely is an aspect of the **complementarity** of location and momentum. The **Heisenberg uncertainty principle** expresses this complementarity quantitatively: it states that if the location of a particle is known to within an uncertainty Δx, then the linear momentum parallel to the x-axis can be known only to within an uncertainty Δp, where

$$\Delta p \Delta x \geq \tfrac{1}{2}\hbar \qquad (7)$$

The symbol $\hbar$, which is read "h bar," means $h/2\pi$, a useful combination that occurs widely in quantum mechanics. From inside the back cover, we see that $\hbar = 1.054\,57 \times 10^{-34}$ J·s. Equation 7 tells us that if the uncertainty in position is very small (Δx very small), then the uncertainty in linear momentum must be large, and vice versa (Fig. 1.13). The uncertainty principle has negligible practical consequences for macroscopic objects, but it is of profound importance for electrons in atoms and for a scientific understanding of the nature of the world.

The location and momentum of a particle are complementary; that is, both the location and momentum cannot be known simultaneously with arbitrary precision. The quantitative relation between the precision of each measurement is described by the Heisenberg uncertainty principle, Eq. 7.

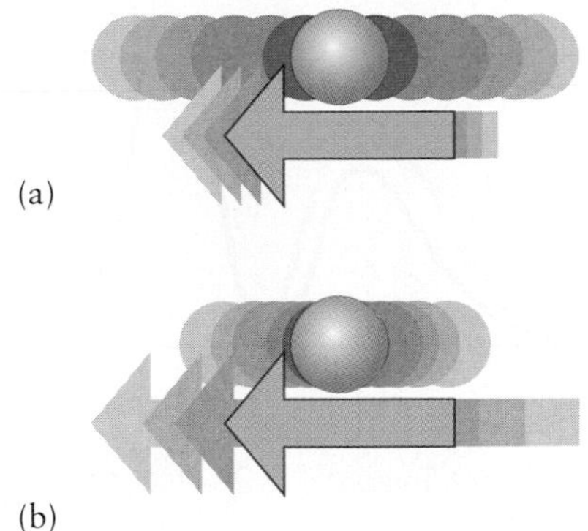

FIGURE 1.13 A representation of the uncertainty principle. (a) The location of the particle is ill defined, so the momentum of the particle (represented by the arrow) can be specified reasonably precisely. (b) The location of the particle is well defined, so the momentum cannot be specified very precisely.

EXAMPLE 1.3 Using the uncertainty principle

Estimate the minimum uncertainty in (a) the position of a marble of mass 1.0 g given that its speed is known to within ± 1.0 mm·s^{-1} and (b) the speed of an electron confined to within the diameter of a typical atom (200 pm).

STRATEGY We expect the uncertainty in the position of an object as heavy as a marble to be very small but the uncertainty in the speed of an electron, which is very light and confined to a small region, to be very large. In each case, be sure to use units that match the units of $\hbar$. (a) We calculate the uncertainty Δp from $m\Delta v$, where Δv is the uncertainty in the speed; then we use Eq. 7 to estimate the minimum uncertainty in position, Δx, along the direction of its travel by assuming $\Delta p \Delta x = \tfrac{1}{2}\hbar$ (which gives the smallest value of the uncertainty). Because the reported uncertainty in speed is given to plus or minus a stated value, the value of Δv is equal to twice that value. (b) We assume Δx to be the diameter of the atom and use Eq. 7 to estimate Δp; by using the mass of the electron in Table B.1, we find Δv from $m\Delta v = \Delta p$.

SOLUTION (a) First we convert mass and speed to SI base units. The mass, m, is 1.0×10^{-3} kg and the uncertainty in the speed, Δv, is $2 \times (1.0 \times 10^{-3}$ m·s$^{-1})$. The minimum uncertainty in position, Δx, is

$$\Delta x = \frac{\hbar}{2m\Delta v} = \frac{1.054\,57 \times 10^{-34}\ \text{J·s}}{2 \times (1.0 \times 10^{-3}\ \text{kg}) \times (2.0 \times 10^{-3}\ \text{m·s}^{-1})}$$
$$= 2.6 \times 10^{-29}\ \text{m}$$

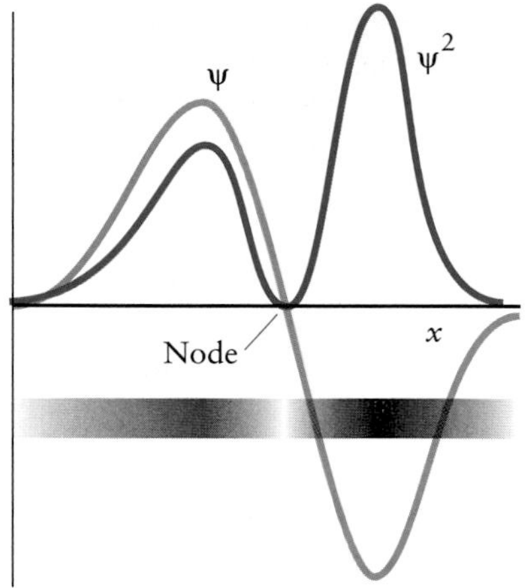

FIGURE 1.14 The Born interpretation of the wavefunction. The probability density is given by the square of the wavefunction (the blue line) and depicted by the density of shading in the band beneath. Note that the probability density is 0 at a node. A node is a point where the wavefunction passes through 0, not merely approaches 0.

As predicted, this uncertainty is very small. (b) The mass of the electron in Table B.1 is 9.109×10^{-31} kg; the diameter of the atom is 200×10^{-12} m, or 2.00×10^{-10} m. The uncertainty in the speed, Δv, is equal to $\Delta p/m$:

$$\Delta v = \frac{\Delta p}{m} = \frac{\hbar}{2m\Delta x} = \frac{1.054 \times 10^{-34}\ \text{J·s}}{2 \times (9.109 \times 10^{-31}\ \text{kg}) \times (2.00 \times 10^{-10}\ \text{m})}$$
$$= 2.89 \times 10^{5}\ \text{m·s}^{-1}$$

As predicted, the uncertainty in the speed of the electron is very large.

Self-Test 1.6A A proton is accelerated in a cyclotron to a very high speed that is known to within 3.0×10^{2} km·s^{-1}. What is the minimum uncertainty in its position?

[*Answer:* 0.10 pm]

Self-Test 1.6B The police are monitoring an automobile of mass 2.0 t ($1\ \text{t} = 10^{3}$ kg) speeding along a highway. They are certain of the location of the vehicle only to within 1 m. What is the minimum uncertainty in the speed of the vehicle?

1.5 Wavefunctions and Energy Levels

Twentieth-century scientists had to revise the nineteenth-century description of matter to take into account wave-particle duality. One of the first people to formulate a successful theory (in 1927) was the Austrian scientist Erwin Schrödinger, and we shall concentrate on his approach.

Because particles have wavelike properties, we cannot expect them to behave like pointlike objects moving along precise trajectories. Schrödinger's approach was to replace the precise trajectory of a particle by a **wavefunction**, denoted by ψ (the Greek letter psi), a mathematical function with values that vary with position. The German physicist Max Born suggested how this function should be interpreted physically. According to the **Born interpretation** of the wavefunction, the probability of finding the particle in a region is proportional to ψ^2 (Fig. 1.14). To be precise, ψ^2 is a **probability density** in the sense that to calculate the probability that the particle is in a small region of space, we multiply ψ^2 by the volume of the region. For instance, if $\psi^2 = 0.1\ \text{pm}^{-3}$ at a point, then the probability of finding the particle in a region of volume 2 pm^3 located at that point would be $(0.1\ \text{pm}^{-3}) \times (2\ \text{pm}^{3}) = 0.2$, or 1 chance in 5. According to the Born interpretation, wherever ψ^2 is large, there is a high probability density for the particle; and wherever ψ^2 is small, there is only a low probability density for the particle.

FIGURE 1.15 Erwin Schrödinger (1887–1961), pondering the Schrödinger equation.

Because the square of any number is positive, we don't have to worry about ψ having a negative sign in some regions of space: a probability density is never negative. Wherever ψ, and hence ψ^2, is 0, there is zero probability density for the particle. A location where ψ passes through 0 is called a **node** of the wavefunction, so we can say that a particle has zero probability density wherever the wavefunction has nodes.

The **Schrödinger equation** is an equation for calculating the wavefunction. You can see what the equation looks like, for Fig. 1.15 shows it emerging from Schrödinger's head. More formally, the equation for a particle of mass m moving in a region where the potential energy is $V(x)$ is

The symbol V is commonly used to denote potential energy, rather than E_p, in this context.

$$-\frac{\hbar^2}{2m}\frac{d^2\psi}{dx^2} + V(x)\psi = E\psi \quad (8a)$$

As we see, the Schrödinger equation is a "differential equation," an equation that relates derivatives of a function (in this case, a second derivative of ψ, $d^2\psi/dx^2$) to the value of the function at each point. It is impossible to solve exactly except in certain simple cases. However, in this text, we need only the form of some of its solutions, not how those solutions are found. The left-hand side of the Schrödinger equation is commonly written $H\psi$, where H is called the **hamiltonian** for the system, then the equation takes the deceptively simple form

Derivatives are reviewed in Appendix 1F.

$$H\psi = E\psi \quad (8b)$$

One of the simplest examples of an actual wavefunction is that for a particle of mass m confined between two rigid walls a distance L apart, the so-called particle in a box (Fig. 1.16). The shapes of the wavefunctions for this one-dimensional system, some of which are shown in the illustration, make sense when we view the particle as a wave. Only certain wavelengths can exist in the box, just as a stretched string can support only certain wavelengths. Think of a guitar string: because it is tied down at each end, it can support only shapes like the ones shown in Fig. 1.16. The shapes of the wavefunctions are identical to the vibrations of a stretched string, and their mathematical form is

$$\psi_n(x) = \left(\frac{2}{L}\right)^{1/2} \sin\left(\frac{n\pi x}{L}\right) \qquad n = 1, 2, \ldots \quad (9)$$

The integer n labels the wavefunctions and is called a "quantum number." In general, a **quantum number** is an integer (sometimes, a half-integer) that labels a wavefunction, specifies a state, and can be used to calculate the value of a property of the system. For example, we can use n to find an expression for the energy corresponding to each wavefunction.

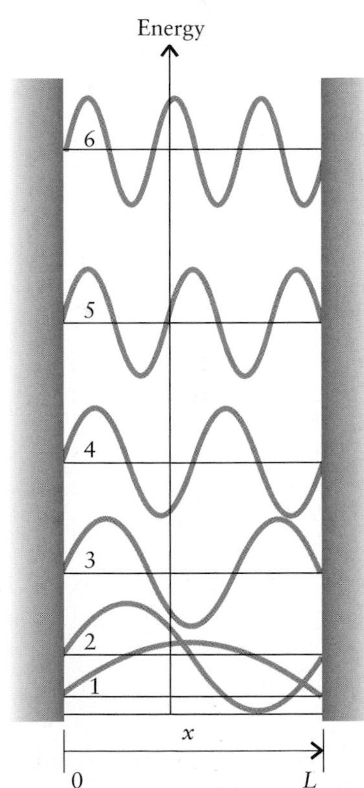

FIGURE 1.16 The arrangement known as "a particle in a box," in which a particle of mass m is confined between two impenetrable walls a distance L apart. The first six wavefunctions and their energies are shown. The numbers on the left are values of the quantum number n. Note that there is a zero-point energy because n cannot be 0.

Media Link **Living graph:** Fig. 1.16, Particle-in-a-box wavefunctions

How do we do that (level 1)?

The best way to find the allowed energies of a particle in a box is to solve the Schrödinger equation directly (see "level 2" below), but there is a simpler way. The kinetic energy of a particle of mass m is related to its speed, v, by $E_K = \frac{1}{2}mv^2$. We can relate this energy to the wavelength of the particle by noting that the linear momentum is $p = mv$ and then using the de Broglie relation (Eq. 6b):

$$E_K = \tfrac{1}{2}mv^2 = \frac{p^2}{2m} = \frac{h^2}{2m\lambda^2}$$

The potential energy of the particle is 0 everywhere inside the box, so the total energy, E, is given by the expression for E_K alone. At this point, we recognize that only whole-number multiples of half-wavelengths can fit into the box (see Fig. 1.16; the waves have one bulge, two bulges, three bulges, and so forth, with each "bulge" a half-wavelength). That is,

$$L = \tfrac{1}{2}\lambda, \lambda, \tfrac{3}{2}\lambda, \ldots = n \times \tfrac{1}{2}\lambda, \quad \text{with } n = 1, 2, \ldots$$

Therefore, the allowed wavelengths are

$$\lambda = 2L/n, \quad \text{with } n = 1, 2, \ldots$$

When this expression for λ is inserted into the expression for the energy, we obtain

$$E_n = \frac{n^2h^2}{8mL^2} \qquad (10)$$

How do we do that (level 2)?

The more sophisticated—and more general—way of finding the energy levels of a particle in a box is to solve the Schrödinger equation. First, we note that the potential energy of the particle is 0 everywhere inside the box, so $V(x) = 0$, and the equation we have to solve is

$$-\frac{\hbar^2}{2m}\frac{d^2\psi}{dx^2} = E\psi$$

This equation has the solutions

$$\psi(x) = A \sin kx + B \cos kx$$

with A, B, and k constants, as may be verified by substituting the solutions into the differential equation and using $d(\sin kx)/dx = k \cos kx$ and $d(\cos kx)/dx = -k \sin kx$. Indeed, because

$$\frac{d^2\psi}{dx^2} = \frac{d^2}{dx^2}(A \sin kx + B \cos kx)$$
$$= -k^2(A \sin kx + B \cos kx) = -k^2\psi$$

we can conclude, by substituting this relation into the Schrödinger equation, that

$$E = \frac{k^2\hbar^2}{2m} = \frac{k^2h^2}{8\pi^2m}$$

Now we have to find the constants A, B, and k. First, we note that the wavefunction is 0 outside the box, but must be continuous. Therefore, $\psi(x)$ must be 0 at the edges of the box, at $x = 0$ and $x = L$. We call these constraints **boundary conditions**. If we set $x = 0$, and use $\sin 0 = 0$ and

$\cos 0 = 1$ in $\psi(x) = A \sin kx + B \cos kx$, we find $\psi(0) = B$. However, $\psi(0) = 0$, so $B = 0$ and the wavefunction is

$$\psi(x) = A \sin kx$$

This expression already resembles that in Eq. 9. To find the value of k, we use the second boundary condition, that $\psi(L) = 0$; that is,

$$\psi(L) = A \sin kL = 0$$

Setting $A = 0$ is unacceptable, because then $\psi(x)$ would be 0 everywhere, which would mean that no particle could be present. Therefore, to guarantee that $\psi(L) = 0$, we use the relation $\sin n\pi = 0$ and set $kL = n\pi$, with $n = 1, 2, \ldots$ We now have

$$\psi(x) = A \sin \frac{n\pi x}{L} \qquad n = 1, 2, \ldots$$

To find the one remaining constant, A, we use the fact that the probability of finding the particle in a region dx at x is $\psi(x)^2 dx$, and therefore the total probability of finding the particle between $x = 0$ and $x = L$ is the sum (integral) of these probabilities and must be equal to 1 (the particle must be somewhere in the box). Therefore,

$$\int_0^L \psi(x)^2\, dx = A^2 \int_0^L \sin^2 \frac{n\pi x}{L}\, dx = 1$$

The integral evaluates to $\frac{1}{2}L$, so $A = (2/L)^{1/2}$ and the final form of the wavefunction is that given in Eq. 9. Moreover, because k is limited to the values $n\pi/L$, it follows that the energy of the particle is limited to

$$E = \frac{(n\pi/L)^2 h^2}{8\pi^2 m} = \frac{n^2 h^2}{8mL^2}$$

exactly as in Eq. 10. We see that the quantization of energy (see below) stems directly from the imposition of boundary conditions on the wavefunction. The origin of quantization from the imposition of boundary conditions on the wavefunction is a general result for any system.

One of the most striking implications of Eq. 8 is that the energy of the particle is **quantized**, meaning that it is restricted to a series of discrete values, called **energy levels**. According to classical mechanics, an object can have any total energy—high, low, and anything in between. For example, a particle in a box could bounce from wall to wall with any speed and hence any kinetic energy. According to quantum mechanics, however, energy is quantized: only certain wavelengths fit into the box. Quantization is a little like pouring water into a bucket. Water seems to be a continuous fluid, and it seems that any amount can be transferred. However, the smallest amount of water that we can transfer is one H_2O

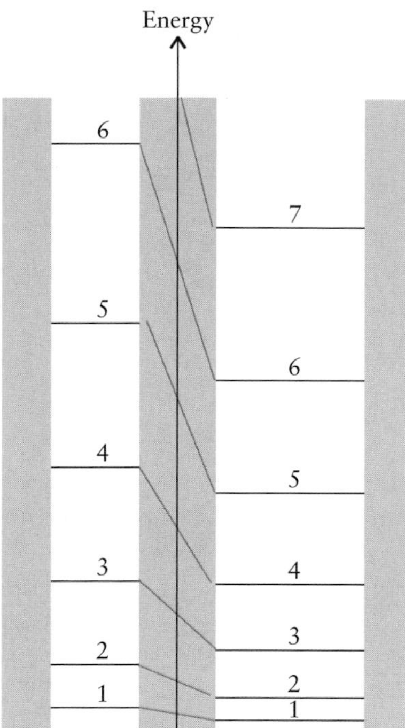

FIGURE 1.17 As the length of the box increases (compare boxes on left and right), the energy levels fall and move closer together.

molecule. Energy seems to be unrestricted on the macroscopic scale; but on the subatomic scale, we can transfer only discrete amounts to a particle in a box.

We can use Eq. 10 to calculate the energy separation between two neighboring levels with quantum numbers n and $n + 1$:

$$E_{n+1} - E_n = \frac{(n+1)^2h^2}{8mL^2} - \frac{n^2h^2}{8mL^2} = \frac{(2n+1)h^2}{8mL^2} \quad (11)$$

We see that as L or m increases, the separation between neighboring energy levels decreases (Fig. 1.17). That is why no one noticed that energy is quantized until they investigated very small systems such as an electron in a hydrogen atom: the separation between levels is so small for ordinary particles in ordinary-sized vessels that it is completely undetectable. We can, in fact, ignore the quantization of the motion of the atoms of a gas in a typical flask.

EXAMPLE 1.4 Calculating the energies of a particle in a box

Treat a hydrogen atom as a one-dimensional box of length 150 pm (the approximate diameter of the atom) containing an electron and predict the wavelength of the radiation emitted when the electron falls to the lowest energy level from the next higher energy level.

STRATEGY The lowest energy level has $n = 1$, so we can use Eq. 11 with $n = 1$ and $m = m_e$, the mass of the electron. The energy difference is carried away as a photon of radiation, so we set the energy difference equal to $h\nu$ and express ν in terms of the corresponding wavelength by using Eq. 1 ($\lambda = c/\nu$).

SOLUTION According to Eq. 11, with $n = 1$:

$$h\nu = E_2 - E_1 = \frac{[(2 \times 1) + 1]h^2}{8m_eL^2} = \frac{3h^2}{8m_eL}$$

where m_e is the mass of the electron. It follows that

$$\nu = \frac{3h}{8m_eL^2}$$

and therefore

$$\begin{aligned}\lambda &= \frac{c}{\nu} = \frac{8m_ecL^2}{3h}\\ &= \frac{8 \times (9.109 \times 10^{-31}\ \text{kg}) \times (2.998 \times 10^{8}\ \text{m·s}^{-1}) \times (1.50 \times 10^{-10}\ \text{m})^2}{3 \times (6.626 \times 10^{-34}\ \text{J·s})}\\ &= 2.47 \times 10^{-8}\ \frac{\text{kg} \times (\text{m·s}^{-1}) \times \text{m}^2}{(\text{kg·m}^2\text{·s}^{-2}) \times \text{s}} = 2.47 \times 10^{-8}\ \text{m}\end{aligned}$$

This wavelength corresponds to 24.7 nm. The experimental value for an electron in a hydrogen atom is 122 nm. Although there is a big discrepancy, the fact that the predicted wavelength has nearly the same order of magnitude as the actual value suggests that a quantum theory of the atom, based on a more realistic three-dimensional model, should give good agreement.

Self-Test 1.7A Use the same model for helium but suppose that the box is 100 pm long, because the atom is smaller. Estimate the wavelength of the same transition.

[*Answer:* 11.0 nm]

Self-Test 1.7B Use the same model for hydrogen and estimate the wavelength for the transition from the $n = 3$ energy level to the $n = 2$ level.

A surprising implication of Eq. 10 is that *a particle in a container cannot have zero energy*. Because the lowest value of n is 1 (corresponding to a wave of one-half wavelength), the lowest energy is $E_1 = h^2/8mL^2$. This lowest possible energy is called the **zero-point energy**. The existence of a zero-point energy means that, according to quantum mechanics, a particle can never be perfectly still when it is confined between two walls: it must always possess an energy of at least $h^2/8mL^2$. This result is consistent with the uncertainty principle. When a particle is confined between two walls, the uncertainty in its position cannot be larger than the distance between the two walls. Because the position is not *completely* uncertain, the linear momentum must be uncertain, too, so we cannot say that the particle is completely still. The particle must therefore have some kinetic energy. The zero-point energy is a purely quantum mechanical phenomenon and is very small for macroscopic systems. For example, a billiard ball on a pool table has a completely negligible zero-point energy of about 10^{-67} J.

Finally, the shapes of the wavefunctions of a particle in a box also reveal some interesting information. Let's look at the two lowest energy wavefunctions, corresponding to $n = 1$ and $n = 2$. Figure 1.18 shows, by the density of shading, the likelihood of finding a particle: we see that when a particle is described by the wavefunction ψ_1 (and has energy $h^2/8mL^2$), it is most likely to be found in the center of the box. Conversely, if the particle is described by the wavefunction ψ_2 (and has energy $h^2/2mL^2$), it is most likely to be found in regions between the center and the walls. Remember that the wavefunction itself does not have any direct physical significance: we have to take the square of ψ before we can interpret it in terms of the probability of finding a particle somewhere. Therefore, the fact that ψ_2 is negative in one-half of the box is not physically significant: when we take the squares of the wavefunctions, the result is not negative anywhere. A similar interpretation can be given for wavefunctions corresponding to other values of n.

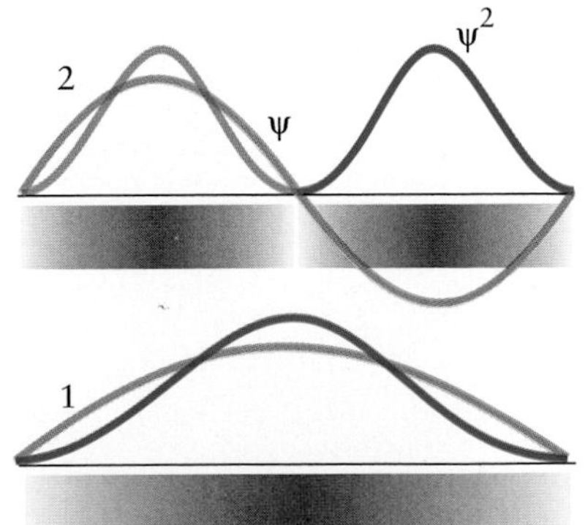

FIGURE 1.18 The two lowest energy wavefunctions (ψ, orange) for a particle in a box and the corresponding probability densities (ψ^2, blue). The probability densities are also shown by the density of shading of the bands beneath each wavefunction.

MEDIA EXAMPLE 1.5 Plotting the wavefunctions for a particle in a box

(a) Using the Living Graphs on the Web site for this book, plot the particle-in-a-box wavefunction for $n = 2$ and $L = 1$ m. (b) How many nodes does the wavefunction have? Where do these nodes occur? (c) Repeat parts (a) and (b) for $n = 3$. (d) What general conclusion can you draw about the relation between n and the number of nodes present in a wavefunction? (e) Convert the $n = 2$ plot to a probability density distribution: at what values of x is it most likely to find the particle? (f) Repeat part (e) for $n = 3$.

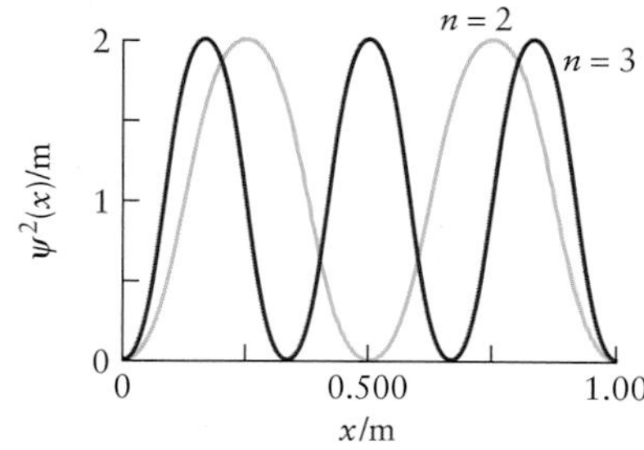

Media Link Example 1.5: Particle-in-a-box wavefunctions

STRATEGY The wavefunctions for a particle in a box are given by Eq. 9. Use the Living graph "Particle-in-a-box wavefunctions" on the Web site for this book to plot them. (a) To graph the wavefunction, enter the parameters $L = 1$ m and $n = 2$, set the range for the entire box ($x = 0$ to 1 m), check the $f(x)$ box, and hit New Plot. (b) The nodes are found where the curve passes through 0. To find the value of x at a point, click on the plot at that point. (c) Create a new plot as before, but set $n = 3$. (d) Look for the patterns in node number and n. (e) Repeat with the $f^2(x)$ box checked and look for the maxima in the probability density. (f) Repeat part (e) for $n = 3$.

SOLUTION The plots are shown on the left. (b) The plot for $n = 2$ has one node at $x = 0.50$ m. (c) The plot for $n = 3$ has two nodes at $x = 0.33$ and 0.67 m. (d) The number of nodes is equal to $n - 1$. (c) Maxima occur at $x = 0.25$ and 0.75 m. (f) Maxima occur at $x = 0.17$, 0.50, and 0.83 m.

Self-Test 1.8A Verify the conclusion in part (d) by plotting the wavefunction for $n = 4$ and determining the number of nodes.

[*Answer:* 3 nodes]

Self-Test 1.8B Verify the conclusion in part (d) by plotting the wavefunction for $n = 5$ and determining the number of nodes.

The probability density for a particle at a location is proportional to the square of the wavefunction at that point; the wavefunction is found by solving the Schrödinger equation for the particle. When the equation is solved, it is found that the particle can possess only certain discrete energies.

1.6 Atomic Spectra and Energy Levels

Compelling evidence for the validity of quantum mechanics came from its ability to explain atomic spectra. When an electric current is passed through a low-pressure sample of hydrogen gas, the sample emits light. The electric current, which is like a storm of electrons, breaks up the H_2 molecules and excites the free hydrogen atoms to higher energies. These excited atoms quickly discard their excess energy by giving off electromagnetic radiation; then they combine to form H_2 molecules again.

When white light is passed through a prism, a continuous spectrum of light results (Fig. 1.19a). However, when the light emitted by excited

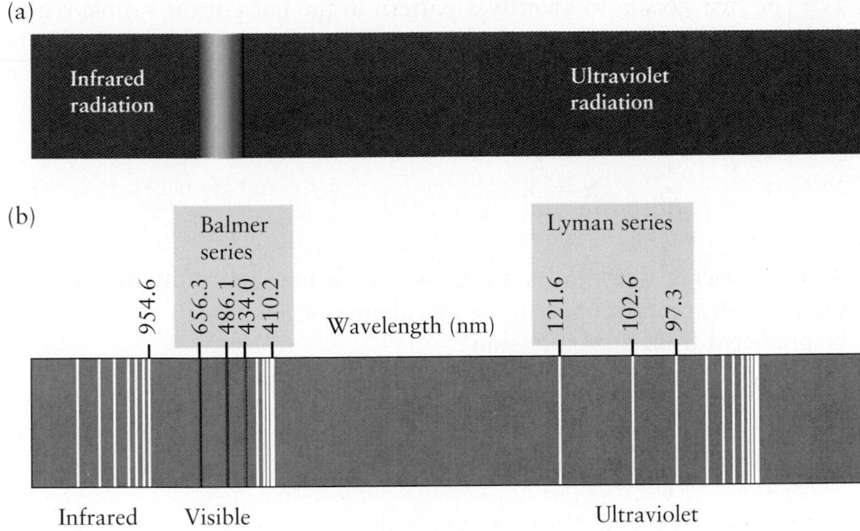

FIGURE 1.19 (a) The visible spectrum. (b) The complete spectrum of atomic hydrogen. The spectral lines have been assigned to various groups called series, two of which are shown with their names.

hydrogen atoms is passed through a prism, it is found that the radiation consists of a number of components, or **spectral lines** (Fig. 1.19b). The brightest line (at 656 nm) is red, and the excited atoms in the gas glow with this red light. Excited hydrogen atoms also emit ultraviolet and infrared radiation, which are invisible to the eye but can be detected electronically and photographically.

The components of different frequency or wavelength are called "lines" because, in the early spectroscopic experiments, the radiation from the sample was passed through a slit and then through a prism; the image of the slit was then focused on a photographic plate, where it appeared as a line.

The series of discrete lines making up the spectrum of atomic hydrogen was highly puzzling. How, the early spectroscopists wondered, can an atom emit only particular frequencies of electromagnetic radiation and not all possible frequencies? The answer must be that an atom can lose energy only in certain discrete amounts. This answer, in turn, suggests that the electron in the atom can exist *only in a series of discrete levels*, just as quantum mechanics predicts. When an electron makes a **transition**, a change of state, its energy changes from one of these levels to another, and the difference in energy, $\Delta E = E_{\text{upper}} - E_{\text{lower}}$, is carried away as a photon. Because the energy of a photon is $h\nu$, where h is the Planck constant, the frequency of an individual line in a spectrum, ν, is related to the energy difference between two energy levels (Fig. 1.20):

$$h\nu = E_{\text{upper}} - E_{\text{lower}} \tag{12}$$

This relation is called the **Bohr frequency condition**. Each spectral line arises from a specific transition. By analyzing the appearance of the spectrum, we can build a picture of the ladder of energy levels for the atom (Fig. 1.21).

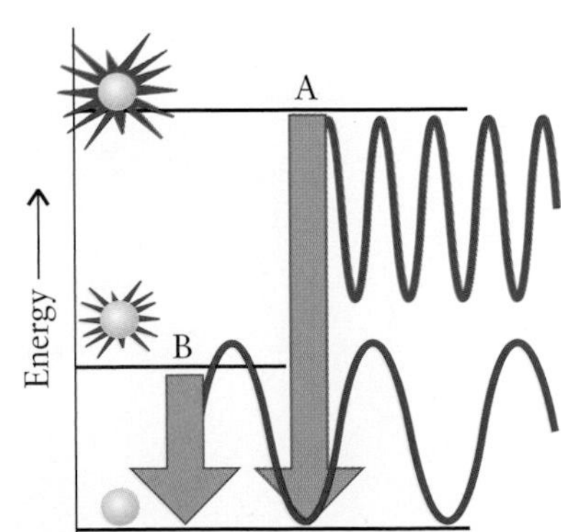

FIGURE 1.20 When an atom undergoes a transition from a state of higher energy to one of lower energy, it loses energy that is carried away as a photon. The greater the energy loss, the higher the frequency (and the shorter the wavelength) of the radiation emitted.

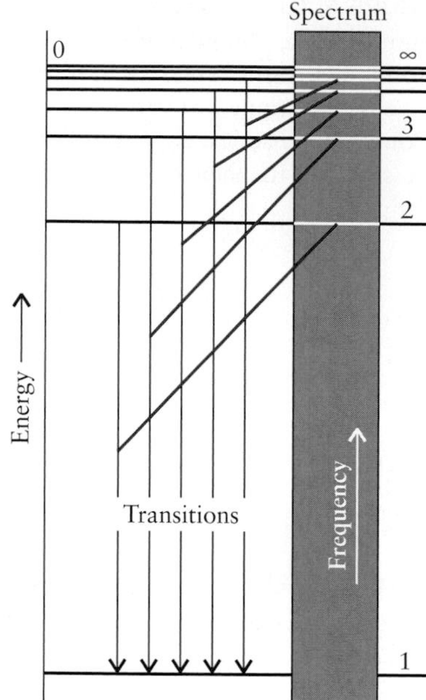

FIGURE 1.21 The spectrum of atomic hydrogen (reproduced on the right) tells us the arrangement of the energy levels of the atom because the frequency of the radiation emitted in a transition is proportional to the energy difference between the two energy levels involved. The zero of energy corresponds to the completely separated proton and electron. The numbers on the right label the energy levels: they are examples of quantum numbers.

The first person to identify a pattern in the lines of the visible region of the spectrum was Joseph Balmer, a Swiss schoolteacher. In 1885, he noticed that the frequencies of all the lines then known could be generated by the expression

$$\nu \propto \frac{1}{2^2} - \frac{1}{n^2} \qquad n = 3, 4, \ldots \tag{13}$$

As experimental techniques advanced, more lines were discovered, and the Swedish spectroscopist Johann Rydberg noticed that all of them could be predicted from the expression

$$\nu = \mathcal{R}\left(\frac{1}{{n_1}^2} - \frac{1}{{n_2}^2}\right) \qquad n_1 = 1, 2, \ldots, \; n_2 = n_1 + 1, n_1 + 2, \ldots \tag{14}$$

Here $\mathcal{R}$ is an empirical (experimentally determined) constant now known as the **Rydberg constant**; its value is 3.29×10^{15} Hz. This empirical formula for the lines, together with the Bohr frequency condition, strongly suggests that the energy levels themselves are proportional to $\mathcal{R}/n^2$. The **Balmer series** is the set of lines with $n_1 = 2$ (and $n_2 = 3, 4, \ldots$). The **Lyman series**, a set of lines in the ultraviolet region of the spectrum, has $n_1 = 1$ (and $n_2 = 2, 3, \ldots$) (Fig. 1.22).

Another consequence of quantization is that an atom can *absorb* only certain frequencies of radiation. The energy of the incoming photon excites electrons from one quantum level to another. If we pass light through a vapor composed of the atoms of an element, we see its *absorption spectrum*, a series of dark lines on an otherwise continuous spectrum

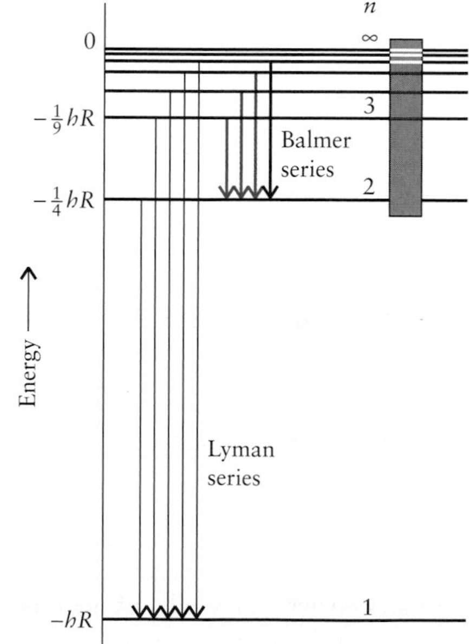

FIGURE 1.22 The permitted energy levels of a hydrogen atom as calculated from Eq. 14. The levels are labeled with the quantum number n, which ranges from 1 (for the lowest state) to infinity (for the separated proton and electron).

(Fig 1.23). The absorption lines have the same frequencies as the lines in the emission spectrum. Absorption spectra are used by astronomers to identify elements in the outer layers of stars.

The observation of discrete spectral lines suggests that an electron in an atom can have only certain energies. Transitions between these energy levels generate or absorb photons in accord with the Bohr frequency condition.

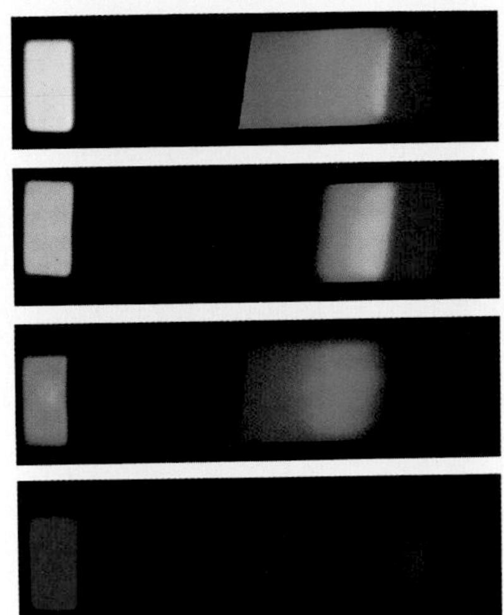

FIGURE 1.23 Spectroscopic observations are also used to study molecules and ions in solution. These four panels show (top to bottom) white light shining through an aperture and through three solutions (potassium dichromate, bromophenol blue, phenolphthalein). The appearance of each solution is shown on the left and the spectrum of light emerging from the solution is shown on the right. Individual lines are not observed because the spectra are much more complex than those of atoms and are affected by the solvent.

EXAMPLE 1.6 Identifying a line in the hydrogen spectrum

Calculate the wavelength of the radiation emitted by a hydrogen atom when an electron makes a transition between the levels with $n_2 = 3$ and $n_1 = 2$. Identify in Fig. 1.19b the spectral line produced by this transition.

STRATEGY The frequency of a line arising from a transition is given by Eq. 14. Convert frequency to a wavelength by using Eq. 1. In the last step, use $1\ \text{Hz} = 1\ \text{s}^{-1}$. The wavelength should match one of the lines in the Balmer series in Fig. 1.19b.

SOLUTION The frequency of the transition from a level with $n_2 = 3$ to a level with $n_1 = 2$ is

$$\nu = \mathcal{R}\left(\frac{1}{2^2} - \frac{1}{3^2}\right) = \frac{5}{36}\mathcal{R}$$

The wavelength of the radiation is therefore

$$\lambda = \frac{c}{\nu} = \frac{c}{(5/36)\mathcal{R}} = \frac{36c}{5\mathcal{R}}$$
$$= \frac{36 \times (3.00 \times 10^8\ \text{m·s}^{-1})}{5 \times (3.29 \times 10^{15}\ \text{s}^{-1})} = 6.57 \times 10^{-7}\ \text{m}$$

This wavelength, 657 nm, corresponds to the red line in the Balmer series of lines in the spectrum. Light of this wavelength is absorbed when the electron is promoted from the $n = 2$ level to the $n = 3$ level and is given off when it falls back to the $n = 2$ level.

Self-Test 1.9A Repeat the calculation for the transition from the state with $n = 4$ to that with $n = 2$ and identify the spectral line in Fig. 1.19b.

[*Answer:* 486 nm; blue line]

Self-Test 1.9B Repeat the calculation for the transition from the state with $n = 5$ to that with $n = 2$ and identify the spectral line in Fig. 1.19b.

MODELS OF ATOMS

Our next task is to combine the wavelike properties of electrons with the nuclear model of the atom and hope to explain the strange ladder of energy levels observed experimentally in a hydrogen atom. In doing so, we shall greatly enrich our model of the internal structure of an atom and transform Rutherford's primitive nuclear model into today's sophisticated version. We proceed in two steps. First, we develop the quantum mechanical model of a hydrogen atom. Then we use that model to account for the structures of atoms with more than one electron and see how a few simple ideas account not only for atomic spectra but also for the structure of the periodic table.

1.7 The Principal Quantum Number

To find the energy levels of an electron in a hydrogen atom, we have to solve the appropriate Schrödinger equation. An electron in an atom is like a particle in a box, in the sense that it is confined within the atom by the pull of the nucleus. We can therefore expect the electron's wavefunctions to obey certain boundary conditions and hence have quantized energy levels. For an electron in a hydrogen atom, the Coulomb potential energy of the electron, V, varies with distance, r, from the nucleus:

$$V(r) = \frac{(-e)(+e)}{4\pi\epsilon_0 r} = -\frac{e^2}{4\pi\epsilon_0 r} \qquad (15)$$

In this expression, $-e$ is the charge of the electron, $+e$ is the charge of the nucleus, and ϵ_0, the "vacuum permittivity," is another fundamental constant; its value is given inside the back cover. This form of potential energy is characteristic of the interaction of all charged particles (see Section A). Solving the Schrödinger equation for a particle with this potential energy is difficult, but Schrödinger himself achieved it in 1927 and found that the allowed energy levels are

$$E_n = -\frac{h\mathcal{R}}{n^2} \qquad \mathcal{R} = \frac{m_e e^4}{8h^3\epsilon_0^2} \qquad n = 1, 2, \ldots \qquad (16)$$

These energy levels have exactly the form suggested by the wavelengths of the lines measured spectroscopically. When the fundamental constants are inserted into the expression for $\mathcal{R}$, the value obtained is 3.29×10^{15} Hz, the same as the experimental value of the Rydberg constant. This agreement is a triumph for Schrödinger's theory and for quantum mechanics.

Figure 1.22 shows the energy levels calculated from Eq. 16. Notice that the energy of an electron bound in a hydrogen atom is always lower than that of a free electron (note the minus signs in Fig. 1.22 and in Eq. 16). The **principal quantum number**, n, is an integer that labels these energy levels, from $n = 1$ for the first level, $n = 2$ for the second, continuing up to infinity. The lowest energy possible for an electron in a hydrogen atom, $-h\mathcal{R}$, is obtained when $n = 1$. This lowest energy state is called the **ground state** of the atom. A hydrogen atom is normally found in its ground state, with its electron in the level where $n = 1$. The energy of the bound electron climbs up the ladder of levels as n increases. It reaches the top of the ladder, corresponding to $E = 0$ and freedom, when n reaches infinity. At that point, the electron has left the atom. This process is called **ionization**. The difference in energy between the ground state and that of the ionized state is the energy required to remove an electron from the neutral atom in its ground state.

The ionization energy is discussed further in Section 1.16.

The energy levels of a hydrogen atom, Eq. 16, are defined by the principal quantum number, $n = 1, 2, \ldots$, and form a converging ladder, as shown in Fig. 1.22.

1.8 Atomic Orbitals

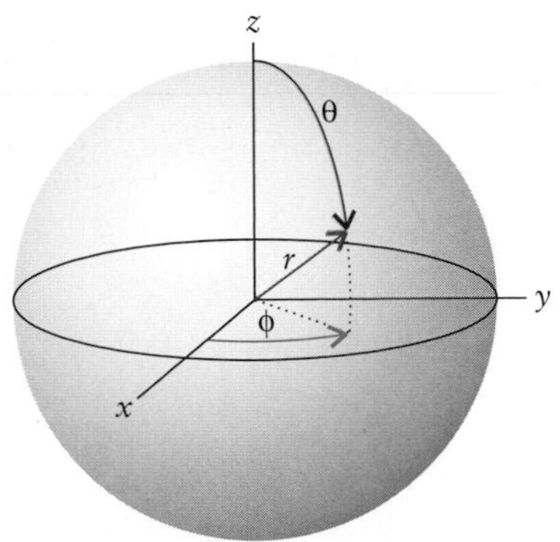

FIGURE 1.24 The spherical polar coordinates: r is the radius, θ is the colatitude, and ϕ, the "longitude," is the azimuth.

Wavefunctions of electrons in atoms are called **atomic orbitals.** The name was chosen to suggest something less definite than an "orbit" of an electron around a nucleus and to take into account the wave nature of the electron. The mathematical expressions for atomic orbitals—which are obtained as solutions of the Schrödinger equation—are more complicated than the sine functions for the particle in a box, but their essential features are quite simple. Moreover, we must never lose sight of their interpretation, that the *square* of a wavefunction tells us the probability density of an electron at each point. To visualize this probability density, imagine a cloud centered on the nucleus. The density of the cloud at each point represents the probability of finding an electron there. Denser regions of the cloud therefore represent locations where the electron is more likely to be found.

To interpret the information in each atomic orbital, we need a way to identify the location of each point around a nucleus. It is most convenient to describe these locations in terms of **spherical polar coordinates,** in which each point is labeled with three coordinates: r, which gives its distance from the center of the atom; θ (theta), the angle from the z-axis (the "pole"), which can be thought of as the "latitude"; and ϕ (phi), the angle about the z-axis, the "longitude," as shown in Fig. 1.24. Each wavefunction, which in general varies in amplitude from point to point, can be written as a function of these three coordinates, $\psi(r,\theta,\phi)$. It turns out, however, that all the wavefunctions can be written as the product of a function R that depends only on r and another function Y that depends only on the angles θ and ϕ. That is,

Geographical latitudes are measured from the equator, not the pole.

$$\psi(r,\theta,\phi) = R(r)Y(\theta,\phi) \tag{17}$$

The function $R(r)$ is called the **radial wavefunction,** and the function $Y(\theta,\phi)$ is called the **angular wavefunction.** For example, the wavefunction corresponding to the ground state of the hydrogen atom is

$$\psi(r,\theta,\phi) = \overbrace{\frac{2e^{-r/a_0}}{a_0^{3/2}}}^{R(r)} \times \overbrace{\frac{1}{2\pi^{1/2}}}^{Y(\theta,\phi)} = \frac{e^{-r/a_0}}{(\pi a_0^{\,3})^{1/2}} \qquad a_0 = \frac{4\pi\epsilon_0\hbar^2}{m_e e^2} \tag{18}$$

In an early model of the hydrogen atom proposed by Niels Bohr, the electron traveled in a circular orbit of radius a_0 around the nucleus. The uncertainty principle rules out this model.

The quantity a_0 is called the **Bohr radius;** when the values of the fundamental constants are inserted, we find $a_0 = 52.9$ pm. For this wavefunction, the angular wavefunction Y is a constant, $1/2\pi^{1/2}$, independent of the angles, and the radial wavefunction decays exponentially toward 0 as r increases. In contrast, the wavefunction for the $2p_x$-orbital of the hydrogen atom is

$$\psi(r,\theta,\phi) = \overbrace{\frac{1}{2\sqrt{6}}\left(\frac{1}{a_0}\right)^{5/2} re^{-r/2a_0}}^{R(r)} \times \overbrace{\left(\frac{3}{4\pi}\right)^{1/2} \sin\theta\cos\phi}^{Y(\theta,\phi)}$$

$$= \frac{1}{4}\left(\frac{1}{2\pi a_0^{\,5}}\right)^{1/2} re^{-r/2a_0} \sin\theta\cos\phi$$

Notice how the wavefunction for a p_x-orbital depends on the angles θ and ϕ. The expressions for a number of other atomic orbitals are shown in Table 1.2.

In general, the wavefunction of an electron in a hydrogen atom is specified by giving the values of three quantum numbers. First is the *principal quantum number, n*, which also specifies the energy of the orbital (through Eq. 16). In a hydrogen atom, all atomic orbitals with the same value of n have the same energy and are said to belong to the same **shell** of the atom.

The second quantum number needed to specify an orbital is l, the **orbital angular momentum quantum number.** This quantum number can take the values $l = 0, 1, 2, \ldots, n - 1$. There are n different values of l for a given value of n. The orbitals of a shell with principal quantum number n therefore fall into n groups (each identified by a different value of l), which are called **subshells** of the shell. There is only one subshell in the $n = 1$ level ($l = 0$), two in the $n = 2$ level ($l = 0$ and 1), three in the $n = 3$ level ($l = 0$, 1, and 2), and so on. In the hydrogen atom, the orbitals of different subshells of a given shell are **degenerate,** which means they all have the same energy. All orbitals with $l = 0$ are called ***s*-orbitals,** those with $l = 1$ are called ***p*-orbitals,** those with $l = 2$ are called ***d*-orbitals,** and those with $l = 3$ are called ***f*-orbitals.** Although higher values of l (corresponding to $g, h, \ldots$ orbitals) are possible, the lower values of l (0, 1, 2, and 3) are the only ones that chemists need in practice.

Spectroscopic lines were once classified as *s*harp, *p*rincipal, *d*iffuse, and *f*undamental.

TABLE 1.2 Hydrogen Wavefunctions (Atomic Orbitals)

Orbital	Wavefunction
1s	$\psi = \left(\frac{1}{\pi a_0^3}\right)^{1/2} e^{-r/a_0}$
2s	$\psi = -\frac{1}{4}\left(\frac{1}{2\pi a_0^3}\right)^{1/2}\left(2 - \frac{r}{a_0}\right)e^{-r/2a_0}$
2p	$\psi = \frac{1}{4}\left(\frac{1}{2\pi a_0^5}\right)^{1/2} re^{-r/2a_0}\begin{cases}\cos\theta & (2p_z)\\ \sin\theta\cos\phi & (2p_x)\\ \sin\theta\sin\phi & (2p_y)\end{cases}$
3s	$\psi = \frac{1}{9}\left(\frac{1}{3\pi a_0^3}\right)^{1/2}\left(3 - \frac{2r}{a_0} + \frac{2r^2}{9a_0^2}\right)e^{-r/3a_0}$
3p	$\psi = -\frac{2}{27}\left(\frac{1}{2\pi a_0^5}\right)^{1/2}\left(2 - \frac{r}{3a_0}\right)re^{-r/3a_0}\begin{cases}\cos\theta & (3p_z)\\ \sin\theta\cos\phi & (3p_x)\\ \sin\theta\sin\phi & (3p_y)\end{cases}$
3d	$\psi = \frac{1}{81}\left(\frac{1}{6\pi a_0^7}\right)^{1/2} r^2e^{-r/3a_0}\begin{cases}(3\cos^2\theta - 1) & (3d_{z^2})\\ 2\sqrt{3}\sin\theta\cos\theta\cos\phi & (3d_{zx})\\ 2\sqrt{3}\sin\theta\cos\theta\sin\phi & (3d_{yz})\\ \sqrt{3}\sin^2\theta\cos 2\phi & (3d_{x^2-y^2})\\ \sqrt{3}\sin^2\theta\sin 2\phi & (3d_{xy})\end{cases}$

Note: In each case, $a_0 = 4\pi\epsilon_0\hbar^2/m_e e^2$, or close to 52.9 pm.

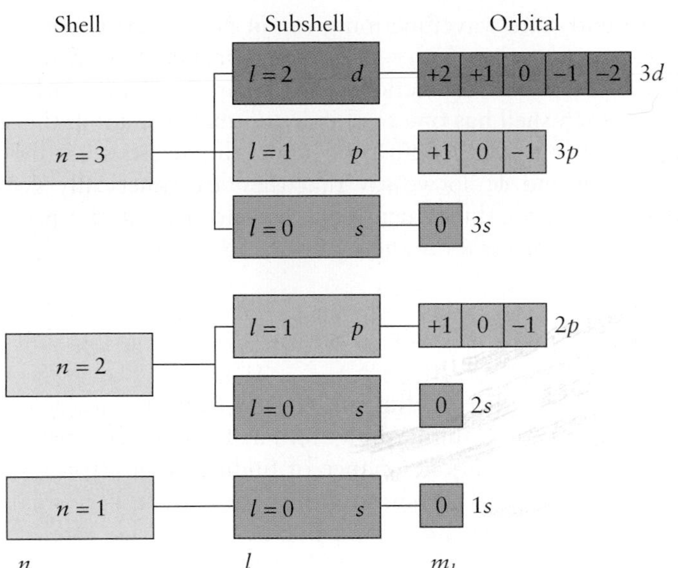

FIGURE 1.25 A summary of the arrangement of shells, subshells, and orbitals in an atom and the corresponding quantum numbers. Note that the quantum number m_l is an alternative label for the individual orbitals: in chemistry, it is more common to use x, y, and z as labels, as shown in Figs. 1.30–1.32.

The third quantum number we need to specify an orbital is m_l, the **magnetic quantum number**, which distinguishes the individual orbitals within a subshell. This quantum number can take the values $l, l-1, \ldots, -l$, giving $2l+1$ different values for a given value of l. For example, when $l = 1$, $m_l = +1, 0, -1$. The magnetic quantum number labels the individual orbitals in a given subshell and determines how the atom behaves in a magnetic field. It follows that there are $2l+1$ orbitals in a subshell of quantum number l.

The hierarchy of shells, subshells, and orbitals is summarized in Fig. 1.25 and Table 1.3. Each possible combination of the three quantum numbers specifies an orbital. For example, an electron in the ground state of a hydrogen atom has the specification $n = 1, l = 0, m_l = 0$. As we shall see, n is related to the *size* of the orbital, l is related to its shape, and m_l is related to its orientation in space.

TABLE 1.3 Quantum Numbers for Electrons in Atoms

Name	Symbol	Values	Labels	Indicates
principal	n	$1, 2, \ldots$	shell	size
orbital angular momentum*	l	$0, 1, \ldots, n-1$	subshell: $l = 0, 1, 2, 3, 4, \ldots$ $s, p, d, f, g, \ldots$	shape
magnetic	m_l	$l, l-1, \ldots, -l$	orbitals of subshell	orientation
spin magnetic	m_s	$+\frac{1}{2}, -\frac{1}{2}$	spin state	spin direction

* Also called the *azimuthal quantum number*.

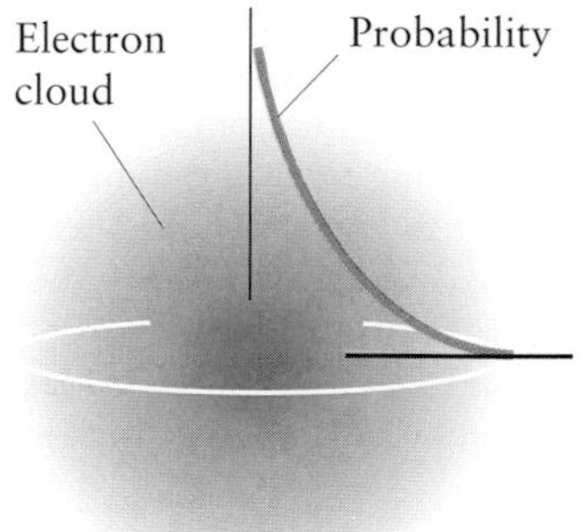

FIGURE 1.26 The three-dimensional electron cloud corresponding to an electron in a 1*s*-orbital of hydrogen. The density of shading represents the probability of finding the electron at any point. The superimposed graph shows how the probability varies with the distance of the point from the nucleus along any radius.

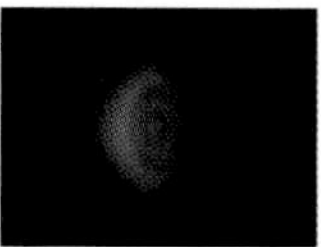

View an animation of an *s*-orbital.

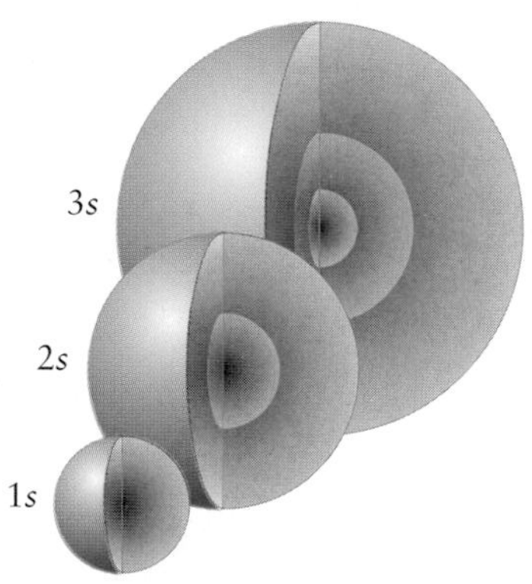

FIGURE 1.27 The three *s*-orbitals of lowest energy. The simplest way of drawing an atomic orbital is as a boundary surface, a surface within which there is a high probability (typically 90%) of finding the electron. We shall use blue to denote *s*-orbitals, but that color is only an aid to their identification. The interior circles in the 2*s*- and 3*s*-orbitals indicate the positions of the radial nodes.

Media Link **Animation:** Fig. 1.27, *s*-Orbitals

The ground-state wavefunction given in Eq. 16 corresponds to $n = 1$ and therefore to $l = 0$ and $m_l = 0$ (the only values allowed when $n = 1$). Because $l = 0$, the wavefunction is an example of an *s*-orbital and is denoted 1*s*. Each shell has one *s*-orbital, and the *s*-orbital in the shell with quantum number *n* is called an ***ns*-orbital.** All *s*-orbitals are independent of the angles θ and ϕ, so we say that they are **spherically symmetrical** (Fig. 1.26). The probability density of an electron at the point *r*, θ, ϕ when it is in a 1*s*-orbital is found from Eq. 18:

$$\psi^2(r, \theta, \phi) = \frac{e^{-2r/a_0}}{\pi a_0^{\,3}} \qquad (19)$$

In principle, the cloud never thins to exactly 0, no matter how large the value of *r*. So, we could think of an atom as being bigger than the Earth! However, there is virtually no chance of finding an electron farther from the nucleus than about 100 pm, so atoms are, in fact, very small. As you can see from the high density of the cloud at the nucleus, an electron in an *s*-orbital has a nonzero probability of being found right at the nucleus.

How can we use this information . . . to calculate the probability of finding an electron at a certain location?

Suppose we wanted to know the probability of finding the electron in a small region a distance a_0 from the nucleus relative to the probability of finding it in the same small region located at the nucleus. We would calculate the following ratio of the squares of the wavefunction at the two points:

$$\frac{\text{Probability density at } r = a_0}{\text{Probability density at } r = 0} = \frac{\psi^2(a_0, \theta, \phi)}{\psi^2(0, \theta, \phi)} = \frac{e^{-2a_0/a_0}/\pi a_0^{\,3}}{1/\pi a_0^{\,3}}$$
$$= \frac{e^{-2}/\pi a_0^{\,3}}{1/\pi a_0^{\,3}} = e^{-2} = 0.14$$

That is, the probability of finding the electron in a small region at a distance a_0 from the nucleus is only 14% that of finding the electron in a similar region located at the nucleus.

Self-Test 1.10A Calculate the same ratio, but for the more distant point at $r = 2a_0$, twice as far from the nucleus.

[*Answer:* 0.018]

Self-Test 1.10B Calculate the same ratio, but for a point at $\frac{1}{2}a_0$ from the nucleus.

Instead of drawing the *s*-orbital as a cloud, chemists usually draw its **boundary surface,** a surface that encloses the densest regions of the cloud. The electron is likely to be found only inside the boundary surface of the orbital. An *s*-orbital has a spherical boundary surface (Fig. 1.27), because the electron cloud is spherical. *s*-Orbitals with higher energies have spherical boundary surfaces of greater diameter. They also have a more complicated radial variation with nodes at locations that can be found by examining the wavefunctions (Fig. 1.28).

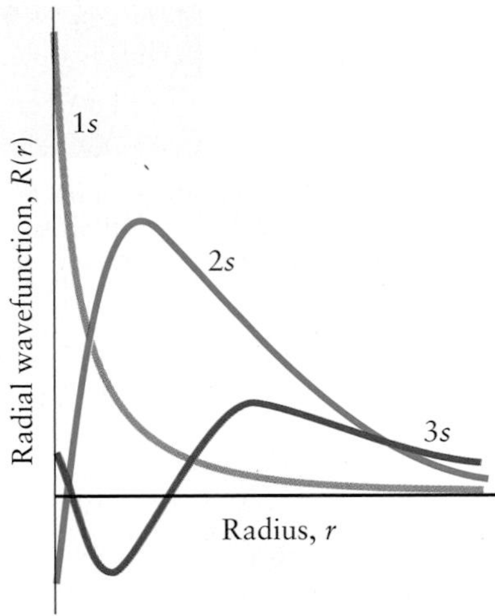

FIGURE 1.28 The radial wave-functions of the first three *s*-orbitals of a hydrogen atom. Note that the number of radial nodes increases (as $n - 1$), as does the average distance of the electron from the nucleus (compare with Fig. 1.27). Because the probability density is given by ψ^2, all *s*-orbitals correspond to a nonzero probability density at the nucleus.

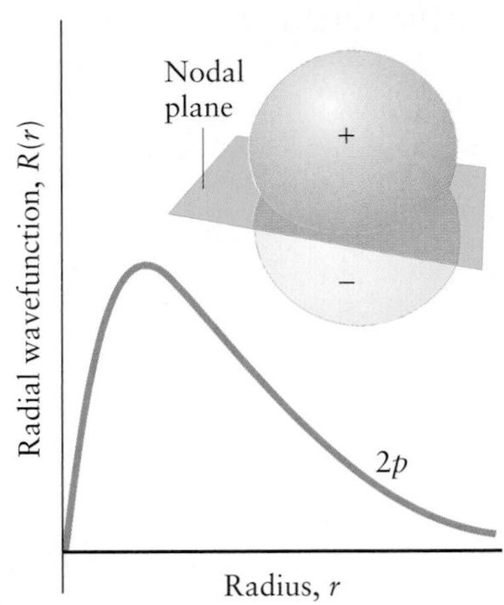

FIGURE 1.29 The boundary surface and the radial variation of a 2*p*-orbital. All *p*-orbitals have boundary surfaces with similar shapes, including one nodal plane. Note that the orbital has opposite signs (as depicted by the depth of color) on each side of the nodal plane.

Media Link Living graph: Fig. 1.28, Hydrogenic atomic orbitals

A subshell with $l = 1$ consists of three *p*-orbitals. A *p*-orbital has two lobes of opposite sign (Fig. 1.29). The two lobes are separated by a planar region called a **nodal plane**, which cuts through the nucleus and on which $\psi = 0$. A *p*-electron will never be found on this plane, so a *p*-electron is never found at the nucleus. This difference from *s*-orbitals will prove to be of major importance for understanding the structure of the periodic table. There are three *p*-orbitals in each subshell, corresponding to the quantum numbers $m_l = +1, 0, -1$. However, chemists commonly refer to the orbitals according to the axes along which the lobes lie; hence, we refer to p_x, p_y, and p_z orbitals (Fig. 1.30).

There is no *direct* relation between the values of m_l and the x, y, z designation of the orbitals: the orbitals labeled with the axes are combinations of the orbitals labeled with the quantum number m_l.

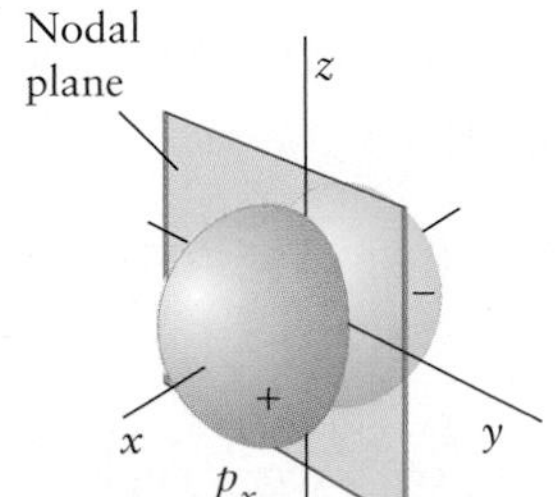

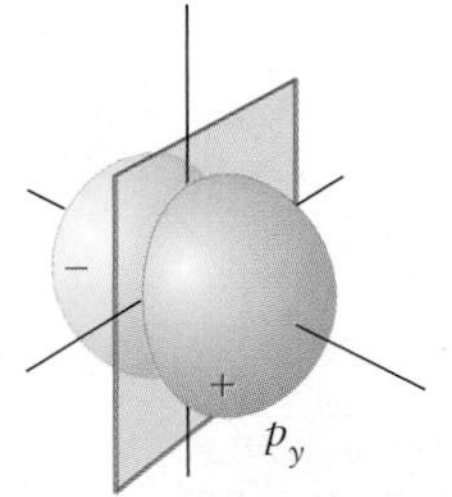

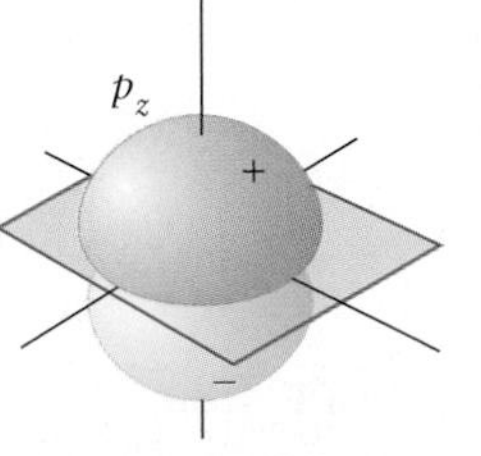

FIGURE 1.30 There are three *p*-orbitals of a given energy, and they lie along three perpendicular axes. We shall use yellow to indicate *p*-orbitals: dark yellow for the positive lobe and light yellow for the negative lobe.

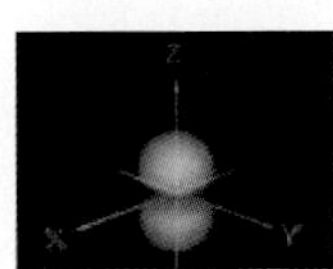

View an animation of the three *p*-orbitals.

Media Link Animation: Fig. 1.30, *p*-Orbitals

FIGURE 1.31 The boundary surface of a d-orbital is more complicated than that of an s- or a p-orbital. There are, in fact, five d-orbitals of a given energy; four of them have four lobes, one is slightly different. In each case, an electron that occupies a d-orbital will not be found at the nucleus. We shall use orange to indicate d-orbitals: dark orange for the positive lobes, and light orange for the negative lobes.

Media Link Animation: Fig. 1.31, *d*-Orbitals

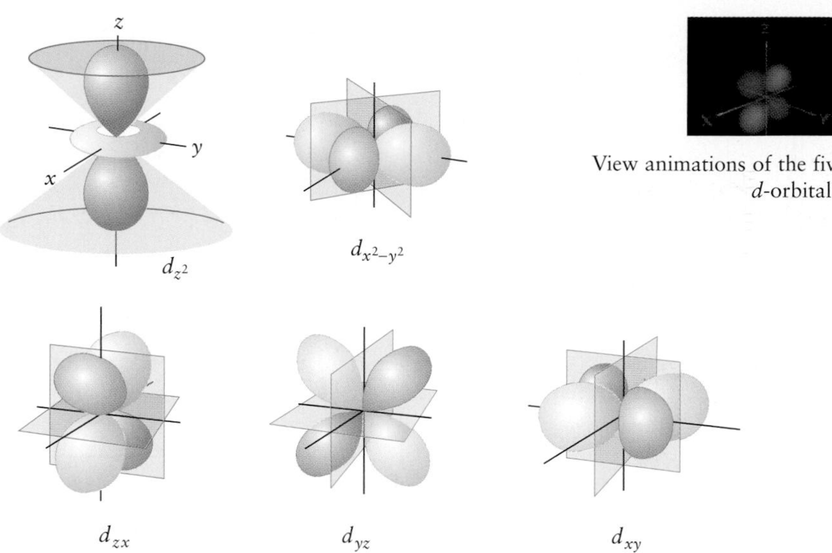

View animations of the five d-orbitals.

A subshell with $l = 2$ consists of five d-orbitals. Each d-orbital has four lobes, except for the orbital designated d_{z^2}, which has a more complicated shape (Fig. 1.31). A shell with $l = 3$ consists of seven f-orbitals with even more complicated shapes (Fig. 1.32). We can find the detailed characteristics of both p-orbitals and d-orbitals by examining their wavefunctions.

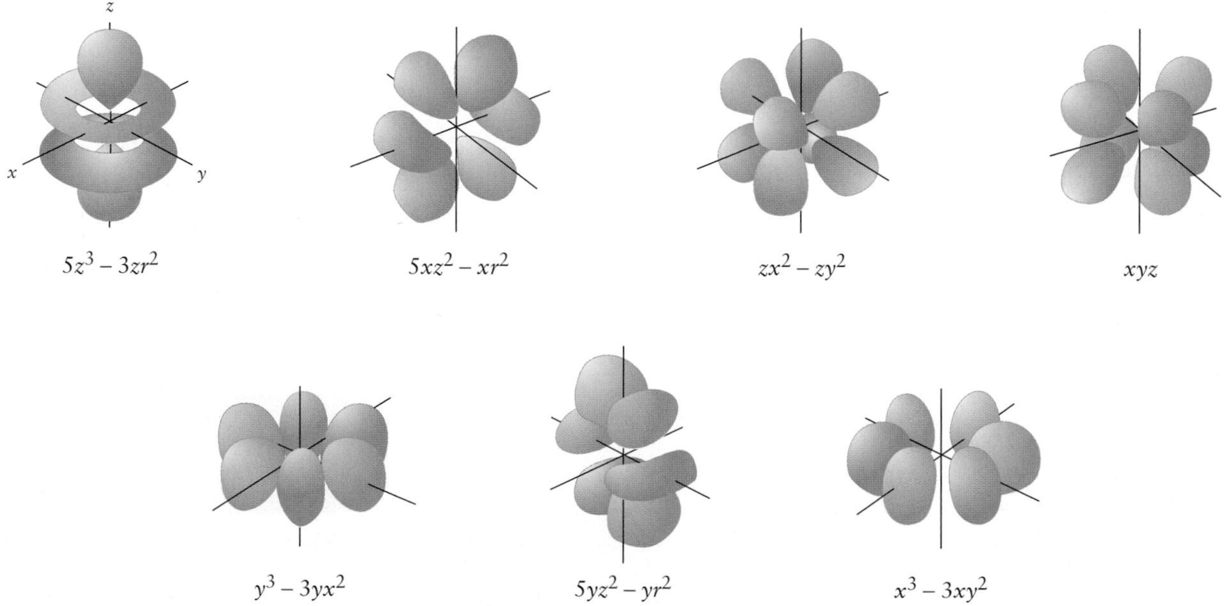

FIGURE 1.32 The seven f-orbitals of a shell (with $n > 3$) have a very complex appearance. Their detailed form will not be used again in this text. However, their existence is important for understanding the periodic table, the presence of the lanthanides and actinides, and the properties of the later d-block elements. A darker color denotes a positive lobe, a lighter color a negative lobe.

It is not immediately obvious from the angular form of the wavefunctions of atomic orbitals, but when there is one electron in each orbital of a given type, then the electron distribution is spherically symmetrical. Thus, one electron in each of the three *p*-orbitals of a shell gives a spherical distribution, and the same is true of one electron in each of the five *d*-orbitals or seven *f*-orbitals of a shell. We get a similar spherical electron cloud when two electrons occupy each orbital of a given type. We should think of atoms as spherical objects even though their electrons occupy orbitals with weird shapes.

The location of an electron in an atom is described by a wavefunction known as an atomic orbital; atomic orbitals are designated by the quantum numbers n, l, and m_l and fall into shells and subshells.

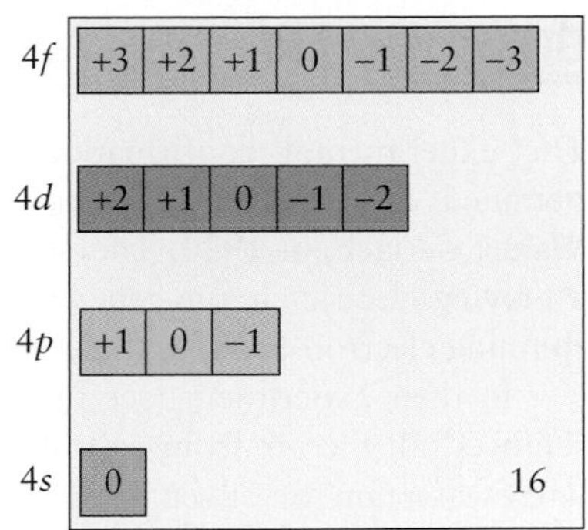

FIGURE 1.33 There are 16 orbitals in the shell with $n = 4$, each of which can hold two electrons, for a total of 32 electrons.

How can we use this information . . . to count the orbitals in a shell?

Suppose we want to know how many orbitals there are in the shell with $n = 4$. Recall that l has integer values from 0 up to $n - 1$ and that the number of orbitals in a subshell for a given value of l is $2l + 1$. Therefore, for $n = 4$, there are four subshells with $l = 0, 1, 2, 3$, consisting of one *s*-orbital, three *p*-orbitals, five *d*-orbitals, and seven *f*-orbitals, respectively. There are therefore $1 + 3 + 5 + 7 = 16$ orbitals in the shell with $n = 4$ (Fig. 1.33).

Self-Test 1.11A Calculate the total number of orbitals in a shell with $n = 6$.
[*Answer:* 36]

Self-Test 1.11B What is the general formula for the number of orbitals in a shell with principal quantum number n?

1.9 Electron Spin

Schrödinger's calculation of the energies of the hydrogen orbitals was a milestone in the development of modern atomic theory. Yet the observed spectral lines did not have exactly the frequencies he predicted. In 1925, two Dutch-American physicists, Samuel Goudsmit and George Uhlenbeck, proposed an explanation for the tiny deviations that had been observed. They suggested that an electron behaves in some respects like a spinning sphere, something like a planet rotating on its axis. This property is called **spin**.

According to quantum mechanics, an electron has two spin states, represented by the arrows ↑ and ↓ or the Greek letters α and β. We can think of an electron as being able to spin counterclockwise at a certain rate (the ↑ state) or clockwise at exactly the same rate (the ↓ state). These two spin states are distinguished by a fourth quantum number, the **spin magnetic quantum number**, m_s. This quantum number can have only two values: $+\frac{1}{2}$ indicates an ↑ electron and $-\frac{1}{2}$ indicates a ↓ electron (Fig. 1.34). Box 1.1 describes how spin explains the results of an important experiment.

An electron has the property of spin; the spin is described by the quantum number m_s, which may have one of two values.

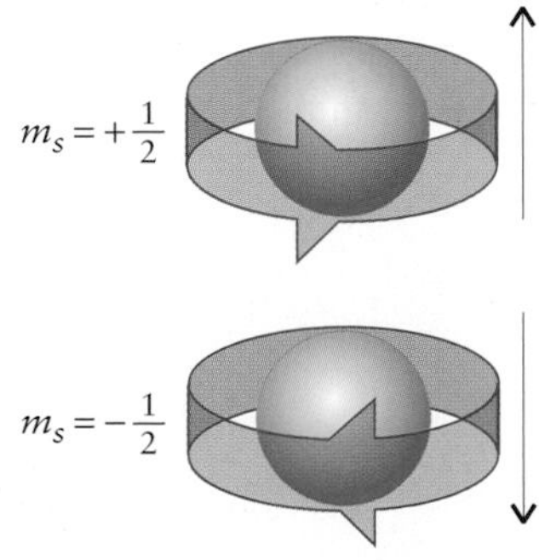

FIGURE 1.34 The two spin states of an electron can be represented as clockwise or counterclockwise rotation around an axis passing through the electron. The two states are labeled by the quantum number m_s and depicted by the arrows shown on the right.

BOX 1.1 HOW DO WE KNOW . . . THAT AN ELECTRON HAS SPIN?

The experimental confirmation of electron spin was obtained by two German scientists, Otto Stern and Walter Gerlach, in 1920. They made use of the fact that a moving electric charge generates a magnetic field, so a spinning electron behaves like a tiny bar magnet.

In their experiment (see figure), Stern and Gerlach removed all the air from a container. They then shot a narrow stream of silver atoms through the container toward a detector at the other end and set up a highly nonuniform magnetic field across the container. A silver atom has one unpaired electron, whereas all the rest of its 47 electrons are paired. The atom therefore behaves like a single unpaired electron riding on a heavy platform, the rest of the atom.

If the electron did not spin, it would not be affected by the applied field and only one spot would be detected where the undeflected atoms hit the screen. If a spinning electron behaved like a spinning ball, we would expect its spin to be able to point in any direction. It would behave like a bar magnet that could have any orientation to the applied magnetic field. In that case, Stern and Gerlach should have observed a broad band of silver atoms arriving at the detector. The field would push the silver atoms by different amounts according to the orientation of the spin. Indeed, that is exactly what they observed when they first carried out the experiment.

However, the first results were misleading. It is a difficult experiment to do properly, because the atoms collide with one another in the beam. An atom moving in one direction might easily be knocked by its neighbors into a different direction. When Stern and Gerlach repeated their experiment, they used a much less dense beam of atoms, thereby reducing the number of collisions between the atoms. They now saw only two narrow bands. One band consisted of atoms flying through the magnetic field with one orientation of their spin; the other band consisted of the atoms with opposite spin. The two narrow bands confirmed not only that an electron has spin but also that it can have only two orientations.

Electron spin is the basis of the experimental technique called *electron paramagnetic resonance* (EPR), which is used to study the structures and motions of molecules and ions that have unpaired electrons. This technique is based on the observation of an electron flipping between its two spin orientations. Like Stern and Gerlach's experiment, it works only with ions or molecules that have an unpaired electron.

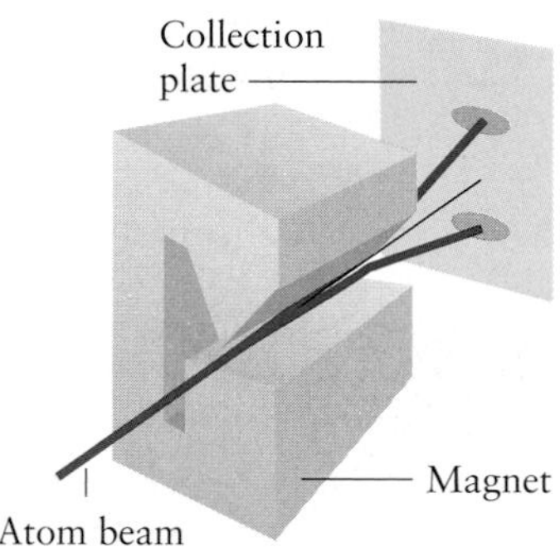

A schematic diagram of the apparatus used by Stern and Gerlach. In the experiment, a stream of atoms splits into two as it passes between the poles of a magnet. The atoms in one stream have an odd ↑ electron, and those in the other an odd ↓ electron.

1.10 The Electronic Structure of Hydrogen

Let's review what we now know about the hydrogen atom by imagining what happens to its electron as the atom acquires energy. Initially, the electron is in the lowest energy level, the ground state of the atom, with $n = 1$. The only orbital with this energy is the 1*s*-orbital; we say that the electron **occupies** a 1*s*-orbital or that it is a "1*s*-electron." The electron in the ground state of a hydrogen atom is described by the following values of the four quantum numbers:

$$n = 1 \quad l = 0 \quad m_l = 0 \quad m_s = +\tfrac{1}{2} \text{ or } -\tfrac{1}{2}$$

The electron can have either spin state.

When the atom acquires enough energy (by absorbing a photon of radiation, for instance) for its electron to reach the shell with $n = 2$, it can occupy any of the four orbitals in that shell. There are one 2*s*- and three 2*p*-orbitals in this shell; and in hydrogen, they all have the same energy. When an electron is described by one of these wavefunctions, we say that it occupies a 2*s*-orbital or one of the 2*p*-orbitals, or that it is a 2*s*- or 2*p*-electron. The average distance of an electron from the nucleus when it occupies any of the orbitals in the shell with $n = 2$ is greater than when $n = 1$, so we can think of the atom as swelling as it is excited energetically. When the atom acquires even more energy, the electron moves into the shell with $n = 3$; the atom is now even larger. In this shell, the electron can occupy any of nine orbitals (one 3*s*-, three 3*p*-, and five 3*d*-orbitals). Still more energy moves the electron still further from the nucleus to the $n = 4$ shell, where 16 orbitals are available (one 4*s*-, three 4*p*-, five 4*d*-, and seven 4*f*-orbitals).

The state of an electron in a hydrogen atom is defined by the four quantum numbers n, l, m_l, and m_s; as the value of n increases, the size of the atom increases.

Self-Test 1.12A The three quantum numbers for an electron in a hydrogen atom in a certain state are $n = 4$, $l = 2$, and $m_l = -1$. In what type of orbital is the electron located?

[*Answer:* 4*d*]

Self-Test 1.12B The three quantum numbers for an electron in a hydrogen atom in a certain state are $n = 3$, $l = 1$, and $m_l = -1$. In what type of orbital is the electron located?

THE STRUCTURES OF MANY-ELECTRON ATOMS

All neutral atoms other than hydrogen have more than one electron. The helium atom ($Z = 2$) has two, the lithium atom ($Z = 3$) has three, and a neutral atom of an element with atomic number Z has Z electrons. All these atoms are examples of **many-electron atoms**, or atoms with more than one electron. In the next three sections, we build on what we have learned about the hydrogen atom to see how the presence of more than one electron affects the energies of atomic orbitals. The resulting electronic structures are the key to the form of the periodic table, the periodic properties of the elements, and the abilities of atoms to form chemical bonds.

Many-electron atoms are also called "polyelectron atoms."

1.11 Orbital Energies

The electrons in a many-electron atom occupy orbitals like those of hydrogen. However, the energies of these orbitals are not the same as for a hydrogen atom. The nucleus of a many-electron atom is more highly charged than the hydrogen nucleus, and the greater charge attracts electrons more strongly and hence lowers their energy. However, the electrons also repel one another; this repulsion opposes the nuclear attraction and raises the energies of the orbitals. In a helium atom, for instance, with two electrons, the charge of the nucleus is $2e$ and the total potential energy is given by three terms:

$$V = -\overbrace{\frac{2e^2}{4\pi\epsilon_0 r_1}}^{\text{Attraction}} - \overbrace{\frac{2e^2}{4\pi\epsilon_0 r_2}}^{\text{Attraction}} + \overbrace{\frac{e^2}{4\pi\epsilon_0 r_{12}}}^{\text{Repulsion}} \quad (20)$$

where r_1 is the distance of electron 1 from the nucleus, r_2 is the distance of electron 2 from the nucleus, and r_{12} is the distance between the two electrons. The two terms with negative signs (indicating that the energy falls as r_1 or r_2 decreases) represent the attractions between the nucleus and the two electrons. The term with a positive sign (indicating an increase in energy as r_{12} decreases) represents the repulsion between the two electrons. The Schrödinger equation based on this potential energy is impossibly difficult to solve exactly, but highly accurate numerical solutions can be obtained by using computers.

Along with code-breakers and weather forecasters, chemists are now among the heaviest users of computers, which they use to calculate the detailed electronic structures of atoms and molecules.

Although chemists can solve the Schrödinger equation by using computers, they look for a physical understanding of the properties they calculate. Let's see how chemists think *qualitatively* about the structures of many-electron atoms. The starting point is the hydrogen atom, in which there are no electron-electron repulsions; we have seen that all the orbitals of a given shell in the hydrogen atom are degenerate. For instance, the 2*s*-orbital and all three 2*p*-orbitals have the same energy. In many-electron atoms, however, spectroscopic experiments and calculations show that

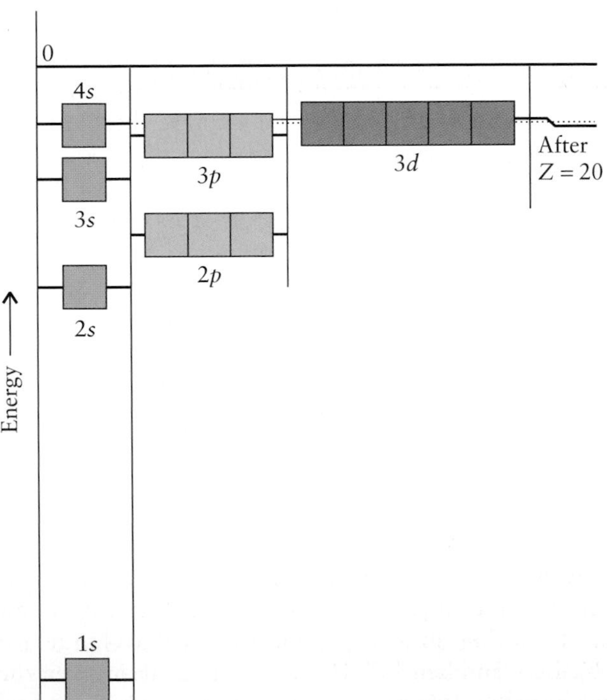

FIGURE 1.35 The relative energies of the shells, subshells, and orbitals in a many-electron atom. Each of the boxes can hold up to two electrons. Note the change in the order of energies of the 3*d*- and 4*s*-orbitals after $Z = 20$.

electron-electron repulsions cause the energy of a 2*p*-orbital to be higher than that of a 2*s*-orbital. Similarly, in the $n = 3$ shell, the three 3*p*-orbitals lie higher than the 3*s*-orbital, and the five 3*d*-orbitals lie higher still (Fig. 1.35). How can we explain these energy differences?

We have to examine the balance of attractions and repulsions in the atom to explain why subshells of a given shell have different energies. As well as being attracted by the nucleus, each electron in a many-electron atom is repelled by the other electrons present. As a result, it is less tightly bound to the nucleus than it would be if those other electrons were absent. We say that each electron is **shielded** from the full attraction of the nucleus by the other electrons in the atom. The shielding effectively reduces the pull of the nucleus on an electron. The **effective nuclear charge,** $Z_{\text{eff}}e$, experienced by the electron is always less than the actual nuclear charge, Ze, because the electron-electron repulsions work against the pull of the nucleus. Note that the other electrons do not "block" the influence of the nucleus: they simply provide additional repulsive coulombic interactions that partly counteract the pull of the nucleus. For example, the pull of the nucleus on an electron in the helium atom is less than its charge of +2 would exert but greater than the net charge of +1 that we would expect if each electron balanced one positive charge exactly.

An *s*-electron of any shell can be found very close to the nucleus (remember that ψ^2 for an *s*-orbital is nonzero at the nucleus), so we say that it can **penetrate** through the inner shells. A *p*-electron penetrates much less (Fig. 1.36). We have seen that its wavefunction has a node passing through the nucleus, so there is zero probability density for finding a *p*-electron at the nucleus. Because a *p*-electron penetrates less than an *s*-electron through the inner shells of the atom, it is more effectively shielded from the nucleus and hence experiences a smaller effective nuclear charge than an *s*-electron does. In other words, an *s*-electron is bound more tightly than a *p*-electron and has a slightly lower (more negative) energy. A *d*-electron is bound less tightly than a *p*-electron of the same shell because it is even less able to penetrate through to the nucleus. That is, *d*-electrons are higher in energy than *p*-electrons of the same shell, which are in turn higher in energy than *s*-electrons of that shell.

The effects of penetration and shielding can be large. A 4*s*-electron may have much lower energy than a 4*p*- or 4*d*-electron; it may even have lower energy than a 3*d*-electron of the same atom (see Fig. 1.35). The precise ordering of orbitals depends on the number of electrons in the atom, as we shall now see.

In a many-electron atom, because of the effects of penetration and shielding, the order of energies of orbitals in a given shell is typically $s < p < d < f$.

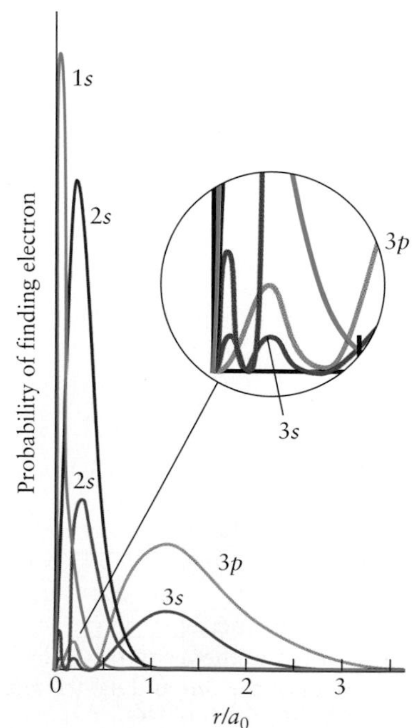

FIGURE 1.36 The probability of finding an electron in an *s*- or *p*-orbital in the shells with $n = 1, 2,$ or 3, at a given radius regardless of direction. Note that the probability maxima for the 3*s*- and 3*p*-orbitals are close to each other, as expected for two orbitals in the same shell. However, note in the inset that an electron in a 3*s*-orbital has a higher probability of being found close to the nucleus.

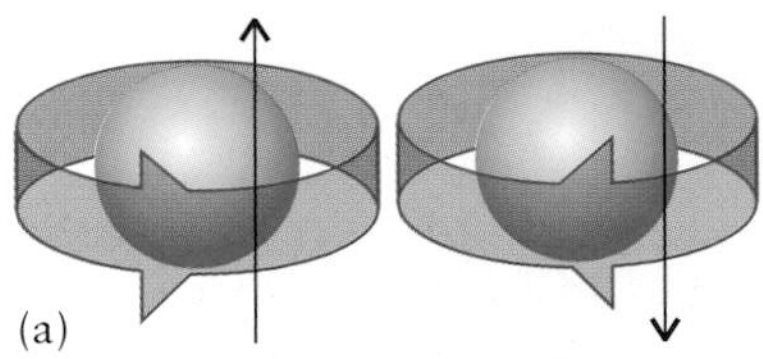

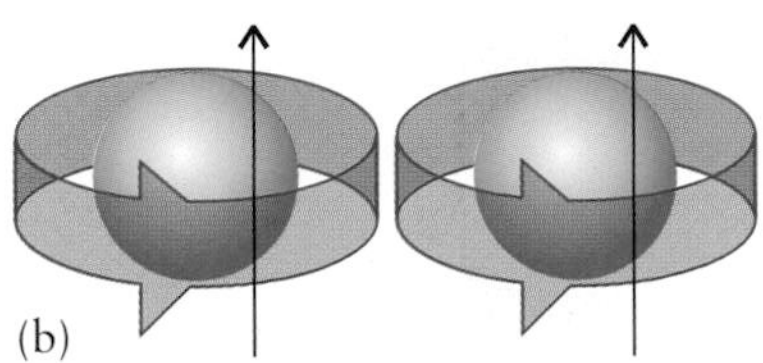

FIGURE 1.37 (a) Two electrons are said to be paired if they have opposite spins (one clockwise, the other counterclockwise). (b) Two electrons are classified as having parallel spins if their spins are in the same direction; in this case, both ↑.

1.12 The Building-Up Principle

The electronic structure of an atom determines its chemical properties, so we need to be able to describe that structure. To do so, we write the **electron configuration** of the atom, a list of all its occupied orbitals, with the numbers of electrons that each one contains. In the ground state of a many-electron atom, the electrons occupy atomic orbitals in such a way that the total energy of the atom is a minimum. At first sight, we might expect an atom to have its lowest energy when all its electrons are in the lowest energy orbital (the 1*s*-orbital), but except for hydrogen and helium, that can never happen. In 1925, the Austrian scientist Wolfgang Pauli discovered why—the **Pauli exclusion principle** states that

- No more than two electrons may occupy any given orbital. When two electrons do occupy one orbital, their spins must be paired.

The spins of two electrons are said to be **paired** if one is ↑ and the other ↓ (Fig. 1.37). Paired spins are denoted ↓↑, and the electrons have spin magnetic quantum numbers of opposite sign. Because an atomic orbital is designated by three quantum numbers (n, l, and m_l) and the two spin states are specified by a fourth quantum number, m_s, another way of expressing the Pauli exclusion principle for atoms is

- No two electrons in an atom can have the same set of four quantum numbers.

The exclusion principle implies that each orbital in the energy level diagram in Fig. 1.35 can hold no more than two electrons.

The hydrogen atom in its ground state has one electron in the 1*s*-orbital. To show this structure, we place a single arrow in the 1*s*-orbital in a "box diagram," which shows each orbital as a box that can hold up to two electrons (see diagram **1**, which is a fragment of Fig. 1.35) and report its configuration as $1s^1$ ("one s one"). The box in this diagram represents one orbital, and it can hold up to two electrons. In the ground state of a helium atom ($Z = 2$), both electrons are in a 1*s*-orbital, which is reported as $1s^2$ ("one s two"). As we see in (**2**), the two electrons are paired.

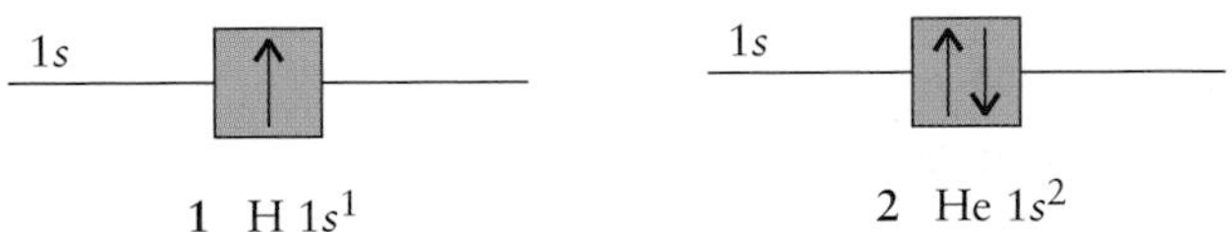

1 H $1s^1$ **2** He $1s^2$

At this point, the 1*s*-orbital holds its maximum number of electrons. The shell with $n = 1$ is now complete, and no more electrons can occupy it. We say that the helium atom has a **closed shell**, a shell containing the maximum number of electrons allowed by the exclusion principle.

Lithium ($Z = 3$) has three electrons. Two electrons occupy the 1*s*-orbital and complete the $n = 1$ shell. The third electron must occupy the next available orbital up the ladder of energy levels, which according to Fig. 1.35 is the 2*s*-orbital. The ground state of a lithium atom is therefore $1s^22s^1$ (**3**). We can think of this atom as consisting of a core made up of the inner heliumlike 1*s* closed shell surrounded by an outer shell

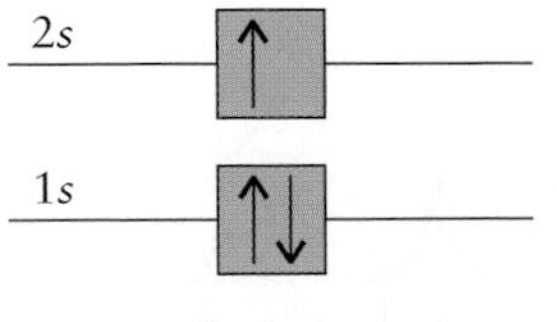

3 Li $1s^22s^1$, [He]$2s^1$

containing a higher energy electron, and denote the configuration [He]$2s^1$. Electrons in the outermost shell are called **valence electrons**. In general, only valence electrons can be lost in chemical reactions, because core electrons are too tightly bound. Thus, lithium will lose only one electron when it forms compounds; it will form Li^+ ions, rather than Li^{2+} or Li^{3+} ions.

The outermost electrons are used in the formation of chemical bonds (Chapter 2), and the theory of bond formation is called "valence theory"; hence the name of these electrons.

The element with $Z = 4$ is beryllium, Be, with four electrons. The first three electrons form the configuration $1s^22s^1$, like Li. The fourth electron pairs with the 2*s*-electron, giving $1s^22s^2$, or more simply [He]$2s^2$ (**4**). A Be atom therefore has a heliumlike core surrounded by a valence shell of two paired electrons. Like lithium—and for the same reason—a Be atom can lose only its valence electrons in chemical reactions. Thus, it loses both 2*s*-electrons to form a Be^{2+} ion.

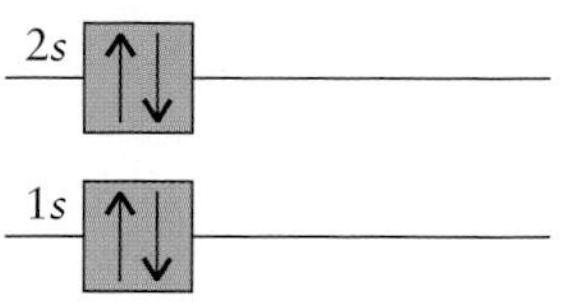

4 Be $1s^22s^2$, [He]$2s^2$

Boron ($Z = 5$) has five electrons. Two enter the 1*s*-orbital and complete the $n = 1$ shell. Two enter the 2*s*-orbital. The fifth electron occupies an orbital of the next available subshell, which Fig. 1.35 shows is a 2*p*-orbital. This arrangement of electrons is reported as the configuration $1s^22s^22p^1$ or [He]$2s^22p^1$ (**5**).

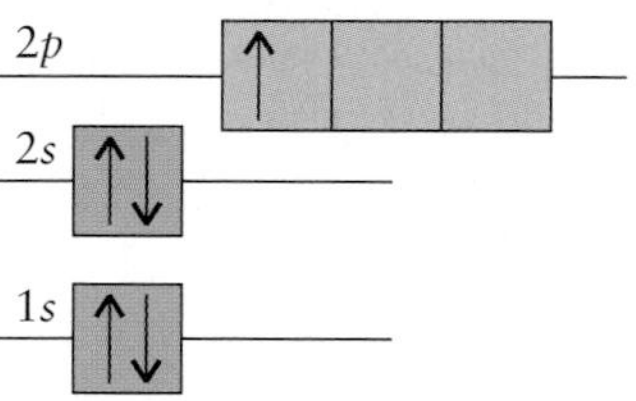

5 B $1s^22s^22p^1$, [He]$2s^22p^1$

We need to make another decision at carbon: does the sixth electron join the one already in the 2*p*-orbital or does it enter a different 2*p*-orbital? (Remember, there are three *p*-orbitals in the subshell, all having the same energy) To answer this question, we note that electrons that occupy different *p*-orbitals are farther from each other than when they occupy the same orbital. Therefore, they repel each other less than if they occupied the same orbital. So the sixth electron goes into an empty *p*-orbital, and the ground state of carbon is $1s^22s^22p_x^{\ 1}2p_y^{\ 1}$ (**6**). We write out the individual orbitals like this only when we need to emphasize that electrons occupy different orbitals within a subshell. In most cases, we can write the shorter form, such as [He]$2s^22p^2$.

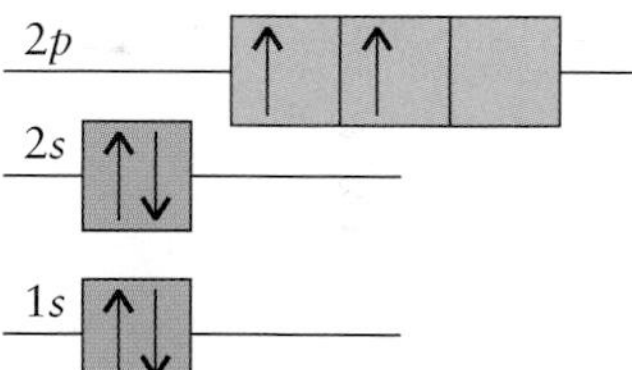

6 C $1s^22s^22p^2$, [He]$2s^22p^2$

Note that in the orbital diagram we have drawn the two 2*p*-electrons with **parallel spins**, (↑↑), indicating that they have the same spin magnetic quantum numbers. For reasons based in quantum mechanics, this arrangement has slightly lower energy than a paired arrangement. However, it is allowed only when the electrons occupy different orbitals.

The procedure we have been using is called the **building-up principle**. It can be summarized by two rules. To predict the ground-state configuration of an element with atomic number *Z*:

Step 1 Add *Z* electrons, one after the other, to the orbitals in the order shown in Fig. 1.38, but with no more than two electrons in any one orbital.

Step 2 If more than one orbital in a subshell is available, add electrons with parallel spins to different orbitals of that subshell rather than pairing two electrons in one of the orbitals.

The first step takes into account the Pauli exclusion principle. The second step is called **Hund's rule**, for the German spectroscopist Friedrich Hund, who first proposed it. This procedure gives the configuration of the atom that corresponds to the lowest total energy, allowing for the attraction of the electrons to the nucleus and their repulsion by one another. An atom with electrons in energy states higher than predicted by the building-up principle is said to be in an **excited state**. For example, the electron

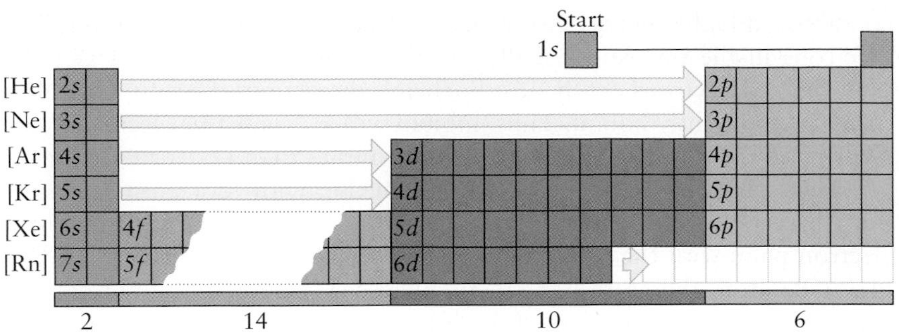

FIGURE 1.38 The order in which atomic orbitals are occupied according to the building-up principle. When we add an electron, we move one place to the right until all the electrons (Z electrons for an element of atomic number Z) have been accommodated. At the end of a row, move to the beginning of the next row down.

2p
2s
1s

7 N $1s^22s^22p^3$, [He]$2s^22p^3$

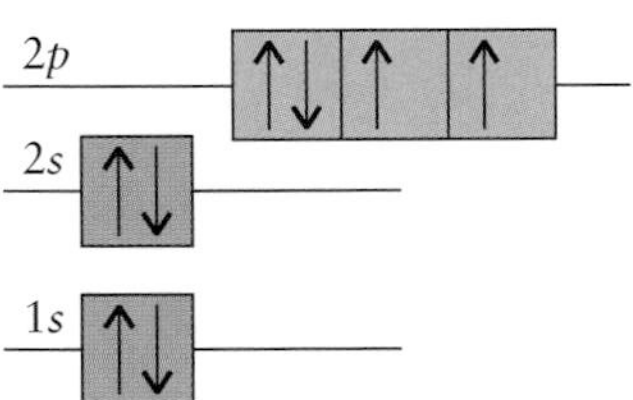

8 O $1s^22s^22p^4$, [He]$2s^22p^4$

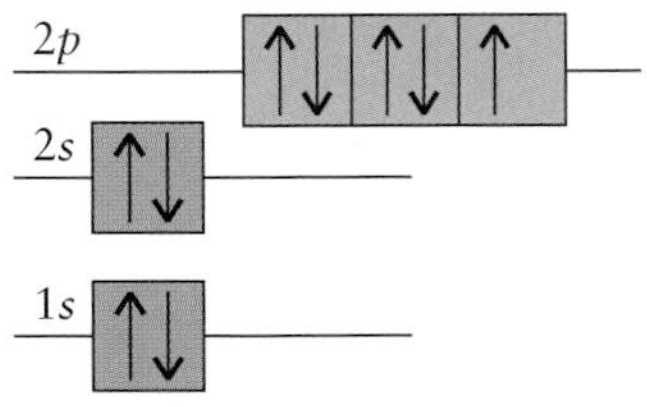

9 F $1s^22s^22p^5$, [He]$2s^22p^5$

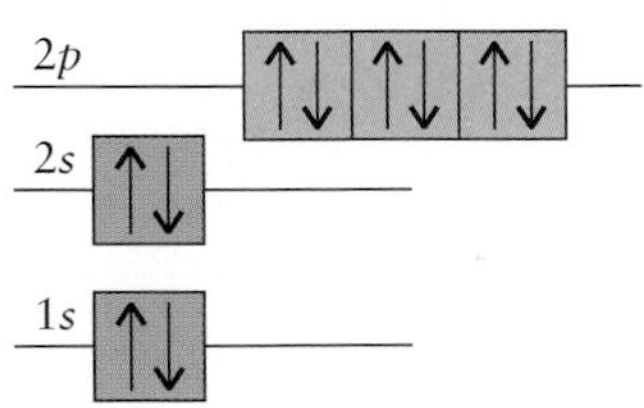

10 Ne $1s^22s^22p^6$, [He]$2s^22p^6$

configuration [He]$2s^12p^3$ represents an excited state for a carbon atom. An excited state is unstable and emits a photon as the electron returns to an orbital that restores the atom to a lower energy.

In general, we can think of an atom of any element as having a noble-gas core surrounded by a number of electrons in the **valence shell**, the outermost occupied shell. The valence shell is the occupied shell with the largest value of n. With that idea in mind, the underlying organization of the periodic table described in Section B begins to unfold. All the atoms in a given period (a horizontal row of the periodic table) have a common valence shell, and the principal quantum number of that shell is equal to the period number. For example, the valence shell of elements in Period 2 (lithium to neon) is the shell with $n = 2$. All the atoms in one period have the same type of core. Thus the atoms of Period 2 elements all have a heliumlike $1s^2$ core, and those of Period 3 elements have a neonlike $1s^22s^22p^6$ core, denoted [Ne]. All the atoms of a given group (in the main groups, particularly) have analogous valence electron configurations that differ only in the value of n. For instance, all the members of Group 1 have the valence configuration ns^1; and all the members of Group 14/IV have the valence configuration ns^2np^2.

With these points in mind, let's continue building up the electron configurations across Period 2. Nitrogen has $Z = 7$ and one more electron than carbon, giving [He]$2s^22p^3$. Each p-electron occupies a different orbital, and the three have parallel spins (7). Oxygen has $Z = 8$ and one more electron than nitrogen; therefore, its configuration is [He]$2s^22p^4$ (8) and two of its $2p$-electrons are paired. Similarly, fluorine, with $Z = 9$ and one more electron than oxygen, has the configuration [He]$2s^22p^5$ (9), with only one unpaired electron. Neon, with $Z = 10$, has one more electron than fluorine. This electron completes the $2p$-subshell, giving [He]$2s^22p^6$ (10). According to Figs. 1.35 and 1.38, the next electron enters the $3s$-orbital, the lowest energy orbital of the next shell. The configuration of sodium is therefore [He]$2s^22p^63s^1$, or more briefly, [Ne]$3s^1$, where [Ne] denotes the neonlike core.

Self-Test 1.13A Predict the ground-state configuration of a magnesium atom.

[***Answer:*** $1s^22s^22p^63s^2$, or [Ne]$3s^2$]

Self-Test 1.13B Predict the ground-state configuration of an aluminum atom.

If we need to know the actual quantum numbers of the electrons in a many-electron atom, we can use the electron configuration or box diagram of the element to find them. For example, the electrons in a sodium atom, with the electron configuration $1s^22s^22p^63s^1$, have the following quantum numbers:

Orbital	Number of electrons	n	l	m_l	m_s
$1s$	2	1	0	0	$+\frac{1}{2}, -\frac{1}{2}$
$2s$	2	2	0	0	$+\frac{1}{2}, -\frac{1}{2}$
$2p$	2	2	1	-1	$+\frac{1}{2}, -\frac{1}{2}$
	2	2	1	0	$+\frac{1}{2}, -\frac{1}{2}$
	2	2	1	$+1$	$+\frac{1}{2}, -\frac{1}{2}$
$3s$	1	3	0	0	$+\frac{1}{2}, -\frac{1}{2}$

The *s*- and *p*-orbitals of the shell with $n = 3$ are full by the time we get to argon, $[\text{Ne}]3s^23p^6$, which is a colorless, odorless, unreactive gas resembling neon. Argon completes the third period. From Fig. 1.35, we see that the energy of the 4*s*-orbital is slightly lower than that of the 3*d*-orbitals. As a result, instead of electrons entering the 3*d*-orbitals, the fourth period now begins by filling the 4*s*-orbitals (see Fig. 1.38). Hence, the next two electron configurations are $[\text{Ar}]4s^1$ for potassium and $[\text{Ar}]4s^2$ for calcium, where [Ar] denotes the argonlike core, $1s^22s^22p^63s^23p^6$. At this point, however, the 3*d*-orbitals begin to be occupied, and there is a change in the rhythm of the periodic table.

According to the pattern of increasing energy of the orbitals (see Fig. 1.35), the next 10 electrons (for scandium, with $Z = 21$, through zinc, with $Z = 30$) enter the 3*d*-orbitals. The ground-state electron configuration of scandium, for example, is $[\text{Ar}]3d^14s^2$, and that of its neighbor titanium is $[\text{Ar}]3d^24s^2$. Note that, beginning at scandium, we write the 4*s*-electrons after the 3*d*-electrons: once they contain electrons, the 3*d*-orbitals lie lower in energy than the 4*s*-orbital (recall Fig. 1.35; the same relation holds true for *nd*- and $(n + 1)s$-orbitals in subsequent periods). Successive electrons are added to the *d*-orbitals as *Z* increases: $[\text{Ar}]3d^14s^2$ for scandium, $[\text{Ar}]3d^24s^2$ for titanium, and so on. However, there are two exceptions: the half-complete subshell configuration d^5 and the complete subshell configuration d^{10} turn out experimentally to have a lower energy than simple theory suggests. As a result, a lower energy may be achieved if an electron enters a 3*d*-orbital instead of the expected 4*s*-orbital, if that arrangement completes a half-subshell or a full subshell. For example, the experimental electron configuration of chromium is $[\text{Ar}]3d^54s^1$ instead of $[\text{Ar}]3d^44s^2$ and that of copper is $[\text{Ar}]3d^{10}4s^1$ instead of $[\text{Ar}]3d^94s^2$. Other exceptions to the building-up principle can be found in Appendix 2C.

As we can anticipate from the structure of the periodic table (see Fig. 1.38), electrons occupy 4*p*-orbitals once the 3*d*-orbitals are full. The configuration of germanium, $[\text{Ar}]3d^{10}4s^24p^2$, for example, is obtained by adding two electrons to the 4*p*-orbitals outside the completed 3*d*-subshell. The fourth period of the table contains 18 elements, because the 4*s*- and 4*p*-orbitals together can accommodate a total of 8 electrons and the 3*d*-orbitals can accommodate 10. Period 4 is the first **long period** of the periodic table.

In this book, the $(n - 1)d$-electrons of *d*-block elements are considered valence electrons, but some chemists exclude these electrons. Take the advice of your instructor.

Next in line for occupation at the beginning of Period 5 is the 5*s*-orbital, followed by the 4*d*-orbitals. As in Period 4, the energy of the 4*d*-orbitals falls below that of the 5*s*-orbital after two electrons have been accommodated in the 5*s*-orbital. A similar effect is seen in Period 6, but now another set of inner orbitals, the 4*f*-orbitals, begins to be occupied. Cerium, for example, has the configuration $[Xe]4f^1 5d^1 6s^2$. Electrons then continue to occupy the seven 4*f*-orbitals, which are complete after 14 electrons have been added, at ytterbium, $[Xe]4f^{14} 6s^2$. Next, the 5*d*-orbitals are occupied. The 6*p*-orbitals are occupied only after the 6*s*-, 4*f*-, and 5*d*-orbitals are filled at mercury; thallium, for example, has the configuration $[Xe]4f^{14} 5d^{10} 6s^2 6p^1$.

We account for the ground-state electron configuration of an atom by using the building-up principle in conjunction with Fig. 1.38, the Pauli exclusion principle, and Hund's rule.

TOOLBOX 1.1 How to predict the ground-state electron configuration of an atom

Conceptual Basis

Electrons occupy orbitals in such a way as to minimize the total energy of an atom in accord with the Pauli exclusion principle and Hund's rule.

Procedure

We use the building-up principle to assign a ground-state configuration to an element with atomic number *Z*:

Step 1 Add *Z* electrons, one after the other, to the orbitals in the order shown in Figs. 1.35 and 1.38, but with no more than two electrons in any one orbital (the Pauli exclusion principle).

Step 2 If more than one orbital in a subshell is available, add electrons to different orbitals of the subshell before doubly occupying any of them (Hund's rule).

Step 3 Write the orbitals in order of increasing energy. The configuration of a filled shell is represented by the symbol of the noble gas having that configuration, as in [He] for $1s^2$.

Step 4 When drawing a box diagram, show the electrons in different orbitals of the same subshell with parallel spins; electrons sharing an orbital have paired spins.

This procedure gives the ground-state electron configuration of an atom. Any other arrangement corresponds to an excited state of the atom. Note that we can use the structure of the periodic table to predict the electron configurations of most elements once we realize which orbitals are being filled in each block of the periodic table (see Fig. 1.38).

EXAMPLE 1.7 Predicting the ground-state electron configuration of a heavy atom

Predict the ground-state electron configuration of (a) a titanium atom and (b) a lead atom.

STRATEGY (a) Use the building-up principle as outlined in Toolbox 1.1. Because titanium is the second member of the *d* block in Period 4, we predict it will have two 3*d*-electrons. (b) There is a shortcut that is particularly useful for elements with large numbers of electrons. First, note the group number, which tells us the number of valence electrons in the ground state of the atom, and the period number, which tells us the value of the principal quantum number of the valence shell. The core consists of the preceding noble-gas configuration together with any completed *d*- and *f*-subshells. Because lead is in the same group as carbon (Group 14/IV), we anticipate that it will have an analogous valence configuration of the form ns^2np^2.

SOLUTION (a) Titanium is in Period 4, so it has an argon core. The 4*s*-orbitals fill next, then the last two electrons fill two separate 3*d*-orbitals. The electron configuration is $[Ar]3d^24s^2$. (b) Lead belongs to Group 14/IV and Period 6. It therefore has four electrons in its valence shell, two in a 6*s*-orbital and two in different 6*p*-orbitals. The period has complete 5*d*- and 4*f*-subshells, and the preceding noble gas is xenon. The electron configuration of lead is therefore $[Xe]4f^{14}5d^{10}6s^26p^2$.

Self-Test 1.14A Write the ground-state configuration of a bismuth atom.
[*Answer:* $[Xe]4f^{14}5d^{10}6s^26p^3$]

Self-Test 1.14B Write the ground-state configuration of an arsenic atom.

1.13 Electronic Structure and the Periodic Table

The periodic table (Section B) was formulated long before the structures of atoms were known, by noting trends in experimental data (Box 1.2). However, to understand the organization of the periodic table, we need to consider the electron configurations of the elements. The table is divided into *s*, *p*, *d*, and *f* blocks, named for the last subshell that is occupied according to the building-up principle (Fig. 1.39). Two elements are exceptions. Strictly, helium belongs in the *s* block, but it is shown in the *p* block. It is a gas with properties matching those of the noble gases in Group 18/VIII, rather than the reactive metals in Group 2. Its place in Group 18/VIII is justified because it has a filled valence shell, like all the other Group 18/VIII elements. Hydrogen occupies a unique position in the periodic table. It has one *s*-electron, so it belongs in Group 1. But it is also one electron short of a noble-gas configuration, so it can act like a member of Group 17/VII.

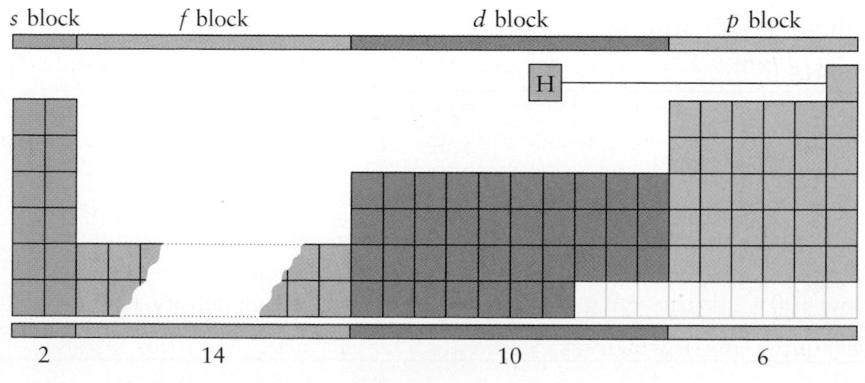

FIGURE 1.39 The names of the blocks of the periodic table are based on the last subshell being occupied in an atom of an element according to the building-up principle. The numbers of electrons that each type of orbital can accommodate are shown by the numbers across the bottom of the table. The colors of the blocks match the colors we are using for the corresponding orbitals.

BOX 1.2 HOW DO WE KNOW . . . THE STRUCTURE OF THE PERIODIC TABLE?

The periodic table is one of the most important achievements of chemistry. Its development is an example of how scientific discoveries can be made by using insight to organize data collected by a large number of scientists over a period of many years. There are often simultaneous discoveries in science, because discoveries tend to be made when enough of the right kind of data has been collected.

In 1860, the Congress of Karlsruhe brought together many prominent chemists in an attempt to come to some agreement on issues such as the existence of atoms, the correct atomic masses, and how the elements are related to one another. No agreement was reached, but many attempts to explain new experimental data presented at the meeting were vigorously debated. One idea discussed was Avogadro's principle—that the number of molecules of different gases in samples of equal volume, pressure, and temperature is the same. This principle allowed the relative atomic masses of the gases to be determined. Two scientists attending the congress were the German Lothar Meyer and the Russian Dmitri Mendeleev, both of whom left with copies of Avogadro's paper. In 1869, Meyer and Mendeleev discovered independently that a regular repeating pattern of properties could be observed when the elements were arranged in order of increasing atomic mass. Mendeleev called this observation the *periodic law*.

Mendeleev's chemical insight led him to leave gaps for elements that would be needed to complete the pattern but were unknown at the time. When they were discovered later, he turned out to be strikingly correct. For example, his pattern required an element he named "eka-silicon" below silicon and between gallium and arsenic. He predicted that the element would have a relative atomic mass of 72 (taking the mass of hydrogen as 1) and properties similar to those of silicon. This prediction spurred the German chemist Clemens Winkler in 1886 to search for eka-silicon, which he eventually discovered and named germanium. It has a relative atomic mass of 72.59 and properties similar to those of silicon, as shown in the accompanying table.

One problem with Mendeleev's table was that some elements seemed to be out of place. For example, when argon was isolated, it did not seem to have the correct mass for its location. Its relative atomic mass of 40 is the same as that of calcium, but argon is an inert gas and calcium a reactive metal. Such anomalies led scientists to question the use of relative atomic mass as the basis for organizing the elements. When Henry Moseley examined x-ray spectra of the elements in the early twentieth century, he realized that all atoms of the same element had identical nuclear charge and, therefore, the same number of protons, which gives the element's atomic number. It was soon discovered that elements fall into the uniformly repeating pattern of the periodic table if they are organized according to atomic number, rather than atomic mass.

Dmitri Ivanovich Mendeleev (1834–1907).

Mendeleev's Predictions for Eka-Silicon (Germanium)

Property	Eka-Silicon (E)	Germanium
molar mass/g·mol^{-1}	72	72.59
density/g·cm^{-3}	5.5	5.32
melting point/°C	high	937
appearance	dark gray	gray-white
oxide	EO_2; white solid; amphoteric; density 4.7 g·cm^{-3}	GeO_2; white solid; amphoteric; density 4.23 g·cm^{-3}
chloride	ECl_4; boils below 100°C; density 1.9 g·cm^{-3}	$GeCl_4$; boils at 84°C; density 1.84 g·cm^{-3}

The s and p blocks form the main groups of the periodic table. The similar electron configurations for the elements in a main group are the reason for the similar properties of these elements. The group number tells us how many valence-shell electrons are present. In the s block, the group number (1 or 2) is the same as the number of valence electrons. This relation is also true for all main groups when the Roman numerals (I–VIII) are used to label the groups. However, when the group labels 1–18 are used, in the p block we have to subtract 10 (for the 10 d-electrons) from the group number to find the number of valence electrons. For example, fluorine in Group 17/VII has seven valence electrons.

Each new period corresponds to the occupation of a shell with a higher principal quantum number. This correspondence explains the different lengths of the periods. Period 1 consists of only two elements, H and He, in which the single 1s-orbital of the $n = 1$ shell is being filled with its two electrons. Period 2 consists of the eight elements Li through Ne, in which the one 2s- and three 2p-orbitals are being filled with eight more electrons. In Period 3 (Na through Ar), the 3s- and 3p-orbitals are being occupied by eight additional electrons. In Period 4, not only are the 8 electrons of the 4s- and 4p-orbitals being added, but so are the 10 electrons of the 3d-orbitals. Hence there are 18 elements in Period 4.

Period 5 elements add another 18 electrons as the 5s-, 4d-, and 5p-orbitals are filled. In Period 6, a total of 32 electrons are added, because 14 electrons are also being added to the seven 4f-orbitals. The f-block elements have very similar properties, because their electron configurations differ only in the population of inner f-orbitals, and electrons in these orbitals do not participate much in bond formation.

The blocks of the periodic table are named for the last orbital to be occupied according to the building-up principle. The periods are numbered according to the principal quantum number of the valence shell.

THE PERIODICITY OF ATOMIC PROPERTIES

Section B introduced some aspects of the periodic table, such as the correlation between the location of an element and whether it is a metal or a nonmetal, and the periodicity of the elements. However, the periodic table summarizes a much wider range of properties than mentioned there, many of which are crucial for understanding chemical bonding (Chapters 2 and 3). As we describe these properties in the following sections, keep in mind the variation of effective nuclear charge through the periodic table, for it plays an important role in the explanation of periodic trends. Figure 1.40 shows the variation for the first three periods. The effective charge increases from left to right across a period and falls back sharply on going to the following period.

Many properties of the elements are listed in Appendix 2D.

1.14 Atomic Radius

Electron clouds do not have sharp boundaries, so we cannot really speak of the radius of an atom. However, when atoms pack together in solids and bond together to form molecules, their centers are found at definite

FIGURE 1.40 The variation of the effective nuclear charge for the outermost valence electron with atomic number. Notice that the effective nuclear charge increases from left to right across a period but drops when the outer electrons occupy a new shell.

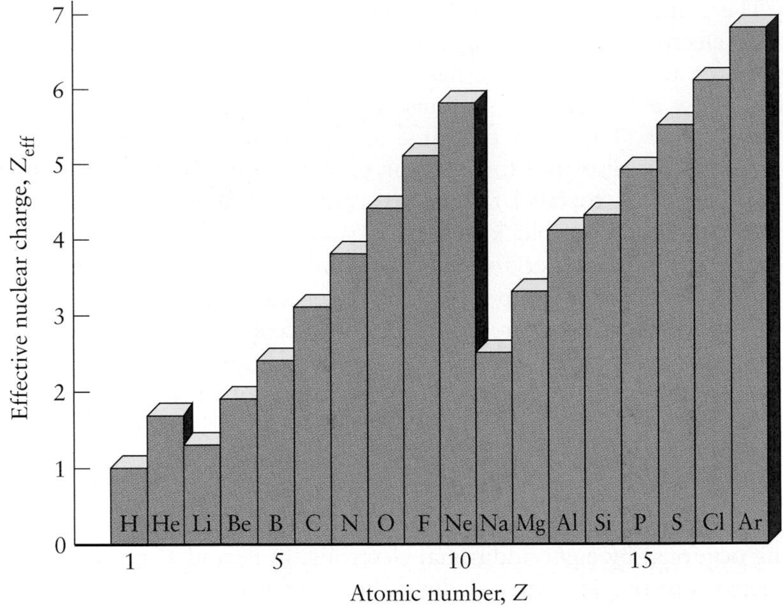

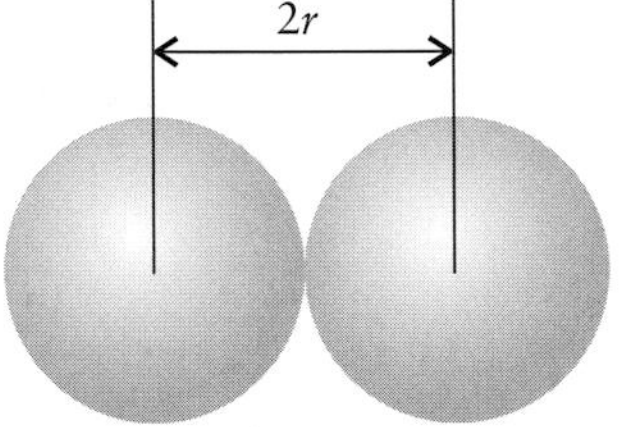

11 Atomic radius

Van der Waals was a Dutch scientist who studied the interactions between molecules; see Chapter 4.

distances from one another. The **atomic radius** of an element is defined as half the distance between the nuclei of neighboring atoms (**11**). If the element is a metal, its atomic radius is taken to be half the distance between the centers of neighboring atoms in a solid sample. For instance, because the distance between neighboring nuclei in solid copper is 256 pm, the atomic radius of copper is 128 pm. If the element is a nonmetal or a metalloid, we use half the distance between the nuclei of atoms joined by a chemical bond; this radius is also called the **covalent radius** of the element. For instance, the distance between the nuclei in a Cl_2 molecule is 198 pm, so the covalent radius of chlorine is 99 pm. If the element is a noble gas, we use the **van der Waals radius**, which is half the distance between the centers of neighboring atoms in a sample of the solidified gas. The atomic radii of the noble gases listed in Appendix 2D are all van der Waals radii. Because the atoms in a sample of a noble gas are not chemically bonded together, van der Waals radii are generally much larger than covalent radii and are best not included in the discussion of trends.

Figure 1.41 shows some atomic radii, and Fig. 1.42 shows the variation in atomic radius with atomic number. Note the periodic, sawtooth pattern in the latter plot. *Atomic radius generally decreases from left to right across a period and increases down a group.*

The increase down a group, like that from Li to Cs, makes sense: with each new period, the outermost electrons occupy shells that lie farther from the nucleus. But the decrease across a period, like that from Li to Ne, is surprising at first, because the number of electrons is increasing along with the number of protons. The explanation is that the new electrons are in the same shell of the atom and about as close to the nucleus as other electrons in the same shell. Because they are spread out, they do

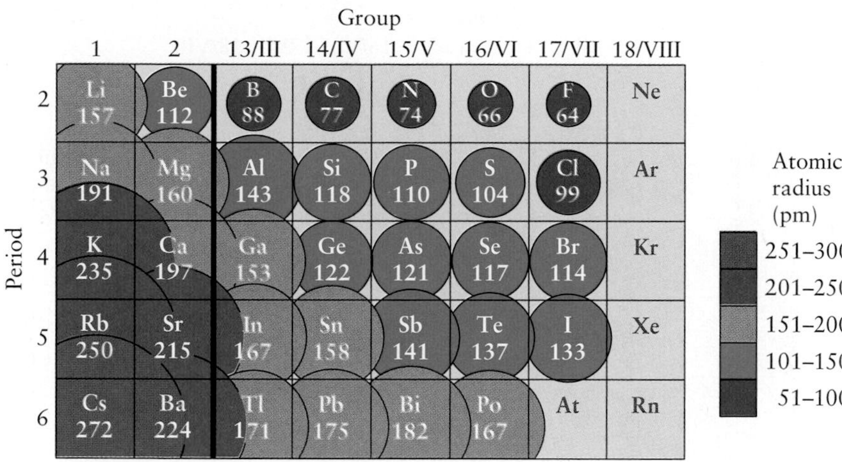

FIGURE 1.41 The atomic radii (in picometers, 1 pm = 10^{-12} m) of the main-group elements. The radii decrease from left to right in a period and increase down a group. The colors used here and in subsequent charts represent the general magnitude of the property, as indicated by the scale on the right. Atomic radii, including those of the *d*-block elements, are listed in Appendix 2D.

not shield one another well from the nuclear charge, so the effective nuclear charge increases across the period. The increasing effective nuclear charge draws the electrons in; and, as a result, the atom is more compact.

Atomic radii generally decrease from left to right across a period and increase down a group.

1.15 Ionic Radius

The **ionic radius** of an element is its share of the distance between neighboring ions in an ionic solid (**12**). The distance between the centers of a neighboring cation and anion is the sum of the two ionic radii. In practice, we take the radius of the oxide ion to be 140 pm and calculate the radii of other ions on the basis of that value. For example, because the distance between the centers of neighboring Mg^{2+} and O^{2-} ions in

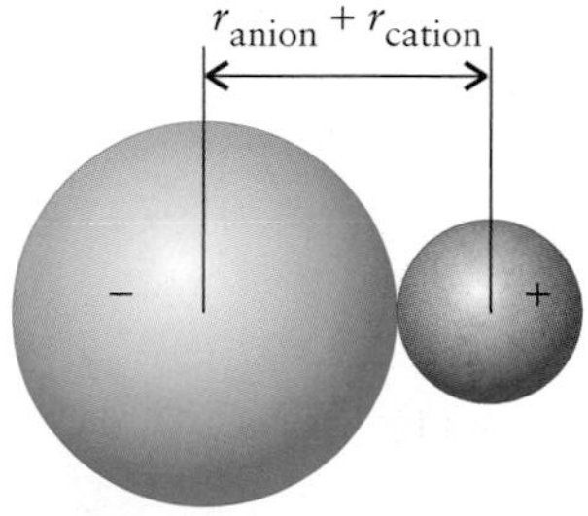

12 Ionic radius

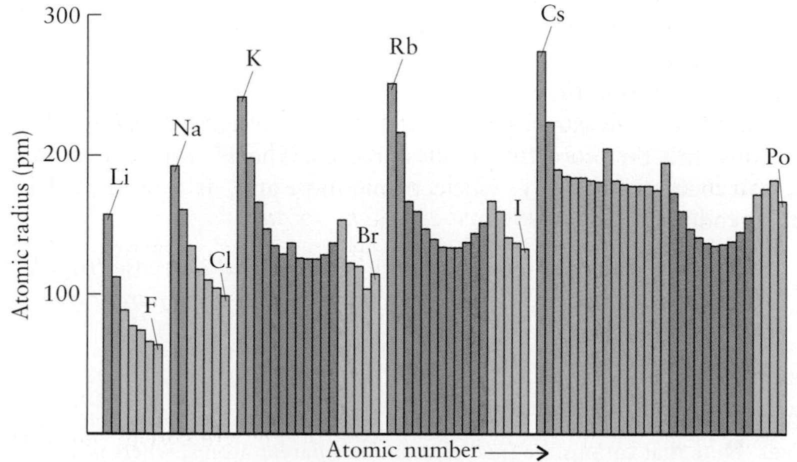

FIGURE 1.42 The periodic variation of the atomic radii of the elements. The variation across a period can be explained in terms of the effect of increasing effective nuclear charge; that down a group by the occupation of shells with increasing principal quantum number as well as the increase in nuclear charge.

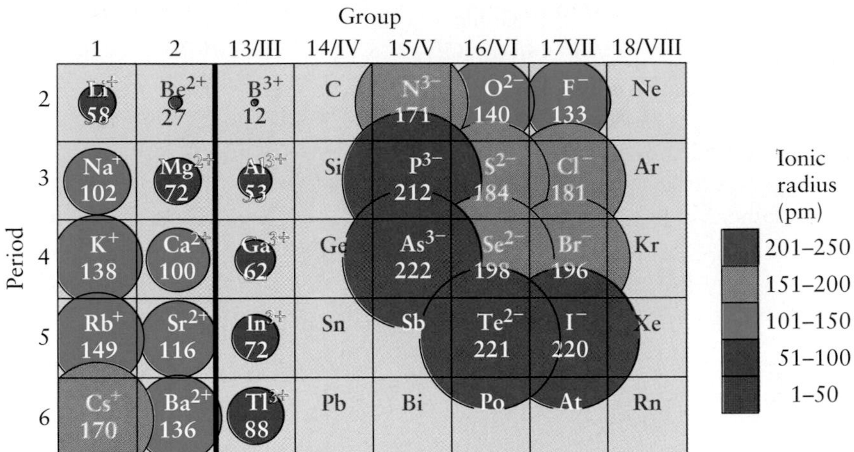

FIGURE 1.43 The ionic radii (in picometers) of the ions of the main-group elements. Note that cations are typically smaller than anions—and, in some cases, very much smaller.

magnesium oxide is 212 pm, the radius of the Mg^{2+} ion is reported as 212 pm − 140 pm = 72 pm.

Figure 1.43 illustrates the trends in ionic radii, and Fig. 1.44 shows the relative sizes of some ions and their parent atoms. All cations are smaller than their parent atoms, because the atom loses one or more electrons to form the cation and exposes its core, which is generally much smaller than the parent atom. For example, the atomic radius of Li, with the configuration $1s^22s^1$, is 157 pm, but the ionic radius of Li^+, the bare heliumlike $1s^2$ core of the parent atom, is only 58 pm. This size difference is comparable to that between a cherry and its pit. Like atomic radii, cation radii increase down each group because electrons are occupying shells with higher principal quantum numbers.

Figure 1.44 shows that anions are larger than their parent atoms. The reason can be traced to the increased number of electrons in the valence shell of the anion and the repulsive effects the electrons exert on one another. The variation in the radii of anions shows the same diagonal trend as that for atoms and cations, with the smallest at the upper right of the periodic table, close to fluorine.

Atoms and ions with the same number of electrons are called **isoelectronic.** For example, Na^+, F^-, and Mg^{2+} are isoelectronic. All three ions have the same electron configuration, $[He]2s^22p^6$, but their radii differ because they have different nuclear charges (see Fig. 1.43). The Mg^{2+} ion has the largest nuclear charge, so it has the strongest attraction for the electrons and therefore the smallest radius. The F^- ion has the lowest nuclear charge of the three isoelectronic ions; and, as a result, it has the largest radius.

Ionic radii generally increase down a group and decrease from left to right across a period. Cations are smaller than their parent atoms and anions are larger.

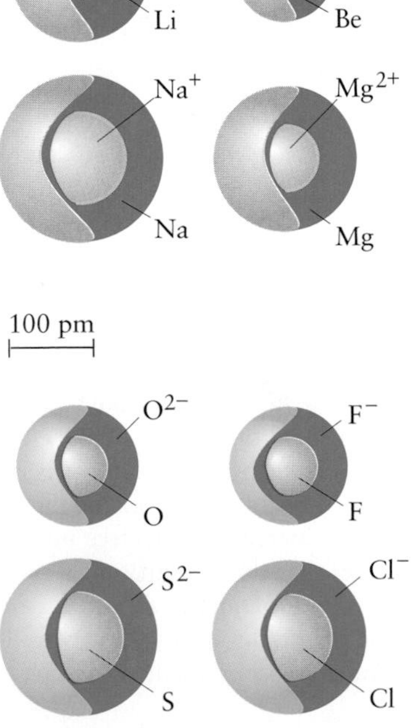

FIGURE 1.44 The relative sizes of some cations and anions with their parent atoms. Note that cations are smaller than their parent atoms, whereas anions are larger.

How can we use this information . . . to decide which is the bigger ion?

We can arrange each of the pairs of ions (a) Mg^{2+} and Ca^{2+}, and (b) O^{2-} and F^-, in order of increasing ionic radius. The smaller member of a pair of ions with the same number of electrons will be an ion of an element that lies further to the right in a period. If the two ions are in the same group, the smaller ion will be the one that lies higher in the group. (a) Because Mg lies above Ca in Group 2, Mg^{2+} will have the smaller ionic radius (the actual values are 72 pm and 100 pm, respectively). (b) Because F lies to the right of O in Period 2, F^- will have the smaller ionic radius (the actual values are 133 pm for F^- and 140 pm for O^{2-}.

Self-Test 1.15A Arrange each pair of ions, (a) Mg^{2+} and Al^{3+}, and (b) O^{2-} and S^{2-}, in order of increasing ionic radius.

[*Answer:* (a) $r(Al^{3+}) < r(Mg^{2+})$; (b) $r(O^{2-}) < r(S^{2-})$]

Self-Test 1.15B Arrange each pair of ions, (a) Ca^{2+} and K^+, and (b) S^{2-} and Cl^-, in order of increasing ionic radius.

1.16 Ionization Energy

We shall see in Chapter 2 that the formation of a bond in an ionic compound depends on the removal of one or more electrons from one atom and their transfer to another atom. The energy needed to remove electrons from atoms is therefore of central importance for understanding their chemical properties. The **ionization energy** is the energy needed to remove an electron from an atom in the gas phase. For the **first ionization energy**, I_1, we start with the neutral atom. For example, for copper,

$$Cu(g) \rightarrow Cu^+(g) + e^-(g) \qquad \text{energy required} = I_1 \qquad (785\ kJ{\cdot}mol^{-1})$$

The **second ionization energy**, I_2, of an element is the energy needed to remove an electron from a singly charged gas-phase cation. For copper,

$$Cu^+(g) \rightarrow Cu^{2+}(g) + e^-(g) \qquad \text{energy required} = I_2 \qquad (1955\ kJ{\cdot}mol^{-1})$$

Figure 1.45 shows that first ionization energies generally decrease down a group. The decrease means that it takes less energy to remove an electron from a cesium atom, for instance, than from a sodium atom. Figure 1.46 shows the variation of first ionization energy with atomic number, and we see a periodic sawtooth pattern like that shown by atomic radii. With few exceptions, the first ionization energy rises from left to right across a period. Then it falls back to a lower value at the start of the following period. The lowest values occur at the bottom left of the periodic table (near cesium) and the highest at the upper right (near helium). In other words, less energy is needed to remove an electron from atoms of elements near cesium and more energy is needed to remove an electron from atoms of elements near helium. Elements with low ionization energies can be expected to form cations readily and to conduct electricity in their solid forms. Elements with high ionization energies are unlikely to form cations and are unlikely to conduct electricity.

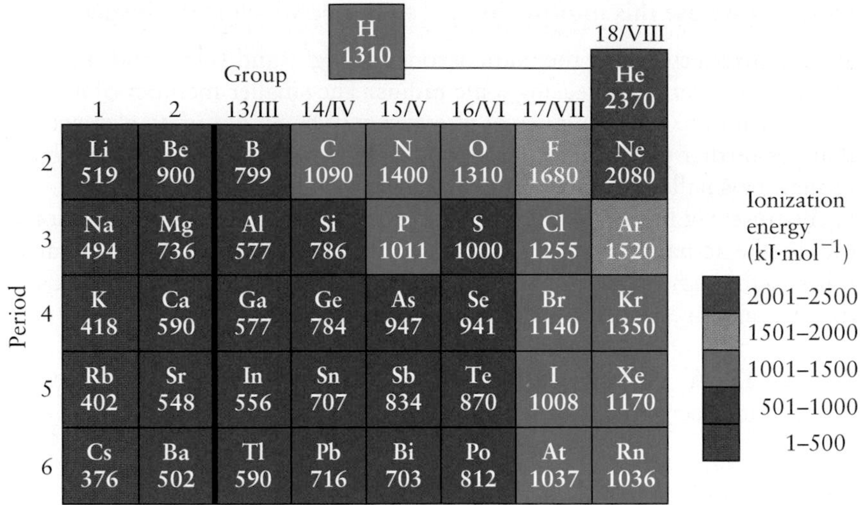

FIGURE 1.45 The first ionization energies of the main-group elements, in kilojoules per mole. In general, low values are found toward the lower left of the table and high values are found to the upper right.

Ionization energy typically decreases down a group because, in successive periods, the outermost electron occupies a shell that is farther from the nucleus and it is therefore less tightly bound. However, as we go from left to right across a given period, the effective nuclear charge increases. As a result, the outermost electron is gripped more tightly and the ionization energies generally increase. The small departures from these trends can usually be traced to repulsions between electrons, particularly electrons occupying the same orbital.

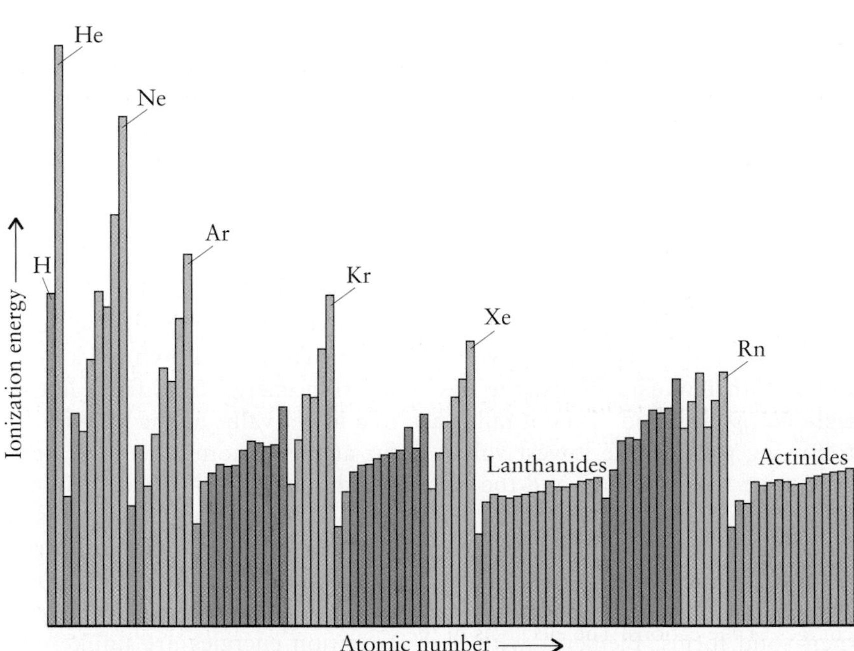

FIGURE 1.46 The periodic variation of the first ionization energies of the elements.

Figure 1.47 shows that the second ionization energy of an element is always higher than its first ionization energy. It takes more energy to remove an electron from a positively charged ion than from a neutral atom. For the Group 1 elements, the second ionization energy is considerably larger than the first; but in Group 2, the two ionization energies have similar values. This difference makes sense, because the Group 1 elements have an ns^1 valence-shell electron configuration. Although the removal of the first electron requires only a small energy, the second electron must come from the noble-gas core. The core electrons have lower principal quantum numbers and are much closer to the nucleus. They are strongly attracted to it and a lot of energy is needed to remove them.

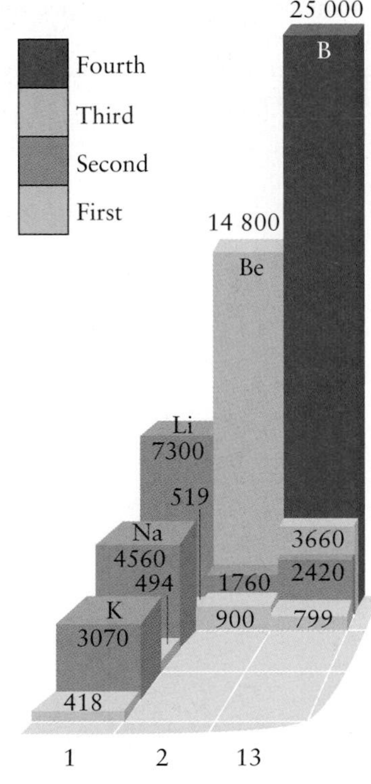

FIGURE 1.47 The successive ionization energies of a selection of main-group elements. Note the great increase in energy required to remove an electron from an inner shell.

Self-Test 1.16A Account for the slight decrease in first ionization energy between beryllium and boron.

[*Answer:* Boron loses an electron from a higher energy subshell than beryllium.]

Self-Test 1.16B Account for the large decrease in third ionization energy between beryllium and boron.

The low ionization energies of elements at the lower left of the periodic table account for their metallic character. A block of metal consists of a collection of cations of the element surrounded by a sea of valence electrons that the atoms have lost (Fig. 1.48). For example, a piece of copper consists of a stack of Cu^+ ions held together by a sea of electrons, each of which comes from one of the atoms in the sample. Only elements with low ionization energies—the members of the *s* block, the *d* block, the *f* block, and the lower left of the *p* block—can form metallic solids, because only they can lose electrons easily.

The elements at the upper right of the periodic table have high ionization energies, so they do not readily lose electrons and, therefore, are not metals. Note that our knowledge of electronic structure has helped us to understand a major feature of the periodic table—in this case, why the metals are found toward the lower left and the nonmetals are found toward the upper right.

The first ionization energy is highest for elements close to helium and is lowest for elements close to cesium. Second ionization energies are higher than first ionization energies (of the same element), and very much higher if the electron is to be removed from a closed shell. Metals are found toward the lower left of the periodic table because these elements have low ionization energies and can readily lose their electrons.

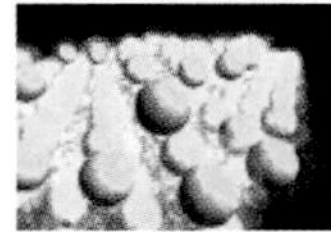

View an animation of bonding in copper metal.

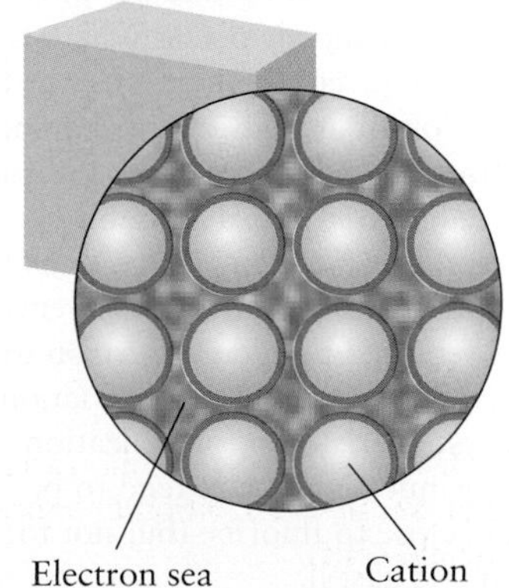

FIGURE 1.48 A block of metal consists of an array of cations (the spheres) surrounded by a sea of electrons. The charge of the electron sea cancels the charges of the cations. The electrons of the sea are mobile and can move past the cations quite easily and hence conduct an electric current.

In some books, you will see electron affinity defined with an opposite sign convention. Those are actually the electron-gain enthalpies.

1.17 Electron Affinity

The ionization energy of an element is a measure of the energy needed to remove an electron from an atom. To predict some chemical properties, we also need to know how the energy changes when an electron *attaches* to an atom. The **electron affinity,** E_{ea}, of an element is the energy *released* when an electron is added to a gas-phase atom. A high electron affinity means that a lot of energy is released when an electron attaches to an atom. A negative electron affinity means that energy must be supplied to push an electron on to an atom. This convention matches the everyday meaning of the term "affinity." More formally, the electron affinity of an element X is defined as

$$E_{ea} = E(X) - E(X^-) \tag{20}$$

where $E(X)$ is the energy of a gas-phase X atom and $E(X^-)$ is the energy of the gas-phase anion.

Figure 1.49 shows the variation in electron affinity in the main groups of the periodic table. It is much less periodic than the variation in radius and ionization energy. However, one broad trend is clearly visible: electron affinities are highest toward the upper right of the periodic table close to oxygen, sulfur, and the halogens. In these atoms, the incoming electron occupies a *p*-orbital close to a highly charged nucleus and can experience its attraction quite strongly. However, the noble gases have negative electron affinities because any electron that is pushed on to them must occupy an orbital outside a closed shell and far from the nucleus.

Once an electron has entered the single vacancy in the valence shell of a Group 17/VII atom, the shell is complete and any additional electron would have to begin a new shell. In that shell, it would not only be farther from the nucleus but would also feel the repulsion of the negative charge already present. As a result, the second electron affinity of fluorine is

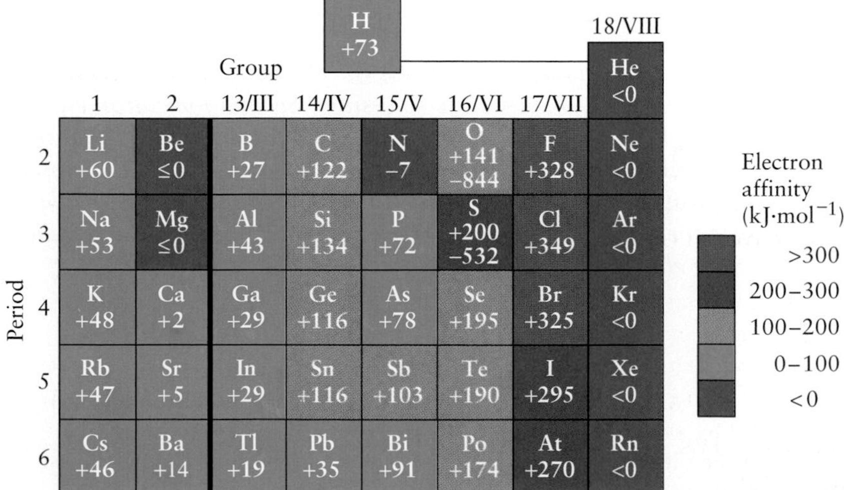

FIGURE 1.49 The variation of electron affinity in kilojoules per mole of the main-group elements. Where two values are given, the first refers to the formation of a singly charged anion and the second is the additional energy needed to produce a doubly charged anion. The negative signs of the second values indicate that energy is required to add an electron to a singly charged anion. The variation is less systematic than for ionization energy, but high values tend to be found close to fluorine (but not for the noble gases).

strongly negative, meaning that a lot of energy has to be expended to form F^{2-} from F^-. Ionic compounds of the halogens are therefore built from singly charged ions such as F^- and never from doubly charged ions such as F^{2-}.

A Group 16/VI atom, such as O or S, has two vacancies in its valence-shell *p*-orbitals and can accommodate two additional electrons. The first electron affinity is positive. However, attachment of the second electron *requires* energy because of the repulsion by the negative charge already present in O^- or S^-. Unlike a halogen, however, the valence shell of the anion has room to accommodate the additional electron. Therefore, we expect less energy to be needed to make O^{2-} from O^- than to make F^{2-} from F^-, where no such vacancy exists (compare diagrams **8** and **9**). In fact, 141 $kJ \cdot mol^{-1}$ is released when the first electron adds to the neutral atom to form O^-, but 844 $kJ \cdot mol^{-1}$ must be supplied to add a second electron to form O^{2-}, so the total energy required to make O^{2-} from O is 703 $kJ \cdot mol^{-1}$. As we shall see in Chapter 2, this energy can be achieved in chemical reactions, and O^{2-} ions are typical of metal oxides. However, no doubly charged anions are stable unless they are surrounded by cations or solvent molecules.

Elements with the highest electron affinities are those in Groups 16/VI and 17/VII. Group 17/VII atoms can acquire one electron with a release of energy and Group 16/VI atoms can accept two electrons with chemically attainable energies; so the halogens typically form X^- ions and Group 16/VI elements typically form X^{2-} ions.

How can we use this information . . . to predict the trend in electron affinity?

From our knowledge of periodic trends, we can suggest a reason for the decrease in electron affinity between carbon and nitrogen. We expect more energy to be released when an electron enters the N atom, because an N atom is smaller than a C atom and its nucleus is more highly charged. That the opposite is observed suggests that electron repulsion in the resulting anions may be important. On forming C^- from C, the additional electron occupies an empty 2*p*-orbital (see **6**). On forming N^- from N, the additional electron must occupy a 2*p*-orbital that is already half full (see 7). Although the nuclear charge of nitrogen is greater than that of carbon, the effective nuclear charge attracting the incoming electron is reduced by repulsion from the electron already in the 2*p*-orbital; hence less energy is released when N^- is formed, and the electron affinity of nitrogen is lower than that of carbon.

Self-Test 1.17A Account for the large decrease in electron affinity between lithium and beryllium.

[*Answer:* The additional electron enters a 2*s*-orbital in Li but a 2*p*-orbital in Be, and a 2*s*-electron is more tightly bound than a 2*p*-electron.]

Self-Test 1.17B Account for the large decrease in electron affinity between fluorine and neon.

FIGURE 1.50 When tin(II) oxide is heated in air, it becomes incandescent as it reacts to form tin(IV) oxide. Even without being heated, it smolders and can ignite.

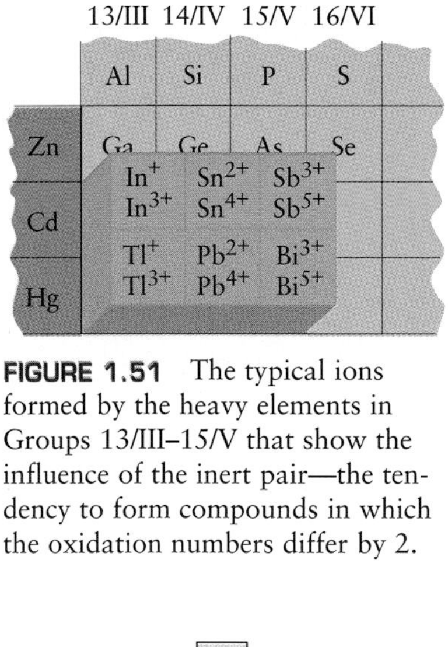

FIGURE 1.51 The typical ions formed by the heavy elements in Groups 13/III–15/V that show the influence of the inert pair—the tendency to form compounds in which the oxidation numbers differ by 2.

1.18 The Inert-Pair Effect

Although aluminum and indium are both in Group 13/III, aluminum forms Al^{3+} ions, whereas indium forms both In^{3+} and In^+ ions. The tendency to form ions two units lower in charge than expected from the group number is called the **inert-pair effect**. Another example of the inert-pair effect is found in Group 14/IV: tin forms tin(IV) oxide when heated in air, but the heavier lead atom loses only its two *p*-electrons and forms lead(II) oxide. Tin(II) oxide can be prepared, but it is readily oxidized to tin(IV) oxide (Fig. 1.50).

The inert-pair effect is due in part to the relative energies of the valence *p*- and *s*-electrons. In the later periods of the periodic table, valence *s*-electrons are very low in energy because of their good penetration and the low shielding ability of the *d*-electrons. The valence *s*-electrons may therefore remain attached to the atom during ion formation. The inert-pair effect is most pronounced among the heaviest members of a group, where the difference in energy between *s*- and *p*-electrons is greatest (Fig. 1.51). Even so, the pair of *s*-electrons can be removed from the atom under sufficiently vigorous conditions. An inert pair would be better called a "lazy pair" of electrons.

The inert-pair effect is the tendency to form ions two units lower in charge than expected from the group number; it is most pronounced for heavy elements in the p block.

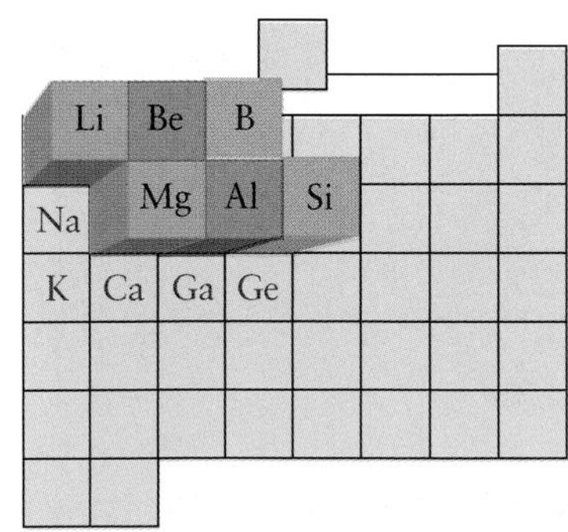

FIGURE 1.52 The pairs of elements represented by colored boxes show a strong diagonal relationship to each other.

1.19 Diagonal Relationships

A **diagonal relationship** is a similarity in properties between diagonal neighbors in the main groups of the periodic table (Fig. 1.52). A part of the reason for this similarity can be seen in Figs. 1.41 and 1.45 by concentrating on the colors that show the general trends in atomic radius and ionization energy. The colored bands of similar values lie in diagonal stripes across the table. Because these characteristics affect the chemical properties of an element, it is not surprising to find that the elements within a diagonal band show similar chemical properties (Fig. 1.53). Diagonal relationships are helpful for making predictions about the properties of elements and their compounds.

The diagonal band of metalloids dividing the metals from the nonmetals is one example of a diagonal relationship (Section B). So is the chemical similarity of lithium and magnesium and of beryllium and aluminum. For example, both lithium and magnesium react directly with nitrogen to form nitrides. Like aluminum, beryllium reacts with both acids and bases. We shall see many examples of this diagonal similarity when we look at the main-group elements in detail (in Chapters 14 and 15).

Diagonally related pairs of elements often show similar chemical properties.

FIGURE 1.53 Boron (top) and silicon (bottom) have a diagonal relationship. They are both brittle solids with high melting points. They also show a number of chemical similarities.

THE IMPACT ON MATERIALS

We are now at the point where we can begin to use the periodic table as chemists and materials scientists do—to predict the properties of elements and see how they can be used to create the materials around us and to design new materials for tomorrow's technologies.

1.20 The Main-Group Elements

An *s*-block element has a low ionization energy, which means its outermost electrons can easily be lost. A Group 1 element is likely to form +1 ions, such as Li^+, Na^+, and K^+. Group 2 elements similarly form +2 ions, such as Mg^{2+}, Ca^{2+}, and Ba^{2+}. An *s*-block element is likely to be a reactive metal with all the characteristics that the name metal implies (Fig. 1.54, Table 1.4). Because ionization energies are lowest at the bottom of

TABLE 1.4 Characteristics of Metals and Nonmetals

Metals	Nonmetals
Physical properties	
good conductors of electricity	poor conductors of electricity
malleable	not malleable
ductile	not ductile
lustrous	not lustrous
typically:	typically:
solid	solid, liquid, or gas
high melting point	low melting point
good conductors of heat	poor conductors of heat
Chemical properties	
react with acids	do not react with acids
form basic oxides	form acidic oxides
(which react with acids)	(which react with bases)
form cations	form anions
form ionic halides	form covalent halides

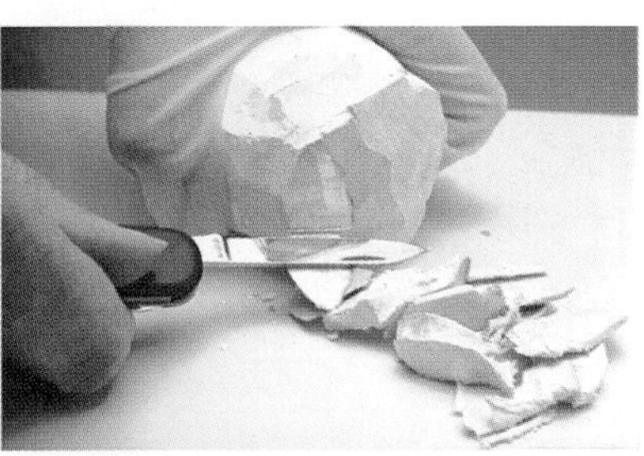

FIGURE 1.54 All the alkali metals are soft, reactive, silvery metals. Sodium is kept under paraffin oil to protect it from air, and a freshly cut surface soon becomes covered with the oxide.

Media Link Lab video: Fig. 1.54, Properties of sodium

FIGURE 1.55 The Group 14/IV elements. From left to right: carbon (as graphite), silicon, germanium, tin, and lead.

FIGURE 1.56 The Group 16/VI elements. From left to right: oxygen, sulfur, selenium, and tellurium. Note the trend from nonmetal to metalloid.

each group, and the elements there lose their valence electrons most easily, the heavy elements cesium and barium react most vigorously of all *s*-block elements. They have to be kept stored out of contact with air and water. The alkali metals have few direct uses as materials but are enormously important as compounds.

Elements on the left of the *p* block, especially the heavier elements, have ionization energies that are low enough for these elements to have some of the metallic properties of the members of the *s* block. However, the ionization energies of the *p*-block metals are quite high, and they are less reactive than those in the *s* block. The useful elements aluminum, tin, and lead all lie in this region of the table. Tin and lead are metals, but, although they are malleable and conduct electricity, they are not nearly as reactive as the *s*-block elements and many *d*-block elements (Fig. 1.55).

Elements at the right of the *p* block have characteristically high electron affinities: they tend to gain electrons to complete closed shells. Except for the metalloids tellurium and polonium, the members of Groups 16/VI and 17/VII are nonmetals (Fig. 1.56). They typically form molecular compounds with each other. They react with metals to form the anions in ionic compounds, and hence many of the minerals that surround us, such as limestone and granite, contain anions formed from nonmetals, such as S^{2-}, CO_3^{2-}, and SO_4^{2-}. Much of the metal industry is concerned with the problem of extracting metals from their combinations with nonmetals.

The elements in the s block are all reactive metals that form basic oxides. The p-block elements tend to gain electrons to complete closed shells; they range from metals through metalloids to nonmetals.

1.21 The Transition Metals

All *d*-block elements are metals (Fig. 1.57). Their properties are transitional between the *s*- and *p*-block elements, which accounts for their alternative name, the *transition metals*. Because transition metals in the same period differ mainly in the number of *d*-electrons, and these electrons are in inner shells, their properties are very similar.

When a *d*-metal atom loses electrons to form a cation, it first loses its outer *s*-electrons. However, most transition metals form ions with different oxidation states, because the *d*-electrons have similar energies and a variable number can also be lost when they form compounds. Iron, for instance, forms Fe^{2+} and Fe^{3+}; copper forms Cu^{+} and Cu^{2+}. The reason for the difference between copper and potassium, which forms only K^{+}, can be understood by comparing their second ionization energies, which are 1958 kJ·mol^{-1} and 3051 kJ·mol^{-1}, respectively. To form Cu^{2+}, an electron is removed from the *d*-subshell of [Ar]$3d^{10}$; but to form K^{2+}, the electron would have to be removed from potassium's argonlike core. Because such huge amounts of energy are not readily available in chemical reactions, a potassium atom can lose only its 4*s*-electron.

The availability of *d*-orbitals, variable valence, and the similarity of the atomic radii of the *d*-block metals have a significant impact on many areas of our lives. The availability of *d*-orbitals is in large measure responsible for the action of transition metals and their compounds as catalysts (substances that accelerate reactions, but are not themselves consumed) throughout the chemical industry. Thus, iron is used in the manufacture of ammonia, nickel is used in the conversion of vegetable oils to shortening, platinum in the manufacture of nitric acid, vanadium(V) oxide in the manufacture of sulfuric acid, and titanium compounds in the manufacture of polyethylene. Variable valence plays a role in catalysis, but it is also very important for facilitating the very subtle changes that take place in organisms. For instance, iron occurs as iron(II) in hemoglobin, the

FIGURE 1.57 The elements in the first row of the *d* block. Top row (left to right): scandium, titanium, vanadium, chromium, and manganese. Bottom row: iron, cobalt, nickel, copper, and zinc.

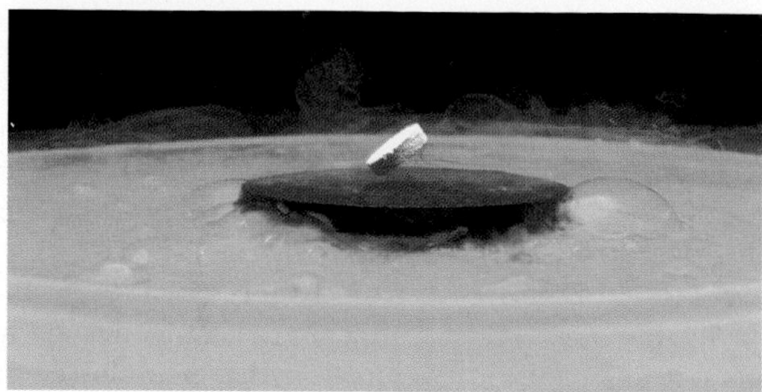

FIGURE 1.58 A sample of a high-temperature superconductor, material first produced in 1987, cooled by liquid nitrogen. Here, a small magnet is levitated by the superconductor. If the assembly were turned over, the magnet would hang at about the same distance below the superconductor.

oxygen-transport protein in mammalian blood; copper is present in the proteins responsible for electron transport; and manganese is present in the proteins responsible for photosynthesis. The similarity of their atomic radii is largely responsible for the ability of transition metals to form the mixtures we know as *alloys*, especially the wide variety of steels that make modern construction and engineering possible.

The lanthanides are rather rare on Earth and so were not used much in earlier technologies. However, today they are studied intensely, because superconducting materials often contain lanthanides (Fig. 1.58). The actinides are all radioactive elements. None of the elements following plutonium occurs naturally on Earth in any significant amount. Because they can be made only in nuclear reactors or particle accelerators, they are available only in small quantities. For example, only a few atoms of element 111 have ever been made.

All d-block elements are metals with properties between those of s-block and p-block metals. Many d-block elements form cations in more than one oxidation state.

SKILLS YOU SHOULD HAVE MASTERED

1 Calculate the wavelength or frequency of light from the relation $\lambda\nu = c$, Section 1.1.

2 Use Wien's law to estimate a temperature, Section 1.2.

3 Use the relation $E = h\nu$ to calculate the energy, frequency, or number of photons emitted from a light source, Example 1.1.

4 Estimate the wavelength of a particle, Example 1.2.

5 Estimate the uncertainty in the location or speed of a particle, Example 1.3.

6 Calculate the energies and describe the wave functions of a particle in a box, Examples 1.4 and 1.5.

7 Explain the origin of the lines in the spectrum of an element and correlate them with specific energy transitions, Example 1.6.

8 Assess the relative probability of finding an electron at a given distance from the nucleus of an atom, Section 1.8.

9 Name and explain the relationship of each of the

four quantum numbers to the properties and relative energies of atomic orbitals, Sections 1.8–1.10.

10 Describe the factors affecting the energy of an electron in a many-electron atom, Section 1.11.

11 Write the ground-state electron configuration for an element, Toolbox 1.1 and Example 1.7.

12 Account for periodic trends in the atomic radii, ionization energies, and electron affinities, Sections 1.14–1.19.

EXERCISES

Observing Atoms

1.1 Arrange the following types of photons of electromagnetic radiation in order of increasing *wavelength*: radio waves, infrared radiation, visible light, ultraviolet radiation.

1.2 Arrange the following types of photons of radiation in order of increasing *speed*: infrared radiation, visible light, x-rays, microwaves.

1.3 Arrange the following types of photons of electromagnetic radiation in order of increasing *energy*: gamma rays, x-rays, visible light, ultraviolet radiation, microwaves.

1.4 Arrange the following types of photons of electromagnetic radiation in order of increasing *frequency*: radio waves, infrared radiation, visible light, ultraviolet radiation.

1.5 (a) Infrared radiation has wavelengths ranging from about 800 nm to 1 mm. What is the frequency of 925-nm radiation? (b) Microwaves, such as those used for radar and to heat food in a microwave oven, have wavelengths greater than about 3 mm. What is the frequency of 4.15-mm radiation?

1.6 (a) Light with a frequency of 7.1×10^{14} Hz lies in the violet region of the visible spectrum. What is the wavelength (in nanometers) of this frequency of light? (b) When an electron beam strikes a block of copper, x-rays with a frequency of 2.0×10^{18} Hz are emitted. What is the wavelength (in picometers) of these x-rays?

1.7 The temperature of molten iron can be monitored by using Wien's law. If the melting point of iron is 1540°C, what will be the wavelength (in nanometers) corresponding to maximum intensity when a piece of iron melts?

1.8 An astronomer discovers a new red star and finds that the maximum intensity occurs at $\lambda = 715$ nm. What is the temperature of the surface of the star?

1.9 Sodium vapor lamps, used for public lighting, emit 589-nm yellow light. How much energy is emitted by (a) an excited sodium atom when it generates a photon; (b) 5.00 mg of sodium atoms emitting light at this wavelength; (c) 1.00 mol of sodium atoms emitting light at this wavelength?

1.10 When an electron beam strikes a block of copper, x-rays with a frequency of 2.0×10^{18} Hz are emitted. How much energy is emitted at this wavelength by (a) an excited copper atom when it generates an x-ray photon; (b) 1.00 mol of excited copper atoms; (c) 1.00 g of copper atoms?

1.11 Most of the gamma radiation emitted by the nuclear decay of a technetium-99 atom used in radiopharmaceuticals has an energy of 140.511 keV. Calculate the wavelength of one photon of this gamma radiation.

1.12 A mixture of argon and mercury vapor used in blue advertising signs emits 470-nm light. Calculate the energy change resulting from the emission of 1.00 mol of photons at this wavelength.

1.13 Consider the following statements about electromagnetic radiation and determine whether they are true or false. If they are false, state why they are false. (a) The total intensity of radiation emitted from a black body at absolute temperature T is directly proportional to the temperature. (b) As the temperature of a black body increases, the wavelength at which the maximum intensity is found decreases. (c) Photons of radio-frequency radiation are higher in energy than photons of ultraviolet radiation.

1.14 Consider the following statements about electromagnetic radiation and determine whether they are true or false. If they are false, state why they are false. (a) Photons of ultraviolet radiation have less energy than photons of infrared radiation. (b) The kinetic energy of an electron ejected from a metal surface when the metal is irradiated with ultraviolet radiation is independent of the frequency of the radiation. (c) The energy of a photon is inversely proportional to the wavelength of the radiation.

1.15 The velocity of an electron that is emitted from a metallic surface by a photon of electromagnetic radiation is 2.2×10^3 km·s^{-1}. (a) What is the wavelength of the electron? (b) No electrons are emitted from the surface of the metal until the frequency of the radiation reaches 1.00×10^{15} Hz. How much energy is required to remove the electron from the metal surface? (c) What was the wavelength of the photon that caused the electron moving at 2.2×10^3 km·s^{-1} to be ejected from the metal? (d) What is the electromagnetic radiation class to which that photon belongs?

1.16 The work function for chromium metal is 4.37 eV. What wavelength of radiation must be supplied to metallic chromium to release an electron with a velocity of 1.5×10^3 km·s^{-1}?

1.17 A baseball must weigh between 5.00 and 5.25 ounces (1 ounce = 28.3 g). What is the wavelength of a 5.15-ounce baseball thrown at 92 mph?

1.18 On a not-so-unusual day, an automobile weighing 1550 kg travels on a German autobahn at 200 km·hr^{-1}. What is the wavelength of the automobile?

1.19 What is the velocity of a neutron that has a wavelength of 100 pm?

1.20 The average speed of a helium atom at 25°C is 1.23×10^3 m·s^{-1}. What is the wavelength associated with the motion of a helium atom at this temperature?

1.21 What is the energy change that occurs when a sodium atom emits a yellow photon with a wavelength of 589 nm? This is commonly called the sodium "D line."

1.22 A particular transition in a mercury atom emits a photon of blue light with a wavelength of 435.8 nm. What is the energy change of the atom?

1.23 (a) Use the Rydberg formula for atomic hydrogen to calculate the wavelength for the transition from $n = 6$ to $n = 2$. (b) What is the name given to the spectroscopic series to which this transition belongs? (c) Use Table 1.1 to determine the region of the spectrum in which the transition occurs. If the change occurs in the visible region of the spectrum, what color will be emitted?

1.24 (a) Use the Rydberg formula for atomic hydrogen to calculate the wavelength for the transition from $n = 5$ to $n = 1$. (b) What is the name given to the spectroscopic series to which this transition belongs? (c) Use Table 1.1 to determine the region of the spectrum in which the transition occurs. If the change occurs in the visible region of the spectrum, what color will be emitted?

1.25 (a) Using the particle-in-the-box model for the hydrogen atom, which considers it to be a one-dimensional box with length 150 pm, predict the wavelength of radiation that will be emitted when the electron falls from the $n = 3$ to $n = 2$ level. (b) Repeat the calculation for an electron moving from the $n = 4$ to $n = 2$ levels.

1.26 (a) What is the highest energy photon that can be absorbed by the hydrogen atom without causing ionization? (b) What is the wavelength of this photon? (c) To what part of the electromagnetic spectrum does this photon belong?

1.27 In the spectrum of atomic hydrogen, several lines are generally classified together as belonging to a series (e.g., Balmer series, Lyman series, Paschen series), as shown in Fig. 1.22. What is common to the lines within a series that makes it logical to group them together?

1.28 In the spectrum of atomic hydrogen, a violet line is observed at 434 nm. Determine the beginning and ending energy levels of the electron during the emission of energy that leads to this spectral line.

Models of Atoms

1.29 (a) Sketch the shape of the boundary surfaces corresponding to the simplest *s*-, *p*-, and *d*-orbitals (1*s*, 2*p*, 3*d*). (b) What is a node as it applies to atomic orbitals? (c) How many *nodal surfaces* does each orbital in (a) have? (d) On the basis of your answer to (c), predict the number of nodal planes expected for a 4*f*-orbital. You may wish to refer to the animation of the atomic orbitals found on the Web site for this text.

1.30 Figure 1.29 showing the *p*-orbital illustrates that a wavefunction changes sign when it crosses a nodal plane. On the basis of this analogy, draw a d_{xy}-orbital and indicate what the signs on the lobes should be.

1.31 Describe the orientation of the lobes of the p_x-, p_y-, and p_z-orbitals with respect to the reference Cartesian axes. You may wish to refer to the animation of the atomic orbitals found on the Web site for this text.

1.32 Describe the difference in orientation of the d_{xy}- and the $d_{x^2-y^2}$-orbitals with respect to the reference

Cartesian axes. You may wish to refer to the animation of the atomic orbitals found on the Web site for this text.

1.33 Evaluate the probability of finding an electron in a small region of a hydrogen 1*s*-orbital at a distance $0.35a_0$ from the nucleus relative to finding it in the same small region located at the nucleus.

1.34 Evaluate the probability of finding an electron in a small region of a hydrogen 1*s*-orbital at a distance $0.65a_0$ from the nucleus relative to finding it in the same small region located at the nucleus.

1.35 How many subshells are there for $n =$ (a) 2; (b) 3? (c) What are the permitted values of l when $n = 3$?

1.36 (a) How many subshells are there for $n = 6$? (b) Identify the subshells by giving the labels associated with them (that is, 6*s*, etc.). (c) How many orbitals (total) are there in the shell with $n = 6$?

1.37 How many orbitals are in subshells with l equal to (a) 0; (b) 2; (c) 1; (d) 3?

1.38 (a) How many *subshells* are there for principal quantum number $n = 5$? (b) Identify the subshells in the form 5*s*, ... (c) How many *orbitals* are there in the shell with $n = 5$?

1.39 (a) How many values of the quantum number l are possible when $n = 7$? (b) How many values of m_l are allowed for an electron in a 6*d*-subshell? (c) How many values of m_l are allowed for an electron in a 3*p*-subshell? (d) How many subshells are there in the shell with $n = 4$?

1.40 (a) How many values of the quantum number l are possible when $n = 6$? (b) How many values of m_l are allowed for an electron in a 5*f*-subshell? (c) How many values of m_l are allowed for an electron in a 2*s*-subshell? (d) How many subshells are there in the shell with $n = 3$?

1.41 What are the principal and orbital angular momentum quantum numbers for each of the following orbitals: (a) 7*s*; (b) 5*f*; (c) 3*d*; (d) 2*p*?

1.42 What are the principal and orbital angular momentum quantum numbers for each of the following orbitals: (a) 2*s*; (b) 6*f*; (c) 4*d*; (d) 5*p*?

1.43 For the orbitals listed in Exercise 1.41, give the possible values for the magnetic quantum number.

1.44 For the orbitals listed in Exercise 1.42, give the possible values for the magnetic quantum number.

1.45 How many electrons total can occupy (a) the 4*p*-orbitals? (b) the 3*d*-orbitals? (c) the 1*s*-orbital? (d) the 4*f*-orbitals?

1.46 How many electrons can occupy a subshell with $l =$ (a) 0; (b) 1; (c) 2; (d) 3?

1.47 Write the subshell notation (3*d*, for instance) and the number of orbitals having these quantum numbers: (a) $n = 5, l = 2$; (b) $n = 1, l = 0$; (c) $n = 6, l = 3$; (d) $n = 2, l = 1$.

1.48 Write the subshell notation (3*d*, for instance) and the number of electrons that can have these quantum numbers if all the orbitals of that subshell are filled: (a) $n = 3; l = 2$; (b) $n = 5, l = 0$; (c) $n = 7, l = 1$; (d) $n = 4; l = 3$.

1.49 Identify the values of the principal quantum number, n, and the orbital angular momentum quantum number, l, for the following subshells: (a) 2*p*; (b) 5*f*; (c) 3*s*; (d) 4*d*.

1.50 Identify the values of the principal quantum numbers, n, and the orbital angular momentum quantum number, l, for the following subshells: (a) 4*s*; (b) 3*d*; (c) 6*p*; (d) 7*f*.

1.51 How many electrons can have the following quantum numbers in an atom: (a) $n = 2, l = 1$; (b) $n = 4, l = 2, m_l = -2$; (c) $n = 2$; (d) $n = 3, l = 2, m_l = +1$?

1.52 How many electrons can have the following quantum numbers in an atom: (a) $n = 3, l = 1$; (b) $n = 5, l = 3, m_l = -1$; (c) $n = 2, l = 1, m_l = 0$; (d) $n = 7$?

1.53 Which of the following subshells cannot exist in an atom: (a) 2*d*; (b) 4*d*; (c) 4*g*; (d) 6*f*?

1.54 Which of the following subshells cannot exist in an atom: (a) 1*p*; (b) 3*f*; (c) 7*p*; (d) 5*d*?

The Structures of Many-Electron Atoms

1.55 (a) Write an expression for the total Coulombic potential energy for a lithium atom. (b) What do the individual terms represent?

1.56 (a) Write an expression for the total Coulombic potential energy for a beryllium atom. (b) If N denotes the number of electrons present in an atom, write a general expression to represent the total number of terms that will be present in the total Coulombic potential energy equation.

1.57 Which of the following statements are true? If false, why are they false? (a) The effective nuclear charge Z_{eff} is independent of the number of electrons present in an atom. (b) Electrons in an *s*-orbital are more effective at shielding electrons in other orbitals from the nuclear charge because an electron in an *s*-orbital can penetrate to the nucleus of the atom. (c) Electrons with $l = 2$ are better at shielding than electrons having $l = 1$. (d) Z_{eff} for an electron in a *p*-orbital is lower than for an electron in an *s*-orbital in the same shell.

1.58 For the electrons on a carbon atom in the ground state, decide which of the following statements are true. If they are false, explain why. (a) Z_{eff} for an electron in a 1*s*-orbital is the same as Z_{eff} for an electron in a 2*s*-orbital. (b) Z_{eff} for an electron in a 2*s*-orbital is the same as Z_{eff} for an electron in a 2*p*-orbital. (c) An electron in the 2*s*-orbital has the same energy as an electron in the 2*p*-orbital. (d) The electrons in the 2*p*-orbitals have spin quantum numbers m_s of opposite sign. (e) The electrons in the 2*s*-orbital have the same value for the quantum number m_s.

1.59 Determine whether the following electron configurations represent the ground state or an excited state of the atom given.

(a) C [↑↓] [↑↓] [↑↓| |] (1*s* 2*s* 2*p*)

(b) N [↑↓] [↑↓] [↑|↓|↑] (1*s* 2*s* 2*p*)

(c) Be [↑↓] [↑] [↑| |] (1*s* 2*s* 2*p*)

(d) O [↑↓] [↑↓] [↑↓|↑|↑] (1*s* 2*s* 2*p*)

1.60 The following *valence-shell configurations* are possible for a neutral atom. What is the element and which configuration represents the ground state?

(a) [↑↓] [↑|↓|] (4*s* 4*p*)

(b) [↑↓] [↑↓| |] (4*s* 4*p*)

(c) [↑] [↑|↑|↑] (4*s* 4*p*)

(d) [↑↓] [↑|↑|] (4*s* 4*p*)

1.61 Of the following sets of four quantum numbers $\{n, l, m_l, m_s\}$, identify the ones that cannot exist for an electron in an atom and explain why not: (a) $\{4, 2, -1, +\frac{1}{2}\}$; (b) $\{5, 0, -1, +\frac{1}{2}\}$; (c) $\{4, 4, -1, +\frac{1}{2}\}$.

1.62 Of the following sets of four quantum numbers $\{n, l, m_l, m_s\}$, identify the ones that cannot exist for an electron in an atom and explain why not: (a) $\{2, 2, -1, +\frac{1}{2}\}$; (b) $\{6, 0, 0, +\frac{1}{2}\}$; (c) $\{5, 4, +5, +\frac{1}{2}\}$.

1.63 The elements Ga, Ge, As, Se, and Br lie in the same period in the periodic table. Write the electron configuration expected for the ground-state atoms of these elements and predict how many unpaired electrons, if any, each element has.

1.64 The elements B, Al, Ga, In, and Tl belong to the same group in the periodic table. Write the electron configuration expected for the ground-state atoms of these elements and predict how many unpaired electrons, if any, each element has.

1.65 What is the ground-state electron configuration expected for each of the following elements: (a) silver; (b) beryllium; (c) antimony; (d) gallium; (e) tungsten; (f) iodine?

1.66 What is the ground-state electron configuration expected for each of the following elements: (a) sulfur; (b) cesium; (c) polonium; (d) palladium; (e) rhenium; (f) vanadium?

1.67 Which elements are predicted to have the following ground-state electron configurations: (a) $[Ar]3d^6 4s^2$; (b) $[Ne]3s^2 3p^3$; (c) $[Xe]4f^{14} 5d^{10} 6s^2 6p^5$; (d) $[Rn]5f^6 7s^2$?

1.68 Which elements are predicted to have the following ground-state electron configurations: (a) $[Ar]3d^{10} 4s^2 4p^1$; (b) $[Ne]3s^1$; (c) $[Kr]5s^2$; (d) $[Xe]4f^7 6s^2$?

1.69 For each of the following ground-state atoms, predict the type of orbital (1*s*, 2*p*, 3*d*, 4*f*, etc.) that an electron will be removed from in order to form the +1 ions: (a) Ge; (b) Mn; (c) Ba; (d) Au.

1.70 For each of the following ground-state atoms, predict the type of orbital (1*s*, 2*p*, 3*d*, 4*f*, etc.) that an electron will be removed from in order to form the +1 ions: (a) Zn; (b) Cl; (c) Al; (d) Cu.

1.71 Predict the number of valence electrons present in each of the following atoms (include the outermost *d*-electrons: (a) P; (b) Al; (c) Te; (d) Fe.

1.72 Predict the number of valence electrons present in each of the following atoms (include the outermost *d*-electrons): (a) Bi; (b) Ba; (c) Mn; (d) Zn.

1.73 How many *unpaired* electrons are predicted for the ground-state configuration of each of the following atoms: (a) Bi; (b) Si; (c) Ta; (d) Ni?

1.74 How many *unpaired* electrons are predicted for the ground-state configuration of each of the following atoms: (a) Pb; (b) Ir; (c) Y; (d) Cd?

1.75 Give the notation for the valence-shell configuration (including the outermost *d*-electrons) of (a) the alkali metals; (b) Group 13/III elements; (c) Group 5 transition metals; (d) "coinage" metals (Cu, Ag, Au) (Group 11).

1.76 Give the notation for the valence-shell configuration (including the outermost *d*-electrons) of (a) the halogens; (b) the chalcogens (the Group 16/VI elements); (c) the transition metals in Group 5; (d) the Group 14/IV elements.

The Periodicity of Atomic Properties

1.77 On the basis of your knowledge of periodicity, place the following sets of elements in order of *decreasing* ionization energy. Explain your choices. (a) selenium, oxygen, tellurium; (b) gold, tantalum, osmium; (c) lead, barium, cesium.

1.78 (a) Generally, the first ionization energies for a *period* of elements increases on going to higher atomic number. Why? (b) Examine the data for the *p*-block elements given in Fig. 1.45. Note any exceptions to the rule given in (a). How are these exceptions explained?

1.79 Look up in Appendix 2D the values for the atomic radii for the first row of transition metals. Explain the trends that are observed as on going from left to right along the period.

Answer the following questions by using the element's location in the periodic table and your knowledge of electronic structure/property relationships.

1.80 Arrange the elements in the following sets in order of *decreasing* atomic radius: (a) sulfur, chlorine, silicon; (b) cobalt, titanium, chromium; (c) zinc, mercury, cadmium; (d) antimony, bismuth, phosphorus.

1.81 Place the following ions in order of *increasing* ionic radius: S^{2-}, Cl^{-}, P^{3-}.

1.82 Which ion of each set has the *larger* radius: (a) Ca^{2+}, Ba^{2+}; (b) As^{3-}, Se^{2-}; (c) Sn^{2+}, Sn^{4+}?

1.83 Which element of each pair has the *higher* electron affinity: (a) oxygen or fluorine; (b) nitrogen or carbon; (c) chlorine or bromine; (d) lithium or sodium?

1.84 Which element of each of the following pairs has the *higher* electron affinity: (a) aluminum or indium; (b) bismuth or antimony; (c) silicon or lead?

1.85 (a) What is a diagonal relationship? (b) How does it arise? (c) Find two examples to illustrate the concept.

1.86 Use Appendix 2D to find the values for the atomic radii of germanium and antimony as well as the ionic radii for Ge^{2+} and Sb^{3+}. What does this suggest about the chemical properties of these two ions?

1.87 Which of the following pairs of elements exhibit a diagonal relationship: (a) N and S; (b) Li and Mg; (c) Ca and Al; (d) F and S?

1.88 Which of the following pairs of elements do *not* exhibit a diagonal relationship: (a) Ga and Si; (b) Be and Al; (c) As and Sn; (d) Na and Ca?

The Impact on Materials

1.89 Why are *s*-block metals more reactive than *p*-block metals?

1.90 Which of the following are transition metals: (a) radium; (b) radon; (c) hafnium; (d) niobium?

1.91 Identify the following elements as metals, nonmetals, or metalloids: (a) lead; (b) sulfur; (c) zinc; (d) silicon; (e) antimony; (f) cadmium.

1.92 Identify the following elements as metals, nonmetals, or metalloids: (a) aluminum; (b) carbon; (c) germanium; (d) arsenic; (e) selenium; (f) tellurium.

INTEGRATED EXERCISES

1.93 Infrared spectroscopy is an important tool for studying vibrations of molecules. Just as an atom can absorb a photon of suitable energy, such as visible light or ultraviolet radiation, to move an electron from one electronic state to another, a molecule can absorb a photon of electromagnetic radiation in the infrared region to move from one *vibrational* energy level to another. In infrared spectroscopy, it is common to express energy in terms of ν/c, with the units cm^{-1} (read as *reciprocal centimeters*). (a) If an absorption occurs in the infrared spectrum at 2143 cm^{-1}, what is the frequency of radiation that corresponds to that absorption? (b) What is the energy, in joules (J), of that absorption? (c) How much energy would be absorbed by 1.00 mol of molecules absorbing at 2143 cm^{-1}?

1.94 *Diffraction* of electromagnetic radiation by atoms and molecules occurs when the wavelength of the electromagnetic radiation is similar to the size of the particle that causes the diffraction, in this case, atoms or molecules. (a) Using 2.0×10^2 pm as the diameter of an atom, decide what type(s) of electromagnetic radiation would give rise to diffraction when passed through a sample of atoms or molecules. Beams of electrons and neutrons can also be used in diffraction experiments because of their high speeds and the de Broglie relation. Calculate the speed of (b) an electron and (c) a neutron that would be necessary to generate wavelengths comparable to the diameter of an atom.

1.95 The ground-state electron configurations for Cr and Cu are not those predicted by the building-up principle. Give their actual electron configurations and explain what causes them to deviate from the anticipated configuration.

1.96 Ionization energies usually increase on going from left to right across the periodic table. The ionization energy for oxygen, however, is lower than that of either nitrogen or fluorine. Explain this anomaly.

Media Link

1.97 Thallium is the heaviest member of Group 13/III. Aluminum is also a member of that group, and its chemical properties are dominated by the +3 oxidation state. Thallium, however, is found most usually in the +1 oxidation state. Examine this issue by plotting the first, second, and third ionization energies for the Group 13/III elements against atomic number (see Appendix 2D or the periodic table data found on the Web site for this text). You may wish to use a graphing program such as the *wfit* routine also on the Web site. Explain any trends you observe.

1.98 The German physicist Lothar Meyer observed a periodicity in the physical properties of the elements at about the same time as Mendeleev was working on their chemical properties. Some of Meyer's observations can be reproduced by examining the molar volume for the solid element as a function of atomic number. Calculate the molar volumes for the elements in Periods 2 and 3 from the densities of the elements found in Appendix 2D and the following solid densities ($g\cdot cm^{-3}$): nitrogen, 0.88; fluorine, 1.11; neon, 1.21. Plot your results as a function of atomic number and discuss any variations you observe.

1.99 The energy required to break a C–C bond in a molecule is 348 $kJ\cdot mol^{-1}$. Will visible light be able to break this bond? If yes, what is the color of that light? If not, what type of electromagnetic radiation will be suitable?

1.100 (a) Explain in very simple terms how light is absorbed by a neutral or ionized atom. (b) What is the phenomenon known as *emission?* (c) The absorption of light by molecules is very similar to that of atoms. Using this knowledge and your answer to (a), suggest how an eye perceives color. Check your suggestions by finding the explanation for color blindness in the literature. Many standard biology, anatomy, physiology, and biochemistry textbooks have explanations of vision, which may be useful. (d) Does your explanation in (c) accommodate the fact that different animals perceive different types of electromagnetic radiation differently, for example, that bees can see ultraviolet radiation? If so, how?

1.101 In the spectroscopic technique known as photoelectron spectroscopy (PES), ultraviolet radiation is directed at an atom or molecule. Electrons are ejected from the valence shell, and their kinetic energies are measured. Because the energy of the incoming ultraviolet photon is known and the kinetic energy of the outgoing electron is measured, the ionization energy, I, can be deduced from the fact that the total energy is conserved. (a) Show that the speed v of the ejected electron and the frequency ν of the incoming radiation are related by

$$E = h\nu = I + \tfrac{1}{2}m_e v^2$$

(b) Use this relation to calculate the ionization energy of a rubidium atom, given that radiation of wavelength 58.4 nm produces electrons with a speed of 2450 $km\cdot s^{-1}$; recall that $1\ J = 1\ kg\cdot m^2\cdot s^{-2}$.

1.102 In the heavier transition series elements, especially the lanthanides and actinides, there are numerous exceptions to the regular filling order predicted by the building-up principle. Suggest an explanation for why more exceptions would be noted for these elements.

1.103 Modern periodic tables sometimes differ in which elements are placed immediately to the right of barium and radium. In some cases, the elements are lanthanum and actinium, whereas others use lutetium and lawrencium. Why?

1.104 Atomic orbitals may be combined to form molecular orbitals. In such orbitals, there is a nonzero probability of finding an electron on any of the atoms that contribute to that molecular orbital. Consider an electron that is confined in a molecular orbital that extends over two adjacent carbon atoms. The electron can move freely between the two atoms. The C—C distance is 139 pm. (a) Using the particle-in-the-box model, calculate the energy required to promote an electron from the $n = 1$ to the $n = 2$ level, assuming that the length of the box is determined by the distance between two carbon atoms. (b) To what wavelength of radiation does this correspond? (c) Repeat the calculation for a linear chain of 1000 carbon atoms. (d) What can you conclude about the energy separation between energy levels as the size of the atom chain increases?

Media Link

1.105 (a) Evaluate the probability (relative to that of the same small region located at the nucleus) of finding an electron in a small region of a hydrogen 2*s*-orbital at distances from 0 to $3a_0$ in increments of $0.10a_0$. (b) With the help of a computer graphing or a spreadsheet program, such as the Function Plotter found on the Web site for this text, plot the data calculated in part (a). (c) Where does the radial node occur?

1.106 The uncertainty principle is negligible for macroscopic objects. Electronic devices, however, are being manufactured on a smaller and smaller scale so that the properties of *nanoparticles*, particles with sizes that range from a few to several hundred nanometers, may be different from those of larger particles as a result of quantum mechanical phenomena. (a) Calculate the minimum uncertainty in the speed of an electron confined in a nanoparticle of diameter 200 nm and compare that uncertainty to the uncertainty in speed of an electron confined to a wire of length 1.00 mm. (b) Calculate the minimum uncertainty in the speed of a Li^+ ion confined in a nanoparticle that has a diameter of 200 nm and is composed of a lithium compound through which the lithium ions can move at elevated temperatures (ionic conductor). (c) Which could be measured more accurately in a nanoparticle, the speed of an electron or the speed of a Li^+ ion?

1.107 Wavefunctions are "normalized." This term means that the total probability of finding an electron in the system is 1. Show that this is true for a particle-in-the-box wavefunction, Eq. 9.

CHAPTER 2

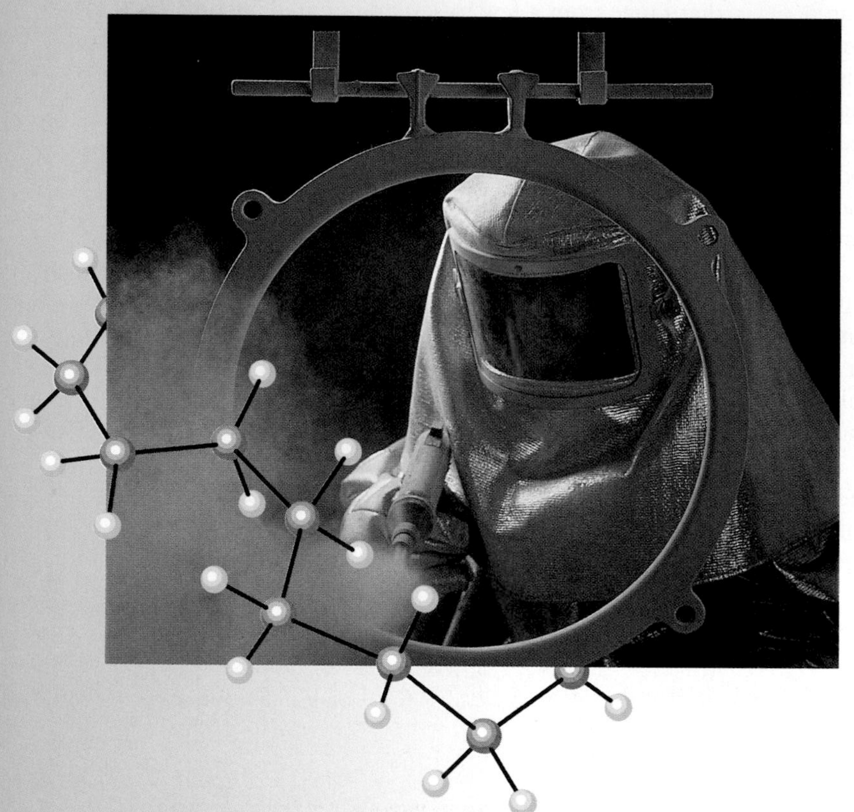

This polytetrafluoroethylene coating is being applied to a gasket that will be used in an industrial chemical reactor. Partly because it has strong bonds, the polymer is very resistant to chemical attack. Only a tiny fragment of the long chain of this polymer molecule is shown here; the actual polymer consists of thousands of $-CF_2CF_2-$ units.

Why Do We Need to Know This Material?

The hundred or so elements in our universe form millions of different compounds by linking to one another in different ways. Chemical bonds are therefore central to chemistry and chemical reactions. In this chapter, we continue our study of individual atoms and learn how they combine. By seeing how bonds form, we come to understand the properties of compounds and see how chemists design new materials. Artificial blood, new pharmaceuticals, agricultural chemicals, and the polymers used to make modern artifacts such as CDs, cellular phones, and synthetic fibers have all become possible because chemists understand how atoms link together in specific ways.

What Do We Need to Know Already?

This chapter assumes that we know about atomic structure and electron configurations (Chapter 1) and the fundamentals of energy (Section A). It is also helpful to be familiar with the nomenclature of compounds (Section D) and oxidation numbers (Section K).

CHEMICAL BONDS

A **chemical bond** is a link between atoms. Chemists understand the properties of matter in terms of the types of bonds that hold atoms together. Why, for example, is calcium phosphate so rigid that nature has adopted it as the inorganic component of bones? Why is the bond in N_2 so strong that it is difficult to make compounds from the nitrogen in air? How can we make lighter airplanes and rockets? Answers to questions like these depend on understanding the different types of bonds and how they can give compounds different properties.

A chemical bond forms between two atoms if the resulting arrangement of the two nuclei and their electrons has a lower energy than the total energy of the separate atoms. If the lowest energy can be achieved by the *complete transfer* of one or more electrons from one atom to another, then ions form and the compound is held together by the attraction between these ions. This attraction is called an **ionic bond.** If the lowest energy can be achieved by *sharing* electrons, then the atoms link through a **covalent bond** and discrete molecules are formed. Sodium and chlorine atoms bond together as ions because solid sodium chloride, which consists of Na^+ and Cl^- ions, has a lower energy than a gas of widely separated sodium and chlorine atoms. Hydrogen and nitrogen atoms bond together to form molecules of ammonia, NH_3, because a gas consisting of NH_3 molecules, in which the nitrogen and hydrogen atoms share electrons, has a lower energy than a gas consisting of the same number of widely separated nitrogen and hydrogen atoms.

Changes in energy occur when bonds form because the **valence electrons** of atoms, the electrons in the outermost shells, are found in new locations. We can therefore expect to explain bond formation in terms of the electronic structures of atoms discussed in Chapter 1. Because the electronic structure of an atom is related to the element's location in the periodic table, we can also expect to be able to predict the type and number of bonds an atom of the element can form from its group and period.

Our current theories of chemical bonding stem from the work of G. N. Lewis, one of the greatest of all chemists. It is even more remarkable that his insights were developed in the early years of the twentieth century, before the electronic structure of atoms was understood.

IONIC BONDS

An ionic bond results from the electrostatic attraction of oppositely charged ions, so we need to understand the energy changes that accompany ion formation and the interactions between ions. Once we know what ions an element is likely to form, we shall be able to predict the formulas of its compounds and explain some of their properties.

2.1 The Formation of Ionic Bonds

Let's consider why a sodium chloride crystal has a lower energy than a gas of widely separated sodium and chlorine atoms. We can picture the formation of the solid as taking place in three steps: sodium atoms release electrons, these electrons attach to chlorine atoms, and then the resulting cations and anions clump together as a crystal. We shall see that chemists often analyze complex processes by breaking them down into simpler steps—and that the steps may be hypothetical (not steps that actually occur in nature).

Sodium is in Group 1 of the periodic table and has the electron configuration [Ne]$3s^1$. This configuration, with one electron outside a closed shell, suggests that a sodium atom will form a +1 ion. However, the valence electron is tightly held by the effective nuclear charge—it does not just fall off: in fact, the ionization energy of sodium is 494 $kJ\cdot mol^{-1}$ (see Fig. 1.45), so we must supply that much energy to form the cations in the process

$$Na(g) \longrightarrow Na^+(g) + e^-(g) \qquad \text{energy required} = 494\ kJ\cdot mol^{-1}$$

A positive electron affinity signifies a release of energy when an electron attaches to a gas-phase neutral atom or ion (Section 1.17).

The electron affinity of chlorine atoms is +349 $kJ\cdot mol^{-1}$ (see Fig. 1.49), so we also know that 349 $kJ\cdot mol^{-1}$ of energy is *released* when electrons attach to chlorine atoms to form anions in the process

$$Cl(g) + e^-(g) \longrightarrow Cl^-(g) \qquad \text{energy released} = 349\ kJ\cdot mol^{-1}$$

At this stage, the net change in energy (energy required − energy released) is 494 − 349 $kJ\cdot mol^{-1}$ = +145 $kJ\cdot mol^{-1}$, an *increase* in energy. We can conclude that a gas of widely separated sodium and chloride ions has a higher energy than a gas of sodium and chlorine atoms. So far, there is no inducement for NaCl to form.

Coulomb potential energy was introduced in Section A.

The missing contribution is the strong **coulombic attraction** (electrostatic attraction) between oppositely charged ions in the solid. When sodium ions and chloride ions come together to form a crystalline solid, their mutual attraction releases a large amount of energy. Experimentally, it is found that

$$Na^+(g) + Cl^-(g) \longrightarrow NaCl(s) \qquad \text{energy released} = 787\ kJ\cdot mol^{-1}$$

Therefore, the net change in energy for the overall process

$$Na(g) + Cl(g) \longrightarrow NaCl(s)$$

is 145 − 787 $kJ\cdot mol^{-1}$ = −642 $kJ\cdot mol^{-1}$ (Fig. 2.1), a huge *decrease* in energy. We conclude that a solid composed of Na^+ and Cl^- ions has a lower energy than a gas of Na and Cl atoms.

We can now begin to understand the formation of ionic bonds and learn when to expect them. There is a net lowering of energy, provided the attraction between ions is greater than the energy needed to make the ions. In practice, the major contribution to the energy requirement is the ionization energy of the element that provides the cation; and, typically, *only metallic elements have ionization energies that are low enough for monatomic cation formation to be energetically feasible*. Broadly speaking, we shall see that metallic elements can lose their valence electrons to form cations and that nonmetallic elements accommodate those electrons in their valence shells and become anions. This **ionic model**, the description of bonding in terms of ions, is particularly appropriate for describing

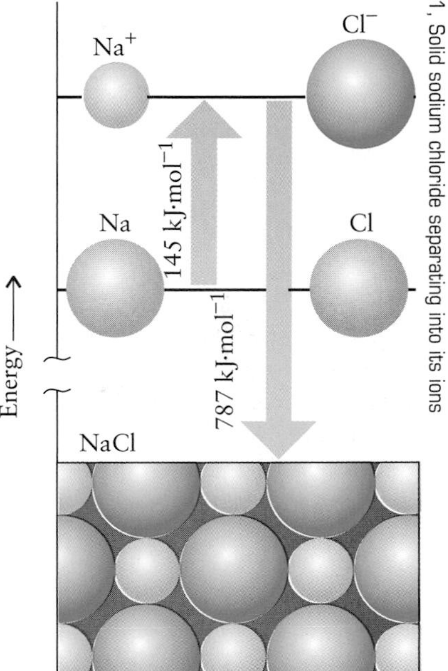

FIGURE 2.1 Considerable energy is needed to produce cations and anions from neutral atoms: the ionization energy of the metal atoms must be supplied, and it is only partly recovered from the electron affinity of the nonmetal atoms. The overall lowering of energy that drags the ionic solid into existence arises from the strong attraction between cations and anions that occurs in the solid.

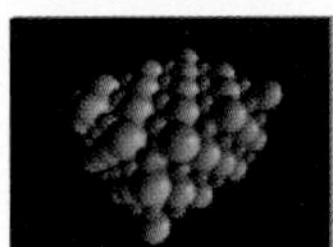

View an animation of a sodium chloride crystal "exploding" into separate sodium and chloride ions.

Media Link Animation: Fig. 2.1, Solid sodium chloride separating into its ions

binary compounds between nonmetallic elements and metallic elements, especially those of the *s*-block metals. An **ionic solid** is an assembly of cations and anions stacked together in a regular array. In the case of sodium chloride, sodium ions alternate with chloride ions, and large numbers of oppositely charged ions are lined up in all three dimensions (Fig. 2.2). Ionic solids are examples of **crystalline solids**, or solids that consist of atoms, molecules, or ions stacked together in a regular pattern.

The energy required for the formation of ionic bonds is supplied largely by the coulombic attraction between oppositely charged ions; the ionic model is a good description of bonding between nonmetals and metals, particularly metals from the s block.

2.2 Interactions Between Ions

We know that ions of opposite charge attract each other and that ions of the same charge repel. Now we express this concept quantitatively and relate the strength of the interaction to the charges on the ions and their radii. Our starting point is the expression for the Coulomb potential energy:

$$E_{P,12} = \frac{(z_1 e) \times (z_2 e)}{4\pi\epsilon_0 r_{12}} = \frac{2.307 z_1 z_2}{r_{12}/\text{nm}} \times 10^{-19}\,\text{J} \qquad (1)$$

In this expression, e is the elementary charge, the absolute value of the charge of an electron (see Section B), z_1 and z_2 are the charge numbers of the two ions (positive for the cation and negative for the anion), r_{12} is the distance between the centers of the ions, and ϵ_0 is the vacuum permittivity (see inside the back cover for the value of this fundamental constant). Our aim is to calculate the **lattice energy** of the solid, the difference in energy between the ions packed together in a solid and widely separated as a gas. The lattice energy is always positive; a high lattice energy indicates that the ions interact strongly with one another to give a tightly bonded solid. In Section 6.20, we shall assess the validity of the ionic model by comparing the calculated lattice energies with experimentally determined values.

FIGURE 2.2 An ionic solid—here we see a fragment of sodium chloride, with the sodium ions represented by pale red spheres and the chloride ions by green spheres—consists of an almost infinite array of cations and anions stacked together to give the lowest energy arrangement. The pattern shown here is repeated throughout the crystal.

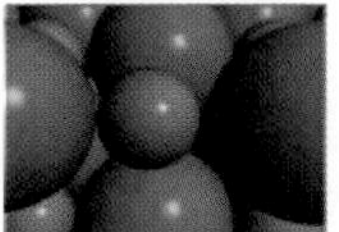

View an animation of a sodium chloride crystal and watch it melt.

Media Link Animation: Fig. 2.2, Sodium chloride crystal melting

Each ion in a solid experiences electrostatic attractions from all the other oppositely charged ions and repulsions from all the other like-charged ions. The total potential energy is the sum of all these contributions. Each cation is surrounded by anions, and there is a large negative (energy lowering) contribution from the attraction of the opposite charges. Beyond those nearest neighbors, there are cations that contribute a positive (repulsive, energy raising) term to the total potential energy of the central cation. There is also a negative contribution from the anions beyond those cations, a positive contribution from the cations beyond them, and so on to the edge of the solid. These repulsions and attractions become progressively weaker as the distance from the central ion increases, but the net outcome of all these contributions is a lowering of energy. We should be able to assess how far the energy is lowered by using the Coulomb potential energy expression in Eq. 1.

How do we do that?

To calculate the change in potential energy when a solid forms from gaseous ions, we consider a simple one-dimensional model of a solid. Imagine a long line of uniformly spaced alternating cations and anions, with d the distance between their centers, the sum of the ionic radii (Fig. 2.3). If the charge numbers of the ions have the same absolute value ($+1$ and -1, or $+2$ and -2, for instance), then $z_1 = +z$, $z_2 = -z$, and $z_1z_2 = -z^2$. The potential energy of the central ion is calculated by summing all the Coulombic potential energy terms, with negative terms representing attractions to oppositely charged ions and positive terms representing repulsions from like-charged ions. For the interaction with ions extending in a line to the right of the central ion, the lattice energy is

$$\begin{aligned} E_P &= \frac{1}{4\pi\epsilon_0} \times \left(-\frac{z^2e^2}{d} + \frac{z^2e^2}{2d} - \frac{z^2e^2}{3d} + \frac{z^2e^2}{4d} - \dots\right) \\ &= -\frac{z^2e^2}{4\pi\epsilon_0 d}\left(1 - \frac{1}{2} + \frac{1}{3} - \frac{1}{4} + \dots\right) \\ &= -\frac{z^2e^2}{4\pi\epsilon_0 d} \times \ln 2 \end{aligned}$$

In the last step, we used the relation $1 - \frac{1}{2} + \frac{1}{3} - \frac{1}{4} + \dots = \ln 2$. Finally, we multiply E_P by 2 to obtain the total energy arising from interactions

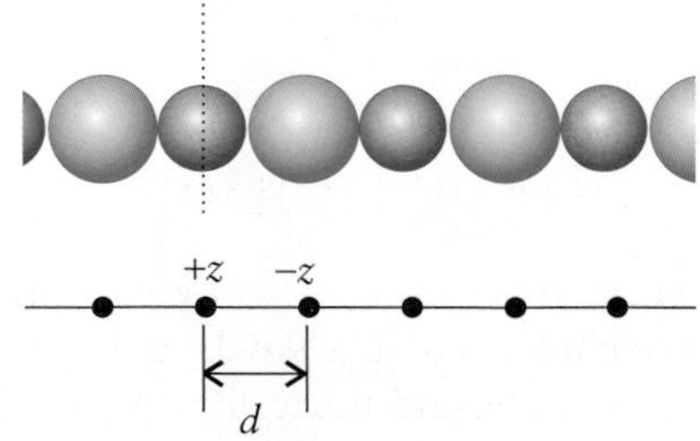

FIGURE 2.3 The arrangement used to calculate the potential energy of an ion in a line of alternating cations and anions. We concentrate on one ion, the "central" ion denoted by the vertical dotted line.

on each side of the ion and then by the Avogadro constant, N_A, to obtain an expression for the lattice energy per mole of ions. The outcome is

$$E_P = -2 \ln 2 \times \frac{z^2 N_A e^2}{4\pi\epsilon_0 d}$$

with $d = r_{cation} + r_{anion}$. This energy is negative, corresponding to a net attraction. The calculation we have just performed can be extended to three-dimensional arrays of ions with different charges:

$$E_P = -A \times \frac{|z_1 z_2| N_A e^2}{4\pi\epsilon_0 d} \quad (2)$$

The symbol $|x|$ means the "absolute value" of x, the value of x without its sign; so $|-2| = 2$.

The factor A is a positive numerical constant called the **Madelung constant**; its value depends on how the ions are arranged about one another. For ions arranged in the same way as in sodium chloride, $A = 1.748$.

Whenever you encounter a new equation, it is good practice to try to interpret what it says, for in that way you can develop insight and deeper understanding. In this case, Eq. 2 tells us that the potential energy is strongly negative when the ions are highly charged (large values of z) and the separation between them is small (small values of d). For example, there is a strong interaction between the Mg^{2+} and O^{2-} ions in magnesium oxide, MgO, because the ions have high charges and small radii. This strong interaction is one reason why magnesium oxide survives at such high temperatures that it can be used for furnace linings. It is an example of a "refractory" material, a substance that can withstand high temperatures.

EXAMPLE 2.1 Estimating the relative lattice energies of solids

The ionic solids NaCl and KCl form the same type of crystal structure. In which solid are the ions bound together more strongly by coulombic interactions?

STRATEGY We can decide in which compound the ions bind together more strongly by taking the ratio of E_P for the two compounds. The two solids have the same crystal structures, so they have the same values of the Madelung constant A. Form the ratio of the two expressions for the potential energy, using Eq. 2.

SOLUTION All the constants cancel when the ratio is taken, including the ionic charges, and we are left with

$$\frac{E_P(\text{NaCl})}{E_P(\text{KCl})} = \frac{d(\text{KCl})}{d(\text{NaCl})}$$

Because the sodium ion has a smaller radius, $d(\text{NaCl}) < d(\text{KCl})$, we can conclude that $|E_P(\text{NaCl})|$ is larger than $|E_P(\text{KCl})|$ and, therefore, that the ions are bound more strongly in NaCl than in KCl.

Self-Test 2.1A The ionic solids CaO and KCl crystallize to form structures of the same type. In which compound are the interactions between the ions stronger?

[*Answer:* CaO, higher charges and smaller radii]

Self-Test 2.1B The ionic solids KBr and KCl crystallize to form structures of the same type. In which compound are the interactions between ions stronger?

The strong coulombic attraction between oppositely charged ions accounts for the typical properties of ionic solids, such as their high melting points and their brittleness. A high temperature is required before the ions are able to move past one another to form a liquid. Ionic solids are brittle because of the same strong attractions and repulsions. We cannot just push a block of ions past another block; instead, when we strike an ionic solid, it shatters into fragments (Fig. 2.4). We can now see why nature has adopted an ionic solid, calcium phosphate, for our skeletons: The doubly charged small Ca^{2+} ions and the triply charged PO_4^{3-} ions attract one another very strongly and clump together tightly to form a rigid, insoluble solid (Fig. 2.5).

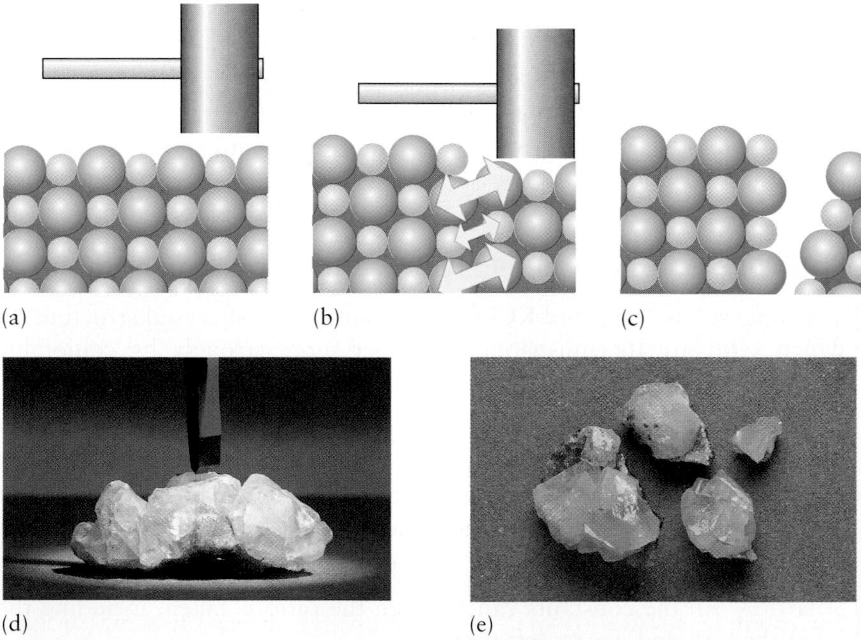

FIGURE 2.4 This sequence of images illustrates why ionic solids are brittle. (a) The original solid consists of an orderly array of cations and anions. (b) A hammer blow can push the ions into positions where cations are next to cations and anions are next to anions; there are now strong repulsive forces acting (as depicted by the double-headed arrows). (c) As a result of these repulsive forces, the solid breaks apart into fragments. (d) This rock consists of several large crystals joined together. (e) The blow of a hammer has shattered the crystal, leaving flat, regular surfaces consisting of planes of ions.

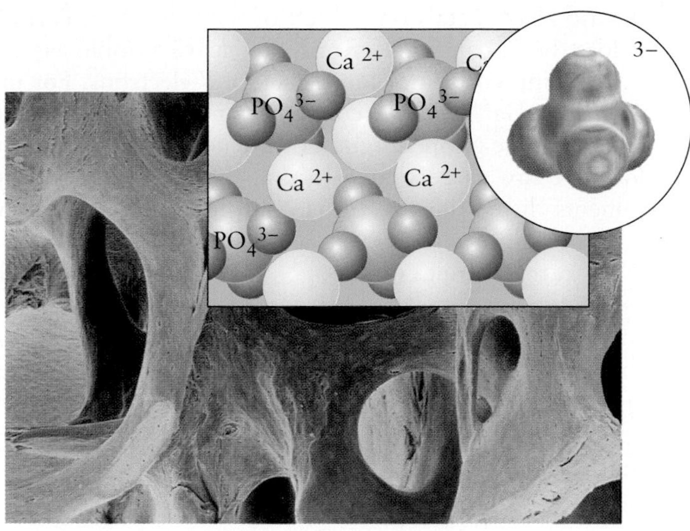

FIGURE 2.5 A micrograph of bone, which owes its rigidity to calcium phosphate. The overlay shows part of the crystal structure of calcium phosphate. Phosphate ions are polyatomic ions; however, as shown in the inset, they are nearly spherical and fit into crystal structures in much the same way as monatomic ions of charge −3.

A conclusion we can draw from the fact that ionic bonds depend on the opposite charges of ions is that *ionic bonds are not directional.* We should think of an ion as bound to all its neighbors by their mutual coulombic attraction; we should not think of an identifiable ion-ion bond between a particular pair of ions. A given chloride ion in sodium chloride is attracted to all the sodium ions around it, not to just one.

Ions stack together in regular crystalline structures. Ionic solids typically have high melting points and are brittle. The coulombic interaction between ions in a solid is large when the ions are small and highly charged.

2.3 The Electron Configurations of Ions

Our next task is to decide on the likely binary formulas of the ionic compounds that may be formed by metals and nonmetals. The key lies in the electronic structures of ions.

An atom becomes a cation when it loses one or more electrons. When an atom of a metallic element from the *s* block forms a cation, it loses electrons down to its noble-gas core (Fig. 2.6). In general, that core has an ns^2np^6 outer electron configuration, which is called an **octet** of electrons. For example, sodium ($[Ne]3s^1$) forms Na^+, which has the same electron configuration as neon ($[He]2s^22p^6$). The Na^+ ions cannot lose more electrons in a chemical reaction because the ionization energies of core electrons are too high to be recovered from ion-ion attractions. Hydrogen loses an electron to form a bare proton. Lithium ($[He]2s^1$) and beryllium ($[He]2s^2$) lose electrons to form a heliumlike **duplet,** a pair of electrons with the configuration $1s^2$, when they become Li^+ and Be^{2+}. Aluminum loses three valence electrons to form the Al^{3+} ion, which has the same electron configuration as neon.

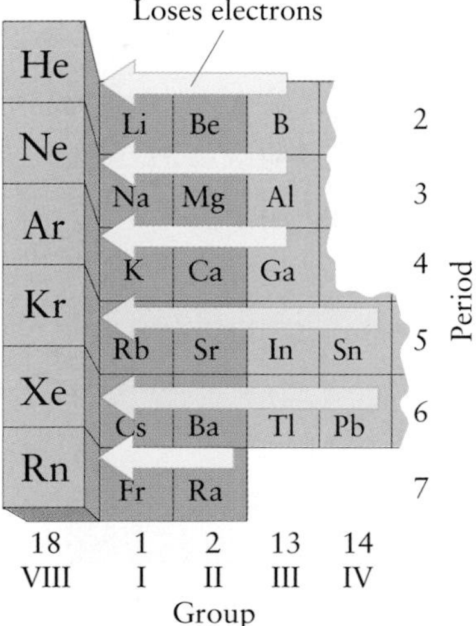

FIGURE 2.6 When a main-group metal atom forms a cation, it loses its valence *s*- and *p*-electrons and acquires the electron configuration of the preceding noble-gas atom. The heavier atoms in Groups 13/III and 14/IV behave similarly, even though their cores consist of the noble-gas configuration and an additional complete subshell of *d*-electrons.

When the atoms of metals on the left of the *p* block in Period 4 and later periods lose their *s*- and *p*-electrons, they leave a noble-gas core surrounded by an additional, complete subshell of *d*-electrons. For instance, gallium forms the ion Ga^{3+} with the configuration $[Ar]3d^{10}$. The *d*-electrons of the *p*-block atoms are gripped tightly by the nucleus and, in most cases, cannot be lost. We saw in Section 1.18 that the inert-pair effect implies that the elements listed in Fig. 1.51 can lose either their valence *p*-electrons alone or all their valence *p*- and *s*-electrons.

In the *d* block, the energies of the $(n-1)d$-orbitals lie below those of the *ns*-orbitals. Therefore, the *ns*-electrons are lost first, followed by a variable number of $(n-1)d$-electrons. For example, to obtain the configuration of the Fe^{3+} ion, we start from the configuration of the Fe atom, which is $[Ar]3d^6 4s^2$, and remove three electrons from it. The first two electrons removed are 4*s*-electrons. The third electron comes from the 3*d*-subshell, giving $[Ar]3d^5$.

1 In $[Kr]4d^{10}5s^2 5p^1$

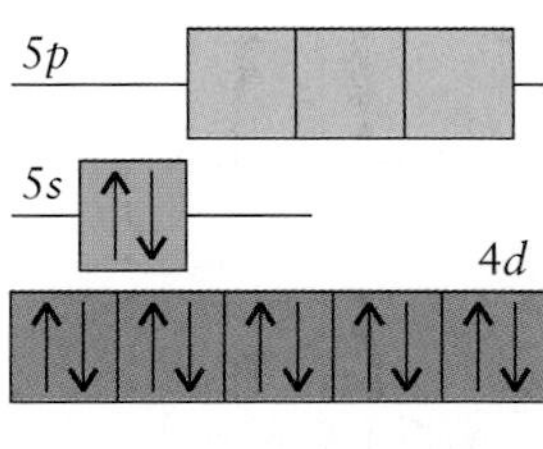

2 In^+ $[Kr]4d^{10}5s^2$

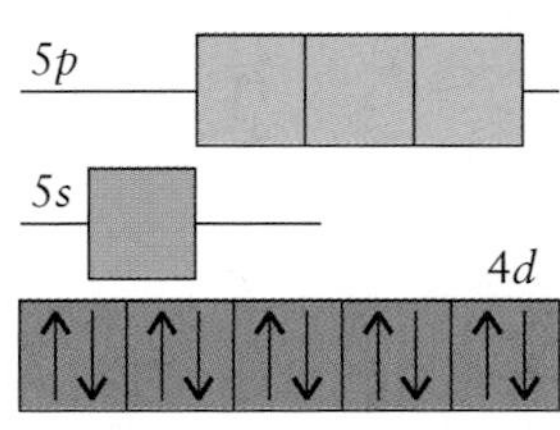

3 In^{3+} $[Kr]4d^{10}$

EXAMPLE 2.2 Writing the electron configurations of cations

Write the electron configurations of the In^+ and In^{3+} ions.

STRATEGY Determine the configuration of the neutral atom. Remove electrons from the valence shell *p*-orbitals first, then from the *s*-orbitals, and finally, if necessary, from the *d*-orbitals in the next lower shell, until the number of electrons removed equals the charge on the ion.

SOLUTION Indium, in Group 13/III, has three valence electrons. It is in Period 5, so its valence shell has $n = 5$. Indium is preceded by the Period 4 noble gas krypton (so indium's noble-gas core is [Kr] and it has a filled 4*d*-subshell). Its ground-state configuration is therefore $[Kr]4d^{10}5s^2 5p^1$ (**1**). One electron can be lost from the 5*p*-orbital, so In^+ is $[Kr]4d^{10}5s^2$ (**2**). The next two electrons are lost from the 5*s*-orbital, so In^{3+} is $[Kr]4d^{10}$ (**3**).

Self-Test 2.2A Write the electron configurations of the copper(I) and copper(II) ions.

[*Answer:* $[Ar]3d^{10}$, $[Ar]3d^9$]

Self-Test 2.2B Write the electron configurations of (a) the manganese(II) ion and (b) the lead(IV) ion.

Nonmetals rarely lose electrons in chemical reactions because their ionization energies are too high. However, they can acquire enough electrons to complete their valence shell and form an anion with an octet corresponding to the configuration of the following noble gas (or $1s^2$ in the case of the hydride ion, H^-), Fig. 2.7. They do not gain more electrons because any additional electrons would have to be accommodated in a higher energy shell. Even the completion of an octet may require considerable energy. For instance, we saw in Section 1.17 that formation of an O^{2-} ion from an O atom requires 703 $kJ \cdot mol^{-1}$ (see Fig. 1.49). The energy needed to achieve a complete octet comes from the favorable ion-ion interactions in the ionic solid. A free O^{2-} ion is unstable and exists in oxides only as a result of its interaction with its positively charged neighbors.

To form monatomic anions, we add enough electrons to complete the valence shell. For example, nitrogen has five valence electrons (**4**), so three more electrons are needed to reach a noble-gas configuration, that of neon. Therefore, the ion will be N^{3-} (**5**).

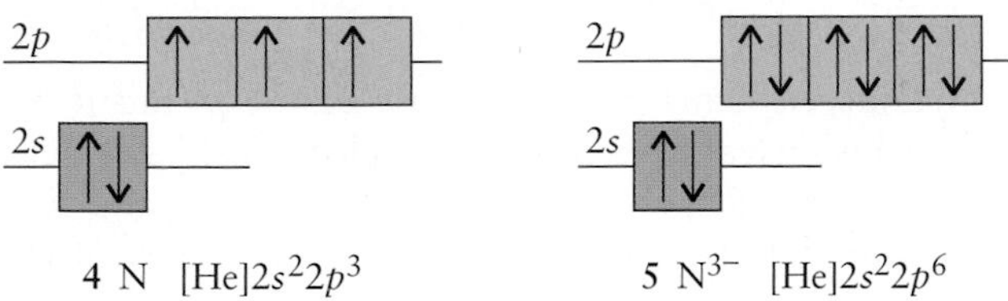

To predict the electron configuration of a monatomic cation, remove outermost electrons in the order np, ns, and (n − 1)d; for a monatomic anion, add electrons until the next noble-gas configuration has been reached. The transfer of electrons results in the formation of an octet (or duplet) of electrons in the valence shell on each of the atoms: metals achieve an octet (or duplet) by electron loss and nonmetals achieve it by electron gain.

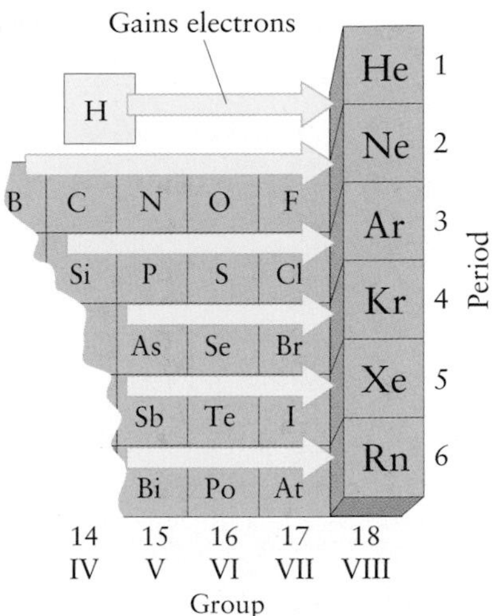

FIGURE 2.7 When nonmetals acquire electrons and form anions, they do so until they have reached the electron configuration of the next noble gas.

Self-Test 2.3A Predict the chemical formula and electron configuration of the phosphide ion.

[*Answer:* P^{3-}, [Ne]$3s^23p^6$]

Self-Test 2.3B Predict the chemical formula and electron configuration of the iodide ion.

2.4 Lewis Symbols

Lewis devised a simple way to keep track of valence electrons when atoms form ionic bonds. He represented each valence electron as a dot and arranged the dots around the symbol of the element. A single dot represents an electron alone in an orbital; a pair of dots represents two paired electrons sharing an orbital. Examples of the **Lewis symbols** of atoms are

H·   He: :N·   ·O·   :Cl·   K·   Mg:

The Lewis symbol for nitrogen, for example, represents the valence electron configuration $2s^22p_x^{\,1}2p_y^{\,1}2p_z^{\,1}$ (see **4**), with two electrons paired in a 2*s*-orbital and three unpaired electrons in different 2*p*-orbitals. The Lewis symbol is a visual summary of the valence-shell electron configuration of an atom and allows us to see what happens to the electrons when an ion forms.

To work out the formula of an ionic compound by using Lewis symbols, we first represent the cation by removing the dots from the symbol for the metal atom. Then we represent the anion by transferring those dots to the Lewis symbol for the nonmetal atom to complete its valence shell. We may need to adjust the numbers of atoms of each kind so that all the dots removed from the metal atom symbols are accommodated by the nonmetal atom symbols. Finally, we write the charge of each ion as a superscript in the normal way. A simple example is the formula of calcium chloride:

$$:\ddot{\underset{..}{Cl}}\cdot + Ca: + :\ddot{\underset{..}{Cl}}\cdot \longrightarrow [:\ddot{\underset{..}{Cl}}:]^- Ca^{2+} [:\ddot{\underset{..}{Cl}}:]^-$$

Many metallic elements, such as those in the *p* and *d* blocks, have atoms that can lose a variable number of electrons. These elements can form different compounds, such as tin(II) oxide, SnO, and tin(IV) oxide, SnO_2, for tin. The ability of an element to form a variety of ions is called **variable valence.**

Formulas of compounds consisting of the monatomic ions of main-group elements can be predicted by assuming that cations have lost all their valence electrons and anions have gained electrons in their valence shells until each ion has an octet of electrons, or a duplet in the case of H, Li, and Be. Atoms of many d-block and p-block metals display variable valence.

COVALENT BONDS

Nonmetallic elements do not typically form monatomic cations because their ionization energies are too high. The nature of bonds between atoms of nonmetals puzzled scientists until 1916, when Lewis published his explanation. With brilliant insight, and before anyone knew about quantum mechanics or orbitals, Lewis proposed that a **covalent bond** is a pair of electrons shared between two atoms. The rest of this chapter and the next develop Lewis's vision of the covalent bond. In this chapter, we consider the types, numbers, and properties of bonds that can be formed by sharing pairs of electrons. In the following chapter, we revisit Lewis's concept and see how to understand it in terms of orbitals.

2.5 The Nature of the Covalent Bond

If you were asked to produce at ordinary temperatures a sample consisting of the free atoms of an element, your only choice would be a Group 18/VIII element, one of the noble gases. All the other elements occur with their atoms linked together in some way. The nonmetallic elements exist as molecules, such as the diatomic species H_2, N_2, O_2, F_2, Cl_2, Br_2, and I_2, and the polyatomic species P_4 and S_8. Elements near the border between metals and nonmetals can form solids with an extended network of atoms, such as crystalline silicon and the graphite or diamond forms of carbon. Covalent bonding is also important in the formation of compounds. There are countless examples of diatomic, polyatomic, and extended network compounds between different nonmetallic elements, including the millions of organic compounds.

Atoms of all metallic elements are held in the solid or liquid state by metallic bonds, which are discussed in Section 5.14.

Lewis thought of the shared pair of electrons as lying between the two atoms and interacting with both their nuclei (**6**). The two atoms are bound together as a result of the coulombic attraction between the two electrons and the nuclei. Neither atom has to release an electron totally, so neither atom has to be supplied with its full ionization energy.

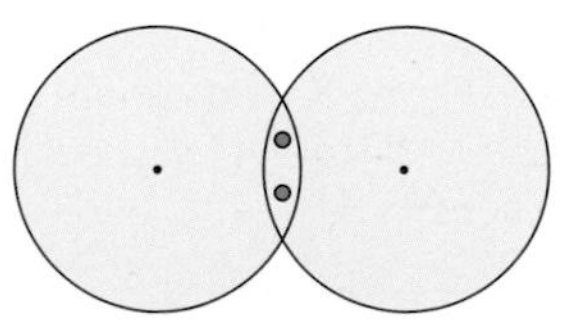

6 Shared electron pair

Nonmetallic elements form covalent bonds to one another by sharing pairs of electrons.

2.6 Lewis Structures

When an ionic bond forms, one atom loses electrons and the other gains them until both atoms have reached a noble-gas configuration—a duplet for elements close to helium and an octet for all other elements. The same idea can be extended to covalent bonds. However, when a covalent bond forms, atoms *share* electrons until they reach a noble-gas configuration. Lewis called this principle the **octet rule:** *in covalent bond formation, atoms go as far as possible toward completing their octets by sharing electron pairs.* For example, nitrogen (:Ṅ̤·) has five valence electrons and needs three more electrons to complete its octet. Chlorine (:C̤̈l·) has seven valence electrons and needs one more electron to complete its octet. Argon (:Ä̤r:) already has a complete octet and has no tendency to share any more electrons. Hydrogen (H·) needs one more electron to reach its heliumlike duplet. Because hydrogen completes its duplet by sharing one pair of electrons, we say that it has a valence of 1 in all its compounds. In general, the **valence** of an atom is the number of bonds that it can form.

We can extend the Lewis symbols we introduced earlier to describe covalent bonding by using a line (—) to represent a shared pair of electrons. For example, the hydrogen molecule formed from two H· atoms is represented by the symbol H—H. A fluorine atom has seven valence electrons and needs one more to complete its octet. It can achieve an octet by accepting a share in an electron supplied by another atom, such as another fluorine atom:

:F̤̈· + ·F̤̈: ⟶ (:F̤̈(:)F̤̈:), or :F̤̈—F̤̈:

The circles drawn around each F atom show how each one gets an octet by sharing one electron pair. The valence of fluorine is therefore 1, the same as that of hydrogen.

A fluorine molecule possesses **lone pairs** of electrons, that is, pairs of valence electrons that are not involved in bonding. The three lone pairs on each of the bonded F atoms repel one another, and this repulsion is almost enough to overcome the favorable attractions of the bonding pair that holds the atoms together in F_2. This repulsion is one of the reasons why fluorine gas is so reactive: the atoms are bound together as F_2 molecules only very weakly. Among the common diatomic molecules, only H_2 has no lone pairs.

The **Lewis structure** of a molecule shows the atoms by their chemical symbols, the covalent bonds by lines, and the lone pairs by pairs of dots. For example, the Lewis structure of HF is H—F̤̈: . We shall see that Lewis structures are a great help in understanding the properties of molecules, including their shapes and the reactions they can undergo.

It is sometimes necessary to write a Lewis structure at the end of a sentence or clause: be careful to distinguish electrons from periods and colons!

Nonmetal atoms share electrons until each has completed its octet (or duplet); a Lewis structure shows the arrangement of electrons as dashes (bonding pairs) and dots (lone pairs).

Self-Test 2.4A Write the Lewis structure for the "interhalogen" compound chlorine monofluoride, ClF, and state how many lone pairs each atom possesses in the compound.

[*Answer:* $:\ddot{\underset{..}{Cl}}-\ddot{\underset{..}{F}}:$; three on each atom]

Self-Test 2.4B Write the Lewis structure for the compound HBr and state how many lone pairs each atom in the compound possesses.

2.7 Lewis Structures for Polyatomic Species

Each atom in a polyatomic molecule completes its octet (or duplet for hydrogen) by sharing pairs of electrons with its immediate neighbors. Each shared pair counts as one covalent bond and is represented by a line between the two atoms. A Lewis structure does not portray the shape of a molecule: it simply displays which atoms are bonded together and which atoms have lone pairs.

Let's write the Lewis structure for the simplest organic molecule, the hydrocarbon methane, CH_4. First we count the valence electrons available from all the atoms in the molecule. For methane, the Lewis symbols of the atoms are

$$:\dot{\underset{\cdot}{C}} \quad H\cdot \quad H\cdot \quad H\cdot \quad H\cdot$$

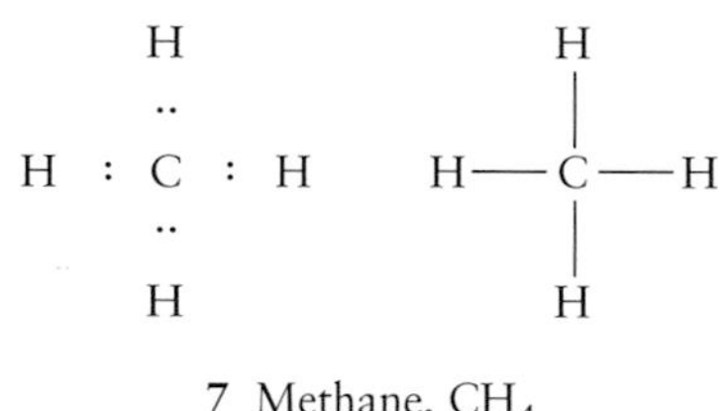

7 Methane, CH_4

so there are eight valence electrons. The next step is to arrange the dots representing the electrons so that the C atom has an octet and each H atom has a duplet. We draw the arrangement shown on the left in (7). The Lewis structure of methane is then drawn as shown on the right in (7). Because the carbon atom is linked by four bonds to other atoms, we say that carbon is *tetravalent*: it has a valence of 4. We emphasize once again that a Lewis structure does not show the shape of the molecule, just the patterns of bonds. In Chapter 3, we shall see that the three-dimensional arrangement of bonds in a methane molecule is, in fact, tetrahedral.

A single shared pair of electrons is called a **single bond.** However, two atoms can share two or three electron pairs. Two electron pairs shared between two atoms are a **double bond,** and three shared electron pairs are a **triple bond.** A double bond, such as C::O is written C=O in a Lewis structure. Similarly, a triple bond, such as C:::C is written C≡C. Double and triple bonds are collectively called **multiple bonds.** As in the case of a single bond, each line represents a pair of electrons. Thus, a double bond involves a total of four electrons; a triple bond involves six electrons. The **bond order** is the number of bonds that link a specific pair of atoms. So, the bond order in H_2 is 1; in the group C=O, it is 2; and for C≡C in a molecule like ethyne, the bond order is 3.

To write a Lewis structure, we need to know which atoms are linked together in a molecule. A "terminal" atom is bonded to only one other atom; an H atom in methane, CH_4, is an example. A "central" atom is bonded to at least two other atoms. An example of a central atom is the O atom in a water molecule, H_2O, or the C atom in methane. The general structure of the molecule and the identity of the central atom is often known in advance. If there is doubt, a good rule of thumb is to *choose as the central atom the element with the lowest ionization energy*. This

In Section 2.14, when we have met the concept of electronegativity, we shall see that another way to express this rule is to say that the central atom is usually the element with lowest electronegativity.

arrangement often leads to the lowest energy because an atom in the central position shares more of its electrons than an atom in a terminal position. The atoms with higher ionization energies are more reluctant to share and are more likely to hold on to their electrons as lone pairs. Except in the strange compounds called the boranes (Section 14.16), hydrogen is never central, because it can form only one bond.

Another rule of thumb for predicting the structure of a molecule is to *arrange the atoms symmetrically around the central atom.* For instance, SO_2 is OSO, not SOO. One common exception to this rule is dinitrogen monoxide, N_2O (nitrous oxide), which has atoms in the asymmetrical arrangement NNO. Another clue is that, in simple chemical formulas, the central atom is often written first, followed by the atoms attached to it. For example, in the compound with the chemical formula OF_2, the arrangement of the atoms is actually FOF, not OFF; and in SF_6, the S atom is surrounded by six F atoms. Water, H_2O, and hydrogen sulfide, H_2S, are two common exceptions, for they are HOH and HSH, respectively. If the compound is an oxoacid, then the acidic hydrogen atoms are attached to oxygen atoms, which in turn are attached to the central atom. For example, sulfuric acid, H_2SO_4, has the structure $(HO)_2SO_2$, and hypochlorous acid, HClO, has the structure HOCl.

Polyatomic ions, such as the ammonium ion, NH_4^+, or the sulfate ion, SO_4^{2-}, consist of atoms held together with covalent bonds. We use the same general procedure to determine the Lewis structures of polyatomic ions. We arrange the atoms in the correct order, count the dots available for bonds and lone pairs, and then construct the Lewis structure. When considering oxoanions, it is usually the case that the atom written first in the chemical formula is the central atom. In SO_4^{2-}, for instance, the S atom is surrounded by four O atoms. Each atom provides the number of dots (electrons) equal to the number of electrons in its valence shell, but we have to adjust the total number of dots to represent the overall charge. For a cation, we subtract one dot for each positive charge. For an anion, we add one dot for each negative charge. The cation and the anion must be treated separately: they are individual ions and are not linked by shared pairs. The Lewis structure of ammonium sulfate, $(NH_4)_2SO_4$, for instance, is written as three bracketed ions (8).

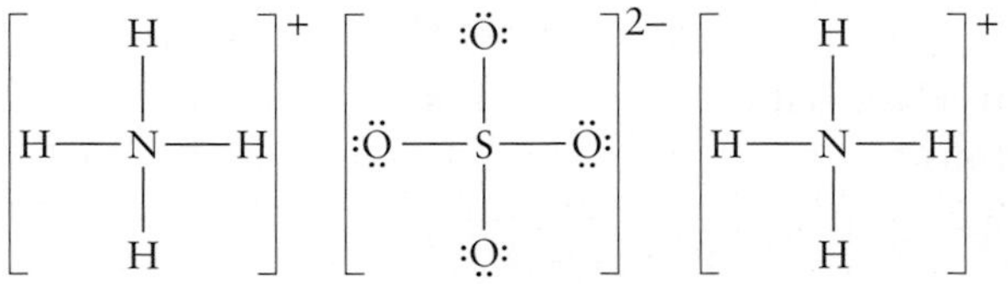

8 Ammonium sulfate, $(NH_4)_2SO_4$

The Lewis structure of a polyatomic species is obtained by using all the valence electrons to complete the octets (or duplets) of the atoms present by forming single or multiple bonds and leaving some electrons as lone pairs.

TOOLBOX 2.1 How to write the Lewis structure of a polyatomic species

Conceptual Basis

We look for ways of using all the valence electrons to complete the octets (or duplets).

Procedure

We take the following steps, which are illustrated for HCN in Fig. 2.8:

Step 1 Count the number of valence electrons on each atom. Divide the total number of valence electrons in the molecule by 2 to obtain the number of electron pairs.

Step 2 Predict the most likely arrangements of atoms by using common patterns and the clues indicated in the text, then write the chemical symbols of the atoms to show their layout in the molecule.

Step 3 Place one electron pair between each pair of bonded atoms.

Step 4 Complete the octet (or duplet, in the case of H) of each atom by placing any remaining electron pairs around the atoms. If there are not enough electron pairs, form multiple bonds.

Step 5 Represent each bonded electron pair by a line.

To check on the validity of a Lewis structure, verify that each atom has an octet or a duplet.

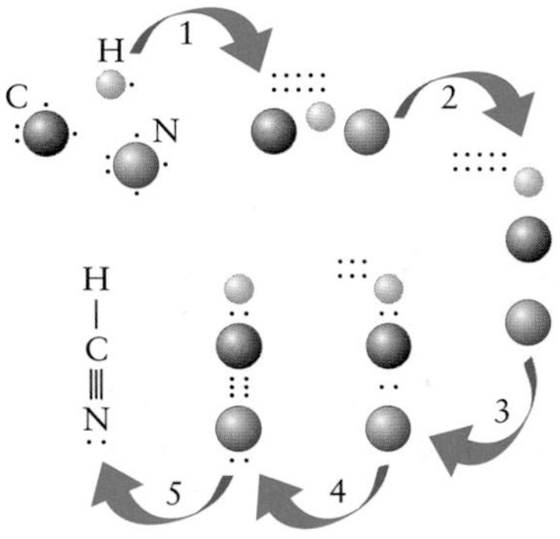

FIGURE 2.8 The five steps used to write the Lewis structure of HCN, as described in this Toolbox.

EXAMPLE 2.3 Writing the Lewis structure of a molecule or an ion

Write the Lewis structures of (a) water, H_2O; (b) methanal, H_2CO; and (c) the chlorite ion, ClO_2^-.

STRATEGY We can write the Lewis structures of molecules and ions by following the steps in Toolbox 2.1 and remembering to adjust the number of electrons for ions. Therefore, we add one electron for the negative charge of ClO_2^-. Find the central atom by inspecting formulas. In methanal, the central atom must be C, because it has a central position and the H atoms can each form only one bond.

SOLUTION

	(a) H_2O	(b) H_2CO	(c) ClO_2^-
1 Number of valence electrons	8	12	20
Number of electron pairs	4	6	10
2 Layout of structure	H O H	H C O H	O Cl O
3 Electron pairs between bonded atoms	H : O : H	H C : O H	O:Cl:O
Electrons not yet located	: :	: : :	: : : : : : : :

4 Lone pairs and multiple bonds	H : Ö : H (lone pairs above and below O)	$H_2C::O$ (H above and below C; lone pairs on O)	:Ö:Cl:Ö: (lone pairs on O, Cl, O)
5 Lines to represent bonds	H—Ö—H	$H_2C{=}O$ (C bonded to H above and below; lone pairs on O)	:Ö—Cl—Ö: (lone pairs on O, Cl, O)

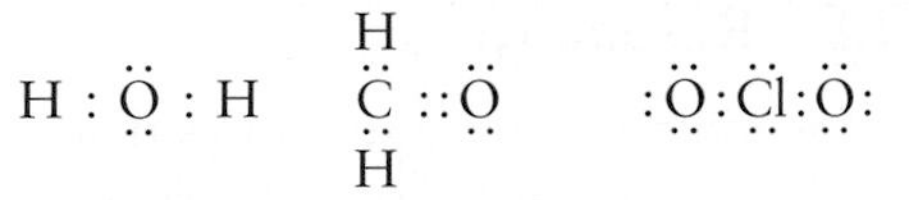

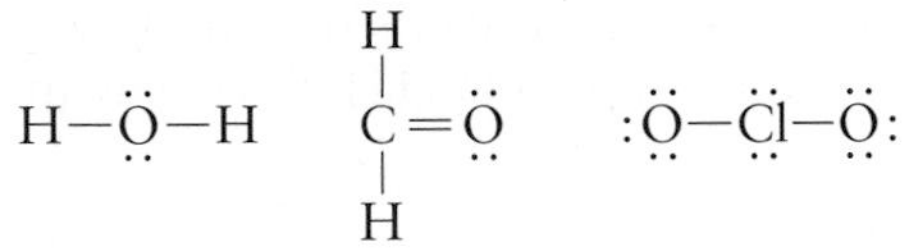

Self-Test 2.5A Write a Lewis structure for the cyanate ion, CNO^-. (In this case, the C atom is in the center.)

[*Answer:* $\ddot{N}{=}C{=}\ddot{O}^-$.]

Self-Test 2.5B Write a Lewis structure for NH_3.

EXAMPLE 2.4 Writing Lewis structures for molecules with more than one "central" atom

Write the Lewis structure for acetic acid, CH_3COOH, a common example of a carboxylic acid. Acetic acid is the acid in vinegar and is formed when the ethanol in wine is oxidized. In the —COOH group, both O atoms are attached to the same C atom, and one of them is bonded to the final H atom. The two C atoms are bonded to each other.

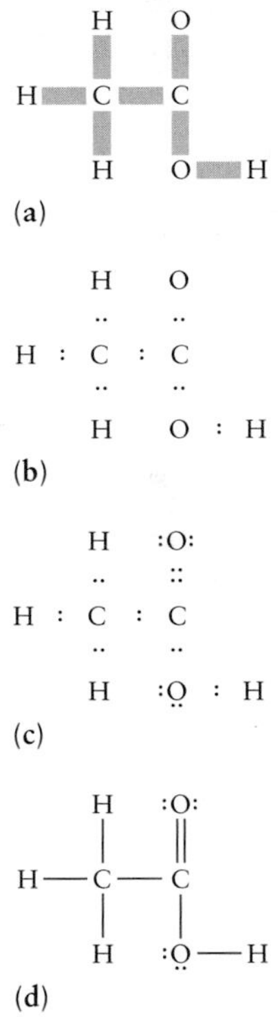

STRATEGY Follow the steps in Toolbox 2.1. The formula for acetic acid suggests that it consists of two groups, with the central C atoms joined together: a CH_3— group and a —COOH group. We anticipate that the CH_3— group, by analogy with methane, will consist of a C atom joined to three H atoms by single bonds.

SOLUTION ***Step 1*** The total number of valence electrons is

C	2×4	=	8
H	4×1	=	4
O	2×6	=	12
Total			24

so the molecule has 12 valence electron pairs. ***Step 2*** The atomic arrangement in the molecule, which is suggested by the way the molecular formula is written, is shown in (**9a**); the linked atoms are indicated by the pale gray rectangles. ***Step 3*** We use seven electron pairs to link neighboring atoms, as shown in (**9b**). Five pairs remain. ***Step 4*** To complete the octets, we arrange electron pairs so that each atom has eight electrons: this can be achieved by adding two lone pairs to each oxygen atom and allowing the terminal oxygen atom to form a double bond to the carbon atom, as depicted in (**9c**). ***Step 5*** The final Lewis structure is shown in (**9d**).

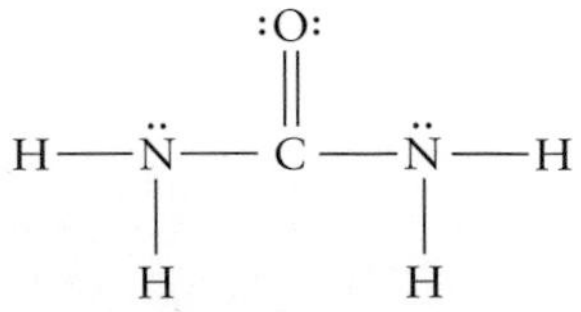

Self-Test 2.6A Write a Lewis structure for the urea molecule, $(NH_2)_2CO$.

[*Answer:* See (**10**).]

Self-Test 2.6B Write a Lewis structure for hydrazine, H_2NNH_2.

(a) H—C(H)(H)—C(O)—O—H (linked atoms shown by gray rectangles)

(b) H : C : C with H above and below the first C, O above the second C, O : H below

(c) H : C : C with H above and below the first C, :O: double-paired above the second C, :Ö : H below

(d) H—C(—H)(—H)—C(=Ö:)—Ö—H

9 Acetic acid, CH_3COOH

H—N̈(—H)—C(=Ö:)—N̈(—H)—H

10 Urea, $(NH_2)_2CO$

2.8 Resonance

Some molecules have structures that are not expressed correctly by a single Lewis structure. For example, consider the nitrate ion, NO_3^-, which, as potassium nitrate, is used in fireworks and fertilizers. The three Lewis structures shown in (**11**) differ only in the position of the double bond.

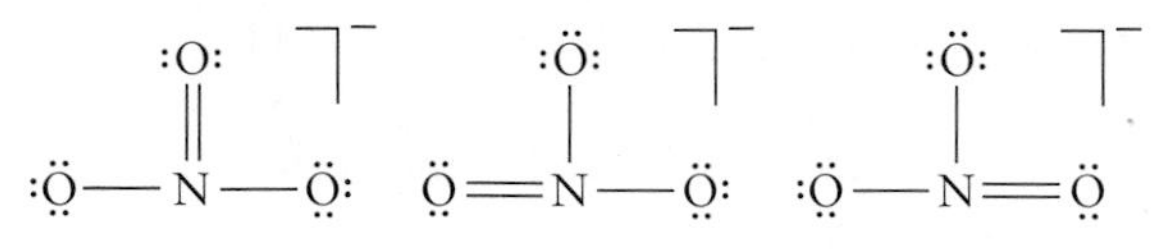

11

All are valid structures and have exactly the same energy. However, measurements of the bond lengths in the nitrate ion show that none alone is the correct structure. A double bond between two atoms is shorter than a single bond between the same types of atoms. If one of the pictured structures were correct, we would expect two long single bonds and one short double bond. However, the experimental evidence is that the bond lengths in a nitrate ion are all the same. At 124 pm, they are longer than a typical N=O double bond (120 pm) but shorter than a typical N—O single bond (140 pm). The bond order in the nitrate ion lies between 1 (a single bond) and 2 (a double bond).

Bond length, the distance between the centers of bonded atoms, is discussed in more detail in Section 2.18.

Because all three bonds are identical, a better model of the nitrate ion is a *blend* of all three Lewis structures. This blending of structures, which is called **resonance,** is depicted in (**12**) by double-headed arrows.

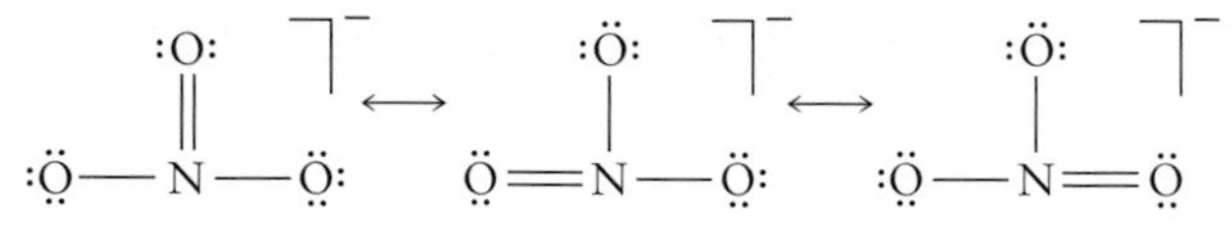

12 Nitrate ion, NO_3^-

The blended structure is a **resonance hybrid** of the contributing Lewis structures, which are called *resonance structures*. Resonance should be thought of as a blend of the individual Lewis structures rather than as the flickering of a molecule between different structures, just as a mule is a blend of a horse and a donkey, not a creature that flickers between the two.

Electrons involved in resonance structures are said to be **delocalized.** Delocalization means that the sharing of an electron pair is distributed over several pairs of atoms and cannot be identified with just one pair of atoms. Because in general we cannot draw one Lewis structure that portrays this delocalization, we depict the structure of a molecule that undergoes resonance by drawing all the contributing Lewis structures.

EXAMPLE 2.5 Writing a resonance structure

Stratospheric ozone, O_3, protects life on Earth from harmful ultraviolet radiation from the Sun. Suggest two Lewis structures that contribute to the resonance structure for the O_3 molecule. Experimental data show that the two bond lengths are the same.

STRATEGY Write a Lewis structure for the molecule, using the method outlined in Toolbox 2.1. Decide whether there is another equivalent structure that results from the interchange of a single bond and a double bond. Write the actual structure as a resonance hybrid of these Lewis structures.

SOLUTION Oxygen is a member of Group 16/VI, so each atom has six valence electrons. The total number of valence electrons in the molecule is $3 \times 6 = 18$. One Lewis structure is therefore $:\ddot{\underset{..}{O}}-\ddot{O}=\ddot{O}$. Interchanging the bonds gives $\ddot{O}=\ddot{O}-\ddot{\underset{..}{O}}:$. The overall structure is the following resonance hybrid:

$$:\ddot{\underset{..}{O}}-\ddot{O}=\ddot{O} \longleftrightarrow \ddot{O}=\ddot{O}-\ddot{\underset{..}{O}}:$$

Self-Test 2.7A Write Lewis structures contributing to the resonance hybrid for the acetate ion, $CH_3CO_2^-$. Recall that the structure of CH_3COOH is described in Example 2.4.

[*Answer:* See (13).]

Self-Test 2.7B Write Lewis structures contributing to the resonance hybrid for the nitrite ion, NO_2^-.

[Structure 13: H–C(H)(H)–C(=O:)(–O:) $\longleftrightarrow$ H–C(H)(H)–C(–O:)(=O:), bracketed with charge −]

13 Acetate ion, $CH_3CO_2^-$

Benzene, C_6H_6, is another molecule best described as a resonance hybrid. It consists of a planar hexagonal ring of six carbon atoms, with a hydrogen atom attached to each one (**14**). One Lewis structure that contributes to the resonance hybrid is shown in (**15**); it is called a **Kekulé structure.** The structure is normally written as a line structure (see Section C), a simple hexagon with alternating double lines (**16**).

The German chemist Friedrich Kekulé first proposed (in 1865) that benzene has a cyclic structure with alternating single and double bonds.

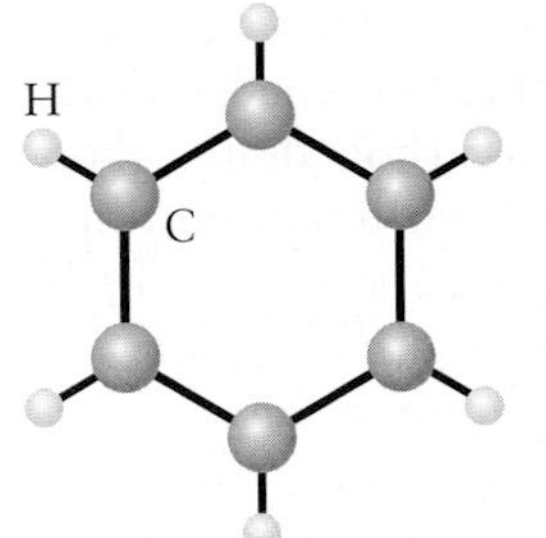

14 Benzene, C_6H_6

[Structure 15: hexagonal ring of six C atoms with alternating single and double bonds, each C bonded to an H]

15 Kekulé structure

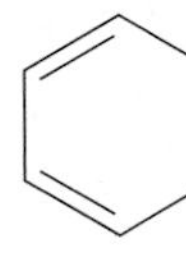

16 Kekulé structure, stick form

The difficulty with a single Kekulé structure is that it does not fit all the evidence that chemists have collected:

- *Reactivity* Benzene does not undergo reactions typical of compounds with double bonds. For example, when ethene, $H_2C{=}CH_2$, is bubbled through a solution of bromine, the solution is decolorized as the Br_2 molecules attack the double bond, attach to the molecule, and form dibromoethane (Fig. 2.9). Benzene does not decolorize bromine.

FIGURE 2.9 When bromine dissolved in a solvent (the brown liquid) is mixed with an alkene (the colorless liquid), the bromine atoms add to the molecule at the double bond, giving a colorless product.

17 Dichlorobenzene

- *Bond lengths* A Kekulé structure suggests that benzene should have two different bond lengths: three longer single bonds (154 pm) and three shorter double bonds (134 pm). Instead, all the bonds are found experimentally to have the same intermediate length (139 pm).
- *Structural evidence* If the Kekulé structure were correct, there would be two distinct dichlorobenzenes in which the chlorine atoms are attached to adjacent carbon atoms, as shown in (**17**). However, only one such compound is known. All six locations on the ring are the same.

18 Benzene resonance structure

We can use the concept of resonance to explain these characteristics of the benzene molecule. There are in fact two Kekulé structures with exactly the same energy: they differ only in the positions of the double bonds (**18**). As a result of resonance between these two structures, the electron density of the C=C double bonds is spread over the whole molecule, thereby giving each bond a length intermediate between that of a single and a double bond. Resonance makes all six C—C bonds identical; this equivalence is implied by representing the resonance hybrid as shown in (**19**). We can see from (**20**) why there can be only one dichlorobenzene with Cl atoms on adjacent C atoms.

Other arrangements may be drawn, but they differ from the one shown only by a rotation of the molecule.

19 Benzene, C_6H_6 **20** 1,2-Dichlorobenzene, $C_6H_4Cl_2$

Resonance stabilization is a quantum mechanical effect: it is discussed further in Section 3.13.

An important consequence of resonance is that it stabilizes a molecule by lowering the total energy. This stabilization makes benzene less reactive than expected for a molecule with three carbon-carbon double bonds. Resonance results in the greatest lowering of energy when the contributing structures have equal energies, as for the two Kekulé structures of benzene. However, in general, a molecule is a blend of all reasonable Lewis structures, including those with different energies. In these cases, the lowest energy structures contribute most strongly to the overall structure.

Resonance occurs only between structures with the same arrangement of atoms. For example, although we might be able to write two hypothetical structures for the dinitrogen oxide molecule, NNO and NON, there is no resonance between them, because the atoms lie in different locations.

Resonance is a blending of structures with the same arrangement of atoms but different arrangements of electrons. It spreads multiple bond character over a molecule and also lowers its energy.

2.9 Formal Charge

Different Lewis structures do not in general make the same contribution to the resonance structure. It is possible to decide which structures are likely to make the major contribution by examining how the electrons are shared in each one and assigning a "formal charge," as defined below, to each atom. The lower the formal charges for a structure, the greater its contribution to a resonance hybrid.

The **formal charge** on an atom in a given Lewis structure is the charge it would have if the bonding were perfectly covalent and the atom had exactly a half-share in the bonding electrons. That is, the formal charge takes into account the number of electrons that an atom can be regarded as "owning" in a molecule. It "owns" all its lone pairs and half of each shared pair. The difference between this number and the number of valence electrons in the free atom is the formal charge:

$$\text{Formal charge} = V - \left(L + \tfrac{1}{2}S\right) \qquad (3)$$

where V is the number of valence electrons in the free atom, L is the number of electrons present as lone pairs, and S is the number of shared electrons. If the atom has more electrons in the molecule than when it is a free, neutral atom, then the atom has a negative formal charge, like a monatomic anion. If the assignment of electrons leaves the atom with fewer electrons than when it is free, then the atom has a positive formal charge, as if it were a monatomic cation.

A Lewis structure typically corresponds to the lowest energy arrangement of atoms and electrons when the formal charges of the individual atoms are closest to 0. A low formal charge indicates that an atom has undergone the least redistribution of electrons relative to the free atom. For example, the formal charge rule suggests that the structure OCO is more likely for carbon dioxide than COO, as shown in (**21**). Similarly, it also suggests that the structure NNO is more likely for dinitrogen monoxide than NON, as shown in (**22**).

0 0 0 / 0 +2 −2
O=C=O   O=O=C

21

−1 +1 0 / −1 +2 −1
N=N=O   N=O=N

22

TOOLBOX 2.2 How to assign formal charge

Conceptual Basis

We need to calculate the "ownership" of the valence electrons when an atom forms a molecule and compare that ownership with the free atom. An atom owns one electron of each bonding pair attached to it and owns its lone pairs completely.

Procedure

Step 1 Decide on the number of valence electrons (V) possessed by each free atom by noting the number of its group in the periodic table.

Step 2 Draw the Lewis structures, showing each electron pair as dots (not as a line).

Step 3 For each bonded atom, count each electron in its lone pairs, plus one electron from each of its bonding pairs.

Step 4 For each bonded atom, subtract the total number of electrons it "owns" from V.

A good check on the calculated formal charges is that their sum is equal to the overall charge of the ion. For an electrically neutral molecule, the sum of the formal charges is 0.

EXAMPLE 2.6 Judging the plausibility of a structure

One test for the presence of iron ions in solution is to add a solution of potassium thiocyanate, KSCN, and obtain a blood-red coloration due to a compound of iron and the thiocyanate ion. Write three Lewis structures with different atomic arrangements for the thiocyanate ion and suggest which one is likely to be the most plausible structure.

STRATEGY We need to calculate the formal charges on the three possible arrangements of atoms and select the one with formal charges closest to 0.

SOLUTION To calculate the formal charges of the Lewis structures corresponding to the three possible arrangements shown in (23), we construct the following table:

	Number of electrons		
Lewis structure	On free atom (V)	Assigned to bonded atom ($L + \frac{1}{2}S$)	Formal charge (FC)
central C	4	4	0
central N	5	4	+1
central S	6	4	+2
=C	4	6	−2
=N	5	6	−1
=S	6	6	0

−1 0 0
N=C=S

−2 +1 0
C=N=S

−2 +2 −1
C=S=N

23

The individual formal charges are closest to 0 in the first structure. That structure is therefore the most likely one for the SCN^- ion.

Self-Test 2.8A Suggest a plausible structure for the poisonous gas phosgene, $COCl_2$. Write its Lewis structure and formal charges.

[*Answer:* See (**24**).]

Self-Test 2.8B Suggest a plausible structure for the oxygen difluoride molecule. Write its Lewis structure and formal charges.

24 Phosgene, $COCl_2$

The formal charge exaggerates the covalent character of the bonds by assuming that the electrons are shared equally. Conversely, oxidation number (Sections D and K) is an exaggeration of the ionic character of bonds. It represents the atoms as ions and all the electrons in a bond are assigned to the more electronegative atom. Thus, the oxidation number of C in CO_2 is +4, because all the bonding electrons are assigned to the oxygen atoms. Formal charges depend on the particular Lewis structure we write; oxidation numbers do not.

The formal charge gives an indication of the extent to which atoms have gained or lost electrons in the process of covalent bond formation; structures with lowest formal charges are likely to have the lowest energy.

EXCEPTIONS TO THE OCTET RULE

The octet rule accounts for the valence of the elements and the structures of many compounds. Carbon, nitrogen, oxygen, and fluorine obey the octet rule rigorously, provided there are enough electrons to go around. However, some compounds have an odd number of electrons. In addition, an atom of phosphorus, sulfur, chlorine, or another nonmetal in Period 3 and later can accommodate more than eight electrons in its valence shell. In the following two sections, we shall learn to recognize exceptions to the octet rule.

2.10 Radicals and Biradicals

Some species have an odd number of valence electrons, so at least one of their atoms cannot have an octet. Species having electrons with unpaired spins are called **radicals**. They are generally highly reactive. One example is the methyl radical, $\cdot CH_3$, which is so reactive that it cannot be stored. It occurs in the flames of burning hydrocarbon fuels.

The older, still widely used term for radicals is *free radicals*.

A single dot is added to the formula of a radical to emphasize that the species contains an unpaired electron.

Atoms bonded together ceaselessly vibrate toward and away from each other. If excited by heat or electromagnetic radiation, the two atoms of a carbon-carbon bond may vibrate so violently that the bond breaks, as happens in one of the reactions that occur when ethane burns:

$$H_3C{-}CH_3 \longrightarrow H_3C\cdot + \cdot CH_3$$

The electron pair that formed the carbon-carbon bond has split into two separate electrons, as indicated by the dot on the C atom in each $\cdot CH_3$. Because the radical has only seven electrons, it is not possible for them all to pair.

Radicals are of crucial importance for the chemical reactions that take place in the upper atmosphere, where they contribute to the formation and decomposition of ozone. Radicals also play a role in our daily lives, sometimes a destructive one. They are responsible for the rancidity of foods and the degradation of plastics in sunlight. Damage from radicals can be delayed by an additive called an **antioxidant**, which reacts rapidly with radicals before the radicals have a chance to do their damage. It is believed that human aging is partly due to the action of radicals, and antioxidants such as vitamins C and E may delay the process (see Box 2.1).

This point is developed further in Box 13.3.

A **biradical** is a molecule with two unpaired electrons. The unpaired electrons are usually on different atoms, as depicted in (**25**). In that biradical, one unpaired electron is on one carbon atom of the chain and the second is on another carbon atom several bonds away. In some cases, though, both electrons are on the same atom. One of the most important examples is the oxygen atom itself. Its electron configuration is $[He]2s^22p_x^{\,2}2p_y^{\,1}2p_z^{\,1}$ and its Lewis symbol is $\cdot\ddot{\underset{..}{O}}\cdot$. The O atom has two unpaired electrons, so it can be regarded as a special case of a biradical.

```
  H    H    H    H    H
  |    |    |    |    |
 ·C————C————C————C————C·
  |    |    |    |    |
  H    H    H    H    H
```

25 A biradical

Although it is not obvious from its Lewis structure, molecular oxygen, O_2, is also a biradical! In fact, experiments have shown that the most plausible Lewis structure, $\ddot{\underset{..}{O}}{=}\ddot{\underset{..}{O}}$ gives a false impression of the arrangement of electrons. In molecular oxygen, two of the electrons that the

BOX 2.1 WHAT HAS THIS TO DO WITH . . . STAYING ALIVE?

Chemical Self-Preservation

In nearly every pharmacy, supermarket, and health food store, you can find bottles of antioxidants and antioxidant-rich natural products, such as fish oils, *Gingko biloba* leaves, and wheat grass. These dietary supplements are intended to help the body control its population of radicals and, as a result, slow aging and degenerative diseases such as heart failure and cancer.

Leaves of the *Gingko biloba* tree, which originated in China. Extracts of these leaves are thought to have antioxidant properties. Some also believe that the extracts improve thinking ability by increasing oxygen flow to the brain.

Radicals occur naturally in the body, partly as a by-product of metabolism. They serve many useful functions but can also cause trouble if they are not eliminated when they are no longer needed. They often contain oxygen atoms and oxidize the *lipid* (fat) molecules that make up cell membranes and other vital tissues. This oxidation changes the structures of the lipid molecules and hence affects the function of the membranes. Cell membranes damaged in this way may not be able to protect cells against disease, and heart and nerve cells may lose their function. Aging is also thought to be accelerated by radicals. When the nucleic acids such as DNA and RNA are attacked by radicals, they cannot replicate properly, an alteration resulting in defective cells with diminished function and a lessened ability to protect themselves against cancer.

The body maintains an antioxidant network consisting of vitamins A, C, and E, antioxidant enzymes, and a group of related compounds called coenzyme Q, for which the general formula is shown. The n represents the number of times that a particular group is repeated; it can be 6, 8, or 10. The coenzyme Q molecules are also called *ubiquinones*, because they are so ubiquitous in the body. *Antioxidants* are molecules that are easily oxidized, so they react readily with radicals before the radicals can react with other compounds in the body. A variety of intricate mechanisms in the body regulate the abundance and locations of these antioxidants.

In a healthy living environment, the antioxidant network controls the radical population efficiently, ensuring

---

Lewis structure implies are responsible for the bonds do not pair with one another. The molecule is really a biradical with an unpaired electron on each O atom. For this reason, its Lewis structure is often written as shown in (**26**). The Lewis model of bonding does not predict the biradical nature of the O_2 molecule, but in Section 3.10 we shall see how modern theories of bonding can predict and explain it.

$\ddot{\underset{\cdot\cdot}{O}}\cdot\!\!-\!\!\cdot\ddot{\underset{\cdot\cdot}{O}}$

26 Dioxygen, O_2

A radical is a species with an unpaired electron; a biradical has two unpaired electrons on either the same or different atoms.

Self-Test 2.9A Write a Lewis structure for the hydrogenperoxyl radical, HOO·, which plays an important role in atmospheric chemistry and which, in the body, has been implicated in the degeneration of neurons.

[*Answer:* See (**27**).]

Self-Test 2.9B Write a Lewis structure for NO_2.

H
|
$:\underset{\cdot\cdot}{O}-\ddot{\underset{\cdot\cdot}{O}}\cdot$

27 Hydrogenperoxyl, $HO_2\cdot$

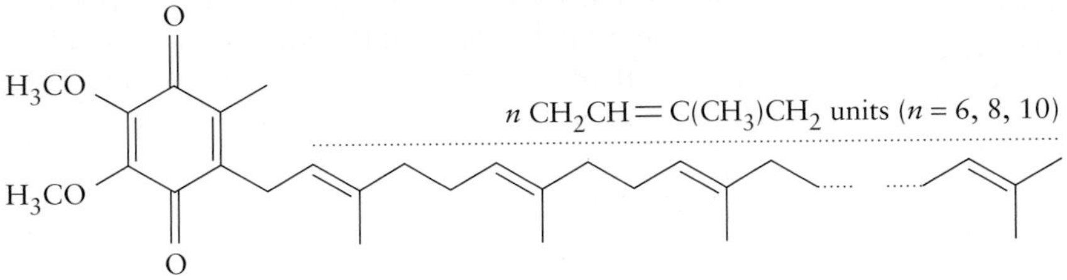

H_3CO, H_3CO, O, O

$n\ CH_2CH{=}C(CH_3)CH_2$ units ($n = 6, 8, 10$)

Molecular structure of coenzyme Q, an antioxidant used by the body to control the level of radicals.

that radicals are present when they are needed but are promptly eliminated before they can cause harm. Many common foods also contain antioxidants, such as green leafy vegetables, orange juice, coffee, tea, and chocolate. However, certain environmental conditions, such as ultraviolet light, ozone in the air we breathe, poor nutrition, and tobacco smoke can cause *oxidative stress*, a condition in which the concentration of radicals has become so high that the body's natural antioxidant network cannot cope. Premature aging of skin that has been overexposed to sunlight and lung cancer in smokers are two possible results. Certain *phytochemicals*, chemicals derived from plant sources and fish oils, in herbal medicines are being investigated as antioxidants that can supplement the diet. Dietary supplementation, unlike our natural antioxidant network, which is highly responsive to cell environment, poses a problem because we do not know how much of the additive to take. In addition, the chemicals consumed will target not only the cells where antioxidants are needed but will be distributed throughout the body.

Unfortunately, there is no convenient marker that can be used to monitor radical or antioxidant levels in the body. Diabetics who are taking insulin to control blood sugar can monitor their blood sugar levels with a simple glucose monitor, but there are no simple tests for antioxidants and therefore no way to know whether the body needs supplementation.

For Further Reading: "Antioxidants are becoming the next big health additive," *Chemical and Engineering News*, April 21, 1997, p. 24. V. Griffith, "The body's fight against aging," *Future*, March 1998, pp. 50–55. J. Raloff, "Panel ups RDAs for some antioxidants," *Science News*, vol. 157 (April 15, 2000), p. 244.

2.11 Expanded Valence Shells

The octet rule tells us that eight electrons fill a shell to give a noble-gas ns^2ns^6 valence-shell configuration. However, when the central atom in a molecule has empty *d*-orbitals, it may be able to accommodate 10, 12, or even more electrons. The electrons in such an **expanded valence shell** may be present as lone pairs or may be used by the central atom to form additional bonds.

The term *expanded octet* is commonly used to denote the expansion of a valence shell to more than eight electrons.

Because the additional electrons must be accommodated in valence orbitals, only nonmetal atoms in Period 3 or later periods can expand their octets. These elements have empty *d*-orbitals in the valence shell. Another factor—possibly the main factor—in determining whether more atoms than allowed by the octet rule can bond to a central atom is the size of that atom. A P atom is big enough for up to six Cl atoms to fit comfortably around it, and PCl_5 is a common laboratory chemical. An N atom, though, is too small, and NCl_5 is unknown.

Elements that can expand their valence shells commonly show **variable covalence**, the ability to form different numbers of covalent bonds. Elements that have variable covalence can form one number of bonds in some compounds and a different number in others. Phosphorus is an example. It reacts directly with a limited supply of chlorine to form the toxic, colorless liquid phosphorus trichloride:

$$P_4(g) + 6\ Cl_2(g) \longrightarrow 4\ PCl_3(l)$$

28 Phosphorus trichloride, PCl_3

The Lewis structure of the PCl_3 molecule is shown in (**28**), and we see that it obeys the octet rule. However, when excess chlorine is present or when phosphorus trichloride reacts with more chlorine (Fig. 2.10), phosphorus pentachloride, a pale yellow crystalline solid, is produced:

$$PCl_3(l) + Cl_2(g) \longrightarrow PCl_5(s)$$

Phosphorus pentachloride is an ionic solid consisting of PCl_4^+ cations and PCl_6^- anions; but at 160°C, it vaporizes to a gas of PCl_5 molecules. The Lewis structures of the polyatomic ions and the molecule are shown in (**29**). In the anion, the P atom has expanded its valence shell to 12 electrons, by making use of two of its 3*d*-orbitals. In PCl_5, the P atom has expanded its valence shell to 10 electrons by using one 3*d*-orbital.

(a) PCl_4^+ (b) PCl_6^- (c) PCl_5

29

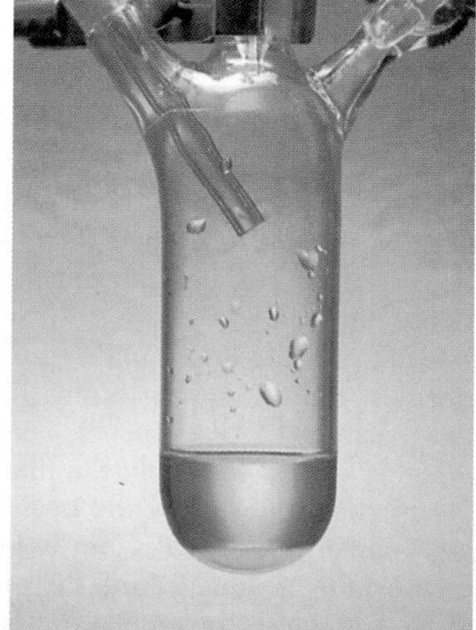

FIGURE 2.10 Phosphorus trichloride is a colorless liquid. When it reacts with chlorine (the pale yellow-green gas in the flask), it forms the very pale yellow solid phosphorus pentachloride (at the bottom of the flask).

EXAMPLE 2.7 Writing a structure with an expanded valence shell

A fluoride of composition SF_4 is formed when a mixture of fluorine and nitrogen gases is passed over a film of sulfur at −75°C in the absence of oxygen and moisture. Write the Lewis structure of sulfur tetrafluoride.

STRATEGY A fluorine atom forms only one bond, a single bond, so we anticipate that the Lewis structure consists of a shared pair between the central S atom and each of the four surrounding F atoms. However, each F atom has three lone pairs and supplies one bonding electron, and the S atom already has six electrons in its valence shell. So, there are two extra electrons. Because sulfur is in Period 3 and has empty 3*d*-orbitals available, it can expand its octet to accept the additional electrons.

SOLUTION Sulfur (·S̤̈·) supplies six valence electrons, and each fluorine atom (:F̤̈·) supplies seven. Hence there are $6 + (4 \times 7) = 34$ electrons, or 17 electron pairs, to accommodate. We write each F atom with three lone pairs and a bonding pair shared with the central S atom, then place the two extra electrons on the S atom, as shown in (**30**).

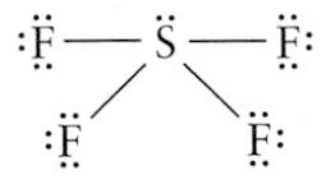

30 Sulfur tetrafluoride, SF_4

Self-Test 2.10A Write the Lewis structure for xenon tetrafluoride, XeF_4, and give the number of electrons in the expanded octet.

[*Answer:* See (**31**); 12 electrons]

Self-Test 2.10B Write the Lewis structure for the I_3^- ion and give the number of electrons in the expanded octet.

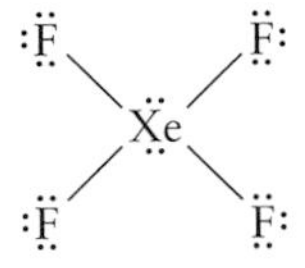

31 Xenon tetrafluoride, XeF_4

Computer calculations show that, for some compounds, a structure with an expanded octet has a lower energy than the octet structure. These calculations are supported by experimental data, including measurements of bond lengths and strengths. They are also supported by formal charge considerations, as in Example 2.8.

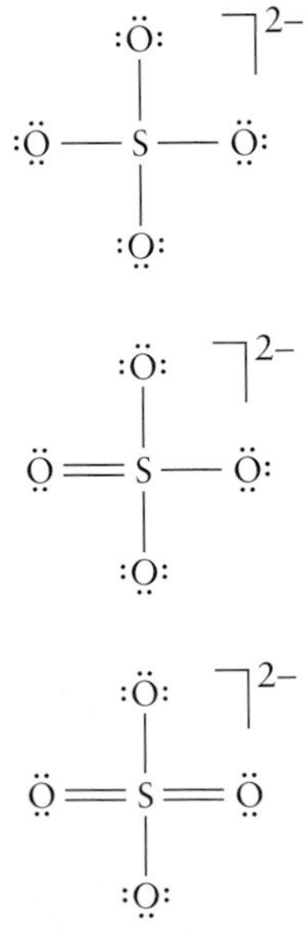

EXAMPLE 2.8 Selecting the favored resonance structure for a molecule

The sulfate ion, SO_4^{2-}, occurs in a number of important minerals, including gypsum ($CaSO_4{\cdot}2H_2O$) and Epsom salts ($MgSO_4{\cdot}7H_2O$). The former is used in cement and the latter as a purgative. Determine the most favored of the three resonance structures of a sulfate ion shown in (**32**) by calculating the formal charges on the atoms in each structure.

STRATEGY Follow the procedure in Toolbox 2.2. We need do only one calculation for equivalent atoms, such as the oxygen atoms in the first diagram, because they all have the same arrangement of electrons and hence the same formal charge.

32 Sulfate ion, SO_4^{2-}

SOLUTION *Step 1* Both O and S belong to Group 16/VI, so the neutral atoms each have six valence electrons and the ion has two additional electrons, for a total of 32 electrons. *Steps 2* and *3* The closed curves that capture the electrons are shown in (**33**).

(a) (b) (c)

33 Sulfate ion, SO_4^{2-}

Step 4 The formal charges are shown in the diagrams. The individual formal charges are closest to 0 in (**33c**), so that structure is the most plausible, even though it involves a valence shell on the S atom expanded to 12 electrons.

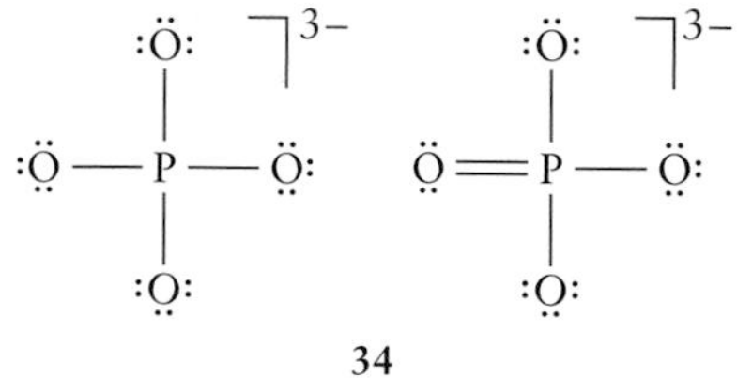

34

Self-Test 2.11A Calculate the formal charges for the two Lewis structures of the phosphate ion shown in (**34**).

[*Answer:* The formal charges are shown in (**35**).]

35

Self-Test 2.11B Calculate the formal charges for the three oxygen atoms in one of the Lewis structures of the ozone resonance structure (Example 2.5).

The Lewis structures with the lowest formal charges usually make the greatest contribution to the overall resonance hybrid, even if they have expanded valence shells. In the case of the sulfate ion, the formal charges are lowest in the structures with double bonds. All three structures contribute to the resonance hybrid, but we can expect the properties of the sulfate ion to indicate the presence of double bonds.

Octet expansion (expansion of the valence shell to more than eight electrons) can occur in elements of Period 3 and later periods and enables them to exhibit variable covalence.

LEWIS ACIDS AND BASES

Boron ([He]$2s^2 2p^1$) and aluminum ([Ne]$3s^2 3p^1$) atoms need five electrons to complete their octets. However, they may be unable to acquire that number of electrons from the atoms to which they are bonded. As a result, the compounds these elements form have special chemical characteristics and take part in an important class of reactions known as "Lewis acid-base reactions," which we introduce here.

2.12 The Unusual Structures of Group 13/III Halides

36 Boron trifluoride, BF_3

Let's investigate the molecular structure of the colorless gas boron trifluoride, BF_3. An important feature of its Lewis structure (**36**) is that the boron atom has an **incomplete octet:** its valence shell consists of only six electrons. We might expect the boron atom to complete its octet by sharing more electrons with fluorine, as depicted in (**37**), but fluorine has such a high ionization energy that this arrangement is unlikely. Experimental evidence, such as the relatively short B—F bond lengths, suggests that the true structure of BF_3 is a resonance hybrid of both types of Lewis structures, with the singly bonded structure making the major contribution.

37

The boron atom in BF_3 can complete its octet if an additional atom or ion with a lone pair of electrons forms a bond by providing *both* electrons. A bond in which both electrons come from one of the atoms is called a **coordinate covalent bond.** For example, the tetrafluoroborate anion, BF_4^- (**38**), forms when boron trifluoride is passed over a metal fluoride. In this anion, the formation of a coordinate covalent bond with a fluoride ion gives the B atom an octet. Another example of a coordinate covalent bond is that formed when boron trifluoride reacts with ammonia:

38 Tetrafluoroborate, BF_4^-

$$BF_3(g) + NH_3(g) \longrightarrow NH_3BF_3(s)$$

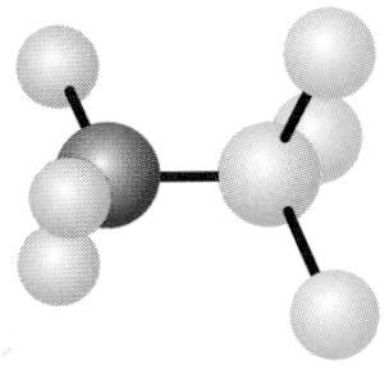

39 NH_3BF_3

A molecular model of the product, a white molecular solid, is shown in (**39**). In this reaction, the lone pair on the nitrogen atom of ammonia can be regarded as completing boron's octet by forming a coordinate covalent bond to give the Lewis structure shown in (**40**).

40 NH_3BF_3

Boron trichloride, a colorless, reactive gas of BCl_3 molecules, behaves chemically like BF_3. However, the trichloride of aluminum, which is in the same group as boron, is a volatile white solid that vaporizes at 180°C to a gas of Al_2Cl_6 molecules (**41**). These molecules survive in the gas up to about 200°C and only then fall apart into $AlCl_3$ molecules. The Al_2Cl_6 molecule exists because a Cl atom in one $AlCl_3$ molecule uses one of its lone pairs to form a coordinate covalent bond to the Al atom in a neighboring $AlCl_3$ molecule (**42**).

Compounds of boron and aluminum may have unusual Lewis structures in which boron and aluminum have incomplete octets or halogen atoms act as bridges.

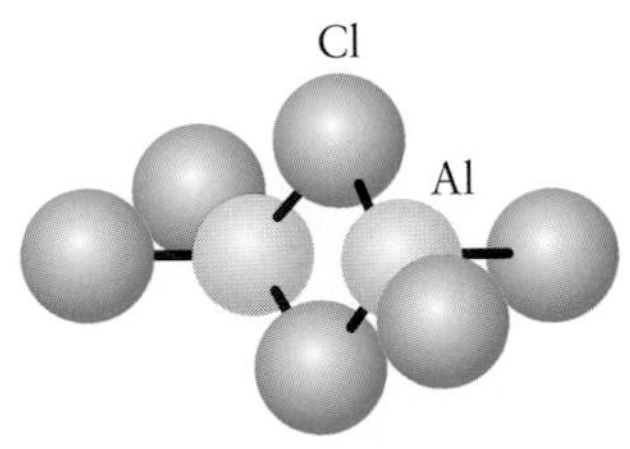

41 Al_2Cl_6

42 Aluminum chloride, Al_2Cl_6

2.13 Lewis Acid-Base Complexes

Coordinate covalent bond formation is common. For reasons that will soon be explained, the species that provides the lone pair is called a **Lewis base,** and the species that accepts it is called a **Lewis acid.** In other words, a Lewis acid is an *electron pair acceptor*, and a Lewis base is an *electron pair donor*. The product of the reaction between a Lewis acid and a Lewis base is called a **complex.** The form of the reaction is therefore

$$\text{Acid} + :\text{base} \longrightarrow \text{complex}$$

For example, in the reaction $BF_3 + F^- \rightarrow BF_4^-$, the BF_3 molecule is the Lewis acid, the F^- ion is the Lewis base, and the BF_4^- ion is the Lewis acid-base complex. The bond formed in a Lewis acid-base reaction is a coordinate covalent bond.

Why do we use the terms *acid* and *base* in this context? Recall from Section J that acids and bases are normally discussed in terms of the behavior of a hydrogen ion, H^+ (a proton), and that, in water, an acid such as HCl donates its proton to a neighboring water molecule. We could imagine this reaction as taking place in two hypothetical steps: first, an HCl molecule releases a hydrogen ion as soon as it dissolves in water and then that hydrogen ion immediately bonds to a neighboring water molecule. The second step is

$$H^+ + :\underset{\cdot\cdot}{\overset{\overset{\displaystyle H}{|}}{O}}-H \longrightarrow \left[H-\underset{\cdot\cdot}{\overset{\overset{\displaystyle H}{|}}{O}}-H\right]^+$$

This step in the reaction has exactly the same form as the Lewis acid-base reaction, with H^+ playing the role of the Lewis acid, H_2O the role of the Lewis base, and H_3O^+ the resulting complex. It is this analogy to the behavior of H^+ (the species typical of acids) that led to the use of the term *acid* to describe an electron pair acceptor like BF_3.

Similarly, just as the hydroxide ion, OH^-, is a typical base in the conventional sense of being a proton acceptor, it acts as an electron pair donor (a Lewis base) in the reaction

$$H^+ + \left[:\ddot{\underset{\cdot\cdot}{O}}-H\right]^- \longrightarrow :\underset{\cdot\cdot}{\overset{\overset{\displaystyle H}{|}}{O}}-H$$

An ammonia molecule is another example of a Brønsted base that has a lone pair to donate (for example, to BF_3, as described earlier), so it can also be regarded as a Lewis base.

> *A Lewis acid is an electron pair acceptor; a Lewis base is an electron pair donor; they react to form a Lewis acid-base complex by forming a coordinate covalent bond.*

Self-Test 2.12A Show that the formation of ammonium and hydroxide ions when ammonia dissolves in water is a Lewis acid-base reaction.

[*Answer:*

$$H_3N: + H-\underset{\cdot\cdot}{\overset{\overset{\displaystyle H}{|}}{O}}: \longrightarrow NH_4^+ + :\ddot{\underset{\cdot\cdot}{O}}-H^-$$

Self-Test 2.12B Show that the formation of Al_2Cl_6 from $AlCl_3$ molecules is a Lewis acid-base reaction.

IONIC VERSUS COVALENT BOND

Ionic and covalent bonding are two extreme models of the chemical bond. Most actual bonds lie somewhere between purely ionic and purely covalent. When describing bonds between nonmetals, covalent bonding is a good model. When a metal and nonmetal are present in a simple compound, ionic bonding is a good model. However, the bonds in many compounds seem to have properties between the two extreme models of bonding. Can we describe these bonds more accurately by improving the two basic models?

2.14 Correcting the Covalent Model: Electronegativity

All bonds can be viewed as resonance hybrids of purely covalent and purely ionic structures. For example, the structure of a Cl_2 molecule can be described as

$$:\underset{\cdot\cdot}{\overset{\cdot\cdot}{\text{Cl}}}-\underset{\cdot\cdot}{\overset{\cdot\cdot}{\text{Cl}}}: \longleftrightarrow \; :\underset{\cdot\cdot}{\overset{\cdot\cdot}{\text{Cl}}}:^{-}\;\underset{\cdot\cdot}{\overset{\cdot\cdot}{\text{Cl}}}:^{+} \longleftrightarrow \; :\underset{\cdot\cdot}{\overset{\cdot\cdot}{\text{Cl}}}^{+}\;:\underset{\cdot\cdot}{\overset{\cdot\cdot}{\text{Cl}}}:^{-}$$

In this case, the ionic structures make only a small contribution to the resonance hybrid, and we regard the bond as almost purely covalent. Moreover, the two ionic structures have the same energy and make equal contributions to the hybrid, so the average charge on each atom is 0. However, in a molecule composed of different elements, such as HCl, the resonance

$$\text{H}-\underset{\cdot\cdot}{\overset{\cdot\cdot}{\text{Cl}}}: \longleftrightarrow \text{H}:^{-}\;\underset{\cdot\cdot}{\overset{\cdot\cdot}{\text{Cl}}}:^{+} \longleftrightarrow \text{H}^{+}\;:\underset{\cdot\cdot}{\overset{\cdot\cdot}{\text{Cl}}}:^{-}$$

has unequal contributions from the two ionic structures. The lower energy ionic structure is $\text{H}^{+}\,:\underset{\cdot\cdot}{\overset{\cdot\cdot}{\text{Cl}}}:^{-}$. Because the chlorine atom has a greater attraction for electrons than the hydrogen atom, the structure with a negative charge on the Cl atom makes a bigger contribution than $\text{H}:^{-}\,\underset{\cdot\cdot}{\overset{\cdot\cdot}{\text{Cl}}}:^{+}$. As a result, there is a small net negative charge on the Cl atom and a small net positive charge on the H atom. The charges on the atoms are called **partial charges.** We show the partial charges on the atoms by writing $^{\delta+}\text{H}-\text{Cl}^{\delta-}$. A bond in which ionic contributions to the resonance result in partial charges is called a **polar covalent bond.** All bonds between atoms of different elements are polar to some extent. The bonds in homonuclear diatomic molecules and ions are nonpolar.

The two atoms in a polar covalent bond form an **electric dipole**, a partial positive charge next to an equal but opposite partial negative charge. A dipole is represented by an arrow that points toward the negative partial charge (**43**). The size of an electric dipole—which is a measure of the magnitude of the partial charges—is reported as the **electric dipole moment**, μ (the Greek letter mu), in units called **debye** (D). The debye is defined so that a single negative charge (an electron) separated by 100 pm from a single positive charge (a proton) has a dipole moment of 4.80 D. The dipole moment associated with a Cl—H bond is about 1.1 D. We can think of this dipole as arising from a partial charge of about 23% of an electron's charge on the Cl atom and an equivalent positive charge on the H atom.

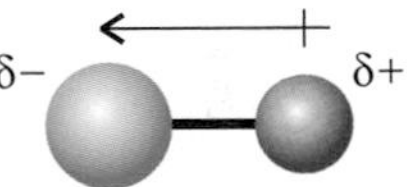

43 Dipole moment

The debye is named for the Dutch chemist Peter Debye, who carried out important studies of dipole moments.

The 4.80 in the definition of the debye arises from an earlier system of units used for electric charge. The SI unit of dipole moment is 1 C·m (1 coulomb-meter). It is the dipole moment of a charge of 1 C separated from a charge of −1 C by a distance of 1 m; 1 D = 3.336×10^{-30} C·m.

A covalent bond is polar if one atom has a greater pulling power than the other atom, because then the electron pair is more likely to be found closer to the former atom. In 1932, the American chemist Linus Pauling proposed a quantitative measure of electron distribution in bonds. The electron-pulling power of an atom when it is part of a molecule is called its **electronegativity**. Electronegativities are denoted χ (the Greek letter chi, pronounced "kye"). An atom of the element with the higher electronegativity has a stronger pulling power on electrons and tends to pull them away from the atom of the element with lower electronegativity (Fig. 2.11). Pauling based his scale on the dissociation energies, D, of the

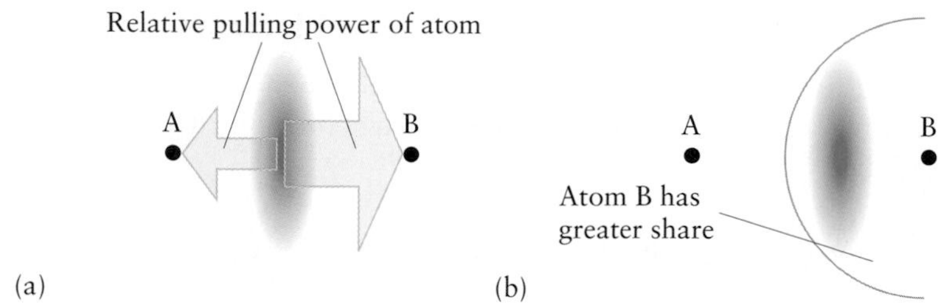

FIGURE 2.11 The electronegativity of an element is its electron-pulling power when it is part of a compound. (a) An atom with a high electronegativity has a strong pulling power on electrons (as represented by the larger arrow), particularly for the electron pair it shares with its neighbor. (b) The outcome of the tug-of-war is that the more electronegative atom has a greater share in the electron pair of the covalent bond.

Dissociation energies, a measure of the strengths of bonds, are discussed in Section 2.16.

A—A, B—B, and A—B bonds, measured in electronvolts. He defined the difference in electronegativity of the two elements A and B as

$$|\chi_A - \chi_B| = 0.102\{D(A\text{–}B) - \tfrac{1}{2}[D(A\text{–}A) + D(B\text{–}B)]\}^{1/2} \qquad (4)$$

A simpler way of setting up a scale of electronegativities was devised by another American chemist, Robert Mulliken. In his approach, the electronegativity is the average of the ionization energy and electron affinity of the element (expressed in electronvolts):

$$\chi = \tfrac{1}{2}(I + E_a) \qquad (5)$$

An atom gives up an electron reluctantly if the ionization energy is high. If the electron affinity is high, it is energetically favorable to attach electrons to an atom. Elements with both these properties are reluctant to lose their electrons and tend to gain them; hence, they are classified as highly electronegative. Conversely, if the ionization energy and the electron affinity are both low, then it takes very little energy for the element to give up its electrons and it has little tendency to gain more; hence, the electronegativity is low.

Figure 2.12 shows the variation of electronegativity for the main-group elements of the periodic table. Because ionization energies and electron affinities are highest at the top right of the periodic table (close to fluorine), it is not surprising to find that nitrogen, oxygen, bromine, chlorine, and fluorine are the elements with the highest electronegativities.

When the two atoms in a bond have only a small electronegativity difference, the partial charges are very small. As the difference in electronegativities increases, so do the partial charges. If the electronegativities are very different, then one atom can acquire the lion's share of the electron pair, and the corresponding ionic structure makes a large contribution to the resonance. Because it has largely robbed the other atom of its share of the electrons, the highly electronegative element resembles an anion and the other atom resembles a cation. We say that such a bond has

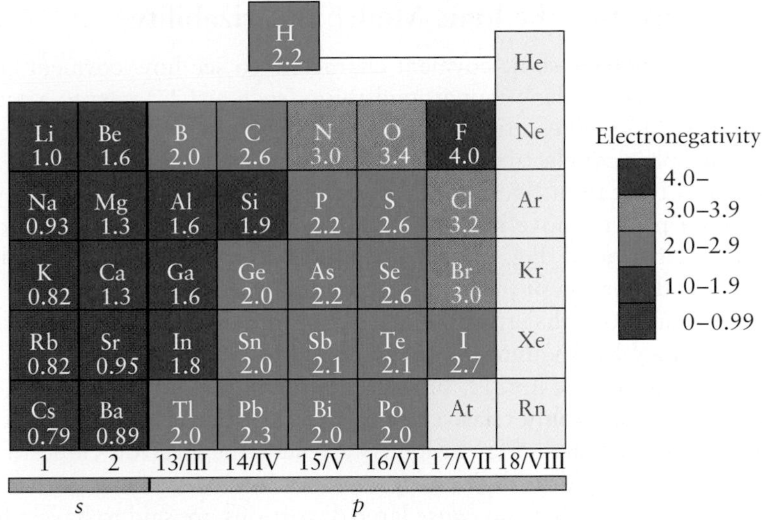

FIGURE 2.12 The variation of the electronegativity of the main-group elements (with the exception of the noble gases). Electronegativity tends to be high toward the upper right corner of the periodic table and low on the lower left. Elements with low electronegativities (such as the *s*-block metals) are often called electropositive. We use Pauling values throughout this text.

considerable **ionic character**. In the limit of very large differences in electronegativity (as in CsF), the ionic contribution dominates the covalent distribution, and it is better to regard the bond as ionic.

There is no hard-and-fast dividing line between ionic and covalent bonding. However, it is a good rule of thumb that an electronegativity difference of about 2 means that there is so much ionic character present that the bond is best regarded as ionic (Fig. 2.13). For electronegativity differences smaller than about 1.5, a covalent description of the bond is probably reasonably reliable. For example, the electronegativities of carbon and oxygen are 2.6 and 3.4 (see Fig. 2.12), an electronegativity difference of 0.8, and C—O bonds are best regarded as polar covalent. However, the electronegativity of calcium is 1.3, and Ca—O bonds, with an electronegativity difference of 2.1, are considered ionic.

Electronegativity is a measure of the pulling power of an atom on the electrons in a bond. A polar covalent bond is a bond between two atoms with partial electric charges arising from their difference in electronegativity. The presence of partial charges gives rise to an electric dipole moment.

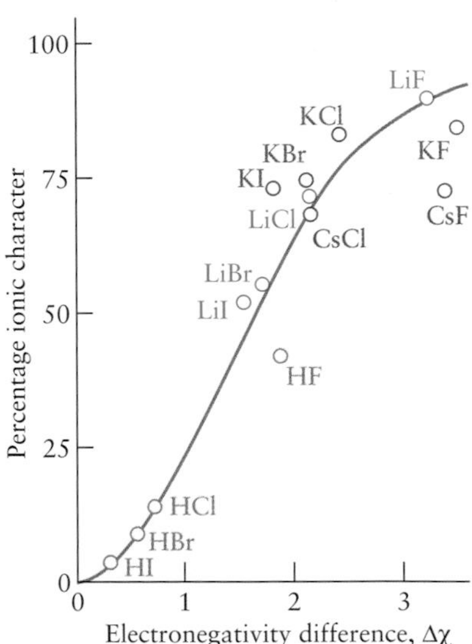

FIGURE 2.13 The dependence of percentage ionic character on the difference in electronegativity, $\Delta\chi$, between two bonded atoms for a number of halides.

Self-Test 2.13A In which of the following compounds do the bonds have greater ionic character: (a) P_4O_{10} or (b) PCl_3?

[*Answer:* (a)]

Self-Test 2.13B In which of the following compounds do the bonds have greater ionic character: (a) CO_2 or (b) NO_2?

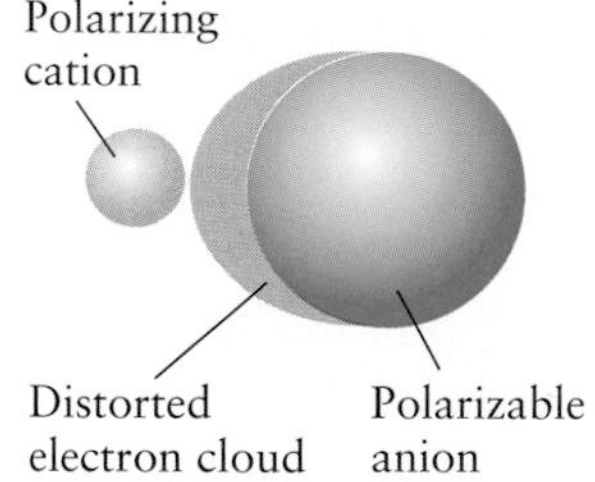

FIGURE 2.14 When a small, highly charged cation is close to a large anion, the electron cloud of the latter is distorted in the process we call polarization. Small, highly charged cations are highly polarizing. Large, electron-rich anions are highly polarizable.

Cations are not significantly polarizable because their electrons are held so tightly.

2.15 Correcting the Ionic Model: Polarizability

All ionic bonds have some covalent character. To see how covalent character can arise, consider a monatomic anion (such as Cl^-) next to a cation (such as Na^+). As the cation's positive charge pulls on the anion's electrons, the spherical electron cloud of the anion becomes distorted in the direction of the cation. We can think of this distortion as the tendency of an electron pair to move into the region between the two nuclei and to form a covalent bond (Fig. 2.14). Ionic bonds acquire more covalent character as the distortion of the electron cloud on the anion increases.

Atoms and ions that readily undergo a large distortion are said to be highly **polarizable.** An anion can be expected to be highly polarizable if it is large, like an iodide ion, I^-. In such a large, highly polarizable ion, the ion's nucleus exerts only relatively weak control over its outermost electrons because they are so far away. As a result, the electron cloud of the large anion is easily distorted.

Atoms and ions that can *cause* large distortions are said to have a high **polarizing power.** A cation can be expected to have a strong polarizing power if it is small and highly charged, like an Al^{3+} cation. A small radius means that the center of charge of a highly charged cation can get very close to the anion, where it can exert a strong pull on the anion's electrons. Compounds composed of a small, highly charged cation and a large, polarizable anion tend to have bonds with considerable covalent character.

The polarizabilities and polarizing powers of ions help to account for the diagonal relationships in the periodic table introduced in Section 1.19 and illustrated in Fig. 1.52. Cations become smaller, more highly charged, and hence more strongly polarizing, from left to right across a period. Thus, Be^{2+} is more strongly polarizing than Li^+, and Mg^{2+} is more strongly polarizing than Na^+. On the other hand, cations become larger and hence less strongly polarizing down a group. Thus, Na^+ is less strongly polarizing than Li^+, and Mg^{2+} is less strongly polarizing than Be^{2+}. Now we can see that, because polarizing power increases from Li^+ to Be^{2+}, but decreases from Be^{2+} to Mg^{2+}, it follows that the polarizing power of the diagonal neighbors Li^+ and Mg^{2+} should be similar. We can expect similarities like these in the properties of other diagonally related neighbors.

Compounds composed of highly polarizing cations and highly polarizable anions have a significant covalent character in their bonding.

Self-Test 2.14A In which of the compounds NaBr and $MgBr_2$ do the bonds have greater covalent character?

[*Answer:* $MgBr_2$]

Self-Test 2.14B In which of the compounds CaS and CaO do the bonds have greater covalent character?

THE STRENGTHS AND LENGTHS OF COVALENT BONDS

The characteristics of a covalent bond between two atoms are due mainly to the properties of the atoms themselves and vary only a little with the identities of the other atoms present in a molecule. Consequently, we can predict some characteristics of a bond with reasonable certainty once we know the identities of the two bonded atoms. For instance, the length of the bond and its strength are approximately the same regardless of the molecule in which it is found. Thus, to understand the properties of a large molecule, such as the resistance of polytetrafluoroethylene (Teflon) to chemical attack, we can study the character of C—F bonds in a much simpler compound, such as tetrafluoromethane, CF_4.

2.16 Bond Strengths

The strength of a chemical bond is measured by its **dissociation energy,** D, the energy required to separate the bonded atoms. An example is the dissociation

$$\text{H—Cl(g)} \longrightarrow \text{H(g)} + \text{Cl(g)}$$

A high dissociation energy indicates a strong bond because a lot of energy has to be supplied to break the bond. The strongest known bond between two nonmetal atoms is the triple bond in carbon monoxide, for which the dissociation energy is 1062 $kJ\cdot mol^{-1}$. One of the weakest bonds is that between the iodine atoms in molecular iodine, for which the dissociation energy is only 139 $kJ\cdot mol^{-1}$.

Carbon-hydrogen bonds, such as those in hydrocarbons like methane and ethane, have dissociation energies close to 400 $kJ\cdot mol^{-1}$, whereas single bonds between carbon and fluorine have dissociation energies close to 500 $kJ\cdot mol^{-1}$. The greater strength of a carbon-fluorine bond helps to explain why fluorocarbon polymers are very resistant to chemical attack. They are used to construct valves for corrosive gases and to line the interiors of chemical reactors.

The strength of a bond between two atoms is measured by its dissociation energy: the greater the dissociation energy, the stronger the bond.

2.17 The Variation of Bond Strength

Tables 2.1 and 2.2 list a selection of typical dissociation energies. The values given in Table 2.2 are *average* dissociation energies for a number of different molecules. For instance, the strength quoted for a C—O single bond is the average strength of such bonds in a selection of organic molecules, such as methanol (CH_3—OH), ethanol (CH_3CH_2—OH), and dimethyl ether (CH_3—O—CH_3). The values should therefore be regarded as typical rather than as accurate values for a particular molecule.

The trends in bond strengths shown in Table 2.1 are explained in part by the Lewis structures for the molecules. Consider, for example, the diatomic molecules of nitrogen, oxygen, and fluorine (Fig. 2.15). Note the decline in bond strength as the bond order decreases—from three in N_2 to one in F_2. The triple bond in nitrogen is the origin of the inertness we mentioned at the beginning of the chapter. A multiple bond is almost always stronger than a single bond because more electrons bind the multiply bonded atoms. A triple bond between two atoms is always stronger

TABLE 2.1 Bond Dissociation Energies of Diatomic Molecules, $kJ\cdot mol^{-1}$

Molecule	Bond dissociation energy
H_2	424
N_2	932
O_2	484
CO	1062
F_2	146
Cl_2	230
Br_2	181
I_2	139
HF	543
HCl	419
HBr	354
HI	287

We shall not distinguish between the average dissociation energy and the average dissociation enthalpy, a concept to be introduced in Section 6.21. The two quantities differ by only a few kilojoules per mole.

TABLE 2.2 Average Bond Dissociation Energies, kJ·mol^{-1}

Bond	Average bond dissociation energy	Bond	Average bond dissociation energy
C—H	412	C—I	238
C—C	348	N—H	388
C=C	612	N—N	163
C⋯C*	518	N=N	409
C≡C	837	N—O	210
C—O	360	N=O	630
C=O	743	N—F	195
C—N	305	N—Cl	381
C—F	484	O—H	463
C—Cl	338	O—O	157
C—Br	276		

* In benzene.

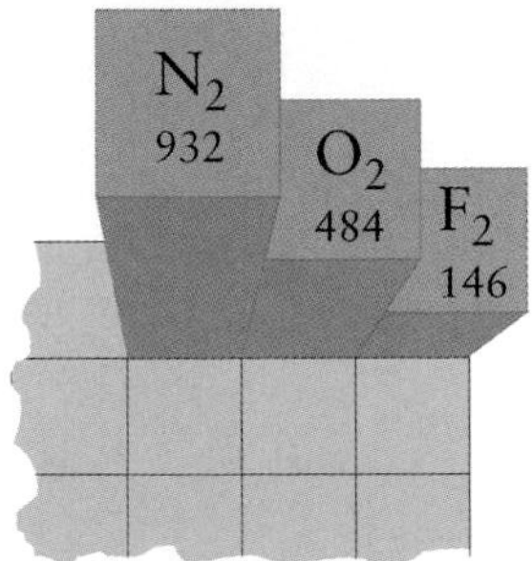

FIGURE 2.15 The bond dissociation energies, in kilojoules per mole of nitrogen, oxygen, and fluorine molecules. Note how the bonds weaken as they change from a triple bond in N_2 to a single bond in F_2.

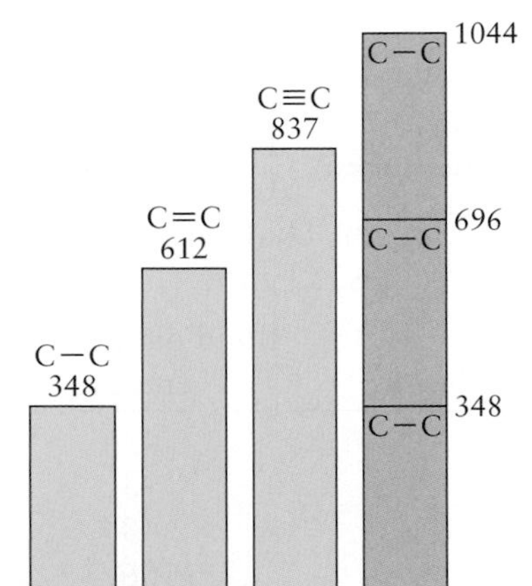

FIGURE 2.16 The strengths (in kilojoules per mole) of single and multiple bonds between two carbon atoms. Note that, for bonds between carbon atoms, a double bond is not twice as strong as a single bond and a triple bond is not three times as strong as a single bond.

than a double bond between the same two atoms, and a double bond is always stronger than a single bond between the same two atoms. However, a double bond between two carbon atoms is not twice as strong as a single bond, and a triple bond is a lot less than three times as strong. We see that the average dissociation energy of a C=C double bond is 612 kJ·mol^{-1}, whereas it takes 696 kJ·mol^{-1} to break two C—C single bonds; similarly, the average dissociation energy of a C≡C triple bond is 837 kJ·mol^{-1}, but it takes 1044 kJ·mol^{-1} to break three C—C single bonds (Fig. 2.16). The origin of these differences is in part the repulsions between the electron pairs in a multiple bond, so each pair is not quite as effective at bonding as a pair of electrons in a single bond.

The values in Table 2.2 show how resonance affects the strengths of bonds. For example, the strength of a carbon-carbon bond in benzene is intermediate between that of a single and a double bond. Resonance spreads multiple bond character over the bonds between atoms: as a result, what were single bonds are strengthened and what were double bonds are weakened. The net effect overall is a stabilization of the molecule because the carbon-carbon bond strength in benzene is closer to that of a double bond than to that of a single bond.

The presence of lone pairs may also influence the strengths of bonds. Lone pairs repel each other; and if they are on neighboring atoms, that repulsion can weaken the bond. This repulsion between lone pairs helps to explain why the bond in F_2 is weaker than the bond in H_2, for the latter molecule has no lone pairs.

Variations in atomic radii also contribute to the trend in bond strengths. If the nuclei of the bonded atoms cannot get very close to the electron pair lying between them, the two atoms will be only weakly bonded together. For example, the bond strengths of the hydrogen halides decrease from HF to HI, as shown in Fig. 2.17.

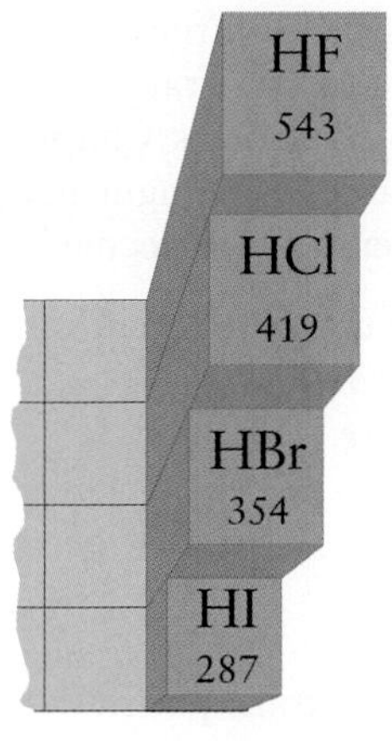

FIGURE 2.17 The bond dissociation energies of the hydrogen halide molecules in kilojoules per mole of molecules. Note how the bonds weaken as the halogen atom becomes larger.

The strength of the bond between hydrogen and a Group 14/IV element also decreases down the group, from carbon (412 kJ·mol^{-1}) to lead (205 kJ·mol^{-1}), as shown in Fig. 2.18. This weakening of the bond correlates with a decrease in the stability of the hydrides down the group. Methane, CH_4, can be kept indefinitely in air at room temperature. Silane, SiH_4, bursts into flame on contact with air. Stannane, SnH_4, decomposes into tin and hydrogen. Plumbane, PbH_4, has never been prepared, except perhaps in trace amounts.

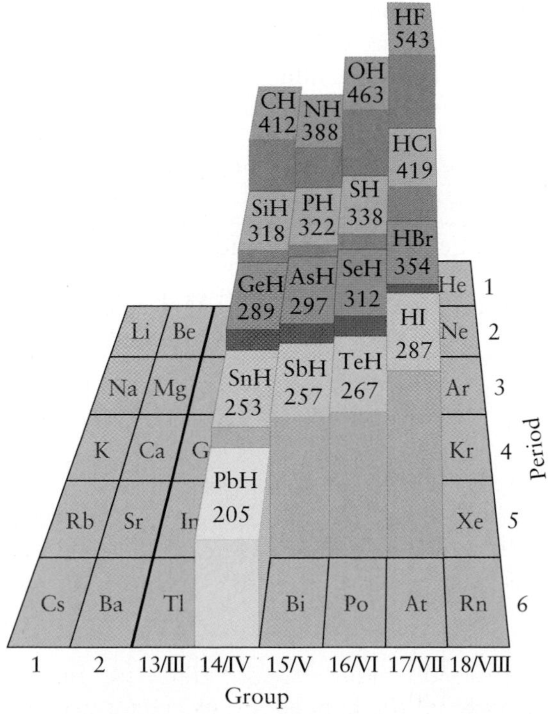

FIGURE 2.18 The dissociation energies for bonds between hydrogen and the *p*-block elements. The bond strengths decrease down each group as the atoms increase in size.

Differences in the characteristics of bonds, such as their stiffness (their resistance to stretching and compressing), are studied by infrared (IR) spectroscopy, as described in Major Technique 1, which follows Chapter 2. In an IR spectrum, different groups of atoms can be distinguished, allowing chemists to determine which groups are present in a molecule.

The bond strength increases as the multiplicity of a bond increases, decreases as the number of lone pairs on neighboring atoms increases, and decreases as the atomic radii increase.

2.18 Bond Lengths

A **bond length** is the distance between the centers of two atoms joined by a covalent bond. Bond lengths help determine the overall size and shape of a molecule. The transmission of hereditary information in DNA, for instance, depends on bond lengths because the two strands of the double helix must fit together like pieces of a jigsaw puzzle (Fig. 2.19). Bond lengths are also crucial to the action of enzymes because only a molecule of the right size and shape will fit into the active site of the enzyme molecule (Section 13.15). As we see from Table 2.3, the lengths of bonds between Period 2 elements typically lie in the range 100 pm to 150 pm. Bond lengths are determined experimentally by using either spectroscopy or x-ray diffraction (Box 2.2).

Bonds between heavy atoms tend to be longer than those between light atoms because heavier atoms have larger radii than lighter ones (Fig. 2.20). *Multiple bonds are shorter than single bonds* between the same two elements because the additional bonding electrons attract the nuclei more strongly and pull the atoms closer together: compare the lengths of the various carbon-carbon bonds in Table 2.3. We can also see the averaging effect of resonance. The length of the carbon-carbon bond in benzene is

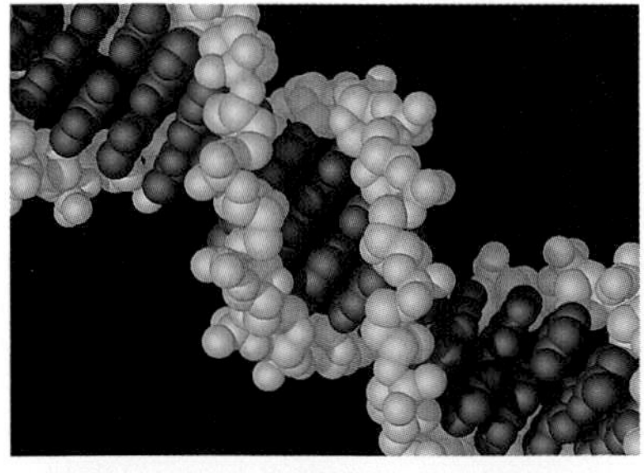

FIGURE 2.19 DNA molecules consist of two very long chains of four units that repeat in a specific order that determines our genetic makeup. The shapes of the units and their charge distributions cause the chains to wind around each other to form a double helix.

TABLE 2.3 Average and Actual Bond Lengths

Bond	Average bond length, pm	Molecule Molecule	Bond length, pm
C—H	109	H_2	74
C—C	154	N_2	110
C=C	134	O_2	121
C┅C*	139	F_2	142
C≡C	120	Cl_2	199
C—O	143	Br_2	228
C=O	112	I_2	268
O—H	96		
N—H	101		
N—O	140		
N=O	120		

* In benzene.

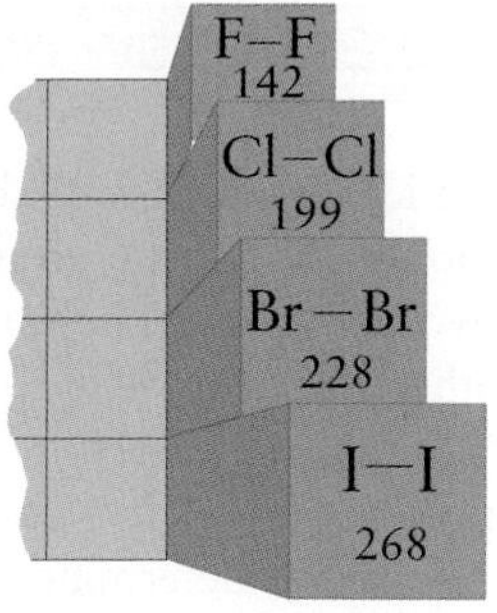

FIGURE 2.20 Bond lengths (in picometers) of the diatomic halogen molecules. Notice how the bond lengths increase down the group as the atomic radii become larger.

intermediate between the lengths of the single and double bonds of a Kekulé structure (but closer to that of a double bond). For bonds between the same pair of atoms, *the stronger the bond, the shorter it is*. Thus, a C≡C triple bond is both stronger and shorter than a C=C double bond. Similarly, a C=O double bond is both stronger and shorter than a C—O single bond.

Each atom makes a characteristic contribution, called its **covalent radius,** to the length of a bond (Fig. 2.21). A bond length is approximately the sum of the covalent radii of the two atoms (**44**). The O—H bond length in ethanol, for example, is the sum of the covalent radii of H and O, 34 + 74 pm = 111 pm. We also see from Fig. 2.21 that the covalent radius of an atom taking part in a multiple bond is smaller than that for a single bond of the same atom.

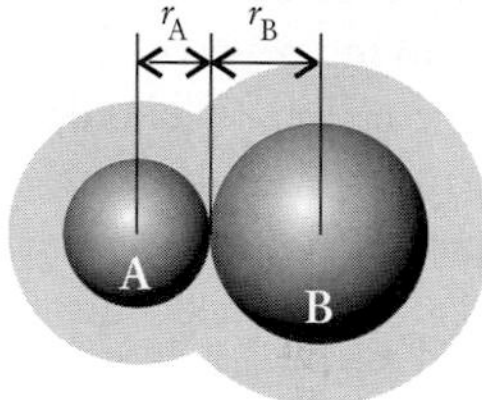

44 Covalent radii

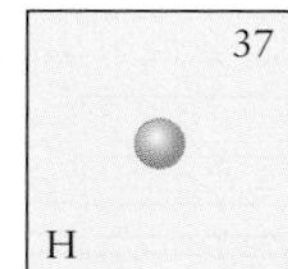

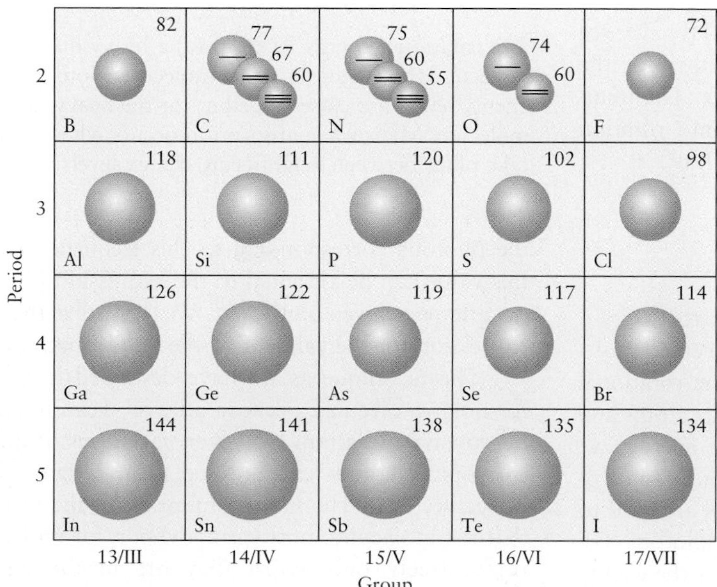

FIGURE 2.21 Covalent radii of hydrogen and the *p*-block elements (in picometers). Where more than one value is given, they refer to single, double, and triple bonds. Covalent radii tend to become smaller toward fluorine. A bond length is approximately the sum of the covalent radii of the two atoms involved.

BOX 2.2 HOW DO WE KNOW . . . THE LENGTH OF A CHEMICAL BOND?

A chemical bond is only about 10^{-10} m long, so special techniques have to be used to measure its length. There are two principal techniques: one is used for solids and the other for gases. The technique used for solids, x-ray diffraction, is described in Major Technique 3, following Chapter 5. *Microwave spectroscopy*, which we discuss here, is used to determine bond lengths in gas molecules. This branch of spectroscopy makes use of the ability of rotating molecules to absorb microwave radiation, radiation having a wavelength close to 1 cm.

According to classical physics, a solid body like a ball can rotate with any energy. According to quantum mechanics, however, rotational energy is quantized, and a body can rotate only with certain energies—that is, only at certain speeds. Let's see what that means for a diatomic molecule AB, with atomic masses m_A and m_B and bond length R. The molecule can rotate only with the following energies:

$$E = \frac{h^2 J(J+1)}{8\pi^2 \mu R^2} \qquad J = 0, 1, 2, \ldots$$

where $\mu = m_A m_B/(m_A + m_B)$, h is the Planck constant, and J is a quantum number. These energy levels are shown in the figure for two kinds of molecules, one with heavy atoms and a long bond (such as ICl) and the other with light atoms and a short bond (such as HF). We see that the energy levels are much closer together for the heavy molecule than for the lighter molecule. The minimum energy needed to excite a molecule into rotation from rest (corresponding to $J = 0$ and $E = 0$) is

$$\Delta E = \frac{h^2}{4\pi^2 \mu R^2}$$

Perhaps surprisingly, less energy is needed to excite a large heavy molecule than a small light molecule.

Because the energy needed to change the rotational state of a molecule depends on the masses of its atoms and the bond length, if we could measure the energy, we would have a way to measure the bond length. When typical values of molar masses and bond lengths are used to calculate rotational energies, it is found that photons corresponding to microwave radiation can supply the energy needed to excite the molecule to higher rotational energies. Therefore, to determine a bond length, we pass a beam of microwave radiation through a gaseous sample, vary its frequency, and determine the frequency, ν (nu), that results in a strong absorption. The energy of the photons corresponding to this frequency is $h\nu$, and this value can be set equal to the expression for the energy difference given on the left. We then solve the resulting expression for the value of R, the bond length.

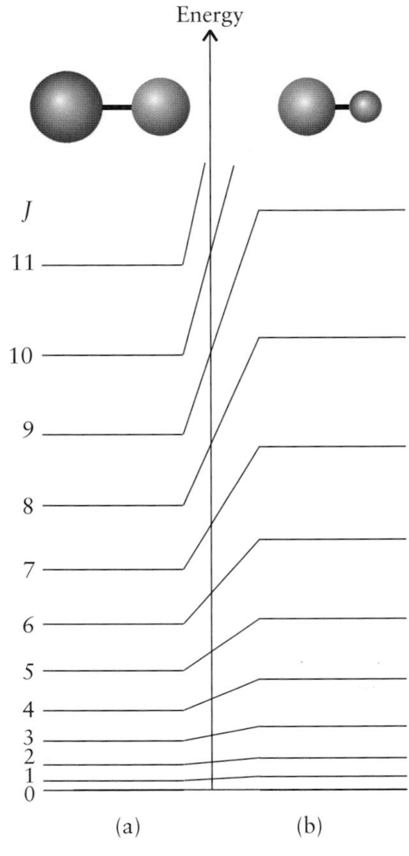

The rotational energy levels of (a) a heavy diatomic molecule, (b) a light diatomic molecule. Note that the energy levels are closer together for the heavy diatomic molecule. Microwave absorption occurs when transitions take place between neighboring energy levels.

The technique as we have described it works only for polar molecules, because only they can interact with microwave radiation. However, variations of these spectroscopic methods can be used to investigate nonpolar molecules, too. The major limitation of the technique is that the molecules must form a vapor, for molecules can rotate freely only when they are in the gas phase. Another limitation is that microwave spectra are very complicated for molecules consisting of more than a few atoms, so the technique is confined to simple molecules. For complex molecules, we use solid samples and x-ray diffraction techniques.

Covalent radii typically decrease from left to right across a period. The reason is the same as for atomic radii (Section 1.14): the increasing effective nuclear charge draws in the electrons and makes the atom more compact. Like atomic radii, covalent radii increase down a group because, in successive periods, the valence electrons occupy shells that are more distant from the nucleus and are better shielded by the inner core of electrons. Such fatter atoms cannot approach their neighbors very closely; hence, they form long, weak bonds.

The covalent radius of an atom is the contribution it makes to the length of a covalent bond; covalent radii are added together to estimate the lengths of bonds in molecules.

SKILLS YOU SHOULD HAVE MASTERED

1 Compare the relative lattice energies of two ionic compounds, Example 2.1.

2 Write the electron configuration for an ion, Example 2.2 and Self-Test 2.3.

3 Draw the Lewis structures of molecules, radicals, and ions, Toolbox 2.1 and Examples 2.3, 2.4, and 2.7.

4 Write the resonance structures for a molecule, Example 2.5.

5 Use formal charge calculations to evaluate alternative Lewis structures, Toolbox 2.2 and Examples 2.6 and 2.8.

6 Explain the characteristics of Lewis acids and bases and how they form bonds, Section 2.13.

7 Predict which of two bonds has greater ionic or covalent character, Self-Tests 2.13 and 2.14.

8 Predict and explain periodic trends in the polarizability of anions and the polarizing power of cations, Section 2.15.

9 Predict and explain relative bond strengths and lengths, Sections 2.16–2.18.

EXERCISES

Ionic Bonds

2.1 Using data in Appendix 2D, predict which of the following pairs of ions would have the greater coulombic attraction in a solid compound: (a) K^+, O^{2-}; (b) Ga^{3+}, O^{2-}; (c) Ca^{2+}, O^{2-}.

2.2 Using data from Appendix 2D, predict which of the following pairs of ions would have the greater coulombic attraction in a solid compound: (a) Mg^{2+}, S^{2-}; (b) Mg^{2+}, Se^-; (c) Mg^{2+}, O^{2-}.

2.3 Explain why the lattice energy of lithium chloride ($861\ kJ{\cdot}mol^{-1}$) is greater than that of rubidium chloride ($695\ kJ{\cdot}mol^{-1}$), given that they have similar arrangements of ions in the crystal lattice. See Appendix 2D.

2.4 Explain why the lattice energy of silver bromide ($903\ kJ{\cdot}mol^{-1}$) is greater than that of silver iodide ($887\ kJ{\cdot}mol^{-1}$). See Appendix 2D.

Electron Configurations of Ions

2.5 Give the number of valence electrons for each of the following elements: (a) Sb; (b) Si; (c) Mn; (d) B.

2.6 Give the number of valence electrons for each of the following elements: (a) I; (b) Ni; (c) Re; (d) Sr.

2.7 Give the ground-state electron configuration expected for each of the following ions: (a) S^{2-}; (b) As^{3+}; (c) Ru^{3+}; (d) Ge^{2+}.

2.8 Give the ground-state electron configuration expected for each of the following ions: (a) Os^{4+}; (b) Pb^{2+}; (c) Tl^+; (d) I^{5+}.

2.9 Give the ground-state electron configuration expected for each of the following ions: (a) Sr^{2+}; (b) Bi^{3+}; (c) Ga^{3+}; (d) Tl^{3+}.

2.10 Give the ground-state electron configuration predicted for each of the following ions: (a) Al^{3+}; (b) Tc^{4+}; (c) Ra^{2+}; (d) I^-.

2.11 Predict the ground-state electron configuration of each of the following metal ions: (a) Pt^{2+}; (b) Cr^{3+}; (c) Cu^{2+}; (d) V^{3+}.

2.12 Predict the ground-state electron configuration of each of the following metal ions: (a) Ir^{+}; (b) Ru^{4+}; (c) Mo^{2+}; (d) Co^{3+}.

2.13 Which M^{2+} ions (M = metal) are predicted to have the following ground-state electron configurations: (a) $[Ar]3d^7$; (b) $[Ar]3d^6$; (c) $[Kr]4d^4$; (d) $[Kr]4d^3$?

2.14 Which E^{3+} (E = element) ions are predicted to have the following ground-state electron configurations: (a) $[Xe]4f^{14}5d^8$; (b) $[Xe]4f^{14}5d^5$; (c) $[Kr]4d^{10}5s^25p^2$; (d) $[Ar]3d^{10}4s^2$?

2.15 Which M^{3+} ions (M = metal) are predicted to have the following ground-state electron configurations: (a) $[Ar]3d^6$; (b) $[Ar]3d^5$; (c) $[Kr]4d^5$; (d) $[Kr]4d^3$?

2.16 Which M^{2+} ions are predicted to have the following ground-state electron configurations: (a) $[Ar]3d^7$; (b) $[Kr]4d^7$; (c) $[Kr]4d^{10}5s^2$; (d) $[Xe]4f^{14}5d^{10}$?

2.17 For each of the following ground-state atoms, predict the type of orbital (1*s*, 2*p*, 3*d*, 4*f*, etc.) from which an electron will need to be removed in order to form the +1 ions: (a) Zn; (b) Cl; (c) Al; (d) Cu.

2.18 For each of the following ground-state ions, predict the type of orbital (1*s*, 2*p*, 3*d*, 4*f*, etc.) from which an electron will need to be removed in order to form the ions of one greater positive charge: (a) Ti^{3+}; (b) In^{+}; (c) Te^{2-}; (d) Ag^{+}.

2.19 Write the most likely charge for the ions formed by each element: (a) Br; (b) Te; (c) Cs; (d) Ga; (e) Tl.

2.20 Write the most likely charge for the ions formed by each element: (a) Sr; (b) Pb; (c) Ge; (d) I; (e) O.

2.21 Predict the number of valence electrons present for each of the following ions: (a) Mn^{4+}; (b) Rh^{3+}; (c) Co^{3+}; (d) P^{3+}.

2.22 Predict the number of valence electrons present for each of the following ions: (a) In^{+}; (b) Tc^{2+}; (c) Ta^{2+}; (d) Re^{+}.

2.23 Give the ground-state electron configuration and number of unpaired electrons expected for each of the following ions: (a) Sb^{3+}; (b) Sn^{4+}; (c) W^{2+}; (d) Br^{-}; (e) Ni^{2+}.

2.24 Give the ground-state electron configuration and number of unpaired electrons expected for each of the following ions: (a) Ga^{+}; (b) Cu^{2+}; (c) Pb^{2+}; (d) Se^{2-}.

2.25 For each of the following ground-state ions, predict the type of orbital (1*s*, 2*p*, 3*d*, 4*f*, etc.) that the electrons of highest energy will occupy: (a) Ti^{3+}; (b) In^{+}; (c) Te^{2-}; (d) Ag^{+}.

2.26 For each of the following ground-state ions, predict the type of orbital (1*s*, 2*p*, 3*d*, 4*f*, etc.) that the electrons of highest energy will occupy: (a) Fe^{2+}; (b) Bi^{3+}; (c) As^{3+}; (d) Os^{+}.

2.27 Iodine can exist in both positive and negative oxidation states. What is the maximum (a) positive and (b) negative oxidation number that iodine can have? (c) Write the electron configuration for each of these states. (d) Explain how you arrived at these values.

2.28 Sulfur can exist in both positive and negative oxidation states. What is the maximum (a) positive and (b) negative oxidation number that sulfur can have? (c) Write the electron configuration for each of these states. (d) Explain how you arrived at these values.

2.29 On the basis of the expected charges on the monatomic ions, give the chemical formula of each of the following compounds: (a) magnesium arsenide; (b) indium(III) sulfide; (c) aluminum hydride; (d) hydrogen telluride; (e) bismuth(III) fluoride.

2.30 On the basis of the expected charges on the monatomic ions, give the chemical formula of each of the following compounds: (a) sodium sulfide; (b) lithium nitride; (c) calcium chloride; (d) gallium arsenide; (e) cobalt(III) oxide.

2.31 On the basis of the expected charges on the monatomic ions, give the chemical formula of each of the following compounds: (a) gallium arsenide; (b) magnesium oxide; (c) aluminum telluride; (d) ruthenium(IV) oxide; (e) vanadium(V) oxide.

2.32 On the basis of the expected charges on the monatomic ions, give the chemical formula of each of the following compounds: (a) manganese(II) telluride; (b) barium arsenide; (c) silicon nitride; (d) lithium bismuthide; (e) zirconium(IV) chloride.

2.33 On the basis of the expected charges of the monatomic ions, give the chemical formula of each of the following compounds: (a) bismuth(III) oxide; (b) lead(IV) oxide; (c) thallium(III) oxide.

2.34 On the basis of the expected charges of the monatomic ions, give the chemical formula of each of the following compounds: (a) iron(II) sulfide; (b) cobalt(III) chloride; (c) lithium phosphide.

Covalent Bonds

2.35 Write the Lewis structure of (a) CCl_4; (b) $COCl_2$; (c) ONF; (d) NF_3.

2.36 Write the Lewis structure of (a) SCl_2; (b) AsH_3; (c) $GeCl_4$; (d) $SnCl_2$.

2.37 Write the Lewis structure of (a) tetrahydridoborate ion, BH_4^-; (b) hypobromite ion, BrO^-; (c) amide ion, NH_2^-.

2.38 Write the Lewis structure of (a) nitronium ion, ONO^+; (b) chlorite ion, ClO_2^-; (c) peroxide ion, O_2^{2-}; (d) formate ion, HCO_2^-.

2.39 Write the complete Lewis structure for each of the following compounds: (a) ammonium chloride; (b) potassium phosphide; (c) sodium hypochlorite.

2.40 Write the complete Lewis structure for each of the following compounds: (a) zinc cyanide; (b) potassium tetrafluoroborate; (c) barium peroxide (the peroxide ion is O_2^{2-}).

2.41 Write the complete Lewis structure for each of the following compounds: (a) formaldehyde, HCHO, which as its aqueous solution "formalin" is used to preserve biological specimens; (b) methanol, CH_3OH, the toxic compound also called wood alcohol; (c) glycine, $CH_2(NH_2)COOH$, the simplest of the amino acids, the building blocks of proteins.

2.42 Write the Lewis structure of each of the following organic compounds: (a) ethanol, CH_3CH_2OH, which is also called ethyl alcohol or grain alcohol; (b) methylamine, CH_3NH_2, a putrid-smelling substance formed when flesh decays; (c) formic acid, HCOOH, a component of the venom injected by ants.

2.43 Anthracene has the formula $C_{14}H_{10}$. It is similar to benzene but has three six-membered rings that share common C—C bonds, as shown below. Complete the structure by drawing in multiple bonds to satisfy the octet rule at each carbon atom. Resonance structures are possible. Draw as many as you can find.

```
      H     H     H
      |     |     |
 H    C     C     C    H
   C     C     C     C
   |     |     |     |
   C     C     C     C
 H    C     C     C    H
      |     |     |
      H     H     H
```

2.44 Write the Lewis structures that contribute to the resonance hybrid of the guanadinium ion, $C(NH_2)_3^+$.

2.45 Draw the Lewis structures that contribute to the resonance hybrid of nitryl chloride, $ClNO_2$ (N is the central atom).

2.46 Do H—C≡N and H—N≡C form a pair of resonance structures? Explain your answer.

2.47 Draw the Lewis structure and determine the formal charge on each atom in (a) NO^+; (b) N_2; (c) CO; (d) C_2^{2-}; (e) CN^-.

2.48 Using only Lewis structures that obey the octet rule, draw the Lewis structures and determine the formal charge on each atom in (a) CH_3^+; (b) OCl^-; (c) BF_4^-.

2.49 Two contributions to the resonance structure are shown below for each species. Determine the formal charge on each atom and then, if possible, identify the Lewis structure of lowest energy for each species.

(a) $[$:O—S(H)(=O:)—O:$]^-$ $[$:O—S(H)(—O:)—O:$]^-$

(b) $[$:O—S(—O:)(H)(=O:)—O:$]^-$ $[$O=S(—O:)(H)(=O:)—O:$]^-$

2.50 Two Lewis structures are shown below for each species. Determine the formal charge on each atom and then, if appropriate, identify the Lewis structure of lower energy for each species.

(a) :O—S=O   O=S=O

(b) O=S(=O:)=O   :O—S(—O:)—O:

2.51 Determine the formal charge on each atom in the following molecules. Identify the structure of lower energy in each pair.

(a) O=Cl(=O:)—O:—H   :O—Cl(—O:)—O:—H

(b) O=C=S   :O—C≡S:

(c) H—C≡N: H—C=N:

2.52 Determine the formal charge on each atom in the following ions. Identify the structure of lowest energy in each case.

(a) $[$N=C=N$]^{2-}$ $[$:N≡C—N:$]^{2-}$

(b) $[$:O—As(—O:)(—O:)—O:$]^{3-}$ $[$:O—As(=O:)(—O:)—O:$]^{3-}$ $[$O=As(=O:)(=O:)=O$]^{3-}$

(c)

```
[   :O:   ]-    [   :O:   ]-    [   :O:   ]-    [   :O:   ]-
[    ||   ]     [    ||   ]     [    |    ]     [    |    ]
[ O = I = O]    [ O = I - O:]   [ O = I = O:]   [:O - I - O:]
[    |    ]     [    |    ]     [    |    ]     [    |    ]
[   :O:   ]     [   :O:   ]     [   :O:   ]     [   :O:   ]
```

2.53 Select from each of the following pairs of Lewis structures the one that is likely to make the dominant contribution to a resonance hybrid. Explain your selection.

(a) :F—Xe—F: :F—Xe=F

(b) O=C=O :O—C≡O:

2.54 Select from each of the following pairs of Lewis structures the one that is likely to make the dominant contribution to a resonance hybrid. Explain your selection.

(a) N=N=O :N≡N—O:

(b)

```
[   :O:   ]3-    [   :O:   ]3-
[    |    ]      [    |    ]
[ O = P - O:]    [ O = P = O ]
[    |    ]      [    |    ]
[   :O:   ]      [   :O:   ]
```

2.55 Write the Lewis structure, including typical contributions to the resonance structure (where appropriate, allow for the possibility of octet expansion, including double bonds in different positions), for (a) sulfite ion; (b) hydrogen sulfite ion; (c) perchlorate ion; (d) nitrite ion.

2.56 Write the Lewis structure, including typical contributions to the resonance structure (where appropriate, allow for the possibility of octet expansion), for (a) dihydrogen phosphate ion; (b) chlorite ion; (c) chlorate ion; (d) nitrate ion.

Exceptions to the Octet Rule

2.57 Which of the following species are radicals? (a) NO_2^-; (b) CH_3; (c) OH; (d) CH_2O.

2.58 Which of the following species are radicals? (a) NO; (b) CH_3^+; (c) BF_4^-; (d) BrO.

2.59 Write the Lewis structure of each of the following reactive species found to contribute to the destruction of the ozone layer and indicate which are radicals: (a) chlorine monoxide, ClO; (b) dichloroperoxide, Cl—O—O—Cl; (c) chlorine nitrate, $ClONO_2$ (the central O atom is attached to the Cl atom and to the N atom of the NO_2 group); (d) chlorine peroxide, Cl—O—O.

2.60 Write the Lewis structure of each of the following species and indicate which are radicals: (a) the superoxide ion, O_2^-; (b) the methoxy group, CH_3O; (c) XeO_4; (d) $HXeO_4^-$.

2.61 Write the Lewis structure for each of the following molecules or ions and give the number of electrons about the central atom: (a) SF_6; (b) XeF_2; (c) AsF_6^-; (d) $TeCl_4$.

2.62 Write the Lewis structure for each of the following molecules or ions and give the number of electrons about the central atom: (a) SiF_6^{2-}; (b) IF_7; (c) ClF_3; (d) BrF_2^+.

2.63 Write the Lewis structure and state the number of lone pairs on xenon, the central atom of each of the following compounds: (a) $XeOF_2$; (b) XeF_4; (c) $XeOF_4$.

2.64 Write the Lewis structure and state the number of lone pairs on the central atom of each of the following compounds: (a) ClF_3; (b) AsF_5; (c) SF_4.

Lewis Acids and Bases

2.65 Identify the following species as Lewis acids or Lewis bases: (a) $N(C_2H_5)_3$; (b) SbF_5; (c) CN^-; (d) Fe^{2+}.

2.66 Identify the following species as Lewis acids or Lewis bases: (a) $Al(CH_3)_3$; (b) In^{3+}; (c) HCO_2^-; (d) CO_3^{2-}.

2.67 Identify the Lewis acid and the Lewis base and then write the product (a complex) for the following Lewis acid-base reactions:

(a) $PF_5 + F^- \rightarrow ?$

(b) $SO_2 + Cl^- \rightarrow ?$

(c) $OH^- + SO_3 \rightarrow ?$

2.68 Identify the Lewis acid and the Lewis base and then write the product (a complex) for the following Lewis acid-base reactions:

(a) $GaCl_3 + Cl^- \rightarrow ?$

(b) $CO_2 + OH^- \rightarrow ?$

(c) $SiF_4 + 2F^- \rightarrow ?$

2.69 Which of the following members of each pair is the stronger Lewis base? Explain your reasoning. (a) CH_3^- or CH_4; (b) H_2O or H_2S; (c) NH_3 or NH_2^-.

2.70 Which of the following members of each pair is the stronger Lewis acid? Explain your reasoning. (a) BF_3 or NF_3; (b) Al^{3+} or K^+; (c) Mg^{2+} in MgF_2 or Mg^{2+} in $MgCl_2$.

Ionic Versus Covalent Bonds

2.71 List the halogens in order of increasing electronegativity.

2.72 Just as some elements are electronegative, others can be *electropositive*, a term meaning that the element has a strong tendency to give up electrons. Elements that are least electronegative are the most electropositive. What trend exists among the alkali metals and the alkaline earth metals with respect to their electropositive character?

2.73 Place the following elements in order of increasing electronegativity: antimony, tin, selenium, indium.

2.74 Place the following elements in order of increasing electronegativity: oxygen, carbon, nitrogen, fluorine, silicon, phosphorus, sulfur.

2.75 For each pair, determine which compound has bonds with greater ionic character: (a) HCl or HI; (b) CH_4 or CF_4; (c) CO_2 or CS_2.

2.76 For each pair, determine which compound has bonds with greater ionic character: (a) PH_3 or NH_3; (b) SO_2 or NO_2; (c) SF_6 or IF_5.

2.77 Arrange the cations Rb^+, Be^{2+}, and Sr^{2+} in order of increasing polarizing power. Give an explanation of your arrangement.

2.78 Arrange the cations K^+, Mg^{2+}, Al^{3+}, and Cs^+ in order of increasing polarizing power. Give an explanation of your arrangement.

2.79 Arrange the ions Cl^-, Br^-, N^{3-}, and O^{2-} in order of increasing polarizability, giving reasons for your decisions.

2.80 Arrange the ions N^{3-}, P^{3-}, I^-, and At^- in order of increasing polarizability, giving reasons for your decisions.

The Strengths and Lengths of Covalent Bonds

2.81 On the basis of their Lewis structures, place the following molecules or ions in order of *decreasing* bond length: (a) the CO bond in CO, CO_2, or CO_3^{2-}; (b) the SO bond in SO_2, SO_3, or SO_3^{2-}; (c) the CN bond in HCN, CH_2NH, or CH_3NH_2.

2.82 On the basis of their Lewis structures, place the following molecules or ions in order of *decreasing* bond order: (a) N—O bond in NO, NO_2, or NO_3^-; (b) CC bond in C_2H_2, C_2H_4, or C_2H_6; (c) CO bond in CH_3OH, CH_2O, or CH_3OCH_3.

2.83 Use the information in Fig. 2.21 to estimate the bond length of (a) the NN bond in hydrazine, H_2NNH_2; (b) the CO bond in CO_2; (c) the CO and CN bonds in urea, $OC(NH_2)_2$; (d) the NN bond in nitrogen hydride, HNNH.

2.84 Use the information in Figure 2.21 to estimate the indicated bond length of (a) the CO bond in formaldehyde, H_2CO; (b) the CO bond in dimethyl ether, CH_3OCH_3; (c) the CO bond in methanol, CH_3OH; (d) the CS bond in methanethiol, CH_3SH.

2.85 Use covalent radii in Fig. 2.21 to calculate the bond lengths in the following molecules. Account for the trends in the values you calculate. (a) CF_4; (b) SiF_4; (c) SnF_4.

2.86 Which do you predict to have the strongest CN bond: (a) $NHCH_2$, (b) NH_2CH_3, or (c) HCN?

INTEGRATED EXERCISES

2.87 Write the Lewis structure, including resonance structures where appropriate, for (a) the oxalate ion, $C_2O_4^{2-}$ (there is a C—C bond with two oxygen atoms attached to each carbon atom); (b) BrO^+; (c) the acetylide ion, C_2^{2-}. Assign formal charges to each atom.

2.88 Draw resonance structures for the trimethylenemethane anion $C(CH_2)_3^{2-}$ in which a central carbon atom is attached to three CH_2 group (CH_2 groups are commonly referred to as methylene).

2.89 Show how resonance can occur in the following organic ions: (a) acetate ion, $CH_3CO_2^-$; (b) enolate ion, $CH_2COCH_3^-$ which has one resonance structure with a C=C double bond and an —O$^-$ group on the central carbon atom; (c) allyl cation, $CH_2CHCH_2^+$; (d) amidate ion, CH_3CONH^- (the O and the N atoms are both bonded to the second C atom).

2.90 For a given formula, it is possible to have a number of reasonable Lewis structures where the atoms have different connectivities (a term meaning that different atoms are bonded to each other to form the various structures). Compounds with the same formula but different connectivities are known as *isomers*. Hydrogen atoms in benzene can be replaced by other atoms such as the halogens. Draw Lewis structures that obey the octet rule but have different arrangements of atoms for the compound dichlorobenzene, $C_6H_4Cl_2$.

2.91 A highly toxic gas that has been used in chemical warfare gives the following elemental analysis figures:

12.1% carbon, 16.2% oxygen, and 71.7% chlorine by mass, and its molar mass is 98.9 g·mol^{-1}. Write the Lewis structure of this compound.

2.92 White phosphorus is composed of tetrahedral molecules of P_4 in which each P atom is connected to three other P atoms. Draw the Lewis structure for this molecule. Does it obey the octet rule?

2.93 Nitrogen, oxygen, phosphorus, and sulfur exist as N_2, O_2, tetrahedral P_4, and cyclic S_8 molecules. Explain this fact in terms of the abilities of the atoms to form different types of bonds with one an other.

2.94 Limestone (calcium carbonate) is used to remove impurities from iron ore in a blast furnace. When limestone is heated, it decomposes into calcium oxide and carbon dioxide, and the calcium oxide combines with silica, SiO_2, impurities to form calcium silicate, $CaSiO_3$. The latter is commonly called "slag."
(a) Write the chemical equation for the Lewis acid-base decomposition of carbonate ions (the complex) and a Lewis acid-base reaction for the formation of calcium silicate. Identify the Lewis acid and Lewis base in each reaction. (b) The SiO_3^{2-} ion can be drawn with a Lewis diagram similar to that of the carbonate ion. Silicon does not readily form double bonds to oxygen, and as a consequence, other means are used to achieve an octet of electrons about the silicon atom and silicate minerals are generally more complex than the formula SiO_3^{2-} would suggest. Go to the literature and find examples of silicate minerals that have a chemical formula based on SiO_3^{2-}. Draw Lewis structures for the silicate part of the structure and discuss how these structures are able to satisfy the octet rule.

2.95 An important principle in chemistry is the *isolobal analogy*. This very simple principle states that chemical fragments with similar valence orbital structures can replace each other in molecules. For example ·Ċ—H and ·Ṡi—H are isolobal fragments, each having three electrons with which to form bonds in addition to the bond to H. An isolobal series of molecules would be HCCH, HCSiH, HSiSiH. Similarly, a lone pair of electrons can be used to replace a bond so that ·Ṅ: is isolobal with ·Ċ—H, with the lone pair taking the place of the C—H bond. The isolobal set here is HCCH, HCN, NN. (a) Draw the Lewis structures for the molecules HCCH, HCSiH, HSiSiH, HCN, and NN. (b) Using the isolobal principle, draw Lewis structures for molecules based on the structure of benzene, C_6H_6, in which one or more CH groups are replaced with N atoms.

2.96 The cyclopentadienide ion, $C_5H_5^-$, is a common organic anion that forms very stable complexes with metal cations. The anion is derived by removing a proton from cyclopentadiene, C_5H_6, with strong base. The molecule has a five-membered ring of carbon atoms, with four carbon atoms attached to only one proton and one carbon atom bonded to two. Draw the Lewis structure of cyclopentadiene. Are resonance forms possible for this molecule? The ion $C_5H_5^-$ consists of a planar five-membered ring of carbon atoms, with one hydrogen atom attached to each carbon atom. Draw the Lewis structure for $C_5H_5^-$. How many resonance structures can you draw for this anion?

2.97 The bond energy in NO is 632 kJ·mol^{-1} and that of each N—O bond in NO_2 is 469 kJ·mol^{-1}. Using Lewis structures and average bond energies in Table 2.2, explain (a) the difference in bond energies between the two molecules and (b) the fact that the bond energies of the two bonds in NO_2 are the same.

2.98 (a) Show that lattice energies are inversely proportional to the distance between the ions in MX (M = alkali metal, X = halide ion) by plotting the lattice energies of KF, KCl, and KI against the internuclear distances $d_{M\text{-}X}$. The lattice energies of KF, KCl, and KI are 826, 717, and 645 kJ·mol^{-1}, respectively. Use the ionic radii found in Appendix 2D to calculate $d_{M\text{-}X}$. How good is the correlation? You should use a standard graphing program to make the plot that will generate an equation for the line and calculate a correlation coefficient for the fit (see the Web site for this book). (b) Estimate the lattice energy of KBr from your graph. (c) Find an experimental value for the lattice energy of KBr in the chemical literature and compare that value to that calculated in (b). How well do they agree?

2.99 (a) Show that the lattice energies of the alkali metal iodides are inversely proportional to the distances between the ions in MI (M = alkali metal) by plotting the lattice energies given below versus the internuclear distances $d_{M\text{-}I}$.

Alkali metal iodide	Lattice energy (kJ·mol^{-1})
LiI	759
NaI	700
KI	645
RbI	632
CsI	601

Use the ionic radii found in Appendix 2D to calculate $d_{M\text{-}X}$. How good is the correlation? You should use a standard graphing program to make the plot that will generate an equation for the line and calculate a correlation coefficient for the fit (see the Web site for this book). (b) From the ionic radii given in Appendix 2D and the plot given in part (a), estimate the lattice energy for silver iodide. (c) Compare your results from part (b) to the experimentally determined value of 886 $kJ\cdot mol^{-1}$. If they do not agree, provide an explanation for the deviation.

2.100 The framework for the tropylium cation, $C_7H_7^+$, is a seven-membered ring of carbon atoms with a hydrogen atom attached to each carbon atom. Complete the structural drawing by adding the multiple bonds as appropriate. Resonance structures are possible. Draw as many as you can find. Determine the C—C bond order.

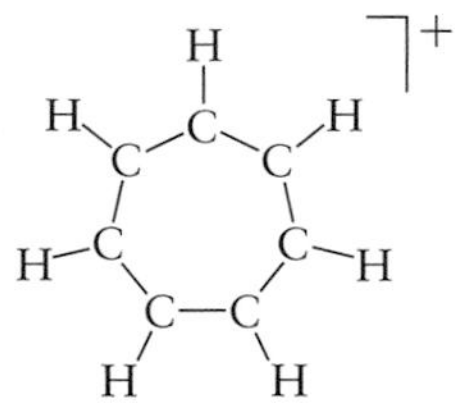

2.101 Quinone, $C_6H_4O_2$, is an organic molecule that can be reduced to give a dianion: $[C_6H_4O_2]^{2-}$. (a) Draw the Lewis structure of the reduced product. (b) On the basis of formal charges derived from the Lewis structure, predict which atoms in the molecule are most negatively charged. (c) If two protons are added to the reduced product, where are they most likely to bond?

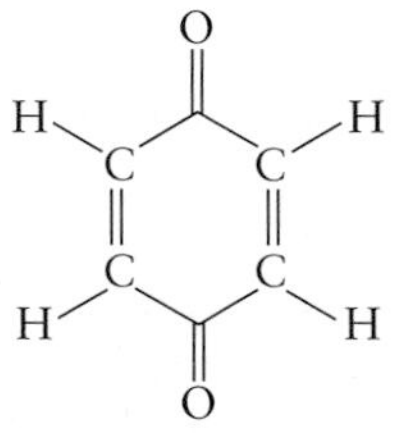

Quinone

2.102 In air, the NO radical can react with the NO_2 radical. What is the most likely product of the reaction? Answer this question by drawing Lewis structures of the reactants and products.

2.103 A common biologically active radical is the pentadienyl radical, RCHCHCHCHCHR′, where the carbons form a long chain with R and R′, which can be a number of different organic groups, at each end. Draw three resonance structures for this compound that maintain carbon's valence of four.

2.104 Hydrofluoric acid is a weak acid; however, a mixture of SbF_5 and HF is a *superacid*. Superacids are extremely strong acids and are very important chemicals, causing a number of important molecular transformations. George Olah won a Nobel prize in part for his research on the formation of carbonium ions in superacid solvents. (a) On the basis of your present knowledge of bonding, describe how the interaction of SbF_5 and HF can form a superacid. (b) Propose some other combinations of molecules that could lead to strong acids.

2.105 An important aspect of the structures of chemical compounds is whether the atoms within a molecule are equivalent or different. For example, all the hydrogen atoms in methane are equivalent: they all exist in an identical environment. On the other hand, the fluorine atoms in a molecule such as PF_5 are not all equivalent. The two fluorine atoms that lie in the axial positions are equivalent to each other but are different from the three fluorine atoms that lie in the equatorial plane. Those three fluorine atoms are all equivalent to each other. The property of the equivalence (or nonequivalence) of atoms in molecules is known as molecular symmetry. Using your knowledge of Lewis structures, predict how many different types of hydrogen atoms will be found in the following molecules: (a) C_2H_2; (b) C_2H_4; (c) C_2H_3Cl; (d) *cis*-$C_2H_2Cl_2$; (e) *trans*-$C_2H_2Cl_2$. (f) In the molecule C_2H_5Cl, the hydrogen atoms could all be different; yet by a variety of experimental techniques, only two types of hydrogen atoms are found. Propose an explanation.

2.106 Predict which bond will absorb light of shorter wavelength and explain why: C—H or C—Cl. Refer to Major Technique 1 on infrared spectroscopy, which follows these exercises.

2.107 Infrared spectra show absorption due to C—H bond stretching at 3.38 μm for a methyl (—CH_3) group and at 3.1 μm for an alkyne (—C≡C—H) group. Which C—H bond is stiffer (has the larger force constant k), assuming that the vibrating atoms have the same effective mass? Refer to Major Technique 1 on infrared spectroscopy, which follows these exercises.

2.108 Vibrational spectra are often so complicated that assignment of a particular absorption to a given bond is difficult. One way to confirm that an assignment is correct is to carry out selective isotopic substitution. For example, we can replace a hydrogen atom with a deuterium atom. If an iron-hydride (Fe—H) stretch occurs at 1950 cm^{-1}, at what energy will this stretch occur, approximately, for a compound that has deuterium in place of the hydrogen?

When we feel the warmth of the Sun on our faces, we are responding to infrared radiation that has traveled nearly 150 million kilometers through space. That radiation stimulates the molecules in our skin to vibrate, and specially adapted nerve cells detect the vibration and interpret it as "warmth." The same type of radiation is used to identify molecules in samples and to determine the stiffness of bonds in molecules.

1 The Technique

Infrared radiation is electromagnetic radiation lying at longer wavelengths (lower frequencies) than red light; a typical wavelength is about 1000 nm. That wavelength is one-thousandth of a millimeter, so it is just possible to imagine its scale. A wavelength of 1000 nm corresponds to a frequency of about 3×10^{14} Hz, which is comparable to the frequency at which molecules vibrate. Therefore, molecules can absorb infrared radiation and become vibrationally excited.

Any bond between two atoms vibrates as the atoms move closer to each other and away again. This type of motion is called a "stretching" mode. Polyatomic molecules can also undergo "bending" vibrations in which bond angles periodically increase and decrease. The frequency at which a molecule vibrates depends on the masses of its atoms and the stiffness of its bonds: a molecule made up of light atoms joined by stiff bonds has a higher vibrational frequency than one made up of heavy atoms joined by loose bonds. The former molecule will therefore absorb higher frequency radiation than the latter. Bending motions of molecules tend to be less stiff than stretching motions, so bending vibrations typically absorb at lower frequencies than stretching vibrations.

The stiffness of a bond is measured by its *force constant, k*. This is the same constant that occurs in Hooke's law for the restoring force of a spring: Hooke observed that the restoring force is proportional to the displacement of the spring from its resting position, and wrote

$$\text{Force} = -k \times \text{displacement} \qquad (1)$$

A stiff bond (like a stiff spring) experiences a strong restoring force, even for quite small displacements, so in this case k is large. A loose bond (like a weak spring) experiences only a weak restoring force, even for quite large displacements, so its associated k is small. In general, the force constant is larger for stretching displacements of molecules than for bending motions.

The vibrational frequency, ν (nu), of a bond between two atoms A and B of masses m_A and m_B is given by the expression

$$\nu = \frac{1}{2\pi}\sqrt{\frac{k}{\mu}} \qquad \mu = \frac{m_A m_B}{m_A + m_B} \qquad (2)$$

The quantity μ (mu) is called the *effective mass* of the molecular vibration (some people call it the "reduced mass"). As we anticipated, the frequency is higher for stiff bonds (large k) and low atomic masses (low μ). We see that by measuring the vibrational frequency of a molecule, which involves measuring the frequency (or wavelength) at which it absorbs infrared radiation, we can measure the stiffness of its bond.

In practice, the vibrational absorption spectrum of a molecule is measured by using an infrared spectrometer. The source of infrared radiation is a hot filament, and the wavelength is selected by diffracting the radiation from a grating: constructive interference* results in intense radiation being obtained in a given direction for a given angle of the grating; and, as that angle is changed, radiation of varying wavelength is passed through the sample. The beam is divided, one beam passing through the sample and the other beam passing through a blank; the intensities of the two beams are then compared at the detector and the reduction in intensity monitored. The result is a spectrum in which dips occur at the wavelengths of radiation absorbed by the sample.

2 Normal Modes and Polyatomic Molecules

A nonlinear molecule consisting of N atoms can vibrate in $3N - 6$ different ways, and a linear molecule can vibrate in $3N - 5$ different ways. The number of ways a molecule can vibrate increases rapidly with the number of atoms: a water molecule, with $N = 3$, can vibrate in three ways, but a benzene molecule, with $N = 12$, can vibrate in 30 different ways. Some of the vibrations of benzene correspond to expansion and contraction of the ring, some to its elongation, and others to flexing and bending. Each way in which a molecule can vibrate is called a *normal mode*, so we say that benzene has 30 normal modes of vibration.

* Interference is discussed more fully in Major Technique 3 after Chapter 5.

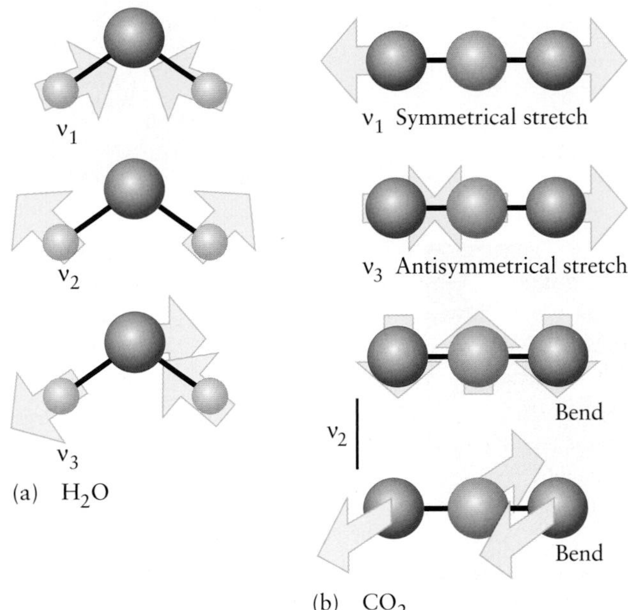

FIGURE 1 (a) The three normal vibrational modes of H_2O. Two of these modes involve principally stretching motions of the bonds, but mode ν_2 is primarily bending. (b) The four normal vibrational modes of CO_2. The first two are symmetrical and antisymmetrical stretching motions, and the last two are perpendicular bending motions.

Each normal mode corresponds to a characteristic, collective motion of the atoms, and each one has a frequency that depends in a complicated way on the masses of the atoms that move during the vibration and the force constants associated with whatever bending or stretching motions are involved (Fig. 1).

Except in simple cases, it is very difficult to predict the infrared absorption spectrum of a polyatomic molecule, because each of the modes has its characteristic absorption frequency rather than just the single frequency of a diatomic molecule. However, certain groups, such as a benzene ring or a carbonyl group, have characteristic frequencies, and their presence can often be detected in a spectrum. Thus, an infrared spectrum can be used to identify the species present in a sample by looking for the characteristic absorption bands associated with various groups. An example and its analysis is shown in Fig. 2.

A further aid to identification comes from the fact that a molecule commonly has complex series of absorptions spanning a range of wavelengths. This *fingerprint region* of the spectrum may be too difficult to analyze in detail, but its presence enables us to recognize the substance by comparing the spectrum to an atlas of spectra.

Related Exercises: 1.93, 2.106, and 2.107.

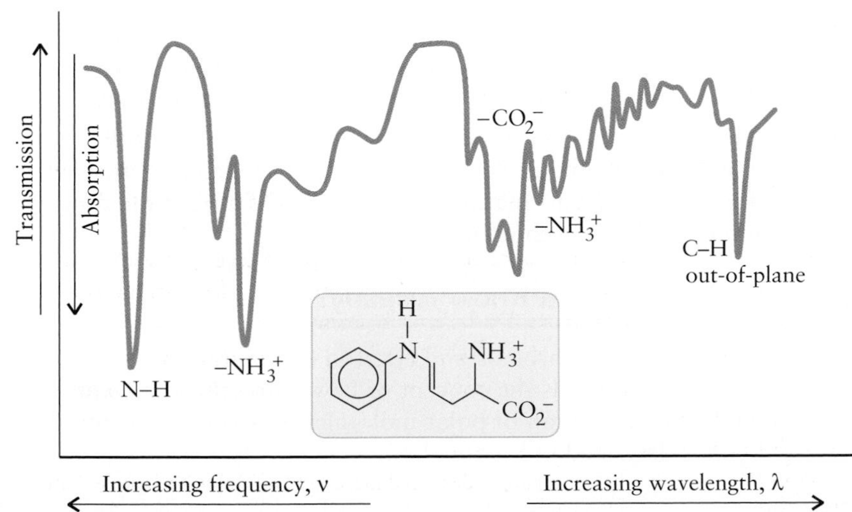

FIGURE 2 The infrared spectrum of an amino acid, with the groups contributing to some of the peaks identified. Notice that the spectrum displays the intensity of absorption.

CHAPTER

The colors of leaves in the fall are due to the absorption of light by large molecules related to carotene, a fragment of which is shown here. The electrons in carotene occupy orbitals that spread over the entire molecule. They are close together in energy, and an incoming photon of visible light can excite an electron from one orbital to another. As a result, the molecules look colored.

Why Do We Need to Know This Material?

The shapes of molecules determine their odors, their tastes, and their actions as drugs. Molecular shape governs the reactions that occur throughout our bodies and are necessary for life. It also affects the properties of the materials around us, including their physical states and their solubilities. Perception, thinking, and learning depend on the shapes of molecules and how they change. Modern theories of the electronic structures of molecules can be extended to describe metals and semiconductors, so the information in this chapter is also at the root of the development of new technologies.

What Do We Need to Know Already?

This chapter uses atomic orbitals and electron configurations (Chapter 1). It also extends the concept of Lewis structures introduced in Chapter 2. The discussion of polar molecules develops the material on polar bonds described in Section 2.15.

MOLECULAR SHAPE AND STRUCTURE

The bright red of fruit, the colors of flowers, and the varied hues of autumn leaves have always been a cause for delight, but it was not until the twentieth century that chemists discovered that the colors arise from the presence of organic compounds with common structural features. They also discovered that small differences in the structures of the molecules of these compounds can enhance photosynthesis, produce an important vitamin, and attract pollinating bees to flowers. A knowledge of the shapes of molecules and the orbitals that their electrons occupy will allow us to begin to understand the properties of these compounds and even the processes occurring in our eyes that allow us to see them.

The impact of modern theories of bonding, however, goes far beyond understanding the colors around us. Growing knowledge of the electronic structures of the atoms and molecules in polymers and semiconductors has led to the development of new technologies. The pharmaceutical industry is increasingly turning to computer-aided design—in which molecular shape and the distribution of electrons play a crucial role—to discover new, more potent drugs. We shall see the basis of that approach in this chapter (Box 3.1).

Molecular shape, bond angles, and bond lengths are determined experimentally, but they can also be predicted by calculations based on the Schrödinger equation. These calculations sometimes are based partly on experimental information, when they are called **semiempirical methods**, and sometimes are purely theoretical predictions, when they are called ***ab initio* methods**. We shall see some of their output later in the chapter.

This chapter is another example of how chemists develop and refine models. Here we meet three models. The first, the most primitive, considers molecular shape to be a consequence merely of the coulombic interaction between pairs of electrons. The other two models are more sophisticated: they both seek to describe the distribution of electrons and their involvement in the determination of molecular shape in terms of simplified descriptions of the available orbitals. Both represent the actual wavefunctions of the electrons as approximations, but they construct these approximations in two different ways.

Loosely translated, *ab initio* means "from scratch."

THE VSEPR MODEL

In this section, we construct our model "empirically," which means that we base it on rules suggested by experimental observations rather than on more fundamental principles. Then, once we have identified rules that work for a number of molecules, we summarize the rules as a model of molecular shape and use it to make predictions about other molecules for which experimental data might not be available. We proceed in three steps. First, we set up the basic model for simple molecules without lone pairs on the central atom. Then we include the effects of lone pairs. Finally, we explore some of the consequences of molecular shape for the properties of molecules.

3.1 The Basic VSEPR Model

A Lewis structure shows the approximate locations of bonding electrons and lone pairs in a molecule. However, it is only a two-dimensional representation of the links between atoms and does not attempt to depict the arrangement of atoms in space. For example many simple molecules (molecules that consist of one central atom to which all the other atoms are attached) have the shapes of the geometrical figures shown in Fig. 3.1; thus, CH_4 (**1**) is tetrahedral, SF_6 (**2**) is octahedral, and PCl_5 (**3**) is trigonal bipyramidal.

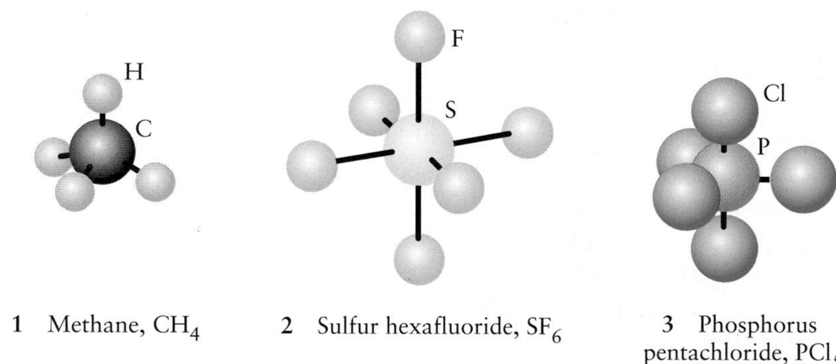

1 Methane, CH_4 **2** Sulfur hexafluoride, SF_6 **3** Phosphorus pentachloride, PCl_5

To describe the shapes of these and less symmetrical molecules, we report each **bond angle**, the angle between the straight lines that join the atom

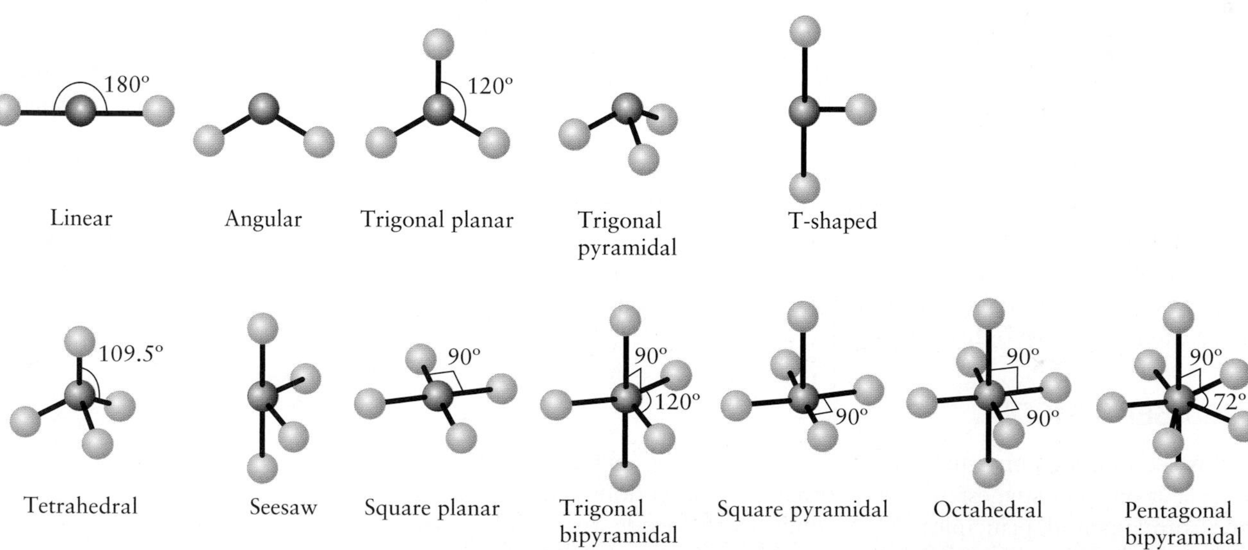

FIGURE 3.1 The names of the shapes of simple molecules and their bond angles. Lone pairs are not shown, because they are not included when identifying molecular shapes.

centers. The bond angle in H_2O, for instance, is the angle (104.5°) between the two O—H bonds. The **valence-shell electron-pair repulsion model** (VSEPR model) extends Lewis's theory of bonding to account for molecular shapes by adding additional rules that account for these bond angles.

This idea was first explored by the British chemists Nevil Sidgwick and Herbert Powell and has been developed by the Canadian chemist Ronald Gillespie. Some people pronounce VSEPR as "vesper."

The first rule suggested by inspection of experimentally determined bond angles in simple molecules is that *regions of high electron concentration repel one another*. In other words, bonding electrons and lone pairs take up positions as far from one another as possible, for then they repel one another the least.

In the VSEPR model, we focus attention on the central atom of a molecule, such as the B atom in BF_3 or the C atom in CO_2. The electrons involved in bonds to the central atom and lone pairs of electrons belonging to that atom are all regions of high electron concentration, so they repel one another. We assume that, to minimize their repulsions, these regions move as far apart as possible while maintaining the same distance from the central atom (Fig. 3.2). Once we have identified the arrangement of the "most distant" locations of these regions, which is called the **electron arrangement** in the molecule, we note where the *atoms* lie and identify the shape of the molecule by giving it the name of the corresponding shape in Fig. 3.1. Notice that, in naming the molecular shape, we consider only the positions of atoms, not any lone pairs that may be present.

A molecule with only two atoms attached to the central atom is $BeCl_2$. The Lewis structure is $:\ddot{\underset{..}{Cl}}-Be-\ddot{\underset{..}{Cl}}:$, and there are no lone pairs on the central atom. To be as far apart as possible, the bonding pairs lie on opposite sides of the Be atom, so the electron arrangement is linear. Because a Cl atom is attached by each bonding pair, the $BeCl_2$ molecule is expected to be linear, with a bond angle of 180° (4). That shape is confirmed by experiment.

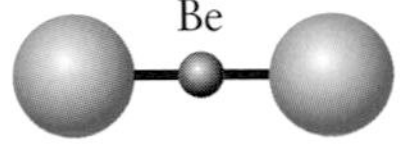

4 Beryllium chloride, $BeCl_2$

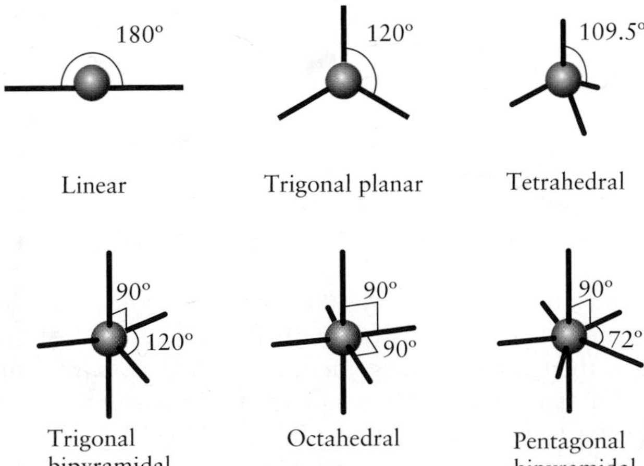

FIGURE 3.2 The positions of from two to seven regions of high electron concentration (atoms *and* lone pairs) around a central atom. The regions are denoted by the straight lines sticking out of the central atom. Use this chart to identify the electron arrangement of a molecule, and then use Fig. 3.1 to identify the shape of the molecule from the locations of its atoms.

A boron trifluoride molecule, BF_3, has the Lewis structure shown in (5). There are three bonding pairs attached to the central atom and no lone pairs. According to the VSEPR model, to be as far apart as possible, the three bonding pairs must lie at the corners of an equilateral triangle. The electron arrangement is trigonal planar, and because an F atom is attached to each bonding pair, the BF_3 molecule is also trigonal planar (see Fig. 3.2), and all three F—B—F angles are 120°, as determined experimentally (**6**).

5 Boron trifluoride, BF_3

6 Boron trifluoride, BF_3

Methane, CH_4, has four bonding pairs on the central atom, each connecting a hydrogen atom to the carbon atom. To be as far apart as possible, the four bonding pairs must take up positions at the corners of a tetrahedron with the C atom at the center. Because the electron arrangement is tetrahedral and an H atom is attached to each bonding pair, we expect the molecule to be tetrahedral (see **1**), with bond angles of 109.5°. That is the shape found experimentally.

In a phosphorus pentachloride molecule, PCl_5 (7), there are five bonding pairs and no lone pairs on the central atom. According to the VSEPR model, the five pairs and the atoms they carry are farthest apart in a trigonal bipyramidal arrangement (see Fig. 3.2). In this arrangement, three atoms lie at the corners of an equilateral triangle and the other two atoms lie above and below the plane of the triangle (**8**). This structure has two different bond angles. Around the "equator" formed by the triangle, the bond angles are 120°. The angle between the axis and the equatorial atoms is 90°. This structure is also confirmed experimentally.

7 Phosphorus pentachloride, PCl_5

8 Phosphorus pentachloride, PCl_5

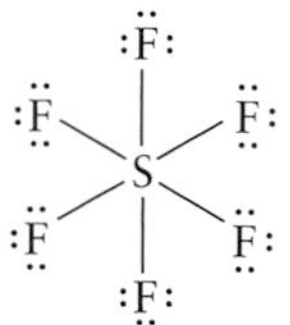

9 Sulfur hexafluoride, SF_6

All six terminal atoms are equivalent in a regular octahedral molecule.

A sulfur hexafluoride molecule, SF_6, has six atoms attached to the central S atom and no lone pairs on that atom (**9**). According to the VSEPR model, the electron arrangement is octahedral, with four pairs at the corners of a square around the equator and the remaining two pairs above and below the plane of the square (see Fig. 3.2). An F atom is attached to each electron pair, so the molecule is predicted to be octahedral. All its bond angles are either 90° or 180°, and all the F atoms are equivalent.

When using the VSEPR model to predict molecular shapes, we do not distinguish between single and multiple bonds. A multiple bond is treated as a single region of high electron concentration. The two electron pairs in a double bond stay together and repel other bonds or lone pairs as a unit. The three electron pairs in a triple bond also stay together and act like a single region of high electron concentration. For instance, a carbon dioxide molecule, O=C=O, has a linear structure similar to that of $BeCl_2$, except for the presence of double bonds (**10**). One of the Lewis structures of a carbonate ion, CO_3^{2-}, is shown in (**11**). The two pairs of electrons in the double bond are treated as a unit, and the resulting shape (**12**) is trigonal planar.

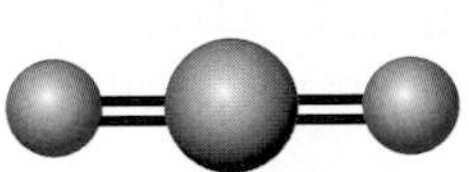

10 Carbon dioxide, CO_2

:O:]2−
‖
:Ö—C—Ö:

11 Carbonate ion, CO_3^{2-}

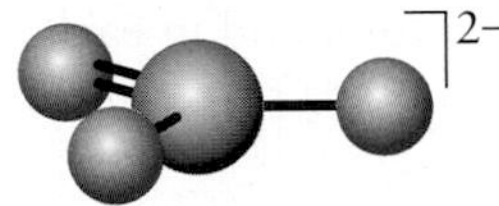

12 Carbonate ion, CO_3^{2-}

When there is more than one central atom, we consider the bonding about each atom independently. For example, let's predict the shape of an ethene (ethylene) molecule, CH_2=CH_2. There are two centers in ethene to consider: the two carbon atoms. The first step is to write the Lewis structure (**13**). Each carbon atom has three regions of electron concentration on it: two single bonds and one double bond. There are no lone pairs. The arrangement around each carbon atom is therefore trigonal planar. We predict that the H—C—H and H—C—C angles will both be 120° (**14**); this prediction is confirmed experimentally.

A limitation of the VSEPR model is that it cannot explain why both CH_2 groups lie in the same plane. We shall see in Section 3.8 that a double bond between two atoms always results in a planar arrangement of the two atoms and the atoms directly attached to them.

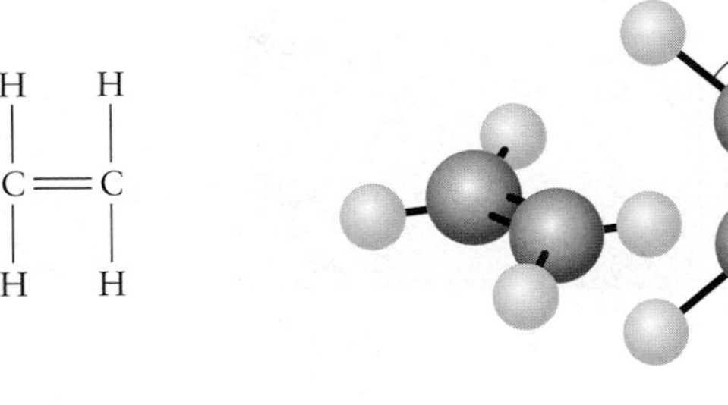

13 Ethene, C_2H_4

14 Ethene, C_2H_4

EXAMPLE 3.1 Predicting the shape of a molecule with no lone pairs on the central atom

Suggest a shape for the ethyne (acetylene) molecule, HC≡CH.

STRATEGY Write down the Lewis structure and identify the electron arrangement around each "central" atom (two C atoms, in this case). Treat each multiple bond as a single unit. Then identify the overall shape of the molecule (refer to Fig. 3.1, if necessary).

SOLUTION The Lewis structure of the molecule is H—C≡C—H. Each C atom is attached to two other atoms (one H atom and one C atom), and there are no lone pairs. The arrangement of atoms around each C atom is therefore predicted to be linear (**15**).

15 Ethyne, C_2H_2

Self-Test 3.1A Predict the shape of an arsenic pentafluoride molecule, AsF_5.
[*Answer:* Trigonal bipyramidal]

Self-Test 3.1B Predict the shape of a formaldehyde molecule, CH_2O.

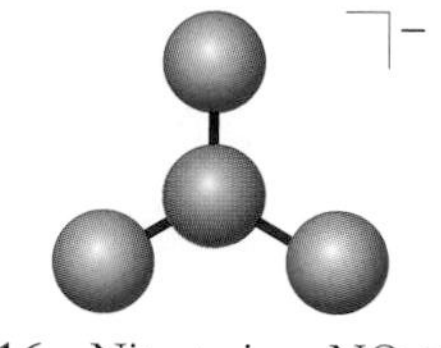

16 Nitrate ion, NO_3^-

Because we treat single bonds and multiple bonds as equivalent in the VSEPR model, it does not matter which of the Lewis structures contributing to the resonance structure we consider. For example, although we can write several different Lewis structures for a nitrate ion, all of them have three regions of electron concentration around the central N atom, and in each case we expect a trigonal planar structure (**16**). A computer graphics image of the nitrate ion shows three equivalent regions of

BOX 3.1 FRONTIERS OF CHEMISTRY: DRUGS BY DESIGN AND DISCOVERY

The search for new drugs relies on the skills of synthetic organic chemists. Because there are so many millions of compounds, it would take too long to start with the elements, combine them in different ways, then test them. Instead, chemists usually start either by *drug discovery*, the identification and possibly the modification of promising medicines that already exist, or by *rational drug design*, the identification of characteristics of a target bacterium or parasite and the design of new compounds to react with it.

In drug discovery, a chemist usually begins by investigating compounds that have already shown medicinal value. A fruitful path is to find a *natural product*, a substance found in nature, that has been shown to have healing characteristics. Nature is the best of all synthetic chemists, with billions of chemicals that fulfill as many different needs. The challenge is to find compounds that have curative powers. These substances are found in different ways: random or "blind" collection of samples that are then tested or collection of specific samples identified by native healers as medically effective.

Observation of the properties of plants and animals can help to guide a random search. For example, if certain types of fruits remain fresh while others rot or become moldy, we might expect them to contain antifungal agents. One example of this type of collection is the gathering of tunicates and sponges in the Caribbean. The chemists harvest the samples by diving from research vessels (see Fig. F.1). The samples are tested for antiviral and antitumor activity in a chemical laboratory on the ship. The antiviral drug didemnin-C and the anticancer drug bryostatin 1 were discovered in marine organisms.

A field biologist examines a plant in a South American rainforest. The plant produces chemicals that will be investigated for their medicinal value.

The guided route involves the testing of fewer samples because the chemist works with a native healer, the ancient lore guiding the modern chemistry. Often an ethnobotanist, a specialist in plants used for native healing, joins the team. This approach saves time for the scientists and can provide an economic benefit for the healers and their nations as well. Drugs that have been discovered in this way include a variety of anticancer and antimalarial drugs, blood-clotting agents, antibiotics, and medicines for the heart and digestive system.

electron density (**17**). The image is an "electrostatic potential plot," in which regions of net negative charge are shown as red and regions of net positive charge are shown as blue (see Section C). Broadly speaking, red regions represent locations where there is a high density of electrons and blue regions represent locations where the electron density is relatively low and the nuclear charge shows through. The equivalence of the three bonds is what we would expect in a resonance structure.

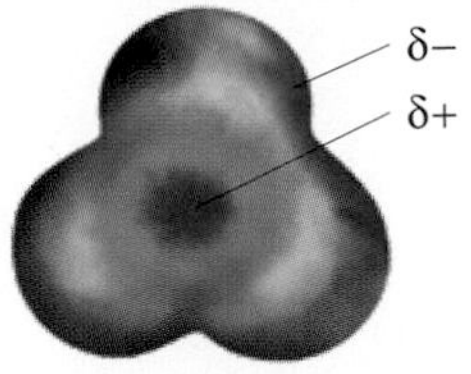

17 Nitrate ion

According to the VSEPR model, regions of high electron concentration take up positions that maximize their separations; electron pairs in a multiple bond are treated as a single unit. The shape of the molecule is then identified from the relative locations of its atoms.

Once the empirical and molecular formulas of the active compounds are determined, then their structural formulas are sought. At that point, synthetic work can begin. The chemist can identify compounds in the material that have medicinal value and find a way to *synthesize*, or prepare them in the laboratory, so that they can be made available in large quantities.

In rational drug design, the chemist begins with the tumor or organism the drug is intended to eradicate. Virtually all processes in living cells depend on specific *enzymes*, types of proteins with very large molecules having specific shapes. Usually there is an *active site* on the enzyme into which only specific molecules can fit and react. If the enzymes that control the growth of parasites or bacteria can be identified and their shapes known, compounds that fit into the active sites and block the reactions can be designed. The chemist taking this route begins by identifying key enzymes in the bacterium or parasite. Then the molecular structure of the enzyme is determined. A computer program is used to design molecules with structures that fit into the active site. The new compounds are synthesized, and their effects and side effects tested.

Suramin

Suramin, which was used against trypanosomes, was developed by Paul Ehrlich. It is one of the first drugs to be developed by the discovery method.

HOW MIGHT YOU CONTRIBUTE?

Despite all the existing medicines, there is still a great need for specific chemotherapeutic agents with few side effects. As an example of an undesirable side effect, one of the drugs used to treat African sleeping sickness fatally damages the brains of about 10% of the patients who use it. You may also be involved in the design of agents that can repair the nucleic acids in human cells and so prevent or cure cancer and congenital diseases. We are beginning to see the investigation of a wide array of natural products and the rational drug design of new therapeutic agents. The selection of good candidates for drug development from these large numbers of compounds needs to be optimized.

For Further Reading: David Hart, "Designer drugs," *Science Spectra*, Number 8, 1997, p. 52. A. Maureen Rouhi, "Seeking drugs in natural products," *Chemical and Engineering News*, April 7, 1997, p. 14. Robert A. Lipper, "E pluribus product," *Modern Drug Discovery*, 1999, 2(1), 55–60.

3.2 Molecules with Lone Pairs on the Central Atom

Lone pairs of electrons on a central atom are regions of electron density and must be considered when identifying a molecular shape. To help us predict the shapes of molecules, we use a generic "VSEPR formula" AX_nE_m to identify the different combinations of atoms and lone pairs attached to the central atom. We let A represent a central atom, X an attached atom, and E a lone pair. For example, the BF_3 molecule, with three attached fluorine atoms and no lone pairs on B, is an example of an AX_3 species. The sulfite ion, SO_3^{2-} (**18**), is an example of an AX_3E species.

:Ö—S̈—Ö: $]^{2-}$
|
:Ö:

18 Sulfite ion, SO_3^{2-}

When deciding on the electron arrangement in an AX_nE_m molecule, we do not distinguish between bonding pairs and lone pairs: all bonds and lone pairs are regions where electrons are concentrated. If there are no lone pairs on the central atom (an AX_n molecule), the molecular shape is the same as the electron arrangement. However, if lone pairs are present, the molecular shape will differ from the electron arrangement because only the positions of the atoms are considered when naming the shape. For example, the four regions of high electron concentration in SO_3^{2-} are farthest apart if they adopt a tetrahedral arrangement (see Fig. 3.2). However, the *shape* of the ion is described by the locations of atoms only. Because three of the tetrahedral locations are occupied by atoms, and one is occupied by a lone pair, an SO_3^{2-} ion is trigonal pyramidal (**19**). In short, all electron pairs, whether bonding or not, are included in a description of the arrangement of electrons. However, *only the positions of atoms are considered when reporting the shape of a molecule.*

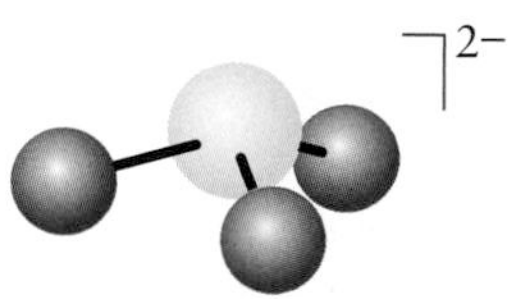

19 Sulfite ion, SO_3^{2-}

An unpaired electron on the central atom also is a region of high electron density and is treated like a lone pair when determining molecular shape. For example, radicals such as NO_2 have a single nonbonding electron. Thus, NO_2 (**20**) has a trigonal planar electron arrangement (including the unpaired electron on N), but an angular molecular shape (**21**).

20 Nitrogen dioxide, NO_2

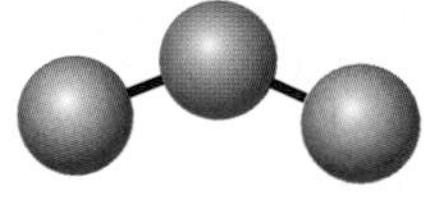

21 Nitrogen dioxide, NO_2

EXAMPLE 3.2 **Predicting the shape of a molecule with lone pairs**

Predict (a) the electron arrangement and (b) the shape of a nitrogen trifluoride molecule, NF_3.

STRATEGY (a) For the electron arrangement, draw the Lewis structure and then decide how the bonding pairs and lone pairs are arranged around the central (nitrogen) atom (consult Fig. 3.2 if necessary). (b) Identify the molecular shape from the layout of atoms, as in Fig. 3.1.

:F̤—N̈—F̤:
|
:F̤:

22 Nitrogen trifluoride, NF_3

SOLUTION (a) The Lewis structure of nitrogen trifluoride is shown in (**22**): we see that the central N atom has four electron pairs, of which three are bonds. According to the VSEPR model, these four electron-rich regions adopt a tetrahedral electron arrangement. (b) Because one of the pairs is a lone pair, the molecular shape is expected to be trigonal pyramidal (**23**). Spectroscopic measurements confirm this prediction.

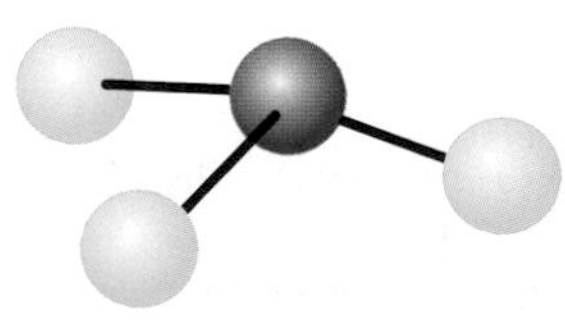

23 Nitrogen trifluoride, NF_3

Self-Test 3.2A Predict (a) the electron arrangement and (b) the shape of an IF_5 molecule.

[*Answer:* (a) Octahedral; (b) square pyramidal]

Self-Test 3.2B Predict (a) the electron arrangement and (b) the shape of an SO_2 molecule.

So far we have regarded lone pairs as equivalent to bonds, but is that really the case? We have predicted, for instance, that the electron arrangement of an SO_3^{2-} ion is tetrahedral, so we might expect OSO angles of 109.5°. However, experiments have shown that, although the sulfite ion is indeed trigonal pyramidal, its bond angle is only 106° (**24**). Experimental evidence like this tells us that the VSEPR model as we have described it is incomplete and needs to be refined.

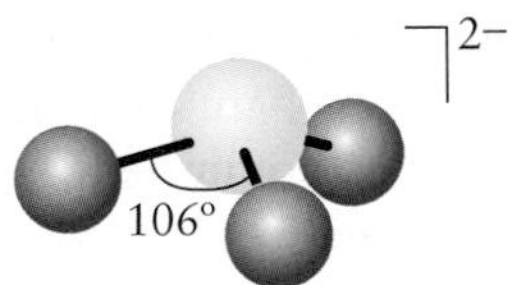

24 Sulfite ion, SO_3^{2-}

Because bond angles in molecules with lone pairs are typically smaller than expected, lone pairs are treated in the VSEPR model as having a more strongly repelling effect than bonding pairs. No one is really sure why this is so. One possible explanation is that the electron cloud of a bonding pair cannot spread over as large a volume as a lone pair, because a bonding pair is pinned down by two atoms, not one (Fig. 3.3). Whatever the true reason, the VSEPR model provides better predictions if we adopt the rule that the strengths of repulsions are in the order

Lone pair-lone pair >
lone pair-bonding pair > bonding pair-bonding pair

The lowest energy is achieved when lone pairs are as far from each other as possible. The energy is also lowest if the atoms bonded to the central atom are far from lone pairs, even though that might bring the atoms closer to other atoms.

Our improved model helps to account for the bond angle of the AX_3E sulfite ion. The electron pairs adopt a tetrahedral arrangement around the S atom. However, the lone pair exerts a strong repulsion on the bonding electrons, forcing them to move together slightly. As a result of this adjustment in positions, the OSO angle is reduced from the 109.5° of the regular tetrahedron to the 106° observed experimentally.

We can now state the revised VSEPR model as follows:

Regions of high electron concentration take up positions that maximize their separations. Lone pairs repel more strongly than bonding pairs and so tend to compress bond angles.

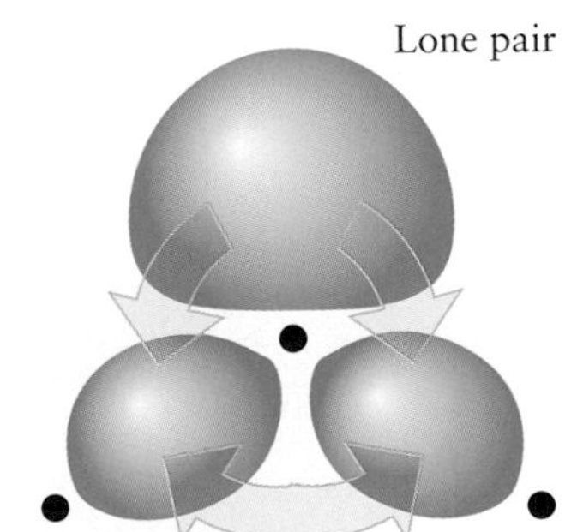

FIGURE 3.3 One possible explanation of why lone pairs have a greater repelling effect than bonding electrons. A lone pair is less restrained than bonding pairs and so takes up more space; the bonding pairs and their atoms move away from the lone pair in an attempt to lower the repulsion they experience.

Note that, although the VSEPR model can predict the direction of the distortion, it cannot predict its extent. We can predict that, in any AX_3E species, the XAX angle will be less than 109.5°, but we cannot predict its actual value; we must measure it experimentally or calculate it by solving the Schrödinger equation numerically on a computer.

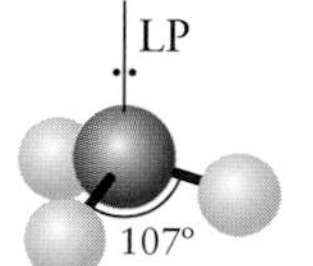

25 Ammonia, NH_3

Self-Test 3.3A (a) Give the VSEPR formula of an NH_3 molecule. Predict (b) its electron arrangement and (c) its shape.

[*Answer:* (a) AX_3E; (b) tetrahedral; (c) trigonal pyramidal (**25**), HNH angle less than 109.5°]

Self-Test 3.3B (a) Give the VSEPR formula of a ClO_2^- ion. Predict (b) its electron arrangement and (c) its shape.

The VSEPR model also allows us to predict the position in which a lone pair will be found. For example, the electron arrangement in an AX_4E molecule or ion, such as IF_4^+, is trigonal bipyramidal, but there are two possible locations for the lone pair. An **axial lone pair** lies on the axis of the molecule. An **equatorial lone pair** lies on the molecule's equator, on the plane perpendicular to the molecular axis. An axial lone pair repels three electron pairs strongly, but an equatorial lone pair repels only two strongly (Fig. 3.4). Therefore, the lowest energy is achieved when a lone pair is equatorial, producing a seesaw-shaped molecule. An AX_3E_2 molecule, such as ClF_3, also has a trigonal bipyramidal arrangement of electron pairs, but two of the pairs are lone pairs. These two pairs are farthest apart if they occupy two of the three equatorial positions, but move away from each other slightly. The result is a T-shaped molecule (Fig. 3.5).

Now consider an AX_4E_2 molecule, which has an octahedral arrangement of electron pairs, two of which are lone pairs. The two lone pairs are farthest apart when they lie opposite each other, so the molecule has a square planar shape (Fig. 3.6).

Molecules with the same VSEPR formula all have the same general shape, although their bond angles generally differ slightly. For example, O_3 is an AX_2E species ($:\ddot{\underset{..}{O}}-\ddot{O}=\ddot{\underset{..}{O}}$); it has a trigonal planar electron

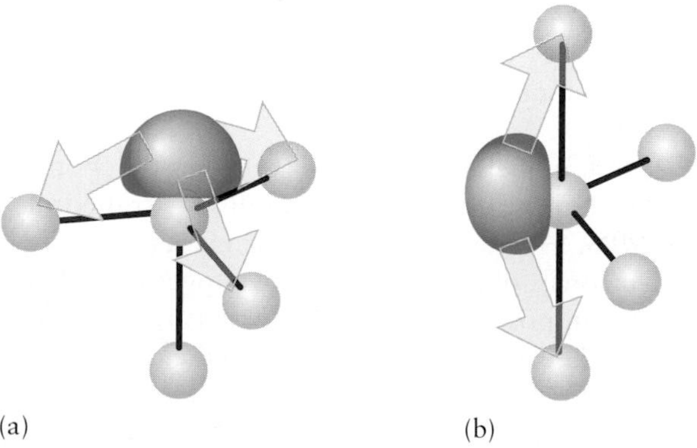

FIGURE 3.4 (a) A lone pair in an axial position is close to three equatorial atoms, whereas (b) in an equatorial position, it is close to only two atoms. The latter arrangement is therefore more favorable.

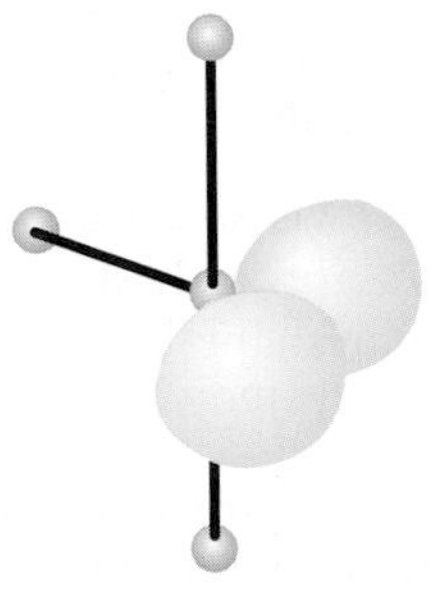

FIGURE 3.5 Two lone pairs in an AX_3E_2 molecule adopt equatorial positions and move away from each other slightly. As a result, the molecule is approximately T-shaped.

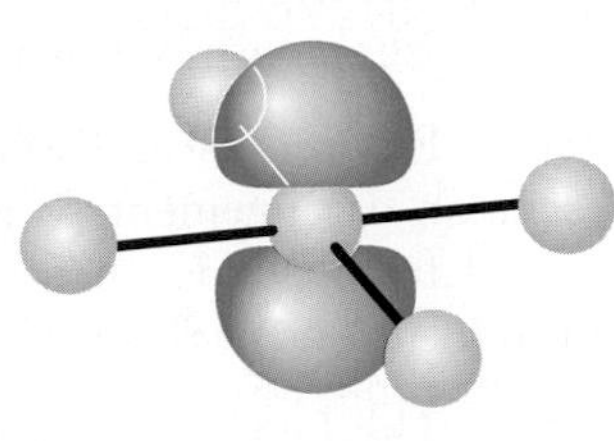

FIGURE 3.6 The square planar arrangement of atoms taken up in AX_4E_2 molecules: the two lone pairs are farthest apart when they are on opposite sides of the central atom.

arrangement and an angular molecular shape (**26**). The nitrite ion, NO_2^-, has the same general formula ($[:\ddot{\text{O}}{-}\ddot{\text{N}}{=}\ddot{\text{O}}]^-$) and the same shape (**27**); so too does sulfur dioxide, SO_2 ($:\ddot{\text{O}}{-}\ddot{\text{S}}{=}\ddot{\text{O}}$, **28**).

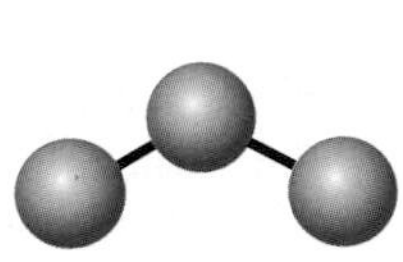

26 Ozone, O_3

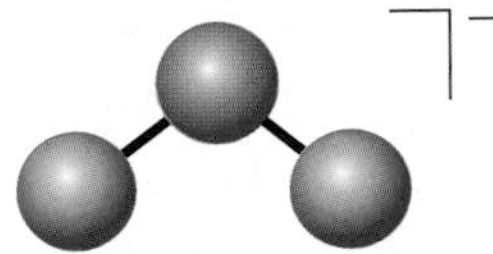

27 Nitrite ion, NO_2^-

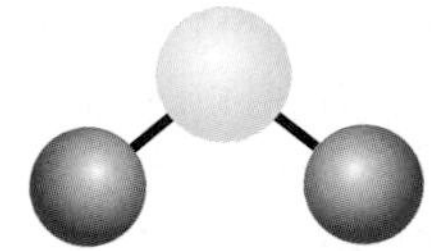

28 Sulfur dioxide, SO_2

In a molecule that has lone pairs or a single electron on the central atom, the valence electrons contribute to the electron arrangement that determines the shape of the molecule but are ignored when identifying the shape. Lone pairs distort the shape of a molecule to reduce lone pair-bonding pair repulsions.

TOOLBOX 3.1 How to use the VSEPR model

Conceptual Basis

Regions of high electron concentration—bonds and lone pairs attached to a central atom in a molecule—arrange themselves in such a way as to minimize mutual repulsions.

Procedure

The general procedure for predicting the shape of a molecule is as follows:

Step 1 Decide how many bonds and lone pairs are present on the central atom by writing a Lewis structure for the molecule.

Step 2 Identify the electron arrangement, treating a multiple bond as the equivalent of a single electron pair (see Fig. 3.2).

Step 3 Locate the atoms and classify the molecular shape (according to Fig. 3.1).

Step 4 Allow the molecule to distort so that lone pairs are as far from one another and from bonding pairs as possible. The repulsions are in the order

Lone pair-lone pair > lone pair-bonding pair > bonding pair-bonding pair

Example 3.3 shows how this procedure is used.

EXAMPLE 3.3 Predicting a molecular shape

Predict the shape of a sulfur tetrafluoride molecule, SF_4.

STRATEGY Proceed as in Toolbox 3.1: write the Lewis structure, identify the electron arrangement in which the lone pairs are farthest from each other and from bonds, then identify the molecular shape by considering only the locations of the atoms.

SOLUTION The Lewis structure of SF_4 is shown in (**29a**) (recall Example 2.7). The molecule is an example of an AX_4E species, with four atoms on the central S atom and one lone pair, so the five electron pairs adopt a trigonal bipyramidal arrangement. To minimize electron pair repulsions, the lone pair occupies an equatorial location. The energy is lowered further if the atoms move away from the lone pair (**29b**). This shape (resembling a slightly bent seesaw) is the one found experimentally.

(a) (b)

29 Sulfur tetrafluoride, SF_4

Self-Test 3.4A Predict the shape of an I_3^- ion.

[*Answer:* Linear]

Self-Test 3.4B Predict the shape of a xenon tetrafluoride molecule, XeF_4.

3.3 Polar Molecules

In Section 2.14, we saw that a polar covalent bond in which electrons are not evenly distributed has a nonzero dipole moment. A **polar molecule** is a molecule with a nonzero dipole moment. All *diatomic* molecules are polar if their bonds are polar. An HCl molecule, with its polar covalent bond ($^{\delta+}H{-}Cl^{\delta-}$), is a polar molecule, and its dipole moment is 1.1 D. That is a typical value for a polar diatomic molecule (Table 3.1). All diatomic molecules that are composed of atoms of different elements are at least slightly polar. A **nonpolar molecule** is a molecule that has zero

TABLE 3.1 Dipole Moments of Selected Molecules

Molecule	Dipole moment, D	Molecule	Dipole moment, D
HF	1.91	PH_3	0.58
HCl	1.08	AsH_3	0.20
HBr	0.80	SbH_3	0.12
HI	0.42	O_3	0.53
CO	0.12	CO_2	0
ClF	0.88	BF_3	0
NaCl*	9.00	CH_4	0
CsCl*	10.42	*cis*-CHCl=CHCl	1.90
H_2O	1.85	*trans*-CHCl=CHCl	0
NH_3	1.47		

* The species consists of pairs of ions in the gas phase.

electric dipole moment. All **homonuclear diatomic molecules**, diatomic molecules containing atoms of only one element, such as O_2, N_2, and Cl_2, are nonpolar, because their bonds are nonpolar.

For polyatomic molecules, it is important to distinguish between a polar *molecule* and a polar *bond*. Because dipole moments are directional, we can treat the moment associated with each bond as a vector. Although each bond in a polyatomic molecule may be polar, the molecule *as a whole* will be nonpolar if the vector sum of the dipole moments of the individual bonds is 0. For example, the two $^{\delta+}C{-}O^{\delta-}$ dipole moments in carbon dioxide, a linear molecule, point in opposite directions, so they cancel each other (30). As a result, CO_2 is a nonpolar molecule even though its bonds are polar. The electrostatic potential diagram (31) illustrates this conclusion. In contrast, the two $^{\delta-}O{-}H^{\delta+}$ dipole moments in H_2O lie at 104.5° to each other and do not cancel, so H_2O is a polar molecule (32). This polarity is part of the reason why water is such a good solvent for ionic compounds.

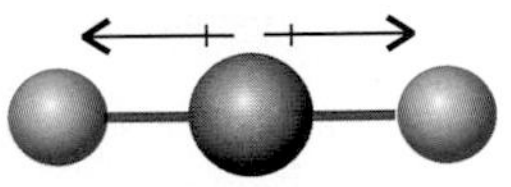

30 Carbon dioxide, CO_2

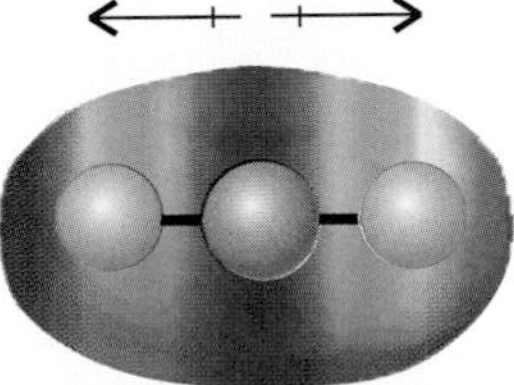

31 Carbon dioxide, CO_2

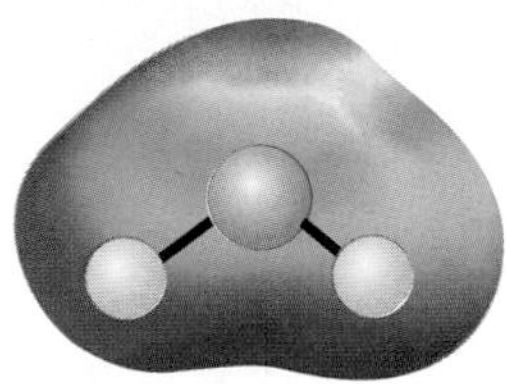

32 Water, H_2O

The shape of a molecule affects whether or not it is polar. The atoms and bonds are the same in *cis*-dichloroethene (33) and *trans*-dichloroethene (34); but in the latter, the C—Cl bonds point in opposite directions and the dipoles (which point along the C—Cl bonds) cancel. Thus, whereas *cis*-dichloroethene is polar, *trans*-dichloroethene is nonpolar.

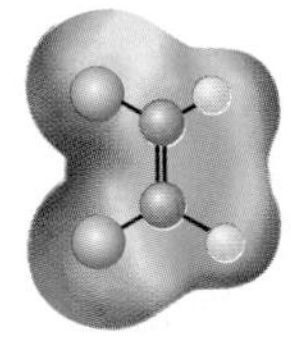

33 *cis*-Dichloreothene, $C_2H_2Cl_2$

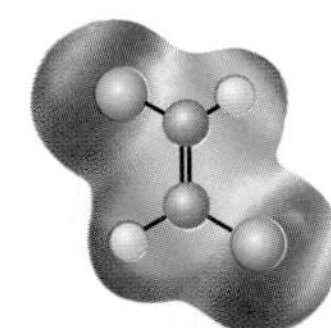

34 *trans*-Dichloroethene, $C_2H_2Cl_2$

Recall that red areas are regions of net negative charge and blue areas are regions of net positive charge.

If the four atoms attached to the central atom in a tetrahedral molecule are the same, as in tetrachloromethane (carbon tetrachloride), CCl_4 (35), the dipole moments cancel and the molecule is nonpolar. We can think of the four dipole moments associated with the bonds as vectors pointing at the more electronegative terminal atoms (Cl) and the molecular dipole moment as the three-dimensional vector sum of these dipole moments. However, if one or more of the atoms are replaced by different atoms, as in trichloromethane (chloroform), $CHCl_3$, or by lone pairs, as in NH_3, then the dipole moments associated with the bonds are not all the same, so they do not cancel. Thus, the $CHCl_3$ molecule is polar (36).

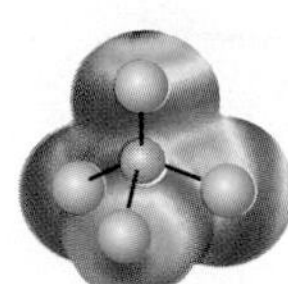

35 Tetrachloromethane, CCl_4

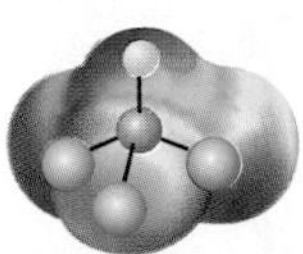

36 Trichloromethane, $CHCl_3$

In general, molecules in which the vector sum of the dipole moments cancel are nonpolar. Such molecules have highly symmetrical arrangements of polar bonds (Fig. 3.7).

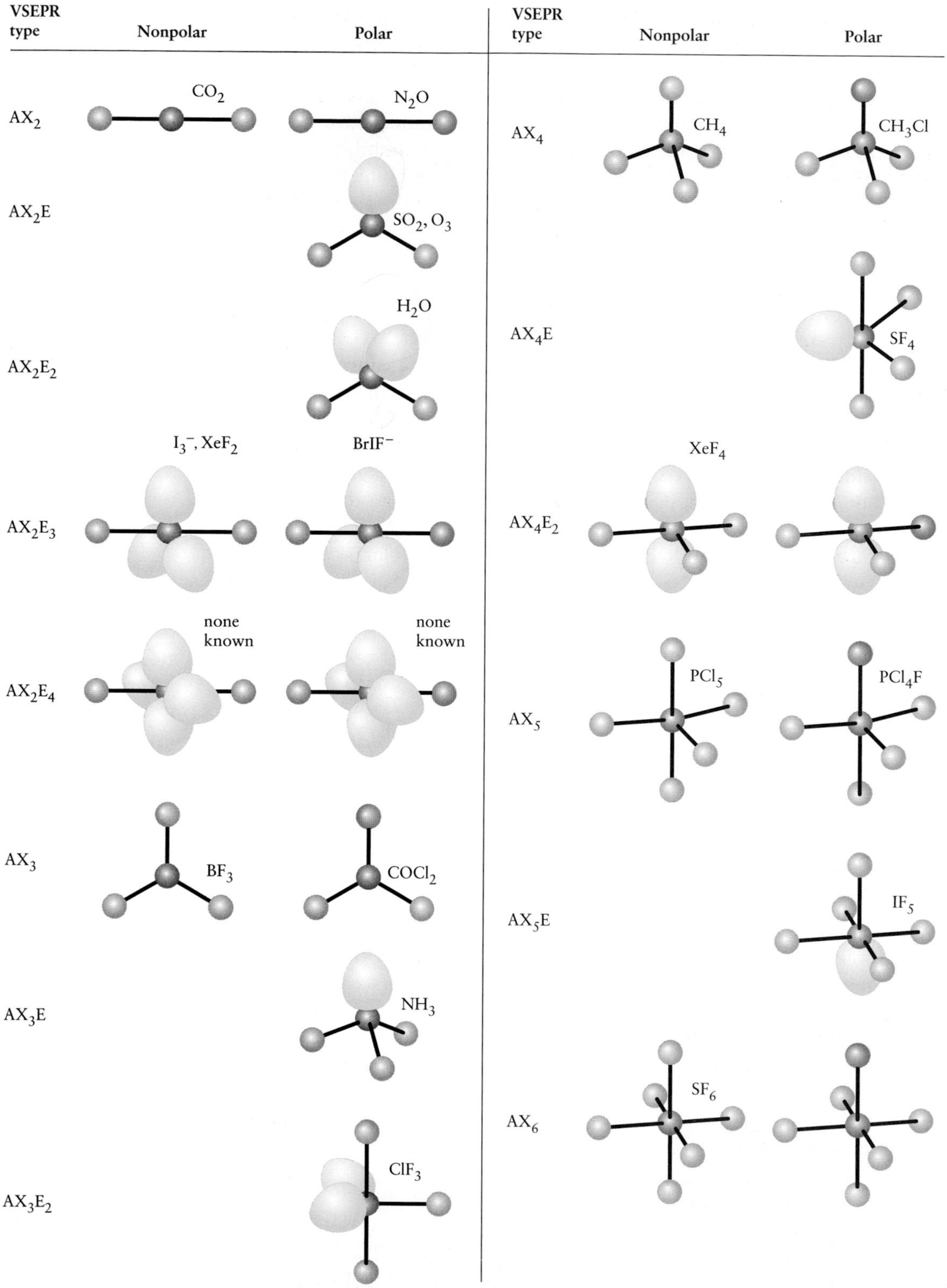

VSEPR type	Nonpolar	Polar
AX_2	CO_2	N_2O
AX_2E		SO_2, O_3
AX_2E_2		H_2O
AX_2E_3	I_3^-, XeF_2	$BrIF^-$
AX_2E_4	none known	none known
AX_3	BF_3	$COCl_2$
AX_3E		NH_3
AX_3E_2		ClF_3

VSEPR type	Nonpolar	Polar
AX_4	CH_4	CH_3Cl
AX_4E		SF_4
AX_4E_2	XeF_4	
AX_5	PCl_5	PCl_4F
AX_5E		IF_5
AX_6	SF_6	

FIGURE 3.7 The arrangements of atoms that give rise to polar and nonpolar molecules. In the notation, A stands for a central atom, X for an attached atom, and E for a lone pair. Atoms of the same color are identical; X atoms colored differently belong to different elements. The green lobes represent lone pairs of electrons.

A diatomic molecule is polar if its bond is polar. A polyatomic molecule is polar if it has polar bonds arranged in space in such a way that their dipoles do not cancel.

EXAMPLE 3.4 Predicting the polar character of a molecule

Predict whether (a) a boron trifluoride molecule, BF_3, and (b) an ozone molecule, O_3, are polar.

STRATEGY In each case, we must decide on the shape of the molecule by using the VSEPR model and then decide whether the symmetry of the molecule results in the cancellation of the dipole moments associated with the bonds. If necessary, refer to Fig. 3.7.

SOLUTION (a) According to the VSEPR model, BF_3 is a trigonal planar AX_3 molecule. This shape has a symmetry in which all three B—F dipoles cancel. It is listed in Fig. 3.7 and classified there as nonpolar (see also **6**). (b) According to the VSEPR model, the O_3 molecule is angular, with one lone pair on the central O atom (**37**). Two positions around that central O atom are linked to O atoms, but the third is a lone pair, making it an AX_2E molecule. The dipoles do not cancel; as a result, the molecule is polar. This example shows that a homonuclear *poly*atomic molecule can be polar.

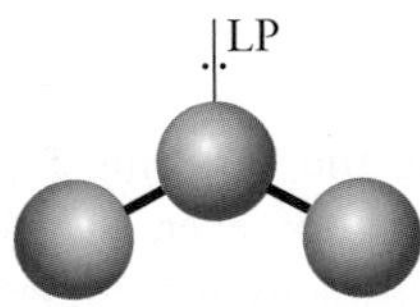

37 Ozone, O_3

Self-Test 3.5A Predict whether (a) SF_4, (b) SF_6 is polar or nonpolar.
[*Answer:* (a) Polar; (b) nonpolar]

Self-Test 3.5B Predict whether (a) PCl_5, (b) IF_5 is polar or nonpolar.

VALENCE-BOND THEORY

The Lewis model of the chemical bond assumes that each bonding electron pair is located between the two bonded atoms—it is a *localized electron model.* However, we know from the wave-particle duality of the electron (Sections 1.3–1.5) that the location of an electron in an atom cannot be described in terms of a precise position, but only in terms of the *probability* of finding it somewhere in a region of space defined by its orbital. The same principles apply to electrons in molecules, except that the volume over which the electrons are distributed is larger. The first description of covalent bonding to be devised in terms of atomic orbitals is called **valence-bond theory** (VB theory). This theory is a quantum mechanical description of the distribution of electrons in bonds that goes beyond Lewis theory and the VSEPR model by providing a way of calculating the numerical values of bond angles and bond lengths. We shall not go into the actual calculations, which are very complicated, but shall concentrate instead on some of the qualitative concepts. These bonding concepts pervade the language of chemistry.

3.4 Sigma and Pi Bonds

Some of the central ideas of valence-bond theory are encountered even in simple molecules. Let's begin with H_2, the simplest molecule of all, and start by thinking about the two hydrogen atoms from which it is formed. The electron configuration of a hydrogen atom in its ground state consists of an electron in a 1*s*-orbital. In valence-bond theory, we suppose that, as two H atoms come together, their 1*s*-electrons pair (denoted ↑↓, as in the discussion of atomic structure), and the atomic orbitals merge together (Fig. 3.8). The resulting sausage-shaped distribution of electrons is called a **σ-bond** (a sigma bond). Notice that in a σ-bond the electron density is concentrated between the two nuclei but that the density is still highest at the nuclei themselves. A hydrogen molecule is held together by a σ-bond. The merging of the two atomic orbitals is called the **overlap** of orbitals. A general point to keep in mind throughout this section is that the greater the extent of orbital overlap, the stronger the bond.

As explained in Section 1.9, the symbols ↑ and ↓ represent the spin states of electrons.

The Greek letter sigma, σ, is the equivalent of our letter *s*. It reminds us that, looking along the internuclear axis, the electron distribution resembles that of an *s*-orbital. Formally, a σ-bond has no nodal plane when viewed along the internuclear axis. All σ-bonds will be colored blue.

Much the same kind of σ-bond formation ("σ-bonding") occurs in the hydrogen halides. For example, in hydrogen fluoride, the unpaired electron on the fluorine atom occupies a $2p_z$-orbital, and the unpaired electron on the hydrogen atom occupies a 1*s*-orbital. These two electrons are the ones that pair to form a bond (**38**). They pair as the orbitals they occupy overlap and merge into a cloud that spreads over both atoms (Fig. 3.9). When viewed from the side, the resulting bond has a more complicated shape than the σ-bond in H_2; however, the bond looks much the same—formally, it has no nodal planes—when viewed along the internuclear (z) axis; hence it too is a σ-bond. All *single* covalent bonds are σ-bonds.

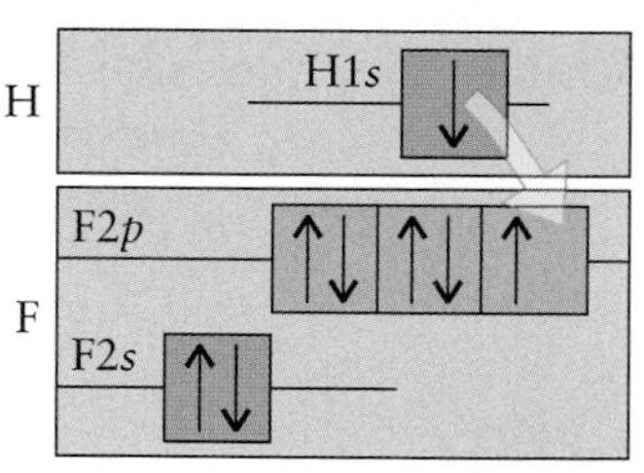

38 Hydrogen fluoride, HF

We encounter a different type of bond when we consider the structure of a nitrogen molecule, N_2. There is a single electron in each of the three

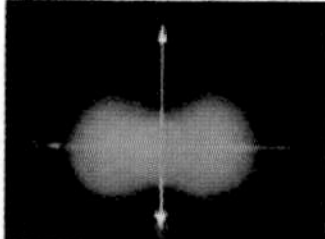

View an animation of the formation of H_2.

Media Link Animation: Fig. 3.8, Formation of the hydrogen molecule

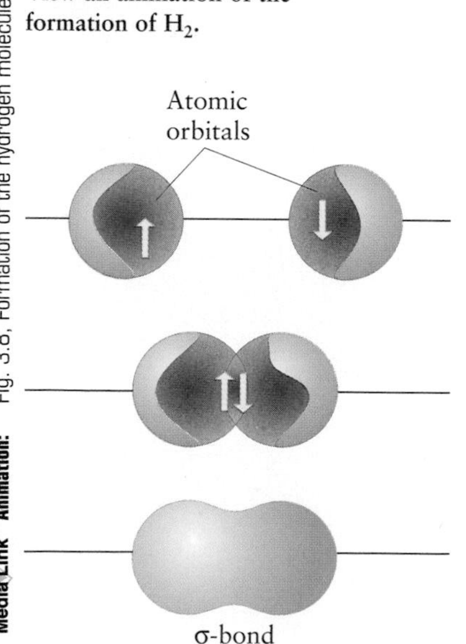

FIGURE 3.8 When electrons with opposite spins (depicted as ↑ and ↓) in two hydrogen 1*s*-orbitals pair and the *s*-orbitals overlap, they form a σ-bond, which is depicted here by the boundary surface of the electron cloud. The cloud has cylindrical symmetry around the internuclear axis and spreads over both nuclei.

2*p*-orbitals on each atom. However, when we try to pair them and form three bonds, only one of the three orbitals on each atom can overlap end to end to form a σ-bond (**39**, Fig. 3.10).

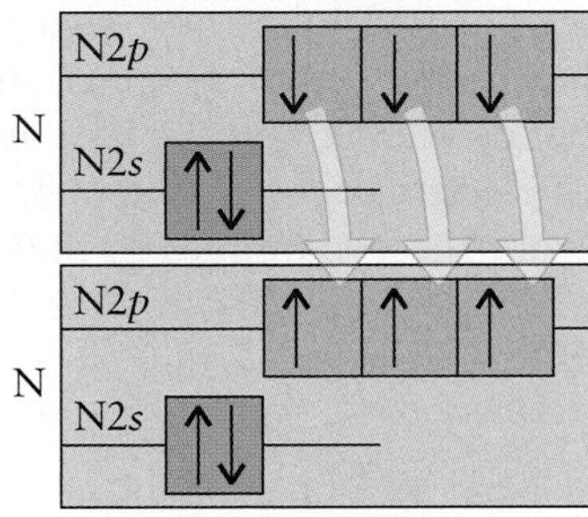

39 Nitrogen, N_2

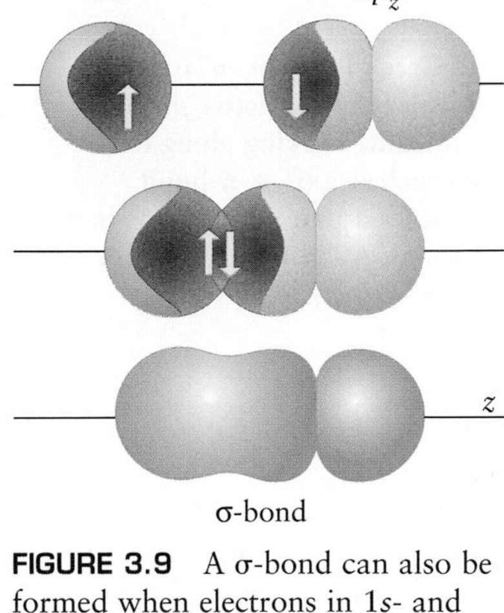

FIGURE 3.9 A σ-bond can also be formed when electrons in 1*s*- and $2p_z$-orbitals pair (where *z* is the direction along the internuclear axis). The two electrons in the bond are spread over the entire region of space enclosed by the boundary surface.

Two of the 2*p*-orbitals on each atom ($2p_x$ and $2p_y$) are perpendicular to the internuclear axis, and each one contains an unpaired electron (Fig. 3.11, top). When the electrons in these orbitals pair, the orbitals can overlap

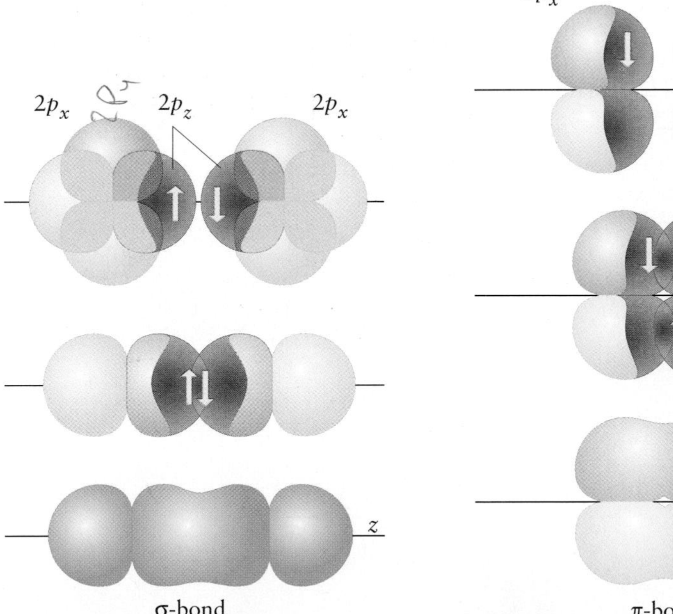

FIGURE 3.10 A σ-bond is formed by the pairing of electron spins in two $2p_z$-orbitals on neighboring atoms. At this stage, we are ignoring the interactions of any $2p_x$- (and $2p_y$-) orbitals that also contain unpaired electrons because they cannot form σ-bonds. The electron pair may be found anywhere within the boundary surface shown in the lowest diagram. Notice that the nodal plane of the p_z-orbital survives in the σ-bond.

$2p_x$ $2p_x$

π-bond

FIGURE 3.11 A π-bond is formed when electrons in two 2*p*-orbitals pair and overlap side by side. The middle diagram shows the density of the resulting electron cloud, and the bottom diagram shows the corresponding boundary surface. Even though the bond has a complicated shape, with two lobes, it is occupied by one pair of electrons and counts as one bond.

The Greek letter pi, π, is the equivalent of our letter *p*. When we imagine looking along the internuclear axis, a π-bond resembles a pair of electrons in a *p*-orbital. All π-bonds will be colored yellow in this text.

only in a side-by-side arrangement. This overlap results in a **π-bond**, a bond in which the two electrons lie in two lobes, one on each side of the internuclear axis (Fig. 3.11, bottom). Although a π-bond has electron density on each side of the internuclear axis, it is only one bond, with the electron cloud in the form of two lobes, just as a *p*-orbital is one orbital with two lobes. In a molecule with two π-bonds, such as N_2, the electron densities in the two π-bonds merge, and the two atoms appear to be surrounded by a cylinder of electron density (Fig. 3.12).

We can generalize from these examples to the description of a multiply bonded species according to valence-bond theory:

A **single bond** is a σ-bond.

A **double bond** is a σ-bond plus one π-bond.

A **triple bond** is a σ-bond plus two π-bonds.

In valence-bond theory, we assume that bonds form when unpaired electrons in valence-shell atomic orbitals pair; the atomic orbitals overlap end to end to form σ-bonds or side by side to form π-bonds.

Self-Test 3.6A How many σ-bonds and how many π-bonds are there in (a) CO_2? (b) CO?

[*Answer:* (a) Two σ, two π; (b) one σ, two π]

Self-Test 3.6B How many σ-bonds and how many π-bonds are there in (a) NH_3? (b) HCN?

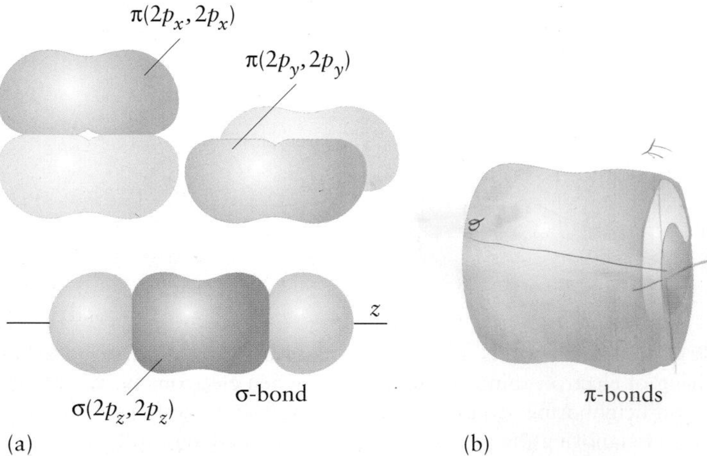

FIGURE 3.12 The bonding pattern in a nitrogen molecule, N_2. (a) The two atoms are bonded together by one σ-bond (blue) and two perpendicular π-bonds (yellow). (b) When the three bonds are put together, it turns out that the two π-bonds result in a long doughnut-shaped cloud surrounding the σ-bond cloud, so the overall structure resembles a cylindrical hot dog.

3.5 Hybridization of Orbitals

Valence-bond theory, as described so far, cannot account for bonding in polyatomic molecules like methane, CH_4, nor for their bond angles. For example, if we tried to apply the theory to methane, we would note that the carbon atom has four valence electrons (**40**). However, two valence electrons are already paired. Only the two half-filled 2*p*-orbitals on the carbon atom appear to be available for bonding, and they are perpendicular to each other. It looks as though a carbon atom should form only two bonds and have a valence of 2, but carbon almost always has a valence of 4. We need to revise valence-bond theory to account for the number of bonds formed.

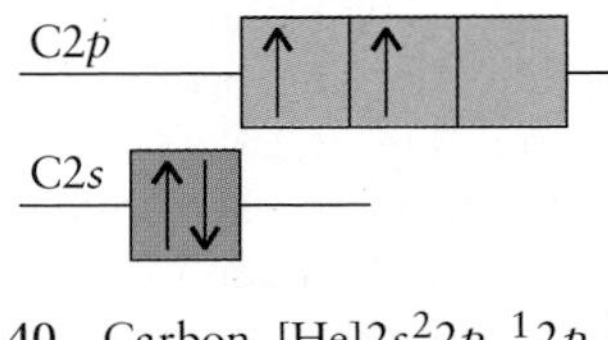

40 Carbon, $[He]2s^2 2p_x^1 2p_y^1$

The carbon atom can have four unpaired electrons if we assume that an electron has been **promoted**, that is, has acquired enough energy to raise it to a higher energy orbital. When we promote a 2*s*-electron into an empty 2*p*-orbital, we get the configuration shown in (**41**). Without promotion, a carbon atom can form only two bonds; after promotion, it can form four bonds. Each bond releases energy as it forms. Therefore, despite the energy cost of promoting the electron, the overall energy of the CH_4 molecule is lower than it would be if carbon formed only two C—H bonds.

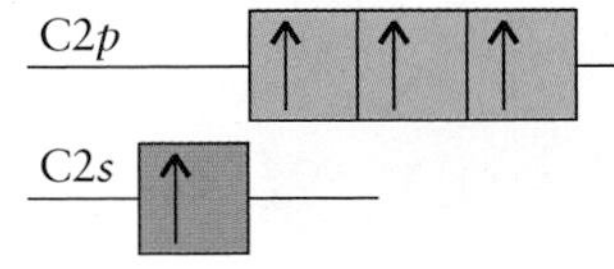

41 Carbon, $[He]2s^1 2p_x^1 2p_y^1 2p_z^1$

The characteristic tetravalence of carbon is due to the small promotion energy of a carbon atom. The promotion energy is small because a 2*s*-electron is transferred from an orbital that it shares with another electron to an empty 2*p*-orbital. Although the promoted electron enters an orbital of higher energy, it experiences less repulsion from other electrons than before it was promoted. As a result, only a small energy is needed to promote the electron. That energy is more than recovered from the atom's ability to form four bonds. Nitrogen, carbon's neighbor, cannot use promotion to increase the number of bonds it can form, because promotion does not increase the number of unpaired electrons it possesses (**42**). The same is true of oxygen and fluorine. Promotion of an electron is possible if the overall change, taking account of all contributions to the energy and especially the greater number of bonds that can thereby be formed, is toward lower energy.

Carbon monoxide, CO, is the only common exception to the tetravalence of carbon.

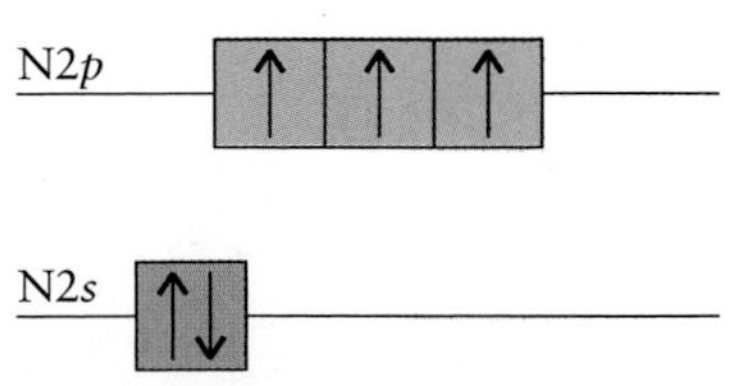

42 Nitrogen, $[He]2s^2 2p_x^1 2p_y^1 2p_z^1$

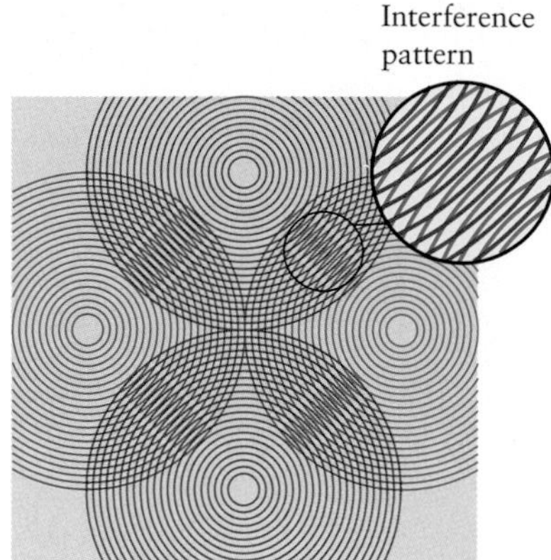

FIGURE 3.13 When waves interfere, new patterns arise.

At this stage, it looks as though electron promotion should result in two different types of bonds, one from the overlap of a hydrogen 1*s*-orbital and a carbon 2*s*-orbital, and three more bonds from the overlap of hydrogen 1*s*-orbitals with each of the three carbon 2*p*-orbitals. The overlap with the 2*p*-orbitals should result in three σ-bonds at 90° to one another. However, that is inconsistent with the known structure of methane.

To improve our model still further and account for the tetrahedral structure of methane, we note that *s*- and *p*-orbitals are waves of electron density centered on the nucleus of an atom. We imagine that the four orbitals interfere with one another and produce new patterns where they intersect, like waves in water (Fig. 3.13). These new patterns are called **hybrid orbitals.** The four identical hybrid orbitals point toward different corners of a tetrahedron (Fig. 3.14). Each orbital has a node close to the nucleus and a small "tail" on the other side where the *s*- and *p*-orbitals do not completely cancel. These four hybrid orbitals are called ***sp*^{3} hybrids** because they are formed from one *s*-orbital and three *p*-orbitals. In an orbital-energy diagram, we represent the hybridization as the formation of four orbitals of equal energy intermediate between the energies of the

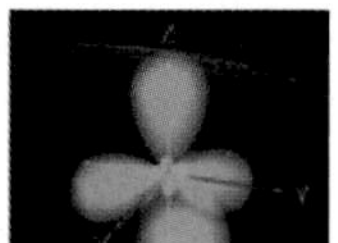

Construct an *sp*^{3} orbital.

Media Link Animation: Fig. 3.14, Hybridization of orbitals

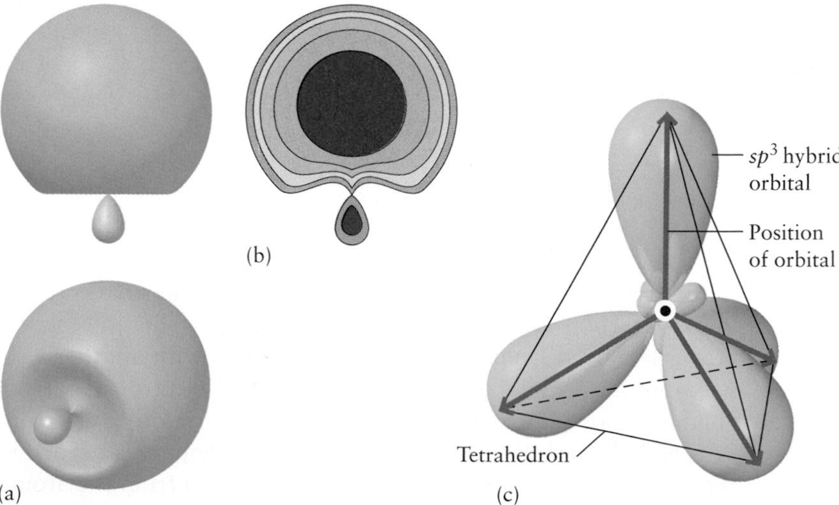

FIGURE 3.14 (a) One *s*-orbital and three *p*-orbitals blend into four *sp*^{3} hybrid orbitals, each of this shape (shown from two viewpoints). (b) These contours indicate the amplitude of the hybrid orbital wavefunction in a plane that bisects it and passes through the nucleus. (c) Each hybrid orbital points toward the vertices of a tetrahedron, as indicated by the orientations of the arrows. Here and elsewhere, the hybrid orbitals are shown as narrow lobes to emphasize their spatial arrangement.

s- and *p*-orbitals from which they are constructed (**43**). The hybrids are colored green to remind us that they are a blend of (blue) *s*-orbitals and (yellow) *p*-orbitals. The fact that hybrid orbitals have their amplitudes concentrated on one side of the nucleus allows them to overlap more effectively with other orbitals, and as a result the bonds they form are stronger than in the absence of hybridization. This stronger bonding is another contribution that makes promotion worthwhile.

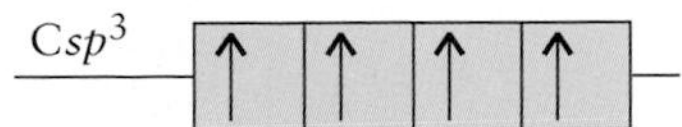

43 sp^3 hybridized carbon

We are now ready to account for the bonding in a methane molecule. An unpaired electron occupies each of carbon's sp^3 hybrid orbitals. Each of these four electrons can pair with an electron in a hydrogen 1*s*-orbital. Their overlapping orbitals form four σ-bonds that point toward the corners of a tetrahedron (Fig. 3.15). All four bonds are identical because they are formed from the same blend of atomic orbitals. The valence-bond description is now consistent with experiment.

Hybrid orbitals are constructed on an atom to reproduce the electron arrangement characteristic of the experimentally determined shape of a molecule.

3.6 Hybridization in More Complex Molecules

To use hybridization to describe bonds in molecules in which there is more than one central atom, we concentrate on each atom in turn. For example, in ethane (**44**), the two carbon atoms are both "central" atoms. According to the VSEPR model, the four electron pairs around each carbon atom take up a tetrahedral arrangement. This arrangement suggests sp^3 hybridization of the carbon atoms, as shown in Fig. 3.14. Each C atom has one unpaired electron in each of its four sp^3 hybrid orbitals and can therefore form four σ-bonds, that point toward the corners of a regular tetrahedron. The C—C bond is formed by spin-pairing of the electrons

44 Ethane, CH_3CH_3

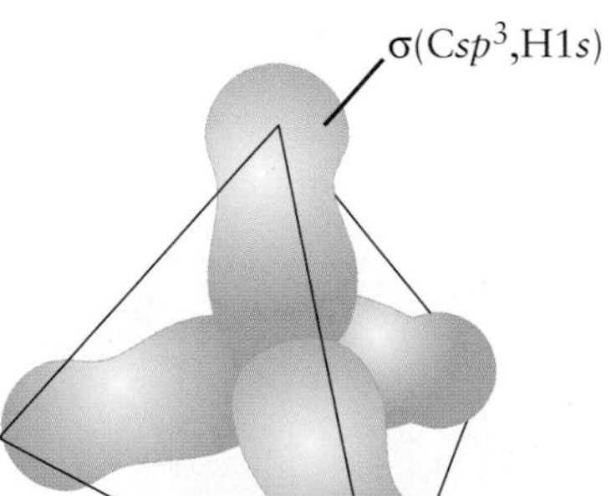

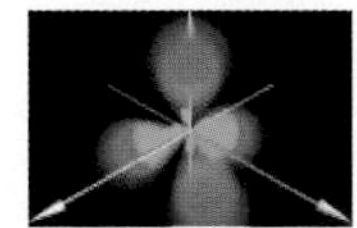

View an animation of the formation of bonds in methane.

Media Link Animation: Fig. 3.15, Methane formation

FIGURE 3.15 Each C–H bond in methane is formed by the pairing of an electron in a hydrogen 1*s*-orbital and an electron in one of the four sp^3 hybrid orbitals of carbon. Therefore, valence bond theory predicts four equivalent σ-bonds in a tetrahedral arrangement, which is consistent with experimental results.

in one sp^3 hybrid orbital of each C atom. We label this bond $\sigma(C2sp^3,C2sp^3)$ to show its composition: $C2sp^3$ denotes an sp^3 hybrid orbital composed of 2*s*- and 2*p*-orbitals on a carbon atom, and the parentheses show which orbitals overlap (Fig. 3.16). Each C—H bond is formed by spin-pairing of an electron in one of the remaining sp^3 hybrid orbitals with an electron in a 1*s*-orbital of an H atom (denoted H1*s*). These bonds are denoted $\sigma(C2sp^3,H1s)$.

Some molecules, such as ammonia, have a lone pair of electrons on the central atom. According to the VSEPR model, the four electron pairs in NH_3 take up a tetrahedral electron arrangement, so we describe the nitrogen atom in terms of four sp^3 hybrid orbitals. Because nitrogen has five valence electrons, one of these hybrid orbitals is already doubly occupied (**45**). The 1*s*-electrons of the three hydrogen atoms pair with the three unpaired electrons in the remaining sp^3 hybrid orbitals. This pairing and overlap result in the formation of three N—H σ-bonds.

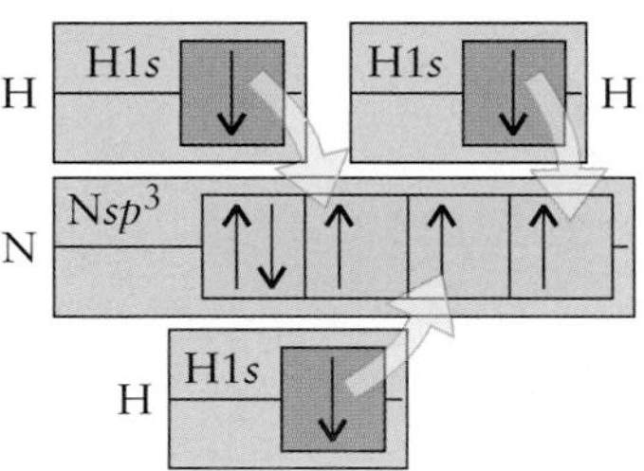

45 Ammonia, NH_3

We use different hybridization schemes to describe other arrangements of electron pairs (Fig. 3.17). For example, to explain a trigonal planar electron arrangement, we mix one *s*-orbital with two *p*-orbitals and so produce three **sp^2 hybrid orbitals.** They point toward the corners of an equilateral triangle. A linear arrangement of pairs requires two hybrid orbitals, so we mix an *s*-orbital with a *p*-orbital to obtain two ***sp* hybrid**

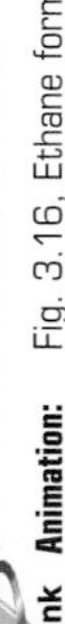

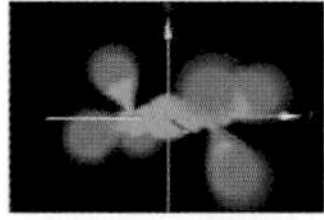

View an animation of the formation of bonds in ethane.

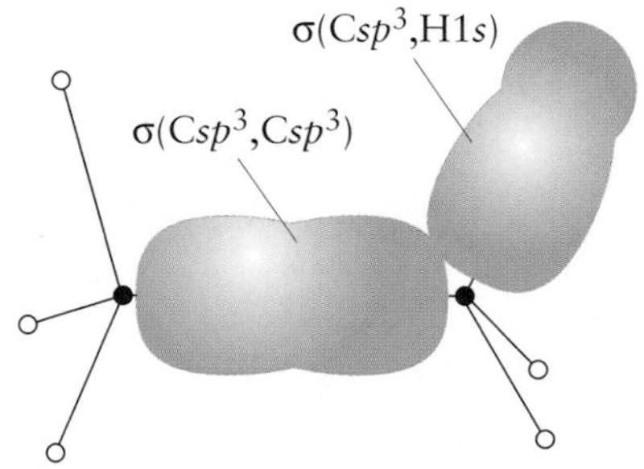

FIGURE 3.16 The valence-bond description of bonding in an ethane molecule, C_2H_6. The boundary surfaces of only two of the bonds are shown. Each pair of neighboring atoms is linked by a σ-bond formed by the pairing of electrons in either H1*s*-orbitals or $C2sp^3$ hybrid orbitals. All the bond angles are close to 109.5° (the tetrahedral angle).

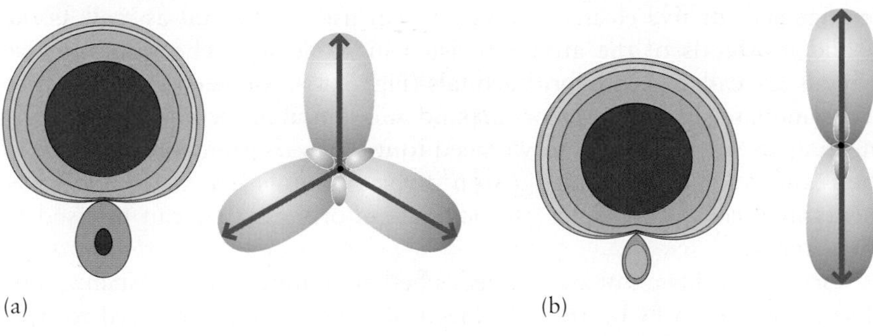

(a) (b)

FIGURE 3.17 Two other common hybridization schemes shown as contours of the amplitude of the wavefunction and in terms of the orientations of the hybrid orbitals. (a) An *s*-orbital and two *p*-orbitals can blend together to give three sp^2 hybrid orbitals that point to the corners of an equilateral triangle. (b) An *s*-orbital and a *p*-orbital hybridize into two *sp* hybrid orbitals that point in opposite directions.

Construct sp^2 and sp hybrid orbitals.

Media Link Animation: Fig. 3.17, Hybrid orbitals

orbitals. Table 3.2 summarizes the relationship between electron arrangement and hybridization type. No matter how many atomic orbitals we mix together, the number of hybrid orbitals is always the same as the number of atomic orbitals we started with, so *N* atomic orbitals produce *N* hybrid orbitals.

Self-Test 3.7A Suggest a structure in terms of hybrid orbitals for BF_3.
[*Answer:* Three σ-bonds formed from B2sp^2 hybrids and F2p_z-orbitals in a trigonal planar arrangement]

Self-Test 3.7B Suggest a structure in terms of hybrid orbitals for carbon dioxide, CO_2.

Some of the hybridization schemes in Table 3.2 include *d*-orbitals. We use *d*-orbitals when a central atom has to accommodate five or more electron pairs, such as in PCl_5. To account for a trigonal bipyramidal

TABLE 3.2 Hybridization and molecular shape*

Electron arrangement	Number of atomic orbitals	Hybridization of the central atom	Number of hybrid orbitals
linear	2	sp	2
trigonal planar	3	sp^2	3
tetrahedral	4	sp^3	4
trigonal bipyramidal	5	sp^3d	5
octahedral	6	sp^3d^2	6

* Other combinations of *s*-, *p*-, and *d*-orbitals can give rise to the same or different shapes, but these combinations are the most common.

arrangement of five electron pairs, we can use a *d*-orbital as well as the *s*- and *p*-orbitals of the atom, as listed in Table 3.2. The resulting five orbitals are called ***sp*^{3}*d* hybrid orbitals** (Fig. 3.18). We need six orbitals to accommodate six electron pairs around an atom in an octahedral arrangement, as in SF_6 and XeF_4, so we need to use two *d*-orbitals in addition to the *s*- and *p*-orbitals to form six *sp*^{3}*d*^{2} hybrid orbitals (Fig. 3.19). Only elements in Period 3 and later periods have *d*-orbitals that can be used to form hybrids.

So far, we have not considered whether terminal atoms, such as the chlorine atoms in PCl_5, are hybridized. Because they are bonded to only one other atom, we cannot use bond angles to predict a hybridization scheme. However, spectroscopic data and calculation suggest that both *s*- and *p*-orbitals of terminal atoms are involved in bond formation, so it is reasonable to suppose that their orbitals are hybridized. The simplest model is to suppose that the three lone pairs and the bonding pair are arranged tetrahedrally and, therefore, that the chlorine atoms bond to the phosphorus atom by using *sp*^{3} hybrid orbitals.

Toolbox 3.2 summarizes how to assign a hybridization scheme to an atom in a molecule.

A hybridization scheme is adopted to match the electron arrangement of the molecule. Octet expansion corresponds to the involvement of d-orbitals.

Media Link Animation: Fig. 3.18 and Fig. 3.19, Hybrid orbitals

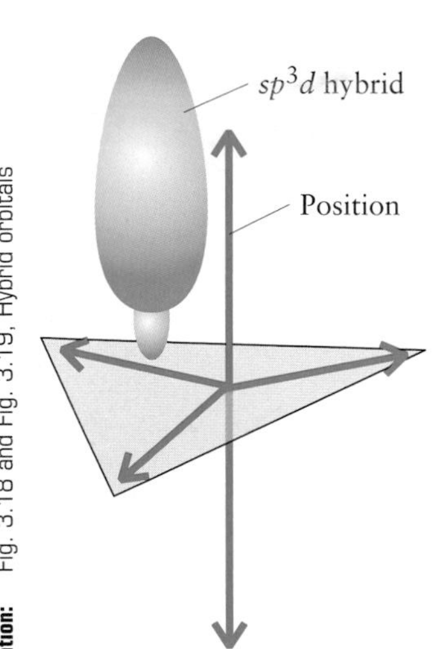

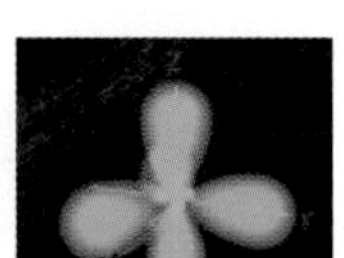

Construct an *sp*^{3}*d* orbital.

FIGURE 3.18 One of the five *sp*^{3}*d* hybrid orbitals, and their five directions, that account for a trigonal bipyramidal arrangement of electron pairs. The *sp*^{3}*d* hybridization scheme can be applied only when *d*-orbitals are available on the central atom.

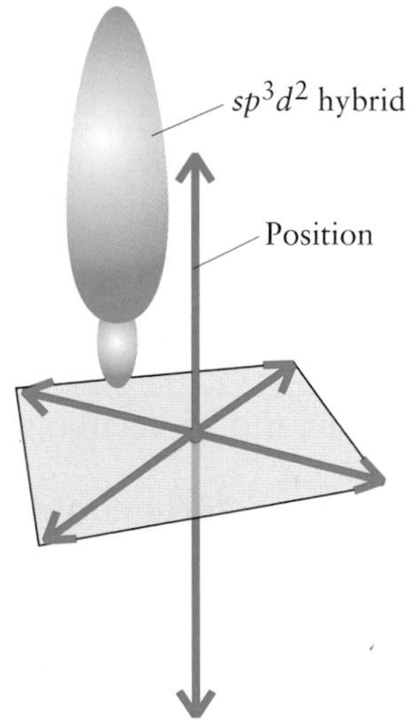

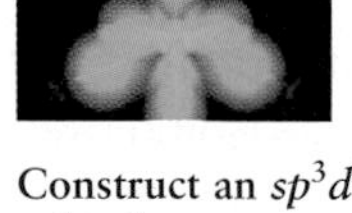

Construct an *sp*^{3}*d*^{2} orbital.

FIGURE 3.19 One of the six *sp*^{3}*d*^{2} hybrid orbitals, and their six directions, that may be formed when *d*-orbitals are available and we need to reproduce an octahedral arrangement of electron pairs.

TOOLBOX 3.2 How to use hybridization

Conceptual Basis

We can explain the shapes of molecules if we assume that atomic orbitals centered on the same nucleus interfere with one another to produce new patterns of electron density. The overlap between these hybrid orbitals and the orbitals of neighboring atoms is enhanced beyond what is possible with unhybridized orbitals.

Procedure

The aim is to select a hybridization scheme that reproduces the shape of the molecule.

Step 1 Draw the Lewis structure and determine the electron arrangement about the central atom.

Step 2 The number of orbitals required for the electron arrangement determines the hybridization scheme used, as shown in Table 3.2.

Step 3 Construct the hybrid orbitals from atomic orbitals, using the same number of atomic orbitals as hybrid orbitals needed to account for the electron arrangement. Start with the *s*-orbital, and add *p*-orbitals, then *d*-orbitals as needed.

More than one scheme may correspond to the same molecular shape: they can be distinguished by considerations of availability and energy. The most common schemes, the only ones that need be considered in most cases, involve *s*- and *p*-orbitals, followed by further inclusion of *d*-orbitals.

How can we use this information . . . to assign a hybridization scheme?

According to Example 3.3, SF_4 has a trigonal bipyramidal arrangement of electron pairs. We can devise a hybridization scheme for the molecule by using the procedure in Toolbox 3.2. A trigonal bipyramidal electron arrangement implies the presence of five bonding orbitals on the S atom. We construct five equivalent hybrid orbitals by mixing five atomic orbitals: an *s*-orbital with three *p*-orbitals and one *d*-orbital. The result is five sp^3d orbitals.

Self-Test 3.8A Describe (a) the electron arrangement, (b) the molecular shape, and (c) the hybridization of the central chlorine atom in chlorine trifluoride.

[*Answer:* (a) Trigonal bipyramidal; (b) T-shaped; (c) sp^3d]

Self-Test 3.8B Describe (a) the electron arrangement, (b) the molecular shape, and (c) the hybridization of the central atom in BrF_4^-.

3.7 Bonding in Hydrocarbons

There are more compounds of carbon than there are of all the other elements put together. Although carbon is not unique in being able to form multiple bonds, it is unique in the extent to which it can link to itself and form chains and rings of carbon atoms that serve as the backbones of organic compounds. Carbon atoms can form four strong, covalent bonds. The other Group 14/IV elements can also form four bonds, but their atoms cannot form multiple bonds with one another as easily as carbon atoms do because their atomic radii are too big for π overlap to be significant. Hence, they are less versatile in their bonding.

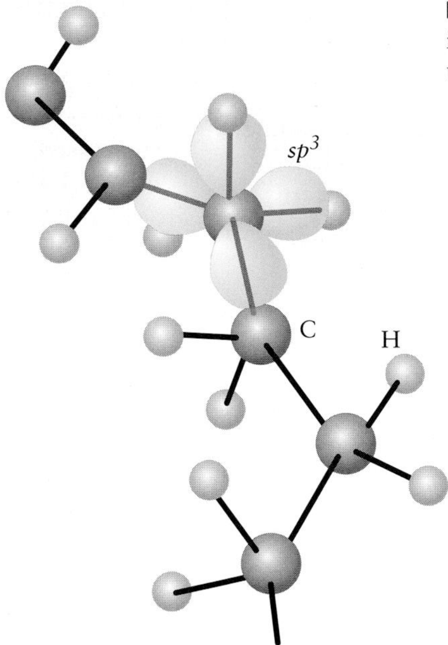

FIGURE 3.20 A polyethylene molecule is a very long chain of $-CH_2-CH_2-$ units. Each C atom is sp^3 hybridized.

We have seen how valence-bond theory accounts for bonding in the hydrocarbons methane and ethane. The same pattern of σ-bonds occurs throughout the alkanes (Section D) and the hydrocarbon polymers like polyethylene (Fig. 3.20). Bond angles in alkanes are very close to the tetrahedral angle of 109.5°, corresponding to sp^3 hybridization. Because σ-bonds are formed by end-to-end overlap, two atoms linked by a σ-bond, along with their attached groups, can rotate with respect to each other, with any methyl groups ($-CH_3$ groups) rotating like tiny propellers.

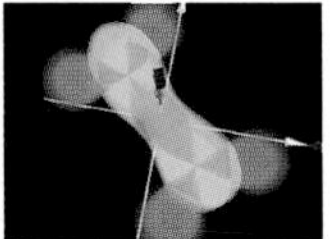

View an animation of the formation of bonds in ethene.

To describe carbon-carbon double bonds in alkenes, we use the pattern provided by ethene, $CH_2{=}CH_2$. We know from experimental data that all six atoms in ethene lie in the same plane, with H—C—H and C—C—H bond angles of 120°. This angle suggests a trigonal planar electron arrangement and sp^2 hybridization for each C atom (**46**). Each of the three hybrid orbitals on the C atom has one electron available for bonding; the fourth valence electron of each C atom occupies the unhybridized 2*p*-orbital, which lies perpendicular to the plane formed by the hybrids. The two carbon atoms form a σ-bond by overlap of an sp^2 hybrid orbital on each atom. The H atoms form σ-bonds with the remaining lobes of the sp^2 hybrids. This arrangement of orbitals leaves the electrons in the two unhybridized 2*p*-orbitals free to pair and to form a π-bond by side-by-side overlap. Figure 3.21 shows that the electron density in the π-bond lies above and below the axis of the C—C σ-bond.

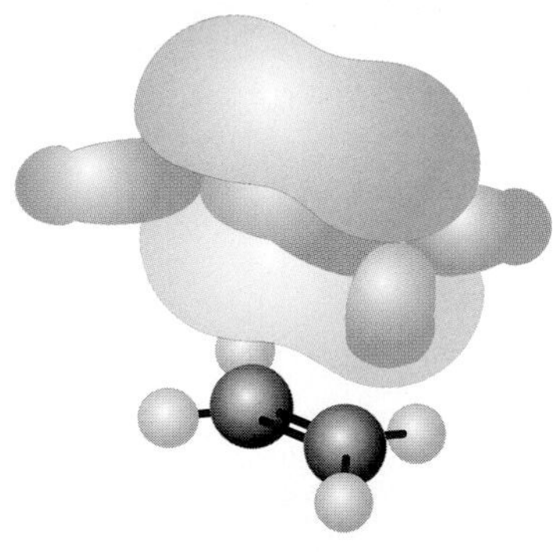

FIGURE 3.21 A view of the bonding pattern in ethene (ethylene), showing the framework of σ-bonds and the single π-bond formed by side-to-side overlap of unhybridized C2*p*-orbitals. The double bond is resistant to twisting because twisting would reduce the overlap between the two C2*p*-orbitals and weaken the π-bond.

Media Link **Animation:** Fig. 3.21, Ethene formation

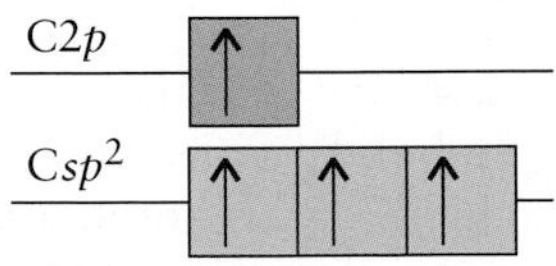

46 sp^2 hybridized carbon

In the benzene molecule, the C atoms and their attached H atoms all lie in the same plane, with the C atoms forming a hexagonal ring. To describe the bonding in the Kekulé structures of benzene (Section 2.8), we need hybrid orbitals that match the 120° bond angles of the hexagonal ring. Therefore, we take each carbon atom to be sp^2 hybridized, as in ethene (Fig. 3.22). There is one electron in each of the three hybrid orbitals and one electron in an unhybridized $2p$-orbital perpendicular to the plane of the hybrids. The ring formed by overlap of the Csp^2 hybrid orbitals brings the six unhybridized carbon $2p$-orbitals close to one another. If only one of the Kekulé structures were correct, the $2p$-orbital on each C atom would overlap side by side with one of its neighbors to form a π-bond to that atom, with only a σ-bond to the other neighboring carbon atom (Fig. 3.23). However, the actual structure of the benzene molecule is a resonance hybrid of the two alternative bonding patterns in the two Kekulé structures. The six carbon-carbon bonds making up the ring are identical, and the electrons in the π-bonds are spread around the entire ring (Fig. 3.24).

Now consider the alkynes, hydrocarbons with triple bonds. The Lewis structure of the linear molecule ethyne (acetylene) is H—C≡C—H. To describe the bonding in a linear molecule, we need a hybridization scheme that produces two equivalent orbitals at 180° from each other: this is sp hybridization. Each C atom has one electron in each of its two sp hybrid orbitals and one electron in each of its two perpendicular unhybridized $2p$-orbitals (**47**). The electrons in the sp hybrid orbitals on the

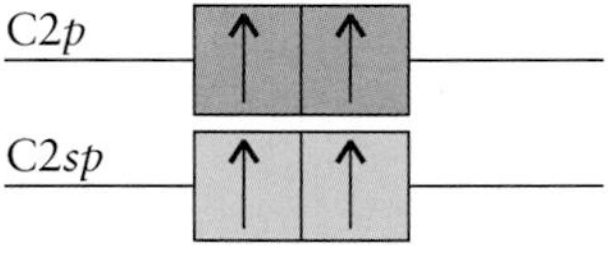

47 *sp* hybridized carbon

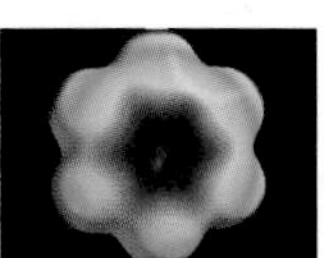

View and manipulate different representions of the structure of benzene.

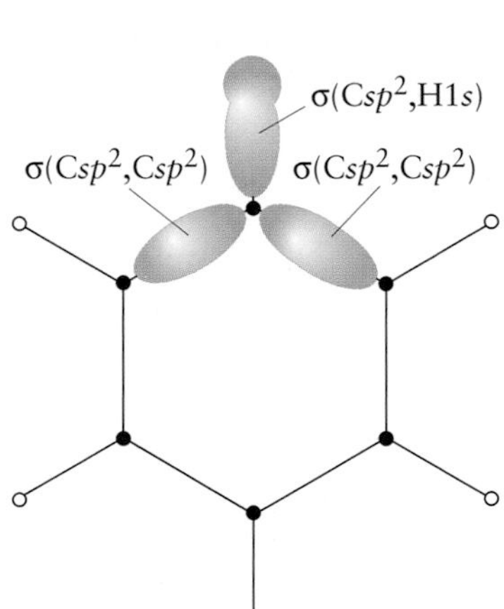

FIGURE 3.22 The framework of σ-bonds in benzene: each carbon atom is sp^2 hybridized, and the array of hybrid orbitals matches the bond angles (of 120°) in the hexagonal molecule. Bonding is shown explicitly around only one carbon atom all the others are the same.

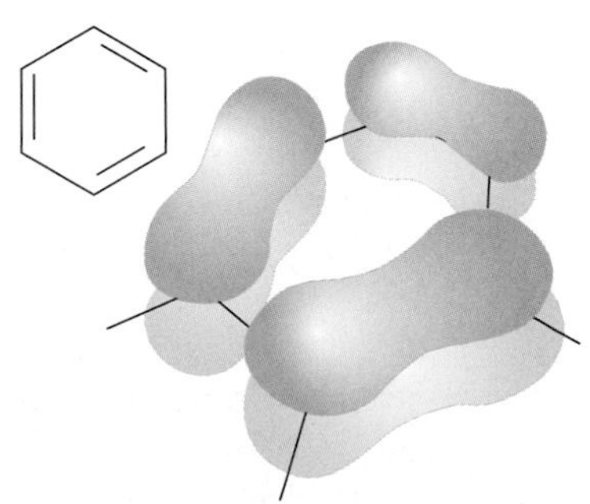

FIGURE 3.23 Unhybridized carbon $2p$-orbitals can form a π-bond with either of their immediate neighbors. Two arrangements are possible, each one corresponding to a different Kekulé structure. One Kekulé structure and the corresponding π-bonds are shown here.

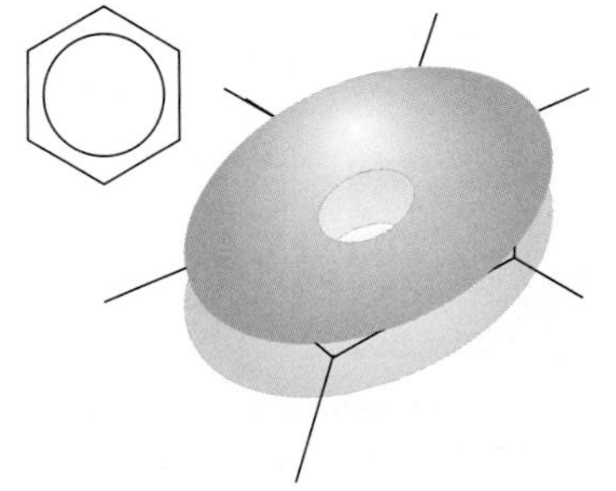

FIGURE 3.24 As a result of resonance between two structures like the one shown in Fig. 3.23 (corresponding to resonance of the two Kekulé structures), the π-electrons form a double doughnut-shaped cloud above and below the plane of the ring.

two C atoms pair and form a carbon-carbon σ-bond. The electrons in the remaining *sp* hybrid orbitals pair with hydrogen 1*s*-electrons to form two carbon-hydrogen σ-bonds. The electrons in the two perpendicular sets of 2*p*-orbitals pair with a side-by-side overlap, forming two π-bonds at 90° to each other. As in the N_2 molecule, the electron density in the π-bonds forms a cylinder about the C—C bond axis. The resulting bonding pattern is shown in Fig. 3.25.

Multiple bonds are formed when an atom forms a σ-bond by using an sp or sp^2 hybrid orbital and one or more π-bonds by using unhybridized p-orbitals.

3.8 Characteristics of Double Bonds

A carbon-carbon double bond is stronger than one carbon-carbon single bond but weaker than the sum of two single bonds (Section 2.17). A carbon-carbon triple bond is also weaker than the sum of three carbon-carbon single bonds. One reason is that, whereas a single bond is a σ-bond, the additional bonds in a multiple bond are π-bonds. Carbon-carbon π-bonds are about 84 $kJ\cdot mol^{-1}$ weaker than carbon-carbon σ-bonds because the side-by-side overlap of the former puts their electrons in less favorable positions for holding the atoms together and increases the electron-electron repulsion between the two bonding pairs.

Atoms of the Period 2 elements C, N, and O readily form double bonds with one another, with themselves, and (especially for oxygen) with atoms of elements from later periods. However, double bonds are rarely found between atoms of elements in Period 3 and higher, because the atoms are so large that it is difficult for their *p*-orbitals to take part in effective side-by-side overlap. Although silicon atoms can also form long chains, it is difficult for them to form double bonds with each other; therefore, they cannot form as large a variety of compounds as carbon atoms can.

Media Link Animation: Fig. 3.25, Ethyne formation

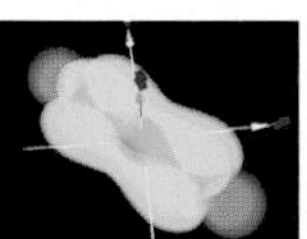

View an animation of the formation of bonds in ethyne.

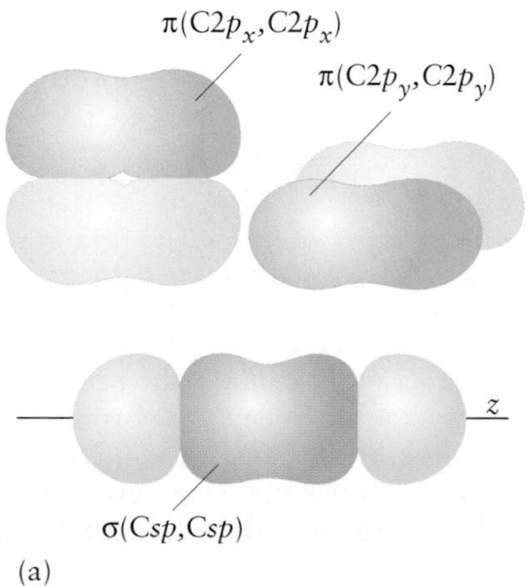

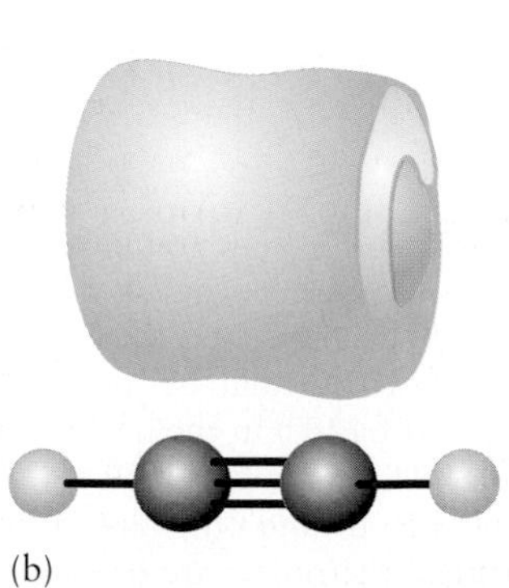

FIGURE 3.25 The pattern of bonding in ethyne (acetylene). The carbon atoms are *sp* hybridized, and the two remaining *p*-orbitals on each C atom form two π-bonds. The resulting pattern is very similar to that for N_2 (Fig. 3.12), but two C—H groups replace the N atoms.

The presence of a double bond has a strong influence on the shape of a molecule because it prevents one part of a molecule from rotating relative to another part. The double bond of ethene, for example, holds the entire molecule flat. Figure 3.21 shows that the two 2*p*-orbitals overlap best if the two CH_2 groups lie in the same plane. The π-bond could not form if the *p*-orbitals on the carbon atoms, which are perpendicular to the CH_2 groups, were twisted about the C—C bond.

Double bonds are vitally important for living organisms. For instance, they help you read these words. Vision depends on the shape of the molecule retinal in the retina of the eye. *cis*-Retinal is held rigid by its double bonds (**48**). When light enters the eye, it excites an electron out of the π-bond marked by the arrow. The double bond is now weaker, and the molecule is free to rotate around the remaining σ-bond. When the excited electron falls back, the molecule is trapped in its new shape (**49**). This change of shape triggers a signal along the optic nerve and is interpreted by the brain as the sensation of vision.

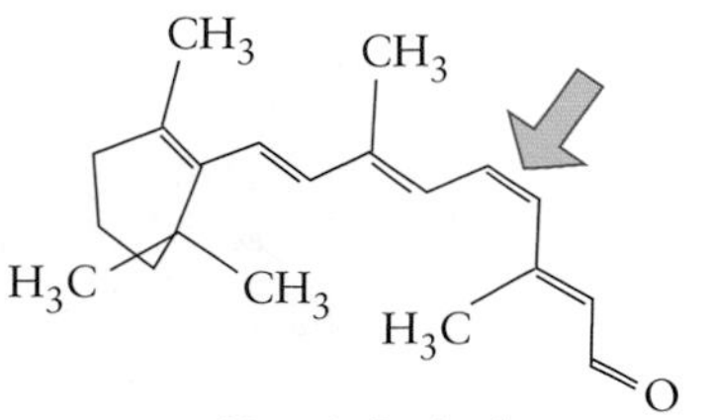

CH_3 CH_3 H_3C CH_3 H_3C O

48 *cis*-Retinal

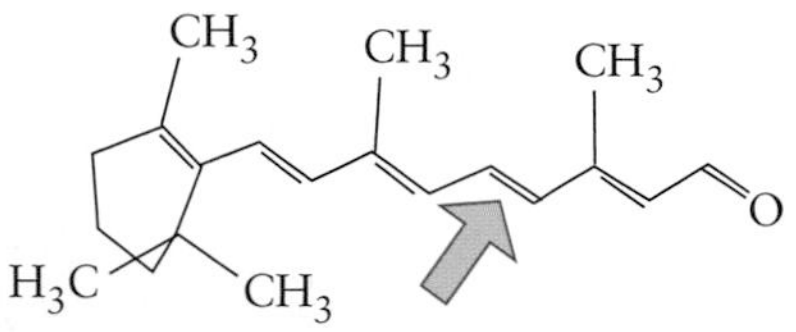

CH_3 CH_3 CH_3 O H_3C CH_3

49 *trans*-Retinal

The side-by-side overlap that forms a π-bond makes a molecule resistant to twisting, results in bonds weaker than σ-bonds, and prevents atoms with large radii from forming multiple bonds.

EXAMPLE 3.5 Accounting for the structure of a molecule with multiple bonds

Account for the structure of a formic acid molecule, HCOOH, in terms of hybrid orbitals, bond angles, and σ- and π-bonds. The C atom is attached to a terminal O atom and an —OH group.

STRATEGY In valence-bond theory, we determine the hybridization of a central atom by noting the number of bonds it forms and the bond angles. The type of bond formed is determined by noting the multiplicity of the bond: single bonds are all σ-bonds and multiple bonds are composed of a σ-bond and one or more π-bonds. Because the C atom is attached to three atoms, we anticipate that its hybridization scheme is sp^2.

SOLUTION ***Step 1*** The Lewis structure of formic acid is shown in (**50**). According to the VSEPR model, the three bonds around the carbon atom take up a trigonal planar arrangement at 120°. The O atom in the —OH group forms a single bond to the C atom and a single bond to an H atom. It also has two lone pairs of electrons, for a total of four regions of high electron concentration; thus, it has a tetrahedral electron arrangement. ***Step 2*** The presence of three bonds around the C atom suggests sp^2 hybridization, and the tetrahedral arrangement around the O atom of the —OH group suggests sp^3 hybridization. ***Step 3*** Carbon's fourth valence electron is in a *p*-orbital perpendicular to the triangle. It forms a π-bond by overlap of *p*-orbitals on the C atom and the terminal O atom.

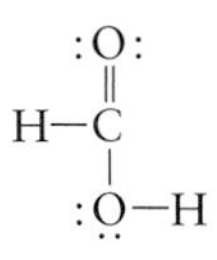

:O:
H—C
:O—H

50 Formic acid, HCOOH

On the basis of this analysis, we can describe the structure of formic acid as follows: the carbon atom is sp^2 hybridized, and the oxygen atom in the —OH group is sp^3 hybridized. There are four σ-bonds and one π-bond in the molecule. All the atoms except for the H of the —OH group are in the same plane. The pattern of bonds is illustrated in Fig. 3.26.

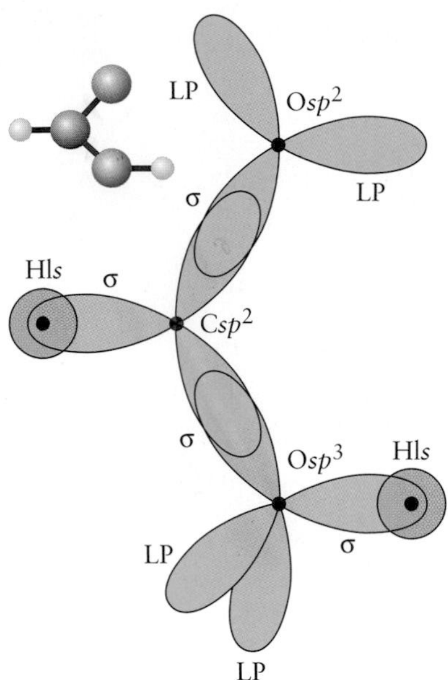

FIGURE 3.26 A schematic diagram showing the pattern of σ-bonds in formic acid (methanoic acid). The π-bond between the C atom and the sp^2 hybridized O atom is not shown.

Self-Test 3.9A Describe the structure of the carbon suboxide molecule, C_3O_2, in terms of hybrid orbitals, bond angles, and σ- and π-bonds. The atoms lie in the order OCCCO.

[*Answer:* Linear; bond angles all 180°; each C atom is *sp* hybridized; it forms one σ-bond and one π-bond to each adjacent C or O atom.]

Self-Test 3.9B Describe the structure of the propene molecule, $CH_3{-}CH{=}CH_2$, in terms of hybrid orbitals, bond angles, and σ- and π-bonds.

MOLECULAR ORBITAL THEORY

Lewis's theory of the chemical bond was brilliant, but it was little more than guesswork inspired by insight. Lewis had no way of knowing why an electron pair was so important for the formation of covalent bonds. Valence-bond theory explained the importance of the electron pair and provided a basis for calculating electron distributions in molecules, but it could not explain the properties of some molecules. Molecular orbital theory, which is also based on quantum mechanics, has proved to be the most successful theory of the chemical bond: it overcomes all the deficiencies of Lewis's theory and is easier to use in calculations than valence-bond theory.

3.9 The Limitations of Lewis's Theory

In the Lewis approach, we would describe bonding in O_2 as $\ddot{\underset{..}{O}}{=}\ddot{\underset{..}{O}}$, with all the electrons paired. However, elemental oxygen is a paramagnetic substance (Fig. 3.27 and Box 3.2), and paramagnetism is a property of *unpaired* electrons. The magnetism of O_2 therefore contradicts both the Lewis structure and the valence-bond description of the molecule.

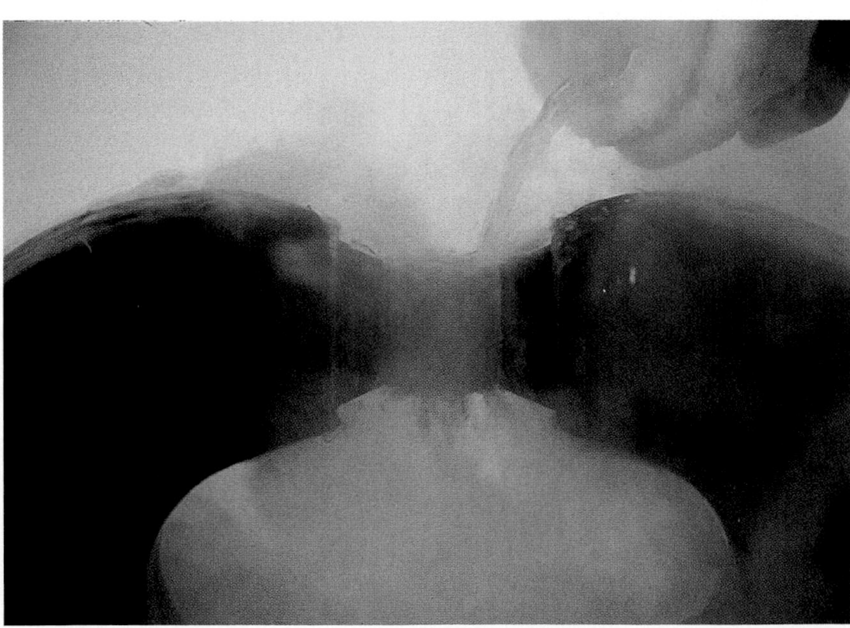

FIGURE 3.27 The paramagnetic properties of oxygen are evident when liquid oxygen is poured between the poles of a magnet. The liquid sticks to the magnet instead of flowing past it.

BOX 3.2 HOW DO WE KNOW . . . THAT ELECTRONS ARE NOT PAIRED?

Most common materials are *diamagnetic*, which means that a sample of the substance tends to move out of a magnetic field. The effect is quite small, but it can be detected by hanging a long thin sample from the pan of a balance and letting it lie between the poles of an electromagnet. This arrangement, which was once the primary technique used to measure the magnetic properties of samples, is called a *Gouy balance*. When the electromagnet is turned on, the sample tends to move upward, out of the field, so it appears to weigh less than in the absence of the field. The diamagnetism arises from the effect of the magnetic field on the electrons in the molecule: the field forces the electrons to circulate through the nuclear framework. Because electrons are charged particles, this circulation corresponds to an electric current circulating within the molecule. That current gives rise to its own magnetic field, which opposes the applied field. The sample tends to move out of the field so as to minimize this opposing field. The modern approach to measuring magnetic properties is to use a superconducting quantum interference device (a SQUID), which is highly sensitive to small magnetic fields and can make very precise measurements on small samples.

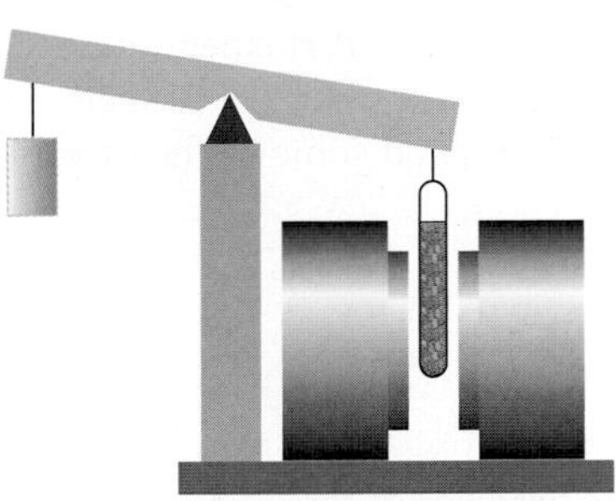

A Gouy balance is used to observe the magnetic character of a sample by detecting the extent to which it is drawn into or driven out of a magnetic field.

Compounds with unpaired electrons are *paramagnetic*. They tend to move into a magnetic field and can be identified because they seem to weigh more in a Gouy balance when a magnetic field is applied than when it is absent. Paramagnetism arises from the electron spins, which behave like tiny bar magnets that tend to line up with the applied field. The more that can line up in this way, the greater the lowering of energy, so the sample tends to move into the applied field. Oxygen is a paramagnetic substance because it has two unpaired electrons: this property is used to detect the concentration of oxygen in incubators. All radicals are paramagnetic. Many compounds of the *d*-block elements are paramagnetic because they have various numbers of unpaired *d*-electrons.

In a paramagnetic substance, the unpaired electron spins become randomly oriented as soon as the applied field is removed. However, in a *ferromagnetic* material such as iron, cobalt, or nickel, the spins retain their alignment when the field is removed. In fact, large numbers of spins align with one another over extensive "domains" of the solid and lock together. Not only does

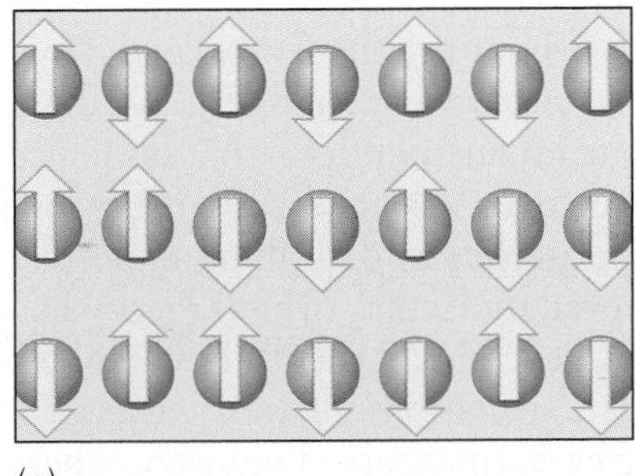

(a)

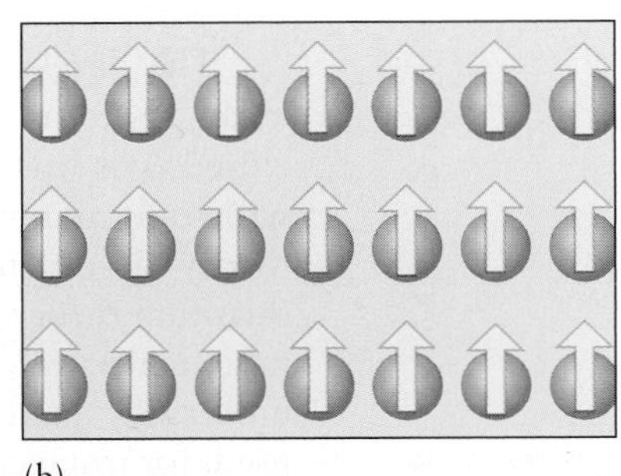

(b)

(a) The spins of electrons in paramagnetic substances return to a random orientation after an applied magnetic field is removed. (b) The spins of electrons in a ferromagnetic substance retain their alignment after a magnetic field is removed.

(continued)

Box 3.2 How Do We Know . . . That Electrons Are Not Paired? *(continued)*

this locking together help the alignment to survive, it gives rise to a powerful magnetic field from the myriad aligned electron spins. Permanent magnets are made from ferromagnetic materials, which include iron, cobalt, their alloys, and some of their compounds.

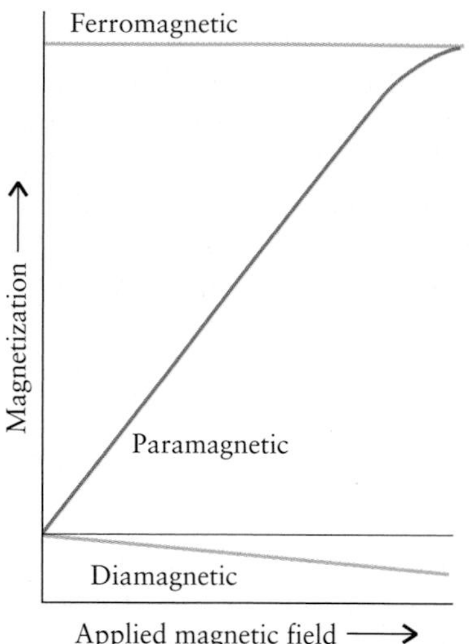

The magnetic behavior of three classes of material. The magnetization of a paramagnetic substance decreases as the applied field is reduced but that of a ferromagnetic material does not. Diamagnetic materials typically have a very weak magnetization at normal field strengths.

Lewis's theory also failed to account for the compound diborane, B_2H_6, a colorless gas that bursts into flame on contact with air. The problem is that diborane has only 12 valence electrons (three from each B atom, one from each H atom), but for a Lewis structure, it needs at least 7 bonds, and therefore 14 electrons, to bind the 8 atoms together! Diborane is an example of an **electron-deficient compound**, a compound with too few valence electrons to be assigned a valid Lewis structure. Valence-bond theory can account for the structures of electron-deficient compounds in terms of resonance, but the explanation is not straightforward.

The development of **molecular orbital theory** in the late 1920s overcame all these difficulties. It explains why the electron pair is so important for bond formation and predicts that oxygen is paramagnetic. It accommodates electron-deficient compounds such as the boron hydrides just as naturally as it deals with methane and water. Furthermore, molecular orbital theory can be extended to account for the structures and properties of metals and semiconductors. It can also be used to account for the electronic spectra of molecules, which arise when an electron makes a transition from an occupied molecular orbital to a vacant molecular orbital.

The development of bonding theory beyond Lewis's model does not mean his ideas were wrong. In fact, modern bonding theory shows how close Lewis was to the truth and confirms that his concepts can still be used with only minor refinement. Valence-bond theory and molecular orbital theory are different approaches to describing molecular structure by using quantum theory. However, molecular orbital theory has been developed far more extensively than valence-bond theory has and is the procedure almost universally used in the calculations of molecular structures. It is heavily used, for instance, in the pharmaceutical industry, where predicting the likely activity of a new drug from molecular orbital calculations is much cheaper than going to the trouble of making one and carrying out expensive tests.

Lewis's theory cannot account for the existence of electron-deficient compounds or the paramagnetism of oxygen.

3.10 Molecular Orbitals

A major difference between the valence-bond theory of chemical bonding and molecular orbital theory is that the former assumes, like the Lewis approach, that the electrons in a bond are localized between the two bonding atoms, whereas molecular orbital theory assumes that electrons belong to the molecule as a whole. In molecular orbital theory, electrons occupy orbitals called **molecular orbitals** that spread throughout the entire molecule. In other words, in the Lewis and valence-bond models of molecular structure, the electrons are localized on or between pairs of atoms, whereas in molecular orbital theory, electrons are **delocalized** over the whole molecule, not confined to individual bonds.

In this section we start, as in valence-bond theory, with the simplest molecule, H_2, and in the following sections extend the same principles to more complex molecules and solids. In every case, molecular orbitals are built by adding together—the technical term is "superimposing"—atomic orbitals belonging to the valence shells of the atoms in the molecule. For example, a molecular orbital for H_2 is

$$\psi = \psi_{A1s} + \psi_{B1s} \qquad (1)$$

where ψ_{A1s} is a 1*s*-orbital centered on one atom (A) and ψ_{B1s} is a 1*s*-orbital centered on the other atom (B). The molecular orbital ψ is called a **linear combination of atomic orbitals** (LCAO). Any molecular orbital formed from a superposition of atomic orbitals is called an **LCAO-MO**.

The LCAO-MO in Eq. 1 has a lower energy than either of the atomic orbitals used in its construction. The two atomic orbitals are like waves centered on different nuclei. Between the nuclei, the waves interfere constructively with each other in the sense that the total amplitude of the wavefunction is increased where they overlap (Fig. 3.28). The increased amplitude in the internuclear region means that there is a greater probability of finding the electron in the orbital in locations where it can interact with both nuclei. Because an electron that occupies the molecular orbital is attracted to both nuclei, it has a lower energy than when it is confined to an atomic orbital on one atom. Moreover, because the electron now occupies a greater volume, it has a lower kinetic energy, just like a particle confined to a bigger box (Section 1.5). A combination of atomic orbitals that results in an overall lowering of energy is called a **bonding orbital**.

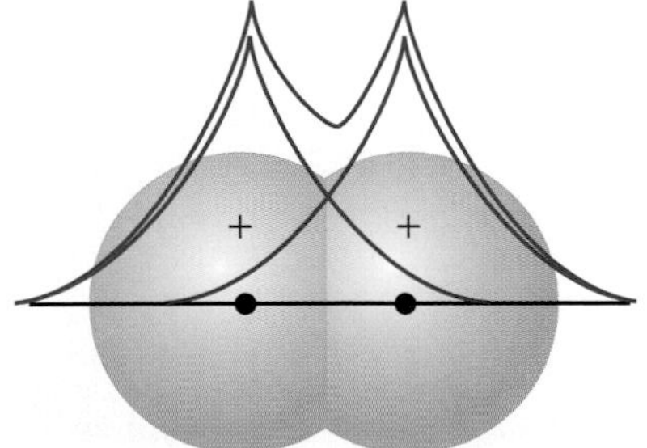

FIGURE 3.28 When two 1*s*-orbitals overlap in the same region of space in such a way that their wavefunctions have the same signs in that region, their wavefunctions (red lines) interfere constructively and give rise to a region of enhanced amplitude between the two nuclei (blue line).

When *N* atomic orbitals overlap, they form *N* molecular orbitals. In molecular hydrogen, where we are building LCAO-MOs from two atomic orbitals, we expect *two* molecular orbitals. In the second molecular orbital, the two atomic orbitals interfere *destructively* where they overlap. This orbital has the form

$$\psi = \psi_{A1s} - \psi_{B1s} \qquad (2)$$

The negative sign indicates that the amplitude of ψ_{B1s} *subtracts* from the amplitude of ψ_{A1s} where they overlap (Fig. 3.29). A nodal surface occurs where the atomic orbitals cancel completely. In a hydrogen molecule, the nodal surface is a plane that lies halfway between the two nuclei. The

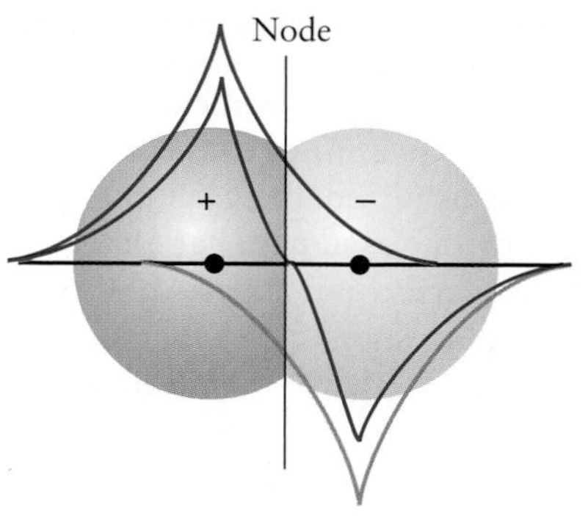

FIGURE 3.29 When two 1*s*-orbitals overlap in the same region of space in such a way that their wavefunctions have opposite signs, the wavefunctions (red and orange lines) interfere destructively and give rise to a region of diminished amplitude and a node between the two nuclei (blue line).

probability density of an electron is 0 on the nodal surface of a molecular orbital that it occupies. Because the electron is excluded from the region where it would have a strong interaction with both nuclei, the energy of the linear combination in Eq. 2 is higher than that of the atomic orbitals alone. A combination of atomic orbitals that results in an overall raising of energy is called an **antibonding orbital.**

The relative energies of the original atomic orbitals and the bonding and antibonding molecular orbitals are shown in a **molecular orbital energy-level diagram** like that in Fig. 3.30. The increase in energy that occurs when an antibonding orbital is occupied is about equal to or a little greater than the decrease in energy that occurs when the corresponding bonding orbital is occupied.

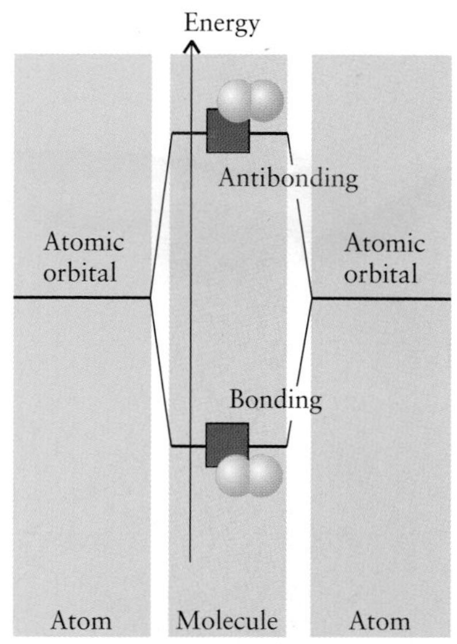

FIGURE 3.30 A molecular orbital energy-level diagram for the bonding and antibonding molecular orbitals that can be built from two *s*-orbitals. Different signs of the *s*-orbitals (reflecting how they are combined together to form the molecular orbital) are depicted by the different shades of blue.

Molecular orbitals are formed by combining atomic orbitals: when atomic orbitals interfere constructively, they give rise to bonding orbitals; when they interfere destructively, they give rise to antibonding orbitals. N atomic orbitals combine to give N molecular orbitals.

3.11 The Electron Configurations of Diatomic Molecules

In the molecular orbital description of homonuclear diatomic molecules, we first build all possible molecular orbitals from the available valence-shell atomic orbitals. Then we accommodate the valence electrons in molecular orbitals by using the same procedure we used in the building-up principle for atomic orbitals (Section 1.12). That is,

Step 1 Electrons are accommodated first in the lowest energy molecular orbital, then in orbitals of increasingly higher energy.

Step 2 According to the Pauli exclusion principle, each orbital can accommodate up to two electrons. If two electrons are present in one orbital, they must be paired.

Step 3 If more than one orbital of the same energy is available, the electrons enter them singly and adopt parallel spins (Hund's rule).

We shall illustrate these rules first with H_2 and then with other diatomic molecules. The same principles apply to polyatomic molecules, but their molecular orbitals are more complicated and their energies are harder to predict. Mathematical software for calculating the molecular orbitals and their energies is now widely available, and we shall show some of the results it provides.

In H_2, two atomic 1*s*-orbitals (one on each atom) merge to form two molecular orbitals. We denote the bonding orbital σ_{1s} and the antibonding orbital $\sigma_{1s}{}^*$; the 1*s* in the label shows the atomic orbitals from which the molecular orbitals are formed. The σ indicates that we have built a **σ-orbital,** a sausage-shaped orbital with no nodal plane when viewed along the internuclear axis. Two electrons, one from each H atom, are available. Both occupy the bonding orbital (the lower energy orbital) and result in the configuration $\sigma_{1s}{}^2$ (Fig. 3.31). Because only the bonding orbital is occupied, the energy of the molecule is lower than that of the separate atoms, and hydrogen exists as H_2 molecules. Two electrons in a σ-orbital form a σ-bond.

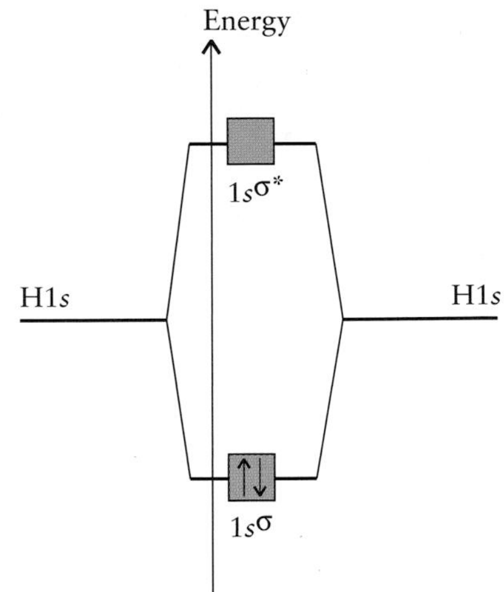

FIGURE 3.31 The two electrons in an H_2 molecule occupy the lower energy (bonding) molecular orbital and result in a stable molecule.

We can now explain something that puzzled Lewis—the fact that a bond normally consists of a pair of electrons. The Pauli exclusion principle allows only *two* electrons (with paired spins) to occupy any one molecular orbital. Hence, a single bond between two atoms consists of two paired electrons in a bonding orbital.

Now let's see how to extend these ideas to other homonuclear diatomic molecules of Period 2 elements. The first step is to build up the molecular orbital energy-level diagram from the valence-shell atomic orbitals provided by the atoms. Because Period 2 atoms have 2*s*- and 2*p*-orbitals in their valence shells, we form molecular orbitals from the overlap of these atomic orbitals. There are a total of eight atomic orbitals (one 2*s*- and three 2*p*-orbitals on each atom), so we can expect to build eight molecular orbitals. The two 2*s*-orbitals overlap to form two sausage-shaped molecular orbitals, one bonding (the σ_{2s} orbital) and the other antibonding (the $\sigma_{2s}{}^*$ orbital); these orbitals resemble the σ_{1s} and $\sigma_{1s}{}^*$ orbitals in H_2. The three 2*p*-orbitals on each neighboring atom can overlap in two distinct ways. The two 2*p*-orbitals that are directed toward each other along the internuclear axis form a bonding σ-orbital (σ_{2p}) and an antibonding σ*-orbital ($\sigma_{2p}{}^*$) where they overlap (Fig. 3.32). Two 2*p*-orbitals that are perpendicular to the internuclear axis overlap

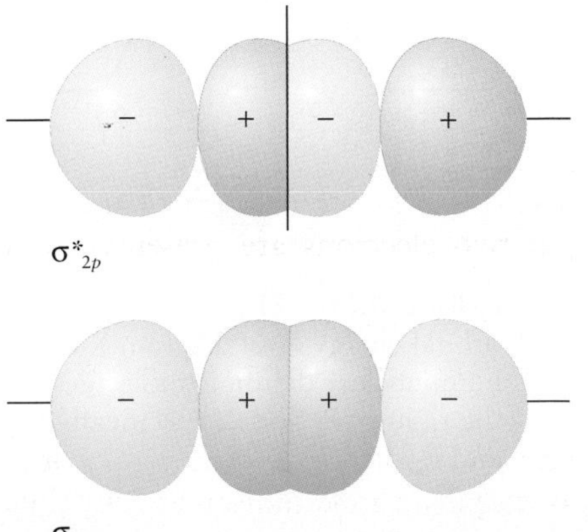

FIGURE 3.32 Two *p*-orbitals can overlap to give bonding (lower) and antibonding (upper) σ-orbitals. Note that the antibonding combination has a node between the two nuclei. Both orbitals have nodes passing through the nuclei.

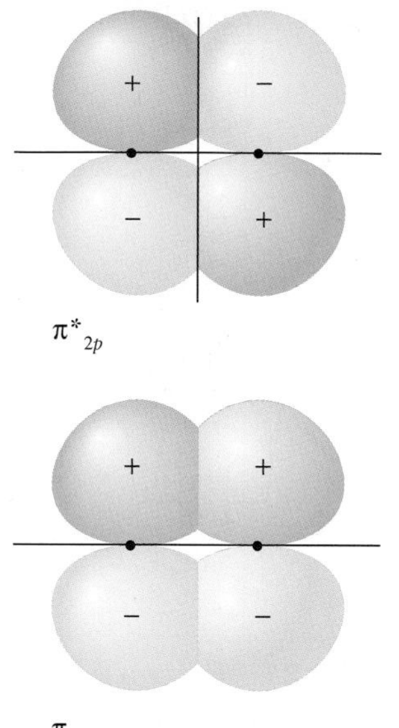

FIGURE 3.33 Two *p*-orbitals can overlap side by side to give bonding (lower) and antibonding (upper) π-orbitals. Note that the latter has a nodal plane between the two nuclei. Both orbitals have a nodal plane through the two nuclei and look like *p*-orbitals when viewed along the internuclear axis.

side by side to form bonding and antibonding **π-orbitals** (Fig. 3.33). A π-orbital—a molecular orbital with a nodal plane when viewed along the internuclear axis—is like the "ghost" of the valence-bond description of a π-bond, but it does not contain electrons at this stage. Because there are two perpendicular 2*p*-orbitals on each atom, four molecular orbitals—two bonding π_{2p}-orbitals and two antibonding $\pi_{2p}{}^*$-orbitals—are formed by their overlap.

Detailed calculation results in the energy-level diagram shown in Fig. 3.34 (Box 3.3). There are some small differences in the order of energy levels from molecule to molecule. The relative energies of the bonding σ_{2p}- and π_{2p}-orbitals, for instance, are interchanged for both O_2 and F_2 compared with their energies for the other Period 2 elements (Fig. 3.35). The order of energy levels is easy to understand in these two molecules, for the end-to-end overlap of 2*p*-orbitals (to give σ_{2p}- and $\sigma_{2p}{}^*$-orbitals) is more effective than the side-by-side overlap (to give π_{2p}- and $\pi_{2p}{}^*$-orbitals). The differences in ordering of the orbital energy levels stem from differences in the energies of the atomic orbitals used to build the molecular orbitals and the extent to which they overlap.

Once we know what molecular orbitals are available, we can construct the ground-state electron configurations of the molecules by using the building-up principle. For example, consider N_2. Because nitrogen

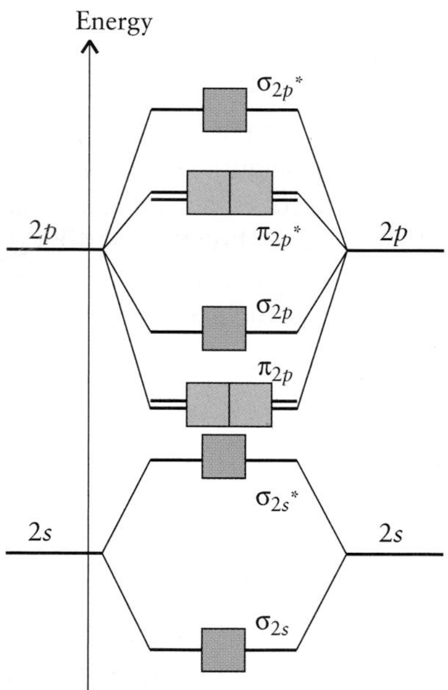

FIGURE 3.34 A typical molecular orbital energy-level diagram for the homonuclear diatomic molecules Li_2 through N_2. Each box represents one molecular orbital and can accommodate up to two electrons.

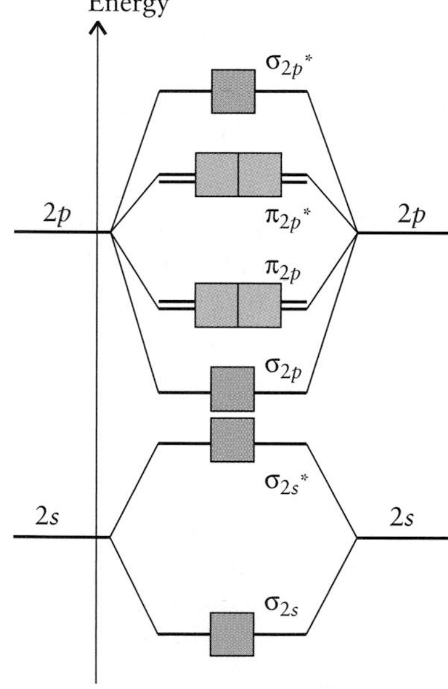

FIGURE 3.35 The molecular orbital energy-level diagram for the homonuclear diatomic molecules to the right of Period 2, specifically O_2 and F_2.

belongs to Group 15/V, each atom supplies five valence electrons. A total of 10 electrons must therefore be assigned to the molecular orbitals shown in Fig. 3.34. Two fill the σ_{2s}-orbital. The next two fill the $\sigma_{2s}{}^*$-orbital. Next in line for occupation are the two π_{2p}-orbitals, which can hold a total of four electrons. The last two electrons then enter the σ_{2p}-orbital. The ground configuration is therefore

$$N_2\colon\ \sigma_{2s}{}^2\sigma_{2s}{}^{*2}\pi_{2p}{}^4\sigma_{2p}{}^2$$

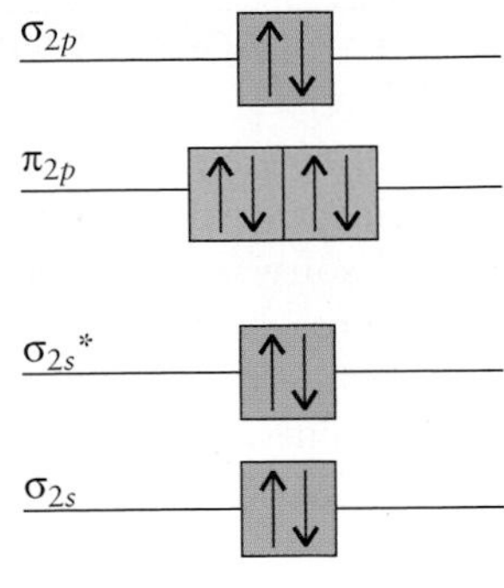

51 Nitrogen, N_2

This configuration is shown as (**51**), where the boxes now represent *molecular* orbitals.

At first sight, the molecular orbital description of N_2 looks quite different from the Lewis description (:N≡N:). However, it is, in fact, very closely related. We can see their similarity by defining the *bond order* (BO) in molecular orbital theory as the net number of bonds, allowing for the cancellation of bonds by antibonds:

$$\text{Bond order} = \tfrac{1}{2} \times (\text{number of electrons in bonding orbitals} - \text{number of electrons in antibonding orbitals}) \quad (3)$$

$$\text{BO} = \tfrac{1}{2}(N - N^*)$$

Here N is the number of electrons in bonding molecular orbitals and N^* is the number of electrons in antiboding molecular orbitals. This definition is the generalization of the one given in Section 2.7, where we simply counted shared pairs. In N_2, there are eight electrons in bonding orbitals and two in antibonding orbitals, so the bond order is $\frac{1}{2}(8 - 2) = 3$. Because its bond order is 3, N_2 effectively has three bonds between the N atoms, just as the Lewis structure suggests.

TOOLBOX 3.3 How to determine the electron configuration and bond order of a diatomic species

Conceptual Basis

When N valence atomic orbitals overlap, they form N molecular orbitals. The ground-state electron configuration is the electron arrangement that corresponds to lowest total energy; it may be deduced by using the building-up principle to accommodate all the valence electrons in the available molecular orbitals. The bond order is the net number of bonds that hold the molecule together.

Procedure

Step 1 Identify *all* the orbitals in the valence shells, ignoring how many electrons they contain. Atoms of Period 1 elements have one atomic orbital each, and atoms of Period 2 elements have four valence-shell atomic orbitals each (one 2*s* and three 2*p*).

Step 2 Use each pair of valence-shell atomic orbitals to build a bonding and an antibonding molecular orbital (still ignoring the electrons) and draw the resulting molecular orbital energy-level diagram.

Step 3 Note how many electrons are present in the valence shells. If the species is an ion, adjust the number of electrons to account for the charge.

Step 4 Accommodate the electrons in the molecular orbitals according to the building-up principle.

Step 5 To determine the bond order, subtract the number of electrons in antibonding orbitals from the number in bonding orbitals and divide the result by 2 (Eq. 3).

BOX 3.3 HOW DO WE KNOW . . . THE ENERGIES OF MOLECULAR ORBITALS?

The energies of orbitals are calculated today by solving the Schrödinger equation with computer software. The commercial software available is now so sophisticated that this approach can be as easy as typing in the name of the molecule or drawing it on screen. But these are theoretical values. How do we verify orbital energies *experimentally*?

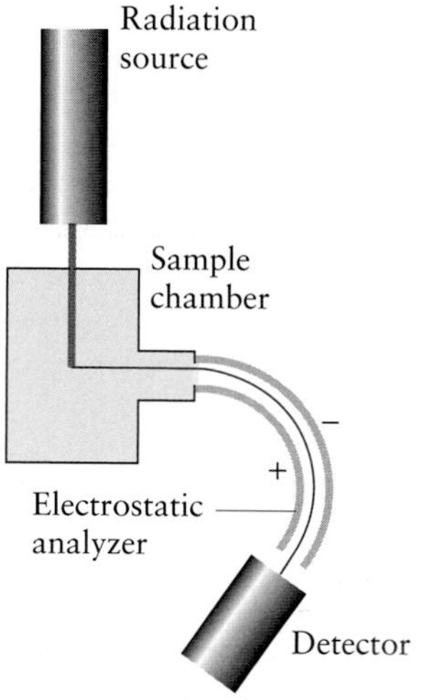

A photoelectron spectrometer.

One of the most direct methods is *photoelectron spectroscopy* (PES), an adaptation of the photoelectric effect (Section 1.2). A photoelectron spectrometer (see illustration) contains a source of high-frequency, short-wavelength radiation. Ultraviolet radiation is used most often for molecules, but x-rays are used to explore orbitals buried deeply inside solids. Photons in both frequency ranges have so much energy that they can eject electrons from the molecular orbitals they occupy.

Let's suppose that the frequency of the radiation is ν (nu), so each photon has an energy $h\nu$. An electron occupying a molecular orbital lies at an energy E_{orbital} below the zero of energy (corresponding to an electron far removed from the molecule), so a photon that collides with the electron can eject it from the molecule if the photon can supply at least that much energy. The remaining energy of the photon, $h\nu - E_{\text{orbital}}$, then appears as the kinetic energy, E_K, of the ejected electron:

$$h\nu - E_{\text{orbital}} = E_K$$

We know the frequency of the radiation being used to bombard the molecules, so if we could measure the kinetic energy of the ejected electron, we could solve this expression to find the orbital energy.

The kinetic energy of an ejected electron depends on its speed, v, because $E_K = \frac{1}{2}m_e v^2$ (Section A). Mass

EXAMPLE 3.6 Deducing the ground-state electron configuration of a diatomic molecule

Deduce the ground-state electron configuration of the fluorine molecule and calculate its bond order.

STRATEGY The Lewis structure of F_2 is $:\ddot{\underset{..}{F}}-\ddot{\underset{..}{F}}:$, so we anticipate that the bond order is 1. To calculate the bond order formally, we need to know the numbers of electrons in bonding and antibonding orbitals in the molecule. Proceed as in Toolbox 3.3.

SOLUTION There are $2 \times 7 = 14$ valence electrons to accommodate. The first 10 electrons repeat the N_2 configuration (apart from the change in order of the σ_{2p}- and π_{2p}-orbitals). The remaining four can enter the two antibonding $\pi_{2p}{}^*$-orbitals. The configuration is therefore

$$F_2\text{: } \sigma_{2s}{}^2\sigma_{2s}{}^{*2}\sigma_{2p}{}^2\pi_{2p}{}^4\pi_{2p}{}^{*4}$$

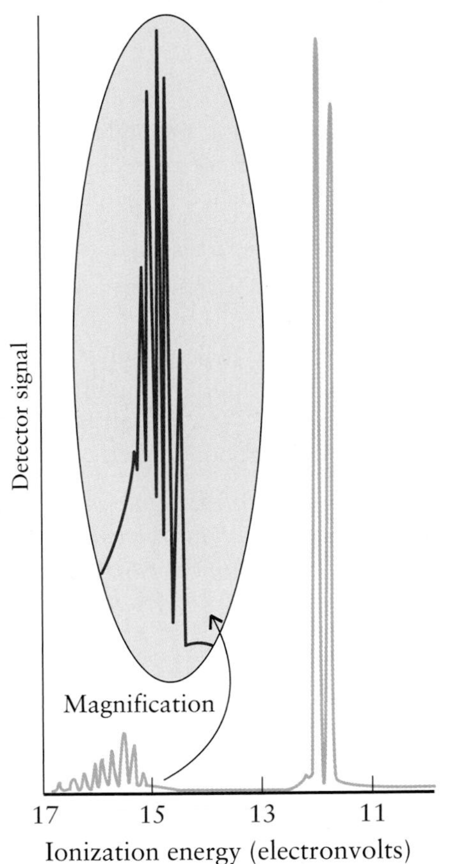

spectrometry can be used to measure v for electrons just as it can for ions, as we saw in Section B. In this method, the electrons pass through a region of electric or magnetic field, which deflects their path. As the strength of the field is changed, the paths of the electrons change, too, until they fall on a detector and generate a signal. From the strength of the field required to obtain a signal, we can work out the speed of electrons ejected from a given orbital. From the speed, we can calculate the kinetic energy of the electrons—and so obtain the energy of the orbital from which they came.

The *photoelectron spectrum* of nitrogen is shown in the illustration. There are several peaks, corresponding to electrons being ejected from orbitals of different energy. A detailed analysis shows that the spectrum is a good portrayal of the qualitative arrangement in structure (**51**).

The photoelectron spectrum of nitrogen has several peaks, a pattern indicating that electrons can be found in several energy levels in the molecule. Each main group of lines corresponds to the energy of a molecular orbital. The additional "fine structure" on some of the groups of lines is due to the excitation of molecular vibration when an electron is expelled.

This configuration is illustrated in (**52**). The bond order is

$$\text{BO} = \tfrac{1}{2} \times [(2 + 2 + 2 + 2) - (2 + 2 + 2)] = 1$$

so, F_2 is a singly bonded molecule, in agreement with the Lewis structure.

Self-Test 3.10A Deduce the electron configuration and bond order of the carbide ion (C_2^{2-}).

[*Answer:* $\sigma_{2s}{}^2\sigma_{2s}{}^{*2}\pi_{2p}{}^4\sigma_{2p}{}^2$, BO = 3]

Self-Test 3.10B Suggest a configuration for the nitrogen monoxide molecule, NO, and state its bond order.

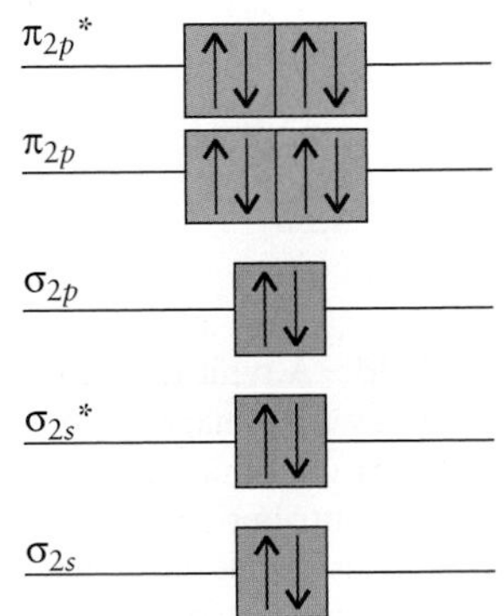

52 Fluorine, F_2

We find the ground-state electron configuration of O_2 by feeding its 12 valence electrons into the molecular orbitals shown in Fig. 3.35. The first 10 repeat the N_2 configuration (apart from the change in order of the

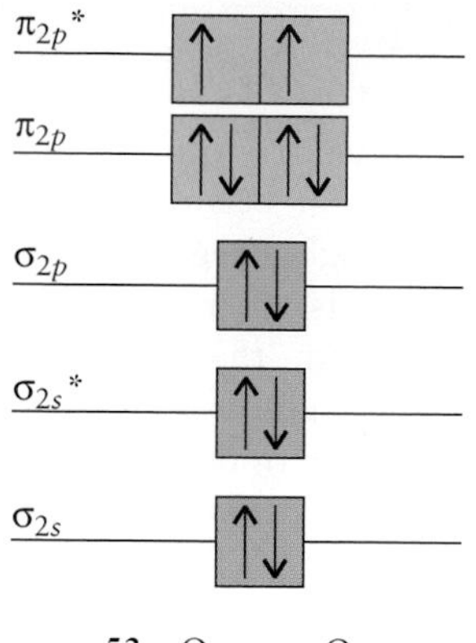

53 Oxygen, O_2

σ_{2p}- and π_{2p}-orbitals). According to the building-up principle, the last two electrons occupy the two separate $\pi_{2p}{}^*$-orbitals with parallel spins. The configuration is therefore

$$O_2\text{: } \sigma_{2s}{}^2\sigma_{2s}{}^{*2}\sigma_{2p}{}^2\pi_{2p}{}^4\pi_{2p}{}^{*1}\pi_{2p}{}^{*1}$$

as shown in (53). This conclusion is a minor triumph for molecular orbital theory: because the last two spins are not paired, their magnetic fields do not cancel, and the molecule is predicted to be paramagnetic—exactly as observed. The bond order of O_2 is

$$BO = \tfrac{1}{2} \times [(2 + 2 + 2 + 2) - (2 + 1 + 1)] = 2$$

This bond order is consistent with the Lewis structure $\ddot{\underset{..}{O}}=\ddot{\underset{..}{O}}$. However, the Lewis structure conceals the fact that the double bond in this molecule is actually a σ-bond plus two "*half* π-bonds"—each "half bond" consisting of a pair of bonding electrons and one antibonding electron.

The ground-state electron configurations of diatomic molecules are deduced by forming molecular orbitals from all the valence-shell atomic orbitals of the two atoms and adding the valence electrons to the molecular orbitals in order of increasing energy, in accord with the building-up principle.

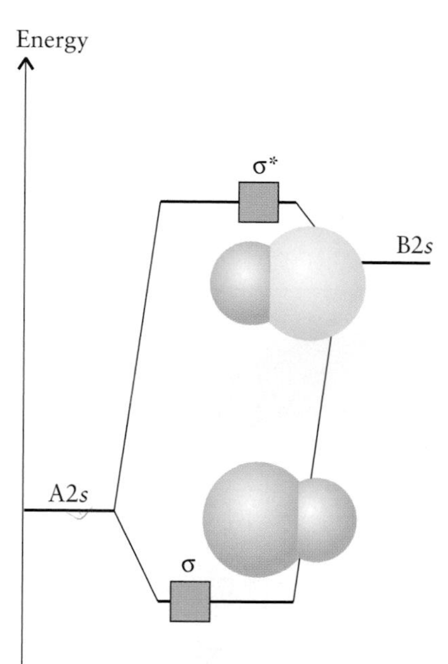

FIGURE 3.36 A typical σ molecular orbital energy-level diagram for a heteronuclear diatomic molecule AB; the relative contributions of the atomic orbitals to the molecular orbitals are represented by the relative sizes of the spheres and the horizontal position of the boxes. In this case, A is the more electronegative of the two elements.

3.12 Bonding in Heteronuclear Diatomic Molecules

The bond in a **heteronuclear diatomic molecule**, a diatomic molecule built from atoms of more than one element, is polar, with the electrons shared unequally by the two atoms. We therefore rewrite Eq. 1 as

$$\psi = c_A\psi_A + c_B\psi_B \tag{4}$$

where the coefficients c_A and c_B are not equal. Because we always take the squares of wavefunctions when interpreting them in terms of probabilities, according to molecular orbital theory the proportion of ψ due to ψ_A is $c_A{}^2$ and the proportion due to ψ_B is $c_B{}^2$. The values of c_A and c_B determine the type of bond. Consider the three types of bonds. In a nonpolar covalent bond, $c_A = c_B$ and the electron pair is shared equally between the two atoms. In an ionic bond, one of the coefficients is 0, because one of the ions captures all the electron density. In a polar covalent bond, the atomic orbital belonging to the more electronegative atom has the lower energy, so it makes the larger contribution to the lowest energy molecular orbital (Fig. 3.36). Conversely, the contribution to the highest energy (most antibonding) orbital is greater for the higher energy atomic orbital, which belongs to the less electronegative atom.

To find the ground-state electron configurations of heteronuclear diatomic molecules, we use the same approach we used for homonuclear diatomic molecules; but first we must modify the energy-level diagrams. For example, consider the HF molecule. The σ-bond in this molecule consists of an electron pair in a σ-orbital built from the F2p_z-orbital and the H1s-orbital. Because the electronegativity of fluorine is 4.0 and that of

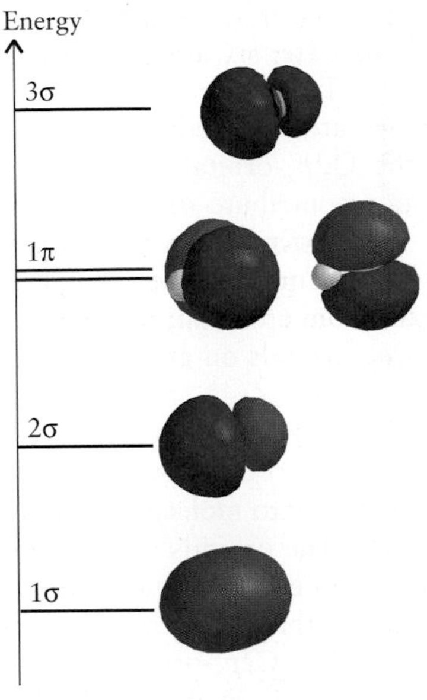

FIGURE 3.37 A schematic diagram of the molecular orbitals of HF. The orbitals are labeled using a standard convention. The software used to generate these images depicts opposite signs by red and blue. The two 1π-orbitals are nonbonding; the three σ-orbitals are progressively more antibonding (1σ, bonding; 3σ, antibonding).

hydrogen is 2.2, we can assume that the bonding σ-orbital is mainly F2p_z and that the antibonding σ-orbital is mainly H1*s* in character. These compositions are confirmed by calculation (Fig. 3.37). Because the two electrons in the bonding orbital are more likely to be found in the F2p_z-orbital than in the H1*s*-orbital, there is a partial negative charge on the F atom and a partial positive charge on the H atom.

Bonding in heteronuclear diatomic molecules involves an unequal sharing of the bonding electrons. The more electronegative element contributes more strongly to the bonding orbitals, whereas the less electronegative element contributes more strongly to the antibonding orbitals.

3.13 Orbitals in Polyatomic Molecules

The molecular orbital theory of polyatomic molecules follows the same principles as those outlined for diatomic molecules, but the molecular orbitals spread over *all* the atoms in the molecule. An electron pair in a bonding orbital helps to bind together the *whole* molecule, not just an individual pair of atoms. The energies of molecular orbitals in polyatomic molecules can be studied experimentally by using ultraviolet and visible spectroscopy (see Major Technique 2, following this chapter).

The description of bonding in polyatomic molecules can be quite complex. However, we can illustrate qualitatively how to use molecular orbital theory to describe bonding in one of the simplest but most important polyatomic molecules, water. A water molecule has six atomic orbitals (one O2*s*, three O2*p*, and two H1*s*). These six orbitals are used to build six molecular orbitals in which the degree of net bonding character is related to the number of internuclear nodes (Fig. 3.38). The molecular orbital with no nodes between neighboring atoms is fully bonding;

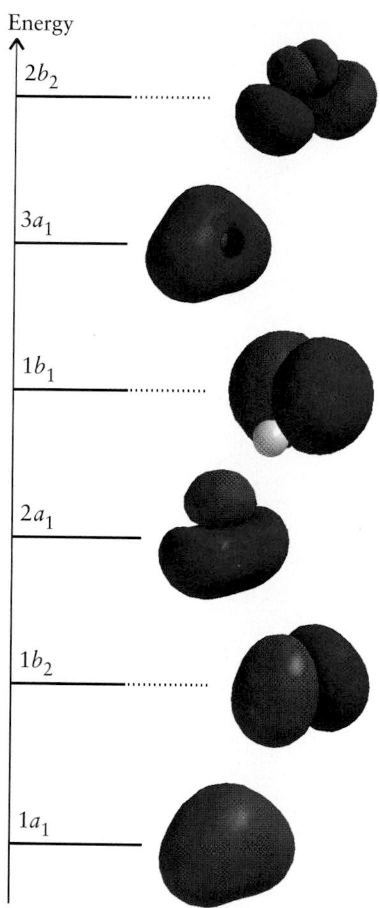

FIGURE 3.38 A schematic diagram of the molecular orbitals of H_2O. The orbitals are labeled according to the conventional notation for angular triatomic molecules. The orbitals become progressively more antibonding ($1a_1$, bonding; $2b_2$, antibonding).

Whereas in linear molecules z is taken to lie along the molecular axis, in planar molecules it is commonly taken to be perpendicular to the plane.

the one with a node between all neighboring pairs of atoms is fully antibonding. There are eight electrons to accommodate: six are provided by the O atom and one comes from each H atom. Two electrons in the lowest energy, most bonding orbital pull all three atoms together. Two electrons in the orbital composed purely of the O2p_x-orbital are localized completely on the oxygen atom and do not contribute directly to the bonding. An orbital such as this one that, when occupied, contributes to neither bonding nor antibonding is classified as a **nonbonding orbital**. In some cases, a nonbonding orbital is a single atomic orbital; in others, it may consist of a linear combination of atomic orbitals on atoms that are not neighbors and so overlap to a negligible extent.

Another important polyatomic molecule is benzene, the parent of the aromatic compounds. In the molecular orbital (MO) description of benzene, all 30 C2s-, C2p-, and H1s-orbitals contribute to molecular orbitals spreading over all 12 atoms (six C plus six H). The orbitals in the plane of the ring (the C2s-, C2p_x-, and C2p_y-orbitals on each carbon atom and all six H1s-orbitals) form delocalized σ-orbitals that bind the C atoms together and link the H atoms to the C atoms. The C2p_z-orbitals, which lie perpendicular to the ring, contribute to delocalized π-orbitals that spread all the way around the ring. However, chemists (other than those carrying out detailed calculations, who do use such a pure MO scheme) commonly mix MO and valence-bond (VB) descriptions when discussing organic molecules. Because the language of hybridization and orbital overlap is well suited to the description of σ-bonds, chemists typically express the σ-framework of molecules in VB terms. Thus, they think of the σ-framework of benzene as formed by the overlap of sp^2 hybridized orbitals on neighboring atoms. Then, because delocalization is such an important feature of the π-bonding component of *conjugated* double bonds, which have alternating double and single bonds, as in —C=C—C=C—C=C—, they treat such π-bonds in terms of MO theory. We shall do the same.

First, the VB part of the description of benzene. Each C atom is sp^2 hybridized, with one electron in each hybrid orbital. Each C atom has a p-orbital perpendicular to the plane defined by the hybrid orbitals, and it contains one electron. Two sp^2 hybrid orbitals on each C atom overlap and form σ-bonds with similar orbitals on the two neighboring C atoms, forming the 120° internal angle of the benzene hexagon. The third, outward-pointing sp^2 orbital on each C atom forms a σ-bond with a hydrogen atom. The resulting σ-framework is the same as that illustrated in Fig. 3.22.

Now for the MO part of the description. From the six C2p_z-orbitals, we form six delocalized π-orbitals: their shapes are shown in Fig. 3.39 and their energies are shown in Fig. 3.40. The character of the orbitals switches from net bonding to net antibonding as the number of internuclear nodes increases from none (most bonding) to six (most antibonding).

Each carbon atom provides one electron for the π-orbitals. Two electrons occupy the lowest energy, most bonding orbital; and the remaining four electrons occupy the next higher energy orbitals (two orbitals of the same energy). The diagram shows one of the reasons for benzene's great stability: the π-electrons occupy only orbitals with a net bonding effect; none of the destabilizing antibonding orbitals is occupied.

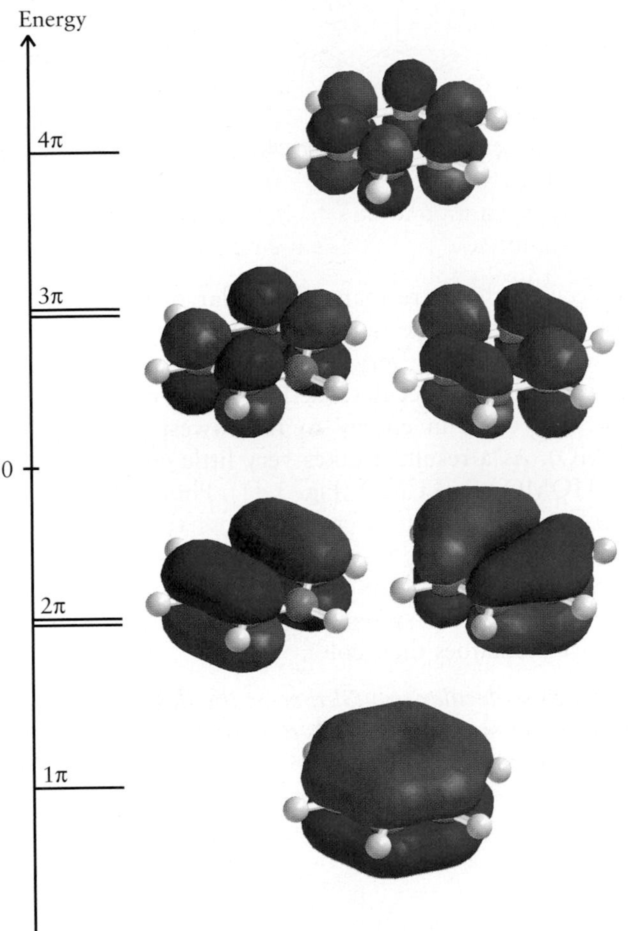

FIGURE 3.39 The six π-orbitals of benzene. Note that they range from fully bonding (no internuclear nodes) to fully antibonding (six internuclear nodes). The zero of energy corresponds to the total energy of the separated atoms. The three orbitals with energies below 0 have net bonding character.

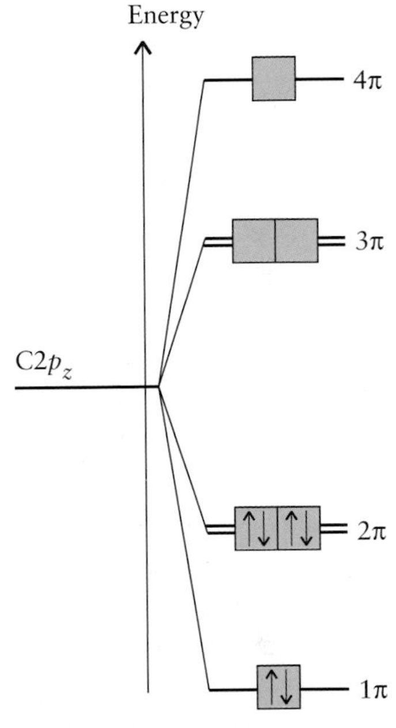

FIGURE 3.40 The molecular orbital energy-level diagram for the π-orbitals of benzene. In the ground state of the molecule, only the net bonding orbitals are occupied.

The delocalization of the bonding influence of electrons also accounts for the existence of electron-deficient molecules. Because the bonding influence of an electron pair is spread over all the atoms in the molecule, there is no need to provide one pair of electrons for each pair of atoms. A smaller number of pairs of electrons spread throughout the molecule may be able to bind all the atoms together, particularly if the nuclei are not highly charged and so do not repel each other strongly. This is the case in B_2H_6, where six electron pairs can hold the eight nuclei together.

Diborane is treated in more detail in Chapter 14.

Finally, we return to the opening remarks of the chapter and see how molecular orbital theory explains the colors of vegetation. The presence of highly delocalized electrons in the large molecules found in the petals of flowers and in fruit and vegetables is largely responsible for their colors. Because many carbon atoms contribute *p*-orbitals to the π system

54 β-Carotene, $C_{40}H_{56}$

55 Lycopene, $C_{40}H_{58}$

of these molecules, there are many molecular orbitals. An electron in a π system of such a molecule is like a particle in a vast one-dimensional box. Because the "box" is very large, the energy levels are very close together. In these large molecules, the **highest occupied molecular orbital** (HOMO) is very close in energy to the **lowest unoccupied molecular orbital** (LUMO). As a result, it takes very little energy to excite an electron from a HOMO to a LUMO (Fig. 3.41). Photons of visible light have enough energy to excite the electrons across this energy gap, and the absorption of these photons results in the colors that we perceive. The coloring agent of carrots and the precursor of vitamin A, β-carotene (**54**), has a highly delocalized π system, as does lycopene (**55**), the red compound that gives tomatoes their color.

According to molecular orbital theory, the delocalization of electrons in a polyatomic molecule spreads the bonding effects of electrons over the entire molecule.

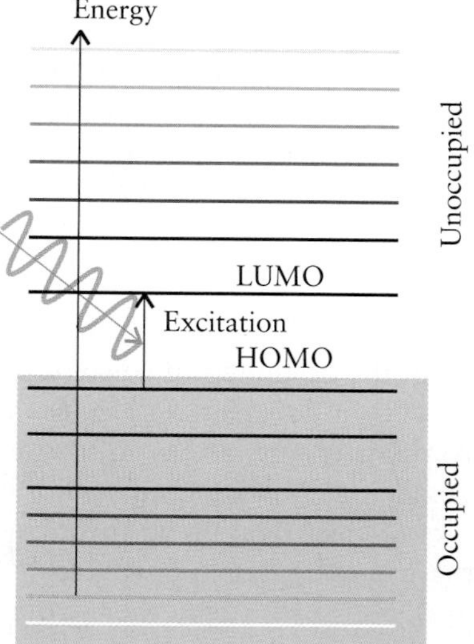

FIGURE 3.41 In large molecules, there are many closely spaced energy levels and the HOMO-LUMO gap is quite small. Such molecules are often colored because photons of visible light can be absorbed when electrons are excited from the HOMO to the LUMO.

3.14 The Impact on Materials: The Band Theory of Solids

Molecular orbital theory explains the electrical properties of metals and semiconductors. Both types of materials are **electronic conductors** in which an electric current is carried by delocalized electrons. A **metallic conductor** is an electronic conductor in which the electrical conductivity *decreases* as the temperature is raised. A **semiconductor** is an electronic conductor in which the electrical conductivity *increases* as the temperature is raised. In most cases, a metallic conductor has a much higher electrical conductivity than a semiconductor, but it is the temperature dependence of the conductivity that distinguishes the two types of conductors. An **insulator** does not conduct electricity. A **superconductor** is a solid that has zero resistance to an electric current. Some metals become superconductors at very low temperatures, at about 20 K or less, and some compounds also show superconductivity. "High-temperature superconductors" (HTSC) are compounds such as $YBa_2Cu_3O_{9-x}$ that turn into superconductors at relatively high temperatures (around 100 K). High-temperature superconductors have enormous potential, because they offer the prospect of more efficient power transmission and the generation of high magnetic fields for use in transport systems (Fig. 3.42).

FIGURE 3.42 Superconductors have the ability to levitate vehicles with embedded magnets. This picture shows an experimental zero-friction train in Japan, built to use helium-cooled metal superconductors. Currently, research is being conducted at this facility on the replacement of the metal superconductors by high-temperature superconductors, which will allow large-scale practical application of the technology.

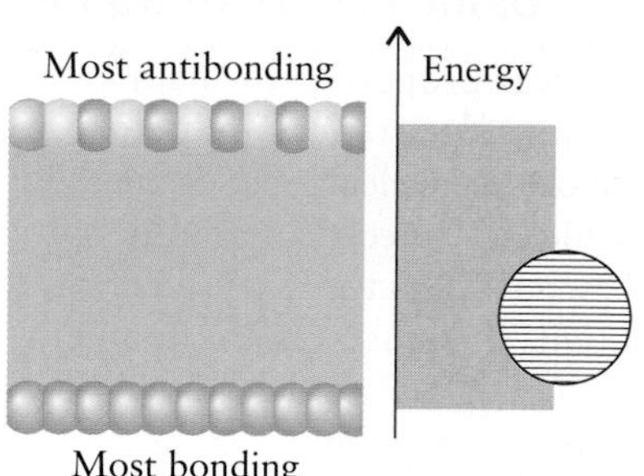

FIGURE 3.43 A line of atoms gives rise to an almost continuous band of molecular orbital energies. At the lower edge of the band, the molecular orbitals are fully bonding; at the upper edge, the molecular orbitals are fully antibonding. The inset shows that, although the band of allowed energies appears to be continuous, it is in fact composed of discrete, closely spaced levels.

Electrical conduction in metals and semiconductors can be explained in terms of molecular orbitals that spread throughout the solid. We have already seen that when N atomic orbitals merge together in a molecule, they form N molecular orbitals. The same is true of a metal; but in the case of a metal, N is enormous (around 10^{23} for 10 g of copper, for example). Instead of the few molecular orbitals with widely spaced energies typical of small molecules, the huge number of molecular orbitals in a metal are so close together in energy that they form a nearly continuous band (Fig. 3.43).

Consider a metal such as sodium. Each atom contributes one valence orbital (the 3*s*-orbital in this case) and one valence electron. If there are N atoms in the sample, then the N 3*s*-orbitals merge to form a band of N molecular orbitals, of which half are net bonding and half are net antibonding. (We say "net" bonding or antibonding, because, in general, a molecular orbital is bonding between some neighbors and antibonding between others, depending on where its internuclear nodes lie.) Because each of the N atoms supplies one valence electron, a total of N electrons must be accommodated in the orbitals. These N electrons occupy the orbitals according to the building-up principle. Because two electrons can occupy each orbital, the N electrons occupy the lower $\frac{1}{2}N$ bonding orbitals. An empty or incompletely filled band of molecular orbitals is called a **conduction band**. Because neighboring orbitals lie so close together in energy, it takes very little additional energy to excite the electrons from the topmost filled molecular orbitals to the empty orbitals just above. These electrons can move freely through the solid, so they can carry an electric current. The resistance of the metal increases with temperature because, when it is heated, the atoms vibrate more vigorously. The passing electrons collide with the vibrating atoms and do not pass through the solid so easily.

In an insulator, the valence electrons fill all the available molecular orbitals to give a full band. A full band is called a **valence band**. There is a substantial **band gap**, a range of energies for which there are no orbitals, before the next band, the empty orbitals that form the conduction band, begins (Fig. 3.44). The electrons in the valence band can be excited into the conduction band only by a very large injection of energy. Because the valence band is full and because the conduction band is separated from it by a large energy gap, the electrons are not mobile and the solid does not conduct electricity.

In a semiconductor, an empty conduction band lies close in energy to a full valence band. As a result, as the solid is warmed, electrons can be excited from the valence band into the conduction band, where they can travel throughout the solid. Hence, the resistance of a semiconductor decreases as its temperature is raised.

The ability of a semiconductor to carry an electric current can also be enhanced by adding electrons to the conduction band or by removing some from the valence band. This modification is carried out chemically by **doping** the solid, or spreading small amounts of impurities throughout it. In one example of an **n-type semiconductor**, a minute amount of a Group 15/V element such as arsenic is added to very pure silicon. The arsenic increases the number of electrons in the solid: each Si atom (Group 14/IV) has four valence electrons, whereas each As atom (Group 15/V) has five. The additional electrons enter the upper, normally empty conduction band of silicon and allow the solid to conduct (Fig. 3.45). This type of material is called an n-type semiconductor because it contains excess *n*egatively charged electrons. When silicon (Group 14/IV) is doped with indium (Group 13/III) instead of arsenic, the solid has fewer valence electrons than pure silicon, so the valence band is no longer completely full (Fig. 3.46). We say that the valence band now contains "holes." Because the valence band is no longer full, it has been turned into a conduction band, and an electric current can flow. This type of

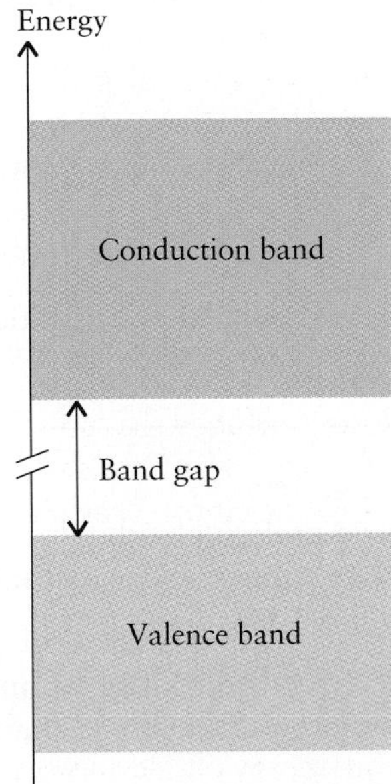

FIGURE 3.44 In a typical insulating solid, a full valence band is separated by a substantial energy gap from the empty conduction band. Note the break in the vertical scale.

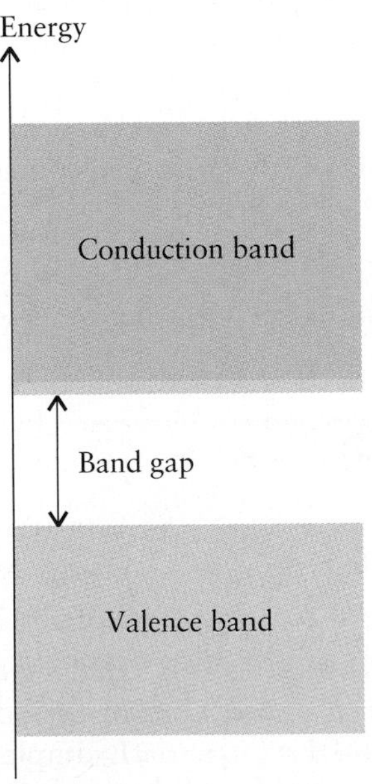

FIGURE 3.45 In an n-type semiconductor, the additional electrons supplied by the electron-rich dopant atoms enter the conduction band (pink band at bottom of conduction band) and act as carriers for the current.

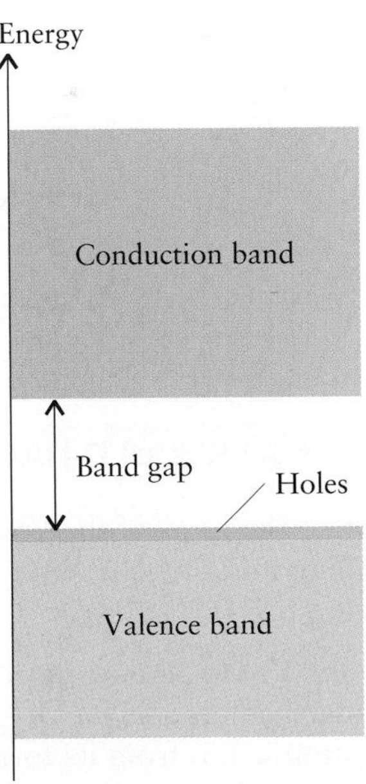

FIGURE 3.46 In a p-type semiconductor, the electron-poor dopant atoms effectively remove electrons from the valence band, and the "holes" that result enable the remaining electrons to become mobile and conduct electricity through the valence band.

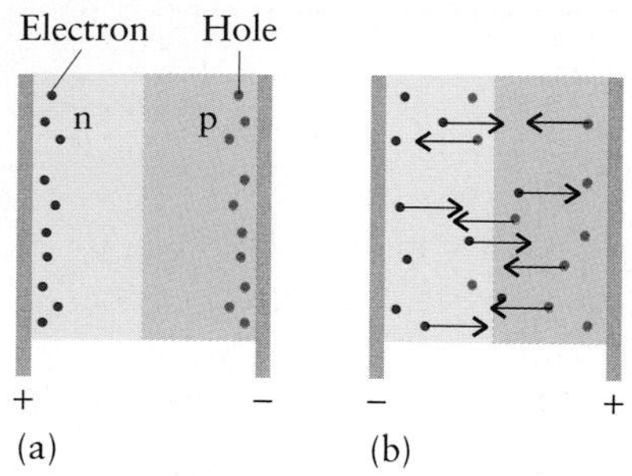

FIGURE 3.47 The structure of a p-n junction allows an electric current to flow in only one direction. (a) In reverse bias, the negative electrode is attached to the p-type semiconductor. The holes in the p-type semiconductor are attracted to the negative electrode, the electrons in the n-type semiconductor are attracted to the positive electrode, and current does not flow. (b) In forward bias, the electrodes are reversed to allow charge carriers to be regenerated. The negative electrode is attached to the n-type semiconductor and, as a result, the charge carriers flow the other way: electrons flow from the n-type semiconductor through the p-type semiconductor toward the positive electrode.

semiconductor is called a **p-type semiconductor**, because the absence of negatively charged electrons is equivalent to the presence of *positively* charged holes. The doped solids are overall electrically neutral, because the nuclei of the doping atoms have a charge that matches the number of their electrons.

Solid-state electronic devices such as diodes, transistors, and integrated circuits contain **p-n junctions** in which a p-type semiconductor is in contact with an n-type semiconductor (Fig. 3.47).

Bonding in solids may be described in terms of bands of molecular orbitals. In metals, the conduction bands are incompletely filled orbitals that allow electrons to flow. In insulators, the valence bands are full and the large band gap prevents promotion of electrons to empty orbitals. In semiconductors, empty levels are close in energy to filled levels.

Self-Test 3.11A Which type of semiconductor is germanium doped with arsenic?

[*Answer:* n-type]

Self-Test 3.11B Which type of semiconductor is antimony doped with tin?

SKILLS YOU SHOULD HAVE MASTERED

1 Explain the basis of the VSEPR model of bonding in terms of repulsions between electrons, Section 3.1.

2 Use the VSEPR model to predict the electron arrangement and shape of a molecule or polyatomic ion from its formula, Toolbox 3.1, and Examples 3.1, 3.2, and 3.3.

3 Predict the polar character of a molecule, Example 3.4.

4 Account for the structure of a molecule in terms of hybrid orbitals and σ- and π-bonds, Toolbox 3.2 and Example 3.5.

5 Construct and interpret a molecular orbital energy-level diagram for a homonuclear diatomic species, Sections 3.9 and 3.10.

6 Deduce the ground-state electron configurations of Period 2 diatomic molecules, Toolbox 3.3 and Example 3.6.

7 Use molecular orbital theory to describe bonding in polyatomic molecules such as the benzene molecule, Section 3.13.

8 Use molecular orbital theory to account for the differences between metals, insulators, and semiconductors, Section 3.14.

EXERCISES

The Shapes of Molecules and Ions

3.1 (a) What is the shape of the thionyl chloride molecule, $SOCl_2$? Sulfur is the central atom. (b) How many different O—S—Cl bond angles are there in this molecule? (c) What values are expected for the O—S—Cl and Cl—S—Cl bond angles?

3.2 (a) What is the shape of the ClO_3^- ion? (b) What values are expected for the O—Cl—O bond angles?

3.3 (a) What is the shape of the ClO_2^+ ion. (b) What is the O—Cl—O bond angle?

3.4 (a) What is the shape of the XeF_5^+ ion? (b) How many different F—Xe—F bond angles are there in this molecule? (c) What values are expected for the F—Xe—F bond angles?

3.5 (a) What is the shape of the carbonate ion, CO_3^{2-}? (b) How many different O—C—O bond angles are there in this molecule? (c) What values are expected for the O—C—O bond angles?

3.6 (a) Describe the shape about each carbon atom of the acetate ion, $CH_3CO_2^-$. (b) How many different C—C—O bond angles are there in this molecule? (c) What values are expected for the C—C—O angles?

3.7 (a) What is the shape of an NSF molecule (sulfur is the central atom)? (b) What value is expected for the N—S—F angle?

3.8 (a) What is the shape of an SbF_5^{2-} ion? (b) How many different F—Sb—F bond angles are there in this molecule? (c) What values are expected for the F—Sb—F angles?

3.9 Using Lewis structures and VSEPR theory, predict the shape of each of the following species: (a) sulfur tetrachloride; (b) iodine trichloride; (c) IF_4^-; (d) xenon trioxide. Give the VSEPR AX_nE_m designation for each.

3.10 Using Lewis structures and VSEPR theory, predict the shape of each of the following species: (a) PF_4^-; (b) ICl_4^+; (c) phosphorus pentafluoride; (d) xenon tetrafluoride. Give the VSEPR AX_nE_m designation for each.

3.11 Predict the molecular shape and bond angles in each of the following: (a) I_3^-; (b) IF_3; (c) IO_4^-; (d) TeF_6. Give the VSEPR AX_nE_m designation for each.

3.12 Predict the molecular shape and bond angles in each of the following: (a) I_3^+; (b) PCl_3; (c) SeO_3^{2-}; (d) GeH_4. Give the VSEPR AX_nE_m designation for each.

3.13 Write the Lewis structure and the VSEPR formula, list the shape, and predict the approximate bond angles in (a) CF_3Cl; (b) GaI_3; (c) $XeOF_4$; (d) CH_3^-. Give the VSEPR AX_nE_m designation for each.

3.14 Write the Lewis structure and the VSEPR formula, list the shape, and predict the approximate bond angles in (a) PCl_3F_2; (b) SnF_4; (c) SnF_6^{2-}; (d) IF_5; (e) XeO_4. Give the VSEPR AX_nE_m designation for each.

3.15 The compound 2,4-pentanedione is acidic and can be deprotonated. The anion forms complexes with metals that are used in gasoline additives, lubricants, insecticides, and fungicides. (a) Estimate the bond angles marked with arcs and lowercase letters in 2,4-pentanedione (also known as acetylacetone or acac) and in the acetylacetonate ion. (b) What are the differences, if any?

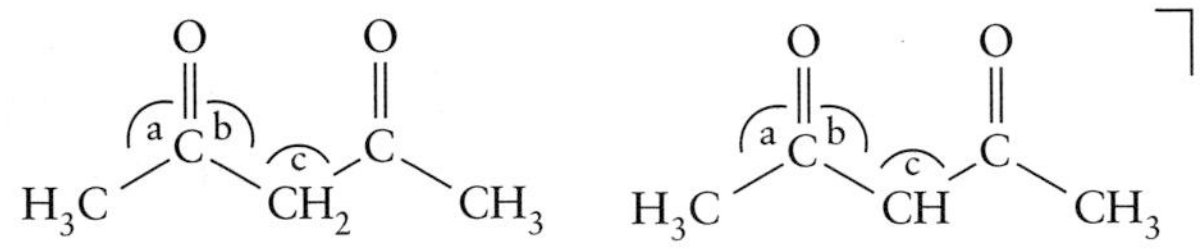

2,4-Pentanedione (acetylacetone) Acetylacetonate ion (one resonance form)

3.16 Estimate the bond angles marked with arcs and lowercase letters in peroxyacetylnitrate, an eye irritant in smog:

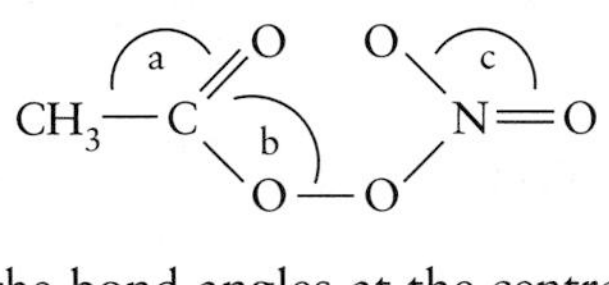

3.17 Predict the bond angles at the central atom of the following molecules and ions: (a) ozone, O_3; (b) azide ion, N_3^-; (c) cyanate ion, CNO^-; (d) hydronium ion, H_3O^+.

3.18 Predict the bond angles at the central atom of the following molecules: (a) OF_2; (b) ClO_2^-; (c) NO_2^-; (d) $SeCl_2$.

3.19 Write its Lewis structure and predict whether each of the following molecules is polar or nonpolar: (a) CH_2Cl_2; (b) CCl_4; (c) CS_2; (d) SF_4.

3.20 Write its Lewis structure and predict whether each of the following molecules is polar or nonpolar: (a) PCl_5; (b) PCl_3; (c) XeF_4; (d) H_2S.

3.21 Predict whether each of the following molecules is likely to be polar or nonpolar: (a) C_6H_5N (pyridine, a molecule like benzene with one —CH— group replaced with nitrogen); (b) C_2H_6 (ethane); (c) $CHCl_3$ (trichloromethane, also known as chloroform, a common organic solvent and anesthetic).

3.22 Predict whether each of the following molecules is likely to be polar or nonpolar: (a) CCl_4 (tetrachloromethane); (b) $CH_3CHOHCH_3$ (2-propanol, rubbing alcohol); (c) CH_3COCH_3 (2-propanone acetone, a common organic solvent used in fingernail polish remover).

3.23 There are three different dichlorobenzenes, $C_6H_4Cl_2$, which differ in the relative positions of the chlorine atoms on the benzene ring. (a) Which of the three forms, if any, are polar? (b) Which has the largest dipole moment?

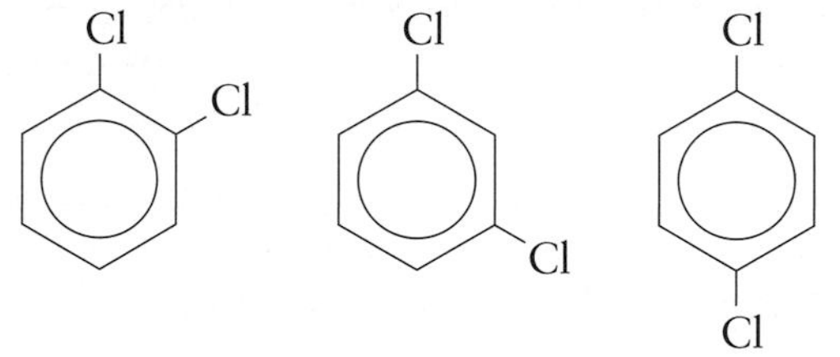

1 2 3

3.24 There are three different difluoroethenes, $C_2H_2F_2$, which differ in the locations of the flourine atoms. (a) Which of the forms are polar and which are nonpolar? (b) Which has the largest dipole moment?

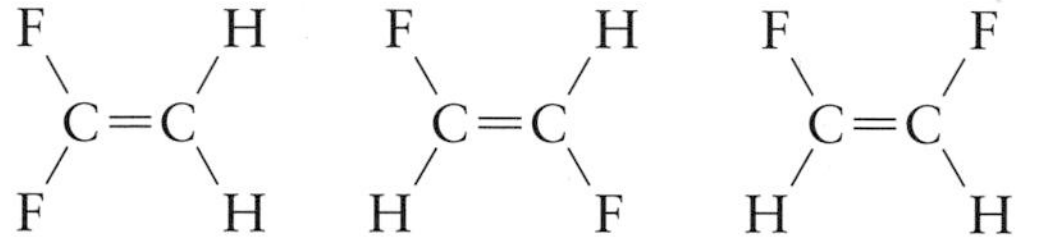

1 2 3

3.25 Acrylonitrile, CH_2CHCN, is used in the synthesis of acrylic fibers (polyacrylonitriles), such as Orlon. Write the Lewis structure of acrylonitrile and describe the hybrid orbitals on each carbon atom. What are the approximate values of the bond angles?

3.26 Xenon forms XeO_3, XeO_4, and XeO_6^{4-}, all of which are powerful oxidizing agents. Give their Lewis structures, their bond angles, and the hybridization of the xenon atom. Which would be expected to have the longest Xe—O distances?

Valence-Bond Theory

3.27 State the orientations of the following hybrid orbitals: (a) sp^3; (b) sp; (c) sp^3d^2; (d) sp^2.

3.28 The orientation of bonds on a central atom of a molecule that has no lone pairs can be any of those listed below. What is the hybridization of the orbitals used by each central atom for its bonding pairs: (a) tetrahedral; (b) trigonal bipyramidal; (c) octahedral; (d) linear?

3.29 State the hybridization of the atom in bold type in the following molecules: (a) $\mathbf{S}F_4$; (b) $\mathbf{B}Cl_3$; (c) $\mathbf{N}H_3$; (d) $(CH_3)_2\mathbf{Be}$.

3.30 State the hybridization of the atom in bold type in the following molecules: (a) $\mathbf{S}F_6$; (b) $O_3\mathbf{Cl}—O—ClO_3$; (c) $H_2\mathbf{N}—CH_2—COOH$ (glycine); (d) $O\mathbf{C}(NH_2)_2$ (urea).

3.31 Identify the hybrid orbitals used by the atoms in bold type in the following species: (a) $\mathbf{B}F_3$; (b) $\mathbf{As}F_3$; (c) $\mathbf{Br}F_3$; (d) $\mathbf{Se}F_3^+$.

3.32 Identify the hybrid orbitals used by the atoms in bold type in the following molecules: (a) $CH_3\mathbf{CC}CH_3$; (b) $CH_3\mathbf{NN}CH_3$; (c) $(CH_3)_2\mathbf{CC}(CH_3)_2$; (d) $(CH_3)_2\mathbf{NN}(CH_3)_2$.

3.33 Identify the hybrid orbitals used by the phosphorus atoms in the following species: (a) PCl_4^+; (b) PCl_6^-; (c) PCl_5; (d) PCl_3.

3.34 Identify the hybrid orbitals used by the atoms in bold type in the following molecules: (a) H_2CCCH_2; (b) H_2CCCH_2; (c) CH_3NNN; (d) CH_3NNN.

3.35 Predict the shapes and estimate the bond angles of (a) the thiosulfate ion, $S_2O_3^{2-}$; (b) $(CH_3)_2Be$; (c) BH_2^-; (d) $SnCl_2$.

3.36 For each molecule or ion, write the Lewis structure, list the number of lone pairs on the central atom, identify the shape, and estimate the bond angles: (a) XeF_5^+; (b) $XeOF_2$; (c) SF_5^+; (d) TeH_2; (e) BrO_3^-.

3.37 Write the Lewis structure and give the approximate bond angles of (a) C_2H_4; (b) ClCN; (c) $OPCl_3$; (d) N_2H_4.

3.38 Write the Lewis structure and predict the shape of (a) TeF_4; (b) NH_2^-; (c) NO_2^+; (d) NH_4^+; (e) SnH_4; (f) OCS.

3.39 Write the Lewis structure and predict the shape of (a) $O\mathbf{Sb}Cl_3$; (b) $\mathbf{S}O_2Cl_2$; (c) $\mathbf{I}O_2F_2^-$. The atom in bold type is the central atom.

3.40 Write the Lewis structure and predict the shape of (a) $OCCl_2$; (b) $OSCl_2$; $OSbCl_2^-$. The atom in bold type is the central atom.

Molecular Orbital Theory

3.41 Draw a molecular orbital energy-level diagram and determine the bond order expected for each of the following diatomic molecules: (a) Li_2; (b) Li_2^+; (c) Li_2^-. State whether each molecule or ion will be paramagnetic or diamagnetic. If it is diamagnetic, give the number of unpaired electrons.

3.42 Draw a molecular orbital energy-level diagram and determine the bond order expected for the following diatomic molecules: (a) B_2; (b) B_2^-; (c) B_2^+. State whether each molecule or ion will be paramagnetic or diamagnetic. If it is diamagnetic, give the number of unpaired electrons.

3.43 (a) On the basis of the configuration of the neutral molecule O_2, write the molecular orbital configuration of the valence molecular orbitals for (1) O_2^-; (2) O_2^+; (3) O_2^{2-}. (b) For each species, give the expected bond order. (c) Which of these are paramagnetic, if any? (d) Is the highest energy orbital that contains an electron σ or π in character?

3.44 (a) On the basis of the configuration for the neutral molecule N_2, write the molecular orbital configuration of the valence molecular orbitals for (1) N_2^+; (2) N_2^{2+}; (3) N_2^{2-}. (b) For each species, give the expected bond order. (c) Which of these are paramagnetic, if any? (d) Is the highest energy orbital that contains an electron σ or π in character?

3.45 Using molecular orbital theory, explain why the bond energy of an N=N (double) bond is not equal to twice the energy of an N—N (single) bond.

3.46 Lewis structures invoke the concept of resonance in order to explain certain physical properties of molecules such as why two bond lengths are equal. Molecular orbital theory does not use the concept of resonance. Using the azide ion, N_3^-, as an example, draw the π molecular orbital that leads to bonding between the three nitrogen atoms.

3.47 (a) Draw the molecular orbital energy-level diagram for N_2 and label the energy levels according to what type of orbitals they are made from, whether or not they are σ or π orbitals, and whether they are bonding or antibonding. (b) The heteronuclear diatomic ion CN^- has a similar orbital structure to N_2. How will the fact that C has a different electronegativity from N affect the molecular-orbital energy-level diagram of CN^- compared with that of N_2? Draw the energy-level diagram for CN^-, using this information. (c) In the molecular orbitals, will the electrons have a higher probability of being at C or at N? Why?

3.48 (a) How does molecular orbital theory make it possible to explain both ionic *and* covalent bonding? (b) The degree of ionic character in bonding was related to electronegativity in Chapter 2. How does electronegativity affect the molecular orbital diagram so that bonds become ionic?

3.49 Write the valence-shell electron configurations and bond orders of (a) C_2; (b) Be_2; (c) Ne_2.

3.50 Write the valence-shell electron configurations and bond order of (a) NO^+; (b) N_2^+; (c) C_2^{2+}. Use Fig. 3.35.

3.51 Which of the following species is paramagnetic: (a) O_2; (b) O_2^-; (c) O_2^+? If the species is paramagnetic, how many unpaired electrons does it possess?

3.52 Which of the following species is paramagnetic: (a) N_2^-; (b) F_2^+; (c) O_2^{2+}? If the species is paramagnetic, how many unpaired electrons does it possess?

3.53 Determine the bond orders and use them to predict which species of each pair has the stronger bond: (a) F_2 or F_2^-; (b) B_2 or B_2^+.

3.54 Determine the bond orders and use them to predict which species of each pair has the stronger bond: (a) C_2^+ or C_2; (b) O_2 or O_2^+.

3.55 How does the change in conductivity of a semiconductor differ from that of a metal as temperature is increased?

3.56 Normally, in conducting materials, we think of current as being carried by electrons as they move through a solid. In semiconductors, it is also common to talk about the current being carried by the "holes" in the valence band. (a) Explain how holes move through a solid material. (b) If, in a p-type semiconductor device, electric current is moving from left to right, in which direction will the holes be moving?

3.57 Germanium is a semiconductor. Including small amounts of impurities of which of these elements—In, P, Sb, and Ga—will make germanium into (a) a p-type semiconductor; (b) an n-type semiconductor?

3.58 Gallium arsenide is a semiconducting material. If we wish to modify the sample by replacing a small amount of the arsenic with an element to produce an n-type semiconductor, which element would we choose, selenium, phosphorus, or silicon? Why?

INTEGRATED EXERCISES

3.59 For each of the following species, write the Lewis structure, predict the shape and hybridization about each central atom, give the bond angles, and state whether it is polar or nonpolar: (a) $GaCl_4^-$; (b) TeF_4; (c) $SbCl_4^-$; (d) $SiCl_4$.

3.60 For each of the following species, write the Lewis structure, predict the shape and hybridization about each central atom, give the bond angles, and state whether it is polar or nonpolar: (a) $SnCl_3^-$; (b) TeO_3; (c) NO_3^-; (d) ICl_3.

3.61 For each of the following species, write the Lewis structure, predict the shape and hybridization about each central atom, give the bond angles, and state whether it is polar or nonpolar: (a) $InCl_4^-$; (b) PCl_4^+; (c) ICl_4^+; (d) $SeOF_2$.

3.62 For each of the following species, write the Lewis structure, predict the shape and hybridization about each central atom, give the bond angles, and state whether it is polar or nonpolar: (a) CH_3^+; (b) H_3O^+; (c) SO_3; (d) $ClOF_2$; (e) $ClOF_2^+$.

3.63 Polar molecules attract other polar molecules through dipole-dipole intermolecular forces. Polar solutes tend to have higher solubilities in polar solvents than nonpolar solutes. Which of the following compounds would be expected to have the higher solubility in hexafluorobenzene, C_6F_6: (a) SiF_4 or PF_3; (b) SF_6 or SF_4; (c) IF_5 or AsF_5?

3.64 The halogens form compounds among themselves. These compounds, called the interhalogens, have the formulas XX′, XX′$_3$, and XX′$_5$, where X is the heavier halogen atom. (a) Predict their structures and bond angles. (b) Which of them are polar? (c) Why is the lighter halogen atom not the central atom of such molecules?

3.65 An organic compound distilled from wood was found to have a molar mass of 32.04 $g{\cdot}mol^{-1}$ and the following composition by mass: 37.5% C, 12.6% H, and 49.9% O. (a) Write the Lewis structure of the compound and determine the bond angles about the carbon and oxygen atoms. (b) Give the hybridization of the carbon and oxygen atoms. (c) Predict whether the molecule is polar or not.

3.66 Draw Lewis structures for each of the following species and predict the hybridization at each carbon atom: (a) H_2CCH^+; (b) $H_2CCH_3^+$; (c) $H_3CCH_2^-$.

3.67 (a) Draw the bonding and antibonding orbitals that correspond to the σ bond in H_2. (b) Repeat this procedure for HF. (c) How do these orbitals compare?

3.68 (a) Consider a hypothetical species "HeH." What charge (magnitude and sign), if any, should be present on this combination of atoms to produce the most stable molecule or ion possible? (b) What is the maximum bond order that such a molecule or ion could have? (c) If the charge on this species were increased or decreased by one, what would be the effect on bonding the molecule?

3.69 (a) Order the following molecules according to increasing C—F bond length: CF^+, CF, CF^-. (b) Which of these molecules will be diamagnetic, if any? Explain your reasoning.

3.70 Describe as completely as you can the structure and bonding in carbamate ion, $CO_2(NH_2)^-$. The C—O bond lengths are both 128 pm, and the C—N bond length is 136 pm.

3.71 Borazine, $B_3N_3H_6$, a compound that has been called "inorganic benzene" because of its similar hexagonal structure (but with alternating B and N atoms in place of C atoms), is the basis of a large class of boron-nitrogen compounds. Write its Lewis structure and predict the composition of the hybrid orbitals used by each B and N atom.

3.72 Knowing that carbon has a valence of four in nearly all its compounds and can form chains and rings of C atoms, (a) draw any two of the three possible structures for C_3H_4; (b) determine all bond angles in each structure; (c) determine the hybridization of each carbon atom in the two structures; (d) ascertain whether the two structures are resonance structures or not and explain your reasoning.

3.73 (a) Draw a Lewis structure for each of the following species: CH_3^+; CH_4; CH_3^-; CH_2; CH_2^{2+}; CH_2^{2-}. (b) Label each as to whether or not it is a radical. (c) Rank them in order of increasing H—C—H bond angles. Explain your choices.

3.74 Consider the molecules H_2CCH_2, H_2CCCH_2, and H_2CCCCH_2. (a) Draw Lewis structures for these molecules. (b) What is the hybridization at each C atom? (c) What type of bond connects the carbon atoms (single, double, etc.)? (d) What are the H—C—H, C—C—H, and C—C—C angles in these molecules? (e) Do all the hydrogen atoms lie in the same plane? (f) A generalized formula for molecules of this type is $H_2C(C)_xCH_2$, where x is 0, 1, 2, etc. What can be said, if anything, about the orientation of the H atoms at the ends of the chain as a function of x?

3.75 Acetylene, C_2H_2 (systematic name, ethyne), can be polymerized. (a) Draw the Lewis structure for acetylene and draw a Lewis structure for the polymer that results when acetylene is polymerized. The polymer has formula $(CH)_n$, where n is large. (b) Consider the polymers polyacetylene and polyethylene. The latter has the formula $(CH_2)_n$ and is an insulating material (plastic wrap is made of polyethylene), whereas polyacetylene is a darkly colored material that can conduct electricity when properly treated. On the basis of your answer to (a), suggest an explanation for the difference in the two polymers.

3.76 Dye molecules are of commercial importance because they are very intensely colored. Most dye molecules possess many multiple bonds and are often aromatic. Why is this important to the properties of the dye molecule?

3.77 In addition to forming σ and π types of bonds similar to p-orbitals, d-orbitals may overlap in δ-type bonds. (a) Draw overlap diagrams showing three different ways in which d-orbitals can combine to form bonds. (b) Place the three types of d-d bonds—σ, π, and δ—in order of strongest to weakest.

3.78 An s-orbital and a p-orbital on different atoms may also overlap to form molecular orbitals. One of these interactions forms a bonding σ-orbital and the other is nonbonding. Draw diagrams to represent the two types of orbital overlap that give rise to σ-bonding and nonbonding orbitals.

3.79 (a) Describe the changes in bonding that would occur in benzene if two electrons were removed from the HOMO (highest occupied molecular orbital). This removal would correspond to an oxidation of benzene to $C_6H_6^{2+}$. (b) Describe the changes in bonding that would occur if two electrons were added to the LUMO (lowest unoccupied molecular orbital). This addition would correspond to a reduction of benzene by two electrons to give $C_6H_6^{2-}$. Do you expect these ions to be diamagnetic or paramagnetic?

3.80 Confirm, using trigonometry, that the dipoles of the three bonds in a trigonal planar AB_3 molecule cancel, resulting in a nonpolar molecule. Go on to show that the dipoles of the four bonds in a tetrahedral AB_4 molecule cancel and the molecule is nonpolar.

3.81 Benzyne, C_6H_4, is a highly reactive molecule that is detected only at low temperatures. It is related to benzene in that it has a six-membered ring of carbon atoms; but instead of three double bonds, the structure is normally drawn with two double bonds and a triple bond. (a) Draw a Lewis structure of the benzyne molecule. Indicate on the structure the hybridization at each carbon atom. (b) On the basis of your understanding of bonding, explain why this molecule might be highly reactive.

3.82 An important technique for determining the structures of chemical compounds is x-ray diffraction. By careful analysis of the data, we can determine the relative positions of atoms in a crystalline lattice. From these data, we can calculate bond distances and angles. (a) Go to the chemical literature and find at least two publications that report x-ray diffraction data for the S—O bond lengths and O—S—O bond angles in a salt containing the sulfate ion, SO_4^{2-}. A number of journals routinely report structural data. One journal that reports such data is *Acta Crystallographica*. Your instructor can suggest other journals that would be good sources of structural data. Be sure to record the full literature citation (authors, journal title, year, volume, and pages), including the formula of the compound. (b) What is the error, given in terms of standard deviation, associated with each number? In standard structural reports, the standard deviation for the last places reported will be given in parentheses following the number. For example, a value of 249.2(1) pm would be 249.2 pm with a standard deviation of 0.1 pm. (*Note:* Most structural data are reported in ångströms (Å) rather than picometers, with 1 Å = 100 pm.) (c) Differences in experimental numbers that are less than three times the standard deviation are not considered to be statistically meaningful. On the basis of this criterion, are the S—O bond distances and angles equivalent to one another? (d) Compare the numbers you found to those found by other students in the class. (e) Discuss the factors that lead to differences in the S—O bond distances and O—S—O angles.

3.83 ClO_2F reacts with AsF_5 in a 1:1 stoichiometry in a simple Lewis acid-base reaction. (a) Predict the product. (b) Draw the complete Lewis structure of the product and describe the formal charges, geometries, bond angles, and hybridizations of the Cl and As atoms (for the most favored resonance structure, if resonance is likely). *Hint:* The reaction involves the central atoms. Consider which could act as the Lewis base.

3.84 The Lewis structure of caffeine, $C_8H_{10}N_4O_2$, a common stimulant, is shown below. (a) Give the hybridization of each atom other than hydrogen and predict the bond angles about that atom. (b) On the basis of your answers in part (a), estimate the bond angles around each carbon and nitrogen atom. (c) Search the chemical literature for the structure of caffeine, and compare the observed structured parameters with your predictions.

3.85 Cyclopropane, C_3H_6, is a hydrocarbon composed of a three-membered ring of carbon atoms.

(a) Determine the hybridization of the carbon atoms. (b) Predict the C—C—C and H—C—H bond angles at each carbon atom on the basis of your answer to part (a). (c) What must the "real" C—C—C bond angles in cyclopropane be? (d) What is the defining characteristic of a σ-bond, as opposed to a π-bond, for example? (e) How do the C—C σ-bonds in cyclopropane extend the definition of conventional σ-bonds? (f) Draw a picture depicting the molecular orbitals to illustrate your answer.

3.86 The variation in bond angles from CH_4 to NH_3 to H_2O was rationalized by using electrostatic arguments asserting that repulsions are different between the less localized lone pair electrons and electrons in bonds. Suggest an explanation based on molecular orbitals.

3.87 The size as well as the shape of a molecule may have a great impact on its reactivity. Many modern industrial catalysts based on zeolites, which are aluminosilicate structures that have well-defined channels and pores, make use of the size and shape differences between molecules to carry out separations and selective transformations. Several zeolite structures may be found on the Web site for this text. (a) On the basis of the average bond lengths given in Chapter 2, calculate the minimum volume of a sphere that would be large enough to encapsulate a methane molecule. Calculate the minimum dimensions of a cylinder that could encapsulate (b) an ethane molecule; (c) a benzene molecule.

3.88 (a) The diameter of a C_{60} molecule is approximately 700 pm. (b) Could more than one lanthanum atom occupy the center of a C_{60} molecule? (c) Because it is possible for C_{60} to undergo reduction (six stepwise reductions to give $C_{60}{}^{6-}$ have been reported), it is also possible for a lanthanum ion, La^{3+}, to exist inside the C_{60} molecule. C_{60} would then be considered to be a -3 ion, or, if two lanthanum ions were present, a -6 ion. Could two La^{3+} ions be placed inside a C_{60} molecule?

3.89 Phenolphthalein is an indicator used to detect the stoichiometric point in acid-base reactions because it is colorless in acidic solution and bright pink in basic solution. The structures of the two forms of phenolphthalein are shown below. (a) Redraw them as Lewis resonance structures and show each atom, each bond, and each lone pair. (b) Give the hybridization of the carbon atom marked with an *. (c) Suggest an explanation for why the base form is colored but the acid form is not.

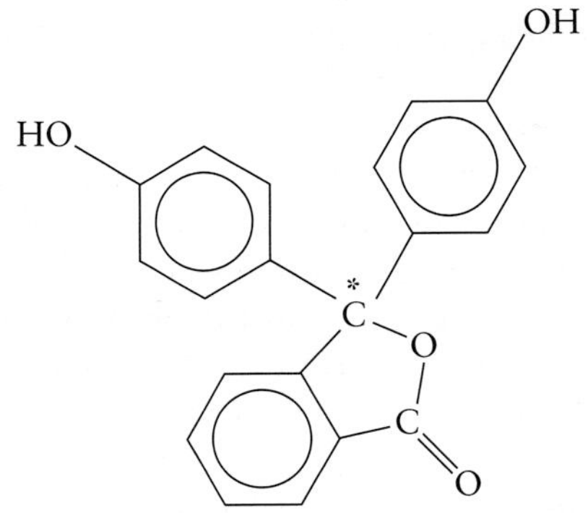

Phenolphthalein (acid form)

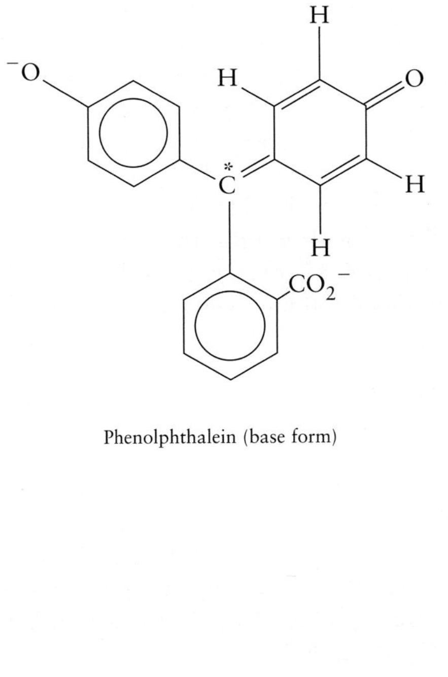

Phenolphthalein (base form)

3.90 Aqueous solutions of compounds containing the Cu^{2+} ion are blue as a result of the presence of the $Cu(H_2O)_6^{2+}$ complex. Does this complex absorb in the visible region? Suggest an explanation. Refer to Major Technique 2 on ultraviolet and visible spectroscopy that follows these exercises.

3.91 In which of the following molecules could there be an n-to-π^* transition? Explain your choices. (a) Formic acid, HCOOH; (b) ethyne, C_2H_2; (c) methanol, CH_3OH; (d) hydrogen cyanide, HCN. Refer to Major Technique 2 on ultraviolet and visible spectroscopy that follows these exercises.

All the colors of vegetation around us arise from the selective absorption and reflection of visible light. If our eyes were sensitive to ultraviolet radiation—the electromagnetic radiation at shorter wavelengths (higher frequencies) than violet light—we would see an even richer tapestry of color, because substances also absorb and reflect selectively in the ultraviolet region. Honeybees do have this ability, and they can "see" radiation in the ultraviolet as well as in the visible range. This selective absorption and transmittance of visible light and ultraviolet radiation is the basis of a technique for identifying compounds and for determining their concentration in samples.

1 The Technique

When electromagnetic radiation falls on a molecule, the electrons in the molecule can be excited to a higher energy state. If the frequency of the radiation is ν (nu), it can raise the molecule to a state that differs in energy from the first state by an amount ΔE, where

$$\Delta E = h\nu \tag{1}$$

This is the *Bohr frequency condition*, which was encountered in Section 1.6; h is the Planck constant. For ultraviolet wavelengths of 300 nm and less (corresponding to a frequency of about 10^{15} Hz or more), each photon brings enough energy to excite the electrons in a molecule into a different distribution. Provided an empty orbital exists at the right energy, the incoming radiation can excite an electron into it, and hence be absorbed. Therefore, the study of visible and ultraviolet absorption gives us information about the electronic energy levels of molecules.

Visible and ultraviolet (UV) absorption spectra are measured in an absorption spectrometer. The source gives out intense visible light or ultraviolet radiation. The wavelengths can be selected with a glass prism for visible light and with a quartz prism or a diffraction grating for ultraviolet radiation (which is absorbed by glass). A typical absorption spectrum, that of chlorophyll, is shown in Fig. 1. Note that chlorophyll absorbs red and blue light, leaving the green light present in white light to be reflected. That is why most vegetation looks green. The spectrum can help us to assess the absorption quantitatively and to make a precise analysis of the energy-capturing power of the molecule.

2 Chromophores

The presence of certain absorption bands in visible and ultraviolet spectra can often be traced to the presence of characteristic groups of atoms in the molecules. These groups of atoms are called *chromophores*, from the Greek words for "color bringer."

One important chromophore is a carbon-carbon double bond. In terms of the language of molecular orbital theory, the electronic transition that occurs when energy is absorbed is the excitation of an electron from a bonding π-orbital to the corresponding antibonding orbital, π^*; this transition is therefore known as a *π-to-π^* (pi to pi star) transition* (Fig. 2); it occurs at about 160 nm, which is in the ultraviolet.

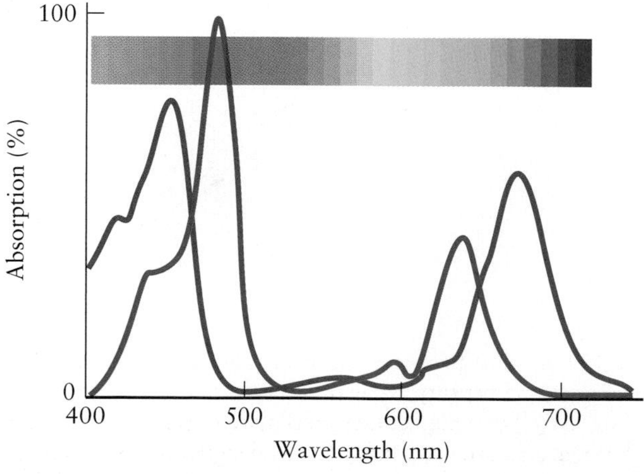

FIGURE 1 The optical absorption spectrum of chlorophyll as a plot of percentage absorption against wavelength. The spectrum of chlorophyll *a* is shown in red, that of chlorophyll *b* in blue.

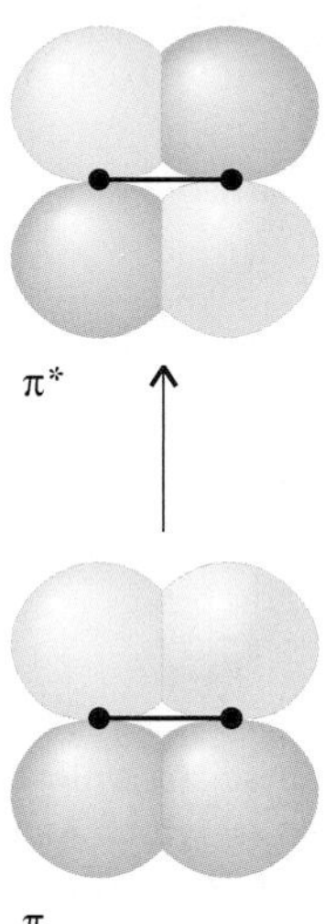

FIGURE 2 In a π-to-π^* transition, an electron in a bonding π-orbital is excited into an empty antibonding π^*-orbital.

The transition can be brought into the visible region—by decreasing the separation between bonding and antibonding orbitals—if the molecule has a chain of alternating single and double bonds. Such "conjugated" double bonding occurs in the compound carotene (structure **54** in Chapter 3), which is partly responsible for the color of carrots, mangoes, and persimmons. Related compounds account for the colors of shrimps and flamingoes. This transition is also responsible for the primary act of vision, as explained in Section 3.8.

Another important chromophore is the carbonyl group, >C=O, which absorbs at about 280 nm. The transition responsible for the absorption is the excitation of a lone-pair electron on the oxygen atom (a "nonbonding" electron, denoted n) into the empty antibonding π^*-orbital of the C=O double bond (Fig. 3). This transition is therefore called an *n-to-π^* (n to pi star) transition.*

A d-metal ion may also be responsible for color, as is apparent from the varied colors of many d-metal complexes (see Chapter 16). Two types of transitions may be involved. In one, which is called a *d-to-d transition*, an electron is excited from a d-orbital of one energy to a d-orbital of higher energy. Because the energy differences between d-orbitals are quite small, visible light brings enough energy to cause this excitation, so colors are absorbed from white light and the sample takes on colors complementary to those observed (Fig. 4). A second type of transition is called a *charge-transfer transition*, in which electrons migrate from the atoms attached to the central metal atom into the latter's d-orbitals, or vice versa. This transfer of charge can result in very intense absorption; it is responsible, for instance, for the deep purple of permanganate ions, MnO_4^-.

Related Exercises: 3.90 and 3.91

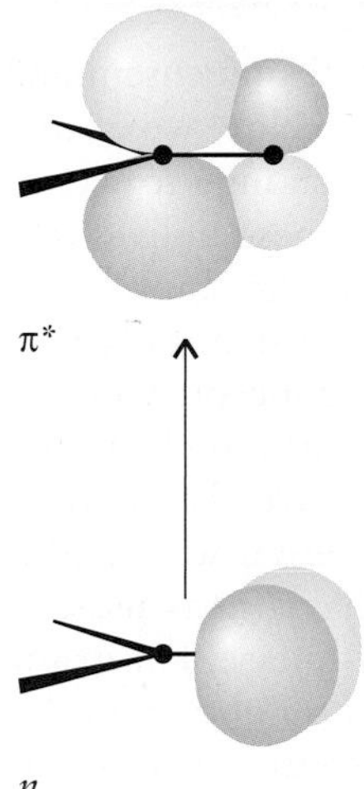

FIGURE 3 In an n-to-π^* transition of a carbonyl group, an electron in a nonbonding orbital (one localized wholly on the oxygen atom) is excited into an antibonding π^* orbital spread over both atoms.

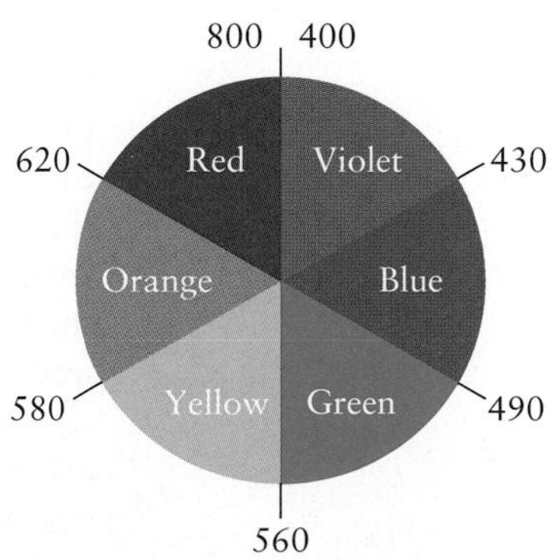

FIGURE 4 The perceived color of a complex in white light is the complementary color of the light it absorbs. In this color wheel, complementary colors are opposite each other. The numbers are approximate wavelengths in nanometers.

CHAPTER

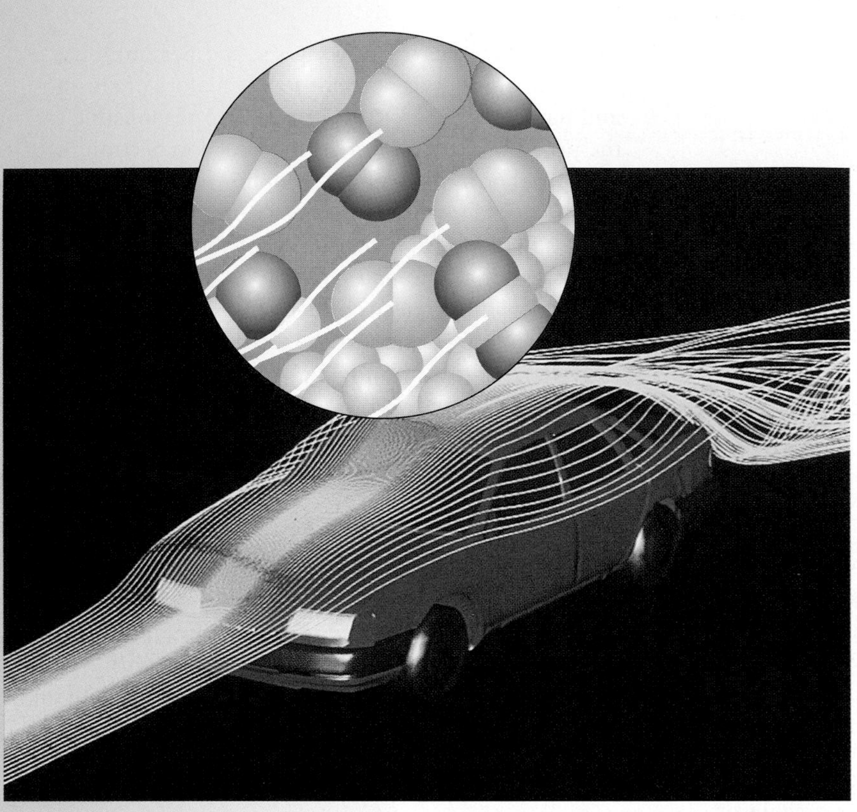

These streamlines show how air, a gas in which molecules are far apart, flows over the solid body of a vehicle, in which the atoms are held in place and cannot move. A supercomputer has been used to simulate the motion of the air. When the actual vehicle is being driven along a highway, the streams of air will be real streams of molecules passing over the atoms and molecules that make up the surface of the car.

Why Do We Need to Know This Material?

We have seen how matter is composed of immense numbers of infinitesimal atoms, ions, and molecules. Now we examine the properties of **bulk matter**, samples of matter large enough to see, and see how these properties emerge from the behavior of these individual particles. Gases are the simplest state of matter, so the connections between the properties of individual molecules and those of bulk matter are relatively easy to identify. In later chapters, these concepts will be used to study thermodynamics, equilibrium, and the rates of chemical reactions. We spend all our lives immersed in a gas, so we also need to understand the processes that go on in the atmosphere if we are to have a full understanding of life on Earth. Gases have many uses, and many chemical reactions take place in the gas phase. For example, gases are used in medicine as anesthetics, in firefighting, in the chemical engineering industry, as fuels, and for sports such as scuba diving.

What Do We Need to Know Already?

We need to be familiar with SI units (Appendix 1B) and the concepts of force and energy (Section A). This chapter also develops the techniques of reaction stoichiometry (Section L) by extending them to gases.

THE PROPERTIES OF GASES

The atmosphere is a precious layer of gas held by gravity to the surface of the Earth. Half the mass of the atmosphere lies below an altitude of 5.5 km. From far out in space where the Earth appears to be the size of a basketball, the atmosphere looks as if it were only 1 mm thick (Fig. 4.1). Yet this thin, delicate layer is vital to life: it shields us from harmful radiation and supplies substances needed for life, such as oxygen, nitrogen, carbon dioxide, and water.

The gases in the atmosphere are not the only ones we are likely to encounter. Eleven elements are gases under normal conditions (Fig. 4.2), as are many compounds with low molar masses. All substances that are gases at ordinary temperatures are molecular, except the six noble gases, which are monatomic (consist of single atoms). Many low-molar-mass organic compounds are gases at room temperature. They include the methane, CH_4, of natural gas and the propane, C_3H_8, and butane, C_4H_{10}, of camping fuel.

As we proceed through this chapter, we follow a path that scientists often employ. First, we collect experimental observations on the properties of gases and summarize these observations mathematically. We then present a simple qualitative molecular model of a gas suggested by these observations. Next, we introduce a more precise molecular model that can be expressed mathematically and used to make quantitative predictions. Finally, we use more detailed experimental observations to refine the model so that it accounts for the properties of real gases.

THE NATURE OF GASES

A remarkable characteristic of gases is that many of their physical properties are very similar, particularly at low pressures. As a result, instead of having to describe the properties of each gas individually, we can describe all gases simultaneously. So, our first task is to discover and describe these shared properties. To do so, we need to establish the meaning of the term

FIGURE 4.1 The delicate film of the Earth's atmosphere as seen from space.

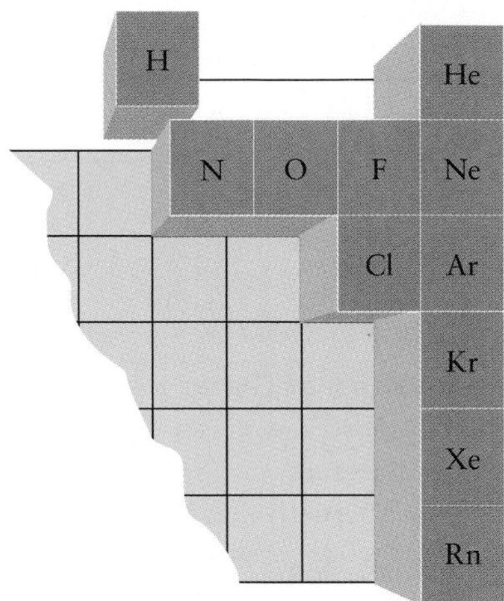

FIGURE 4.2 The 11 elements that are gases under normal conditions. Note how they lie toward the upper right of the periodic table.

"pressure," report its units, and then use the concept of pressure to formulate the laws that gases obey. Finally, we can ask why all gases behave similarly and attempt to build a model of this state of matter in terms of the behavior of individual molecules.

4.1 Observing Gases

We can begin to build an initial model of a gas from our observations of gases in our daily lives. For example, when we push on a bicycle pump, we see that air is **compressible,** that is, it can be confined into a smaller volume. The observation that gases are more compressible than solids and liquids suggests that there is a lot of space between the molecules of gases. We also know from everyday experience, such as filling balloons, that a gas expands rapidly to fill the space available to it. This observation suggests that the molecules are always moving rapidly and hence can respond quickly to changes in the volume available to them. Our first primitive picture of a gas could therefore be as a collection of widely spaced molecules that are in ceaseless rapid random motion. We shall use more quantitative observations to refine this simple model as we proceed through the chapter.

Gases are readily compressible and expandable, always filling the space available to them. These observations suggest that molecules of gases are widely separated and in ceaseless chaotic motion.

4.2 Pressure

If you have ever pumped up a bicycle tire or squeezed an inflated balloon, you have experienced an opposing force arising from the confined air. The **pressure,** P, of a gas is the force, F, exerted by the gas divided by the area, A, on which the force is exerted:

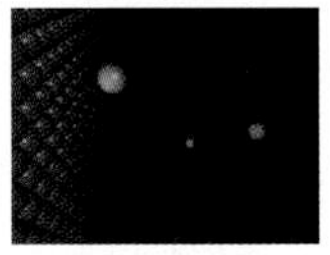

View an animation of how gas molecules generate pressure by striking the walls of their container.

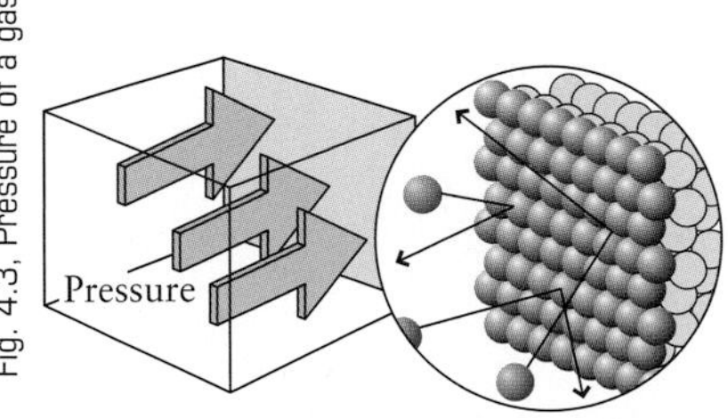

Media Link Animation Fig. 4.3, Pressure of a gas

FIGURE 4.3 The pressure of a gas arises from the collisions that its molecules make with the walls of the container. The storm of collisions exerts an almost steady force on the walls.

$$\text{Pressure} = \frac{\text{force}}{\text{area}} \qquad \text{or, more simply,} \qquad P = \frac{F}{A} \qquad (1)$$

In terms of the model we are developing, the pressure a gas exerts on the walls of its container results from the collisions of its molecules with the container's surface (Fig. 4.3). A vigorous, chaotic storm of molecules on a surface results in a strong force and hence a high pressure. A gentle storm of molecules on the same area results in a low pressure. The pressure of the atmosphere arises from the ceaseless motion of the molecules in air. Any object on the surface of the Earth stands in an invisible storm of molecules that beat on it incessantly and exert a force all over its surface. Even on an apparently calm day, we are in the midst of a molecular storm.

The pressure exerted by the atmosphere can be measured by various means, most commonly by using a pressure gauge. When we measure the pressure of air in a tire with a pressure gauge, we are actually measuring the **gauge pressure,** the difference between the pressure inside the tire and the atmospheric pressure. A flat tire registers a gauge pressure of 0, because the pressure inside the tire is the same as the pressure of the atmosphere. A pressure gauge attached to a piece of scientific apparatus, however, measures the *actual* pressure in the apparatus.

The pressure of the atmosphere is measured with a **barometer**, an instrument invented in the seventeenth century by Evangelista Torricelli, a student of Galileo. Torricelli (whose name coincidentally means "little tower" in Italian) formed a little tower of liquid mercury, which is very dense. He sealed a long glass tube at one end, filled it with mercury, and inverted it into a beaker (Fig. 4.4). The column of mercury fell until the pressure it exerted at its base matched the pressure exerted by the atmosphere. To interpret measurements with a barometer, we need to find how the height of the column of mercury depends on the atmospheric pressure.

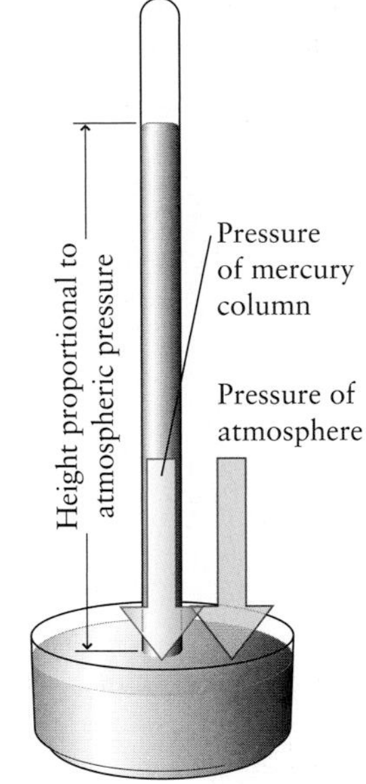

FIGURE 4.4 A barometer is used to measure the pressure of the atmosphere. The pressure of the atmosphere is balanced by the pressure exerted by the column of mercury. The height of the column is proportional to the atmospheric pressure. Therefore, by measuring the height of the column, we can monitor the pressure of the atmosphere.

How do we do that?

We wish to find the relation between the height, h, of the column of mercury and the atmospheric pressure, P. Suppose the cross-sectional area of the column is A. The volume of mercury in the column is the height of the cylinder times this area, or hA. The mass, m, of this volume of mercury is the product of mercury's density, d, and the volume, so $m = dV = dhA$. This mass of mercury is pulled down by the force of gravity; and the total force it exerts at its base is the product of the mass and the acceleration of free fall (the acceleration due to gravity), g: $F = mg$. Therefore, the pressure at the base of the column, the force divided by the area, is

$$P = \frac{F}{A} = \frac{mg}{A} = \frac{dhAg}{A} = dhg \qquad (2)$$

This equation shows that the pressure exerted by the column of mercury, P, is proportional to the height of the column. Because this pressure balances the atmospheric pressure, the height of the column is also proportional to the atmoseric pressure.

How can we use this information . . . to calculate the atmospheric pressure?

Suppose the height of the column of mercury in a barometer is 760 mm. Given that the density of mercury at 20°C is 13.546 $g{\cdot}cm^{-3}$ (corresponding to 13 546 $kg{\cdot}m^{-3}$) and the acceleration of free fall at the surface of the Earth is 9.806 65 $m{\cdot}s^{-2}$, we can conclude that the atmospheric pressure is

$$\begin{aligned} P &= dhg = (13\,546\ kg{\cdot}m^{-3}) \times (0.760\ m) \times (9.806\,65\ m{\cdot}s^{-2}) \\ &= 1.01 \times 10^5\ kg{\cdot}m^{-1}{\cdot}s^{-2} \end{aligned}$$

Self-Test 4.1A What is the atmospheric pressure when the height of the mercury column in a barometer is 756 mm?

[*Answer:* $1.00 \times 10^5\ kg{\cdot}m^{-1}{\cdot}s^{-2}$]

Self-Test 4.1B The density of water at 20°C is 0.998 $g{\cdot}cm^{-3}$. What height would the column of liquid be in a water barometer when the atmospheric pressure corresponds to 760 mm of mercury?

The unit is named after the French philosopher and mathematician Blaise Pascal.

The SI unit of pressure is the **pascal,** Pa:

$$1 \text{ Pa} = 1 \text{ kg·m}^{-1}\text{·s}^{-2}$$

As we have seen, when the height of a barometer column is 760 millimeters of mercury (written 760 mmHg) the atmospheric pressure is 1.01×10^5 kg·m^{-1}·s^{-2}, so this pressure can be reported as 1.01×10^5 Pa, or 101 kPa, where 1 kPa = 10^3 Pa. A shorter column of mercury corresponds to a lower atmospheric pressure, and a taller column corresponds to a higher pressure.

Although a pressure gauge is more commonly used to measure the pressure inside a laboratory vessel, a **manometer** is sometimes used (Fig. 4.5). It consists of a U-shaped tube connected to the experimental system. The other end of the tube may be either open to the atmosphere or sealed. For an open-tube manometer (like that shown in Fig. 4.5a), the pressure in the system is equal to that of the atmosphere when the levels of the liquid in each arm of the U-tube are the same. If the level of mercury on the system side of an open manometer is above that of the atmosphere side, the pressure in the system is lower than the atmospheric pressure. In a closed-tube manometer (like that shown in Fig. 4.5b), one side is connected to a closed flask (the system) and the other side is vacuum. The difference in heights of the two columns is proportional to the pressure in the system.

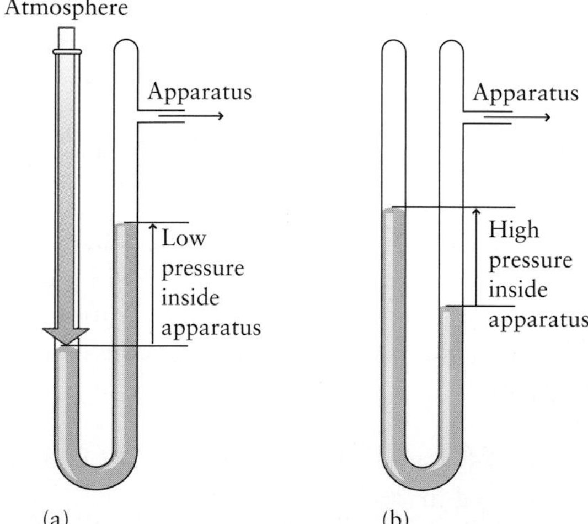

FIGURE 4.5 (a) An open-tube manometer. The pressure inside the apparatus to which the narrow horizontal tube is connected pushes against the external pressure. In this instance, the pressure inside the system is lower than the atmospheric pressure by an amount proportional to the difference in heights of the liquid in the two arms. (b) A closed-tube manometer. The pressure in the adjoining apparatus is proportional to the difference in heights of the liquid in the two arms. The space inside the enclosed end is a vacuum.

How can we use this information . . . to interpret a manometer reading?

The height of the mercury in the system-side column of an open-tube mercury manometer was 10 mm above that of the open side when the atmospheric pressure corresponded to 756 mm of mercury. The pressure in the system therefore corresponds to 756 mm − 10 mm = 746 mm of mercury (746 mmHg). To express the pressure in pascals, we write

$$\begin{aligned} P &= (13\ 546\ \text{kg·m}^{-3}) \times (0.746\ \text{m}) \times (9.806\ 65\ \text{m·s}^{-2}) \\ &= 9.91 \times 10^4\ \text{kg·m}^{-1}\text{·s}^{-2} = 9.91 \times 10^4\ \text{Pa} \end{aligned}$$

or 99.1 kPa.

Self-Test 4.2A What is the pressure in a system when the mercury level in the system-side column in an open-tube mercury manometer is 25 mm lower than the mercury level in the atmosphere-side column and the atmospheric pressure corresponds to 760 mmHg?

[*Answer:* 785 mmHg, 104 kPa]

Self-Test 4.2B What is the pressure in millimeters of mercury inside a system when a *closed* mercury manometer shows a height difference of 10 cm (higher on the closed side)?

The pressure of a gas, the force it exerts divided by the area subjected to the force, arises from the impact of its molecules.

4.3 Alternative Units of Pressure

Although the SI unit of pressure is the pascal, Pa, there are several other units in common use. It is important to be familiar with them all (Table 4.1).

The height of the mercury column in a barometer on a typical day at sea level is about 760 mm, which we have seen corresponds to 101 kPa. This pressure could be reported as 760 mmHg. However, the unit 1 mmHg has been largely replaced by the unit 1 Torr (named for Torricelli), with 1 Torr = 1 mmHg, so normal atmospheric pressure at sea level is about 760 Torr. Atmospheric pressure varies with altitude and weather. The pressure of the atmosphere at the cruising height of a commercial jetliner (10 km) is only about 200 Torr, so airplane cabins must be pressurized.

The two units are not *exactly* the same. 1 Torr = (1/760) atm exactly; 1 mmHg is the pressure exerted by a column of mercury 1 mm high in a particular gravitational field and at a given temperature. In practice, the units are interchangeable, but Torr is now preferred.

Because we commonly deal with substances close to atmospheric pressure, it is often convenient to use the unit **atmosphere** (atm), where 1 atm = 760 Torr exactly. The *exact* conversion relation between pascals and atmospheres is

$$1\ \text{atm} = 1.013\ 25 \times 10^5\ \text{Pa}$$

A very low pressure atmospheric region, like an area of low pressure on the weather chart in Fig. 4.6, typically has a pressure of about 0.98 atm at sea level. The pressure in the eye of a hurricane might be as low as 0.90 atm. A typical region of high pressure is about 1.03 atm. The relation between the atmosphere and pascals is numerically awkward, so scientists also use a unit called the **bar.**

The name *bar* is derived from the Greek word for "heavy."

$$1\ \text{bar} = 10^5\ \text{Pa}$$

TABLE 4.1 Pressure Units*

SI unit: pascal (Pa)
$1 \text{ Pa} = 1 \text{ kg}\cdot\text{m}^{-1}\cdot\text{s}^{-2} = 1 \text{ N}\cdot\text{m}^{-2}$

Conventional units
$1 \text{ bar} = \mathbf{10^5} \textbf{ Pa} = \mathbf{100} \textbf{ kPa}$
$1 \text{ atm} = \mathbf{1.013\,25 \times 10^5} \textbf{ Pa}$
$\quad\quad\;\; = \mathbf{101.325} \text{ kPa}$
$1 \text{ atm} = \mathbf{760} \text{ Torr}$
$1 \text{ atm} = 14.7 \text{ lb}\cdot\text{inch}^{-2} \text{ (psi)}$

*Figures in bold are exact. See inside back cover for more relations. N, newton.

FIGURE 4.6 A typical weather map showing Europe and the North Atlantic. The closed curves are called isobars and are contours of constant atmospheric pressure. Regions of low pressure (L) are called cyclones and regions of high pressure (H) are called anticyclones.

This relation is exact; it implies that 1 bar is about 1% less than 1 atm.

The units for reporting pressure are Torr and atmospheres; the SI unit is pascal. Pressure units can be interconverted by using the information in Table 4.1.

Self-Test 4.3A Express a pressure of 500 Torr in atmospheres.

[*Answer:* 0.658 atm]

Self-Test 4.3B The atmospheric pressure in Denver, Colorado, on a certain day was 630 Torr. Express this pressure in pascals.

THE GAS LAWS

Summaries of the properties of gases, particularly the variation of pressure with volume and temperature, are known as the "gas laws." The first reliable measurements of the properties of gases were made by the Anglo-Irish scientist Robert Boyle in 1662 when he examined the effect of pressure on volume. A century and a half later, a new pastime, hot-air ballooning, motivated two French scientists, Jacques Charles and Joseph-Louis Gay-Lussac, to formulate additional gas laws. Charles and Gay-Lussac measured how the temperature of a gas affects its pressure, volume, and density.

4.4 Boyle's Law

Boyle took a long tube of glass curved into a J-shape, with the short end sealed (Fig. 4.7). He then poured mercury into the tube, trapping air in the short end of the J. The more mercury he added, the more the gas was compressed. He concluded that the volume of a fixed amount of gas decreases as the pressure on it increases. Figure 4.8 shows a graph of the dependence. The curve shown there is called an **isotherm,** which is a general term for a plot showing the variation of a property at constant temperature. Scientists, however, often look for ways of plotting experimental data in a manner that gives straight lines, because such graphs are easier to identify, analyze, and interpret. Boyle's data gives a straight line when we plot the pressure against 1/volume (Fig. 4.9). The implication of this result is that *for a fixed amount of gas at constant temperature, volume is inversely proportional to pressure.*

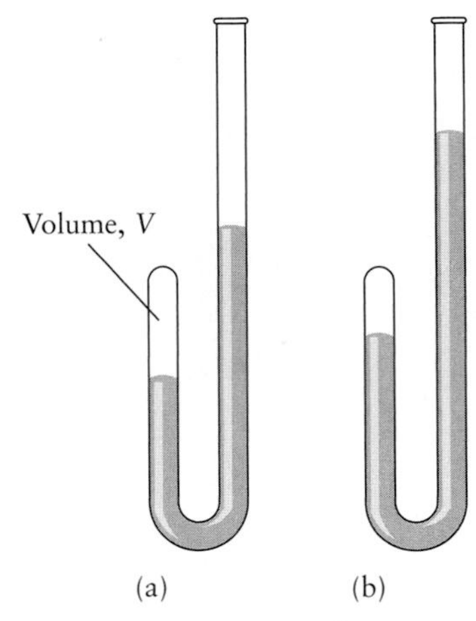

FIGURE 4.7 (a) In Boyle's experiment, a gas was trapped by mercury inside the closed end of the J-shaped tube. (b) The volume of the trapped gas decreased as the pressure on it was increased by adding more mercury to the open end of the tube.

$$\text{Volume} \propto \frac{1}{\text{pressure}} \qquad \text{or, more simply,} \qquad V \propto \frac{1}{P} \tag{3a}$$

Equivalently, the pressure exerted by a given amount of gas at constant temperature is inversely proportional to the volume to which it is confined:

$$\text{Pressure} \propto \frac{1}{\text{volume}} \qquad \text{or, more simply,} \qquad P \propto \frac{1}{V} \tag{3b}$$

These two relations are known as **Boyle's law.** The law implies that if we compress a gas isothermally (at constant temperature) into half its initial

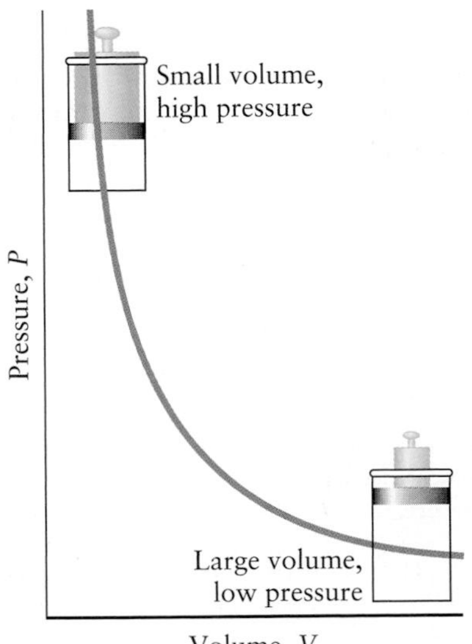

FIGURE 4.8 Boyle's law summarizes the effect of pressure on the volume of a fixed amount of gas at constant temperature. As the pressure of a gas sample is increased, the volume of the gas decreases.

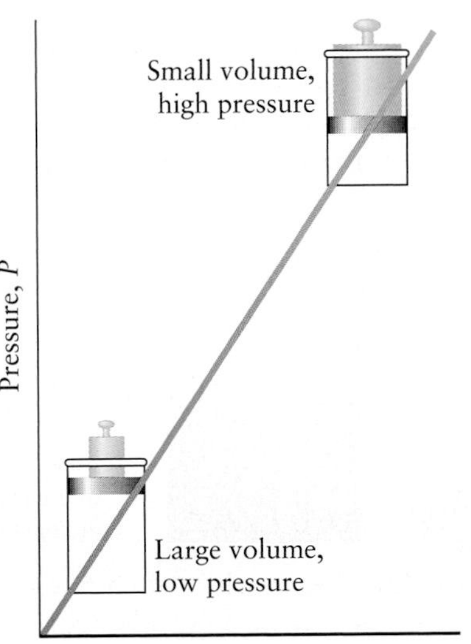

FIGURE 4.9 When the pressure is plotted against 1/volume, a straight line is obtained. Boyle's law breaks down at high pressures, and a straight line is not obtained in these regions (not shown).

Media Link **Living graph:** Fig. 4.8, Boyle's Law

volume, from 1 L to 0.5 L, for example, then the pressure of the gas will double.

We can write Boyle's law as P = constant/V and then rearrange it into

$$PV = \text{constant} \tag{4}$$

This relation in turn tells us that the isotherm in Fig. 4.8 is in fact a *hyperbola*, the name given to a graph of y against x when x and y are related by xy = constant. Note that it would have been difficult to conclude from inspection that the curve in Fig. 4.8 is a hyperbola without first going through the step of finding a way of plotting the data that gave a straight line.

Our molecular model of a gas is consistent with Boyle's law. As a gas is compressed, its molecules are confined to a smaller volume. The interior of the container becomes more crowded and more of the molecules strike the walls in a given time (Fig. 4.10). As a result, they exert a higher total force on the walls, so giving rise to a higher pressure.

The pressure of a fixed amount of gas at constant temperature is inversely proportional to the volume; PV = constant.

EXAMPLE 4.1 Using Boyle's law

Assume that when you press in the piston of a bicycle pump, the volume inside the pump is decreased from about 100 cm^3 to 20 cm^3 before the compressed air flows into the tire. Suppose that the compression is isothermal; estimate the final pressure of the compressed air in the pump, given an initial pressure of 1.00 atm.

STRATEGY In calculations of this kind, it is a good idea to summarize in a table the data supplied and the information required. We expect the final pressure, P_2, to be higher than the initial pressure, P_1, because the volume occupied by the air has been decreased. Because the product of pressure and volume is a constant, we can use Eq. 4 to write

$$P_2V_2 = P_1V_1 \quad \text{or} \quad \frac{P_2}{P_1} = \frac{V_1}{V_2}$$

Media Link Animation: Fig. 4.10, Effect of compression on the pressure of a gas

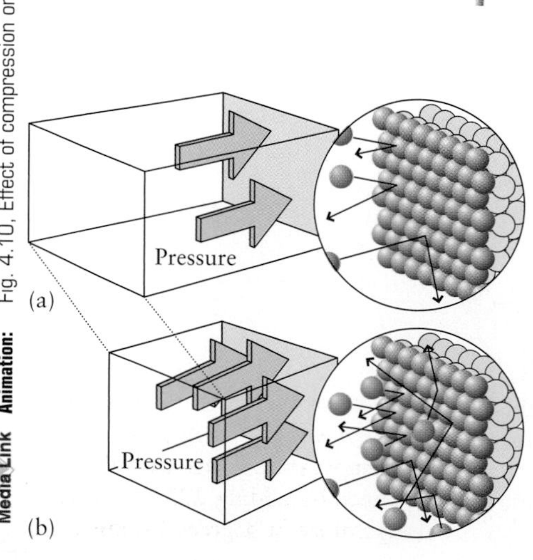

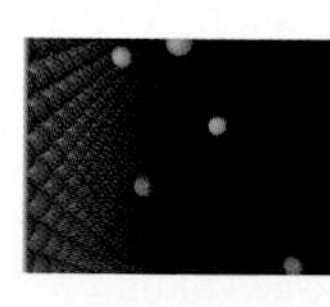

View an animation of the molecules of a gas as its volume is decreased at constant temperature.

FIGURE 4.10 (a) As we have seen, the pressure of a gas arises from the impact of its molecules on the walls of the container. (b) When the volume of the sample is decreased, there are more molecules in a given volume, so there are more collisions with the walls. Because the impact on the walls is now greater, so is the pressure. In other words, decreasing the volume occupied by a given amount of a gas increases the pressure.

where V_1 is the initial volume and V_2 is the final volume. We then rearrange this expression into one for P_2 in terms of P_1 and the ratio of volumes, and substitute the data.

SOLUTION The information given and required is summarized in the following table:

	Pressure, P	Volume, V	Temperature, T	Amount, n
1. Initial	1.00 atm	100 cm^3	same	same
2. Final	?	20 cm^3	same	same

We rearrange the equation given on the previous page and substitute the data:

$$P_2 = \frac{V_1}{V_2} \times P_1 = \frac{100\ \text{cm}^3}{20\ \text{cm}^3} \times 1.00\ \text{atm} = 5.0\ \text{atm}$$

The final pressure is higher, as expected.

Self-Test 4.4A A sample of neon of volume 1.00×10^{-2} L at 200 Torr is allowed to expand isothermally into an evacuated tube with a volume of 0.200 L. What is the final pressure of the neon in the tube?

[*Answer:* 10.0 Torr]

Self-Test 4.4B A sample of gas in the cylinder of a test engine at 80 cm^3 and 1.00 atm is compressed isothermally by pushing a piston into the cylinder. The final pressure is 3.20 atm. What is the final volume of the sample?

4.5 Charles's Law

Charles and Gay-Lussac were both inspired by the problems associated with the new technology of balloon flight, and they independently carried out a number of experiments with the hope of improving the performance of their balloons. They found that, provided the pressure is kept constant, the volume of a gas increases as its temperature is raised. When we plot the volume against the temperature, we get a straight-line graph (Fig. 4.11). We can conclude that, *for a fixed amount of gas under constant pressure, the volume varies linearly with the temperature.* This relation is called **Charles's law.**

When we plot volume as a function of temperature, using data obtained from measurements on different gases and at different pressures, we find that the straight lines all meet at one point (Fig. 4.12). The extension of a graph beyond the data is called **extrapolation,** and we say that, in this case, the straight lines extrapolate to the same point. For Charles's law, this unique point corresponds to zero volume and −273.15°C. Because a volume cannot be negative, this temperature must be the lowest possible temperature. It is the value corresponding to 0 on the Kelvin scale. It follows that if we use temperature, T, on the Kelvin scale, then we can write Charles's law as

The Kelvin (absolute) temperature scale is described in Appendix 1B: the temperature in kelvins (K) is obtained by adding 273.15 to the temperature in degrees Celsius (°C).

$$\text{Volume} \propto \text{absolute temperature} \qquad \text{or, more simply,} \quad V \propto T \qquad (5)$$

Media Link Living graph: Fig. 4.11, Charles's law

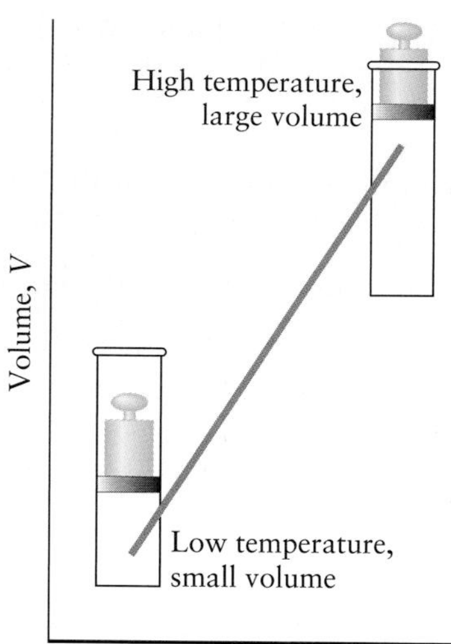

FIGURE 4.11 When the temperature of a gas is increased and its volume is free to change at constant pressure (as depicted by the constant weight acting on the piston), the volume increases. A graph of volume against temperature is a straight line.

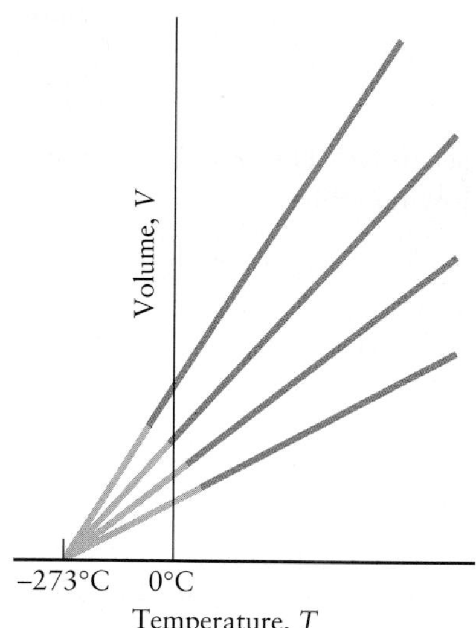

FIGURE 4.12 The extrapolation of data like that in Fig. 4.11 for a number of gases suggests that the volume of all gases should become 0 at $T = 0$ (−273°C). In practice, all gases condense to liquids well before that temperature is reached.

Media Link Living graph: Fig. 4.13, Temperature dependence of the pressure of an ideal gas

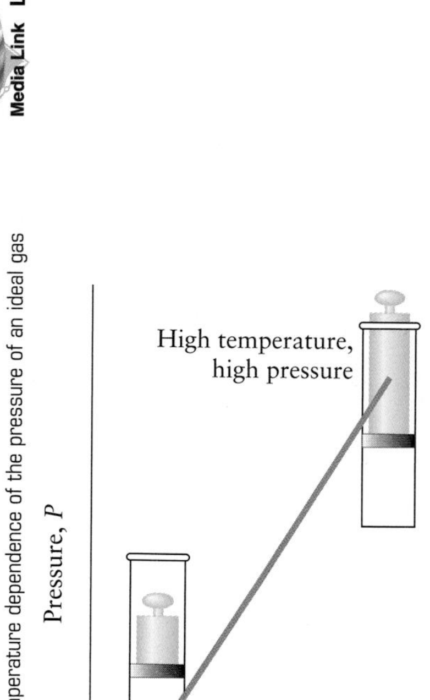

FIGURE 4.13 The pressure of a fixed amount of gas in a vessel of constant volume is proportional to the absolute temperature. The pressure extrapolates to 0 at $T = 0$.

This expression tells us that if the (absolute) temperature of a fixed amount of gas at constant pressure is doubled, then the volume doubles. For example, if a gas is heated from 300 K (27°C) to 600 K (327°C) in a container fitted with a piston, then its volume doubles.

A similar expression summarizes the variation in pressure of a sample of gas when it is heated in a container of fixed volume. Once again, it is found experimentally that the pressure varies linearly with temperature and that the data extrapolate to zero pressure at −273.15°C (Fig. 4.13). Therefore, we can write

$$\text{Pressure} \propto \text{absolute temperature} \qquad \text{or, more simply,} \qquad P \propto T \quad (6)$$

It follows that doubling the absolute temperature doubles the pressure of a gas, provided the amount and volume are constant.

Self-Test 4.5A An oxygen tank stored outside a building has a pressure of 20.00 atm at 6:00 AM, when the temperature is 10°C. What will be the pressure in the tank at 6:00 PM, when the temperature is 30°C?

[*Answer:* 21.4 atm]

Self-Test 4.5B A sample of hydrogen gas at 760 mmHg and 20°C is heated to 300°C in a container of constant volume. What is the final pressure of the sample?

We can explain the effect of temperature on the pressure of a gas by adding a new feature to our molecular model: *as the temperature of a gas is raised, the average speed of the molecules increases.* As a result of this increase in average speed, each molecule strikes the walls more often and with greater force. Therefore, the gas exerts a greater pressure as the temperature is increased at constant volume. We can use this same model of a gas to explain the effect of temperature on the volume. To maintain constant pressure, the walls need to move out and increase the volume available to the gas; so the volume of a gas increases as the temperature is increased at constant pressure. Notice how experimental results are helping us to improve our original model of a gas. The same procedure—the progressive enrichment of a model in the light of experimental information—is widely used in science.

The volume of a fixed amount of gas in a container at constant pressure is directly proportional to the absolute temperature: $V \propto T$. The pressure of a fixed amount of gas in a container of constant volume is directly proportional to the absolute temperature: $P \propto T$.

4.6 Avogadro's Principle

A third contribution to our understanding of gases was made by the Italian scientist Amedeo Avogadro, whose work enabled scientists to accept the reality of atoms. The **molar volume,** V_m, of a substance—any substance, not only a gas—is the volume it occupies per mole of its molecules. Specifically, if V is the volume of a sample and n is the amount (in moles) of molecules the sample contains, then the molar volume is

$$V_m = \frac{V}{n} \qquad (7)$$

Figure 4.14 summarizes some measurements of the molar volumes of gases at the same pressure and temperature. We see that all the molar volumes are very similar. Results like these suggest that *under the same*

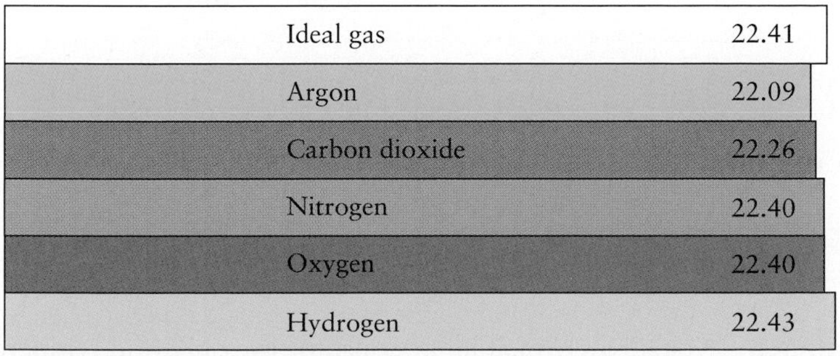

Gas	Molar volume
Ideal gas	22.41
Argon	22.09
Carbon dioxide	22.26
Nitrogen	22.40
Oxygen	22.40
Hydrogen	22.43

FIGURE 4.14 The molar volumes (in liters per mole) of various gases at 0°C and 1 atm. The values are all very similar and close to the molar volume of an ideal gas under these conditions, 22.41 $L\cdot mol^{-1}$ (Section 4.8).

conditions of temperature and pressure, a given number of gas molecules occupy the same volume regardless of their chemical identity. This observation is now known as **Avogadro's principle.**

It follows from the definition of the molar volume that

$$V = nV_m \tag{8}$$

Because V_m is a constant at constant temperature and pressure, this expression tells us that doubling the number of moles of molecules at constant temperature and pressure doubles the volume occupied by a gas. This behavior is consistent with our model of gases. In our model, gas molecules are in constant motion and strike the walls of their container. To keep the pressure constant as more molecules are added, the size of the container must increase.

The volume occupied by a sample of gas at constant pressure and temperature is directly proportional to the number of moles of molecules present; $V \propto n$.

Self-Test 4.6A A helium weather balloon was filled at $-20°C$ to a volume of 2.5×10^4 L with 1.2×10^3 mol He. What is the molar volume of helium at $-20°C$?

[*Answer:* $21\ L{\cdot}mol^{-1}$]

Self-Test 4.6B A large natural gas storage tank contains 200 mol $CH_4(g)$ at 1.20 atm. An additional 100 mol $CH_4(g)$ is pumped into the tank at constant temperature. What is the final pressure in the tank?

4.7 The Ideal Gas Law

At this point, we have three different laws. Can we combine them into a single expression relating pressure (P), volume (V), temperature (T), and amount (n) of a gas?

How do we do that?

To combine the three laws into one, we first note that when the temperature and amount of gas are constant, Boyle's law, Eq. 4, has the form

$$PV = \text{constant}_1$$

When the pressure and amount are constant, Charles's law, Eq. 5, tells us that

$$V = \text{constant}_2 \times T$$

According to Avogadro's principle, at constant pressure and temperature, the volume is proportional to n, so, from Eq. 8, we can also write

$$V = \text{constant}_3 \times n$$

An expression that combines these three laws into one is

$$PV = \text{constant} \times nT$$

When the constant of proportionality for the laws is written R, the combined expression becomes the **ideal gas law:**

$$PV = nRT \quad (9)$$

The constant R in Eq. 9 is called the **gas constant** and has the same value for all gases; because R is independent of the identity of the gas, we say that it is a "universal constant." The value of the gas constant can be found by measuring P, V, n, and T and substituting their values into

$$R = \frac{PV}{nT} \quad (10)$$

Table 4.2 lists the values of R for different units of pressure; in SI units, $R = 8.314\ 51\ \text{J·K}^{-1}\text{·mol}^{-1}$. When doing calculations that involve R, select from the table the value that has units matching the ones you need. Alternatively, if you prefer to memorize one value of R and use it consistently, convert the pressure and volume units given into the units corresponding to the chosen value of R.

TABLE 4.2 The Gas Constant, R

$8.205\ 78 \times 10^{-2}\ \text{L·atm·K}^{-1}\text{·mol}^{-1}$
$8.314\ 51 \times 10^{-2}\ \text{L·bar·K}^{-1}\text{·mol}^{-1}$
$8.314\ 51\ \text{L·kPa·K}^{-1}\text{·mol}^{-1}$
$8.314\ 51\ \text{J·K}^{-1}\text{·mol}^{-1}$
$62.364\ \text{L·Torr·K}^{-1}\text{·mol}^{-1}$

The ideal gas law is an example of an **equation of state,** an expression showing how the pressure of a substance—in this case, a gas—is related to the temperature, volume, and amount of substance in the sample. A hypothetical gas that obeys the ideal gas law under all conditions is called an **ideal gas.** All actual gases are found to obey Eq. 9 with increasing accuracy as the pressure is reduced toward 0 (which we write $P \rightarrow 0$). Therefore, the ideal gas law is an example of a **limiting law,** a law that is strictly valid only in some limit, in this case, as $P \rightarrow 0$. Although the ideal gas law is a limiting law, it is in fact reasonably reliable at normal pressures, so we can use it to describe the behavior of most gases under normal conditions.

The ideal gas law, $PV = nRT$, is an equation of state that summarizes all the relations describing the response of an ideal gas to changes in pressure, volume, temperature, and amount of molecules; it is an example of a limiting law.

4.8 Applications of the Ideal Gas Law

The ideal gas law has many uses in chemistry, some of which we meet later in this text. To begin to see how useful the law can be, recall that we have seen how to use the individual laws to make predictions when only one variable is changed, such as heating a fixed amount of gas at constant volume. The ideal gas law enables us to make predictions when two or more variables are changed. For example, when we pump up an actual bicycle tire, the temperature of the gas in the pump increases as we press in the piston, so the compression is not strictly isothermal as we assumed in Example 4.1.

To do calculations like this, we note that, if the initial conditions of a gas are P_1, V_1, and T_1, then from Eq. 9 we can write $P_1V_1 = nRT_1$, or $P_1V_1/T_1 = nR$, where n is the amount of gas. After the change, the conditions become P_2, V_2, and T_2; and because the ideal gas law still applies, we know that $P_2V_2 = nRT_2$, or $P_2V_2/T_2 = nR$. Because the amount of gas remains the same, nR is a constant, so we can equate P_1V_1/T_1 and P_2V_2/T_2, and hence obtain an expression that combines Boyle's law and Charles's law:

$$\frac{P_1V_1}{T_1} = \frac{P_2V_2}{T_2} \tag{11}$$

This expression is called the **combined gas law.** However, it is a direct consequence of the ideal gas law and is not a new law.

TOOLBOX 4.1 How to use the ideal gas law

Conceptual Basis

The ideal gas law, Eq. 9 ($PV = nRT$), is a quantitative expression of the relations between the properties of an ideal gas that is reliable for real gases at low pressures.

Procedure

To calculate the pressure (or other property) of a given sample

Step 1 To use Eq. 9 directly, rearrange the equation $PV = nRT$ to give the desired quantity on the left and all other quantities on the right.

Step 2 Substitute the data, if necessary converting from mass of gas to the amount in moles. Note that the temperature must be in kelvins.

Step 3 Choose the value of R from Table 4.2 that matches the units of pressure and volume you need to use. Alternatively, convert the pressure units to match the value of R you prefer to use.

The procedure is illustrated in Example 4.2.

To calculate the response of a gas to a change in conditions

The initial and final conditions for a fixed amount of gas are related by Eq. 11. To use this relation:

Step 1 Construct a table summarizing the data, which may include changes in any of the four variables pressure, volume, amount, and temperature. Express the temperatures in kelvins and the amount in moles.

Step 2 Rearrange the relation so that the quantity required is on the left and all other quantities are on the right. Cancel any quantities that are unchanged.

Step 3 Substitute the data and check to see whether the answer agrees with your predictions.

The procedure is illustrated in Example 4.3.

EXAMPLE 4.2 Calculating the pressure of a given sample

Have you ever wondered what the pressure is inside a television picture tube? Estimate the pressure (in atmospheres), given that the volume of the tube is 5.0 L, its temperature is 23°C, and it contains 0.010 mg of nitrogen gas.

STRATEGY A television picture tube is a form of cathode ray tube in which the beam of electrons is directed toward colored phosphors on the screen. A very low pressure is required to minimize the collisions between the electrons in the beam and the gas molecules. Collisions and the resulting deflections of the electrons would give a blurred, dim picture. Therefore, expect a very low pressure. Proceed as in Toolbox 4.1, and rearrange Eq. 9 into

$$P = \frac{nRT}{V}$$

SOLUTION The molar mass of N_2 is 28.02 g·mol^{-1}, and the temperature, 23°C, corresponds to $T = (273.15 + 23)$ K = 296 K. We want the pressure in atmospheres, so we take from Table 4.2 the value of R that is expressed in those units. Therefore, the pressure is

$$\begin{aligned} P &= (1.0 \times 10^{-5}\ \text{g}) \times \left(\frac{1\ \text{mol}}{28.02\ \text{g}}\right) \\ &\quad \times \frac{(8.206 \times 10^{-2}\ \text{L·atm·K}^{-1}\text{·mol}^{-1}) \times (296\ \text{K})}{5.0\ \text{L}} \\ &= 1.7 \times 10^{-6}\ \text{atm} \end{aligned}$$

The pressure is very low, as anticipated.

Self-Test 4.7A Calculate the pressure (in kilopascals) exerted by 1.0 g of carbon dioxide in a flask of volume 1.0 L at 300°C.

[*Answer:* 1.1×10^2 kPa]

Self-Test 4.7B An idling, badly tuned automobile engine can release as much as 1.00 mol CO per minute into the atmosphere. At 27°C, what volume of CO, adjusted to 1.00 atm, is emitted per minute?

EXAMPLE 4.3 Using the combined gas law

In an experiment to investigate the properties of the coolant gas used in an air-conditioning system, a sample of volume 500 mL at 28.0°C was found to exert a pressure of 92.0 kPa. What pressure will the sample exert when it is compressed to 300 mL and cooled to −5.0°C?

STRATEGY Compression increases the pressure but cooling lowers the pressure, so it is not easy to predict the outcome without doing a calculation. Follow the procedure set out in Toolbox 4.1.

SOLUTION After conversion of the temperatures to kelvins, the data are summarized in the following table:

	Pressure, P	Volume, V	Temperature, T
1. Initial	92.0 kPa	500 mL	301.2 K
2. Final	?	300 mL	268.2 K

We now rearrange Eq. 11:

$$\begin{aligned} P_2 &= P_1 \times \frac{V_1}{V_2} \times \frac{T_2}{T_1} \\ &= (92.0\ \text{kPa}) \times \frac{500\ \text{mL}}{300\ \text{mL}} \times \frac{268.2\ \text{K}}{301.2\ \text{K}} = 137\ \text{kPa} \end{aligned}$$

The net outcome is an increase in pressure, so, in this instance, compression has a greater effect than cooling.

Self-Test 4.8A A parcel of air of volume 1.00×10^3 L at 20°C and 1.00 atm rises up the side of a mountain range. At the summit, where the pressure is 0.750 atm, the parcel of air has cooled to −10°C. What is the volume of the parcel at that point?

[*Answer:* 1.20×10^3 L]

Self-Test 4.8B A weather balloon is filled with helium gas at 20°C and 1.00 atm. The volume of the balloon is 250 L. When the balloon rises to a layer of air where the temperature is −30°C, it has expanded to 800 L. What is the pressure of the atmosphere at that point?

The ideal gas law can also be used to predict the molar volume of an ideal gas under any conditions of temperature and pressure. To do so, we combine Eq. 7 and Eq. 9 by writing

$$V_m = \frac{V}{n} = \frac{nRT/P}{n} = \frac{RT}{P} \tag{12}$$

At 1 bar and 25.00°C (298.15 K), the temperature that is commonly used for chemistry calculations, the molar volume of an ideal gas, V_m, is 24.79 L·mol^{-1}, which is about the volume of a cube 1 ft on a side (Fig. 4.15). The expression **standard temperature and pressure** (STP) means 0°C and 1 atm. At STP, the molar volume of an ideal gas is 22.41 L·mol^{-1}. Note the slightly smaller value: the temperature is lower and the pressure is slightly higher, so the same amount of gas molecules occupies a smaller volume.

Table 4.3 gives values of the molar volume under common conditions. To obtain the volume of a known amount of gas at a specified temperature and pressure, we simply multiply the molar volume *at that temperature and pressure* by the amount in moles. Alternatively, we can use the ideal gas law to calculate the volume.

Standard temperature and pressure (STP) are 0°C and 1 atm. The molar volume of an ideal gas under various conditions is given in Table 4.3.

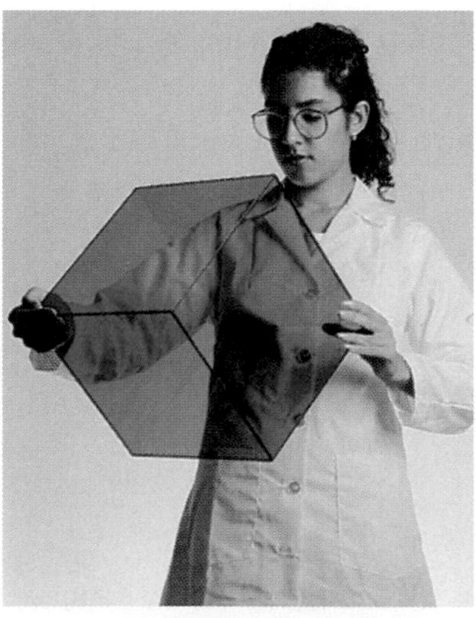

FIGURE 4.15 The blue cube is the volume (25 L) occupied by 1 mol of ideal gas molecules at 25°C and 1 bar.

TABLE 4.3 Molar Volume of an Ideal Gas

Temperature	Pressure	Molar volume, $L\cdot mol^{-1}$
0 K	0	0
0°C	1 atm	22.4141
0°C	1 bar	22.7111
25°C	1 atm	24.4655
25°C	1 bar	24.7897

Self-Test 4.9A Calculate the volume occupied by 1.0 kg of hydrogen at 25°C and 1.0 atm.

[*Answer:* 1.2×10^4 L]

Self-Test 4.9B Calculate the volume occupied by 2.0 g of helium at 25°C and 1.0 atm.

4.9 Gas Density

The molar concentration of a gas and a related quantity, the density of a gas, are both of great interest to meteorologists as well as to engineers. As we saw in Section G, the molar concentration is the amount of molecules (n, in moles) divided by the volume of the sample (V). It follows from the ideal gas law that, for a gas behaving ideally (so we can write $n = PV/RT$),

$$\text{Molar concentration} = \frac{n}{V} = \frac{P}{RT} \qquad (13)$$

This expression shows that, for a given pressure and temperature, the molar concentration is the same for every gas.

The density, d, of a gas, like that of any substance, is the mass of the sample divided by its volume, $d = m/V$. It is usually measured in grams per liter ($g\cdot L^{-1}$). Because the mass of the sample is equal to the amount in moles times the molar mass, $m = nM$, and $n = PV/RT$, it follows that

$$d = \frac{m}{V} = \frac{nM}{V} = \frac{MP}{RT} \qquad (14)$$

We see that, for a given pressure and temperature, the greater the molar mass of the gas, the greater its density. Equation 14 also shows that, at constant temperature, the density of a gas increases with pressure. When we compress a gas, its pressure and therefore its density increase because the same number of molecules are being confined into a smaller volume. Similarly, heating a gas that is free to expand at constant pressure increases the volume occupied by the gas and therefore *reduces* its density. The effect of temperature on density is the principle behind hot-air balloons:

the hot air inside the envelope has a lower density than the surrounding cool air. Equation 14 is also the basis for using density measurements to determine the molar mass of a gas or vapor.

How can we use this information . . . to calculate the molar mass of a gas, given its density?

The volatile organic compound geraniol, a component of oil of roses, is used in perfumery. The density of the vapor at 260°C and 103 Torr is 0.480 $g \cdot L^{-1}$. To find the molar mass of geraniol, we rearrange Eq. 14 and substitute the data. For this problem, $d = 0.480\ g \cdot L^{-1}$, $P = 103$ Torr, and $T = (273.15 + 260)\ K = 533\ K$:

$$M = \frac{dRT}{P} = \frac{(0.480\ g \cdot L^{-1}) \times (62.364\ L \cdot Torr \cdot K^{-1} \cdot mol^{-1}) \times (533\ K)}{103\ Torr}$$
$$= 155\ g \cdot mol^{-1}$$

Self-Test 4.10A The oil produced from eucalyptus leaves contains the volatile organic compound eucalyptol. At 190°C and 60.0 Torr, a sample of eucalyptol vapor had a density of 0.320 $g \cdot L^{-1}$. Calculate the molar mass of eucalyptol.

[*Answer:* 154 $g \cdot mol^{-1}$]

Self-Test 4.10B The Codex Ebers, an Egyptian medical papyrus, describes the use of garlic as an antiseptic. Chemists today have discovered that the oxide of diallyl disulfide (the volatile compound responsible for garlic odor) is a powerful antibacterial agent. At 177°C and 200 Torr, a sample of diallyl disulfide vapor has a density of 1.04 $g \cdot L^{-1}$. What is the molar mass of diallyl disulfide?

BOX 4.1 WHAT HAS THIS TO DO WITH . . . THE ENVIRONMENT?

Gas Laws and the Weather

One of the most graceful flights is the dance of the white pelican as flocks of the birds circle through the air, spiraling higher and higher without apparent effort. Like glider pilots, the birds are taking advantage of a *thermal*, a parcel (three-dimensional region) of warm air that is itself rising. The atmosphere is so large that individual parcels of air can maintain uniform conditions within themselves. These immense air masses are affected by their surroundings, but slowly. As a result, the atmosphere is not uniform in composition, temperature, and density; but these differences circulate air, create wind, and disperse pollutants, thereby making it difficult to contain pollution.

The atmosphere is divided vertically into several regions, as shown in the diagram. The lowest region is called the *troposphere*. In this region, the temperature decreases with increasing altitude. Between about 11 and 16 km lies the *tropopause*, a region where the temperature remains constant at about −55°C. Above about 16 km, the *stratosphere* begins. Here the temperature rises, until it reaches about 0°C in the *stratopause*, a region lying about 45 km above sea level. Above that altitude, in the *mesosphere*, the temperature drops again, until the *mesopause* is reached. After the mesopause, in the thermosphere, the temperature rises; and at very high altitudes, it reaches over 1000°C.

Notice how the temperature of the atmosphere changes with altitude. Close to the surface of the Earth, the temperature is about 20°C. The temperature falls to about −55°C at 15 km and then rises again at higher altitudes. One reason that the stratosphere is warmer than lower regions of the atmosphere is that solar radiation causes different chemical reactions at different altitudes. For example, these reactions result in a relatively stable concentration of ozone we call the ozone layer in the stratosphere. The reactions that produce ozone also release energy; and, as a result, the temperature rises with altitude.

TABLE 4.4 Composition of Dry Air at Sea Level

Constituent	Molar mass,* $g \cdot mol^{-1}$	Composition, % Volume	Composition, % Mass
nitrogen, N_2	28.02	78.09	75.52
oxygen, O_2	32.00	20.95	23.14
argon, Ar	39.95	0.93	1.29
carbon dioxide, CO_2	44.01	0.03	0.05

*The average molar mass of molecules in the air, allowing for their different abundances, is 28.97 $g \cdot mol^{-1}$. The percentage of water vapor in ordinary air varies with the humidity.

The density of the atmosphere varies greatly from place to place, as does its composition and temperature. The average composition of dry air (air from which water vapor has been removed) is shown in Table 4.4. One reason for the nonuniformity of air is the effect of solar radiation, which causes different chemical reactions at different altitudes (Box 4.1). The density of air also varies with altitude, partly because the temperature varies, but also as a result of the compressibility of air. The weight of air above compresses the air at sea level, causing it to be denser than the air at higher altitudes. The air outside an airplane cruising at 10 km is only 25% as dense as air at sea level.

The molar concentrations and densities of gases increase with decreasing volume.

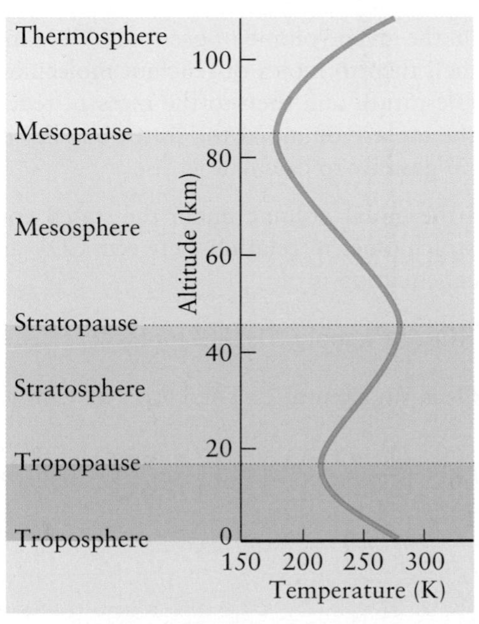

Because the stratosphere is warmer than lower regions, convection mixing between the stratosphere and troposphere is slow. There is little driving force for mixing because the higher layers are lower in density and the higher temperature means that local regions of higher density that might promote mixing are unlikely to develop. An important implication of this small degree of mixing is that it takes a long time for chemicals to diffuse into the stratosphere. However, once they are there, they remain for a very long time, and pollutants that threaten the ozone layer can act over a period of many years.

You can learn more about the ozone layer in Box 13.3.

Plot of temperature as a function of altitude.

4.10 The Stoichiometry of Reacting Gases

Chemists sometimes need to calculate the volume of gas consumed or produced in a reaction. For example, we might need to know the volume of carbon dioxide produced when a fuel burns or the volume of oxygen needed to react with a given mass of hemoglobin in the red cells of our blood. To answer this kind of question, we have to combine the mole-to-mole calculations of the type described in Section L with the conversion of moles of gas molecules to the volume they occupy (**1**).

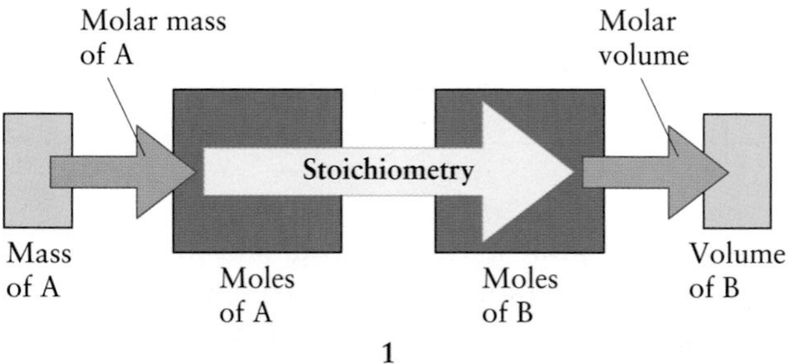

1

EXAMPLE 4.4 Calculating the mass of reagent needed to react with a specified volume of gas

The carbon dioxide generated by the personnel in the artificial atmosphere of submarines and spacecraft must be removed from the air and the oxygen recovered. Submarine design teams have investigated the use of potassium superoxide, KO_2, as an air purifier because this compound reacts with carbon dioxide and releases oxygen (Fig. 4.16):

$$4\,KO_2(s) + 2\,CO_2(g) \longrightarrow 2\,K_2CO_3(s) + 3\,O_2(g)$$

Calculate the mass of KO_2 needed to react with 50 L of carbon dioxide at 25°C and 1.0 atm.

FIGURE 4.16 When carbon dioxide is passed over potassium superoxide (the yellow solid), it reacts to form colorless potassium carbonate and oxygen gas. The reaction is used to remove carbon dioxide from the air in a closed-system breathing environment.

STRATEGY We convert from the given volume of gas to moles of molecules (by using the molar volume), then to moles of reactant molecules or formula units (by using a mole ratio), and then to the mass of reactant (by using its molar mass). If the molar volume at the stated conditions is not available, then use the ideal gas law to calculate it.

SOLUTION Table 4.3 gives the molar volume under the stated conditions as 24.47 $L{\cdot}mol^{-1}$. The stoichiometric relation between CO_2 and KO_2 from the balanced chemical equation is

$$2\text{ mol } CO_2 \simeq 4\text{ mol } KO_2, \text{ or } 1\text{ mol } CO_2 \simeq 2\text{ mol } KO_2$$

and the molar mass of KO_2 is 71.1 $g{\cdot}mol^{-1}$. Therefore, we can write

$$\text{Mass of } KO_2 = (50\text{ L}) \times \left(\frac{1\text{ mol } CO_2}{24.79\text{ L}}\right) \times \left(\frac{2\text{ mol } KO_2}{1\text{ mol } CO_2}\right) \times \left(\frac{71.1\text{ g}}{1\text{ mol } KO_2}\right) = 2.9 \times 10^2\text{ g}$$

Self-Test 4.11A Calculate the volume of carbon dioxide, adjusted to 25°C and 1.0 atm, that plants need to make 1.00 g of glucose, $C_6H_{12}O_6$, by photosynthesis in the reaction $6\ CO_2(g) + 6\ H_2O(l) \longrightarrow C_6H_{12}O_6(s) + 6\ O_2(g)$.

[*Answer:* 0.82 L]

Self-Test 4.11B The reaction of H_2 and O_2 gases to produce liquid H_2O is used in fuel cells on the space shuttles to provide electricity. What mass of water is produced in the reaction of 100.0 L of oxygen stored at 25°C and 1.00 atm?

There may be over a thousandfold increase in volume when liquids or solids react to form a gas. The molar volumes of gases are close to 25 $L{\cdot}mol^{-1}$ under normal conditions, whereas liquids and solids occupy only about a few tens of milliliters per mole. The molar volume of liquid water, for instance, is only 18 $mL{\cdot}mol^{-1}$. In other words, 1 mol of gas molecules at 25°C and 1 atm occupies as much as 1000 times the volume of 1 mol of molecules in a typical liquid or solid.

The increase in volume as gaseous products are formed in a chemical reaction is even larger if several gas molecules are produced from each reactant molecule, such as the formation of CO and CO_2 from a solid fuel (Fig. 4.17). The explosive action of ammonium nitrate, which was used in the Oklahoma City federal building bomb in 1995, illustrates the point. Ammonium nitrate decomposes in various ways, including

$$NH_4NO_3(s) \longrightarrow N_2O(g) + 2\ H_2O(g)$$

In this reaction, 1 mol $NH_4NO_3(s)$, corresponding to a little over 46 cm^3 of the solid, produces 3 mol of gas molecules, giving a total of about 75 L of gas under normal conditions. The pressure wave from the sudden 1300-fold expansion is the destructive shock of the explosion. The reaction needs to be initiated by a short, sharp shock from a detonator, an easily initiated explosive. One detonator, lead azide, $Pb(N_3)_2$, releases a large volume of nitrogen gas when it is struck:

$$Pb(N_3)_2(s) \longrightarrow Pb(s) + 3\ N_2(g)$$

FIGURE 4.17 An explosion caused by the ignition of coal dust. A shock wave is created by the tremendous expansion of volume as large numbers of gas molecules form.

An explosion of the same kind, but using sodium azide, NaN_3, is used in air bags in automobiles (Fig.4.18). The explosive release of nitrogen is detonated electrically when the vehicle decelerates abruptly in a collision.

The molar volume (at the specified temperature and pressure) is used to convert the amount of a reactant or product in a chemical reaction into a volume of gas.

4.11 Mixtures of Gases

Many of the gases we meet in chemistry—and everyday life—are mixtures. The atmosphere, for instance, is a mixture of nitrogen, oxygen, argon, carbon dioxide, and many other gases. The air we inhale is one mixture; the air we exhale is another mixture. Many gaseous anesthetics are carefully controlled mixtures. An important industrial chemical reaction is the synthesis of ammonia, in which a mixture of nitrogen and hydrogen is fed into a reaction vessel. We need to extend the model of a gas we have been developing to include the properties of mixtures of gases.

To extend our model we should note that, at low pressures at least, all gases respond in the same way to changes in pressure, volume, and temperature. Therefore, for calculations of the type we are doing in this chapter, it does not matter whether all the molecules in a sample are the same or not. *A mixture of gases that do not react with one another behaves like a single pure gas.* For instance, we can treat air as a single gas when we want to use the ideal gas law to predict its properties.

This is the same Dalton whose contribution to atomic theory we met in Section B.

John Dalton was the first to see how to calculate the pressure of a mixture of gases. His reasoning was something like this. Imagine that we introduce a certain amount of oxygen into a container and end up with a pressure of 0.60 atm. Then we evacuate the container to empty it of all gas. Now we introduce into the container enough nitrogen gas to give a pressure of 0.40 atm at the same temperature. Dalton wondered what the total pressure would be if these same amounts of the two gases were present in the container simultaneously. From some fairly crude measurements, he concluded that the total pressure resulting from the pressure of both gases in the same container would be 1.00 atm, the sum of the individual pressures.

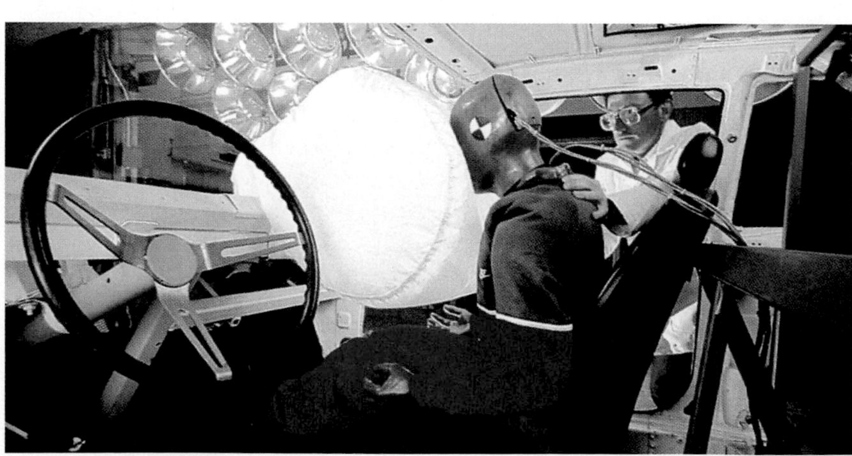

FIGURE 4.18 The rapid decomposition of sodium azide, NaN_3, results in the formation of a large volume of nitrogen gas. The reaction is triggered electrically in this air bag.

Dalton summarized his observations in terms of the **partial pressure** of each gas, the pressure the gas would exert if it occupied the container alone. In our example, the partial pressures of oxygen and nitrogen in the mixture are 0.60 atm and 0.40 atm, respectively, because those are the pressures the gases exert when each one is in the container alone. Dalton then described the behavior of gaseous mixtures by his **law of partial pressures:**

> The total pressure of a mixture of gases is the sum of the partial pressures of its components.

If we denote the partial pressures of the gases A, B, . . . as P_A, P_B, . . . , and the total pressure of the mixture as P, then we can write Dalton's law as follows:

$$P = P_A + P_B + \ldots \qquad (15)$$

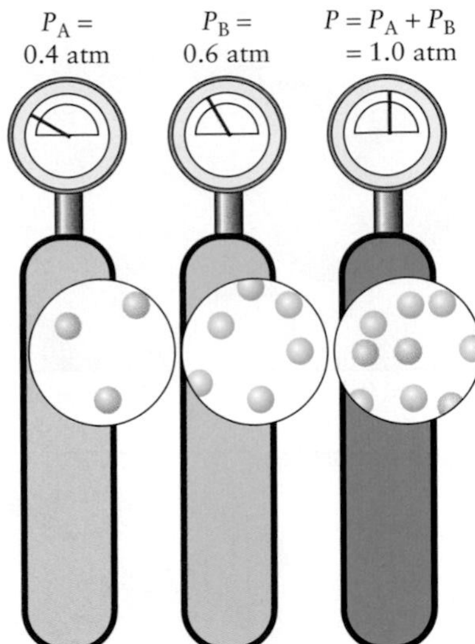

FIGURE 4.19 According to Dalton's law, the total pressure, P, of a mixture of gases is the sum of the partial pressures P_A and P_B of gases A and B. These partial pressures are the pressures the gases would exert if they were alone in the container (at the same temperature).

The law is illustrated in Fig. 4.19. It is exactly true only for gases that behave ideally, but it is a good approximation for nearly all gases under normal conditions.

Dalton's law is consistent with our picture of a gas and adds a little more information. The total pressure of a gas arises, as we have seen, from the battering of the molecules on the walls of the container. That battering is due to all the molecules in a mixture. The molecules of gas A exert a pressure, the molecules of gas B do too, and the total pressure is the sum of these individual pressures. The additional piece of information we can use to expand our model is that, for the total pressure to be the sum of the individual contributions, the A molecules must be blind to the presence of the B molecules. That is, there are no interactions—neither attractions nor repulsions—between the two kinds of molecules. Later we shall see that the absence of any interactions at all is a characteristic feature of an ideal gas.

How can we use this information . . . to find the partial pressure of a gas in a mixture?

The total pressure of the damp air in our lungs is

$$P = P_{\text{dry air}} + P_{\text{water vapor}}$$

In a closed container, to which a lung is a good first approximation, water vaporizes until its partial pressure has reached a certain value, called its *vapor pressure*. The vapor pressure of water at normal body temperature is 47 Torr. The partial pressure of the air itself in our lungs is therefore

Vapor pressure is treated in more detail in Section 8.1.

$$P_{\text{dry air}} = P - P_{\text{water vapor}} = P - 47\ \text{Torr}$$

On a typical day, the total pressure at sea level is 760 Torr, so the pressure in our lungs due to all the gas except the water vapor is 760 − 47 Torr = 713 Torr.

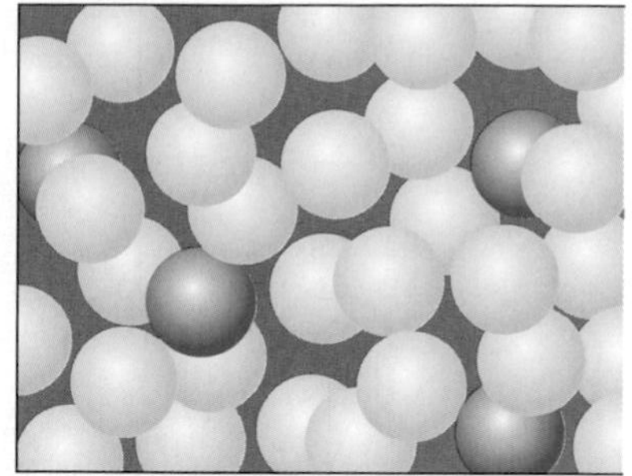
$x_{RED} = 0.1$

$x_{RED} = 0.5$

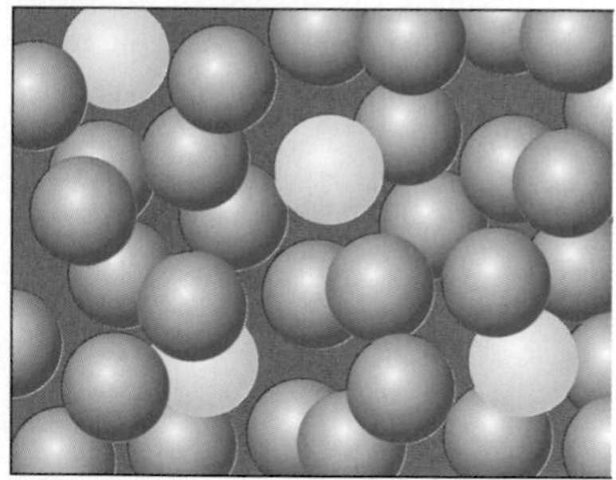
$x_{RED} = 0.9$

FIGURE 4.20 A mole fraction, x, tells us the fraction of molecules of a particular kind in a mixture of two or more kinds of molecules. In this illustration, the mole fraction of A molecules, which are colored red, is given below each mixture and the B molecules are colored yellow.

Self-Test 4.12A As a sample of oxygen was collected over water at 24°C and 745 Torr, it became saturated with water vapor. At this temperature, the vapor pressure of water is 24.38 Torr. What is the partial pressure of the oxygen?

[*Answer:* 721 Torr]

Self-Test 4.12B Students collecting hydrogen and oxygen gases by electrolysis of water failed to separate the two gases. If the total pressure of the dry mixture is 720 Torr, what is the partial pressure of each gas?

The easiest way to express the relation between the total pressure of a mixture and the partial pressures of its components is to introduce the **mole fraction,** x_J, of each component, the number of moles of molecules of the gas J expressed as a fraction of the total number of moles of molecules in the sample. If the amounts of gas molecules present are n_A, n_B, and so forth, the mole fraction of any one of the gases is

$$x_J = \frac{n_J}{n_A + n_B + \ldots} \quad (16)$$

A similar expression is used to calculate the mole fraction of each component of the mixture. In a binary (two-component) mixture of gases A and B,

$$x_A + x_B = 1 \quad (17)$$

When $x_A = 1$, the mixture is pure A; when $x_B = 1$, the mixture is pure B. When $x_A = x_B = \frac{1}{2}$, half the molecules are A and half are B (Fig. 4.20). Our task is to find the relation between the partial pressure of a gas and its mole fraction.

How do we do that?

To find the relation between the partial pressure of a gas in a mixture and its mole fraction, we first express the partial pressure, P_J, of a gas J in terms of the amount of J molecules present, n_J, the volume, V, occupied by the mixture, and the temperature, T:

$$P_J = \frac{n_J RT}{V}$$

Next, we express the total pressure in terms of n, the total number of moles of molecules present:

$$P = \frac{nRT}{V} = (n_A + n_B + \ldots)\frac{RT}{V}$$

We can rearrange this relation to

$$\frac{RT}{V} = \frac{P}{n_A + n_B + \ldots}$$

and then substitute this expression for RT/V in the first equation of this derivation ($P_J = n_J RT/V$) to obtain

$$P_J = \frac{n_J P}{n_A + n_B + \ldots} = x_J P$$

The relation we have derived is

$$P_J = x_J P \qquad (18)$$

where P is the total pressure. This expression is useful because it allows us to calculate the mole fraction from masses or amounts of molecules in the sample and the total pressure. In more advanced applications, Eq. 18 is regarded as the *definition* of the partial pressure of a gas, either ideal or real.

The partial pressure of a gas is the pressure it would exert if it were alone in the container; the total pressure of a mixture of gases is the sum of the partial pressures of the components; the partial pressure of a gas is related to the total pressure by the mole fraction: $P_J = x_J P$.

EXAMPLE 4.5 Calculating partial pressures

Air is a source of reactants for many chemical and physical processes: for example, oxygen is essential for combustion and respiration, and nitrogen is a starting material for the production of ammonia. To determine how much air is needed for these applications, we need to know the composition of air and, in some applications, the partial pressures of the components. A certain sample of dry air of total mass 1.00 g consists almost entirely of 0.76 g of nitrogen and 0.24 g of oxygen. Calculate the partial pressures of these gases when the total pressure is 1.00 atm.

STRATEGY To use Eq. 18, we need the total pressure (given) and the mole fraction of each component. The first step is to calculate the amount (in moles) of each gas present and the total amount (the sum of the individual amounts, the total number of moles). Then we calculate the mole fractions from Eq. 16. To obtain the partial pressures of the gases, we multiply the total pressure by the mole fractions of the gases in the mixture (Eq. 18).

SOLUTION The molar masses of N_2 and O_2 are 28.02 and 32.00 g·mol^{-1}, respectively. The numbers of moles of the gas molecules in 1.00 g of dry air are

$$n_{N_2} = (0.76\ \text{g}\ N_2) \times \left(\frac{1\ \text{mol}\ N_2}{28.02\ \text{g}\ N_2}\right) = \frac{0.76}{28.02}\ \text{mol}\ N_2$$

$$n_{O_2} = (0.24\ \text{g}\ O_2) \times \left(\frac{1\ \text{mol}\ O_2}{32.00\ \text{g}\ O_2}\right) = \frac{0.24}{32.00}\ \text{mol}\ O_2$$

The mole fractions are therefore

$$x_{N_2} = \frac{0.76 / 28.02}{0.76 / 28.02 + 0.24 / 32.00} = 0.78$$
$$x_{O_2} = \frac{0.24 / 32.00}{0.76 / 28.02 + 0.24 / 32.00} = 0.22$$

To find the partial pressures, we multiply each mole fraction by the total pressure, 1.00 atm, and obtain

$$P_{N_2} = 0.78 \times (1.00 \text{ atm}) = 0.78 \text{ atm} \qquad P_{O_2} = 0.22 \text{ atm}$$

To check this answer, we calculate the total pressure, the sum $P_{N_2} + P_{O_2}$. It is 1.00 atm, as in the data.

Self-Test 4.13A A baby with a severe bronchial infection is in respiratory distress. The anesthetist administers heliox, a mixture of oxygen and helium with 92.3% by mass O_2. What is the partial pressure of oxygen being administered to the baby if atmospheric pressure is 730 Torr?

[*Answer:* 4.4×10^2 Torr]

Self-Test 4.13B Divers exploring a shipwreck and wishing to avoid the narcosis associated with breathing nitrogen under high pressure switch to a neon-oxygen gas mixture containing 141.2 g of oxygen and 335.0 g of neon. The pressure in the gas tanks is 50.0 atm. What is the partial pressure of oxygen in the tanks?

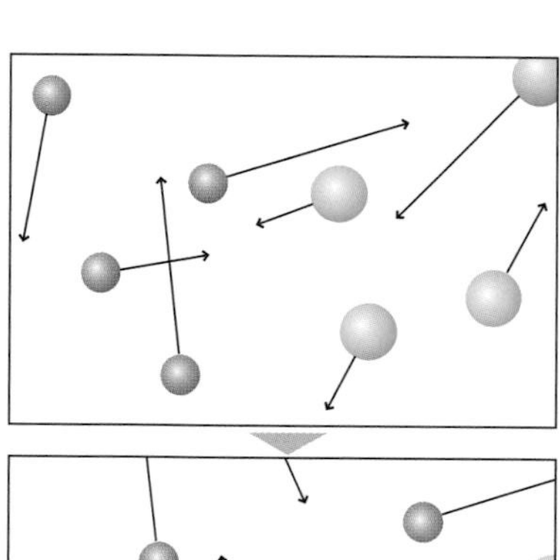

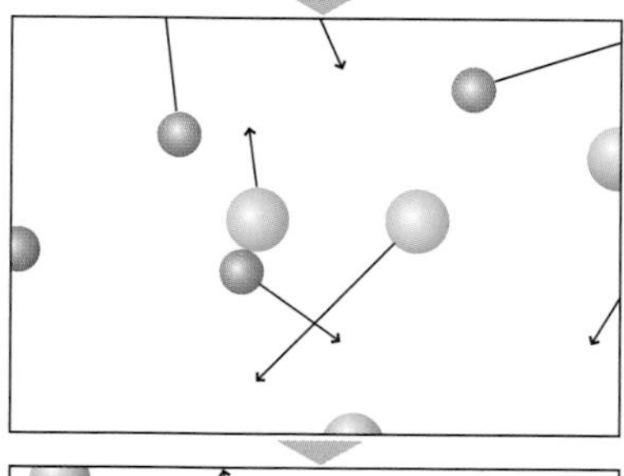

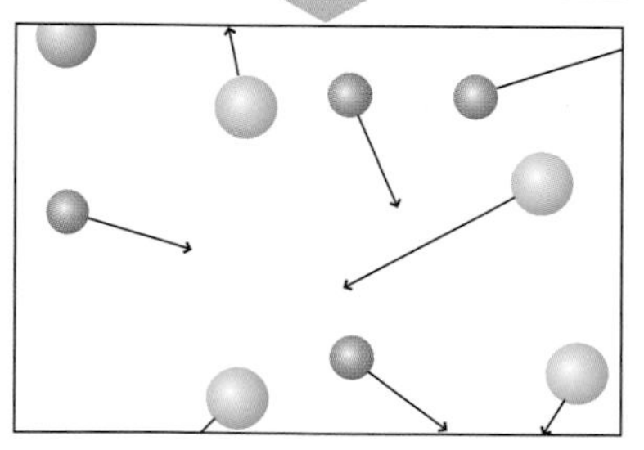

FIGURE 4.21 In diffusion, the molecules of one substance spread into the region occupied by molecules of another in a series of random steps, undergoing collisions as they move.

MOLECULAR MOTION

The empirical results summarized by the gas laws have suggested a model in which an ideal gas consists of widely spaced, noninteracting molecules undergoing ceaseless motion, with average speeds that increase with temperature. Can we refine this model still further? Can we also infer, for example, how molecular speeds depend on temperature?

In this section, we refine our model in two steps. First, we see how experimental measurements of the rate at which gases spread from one region to another give valuable insight into the *average* speeds of molecules. Then we use these average speeds to express our model of an ideal gas quantitatively, and check that it is in agreement with the gas laws. We shall see, too, how this quantitative treatment can be extended to derive detailed information about the proportion of molecules having any specified speed.

4.12 Diffusion and Effusion

Two kinds of processes, diffusion and effusion, provide data that show how the average speeds of gas molecules are related to molar mass and temperature. **Diffusion** is the gradual dispersal of one substance through another substance, such as krypton dispersing through a neon atmosphere (Fig. 4.21). Diffusion explains the spread of perfumes and pheromones, which are chemical signals exchanged by animals, through air. It also helps to keep the composition of the atmosphere approximately constant, because abnormally high concentrations of one gas diffuse away and disperse. One reason why chlorofluorocarbons (CFCs) are a danger to the ozone layer is that molecules that diffuse into the stratosphere stay there for long periods because the stratosphere is warmer than the layers below it and there is little mixing between them (see Box 4.1). **Effusion** is the escape of a gas through a small hole into a vacuum (Fig. 4.22). Effusion occurs whenever a gas is separated from a vacuum by a porous barrier—a barrier that contains microscopic holes—or a single pinhole. A gas escapes through a pinhole because there are more "collisions" with the hole on the high-pressure side than on the low-pressure side, so more molecules pass from the high-pressure region into the low-pressure region than pass in the opposite direction.

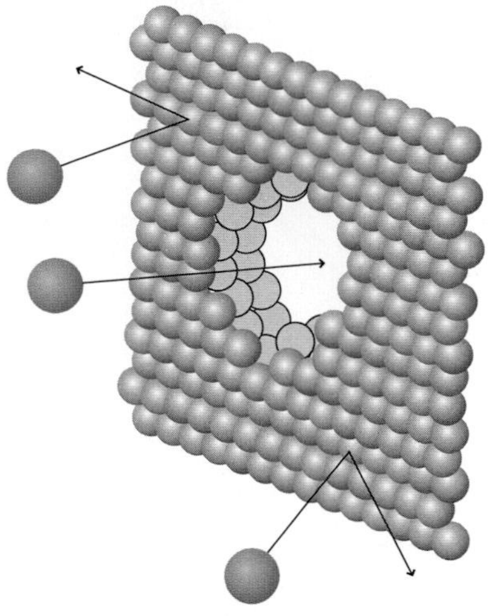

FIGURE 4.22 In effusion, the molecules of one substance escape through a small hole in a barrier into a vacuum or a region of low pressure.

The nineteenth-century Scottish chemist Thomas Graham carried out a series of experiments on the rates of effusion of gases. He found that *at constant temperature, the rate of effusion of a gas is inversely proportional to the square root of its molar mass*:

$$\text{Rate of effusion} \propto \frac{1}{\sqrt{\text{molar mass}}}$$

This observation is now known as **Graham's law of effusion.** If follows that, for two gases A and B, because the constants of proportionality cancel,

$$\frac{\text{Rate of effusion of A}}{\text{Rate of effusion of B}} = \sqrt{\frac{\text{molar mass of B}}{\text{molar mass of A}}} \quad (19)$$

The diffusion of a gas has also been found to follow an approximate inverse square-root dependence of the same kind, with heavy molecules diffusing more slowly than light molecules. However, because effusion involves the motion of only one gas, it is simpler to study, and we shall concentrate on it.

What does Graham's law add to our model? It is physically plausible that the rate of effusion is proportional to the average speed of the molecules in the gas, just as the rate at which cars pass beneath a bridge is proportional to their average speed. Therefore, Graham's law strongly suggests that *the average speed of molecules in a gas is inversely proportional to the square root of their molar mass*. For two gases A and B,

$$\frac{\text{Average speed of A molecules}}{\text{Average speed of B molecules}} = \sqrt{\frac{\text{molar mass of B}}{\text{molar mass of A}}} \quad (20)$$

This relation can be used to estimate the molar mass of a substance, but mass spectrometry has replaced it.

Self-Test 4.14A It takes 30 mL of argon 40 s to effuse through a porous barrier. The same volume of vapor of a volatile compound extracted from Caribbean sponges takes 120 s to effuse through the same barrier under the same conditions. What is the molar mass of this compound?

[*Answer:* $3.6 \times 10^2 \text{ g}\cdot\text{mol}^{-1}$]

Self-Test 4.14B It takes a certain amount of helium atoms 10 s to effuse through a porous barrier. How long does it take the same amount of methane, CH_4, molecules under the same conditions?

When we carry out effusion experiments at different temperatures, we find that the rate of effusion increases as the temperature is raised. Specifically, for a given gas, *the rate of effusion increases as the square root of the temperature:*

$$\frac{\text{Rate of effusion at } T_2}{\text{Rate of effusion at } T_1} = \sqrt{\frac{T_2}{T_1}} \qquad (21)$$

Because the rate of effusion is proportional to the average speed of the molecules, we can infer that *the average speed of molecules in a gas is proportional to the square root of the temperature:*

$$\frac{\text{Average speed of molecules at } T_2}{\text{Average speed of molecules at } T_1} = \sqrt{\frac{T_2}{T_1}} \qquad (22)$$

This very important relation will help us refine our model of a gas. It also begins to reveal the significance of one of the most elusive concepts in science: the nature of temperature. We see that, when referring to a gas, *the temperature is an indication of the average speed of the molecules*, and the higher the temperature, the higher the average speed of the molecules.

We can combine the two relations we have found. Because the average speed of molecules in a gas is proportional to the square root of the temperature and inversely proportional to the square root of the molar mass, we can write

$$\text{Average speed of molecules in a gas} \propto \sqrt{\frac{T}{M}} \qquad (23)$$

That is, the higher the temperature and the lower the molar mass, the higher is the average speed of molecules in a gas.

> *The average speed of molecules in a gas is directly proportional to the square root of the temperature and inversely proportional to the square root of the molar mass: average speed* $\propto (T/M)^{1/2}$.

4.13 The Kinetic Model of Gases

We now have enough information to turn our qualitative ideas about a gas into a quantitative model that can be used to make numerical predictions. The **kinetic model** of a gas is based on four assumptions (Fig. 4.23):

In this context, "molecules" includes all types of particles, whether atoms, ions, or molecules.

1 A gas consists of a collection of molecules in continuous random motion.

2 Gas molecules are infinitesimally small points.

3 The molecules move in straight lines until they collide.

4 The molecules do not influence one another except during collisions.

The fourth assumption means that we are proposing, as part of the model, that there are no attractive or repulsive forces between ideal gas molecules except during collisions.

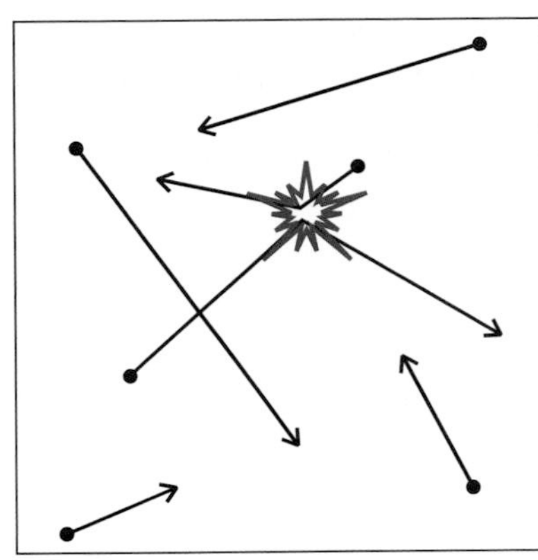

FIGURE 4.23 In the kinetic model of gases, the molecules are regarded as infinitesimal points that travel in straight lines until they undergo instantaneous collisions.

Let's express the model more informally. In the kinetic model of gases, we picture the molecules as widely separated for most of the time and in ceaseless random motion. They zoom from place to place, always in straight lines, changing direction only when they collide with a wall of the container or another molecule. The collisions change the speed and direction of the molecules, just like balls in a three-dimensional cosmic game of pool.

In Section 4.5, we used a molecular model of a gas to explain why the pressure of a gas rises as the temperature is increased: as a gas is heated, its molecules move faster and strike the walls of their container more often. The kinetic model of a gas allows us to derive the *quantitative* relation between pressure and the speeds of the molecules.

How do we do that?

The following calculation of the pressure of a gas based on the kinetic model may seem long, but it breaks down into many small steps. We begin with the gas sample illustrated in Fig. 4.24 and initially suppose that all the molecules are traveling at the same speed; we remove that condition later. The molecules strike the wall on the right. If we know how often these impacts occur and what force they exert on the wall, we can calculate the pressure that results. To calculate the force exerted, we use Newton's second law of motion: force is equal to the rate of change of momentum of a particle (Section A). **Momentum** is the product of mass and velocity, so if a molecule of mass m is traveling with a velocity v_x parallel to the side of the box we are calling x, then its linear momentum before it strikes the wall on the right is mv_x. Immediately after the collision, the momentum of the molecule is $-mv_x$ because the velocity has changed from v_x to $-v_x$. The change in momentum of the molecule is therefore $2mv_x$ on each collision.

Next, we calculate the total rate of change of momentum of all the molecules. For this step, we consider a time interval Δt and begin by calculating the total number of collisions in that interval. A molecule with velocity v_x can travel a distance $v_x\Delta t$ in an interval Δt. Therefore, all the molecules within a distance $v_x\Delta t$ of the wall and traveling toward it will strike the wall during the interval Δt. If the wall has area A, all the particles in a volume $Av_x\Delta t$ will reach the wall if they are traveling toward it (Fig. 4.25).

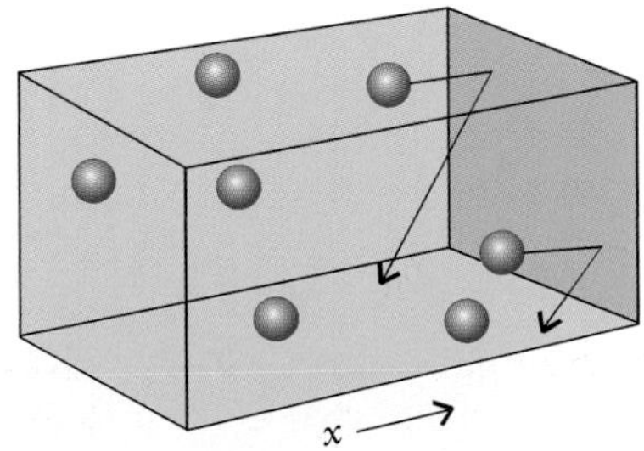

FIGURE 4.24 In the kinetic model of gases, the pressure arises from the force exerted on the walls of the container when the impacting molecules are deflected. We need to know the force of each impact and the number of impacts in a given interval.

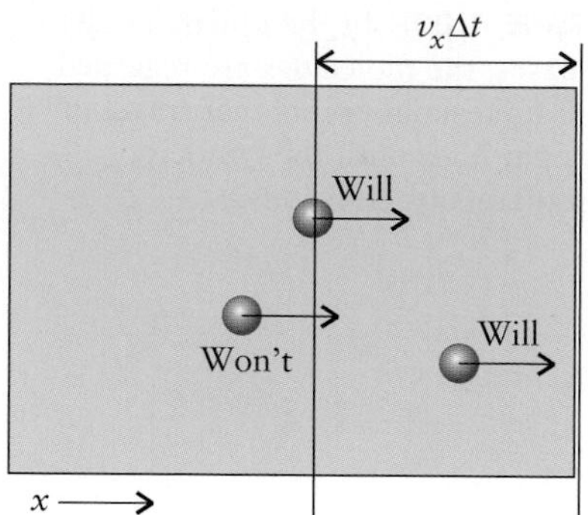

FIGURE 4.25 A molecule moving to the right with speed v_x in that direction will reach the wall in an interval Δt only if it lies within a distance $v_x\Delta t$ of the wall.

Suppose that the total number of particles in the container is N and that the volume of the container is V. Because a gas fills its container, we know that the molecules are spread evenly throughout the container. Therefore, the number of molecules in the volume $Av_x\Delta t$ is the fraction of the total volume represented by the volume $Av_x\Delta t$, multiplied by the total number of molecules in the volume V:

$$\text{Number of molecules} = \frac{Av_x\Delta t}{V} \times N = \frac{NAv_x\Delta t}{V}$$

Half the molecules in the box are moving to the right and half are moving to the left. Therefore, the average number of collisions with the wall during the interval Δt is half the number in the volume $Av_x\Delta t$, or

$$\text{Number of collisions} = \frac{NAv_x\Delta t}{2V}$$

We have calculated the momentum of a molecule and the number of collisions during the interval Δt. Now we bring the parts of the calculation together. The total momentum change in that interval is the change $2mv_x$ that an individual molecule undergoes multiplied by the total number of collisions:

$$\text{Total momentum change} = \frac{NAv_x\Delta t}{2V} \times 2mv_x = \frac{NmAv_x^2\Delta t}{V}$$

At this point, we calculate the rate of change of momentum by dividing this total momentum change by the interval Δt during which it occurs:

$$\text{Rate of change of momentum} = \frac{NmAv_x^2\Delta t}{V\Delta t} = \frac{NmAv_x^2}{V}$$

Now we use Newton's second law, that the rate of change of momentum is equal to the force, and hence find that the force exerted on the wall is

$$\text{Force} = \frac{NmAv_x^2}{V}$$

It follows that the pressure, the force divided by the area A on which the force acts, is

$$\text{Pressure} = \frac{NmAv_x^2}{VA} = \frac{Nmv_x^2}{V}$$

Not all the molecules in the sample are traveling at the same speed. To obtain the detected pressure, P, we need to use the *average* value of v_x^2 in place of v_x^2 for each individual molecule. Averages are commonly denoted by angular brackets, so we write

$$P = \frac{Nm\langle v_x^2\rangle}{V}$$

where $\langle v_x^2\rangle$ is the average value of v_x^2 for all the molecules in the sample.

At this stage, we relate $\langle v_x^2\rangle$ to the *root mean square speed*, $c = \langle v^2\rangle^{1/2}$, the square root of the average of the squares of the molecular speeds (this quantity is explained in more detail in the text following this derivation). The root mean square speed occurs frequently in expressions derived from kinetic molecular theory. First, we note that the speed of a single molecule, v, is related to the velocity parallel to the x, y, and z directions by

Its importance stems from the fact that c^2 is proportional to the average kinetic energy of the molecules, $\langle E_K\rangle = \frac{1}{2}m\langle v^2\rangle = \frac{1}{2}mc^2$.

$$v^2 = v_x^2 + v_y^2 + v_z^2$$

Therefore, mean square speed is

$$c^2 = \langle v^2\rangle = \langle v_x^2 + v_y^2 + v_z^2\rangle = \langle v_x^2\rangle + \langle v_y^2\rangle + \langle v_z^2\rangle$$

However, because the particles are moving randomly, the average of v_x^2 is the same as the average of v_y^2 and the average of v_z^2, the analogous quantities in the y and z directions. Because $\langle v_x^2\rangle$, $\langle v_y^2\rangle$, and $\langle v_z^2\rangle$ are all equal, we know that $\langle v^2\rangle = 3\langle v_x^2\rangle$; therefore, $\langle v_x^2\rangle = \frac{1}{3}c^2$. It follows that

$$P = \frac{Nmc^2}{3V}$$

The total number of molecules, N, is the product of the amount, n, and the Avogadro constant, N_A, so the last equation becomes

$$P = \frac{nN_A mc^2}{3V} = \frac{nMc^2}{3V}$$

where $M = mN_A$ is the molar mass of the molecules.

We have deduced that the pressure of a gas and the volume are related by

$$PV = \tfrac{1}{3}nMc^2 \qquad (24)$$

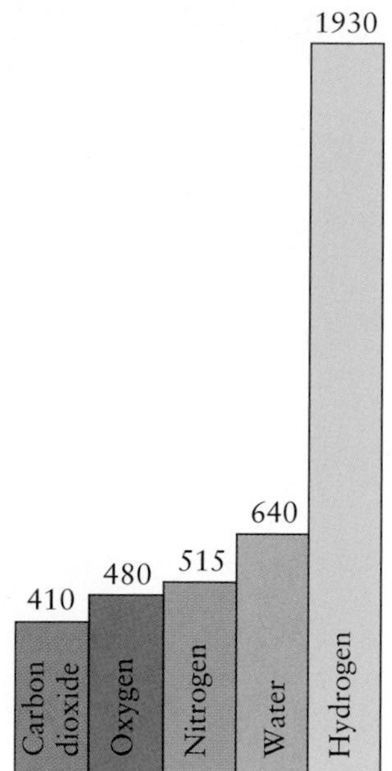

FIGURE 4.26 The average molecular speeds of five gases at 25°C, in meters per second. The gases are some of the components of air; hydrogen is included to show that the average speed of light molecules is much greater than that of heavy molecules.

where n is the amount (in moles) of gas molecules, M is their molar mass, and c is the root mean square speed of the molecules. As remarked in the derivation above, the **root mean square speed,** c, is the square root of the average value of the squares of the molecular speeds. That may seem a bit of a mouthful in words: it is easier—and more explicit—to express it in symbols. If there are N molecules in the sample, and the speeds of these molecules at some instant are $v_1, v_2, \ldots, v_N$, then the root mean square speed is

$$c = \left(\frac{v_1{}^2 + v_2{}^2 + \cdots + v_N{}^2}{N}\right)^{1/2} \tag{25}$$

Equation 24 already looks very much like the ideal gas law on the left (the PV term), so the kinetic model is consistent with observation. Because we know that $PV = nRT$ for an ideal gas, we can set the right-hand side of Eq. 24 equal to nRT and rearrange the resulting expression ($\frac{1}{3}nMc^2 = nRT$) into

$$c = \left(\frac{3RT}{M}\right)^{1/2} \tag{26}$$

This important result allows us to calculate the root mean square speed of the molecules of any gas at any temperature, simply by substituting the values of M and T (Fig. 4.26).

How can we use this information . . . to find the root mean square speed of gas molecules?

The molar mass of N_2 is 28.02 $g{\cdot}mol^{-1}$ (corresponding to 2.802×10^{-2} $kg{\cdot}mol^{-1}$), so the root mean square speed of nitrogen molecules in air at 20°C (293 K) is

$$c = \left(\frac{3 \times (8.3145\ \text{J}\cdot\text{K}^{-1}\cdot\text{mol}^{-1}) \times (293\ \text{K})}{2.802 \times 10^{-2}\ \text{kg}\cdot\text{mol}^{-1}}\right)^{1/2} = 511\ \text{m}\cdot\text{s}^{-1}$$

This result means that nitrogen molecules are zooming about your head at about 1140 miles per hour. Note the values chosen: first, we expressed the temperature in kelvins; second, we have used R in its fundamental SI form. Third, we have expressed the molar mass in SI base units to be consistent with the choice for R. Fourth, the conversion of the units has made use of the relation $1\ \text{J} = 1\ \text{kg}\cdot\text{m}^2\cdot\text{s}^{-2}$.

Self-Test 4.15A Estimate the root mean square speed of water molecules in the vapor above boiling water at 100°C.

[***Answer:*** 719 $m{\cdot}s^{-1}$]

Self-Test 4.15B Estimate the root mean square speed of CH_4 molecules at 25°C.

We can rewrite Eq. 26 to emphasize that, for a gas, the temperature is a measure of mean molecular speed. From $c^2 = 3RT/M$, it follows that

$$T = \frac{Mc^2}{3R} \tag{27}$$

That is, *the temperature is proportional to the mean square speed of the molecules.* We can also use Eq. 26 to obtain an expression for the mean kinetic energy of a gas molecule. Because the kinetic energy of a particle of mass m traveling at a speed v is equal to $\frac{1}{2}mv^2$ (Section A), the mean kinetic energy, E_K, of a molecule of mass m is

$$E_K = \tfrac{1}{2}mc^2 = \tfrac{1}{2}m \times \frac{3RT}{M} = \frac{3RT}{2N_A}$$

We have used $M = mN_A$ in the last step. The *molar* kinetic energy of an ideal gas is the mean energy of a molecule (the quantity we have just calculated) multiplied by the number of molecules per mole (the Avogadro constant). Therefore, for an ideal gas,

$$\text{Molar kinetic energy} = \frac{3RTN_A}{2N_A} = \frac{3}{2}RT \tag{28}$$

We see that the molar kinetic energy increases in proportion to temperature and that it is independent of the molecular masses.

The kinetic model of gases is consistent with the ideal gas law and provides an expression for the root mean square speed of the molecules: $c = (3RT/M)^{1/2}$. The molar kinetic energy of a gas is proportional to the temperature.

4.14 The Maxwell Distribution of Speeds

Useful as it is, Eq. 26 gives only the root mean square speed of gas molecules. Like cars in traffic, individual molecules have speeds that vary over a wide range. Moreover, like a car in a head-on collision, a molecule might be brought almost to a standstill when it collides with another. In the next instant (but now unlike a colliding car), it might be struck by another molecule and move off at the speed of sound. An individual molecule undergoes several billion changes of speed and direction each second.

The formula for calculating the fraction of gas molecules having a given speed, v, at any instant was first derived from the kinetic model by the Scottish scientist James Maxwell. He derived the expression

$$\Delta N = Nf(v)\Delta v \qquad \text{with } f(v) = 4\pi N\left(\frac{M}{2RT}\right)^{1/2} v^2 e^{-Mv^2/2RT} \tag{29}$$

where ΔN is the number of molecules with speeds in the narrow range between v and $v + \Delta v$, N is the total number of molecules in the sample, M is the molar mass, and R is the gas constant. This expression for $f(v)$ is the **Maxwell distribution of speeds** (Box 4.2).

Figure 4.27 shows a plot of the Maxwell distribution against speed for several different gases. We see that heavy molecules (such as CO_2) travel with speeds close to their average values. Light molecules (such as H_2) not only have a higher average speed but also a wider range of speeds. Some molecules of gases with low molar masses have such high speeds that they can escape from the gravitational pull of small planets and go off into space. As a consequence, hydrogen molecules and helium atoms, which are both very light, are rare in the Earth's atmosphere, although they are abundant on massive planets like Jupiter.

A plot of the Maxwell distribution for the same gas at several different temperatures shows that the average speed increases as the temperature is raised (Fig 4.28). We knew that already (Section 4.12); but what the curves also show is that the spread of speeds widens as the temperature increases. At low temperatures, most molecules of a gas have speeds close to the average speed. At high temperatures, a high proportion have speeds widely different from their average speed. Because the kinetic energy of a molecule in a gas is proportional to the square of its speed, the distribution of molecular kinetic energies follows the same trends.

The molecules of all gases have a wide range of speeds. As the temperature increases, the root mean square speed and the range of speeds both increase. The range of speeds is described by the Maxwell distribution, Eq. 29.

Media Link **Living graphs:** Fig. 4.27 and Fig. 4.28, The Maxwell distribution

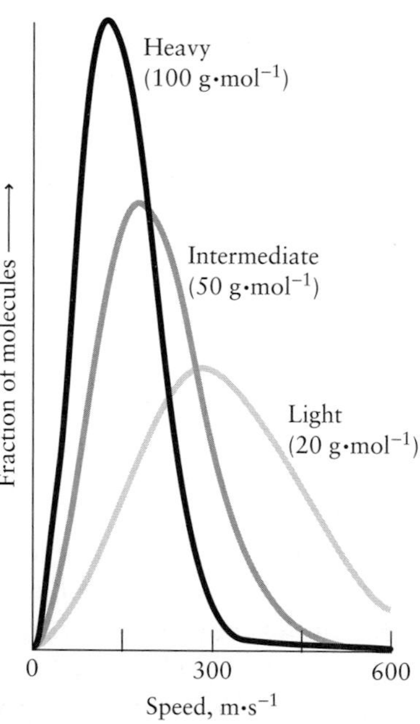

FIGURE 4.27 The range of molecular speeds for three gases, as given by the Maxwell distribution. All the curves correspond to the same temperature. The greater the molar mass, the lower the average speed and the narrower the spread of speeds.

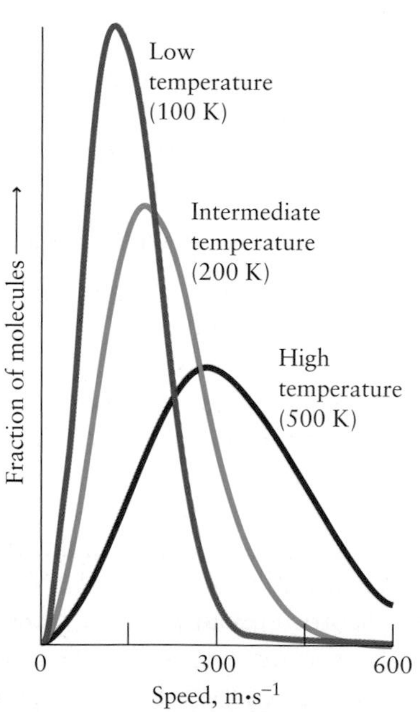

FIGURE 4.28 The Maxwell distribution again, but now the curves correspond to the speeds of a single substance at different temperatures. The higher the temperature, the higher the average speed and the broader the spread of speeds.

BOX 4.2 HOW DO WE KNOW . . . THE DISTRIBUTION OF MOLECULAR SPEEDS?

The distribution of molecular speeds in a gas can be determined experimentally. The gas is heated to the required temperature in an oven. The molecules then stream out of the oven through a small hole into an evacuated region. To ensure that the molecules form a beam, they may also pass through a series of slits, and the pressure must be kept very low so that collisions within the beam do not cause spreading.

The molecular beam passes through a series of rotating disks (see figure). Each disk contains a slit that is offset by a certain angle from its neighbor. A molecule that passes through the first slit will pass through the slit in the next disk only if the time it takes to pass between the disks is the same as the time required for the slit in the second disk to move into the orientation originally occupied by the first disk. The two times must continue to match if the same molecule is to pass through subsequent disks. Therefore, a particular rate of rotation of the disks allows the passage of only those molecules having the speed that carries them through the slits. To determine the distribution of molecular speeds, all we have to do is make a series of measurements at different rotational rates of the disks and measure the intensity of the beam of molecules arriving at the detector. A typical result is shown in the graph, and we see that it is a good match to the theoretical Maxwell expression (see Fig. 4.27).

Molecular beams are an important and current area of research in chemistry (see Box 13.2) because they enable chemists to explore the reactions that take place between individual molecules that are aimed at each other and collide at a speed that can be controlled.

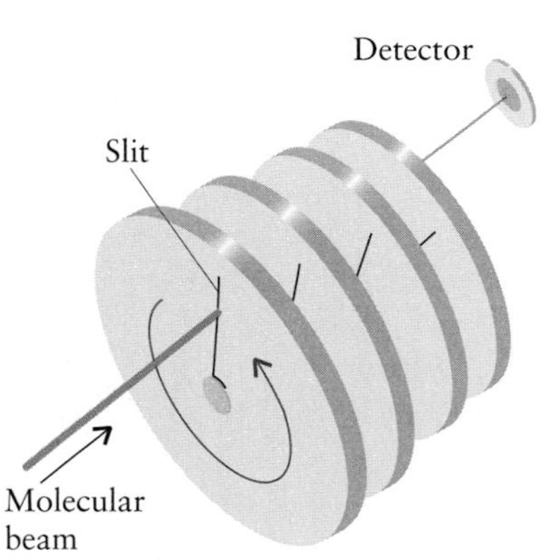

Schematic diagram of the rototating disks that serve as a velocity selector in a molecular beam apparatus.

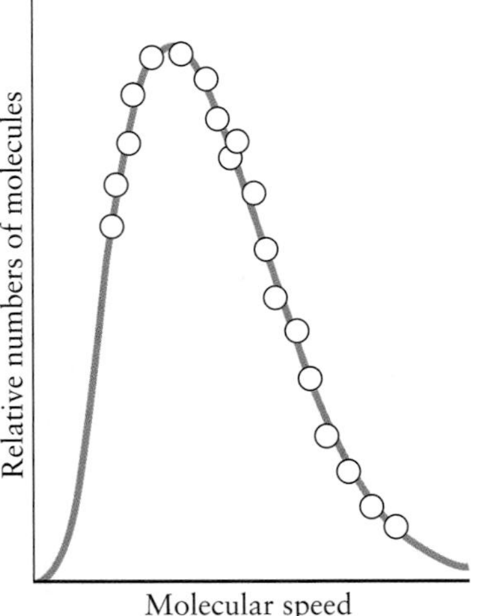

A typical result of a molecular speed distribution measurement.

THE IMPACT ON MATERIALS: REAL GASES

In industry and in many research laboratories, gases must be used under high pressures, when the ideal gas law is not followed closely. Recall that the ideal gas law is a limiting law, valid only as $P \rightarrow 0$. All actual gases, which are called **real gases,** have properties that differ from those predicted by the ideal gas law. These differences are significant at high pressures and low temperatures.

Experimental data on real gases show that they behave differently from an ideal gas. We first explore those differences, then relate the behavior of real gases to the properties of their molecules and so come to the final, refined version of our model of gases.

4.15 Deviations from Ideality

Two types of observations show that our model of a gas must be refined. The qualitative observation is that gases can be condensed to liquids when cooled or compressed. This property strongly suggests that, contrary to the assumptions of the kinetic model, gas molecules must attract one another for otherwise they would not cohere. In addition, liquids are very difficult to compress. This observation strongly suggests that there must be powerful repulsive forces opposing molecules being squashed together into a tiny volume. Once again, the presence of repulsive forces implies that the kinetic model must be refined.

The quantitative observation of deviations from ideal behavior shows up when we compare the behavior of real gases with that expected of an ideal gas: we find deviations from Boyle's and Charles's laws. One of the best ways of exhibiting these deviations is to measure the **compression factor,** Z, the ratio of the actual molar volume of the gas to the molar volume of an ideal gas under the same conditions:

$$Z = \frac{V_m}{V_m^{ideal}} \tag{30}$$

The compression factor of an ideal gas is 1, so deviations from $Z = 1$ are a sign of nonideality. Figure 4.29 shows the experimental variation of Z for a number of gases. We see that all gases deviate from $Z = 1$ as the pressure is raised. A few gases, such as hydrogen, have $Z > 1$ at all pressures. Our refined model of a gas needs to accommodate these variations.

Media Link **Living graph:** Fig. 4.29, Van der Waals isotherms

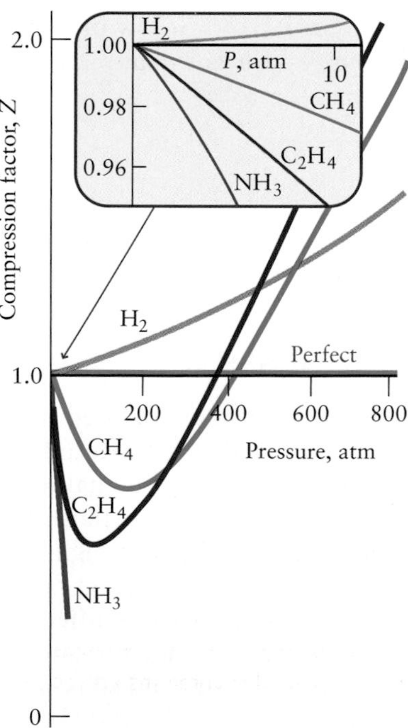

FIGURE 4.29 A plot of the compression factor, Z, as a function of pressure for a variety of gases. An ideal gas has $Z = 1$ for all pressures. For a few real gases with very weak intermolecular attractions, Z is always greater than 1. For most gases, at low pressures the attractive forces are dominant and $Z < 1$ (see inset). At high pressures, repulsive forces become dominant and $Z > 1$ for all gases.

All deviations from ideal behavior can be traced to the presence of **intermolecular forces,** or attractions and repulsions between molecules. Sections 5.1–5.5 describe the origin of intermolecular forces. Here we need only know that all molecules attract one another when they are a few molecular diameters apart but (provided they do not react) repel one another as soon as their electron clouds come into contact. Figure 4.30 shows how the potential energy of a molecule varies with its distance from a second molecule. At moderate separations, its potential energy is lower than when it is infinitely far away: attractions always lower the potential energy of an object. As the molecules come into contact, the potential energy starts to rise, because repulsions always increase the potential energy of an object.

The presence of attractive intermolecular forces accounts for the condensation of gases to liquids when they are compressed or cooled. Compression reduces the average separation of the molecules, and neighboring molecules can be captured by each other's attraction, providing they are traveling slowly enough (that is, have a low enough temperature). The low compressibility of liquids and solids is consistent, as we have seen, with the presence of strong repulsive forces when molecules are pushed together. Another way to describe the repulsive forces between molecules is to say that molecules have definite volumes. When you touch a solid object, you feel its size and shape because your fingers cannot penetrate into it. The resistance offered by the solid is due to the repulsive forces its atoms exert on the atoms in your fingers. When you dip your finger into a liquid, your molecules repel the molecules of the liquid and push them aside.

The presence of intermolecular forces also accounts for the variation of the compression factor. Thus, for gases that have $Z > 1$, the repulsions are more important than the attractions. Their molar volumes are greater than expected for an ideal gas because repulsions tend to drive the molecules apart. For gases such as ammonia, that have $Z < 1$ at low pressures (see inset in Fig. 4.29), the attractions are more important than the repulsions, and the molar volume is smaller than for an ideal gas because attractions tend to draw molecules together.

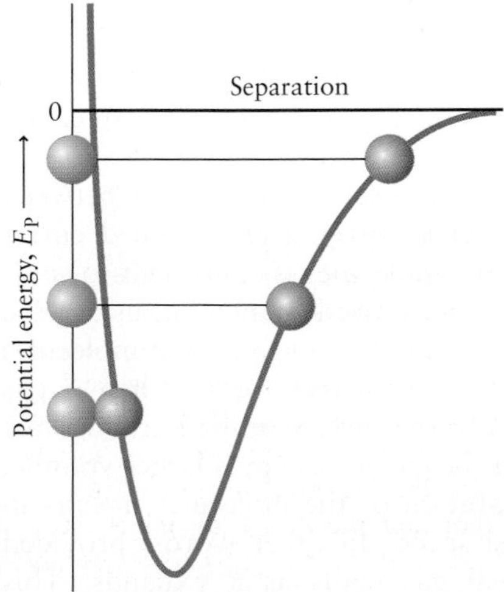

FIGURE 4.30 The variation of the potential energy of a molecule as it approaches another molecule. The potential energy rises sharply once the two molecules come into direct contact.

Media Link Living graph: Fig. 4.30, Lennard-Jones potential

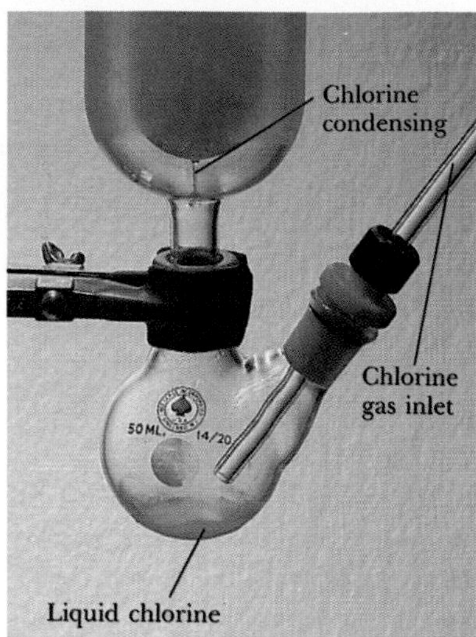

FIGURE 4.31 Chlorine can be condensed to a liquid at atmospheric pressure by cooling it to −35°C or lower. The upper flask contains a "cold-finger," a smaller tube filled with dry ice in acetone at −78°C.

Real gases consist of atoms or molecules with intermolecular attractions and repulsions. Attractions have a longer range than repulsions. The compression factor is a measure of the strength and type of intermolecular forces. When $Z > 1$, intermolecular repulsions are dominant; when $Z < 1$, attractions are dominant.

4.16 The Liquefaction of Gases

As described in Section 4.15, at low temperatures, molecules of a real gas move so slowly that intermolecular attractions may result in one molecule being captured by others and sticking to them instead of moving freely. When the temperature is reduced to below the boiling point of the substance, the gas condenses to a liquid. At and below this temperature, the molecules move too slowly to escape from one another, and the sample condenses to a jostling crowd of molecules held together by intermolecular attractions.

The simplest method of liquefying a gas is to immerse a sample in a bath kept at a temperature lower than the boiling point of the substance (Fig. 4.31). Temperatures as low as about 195 K (−78°C) can be reached by adding chips of solid carbon dioxide ("dry ice") to a low-freezing-point liquid, such as acetone.

Gases can also be liquefied by making use of the relation between temperature and molecular speed. Because lower average speed corresponds to lower temperature, slowing the molecules is equivalent to cooling the gas. The molecules of a gas can be slowed by making use of the attractions between them and allowing the gas to expand: the molecules have to climb away from one another's attractive forces when a gas expands, rather like a ball rising up above the surface of the Earth against the pull of gravity. Therefore, allowing the gas to occupy a larger volume, and hence increasing the average separation of the molecules, results in the molecules having a lower average speed. In other words, provided attractive effects are dominant, a real gas cools as it expands. This

Exceptions are helium and hydrogen, which have very weak attractive interactions and relatively strong repulsive interactions; at room temperature, they get warmer as they expand.

observation is called the **Joule-Thomson effect** in honor of the scientists who first studied it, James Joule and William Thomson (later to become Lord Kelvin), the inventor of the absolute temperature scale (Fig. 4.32).

The Joule-Thomson effect is used in some commercial refrigerators to liquefy gases. The gas to be liquefied is compressed and then allowed to expand through a small hole, called the throttle. The gas cools as it expands, and the cooled gas circulates past the incoming compressed gas (Fig. 4.33). This contact cools the incoming gas before it expands and cools still further. As the gas is continually recompressed and recirculated, its temperature progressively falls until finally it condenses to liquid. If the gas is a mixture, like air, then the liquid it forms can later be distilled to separate its components. This technique is used for harvesting nitrogen, oxygen, neon, argon, krypton, and xenon from the atmosphere.

Many gases can be liquefied by making use of the Joule-Thomson effect, cooling brought about by expansion.

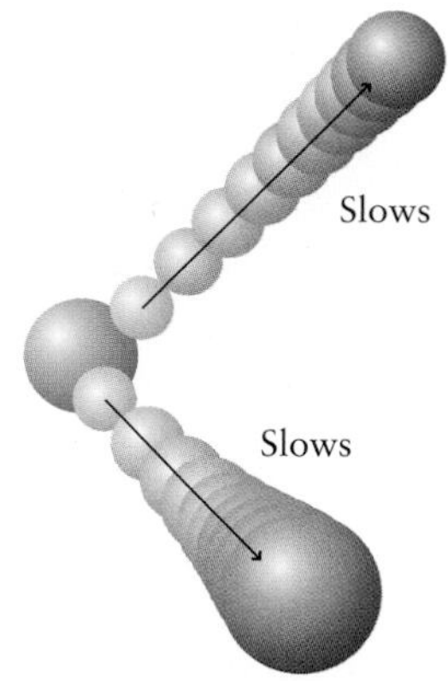

FIGURE 4.32 Cooling by the Joule-Thomson effect can be visualized as a slowing of the molecules as they climb away from each other against the force of attraction between them.

4.17 Equations of State of Real Gases

How do we describe the behavior of real gases that do not obey the ideal gas law? A common procedure in chemistry is to suppose that the term on the right-hand side of an equation (such as the nRT in $PV = nRT$ of the ideal gas law) is just the leading (and predominant) term of a more complicated expression. Thus, to extend the ideal gas law to real gases, one procedure is to write.

$$PV = nRT\left(1 + \frac{B}{V_m} + \frac{C}{V_m^2} + \ldots\right) \tag{31}$$

This expression is called the **virial equation.** The coefficients B, C, . . . , are called the **second virial coefficient, third virial coefficient,** and so on.

The name *virial* comes from the Latin word meaning "force."

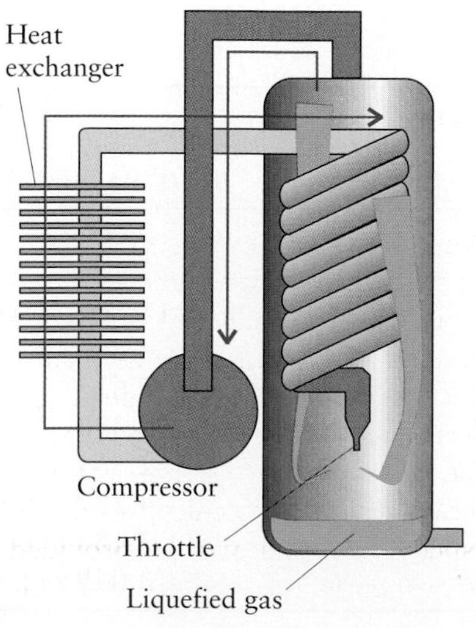

FIGURE 4.33 A Linde refrigerator for liquefying gases. The compressed gas gives up heat to the surroundings in the heat exchanger (left) and is cooled further as it passes through the coil (right) past which cooled gas circulates. The gas is cooled still further by the Joule-Thomson effect as it emerges through the throttle. This gas cools the incoming gas and is recirculated through the system. Eventually, the temperature of the incoming gas is so low that it condenses to a liquid.

The virial coefficients, which depend on the temperature, are found by fitting experimental data to the virial equation.

Although the virial equation can be used to make accurate predictions about the properties of a real gas, providing the virial coefficients are known for the temperature of interest, it does not give much insight without a lot of advanced analysis. An equation that is less accurate but easier to interpret was proposed by the Dutch scientist Johannes van der Waals. The **van der Waals equation** is

$$\left(P + a\frac{n^2}{V^2}\right)(V - nb) = nRT \tag{32}$$

The temperature-independent **van der Waals parameters** a and b are unique for each gas and are determined experimentally (Table 4.5). The parameter a represents the role of *attractions*, so it is relatively large for molecules that attract each other strongly. The parameter b represents the role of *repulsions*; it can be thought of as representing the volume of an individual molecule. The roles of the parameters become clearer when we rearrange the van der Waals equation into

$$P = \frac{nRT}{V - nb} - a\frac{n^2}{V^2} \tag{33}$$

and then write the compression factor (Eq. 30) as

$$\begin{aligned} Z &= \frac{V_{\text{m}}}{(RT/P)} = \frac{PV}{nRT} = \frac{V}{V - nb} - \frac{an}{RTV} \\ &= \frac{1}{1 - (nb/V)} - \frac{an}{RTV} \end{aligned} \tag{34}$$

For an ideal gas, a and b are both 0, and $Z = 1$. We see that $Z > 1$ when the attractive contribution (a) is small and the repulsive contribution (b) is appreciable. Conversely, $Z < 1$ when the repulsive contribution

TABLE 4.5 Van der Waals Parameters

Gas	a, $L^2 \cdot atm \cdot mol^{-2}$	b, $10^{-2}\ L \cdot mol^{-1}$
ammonia	4.225	3.707
argon	1.363	3.219
benzene	18.24	11.54
carbon dioxide	3.640	4.267
chlorine	6.579	5.622
ethane	5.562	6.380
hydrogen	0.2476	2.661
hydrogen sulfide	4.490	4.287
oxygen	1.378	3.183
water	5.536	3.049

is weak and the attractive interaction is strong. The values of a and b for a gas are found experimentally by fitting this expression for Z to curves like those in Fig. 4.29. Once the parameters have been determined, they can be used in the van der Waals equation to predict the pressure of the gas under the conditions of interest.

How can we use this information . . . to calculate the pressure of a gas?

Investigators studying the physical properties of a gas to be used as a refrigerant in an air-conditioning unit need to estimate its pressure when a 1.50-mol sample is confined to 5.00 L at 0°C. From tables of van der Waals parameters, they know that $a = 16.2\ L^2\cdot atm\cdot mol^{-2}$ and $b = 8.4 \times 10^{-2}\ L\cdot mol^{-1}$. To find the pressure of the gas, we substitute the data into Eq. 33 after converting the temperature to the Kelvin scale. We use R in units that match those required and given (atmospheres and liters).

$$\begin{aligned}
P &= \frac{nRT}{V - nb} - a\frac{n^2}{V^2} \\
&= \frac{(1.50\ \text{mol}) \times (0.082\ 06\ \text{L}\cdot\text{atm}\cdot\text{K}^{-1}\cdot\text{mol}^{-1}) \times (273\ \text{K})}{5.00\ \text{L} - (1.50\ \text{mol}) \times (8.4 \times 10^{-2}\ \text{L}\cdot\text{mol}^{-1})} \\
&\quad - (16.2\ \text{L}^2\cdot\text{atm}\cdot\text{mol}^{-2}) \times \frac{(1.50\ \text{mol})^2}{(5.00\ \text{L})^2} \\
&= \frac{1.50 \times (0.082\ 06\ \text{atm}) \times 273}{5.00 - 1.50 \times 8.4 \times 10^{-2}} - (16.2\ \text{atm}) \times \frac{(1.50)^2}{(5.00)^2} \\
&= 5.44\ \text{atm}
\end{aligned}$$

Between the second and third equalities, we have canceled the units to clarify the numerical calculation. An ideal gas under the same circumstances has a pressure of 6.72 atm.

Self-Test 4.16A A 10.0-L tank containing 25 mol O_2 is stored in a diving supply shop at 25°C. Use the data in Table 4.5 and the van der Waals equation to estimate the pressure in the tank.

[*Answer:* 58 atm]

Self-Test 4.16B The properties of carbon dioxide gas are well known in the bottled beverage industry. In an industrial process, a tank of volume 100 L at 20°C contains 20 mol CO_2. Use the data in Table 4.5 and the van der Waals equation to estimate the pressure in the tank.

We have arrived at our final model. By drawing on a wide range of experimental information, we now understand a gas as consisting of a large number of molecules in ceaseless, random motion. The average speed—and the spread of speeds—of the molecules increases with temperature and decreases with molar mass. Because the average separation of the molecules is so great, intermolecular forces have only a weak effect, and the molecules travel in straight lines until they collide with other molecules or with the walls of the container. However, the intermolecular forces are not completely negligible, and repulsions increase the molar volume (at a given temperature and pressure) whereas attractions decrease the molar volume. Which effect dominates in a real gas depends on the relative strengths of the two kinds of interactions.

The virial equation is a general equation for describing real gases. The van der Waals equation is an approximate equation of state for a real gas; the parameter a represents the role of attractive forces and b represents the role of repulsive forces.

SKILLS YOU SHOULD HAVE MASTERED

1 Calculate the pressure at the foot of a column of liquid, Self-Test 4.1.

2 Interpret a manometer reading, Self-Test 4.2.

3 Use the gas laws to calculate P, V, T, or n for given conditions or after a change in conditions, Toolbox 4.1 and Examples 4.1–4.3.

4 Determine molar mass from gas density and vice versa, Section 4.9.

5 Calculate the mass or volume of reactant needed to react with a specified volume of gas, Example 4.4.

6 Calculate the partial pressures of gases in a mixture and the total pressure of the mixture, Example 4.5.

7 Use Graham's law to account for relative rates of effusion, Self-Test 4.14.

8 Describe the effect of temperature on average speed, Section 4.12

9 List and explain the assumptions of the kinetic model of gases, Section 4.13.

10 Calculate the root mean square speed of molecules in a sample of gas, Self-Test 4.15.

11 Describe the effect of molar mass and temperature on the Maxwell distribution of molecular speeds, Section 4.14.

12 Explain how real gases differ from ideal gases, Sections 4.15–4.17.

13 Use the van der Waals equation to estimate the pressure of a gas, Self-Test 4.16.

EXERCISES

The Nature of Gases

4.1 The pressure needed to make synthetic diamonds from graphite is 8×10^4 atm. Express this pressure in (a) Pa; (b) kbar; (c) Torr; (d) lb·inch^{-2}.

4.2 An argon gas cylinder measures a pressure of 22.5 lb·inch^{-2}. Convert this pressure to (a) kPa; (b) Torr; (c) bar; (d) atm.

4.3 A student attaches a glass bulb containing neon gas to an open-tube manometer (refer to Fig. 4.5) and calculates the pressure of the gas to be 1.200 atm. (a) If the atmospheric pressure is 754 Torr, what height difference between the two sides of the mercury in the manometer did the student find? (b) Which side is higher, the side of the manometer attached to the bulb or the side open to the atmosphere? (c) If the student had mistakenly switched the numbers for the sides of the manometer when recording the data in the laboratory notebook, what would have been the actual pressure in the gas bulb?

4.4 A reaction is performed in a vessel attached to a closed-tube manometer. Before the reaction, the levels of mercury in the two sides of the manometer were at the same height. As the reaction proceeds, a gas is produced. At the end of the reaction, the height of the mercury column on the vacuum side of the manometer has risen 35.96 cm and the height on the side of the manometer connected to the flask has fallen by the same amount. What is the pressure in the apparatus at the end of the reaction expressed in (a) Torr; (b) atm; (c) Pa; (d) bar?

4.5 Suppose you were marooned on a tropical island and had to use seawater (density 1.10 $g \cdot cm^{-1}$) to make a primitive barometer. What height would the water reach in your barometer when a mercury barometer would reach 73.5 cm? The density of mercury is 13.6 $g \cdot cm^{-1}$.

4.6 An unknown liquid is used to fill a closed-tube manometer. The atmosphere is found to produce a height difference of 4.02 m in this manometer at the same time that a mercury manometer gives a displacement of 753.2 mm. What is the density of the unknown liquid?

4.7 Assume that the width of your body (across your shoulders) is 2.0×10 inches and the depth of your body (chest to back) is 1.0×10 inches. If atmospheric pressure is 14.7 $lb \cdot inch^{-2}$, what mass of air does your body support when you are in an upright position?

4.8 Low-pressure gauges in research laboratories are occasionally calibrated in inches of water (inH_2O). Given that the density of mercury at 15°C is 13.5 $g \cdot cm^{-3}$ and the density of water at that temperature is 1.0 $g \cdot cm^{-3}$, what is the pressure (in Torr) inside a gas cylinder that reads 1.5 inH_2O at 15°C?

The Gas Laws

4.9 (a) Using the Living Graphs on the Web site for this text, prepare and print a single plot of the pressure against temperature for 1.00 mol of gas, showing the curves at volumes of 0.01 L to 0.05 L in increments of 0.01 L for $T = 0$ to 400 K. This plot can be made by using the pull-down menu item *Set Variable* under the *Options* menu. Be sure to click on the *Add to existing graph* button in order to have all the plots appear on one graph. (b) What is the expression for the slope of each of these lines? (c) What is each intercept?

4.10 (a) Using the Living Graphs on the Web site for this text, prepare and print a single plot of the volume against temperature graph for 1.00 mol of gas, showing the curves at pressures of 11 000 atm to 15 000 atm in increments of 1000 atm for $T = 0$ to 400 K. This plot can be made by using the pull-down menu item *Set Variable* under the *Options* menu. Be sure to click on the *Add to existing graph* button in order to have all the plots appear on one graph. (b) What is the expression for the slope of each of these lines? (c) What is each intercept?

4.11 Determine the final pressure when (a) 7.36 mL of krypton at 4.9×10^5 kPa is transferred to a 1.0-L vessel; (b) 67.3 cm^3 of O_2 at 537 Torr is compressed to 9.5 cm^3. Assume constant temperature.

4.12 (a) 3.95 L of methane at a pressure of 1075 Torr is transferred to a 6.35-L vessel. What is the final pressure of methane if the change occurs at constant temperature? (b) A fluorinated organic gas in a cylinder is compressed from an initial volume of 854 mL at 219 Pa to 235 mL at the same temperature. What is the final pressure?

4.13 A 250-mL aerosol can at 25°C and 1.10 atm was thrown into an incinerator. When the temperature in the can reached 625°C, it exploded. What was the pressure in the can just before it exploded, assuming it reached the maximum pressure possible at that temperature?

4.14 A helium balloon has a volume of 22.5 L when the pressure is 0.951 atm and the temperature is 18°C. The balloon is cooled at a constant pressure until the temperature is −15°C. What is the volume of the balloon at this stage?

4.15 An outdoor storage vessel for hydrogen gas with a volume of 300 m^3 is at 1.5 atm and 10°C at 2:00 AM. By 2:00 PM, the temperature has risen to 30°C. What is the new pressure of the hydrogen in the vessel?

4.16 A chemist prepares a sample of hydrogen bromide and finds that it occupies 255 mL at 85°C and 600 Torr. What volume would it occupy at 0°C at the same pressure?

4.17 A chemist prepares 0.100 mol Ne gas at a certain pressure and temperature in an expandable container. Another 0.010 mol Ne is then added to the same container. How must the volume be changed to keep the pressure and temperature the same?

4.18 A chemist prepares a sample of helium gas at a certain pressure, temperature, and volume, and then removes half the gas molecules. How must the temperature be changed to keep the pressure and volume the same?

4.19 35.5 mL of xenon exerts a pressure of 0.255 atm at −45°C. (a) What volume does the sample occupy at 1.00 atm and 298 K? (b) What pressure would it exert if it were transferred to a 12.0-mL flask at 20°C? (c) Calculate the temperature needed for the xenon to exert a pressure of 5.00×10^2 Torr in the 12.0-mL flask.

4.20 A lungful of air (355 cm^3) is exhaled into a machine that measures lung capacity. If the air is exhaled from the lungs at a pressure of 1.08 atm at 37°C but the machine is at ambient conditions of 0.958 atm and 23°C, what is the volume of air measured by the machine?

4.21 (a) A 350.0-mL flask contains 0.1500 mol Ar at 24°C. What is the pressure of the gas in kilopascals? (b) 23.9 mg of bromine trifluoride exerts a pressure of 10.0 Torr at 100°C. What is the volume of the container in milliliters? (c) A 100.0-mL flask contains sulfur dioxide at 0.77 atm and 30°C. What mass of gas is present? (d) A 6.00×10^3-m^3 storage tank contains methane at 129 kPa and 14°C. How many moles of CH_4 are present? (e) A 1.0-μL ampoule of helium has a pressure of 2.00 kPa at −115°C. How many He atoms are present?

4.22 (a) A 125-mL flask contains argon at 1.30 atm and 77°C. What amount of Ar is present (in moles)? (b) A 120-mL flask contains 2.7 μg of O_2 at 17°C. What is the pressure (in Torr)? (c) A 20.0-L flask at 215 K and 20 Torr contains nitrogen. What mass of nitrogen is present (in grams)? (d) 16.7-g of krypton exerts a pressure of 1.00×10^2 mTorr at 44°C. What is the volume of the container (in liters)? (e) A 2.6-μL ampoule of xenon has a pressure of 2.00 Torr at 15°C. How many Xe atoms are present?

4.23 Calculate the number of molecules in (a) 74.57 nL of nitrogen; (b) a spherical flask of ammonia with a diameter of 15.5 cm; (c) an ampoule of krypton that is in the shape of a cylinder with diameter 2.50 mm and length 10.0 mm. (All measurements are at 1.00 atm and 298 K.)

4.24 Calculate the mass of gas in (a) 7.45 L of O_2; (b) 25.76 mL of SO_3; (c) 5.500×10^3 L of CH_4; (d) 4.76 mL of CO_2 (all measured at 1.00 atm and 298 K).

4.25 What is the molar volume of an ideal gas at atmospheric pressure and (a) 500°C; (b) at the boiling point of liquid nitrogen (−196°C)?

4.26 What is the resultant pressure if 1.00 mol of ideal gas at 273 K and 1.00 atm in a closed container of constant volume is heated to (a) 373 K; (b) 500 K?

4.27 Originally at −20°C and 759 Torr, a 1.00-L air sample is heated to 235°C. Next, the pressure is increased to 765 Torr. Then it is heated to 1250°C, and finally the pressure is decreased to 252 Torr. What is the final volume of the air?

4.28 A domestic water-carbonating kit uses steel cylinders of carbon dioxide, each with a volume of 250 mL. Each weighs 1.04 kg when full and 0.74 kg when empty. What is the pressure of gas (in bar) in a full cylinder at 20°C?

4.29 The effect of high pressure on organisms, including humans, is studied to gain information about deep-sea diving and anesthesia. A sample of air occupied 1.00 L at 25°C and 1.00 atm. What pressure (in atm) is needed to compress it to 239 cm^3 at this temperature?

4.30 To what temperature must a sample of helium gas be cooled from 131.0°C to reduce its volume from 2.97 L to 0.345 L at constant pressure?

4.31 Nitrogen monoxide, NO, has been found to act as a neurotransmitter. To prepare to study its effect, a sample was collected in a 250.0-mL container. At 19.5°C, its pressure is found to be 24.5 kPa. What amount (in moles) of NO has been collected?

4.32 Hot-air balloons gain their lift from the reduction in the density of air that occurs when the air in the envelope is heated. To what temperature should you heat a sample of air, initially at 340 K, to increase its volume by 14%?

4.33 At sea level, where the pressure was 104 kPa and the temperature 21.1°C, a certain mass of air occupied 2.0 m^3. To what volume will the air mass expand when it has risen to an altitude where the pressure and temperature are (a) 52 kPa, −5.0°C; (b) 880 Pa, −52.0°C?

4.34 A meteorological balloon had a radius of 1.0 m when released at sea level at 20°C. It expanded to a radius of 3.0 m when it had risen to its maximum altitude, where the temperature was −20°C. What is the pressure inside the balloon at that altitude?

4.35 How much *additional* pressure would you need to exert on a sample of helium at 765 Torr to compress it from 555 mL to 125 mL at constant temperature?

4.36 At the cruising altitude of an airplane (10 km), the atmospheric pressure is about 25 kPa. Disregarding the effect of temperature, calculate the volume of a sample of air at that altitude, which at the surface of the Earth would occupy 1.0 L.

Gas Density

4.37 A 2.00-mg sample of argon is confined to a 0.0500-L vial at 20°C; 2.00-mg of krypton is confined to a different 0.0500-L vial. What must the temperature of the krypton be if it is to have the same pressure as the argon?

4.38 What mass of ammonia will exert the same pressure as 12 mg of hydrogen sulfide, H_2S, in the same container under the same conditions?

4.39 Order the following gases according to increasing density: CO; CO_2; H_2S. The temperature and pressure are the same for all three samples.

4.40 Order the following gases according to increasing density: N_2; NH_3; NO_2. The temperature and pressure are the same for all three samples.

4.41 What is the density (in $g \cdot L^{-1}$) of chloroform, $CHCl_3$, vapor at (a) 2.00×10^2 Torr and 298 K; (b) 100°C and 1.00 atm?

4.42 What is the density (in $g \cdot L^{-1}$) of hydrogen sulfide (H_2S) at (a) 1.00 atm and 298 K; (b) 35.0°C and 0.962 atm?

4.43 A gaseous fluorinated methane compound has a density of 8.0 $g \cdot L^{-1}$ at 2.81 atm and 300 K. (a) What is the molar mass of the compound? (b) What is the formula of the compound if it is composed solely of C, H, and F? (c) What is the density of the gas at 1.00 atm and 298 K?

4.44 The density of a gaseous compound of phosphorus is 0.943 $g \cdot L^{-1}$ at 420 K when its pressure is 727 Torr. (a) What is the molar mass of the compound? (b) Assuming that the compound remains gaseous, what would be its density at 1.00 atm and 298 K?

4.45 A compound used in the manufacture of Saran is 24.7% C, 2.1% H, and 73.2% Cl by mass. The storage of 3.557 g of the gaseous compound in a 755-mL vessel at 0°C results in a pressure of 1.10 atm. What is the molecular formula of the compound?

4.46 The analysis of a hydrocarbon revealed that it was 85.7% C and 14.3% H by mass. When 1.77 g of the gas was stored in a 1.500-L flask at 17°C, it exerted a pressure of 508 Torr. What is the molecular formula of the hydrocarbon?

4.47 The density of a gaseous compound was found to be 1.23 $g \cdot L^{-1}$ at 330 K and 25.5 kPa. What is the molar mass of the compound?

4.48 A 115-mg sample of eugenol, the compound responsible for the odor of cloves, was placed in an evacuated flask with a volume of 500.0 mL at 280.0°C. The pressure eugenol exerted in the flask under those conditions was found to be 48.3 Torr. In a combustion experiment, 18.8 mg of eugenol burned to give 50.0 mg of carbon dioxide and 12.4 mg of water. What is the molecular formula of eugenol?

The Stoichiometry of Reacting Gases

4.49 The Haber process for the synthesis of ammonia is one of the most significant industrial processes for the well-being of humanity. It is used extensively in the production of fertilizers as well as polymers and other products. (a) What volume of hydrogen at 15.00 atm and 350°C must be supplied to produce 1.0 metric tonne (1 tonne = 10^3 kg) of NH_3? (b) What volume of hydrogen is needed in (a) if it is supplied at 3.76×10^2 atm and 250°C?

4.50 Nitroglycerin is a shock-sensitive liquid that detonates by the reaction

$$4\ C_3H_5(NO_3)_3(l) \longrightarrow 6\ N_2(g) + 10\ H_2O(g) + 12\ CO_2(g) + O_2(g)$$

Calculate the total volume of product gases at 215 kPa and 275°C from the detonation of 1.00 lb (454 g) of nitroglycerin.

4.51 Which starting condition would produce the larger volume of carbon dioxide by combustion of $CH_4(g)$ with an excess of oxygen gas to produce carbon dioxide and water: (a) 2.00 L of $CH_4(g)$; (b) 2.00 g of $CH_4(g)$? Justify your answer. The system is maintained at a temperature of 75°C and 1.00 atm.

4.52 Which starting condition would produce the larger volume of carbon dioxide by combustion of $C_2H_4(g)$ with an excess of oxygen gas to produce carbon dioxide and water: (a) 1.00 L of $C_2H_4(g)$; (b) 1.20 g of $C_2H_4(g)$? Justify your answer. The system is maintained at a temperature of 45°C and 2.00 atm.

Mixtures of Gases

4.53 Through a series of enzymatic steps, carbon dioxide and water undergo photosynthesis to produce glucose and oxygen according to the equation

$$6\ CO_2(g) + 6\ H_2O(l) \longrightarrow C_6H_{12}O_6(s) + 6\ O_2(g)$$

Given that the partial pressure of carbon dioxide in the troposphere is 0.26 Torr and that the temperature is 25°C, calculate the volume of air needed to produce 10.0 g of glucose.

4.54 The naturally occurring compound urea was first synthesized by Friedrich Wöhler in Germany in 1828. This was a significant event because it was the first time that an organic compound had been produced from an inorganic substance (thermal decomposition of ammonium cyanate). The compound has a formula of $CO(NH_2)_2$ and is used as a fertilizer. It may also be made by the reaction of carbon dioxide and ammonia:

$$CO_2(g) + 2\ NH_3(l) \longrightarrow CO(NH_2)_2(s) + H_2O(g)$$

What volumes of CO_2 and NH_3 at 200 atm and 450°C are needed to produce 2.50 kg of urea, assuming the reaction goes to completion?

4.55 When titanium tetrachloride is mixed with water, voluminous clouds of a white smoke of titanium dioxide particles are produced by the reaction

$$TiCl_4(g) + 2\ H_2O(l) \longrightarrow TiO_2(s) + 4\ HCl(g)$$

(a) What mass of TiO_2 is produced from the reaction of 1.00 L of $TiCl_4(g)$, measured at 11 Torr and 25°C? (b) Determine the volume of hydrogen chloride gas produced at 1.00 atm and 298 K.

4.56 The first stage in the production of nitric acid by the Ostwald process is the oxidation of ammonia:

$$4\ NH_3(g) + 5\ O_2(g) \longrightarrow 4\ NO(g) + 6\ H_2O(g)$$

(a) Calculate the mass of nitric oxide that will be produced from the reaction of 65.5 L of ammonia at 9.78 atm and 225°C with an excess of oxygen, assuming that all the ammonia is converted into NO. (b) If the water produced in (a) is condensed to a liquid with a density of 1.00 $g \cdot mL^{-1}$, what volume will it occupy?

4.57 A 15.0-mL sample of ammonia gas at 1.00×10^2 Torr and 30°C is mixed with 25.0 mL of hydrogen chloride gas at 1.50×10^2 Torr and 25°C, and the reaction

$$NH_3(g) + HCl(g) \longrightarrow NH_4Cl(s)$$

occurs. (a) Calculate the mass of NH_4Cl that forms. (b) Identify the gas in excess and determine the pressure of the excess gas at 27°C after the reaction is complete (in the combined volume of the original two flasks).

4.58 A 2.00-L sample of ethene gas, C_2H_4, at 1.00 atm and 298 K is burned in 2.00 L of oxygen gas to form carbon dioxide gas and liquid water at the same pressure and temperature. Ignoring the volume of water, what is the final volume of the reaction mixture at 1.00 atm and 298 K if the reaction goes to completion?

4.59 A vessel of volume 22.4 L contains 2.0 mol $H_2(g)$ and 1.0 mol $N_2(g)$ at 273.15 K. Calculate (a) their partial pressures and (b) the total pressure.

4.60 An apparatus consists of a 4.0-L flask containing nitrogen gas at 25°C and 803 kPa, joined by a valve to a 10.0-L flask containing argon gas at 25°C and 47.2 kPa. The valve is opened and the gases mix. (a) What is the partial pressure of each gas after mixing? (b) What is the total pressure of the gas mixture?

4.61 The total pressure of a mixture of sulfur dioxide and nitrogen gases at 25°C in a 5.00×10^2-mL vessel is 1.09 atm. The mixture is passed over warm calcium oxide powder, which removes the sulfur dioxide by the reaction $CaO(s) + SO_2(g) \longrightarrow CaSO_3(s)$ and is then transferred to a 1.50×10^2-mL vessel, where the pressure is 1.09 atm at 50°C. (a) What was the partial pressure of the SO_2 in the initial mixture? (b) What is the mass of SO_2 in the initial mixture?

4.62 A gas mixture being used to simulate the atmosphere of another planet consists of 320 mg of methane, 175 mg of argon, and 225 mg of nitrogen. The partial pressure of nitrogen at 300 K is 15.2 kPa. Calculate (a) the total pressure of the mixture and (b) the volume .

4.63 During the course of the electrolysis of water, hydrogen gas was collected at one electrode over water at 20°C when the external pressure was 756.7 Torr. The vapor pressure of water at 20°C is 17.54 Torr. The volume of the gas was measured to be 0.220 L. (a) What is the partial pressure of hydrogen? (b) The other product of the electrolysis of water is oxygen gas. Write a balanced equation for the electrolysis of water into H_2 and O_2. (c) What mass of oxygen was also produced in the reaction?

4.64 Dinitrogen oxide, N_2O, gas was generated from the thermal decomposition of ammonium nitrate and collected over water. The wet gas occupied 126 mL at 21°C when the atmospheric pressure was 755 Torr. What volume would the same amount of *dry* dinitrogen oxide have occupied if collected at 755 Torr and 21°C? The vapor pressure of water is 18.65 Torr at 21°C.

Molecular Motion

4.65 What is the molecular formula of a compound of empirical formula CH that diffuses 1.24 times more slowly than krypton at the same temperature and pressure?

4.66 What is the molar mass of a compound that takes 2.7 times longer to effuse through a porous plug than it did for the same amount of XeF_2 at the same temperature and pressure?

4.67 A sample of argon gas effuses through a porous plug in 147 s. Calculate the time required for the same number of moles of (a) CO_2: (b) C_2H_4; (c) H_2; (d) SO_2 to effuse under the same conditions of pressure and temperature.

4.68 Two identical flasks are each filled with a gas at 0°C. One flask contains 1 mol CO_2 and the other 1 mol Ne. In which flask do the particles have more collisions per second with the walls of the container?

4.69 A hydrocarbon of empirical formula C_2H_3 takes 349 s to effuse through a porous plug; under the same conditions of temperature and pressure, it took 210 s for the same number of molecules of argon to effuse. What is the molar mass and molecular formula of the hydrocarbon?

4.70 A compound used to make polyvinyl chloride (PVC) has composition 38.4% C, 4.82% H, and 56.8% Cl by mass. It took 7.73 min for a given volume of the compound to effuse through a porous plug, but it took only 6.18 min for the same amount of Ar to diffuse at the same temperature and pressure. What is the molecular formula of the compound?

4.71 Calculate the average kinetic energy (in joules) of 1 mol Kr(g) at (a) 55.85°C and (b) 54.85°C. (c) The molar energy difference between the answers in (a) and (b) is the energy it takes to raise the temperature of 1 mol Kr(g) by 1.00°C. The quantity is known as the heat capacity. What is its value?

4.72 Calculate the average kinetic energy (in joules) of a sample of 1 mol Ne(g) at (a) 25.00°C and (b) 26.00°C. (c) The energy difference between the answers in (a) and (b) is the energy it takes to raise the temperature of 1 mole of Ne(g) by 1°C. The quantity is known as the heat capacity. What is its value?

4.73 Calculate the root mean square speeds of (a) methane, (b) ethane, and (c) propane molecules, all at −20°C.

4.74 Calculate the root mean square speed of (a) fluorine gas, (b) chlorine gas, and (c) bromine gas molecules, all at 350°C.

4.75 Consider the Maxwell distribution of speeds found in Fig. 4.28. (a) From the graph, find the location that represents the most probable speed of the molecules. (b) What happens to the percentage of molecules having the most probable speed as the temperature increases?

Media Link **4.76** (a) Using the Living Graphs program on the Web site for this text, create and print a plot of the Maxwell distribution of molecular speeds for the noble gases He, Ne, Ar, Xe, and Kr from $s = 0$ to 2000 $m \cdot s^{-1}$. (b) Describe the trends observed in the graph in part (a). (c) Again using the Living Graphs program, create graphs for He and for Kr at 100 K and at 300 K. (d) Describe the trends observed in part (c).

Real Gases

4.77 The pressure of a sample of hydrogen fluoride is lower than expected and, as the temperature is increased, rises more quickly than the ideal gas law predicts. Suggest an explanation.

4.78 Under what conditions would you expect a real gas to be (a) more compressible than an ideal gas; (b) less compressible than an ideal gas?

4.79 Using the ideal gas equation, calculate the pressure at 298 K exerted by 1.00 mol CO_2(g) when confined in a volume of (a) 15.0 L; (b) 0.500 L; (c) 50.0 mL. Repeat these calculations, using the van der Waals equation. What do these calculations indicate about the accuracy of the ideal gas law?

4.80 Using the ideal gas equation, calculate the pressure at 298 K exerted by 1.00 mol H_2(g) when confined in a volume of (a) 30.0 L; (b) 1.00 L; (c) 50.0 mL. Repeat these calculations, using the van der Waals equation. What do these calculations indicate about the accuracy of the ideal gas law?

4.81 For CO_2(g) confined in a 1.00-L vessel at a constant temperature of 27°C, calculate the pressure of the gas, using the ideal gas law and the van der Waals equation for 0.100 mol to 0.500 mol CO_2 at 0.100-mol increments. (a) Calculate the percentage deviation of the ideal value from the "real" value at each point. (b) Under these conditions, which term has the larger effect on the real pressure of CO_2, the intermolecular attractions or the molar volume? (c) If we consider to be ideal those gases for which the observed pressure differs by no more than 5% from the ideal value, at what pressure does CO_2 become a "real" gas?

4.82 Calculate the pressure exerted by 1.00 mol C_2H_6(g) behaving as (a) an ideal gas; (b) a van der Waals gas when it is confined under the following conditions: (1) at 273.15 K in 22.414 L; (2) at 1.00×10^3 K in 0.100 L.

4.83 Plot pressure against volume for 1 mol of (a) an ideal gas, (b) ammonia, and (c) oxygen gas, for the range V = 0.05 L to 1.0 L at 298 K. Use a spreadsheet or the graphical plotter on the Web site for this book. Compare the plots and explain the origin of any differences you find.

4.84 Plot pressure against volume for 1 mol of (a) an ideal gas, (b) carbon dioxide, (c) ammonia, and (d) benzene for the range V = 0.1 L to 1.0 L at 298 K. Use the van der Waals equation to determine the pressures of the real gases. Use a spreadsheet or the graphical plotter on the Web site for this book. Compare the shapes of the curves and explain the differences.

INTEGRATED EXERCISES

4.85 A gaseous air-fuel mixture in an internal combustion engine occupies 545 cm^3 at 764 Torr. (a) If the volume is reduced to 58.8 cm^3 at the same temperature, what is the resulting pressure of the mixture prior to ignition? The fuel does not condense under these conditions. (b) Assume that the fuel is octane and that the gases in air other than nitrogen and oxygen can be neglected. At the temperature of the engine, octane exists completely in the gaseous state. If the air-fuel mixture is 97.5% by volume air, what is the mole fraction of octane present? (c) What mole fraction of oxygen is present? (d) If the combustion of octane takes place, what will be the final values for the mole fractions of the gases in the cylinder? Assume all products are gaseous for this calculation and that the nitrogen and other gases present in the cylinder do not react.

4.86 A 0.297-g sample of an unidentified gas occupies 0.250 L at 300 K and 670 Torr. (a) What is the density of the gas at 1.00 atm and 298 K? (b) What is the density of the gas if it is transferred to a vessel where its pressure is 170 kPa and its temperature is 70.0°C? (c) What is the molar mass of the gas?

4.87 A 5.00-L flask is evacuated and 43.78 g of colorless $N_2O_4(g)$ introduced. At −196°C, the gas solidifies on the walls of the flask. The sample is then warmed to 25°C, during which time the N_2O_4 vaporizes and some of it dissociates to form brown NO_2. The pressure slowly increases until it stabilizes at 2.96 atm. (a) Write a balanced equation for the reaction. (b) If the gas in the flask at 25°C were all N_2O_4, what would the pressure be? (c) If the gas in the flask all converts to NO_2, what would the pressure be? (d) What are the mole fractions of N_2O_4 and NO_2 once the pressure stabilizes at 2.96 atm?

4.88 Small quantities of hydrogen gas can be generated in the laboratory by the action of dilute hydrochloric acid on zinc metal. When 0.40 g of impure zinc reacted with an excess of hydrochloric acid, 127 mL of hydrogen was collected over water at 10°C. The external pressure was 737.7 Torr. (a) What volume would the dry hydrogen occupy at 1.00 atm and 298K? (b) What amount (in moles) of H_2 was collected? (c) What is the percentage purity of the zinc, assuming that all the zinc present reacted completely with HCl and that the impurities did not react with HCl to produce hydrogen? The vapor pressure of water at 10°C is 9.21 Torr.

4.89 A 0.473-g sample of unknown gas that occupies 200 mL at 1.81 atm and 25°C was analyzed and found to contain 0.414 g of nitrogen and 0.0591 g of hydrogen. (a) What is the molecular formula of the compound? (b) Draw the Lewis structure of the molecule. (c) If ammonia effuses through a small opening in a glass apparatus at the rate of 3.5×10^{-4} mol in 15.0 min at 200°C, how much of the compound would effuse through the same opening in 25.0 min at 200°C?

4.90 When a 2.36-g sample of phosphorus was burned in chlorine, the product was 10.5 g of a phosphorus chloride. Its vapor took 1.77 times longer to effuse than the same amount of CO_2 under the same conditions of temperature and pressure. What is the molar mass and molecular formula of the phosphorus chloride?

4.91 Sulfur is a yellow solid consisting of S_8 molecules that each have eight sulfur atoms in a ring. (a) Draw a Lewis structure for the S_8 molecule. (b) What is the geometry of sulfur atoms in S_8? (c) Do the sulfur atoms obey the octet rule? (d) When a sample of sulfur was vaporized, it gave rise to a pressure that was about three times larger than that expected; and as heating continued, the pressure rose more quickly than the ideal gas law predicts. Suggest an explanation.

4.92 In a gaseous mixture containing equal masses of CH_4 and CO_2, the partial pressure of CH_4 is 365 Torr. What is the partial pressure and mole fraction of CO_2 in this mixture?

4.93 A graduate student needs to prepare isotopically labeled phosgene gas, $^{13}COCl_2(g)$, for use in the synthesis of an organic compound. The student places 3.59 atm of $^{13}CO(g)$ and 2.75 atm of $Cl_2(g)$ in a steel cylinder at 25°C, seals the cylinder, and heats it to 227°C. After the cylinder is kept at the elevated temperature for a week, the pressure inside the cylinder is found to be 9.75 atm while still at 227°C. (a) What are the mole fractions of ^{13}CO, Cl_2, and $^{13}COCl_2$ present in the vessel at the end of the week? (b) What is the gas density of this mixture?

4.94 On an as-yet-undiscovered planet, the atmosphere is composed of a mixture of 79% argon and 20% oxygen (volume percentage) along with trace amounts of other gases, including CO_2, Kr, and Ne. An enterprising student on that planet wishes to sell the equivalent of helium balloons to small children. Helium is not available. Which of the following gases would be an acceptable replacement so that the balloons will float in that atmosphere: (a) krypton; (b) carbon dioxide; (c) methane; (d) propane; (e) nitrogen; (f) carbon monoxide? If the student is concerned about safety, which of the possible choices of gases would be the most appropriate?

4.95 A finely powdered solid sample of an osmium oxide (which melts at 40°C and boils at 130°C) with a mass of 1.509 g is placed into a cylinder with a movable piston that can expand against the atmospheric pressure of 745 Torr. Assume that the amount of residual air initially present in the cylinder is negligible. When the sample is heated to 200°C, it is completely vaporized and the volume of the cylinder expands by 235 mL. What is the molar mass of the oxide? Assuming that the oxide is OsO_x, what is the value of x?

4.96 Iron pyrite, FeS_2, is the form in which much of the sulfur occurs in coal. In the combustion of the coal, oxygen reacts with iron pyrite to produce iron(III) oxide and sulfur dioxide, which is a major source of air pollution and a substantial contributor to acid rain. (a) Write a balanced equation for the burning of FeS_2 in air to produce iron(III) oxide and sulfur dioxide. (b) Calculate the mass of Fe_2O_3 that is produced from the reaction of 75.0 L of oxygen at 2.33 atm and 150°C with an excess of iron pyrite. (c) If the sulfur dioxide that is generated in (b) is dissolved to form 5.00 L of aqueous solution, what is the molar concentration of the resulting sulfurous acid, H_2SO_3, solution? (d) What mass of SO_2 is produced in the burning of 1 metric ton (10^3 kg) of high-sulfur coal, if the coal contains 5% pyrite by mass? (e) What is the volume of the SO_2 at 1.00 atm and 25°C?

4.97 A group of chemistry students have injected 32.5 g of a gas into an evacuated, constant-volume container at 22°C and atmospheric pressure. Now the students want to heat the gas at constant pressure by allowing some of the gas to escape during the heating. What mass of gas must be released if the temperature is raised to 212°C?

4.98 Using the kinetic model of gases, the root mean square speed c was derived in Section 4.13. Using the Maxwell distribution of speeds, we can also calculate the mean speed and most probable speed of a collection of molecules. The equations used to calculate these two quantities are $c_{\text{average}} = (8RT/\pi M)^{1/2}$ and $c_{\text{most probable}} = (2RT/M)^{1/2}$, respectively. These values have a fixed relation to one another. (a) Place these three quantities in order of increasing magnitude. (b) Show that the relative magnitudes are independent of the molar mass of the gas. (c) Using the smallest speed as the reference point, determine the ratio of the larger values to the smallest.

4.99 (a) The van der Waals parameters for He are $a = 3.412 \times 10^{-2}$ $L^2 \cdot atm \cdot mol^{-2}$, $b = 2.370 \times 10^{-2}$ $L \cdot mol^{-1}$. Calculate the apparent volume (pm^3) and radius (pm) of a helium atom as determined from the van der Waals parameters. (b) Estimate the volume of a helium atom on the basis of its atomic radius. (c) How do these quantities compare? Should they be same? Discuss.

4.100 A student is given a gaseous sample of an unknown hydrocarbon. To measure the molar mass of the compound, the student uses an effusion apparatus. When methane was placed in the apparatus, it was found that 0.9560 g effused over a period of 2.5 h at room temperature. Under identical conditions, 2.292 g of the unknown effused. (a) What is the molar mass of the unknown? (b) Assuming that this molar mass is accurate and that it contains only C and H, what is the most likely formula for this compound? (c) Draw Lewis structures of at least four possible compounds with the formula predicted in (b).

CHAPTER

Quartz crystals are composed of silica, SiO_2. When the silica is pure, like that shown here, it is colorless. The internal structure of quartz is highly complex but can be pictured in terms of tetrahedral SiO_4 units, linked through shared oxygen atoms.

Why Do We Need to Know This Material?

The science of materials may have begun in the blacksmith's forge, but the materials of tomorrow will be formulated by understanding how the properties of matter are determined by the arrangements of its atoms and molecules. Scientists understand and design new materials by considering the properties and interactions of individual particles and predicting how those properties translate into bulk properties. This chapter continues the important task begun in Chapter 4: relating atomic and molecular properties to the structure and properties of bulk matter.

What Do We Need to Know Already?

This chapter assumes that we know about potential energy (Section A), coulombic interactions (Section 2.2), polar molecules and dipoles (Section 3.3), and intermolecular forces in gases (Section 4.14).

LIQUIDS AND SOLIDS

Molecules attract one another. From that simple fact spring fundamentally important consequences. Rivers, lakes, and oceans exist because water molecules attract one another and form a liquid. Without that liquid, there would be no life. Without forces between molecules, our flesh would drip off our bones and the oceans would be gas. Less dramatically, the forces between molecules govern the physical properties of bulk matter and help to account for the differences in the substances around us. They explain why carbon dioxide is a gas that we exhale, why wood is a solid we can stand on, and why ice floats on water.

In Chapter 4, we discussed gases, in which intermolecular forces play only a minor role. Here we deal with liquids and solids, in which the forces that hold molecules together are of crucial importance for determining the physical properties of bulk samples (large collections of molecules). Individual water molecules, for instance, are not wet, but bulk water is wet because water molecules are attracted to each other and to other substances. Individual water molecules neither freeze nor boil, but bulk water does, because in the process of freezing molecules stick together and in boiling they separate. We have to refine our atomic and molecular model of matter to see how properties like these, which we observe when we examine samples consisting of huge numbers of molecules, can be interpreted in terms of the properties of individual molecules, such as their size, shape, and polarity.

We begin this chapter by exploring the forces responsible for the formation of liquids and solids. These are the **intermolecular forces**, the forces between molecules, as distinct from the forces responsible for the formation of chemical bonds between atoms (chemical bonds are due to *intra*molecular forces). Then we discuss the physical properties of liquids and the structures and physical properties of solids. Elemental metals typically have very simple structures because they consist of the same atoms stacked together in definite patterns. Although the structures of ionic and molecular solids are much more varied, in many cases they are related to the structures of these metallic solids. As we work through the chapter, we encounter solid structures of increasing complexity, but the relations between them should remain clear.

INTERMOLECULAR FORCES

The simplest state of matter is a gas. We can understand many of the bulk properties of a gas—its pressure, for instance—in terms of the kinetic model introduced in Chapter 4, in which the molecules do not interact with one another except during collisions. We have also seen that this model can be improved and used to account for the properties of real gases, by taking into account the fact that molecules attract and repel one another. But what is the origin of these attractive and repulsive forces? How does their strength relate to the properties of the molecules? These are some of the questions we tackle in the next five sections.

5.1 The Formation of Condensed Phases

A "vapor" usually means the gaseous state of a substance that is a liquid or a solid under ordinary conditions.

Intermolecular forces are responsible for the existence of several different "phases" of matter. A **phase** is a form of matter that is uniform throughout in both chemical composition and physical state. The phases of matter include the three common physical states, solid, liquid, and gas (or vapor), introduced in Section A. Many substances have more than one solid phase, with different arrangements of their atoms or molecules. For instance, carbon has several solid phases; one is diamond and another is graphite. A **condensed phase** means simply a solid or liquid phase.

The temperature at which a gas condenses to a liquid or a solid depends on both the pressure and the strength of the attractive forces between its molecules. In the gas phase, molecules are far apart for most of the time, so the properties of the substance are dominated by the nearly free motion of the molecules. In condensed phases, however, the molecules are very close to one another all the time and the properties of the phases are dominated by the intermolecular forces. The molecules of a solid are held in place by strong attractive forces. In the liquid phase, molecules have more energy and can move about, but they are still held close together by intermolecular forces. When the temperature of a gas is lowered, intermolecular forces pull the molecules together and, provided the temperature is low enough, produce a condensed phase. Note that molecules remain intact throughout all phase changes.

Sections 5.2–5.5 describe the origins of some of the more important forces between molecules. Table 5.1 summarizes the interactions and compares their energies with the typical energies of ionic bonds, which are much stronger than intermolecular interactions. So that we can talk about all liquids—solutions as well as pure liquids—we also describe the forces that can act when ions are present in a molecular liquid like water.

TABLE 5.1 Interionic and Intermolecular Forces*

Type of interaction	Typical energy, $kJ{\cdot}mol^{-1}$	Interacting species
ion-ion	250	ions only
ion-dipole	15	ions and polar molecules
dipole-dipole	2	stationary polar molecules
	0.3	rotating polar molecules
dipole-induced dipole	2	at least one molecule must be polar
London (dispersion)†	2	all types of molecules
hydrogen bonding	20	N, O, F; the link is a shared H atom

* The total force experienced by a species is the sum of all the forces in which it can participate.
† Also known as induced dipole-induced dipole interaction.

Almost all intermolecular interactions, as well as those between ions, can be traced to the coulombic interaction between two charges (Section 2.2). Recall from Section A that the potential energy E_P of the coulombic interaction between two charges q_1 and q_2 separated by a distance r has the form

$$E_P \propto \frac{q_1 q_2}{r} \quad (1)$$

Figure 5.1 depicts the distance dependence of this potential energy and that of other interactions described later.

Condensed phases form when attractive coulombic forces between particles pull them together.

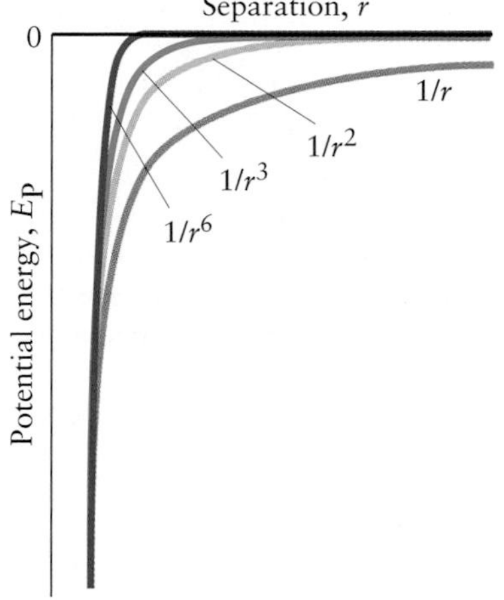

FIGURE 5.1 The distance dependence of the potential, energy of the interaction between ions (orange), ions and dipoles (yellow), stationary dipoles (green), and rotating dipoles (blue).

Media Link Living graph: Fig. 5.1, Variation of the Coulomb interaction with distance

5.2 Ion-Dipole Forces

If we could follow the individual ions in an electrolyte solution, we would see each ion jostling through the solvent, meeting another ion only occasionally, but sometimes lingering near an ion of opposite charge, then moving off again (Fig. 5.2). Each ion has a number of water molecules closely associated with it, and this coating of water molecules helps to prevent oppositely charged ions from coming together to form a solid.

What would we see if we could observe a cation dissolved in water? Recall that H_2O is a polar molecule (**1**). The partial negative charge on the O atom is likely to be found close to the cation, and the partial positive charges of the H atoms are likely to be found as far away as possible. Therefore, the water molecules can be expected to cluster around the cation, with the O atoms pointing inward and the H atoms pointing outward (Fig. 5.3a). We expect the reverse arrangement around an anion: the H atoms have partial positive charges, so they are attracted to the anion's negative charge (Fig. 5.3b). The attachment of water molecules to ions is called **hydration**. Because hydration arises from the interaction between the ion and the partial charges on the polar water molecule, it is a type of **ion-dipole interaction**. This interaction is characteristic of ions dissolved in polar solvents, such as water and liquid ammonia.

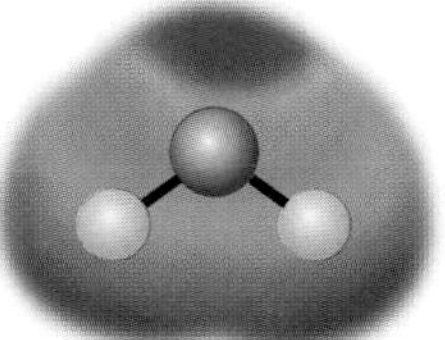

1 Water, H_2O

For solvents other than water, the analogous process is called *solvation*.

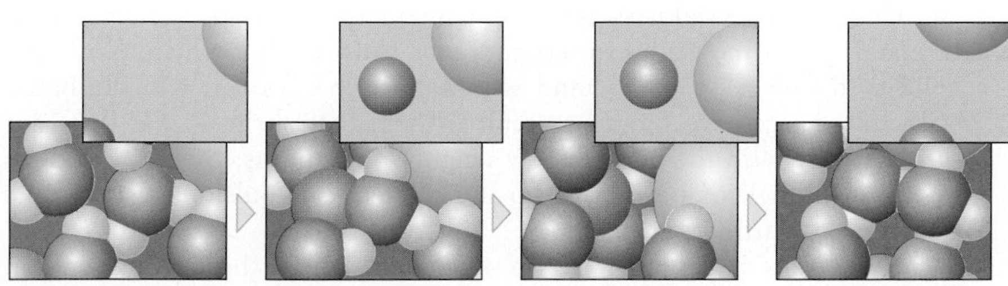

FIGURE 5.2 This animation sequence shows a series of events in a solution of sodium chloride. A sodium ion and a chloride ion move together, linger near each other for a time, attracted by their opposite charges, and then move apart; later, they will meet other partners elsewhere. The loose, transient association of oppositely charged ions is called an ion pair.

View an animation of ion-dipole forces acting as sodium chloride dissolves in water.

Media Link Animation: Fig. 5.2, Sodium chloride dissolving

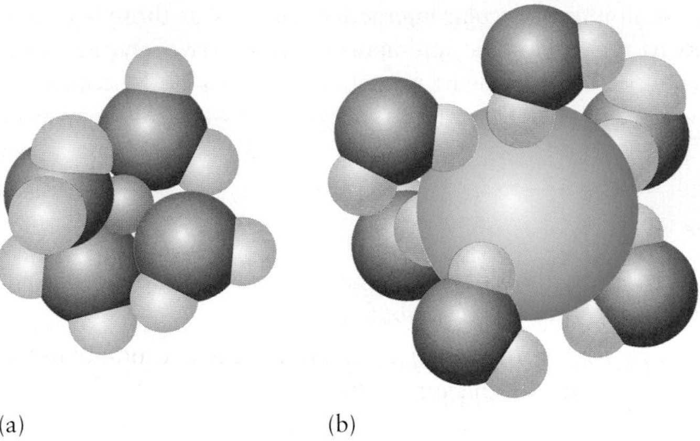

FIGURE 5.3 In water, ions are hydrated. (a) A cation is surrounded by water molecules, with water's oxygen atoms closest to the ion. (b) An anion is surrounded by water molecules that direct their hydrogen atoms toward the ion.

The ion-dipole interaction lowers the potential energy of an ion in a solvent relative to its value for the same ion in a vacuum. When we use Eq. 1 to calculate the net potential energy of the interaction between the full charge of an ion and the two partial charges of a polar molecule, we find that

$$E_P \propto -\frac{|z|\mu}{r^2} \tag{2}$$

Here z is the charge number of the ion and μ is the electric dipole moment of the polar molecule. The negative sign means that the potential energy of the ion is *lowered* by its interaction with the polar solvent. The dependence on $1/r^2$ means that the interaction between an ion and a dipole depends more strongly on distance than the interaction between two ions does (see Fig. 5.1), and therefore that it acts over a shorter range. Consequently, polar molecules need to be very close to an ion—almost in contact—before the interaction is significant.

Ion-dipole interactions are much weaker than ion-ion interactions (see Table 5.1). One reason for their weakness is that the polar molecule has only partial charges. Another reason is that the attraction of the ion to the partial charge at one end of a molecule is nearly canceled by the repulsion from the opposite partial charge at the other end of the molecule.

Despite their weakness, ion-dipole interactions may be strong enough to enable water molecules to separate ions from a solid, which allows the solid to dissolve. To understand why partial charges can win the battle with full charges, we should remember that a cation in a solid does not interact only with the immediately neighboring anions: there are cations in the next shell of neighbors, so it experiences repulsions that almost overcome the attractions.

This point was discussed in Section 2.2 in connection with ionic bonding.

When salts crystallize from an aqueous solution, the ions may retain some of the water molecules that surround them in solution and form hydrates such as $Na_2CO_3 \cdot 10H_2O$ and $CuSO_4 \cdot 5H_2O$. Both the size of the ion and its charge control the extent of hydration. The strength of the ion-dipole interaction is greater the closer the dipole can come to the center of the ion (that is, the smaller the value of r in Eq. 2). Because of their stronger ion-dipole interactions, small cations attract the polar H_2O molecules more strongly than large cations do. As a result, small cations are more extensively hydrated than large cations. In fact, lithium and sodium commonly form hydrated salts, whereas the heavier Group 1 elements that have bigger cations—potassium, rubidium, and cesium—do not. Ammonium salts are usually **anhydrous**, or water-free, for a similar reason:an NH_4^+ ion has about the same radius (143 pm) as an Rb^+ ion (149 pm).

We can see the effect of charge on the extent of hydration by comparing barium and potassium cations, which have similar radii (136 pm for Ba^{2+} and 138 pm for K^+). Potassium salts are not hydrated to any appreciable extent, but barium salts are often hydrated. The difference can be traced to the barium ion's higher charge. For example, barium chloride is found as $BaCl_2 \cdot 2H_2O$, but potassium chloride is anhydrous. Lanthanum, barium's neighbor, is both smaller (122 pm) and more highly charged (La^{3+}), so we can expect it to have strong ion-dipole interactions and its compounds to be extensively hydrated. In fact, its salts include $La(NO_3)_3 \cdot 6H_2O$ and $La_2(SO_4)_3 \cdot 9H_2O$.

Ion-dipole interactions are strong for small, highly charged ions; one consequence is that small, highly charged cations are often hydrated in compounds.

5.3 Dipole-Dipole Forces

Now suppose that we can observe a polar molecule surrounded by other polar molecules in a solid. What would we see? An example of a polar molecule is chloromethane, CH_3Cl, with the partial negative charge on the Cl atom and the partial positive charge spread over the H atoms (**2**). For molecules that are held in place in a solid, the partial negative charge of a Cl atom on one molecule is most likely to lie close to the partial positive charge on the H atoms on a neighboring molecule (Fig. 5.4). So, we would expect to see the surrounding molecules lined up as much as possible, with opposite partial charges on neighboring molecules as close together as possible. The interaction between dipoles is called the **dipole-dipole interaction**, and the resulting potential energy—which is based on Eq. 1 but takes into account the interactions between all four partial charges on the two molecules—varies as the cube of the separation:

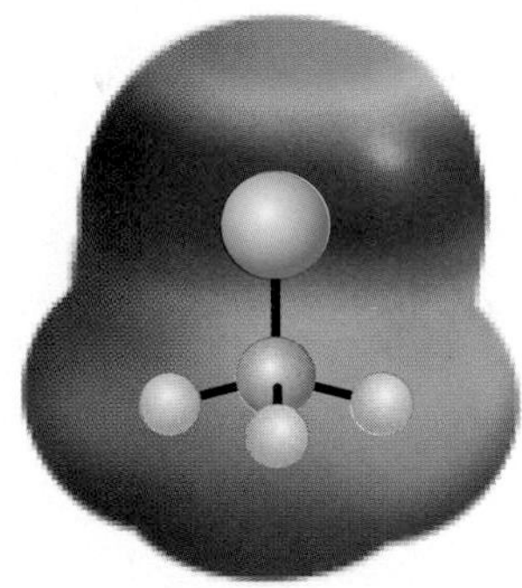

2 Chloromethane, CH_3Cl

$$E_P \propto -\frac{\mu_1\mu_2}{r^3} \tag{3}$$

In this expression, μ_1 and μ_2 are the dipole moments of the interacting molecules. The greater the polarity of the molecules, the stronger are the interactions. Equation 3 also shows that doubling the distance between the molecules reduces the strength of their interaction by a factor of $2^3 = 8$. That strong dependence on distance means that the potential energy falls off much faster than the energies of ion-ion or ion-dipole forces (see Fig. 5.1). Another reason why the interaction is relatively weak is that the charges on dipoles are partial, not full, charges.

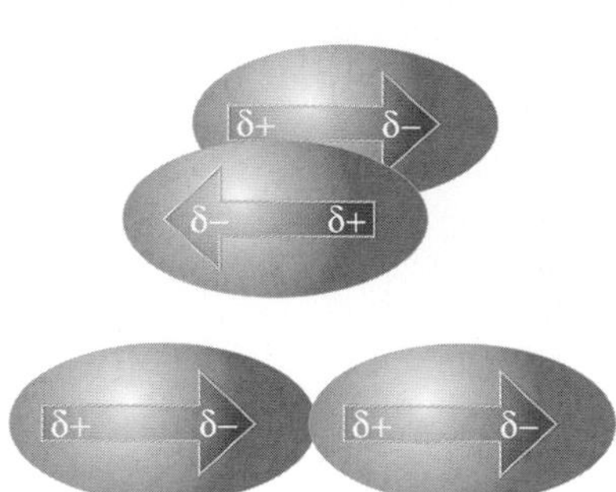

FIGURE 5.4 Polar molecules attract each other by the interaction between the partial charges of their electric dipoles (represented by the arrows). Both the relative orientations shown (end to end or side by side) result in a lower energy.

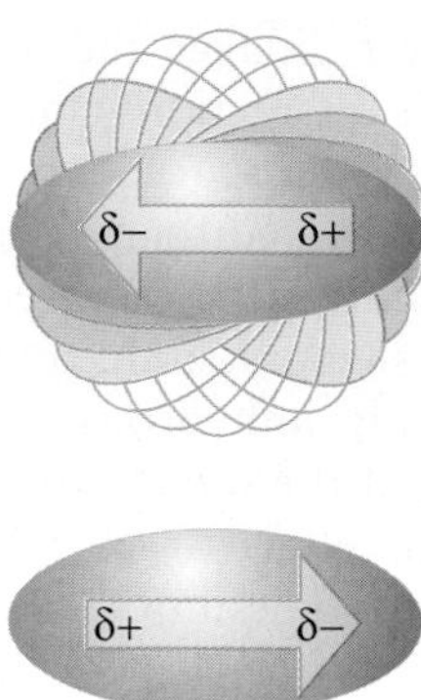

FIGURE 5.5 A polar molecule rotating near another polar molecule spends more time in the low-energy orientations (shown shaded), so the net interaction is attractive.

It takes about 1 ps for a small molecule to make a complete revolution in the gas phase.

Now imagine that we can observe the same chloromethane molecule in a gas. The molecule and its neighbors are all rotating rapidly (Fig. 5.5). This rotation brings like partial charges together some of the time, opposing the attraction of the unlike partial charges. However, the neighbors linger slightly in the energetically favorable orientations (their Cl atom close to the H atoms of the central molecule and vice versa). Although the lingering is only very fleeting, it is enough to upset the averaging slightly, and the attractive interactions between opposite partial charges slightly outweigh the repulsive interactions between like partial charges. As a result, there is a weak net attraction between rotating polar molecules in the gas phase. This type of dipole-dipole interaction decreases very strongly with separation, and it turns out that

$$E_P \propto -\frac{\mu_1^2\mu_2^2}{r^6} \qquad (4)$$

The potential energy varies as the *sixth* power of the distance between the molecules (see Fig. 5.1 again). Doubling the separation of polar molecules reduces the strength of the interaction by a factor of $2^6 = 64$, so dipole-dipole interactions between rotating molecules have a significant effect only when the molecules are very close. We can now start to understand why the kinetic model accounts for the properties of gases so well: gas molecules rotate and are far apart for most of the time, so any intermolecular interactions between them are very weak, the potential energy makes little contribution to the total energy of the gas, and a "kinetic energy only" model is a good first approximation.

Boiling points are discussed more fully in Section 8.4. For the time being, we take a boiling point to be the temperature above which a liquid changes into a vapor when the pressure is 1 atm and a melting point to be the temperature above which a solid changes to a liquid.

Molecules also rotate in liquids, so Eq. 4 describes their dipole-dipole interactions, too. However, the molecules are much closer together, and the interaction is much stronger than in a gas. The stronger the intermolecular forces in a liquid, the greater the energy required to separate the molecules. Therefore, substances with strong intermolecular forces have high melting points and boiling points.

Polar molecules take part in dipole-dipole interactions, the attraction between the partial charges of their molecules. Dipole-dipole interactions are weaker than forces between ions and fall off rapidly with distance, especially in the liquid and gas phases, where the molecules rotate.

EXAMPLE 5.1 **Predicting relative boiling points on the basis of dipole-dipole interactions**

Which would you expect to have the higher boiling point, *p*-dichlorobenzene (3) or *o*-dichlorobenzene (4)?

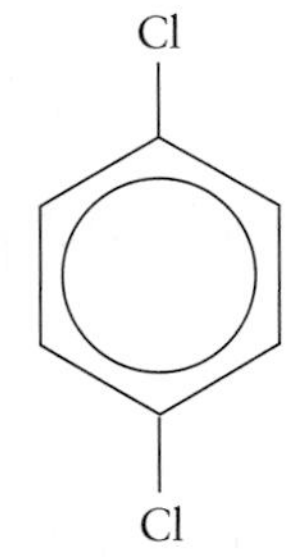

3 *p*-Dichlorobenzene

4 *o*-Dichlorobenzene

STRATEGY When two compounds have different dipole moments but are otherwise very similar, we expect the molecules with the larger electric dipole moment to interact more strongly. Therefore, assign the higher boiling point to the more strongly polar compound. To decide whether a molecule is polar, judge whether or not the dipole moments of the bonds cancel each other, as explained in Section 3.3.

SOLUTION The two C—Cl bonds in *p*-dichlorobenzene lie directly across the ring and their dipole moments cancel, thereby producing a nonpolar molecule. An *o*-dichlorobenzene molecule is polar because the two C—Cl bond dipoles do not cancel. We therefore predict that *o*-dichlorobenzene will have a higher boiling point than *p*-dichlorobenzene. The experimental values are found to be 180°C for *o*-dichlorobenzene and 174°C for *p*-dichlorobenzene.

Self-Test 5.1A Which will have the higher boiling point, *cis*-dichloroethene (5) or *trans*-dichloroethene (6)?

5 *cis*-Dichloroethene

6 *trans*-Dichloroethene

[*Answer:* *cis*-Dichloroethene]

Self-Test 5.1B Which will have the higher boiling point, 1,1-dichloroethene (7) or *trans*-dichloroethene?

7 1,1-Dichloroethene

5.4 London Forces

Finally, suppose we could observe a nonpolar molecule, with no electric dipole moment. Would we see the molecule interacting with its neighbors or would they glide past without being affected? The evidence for an attractive interaction between nonpolar molecules is that the noble gases—which, because they are monatomic, are necessarily nonpolar—can be liquefied, and many nonpolar compounds, such as the hydrocarbons that make up gasoline, are liquids.

At first sight, there seem to be no means by which two nonpolar molecules could attract each other. However, we know they do, so there must be an explanation. To find it, we must refine our model of the electron distribution in a molecule. First, recall that all the representations of electron distributions and charge distributions in the computer graphics we have seen are *average* values. In fact, at any instant, the electron clouds of atoms and molecules are not uniform: they are more like a swirling fog. If we could take a snapshot of a molecule at a given instant, the electron distribution would look deformed. Electrons may pile up somewhere in a molecule, leaving a nucleus elsewhere partially exposed. As a result, one region of the molecule will have a fleeting partial negative charge and another region will have a fleeting partial positive charge (Fig. 5.6).

Only a movie filmed at the impossible rate of 10^{16} frames a second would show the swirling motion.

Now that we know that any molecule—even a nonpolar molecule—may have *instantaneous* partial charges and a corresponding *instantaneous* dipole moment, let's explore the consequences. An instantaneous partial positive charge on one molecule attracts the electrons on the second molecule and gives rise to a region of partial negative charge on the second molecule: the two partial charges attract each other. An instant later, the partial positive charge of the first molecule might be elsewhere, but it can induce a partial negative charge elsewhere in the second molecule, and still the two molecules attract each other. That is, although there is a constant flickering of the dipole moment of one molecule from one orientation to another, the **induced dipole moment** of the second molecule follows it faithfully, and there is a net attraction between the two molecules. This interaction is called the **London force.** It acts between *all* molecules; for nonpolar molecules, it is the only force that acts between them.

Fritz London was the German-American physicist who first explained the force. The London force is also called the *dispersion force.*

The London force is always attractive. It holds the molecules of hydrocarbons together; as a result, gasoline is a liquid at normal temperatures. It holds I_2 molecules together as a solid at ordinary temperatures and N_2 molecules as a solid at very low temperatures. Even noble gas atoms can have instantaneous dipole moments and so condense to a liquid.

The strength of the London interaction depends on the **polarizability,** α (alpha), of the molecules, that is, the ease with which their electron clouds can be distorted. This dependence is reasonable, because the nuclei in highly polarizable molecules have only weak control over the surrounding electrons; consequently, there can be big fluctuations in electron density and hence large instantaneous partial charges. Detailed calculations show that the potential energy of the London interaction varies as the sixth power of the separation of two molecules:

$$E_P \propto -\frac{\alpha_1\alpha_2}{r^6} \qquad (5)$$

Like the potential energy of dipole-dipole interactions between rotating molecules, the potential energy of the London interaction also decreases very rapidly with distance (as $1/r^6$; see Fig. 5.1).

It is always sensible when presented with an equation to try to work out what information it contains, so let's see whether we can infer from Eq. 5 any other features of the London interaction. In this case, we see that the strength of the interaction increases with the polarizability of the interacting molecules, so we need to know when to expect large polarizabilities. The key is size: a large molecule with many electrons is more

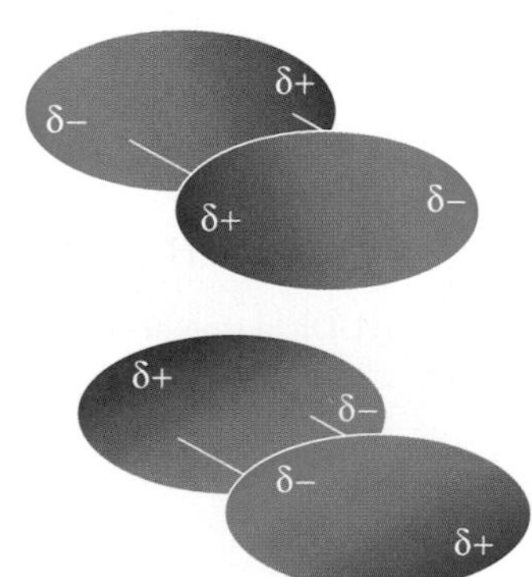

FIGURE 5.6 A fluctuation in the electron distribution in two neighboring molecules results in two instantaneous electric dipole moments that attract each other. The fluctuations move, but each new arrangement results in attraction.

polarizable than a small molecule with only a few electrons. The reason for this difference is the low degree of control that the nuclei have over the electrons when many are present and the molecules are large. In other words, we can expect molecules with large numbers of electrons to be more polarizable, and therefore to have stronger London interactions, than small molecules. Because the number of electrons usually increases with molar mass, we can also expect London interactions to increase with molar mass (Fig. 5.7). However, high molar mass is not *directly* the source of strong interactions.

The correlation between number of electrons and strength of London interactions accounts for many properties. For instance, an H_2 molecule has only two electrons, so we can expect it to interact very weakly with other H_2 molecules. Indeed, at 1 atm, hydrogen gas condenses to a liquid only if the temperature is lowered to 20 K (−253°C). We can now see why the halogens range from gases (F_2 and Cl_2), through a liquid (Br_2), to a solid (I_2) at room temperature: the molecules have an increasing number of electrons, so their polarizabilities and hence their London interactions increase on going down the group. The increase in strength of the London interaction is striking when heavier atoms are substituted for hydrogen atoms. Methane boils at −161°C, but tetrachloromethane (carbon tetrachloride, CCl_4) has many more electrons and is a liquid that boils at 77°C. Tetrabromomethane, CBr_4, has even more electrons: it is a solid at room temperature. Most covalent fluorides boil at lower temperatures than the corresponding covalent chlorides do, because fluorides have fewer electrons. Compare, for example, the values for CF_4 and CCl_4 in Table 5.2.

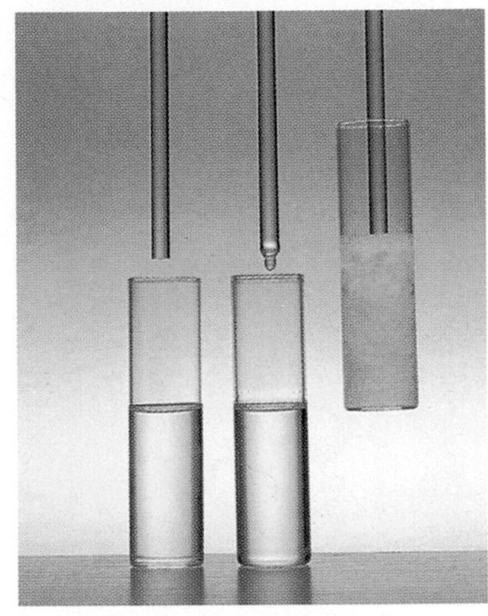

FIGURE 5.7 Hydrocarbons show how London forces increase in strength with molar mass. Pentane is a mobile fluid (left); pentadecane, $C_{15}H_{32}$, a viscous liquid (middle); and octadecane, $C_{18}H_{38}$, a waxy solid (right). To some extent, the effect of increasing intermolecular forces is increased by the ability of long-chain molecules to tangle with one another.

TABLE 5.2 Melting and Boiling Points of Substances

Substance	Melting point, °C	Boiling point, °C	Substance	Melting point, °C	Boiling point, °C
Noble gases			**Small inorganic species**		
He	−270 (3.5 K)*	−269 (4.2 K)	H_2	−259	−253
Ne	−249	−246	N_2	−210	−196
Ar	−189	−186	O_2	−218	−183
Kr	−157	−153	H_2O	0	100
Xe	−112	−108	H_2S	−86	−60
Halogens			NH_3	−78	−33
F_2	−220	−188	CO_2	—	−78s†
Cl_2	−101	−34	SO_2	−76	−10
Br_2	−7	59	**Organic compounds**		
I_2	114	184	CH_4	−182	−162
Hydrogen halides			CF_4	−150	−129
HF	−93	20	CCl_4	−23	77
HCl	−114	−85	C_6H_6	6	80
HBr	−89	−67	CH_3OH	−94	65
HI	−51	−35	glucose	142	d†
			sucrose	184d†	—

* Under pressure.
† s, solid sublimes; d, solid decomposes.

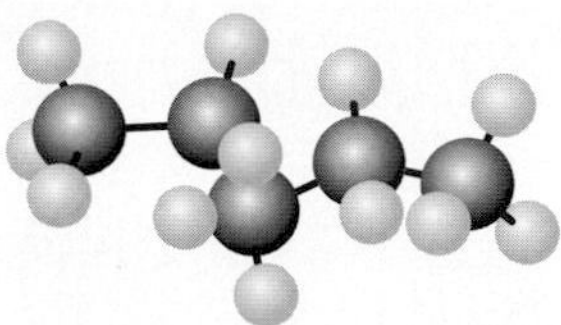

8 Pentane, C_5H_{12}

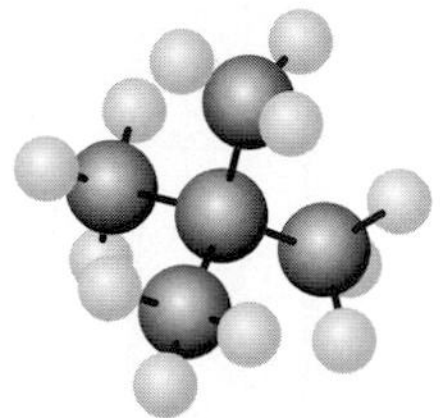

9 2,2-Dimethylpropane

Butane gas is a familiar camp stove fuel.

Acetone is a common laboratory solvent and the primary ingredient in many brands of nail polish remover.

The strength of the London interaction also depends on the shapes of the molecules. Both pentane (8) and 2,2-dimethylpropane (9), for instance, have the molecular formula C_5H_{12}, so they each have the same number of electrons. However, they have different boiling points (36° and 10°C, respectively). Molecules of pentane are relatively long and rod-shaped. The instantaneous partial charges on adjacent rod-shaped molecules can interact strongly. In contrast, the instantaneous partial charges on more spherical molecules like 2,2-dimethylpropane cannot get so close to one another because only a small region of each molecule can be in contact (Fig. 5.8). As a result of the strong dependence on distance ($1/r^6$ in Eq. 5), the interactions between rod-shaped molecules are stronger than those between spherical molecules with the same number of electrons.

All molecules, including polar molecules, interact by the London mechanism. However, if the molecules are polar, they may also interact by the dipole-dipole mechanism, and the energy of interaction is the sum of both contributions. Dipole-dipole interactions are particularly important when the molecules are in a condensed phase and do not rotate freely (the interaction then has a longer range because it varies as $1/r^3$ rather than $1/r^6$). For instance, butane, C_4H_{10}, consists of nonpolar molecules and is a gas that boils at −0.5°C. The liquid condenses as a result of London interactions. However, although 2-propanone (acetone), C_3H_6O, has approximately the same number of electrons as butane, it is a liquid that boils at 56°C. For 2-propanone, the interactions between the polar carbonyl groups occur in addition to the London interaction, so the total interaction between the molecules is stronger than for butane.

Closely related to the London interaction is the **dipole–induced-dipole interaction**, in which a polar molecule interacts with a nonpolar molecule. Like the London interaction, the dipole–induced-dipole interaction arises from the ability of one molecule to induce a dipole moment in the other. However, in this case, the molecule that induces the dipole moment has a permanent moment. The potential energy of the interaction is

$$E_P \propto -\frac{\mu_1^2 \alpha_2}{r^6} \qquad (6)$$

Like many of the other interactions we have considered, this potential energy is also inversely proportional to the sixth power of the separation.

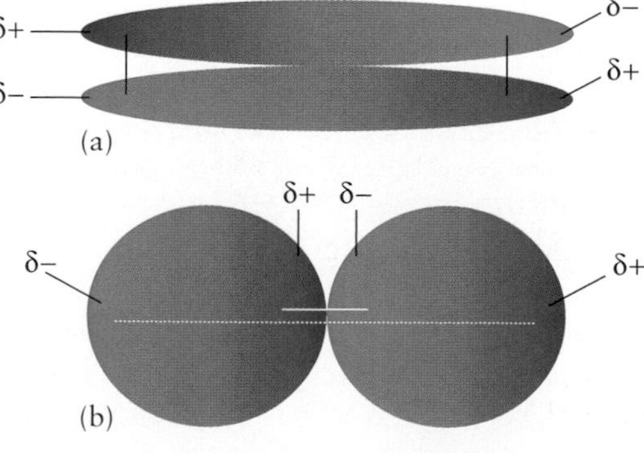

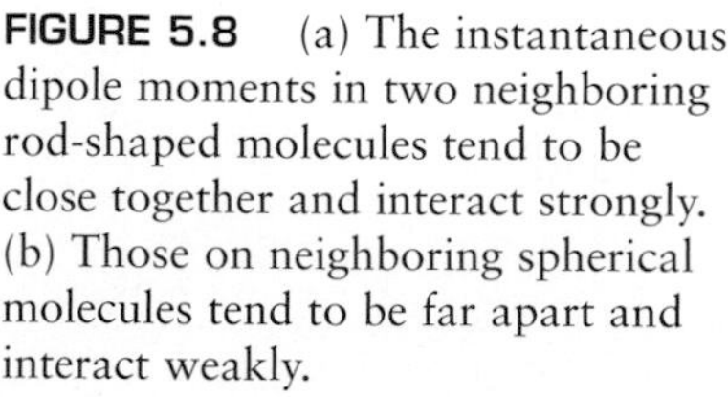

FIGURE 5.8 (a) The instantaneous dipole moments in two neighboring rod-shaped molecules tend to be close together and interact strongly. (b) Those on neighboring spherical molecules tend to be far apart and interact weakly.

The London force arises from the attraction between instantaneous electric dipoles on neighboring molecules and acts between all types of molecules; its strength increases with the number of electrons in a molecule and occurs in addition to any dipole-dipole interactions. Polar molecules also attract nonpolar molecules by weak dipole–induced-dipole interactions.

EXAMPLE 5.2 Accounting for a trend in boiling points

Explain the trend in the boiling points of the hydrogen halides: HCl, −85°C; HBr, −67°C; HI, −35°C.

STRATEGY Stronger intermolecular forces result in higher boiling points. The dipole moments, and therefore the strength of the dipole-dipole interactions, increase with the polarity of the H—X bond and therefore with the difference in electronegativity between the hydrogen and halogen atoms. The strength of London forces increases with the number of electrons. Use the data to decide which is the dominant effect.

SOLUTION The data in Fig. 2.12 show that electronegativity differences decrease from HCl to HI, so the dipole moments decrease as well. Therefore, dipole-dipole forces decrease, too, a trend suggesting that the boiling points should decrease from HCl to HI. This prediction conflicts with the data, so we examine the London forces. The number of electrons in a molecule increases from HCl to HI, so the London interaction increases, too. Therefore, the boiling points should increase from HCl to HI, in accord with the data. This analysis suggests that London forces dominate dipole-dipole interactions for these molecules.

Self-Test 5.2A Account for the trend in boiling points of the noble gases, which increase from helium to xenon.

[*Answer:* London interactions increase as the number of electrons increases.]

Self-Test 5.2B Suggest a reason why trifluoromethane, CHF_3, has a higher boiling point than tetrafluoromethane, CF_4.

5.5 Hydrogen Bonding

London forces are "universal" in the sense that they apply to all molecules regardless of their chemical identity. The dipole-dipole interaction depends only on the polarity of the molecule, regardless of its chemical identity. However, there is another very strong interaction between molecules that is *specific* to certain types of molecule.

We might suspect the existence of a special interaction when we plot the boiling points of the common binary hydrogen compounds of the elements in Groups 14/IV to 17/VII (Fig. 5.9). The trend in Group 14/IV is what we expect for similar compounds that differ in their number of electrons—the boiling points increase down the group because the strengths of the London interactions increase. However, ammonia, water, and hydrogen fluoride all show anomalous behavior. Their exceptionally high boiling points suggest that there are unusually strong attractive forces between their molecules.

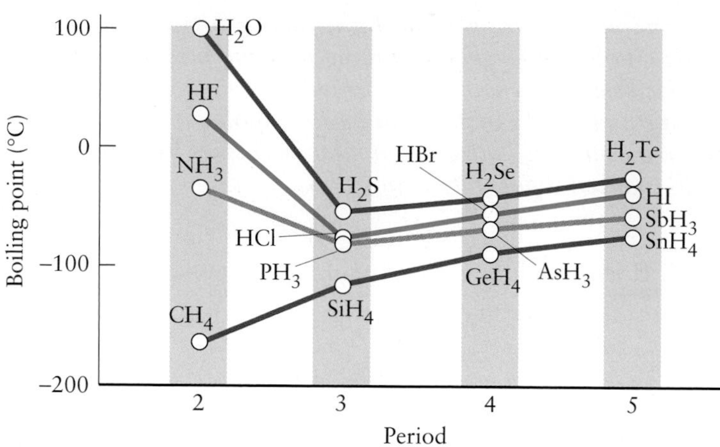

FIGURE 5.9 The boiling points of the hydrogen compounds of the *p*-block elements show a smooth variation and increase with molar mass in each group. However, three compounds—ammonia, water, and hydrogen fluoride—are strikingly out of line.

The strong interaction responsible for the high boiling points of these and certain other substances is due to a **hydrogen bond**. A hydrogen bond consists of a hydrogen atom lying between two small, strongly electronegative atoms with lone pairs of electrons, specifically N, O, or F. One molecule must contain an O—H, N—H, or H—F group to provide the H atom, and the other molecule the O, N, or F atom that provides the lone pair of electrons.

To understand how a hydrogen bond forms, let's imagine what happens when one water molecule comes close to another. Each O—H bond is polar. The electronegative O atom exerts a strong pull on the electrons in the bond, and the proton is almost completely unshielded. Because it is so small, the hydrogen atom, with its partial positive charge, can get very close to one of the lone pairs of electrons on the O atom of another water molecule. The lone pair and the partial positive charge attract each other strongly and form a bond. Hydrogen bonding is strongest when the hydrogen atom is on a straight line between the two oxygen atoms (**10**). A hydrogen bond between two O atoms, for instance, is denoted O—H···O. The O—H distance in the actual bond is shorter than the H···O distance in the hydrogen bond. In ice, for instance, the two distances are close to 101 pm and 175 pm, respectively.

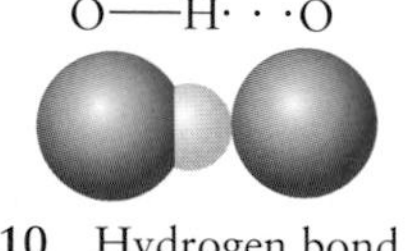

10 Hydrogen bond

Self-Test 5.3A Which of the following intermolecular links can be made by hydrogen bonds: (a) CH_3NH_2 to CH_3NH_2; (b) CH_3—O—CH_3 to CH_3—O—CH_3; (c) HBr to HBr?

[*Answer:* Only (a) has H bonded directly to N, O, or F.]

Self-Test 5.3B Which of the following molecules can take part in hydrogen bonding in the pure state: (a) CH_3OH; (b) PH_3; (c) HClO (which has the structure Cl—O—H)?

When it can occur, a hydrogen bond is strong enough to dominate all the other types of intermolecular interactions. Hydrogen bonding is so strong that it survives even in the vapor of some substances. Liquid hydrogen fluoride, for instance, contains zigzag chains of HF molecules

(**11**), and the vapor contains short fragments of the chains and $(HF)_6$ rings. Acetic acid vapor contains **dimers**, or pairs of molecules, linked by two hydrogen bonds (**12**).

11 Hydrogen fluoride, $(HF)_n$

12 Acetic acid dimer

FIGURE 5.10 Vegetation, particularly huge trees, is held upright by the powerful intermolecular hydrogen bonds that exist between the ribbonlike cellulose molecules that form much of its bulk.

Hydrogen bonds play a vital role in maintaining the shapes of biological molecules. The shape of a protein molecule is governed partly by hydrogen bonds; and once the bonds are broken, the delicately organized protein molecule loses its function. When we cook an egg, the clear albumen becomes milky white, because the heat breaks the hydrogen bonds in its molecules, which collapse into a random jumble. Trees are held upright by hydrogen bonds (Fig. 5.10). Cellulose molecules (which have many —OH groups) can form many hydrogen bonds with one another, and the strength of wood is due in large part to the strength of the hydrogen bonds between neighboring ribbonlike cellulose molecules. Hydrogen bonding also binds the two chains of a DNA molecule together, so hydrogen bonding is a key to understanding reproduction (see Section 19.15). The hydrogen bonds are strong enough to keep the two chains of a DNA molecule together but so much weaker than typical covalent bonds that they readily give way during cell division without affecting the covalent bonds in DNA.

Hydrogen bonding, which occurs when hydrogen atoms are bonded to oxygen, nitrogen, and fluorine atoms, is the strongest type of intermolecular force.

LIQUID STRUCTURE

Now that we have seen the origin of intermolecular forces, we can move on to consider how those forces affect the properties of liquids. The molecules in a liquid are in contact with their neighbors, but they can move past and tumble over one another (Fig. 5.11). When we imagine a liquid,

Media Link Animation: Fig. 5.11, Liquid water

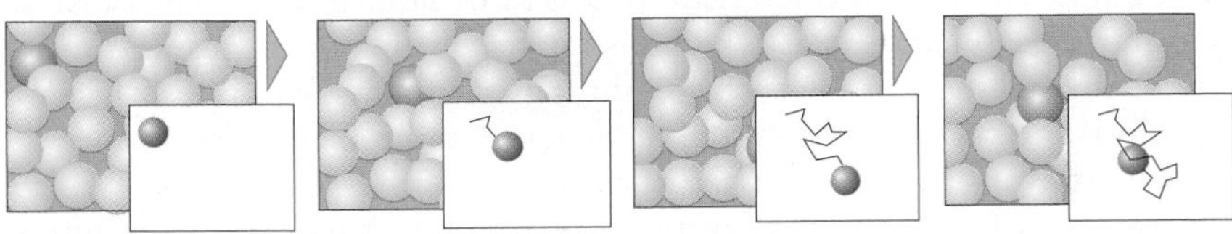

FIGURE 5.11 The structure of a liquid. Although the molecules (represented by the spheres in this series of diagrams) remain in contact with their neighbors, they have enough energy to push through to a new neighborhood, and the entire substance is fluid. One sphere is slightly darker so that you can follow its motion.

View an animation of the liquid phase of water

FIGURE 5.12 A liquid is characterized by some short-range order but no long-range order. Here we see an ordered structure within the first sphere of neighbors, but a less ordered arrangement within the next sphere of neighbors.

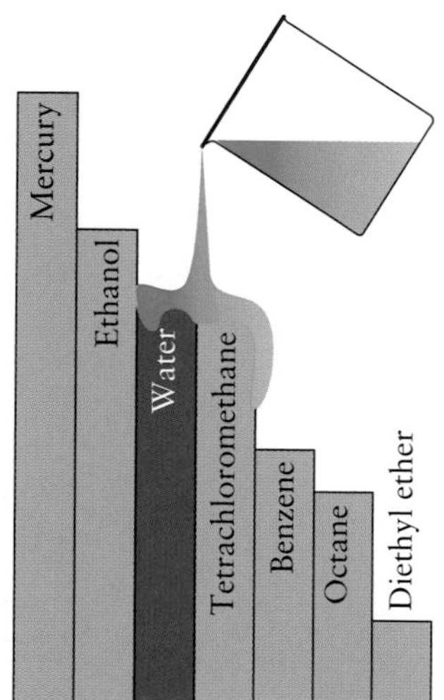

FIGURE 5.13 The viscosities of several liquids relative to that of water. Liquids composed of molecules that cannot form hydrogen bonds are generally less viscous than those that can form hydrogen bonds. Mercury is an exception: its atoms stick together by a kind of metallic bonding, and its viscosity is high.

we can think of a jostling crowd of molecules, each one constantly changing places with its neighbors. A liquid at rest is like a crowd of people milling about in a stadium; a flowing liquid is like a crowd of people leaving a stadium.

5.6 Order in Liquids

The liquid phase of matter is the most difficult to visualize. We have seen that a gas-phase molecule moves almost completely freely. The intermolecular forces from other molecules are minimal, and movement is highly disordered. In the solid phase, a molecule cannot easily move from its position: it is locked in place by intermolecular forces and can only oscillate around an average location. The liquid phase lies between the extremes of the gas and solid phases. The molecules are mobile, but they cannot escape from one another completely.

A crystalline solid has **long-range order**. In other words, its atoms or molecules lie in an orderly arrangement that is repeated over long distances. When the crystal melts, that long-range order is lost. In the liquid, the kinetic energy of the molecules can overcome the intermolecular forces, so the molecules are able to move past one another. However, the molecules still experience strong attractions to one another. In water, for instance, only about 10% of the hydrogen bonds are broken on melting. The rest are continually broken and reformed with different water molecules. At any given moment, the immediate vicinity of a molecule in the liquid is like that in the solid, but the ordering does not extend very far past the nearest neighbors (Fig. 5.12). This local ordering is called **short-range order**. In water, each molecule is surrounded by an approximately tetrahedral arrangement of other molecules as a result of hydrogen bonds. These bonds are strong enough to affect the local structure of molecules up to the boiling point. In some cases, this short-range order can be detected by diffraction (see Major Technique 3, following this chapter).

In the liquid phase, the molecules have short-range order but not long-range order.

5.7 Viscosity and Surface Tension

The viscosity and surface tension of a liquid are clues to the strength of its intermolecular forces. The **viscosity** of a liquid is its resistance to flow: the higher the viscosity of the liquid, the more sluggish the flow. A liquid with a high viscosity, like molasses at room temperature, or molten glass, is said to be "viscous." Figure 5.13 compares the relative viscosities of several liquids. Another characteristic property of liquids is their possession of a sharply defined surface, which distinguishes them from gases. A measure of the tightness of the surface film that seems to cover the surface of a liquid is the **surface tension**, the tendency of molecules at the surface to be pulled inward: the greater the surface tension, the tighter the surface film seems to be. Because viscosity and surface tension arise from intermolecular forces, we ought to be able to relate these two bulk properties to the properties of individual molecules.

We can understand intuitively how the viscosity of a liquid arises from the forces between its molecules: strong intermolecular forces hold molecules together and do not let them move past one another easily. With this connection in mind, and noting that hydrogen bonding is the strongest kind of intermolecular force, we can explain why water has a greater viscosity than benzene: benzene molecules can easily slip past one another in the liquid, but for water molecules to move, hydrogen bonds have to be wrenched apart. Phosphoric acid, H_3PO_4, and glycerol, $C_3H_8O_3$, are very viscous at room temperature because of the numerous hydrogen bonds their molecules can form. The long chains in hydrocarbon oils and greases are tangled together like cooked spaghetti (Fig. 5.14), and the molecules move past one another only with difficulty.

FIGURE 5.14 Heavy hydrocarbon oils, which consist of long hydrocarbon chains, tend to get tangled together like a plate of cooked spaghetti. As a result, the molecules do not move past one another very readily, and the liquid is very viscous.

Viscosity usually decreases as the temperature rises. Molecules have more energy at high temperatures and can wriggle past their neighbors more readily. The viscosity of water at 100°C, for instance, is only one-sixth of its value at 0°C, so the same amount of water flows through a tube six times faster at the higher temperature. In some cases, though, a change in molecular structure takes place during heating, and the viscosity increases. For example, just above 113°C, rhombic sulfur is a mobile, straw-colored liquid. Its S_8 rings can move past one another readily. The viscosity increases when the liquid is heated further because the S_8 rings break open into chains that become tangled. The viscosity falls again at still higher temperatures, and the color changes to deep red-brown: now the S_8 chains are breaking up into smaller, more mobile, highly colored, S_2 and S_3 molecules.

Now let's think about the surface tension responsible for the smooth surface of a liquid. The surface of a liquid is smooth because intermolecular forces tend to pull the molecules together and inward (Fig. 5.15); the surface tension is the net inward pull. Once again, we can expect liquids composed of molecules with strong intermolecular forces to have high surface tensions, because the inward pull at the surface will be strong. Indeed, the surface tension of water is about three times higher than that of most other common liquids, because of its strong hydrogen bonds (Table 5.3). The surface tension of mercury is higher still—more than six times that of water. Its high surface tension suggests that there are strong bonds between mercury atoms in the liquid.

Surface tension gives us insight into a number of everyday phenomena. For example, a droplet of liquid suspended in air or on a waxy surface

TABLE 5.3 Surface Tensions of Liquids at 25°C

	Surface tension, γ, $mN \cdot m^{-1}$
benzene	28.88
carbon tetrachloride	27.0
ethanol	22.8
hexane	18.4
mercury	472
methanol	22.6
water	72.75
	58.0 at 100°C

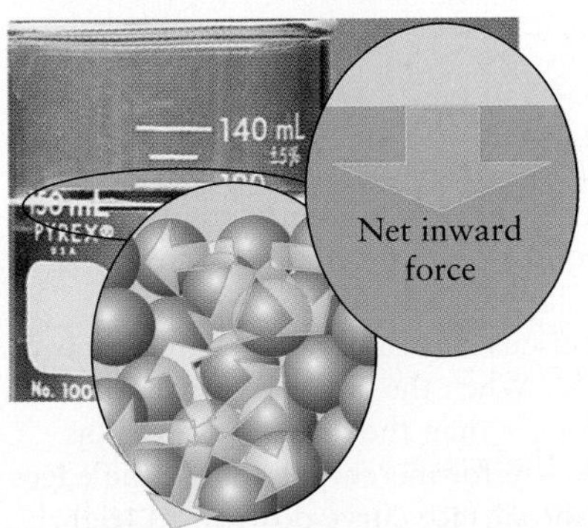

FIGURE 5.15 Surface tension arises from the attractive forces acting on the molecules at the surface. A molecule within the liquid experiences attractive forces from all directions, but a molecule at the surface experiences a net inward force.

FIGURE 5.16 The nearly spherical shape of these beads of water on a waxed surface arises from the effect of surface tension. The beads are flattened slightly by the effect of the Earth's gravity.

is spherical because the surface tension pulls the molecules into the most compact shape, a sphere (Fig. 5.16). The attractive forces between water molecules are greater than those between water and wax, which is largely hydrocarbon. On Earth, the droplets are slightly flattened by gravity; but in the gravity-free environment of an orbiting space shuttle, the shape of droplets is governed by surface tension alone, and tiny (0.01-mm diameter), perfect spheres of polymers that have been formed in space are now commercially available (Fig. 5.17). The spheres are used to calibrate particle sizes for powdered pharmaceuticals.

Water has strong interactions with paper, wood, and cloth because the molecules in their surfaces form hydrogen bonds that can replace some of the hydrogen bonds between water molecules. As a result, water maximizes its contact with these materials by spreading over them; in other words, water *wets* them. We now see that water is wet because of the hydrogen bonds its molecules can form.

Capillary action, the rise of liquids up narrow tubes, occurs when there are favorable attractions between the molecules of the liquid and the tube's inner surface. These are forces of **adhesion**, forces that bind a substance to a surface, as distinct from the forces of **cohesion**, the forces that bind the molecules of a substance together to form a bulk material.

The **meniscus** of a liquid is the curved surface it forms in a narrow tube (Fig. 5.18). The meniscus of water in a glass capillary is curved upward at the edges (forming a concave shape) because the adhesive forces between water molecules and the oxygen atoms and —OH groups that are present on a typical glass surface are stronger than the cohesive forces between water molecules. The water therefore tends to spread over the greatest possible area of glass. The meniscus of mercury curves downward in glass (forming a convex shape). This shape is a sign that the cohesive forces between mercury atoms are stronger than the forces between mercury atoms and the glass, so the liquid tends to reduce its contact with the glass.

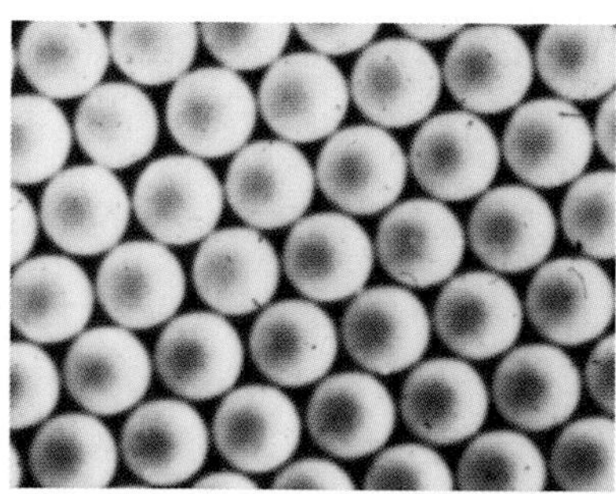

FIGURE 5.17 Latex spheres produced on one of the space shuttle missions. Each one is a perfect sphere of diameter 0.01 mm, formed by surface tension in the absence of gravity.

FIGURE 5.18 When the adhesive forces between a liquid and glass are stronger than the cohesive forces within the liquid, the liquid forms the meniscus shown here for water in glass (left). When the cohesive forces are stronger than the adhesive forces (as they are for mercury in glass), the edges of the surface curve downward (right).

The greater the viscosity of a liquid, the more slowly it flows. Viscosity usually decreases with increasing temperature. Surface tension arises from the imbalance of intermolecular forces at the surface of a liquid. It is responsible for the tendency of liquids to form droplets. Intermolecular forces are also responsible for capillary action.

FIGURE 5.19 Crystalline solids have well-defined faces and an orderly internal structure. Each face is the edge of a stack of atoms, molecules, or ions.

SOLID STRUCTURES

When the temperature is so low that the molecules in a liquid are moving too slowly to overcome the intermolecular forces, they tend to solidify. The nature of the solid depends on the types of forces that hold the atoms, ions, or molecules together. Most of the objects we encounter every day—clothes, food, furniture, vehicles, and mountains—are solids, so an understanding of solids in terms of the properties of their atoms, ions, and molecules will greatly enrich our insight into the everyday world.

5.8 Classification of Solids

A **crystalline solid** is a solid in which the atoms, ions, or molecules lie in an orderly array (Fig. 5.19). A crystalline solid has long-range order. An **amorphous solid** is one in which the atoms, ions, or molecules lie in a random jumble, as in butter, rubber, and glass (Fig. 5.20). An amorphous solid has a structure like that of a frozen instant in the life of a liquid, with only short-range order. Crystalline solids typically have flat, well-defined planar surfaces called **crystal faces**, which lie at definite angles to one another. These faces are formed by orderly layers of atoms (Box 5.1). Amorphous solids do not have well-defined faces unless they have been molded or cut.

The word *amorphous* comes from the Greek words for "without form."

FIGURE 5.20 (a) Quartz is a crystalline form of silica, SiO_2, with the atoms in an orderly network, represented here in two dimensions. (b) When molten silica solidifies, it becomes glass. Now the atoms form a disorderly network.

Alloys are discussed in Section 5.15.

Each pure element forms a crystalline solid when it freezes. In these crystals, the atoms of the elements are stacked in regular arrays that form a pattern characteristic of the particular element. Molecular compounds (such as carbon dioxide, water, and glucose) and metal alloys (such as brass) are also crystalline when solid. Ionic compounds (such as sodium chloride and potassium nitrate) crystallize into structures that allow cations and anions to stack together in energetically favorable arrangements, sometimes with hydrated water molecules (as in $CuSO_4{\cdot}5H_2O$). The arrangement of atoms, ions, and molecules within a crystal is determined by x-ray diffraction (Major Technique 3, which follows this chapter), one of the most useful techniques for determining the structures of solids.

We classify crystalline solids according to the bonds that hold their atoms, ions, or molecules in place:

- **Metallic solids**, also called simply **metals**, consist of cations held together by a sea of electrons.

BOX 5.1 HOW DO WE KNOW . . . WHAT A SURFACE LOOKS LIKE?

Even with the most powerful optical microscopes available, we cannot make out the individual atoms that make up the surface of a solid. However, recent developments have allowed us to overcome this restriction and to resolve surfaces on an atomic scale. The first two images in this box illustrate the kind of detail we can now obtain. One shows individual nickel atoms lined up in rank upon rank to form a clean surface. There could hardly be more direct evidence for the existence of atoms! Although these images are not photographs in the normal sense, we can make out the close packing of the atoms. The other image shows a tiny sodium iodide crystal lying on a copper surface.

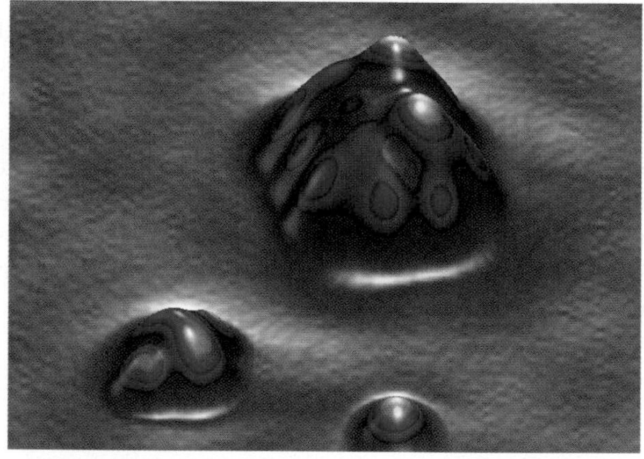

An experimenter attempted to create a two-dimensional NaI crystal on a copper surface, but the ions spontaneously rearranged themselves into a tiny three-dimensional crystal.

The surface of a single crystal of nickel, showing the regularity of its cubic close-packed structure.

The technique used to obtain these images is called *scanning tunneling microscopy* (STM). A scanning tunneling microscope works on a principle entirely different from that on which optical microscopes and their electronic cousins, electron microscopes, operate. The key idea we need for understanding STM is that, because electrons have wavelike properties (Section 1.3), electrons can penetrate into and through regions where classical mechanics would forbid them to be. This penetration is called *tunneling*. The effect is used in STM

- **Ionic solids** are built from the mutual attractions of cations and anions.
- **Molecular solids** are assemblies of discrete molecules held in place by intermolecular forces.
- **Network solids** consist of atoms bonded to their neighbors covalently throughout the extent of the solid.

Examples of each type of solid and their typical characteristics are listed in Table 5.4.

In this section, we build up an understanding of solids of increasing complexity. We begin with metallic solids, which in the case of elements are built from identical atoms stacked together in orderly arrays. These arrays turn out to be a convenient basis for the discussion of the other kinds of solids. The structures of ionic solids are based on the same set of arrays but are complicated by the need to take into account the presence of ions of opposite charges and different sizes.

(hence the name) by bringing a sharp needle point up to a surface and monitoring the current that flows through the gap between the needle point and the surface. The current flows, by tunneling, even though the two parts of the circuit are not in actual contact. The magnitude of the current, like the tunneling itself, is very sensitive to the separation of the tip and the surface, and even variations the size of an atom's dimensions affect it.

To obtain an STM image, a fine tip is moved back and forth across the surface in a series of closely spaced lines—hence the "scanning" in the name. The tip is extremely fine: it ends in a single atom (see diagram). As the tip moves across the surface at a constant height, the tunneling ebbs and flows, and the current varies through the circuit correspondingly. The image is a portrayal of the current on each scan.

In a modification, the current is maintained at a constant level by varying the height of the probe tip above the surface and monitoring its height. That degree of control of height can be achieved by using a special property of the support to which the tip is attached: a *piezoelectric* substance changes dimensions according to the electric potential difference applied to it. So, by measuring the voltage that must be applied to the piezoelectric support to maintain a constant current through the tip, the height of the tip can be inferred and plotted on a computer screen.

Yet another modification is *atomic force microscopy*, where a fine tip attached to a tiny beam is scanned across the surface. The atom at the end of the tip experiences a force that pulls it toward or pushes it away from the atoms on the surface. The deflection of the beam, which shows the shape of the surface, can be monitored by using light from a laser.

These images are computer-generated representations, not true photographs. However, they have opened our eyes to the appearance of surfaces in the most extraordinary ways.

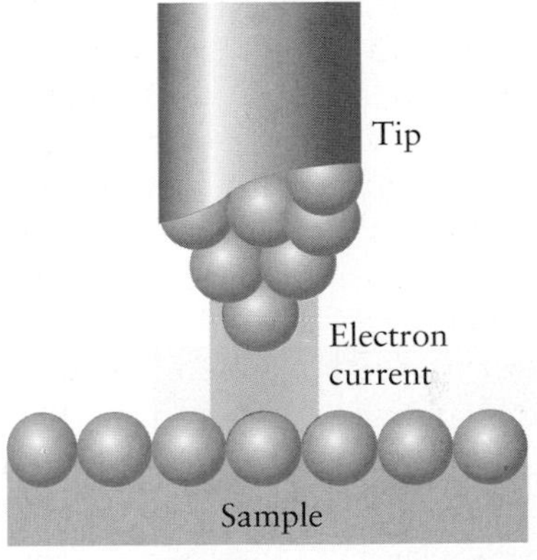

The tip of a scanning tunneling microscope above a surface. Because the tip is too close to the surface for other molecules to interfere, STM devices can be used in a gaseous atmosphere or even in liquids.

TABLE 5.4 Typical Characteristics of Solids

Class	Examples	Characteristics
metallic	*s*- and *d*-block elements	malleable, ductile, lustrous, electrically and thermally conducting
ionic	NaCl, KNO_3, $CuSO_4 \cdot 5H_2O$	hard, rigid, brittle; high melting and boiling points; those soluble in water give conducting solutions
network	B, C, black P, BN, SiO_2	hard, rigid, brittle; very high melting points; insoluble in water
molecular	$BeCl_2$, S_8, P_4, I_2, ice, glucose, naphthalene	relatively low melting and boiling points; brittle if pure

Crystalline solids have a regular internal arrangement of atoms or ions; solids are classified as metallic, ionic, molecular, or network.

5.9 Metallic Solids

The cations in a metal are bound together by their interaction with the sea formed by the electrons they have lost. Because this interaction is the same in any direction, the arrangement of the cations can be modeled by thinking of them as hard spheres and stacking them together. We can explain the structures of many metals if we suppose that the spheres representing the cations adopt a **close-packed structure**, in which the spheres stack together with the least waste of space, like oranges in a display (Fig. 5.21). We can also start to understand some of the physical properties of metals in terms of these models.

FIGURE 5.21 The stacks of apples, oranges, and other produce in a grocery display illustrate how atoms stack together in metals to give single crystals with flat faces.

To see how to stack identical spheres together to give a close-packed structure, look at Fig. 5.22. The spheres in the first layer (A) touch their six neighbors. The spheres of the second (upper) layer (B) lie in the dips of the first layer. The third layer of spheres will lie in the dips of the second layer, with the pattern repeating over and over again.

The third layer of spheres may be added in either of two ways. In the first arrangement, the spheres lie in the dips that are directly above the spheres of the first layer (Fig. 5.23). The third layer then duplicates layer A, the next layer duplicates B, and so on. This procedure results in an ABABAB . . . pattern of layers called a **hexagonal close-packed structure** (hcp). The hexagonal pattern in the arrangement of atoms can be seen in Fig. 5.24. Magnesium and zinc are examples of metals that crystallize in this way. As can be seen from the figure, each sphere has three nearest neighbors in the plane below, six in its own plane, and three in the plane above, giving 12 in all. This arrangement is reported by saying that the **coordination number**, the number of nearest neighbors of each atom, in the solid is 12. It is impossible to pack identical spheres together with a coordination number greater than 12.

In the second arrangement, the spheres of the third layer lie in the dips of the second layer that do not lie directly over the atoms of the first

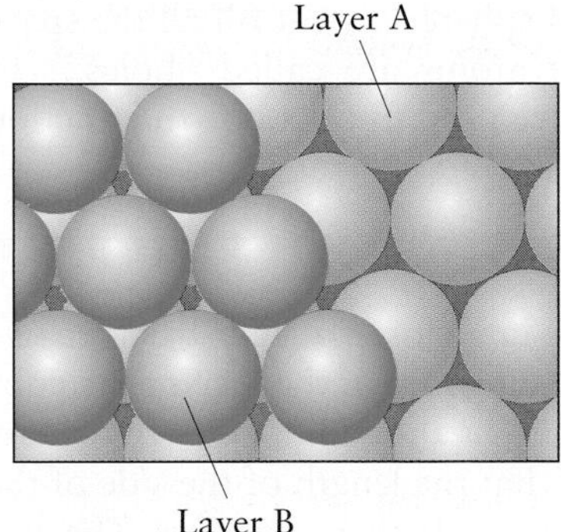

FIGURE 5.22 A close-packed structure can be built up in stages. The first layer (A) is laid down with minimum waste of space, and the second layer (B) lies in the dips—the depressions—between the spheres of the first layer. Each sphere is touching six other spheres in its layer, three in the layer below, and three in the layer above.

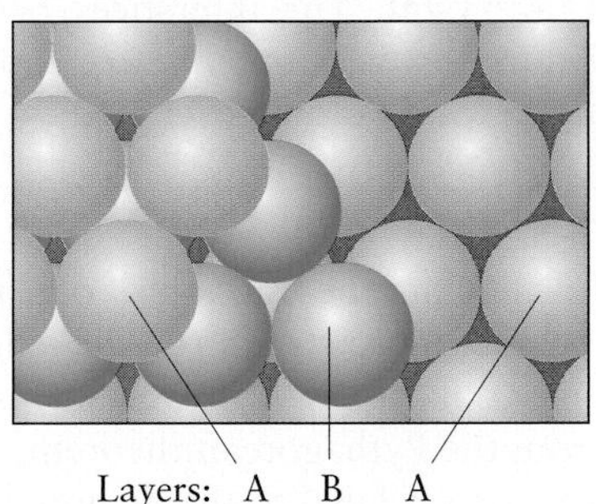

FIGURE 5.23 The third layer of spheres can lie above the atoms of the first layer to give an ABABAB . . . structure.

FIGURE 5.24 A fragment of the structure formed as described in Fig. 5.23 shows the hexagonal symmetry of the arrangement—and the origin of its name, "hexagonal close-packed."

layer (Fig. 5.25). If we call this third layer C, the resulting structure has an ABCABC . . . pattern of layers to give a **cubic close-packed structure** (ccp). The name acknowledges that the atoms in a ccp structure form a cubic pattern. The pattern is called "face-centered cubic" because it can be broken down into cubes with an atom at each corner and one in the center of each face (Fig. 5.26). The coordination number is also 12: each sphere has three nearest neighbors in the layer below, six in its own layer, and three in the layer above. Aluminum, copper, silver, and gold are examples of metals that crystallize in this way.

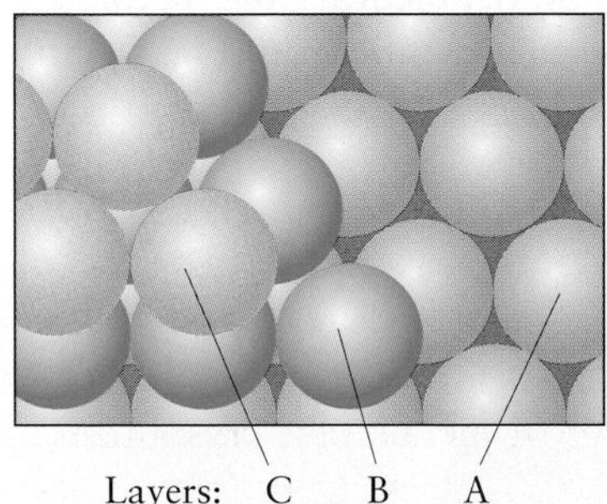

FIGURE 5.25 As an alternative to the scheme shown in Fig. 5.23, the spheres of the third layer can lie in the dips in the second layer that are directly above the dips in the first layer, to give an ABCABC . . . arrangement of layers.

FIGURE 5.26 A fragment of the structure constructed as described in Fig. 5.25. This fragment shows the origin of the names "cubic close-packed" or "face-centered cubic" for this arrangement.

The holes are not really empty in a metal, because atoms and ions do not have sharp edges. The electron sea extends into the holes.

Even in a close-packed structure, hard spheres do not fill all the space in a crystal. The interstices between the atoms are called "holes." To deduce just how much space is occupied, we need to calculate the fraction of the total volume occupied by the spheres.

How do we do that?

The Pythagorean theorem states that the square of the hypotenuse of a right-angled triangle is equal to the sums of the squares of the other two sides. That is, if the hypotenuse is c and the other two sides are a and b, then $a^2 + b^2 = c^2$.

To calculate the fraction of occupied space in a close-packed structure, we consider a ccp structure. The length of the diagonal of the face of the cube shown in Fig. 5.27 (lower left) is $4r$, where r is the radius of each sphere. From the Pythagorean theorem, we know that the length of the side of the face, a, is related to the diagonal by $a^2 + a^2 = (4r)^2$, so $a = 8^{1/2}r$. The volume of the cube is therefore $a^3 = 8^{3/2}r^3$. If we look at the diagram carefully, we see that there are eight spheres at the corners, but only $\frac{1}{8}$ of each sphere projects into the cube, so these spheres contribute $\frac{1}{8} \times 8 = 1$ sphere to the cube. There is half a sphere on each of the six faces, so these spheres contribute $\frac{1}{2} \times 6 = 3$ spheres, giving four spheres in all. The volume of each sphere is $\frac{4}{3}\pi r^3$, so the total volume of the spheres inside the cube is $4 \times \frac{4}{3}\pi r^3$. The ratio of this occupied volume to the total volume of the cube is therefore

$$\frac{\text{Total volume of spheres}}{\text{Total volume of cube}} = \frac{(16/3)\pi r^3}{8^{3/2}r^3} = \frac{16\pi}{3 \times 8^{3/2}} = \frac{\pi}{3 \times 2^{1/2}} = 0.74$$

Therefore, 74% of the crystal's space is occupied by spheres and 26% is considered to be empty. The hcp structure has the same coordination number of 12, so it must have the same fraction of occupied space.

The holes in the close-packed structure of a metal can be filled with smaller atoms to form alloys (alloys are described in more detail in Section 5.15). If a dip between three atoms is directly covered by another atom, we obtain a **tetrahedral hole**, because it is formed by four atoms at the corners of a regular tetrahedron (Fig. 5.28a). There are two tetrahedral holes per atom in a close-packed lattice. When a dip in a layer coincides with a dip in the next layer, we obtain an **octahedral hole**, because it is formed by six atoms at the corners of a regular octahedron (Fig. 5.28b).

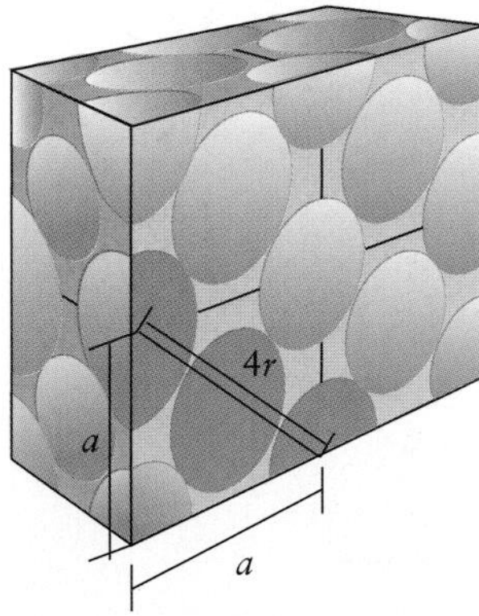

FIGURE 5.27 The relation of the dimensions of a face-centered cubic unit cell to the radius, r, of the spheres. The spheres are in contact along a face diagonal. Four unit cells are shown here.

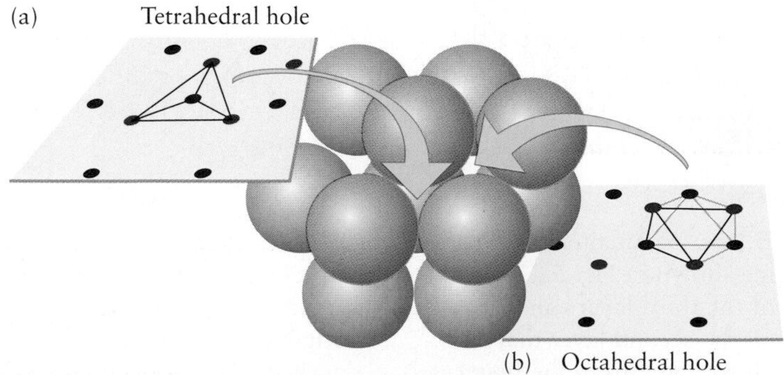

FIGURE 5.28 The locations of (a) tetrahedral and (b) octahedral holes. Note that both types of holes are defined by two neighboring close-packed layers, so they occur with equal density in both hcp and ccp structures.

FIGURE 5.29 The body-centered cubic (bcc) structure. This structure is not packed as closely as the others we have illustrated. It is less common among metals than the close-packed structures. Some ionic structures are based on it.

FIGURE 5.30 A primitive cubic unit cell has an atom at each corner.

There is one octahedral hole for each atom in the lattice. Note that, because holes are formed by two adjacent layers, and neighboring close-packed layers are identical in hcp and ccp, the numbers of holes are the same for both close-packed structures.

Which close-packed structure—if either—a metal adopts depends on which gives the lower energy, and that in turn depends on details of its electronic structure. In fact, some elements achieve a lower energy by adopting a different arrangement altogether. In a **body-centered cubic** (bcc) **structure**, for instance, a single atom lies at the center of a cube formed by eight other atoms (Fig. 5.29). This structure is not close packed, and metals that have a body-centered cubic structure can often be forced under pressure into a close-packed form. Iron, sodium, and potassium are examples of metals that crystallize in bcc lattices. The **primitive cubic structure** has an atom at each corner of a cube. The atoms touch along the edges (Fig. 5.30). This structure is known for only one element, the metalloid polonium, a fact suggesting that covalent forces are so strong in this element that they overcome the tendency toward close packing characteristic of metallic bonding.

Many metals have close-packed structures, with the atoms stacked in either a hexagonal or a cubic arrangement; close-packed atoms have a coordination number of 12. Some metals have body-centered cubic structures, whereas primitive cubic structures are rare.

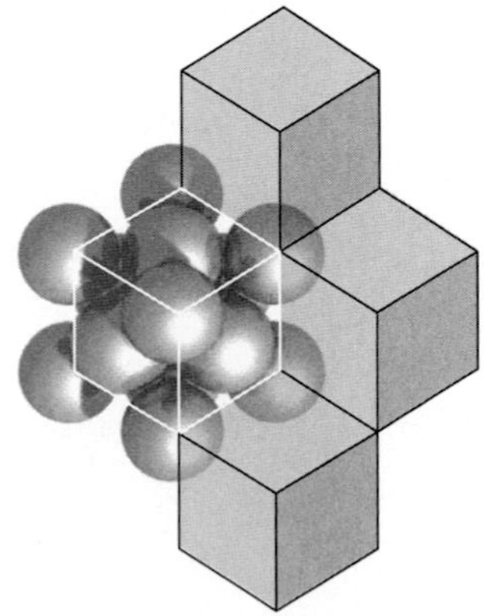

FIGURE 5.31 The entire crystal structure is constructed from a single type of unit cell by stacking the cells together without any gaps.

Self-Test 5.4A What is the coordination number of an atom in a body-centered cubic (bcc) structure?

[*Answer:* 8]

Self-Test 5.4B What is the coordination number of an atom in a primitive cubic structure? (*Hint*: Take into account the atoms in the neighboring cubes as shown for a ccp structure in Fig. 5.31.)

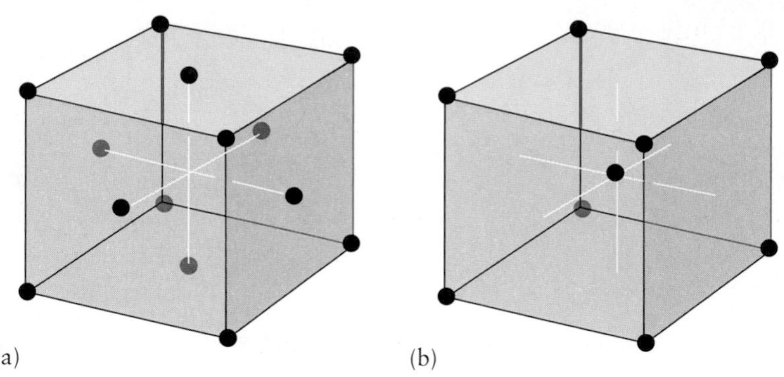

FIGURE 5.32 The unit cells of the (a) ccp (or fcc) and (b) bcc structures in which the locations of the centers of the spheres are marked by dots.

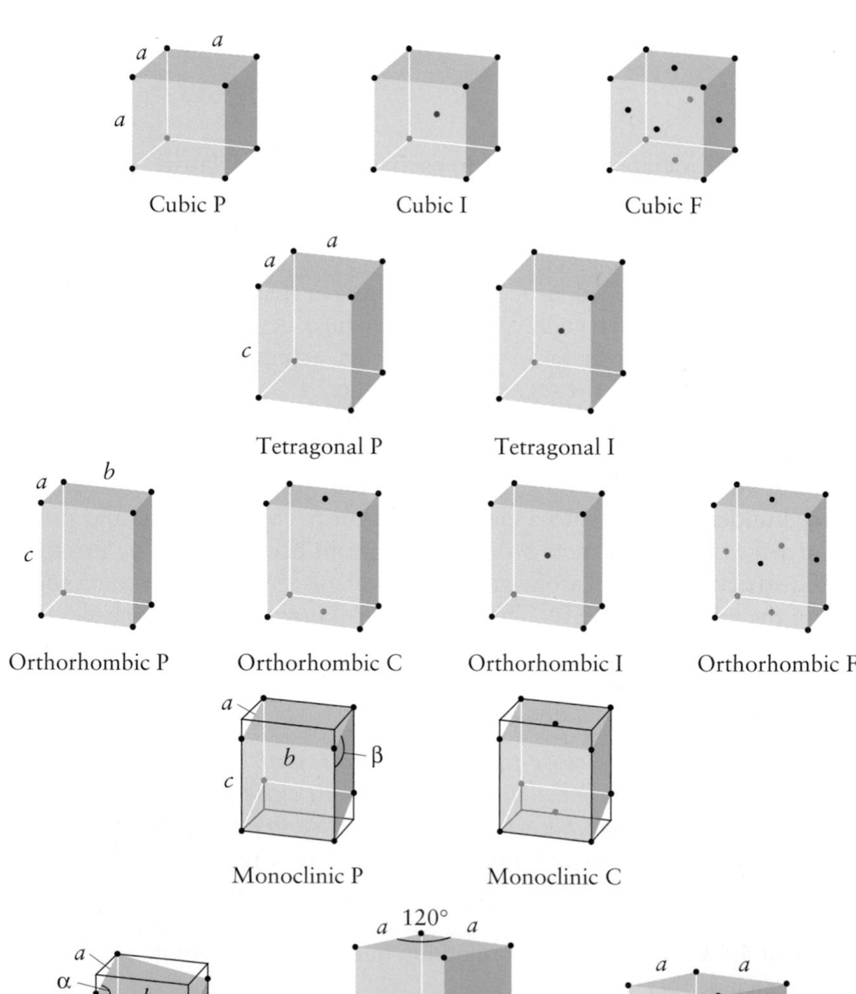

FIGURE 5.33 The 14 Bravais lattices. P denotes primitive; I, body-centered; F, face-centered; C, with a lattice point on two opposite faces; and R, rhombohedral (a rhomb is an oblique equilateral parallelogram).

5.10 Unit Cells

We do not have to draw the complete crystal lattice every time we want to convey the structure of a solid. Instead, we can focus our attention on a small region of the crystal that is representative of the entire crystal. Each of the small units highlighted in Figs. 5.26, 5.29–5.31 is called a **unit cell**, the smallest unit that, when stacked together repeatedly without any gaps, can reproduce the entire crystal (as suggested by Fig. 5.31). Unit cells are drawn by representing each atom by a dot that marks the location of the atom's center (Fig. 5.32). A cubic close-packed unit cell has an atom at the center of each face of the unit cell; for this reason, it is also called a **face-centered cubic** (fcc) **structure**.

At first thought, we might expect that there are limitless ways of stacking atoms in regular three-dimensional arrays to give crystals. However, scientists have discovered that all crystal structures can be related to only 14 basic patterns. When we stack together a large number of one of these basic types of unit cells, we get one of the 14 possible **Bravais lattices** (Fig. 5.33). We can begin to appreciate why the number of types of unit cells is limited by recognizing that regular pentagonal shapes are absent from the Bravais lattices: regular pentagons cannot cover space without gaps (Fig. 5.34). Similarly, regular heptagonal (seven-sided) and higher shapes cannot be stacked together to cover all space; hence they do not occur among the Bravais lattices.

It is equally amazing that scientists have also found that there are only 17 basic ways in which patterns can be repeated on wallpaper—the two dimensional analogue of crystals.

The number of atoms in a unit cell is counted by noting how they are shared between neighboring cells. For example, an atom at the center of a cell belongs entirely to that cell, but one on a face is shared between two cells and counts as one-half an atom. As noted earlier, for an fcc structure, the eight corner atoms contribute $8 \times \frac{1}{8} = 1$ atom to the cell. The six atoms at the centers of faces contribute $6 \times \frac{1}{2} = 3$ atoms (Fig. 5.35). The total number of atoms in an fcc unit cell is therefore $1 + 3 = 4$, and the mass of the unit cell is four times the mass of one atom.

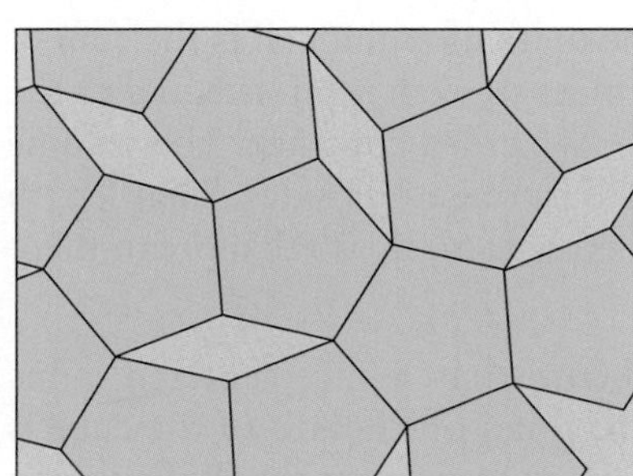

FIGURE 5.34 A plane surface cannot be covered by regular pentagons without leaving gaps. The same is true of regular heptagons.

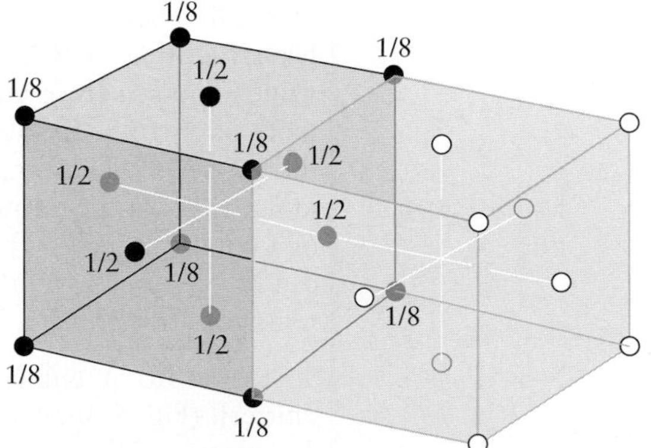

FIGURE 5.35 The calculation of the net number of atoms in a face-centered cubic cell; one of the cell's neighbors is also shown.

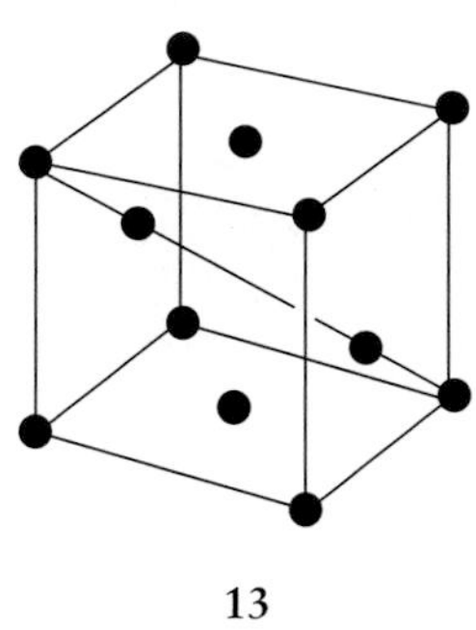

13

How can we use this information . . . to count the atoms in a unit cell?

To count the number of atoms in the bcc cell shown in Fig. 5.29, we count 1 for the atom at the center and $\frac{1}{8}$ for each of the eight atoms at the vertices, giving $1 + (8 \times \frac{1}{8}) = 2$ overall.

Self-Test 5.5A How many atoms are there in a primitive cubic cell (see Fig. 5.30)?

[*Answer:* 1]

Self-Test 5.5B How many atoms are there in the structure made up of unit cells like the one shown in (**13**), which has an atom at each corner, two on opposite faces, and two inside the cell on a diagonal?

The best way to determine the type of unit cell adopted by a metal is x-ray diffraction, which gives characteristic diffraction patterns for each of the types of unit cells (see Major Technique 3). However, a simpler procedure that can be used to distinguish between close-packed and other structures is to measure the density of the metal. The densities of the candidate unit cells are calculated and compared with experiment to determine which accounts for the observed density. Because density is an intensive property, it is the same for a unit cell or a bulk sample. Hexagonal and cubic close packing cannot be distinguished in this way, because they have the same densities.

Remember that an intensive property is a property that is independent of the size of the sample (Section A).

All crystal structures are derived from the 14 Bravais lattices. The atoms in a unit cell are counted by determining what fraction of each atom resides within the cell.

EXAMPLE 5.3 Deducing the structure of a metal from its density

The density of copper is 8.93 $g\cdot cm^{-3}$ and its atomic radius is 128 pm. Is the metal more likely to be (a) close packed or (b) body-centered cubic?

STRATEGY We first calculate the density of the metal for a close-packed unit cell. Then we calculate the density by assuming that the metal has a bcc structure. Finally, we compare the two values with the experimental value. The structure with the density closer to the experimental value is more likely to be the actual structure. The mass of a unit cell is the sum of the masses of the atoms it contains. The mass of each atom is equal to the molar mass of the element divided by the Avogadro constant. The volume of a cubic unit cell is the cube of the length of one of its sides. That length can be obtained from the radius of the metal atom, the Pythagorean theorem, and the geometry of the cell.

SOLUTION (a) A cubic close-packed structure has a face-centered cubic unit cell (Fig. 5.36c). The length of the diagonal of the face of the cube is $4r$, where r is the radius of each sphere. Therefore, from the Pythagorean theorem,

$$a^2 + a^2 = (4r)^2 = 16r^2$$

It follows that

$$a^2 = \tfrac{1}{2}(16r^2) = 8r^2$$

Taking the square root of both sides gives $a = 8^{1/2}r$. As noted earlier, there are four atoms in a face-centered cubic unit cell, so the mass of one unit cell is four times the mass of one atom (M/N_A). The density is therefore

$$d = \frac{m}{a^3} = \frac{4M/N_A}{(8^{1/2}r)^3} = \frac{4M}{8^{3/2}N_A r^3}$$

A radius of 128 pm corresponds to 1.28×10^{-8} cm and the molar mass of copper (from the periodic table on the inside front cover) is 63.55 g·mol^{-1}. The density is therefore

$$d = \frac{4 \times (63.55\ \text{g·mol}^{-1})}{8^{3/2} \times (1.28 \times 10^{-8}\ \text{cm})^3 \times (6.022 \times 10^{23}\ \text{mol}^{-1})}$$
$$= 8.90\ \text{g·cm}^{-3}$$

(b) To calculate the density of a bcc unit cell, we let the length of the diagonal of a face of the cell be f and the length of the diagonal through the body of the cell be b. Then, from Fig. 5.36b and the Pythagorean theorem,

$$a^2 + f^2 = b^2 = (4r)^2$$

The Pythagorean theorem also tells us that $f^2 = 2a^2$, so

$$a^2 + f^2 = a^2 + a^2 + a^2 = 3a^2$$

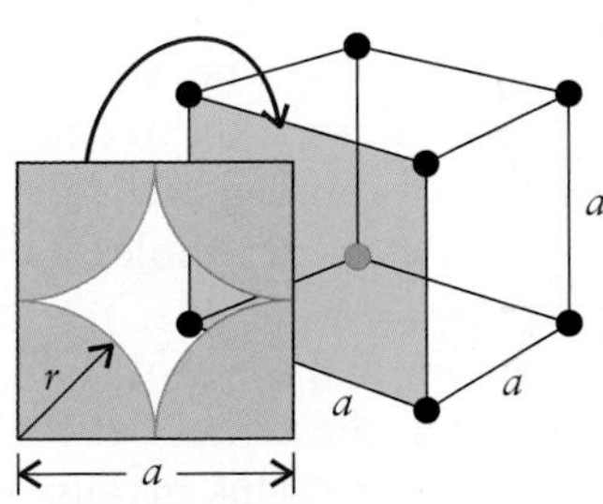

(a) Primitive cubic

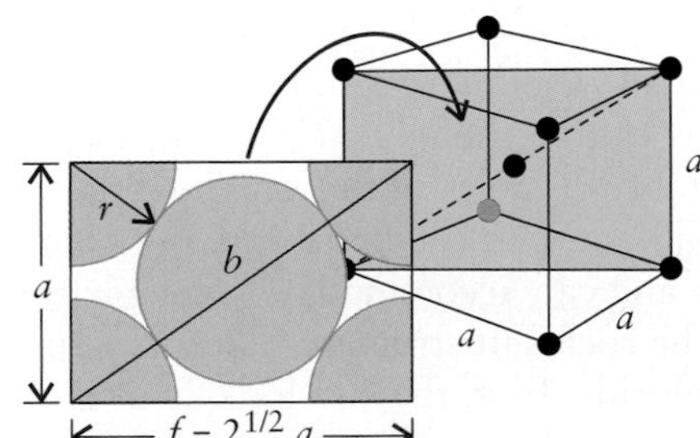

(b) Body-centered cubic

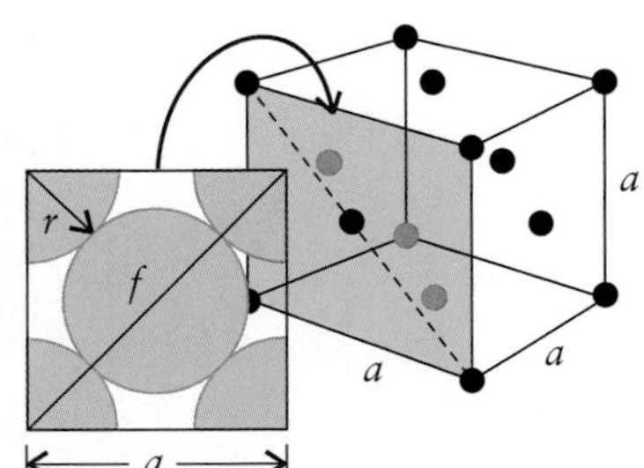

(c) Face-centered cubic (ccp)

FIGURE 5.36 The geometries of three cubic unit cells, showing the relation of the dimensions of each cell to the radius, r, of a sphere representing an atom or ion. The side of a cell is a, the diagonal of the body of a cell, b, and the diagonal of a face, f.

It follows that $3a^2 = (4r)^2$ and therefore that $a = 4r/3^{1/2}$. Each unit cell contains $8 \times \frac{1}{8} + 1 = 2$ spheres, so the total mass of a body-centered cubic unit cell is $2M/N_A$. Therefore,

$$d = \frac{m}{a^3} = \frac{2M/N_A}{(4r/3^{1/2})^3} = \frac{3^{3/2}M}{32N_A r^3}$$

When we insert numerical values, we obtain

$$d = \frac{3^{3/2} \times (63.55 \text{ g·mol}^{-1})}{32 \times (6.022 \times 10^{23} \text{ mol}^{-1}) \times (1.28 \times 10^{-8} \text{ cm})^3}$$
$$= 8.17 \text{ g·cm}^{-3}$$

The value is further from 8.93 g·cm^{-3}, the experimental value, than that for a close-packed structure, 8.90 g·cm^{-3}, so we conclude that copper has a close-packed structure.

Self-Test 5.6A The atomic radius of silver is 144 pm and its density is 10.5 g·cm^{-3}. Is the structure close packed or body-centered cubic?

[*Answer:* Close packed]

Self-Test 5.6B The atomic radius of iron is 124 pm and its density is 7.87 g·cm^{-3}. Is this density consistent with a close-packed or a body-centered cubic structure?

5.11 Ionic Structures

As we have seen, we can model elemental metals by spheres of the same radius. However, to model ionic solids, we need to pack together spheres of different radii and opposite charges. Sodium chloride, for instance, is modeled by stacking together positively charged spheres of radius 102 pm, representing the Na^+ ions, and negatively charged spheres of radius 181 pm, representing the Cl^- ions. Because the crystal is electrically neutral overall, each unit cell must reflect the stoichiometry of the compound and itself be electrically neutral.

In many cases, the structure of an ionic solid can be related to a close-packed structure. The big anions can be thought of as forming a slightly expanded version of one of the close-packed structures. The smaller cations can be thought of as occupying some of the enlarged tetrahedral or octahedral holes in the expanded lattice. A tetrahedral hole is relatively small and can accommodate only small cations. Octahedral holes are larger and can accommodate somewhat bigger cations.

The **rock-salt structure** takes its name from the mineral form of sodium chloride. In it, the Cl^- ions lie at the corners and in the centers of the faces of a cube, forming a face-centered cube (Fig. 5.37). This arrangement is like an expanded cubic close-packed arrangement: the expansion keeps the anions out of contact with one another, thereby reducing their repulsion, and opens up holes that are big enough to accommodate the Na^+ ions. These ions fit into the octahedral holes between the Cl^- ions. There is one octahedral hole for each sphere in a close-packed array, so all the octahedral holes are occupied. If we look carefully at the structure, we

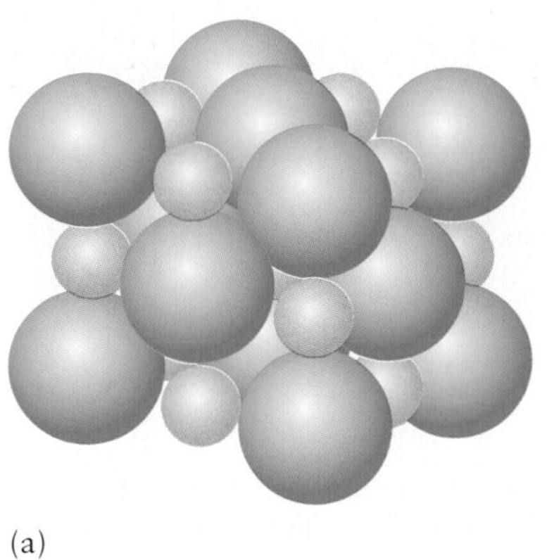

(a)

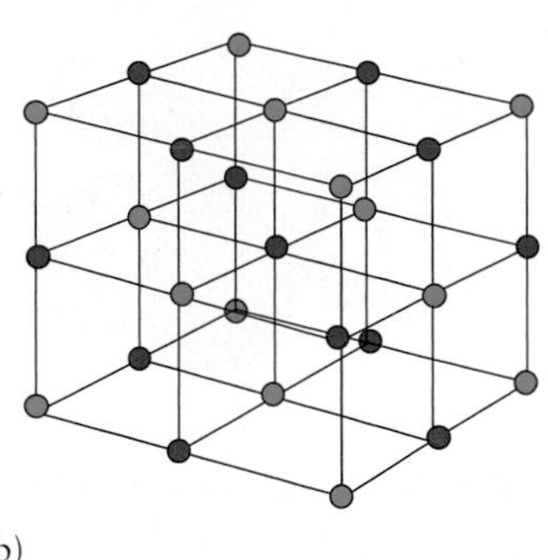

(b)

View an animation of a crystal of sodium chloride

Media Link Animation: Fig. 5.37, Sodium chloride

FIGURE 5.37 The arrangement of ions in the rock-salt structure. (a) The unit cell, showing the packing of the individual ions, and (b) a representation of the structure by dots that identify the centers of the ions.

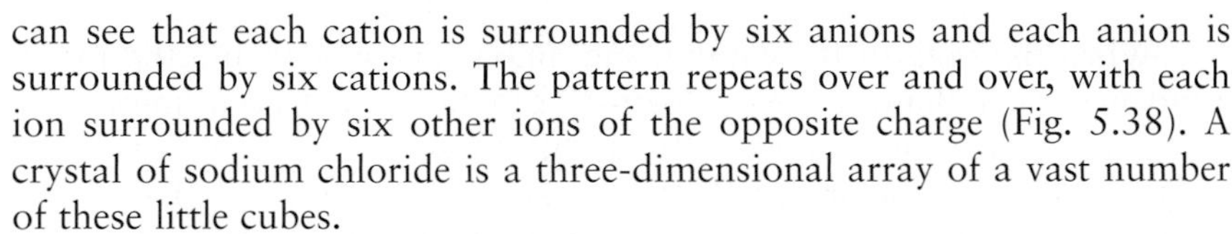

can see that each cation is surrounded by six anions and each anion is surrounded by six cations. The pattern repeats over and over, with each ion surrounded by six other ions of the opposite charge (Fig. 5.38). A crystal of sodium chloride is a three-dimensional array of a vast number of these little cubes.

In an ionic solid, the "coordination number" means the number of ions of *opposite* charge immediately surrounding it. In the rock-salt structure, the coordination numbers of the cations and the anions are both 6, and the structure overall is described as having *(6,6)-coordination.* In this notation, the first number is the cation coordination number and the second is that of the anion. The rock-salt structure is found for a number of other minerals with ions of the same charge number, including KBr, RbI, MgO, CaO, and AgCl. It is common whenever the cations and anions have very different radii, in which case the cations can fit into the octahedral holes in a face-centered cubic array of anions. Specifically, although there are many exceptions, a rock-salt structure is generally expected when the **radius ratio**, the radius of the smaller ion divided by that of the larger ion, is in the range 0.4 to 0.7. For sodium chloride,

$$\text{Radius ratio} = \frac{102\ \text{pm}}{181\ \text{pm}} = 0.564$$

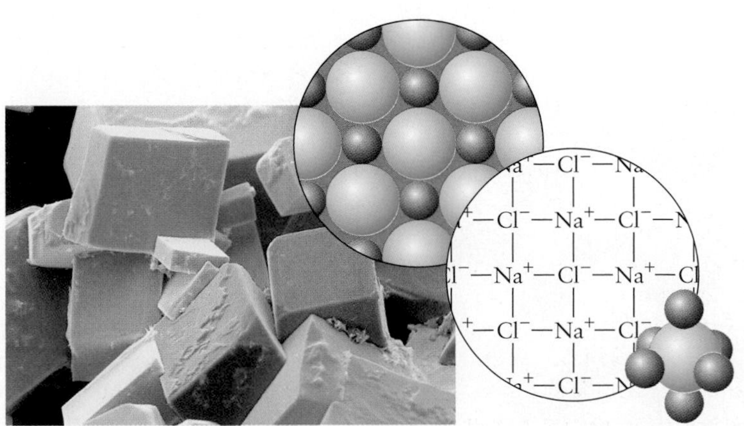

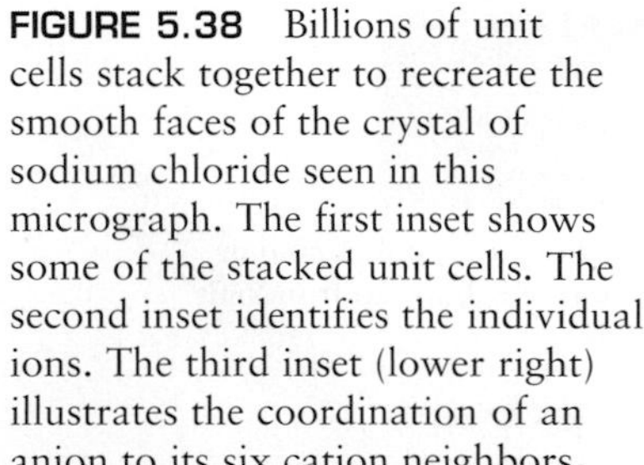

FIGURE 5.38 Billions of unit cells stack together to recreate the smooth faces of the crystal of sodium chloride seen in this micrograph. The first inset shows some of the stacked unit cells. The second inset identifies the individual ions. The third inset (lower right) illustrates the coordination of an anion to its six cation neighbors.

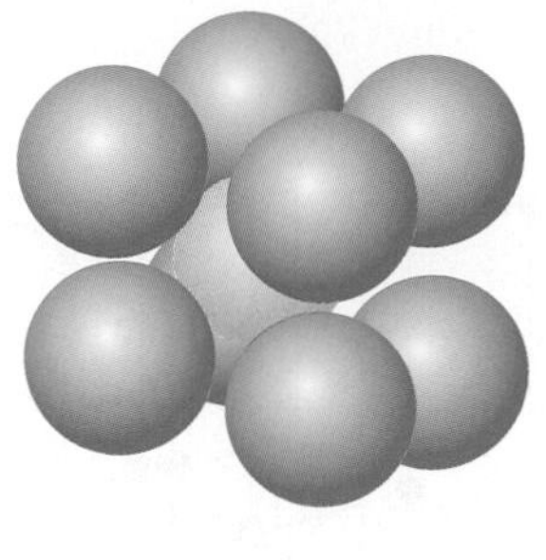
(a)

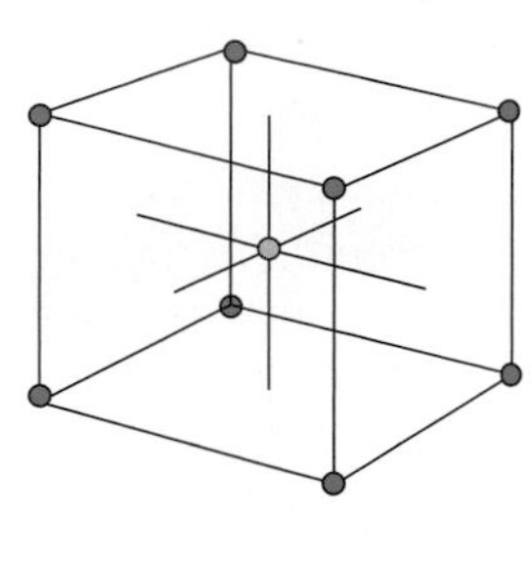
(b)

FIGURE 5.39 The cesium-chloride structure: (a) the unit cell and (b) the location of the centers of the ions. The pink spheres are cations and the green spheres are anions.

When the radii of the cations and anions are similar and the radius ratio is greater than 0.7, more cations can fit around each anion, and more anions can fit around each cation. Now the ions may adopt the **cesium-chloride structure** typified by cesium chloride, CsCl (Fig. 5.39). The radius of a Cs^+ ion is 170 pm and that of a Cl^- ion is 181 pm, giving a radius ratio of 0.939, so the two ions are almost the same size. In this structure, the anions form an expanded primitive cubic array, with a Cl^- ion at the eight corners of each cubic unit cell. There is a large "cubic" hole at the center of the cell, and the Cs^+ ion fits into it. Equivalently, each Cl^- ion can be regarded as being at the center of a cubic unit cell with eight Cs^+ ions at its corners, so overall the crystal can be thought of as built from two interpenetrating primitive cubic lattices (Fig. 5.40). The coordination number of each type of ion is 8, and overall the structure has *(8,8)-coordination*. The cesium-chloride structure is much less common than the rock-salt structure, but it is found for CsI as well as CsCl.

Self-Test 5.7A How many Na^+ ions are present in the NaCl unit cell? Account for each one.

[*Answer:* 4; 1 center atom and $\frac{1}{4} \times 12$ (1 on each of 12 edges)]

Self-Test 5.7B How many Cl^- ions are there in one unit cell of CsCl?

Media Link Animation: Fig. 5.40, Cesium chloride

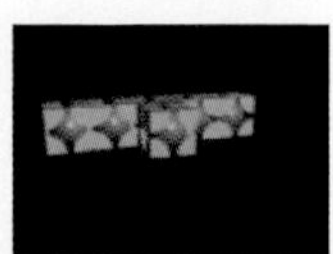

See how a crystal of cesium chloride is built up from unit cells.

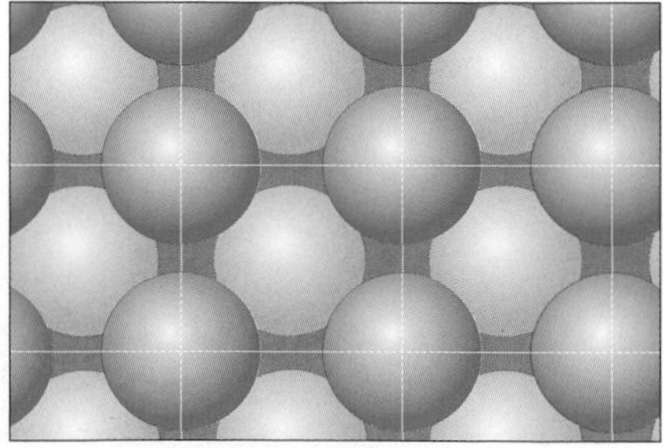

FIGURE 5.40 The repetition of the cesium-chloride unit cell recreates the entire crystal. This view is from one side of the crystal and shows six unit cells.

EXAMPLE 5.4 Estimating the density of an ionic solid

Estimate the density of cesium chloride from its crystal structure.

STRATEGY As for metals, the density of the bulk solid is the same as that of the unit cell. The volume of the cesium-chloride unit cell is the cube of the length of one of its sides, and that length can be worked out from the Pythagorean theorem. The mass of the unit cell is obtained by counting the net number of ions of each kind and using the mass of each ion (its molar mass divided by the Avogadro constant). The density is obtained by dividing the mass by the volume. Assume cations and anions touch either along diagonals (as in cesium chloride) or along edges (as in rock salt).

SOLUTION The radius of the Cs^+ ion is 170 pm and that of the Cl^- is 181 pm. Therefore, the length of the diagonal of the unit cell is $b = 170 + 2(181) + 170$ pm $= 702$ pm (Fig. 5.41), corresponding to 7.02×10^{-8} cm. The length of the side, a, is related to b by $a = b/3^{1/2}$ (recall Example 5.3). The volume of the unit cell is therefore $a^3 = (b/3^{1/2})^3$. Each unit cell contains one Cs^+ ion (of molar mass 132.91 g·mol^{-1}) and one Cl^- ion (of molar mass 35.45 g·mol^{-1}). The total mass of the unit cell is the sum of these two masses divided by the Avogadro constant, N_A. The density is therefore

$$d = \frac{M}{N_A(b/3^{1/2})^3} = \frac{(132.91 + 35.45)\ \text{g·mol}^{-1}}{(6.022 \times 10^{23}\ \text{mol}^{-1}) \times (7.02 \times 10^{-8}\ \text{cm}/3^{1/2})^3}$$
$$= 4.20\ \text{g·cm}^{-3}$$

The experimental value is 3.99 g·cm^{-3}.

Self-Test 5.8A Estimate the density of sodium chloride from its crystal structure.

[*Answer:* 2.14 g·cm^{-3}]

Self-Test 5.8B Estimate the density of cesium iodide from its crystal structure.

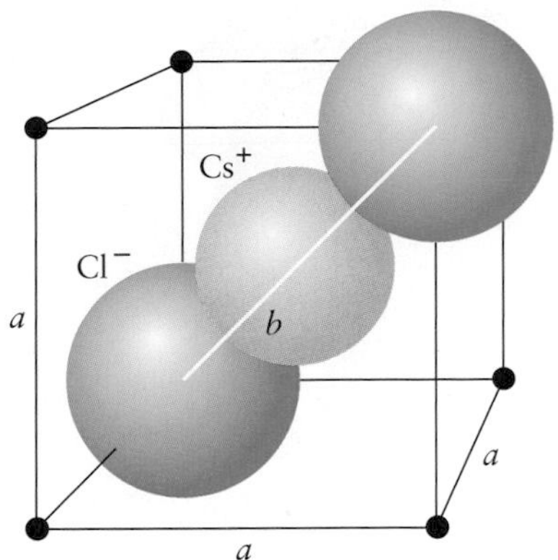

FIGURE 5.41 The relation between the radii of the ions in a cesium-chloride unit cell and the length of the sides of the cell.

When the radius ratio of an ionic compound is less than about 0.4, tetrahedral holes may be filled. An example of such a structure is the **zinc-blende structure** (or sphalerite structure), which is named after a form of the mineral ZnS (Fig. 5.42). This structure is based on an expanded cubic close-packed lattice for the anions, with the cations occupying half the tetrahedral holes. Each Zn^{2+} ion is surrounded by four S^{2-} ions and each S^{2-} ion is surrounded by four Zn^{2+} ions, so the zinc-blende structure has *(4,4)-coordination.*

In these and many other ionic structures, we treat ions as spheres of the appropriate radius and stack them together in the arrangement that has the lowest total energy. However, the simple sphere-packing arrangement may break down if the bonding is not purely ionic. In cases where the bonding has a significant amount of covalent character, the ions may have to lie in specific positions around one another. An example is nickel arsenide, Ni_3As_2. In this solid, the small Ni^{2+} cations polarize the big

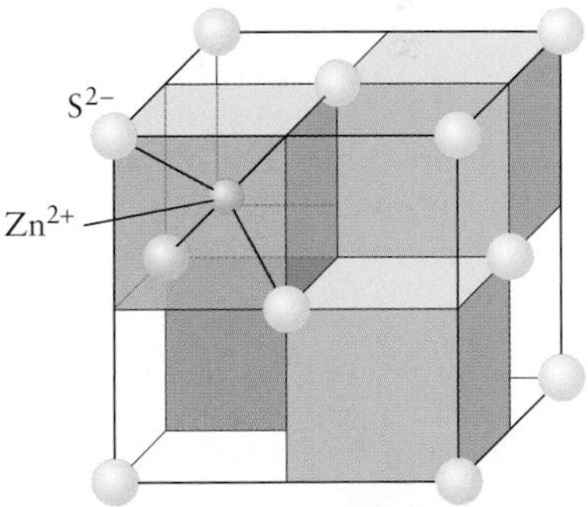

FIGURE 5.42 The zinc-blende, or sphalerite, structure. The four zinc ions (pink) form a tetrahedron within a face-centered cubic unit cell composed of sulfide ions (yellow). The zinc ions occupy half the tetrahedral holes between the sulfide ions and the parts of the unit cell occupied by zinc ions are shaded gold. The detail shows how each zinc ion is surrounded by four sulfide ions; each sulfide ion is similarly surrounded by four zinc ions.

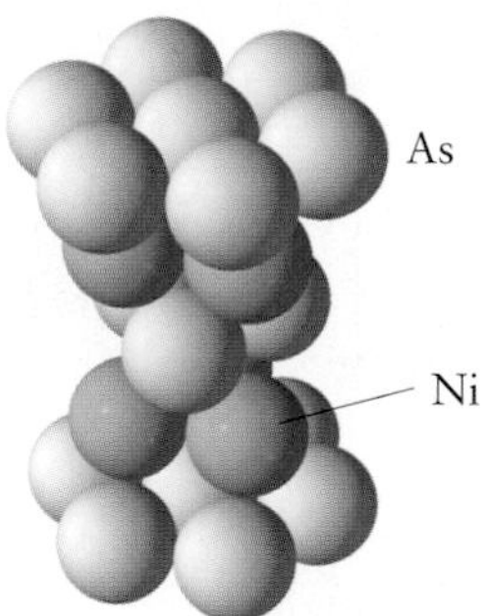

FIGURE 5.43 The structure of nickel arsenide, Ni_3As_2. Structures like this are often found when covalent bonding is starting to become important and the ions have to take up specific positions relative to one another to maximize their bonding.

As^{3-} anions, and the bonds have some covalent character. The ions pack together in an arrangement that is quite different from a purely ionic, sphere-packing model (Fig. 5.43).

Ions stack together in the regular crystalline structure corresponding to lowest energy. The structure adopted depends on the radius ratio of cation and anion. Covalent character in an ionic bond imposes a directional character on the bonding.

FIGURE 5.44 Titanium tetrachloride, $TiCl_4$, is a liquid that reacts with water to form solid titanium dioxide, TiO_2. When squirted out of appropriate containers, it can be used for producing smoke screens or, as here, for skywriting.

5.12 Molecular Solids

Molecular solids are composed of individual molecules that are not spheres and are held together by relatively weak intermolecular forces. They often have low melting points: only a little thermal motion is needed to overcome the relatively weak intermolecular forces. For example, nonpolar carbon dioxide is a gas at ordinary temperatures. Titanium tetrachloride ($TiCl_4$), with many more electrons, is a liquid that boils at 136°C and freezes at −24°C to a molecular solid (Fig. 5.44). The solid is composed of tetrahedral $TiCl_4$ molecules held together largely by London forces between the electron-rich chlorine atoms. Most organic materials, such as polymers, waxes, and lubricants, are molecular solids or liquids.

The physical properties of molecular solids depend on the strengths of their intermolecular forces. Amorphous molecular solids may be as soft as paraffin wax, which is a mixture of long-chain hydrocarbons. These molecules lie together in a disorderly way, and the forces between them are so weak that they can be pushed past one another

Travel through a crystal of ice.

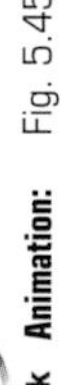

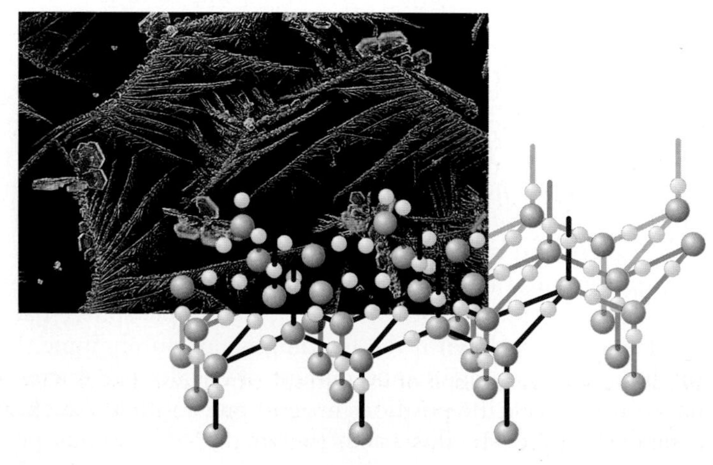

FIGURE 5.45 Ice is made up of water molecules that are held together by hydrogen bonds in a relatively open structure. Each O atom is surrounded tetrahedrally by four hydrogen atoms, two of which are σ-bonded to it and two of which are hydrogen-bonded to it.

FIGURE 5.46 The variation of the densities of water and tetrachloromethane with temperature. Note that ice is less dense than liquid water at its freezing point and that water has its maximum density at 4°C.

View an animation of ice melting

Media Link Animation: Fig. 5.46, Ice melting

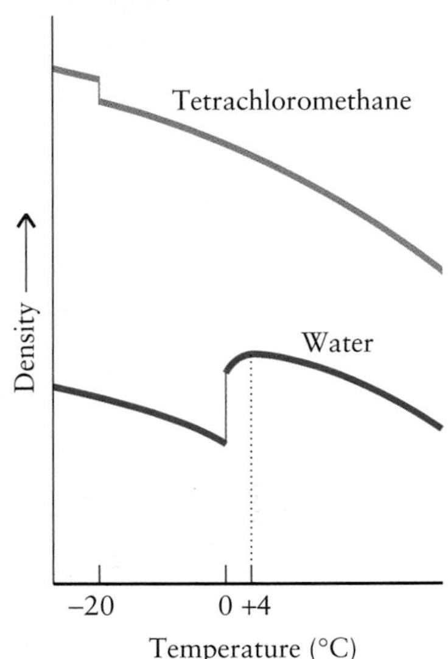

very easily. Many other molecular solids are brittle and hard. For example, sucrose molecules, $C_{12}H_{22}O_{11}$, are held together by hydrogen bonds between their numerous —OH groups. Hydrogen bonding between sucrose molecules is so strong that by the time the melting point has been reached (at 184°C), the molecules themselves have started to decompose. The partly decomposed mixture of products, called caramel, is used to add flavor and color to food.

Because molecules have such widely varying shapes, they stack together in a wide variety of different ways. In ice, for example, each O atom is surrounded by four H atoms. Two of these H atoms are linked to the O atom through σ-bonds. The other two belong to neighboring H_2O molecules and are linked to the O atom by hydrogen bonds. As a result, the structure of ice is an open network of H_2O molecules held in place by hydrogen bonds (Fig. 5.45). Some of the hydrogen bonds break when ice melts; and as the orderly arrangement collapses, the molecules pack less uniformly but more densely (Fig. 5.46). The openness of the network in ice compared with that in the liquid explains why it has a lower density than liquid water (0.92 $g\cdot cm^{-3}$ and 1.00 $g\cdot cm^{-3}$, respectively, at 0°C). Solid benzene and solid carbon dioxide, in contrast, have higher densities than their liquids. Their molecules are held in place by London forces, which are much less directional than hydrogen bonds, so they can pack together more closely in the solid than in the liquid (Fig. 5.47).

Molecular solids are typically less hard than ionic solids and melt at lower temperatures.

5.13 Network Solids

Whereas molecular solids consist of molecules held together by relatively weak intermolecular forces, the atoms in network solids are joined to their neighbors by strong covalent bonds that form a network extending throughout the crystal. Network solids exhibit the strength of the covalent bonds that bind them by being very hard, rigid materials with high melting and boiling points.

Diamond and graphite are elemental network solids. These two forms of carbon are **allotropes**, meaning forms of an element that differ in the way the atoms are linked. Each C atom in diamond is covalently bonded

FIGURE 5.47 As a result of its open structure, ice is less dense than liquid water, and floats in it (left). Solid benzene is denser than liquid benzene and "benzenebergs" sink in liquid benzene (right).

to four neighbors through sp^3 hybrid σ-bonds (Fig. 5.48). The tetrahedral framework extends throughout the solid like the steel framework of a large building. This structure accounts for the great hardness of the solid. Diamond is so hard that it is used to protect drill bits and as a long-lasting abrasive. Because diamond is also one of the best conductors of heat, thin films of diamond are used as a base for some integrated circuits so that they do not overheat. The vigorous vibration of an atom in a hot part of the crystal is rapidly transmitted to distant, cooler parts through the covalent bonds, rather like the effect of slamming a door in a steel-framed building.

Graphite, the most important component of the "lead" of pencils, is a black, lustrous, electrically conducting, slippery solid that vaporizes at 3700°C. It consists of flat sheets of sp^2 hybridized carbon atoms bonded covalently into hexagons like chicken wire (Fig. 5.49). There are also weak bonds between the sheets. In the commercially available forms of graphite, there are many impurity atoms trapped between the sheets; these atoms weaken the already weak intersheet bonds and let the sheets of atoms slide over one another. So, in contrast to the hardness of diamonds, graphite is soft and, when impure, slippery. When we write with a pencil, the mark left on the paper consists of rubbed-off layers of graphite. Electrons can move within the sheets of graphite but much less readily from one sheet to another. Hence, graphite conducts electricity better parallel to the sheets than perpendicular to them.

Ceramic materials are often noncrystalline inorganic oxides with a network structure and include many silicate minerals, such as quartz (silicon dioxide), and high-temperature superconductors (see Box 5.2). Ceramic materials have great strength and stability, because covalent bonds must be broken to cause any deformation in the crystal. As a result, ceramic materials under physical stress tend to shatter rather than bend. Section 14.24 contains further information on the properties of ceramic materials.

Network solids are typically hard and rigid; they have high melting and boiling points. Ceramic materials tend to be network solids.

Media Link Animation: Fig. 5.48, Diamond

View the internal structure of diamond.

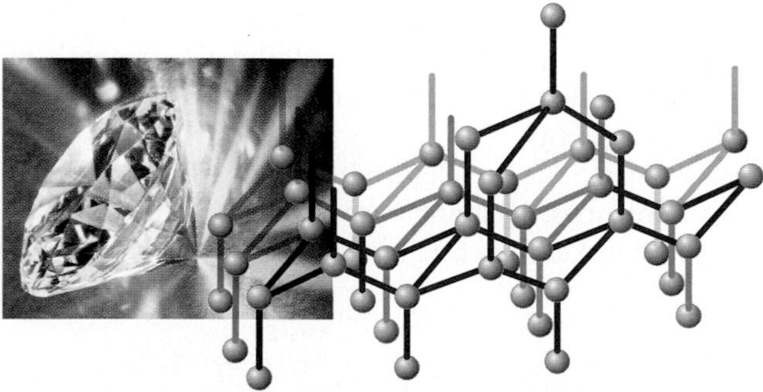

FIGURE 5.48 The structure of diamond. Each dot represents the location of the center of a carbon atom. Each atom forms an sp^3 hybrid covalent bond to each of its four neighbors.

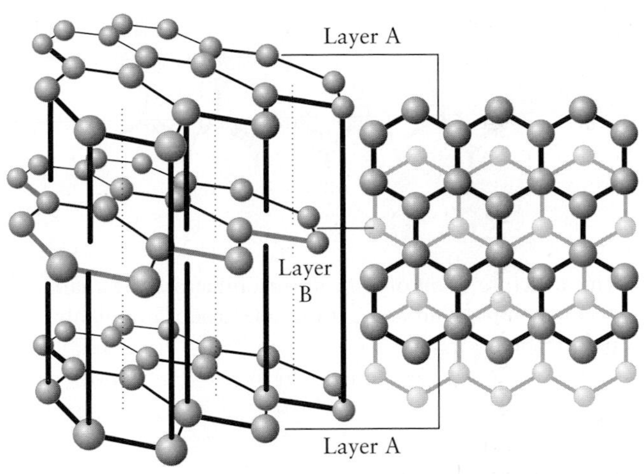

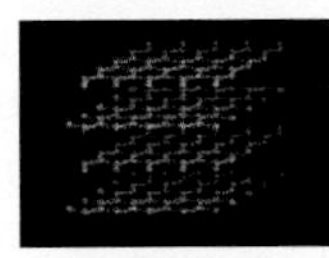

View the internal structure of graphite.

Media Link Animation: Fig. 5.49, Graphite

FIGURE 5.49 Graphite consists of layers of hexagonal rings of sp^2 hybridized carbon atoms. The slipperiness of graphite results from the ease with which the layers can slide over one another when there are impurity atoms lying between the planes.

BOX 5.2 FRONTIERS OF CHEMISTRY: HIGH-TEMPERATURE SUPERCONDUCTORS*

Three of the biggest problems facing scientists and engineers in the future will be providing economical and energy-efficient transportation, improving the capacity and quality of communication systems, and reducing the loss of scarce power resources during transmission. Currently, only about 37% of the energy generated in an electrical power plant actually reaches our homes. Much of the energy loss occurs during transmission. Scientists and engineers believe that all three of these problems can be resolved with the development and refinement of appropriate high-temperature superconductors.

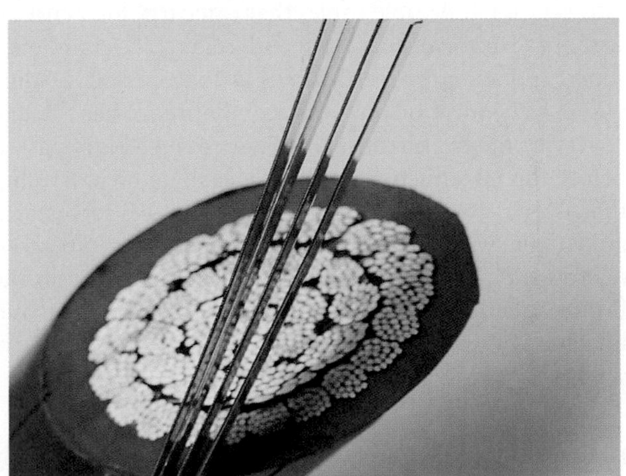

Twenty-five kilograms of this experimental superconducting wire can carry as much current as 1800 kg of the bulky cable shown behind the wire.

Superconductivity is the loss of all electrical resistance when a substance is cooled below a certain characteristic transition temperature (T_s). It was first observed in 1911 in mercury, for which $T_s = 4$ K. Over the years, many other superconducting materials were identified, but, before 1986, none for which the transition temperature was greater than 23 K. It is thought that the low temperatures are required to reduce the effect of the vibrations of the atoms in their crystalline lattice. One problem is that these low-temperature superconductors are expensive to use: they need to be cooled with liquid helium, which costs about \$10 per liter. To put superconducting devices to use on a large scale, higher temperatures would be required.

Research chemists found that they could modify the conducting properties of solids by doping them, a process commonly used to control the properties of semiconductors (see Section 3.14). In 1986, a record-high T_s of 35 K was observed, surprisingly not for a metal, but for a ceramic material (Section 14.24), a lanthanum-copper oxide doped with barium. Then early in 1987, a new record T_s of 93 K was set with yttrium-barium-copper oxide and a series of related compounds. In 1988, two more oxide series of bismuth-strontium-calcium-copper and thallium-barium-calcium-copper

* This box is an adaption of a contribution by C. W. Chu, Texas Center for Superconductivity, University of Houston.

(continued)

Box 5.2 Frontiers of Chemistry: High-Temperature Superconductors *(continued)*

exhibited transition temperatures of 110 K and 125 K, respectively. These temperatures can be reached by cooling the materials with liquid nitrogen, which costs only about $0.20 per liter. Suddenly, superconducting devices became economically viable.

Almost all high-temperature superconductors are hard, brittle oxides that have sheets of copper and oxygen atoms sandwiched between layers of either cations or a combination of cations and oxide ions, and all are derived from their respective parent insulators by doping. Because of their layered structure, their electrical and magnetic properties are strongly anisotropic (Section 5.16). The electric current flows easily along the planes of the copper-oxygen sheets, but only weakly perpendicular to them. Thus, the materials must be positioned with the copper-oxygen planes in a favorable orientation. For example, one challenge to the use of superconductors for transmitting electrical power is the difficulty in making electrical wires from a brittle ceramic material. One solution has been to deposit the superconducting material on the surface of wire or tape made of a metal such as silver or palladium. The metal is usually prepared with a textured surface that helps to align the crystal grains in the desired direction.

An interesting property of superconductors is their ability to create a mirror image of a magnetic field within themselves. As a result, they trap magnetic fields and can even levitate objects (see Fig. 1.58). This property has led to research on the possibility of levitating railroad trains and other vehicles fitted with superconductors over magnetic tracks (see Fig. 3.42).

The high-temperature ceramic superconductors are ionic compounds, although the arrangement of ions suggests that some covalent character is present. Molecular

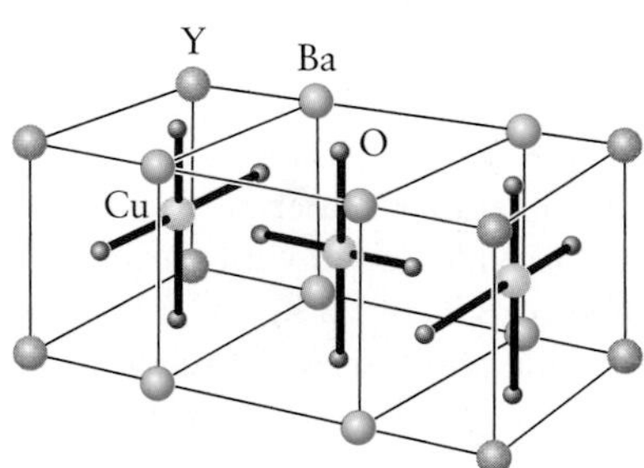

The structure of a 123 superconductor, a ceramic that has the variable formula $YBa_2Cu_3O_{6.5-7.0}$. The numbers 1, 2, and 3 refer to the implied or specified subscripts on the first three elements in the formula.

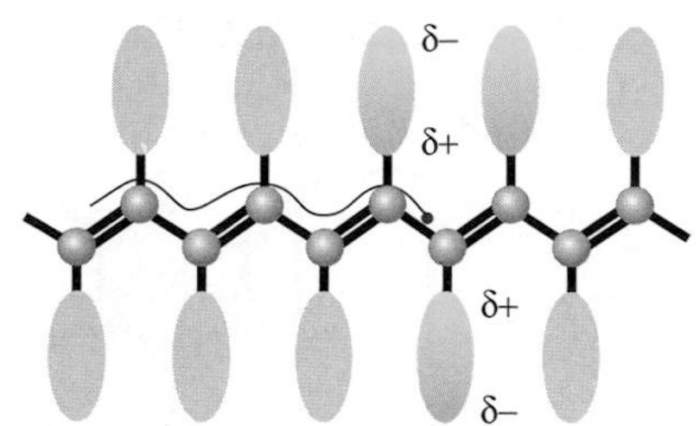

The structure of an organic superconductor. The chain becomes superconducting if the side groups are highly polarizable.

superconductors are also being studied. Conducting polymers (Box 19.2) have long chains of carbon atoms with alternating double bonds. Electrons become delocalized along the chain, thus allowing conductivity approaching that of metals. When highly polarizable side groups are attached all along the chain, they may become superconducting. As electrons travel along the chain, they repel the easily polarized electrons on the side groups, thereby creating transient partial charges that attract other electrons.

HOW MIGHT YOU CONTRIBUTE?

Many technical problems remain to be solved before the promise of high-temperature superconductivity can be realized. Superconducting materials need to be formed into very long, flexible cables that can carry high current without excessive heat loss and that can retain their superconducting properties over a long period of time without chemical or physical degradation. After the discovery of the transistor in 1947, almost 40 years passed before the introduction of the 1-megabyte memory chip that is vital to today's computers. If history serves as a guide, the wonderland of high-temperature superconductor applications is destined to become reality in the foreseeable future, given determination, persistence, and imaginative experimentation by scientists and engineers with vision.

For Further Reading: P. Ball, "Metals from molecules," *Designing the Molecular World* (Princeton, NJ: Princeton University Press, 1994). B. Schechter, "Engineering superconductivity: No resistance," *Scientific American*, August, 2000, pp. 32–33.

THE IMPACT ON MATERIALS

Metals are the mainstays of technology, and it is hardly surprising that so much effort has gone into the development and understanding of their properties. We dealt with their electrical properties in Section 3.14. Here we explore some of their other physical properties.

5.14 The Properties of Metals

In a metallic solid, cations lie in a regular array and are surrounded by a sea of electrons. This structure results in unique properties. The characteristic luster of metals is due to the mobility of their electrons (Section 3.14). An incident light wave is an oscillating electromagnetic field. When it strikes the surface, the electric field of the radiation pushes the mobile electrons backward and forward. These oscillating electrons radiate light, and we see it as a luster—essentially a reflection of the incident light (Fig. 5.50). The electrons oscillate in step with the incident light, so they generate light of the same frequency. In other words, red light reflected from a metallic surface is red and blue light is reflected as blue. That is why an image in a mirror—a thin metallic coating on glass—is a faithful portrayal of the reflected object.

The mobility of their electrons explains a metal's **malleability**, the ability to be hammered into shape, and its **ductility**, the ability to be drawn into wires. Because the cations are surrounded by an electron sea, there is very little directional character in the bonding. As a result, a cation can be pushed past its neighbors without too much effort. A blow from a hammer can drive large numbers of cations past their neighbors. The electron sea immediately adjusts to ensure that the atoms are not simply broken off but remain attached in their new positions (Fig. 5.51). The

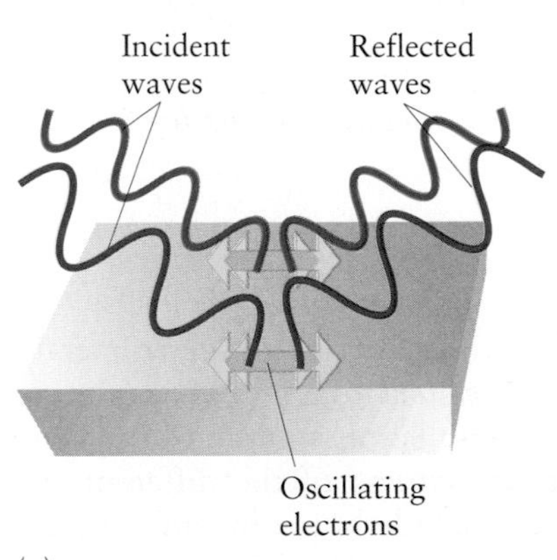

(a)

(b)

FIGURE 5.50 (a) When light of a particular color shines on the surface of a metal, the electrons at the surface oscillate in step. This oscillating motion gives rise to an electromagnetic wave that we perceive as the reflection of the source. (b) Each of these solar mirrors at Sandia National Laboratories in California is positioned at the best angle to reflect sunlight into a collector that uses the incident energy to generate electricity.

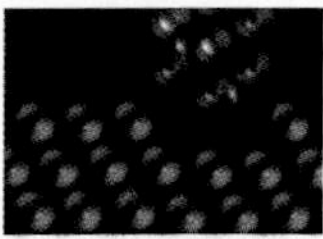

Watch what happens to the atoms inside a metal when it is struck by a hammer.

Media Link Animation: Fig. 5.51, Malleability

(a)

(b)

FIGURE 5.51 (a) A metal is malleable because, when cations are displaced by a blow from a hammer, the mobile electrons can immediately respond and follow the cations to their new positions. (b) This piece of lead has been flattened by a hammer, whereas crystals of the orange-colored ionic compound lead(II) oxide have shattered.

metallic bond is relatively strong. As a result, most metals have high melting points and serve as tough, strong materials for construction.

The differing brittleness of metals can be traced to their crystal structures. The crystal structure of a metal typically has **slip planes**, which are planes of atoms that under stress may slip or slide relative to one another. The slip planes of a ccp structure are the close-packed planes, and careful inspection of a unit cell shows that there are eight sets of slip planes in different directions. As a result, metals with cubic close-packed structures, like copper, are malleable: they can easily be bent, flattened, or pounded into shape. In contrast, a hexagonal close-packed structure has only one set of slip planes; and metals with hexagonal close packing, like zinc or cadmium, tend to be brittle.

The mobility of the valence electrons in a metal accounts for its luster, malleability, and ductility.

5.15 Alloys

Alloys are metallic materials that are mixtures of two or more metals. They are used for many purposes, such as construction, and are central to the transportation and electronics industries. Some common alloys are listed in Table 5.5. In **homogeneous alloys**, atoms of the different elements are distributed uniformly. Examples include brass, bronze, and the coinage alloys. **Heterogeneous alloys**, such as tin-lead solder and the mercury amalgam used to fill teeth, consist of a mixture of crystalline phases with different compositions.

Alloys are discussed further in Sections 16.15 and 16.16

The structures of alloys are more complicated than those of pure metals because we have to pack together two or more types of metal atoms with different radii. The packing problem is now like that of a storekeeper trying to stack oranges and grapefruit in the same pile.

TABLE 5.5 Compositions of Typical Alloys

Alloy	Mass percentage composition
brass	up to 40% zinc in copper
bronze	a metal other than zinc or nickel in copper (casting bronze: 10% Sn and 5% Pb)
cupronickel	nickel in copper (coinage cupronickel: 25% Ni)
pewter	6% antimony and 1.5% copper in tin
solder	tin and lead
stainless steel*	over 12% chromium in iron

* For more detailed information on steels, see Tables 16.2 and 16.6.

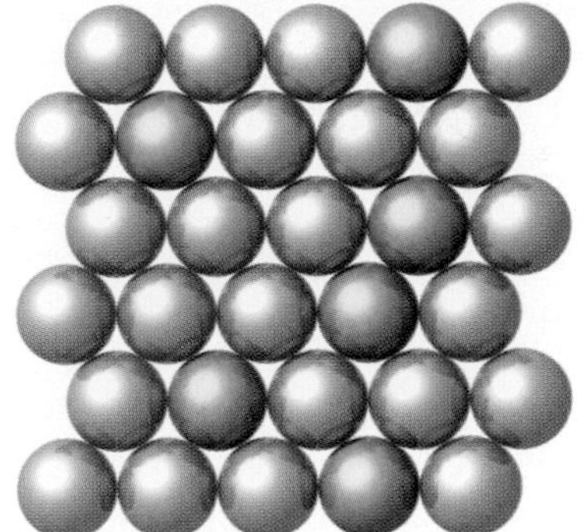

FIGURE 5.52 In a substitutional alloy, the positions of some of the atoms of one metal are taken by atoms of another metal. The two elements must have similar atomic radii.

Because the metallic radii of the *d*-block elements are all similar, for them the stacking problem is easily solved. The elements form an extensive range of alloys with one another because one type of atom can take the place of another with little distortion of the original crystal structure. One example is the copper-zinc alloy used for some "copper" coins. Because zinc atoms are nearly the same size as copper atoms and have similar electronic properties, they can take the place of some of the copper atoms in the crystal. An alloy in which atoms of one metal are substituted for atoms of another metal is called a **substitutional alloy** (Fig. 5.52). Elements that can form substitutional alloys have atoms with atomic radii that differ by no more than about 15%. Because there are slight differences in size and electronic structure, the solute atoms in a substitutional alloy distort the shape of the lattice and hinder the flow of electrons. Because the lattice is distorted, it is harder for one plane of atoms to slip past another. Therefore, although a substitutional alloy has lower electrical and thermal conductivity than the pure element, it is harder and stronger.

Steel is an alloy of about 2% or less carbon in iron. Carbon atoms are much smaller than iron atoms, so they cannot substitute for iron in the crystal lattice. Indeed, they are so small that they can fit into the **interstices**, or holes, in the iron lattice. The resulting material is called an **interstitial alloy** (Fig. 5.53). For two elements to form an interstitial alloy, the atomic radius of the solute element must be less than about 60% of the atomic radius of the host metal. The interstitial atoms interfere with electrical conductivity and with the movement of the atoms forming the lattice. This restricted motion makes the alloy harder and stronger than the pure host metal would be.

Although carbon is not a metal, it is usual to regard steel as an alloy: all commercial steels contain other metals as well as iron and carbon.

Some alloys are softer than the component metals. The presence of big bismuth atoms helps to soften a metal and lower its melting point, much as melons would destabilize a stack of oranges because they just do not fit together well. A low-melting-point alloy of lead, tin, and bismuth is employed to control water sprinklers used in certain fire-extinguishing systems. The heat of the fire melts the alloy, which activates the sprinklers before the fire can spread.

Media Link **Animation:** Fig.5.53, Stainless steel

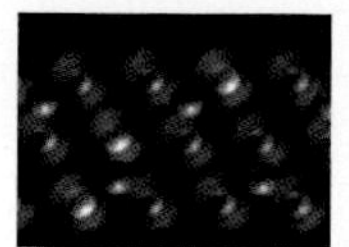

See how the atoms are arranged inside stainless steel.

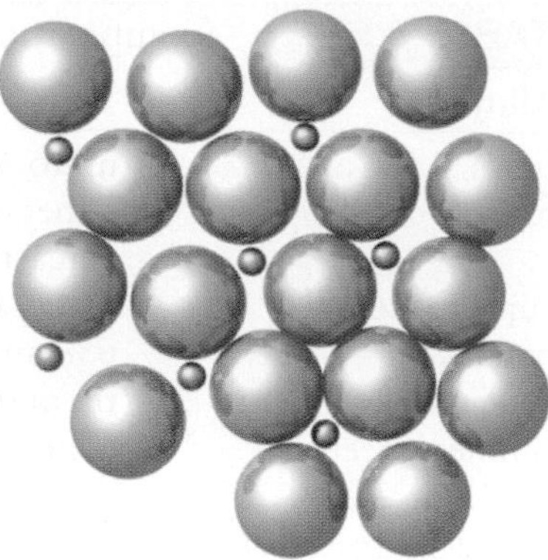

FIGURE 5.53 In an interstitial alloy, the atoms of one metal lie in the gaps between the atoms of another metal. The two elements need to have markedly different atomic radii.

Alloys of metals tend to be harder and have lower electrical conductivity than pure metals. In substitutional alloys, atoms of the solute metal take the place of some atoms of a metal of similar atomic radius. In interstitial alloys, atoms of the solute element fit into the interstices in a lattice formed by atoms of a metal with a larger atomic radius.

5.16 Liquid Crystals

One type of material that has transformed electronic displays is neither a solid nor a liquid, but something intermediate between these two states of matter. **Liquid crystals** are substances that flow like viscous liquids, but their molecules lie in a moderately orderly array, like those in a crystal. They are examples of a **mesophase**, an intermediate state of matter with the fluidity of a liquid and some of the molecular order of a solid. Liquid crystalline materials are finding many applications in the electronics industry because they are responsive to changes in temperature and electric fields.

The word *mesophase* comes from the Greek words for "intermediate appearance."

A typical liquid crystal molecule, such as *p*-azoxyanisole, is long and rodlike (**14**). The rodlike shape causes the molecules to stack together like dry, uncooked spaghetti: they lie parallel but are free to slide past one another along their long axes. Liquid crystals are anisotropic because of this ordering. **Anisotropic** materials have properties that depend on the direction of measurement. The viscosity of liquid crystals is least in the direction parallel to the long axis of the molecules: it is easier for the long rod-shaped molecules to slip past one another along their axes than to move sideways. **Isotropic** materials have properties that do not depend on the direction of measurement. Ordinary liquids are isotropic: their viscosities are the same in every direction.

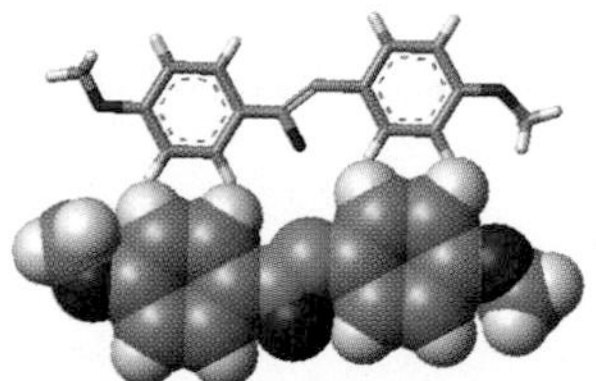

14 *p*-Azoxyanisole

There are three classes of liquid crystals, which differ in the arrangement of their molecules. In the **nematic phase**, the molecules lie together, all in the same direction, but staggered, like cars on a busy multilane

Nematic comes from the Greek word for "thread"; *smectic* comes from the Greek word for "soapy"; *cholesteric* is related to the word *cholesterol*, which comes from the Greek words for "bile solid."

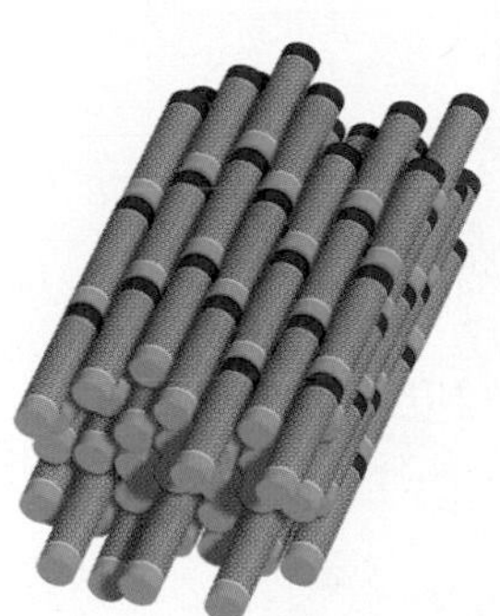

FIGURE 5.54 A representation of the nematic phase of a liquid crystal. The long molecules lie parallel to one another.

FIGURE 5.55 The smectic phase of a liquid crystal. Not only do the molecules lie parallel to one another, they also form sheets.

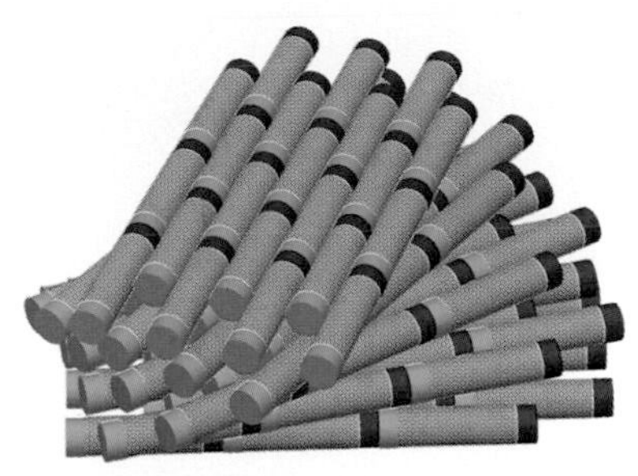

FIGURE 5.56 The cholesteric phase of a liquid crystal. In this phase, sheets of parallel molecules are rotated relative to their neighbors and form a helical structure.

highway (Fig. 5.54). In the **smectic phase**, the molecules line up like soldiers on parade and form layers (Fig. 5.55). Cell membranes are composed mainly of smectic liquid crystals. In the **cholesteric phase**, the molecules form ordered layers, but neighboring layers have molecules at different angles, so the liquid crystal has a helical arrangement of molecules (Fig. 5.56).

We can also classify liquid crystals by their mode of preparation. **Thermotropic liquid crystals** are made by melting the solid phase. The highly viscous liquid crystal phase exists over a short temperature range between the solid and liquid states. Thermotropic liquid crystals become isotropic liquids when they are heated above a characteristic temperature, because then the molecules have enough energy to overcome the attractions that restrict their movement. *p*-Azoxyanisole is a thermotropic liquid crystal. Thermotropic liquid crystals are used in applications such as watches, computer screens, and thermometers. **Lyotropic liquid crystals** are layered structures that result from the action of a solvent on a solid or liquid. Examples are cell membranes and aqueous solutions of detergents and lipids (fats). These molecules, like the detergent sodium lauryl sulfate, have long, nonpolar hydrocarbon chains attached to polar heads (**15**). When the lipids that form cell membranes are mixed with water, they spontaneously form sheets in which the molecules are aligned in rows, forming a double layer, with their polar heads facing outward on each side of the sheet. These sheets form the protective membranes of the cells that make up living tissues.

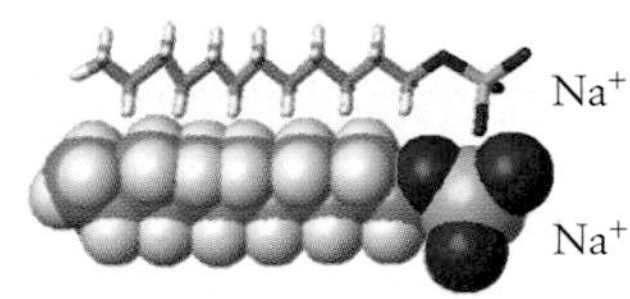

15 Sodium lauryl sulfate

Electronic displays make use of the changes in orientation of liquid crystal molecules caused by an electric field (Fig. 5.57). In a liquid crystal display, layers of a liquid crystal in a nematic phase lie between the surfaces of two glass or plastic plates. The long axis of the molecules in each layer is parallel to the plates but is oriented in such a way that polarized light (see Section 16.7) is transmitted. However, where a potential difference is applied, the molecules become oriented with the electrical field and become opaque to the polarized light, hence forming a dark spot on

FIGURE 5.57 Liquid crystals are used in flat-screen displays for laptop computers and, increasingly, for flat displays for desktop computers.

the screen. Cholesteric liquid crystals are also of interest because the helical structure unwinds slightly as the temperature is changed. Because the twist of the helical structure affects the optical properties of the liquid crystal, such as its color, these properties change with temperature. The effect is utilized in liquid crystal thermometers.

Liquid crystals have a degree of order characteristic of solid crystals, but they can flow like liquids. They are mesophases, intermediate between solids and liquids; their properties can be modified by electric fields and changes in temperature.

SKILLS YOU SHOULD HAVE MASTERED

1 Predict the relative strengths of ion-dipole interactions, Section 5.2.

2 Explain how London forces arise and how they vary with the polarizability of an atom and the size and shape of a molecule, Section 5.4.

3 Predict the relative order of the boiling points of two substances from the strengths of their intermolecular forces, Examples 5.1 and 5.2.

4 Identify molecules that can experience hydrogen bonding, Self-Test 5.3.

5 Describe the structure of a liquid and explain how viscosity and surface tension vary with temperature and the strength of intermolecular forces, Sections 5.6 and 5.7.

6 Distinguish metallic solids, ionic solids, network solids, and molecular solids by their structures and by their properties, Sections 5.8 and 5.14.

7 Determine the fraction of occupied space in a given crystal lattice, Section 5.9.

8 Give the coordination number of an atom or ion in a given crystal lattice, Self-Test 5.4.

9 Find the number of atoms or ions in a given unit cell, Self-Test 5.7.

11 Deduce the crystal structure of a metal from its density, Example 5.3.

10 Predict the density of an ionic solid from its crystal structure, Example 5.4.

12 Identify the different types of liquid crystals, Section 5.16.

EXERCISES

Intermolecular Forces

5.1 Identify the kinds of intermolecular forces that might arise between molecules of the following substances: (a) NO_2; (b) N_2H_4; (c) HF; (d) CI_4.

5.2 Identify the kinds of intermolecular forces that might arise between molecules of the following substances: (a) P_4; (b) *cis*-CHCl=CHCl; (c) *trans*-CHCl=CHCl; (d) Br_2.

5.3 For which of the following molecules will dipole-dipole interactions be important: (a) CH_4; (b) CH_3Cl; (c) CH_2Cl_2; (d) $CHCl_3$; (e) CCl_4?

5.4 For which of the following molecules will dipole-dipole interactions be important: (a) O_2; (b) O_3; (c) CO_2; (d) SO_2?

5.5 Place the following types of molecular and ion interactions in order of increasing magnitude: (a) ion-dipole; (b) induced dipole-induced dipole; (c) dipole-dipole in the gas phase; (d) ion-ion; (e) dipole-dipole in the solid phase.

5.6 Explain why ionic solids like NaCl have high melting points yet dissolve readily in water, whereas network solids like diamond have very high melting points and do not dissolve in solvents.

5.7 Which of the following molecules are likely to form hydrogen bonds: (a) H_2S; (b) CH_4; (c) H_2SO_3; (d) PH_3?

5.8 Which of the following molecules are likely to form hydrogen bonds: (a) CH_3OCH_3; (b) CH_3COOH; (c) CH_3CH_2OH; (d) CH_3CHO?

5.9 Suggest, giving reasons, which substance in each pair is likely to have the higher normal melting point (Lewis structures may help your arguments): (a) HCl or NaCl; (b) $C_2H_5OC_2H_5$ (diethyl ether) or C_4H_9OH (butanol); (c) HF or HCl; (d) H_2O or CH_3OH.

5.10 Suggest, giving reasons, which substance in each pair is likely to have the higher normal boiling point: (a) H_2S or H_2O; (b) NH_3 or PH_3; (c) KBr or CH_3Br; (d) CH_4 or SiH_4.

5.11 Using VSEPR theory, predict the shapes of each of the following molecules. Which of each pair should have the higher boiling point: (a) PF_3 or PCl_3; (b) SO_2 or CO_2; (c) BF_3 or BCl_3; (d) AsF_3 or AsF_5?

5.12 Using VSEPR theory, predict the shapes of each of the following molecules. Which of each pair should have the higher boiling point: (a) SF_4 or SF_6; (b) BF_3 or ClF_3; (c) SF_4 or CF_4; (d) *cis*-CHCl=CHCl or *trans*-CHCl=CHCl?

5.13 Calculate the ratio of the potential energies for the interaction of a water molecule with an Al^{3+} ion and with a Be^{2+} ion. Which will attract a water molecule more strongly?

5.14 Calculate the ratio of the potential energies for the interaction of a water molecule with a Ca^{2+} ion and with an In^{3+} ion. Which will attract a water molecule more strongly?

5.15 Calculate the ratio of the potential energies for the interaction of a water molecule with an Al^{3+} ion and with a Ga^{3+} ion. Which will attract a water molecule more strongly?

5.16 Calculate the ratio of the potential energies for the interaction of a water molecule with an Na^+ ion and with a K^+ ion. Which will attract a water molecule more strongly?

5.17 Account for the following observations in terms of the type and strength of intermolecular forces: (a) The melting point of solid xenon is −112°C and that of solid argon is −189°C. (b) The vapor pressure of diethyl ether ($C_2H_5OC_2H_5$) is greater than that of water. (c) The boiling point of pentane, $CH_3(CH_2)_3CH_3$, is 36.1°C whereas that of 2,2-dimethylpropane (also known as neopentane) is 9.5°C.

5.18 Classify the following solids as ionic, network, or molecular: (a) quartz, SiO_2; (b) limestone, $CaCO_3$; (c) dry ice, CO_2; (d) sucrose, $C_{12}H_{22}O_{11}$; (e) polyethylene, a polymer of repeating $-CH_2CH_2-$ units.

Liquid Structure

5.19 Predict which liquid has the greater surface tension: (a) *cis*-dichloroethene or *trans*-dichloroethene (see structures **5** and **6**); (b) benzene at 20°C or benzene at 60°C.

5.20 Predict which substance has the greater viscosity in its liquid form at 0°C: (a) ethanol, CH_3CH_2OH, or dimethyl ether, CH_3OCH_3; (b) butane, C_4H_{10}, or propanone, $CH_3C(=O)CH_3$.

5.21 Place the following molecules in order of increasing viscosity at 50°C: C_6H_5SH, C_6H_5OH, C_6H_6.

5.22 The surface of glass contains many —OH groups that are bonded to the silicon atoms of SiO_2, which is the major component of glass. If the glass is treated with $Si(CH_3)_3Cl$ (chlorotrimethylsilane), a reaction takes place to eliminate HCl and attach the Si atom to the oxygen atom:

$$\text{(surface of glass)—OH} + Si(CH_3)_3Cl \longrightarrow \text{(surface of glass)—OSi(CH}_3)_3 + HCl$$

How will this rearrangement affect the interaction of liquids with the glass surface?

5.23 The height, h, of a column of liquid in a capillary tube can be estimated by using $h = 2\gamma/gdr$, where γ is the surface tension, d is the density of the liquid, g is the acceleration of free fall, and r is the radius of the tube. Which will rise higher in a tube that is 0.15 mm in diameter at 25°C, water or ethanol? The density of water is 0.997 $g\cdot cm^{-3}$ and that of ethanol is 0.79 $g\cdot cm^{-3}$. See Table 5.3.

5.24 Calculate the height to which liquid water will rise in a capillary tube that is 0.30 mm in diameter at (a) 25°C and (b) 100°C. The density of water at 25°C is 0.997 $g\cdot cm^{-3}$, and at 100°C, it is 0.958 $g\cdot cm^{-3}$. See Table 5.3 and Exercise 5.23.

Solid Structures

5.25 Potassium crystallizes in a bcc structure. The atomic radius of potassium is 235 pm. Determine (a) the number of atoms per unit cell; (b) the coordination number of the lattice; (c) the length of the side of the unit cell.

5.26 The metal polonium (which was named by Marie Curie after her homeland, Poland) crystallizes in a primitive cubic structure, with an atom at each corner of a cubic unit cell. The atomic radius of polonium is 167 pm. Sketch the unit cell and determine (a) the number of atoms per unit cell; (b) the coordination number of an atom of polonium; (c) the length of the side of the unit cell.

5.27 Calculate the density of each of the following metals from the data: (a) platinum, fcc structure, atomic radius 139 pm; (b) cesium, bcc structure, atomic radius 272 pm.

5.28 Determine the density of each of the following metals from the data given: (a) nickel, fcc structure, atomic radius 125 pm; (b) rubidium, bcc structure, atomic radius 250 pm.

5.29 Calculate the atomic radius of each of the following elements from the data given: (a) platinum, a rare metal used in jewelry and catalysis, fcc structure, density 21.450 $g\cdot cm^{-3}$; (b) tantalum, bcc structure, density 16.654 $g\cdot cm^{-3}$.

5.30 Calculate the atomic radius of the following elements from the data given: (a) silver, fcc structure, density 10.500 $g\cdot cm^{-3}$; (b) chromium, bcc structure, density 7.190 $g\cdot cm^{-3}$.

5.31 Aluminum metal has a density of 2.70 $g\cdot cm^{-3}$ and crystallizes in a cubic lattice with a unit cell edge of 404 pm. (a) Which type of cubic unit cell is formed by aluminum? (b) What is the coordination number of aluminum?

5.32 If the edge length of a fcc unit cell of RbI is 732.6 pm, how long would the edge of a cubic single crystal of RbI be that contains 1.00 mol RbI?

5.33 The edge length of the fcc unit cell of NaCl is 562.8 pm. (a) How many unit cells are present in a single crystal of NaCl (table salt) that is a cube with edges 1.00 mm in length? (b) How many moles of NaCl are present in this crystal?

5.34 Krypton crystallizes with a face-centered cubic unit cell of edge 559 pm. (a) What is the density of solid krypton? (b) What is the atomic radius of a krypton atom? (c) What is the volume of one krypton atom? (d) What percentage of the unit cell is empty space if each atom is treated as a hard sphere?

Nonmetallic Solids

5.35 Calculate the number of cations, anions, and formula units per unit cell in the following solids: (a) the cesium-chloride unit cell shown in Fig. 5.39; (b) the rutile (TiO_2) unit cell shown below. (c) What are the coordination numbers of the ions in rutile?

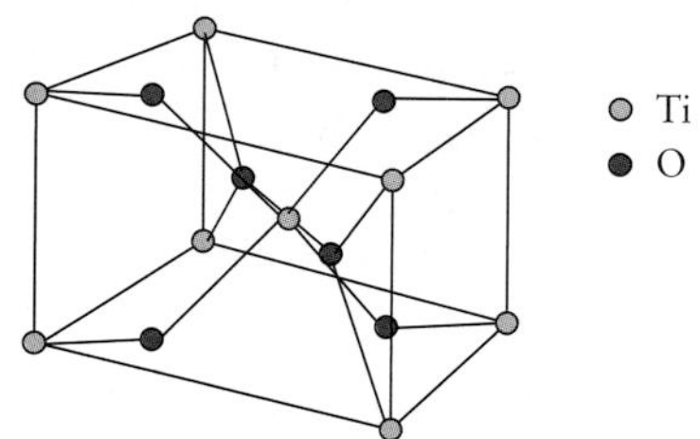

5.36 Calculate the number of cations, anions, and formula units per unit cell in the following solids: (a) the rock-salt unit cell shown in Fig. 5.37; (b) the fluorite (CaF_2) unit cell shown below. (c) What are the coordination numbers of the ions in fluorite?

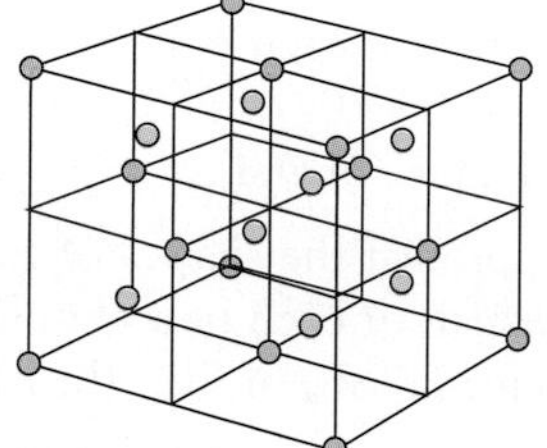

5.37 A unit cell of one of the new high-temperature superconductors is shown below. What is its formula?

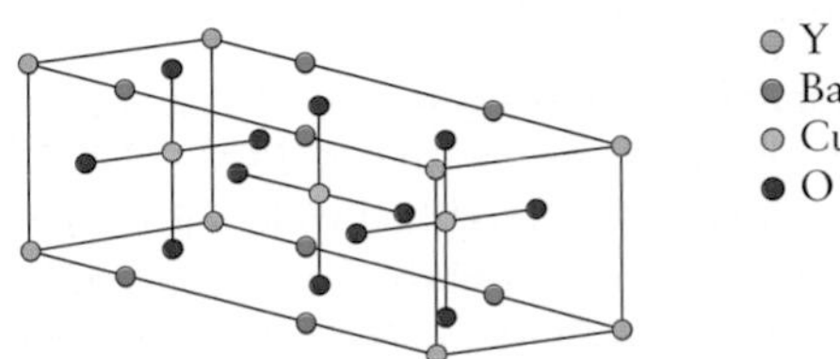

5.38 A unit cell of the mineral perovskite is drawn below. What is its formula?

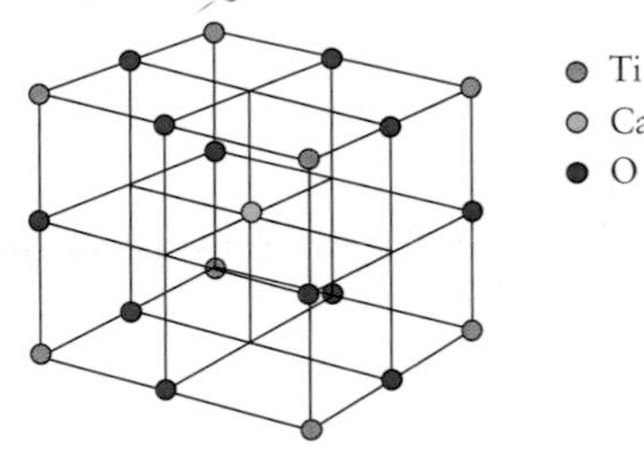

5.39 Use radius ratios to predict the coordination number of the cation in (a) RbF; (b) MgO; (c) NaBr. (See Fig. 1.43.)

5.40 Use radius ratios to predict the coordination number of the cation in (a) KBr; (b) LiBr; (c) BaO. (See Fig. 1.43.)

5.41 Estimate the density of the following solids from the atomic radii of the ions given in Fig. 1.43: (a) calcium oxide (rock-salt structure, Fig. 5.37); (b) cesium bromide (cesium-chloride structure, Fig. 5.39).

5.42 Calculate the density of the following solids: (a) sodium iodide (rock-salt structure, Fig. 5.37), distance between centers of Na^+ and I^- ions is 322 pm; (b) cesium iodide (cesium-chloride structure, Fig. 5.39), distance between the centers of the Cs^+ and I^- ions is 356 pm.

5.43 Glucose, benzophenone ($C_6H_5C(=O)C_6H_5$), and methane are examples of compounds that form molecular solids. The structures of glucose and benzophenone are given below. (a) What types of forces hold these molecules in a molecular solid? (b) Place the solids in order of increasing melting point.

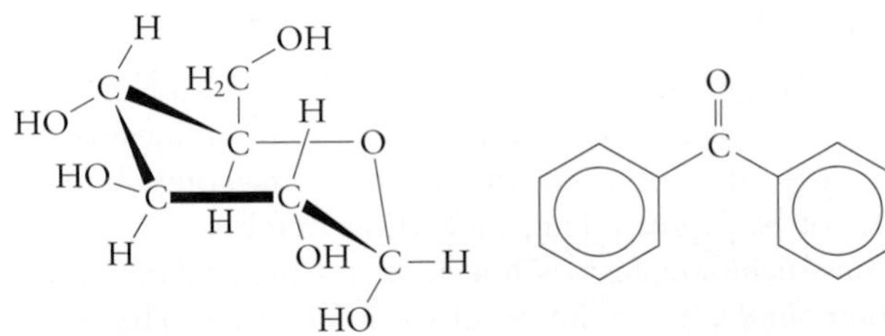

5.44 Chloromethane (CH_3Cl), methane, and acetic acid (CH_3COOH) form molecular solids. (a) What types of forces hold these molecules in a molecular solid? (b) Place the solids in order of increasing melting point.

5.45 Name two compounds that exist as network solids at room temperature and pressure.

5.46 Name four elements that, in at least one allotropic form, are network solids.

The Impact on Materials: Metals

5.47 When iron surfaces are exposed to ammonia at high temperatures, "nitriding"—the incorporation of nitrogen into the iron lattice—occurs. The atomic radius of iron is 124 pm. (a) Is the alloy interstitial or substitutional? Justify your answer. (b) How do you expect nitriding to change the properties of the iron?

5.48 Silicon can be doped with small amounts of phosphorus to create a semiconductor used in transistors. (a) Is the alloy interstitial or substitutional? Justify your answer. (b) How do you expect the properties of the doped material to differ from those of pure silicon?

5.49 The electrical conductivity of graphite parallel to its planes is different from that perpendicular to them. Parallel to the planes, the conductivity decreases as the temperature is raised; but perpendicular to them, it rises. In what sense is graphite a metallic conductor or a semiconductor?

5.50 Zinc oxide is a semiconductor that loses a small percentage of the oxygen atoms in a vacuum but regains these atoms when heated in oxygen. The electrical conductivity increases as the oxygen atoms are lost in the vacuum but decreases when they are regained upon heating in oxygen. Account for these observations.

5.51 Calculate the relative number of atoms of each element contained in each alloy: (a) coinage cupronickel, which is 25% Ni by mass in copper; (b) a type of pewter that is about 7% antimony and 3% copper by mass in tin.

5.52 Calculate the relative number of atoms of each element contained in each alloy: (a) Wood's metal, which is a low-melting-point alloy used to trigger automatic sprinkler systems and is 12.5% tin, 12.5% cadmium, and 24% lead by mass in bismuth; (b) a steel that is 1.75% by mass carbon in iron.

5.53 Why do long hydrocarbon molecules that do not have multiple bonds, such as decane, $CH_3(CH_2)_8CH_3$, not form liquid crystals?

5.54 Consider the structure of *p*-azoxyanisole, **14**. (a) Using simple VSEPR rules, draw a picture that represents the shape of the molecule and predict the C—N—N bond angles. (b) What features of the bonding of this molecule give rise to its rodlike nature? (c) Use your conclusions in part (b) to design another simple organic molecule that is composed only of C, H, N, and O and might be likely to form rodlike, liquid crystalline materials.

5.55 When long surfactant molecules having a polar headgroup and a nonpolar "tail" are placed into water, *micelles* are formed in which the nonpolar tails aggregate with the polar headgroups pointing out toward the solvent. *Inverse micelles* are similar but have the nonpolar regions pointing outward. How can inverse micelles be produced?

5.56 The molecule *p*-azoxyanisole (**14**) has a liquid crystalline range of approximately 117°–137°C. How might this molecule be modified to lower its melting point and make it more suitable for lower temperature applications (near room temperature, for example)? Remember that substances exhibit liquid crystalline properties over a short temperature range just above the melting point of the solid.

INTEGRATED EXERCISES

5.57 (a) Calculate the surface areas of the isomers 2,2-dimethylpropane and pentane. Assume that 2,2-dimethylpropane is spherical with a radius of 254 pm and that pentane can be approximated by a rectangular prism with dimensions 295 pm × 692 pm × 766 pm.
(b) Which has the larger surface area?
(c) Which do you expect to have the higher boiling point?

Media Link 5.58 Using the Molecules Database on the Web site for this book examine the structure of tristearin, formed by the condensation of three molecules of stearic acid, $C_{17}H_{35}COOH$, with glycerol, $C_3H_5(OH)_3$. (a) Draw the Lewis structure of tristearin. (b) What types of forces are responsible for the molecule adopting the shape shown? (c) Do you expect this molecule to be soluble or insoluble in water? For comparison, glycerol is very soluble, but stearic acid is only slightly soluble in water.

5.59 Using the Molecules Database on the Web site for this book, determine the chemical formula and write the Lewis structure of (a) anthracene; (b) phosgene; (c) glutamic acid. For each molecule, list the types of intermolecular forces expected to be significant.

5.60 All the alkali metals crystallize with bcc structures. (a) Find a general equation relating the metallic radius to the density of a bcc solid of an element in terms of its molar mass and use it to deduce the atomic radii of the elements, given the following densities (in $g \cdot cm^{-3}$): Li, 0.53; Na, 0.97; K, 0.86; Rb, 1.53; Cs, 1.87. (b) Find a factor for converting a bcc density to a cubic close-packed density of the same element. (c) Calculate what the densities of the alkali metals would be if they were cubic close-packed. (d) Which, if any, would float on water?

5.61 All noble gases except helium crystallize with cubic close-packed structures at very low temperatures. Find an equation relating the atomic radius to the density of a cubic close-packed solid of given molar mass and apply it to deduce the atomic radii of the noble gases, given the following densities (in $g \cdot cm^{-3}$): Ne, 1.20; Ar, 1.40; Kr, 2.16; Xe, 2.83; Rn, 4.4 (estimated).

5.62 (a) Calculate the ratio of potential energies for an ion-ion interaction for Li^+ and for K^+ with the same anion (b) Repeat the ratio calculation, but for an ion-dipole interaction between the Li^+ and K^+ and a water molecule. (c) What do these numbers indicate about the relative importance of hydration of salts of lithium and potassium?

5.63 Metals with bcc structures, like tungsten, are not close packed. Therefore, their densities would be greater if they were to change to a cubic close-packed structure (under pressure, for instance). What would the density of tungsten be if it had a cubic close-packed structure rather than bcc? Its actual density is 19.3 $g \cdot cm^{-3}$.

5.64 An oxide of niobium has a unit cell in which there are oxide ions at the middle of each edge and niobium ions at the center of each face. What is the empirical formula of this oxide?

5.65 A number of metal oxides are known to form *nonstoichiometric* compounds, a description meaning that the ratios of atoms that make up the compound cannot be expressed in small whole numbers. Nonstoichiometry does not occur for molecular compounds, but it can occur for ionic solids. The regular arrangement of the atoms in a crystal of a nonstoichiometric compound is not perfect and some places where one would have expected to find atoms are vacant. Transition metals most easily form nonstoichiometric compounds because of the number of oxidation states they can have. For example, a titanium oxide with formula $TiO_{1.18}$ is known. (a) Calculate the average oxidation state of titanium in this compound. (b) If the compound has both Ti^{2+} and Ti^{3+} ions present, what fraction of the titanium ions will be in each oxidation state?

5.66 Uranium dioxide, UO_2, can be further oxidized to give a nonstoichiometric compound UO_{2+x}, where $0 < x < 0.25$. See Exercise 5.65 for a description of nonstoichiometric compounds. (a) What is the average oxidation state of uranium in a compound with composition $UO_{2.17}$? (b) If we assume that the uranium exists in either the +4 or the +5 oxidation state, what is the fraction of uranium ions in each?

5.67 Indicate whether the following statements are true or false: (a) If there is an atom present at the corner of a unit cell, there must be the same type of atom at all the corners of the unit cell. (b) A unit cell must be defined so that there are atoms at the corners. (c) If one face of a unit cell has an atom in its center, then the face opposite that face must also have an atom at its center. (d) If one face of a unit cell has an atom in its center, all the faces of the unit cell must also have atoms at their centers.

5.68 Indicate whether the following statements are true or false: (a) Because cesium chloride has chloride ions at the corners of the unit cell and a cesium ion at the center of the unit cell, it is classified as having a body-centered unit cell. (b) The density of the unit cell must be the same as the density of the bulk material. (c) When x-rays are passed through a single crystal of a compound, the x-ray beam will be diffracted because it interacts with the electrons in the atoms of the crystal. (d) All the angles of a unit cell must be equal to 90°.

5.69 The unit cell of rutile, TiO_2, shown in Exercise 5.35 is tetragonal. If the unit cell dimensions are $a = b = 459$ pm and $c = 296$ pm, what is the density of TiO_2? You will need to use the answer from Exercise 5.35 (b) to answer this question.

5.70 As can be seen in Fig. 5.33, not all unit cells are cubic. Other types of unit cells have different restrictions placed on the lattice parameters (edge lengths and angles). Calculations for unit cell properties, such as cell volume, density, and distances between atoms, are similar to those for cubic unit cells, but the geometry is more complex. (a) Calculate the distance between a corner atom and the atom at the body center of a tetragonal unit cell that has $a = b = 549$ pm and $c = 769$ pm. (b) What is the volume of this unit cell?

5.71 For the two-dimensional arrays shown below, draw unit cells that, when repeated, generate the entire two-dimensional lattice.

(a)
✻ ✻ ✻ ✻
✻ ✻ ✻ ✻
✻ ✻ ✻ ✻
✻ ✻ ✻ ✻

(b)
✻ ✻ ✻ ✻
  ✻ ✻ ✻ ✻
✻ ✻ ✻ ✻
  ✻ ✻ ✻ ✻

5.72 For the two-dimensional arrays shown below, draw unit cells that, when repeated, generate the entire two-dimensional lattice.

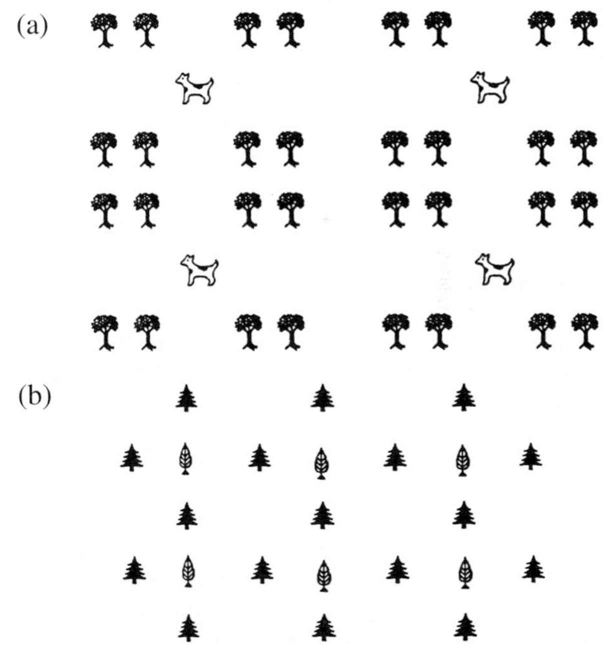

5.73 Salts of organic molecules such as acetic acid and methanol can be prepared. For example, you could prepare sodium acetate, $NaCH_3CO_2$, and sodium methoxide, $NaCH_3O$. How do you expect the forces that hold these compounds together in the solid state to differ from those that hold together salts like sodium chloride or sodium bromide?

5.74 (a) From standard reference sources, find the boiling points of the following substances: methane (CH_4), ethane (C_2H_6), propane ($CH_3CH_2CH_3$), butane

($CH_3CH_2CH_2CH_3$, also called *n*-butane), pentane ($CH_3CH_2CH_2CH_2CH_3$, also called *n*-pentane), 2-methylpropane ($(CH_3)_3CH$, also called isobutane), 2-methylbutane ($CH_3CH_2CH(CH_3)_2$, also known as isopentane), and 2,2-dimethylpropane ($C(CH_3)_4$, also known as neopentane). (b) How do the boiling points vary with molecular size? (c) Is there a correlation between molecular shape and boiling point?

5.75 Glass is composed mainly of silica, SiO_2, with various oxides added to change the properties. The property differences are substantial, and different amounts of added oxide compounds produce glasses with very different applications derived from these properties. Often the added material is B_2O_3 and the products are called borosilicate glasses. From standard reference sources, determine how the properties of glass change as the amount of B_2O_3 is increased.

5.76 In an aqueous solution, a certain amount of time is required for solute molecules or ions to migrate through the solution. The rate at which this migration takes place sets an upper limit on how fast reactions can occur, because no reaction can occur faster than the ions can be supplied. This limit is known as the *diffusion-controlled rate*. It has been found that the diffusion rate for hydrogen ions is about three times faster than that for other ions in aqueous solution. Explain why this is so.

5.77 A commonly occurring mineral has a cubic unit cell in which the metal cations M occupy the corners and face centers. Inside the unit cell, there are anions A that occupy all the tetrahedral holes created by the cations. What is the empirical formula of the M_mA_a compound?

5.78 Tetrahedral and octahedral interstitial holes are formed by the vacancies left when anions pack in a cubic close-packed array. (a) Which hole can accommodate the larger ions? (b) What is the size ratio of the largest metal cation that can occupy an octahedral hole to the largest that can occupy the tetrahedral hole while maintaining the close-packed nature of the anion lattice? (c) If half the tetrahedral holes are occupied, what will be the empirical formula of the compound M_xA_y, where M represents the cations and A the anions?

5.79 The density of cesium chloride is 3.988 $g \cdot cm^{-3}$. Calculate the amount of empty space in a cesium chloride lattice with the ions treated as hard spheres.

5.80 Use information from Exercise 5.36 to calculate the amount of empty space in a calcium fluoride lattice with the ions treated as hard spheres. The density of CaF_2 is 3.180 $g \cdot cm^{-3}$.

5.81 Lanthanum exists in three different solid-state phases. The form normally encountered is hexagonal, whereas the other two phases are cubic. One is face-centered cubic and the other body-centered cubic. The lattice constants are $a = 529.6$ pm for the fcc phase and $a = 426$ pm for the bcc phase. (a) Which of these two phases would you expect to be more stable under high pressures, the face-centered cubic or the body-centered cubic form? Substantiate your answer with the appropriate calculations. (b) The experimental density for the hexagonal phase is 6.145 $g \cdot cm^{-3}$. Is this what you would have expected on the basis of your calculations in part (a)? Explain.

5.82 Many ionic compounds are considered to pack in such as way that the anions form a close-packed lattice in which the metal cations fill holes or interstitial sites left between the anions. These lattices, however, may not necessarily be as tightly packed as the label "close-packed" implies. The radius of an F^- ion is approximately 133 pm. The edge distances of the cubic unit cells of LiF, NaF, KF, RbF, and CsF, which all pack in the rock-salt structure, are 568 pm, 652 pm, 754 pm, 796 pm, and 850 pm, respectively. Which of these lattices, if any, can be thought to be based on close-packed arrays of F^- ions? Discuss.

5.83 Using the density of calcium fluoride (CaF_2, found in nature as the mineral fluorite), which is known to be 3.180 $g \cdot cm^{-3}$, and the information given in Exercise 5.36, calculate (a) the edge length a of the unit cell and (b) the Ca-F separation in fluorite. (c) Compare this value to that expected from the ionic radii tabulated in Fig. 1.43.

5.84 (a) If a pure element crystallizes with a primitive cubic lattice, what percentage of the unit cell is empty space? (b) How does this compare with the percentage of empty space in a fcc unit cell?

5.85 Consider a metallic element that crystallizes in a cubic close-packed lattice. The edge length a of the unit cell is 408 pm. If close-packed layers are deposited on a flat surface to a depth (of metal) of 0.125 mm, how many close-packed layers are present?

5.86 Prove that the percentage of occupied space for a hexagonal close-packed lattice is the same as that for a cubic close-packed lattice. Refer to Fig. 5.24.

5.87 The diagram below shows two lattice planes from which two parallel x-rays are diffracted. If the two incoming x-rays are in phase, show that the Bragg equation $2d \sin 2 = n\lambda$ is true when n is an integer. Refer to Major Technique 3 on x-ray diffraction, which follows this set of exercises.

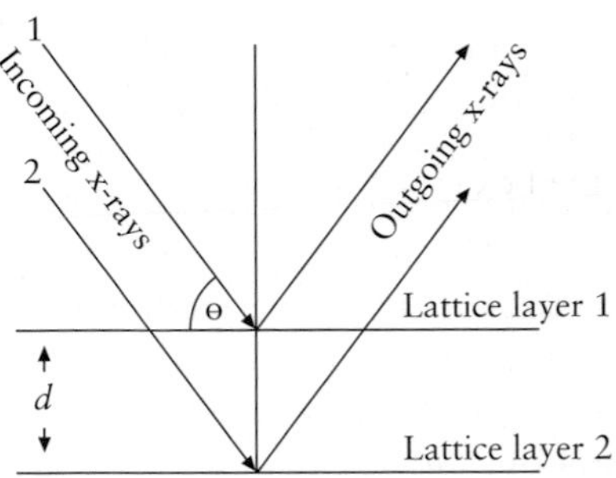

5.88 X-rays generated from a copper target have a wavelength of 154 pm. If an x-ray beam of this type is passed through a single crystal of NaBr, a diffracted beam is observed when the incident x-ray beam is at an angle of 7.42° to the surface of the crystal. What is the minimum spacing between the planes in the crystal that give rise to this diffracted beam? Refer to Major Technique 3 on x-ray diffraction, which follows this set of exercises.

5.89 The unit cell edge length of lithium fluoride is 401.8 pm. What is the smallest angle at which the x-ray beam generated from a Mo target ($\lambda = 71.07$ pm) must strike the planes making up the faces of the unit cell in order for it to be diffracted from those planes. Refer to Major Technique 3 on x-ray diffraction, which follows this set of exercises.

5.90 A reflection from a cubic crystal was observed at a glancing angle of 12.1° when x-rays of wavelength 152 pm were used. (a) What is d, the separation of layers causing this reflection? (b) At what glancing angle would you detect the reflection arising from planes at twice the layer separation? Refer to Major Technique 3 on x-ray diffraction, which follows this set of exercises.

5.91 "Graphite bisulfates" are formed by heating graphite with a mixture of sulfuric and nitric acids. In the reaction, the graphite planes are partially oxidized. There is approximately one positive charge for every 24 carbon atoms, and the HSO_4^- anions are distributed between the planes. (a) What effect should this have on the electrical conductivity? (b) What effect would you expect this to have on the x-ray diffraction pattern observed for this material? Refer to Major Technique 3 on x-ray diffraction, which follows this set of exercises.

MAJOR TECHNIQUE 3 X-RAY DIFFRACTION

A great deal of our knowledge about the interior of solids has come from *x-ray diffraction*. This important technique is used to determine the arrangement of atoms in solid compounds and to determine bond lengths and angles. Moreover, almost all recent advances in molecular biology have stemmed from the application of this technique to determine the structures of molecules such as proteins and nucleic acids.

1 The Technique

The phenomenon called *interference* occurs when two or more waves pass through the same region of space; interference is observed as an increase or a decrease in the total amplitude of the wave (Fig. 1). *Constructive interference* occurs when the peaks of one wave coincide with the peaks of another wave: it results in an increase in the total amplitude of the wave. If the waves are electromagnetic radiation, the increased amplitude corresponds to an increased intensity of the radiation because the intensity of a wave of electromagnetic radiation is proportional to the square of the amplitude of the wave. *Destructive interference* occurs when the peaks of one wave coincide with the troughs of the other wave: it produces a decrease in the total amplitude of the wave and a reduction in intensity (Fig 2).

The phenomenon of *diffraction* is interference between waves that arises when there is an object in their path. One of the earliest demonstrations of interference was the *Young's slit experiment*, in which light passes through two slits and gives rise to a pattern on a screen (Fig. 3). Each location on the screen corresponds to slightly different path lengths for the two routes from the slits to the location; some of the differences in path length result in constructive interference and others result in destructive interference, thereby giving rise to the pattern of dark and light stripes. If you were presented with only the pattern and were told the wavelength of the light and the distance of the detection screen from the screen containing the slits, it would be possible to work out the spacing of the two slits.

An x-ray diffraction experiment is a very elaborate version of the Young's slit experiment. The regular layers of atoms in a crystal act as a three-dimensional collection of slits and give rise to a diffraction pattern that varies as the crystal is rotated and the "slits" are brought into a new arrangement. The task for the x-ray crystallographer is to use the diffraction pattern to determine the spacing and arrangement of the "slits"

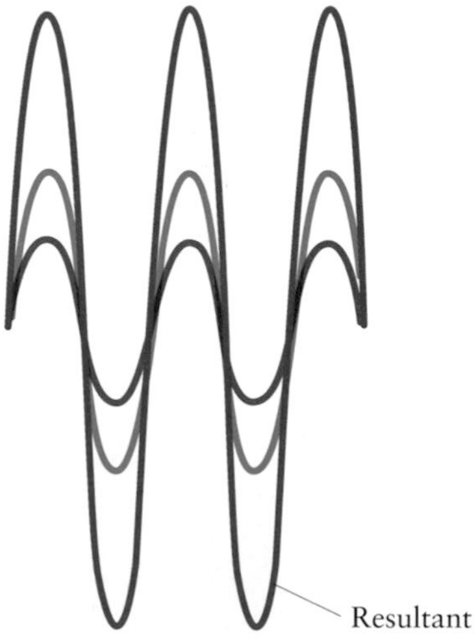

Resultant

FIGURE 1 In constructive interference, two waves (red and green) pass through the same region of space and their amplitudes add together to give a resultant with an enhanced amplitude (blue). The intensity of the radiation is proportional to the square of the amplitude, so constructive interference increases the intensity.

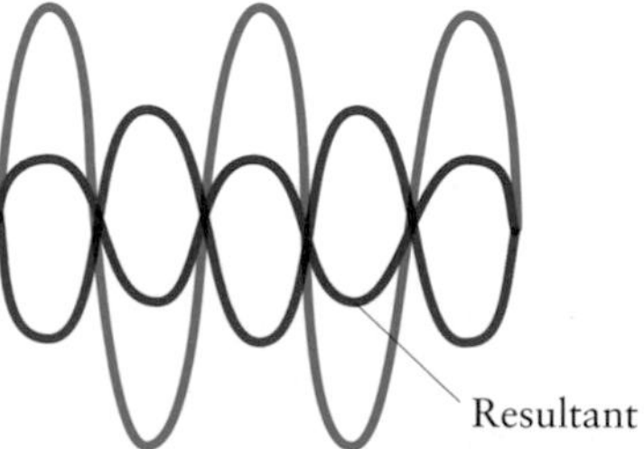

FIGURE 2 In destructive interference, the amplitudes of two waves (red and green) combine to give a resultant with a reduced amplitude (blue). Destructive interference decreases the intensity.

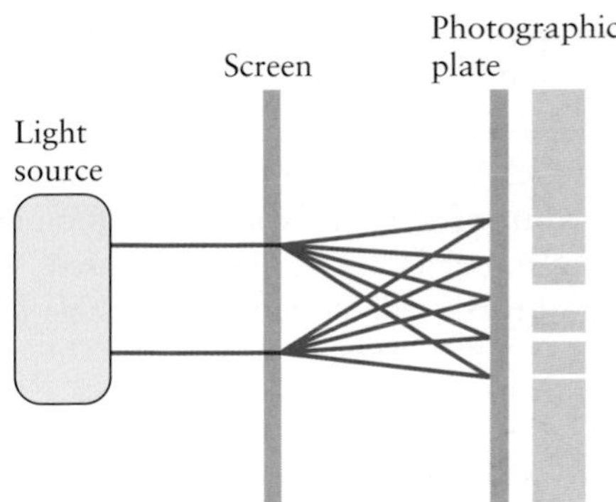

FIGURE 3 A "Young's slit" experiment demonstrates diffraction caused by a screen with two slits. The light rays passing through the slits interfere either constructively or destructively, depending on their relative path lengths, and bands of high and low intensity are recorded photographically.

that gave rise to it. This task is enormously complex, and although it was carried out by hand in the early days of the technique, it is now universally carried out on computers integrated into the diffraction equipment.

Why x-rays? Diffraction occurs when the wavelength of the radiation is comparable to characteristic spacings within the object causing the diffraction. Therefore, to obtain diffraction patterns from layers of atoms, we need to use radiation with a wavelength comparable to the spacing of the layers. The separation between layers of atoms in a crystal is of the order of 100 pm, so electromagnetic radiation of that wavelength, which corresponds to the x-ray region, must be used.

2 Experimental Techniques

X-rays are generated by accelerating electrons to very high speeds and then letting them plunge into a metal target. This technique generates two types of x-radiation. One type comes from the bombarding electrons themselves. Because accelerating and decelerating charges emit electromagnetic radiation, the electrons generate radiation as they are brought to a standstill in the metal. The radiation covers a wide range of frequencies, including x-rays. However, for x-ray diffraction, we need a well-defined wavelength. Such radiation is also generated when the electrons plunge into the target, but by a second mechanism. The fast electrons hit electrons occupying orbitals in the inner shells of atoms and drive them out of the atom. That collision leaves a gap in the atom, which is filled when an electron in another shell falls into the vacancy. The difference in energy is given out as a photon. Because the energy difference between shells is so great, the photon has a very high energy, corresponding to the x-ray region. When copper is used as the target, for example, the x-radiation has a wavelength of 154 pm.

An increasing amount of x-ray diffraction is now carried out by using the intense radiation from a *synchrotron source*, in which electrons are accelerated in a circle by electromagnetic fields. However, such sources are huge and are available only at national facilities.

In the *powder diffraction* technique, a monochromatic (single-frequency) beam of x-rays is directed at a powdered sample spread on a support, and the diffraction intensity is measured as the detector is moved to different angles (Fig. 4). The pattern obtained is characteristic of the material in the sample, and it can be identified by comparison with a database

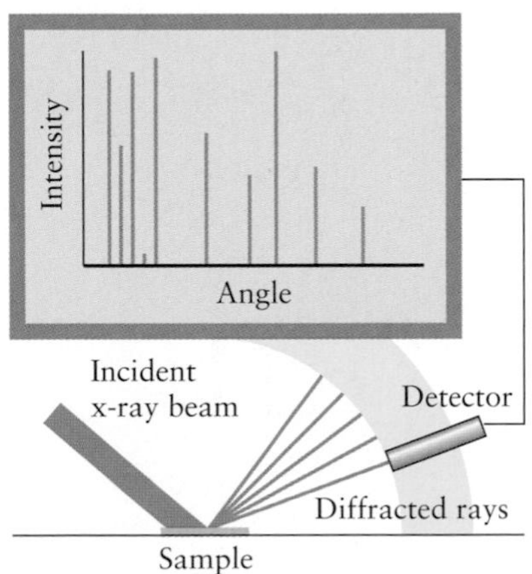

FIGURE 4 In the powder diffraction technique, a sample is spread on a flat plate and exposed to a beam of monochromatic (single-frequency) x-rays. The diffraction pattern (inset) is recorded by moving the detector to different angles.

of patterns. In effect, powder x-ray diffraction takes a fingerprint of the sample. It can also be used to identify the size and shape of the unit cell by measuring the spacing of the lines in the diffraction pattern. The central equation for analyzing the results of a powder diffraction experiment is the *Bragg equation*:

$$2d \sin \theta = \lambda$$

which relates the angles, θ (theta), at which constructive interference occurs to the spacing, d, of layers of atoms in the sample for x-rays of wavelength λ (lambda).

How can we use this information . . . to determine the spacing between layers of atoms in a crystal?

In an x-ray diffraction experiment on a single crystal of sodium chloride, using radiation from a copper source (λ = 154 pm), constructive interference was observed at θ = 11.2°. The spacing of the layers of atoms responsible for this interference is therefore

$$d = \frac{\lambda}{2 \sin \theta} = \frac{154\ \text{pm}}{2 \sin 11.2^\circ} = 396\ \text{pm}$$

Self-Test MT3.1A Constructive interference at 7.23° was observed from a crystal when x-radiation from a molybdenum source (λ = 71.0 pm) was used. What is the spacing of the layers of atoms responsible for the diffraction?

[*Answer:* 282 pm]

Self-Test MT3.1B Constructive interference at 12.1° was observed from a cubic crystal when x-radiation of wavelength 152 pm was used. What is the spacing of the layers of atoms responsible for the diffraction?

The x-ray diffraction pattern from a liquid resembles that from a powdered sample, but the lines are diffuse rather than sharp. This pattern indicates that the molecules of a liquid have a degree of short-range order but that the distances between molecules vary; this variation produces the diffuse lines.

The *single-crystal diffraction* technique is much more elaborate and gives much richer information. The first task is to grow a perfect single crystal of the sample. Whereas that task is usually straightforward for simple inorganic solids, it is often one of the most challenging parts of the investigation for the huge molecules characteristic of biologically important compounds such as proteins. Only a tiny crystal is needed, about 0.1 mm on a side, but the task of growing one can be very challenging. Once the crystal has been grown, it is placed at the center of a *four-circle diffractometer* (Fig. 5), a device for rotating the crystal and the detector so that the entire diffraction pattern can be recorded under the control of a computer.

The raw data consist of the intensities of the x-rays at all the settings of the angles in the diffractometer. The computational task is to analyze these measurements and convert them into the locations of the atoms. The process of conversion is called *Fourier synthesis* and involves lengthy calculations carried out on a computer that is an integral part of the diffractometer. The end product is a detailed description of the location of all the atoms in the molecule, the bond lengths, and the bond angles. The most spectacular achievement of this kind was the elucidation of the functioning of the genetic messenger, DNA, where x-rays gave us insight into the workings of life itself.

Related Exercises: 5.87–5.91

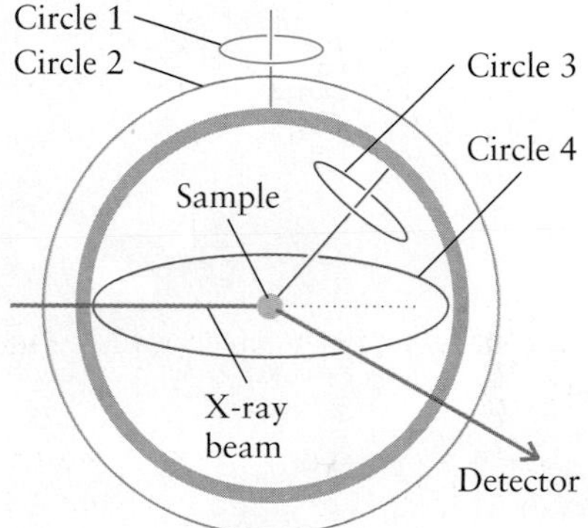

FIGURE 5 A four-circle diffractometer is used to obtain highly detailed information about x-ray diffraction patterns from single crystals. The diffraction pattern is monitored as the orientations around each of the four axes are changed.

CHAPTER

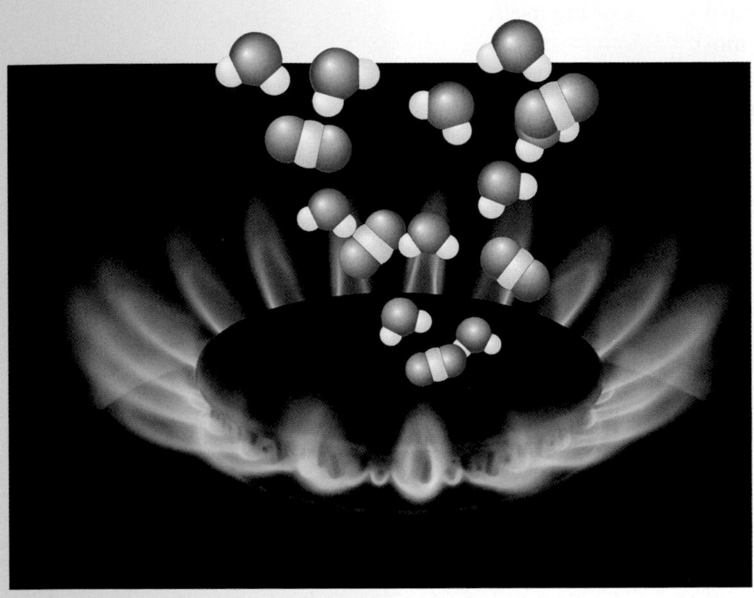

When natural gas, which is largely methane, CH_4, burns, it forms carbon dioxide and water molecules. The rearrangement of bonds—the breaking of C—H and O—O bonds and the formation of H—O and C—O bonds—releases a large quantity of energy as heat.

Why Do We Need to Know This Material?

Because energy is the basis of all chemical change, thermodynamics—the study of the transformations of energy—is central to chemistry. Thermodynamics explains why reactions occur at all. It also lets us predict how much heat they release or require. Heat output is a critical property of a fuel, and the first law of thermodynamics allows us to measure and report heat output systematically. The material in this chapter provides the foundation for the following chapters, in particular Chapter 7, which deals with the driving force of chemical reactions—why they occur and in which direction they can be expected to go.

What Do We Need to Know Already?

This chapter assumes that we know about energy (Section A) and the ideal gas law (Chapter 4). Some of the explanations refer to intermolecular forces (Sections 4.13 and 5.1–5.5). The material here builds on the introductions to ionic bonding and bond strengths in Sections 2.1, 2.2, 2.16, and 2.17.

THERMODYNAMICS: THE FIRST LAW

Thermodynamics, the study of the transformation of energy, falls naturally into two parts. The *first law* is concerned with keeping track of energy changes and allows us to calculate, for instance, how much heat a reaction produces. The *second law*, which is the subject of Chapter 7, explains why some chemical reactions occur but others do not. Both laws are summaries of experience with bulk matter and are independent of any model we might have of its microscopic structure—you can do thermodynamics even if you don't believe in the existence of atoms! However, we can interpret both laws in terms of models based on the behavior of atoms and molecules, and thereby deepen our understanding of the properties of matter. The link between the properties of atoms and those of bulk matter is **statistical thermodynamics**, the interpretation of the laws of thermodynamics in terms of the *average* behavior of the large numbers of atoms and molecules that make up a typical sample. We shall move backward and forward between these levels of explanation, because a knowledge of how molecular behavior contributes to bulk properties enriches our understanding of thermodynamics and develops our insight into the properties of matter.

This chapter introduces the first law of thermodynamics and its applications in three main parts. The first part introduces the basic concepts of thermodynamics—including the concepts of system, surroundings, heat, work, and energy—and the experimental basis of the first law. The second part introduces the enthalpy, a quantity that is widely used in chemical thermodynamics, and applies it to characterize physical changes. The third part applies the concept of enthalpy to a wide variety of chemical changes.

SYSTEMS, STATES, AND ENERGY

At one time, heat was thought to be a fluid called *caloric*, which flowed from a hot substance to a cooler one. The French engineer Sadi Carnot (Fig. 6.1), who helped to lay the foundations of thermodynamics, believed that work resulted from the flow of caloric, just as water turns a water wheel. Some of Carnot's work survives, but we now know that caloric is a myth. Not long after Carnot proposed his ideas in the early nineteenth century, the English physicist James Joule changed the course of science with the suggestion that heat is a form of energy (Fig. 6.2). Indeed, he went much further: because he found that he could raise the temperature of substances either by heating them or by doing work on them (for instance, by stirring them vigorously), he proposed that heat and work are both forms of energy and that one form can be changed into another. We first examine what these and other fundamental terms mean and how they are used, for they lie at the heart of thermodynamics.

FIGURE 6.1 Nicolas Leonard Sadi Carnot (1796–1832).

FIGURE 6.2 James Prescott Joule (1818–1889).

6.1 Systems

Thermodynamics is based on experiments in which energy is transformed from one form into another and transferred from one place to another. For instance, electricity may be generated in a power station but used in a factory a great distance away; food may be digested in your stomach, but the energy released may be used in your head. To keep track of energy, we divide the world into two parts. The region in which we are interested, such as a flask of gas, a beaker of acid, or a reaction mixture, is called a **system** (Fig. 6.3). Everything else, such as the water bath in which a reaction mixture may be immersed, is called the **surroundings.** The surroundings are where we make observations on the energy transferred into or out of the system.

A system can be open, closed, or isolated (Fig. 6.4). An **open system** can exchange both matter and energy with the surroundings. Examples of open systems are automobile engines and the human body. A **closed system** has a fixed amount of matter, but it can exchange energy with the surroundings. Examples of closed systems are the cold packs used for athletic injuries. An **isolated system** has no contact with its surroundings. We can think of an isolated system as sealed inside rigid, thermally insulated walls. A good approximation to an isolated system is hot coffee inside a sealed vacuum flask.

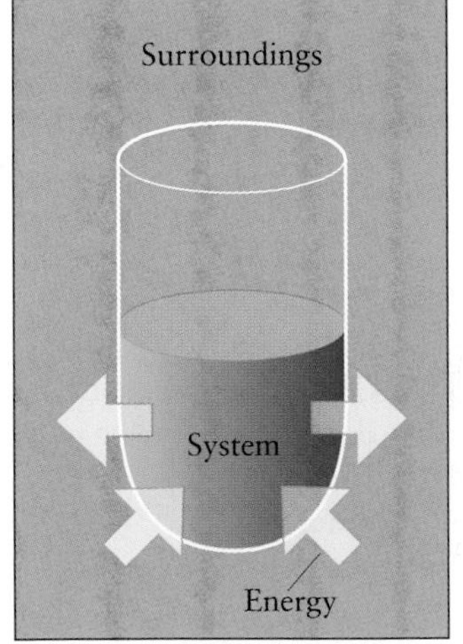

FIGURE 6.3 The system is the sample or reaction mixture in which we are interested. Outside the system are the surroundings. The system plus its surroundings is sometimes called the universe.

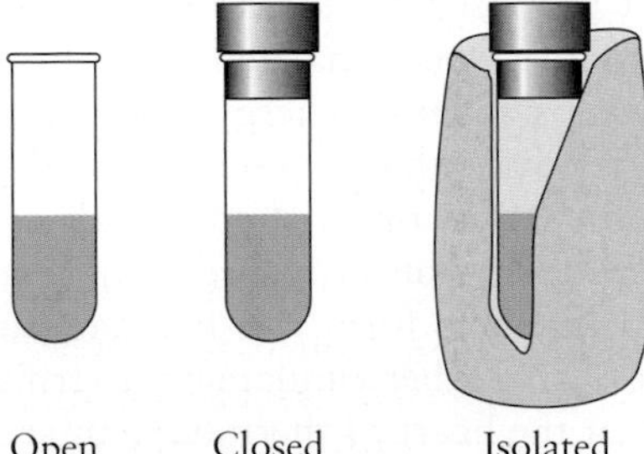

FIGURE 6.4 We classify any system as one of three kinds. An open system can exchange matter and energy with its surroundings. A closed system can exchange energy but not matter. An isolated system can exchange neither matter nor energy.

In thermodynamics, the world is divided into a system and its surroundings. An open system can exchange both matter and energy with the surroundings; a closed system can exchange only energy; an isolated system can exchange nothing.

6.2 Work and Energy

The most fundamental property in thermodynamics—in the sense that it provides a basis for defining the principal concepts—is **work,** or movement against an opposing force. Work is done when a weight is raised against the pull of gravity. The reaction in a battery does work when it pushes an electric current through a circuit. Gas in a cylinder—perhaps the hot gas mixture in the cylinder of an automobile engine—does work when it pushes back a piston. We can identify a process that does work by noting whether, in principle at least, it can be harnessed to raise a weight. The expansion of a gas, for example, can be used to raise a weight because the piston can be connected to a weight. An electric current produced by a battery can be used to raise a weight if the circuit includes an electric motor.

The work to be done to move an object a certain distance against an opposing force is calculated by multiplying the force by the distance:

$$\text{Work} = \text{force} \times \text{distance} \qquad (1)$$

As pointed out and illustrated in Section A, the unit in which work, and therefore energy, is reported is the *joule*, J, with

$$1\ \text{J} = 1\ \text{kg}\cdot\text{m}^2\cdot\text{s}^{-2} \qquad (2)$$

In thermodynamics, the *total* capacity of a system to do work is called its **internal energy,** U. A compressed gas has more internal energy before it is allowed to expand than after the expansion. A wound spring has more internal energy than an unwound spring. A charged battery has more internal energy than a discharged battery. Steam has a higher internal energy than the same mass of cold water.

We cannot measure the total energy of a system: all we can do is measure *changes* in energy. If a system does 15 J of work, it has used up some of its store of energy, and we say that its internal energy has fallen by 15 J. To denote this change, we write $\Delta U = -15$ J. Throughout thermodynamics, the symbol ΔX means a difference in a property X:

Einstein would have disagreed: he would calculate the total internal energy from mc^2, where m is the mass of the system. He would have agreed, though, that in practice in chemistry we can only measure energy changes.

$$\Delta X = X_{\text{final}} - X_{\text{initial}} \qquad (3)$$

A *negative* value of ΔX, as in $\Delta U = -15$ J, signifies that the value of X has decreased during the change and a *positive* value signifies that X has increased.

When we do work *on* a system, the internal energy of the system is increased. Compressing a gas inside a thermally insulated container increases its internal energy, because the compressed gas can do more work than an uncompressed gas. Winding a spring transfers energy to the

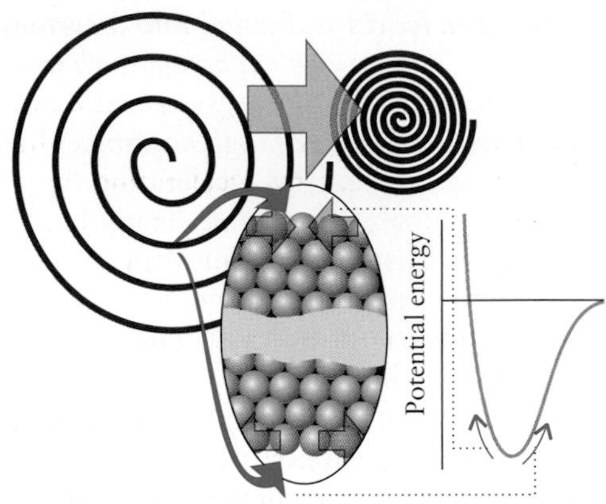

FIGURE 6.5 When we wind a spring, the potential energy of the atoms changes because they are squashed together on the inside of the curve and pulled apart on the outside. The internal energy of the spring rises as a result of this increase in potential energy. The graph shows how the potential energy of an atom in the spring changes with distance from its neighbors. The lowest point of the graph corresponds to the unwound, relaxed spring.

spring: when the spring is fully wound, it can do more work than before it was wound (Fig. 6.5). Doing work on a system also includes passing an electric current through it, as we do when we charge a rechargeable battery in a laptop computer. If we do 15 J of work on a system, we increase the capacity of the system to do work by 15 J. In other words, the internal energy of the system is increased by 15 J, and we write $\Delta U = +15$ J.

We use the symbol w to denote the energy transferred to a system by doing work, so

$$\Delta U = w \tag{4}$$

provided no other type of transfer of energy is taking place. If energy is transferred *to* a system as work (for example, by compressing a gas), w is positive; if energy *leaves* the system as work (for example, by allowing a gas to expand), w is negative. The internal energy of a system increases when work is done on it (w is positive) and decreases when it does work (w is negative).

Work is the transfer of energy to a system by a process that is equivalent to raising or lowering a weight. The internal energy of a system may be changed by doing work: $\Delta U = w$.

6.3 The Molecular Origin of Internal Energy

The internal energy is a *bulk* property of a system. How can we describe internal energy in terms of the properties of the individual atoms and molecules that make up the system? Internal energy is energy stored in a system as kinetic energy and potential energy. We saw in Section A that kinetic energy is energy due to motion, and the faster a molecule travels, the greater its kinetic energy. When we heat a gas, the average speed of the molecules increases. When we do work on a gas in an insulated container, the molecules also move faster. For example, when we push a piston in against the opposing pressure of a gas, the piston acts like a very slowly moving bat. It does work on the gas by accelerating the gas molecules to higher speeds, just as you accelerate a table-tennis ball when you hit it with a paddle. However, the piston in our example is pelted with billions of balls each second. The increase in the average speed of the gas molecules corresponds to an increase in the total kinetic energy of the molecules, and therefore to an increase in the internal energy of the gas. We have already seen (in Section 4.13) that the average speed of gas molecules is an indication of temperature, so this increase in internal energy also corresponds to an increase in temperature. It is always the case that *a system at high temperature has a greater internal energy than the same system at a lower temperature.*

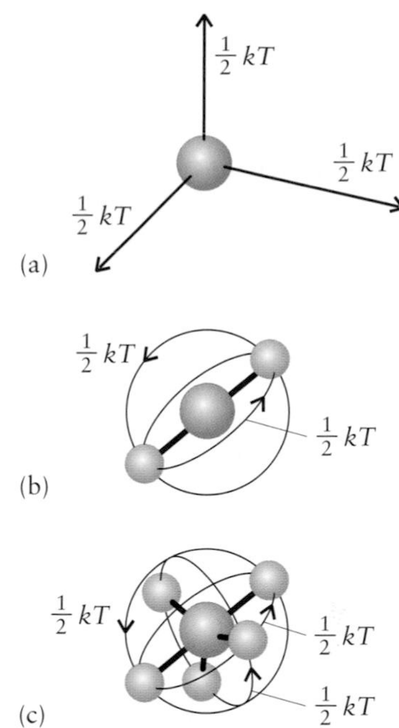

FIGURE 6.6 The translational and rotational modes of atoms and molecules and the corresponding average energies at a temperature T. (a) An atom or molecule can undergo translational motion in three dimensions. (b) A linear molecule can also rotate about two axes perpendicular to the line of atoms. (c) A nonlinear molecule can rotate about three perpendicular axes.

Molecules can move in different ways in a gas (Fig. 6.6). The kinetic energy of an atom or molecule as it moves bodily through space is called its **translational kinetic energy.** Molecules (in contrast to atoms) can also store energy as **rotational kinetic energy,** the kinetic energy arising from their rotational motion. A third store of kinetic energy for molecules is in the oscillation of their atoms relative to one another: this contribution is called **vibrational kinetic energy.** During a vibration, the potential energy of the atoms also changes as they move toward and away from each other, so energy can also be stored as potential energy of molecular vibration.

Even though pushing in a piston affects only the translational energy of the gas molecules directly, molecular collisions cause the molecules to rotate more rapidly. That is, rotations are excited into states of higher energy, so the molecules also store internal energy in their rotational modes of motion. Most molecular vibrations are not excited at temperatures close to room temperature because they require too much energy. The contribution of each different mode of motion can be estimated from the temperature.

How do we do that?

To estimate the contribution of a mode of motion, we use a fundamental theorem of classical mechanics. The **equipartition theorem** (which we shall not derive here) states that *the average energy of each mode of motion of a molecule in a sample at a temperature T is equal to $\frac{1}{2}kT$.* In this simple expression, k is the Boltzmann constant, a fundamental constant with the value $1.380\,66 \times 10^{-23}$ J·K^{-1}. The Boltzmann constant is related to the gas constant by $R = N_A k$, where N_A is the Avogadro constant. The equipartition theorem is a result from classical mechanics, so we can use it for translational and rotational motion of molecules at room temperature and above, where quantization is unimportant, but we cannot use it reliably for vibrational motion, except at high temperatures. The following remarks therefore apply only to translational and rotational motion.

A molecule can move through space along any of three dimensions, so it has *three* translational modes of motion. It follows from the equipartition theorem that the average translational kinetic energy of a molecule in a sample at a temperature T is $3 \times \frac{1}{2}kT = \frac{3}{2}kT$. The contribution to the molar internal energy is therefore N_A times this value, or

We derived essentially the same expression by a different argument in Section 4.13 in the discussion of the kinetic theory of gases.

$$U_m\text{(translation)} = \tfrac{3}{2}N_A kT = \tfrac{3}{2}RT$$

Because $RT = 2.48\ \text{kJ·mol}^{-1}$ at 25°C, the translational motion of gas molecules contributes about $3.72\ \text{kJ·mol}^{-1}$ to the internal energy of the sample at 25°C. Apart from the energies of the atoms themselves, this is the only motional contribution to the internal energy of a monatomic gas, such as argon or any other noble gas.

A *linear molecule*, such as any diatomic molecule, carbon dioxide, and ethyne (acetylene), can rotate around two axes perpendicular to the line of atoms, so it has two rotational modes of motion. Its average rotational energy is therefore $2 \times \frac{1}{2}kT = kT$, and the contribution to the molar internal energy is N_A times this value:

$$U_m\text{(rotation, linear)} = RT$$

or about $2.48\ \text{kJ·mol}^{-1}$ at 25°C. The total motional contribution to the internal energy of a gas of linear molecules is the sum of the translational and rotational contributions, or $\frac{5}{2}RT$, which corresponds to $6.20\ \text{kJ·mol}^{-1}$ at 25°C.

A nonlinear molecule, such as water, methane, or benzene, can rotate around any of three perpendicular axes, so it has three rotational modes of motion. The average rotational energy of such a molecule is therefore $3 \times \frac{1}{2}kT = \frac{3}{2}kT$. The contribution of rotation to the molar internal energy of a gas of nonlinear molecules is therefore

$$U_m\text{(rotation, nonlinear)} = \tfrac{3}{2}RT$$

At 25°C, the contribution of rotation to the energy is $3.72\ \text{kJ·mol}^{-1}$ (the same as for the translational motion), and the total motional contribution for nonlinear molecules is therefore $3RT$, or about $7.44\ \text{kJ·mol}^{-1}$ at 25°C.

We saw in Section A that potential energy is energy due to position.

Changes in internal energy also occur when the potential energies of atoms and molecules change. A molecule of an ideal gas can be taken to have zero potential energy because its energy is not affected by how close it is to any of the other molecules in the sample. Therefore, compressing or expanding an ideal gas does not change the potential energy of the molecules from 0; so, *for an ideal gas, the internal energy is independent of the volume; it depends only on the temperature.* A molecule in a real gas, a liquid, or a solid, however, does interact with its neighbors, and the potential energy does contribute to the internal energy. When we squeeze a rubber ball, we push the molecules closer together, their potential energy rises, and the internal energy of the ball increases. In other words, when we do work on the system (compressing the rubber), its internal energy (its capacity to do work, by springing back into its original shape) increases.

The internal energy is an extensive property, so it depends on the amount of substance. For example, a 50-g sample of a substance has twice as much internal energy as 25 g of the same substance under the same conditions.

Recall from Section A that extensive properties are properties that depend on the mass of the sample; intensive properties are independent of the mass.

Internal energy is stored as molecular kinetic and potential energy. The equipartition theorem can be used to estimate the translational and rotational contributions to the internal energy of an ideal gas.

6.4 Heat

In addition to work, the internal energy of a system can also be changed by transferring energy to or from the surroundings as heat. *Heat* is a common term in everyday conversation, but in thermodynamics it has a special meaning. In thermodynamics, **heat** is energy transferred as a result of a temperature difference. Energy flows as heat from a high-temperature region to a low-temperature region. Therefore, provided the walls of the system are not thermally insulating, if the system is cooler than its surroundings, energy flows into the system from the surroundings and the internal energy of the system increases. The verb *to heat* means to transfer energy as heat; that is, by making use of a temperature difference. When we heat water in a kettle, energy flows from the heating element into the cooler water and stimulates the water molecules into more vigorous motion.

The energy transferred as heat is measured in joules, J. However, a unit of energy still widely used in biochemistry and related fields is the **calorie,** cal. The original definition of 1 cal stated that it is the energy needed to raise the temperature of 1 g of water by 1°C. The modern definition is

This exact relation defines the calorie in terms of the joule; the joule is the fundamental unit.

$$1 \text{ cal} = 4.184 \text{ J (exactly)} \tag{5}$$

We can use the measurement of heat produced or absorbed by a process to monitor the change in internal energy of a system. However, before we see how this is done, we have to distinguish between two types of closed systems. One type of closed system we shall study has thermally insulating walls. The walls of a vacuum flask are thermally insulating because the vacuum between them does not allow energy to be carried from one wall to another by molecules and because their silvered surfaces cut down the transfer of energy by radiation. The technical term for a thermally insulating wall is an **adiabatic** wall. So, a vacuum flask is a good approximation to an adiabatic container. There is no flow of energy as heat into or out of a system in an adiabatic container, even if there is a temperature difference between the system and its surroundings. However, an adiabatic system is not an isolated system: energy may be tranferred to or from a system in an adiabatic container, but only as work done by or on the system, not by the transfer of energy as heat. In an adiabatic system, $\Delta U = w$.

The word *adiabatic* comes from the Greek words for "not passing through."

The second type of closed system we consider has *nonadiabatic*, or **diathermic,** walls, which permit the transfer of energy as heat. An influx of energy through the walls of diathermic container might raise the temperature of the system, so monitoring the temperature change is a way to monitor the heat transferred. The newly heated contents of the system can then do more work than they could do initially, so the internal energy of the system has been increased. For example, suppose we placed a sample of gas into a sealed rigid container in contact with a heater. Energy flows from the heater into the gas; the molecules store that energy as increased kinetic energy, and, as a consequence, they travel faster. That increase in average speed corresponds to an increase in temperature of the gas. An increase in temperature, in turn, results in an increase in the pressure of the confined gas, and a gas at high pressure can do more work than a gas at low pressure. Therefore, because the sample of gas can now do more work, the internal energy of the gas has been increased.

We write the energy transferred *to* a system as heat as q. Therefore, when the internal energy of the system is changed by transferring energy only as heat,

$$\Delta U = q \qquad (6)$$

If energy enters a system as heat, q is positive; if energy leaves the system as heat, q is negative. Thus, if 10 J of energy enters the system as a result of a temperature difference, we write $q = +10$ J. If 10 J leaves the system, we write $q = -10$ J. In each case, $\Delta U = q$. In the former case, the internal energy increases by 10 J ($\Delta U = +10$ J), whereas, in the latter case, the internal energy decreases by 10 J ($\Delta U = -10$ J).

Heat is the transfer of energy as a result of a temperature difference. When the only transfer of energy is as heat, $\Delta U = q$.

6.5 The First Law

It is often the case that the internal energy of a system can change as a result of both work and heat. For example, the hot engine of a moving automobile transfers energy to its surroundings as both work and heat. In general, therefore, the change in internal energy of a system is the net result of both kinds of transfers, so we combine Eqs. 4 and 6 and write

$$\Delta U = q + w \qquad (7)$$

This expression summarizes the experimental fact that heat and work are equivalent means of transferring energy and thereby changing the internal energy of a system.

Self-Test 6.1A An automobile engine does 520 kJ of work and loses 220 kJ of energy as heat. What is the change in internal energy of the engine? Treat the engine, fuel, and exhaust gases as a closed system.

[*Answer:* −740 kJ]

Self-Test 6.1B A system was heated by using 300 J of heat, yet it was found that its internal energy decreased by 150 J (so $\Delta U = -150$ J). Calculate w. Was work done on the system or did the system do work?

Equation 7 is one form of the first law of thermodynamics. It is an experimental fact—a fact supported by thousands of experiments—that we cannot use a system to do work, leave it isolated for a while, and then return to it to find its internal energy restored to its original value. This observation is summarized as the **first law of thermodynamics:**

- *The internal energy of an isolated system is constant.*

The first law is closely related to the conservation of energy (Section A), but goes beyond it in the sense that the concept of heat does not apply to single particles.

- *The first law of thermodynamics states that the internal energy of an isolated system is constant; for a system that is not isolated, $\Delta U = q + w$.*

6.6 State Functions

A **state function** is a property with a value that depends only on the current state of the system and is independent of the manner in which the state was prepared. The density of a substance is a state function, because a sample will have the same density (for a given set of conditions) regardless of its history. Internal energy is a state function; so the internal energy of a beaker of water at 25°C is the same regardless of its history. State functions may be either intensive or extensive: density is an intensive state function; internal energy is an extensive state function.

The importance of state functions in thermodynamics is that if the system *is changed from one state to another, the change in a state function is independent of how that change was brought about.* A state function is like altitude on a mountain (Fig. 6.7). We may take any number of different paths between two huts on the mountain, but the change in altitude between the two huts is the same regardless of the path. Similarly, if we took 100 g of water at 25°C and raised its temperature to 60°C, its internal energy would change by a certain amount. However, if we took the same mass of water, heated it to boiling, vaporized it, condensed the vapor, and then allowed the water to cool to 60°C, the *net* change in internal energy of the water would be exactly the same as before.

The work done by a system is *not* a state function: it depends on how the change is brought about. For example, we could let a gas at 25°C expand at constant temperature through 100 cm^3 by two different paths. In the first experiment, the gas might push on a piston and do a certain amount of work against an external force. In the second experiment, the gas might push a piston into a vacuum; it then does no work because there is no external opposing force (Fig. 6.8). The change in the state of the gas is the same in each case, but the work done by the system is different: in the first case, w is nonzero; in the second case, $w = 0$. Indeed, even everyday language suggests that work is not a state function, because we never speak of a system as possessing a certain amount of "work." For just this reason, we write the work done as w, not Δw, for the latter would imply a difference of two "works."

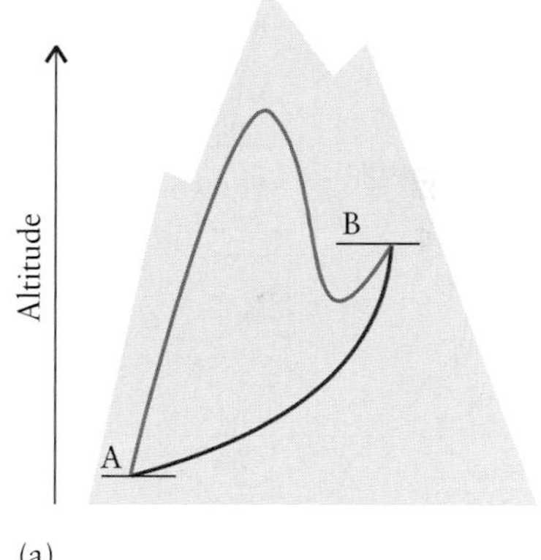

(a)

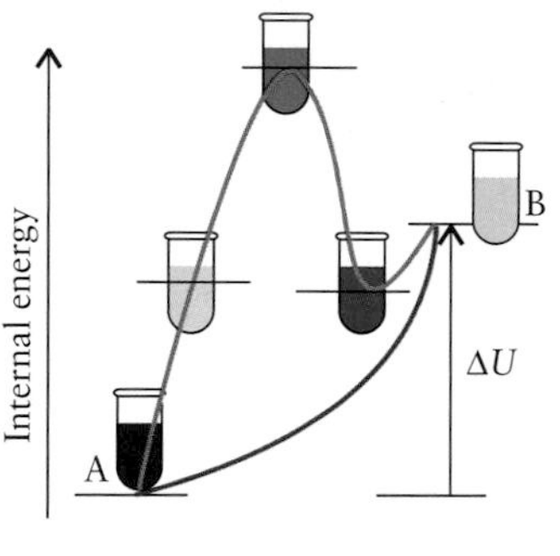

(b)

FIGURE 6.7 (a) The altitude of a location on a mountain is like a state function: it does not matter what route you take between two points, the net change in altitude is the same. (b) Internal energy is a state function: if a system changes from state A to state B (as depicted highly diagrammatically here), the net change in internal energy is the same whatever the route—the sequence of chemical or physical changes—between the two states.

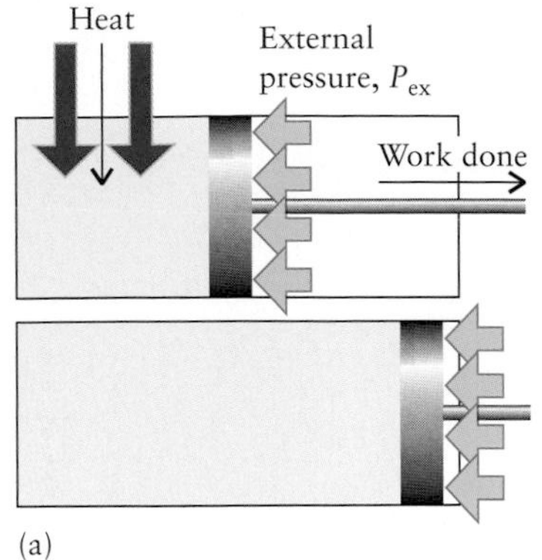

(a)

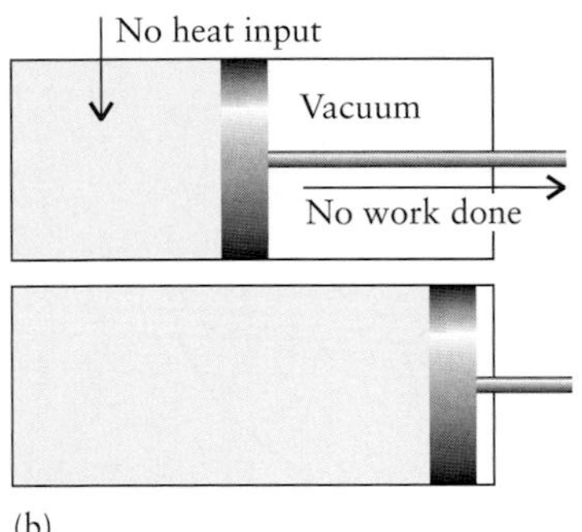

(b)

FIGURE 6.8 Two different paths between the same initial and final states of an ideal gas. (a) The gas does work as it expands isothermally, but heat flows in to restore the energy lost as work. (b) The gas does no work as it expands isothermally into a vacuum.

Similarly, heat is not a state function. The energy transferred as heat during a change in the state of a system depends on how the change is brought about. For example, suppose we want to raise the temperature of 100 g of water from 25°C to 30°C. One way would be to supply energy as heat by using an electric heater. Another way would be to stir the water vigorously with paddles. In the latter case, all the energy required is transferred as work, and none is supplied as heat. So, in the first case, q is nonzero; and in the second case, q is 0. Because heat is not a state function, we should not speak of a system as possessing a certain amount of "heat." Just as we do for work, we write the energy transferred as heat as q, not Δq.

Because internal energy is a state function, we can choose any convenient path between the given initial and final states of a system and calculate ΔU for that path. The result will have the same value of ΔU as the actual path between the same two states, even if the actual path is so complicated that ΔU cannot be calculated.

A state function depends only on the current state of a system. The change in a state function between two states is independent of the path between them. Internal energy is a state function; work and heat are not.

6.7 Expansion Work

Energy may flow as work as well as heat between the system and its surroundings. A system can do two kinds of work. The first type is **expansion work,** which involves a change in volume of the system against an external pressure. For example, a gas expanding in a cylinder fitted with a piston pushes out against the atmosphere and thus does work on it. The second type of work is **nonexpansion work,** work that does not involve a change in volume. Table 6.1 lists some of the kinds of work a system can

TABLE 6.1 Varieties of Work

Type of work	w	Comment	Units*
expansion	$-P_{ex}\Delta V$	P_{ex} is the external pressure ΔV is the change in volume	Pa m^3
extension	$f\Delta l$	f is the tension Δl is the change in length	N m
weight lifting	$mg\Delta h$	m is the mass g is the acceleration of free fall Δh is the change in height	kg $m{\cdot}s^{-2}$ m
electrical	$\phi\Delta q$	ϕ is the electrical potential Δq is the change in charge	V C
surface expansion	$\gamma\Delta A$	γ is the surface tension ΔA is the change in area	$N{\cdot}m^{-1}$ m^2

* For work in joules (J). Note that 1 N·m = 1 J and 1 V·C = 1 J.

do. For example, a chemical reaction can do work by causing an electrical current to flow, and our bodies do work by moving about.

In some chemical reactions, no work is done. For example, suppose the chemical reaction takes place inside a sealed, rigid container. We could think of heating a lump of calcium carbonate inside a container until it decomposed into calcium oxide and carbon dioxide. A gas is produced, but, because the walls are rigid, the system cannot do work by pushing back the walls like a piston. That is, *a system maintained at constant volume can do no expansion work*. Let's also suppose that the system can do no other types of work—for instance, the reaction is not taking place inside an electrical cell, so it cannot do electrical work either. Because the system can do neither expansion work nor any other kind of work, we know that $w = 0$. Therefore, according to the first law, and specifically Eq. 7, any change in internal energy of the system must arise from the transfer of energy as heat between the system and the surroundings, and we can write

$$\text{At constant volume: } \Delta U = q \qquad (8)$$

This equation means that we can measure the change in internal energy of a constant-volume system that does no work of any kind simply by monitoring the transfer of energy to it as heat. For instance, if the decomposition of a few grams of calcium carbonate in a closed, rigid container absorbs 16 kJ of energy as heat from the surroundings ($q = +16$ kJ), we can conclude that $\Delta U = +16$ kJ.

Notice that the + sign is written explicitly for positive values.

Now suppose that the walls are movable, so the system can expand when the reaction takes place. At this stage, let's suppose that the external pressure is constant, as would be the case if the piston were to press against the atmosphere (Fig. 6.9). The external pressure acting on the outer face of the piston provides the force opposing expansion. We need to find how the work done when the system expands through a volume ΔV is related to the external pressure P_{ex}.

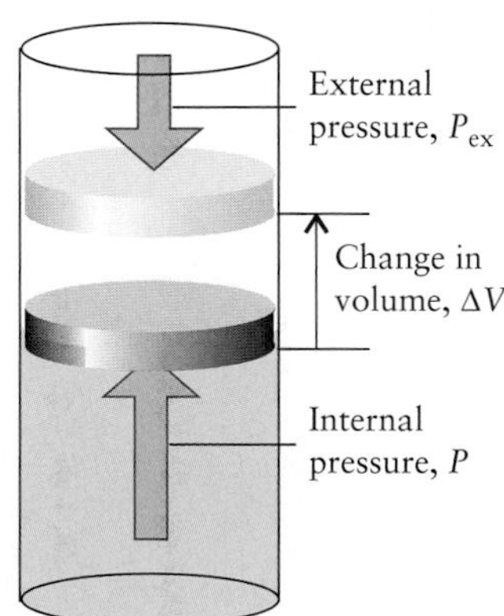

FIGURE 6.9 A system does work when it expands against an external pressure. Here we see a gas that pushes a piston out against a pressure P_{ex}. The work done is proportional to P_{ex} and the change in volume, ΔV, that the system undergoes.

How do we do that?

To calculate the work done when a system expands against constant pressure, we need to return to the basic definition of work: Eq. 1 gives the formula for calculating work (force times distance). Because pressure is force divided by area ($P = f/A$), the force opposing expansion is the product of the pressure acting on the outside of the piston and the area of the piston. Therefore, if the external pressure is P_{ex} and the area of the piston is A, the force opposing expansion is $f = P_{ex}A$. Suppose the piston is driven out through a distance d; then the work the system does is

$$\text{Work} = d \times P_{ex}A$$

However, the product of area and distance moved is equal to the change in volume of the sample:

$$d \times A = \Delta V$$

Therefore, the work done by the expanding gas is $P_{ex}\Delta V$. At this point, we match the signs to our convention. When a system expands, it loses energy as work, so when ΔV is positive, w is negative. We can therefore write

$$w = -P_{ex}\Delta V \tag{9}$$

This expression confirms that the work done is proportional to the *external* pressure; it also shows that the work done is proportional to the change in volume.

Equation 9 applies to all systems. A gas is easiest to visualize, but the expression also applies to an expanding liquid or solid. However, Eq. 9 applies *only when the external pressure is constant* during the expansion.

If the external pressure is 0 (a vacuum), it follows from Eq. 9 that $w = 0$: *a system does no expansion work when it expands into a vacuum*, because there is no opposing force. You do no work by pushing if there is nothing to push against. Expansion against zero pressure is called **free expansion.** Free expansion does no work.

If we are using SI units, the external pressure will be expressed in pascals (1 Pa = 1 $\text{kg·m}^{-1}\text{·s}^{-2}$, Section 4.2) and the change in volume in cubic meters (m^3). The product of 1 Pa and 1 m^3 is

$$1\ \text{Pa·m}^3 = 1\ \text{kg·m}^{-1}\text{·s}^{-2} \times 1\ \text{m}^3 = 1\ \text{kg·m}^2\text{·s}^{-2} = 1\ \text{J} \tag{10}$$

Therefore, if we work in pascals and cubic meters, the work is obtained in joules. However, we might have expressed the pressure in atmospheres and the volume in liters. In this case, we may need to convert the answer (in liter-atmospheres) into joules. The conversion factor is obtained by noting that 1 L = 10^{-3} m^3 and 1 atm = 101 325 Pa exactly; therefore

$$\begin{aligned} 1\ \text{L·atm} &= 10^{-3}\ \text{m}^3 \times 101\ 325\ \text{Pa} \\ &= 101.325\ \text{Pa·m}^3 = 101.325\ \text{J (exactly)} \end{aligned} \tag{11}$$

How can we use this information . . . to calculate the work done when a gas expands?

Suppose a gas expands by 500 mL (0.500 L) against a pressure of 1.20 atm. The work done is

$$w = -P_{ex}\Delta V = -(1.20\ \text{atm}) \times (0.500\ \text{L}) = -0.600\ \text{L· atm}$$

and the work in joules is

$$w = -(0.600\ \text{L·atm}) \times \left(\frac{101.325\ \text{J}}{1\ \text{L·atm}}\right) = -60.8\ \text{J}$$

The negative sign means that the internal energy decreases by 60.8 J when the gas expands and does 60.8 kJ of work on its surroundings.

Self-Test 6.2A Water expands when it freezes. How much work does 100 g of water do when it freezes at 0°C and bursts a water pipe with an opposing pressure of 1070 atm? The densities of water and ice at 0°C are 1.00 and 0.92 $g\cdot cm^{-3}$, respectively.

[*Answer:* $w = -0.9$ kJ]

Self-Test 6.2B The gases in the four cylinders of an automobile engine expand from 0.22 L to 2.2 L during one ignition cycle. Assuming that the gear train maintains a steady pressure of 9.60 atm on the gases, how much work can the engine do in one cycle?

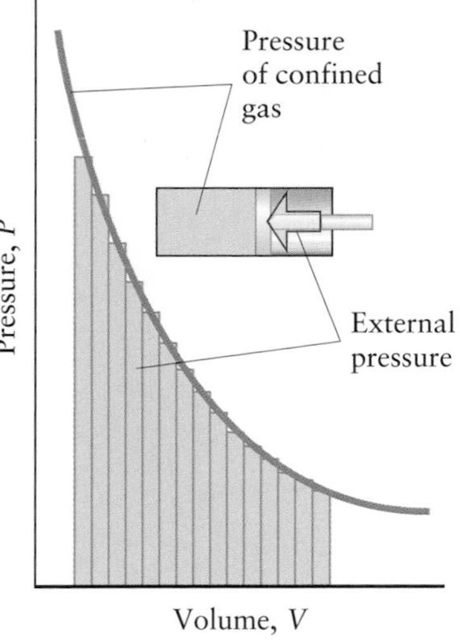

FIGURE 6.10 When a gas expands reversibly, the external pressure is matched to the pressure of the gas at every stage of the expansion. This arrangement (when the steps corresponding to the increase in volume are infinitesimal) achieves maximum work.

So far, we have considered systems for which the external pressure was constant. Now let's consider the case of a gas that expands against a *changing* external pressure. In particular, we consider the very important case of **reversible, isothermal expansion** of an ideal gas. The term *reversible*, as we shall see in more detail shortly, means that *the external pressure is matched to the pressure of the gas at every stage of the expansion.* The term *isothermal* means that the expansion takes place at constant temperature. In an isothermal expansion, the pressure of the gas falls as it expands; so to achieve reversible, isothermal expansion, the external pressure must also be gradually reduced in step with the change in volume (Fig. 6.10). To calculate the work, we have to take into account the gradual reduction in external pressure.

How do we do that?

To calculate the work of reversible, isothermal expansion of a gas, we have to use calculus, starting at Eq. 9 written for an infinitesimal change in volume:

$$dw = -P_{ex}dV$$

The overall amount of work done is the sum of these infinitesimal contributions as the volume changes from its initial value to its final value. That is, the work is the following integral:

$$w = -\int_{V_{initial}}^{V_{final}} P_{ex}\, dV$$

Because the integral of a function is equal to the area beneath the graph of the function, *the work done is equal to the (negative of the) area beneath a graph of P_{ex} as a function of V.* We see an example of this interpretation shortly.

When the external pressure is matched to the pressure, P, of the gas, the expression for dw becomes

$$dw = -PdV$$

The pressure of the gas changes as it expands; but at any stage of the expansion, the pressure is related to the volume by the ideal gas law, $PV = nRT$. Therefore, we can write

$$dw = -\frac{nRT\,dV}{V}$$

The total work done is the sum of these infinitesimal contributions as the volume changes from its initial value to its final value. That is, the work done is the following integral:

$$\begin{aligned} w &= -\int_{V_{\text{initial}}}^{V_{\text{final}}} \frac{nRT\,dV}{V} = -nRT\int_{V_{\text{initial}}}^{V_{\text{final}}} \frac{dV}{V} \\ &= -nRT \ln \frac{V_{\text{final}}}{V_{\text{initial}}} \end{aligned}$$

Apart from the sign, this integral is equal to the area below the ideal gas isotherm between the initial and final volumes. The temperature T has been taken outside the integral because the expansion is isothermal, so T is a constant during the expansion. The final line has made use of the standard integral

$$\int \frac{dx}{x} = \ln x + \text{constant}$$

We have found that, provided the gas is ideal, the work of reversible, isothermal expansion from the volume V_{initial} to the volume V_{final} is

$$w = -nRT \ln \frac{V_{\text{final}}}{V_{\text{initial}}} \tag{12}$$

where n is the amount of gas (in moles) in the container and T is the temperature. In some cases, we know only the initial and final pressures of the gas, not the initial and final volumes. In such a case, we can use Boyle's law (Eq. 4 in Section 4.4) to write $V_{\text{final}}/V_{\text{initial}} = P_{\text{initial}}/P_{\text{final}}$; then Eq. 12 becomes

$$w = -nRT \ln \frac{P_{\text{initial}}}{P_{\text{final}}} \tag{13}$$

Before using these equations to calculate the work of a reversible process, let's examine the meaning of reversible in this context. In everyday language, a reversible process is one that can take place in either direction. This common usage is refined in science: in thermodynamics, a **reversible process** is one that can be reversed by an *infinitesimal* change in a variable. For example, if the external pressure exactly matches the pressure of the gas in the system, the piston moves in neither direction. If the external pressure is increased by an infinitesimal amount, the piston moves in. If, instead, the external pressure is reduced by an infinitesimal amount, the piston moves out.

Expansion against an external pressure that differs by a finite (measurable) amount from the pressure of the system is an **irreversible process** in the sense that an infinitesimal change in the external pressure does not reverse the direction of travel of the piston. For instance, if the pressure of the system is 2.0 atm at some stage of the expansion and the external pressure is 1.0 atm, then an infinitesimal change in the latter does not convert expansion into compression. Reversible processes are of the greatest importance in thermodynamics, and we shall encounter them many times.

The work done by any system on its surroundings during expansion against a constant pressure is calculated from Eq. 9; for an isothermal, reversible expansion of an ideal gas, the work is calculated from Eqs. 12 or 13. A reversible process is a process that can be reversed by an infinitesimal change in a variable.

EXAMPLE 6.1 **Calculating the change in internal energy due to expansion**

Suppose that 1.00 mol of an ideal gas at 292 K and 3.00 atm expands from 8.00 L to 20.00 L and a final pressure of 1.20 atm by two different paths. (a) Path A is an isothermal, reversible expansion. (b) Path B has two parts. In step 1, the gas is cooled at constant volume until its pressure has fallen to 1.20 atm. In step 2, it is heated and allowed to expand against a constant pressure of 1.20 atm until its volume is 20.00 L and $T = 292$ K. Determine for each path the work done, the heat transferred, and the change in internal energy (w, q, and ΔU).

STRATEGY A good first step is to draw the graphs of pressure as a function of volume for the two paths (Fig. 6.11). We use the ideal gas law to plot the isothermal expansion. Recall that the work done by the system during expansion is equal to the area underneath a curve showing how the external pressure is varied as the volume is changed. (a) For an isothermal, reversible expansion, we use Eq. 12 to calculate w. We expect w to be negative, because energy is lost as work is done. There are no intermolecular forces in an ideal gas, so the internal energy does not change when it expands isothermally: $\Delta U = 0$. (b) In step 1, the volume does not change, so no work is done ($w = 0$). Step 2 is a constant-pressure process, so we use Eq. 9 to calculate w. Because internal energy is a state function, and the initial and final states are the same in both paths, ΔU for path B is the same as for path A. In each case, we find q for the overall path from $\Delta U = q + w$. Use 1 L·atm = 101.325 J to convert liter-atmospheres to joules.

SOLUTION (a) Use Eq. 12 to find w:

$$\begin{aligned} w &= -nRT \ln \frac{V_{\text{final}}}{V_{\text{initial}}} \\ &= -(1.00 \text{ mol}) \times (8.3145 \text{ J·K}^{-1}\text{·mol}^{-1}) \times (292 \text{ K}) \times \ln\left(\frac{20.00}{8.00}\right) \\ &= -2.22 \times 10^3 \text{ J} = -2.22 \text{ kJ} \end{aligned}$$

The value of w is negative, as we expected. Because $\Delta U = 0$ for the isothermal expansion of an ideal gas, and $\Delta U = q + w$, it follows that $q = -w$. Therefore, $q = +2.22$ kJ. This influx of energy as heat counteracts the loss of energy as work and maintains the kinetic energy, and hence the temperature, at a constant value.

(b) Because the volume did not change in step 1, no work is done, so $w = 0$. In step 2, from Eq. 9:

$$w = -P_{ex}\Delta V = -(1.20 \text{ atm}) \times (20.00 - 8.00) \text{ L} = -14.4 \text{ L·atm}$$

Because the work in the first step is 0, the work done in step 2 is the total work for path B. We convert it to joules:

$$w = -(14.4 \text{ L· atm}) \times \left(\frac{101.325 \text{ J}}{1 \text{ L·atm}}\right) = -1.46 \times 10^3 \text{ J} = -1.46 \text{ kJ}$$

Less work is done on path B than on path A, a result confirming that work is not a state function. Overall, the change of state that occurs in path B is the same as that in path A, and we know that $\Delta U = 0$ regardless of path, provided the initial and final states have the same temperature and the gas is ideal. Therefore, $\Delta U = 0$ for path B. Moreover, because $\Delta U = q + w$, we can once again write $q = -w$; and we can conclude that $q = +1.46$ kJ. This quantity is different from q for path A, a result confirming that heat transferred is not a state function. In summary,

	ΔU	q	w
For the reversible path:	0	+2.22 kJ	−2.22 kJ
For the irreversible path:	0	+1.46 kJ	−1.46 kJ

The difference in the work done in the two expansions is illustrated in Fig. 6.11.

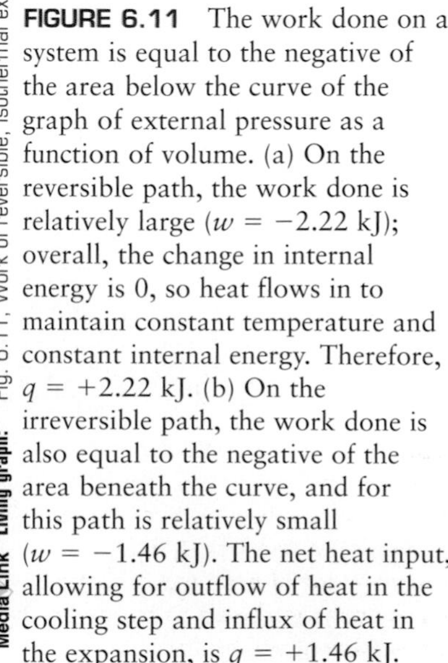

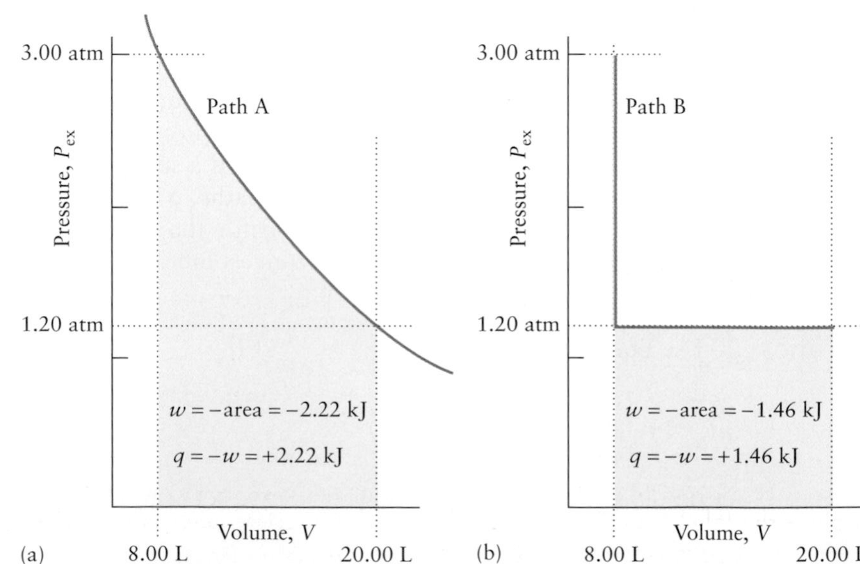

FIGURE 6.11 The work done on a system is equal to the negative of the area below the curve of the graph of external pressure as a function of volume. (a) On the reversible path, the work done is relatively large ($w = -2.22$ kJ); overall, the change in internal energy is 0, so heat flows in to maintain constant temperature and constant internal energy. Therefore, $q = +2.22$ kJ. (b) On the irreversible path, the work done is also equal to the negative of the area beneath the curve, and for this path is relatively small ($w = -1.46$ kJ). The net heat input, allowing for outflow of heat in the cooling step and influx of heat in the expansion, is $q = +1.46$ kJ.

Self-Test 6.3A Suppose that 2.00 mol CO_2, treated as an ideal gas, at 2.00 atm and 300 K is compressed isothermally and reversibly to half its original volume prior to being used to produce soda water. Calculate w, q, and ΔU.

[*Answer:* $w = +3.46$ kJ, $q = -3.46$ kJ, $\Delta U = 0$]

Self-Test 6.3B Suppose that 1.00 kJ of energy is transferred as heat to oxygen in a cylinder fitted with a piston; the external pressure is 2.00 atm. The oxygen expands from 1.00 L to 3.00 L against this constant pressure. Calculate w and ΔU for the entire process.

6.8 The Measurement of Heat: Calorimetry

Many chemical reactions release heat into the surroundings. Almost all our heating, transport, and electricity production depend on reactions of this kind. A process that releases heat into the surroundings is called an **exothermic process.** Most common chemical reactions—and all combustions—are exothermic (Fig. 6.12). Less familiar are chemical reactions that absorb heat from the surroundings. A process that absorbs heat is called an **endothermic process** (Fig. 6.13). A number of common physical processes are endothermic. For instance, vaporization is endothermic, because heat must be supplied to drive molecules of a liquid apart from one another. The dissolution of ammonium nitrate in water is endothermic; in fact, this process is used in instant cold packs for sports injuries. Our task in this section is to see how to measure the energy released or absorbed as heat by a chemical or physical process.

Media Link Lab video: Fig. 6.12, Thermite reaction

FIGURE 6.12 The thermite reaction is a highly exothermic reaction—one that can melt the metal it produces. In this reaction, sluminum metal is reacting with iron(III) oxide, Fe_2O_3, causing a shower of molten iron sparks.

Any energy that leaves a system as heat, such as the energy released when a fuel burns, must end up in the surroundings, and vice versa. Therefore, if we use q to denote the energy transferred as heat from the surroundings to the system and q_{surr} to denote the energy transferred as heat from the system to the surroundings for the same process, the two quantities are related by $q = -q_{surr}$. For instance, if the energy transferred to the system is 10 kJ, we write $q = +10$ kJ; this energy must have come from the surroundings as heat, so we write $q_{surr} = -10$ kJ. It follows that, by measuring the heat that enters or leaves the surroundings, we can determine the heat that leaves or enters the system.

Transfers of energy as heat are measured with a **calorimeter,** a device in which the heat transferred is monitored by observing changes in temperature. A simple calorimeter consists of a reaction vessel containing the reaction mixture, a stirrer, and a thermometer (Fig. 6.14). Reactions in such a calorimeter are studied at atmospheric pressure, and therefore under conditions of constant pressure. However, much more elaborate and precise calorimeters are available.

Media Link Lab video: Fig. 6.13, Endothermic reaction between solids

FIGURE 6.13 The reaction between ammonium thiocyanate, NH_4SCN, and barium hydroxide octahydrate, $Ba(OH)_2{\cdot}8H_2O$, absorbs a lot of heat and can cause water vapor in the air to freeze on the outside of the beaker.

A common type of calorimeter in which measurements are made at constant volume is a *bomb calorimeter* (Fig. 6.15). The reaction takes place inside the sturdy sealed metal vessel (the bomb), which is immersed in water, and the rise in temperature of the entire assembly is monitored. In calorimetry, the surroundings are taken to be the entire calorimeter assembly—the reaction vessel, its contents, including the solvent and, when one is present, the water bath. When discussing calorimetry, we replace the label "surr" with the label "cal" for calorimeter; so, in place of the expression $q = -q_{surr}$, we write $q = -q_{cal}$.

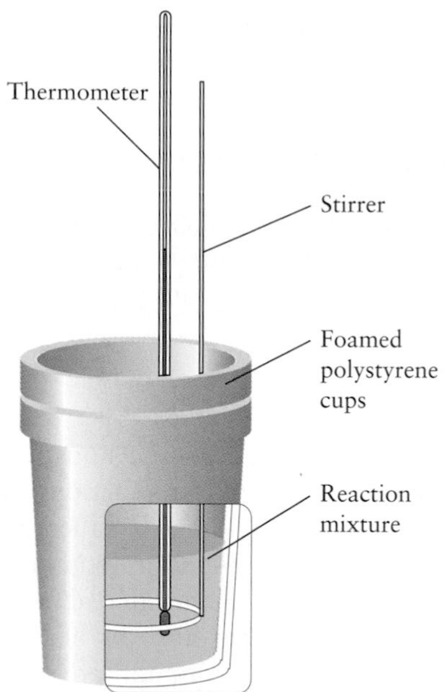

FIGURE 6.14 The quantity of heat released or absorbed by a reaction can be measured in this primitive calorimeter. The outer polystyrene cup acts as an extra layer of insulation to ensure that no heat enters or leaves the inner cup. The quantity of heat released or absorbed is proportional to the change in temperature of the calorimeter.

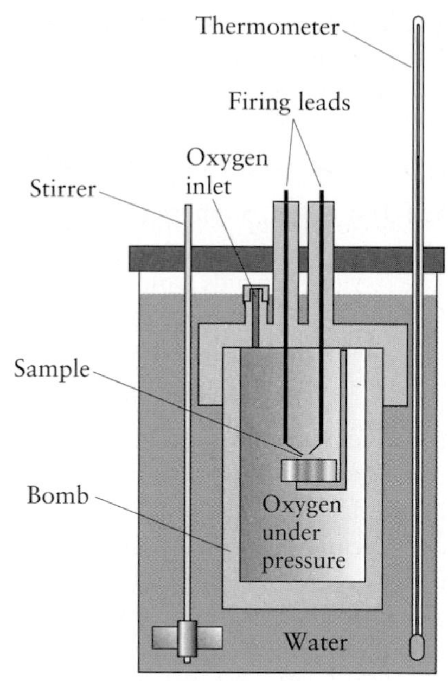

FIGURE 6.15 A bomb calorimeter. The combustion is started with an electrically ignited fuse wire in contact with the sample. Once the combustion has begun, energy is released as heat that spreads through the walls of the bomb into the water. The heat released is proportional to the temperature change of the entire assembly.

The heat released or absorbed in a calorimeter is monitored by measuring the change of temperature it causes. If the reaction is exothermic, the energy transferred as heat to the calorimeter results in a rise in temperature, ΔT. The temperature increase is proportional to the energy transferred, and the constant of proportionality is called the **heat capacity** of the calorimeter, C_{cal}:

$$q_{cal} = C_{cal}\Delta T \qquad (14)$$

The heat capacity tells us how much heat an object absorbs per kelvin or per degree Celsius rise in temperature. The heat capacity of a calorimeter is an empirical quantity and is measured in a separate experiment.

How can we use this information . . . to measure the heat transferred in a calorimeter?

First, we calibrate the calorimeter to determine its heat capacity. Suppose we supply 80.0 kJ of heat to a calorimeter by heating it electrically and observe that the temperature of the calorimeter increases by 8.40°C; then

$$C_{cal} = \frac{\text{heat supplied}}{\text{temperature increase}} = \frac{80.0\ \text{kJ}}{8.40°\ \text{C}} = 9.52\ \text{kJ}\cdot(°\text{C})^{-1}$$

Now we use the calibrated calorimeter. Suppose, under identical conditions, that a reaction taking place in the calorimeter results in a rise in temperature from 20.0°C to 25.2°C (a change of +5.2°C), then

$$q_{cal} = C_{cal}\Delta T = (9.52\ \text{kJ}\cdot(°\text{C})^{-1}) \times (5.2°\text{C}) = +50\ \text{kJ}$$

Because $q = -q_{cal}$, we now know that $q = -50$ kJ. Notice that the reaction has released energy as heat.

Self-Test 6.4A A reaction took place in a calorimeter of heat capacity 2.0 kJ·(°C)$^{-1}$. The temperature of the calorimeter rose by 3.25°C. What is the value of q for the reaction mixture?

[*Answer:* −6.5 kJ]

Self-Test 6.4B A reaction took place in a calorimeter of heat capacity 3.0 kJ·(°C)$^{-1}$. The temperature of the calorimeter rose by 5.1°C. What is the value of q for the reaction mixture?

Heat capacities, C, are also reported for pure substances, not just for the complicated assembly of substances that makes up a typical calorimeter. For instance, we can report the heat capacity of water or of copper. More heat is needed to raise the temperature of a large sample of a substance by a given amount than is required to raise the temperature of a small sample by the same amount, so heat capacity is an extensive property: the larger the sample, the greater its heat capacity (Fig. 6.16). It is therefore common to report either the **specific heat capacity** (often called just "specific heat"), C_s, which is the heat capacity divided by the mass of the sample ($C_s = C/m$), or the **molar heat capacity,** C_m, the heat capacity divided by the number of moles in the sample ($C_m = C/n$). Specific heat capacities and molar heat capacities are intensive properties.

The specific and molar heat capacities of some common substances are listed in Table 6.2. Note that, although the values of the specific heat capacities are given in joules per degree Celsius per gram (J·(°C)$^{-1}$·g^{-1}), they could equally well be reported in joules per kelvin per gram (J·K^{-1}·g^{-1}) with the same numerical values, because the Celsius degree is the same size as the kelvin. We can calculate the heat capacity of a sample of a substance from its mass and its specific heat capacity by rearranging the definition $C_s = C/m$ into $C = mC_s$. Then we can use

$$q = C\Delta T = mC_s\Delta T \qquad (15)$$

to calculate the heat required to raise the temperature of a system by a given amount. One word of warning: for most substances, heat capacities in general depend on the temperature, so using a heat capacity at a single temperature gives only an approximate indication of the heat required to achieve a given increase in temperature.

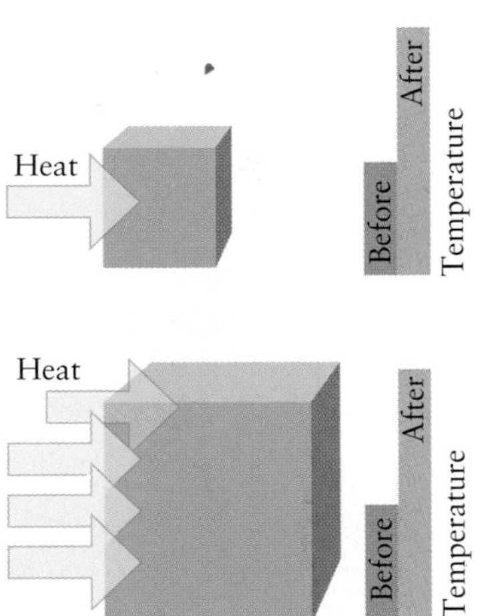

FIGURE 6.16 The heat capacity of an object determines the change in temperature brought about by a given quantity of heat: an object with a large heat capacity requires a lot of heat to bring about a given rise in temperature. Heat capacity is an extensive property, so a large object (bottom) has a larger heat capacity than a small object (top) made of the same material. Heat capacities also, in general, depend on temperature.

TABLE 6.2 Specific and Molar Heat Capacities of Common Materials*

Material	Specific heat capacity, $J\cdot(°C)^{-1}\cdot g^{-1}$	Molar heat capacity, $J\cdot K^{-1}\cdot mol^{-1}$
air	1.01	—
benzene	1.05	136
brass	0.37	—
copper	0.38	33
ethanol	2.42	111
glass (Pyrex)	0.78	—
granite	0.80	—
marble	0.84	—
polyethylene	2.3	—
stainless steel	0.51	—
water: solid	2.03	37
liquid	4.184	75
vapor	2.01	34

* More values are available in Appendices 2A and 2D; values assume constant pressure. Specific heat capacities commonly use degrees Celsius in their units, whereas molar heat capacities commonly use kelvins. All values except that for ice are for 25°C.

Molar heat capacities increase with increasing molecular complexity, because more ways of storing the energy become available as the number of atoms increases. Notice that, in Table 6.2, molecular solids have greater molar heat capacities than atomic solids like metals: molecules provide more modes of energy storage in the form of vibrational motion. Notice also that the molar heat capacities of molecular liquids are greater than those of their solid forms. In the liquid state, the molecules are free to translate (move through space) and rotate as well as vibrate. The molar heat capacities of liquid and solid sodium, both of which consist of atoms (which cannot rotate), differ by only about 3 $J\cdot K^{-1}\cdot mol^{-1}$. In contrast, the molar heat capacities of liquid and solid water, both of which consist of molecules (which can rotate in the liquid phase), differ by about 38 $J\cdot K^{-1}\cdot mol^{-1}$. A consequence of the high heat capacity of water is that lakes and oceans are slow to freeze because so much energy has to be extracted from them to lower their temperature to the freezing point. The great heat capacity of the oceans ensures that the coldest days of winter do not occur until well after the winter solstice and that the hottest days of summer do not occur until well after the summer solstice.

Heat transfers are measured by using a calibrated calorimeter. The heat capacity of an object is the ratio of the heat supplied to the temperature rise produced. Molar heat capacities of liquids are generally greater than those of the solid phase of the same substance. Molar heat capacities increase as molecular complexity increases.

EXAMPLE 6.2 Measuring specific heat capacity at constant pressure with a simple calorimeter

The constant-pressure heat capacity of a metal can be measured by a simple experiment. Add a known mass of water to a calorimeter like that in Fig. 6.14 and measure the temperature. Then place a piece of metal heated to a known temperature into the calorimeter and measure the final temperature. In one experiment, 21 g of iron at 90.2°C is added to 50 g of water at 20.0°C. The final temperature of the mixture is 23.2°C. What is the specific heat capacity of iron?

STRATEGY All the heat lost by the iron is transferred to the water and the rest of the calorimeter. The heat lost or gained by each substance is equal to its heat capacity multiplied by the temperature rise or fall; the heat capacity is the specific heat capacity multiplied by the mass. The heat capacity of the polystyrene cup is so small that we assume it can be neglected. Therefore, we can use Eq. 15 to write

$$\begin{aligned} &m(\text{metal})C_s(\text{metal}) \times (\text{fall in temperature})_{\text{metal}} \\ &\quad = m(\text{water})C_s(\text{water}) \times (\text{rise in temperature})_{\text{water}} \end{aligned}$$

There is enough information to find the one unknown, the specific heat capacity of the metal.

SOLUTION We substitute the data and solve for the specific heat capacity of the metal. Note that the temperature of the iron falls from 90.2°C to 23.2°C, or by 67.0°C, and the temperature of the water rises from 20.0°C to 23.2°C, or by 3.2°C. Therefore,

$$(21\text{ g}) \times C_s(\text{metal}) \times (67.0°\text{C}) = (50\text{ g}) \times (4.184\text{ J}\cdot(°\text{C})^{-1}\cdot\text{g}^{-1}) \times (3.2°\text{C})$$

That is,

$$\begin{aligned} C_s(\text{metal}) &= \frac{(50\text{ g}) \times (4.184\text{ J}\cdot(°\text{C})^{-1}\cdot\text{g}^{-1}) \times (3.2°\text{C})}{(21\text{ g}) \times (67.0°\text{C})} \\ &= 0.48\text{ J}\cdot(°\text{C})^{-1}\cdot\text{g}^{-1} \end{aligned}$$

Self-Test 6.5A A piece of copper of mass 19.0 g was heated to 87.4°C and then placed in a calorimeter that contained 55.5 g of water at 18.3°C. The temperature of the water rose to 20.4°C. What is the specific heat capacity of copper?

[*Answer:* $0.38\text{ J}\cdot(°\text{C})^{-1}\cdot\text{g}^{-1}$]

Self-Test 6.5B An alloy of mass 25.0 g was heated to 88.6°C and then placed in a calorimeter that contained 61.2 g of water at 19.6°C. The temperature of the water rose to 21.3°C. What is the specific heat capacity of the alloy?

EXAMPLE 6.3 Determining the heat output of a reaction

A reaction known to release 1.78 kJ of heat took place in a calorimeter like that shown in Fig. 6.14. The calorimeter contained 0.100 L of solution and the temperature rose by 3.65°C. Next, 50 mL of 0.20 M HCl(aq) and 50 mL of 0.20 M NaOH(ag) were mixed in the same calorimeter and the temperature rose by 1.26°C. What is the heat output of the neutralization reaction?

STRATEGY First, calibrate the calorimeter by calculating the heat capacity of the calorimeter from the information on the first reaction:

$$C_{cal} = \frac{q_{cal}}{\Delta T}$$

Then use the value of that heat capacity to convert the temperature rise caused by the neutralization reaction into the heat output of the reaction. For this step, use the same equation rearranged to $q_{cal} = C_{cal}\Delta T$, but with ΔT now the change in temperature observed during the reaction, and then convert the value of q_{cal} to q for the reaction mixture, $q = -q_{cal}$. So, overall, we use

$$q = -q_{cal} = -C_{cal}\Delta T$$

Note that the calorimeter contains the same volume of liquid in both cases. Because dilute aqueous solutions have approximately the same heat capacities as pure water, the heat capacity is the same as in the calibration.

SOLUTION The heat capacity of the calorimeter is

$$C_{cal} = \frac{1.78 \text{ kJ}}{3.65°\text{C}} = \frac{1.78}{3.65} \text{ kJ}\cdot(°\text{C})^{-1}$$

It follows that, for the neutralization reaction,

$$q = -\left(\frac{1.78}{3.65} kJ\cdot(°\text{C})^{-1}\right) \times (1.26°\text{C}) = -0.614 \text{ kJ}$$

The negative sign indicates that the reaction is exothermic.

Self-Test 6.6A A small piece of calcium carbonate was placed in the same calorimeter, and 0.100 L of dilute hydrochloric acid was poured over it. The temperature of the calorimeter rose by 3.57°C. What is the value of q?

[*Answer:* −1.74 kJ]

Self-Test 6.6B A calorimeter was calibrated by mixing two aqueous solutions together, each of volume 0.100 L. The heat output of the reaction that occurred was known to be −4.16 kJ, and the temperature of the calorimeter rose by 3.24°C. Calculate the heat capacity of this calorimeter when it contains 0.200 L of water.

ENTHALPY

If there is no change in the volume of a system and no nonexpansion work is done, then the change in the internal energy of the system is equal to the energy supplied to it as heat ($\Delta U = q$, at constant volume). In chemistry, though, we are more concerned with heat transfers at constant pressure. Many chemical reactions take place in containers open to the atmosphere and therefore occur at a constant pressure of about 1 atm. Such systems are free to expand or contract. For example, when calcium carbonate is heated, it decomposes and the carbon dioxide produced spreads away. Although no piston is involved, the expanding gas still needs to make room for itself and has to drive back the surrounding atmosphere.

If in some process we supply 100 J of heat to a system ($q = +100$ J) and the system does 20 J of work ($w = -20$ J) as it expands at constant pressure, the internal energy of the system rises by only 80 J. Clearly, in this case, we cannot equate the heat supplied to a system with the rise in internal energy. Is there another state function that we can use to keep track of energy changes occurring at constant pressure?

6.9 Heat Transfers at Constant Pressure

The state function that allows us to keep track of energy changes at constant pressure is called the **enthalpy,** H:

$$H = U + PV \tag{16}$$

where U, P, and V are the internal energy, pressure, and volume of the system. What we now need to show is that it follows from this definition and the first law of thermodynamics that *a change in the enthalpy of a system is equal to the heat released or absorbed at constant pressure.*

How do we do that?

To show that the change in the enthalpy of a system is equal to the heat released or absorbed at constant pressure, we consider a process at constant pressure for which the change in internal energy is ΔU and the change in volume is ΔV. It then follows from the definition of enthalpy in Eq. 16 that the change in enthalpy is

$$\Delta H = \Delta U + P\Delta V$$

Now we use the first law in the form $\Delta U = q + w$, where q is the energy supplied to the system as heat and w is the energy supplied as work. Therefore,

$$\Delta H = q + w + P\Delta V$$

At this stage, we use the expression for expansion work, Eq. 9 ($w = -P_{ex}\Delta V$), to write

$$\Delta H = q - P_{ex}\Delta V + P\Delta V$$

Finally, because the system is open to the atmosphere, the pressure of the system is the same as the external pressure, so $P_{ex} = P$, and the last two terms cancel to leave

$$\text{At constant pressure: } \Delta H = q \tag{17}$$

The implication of Eq. 17 is that, because chemical reactions typically take place at constant pressure in vessels open to the atmosphere, the heat they release or require can be equated to the change in enthalpy of the system. Equations 8 and 17 have the following implications for calorimetry:

- The heat transfer in a constant-volume calorimeter can be equated to the change in internal energy.
- The heat transfer in a constant-pressure calorimeter can be equated to the change in enthalpy.

When we transfer energy to a constant-pressure system as heat, the enthalpy of the system *increases* by that amount. When energy leaves a constant-pressure system as heat, the enthalpy of the system *decreases* by that amount. For example, the formation of zinc iodide from its elements is an exothermic reaction that (at constant pressure) releases 208 kJ of heat to the surroundings for each mole of ZnI_2 formed:

$$Zn(s) + I_2(s) \longrightarrow ZnI_2(s)$$

We can therefore report that $\Delta H = -208$ kJ because the enthalpy of the reaction mixture decreases by 208 kJ in this reaction (Fig. 6.17). An endothermic process absorbs heat, so when ammonium nitrate dissolves in water—as occurs when a cold-pack is used—the enthalpy of the system increases (Fig. 6.18). Note that $\Delta H < 0$ for exothermic reactions, whereas $\Delta H > 0$ for endothermic reactions.

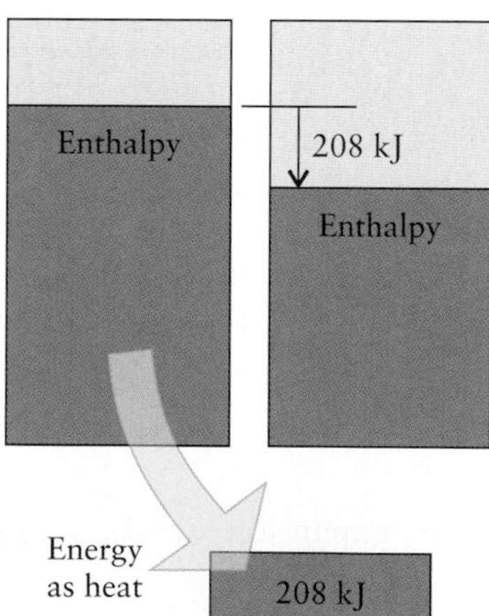

FIGURE 6.17 The enthalpy of a system is like a measure of the height of water in a reservoir. When a reaction releases 208 kJ of heat at constant pressure, the "reservoir" falls by 208 kJ and $\Delta H = -208$ kJ.

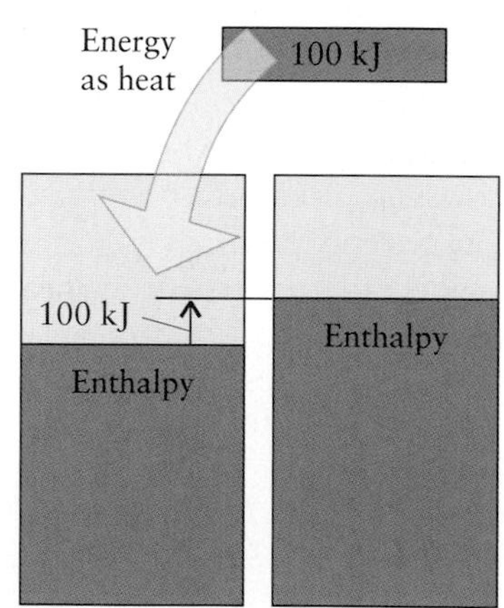

FIGURE 6.18 If an endothermic reaction absorbs 100 kJ of heat at constant pressure, the height of the enthalpy "reservoir" rises by 100 kJ and $\Delta H = +100$ kJ.

We can see that *enthalpy is a state function* by noting that U, P, and V are all state functions, so $H = U + PV$ must be a state function, too. That U is a state function is a consequence of the first law of thermodynamics, so the fact that H is a state function is also a consequence of the first law. It then follows that, because the enthalpy is a state function, we can calculate the value of ΔH for any change of state by choosing the most convenient path between the specified initial and final states.

The change in enthalpy of a system is equal to the heat supplied to the system at constant pressure. For an endothermic process, $\Delta H > 0$; for an exothermic process, $\Delta H < 0$.

Self-Test 6.7A In a certain exothermic reaction at constant pressure, 50 kJ of energy left the system as heat and 20 kJ of energy left the system as expansion work to make room for the products. What are the values of (a) ΔH and (b) ΔU for this process?

[*Answer:* (a) −50 kJ; (b) −70 kJ]

Self-Test 6.7B In a certain endothermic reaction at constant pressure, 30 kJ of energy entered the system as heat. The products took up less volume than the reactants, and 40 kJ of energy entered the system as work as the outside atmosphere pressed down on it. What are the values of (a) ΔH and (b) ΔU for this process?

6.10 Heat Capacities at Constant Volume and Constant Pressure

As explained in Section 6.8, the heat capacity of a substance is a measure of the temperature rise that occurs when the substance is heated. The rise in temperature and therefore the heat capacity depend on the conditions under which the heating takes place because, at constant pressure, some of the heat is used to do expansion work rather than to raise the temperature of the system.

With rare exceptions, solids and liquids expand when heated, but to a much smaller extent than gases.

We can express these ideas mathematically. First, the definition of heat capacity, C, is the ratio of heat supplied, q, to the temperature rise the heat produces, ΔT:

$$C = \frac{q}{\Delta T} \qquad (18)$$

We have already seen (from the first law) that heat transferred at constant volume can be identified with the change in internal energy, ΔU. Therefore, the **heat capacity at constant volume,** C_V, is

$$C_V = \frac{\Delta U}{\Delta T} \qquad (19)$$

We have also seen that the heat transferred at constant pressure can be identified with the change in enthalpy, ΔH. It follows that the **heat capacity at constant pressure,** C_P, is

$$C_P = \frac{\Delta H}{\Delta T} \quad (20)$$

The molar heat capacities are these quantities divided by the amount of substance and are denoted $C_{V,\mathrm{m}}$ and $C_{P,\mathrm{m}}$.

We can use the definition of enthalpy and the ideal gas law to find a simple quantitative relation between C_P and C_V for an ideal gas.

How do we do that?

To find the relation between the two heat capacities, we start with the definition of enthalpy in Eq. 16. For an ideal gas, the PV in $H = U + PV$ can be replaced by nRT, so

$$H = U + nRT$$

When a sample of an ideal gas is heated, the enthalpy, internal energy, and temperature all change, and it follows that

$$\Delta H = \Delta U + nR\Delta T$$

The heat capacity at constant pressure can therefore be expressed as

$$C_P = \frac{\Delta H}{\Delta T} = \frac{\Delta U + nR\Delta T}{\Delta T} = \frac{\Delta U}{\Delta T} + nR = C_V + nR$$

It follows that the two molar heat capacities of an ideal gas are related by

$$C_{P,\mathrm{m}} = C_{V,\mathrm{m}} + R \quad (21)$$

As an example, the molar constant-volume heat capacity of argon is $12.8\ \mathrm{J{\cdot}K^{-1}{\cdot}mol^{-1}}$, so the corresponding constant-pressure value is $12.8 + 8.3\ \mathrm{J{\cdot}K^{-1}{\cdot}mol^{-1}} = 21.1\ \mathrm{J{\cdot}K^{-1}{\cdot}mol^{-1}}$, a difference of 65%.

The molar heat capacity of an ideal gas at constant pressure is greater than that at constant volume; the two quantities are related by Eq. 21.

6.11 The Molecular Origin of the Heat Capacities of Gases

Heat capacities are bulk properties and are determined experimentally. However, just as we enriched our understanding of work and heat by interpreting them in terms of molecules, so we can relate the values of heat capacities to molecular properties. First, let's see how to account for the molar heat capacity of a monatomic ideal gas, such as argon. We saw in Section 6.3 that the molar internal energy of a monatomic ideal gas at a temperature T is $\frac{3}{2}RT$. We can conclude that the change in molar internal energy when the temperature is changed by ΔT is

$$\Delta U_m = \tfrac{3}{2}R\Delta T$$

It follows from Eq. 19 that the molar heat capacity at constant volume is

$$C_{V,m} = \frac{\Delta U_m}{\Delta T} = \frac{\frac{3}{2}R\Delta T}{\Delta T} = \tfrac{3}{2}R \qquad (22a)$$

or about 12.5 $J\cdot K^{-1}\cdot mol^{-1}$, in agreement with the value found experimentally. Note that, according to Eq. 22a, the molar heat capacity of a monatomic ideal gas is a constant and so independent of temperature and pressure.

The molar heat capacities of gases composed of molecules (as distinct from atoms) are higher than those of monatomic gases because the molecules can store energy as rotational kinetic energy as well as translational kinetic energy. We saw in Section 6.3 that the rotational motion of linear molecules contributes another RT to the molar internal energy, to give a total of $\frac{5}{2}RT$. For nonlinear molecules, the rotational contribution is $\frac{3}{2}RT$, for a total contribution of $3RT$. Therefore, by the same argument used earlier,

$$\text{Linear molecules:} \quad C_{V,m} = \tfrac{5}{2}R \qquad C_{P,m} = \tfrac{7}{2}R \qquad (22b)$$

$$\text{Nonlinear molecules:} \quad C_{V,m} = 3R \qquad C_{P,m} = 4R \qquad (22c)$$

(In each case, $C_{P,m}$ has been calculated from $C_{P,m} = C_{V,m} + R$.) Note that the molar heat capacity of nonlinear molecules is higher than that of linear molecules because nonlinear molecules can rotate around three rather than only two axes (recall Fig. 6.6).

The graph in Fig. 6.19 shows how $C_{V,m}$ for iodine vapor varies with temperature. Although $C_{V,m}$ is constant at $\frac{3}{2}R$ at low temperatures, it rises to $\frac{5}{2}R$ as the temperature is raised and to even higher values at high temperatures. The increase from $\frac{3}{2}R$ to $\frac{5}{2}R$ is due to the onset of molecular rotation. At still higher temperatures, molecular vibrations start to absorb energy and the heat capacity rises to even higher values.

The constant-pressure heat capacity of a substance is greater than its constant-volume heat capacity. Rotation requires energy and leads to higher heat capacities for complex molecules; the equipartition theorem can be used to estimate the molar heat capacities of gas-phase molecules, Eq. 22.

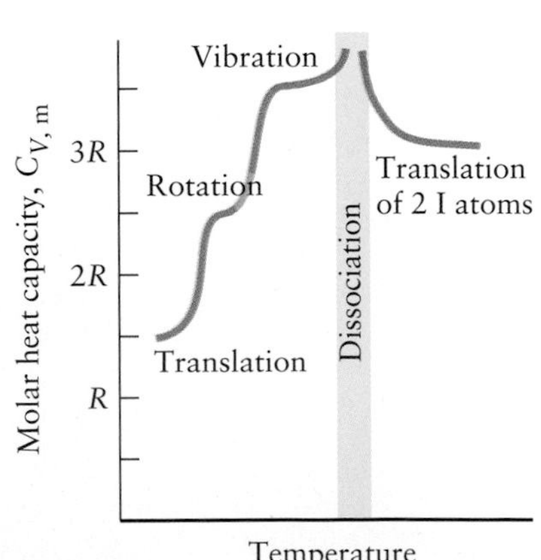

FIGURE 6.19 The variation of the molar heat capacity of iodine vapor at constant volume. Translation always contributes; rotation contributes except at very low temperature, and vibrations of the molecule contribute only at very high temperatures. When the molecules dissociate, the heat capacity becomes very large during dissociation, but then settles down to a value characteristic of 2 mol I atoms undergoing only translational motion.

EXAMPLE 6.4 Calculating the energy changes when heating an ideal gas

Calculate the final temperature and the change in internal energy when 500 J of energy is transferred as heat to 0.900 mol O_2 at 298 K and 1.00 atm at (a) constant volume; (b) constant pressure. Treat the gas as ideal.

STRATEGY We expect the temperature to rise more as a result of heating at constant volume than at constant pressure because, in the latter case, some of the energy is used to expand the system. The heat capacities of oxygen can be estimated from the equipartition principle, as stated in Eq. 22; then we can use $q = nC_m\Delta T$ to find the final temperature. The internal energy change at constant volume is equal to the heat supplied. At constant pressure, find the internal energy change by considering the process as taking place in two steps, the first heating to the final temperature at constant volume, then isothermal expansion.

SOLUTION The molar heat capacities of oxygen, a linear molecule, are

$$C_{V,\mathrm{m}} = \tfrac{5}{2}R = 20.79\ \mathrm{J \cdot K^{-1} \cdot mol^{-1}}$$
$$C_{P,\mathrm{m}} = C_{V,\mathrm{m}} + R = \tfrac{7}{2}R = 29.10\ \mathrm{J \cdot K^{-1} \cdot mol^{-1}}$$

(a) The increase in temperature when the gas is heated at constant volume is

$$\Delta T = \frac{q}{nC_{V,\mathrm{m}}} = \frac{500\ \mathrm{J}}{(0.900\ \mathrm{mol}) \times (20.79\ \mathrm{J \cdot K^{-1} \cdot mol^{-1}})} = +26.7\ \mathrm{K}$$

The final temperature is therefore 298 + 26.7 K = 325 K, or 52°C. The increase in internal energy when heating takes place at constant volume is equal to the heat input ($\Delta U = q$ at constant volume), so $\Delta U = +500$ J.
(b) The increase in temperature when the gas is heated at constant pressure is

$$\Delta T = \frac{q}{nC_{P,\mathrm{m}}} = \frac{500\ \mathrm{J}}{(0.900\ \mathrm{mol}) \times (29.10\ \mathrm{J \cdot K^{-1} \cdot mol^{-1}})} = +19.1\ \mathrm{K}$$

The final temperature is therefore 298 + 19.1 K = 317 K, or 44°C. For heating at constant pressure, the overall process can be regarded as taking place in two steps.

Step 1 Enough energy is transferred as heat at constant volume to raise the temperature to its final value (317 K). This step requires

$$q = (0.900\ \mathrm{mol}) \times (20.79\ \mathrm{J \cdot K^{-1} \cdot mol^{-1}}) \times (19.1\ \mathrm{K}) = +357\ \mathrm{J}$$

Step 2 The sample is allowed to expand isothermally to its final volume. This step involves no change in internal energy, because U is independent of volume for an ideal gas.

Therefore, overall, $\Delta U = +357$ J, a value lower than that for the constant-volume case, as predicted.

Self-Test 6.8A Calculate the final temperature and the change in internal energy when 500 J of energy is transferred as heat to 0.900 mol Ne(g) at 298 K and 1.00 atm (a) at constant volume; (b) at constant pressure. Treat the gas as ideal.

[*Answer:* (a) 343 K, 500 J; (b) 325 K, 300 J]

Self-Test 6.8B Calculate the final temperature and the change in internal energy when 1.20 kJ of energy is transferred as heat to 1.00 mol $H_2(g)$ at 298 K and 1.00 atm (a) at constant volume; (b) at constant pressure. Treat the gas as ideal.

6.12 The Enthalpy of Physical Change

Because no change of substance is involved, the simplest processes to consider in terms of their energy requirements are physical changes, which include vaporization, condensation, and freezing. Each of these phase transitions involves a change in the energy of the substance as molecules separate from one another (in vaporization), stick together (in condensation), or settle into new arrangements (in phase transitions between solid forms). A phase change that results in greater separation of the molecules requires energy and is therefore endothermic. Vaporization, for instance, is endothermic. Phase changes that increase molecular contact, such as freezing, are exothermic. Because phase transitions typically take place at constant pressure, we identify the heat transfer due to a phase change with the change in enthalpy of the substance.

The difference in molar enthalpy between the vapor and liquid states of a substance is called the **enthalpy of vaporization, ΔH_{vap}**.

$$\Delta H_{vap} = H_m(\text{vapor}) - H_m(\text{liquid}) \qquad (23)$$

The enthalpy of vaporization of most substances changes little with temperature. For water at its boiling point, 100°C, $\Delta H_{vap} = 40.7\ \text{kJ}\cdot\text{mol}^{-1}$; but at 25°C, $\Delta H_{vap} = 44.0\ \text{kJ}\cdot\text{mol}^{-1}$. This latter value means that to vaporize 1.00 mol $H_2O(l)$, corresponding to 18.02 g of water, at 25°C and constant pressure, we have to supply 44.0 kJ of energy as heat.

Self-Test 6.9A A sample of benzene, C_6H_6, was heated to 80°C, its boiling point. The heating was continued until 28 kJ had been supplied; as a result, 71 g of boiling benzene was vaporized. What is the enthalpy of vaporization of benzene at its boiling point?

[*Answer:* $31\ \text{kJ}\cdot\text{mol}^{-1}$]

Self-Test 6.9B The same heater was used to heat a sample of ethanol, C_2H_5OH, of mass 23 g to its boiling point. It was found that 22 kJ was required to vaporize all the ethanol. What is the enthalpy of vaporization of ethanol at its boiling point?

Enthalpies of vaporization for various substances are listed in Table 6.3. All enthalpies of vaporization are positive, so it is not necessary to report their sign. Notice that compounds that can form hydrogen bonds tend to have the highest enthalpies of vaporization. That is easy to explain, because the enthalpy of vaporization is a measure of the energy

TABLE 6.3 Standard Enthalpies of Physical Change*

Substance	Formula	Freezing point, K	$\Delta H_{fus}°$, kJ·mol^{-1}	Boiling point, K	$\Delta H_{vap}°$, kJ·mol^{-1}
acetone	CH_3COCH_3	177.8	5.72	329.4	29.1
ammonia	NH_3	195.4	5.65	239.7	23.4
argon	Ar	83.8	1.2	87.3	6.5
benzene	C_6H_6	278.6	10.59	353.2	30.8
ethanol	C_2H_5OH	158.7	4.60	351.5	43.5
helium	He	3.5	0.021	4.22	0.084
mercury	Hg	234.3	2.292	629.7	59.3
methane	CH_4	90.7	0.94	111.7	8.2
methanol	CH_3OH	175.2	3.16	337.8	35.3
water	H_2O	273.2	6.01	373.2	40.7 (44.0 at 25°C)

* Values correspond to the temperature of the phase change. The superscript ° signifies that the change takes place at 1 bar and that the substance is pure (that is, the values are for standard states; see Section 6.16).

needed to separate molecules from their attractions in the liquid state into a free state in the vapor. In plots of the potential energy arising from intermolecular interactions, like that shown in Fig. 6.20, the enthalpy of the substance in the liquid state, where molecular interactions are strong, is related to the depth of the bottom of the well in the curve. The enthalpy of the vapor state, where the interactions are almost insignificant, is related to the horizontal part of the curve to the right. A substance with a high molar enthalpy of vaporization has a deep intermolecular potential well, indicating strong intermolecular attractions.

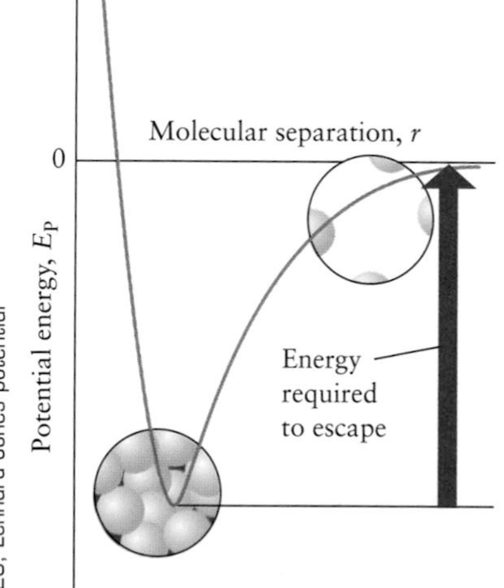

Media Link Living graph: Fig. 6.20, Lennard Jones potential

FIGURE 6.20 The potential energy of molecules decreases as they approach each other, and then it rises again as they are pressed together. To vaporize a liquid, molecules must be raised from the bottom of the well to the energy of the horizontal part of the curve on the right.

The enthalpy change that accompanies melting per mole of molecules is called the **enthalpy of fusion,** ΔH_{fus}, of the substance:

$$\Delta H_{fus} = H_m(\text{liquid}) - H_m(\text{solid}) \quad (24)$$

Melting is endothermic, so all enthalpies of fusion are positive (see Table 6.3). The enthalpy of fusion of water at 0°C is 6.0 kJ·mol^{-1}: to melt 1.0 mol $H_2O(s)$ (18 g of ice) at 0°C, we have to supply 6.0 kJ of heat. Vaporizing the same amount of water takes much more energy (over 40 kJ), because when water is vaporized to a gas, its molecules must be separated completely. In melting, the molecules stay close together, so the forces of attraction and repulsion are nearly as strong as those experienced in the solid (Fig. 6.21).

The **enthalpy of freezing** is the change in enthalpy per mole when a liquid turns into a solid. For water at 0°C, the enthalpy of freezing is −6.0 kJ·mol^{-1}, because 6.0 kJ of heat is *released* when 1 mol $H_2O(l)$ freezes. Because enthalpy is a state property, the enthalpy of freezing of a substance is the negative of its enthalpy of fusion. The enthalpy of water must be the same after being frozen and then melted as it was before it was frozen. Therefore, the same quantity of heat is released in freezing as

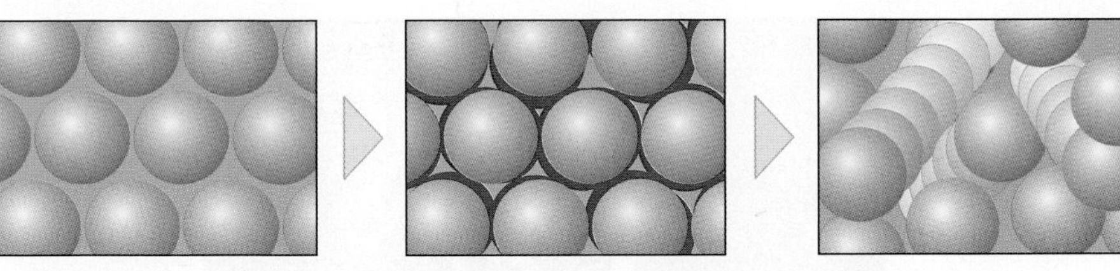

FIGURE 6.21 Melting (fusion) is an endothermic process. As molecules acquire energy, they begin to struggle past their neighbors. Finally the sample changes from a solid with ordered molecules (left) to a liquid with disordered, mobile molecules (far right).

View an animation of a solid melting.

Media Link Animation: Fig. 6.21, Melting

is absorbed in melting. In general, to obtain the enthalpy change for the reverse of any process, we just take the negative of the enthalpy change for the original process:

$$\Delta H_{\text{reverse process}} = -\Delta H_{\text{forward process}} \tag{25}$$

This relation, which we can trace back to the first law of thermodynamics, is illustrated in Fig. 6.22. If we find, for example, that the enthalpy of vaporization of mercury is 59 $\text{kJ}\cdot\text{mol}^{-1}$ at its boiling point, then we know that the enthalpy change occurring when mercury vapor condenses at that temperature is $-59\ \text{kJ}\cdot\text{mol}^{-1}$. This value tells us that 59 kJ of heat is released when 1 mol Hg(g) condenses to a liquid.

Sublimation is the direct conversion of a solid into its vapor. Frost disappears on a cold, dry morning as the ice sublimes directly into water vapor. Solid carbon dioxide also sublimes, which is why it is called dry ice. Each winter on Mars, solid carbon dioxide is deposited as polar frost, which sublimes when the feeble summer arrives (Fig. 6.23). The **enthalpy of sublimation, ΔH_{sub},** is the enthalpy change per mole when a solid sublimes:

$$\Delta H_{\text{sub}} = H_{\text{m}}(\text{vapor}) - H_{\text{m}}(\text{solid}) \tag{26}$$

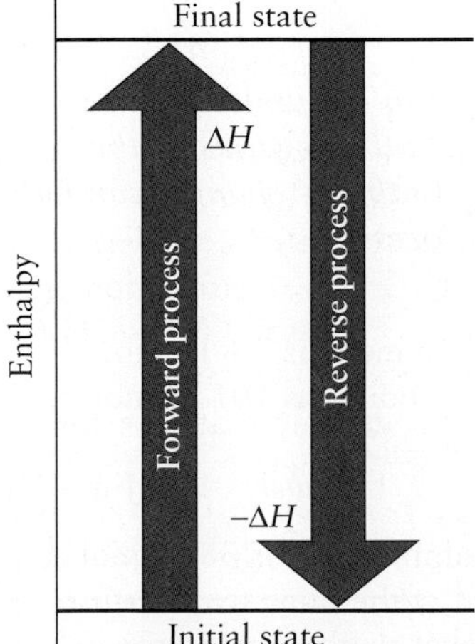

FIGURE 6.22 The enthalpy change for a reverse process is the negative of the enthalpy change for the forward process at the same temperature.

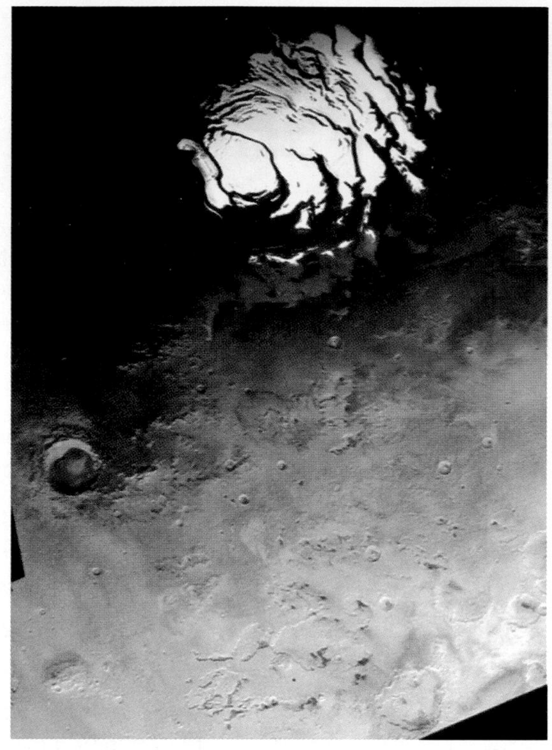

FIGURE 6.23 The polar ice caps on Mars extend and recede with the seasons. They are solid carbon dioxide and form by direct conversion of the gas to a solid. They disappear by sublimation. Although some water ice is also present in the polar caps, the temperature on Mars never becomes high enough to melt it. On Mars, ice is just another rock.

Because enthalpy is a state function, the enthalpy of sublimation of a substance is the same whether the transition takes place in one step, directly from solid to gas, or in two steps, first from solid to liquid, then from liquid to gas. The enthalpy of sublimation of a substance must therefore be equal to the sum of the enthalpies of fusion and vaporization measured at the same temperature (Fig. 6.24):

$$\Delta H_{\text{sub}} = \Delta H_{\text{fus}} + \Delta H_{\text{vap}} \tag{27}$$

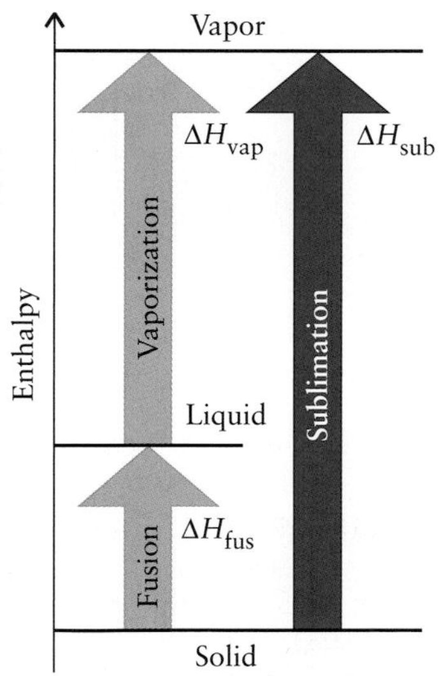

FIGURE 6.24 Because enthalpy is a state property, the enthalpy of sublimation can be expressed as the sum of the enthalpies of fusion and vaporization at the same temperature.

Note that we can add enthalpy changes like this only if they correspond to the same temperature.

Enthalpies of phase transitions are reported in kilojoules per mole. The enthalpy change for a reverse reaction is the negative of the enthalpy change for the forward reaction. Enthalpy changes can be added to obtain the value for an overall process.

Self-Test 6.10A The enthalpy of fusion of sodium metal is 2.6 kJ·mol^{-1} at 25°C, and the enthalpy of sublimation of solid sodium is 101 kJ·mol^{-1}. What is the enthalpy of vaporization of sodium at 25°C?

[*Answer:* 98 kJ·mol^{-1}]

Self-Test 6.10B The enthalpy of vaporization of methanol is 38 kJ·mol^{-1} at 25°C and the enthalpy of fusion is 3 kJ·mol^{-1} at the same temperature. What is the enthalpy of sublimation of methanol at that temperature?

6.13 Heating Curves

The enthalpies of fusion and vaporization affect the appearance of the **heating curve** of a substance; a heating curve is the graph showing the variation of the temperature of a sample as it is heated at a constant rate at constant pressure and therefore at a constant rate of increase in enthalpy (Box 6.1).

Consider what happens when we heat a sample of very cold ice. As we see in Fig. 6.25, at first its temperature rises steadily. Although the molecules are still locked in a solid mass, they are oscillating more and more vigorously around their mean positions. However, once the temperature has reached the melting point, the molecules have enough energy to move past one another. At this temperature, all the added energy is used to overcome the attractive forces between molecules. Thus, although heating continues, the temperature remains constant at the melting point until all the ice has melted. Only then does the temperature rise again, and the rise continues right up to the boiling point. At the boiling point, the temperature rise again comes to a halt. Now the water molecules have enough energy to escape into the vapor state, and all the heat supplied is used to form the vapor. After the sample has evaporated and heating continues, the temperature of the vapor rises again.

The slope of a heating curve is greater for samples that have a low heat capacity than for those with a high heat capacity. In the heating curve for water, for instance, the heating curve of the liquid has a smaller slope than those of the solid and vapor; therefore, we know that the liquid has a greater heat capacity than either the solid or the gas phase does. The high heat capacity of liquid water results partly from the fact that strong hydrogen bonds are being broken as the sample is heated.

The temperature of a sample is constant at its melting and boiling points, even though heat is being supplied.

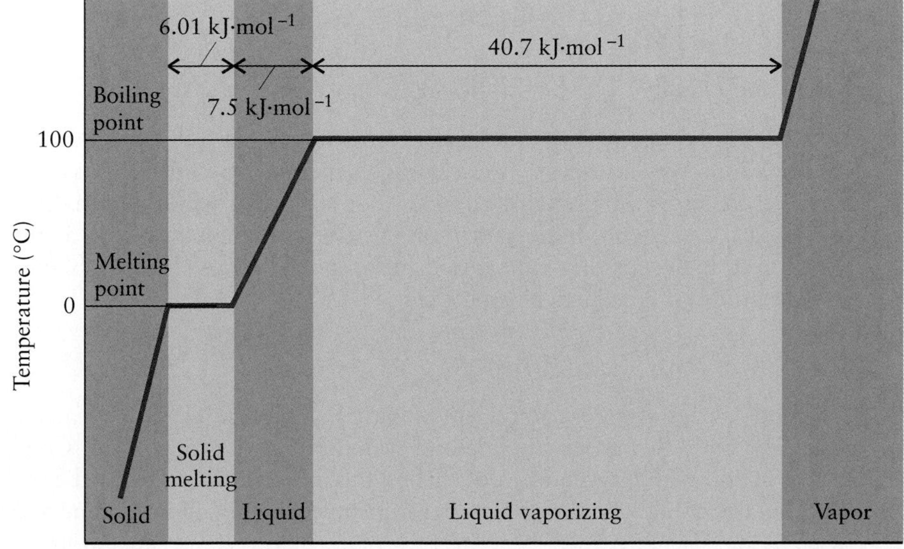

FIGURE 6.25 The heating curve of water. The temperature of a solid rises as heat is supplied. At the melting point, the temperature remains constant and the heat is used to melt the sample. When enough heat has been supplied to melt all the solid, the temperature begins to rise again. A similar pause occurs at the boiling point.

BOX 6.1 HOW DO WE KNOW . . . THE SHAPE OF A HEATING CURVE?

The heating curve of a substance, like that in Fig. 6.25, shows how its temperature changes as heat is supplied at a constant rate, usually at constant pressure. The curve contains a lot of useful information about a substance, including the temperature and enthalpy of each phase transition and the heat capacity of each phase. Simple laboratory heaters can be used to obtain a crude estimate of a heating curve. However, for accuracy, one of two related techniques is normally used.

In *differential thermal analysis* (DTA), equal masses of a sample and a reference material that will not undergo any phase changes, such as Al_2O_3 (which melts at a very high temperature), are inserted into two separate sample wells in a large steel block that acts as a heat sink (see illustration). Because the steel block is so large, it is possible to heat the sample and reference at the same slow and accurately monitored rate. Thermocouples are placed in each well and in the block itself. Then the block is gradually heated and the temperatures of the sample and the reference compared. An electrical signal is generated if the temperature of the sample suddenly stops rising while that of the reference continues to rise. Such an event signals an endothermic process in the sample, such as a phase change. The output of a DTA analysis is a *thermogram*, which shows the temperatures of phase changes as heat absorption peaks at the transition temperatures.

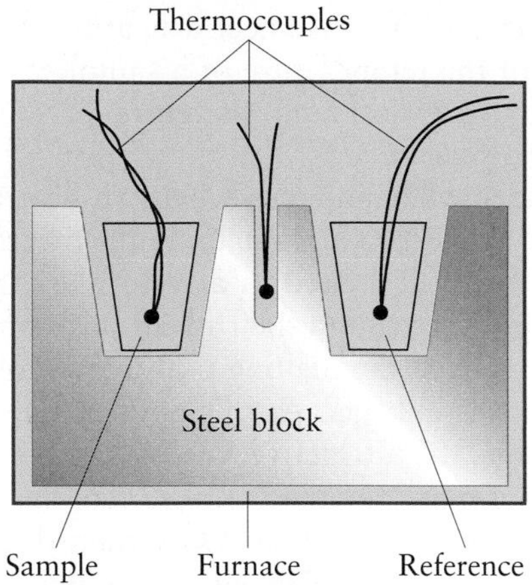

In differential thermal analysis, a sample and reference material are placed in the same large metal heat sink. Changes in the heat capacity of the sample are measured by changes in temperature between the sample and reference materials as they are heated at the same rate.

In *differential scanning calorimetry* (DSC), quantitative results can be obtained and heat capacities can also be measured. The apparatus is similar to that for a DTA analysis, with the primary difference being that the sample and reference are in separate heat sinks that are heated by individual heaters (see illustration). The temperatures of the two samples are kept the same by differential heating. Even slight temperature differences between

THE ENTHALPY OF CHEMICAL CHANGE

We have seen that enthalpy changes accompany physical changes, such as vaporization. We now apply these same principles to chemical changes and to the energy and enthalpy changes that accompany them. Enthalpies of chemical change are important in many areas of chemistry, such as the selection of materials that make good fuels, the design of chemical plants, and the study of biochemical processes.

6.14 Reaction Enthalpies

When methane, the major component of natural gas, burns in air, the reaction produces, not only carbon dioxide and water, but also a lot of heat. Complete reaction with oxygen to form carbon dioxide and water is called **combustion.** The combustion of methane, for instance, is the following reaction:

$$CH_4(g) + 2\,O_2(g) \longrightarrow CO_2(g) + 2\,H_2O(l) \qquad \textbf{(A)}$$

the reference and sample will trigger a switch that sends more or less power to the sample to maintain the constant temperature. If the heat capacity of the sample is higher than the heat capacity of the reference, energy must be supplied at a greater rate to the sample cell. If a phase transition occurs in the sample, a great deal of energy must be supplied to the sample until the phase transition is complete and the temperature begins to rise again.

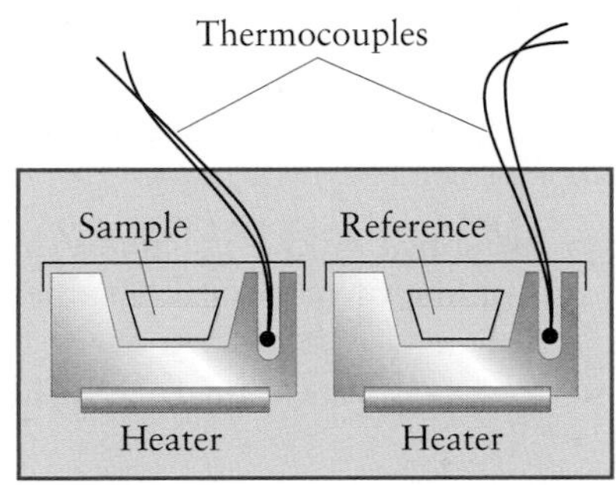

In a differential scanning calorimeter, a sample and reference material are heated in separate, but identical, metal heat sinks. The temperatures of the sample and reference material are kept the same by varying the power supplied to the two heaters. The output is the difference in power as a function of heat added.

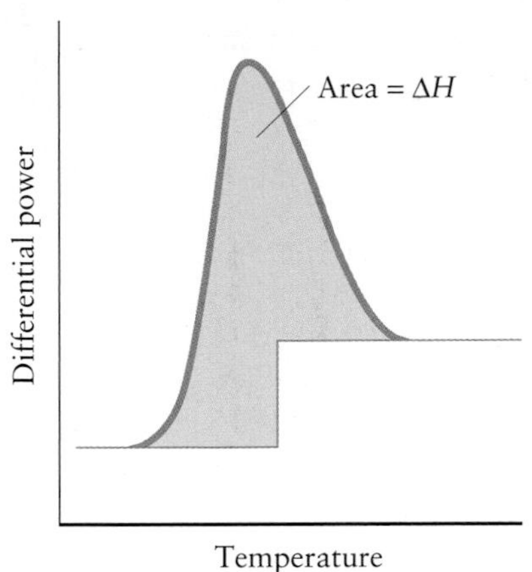

A thermogram from a differential scanning calorimeter. The peak indicates a phase change in the sample, and the difference in base line before and after the phase transition reflects the difference in heat capacities of the two phases.

The output of a differential scanning calorimeter is a measure of the power (the rate of energy supply) supplied to the sample cell. The thermogram illustrated shows a peak that signals a phase change. The thermogram does not look much like a heating curve, but it contains all the necessary information and is easily transformed into the familiar shape.

Calorimetry shows that burning 1 mol $CH_4(g)$ produces 890 kJ of heat at 298 K and 1 bar. To report this value, we write

$$CH_4(g) + 2\ O_2(g) \longrightarrow CO_2(g) + 2\ H_2O(l) \qquad \Delta H = -890\ \text{kJ} \qquad (A)$$

Note that, although burning takes place at much higher temperatures than 298 K, the value of ΔH is determined by the difference in enthalpies of the products and reactants, both measured at 298 K.

This entire expression is a **thermochemical equation,** a chemical equation together with the corresponding enthalpy change. The stoichiometric coefficients in a thermochemical equation are interpreted as the number of moles that react to give the reported change in enthalpy. Therefore, in this case, the enthalpy change is that resulting from the complete reaction of 1 mol CH_4 and 2 mol O_2. The enthalpy change in a thermochemical equation refers to the chemical equation exactly as written. If the same reaction is written with the coefficients all multiplied by 2, then the change in enthalpy will be twice as great:

$$2\ CH_4(g) + 4\ O_2(g) \longrightarrow 2\ CO_2(g) + 4\ H_2O(l) \qquad \Delta H = -1780\ \text{kJ}$$

This result makes sense because the equation now represents the burning of twice as much methane.

The **enthalpy of reaction,** ΔH_r (or "reaction enthalpy"), has the same numerical value as the quantity ΔH that appears in a thermochemical equation, but its units are kilojoules per mole ($kJ \cdot mol^{-1}$) rather than kilojoules. We interpret the "per mole" to mean per number of moles of each substance as indicated by its stoichiometric coefficient in the chemical equation. For example, the oxidation of methane described by reaction A signifies that 890 kJ of heat is released per mole of CH_4 molecules consumed or per two moles of H_2O molecules formed. It may seem unnecessary to have two ways of reporting the same information, but we shall see later that they are useful in different contexts.

We have seen in Section 6.12 that the first law of thermodynamics implies that the enthalpy change for the reverse of a process is the negative of the enthalpy change of the forward process. The same relation applies to forward and reverse chemical reactions. For the reverse of reaction A, for instance, we can write

$$CO_2(g) + 2\ H_2O(l) \longrightarrow CH_4(g) + 2\ O_2(g) \qquad \Delta H = +890\ kJ$$

Once we know the reaction enthalpy, we can calculate the enthalpy change for any amount, mass, or volume of reactant consumed or product formed. As shown in the following example, we carry out a stoichiometry calculation like those described in Section L, as if heat were a reactant or product.

A thermochemical equation is a statement of a chemical equation and the corresponding change in enthalpy for the stoichiometric numbers of moles of substances. The reaction enthalpy is the change in enthalpy per mole of molecules as expressed by the stoichiometric numbers in the chemical equation.

EXAMPLE 6.5 Determining a reaction enthalpy from experimental data

When 0.113 g of benzene, C_6H_6, burns in excess oxygen in a calibrated constant-pressure calorimeter with a heat capacity of 551 $J \cdot (°C)^{-1}$, the temperature of the calorimeter rises by 8.60°C. Calculate the reaction enthalpy for

$$2\ C_6H_6(l) + 15\ O_2(g) \longrightarrow 12\ CO_2(g) + 6\ H_2O(l)$$

and write the corresponding thermochemical equation.

STRATEGY First, we note whether the temperature rises or falls. If it rises, heat is given off, the reaction is exothermic, and ΔH is negative. If the temperature falls, heat is absorbed, the reaction is endothermic, and ΔH is positive. The enthalpy change is calculated from the temperature change times the heat capacity of the calorimeter. Once we know the enthalpy change for the masses of reactants used, we use the molar mass of one species to convert to the reaction enthalpy for moles of reactants equal to the stoichiometric coefficients.

SOLUTION The temperature of the calorimeter rises, so we know that the reaction is exothermic and ΔH is negative. Because the heat capacity of the calorimeter is 551 $J\cdot(°C)^{-1}$, it follows that

$$q_{cal} = (8.60°C) \times (551\ J\cdot(°C)^{-1}) = 8.60 \times 551\ J$$

Then, because $q = -q_{cal}$, we know that $q = -(8.60 \times 551)$ J. Because the molar mass of benzene is 78.12 $g\cdot mol^{-1}$, the amount of C_6H_6 that reacts is

$$n(C_6H_6) = \frac{0.113\ g}{78.12\ g\cdot mol^{-1}} = \frac{0.113}{78.12}\ mol$$

The reaction enthalpy for the equation as written, the change in enthalpy per 2 mol C_6H_6 that reacts, is therefore

$$\Delta H_r = \frac{2q}{n(C_6H_6)} = \frac{2 \times (-8.60 \times 551)\ J}{(0.113\,/\,78.12)\ mol}$$
$$= -6.55 \times 10^6\ J\cdot mol^{-1} = -6.55 \times 10^3\ kJ\cdot mol^{-1}$$

The thermochemical equation is therefore

$$2\ C_6H_6(l) + 15\ O_2(g) \longrightarrow 12\ CO_2(g) + 6\ H_2O(l), \qquad \Delta H = -6.55 \times 10^3\ kJ$$

Note that we could write $\Delta H = -6.55$ MJ.

Self-Test 6.11A When 0.231 g of phosphorus reacts with chlorine to form phosphorus trichloride, PCl_3, in a constant-pressure calorimeter of heat capacity 216 $J\cdot(°C)^{-1}$, the temperature of the calorimeter rises by 11.06°C. Write the thermochemical equation for the reaction.

[*Answer:* $2\ P(s) + 3\ Cl_2(g) \rightarrow 2\ PCl_3(l)$, $\Delta H = -641$ kJ]

Self-Test 6.11B When 0.338 g of pentane, C_5H_{12}, burns in an excess of oxygen to form carbon dioxide and liquid water in the same calorimeter used in Self-Test 6.11A, the temperature rises by 76.7°C. Write the thermochemical equation for the reaction.

6.15 The Relation Between ΔH and ΔU

We saw in Section 6.9 that a constant-pressure calorimeter and a bomb calorimeter measure slightly different quantities: at constant volume, the heat transfer is interpreted as ΔU; and at constant pressure, it is interpreted as ΔH. However, we may need to convert the measured value of ΔU to ΔH. For example, it is easy to measure the heat released by the combustion of glucose in a bomb calorimeter, which has a constant volume; but to use that information in assessing energy changes in metabolism, which take place at constant pressure, we need the enthalpy of reaction.

For reactions that do not involve gases, the difference between ΔH and ΔU is negligible, so we can assume that $\Delta H = \Delta U$. However, if a gas is formed in the reaction, so much work is done to make room for the gaseous products that the difference can be significant. Once again, we can use the ideal gas law to relate the values of ΔH and ΔU for gases that behave ideally.

How do we do that?

To calculate the relation between ΔH and ΔU, we suppose that the initial amount of ideal gas molecules is n_{initial} (as reactants). The initial enthalpy is

$$H_{\text{initial}} = U_{\text{initial}} + PV_{\text{initial}} = U_{\text{initial}} + n_{\text{initial}}RT$$

For the second equality, we have used the ideal gas law to replace PV by nRT. After the reaction is complete, the amount of ideal gas molecules is n_{final} (as products). The enthalpy is then

$$H_{\text{final}} = U_{\text{final}} + PV_{\text{final}} = U_{\text{final}} + n_{\text{final}}RT$$

The difference is

$$\Delta H = H_{\text{final}} - H_{\text{initial}} = \Delta U + (n_{\text{final}} - n_{\text{initial}})RT$$

If we write $\Delta n_{\text{gas}} = n_{\text{final}} - n_{\text{initial}}$, we obtain

$$\Delta H = \Delta U + \Delta n_{\text{gas}}RT \qquad (28)$$

where Δn_{gas} is the change in the amount of gas molecules in the reaction (positive for net gas formation, negative for net gas consumption). Notice that ΔH is less negative than ΔU for exothermic reactions that generate gases; for reactions with no change in the amount of gas, the two quantities are about the same.

The reaction enthalpy is less negative than the reaction internal energy for reactions that generate gases; for reactions with no change in the amount of gas, the two quantities are almost the same.

EXAMPLE 6.6 **Relating the enthalpy change and internal energy change for a chemical reaction**

A constant-volume calorimeter showed that the heat loss accompanying the combustion of 1.000 mol glucose molecules in the reaction

$$C_6H_{12}O_6(s) + 6\ O_2(g) \longrightarrow 6\ CO_2(g) + 6\ H_2O(g)$$

is 2559 kJ at 298 K, so $\Delta U = -2559$ kJ. What is the change in enthalpy for the same reaction?

STRATEGY We need to identify the change in the number of moles of gas molecules and then use Eq. 28. For R, use the value in joules per kelvin per mole.

SOLUTION There are 6 mol of gas molecules on the left as reactants and 12 mol on the right as products, so $\Delta n_{\text{gas}} = +6$ mol. It then follows from Eq. 28 that

$$\begin{aligned}\Delta H &= \Delta U + \Delta n_{\text{gas}}RT \\ &= -2559\text{ kJ} + [(6\text{ mol}) \times (8.3145\text{ J}\cdot\text{K}^{-1}\cdot\text{mol}^{-1}) \times (298\text{ K})] \\ &= -2559\text{ kJ} + 1.49 \times 10^4\text{ J} = -2559\text{ kJ} + 14.9\text{ kJ} = -2544\text{ kJ}\end{aligned}$$

Note that the change in enthalpy is less negative than the change in internal energy. Less energy can be obtained as heat at constant pressure than at constant volume because the system must use some energy to expand to make room for the reaction products.

Self-Test 6.12A The thermochemical equation for the combustion of cyclohexane, C_6H_{12}, is $C_6H_{12}(l) + 9\ O_2(g) \rightarrow 6\ CO_2(g) + 6\ H_2O(l)$, $\Delta H = -3920$ kJ at 298 K. What is the change in internal energy for the combustion of 1.00 mol $C_6H_{12}(l)$ at 298 K?

[*Answer:* -3.91×10^3 kJ]

Self-Test 6.12B The reaction $4\ Al(s) + 3\ O_2(g) \rightarrow 2\ Al_2O_3(s)$ was studied as part of an investigation into using aluminum powder as a rocket fuel (Fig. 6.26). It was found that 1.00 mol Al produced 3378 kJ of heat under constant-pressure conditions at 1000°C. What is the change in internal energy for the combustion of 1.00 mol Al?

FIGURE 6.26 In this preparation of rocket fuel for the space shuttle, powdered aluminum is mixed with an oxidizing agent in a liquid polymer base that hardens inside the booster rocket shell.

6.16 Standard Reaction Enthalpies

The heat released or absorbed by a reaction depends on the physical states of the reactants and products. This dependence suggests that we should introduce a standard set of conditions for reporting reaction enthalpies. For example, let's consider again the combustion of methane. We can write two different thermochemical equations, for two different sets of products:

$$CH_4(g) + 2\ O_2(g) \longrightarrow CO_2(g) + 2\ H_2O(g) \qquad \Delta H = -802 \text{ kJ} \qquad \text{(B)}$$

$$CH_4(g) + 2\ O_2(g) \longrightarrow CO_2(g) + 2\ H_2O(l) \qquad \Delta H = -890 \text{ kJ} \qquad \text{(C)}$$

In the first reaction, all the water is produced as a vapor; in the second, it is produced as a liquid. The heat produced is different in each case. We have already seen that the enthalpy of water vapor is 44 kJ·mol^{-1} higher than that of liquid water at 25°C (see Table 6.3). As a result, an additional 88 kJ (for 2 mol H_2O) remains stored in the system if water vapor is formed (Fig. 6.27). If the 2 mol $H_2O(g)$ subsequently condenses, an additional 88 kJ is given off as heat.

All the tables in this book list data for reactions in which each reactant and product is in its **standard state,** its pure form at a pressure of exactly 1 bar. The standard state of liquid water is pure water at 1 bar. The standard state of ice is pure ice at 1 bar. A solute in a liquid solution is in its standard state when its concentration is 1 mol·L^{-1}. The standard value of a property X (that is, the value of X for the standard state of the substance) is denoted $X°$.

A reaction enthalpy based on standard states is called a *standard reaction enthalpy*, $\Delta H_r°$. The **standard reaction enthalpy** is the reaction enthalpy when reactants in their standard states change to products in their standard states. For example, for reaction C, the value $\Delta H_r° = -890$ kJ·mol^{-1} signifies that the heat output is 890 kJ per mole of CH_4, when pure methane gas at 1 bar is allowed to react with pure oxygen gas at 1 bar, giving pure carbon dioxide gas and pure liquid water, both

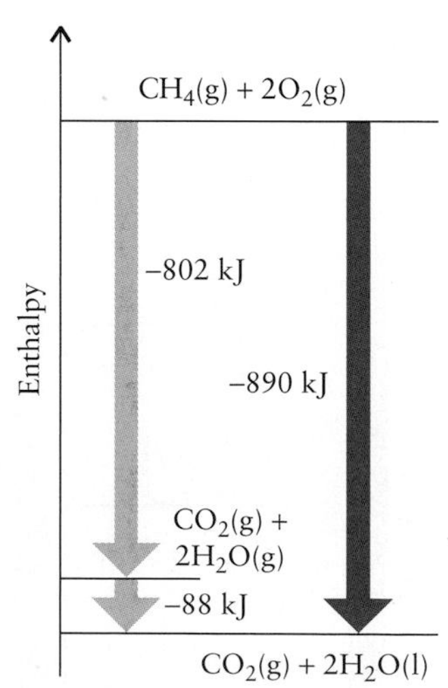

FIGURE 6.27 The enthalpy changes when methane burns to give carbon dioxide and either gaseous or liquid water.

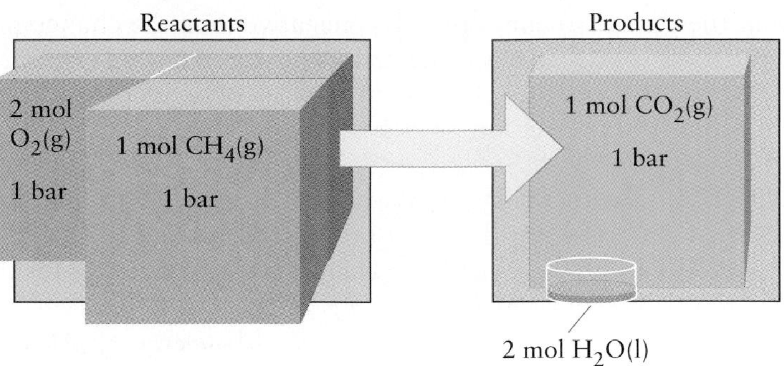

FIGURE 6.28 The standard reaction enthalpy is the difference in enthalpy between the pure products, each at 1 bar, and the pure reactants, each at 1 bar, and the specified temperature (which is commonly but not necessarily 298 K). The scheme here is for the combustion of methane to carbon dioxide and liquid water.

The standard state used to be defined for 1 atm rather than for 1 bar; however, 1 bar is now the accepted standard. The small change in standard pressure makes a negligible difference to most numerical values, so it is normally safe to use tables of data compiled for 1 atm.

at 1 bar (Fig. 6.28). Reaction enthalpies do not change very much with pressure, so the standard value gives a good indication of the change in enthalpy for pressures near 1 bar.

Most thermochemical data are reported for 25°C (more precisely, for 298.15 K). Temperature is not part of the definition of standard states: we can have a standard state at any temperature; 298.15 K is simply the most common temperature used in tables of data. All reaction enthalpies used in this text are for 298.15 K unless another temperature is indicated.

Standard reaction enthalpies refer to reactions in which the reactants and products are in their standard states; they are usually reported for a temperature of 298.15 K.

6.17 Combining Reaction Enthalpies: Hess's Law

The enthalpy is a state function; therefore, the value of ΔH is independent of the path between given initial and final states. We saw an application of this approach in Section 6.12, where we calculated the enthalpy change for an overall physical process as the sum of the enthalpy changes for a series of two individual steps. The same rule applies to chemical reactions. In this context, the rule is known as **Hess's law:** *the overall reaction enthalpy is the sum of the reaction enthalpies of the steps into which the reaction can be divided.* Hess's law applies even if the intermediate reactions or the overall reaction cannot actually be carried out. Provided the equation for each step balances and the individual equations add up to the equation for the reaction of interest, a reaction enthalpy can be calculated from any convenient sequence of reactions (Fig. 6.29). As we shall see, Hess's law also allows us to use readily obtainable combustion data to compile information on a number of reactions.

As an example of Hess's law, let's consider the oxidation of carbon as graphite, denoted C(gr), to carbon dioxide:

$$C(gr) + O_2(g) \longrightarrow CO_2(g)$$

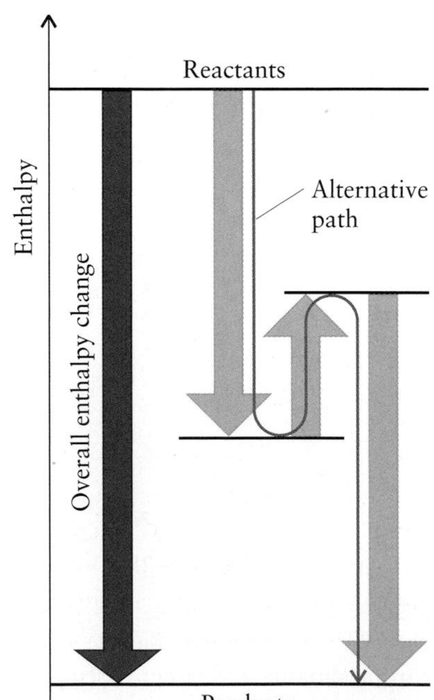

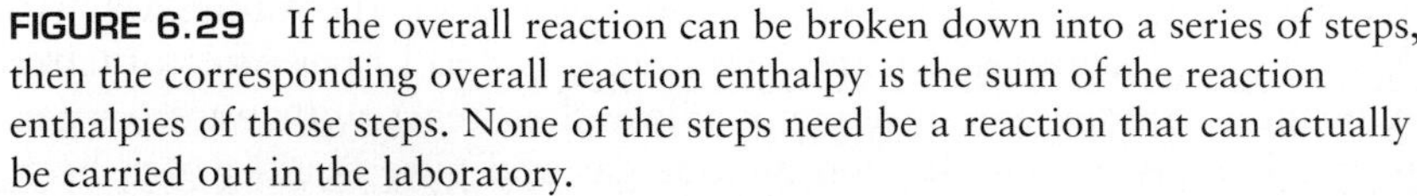

FIGURE 6.29 If the overall reaction can be broken down into a series of steps, then the corresponding overall reaction enthalpy is the sum of the reaction enthalpies of those steps. None of the steps need be a reaction that can actually be carried out in the laboratory.

This reaction can be thought of as the outcome of two steps. One step is the oxidation of carbon to carbon monoxide:

$$C(gr) + \tfrac{1}{2} O_2(g) \rightarrow CO(g) \qquad \Delta H^\circ = -110.5\ kJ$$

The second step is the oxidation of carbon monoxide to carbon dioxide:

$$CO(g) + \tfrac{1}{2} O_2(g) \rightarrow CO_2(g) \qquad \Delta H^\circ = -283.0\ kJ$$

This two-step process is an example of a **reaction sequence,** a series of reactions in which the products of one reaction take part as reactants in another reaction. The equation for the overall reaction, the net outcome of the sequence, is the sum of the equations for the intermediate steps:

$$\begin{array}{llr} C(gr) + \tfrac{1}{2} O_2(g) \longrightarrow CO(g) & \Delta H^\circ = -110.5\ kJ & (a) \\ \underline{CO(g) + \tfrac{1}{2} O_2(g) \longrightarrow CO_2(g)} & \underline{\Delta H^\circ = -283.0\ kJ} & (b) \\ C(gr) + O_2(g) \longrightarrow CO_2(g) & \Delta H^\circ = -393.5\ kJ & (a + b) \end{array}$$

The same procedure is used to predict the enthalpies of reactions that we cannot measure directly in the laboratory. The procedure is described in Toolbox 6.1.

Thermochemical equations for the individual steps of a reaction sequence may be combined to give the thermochemical equation for the overall reaction.

TOOLBOX 6.1 How to use Hess's law

Conceptual Basis

Because enthalpy is a state function, the enthalpy change of a system depends only on its initial and final states. Therefore, we can carry out a reaction in one step or visualize it as the sum of several steps; the reaction enthalpy is the same in each case.

Procedure

To use Hess's law, we need to find a sequence of reactions that adds up to the reaction of interest. In many cases (as in the oxidation of carbon mentioned in the text), we can see at a glance what reactions to use. A more systematic procedure is often helpful when the sequence is not obvious.

Step 1 Select one of the reactants in the overall reaction and write down a chemical equation in which it also appears as a reactant.

Step 2 Select one of the products in the overall reaction and write down a chemical equation in which it also appears as a product. Add this equation to the equation written in step 1.

Step 3 Cancel unwanted species in the sum obtained in step 2 by adding an equation that has the same substance or substances on the opposite side of the arrow.

Step 4 Once the sequence is complete, combine the standard reaction enthalpies.

In each step, we may need to reverse the equation or multiply it by a factor. Recall from Eq. 25 that if we want to reverse a chemical equation, we have to change the sign of the reaction enthalpy. If we multiply the stoichiometric coefficients by a factor, we must multiply the reaction enthalpy by the same factor.

EXAMPLE 6.7 Using Hess's law

Consider the synthesis of propane, C_3H_8, a gas used as camping fuel:

$$3\ C(gr) + 4\ H_2(g) \longrightarrow C_3H_8(g)$$

It is difficult to measure the enthalpy change of this reaction. However, standard enthalpies of combustion reactions are easy to measure. We have the following experimental data:

$$C_3H_8(g) + 5\ O_2(g) \longrightarrow 3\ CO_2(g) + 4\ H_2O(l) \quad \Delta H° = -2220\ kJ \quad (a)$$

$$C(gr) + O_2(g) \longrightarrow CO_2(g) \quad \Delta H° = -394\ kJ \quad (b)$$

$$H_2(g) + \tfrac{1}{2}\ O_2(g) \longrightarrow H_2O(l) \quad \Delta H° = -286\ kJ \quad (c)$$

Step 1 Only (b) and (c) have at least one of the reactants. In both cases, they are on the correct side of the arrow in the overall chemical equation. We select (b) and multiply it through by 3 to give carbon the coefficient it will have in the final equation:

$$3\ C(gr) + 3\ O_2(g) \longrightarrow 3\ CO_2(g) \quad \Delta H° = 3 \times (-394\ kJ) = -1182\ kJ$$

Step 2 To obtain C_3H_8 on the right, we reverse equation (a), changing the sign of its reaction enthalpy, and add it to the equation from step 1:

$$3\ C(gr) + 3\ O_2(g) \longrightarrow 3\ CO_2(g) \quad \Delta H° = -1182\ kJ$$

$$3\ CO_2(g) + 4\ H_2O(l) \longrightarrow C_3H_8(g) + 5\ O_2(g) \quad \Delta H° = +2220\ kJ$$

The sum of these two equations is

$$3\ C(gr) + 3\ O_2(g) + 3\ CO_2(g) + 4\ H_2O(l) \longrightarrow C_3H_8(g) + 5\ O_2(g) + 3\ CO_2(g) \quad \Delta H° = +1038\ kJ$$

This equation simplifies to

$$3\ C(gr) + 4\ H_2O(l) \longrightarrow C_3H_8(g) + 2\ O_2(g) \quad \Delta H° = +1038\ kJ$$

Step 3 To cancel the unwanted reactant H_2O and product O_2, we add equation (c) after multiplying it by 4:

$$3\ C(gr) + 4\ H_2O(l) \longrightarrow C_3H_8(g) + 2\ O_2(g) \quad \Delta H° = +1038\ kJ$$

$$4\ H_2(g) + 2\ O_2(g) \longrightarrow 4\ H_2O(l) \quad \Delta H° = 4 \times (-286\ kJ) = -1144\ kJ$$

Step 4 The sum of these two reactions is

$$3\ C(gr) + 4\ H_2(g) + 4\ H_2O(l) + 2\ O_2(g) \longrightarrow C_3H_8(g) + 2\ O_2(g) + 4\ H_2O(l) \qquad \Delta H° = -106\ kJ$$

which simplifies to

$$3\ C(gr) + 4\ H_2(g) \longrightarrow C_3H_8(g) \qquad \Delta H° = -106\ kJ$$

giving us the quantity we require.

Self-Test 6.13A Gasoline, which contains octane as one component, may burn to carbon monoxide if the air supply is restricted. Derive the standard reaction enthalpy for the incomplete combustion of liquid octane in air to carbon monoxide gas and liquid water from the standard reaction enthalpies for the combustions of octane and carbon monoxide:

$$2\ C_8H_{18}(l) + 25\ O_2(g) \longrightarrow 16\ CO_2(g) + 18\ H_2O(l) \qquad \Delta H° = -10\ 942\ kJ$$

$$2\ CO(g) + O_2(g) \longrightarrow 2\ CO_2(g) \qquad \Delta H° = -566.0\ kJ$$

[*Answer:* $2\ C_8H_{18}(l) + 17\ O_2(g) \longrightarrow 16\ CO(g) + 18\ H_2O(l)$, $\Delta H° = -6414\ kJ$]

Self-Test 6.13B Methanol is a clean-burning liquid fuel being considered as a replacement for gasoline. It can be produced by the controlled reaction of the oxygen in air with methane in natural gas. Find the standard reaction enthalpy for the formation of 1 mol $CH_3OH(l)$ from methane and oxygen, given the following information:

$$CH_4(g) + H_2O(g) \longrightarrow CO(g) + 3\ H_2(g) \qquad \Delta H° = +206.10\ kJ$$

$$2\ H_2(g) + CO(g) \longrightarrow CH_3OH(l) \qquad \Delta H° = -128.33\ kJ$$

$$2\ H_2(g) + O_2(g) \longrightarrow 2\ H_2O(g) \qquad \Delta H° = -483.64\ kJ$$

6.18 The Heat Output of Reactions

Combustion reactions supply most of our energy needs, producing energy by the burning of fossil fuels (coal, petroleum, and natural gas). Because the reserves of these fuels are limited, alternatives are being sought (Box 6.2); but even these new fuels will be burned.

The **standard enthalpy of combustion,** $\Delta H_c°$, is the change in enthalpy per mole of a substance that is burned in a combustion reaction under standard conditions. The products of the combustion of an organic compound are carbon dioxide gas and liquid water; any nitrogen present is released as N_2, unless we specify that nitrogen oxides are formed.

Standard enthalpies of combustion are listed in Table 6.4 and Appendix 2A. We have seen in Toolbox 6.1 how to use enthalpies of combustion

TABLE 6.4 Standard Enthalpies of Combustion at 25°C*

Substance	Formula	$\Delta H_c°$, $kJ \cdot mol^{-1}$
benzene	$C_6H_6(l)$	−3268
carbon	C(s, graphite)	−394
ethanol	$C_2H_5OH(l)$	−1368
ethyne (acetylene)	$C_2H_2(g)$	−1300
glucose	$C_6H_{12}O_6(s)$	−2808
hydrogen	$H_2(g)$	−286
methane	$CH_4(g)$	−890
octane	$C_8H_{18}(l)$	−5471
propane	$C_3H_8(g)$	−2220
urea	$CO(NH_2)_2(s)$	−632

* In a combustion, carbon is converted to carbon dioxide, hydrogen to liquid water, and nitrogen to nitrogen gas. More values are given in Appendix 2A.

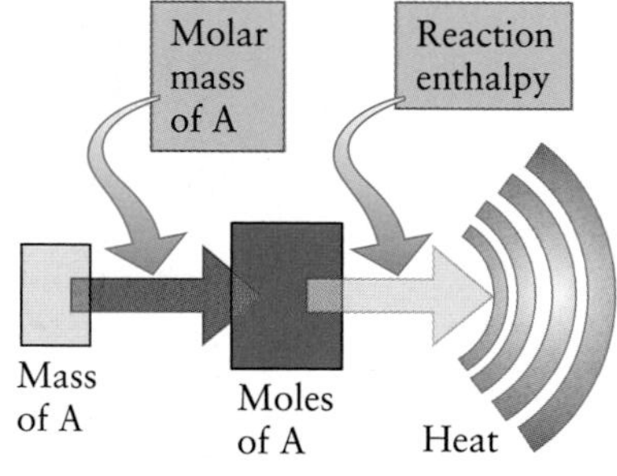

1

to obtain the standard enthalpies of reactions. Here we consider another practical application—the choice of a fuel. For example, suppose we want to know the heat output from the combustion of 150 g of methane. The thermochemical equation allows us to write the following relation

$$1 \text{ mol } CH_4 \simeq 890 \text{ kJ}$$

and to use it as a conversion factor in the usual way (**1**).

The heat absorbed or given off by a reaction can be treated like a reactant or product in a stoichiometric relation.

EXAMPLE 6.8 Calculating the heat output of a fuel

How much propane should a backpacker carry: do we really need to carry a kilogram of gas? Calculate the mass of propane that you would need to burn to obtain 350 kJ of heat, which is just enough energy to heat 1 L of water from room temperature (20°C) to boiling at sea level (if we ignore all heat losses). The thermochemical equation is

$$C_3H_8(g) + 5\,O_2(g) \longrightarrow 3\,CO_2(g) + 4\,H_2O(l) \qquad \Delta H° = -2220 \text{ kJ}$$

STRATEGY The first step is to convert heat output required to moles of fuel molecules by using the thermochemical equation. Then convert from moles of fuel molecules to grams by using the molar mass of the fuel.

SOLUTION The thermochemical equation tells us that

$$2220 \text{ kJ} \simeq 1 \text{ mol } C_3H_8$$

BOX 6.2 WHAT HAS THIS TO DO WITH . . . THE ENVIRONMENT?

Alternative Fuels

Our complex modern lifestyle depends on the constant availability of inexpensive power. In fact, our way of life was made possible by the discovery and refining of fossil fuels, fuels that are the result of the decay of organic matter laid down millions of years ago. The natural gas that heats our homes, the gasoline that powers our automobiles, and the coal that provides much of our electrical power are fossil fuels. Vast reserves of petroleum, the source of liquid hydrocarbon fuels such as gasoline, and coal exist in many areas of the world. However, although large, these reserves are limited, and we are using them up at a much faster rate than they can be replaced.

In addition, the burning of fossil fuels harms our environment, and the carbon dioxide produced as a combustion product contributes to global warming (Box 14.2). Alternative and sustainable power generation methods, such as hydroelectric power, wind power, solar power, and alternative fuels are being sought to reduce the demand for fossil fuels.

Three of the most promising alternative fuels are hydrogen, ethanol, and methane. Hydrogen is extracted from ocean water by electrolysis. Ethanol is obtained by fermenting biomass, the name given to plant materials that can be burned or reacted to produce fuels. Methane is generated by bacterial digestion of wastes such as sewage and agricultural wastes. In each case, the fuel is renewable. That means the source of the fuel is replenished every year by the Sun. The use of hydrogen as a fuel is discussed in Section 14.5. Here we look at ethanol and methane.

These tanks at a water treatment facility are used to store the methane generated by anaerobic digestion of sewage. The methane produced provides much of the power needed to run the facility.

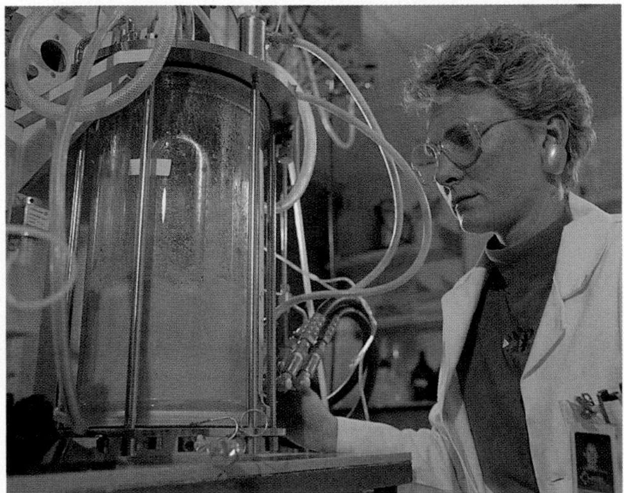

The cellulose-based waste biomass in this reactor is being digested by special enzymes that break it down into ethanol. These enzymes are being intensively studied to improve the efficiency of conversion.

Ethanol, CH_3CH_2OH, is produced from the biological fermentation of the starches in grains, mainly corn. It currently makes up about 10% by volume of gasoline in the United States, thereby reducing pollution as well as usage of petroleum. The oxygen atom in the ethanol molecule reduces emissions of carbon monoxide and hydrocarbons by helping to ensure complete combustion. One bushel of corn (about 30 L) can produce nearly 10 L of ethanol. A problem with ethanol as a fuel is that the sugars and starches fermented to produce it are expensive. However, the cellulose in straw and cornstalks left behind as rubble when grains are harvested is now attracting attention. Cellulose is the structural material in plants. It is made up of simple sugars, just

(continued)

Box 6.2 What Has This to Do with . . . the Environment? *(continued)*

as starches are, but the bacteria that ferment starches cannot digest cellulose. Research is now being conducted on enzymes that break down cellulose into sugars that can be digested. This process would greatly increase the biomass available for fuel production, because straw, wood, grasses, and in fact, nearly any plant materials could be used for fuel. In fact, the amount of ethanol that could be obtained from biomass in this way would be sufficient to replace all the gasoline now used. The disadvantage is that the biomass would then not be available for soil enrichment.

Methane, CH_4, is found in underground reserves as the primary component of natural gas, but it is also obtained from biological materials. The digestion is anaerobic, which means that it takes place in the absence of oxygen. Currently, many sewage treatment plants have anaerobic digesters that produce the methane used to operate the plants. To generate methane by anaerobic digestion on a large-scale basis, additional materials, such as sugars from the enzymatic breakdown of biomass, would need to be used. Methane would be less useful than ethanol as a transportation fuel, because it is a gas and so has a low *enthalpy density*, the enthalpy of combustion per liter. However, it can be used wherever natural gas is used.

Methane and ethanol do produce carbon dioxide when burned and thus contribute to the greenhouse effect and global warming. However, they generate less carbon dioxide per gram than gasoline and can be renewed every year, as long as the Sun shines and produces green plants.

For Further Reading: *Biofuels for Sustainable Transportation*, www.ott.doe.gov/biofuels/ (U.S. Department of Energy, 2000). Energy Information Administration, *Alternative Fuels Information at a Glance*, www.eia.doe.gov/cneaf/alternate/page/fuelalternate.html (Washington, DC: U.S. Department of Energy, 2000). *Alternative Fuel Vehicles*, www.energy.ca.gov/afvs/index.html (California Energy Commission, 2000). C. Wu, "Fill 'er up . . . with veggie oil," *Science News* **154**, 364–366 (1988).

The molar mass of propane is 44.09 g·mol^{-1}. It follows that

$$\begin{aligned}\text{Mass of } C_3H_8 \text{ required} &= (350\text{ kJ}) \times \left(\frac{1\text{ mol } C_3H_8}{2220\text{ kJ}}\right) \\ &\quad \times (44.09\text{ g}\cdot\text{mol}^{-1}) \\ &= 6.95\text{ g } C_3H_8\end{aligned}$$

That is, just under 7 g of propane is needed to boil the water (or more, if we allow for heat losses).

Self-Test 6.14A The thermochemical equation for the combustion of butane is

$$2\ C_4H_{10}(g) + 13\ O_2(g) \longrightarrow 8\ CO_2(g) + 10\ H_2O(l) \qquad \Delta H^\circ = -5756\text{ kJ}$$

What mass of butane must be burned to supply 350 kJ of heat? Would it be easier to pack butane rather than propane?

[*Answer:* 7.07 g. Propane would be very slightly lighter to carry.]

Self-Test 6.14B Ethanol trapped in a gel is another common camping fuel. What mass of ethanol must be burned to supply 350 kJ of heat? The thermochemical equation for the combustion of ethanol is

$$C_2H_5OH(l) + 3\ O_2(g) \longrightarrow 2\ CO_2(g) + 3\ H_2O(l) \qquad \Delta H^\circ = -1368\text{ kJ}$$

6.19 Standard Enthalpies of Formation

There are millions of possible reactions, and it is impractical to list every one with its standard reaction enthalpy. However, chemists have devised an ingenious alternative. There are two stages. First, they report the standard enthalpy of formation of substances. Then they combine these quantities to obtain the standard enthalpy of reaction needed. Let's look at these two stages in turn.

The **standard enthalpy of formation,** $\Delta H_f°$, of a substance is the standard reaction enthalpy for the formation of a substance from its elements in their *most stable form.* (Phosphorus is an exception: white phosphorus is used because it is much easier to obtain pure than the other, more stable allotropes.) Standard enthalpies of formation are expressed in kilojoules per mole of the substance ($kJ \cdot mol^{-1}$). We obtain $\Delta H_f°$ for ethanol, for instance, from the thermochemical equation for its formation from graphite (the most stable form of carbon) and gaseous hydrogen and oxygen:

$$2\,C(gr) + 3\,H_2O(g) + \tfrac{1}{2}\,O_2(g) \longrightarrow C_2H_5OH(l) \qquad \Delta H° = -277.69\text{ kJ}$$

Notice that the thermochemical equation is written for the formation of 1 mol $C_2H_5OH(l)$; so a fractional coefficient is necessary for O_2. We report that $\Delta H_f°(C_2H_5OH, l) = -277.69\ kJ \cdot mol^{-1}$. Note also how the substance and its state are used to label the enthalpy change, so that we know which species and which form of that species we are talking about.

The standard enthalpy of formation of an element in its most stable form is defined as 0. However, the enthalpy of formation of an element in a form other than its most stable one is nonzero. For example, the conversion of carbon from graphite to diamond is endothermic:

$$C(gr) \longrightarrow C(s, \text{diamond}) \qquad \Delta H° = +1.9\text{ kJ}$$

The standard enthalpy of formation of diamond is $+1.9\ kJ \cdot mol^{-1}$. When this change takes place, $1.9\ kJ \cdot mol^{-1}$ of energy must be supplied as heat to change the bonding arrangement in graphite into the bonding arrangement in diamond. When the bonding arrangement in diamond collapses back into that characteristic of graphite, $1.9\ kJ \cdot mol^{-1}$ of energy is released as heat.

When a compound cannot be synthesized from its elements directly (or the reaction is too difficult to study), its enthalpy of formation can still be found from its enthalpy of combustion (see Example 6.7). Some values are listed in Table 6.5 and Appendix 2A.

Now let's see how to combine standard enthalpies of formation to calculate a standard reaction enthalpy. First, we calculate the reaction enthalpy for the formation of all the products from their elements. For this step, we use the enthalpies of formation of the products. Then we calculate the reaction enthalpy for the formation of all the reactants from their elements. The difference between these two totals is the standard enthalpy of the reaction (Fig. 6.30):

$$\Delta H_r° = \sum n\Delta H_f°(\text{products}) - \sum n\Delta H_f°(\text{reactants}) \qquad (29)$$

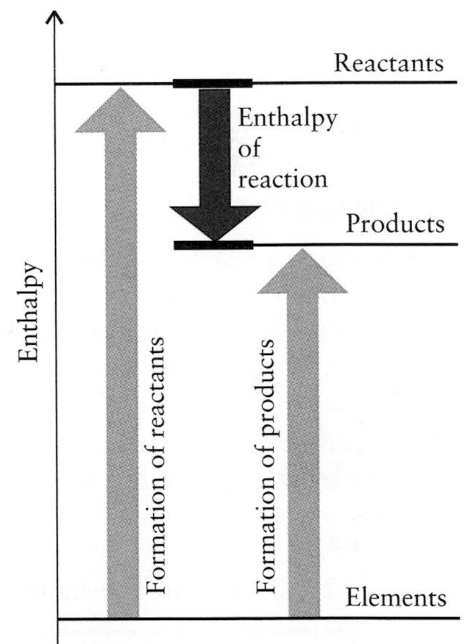

FIGURE 6.30 The reaction enthalpy can be constructed from enthalpies of formation by imagining forming the reactants and products from their elements. Then the reaction enthalpy is the difference between the enthalpies of products and reactants.

TABLE 6.5 Standard Enthalpies of Formation at 25°C*

Substance	Formula	$\Delta H_f°$, kJ·mol^{-1}	Substance	Formula	$\Delta H_f°$, kJ·mol^{-1}
Inorganic compounds			**Organic compounds**		
ammonia	$NH_3(g)$	−46.11	benzene	$C_6H_6(l)$	+49.0
carbon dioxide	$CO_2(g)$	−393.51	ethanol	$C_2H_5OH(l)$	−277.69
carbon monoxide	$CO(g)$	−110.53	ethyne (acetylene)	$C_2H_2(g)$	+226.73
dinitrogen tetroxide	$N_2O_4(g)$	+9.16	glucose	$C_6H_{12}O_6(s)$	−1268
hydrogen chloride	$HCl(g)$	−92.31	methane	$CH_4(g)$	−74.81
hydrogen fluoride	$HF(g)$	−271.1			
nitrogen dioxide	$NO_2(g)$	+33.18			
nitric oxide	$NO(g)$	+90.25			
sodium chloride	$NaCl(s)$	−411.15			
water	$H_2O(l)$	−285.83			
	$H_2O(g)$	−241.82			

* A much longer list is given in Appendix 2A.

In this expression, the n are the stoichiometric coefficients in the chemical equation and the symbol Σ (sigma) means a sum. The first sum is the total enthalpy of formation of the products. The second sum is the similar total for the reactants. Toolbox 6.2 explains in more detail how to use this expression.

TOOLBOX 6.2 How to use standard enthalpies of formation

Conceptual Basis

We imagine a reaction in which the products are formed from their elements, and another reaction in which the reactants are formed from the same elements. We calculate the standard reaction enthalpies of both hypothetical reactions and then calculate the standard enthalpy of the reaction we require by taking the difference of these two values.

Procedure

Use Eq. 29 to calculate the standard reaction enthalpy:

Step 1 Write down the chemical equation for the reaction of interest.

Step 2 Add up the individual standard enthalpies of formation of the products. Each value is multiplied by the appropriate stoichiometric coefficient. Remember that the standard enthalpy of formation of an element in its most stable form is 0.

Step 3 Calculate the total standard enthalpy of formation of the reactants in the same way.

Step 4 Subtract the second sum from the first.

EXAMPLE 6.9 Using standard enthalpies of formation to calculate a standard enthalpy of reaction

Amino acids are the building blocks of protein molecules, which are long chainlike molecules. They are oxidized in the body to urea, carbon dioxide, and water. Is this reaction a source of heat for the body? Use the information in Appendix 2A to predict the standard enthalpy of reaction for the oxidation of the simplest amino acid, glycine, NH_2CH_2COOH, a solid, to solid urea, H_2NCONH_2, carbon dioxide gas, and water. Report the enthalpy change as enthalpy per mole of glycine molecules, and compare this value with the enthalpy of combustion of glucose in Table 6.4.

STRATEGY We expect a strongly negative value because all combustions are exothermic and, in this case, the oxidation is like an incomplete combustion. Proceed as suggested in the Toolbox.

SOLUTION ***Step 1*** The reaction to consider is

$$2\ NH_2CH_2COOH(s) + 3\ O_2(g) \longrightarrow H_2NCONH_2(s) + 3\ CO_2(g) + 3\ H_2O(l)$$

Step 2 The total enthalpy of formation of the products is

$$\begin{aligned}\textstyle\sum n\Delta H_f^\circ(\text{products}) &= \Delta H_f^\circ(H_2NCONH_2, s) \\ &\quad + 3\Delta H_f^\circ(CO_2, g) + 3\Delta H_f^\circ(H_2O, l) \\ &= -333.51\ \text{kJ·mol}^{-1} + 3(-393.51\ \text{kJ·mol}^{-1}) \\ &\quad + 3(-285.83\ \text{kJ·mol}^{-1}) \\ &= -333.51 + (-1180.53) + (-857.49)\ \text{kJ·mol}^{-1} \\ &= -2371.53\ \text{kJ·mol}^{-1}\end{aligned}$$

Step 3 The total enthalpy of formation of the reactants is

$$\begin{aligned}\textstyle\sum n\Delta H_f^\circ(\text{reactants}) &= 2\Delta H_f^\circ(NH_2CH_2COOH, s) + 3\Delta H_f^\circ(O_2, g) \\ &= 2(-532.9\ \text{kJ·mol}^{-1}) + 3(0) \\ &= -1065.8\ \text{kJ·mol}^{-1}\end{aligned}$$

Step 4 The difference between the two totals is

$$\Delta H_r^\circ = -2371.53 - (-1065.8)\ \text{kJ·mol}^{-1} = -1305.7\ \text{kJ·mol}^{-1}$$

This value is the enthalpy of reaction per 2 mol glycine molecules. To report the enthalpy change per mole of glycine molecules, we must divide by 2:

$$\Delta H_r^\circ = \frac{-1305.7\ \text{kJ·mol}^{-1}}{2} = -652.8\ \text{kJ·mol}^{-1}$$

Because reactions in the body take place in aqueous solution, this value is not the same as the enthalpy change for the reaction in the body. However, the two values are fairly close. The enthalpy change per mole of glycine molecules is much less than the standard enthalpy of combustion of glucose (-2808 kJ·mol^{-1}), partly because the oxidation of glycine is incomplete.

Self-Test 6.15A Calculate the standard enthalpy of combustion of glucose from the information in Table 6.5 and Appendix 2A.

[*Answer:* -2808 kJ·mol^{-1}]

Self-Test 6.15B You have an inspiration: maybe diamonds would make a great fuel! Calculate the standard enthalpy of combustion of diamonds from the information in Appendix 2A.

Standard enthalpies of formation are commonly determined from combustion data by using Eq. 29. The procedure is the same, but the standard reaction enthalpy is known and the unknown value is one of the standard enthalpies of formation.

Standard enthalpies of formation can be combined to obtain the standard enthalpy of any reaction.

EXAMPLE 6.10 Using the enthalpy of reaction to calculate an enthalpy of formation

Use the information in Table 6.5 and the enthalpy of combustion of propane gas to calculate the enthalpy of formation of propane, a gas commonly used for camping stoves and outdoor barbecues.

STRATEGY It is hard to anticipate whether the enthalpy of formation of a hydrocarbon is positive or negative. Use the procedure set out in Toolbox 6.2.

SOLUTION ***Step 1*** The combustion reaction to consider is

$$C_3H_8(g) + 5\,O_2(g) \longrightarrow 3\,CO_2(g) + 4\,H_2O(l) \quad \Delta H° = -2220\text{ kJ}$$

Step 2 The total enthalpy of formation of the products is

$$\begin{aligned}\sum n\Delta H_f°(\text{products}) &= 3\Delta H_f°(CO_2, g) + 4\Delta H_f°(H_2O, l)\\ &= 3(-393.51\text{ kJ·mol}^{-1}) + 4(-285.83\text{ kJ·mol}^{-1})\\ &= -1180.53 - 1143.32\text{ kJ·mol}^{-1}\\ &= -2323.85\text{ kJ·mol}^{-1}\end{aligned}$$

Step 3 The total enthalpy of formation of the reactants is

$$\begin{aligned}\sum n\Delta H_f°(\text{reactants}) &= \Delta H_f°(C_3H_8, g) + 5\Delta H_f°(O_2, g)\\ &= \Delta H_f°(C_3H_8, g)\end{aligned}$$

because $\Delta H_f°(O_2, g) = 0$.

Step 4 The difference between the two totals is equal to the standard enthalpy of the reaction

$$-2323.85\text{ kJ·mol}^{-1} - \Delta H_f°(C_3H_8, g) = -2220\text{ kJ·mol}^{-1}$$

It follows that

$$\Delta H_f°(C_3H_8, g) = -104\text{ kJ·mol}^{-1}.$$

Self-Test 6.16A Calculate the standard enthalpy of formation of ethyne, the fuel used in oxyacetylene welding torches, from the information in Tables 6.4 and 6.5.

[*Answer:* $+227\ \text{kJ}\cdot\text{mol}^{-1}$]

Self-Test 6.16B Calculate the standard enthalpy of formation of urea, $CO(NH_2)_2$, a by-product of the metabolism of proteins, from the information in Tables 6.4 and 6.5.

6.20 The Born-Haber Cycle

We saw in Section 2.2 that the energy changes accompanying the formation of a solid could be estimated on the basis of a model—the ionic model—in which the principal contribution to the lattice energy was the Coulomb interaction between ions. However, a model can give only an estimate: we need a way to measure the energy change. If the measured and calculated energies are similar, we can conclude that the ionic model is reliable for the particular substance. If the two energies are markedly different, then we know that the ionic model must be improved or even discarded.

For a given solid, the difference in molar enthalpy between the solid and a gas of widely separated ions is called the **lattice enthalpy** of the solid, ΔH_L:

$$\Delta H_L = H_m(\text{ions, g}) - H_m(\text{solid}) \qquad (30)$$

The lattice enthalpy can be identified with the heat required to vaporize the solid at constant pressure; and the greater the lattice enthalpy, the greater the heat required. Heat equal to the lattice enthalpy is released when the solid forms from gaseous ions.

The lattice enthalpy of a solid cannot be measured directly. However, we can obtain it by another path, one that combines other measurements. This approach takes advantage of the first law of thermodynamics and, in particular, the fact that enthalpy is a state function. The procedure uses a **Born-Haber cycle**, a closed path of steps, one of which is the formation of a solid lattice from the gaseous ions (Fig. 6.31). The cycle traces the enthalpy changes that occur when we begin with the pure elements, atomize them to form gaseous atoms, ionize the atoms to form gaseous ions, allow the ions to form an ionic solid, then convert the solid back into the pure elements. The only unknown enthalpy is that of the step in which the ionic solid forms from gaseous ions. The Born-Haber cycle allows us to calculate this value, which is the negative of the lattice enthalpy, as explained in Toolbox 6.3.

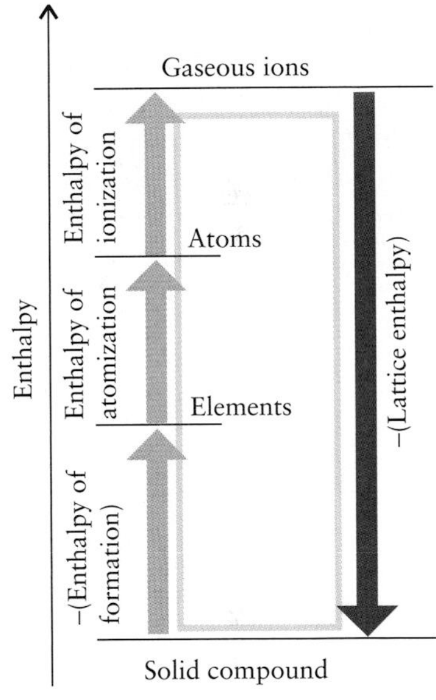

FIGURE 6.31 In a Born-Haber cycle, we select a sequence of steps that starts and ends at the same point (the elements, for instance). The step giving the enthalpy change for the formation of the solid from a gas of ions corresponds to the negative of the lattice enthalpy. The sum of the enthalpy changes for the complete cycle is 0 because enthalpy is a state function.

TOOLBOX 6.3 How to use a Born-Haber cycle

Conceptual Basis

Because the enthalpy is a state function, the sum of the enthalpy changes around a cycle is 0: the enthalpy at the end of a cycle must be the same as the enthalpy at the start because the system is in the same state. Therefore, if we know all the enthalpy changes except the lattice enthalpy, we can determine it from the enthalpies of the other steps.

Procedure

Start with the elements—typically a metal and a nonmetal—in the appropriate amounts to form the compound, and then make the following changes, adding an arrow to the diagram for each change. For an endothermic process, draw the arrow pointing up; for an exothermic process, draw the arrow pointing down.

Step 1 Atomize the metal and the nonmental and write the corresponding enthalpies of formation of the atoms (Appendix 2A).

Step 2 Form the gaseous cation of the metal. This step requires the ionization energy of the element and possibly the sum of the first and higher ionization energies.

Step 3 Form the gaseous anion of the nonmetal. This step releases an energy equal to the electron affinity of the element. If the electron affinity is positive, energy is released (so ΔH is negative). If it is negative, energy must be supplied (so ΔH is positive).

Step 4 Let the gas of ions form the solid compound. This step is the reverse of the formation of the ions from the solid, so its enthalpy change is the negative of the lattice enthalpy, $-\Delta H_L$.

Step 5 Complete the cycle by forming the elements from the compound: the enthalpy change for this step is the negative of the enthalpy of formation of the compound.

Step 6 Finally, calculate ΔH_L from the fact that the sum of all the enthalpy changes for the complete cycle is 0.

The procedure is illustrated in the following example. Lattice enthalpies of other compounds obtained in this way are listed in Table 6.6.

The strength of interaction between ions is determined by the lattice enthalpy, which can be calculated by using a Born-Haber cycle.

EXAMPLE 6.11 Using a Born-Haber cycle to calculate a lattice enthalpy

Devise and use a Born-Haber cycle to calculate the lattice enthalpy of potassium chloride.

STRATEGY We follow the strategy set out in Toolbox 6.3. Enthalpies of formation are listed in Appendix 2A. Ionization energies and electron affinities are given in Figs. 1.45 and 1.49, respectively; more extensive ionization energy data are in Appendix 2D. All lattice enthalpies are positive.

TABLE 6.6 Lattice Enthalpies at 25°C, $kJ\cdot mol^{-1}$

Halides							
LiF	1046	LiCl	861	LiBr	818	LiI	759
NaF	929	NaCl	787	NaBr	751	NaI	700
KF	826	KCl	717	KBr	689	KI	645
AgF	971	AgCl	916	AgBr	903	AgI	887
$BeCl_2$	3017	$MgCl_2$	2524	$CaCl_2$	2260	$SrCl_2$	2153
		MgF_2	2961	$CaBr_2$	1984		
Oxides							
MgO	3850	CaO	3461	SrO	3283	BaO	3114
Sulfides							
MgS	3406	CaS	3119	SrS	2974	BaS	2832

SOLUTION The Born-Haber cycle for KCl is shown in Fig. 6.32. The sum of the enthalpy changes for the complete cycle is 0, so we can write

$$\{89 + 122 + 418 - 349 - (-437)\}\ \text{kJ}\cdot\text{mol}^{-1} - \Delta H_L = 0$$

hence

$$\Delta H_L = (89 + 122 + 418 - 349 + 437)\ \text{kJ}\cdot\text{mol}^{-1} = +717\ \text{kJ}\cdot\text{mol}^{-1}$$

Therefore, the lattice enthalpy of calcium chloride is 717 kJ·mol^{-1}.

Self-Test 6.17A Calculate the lattice enthalpy of calcium chloride, $CaCl_2$, by using the data in Appendices 2A and 2D.

[*Answer:* 2259 kJ·mol^{-1}]

Self-Test 6.17B Calculate the lattice enthalpy of magnesium bromide, $MgBr_2$.

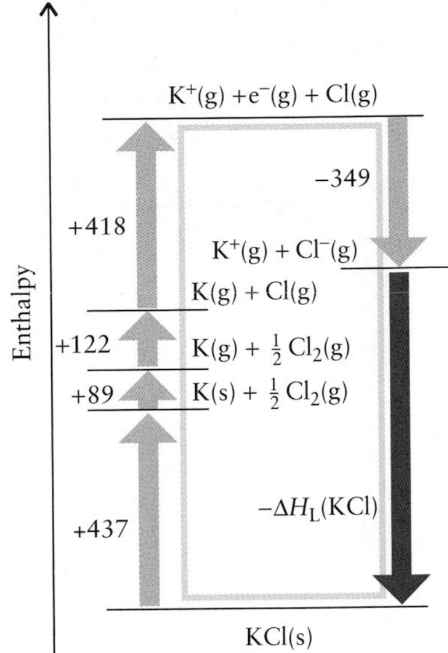

FIGURE 6.32 The Born-Haber cycle used to determine the lattice enthalpy of potassium chloride (see Example 6.11). The enthalpy changes are in kilojoules per mole.

6.21 Bond Enthalpies

In a chemical reaction, old bonds are broken and new ones formed. We ought to be able to estimate reaction enthalpies if we know the enthalpy changes that accompany the breaking and making of bonds. The strength of a chemical bond is measured by the **bond enthalpy, ΔH_B**, the difference between the standard molar enghalpies ($H_m°$) of the molecule, X—Y (for instance, H_3C—OH) and its fragments X and Y (such as CH_3 and OH):

$$\Delta H_B(\text{X—Y}) = \{H_m°(\text{X, g}) + H_m°(\text{Y, g})\} - H_m°(\text{XY, g}) \quad (31)$$

For example, the bond enthalpy of H_2 is derived from the thermochemical equation

$$H_2(g) \longrightarrow 2\ H(g) \qquad \Delta H° = +436\ \text{kJ}$$

We write ΔH_B(H—H) = 436 kJ·mol^{-1} to report this value.

All bond enthalpies are positive because heat must be supplied to break a bond. In other words, bond breaking is always endothermic and bond formation is always exothermic. To understand this conclusion in more detail, we can plot the energy of the two atoms of a diatomic molecule against internuclear separation to see how the energy of the molecule depends on the distance between the two nuclei (Fig. 6.33). The interatomic separation corresponding to the bottom of the energy well is the distance between the atoms when they have formed a bond. The energy climbs steeply when the nuclei approach too closely, but it also climbs when the atoms are pulled apart. A high bond enthalpy means that the energy well is deep and that a lot of heat is required to break the bond. Table 6.7 lists bond enthalpies of some diatomic molecules.

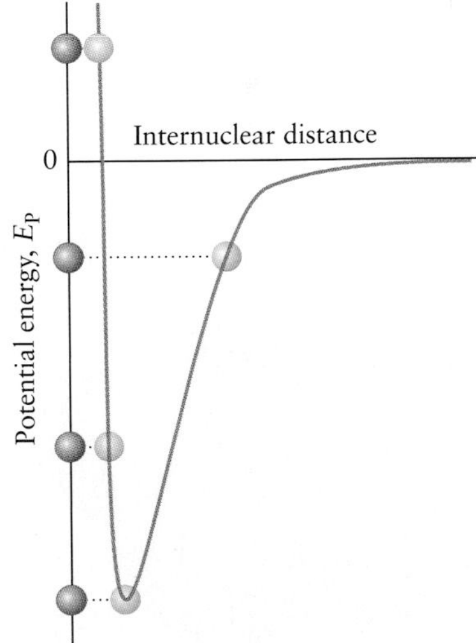

FIGURE 6.33 The variation of the energy of two atoms as the bond length changes. The equilibrium bond length is at the position of minimum energy and the depth of the well determines the bond enthalpy. This curve resembles that in Fig. 6.20, but the well is much deeper and narrower.

In a polyatomic molecule, all the atoms in the molecule exert a pull—through their electronegativities—on all the electrons in the molecule (Fig. 6.34). As a result, the bond strength between a given pair of atoms varies to a small extent from one compound to another. For example, the O—H bond enthalpy in HO—H (492 kJ·mol^{-1}) is somewhat different from that of the same bond in CH_3O—H (437 kJ·mol^{-1}). However, these variations

TABLE 6.7 Bond Enthalpies of Diatomic Molecules, kJ·mol^{-1}

Molecule	ΔH_B
H_2	436
N_2	944
O_2	496
CO	1074
F_2	158
Cl_2	242
Br_2	193
I_2	151
HF	565
HCl	431
HBr	366
HI	299

in bond enthalpy are not very great, so the **mean bond enthalpy**, which we also denote ΔH_B, is a guide to the strength of a bond in any molecule containing the bond (Table 6.8). The tabulated values are for gaseous samples.

Reaction enthalpies can be estimated by using mean bond enthalpies to determine the total energy required to break the reactant bonds and form the product bonds. Finding a reaction enthalpy by using bond enthalpies is equivalent to imagining that all the reactants separate into atoms and that these atoms combine to form the products. In practice, only the bonds that change are treated. To use the tabulated values of bond enthalpies, substances must all be gases or converted to the gas phase.

A mean bond enthalpy is the average molar enthalpy change accompanying the dissociation of a given type of bond.

EXAMPLE 6.12 **Using mean bond enthalpies to estimate the enthalpy of a reaction**

Estimate the enthalpy change of the reaction between gaseous iodoethane and water vapor:

$$CH_3CH_2I(g) + H_2O(g) \longrightarrow CH_3CH_2OH(g) + HI(g)$$

STRATEGY Decide which bonds are broken and which bonds are formed. Use the mean bond enthalpies in Table 6.8 to calculate the change in enthalpy when the bonds are broken in the reactants and the change in enthalpy when the new bonds are formed to produce the products. For diatomic molecules, use the information in Table 6.7 for the specific molecule. Finally, add the enthalpy change required to break the reactant bonds (a positive value) to the enthalpy change that occurs when the product bonds form (a negative value).

TABLE 6.8 Mean Bond Enthalpies, kJ·mol^{-1}

Bond	Mean bond enthalpy	Bond	Mean bond enthalpy
C—H	412	C—I	238
C—C	348	N—H	388
C=C	612	N—N	163
C⋯C*	518	N=N	409
C≡C	837	N—O	210
C—O	360	N=O	630
C=O	743	N—F	195
C—N	305	N—Cl	381
C—F	484	O—H	463
C—Cl	338	O—O	157
C—Br	276		

* In benzene.

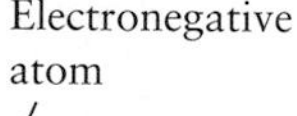

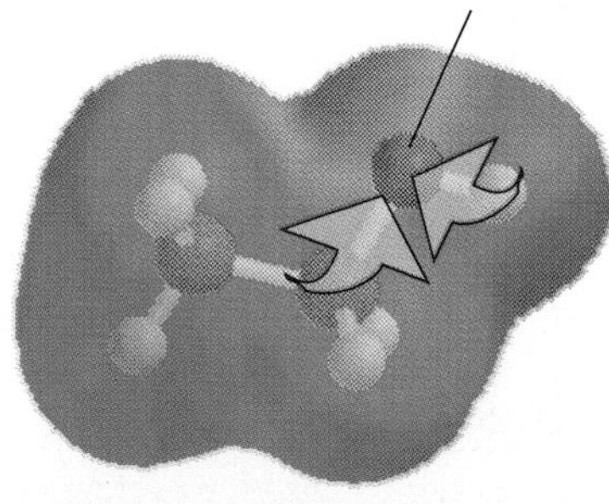

FIGURE 6.34 An electrostatic potential diagram of ethanol, C_2H_5OH. An electronegative atom (here, the O atom) can pull electrons toward itself from more distant parts of the molecule. Hence it can influence the strengths of bonds even between atoms to which it is not directly attached.

SOLUTION In the reactant molecules, we need to break a C—I bond in CH_3CH_2I (average value 238 kJ·mol^{-1}) and an O—H bond in H_2O (average value 463 kJ·mol^{-1}), so the total reaction enthalpy for the dissociation is

$$\Delta H° = 238 + 463\ \text{kJ} = +701\ \text{kJ}$$

To generate product molecules, a C—O bond and an H—I bond must be formed. To break 1 mol C—O bonds (average value 360 kJ·mol^{-1}) and 1 mol H—I bonds (299 kJ·mol^{-1}) corresponds to

$$\Delta H° = 360 + 299\ \text{kJ} = +659\ \text{kJ}$$

The enthalpy change when the product bonds are *formed* is therefore −659 kJ. The overall enthalpy change is the sum of these two changes:

$$\Delta H° = 701 + (-659)\ \text{kJ} = +42\ \text{kJ}$$

Therefore, the reaction is endothermic, primarily because a relatively large energy is needed to break an O—H bond in a water molecule.

Self-Test 6.18A Estimate the standard enthalpy of the reaction $CCl_3CHCl_2(g) + 2\ HF(g) \longrightarrow CCl_3CHF_2(g) + 2\ HCl(g)$.

[*Answer:* −24 kJ·mol^{-1}]

Self-Test 6.18B Estimate the standard enthalpy of the reaction in which 1.00 mol of gaseous CH_4 reacts with gaseous F_2 to form gaseous CH_2F_2 and HF.

6.22 The Variation of Reaction Enthalpy with Temperature

Suppose we know the reaction enthalpy at one temperature but require it at another temperature. For instance, the temperature of the human body is about 37°C, but the data in Appendix 2A are for 25°C. Does an increase of 12°C make much difference to the reaction enthalpy accompanying a metabolic process? Suppose we are chemical engineers and need to find the reaction enthalpy for the synthesis of ammonia at 500°C, using only the data in this text? Does an increase of over 400°C make much difference? To obtain *reliable* information on these questions, we should really make measurements at the temperature of interest. However, it is possible to make reasonable estimates without having to do more experiments.

The enthalpies of reactants and products increase with temperature. If the total enthalpy of the reactants increases more than that of the products, then the reaction enthalpy of an exothermic reaction becomes more negative as the temperature is raised (Fig. 6.35). On the other hand, if the enthalpy of the products increases more with an increase in temperature than that of the reactants, then the reaction enthalpy will become less negative. The increase in enthalpy of a substance when the temperature is raised depends on its heat capacity at constant pressure (Eq. 20), so we ought to be able to predict the change in reaction enthalpy from the heat capacities of all the reactants and products.

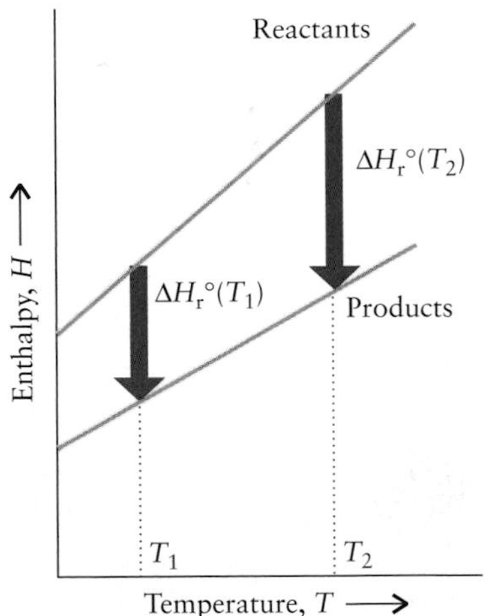

FIGURE 6.35 If the heat capacity of the reactants is larger than that of the products, the enthalpy of the reactants will increase more sharply with increasing temperature. If the reaction is exothermic, as here, the reaction enthalpy will become more negative. If the reaction is endothermic, the reaction enthalpy will become less positive and may even become negative.

How do we do that?

To find the effect of temperature on the reaction enthalpy, we note first from Eq. 20 that, if we raise the temperature of a substance from T_1 to T_2, then the enthalpy of the substance increases from H_1 to H_2, where

$$H_2 - H_1 = C_P(T_2 - T_1)$$

Therefore, the enthalpy of any reactant or product at the final temperature is related to the enthalpy at the initial temperature by

$$H_2 = H_1 + C_P(T_2 - T_1)$$

This expression applies to all the substances involved in the reaction, so

$$\begin{aligned}\Delta H_{r,2}° &= H_{m,2}°(\text{products}) - H_{m,2}°(\text{reactants}) \\ &= \{H_{m,1}°(\text{products}) + C_{P,m}(\text{products})\,(T_2 - T_1)\} \\ &\quad - \{H_{m,1}°(\text{reactants}) + C_{P,m}(\text{reactants})\,(T_2 - T_1)\} \\ &= H_{m,1}°(\text{products}) - H_{m,1}°(\text{reactants}) \\ &\quad + \{C_{P,m}(\text{products}) - C_{P,m}(\text{reactants})\}\,(T_2 - T_1)\} \\ &= \Delta H_{r,1}° + \Delta C_P(T_2 - T_1)\end{aligned} \quad (32)$$

This relation is called **Kirchhoff's law.** To use it, we need to know ΔC_P, the difference between the molar constant-pressure heat capacities of the products and reactants:

$$\Delta C_P = \sum nC_{P,m}(\text{products}) - \sum nC_{P,m}(\text{reactants}) \quad (33)$$

The temperature variation of the standard reaction enthalpy is given by Kirchhoff's law, Eq. 32, in terms of the difference in molar heat capacities at constant pressure between the products and the reactants.

EXAMPLE 6.13 Predicting the reaction enthalpy at a different temperature

In the synthesis of ammonia by the reaction $N_2(g) + 3\,H_2(g) \longrightarrow 2\,NH_3(g)$, which is crucial to the economies and well-being of all nations, the standard enthalpy of reaction is -92.22 kJ·mol^{-1} at 298 K. However, the industrial synthesis is run at 450°C. What is the standard reaction enthalpy at the latter temperature?

STRATEGY To use Kirchhoff's law, we need to know the molar heat capacities of the reactants and products: these can be found in Appendix 2A. Combine them in the same way that enthalpies of formation would be combined to calculate a reaction enthalpy and then substitute in Eq. 32. Note that the temperatures must be expressed in kelvins.

SOLUTION The difference in molar heat capacities we need to calculate is

$$\begin{aligned}\Delta C_P &= 2C_{P,m}(NH_3, g) - \{C_{P,m}(N_2, g) + 3C_{P,m}(H_2, g)\}\\ &= 2(35.06\ J\cdot K^{-1}\cdot mol^{-1}) - \{(29.12\ J\cdot K^{-1}\cdot mol^{-1})\\ &\quad + 3(28.82\ J\cdot K^{-1}\cdot mol^{-1})\}\\ &= -45.46\ J\cdot K^{-1}\cdot mol^{-1}\end{aligned}$$

The temperature change is

$$T_2 - T_1 = (450 + 273\ K) - 298\ K = +425\ K$$

Therefore, from Eq. 32,

$$\begin{aligned}\Delta H_{r,2}° &= -92.22\ kJ\cdot mol^{-1} + \{(-45.46\ J\cdot K^{-1}\cdot mol^{-1}) \times (425\ K)\}\\ &= -92.22\ kJ\cdot mol^{-1} - 19.32\ kJ\cdot mol^{-1} = -111.54\ kJ\cdot mol^{-1}\end{aligned}$$

Note how this reaction is more exothermic at the higher temperature than at the lower.

Self-Test 6.19A The reaction enthalpy of $4\ Al(s) + 3\ O_2(g) \rightarrow 2\ Al_2O_3(s)$ is $-3351\ kJ\cdot mol^{-1}$ at 298 K. Estimate its value at 1000°C.

[*Answer:* $-3378\ kJ\cdot mol^{-1}$]

Self-Test 6.19B The standard enthalpy of formation of ammonium nitrate is $-365.56\ kJ\cdot mol^{-1}$ at 298.15 K. Estimate its value at 250°C.

SKILLS YOU SHOULD HAVE MASTERED

1 Calculate the change in internal energy due to heat and work, Self-Test 6.2.

2 Calculate the work, heat, and internal energy changes in a gas due to expansion, Example 6.1.

3 State and explain the implications of the first law of thermodynamics, Section 6.5.

4 Interpret the calorimetric measurement of the specific heat capacity of a substance, Example 6.2.

5 Calculate enthalpy changes from calorimetry data and write a thermochemical equation, Examples 6.3 and 6.5.

6 Calculate the change in internal energy when an ideal gas is heated, Example 6.4.

7 Determine the enthalpy of vaporization, Self-Test 6.10.

8 Interpret the heating curve of a substance, Section 6.13.

9 Distinguish between the internal energy and enthalpy of a process and describe how each is measured, Section 6.15.

10 Determine the enthalpy change of a reaction, given the change in internal energy, and vice versa, Self-Test 6.8 and Example 6.6.

11 Calculate an overall reaction enthalpy from the enthalpies of the reactions in a reaction sequence by using Hess's law, Toolbox 6.1 and Example 6.7.

12 Calculate the heat output of a fuel, Example 6.8.

13 Use standard enthalpies of formation to calculate the standard enthalpy of a reaction, and vice versa, Toolbox 6.2 and Examples 6.9 and 6.10.

14 Calculate a lattice enthalpy by using the Born-Haber cycle, Toolbox 6.3 and Example 6.11.

15 Use average bond enthalpies to estimate the standard enthalpy of a reaction, Example 6.12.

16 Predict the reaction enthalpy at a temperature different from that of the data, Example 6.13.

EXERCISES

Systems

6.1 Identify the following systems as open, closed, or isolated: (a) coffee in a very high quality thermos bottle; (b) coolant in a refrigerator coil; (c) a bomb calorimeter in which benzene is burned; (d) gasoline burning in an automobile engine; (e) mercury in a thermometer; (f) a living plant.

6.2 (a) Describe three ways in which you could increase the internal energy of an open system. (b) Which of these methods could you use to increase the internal energy of a closed system? (c) Which, if any, of these methods could you use to increase the internal energy of an isolated system?

Energy and Work

6.3 A gas sample is heated in a cylinder, using 437 kJ of heat. At the same time, a piston compresses the gas, doing 295 kJ of work. What is the change in internal energy of the gas during this process?

6.4 A gas sample in a piston assembly expands, doing 515 kJ of work on its surrounding at the same time that 325 kJ of heat is added to the gas. (a) What is the change in internal energy of the gas during this process? (b) Will the pressure of the gas be higher or lower when these changes are completed?

6.5 The internal energy of a system increased by 982 J when it absorbed 492 J of heat. (a) Was work done by or on the system? (b) How much work was done?

6.6 (a) Calculate the work for a system that releases 346 kJ of heat in a process for which the decrease in internal energy is 125 kJ. (b) Is work done on or by the system during this process?

6.7 A gas in a cylinder was placed in a heater and gained 5500 kJ of heat. If the cylinder increased in volume from 345 mL to 1846 mL against an atmospheric pressure of 750 Torr during this process, what is the change in internal energy of the gas in the cylinder?

6.8 A 100-W electric heater ($1\ W = 1\ J \cdot s^{-1}$) operates for 20 min to heat the gas in a cylinder. At the same time, the gas expands from 2.00 L to 52.50 L against a constant atmospheric pressure of 0.975 atm. What is the change in internal energy of the gas?

6.9 In a combustion cyclinder, the total internal energy change produced from the burning of a fuel is −2573 kJ. The cooling system that surrounds the cylinder absorbs 947 kJ as heat. How much work can be done by the fuel in the cylinder?

6.10 The change in internal energy for the combustion of 1.00 mol $CH_4(g)$ in a cylinder according to the reaction $CH_4(g) + 2\ O_2(g) \rightarrow CO_2(g) + 2\ H_2O(g)$ is −892.4 kJ. If a piston connected to the cylinder performs 492 kJ of expansion work due to the combustion, how much heat is lost from the system (the reaction mixture)?

6.11 For a certain reaction at constant pressure, $\Delta H = -15$ kJ, and 22 kJ of expansion work is done on the system. What is ΔU for this process?

6.12 For a certain reaction at constant pressure, $\Delta U = -95$ kJ, and 56 kJ of expansion work is done by the system. What is ΔH for this process?

6.13 Calculate the work for each of the following processes beginning with a gas sample in a piston assembly with $T = 305$ K, $P = 1.79$ atm, and $V = 4.29$ L: (a) irreversible expansion against a constant external pressure of 1.00 atm to a final volume of 6.52 L; (b) isothermal, reversible expansion to a final volume of 6.52 L.

6.14 A sample of gas in a cylinder of volume 3.42 L at 298 K and 2.57 atm expands to 7.39 L by two different pathways. Path A is an isothermal, reversible expansion. Path B involves two steps. In the first, the gas is cooled at constant volume to 1.19 atm. In the second step, the gas is heated and allowed to expand against a constant external pressure of 1.19 atm until the final volume is 7.39 L. Calculate the work for each path.

6.15 (a) Near room temperature, the specific heat capacity of ethanol is $2.42\ J \cdot (°C)^{-1} \cdot g^{-1}$. (a) Calculate the heat that must be removed to reduce the temperature of 150.0 g of $C_2H_5OH(l)$ from 50.0°C to 16.6°C. (b) What is the heat that must be supplied to raise the temperature of 1.00 L of ethanol from 25°C to its boiling point at 78°C?

6.16 Near room temperature, the specific heat capacity of benzene is $1.05\ J \cdot (°C)^{-1} \cdot g^{-1}$. (a) Calculate the heat needed to raise the temperature of 50.0 g of benzene from 25.3°C to 37.2°C. (b) How much heat must be removed to lower the temperature of 1.00 mol $C_6H_6(l)$ from its boiling point (353.2 K) to its freezing point (278.6 K)?

6.17 (a) Calculate the molar heat capacity of copper.

(b) What mass of copper can be heated from 15°C to 205°C when 425 kJ of energy is available?

6.18 (a) Calculate the molar heat capacity of aluminum. (b) A 1.0-kg block of aluminum is supplied with 490 kJ of heat. What is the temperature change of the aluminum? The specific heat capacity of aluminum is 0.90 $J\cdot(°C)^{-1}\cdot g^{-1}$.

6.19 (a) Calculate the heat that must be supplied to a 500.0-g copper kettle containing 450.0 g of water to raise its temperature from 25°C to the boiling point of water, 100°C. (b) What percentage of the heat is used to raise the temperature of the water?

6.20 (a) Calculate the heat that must be supplied to a 500.0-g stainless steel vessel containing 450.0 g of water to raise its temperature from 25.0°C to the boiling point of water, 100.0°C. (b) What percentage of the heat is used to raise the temperature of the water? (c) Compare these answers with those of the preceding exercise.

6.21 A piece of copper of mass 20.0 g at 100.0°C is placed in a vessel of negligible heat capacity but containing 50.7 g of water at 22.0°C. Calculate the final temperature of the water. Assume that there is no energy lost to the surroundings.

6.22 A piece of metal of mass 20.0 g at 100.0°C is placed in a calorimeter containing 50.7 g of water at 22.0°C. The final temperature of the mixture is 25.7°C. What is the specific heat capacity of the metal? Assume that there is no energy lost to the surroundings.

6.23 A calorimeter was calibrated with an electric heater, which supplied 22.5 kJ of energy to the calorimeter and increased the temperature of the calorimeter and its water bath from 22.45°C to 23.97°C. What is the heat capacity of the calorimeter?

6.24 The enthalpy of combustion of benzoic acid, C_6H_5COOH, which is often used to calibrate calorimeters, is $-3227\ kJ\cdot mol^{-1}$. When 1.236 g of benzoic acid was burned in a calorimeter, the temperature increased by 2.345°C. What is the heat capacity of the calorimeter?

6.25 Calculate the amount of heat needed to raise the temperature of 0.325 mol of a monatomic ideal gas from a temperature of −25°C to +50°C at (a) constant volume and (b) constant pressure. (c) Why is more heat needed to raise the temperature at constant pressure than at constant volume?

6.26 A calorimeter has a measured heat capacity of 6.27 $kJ\cdot(°C)^{-1}$. The combustion of 1.84 g of magnesium in this calorimeter led to a temperature change from 21.30°C to 28.56°C. Calculate the enthalpy of reaction for $2\ Mg(s) + O_2(g) \rightarrow 2\ MgO(s)$.

6.27 Which molecular substance do you expect to have the higher molar heat capacity, NO or NO_2? Why?

6.28 Explain why the heat capacities of methane and ethane differ from the values expected for an ideal monatomic gas and from each other. The values are 35.309 $J\cdot mol^{-1}\cdot K^{-1}$ for CH_4 and 52.63 $J\cdot mol^{-1}\cdot K^{-1}$ for C_2H_6.

6.29 Calculate the heat released by 1.078 g of Kr(g) at 0.0572 atm as it cools from 97.6°C to 25.0°C at (a) constant pressure and (b) constant volume. Assume that krypton behaves as an ideal gas.

6.30 Calculate the heat one must supply to 10.35 g of Ne(g) at 0.150 atm to raise its temperature from 25.0°C to 50°C at (a) constant pressure and (b) constant volume. Assume that neon behaves as an ideal gas.

6.31 Predict the contribution to the heat capacity $C_{V,m}$ made by molecular motions for the following molecules: (a) HCN; (b) C_2H_6; (c) Ar; (d) HBr.

6.32 Predict the contribution to the heat capacity $C_{V,m}$ made by molecular motions for the following molecules: (a) BF_3; (b) N_2O; (c) HCl; (d) SO_2.

Enthalpies of Phase Changes

6.33 (a) At its boiling point, the vaporization of 0.579 mol CH_4(l) requires 4.76 kJ of heat. What is the enthalpy of vaporization of methane? (b) An electric heater was immersed in a flask of boiling ethanol, C_2H_5OH, and 22.45 g of ethanol was vaporized when 21.2 kJ of energy was supplied. What is the enthalpy of vaporization of ethanol?

6.34 (a) When 25.23 g of methanol, CH_3OH, froze, 4.01 kJ of heat was released. What is the enthalpy of fusion of methanol? (b) A sample of benzene was vaporized at 25°C. When 37.5 kJ of heat was supplied, 95 g of the liquid benzene vaporized. What is the enthalpy of vaporization of benzene at 25°C?

6.35 Use the information in Table 6.3 to calculate the enthalpy change for (a) the vaporization of 100.0 g of water at 373.2 K; (b) the melting of 612 g of solid ammonia at its freezing point (195.4 K).

6.36 Use the information in Table 6.3 to calculate the enthalpy change that occurs when (a) 207 g of methanol condenses at its boiling point (337.8 K); (b) 17.7 g of acetone, CH_3COCH_3, solidifies at it freezing point (175.2 K).

6.37 How much heat is needed to convert 50.0 g of ice at 0°C to liquid water at 25°C (see Tables 6.2 and 6.3)?

6.38 If we start with 155 g of water at 30°C, how much heat must we add to convert all the liquid to vapor at 100°C (see Tables 6.2 and 6.3)?

6.39 A 50.0-g ice cube at 0°C is added to a glass containing 400 g of water at 45°C. What is the final temperature of the system? Assume that no heat is lost to the surroundings.

6.40 When 25.0 g of a metal at 90.0°C is added to 50.0 g of water at 25.0°C, the temperature of the water rises to 29.8°C. What is the specific heat capacity of the metal?

Reaction Enthalpies

6.41 Carbon disulfide can be prepared from coke (an impure form of carbon) and elemental sulfur:

$$4\,C(s) + S_8(s) \longrightarrow 4\,CS_2(l) \qquad \Delta H° = +358.8\text{ kJ}$$

(a) How much heat is absorbed in the reaction of 1.25 mol S_8? (b) Calculate the heat absorbed in the reaction of 197 g of carbon with an excess of sulfur. (c) If the heat absorbed in the reaction was 415 kJ, how much CS_2 was produced?

6.42 The oxidation of nitrogen in the hot exhaust of jet engines and automobiles occurs by the reaction

$$N_2(g) + O_2(g) \longrightarrow 2\,NO(g) \qquad \Delta H° = +180.6\text{ kJ}$$

(a) How much heat is absorbed in the formation of 1.55 mol NO? (b) How much heat is absorbed in the oxidation of 5.45 L of nitrogen measured at 1.00 atm and 273 K? (c) When the oxidation of N_2 to NO was completed in a bomb calorimeter, the heat absorbed was measured as 492 J. What mass of nitrogen gas was oxidized?

6.43 The combustion of octane is expressed by the thermochemical equation

$$C_8H_{18}(l) + \tfrac{25}{2}\,O_2(g) \longrightarrow 8\,CO_2(g) + 9\,H_2O(l) \qquad \Delta H° = -5471\text{ kJ}$$

(a) Estimate the mass of octane that would need to be burned to produce enough heat to raise the temperature of the air in a 12 ft × 12 ft × 8 ft room from 40°F to 78°F on a mild winter's day. Use the normal composition of air to determine its density and assume a pressure of 1.00 atm. (b) How much heat will be evolved from the combustion of 1.0 gal of gasoline (assumed to be exclusively octane)? The density of octane is 0.70 $g{\cdot}mL^{-1}$.

6.44 Suppose that coal of density 1.5 $g{\cdot}cm^{-3}$ is carbon (it is, in fact, much more complicated, but this is a reasonable first approximation). The combustion of carbon is described by the equation

$$C(s) + O_2(g) \longrightarrow CO_2(g) \qquad \Delta H° = -394\text{ kJ}$$

(a) Calculate the heat produced when a lump of coal of size 7.0 cm × 6.0 cm × 5.0 cm is burned. (b) Estimate the mass of water that can be heated from 15°C to 100°C by burning this piece of coal.

6.45 (a) If a 70-kg man uses 2000 kJ of energy while jogging for one hour, what mass of fat would be consumed, given that the typical standard energy of combustion of fat is about 38 $kJ{\cdot}g^{-1}$? (b) How many hours would he need to jog if he wished to lose 1.0 lb of fat?

6.46 Calculate the heat evolved from a reaction mixture of 13.4 L of sulfur dioxide at 1.00 atm and 273 K and 15.0 g of oxygen in the reaction

$$2\,SO_2(g) + O_2(g) \longrightarrow 2\,SO_3(g) \qquad \Delta H° = -198\text{ kJ}$$

6.47 Use the information in Tables 6.3, 6.7, and 6.8 to estimate the enthalpy of formation of the following compounds in the *liquid* state. The standard enthalpy of sublimation of carbon is +717 $kJ{\cdot}mol^{-1}$: (a) H_2O; (b) methanol, CH_3OH; (c) benzene, C_6H_6 (without resonance); (d) benzene, C_6H_6 (with resonance).

6.48 Use the information in Tables 6.3, 6.7, and 6.8 to estimate the enthalpy of formation of the following compounds in the *liquid* state, given that the standard enthalpy of sublimation of carbon is +717 $kJ{\cdot}mol^{-1}$: (a) NH_3; (b) ethanol, CH_3CH_2OH; (c) acetone, CH_3COCH_3.

Hess's Law

6.49 The standard enthalpies of combustion of graphite and diamond are −393.51 $kJ{\cdot}mol^{-1}$ and −395.41 $kJ{\cdot}mol^{-1}$, respectively. Calculate the enthalpy of the graphite → diamond transition.

6.50 Elemental sulfur occurs in several forms, with rhombic sulfur the most stable under normal conditions and monoclinic sulfur slightly less stable. The standard enthalpies of combustion of the two forms (to sulfur dioxide) are $-296.83\ \text{kJ·mol}^{-1}$ and $-297.16\ \text{kJ·mol}^{-1}$, respectively. Calculate the enthalpy of the rhombic → monoclinic transition.

6.51 Two successive stages in the industrial manufacture of sulfuric acid are the combustion of sulfur and the oxidation of sulfur dioxide to sulfur trioxide. From the standard reaction enthalpies

$$S(s) + O_2(g) \longrightarrow SO_2(g) \qquad \Delta H° = -296.83\ \text{kJ}$$

$$2\ S(s) + 3\ O_2(g) \longrightarrow 2\ SO_3(g) \qquad \Delta H° = -791.44\ \text{kJ}$$

calculate the reaction enthalpy for the oxidation of sulfur dioxide to sulfur trioxide in the reaction $2\ SO_2(g) + O_2(g) \rightarrow 2\ SO_3(g)$

6.52 In the manufacture of nitric acid by the oxidation of ammonia, the first product is nitric oxide, which is then oxidized to nitrogen dioxide. From the standard reaction enthalpies

$$N_2(g) + O_2(g) \longrightarrow 2\ NO(g) \qquad \Delta H° = +180.5\ \text{kJ}$$

$$N_2(g) + 2\ O_2(g) \longrightarrow 2\ NO_2(g) \qquad \Delta H° = +66.4\ \text{kJ}$$

calculate the standard reaction enthalpy for the oxidation of nitric oxide to nitrogen dioxide:

$$2\ NO(g) + O_2(g) \longrightarrow 2\ NO_2(g)$$

6.53 Calculate the enthalpy of the reaction $P_4(s) + 10\ Cl_2(g) \rightarrow 4\ PCl_5(s)$ from the reactions

$$P_4(s) + 6\ Cl_2(g) \longrightarrow 4\ PCl_3(l) \qquad \Delta H° = -1278.8\ \text{kJ}$$

$$PCl_3(l) + Cl_2(g) \longrightarrow PCl_5(s) \qquad \Delta H° = -124\ \text{kJ}$$

6.54 Calculate the reaction enthalpy for the reduction of hydrazine to ammonia, $N_2H_4(l) + H_2(g) \rightarrow 2\ NH_3(g)$, from the following data:

$$N_2(g) + 2\ H_2(g) \longrightarrow N_2H_4(l) \qquad \Delta H° = +50.63\ \text{kJ}$$

$$N_2(g) + 3\ H_2(g) \longrightarrow 2\ NH_3(l) \qquad \Delta H° = -92.22\ \text{kJ}$$

6.55 Determine the reaction enthalpy for the hydrogenation of ethyne to ethane, $C_2H_2(g) + 2\ H_2(g) \rightarrow C_2H_6(g)$, from the following data: $\Delta H_c°(C_2H_2, g) = -1300\ \text{kJ·mol}^{-1}$, $\Delta H_c°(C_2H_6, g) = -1560\ \text{kJ·mol}^{-1}$, $\Delta H_c°(H_2, g) = -286\ \text{kJ·mol}^{-1}$.

6.56 Determine the enthalpy for the partial combustion of methane to carbon monoxide, $2\ CH_4(g) + 3\ O_2(g) \rightarrow 2\ CO(g) + 4\ H_2O(l)$, from $\Delta H_c°(CH_4, g) = -890\ \text{kJ·mol}^{-1}$ and $\Delta H_c°(CO, g) = -283.0\ \text{kJ·mol}^{-1}$.

6.57 Use the data provided in Appendix 2A to calculate the standard reaction enthalpy for the reaction of pure nitric acid with hydrazine:

$$4\ HNO_3(l) + 5\ N_2H_4(l) \longrightarrow 7\ N_2(g) + 12\ H_2O(l)$$

6.58 Use the data provided in Appendix 2A to calculate the standard reaction enthalpy for the reaction of calcite with hydrochloric acid:

$$CaCO_3(s) + 2\ HCl(aq) \longrightarrow CaCl_2(aq) + H_2O(l) + CO_2(g)$$

6.59 Calculate the reaction enthalpy for the synthesis of hydrogen chloride gas, $H_2(g) + Cl_2(g) \rightarrow 2\ HCl(g)$, from the following data:

$$NH_3(g) + HCl(g) \longrightarrow NH_4Cl(s) \qquad \Delta H° = -176.0\ \text{kJ}$$

$$N_2(g) + 3\ H_2(g) \longrightarrow 2\ NH_3(g) \qquad \Delta H° = -92.22\ \text{kJ}$$

$$N_2(g) + 4\ H_2(g) + Cl_2(g) \longrightarrow 2\ NH_4Cl(s) \qquad \Delta H° = -628.86\ \text{kJ}$$

6.60 Calculate the reaction enthalpy for the formation of anhydrous aluminum chloride, $2\ Al(s) + 3\ Cl_2(g) \rightarrow 2\ AlCl_3(s)$, from the following data:

$$2\ Al(s) + 6\ HCl(aq) \longrightarrow 2\ AlCl_3(aq) + 3\ H_2(g) \qquad \Delta H° = -1049\ \text{kJ}$$

$$HCl(g) \longrightarrow HCl(aq) \qquad \Delta H° = -74.8\ \text{kJ}$$

$$H_2(g) + Cl_2(g) \longrightarrow 2\ HCl(g) \qquad \Delta H° = -185\ \text{kJ}$$

$$AlCl_3(s) \longrightarrow AlCl_3(aq) \qquad \Delta H° = -323\ \text{kJ}$$

Standard Enthalpies of Formation

6.61 Write the thermochemical equations that give the values of the standard enthalpies of formation of (a) $CH_3CHO(l)$, ethanal; (b) $CuSO_4{\cdot}5H_2O(s)$, copper(II) sulfate pentahydrate; (c) $NH_4ClO_4(s)$, ammonium perchlorate.

6.62 Write the thermochemical equations that give the values of the standard enthalpies of formation of (a) $Ca_5(PO_4)_3F(s)$, calcium fluorapatite; (b) $CO(NH_2)_2(s)$, urea; (c) $C_6H_{12}O_6(s)$, glucose.

6.63 Calculate the standard enthalpy of formation of dinitrogen pentoxide from the data

$$2\,NO(g) + O_2(g) \longrightarrow 2\,NO_2(g) \qquad \Delta H^\circ = -114.1\text{ kJ}$$

$$4\,NO_2(g) + O_2(g) \longrightarrow 2\,N_2O_5(g) \qquad \Delta H^\circ = -110.2\text{ kJ}$$

and the standard enthalpy of formation of nitric oxide, NO (see Appendix 2A).

6.64 An important reaction that occurs in the atmosphere is $NO_2(g) \longrightarrow NO(g) + O(g)$, which is brought about by sunlight. How much energy must be supplied by the Sun to cause it? Calculate the standard enthalpy of the reaction from the following information:

$$O_2(g) \longrightarrow 2\,O(g) \qquad \Delta H^\circ = +498.4\text{ kJ}$$

$$NO(g) + O_3(g) \longrightarrow NO_2(g) + O_2(g) \qquad \Delta H^\circ = -200\text{ kJ}$$

and additional information from Appendix 2A.

6.65 Calculate the standard enthalpies of formation of $PCl_5(s)$ from the standard enthalpy of formation of $PCl_3(l)$ (see Appendix 2A) and $PCl_3(l) + Cl_2(g) \longrightarrow PCl_5(s)$, $\Delta H^\circ = -124$ kJ.

6.66 Using standard enthalpies of formation from Appendix 2A, calculate the standard reaction enthalpy for each reaction:
(a) the final stage in the production of nitric acid, when nitrogen dioxide dissolves in and reacts with water:

$$3\,NO_2(g) + H_2O(l) \longrightarrow 2\,HNO_3(aq) + NO(g)$$

(b) the formation of boron trifluoride, which is widely used in the chemical industry:

$$B_2O_3(s) + 3\,CaF_2(s) \longrightarrow 2\,BF_3(g) + 3\,CaO(s)$$

(c) the formation of a sulfide by the action of hydrogen sulfide on an aqueous solution of a base:

$$H_2S(aq) + 2\,KOH(aq) \longrightarrow K_2S(aq) + 2\,H_2O(l)$$

The Born-Haber Cycle

6.67 Calculate the lattice enthalpy of calcium sulfide from the data in Appendix 2D, Fig. 1.49, and the following information:

Enthalpy of formation of Ca(g)	$+178\text{ kJ·mol}^{-1}$
Enthalpy of formation of S(g)	$+279\text{ kJ·mol}^{-1}$
Enthalpy of formation of CaS(s)	-482 kJ·mol^{-1}
Second ionization energy of Ca(g)	$+1150\text{ kJ·mol}^{-1}$

6.68 Calculate the lattice enthalpy of silver fluoride from the data in Fig. 1.49 and the following information:

Enthalpy of formation of Ag(g)	$+284\text{ kJ·mol}^{-1}$
First ionization energy of Ag(g)	$+731\text{ kJ·mol}^{-1}$
Enthalpy of formation of F(g)	$+79\text{ kJ·mol}^{-1}$
Enthalpy of formation of AgF(s)	-205 kJ·mol^{-1}

6.69 Use Fig. 1.49, Appendix 2A, Appendix 2D, and the following data to calculate the lattice enthalpy of Na_2O. $\Delta H_f^\circ(Na_2O) = -409\text{ kJ·mol}^{-1}$; $\Delta H_f^\circ(O, g) = +249\text{ kJ·mol}^{-1}$.

6.70 Use Fig. 1.49, Appendix 2A, Appendix 2D, and the following data to calculate the lattice enthalpy of $AlCl_3$. $\Delta H_f^\circ(Al, g) = +326\text{ kJ·mol}^{-1}$.

6.71 Complete the table below (all values are in kilojoules per mole).

Compound MX	ΔH_f° M(g)	Ionization energy M	ΔH_f° X(g)	Electron affinity X	ΔH_L MX	ΔH_f° MX(s)
(a) NaCl	108	494	122	+349	787	?
(b) KBr	89	418	97	+325	?	−394
(c) RbF	?	402	79	+328	774	−558

6.72 By assuming that the lattice enthalpy of $NaCl_2$ is the same as that of $MgCl_2$, use enthalpy arguments based on data in Appendix 2A, Appendix 2D, and Fig. 1.49 to explain why $NaCl_2$ is an unlikely compound.

Bond Enthalpies

6.73 Use the bond enthalpies in Tables 6.8 and 6.9 to estimate the reaction enthalpy for
(a) $3\ C_2H_2(g) \rightarrow C_6H_6(l)$
(b) $CH_4(g) + 4\ Cl_2(g) \rightarrow CCl_4(g) + 4\ HCl(g)$
(c) $CH_4(g) + CCl_4(l) \rightarrow CHCl_3(l) + CH_3Cl(g)$

6.74 Use the bond enthalpies in Tables 6.8 and 6.9 to estimate the reaction enthalpy for
(a) $HCl(g) + F_2(g) \rightarrow HF(g) + ClF(g)$, given that $\Delta H_B(\text{Cl—F}) = 256\ \text{kJ·mol}^{-1}$
(b) $C_2H_4(g) + HCl(g) \rightarrow CH_3CH_2Cl(g)$
(c) $C_2H_4(g) + H_2(g) \rightarrow CH_3CH_3(g)$

6.75 Use the bond enthalpies in Tables 6.8 and 6.9 to estimate the reaction enthalpy for
(a) $N_2(g) + 3\ F_2(g) \rightarrow 2\ NF_3(g)$
(b) $CH_3CH{=}CH_2(g) + H_2O(g) \rightarrow CH_3CH(OH)CH_3(g)$
(c) $CH_4(g) + Cl_2(g) \rightarrow CH_3Cl(g) + HCl(g)$

6.76 The bond enthalpy in NO is 632 kJ·mol^{-1} and that of each N—O bond in NO_2 is 469 kJ·mol^{-1}. Using Lewis structures and the mean bond enthalpies given in Table 6.8, explain (a) the difference in bond enthalpies between the two molecules; (b) the fact that the bond enthalpies of the two bonds in NO_2 are the same.

6.77 Benzene is more stable and less reactive than would be predicted from its Kekulé structures. Use the mean bond enthalpies in Table 6.8 to calculate the lowering in molar energy that occurs when resonance is allowed between the Kekulé structures of benzene.

6.78 Draw the Lewis structure for the hypothetical molecule N_6 consisting of a six-membered ring of nitrogen atoms. Using bond enthalpies, calculate the enthalpy of reaction for the decomposition of N_6 to $N_2(g)$. Do you expect N_6 to be a stable molecule?

The Variation of Reaction Enthalpy with Temperature

6.79 (a) From the data in Appendix 2A, calculate the enthalpy of vaporization of benzene (C_6H_6) at 298.2 K. The standard enthalpy of formation of gaseous benzene is 82.93 kJ·mol^{-1}. (b) Given that for liquid benzene, $C_{P,\text{m}} = 136.1\ \text{J·mol}^{-1}\text{·K}^{-1}$ and that for gaseous benzene $C_{P,\text{m}} = 81.67\ \text{J·mol}^{-1}\text{·K}^{-1}$, calculate the enthalpy of vaporization of benzene at its boiling point (353.2 K). (c) Compare the value obtained in (b) to that found in Table 6.3. What is the source of difference between these numbers?

6.80 (a) From the data in Appendix 2A, calculate the enthalpy required to vaporize 1 mol $CH_3OH(l)$ at 298.2 K. (b) Given that the molar heat capacity, $C_{P,\text{m}}$, of liquid methanol is 81.6 $\text{J·mol}^{-1}\text{·K}^{-1}$ and that of gaseous methanol is 43.89 $\text{J·mol}^{-1}\text{·K}^{-1}$, calculate the enthalpy of vaporization of methanol at its boiling point (64.7°C). (c) Compare the value obtained in (b) to that found in Table 6.3. What is the source of difference between these values?

INTEGRATED EXERCISES

6.81 How much heat is required to convert a 27.96-g block of ice at −30.27°C to water vapor at 150.35°C?

6.82 A 155.7-g piece of stainless steel was heated to 475°C and quickly added to 25.34 g of water at −24°C in a well-insulated flask that was immediately sealed. (a) Assuming no loss of energy to the surroundings, what will be the final temperature of this system? (b) What phases of water will be present and in what quantities when the system reaches its final temperature?

6.83 The heat capacity of liquid iodine is 80.7 $\text{J·K}^{-1}\text{·mol}^{-1}$ and the enthalpy of vaporization of iodine is 41.96 kJ·mol^{-1} at its boiling point (184.3°C). Using these facts and information in Appendix 2A, calculate the enthalpy of fusion of iodine at 25°C.

6.84 A typical bathtub can hold about 100 gallons of water. (a) Calculate the mass of natural gas that would need to be burned to heat the water for a tub of this size from 70°F to 100°F. Assume that the natural gas is pure methane, CH_4. (b) What volume of natural gas does this correspond to at 25°C and 1.00 atm? See Table 6.4.

Media Link

6.85 (a) Using the Living Graphs, plot the work done by a gas as it expands isothermally and reversibly to 10 times its original volume at 100 K, 200 K and 300 K. (b) Is more or less work done for this expansion as the temperature is raised? (c) Compare the work done by a gas as it expands reversibly and isothermally from 1.00 L to 5.00 L to that done when it expands reversibly from 5.00 L to 9.00 L at any given temperature.

6.86 (a) Calculate the work associated with the isothermal, reversible expansion of 0.750 mol of an ideal gas from 6.97 L to 18.76 L at 50.0°C. (b) Calculate the work associated with the irreversible expansion of the same sample of gas in (a) against a constant atmospheric pressure of 743 Torr adiabatically. (c) How will the temperature of the gas in part (b) compare to that in part (a) after the expansion?

6.87 Strong sunshine bombards the Earth with about $1\ kJ\cdot m^{-2}$ in 1 s. Calculate the maximum mass of pure ethanol that can be vaporized in 10 min from a beaker left in strong sunshine, assuming the surface area of the ethanol to be $50\ cm^2$. Assume all the heat is used for vaporization, not to increase the temperature.

6.88 The following reaction can be used for the production of manganese:

$$3\ MnO_2(s) + 4\ Al(s) \longrightarrow 2\ Al_2O_3(s) + 3\ Mn(s)$$

(a) Use the information in Appendix 2A and the enthalpy of formation of $MnO_2(s)$, which is $-521\ kJ\cdot mol^{-1}$, to calculate the standard enthalpy of reaction. (b) What is the enthalpy change in the production of 10.0 g of manganese?

6.89 Aniline, $C_6H_5NH_2(l)$, is a derivative of benzene in which a hydrogen atom has been replaced by an NH_2 group. (a) Write the balanced equation for the combustion of aniline. (b) What is the mass of each product when 0.1754 g of aniline is burned in excess oxygen? (c) If the bomb calorimeter in which this reaction was carried out had a volume of 355 mL, what minimum pressure of oxygen at 23°C must have been used to ensure complete combustion? Assume that the volume of the aniline is negligible.

6.90 The standard enthalpy of formation of solid ammonium acetate is $-616.14\ kJ\cdot mol^{-1}$. Calculate its standard enthalpy of combustion.

6.91 (a) Is the production of water gas (a cheap, low-grade industrial fuel) exothermic or endothermic? The reaction is

$$C(s) + H_2O(g) \longrightarrow CO(g) + H_2(g).$$

(b) Calculate the enthalpy change for the production of 200 L of hydrogen at 500 Torr and 65°C by this reaction.

6.92 (a) Acetic acid can be produced by the reaction of carbon monoxide with methanol in the presence of a catalyst:

$$CO(g) + CH_3OH(l) \longrightarrow CH_3COOH(l).$$

Use the information in Appendix 2A to determine whether this reaction is exothermic or endothermic. (b) Acetic acid can also be formed by the oxidation of ethanol:

$$C_2H_5OH(l) + O_2(g) \longrightarrow CH_3COOH(l) + H_2O(l).$$

This reaction occurs when wine goes sour—the product being vinegar. Calculate the standard reaction enthalpy for this reaction.

6.93 (a) Hydrogen is being extensively investigated as a fuel for automobiles because it is the cleanest burning fuel available (the only product is water). How does the enthalpy of combustion of hydrogen compare with that of an organic hydrocarbon such as octane on a per gram basis? (b) What problems would be associated with the use of hydrogen as a fuel in automobiles?

6.94 The standard enthalpy of combustion of hexane, $C_6H_{14}(l)$, is $-4163\ kJ\cdot mol^{-1}$. Use this fact and the values for the enthalpies of combustion of various organic compounds found in Appendix 2A to compare benzene, ethanol, hexane, and octane as possible fuels by investigating the following problems. (a) Calculate the heat produced per gram of each of these substances. (b) From standard reference sources, find the density of each liquid and determine the heat produced per liter of liquid. (c) Using a current chemical catalogue, rank the organic liquids in increasing cost for samples of similar purity and quantity. (d) On the basis of this information, which would you choose as a fuel?

6.95 (a) Calculate the work that must be done against the atmosphere for the expansion of the gaseous products in the combustion of $C_6H_6(l)$ at 25°C and 1 bar. (b) Using data in Appendix 2A, calculate the standard enthalpy of the reaction. (c) Calculate the change in internal energy, $\Delta U°$, of the system.

6.96 A system undergoes a two-step process. In step 1, it absorbs 50 J of heat at constant volume. In step 2, it gives off 5 J of heat at 1.00 atm as it is returned to the same internal energy it had originally. Find the change in volume of the system during the second step and identify it as an expansion or compression.

6.97 50.0 mL of 0.500 M NaOH(aq) and 50.0 mL of 0.500 M $HNO_3(aq)$, both initially at 18.6°C, were mixed and stirred in a calorimeter having a heat capacity of $525.0\ J\cdot(°C)^{-1}$ when empty. The temperature of the mixture rose to 20.0°C. (a)What is the change in enthalpy for the neutralization reaction? (b) What is the change in enthalpy for the neutralization in kilojoules per mole of HNO_3?

6.98 The heat capacity of a certain calorimeter is 488.1 $J\cdot(°C)^{-1}$. When 25.0 mL of 0.700 M NaOH(aq) was mixed in that calorimeter with 25.0 mL of 0.700 M HCl(aq), both initially at 20.00°C, the temperature increased to 21.34°C. Calculate the enthalpy of neutralization in kilojoules per mole of HCl.

6.99 Consider the hydrogenation of benzene to cyclohexane, which occurs by a stepwise addition of two H atoms per step:

$$(1)\ C_6H_6(l) + H_2(g) \longrightarrow C_6H_8(l) \qquad \Delta H° = ?$$

$$(2)\ C_6H_8(l) + H_2(g) \longrightarrow C_6H_{10}(l) \qquad \Delta H° = ?$$

$$(3)\ C_6H_{10}(l) + H_2(g) \longrightarrow C_6H_{12}(l) \qquad \Delta H° = ?$$

(a) Draw Lewis structures for the products of the hydrogenation of benzene. If resonance is possible, show only one resonance structure. (b) Use bond enthalpies to estimate the enthalpy changes of each step and the total hydrogenation. Ignore delocalization of electrons in this calculation. (c) Use data from Appendix 2A to calculate the enthalpy of the complete hydrogenation of benzene to cyclohexane. (d) Compare the value obtained in (c) to that obtained in (d). Explain any difference.

6.100 A natural gas mixture is burned in a furnace at a power-generating station at a rate of 13.0 mol per minute. (a) If the fuel consists of 9.3 mol CH_4, 3.1 mol C_2H_6, 0.40 mol C_3H_8, and 0.20 mol C_4H_{10}, what mass of CO_2(g) is produced per minute? (b) How much heat is released per minute?

6.101 Calculate the lattice enthalpy of solid potassium bromide from the following information:

$\Delta H_f°$ (KBr, s) = −394 kJ.mol^{-1}

$\Delta H_{atomization}°$ of K(s) = +89.2 kJ·mol^{-1}

First ionization energy of K(g) = +425.0 kJ·mol^{-1}

$\Delta H_{vap}°$ (Br_2, l) = +30.9 kJ·mol^{-1}

Br—Br bond dissociation enthalpy = +192.9 kJ·mol^{-1}

Electron attachment to Br(g):

$$Br(g) + e^- \longrightarrow Br^-(g), \qquad \Delta H° = -331.0\ kJ$$

6.102 Use the reactions (a), (b), and (c) to determine the enthalpy change of this reaction:

$$CH_4(g) + \tfrac{3}{2}\,O_2(g) \longrightarrow CO(g) + 2\,H_2O(g)$$

(a) $CH_4(g) + 2\,O_2(g) \longrightarrow CO_2(g) + 2\,H_2O(g)$ $\quad \Delta H° = -802$ kJ

(b) $CH_4(g) + CO_2(g) \longrightarrow 2\,CO(g) + 2\,H_2(g)$ $\quad \Delta H° = +206$ kJ

(c) $CH_4(g) + H_2O(g) \longrightarrow CO(g) + 3\,H_2(g)$ $\quad \Delta H° = +247$ kJ

6.103 Robert Curl, Richard Smalley, and Harold Kroto were awarded the Nobel prize in chemistry in 1996 for the discovery of the soccer ball-shaped molecule C_{60}. This fundamental molecule represented the first of a new series of *molecular* allotropes of carbon (in addition to the known forms graphite, diamond, and lonsdalite; lonsdalite is similar to diamond in that it has sp^3 hybridized carbon, with each C atom bound to four other C atoms, but dissimiliar in that the unit cell of lonsdalite is hexagonal rather than cubic like that of diamond). This new series of compounds has come to be known as the fullerenes in honor of Buckminster Fuller, who created architectural frameworks similar to the arrangement of atoms in C_{60}. The enthalpy of combustion of C_{60} is −25 937 kJ·mol^{-1}, and its enthalpy of sublimation is +233 kJ·mol^{-1}. There are 90 bonds in C_{60}, of which 60 are single bonds and 30 are double bonds. C_{60}, like benzene, has a set of multiple bonds for which resonance structures may be drawn. (a) Determine the enthalpy of formation of C_{60} from its enthalpy of combustion. (b) Calculate the expected enthalpy of formation of C_{60} from bond enthalpies, assuming the bonds to be isolated double and single bonds. (c) Is C_{60} more or less stable than predicted based on the isolated bond model? (d) Quantify the answer in (c) by dividing the difference between the enthalpy of formation calculated from the combustion data and that obtained from the bond enthalpy calculation by 60 in order to obtain a per carbon value. (e) How does the number in (d) compare with the per carbon resonance stabilization energy of benzene (the total resonance stabilization energy of benzene is approximately 150 kJ·mol^{-1})? (f) Why might these values differ? The enthalpy of atomization of C(gr) is +717 kJ·mol^{-1}.

6.104 Normally when sulfur is oxidized, the product is SO_2, but SO_3 may also be formed. When 0.6192 g of sulfur was burned, using ultrapure oxygen in a bomb calorimeter with a heat capacity of 5.270 $kJ\cdot(°C)^{-1}$, the temperature rose 1.140°C. Assuming that all the sulfur was consumed in the reaction, what was the ratio of sulfur dioxide to sulfur trioxide produced?

6.105 Speculate as to whether the oxidation of 1,3,5-trinitrobenzene, $C_6H_3(NO_2)_3$, or the oxidation of 1,3,5-triaminobenzene, $C_6H_3(NH_2)_3$, would be more exothermic. Support your answer with appropriate calculations.

CHAPTER

The natural direction of change is for sugars like glucose to combine with oxygen in the process of combustion and to form carbon dioxide and water. This natural direction of change can be driven in the reverse direction if sufficient energy is provided in the appropriate form. The inset shows a schematic and highly simplified version of the photosynthetic center that is responsible for harvesting the energy provided by the Sun and using it to form glucose from carbon dioxide and water. The energy stored in carbohydrates is used to power the animals and people who eat them and use the energy to grow, learn, and think.

Why Do We Need to Know This Material?

The second law of thermodynamics summarizes the origin of all chemical change. So, if we want to know why any chemical change occurs, we need to understand the second law and two very important concepts to which it leads, namely, entropy and free energy. The third law of thermodynamics is the basis of the numerical scale of these two quantities. The second and third laws jointly provide a way to predict the effects of changes in temperature and pressure on physical and chemical processes. This chapter also lays the thermodynamic foundations for discussing chemical equilibrium, which the following chapters explore in detail.

What Do We Need to Know Already?

The discussion draws on concepts related to the first law of thermodynamics, particularly enthalpy (Section 6.9) and work (Section 6.2).

THERMODYNAMICS: THE SECOND AND THIRD LAWS

Some things happen naturally; some things don't. Water flows downhill naturally; we have to pump it uphill. We grow old as the chemical reactions of life take their toll; we do not grow young again. A spark is enough to initiate a vast forest fire; it requires the ceaseless input of energy from the Sun to rebuild the forest from carbon dioxide and water. Decay is natural; construction needs to be contrived. Anyone who thinks about the world may wonder what determines the *natural* direction of change. What drives events forward? What drives a single reaction toward products, and, on a larger scale, what drives the great and intricate network of reactions in biological systems forward and gives rise to the extraordinary phenomenon we call life? As this chapter unfolds, we shall see that a single quantity, "entropy," provides an elegant and quantitative answer to all these questions. We shall see that the concept of entropy accounts both for physical change and for the more intricate changes that accompany chemical reactions. We shall also see that, although entropy is a thermodynamic concept, in the sense that it is a property of bulk samples of matter, it has a very straightforward molecular interpretation.

ENTROPY

The first law of thermodynamics tells us only how much energy is transferred in a reaction. It reveals nothing about the conditions necessary for the reaction to take place, nor why it goes in a particular direction. All the first law tells us is that *if* a reaction takes place, then the total energy of the reaction system and its surroundings remains unchanged. But what lies behind the *if*? Why do some reactions have a tendency to occur whereas others do not? Why does anything happen at all? To answer these deeply important questions about the world around us, we need to take a further step into thermodynamics and learn more about energy beyond the fact that it is conserved.

7.1 Spontaneous Change

Natural change occurs in only one direction. Water does not flow uphill, rotten apples do not become fresh, and people do not grow younger. The technical term for a natural change is a **spontaneous change,** which is a change that tends to occur without needing to be driven by an external influence. A simple example is the cooling of a block of hot metal to the temperature of its surroundings (Fig. 7.1). The reverse change, a block of metal spontaneously growing hotter than its surroundings, has never been observed. The expansion of a gas into a vacuum is also spontaneous (Fig. 7.2): a gas has no tendency to contract spontaneously into one part of a container.

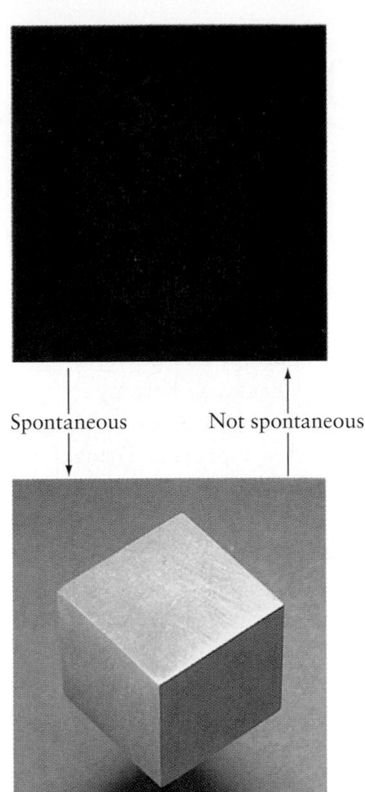

FIGURE 7.1 A hot block of metal (top) spontaneously cools to the temperature of its surroundings (bottom). The reverse process, in which a block at the same temperature as its surroundings spontaneously become hotter, does not occur. Observations like these are evidence for the claim that spontaneous change is directional.

Nonspontaneous changes can be made to happen in the "unnatural" direction. For example, we can force an electric current through a block of metal to heat it to a temperature higher than that of its surroundings. We can drive a gas into a smaller volume by pushing in a piston. In every case, though, to bring about a nonspontaneous change in a system, we have to contrive a way of *forcing* it to happen by applying an influence from outside the system. In short, *a nonspontaneous change can be brought about only by doing work.*

It is important to keep in mind that a spontaneous change need not be fast. Molasses has a spontaneous tendency to flow out of an overturned can; but at low temperatures, that flow may be very slow. Hydrogen and oxygen have a tendency to react to form water—the reaction is spontaneous in the thermodynamic sense—but a mixture of the two gases can be kept safely for centuries, provided we do not ignite it with a spark. Diamonds have a natural tendency to turn into graphite, but diamonds last unchanged for countless years—diamonds are, in practice, forever. *A spontaneous process has a natural tendency to occur; it does not necessarily take place rapidly.*

A spontaneous change is a change that has a tendency to occur without being driven by an external influence; spontaneous changes need not be fast.

7.2 Entropy and Disorder

When we study large numbers of spontaneous processes, we find a single underlying pattern: *energy and matter tend to become more disordered.* For example, a hot block of metal cools because the energy of its vigorously vibrating atoms tends to spread into the surroundings. The reverse change is very unlikely because it is very improbable that energy will gather from the surroundings and concentrate in a small block of metal. Similarly, it is natural for randomly moving gas molecules to spread out all over their container but very unlikely that their random motion will bring them all simultaneously back into one corner. We can use this idea of increasing disorder to account for any spontaneous change, even the complicated changes that occur in chemical reactions.

In thermodynamics, a measure of disorder is **entropy,** S. Low entropy means little disorder; high entropy means great disorder. We shall need to express the concept of entropy quantitatively; but even at this early stage, we can see that the tendency of energy and matter to become more disordered can be expressed as

The entropy of an isolated system increases in the course of any spontaneous change.

This is one statement of the **second law of thermodynamics.** The natural progression of the universe—or, more prosaically, a small isolated system within the universe, such as a thermally insulated, sealed flask—is from order to disorder, from lower to higher entropy.

In short, things get worse.

We need a quantitative definition of entropy before we can measure its value and use it to make predictions. Chemists have found that, provided the temperature is constant, a *change* in the entropy of a system can be calculated from the following expression:

$$\Delta S = \frac{q_{rev}}{T} \quad (1)$$

We generalize the definition in the following section to changes in which the temperature is not constant.

where q is the energy transferred as heat and T is the temperature. The subscript *rev* on q signifies that the energy must be transferred *reversibly* (see Section 6.7). For a reversible transfer of energy as heat, the temperature of the surroundings must be the same as that of the system. The transfer of heat is then reversible because an infinitesimal increase in the temperature of the surroundings results in a flow of energy into the system but an infinitesimal decrease in temperature results in an outflow of energy from the system.

How can we use this information . . . to calculate the change in entropy when heat is transferred to a system?

A large flask of water was placed on a heater and 100 J of energy was transferred reversibly to the water at 25°C. From Eq. 1, the change in entropy of the water is

$$\Delta S = \frac{100\ \text{J}}{(273.15 + 25)\ \text{K}} = \frac{1.00 \times 10^2\ \text{J}}{298\ \text{K}} = +0.336\ \text{J}\cdot\text{K}^{-1}$$

Note that the entropy of the water increases as a result of the flow of heat into it and that the units of entropy are joules per kelvin ($\text{J}\cdot\text{K}^{-1}$). We refer to a "large flask of water" to ensure that the temperature remains virtually constant as the heat is transferred: Eq. 1 applies only at constant temperature.

Self-Test 7.1A Calculate the change in entropy of a large block of ice when 50 J of energy as heat is removed reversibly from it at 0°C in a freezer.

[*Answer:* $-0.18\ \text{J}\cdot\text{K}^{-1}$]

Self-Test 7.1B Calculate the change in entropy of a large vat of molten copper when 50 J of energy as heat is removed reversibly from it at 1100°C.

Although the definition of ΔS in Eq. 1 might look arbitrary, it makes sense if we think about the physical meaning of each term. If a lot of energy is transferred as heat (large q_{rev}), a lot of disorder is stirred up in the system and we expect a correspondingly big increase in entropy. This reasoning suggests that ΔS should be proportional to q_{rev}. Moreover, for a given transfer of energy, we expect a greater change in disorder when the temperature is low than when it is high. The molecules of a cool system have little thermal motion, so the arriving energy stirs them up more than if the system is already hot and the molecules are moving vigorously. Think of how sneezing (corresponding to the transfer of energy) in a noisy street (a system with a lot of molecular motion, equivalent to a high temperature) may pass unnoticed, but sneezing in a quiet library (little molecular motion, corresponding to low temperature) will not.

FIGURE 7.2 A gas spontaneously fills its container. The reverse process, in which a gas that already fills its container collects spontaneously in a small region of the container, does not occur. A glass cylinder containing the brown gas nitrogen dioxide (upper piece of glassware in the top illustration) is attached to an evacuated flask. When the stopcock between them is opened, the gas spontaneously fills both upper and lower vessels (bottom illustration).

One very important characteristic of entropy that is not immediately obvious from Eq. 1 but can be proved by using thermodynamics is that *entropy is a state function*. This property is consistent with entropy being a measure of disorder, for the disorder of a system depends only on its current state and is independent of how that state was achieved.

Because entropy is a state function, the change in entropy of a system is independent of the path between its initial and final states. Note that, even though entropy is a state function, we cannot just disregard the *rev* subscript in Eq. 1 and calculate ΔS directly from q/T for any path. If we want to calculate the entropy difference between a pair of states joined by an irreversible path—such as between the initial and final states of an ideal gas that undergoes free expansion—then we can calculate that entropy difference from a *reversible* path between the same two states. The values of q and w will differ for the two paths, but ΔS for the reversible path will be the same as that for the irreversible path because entropy is a state function.

Entropy is a measure of disorder; according to the second law of thermodynamics, the entropy of an isolated system increases in any spontaneous process. Entropy is a state function.

7.3 Changes in Entropy

To understand the direction of spontaneous change, we must identify the factors that increase the entropy. Qualitatively, we can expect disorder to increase when the temperature is raised, because an increase in temperature corresponds to an increase in the thermal motion of the molecules. Heating increases the **thermal disorder,** the disorder arising from the thermal motion of the molecules. We can also expect the entropy to increase when a given amount of matter spreads into a greater volume or is mixed with another substance. These processes spread the molecules of the substance into additional locations and increase the **positional disorder,** the disorder related to the locations of the molecules.

Equation 1 allows us to express these changes quantitatively. First, let's find out how to calculate the entropy change that results when a substance is heated.

How do we do that?

To derive an expression for the change in entropy when a system is heated, we first note that Eq. 1 applies only when the temperature remains constant as heat is supplied to a system. Except in special cases, that can be true only for infinitesimal transfers of heat, so we have to break down the calculation into an infinite number of infinitesimal steps, with each step taking place at a constant but slightly different temperature. In other words, we have to use calculus. For an infinitesimal transfer dq at the temperature T, the increase in entropy is also infinitesimal; and instead of Eq. 1, we write

These special cases include any isothermal process, such as isothermal expansion of a gas, and transfers of energy at freezing or boiling points (see Section 7.4).

$$dS = \frac{dq_{rev}}{T}$$

When an infinitesimal quantity of energy dq is supplied as heat, the temperature rises by an infinitesimal amount dT. Because the rise in temperature is proportional to the energy transferred as heat (Eq. 15 of Chapter 6), we can write

$$dq = CdT$$

where C is the heat capacity of the system. If the change in temperature is carried out at constant volume, we use the constant-volume heat capacity, C_V. If the change is carried out at constant pressure, we use the constant-pressure heat capacity, C_P. This formula also applies when the transfer is reversible; so, by combining the last two equations, we can write

$$dS = \frac{CdT}{T}$$

Now suppose that the temperature of a sample is increased from T_1 to T_2. The overall change in entropy is the sum (integral) of all these infinitesimal changes:

$$\Delta S = \int_{T_1}^{T_2} \frac{CdT}{T} = C\int_{T_1}^{T_2} \frac{dT}{T} = C \ln \frac{T_2}{T_1} \qquad (2a)$$

We have assumed that the heat capacity is independent of temperature in the temperature range of interest, so C could be taken outside the integral. To evaluate the integral, we have used the standard result in Section 6.7.

When the pressure is kept constant, the entropy change that occurs when the temperature of a substance is increased from T_1 to T_2 is given by Eq. 2a, using C_P, the heat capacity at constant pressure:

$$\Delta S = C_P \ln \frac{T_2}{T_1} \qquad (2b)$$

Note that if T_2 is higher than T_1, then ΔS is positive, corresponding to the expected increase in entropy as the temperature is raised. Equation 2 applies to all types of substances and states of matter, provided the heat capacity is constant over the temperature range of interest.

EXAMPLE 7.1 Calculating the entropy change due to an increase in temperature

The temperature of a sample of nitrogen gas of volume 20.0 L at 5.00 kPa is increased from 20°C to 400°C at constant volume. What is the change in the entropy of the nitrogen? The molar heat capacity of nitrogen at constant volume, $C_{V,m}$, is 20.81 $J \cdot K^{-1} \cdot mol^{-1}$. Assume ideal behavior.

STRATEGY We expect a positive entropy change because the disorder in a system increases as the temperature is raised. We use Eq. 2a, with the heat capacity at constant volume, C_V. Find the amount (in moles) of gas molecules by using the ideal gas law, $PV = nRT$, and the initial conditions;

remember to express temperature in kelvins. Because the data are liters and kilopascals, use R expressed in those units. As always, avoid rounding errors by delaying the numerical calculation to the last possible stage.

SOLUTION The initial temperature is 293 K; the amount of N_2 is

$$\begin{aligned} n &= \frac{PV}{RT} \\ &= \frac{(5.00 \text{ kPa}) \times (20.0 \text{ L})}{(8.314\,51 \text{ L·kPa·K}^{-1}\text{·mol}^{-1}) \times (293 \text{ K})} \\ &= \frac{5.00 \times 20.0}{8.314\,51 \times 293} \text{ mol} \end{aligned}$$

It then follows from Eq. 2a that, when the temperature is raised from 293 K to 673 K, the change in entropy is

$$\begin{aligned} \Delta S &= C_V \ln \frac{T_2}{T_1} = nC_{V,\text{m}} \ln \frac{T_2}{T_1} \\ &= \left(\frac{5.00 \times 20.0}{8.314\,51 \times 293} \text{ mol}\right) \times (20.81 \text{ J·K}^{-1}\text{·mol}^{-1}) \times \ln \frac{673 \text{ K}}{293 \text{ K}} \\ &= +0.710 \text{ J·K}^{-1} \end{aligned}$$

Self-Test 7.2A The temperature of 1.00 mol He(g) is increased from 25°C to 300°C at constant pressure. What is the change in the entropy of the helium? Assume ideal behavior and use $C_{P,\text{m}} = \frac{5}{2}R$.

[*Answer:* $+13.6$ J·K^{-1}]

Self-Test 7.2B The temperature of 5.5 g of stainless steel is increased from 20°C to 100°C. What is the change in the entropy of the stainless steel? The specific heat capacity of stainless steel is 0.51 J·(°C)$^{-1}$·g^{-1}.

Some spontaneous changes are accompanied by a change in volume, so we also need to know how entropy depends on volume. Using our intuition about disorder once again, we can expect the entropy to increase as the volume occupied by a sample is increased. We can use Eq. 1 to turn this intuitive idea into a quantitative expression of the entropy change for the isothermal expansion of an ideal gas.

We choose an isothermal expansion because both temperature and volume affect the entropy of a substance and at this stage we do not want to have to consider changes in both temperature and volume.

How do we do that?

To derive an expression for the dependence of the entropy of an ideal gas on its volume, we note that, because the expansion is isothermal, implying that T is constant, we can use Eq. 1 directly. However, the formula requires q for a reversible, isothermal expansion. We find the value of q_{rev} from the first law, $\Delta U = q + w$. We know (Section 6.3) that $\Delta U = 0$ for the expansion of an ideal gas at constant temperature, so we can conclude that $q + w = 0$ and therefore that $q = -w$. The same relation applies if the change is carried out reversibly, so we can write $q_{\text{rev}} = -w_{\text{rev}}$. Therefore, to calculate q_{rev}, we calculate the work done when an ideal gas expands reversibly and isothermally, and then change the sign. For that calculation, we can use the following expression, which we derived in Section 6.7:

$$w_{rev} = -nRT \ln \frac{V_2}{V_1}$$

It follows that

$$\Delta S = \frac{q_{rev}}{T} = -\frac{w_{rev}}{T} = nR \ln \frac{V_2}{V_1} \qquad (3)$$

where n is the amount of gas molecules (in moles) and R is the gas constant (in joules per kelvin per mole). As expected, when the final volume is greater than the initial volume ($V_2 > V_1$), the change in entropy is positive, corresponding to an increase in entropy.

How can we use this information . . . to calculate the change in entropy when an ideal gas expands isothermally?

Suppose 1.00 mol N_2(g) expands isothermally from 22.0 L to 44.0 L. The change in entropy is

$$\Delta S = (1.00\ \text{mol}) \times (8.314\ 51\ \text{J·K}^{-1}\text{·mol}^{-1}) \times \ln \frac{44.0\ \text{L}}{22.0\ \text{L}}$$
$$= +5.76\ \text{J·K}^{-1}$$

As expected, the entropy increases as the space available to the gas expands.

Self-Test 7.3A Calculate the change in molar entropy (the entropy per mole) of an ideal gas when it is compressed isothermally to one-half its initial volume.

[*Answer:* $-5.76\ \text{J·K}^{-1}\text{·mol}^{-1}$]

Self-Test 7.3B Calculate the change in molar entropy of carbon dioxide that is allowed to expand isothermally to 10 times its initial volume (treat the carbon dioxide as an ideal gas).

Some reversible changes are accompanied by a change in pressure, so we also need to know how entropy depends on pressure. The simplest and most important case is once again that of an ideal gas undergoing isothermal expansion. If we know the initial and final pressures, we can use the ideal gas law, specifically Boyle's law, to express the ratio of volumes in Eq. 3 in terms of the ratio of the initial and final pressures. Because pressure is inversely proportional to volume (Boyle's law), we know that at constant temperature

$$\frac{V_2}{V_1} = \frac{P_1}{P_2}$$

Therefore, it follows from Eq. 3 that the change in entropy at constant temperature is

$$\Delta S = nR \ln \frac{P_1}{P_2} \qquad (4)$$

where P_1 is the initial pressure and P_2 is the final pressure.

The entropy of a system increases when its temperature increases and during an isothermal expansion. Equations 2, 3, and 4 are used to calculate the value of ΔS for these processes.

How can we use this information . . . to calculate the change in entropy when the pressure of an ideal gas is changed?

To calculate the change in entropy when the pressure of 0.321 mol $O_2(g)$ is increased from 0.300 atm to 12.00 atm at a constant temperature of 298 K, we use Eq. 4 and substitute the data:

$$\begin{aligned}\Delta S &= nR \ln \frac{P_1}{P_2} \\ &= (0.321 \text{ mol}) \times (8.314\ 51 \text{ J·K}^{-1}\text{·mol}^{-1}) \times \ln \frac{0.300 \text{ atm}}{12.00 \text{ atm}} \\ &= -9.85 \text{ J·K}^{-1}\end{aligned}$$

The entropy decreases, as expected for a sample that has been compressed at constant temperature.

Self-Test 7.4A Calculate the change in entropy when the pressure of 1.50 mol Ne(g) is increased isothermally from 5.00 bar to 20.00 bar. Assume ideal behavior.

[*Answer:* -17.3 J·K^{-1}]

Self-Test 7.4B Calculate the change in entropy when the pressure of 70.9 g of chlorine gas is increased isothermally from 3.00 kPa to 24.00 kPa. Assume ideal behavior.

7.4 Entropy Changes Accompanying Changes of Physical State

The simplest types of spontaneous changes are physical changes, such as melting and vaporizing, because the identity of the substance remains the same. The entropy changes accompanying changes of state are easy to explain in terms of the rearrangement of molecules. We can expect the entropy of a substance to increase when the solid form melts and its molecules become more disordered. Similarly, we can expect an even greater increase in entropy when the liquid form vaporizes, for its molecules then occupy a much greater volume and their motion is highly chaotic.

To calculate the entropy change for a substance undergoing a transition from one phase to another, we need to note three facts:

1 At the transition temperature (such as the boiling point for vaporization), the temperature of the substance remains constant as heat is supplied.

All the energy supplied is used to drive the phase transition, such as converting liquid to vapor, rather than to raise the temperature. The T in the denominator of Eq. 1 is therefore a constant and may be set equal to the transition temperature (in kelvins), such as the **normal melting point,** T_f, the temperature at which a solid melts when the pressure is 1 atm, or the **normal boiling point,** T_b, the temperature at which a liquid boils when the pressure is 1 atm.

2 At the temperature of a phase transition, the transfer of heat is reversible.

Provided the external pressure is fixed (at 1 atm, for instance), raising the temperature of the surroundings an infinitesimal amount results in complete vaporization and lowering the temperature causes complete condensation.

3 Because the transition takes place at constant pressure (for instance, 1 atm), we can identify the heat supplied with the change in enthalpy of the substance (Section 6.9).

Now we bring these three points together to derive an expression for the entropy of a phase change in terms of its enthalpy and temperature.

The **entropy of vaporization,** ΔS_{vap}, is the change in entropy per mole of molecules when a substance changes from a liquid to a vapor. The heat required to vaporize the liquid at constant pressure is the enthalpy of vaporization (ΔH_{vap}, Section 6.12). It then follows from Eq. 1, by setting $q_{rev} = \Delta H_{vap}$, that the entropy of vaporization at the boiling temperature is

$$\Delta S_{vap} = \frac{\Delta H_{vap}}{T_b} \qquad (5)$$

where T_b is the boiling temperature. When the liquid and the vapor are in their standard states (both pure, both at 1 bar), we refer to the **standard entropy of vaporization** and write it $\Delta S_{vap}°$. All entropies of vaporization are positive, so they are normally reported without their positive sign.

Strictly, the boiling temperature should be for 1 bar; but the normal boiling point, defined for 1 atm, is commonly used and makes negligible difference.

How can we use this information . . . to calculate the standard entropy of vaporization of a liquid?

To calculate the standard entropy of vaporization of acetone at its boiling point of 329.4 K (corresponding to 56.2°C), we note from Table 6.3 that its standard enthalpy of vaporization at its boiling point is 29.1 $kJ{\cdot}mol^{-1}$. Therefore, from Eq. 5, its standard entropy of vaporization at its boiling point is

$$\Delta S_{vap}{}^{\circ} = \frac{2.91 \times 10^{4}\ J{\cdot}mol^{-1}}{329.4\ K} = 88.3\ J{\cdot}K^{-1}{\cdot}mol^{-1}$$

Self-Test 7.5A Calculate the standard entropy of vaporization of argon at its boiling point (see Table 6.3).

[*Answer:* 74 $J{\cdot}K^{-1}{\cdot}mol^{-1}$]

Self-Test 7.5B Calculate the standard entropy of vaporization of water at its boiling point (see Table 6.3).

Table 7.1 lists the standard entropies of vaporization of a number of liquids. These and other data show a striking trend: many values are close to 85 $J{\cdot}K^{-1}{\cdot}mol^{-1}$. This observation is called **Trouton's rule.** The explanation of Trouton's rule is that approximately the same increase in disorder occurs when any liquid is converted into vapor, so we can expect the change in entropy to be the same. Substantial departures from Trouton's rule normally indicate that the molecules have a more orderly arrangement in the liquid than is typical of most liquids, so there is a greater increase in disorder when such liquids vaporize. The molecules in water are in a quite orderly arrangement as a result of the hydrogen bonds they form. Liquids such as water that can take part in hydrogen bonding typically have entropies of vaporization greater than 85 $J{\cdot}K^{-1}{\cdot}mol^{-1}$.

Self-Test 7.6A Use Trouton's rule to estimate the enthalpy of vaporization of liquid bromine, which boils at 59°C.

[*Answer:* 28 $kJ{\cdot}mol^{-1}$ (experimentally, 29.45 $kJ{\cdot}mol^{-1}$)]

Self-Test 7.6B Use Trouton's rule to estimate the enthalpy of vaporization of diethyl ether, $C_4H_{10}O$, which boils at 34.5°C.

TABLE 7.1 Standard Entropy of Vaporization at the Boiling Point

Liquid	Boiling point, K	$\Delta S_{vap}{}^{\circ}$, $J{\cdot}K^{-1}{\cdot}mol^{-1}$
acetone	329.4	88.3
ammonia	239.7	97.6
argon	87.3	74
benzene	353.2	87.2
ethanol	351.5	124
helium	4.22	20
mercury	629.7	94.2
methane	111.7	73
methanol	337.8	105
water	373.2	109

A smaller increase in entropy occurs when solids melt, because a liquid is only slightly more disordered than a solid (Fig. 7.3). By applying the same argument used for vaporization to the standard entropy of fusion of a substance at its melting (or freezing) point, we can write

$$\Delta S_{\text{fus}}{}^{\circ} = \frac{\Delta H_{\text{fus}}{}^{\circ}}{T_{\text{f}}} \qquad (6)$$

Here, $\Delta H_{\text{fus}}{}^{\circ}$ is the enthalpy of fusion and T_{f} is the melting point. All entropies of fusion are positive, so they are normally reported without their positive sign.

The entropy of a substance increases both when it melts and when it vaporizes.

How can we use this information . . . to calculate the standard entropy of fusion of a solid?

To calculate the standard entropy of fusion of water at its melting point, we note (from Table 6.3) that the enthalpy of fusion is 6.01 $\text{kJ}\cdot\text{mol}^{-1}$ and the melting temperature is 273.15 K; then, from Eq. 6,

$$\Delta S_{\text{fus}}{}^{\circ} = \frac{6.01 \times 10^{3}\ \text{J}\cdot\text{mol}^{-1}}{273.15\ \text{K}} = 22.0\ \text{J}\cdot\text{K}^{-1}\cdot\text{mol}^{-1}$$

Self-Test 7.7A Calculate the standard entropy of fusion of mercury at its freezing point (see Table 6.3).

[*Answer:* $9.782\ \text{J}\cdot\text{K}^{-1}\cdot\text{mol}^{-1}$]

Self-Test 7.7B Calculate the standard entropy of freezing of benzene at its freezing point (see Table 6.3).

View the increasing disorder as a solid melts.

Media Link Animation: Fig. 7.3, Fusion

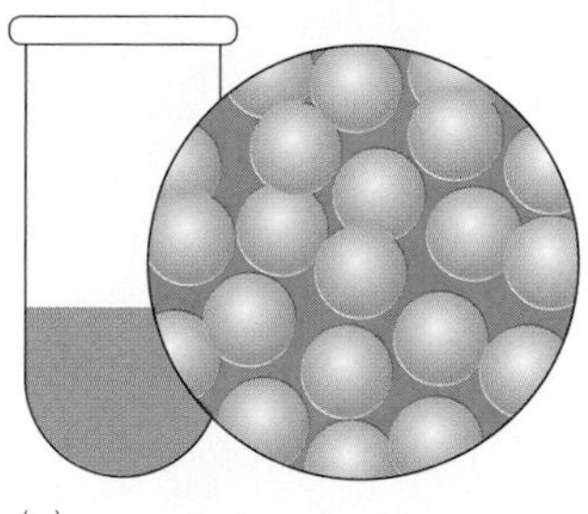

(a)

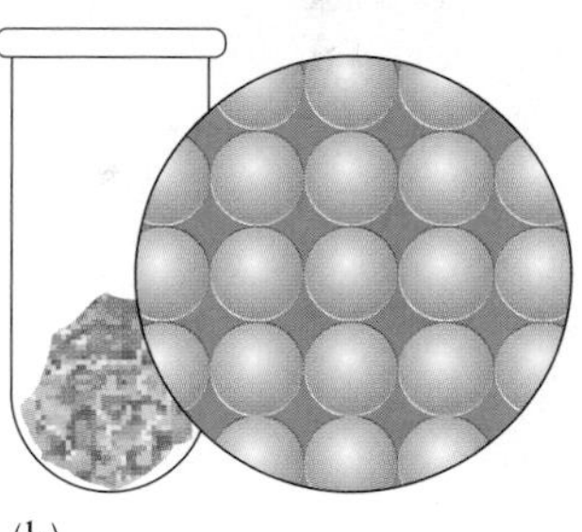

(b)

FIGURE 7.3 A representation of the arrangement of molecules in (a) a liquid and (b) a solid. When the solid melts, there is an increase in the disorder of the system and hence a rise in entropy.

7.5 A Molecular Interpretation of Entropy

We have seen that it is not possible to determine absolute values of the enthalpy and, so far, we have calculated only *changes* in the entropy of a substance. However, entropy is a measure of disorder, and it is possible to imagine a perfectly orderly state of matter with neither positional nor, at $T = 0$, thermal disorder. This idea is summarized by the **third law of thermodynamics:**

> The entropies of all perfect crystals are the same at the absolute zero of temperature.

It is conventional to take the entropy of all perfect crystals as 0 at $T = 0$. The "perfect crystal" part of this statement of the third law refers to a substance in which all the atoms are in a perfectly orderly array. As the temperature of a substance is raised above $T = 0$, more orientations become available to the molecules and the entropy of the substance increases. Thus we can expect the entropy of any substance at room temperature to be greater than 0.

As we shall see, the molecular interpretation of entropy justifies this convention.

FIGURE 7.4 Ludwig Boltzmann (1844–1906). His formula for entropy (using an earlier notation for natural logarithms) became his epitaph.

In 1877, the Austrian physicist Ludwig Boltzmann proposed an alternative definition of entropy that enables us to calculate its *absolute* value at any temperature (Fig. 7.4). He was able to show that his definition of entropy had all the characteristics of the thermodynamic definition, provided a way of calculating the entropy when measurements could not be made, and deepened our insight into the meaning of entropy at the molecular level.

The **Boltzmann formula** for the entropy is

$$S = k \ln W \tag{7}$$

We met the Boltzmann constant $k = 1.3807 \times 10^{-23}\ \mathrm{J \cdot K^{-1}}$ in Section 6.3 in connection with the equipartition theorem: it is related to the gas constant by $R = kN_A$. The quantity W is the *number of ways that the atoms or molecules in the sample can be arranged and yet still give rise to the same total energy*. Each arrangement of the molecules in a sample is a different **microstate;** so W is equal to the number of microstates that correspond to the same energy. A microstate lasts only for an instant, so when we measure the bulk properties of a system, we are measuring an average taken over the many microstates the system occupied during the measurement. The entropy calculated from the Boltzmann formula is sometimes called the **statistical entropy.**

To understand Eq. 7, imagine that we have a large number of replicas of the system we are investigating. This collection is called an **ensemble.** Each replica has the same energy but may have a different distribution of molecules over the available energy levels. We can imagine that all the replicas are placed in a box, and we select one of the replicas blindly. If $W = 1$, which is the case if there is only one way of achieving a given energy, then every replica we select will be identical. Such a system has zero entropy ($\ln 1 = 0$) and zero disorder because we can be quite sure that we will find the system in a single given microstate. However, if there is more than one way of arranging the molecules in the sample to achieve the same energy, then not every replica we draw will be the same. For instance, if a certain energy can be achieved by 1000 different ways of arranging the molecules ($W = 1000$), then there is only 1 chance in 1000 of drawing a *specific* microstate. The entropy of the system, and its disorder, is higher for this state than for that with $W = 1$.

We are now in a position to understand the statistical basis of the second law of thermodynamics. Changes occur in the direction of increased entropy because the more disordered a state, the more likely it is that the system will be found in it. That is, *spontaneous change occurs toward more probable states.*

How can we use this information . . . to calculate the entropy of a system with a specified degree of disorder?

Let's calculate the entropy of a tiny solid made up of four diatomic molecules of a binary compound such as carbon monoxide, CO. Suppose the four molecules have formed a perfectly ordered crystal in which all molecules are aligned with their C atoms on the left. Because $T = 0$, all motion has ceased (Fig. 7.5). We expect the sample to have zero entropy, because there is no disorder in either location or energy.

This value is confirmed by the Boltzmann formula: because there is only one way of arranging the molecules in the perfect crystal, $W = 1$ and $S = k \ln 1 = 0$. In this case, every member of the ensemble is identical, and we can be sure that any selection results in the same microstate. Now imagine a different arrangement in which the molecules lie with their C atoms on either side yet still have the same total energy (Fig. 7.6). The total number of ways of arranging the four molecules in this way is

$$W = 2 \times 2 \times 2 \times 2 = 2^4$$

or 16 different possible arrangements corresponding to the same total energy. Now there is only 1 chance in 16 of selecting a given microstate from the ensemble, so we can be less sure about the actual state of the system. The entropy of this tiny solid is therefore

$$\begin{aligned} S &= k \ln 2^4 = (1.3807 \times 10^{-23}\ \text{J·K}^{-1}) \times (\ln 16) \\ &= 3.8281 \times 10^{-23}\ \text{J·K}^{-1} \end{aligned}$$

As expected, the entropy of the disordered solid is higher than that of the perfectly ordered solid.

Self-Test 7.8A Calculate the entropy of a sample of a solid in which the molecules can take any one of three orientations with the same energy. Suppose there are 30 molecules in the sample.

[*Answer:* 4.5×10^{-22} J·K^{-1}]

Self-Test 7.8B Calculate the entropy of a sample of a solid in which it is supposed that a substituted benzene molecule, C_6H_5F, can take any one of six orientations with the same energy. Suppose there is 1.0 mol of molecules in the sample.

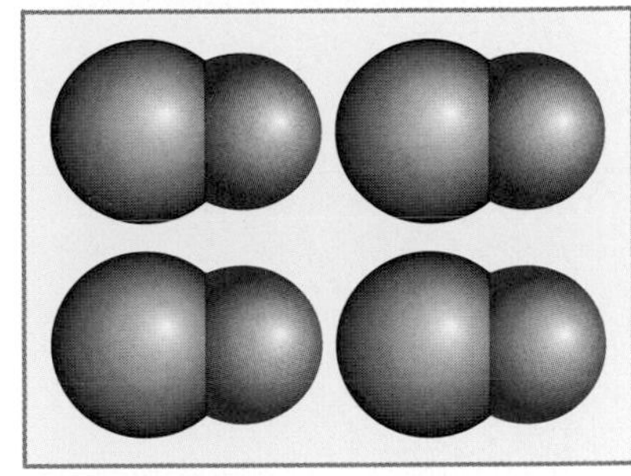

FIGURE 7.5 A tiny sample of solid carbon monoxide consisting of four molecules. When there is only one way of arranging the molecules so that all the molecules point in one direction, the entropy of the solid is 0.

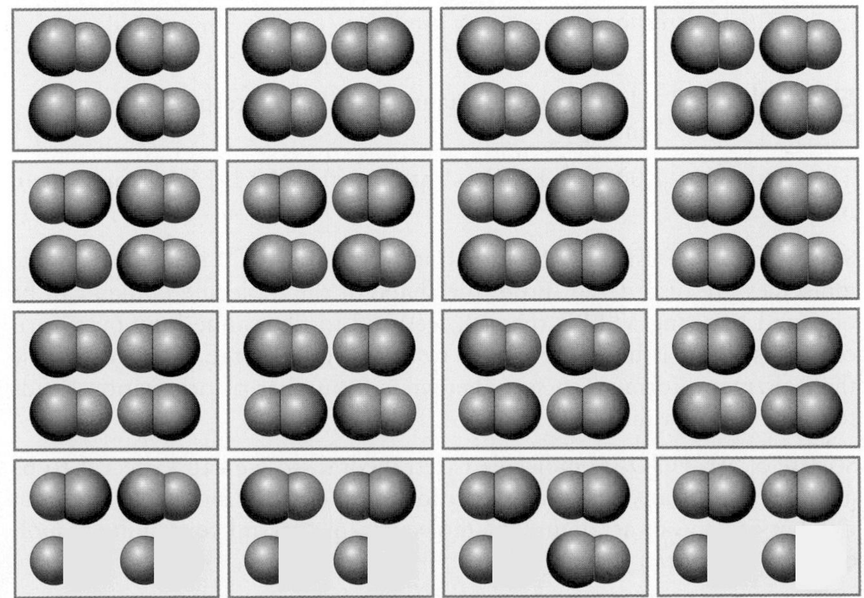

FIGURE 7.6 All 16 ways of arranging four CO molecules, assuming that each orientation is equally likely. The entropy of such a solid is greater than 0.

For a more realistic sample size than that in the example, one that contains 1.00 mol CO, corresponding to 6.02×10^{23} CO molecules, each of which could be oriented in either of two ways, there are $2^{6.02 \times 10^{23}}$ (an astronomically large number) different microstates, and a chance of only 1 in $2^{6.02 \times 10^{23}}$ of drawing a given microstate in a blind selection. We can expect the entropy of the solid to be high and calculate that

$$\begin{aligned} S &= k \ln 2^{6.02 \times 10^{23}} \\ &= (1.3807 \times 10^{23}\ \text{J}\cdot\text{K}^{-1}) \times (6.02 \times 10^{23} \ln 2) = 5.76\ \text{J}\cdot\text{K}^{-1} \end{aligned}$$

When the entropy of 1.00 mol CO is actually measured at temperatures close to $T = 0$, the value found is 4.6 J·K^{-1}. This value, which is called the **residual entropy** of the sample, the entropy of a sample at $T = 0$ arising from molecular disorder surviving at that temperature, is close enough to 5.76 J·K^{-1} to suggest that in the crystal the molecules are indeed arranged nearly randomly. The reason for this randomness is that the electric dipole moment of a CO molecule is very small, so there is little energy advantage in the molecules lying head-to-tail, head-to-head, or tail-to-tail; thus the molecules lie in either direction. For solid HCl, the same experimental measurements give $S \approx 0$ close to $T = 0$. This value indicates that the bigger dipole moments of the HCl molecules pull them into an orderly arrangement, so they lie strictly head-to-tail.

As we see from these examples, the Boltzmann formula justifies the convention adopted for the thermodynamic entropy that $S(0) = 0$. The third law simply implies that all perfect crystals have the same entropy: convention and convenience then suggest that we set that common entropy to 0. However, the "statistical" approach sets $W = 1$ for a perfect crystal (no spatial disorder) at $T = 0$ (no thermal disorder). Therefore, according to Eq. 7, $S(0) = 0$ for all perfect crystals.

The Boltzmann formula interprets the entropy of a substance in terms of the number of arrangements of molecules that result in the same energy: when many energy levels are accessible, this number and the corresponding entropy are large.

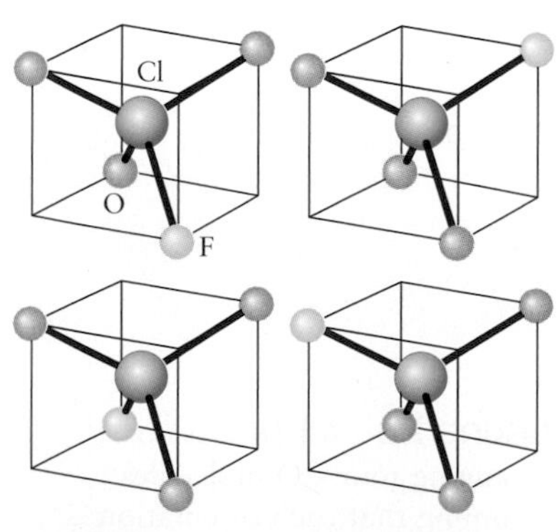

FIGURE 7.7 The four possible orientations of a tetrahedral $FClO_3$ molecule in a solid.

EXAMPLE 7.2 Using the Boltzmann formula to interpret a residual entropy

The entropy of 1.00 mol $FClO_3(s)$ at $T = 0$ is 10.1 J·K^{-1}. Suggest an interpretation.

STRATEGY The existence of residual entropy at $T = 0$ suggests that the molecules are disordered. From the shape of the molecule (which can be obtained by using VSEPR theory), we need to determine how many orientations, W, it is likely to be able to adopt in a crystal; then we can use the Boltzmann formula to see whether that number of orientations leads to the observed value of S.

SOLUTION An $FClO_3(s)$ molecule is tetrahedral, so we can expect it to be able to take up any of four orientations in a crystal (Fig. 7.7). The total number of ways of arranging the molecules in a crystal containing N molecules is therefore

$$W = (4 \times 4 \times 4 \times \ldots \times 4)_{N\ \text{factors}} = 4^N$$

The entropy is then

$$S = k \ln 4^N = k \times (N \ln 4)$$

Because in 1.00 mol $FClO_3$ there are 6.02×10^{23} molecules,

$$S = (1.3807 \times 10^{-23}\ J{\cdot}K^{-1}) \times (6.02 \times 10^{23} \times \ln 4) = 11.5\ J{\cdot}K^{-1}$$

This value is reasonably close to the experimental value of $10.1\ J{\cdot}K^{-1}$, a result suggesting that, at $T = 0$, the molecules lie randomly in any of four possible orientations.

Self-Test 7.9A Explain the observation that, at $T = 0$, the entropy of 1 mol $N_2O(s)$ is $6\ J{\cdot}K^{-1}$.

[*Answer:* In the crystal, the orientations NNO and ONN are equally likely.]

Self-Test 7.9B Suggest a reason why the entropy of ice is nonzero at $T = 0$; think about the hydrogen bonds.

7.6 The Equivalence of Statistical and Thermodynamic Entropies

Although the Boltzmann formula for the "statistical" entropy, Eq. 7, and the formula for the "thermodynamic" entropy, Eq. 1, look completely different, they refer to the same property. To verify their equivalence, we need to show that the entropy changes predicted by Boltzmann's formula are the same as those deduced from thermodynamics by using Eq. 1. First, we show that the Boltzmann formula predicts the volume-dependence of the entropy of an ideal gas (Eq. 3).

We regard the gas as consisting of a large number of molecules that occupy the energy levels characteristic of a particle in a box (Fig. 7.8).

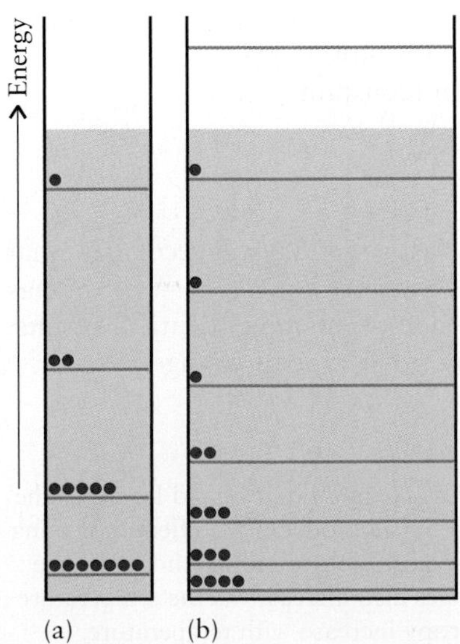

FIGURE 7.8 The energy levels of a particle in a box become closer together as the length of the box is increased. As a result, the number of levels accessible to the particles in the box increases, and the entropy of the system increases accordingly. The range of thermally accessible levels is shown by the tinted band. The change from (a) to (b) is a model of the isothermal expansion of an ideal gas. The total energy of the particles is the same in each case.

TABLE 7.2 Standard Molar Entropy of Water at Various Temperatures

Phase	Temperature, °C	$S_m°$, $J·K^{-1}·mol^{-1}$
solid	−273 (0 K)	3.4
	0	43.2
liquid	0	65.2
	20	69.6
	50	75.3
	100	86.8
vapor	100	196.9
	200	204.1

This analysis can also be applied to a gas in a three-dimensional container, but for simplicity we consider here a one-dimensional box like that treated in Section 1.5. At $T = 0$, only the lowest energy level is occupied, so $W = 1$ and the entropy is 0. At any temperature above $T = 0$, the molecules occupy large numbers of energy levels like those shown in Fig. 7.8, W is larger than 1, and the entropy (which is proportional to ln W) is greater than 0.

When the length of the box is increased (corresponding to an increase in volume in three dimensions) at constant temperature, more energy levels become accessible to the molecules because the levels now lie closer together. Therefore, the value of W increases as the box is lengthened and, by the Boltzmann formula, the entropy increases, too. The same argument applies to a three-dimensional box: as the volume of the box increases, the number of accessible states increases, too. The logarithm of the volume in Eq. 3 can now be seen to stem from the logarithm of W in the Boltzmann formula.

We can also use the Boltzmann formula to interpret the increase in entropy of a substance as its temperature is raised (Eq. 2 and Table 7.2). We use the same particle-in-a-box model of a gas, but this reasoning also applies to liquids and solids, even though their energy levels are much more complicated. At low temperatures, the molecules of a gas can occupy only a few of the energy levels, so W is small and the entropy is low. As the temperature is raised, the molecules have access to larger numbers of energy levels (Fig. 7.9), so W rises and the entropy increases, too.

We can also show that the definitions of entropy in Eqs. 1 and 7 are quantitatively equivalent, even though they look totally different. First, we have seen that the thermodynamic entropy of an ideal gas increases when it expands isothermally (Eq. 3). If we suppose that the number of microstates available to a single molecule is proportional to the volume available to it, we can write W = constant × V. For N molecules, the number of microstates is proportional to the Nth power of the volume:

$$\begin{aligned} W &= (\text{constant} \times V) \times (\text{constant} \times V) \ldots N \text{ times} \\ &= (\text{constant} \times V)^N \end{aligned}$$

Therefore, the change in the statistical entropy when a sample expands isothermally from volume V_1 to a volume V_2 is

$$\begin{aligned} \Delta S &= k \ln (\text{constant} \times V_2)^N - k \ln (\text{constant} \times V_1)^N \\ &= Nk \ln (\text{constant} \times V_2) - Nk \ln (\text{constant} \times V_1) \\ &= Nk \ln \frac{\text{constant} \times V_2}{\text{constant} \times V_1} = Nk \ln \frac{V_2}{V_1} \end{aligned}$$

This expression is identical to Eq. 3 as soon as we recognize that $N = nN_A$ (where N_A is the Avogadro constant) and $N_A k = R$, the gas constant. In this case, at least, the two definitions give identical results. We can show that the same is true of any other type of change.

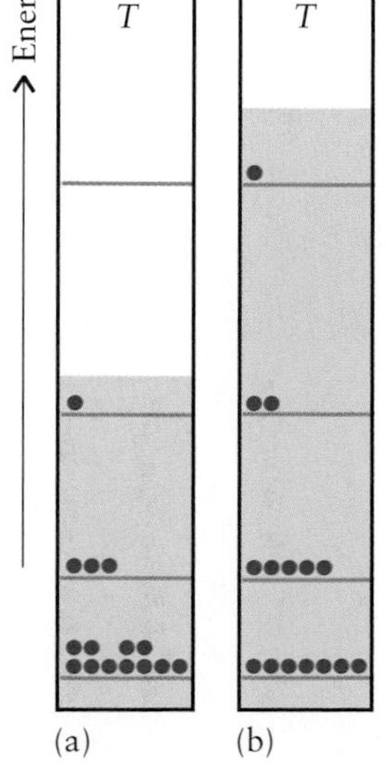

FIGURE 7.9 More energy levels become accessible in a box of fixed length as the temperature is raised. The change from (a) to (b) is a model of the effect of heating an ideal gas at constant volume. The thermally accessible levels are shown by the tinted band. The average energy of the molecules also increases as the temperature is raised: that is, both internal energy and entropy increase with temperature.

How do we do that?

To show in general that the statistical and thermodynamic definitions of entropy are essentially the same, we note that the number of microstates accessible by the system depends on only its current state, not its past history. Therefore W, and hence the statistical entropy, is a state function (like the thermodynamic entropy). Second, if we double the number of molecules in the system, then the number of microstates increases as the second power, and the entropy changes from $k \ln W$ to $k \ln W^2$, or $2k \ln W$. Therefore, the statistical entropy is an extensive property (like the thermodynamic entropy). Next, in any irreversible change, the system and its surroundings become more disordered, which means that the number of microstates increases. If W increases, then so does $\ln W$, and the statistical entropy increases, too. Finally, if the temperature of the system increases, then more microstates become accessible, so the statistical entropy increases, just as the thermodynamic entropy does.

The equations used to calculate changes in the statistical entropy and the thermodynamic entropy lead to the same result. A system is more likely to be found in a condition that corresponds to more microstates.

7.7 Standard Molar Entropies

To *calculate* the entropy, we use Boltzmann's formula, but the calculations are sometimes very difficult and involve a lot of manipulation of Eq. 7. To *measure* the entropy, we use the thermodynamic definition, Eq. 1, in combination with the third law. Because the third law tells us that $S(0) = 0$ and Eq. 2 can be used to calculate the change in entropy as a substance is heated to the temperature of interest, we can write

$$S(T) = S(0) + \Delta S(\text{heating from 0 to } T) = \Delta S(\text{heating from 0 to } T)$$

The change in entropy on heating is calculated from the heat capacity and the temperature, as we did in Example 7.1. However, we cannot use Eq. 2 directly because its derivation was based on the assumption that heat capacities are constant over the temperature range of interest. It is known experimentally that heat capacities vary with temperature and fall to 0 as T approaches 0. Therefore, we have to adopt a different procedure.

How do we do that?

To take into account the possibility that the heat capacity changes with temperature, we go back to the derivation of Eq. 2. There we assumed (in the last line) that the heat capacity is constant; but here we cannot make that assumption, so we write

$$\Delta S(\text{heating from } T_1 \text{ to } T_2) = \int_{T_1}^{T_2} \frac{C\mathrm{d}T}{T}$$

We set $T_1 = 0$ and $T_2 = T$, the temperature of interest, in the limits of the integral. Heating commonly takes place at constant pressure, so we replace C by C_P:

$$\Delta S(\text{heating from 0 to } T) = \int_0^T \frac{C_P}{T}\,dT$$

Therefore, by invoking the third law as explained in the text,

$$S(T) = \int_0^T \frac{C_P}{T}\,dT \qquad (8)$$

An integral of a function—in this case, the integral of C_P/T—is the area under the graph of the function. Therefore, to measure the entropy of a substance at a temperature above 0, we need to measure the heat capacity (typically the constant-pressure heat capacity) at all temperatures from $T = 0$ up to the temperature of interest. Then the entropy of the substance is obtained by plotting C_P/T against T and measuring the area under the curve (Fig. 7.10).

If a phase transition takes place between $T = 0$ and the temperature of interest, we also have to include the corresponding entropy of transition by using Eq. 5 or 6 (Fig. 7.11). For instance, if we wanted the entropy of liquid water at 25°C, we would measure the heat capacity of ice from $T = 0$ (or as close to it as we can get), up to $T = 273.15$ K, determine the entropy of fusion at that temperature from the enthalpy of fusion, then measure the heat capacity of liquid water from $T = 273.15$ K up to $T = 298.15$ K. Table 7.3 gives selected values of the **standard molar entropy,** $S_m°$, the molar entropy of the pure substance at 1 bar. Note that all the values in the table refer to 298 K. They are all positive, which is consistent with all substances being more disordered at 298 K than at $T = 0$.

A longer list is given in Appendix 2A.

As remarked previously, in some texts you will see 1 atm used in the definition of standard states. The modern definition is 1 bar. The change in standard pressure has little effect on the tabulated values.

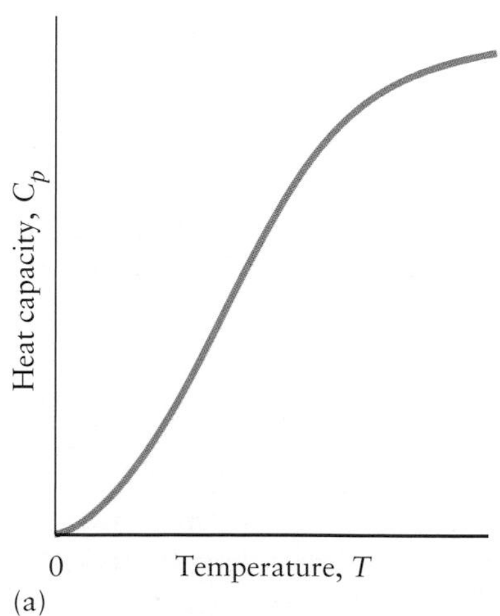

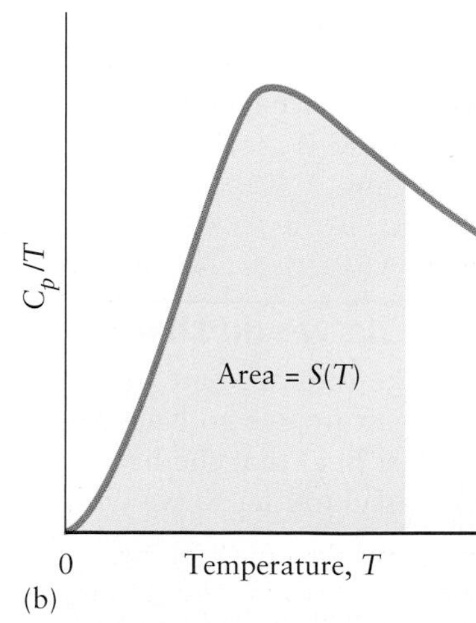

FIGURE 7.10 The two steps in the experimental determination of entropy. (a) The heat capacity (at constant pressure in this instance) of the substance is determined from close to absolute zero up to at least the temperature of interest. (b) The area under the plot of C_P/T against T is determined up to the temperature of interest. This area is the entropy of the substance at that temperature.

Media Link Living graph Fig. 7.10, The entropy from the heat capcatiy

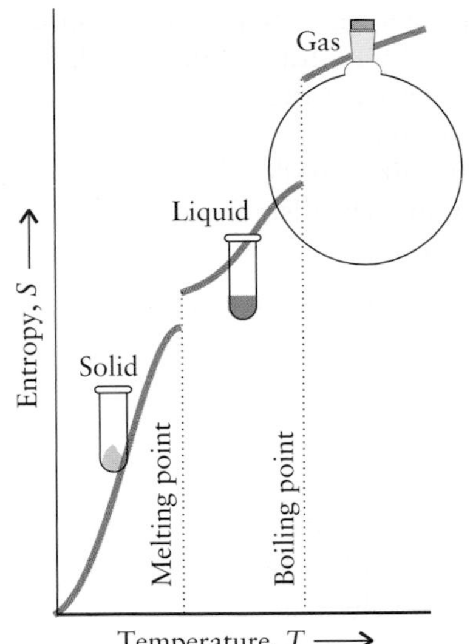

FIGURE 7.11 The entropy of a solid increases as its temperature is raised. The entropy increases sharply when the solid melts to form the more disordered liquid, and then gradually increases again up to the boiling point. A second, larger jump in entropy occurs when the liquid vaporizes.

TABLE 7.3 Standard Molar Entropies at 25°C*

Substance	$S_m°$, $J{\cdot}K^{-1}{\cdot}mol^{-1}$
Gases	
ammonia, NH_3	192.4
carbon dioxide, CO_2	213.7
hydrogen, H_2	130.7
nitrogen, N_2	191.6
oxygen, O_2	205.1
Liquids	
benzene, C_6H_6	173.3
ethanol, C_2H_5OH	160.7
water, H_2O	69.9
Solids	
calcium oxide, CaO	39.8
calcium carbonate, $CaCO_3$†	92.9
diamond, C	2.4
graphite, C	5.7
lead, Pb	64.8

* Additional values are given in Appendix 2A.
† Calcite.

We can use the molecular interpretation of entropy to understand why some substances have high molar entropies and others have low molar entropies. For example, let's compare the molar entropy of diamond, 2.4 $J{\cdot}K^{-1}{\cdot}mol^{-1}$, with the much higher value for lead, 64.8 $J{\cdot}K^{-1}{\cdot}mol^{-1}$. The low entropy of diamond is what we should expect for a solid that has rigid bonds: at room temperature, its atoms are not able to jiggle around as much as the atoms of lead, which have less directional bonds. Lead atoms are also much heavier than carbon atoms, and heavier atoms have more vibrational energy levels available than do lighter ones. This difference is easier to see for gas molecules and the particle in a box model, for the energy levels are closer together the greater the mass of the particles (Fig. 7.12), but it is also true of atoms vibrating in a solid. The larger number of accessible energy levels means that more microstates are possible for heavier atoms, so we expect molar entropy to increase with molecular mass (compare also H_2, N_2, and O_2 in Table 7.3). We also see that

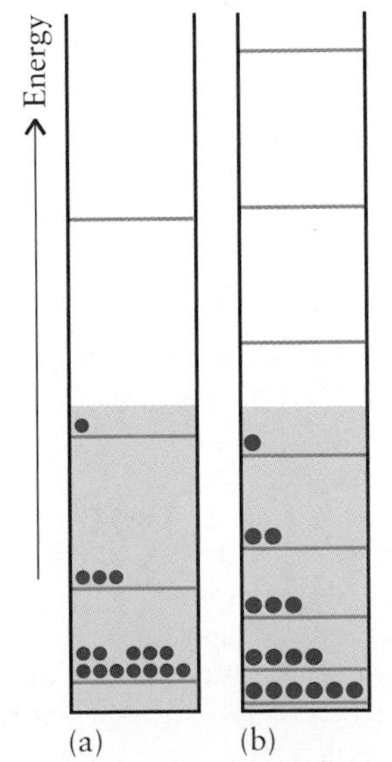

FIGURE 7.12 The energy levels of a particle in a box are more widely spaced for (a) light molecules than for (b) heavy molecules. As a result, the number of thermally accessible levels, as shown by the tinted band, is greater for heavy molecules than for light molecules at the same temperature, and the entropy of the substance with heavy molecules is correspondingly greater.

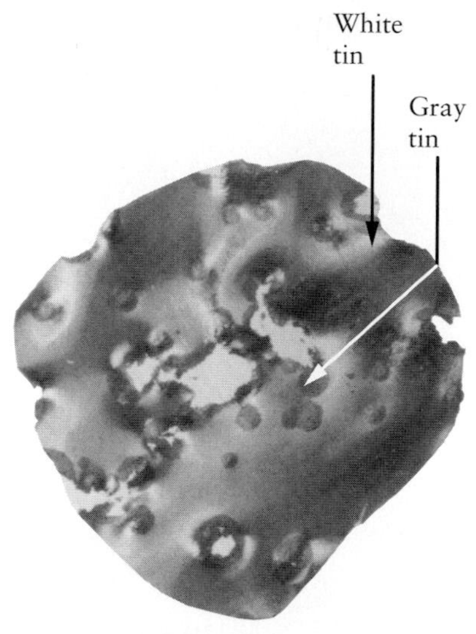

FIGURE 7.13 Gray tin and white tin are two solid forms of tin. The denser white metallic form is the more stable phase above 13°C, and the powdery gray form is more stable below that temperature.

large, complex species have higher molar entropies than smaller, simpler ones (compare $CaCO_3$ with CaO or C_2H_5OH with H_2O). Liquids have higher molar entropies than solids have because the greater freedom of movement of the molecules in a liquid results in a less ordered state of matter. The molar entropies of gases, in which molecules occupy much larger volumes and undergo almost completely disordered motion, are substantially higher than those of the corresponding liquids.

Standard molar entropies increase as the complexity of a substance increases. The standard molar entropies of gases are higher than those of comparable solids and liquids at the same temperature.

Self-Test 7.10A Which substance in each pair has the higher molar entropy: (a) 1 mol CO_2 at 25°C and 1 bar or 1 mol CO_2 at 25°C and 3 bar; (b) 1 mol He at 25°C in a 20-L container or 1 mol He at 100°C in a 20-L container; (c) $Br_2(l)$ or $Br_2(g)$? Explain your conclusions.

[*Answer:* (a) CO_2 at 1 bar, because disorder increases with volume; (b) He at 100°C, because disorder increases with temperature; (c) $Br_2(g)$, because the vapor state has greater disorder than the liquid state.]

Self-Test 7.10B Use the information in Table 7.3 or Appendix 2A to determine which allotrope is the more ordered form and predict the sign of ΔS for each transition: (a) white tin (Fig. 7.13) changes to gray tin at 25°C; (b) diamond changes to graphite at 25°C.

7.8 Standard Reaction Entropies

One of the main reasons for the importance of entropy in chemistry is that we can use it to predict the spontaneity of a reaction. However, we need to know that there are two contributions to the change in entropy during a reaction: not only does the entropy of the reaction system change as reactants form products, but the entropy of the surroundings changes as the heat produced or absorbed by the reaction enters or leaves them. We explore the contribution of the system in this section and the contribution of the surroundings in the next section.

In some cases, we can predict the sign of the entropy change of the system without doing any calculation. Because the molar entropy of a gas is so much greater than that of solids and liquids, a change in the amount of gas molecules normally dominates any other entropy change in a reaction. A net increase in the amount of gas usually results in a positive change in entropy. Conversely, a net consumption of gas usually results in a negative change. Entropy changes are much more finely balanced for reactions in which there is no net change in the amount of gas or in which gases are not involved, and for these reactions we have to use numerical data to predict the sign of a reaction entropy. We also need tables of numerical data (such as Table 7.3 and Appendix 2A) to predict the numerical value of any reaction entropy, whether or not gases are involved.

To calculate the change in entropy that accompanies a reaction, we need to know the molar entropies of all the substances involved; then, we determine the difference between the entropies of the products and the reactants. More specifically, the **standard reaction entropy,** $\Delta S_r°$, is the difference between the standard molar entropies of the products and those of the reactants, taking into account their stoichiometric coefficients:

$$\Delta S_r° = \sum nS_m°(\text{products}) - \sum nS_m°(\text{reactants}) \quad (9)$$

The first term on the right is the total standard molar entropy of the products and the second term is that of the reactants; n denotes the various stoichiometric coefficients in the chemical equation. Example 7.3 illustrates how this expression is used.

The standard reaction entropy is positive (an increase in entropy) if there is a net production of gas in a reaction; it is negative (a decrease) if there is a net consumption of gas.

EXAMPLE 7.3 Calculating the standard reaction entropy

Calculate the standard entropy for the reaction $N_2(g) + 3\ H_2(g) \rightarrow 2\ NH_3(g)$.

STRATEGY We expect a decrease in entropy because there is a net reduction in the amount of gas. To find the numerical value, we use the chemical equation to write an expression for $\Delta S_r°$, as shown in Eq. 9, and then substitute values from Table 7.3 or Appendix 2A.

SOLUTION From the chemical equation, we can write

$$\begin{aligned}\Delta S_r° &= 2S_m°(NH_3, g) - \{S_m°(N_2, g) + 3S_m°(H_2, g)\} \\ &= (2 \times 192.4\ J{\cdot}K^{-1}{\cdot}mol^{-1}) - \\ &\quad \{(191.6\ J{\cdot}K^{-1}{\cdot}mol^{-1}) + (3 \times 130.7\ J{\cdot}K^{-1}{\cdot}mol^{-1})\} \\ &= -198.9\ J{\cdot}K^{-1}{\cdot}mol^{-1}\end{aligned}$$

Because the value of $\Delta S_r°$ is negative, the product is less disordered than the reactants—in part, because it occupies a smaller volume—just as we expected.

Self-Test 7.11A Use data from Appendix 2A to calculate the standard entropy of the reaction $N_2O_4(g) \rightarrow 2\ NO_2(g)$ at 25°C.

[*Answer:* $+176.03\ J{\cdot}K^{-1}{\cdot}mol^{-1}$]

Self-Test 7.11B Use data from Appendix 2A to calculate the standard entropy of the reaction $C_2H_4(g) + H_2(g) \rightarrow C_2H_6(g)$ at 25°C.

GLOBAL CHANGES IN ENTROPY

Some processes seem to defy the second law. For example, water freezes to ice at low temperatures, and cold packs for athletic injuries spontaneously becomes ice-cold even in warm air when the ammonium nitrate inside dissolves in the water. Life itself appears to go against the second law. Every cell of a living being is organized to an extraordinary extent. Thousands of different compounds, each one having a specific function to perform, move in the intricately choreographed dance we call life. How can the molecules in our bodies form such highly organized, complex structures from slime, ooze, or gas? We are examples of systems with very low entropy. In fact, our existence seems at first sight to be a contradiction of the second law of thermodynamics.

The dilemma is resolved when we realize that the second law refers to an *isolated* system. That is, we must consider any system as part of a larger isolated system that includes the surroundings of the system of interest. If we want to determine whether a change in the system of interest is spontaneous or not, we must consider the total change in entropy, not only of the system itself, but also of the surroundings with which it can exchange energy. Only if the *total* change in entropy, the sum of the changes in the system and the surroundings, is positive is the process spontaneous. Our task, therefore, is first to see how to calculate the change in entropy of the surroundings, then to combine that change with the change in entropy of the system (which we have already seen how to calculate). Once we know the total entropy change, we shall be able to predict whether a reaction is spontaneous, whether its reverse is spontaneous, or whether neither the forward nor the reverse reaction is spontaneous and the system is at equilibrium.

7.9 The Surroundings

The actual system and its surroundings constitute the "isolated system" to which the second law refers (Fig. 7.14). We have seen how to calculate the entropy change of the system; now we turn our attention to the entropy changes in its surroundings. Only if the *total* entropy change,

$$\Delta S_{\text{tot}} = \Delta S + \Delta S_{\text{surr}} \tag{10}$$

is positive will the process be spontaneous. As is customary, properties of the system are written without subscripts, so here ΔS is the entropy change of the system and ΔS_{surr} that of the surroundings. It is critically important to understand that a process in which ΔS decreases may occur spontaneously, provided the entropy increases so much in the surroundings that ΔS_{tot} increases.

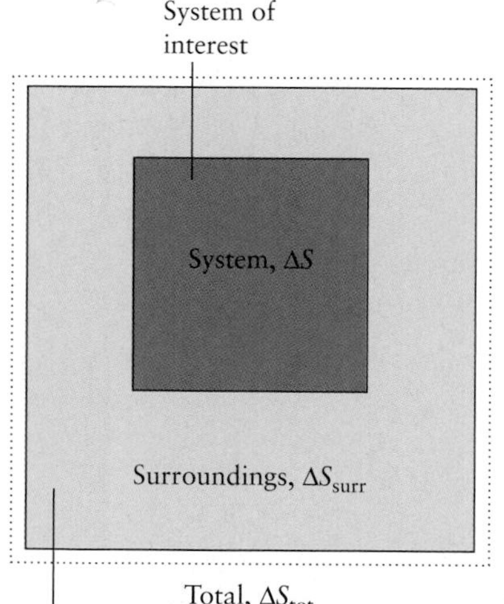

FIGURE 7.14 The global isolated system that models events taking place in the world consists of a smaller system together with its immediate surroundings. The only events that can take place spontaneously in the global isolated system are those corresponding to an increase in the total entropy of both the smaller system and its immediate surroundings.

A good example of the role of the surroundings in determining the direction of a process is the freezing of water. We can see from Table 7.2 that, at 0°C, the molar entropy of liquid water is 22.0 $\text{J·K}^{-1}\text{·mol}^{-1}$ higher than that of ice at the same temperature. This difference makes sense, because the molecules in liquid water are more disordered than in ice. It follows that, when water freezes at 0°C, its entropy *decreases* by 22.0 $\text{J·K}^{-1}\text{·mol}^{-1}$. Entropy changes do not vary much with temperature; so just below 0°C, we can expect almost the same decrease. Yet we know from everyday experience that water freezes spontaneously below 0°C. Clearly, the surroundings must be playing a deciding role: if we can show that their entropy increases by more than 22.0 $\text{J·K}^{-1}\text{·mol}^{-1}$ when water freezes, then the total entropy change will be positive.

We can expect the entropy of the surroundings to increase when water freezes because freezing is an exothermic process, and the heat released passes into the surroundings. This heat stirs up the thermal motion of the atoms in the surroundings, which increases their disorder and hence their entropy (Fig. 7.15).

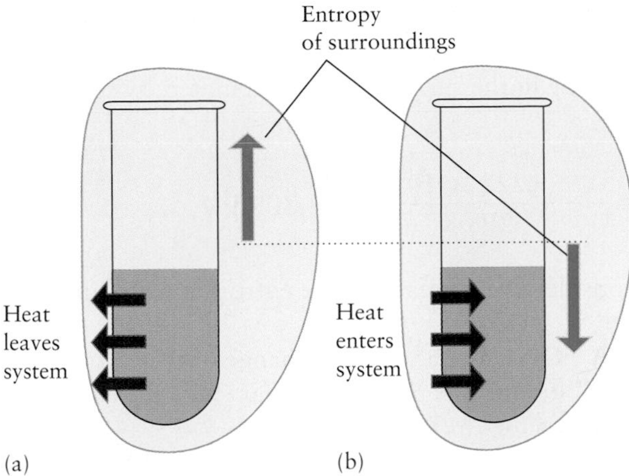

FIGURE 7.15 (a) In an exothermic process, heat escapes into the surroundings and increases their entropy. (b) In an endothermic process, the entropy of the surroundings decreases. The red arrows represent the transfer of heat between system and surroundings, and the green arrows indicate the entropy change of the surroundings.

We can use Eq. 1 to calculate the entropy change of the surroundings, provided we assume that the surroundings are so large that their temperature and pressure remain constant. If the enthalpy change of the system is ΔH, then, for heat transfers at constant pressure, $q_{\text{surr}} = -\Delta H$. We can now use Eq. 1 to write

$$\Delta S_{\text{surr}} = -\frac{\Delta H}{T} \qquad (11)$$

This important formula, which can be derived more formally from the laws of thermodynamics, applies when any change takes place at constant pressure and temperature. Notice that, for a given enthalpy change of the system (that is, a given output of heat), the entropy of the surroundings increases more if their temperature is low than if it is high (Fig. 7.16). The explanation is the "sneeze in the street" analogy mentioned in Section 7.2. Because ΔH is independent of path, Eq. 11 is applicable whether the process occurs reversibly or irreversibly.

How can we use this information . . . to calculate the change of entropy in the surroundings?

The synthesis of NH_3 is exothermic:

$$N_2(g) + 3\,H_2(g) \longrightarrow 2\,NH_3(g) \qquad \Delta H^\circ = -92.22\text{ kJ}$$

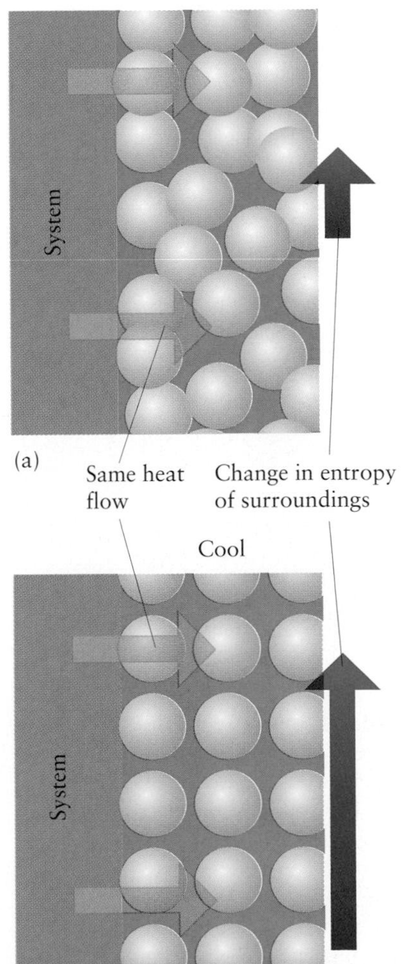

FIGURE 7.16 (a) When a given quantity of heat flows into hot surroundings, it produces very little additional chaos and the increase in entropy is quite small. (b) When the surroundings are cool, however, the same quantity of heat can make a considerable difference to the disorder, and the change in entropy of the surroundings is correspondingly large.

so we expect the entropy of the surroundings to increase as heat spreads into them. In fact, if the reaction were to go to completion at 298 K, the change in entropy in the surroundings, calculated by using Eq. 11, would be

$$\Delta S_{surr} = -\frac{(-9.222 \times 10^4 \text{ J})}{298 \text{ K}} = +309 \text{ J·K}^{-1}$$

which is a considerable increase in the entropy of the surroundings.

Self-Test 7.12A Calculate the entropy change of the surroundings when 1.00 mol $H_2O(l)$ vaporizes at 90°C and 1 bar. Take the enthalpy of vaporization of water as 40.7 kJ·mol^{-1}.

[*Answer:* -112 J·K^{-1}]

Self-Test 7.12B Calculate the entropy change in the surroundings when 1.00 mol $NH_3(g)$ condenses to a liquid at -30°C. Take the enthalpy of vaporization of ammonia as 23.4 kJ·mol^{-1}.

The entropy change of the surroundings due to a process taking place at constant pressure and temperature is equal to $-\Delta H/T$.

7.10 The Overall Change in Entropy

As we have already emphasized, *to use the entropy to judge the direction of spontaneous change, we must consider changes in the entropy of the system plus the entropy change in the surroundings.* Then, if the total entropy change is positive (an increase), the process is spontaneous. If the total entropy change is negative (a decrease), then the reverse process is spontaneous.

In an exothermic reaction, such as the synthesis of ammonia or a combustion reaction, the heat released by the reaction increases the disorder of the surroundings. In some cases, the entropy of the system may decrease, as in the combustion of magnesium, where a gas is converted to a solid:

$$2\ \text{Mg(s)} + \text{O}_2\text{(g)} \longrightarrow 2\ \text{MgO(s)} \qquad \Delta S° = -217 \text{ J·K}^{-1} \qquad \Delta H° = -1202 \text{ kJ}$$

However, provided the reaction is strongly exothermic, the increased disorder of the surroundings is so great that overall the total entropy increases.

How can we use this information . . . to calculate the total change in entropy for a process?

Let's assess whether the combustion of magnesium is spontaneous at 25°C by calculating the total change in entropy for the reaction. The change in entropy of the system is the standard reaction entropy, -217 J·K^{-1}, for the combustion of 2 mol Mg(s) (as shown in the preceding equation) and $\Delta H° = -1202$ kJ. The change in entropy of the surroundings is

$$\Delta S_{surr} = -\frac{\Delta H°}{T} = -\frac{(-1.202 \times 10^6 \text{ J})}{298 \text{ K}} = +4.03 \times 10^3 \text{ J·K}^{-1}$$

The total change is

$$\Delta S_{tot} = -217\ J{\cdot}K^{-1} + (4.03 \times 10^3\ J{\cdot}K^{-1}) = +3.81 \times 10^3\ J{\cdot}K^{-1}$$

Which is positive, so the reaction is spontaneous.

Self-Test 7.13A Is the formation of hydrogen fluoride from its elements in their most stable forms spontaneous at 25°C?

$$H_2(g) + F_2(g) \longrightarrow 2\ HF(g) \qquad \Delta H° = -542.2\ kJ \qquad \Delta S° = +14.1\ J{\cdot}K^{-1}$$

[*Answer:* $\Delta S_{surr}° = +1936\ J{\cdot}K^{-1}$; therefore, $\Delta S_{tot}° = +1950\ J{\cdot}K^{-1}$; spontaneous]

Self-Test 7.13B Is the formation of benzene from its elements in their most stable forms spontaneous at 25°C?

$$6\ C(s) + 3\ H_2(g) \longrightarrow C_6H_6(l) \qquad \Delta H° = +49.0\ kJ \qquad \Delta S° = -253.18\ J{\cdot}K^{-1}$$

The highly positive ΔS_{tot} for the burning of magnesium shows why exothermic reactions are so common. Provided ΔH has a large numerical value, the release of energy as heat into the surroundings increases their entropy so much that it dominates the overall change in entropy. So, even if the entropy of the system decreases (as it does in the magnesium combustion), *overall* the entropy increases and the reaction is spontaneous (Fig. 7.17).

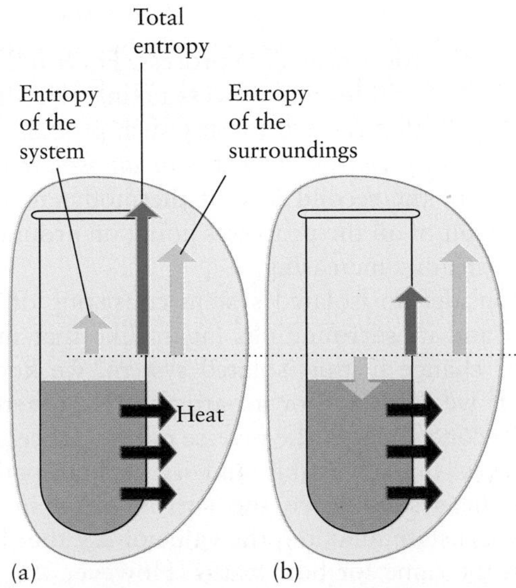

FIGURE 7.17 In an exothermic reaction, (a) the overall entropy change is certainly positive when the entropy of the system increases. (b) The overall entropy change may also be positive when the entropy of the system decreases. The reaction is spontaneous in both cases.

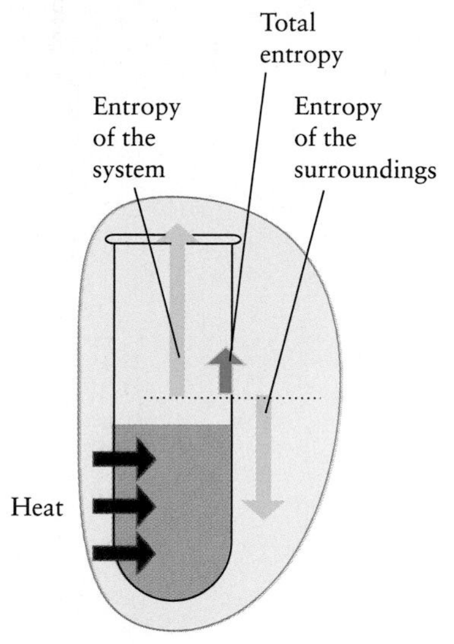

FIGURE 7.18 An endothermic reaction is spontaneous only if the entropy of the system increases enough to overcome the decrease in entropy of the surroundings.

Spontaneous endothermic reactions were a puzzle for nineteenth-century chemists, who believed that reactions ran only in the direction of decreasing energy of the system. It seemed to them that, in endothermic reactions, reactants were spontaneously rising to higher energies, like a weight suddenly leaping up from the floor on to a table. However, *the criterion for spontaneity is an increase in the total entropy, not a decrease in the energy of the system.* In an endothermic reaction, the entropy of the surroundings certainly decreases as heat flows from them into the system. Nevertheless, there can still be an overall increase in entropy if the disorder of the system increases enough. Every endothermic reaction must be accompanied by increased disorder of the system if it is to be spontaneous (Fig. 7.18), but this constraint does not apply to exothermic reactions.

By considering the total energy change, we can draw some far-reaching conclusions about processes going on in the universe. For instance, we saw in Section 6.7 that the maximum expansion work is achieved if the expansion is carried out reversibly, by matching the external pressure to the pressure of the system at every stage. That relation is always true: *a process produces maximum work if it takes place reversibly*, so w_{rev} is more negative than w_{irrev}. However, because the internal energy is a state function, ΔU is the same for any path between the same two states. Therefore, because $\Delta U = q + w$, it follows that q_{rev}, the heat absorbed along the reversible path, must be more positive than q_{irrev}, the heat absorbed along any other path, for only then can the sums of q and w be the same for the two types of processes. If now we replace q_{rev} in Eq. 1 by the smaller quantity q_{irrev}, we get a quantity q_{irrev}/T that is smaller than q_{rev}/T and therefore smaller than ΔS. That is, in general,

$$\Delta S \geq \frac{q}{T} \qquad (12)$$

with the equality valid for a reversible process. For a fully isolated system like that in Fig. 7.14, $q = 0$ for any process taking place inside the system. It follows from Eq. 12 that $\Delta S \geq 0$ for any such process. That is, we have shown that *the entropy cannot decrease in an isolated system.* This is another statement of the second law of thermodynamics. It tells us, in effect, that as a result of all the processes going on around us, the entropy of the universe is steadily increasing.

Now let's consider an isolated system consisting of both the system that interests us and its surroundings (again like that in Fig. 7.14). For any spontaneous change in this isolated system, we know from Eq. 12 that $\Delta S_{tot} > 0$. If we calculate for a particular hypothetical process that $\Delta S_{tot} < 0$, we can conclude that the reverse of that process is spontaneous.

A final point is that reversible and irreversible paths between two given states of the system leave the surroundings in different states. Because entropy is a state function, the value of ΔS, the change in entropy of the system, is the same for both paths. However, ΔS_{tot} is different for the two paths because the entropy of the surroundings changes by a different amount in each case. For example, the isothermal expansion of an ideal gas always results in the change in entropy of the system given by Eq. 3. However, the change in entropy of the surroundings is different for

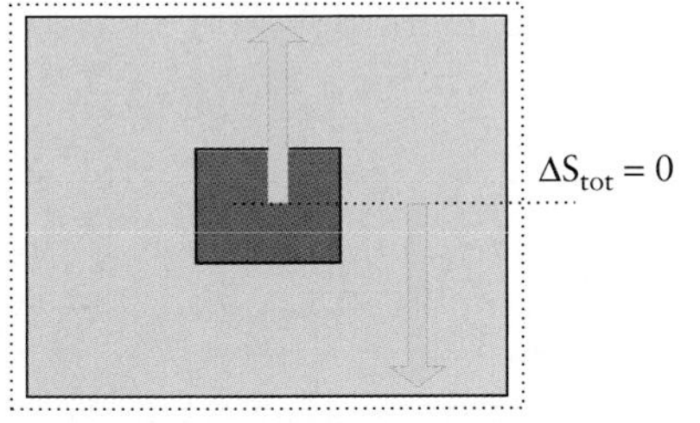

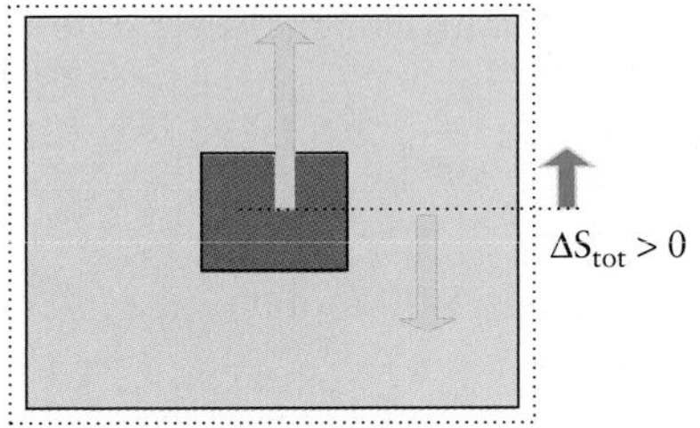

(a) Reversible process (b) Irreversible process

FIGURE 7.19 (a) When a process takes place reversibly—that is, when the system is at equilibrium with its surroundings—the change in entropy of the surroundings is the negative of the change in entropy of the system and the overall change is 0. (b) For an irreversible change between the same two states of the system, the final state of the surroundings is different from the final state in (a), and the change of the system is not canceled by the entropy change of the surroundings. Overall, there is an increase in entropy.

reversible and irreversible paths because the surroundings are left in different states in each case (Fig. 7.19). The characteristics of reversible and irreversible processes are listed in Table 7.4.

A process is spontaneous if it is accompanied by an increase in the total entropy of the system and the surroundings.

EXAMPLE 7.4 Calculating the total entropy change for a process

Calculate ΔS, ΔS_{surr}, and ΔS_{tot} for (a) the isothermal, reversible expansion and (b) the isothermal, irreversible free expansion (expansion against zero pressure) of 1.00 mol of ideal gas molecules from 8.00 L to 20.00 L atm at 292 K. Explain any differences between the two paths.

STRATEGY Because entropy is a state function, the change in entropy of the system is the same regardless of the path between two states, so we can use Eq. 3 to calculate ΔS for (a) and (b). For the entropy of the surroundings, we need to find the heat transferred to the surroundings. In each case, we can combine the fact that $\Delta U = 0$ for an isothermal expansion of an ideal gas with $\Delta U = w + q$ and conclude that $q = -w$. We then use Eq. 12 from Chapter 6 to calculate the work done in an isothermal, reversible expansion of an ideal gas and Eq. 12 from this chapter to find the sum. The changes we calculate are summarized in Fig. 7.20.

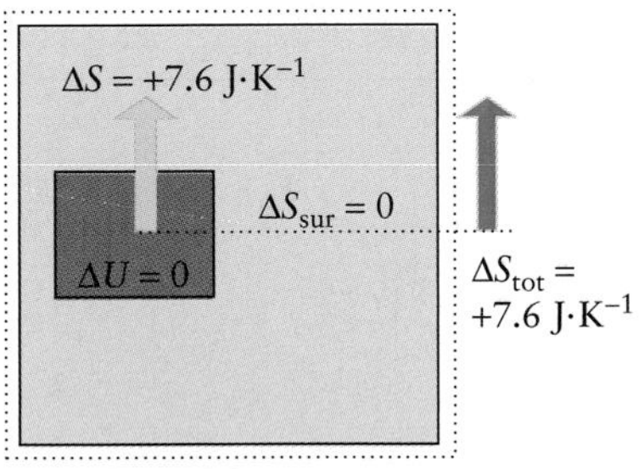

(a) Spontaneous process

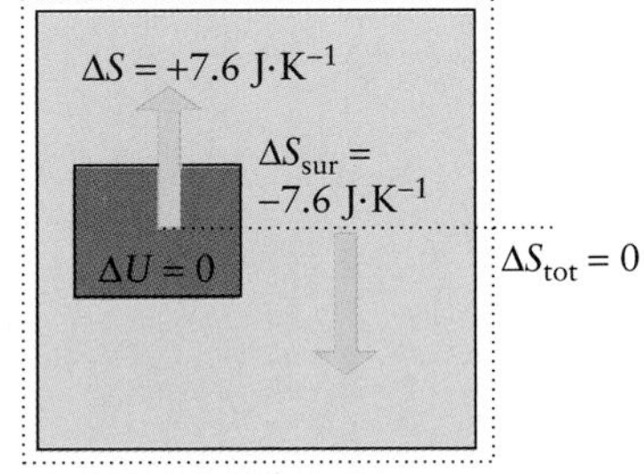

(b) Nonspontaneous process

FIGURE 7.20 The changes in entropy and internal energy when an ideal gas undergoes (a) irreversible and (b) reversible changes between the same two states, as described in Example 7.4.

TABLE 7.4 Reversible and Irreversible Processes

ΔS	ΔS_{surr}	ΔS_{tot}	Character
> 0	> 0	> 0	spontaneous
< 0	< 0	< 0	not spontaneous; reverse change is spontaneous
> 0	< 0		spontaneous if ΔS greater than $-\Delta S_{surr}$
< 0	> 0		spontaneous if ΔS_{surr} greater than $-\Delta S$

SOLUTION (a) For the reversible path, the change in entropy is

$$\begin{aligned}\Delta S &= nR \ln \frac{V_2}{V_1} \\ &= (1.00 \text{ mol}) \times (8.314\ 51 \text{ J·K}^{-1}\text{·mol}^{-1}) \times \ln \frac{20.0 \text{ L}}{8.00 \text{ L}} \\ &= +7.6 \text{ J·K}^{-1}\end{aligned}$$

The change in the entropy of the gas is positive, as expected. Because $\Delta U = 0$, $q = -w$. Therefore, because the heat that flows into the surroundings is equal to the heat that flows *out of* the system

$$q_{\text{surr}} = -q = w = -nRT \ln \frac{V_2}{V_1}$$

The change in entropy of the surroundings, Eq. 12, is therefore

$$\Delta S_{\text{surr}} = \frac{q_{\text{surr}}}{T} = -nR \ln \frac{V_2}{V_1}$$

This expression is the negative of the expression for ΔS, so we know that $\Delta S_{\text{surr}} = -7.6$ J·K^{-1} and therefore that $\Delta S_{\text{tot}} = 0$. This value is in accord with the statement that the process is reversible. (b) For the irreversible process, ΔS is the same, at $+7.6$ J·K^{-1}. No work is done in free expansion (Section 6.7), so $w = 0$. Because $\Delta U = 0$, it follows that $q = 0$. Therefore, no heat is transferred into the surroundings, and their entropy is unchanged: $\Delta S_{\text{surr}} = 0$. The total change in entropy is therefore $\Delta S_{\text{tot}} = +7.6$ J·K^{-1}. The positive value is consistent with an irreversible expansion.

Self-Test 7.14A Determine ΔS, ΔS_{surr}, and ΔS_{tot} for (a) the reversible, isothermal expansion and (b) the isothermal free expansion of 1.00 mol of ideal gas molecules from 10.00 atm and 0.200 L to 1.00 atm and 2.00 L at 298 K.

[*Answer:* (a) $\Delta S = +19.1$ J·K^{-1}; $\Delta S_{\text{surr}} = -19.1$ J·K^{-1}; $\Delta S_{\text{tot}} = 0$; (b) $\Delta S = +19.1$ J·K^{-1}; $\Delta S_{\text{surr}} = 0$; $\Delta S_{\text{tot}} = +19.1$ J·K^{-1}]

Self-Test 7.14B Determine ΔS, ΔS_{surr}, and ΔS_{tot} for the reversible, isothermal compression of 2.00 mol of ideal gas molecules from 1.00 atm and 4.00 L to 20.00 atm and 0.200 L at 298 K.

7.11 Equilibrium

Although this section is short, the principles presented here are critically important, for the tendency of reactions to proceed toward equilibrium is the basis for much of chemistry. The material in this section lays the foundation for the next five chapters, including phase changes, the reactions of acids and bases, and redox reactions.

A system at **equilibrium** has no tendency to change in either direction (forward or reverse), such a system remains in its state until it is disturbed from outside the system. We shall be concerned with **dynamic equilibria,** which are equilibria in which forward and reverse processes still continue, but at matching rates. For example, when a block of metal is at the same temperature as its surroundings, it is in **thermal equilibrium** with them, and energy has no *net* tendency to flow into or out of the block as heat. Energy does flow in both directions, but there is no *net* flow. When a gas confined to a cylinder by a piston has the same pressure as the surroundings, the system is in **mechanical equilibrium** with the surroundings and the gas has no tendency to expand or contract (Fig. 7.21). The pressure inside is causing the piston to move outward, but the pressure outside is causing the piston to move inward to the same extent, and there is no *net* change.

When a solid, such as ice, is in contact with its liquid form, such as water, at certain conditions of temperature and pressure (at 0°C and 1 atm for water), the two states of matter are in **physical equilibrium** with each other, and there is no tendency for one form of matter to change into the other form. When solid and liquid water are at equilibrium, water molecules continually leave solid ice to form liquid water, and water molecules continually leave the liquid phase to form ice. However, there is no *net* change, because these processes balance.

Physical equilibria, which include vaporization sublimation; and dissolving as well as melting, are discussed in detail in Chapter 8.

When a chemical reaction mixture reaches a certain composition, the reaction seems to come to a halt. A mixture of substances at **chemical equilibrium** has no tendency either to produce more products or to revert to reactants. At equilibrium, reactants are still forming products, but products are decaying at a matching rate into reactants and there is no *net* change of composition.

The common characteristic of any kind of dynamic equilibrium is the continuation of processes at the microscopic level but the absence of a *net* tendency for the system to change in either the forward or reverse direction. That is, neither the forward nor the reverse process is spontaneous. Expressed thermodynamically,

$$\text{At equilibrium: } \Delta S_{\text{tot}} = 0 \tag{13}$$

This condition is of the greatest importance. For instance, if we can find how the total entropy varies with the composition of a reaction mixture, then, by looking for the composition at which $\Delta S_{\text{tot}} = 0$, we can predict the equilibrium composition of the reaction. At this composition, the reaction has no tendency to form more products nor decompose into reactants.

The criterion for equilibrium in thermodynamics is $\Delta S_{tot} = 0$.

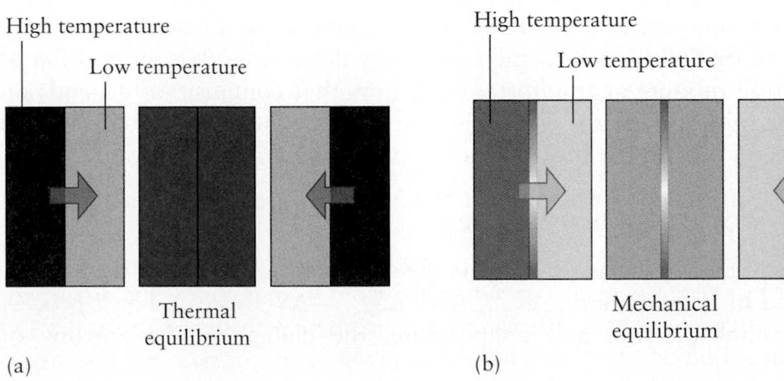

FIGURE 7.21 (a) Energy flows as heat from high temperature (dark red) to low temperature (light red); a system is in thermal equilibrium with its surroundings when the temperatures on both sides of the thermally conducting wall are the same (center). (b) A piston moves from the high-pressure region (dark) to the low-pressure region (light); the system and its surroundings are in mechanical equilibrium when the pressures on both sides of the piston are the same (center).

How can we use this information . . . to decide whether two phases are in equilibrium?

Let's confirm that ice and water are in equilibrium at 0°C. First, we know from Table 7.2 that the entropy of fusion for the process $H_2O(s) \rightarrow H_2O(l)$ is 22.0 $J{\cdot}K^{-1}{\cdot}mol^{-1}$; this value is the entropy change for the system. We can use the enthalpy of fusion of ice in Table 6.3 and Eq. 11 to calculate that the entropy change of the surroundings is

$$\Delta S_{surr} = -\frac{6.01 \times 10^3\ J{\cdot}mol^{-1}}{273.15\ K} = -22.0\ J{\cdot}K^{-1}{\cdot}mol^{-1}$$
$$\Delta S_{tot} = \Delta S + \Delta S_{surr} = 22.0 + (-22.0)\ J{\cdot}K^{-1}{\cdot}mol^{-1} = 0.0\ J{\cdot}K^{-1}{\cdot}mol^{-1}$$

Therefore, the total change in entropy per mole of H_2O that melts at 0°C is 0, and the ice and the water are in equilibrium at that temperature.

Self-Test 7.15A Confirm that liquid water and water vapor are in equilibrium when the temperature is 100°C and the pressure is 1 atm. Data are available in Tables 6.3 and 7.2.

[*Answer:* $\Delta S_{tot} = 0$ at 100°C]

Self-Test 7.15B Confirm that liquid benzene and benzene vapor are in equilibrium at the normal boiling point of benzene, 80.1°C, and 1 atm pressure. The enthalpy of vaporization of benzene at its boiling point is 30.8 $kJ{\cdot}mol^{-1}$ and its entropy of vaporization is 87.2 $J{\cdot}K^{-1}{\cdot}mol^{-1}$.

FREE ENERGY

One of the problems with using the second law to judge whether a reaction is spontaneous is that there are three steps. First, we have to calculate the entropy change of the system from information like that in Table 7.3. Then, we have to calculate the entropy change of the surroundings by using enthalpy data like that in Table 6.3. Finally, we have to add the two changes together to get the total entropy change. We could avoid this lengthy business if a single property combined the entropy calculations for the system and the surroundings. We could then carry out the calculation in one step, using a single table of data. This simplification is achieved by introducing a new and enormously useful state function, the "free energy." Not only does the free energy allow us to assess whether a reaction is spontaneous, it also tells us how much work we can get from the system—a central issue in the study of foods and fuels—and how a change in temperature can change the spontaneity of a reaction. In Chapter 9, we shall also see how the free energy determines the composition of a reaction mixture at equilibrium and how that composition depends on the conditions. The free energy is probably the single most widely used and useful quantity in the application of thermodynamics to chemistry.

7.12 Focusing on the System

The total entropy change, ΔS_{tot}, is the sum of the changes in the system, ΔS, and its surroundings, ΔS_{surr}, with $\Delta S_{tot} = \Delta S + \Delta S_{surr}$. For a process at constant pressure and temperature, the change in the entropy of the surroundings is given by Eq. 11, $\Delta S_{\text{surr}} = -\Delta H/T$. Therefore, under these conditions,

$$\Delta S_{\text{tot}} = \Delta S - \frac{\Delta H}{T} \tag{14}$$

This equation shows how to calculate the total entropy change from information about the system alone. The limitation is that the equation is valid only at constant pressure and temperature.

To take the next step, we introduce the **Gibbs energy,** G, which is defined as

$$G = H - TS \tag{15}$$

The Gibbs energy, G, is named for Josiah Willard Gibbs (Fig. 7.22), the nineteenth-century American physicist who was responsible for turning thermodynamics from an abstract theory into a subject of great usefulness.

This quantity, which is commonly known as the *free energy*, is defined solely in terms of state functions, so G is also a state function. As for enthalpy, the absolute value of G for a substance is not measurable. Throughout chemistry, we use only changes in G rather than its absolute value; and for a process occurring at constant temperature, the change in free energy is

$$\Delta G = \Delta H - T\Delta S \tag{16}$$

By comparing this expression with Eq. 14, we see that, at constant temperature and pressure,

$$\Delta G = -T\Delta S_{\text{tot}} \tag{17}$$

The minus sign in this equation means that an increase in total entropy corresponds to a decrease in free energy. Therefore, at constant pressure and temperature, *the direction of spontaneous change is the direction of decreasing free energy* (Fig. 7.23). The great importance of the introduction of free energy is that, *providing we can accept the limitation to changes at constant pressure and temperature, we can discuss the spontaneity of a process solely in terms of the thermodynamic properties of the system.*

FIGURE 7.22 Josiah Willard Gibbs (1839–1903).

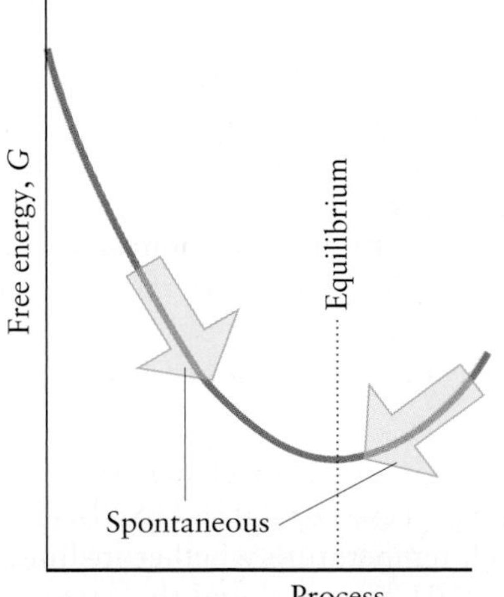

FIGURE 7.23 At constant temperature and pressure, the direction of spontaneous change is toward lower free energy. The equilibrium state of a system corresponds to the lowest point on the curve.

TABLE 7.5 Factors That Favor Spontaneity

Enthalpy change	Entropy change	Spontaneous?
exothermic ($\Delta H < 0$)	increase ($\Delta S > 0$)	yes, $\Delta G < 0$
exothermic ($\Delta H < 0$)	decrease ($\Delta S < 0$)	yes, if $\|T\Delta S\| < \|\Delta H\|$, $\Delta G < 0$
endothermic ($\Delta H > 0$)	increase ($\Delta S > 0$)	yes, if $T\Delta S > \Delta H$, $\Delta G < 0$
endothermic ($\Delta H > 0$)	decrease ($\Delta S < 0$)	no, $\Delta G > 0$

Equation 16 summarizes the factors that determine the direction of spontaneous change at constant temperature and pressure: we look for values of ΔH, ΔS, and T that result in a negative value of ΔG (Table 7.5). One condition that may result in a negative ΔG is a large negative ΔH, such as that for a combustion reaction. As we have seen, a large negative ΔH corresponds to a large increase in the entropy of the surroundings. However, a negative ΔG can occur even if ΔH is positive (an endothermic reaction), provided $T\Delta S$ is large and positive. In this case, as we have already seen, the driving force of the reaction is the increase in entropy of the system.

Self-Test 7.16A Can a nonspontaneous process with a negative ΔS become spontaneous if the temperature is increased?

[*Answer:* No]

Self-Test 7.16B Can a nonspontaneous process with a positive ΔS become spontaneous if the temperature is increased?

We have seen that the criterion for equilibrium is $\Delta S_{tot} = 0$. It follows from Eq. 17 that, for a process at constant temperature and pressure, the condition for equilibrium is

$$\Delta G = 0 \text{ at constant temperature and pressure} \qquad (18)$$

If, when we calculate ΔG for a process, such as the conversion of ice to water at a particular temperature, we find that $\Delta G = 0$, then we know at once that the system is at equilibrium (in this example, that the ice and water are in equilibrium with each other). The following chapters build on this enormously important relation.

How can we use this information . . . to decide whether a process is spontaneous?

Let's calculate the change in molar free energy, ΔG_m, the change in free energy per mole of molecules, for the process $H_2O(s) \rightarrow H_2O(l)$ at (a) 10°C, and (b) 0°C, and decide for each temperature whether melting is spontaneous. The enthalpy of fusion is 6.01 $kJ{\cdot}mol^{-1}$ and the entropy of fusion is 22.0 J·K^{-1}·mol^{-1}. These values are almost independent of temperature over the temperature range considered. From Eq. 16 ($\Delta G = \Delta H - T\Delta S$), it follows that

$$\text{(a) } \Delta G_m = 6.01 \text{ kJ·mol}^{-1} - \{(283 \text{ K}) \times (22.0 \times 10^{-3} \text{ kJ·K}^{-1}\text{·mol}^{-1})\}$$
$$= 6.01 \text{ kJ·mol}^{-1} - 6.23 \text{ kJ·mol}^{-1} = -0.22 \text{ kJ·mol}^{-1}$$

$$\text{(b) } \Delta G_m = 6.01 \text{ kJ·mol}^{-1} - \{(273 \text{ K}) \times (22.0 \times 10^{-3} \text{ kJ·K}^{-1}\text{·mol}^{-1})\}$$
$$= 6.01 \text{ kJ·mol}^{-1} - 6.01 \text{ kJ·mol}^{-1} = 0.00 \text{ kJ·mol}^{-1}$$

It follows that melting is spontaneous at 10°C, but that, at 0°C, ice and water are in equilibrium.

Self-Test 7.17A Calculate the change in molar free energy for the process $H_2O(l) \rightarrow H_2O(g)$ at (a) 95°C; (b) 105°C. The enthalpy of vaporization is 40.7 kJ·mol^{-1} and the entropy of vaporization is +109.1 J·K^{-1}·mol^{-1}. In each case, indicate whether vaporization would be spontaneous or not.

[*Answer:* (a) +0.6 kJ·mol^{-1}, not spontaneous; (b) −0.5 kJ·mol^{-1}, spontaneous]

Self-Test 7.17B Calculate the change in molar free energy for the process $Hg(l) \rightarrow Hg(g)$ at (a) 350°C; (b) 370°C. The enthalpy of vaporization is 59.3 kJ·mol^{-1} and the entropy of vaporization is 94.2 J·K^{-1}·mol^{-1}. In each case, indicate whether vaporization would be spontaneous or not.

The free energy of a substance decreases as the temperature is raised at constant pressure. This conclusion follows from the definition $G = H - TS$ and the fact that the entropy of a pure substance is invariably positive; as T increases, TS increases, too, and a larger quantity is subtracted from H. Another important conclusion is that the free energy decreases more sharply with temperature for the gas phase of a substance than for the liquid phase. Similarly, the free energy of the liquid phase decreases more sharply than that of the solid phase. These conclusions follow because the molar entropy is greatest for the gas phase, less for the liquid phase, and least for the solid phase of a substance. As a result, we can anticipate that, at constant pressure, the free energies of these three phases of a substance vary with temperature (Fig. 7.24).

We can now see the thermodynamic origin of melting, vaporization, and their opposites. At low temperatures, the molar free energy of the solid phase lies lowest, so there is a spontaneous tendency for a liquid to freeze and thereby reduce its free energy. Above a certain temperature, the molar free energy of the liquid falls below that of the solid, and the substance has a spontaneous tendency to melt. At an even higher temperature, the molar free energy of the gas phase plunges down below the liquid line, and the substance has a spontaneous tendency to vaporize. The temperature of each phase change corresponds to the point of intersection of the lines for the two phases, as can be seen in Fig. 7.24.

This discussion also helps to illuminate why some substances, such as solid carbon dioxide, sublime (convert directly to vapor). The relative positions of the three lines shown in Fig. 7.24 are different for each substance. One possibility—which depends on the strength of intermolecular

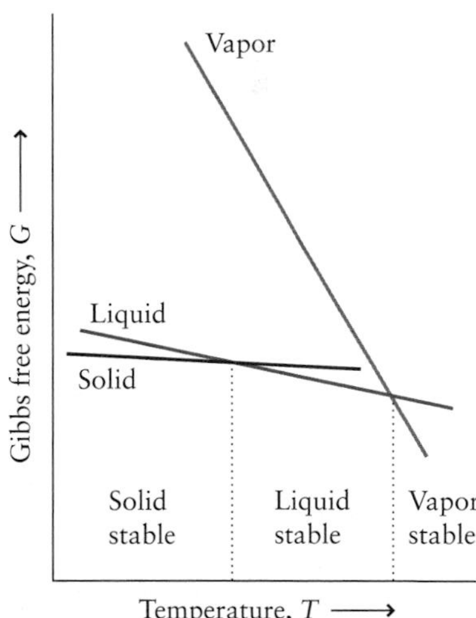

FIGURE 7.24 The variation of the (molar) Gibbs free energy with temperature for three phases of a substance at a given pressure. The most stable phase is the phase with lowest molar free energy. We see that as the temperature is raised, the solid, liquid, and vapor phases in succession become the most stable.

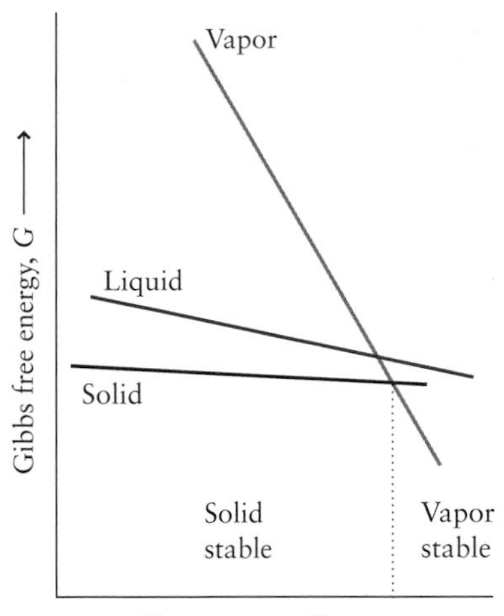

FIGURE 7.25 For some substances and at certain pressures, the molar free energy of the liquid phase might never lie lower than that of the other two phases. For such substances, the liquid is never the stable phase, and at constant pressure the solid sublimes when the temperature is raised to the point of intersection of the solid and vapor lines.

interactions in the condensed phases—is for the liquid line to lie in the position shown in Fig. 7.25. In this case, the liquid line is never the lowest line, at any temperature. As soon as the temperature has been raised above the point corresponding to the intersection of the solid and gas lines, the substance makes a spontaneous transition to the gas phase without first becoming a liquid.

The free energy change is a measure of the change in the total entropy of a system and its surroundings at constant temperature and pressure. Spontaneous processes at constant temperature and pressure are accompanied by a decrease in free energy.

7.13 Reaction Free Energy

The decrease in free energy as a signpost of spontaneous change and $\Delta G = 0$ as a criterion of equilibrium are applicable to any kind of process, provided it is occurring at constant temperature and pressure. Because chemical reactions are our principal interest in chemistry, we now concentrate on them and look for a way to calculate ΔG for a reaction.

The thermodynamic function used as the criterion of spontaneity for a chemical reaction is the **reaction Gibbs energy,** ΔG_r (which is commonly referred to as the "reaction free energy"). This quantity is defined as the difference in molar free energies, G_m, of the products and the reactants:

$$\Delta G_r = \sum nG_m(\text{products}) - \sum nG_m(\text{reactants}) \qquad (19)$$

The n are the stoichiometric coefficients in the chemical equation. For example, for the formation of ammonia, $N_2(g) + 3\,H_2(g) \rightarrow 2\,NH_3(g)$:

$$\Delta G_r = 2G_m(NH_3) - \{G_m(N_2) + 3G_m(H_2)\}$$

If $\Delta G_r < 0$ at a certain composition of the reaction mixture, then the reaction is spontaneous. If $\Delta G_r > 0$ at a composition, then the reverse reaction, the decomposition of ammonia, is spontaneous.

The **standard reaction Gibbs energy,** $\Delta G_r°$, is defined like the reaction free energy but in terms of the *standard* molar free energies of the reactants and products:

$$\Delta G_r° = \sum nG_m°(\text{products}) - \sum nG_m°(\text{reactants}) \qquad (20)$$

Standard states are defined in Section 6.16.

That is, the standard reaction free energy is the difference in free energy between the pure products in their standard states and the pure reactants in their standard states (at the specified temperature). There is a very subtle relation between the sign of $\Delta G_r°$ and the conditions under which a reaction is spontaneous: we explore this point in Chapter 9.

Equations 19 and 20 are not very useful in practice because, as remarked earlier, we do not know the *absolute* values of the molar free energies of substances. However, we can use the same technique we used to find the standard reaction enthalpy in Section 6.19, where we assigned each compound a standard enthalpy of formation, $\Delta H_f°$. We can also tabulate free energies of formation of substances and then use them to calculate

$\Delta G_r°$. The **standard Gibbs energy of formation,** $\Delta G_f°$ (the "standard free energy of formation") of a substance is the *standard reaction free energy per mole for the formation of a compound from its elements in their most stable form.* The most stable form of an element is the state with the lowest free energy (Table 7.6). For example, the standard free energy of formation of hydrogen iodide gas at 25°C is $\Delta G_f° = -1.70\ kJ{\cdot}mol^{-1}$. It is the standard reaction free energy for $\frac{1}{2}\,H_2(g) + \frac{1}{2}\,I_2(s) \rightarrow HI(g)$. The standard free energies of formation of elements in their most stable forms are 0.

Standard free energies of formation can be determined in various ways. One straightforward way is to combine enthalpy and entropy data from tables such as Tables 6.3 and 7.3. A list of values for several common substances is given in Table 7.7, and a more extensive one appears in Appendix 2A.

TABLE 7.6 Examples of the Most Stable Forms of Elements

Element	Most stable form at 25°C and 1 bar
H_2, O_2, Cl_2, Xe	gas
Br_2, Hg	liquid
C	graphite
Na, Fe, I_2	solid

EXAMPLE 7.5 **Calculating a standard free energy of formation from enthalpy and entropy data**

Calculate the standard free energy of formation of HI(g) at 25°C from its standard entropy and standard enthalpy of formation.

STRATEGY First, we write the chemical equation for the formation of HI(g) and calculate the standard reaction free energy from $\Delta G_r° = \Delta H_r° - T\Delta S_r°$. It is best to write the equation with a stoichiometric coefficient of 1 for the compound of interest, for then $\Delta G_r° = \Delta G_f°$. The standard enthalpy of formation is found in Appendix 2A. The standard reaction entropy is found as shown in Example 7.3, by using the data from Table 7.3 or Appendix 2A. If necessary, convert the units of $\Delta S_r°$ from joules to kilojoules.

SOLUTION The chemical equation is $\frac{1}{2}\,H_2(g) + \frac{1}{2}\,I_2(s) \rightarrow HI(g)$. From data in Appendix 2A,

$$\Delta H_r° = \Delta H_f°(\text{HI, g}) = +26.48\ \text{kJ}{\cdot}\text{mol}^{-1}$$

From the data in Appendix 2A,

$$\begin{aligned}\Delta S_r° &= S_m°(\text{HI, g}) - \left\{\tfrac{1}{2}\,S_m°(\text{H}_2,\text{g}) + \tfrac{1}{2}\,S_m°(\text{I}_2,\text{s})\right\}\\ &= 206.6 - \left\{\tfrac{1}{2}(130.7) + \tfrac{1}{2}(116.1)\right\}\ \text{J}{\cdot}\text{K}^{-1}{\cdot}\text{mol}^{-1}\\ &= +83.2\ \text{J}{\cdot}\text{K}^{-1}{\cdot}\text{mol}^{-1} = +83.2 \times 10^{-3}\ \text{kJ}{\cdot}\text{K}^{-1}\end{aligned}$$

Therefore, because $T = 298$ K (corresponding to 25°C),

$$\begin{aligned}\Delta G_r° &= \Delta H_r° - T\Delta S_r°\\ &= 26.48\ \text{kJ}{\cdot}\text{mol}^{-1} - \{(298\ \text{K}) \times (83.2 \times 10^{-3}\ \text{kJ}{\cdot}\text{K}^{-1}{\cdot}\text{mol}^{-1})\}\\ &= +1.69\ \text{kJ}{\cdot}\text{mol}^{-1}\end{aligned}$$

The standard free energy of formation of HI is therefore $+1.69\ kJ{\cdot}mol^{-1}$, in good agreement with the value of $+1.70\ kJ{\cdot}mol^{-1}$ quoted in the text. Note that, because this value is positive, the formation of pure HI from the elements is not spontaneous. Pure HI has a tendency to decompose into its elements at 25°C when the pressure of each substance is 1 bar.

TABLE 7.7 Standard Free Energies of Formation at 25°C*

Substance	$\Delta G_f°$, $kJ{\cdot}mol^{-1}$
Gases	
ammonia, NH_3	−16.45
carbon dioxide, CO_2	−394.4
nitrogen dioxide, NO_2	+51.3
water, H_2O	−228.6
Liquids	
benzene, C_6H_6	+124.3
ethanol, C_2H_5OH	−174.8
water, H_2O	−237.1
Solids	
calcium carbonate, $CaCO_3$†	−1128.8
silver chloride, AgCl	−109.8

* Additional values are given in Appendix 2A.
† Calcite.

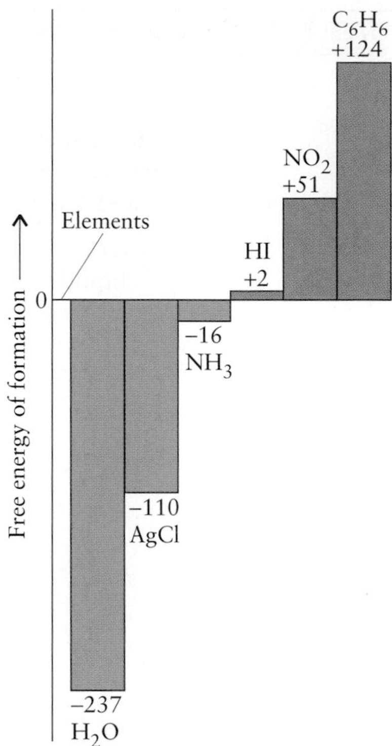

FIGURE 7.26 The standard free energies of formation of compounds are defined as the standard reaction free energy for their formation from the elements. They represent a "thermodynamic altitude" with respect to the elements at "sea level." The numerical values are in kilojoules per mole.

Self-Test 7.18A Calculate the standard free energy of formation of $NH_3(g)$ at 25°C from the enthalpy of formation and the molar entropies of the species involved in its formation.

[*Answer:* $-16.5\ kJ{\cdot}mol^{-1}$]

Self-Test 7.18B Calculate the standard free energy of formation of $C_3H_6(g)$, cyclopropane, at 25°C.

The standard free energy of formation at a given temperature is an indication of a compound's stability relative to its elements under standard conditions; that is, when the pressure of any gas is 1 bar and the concentration of any solute is 1 $mol{\cdot}L^{-1}$. The data in Appendix 2A are for 298.15 K, the conventional temperature for reporting thermodynamic data. If $\Delta G_f° > 0$ at a certain temperature, the pure compound has a lower free energy than its pure elements and the elements are poised to change spontaneously into the compound at that temperature (Fig. 7.26). We say that the compound is "more stable" under standard conditions than the elements. If $\Delta G_f° > 0$, the compound has a higher free energy than its elements and the compound is poised to change spontaneously into the pure elements. In this case, we say that the pure elements are "more stable" than the pure compound. For example, the standard free energy of formation of benzene is $+124\ kJ{\cdot}mol^{-1}$ at 25°C, and benzene is unstable with respect to its elements under standard conditions at 25°C.

A **thermodynamically stable compound** is a compound with a negative standard free energy of formation (water is an example). A **thermodynamically unstable compound** is a compound with a positive standard free energy of formation (benzene is an example). Such a compound is poised to decompose spontaneously into its elements. However, that tendency may not be realized in practice because the decomposition may be very slow. Benzene can, in fact, be kept indefinitely without decomposing at all. Substances that are thermodynamically unstable but survive for long periods are called **nonlabile** or even **inert.** For example, benzene is thermodynamically unstable but nonlabile. Substances that do decompose or react rapidly are called **labile.** Most radicals are labile. It is important to keep in mind the distinction between stability and lability:

The rates of reactions are discussed in Chapter 13.

Stable and *unstable* are terms that refer to the thermodynamic tendency of a substance to decompose into its elements.

Labile, nonlabile, and *inert* are terms that refer to the rate at which that tendency is realized.

Self-Test 7.19A Is glucose stable relative to its elements at 25°C and under standard conditions? Explain your answer.

[*Answer:* Yes; for glucose, $\Delta G_f° = -910\ kJ{\cdot}mol^{-1}$, a negative value]

Self-Test 7.19B Is methylamine, CH_3NH_2, stable relative to its elements at 25°C and under standard conditions? Explain your answer.

Just as we can combine standard enthalpies of formation to obtain standard reaction enthalpies, we can also combine standard free energies of formation to obtain standard reaction free energies:

$$\Delta G_r^\circ = \sum n\Delta G_f^\circ(\text{products}) - \sum n\Delta G_f^\circ(\text{reactants}) \quad (21)$$

where the n are the stoichiometric coefficients in the chemical equation.

The standard free energy of formation of a substance is the standard free energy of reaction per mole of compound when it is formed from its elements in their most stable forms. The sign of ΔG_f° tells us whether a compound is stable or unstable with respect to its elements. Standard free energies of formation are used to calculate standard reaction free energies by using Eq. 21.

How can we use this information . . . to calculate the standard free energy of a reaction?

For the oxidation of ammonia,

$$4\ NH_3(g) + 5\ O_2(g) \longrightarrow 4\ NO(g) + 6\ H_2O(g)$$

we have (from Appendix 2A)

$$\begin{aligned}\Delta G_r^\circ &= \{4\Delta G_f^\circ(NO, g) + 6\Delta G_f^\circ(H_2O, g)\} - \\ &\quad \{4\Delta G_f^\circ(NH_3, g) + 5\Delta G_f^\circ(O_2, g)\} \\ &= \{4(+86.55) + 6(-228.57)\} - \{4(-16.45) + 0\}\ \text{kJ·mol}^{-1} \\ &= -959.42\ \text{kJ·mol}^{-1}\end{aligned}$$

We can infer that the oxidation of ammonia is spontaneous at 25°C under standard conditions.

Self-Test 7.20A Calculate the standard reaction free energy of $2\ CO(g) + O_2(g) \rightarrow 2\ CO_2(g)$ from standard free energies of formation.

[*Answer:* $\Delta G_r^\circ = -514.38$ kJ·mol^{-1}]

Self-Test 7.20B Calculate the standard reaction free energy for $6\ CO_2(g) + 6\ H_2O(l) \rightarrow C_6H_{12}O_6$ (s, glucose) $+ 6\ O_2(g)$ from standard free energies of formation.

7.14 Free Energy and Nonexpansion Work

The free energy change of a system is more than just a disguised form of the total entropy change of the system and its surroundings: once we know its value, we can predict the maximum *nonexpansion work* that a process can do at constant temperature and pressure. **Nonexpansion work,** w_e, is any kind of work other than that due to expansion against an opposing pressure, so it includes electrical work and mechanical work (like stretching a spring or carrying a weight up a slope). The fact that nonexpansion work includes electrical work—the work of pushing electrons through an

The subscript e in w_e stands for "extra."

electrical circuit—is the basis of the chemical generation of electrical power: we pursue that idea in Chapter 12. Nonexpansion work also includes the work of muscular contraction, the work involved in linking amino acids together to form protein molecules, and the work of sending nerve signals through neurons, so a knowledge of changes in free energy is central to an understanding of **bioenergetics**, the deployment and utilization of energy in living cells.

The challenge we now face is to justify these remarks and to derive a *quantitative* relation between the free energy and the maximum nonexpansion work a system can do.

How do we do that?

To derive the relation between the free energy and the maximum nonexpansion work a system can do at constant temperature and pressure, we start with Eq. 16 for an infinitesimal change:

$$dG = dH - TdS$$

We then express the enthalpy change at constant pressure in terms of the internal energy ($dH = dU + PdV$):

$$dG = dU + PdV - TdS$$

Now we use Eq. 7 from Chapter 6 ($dU = dw + dq$) to replace the change in internal energy by $dw + dq$. When we substitute this expression into the preceding one, we obtain

$$dG = dw + dq + PdV - TdS$$

The only constraints at this stage are those of constant temperature and pressure. We are interested in the maximum work that a process can do, which means that the process has to occur reversibly (recall Section 6.7). For a reversible change,

$$dG = dw_{\text{rev}} + dq_{\text{rev}} + PdV - TdS$$

We can use Eq. 1, $dS = dq_{\text{rev}}/T$, to replace dq_{rev} by TdS:

$$dG = dw_{\text{rev}} + TdS + PdV - TdS = dw_{\text{rev}} + PdV$$

At this point, we recognize that the system may do both expansion work and nonexpansion work. Reversible expansion work (achieved by matching the external pressure to the internal) is given by

$$dw_{\text{rev, expansion}} = -PdV$$

(Eq. 9 of Chapter 6), so

$$dw_{\text{rev}} = dw_{\text{rev, expansion}} + dw_{\text{rev, e}} = -PdV + dw_{\text{rev, e}}$$

When we substitute this expression into the last line of the expression for dG, the PdV terms cancel and we are left with

$$dG = dw_{\text{rev, e}}$$

If we recognize that $dw_{\text{rev, e}}$ is the *maximum* nonexpansion work that the system can do (because it is achieved reversibly), we obtain

$$dG = dw_{\text{e,max}} \tag{22a}$$

When we consider a measurable change, this equation becomes

$$w_{\text{e,max}} = \Delta G \text{ at constant temperature and pressure} \tag{22b}$$

The important relation in Eq. 22 tells us that if we know the change in free energy of a process taking place at constant temperature and pressure, then we immediately know how much nonexpansion work it can do. If $\Delta G = -100$ kJ, for instance, then we know that $w_{\text{e,max}} = -100$ kJ, which means—because of the negative sign—that 100 kJ of energy can leave the system as nonexpansion work.

How can we use this information . . . to calculate the work available from the combustion of a compound?

Suppose we are investigating a problem in bioenergetics and need to know how much energy is available from the oxidation of 1.00 mol $C_6H_{12}O_6$ (glucose) for powering muscular contraction. First, we calculate the standard reaction free energy for

$$C_6H_{12}O_6(s) + 6\ O_2(g) \longrightarrow 6\ CO_2(g) + 6\ H_2O(l)$$

From tables of standard free energies of formation, we find $\Delta G_r^\circ = -2879$ kJ·mol^{-1}. Therefore, the maximum nonexpansion work obtainable from 1.00 mol $C_6H_{12}O_6(s)$ is 2.88×10^3 kJ. About 17 kJ of work must be done to build 1 mol of peptide links (a link between amino acids) in a protein, so the oxidation of 1 mol (180 g) of glucose can be used to build up to about 170 mol of such links. In other words, the oxidation of one glucose molecule is needed to build about 170 peptide links. In practice, biosynthesis occurs indirectly, there are energy losses, and only about 10 such links can be built. A typical protein has several hundred peptide links, so several dozen glucose molecules must be sacrificed to build one protein molecule.

Self-Test 7.21A The energy needed for a person of mass m to climb through a height h on the surface of Earth is equal to mgh (see Section A). What is the minimum mass of sucrose a person of mass 60 kg must metabolize to provide the energy needed to climb through 3.0 m? The free energy of combustion of sucrose is -5796 kJ·mol^{-1}.

[*Answer:* 0.10 g]

Self-Test 7.21B What is the maximum amount of electrical energy per gram of hydrogen that can be generated under standard conditions in a fuel cell like the one used on the space shuttle, in which oxygen and hydrogen gases react to form liquid water?

The change in free energy for a process is equal to the maximum nonexpansion work that the system can do at constant temperature and pressure.

7.15 The Effect of Temperature

The values of $\Delta H_r°$ and $\Delta S_r°$ do not change much with temperature because the enthalpies and entropies of both reactants and products are affected by the temperature rise. However, $\Delta G_r°$ does depend on temperature (remember the T in $\Delta G_r° = \Delta H_r° - T\Delta S_r°$) and might change sign as the temperature is changed.

For an exothermic reaction with a negative reaction entropy (so $\Delta H_r°$ and $\Delta S_r°$ are both negative), the enthalpy contributes a negative term to $\Delta G_r°$ (recall Table 7.5). However, $-T\Delta S_r°$ contributes a positive term, and there is a temperature above which $\Delta G_r°$ becomes positive and the reactants are no longer poised to form products spontaneously. For such a reaction, $\Delta G_r°$ is negative (and the reactants poised to form products spontaneously) at low temperatures, but it may become positive (and the *reverse* reaction spontaneous) at higher temperatures (Fig. 7.27a).

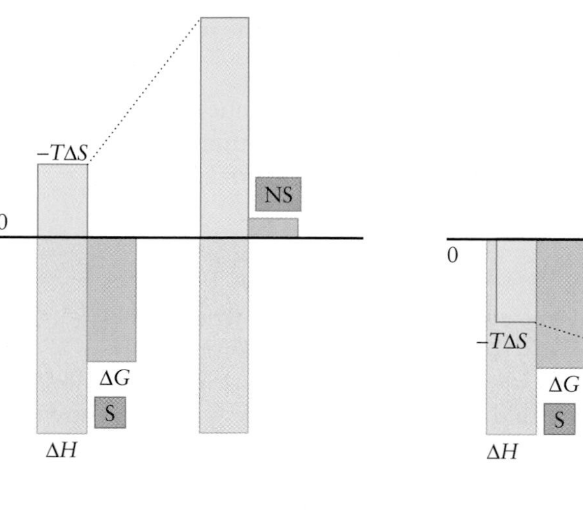

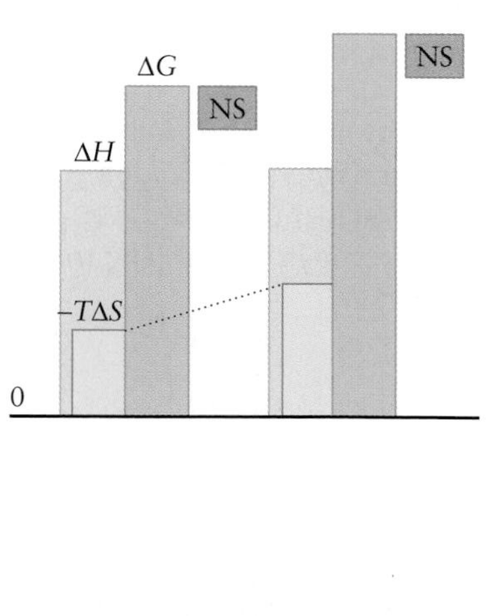

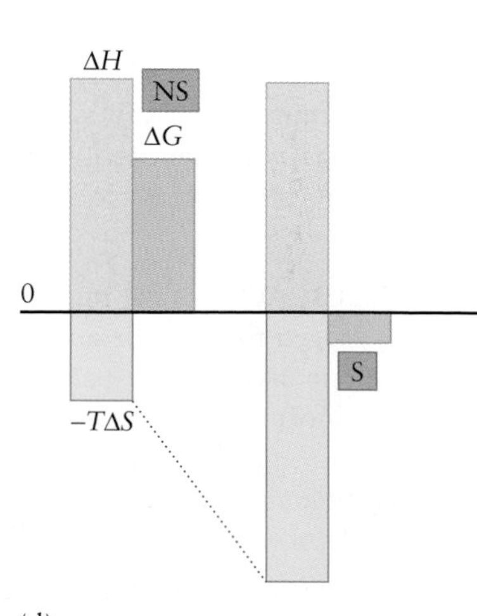

FIGURE 7.27 The effect of an increase in temperature on the spontaneity of a reaction. In each case, S denotes spontaneous and NS nonspontaneous. (a) A spontaneous exothermic reaction with negative reaction entropy may become nonspontaneous; (b) a spontaneous exothermic reaction with positive reaction entropy remains spontaneous; (c) a nonspontaneous endothermic reaction with negative reaction entropy remains nonspontaneous; (d) a nonspontaneous endothermic reaction with a positive reaction entropy may become spontaneous. In each case, the set of bars on the right correspond to higher temperature.

The reverse is true for an endothermic reaction ($\Delta H_r°$ positive) with a positive reaction entropy (Fig. 7.27d). In this case, $\Delta G_r°$ is positive at low temperatures but may become negative when the temperature is raised to the point that $T\Delta S_r°$ is larger than $\Delta H_r°$. The reactants in such a reaction may become poised to form products spontaneously when the temperature is raised sufficiently.

If $\Delta H_r°$ is positive and the reaction entropy is negative, then both terms are positive, $\Delta G_r°$ is positive and the reactants do not form products spontaneously at any temperature (Fig. 7.27c). The reverse is true for a reaction for which $\Delta H_r°$ is negative and the reaction entropy positive: $\Delta G_r°$ for such a process is negative and the reactants are poised to form products spontaneously at all temperatures (Fig. 7.27b).

The free energy increases with temperature for reactions with a negative $\Delta S_r°$ and decreases with temperature for reactions with a positive $\Delta S_r°$.

EXAMPLE 7.6 Calculating the temperature at which an endothermic reaction becomes spontaneous

Estimate the temperature at which it is thermodynamically possible for carbon to reduce iron(III) oxide to iron by the endothermic reaction $2\ Fe_2O_3(s) + 3\ C(s) \rightarrow 4\ Fe(s) + 3\ CO_2(g)$. In other words, find the temperature at which the reaction free energy, $\Delta G_r° = \Delta H_r° - T\Delta S_r°$, becomes negative.

STRATEGY If a reaction is endothermic, it becomes spontaneous at temperatures above which $\Delta G_r° = 0$; that is, the reaction is spontaneous for temperatures above which $\Delta H_r° - T\Delta S_r° = 0$. If it is exothermic, it becomes spontaneous at temperatures below which $\Delta G_r° = 0$. The reaction free energy changes sign at

$$T = \frac{\Delta H_r°}{\Delta S_r°}$$

SOLUTION From data in Appendix 2A, and recalling that the standard enthalpies of formation of the elements are 0,

$$\begin{aligned}\Delta H_r° &= 3\Delta H_f°(CO_2, g) - 2\Delta H_f°(Fe_2O_3, s)\\ &= 3(-393.5) - 2(-824.2)\ \text{kJ·mol}^{-1} = +467.9\ \text{kJ·mol}^{-1}\end{aligned}$$

Therefore, the reaction is endothermic. Similarly, but noting that the standard molar entropies of elements are not 0,

$$\begin{aligned}\Delta S_r° &= \{4S_m°(Fe, s) + 3S_m°(CO_2, g)\} - \{2S_m°(Fe_2O_3, s) + 3S_m°(C, s)\}\\ &= \{4(27.3) + 3(213.7)\} - \{2(87.4) + 3(5.7)\}\ \text{J·K}^{-1}\text{·mol}^{-1}\\ &= +558.4\ \text{J·K}^{-1}\text{·mol}^{-1}\end{aligned}$$

The temperature at which $\Delta G_r°$ changes from positive to negative is

$$T = \frac{\Delta H_r°}{\Delta S_r°} = \frac{467.9 \times 10^3\ \text{J·mol}^{-1}}{558.4\ \text{J·K}^{-1}\text{·mol}^{-1}} = 838\ \text{K}$$

Because the reaction is endothermic, the minimum temperature at which reduction occurs is about 565°C.

Self-Test 7.22A What is the minimum temperature at which magnetite, Fe_3O_4, can be reduced to iron by using carbon (to produce CO_2)?

[*Answer:* 943 K]

Self-Test 7.22B Estimate the temperature at which magnesium carbonate can be expected to decompose to magnesium oxide and carbon dioxide.

7.16 Free Energy Changes in Biological Systems

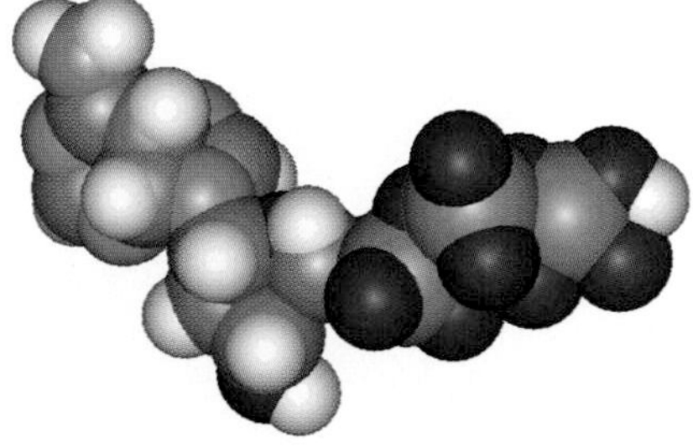

1 Adenosine triphosphate, ATP

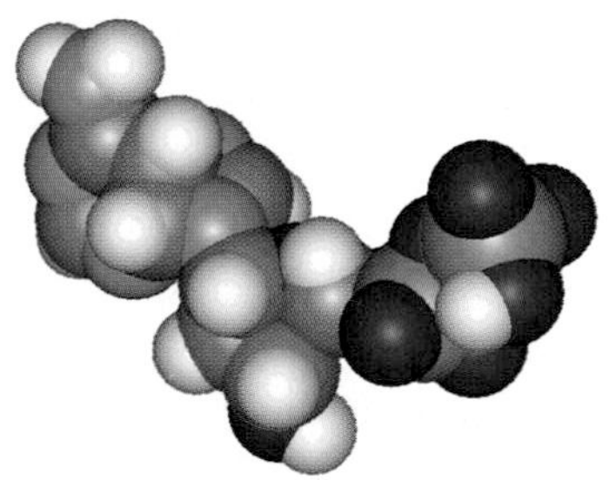

2 Adenosine diphosphate, ADP

Many biological reactions are accompanied by a decrease in entropy of the system. So how can they occur at all? The answer lies in sunlight. We get our energy, not directly from the Sun, but from chemicals—food—that have stored the energy of sunlight. In each case, one biochemical process, which generates a lot of entropy as it runs in its spontaneous direction, drives other reactions in a nonspontaneous direction, perhaps to build a protein or contribute to the construction of a DNA molecule. In other words, biochemical processes are *coupled*: one may be driven uphill in free energy by another reaction that rolls downhill. Staying alive is very much like the effect of a heavy weight tied to another weight by a string that passes over a pulley (Fig. 7.28). The lighter weight could never fly up into the air on its own. However, when it is connected to a heavier weight falling downward on the other side of a pulley, the light weight can soar upward.

The hydrolysis of adenosine triphosphate, ATP (**1**), to adenosine diphosphate, ADP (**2**), is the reaction used most frequently by biological organisms to couple with and drive nonspontaneous reactions. This hydrolysis is the key metabolic reaction by which free energy is stored and used in living systems. The value of $\Delta G_r°$ for the hydrolysis of ATP is about -30 kJ·mol^{-1}. To restore ADP back to ATP, which involves a free energy change of $+30$ kJ·mol^{-1}, an ADP molecule and a phosphate group must be linked by coupling them to another reaction for which the free energy change is more negative than -30 kJ·mol^{-1}. That is one reason why we have to eat. When we eat food containing glucose, we consume a fuel. Like all fuels, it has a spontaneous tendency to form combustion

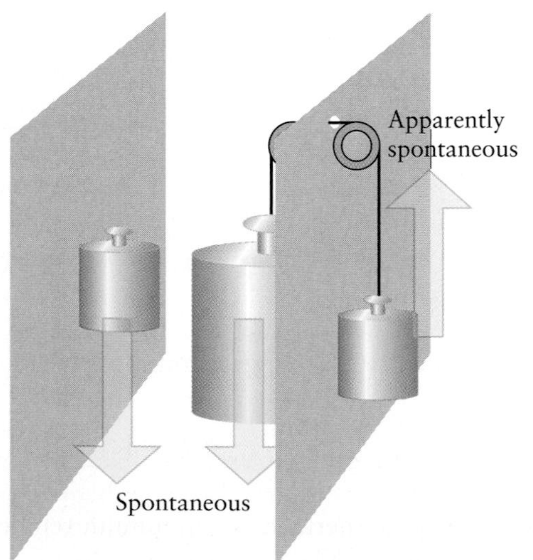

FIGURE 7.28 A natural process can be represented as a falling weight (left). A weight rising spontaneously would be regarded as highly unusual until we recognize that it is actually part of a natural process overall (right). The natural fall of the heavier weight causes the "unnatural" rise of the smaller weight.

products. If we were simply to burn glucose in an open container, it would do no work other than pushing back the atmosphere, and it would give off a lot of heat. However, in our body, the "combustion" is a highly controlled and sophisticated version of burning. In such a controlled reaction, the nonexpansion work the process can do approaches 2500 $kJ \cdot mol^{-1}$, which is enough to "recharge" about 80 mol ADP molecules.

When we die, we will no longer ingest the second-hand sunlight stored in molecules of carbohydrate, protein, and fat. Then the natural direction of change will become dominant, and our intricate molecules will start to decompose into the ooze and slime that we managed to avoid becoming during our lives. Life is a constant battle to generate enough entropy in our surroundings to go on building and maintaining our intricate interiors. As soon as we stop the battle, we stop generating that external entropy, and our bodies decay.

Reactions that are nonspontaneous can be made to occur by coupling them with spontaneous reactions. This coupling is used extensively in biological systems.

SKILLS YOU SHOULD HAVE MASTERED

1 Calculate the entropy change for reversible heat transfer, Section 7.2.

2 Calculate the entropy change when the temperature of a substance is changed, Example 7.1.

3 Determine the entropy change for the isothermal expansion or compression of an ideal gas, Section 7.3.

4 Calculate the standard entropy of a phase change, Section 7.4.

5 Use the Boltzmann formula to calculate and interpret the entropy of a substance, Self-Test 7.8 and Example 7.2.

6 Predict which of two systems has the greater entropy, given their compositions and conditions, Self-Test 7.10.

7 Calculate the standard reaction entropy from standard molar entropies, Example 7.3.

8 Estimate the change in entropy of the surroundings due to heat transfer at constant pressure and temperature, Self-Test 7.12.

9 Predict the spontaneity of a reaction from the total change in entropy, Self-Test 7.13 and Example 7.4.

10 Use the total entropy change to determine whether a system is at equilibrium, Section 7.11.

11 Use the change in free energy to determine the spontaneity of a process at a given temperature, Section 7.12.

12 Calculate a standard free energy of formation from enthalpy and entropy data, Example 7.5.

13 Calculate the standard reaction free energy from standard free energies of formation, Section 7.13.

14 Estimate the maximum nonexpansion work that can be done by a process, Section 7.14.

15 Predict the minimum temperature at which an endothermic reaction can occur spontaneously, Example 7.6.

EXERCISES

Entropy

7.1 A human body generates heat at the rate of about 100 W ($1\ W = 1\ J \cdot s^{-1}$). (a) At what rate does your body heat generate entropy in your surroundings, taken to be at 20°C? (b) How much entropy do you generate each day? (c) Would the entropy generated be greater or less if you were in a room kept at 30°C? Explain your answer.

7.2 An electric heater is rated at 2 kW ($1\ W = 1\ J \cdot s^{-1}$). (a) At what rate does it generate entropy in a room maintained at 28°C? (b) How much entropy does it generate in the course of a day? (c) Would the entropy generated be greater or less if the room were maintained at 25°C? Explain your answer.

7.3 (a) Calculate the change in entropy of a block of copper at 25°C that absorbs 65 J of energy from a heater. (b) If the block of copper is at 100°C and it absorbs 65 J of energy from the heater, what is its entropy change? (c) Explain any difference in entropy change.

7.4 (a) Calculate the change in entropy of 1.0 L of water at 0°C when it absorbs 235 J of energy from a heater. (b) If the 1.0 L of water is at 99°C, what is its entropy change? (c) Explain any difference in entropy change.

7.5 Assuming that the heat capacity of an ideal gas is independent of temperature, calculate the entropy change associated with raising the temperature of 1.00 mol of monatomic ideal gas reversibly from 37.6°C to 157.9°C at (a) constant pressure and (b) constant volume.

7.6 Assuming that the heat capacity of an ideal gas is independent of temperature, calculate the entropy change associated with lowering the temperature of 2.92 mol of monatomic ideal gas from 107.35°C to −52.39°C at (a) constant pressure and (b) constant volume.

7.7 Calculate the entropy change associated with the isothermal expansion of 7.29 mol of monatomic ideal gas from 2.475 L to 10.589 L.

7.8 Calculate the entropy change associated with the isothermal compression of 6.32 mol of monatomic ideal gas from 6.72 atm to 13.44 atm.

7.9 Use data from Table 6.3 or Appendix 2A to calculate the entropy change for (a) the freezing of 1.00 mol H_2O at 0°C; (b) the vaporization of 50.0 g of ethanol, C_2H_5OH, at 351.5 K.

7.10 Use data from Table 6.3 or Table 7.2 to calculate the entropy change for (a) the vaporization of 1.00 mol H_2O at 100°C and 1 atm; (b) the freezing of 3.33 g of NH_3 at 195.3 K.

7.11 (a) Using data provided in Appendix 2A, estimate the boiling point of ethanal, $CH_3CHO(l)$. (b) From standard reference sources, find the actual boiling point of ethanal. (c) How do these values compare? (d) What gives rise to the differences?

7.12 Using data provided in Appendix 2A, estimate the boiling point of $Br_2(l)$. (b) From standard reference sources, find the actual boiling point of bromine. (c) How do these values compare?

7.13 (a) Using Trouton's rule, estimate the boiling point of dimethyl ether, CH_3OCH_3, given that $\Delta H_{vap}° = 21.51\ kJ{\cdot}mol^{-1}$. (b) Using standard reference sources available in your library, find the actual boiling point of dimethyl ether and compare this value to the value obtained by using Trouton's rule.

7.14 (a) Using Trouton's rule, estimate the boiling point of chloroethane, CH_3CH_2Cl, given that $\Delta H_{vap}° = 25\ kJ{\cdot}mol^{-1}$. (b) Using standard reference sources available in your library, find the actual boiling point of chloroethane and compare this value to the value obtained by using Trouton's rule.

7.15 The entropy of vaporization of benzene is approximately 85 $J{\cdot}mol^{-1}{\cdot}K^{-1}$. (a) Estimate the enthalpy of vaporization of benzene at its normal boiling point of 80°C. (b) What is the entropy change of the surroundings when 10 g of benzene, C_6H_6, vaporizes at its normal boiling point?

7.16 The entropy of vaporization of acetone is approximately 85 $J{\cdot}mol^{-1}{\cdot}K^{-1}$. (a) Estimate the enthalpy of vaporization of acetone at its normal boiling point of 56.2°C. (b) What is the entropy change of the surroundings when 10 g of acetone, CH_3COCH_3, condenses at its normal boiling point?

7.17 Which would you expect to have a higher molar entropy at $T = 0$, single crystals of BF_3 or of COF_2? Why?

7.18 On the basis of the structures predicted for each of these molecules, predict which ones would be most likely to have a residual entropy in their crystal forms at $T = 0$: (a) CO_2; (b) NO; (c) N_2O; (d) Cl_2.

7.19 If SO_2F_2 adopts a disordered arrangement in its crystal form, what would its residual molar entropy be?

7.20 What would you expect the residual molar entropy to be for PH_2F if it were to adopt a disordered arrangement in its crystal form?

7.21 Which substance in the pair has the higher molar entropy at 298 K: (a) HBr(g), HF(g); (b) $NH_3(g)$, Ne(g); (c) $I_2(s)$, $I_2(l)$; (d) 1.0 mol Ar(g) at 1.00 atm or 1.0 mol Ar(g) at 2.00 atm?

7.22 Which substance in each pair has the higher molar entropy? (Assume the temperature is 298 K unless otherwise specified.) (a) $CH_4(g)$, $C_3H_8(g)$; (b) KCl(aq), KCl(s); (c) He(g), Kr(g); (d) $O_2(g)$ at 273 K and 1.00 atm, $O_2(g)$ at 450 K and 1.00 atm.

7.23 List the following substances in order of increasing molar entropy at 298 K: $H_2O(l)$, $H_2O(g)$, $H_2O(s)$, C(s, diamond). Explain your reasoning.

7.24 List the following substances in order of increasing molar entropy at 298 K: $CO_2(g)$, Ar(g), $H_2O(l)$, Ne(g). Explain your reasoning.

7.25 Which of the following pairs of molecules would

you expect to have the higher standard molar entropy at 298 K? Explain your reasoning. (a) Iodine or bromine; (b) the two liquids cyclopentane and 1-pentene;

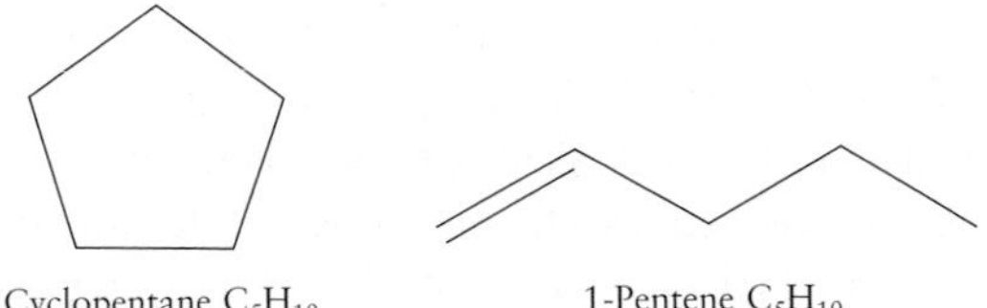

Cyclopentane C_5H_{10} 1-Pentene C_5H_{10}

(c) ethene (also known as ethylene) or an equivalent mass of polyethylene, a substance formed by the polymerization of ethylene.

7.26 Which of the following pairs of molecules would you expect to have the higher standard molar entropy at 298 K? (a) H_2O or H_2S; (b) CH_4 or C_2H_6; (c) the gases cyclopropane and propene (also known as propylene):

CH_2

H_2C—CH_2

Cyclopropane

H

C

H_2C CH_3

Propene

7.27 The standard molar entropy of Cl(g) is 165.2 $J\cdot K^{-1}\cdot mol^{-1}$ and that of $Cl_2(g)$ is 223.07 $J\cdot mol^{-1}\cdot K^{-1}$. Suggest a reason why the standard molar entropy of the diatomic molecules is greater than that of the atoms at 25°C.

7.28 Explain why the standard molar entropy of liquid benzene is less than that of liquid cyclohexane.

7.29 Without performing any calculations, state whether the entropy of the system increases or decreases during each of the following processes: (a) oxidation of nitrogen: $N_2(g) + 2\ O_2(g) \rightarrow 2\ NO_2(g)$; (b) the sublimation of dry ice: $CO_2(s) \rightarrow CO_2(g)$; (c) the cooling of water from 50°C to 4°C. Explain your reasoning.

7.30 Without performing any calculations, state whether the entropy of the system increases or decreases during each of the following processes: (a) the dissolution of table salt: $NaCl(s) \rightarrow NaCl\ (aq)$; (b) the photosynthesis of glucose: $6\ CO_2(g) + 6\ H_2O(l) \rightarrow C_6H_{12}O_6(s) + 6\ O_2(g)$; (c) the evaporation of water from damp clothes. Explain your reasoning.

7.31 Without performing any calculations, predict whether an increase or a decrease in entropy occurs for each of the following processes:
(a) $Cl_2(g) + H_2O(l) \rightarrow HCl(aq) + HClO(aq)$
(b) $Cu_3(PO_4)_2(s) \rightarrow 3\ Cu^{2+}(aq) + 2\ PO_4^{3-}(aq)$
(c) $SO_2(g) + Br_2(g) + 2\ H_2O(l) \rightarrow H_2SO_4(aq) + 2\ HBr(aq)$

7.32 Without performing any calculations, predict whether an increase or a decrease in entropy occurs for each of the following processes: (a) combustion of methane; (b) the formation of carbon dioxide gas; (c) the coiling of two strands of DNA to form a double helix.

7.33 Use data from Table 7.3 or Appendix 2A to calculate the standard reaction entropy for each of the following reactions at 25°C. For each reaction, interpret the sign and magnitude of the reaction entropy. (a) the formation of 1 mol $H_2O(l)$ from the elements in their most stable state at 298 K; (b) the oxidation of 1 mol CO(g) to carbon dioxide; (c) the decomposition of calcite, $CaCO_3(s)$, to carbon dioxide gas and solid calcium oxide; (d) the decomposition of potassium chlorate: $4\ KClO_3(s) \rightarrow 3\ KClO_4(s) + KCl(s)$.

7.34 Use data from Table 7.3 or Appendix 2A to calculate the standard entropy change for each of the following reactions at 25°C. For each reaction, interpret the sign and magnitude of the reaction entropy.
(a) the synthesis of carbon disulfide from natural gas (methane):

$$CH_4(g) + 4\ S(s,\ \text{rhombic}) \longrightarrow CS_2(l) + 2\ H_2S(g)$$

(b) the production of acetylene from calcium carbide and water:

$$CaC_2(s) + 2\ H_2O(l) \longrightarrow C_2H_2(g) + Ca(OH)_2(s)$$

(c) the oxidation of ammonia, which is the first step in the commercial production of nitric acid:

$$4\ NH_3(g) + 5\ O_2(g) \longrightarrow 4\ NO(g) + 6\ H_2O(l)$$

(d) the industrial synthesis of urea, a common fertilizer:

$$CO_2(g) + 2\ NH_3(g) \longrightarrow CO(NH_2)_2(s) + H_2O(l)$$

Global Changes in Entropy

7.35 Calculate the change in the entropy of the surroundings when (a) 1 mJ is released to the surroundings at 2×10^{-7} K; (b) 1 J, the energy of a single heartbeat, is released into the surroundings at 37°C (normal body temperature); (c) 20 J, the energy released when 1 mol He(g) freezes at 3.5 K under pressure, is absorbed by the surroundings.

7.36 Calculate the change in the entropy of the surroundings when (a) 120 kJ is released into the surroundings at 25°C; (b) 120 kJ is released into the surroundings at 100°C; (c) 100 J is absorbed from the surroundings at 50°C.

7.37 Use the information in Table 6.3 to calculate the changes in entropy of the surroundings and of the system for (a) the vaporization of 1.00 mol $CH_4(l)$ at its normal boiling point; (b) the melting of 1.00 mol

$C_2H_5OH(s)$ at its normal melting point; (c) the freezing of 1.00 mol $C_2H_5OH(l)$ at its normal freezing point.

7.38 Use the information in Table 6.3 to calculate the changes in entropy of the surroundings and of the system for (a) the melting of 1.00 mol $NH_3(s)$ at its normal melting point; (b) the freezing of 1.00 mol $CH_3OH(l)$ at its normal freezing point; (c) the vaporization of 1.00 mol $H_2O(l)$ at its normal boiling point.

7.39 Initially a sample of ideal gas at 323 K occupies 1.67 L at 4.95 atm. The gas is allowed to expand to 7.33 L via two pathways: (a) isothermal, reversible expansion; (b) isothermal, irreversible free expansion. Calculate ΔS_{tot}, ΔS, and ΔS_{surr} for each pathway.

7.40 Initially a sample of ideal gas at 412 K occupies 12.62 L at 0.6789 atm. The gas is allowed to expand to 19.44 L via two pathways: (a) isothermal, reversible expansion and (b) isothermal, irreversible free expansion. Calculate ΔS_{tot}, ΔS, and ΔS_{surr} for each pathway.

Free Energy

7.41 Consider the reaction for the production of formaldehyde:

$$H_2(g) + CO(g) \longrightarrow H_2CO(g)$$
$$\Delta H° = +1.96 \text{ kJ}, \Delta S° = -109.58 \text{ J·K}^{-1}$$

Calculate the change in entropy of the surroundings and predict the spontaneity of the reaction under standard conditions at 25°C.

7.42 Consider the reaction in which deuterium exchanges with ordinary hydrogen in water:

$$D_2(g) + H_2O(l) \longrightarrow D_2O(l) + H_2(g)$$
$$\Delta H° = -8.77 \text{ kJ}, \Delta S° = -8.25 \text{ J·K}^{-1}$$

Calculate the change in entropy of the surroundings and predict the spontaneity of the reaction under standard conditions at 25°C.

7.43 Why are so many exothermic reactions spontaneous?

7.44 Explain how an endothermic reaction can be spontaneous.

7.45 Calculate the standard enthalpy change, entropy change, and free energy change at 298 K for each reaction from data in Appendix 2A. For each case, confirm that the value obtained from the free energies of formation is the same as that obtained by using the relation $\Delta G_r° = \Delta H_r° - T\Delta S_r°$.
(a) the oxidation of magnetite to hematite:

$$2 \, Fe_3O_4(s) + \tfrac{1}{2} O_2(g) \longrightarrow 3 \, Fe_2O_3(s)$$

(b) the dissolution of CaF_2 in water:

$$CaF_2(s) \longrightarrow CaF_2(aq)$$

(c) the dimerization of NO_2:

$$2 \, NO_2(g) \longrightarrow N_2O_4(g)$$

7.46 Calculate the standard enthalpy change, entropy change, and free energy change for each reaction at 298 K, using data from Appendix 2A. For each case, confirm that the value obtained from the free energies of formation is the same as that obtained by using the relation $\Delta G_r° = \Delta H_r° - T\Delta S_r°$.
(a) the hydration of copper sulfate:

$$CuSO_4(s) + 5 \, H_2O(l) \longrightarrow CuSO_4{\cdot}5H_2O(s)$$

(b) the reaction of $H_2SO_4(l)$, an effective dehydrating agent, with water to form aqueous sulfuric acid:

$$H_2SO_4(l) \longrightarrow H_2SO_4(aq)$$

(c) the reaction of calcium oxide with water to form calcium hydroxide:

$$CaO(s) + H_2O(l) \longrightarrow Ca(OH)_2(s)$$

(d) Which of the above reagents would you expect to be the most effective at removing water from a substance?

7.47 Write a balanced chemical equation for the formation reaction of (a) $NH_3(g)$; (b) $H_2O(g)$; (c) $CO(g)$; (d) $NO_2(g)$. For each, calculate $\Delta H_r°$, $\Delta S_r°$, and $\Delta G_r°$ from data available in Appendix 2A.

7.48 Write a balanced chemical equation for the formation reaction of (a) $SO_3(g)$; (b) $C_6H_6(l)$; (c) $C_2H_5OH(l)$; (d) $CaCO_3$(s, calcite). For each reaction calculate $\Delta H_r°$, $\Delta S_r°$, and $\Delta G_r°$ from data available in Appendix 2A.

7.49 Use the standard free energies of formation in Appendix 2A to calculate $\Delta G_r°$ for each of the following reactions at 25°C. Comment on the spontaneity of each reaction under standard conditions at 25°C.
(a) $2 \, SO_2(g) + O_2(g) \rightarrow 2 \, SO_3(g)$
(b) $CaCO_3$ (s, calcite) $\rightarrow CaO(s) + CO_2(g)$
(c) $2 \, C_8H_{18}(l) + 25 \, O_2(g) \rightarrow 16 \, CO_2(g) + 18 \, H_2O(l)$

7.50 Use the standard free energies of formation in Appendix 2A to calculate $\Delta G_r°$ for each of the following reactions at 25°C. Comment on the spontaneity of each reaction at 25°C.
(a) $NH_4Cl(s) \rightarrow NH_3(g) + HCl(g)$
(b) $H_2(g) + D_2O(l) \rightarrow D_2(g) + H_2O(l)$
(c) $N_2(g) + NO_2(g) \rightarrow NO(g) + N_2O(g)$
(d) $2 \, CH_3OH(g) + 3 \, O_2(g) \rightarrow 2 \, CO_2(g) + 4 \, H_2O(l)$

7.51 Determine which of the following compounds are stable with respect to decomposition into their elements under standard conditions at 25°C (see Appendix 2A): (a) $PCl_5(g)$; (b) $HCN(g)$; (c) $NO(g)$; (d) $SO_2(g)$.

7.52 Determine which of the following compounds are stable with respect to decomposition into their elements under standard conditions at 25°C (see Appendix 2A): (a) $CuO(s)$; (b) $C_6H_{12}(l)$, cyclohexane; (c) $PCl_3(g)$; (d) $N_2H_4(l)$.

7.53 Does potassium chlorate have a thermodynamic tendency to decompose into potassium perchlorate and potassium chloride at 25°C? Would the tendency be stronger at a higher or lower temperature?

7.54 Does methanol have a thermodynamic tendency to decompose into carbon monoxide and hydrogen at 25°C? Would the tendency be stronger at a higher or lower temperature?

7.55 Which of the following compounds become less stable with respect to the elements as the temperature is raised: (a) $PCl_5(g)$; (b) $HCN(g)$; (c) $NO(g)$; (d) $SO_2(g)$?

7.56 Which of the following compounds become less stable with respect to their elements as the temperature is raised: (a) $CuO(s)$; (b) C_6H_{12}(l, cyclohexane); (c) $PCl_3(g)$; (d) $N_2H_4(l)$?

7.57 Calculate the standard reaction entropy, enthalpy, and free energy for each of the following reactions from data found in Appendix 2A:
(a) the decomposition of hydrogen peroxide:

$$2\,H_2O_2(l) \longrightarrow 2\,H_2O(l) + O_2(g)$$

(b) the preparation of hydrofluoric acid from fluorine and water:

$$2\,F_2(g) + 2\,H_2O(l) \longrightarrow 4\,HF(aq) + O_2(g)$$

7.58 Calculate the standard reaction entropy, enthalpy, and free energy for each of the following reactions from data found in Appendix 2A:
(a) the production of "synthesis gas," a low-grade industrial fuel:

$$CH_4(g) + H_2O(g) \longrightarrow CO(g) + 3\,H_2(g)$$

(b) the thermal decomposition of ammonium nitrate:

$$NH_4NO_3(s) \longrightarrow N_2O(g) + 2\,H_2O(g)$$

7.59 Assuming that $\Delta H_r°$ and $\Delta S_r°$ are independent of temperature, use data available in Appendix 2A to calculate $\Delta G_r°$ for the following reactions at 80°C. Over what temperature range will each reaction be spontaneous under standard conditions?
(a) $B_2O_3(s) + 6\,HF(g) \rightarrow 2\,BF_3(g) + 3\,H_2O(l)$
(b) $CaC_2(s) + 2\,HCl\,(aq) \rightarrow CaCl_2\,(aq) + C_2H_2(g)$
(c) C(s, graphite) → C(s, diamond)

7.60 Assuming that $\Delta H_r°$ and $\Delta S_r°$ are independent of temperature, use data available in Appendix 2A to calculate $\Delta G_r°$ for the following reactions at −50°C. Over what temperature range will each reaction be spontaneous under standard conditions?
(a) $HCN(g) + 2\,H_2(g) \rightarrow CH_3NH_2(g)$
(b) $2\,Cu^+\,(aq) \rightarrow Cu(s) + Cu^{2+}(aq)$
(c) $CaCl_2(s) + 2\,HF(g) \rightarrow CaF_2(s) + 2\,HCl(g)$

INTEGRATED EXERCISES

7.61 (a) Which substance would you expect to have the higher molar entropy in the liquid phase, 2-propanone (acetone, CH_3COCH_3) or 1-propanol ($CH_3CH_2CH_2OH$)? (b) Would your answer be different if these substances were in the gas phase?

7.62 (a) Before checking the numbers, which substance would you expect to have the higher standard molar entropy, $CH_3COOH(l)$ or $CH_3COOH(aq)$? Having made this prediction, examine the values in Appendix 2A and explain your findings.

7.63 Assuming statistical disorder, would you expect a crystal of octahedral *cis*-MX_2Y_4 to have the same, higher, or lower residual entropy that the corresponding trans isomer? Explain.

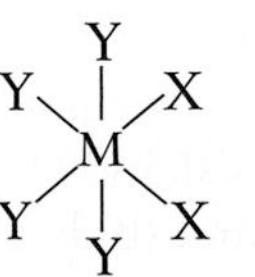

cis-MX_2Y_4

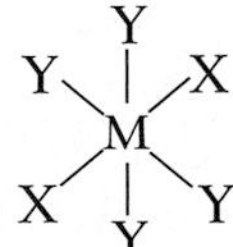

trans-MX_2Y_4

7.64 There are three different substituted benzenes with the formula $C_6H_4F_2$. Assume that the benzene rings pack similarly into their crystal lattices. If the positions of the H and F atoms are statistically disordered in the solid state, which isomer will have the *least* residual molar entropy?

7.65 Consider the enthalpies of fusion and the melting points of the following elements: Pb, 5.10 kJ·mol^{-1},

327°C; Hg, 2.29 $kJ\cdot mol^{-1}$, −39°C; Na, 2.64 $kJ\cdot mol^{-1}$, 98°C. Given these data, determine whether a relationship similar to Trouton's rule can be obtained for the entropy of fusion of the metallic elements.

7.66 Determine whether titanium dioxide can be reduced by carbon at 1000 K in the reactions:
(a) $TiO_2(s) + 2\ C(s) \rightarrow Ti(s) + 2\ CO(g)$
(b) $TiO_2(s) + C(s) \rightarrow Ti(s) + CO_2(g)$
given that at 1000 K, $\Delta G_f°$ (CO, g) = −200 $kJ\cdot mol^{-1}$, $\Delta G_f°(CO_2, g)$ = −396 $kJ\cdot mol^{-1}$; $\Delta G_f°(TiO_2, s)$ = −762 $kJ\cdot mol^{-1}$.

7.67 Which is the thermodynamically more stable iron oxide in air, $Fe_3O_4(s)$ or $Fe_2O_3(s)$? Justify your selection.

7.68 For the reaction $PCl_3(g) + Cl_2(g) \rightarrow PCl_5(g)$ at 25°C, $\Delta G_r° = -37.2\ kJ\cdot mol^{-1}$ and $\Delta H_r° = -87.9\ kJ\cdot mol^{-1}$. Calculate $\Delta S_r°$ for the reaction at 25°C. (b) Does the sign of the entropy change agree with your expectations? Explain your conclusion. (c) Calculate $\Delta G_r°$ for the reaction at −25°C.

7.69 The reaction for the production of the synthetic fuel "water-gas" from coal is

$$C(s, graphite) + H_2O(g) \rightarrow CO(g) + H_2(g)$$

(a) Calculate the standard reaction free energy at 25°C. (b) Assuming $\Delta H_r°$ and $\Delta S_r°$ are unaffected by temperature changes, calculate the temperature at which $\Delta G_r° = 0$.

7.70 (a) Calculate $\Delta H_r°$ and $\Delta S_r°$ at 25°C for the reaction $C_2H_2(g) + 2\ H_2(g) \rightarrow C_2H_6(g)$. (b) Calculate the standard reaction free energy from the equation $\Delta G_r° = \Delta H_r° - T\Delta S_r°$. (c) Interpret the calculated values for $\Delta H_r°$ and $\Delta S_r°$. (d) Assuming $\Delta H_r°$ and $\Delta S_r°$ are unaffected by temperature changes, determine the temperature at which $\Delta G_r° = 0$. (e) What is the significance of this temperature?

7.71 (a) Would you expect pure FeS(s) to convert into Fe(s) and $FeS_2(s)$ if heated? (b) Would you expect pure $FeS_2(s)$ to convert into FeS(s) and S(s) if heated? (c) Would your answers to (a) and (b) have been different if iron or sulfur, respectively, had also been present in addition to the compound?

7.72 (a) Calculate the work that must be done at standard temperature and pressure against the atmosphere for the expansion of the gaseous products in the combustion of 2.00 mol $C_6H_6(l)$ to give carbon dioxide and water vapor. Using data from Appendix 2A, calculate the standard enthalpy of the reaction. (c) Calculate the change in internal energy, $\Delta U°$, of the system.

7.73 Potassium nitrate dissolves readily in water and its enthalpy of solution is +34.9 $kJ\cdot mol^{-1}$. (a) How does the enthalpy of solution favor or not favor the dissolving process? (b) Is the entropy change of the system positive or negative when the salt dissolves? (c) Is the entropy change of the system primarily a result of changes in positional disorder or thermal disorder? (d) Is the entropy change of the surroundings primarily a result of changes in positional disorder or thermal disorder? (e) What is the driving force for the dissolution of KNO_3?

7.74 Explain why each of the following statements is false: (a) Reactions with negative reaction free energies occur spontaneously and rapidly. (b) Every sample of a pure element, regardless of its physical state, is assigned zero free energy of formation. (c) An exothermic reaction producing more moles of gas molecules than are consumed has a positive standard reaction free energy.

7.75 Consider the conversion of graphite to diamond. This reaction is significant, not only because diamond is a valued gemstone, but also because diamond is very hard and serves as a coating material to prolong the life of drilling tools and other objects used in demanding environments. Using the data in Appendix 2A, calculate the standard enthalpy, entropy, and free energy for the conversion of graphite to diamond. Rationalize the signs of the values that are obtained on the basis of what you know about the structures of diamond and graphite.

7.76 According to current theories of biological evolution, complex amino and nucleic acids were produced from randomly occurring reactions involving compounds thought to be present in the Earth's early atmosphere. These simple molecules then assembled into more and more complex molecules, such as DNA and RNA. Is this process consistent with the second law of thermodynamics? Discuss.

7.77 Some entries for $S°$ in Appendix 2A are negative. What is common about these substances and why would the entropy be negative?

7.78 Acetic acid, $CH_3COOH(l)$, could be produced from (a) the reaction of methanol with carbon monoxide; (b) the oxidation of ethanol; (c) the reaction of carbon dioxide with methane. Write balanced chemical equations for each process. Carry out a thermodynamic analysis of the three possibilities and decide which you would expect to be the easiest to accomplish.

7.79 Consider the reaction of $InCl_3$ with $InBr_3$ to form $InCl_2Br$:

$$2\ InCl_3(s) + InBr_3(s) \xrightarrow{\Delta} 3\ InCl_2Br(s)$$

Without performing any calculations, consider the above reaction and predict whether $\Delta G_r°$ should be positive or negative. Explain your reasoning.

7.80 The major uses of ammonia are in the manufacture of fertilizers, plastics, and explosives. (a) Write a balanced equation for the production of 1 mol NH_3(g) from nitrogen and hydrogen. (b) Calculate the standard enthalpy and entropy changes of this reaction at 298 K. (c) Calculate the standard free energy change of this reaction at 150 K, 298 K, and 350 K, assuming that $\Delta H_r°$ and $\Delta S_r°$ are constant with temperature. (d) The synthesis of ammonia at room temperature is slow. Is raising the temperature in order to speed up the reaction a good idea? Explain your conclusion.

7.81 Using values found in Appendix 2A, calculate the standard free energy for the vaporization of water at 25.0°C, 100.0°C, and 150°C. (b) What should the value at 100.0°C be? (c) Why is there a discrepancy?

7.82 Octane, C_8H_{18}, a hydrocarbon used in gasoline, has molecules that consist of long, flexible chains of eight carbon atoms with hydrogen atoms attached to satisfy the remaining valence of each carbon atom. It does not burn smoothly in automobile engines, thus causing a noisy condition known as "knocking." Isooctane, a less flexible, branched isomer of octane with the formal name 2,2,4-trimethylpentane, burns more smoothly. The octane rating of isooctane is 100; that of octane is close to 0. To improve the octane rating of gasoline, some of the octane is converted into isooctane. (a) Draw Lewis structures for octane and 2,2,4-trimethylpentane, $(CH_3)_3CCH_2CH(CH_3)_2$. (b) Calculate the change in entropy, the change in enthalpy, and the change in free energy that accompany the conversion of octane into isooctane. (c) Which hydrocarbon is thermodynamically favored? (d) Is the specific enthalpy (the enthalpy of combustion per gram of substance) of the hydrocarbon increased or decreased in the conversion? For isooctane $\Delta H_f° = -225.0\ kJ\cdot mol^{-1}$; $\Delta G_f° = +12.8\ kJ\cdot mol^{-1}$; $S_m° = 423.0\ J\cdot K^{-1}\cdot mol^{-1}$. For octane, see Appendix 2A.

7.83 One technique that is used to overcome the unfavorable thermodynamics of one reaction is to "couple" that reaction with another process that is thermodynamically favored. For instance, the dehydrogenation of cyclohexane to form benzene and hydrogen gas is not spontaneous. Show that if another molecule such as ethene is present to act as a hydrogen acceptor (that is, the ethene reacts with the hydrogen produced to form ethane), then the process can be made spontaneous.

7.84 Adenosine triphosphate (ATP) is an extremely important molecule in biological systems. Consult standard reference sources in your library to determine how this molecule is used in energy transfer to facilitate nonspontaneous processes necessary for life.

7.85 There are three isomeric *alkenes* that have the formula C_4H_8 (see the table below). (a) Draw Lewis structures of these compounds. (b) Calculate $\Delta G°$, $\Delta H°$, and $\Delta S°$ for the three reactions that interconvert each pair of compounds. (c) Which isomer is the most stable? (d) Place the isomers in order of decreasing $S_m°$.

Compound	$\Delta H_f°$	$\Delta G_f°$
2-methylpropene	−16.90	+58.07
cis-2-butene	−6.99	+65.86
trans-2-butene	−11.17	+62.97

7.86 From standard reference sources available in your library, find the standard enthalpies and free energies of formation of 1-butene ($CH_2{=}CHCH_2CH_3$), 1-pentene ($CH_2{=}CHCH_2CH_2CH_3$), styrene ($C_6H_5CH{=}CH_2$), and ethyl benzene ($C_6H_5CH_2CH_3$). For comparison purposes, use values for the gases in all cases. Using these data and the data available in Appendix 2A, calculate the standard enthalpy change, free energy change, and entropy change of the hydrogenation of the double bond in each of these molecules as well as those of ethene and propene. How similar are these values? What does this indicate about the use of bond enthalpies to calculate standard enthalpy changes for reactions?

7.87 Heating organic carboxylic acids may lead to decarboxylation (loss of CO_2). Using values available in Appendix 2A, calculate the temperature at which the decarboxylation of (a) formic acid, (b) acetic acid, and (c) benzoic acid become spontaneous. What is the driving force for getting these reactions to proceed, the enthalpy change or the entropy change?

7.88 (a) At 25.0°C, the enthalpies of hydrogenation of the gases *cis*-2 butene and *trans*-2-butene are $-120.4\ kJ\cdot mol^{-1}$ and $-115.5\ kJ\cdot mol^{-1}$, respectively. In both cases, the product of hydrogenation is butane. Calculate the enthalpies of formation of *cis*- and *trans*-2-butene from their enthalpies of hydrogenation. (b) Which is more stable, *cis*- or *trans*-2-butene? (c) Calculate $\Delta G_f°$ for *cis*- and *trans*-2-butene. (d) Suggest an explanation for the relative stability found in part (b). For *cis*-2-butene, $S_m° = 300.8\ J\cdot K^{-1}\cdot mol^{-1}$, and for *trans*-2-butene, $S_m° = 296.5\ J\cdot K^{-1}\cdot mol^{-1}$.

CHAPTER

Although the bald eagle perched on this iceberg in Tongass National Forest, Alaska, is not aware of it, the melting of the ice is accompanied by the partial loss of the open structure of the solid, in which hydrogen bonds hold water molecules apart as well as together. This structure collapses when ice melts and results in the formation of a denser liquid.

Why Do We Need This Material?

In earlier chapters, we investigated the nature of the solid, liquid, and gaseous states of matter; in this chapter, we extend the discussion and look at transformations between these states. The discussion introduces the concept of equilibrium between different phases of a substance. This concept will prove to be of the greatest importance when, in the following chapter, we consider chemical transformations. We also take a deeper look at solutions in this chapter. The presence of solutes alters the physical properties of the solvent. These changes have important consequences, such as the control of the flow of nutrients into and out of living cells and the refining of fuels.

What Do We Need to Know Already?

This chapter develops the concepts of physical equilibria introduced in the context of thermodynamics (Chapters 6 and 7). The material also assumes an acquaintance with the nature of intermolecular forces (Sections 5.1–5.5). The compositions of some of the solutions we discuss are expressed in terms of mole fraction (Section 4.11).

Physical Equilibria

We are surrounded by substances that change from one form to another but retain their chemical identity. We call some of these changes "the weather." As the oceans evaporate, water changes into vapor and adds to the moisture of the air. That vapor condenses to droplets of liquid to form clouds and later falls back to Earth as rain, perhaps to freeze to ice. In this chapter, we see how to use thermodynamic principles, like those introduced in the preceding chapters, to deepen our understanding, not only of these familiar processes, but also of all kinds of physical change.

Many of the liquids we encounter are solutions, and the solutes present in them affect their physical properties. The largest solutions we encounter on Earth are the oceans, which account for 1.4×10^{12} kg of the Earth's surface water. That mass amounts to nearly 300 000 tons of water for each inhabitant, more than enough to supply all the drinking water needed for a thirsty world. However, seawater cannot be drunk directly, because of the high concentrations of dissolved salts. The main solutes in seawater are Na^+ and Cl^- ions (Table 8.1), but seawater also contains huge amounts of other substances and at least a trace of every naturally occurring element. Transforming seawater into potable water is a major challenge to water engineers. The material in this chapter suggests how this purification can be achieved.

Another solution of great importance to our lives is the plasma that carries the erythrocytes (red blood cells) through our bodies. These cells can cease to function if the concentrations of solutes in the plasma drift from their proper values. This chapter explains why such changes occur and guides us through building a model for the behavior of solutions.

In this chapter, we consider the phase transitions of pure materials and see how to describe them precisely, often in terms of thermodynamics. Then we move on to the physical properties of mixtures. The first step is to understand the nature of dissolving and the characteristics of solutes and solvents that determine the extent to which one substance dissolves in another. Thermodynamic concepts will help us do that. Next, we consider how the presence of a solute can affect the physical properties of the solvent. Once again, we shall see that thermodynamics provides a consistent foundation that helps us to explain the consequences of the presence of a solute on the solvent. Finally, we turn to the special case of a solution in which the solute and the solvent are both liquids. The principal feature of the binary liquid mixtures we consider is the task of separating one liquid from another, and we see how that enormously important practical process can be described and controlled.

The underlying unifying theme of this chapter is *equilibrium*. At the transition temperature, transitions from one phase to another are equilibrium processes; for example, the solid and liquid phases are in equilibrium at the melting point. When a solvent has dissolved the maximum amount of solute, the dissolved and undissolved solute are in equilibrium.

TABLE 8.1 Principal Ions Found in Seawater

Element	Principal form	Concentration, $g \cdot L^{-1}$
Cl	Cl^-	19.0
Na	Na^+	10.5
Mg	Mg^{2+}	1.35
S	SO_4^{2-}	0.89
Ca	Ca^{2+}	0.40
K	K^+	0.38
Br	Br^-	0.065
C	HCO_3^-, H_2CO_3, CO_2	0.028

The principal physical properties that we consider are modifications of the conditions, such as the temperature, needed to bring two phases into equilibrium. The problem of separating one liquid from another depends on being able to make use of the equilibrium between the liquid mixture and its vapor. Because we know that equilibrium (at constant temperature and pressure) is characterized by $\Delta G = 0$, the *single* unifying theme of this chapter is the exploration of the consequences of this thermodynamic condition. In Chapter 9, we shall see how these same ideas can be applied to chemical equilibria.

PHASES AND PHASE TRANSITIONS

First, we consider changes of state of pure substances. As we saw in Section 5.1, a single substance can exist in different *phases*, or different physical forms. The phases of a substance include its solid, liquid, and gaseous forms and its different solid forms, such as the diamond and graphite phases of carbon. In one case—helium—there are two liquid phases of the same substance. There are several phases of ice, which differ from one another in the way the water molecules pack together when high pressures are applied. The conversion of a substance from one phase to another, such as the melting of ice, the vaporization of water, or the conversion of graphite into diamond, is called a **phase transition.** Phase transitions take place at specific temperatures and pressures that depend on the purity of the substance. Seawater, for instance, freezes at a lower temperature than pure water does.

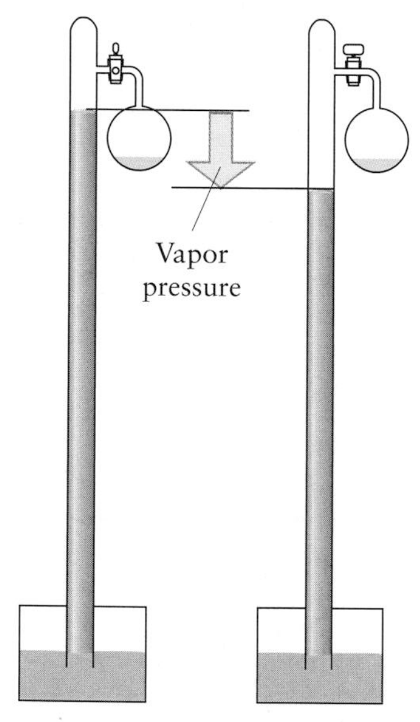

FIGURE 8.1 The apparatus on the left shows a mercury barometer, with a vacuum over the mercury column and a flask containing a liquid and its vapor. The stopcock between the top of the barometer and flask is closed. In the apparatus on the right, the stopcock has been opened and the vapor pressure exerted by the liquid in the flask is proportional to the distance by which the column of mercury is lowered. The vapor pressure is the same, however much liquid is present in the flask.

8.1 Vapor Pressure

In this section, we investigate the transition between liquid and vapor, such as the vaporization of water and its reverse, the condensation of water vapor to liquid. A simple experiment shows that, in a closed vessel, the two phases reach equilibrium with each other.

Suppose we have a long glass tube full of mercury (like a barometer) and upend it in a dish of mercury. As we saw in Section 4.2, the mercury inside the tube falls to a height proportional to the external atmospheric pressure, leaving it about 76 cm high at sea level on Earth. The space above the mercury is almost a vacuum. Now suppose we connect a bulb containing a small drop of water to the space above the mercury. The water immediately evaporates and fills the space with vapor. This vapor exerts a pressure that pushes the surface of the mercury down a few centimeters. The pressure exerted by the vapor—as measured by the change in the height of mercury—depends on the amount of water added. However, suppose we add so much water that a little liquid remains in the bulb. The pressure of the vapor now remains constant, no matter how much liquid water is present (Fig. 8.1). We conclude that, *at a fixed temperature, the vapor exerts a characteristic pressure regardless of the amount of liquid water present.* For example, at 20°C, the mercury falls 18 mm, so the pressure exerted by the vapor is 18 Torr. The pressure of the water vapor is the same whether there is 0.1 mL or 1 mL of liquid water present. This characteristic pressure is the "vapor pressure" of the liquid at the temperature of the experiment (Table 8.2).

Liquids with high vapor pressures at ordinary temperatures are said to be **volatile.** Diethyl ether is highly volatile; mercury is not. Solids also exert a vapor pressure, but their vapor pressures are usually much lower than those of liquids because the molecules are gripped more tightly in a solid than they are in a liquid. Nevertheless, solids vaporize in the process called **sublimation,** which we can observe in the presence of some pungent solids—such as menthol and mothballs.

TABLE 8.2 Vapor Pressures at 25°C

Substance	Vapor pressure, Torr
benzene	94.6
ethanol	58.9
mercury	0.0017
methanol	122.7
water*	23.8

* For values at other temperatures, see Table 8.3.

The fact that at a given temperature a condensed phase gives rise to a characteristic vapor pressure is a sign that the condensed phase and the vapor are in equilibrium with each other. We can begin to build a molecular model of this equilibrium by considering what happens when a liquid vaporizes and reaches equilibrium with its vapor. Let's suppose that the liquid is introduced into a closed container. The vapor forms as molecules leave the surface of the liquid. Most evaporation takes place from the surface of the liquid because the molecules there are least strongly bound and can escape more easily than those in the bulk. As the number of molecules in the vapor increases, more of them are available to strike the surface of the liquid, stick to it, and become part of the liquid again. Eventually the number of molecules returning to the liquid each second matches the number escaping (Fig. 8.2). The vapor is now condensing as fast as the liquid is vaporizing. We say that the liquid and vapor are in **dynamic equilibrium,** a condition in which a forward process and its reverse take place at equal rates. We write

It is important to note that all these events are taking place in a closed container. We shall see in Section 8.4 what happens when the container is open.

Rate of evaporation = rate of condensation

The word *dynamic* is important: it implies continuous activity, unlike the static equilibrium of a ball resting at the foot of a hill. A liquid in dynamic equilibrium with its vapor is like a busy store that has customers continually arriving and leaving, with the number of customers inside the store remaining constant.

The dynamic equilibrium between liquid water and its vapor is denoted

$$H_2O(l) \rightleftharpoons H_2O(g)$$

View an animation of the dynamic equilibrium between liquid water and water vapor.

Media Link **Animation:** Fig. 8.2, Liquid/vapor equilibrium.

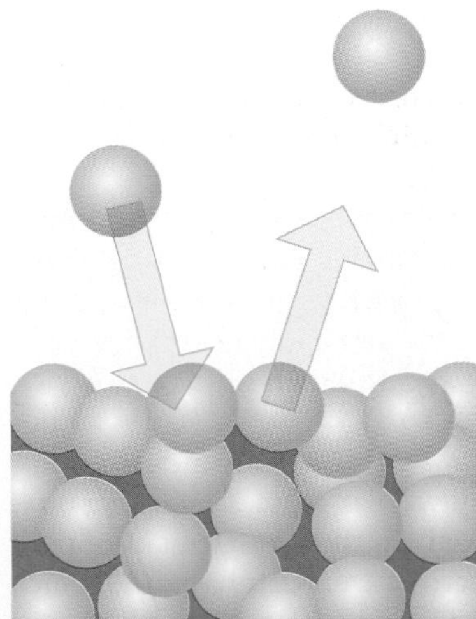

FIGURE 8.2 When a liquid and its vapor are in dynamic equilibrium inside a closed container, the rate at which the molecules leave the liquid is equal to the rate at which they return.

Whenever we see the symbol $\rightleftharpoons$, it means that the species on both sides are in dynamic equilibrium with each other. Although "products" (water molecules in the gas phase) are being formed from "reactants" (water molecules in the liquid phase), the products are changing back into reactants at a matching rate. With this picture in mind, we can define the **vapor pressure** of a liquid (or a solid) as the pressure exerted by its vapor when the vapor and the liquid (or the solid) are in dynamic equilibrium with each other.

Dynamic equilibrium implies the simultaneous continuous action of opposing processes. The vapor pressure is the pressure exerted by the vapor of a substance when the vapor and its condensed phase are in dynamic equilibrium.

8.2 Volatility and Intermolecular Forces

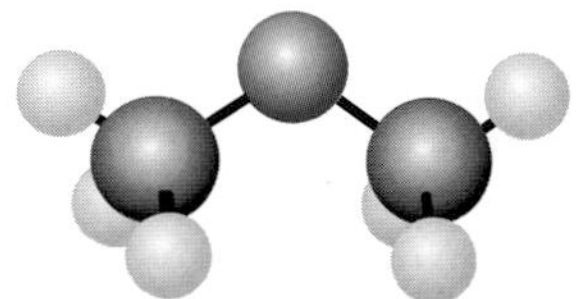

1 Dimethyl ether, CH_3OCH_3

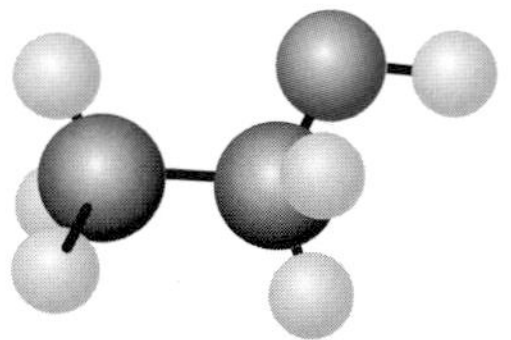

2 Ethanol, CH_3CH_2OH

We can expect high vapor pressure when the molecules of a liquid are held together only by weak intermolecular forces in the liquid and low vapor pressure when the intermolecular forces are strong. We therefore expect liquids composed of molecules capable of forming hydrogen bonds to be less volatile than others. We can see the effect of hydrogen bonding clearly when we compare dimethyl ether (**1**) and ethanol (**2**), both of which have the molecular formula C_2H_6O. Because the two compounds have the same numbers of electrons, we can expect them to have similar London interactions and therefore similar vapor pressures. However, an ethanol molecule has an —OH group and can form a hydrogen bond to another ethanol molecule. Ether molecules cannot form a hydrogen bond to one another because all their hydrogen atoms are attached to carbon atoms. As a result of this difference, ethanol is a liquid at room temperature, whereas dimethyl ether is a gas. Even diethyl ether, $(C_2H_5)_2O$, which has more electrons than ethanol, and much stronger London interactions, has a higher vapor pressure than ethanol. At ordinary temperatures, benzene also has a vapor pressure slightly higher than that of ethanol: although it has more electrons than ethanol and its London forces are relatively strong, benzene cannot form hydrogen bonds.

EXAMPLE 8.1 Predicting the relative vapor pressures of substances

Which liquid do you expect to have the higher vapor pressure at room temperature, *cis*-dibromoethene (**3**) or *trans*-dibromoethene (**4**)?

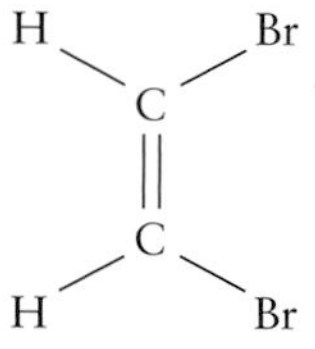

3 *cis*-Dibromoethene

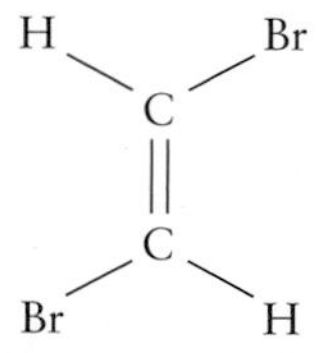

4 *trans*-Dibromoethene

STRATEGY The two molecules have the same molar masses (and therefore the same numbers of electrons) and differ only in the arrangement of their atoms. Decide which intermolecular interactions each can have and which has the stronger interactions. That substance is likely to have the lower vapor pressure.

SOLUTION Both substances can interact by London forces, and the strengths will be about the same in each case because they are composed of the same atoms. However, only the cis compound is polar, so only it can have dipole-dipole interactions. Therefore, we can expect *cis*-dibromoethene to have stronger intermolecular forces than *trans*-dibromoethene. It follows that we should expect *trans*-dibromoethene to be the more volatile.

Self-Test 8.1A Which do you expect to have the higher vapor pressure at room temperature, tetrabromomethane, CBr_4, or tetrachloromethane, CCl_4? Give reasons.

[*Answer:* CCl_4; weaker London forces]

Self-Test 8.1B Which do you expect to have the higher vapor pressure at $-50°C$, CH_3CHO or $CH_3CH_2CH_3$?

To predict the numerical value of the vapor pressure, we need to express these ideas more quantitatively. We can use the thermodynamic relations we met in Chapter 7 to find an expression for the vapor pressure in terms of thermodynamic properties.

How do we do that?

The starting point for setting up an expression for the vapor pressure is to note that when a liquid and its vapor are in equilibrium, there is no tendency for one phase to change into the other. In the language of Section 7.12, the free energy change for the formation of vapor from liquid is 0 when the two phases are in equilibrium:

$$\Delta G_{vap} = G_m(g) - G_m(l) = 0 \qquad (1)$$

The molar free energy of a liquid is almost independent of pressure, so we can replace $G_m(l)$ by its standard value (its value at 1 bar), $G_m°(l)$. The molar free energy of an ideal gas does vary with pressure, and thermodynamic arguments show that, for an ideal gas,

$$G_m(g,P) = G_m°(g) + RT \ln P \qquad (2)$$

where P is the numerical value of the pressure in bars and $G_m°$ is the standard molar free energy of the gas (its molar free energy at 1 bar). This pressure dependence is illustrated in Fig. 8.3. It follows from these remarks that, at equilibrium, when the pressure, P, is the vapor pressure of the liquid,

$$\begin{aligned}\Delta G_{vap} &= \{G_m°(g) + RT \ln P\} - G_m°(l)\\ &= \{G_m°(g) - G_m°(l)\} + RT \ln P\\ &= \Delta G_{vap}° + RT \ln P = 0\end{aligned}$$

$\Delta G_{vap}°$ is the **standard free energy of vaporization,** the change in molar (Gibbs) free energy when a liquid at 1 bar changes to a vapor at 1 bar. Therefore, the vapor pressure is given by

$$\ln P = -\frac{\Delta G_{vap}°}{RT}$$

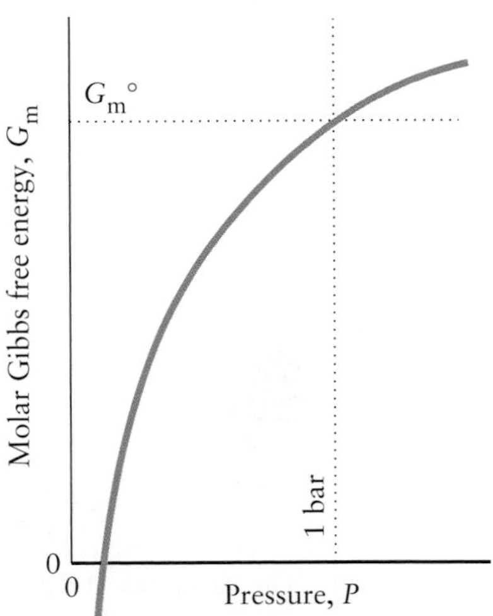

FIGURE 8.3 The variation of the molar free energy of an ideal gas with pressure. The free energy has its standard value when the pressure of the gas is 1 bar. The free energy tends to minus infinity as the pressure falls to 0.

When we insert $\Delta G_{vap}° = \Delta H_{vap}° - T\Delta S_{vap}°$, we obtain

$$\ln P = -\frac{\Delta H_{vap}°}{RT} + \frac{\Delta S_{vap}°}{R} \quad (3)$$

where $\Delta H_{vap}°$ and $\Delta S_{vap}°$ are the standard enthalpy and entropy of vaporization, respectively. To a good approximation, the entropies of vaporization of all liquids are about the same because they undergo nearly the same expansion when they vaporize. Therefore, most of the variation in vapor pressure stems from the first term on the right. We see that if the enthalpy of vaporization is high—which is likely to be the case when there are strong intermolecular forces in the liquid—then ln P, and therefore the vapor pressure, P, can be expected to be low, just as we anticipated.

The same argument was used in the discussion of Trouton's rule, Section 7.4.

The vapor pressure of a liquid at a given temperature is expected to be low if the forces acting between its molecules are strong.

8.3 The Variation of Vapor Pressure with Temperature

The vapor pressure of a liquid is a measure of how readily the molecules in the liquid can escape from the forces that hold them together. More energy to overcome these attractions is available at higher temperatures than at low, so we can expect the vapor pressure of a liquid to rise with increasing temperature. Figure 8.4 shows the temperature dependence of the vapor pressure of several liquids, and we see that, as expected, in each case the vapor pressure does indeed increase with temperature. Table 8.3 shows the temperature dependence of the vapor pressure of water in more detail.

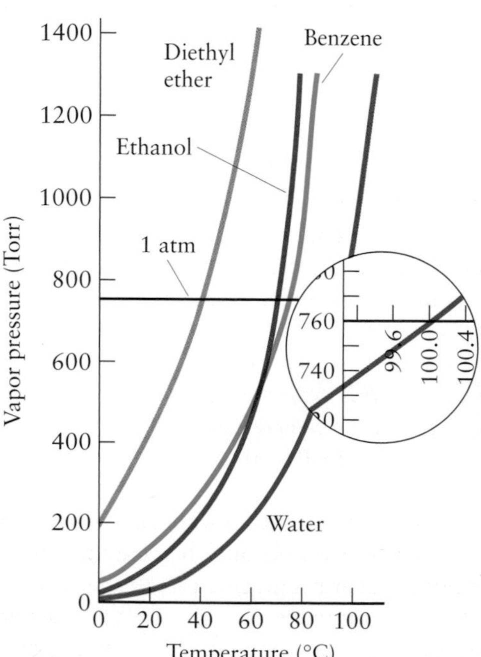

FIGURE 8.4 The variation of the vapor pressure of liquids with temperature, for diethyl ether (orange), ethanol (red), benzene (green), and water (blue). The normal boiling point is the temperature at which the vapor pressure is 1 atm (760 Torr).

We can use Eq. 3 to obtain a quantitative expression for the temperature variation of the vapor pressure. It is known experimentally that neither the enthalpy nor the entropy of vaporization varies much with temperature; so, for a given substance, $\Delta S_{vap}°$ and $\Delta H_{vap}°$ can both be treated as approximately constant. It follows that the vapor pressures P_1 and P_2 at any two temperatures T_1 and T_2 are related by writing Eq. 3 for two temperatures and subtracting one from the other. In the process, the entropy terms cancel:

$$\ln P_2 - \ln P_1 = \left(-\frac{\Delta H_{vap}°}{RT_2} + \frac{\Delta S_{vap}°}{R}\right) - \left(-\frac{\Delta H_{vap}°}{RT_1} + \frac{\Delta S_{vap}°}{R}\right)$$
$$= \frac{\Delta H_{vap}°}{R}\left(\frac{1}{T_1} - \frac{1}{T_2}\right)$$

Then, because $\ln a - \ln b = \ln(a/b)$, we obtain the **Clausius-Clapeyron equation**

$$\ln\frac{P_2}{P_1} = \frac{\Delta H_{vap}°}{R}\left(\frac{1}{T_1} - \frac{1}{T_2}\right) \quad (4)$$

We can use this equation to predict the vapor pressure at any temperature of interest, given the vapor pressure at some other temperature as described in Toolbox 8.1 and Example 8.2. The curves in Fig. 8.4 show the dependence of vapor pressure on temperature predicted by this equation. We also see that the higher the enthalpy of vaporization, the more sharply the vapor pressure increases with temperature.

The vapor pressure of a liquid increases as the temperature increases. The Clausius-Clapeyron equation gives the quantitative dependence of the vapor pressure of a liquid on temperature.

TABLE 8.3 Vapor Pressure of Water

Temperature, °C	Vapor pressure, Torr
0	4.58
10	9.21
20	17.54
21	18.65
22	19.83
23	21.07
24	22.38
25	23.76
30	31.83
37*	47.08
40	55.34
60	149.44
80	355.26
100	760.00

* Body temperature.

8.4 Boiling

So far, we have considered a liquid in equilibrium with its vapor inside a closed container. Now let's consider what happens when we heat a liquid in a container that is open to the atmosphere—water heated in a kettle is an example. As the vapor forms, it spreads away from the liquid and the rate of condensation never matches the rate of vaporization. The liquid and vapor never reach dynamic equilibrium and, given enough time, the liquid vaporizes completely.

When the temperature is raised to the point at which the vapor pressure is equal to the atmospheric pressure (for instance, when water is heated to 100°C and the external pressure is 1 atm), vaporization occurs *throughout* the liquid, not just from its surface. At this temperature, any vapor formed can drive back the atmosphere and make room for itself. Thus, bubbles of vapor form in the liquid and rise to the surface. This rapid vaporization taking place throughout the bulk is called **boiling**. The **normal boiling point,** T_b, of a liquid is the temperature at which a liquid boils when the atmospheric pressure is 1 atm. In other words, *the normal boiling point is the temperature at which the vapor pressure of the liquid* *is 1 atm*. We can find the boiling points of the compounds in Fig. 8.4 by drawing a horizontal line at $P = 1$ atm and noting the temperatures at which it intercepts the curves.

Boiling occurs at a temperature higher than the normal boiling point when the pressure is greater than 1 atm, as it is in a pressure cooker. A higher temperature is needed to raise the vapor pressure of the liquid up to the prevailing atmospheric pressure. Boiling occurs at a lower temperature when the pressure is less than 1 atm, because now the vapor pressure matches the external pressure at a lower temperature. At the summit of Mt. Everest—where the pressure is about 240 Torr—water boils at only 71°C.

A high normal boiling point is a sign of strong intermolecular forces. Unusually high boiling points of liquids are in many cases due to hydrogen bonding, because high temperatures are needed to overcome the forces and raise the vapor pressure to 1 atm. For example, water has a boiling point much higher than that of hydrogen sulfide, which does not form hydrogen bonds.

Boiling occurs when the vapor pressure of a liquid is equal to the atmospheric pressure. Strong intermolecular forces usually lead to high normal boiling points.

TOOLBOX 8.1 How to use the temperature dependence of vapor pressure

Conceptual Basis

We expect the vapor pressure of a condensed phase (a liquid or a solid) to be high when the temperature is high (because there is then more energy available to help the molecules to escape) and when the intermolecular forces are weak.

Procedure

The temperature dependence of the vapor pressure is expressed by the Clausius-Clapeyron equation, Eq. 4. To use this expression, take the standard enthalpy of vaporization from Table 6.3 and assume that its value is independent of temperature. Note that the equation can be expressed in the form

$$\ln P = A + \frac{B}{T}$$

where B is proportional to the enthalpy of vaporization. Therefore, a plot of $\ln P$ against $1/T$ should be a straight line (over a limited range of temperatures). The slope is steeper, and the vapor pressure more responsive to changes in temperature, for substances with high enthalpies of vaporization.

If the vapor pressure of a liquid at one temperature is known, use the Clausius-Clapeyron equation, Eq. 4, to calculate the vapor pressure of the liquid at a different temperature, as in Example 8.2.

The normal boiling point of a liquid is the temperature at which its vapor pressure is 1 atm. To find this temperature, rearrange the Clausius-Clapeyron equation into

$$\frac{1}{T_1} = \frac{1}{T_2} + \frac{R}{\Delta H_{vap}{}^{\circ}} \ln \frac{P_2}{P_1}$$

and calculate the temperature at which the vapor pressure of the liquid is equal to 1 atm = 101.325 kPa, given its vapor pressure P_2 at another temperature T_2 (see Example 8.3).

EXAMPLE 8.2 Estimating the vapor pressure of a liquid from its value at a different temperature

Tetrachloromethane, CCl_4, which is now known to be carcinogenic, was once used as a dry cleaning solvent and to treat patients infested with hookworms. The enthalpy of vaporization of CCl_4 is 33.05 kJ·mol^{-1}, and its vapor pressure at 57.8°C is 405 Torr. What is the vapor pressure of tetrachloromethane at 25.0°C?

STRATEGY We expect the vapor pressure of CCl_4 to be lower at 25.0°C than at 57.8°C. The Clausius-Clapeyron equation gives the quantitative relation between vapor pressure and temperature, so we can substitute the temperatures and the enthalpy of vaporization to find the ratio of vapor pressures. Then we can substitute the known vapor pressure to find the desired one. To use the equation, we convert the enthalpy of vaporization into joules per mole and all temperatures into kelvins.

SOLUTION The enthalpy of vaporization (33.05 kJ·mol^{-1}) corresponds to 3.305×10^4 J·mol^{-1}; 25.0°C corresponds to $T_2 = 298.2$ K, and 57.8°C corresponds to $T_1 = 331.0$ K. We substitute these values into the Clausius-Clapeyron equation and obtain

$$\begin{aligned}\ln\frac{P_2}{P_1} &= \frac{\Delta H_{\text{vap}}^\circ}{R}\left(\frac{1}{T_1}-\frac{1}{T_2}\right)\\ &= \frac{3.305\times10^4\ \text{J·mol}^{-1}}{8.3145\ \text{J·K}^{-1}\text{·mol}^{-1}}\left(\frac{1}{331.0\ \text{K}}-\frac{1}{298.2\ \text{K}}\right)\\ &= \frac{3.305\times10^4}{8.3145}\left(\frac{1}{331.0}-\frac{1}{298.2}\right)\\ &= -1.32\end{aligned}$$

Now we can solve for P_2, using $P_1 = 405$ Torr:

$$P_2 = (405\ \text{Torr}) \times e^{-1.32} = 108\ \text{Torr}$$

To avoid rounding errors, it is best to leave the numerical calculation until the end, especially when dealing with exponentials.

Self-Test 8.2A The vapor pressure of water at 25°C is 23.76 Torr and its enthalpy of vaporization at that temperature is 44.0 kJ·mol^{-1}. Estimate the vapor pressure of water at 35°C.

[*Answer:* 42 Torr]

Self-Test 8.2B The vapor pressure of benzene at 25°C is 94.6 Torr and its enthalpy of vaporization is 30.8 kJ·mol^{-1}. Estimate the vapor pressure of benzene at 35°C.

EXAMPLE 8.3 Estimating the boiling point of a liquid

The vapor pressure of ethanol at 34.9°C is 13.3 kPa. Estimate the normal boiling point of ethanol.

STRATEGY The normal boiling point of a substance is the temperature at which its vapor pressure is equal to 1 atm, and 1 atm = 101.325 kPa exactly. The Clausius-Clapeyron equation provides the relation we require. The first step is to rearrange the equation to solve for the desired temperature; second, we substitute the data. For the latter step, we need the enthalpy of vaporization of the liquid: see Table 6.3.

SOLUTION The enthalpy of vaporization of ethanol is 43.5 kJ·mol^{-1} (Table 6.3). Therefore, with P_2 = 13.3 kPa, T_2 = 308.0 K, P_1 = 101.325 kPa, and $\Delta H_{vap}°$ = 43.5 kJ·mol^{-1}, we find

$$\frac{1}{T_1} = \frac{1}{308.0\text{ K}} + \frac{8.3145\text{ J}\cdot\text{K}^{-1}\cdot\text{mol}^{-1}}{4.35\times10^4\text{ J}\cdot\text{mol}^{-1}}\ln\left(\frac{13.3\text{ kPa}}{101.325\text{ kPa}}\right)$$
$$= \frac{1}{308.0\text{ K}} + \frac{8.3145}{4.35\times10^4\text{ K}}\ln\left(\frac{13.3}{101.325}\right)$$

When we evaluate and then take the reciprocal of the right-hand side, we obtain T_1 = 350 K, which corresponds to 77°C. The actual boiling point of ethanol is 78°C. The small error probably comes from assuming that the enthalpy of vaporization is a constant over the temperature range of the question and assuming that the vapor behaves like an ideal gas.

Self-Test 8.3A The vapor pressure of acetone, C_3H_6O, at 7.7°C is 13.3 kPa and its enthalpy of vaporization is 29.1 kJ·mol^{-1}. Estimate the normal boiling point of acetone.

[*Answer:* 62.3°C (actual: 56.2°C)]

Self-Test 8.3B The vapor pressure of methanol, CH_3OH, at 49.9°C is 400 Torr and its enthalpy of vaporization is 35.3 kJ·mol^{-1}. Estimate the normal boiling point of methanol.

8.5 Freezing and Melting

Two common phase transitions are freezing and its reverse, melting (or fusion). A liquid solidifies when its molecules have such low energies that they are unable to wriggle past their neighbors. In the solid, the molecules vibrate about their average positions but rarely move from place to place. The **freezing temperature,** the temperature at which the solid and liquid phases are in dynamic equilibrium, varies slightly as the pressure is changed, and the **normal freezing point,** T_f, of a liquid is the temperature at which it freezes at 1 atm. In practice, a liquid sometimes does not freeze until the temperature is a few degrees below its freezing point. A liquid that survives below its freezing point is said to be **supercooled.**

The freezing point of a liquid is the same as the melting point of the solid.

For most substances, the solid phase of a substance is denser than the liquid phase, because the molecules pack together more closely in the solid phase. Applied pressure helps to hold the molecules together, so a high temperature must be reached before the molecules of a liquid at high pressure can break apart. As a result, most liquids freeze at higher temperatures when subjected to pressure. However, except at very high pressures, the effect of pressure is usually quite small. Iron at 1 atm, for example, melts at 1800 K, but it melts at only a few degrees higher when the pressure is a thousand times greater. At the center of the Earth, however, the pressure is high enough for iron to be solid despite the high temperatures there, so the Earth's inner core is believed to be solid.

Water is highly unusual in that at 0°C the density of the liquid is greater than that of ice. We already know this fact from a common observation: ice floats on water. Therefore, at the melting point, the molar volume of liquid water is less than that of ice. This anomalous behavior is due to the hydrogen bonds in water, which result in a very open structure (see Section 5.12). When ice melts, many of the hydrogen bonds collapse, thereby allowing the water molecules to pack together more closely. As a result, at a high pressure, ice melts at a slightly lower temperature, and the melting point of water *decreases* as the pressure is increased.

Most liquids freeze at a higher temperature under pressure. Water's hydrogen bonds make it anomalous: it freezes at a lower temperature under pressure.

8.6 Phase Diagrams

Which phase a substance adopts—ice, liquid water, or water vapor, for instance—depends on the temperature and pressure of the substance. For example, at 1 atm, water exists as ice below 0°C, as liquid between 0°C and 100°C, and as vapor above 100°C. If we were to increase the pressure, we would find that water survived as a liquid to temperatures a little below 0°C (the melting point barely changes with pressure) but to temperatures well above 100°C (the boiling point depends strongly on pressure).

A **phase diagram** is a map showing which phase is the most stable at different pressures and temperatures. The phase diagram for water is shown in Fig. 8.5, and that for carbon dioxide is shown in Fig. 8.6. These diagrams are examples of phase diagrams for a single substance (water or carbon dioxide) and hence are known as **one-component phase diagrams.** Any point in the region marked "solid" (or, specifically, ice in the case of water) corresponds to conditions for which the solid phase of the substance is the most stable, and similarly for the regions marked "liquid" and "vapor" (or "gas"). We can see from the phase diagram for carbon dioxide, for instance, that a sample of carbon dioxide at 10°C and 2 atm will be a gas, but if the pressure is increased at constant temperature to 10 atm, then carbon dioxide will condense to a liquid. Sulfur has two solid phases (Fig. 8.7), corresponding to the two ways in which its crown-like S_8 molecules can stack together. Many substances have several solid phases: even water forms at least seven kinds of ice, depending on how the H_2O molecules fit together, but only one of them is stable at ordinary pressures. Ice-VIII, for instance, is stable only above 20 000 atm and 100°C.

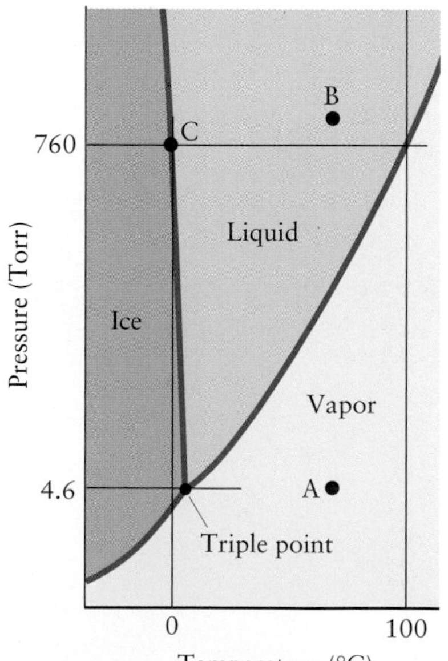

FIGURE 8.5 The phase diagram for water (not to scale). The solid lines define the boundaries of the regions of pressure and temperature at which each phase is the most stable. Note that the freezing point decreases slightly with increasing pressure. The letters A and B are referred to in Example 8.4.

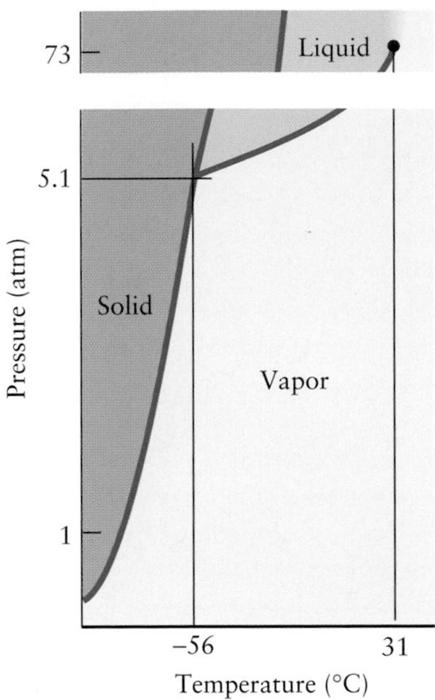

FIGURE 8.6 The phase diagram for carbon dioxide (not to scale). The liquid can exist only at pressures above 5.1 atm. Note the slope of the boundary between the solid and liquid phases; it shows that the freezing point rises as pressure is applied.

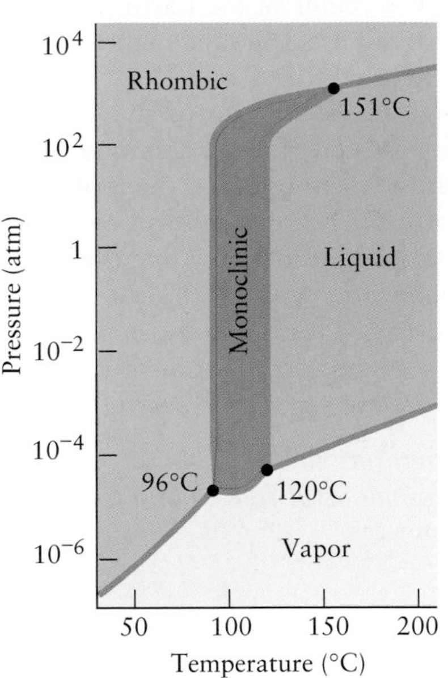

FIGURE 8.7 The phase diagram for sulfur. Notice that there are two solid phases and three triple points. The pressure scale, which is logarithmic, covers a very wide range of values.

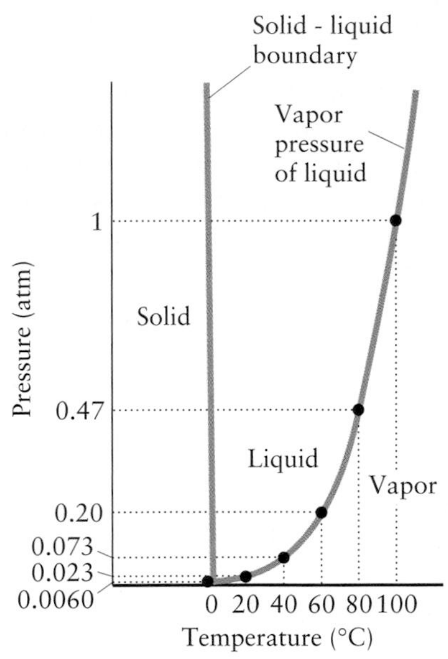

FIGURE 8.8 The liquid-vapor boundary curve is a plot of the vapor pressure of the liquid (in this case, water) as a function of temperature. The liquid and its vapor are in equilibrium at each point on the curve.

The lines separating the regions in a phase diagram are called **phase boundaries.** At any point on a boundary between two regions, the two neighboring phases coexist in dynamic equilibrium. For example, the point at 80°C and 0.47 atm in the phase diagram for water lies on the phase boundary between liquid and vapor (Fig. 8.8), so we know that, at 80°C, liquid water is in equilibrium with its vapor when the pressure of the latter is 0.47 atm. In other words, the point shows that the vapor pressure of water at 80°C is 0.47 atm. The liquid-vapor phase boundary is simply a plot of the vapor pressure of the liquid as a function of temperature. Similarly, the solid-vapor phase boundary shows how the vapor pressure of the solid varies with temperature (see Fig. 8.5).

The solid-liquid boundary, the almost vertical line in Figs. 8.5 and 8.8, shows the pressures and temperatures at which solid and liquid water coexist in equilibrium. In other words, it shows how the melting point varies with pressure. To show the slopes more clearly, the plots are not to scale. The steepness of the lines shows that even large changes in pressure result in quite small variations in melting point.

A **triple point** is a point where three phase boundaries meet. For water, the triple point for the solid, liquid, and vapor phases lies at 4.6 Torr and 0.01°C (see Fig. 8.5). At the triple point, all three phases (ice, liquid, and vapor) coexist in dynamic equilibrium. Under these conditions, water molecules leave ice to become liquid and return to form ice at the same rate; liquid vaporizes and vapor condenses at the same rate; and ice sublimes and vapor condenses directly to ice again at the same rate. The location of a triple point of a substance is a fixed property of that substance and cannot be changed by changing the conditions. The triple point of water is used to define the size of the kelvin: by definition, there are exactly 273.16 kelvins between absolute zero and the triple point of water. The normal freezing point of water is found to lie 0.01 K below the triple point, so 0°C corresponds to 273.15 K.

A phase diagram summarizes the regions of pressure and temperature at which each phase of a substance is most stable. The phase boundaries show the conditions under which two phases can coexist in dynamic equilibrium with each other. Three phases coexist in mutual equilibrium at a triple point.

Self-Test 8.4A From the phase diagram for carbon dioxide (Fig. 8.6), predict which is more dense, the solid or the liquid phase. Explain your conclusion.

[*Answer:* The solid, because it becomes more stable at higher pressures.]

Self-Test 8.4B From the phase diagram for sulfur (Fig. 8.7), predict which phase is more dense, liquid sulfur or monoclinic sulfur? Explain your conclusion.

8.7 The Phase Rule

In the range of pressures and temperatures shown in Fig. 8.7, we see that sulfur can exist in any of four phases: two solid phases (rhombic and monoclinic sulfur), one liquid phase, and one vapor phase. There are three triple points in the diagram, where various combinations of these phases, such as monoclinic solid, liquid, and vapor or monoclinic solid, rhombic solid, and liquid, can coexist.

The simultaneous existence of *four* phases in mutual equilibrium in a one-component system has never been observed and is ruled out by a famous result from thermodynamics. The same Gibbs who introduced the concept of free energy derived this result by considering a system consisting of c components (individual substances) and p phases in mutual equilibrium. For example, a one-component system such as water has $c = 1$; and if the two phases ice and liquid water are in equilibrium, $p = 2$. He was able to show that the number of **degrees of freedom,** f, or number of intensive variables (such as the pressure or temperature) that can be changed independently without disturbing the number of phases in equilibrium, is given by the **phase rule:**

$$f = c - p + 2 \qquad (5)$$

In this section, we consider only one-component systems, such as pure water, sulfur, or carbon dioxide. Therefore, we set $c = 1$ and obtain

$$f = 3 - p \qquad (6)$$

Because the phase rule is concerned with the number of phases in equilibrium, we should be able to justify it thermodynamically.

How do we do that?

Although we can't be sure, Gibbs probably thought about a one-component system first and then generalized his conclusions to an arbitrary number of components. Let's try to imagine how Gibbs took the first step in his derivation of the phase rule.

When two phases of a substance are in equilibrium, their molar free energies are equal. Each molar free energy depends on the pressure and the temperature, so the condition for equilibrium of two phases α and β at a given pressure and temperature is

$$G_m(\alpha; P,T) = G_m(\beta; P,T)$$

Because the values of G_m depend on P and T, this expression can be regarded as an equation relating P and T. That is, we have an equation relating two unknowns (P and T); so if P is set, T must have a corresponding value. Only one of the variables, P or T, is independent, so $f = 1$, as given by the phase rule in Eq. 6. Now consider three phases α, β, and γ in equilibrium. Three equations must be satisfied:

$$G_m(\alpha; P,T) = G_m(\beta; P,T) \qquad G_m(\beta; P,T) = G_m(\gamma; P,T)$$
$$G_m(\gamma; P,T) = G_m(\alpha; P,T)$$

The first two equalities imply the third (if $a = b$ and $b = c$, then necessarily $a = c$); so there are actually only two equations to satisfy. We can think of the two equations as two simultaneous equations in two unknowns (P and T). Such a pair of equations is satisfied for fixed values of the two unknowns (just as $x + y = 1$ and $3x + 2y = 4$ have the single solution $x = 2$ and $y = -1$). That is, we have no freedom to choose our own values of either P or T, so $f = 0$, just as the phase rule predicts.

Let's see how Eq. 6 applies to the phase diagrams we have encountered so far. When two phases are in equilibrium, such as liquid water and water vapor, we set $p = 2$ and obtain $f = 3 - 2 = 1$. One degree of freedom means that we are free to vary *either* the pressure *or* the temperature but not both. To preserve the equilibrium when we vary the pressure, we have to adjust the temperature by an appropriate amount; to preserve the equilibrium when we vary the temperature, the pressure must change appropriately. That is, the pressures and temperatures at which two phases are present in a one-component system are not independent of each other, and the relation between them is represented by a *line* on the phase diagram. As the temperature is changed, the pressure changes as indicated by the phase boundary. When three phases are in mutual equilibrium, we set $p = 3$ and find that $f = 3 - 3 = 0$. The condition is "invariant": neither temperature nor pressure can be varied. That is why the condition for the mutual existence of three phases in equilibrium is represented by a *point*, the triple point.

Now let's see what the phase rule tells us about a one-component system in which there are four phases in equilibrium (such as both forms of sulfur, sulfur liquid, and sulfur vapor). We set $p = 4$ and obtain $f = -1$. It is impossible to have a negative number of variables, so we have to conclude that *four phases cannot be in mutual equilibrium in a one-component system*. There are no points in a one-component phase diagram where four lines meet.

We can use a phase diagram and the phase rule to explain the changes that occur when we reduce the pressure on a liquid. Imagine that we have a sample of water in a cylinder fitted with a piston, that the temperature is constant at 50°C, and that the weights on the piston exert a pressure of 1.0 atm (Fig. 8.9). The piston presses on the surface of the liquid, as in (a). Now we gradually reduce the pressure by removing some of the weights (b). At first nothing seems to happen. The high pressure is keeping all the water molecules in the liquid state and the volume of a liquid changes very little with pressure. However, when so many weights have been removed that the pressure has fallen to 0.12 atm (93 Torr, the vapor pressure of water at that temperature), vapor begins to appear (c). We are now at the vapor-liquid boundary on the phase diagram. Because $f = 1$ for this one-component system and we have fixed the temperature, the pressure must remain constant as long as the liquid and vapor phases are present in equilibrium. As we pull the piston up by an arbitrary extent (d), enough water evaporates to maintain the pressure at 0.12 atm. The sample stays on the liquid-vapor boundary until no more liquid is left. Now, as weights are removed from the piston, we move into the "vapor" region (e). The liquid has evaporated completely and the vapor drives the piston out in accordance with the ideal gas law (f). Because only one phase, the vapor, is now present, p (the number of phases) falls to 1 and f rises to 2. Even though the temperature is maintained at a constant value, the pressure now has the freedom to fall again.

The phase rule, $f = c - p + 2$, implies that, for a one-component system, a line represents the equilibrium of two phases, and a point represents the mutual equilibrium of three phases.

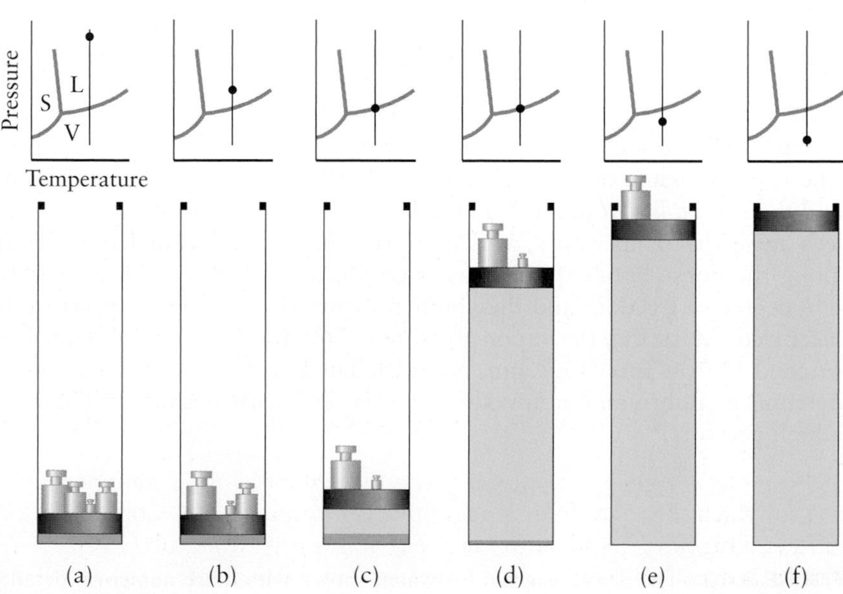

FIGURE 8.9 The changes undergone by a liquid as its pressure is decreased at constant temperature. The dot on the vertical line traces the path taken by the system, which is described in the text. The blue region is the liquid and the light green region is the vapor.

EXAMPLE 8.4 **Interpreting a phase diagram**

Use the phase diagram in Fig. 8.5 to describe the physical states and phase changes of water as the pressure on it is increased from 5 Torr to 800 Torr at 70°C.

STRATEGY First, locate the points corresponding to the initial and final conditions on the phase diagram. The region in which each point lies shows the stable phase of the sample under those conditions. If a point lies on one of the curves, then both phases are present in mutual equilibrium.

SOLUTION Although the phase diagram in Fig. 8.5 is not to scale, we can find the approximate locations of the points we need. Point A is at 5 Torr and 70°C, so it lies in the vapor region. Increasing the pressure takes the system to the liquid-vapor phase boundary, at which point liquid begins to form. At this pressure, liquid and vapor are in equilibrium and the pressure remains constant until all the vapor has condensed. The pressure is increased further to 800 Torr, which takes it to point B, in the liquid region. The sample is a liquid in this region because the applied pressure is greater than the vapor pressure and all the molecules have been pushed back into the condensed phase.

Self-Test 8.5A The phase diagram for carbon dioxide is shown in Fig. 8.6. Describe the physical states and phase changes of carbon dioxide as it is heated at 2 atm from −155°C to 25°C.

[*Answer:* Solid CO_2 is heated until it begins to sublime at the solid-vapor boundary. The temperature remains constant until all the CO_2 has vaporized. The vapor is then heated to 25°C.]

Self-Test 8.5B Describe what happens when liquid carbon dioxide in a container at 60 atm and 25°C is released into a room at 1 atm and the same temperature.

8.8 Critical Properties

A surprising feature of the phase diagram in Fig. 8.10 (and, in fact, in all phase diagrams) is that the liquid-vapor boundary comes to an end at point C. To see what happens at that point, suppose that the tube shown in Fig. 8.11 contains liquid water and water vapor at 25°C and 24 Torr (the vapor pressure of water at 25°C). The two phases are in equilibrium, and the system lies at point A on the liquid-vapor curve in Fig. 8.10. Now let's raise the temperature, which moves the system from left to right along the phase boundary. Because two phases are present, there is only one degree of freedom and the vapor pressure rises as the temperature is increased. At 100°C, the vapor pressure is 760 Torr; and at 200°C, it has reached 11 700 Torr (15.4 atm, point B). The liquid and vapor are still in dynamic equilibrium, but now the vapor is very dense because it is at such a high pressure.

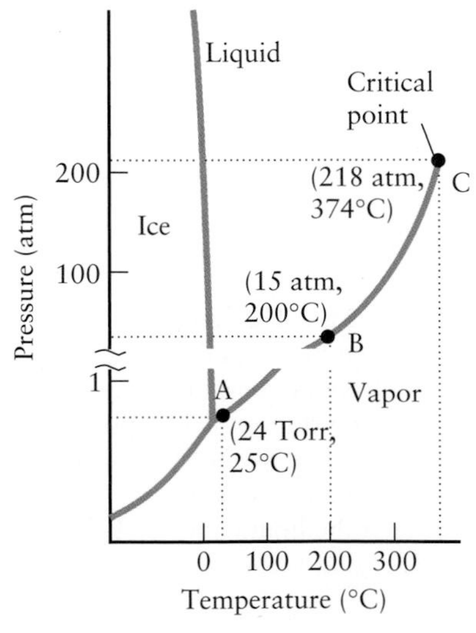

FIGURE 8.10 The phase diagram for water, shown with more numerical detail. C is the critical point. Pressures are in atmospheres, except for the triple point.

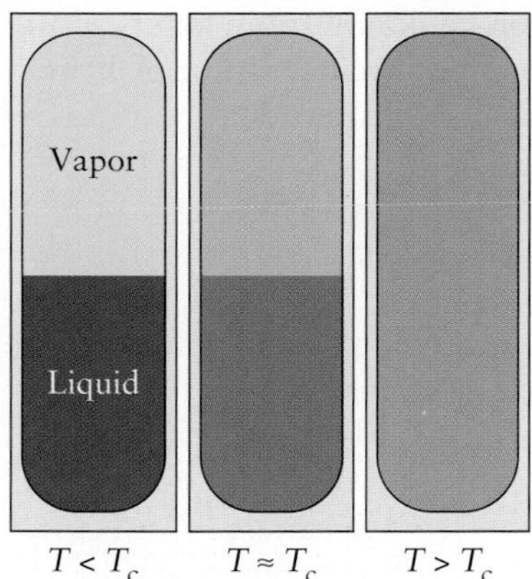

FIGURE 8.11 When the temperature of a liquid in a sealed, constant-volume container (left) is raised, the density of the liquid decreases and the density of the vapor increases (center). At the critical temperature, T_c, the density of the vapor becomes the same as the density of the liquid; at and above that temperature, a single uniform phase fills the container (right).

TABLE 8.4 Critical Temperatures and Pressures of Selected Substances

Substance	Critical temperature, °C	Critical pressure, atm
He	−268 (5.2 K)	2.3
Ne	−229	27
Ar	−123	48
Kr	−64	54
Xe	17	58
H_2	−240	13
O_2	−118	50
H_2O	374	218
N_2	−147	34
NH_3	132	111
CO_2	31	73
CH_4	−83	46
C_6H_6	289	49

When the temperature has been increased to 374°C (point C), the vapor pressure has reached 218 atm—the container must be very strong! The density of the vapor is now so great that it is equal to that of the remaining liquid. At this stage, the surface separating the liquid from its vapor vanishes; and from this point on, we can no longer identify the liquid phase. Instead, a single uniform phase fills the container. Because a substance that fills any container it occupies is by definition a gas, we conclude that we have reached the **critical temperature,** T_c, of water, the temperature at and above which it cannot be condensed to a liquid or a solid. The critical temperature of water is 374°C; that of carbon dioxide is only 31°C (Table 8.4). The vapor pressure at the critical point is called the **critical pressure,** P_c, of the substance. The critical pressure of water is 218 atm; that of carbon dioxide is 73 atm.

A gas can be liquefied by applying pressure only if it is below its critical temperature (Fig. 8.12). For example, carbon dioxide can be liquefied by applying pressure only if its temperature is lower than 31°C. According to Table 8.4, the critical temperature of oxygen is −118°C, so we know that it cannot exist as a liquid at room temperature whatever the

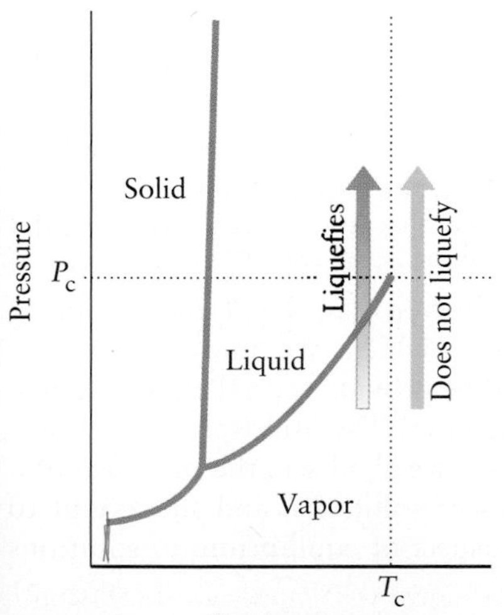

FIGURE 8.12 When pressure is applied to a vapor below its critical temperature, the vapor may be liquefied. However, if the temperature is above the critical temperature (the vertical dotted line), pressure does not take the vapor into the liquid region. At high pressures above its critical temperature, the substance becomes a supercritical fluid.

pressure. We know, therefore, that oxygen tanks at room temperature contain only gaseous oxygen. Because the critical temperature of helium is 5.2 K, the gas must be cooled almost to absolute zero before it can be liquefied. The distinction between gases and vapor is sometimes made on the basis of their critical temperatures: a vapor is a gaseous form of matter that can be liquefied by pressure alone; a gas is a form of matter that cannot be liquefied by pressure alone.

The dense fluid that exists above the critical temperature and pressure of a substance is called a **supercritical fluid.** It may be so dense that, although it is formally a gas, it is as dense as a liquid phase and can act as a solvent for liquids and solids. Supercritical carbon dioxide, for instance, can dissolve organic compounds. It is used to remove caffeine from coffee beans and to extract perfumes from flowers without contaminating the extracts with potentially harmful solvents. The carbon dioxide is easily separated by allowing it to evaporate at a lower pressure. Supercritical hydrocarbons are used to dissolve coal and separate it from ash, and they have been proposed for extracting oil from oil-rich tar sands.

A substance can be converted to a liquid by the application of pressure only if it is below its critical temperature.

Self-Test 8.6A Identify trends in the data in Table 8.4 that indicate the effect of increasing London forces on the critical temperature.

[*Answer:* Noble gases have higher critical temperatures as their atomic number increases, so critical temperature increases with the strength of London forces.]

Self-Test 8.6B Identify trends in the data in Table 8.4 that indicate the effect of hydrogen bonding on the critical temperature.

SOLUBILITY

So far we have considered equilibria in pure, one-component systems. When another substance dissolves in such a system, it produces a solution—a homogeneous mixture of two or more components—which can be solid, liquid, or gas. The solute affects the physical properties of the solvent. Solid solutions of solutes in metalloid and nonmetal solvents, such as silicon "doped" with a tiny amount of phosphorus as solute, exhibit electrical properties that make them the primary materials of the electronics industry (Section 3.14). When we spread salt on icy sidewalks, we are creating a mixture of salt and water that lowers the freezing point of water. If that temperature is lower than the ambient temperature, the ice melts. Gaseous solutions—which are more commonly referred to simply as mixtures—are used to provide special environments for deep-sea divers and food storage facilities. The atmosphere is another example of a huge gaseous solution. As we shall see, the presence of a solute affects the equilibrium properties of solutions, and the extent to which a substance dissolves is the first aspect of equilibrium in solutions that we consider.

8.9 The Molecular Nature of Dissolving

What would we see at the molecular level if we could follow the events that occur when a solution forms? Suppose, for example, we add a crystal of glucose to some water. At the surface of the crystal, the glucose molecules are in contact with water molecules, and hydrogen bonds begin to form between the two. The surface glucose molecules are pulled into solution by water molecules but are held back by other glucose molecules. When the water molecules are successful, the surface glucose molecules break away from the crystal and drift off into the solvent, surrounded by water molecules. A similar process occurs when an ionic solid dissolves. The polar water molecules hydrate the ions (surround them in a closely held "shell") and pry them away from the crystal lattice (Fig. 8.13). Stirring and shaking speed the process because they bring more free water molecules to the surface of the solid and sweep the hydrated ions away.

We can get more insight into the process if we think about what happens as we try to dissolve more and more solute. If we add only a small amount—20 g, say—of glucose to 100 mL of water at room temperature, it all dissolves. However, if we add 200 g, some glucose remains undissolved (Fig. 8.14). A solution is said to be **saturated** when the solvent has dissolved all the solute it can and some undissolved solute remains. The concentration of solid solute in a saturated solution has reached its greatest value and no more can dissolve. In other words, a saturated solution represents the limit of a solute's ability to dissolve in a given quantity of solvent.

The **molar solubility,** *s*, of a substance is its molar concentration in a saturated solution. In a saturated solution, any solid solute present still continues to dissolve, but the rate at which it dissolves exactly matches the rate at which the solute returns to the solid (Fig. 8.15). In other words,

View an animation of sodium chloride dissolving.

Media Link Animation: Fig. 8.13, NaCl dissolving.

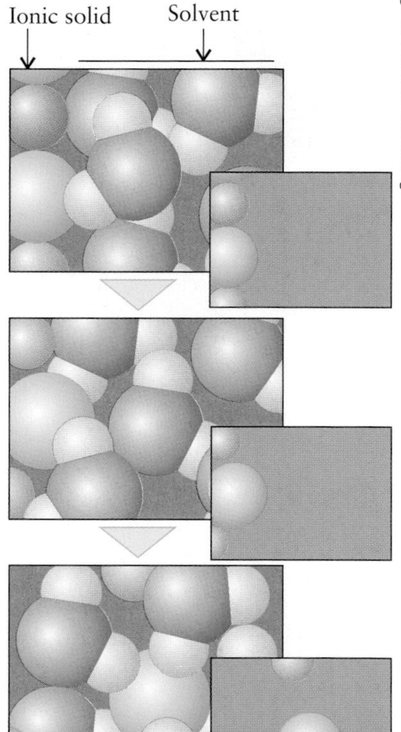

FIGURE 8.13 The events that take place at the interface of a solid solute and a solvent (water). The ions at the surface of the solid become hydrated, and as they do so, they move off into the solution.

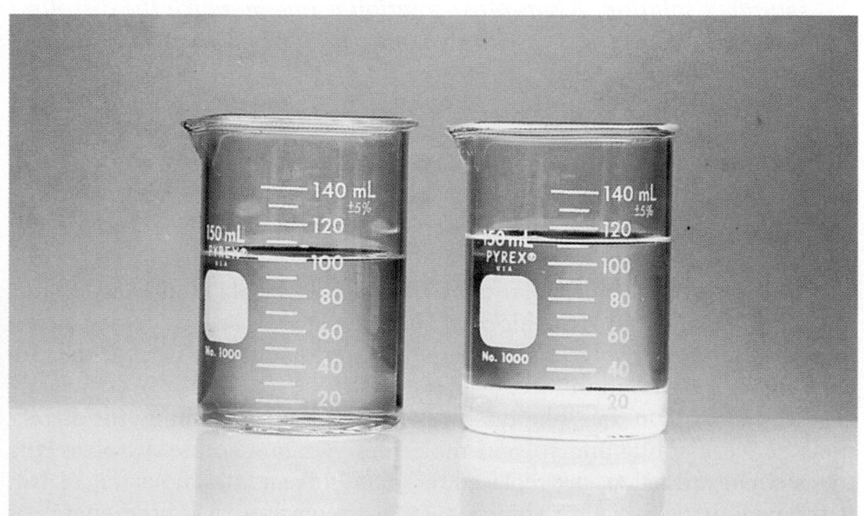

FIGURE 8.14 When a little glucose is stirred into 100 mL of water, it all dissolves (left). However, when a large amount is added, some undissolved glucose remains and the solution is saturated with glucose (right).

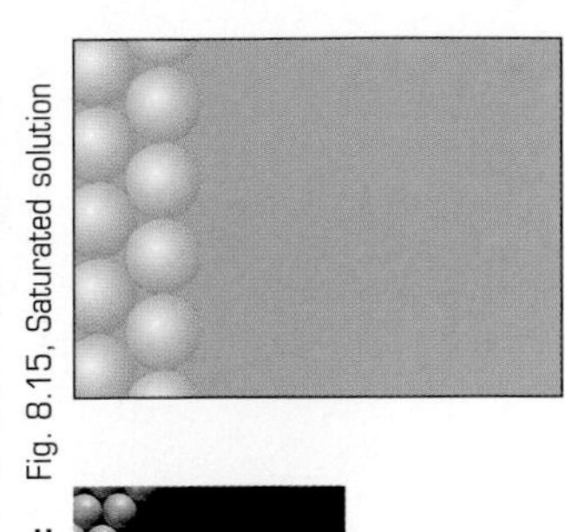

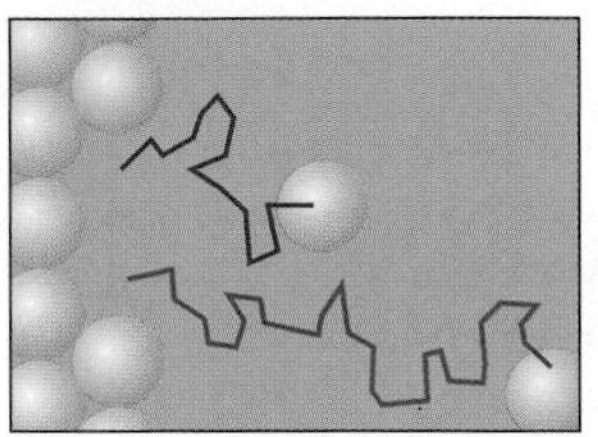

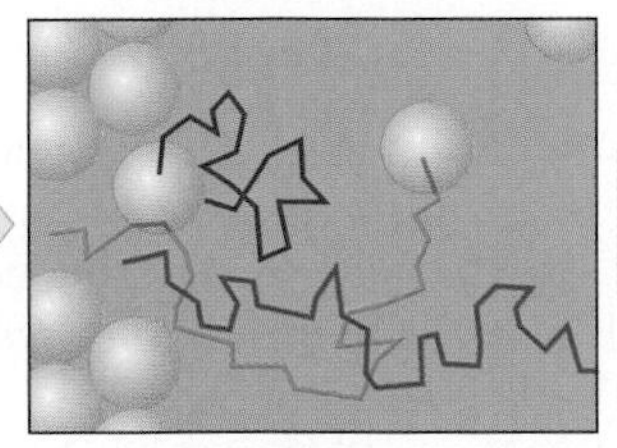

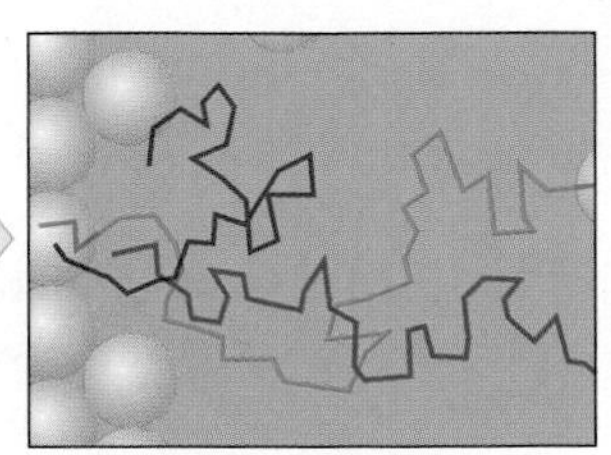

FIGURE 8.15 The solute in a saturated solution is in dynamic equilibrium with the undissolved solute. If we could follow the history of the solute particles (yellow spheres), we would sometimes find them in solution and sometimes back within the solid. Red, green, and blue lines represent the paths of individual solute particles. The solvent molecules are not shown.

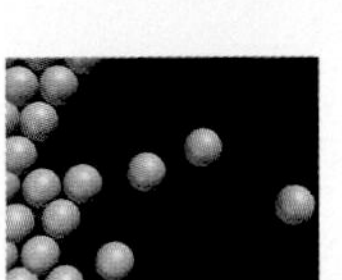

Media Link **Animation:** Fig. 8.15, Saturated solution

View an animation of the dynamic equilibrium in a saturated solution.

in a saturated solution, the dissolved and undissolved solute are in dynamic equilibrium with each other. For example, in a saturated solution of the ionic compound calcium hydroxide, the dynamic equilibrium is

$$Ca(OH)_2(s) \rightleftharpoons Ca^{2+}(aq) + 2\ OH^-\ (aq)$$

with the ions leaving and returning to the surface of the solid at matching rates.

The results of a simple experiment confirm that the equilibrium in a saturated solution is dynamic. First, we prepare a saturated solution of silver chloride. Then we add a little solid silver chloride containing radioactive ^{112}Ag to the solution. After a while, the solution itself is found to be radioactive. We can conclude that exchange of silver ions is still taking place between the solid and the solution even though the solution is saturated.

The molar solubility of a substance is its molar concentration in a saturated solution. A saturated solution is one in which the dissolved and undissolved solute are in dynamic equilibrium with each other.

8.10 The Like-Dissolves-Like Rule

Some fundamental principles of science underlie the answers to simple, everyday questions. Suppose, for instance, we need to remove some spilled wax. How can we know what solvent to use? A good guide is the rule that *like dissolves like*. A polar liquid, such as water, is generally the best solvent for ionic and polar compounds. Conversely, nonpolar liquids, including hexane and tetrachloroethene, $Cl_2C{=}CCl_2$, used for dry cleaning, are often better solvents for nonpolar compounds like wax, which are held together by London forces.

We can explain the "like-dissolves-like" rule by examining the attractions between solute and solvent molecules. When a solute dissolves, the solute-solute attractions are replaced by solute-solvent attractions (Fig. 8.16). If the new attractions are similar to those replaced, very little energy is required for the solution to form. For example, when the main cohesive forces in a solute are hydrogen bonds, the solute is more likely to dissolve in a hydrogen-bonded solvent than in other solvents. The molecules can

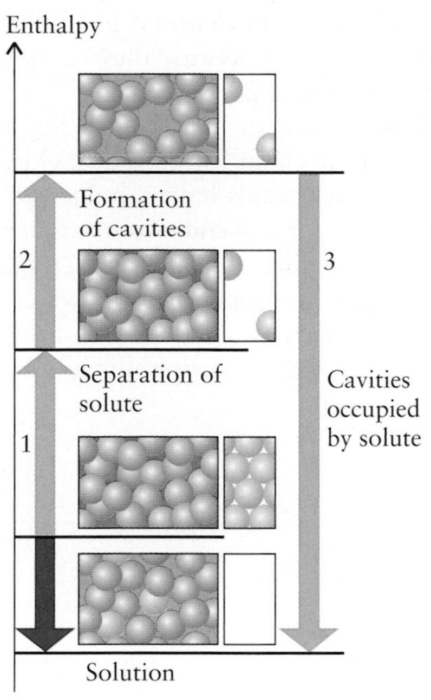

FIGURE 8.16 A molecular representation of the changes in molecular interactions and energy associated with the formation of a dilute solution. For the first step, solute molecules are separated from one another. Then (step 2) some of the solvent molecules move apart and leave cavities in the solvent. In the third step, the solute molecules occupy the cavities in the solvent and release energy. The overall energy change for the process is shown by the dark red arrow. In the actual dissolving process, these steps do not occur independently.

slip into solution only if they can replace their solute-solute hydrogen bonds with solute-solvent hydrogen bonds. Glucose, for example, has hydrogen bond-forming —OH groups and dissolves readily in water but not in hexane.

If the principal cohesive forces between solute molecules are London forces, then the best solvent is likely to be one that can mimic those forces. For example, a good solvent for nonpolar substances is the nonpolar liquid carbon disulfide, CS_2. It is a far better solvent than water for sulfur because solid sulfur is a molecular solid of S_8 molecules held together by London forces (Fig. 8.17). The sulfur molecules slip easily

FIGURE 8.17 The molecular solid sulfur does not dissolve in water (left), but it does dissolve in carbon disulfide (right), with which the S_8 molecules have favorable London interactions.

between the nonpolar carbon disulfide molecules. They cannot penetrate into the strongly hydrogen-bonded structure of water because they cannot replace those bonds with interactions of similar strength.

Soaps and detergents are practical applications of the like-dissolves-like rule. Soaps are the sodium salts of long-chain carboxylic acids, including sodium stearate (**5**). The anions of these acids have a polar carboxylate ($-CO_2^-$) group, called the *head group*, at one end of a nonpolar hydrocarbon chain. The head group is **hydrophilic,** or water-attracting, whereas the nonpolar hydrocarbon tails are **hydrophobic,** or water-repelling. The like-dissolves-like rule suggests that the hydrophilic head group of the anion ought to dissolve in water and the hydrophobic hydrocarbon tails ought to dissolve in grease.

Na^+ ^-O O

5 Sodium stearate, $NaCH_3(CH_2)_{16}CO_2$

This dual character makes soap very effective at removing grease. The hydrocarbon tails sink into the blob of grease up to the head groups. The hydrophilic head groups remain on the surface of the blob and enable the entire unit to dissolve in water (Fig. 8.18). The grease is carried away in a casing of soap molecules called a **micelle.** The hydrophobic tail of each soap molecule sticks into the grease blob at the center and its hydrophilic head points out, toward the surrounding water molecules.

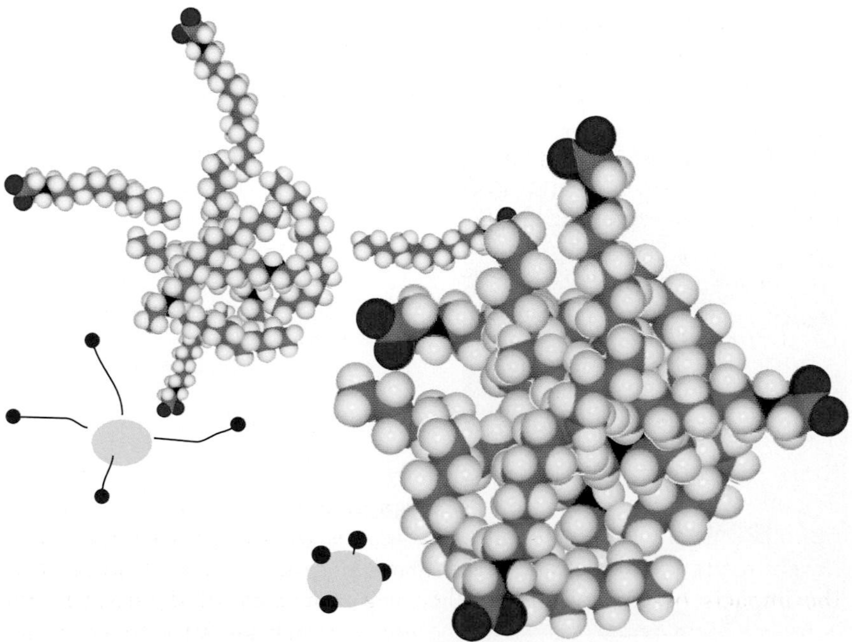

FIGURE 8.18 The hydrocarbon tails of a soap or surfactant molecule beginning to dissolve in grease (left). The water-attracting head groups remain on the surface where they can interact favorably with water (right). As more surfactant molecules dissolve in the grease, the whole blob, called a micelle, dissolves in water and is washed away.

Early soaps were made by heating sodium hydroxide with beef fat, which contains an *ester* formed between glycerol and stearic acid (see Section 19.7). The sodium hydroxide attacks the ester and forms the soluble soap sodium stearate. Soaps, however, form a scum in hard water. The scum is a precipitate of calcium stearate, which forms because calcium salts are less soluble than sodium salts. One way to avoid the formation of scum is to precipitate the Ca^{2+} ions from the solution before adding the soap. For example, sodium carbonate (washing soda) can be added to water to precipitate Ca^{2+} ions as calcium carbonate:

Hard water is discussed in Section 8.14.

$$Ca(HCO_3)_2(aq) + Na_2CO_3(aq) \longrightarrow CaCO_3(s) + 2\ NaHCO_3(aq)$$

Modern commercial detergents are mixtures. Their most important component is a **surfactant**, or *surface-active agent*, which takes the place of the soap. Surfactant molecules are organic compounds that resemble the one shown as (**6**). Like the stearate ion, they have a hydrophilic head group and a hydrophobic tail and act similarly. Most commercial detergents also contain additives such as fillers (to generate bulk), foam-reducing agents, and bleach. Often, additives that *fluoresce* (absorb ultraviolet radiation and then emit visible light) are added to detergents to give the impression of greater cleanliness in sunlight, which contains some ultraviolet radiation.

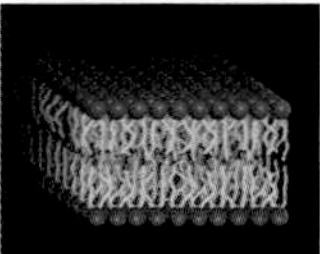

View an animation of a bilayer membrane in action.

Media Link Animation: Fig. 8.19, Bilayer membranes.

SO_3^-

6 A typical surfactant ion

The membranes of living cells are double layers of surfactant molecules called *phospholipids* that line up with their hydrocarbon tails pointing into the membrane and the polar head groups forming the membrane surface. It is possible to coax phospholipids to form artificial membranes that close up to form tiny bags, called *liposomes*, that are much smaller than living cells (Fig. 8.19). If the liposomes are formed in a solution containing a drug, some drug molecules become encapsulated. This behavior has led to the study of liposomes as a way of delivering drugs into different organs and regions of the body (Box 8.1).

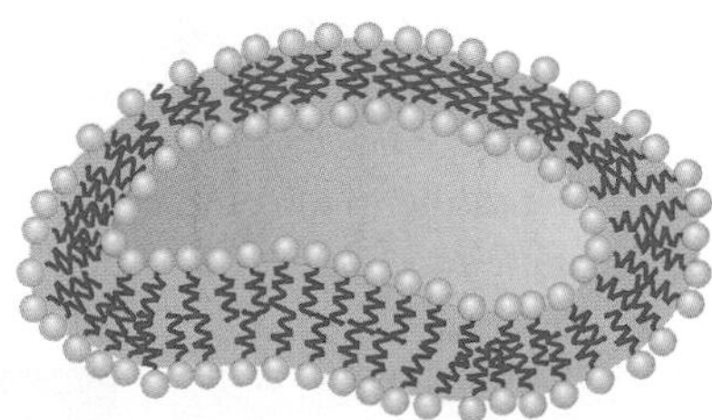

FIGURE 8.19 A cross-section of a liposome, in which a bilayer membrane is formed from surfactantlike phospholipid molecules.

A general guide to the suitability of a solvent is the rule that like dissolves like. Soaps and detergents contain surfactant molecules that have both hydrophobic and hydrophilic regions.

8.11 Pressure and Gas Solubility: Henry's Law

Every fish we see reminds us that gases dissolve in liquids, for almost all aquatic organisms rely on the presence of dissolved oxygen for respiration. The extent to which oxygen dissolves depends on its pressure. We have already seen (in Section 4.2) that the pressure of a gas arises from the impacts of its molecules. When a gas is introduced into the same container as a liquid, the gas molecules can burrow into the liquid like

BOX 8.1 FRONTIERS OF CHEMISTRY: DRUG DELIVERY

Providing drugs to ease disease and chronic, severe pain or to provide benefits such as hormone replacement therapy is difficult because drugs taken orally may lose much of their potency in the harsh conditions of the digestive system. In addition, they are distributed throughout the entire body, not just where they are needed, and side effects can be significant. Recently, however, new techniques have been developed to deliver drugs gradually over time, to the exact location in the body where they are needed, and even at the time when they are needed.

Transdermal patches are applied to the skin. The drug is mixed with the adhesive for the patch, so it lies next to the skin. The skin can readily absorb many chemicals and so can absorb drugs such as nitroglycerin (for heart disease), morphine derivatives (for constant, severe pain), estrogen (for hormone replacement therapy), or nicotine (for easing symptoms that result when a patient stops smoking). Some patches must be renewed daily, but others can last for up to seven days. The patches are thin, lie flat against the skin, and allow gases to flow and moisture to evaporate from the skin.

Implants provide a means of delivering drugs over a longer period of time at a controlled rate inside the body. Subcutaneous implants are used to provide appropriate doses of psychoactive medications, birth-control drugs, painkillers, and other medications that must be administered frequently. The implants last for as long as a month and can easily be replaced or renewed. When the location of the drug release is critical, implants can be placed deeper into the body. For example, implants can be introduced into the brain or spinal column to provide effective relief from pain or to protect neurons from degeneration. The implant is contained inside a cylinder of porous foam through which the drug is released. Usually a long, thin plastic tail is attached as a tether to allow easy retrieval of the implant. Some implants contain living animal cells that have been engineered to produce natural hormones or painkillers that are released to the body as they are produced. In other types of implants, special membranes allow a gradual release of the drug.

Controlled-release drug delivery systems can also be fashioned from soft materials. Molecules called lipids are found in fats and also form the membranes of living cells. These molecules are similar in structure to the surfactant molecules in detergents: they have a polar head and a nonpolar hydrocarbon tail. Some of these lipids are found to form liquid crystal structures spontaneously in water (Section 15.13); in these structures, sheets consisting of rows of molecules are lined up next to one another. The sheets can be coaxed into forming liposomes (see Fig. 8.19). Liposomes are similar to micelles, but are formed from a double layer of molecules, like the membrane of a living cell. When a drug is present in the aqueous solution in which liposomes are being formed,

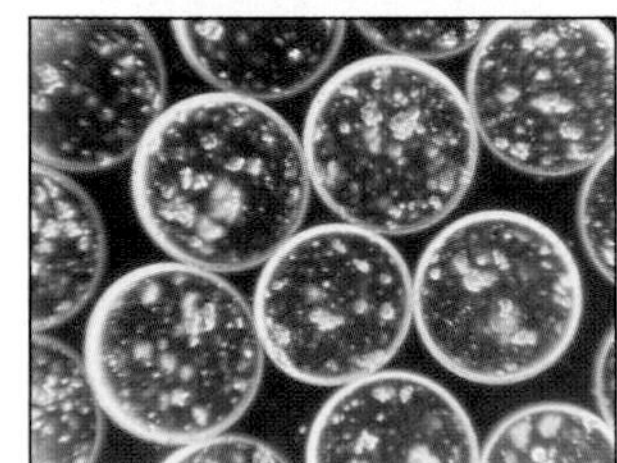

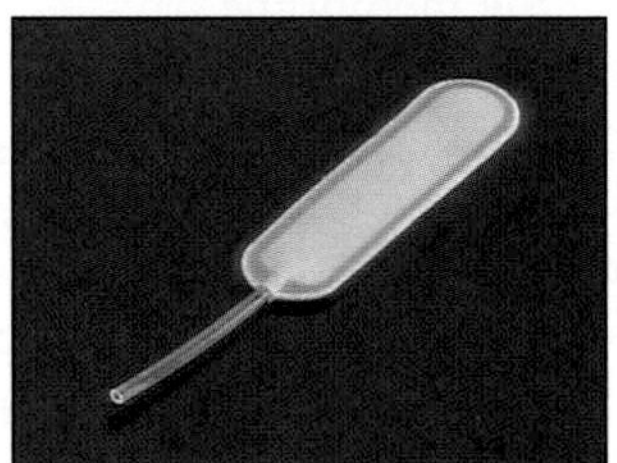

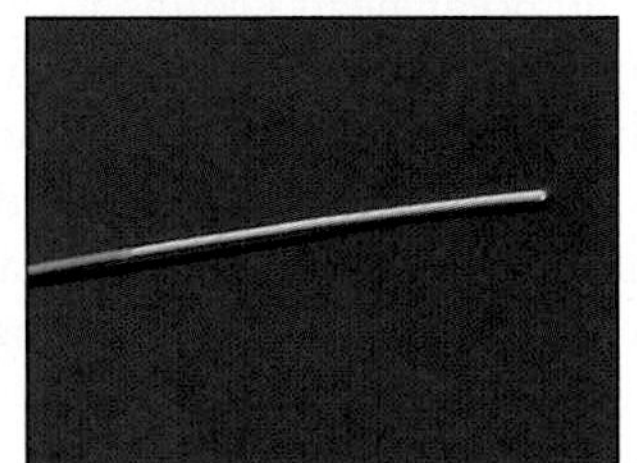

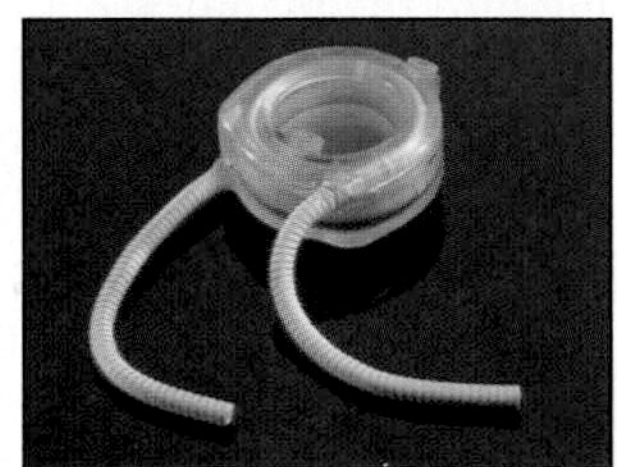

Examples of implants used to insert living cells into the body. The cells continually produce enzymes, hormones, or painkillers needed by the body.

meteorites plunging into the ocean. Because the number of impacts increases as the pressure of a gas increases, we should expect the solubility of the gas—its molar concentration when the dissolved gas is in dynamic equilibrium with the free gas—to increase as its pressure increases. If the gas above the liquid is a mixture (like air), then the solubility of each component depends on that component's partial pressure (Fig. 8.20).

some of the drug is encapsulated inside each liposome, thereby forming a tiny container for the drug. The liposomes can then be injected into the body, where they stick only to certain types of cells—cancer cells, for instance. By this means, the drug is delivered to target cells. A smaller dosage is required and side effects are greatly reduced. Liposome drug delivery has already been used to treat patients with leukemia.

Smart gels (see Box 15.1) are also actively being investigated for drug delivery. The use of these gels in medicine is promising because the molecules making up the gels are similar to biological molecules. For example, the delivery of insulin to diabetics is fraught with uncertainty. The amount of insulin a nondiabetic person needs is delivered by the body only when blood sugar rises. However, a diabetic person must take insulin at specified times of the day and the same amount each time. If the blood sugar is already low, a hypoglycemic reaction, and possibly coma, can result. An insulin delivery system that would be responsive to blood sugar levels and provide the correct dosage at the time it is needed is now being investigated. The system makes use of a smart gel in which molecules of insulin have been trapped. The gel incorporates into its structure molecules of phenylboronic acid, to which glucose (blood sugar) molecules are attracted. If the blood sugar is high, more and more glucose molecules stick to the gel and cause it to swell. When the level of glucose rises above a certain concentration, the gel swells so much that it becomes porous, releasing insulin into the blood.

A cross section through one of the two layers of a liposome membrane. The long, narrow molecules are aligned with their polar heads toward the surfaces of the membrane.

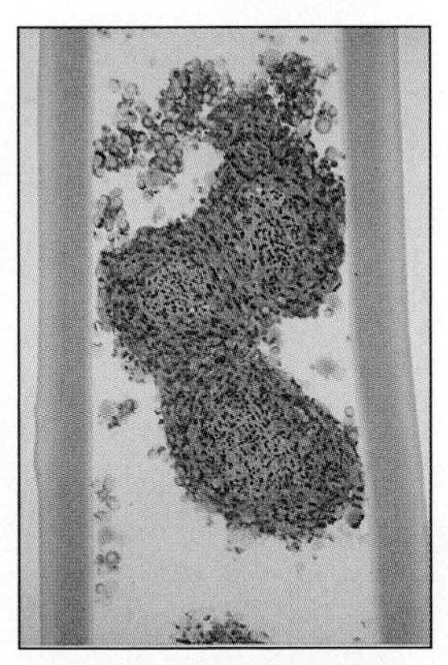

This implant containing live hamster cells was inserted into the spinal column of a patient for 17 weeks. After removal, the cells were still alive and secreting the hormone required to keep the patient healthy.

HOW MIGHT YOU CONTRIBUTE?

Drug delivery systems must be designed for each specific drug. Therefore, a great deal of research will be required before new drug delivery systems are available for all candidate drugs. The length of time a drug delivery system can remain active inside the body and the stability of the system need to be increased. New drug delivery systems also need rigorous testing for safety and effectiveness.

For Further Reading: C. M. Henry, "Special delivery," *Chemical and Engineering News* (September 18, 2000), pp. 49–64. M. J. Lysaght and P. Aebischer, "Encapsulated cells as therapy," *Scientific American* (April 1999), pp. 76–82. N. Platé and I. L. Valuev, "Drug delivery by 'smart' polymer gels," *Science Spectra* (Issue 22, 2000), pp. 60–69.

Deep-sea divers breathe compressed air. Nitrogen is not very soluble in blood at normal pressures; but at great depths, when the divers' bodies are exposed to very high pressures, the nitrogen becomes more soluble. The dissolved nitrogen comes out of solution rapidly when the divers return to the surface, and numerous small bubbles form in the bloodstream. These bubbles can burst the capillaries—the narrow vessels that

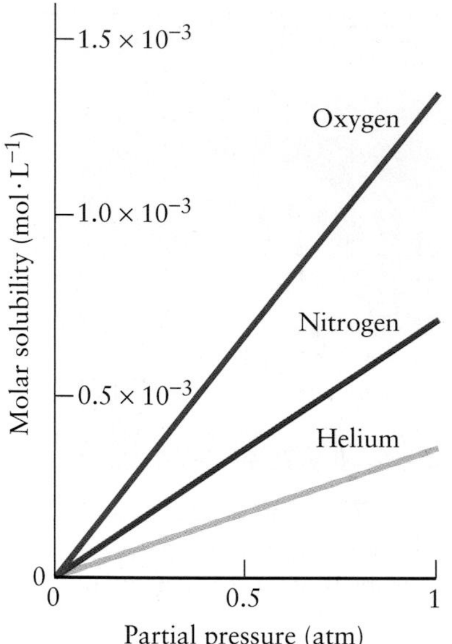

FIGURE 8.20 The variation of the molar solubilities of oxygen, nitrogen, and helium gases with partial pressure. Note that the solubility of each gas doubles when its partial pressure is doubled.

distribute the blood—or block them and starve the tissues of oxygen (Fig. 8.21). This condition is called "the bends." In serious cases, the bends can be fatal. The risk is reduced if helium is used instead of nitrogen to dilute the diver's oxygen supply, because helium is less soluble in plasma than nitrogen. Moreover, because helium atoms are so small, they can pass more readily than nitrogen through cell walls without damaging them.

The straight lines in Fig. 8.20 show that *the solubility of a gas is directly proportional to its partial pressure, P*. This observation was first made in 1801 by the English chemist William Henry and is now known as **Henry's law.** The law is normally written

$$s = k_H P \qquad (7)$$

The constant k_H, which is called **Henry's constant,** depends on the gas, the solvent, and the temperature (Table 8.5). The law implies that, at constant temperature, doubling the partial pressure of a gas doubles its solubility.

The solubility of a gas is proportional to its partial pressure, because an increase in pressure corresponds to an increase in the rate at which gas molecules strike the surface of the solvent.

How can we use this information . . . to estimate the solubility of a gas in a liquid?

We can use Henry's law to verify that the concentration of oxygen in lake water is normally adequate to sustain aquatic life, which requires a concentration of at least 1.3×10^{-4} mol·L^{-1}. The partial pressure of oxygen is 0.21 atm at sea level, so the molar solubility of oxygen is

$$s = (1.3 \times 10^{-3}\ \text{mol·L}^{-1}\text{·atm}^{-1}) \times (0.21\ \text{atm}) = 2.7 \times 10^{-4}\ \text{mol·L}^{-1}$$

This molar concentration corresponds to about 8.6 mg of oxygen per liter of water (8.6 mg·L^{-1}), which is more than adequate to sustain life.

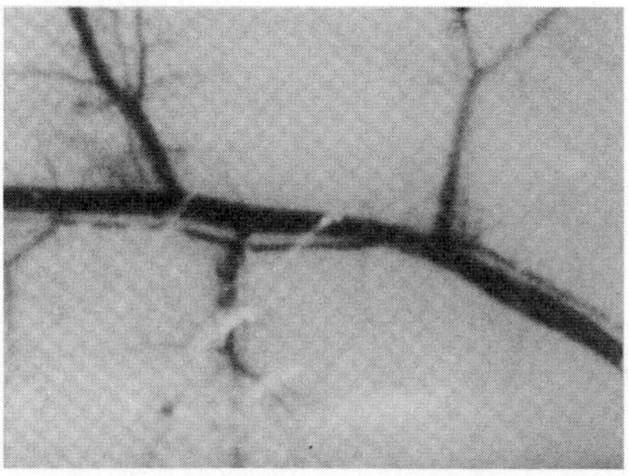

FIGURE 8.21 Small bubbles of nitrogen are responsible for the bends. (a) Normal blood vessels; (b) catastrophic collapse as bubbles escape from solution in the blood plasma.

Self-Test 8.7A At the elevation of Bear Lake in Rocky Mountain National Park, 2900 m, the partial pressure of oxygen is 0.14 atm. What is the molar solubility of oxygen in Bear Lake at 20°C?

[*Answer:* 1.8×10^{-4} mol·L^{-1}]

Self-Test 8.7B Use the information in Table 8.5 to calculate the amount (in moles) of CO_2 that will dissolve in enough water to form 900 mL of solution at 20°C if the partial pressure of CO_2 is 1.00 atm.

TABLE 8.5 Henry's Constants for Gases in Water at 20°C

Gas	k_H, mol·L^{-1}·atm^{-1}
air	7.9×10^{-4}
argon	1.5×10^{-3}
carbon dioxide	2.3×10^{-2}
helium	3.7×10^{-4}
hydrogen	8.5×10^{-4}
neon	5.0×10^{-4}
nitrogen	7.0×10^{-4}
oxygen	1.3×10^{-3}

8.12 Temperature and Solubility

Most substances dissolve more quickly at higher temperatures. However, that does not necessarily mean that they are more soluble—that is, reach a higher concentration of solute—at higher temperatures. In a number of cases, the solubility turns out to be lower at higher temperatures. We must always distinguish the effect of temperature on the *rate* of a process from its effect on the final outcome.

Most gases become less soluble as the temperature is raised. The lower solubility of gases in warm water is responsible for the tiny bubbles that appear when cool water from the faucet is left to stand in a warm room. The bubbles consist of air that dissolved when the water was cooler; it comes out of solution as the temperature rises. Most ionic and molecular solids are more soluble in hot water than in cold (Fig. 8.22). We make use of this characteristic in the laboratory to dissolve a substance and to grow crystals by letting a saturated solution cool slowly. However, a few solids, such as lithium sulfate, are less soluble at high temperatures than at low. A smaller number of compounds show a mixed behavior. For example, the solubility of sodium sulfate increases up to 32°C but then decreases as the temperature is raised further.

The solubility of most gases is lower at higher temperatures; solids show a more varied behavior.

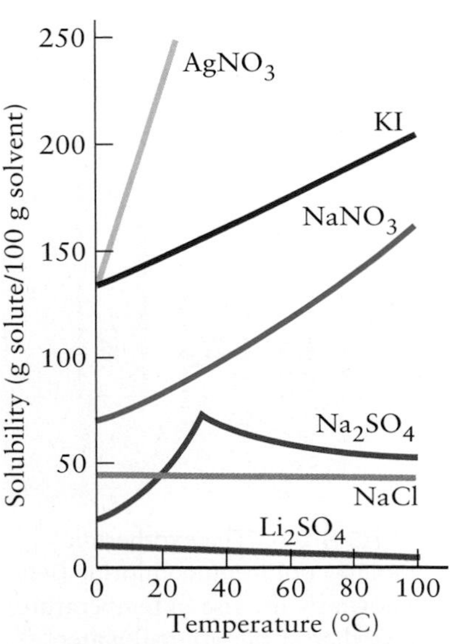

FIGURE 8.22 The variation with temperature of the solubilities of six substances in water.

8.13 The Enthalpy of Solution

"Like dissolves like" is a good rule because good solvents mimic the surroundings of the solute molecules: partial charges on polar solvent molecules play the role of ionic charges, hydrogen bonds replace hydrogen bonds, and London forces exerted by a solvent mimic those of the solute molecules themselves. But how do we assess these effects *quantitatively*? As always when considering equilibria, we should focus on the change in free energy that accompanies the process—in this case, dissolving. If $\Delta G < 0$, dissolving is spontaneous; if $\Delta G > 0$, the reverse of dissolving—precipitation or crystallization—is spontaneous. Because $\Delta G = \Delta H - T\Delta S$, we need to examine the changes in both enthalpy and entropy that occur when a substance dissolves to decide whether ΔG is positive or negative.

The change in enthalpy per mole of formula units when a substance dissolves is called the **enthalpy of solution, ΔH_{sol}**. The change in enthalpy can be measured calorimetrically from the heat released or absorbed when the substance dissolves at constant pressure. However, in concentrated solutions, solute particles interact with one another, and this interaction

TABLE 8.6 Limiting Enthalpies of Solution, ΔH_{sol}, at 25°C, in Kilojoules per Mole*

	Anion							
Cation	fluoride	chloride	bromide	iodide	hydroxide	carbonate	nitrate	sulfate
lithium	+4.9	−37.0	−48.8	−63.3	−23.6	−18.2	−2.7	−29.8
sodium	+1.9	+3.9	−0.6	−7.5	−44.5	−26.7	+20.4	−2.4
potassium	−17.7	+17.2	+19.9	+20.3	−57.1	−30.9	+34.9	+23.8
ammonium	−1.2	+14.8	+16.0	+13.7	—	—	+25.7	+6.6
silver	−22.5	+65.5	+84.4	+112.2	—	+41.8	+22.6	+17.8
magnesium	−12.6	−160.0	−185.6	−213.2	+2.3	−25.3	−90.9	−91.2
calcium	+11.5	−81.3	−103.1	−119.7	−16.7	−13.1	−19.2	−18.0
aluminum	−27	−329	−368	−385	—	—	—	−350

* The value for silver iodide, for example, is the entry found where the row labeled "silver" intersects the column labeled "iodide."

affects their enthalpies. To avoid this complication, data are normally reported as the **limiting enthalpy of solution**, the enthalpy change accompanying the formation of a very dilute solution (Table 8.6). The data show that some solids, like lithium chloride, dissolve exothermically, with a release of heat (Fig. 8.23). Others, like ammonium nitrate, dissolve endothermically.

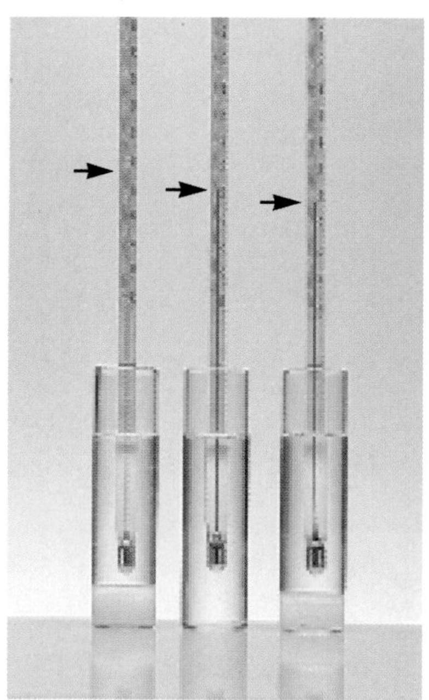

FIGURE 8.23 The exothermic dissolution of lithium chloride (left) is shown by the rise in temperature above that of the original water (center); in contrast, ammonium nitrate (right) dissolves endothermically.

To understand the values in Table 8.6, we can think of dissolving as a two-step process (Fig. 8.24). In the first hypothetical step, we imagine the ions separating from the solid to form a gas of ions. The change in enthalpy accompanying this highly endothermic step is the lattice enthalpy, ΔH_L, of the solid, which was introduced in Section 6.20 (see Table 6.6 for values). The lattice enthalpy of sodium chloride (787 kJ·mol^{-1}), for instance, is the molar enthalpy change for the process

$$NaCl(s) \longrightarrow Na^+(g) + Cl^-(g)$$

We saw in Section 6.20 that compounds formed from small, highly charged ions (such as Mg^{2+} and O^{2-}) cohere together strongly and that a lot of energy is needed to break up the lattice. Such compounds have high lattice enthalpies. Compounds formed from large ions with low charges, such as potassium iodide, have weak attractive forces and correspondingly low lattice enthalpies.

In the second hypothetical step, we imagine the gaseous ions plunging into water and forming the final solution. The molar enthalpy of this step is called the **enthalpy of hydration**, ΔH_{hyd}, of the compound (Table 8.7). Enthalpies of hydration are large and comparable to the lattice enthalpies of the compounds. For sodium chloride, for instance, the enthalpy of hydration, the enthalpy change for the process

$$Na^+(g) + Cl^-(g) \longrightarrow Na^+(aq) + Cl^-(aq)$$

is −784 kJ·mol^{-1}. That is enough energy for 1 g of NaCl (as a gas of Na^+ and Cl^- ions) to raise the temperature of 100 mL of water by nearly 50°C.

TABLE 8.7 Enthalpies of Hydration, ΔH_{hyd}, at 25°C, of Some Halides, in Kilojoules per Mole*

Cation	Anion F^-	Cl^-	Br^-	I^-
H^+	−1613	−1470	−1439	−1426
Li^+	−1041	−898	−867	−854
Na^+	−927	−784	−753	−740
K^+	−844	−701	−670	−657
Ag^+	−993	−850	−819	−806
Ca^{2+}	—	−2337	—	—

* The entry where the row labeled Na^+ intersects the column labeled Cl^-, for instance, is the enthalpy change, −784 kJ·mol^{-1}, for the process $Na^+(g) + Cl^-(g) \rightarrow Na^+(aq) + Cl^-(aq)$; the values here apply only when the resulting solution is very dilute.

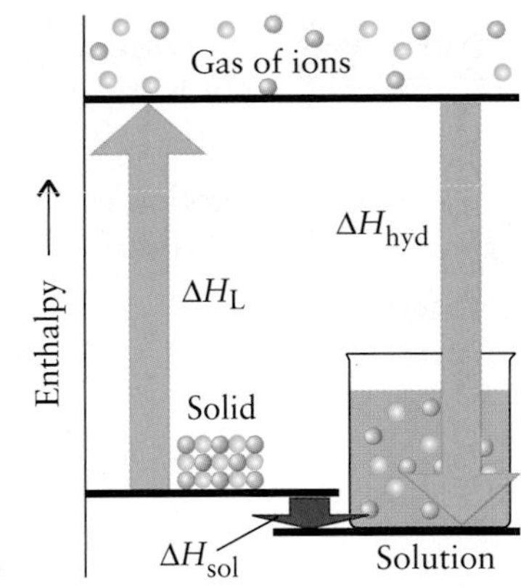

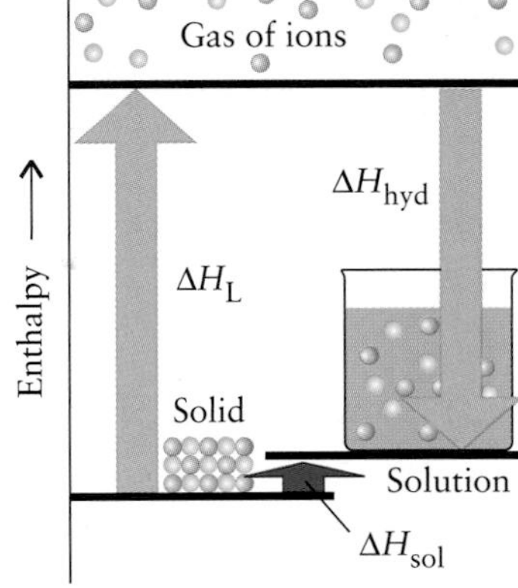

FIGURE 8.24 The enthalpy of solution, ΔH_{sol}, is the sum of the enthalpy change required to separate the molecules or ions of the solute, the lattice enthalpy, ΔH_L (step 1 in Fig. 8.16), and the enthalpy change accompanying their hydration, ΔH_{hyd} (steps 2 and 3 in Fig. 8.16). The outcome is finely balanced: (a) in some cases, it is exothermic; (b) in others, it is endothermic. For gaseous solutes, the lattice enthalpy is 0 because the molecules are already widely separated.

Hydration is always exothermic for ionic compounds because of the formation of attractive ion-dipole interactions between the water molecules and the ions. It is also exothermic for molecules that can form hydrogen bonds with water, such as sucrose, glucose, acetone, and ethanol.

Now we bring the two steps of the dissolving process together and calculate the enthalpy for the overall change. When we include the data, Fig. 8.24b implies that the limiting enthalpy of solution of sodium chloride, the enthalpy change for the process

$$NaCl(s) \longrightarrow Na^+(aq) + Cl^-(aq)$$

is

$$\Delta H_{sol} = \underset{\text{Lattice enthalpy}}{787 \text{ kJ·mol}^{-1}} - \underset{\text{Enthalpy of hydration}}{784 \text{ kJ·mol}^{-1}} = \underset{\text{Enthalpy of solution}}{+3 \text{ kJ·mol}^{-1}}$$

Because the enthalpy of solution is positive, there is a net inflow of energy as heat when the solid dissolves. Sodium chloride therefore dissolves endothermically, but only to the extent of 3 kJ·mol^{-1}. As this example shows, the overall change in enthalpy depends on a very delicate balance between the lattice enthalpy and the enthalpy of hydration.

Enthalpies of solution in dilute solutions can be expressed as the sum of the lattice enthalpy and the enthalpy of hydration of the compound.

8.14 Individual Ion Hydration Enthalpies

We can analyze the process of hydration still further and get a deeper understanding of the energy changes that take place during the formation of a solution. The entries in Table 8.7 are for gaseous cations and anions jointly entering solution. It would be more revealing to know the enthalpy changes for the cations and anions separately, for then we could predict

TABLE 8.8 Individual Ion Hydration Enthalpies

Ion	Hydration enthalpy, $kJ\cdot mol^{-1}$
Cations	
H^+	−1130
Li^+	−558
Na^+	−444
K^+	−361
Ag^+	−510
Be^{2+}	−1435
Mg^{2+}	−2003
Ca^{2+}	−1657
Sr^{2+}	−1524
Al^{3+}	−2537
Anions	
F^-	−483
Cl^-	−340
Br^-	−309
I^-	−296

the enthalpy of hydration of NaCl(g), for instance, by adding together the values for Na^+(g) and Cl^-(g), and could analyze the individual values in terms of the characteristics of the individual ions. Unfortunately, except in one special case, we cannot measure enthalpies of hydration of cations without anions being present, too. The exception is the hydrogen ion, for which the enthalpy of hydration has been measured by a special technique based on mass spectrometry and found to have the value $-1130\ kJ\cdot mol^{-1}$. That is, when 1 mol H^+(g) dissolves in a very large sample of water, the system releases 1130 kJ of energy as heat. This large value is the same as the energy obtained by burning about 50 mL of gasoline. With the value for the hydrogen ion determined, we can use the entries in Table 8.7 to obtain the individual hydration enthalpies for other ions (Table 8.8): all the values are negative because hydration is always exothermic.

How can we use this information . . . to determine the enthalpy of hydration of an ion?

The enthalpy of hydration for

$$H^+(g) + Cl^-(g) \longrightarrow H^+(aq) + Cl^-(aq)$$

is given in Table 8.7 as $-1470\ kJ\cdot mol^{-1}$. Because $-1130\ kJ\cdot mol^{-1}$ of this value is due to the hydration of H^+, the enthalpy of hydration of the Cl^- ion is the difference:

$$(-1470\ kJ\cdot mol^{-1}) - (-1130\ kJ\cdot mol^{-1}) = -340\ kJ\cdot mol^{-1}$$

Self-Test 8.8A Use the data for the hydration of calcium chloride in Table 8.7 and the information above to derive the enthalpy of hydration of Ca^{2+}.

[*Answer:* $-1657\ kJ\cdot mol^{-1}$]

Self-Test 8.8B Use the data for the hydration of I^- in Table 8.8 and the enthalpy of hydration of ZnI_2, $-2638\ kJ\cdot mol^{-1}$, to calculate the enthalpy of hydration of Zn^{2+} ions.

Numerical data are much more interesting—and lead to deeper understanding—when we can find patterns in the values. When we examine the data in Table 8.8, we see that *the transfer of an ion from the gas phase into water is more exothermic the greater the charge of the ion and the smaller its radius.* There are exceptions: the radius of the Ag^+ ion (113 pm) is larger than that of the Na^+ ion (102 pm), but its hydration is more exothermic. The reason for this anomaly may be that the Ag^+ ion can form covalent bonds with the hydrating water molecules. The low solubility of many silver salts supports this explanation.

High charge and small radius both contribute to a high enthalpy of hydration. However, the same properties also contribute to high lattice enthalpy (Section 6.20). It is therefore very difficult to make reliable predictions about solubilities on the basis of ion charge and radius. The best we can do is to use these properties to rationalize what is observed. With that limitation in mind, we can begin to understand the behavior of some everyday substances and the properties of some minerals. For example, nitrates have big, singly charged anions and hence low lattice enthalpies. Their hydration enthalpies, though, are quite large because water can form hydrogen bonds with the nitrate anions. As a result, they are rarely found in mineral deposits because they are soluble in groundwater, the water that trickles through the ground and washes away soluble substances. Carbonate ions are even more bulky than nitrate ions, but they are doubly charged. As a result, carbonates commonly have bigger lattice enthalpies than nitrates, and it is much harder to break the ions out of solids such as limestone (calcium carbonate). Hydrogen carbonate ions (bicarbonates, HCO_3^-) have a single charge and their salts are more soluble than carbonates.

The difference in solubility between carbonates and hydrogen carbonates is responsible for the behavior of "hard water," which is water that contains dissolved calcium and magnesium salts. Hard water originates as rainwater, which dissolves carbon dioxide from the air and forms a very dilute solution of carbonic acid.

Although we write the reaction as leading to the formation of H_2CO_3, a high proportion of the dissolved carbon dioxide remains unreacted.

$$CO_2(g) + H_2O(l) \longrightarrow H_2CO_3(aq)$$

As the water runs over and through the ground, the carbonic acid reacts with the calcium carbonate of limestone or chalk and forms the more soluble hydrogen carbonate:

$$CaCO_3(s) + H_2CO_3(aq) \longrightarrow Ca^{2+}(aq) + 2\ HCO_3^-(aq)$$

These two reactions are reversed when water containing $Ca(HCO_3)_2$ is heated in a kettle or furnace:

$$Ca^{2+}(aq) + 2\ HCO_3^-(aq) \longrightarrow CaCO_3(s) + H_2O(l) + CO_2(g)$$

The carbon dioxide is driven off and the calcium carbonate is deposited as the hard, insoluble deposit known as *scale*.

Small, highly charged ions have large negative enthalpies of hydration.

Self-Test 8.9A Why is the hydration enthalpy of Cl^- more negative than that of I^-?

[*Answer:* Cl^- has the smaller radius.]

Self-Test 8.9B Would you expect Rb^+ to have a more negative or a less negative hydration enthalpy than Sr^{2+}? Explain your answer.

8.15 The Free Energy of Solution

The enthalpy of solution of a substance is only half the story when considering solubility. A negative enthalpy of solution tells us that energy is released as heat when a substance dissolves. However, to judge whether dissolving is spontaneous, we need to consider the change in free energy, $\Delta G = \Delta H - T\Delta S$, not just the change in enthalpy. In other words, we have to consider changes in the entropy of the system, too.

Remember that ΔG is a measure of the overall change in entropy: at constant temperature and pressure, ΔS is the change in entropy of the system and $-\Delta H/T$ is the change in entropy of the surroundings (Section 7.12).

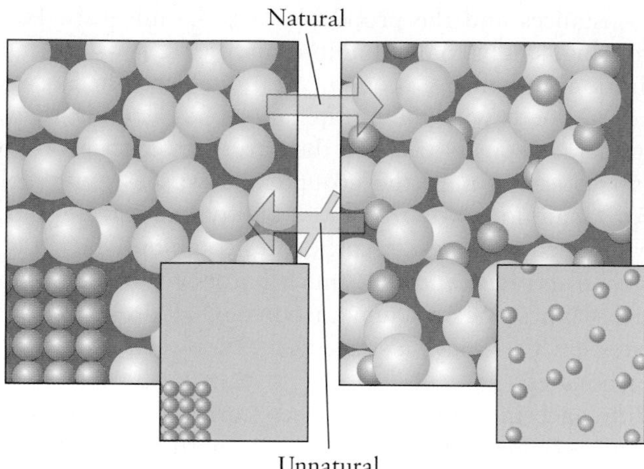

FIGURE 8.25 The dissolution of a solid is usually accompanied by an increase in disorder. However, to decide whether the dissolution is spontaneous, the accompanying changes in the surroundings must be considered, too. The insets show the solute particles alone.

Disorder normally increases when an orderly solid dissolves (Fig. 8.25). Therefore, in most cases, we can expect the entropy of the system to increase when a solution forms. Because $T\Delta S$ is positive, this increase in disorder makes a negative contribution to ΔG. If ΔH is also negative we can be confident that ΔG is negative overall; so we can expect most substances with negative enthalpies of solution to be soluble (Fig. 8.26).

Although it is reasonable to expect dissolving to be accompanied by an increase in entropy of the solute, we must not forget the effect of the solute on the solvent. In some cases, the presence of a solute causes the solvent molecules to take up a more organized arrangement than they had in the pure liquid (Fig. 8.27). In such cases, the entropy of the system is lowered when the solution forms because the solvent molecules form cagelike structures around the solute molecules. As a result, ΔS is negative

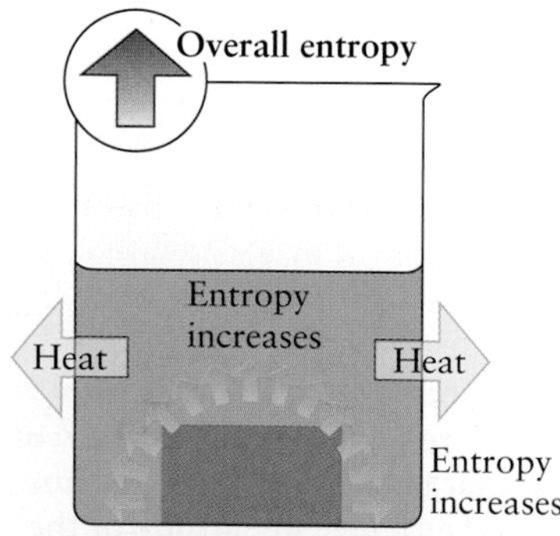

FIGURE 8.26 The changes in entropy accompanying the exothermic dissolution of a solid.

FIGURE 8.27 When a nonpolar compound (the yellow sphere) dissolves in water, the water molecules may become organized around it. As a result, the entropy of the solvent is reduced when the solution forms.

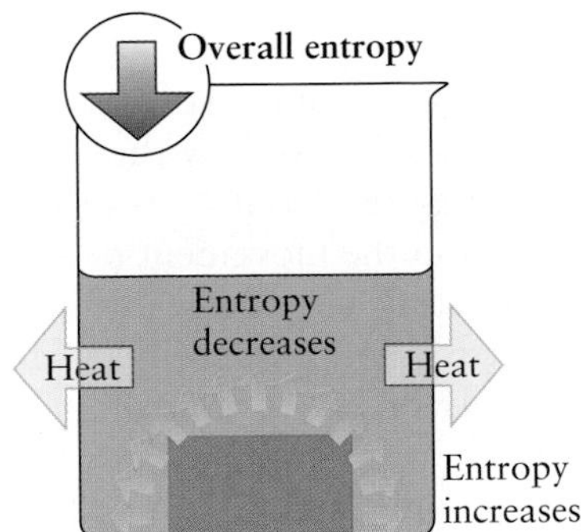

FIGURE 8.28 Even though dissolving may be weakly exothermic, and hence increase the entropy of the surroundings, the reduced entropy of the solvent illustrated in Fig. 8.27 might cancel the increase. As a result, dissolution may not be spontaneous.

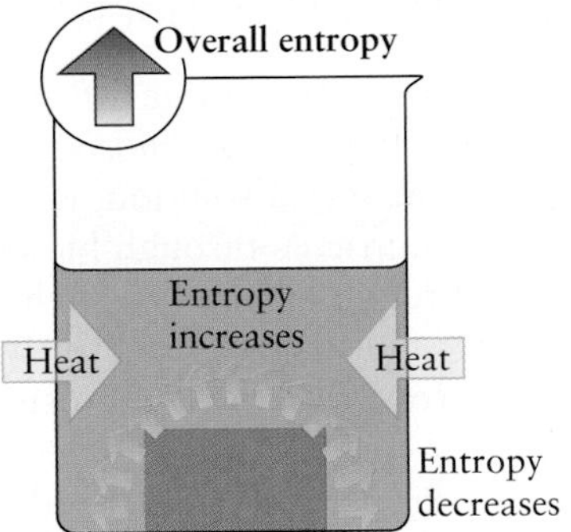

FIGURE 8.29 If dissolution is endothermic, it will occur only if the disorder caused by the solute dispersing is great enough to produce an overall increase in entropy.

and the $-T\Delta S$ term makes a positive contribution to ΔG. Even if energy is released to the surroundings, the increase in the entropy of the surroundings may not be enough to overcome the decrease in entropy of the solution (Fig. 8.28). Then ΔG is positive and the substance does not dissolve. For this reason, some hydrocarbons, such as heptane, are insoluble in water even though they have weakly negative enthalpies of solution.

The increase in entropy of the system is responsible for the solubilities of substances that dissolve endothermically. The endothermic character of dissolving means that energy flows from the surroundings into the solution as the solute dissolves, with the result that the entropy of the surroundings decreases. However, provided the entropy of the system—the solution—increases sufficiently, there may still be a net increase in total entropy (Fig. 8.29). Whether or not dissolving is spontaneous depends on the balance of these two effects. A substance with a strongly positive enthalpy of solution is likely to be insoluble, because the entropy of the surroundings may decrease so much that dissolving corresponds to an overall decrease in disorder.

We can also now begin to see why a substance has a characteristic solubility in water. The solubility of a substance is very sensitive to the enthalpy of hydration partly because there are fewer water molecules to attach to the solute particles at higher concentrations and because the solute molecules interact with one another more strongly. These effects modify both the enthalpy and entropy of solution. Although ΔG for the dissolving of a solute might be negative at low concentrations, it may become positive at high concentrations (Fig. 8.30). Spontaneous dissolving occurs only up to the stage at which $\Delta G = 0$.

Dissolving depends on the balance between the increase in entropy of the solution and the increase in entropy of the surroundings.

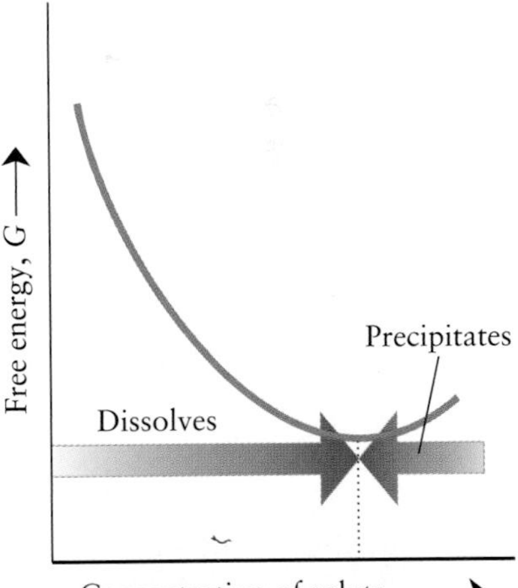

FIGURE 8.30 At low concentrations of solute, dissolving is accompanied by a decrease in free energy; so it is spontaneous. At high concentrations, dissolving is accompanied by an increase in free energy; so the reverse process, precipitation, is spontaneous. The concentration of a saturated solution corresponds to the state of lowest free energy at the temperature of the experiment.

COLLIGATIVE PROPERTIES

The presence of a solute affects the physical properties of the solvent. One of these properties, *osmosis*, the tendency of a solvent to flow through a membrane into a solution, is essential to life because it contributes to the flow of nutrients through biological cell walls and the movement of solutions through plants. When chemists started to study the physical properties of solutions quantitatively, they came across a very peculiar fact: some properties, including osmosis, depend only on the relative amounts of solute and solvent and are independent of the chemical identity of the solute. Properties that depend on the relative numbers of solute and solvent molecules and not on the chemical identity of the solute are called **colligative properties.** Four colligative properties of major importance are the lowering of the vapor pressure of the solvent, the raising of its boiling point, the lowering of its freezing point, and osmosis. All four properties involve an equilibrium between two phases of a solvent or (for osmosis) between two solutions with different concentrations.

The meaning of the word *colligative* is "depending on the collection."

8.16 Molality

Because colligative properties depend on the relative numbers of molecules of solute and solvent, the two measures of concentration necessary for the study of colligative properties are mole fraction and molality. We first met the *mole fraction, x*, in Section 4.11, where we saw that it is the ratio of the number of moles of a species to the total number of moles of all the species present in a mixture. The **molality** of a solute is the amount of solute species (in moles) in a solution divided by the mass of the solvent (in kilograms):

$$\text{Molality of solute} = \frac{\text{amount of solute (mol)}}{\text{mass of solvent (kg)}} \qquad (8)$$

The units of molality are moles of solute per kilogram of solvent ($\text{mol}\cdot\text{kg}^{-1}$); these units are often denoted m (for example, a 1 m $NiSO_4(aq)$ solution) and read "molal." Note the emphasis on solvent in the definition. Therefore, to prepare a 1 m $NiSO_4(aq)$ solution, we dissolve 1 mol $NiSO_4$ in 1 kg of water (Fig. 8.31).

How can we use this information . . . to calculate the molality of a solute?

To calculate the molality of a solution prepared by dissolving 10.5 g of sodium chloride in 250 g of water, we convert the mass of sodium chloride to moles of NaCl (by dividing the mass by the molar mass) and divide that amount by the mass of water in kilograms:

$$\text{Molality of NaCl} = \frac{(10.5\text{ g}) / (58.44\text{ g}\cdot\text{mol}^{-1})}{0.250\text{ kg}} = 0.719\text{ mol}\cdot\text{kg}^{-1}$$

We report this molality as 0.719 m NaCl(aq).

FIGURE 8.31 The steps taken to prepare a solution of a given molality. First (left), the required masses of solute and solvent are measured out. Then (right), the solute is dissolved in the solvent. Compare this procedure with that for preparing a solution of given molarity (see Fig. G.8).

Self-Test 8.10A Calculate the molality of $ZnCl_2$ in a solution prepared by dissolving 4.11 g $ZnCl_2$ in 150 g of water.

[*Answer:* 0.201 $mol \cdot kg^{-1}$]

Self-Test 8.10B Calculate the molality of $KClO_3$ in a solution prepared by dissolving 7.36 g $KClO_3$ in 200.0 g of water.

Molality differs from molarity in that molality is defined in terms of the mass of *solvent* used to make up the solution; molarity is defined in terms of the volume of the resulting *solution* (not the volume of solvent used to make the solution). As a result, when we want to emphasize the relative numbers of molecules of the components of a mixture we use either molality or mole fraction to express the concentration.

The molality of a solute in a solution is the amount (in moles) of solute divided by the mass (in kilograms) of solvent used to prepare the solution.

TOOLBOX 8.2 How to use the molality

Conceptual Basis

The molality of a solute is the concentration of solute in moles per kilogram of solvent. The mole fraction of a species is the number of moles of a species expressed as a fraction of the total number of moles of ions and molecules in the solution. These quantities are calculated from measurements made before mixing. The molarity of a solute is the amount of solute per liter of solution. It is calculated by measuring the mass of solute before mixing and the volume of solution *after* mixing.

Procedure

(1) Calculating the mass of solute from molality and mass of solvent

Step 1 Calculate the amount of solute molecules present in a given mass of solvent by rearranging the equation defining molality (Eq. 8) into

$$\text{Amount of solute (mol)} = \text{molality} \times \text{mass of solvent (kg)}$$

Step 2 Use the molar mass of the solute to find the mass of the amount of solute calculated in step 1.

(2) Calculating the molality from a mole fraction

The mole fraction tells us that in a total of 1 mol of molecules in a sample, there is x_{solute} mol of solute molecules and $(1 - x_{\text{solute}})$ mol of solvent molecules.

Step 1 Use the molar mass, *M*, of the solvent to convert the amount of solvent molecules, $(1 - x_{\text{solute}})$ mol, to mass of solvent in kilograms:

$$\text{Mass of solvent} = (1 - x_{\text{solute}})\text{ mol} \times M \times \frac{1\text{ kg}}{10^3\text{ g}}$$

At this point, we know both the amount of solute molecules (x_{solute} mol) and the mass of solvent in a sample that contains a total of 1 mol of molecules.

Step 2 Calculate the molality of the solute by dividing the amount of solute molecules by the mass of solvent (m_{solvent}, in kilograms):

$$\text{Molality} = \frac{n_{\text{solute}}\text{ (mol)}}{m_{\text{solvent}}\text{ (kg)}}$$

Example 8.5 shows how these steps are applied.

(3) Calculating the mole fraction from the molality

Consider a solution in which exactly 1 kg of solvent is present.

Step 1 Convert that mass of solvent to an amount of solvent molecules (n_{solvent}, in moles) by using the molar mass, *M*, of the solvent:

$$n_{\text{solvent}} = \frac{1\text{ kg}}{M} = \frac{10^3\text{ g}}{M}$$

We already know the amount of solute molecules, n_{solute}, in the solution (from the molality).

$$n_{\text{solute}} = \text{molality} \times (1\text{ kg})$$

Step 2 Calculate the mole fractions from the amounts of solute and solvent molecules:

$$x_{\text{solute}} = \frac{n_{\text{solute}}}{n_{\text{solute}} + n_{\text{solvent}}}$$

(4) Calculating the molality, given the molarity

Molality is defined as amount of solute divided by the mass of *solvent*, whereas molarity is defined as amount of solute divided by the volume of *solution*. To find the mass of solvent in the solution of given molarity, we need the density of the solution, which must either be provided as data or be measured.

Step 1 Calculate the total mass of exactly 1 L of solution by using the density of the solution.

Step 2 Use the molarity to calculate the mass of solute in 1 L of solution.

Step 3 Subtract the mass of solute from the total mass to find the mass of solvent in 1 L of solution and convert the mass of solvent into kilograms.

Step 4 The amount of solute present in 1 L of solution is numerically equal to the molarity. Divide the amount of solute (in moles) by the mass of the solvent in kilograms to obtain the molality.

See Example 8.6.

EXAMPLE 8.5 Calculating a molality from a mole fraction

What is the molality of benzene, C_6H_6, dissolved in toluene, $C_6H_5CH_3$, in which the mole fraction of benzene is 0.150?

STRATEGY Follow procedure (2) in Toolbox 8.2.

SOLUTION From the definition of mole fraction, 1 mol of solution molecules consists of 0.150 mol C_6H_6 and 1 − 0.150 mol = 0.850 mol $CH_3C_6H_5$. ***Step 1*** The mass of this amount of toluene (of molar mass 92.13 g·mol^{-1}) is

$$\begin{aligned}\text{Mass of toluene (kg)} &= (0.850\text{ mol}) \times \left(\frac{92.13\text{ g}}{1\text{ mol}}\right) \times \left(\frac{1\text{ kg}}{10^3\text{ g}}\right)\\ &= 0.850 \times 92.13 \times 10^{-3}\text{ kg}\end{aligned}$$

Step 2 The molality of benzene in the solution is therefore

$$\text{Molality of } C_6H_6 = \frac{0.150 \text{ mol}}{0.850 \times 92.13 \times 10^{-3} \text{ kg}} = 1.92 \text{ mol·kg}^{-1}$$

Self-Test 8.11A Calculate the molality of toluene dissolved in benzene, given that the mole fraction of toluene is 0.150.

[*Answer:* 2.26 mol·kg^{-1}]

Self-Test 8.11B Calculate the molality of methanol in aqueous solution, given that the mole fraction of methanol is 0.250.

EXAMPLE 8.6 Converting from molarity to molality

Find the molality of sucrose, $C_{12}H_{22}O_{11}$, in 1.06 M $C_{12}H_{22}O_{11}$(aq) which is known to have density 1.14 g·mL^{-1}.

STRATEGY First find the mass of solvent in 1 L of solution, then convert that mass to kilograms, as in the last procedure in Toolbox 8.2.

SOLUTION *Step 1* The mass of exactly 1 L of solution is

$$\text{Mass of solution} = (1.14 \text{ g·mL}^{-1}) \times (1 \times 10^3 \text{ mL}) = 1.14 \times 10^3 \text{ g}$$

Step 2 The amount of solute molecules in 1.00 L of the 1.06 M solution is 1.06 mol, and the molar mass of sucrose is 342.3 g·mol^{-1}. The mass of sucrose in 1.00 L of solution is therefore

$$\text{Mass of sucrose} = (1.06 \text{ mol}) \times (342.3 \text{ g· mol}^{-1}) = 363 \text{ g}$$

Step 3 The mass of water present in 1.00 L of solution is

$$\text{Mass of water} = 1.14 \times 10^3 - 363 \text{ g} = 7.8 \times 10^2 \text{ g, or } 0.78 \text{ kg}$$

Step 4 Because 1.06 mol $C_{12}H_{22}O_{11}$ is dissolved in 0.78 kg of water, we conclude that the molality of the solution is

$$\text{Molality of } C_{12}H_{22}O_{11} = \frac{1.06 \text{ mol}}{0.78 \text{ kg}} = 1.4 \text{ mol·kg}^{-1}$$

Self-Test 8.12A Battery acid is 4.27 M H_2SO_4(aq) and has a density of 1.25 g·cm^{-3}. What is the molality of H_2SO_4 in the solution?

[*Answer:* 5.14 *m* H_2SO_4(aq)]

Self-Test 8.12B The density of 1.83 M NaCl(aq) is 1.07 g·cm^{-3}. What is the molality of NaCl in the solution?

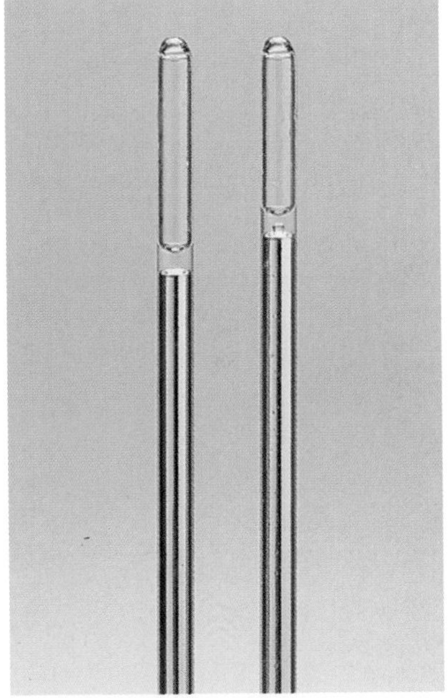

FIGURE 8.32 The vapor pressure of a solvent is lowered by a nonvolatile solute. The barometer tube on the left has a small volume of pure water floating on the mercury. That on the right has a small volume of 10 *m* NaCl(aq), and a lower vapor pressure. Note that the column on the right is depressed less by the vapor in the space above the mercury than the one on the left, showing that the vapor pressure is lower when solute is present.

8.17 Vapor-Pressure Lowering

Saltwater lakes evaporate more slowly than freshwater lakes. The slower evaporation of salt lakes implies that the presence of the salt has reduced the vapor pressure. We see the same phenomenon in the laboratory: the vapor pressure of a solvent is found to be lower when a nonvolatile solute is present (Fig. 8.32).

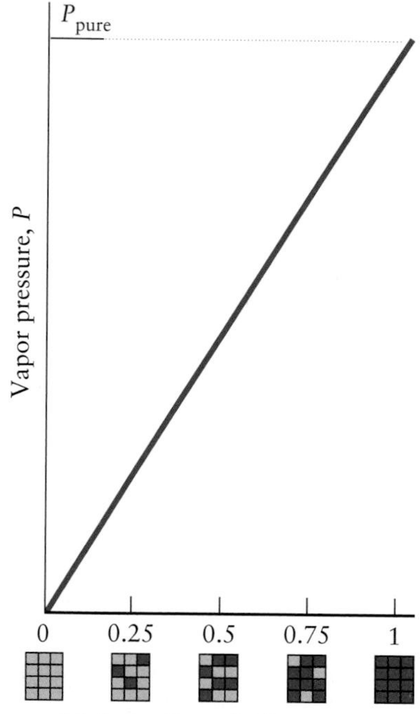

FIGURE 8.33 Raoult's law predicts that the vapor pressure of a solvent in a solution should be proportional to the mole fraction of the solvent molecules. The horizontal axis shows the mole fraction of the red molecules in the pure solute, three different solutions, and pure solvent, pictured below.

The French scientist François-Marie Raoult, who spent much of his life measuring vapor pressures, discovered that *the vapor pressure of a solvent is proportional to its mole fraction in a solution.* This statement, which is called **Raoult's law,** is normally written

$$P = x_{solvent}P_{pure} \tag{9}$$

P_{pure} is the vapor pressure of the pure solvent, $x_{solvent}$ is the mole fraction of the solvent, and P is the vapor pressure of the solvent in the solution (Fig. 8.33). According to this equation, at any given temperature, the vapor pressure of the solvent is directly proportional to the mole fraction of solvent molecules in the solution. For instance, if only 9 in 10 of the molecules present in a solution are solvent molecules (the others being solute molecules or ions), then the vapor pressure of the solvent is only nine-tenths that of the pure solvent.

EXAMPLE 8.7 Using Raoult's law

Calculate the vapor pressure of water at 20°C in a solution prepared by dissolving 10.00 g of the nonelectrolyte sucrose, $C_{12}H_{22}O_{11}$, in 100.0 g of water.

STRATEGY Expect a lower vapor pressure when the solute is present. Once we know the mole fraction of the solvent (water) in the solution, we can apply Raoult's law. To calculate the mole fractions, convert masses to amounts in moles, then divide each value by the total number of moles. To use Raoult's law, we need the vapor pressure of the pure solvent, Table 8.2 or 8.3.

SOLUTION The molar mass of sucrose is 342.3 g·mol^{-1} and that of water is 18.02 g·mol^{-1}. Therefore, the amount of each species is

$$\text{Amount of } C_{12}H_{22}O_{11} = \frac{10.00\text{ g}}{342.3\text{ g·mol}^{-1}} = \frac{10.00}{342.3}\text{ mol}$$
$$\text{Amount of } H_2O = \frac{100.0\text{ g}}{18.02\text{ g·mol}^{-1}} = \frac{100.0}{18.02}\text{ mol}$$

The two mole fractions are

$$x_{sucrose} = \frac{10.00 / 342.3\text{ mol}}{(10.00 / 342.3 + 100.0 / 18.02)\text{ mol}} = 0.005\ 24$$
$$x_{water} = \frac{100.0 / 18.02\text{ mol}}{(10.00 / 342.3 + 100.0 / 18.02)\text{ mol}} = 0.995$$

Therefore, because P_{pure} = 17.54 Torr (Table 8.3), the vapor pressure of the water in the solution is

$$P_{water} = \frac{100.0/18.02}{10.00/342.3 + 100.0/18.02} \times (17.54\text{ Torr}) = 17.45\text{ Torr}$$

Self-Test 8.13A Calculate the vapor pressure of water at 90°C for a solution prepared by dissolving 5.00 g of glucose ($C_6H_{12}O_6$) in 100 g of water. The vapor pressure of pure water at 90°C is 524 Torr.
[*Answer:* 521 Torr]

Self-Test 8.13B Calculate the vapor pressure of ethanol in kilopascals (kPa) at 19°C for a solution prepared by dissolving 2.00 g of cinnamaldehyde, C_9H_8O, in 50.0 g of ethanol, C_2H_5OH. The vapor pressure of pure ethanol at that temperature is 5.3 kPa.

To understand why a solute lowers the vapor pressure, we need to look at the thermodynamic properties of the solution. We saw in Section 8.2, specifically Eq. 1, that at equilibrium, and in the absence of any solute, the molar free energy of the vapor is equal to that of the pure solvent. We now need to think about how the molar free energies of the solvent and the vapor are affected by the presence of a solute. At this point, we shall consider only nonvolatile solutes, which do not appear in the vapor phase, and limit the discussion to ideal solutions.

A hypothetical solution that obeys Raoult's law exactly at all concentrations is called an **ideal solution.** In an ideal solution, the interactions between solute and solvent molecules are the same as the interactions between solvent molecules, so the solute molecules mingle freely with the solvent molecules. That is, in an ideal solution, the enthalpy of solution is 0. Solutes that form nearly ideal solutions are often similar in composition and structure to the solvent molecules. For instance, methylbenzene (toluene) forms nearly ideal solutions with benzene.

Even in an ideal solution, a solute increases the entropy of the liquid phase: we can no longer be sure that a molecule picked at random from the solution will be a solvent molecule. Because the entropy of the liquid phase is raised by a solute but the enthalpy is left unchanged, overall there is a decrease in the molar free energy of the solvent.

Now consider the vapor. At equilibrium, the free energy of the vapor must equal that of the solvent. Because the free energy of the solvent has been lowered, for the two phases to remain in equilibrium, the free energy of the vapor must be lower, too. Because the free energy of a gas depends on its pressure (Eq. 2 and Fig. 8.3), it follows that the vapor pressure must also be lower. In other words, *the vapor pressure of the solvent is lower in the presence of the solute* (Fig. 8.34).

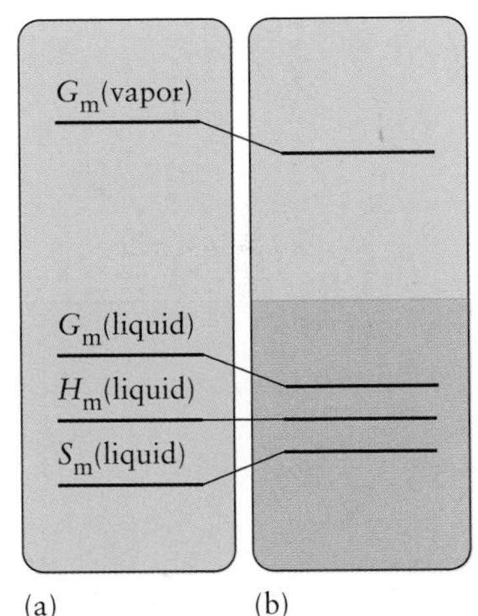

FIGURE 8.34 (a) The molar enthalpy, entropy, and free energy of a pure liquid solvent and its vapor. At equilibrium, the molar free energies of liquid and vapor are equal (separate scales are used for liquid and vapor). (b) When a solute is added to form an ideal solution, the molar entropy of the solvent rises, so the free energy of the solvent falls. For the vapor to remain in equilibrium, its molar free energy must also fall: that reduction is achieved when some of the vapor condenses, which results in a decrease of the vapor pressure.

Real solutions do not obey Raoult's law at all concentrations, but the lower the solute concentration, the more closely they resemble ideal solutions. Raoult's law is another example of a limiting law (Section 4.7), which in this case becomes increasingly valid as the concentration of the solute approaches 0. A solution that does not obey Raoult's law at a particular solute concentration is called a **nonideal solution.** Real solutions are approximately ideal at solute concentrations below about 10^{-1} M for nonelectrolyte solutions and 10^{-2} M for electrolyte solutions. The problem with electrolyte solutions is that interactions between ions occur over a long distance and hence have a pronounced effect. Unless stated otherwise, we shall assume that all the solutions we meet are ideal.

The vapor pressure of a solvent is reduced by the presence of a nonvolatile solute: in an ideal solution, the vapor pressure of the solvent is proportional to the mole fraction of solvent.

8.18 Boiling-Point Elevation and Freezing-Point Depression

When a nonvolatile solute is present in a solution (such as the sodium chloride in seawater), the lowering of solvent vapor pressure results in an increase in the boiling point. This increase is called **boiling-point elevation.** The origin of the boiling-point elevation can be explained in terms of the effect of the solute on the entropy of the solvent.

Figure 8.35a shows how the standard molar free energies of the liquid and vapor phases of a pure solvent vary with temperature. Above the boiling point, where the free energy of the vapor phase is lower than that of the liquid, the vapor is the more stable phase. We have already seen that the presence of a solute in the liquid phase of the solvent increases the entropy of the solute and therefore lowers its free energy. As shown in Fig. 8.35b, the lines representing the free energies of the liquid solution and the vapor intersect at a higher temperature than they do for the pure solvent, so the boiling point is higher in the presence of the solute. The increase is usually quite small and is of little practical importance. A 0.1 m aqueous sucrose solution, for instance, boils at 100.05°C.

Of greater practical significance is **freezing-point depression,** the lowering of the freezing point of the solvent caused by a solute. For example, seawater freezes about 1°C lower than fresh water. People living in regions with cold winters make use of this effect when they spread salt on walkways and roads to melt ice. In the laboratory, chemists make use of the depression of the freezing point to judge the purity of a solid compound: if impurities are present, the compound's melting point is lower than the accepted value.

Figure 8.36a shows how the standard molar free energies of the liquid and solid phases of a pure solvent vary with temperature. The solid is the more stable phase below the freezing point, where the free energy of the solid is lower than that of the liquid. As before, the presence of a solute in the liquid phase of the solvent raises the entropy of the solvent and hence lowers its free energy (Fig. 8.36b) without affecting the free energy of the

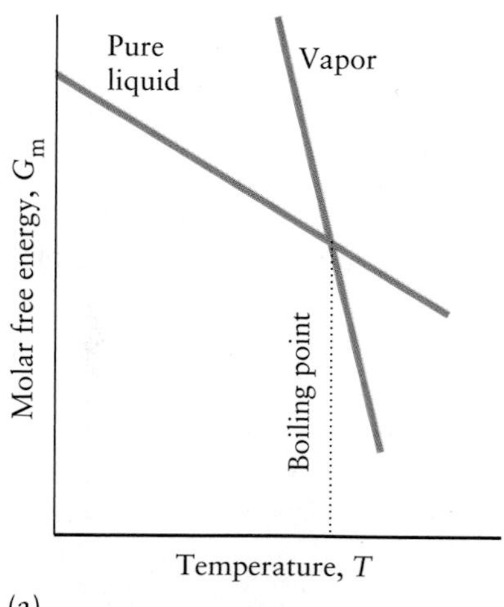

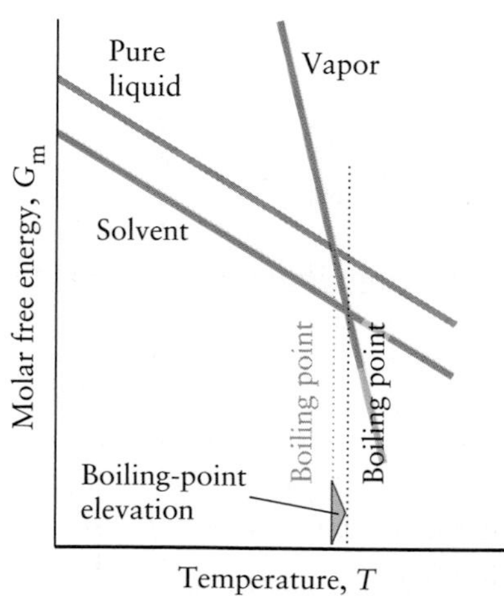

FIGURE 8.35 (a) The molar free energy of a liquid and its vapor both decrease with increasing temperature, but that of the vapor decreases more sharply. The vapor is the stable phase at temperatures higher than the point of intersection of the two lines (the boiling point). (b) When a nonvolatile solute is present, the molar free energy of the solvent is lowered (an entropy effect), but that of the vapor is left unchanged. The point of intersection of the lines moves to a slightly higher temperature.

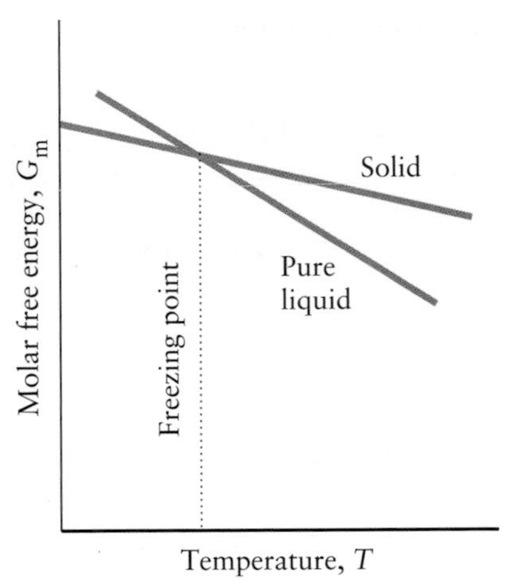

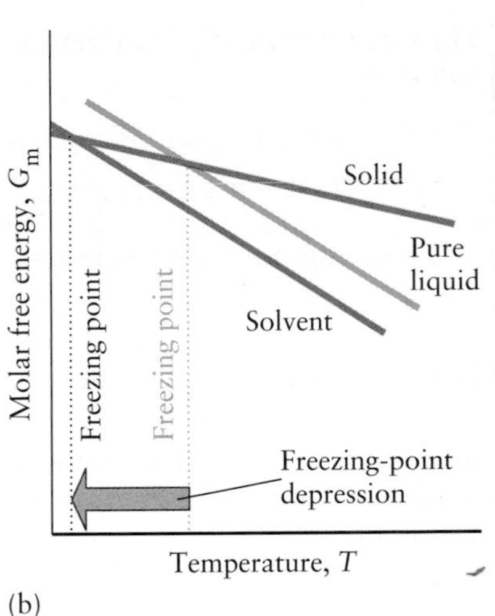

FIGURE 8.36 (a) The molar free energy of a solid and its liquid phase both decrease with increasing temperature, but that of the liquid decreases slightly more steeply. The liquid is the stable phase at temperatures higher than the point of intersection of the two lines. (b) When a solute is present, the molar free energy of the solvent is lowered (an entropy effect), but that of the solid is left unchanged. The point of intersection of the lines moves to a lower temperature.

solid phase of the solvent (because the solute is insoluble in the solid solvent). The lines representing the free energies of the liquid and solid phases of the solvent intersect at a lower temperature than they do for the pure solvent, so the freezing point is lower in the presence of the solute.

It is found empirically (and can be justified thermodynamically) that the freezing-point depression for an ideal solution is proportional to the molality of the solute. For a nonelectrolyte solution,

$$\text{Freezing-point depression} = k_f \times \text{molality} \quad (10)$$

The constant k_f is called the **freezing-point constant** of the solvent; it is different for each solvent and must be determined experimentally (Table 8.9). A similar expression (boiling-point elevation = $k_b \times$ molality) is used to relate the elevation of boiling point to the molality of the solute.

TABLE 8.9 Boiling-Point and Freezing-Point Constants

Solvent	Freezing point, °C	k_f, K·kg·mol^{-1}	Boiling point, °C	k_b, K·kg·mol^{-1}
acetone	−95.35	2.40	56.2	1.71
benzene	5.5	5.12	80.1	2.53
camphor	179.8	39.7	204	5.61
carbon tetrachloride	−23	29.8	76.5	4.95
cyclohexane	6.5	20.1	80.7	2.79
naphthalene	80.5	6.94	217.7	5.80
phenol	43	7.27	182	3.04
water	0	1.86	100.0	0.51

How can we use this information . . . to estimate the freezing point of a solution?

The freezing-point depression for a 0.1 *m* $C_{12}H_{22}O_{11}(aq)$ (sucrose) solution is

$$\begin{aligned}\text{Freezing-point depression} &= (1.86\ \text{K·kg·mol}^{-1}) \times (0.1\ \text{mol·kg}^{-1}) \\ &= 0.2\ \text{K}\end{aligned}$$

Hence the water in the solution freezes at −0.2°C.

Self-Test 8.14A Use the data in Table 8.9 to determine at what temperature a 0.20 mol·kg^{-1} solution of the analgesic codeine, $C_{18}H_{21}NO_3$, in benzene will freeze.

[*Answer:* 4.5°C]

Self-Test 8.14B Use the data in Table 8.9 to determine at what temperature a 0.050 mol·kg^{-1} solution of the insecticide malathion, $C_{10}H_{19}O_6PS_2$, in camphor will freeze.

In an electrolyte solution, each formula unit contributes two or more ions. Sodium chloride, for instance, dissolves to give Na^+ and Cl^- ions, and both kinds of ions contribute to the depression of the freezing point. The cations and anions contribute nearly independently in very dilute solutions, so the total solute molality is twice the molality of NaCl formula units. In more concentrated solutions, the ions do not move independently. For instance, some stick together to form ion pairs and other aggregates of small numbers of ions. The effect of the solute on the freezing point in these solutions is very difficult to predict. We can write

$$\text{Freezing-point depression} = ik_f \times \text{molality} \tag{11}$$

Here, i, the **van 't Hoff *i* factor,** is determined experimentally. In a very dilute solution (less than about 10^{-3} mol·L^{-1}), when all ions are independent, $i = 2$ for MX salts such as NaCl, $i = 3$ for MX_2 salts such as $CaCl_2$, and so on. For dilute nonelectrolyte solutions, $i = 1$. The i factor is so unreliable, however, that it is best to confine quantitative calculations of freezing-point depression to nonelectrolyte solutions. Even these solutions must be dilute enough to be approximately ideal. However, the i factor can be used to help determine the extent to which a substance is dissociated into ions in solution. For example, in dilute solution, HCl has an i factor of 1 in toluene and 2 in water. These values suggest that HCl retains its molecular form in toluene but is fully deprotonated in water.

How can we use this information . . . to find the *i* factor for a solution of a weak acid?

A weak acid that is 5% deprotonated in water would have an i factor of 1.05. Because each deprotonated molecule produces two ions, $i = (0.05 \times 2) + 0.95 = 1.05$.

Percentage deprotonation is a mole percentage, not a mass percentage.

Self-Test 8.15A How many moles of ions are present in a solution containing 0.10 mol Na_2SO_4, assuming complete dissociation (separation of ions)? Estimate the *i* factor.

[*Answer:* 0.30 mol (0.20 mol Na^+ ions and 0.10 mol SO_4^{2-} ions); $i = 3$]

Self-Test 8.15B How many moles of ions are present in a solution containing 0.25 mol $CoCl_3$, assuming complete dissociation? Estimate the *i* factor.

Cryoscopy is the determination of the molar mass of a solute by measuring the depression of freezing point it causes when dissolved in a solvent. Camphor is often used as the solvent for organic compounds because it has a large freezing-point constant; consequently, solutes depress its freezing point significantly. However, this procedure is rarely used in modern laboratories because techniques like mass spectrometry give far more reliable results (see Major Technique 5, following Chapter 18). The procedure is described in Toolbox 8.2 at the end of Section 8.19.

The presence of a solute lowers the freezing point of a solvent; if the solute is nonvolatile, the boiling point is also raised. The freezing-point depression can be used to calculate the molar mass of the solute. If the solute is an electrolyte, the extent of its dissociation, protonation, or deprotonation must also be taken into account.

8.19 Osmosis

Osmosis is the flow of solvent through a membrane into a more concentrated solution. The phenomenon can be demonstrated in the laboratory by separating a solution and the pure solvent by a **semipermeable membrane,** a membrane that permits only certain types of molecules or ions to pass through (Fig. 8.37). Cellulose acetate, for instance, allows water molecules to pass through it, but not solute molecules or ions with their bulky coating

The name *osmosis* comes from the Greek word for "push."

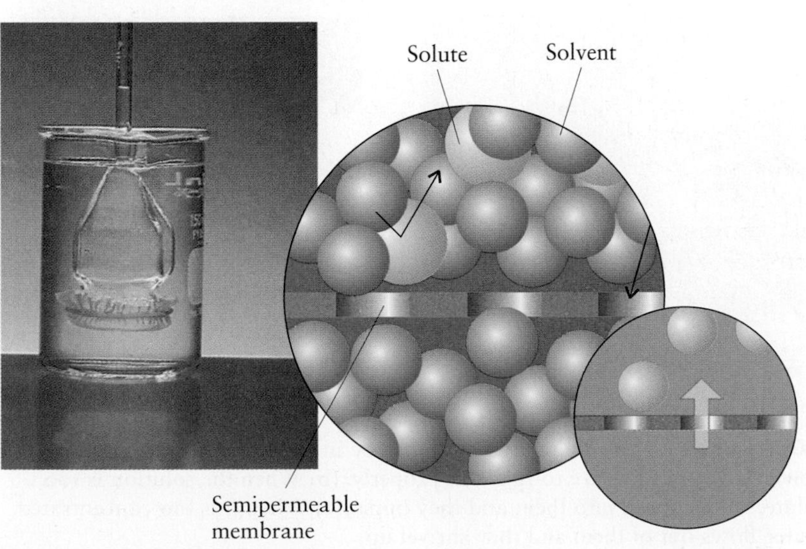

FIGURE 8.37 An experiment to illustrate osmosis. Initially, the tube contained a sucrose solution and the beaker contained pure water: the initial heights of the two liquids were the same. At the stage shown here, water has passed into the solution through the membrane by osmosis, and the level of solution in the tube has risen above that of the pure water. The large inset shows the molecules in the pure solvent (below the membrane) tending to join those in the solution (above the membrane) because the presence of solute molecules there have led to increased disorder. The small inset shows just the solute molecules; the arrow shows the direction of flow of solvent molecules.

of hydrating water molecules. Initially, the heights of the solution and the pure solvent shown in the illustration are the same. However, the level of the solution inside the tube begins to rise as pure solvent passes through the membrane into the solution. At equilibrium, the pressure exerted by the column of solution stops the net flow of molecules through the membrane. The pressure needed to stop the flow of solvent is called the **osmotic pressure,** Π (the Greek uppercase letter pi). The greater the osmotic pressure, the greater the height of the solution needed to stop the net flow.

The pressure exerted by a vertical column of liquid is proportional to its height: see Eq. 2 in Section 4.2.

Life depends on osmosis. Biological cell walls act as semipermeable membranes that allow water, small molecules, and hydrated ions to pass (Fig. 8.38). However, they block the passage of the enzymes and proteins that have been synthesized within the cell. The difference in concentrations of solute inside and outside a cell gives rise to an osmotic pressure, and water passes into the more concentrated solution in the interior of the cell, carrying small nutrient molecules with it. This influx of water also keeps the cell turgid (swollen). When the water supply is cut off, the turgidity is lost and the cell becomes dehydrated. In a plant, this dehydration results in wilting. Salted meat is preserved from bacterial attack by osmosis. In this case, the concentrated salt solution dehydrates—and kills—the bacteria by causing water to flow out of them.

Osmosis is a thermodynamic property, so we can expect it to be related to enthalpy and entropy changes brought about by the solute. We have already seen several times that a solute lowers the molar free energy of the solution below that of the pure solvent, and solvent therefore has a spontaneous tendency to pass into the solution (Fig. 8.39).

The same van 't Hoff responsible for the i factor showed that the osmotic pressure of a nonelectrolyte solution is related to the molarity, c, of the solute in the solution:

$$\Pi = iRTc \tag{12}$$

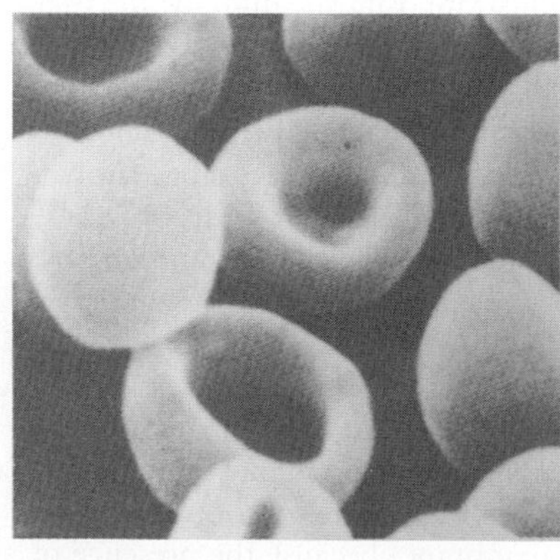
(a)

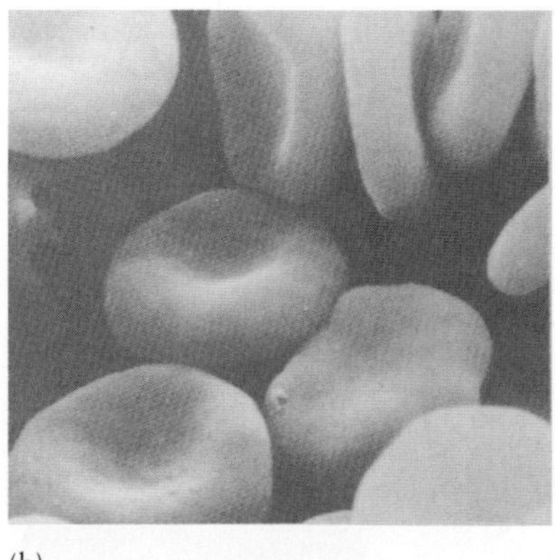
(b)

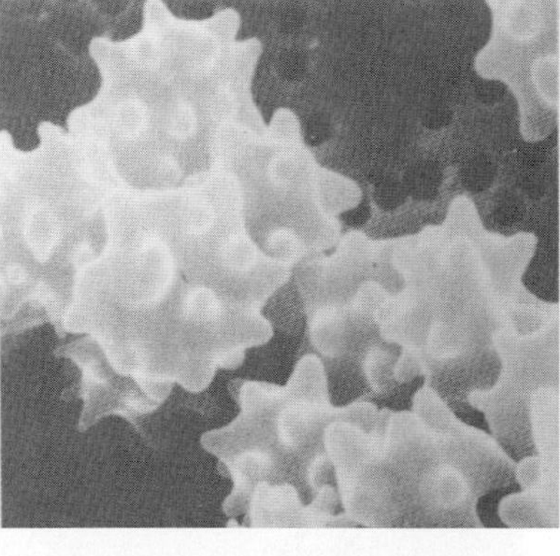
(c)

FIGURE 8.38 (a) Red blood cells need to be in a solution of the correct solute concentration if they are to function properly. (b) When the solution is too dilute, water passes into them and they burst. (c) When it is too concentrated, water flows out of them and they shrivel up.

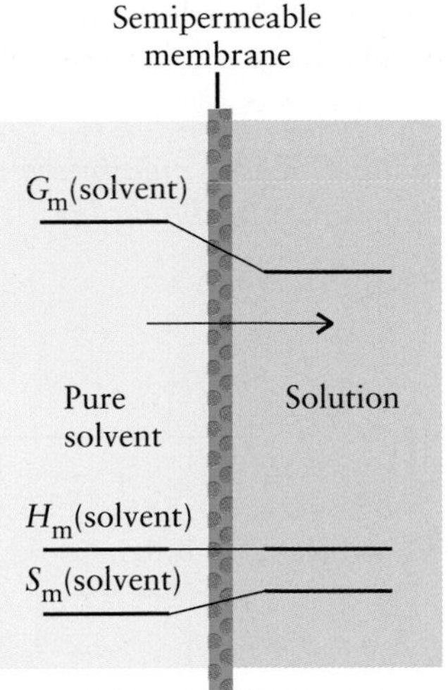

FIGURE 8.39 On the left of the semipermeable membrane is the pure solvent with its characteristic molar enthalpy, entropy, and free energy. On the right is the solution. The molar free energy of the solvent is lower in the solution (an entropy effect), so there is a spontaneous tendency for the solvent to flow into the solution.

where i is the i factor, R is the gas constant, and T is the temperature. This expression is now known as the **van 't Hoff equation.** Notice that the osmotic pressure is independent of the identities of both the solute and the solvent. However, the height of the column of solvent depends on its density (Fig. 8.40).

How can we use this information . . . to determine the osmotic pressure of a solution?

To use Eq. 12, we select R in the pressure units required for the osmotic pressure. For example, if we want the pressure in atmospheres, then we use R in units that include atmospheres. So, the osmotic pressure in atmospheres of any ideal 0.010 M nonelectrolyte solution, for which $i = 1$, at 25°C (298 K) is calculated by writing

$$\begin{aligned}\Pi &= (1) \times (0.0821\ \text{L·atm·K}^{-1}\text{·mol}^{-1}) \times (298\ \text{K}) \times (0.010\ \text{mol·L}^{-1})\\ &= 0.24\ \text{atm}\end{aligned}$$

This pressure is enough to push a column of water to a height of over 2 m (the height needed for the column to exert a pressure of 0.24 atm).

Self-Test 8.16A What is the osmotic pressure of 0.0100 M KCl(aq) at 298 K? (Assume $i = 2$.)

[*Answer:* 0.49 atm]

Self-Test 8.16B What is the osmotic pressure of a 0.120 M sucrose solution at 298 K?

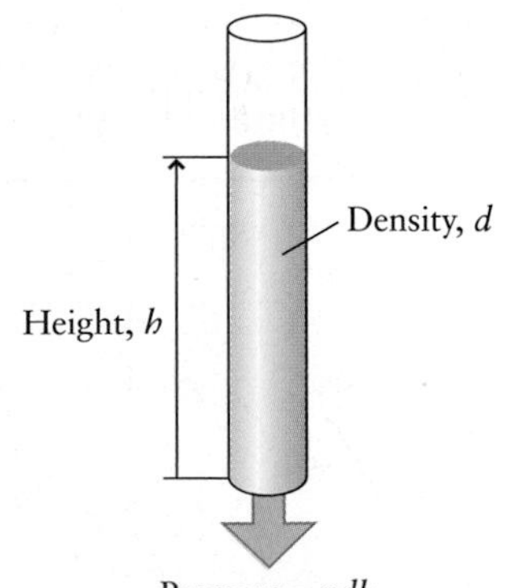

FIGURE 8.40 The pressure at the base of a column of fluid is equal to the product of the height, h, of the column, the density, d, of the liquid, and the acceleration of free fall, g.

The van 't Hoff equation is used to determine the molar mass of a solute from osmotic pressure measurements. This technique, which is called **osmometry,** involves the determination of the osmotic pressure of a solution prepared by using a known mass concentration of solute with an unknown molar mass. Osmometry is very sensitive, even at low concentrations, and is commonly used to determine very large molar masses, such as those of polymers.

Osmosis is the flow of solvent through a semipermeable membrane into a solution; the osmotic pressure is proportional to the molar concentration of the solute in the solution.

TOOLBOX 8.3 How to use colligative properties to determine molar mass

Conceptual basis

The presence of a solute lowers the free energy of a solvent, so it freezes at a lower temperature and has a characteristic osmotic pressure. Both properties depend on the relative numbers of solute and solvent molecules in the solution and can be used to determine the molar mass of a solute.

Procedure

(1) Cryoscopy

Step 1 Convert the observed freezing-point depression into solute molality by writing Eq. 10 in the form

$$\text{Molality} = \frac{\text{freezing-point depression}}{k_f}$$

Take the freezing-point constant from Table 8.9.

Step 2 Use this molality to calculate the moles of solute in the sample by multiplying it by the mass of solvent in kilograms.

Step 3 Determine the molar mass of the solute by dividing the given mass of solute by the number of moles calculated in step 2.

This procedure is illustrated in Example 8.8.

(2) Osmometry

Step 1 Convert the observed osmotic pressure into solute molarity (not molality) by writing Eq. 12 in the form

$$c = \frac{\Pi}{iRT}$$

In some cases, it may be necessary to calculate the osmotic pressure from the height, h, of the solution (in an apparatus like that in Fig. 8.37) by using $\Pi = gdh$, where d is the density of the solution and g is the acceleration of free fall (see inside back cover).

Step 2 Use this molarity to calculate the amount of solute in the stated volume of solution.

Step 3 Determine the molar mass of the solute by dividing the given mass of solute by the amount (in moles) calculated in step 2.

This procedure is illustrated in Example 8.9.

EXAMPLE 8.8 Determining molar mass cryoscopically

The addition of 0.24 g of sulfur to 100 g of the solvent carbon tetrachloride lowers the solvent's freezing point by 0.28°C. What is the molar mass and molecular formula of sulfur?

STRATEGY We have already remarked that sulfur occurs as S_8 molecules, but this example illustrates one way of reaching that conclusion. First, determine the molar mass of sulfur, using the cryoscopy procedure in Toolbox 8.3. Then, for the molecular formula, decide how many atoms of sulfur are needed in each molecule to account for the molar mass.

SOLUTION From the data given here and in Table 8.9,

$$\text{Molality of solute} = \frac{0.28\ \text{K}}{29.8\ \text{K}\cdot\text{kg}\cdot\text{mol}^{-1}} = \frac{0.28}{29.8}\ \text{mol}\cdot\text{kg}^{-1}$$

Because the mass of carbon tetrachloride is 100 g (0.100 kg), we find

$$\text{Moles of S}_x = (0.100\ \text{kg}) \times \left(\frac{0.28}{29.8}\ \text{mol}\cdot\text{kg}^{-1}\right) = \frac{0.100 \times 0.28}{29.8}\ \text{mol}$$

It then follows that

$$\text{Molar mass of S}_x = \frac{0.24\ \text{g}}{(0.100 \times 0.28/29.8)\ \text{mol}} = 2.6 \times 10^2\ \text{g}\cdot\text{mol}^{-1}$$

Because the molar mass of atomic sulfur is 32.1 $\text{g}\cdot\text{mol}^{-1}$, we can conclude that the value of x in the molecular formula S_x is

$$x = \frac{2.6 \times 10^2\ \text{g}\cdot\text{mol}^{-1}}{32.1\ \text{g}\cdot\text{mol}^{-1}} = 8.1$$

Elemental sulfur is therefore composed of S_8 molecules.

Self-Test 8.17A When 250 mg of eugenol, the compound responsible for the odor of oil of cloves, was added to 100 g of camphor, it lowered the freezing point of camphor by 0.62°C. Calculate the molar mass of eugenol.
[*Answer:* 1.6×10^2 $\text{g}\cdot\text{mol}^{-1}$ (actual: 164.2 $\text{g}\cdot\text{mol}^{-1}$)]

Self-Test 8.17B When 200 mg of linalool, a fragrant compound extracted from Ceylon cinnamon oil, was added to 100 g of camphor, it lowered the freezing point of camphor by 0.51°C. What is the molar mass of linalool?

EXAMPLE 8.9 Using osmometry to determine molar mass

The osmotic pressure due to 2.20 g of polyethylene (PE) dissolved in enough benzene to produce 100 mL of solution was 1.10×10^{-2} atm at 25°C. Calculate the molar mass of the polymer, which is a nonelectrolyte.

STRATEGY Polymers commonly have very high molar masses (of the order of kilograms per mole). Proceed as in Toolbox 8.3 for osmometry. Because polyethylene is a nonelectrolyte, $i = 1$. Use R in the units that match the data: in this case, liters and atmospheres. To avoid rounding errors, do the numerical calculation at the end.

SOLUTION From the data,

$$c = \frac{1.10 \times 10^{-2}\ \text{atm}}{(0.0821\ \text{L}\cdot\text{atm}\cdot\text{K}^{-1}\cdot\text{mol}^{-1}) \times (298\ \text{K})}$$
$$= \frac{1.10 \times 10^{-2}}{0.0821 \times 298}\ \text{mol}\cdot\text{L}^{-1}$$

It follows that, in 0.100 L of solution, the amount of PE molecules, n_{PE}, is

$$n_{PE} = (0.100\ \text{L}) \times \left(\frac{1.10 \times 10^{-2}}{0.0821 \times 298}\ \text{mol·L}^{-1}\right) = \frac{0.100 \times 1.10 \times 10^{-2}}{0.0821 \times 298}\text{mol}$$

Because this amount corresponds to 2.20 g, we conclude that

$$\text{Molar mass of PE} = \frac{2.20\ \text{g}}{\left(\dfrac{0.100 \times 1.10 \times 10^{-2}}{0.0821 \times 298}\right)\text{mol}} = 4.89 \times 10^{4}\ \text{g·mol}^{-1}$$

We report this molar mass as 48.9 kg·mol^{-1}.

Self-Test 8.18A The osmotic pressure of 3.0 g of polystyrene dissolved in enough benzene to produce 150 mL of solution was 1.21 kPa at 25°C. Calculate the average molar mass of the sample of polystyrene.

[***Answer:*** 41 kg·mol^{-1}]

Self-Test 8.18B The osmotic pressure of 1.50 g of polymethyl methacrylate dissolved in enough methylbenzene to produce 175 mL of solution was 2.11 kPa at 20°C. Calculate the average molar mass of the sample of polymethyl methacrylate.

In **reverse osmosis**, a pressure greater than the osmotic pressure is applied to the solution side of the semipermeable membrane. This application of pressure increases the rate at which solvent molecules leave the solution and thus reverses the flow of solvent, forcing it to flow from the solution to pure solvent. Reverse osmosis is used to remove salts from seawater to produce fresh water for drinking and irrigation. The water is almost literally squeezed out of the salt solution through the membrane. The technological challenge is to fabricate new membranes that are strong enough to withstand high pressures and do not easily become clogged. Commercial plants use cellulose acetate membranes at pressures of up to 70 atm. Each cubic meter of membrane can produce about 250 000 L of pure water a day.

Osmometry is used to determine the molar masses of polymers and natural macromolecules; osmosis helps to transport nutrients in plants; reverse osmosis is used in water purification.

BINARY LIQUID MIXTURES

Distillation, which we first encountered in Section G, is the foundation of the petroleum industry. It is used to separate the many compounds that make up petroleum and to purify alternative fuels such as ethanol and methanol. The separation of a liquid from solids by distillation was described in Section G, but now we can consider how two liquids can be separated by distillation. We have seen how the presence of a nonvolatile solute affects the vapor pressure of a liquid. The final sections of this chapter examine how the vapor pressure varies with composition when both components of a mixture are volatile and how that information can be used to separate them. A mixture of two liquids is formally a solution (a homogeneous mixture), but the term "mixture" is more commonly used.

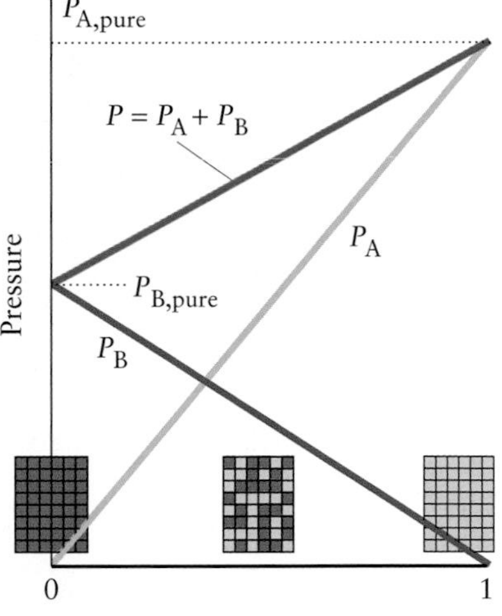

FIGURE 8.41 The vapor pressures of the two components of an ideal binary mixture obey Raoult's law. The total vapor pressure is the sum of the two partial vapor pressures (Dalton's law).

8.20 The Vapor Pressure of a Binary Liquid Mixture

Consider a simple system consisting of an ideal binary mixture of the volatile liquids A and B. We could think of A as benzene and B as toluene (methylbenzene), for example, because these two compounds have similar molecular structures and so form nearly ideal solutions. Because the mixture is ideal, each component has a vapor pressure given by Raoult's law:

$$P_A = x_{A,liquid} P_{A,pure} \qquad P_B = x_{B,liquid} P_{B,pure} \qquad (13)$$

In these equations, $x_{A,liquid}$ is the mole fraction of A in the liquid mixture and $P_{A,pure}$ is the vapor pressure of pure A; similarly, $x_{B,liquid}$ is the mole fraction of B in the liquid and $P_{B,pure}$ is the vapor pressure of pure B. Dalton's law (Section 4.11) tells us that the total pressure of the vapor is the sum of these two partial pressures (Fig. 8.41):

$$\begin{aligned}\text{Total vapor pressure} &= P_{total} = P_A + P_B \\ &= x_{A,liquid}P_{A,pure} + x_{B,liquid}P_{B,pure}\end{aligned} \qquad (14)$$

How can we use this information . . . to predict the vapor pressure of a mixture of two liquids?

The vapor pressures of benzene and toluene at 25°C are 94.6 and 29.1 Torr, respectively. Thus, the vapor pressure of each component and the total vapor pressure of a mixture in which one-third of the molecules are benzene (so $x_{benzene,liquid} = \frac{1}{3}$ and $x_{toluene,liquid} = \frac{2}{3}$) are

$$\begin{aligned}P_{benzene} &= \tfrac{1}{3}(94.6 \text{ Torr}) = 31.5 \text{ Torr} \\ P_{toluene} &= \tfrac{2}{3}(29.1 \text{ Torr}) = 19.4 \text{ Torr} \\ P_{total} &= 31.5 + 19.4 \text{ Torr} = 50.9 \text{ Torr}\end{aligned}$$

Self-Test 8.19A What is the total vapor pressure at 25°C of a mixture of 3.00 mol $C_6H_6(l)$ and 2.00 mol $CH_3C_6H_5(l)$?

[*Answer:* 68.4 Torr]

Self-Test 8.19B What is the total vapor pressure at 25°C of a mixture of equal masses of benzene and toluene?

The vapor of the mixture is richer in the more volatile component (the component with the greater vapor pressure). Benzene, for instance, is more volatile than toluene, so we can expect relatively more of it in the vapor than in the liquid mixture. If we could express the composition of the vapor in terms of the composition of the liquid, we could confirm that the vapor is richer than the liquid in the more volatile component.

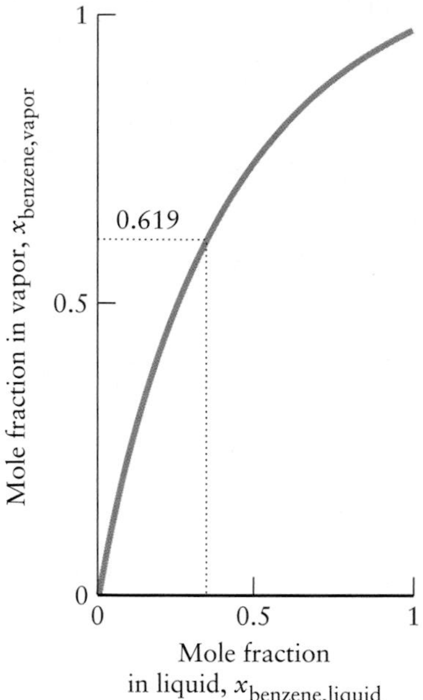

FIGURE 8.42 The composition of the vapor in equilibrium with a mixture of two volatile liquids (here, benzene and toluene) and its variation with the composition of the liquid. Note that the vapor is richer in benzene than the liquid mixture at each composition. For instance, when the mole fraction of benzene is 0.333 in the liquid, in the vapor it is 0.619.

How do we do that?

To express the composition of the vapor in equilibrium with the liquid phase, we first note that the definition of partial pressure ($P_J = x_J P$) and Dalton's law ($P = P_A + P_B$) allow us to express the composition of the vapor of a mixture of liquids A and B in terms of the partial pressures of the components:

$$x_{\text{A,vapor}} = \frac{P_A}{P} = \frac{P_A}{P_A + P_B} \tag{15}$$

and likewise for $x_{\text{B,vapor}}$. However, we have seen, in Eq. 13, how to express the vapor pressure of the components in terms of the composition of the liquid (by using Raoult's law); we can therefore write

$$x_{\text{A,vapor}} = \frac{x_{\text{A,liquid}}P_{\text{A,pure}}}{x_{\text{A,liquid}}P_{\text{A,pure}} + x_{\text{B,liquid}}P_{\text{B,pure}}} \tag{16}$$

This expression relates the composition of the vapor (in terms of the mole fraction of A in the vapor) to the composition of the liquid (in terms of the mole fraction of A in the liquid, remembering that $x_B = 1 - x_A$). It is plotted for benzene and toluene in Fig. 8.42, and we see that $x_{\text{benzene,vapor}} > x_{\text{benzene,liquid}}$, just as we anticipated.

The vapor of an ideal mixture of two volatile liquids is richer than the liquid in the more volatile component. The contribution of each component to the total vapor pressure and its mole fraction in the vapor can be calculated by combining Raoult's law and Dalton's law.

How can we use this information . . . to predict the composition of the vapor in equilibrium with a mixture of liquids?

In a solution of benzene in toluene in which one-third the molecules in the liquid are benzene, the mole fraction of benzene in the vapor at 25°C is found to be

$$x_{\text{benzene, vapor}} = \frac{0.333 \times 94.6\ \text{Torr}}{0.333 \times 94.6\ \text{Torr} + 0.667 \times 29.1\ \text{Torr}} = 0.619$$

Note that the mole fraction of benzene in the vapor is nearly twice that in the liquid.

Self-Test 8.20A (a) Determine the vapor pressure at 25°C of a solution of toluene in benzene in which the mole fraction of benzene is 0.900. (b) Calculate the mole fractions of benzene and toluene in the vapor.

[*Answer:* (a) 88.0 Torr; (b) 0.967 and 0.033]

Self-Test 8.20B (a) Determine the vapor pressure at 25°C of a solution of benzene in toluene in which the mole fraction of benzene is 0.500. (b) Calculate the mole fractions of benzene and toluene in the vapor.

8.21 Distillation

The normal boiling point of a binary liquid mixture is the temperature at which the *total* vapor pressure is equal to 1 atm. Because the vapor pressure varies with composition, the boiling point also varies as the more volatile component of the mixture boils away. If we were to heat a sample of pure benzene at a constant pressure of 1 atm, it would begin to boil at 80.1°C. Similarly, pure toluene boils at 110.6°C. Because, at a given temperature, the vapor pressure of a mixture of benzene and toluene is intermediate between that of toluene and benzene, the boiling point of the mixture will be intermediate between that of the two pure liquids. In Fig. 8.43, which is called a **temperature-composition diagram,** the lower line shows how the normal boiling point of the mixture varies with the composition. The graph, which has been constructed empirically, allows us to predict the boiling point at any composition of the liquid mixture.

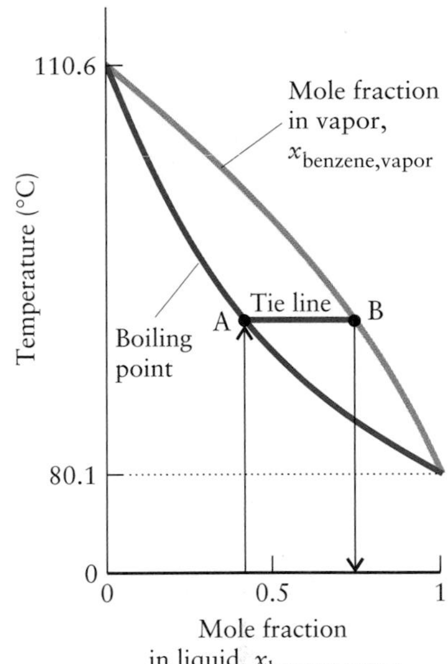

FIGURE 8.43 A temperature-composition diagram for benzene and toluene. The lower, blue line shows the boiling point of the mixture as a function of composition. The upper, orange line shows the composition of the vapor in equilibrium with the liquid at each boiling point. Thus, point B shows the vapor composition for a mixture that boils at A.

The upper curve in Fig. 8.43 shows the composition of the vapor in equilibrium with the liquid mixture at the boiling point. To find the composition of the vapor, we simply look along the **tie line,** the horizontal line at the boiling point, and see where it cuts through the upper line. So, if we heat a liquid mixture with the composition given by the vertical line through A in Fig. 8.43 ($x_{\text{benzene, liquid}} = 0.45$) at a constant pressure of 1 atm, the mixture boils at the temperature corresponding to point A. At this temperature, the composition of the vapor in equilibrium with the liquid is given by point B ($x_{\text{benzene, vapor}} = 0.73$).

We can put these ideas to work and see how to use distillation to separate two volatile liquids. When a mixture of benzene and toluene molecules with $x_{\text{benzene}} = 0.20$ begins to boil (point A in Fig. 8.44), the initial composition of the vapor formed is given by point B ($x_{\text{benzene}} = 0.45$). If we were to cool the vapor until it condenses, the first drop of condensed vapor, the **distillate,** would be richer in benzene than the original mixture

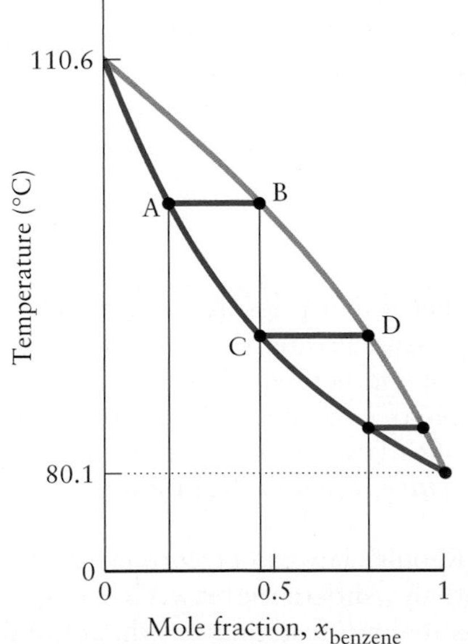

FIGURE 8.44 Some of the steps that represent fractional distillation of a mixture of two volatile liquids (benzene and toluene). The original mixture boils at A, and its vapor has composition B. After condensation of the vapor, the resulting liquid boils at C, and the vapor has composition D, and so on.

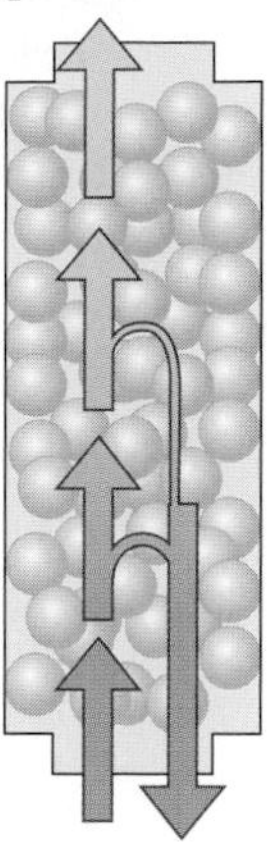

FIGURE 8.45 A schematic illustration of the process of fractional distillation. The temperature in the fractionating column decreases with height. The condensations and reboilings illustrated in Fig. 8.44 occur at increasing heights in the column, until pure benzene begins to distill. The less volatile component returns to the flask beneath the fractionating column, and the more volatile component escapes from the top, to be condensed and collected.

and the liquid remaining in the pot would be richer in toluene. If we take that drop of distillate and reheat it, it will boil at the temperature represented by the point C, and the vapor above the boiling solution will have the composition D ($x_{benzene} = 0.73$), as indicated by the tie line. Note that the distillate from this second stage of distillation is even richer in benzene than the distillate from the first stage. If we continued these steps of boiling, condensation, and boiling again, we would eventually obtain a tiny amount of nearly pure benzene.

The process called **fractional distillation** uses a method of continuous redistillation to separate mixtures of liquids with similar boiling points, such as benzene and toluene: the mixture is heated and the rising vapor passes up through a tall column packed with material having a high surface area, such as glass beads (Fig. 8.45). The vapor begins to condense on the beads near the bottom of the column. However, as heating continues, the vapor condenses and vaporizes over and over as it rises; the liquid drips back into the boiling mixture. The vapor becomes richer in the component with the lower boiling point as it continues to rise through the column and passes out into the condenser. Therefore, the final distillate is nearly pure benzene, the more volatile of the components, whereas the liquid in the pot is nearly pure toluene.

If the original sample consists of several volatile liquids, the component liquids appear in the distillate in succession in a series of **fractions,** or samples of the distillate, that boil in specific temperature ranges. The most volatile (lowest boiling point) component distills first, and the least volatile (highest boiling point) component is obtained as the final fraction.

Giant fractionating columns are used in industry to separate complex mixtures such as crude petroleum. We use the volatile fractions as natural gas (boiling below 0°C), gasoline (boiling in the range 30°C to 200°C), and kerosene (180°C to 325°C). The less volatile fractions are used as diesel fuel (above 275°C). The residue that remains behind after distillation is asphalt, which is used for surfacing highways.

Volatile liquids can be separated by fractional distillation. Liquid and vapor are in equilibrium at each point in the fractionating column, but their compositions vary with height. As a result, the lowest-boiling-point component can be removed from the top of the column before the next-higher-boiling-point component distills.

8.22 Azeotropes

Most liquid mixtures are not ideal, so their vapor pressures do not follow Raoult's law. A mixture with a vapor pressure greater than that predicted by Raoult's law, like mixtures of ethanol and benzene, is said to show a *positive deviation* from the law (Fig. 8.46a). Mixtures with a vapor pressure lower than that predicted by Raoult's law, like mixtures of acetone and chloroform, are said to show a *negative deviation* from Raoult's law (Fig. 8.46b).

The enthalpy of mixing of two liquids is the same as the enthalpy of solution of one liquid in the other, but the former term is used more commonly.

The direction of a deviation from Raoult's law can be correlated with the **enthalpy of mixing,** ΔH_{mix}, the enthalpy difference between the mixture and the pure components. The enthalpy of mixing of ethanol and

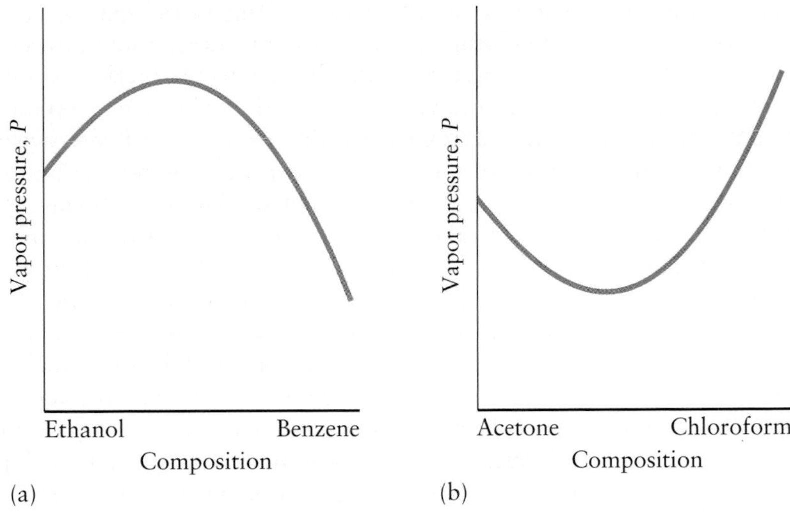

FIGURE 8.46 A graphical illustration of the variation of the vapor pressures of a mixture of (a) ethanol and benzene, (b) acetone and chloroform. Note that the mixture in (a) shows a vapor-pressure maximum, and therefore displays a positive deviation from Raoult's law. The one in (b) shows a minimum, and hence displays a negative deviation from Raoult's law.

benzene is positive—the mixing process is endothermic—and this mixture also shows a positive deviation from Raoult's law. The enthalpy of mixing of acetone and chloroform is negative—the mixing process is exothermic—and this mixture shows a negative deviation from Raoult's law.

A negative enthalpy of mixing indicates that the interactions between molecules of acetone and chloroform are more favorable than those between the molecules in the unmixed liquids: energy is released as the mixture is formed. We can understand the reason for this by looking at the molecular structures. Chloroform (trichloromethane, $CHCl_3$) is a polar molecule with a relatively high dipole moment (7): the H atom has a partial positive charge. When an acetone molecule (**8**) approaches, the H atom of chloroform is attracted to the lone pairs of electrons on the O atom in acetone. The attraction is so strong that it can be regarded as a hydrogen bond. On the other hand, the positive enthalpy of mixing of ethanol and benzene implies that the intermolecular forces between the two compounds are less favorable than those in one or both of the pure compounds: energy is absorbed as the mixture is formed. In pure liquid benzene, the molecules are held together by London forces, but ethanol molecules are held together by strong hydrogen bonds. Many of the latter bonds must be broken when ethanol is mixed with benzene.

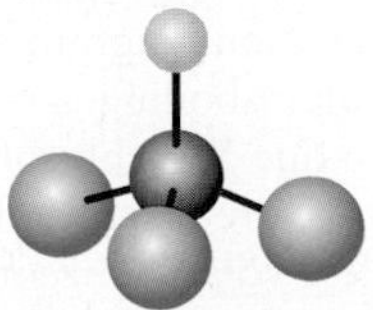

7 Trichloromethane, $CHCl_3$

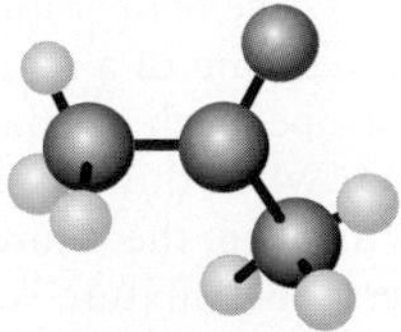

8 Acetone, C_3H_6O

Deviations from Raoult's law can make it difficult to separate liquids by distillation. The temperature-composition diagrams for mixtures of ethanol and benzene and of acetone and chloroform show why. A positive deviation from Raoult's law means that the boiling point of the mixture is lower than that predicted by Raoult's law. For some pairs of components, the boiling point of the mixture is in fact lower than the boiling point of either constituent (Fig. 8.47). A mixture for which the lowest boiling temperature is below that of both the pure components is called a **minimum-boiling azeotrope**, and the components cannot be separated by distillation: in a fractional distillation, a mixture with the azeotropic composition boils over first, not the more volatile pure liquid. The opposite behavior is found for a mixture of acetone in chloroform (Fig. 8.48). This **maximum-boiling azeotrope** boils at a higher temperature than either constituent and is the last fraction to be collected, not the less volatile pure liquid. Pure ethanol cannot be obtained from aqueous solutions in fermentation chambers because it forms a minimum-boiling azeotrope with water; the azeotropic mixture has the composition 95% ethanol.

Solutions in which intermolecular forces are stronger in the solution than in the pure components have negative deviations from Raoult's law and negative enthalpies of mixing; they often form maximum-boiling azeotropes. Solutions in which intermolecular forces are weaker in the solution than in the pure components have positive deviations from Raoult's law and positive enthalpies of mixing; they often form minimum-boiling azeotropes.

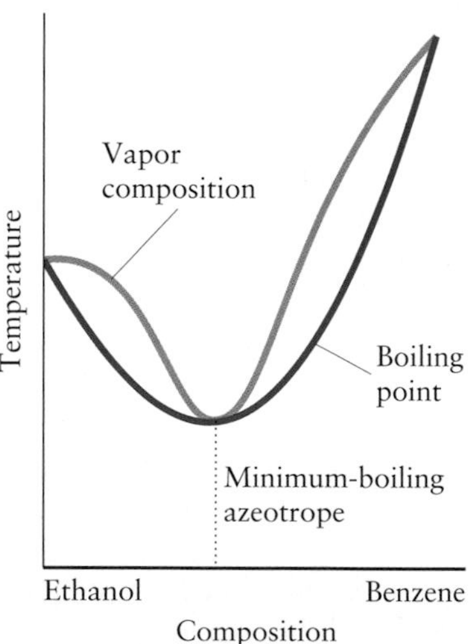

FIGURE 8.47 The temperature-composition diagram of a minimum-boiling azeotrope (such as ethanol and benzene). When this mixture is fractionally distilled, the (more volatile) azeotropic mixture is obtained as the initial distillate.

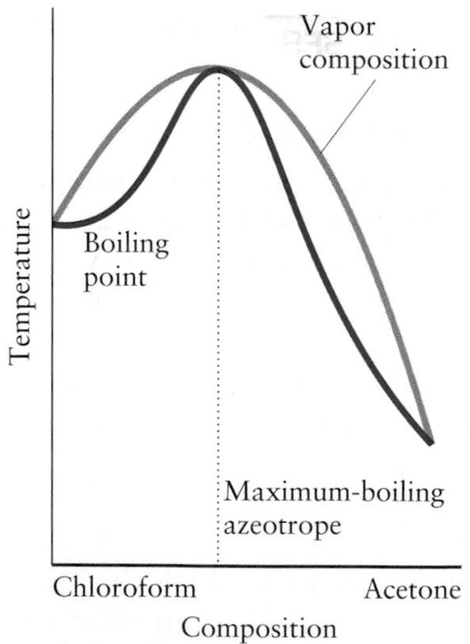

FIGURE 8.48 The temperature-composition diagram showing a maximum-boiling azeotrope (such as acetone and chloroform). When this mixture is fractionally distilled, the (less volatile) azeotropic mixture is left in the flask.

SKILLS YOU SHOULD HAVE MASTERED

1 Predict the relative value of the vapor pressure, Example 8.1.

2 Estimate the vapor pressure of a liquid by using the Clausius-Clapeyron equation, Toolbox 8.1 and Example 8.2.

3 Use the Clausius-Clapeyron equation to estimate the boiling point of a liquid, Toolbox 8.1 and Example 8.3.

4 Interpret a one-component phase diagram, Example 8.4.

5 Predict relative solubilities from molecular polarity, Section 8.10.

6 Use Henry's law to calculate the solubility of a gas, Section 8.11.

7 Estimate the hydration enthalpy of an ion, Sections 8.13 and 8.14.

8 Calculate the molality of a solute, Self-Test 8.10.

9 Convert between molality and mole fraction or molarity, Toolbox 8.2 and Examples 8.5 and 8.6.

10 Calculate the vapor pressure of a solvent in a solution by using Raoult's law, Example 8.7.

11 Estimate a freezing-point depression from the molality of a solute, Section 8.18.

12 Interpret a measured value of the van 't Hoff *i* factor, Self-Test 8.15.

13 Determine the osmotic pressure of a solution of known molarity, Self-Test 8.16.

14 Determine molar mass cryoscopically, Toolbox 8.3 and Example 8.8.

15 Use osmometry to find the molar mass of a solute, Toolbox 8.2 and Example 8.9.

16 Calculate the vapor pressure and vapor composition for a solution of two liquids, Section 8.20.

17 Interpret a two-component phase diagram and discuss fractional distillation, Sections 8.20–8.22.

EXERCISES

Phases and Phase Transitions

8.1 Suppose you were to collect 1.0 L of air by passing it slowly through water at 20°C and into a container. Estimate the mass of water vapor in the collected air, assuming that the air is saturated with it.

8.2 Suppose you were to collect 500.0 mL of nitrogen gas by slowly passing it through ethanol at 25°C, at which temperature ethanol's vapor pressure is 58.9 Torr. Estimate the mass of ethanol vapor in the collected gas, assuming that the nitrogen is saturated with it.

8.3 What mass of water vapor can you expect to find in the air of a bathroom of dimensions 4.0 m by 3.0 m by 3.0 m when water has been left in the bathtub at 40°C with the door shut? The vapor pressure of water at that temperature is 7.4 kPa. Assume that the air is saturated with water vapor.

8.4 A bottle of mercury at 25°C was left unstoppered in a chemical supply room measuring 3.0 m by 2.0 m by 2.0 m. What mass of mercury vapor would be present if the air became saturated with it? The vapor pressure of mercury at 25°C is 0.227 Pa.

8.5 Use the vapor pressure curve in Fig. 8.4 to estimate the boiling point of water when the atmospheric pressure is (a) 735 Torr; (b) 750 Torr.

8.6 Use the vapor pressure curve in Fig. 8.4 to estimate the boiling point of ethanol when the atmospheric pressure is (a) 600 Torr; (b) 400 Torr.

8.7 The vapor pressure of hydrogen azide, HN_3, is 58 Torr at −22.75°C and 512 Torr at 25.0°C. Calculate (a) the standard enthalpy of vaporization; (b) the standard entropy of vaporization; (c) the standard free energy of vaporization; (d) the boiling point of hydrogen azide.

8.8 The vapor pressure of 2-propanone, CH_3COCH_3 (more commonly known as acetone), is 67 Torr at 0.0°C and 222 Torr at 25.0°C. Calculate (a) the standard enthalpy of vaporization; (b) the standard entropy of vaporization; (c) the standard free energy of vaporization; (d) the boiling point of 2-propanone.

8.9 The normal boiling point of hydrogen fluoride is 19.5°C, and its vapor pressure at 0°C is 359 Torr. Calculate (a) the standard enthalpy of vaporization of hydrogen fluoride; (b) the standard entropy of vaporization of hydrogen fluoride; and (c) the vapor pressure of hydrogen fluoride at 8.50°C.

8.10 The normal boiling point of trisilane, Si_3H_8, is 52.9°C, and its vapor pressure at 0°C is 96 Torr. Calculate (a) the standard enthalpy of vaporization of trisilane; (b) the standard entropy of vaporization of trisilane; and (c) the vapor pressure of trisilane at 35.0°C.

8.11 Arsine, AsH_3, is a highly toxic compound used in the electronics industry for the production of semiconductors. Its vapor pressure is 35 Torr at −111.95°C and 253 Torr at −83.6°C. Using these data, calculate (a) the standard enthalpy of vaporization; (b) the standard entropy of vaporization; (c) the standard free energy of vaporization; (d) the normal boiling point of arsine.

8.12 The vapor pressure of chlorine dioxide, ClO_2, is 155 Torr at −22.75°C and 485 Torr at 0°C. Calculate (a) the standard enthalpy of vaporization; (b) the standard entropy of vaporization; (c) the standard free energy of vaporization; (d) the normal boiling point of ClO_2.

8.13 The normal boiling point of iodomethane, CH_3I, is 42.43°C, and its vapor pressure at 0°C is 140 Torr. Calculate (a) the standard enthalpy of vaporization of iodomethane; (b) the standard entropy of vaporization of iodomethane; (c) the vapor pressure of iodomethane at 25.0°C.

8.14 The normal boiling point of trimethylphosphine, $P(CH_3)_3$, is 38.4°C, and its vapor pressure at −45.21°C is 13 Torr. Calculate (a) the standard enthalpy of vaporization of trimethylphosphine; (b) the standard entropy of vaporization of trimethylphosphine; (c) the vapor pressure of trimethylphosphine at 15.0°C.

8.15 Using data from Table 6.3, calculate the vapor pressure of methanol at 25.0°C.

8.16 Using data from Table 6.3, calculate the vapor pressure of ammonia at 215 K.

8.17 Using data from Table 6.3, calculate the vapor pressure of mercury at 175°C.

8.18 Using data from Table 6.3, calculate the vapor pressure of methane at 75 K.

8.19 Use Fig. 8.5 to predict the state of a sample of water under the following sets of conditions: (a) 1 atm, 200°C; (b) 100 atm, 200°C; (c) 3 Torr, 25°C; (d) 218 atm, 375°C.

8.20 Use Fig. 8.6 to predict the state of a sample of CO_2 under the following sets of conditions: (a) 6 atm, −80°C; (b) 1 atm, −56°C; (c) 80 atm, 25°C; (c) 5.1 atm, −56°C.

8.21 The phase diagram for helium is shown below. Use it to answer the following questions: (a) What is the maximum temperature at which superfluid helium-II can exist? (b) What is the minimum pressure at which solid helium-II can exist? (c) What is the normal boiling point of helium-I? (d) Can solid helium sublime?

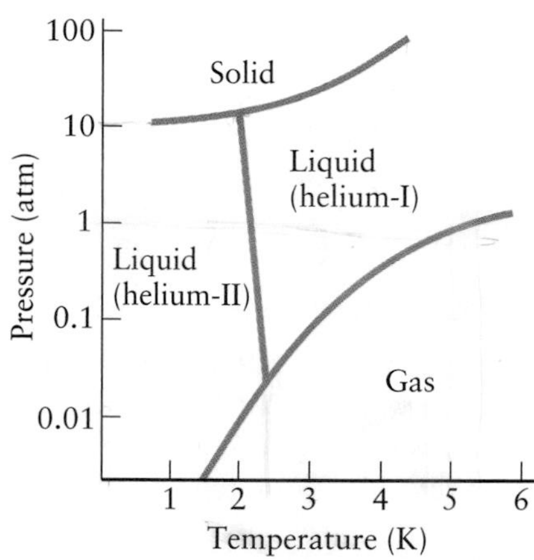

8.22 The phase diagram for carbon is shown on the next page: it indicates the extreme conditions that are needed to form diamonds from graphite. Use the diagram to answer the following questions: (a) At 2000 K, what is the minimum pressure needed before graphite changes to diamond? (b) What is the minimum temperature at which liquid carbon can exist at pressures below 10 000 atm? (c) At what pressure does graphite melt at 3000 K? (d) Are diamonds stable under normal conditions? If not, why is it that people can wear them without having to compress and heat them?

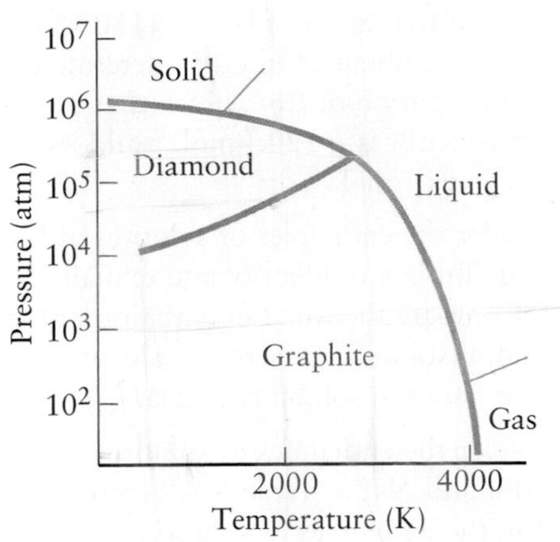

8.23 Use the phase diagram for helium in Exercise 8.21(a) to describe the phases in equilibrium at each of helium's two triple points; (b) to decide which liquid phase is more dense, helium-I or helium-II.

8.24 Use the phase diagram for carbon in Exercise 8.22(a) to describe the phase transitions that carbon would experience if compressed at a constant temperature of 2000 K from 100 atm to 1×10^6 atm; (b) to rank the diamond, graphite, and liquid phases of carbon in order of increasing density.

8.25 Use the phase diagram for carbon dioxide (Fig. 8.6) to predict what would happen to a sample of carbon dioxide gas at −50°C and 1 atm if its pressure were suddenly increased to 73 atm. What would be the final physical state of the carbon dioxide?

8.26 A new substance developed in a laboratory has the following properties: normal melting point, 83.7°C; normal boiling point, 177°C; triple point, 200 Torr and 38.6°C. (a) Sketch the approximate phase diagram and label the solid, liquid, and gaseous phases and the solid-liquid, liquid-gas, and solid-gas phase boundaries. (b) Sketch an approximate cooling curve for a sample at constant pressure, beginning at 500 Torr and 25°C and ending at 200°C.

Solubility

8.27 Which would be the better solvent, H_2O or benzene, for each of the following: (a) KCl; (b) CCl_4; (c) CH_3COOH?

8.28 Which would be the better solvent, H_2O or CCl_4, for each of the following: (a) NH_3; (b) HCl; (c) I_2?

8.29 The following groups are found in some organic molecules. Which are hydrophilic and which are hydrophobic: (a) $-NH_2$; (b) $-CH_3$; (c) $-Br$; (d) $-COOH$?

8.30 The following groups are found in some organic molecules. Which are hydrophilic and which are hydrophobic: (a) $-OH$; (b) $-CH_2CH_3$; (c) $-CONH_2$; (d) $-Cl$?

8.31 State the molar solubility in water of (a) O_2 at 50 kPa; (b) CO_2 at 500 Torr; (c) CO_2 at 0.10 atm. The temperature in each case is 20°C, and the pressures are partial pressures of the gas. Use the information in Table 8.5.

8.32 Calculate the aqueous solubility (in milligrams per liter) of (a) air at 1.0 atm; (b) He at 1.0 atm; (c) He at 25 kPa. The temperature is 20°C in each case, and the pressures are partial pressures of the gas. Use information in Table 8.5.

8.33 A soft drink is made by dissolving CO_2 at 3.00 atm in a flavored solution and sealing the solution in an aluminum can. What volume of CO_2 is released when a 355-mL can is opened to 1.00 atm at 20°C and all the CO_2 is allowed to escape? Assume that the can is completely filled with the solution.

8.34 The volume of blood in the body of a certain deep-sea diver is about 6.00 L. Blood cells make up about 55% of the blood volume, and the remaining 45% is the aqueous solution called plasma. What is the maximum volume of nitrogen measured at 1.00 atm and 37°C that could dissolve in the diver's blood plasma at a depth of 93 m, where the pressure is 10.0 atm? (This is the volume that could come out of solution suddenly, causing the bends, if the diver were to ascend too quickly.) Assume that the Henry's constant for nitrogen at 37°C (body temperature) is 5.8×10^{-4} $mol \cdot L^{-1} \cdot atm^{-1}$.

8.35 The carbon dioxide gas dissolved in a sample of water in a partially filled, sealed container has reached equilibrium with its partial pressure in the air above the solution. Explain what happens to the solubility of the CO_2 if (a) the partial pressure of the CO_2 gas is doubled by the addition of more CO_2; (b) the total pressure of the gas above the liquid is doubled by the addition of nitrogen.

8.36 Explain what happens to the solubility of the CO_2 in Exercise 8.35 if (a) the partial pressure of $CO_2(g)$ is increased by compressing the gas to a third of its original volume; (b) the temperature is raised.

8.37 The minimum mass concentration of oxygen required for fish life is 4 mg·L^{-1}. (a) Assuming the density of lake water to be 1.00 g·mL^{-1}, express this concentration in parts per million (mg·kg^{-1}; see Table 8.5). (b) What is the minimum partial pressure of O_2 that would supply the minimum mass concentration of oxygen in water in order to support fish life at 20°C? (c) What is the minimum atmospheric pressure that would give this partial pressure, assuming oxygen exerts about 21% of the atmospheric pressure?

8.38 The solubility of oxygen in rivers and streams depends on various factors, such as temperature, salt concentration, depth, and biological activity. Its solubility equilibrium is $O_2(g) \rightleftharpoons O_2(aq)$. Compare the dissolved oxygen concentration at 1.0 atm total pressure (0.21 atm partial pressure) with its concentration for pure oxygen at 1.0 atm.

8.39 Describe and explain the trend in the ion hydration enthalpies for the halide ions. See Table 8.8.

8.40 Describe and explain the trend in ion hydration enthalpies for the isoelectronic cations Na^+, Mg^{2+}, and Al^{3+}. See Table 8.8.

8.41 Lithium sulfate dissolves exothermically in water. (a) Is the enthalpy of solution for Li_2SO_4 positive or negative? (b) Write the chemical equation for the dissolving process. (c) Which is larger for lithium sulfate, the lattice enthalpy or the enthalpy of hydration?

8.42 The enthalpy of solution of ammonium nitrate in water is positive. (a) Does NH_4NO_3 dissolve endothermically or exothermically? (b) Write the chemical equation for the dissolving process. (c) Which is larger for NH_4NO_3, the lattice enthalpy or the enthalpy of hydration?

8.43 Calculate the heat evolved or absorbed when 10.0 g of (a) NaCl; (b) NaBr; (c) $AlCl_3$; (d) NH_4NO_3 is dissolved in 100.0 g of water. Assume that the enthalpies of solution in Table 8.6 are applicable.

8.44 Determine the temperature change when 10.0 g of (a) KCl; (b) $MgBr_2$; (c) KNO_3; (d) NaOH is dissolved in 100.0 g of water. Assume that the specific heat capacity of the solution is 4.18 J·K^{-1}·g^{-1} and that the enthalpies of solution in Table 8.6 are applicable.

8.45 Estimate the enthalpy of solution of $SrCl_2$ from the data for the lattice enthalpy and the ion enthalpies of hydration in Tables 6.6 and 8.8.

8.46 (a) Calculate the enthalpy of hydration of Br^- by using the appropriate data from Table 8.7 and the enthalpy of hydration of H^+ (−1130 kJ·mol^{-1}). (b) Use the value obtained in (a) to determine the enthalpy of hydration of Rb^+, given that the enthalpy of solution of RbBr is +22 kJ·mol^{-1} and its lattice enthalpy is 651 kJ·mol^{-1}.

8.47 Consider the enthalpies of solution of LiCl, NaCl, and AgCl in Table 8.6. Identify and explain any trends you find. Compare the trend in enthalpies of solution to the trend in solubilities of these salts and explain any relationship between solubility and ΔH_{sol}.

8.48 Consider the enthalpies of solution of NaF, NaCl, NaBr, and NaI in Table 8.6. Identify and explain the trend in the enthalpies of solution on proceeding down the group of halides.

Colligative Properties

8.49 Calculate the following: (a) the molality of sodium chloride in a solution prepared by dissolving 10.0 g of NaCl in 250.0 g water; (b) the mass (in grams) of NaOH that must be mixed with 345 g of water to prepare 0.22 *m* NaOH(aq); (c) the molality of urea, $CO(NH_2)_2$, in a solution prepared by dissolving 1.94 g urea in 515 mL water.

8.50 Calculate the following: (a) the molality of KOH in a solution prepared from 13.72 g of KOH and 50.0 g of water; (b) the mass (in grams) of ethylene glycol, HOC_2H_4OH, that should be added to 2.0 kg of water to prepare 0.44 *m* $HOC_2H_4OH(aq)$; (c) the molality of an aqueous 4.66% by mass HCl solution.

8.51 A 5.00% by mass K_3PO_4 aqueous solution has a density of 1.043 g·cm^{-3}. Determine (a) the molality; (b) the molarity of potassium phosphate in the solution.

8.52 Calculate the concentrations of the following solutions: (a) the molality of 15.34 g of sucrose, $C_{12}H_{22}O_{11}$, dissolved in 549 mL water; (b) the molality of CsCl in a 10.00% by mass aqueous solution with a density of 1.0818 g·cm^{-3}; (c) the molality of acetone in an aqueous solution with a mole fraction for acetone of 0.235.

8.53 Calculate the concentrations of the following solutions: (a) the molality of chloride ions in an aqueous solution of magnesium chloride for which x_{MgCl_2} is 0.0175; (b) the molality of 7.12 g of sodium hydroxide dissolved in 325 g water; (c) the molality of 15.00 M HCl(aq) with a density of 1.0745 g·cm^{-3}.

8.54 Calculate the concentrations of the following solutions: (a) the molality of chloride ions in an aqueous solution of iron(III) chloride for which x_{FeCl_3} is

0.0231; (b) the molality of hydroxide ions in a solution prepared from 9.25 g of barium hydroxide dissolved in 155 g of water.; (c) the molality of 12.00 M $NH_3(aq)$ with a density of 0.9519 $g{\cdot}cm^{-3}$.

8.55 (a) Calculate the mass of $CaCl_2{\cdot}6H_2O$ needed to prepare 0.175 *m* $CaCl_2(aq)$, using 500 g of water. (b) What mass of $NiSO_4{\cdot}6H_2O$ must be dissolved in 500 g of water to produce 0.33 *m* $NiSO_4(aq)$?

8.56 A 10.0% by mass $H_2SO_4(aq)$ solution has a density of 1.07 $g{\cdot}cm^{-3}$. (a) How many milliliters of solution contain 8.37 g of H_2SO_4? (b) What is the molality of H_2SO_4 in solution? (c) What mass (in grams) of H_2SO_4 is in 250 mL of solution?

8.57 Calculate the vapor pressure of the solvent in each solution listed below. Use Table 8.3 to find the vapor pressure of water in (a) an aqueous solution at 100°C in which the mole fraction of sucrose is 0.100; (b) an aqueous solution at 100°C in which the molality of sucrose is 0.100 $mol{\cdot}kg^{-1}$.

8.58 What is the vapor pressure of the solvent in each of the solutions listed below? Use the data in Table 8.3 for the vapor pressure of water at various temperatures. (a) the mole fraction of glucose is 0.050 in an aqueous solution at 80°C. (b) an aqueous solution at 25°C is 0.10 *m* urea, $CO(NH_2)_2$, a nonelectrolyte.

8.59 (a) Calculate the vapor pressure of a 2.50% by mass aqueous ethylene glycol, HOC_2H_4OH, solution at 0°C. (b) What is the vapor pressure of 0.155 *m* NaOH(aq) at 80°C? (c) Determine the *change* in the vapor pressure of water when 5.95 g of urea, $CO(NH_2)_2$, is dissolved in 100 g of water at 10°C. For data, see Table 8.3.

8.60 (a) Calculate the *change* in vapor pressure from that of pure water for an aqueous solution at 40°C in which the mole fraction of fructose is 0.33. (b) Determine the vapor pressure of a saturated magnesium fluoride solution at 20°C. The solubility of MgF_2 at 20°C is 8 mg/100 g of water. (c) What is the vapor pressure of 0.025 *m* $Fe(NO_3)_3(aq)$ at 0°C? For data, see Table 8.3.

8.61 When 8.05 g of an unknown compound X was dissolved in 100 g of benzene, the vapor pressure of the benzene decreased from 100.0 Torr to 94.8 Torr at 26°C. What is (a) the mole fraction and (b) the molar mass of X?

8.62 The normal boiling point of ethanol is 78.4°C. When 9.15 g of a soluble nonelectrolyte was dissolved in 100 g of ethanol, the vapor pressure of the solution at that temperature was 7.40×10^2 Torr. (a) What are the mole fractions of ethanol and solute? (b) What is the molar mass of the solute?

8.63 Estimate the boiling-point elevation and the normal boiling points of (a) 0.10 *m* $C_{12}H_{22}O_{11}(aq)$; (b) 0.22 *m* NaCl(aq); (c) a saturated solution of LiF at 100°C (the solubility of LiF is 230 mg/100 g of water at 100°C). Assume that the ionic compounds undergo complete dissociation.

8.64 Estimate the boiling-point elevation and the normal boiling points of (a) 0.22 *m* $CaCl_2(aq)$; (b) a saturated solution of Li_2CO_3, of solubility 0.72 g/100 g of water at 100°C; (c) 1.7% by mass $CO(NH_2)_2(aq)$. Assume that the ionic compounds undergo complete dissociation.

8.65 (a) What is the normal boiling point of an aqueous solution that has a vapor pressure of 751 Torr at 100°C? (b) Determine the normal boiling point of a benzene solution that has a vapor pressure of 740 Torr at 80.1°C, the normal boiling point of pure benzene.

8.66 (a) What is the normal boiling point of an aqueous solution that has a freezing point of −1.04°C? (b) The freezing point of a benzene solution is 2.0°C. The normal freezing point of benzene is 5.5°C. What is the expected normal boiling point of the solution? The normal boiling point of benzene is 80.1°C.

8.67 A 1.05-g sample of a molecular compound is dissolved in 100 g of tetrachloromethane (carbon tetrachloride). The normal boiling point of the solution is 61.51°C; the normal boiling point of CCl_4 is 61.20°C. What is the molar mass of the compound?

8.68 When 2.25 g of an unknown compound was dissolved in 150 g of cyclohexane, the boiling point increased by 0.481°C. Determine the molar mass of the compound.

8.69 Estimate the freezing-point depression and the freezing point of each of the solutions in Exercise 8.63 (the solubility of LiF near 0°C is 120 mg/100 g of water).

8.70 Estimate the freezing-point depression and the freezing point of each of the solutions in Exercise 8.64 (the solubility of Li_2CO_3 near 0°C is 1.54 mg/100 g of water).

8.71 A 1.14-g sample of a molecular substance dissolved in 100 g of camphor freezes at 176.9°C. What is the molar mass of the substance?

8.72 When 1.75 g of a nonpolar solute was dissolved in 50.0 g of phenol, the latter's freezing point was lowered by 1.454°C. Calculate the molar mass of the solute.

8.73 (a) What would be the freezing point of a benzene solution that boils at 82.0°C? (b) An aqueous solution freezes at −3.04°C. What is the molality of the solution? Assume there is no dissociation of the solute. (c) The freezing point of an aqueous solution containing a nonelectrolyte dissolved in 200 g of water is −1.94°C. How many moles of the solute are present?

8.74 (a) A benzene solution has a vapor pressure of 740 Torr at 80.1°C. What is the expected freezing point of the solution? (b) How many moles of $CO(NH_2)_2$ are present in 120.0 g of water, given that the freezing point of the solution is −4.02°C? (c) A 1.0-g sample of a protein of molar mass 1.0×10^5 g·mol^{-1} is dissolved in 1.0 kg of water. What is the expected freezing point of the solution?

8.75 A 1.00% by mass NaCl(aq) solution has a freezing point of −0.593°C. (a) Estimate the van 't Hoff i factor from the data. (b) Determine the total molality of all solute species. (c) Calculate the percentage dissociation of NaCl in this solution. (The molality calculated from the freezing-point depression is the sum of the molalities of the undissociated ion pairs, the Na^+ ions, and the Cl^- ions.)

8.76 A 1.00% by mass $MgSO_4$(aq) solution has a freezing point of −0.192°C. (a) Estimate the van 't Hoff i factor from the data. (b) Determine the total molality of all solute species. (c) Calculate the percentage dissociation of $MgSO_4$ in this solution.

8.77 Determine the freezing point of a 0.10 mol·kg^{-1} aqueous solution of a weak electrolyte that is 7.5% dissociated into two ions.

8.78 A 0.124 m CCl_3COOH(aq) solution has a freezing point of −0.423°C. What is the percentage deprotonation of the acid?

8.79 What is the osmotic pressure at 20°C of (a) 0.010 M $C_{12}H_{22}O_{11}$(aq); (b) 1.0 M HCl(aq); (c) 0.010 M$CaCl_2$(aq)?

8.80 Which of the following solutions has the highest osmotic pressure at 50°C: (a) 0.10 M KCl(aq); (b) 0.60 M $CO(NH_2)_2$(aq); (c) 0.30 m K_2SO_4(aq)? Justify your answer by calculating the osmotic pressure of each solution.

8.81 A 0.40-g sample of a polypeptide dissolved in 1.0 L of an aqueous solution at 27°C has an osmotic pressure of 3.74 Torr. What is the molar mass of the polypeptide?

8.82 When 0.10 g of insulin is dissolved in 0.200 L of water, the osmotic pressure is 2.30 Torr at 20°C. What is the molar mass of insulin?

8.83 A 0.10-g sample of a polymer, dissolved in 100 mL of toluene, has an osmotic pressure of 5.4 Torr at 20°C. What is the molar mass of the polymer?

8.84 A solution prepared by adding 0.50 g of a polymer to 200 mL of toluene (methylbenzene, a common solvent) showed an osmotic pressure of 0.582 Torr at 20°C. What is the molar mass of the polymer?

8.85 Calculate the osmotic pressure at 20°C of each of the solutions listed below; assume complete dissociation for any ionic solutes. (a) 0.050 M $C_{12}H_{22}O_{11}$(aq); (b) 0.0010 M NaCl(aq); (c) a saturated aqueous solution of AgCN of solubility 2.3×10^{-5} g/100 g of water.

8.86 Calculate the osmotic pressure at 20°C of the solutions specified below. Assume complete dissociation of ionic compounds. (a) 3.0×10^{-3} M $C_6H_{12}O_6$(aq); (b) 2.0×10^{-3}M $CaCl_2$(aq); (c) 0.010 M K_2SO_4(aq).

8.87 Catalase, a liver enzyme, dissolves in water. A 10-mL solution containing 0.166 g of catalase exhibits an osmotic pressure of 1.2 Torr at 20°C. What is the molar mass of catalase?

8.88 Intravenous medications are often administered in 5.0% glucose, $C_6H_{12}O_6$(aq), by mass. What is the osmotic pressure of such solutions at 37°C (body temperature)?

Binary Liquid Mixtures

8.89 Benzene, C_6H_6, and toluene, $C_6H_5CH_3$, form an ideal solution. The vapor pressure of benzene is 94.6 Torr and that of toluene is 29.1 Torr at 25.0°C. Calculate the vapor pressure of each of the following solutions and the mole fraction of each substance in the vapor phase above those solutions: (a) 1.50 mol C_6H_6 mixed with 0.50 mol $C_6H_5CH_3$; (b) 15.0 g of benzene mixed with 64.3 g of toluene.

8.90 Hexane, C_6H_{14}, and cyclohexane, C_6H_{12}, form an ideal solution. The vapor pressure of hexane is 151 Torr and that of cyclohexane is 98 Torr at 25.0°C. Calculate the vapor pressure of each of the following solutions and the mole fraction of each substance in the vapor phase above those solutions: (a) 0.25 mol C_6H_{14} mixed with 0.65 mol C_6H_{12}; (b) 10.0 g of hexane mixed with 10.0 g of cyclohexane.

8.91 1,1-Dichloroethane, CH_3CHCl_2, has a vapor pressure of 228 Torr at 25°C; at the same temperature, 1,1-dichlorotetrafluoroethane has a vapor pressure of 79 Torr. What mass of 1,1-dichloroethane must be mixed with 100.0 g of 1,1-dichlorotetrafluoroethane to give a solution whose vapor pressure is 157 Torr? Assume ideal behavior.

8.92 2-Butanone, $CH_3CH_2COCH_3$, has a vapor pressure of 100 Torr at 25°C; at the same temperature, 2-propanone, CH_3COCH_3, has a vapor pressure of 222 Torr. What mass of 2-propanone must be mixed with 350.0 g of 2-butanone to give a solution with a vapor pressure of 135 Torr? Assume ideal behavior.

8.93 Which of the following mixtures would you expect to show a positive deviation, negative deviation, or no deviation (that is, form an ideal solution) from Raoult's law? Explain. (a) methanol, CH_3OH, and ethanol, CH_3CH_2OH; (b) HF and H_2O; (c) hexane, C_6H_{14}, and H_2O.

8.94 Which of the following mixtures would you expect to show a positive deviation, negative deviation, or no deviation (that is, form an ideal solution) from Raoult's law? Explain. (a) HBr and H_2O; (b) formic acid, HCOOH, and benzene; (c) cyclopentane, C_5H_{10}, and cyclohexane, C_6H_{12}.

INTEGRATED EXERCISES

8.95 Complete the following statements about the effect of intermolecular forces on the physical properties of a substance: (a) The higher the boiling point of a liquid, the (stronger, weaker) its intermolecular forces. (b) Substances with strong intermolecular forces have (high, low) vapor pressures. (c) Substances with strong intermolecular forces typically have (high, low) surface tensions. (d) The higher the vapor pressure of a liquid, the (stronger, weaker) its intermolecular forces. (e) Because nitrogen, N_2, has (strong, weak) intermolecular forces, it has a (high, low) critical temperature. (f) Substances with high vapor pressures have correspondingly (high, low) boiling points. (g) Because water has a relatively high boiling point, it must have (strong, weak) intermolecular forces and a correspondingly (high, low) enthalpy of vaporization.

8.96 Hydrogen peroxide, H_2O_2, is a syrupy liquid with a vapor pressure lower than that of water and a boiling point of 152°C. Account for the differences between these properties and those of water.

8.97 Explain the effect that an increase in temperature has on each of the following properties: (a) viscosity; (b) surface tension; (c) vapor pressure; (d) evaporation rate.

8.98 Explain how the vapor pressure of a liquid is affected by each of the following changes in conditions: (a) an increase in temperature; (b) an increase in surface area of the liquid; (c) an increase in volume above the liquid; (d) the addition of air to the volume above the liquid.

8.99 The normal boiling point of water is 100°C. Suppose a cyclonic region (a region of low pressure) moves into the area. State and explain what happens to the boiling point of the water.

8.100 A vacuum pump is attached to a flask of water at 0°C and 2 atm, and the pressure on the liquid is decreased to 5 Torr. (a) Explain what would be observed, basing your answer on the phase diagram of water, Fig. 8.5. (b) Explain what would be observed if the water were at 50°C instead of 0°C.

8.101 Using the phase diagrams in Exercises 8.21, 8.22, and Fig. 8.7, answer the following questions. (a) Does solid helium sublime at temperatures below 5 K? (b) At atmospheric pressure, will the rhombic form of sulfur convert into the monoclinic form before the solid melts? (c) Which form of sulfur, rhombic or monoclinic, is more stable at room temperature? (d) Which form of sulfur is more stable at 100°C and 1 atm? (e) Does diamond sublime at high temperatures?

8.102 An air mass is saturated with water vapor at 10°C. Given that at 10°C the vapor pressure of water is 9.21 Torr, explain what happens to the water in the air mass if it meets (a) a warm weather front; (b) a cold front.

8.103 Relative humidity at a particular temperature is defined as

$$\text{Relative humidity} = \frac{\text{partial pressure of water}}{\text{vapor pressure of water}} \times 100\%$$

The vapor pressure of water at various temperatures is given in Table 8.3. (a) What is the relative humidity at 30°C when the partial pressure of water is 25.0 Torr? (b) Explain what would be observed if the temperature of the air were to fall to 25°C.

8.104 The 5:00 PM weather report states that the ambient temperature is 94°F (34°C) and the relative humidity (see Exercise 8.103) is 64%; an overnight low of 77°F (25°C) is expected. Can fog or dew be predicted? Explain your reasoning.

8.105 The density of a 16.0% by mass $C_{12}H_{22}O_{11}(aq)$ solution is 1.0635 $g \cdot cm^{-3}$ at 20°C. (a) What is the molarity of $C_{12}H_{22}O_{11}(aq)$? (b) What is the vapor pressure of the solution at 20°C? (c) Predict the boiling point of the solution. See Tables 8.3 and 8.9.

8.106 Explain in terms of events at a molecular level why a solute raises the boiling point and lowers the freezing point of a solution.

8.107 Water temperatures in the Gulf of Mexico can be as high as 30°C along the coast during the summer. Use Table 8.3 to estimate the vapor pressure of seawater at that temperature and at 100°C and 0°C, assuming that 0.50 *m* NaCl(aq) simulates seawater.

8.108 Organic chemists sometimes use freezing-point and boiling-point measurements to determine the molar mass of compounds they have synthesized. When 0.30 g of a nonvolatile solute is dissolved in 30.0 g of CCl_4, the boiling point of the solution is 77.19°C. What is the molar mass of the compound?

8.109 A 10.0-g sample of an organic compound used as a component of mothballs is dissolved in 80.0 g of benzene. The freezing point of the solution is 1.20°C. (a) What is an approximate molar mass of the organic compound? (b) An elemental analysis of that substance indicated that the empirical formula is C_3H_2Cl. What is its molecular formula? (c) Using the atomic molar masses from the periodic table, calculate a more accurate molar mass of the compound.

8.110 When determining a molar mass from freezing-point depression, it is possible to make each of the following errors (among others). In each case, predict whether the error would cause the reported molar mass to be greater or less than the actual molar mass: (a) There was dust on the balance, causing the mass of solute to appear greater than it actually was. (b) The water was measured by volume, assuming a density of 1.00 $g \cdot cm^{-3}$, but the water was warmer and less dense than assumed. (c) The thermometer was not calibrated accurately, so the temperature of the freezing point was actually 0.5°C higher than recorded. (d) The solution was not stirred sufficiently, so not all the solute dissolved.

8.111 An elemental analysis of epinephrine resulted in the following composition: 59.0% carbon, 26.2% oxygen, 7.15% hydrogen, and 7.65% nitrogen by mass. When 0.64 g of epinephrine was dissolved in 36.0 g of benzene, the freezing point decreased by 0.50°C. (a) Determine the empirical formula of epinephrine. (b) What is the molar mass of epinephrine? (c) Deduce the molecular formula of epinephrine.

8.112 Suggest a reason why saltwater fish die when they are suddenly transferred to a freshwater aquarium.

8.113 Solutions that are injected into the bloodstream (for blood transfusions and intravenous feeding) must be "isotonic" with the blood (that is, have the same osmotic pressure). If the injected solution is too dilute, it is termed "hypotonic" and if too concentrated, "hypertonic." Explain the effects of hypotonic and hypertonic solutions on the cells in the bloodstream.

8.114 Interpret the following verse from the Coleridge's *The Rime of the Ancient Mariner:*

Water, water, every where,
And all the boards did shrink,
Water, water every where,
Nor any drop to drink.

8.115 Diethyl ether, $C_2H_5OC_2H_5$, and ethyl methyl ether, $C_2H_5OCH_3$, form an ideal solution. The vapor pressure of diethyl ether is 185 Torr and that of ethyl methyl ether is 554 Torr at 0°C. Calculate the vapor pressure of each of the following solutions and the mole fraction of each substance in the vapor phase above those solutions at 0°C: (a) 0.75 mol $C_2H_5OC_2H_5$ is mixed with 0.50 mol $C_2H_5OCH_3$; (b) 25.0 g of diethyl ether is mixed with 35.0 g of ethyl methyl ether.

8.116 Bromomethane, CH_3Br, and iodomethane, CH_3I, form an ideal solution. The vapor pressure of bromomethane is 661 Torr and that of iodomethane is 140 Torr at 0.0°C. Calculate the vapor pressure of each of the following solutions and the mole fraction of each substance in the vapor phase above those solutions at 0.0°C: (a) 0.33 mol CH_3Br is mixed with 0.67 mol CH_3I; (b) 35.0 g of bromomethane is mixed with 35.0 g of iodomethane.

8.117 The vapor pressure of ethanol at 25°C is 58.9 Torr. A sample of ethanol vapor at 25°C and 58.9 Torr partial pressure is in equilibrium with a very small amount of liquid ethanol in a 10-L container also containing dry air, at a total pressure of 750 Torr. The volume of the container is then reduced at constant temperature to 5.0 L. (a) What is the partial pressure of ethanol in the smaller volume? Explain your reasoning. (b) What is the total pressure of the mixture?

8.118 The reaction $Ca(HCO_3)_2(aq) \rightarrow CaCO_3(s) + CO_2(g) + H_2O(l)$ takes place during the formation of stalagmites and stalactites in caves. The soluble calcium hydrogen carbonate seeps through the water in the solid until an air space is encountered. Explain, using Henry's law and principles of solubility, the formation of the stalagmites and stalactites, the major component of which is calcium carbonate.

8.119 What is the osmotic pressure of each of the following aqueous solutions at the given temperature? Assume the density of each solution is 1.00 $g{\cdot}cm^{-3}$. (a) a saturated solution of Li_2CO_3, which has a solubility of 1.54 g/100 g H_2O at 0°C; (b) a solution that has a vapor pressure of 751 Torr at 100°C; (c) a solution that has a boiling point of 101°C.

8.120 The height of a column of liquid that can be supported by a given pressure is inversely proportional to its density. An aqueous solution of 0.010 g of a protein in 10 mL of water at 20°C shows a 5.22-cm rise in the apparatus shown in Fig. 8.37. Assume the density of the solution to be 0.998 $g{\cdot}cm^{-3}$ and the density of mercury to be 13.6 $g{\cdot}cm^{-3}$. (a) What is the molar mass of the protein? (b) What is the freezing point of the solution? (c) Which colligative property is best for measuring the molar mass of these large molecules? Give reasons for your answer.

8.121 A 0.020 M $C_6H_{12}O_6(aq)$ solution (glucose) is separated from a 0.050 M $CO(NH_2)_2(aq)$ solution (urea) by a semipermeable membrane at 25°C. (a) Which solution has the higher osmotic pressure? (b) Which solution becomes more dilute with the passage of H_2O molecules through the membrane? (c) To which solution should an external pressure be applied in order to maintain an equilibrium flow of H_2O molecules across the membrane? (d) What external pressure (in atm) should be applied in (c)?

8.122 (a) A solution of a nonvolatile solute in benzene has a vapor pressure of 740 Torr at 80.1°C. What is the osmotic pressure of this solution at 20°C? Assume that the density of the solution is the same as that of benzene, 0.88 $g{\cdot}cm^{-3}$. (b) The freezing point of a different benzene solution was 5.4°C. Determine the osmotic pressure of the solution at 10°C. (c) Determine the height to which osmosis will force the solution in (b) to rise in an arrangement like that shown in Fig. 8.37. The density of mercury is 13.6 $g{\cdot}cm^{-3}$. See Exercise 8.120 for more information.

8.123 (a) From data in Appendix 2A and Table 6.3, derive the Clausius-Clapeyron equation for benzene. (b) Using this equation, plot the appropriate quantities that should give a straight-line relationship between vapor pressure and temperature. (c) Estimate the boiling point of benzene when the external pressure is 0.655 atm. (d) Calculate $S_m°$ for gaseous benzene.

8.124 (a) From data in Appendix 2A, derive the Clausius-Clapeyron equation for methanol. (b) Using this equation, plot the appropriate quantities that should give a straight-line relationship between vapor pressure and temperature. (c) Estimate the vapor pressure of methanol at 0°C. (d) Estimate the normal boiling point of methanol.

8.125 It is standard practice in chemical laboratories to distill high-boiling-point substances under reduced pressure. Trichloroacetic acid has a standard enthalpy of vaporization of 57.814 $kJ{\cdot}mol^{-1}$ and a standard entropy of vaporization of 124 $J{\cdot}K^{-1}{\cdot}mol^{-1}$. Use this information to determine the pressure that one would need to achieve in order to distill trichloroacetic acid at 80°C.

8.126 The vapor pressure of phosphoryl chloride difluoride ($OPClF_2$) has been measured as a function of temperature:

Temperature (K)	Vapor pressure (Torr)
190	3.2
228	68
250	240
273	672

(a) Make a plot of ln P versus T^{-1} (this project may be done with the aid of a computer or a graphing calculator that can calculate a least-squares fit to the data). (b) From the plot (or a linear equation derived therefrom) in (a), determine the standard enthalpy of vaporization of $OPClF_2$; (c) the standard entropy of vaporization of $OPClF_2$; and (d) the normal boiling point of $OPClF_2$. (e) If a sample of $OPClF_2$ is placed under a vacuum such that the pressure is reduced to 15 Torr, at what temperature will the sample boil?

8.127 From standard literature references, find the critical pressures and temperatures of methane, methyl amine (CH_3NH_2), ammonia, and tetrafluoromethane. Discuss the suitability of using each of these solvents for a supercritical extraction at room temperature in an autoclave that can withstand pressures of 100 atm.

8.128 From standard literature sources, find the critical temperatures for the gaseous hydrocarbons methane, ethane, propane, and butane. Explain the trends observed.

8.129 Consider the following apparatus in which A and B are two 1.00 L-flasks joined by a stopcock C. The volume of the stopcock is negligble.

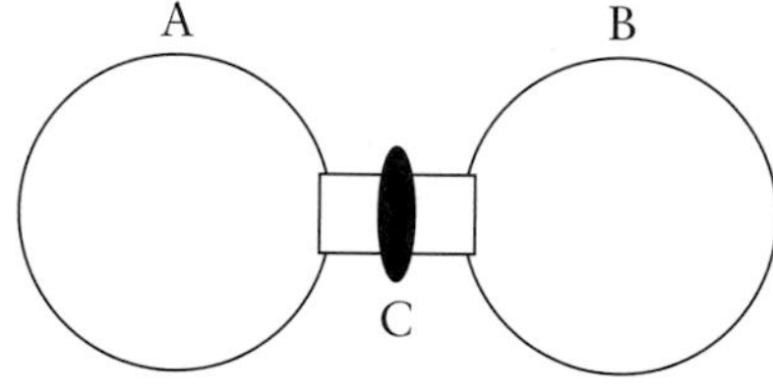

Initially A and B are evacuated, the stopcock C is closed, and 1.50 g of diethyl ether, $C_2H_5OC_2H_5$, is introduced into flask A. The vapor pressure of diethyl ether is 57 Torr at −45°C, 185 Torr at 0°C, 534 Torr at 25°C, and negligible below −86°C. (a) If the stopcock is left closed and the flask is brought to equilibrium at −45°C, what will be the pressure of diethyl ether in flask A? (b) If the temperature is raised to 25°C, what will be the pressure of diethyl ether in the flask? (c) If the temperature of the assembly is returned to −45°C and the stopcock C is opened, what will be the pressure of diethyl ether in the apparatus? (d) If flask A is maintained at −45°C and flask B is cooled with liquid nitrogen (boiling point, −196°C) with the stopcock open, what changes will take place in the apparatus? Assume ideal behavior.

8.130 Consider the apparatus in Exercise 8.129. The apparatus is again evacuated; and 35.0 g of chloroform, $CHCl_3$, is placed in flask A and 35.0 g of acetone, CH_3COCH_3, is placed in flask B. The system is allowed to come to equilibrium at 25°C with the stopcock closed. The vapor pressures of chloroform and acetone at 25°C are 195 Torr and 222 Torr, respectively. (a) What is the pressure in each flask at equilibrium? (b) The stopcock is now opened. What changes will occur? What will be the final composition of the gas and liquid phases in each flask once the system reaches equilibrium, assuming ideal behavior? (c) Acetone and chloroform shows negative deviation from Raoult's law. How will this behavior affect the answers given in part (b)?

8.131 Pentane is a liquid with a vapor pressure of 512 Torr at 25°C; at the same temperature, the vapor pressure of hexane is only 151 Torr. What composition must the liquid phase have if the gas-phase composition is to have equal amounts of pentane and hexane?

8.132 The apparatus in Exercise 8.129 is evacuated. A student places 15.0 mL of a 15 *m* solution of sucrose dissolved in ethanol in flask A. In flask B, the student places 15.0 mL of pure ethanol. (a) Describe qualitatively the changes that take place in this system when the stopcock is opened and the system is allowed to reestablish equilibrium at 25°C (b) What is the final pressure in the assembly when it has reached equilibrium? The vapor pressure of ethanol is 59 Torr at 25°C.

8.133 The apparatus in Exercise 8.129 is evacuated. A student places 15.0 g of a 0.15 *m* solution of sucrose dissolved in ethanol in flask A. In flask B, the student places 15.0 g of 0.050 *m* sucrose in ethanol. (a) Describe qualitatively the changes that take place in this system when the stopcock is opened and the system is allowed to reestablish equilibrium. (b) What are the final concentrations of the solutions on each side of the apparatus and the final pressure in the assembly when it has reached equilibrium? The vapor pressure of ethanol is 58.9 Torr at 25°C and that of sucrose is negligible; the density of ethanol is 0.816 $g \cdot cm^{-3}$.

8.134 Using the Clausius-Clapeyron Equation Living Graph on the Web site for this book, plot on the same set of axes the lines for $\Delta H_{vap} = 15\,000$ J, 20 000 J, 25 000 J, and 30 000 J. Is the vapor pressure of a liquid more sensitive to changes in temperature if ΔH_{vap} is small or large? Explain.

8.135 A pair of amino acids is separated in a column in which the stationary phase is saturated with water and the carrier solvent is methanol, CH_3OH. The more polar the acid, the more strongly it is adsorbed by the stationary phase. The amino acids that were separated in this column are (a) $HOOCCHNH_2CH_2COOH$ and (b) $HOOCCHNH_2CH(CH_3)_2$. Which amino acid would you expect to have the greater value of *k*? Explain your reasoning. Refer to Major Technique 4 on chromatography, which follows these exercises.

8.136 Compounds A and B were extracted from a sample of Martian soil. A mixture of 0.52 mg of A and 2.30 mg of B in 1.00 mL of solution was separated by gas chromatography. The areas of the two peaks were 5.44 cm^2 for A and 8.72 cm^2 for B. A second solution contained an extract with an unknown amount of A. To determine the concentration of A in the solution, 2.00 mg of B was added to 2.0 mL of the solution, which was then introduced into a gas chromatograph. Peak areas of 3.52 cm^2 for A and 7.58 cm^2 for B were measured. What is the concentration of A in the second solution? Refer to Major Technique 4, which follows these exercises.

The scent of a flower may be due to hundreds of different compounds, so how do perfume manufacturers duplicate floral scents? Establishing the identities and relative amounts of the components of a fragrance was impossible until the development of chromatography. Related techniques are used in forensic laboratories to match samples of fluids, by food manufacturers to test product quality, and on Mars to search for evidence of life. All these techniques depend on subtle differences in intermolecular forces to separate compounds.

1 Solvent Extraction

If an aqueous solution of a compound is shaken with another liquid that is *immiscible* (mutually insoluble) with water, some of the compound may dissolve in the other solvent. For example, molecular iodine, I_2, is very slightly soluble in water, but highly soluble in tetrachloromethane, CCl_4, which is immiscible with water. When tetrachloromethane is added to water containing iodine, most of the iodine dissolves in the CCl_4. The solute is said to *partition* itself between the two solvents. *Solvent extraction* is used to obtain plant flavors and aromas from aqueous slurries of the plant that have been crushed in a blender.

In some cases, the solids themselves are subjected to extraction by a solvent. For example, at one time coffee beans were decaffeinated by heating them with hexane, which dissolves caffeine. Unfortunately, hexane, which is a suspected carcinogen, also dissolves some of the flavoring agents in coffee beans and is difficult to remove completely from the beans. Today, the beans are mixed with activated charcoal, and a high-pressure stream of "supercritical carbon dioxide" (carbon dioxide at high pressure and above its critical temperature) is passed over them at approximately 90°C. A supercritical solvent is a highly mobile fluid with a very low viscosity. The carbon dioxide preferentially removes the highly soluble caffeine without extracting the flavoring agents and evaporates without leaving a harmful residue.

2 Liquid Chromatography

In the early part of the twentieth century, the Russian botanist M. S. Tsvet found a way to separate the many pigments in flowers and leaves. He ground up the plant materials and dissolved the pigments, then poured the solution over the top of a vertical tube of ground chalk. The different pigments trickled through the chalk at different rates, producing colored bands in the tube, inspiring the name *chromatography* ("color writing"). The separation occurred because the different pigments were absorbed to different extents by the chalk.

Chromatography is one of the most powerful and widely used means for separating mixtures, because it is often inexpensive and it can be used to provide quantitative as well as qualitative information. The simplest method is *paper chromatography*. A drop of solution is placed near the bottom edge of an absorbant support, such as a strip of paper, which is called the *stationary phase*. A solvent called the *mobile phase* is added below the spot and the solvent is absorbed on the support. As the level of the solvent rises in the support, the materials in the spot also begin to travel upward, the rate depending on how strongly they adsorbed on the support (Fig. 1). In *adsorption*, the solutes adhere to a surface. In a sense, there is a continuous extraction going on between the support and the solvent, which is also called the *carrier solvent*. Those solutes attracted more strongly to the solvent than to the support will move the fastest. Each

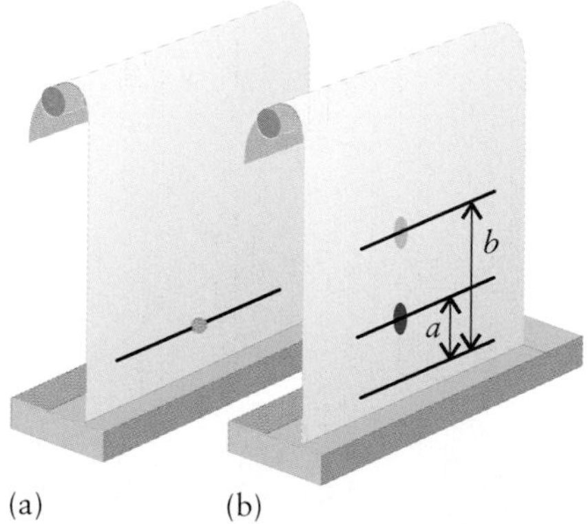

FIGURE 1 Two stages in a paper chromatography separation of a mixture of two components. (a) Before separation; (b) after separation. The relative values of the distances *a* and *b* are used to identify the components.

substance is identified by its *capacity factor, k*, the relative time it spends in each phase:

$$k = \frac{\text{time in stationary phase}}{\text{time in mobile phase}} \quad (1)$$

The more strongly the solute is adsorbed on the stationary phase, the longer it will take to travel up the support, and the larger the value of k. The value of k is normally determined by measuring how far a given solute has traveled relative to the solvent itself.

The same concepts apply to *column chromatography*, where the stationary phase is normally small particles of silica, SiO_2, or alumina, Al_2O_3. These substances are not very reactive and have specially prepared surfaces to increase their ability to adsorb solvents. The column is saturated with solvent, and a small volume of solution containing the solutes is poured on to the top. As soon as it has soaked in, more solvent is added. The solutes travel slowly down the column and are *eluted* (removed as fractions) at the bottom (Fig. 2). If the mobile phase is less polar than the stationary phase, the less polar solutes will be eluted first and the more polar ones last.

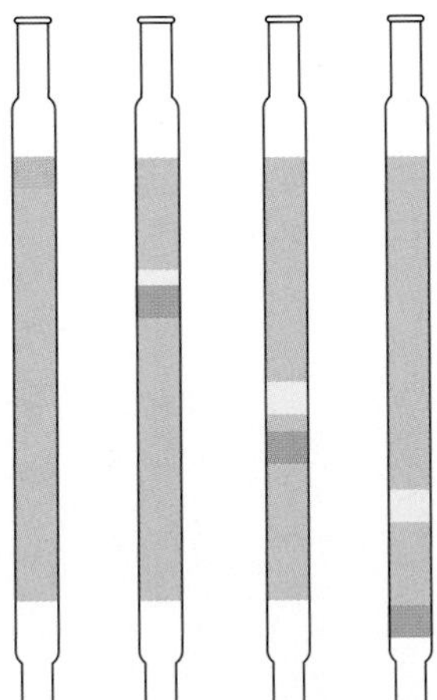

FIGURE 2 A column chromatography experiment. The mixture (left) is washed down the column and separates into bands as the solvent passes through the support.

A difficulty with column chromatography is that, for a good separation, the column needs to be long and elution may take a very long time: days may be required. An improvement is *high-performance liquid chromatography* (HPLC), in which the mobile phase is forced under pressure through a long, narrow column; this technique results in an excellent separation in a relatively short time. HPLC has become the primary means of monitoring the use of therapeutic drugs and in the detection of drug abuse; it is also used to separate the compounds contributing to the fragrance of flowers.

Self-Test M4.1A A pair of amino acids is separated in a column in which the stationary phase is saturated with water and the carrier solvent is methanol, CH_3OH. The more polar the acid, the more strongly it is adsorbed by the stationary phase. The amino acids that were separated in this column are
(a) $HOOC{-}HNH_2{-}CH_2{-}OH$ and
(b) $HOOC{-}CHNH_2{-}CH_3$. Which amino acid would you expect to have the greater value of the capacity factor k? Explain your reasoning.

[*Answer:* Compound (a) has more polar groups, because it has an $-OH$ group, and so will be retained longer on the column.]

Self-Test M4.1B Inorganic cations can be separated by liquid chromatography according to their ability to form complexes with chloride ions. For the separation, the stationary phase is saturated with water and the carrier solvent is a solution of HCl in acetone. The relative solubilities of the following chlorides in concentrated hydrochloric acid are $CuCl_2 > CoCl_2 > NiCl_2$. Predict the relative values of the capacity factor k for the three salts.

3 Gas Chromatography

Volatile compounds can be separated in a *gas chromatograph*, in which the mobile phase is usually a relatively unreactive gas such as helium, nitrogen, or hydrogen. The principles are the same as those for liquid chromatography, but the output is more often a *chromatogram*, rather than a series of eluted samples (Fig. 3). The chromatogram shows when each solute was

(*continued*)

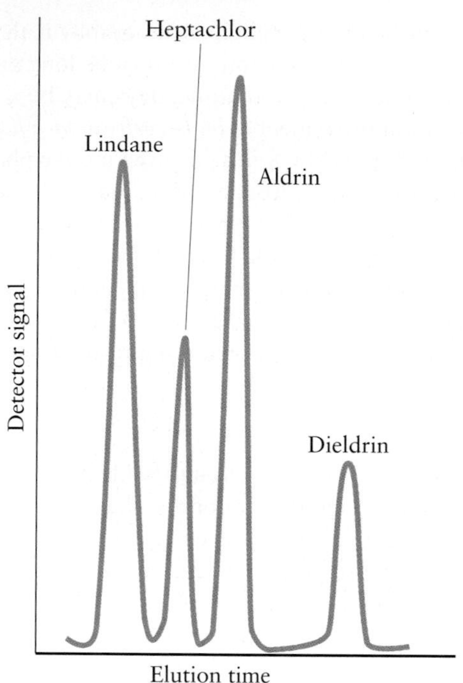

FIGURE 3 A gas chromatogram of a mixture of pesticides from farmland. The relative heights of the peaks indicate the relative abundances of the compounds.

eluted and the peak heights tell how much of each component is present. The identity of the solute producing each peak can be determined by comparing its location against a database of known compounds. The amount of each solute present is determined from the area under the peak.

In *gas-liquid partition chromatography* (GLPC), the stationary phase is a liquid that coats the particles in the tube or the walls of the tube. Often the tube itself is very narrow and long, perhaps 100 m, and has to be coiled (Fig. 4). Solutes are separated as in liquid chromatography, by their relative solubility in the gas and liquid phases. In *gas-solid adsorption chromatography*, solid particles coat the inside of the narrow tube. The solute vapors are separated by their relative attraction for the solid particles. In both cases, it is relative polarity that determines the distances between peaks.

As the vapor leaves the tube, the compounds in the sample are detected by a device such as a thermal conductivity detector. This instrument continuously measures the thermal conductivity (the ability to conduct heat) of the carrier gas, which changes when a solute is present. The detection techniques are very

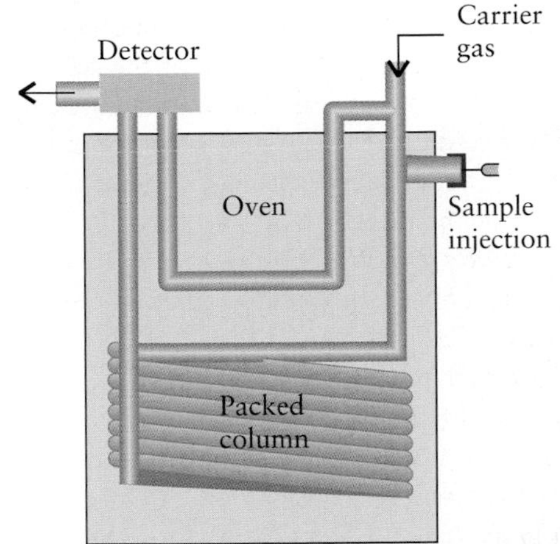

FIGURE 4 A schematic diagram of the arrangement in a gas chromatograph. The coiled column may be up to 100 m in length.

sensitive, allowing tiny amounts of solutes to be detected. Many environmental monitoring and forensic applications have been developed.

4 Gas Chromatography-Mass Spectrometry

Some detectors can give additional information about the solutes. One example is the *gas chromatograph-mass spectrometer* (GC-MS), which produces a mass spectrum of each compound as well as its mass and location in the chromatogram.* This powerful means of detection can be used when standard samples are not available to help determine the identities of the solutes. A beam of ions bombards each compound as it emerges from the chromatograph. The compound breaks up into ions of different masses, providing a spread of narrow peaks instead of one peak for each compound. The relative amount of each fragment is determined and used to help identify the compound.

The GC-MS instrument can be very compact and used in mobile equipment. A GC-MS device was used on the Viking Lander on the first Mars mission to search for organic compounds that would have provided evidence of life.

Related Exercises: 8.135 and 8.136

*Mass spectrometry is the subject of Major Technique 5, following Chapter 18.

CHAPTER

When drops of hydrochloric acid fall into an aqueous solution of silver nitrate, a precipitate of solid silver chloride forms immediately. However, hidden processes continue to occur: silver and chloride ions are continuously moving between the solid and the solution.

Why Do We Need to Know This Material?

The dynamic equilibrium toward which every chemical reaction tends is such an important aspect of the study of chemistry that four chapters of this book are devoted to it. We need to know the composition of a reaction mixture at equilibrium if we are to calculate how much product to expect. To control the outcome of a reaction, we need to understand the relation of equilibrium to thermodynamics and how the position of equilibrium is affected by conditions such as temperature and pressure. The response of equilibria to changes in conditions has considerable economic and biological significance: the regulation of chemical equilibrium affects the yields of products in industrial processes, and living cells struggle to avoid sinking into equilibrium.

What Do We Need to Know Already?

The concepts of chemical equilibrium are related to those of physical equilibrium (Sections 8.1–8.3). Because chemical equilibrium is related to the thermodynamics of chemical reactions, we need to know about the free energy of reaction (Section 7.13) and the enthalpy of reaction (Section 6.14). Chemical equilibrium calculations require a thorough knowledge of molar concentration (Section G), reaction stoichiometry (Section L), and the gas laws (Chapter 4).

Chemical Equilibria

At the opening of the twentieth century, the looming prospect of World War I created a desperate need for nitrogen compounds because nitrates normally used for agriculture were being made into explosives. Almost all the nitrates used for fertilizers and explosives were quarried from deposits in Chile, and the limited supply could not keep up with the demand. Moreover, shipping channels were vulnerable to attack, which threatened to cut off the supply altogether. Although nitrogen is abundant in air, the methods used at the time to convert nitrogen into compounds were too costly to employ on a large scale. Any nation that could develop an economical process to **fix** atmospheric nitrogen, that is, to combine it with other elements, would have all the nitrogen compounds they needed.

A difficulty with the fixation of nitrogen is that N_2 has a strong triple bond and a correspondingly high bond enthalpy (944 $kJ \cdot mol^{-1}$), and, as a result, it is very unreactive. However, if nitrogen could be converted into ammonia, the ammonia could serve as the starting material for other compounds. The average N—H bond enthalpy in ammonia is only 388 $kJ \cdot mol^{-1}$, so ammonia is much more reactive than nitrogen. Microorganisms in the roots of certain plants fix nitrogen, but they use complicated enzymes that chemists are still trying to replicate and that were a complete mystery at the beginning of the twentieth century.

Scientists on both sides of the conflict were urgently seeking better ways to fix nitrogen, using the technological resources of the early twentieth century. Finally, through determination, application, and—as so often happens in research—a bit of luck, the German chemist Fritz Haber found an economical way to harvest the nitrogen of the air and to provide an abundant source of compounds for both agriculture and armaments.

One difficulty Haber faced is that the reactions used to produce compounds from nitrogen go, not to completion, but to a state of equilibrium: some reactants remain. To achieve the greatest conversion of nitrogen to its compounds, Haber had to understand how a reaction approaches and eventually reaches equilibrium and then use that knowledge to improve the yield by changing the conditions of the reaction. **Chemical equilibrium** is the stage in a chemical reaction when there is no further tendency for the composition of the reaction mixture—the concentrations or partial pressures of the reactants and products—to change. Like physical equilibria, all chemical equilibria are *dynamic* equilibria, with the forward and reverse reactions occurring at the same rate. In Chapter 8, we considered several physical processes, including vaporizing and dissolving, that reach dynamic equilibrium. This chapter shows how to apply the same ideas to chemical changes. It also shows how to use thermodynamics to describe equilibria quantitatively, which puts enormous power into our hands—the power to control the direction of a reaction and the yield of products.

In this chapter, we first show that a single quantity, the *equilibrium constant*, characterizes the composition of a reaction mixture at equilibrium. Then we explore how the equilibrium constant arises from thermodynamics and see how it is related to the free energy of the reaction. With the thermodynamic foundations of the equilibrium constant established, we go on to see how to use it to calculate the equilibrium composition of the reaction mixture for arbitrary initial compositions. Finally, we see how to use thermodynamic arguments to deduce how the equilibrium constant and therefore the composition of the reaction mixture at equilibrium respond to changes in the conditions under which the reaction is taking place. The key idea in this chapter is that *dynamic equilibria respond to changes in the conditions*. In other words, a dynamic equilibrium is a responsive, living equilibrium.

REACTIONS AT EQUILIBRIUM

When a reaction has reached equilibrium, its forward and reverse rates are equal and all the properties of the reaction mixture, including the composition, are constant. So, are the three systems in Fig. 9.1 at equilibrium? At first sight, the properties of all three systems appear to be constant. However, when methane, CH_4, burns with a steady flame to form carbon dioxide gas and water, the composition of the reaction mixture is not constant because methane is being consumed and the products are being formed. This combustion reaction is not at equilibrium. The sample of glucose does not change, even if we leave it for a very long time. However, the glucose is not at equilibrium with the products of its combustion (carbon dioxide and water) because the rate of its combustion is immeasurably slow at room temperature. The gas-phase reaction, however, is at equilibrium, because as well as the composition being constant, additional

Media Link **Lab video:** Fig. 9.1, Temperature-dependent equilibrium of a gaseous system

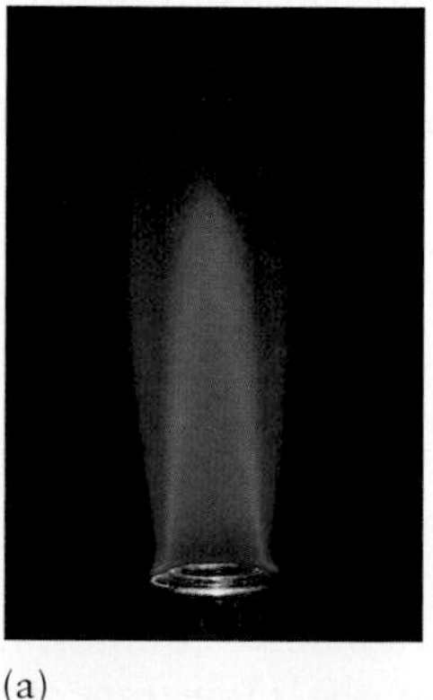
(a)

(b)

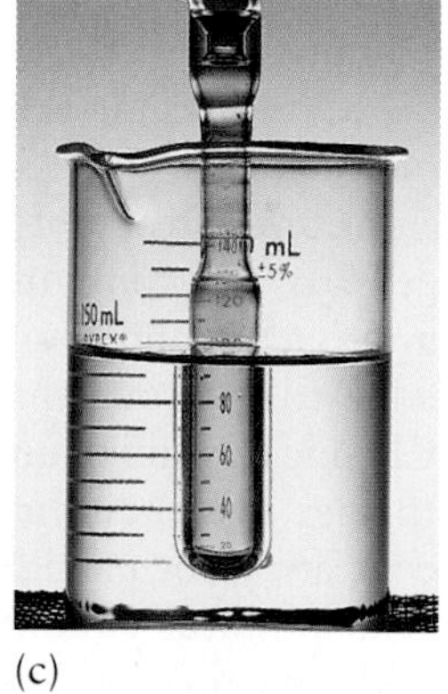
(c)

FIGURE 9.1 (a) Methane burns in air with a steady flame, but because matter is being added and removed, the reaction is not at equilibrium. (b) This sample of glucose in air is unchanging in composition, but it is not in equilibrium with its combustion products because the reaction proceeds far too slowly at room temperature. (c) Nitrogen dioxide (a brown gas) and dinitrogen tetroxide (a colorless volatile solid) are in equilibrium in this vessel. We can tell they are by changing the temperature slightly and seeing the composition adjust to a new value.

experiments show that NO_2 is ceaselessly forming N_2O_4 and that N_2O_4 is decomposing into NO_2 at the same rate.

As these examples show, it is sometimes difficult to tell whether a chemical reaction is at equilibrium. The criteria that identify a chemical equilibrium are:

1 The forward reaction and its reverse are both taking place.

2 They are doing so at equal rates (so there is no net change).

One way to decide whether a reaction is at equilibrium is to change the conditions, such as adding more reactant: the composition will change if the reaction is at equilibrium, and it will change back to its earlier value once the conditions are restored to their initial values.

Now we begin to develop these ideas. First, we establish that reactions can take place in their reverse direction as soon as some products have accumulated. The ability of the reverse reaction to take place is the underlying characteristic of dynamic equilibrium. Then we see how to relate the equilibrium composition to other properties of the system.

9.1 The Reversibility of Reactions

Some reactions, like the explosive reaction of hydrogen and oxygen, appear to proceed to completion, but others seem to stop at an early stage. For example, consider the reaction that occurs when nitrogen and hydrogen are heated together under pressure in the presence of a small amount of iron.

The iron acts as a *catalyst* for this reaction, a substance that helps to make a reaction go faster (Section 9.12).

$$N_2(g) + 3\,H_2(g) \longrightarrow 2\,NH_3(g) \qquad \text{(A)}$$

The standard reaction free energy, $\Delta G_r°$, is $-32.90\ \text{kJ}\cdot\text{mol}^{-1}$, so the reaction is spontaneous when all the gases are at 1 bar. The reaction does produce ammonia rapidly at first, but eventually the reaction seems to stop (Fig. 9.2). As the graph shows, no matter how long we wait, no more product forms. The reaction has reached equilibrium.

As we have indicated, like phase changes, chemical reactions tend toward a *dynamic* equilibrium in which, although there is no net change, the forward and reverse reactions are still occurring, but at matching rates. What actually happens when the formation of ammonia appears to stop is that the rate of the reverse reaction,

$$2\,NH_3(g) \longrightarrow N_2(g) + 3\,H_2(g) \qquad \text{(B)}$$

increases as more ammonia is formed. At equilibrium, the ammonia is decomposing as fast as it is being formed. Just as we did for phase transitions, we express this state of dynamic equilibrium by replacing the arrow in the equation with equilibrium "harpoons":

$$N_2(g) + 3\,H_2(g) \rightleftharpoons 2\,NH_3(g) \qquad \text{(C)}$$

All chemical equilibria are dynamic equilibria. Although there is no further net change at equilibrium, the forward and reverse reactions are still taking place.

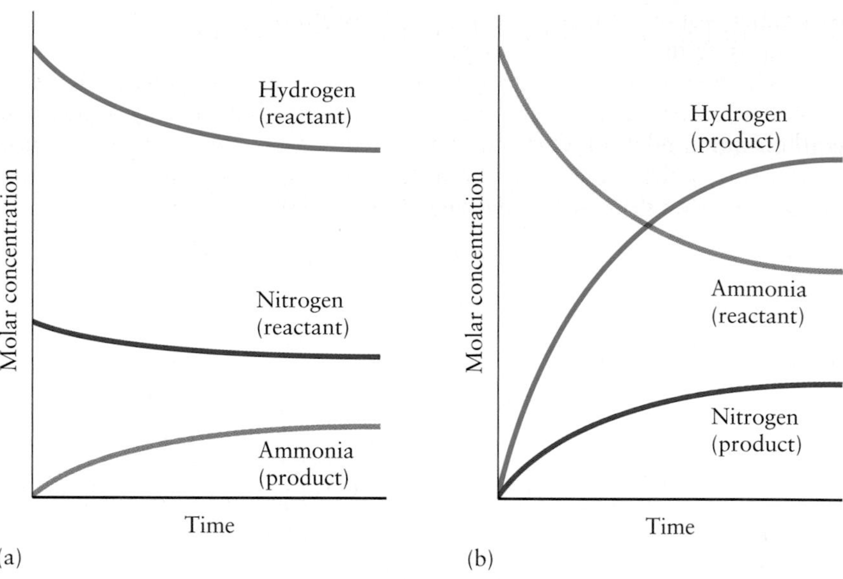

FIGURE 9.2 (a) In the synthesis of ammonia, the concentrations of N_2 and H_2 decrease with time and that of NH_3 increases until they finally settle into values corresponding to a mixture in which all three are present and there is no further net change. (b) If the experiment is repeated with pure ammonia, it decomposes, and the composition settles down into a mixture of ammonia, nitrogen, and hydrogen. (The two graphs correspond to experiments at two different temperatures, so they correspond to different equilibrium compositions.)

To show that the chemical equilibrium is indeed dynamic, we could carry out two ammonia syntheses with the same starting conditions, but using D_2 (where D denotes deuterium) in place of H_2 in one of them (Fig. 9.3). Deuterium is an isotope of hydrogen and has essentially the same chemical properties, so the two reaction mixtures reach equilibrium with almost exactly the same composition, except that D_2 and ND_3 are present in one of the systems instead of H_2 and NH_3. Suppose we now combine the two equilibrium mixtures and leave them for a while. When later we analyze the mixture, we find that the concentration of ammonia has not changed. However, when we analyze the sample with a mass spectrometer, we find that all isotopic forms of ammonia (NH_3, NH_2D, NHD_2, and ND_3) and all isotopic forms of hydrogen (H_2, HD, and D_2) are present. This scrambling of H and D atoms in the molecules must result from a continuation of the forward and reverse reactions in the mixture. If the reactions had simply stopped when they reached equilibrium, there would have been no mixing of isotopes in this way.

Chemical reactions reach a state of dynamic equilibrium in which the rates of forward and reverse reactions are equal and there is no net change in composition.

(a)

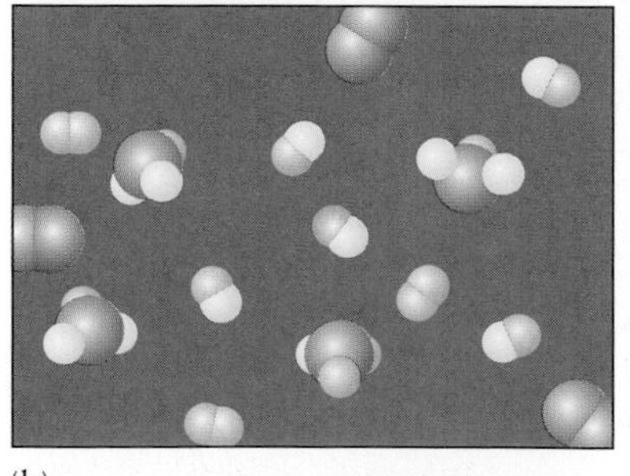
(b)

FIGURE 9.3 (a) In an experiment showing that equilibrium is dynamic, a reaction mixture in which N_2 (pairs of blue spheres), D_2 (pairs of yellow spheres), and ND_3 have reached equilibrium is mixed with one with the same concentrations of N_2, H_2 (pairs of gray spheres), and NH_3. (b) After some time, the concentrations of nitrogen, hydrogen, and ammonia are found to be the same, but the D atoms are distributed among the hydrogen and ammonia molecules.

9.2 Equilibrium and the Law of Mass Action

In 1864, long before Haber began his work, the Norwegians Cato Guldberg (a mathematician) and Peter Waage (a chemist) discovered the mathematical relation that summarizes the composition of a reaction mixture at equilibrium. As an example of their approach, look at the data in Table 9.1 for the reaction between SO_2 and O_2:

$$2\,SO_2(g) + O_2(g) \rightleftharpoons 2\,SO_3(g) \qquad \text{(D)}$$

In the experiments reported in the table, five mixtures with different initial compositions of the three gases were prepared and allowed to reach equilibrium at 1000 K. The compositions of the equilibrium mixtures were then determined. At first, there seemed to be no pattern in the data. However, Guldberg and Waage noticed an extraordinary relation: the value of the following quantity was the same for every experiment, regardless of the initial compositions:

$$K = \frac{(P_{SO_3}/P°)^2}{(P_{SO_2}/P°)^2\,(P_{O_2}/P°)}$$

Here, P_J is the equilibrium partial pressure of gas J and $P° = 1$ bar, the standard pressure. Note that the units cancel, so K is unitless. Within experimental error, the same value of K was obtained in each case. This remarkable result shows that K is characteristic of the composition of the reaction mixture at equilibrium. It is known as the **equilibrium constant** for the reaction. The **law of mass action** summarizes this result: it states that, at equilibrium, *the composition of the reaction mixture can be expressed in terms of an equilibrium constant.*

The equilibrium constant is a simple and powerful quantity that we can use to summarize the equilibrium composition of any reaction mixture. For any reaction between gases that can be treated as ideal, the equilibrium constant has the general form

$$K = \left\{\frac{\text{partial pressures of products relative to the standard pressure}}{\text{partial pressures of reactants relative to the standard pressure}}\right\}_{\text{equilibrium}} \qquad (1a)$$

TABLE 9.1 Equilibrium Data and the Equilibrium Constant for the Reaction $2\,SO_2(g) + O_2(g) \rightleftharpoons 2\,SO_3(g)$ at 1000 K

P_{SO_2}, bar	P_{O_2}, bar	P_{SO_3}, bar	P, bar	K
5.49×10^{-2}	3.24×10^{-2}	2.21×10^{-4}	8.75×10^{-2}	5.00×10^{-4}
3.16×10^{-3}	1.83×10^{-2}	9.49×10^{-6}	2.15×10^{-2}	4.93×10^{-4}
9.15×10^{-3}	9.15×10^{-3}	1.96×10^{-5}	1.83×10^{-2}	5.02×10^{-4}
7.90×10^{-2}	7.32×10^{-2}	4.75×10^{-4}	0.153	4.93×10^{-4}
0.120	0.165	1.08×10^{-3}	0.286	4.89×10^{-4}
			Average:	4.95×10^{-4}

in which each ratio $P_J/P°$ is raised to a power equal to the coefficient of that species in the balanced equation. We will not in general write "equilibrium" on expressions for K but will rely on the context to recognize that K is defined in terms of the equilibrium values. Thus, if we are interested in the reaction

$$aA(g) + bB(g) \rightleftharpoons cC(g) + dD(g) \tag{E}$$

and if all the gases are treated as ideal, we write

$$K = \frac{(P_C/P°)^c\,(P_D/P°)^d}{(P_A/P°)^a\,(P_B/P°)^b} \tag{1b}$$

with the partial pressures being those at equilibrium. This expression is normally written in the less cumbersome form

$$K = \frac{P_C{}^c\,P_D{}^d}{P_A{}^a\,P_B{}^b} \tag{1c}$$

by interpreting each P_J as the numerical value of the partial pressure without the units.

How can we use this information . . . to write the expression for an equilibrium constant?

For the special case of the ammonia equilibrium, reaction C, we write the equilibrium constant with the partial pressure of the product, NH_3, in the numerator, raised to a power equal to its coefficient in the balanced equation. We do the same for the reactants, but place their partial pressures in the denominator:

$$K = \frac{P_{NH_3}{}^2}{P_{N_2}P_{H_2}{}^3} \tag{2}$$

As explained above, each P_J in this expression should be interpreted as $P_J/P°$.

Self-Test 9.1A Write the expression for the equilibrium constant for the reaction $4\ NH_3(g) + 5\ O_2(g) \rightleftharpoons 4\ NO(g) + 6\ H_2O(g)$.

[*Answer:* $K = P_{NO}{}^4P_{H_2O}{}^6/P_{NH_3}{}^4P_{O_2}{}^5$]

Self-Test 9.1B Write the expression for the equilibrium constant for $2\ H_2S(g) + 3\ O_2(g) \rightleftharpoons 2\ SO_2(g) + 2\ H_2O(g)$.

It has been found empirically that similar expressions can be written for the equilibrium constants of reactions that involve species other than gases. Thus, for a species J that forms an ideal solution in a liquid solvent, the partial pressure in the expression for K is replaced by the molarity [J] relative to the standard molarity $[J]° = 1\ mol{\cdot}L^{-1}$. In practice, although an equilibrium constant should be written in terms of the ratios [J]/[J]°, it is common practice to write them in terms of [J] alone and to interpret each [J] as the molarity with the units struck out. It has also been found empirically that pure liquids or solids should not appear in K, so even though $CaCO_3(s)$ and $CaO(s)$ occur in the equilibrium

$$CaCO_3(s) \rightleftharpoons CaO(s) + CO_2(g) \quad \text{(F)}$$

neither appears in the equilibrium constant, which is simply $K = P_{CO_2}/P°$.

We can summarize these empirical rules by introducing the concept of the **activity,** a_J, of a substance J, the partial pressure or concentration of a substance relative to its standard value:

- For an ideal gas, $a_J = P_J/P°$.
- For a solute in a dilute solution, $a_J = [J]/[J]°$.
- For a pure solid or liquid, $a_J = 1$.

In other words, the activity of a pure gas is its partial pressure in bar with the units struck out; that of a solute is its molarity with the units $mol{\cdot}L^{-1}$ struck out. Note that all activities are pure numbers: they are unitless.

At this stage, these activities are purely empirical quantities and are introduced to facilitate writing the expression for K in accord with the empirical rules set out above. Later, though, we shall see that both the activities and the rules can be justified thermodynamically. In terms of activities, the equilibrium constant for any reaction has the general form

$$K = \left\{\frac{\text{activities of products}}{\text{activities of reactants}}\right\}_{\text{equilibrium}} \quad (3)$$

The activity of each species is raised to a power equal to its coefficient in the chemical equation for the reaction.

Because activities are unitless, so too is K.

Chemical equilibria with reactants and products that are all in the same phase are called **homogeneous equilibria.** Equilibria C, D, and E are homogeneous. Equilibria in systems having more than one phase are called **heterogeneous equilibria.** Equilibrium F is heterogeneous; so too is the equilibrium between water vapor and liquid water in a closed system:

$$H_2O(l) \rightleftharpoons H_2O(g) \quad \text{(G)}$$

In this reaction, there is a gas phase and a liquid phase. Likewise, the equilibrium between a solid and its saturated solution is heterogeneous:

$$Ca(OH)_2(s) \rightleftharpoons Ca^{2+}(aq) + 2\ OH^-(aq) \quad \text{(H)}$$

The equilibrium constants for heterogeneous reactions are given by the general expression in Eq. 3: all we have to remember is that the activity of a pure solid or liquid is 1. For instance, for the calcium hydroxide equilibrium (reaction H),

$$K = \frac{a_{Ca^{2+}}a_{OH^-}^{\ 2}}{a_{Ca(OH)_2}} = ([Ca^{2+}]/[Ca^{2+}]°)([OH^-]/[OH^-]°)^2$$

because the calcium hydroxide is a pure solid. As indicated above, for simplicity we would normally write

$$K = [Ca^{2+}][OH^-]^2$$

and interpret each [J] as the molar concentration of J with the units struck out. Similarly, in the equilibrium between solid nickel, gaseous carbon monoxide, and gaseous nickel carbonyl used in the purification of nickel,

$$Ni(s) + 4\,CO(g) \rightleftharpoons Ni(CO)_4(g) \qquad K = \frac{a_{Ni(CO)_4}}{a_{Ni}a_{CO}^{\,4}} = \frac{P_{Ni(CO)_4}}{P_{CO}^{\,4}}$$

The pure solid nickel must be present for the equilibrium to exist, but it does not appear in the expression for the equilibrium constant.

Some reactions in solution involve the solvent as a reactant or product. When the solution is very dilute, the change in solvent concentration due to the reaction is insignificant. In such cases, the solvent is treated as a pure substance and ignored when writing K. In other words,

For a nearly pure solvent: $a_{solvent} = 1$

Self-Test 9.2A Write the equilibrium constant for $Ag_2O(s) + 2\,HNO_3(aq) \rightleftharpoons 2\,AgNO_3(aq) + H_2O(l)$. (*Hint:* Use the net ionic equation.)

[***Answer:*** $K = [Ag^+]^2/[H^+]^2$]

Self-Test 9.2B Write the equilibrium constant K for $P_4(s) + 5\,O_2(g) \rightleftharpoons P_4O_{10}(s)$.

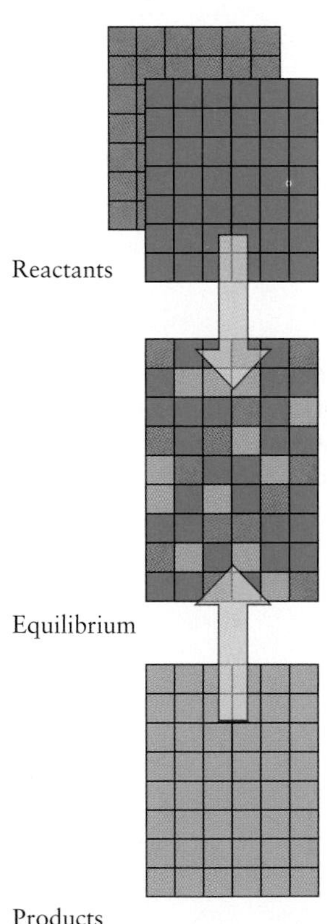

FIGURE 9.4 Whether we start with pure reactants or with pure products, a reaction mixture will always tend toward a mixture of reactants and products that has a composition in accord with the equilibrium constant for the reaction at the temperature of the experiment.

Each reaction has its own characteristic equilibrium constant, with a value that can be changed only by varying the temperature (Table 9.2). Whatever the initial composition of the reaction mixture, the composition adjusts so that at equilibrium it corresponds to the value of K for the reaction. The extraordinary empirical result, which we justify in the next section in terms of thermodynamics, is that *regardless of the initial composition of a reaction mixture, the composition tends to adjust until the activities give rise to the characteristic value of K for the reaction and temperature* (Fig. 9.4)

The equilibrium composition of a reaction mixture is described by the equilibrium constant, which is equal to the activities of the products (raised to powers equal to their stoichiometric coefficients in the balanced chemical equation for the reaction) divided by the activities of the reactants (raised to powers equal to their stoichiometric coefficients).

9.3 The Thermodynamic Origin of Equilibrium Constants

Guldberg and Waage's description of equilibrium was entirely empirical. Now, though, chemists know that the law of mass action is a consequence of the role of free energy in determining the direction of spontaneous change. Let's begin our exploration of the thermodynamic basis for equilibrium by analyzing how changes in free energy determine the tendency of a reaction to go in the forward or reverse directions.

When a reaction mixture has not yet produced enough products to establish equilibrium, the spontaneous direction of change is toward more

TABLE 9.2 Equilibrium Constants for Various Reactions

Reaction	Temperature, K	K	K_c
$H_2(g) + Cl_2(g) \rightleftharpoons 2\ HCl(g)$	300	4.0×10^{31}	4.0×10^{31}
	500	4.0×10^{18}	4.0×10^{18}
	1000	5.1×10^{8}	5.1×10^{8}
$H_2(g) + Br_2(g) \rightleftharpoons 2\ HBr(g)$	300	1.9×10^{17}	1.9×10^{17}
	500	1.3×10^{10}	1.3×10^{10}
	1000	3.8×10^{4}	3.8×10^{4}
$H_2(g) + I_2(g) \rightleftharpoons 2\ HI(g)$	298	794	794
	500	160	160
	700	54	54
$2\ BrCl(g) \rightleftharpoons Br_2(g) + Cl_2(g)$	300	377	377
	500	32	32
	1000	5	5
$2\ HD(g) \rightleftharpoons H_2(g) + D_2(g)$	100	0.52	0.52
	500	0.28	0.28
	1000	0.26	0.26
$F_2(g) \rightleftharpoons 2\ F(g)$	500	3.0×10^{-11}	7.3×10^{-13}
	1000	1.0×10^{-2}	1.2×10^{-4}
	1200	0.27	2.7×10^{-3}
$Cl_2(g) \rightleftharpoons 2\ Cl(g)$	1000	1.0×10^{-5}	1.2×10^{-7}
	1200	1.7×10^{-3}	1.7×10^{-5}
$Br_2(g) \rightleftharpoons 2\ Br(g)$	1000	3.4×10^{-5}	4.1×10^{-7}
	1200	1.7×10^{-3}	1.7×10^{-5}
$I_2(g) \rightleftharpoons 2\ I(g)$	800	2.1×10^{-3}	3.1×10^{-5}
	1000	0.26	3.1×10^{-3}
	1200	6.8	6.8×10^{-2}
$N_2(g) + 3\ H_2(g) \rightleftharpoons 2\ NH_3(g)$	298	6.8×10^{5}	4.2×10^{8}
	400	41	4.5×10^{4}
	500	3.6×10^{-2}	62
$2\ SO_2(g) + O_2(g) \rightleftharpoons 2\ SO_3(g)$	298	4.0×10^{24}	9.9×10^{25}
	500	2.5×10^{10}	1.0×10^{12}
	700	3.0×10^{4}	1.7×10^{6}
$N_2O_4(g) \rightleftharpoons 2\ NO_2(g)$	298	0.98	4.0×10^{-2}
	400	47.9	1.44
	500	1.7×10^{3}	41

products. In terms of the reaction free energy, $\Delta G_r < 0$ for the reaction at constant temperature and pressure. The initial reaction mixture in Haber's process has no ammonia present, ΔG_r at this stage of the reaction is negative, and the formation of ammonia is spontaneous. If we start with a reaction mixture that has excess products (pure ammonia, for instance), the *reverse* reaction is spontaneous. In thermodynamic terms, $\Delta G_r > 0$ for the forward reaction. For the reaction at equilibrium, when nitrogen, hydrogen, and ammonia are all present at their equilibrium concentrations, there is no tendency for

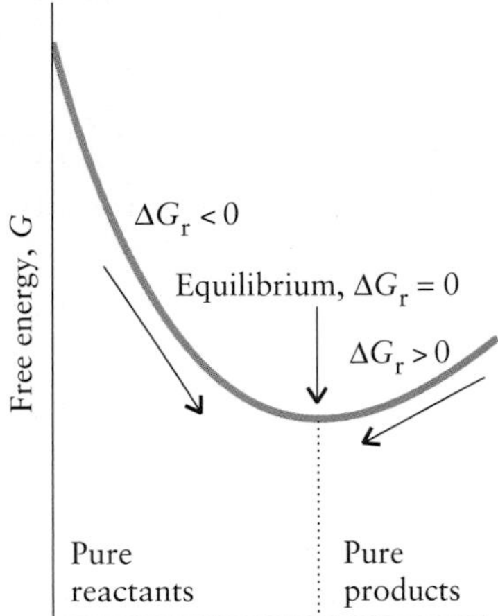

FIGURE 9.5 The variation of free energy of a reaction mixture with composition. A reaction mixture has a spontaneous tendency to change in the direction of decreasing free energy.

spontaneous change in either the forward or the reverse direction and $\Delta G_r = 0$. Clearly, the reaction free energy changes as the proportions of reactants and products change; and to find the composition of the reaction mixture at equilibrium, we need to find the composition at which the reaction free energy is 0 (Fig. 9.5).

The reaction free energy is the difference of the molar free energies of the products and reactants (weighted by the stoichiometric coefficients):

$$\Delta G_r = \sum nG_m°(\text{products}) - \sum nG_m°(\text{reactants})$$

Therefore, to find how ΔG_r changes with composition, we need to know how the molar free energy of each substance varies with its partial pressure, if it is a gas, or with its concentration, if it is a solute. We have already seen (in Section 8.2) that the molar free energy of an ideal gas J is related to its partial pressure, P_J, by

$$G_m(\text{J}) = G_m°(\text{J}) + RT \ln(P_J/P°) \quad (4a)$$

Note that because $\ln 1 = 0$, $G_m(\text{J})$ has its standard value, $G_m°(\text{J})$, when the partial pressure of J, P_J, is 1 bar. Thermodynamic arguments (which we do not reproduce here) show that a similar expression applies to solutes and pure substances. In each case, we can write the molar free energy of a substance J as

$$G_m(\text{J}) = G_m°(\text{J}) + RT \ln a_J \quad (4b)$$

with the activity defined as in Section 9.2. Thus, we see that *an activity is an effective partial pressure or concentration relative to a standard value that, when substituted into Eq. 4b, correctly gives the molar free energy of a species.* For an ideal gas, Eq. 4b is the same as Eq. 4a. For a solute in an ideal solution, Eq. 4b becomes

$$G_m(\text{J}) = G_m°(\text{J}) + RT \ln([\text{J}]/[\text{J}]°) \quad (4c)$$

Note that $G_m(\text{J})$ has its standard value, $G_m°(\text{J})$, when the molarity of J is 1 mol·L^{-1}. For a pure solid or liquid,

$$G_m(\text{J}) = G_m°(\text{J}) \quad (4d)$$

That is, for a pure solid or liquid, the molar free energy always has its standard value. That may seem puzzling at first sight, but it stems from the fact that the free energy of a solid or liquid is almost independent of pressure. As a result, to a very good approximation, the substance can be treated as being in its standard state whatever the pressure (provided the pressure is not too great).

How can we use this information . . . to calculate the change in free energy when the concentration is changed?

To calculate the change in molar free energy when the concentration of methylbenzene (toluene) in benzene is increased from 0.100 mol·L^{-1} to 0.200 mol·L^{-1} at 25.00° C, we write the expression for G_m at each concentration, noting that [J]/[J]° is equal to the molarity of J with the units struck out:

$$G_m(0.200\ \text{mol·L}^{-1}) = G_m°(\text{methylbenzene}) + RT \ln 0.200$$

$$G_m(0.100\ \text{mol·L}^{-1}) = G_m°(\text{methylbenzene}) + RT \ln 0.100$$

The change in free energy when the concentration is increased is the difference

$$\begin{aligned}\Delta G_m &= G_m(0.200\ \text{mol·L}^{-1}) - G_m(0.100\ \text{mol·L}^{-1})\\ &= RT(\ln 0.200 - \ln 0.100) = RT \ln 2.00\end{aligned}$$

At 25.00°C, $RT = 2.479\ \text{kJ·mol}^{-1}$, so $\Delta G_m = +1.72\ \text{kJ·mol}^{-1}$. Note that the molar free energy *increases* as the concentration of solute increases.

Self-Test 9.3A Calculate the change in molar free energy when the partial pressure of nitrogen in a mixture of gases is increased from 0.600 bar to 1.800 bar at 298 K.

[*Answer:* $+2.72\ \text{kJ·mol}^{-1}$]

Self-Test 9.3B Calculate the change in molar free energy when the concentration of ethanol in water is increased from $0.0500\ \text{mol·L}^{-1}$ to $0.250\ \text{mol·L}^{-1}$ at 30°C.

Real gases and real solutions may not behave as predicted by the simple treatment we have described so far because intermolecular interactions affect their properties. However, there is a way to adjust a_J appropriately, so that Eq. 3 also applies to *real* gases and *real* solutions. For real systems, in which molecules interact with one another, we write

- For a gas J, $a_J = \gamma_J P_J/P°$.
- For a solute J, $a_J = \gamma_J [J]/[J]°$.

The dimensionless quantity γ_J is called the **activity coefficient** of the substance J (Fig. 9.6). More advanced techniques in chemistry enable us to relate the activity coefficient to the actual composition. However, for the low-pressure gases and dilute solutions that concern us here, we can ignore the interactions between molecules and set all $\gamma_J = 1$. We have to remember, though, that our expressions are then valid only for gases behaving ideally and for very dilute solutions.

When we follow the changes in free energy that occur during a reaction, we need to distinguish the *standard* reaction free energy, $\Delta G_r°$, from the reaction free energy, ΔG_r:

$\Delta G_r°$ is the difference in molar free energies of the products and reactants in their standard states.

As we have remarked, the standard state of a pure substance is its pure form at a pressure of 1 bar. For a solute, the standard state is for a concentration of $1\ \text{mol·L}^{-1}$. A pure solid may always be regarded as in its standard state.

ΔG_r is the reaction free energy at any definite, fixed composition of the reaction mixture.

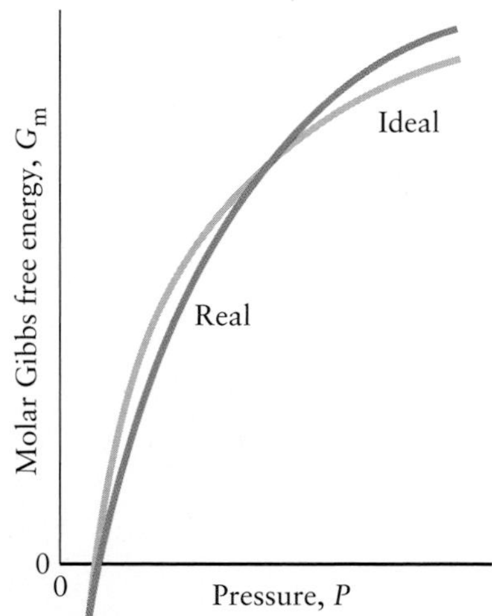

FIGURE 9.6 The variation of the molar free energy of a typical real gas with its partial pressure (orange line) superimposed on the variation for an ideal gas (gray line). The deviation from ideality is taken into account by allowing the activity coefficient to vary from 1.

The value of ΔG_r refers to any chosen composition of the reaction mixture and represents the difference in molar free energy between the products and reactants *at the concentrations present at a specified stage of the reaction.*

To see how ΔG_r, which is a *change* in a quantity, can apply to a system at fixed composition, consider the change in free energy when a tiny fraction of reactant molecules changes into products. Such a small change has a negligible effect on the composition of the reaction mixture. For example, suppose nitrogen and hydrogen react to form ammonia in a large container, with 10 000 mol of each substance present. The reaction free energy is the change in free energy when 1 mol N_2 and 3 mol H_2 are consumed at the partial pressures present in that environment. Because the composition of the reaction mixture is virtually unchanged when such small amounts of reactants are changed into products, we can speak of the change in free energy at a fixed composition.

The actual value of ΔG_r changes as the composition of the reaction mixture changes, as we saw in Fig. 9.5. If the reaction mixture contains relatively more reactants than it would contain at equilibrium, then $\Delta G_r < 0$ at that composition, and the reaction will have a tendency to form more products until the equilibrium composition is reached. If the reaction mixture happens to contain a higher proportion of products than it would contain at equilibrium, then $\Delta G_r > 0$, and the reverse reaction, the formation of reactants, will be spontaneous. However, if the reaction mixture happens to have its equilibrium composition, then $\Delta G_r = 0$, and there is no tendency for the reaction to run in either direction.

We can express these remarks quantitatively in a form that we can use to predict the reaction free energy for any concentrations of the reactants and products. First, we note that the reaction free energy is related to the molar free energies of the reactants and products by the expression first given in Section 7.13:

$$\Delta G_r = \sum nG_m(\text{products}) - \sum nG_m(\text{reactants}) \qquad (5)$$

The n are the stoichiometric coefficients in the chemical equation. To find the value of ΔG_r at any composition of a reaction, all we need to do is substitute Eq. 4 into this expression. For example, to find the reaction free energy for the synthesis of ammonia, reaction C, in terms of the partial pressures of the three gases (taken to be ideal), we use Eq. 4a to write

$$\begin{aligned}\Delta G_r &= 2G_m(NH_3) - \{G_m(N_2) + 3G_m(H_2)\} \\ &= 2(G_m^\circ(NH_3) + RT\ln P_{NH_3}) - \\ &\qquad \{(G_m^\circ(N_2) + RT\ln P_{N_2}) + 3(G_m^\circ(H_2) + RT\ln P_{H_2})\} \\ &= \{2G_m^\circ(NH_3) - G_m^\circ(N_2) - 3G_m^\circ(H_2)\} \\ &\qquad + 2RT\ln P_{NH_3} - RT\ln P_{N_2} - 3RT\ln P_{H_2}\end{aligned}$$

To keep the notation simple, we have written each P_J/P° as P_J. We can simplify this expression further. The combination of the first three terms in the final equation is the *standard* reaction free energy, ΔG_r°:

$$\Delta G_r^\circ = 2G_m^\circ(NH_3) - \{G_m^\circ(N_2) + 3G_m^\circ(H_2)\}$$

We can also tidy up the three logarithmic terms. First, we use $s \ln x = \ln x^s$ to write them as

$$2RT\ln P_{NH_3} - RT\ln P_{N_2} - 3RT\ln P_{H_2} = RT(\ln P_{NH_3}{}^2 - \ln P_{N_2} - \ln P_{H_2}{}^3)$$

Then we use $\ln x - \ln y = \ln(x/y)$ to turn this expression into

$$RT(\ln P_{NH_3}{}^2 - \ln P_{N_2} - \ln P_{H_2}{}^3) = RT\ln\left(\frac{P_{NH_3}{}^2}{P_{N_2}P_{H_2}{}^3}\right)$$

Now we can bring our work together and write:

$$\Delta G_r = \Delta G_r{}^\circ + RT\ln\left(\frac{P_{NH_3}{}^2}{P_{N_2}P_{H_2}{}^3}\right)$$

This expression tells us how the reaction free energy varies with the partial pressures of the reactants and products. The expression in parentheses is called the **reaction quotient, Q**. It has the same form as the expression for K, but the partial pressures refer to any stage of the reaction. For reaction C, the reaction quotient is

$$Q = \frac{P_{NH_3}{}^2}{P_{N_2}P_{H_2}{}^3} \qquad (6)$$

With the introduction of Q, the equation becomes simply

$$\Delta G_r = \Delta G_r{}^\circ + RT \ln Q \qquad (7)$$

For a general reaction, Q has the form

$$Q = \frac{\text{activities of products}}{\text{activities of reactants}} \qquad (8)$$

with each activity (partial pressure or molarity relative to its standard value) raised to a power equal to the stoichiometric coefficient in the chemical equation, just as in the expression for K. Thus, for a gas-phase reaction of the form $a\text{A} + b\text{B} \rightarrow c\text{C} + d\text{D}$,

$$Q = \frac{P_C{}^c P_D{}^d}{P_A{}^a P_B{}^b} \qquad (9)$$

with, as usual, each P_J interpreted as P_J/P°.

The reaction free energy, ΔG_r, is related to the composition of the reaction mixture at any stage of the reaction by Eq. 7.

EXAMPLE 9.1 Calculating the reaction free energy from the reaction quotient

The standard reaction free energy for $2\ SO_2(g) + O_2(g) \rightarrow 2\ SO_3(g)$ is $\Delta G_r{}^\circ = -141.74\ \text{kJ·mol}^{-1}$ at 25.00°C. What is the reaction free energy when the partial pressure of each gas is 100 bar? What is the spontaneous direction of the reaction under these conditions?

STRATEGY Calculate the reaction quotient and substitute it and the standard reaction free energy into Eq. 7. If ΔG_r turns out to be negative, the oxidation of sulfur dioxide is spontaneous at the given composition. If ΔG_r is positive, sulfur trioxide has a spontaneous tendency to decompose into sulfur dioxide and oxygen at the given composition. If ΔG_r is 0, there is no tendency to react in either direction; the reaction is at equilibrium.

SOLUTION The reaction quotient is

$$Q = \frac{P_{SO_3}{}^2}{P_{SO_2}{}^2 P_{O_2}} = \frac{(100)^2}{(100)^2 \times (100)} = 1.00 \times 10^{-2}$$

Therefore, the reaction free energy at this composition is

$$\begin{aligned}\Delta G_r &= \Delta G_r^\circ + RT \ln Q \\ &= (-141.74 \text{ kJ·mol}^{-1}) + (2.479 \text{ kJ·mol}^{-1}) \ln (1.00 \times 10^{-2}) \\ &= -153.16 \text{ kJ·mol}^{-1}\end{aligned}$$

Because the reaction free energy is negative, the formation of products is spontaneous at this composition and temperature.

Self-Test 9.4A The standard reaction free energy for $H_2(g) + I_2(g) \rightarrow 2\ HI(g)$ is $\Delta G_r^\circ = -21.1$ kJ·mol^{-1} at 500 K (when $RT = 4.16$ kJ·mol^{-1}). What is the value of ΔG_r when the partial pressures of the gases are $P_{H_2} = 1.5$ bar, $P_{I_2} = 0.88$ bar, and $P_{HI} = 0.065$ bar? What is the spontaneous direction of the reaction?

[*Answer:* −45 kJ·mol^{-1}; toward products]

Self-Test 9.4B The standard reaction free energy for $N_2O_4(g) \rightarrow 2\ NO_2(g)$ is $\Delta G_r^\circ = +4.73$ kJ·mol^{-1} at 298 K. What is the value of ΔG_r when the partial pressures of the gases are $P_{N_2O_4} = 0.80$ bar and $P_{NO_2} = 2.10$ bar? What is the spontaneous direction of the reaction?

At equilibrium, the partial pressures have their equilibrium values, and Q has a special value that we recognize as the equilibrium constant, K, of the reaction. Thermodynamics has provided a clear interpretation of the puzzling form of K: we see that *K is simply the value of the reaction quotient evaluated at equilibrium*. Moreover, we see that the reaction quotient is the quantity that allows us to calculate the difference between a reaction free energy and its standard value.

Now we are ready to turn Eq. 7 into the most important equation in the whole of chemical thermodynamics. We know that at equilibrium $\Delta G_r = 0$. We also know that $Q = K$ at equilibrium. It follows that, at equilibrium,

$$0 = \Delta G_r^\circ + RT \ln K$$

and therefore that

$$\Delta G_r^\circ = -RT \ln K \qquad (10)$$

This fundamentally important equation links thermodynamic quantities—which are widely available from tables of thermodynamic data—and the composition of a system at equilibrium.

The reaction quotient, Q, has the same form as K, the equilibrium constant, except that Q uses the activities evaluated at an arbitrary stage of the reaction. The equilibrium constant is related to the standard reaction free energy by Eq. 10, $\Delta G_r^\circ = -RT \ln K$.

How can we use this information . . . to predict the value of K?

The value of the standard reaction free energy for $\frac{1}{2} H_2(g) + \frac{1}{2} I_2(s) \rightarrow HI(g)$ is $+ 1.70\ kJ{\cdot}mol^{-1}$ at 25.00°C (see Appendix 2A). Therefore, the equilibrium constant for $\frac{1}{2} H_2(g) + \frac{1}{2} I_2(s) \rightleftharpoons HI(g)$ is obtained from Eq. 10:

$$\begin{aligned} \ln K &= -\frac{\Delta G_r^\circ}{RT} \\ &= -\frac{1.70 \times 10^3\ J{\cdot}mol^{-1}}{(8.3145\ J{\cdot}K^{-1}{\cdot}mol^{-1}) \times (298.15\ K)} = -\frac{1.70 \times 10^3}{8.3145 \times 298.15} \\ &= -0.686 \end{aligned}$$

The equilibrium constant itself is therefore $e^{-0.686}$, which is 0.50. Note that reaction free energies are normally given in kilojoules per mole, so we need to convert them to joules per mole (to match the units of R) when using this expression.

Self-Test 9.5A Use the thermodynamic data in Appendix 2A to calculate K from ΔG_r° for $N_2O_4(g) \rightleftharpoons 2\ NO_2(g)$ at 298 K.

[*Answer:* $K = 0.15$]

Self-Test 9.5B Use the thermodynamic data in Appendix 2A to calculate K from ΔG_r° for $2\ NO(g) + O_2(g) \rightleftharpoons 2\ NO_2(g)$ at 298 K.

We can now begin to get some insight into why some reactions have large equilibrium constants and others have small ones. It follows from $\Delta G_r^\circ = \Delta H_r^\circ - T\Delta S_r^\circ$ and $\Delta G_r^\circ = -RT \ln K$ that

$$\ln K = -\frac{\Delta G_r^\circ}{RT} = -\frac{\Delta H_r^\circ}{RT} + \frac{\Delta S_r^\circ}{R}$$

When we take antilogarithms of both sides, we obtain

$$K = e^{-\Delta H_r^\circ/RT}\, e^{\Delta S_r^\circ/R} \tag{11}$$

We see that K can be expected to be small if ΔH_r° is positive. An endothermic reaction is therefore likely to have $K < 1$ and is unlikely to form much product. Only if ΔS_r° is large and positive can we expect $K > 1$ for an endothermic reaction. Conversely, if a reaction is strongly exothermic, then ΔH_r° is large and negative. Now we can expect $K > 1$ and the products to be favored. In other words, we can expect strongly exothermic reactions to go to completion. K is likely to fall below 1 for exothermic reactions only if the reaction enthalpy is small and the reaction entropy strongly negative.

Strongly exothermic reactions are likely to have large equilibrium constants and thus favor the formation of products.

EQUILIBRIUM CONSTANTS

The value of an equilibrium constant tells us whether we can expect a high or low concentration of product at equilibrium. The constant also allows us to predict the spontaneous direction of reaction when the reactants and products are present at arbitrary concentrations. In the following five sections, we see how to express the equilibrium constant in terms of molar concentrations as well as partial pressures and how to extract *quantitative* information about the equilibrium composition of a reaction mixture, given the value of the equilibrium constant for the reaction. Such information is critical to the success of many industrial processes and is fundamental to the discussion of acids and bases, as we shall see in the following chapters.

9.4 The Equilibrium Constant in Terms of Molar Concentrations of Gases

To emphasize that partial pressures are being used, the equilibrium constant is often denoted K_P.

The equilibrium constant in Eq. 3 is defined in terms of dimensionless activities, and the activities are interpreted in terms of the partial pressures or concentrations. Gases always appear in K as partial pressures, and solutes appear as molarities. Often, however, we want to discuss gas-phase equilibria in terms of molar concentrations (moles of gas molecules divided by the volume of the container in liters), not partial pressures. To do this, we introduce the equilibrium constant K_c, which is defined as

$$K_c = \left\{ \frac{\text{molar concentrations of products relative to standard concentration}}{\text{molar concentrations of reactants relative to standard concentration}} \right\}_{\text{equilibrium}} \quad (12)$$

with each molar concentration raised to a power equal to the stoichiometric coefficient of the species in the chemical equation. For example, for the ammonia synthesis equilibrium, reaction C,

$$K_c = \frac{[NH_3]^2}{[N_2][H_2]^3} \quad (13)$$

As usual, we have replaced $[J]/[J]^\circ$ by $[J]$ itself, which should be interpreted as the numerical value of the molar concentration of the gas J. Thus, if the molar concentration of NH_3 is 0.10 mol·L^{-1}, then in the expression for the equilibrium constant (but not in other applications), we write $[NH_3] = 0.10$. In general, for reaction E, the equilibrium constant K_c is written

$$K_c = \frac{[C]^c[D]^d}{[A]^a[B]^b} \quad (14)$$

with each $[J]$ standing for $[J]/[J]^\circ$.

Self-Test 9.6A Write the equilibrium constants K and K_c for $2\ CO(g) + O_2(g) \rightleftharpoons 2\ CO_2(g)$.

[*Answer:* $K = P_{CO_2}{}^2/P_{CO}{}^2P_{O_2}$; $K_c = [CO_2]^2/[CO]^2[O_2]$]

Self-Test 9.6B Write the equilibrium constants K and K_c for $2\ O_3(g) \rightleftharpoons 3\ O_2(g)$.

Experimentally, we are free to choose either K or K_c to report the equilibrium constant of a reaction. However, it is important to remember that calculations of an equilibrium constant from thermodynamic tables of data (standard free energies of formation, for instance) and Eq. 10 give K, not K_c. In some cases, we need to know K_c after we have calculated K from thermodynamic data, so we need to be able to convert between these two constants.

How do we do that?

The overall strategy for finding the relation between K and K_c is to replace the partial pressures that appear in K by the molar concentrations, and thereby generate K_c. For this calculation, we write activities as $P_J/P°$ and $[J]/[J]°$. Moreover, because $[J]° = 1\ mol{\cdot}L^{-1}$ regardless of the species J, in this calculation we shall write $c° = 1\ mol{\cdot}L^{-1}$ in place of $[J]°$. We shall consider a specific case: the relation between K and K_c for the ammonia equilibrium, reaction C.

First, we need the relation between the partial pressure, P_J, and the molar concentration, $[J] = n_J/V$, of each gas. Assuming ideal gas behavior, we write the ideal gas law $P_JV = n_JRT$ as

$$P_J = \frac{n_JRT}{V} = RT\frac{n_J}{V} = RT[J]$$

When this expression is substituted in Eq. 2, we obtain

$$K = \frac{(P_{NH_3}/P°)^2}{(P_{N_2}/P°)(P_{H_2}/P°)^3} = \frac{P_{NH_3}{}^2(P°)^2}{P_{N_2}P_{H_2}{}^3} = \frac{(RT)^2[NH_3]^2(P°)^2}{RT[N_2](RT)^3[H_2]^3}$$
$$= \left(\frac{P°}{RT}\right)^2\frac{[NH_3]^2}{[N_2][H_2]^3}$$

By K_c for reaction C, we mean

$$K_c = \frac{([NH_3]/[NH_3]°)^2}{([N_2]/[N_2]°)([H_2]/[H_2]°)^3} = \frac{([NH_3]/c°)^2}{([N_2]/c°)([H_2]/c°)^3} = \frac{c°^2[NH_3]^2}{[N_2][H_2]^3}$$

This expression can be rearranged into

$$\frac{[NH_3]^2}{[N_2][H_2]^3} = \frac{K_c}{c°^2}$$

If we substitute this relation into the expression for K, it follows that

$$K = \left(\frac{P°}{RT}\right)^2\frac{K_c}{c°^2} = \left(\frac{P°}{RTc°}\right)^2K_c = \left(\frac{RTc°}{P°}\right)^{-2}K_c$$

A useful way of remembering the general form of this expression, and others like it, is to write it as

$$K = \left(\frac{RTc^{\circ}}{P^{\circ}}\right)^{\Delta n} K_c = \left(\frac{RT \times (\text{standard molarity})}{\text{standard pressure}}\right)^{\Delta n} K_c \qquad \textbf{(15a)}$$

where Δn is the change in the number of gas molecules between reactants and products in the chemical equation, calculated as $\Delta n = n(\text{products}) - n(\text{reactants})$ (so $\Delta n = -2$ for reaction C). If no gases are involved in the reaction, or if $\Delta n = 0$, then $K = K_c$. The same relation holds between Q and Q_c, the reaction quotient in terms of concentrations. Equation 15a is commonly written more succinctly as

$$K = (RT)^{\Delta n} K_c \qquad \textbf{(15b)}$$

but the full version, Eq. 15a, clarifies the units and should be used in calculations. Because P° is 1 bar and c° is 1 mol·L^{-1}, it is sensible to use R expressed in bar and liters, $R = 8.3145 \times 10^{-2}$ L·bar·K^{-1}·mol^{-1}, when using this relation.

Equilibrium constants for reactions involving gases can be written either by using molar concentrations and writing K_c or by using partial pressures and writing K. The two forms of the equilibrium constant are related by Eq. 15.

EXAMPLE 9.2 Converting between K and K_c

At 400°C, the equilibrium constant K for $2\ SO_2(g) + O_2(g) \rightleftharpoons 2\ SO_3(g)$ is 3.1×10^4. What is the value of K_c at that temperature?

STRATEGY First, we identify the value of Δn for the reaction, and then substitute the data into Eq. 15, remembering to convert the temperature to the Kelvin scale, and to rearrange the equation to solve for K_c. Because the standard pressure is 1 bar and the standard molarity is in moles per liter, we use R in units that include bars and liters. Postpone numerical work until the last possible stage to avoid rounding errors. Convert temperatures to kelvins.

SOLUTION First, we calculate the factor in parentheses in Eq. 15:

$$\frac{RTc^{\circ}}{P^{\circ}} = \frac{(8.3145 \times 10^{-2}\ \text{L·bar·K}^{-1}\text{·mol}^{-1}) \times (673\ \text{K}) \times (1\ \text{mol·L}^{-1})}{1\ \text{bar}}$$
$$= 8.3145 \times 6.73$$

Therefore, because $\Delta n = 2 - (2 + 1) = -1$,

$$K_c = \left(\frac{RTc^{\circ}}{P^{\circ}}\right)^{-\Delta n} K = (8.3145 \times 6.73) \times (3.1 \times 10^4) = 1.7 \times 10^6$$

Self-Test 9.7A The equilibrium constant, K, for the ammonia synthesis (reaction C) is 41 at 127°C. What is the value of K_c at that temperature?

[*Answer:* $\Delta n = -2$, so $K_c = 4.5 \times 10^4$]

Self-Test 9.7B At 127°C, the equilibrium constant K for $N_2O_4(g) \rightleftharpoons 2\ NO_2(g)$ is 47.9. What is the value of K_c at that temperature?

EXAMPLE 9.3 Determining the equilibrium constant from equilibrium concentrations

Haber mixed some nitrogen and hydrogen and allowed them to react at 500 K until the mixture reached equilibrium with the product, ammonia. When he analyzed the equilibrium mixture, he found it to consist of 0.796 mol·L^{-1} NH_3, 0.305 mol·L^{-1} N_2, and 0.324 mol·L^{-1} H_2. What is the equilibrium constant K for the reaction?

STRATEGY Begin by writing the chemical equation and the expression for the equilibrium constant. Then substitute the equilibrium concentrations in Eq. 13 to find K_c. Raise each concentration to a power equal to the stoichiometric coefficient of that species in the chemical equation. Equation 15 can then be used to obtain K.

SOLUTION The chemical equation and the expression for the equilibrium constant K_c are given in reaction C and Eq. 13, respectively. We substitute the equilibrium molar concentrations (without their units) into the expression for K_c:

$$K_c = \frac{[NH_3]^2}{[N_2][H_2]^3} = \frac{(0.796)^2}{0.305 \times (0.324)^3} = 61.1$$

To obtain the value of K, we use Eq. 15 with $T = 500$ K. We first write

$$\frac{RTc^\circ}{P^\circ} = \frac{(8.3145 \times 10^{-2}\ \text{L·bar·K}^{-1}\text{·mol}^{-1}) \times (500\ \text{K}) \times (1\ \text{mol·L}^{-1})}{1\ \text{bar}}$$
$$= 8.3145 \times 5.00$$

and then substitute this value into Eq. 15:

$$K = \left(\frac{RTc^\circ}{P^\circ}\right)^{\Delta n} K_c = (8.3145 \times 5.00)^{-2} \times 61.1 = 3.54 \times 10^{-2}$$

Self-Test 9.8A A reaction important in the gasification of coal is $2\ CO(g) + 2\ H_2(g) \rightleftharpoons CH_4(g) + CO_2(g)$. Determine the equilibrium constants K_c and K for this reaction at 298 K, given the following equilibrium concentrations: CO, 4.30×10^{-9} mol·L^{-1}; H_2, 1.15×10^{-8} mol·L^{-1}; CH_4, 5.14×10^{-2} mol·L^{-1}; and CO_2, 4.12×10^{-2} mol·L^{-1}.

[*Answer:* $K_c = 8.66 \times 10^{29}$, $K = 1.41 \times 10^{27}$]

Self-Test 9.8B Determine the equilibrium constants K_c and K for $2\ BrCl(g) \rightleftharpoons Br_2(g) + Cl_2(g)$ at 500 K, given the following equilibrium concentrations: BrCl, 0.131 mol·L^{-1}; Br_2, 3.51 mol·L^{-1}; and Cl_2, 0.156 mol·L^{-1}.

For thermodynamic calculations, gas-phase equilibria are expressed in terms of K, but for practical calculations may be expressed in terms of molar concentrations by using Eq. 15.

9.5 Alternative Forms of the Equilibrium Constant

Equilibrium constants are normally given for chemical equations written with the smallest whole numbers for the stoichiometric coefficients. However, if we change the stoichiometric coefficients in a chemical equation (for instance, by multiplying through by a factor), then we must make sure that the equilibrium constant reflects that change. For example, at 700 K,

$$H_2(g) + I_2(g) \rightleftharpoons 2\,HI(g) \qquad K_{c1} = \frac{[HI]^2}{[H_2][I_2]} = 54$$

If we rewrite the chemical equation by multiplying through by 2, the equilibrium constant becomes

$$2\,H_2(g) + 2\,I_2(g) \rightleftharpoons 4\,HI(g)$$

$$K_{c2} = \frac{[HI]^4}{[H_2]^2[I_2]^2} = (K_{c1})^2 = (54)^2 = 2.9 \times 10^3$$

In general, if we multiply a chemical equation by a factor n, we raise K (and K_c) to the nth power.

Now suppose we reverse the original equation for the reaction:

$$2\,HI(g) \rightleftharpoons H_2(g) + I_2(g)$$

This equation still describes the same equilibrium; but how is the equilibrium constant related to the one we wrote earlier? To find out, we simply write the equilibrium constant for this equation and obtain

$$K_{c3} = \frac{[H_2][I_2]}{[HI]^2} = \frac{1}{K_{c1}} = \frac{1}{54} = 0.019$$

In general, *the equilibrium constant for an equilibrium written in one direction is the reciprocal of the equilibrium constant for the equilibrium written in the opposite direction.*

If a chemical equation can be expressed as the sum of two or more chemical equations, *the equilibrium constant for the overall reaction is the product of the equilibrium constants for the component reactions.* For example, consider the three gas-phase reactions

$$2\,P(g) + 3\,Cl_2(g) \rightleftharpoons 2\,PCl_3(g) \qquad K_1 = \frac{P_{PCl_3}{}^2}{P_P{}^2 P_{Cl_2}{}^3}$$

$$PCl_3(g) + Cl_2(g) \rightleftharpoons PCl_5(g) \qquad K_2 = \frac{P_{PCl_5}}{P_{PCl_3} P_{Cl_2}}$$

$$2\,P(g) + 5\,Cl_2(g) \rightleftharpoons 2\,PCl_5(g) \qquad K_3 = \frac{P_{PCl_5}{}^2}{P_P{}^2 P_{Cl_2}{}^5}$$

TABLE 9.3 Relations Between Equilibrium Constants

Chemical equation	Equilibrium constant
$a\mathrm{A} + b\mathrm{B} \rightleftharpoons c\mathrm{C} + d\mathrm{D}$	K_1
$c\mathrm{C} + d\mathrm{D} \rightleftharpoons a\mathrm{A} + b\mathrm{B}$	$K_2 = \dfrac{1}{K_1} = K_1^{-1}$
$na\mathrm{A} + nb\mathrm{B} \rightleftharpoons nc\mathrm{C} + nd\mathrm{D}$	$K_3 = K_1^{\,n}$

The third reaction is the following sum of the first two reactions:

$$\begin{array}{r}
2\ \mathrm{P(g)} + 3\ \mathrm{Cl_2(g)} \rightleftharpoons 2\ \mathrm{PCl_3(g)} \\
2\ \mathrm{PCl_3(g)} + 2\ \mathrm{Cl_2(g)} \rightleftharpoons 2\ \mathrm{PCl_5(g)} \\
\hline
2\ \mathrm{P(g)} + 5\ \mathrm{Cl_2(g)} \rightleftharpoons 2\ \mathrm{PCl_5(g)}
\end{array}$$

and the equilibrium constant, K_3, of the overall reaction can be written

$$\begin{aligned}
K_3 &= \frac{P_{\mathrm{PCl_5}}{}^2}{P_{\mathrm{P}}{}^2 P_{\mathrm{Cl_2}}{}^5} = \frac{P_{\mathrm{PCl_3}}{}^2}{P_{\mathrm{P}}{}^2 P_{\mathrm{Cl_2}}{}^3} \times \frac{P_{\mathrm{PCl_5}}}{P_{\mathrm{PCl_3}} P_{\mathrm{Cl_2}}} \times \frac{P_{\mathrm{PCl_5}}}{P_{\mathrm{PCl_3}} P_{\mathrm{Cl_2}}} \\
&= K_1 \times K_2 \times K_2 = K_1 K_2{}^2
\end{aligned}$$

Notice that, because we used twice the second reaction in the sum, its equilibrium constant appears twice in the product.

Table 9.3 summarizes the relations between equilibrium constants.

Self-Test 9.9A At 500 K, K for $\mathrm{H_2(g)} + \mathrm{D_2(g)} \rightleftharpoons 2\ \mathrm{HD(g)}$ is 3.6. What is the value of K for $2\ \mathrm{HD(g)} \rightleftharpoons \mathrm{H_2(g)} + \mathrm{D_2(g)}$?

[*Answer:* 0.28]

Self-Test 9.9B At 500 K, K_c for $\mathrm{F_2(g)} \rightleftharpoons 2\ \mathrm{F(g)}$ is 7.3×10^{-13}. What is the value of K_c for $\frac{1}{2}\ \mathrm{F_2(g)} \rightleftharpoons \mathrm{F(g)}$?

9.6 The Extent of Reaction

Product concentrations or partial pressures appear in the numerator of K, and reactant concentrations or partial pressures appear in the denominator. If the products are relatively abundant at equilibrium, the numerator will be large and the denominator small. Therefore, K is large when the equilibrium mixture consists mostly of products. In contrast, when the numerator is small and the denominator large, K is small and equilibrium is reached after very little reaction has occurred.

For instance, consider the reaction

$$\mathrm{H_2(g)} + \mathrm{Cl_2(g)} \rightleftharpoons 2\ \mathrm{HCl(g)} \qquad K_c = \frac{[\mathrm{HCl}]^2}{[\mathrm{H_2}][\mathrm{Cl_2}]}$$

Experiment shows that $K = K_c = 4.0 \times 10^{18}$ at 500 K. Such a large value for K tells us that the system does not reach equilibrium until most of the reactants have been converted to HCl. In fact, this reaction is fast as well as favorable, because it goes explosively as soon as the reactants are mixed and exposed to sunlight. Now consider the equilibrium

$$N_2(g) + O_2(g) \rightleftharpoons 2\,NO(g) \qquad K_c = \frac{[NO]^2}{[N_2][O_2]}$$

Experiment gives $K = K_c = 3.4 \times 10^{-21}$ at 800 K. The very small value of K_c tells us that the reactants N_2 and O_2 are the dominant species in the system at equilibrium.

We can summarize these remarks as follows (Fig. 9.7):

- Large values of K (larger than about 10^3): the equilibrium favors the products.
- Intermediate values of K (approximately in the range 10^{-3} to 10^3): neither reactants nor products are strongly favored at equilibrium.
- Small values of K (smaller than about 1023): the equilibrium favors the reactants.

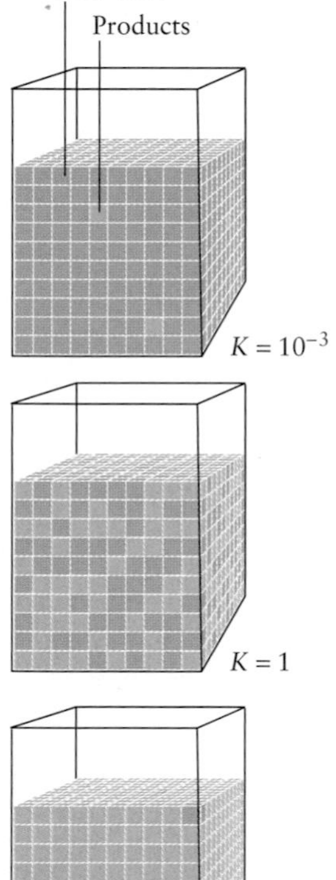

FIGURE 9.7 The size of the equilibrium constant indicates whether the reactants or the products are favored. In this diagram, the reactants are represented by blue cubes and the products by yellow cubes. Note that reactants are favored when K is small (top), products are favored when K is large (bottom), and reactants and products are in equal abundance when $K = 1$.

The following example shows how to use the equilibrium constant to calculate an equilibrium concentration.

If K is large, products are favored at equilibrium; if K is small, reactants are favored.

EXAMPLE 9.4 Calculating an equilibrium concentration

Suppose that in an equilibrium mixture of HCl, Cl_2, and H_2, the concentration of H_2 is 1.0×10^{-11} mol·L^{-1} and that of Cl_2 is 2.0×10^{-10} mol·L^{-1}. What is the equilibrium molar concentration of HCl at 500 K, given $K_c = 4.0 \times 10^{18}$ for $H_2(g) + Cl_2(g) \rightleftharpoons 2\,HCl(g)$?

STRATEGY At equilibrium, the molar concentrations of the reactants and products satisfy the expression for K_c. Therefore, rearrange the expression to give the one unknown concentration, and substitute the data.

SOLUTION The expression for K_c is given in the text; to find [HCl], we rearrange the expression into

$$[HCl] = \{K_c[H_2][Cl_2]\}^{1/2}$$

and substitute the data:

$$\begin{aligned}[HCl] &= \{(4.0 \times 10^{18}) \times (1.0 \times 10^{-11}) \times (2.0 \times 10^{-10})\}^{1/2} \\ &= 0.089\end{aligned}$$

This result means that, at equilibrium, the amount of product in the system is overwhelming relative to the small amounts of reactants.

Self-Test 9.10A Suppose that the equilibrium molar concentrations of H_2 and Cl_2 are both 1.0×10^{-12} mol·L^{-1}. What is the equilibrium molar concentration of HCl at 500 K, given $K_c = 4.0 \times 10^{18}$?

[*Answer:* [HCl] = 2.0×10^{-3} mol·L^{-1}, about 10^9 times greater than the concentrations of H_2 and Cl_2]

Self-Test 9.10B Suppose that the equilibrium molar concentrations of N_2 and O_2 in the reaction $N_2(g) + O_2(g) \rightleftharpoons 2\ NO(g)$ at 800 K are 0.100 mol·L^{-1}. What is the equilibrium molar concentration of NO if $K = 3.4 \times 10^{-21}$ at 800 K?

9.7 The Direction of Reaction

The question we now consider is how to tell whether a reaction mixture with an arbitrary concentration will have a tendency to form more products or to decompose into reactants. To answer this question, we first determine the equilibrium constant experimentally or calculate it from standard free energy data. Then we calculate the reaction quotient, Q, from the actual composition of the reaction mixture, as described in Section 9.3. To predict whether a particular mixture of reactants and products will tend to produce more products or more reactants, we compare Q with K:

- If $Q > K$, the concentrations or partial pressures of the products are too high (or the concentrations or partial pressures of the reactants too low) for equilibrium. Hence, the reaction has a tendency to proceed in the reverse direction, toward reactants.
- If $Q < K$, the reaction tends to go forward and form products.
- If $Q = K$, the mixture has its equilibrium composition and has no tendency to change in either direction.

This pattern is summarized in Fig. 9.8.

The same pattern is seen when we plot the free energy of a reaction mixture against its changing composition, as we did in Fig. 9.5. Reaction tends to proceed toward the composition at the lowest point of the curve, because that is the direction of decreasing free energy. The composition at the lowest point of the curve—the point of minimum free energy—corresponds to equilibrium. For a system at equilibrium, any change—either the forward or the reverse reaction—would lead to an increase in free energy, so neither change is spontaneous. When the free-energy minimum lies very close to pure products, the equilibrium composition strongly favors products and "goes to completion" (Fig. 9.9a). When the free-energy minimum lies very close to the pure reactants, the equilibrium composition strongly favors the reactants and the reaction "does not go" (Fig. 9.9b).

A reaction has a tendency to form products if $Q < K$ and to form reactants if $Q > K$.

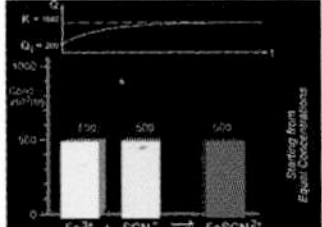

Explore the approach to equilibrium in reaction mixtures having different values of Q.

Media Link **Animation:** Fig. 9.8, Equilibrium simulation

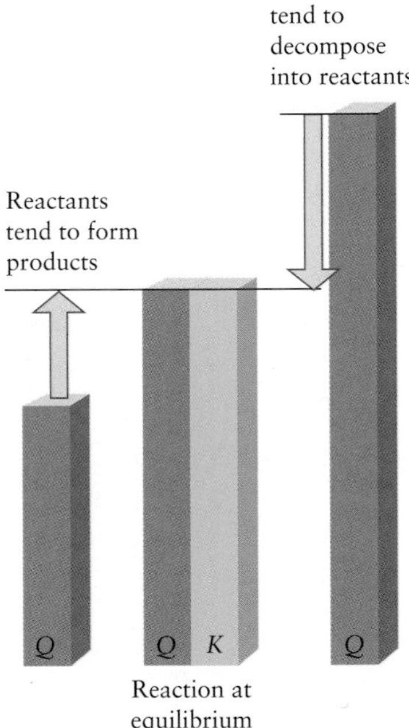

FIGURE 9.8 The relative sizes of the reaction quotient Q and the equilibrium constant K indicate the direction in which a reaction mixture tends to change. The arrows show that when $Q < K$, reactants form products (left) and when $Q > K$, products form reactants (right). There is no tendency to change once the reaction quotient has become equal to the equilibrium constant.

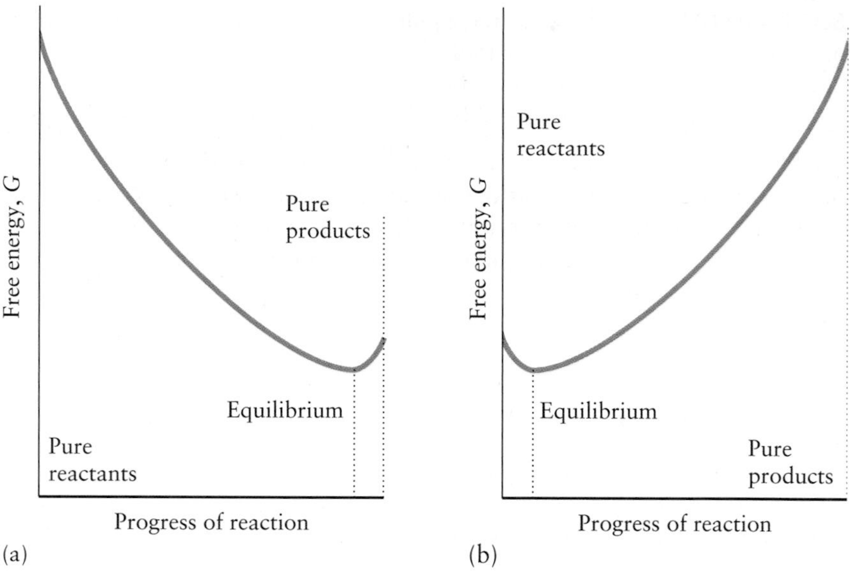

FIGURE 9.9 (a) A reaction that has the potential to go to completion ($K \gg 1$) is one in which the minimum of the free energy curve lies close to pure products. (b) A reaction that has little tendency to form products ($K \ll 1$) is one in which the minimum of the free energy curve lies close to pure reactants.

EXAMPLE 9.5 Predicting the direction of reaction

A mixture of hydrogen, iodine, and hydrogen iodide, each at a concentration of 0.0020 $mol \cdot L^{-1}$, was introduced into a container heated to 783 K. At this temperature, $K = 46$ for $H_2(g) + I_2(g) \rightleftharpoons 2\ HI(g)$. Predict whether or not more HI has a tendency to form.

STRATEGY We need to calculate Q (or Q_c) and compare it with K (or K_c). If $Q > K$, the products need to decompose until their concentrations match K. The opposite is true if $Q < K$: in that case, more products need to form. For this reaction, because $\Delta n = 0$, $Q = Q_c$.

SOLUTION The reaction quotient is

$$Q = Q_c = \frac{[HI]^2}{[H_2][I_2]} = \frac{(0.0020)^2}{(0.0020) \times (0.0020)} = 1.0$$

Because $Q < K$, we conclude that the reaction will tend to form more product and consume reactants.

Self-Test 9.11A A mixture of H_2, N_2, and NH_3 with molar concentrations 3.0×10^{-3}, 1.0×10^{-3}, and 2.0×10^{-3} $mol \cdot L^{-1}$, respectively, was prepared and heated to 500 K, at which temperature $K_c = 62$ for reaction C. Decide whether ammonia tends to form or decompose.

[*Answer:* $Q_c = 1.5 \times 10^5$; tends to decompose]

Self-Test 9.11B For the reaction $N_2O_4(g) \rightleftharpoons 2\ NO_2(g)$ at 298 K, $K = 0.98$. A mixture of N_2O_4 and NO_2 with initial partial pressures of 2.4 and 1.2 bar, respectively, was prepared at 298 K. Which compound will tend to increase its partial pressure?

9.8 Using Equilibrium Constants

The equilibrium constant of a reaction contains information about the equilibrium composition at the given temperature. However, because K is expressed as a ratio of powers of activities (partial pressures or concentrations), we need special techniques to extract the information it contains.

In many cases, we know the initial composition of the reaction mixture and are given apparently incomplete information about the equilibrium composition. In fact, the missing information can usually be inferred by using the reaction stoichiometry. The easiest way to proceed is to draw up an **equilibrium table,** a table showing the initial composition, the changes needed to reach equilibrium in terms of some unknown quantity x, and the final equilibrium composition. The procedure is summarized in Toolbox 9.1 and illustrated in the examples that follow.

To calculate the equilibrium composition of a reaction mixture, set up an equilibrium table in terms of changes in the concentrations of reactants and products, express the equilibrium constant in terms of those changes, and solve the resulting equation.

TOOLBOX 9.1 How to set up and use an equilibrium table

Conceptual Basis

The composition of a reaction mixture tends to adjust until the molar concentrations or partial pressures of gases ensure that $Q_c = K_c$ or $Q = K$. The adjustment of any one composition is linked to changes in the others by the reaction stoichiometry.

Procedure

Begin by writing the balanced chemical equation for the equilibrium and the corresponding expression for the equilibrium constant. Then construct an equilibrium table by working through the following steps.

Step 1 Set up a table as shown below, with columns labeled by the species taking part in the reaction. In the first row, show the initial molar concentration or partial pressure of each species.

	Species 1	Species 2	Species 3 . . .
Initial concentration			
Change in concentration			
Final concentration			

The first step shows how the reaction system is prepared. As always, strike out the units from molar concentrations (in moles per liter) and partial pressures (in bars). Omit pure solids and liquids.

Step 2 In the second row, write the changes in the molar concentrations or partial pressures that are needed for the reaction to reach equilibrium.

It is often the case that we do not know the changes, so write one of them as x or a multiple of x and then use the stoichiometric coefficients in the balanced equation to express the other changes in terms of that x.

Step 3 In the third row, write the equilibrium molar concentrations or partial pressures by adding the change in molar concentration or partial pressure (from step 2) to the initial value for each substance (from step 1).

Although a change in concentration may be positive (an increase) or negative (a decrease), the value of the concentration itself must always be positive.

Step 4 Use the equilibrium constant to determine the value of x, the unknown molar concentration or partial pressure at equilibrium.

You should adopt the good habit of checking the answer by substituting the concentrations or pressures into the expression for K.

An approximation technique can greatly simplify calculations when the change in molar concentration (x) is less than about 5% of the initial concentrations.* To use it, assume that x is negligible when added to or subtracted from a number. Thus, we can replace all expressions like $A - x$ or $A - 2x$, for example, by A. When x occurs on its own (not added to or subtracted from another number), it is left unchanged. So, an expression like $(0.1 - 2x)^2x$ simplifies to $(0.1)^2x$ provided $2x \ll 0.1$ (and specifically, if $x < 0.005$). At the end of the calculation, it is important to verify that the calculated value of x is indeed smaller than about 5% of the initial values. If it is, then the approximation is valid. If it is not, then we must solve the equation without making an approximation. The approximation procedure is illustrated in Example 9.6.

In some cases, the full equation for x is a quadratic equation of the form

$$ax^2 + bx + c = 0$$

When the approximation procedure is invalid, we use the exact solutions of this equation, which are

$$x = \frac{-b \pm \sqrt{b^2 - 4ac}}{2a}$$

We have to decide which of the two solutions given by this expression is valid (the one with the + sign or the one with the − sign in front of the square root) by seeing which solution is chemically possible. See Example 9.7 for an illustration of this procedure.

In other cases, the equation for x in terms of K may be quite complicated. If an approximation is not valid, one approach to solving the equation is to use a graphing calculator, a scientific calculator or mathematical software.

*Why 5%? Because we are ignoring activity coefficients, the equilibrium constants are only approximate, and we are judging 5% to be the typical level of inaccuracy.

EXAMPLE 9.6 Calculating the equilibrium composition by approximation

Under certain conditions, nitrogen and oxygen react to form dinitrogen oxide, N_2O. Suppose that a mixture of 0.482 mol N_2 and 0.933 mol O_2 is placed in a reaction vessel of volume 10.0 L and allowed to form N_2O at a temperature for which $K_c = 2.0 \times 10^{-13}$. What is the composition of the equilibrium mixture?

STRATEGY Proceed as set out in Toolbox 9.1. Use the reaction stoichiometry to express the molar concentration of the product in terms of x. Because K_c is so small, we can expect that very little N_2O is produced, an expectation suggesting that terms like $A - x$ can be approximated by A itself.

SOLUTION The equilibrium is

$$2\,N_2(g) + O_2(g) \rightleftharpoons 2\,N_2O(g) \qquad K_c = \frac{[N_2O]^2}{[N_2]^2[O_2]}$$

The initial molar concentrations are

$$\text{Molar concentration of } N_2 = \frac{0.482\text{ mol}}{10.0\text{ L}} = 0.0482\text{ mol·L}^{-1}$$

$$\text{Molar concentration of } O_2 = \frac{0.933\text{ mol}}{10.0\text{ L}} = 0.0933\text{ mol·L}^{-1}$$

Because there is no N_2O present initially, the equilibrium table, with all molar concentrations in moles per liter, is

	Species		
	N_2	O_2	N_2O
Step 1 Initial molar concentration	0.0482	0.0933	0
Step 2 Change in molar concentration	$-2x$	$-x$	$+2x$
Step 3 Equilibrium molar concentration	$0.0482 - 2x$	$0.0933 - x$	$2x$

The stoichiometry of the reaction implies that if the molar concentration of O_2 decreases by x, then the molar concentration of N_2 decreases by $2x$ and that of N_2O increases by $2x$. The equilibrium values are the sums of the initial concentrations, step 1, and the changes in concentration brought about by reaction, step 2.

Step 4 Substitution of the values in the last line of the table into the expression for K_c gives

$$K_c = \frac{[N_2O]^2}{[N_2]^2[O_2]} = \frac{(2x)^2}{(0.0482 - 2x)^2(0.0933 - x)}$$

When rearranged, this equation is a cubic equation (an equation in x^3). It is difficult to solve cubic equations exactly, but a graphing calculator or mathematical software can be used. If these calculation aids are not available, we can use an approximation procedure to estimate the answer. Because K_c is very small, we suppose that x will turn out to be so small that we can replace $0.0482 - 2x$ by 0.0482 and $0.0933 - x$ by 0.0933. These approximations simplify the equation above to

$$K_c \approx \frac{(2x)^2}{(0.0482)^2(0.0933)}$$

This equation is easily solved by rearranging it to

$$x \approx \left\{\frac{(0.0482)^2 \times 0.0933 \times K_c}{4}\right\}^{1/2}$$

Because $K_c = 2.0 \times 10^{-13}$, we obtain

$$x \approx \left\{\frac{(0.0482)^2 \times 0.0933 \times (2.0 \times 10^{-13})}{4}\right\}^{1/2} = 3.3 \times 10^{-9}$$

The value of x is very small compared with 0.0482 (far smaller than 5% of it), so the approximation is valid. We conclude that, at equilibrium,

$$[N_2] = 0.0482 - 2x \text{ mol·L}^{-1} \approx 0.0482 \text{ mol·L}^{-1}$$

$$[O_2] = 0.0933 - x \text{ mol·L}^{-1} \approx 0.0933 \text{ mol·L}^{-1}$$

$$[N_2O] = 2x \text{ mol·L}^{-1} \approx 6.6 \times 10^{-9} \text{ mol·L}^{-1}$$

When substituted into the full expression for the equilibrium constant, these values give $K_c = 2.0 \times 10^{-13}$, in excellent agreement with the experimental value.

Self-Test 9.12A The initial concentrations of nitrogen and hydrogen are 0.010 and 0.020 mol·L^{-1}, respectively. The mixture is heated to a temperature at which $K_c = 0.11$ for $N_2(g) + 3\ H_2(g) \rightleftharpoons 2\ NH_3(g)$. What is the equilibrium composition of the mixture?

[*Answer:* 0.010 mol·L^{-1} N_2; 0.020 mol·L^{-1} H_2; 9.4×10^{-5} mol·L^{-1} NH_3]

Self-Test 9.12B Hydrogen chloride gas is added to a reaction vessel containing solid iodine until its concentration reaches 0.012 mol·L^{-1}. At the temperature of the experiment, $K_c = 3.5 \times 10^{-32}$ for $2\ HCl(g) + I_2(s) \rightleftharpoons 2\ HI(g) + Cl_2(g)$. Assume that some I_2 remains at equilibrium. What is the equilibrium composition of the mixture?

EXAMPLE 9.7 Calculating the equilibrium composition by using the quadratic equation

Suppose that we place 31.2 g of PCl_5 in a 500-mL reaction vessel and allow the sample to reach equilibrium with its decomposition products phosphorus trichloride and chlorine at 250°C, when $K = 78.3$. All three substances are gases at 250°C. (a) What is the composition of the equilibrium mixture in moles per liter? (b) What is the percentage decomposition of the PCl_5?

STRATEGY (a) Follow the general procedure set out in Toolbox 9.1. Find the amount (in moles) of PCl_5 added from its molar mass. Calculate the molar concentrations by dividing the number of moles of each substance by the volume of the container. To convert K to K_c, determine Δn from the chemical equation and use Eq. 15. (b) The amount of PCl_5 that has reacted is determined from x, the change in concentration. The amount reacted divided by the original amount times 100% gives the percentage decomposition.

SOLUTION (a) The chemical equation is

$$PCl_5(g) \rightleftharpoons PCl_3(g) + Cl_2(g) \qquad K_c = \frac{[PCl_3][Cl_2]}{[PCl_5]} \qquad \text{(I)}$$

The amount of PCl_5 added is

$$\frac{31.2 \text{ g}}{208.24 \text{ g·mol}^{-1}} = 0.150 \text{ mol}$$

The initial molar concentration of PCl_5 is therefore

$$[PCl_5] = \frac{0.150\ mol}{0.500\ L} = 0.300\ mol{\cdot}L^{-1}$$

so we use $[PCl_5] = 0.300$ in the expression for K_c. We suppose that the change in its molar concentration is $-x\ mol{\cdot}L^{-1}$. Using chemical equation I, we construct the following table, with all concentrations in moles per liter:

	Species		
	PCl_5	PCl_3	Cl_2
Step 1 Initial molar concentration	0.300	0	0
Step 2 Change in molar concentration	$-x$	$+x$	$+x$
Step 3 Equilibrium molar concentration	$0.300 - x$	x	x

Step 4 Substitution of these equilibrium values into the expression for the equilibrium constant gives

$$K_c = \frac{[PCl_3][Cl_2]}{[PCl_5]} = \frac{x \times x}{0.300 - x} = \frac{x^2}{0.300 - x}$$

Now convert K to K_c. The value of Δn from the balanced equation is $+1$, so by the technique illustrated in Example 9.2 ($K_c = (RTc°/P°)^{-\Delta n}K$), we find:

$$K_c = \left(\frac{RTc°}{P°}\right)^{-\Delta n} K$$

$$= \left(\frac{(8.3145 \times 10^{-2}\ L{\cdot}bar{\cdot}mol^{-1}) \times (523\ K) \times (1\ mol{\cdot}L^{-1})}{1\ bar}\right)^{-1} \times 78.3$$

$$= 1.80$$

It follows that we must solve

$$\frac{x^2}{0.300 - x} = 1.80$$

or

$$x^2 + 1.80x - 0.540 = 0$$

The solutions follow from the quadratic formula (see Toolbox 9.1):

$$x = \frac{-1.80 \pm \sqrt{(1.80)^2 - 4(1)(-0.540)}}{2} = 0.262 \text{ or } -2.06$$

Because the concentrations must be positive and because (from step 3) x is the molar concentration of each of the products, we select 0.262 as the solution. It follows that at equilibrium

$$[PCl_5] = 0.300 - x \text{ mol·L}^{-1} = 0.300 - 0.262 \text{ mol·L}^{-1} = 0.038 \text{ mol·L}^{-1}$$

$$[PCl_3] = x \text{ mol·L}^{-1} = 0.262 \text{ mol·L}^{-1}$$

$$[Cl_2] = x \text{ mol·L}^{-1} = 0.262 \text{ mol·L}^{-1}$$

That is, the equilibrium concentrations of PCl_5, PCl_3, and Cl_2 are 0.038, 0.262, and 0.262 mol·L^{-1}, respectively. To verify that these concentrations are consistent with the equilibrium constant, we use them to calculate its value:

$$K_c = \frac{[PCl_3][Cl_2]}{[PCl_5]} = \frac{0.262 \times 0.262}{0.038} = 1.8$$

which is the same as the value used. (b) The amount of PCl_5 that has decomposed is equal to x times the volume: 0.262 mol·L^{-1} × 0.500 L = 0.131 mol. The amount initially present was 0.150 mol, so the percentage decomposition is

$$\frac{0.131 \text{ mol}}{0.150 \text{ mol}} \times 100\% = 87.3\%$$

87.3% of the PCl_5 has decomposed at equilibrium.

Self-Test 9.13A Bromine monochloride, BrCl, decomposes into bromine and chlorine and reaches the equilibrium $2\ BrCl(g) \rightleftharpoons Br_2(g) + Cl_2(g)$, for which $K_c = 32$ at 500 K. If initially pure BrCl is present at a concentration of 3.30×10^{-3} mol·L^{-1}, what is its molar concentration in the mixture at equilibrium?

[*Answer:* 3×10^{-4} mol·L^{-1}]

Self-Test 9.13B Chlorine and fluorine react at 2500 K to produce ClF and reach the equilibrium $Cl_2(g) + F_2(g) \rightleftharpoons 2\ ClF(g)$ with $K_c = 20$. If 0.200 mol Cl_2, 0.100 mol F_2, and 0.100 mol ClF are added to a 1.00-L container and allowed to come to equilibrium at 2500 K, what is the molar concentration of ClF in the equilibrium mixture?

THE RESPONSE OF EQUILIBRIA TO CHANGES IN CONDITIONS

Because chemical equilibria are dynamic, they respond to changes in the conditions under which the reaction is taking place. This responsiveness gives us a degree of control over the equilibrium composition. When we add or remove a reactant, the value of ΔG_r is changed and the composition shifts until $\Delta G_r = 0$ again. The composition also might change in response to a change in pressure or temperature. In the following sections, we shall see how Haber used this responsiveness to manipulate the equilibrium between ammonia, nitrogen, and hydrogen to increase the production of ammonia.

It is quite easy to predict how the composition of a reaction mixture at equilibrium tends to change when the conditions are changed. The French chemist Henri Le Chatelier (Fig. 9.10) identified the general principle:

> **Le Chatelier's principle:** When a stress is applied to a system in dynamic equilibrium, the equilibrium tends to adjust to minimize the effect of the stress.

This empirical principle is no more than a rule of thumb. It provides neither an explanation nor a quantitative prediction. However, we shall see that it is consistent with thermodynamics; and as we develop it, we shall see the powerful *quantitative* conclusions that can be drawn from thermodynamics.

FIGURE 9.10 Henri Le Chatelier (1850–1936).

9.9 Adding and Removing Reagents

We start by considering changes in the composition of a system. Let's suppose that the ammonia synthesis reaction has reached the equilibrium in reaction C. Now suppose we pump in more hydrogen gas. According to Le Chatelier's principle, the reaction will tend to minimize the increase in the number of hydrogen molecules. Hydrogen will tend to react with nitrogen. As a result, additional ammonia will be formed. If, instead of hydrogen, we were to add some ammonia, then the reaction would tend instead to form reactants at the expense of the added ammonia (Fig. 9.11).

We can explain these responses thermodynamically by considering the relative sizes of Q and K (Fig. 9.12). When reactants are added, the reaction quotient Q falls below K, because the reactant concentrations appear in the denominator of Q. As we have seen, when $Q < K$, the reaction mixture responds by forming products and Q is restored to K. Likewise, when products are added, Q rises above K, because products appear in the numerator. Then, because $Q > K$, the reaction mixture responds by forming reactants at the expense of products until $Q = K$ again. It is important to understand that K is a constant that is not altered by changing concentrations. Only the value of Q changes.

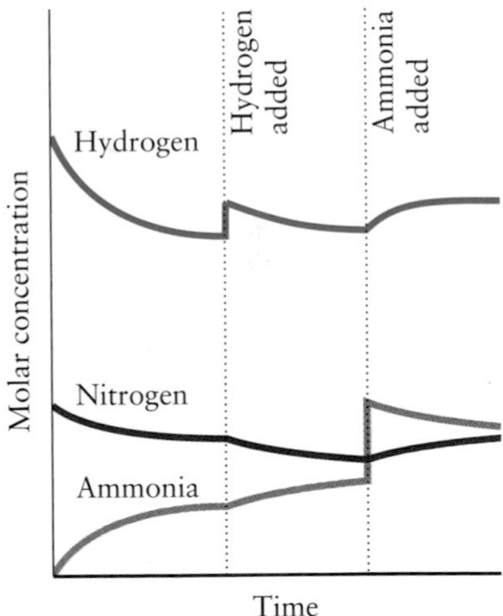

FIGURE 9.11 These graphs show the kinds of changes in composition that can be expected when excess hydrogen and then ammonia are added to an equilibrium mixture of nitrogen, hydrogen, and ammonia. Note that the addition of hydrogen results in the formation of ammonia, whereas the addition of ammonia results in the decomposition of some of the added ammonia. In each case, the mixture settles into a composition in accord with the equilibrium constant of the reaction.

EXAMPLE 9.8 Predicting the effect of adding or removing reactants and products

Consider the equilibrium

$$4\ NH_3\ (g) + 3\ O_2\ (g) \rightleftharpoons 2\ N_2\ (g) + 6\ H_2O\ (g)$$

Predict the effect on each equilibrium concentration of (a) the addition of N_2; (b) the removal of NH_3; (c) the removal of H_2O.

STRATEGY Le Chatelier's principle tells us that an equilibrium mixture tends to minimize the effect of a change. Therefore, when we add a substance, the reaction tends to reduce the concentration of that substance. When we remove a substance, the reaction tends to replace it.

Explore the effect on an equilibrium of changing the concentrations of reactants and products.

SOLUTION (a) The addition of N_2 to the equilibrium mixture causes the reaction to shift in the direction that minimizes the increase in N_2. Therefore, the reaction shifts toward reactant formation. Forming more reactants increases the amounts of NH_3 and O_2 while decreasing the amount of H_2O. The concentration of N_2 remains slightly higher than its original equilibrium value but lower than its concentration immediately after the additional N_2 was supplied. (b) When NH_3 is removed from the system at equilibrium, the reaction shifts to minimize its loss. The reaction tends to increase the amount of O_2 and decrease the amounts of N_2 and H_2O. The concentration of NH_3 will be somewhat lower than its original equilibrium value but not as low as it was immediately after the removal of NH_3. (c) The removal of H_2O causes the equilibrium to shift in favor of products to restore (partially) the amount removed. This shift increases the amount of N_2 while decreasing the amounts of NH_3 and O_2. The concentration of H_2O is somewhat lower than its original value.

Self-Test 9.14A Consider the equilibrium $SO_3(g) + NO(g) \rightleftharpoons SO_2(g) + NO_2(g)$. Predict the effect on the equilibrium of (a) the addition of NO; (b) the removal of SO_2; (c) the addition of NO_2.

[*Answer:* The equilibrium tends to shift toward (a) products; (b) products; (c) reactants.]

Self-Test 9.14B Consider the equilibrium $CO(g) + 2\,H_2(g) \rightleftharpoons CH_3OH(g)$. Predict the effect on the equilibrium of (a) the addition of H_2; (b) the removal of CH_3OH; (c) the removal of CO.

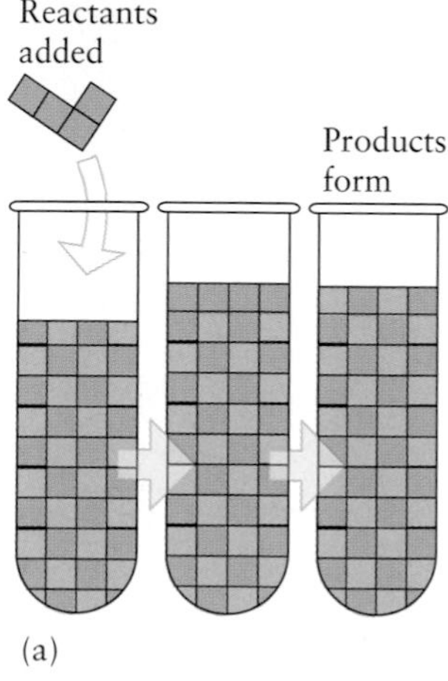

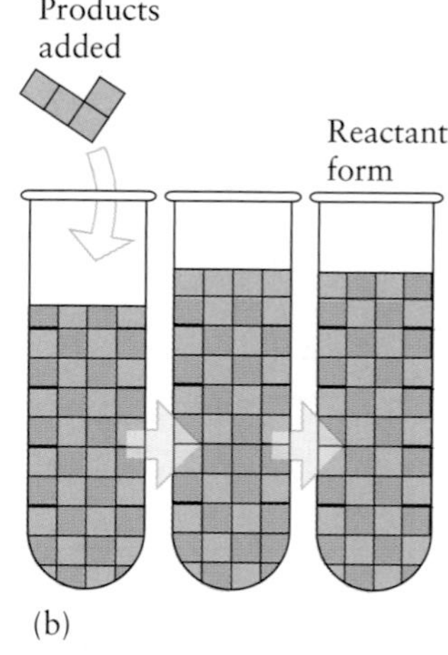

FIGURE 9.12 (a) When a reactant (blue) is added to a reaction mixture at equilibrium, $Q < K$ and products have a tendency to form. (b) When a product (yellow) is added instead, $Q > K$ and reactants tend to be formed. For this reaction, we have used $K = 1$ for an equilibrium of the form blue $\rightleftharpoons$ yellow.

Media Link Animation: Fig. 9.12, Equilibrium simulation

Le Chatelier's principle suggests a good way of ensuring that a reaction goes on generating a substance: simply remove products as they are formed. In its continuing hunt for equilibrium, the reaction generates more product. Industrial processes rarely reach equilibrium, for just this reason. In the commercial synthesis of ammonia, for instance, ammonia can be continuously removed by passing the equilibrium mixture through a refrigeration unit, in which only the ammonia condenses. The hydrogen and nitrogen go on to react to form more product. In aqueous systems, a substance is often removed by chemical reaction. Reactions in living systems also tend to remain far from equilibrium, thanks to the ceaseless exchange of products and reactants with the surroundings.

When a reactant is added to a reaction mixture at equilibrium, the reaction tends to form products. When a reactant is removed, more reactant tends to form. When a product is added, the reaction tends to form reactants. When a product is removed, more product is formed.

EXAMPLE 9.9 Calculating the equilibrium composition after addition of a reagent

Suppose that the equilibrium mixture in Example 9.7 is perturbed by adding 0.100 mol $Cl_2(g)$ to the 0.500-L container; then the system is once again allowed to reach equilibrium. Use this information and data from Example 9.7 to calculate the new composition of the equilibrium mixture.

STRATEGY The general procedure is like that set out in Toolbox 9.1. Write the chemical equation and the expression for its equilibrium constant, then set up an equilibrium table, using the concentrations immediately after addition of the reagent (before any reaction has taken place) as the initial concentrations. The initial concentration of the substance added is the sum of the original and added amounts divided by the volume. Because a product has been added, we expect the reaction to respond by producing more reactants. Obtain the initial equilibrium concentrations, the value of K_c, and the balanced equation from Example 9.7.

SOLUTION The chemical equation is reaction I in Example 9.7. The initial equilibrium molar concentrations of PCl_5 and PCl_3 from Example 9.7 are $[PCl_5] = 0.038\ mol{\cdot}L^{-1}$ and $[PCl_3] = 0.262\ mol{\cdot}L^{-1}$ and the volume of the container is given as 0.500 L. The amount of $Cl_2(g)$ in the container before the addition is $(0.262\ mol{\cdot}L^{-1}) \times 0.500\ L = 0.131$ mol. After the addition of 0.100 mol $Cl_2(g)$ and before any reaction begins, the molar concentration of $Cl_2(g)$ is

$$[Cl_2] = \frac{(0.100 + 0.131)\ mol}{0.500\ L} = 0.462\ mol{\cdot}L^{-1}$$

We suppose that the change in molar concentration of both Cl_2 and PCl_3 is $-x\ mol{\cdot}L^{-1}$, because the concentrations of products are expected to decrease. Using the chemical equation above, we draw up the following table, with all concentrations in moles per liter:

	Species		
	PCl_5	PCl_3	Cl_2
Step 1 Initial molar concentration	0.038	0.262	0.462
Step 2 Change in molar concentration	$+x$	$-x$	$-x$
Step 3 Equilibrium molar concentration	$0.038 + x$	$0.262 - x$	$0.462 - x$

Step 4 Substitution of these equilibrium values into the expression for the equilibrium constant gives

$$K_c = \frac{[PCl_3][Cl_2]}{[PCl_5]} = \frac{(0.262 - x) \times (0.462 - x)}{0.038 + x}$$
$$= \frac{0.121 - 0.724x + x^2}{0.038 + x}$$

From Example 9.7, we know that $K_c = 1.80$. The expression then rearranges to the quadratic equation

$$x^2 - 2.52x + 0.053 = 0$$

The roots of this equation are 2.5 and 0.021. Because $0.262 - x$ mol·L^{-1} is the molar concentration of PCl_3 at equilibrium, and all concentrations are positive, we select $x = 0.021$ as the acceptable solution. It follows that, at equilibrium,

$$[PCl_5] = 0.038 + x \text{ mol·L}^{-1} = 0.038 + 0.021 \text{ mol·L}^{-1} = 0.059 \text{ mol·L}^{-1}$$

$$[PCl_3] = 0.262 - x \text{ mol·L}^{-1} = 0.262 - 0.021 \text{ mol·L}^{-1} = 0.241 \text{ mol·L}^{-1}$$

$$[Cl_2] = 0.4602 - x \text{ mol·L}^{-1} = 0.462 - 0.021 \text{ mol·L}^{-1} = 0.441 \text{ mol·L}^{-1}$$

(*Check:* These concentrations give $K_c = 1.8$, in good agreement with the value of K_c we used.) Notice that the concentration of reactant has increased and the concentrations of the products have decreased from their initial values. Although the concentration of Cl_2 is less than it was after the addition of more Cl_2 (0.462 mol·L^{-1}) it is greater than its original equilibrium concentration (0.262 mol·L^{-1}).

Self-Test 9.15A Suppose that the equilibrium mixture in Example 9.6 is perturbed by adding 3.00 mol N_2(g) and then the system is once again allowed to reach equilibrium. Use this information and data from Example 9.6 to calculate the new composition of the equilibrium mixture.

[*Answer:* 0.3482 mol·L^{-1} N_2; 0.0933 mol·L^{-1} O_2; 4.8×10^{-8} mol·L^{-1} N_2O]

Self-Test 9.15B Nitrogen, hydrogen, and ammonia gases have reached equilibrium in a 1.00-L container at 298 K. Equilibrium partial pressures are 0.080 atm N_2, 0.050 atm H_2, and 2.60 atm NH_3. For $N_2(g) + 3\,H_2(g) \rightleftharpoons 2\,NH_3(g)$, $K = 6.8 \times 10^5$. Calculate the new equilibrium partial pressures if half the NH_3 is removed from the container and equilibrium is reestablished.

9.10 Compressing a Reaction Mixture

A gas-phase equilibrium responds to compression—a reduction in volume—of the reaction vessel. According to Le Chatelier's principle, the composition will tend to change in a way that minimizes the resulting increase in pressure. For instance, in the dissociation of gaseous I_2 to form I atoms, $I_2(g) \rightleftharpoons 2\,I(g)$, 1 mol of reactant molecules produces 2 mol of product molecules. However, the reverse reaction decreases the number of molecules in the container and hence the pressure. It follows that, when the mixture is compressed, the equilibrium composition will tend to shift in favor of the reactant, I_2, because that response minimizes the increase in pressure (Fig. 9.13). Expansion results in the opposite response, a tendency for the I_2 to dissociate to free atoms. In the formation of ammonia, two moles of gas molecules are produced from four moles of gas molecules. Therefore, Haber realized that, to increase the yield of ammonia, he needed to carry out the synthesis with highly compressed gases. The actual industrial process uses pressures of 250 atm and more (Fig. 9.14).

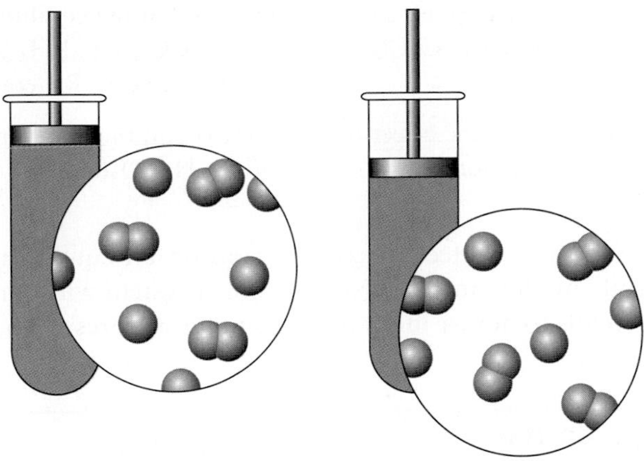

FIGURE 9.13 Le Chatelier's principle predicts that when a reaction at equilibrium is compressed, the number of molecules in the gas phase will tend to decrease. This diagram illustrates the effect of compression on the dissociation equilibrium of a diatomic molecule. Note the increase in the relative concentration of diatomic molecules as the system is compressed.

EXAMPLE 9.10 Predicting the effect of compression on an equilibrium

Predict the effect of compression on the equilibrium composition of the reaction mixtures in which the equilibria (a) $2\ NO_2(g) \rightleftharpoons N_2O_4(g)$ and (b) $H_2(g) + I_2(g) \rightleftharpoons 2\ HI(g)$ have been established.

STRATEGY Reaction will occur in the direction that reduces the increase in pressure. A glance at the chemical equation shows which direction corresponds to a decrease in the number of gas-phase molecules.

SOLUTION (a) In the forward reaction, two NO_2 molecules combine to form one N_2O_4 molecule. Hence, compression favors the formation of N_2O_4. (b) Because neither direction corresponds to a reduction of gas-phase molecules, compressing the mixture should have little effect on the composition of the equilibrium mixture.

FIGURE 9.14 One of the high-pressure vessels used for the catalytic synthesis of ammonia. The vessel must be able to withstand internal pressures of greater than 250 atm.

Self-Test 9.16A Predict the effect of compression on the equilibrium composition of the reaction $CH_4(g) + H_2O(g) \rightleftharpoons CO(g) + 3\,H_2(g)$.
[*Answer:* Reactants favored]

Self-Test 9.16B Predict the effect of compression on the equilibrium composition of the reaction $CO_2(g) + H_2O(l) \rightleftharpoons H_2CO_3(aq)$.

We can justify the effect of compression on an equilibrium mixture mathematically by showing that compressing a system will change Q, and the reaction tends to adjust in the direction that restores Q to K.

How do we do that?

Suppose we want to discover the effect of compression on the equilibrium in Example 9.10a. We write the equilibrium constant in its complete form as

$$2\,NO_2(g) \rightleftharpoons N_2O_4(g) \qquad K_c = \frac{[N_2O_4]/[N_2O_4]^\circ}{([NO_2]/[NO_2]^\circ)^2} = \frac{[N_2O_4]c^\circ}{[NO_2]^2}$$

Next, we express K_c in terms of the volume of the system. The molar concentration of each gas is the amount, n, of that gas divided by the volume, V, of the reaction vessel:

$$[NO_2] = \frac{n_{NO_2}}{V} \qquad [N_2O_4] = \frac{n_{N_2O_4}}{V}$$

Hence, the expression for the equilibrium constant has the form

$$K_c = \frac{(n_{N_2O_4}/V)c^\circ}{(n_{NO_2}/V)^2} = \frac{n_{N_2O_4}}{n_{NO_2}^{\;2}} \times Vc^\circ$$

Because c° is a constant, for this expression to remain constant when the volume (V) of the system is reduced, the ratio $n_{N_2O_4}/n_{NO_2}^{\;2}$ must increase. That is, the amount of NO_2 must decrease and the amount of N_2O_4 must increase. Therefore, as we have seen before, as the volume of the system is decreased, the equilibrium shifts to a smaller total number of gas molecules.

Suppose that we were to increase the total pressure inside a reaction vessel by pumping in argon or some other inert gas without decreasing the volume. The reacting gases continue to occupy the same volume, so their individual molar concentrations and partial pressures remain unchanged despite the presence of an inert gas. In this case, therefore, provided the gases can be regarded as ideal, the equilibrium composition is unaffected.

Compression of a reaction mixture at equilibrium tends to drive the reaction in the direction that reduces the number of gas-phase molecules; increasing the pressure by introducing an inert gas has no effect on the equilibrium composition.

9.11 Temperature and Equilibrium

We can see from Table 9.2 that the equilibrium constant depends on the temperature. It is found experimentally that, for an exothermic reaction, the formation of products is favored by lowering the temperature. Conversely, for an endothermic reaction, the products are favored by an increase in temperature.

Le Chatelier's principle is consistent with these observations. We can imagine the heat generated in an exothermic reaction as helping to offset lowering the temperature. Similarly, we can imagine the heat absorbed in an endothermic reaction as helping to offset an increase in temperature. In other words, raising the temperature of a reaction mixture at equilibrium will shift the reaction in the endothermic direction.

An example is the decomposition of carbonates. A reaction such as

$$CaCO_3(s) \rightleftharpoons CaO(s) + CO_2(g)$$

is strongly endothermic in the forward direction, and an appreciable partial pressure of carbon dioxide is present at equilibrium only if the temperature is high. For instance, at 800°C, the partial pressure is 0.22 atm at equilibrium. If the heating takes place in an open container, this partial pressure is never reached because equilibrium is never reached. The gas drifts away, and the calcium carbonate decomposes completely, leaving a solid residue of CaO—just as water evaporates completely in an open vessel. However, if the surroundings are already so rich in carbon dioxide that its partial pressure exceeds 0.22 atm, then virtually no decomposition occurs: for every CO_2 molecule that is formed, one is converted back to carbonate. This dynamic process is probably what happens on the hot surface of Venus (Fig. 9.15), where the partial pressure of carbon dioxide is about 87 atm. This high pressure has led to speculation that the planet's surface is rich in carbonates, despite its high temperature (about 500°C).

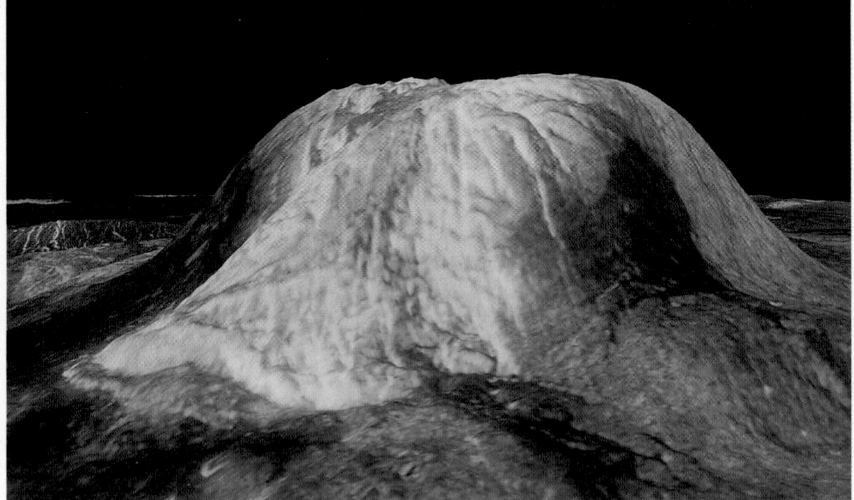

FIGURE 9.15 A radar image of the surface of Venus. Although the rocks are very hot, the partial pressure of carbon dioxide in the atmosphere is so great that carbonates may be abundant.

EXAMPLE 9.11 Predicting the effect of temperature on an equilibrium

One stage in the manufacture of sulfuric acid is the formation of sulfur trioxide by the reaction of SO_2 with O_2 in the presence of a vanadium(V) oxide catalyst. Predict how the equilibrium composition for the sulfur trioxide synthesis will tend to change when the temperature is raised.

STRATEGY Raising the temperature of an equilibrium mixture will tend to shift its composition in the endothermic direction of the reaction. A positive reaction enthalpy indicates that the reaction is endothermic in the forward direction. A negative reaction enthalpy indicates that the reaction is endothermic in the reverse direction. To find the standard reaction enthalpy, use the standard enthalpies of formation given in Appendix 2A.

SOLUTION The chemical equation is

$$2\ SO_2(g) + O_2(g) \rightleftharpoons 2\ SO_3(g)$$

The standard reaction enthalpy of the forward reaction is

$$\begin{aligned}\Delta H_r^\circ &= 2\Delta H_f^\circ(SO_3, g) - 2\Delta H_f^\circ(SO_2, g)\\ &= 2(-395.72\ \text{kJ}\cdot\text{mol}^{-1}) - 2(-296.83\ \text{kJ}\cdot\text{mol}^{-1})\\ &= -197.78\ \text{kJ}\cdot\text{mol}^{-1}\end{aligned}$$

Because the formation of SO_3 is exothermic, the reverse reaction is endothermic. Hence, raising the temperature of the equilibrium mixture favors the decomposition of SO_3 to SO_2 and O_2.

Self-Test 9.17A Predict the effect of raising the temperature on the equilibrium composition of the reaction $N_2O_4(g) \rightleftharpoons 2\ NO_2(g)$. See Appendix 2A for data.

[*Answer:* NO_2 favored]

Self-Test 9.17B Predict the effect of lowering the temperature on the equilibrium composition of the reaction $2\ CO(g) + O_2(g) \rightleftharpoons 2\ CO_2(g)$. See Appendix 2A for data.

The explanation of the effect of temperature on the equilibrium compostion must be found in the fact that the equilibrium constant of a reaction varies with temperature. The relation between the equilibrium constant and the standard reaction free energy in Eq. 10 applies to any temperature. Therefore, we ought to be able to use it to relate the equilibrium constant at one temperature to its value at another temperature. In this way, we obtain a quantitative version of Le Chatelier's principle for the effect of temperature.

How do we do that?

To find the quantitative relation between the equilibrium constants for the same reaction at two temperatures T_1 and T_2, we note that Eq. 10 allows us to write the relation between reaction free energy and the equilibrium constant at the two temperatures:

$$\Delta G_{r,1}^\circ = -RT_1 \ln K_1 \qquad \Delta G_{r,2}^\circ = -RT_2 \ln K_2$$

These two expressions rearrange to

$$\ln K_1 = -\frac{\Delta G_{r,1}°}{RT_1} \qquad \ln K_2 = -\frac{\Delta G_{r,2}°}{RT_2}$$

Subtraction of the second from the first gives

$$\ln K_1 - \ln K_2 = -\frac{1}{R}\left(\frac{\Delta G_{r,1}°}{T_1} - \frac{\Delta G_{r,2}°}{T_2}\right)$$

At this point, we introduce the definition of $\Delta G_r°$ in terms of $\Delta H_r°$ and $\Delta S_r°$:

$$\Delta G_{r,1}° = \Delta H_{r,1}° - T_1\Delta S_{r,1}° \quad \Delta G_{r,2}° = \Delta H_{r,2}° - T_2\Delta S_{r,2}°$$

which gives

$$\begin{aligned}\ln K_1 - \ln K_2 &= -\frac{1}{R}\left(\frac{\Delta H_{r,1}° - T_1\Delta S_{r,1}°}{T_1} - \frac{\Delta H_{r,2}° - T_2\Delta S_{r,2}°}{T_2}\right)\\ &= -\frac{1}{R}\left(\frac{\Delta H_{r,1}°}{T_1} - \frac{\Delta H_{r,2}°}{T_2} - \Delta S_{r,1}° + \Delta S_{r,2}°\right)\end{aligned}$$

It is usually reasonable to assume that $\Delta H_r°$ and $\Delta S_r°$ are both approximately independent of temperature over the range of temperatures of interest. When we make that approximation, the reaction entropies cancel, and we are left with

$$\ln K_1 - \ln K_2 = -\frac{\Delta H_r°}{R}\left(\frac{1}{T_1} - \frac{1}{T_2}\right)$$

We now multiply through by -1, use $\ln a - \ln b = \ln(a/b)$, and obtain the **van 't Hoff equation:**

$$\ln\frac{K_2}{K_1} = \frac{\Delta H_r°}{R}\left(\frac{1}{T_1} - \frac{1}{T_2}\right) \qquad (16)$$

In this expression, K_1 is the equilibrium constant when the temperature is T_1 and K_2 is the equilibrium constant when the temperature is T_2.

This equation is sometimes called the *van 't Hoff isochore*, to distinguish it from van 't Hoff's osmotic pressure equation (Section 8.19). An isochore is the plot of an equation for a constant-volume process.

Let's see how the van 't Hoff equation accounts for Le Chatelier's principle for the effect of temperature on an equilibrium. Suppose that the reaction is endothermic, then $\Delta H_r°$ is positive. If $T_2 > T_1$, then $1/T_2 < 1/T_1$ and the term in parentheses is also positive. Therefore, $\ln(K_2/K_1)$ is positive, which implies that $K_2/K_1 > 1$, and therefore that $K_2 > K_1$. In other words, an increase in temperature favors the formation of product if the reaction is endothermic. We predict the opposite effect for an exothermic reaction because $\Delta H_r°$ is then negative.

One word of warning: when using the van 't Hoff equation for gas-phase reactions, the equilibrium constants must be K, not K_c. If we want a new value for K_c for a gas-phase reaction, we convert from K_c to K, use the van 't Hoff equation to calculate the new value of K, and then convert that K to the new K_c by using Eq. 15.

Raising the temperature of an exothermic reaction favors the formation of reactants; raising the temperature of an endothermic reaction favors the formation of products.

EXAMPLE 9.12 **Predicting the value of an equilibrium constant at a different temperature**

The equilibrium constant K for the synthesis of ammonia (reaction C) is 6.8×10^5 at 298 K. Predict its value at 400 K.

STRATEGY The synthesis of ammonia is exothermic, so we expect the equilibrium constant to be smaller at the higher temperature. To use the van 't Hoff equation, we need the standard reaction enthalpy, which can be calculated from the standard enthalpies of formation in Appendix 2A.

SOLUTION The standard reaction enthalpy for the forward reaction in C is

$$\begin{aligned}\Delta H_r° &= 2\Delta H_f°(NH_3, g) - \Delta H_f°(N_2, g) - 3\Delta H_f°(H_2, g)\\ &= 2(-46.11\ kJ\cdot mol^{-1}) = -92.22\ kJ\cdot mol^{-1}\end{aligned}$$

Therefore, from Eq. 16,

$$\ln\frac{K_2}{K_1} = \left(\frac{-9.222 \times 10^4\ J\cdot mol^{-1}}{8.3145\ J\cdot K^{-1}\cdot mol^{-1}}\right)\left(\frac{1}{298\ K} - \frac{1}{400\ K}\right) = -9.49$$

It follows that

$$K_2 = K_1 e^{-9.49} = (6.8 \times 10^5) \times e^{-9.49} = 51$$

The answer is close to the experimental value of 41 in Table 9.2. It is not exactly the same because $\Delta H_r°$ may change over the temperature range in question.

Self-Test 9.18A The equilibrium constant K for $2\ SO_3(g) \rightleftharpoons 2\ SO_2(g) + O_2(g)$ is 2.5×10^{-25} at 298 K. Predict its value at 500 K.

[*Answer:* 2.5×10^{-11}]

Self-Test 9.18B The equilibrium constant K for $PCl_5(g) \rightleftharpoons PCl_3(g) + Cl_2(g)$ is 78.3 at 523 K. Predict its value at 800 K.

9.12 Catalysts and Haber's Achievement

The role of catalysts is discussed in detail in Section 13.14.

A **catalyst** is a substance that increases the rate of a chemical reaction without being consumed itself. We shall see a lot more of catalysts later, when we discuss reaction rates. However, it is important to be aware at this stage that a catalyst has no effect on the equilibrium composition of a reaction mixture. A catalyst can speed up the rate at which a reaction reaches equilibrium, but it does not affect the composition at equilibrium. It acts by providing a faster route to the same destination.

FIGURE 9.16 Fritz Haber (1868–1934).

FIGURE 9.17 Carl Bosch (1874–1940).

A catalyst speeds up both the forward and reverse reactions by the same amount. Therefore, the dynamic equilibrium is unaffected. The thermodynamic justification of this observation is based on the fact that the equilibrium constant depends only on the temperature and the value of $\Delta G_r°$. A standard reaction free energy depends only on the identities of the reactants and products, and is independent of its rate or the presence of any substances that do not appear in the overall chemical equation for the reaction.

We have seen how Haber's understanding of chemical equilibrium enabled him to increase the *potential* yield of ammonia (Fig. 9.16). He realized that he had to compress the gases and remove the ammonia as it was formed. As we have seen, compression shifts the equilibrium composition in favor of ammonia, which increases the yield of product. Removing the ammonia also encourages more to be formed. In addition, Haber realized that he ought to run the reaction at as low a temperature as possible: low temperature favors the formation of product because the reaction is exothermic. However, nitrogen and hydrogen combine too slowly at low temperatures, so Haber solved this problem by developing an appropriate catalyst. The process Haber devised in collaboration with a chemical engineer, Carl Bosch (Fig. 9.17), is still in use throughout the world. In the United States alone, it accounts for almost the entire annual production of over 1.6×10^{10} kg of ammonia (Fig. 9.18).

A catalyst does not affect the equilibrium composition of a reaction mixture.

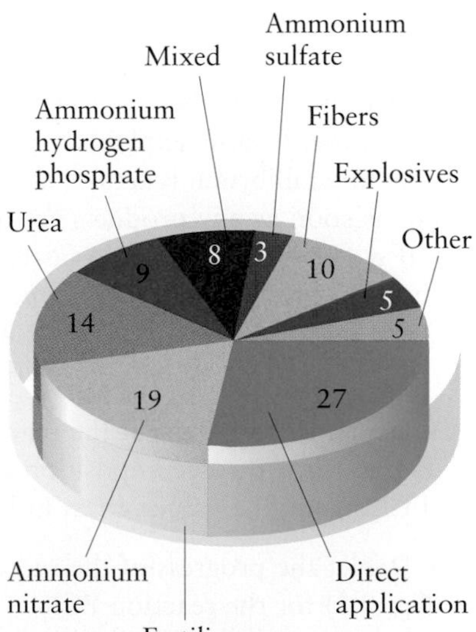

FIGURE 9.18 The Haber process is still used to produce almost all the ammonia manufactured in the world. This pie chart shows how the ammonia is used. The figures are percentages. Note that 80%—as shown by the green band—is used as fertilizer, either directly or after conversion to another compound.

SKILLS YOU SHOULD HAVE MASTERED

1 Distinguish homogeneous and heterogeneous equilibria and write equilibrium constants for both types of reaction from a balanced equation, Self-Tests 9.1, 9.2, and 9.6.

2 Calculate the change in free energy with concentration, Self-Test 9.3.

3 Determine the reaction free energy from the reaction quotient, Example 9.1.

4 Calculate an equilibrium constant from a standard free energy, Self-Test 9.5.

5 Convert between K and K_c, Examples 9.2 and 9.3.

6 Determine the equilibrium constant from equilibrium concentrations, Example 9.3.

7 Calculate the effect on K of reversing a reaction or multiplying the chemical equation by a factor, Self-Test 9.9.

8 Calculate an equilibrium concentration, Example 9.4.

9 Predict the direction of a reaction, given K and the concentrations of reactants and products, Example 9.5.

10 Use an equilibrium table to carry out equilibrium calculations, Toolbox 9.1 and Examples 9.6, 9.7, and 9.9.

11 Use Le Chatelier's principle to predict how the equilibrium composition of a reaction mixture is affected by adding or removing reagents, compressing or expanding the mixture, or changing the temperature, Examples 9.8, 9.10, and 9.11.

12 Predict the value of K at a different temperature, Example 9.12.

EXERCISES

Reactions at Equilibrium

9.1 State whether the following statements are true or false. If false, explain why.
(a) A reaction stops when equilibrium is reached.
(b) An equilibrium reaction is not affected by increasing the concentrations of products.
(c) If one starts with a higher pressure of reactant, the equilibrium constant will be larger.
(d) If one starts with higher concentrations of reactants, the equilibrium concentrations of the products will be larger.

9.2 Determine whether the following statements are true or false. If false, explain why.
(a) In an equilibrium reaction, the reverse reaction will begin as soon as any products are formed.
(b) If we make a reaction go faster, we can increase the amount of product at equilibrium.
(c) The reaction free energy at equilibrium is 0.
(d) The standard reaction free energy at equilibrium is 0.

9.3 Depict the progress of the reaction graphically (as in Fig. 9.2) for the decomposition of PCl_5 to give PCl_3 and Cl_2, using the values given in Example 9.7.

9.4 Depict the progress of the reaction graphically (as in Fig. 9.2) for the reaction $PCl_5(g) \rightarrow PCl_3(g) + Cl_2(g)$, using the conditions described in Example 9.9, starting from the point where additional Cl_2 is added, and continuing until equilibrium is restored.

9.5 A mixture of gases consists of CO(g), $H_2(g)$, and $CH_4(g)$. Calculate the molar free energy change at 298 K (supposing there is no reaction) if (a) the partial pressure of CO(g) is increased from 0.500 bar to 25.00 bar; (b) the partial pressure of $H_2(g)$ is decreased from 10.00 bar to 5.00 bar; (c) the partial pressure of $H_2(g)$ is decreased from 20.00 bar to 10.00 bar; (d) the partial pressure of CO(g) is decreased from 20.00 bar to 10.00 bar.

9.6 Calculate the molar free energy change at 298 K when (a) the partial pressure of NO(g) in a mixture of gases is increased from 5.00 bar to 15.00 bar; (b) the partial pressure of HCN(g) in a mixture of gases is increased from 5.00 bar to 15.00 bar; (c) the partial pressure of $O_2(g)$ in a mixture of gases is decreased from 3.00 bar to 1.50 bar; (d) the partial pressure of $O_2(g)$ in a mixture of gases is decreased from 9.00 bar to 4.50 bar.

9.7 Calculate the molar free energy change at 298 K when (a) 0.450 M $FeCl_3(aq)$ is diluted to 0.350 M; (b) 0.900 M NaCl(aq) is diluted to 0.700 M; (c) the concentration of ethanol in water is increased from 0.175 M to 0.350 M; (d) the concentration of water in ethanol is increased from 0.175 M to 0.350 M.

9.8 Calculate the molar free energy change at 298 K when (a) 1.00 M sucrose(aq) is diluted to 0.250 M; (b) 2.00 M methanol(aq) is diluted to 0.500 M; (c) the concentration of acetone in water is increased from 0.500 M to 0.750 M; (d) the concentration of water in acetone is increased from 0.500 M to 0.750 M.

9.9 Calculate the equilibrium constant at 25°C for each of the following reactions, using data in Appendix 2A:
(a) the combustion of hydrogen:

$$2\,H_2(g) + O_2(g) \rightleftharpoons 2\,H_2O(g)$$

(b) the oxidation of carbon monoxide:

$$2\,CO(g) + O_2(g) \rightleftharpoons 2\,CO_2(g)$$

(c) the decomposition of limestone:

$$CaCO_3(s) \rightleftharpoons CaO(s) + CO_2(g)$$

9.10 Calculate the equilibrium constant at 25°C for each reaction, using data in Appendix 2A:
(a) The synthesis of carbon disulfide from natural gas (methane):

$$2\,CH_4(g) + S_8(s) \rightleftharpoons 2\,CS_2(l) + 4\,H_2S(g)$$

(b) the production of acetylene from calcium carbide:

$$CaC_2(s) + 2\,H_2O(l) \rightleftharpoons Ca(OH)_2(s) + C_2H_2(g)$$

(c) the oxidation of ammonia, the first step in the production of nitric acid:

$$4\,NH_3(g) + 5\,O_2(g) \rightleftharpoons 4\,NO(g) + 6\,H_2O(l)$$

(d) the industrial synthesis of urea, a common fertilizer:

$$CO_2(g) + 2\,NH_3(g) \rightleftharpoons CO(NH_2)_2(s) + H_2O(l)$$

9.11 Calculate the standard free energy of each of the following reactions:
(a) $I_2(g) \rightleftharpoons 2\,I(g)$, $K = 6.8$ at 1200 K
(b) $Ag_2CrO_4(s) \rightleftharpoons 2\,Ag^+(aq) + CrO_4^{2-}(aq)$, $K = 1.1 \times 10^{-12}$ at 298 K

9.12 Calculate the standard free energy for each of the following reactions:
(a) $H_2(g) + I_2(g) \rightleftharpoons 2\,HI(g)$, $K = 54$ at 700 K
(b) $CCl_3COOH(aq) + H_2O(l) \rightleftharpoons CCl_3CO_2^-(aq) + H_3O^+(aq)$, $K = 0.30$ at 298 K

9.13 If $Q = 1.0$ for the reaction $N_2(g) + O_2(g) \rightarrow 2\,NO(g)$ at 25°C, will the reaction have a tendency to form products or reactants, or will it be at equilibrium?

9.14 If $Q = 1.0 \times 10^{50}$ for the reaction $C(s) + O_2(g) \rightarrow CO_2(g)$ at 25°C, will the reaction have a tendency to form products or reactants, or will it be at equilibrium?

9.15 (a) Calculate the reaction free energy of $I_2(g) \rightarrow 2\,I(g)$ at 1200 K ($K_c = 6.8 \times 10^{-2}$) when the concentrations of I_2 and I are 0.026 mol·L^{-1} and 0.0084 mol·L^{-1}, respectively. (b) What is the spontaneous direction of the reaction?

9.16 Calculate the reaction free energy of $PCl_3(g) + Cl_2(g) \rightarrow PCl_5(g)$ at 230°C when the concentrations of PCl_3, Cl_2, and PCl_5 are 0.22 mol·L^{-1}, 0.41 mol·L^{-1}, and 1.33 mol·L^{-1}, respectively. What is the spontaneous direction of change, given that $K_c = 49$ at 230°C?

9.17 (a) Calculate the reaction free energy of $N_2(g) + 3\,H_2(g) \rightarrow 2\,NH_3(g)$ when the partial pressures of N_2, H_2, and NH_3 are 5.7 bar, 1.9 bar, and 15 bar, respectively, and the temperature is 400 K. For this reaction, $K = 41$ at 400 K. (b) Indicate whether this reaction mixture is likely to form reactants, is likely to form products, or is at equilibrium.

9.18 (a) Calculate the reaction free energy of $H_2(g) + I_2(g) \rightarrow 2\,HI(g)$ at 700 K when the concentrations of H_2, I_2, and HI are 0.14 mol·L^{-1}, 0.57 mol·L^{-1}, and 1.07 mol·L^{-1}, respectively. For this reaction, $K_c = 54$ at 700 K. (b) Indicate whether this reaction mixture is likely to form reactants, is likely to form products, or is at equilibrium.

9.19 Write the equilibrium expressions K_c for the following reactions:
(a) $CO(g) + Cl_2(g) \rightleftharpoons COCl(g) + Cl(g)$
(b) $H_2(g) + Br_2(g) \rightleftharpoons 2\,HBr(g)$
(c) $2\,H_2S(g) + 3\,O_2(g) \rightleftharpoons 2\,SO_2(g) + 2\,H_2O(g)$

9.20 Write the equilibrium expressions K for the following reactions:
(a) $2\,NO(g) + O_2(g) \rightleftharpoons 2\,NO_2(g)$
(b) $SbCl_5(g) \rightleftharpoons SbCl_3(g) + Cl_2(g)$
(c) $N_2(g) + 2\,H_2(g) \rightleftharpoons N_2H_4(g)$

9.21 Determine K_c for each of the following equilibria from the value of K:
(a) $2\,NOCl(g) \rightleftharpoons 2\,NO(g) + Cl_2(g)$, $K = 1.8 \times 10^{-2}$ at 500 K
(b) $CaCO_3(s) \rightleftharpoons CaO(s) + CO_2(g)$, $K = 167$ at 1073 K

9.22 Determine K_c for each of the following equilibria from the value of K:
(a) $2\,SO_2(g) + O_2(g) \rightleftharpoons 2\,SO_3(g)$, $K = 3.4$ at 1000 K
(b) $NH_4HS(s) \rightleftharpoons NH_3(g) + H_2S(g)$, $K = 9.4 \times 10^{-2}$ at 24°C

9.23 A 0.10-mol sample of pure ozone, O_3, is placed in a sealed 1.0-L container and the reaction $2\,O_3(g) \rightleftharpoons 3\,O_2(g)$ is allowed to reach equilibrium. A 0.50-mol sample of pure ozone is placed in a second 1.0-L container at the same temperature and allowed to reach equilibrium. Without doing any calculations, predict which of the following will be different in the two containers at equilibrium. Which will be the same? Explain each of your answers. (a) Amount of O_2; (b) concentration of O_2; (c) the ratio of $[O_2]/[O_3]$; (d) the ratio $[O_2]^3/[O_3]^2$; (e) the ratio $[O_3]^2/[O_2]^3$.

9.24 A 0.10-mol sample of $H_2(g)$ and a 0.10-mol sample of $Br_2(g)$ are placed into a 2.0-L container. The reaction $H_2(g) + Br_2(g) \rightleftharpoons 2\ HBr(g)$ is then allowed to come to equilibrium. A 0.20-mol sample of HBr is placed into a second 2.0-L sealed container at the same temperature and allowed to reach equilibrium with H_2 and Br_2. Which of the following will be different in the two containers at equilibrium? Which will be the same? (a) Amount of Br_2; (b) concentration of H_2; (c) the ratio $[HBr]/[H_2][Br_2]$; (d) the ratio $[HBr]/[Br_2]$; (e) the ratio $[HBr]^2/[H_2][Br_2]$; (f) the total pressure in the container. Explain each of your answers.

9.25 For the reaction $N_2(g) + 3\ H_2(g) \rightleftharpoons 2\ NH_3(g)$ at 400 K $K = 41$. Find the value of K for each of the following reactions at the same temperature:

(a) $2\ NH_3(g) \rightleftharpoons N_2(g) + 3H_2(g)$

(b) $\frac{1}{2}\ N_2(g) + \frac{3}{2}\ H_2(g) \rightleftharpoons NH_3(g)$

(c) $2\ N_2(g) + 6\ H_2(g) \rightleftharpoons 4\ NH_3(g)$

9.26 The equilibrium constant for the reaction $2\ SO_2(g) + O_2(g) \rightleftharpoons 2\ SO_3(g)$ has the value $K = 2.5 \times 10^{10}$ at 500 K. Find the value of K for each of the following reactions at the same temperature.

(a) $SO_2(g) + \frac{1}{2}\ O_2(g) \rightleftharpoons SO_3(g)$

(b) $SO_3(g) \rightleftharpoons SO_2(g) + \frac{1}{2}\ O_2(g)$

(c) $3\ SO_2(g) + \frac{3}{2}\ O_2(g) \rightleftharpoons 3\ SO_3(g)$

9.27 Use the following data, which were collected at 460°C and are equilibrium molar concentrations, to determine K_c for the reaction $H_2(g) + I_2(g) \rightleftharpoons 2\ HI(g)$:

$[H_2]$, mol·L^{-1}	$[I_2]$, mol·L^{-1}	[HI], mol·L^{-1}
6.47×10^{-3}	0.594×10^{-3}	0.0137
3.84×10^{-3}	1.52×10^{-3}	0.0169
1.43×10^{-3}	1.43×10^{-3}	0.0100

9.28 Determine K from the following equilibrium data collected at 24°C for the reaction $NH_4HS(s) \rightleftharpoons NH_3(g) + H_2S(g)$:

P_{NH_3}, bar	P_{H_2S}, bar
0.307	0.307
0.364	0.258
0.539	0.174

9.29 Write the reaction quotients Q for

(a) $4\ Bi(s) + 3\ O_2(g) \rightarrow 2\ Bi_2O_3(s)$

(b) $MgSO_4{\cdot}7H_2O(s) \rightarrow MgSO_4(s) + 7\ H_2O(g)$

(c) $N_2O_3(g) \rightarrow NO(g) + NO_2(g)$

9.30 Write the reaction quotients Q_c for

(a) $3\ ClO^-(aq) \rightarrow 2\ Cl^-(aq) + ClO_3^-(aq)$

(b) $Li_2CO_3(s) \rightarrow Li_2O(s) + CO_2(g)$

(c) $CH_3COOH(aq) + H_2O(l) \rightarrow H_3O^+(aq) + CH_3CO_2^-(aq)$

9.31 In a gas-phase equilibrium mixture of H_2, I_2, and HI at 500 K, $[HI] = 2.21 \times 10^{-3}$ mol·L^{-1} and $[I_2] = 1.46 \times 10^{-3}$ mol·L^{-1}. Given the value of the equilibrium constant in Table 9.2, calculate the concentration of H_2.

9.32 In a gas-phase equilibrium mixture of H_2, Cl_2, and HCl at 1000 K, $[HCl] = 1.45 \times 10^{-3}$ mol·L^{-1} and $[Cl_2] = 2.45 \times 10^{-3}$ mol·L^{-1}. Use the information in Table 9.2 to calculate the concentration of H_2.

9.33 In a gas-phase equilibrium mixture of PCl_5, PCl_3, and Cl_2 at 500 K, $P_{PCl_5} = 1.45$ bar, $P_{Cl_2} = 2.35$ bar. What is the partial pressure of PCl_3, given that $K = 25$ for the reaction $PCl_5(g) \rightleftharpoons PCl_3(g) + Cl_2(g)$?

9.34 In a gas-phase equilibrium mixture of $SbCl_5$, $SbCl_3$, and Cl_2 at 500 K, $P_{SbCl_5} = 0.089$ bar and $P_{SbCl_3} = 0.067$ bar. Calculate the equilibrium partial pressure of Cl_2, given that $K = 3.5 \times 10^{-4}$ for the reaction $SbCl_5(g) \rightleftharpoons SbCl_3(g) + Cl_2(g)$.

9.35 For the reaction $H_2(g) + I_2(g) \rightleftharpoons 2\ HI(g)$, $K_c = 160$ at 500 K. An analysis of a reaction mixture at 500 K showed that it had the composition 4.8×10^{-3} mol·L^{-1} H_2, 2.4×10^{-3} mol·L^{-1} I_2, and 2.4×10^{-3} mol·L^{-1} HI. (a) Calculate the reaction quotient. (b) Is the reaction mixture at equilibrium? (c) If not, is there a tendency to form more reactants or more products?

9.36 Analysis of a reaction mixture showed that it had the composition 0.417 mol·L^{-1} N_2, 0.524 mol·L^{-1} H_2, and 0.122 mol·L^{-1} NH_3 at 800 K, at which temperature $K_c = 0.278$ for $N_2(g) + 3\ H_2(g) \rightleftharpoons 2\ NH_3(g)$. (a) Calculate the reaction quotient. (b) Is the reaction mixture at equilibrium? (c) If not, is there a tendency to form more reactants or more products?

9.37 A 0.500-L reaction vessel at 700 K contains 1.20×10^{-3} mol $SO_2(g)$, 5.0×10^{-4} mol $O_2(g)$, and 1.0×10^{-4} $SO_3(g)$. At 700 K, $K_c = 1.7 \times 10^6$ for the equilibrium $2\ SO_2(g) + O_2(g) \rightleftharpoons 2\ SO_3(g)$. (a) Calculate the reaction quotient Q_c. (b) Will more $SO_3(g)$ tend to form?

9.38 Given that $K_c = 62$ for the reaction $N_2(g) + 3\ H_2(g) \rightleftharpoons 2\ NH_3(g)$ at 500 K, calculate whether more ammonia will tend to form when a mixture of composition $2.23 \times 10^{-3}\ mol{\cdot}L^{-1}\ N_2$, $1.24 \times 10^{-3}\ mol{\cdot}L^{-1}\ H_2$, and $1.12 \times 10^{-4}\ mol{\cdot}L^{-1}$ NH_3 is present in a container at 500 K.

9.39 When 0.0172 mol HI is heated to 500 K in a 2.00-L sealed container, the resulting equilibrium mixture contains 1.90 g of HI. Calculate K_c for the decomposition reaction $2\ HI(g) \rightleftharpoons H_2(g) + I_2(g)$.

9.40 When 1.00 g of gaseous I_2 is heated to 1000 K in a 1.00-L sealed container, the resulting equilibrium mixture contains 0.830 g of I_2. Calculate K_c for the dissociation equilibrium $I_2(g) \rightleftharpoons 2\ I(g)$.

9.41 A 25.0-g sample of ammonium carbamate, $NH_4(NH_2CO_2)$, was placed in an evacuated 0.250-L flask and kept at 25°C. At equilibrium, 17.4 mg of CO_2 was present. What is the value of K_c for the decomposition of ammonium carbamate into ammonia and carbon dioxide? The reaction is $NH_4(NH_2CO_2)(s) \rightleftharpoons 2\ NH_3(g) + CO_2(g)$.

9.42 Carbon monoxide and water vapor, each at 200 Torr, were introduced into a 250-mL container. When the mixture reached equilibrium at 700°C, the partial pressure of $CO_2(g)$ was 88 Torr. Calculate the value of K for the equilibrium $CO(g) + H_2O(g) \rightleftharpoons CO_2(g) + H_2(g)$.

9.43 Data for the equilibrium $CH_3COOH(l) + C_2H_5OH(l) \rightleftharpoons CH_3COOC_2H_5(l) + H_2O(l)$ at 100°C were obtained for the reaction in an organic solvent. The initial concentrations of the reactants are reported in columns 1 and 2 in the following table and the equilibrium concentration of $CH_3COOC_2H_5$ is reported in column 3. Calculate $[H_2O]$ and determine the value of K_c for each experiment.

Initial concentrations, $mol{\cdot}L^{-1}$		Equilibrium concentration, $mol{\cdot}L^{-1}$
$[CH_3COOH]$	$[C_2H_5OH]$	$[CH_3COOC_2H_5]$
1.00	0.180	0.171
1.00	1.00	0.667
1.00	8.00	0.966

9.44 A 1.0-L reaction vessel was filled to the partial pressures of N_2O_4 given in column 1 in the following table, and the gas was allowed to decompose into NO_2. At 25°C, the total pressure ($P = P_{N_2O_4} + P_{NO_2}$) of the system at equilibrium is listed in column 2. Write the chemical equation for the reaction and calculate K for each set of data.

$P_{N_2O_4}$, bar	P, bar
0.154	0.212
0.333	0.425

9.45 (a) A sample of 2.0 mmol Cl_2 was sealed into a 2.0-L reaction vessel and heated to 1000 K to study its dissociation into Cl atoms. Use the information in Table 9.2 to calculate the equilibrium composition of the mixture. What is the percentage decomposition of the Cl_2? (b) If 2.0 mmol F_2 was placed into the reaction vessel instead of the chlorine, what would be its equilibrium composition at 1000 K? (c) Use your results from (a) and (b) to determine which is more stable relative to its atoms at 1000 K, Cl_2 or F_2.

9.46 (a) A sample of 5.0 mmol Cl_2 was sealed into a 2.0-L reaction vessel and heated to 1200 K, and the dissociation equilibrium was established. What is the equilibrium composition of the mixture? What is the percentage decomposition of the Cl_2? Use the information in Table 9.2. (b) If 5.0 mol Br_2 was placed into the reaction vessel instead of the chlorine, what would be the equilibrium composition at 1200 K? (c) Use your results from (a) and (b) to determine which is more stable relative to its atoms at 1200 K, Cl_2 or Br_2.

9.47 The initial concentration of HBr in a reaction vessel is $1.2 \times 10^{-3}\ mol{\cdot}L^{-1}$. If the vessel is heated to 500 K, what is the percentage decomposition of the HBr and the equilibrium composition of the mixture? For $2\ HBr(g) \rightleftharpoons H_2(g) + Br_2(g)$, $K_c = 7.7 \times 10^{-11}$ at 500 K.

9.48 The initial concentration of BrCl in a reaction vessel is $1.4 \times 10^{-3}\ mol{\cdot}L^{-1}$. If the vessel is heated to 500 K, what is the percentage decomposition of the BrCl and what is the equilibrium composition of the mixture? See Table 9.2 for data on the reaction.

9.49 The equilibrium constant $K_c = 1.1 \times 10^{-2}$ for the reaction $PCl_5(g) \rightleftharpoons PCl_3(g) + Cl_2(g)$ at 400 K. (a) Given that 1.0 g of PCl_5 is placed in a 250-mL reaction vessel, determine the molar concentrations in the mixture at equilibrium. (b) What percentage of PCl_5 is decomposed at 400 K?

9.50 For the reaction $PCl_5(g) \rightleftharpoons PCl_3(g) + Cl_2(g)$, $K_c = 0.61$ at 500 K. (a) Calculate the equilibrium molar concentrations of the components in the mixture when 2.0 g of PCl_5 is placed in a 300-mL reaction vessel and allowed to come to equilibrium. (b) What percentage of PCl_5 is decomposed at 500 K?

9.51 When solid NH_4HS and 0.400 mol $NH_3(g)$ were placed in a 2.0-L vessel at 24°C, the equilibrium $NH_4HS(s) \rightleftharpoons NH_3(g) + H_2S(g)$, for which $K_c = 1.6 \times 10^{-4}$, was reached. What are the equilibrium concentrations of NH_3 and H_2S?

9.52 When solid NH_4HS and 0.200 mol $NH_3(g)$ were placed into a 2.0-L vessel at 24°C, the equilibrium $NH_4HS(s) \rightleftharpoons NH_3(g) + H_2S(g)$, for which $K_c = 1.6 \times 10^{-4}$, was reached. What are the equilibrium concentrations of NH_3 and H_2S?

9.53 At 760°C, $K_c = 33.3$ for the reaction $PCl_5(g) \rightleftharpoons PCl_3(g) + Cl_2(g)$. If a mixture that consists of 0.200 mol PCl_5 and 0.600 mol PCl_3 is placed in a 4.00-L reaction vessel and heated to 760°C, what is the equilibrium composition of the system?

9.54 At 760°C, $K_c = 33.3$ for the reaction $PCl_5(g) \rightleftharpoons PCl_3(g) + Cl_2(g)$. If a mixture that consists of 0.200 mol PCl_3 and 0.600 mol Cl_2 is placed in an 8.00-L reaction vessel and heated to 760°C, what is the equilibrium composition of the system?

9.55 The equilibrium constant K_c for the reaction $N_2(g) + O_2(g) \rightleftharpoons 2\ NO(g)$ at 1200°C is 1.00×10^{-5}. Calculate the equilibrium molar concentrations of NO, N_2, and O_2 in a 1.00-L reaction vessel that initially held 0.114 mol N_2 and 0.114 mol O_2.

9.56 The equilibrium constant K_c for the reaction $N_2(g) + O_2(g) \rightleftharpoons 2\ NO(g)$ at 1200°C is 1.00×10^{-5}. Calculate the equilibrium molar concentrations of NO, N_2, and O_2 in a 10.00-L reaction vessel that initially held 0.0140 mol N_2 and 0.214 mol O_2.

9.57 A reaction mixture that consisted of 0.400 mol H_2 and 1.60 mol I_2 was introduced into a 3.00-L flask and heated. At equilibrium, 60.0% of the hydrogen gas had reacted. What is the equilibrium constant for the reaction $H_2(g) + I_2(g) \rightleftharpoons 2\ HI(g)$ at this temperature?

9.58 A reaction mixture that consisted of 0.20 mol N_2 and 0.20 mol H_2 was introduced into a 25.0-L reactor and heated. At equilibrium, 5.0% of the nitrogen gas had reacted. What is the value of the equilibrium constant K_c for the reaction $N_2(g) + 3\ H_2(g) \rightleftharpoons 2\ NH_3(g)$ at this temperature?

9.59 The equilibrium constant K_c of the reaction $2\ CO(g) + O_2(g) \rightleftharpoons 2\ CO_2(g)$ is 0.66 at 2000°C. If 0.28 g of CO and 0.032 g of $O_2(g)$ are placed in a 2.0-L reaction vessel and heated to 2000°C, what will the equilibrium composition of the system be? (You may wish to use graphing software or calculator to solve the cubic equation.)

9.60 In the Haber process for ammonia synthesis, $K = 0.036$ for $N_2(g) + 3\ H_2(g) \rightleftharpoons 2\ NH_3(g)$ at 500 K. If a 2.0-L reactor is charged with 0.010 bar of N_2 and 0.010 bar of H_2, what will the equilibrium partial pressures in the mixture be?

9.61 An ester is formed in the reaction of an organic acid with an alcohol. For example, in the reaction of acetic acid, CH_3COOH, with ethanol, C_2H_5OH, in an organic solvent, the ester known as ethyl acetate, $CH_3COOC_2H_5$, and water form. The reaction is

$$CH_3COOH + C_2H_5OH \rightleftharpoons CH_3COOC_2H_5 + H_2O \qquad K_c = 4.0 \text{ at } 100°C$$

If the initial concentrations of CH_3COOH and C_2H_5OH are 0.32 mol·L^{-1} and 6.30 mol·L^{-1}, respectively, and no products are present initially, what will the equilibrium concentration of the ester be?

9.62 In the esterification reaction specified in Exercise 9.61, if the initial concentrations of CH_3COOH and C_2H_5OH are 0.024 mol·L^{-1} and 0.059 mol·L^{-1}, respectively, but in addition the concentration of the water in the initial mixture is 0.015 mol·L^{-1}, what will the equilibrium concentration of the ester be?

9.63 At 500 K, the equilibrium constant for the reaction $Cl_2(g) + Br_2(g) \rightleftharpoons 2\ BrCl(g)$ is $K_c = 0.031$. If the equilibrium composition is 0.495 mol·L^{-1} Cl_2 and 0.145 mol·L^{-1} BrCl, what is the equilibrium concentration of Br_2?

9.64 The equilibrium constant for the reaction $PCl_3(g) + Cl_2(g) \rightleftharpoons PCl_5(g)$ is $K = 3.5 \times 10^4$ at 760°C. At equilibrium, the partial pressure of PCl_5 was 1.3×10^2 bar and that of PCl_3 was 9.56 bar. What was the equlibrium partial pressure of Cl_2?

9.65 A reaction mixture consisting of 2.00 mol CO and 3.00 mol H_2 is placed into a 10.0-L reaction vessel and heated to 1200 K. At equilibrium, 0.478 mol CH_4 was present in the system. Determine the value of K_c for the reaction $CO(g) + 3\ H_2(g) \rightleftharpoons CH_4(g) + H_2O(g)$ at 1200 K.

9.66 A mixture consisting of 1.000 mol $H_2O(g)$ and 1.000 mol CO(g) is placed in a 10.00-L reaction vessel at 800 K. At equilibrium, 0.665 mol $CO_2(g)$ is present as a result of the reaction $CO(g) + H_2O(g) \rightleftharpoons CO_2(g) + H_2(g)$. What are (a) the equilibrium concentrations for all substances and (b) the value of K_c at 800 K?

9.67 A reaction mixture was prepared by mixing 0.100 mol SO_2, 0.200 mol NO_2, 0.100 mol NO, and

0.150 mol SO_3 in a 5.00-L reaction vessel. The reaction $SO_2(g) + NO_2(g) \rightleftharpoons NO(g) + SO_3(g)$ is allowed to reach equilibrium at 460°C, when $K_c = 85.0$. What is the equilibrium concentration of each substance?

9.68 A 0.100-mol sample of H_2S is placed in a 10.0-L reaction vessel and heated to 1132°C. At equilibrium, 0.0285 mol H_2 is present. Calculate the value of K_c for the reaction $2\ H_2S(g) \rightleftharpoons 2\ H_2(g) + S_2(g)$ at 1132°C.

9.69 A mixture of 0.0560 mol O_2 and 0.0200 mol N_2O is placed in a 1.00-L reaction vessel at 25°C. When the reaction $2\ N_2O(g) + 3\ O_2(g) \rightleftharpoons 4\ NO_2(g)$ is at equilibrium, 0.0200 mol NO_2 is present. (a) What are the equilibrium concentrations of O_2 and N_2O? (b) What is the value of K_c at 25°C?

9.70 At 500 K, 1.00 mol NOCl is 9.0% dissociated in a 1.00-L vessel. Calculate the value of K_c for the reaction $2\ NOCl(g) \rightleftharpoons 2\ NO(g) + Cl_2(g)$.

9.71 The equilibrium constant $K_c = 0.56$ for the reaction $PCl_3(g) + Cl_2(g) \rightleftharpoons PCl_5(g)$ at 250°C. Upon analysis, it was found that 1.50 mol PCl_5, 3.00 mol PCl_3,and 0.500 mol Cl_2 were present in a 0.500-L reaction vessel at 250°C. (a) Is the reaction at equilibrium? (b) If not, in which direction does it tend to proceed and (c) what is the equilibrium composition of the reaction system?

9.72 A 1.50-mol sample of PCl_5 is placed in a 500-mL reaction vessel. What is the concentration of each substance when the reaction $PCl_5(g) \rightleftharpoons PCl_3(g) + Cl_2(g)$ has reached equilibrium at 250°C (when $K_c = 1.80$)?

9.73 At 25°C, $K = 3.2 \times 10^{-34}$ for the reaction $2\ HCl(g) \rightleftharpoons H_2(g) + Cl_2(g)$. If a 1.0-L reaction vessel is filled with HCl at 0.22 bar, what are the equilibrium partial pressures of HCl, H_2, and Cl_2?

9.74 If 4.00 L of HCl(g) at 1.00 bar and 273 K and 26.0 g of $I_2(s)$ are transferred to a 12.00-L reaction vessel and heated to 25°C, what will the equilibrium concentration of HCl, HI, and Cl_2 be? $K_c = 1.6 \times 10^{-34}$ at 25°C for $2\ HCl(g) + I_2(s) \rightleftharpoons 2\ HI(g) + Cl_2(g)$.

9.75 A 3.00-L reaction vessel is filled with 0.342 mol CO, 0.215 mol H_2, and 0.125 mol CH_3OH. Equilibrium is reached in the presence of a zinc oxide-chromium(III) oxide catalyst, and at 300°C, $K_c = 1.1 \times 10^{-2}$ for the reaction $CO(g) + 2\ H_2(g) \rightleftharpoons CH_3OH(g)$. (a) As the reaction approaches equilibrium, will the concentration of CH_3OH increase, decrease, or remain unchanged? (b) What is the equilibrium composition of the mixture?

9.76 $K_c = 0.395$ at 350°C for the reaction $2\ NH_3(g) \rightleftharpoons N_2(g) + 3\ H_2(g)$. A 25.6-g sample of NH_3 is placed in a 5.00-L reaction vessel and heated to 350°C. What are the equilibrium concentrations of NH_3, N_2, and H_2?

The Response of Equilibria to Changes in Conditions

9.77 Consider the equilibrium $CO(g) + H_2O(g) \rightleftharpoons CO_2(g) + H_2(g)$. (a) If the partial pressure of CO_2 is increased, what happens to the partial pressure of H_2? (b) If the partial pressure of CO is decreased, what happens to the partial pressure of CO_2? (c) If the concentration of CO is increased, what happens to the concentration of H_2? (d) If the concentration of H_2O is decreased, what happens to the equilibrium constant for the reaction?

9.78 Consider the equilibrium $CH_4(g) + 2\ O_2(g) \rightleftharpoons CO_2(g) + 2\ H_2O(g)$. (a) If the partial pressure of CO_2 is increased, what happens to the partial pressure of CH_4? (b) If the partial pressure of CH_4 is decreased, what happens to the partial pressure of CO_2? (c) If the concentration of CH_4 is increased, what happens to the equilibrium constant for the reaction? (d) If the concentration of H_2O is decreased, what happens to the concentration of CO_2?

9.79 The four gases NH_3, O_2, NO, and H_2O are mixed in a reaction vessel and allowed to reach equilibrium in the reaction $4\ NH_3(g) + 5\ O_2(g) \rightleftharpoons 4\ NO(g) + 6\ H_2O(g)$. Certain changes (see the following table) are then made to this mixture. Considering each change separately, state the effect (increase, decrease, or no change) that the change has on the original equilibrium values of the quantity in the second column (or K_c, if that is specified). The temperature and volume are constant.

Change	Quantity
(a) add NO	amount of H_2O
(b) add NO	amount of O_2
(c) remove H_2O	amount of NO
(d) remove O_2	amount of NH_3
(e) add NH_3	K_c
(f) remove NO	amount of NH_3
(g) add NH_3	amount of O_2

9.80 The four substances HCl, I_2, HI, and Cl_2 are mixed in a reaction vessel and allowed to reach equilibrium in the reaction $2\ HCl(g) + I_2(s) \rightleftharpoons 2\ HI(g) + Cl_2(g)$. Certain changes (which are specified in the first column in the following table) are then made to this mixture. Considering each change separately, state the effect (increase, decrease, or no change) that the change has on the original equilibrium value of the quantity in the second column (or K_c, if that is specified). The temperature and volume are constant.

Change	Quantity
(a) add HCl	amount of HI
(b) add I_2	amount of Cl_2
(c) remove HI	amount of Cl_2
(d) remove Cl_2	amount of HCl
(e) add HCl	K_c
(f) remove HCl	amount of I_2
(g) add I_2	K_c

9.81 State whether reactants or products will be favored by an increase in the total pressure on each of the following equilibria. If no change occurs, explain why that is so.

(a) $2\ O_3(g) \rightleftharpoons 3\ O_2(g)$

(b) $H_2O(g) + C(s) \rightleftharpoons H_2(g) + CO(g)$

(c) $4\ NH_3(g) + 5\ O_2(g) \rightleftharpoons 4\ NO(g) + 6\ H_2O(g)$

(d) $2\ HD(g) \rightleftharpoons H_2(g) + D_2(g)$

(e) $Cl_2(g) \rightleftharpoons 2\ Cl(g)$

9.82 State what happens to the concentration of the indicated substance when the total pressure on each of the following equilibria is increased:

(a) $NO_2(g)$ in $2\ Pb(NO_3)_2(s) \rightleftharpoons 2\ PbO(s) + 4\ NO_2(g) + O_2(g)$

(b) $NO(g)$ in $3\ NO_2(g) + H_2O(l) \rightleftharpoons 2\ HNO_3(aq) + NO(g)$

(c) $HI(g)$ in $2\ HCl(g) + I_2(s) \rightleftharpoons 2\ HI(g) + Cl_2(g)$

(d) $SO_2(g)$ in $2\ SO_2(g) + O_2(g) \rightleftharpoons 2\ SO_3(g)$

(e) $NO_2(g)$ in $2\ NO(g) + O_2(g) \rightleftharpoons 2\ NO_2(g)$

9.83 Consider the equilibrium $4\ NH_3(g) + 5\ O_2(g) \rightleftharpoons 4\ NO(g) + 6\ H_2O(g)$. (a) What happens to the partial pressure of NH_3 when the partial pressure of NO is increased? (b) Does the partial pressure of O_2 decrease when the partial pressure of NH_3 is decreased?

9.84 Consider the equilibrium $2\ SO_2(g) + O_2(g) \rightleftharpoons 2\ SO_3(g)$. (a) What happens to the partial pressure of SO_3 when the partial pressure of SO_2 is decreased? (b) If the partial pressure of SO_2 is increased, what happens to the partial pressure of O_2?

9.85 Predict whether each of the following equilibria will shift toward products or reactants with a temperature increase:

(a) $N_2O_4(g) \rightleftharpoons 2\ NO_2(g)$, $\Delta H° = +57$ kJ

(b) $X_2(g) \rightleftharpoons 2\ X(g)$, where X is a halogen

(c) $Ni(s) + 4\ CO(g) \rightleftharpoons Ni(CO)_4(g)$, $\Delta H° = -161$ kJ

(d) $CO_2(g) + 2\ NH_3(g) \rightleftharpoons CO(NH_2)_2(s) + H_2O(g)$, $\Delta H° = -90$ kJ

9.86 Predict whether each of the following equilibria will shift toward products or reactants with a temperature increase:

(a) $CH_4(g) + H_2O(g) \rightleftharpoons CO(g) + 3\ H_2(g)$, $\Delta H° = +206$ kJ

(b) $CO(g) + H_2O(g) \rightleftharpoons CO_2(g) + H_2(g)$, $\Delta H° = -41$ kJ

(c) $2\ SO_2(g) + O_2(g) \rightleftharpoons 2\ SO_3(g)$, $\Delta H° = -198$ kJ

9.87 A mixture consisting of 2.23×10^{-3} mol N_2 and 6.69×10^{-3} mol H_2 in a 500-mL container was heated to 600 K and allowed to reach equilibrium. Will more ammonia be formed if that equilibrium mixture is then heated to 700 K? For $N_2(g) + 3\ H_2(g) \rightleftharpoons 2\ NH_3(g)$, $K = 1.7 \times 10^{-3}$ at 600 K and 7.8×10^{-5} at 700 K.

9.88 A mixture consisting of 1.1 mmol SO_2 and 2.2 mmol O_2 in a 250-mL container was heated to 500 K and allowed to reach equilibrium. Will more sulfur trioxide be formed if that equilibrium mixture is cooled to 298 K? For the reaction $2\ SO_2(g) + O_2(g) \rightleftharpoons 2\ SO_3(g)$, $K = 2.5 \times 10^{10}$ at 500 K and 4.0×10^{24} at 298 K.

9.89 At 500°C, $K_c = 0.061$ for $N_2(g) + 3\ H_2(g) \rightleftharpoons 2\ NH_3(g)$. If analysis shows that the composition is 3.00 mol·L^{-1} N_2, 2.00 mol·L^{-1} H_2, and 0.500 mol·L^{-1} NH_3, is the reaction at equilibrium? If not, in which direction does the reaction tend to proceed to reach equilibrium?

9.90 At 2500 K, the equilibrium constant is $K_c = 20$ for the reaction $Cl_2(g) + F_2(g) \rightleftharpoons 2\ ClF(g)$. An analysis of a reaction vessel at 2500 K revealed the presence of 0.18 mol·L^{-1} Cl_2, 0.31 mol·L^{-1} F_2, and 0.92 mol·L^{-1} ClF. Will ClF tend to form or to decompose as the reaction proceeds toward equilibrium?

9.91 Calculate the equilibrium constant at 25°C and at 150°C for each of the following reactions, using data available in Appendix 2A:

(a) $NH_4Cl(s) \rightleftharpoons NH_3(g) + HCl(g)$

(b) $H_2(g) + D_2O(l) \rightleftharpoons D_2(g) + H_2O(l)$

9.92 Calculate the equilibrium constant at 25°C and at 100°C for each of the following reactions, using data available in Appendix 2A:

(a) $HgO(s) \rightleftharpoons Hg(l) + \frac{1}{2}\ O_2(g)$

(b) propene (C_3H_6, g) $\rightleftharpoons$ cyclopropane (C_3H_6, g)

INTEGRATED EXERCISES

9.93 A 30.1-g sample of NOCl is placed into a 200-mL reaction vessel and heated to 500 K. The value of K for the decomposition of NOCl at 500 K in the reaction $2\ NOCl(g) \rightleftharpoons 2\ NO(g) + Cl_2(g)$ is 1.13×10^{-3}. (a) What are the equilibrium partial pressures of NOCl, NO, and Cl_2? (b) What is the percentage decomposition of NOCl at this temperature?

9.94 At 25°C, $K_c = 4.66 \times 10^{-3}$ for $N_2O_4(g) \rightleftharpoons 2\ NO_2(g)$. If 2.50 g of N_2O_4 and 0.330 g of NO_2 are placed in a 2.00-L reaction vessel, what are the equilibrium concentrations of N_2O_4 and NO_2?

9.95 The photosynthesis reaction is $6\ CO_2(g) + 6\ H_2O(l) \rightleftharpoons C_6H_{12}O_6(aq) + 6\ O_2(g)$, and $\Delta H° = +2802$ kJ. Suppose that the reaction is at equilibrium. State the effect that each of the following changes will have on the equilibrium composition (tend to shift toward the formation of reactants, tend to shift toward the formation of products, or have no effect). (a) The partial pressure of O_2 is increased. (b) The system is compressed. (c) The amount of CO_2 is increased. (d) The temperature is increased. (e) Some of the $C_6H_{12}O_6$ is removed. (f) Water is added. (g) The partial pressure of CO_2 is decreased.

9.96 The depletion of ozone in the stratosphere can be summarized by the equation $2\ O_3(g) \rightleftharpoons 3\ O_2(g)$. (a) From values in Appendix 2A, determine the standard free energy and the standard entropy for the reaction. (b) What is the equilibrium constant of the reaction at 25°C? What is the significance of your answers with regard to ozone depletion?

9.97 At 1565 K, the equilibrium constants for the reactions (1) $2\ H_2O(g) \rightleftharpoons 2\ H_2(g) + O_2(g)$ and (2) $2\ CO_2(g) \rightleftharpoons 2\ CO(g) + O_2(g)$ are 1.6×10^{-11} and 1.3×10^{-10}, respectively. (a) What is the equilibrium constant for the reaction (3) $CO_2(g) + H_2(g) \rightleftharpoons H_2O(g) + CO(g)$ at that temperature? (b) Show that the manner in which equilibrium constants are calculated is consistent with the manner in which the $\Delta G_r°$ values are calculated when combining two or more equations by determining $\Delta G_r°$ for (1) and (2) and using those values to calculate $\Delta G_r°$ for reaction (3) and K_3.

9.98 Suppose that in an esterification reaction in which an organic acid RCOOH reacts with an alcohol R′OH in an organic solvent to produce an ester RCOOR′

$$RCOOH + R'OH \rightleftharpoons RCOOR' + H_2O$$

A moles of acid and B moles of alcohol are mixed and heated to 100°C. Find an expression for the amount (in moles) of ester (C) that is present at equilibrium, in terms of A, B, and K_c. Evaluate the expression for $A = 1.0$, $B = 0.50$, and $K_c = 3.5$.

9.99 Let α be the fraction of PCl_5 molecules that have decomposed to PCl_3 and Cl_2 in the reaction $PCl_5(g) \rightleftharpoons PCl_3(g) + Cl_2(g)$ in a constant-volume container; so the amount of PCl_5 at equilibrium is $n(1 - \alpha)$, where n is the amount present initially. Derive an equation for K in terms of α and the total pressure P, and solve it for α in terms of P. Calculate the fraction decomposed at 556 K, at which temperature $K = 4.96$, and the total pressure is (a) 0.50 bar; (b) 1.00 bar.

9.100 The three compounds 2-methyl-2-propene, *cis*-2-butene, and *trans*-2-butene are isomers with formula C_4H_8, with $\Delta G_f°$ equal to +58.07, +65.86, and +62.97 kJ·mol^{-1}, respectively. In the presence of a suitable metal catalyst, these three compounds can be interconverted to give an equilibrium gaseous mixture. What will be the percentage of each isomer present at 25°C once equilibrium is established?

9.101 (a) What is the standard free energy of the reaction $CO(g) + H_2O(g) \rightleftharpoons CO_2(g) + H_2(g)$ when $K = 1.00$. (b) From data available in Appendix 2A, estimate the temperature at which $K = 1$. (c) At this temperature, a cylinder is filled with CO(g) at 10.00 bar, $H_2O(g)$ at 10.00 bar, $H_2(g)$ at 5.00 bar, and $CO_2(g)$ at 5.00 bar. What will be the partial pressure of each of these gases when the system reaches equilibrium? (d) If the cylinder were filled instead with CO(g) at 6.00 bar, $H_2O(g)$ at 4.00 bar, $H_2(g)$ at 5.00 bar, and $CO_2(g)$ at 10.00 bar, what would the partial pressures be at equilibrium?

9.102 Consider a hypothetical equilibrium reaction $A \rightleftharpoons 2\ B + 3\ C$ at 25°C, where all the compounds are gases. When A is loaded into a cylinder at 10.0 atm and the system allowed to come to equilibrium, the final pressure is found to be 15.76 atm. What is $\Delta G_r°$ for this reaction?

9.103 (a) Calculate the free energies of formation of the halogen atoms X(g) at 1000 K from data available in Table 9.2. (b) Show how these data correlate with the X—X bond strength by plotting the free energy of formation of the atoms against the bond dissociation energy and atomic number. Rationalize any trends you observe.

9.104 (a) 10.00 g of pure $NH_3(g)$ is introduced into a 4.0-L cylinder at 25°C and then heated to 127°C. What will be the total pressure in the cylinder when it reaches equilibrium? (b) How well does this equilibrium constant compare with that calculated from the thermodynamic data tabulated in Appendix 2A?

9.105 Calculate K at 25°C for the reaction $Br_2(g) \rightleftharpoons 2\ Br(g)$ in two ways: (a) from the equilibrium constant data given in Table 9.2; (b) from the thermodynamic data provided in Appendix 2A. Compare the values. How well do they agree? (c) What is the vapor pressure of liquid bromine? (d) What is the partial pressure of Br(g) above the liquid in a bottle of bromine at 25°C? (e) A student wishes to add 0.0100 mol Br_2 to a reaction. This will be done by filling an evacuated flask with Br_2 vapor from a reservoir that contains only bromine liquid in equilibrium with its vapor. The flask will be sealed and then transferred to the reaction vessel. What volume container should the student use in order to deliver 1.00 mol $Br_2(g)$ at 25°C?

9.106 A reaction vessel is filled with $Cl_2(g)$ at 1.00 bar and $Br_2(g)$ at 1.00 bar, which are allowed to react at 1000 K to form BrCl(g) according to the equation $Br_2(g) + Cl_2(g) \rightleftharpoons 2\ BrCl(g)$, $K = 0.2$. Construct a plot of the free energy of this system as a function of partial pressure of BrCl as the reaction approaches equilibrium.

9.107 A reactor for the production of ammonia by the Haber process is found to be at equilibrium with $P_{N_2} = 3.11$ bar, $P_{H_2} = 1.64$ bar, and $P_{NH_3} = 23.72$ bar. If the partial pressure of N_2 is increased by 1.57 bar, what will be the partial pressure of each gas once equilibrium is reestablished?

9.108 Using data from Table 9.2 and standard graphing software, determine the enthalpy and entropy of the equilibrium $N_2O_4(g) \rightleftharpoons 2\ NO_2(g)$ and estimate the N—N bond enthalpy in N_2O_4. How does this value compare with N—N bond enthalpies in general?

9.109 Using the Living Graph *Variation of Equilibrium Constant* on the Web site, construct a plot from 250 K to 350 K for reactions with standard free energies of 11 $kJ \cdot mol^{-1}$ to 15 $kJ \cdot mol^{-1}$ in increments of 1 $kJ \cdot mol^{-1}$. Which equilibrium constant is most sensitive to changes in temperature?

9.110 (a) Using appropriate graphing software, such as that found on the Web site for this book, and data at 1000 K and 1200 K from Table 9.2, plot the van 't Hoff equations for the dissociation of the diatomic halogens into the atoms, $X_2(g) \rightleftharpoons 2\ X(g)$. (b) From the graphs, determine the enthalpies and entropies of dissociation. (c) Use this data to calculate the molar entropies of the gaseous halogen atoms X(g).

9.111 The reaction $N_2O_4 \rightleftharpoons 2\ NO_2$ is allowed to reach equilibrium in chloroform solution at 25°C. The equilibrium concentrations are 0.405 $mol \cdot L^{-1}$ N_2O_4 and 2.13 $mol \cdot L^{-1}$ NO_2. (a) Calculate K_c for the reaction. (b) An additional 1.00 mol NO_2 is added to 1.00 L of the solution and the system is allowed to reach equilibrium again at the same temperature. Use Le Chatelier's principle to predict the direction of change (increase, decrease, or no change) for N_2O_4, NO_2, and K_c after the addition of NO_2. (c) Calculate the final equilibrium concentrations after the addition of NO_2 and confirm that your predictions in (b) were valid. If they do not agree, check your procedure and repeat it, if necessary.

9.112 The temperature dependence of the equilibrium constant of the reaction $N_2(g) + O_2(g) \rightleftharpoons 2\ NO(g)$, which makes an important contribution to atmospheric nitrogen oxides, can be expressed as $\ln K = 2.5 - (21\ 700\ K)/T$. What is the standard enthalpy of the forward reaction?

9.113 Determine the vapor pressure of heavy water, D_2O, and of normal water at 25°C, using data in Appendix 2A. How do these values compare with each other? Using your knowledge of intermolecular forces, explain the reason for the difference observed.

9.114 Hydrogen, chlorine, and bromine are placed in a flask and heated to 1000 K. (a) What molecules will be present when the reaction comes to equilibrium (neglect atomic species)? Consult Table 9.2. (b) Write all the equilibrium reactions that are possible for this system (neglect reactions that are the reverse of reactions already written and reactions that differ from the others only by a multiplicative factor. Also ignore reactions that are nonproductive (that is, reactions in which the reactants and products are identical). Choose balanced equations that have the lowest possible, whole-number coefficients. (c) Give the numerical values of the equilibrium constants for each of these reactions.

9.115 It is possible to choose a different standard state (after all, molecules don't know what their standard state is) for the pressure or concentration of chemical species. The standard state corresponding to data presented in Appendix 2A is for $P = 1$ bar or

concentration 1 M, as appropriate. Calculate ΔG_f° for the following species if the standard state condition were changed to that listed (the temperature remains at 298 K):
(a) HI(g), standard state corresponds to 1 atm
(b) CO(g), standard state corresponds to 1 atm
(c) HCN(g), standard state corresponds to 1 Torr
(d) CH_4(g), standard state corresponds to 1 Pa

9.116 The boron trihalides can undergo redistribution reactions of the type

$$BCl_3(g) + BBr_3(g) \rightleftharpoons BCl_2Br(g) + BClBr_2(g)$$

(a) From standard reference sources, find the standard free energies of formation of the boron halides given in the equation above (at 298.15 K).
(b) What other equilibrium reactions are possible that will interconvert these species?
(c) Before doing any calculation, predict which compound will be present in the largest amount in an equilibrium system resulting from mixing equal numbers of moles of BCl_3 and BBr_3.
(d) Without doing any calculation, determine whether it is required that the partial pressure of BCl_2Br equal the partial pressure of $BClBr_2$ at equilibrium.
(e) From the standard free energies of formation, calculate the partial pressure of each species present at equilibrium if a vessel is filled initially with BCl_3(g) at 1.00 bar and BBr_3(g) at 1.00 bar. You will need to use a computer program that can solve a set of simultaneous equations.

9.117 The halogen atom in haloalkanes may undergo exchange reactions with halide ions on boron halides:

$$CH_3Cl(g) + BBr_3(g) \rightleftharpoons CH_3Br(g) + BClBr_2(g)$$

(a) From standard reference sources, find the standard free energies of formation of chloromethane gas (CH_3Cl) and bromomethane gas (CH_3Br) at 298.15 K.
(b) From the data in (a) and data obtained for Exercise 9.116, determine the equilibrium constant for the reaction written above.
(c) Given the information in Exercise 9.116, what other species will be present at equilibrium?
(d) If a vessel is filled with CH_3Cl(g) at 1.000 bar and BBr_3(g) at 1.000 bar, calculate the concentration of each compound that will be present at equilibrium. You will need a computer program that can solve a series of simultaneous equations.

CHAPTER

Citrus fruits—like these oranges—contain acidic juices. A characteristic of aqueous acidic solutions is the presence of hydronium ions—one can be seen among the water molecules. There is a continuing dynamic equilibrium between the acid molecules and the solvent in the solution.

Why Do We Need to Know This Material?

Acids and bases are two of the most common types of substances in the laboratory and in the everyday world. We need to know how to recognize them, what their characteristic reactions are, and why they are such important chemicals. This chapter extends to aqueous systems the concepts of equilibrium developed in Chapter 9 for gaseous systems. Equilibria between acids, bases, and water in plant and animal cells are vital for the survival of individual organisms. To sustain human societies, we also need to control—using the concepts introduced in this chapter—the acidity of rain, of natural waters such as lakes and rivers, and of municipal water supplies.

What Do We Need to Know Already?

This chapter builds on the introduction to acids and bases presented in Section J. It also draws on and illustrates the principles of thermodynamics (Chapters 6 and 7) and chemical equilibrium (Chapter 9). To a smaller extent, it uses the concepts of hydrogen bonding (Section 5.5), bond polarity (Section 2.14), and bond strength (Sections 2.16 and 2.17).

Acids and Bases

Acids and bases are found almost everywhere. Inside each living cell is a factory that manufactures the acids and bases that support life and control the composition of our blood and cellular fluids. Acids and bases affect the flavor, quality, and digestion of our food. Acid rain is a concern because it contributes to the stress that transportation and industrial activity impose on the environment. Almost all the consumer products that surround us—synthetic fabrics, dyes, metals, plastics—have made use of acids and bases in the course of their manufacture. In the laboratory, many of the reactions that chemists carry out to study the properties of materials or create new kinds of matter involve acids and bases.

In this chapter, we see why some substances are acids, why some acids are stronger than others, and why acids—like their partners the bases—are so important. We shall see that we can use thermodynamics, particularly equilibrium constants, to discuss the strengths of acids and bases quantitatively, and thereby develop our insight into the behavior of species in solution.

PROTON TRANSFER REACTIONS

Chemists sometimes open up a new field of study by taking a simple definition that has worked well for a certain type of substance or reaction, noticing that many other substances display a similar pattern, and then generalizing the original definition. The reactions of acids and bases are an excellent illustration of this approach. These reactions were first identified by studying solutions of acids and bases in water, which led to the Arrhenius definitions of acids and bases (Section J). However, chemists discovered that reactions between acids and bases also take place in nonaqueous solutions. They can even take place in the absence of solvent. The original definitions had to be discarded and replaced by more general definitions that encompassed this new knowledge. Following the pattern established earlier in this book, we begin by seeing how these new definitions enriched our understanding of the properties of matter and then go on to see how to express that new understanding quantitatively.

10.1 Brønsted-Lowry Acids and Bases

In 1923, the Danish chemist Johannes Brønsted proposed that

An **acid** is a proton donor

A **base** is a proton acceptor

(Fig. 10.1). The term *proton* in these definitions refers to the hydrogen ion, H^+. A species containing an **acidic hydrogen atom**, which is a hydrogen atom that can release its nucleus, a proton, is an acid. The same definitions were proposed independently by the English chemist Thomas Lowry, and the theory based on them is called the **Brønsted-Lowry theory**

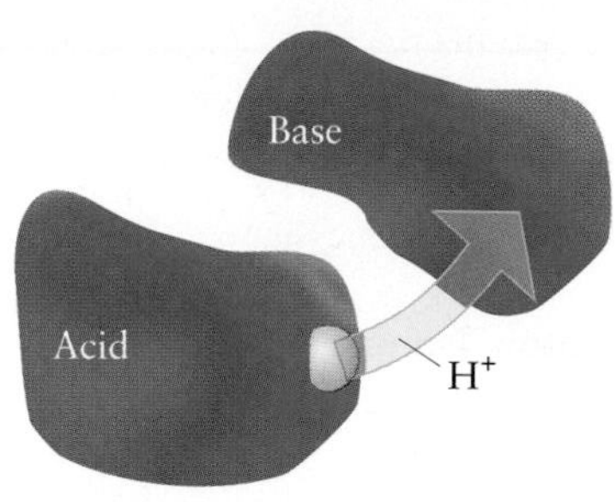

FIGURE 10.1 A Brønsted acid is a proton donor, and a Brønsted base is a proton acceptor. In this context, a proton is an H^+ ion.

of acids and bases. Proton transfer, and the rapid equilibrium established between proton donors and acceptors in solution, is the central theme of this and the following chapter.

Even though a substance can be *classified* as an acid in the absence of a base, it can *act* as an acid only if a base is present. For example, HCl is a Brønsted acid. In the gas phase, an HCl molecule holds on to its proton. However, when hydrogen chloride dissolves in water, the hydrogen atom forms a hydrogen bond to the oxygen atom of a neighboring H_2O molecule and then its proton immediately migrates to the water molecule, which here is acting as a base (Fig. 10.2). This process is a **proton transfer reaction**, a reaction in which a proton is transferred from one species to another. We say that the HCl molecule becomes **deprotonated:**

$$HCl(aq) + H_2O(l) \longrightarrow H_3O^+(aq) + Cl^-(aq)$$

Because essentially all the HCl molecules donate their protons to water, HCl is classified as a *strong acid.* At equilibrium, virtually all the HCl molecules have donated their protons, and the proton transfer reaction has gone to completion. The H_3O^+ ion, which will play a central role in the discussion in this chapter and the next, is called the *hydronium ion*. It is strongly hydrated in solution, and there is some evidence that a better representation of the species is $H_9O_4^+$. However, formulating the hydronium ion as H_3O^+ is adequate for all our purposes and reminds us that proton transfer always takes place between an acid and a base; H^+ itself does not exist in water.

Another example of an acid is hydrogen cyanide, HCN, which also transfers its proton to water when it forms the solution known as hydrocyanic acid, HCN(aq). However, only a small fraction of the HCN molecules donate their protons. At equilibrium, both HCN molecules and CN^- ions are present in hydrocyanic acid. We write the proton transfer equilibrium as

$$HCN(aq) + H_2O(l) \rightleftharpoons H_3O^+(aq) + CN^-(aq)$$

Because only a fraction of the HCN molecules donate their protons, we classify HCN as a *weak acid* in water. As in any chemical reaction, the equilibrium between HCN and its deprotonated form, CN^-, is dynamic. In a molecular picture of hydrocyanic acid, we would think of protons ceaselessly exchanging between HCN and H_2O molecules, such that there is a constant, but low concentration of CN^- and H_3O^+ ions.

The Brønsted definition also includes the possibility that an ion is an acid (an option not allowed by the Arrhenius definition). For instance, a hydrogen carbonate ion, HCO_3^-, one of the species present in natural waters, can act as a weak proton donor. In water, HCO_3^- takes part in a proton transfer equilibrium (Fig. 10.3):

$$HCO_3^-(aq) + H_2O(l) \rightleftharpoons H_3O^+(aq) + CO_3^{2-}(aq)$$

In the Brønsted-Lowry theory, the strength of an acid depends on the extent to which it donates protons to the solvent. We can therefore summarize the distinction between strong and weak acids as follows:

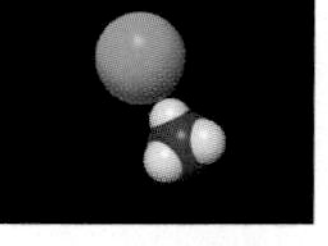

View an animation of the proton transfer reaction between HCl and H_2O molecules.

Media Link **Animation:** Fig. 10.2, Reaction of HCl with water

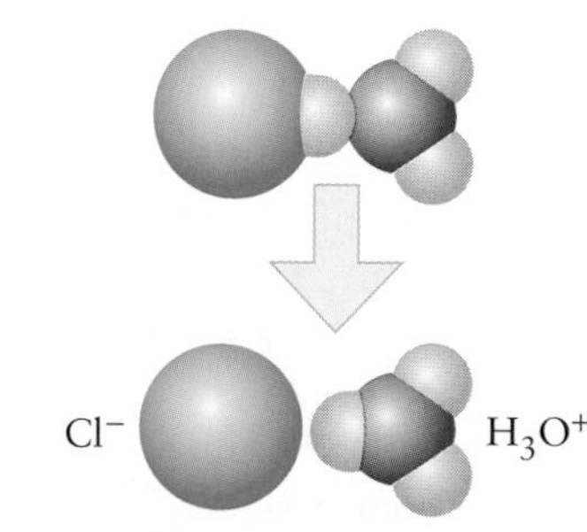

FIGURE 10.2 When an HCl molecule dissolves in water, a hydrogen bond forms between the H atom of HCl and the O atom of a neighboring H_2O molecule. The nucleus of the hydrogen atom is pulled out of the HCl molecule to become part of a hydronium ion.

FIGURE 10.3 Stalactites hang from the roof of a cave and stalagmites grow from the floor. Both are caused by the formation of insoluble calcium carbonate from the more soluble hydrogen carbonate carried into the cave by groundwater.

A strong acid is fully deprotonated in solution.

A weak acid is only partially deprotonated in solution.

In almost all the cases we consider, water is the solvent. However, solvents other than water are often used, and an acid that is strong in water may be weak in another solvent, or vice versa.

A Brønsted base is a proton acceptor, which in most cases means that it possesses a lone pair of electrons to which the proton can bond. An example of a Brønsted base is an oxide ion, such as the oxide ion in calcium oxide, CaO. When CaO dissolves in water, the strong electric field of the small, highly charged O^{2-} ion pulls a proton out of a neighboring H_2O molecule (Fig. 10.4). By accepting a proton, the oxide ion has become **protonated.** Every oxide ion present accepts a proton from water, so O^{2-} is an example of a *strong base* in water, a species that is fully protonated:

$$O^{2-}(aq) + H_2O(l) \longrightarrow 2\ OH^-(aq)$$

Another example of a Brønsted base is ammonia. When ammonia dissolves in water, it is protonated by H_2O:

$$NH_3(aq) + H_2O(l) \rightleftharpoons NH_4^+(aq) + OH^-(aq)$$

The lone pair of electrons on the N atom in the neutral NH_3 molecule has much less proton-pulling power than the double negative charge of the oxide ion. As a result, only a very small proportion of the NH_3 molecules are converted into NH_4^+ ions (Fig. 10.5). Ammonia is therefore an example of a *weak base.* All amines (Section 19.8) are weak bases in water. When we look at an aqueous solution of ammonia, we can imagine the protons ceaselessly exchanging between NH_3 and H_2O molecules, such that there is a constant, low concentration of NH_4^+ and OH^- ions.

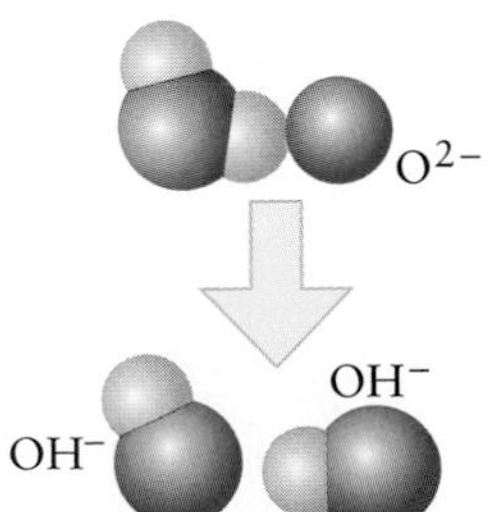

FIGURE 10.4 When an oxide ion is present in water, it exerts such a strong attraction on the nucleus of a hydrogen atom in a neighboring water molecule that the nucleus is pulled out of the molecule. As a result, the oxide ion forms two hydroxide ions.

Media Link Animation: Fig. 10.5, Proton transfer reaction of ammonia and water

View an animation of ammonia reacting with water.

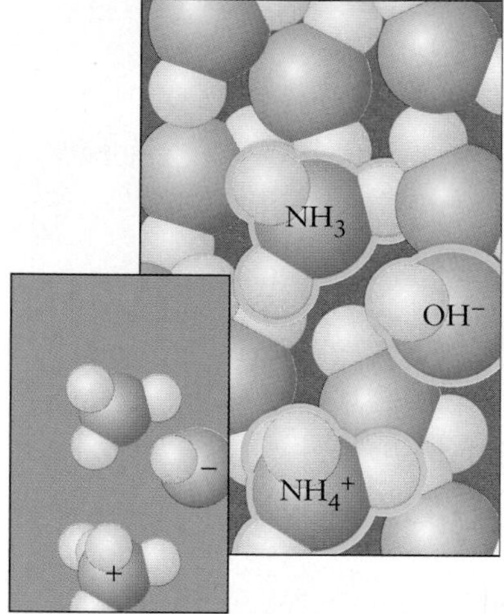

FIGURE 10.5 In this molecular portrayal of the structure of a solution of ammonia in water, we see that there are NH_3 molecules still present because only a small percentage of them have been protonated by transfer of hydrogen ions from water. In a typical solution, only about 1 in 100 NH_3 molecules is protonated. The overlay shows only the solute species.

We can summarize the distinction between strong and weak bases as follows:

A strong base is completely protonated in solution.

A weak base is only partially protonated in solution.

As for acids, this distinction depends on the solvent, and a base that is strong in water may be weak in another solvent, or vice versa.

Proton transfer in solution reaches a dynamic equilibrium very rapidly. Therefore, for all weak acids and bases, it is essential to consider the reverse as well as the forward reaction. For example, the CN^- ion produced when HCN loses a proton to water can accept a proton from a water molecule and form HCN again. Therefore, according to the Brønsted definition, CN^- is a base: it is called the "conjugate base" of HCN. In general, a **conjugate base** is the species left when an acid donates a proton:

$$\text{Acid} \xrightarrow{\text{denotes } H^+} \text{conjugate base}$$

Because HCN is an acid that we can regard as formed by proton transfer to a cyanide ion, it is the "conjugate acid" of the base CN^-. In general, a **conjugate acid** is the species formed when a base accepts a proton:

$$\text{Base} \xrightarrow{\text{accepts } H^+} \text{conjugate acid}$$

How can we use this information . . . to write the formulas of conjugate acids and bases?

The formula of the conjugate base of an acid has one less H^+ ion than that of the acid: CO_3^{2-} is the conjugate base of HCO_3^-, and O^{2-} is the conjugate base of OH^-. The formula of the conjugate acid of a base has one more H^+ ion than the base: HCO_3^- is the conjugate acid of CO_3^{2-} and OH^- is the conjugate acid of O^{2-}.

Self-Test 10.1A What is (a) the conjugate acid of OH^-; (b) the conjugate base of HPO_4^{2-}?

[*Answer:* (a) H_2O; (b) PO_4^{3-}]

Self-Test 10.1B What is (a) the conjugate acid of H_2O; (b) the conjugate base of NH_3?

Whereas the Arrhenius definitions are restricted to aqueous solutions, the Brønsted definitions of acids and bases also apply to species in nonaqueous solvents. For example, when pure acetic acid is added to liquid ammonia, proton transfer takes place and the following equilibrium is reached:

$$CH_3COOH(l) + NH_3(l) \rightleftharpoons CH_3CO_2^-(am) + NH_4^+(am)$$

(The label *am* indicates a species dissolved in liquid ammonia.) Protons can be transferred from acids to bases even in the gas phase, in the absence of a solvent. For example, hydrogen chloride and ammonia gases react by proton transfer when they mingle together. They produce the fine powder of ammonium chloride often seen coating surfaces in chemical laboratories:

$$HCl(g) + NH_3(g) \longrightarrow NH_4Cl(s)$$

An acid is a proton donor and a base is a proton acceptor. The conjugate base of an acid is the base formed when the acid has donated a proton. The conjugate acid of a base is the acid that forms when the base has accepted a proton.

10.2 Proton Transfer as a Lewis Acid-Base Reaction

The Brønsted-Lowry theory focuses on the transfer of a proton from one species to another. However, chemists look for patterns of reactions that are not restricted to a single, specific species but apply to a wide range of species with a common characteristic. That enables them to transfer their experience with one type of substance to a much wider range of substances. We can apply this approach to show that the concepts of acids and bases have a much wider significance than the transfer of protons.

The crucial insight is that proton transfer is a special type of Lewis acid-base reaction. Recall from Sections 2.12 and 2.13 that a Lewis acid is defined as an electron pair acceptor and a Lewis base is defined as an electron pair donor. A proton (H^+) is an electron pair acceptor, and therefore a Lewis acid, because it can attach to a lone pair of electrons on a Lewis base. In other words, a Brønsted acid is a supplier of one particular Lewis acid, a proton. Likewise, a Brønsted base is a special kind of Lewis base, one that can use a lone pair to bond to a proton.

The oxide ion is a Lewis base. It forms a coordinate covalent bond to a proton, a Lewis acid, by supplying both the electrons for the bond; in the process, it becomes a hydroxide ion:

$$:\ddot{\underset{..}{O}}:^{2-} + H{-}\ddot{\underset{..}{O}}{-}H \longrightarrow :\ddot{\underset{..}{O}}{-}H^- + :\ddot{\underset{..}{O}}{-}H^-$$

(The curved arrows show the direction in which the electrons can be thought to move.) Most soluble metal oxides are strong bases in water. Similarly, when the Lewis base ammonia, NH_3, dissolves in water, some of the molecules accept protons from water molecules:

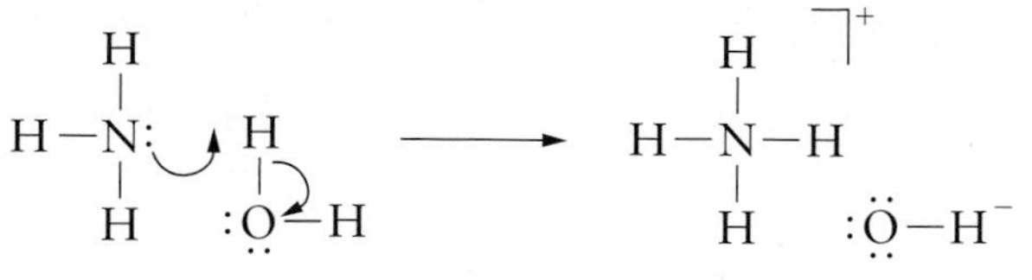

$$H_3N{:} + H{-}\ddot{O}{-}H \longrightarrow [NH_4]^+ + {:}\ddot{O}{-}H^-$$

Whereas soluble metal oxides are strong bases, many nonmetal oxides react with water to give acids:

$$CO_2(g) + H_2O(l) \longrightarrow H_2CO_3(aq)$$

$$SO_2(g) + H_2O(l) \longrightarrow H_2SO_3(aq)$$

$$P_4O_{10}(S) + 6\ H_2O(l) \longrightarrow 4\ H_3PO_4(aq)$$

This acidic character stems from the ability of a nonmetal oxide molecule to act as a Lewis acid:

For clarity, we are showing only one of the lone pairs.

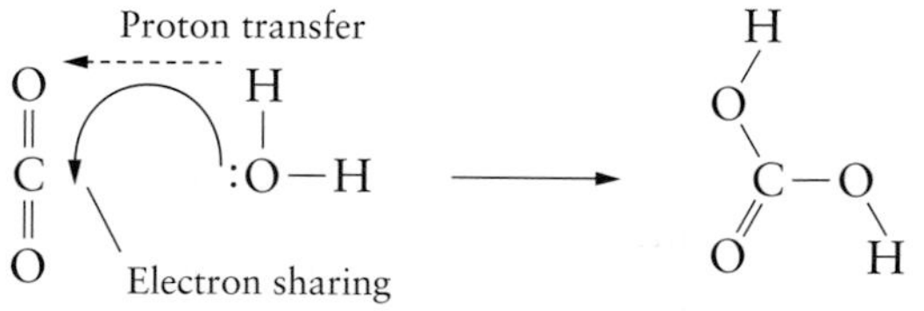

Proton transfer

$$O{=}C{=}O + H{-}\ddot{O}{-}H \longrightarrow (HO)(O{=})C{-}O{-}H$$

Electron sharing

In this reaction, the C atom of CO_2, the Lewis acid, accepts an electron pair from the O atom of a water molecule, the Lewis base, and a proton migrates from an H_2O oxygen atom to a CO_2 oxygen atom. The product, an H_2CO_3 molecule, is a Brønsted acid.

A proton is a Lewis acid that attaches to a lone pair provided by a Lewis base.

Self-Test 10.2A Identify (a) the Brønsted acids and bases in both reactants and products in the proton transfer equilibrium $HNO_2(aq) + HPO_4^{2-}(aq) \rightleftharpoons NO_2^-(aq) + H_2PO_4^-(aq)$. (b) Which species (not necessarily shown explicitly) is a Lewis acid and which are Lewis bases?

[*Answer:* (a) Acids, HNO_2, $H_2PO_4^-$; bases, HPO_4^{2-}, and NO_2^-; (b) Lewis acid, H^+; Lewis bases, HPO_4^{2-} and NO_2^-]

Self-Test 10.2B Identify (a) the Brønsted acids and bases in both reactants and products in the proton transfer equilibrium $HCO_3^-(aq) + NH_4^+(aq) \rightleftharpoons H_2CO_3(aq) + NH_3(aq)$. (b) Which species (not necessarily shown explicitly) is a Lewis acid and which are Lewis bases?

10.3 Acidic, Basic, and Amphoteric Oxides

As illustrated in Section 10.2, some oxides can react with water as Lewis acids or bases. An **acidic oxide** is an oxide that reacts with water to form a Brønsted acid; a **basic oxide** is an oxide that reacts with water to form a Brønsted base. Acidic oxides are molecular compounds that react with bases to give a salt and water; similarly, basic oxides are ionic compounds that react with acids to give a salt and water. For instance, magnesium oxide, a basic oxide, reacts with hydrochloric acid:

$$MgO(s) + 2\,HCl(aq) \longrightarrow MgCl_2(aq) + H_2O(l)$$

In this reaction, the base O^{2-} accepts two protons from the hydronium ions present in hydrochloric acid. Carbon dioxide, an acidic oxide, reacts with sodium hydroxide:

$$2\,NaOH(aq) + CO_2(g) \longrightarrow Na_2CO_3(aq) + H_2O(l)$$

The white crust often seen on pellets of sodium hydroxide is a mixture of sodium carbonate formed in this way and of sodium hydrogen carbonate formed in a similar reaction:

$$NaOH(aq) + CO_2(g) \longrightarrow NaHCO_3(aq)$$

In both these reactions, the Lewis base OH^- present in aqueous sodium hydroxide attacks the Lewis acid CO_2.

Metals tend to form basic oxides and nonmetals tend to form acidic oxides, but what about the elements that lie on the diagonal frontier between the metals and nonmetals? On this frontier, metallic character blends into nonmetallic character, and the oxides of these elements have both acidic and basic character (Fig. 10.6). As you can see, these elements lie in a diagonal band across the table from beryllium to polonium. Substances that react with both acids and bases are classified as **amphoteric,** from the Greek word for "both." For example, aluminum oxide, Al_2O_3, is amphoteric. Its reaction with acids shows that it is basic:

$$Al_2O_3(s) + 6\,HCl(aq) \longrightarrow 2\,AlCl_3(aq) + 3\,H_2O(l)$$

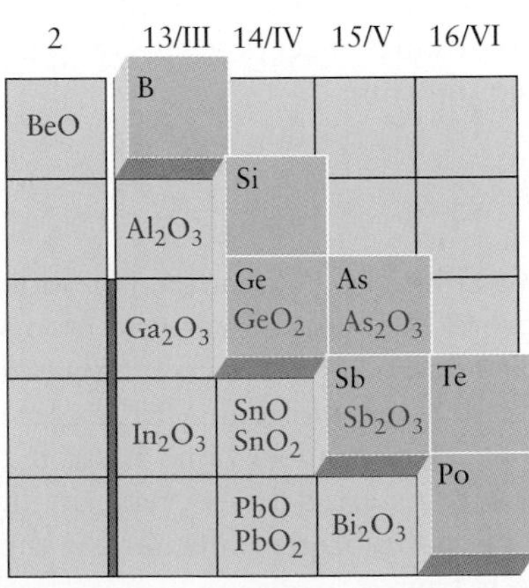

FIGURE 10.6 The elements in and close to the diagonal line of metalloids (shown as blocks) typically form amphoteric oxides, but their ability to do so depends on their oxidation state.

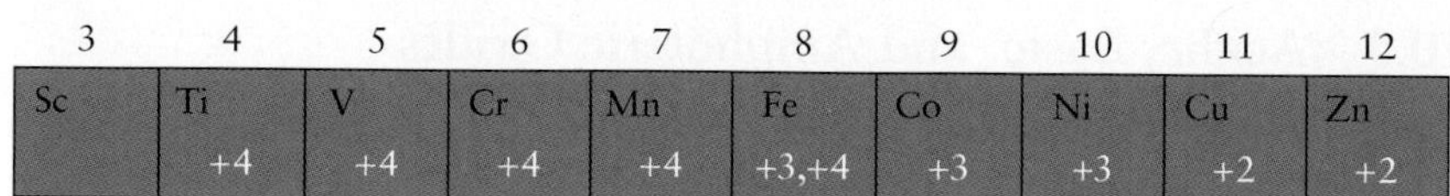

FIGURE 10.7 Certain elements of the *d* block form amphoteric oxides, particularly in oxidation states intermediate in their range (as shown here for the first series).

However, its reaction with bases shows that it is acidic:

$$2\ NaOH(aq) + Al_2O_3(s) + 3\ H_2O(l) \longrightarrow 2\ Na[Al(OH)_4](aq)$$

The product of the second reaction is sodium aluminate, which contains the aluminate ion, $Al(OH)_4^-$. Other main-group elements that form amphoteric oxides are shown in Fig. 10.7. The acidic, amphoteric, or basic character of the oxides of the *d*-block metals depends on their oxidation state (see Chapter 16).

The chemistry of acidic and basic oxides is explored further in Chapters 14–16.

Metals form basic oxides, nonmetals form acidic oxides, and the elements on a diagonal line from beryllium to polonium, and several d-block metals, form amphoteric oxides.

10.4 Proton Exchange Between Water Molecules

One of the most startling implications of the Brønsted definitions of acids and bases is that the same substance can function as both an acid and a base! For example, we have seen that a water molecule accepts a proton from an acid molecule (such as HCl or HCN) to form an H_3O^+ ion. So water is a base. However, a water molecule can donate a proton to a base (such as O^{2-} or NH_3) and become an OH^- ion. So water is also an acid. We describe water as **amphiprotic,** meaning that it can act as both a proton donor and a proton acceptor.

Because water is amphiprotic—because it is both an acid and a base—proton transfer between water molecules occurs even in pure water. The reaction is very fast, and the equilibrium

$$2\ H_2O(l) \rightleftharpoons H_3O^+(aq) + OH^-(aq) \qquad \text{(A)}$$

is always present in water and in aqueous solutions. Whenever you look at a glass of water, you should think of protons from the hydrogen atoms as ceaselessly migrating between the molecules. This type of reaction, in which one molecule transfers a proton to another molecule of the same kind, is called **autoprotolysis** (Fig. 10.8). Because an O–H bond is strong, we should expect the fraction of protons transferred to be very small. Thermodynamics enables us to estimate the equilibrium constant for the reaction and see whether this expectation is true.

Another common term for autoprotolysis is *autoionization.*

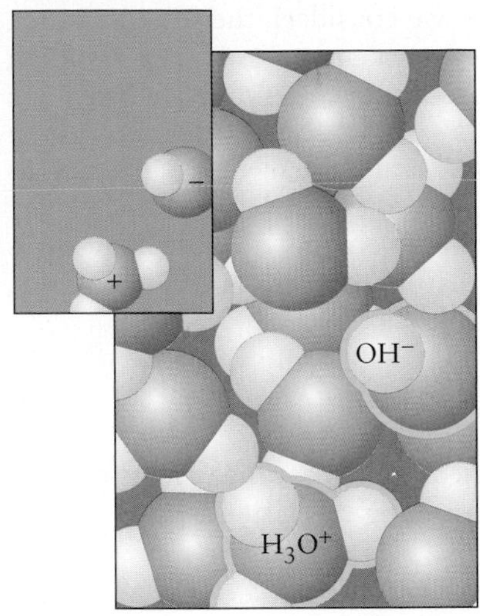

FIGURE 10.8 As a result of autoprotolysis, pure water consists of hydronium ions and hydroxide ions as well as water molecules. The concentration of ions that results from autoprotolysis is only about 10^{-7} mol·L^{-1}, so only about 1 molecule in 200 million is ionized. The overlay shows only the ions.

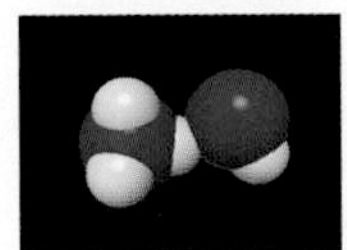

View an animation of proton transfer between water molecules.

Media Link Animation: Fig. 10.8, Autoprotolysis of water

How do we do that?

We can calculate equilibrium constants—and in particular the equilibrium constant for the autoprotolysis reaction—from standard reaction free energies (Section 9.3). From the data in Appendix 2A for reaction A, we can write

$$\begin{aligned}\Delta G_r{}^\circ &= \Delta G_f{}^\circ(H_3O^+, aq) + \Delta G_f{}^\circ(OH^-, aq) - 2\Delta G_f{}^\circ(H_2O, l)\\ &= (-237.13) + (-157.24) - 2 \times (-237.13)\ \text{kJ·mol}^{-1}\\ &= +79.89\ \text{kJ·mol}^{-1}\end{aligned}$$

The reaction has a large positive standard free energy, so we expect the equilibrium constant to be much smaller than 1 and, therefore, for very few molecules of water to be present as ions at equilibrium. Specifically, from Eq. 10 of Chapter 9, at 298 K

$$\begin{aligned}\ln K &= -\frac{\Delta G_r{}^\circ}{RT} = -\frac{7.989 \times 10^4\ \text{J·mol}^{-1}}{(8.3145\ \text{J·K}^{-1}\text{·mol}^{-1}) \times (298\ \text{K})}\\ &= -\frac{7.989 \times 10^4}{8.3145 \times 298}\\ &= -32.24\end{aligned}$$

Taking the natural antilogarithm (e^x) of ln K yields $K = 1.0 \times 10^{-14}$ at 298 K. This very small value indicates that only a tiny proportion of water molecules are present as ions, as we suspected.

Let's see whether the thermodynamic prediction is supported experimentally. The equilibrium constant for reaction A is

$$K = \frac{a_{H_3O^+}\, a_{OH^-}}{a_{H_2O}{}^2}$$

In dilute aqueous solutions (the only ones we consider), the solvent, water, is very nearly pure, so its activity may be taken to be 1. The resulting expression is called the **autoprotolysis constant** of water and is written K_w:

K_w is also widely called the *autoionization constant* and sometimes the *ion product constant* of water.

$$K_w = a_{H_3O^+}\,a_{OH^-} \tag{1a}$$

As we saw in Section 9.3, the activity of a solute J in a dilute solution is approximately equal to the molarity relative to the standard molarity, [J]/[J]°, so a practical form of this expression is

$$K_w = [H_3O^+][OH^-] \tag{1b}$$

where, as in Chapter 9, we have simplified the appearance of the expression by replacing [J]/[J]° by [J] and interpreting the latter as the value of the molarity with the units struck out.

In pure water at 25°C, the molar concentrations of H_3O^+ and OH^- are equal and are known by experiment to be 1.0×10^{-7} mol·L^{-1}, so

$$K_w = (1.0 \times 10^{-7}) \times (1.0 \times 10^{-7}) = 1.0 \times 10^{-14}$$

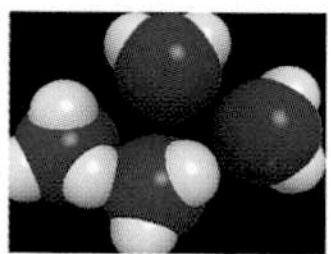

View animations that show how hydronium and hydroxide ions move through water by proton transfer.

This value agrees with the value obtained from thermodynamics. The concentrations of H_3O^+ and OH^- are very low in pure water, which explains why pure water is such a poor conductor of electricity. To imagine the very tiny extent of autoprotolysis, think of each letter in this book as a water molecule. We would need to search through more than 50 books to find one ionized water molecule.

Now we come to a very important point that will be the basis of much of the discussion in this chapter and the next. *Because K_w is an equilibrium constant, the product of the concentrations of H_3O^+ and OH^- ions is always equal to K_w.* We can increase the concentration of H_3O^+ ions by adding acid, but the concentration of OH^- ions will simultaneously decrease to preserve the value of K_w. Alternatively, we can increase the concentration of OH^- ions by adding base, but the concentration of H_3O^+ ions will then decrease. The autoprotolysis equilibrium links the concentrations of H_3O^+ and OH^- ions rather like a seesaw: when one goes up, the other must go down (Fig. 10.9).

In aqueous solutions, the concentrations of H_3O^+ and OH^- ions are related by the autoprotolysis equilibrium; if one concentration is increased, then the other must decrease to maintain the value of K_w.

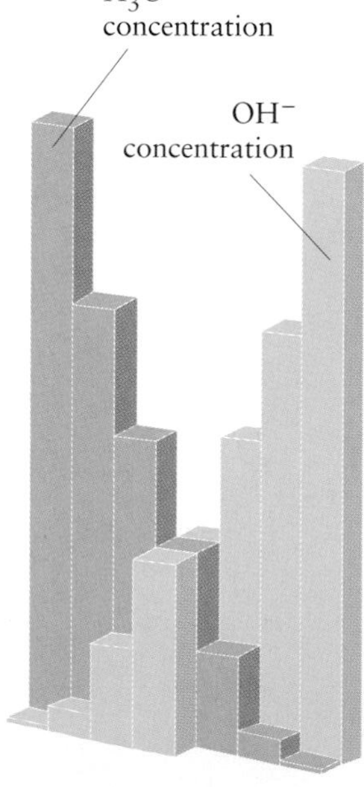

Media Link Animation: Fig. 10.9, Migration of hydronium and hydroxide ions through water

FIGURE 10.9 The product of the concentrations of hydronium and hydroxide ions in water (pure water and aqueous solutions) is a constant. If the concentration of one type of ion increases, the other must decrease to keep the product of concentrations constant.

EXAMPLE 10.1 Calculating the molarities of ions in a solution of a strong base

What are the molarities of H_3O^+ and OH^- in 0.0030 M $Ba(OH)_2$(aq) at 25°C?

STRATEGY The anion of a strong base is virtually entirely present as OH^- in water. First, decide whether the base is strong; if so, decide from the chemical formula how many OH^- ions are provided by each formula unit. Then calculate the molarity of these ions in the solution. To find the molarity of H_3O^+ ions, use the water autoprotolysis constant $K_w = [H_3O^+][OH^-]$.

SOLUTION Barium hydroxide is a hydroxide of an alkaline earth metal, so it is a strong base (see Table J.1). The equation

$$Ba(OH)_2(s) \longrightarrow Ba^{2+}(aq) + 2\ OH^-(aq)$$

tells us that 1 mol $Ba(OH)_2 \simeq$ 2 mol OH^-. Because the molarity of $Ba(OH)_2(aq)$ is 0.0030 mol·L^{-1}, it follows that the molarity of OH^- is twice that value, or 0.0060 mol·L^{-1}. Then, for the molarity of H_3O^+ ions, we write

$$[H_3O^+] = \frac{K_w}{[OH^-]} = \frac{1.0 \times 10^{-14}}{0.0060} = 1.7 \times 10^{-12}$$

That is, the molarity of H_3O^+ ions in the solution is 1.7×10^{-12} mol·L^{-1}.

Self-Test 10.3A Estimate the molarities of (a) H_3O^+ and (b) OH^- at 25°C in 6.0×10^{-5} M HI(aq).

[*Answer:* (a) 6.0×10^{-5} mol·L^{-1}; (b) 1.7×10^{-10} mol·L^{-1}]

Self-Test 10.3B Estimate the molarities of (a) H_3O^+ and (b) OH^- at 25°C in 2.2×10^{-3} M NaOH(aq).

10.5 The pH Scale

So far, we have seen *qualitatively* that hydronium ions are present in water and that, in an aqueous solution of an acid or base, the concentration of H_3O^+ ions depends on the concentration of solute. The time has come for us to express this concentration *quantitatively* and to see how it depends on the concentration of acid or base in solution. However, a minor problem is that, depending on the identity of the solute and its concentration, the molarity of H_3O^+ ion varies over many orders of magnitude: in some solutions, it may be higher than 1 mol·L^{-1} and, in others, it may be lower than 10^{-14} mol·L^{-1}. Chemists avoid the awkwardness of working with a wide range of values by using logarithms, which condense the values into a much smaller and more convenient range (Fig. 10.10). Therefore, the hydronium ion molarity is usually reported in terms of the **pH** of the solution:

The pH scale was introduced by the Danish chemist Søren Sørensen in 1909 in the course of his work on quality control in the brewing of beer and is now used throughout science, medicine, and engineering.

$$pH = -\log a_{H_3O^+} \qquad (2a)$$

where (for solutions so dilute that they may be treated as ideal) $a_{H_3O^+} = [H_3O^+]/[H_3O^+]°$. The logarithm in this definition is a common logarithm, to the base 10. As in Chapter 9, we simplify the appearance of this expression by interpreting $[H_3O^+]$ as the molarity of H_3O^+ with the units struck out, so we can write

$$pH = -\log\,[H_3O^+] \qquad (2b)$$

For example, the pH of pure water, in which the molarity of H_3O^+ ions is 1.0×10^{-7} mol·L^{-1} at 25°C, is

$$pH = -\log\,(1.0 \times 10^{-7}) = 7.00$$

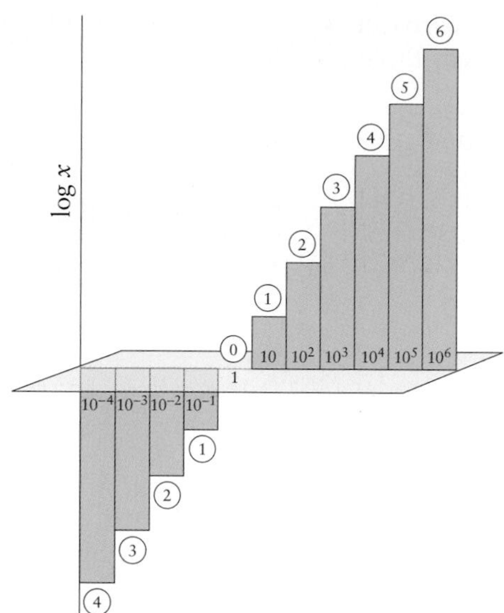

FIGURE 10.10 Very large ranges of numbers are difficult to represent graphically. However, their logarithms span a much smaller range that can be represented easily. Note how the numbers shown here range over 10 orders of magnitude (from 10^{-4} to 10^6), but their logarithms range over 10 units (from -4 to 6).

Most solutions we use in chemistry have a pH in the range 0 to 14, but values outside this range are possible.

The negative sign in the definition of pH means that the higher the H_3O^+ molarity, the lower the pH:

The negative logarithm is used so that most pH values are positive numbers.

- The pH of pure water is 7.
- The pH of an acidic solution is less than 7.
- The pH of a basic solution is greater than 7.

Notice that, as the concentration of an acid increases, the pH decreases. Because pH is proportional to the common logarithm of the concentration, a change of one pH unit means the molarity of H_3O^+ ions has changed by a factor of 10. For example, when the molarity of H_3O^+ increases by a factor of 10, from 10^{-5} mol·L^{-1} to 10^{-4} mol·L^{-1}, the pH decreases from 5 to 4.

EXAMPLE 10.2 Calculating a pH from a molarity

What is the pH of (a) human blood, in which the molarity of H_3O^+ ions is 4.0×10^{-8} mol·L^{-1}; (b) 0.020 M HCl(aq); (c) 0.040 M KOH(aq)?

STRATEGY The pH is calculated from Eq. 2. For strong acids, the molarity of H_3O^+ is equal to the molarity of the acid; for acids, expect pH < 7. For strong bases, first find the molarity of OH^-, then convert that molarity to $[H_3O^+]$ by using $[H_3O^+][OH^-] = K_w$, as described in Example 10.1; for bases, expect pH > 7. The number of digits following the decimal point in a pH value is equal to the number of significant figures in the corresponding molarity, because the digits preceding the decimal point simply report the power of 10 in the data (as in $\log 10^5 = 5$).

SOLUTION (a) For a solution in which the molarity of H_3O^+ ions is 4.0×10^{-8} mol·L^{-1},

$$pH = -\log (4.0 \times 10^{-8}) = 7.40$$

(b) Because HCl is a strong acid, it is completely deprotonated, so the molarity of H_3O^+ is 0.020 mol·L^{-1}. Hence,

$$pH = -\log 0.020 = 1.70$$

(c) Each formula unit of KOH (a strong base) provides one OH^- ion; therefore, the molarity of OH^- is 0.040 mol·L^{-1} and

$$[H_3O^+] = \frac{K_w}{[OH^-]} = \frac{1.0 \times 10^{-14}}{0.040} = 2.5 \times 10^{-13}$$

Hence

$$pH = -\log (2.5 \times 10^{-13}) = 12.60$$

Self-Test 10.4A Calculate the pH of (a) household ammonia, for which the OH^- molarity is about 3×10^{-3} mol·L^{-1}; (b) 6.0×10^{-5} M $HClO_4$(aq).

[*Answer:* (a) 11.5; (b) 4.22]

Self-Test 10.4B Calculate the pH of 0.077 M NaOH(aq).

We can measure the pH of an aqueous solution very quickly with a strip of *universal indicator paper*, which turns different colors at different pH values. More precise measurements are made with a *pH meter* (Fig. 10.11). This instrument consists of a voltmeter connected to two electrodes that dip into the solution. The difference in electrical potential between the electrodes is proportional to the pH (as will be explained in Section 12.11); so once the scale on the meter has been calibrated, the pH can be read directly. In the United States, the Environmental Protection

Indicators are discussed in Section 11.7.

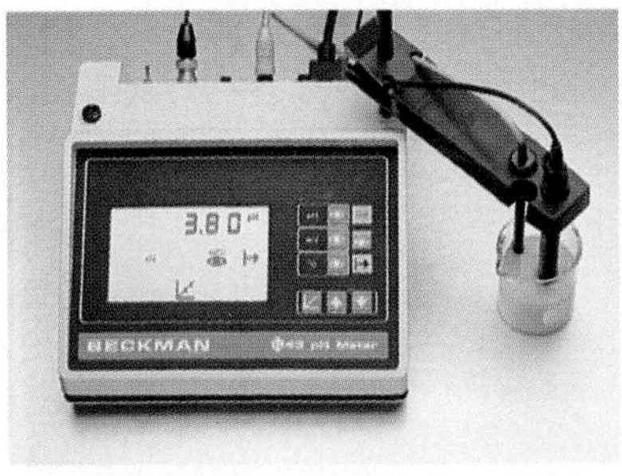

(a)

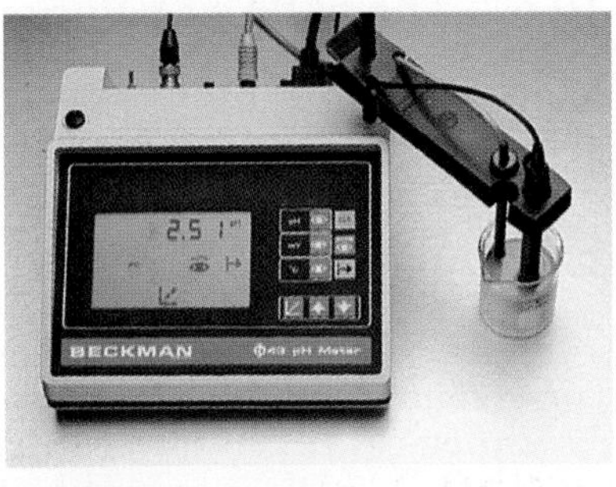

(b)

FIGURE 10.11 A pH meter is a voltmeter that measures the pH electrochemically. The two samples are (a) orange juice and (b) lemon juice. Notice that the latter has a lower pH and hence a higher concentration of hydronium ions.

Agency (EPA) defines waste as "corrosive" if its pH is either lower than 3.0 (highly acidic) or higher than 12.5 (highly basic). Lemon juice, with a pH of about 2.5, would be regarded as corrosive; so would solutions of 1 M NaOH(aq), for which the pH is 14.

To convert from pH to the molarity of H_3O^+, we reverse the sign of the pH and then take its antilogarithm:

$$[H_3O^+] = 10^{-pH}\ mol{\cdot}L^{-1} \quad (3)$$

How can we use this information . . . to calculate the hydronium ion concentration from the pH?

Suppose we need to find the hydronium ion concentration in a solution with pH = 4.83. Because pH < 7, the solution is acidic and we expect that $[H_3O^+] > 10^{-7}\ mol{\cdot}L^{-1}$. To calculate the precise value, we first change the sign of the pH, then take its antilogarithm:

$$[H_3O^+] = 10^{-4.83}\ mol{\cdot}L^{-1} = 1.5 \times 10^{-5}\ mol{\cdot}L^{-1}$$

Self-Test 10.5A The pH of stomach fluids is about 1.7. What is the molarity of H_3O^+ ions in the stomach?

[*Answer:* $2 \times 10^{-2}\ mol{\cdot}L^{-1}$]

Self-Test 10.5B The pH of pancreatic fluids, which help to digest food once it has left the stomach, is about 8.2. What is the approximate molarity of H_3O^+ ions in pancreatic fluids?

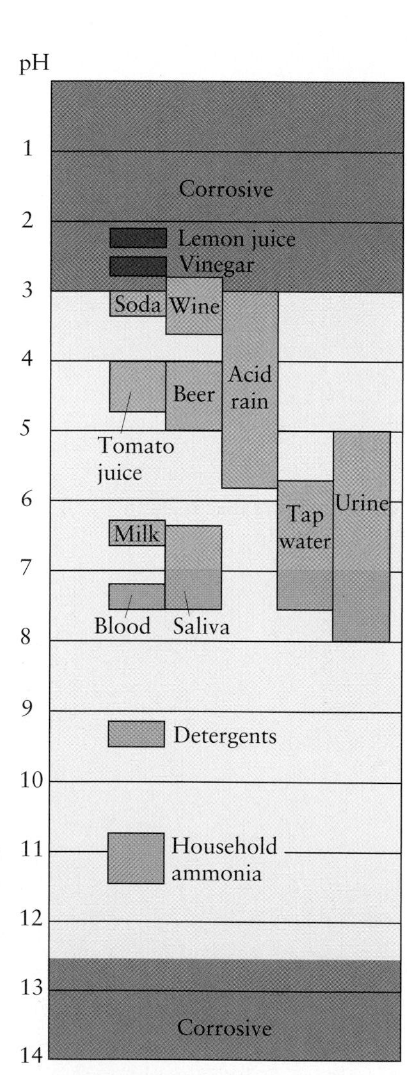

FIGURE 10.12 Typical pH values of common aqueous solutions. The brown regions indicate the pH ranges for liquids regarded as corrosive.

Figure 10.12 shows the results of measuring the pH of a selection of liquids and beverages. Fresh lemon juice, as we have seen, has a pH of 2.5, corresponding to an H_3O^+ molarity of $3 \times 10^{-3}\ mol{\cdot}L^{-1}$. Natural (unpolluted) rain, with an acidity due largely to dissolved carbon dioxide, typically has a pH of about 5.7.

The pH scale is used to report H_3O^+ molarity: $pH = -log[H_3O^+]$. High pH denotes a basic solution, low pH an acidic solution; a neutral solution has pH = 7.

10.6 The pOH of Solutions

Many quantitative expressions relating to acids and bases are greatly simplified and more easily remembered when we use logarithms. The quantity **pX** is a generalization of pH:

$$pX = -\log X \quad (4)$$

For example, **pOH** is defined as

$$pOH = -\log a_{OH^-} \quad (5a)$$

which, for the same reasons as pH, is normally simplified to

$$pOH = -\log[OH^-] \quad (5b)$$

The pOH is convenient for reporting the molarities of OH^- ions in solution. For example, in pure water, where the molarity of OH^- ions is 1.0 $\times 10^{-7}$ mol·L^{-1}, the pOH is 7.00. Similarly, by pK_w, we mean

$$pK_w = -\log K_w = -\log (1.0 \times 10^{-14}) = 14.00$$

By using logarithms, we can also derive a simple relation between the pH and pOH of a solution.

How do we do that?

To derive the relation between pH and pOH, we start by writing down the expression for the autoprotolysis constant of water: $K_w = [H_3O^+][OH^-]$. Then we take logarithms of both sides and use $\log ab = \log a + \log b$. We obtain

$$\log [H_3O^+] + \log [OH^-] = \log K_w$$

Multiplication of both sides of the equation by −1 gives

$$-\log [H_3O^+] - \log [OH^-] = -\log K_w$$

which is the same as

$$pH + pOH = pK_w \quad (6a)$$

Because pK_w = 14.00 at 25°C, at that temperature

$$pH + pOH = 14.00 \quad (6b)$$

Equation 6 shows that the pH and pOH of a solution have complementary values: if one increases, the other decreases such that their sum remains constant (Fig. 10.13).

The pH and pOH of a solution are related by the expression pH + pOH = pK_w.

WEAK ACIDS AND BASES

Solutions with the same concentration of acid may not have the same pH. For instance, when we measure the pH of 0.10 M CH_3COOH(aq), we find a value close to 3; but when we make the same measurement on 0.10 M HCl(aq), we find a pH close to 1. We have to conclude that the H_3O^+ molarity in 0.10 M CH_3COOH(aq) is *lower* than that in 0.10 M HCl(aq). Similarly, we find that the molarity of H_3O^+ is *higher* in 0.10 M NH_3(aq) than it is in 0.10 M NaOH(aq). The explanation must be that CH_3COOH is not fully deprotonated and NH_3 is not fully protonated in water. That is, acetic acid and ammonia are, respectively, a weak acid and a weak base. The incomplete deprotonation of CH_3COOH explains why solutions of HCl and CH_3COOH with the same molarity react with a metal at different rates (Fig. 10.14).

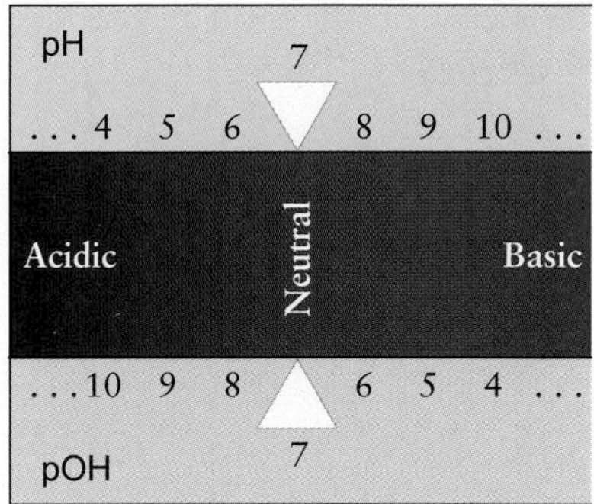

FIGURE 10.13 The numbers along the top of the rectangle are values of pH for a range of solutions and those on the bottom are the pOH values of the same solutions. Neutrality corresponds to a pH of 7; smaller values of pH correspond to acidic solutions, larger values to basic solutions. Most values of pH and pOH lie in the range 1 to 14, but pH and pOH values can lie outside this range and even be negative.

FIGURE 10.14 Equal masses of magnesium metal have been added to solutions of HCl, a strong acid (left), and CH_3COOH, a weak acid (right). Although the acids have the same concentrations, the rate of hydrogen evolution, which depends on the concentration of hydronium ions, is much greater in the strong acid.

Most acids and bases that exist in nature are weak. For example, the natural acidity of river water is generally due to the presence of carbonic acid (H_2CO_3, from dissolved CO_2), hydrogen phosphate ions, HPO_4^{2-} and dihydrogen phosphate ions, $H_2PO_4^-$, from fertilizer runoff, or carboxylic acids arising from the degradation of plant tissues. Similarly, most naturally occurring bases are weak. They often arise from the decomposition in the absence of air of compounds containing nitrogen. For example, the odor of dead fish is due to amines, which are weak bases.

Lemon juice, which contains citric acid, is often added to fish dishes to help eliminate the smell of some of these amines.

In this part of the chapter, we develop a quantitative measure of the strengths of weak acids and bases. We then use this information to explore how acid strength is related to molecular structure.

10.7 Acidity and Basicity Constants

When we imagine the composition of a solution of a weak acid in water, we think of a solution that contains

1 The acid molecules or ions.

2 Small concentrations of H_3O^+ ions and the conjugate base of the acid formed by proton transfer to water molecules.

3 A very, very small concentration of OH^- ions, which maintain the autoprotolysis equilibrium.

All these species are in ceaseless dynamic equilibrium. Similarly, for a solution of a weak base, we imagine

1 The base molecules or ions.

2 Small concentrations of OH^- ions and the conjugate acid of the base formed by proton transfer from water molecules.

3 A very, very small concentration of H_3O^+ ions that maintain the autoprotolysis equilibrium.

Because conjugate acids and bases are in equilibrium in solution, we can express the composition of a solution of an acid or a base in terms of the equilibrium constant for proton transfer between the solute and the solvent. For example, for acetic acid in water,

$$CH_3COOH(aq) + H_2O(l) \rightleftharpoons H_3O^+(aq) + CH_3CO_2^-(aq) \qquad \text{(B)}$$

the equilibrium constant is

$$K = \frac{a_{H_3O^+}\, a_{CH_3CO_2^-}}{a_{CH_3COOH}\, a_{H_2O}} \qquad (7)$$

Because the solutions we are considering are dilute and the water is almost pure, the activity of H_2O can be set equal to 1. The resulting expression is called an **acidity constant,** K_a. If we make the further approximation of replacing the activities of the solute species by the numerical values of their molarities, we can write the acidity constant expression as

K_a is also widely called the *acid ionization constant* or *acid dissociation constant.*

$$K_a = \frac{[H_3O^+][CH_3CO_2^-]}{[CH_3COOH]} \qquad (8)$$

The experimental value of K_a for acetic acid at 25°C is 1.8×10^{-5}. This small value tells us that only a small proportion of CH_3COOH molecules donate their protons when dissolved in water. Depending on the concentration, about 99 out of 100 CH_3COOH molecules remain intact. This value is typical of weak acids in water (Fig. 10.15). Table 10.1 lists the acidity constants of other weak acids.

We can also write an equilibrium constant for the proton transfer equilibrium of a base in water. For aqueous ammonia, for instance,

$$NH_3(aq) + H_2O(l) \rightleftharpoons NH_4^+(aq) + OH^-(aq) \qquad \text{(C)}$$

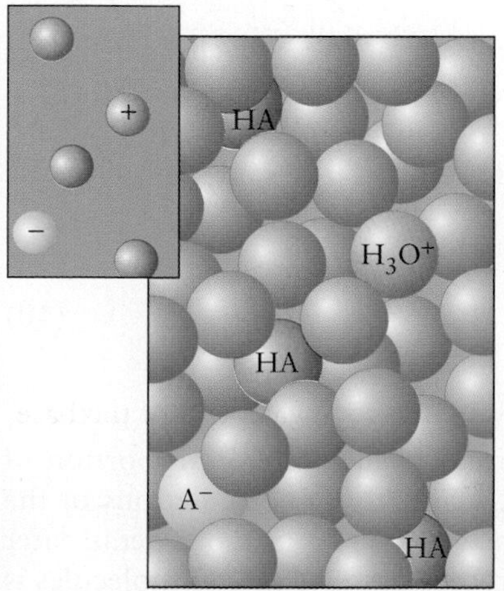

FIGURE 10.15 In a solution of a weak acid, only some of the acidic hydrogen atoms are present as hydronium ions (the red sphere), and the solution contains a high proportion of the original acid molecules (HA, gray spheres). The green sphere represents the conjugate base of the acid. The overlay shows only the solute species.

View an animation of the proton transfer equilibrium of a weak acid in water.

Media Link Animation: Fig. 10.15, Acetic acid in water

TABLE 10.1 Acidity Constants at 25°C*

Acid	K_a	pK_a
trichloroacetic acid, CCl_3COOH	3.0×10^{-1}	0.52
benzene sulfonic acid, $C_6H_5SO_3H$	2.0×10^{-1}	0.70
iodic acid, HIO_3	1.7×10^{-1}	0.77
sulfurous acid, H_2SO_3	1.5×10^{-2}	1.81
chlorous acid, $HClO_2$	1.0×10^{-2}	2.00
phosphoric acid, H_3PO_4	7.6×10^{-3}	2.12
chloroacetic acid, $CH_2ClCOOH$	1.4×10^{-3}	2.85
lactic acid, $CH_3CH(OH)COOH$	8.4×10^{-4}	3.08
nitrous acid, HNO_2	4.3×10^{-4}	3.37
hydrofluoric acid, HF	3.5×10^{-4}	3.45
formic acid, HCOOH	1.8×10^{-4}	3.75
benzoic acid, C_6H_5COOH	6.5×10^{-5}	4.19
acetic acid, CH_3COOH	1.8×10^{-5}	4.75
carbonic acid, H_2CO_3	4.3×10^{-7}	6.37
hypochlorous acid, HClO	3.0×10^{-8}	7.53
hypobromous acid, HBrO	2.0×10^{-9}	8.69
boric acid, $B(OH)_3$†	7.2×10^{-10}	9.14
hydrocyanic acid, HCN	4.9×10^{-10}	9.31
phenol, C_6H_5OH	1.3×10^{-10}	9.89
hypoiodous acid, HIO	2.3×10^{-11}	10.64

*The values for K_a listed here have been calculated from pK_a values with more significant figures than shown so as to minimize rounding errors. Values for polyprotic acids—those capable of donating more than one proton—refer to the first deprotonation.

†The proton transfer equilibrium is $B(OH)_3(aq) + 2\,H_2O(l) \rightleftharpoons H_3O^+(aq) + B(OH)_4^-(aq)$.

the equilibrium constant is

$$K = \frac{a_{NH_4^+}\, a_{OH^-}}{a_{H_2O}\, a_{NH_3}} \qquad (9)$$

K_b **is also widely called the *base ionization constant.***

In dilute solutions, the water is almost pure, and its activity can be set equal to 1. With this approximation, we obtain the **basicity constant,** K_b. If we make the further approximation of replacing the activities of the solute species by the numerical values of their molarities, we can write the basicity constant expression as

$$K_b = \frac{[NH_4^+][OH^-]}{[NH_3]} \qquad (10)$$

The experimental value of K_b for ammonia in water at 25°C is 1.8×10^{-5}. This small value tells us that normally only a small proportion of the NH_3 molecules are present as NH_4^+. Equilibrium calculations of the type we used in Section 9.8 and which we shall do in more detail later confirm this conclusion. They show that only about 1 in 100 molecules is protonated in a typical solution (Fig. 10.16). Table 10.2 lists the basicity constants of some weak bases.

Acidity and basicity constants are commonly reported as their negative logarithms, by defining

$$pK_a = -\log K_a \qquad pK_b = -\log K_b \tag{11}$$

The smaller the value of K_a, the greater the value of pK_a. In other words, *the greater the value of pK_a, the weaker the acid.* For example, the pK_a of trichloroacetic acid is 0.5, whereas that of acetic acid, a much weaker acid, is nearly 5. Similar remarks apply to bases: *the greater the value of pK_b, the weaker the base.* Values of pK_a and pK_b are included in Tables 10.1 and 10.2.

The proton-donating strength of an acid is measured by its acidity constant; the proton-accepting strength of a base is measured by its basicity constant. The smaller the constants, the weaker the respective strengths. The larger the value of pK, the weaker the acid or base.

View an animation of the proton transfer equilibrium of a weak base in water.

Media Link Animation: Fig. 10.16, Ammonia in water

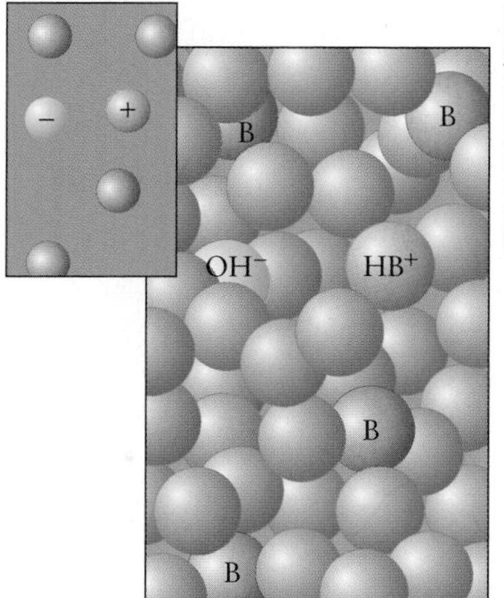

FIGURE 10.16 In a solution of a weak base, only a small proportion of the base molecules (B, represented here by gray spheres) have accepted protons from water molecules (the blue spheres) to form HB^+ ions (the red sphere) and OH^- ions (green sphere). The overlay shows only the solute species.

10.8 The Conjugate Seesaw

Hydrochloric acid is classified as a strong acid because it is fully deprotonated in water. It follows that its conjugate base, Cl^-, must be a very, very weak proton acceptor (weaker, in fact, than H_2O). Conversely, acetic acid is a weak acid. Its conjugate base, the acetate ion, $CH_3CO_2^-$, must be a relatively good proton acceptor, because it readily forms CH_3COOH molecules in water. Similarly, because methylamine, CH_3NH_2, is a stronger

TABLE 10.2 Basicity Constants at 25°C*

Base	K_b	pK_b
urea, $CO(NH_2)_2$	1.3×10^{-14}	13.90
aniline, $C_6H_5NH_2$	4.3×10^{-10}	9.37
pyridine, C_5H_5N	1.8×10^{-9}	8.75
hydroxylamine, NH_2OH	1.1×10^{-8}	7.97
nicotine, $C_{10}H_{14}N_2$	1.0×10^{-6}	5.98
morphine, $C_{17}H_{19}O_3N$	1.6×10^{-6}	5.79
hydrazine, NH_2NH_2	1.7×10^{-6}	5.77
ammonia, NH_3	1.8×10^{-5}	4.75
trimethylamine, $(CH_3)_3N$	6.5×10^{-5}	4.19
methylamine, CH_3NH_2	3.6×10^{-4}	3.44
dimethylamine, $(CH_3)_2NH$	5.4×10^{-4}	3.27
ethylamine, $C_2H_5NH_2$	6.5×10^{-4}	3.19
triethylamine, $(C_2H_5)_3N$	1.0×10^{-3}	2.99

*The values for K_b listed here have been calculated from pK_b values with more significant figures than shown so as to minimize rounding errors.

base than ammonia (see Table 10.2), the conjugate acid of methylamine—the methylammonium ion, $CH_3NH_3^+$—must be a weaker proton donor (and therefore a weaker acid) than NH_4^+. In general,

- *The stronger the acid, the weaker its conjugate base.*
- *The stronger the base, the weaker its conjugate acid.*

These are qualitative conclusions. We ought to be able to use the acidity and basicity constants—which are measures of acid and base strength—to express them quantitatively.

How do we do that?

To express the relative strengths of a conjugate acid and base, we consider the special case of the ammonia proton transfer equilibrium, reaction C, for which

$$K_b = \frac{[NH_4^+][OH^-]}{[NH_3]}$$

Now let's consider the proton transfer equilibrium of ammonia's conjugate acid, NH_4^+, in water:

$$NH_4^+(aq) + H_2O(l) \rightleftharpoons H_3O^+(aq) + NH_3(aq)$$

The acidity constant for the ammonium ion is therefore

$$K_a = \frac{[H_3O^+][NH_3]}{[NH_4^+]}$$

Now multiply these two equilibrium constants together and obtain

$$K_a \times K_b = \frac{[H_3O^+]\cancel{[NH_3]}}{\cancel{[NH_4^+]}} \times \frac{\cancel{[NH_4^+]}[OH^-]}{\cancel{[NH_3]}} = [H_3O^+][OH^-]$$

We recognize K_w on the right; so we can write

$$K_a \times K_b = K_w \tag{12a}$$

When we take logarithms of both sides, we obtain

$$\log K_a + \log K_b = \log K_w$$

Multiplication throughout by -1 turns this expression into

$$pK_a + pK_b = pK_w \tag{12b}$$

In these expressions, K_a is the acidity constant and K_b is the basicity constant of a given conjugate pair of an acid and a base (such as $CH_3COOH/CH_3CO_2^-$ or NH_4^+/NH_3) in water.

Equation 12—in either form—confirms the seesaw relation between the strengths of conjugate acids and bases, for if K_a is high, K_b must be low, or vice versa, to satisfy Eq. 12. That is, if an acid has a large K_a, then its conjugate base has a small K_b. This reciprocal relation is summarized in Fig. 10.17 and Table 10.3.

The stronger the acid, the weaker its conjugate base; the stronger the base, the weaker its conjugate acid.

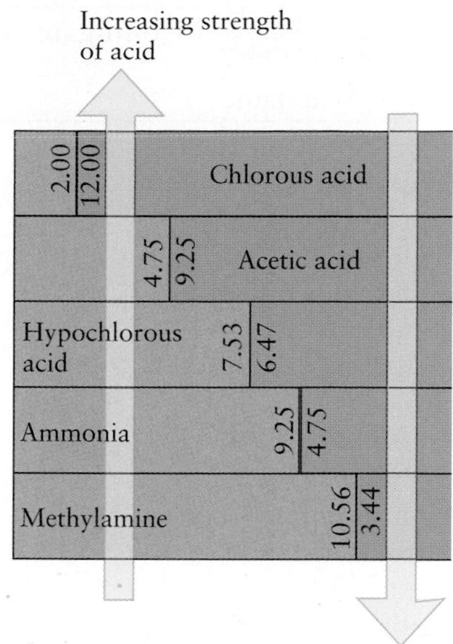

FIGURE 10.17 The sum of the pK_a of an acid (pink) and the pK_b of its conjugate base (blue) is constant, and equal to pK_w, which is 14.00 at 25°C. The values of the constants allow us to arrange all acids and bases in order of strength on a single chart.

How can we use this information . . . to calculate the pK_a of the conjugate acid of a base?

Because the pK_b of NH_3 is 4.75, the pK_a of NH_4^+ at 25°C is

$$pK_a = pK_w - pK_b = 14.00 - 4.75 = 9.25$$

This value shows that NH_4^+ is a weaker acid than acetic acid (pK_a = 4.75) but stronger than hypoiodous acid (HIO, pK_a = 10.64).

Self-Test 10.6A Write the chemical formula for the conjugate acid of the base pyridine, C_5H_5N, and calculate its pK_a.

[*Answer:* $C_5H_5NH^+$, 5.25]

Self-Test 10.6B Write the chemical formula for the conjugate base of the acid HIO_3 and calculate its pK_b.

How can we use this information . . . to decide which of two species is the stronger acid or base?

Suppose we need to decide which member of the following pairs is the stronger acid or base in water: (a) acid: HF or HIO_3; (b) base: NO_2^- or CN^-. We know that the greater the K_a of a weak acid, the stronger the acid and the weaker its conjugate base. Similarly, the greater the K_b of a weak base, the stronger the base and the weaker its conjugate acid. (a) Because $K_a(HIO_3) > K_a(HF)$, it follows that HIO_3 is a stronger acid than HF. (b) Because $K_a(HNO_2) > K_a(HCN)$, and because the stronger acid has the weaker conjugate base, it follows that NO_2^- is a weaker base than CN^-. Hence, CN^- is the stronger base.

Self-Test 10.7A Use Tables 10.1 and 10.2 to decide which species of each pair is the stronger acid or base: (a) acid: HF or HIO; (b) base: $C_6H_5CO_2^-$ or $CH_2ClCO_2^-$; (c) base: $C_6H_5NH_2$ or $(CH_3)_3N$; (d) acid: $C_6H_5NH_3^+$ or $(CH_3)_3NH^+$.

[*Answer:* Stronger acids: (a) HF; (d) $C_6H_5NH_3^+$. Stronger bases: (b) $C_6H_5CO_2^-$; (c) $(CH_3)_3N$]

Self-Test 10.7B Use Tables 10.1 and 10.2 to decide which species of each pair is the stronger acid or base: (a) base: C_5H_5N or NH_2NH_2; (b) acid: $C_5H_5NH^+$ or $NH_2NH_3^+$; (c) acid: HIO_3 or $HClO_2$; (d) base: ClO_2^- or HSO_3^-.

TABLE 10.3 Conjugate Acid-Base Pairs Arranged by Strength

Acid name	Acid formula	Base formula	Base name
Strong acid			**Very weak base**
hydroiodic acid	HI	I^-	iodide ion
perchloric acid	$HClO_4$	ClO_4^-	perchlorate ion
hydrobromic acid	HBr	Br^-	bromide ion
hydrochloric acid	HCl	Cl^-	chloride ion
sulfuric acid	H_2SO_4	HSO_4^-	hydrogen sulfate ion
chloric acid	$HClO_3$	ClO_3^-	chlorate ion
nitric acid	HNO_3	NO_3^-	nitrate ion
hydronium ion	H_3O^+	H_2O	*water*
hydrogen sulfate ion	HSO_4^-	SO_4^{2-}	sulfate ion
hydrofluoric acid	HF	F^-	fluoride ion
nitrous acid	HNO_2	NO_2^-	nitrite ion
acetic acid	CH_3COOH	$CH_3CO_2^-$	acetate ion
carbonic acid	H_2CO_3	HCO_3^-	hydrogen carbonate ion
hydrosulfuric acid	H_2S	HS^-	hydrogen sulfide ion
ammonium ion	NH_4^+	NH_3	ammonia
hydrocyanic acid	HCN	CN^-	cyanide ion
hydrogen carbonate ion	HCO_3^-	CO_3^{2-}	carbonate ion
methylammonium ion	$CH_3NH_3^+$	CH_3NH_2	methylamine
water	H_2O	OH^-	*hydroxide ion*
ammonia	NH_3	NH_2^-	amide ion
hydrogen	H_2	H^-	hydride ion
methane	CH_4	CH_3^-	methide ion
hydroxide ion	OH^-	O^{2-}	oxide ion
Very weak acid			**Strong base**

10.9 Solvent Leveling

The equilibrium composition of a solution of an acid or a base, and therefore the value of the acidity or basicity constant, depends on the identity of the solvent. To illustrate how the solvent plays a role, we can consider the reactions of the strong acid HA and the weak acid HA′ with water:

$$\text{Strong acid: } HA(aq) + H_2O(aq) \longrightarrow H_3O^+(aq) + A^-(aq)$$

Although we have written the reaction with an arrow, the proton transfer is actually an equilibrium with the products in overwhelming abundance, so we could write it as

$$\text{Strong acid: } HA(aq) + H_2O(aq) \rightleftharpoons H_3O^+(aq) + A^-(aq)$$

with *products* dominant. For a weak acid, we write

$$\text{Weak acid: } HA'(aq) + H_2O(aq) \rightleftharpoons H_3O^+(aq) + A'^-(aq)$$

with the *reactants* dominant. Each equilibrium is the outcome of a battle between the proton-donating powers of H_3O^+ and the acid. In the first equilibrium, the proton-donating power of HA is greater than that of

H_3O^+, so the products are favored and the proton transfer effectively goes to completion. As a result, the solution contains only H_3O^+ ions and A^- ions; there are virtually no HA molecules left. In other words, the only acid species present in an aqueous solution of a strong acid is the H_3O^+ ion. Because all strong acids in water behave as though they were solutions of the acid H_3O^+, we say that strong acids are **leveled** in water to the strength of the acid H_3O^+. In the second equilibrium, the proton-donating power of HA′ is less than that of H_3O^+, so the reactants are favored and only a small fraction of HA′ molecules are deprotonated. For instance, if the acid is HCN, a high proportion of HCN molecules survive in solution.

We can summarize these remarks as follows:

- Strong acid: the proton-donating power of HA is greater than that of H_3O^+.
- Weak acid: the proton-donating power of HA is less than that of H_3O^+.

The strength of H_3O^+ as a proton donor therefore marks the frontier between strong and weak acids in water. Any acid that lies above H_3O^+ in Table 10.3 is a strong acid in aqueous solution. Sulfuric acid, H_2SO_4, is an example: water is a stronger proton donor than HSO_4^-. Any acid that lies below H_3O^+ in Table 10.3 is a weak acid. An example is acetic acid.

The common strong acids are listed in Table J.1.

Similar arguments apply to bases. Consider the following two equilibria for the strong base B and the weak base B′:

$$\text{Strong base: } B(aq) + H_2O(aq) \rightleftharpoons BH^+(aq) + OH^-(aq)$$

with products dominant, and

$$\text{Weak base: } B'(aq) + H_2O(aq) \rightleftharpoons B'H^+(aq) + OH^-(aq)$$

with reactants dominant. Each equilibrium is the outcome of a tug-of-war between the proton-accepting powers of OH^- and the base. In the first equilibrium, the proton-accepting power of B is greater than that of OH^-, so the products are favored and the proton transfer effectively goes to completion. As a result, the solution will contain only OH^- ions and BH^+ ions; there are virtually no B molecules left. In other words, the only base species present in an aqueous solution of a strong base is the OH^- ion. In the second equilibrium, the proton-accepting power of B′ is less than that of OH^-, so the reactants are favored and only a small fraction of B′ molecules are protonated.

We can summarize these remarks as follows:

- Strong base: the proton-accepting power of B is greater than that of OH^-.
- Weak base: the proton-accepting power of B is less than that of OH^-.

Once again, we see that the properties of ions that arise from the solvent determine the location of the frontier between weak and strong. Any base that lies below OH^- in Table 10.3 is strong in aqueous solution, and any base that lies above OH^- is weak.

Media Link **Animation:** Fig. 10.18, Acetate ions in water

View an animation of the weak base acetate ion reacting with water.

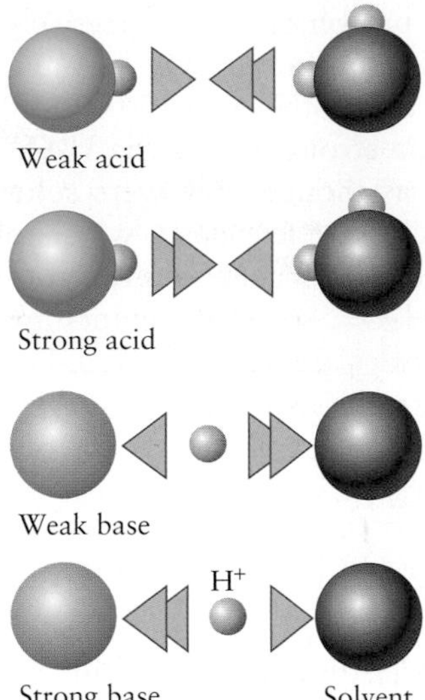

FIGURE 10.18 The ability of a molecule or ion to act as a strong or weak acid (upper left) depends on its proton-donating power (represented by the number of triangles) relative to that of the conjugate acid of the solvent (right). The ability of a molecule or ion to act as a strong or weak base (lower left) depends on its proton-attracting power (represented by the number of triangles) relative to that of the conjugate base of the solvent.

These principles apply to any solvent (Fig. 10.18): *an acid is strong if it is a stronger proton donor than the conjugate acid of the solvent.* In such a case, the tug-of-war for the proton is resolved in favor of the solvent, so the acid is completely deprotonated. *A base is strong if it is a stronger proton acceptor than the conjugate base of the solvent*; if the opposite is true, the base is weak.

A stronger proton donor than H_3O^+ is a strong acid in water; a weaker proton donor than H_3O^+ is a weak acid in water. A weaker proton acceptor than OH^- is a weak base in water; a stronger proton acceptor than OH^- is a strong base in water.

Self-Test 10.8A Use the data in Table 10.3 to label each of the following species as a strong or a weak acid in water: (a) $HClO_4$; (b) NH_4^+; (c) HNO_2.
[*Answer:* (a) Strong acid; (b) weak acid; (c) weak acid]

Self-Test 10.8B Use the data in Table 10.3 to label each of the following species as a strong or a weak base in water: (a) ClO_4^-; (b) NH_2^-; (c) $CH_3CO_2^-$.

10.10 Molecular Structure and Acid Strength

Here is an opportunity for you to come up with a theory.

Chemists commonly interpret tables of data, including tables of acidity and basicity constants, in terms of the structures of the molecules involved. However, there are two reasons why the relative strengths of acids and bases are difficult to predict. First, K_a and K_b are equilibrium constants, so they are related to the free energy of the proton transfer reaction. They therefore depend on considerations of entropy as well as energy. Second, the solvent plays a significant role in the forward and reverse proton transfer reactions, so we cannot expect to relate acid strength solely to the molecule itself. However, although absolute values are difficult to predict, we can look for trends among series of compounds with similar structures. Because acid strength in aqueous solutions involves the breaking of the H—A bond and the formation of an H—OH_2 bond, we might suspect that one factor in determining strength is the ease with which these bonds can be broken and formed.

Let's look first at the strength of the hydrogen bond that HA forms with the O atom of an H_2O molecule. We know that the more polar the H—A bond, the greater the partial positive charge on the H atom, therefore, the stronger the $H_2O\cdots$H—A hydrogen bond and the more readily HA loses its proton. Because the polarity of the H—A bond increases with the electronegativity of A, we can predict that *the greater the electronegativity of A, the stronger the acid HA*. For instance, the electronegativity difference is 0.8 for the N—H bond and 1.8 for F—H; therefore, the F—H bond is markedly more polar than the N—H bond. This difference is consistent with the observation that HF, but not NH_3, is an acid in water. In general, bond polarity dominates the trend of acid strengths for binary acids of elements of the same period.

But what about binary acids of elements in the same group? When we compare the relative strengths of the hydrohalic acids, we find that they lie in the order HF < HCl < HBr < HI. Bond polarities *decrease* down the group, so another factor must be affecting the acidity. In this case, the order of acid strength is consistent with the weakening of the H—A bond down the group (recall Fig. 2.18). The weaker the H—A bond, the easier it is for the proton to leave and the stronger the acid HA. The same trend is also found for the Group 16/VI acids in aqueous solution: the acid strengths lie in the order $H_2O < H_2S < H_2Se < H_2Te$. Because bond strengths and bond polarities both decrease down the group, it appears that bond strength dominates the trend in acid strength in these two sets of binary acids.

Because all the hydrohalic acids except HF are strong acids, they are leveled in water. Therefore, to determine their relative strengths, they must be studied in a solvent that is a poorer proton acceptor than water (such as pure acetic acid).

The more polar or the weaker the H—A bond, the stronger the acid.

10.11 The Strengths of Oxoacids

We can learn more from the oxoacids about the effect of structure on acidity. These acids form series of families, so we can change the central atom without changing the number of O atoms (as in $HClO_3$ and $HBrO_3$). Alternatively, we can change the number of O atoms without changing the central atom (as in $HClO_3$ and $HClO_4$).

The high polarity of the O—H bond is one reason why the proton of an —OH group in an oxoacid molecule is acidic. For example, phosphorous acid, H_3PO_3, has the structure $(HO)_2PHO$ (**1**): it can donate the protons from its two —OH groups but not the proton attached directly to the phosphorus atom. The difference in behavior can be traced to the much lower electronegativity of phosphorus relative to that of oxygen. In nearly all common inorganic oxoacids, however, all the hydrogen atoms are attached to oxygen atoms.

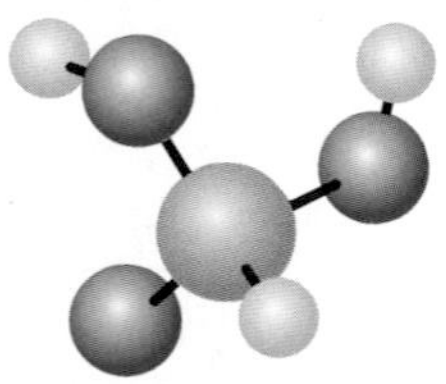

1 Phosphorous acid, H_3PO_3

Let's consider a family of oxoacids in which the number of O atoms is constant, as in the hypohalous acids HClO, HBrO, and HIO. By referring to Table 10.4, we see that *the greater the electronegativity of the halogen, the stronger the oxoacid*. A partial explanation of this trend is that electrons are withdrawn slightly from the O—H bond as the electronegativity of the halogen increases. As these bonding electrons move

TABLE 10.4 Correlation of Acid Strength and Electronegativity

Acid, HXO	Structure*	Electronegativity of atom X	pK_a
hypochlorous acid, HClO	:C̈l—Ö—H	3.2	7.53
hypobromous acid, HBrO	:B̈r—Ö—H	3.0	8.69
hypoiodous acid, HIO	:Ï—Ö—H	2.7	10.64

*The red arrows indicate the direction of the shift of electron density away from the O—H bond.

toward the central atom, the O—H bond becomes more polar, so the molecule becomes a stronger acid. A halogen that has a high electronegativity also weakens the conjugate base by making the electrons in the molecule less accessible to an incoming proton.

Now let's consider a family of oxoacids in which the number of oxygen atoms varies, as in the chlorine oxoacids HClO, $HClO_2$, $HClO_3$, and $HClO_4$, or the sulfur oxoacids H_2SO_3 and H_2SO_4. When we refer to Table 10.5, we see that *the greater the number of oxygen atoms attached to the central atom, the stronger the acid*. Because the oxidation number of the central atom increases as the number of O atoms increases, we can also conclude that *the greater the oxidation number of the central atom, the stronger the acid*.

We can identify the effect of the number of O atoms on the strengths of organic acids by comparing alcohols and carboxylic acids. We have seen (Section D) that alcohols are organic compounds in which an —OH group is attached to a carbon atom, as in ethanol (**2**). Carboxylic acids have two O atoms attached to the same carbon atom: one is attached with a double bond and the other is the O atom of an —OH group, as in acetic acid (**3**). Although carboxylic acids are weak acids, they are much stronger acids than alcohols, partly as a result of the electron-withdrawing power of the second O atom.

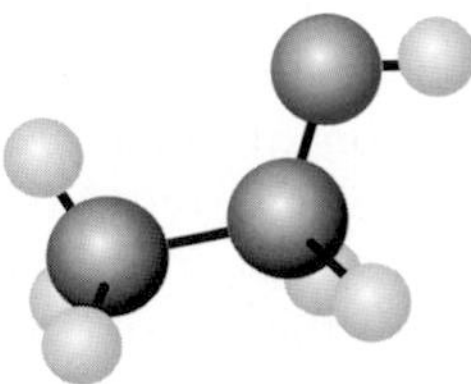

2 Ethanol, CH_3CH_2OH

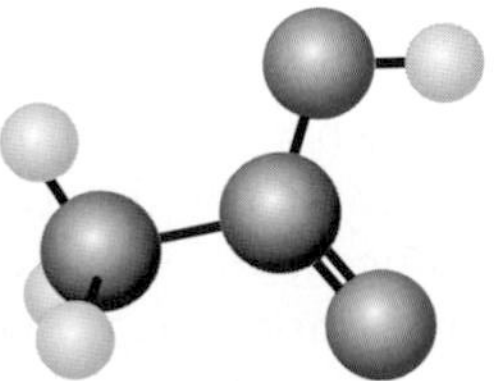

3 Acetic acid, CH_3COOH

The strength of a carboxylic acid is also increased relative to that of an alcohol by electron delocalization in the conjugate base. The second O atom of the carboxyl group provides an additional electronegative atom over which the negative charge of the conjugate base can spread. This electron delocalization stabilizes the anion (**4**). Moreover, because the charge is spread over several atoms, it is less effective at attracting a proton. A carboxylate ion is therefore a weaker base than the conjugate base of an alcohol (for example, the ethoxide ion, $CH_3CH_2O^-$). Hence, carboxylic acids are stronger acids than alcohols. In fact, alcohols have such weak proton-donating power that usually they are not regarded as oxoacids.

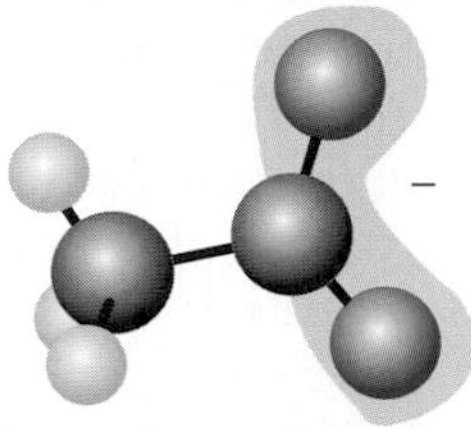

4 Acetate ion, $CH_3CO_2^-$

The strengths of carboxylic acids also vary with total electron-withdrawing power of the atoms bonded to the carboxyl group. Because hydrogen is less electronegative than chlorine, the $-CH_3$ group bonded to $-COOH$ in acetic acid is less electron withdrawing than the $-CCl_3$ group in trichloroacetic acid. Therefore, we expect CCl_3COOH to be a stronger acid than CH_3COOH. In agreement with this prediction, the pK_a of acetic acid is 4.75, whereas that of trichloroacetic acid is 0.52.

TABLE 10.5 Correlation of Acid Strength and Oxidation Number

Acid	Structure*	Oxidation number of chlorine atom	pK_a
hypochlorous acid, HClO	:C̈l—Ö—H	+1	7.53
chlorous acid, $HClO_2$	:O: ‖ :C̈l—Ö—H	+3	2.00
chloric acid, $HClO_3$	:O: ‖ :Cl—Ö—H ‖ :O:	+5	strong
perchloric acid, $HClO_4$	:O: ‖ Ö=Cl—Ö—H ‖ :O:	+7	strong

*The red arrows indicate the direction of the shift of electron density away from the O—H bond. The Lewis structures shown are the ones with the most favorable formal charges.

TABLE 10.6 Correlations of Molecular Structure and Acid Strength

Acid Type	Trend
Binary	1 The more polar the H—A bond, the stronger the acid. This effect is dominant for acids of the same period. 2 The weaker the H—A bond, the stronger the acid. This effect is dominant for acids of the same group.
Oxoacid	1 The greater the number of O atoms attached to the central atom (the greater the oxidation number of the central atom), the stronger the acid. 2 For the same number of O atoms attached to the central atom, then the greater the electronegativity of the central atom, the stronger the acid.
Carboxylic	1 The greater the electronegativities of the groups attached to the carboxyl group, the stronger the acid.

The greater the number of oxygen atoms and the more electronegative the atoms present in a molecule, the stronger the acid. These trends are summarized in Table 10.6.

How can we use this information . . . to decide which of two acids is the stronger?

Let's predict which acid in each of the following pairs is the stronger: (a) H_2S and H_2Se; (b) H_2SO_4 and H_2SO_3; (c) H_2SO_4 and H_3PO_4. We refer to the summary in Table 10.6. (a) Sulfur and selenium are in the same group, and we expect the H—Se bond to be weaker than the H—S bond. Thus, according to rule 2 for binary acids, H_2Se can be expected to be the stronger acid. (b) H_2SO_4 has the greater number of O atoms bonded to the S atom and the oxidation number of sulfur is +6, whereas in H_2SO_3 the sulfur has an oxidation number of only +4. Thus, according to rule 1 for oxoacids, H_2SO_4 is expected to be the stronger acid. (c) Both acids have four O atoms bonded to the central atom; but the electronegativity of sulfur is greater than that of phosphorus, so, according to rule 2 for oxoacids, H_2SO_4 is expected to be the stronger acid.

Self-Test 10.9A In the following pairs, predict which acid is stronger. (a) H_2S and HCl; (b) HNO_2 and HNO_3; (c) H_2SO_3 and $HClO_3$.
[*Answer:* (a) HCl; (b) HNO_3; (c) $HClO_3$]

Self-Test 10.9B List the following carboxylic acids in order of increasing strength: $CHCl_2COOH$, CH_3COOH, and $CH_2ClCOOH$.

THE pH OF SOLUTIONS OF WEAK ACIDS AND BASES

The rest of this chapter is a variation on a theme introduced in Chapter 9: the use of equilibrium constants to calculate the equilibrium composition of solutions of acids, bases, and salts. For acids, we may need to know the extent to which acid molecules have become deprotonated; similarly, for bases, we may need to know the extent to which base molecules have become protonated. We shall see how to calculate both quantities and the pH of solutions of weak acids and bases. We shall also see how to calculate the pH of solutions of salts, which in general differ from 7 because the cations and anions of a salt may themselves be weak acids or bases.

10.12 Solutions of Weak Acids

Our first task is to calculate the pH of a solution of a weak acid, such as acetic acid in water. The **initial concentration** of the acid is its concentration as prepared, as if no acid molecules had donated any protons. If the acid were strong, the H_3O^+ concentration in solution would be the same as the initial concentration of the strong acid, because all the acid molecules would have been deprotonated. However, to find the H_3O^+ molarity in a solution of a weak acid, we have to take into account the equilibrium between the acid HA and its conjugate base A^-:

$$HA(aq) + H_2O(l) \rightleftharpoons H_3O^+(aq) + A^-(aq) \qquad K_a = \frac{[H_3O^+][A^-]}{[HA]} \tag{13}$$

We can expect the pH to lie somewhere between 7, a value indicating no deprotonation, and the value we would calculate for a strong acid, which undergoes complete deprotonation. The technique, which is based on the use of an equilibrium table like those introduced in Chapter 9, is set out in Toolbox 10.1.

The calculation summarized in the toolbox also allows us to predict the **percentage deprotonation,** the percentage of HA molecules that are deprotonated in the solution. To calculate the percentage deprotonation, we use the equality $[H_3O^+] = [A^-]$, which follows from the stoichiometric relation 1 mol $A^- \simeq$ 1 mol H_3O^+ for the deprotonation reaction:

$$\begin{aligned}\text{Percentage protonated} &= \frac{\text{molarity of } A^-}{\text{initial molarity of HA}} \times 100\% \\ &= \frac{[H_3O^+]}{[HA]_{\text{initial}}} \times 100\% \end{aligned} \tag{14}$$

A small percentage of deprotonated molecules indicates that the solute consists primarily of the acid HA. The concentrations of H_3O^+ ions and conjugate base ions must then be very low.

The autoprotolysis of water contributes significantly to the pH when the acid is so dilute or so weak that the calculation predicts an H_3O^+ molarity close to 10^{-7} mol·L^{-1}. In such cases, we must use the procedures described in Sections 10.15 and 10.16. We can ignore the contribution of the autoprotolysis of water only when the calculated H_3O^+ molarity is substantially (about 10 times) higher than 10^{-7} mol·L^{-1}, corresponding to a pH of 6 or less.

To calculate the pH and percentage deprotonation of a solution of a weak acid, set up an equilibrium table and determine the H_3O^+ molarity by using the acidity constant.

TOOLBOX 10.1 How to calculate the pH of a solution of a weak acid

Conceptual Basis

Because a proton transfer equilibrium is established as soon as a weak acid is dissolved in water, we can calculate the hydronium ion concentration from the value of the acidity constant for the acid by using a procedure based on that in Toolbox 9.1.

Procedure

Step 1 Set up a table with columns labeled by the acid HA, its conjugate base A^-, and H_3O^+. In the first row below the headings, show the initial molarities of each species.

	Acid, HA	Conjugate base, A^-	H_3O^+
Initial molarity	$[HA]_{initial}$	0	0

For the initial values, assume that no acid molecules have donated a proton.

Step 2 In the second row, write the changes in the molarities that are needed for the reaction to reach equilibrium.

We do not know the number of acid molecules that lose their protons, so we assume that the molarity of the acid decreases by x mol·L^{-1} as a result of deprotonation. The reaction stoichiometry gives us the other changes in terms of x.

	Acid, HA	Conjugate base, A^-	H_3O^+
Initial molarity	$[HA]_{initial}$	0	0
Change in molarity	$-x$	$+x$	$+x$

Step 3 In the third row, write the equilibrium molarities by adding the change in molarities (step 2) to the initial values for each substance (step 1).

	Acid, HA	Conjugate base, A^-	H_3O^+
Initial molarity	$[HA]_{initial}$	0	0
Change in molarity	$-x$	$+x$	$+x$
Equilibrium molarity	$[HA]_{initial} - x$	x	x

Although a *change* in concentration may be positive (an increase) or negative (a decrease), the value of the concentration must always be positive.

Step 4 Use the value of K_a to calculate the value of x.

The calculation of x can often be simplified, as shown in Toolbox 9.1, by neglecting changes of less than 5% of the initial molarity of the acid. However, at the end of the calculation, we must check that x is consistent with the approximation, by calculating the percentage of acid deprotonated. If this percentage is greater than 5%, then the exact expression for K_a must be solved for x. An exact calculation often requires solving a quadratic equation, as explained in Toolbox 9.1. Although we calculate the pH to the number of significant figures appropriate to the data, the answers are often considerably less reliable than that. One reason for this poor reliability is that we are ignoring interactions between the ions in solution when we use concentrations in place of activities.

EXAMPLE 10.3 Estimating the pH and percentage deprotonation of a weak acid

Calculate the pH and percentage deprotonation of 0.10 M $CH_3COOH(aq)$, given that K_a for acetic acid is 1.8×10^{-5}.

STRATEGY Acetic acid is a weak acid; consequently, we expect the molarity of H_3O^+ ions to be less than 0.10 mol·L^{-1} and, therefore, its pH to be greater than 1.0. To find the actual value, follow the procedure in Toolbox 10.1, with the initial molarity of acid equal to 0.10 mol·L^{-1}. Assume that the presence of acid dominates the pH and therefore that the autoprotolysis of water need not be considered. Use the approximation procedure described in step 4, but verify the assumption that x is small at the end of the calculation.

SOLUTION The proton transfer equilibrium we consider is

$$CH_3COOH(aq) + H_2O(l) \rightleftharpoons H_3O^+(aq) + CH_3CO_2^-(aq)$$

$$K_a = \frac{[H_3O^+][CH_3CO_2^-]}{[CH_3COOH]}$$

and the equilibrium table, with the concentrations in moles per liter, is

	Species		
	CH_3COOH	$CH_3CO_2^-$	H_3O^+
Step 1 Initial molarity	0.10	0	0
Step 2 Change in molarity	$-x$	$+x$	$+x$
Step 3 Equilibrium molarity	$0.10 - x$	x	x

Step 4 Substitute these equilibrium molarities into the expression for the acidity constant:

$$K_a = 1.8 \times 10^{-5} \approx \frac{x \times x}{0.10 - x}$$

If we anticipate that $x \ll 0.1$, we can write an approximation of this expression as

$$1.8 \times 10^{-5} \approx \frac{x^2}{0.10}$$

The solution for x is

$$x \approx \sqrt{(0.10) \times (1.8 \times 10^{-5})} = 1.3 \times 10^{-3}$$

From step 3, $x = [H_3O^+] \approx 1.3 \times 10^{-3}\ mol \cdot L^{-1}$, so

$$pH \approx -\log(1.3 \times 10^{-3}) = 2.89$$

The percentage deprotonated is

$$\text{Percentage deprotonated} = \frac{[CH_3CO_2^-]}{[CH_3COOH]_{initial}} \times 100\%$$

$$= \frac{1.3 \times 10^{-3}}{0.10} \times 100\% = 1.3\%$$

We see that x is less than 5% of 0.10, as assumed, and the approximation is valid. Because the pH < 6, the assumption that the autoprotolysis of water can be ignored is also valid.

Self-Test 10.10A Calculate the pH of 0.50 M lactic acid. See Table 10.1 for K_a. Be sure to check any approximation to see whether it is valid.
[*Answer:* 1.69]

Self-Test 10.10B Calculate the pH of 0.22 M chloroacetic acid. Be sure to check any approximation to see whether it is valid.

EXAMPLE 10.4 **Calculating the K_a of a weak acid from the pH**

The pH of a 0.010 M aqueous solution of the antiseptic mandelic acid is 2.95. What is the K_a of mandelic acid?

STRATEGY We can calculate the hydronium ion concentration from the pH and then calculate the value of K_a from the initial concentration of the acid and the hydronium ion concentration.

SOLUTION The hydronium ion concentration is equal to 10^{-pH}, so $[H_3O^+] = 10^{-2.95}\ mol \cdot L^{-1} = 1.8 \times 10^{-3}\ mol \cdot L^{-1}$ The concentration of the conjugate base is equal to the hydronium ion concentration, and the equilibrium concentration of the acid is $[HA]_{initial} - [H_3O^+]$. Therefore, the value of K_a is

$$K_a = \frac{(1.1 \times 10^{-3})^2}{0.010 - 0.0011} = 1.4 \times 10^{-4}$$

Self-Test 10.11A The pH of a 0.20 M aqueous solution of crotonic acid, which is used in medicinal research, is 2.69. What is the K_a of crotonic acid?
[*Answer:* 2.1×10^{-5}]

Self-Test 10.11B The pH of a 0.50 M aqueous solution of the metabolic intermediate homogentisic acid is 2.35. What is the K_a of homogentisic acid?

10.13 Solutions of Weak Bases

In the treatment of weak acids, we were interested in the percentage deprotonated. Similarly, when describing the strengths of weak bases, it is useful to know the **percentage protonated**, the percentage of base molecules that have been protonated:

$$\text{Percentage protonated} = \frac{\text{molarity of HB}^+}{\text{initial molarity of B}} \times 100\% = \frac{[BH^+]}{[B]_{initial}} \times 100\% \qquad (15)$$

Here $[B]_{initial}$ is the initial molar concentration of base, its molarity assuming that no protonation has occurred.

We calculate the pH of solutions of weak bases in the same way we calculate the pH of solutions of weak acids—by using an equilibrium table. The protonation equilibrium is

$$B(aq) + H_2O(l) \rightleftharpoons HB^+(aq) + OH^-(aq) \qquad K_b = \frac{[HB^+][OH^-]}{[B]} \qquad (16)$$

To calculate the pH of the solution, we first calculate the molarity of OH^- ions at equilibrium, express that molarity as pOH, and then calculate the pH from the relation pH + pOH = 14.00 at 25°C. For very weak or very dilute bases, the autoprotolysis of water must be taken into consideration.

To calculate the pH of a solution of a weak base, set up an equilibrium table to calculate pOH from the value of K_b and convert that pOH to pH by using pH + pOH = 14.00.

TOOLBOX 10.2 How to calculate the pH of a solution of a weak base

Conceptual Basis

Proton transfer equilibrium is established as soon as a weak base is dissolved in water, so we can calculate the hydroxide ion concentration from the initial concentration of the base and the value of its basicity constant. The hydroxide ions are in equilibrium with the hydronium ions arising from the autoprotolysis of water, so we can use the pOH and pK_w to calculate the pH.

Procedure

Step 1 Set up a table with columns labeled by the base B, its conjugate acid BH^+, and OH^-. In the first row, show the initial molarities of each species.

	Base, B	Conjugate acid, BH^+	OH^-
Initial molarity	$[B]_{initial}$	0	0

For the initial values, assume that no base molecules have been protonated.

Step 2 Write the changes in the molarities that are needed for the reaction to reach equilibrium.

We do not know the number of base molecules that acquire protons, so we assume that the molarity of the base decreases by x mol·L^{-1} as a result of protonation. The reaction stoichiometry gives us the other changes in terms of x.

	Base, B	Conjugate acid, BH^+	OH^-
Initial molarity	$[B]_{initial}$	0	0
Change in molarity	$-x$	$+x$	$+x$

Step 3 Write the equilibrium molarities by adding the change in molarities (step 2) to the initial values for each substance (step 1).

	Base, B	Conjugate acid, BH^+	OH^-
Initial molarity	$[B]_{initial}$	0	0
Change in molarity	$-x$	$+x$	$+x$
Equilibrium molarity	$[B]_{initial} - x$	x	x

Although a change in concentration may be positive (an increase) or negative (a decrease), the concentration itself must always be positive.

Step 4 Use the value of K_b to calculate the value of x.

The calculation of x can often be simplified, as explained in Toolbox 10.1. We ignore contributions from the autoprotolysis of water to the hydroxide ion concentration if the concentration of hydroxide ions is greater than 10^{-6} mol·L^{-1}.

Step 5 Calculate the pH of the solution from the pOH by using Eq. 6b.

As in Toolbox 10.1, although we calculate the pH to the number of significant figures appropriate to the data, the answers are often considerably less reliable than that.

EXAMPLE 10.5 Calculating the pH and percentage protonation of a weak base

Calculate the pH and percentage protonation of a 0.20 M aqueous solution of methylamine, CH_3NH_2. The K_b for CH_3NH_2 is 3.6×10^{-4}.

STRATEGY Expect pH > 7, because amines are bases. Assume that the presence of base dominates the pH and therefore that the autoprotolysis of water need not be considered. Proceed as in Toolbox 10.2.

SOLUTION The proton transfer equilibrium is

$$H_2O(l) + CH_3NH_2(aq) \rightleftharpoons CH_3NH_3^+(aq) + OH^-(aq)$$

$$K_b = \frac{[CH_3NH_3^+][OH^-]}{[CH_3NH_2]}$$

The equilibrium table, with all concentrations in moles per liter, is

	Species		
	CH_3NH_2	$CH_3NH_3^+$	OH^-
Step 1 Initial molarity	0.20	0	0
Step 2 Change in molarity	$-x$	$+x$	$+x$
Step 3 Equilibrium molarity	$0.20 - x$	x	x

Step 4 Substitution of the equilibrium molarities into the expression for the basicity constant ($K_b = 3.6 \times 10^{-4}$) yields

$$K_b = 3.6 \times 10^{-4} = \frac{x \times x}{0.20 - x}$$

We now anticipate that x is less than 5% of 0.20 (that is, $x < 0.01$) and approximate this expression by

$$3.6 \times 10^{-4} \approx \frac{x^2}{0.20}$$

Therefore,

$$x \approx \sqrt{0.20 \times (3.6 \times 10^{-4})} = 8.5 \times 10^{-3}$$

which is consistent with the assumption that $x < 0.01$. According to step 3, $[OH^-] = x = 8.5 \times 10^{-3}$ mol·L^{-1}, so

$$\text{pOH} \approx -\log(8.5 \times 10^{-3}) = 2.1$$

Step 5 We conclude that

$$\text{pH} \approx 14.00 - 2.1 = 11.9$$

The percentage of base molecules protonated is

$$\text{Percentage protonated} = \frac{[CH_3NH_3^+]}{[CH_3NH_2]_{\text{initial}}} \times 100\%$$

$$= \frac{8.5 \times 10^{-3}}{0.20} \times 100\% = 4.2\%$$

That is, 4.2% of the methylamine is present as the protonated form, $CH_3NH_3^+$, and the approximation is valid. Because the pH is greater than 8, the assumption that this equilibrium dominates the pH and that autoprotolysis can be ignored is valid.

Self-Test 10.12A Estimate the pH and percentage of protonated base in 0.15 M $NH_2OH(aq)$ (aqueous hydroxylamine).

[*Answer:* 9.61; 0.027%]

Self-Test 10.12B Estimate the pH and percentage of protonated base in 0.012 M $C_{10}H_{14}N_2(aq)$ (nicotine).

10.14 The pH of Salt Solutions

In Section J, we saw that a salt is produced by the neutralization of an acid by a base. However, if we measure the pH of a solution of a salt, we do not in general find the "neutral" value (pH = 7). For instance, if we neutralize 0.3 M acetic acid with 0.3 M sodium hydroxide, the resulting solution of sodium acetate has pH = 9.0. How can this be? The Brønsted-Lowry theory provides the explanation. According to this theory, an ion may be an acid or a base. The acetate ion, for instance, is a base, and the ammonium ion is an acid. Therefore, a solution of a salt is a solution of an acid (usually the cation) and a base (usually the anion), and the pH depends on their relative strengths.

Table 10.7 lists some cations that are acidic in water. All cations that are the conjugate acids of weak bases act as proton donors, so we can

TABLE 10.7 Acidic Character and K_a Values of Common Cations in Water*

Character	Examples	K_a	pK_a
Acidic			
conjugate acids of weak bases	anilinium ion, $C_6H_5NH_3^+$	2.3×10^{-5}	4.64
	pyridinium ion, $C_5H_5NH^+$	5.6×10^{-6}	5.24
	ammonium ion, NH_4^+	5.6×10^{-10}	9.25
	methylammonium ion, $CH_3NH_3^+$	2.8×10^{-11}	10.56
small, highly charged metal cations	Fe^{3+} as $Fe(H_2O)_6^{3+}$	3.5×10^{-3}	2.46
	Cr^{3+} as $Cr(H_2O)_6^{3+}$	1.3×10^{-4}	3.89
	Al^{3+} as $Al(H_2O)_6^{3+}$	1.4×10^{-5}	4.85
	Fe^{2+} as $Fe(H_2O)_6^{2+}$	1.3×10^{-6}	5.89
	Cu^{2+} as $Cu(H_2O)_6^{2+}$	3.2×10^{-8}	7.49
	Ni^{2+} as $Ni(H_2O)_6^{2+}$	9.3×10^{-10}	9.03
Neutral			
Group 1 and 2 cations; metal cations with charge +1	Li^+, Na^+, K^+		
	Mg^{2+}, Ca^{2+}, Ag^+		
Basic	none		

*As in Table 10.1, the experimental pK_a values have more significant figures than shown here, and the K_a values have been calculated from these better data.

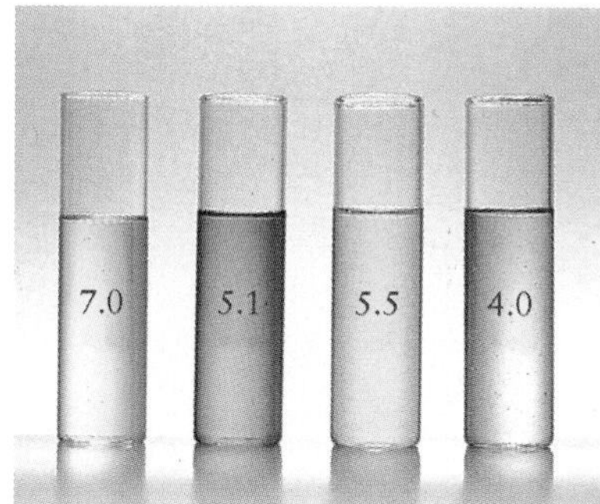

FIGURE 10.19 These four solutions show that hydrated cations can be significantly acidic. The flasks contain, from left to right, pure water, 0.1 M $Al_2(SO_4)_3(aq)$, 0.1 M $Ti_2(SO_4)_3(aq)$, and 0.1 M $CH_3COOH(aq)$. All four tubes contain a few drops of universal indicator, which turns from green in a neutral solution through yellow to red with increasing acidity. The superimposed numbers are the pH of each solution.

expect them to give acidic solutions. The acidic character of monatomic ions is generally due to the ability of some metal cations to form Lewis acid-base complexes with water.

- Small, highly charged metal cations that can act as Lewis acids in water, such as Al^{3+} and Fe^{3+}, produce acidic solutions, even though the cations themselves have no hydrogen ions to donate (Fig. 10.19).

The protons come from the water molecules that hydrate the ions in solution (Fig. 10.20). The water molecules act as Lewis bases and share electrons with the metal cation. This partial loss of electrons weakens the O—H bonds and allows one or more hydrogen ions to be lost. Small, highly charged cations exert the greatest pull on the electrons and so form the most acidic solutions.

- The cations of Group 1 and 2 metals, and those of charge +1 from other groups, are such weak Lewis acids that the hydrated ions do not act as acids.

These metal cations are too large or have too low a charge to have an appreciable polarizing effect on the hydrating water molecules that surround them, so the water molecules do not readily release their protons. These cations are sometimes called "neutral cations," because they have so little effect on the pH.

- Very few anions are acids.

It is difficult for a positively charged proton to leave a negatively charged anion. The few anions that do act as acids include $H_2PO_4^-$ and HSO_4^-.

Table 10.8 summarizes the types of ions that act as bases in solution. Cations cannot readily accept protons because the positive charge of the cation repels the positive charge of the incoming proton; so we need to look for anions that can accept protons:

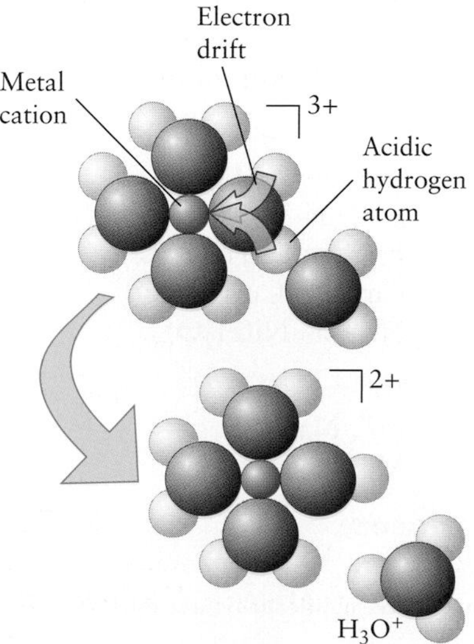

FIGURE 10.20 In water, Al^{3+} cations exist as hydrated ions that can act as Brønsted acids. Although, for clarity, only four water molecules are shown here, metal cations typically have six H_2O molecules attached to them.

TABLE 10.8 Acidic and Basic Character of Common Anions in Water

Character	Examples
Acidic	
very few	HSO_4^-, $H_2PO_4^-$
Neutral	
conjugate bases of strong acids	Cl^-, Br^-, I^-, NO_3^-, ClO_4^-
Basic	
conjugate bases of weak acids	F^-, O^{2-}, OH^-, S^{2-}, HS^-, CN^-, CO_3^{2-}, PO_4^{3-}, NO_2^-, $CH_3CO_2^-$, other carboxylate ions

- All anions that are the conjugate bases of weak acids act as proton acceptors; so we can expect them to give basic solutions.

For example, formic acid, HCOOH, the acid in ant venom, is a weak acid, so the formate ion acts as a base in water:

$$H_2O(l) + HCO_2^-(aq) \rightleftharpoons HCOOH(aq) + OH^-(aq)$$

Acetate ions and the other ions listed in Table 10.8 act as bases in water.

- The anions of strong acids—which include Cl^-, Br^-, I^-, NO_3^-, and ClO_4^-—are such weak bases that they have no significant effect on the pH of a solution.

To determine whether the solution of a salt will be acidic, basic, or neither, we must consider both the cation and anion. If the cation is not acidic, we examine the anion to see whether it is the conjugate base of a weak acid. If the anion is neither acidic nor basic, then we examine the cation to see whether it is an acidic metal ion or the conjugate acid of a weak base. If one ion is an acid and the other a base, as in NH_4F, then the pH is affected by both the reactions of both ions with water and both equilibria must be considered, as in Section 10.16.

Self-Test 10.13A Use Tables 10.7 and 10.8 to decide whether aqueous solutions of the salts (a) $Ba(NO_2)_2$; (b) $CrCl_3$; (c) NH_4NO_3 are acidic, neutral, or basic.

[*Answer:* (a) Basic; (b) acidic; (c) acidic]

Self-Test 10.13B Decide whether aqueous solutions of (a) Na_2CO_3; (b) $AlCl_3$; (c) KNO_3 are acidic, neutral, or basic.

To calculate the pH of a salt solution, we can use the equilibrium table procedure described in Toolboxes 10.1 and 10.2—an acidic cation is treated as a weak acid and a basic anion as a weak base. Toolbox 10.3 describes the procedure.

Salts that contain the conjugate acids of weak bases produce acidic aqueous solutions; so do salts that contain small, highly charged metal cations. Salts that contain the conjugate bases of weak acids produce basic aqueous solutions.

TOOLBOX 10.3 How to calculate the pH of an electrolyte solution

Conceptual Basis

The ions provided by a dissolved salt can act as acids and bases: the cations are typically acidic and the anions are generally basic. We normally assume that the acidic or basic character of one type of ion dominates the other. We then treat the solution as that of a weak acid or a weak base, using an equilibrium table to calculate the composition and hence the pH.

Procedure

The procedure is similar to the one set out in Toolboxes 10.1 and 10.2. The only difference being that the acid or base is now an ion added as a salt. As in Toolboxes 10.1 and 10.2, although we calculate the pH to the number of significant figures appropriate to the data, the answers are often considerably less reliable than that.

A salt with an acidic cation First, write the chemical equation for proton transfer to water and the expression for K_a. Then set up an equilibrium table:

Step 1 The initial molarity of the acidic cation is equal to the molarity of the cation that the salt would produce if the salt were fully dissociated and the cation retained all its acidic protons. The initial molarities of its conjugate base and H_3O^+ are assumed to be 0.

Step 2 Write the change in molarity of the acidic cation as $-x$ mol·L^{-1} and use the reaction stoichiometry to write the corresponding changes for the conjugate base and H_3O^+. Ignore H_3O^+ from the autoprotolysis of water; this approximation is valid if $[H_3O^+]$ is substantially (at least 10 times) greater than 1×10^{-7} mol·L^{-1}.

Step 3 Write the equilibrium molarities of each species in terms of x.

Step 4 Express the acidity constant for the acidic cation in terms of x and solve the equation for x. If K_a is not available, obtain it from the value of K_b for the conjugate base by using $K_a = K_w/K_b$ (Eq. 12a). Because x mol·L^{-1} is the H_3O^+ molarity, the pH of the solution is $-\log x$.

This procedure is illustrated in Example 10.6.

A salt with a basic anion In this case, we expect pH > 7. We follow the same procedure used for an acidic cation, except that now proton transfer to the anion results in the formation of OH^- ions. We therefore need K_b for the basic anion, which we may have to calculate from K_a for its conjugate acid. The equilibrium table leads to a value for pOH, so we convert pOH to pH by using pH + pOH = 14.00. This procedure is illustrated in Example 10.7.

EXAMPLE 10.6 Calculating the pH of a salt solution with an acidic cation

Estimate the pH of 0.15 M $NH_4Cl(aq)$.

STRATEGY Because NH_4^+ is an acid and Cl^- is neutral, we expect pH < 7. Follow the procedure in Toolbox 10.3. The equilibrium to consider is

$$NH_4^+(aq) + H_2O(l) \rightleftharpoons H_3O^+(aq) + NH_3(aq)$$

$$K_a = \frac{[H_3O^+][NH_3]}{[NH_4^+]}$$

SOLUTION We construct the following equilibrium table, with all concentrations in moles per liter:

	Species		
	NH_4^+	NH_3	H_3O^+
Step 1 Initial molarity	0.15	0	0
Step 2 Change in molarity	$-x$	$+x$	$+x$
Step 3 Equilibrium molarity	$0.15 - x$	x	x

Step 4 The acidity constant K_a for NH_4^+ is obtained from the value of K_b for NH_3:

$$K_a = \frac{K_w}{K_b} = \frac{1.0 \times 10^{-14}}{1.8 \times 10^{-5}} = 5.6 \times 10^{-10}$$

Substitution of this value and the information from step 3 into the expression for K_a gives

$$5.6 \times 10^{-10} = \frac{x \times x}{0.15 - x}$$

We now suppose that x is less than 5% of 0.15 and simplify this expression to

$$5.6 \times 10^{-10} \approx \frac{x^2}{0.15}$$

The solution of this equation is

$$x \approx \sqrt{0.15 \times (5.6 \times 10^{-10})} = 9.2 \times 10^{-6}$$

The approximation that x is less than 5% of 0.15 is valid. Moreover, the H_3O^+ molarity (9.2×10^{-6} mol·L^{-1}) is much larger than that generated by the autoprotolysis of water (1.0×10^{-7} mol·L^{-1}), so the neglect of the latter contribution is also valid. The pH of the solution is

$$\text{pH} \approx -\log(9.2 \times 10^{-6}) = 5.04$$

or about 5.0.

Self-Test 10.14A Estimate the pH of 0.10 M $CH_3NH_3Cl(aq)$, aqueous methylammonium chloride; the cation is $CH_3NH_3^+$.

[*Answer:* 5.78]

Self-Test 10.14B Estimate the pH of 0.10 M $NH_4NO_3(aq)$.

EXAMPLE 10.7 Calculating the pH of a salt solution with a basic anion

Estimate the pH of 0.15 M $Ca(CH_3CO_2)_2(aq)$. The proton transfer equilibrium is

$$H_2O(l) + CH_3CO_2^-(aq) \rightleftharpoons CH_3COOH(aq) + OH^-(aq)$$

$$K_b = \frac{[CH_3COOH][OH^-]}{[CH_3CO_2^-]}$$

STRATEGY Because the acetate ion is a base and the calcium ion is a Group 2 cation and therefore neutral, expect a basic solution with pH > 7. Use the procedure in Toolbox 10.3.

SOLUTION The initial molarity of $CH_3CO_2^-$ is $2 \times 0.15\ mol{\cdot}L^{-1} = 0.30\ mol{\cdot}L^{-1}$, because each formula unit of salt provides two $CH_3CO_2^-$ ions.

	Species		
	$CH_3CO_2^-$	CH_3COOH	OH^-
Step 1 Initial molarity	0.30	0	0
Step 2 Change in molarity	$-x$	$+x$	$+x$
Step 3 Equilibrium molarity	$0.30 - x$	x	x

Step 4 Table 10.1 gives the K_a of CH_3COOH as 1.8×10^{-5}, so the K_b of its conjugate base, the $CH_3CO_2^-$ ion, is

$$K_b = \frac{K_w}{K_a} = \frac{1.0 \times 10^{-14}}{1.8 \times 10^{-5}} = 5.6 \times 10^{-10}$$

Next, we insert this value and the information from step 3 into the expression for K_b and obtain

$$5.6 \times 10^{-10} = \frac{x \times x}{0.30 - x}$$

Because K_b is so small, we can anticipate that x is less than 5% of 0.30; hence we simplify this expression to

$$5.6 \times 10^{-10} \approx \frac{x^2}{0.30}$$

It follows that

$$x \approx \sqrt{0.30 \times (5.6 \times 10^{-10})} = 1.3 \times 10^{-5}$$

which is far less than 5% of 0.30. The OH^- molarity ($1.3 \times 10^{-5}\ mol{\cdot}L^{-1}$, from step 3) arising from the proton transfer equilibrium is much larger than that arising from the autoprotolysis of water (1.0×10^{-7} mol·L^{-1}), so it is valid to neglect autoprotolysis. Because the molarity of OH^- ions is 1.3×10^{-5} mol·L^{-1},

$$\text{pOH} = -\log(1.3 \times 10^{-5}) = 4.89$$

and

$$\text{pH} = 14.00 - 4.89 = 9.11$$

or about 9.1. The solution is basic, as expected.

Self-Test 10.15A Estimate the pH of 0.10 M $KC_6H_5CO_2$(aq), potassium benzoate. (The conjugate acid of the benzoate ion is benzoic acid, C_6H_5COOH, $K_a = 6.5 \times 10^{-5}$.)

[*Answer:* 8.59]

Self-Test 10.15B Estimate the pH of 0.020 M KF(aq); see Table 10.1 for data.

AUTOPROTOLYSIS AND pH

Suppose we were asked to estimate the pH of 1.0×10^{-8} M HCl(aq). If we used the techniques of Example 10.2 to calculate the pH from the concentration of the acid itself, we would find a pH of 8. That value, though, is absurd, because it lies on the basic side of neutrality, whereas HCl is an acid! The error stems from there being two sources of hydronium ions, but we have considered only one. At very low acid concentrations, the supply of hydronium ions from autoprotolysis of water is similar to the supply provided by the very low concentration of HCl, and both supplies must be taken into account. The following two sections explain how to take autoprotolysis into account, first for strong acids and bases, and then for weak ones.

10.15 Very Dilute Solutions of Strong Acids and Bases

We have to include the contribution of autoprotolysis to pH only when the concentration of strong acid or base is less than about 10^{-6} mol·L^{-1}. To calculate the pH in such a case, we need to consider all species in solution. As an example, consider a solution of HCl, a strong acid. Other than water, the species present are H_3O^+, OH^-, and Cl^-. There are three unknown concentrations. To find them, we need three equations.

One equation takes **charge balance** into account, the requirement that the solution be electrically neutral overall. That is, the concentration of cations must equal that of anions. Because there is only one type of cation, H_3O^+, the concentration of H_3O^+ ions must equal the sum of the concentrations of the two types of anions, Cl^- and OH^-. The charge balance relation $[H_3O^+] = [Cl^-] + [OH^-]$ then tells us that

$$[OH^-] = [H_3O^+] - [Cl^-]$$

The second equation takes **material balance** into account, the requirement that all the added solute must be accounted for even though it is now present as ions. In this example, material balance means that the concentration of Cl^- ions is equal to the concentration of the HCl added initially (all the HCl molecules are deprotonated). If we denote the numerical value of that initial concentration by $[HCl]_{initial}$, the material balance relation is $[Cl^-] = [HCl]_{initial}$. We can combine it with the preceding equation to write

$$[OH^-] = [H_3O^+] - [HCl]_{initial}$$

The third equation relating the species is the autoprotolysis constant, K_w (Eq. 1). We can substitute the preceding expression for $[OH^-]$ into this equation:

$$K_w = [H_3O^+][OH^-] = [H_3O^+]([H_3O^+] - [HCl]_{initial})$$

and rearrange this expression into a quadratic equation:

$$[H_3O^+]^2 - [HCl]_{initial}[H_3O^+] - K_w = 0 \qquad (17)$$

As shown in Example 10.8, we can use the quadratic formula to solve this equation for the concentration of hydronium ions.

Now consider a very dilute solution of a strong base, such as NaOH. Apart from water, the species present in solution are Na^+, OH^-, and H_3O^+. We can write down three similar equations relating their concentrations by using charge balance, material balance, and the autoprotolysis constant. Because the cations present are hydronium ions and sodium ions, the charge balance relation is

$$[OH^-] = [H_3O^+] + [Na^+]$$

The concentration of sodium ions is the same as the initial concentration of the NaOH, $[NaOH]_{initial}$, so the material balance relation is $[Na^+] = [NaOH]_{initial}$. It follows that

$$[OH^-] = [H_3O^+] + [NaOH]_{initial}$$

The autoprotolysis constant now becomes

$$K_w = [H_3O^+]([H_3O^+] + [NaOH]_{initial}) \qquad (18)$$

This expression gives us a quadratic equation that can be solved in the same way we solved the equation for a strong acid.

In very dilute solutions of strong acids and bases, the pH is significantly affected by the autoprotolysis of water. The pH is determined by solving three simultaneous equations: the charge balance equation, the material balance equation, and the expression for K_w.

EXAMPLE 10.8 Finding the pH of a very dilute aqueous solution of a strong acid

What is the pH of 8.0×10^{-8} M HCl(aq)?

STRATEGY A very dilute acidic solution can be expected to have a pH slightly below 7. Because the concentration of acid is so low, we need to use Eq. 17 to calculate the value. Solve for the hydronium ion concentration by using the quadratic formula. Calculate the pH from $\text{pH} = -\log[H_3O^+]$.

SOLUTION If we let $[\text{HCl}]_{\text{initial}} = 8.0 \times 10^{-8}\ \text{mol}\cdot\text{L}^{-1}$ and $x = [H_3O^+]$, then Eq. 17 becomes

$$x^2 - (8.0 \times 10^{-8})x - (1.0 \times 10^{-14}) = 0$$

Therefore, from the quadratic formula,

$$x = \frac{(8.0 \times 10^{-8}) \pm \sqrt{(-8.0 \times 10^{-8})^2 - 4(1)(-1.0 \times 10^{-14})}}{2}$$

$$= 1.5 \times 10^{-7} \quad \text{or} \quad -6.8 \times 10^{-8}$$

The negative root must be rejected; we are left with $[H_3O^+] = 1.5 \times 10^{-7}$ $\text{mol}\cdot\text{L}^{-1}$, so pH = 6.82. This value is just less than 7, as we anticipated.

Self-Test 10.16A What is the pH of 1.0×10^{-7} M HNO_3(aq)?

[*Answer:* 6.79]

Self-Test 10.16B What is the pH of 2.0×10^{-7} M NaOH(aq)?

10.16 Very Dilute Solutions of Weak Acids

Some acids, such as hypoiodous acid, HIO, are so weak and undergo so little deprotonation that the autoprotolysis of water almost always contributes significantly to the pH. Another indication that autoprotolysis must be taken into account occurs when we use the procedures in Toolboxes 10.1 and 10.2 and find that the pH of a weak acid is greater than 6 or the pH of a weak base is less than 8. We must then carry out a calculation that takes into account the autoprotolysis of water.

How do we do that?

The calculation of pH for very dilute solutions of a weak acid HA is based on the fact that, apart from water, there are four species in solution, namely, HA, A^-, H_3O^+, and OH^-. Because there are four unknowns, we need four equations to find their concentrations. Two relations that we can use are the autoprotolysis constant of the solvent and the acidity constant of the acid HA:

$$K_w = [H_3O^+][OH^-]$$

$$K_a = \frac{[H_3O^+][A^-]}{[HA]}$$

Charge balance provides another equation:

$$[H_3O^+] = [OH^-] + [A^-]$$

Recall that this relation states that the total concentration of (singly charged) cations must be equal to the total concentration of (singly charged) anions, for only then is the solution electrically neutral. Material balance provides the fourth equation: the total concentration of A groups (for instance, F atoms if the acid added is HF) must be equal to the initial concentration of the acid:

$$[HA]_{initial} = [HA] + [A^-]$$

This relation follows from the fact that, at equilibrium, all the weak acid originally present must be in the form of either the acid, HA, or the deprotonated ion, A^-.

Our task is to use these four equations to find an expression for the concentration of hydronium ions in terms of the initial concentration of the acid. First, we use the charge balance relation to express the concentration of A^- in terms of $[H_3O^+]$:

$$[A^-] = [H_3O^+] - [OH^-]$$

Then we express the $[OH^-]$ that occurs in this expression in terms of the hydronium ion concentration by using the autoprotolysis expression:

$$[A^-] = [H_3O^+] - \frac{K_w}{[H_3O^+]}$$

When we substitute this expression into the material balance equation, we obtain

$$\begin{aligned}[HA] &= [HA]_{initial} - [A^-] = [HA]_{initial} - \left([H_3O^+] - \frac{K_w}{[H_3O^+]}\right) \\ &= [HA]_{initial} - [H_3O^+] + \frac{K_w}{[H_3O^+]}\end{aligned}$$

Now we substitute the expressions for [HA] and $[A^-]$ into K_a to obtain:

Don't be alarmed by the complexity of this expression. We would be surprised if you were ever expected to memorize it. We certainly could not write it down from memory.

$$K_a = \frac{[H_3O^+]\left([H_3O^+] - \dfrac{K_w}{[H_3O^+]}\right)}{[HA]_{initial} - [H_3O^+] + \dfrac{K_w}{[H_3O^+]}} \qquad (19)$$

Equation 19 is certainly intimidating, but it can often be simplified. A useful talent to develop as a scientist is the ability to *interpret* an equation. One of the techniques used to interpret an equation is to look for experimental conditions that simplify it. Let's put this idea to work.

In many solutions of weak acids, the concentration of hydronium ions is so high that $[H_3O^+] > 10^{-6}$ mol·L^{-1} (that is, pH < 6). Under these conditions, $K_w/[H_3O^+] < 10^{-8}\ mol{\cdot}L^{-1}$ and we can ignore this term in both the numerator and the denominator of Eq. 19. The expression for K_a simplifies to the familiar one we use when the autoprotolysis of water has an insignificant effect on pH (Section 10.12):

$$K_a = \frac{[H_3O^+]^2}{[HA]_{initial} - [H_3O^+]} \tag{20}$$

This equation can be solved by using the quadratic formula (Toolbox 10.1).

Another simplification is possible if the extent of deprotonation is very small, in the sense that $[H_3O^+] \ll [HA]_{initial}$, and if $K_w/[H_3O^+] \ll [HA]_{initial}$. In this case, replace the denominator of Eq. 19 by $[HA]_{initial}$ alone:

$$K_a = \frac{[H_3O^+]}{[HA]_{initial}}\left([H_3O^+] - \frac{K_w}{[H_3O^+]}\right) \tag{21}$$

If both approximations can be made, we can simplify this relation even further and obtain the expression used in Example 10.3:

$$K_a \approx \frac{[H_3O^+]^2}{[HA]_{initial}} \tag{22}$$

It is always wise to check to see whether a complicated expression can be reduced to a known, simpler expression.

In the worst case, when the acid is so dilute or so weak that it is not even true that $[H_3O^+] > 10^{-6}\ mol{\cdot}L^{-1}$ (that is, when the pH lies between 6 and 7), we have to use the full expression, Eq. 19. To see what this involves, we write $x = [H_3O^+]$ and rearrange Eq. 19 into

$$x^3 + K_a x^2 - (K_w + K_a[HA]_{initial})x - K_a K_w = 0 \tag{23}$$

This is a cubic equation in x. To solve this equation, it is best to use a graphing calculator or mathematical software like that on the Web site of this text.

In aqueous solutions of very weak acids, the autoprotolysis of water must be taken into account if the hydronium ion concentration is less than $10^{-6}\ mol{\cdot}L^{-1}$. The expressions for K_w and K_a are combined with the equations for charge balance and material balance to find the pH.

EXAMPLE 10.9 Estimating the pH of a dilute aqueous solution of a weak acid

Use the information in Table 10.1 to estimate the pH of a 1.0×10^{-3} M aqueous phenol solution.

STRATEGY First, check to see whether approximations can be made so that a simple equation can be used. If they cannot, use the full expression, Eq. 19. Check two assumptions: that $[H_3O^+] > 10^{-6}\ mol{\cdot}L^{-1}$ (which allows Eq. 20 to be used) and that $[H_3O^+] \ll [HA]_{initial}$, which can simplify calculations as in Eq. 21 or Eq. 22. If the assumptions are not justified, solve Eq. 19 with a graphing calculator or the appropriate software.

SOLUTION From Table 10.1, we find that K_a for phenol is 1.3×10^{-10}. First, we make both assumptions described in the strategy, so that we can use the simple expression in Eq. 22 in the form

$$[H_3O^+]^2 = [HA]_{initial}K_a = (1.0 \times 10^{-3}) \times (1.3 \times 10^{-10}) = 1.3 \times 10^{-13}$$

We take square roots of both sides and obtain

$$[H_3O^+] = 3.6 \times 10^{-7}\ mol{\cdot}L^{-1}$$

We see that the assumption that $[H_3O^+] > 10^{-6}\ mol{\cdot}L^{-1}$ is not justified. However, the second set of assumptions, that $[H_3O^+] \ll [HA]_{initial}$ and $K_w/[H_3O^+] \ll [HA]_{initial}$, is justified. In this case, we can simplify Eq. 19 to Eq. 21:

$$K_a \approx \frac{[H_3O^+]}{[HA]_{initial}}\left([H_3O^+] - \frac{K_w}{[H_3O^+]}\right)$$

With $x = [H_3O^+]$, this expression can be rearranged into

$$x^2 \approx K_w + K_a\,[HA]_{initial}$$

and solved for x:

$$x \approx \sqrt{K_w + K_a\,[HA]_{initial}}$$

Now we substitute the values for K_a, K_w, and $[HA]_{initial}$ into the equation:

$$\begin{aligned}[H_3O^+] &= \sqrt{(1.0 \times 10^{-14}) + (1.3 \times 10^{-10})\,(1.0 \times 10^{-3})}\ mol{\cdot}L^{-1} \\ &= 3.7 \times 10^{-7}\ mol{\cdot}L^{-1}\end{aligned}$$

It follows that pH = 6.43. As can be appreciated by the steps taken in this example, it is important to develop a sense of the relative concentrations of all species in solution to help us decide when approximations will be valid.

Self-Test 10.17A Use the information in Table 10.1 to estimate the pH of 2.0×10^{-4} M HCN(aq).

[*Answer:* 6.48]

Self-Test 10.17B Use the information in Table 10.1 to estimate the pH of 1.0×10^{-2} M HIO(aq).

POLYPROTIC ACIDS AND BASES

A **polyprotic acid** is a compound that can donate more than one proton. Many acids encountered in industry, medicine, the environment, and the laboratory are polyprotic. They include sulfuric acid, H_2SO_4, and carbonic acid, H_2CO_3, each of which can donate two protons, and phosphoric acid, H_3PO_4, which can donate three protons. Polyprotic acids are of the greatest significance in biological systems, because many enzymes can be regarded as polyprotic acids that carry out their vital functions by donating one proton after another. A **polyprotic base** is a species that can accept more than one proton. Examples include the CO_3^{2-} and SO_3^{2-} anions, both of which can accept two protons, and the PO_4^{3-} anion, which can accept three protons.

Polyprotic acids can be treated exactly like monoprotic acids. The principal difference is that we have to allow for a succession of deprotonation steps. For example, carbonic acid can lose one proton to form HCO_3^-, then that ion can donate the remaining proton, to form CO_3^{2-}. We need to see when and how to take this succession of deprotonations into account when assessing the pH of the solution of a polyprotic acid or one of its salts. In addition, here we see how the relative concentrations of the ions in solution, such as PO_4^{3-}, HPO_4^{2-}, and $H_2PO_4^-$ depend on the pH of the solution.

10.17 The pH of a Polyprotic Acid Solution

Carbonic acid is an important natural component of the environment because it is formed whenever carbon dioxide dissolves in lake water or seawater. In fact, the oceans provide one of the delicate mechanisms for helping to maintain a constant concentration of carbon dioxide in the atmosphere. Carbonic acid takes part in two successive proton transfer equilibria:

$$H_2CO_3(aq) + H_2O(l) \rightleftharpoons H_3O^+(aq) + HCO_3^-(aq) \qquad K_{a1} = 4.3 \times 10^{-7}$$

$$HCO_3^-(aq) + H_2O(l) \rightleftharpoons H_3O^+(aq) + CO_3^{2-}(aq) \qquad K_{a2} = 5.6 \times 10^{-11}$$

The conjugate base of H_2CO_3 in the first equilibrium, HCO_3^-, acts as an acid in the second equilibrium. That ion produces in turn its own conjugate base, CO_3^{2-}.

Protons are donated successively by polyprotic acids, with the acidity constant decreasing significantly, usually by a factor of about 10^3, with each proton lost (Table 10.9):

$$K_{a1} \gg K_{a2} \gg K_{a3} \gg \cdots$$

The decrease is reasonable: because of the attraction between opposite charges, it is harder to lose a positively charged proton from a negatively charged ion (such as HCO_3^-) than from the original uncharged molecule (H_2CO_3). Sulfuric acid, for example, is a strong acid, and H_2SO_4 loses its first proton to give its conjugate base, the hydrogen sulfate ion, HSO_4^-; this ion, however, is a weak acid.

EXAMPLE 10.10 Calculating the pH of a solution of sulfuric acid

Calculate the pH of 0.010 M H_2SO_4(aq) at 25°C. Use the information in Table 10.9.

TABLE 10.9 Acidity Constants of Polyprotic Acids

Acid	K_{a1}	pK_{a1}	K_{a2}	pK_{a2}	K_{a3}	pK_{a3}
sulfuric acid, H_2SO_4	strong		1.2×10^{-2}	1.92		
oxalic acid, $(COOH)_2$	5.9×10^{-2}	1.23	6.5×10^{-5}	4.19		
sulfurous acid, H_2SO_3	1.5×10^{-2}	1.81	1.2×10^{-7}	6.91		
phosphorous acid, H_3PO_3	1.0×10^{-2}	2.00	2.6×10^{-7}	6.59		
phosphoric acid, H_3PO_4	7.6×10^{-3}	2.12	6.2×10^{-8}	7.21	2.1×10^{-13}	12.68
tartaric acid, $C_2H_4O_2(COOH)_2$	6.0×10^{-4}	3.22	1.5×10^{-5}	4.82		
carbonic acid, H_2CO_3	4.3×10^{-7}	6.37	5.6×10^{-11}	10.25		
hydrosulfuric acid, H_2S	1.3×10^{-7}	6.89	7.1×10^{-15}	14.15		

STRATEGY Sulfuric acid is the only common polyprotic acid for which the first deprotonation is complete. The second deprotonation adds to the H_3O^+ molarity slightly, so the overall pH will be slightly less than that due to the first deprotonation alone. To find the pH, we construct the equilibrium table in the usual way. For the initial H_3O^+ and HSO_4^- molarities, we suppose that every H_2SO_4 molecule has lost its first proton completely. Because K_{a2} is relatively large (0.012), there are no shortcuts: it will be necessary to solve a quadratic equation.

SOLUTION The first deprotonation is complete:

$$H_2SO_4(aq) + H_2O(l) \longrightarrow H_3O^+(aq) + HSO_4^-(aq)$$

It results in an H_3O^+ molarity equal to the original molarity of the acid before deprotonation, 0.010 mol·L^{-1}. This value corresponds to pH = 2.0. The second proton transfer equilibrium is

$$HSO_4^-(aq) + H_2O(l) \rightleftharpoons H_3O^+(aq) + SO_4^{2-}(aq) \qquad K_{a2} = 0.012$$

The equilibrium table, with all concentrations in moles per liter, is

	Species		
	HSO_4^-	SO_4^{2-}	H_3O^+
Step 1 Initial molarity	0.010	0	0.010
Step 2 Change in molarity	$-x$	$+x$	$+x$
Step 3 Equilibrium molarity	$0.010 - x$	x	$0.010 + x$

Step 4 Now substitute the molarities in step 3 into the expression for the second acidity constant and use $K_{a2} = 0.012$, which gives

$$0.012 = \frac{(0.010 + x) \times x}{0.010 - x}$$

To find x, rearrange this expression into a quadratic equation:

$$x^2 + 0.022x - (1.2 \times 10^{-4}) = 0$$

The quadratic formula yields $x = 4.5 \times 10^{-3}$. The total H_3O^+ molarity is therefore

$$\begin{aligned}[H_3O^+] &= 0.010 + x \text{ mol·L}^{-1} = 0.010 + (4.5 \times 10^{-3}) \text{ mol·L}^{-1} \\ &= 1.4 \times 10^{-2} \text{ mol·L}^{-1}\end{aligned}$$

so

$$\text{pH} = -\log (1.4 \times 10^{-2}) = 1.9$$

The pH of the solution is less than 2.0, as we expected.

Self-Test 10.18A Estimate the pH of 0.050 M H_2SO_4(aq) at 25°C.
[*Answer:* 1.2]

Self-Test 10.18B Estimate the pH of 0.10 M H_2SO_4(aq) at 25°C.

When we turn to polyprotic acids other than sulfuric, we find that the parent acid is weak and the acidity constants of successive deprotonation steps are normally widely different. As a result, we can treat a polyprotic acid or the salt of any anion derived from it as the *only significant species* in solution. This approximation leads to a major simplification: to calculate the pH of a polyprotic acid, we just use K_{a1} and take only the first deprotonation into account; that is, we treat the acid as a monoprotic weak acid. Subsequent deprotonations do take place, but they do not affect the pH significantly and can be ignored.

Estimate the pH of a polyprotic acid for which all deprotonations are weak by using only the first deprotonation equilibrium and assuming that further deprotonation is insignificant. An exception is sulfuric acid, the only common polyprotic acid that is a strong acid in its first deprotonation.

EXAMPLE 10.11 **Calculating the pH of a solution of a polyprotic acid**

Use the information in Table 10.9 to calculate the pH of 0.20 M H_2S(aq).

STRATEGY The primary solute species in a solution of a polyprotic acid (other than H_2SO_4) is the parent acid itself, in this case, H_2S. We can find the pH of the solution by assuming that the acid loses only one proton and treat it as a monoprotic weak acid. Set up an equilibrium table as described in Toolbox 10.1 and determine the H_3O^+ molarity by using the first acidity constant, K_{a1}.

SOLUTION The primary proton transfer equilibrium is

$$H_2S(aq) + H_2O(l) \rightleftharpoons H_3O^+(aq) + HS^-(aq)$$

and the equilibrium table, with the concentrations in moles per liter, is

	Species		
	H_2S	HS^-	H_3O^+
Step 1 Initial molarity	0.20	0	0
Step 2 Change in molarity	$-x$	$+x$	$+x$
Step 3 Equilibrium molarity	$0.20 - x$	x	x

Step 4 Substitute the equilibrium molarities into the acidity constant expression:

$$K_{a1} = 1.3 \times 10^{-7} = \frac{x^2}{0.20 - x}$$

Because of the small value of the acidity constant, we anticipate that $x \ll 0.20$ and therefore approximate the expression to

$$1.3 \times 10^{-7} \approx \frac{x^2}{0.20}$$

Solving for x gives

$$x \approx \sqrt{0.20 \times (1.3 \times 10^{-7})} = 1.6 \times 10^{-4}$$

Because the concentration of HS^- is also 1.6×10^{-4} and its acidity constant only 7.1×10^{-15}, the deprotonation of HS^- will not significantly affect the pH. The pH of the solution is therefore $-\log(1.6 \times 10^{-4}) = 3.8$.

Self-Test 10.19A Use the information in Table 10.9 to calculate the pH of a 0.010 M aqueous oxalic acid solution.

[*Answer:* 2.06]

Self-Test 10.19B Use the information in Table 10.9 to calculate the pH of 0.50 M $H_3PO_4(aq)$.

10.18 Solutions of Salts of Polyprotic Acids

The conjugate base of a polyprotic acid is amphiprotic: it can act as either an acid or a base because it can either donate its remaining acidic hydrogen atom or accept an acidic hydrogen atom and revert to the original acid. For example, a hydrogen sulfide ion, HS^-, in water acts as both an acid and a base:

Exceptions are HSO_4^-, which is an extremely weak base, and HPO_3^{2-}, which does not act as an acid because its proton is not acidic (Section 10.11).

$$HS^-(aq) + H_2O(l) \rightleftharpoons H_3O^+(aq) + S^{2-}(aq) \qquad K_{a2} = 7.1 \times 10^{-15},\ pK_{a2} = 14.15$$

$$HS^-(aq) + H_2O(l) \rightleftharpoons H_2S(aq) + OH^-(aq) \qquad K_{b1} = K_w/K_{a1} = 7.7 \times 10^{-8}$$

In the second line, we have obtained K_{b1} from the relation $K_aK_b = K_w$ and the acidity constant for H_2S, the conjugate acid of HS^-:

$$H_2S(aq) + H_2O(l) \rightleftharpoons H_3O^+(aq) + HS^-(aq)$$
$$K_{a1} = 1.3 \times 10^{-7}, pK_{a1} = 6.89$$

The question that arises is whether a solution of a salt of HS^-, such as NaHS, is acidic as a result of the acid character of HS^- or basic as a result of that ion's basic character. We can suspect that if pK_{a2} is large (corresponding to HS^- being a very weak acid), then the solution will have a high pH because the basic character of HS^- will dominate. Furthermore, the basic character of HS^- is likely to be dominant if pK_{a1} is large (and hence K_{a1} small), for then HS^- is a relatively "strong" weak base (in the sense that K_{b1} is not very small). This reasoning suggests that the pH will be high if both pK_{a1} and pK_{a2} are large. Indeed, if we make some reasonable assumptions and carry through some intricate algebra, it turns out that the pH of the solution is

$$\text{pH} = \tfrac{1}{2}(\text{p}K_{a1} + \text{p}K_{a2}) \qquad (24)$$

independent of the concentration of the salt.

The pH of the aqueous solution of an amphiprotic salt is equal to the average of the pK_as of the salt and its conjugate acid.

EXAMPLE 10.12 Estimating the pH of a solution of an amphiprotic salt

Estimate the pH of (a) 0.20 M $NaH_2PO_4(aq)$; (b) 0.20 M $Na_2HPO_4(aq)$.

STRATEGY We use Eq. 24 to determine the pH of an amphiprotic salt. This equation is used for salts of the diprotic conjugate base (H_2A^-) of a triprotic acid (H_3A) and the monoprotic conjugate base (HA^-) of a diprotic acid (H_2A). However, when the solute is a salt of an anion that has lost two protons, such as HPO_4^{2-}, we must adjust the expression to use the appropriate neighboring pK_as.

SOLUTION (a) For H_3PO_4, $K_{a1} = 7.6 \times 10^{-3}$ and $K_{a2} = 6.2 \times 10^{-8}$, so for the solution of $H_2PO_4^-$,

$$\text{pH} = \tfrac{1}{2}(2.12 + 7.21) = 4.66$$

(b) For H_3PO_4, $K_{a2} = 6.2 \times 10^{-8}$ and $K_{a3} = 2.1 \times 10^{-13}$, so for the solution of $Na_2HPO_4(aq)$,

$$\text{pH} = \tfrac{1}{2}(7.21 + 12.68) = 9.94$$

Self-Test 10.20A Estimate the pH of 0.10 M NaHS(aq).

[*Answer:* 10.52]

Self-Test 10.20B Estimate the pH of 0.50 M $KH_2PO_4(aq)$.

Suppose we need to estimate the pH of an aqueous solution of the anion remaining after all protons have been stripped away from a polyprotic acid. An example is a solution of sodium sulfide, in which sulfide ions, S^{2-}, are present; another example is a solution of potassium phosphate, which contains PO_4^{3-} ions. In such a solution, the anion acts as a base: it accepts protons from water. For such a solution, we can use the techniques for calculating the pH of a basic anion described in Toolbox 10.2. The K_a to use in the calculation is for the deprotonation that produces the ion being studied. For S^{2-}, we would use K_{a2} for H_2S; and for PO_4^{3-}, we would use K_{a3} for H_3PO_4.

The pH of a solution of a salt of the final conjugate base of a polyprotic acid is found from the reaction of the anion with water.

10.19 The Concentrations of Solute Species

Environmental chemists studying the pollution caused by fertilizers in runoff from fields or mineralogists studying the formation of sedimentary rocks as groundwater trickles through rock formations may need to know the concentrations of each of the ions present in the solution. For example, they may need to know the concentration of sulfite ion in a solution of sulfurous acid or the concentrations of phosphate and hydrogen phosphate ions in a solution of phosphoric acid. The calculations in Example 10.12 tell us the pH—the hydrogen ion molarity—but they do not give the concentrations of all the solute species in solution, which may include H_3PO_4, $H_2PO_4^-$, HPO_4^{2-}, and PO_4^{3-}. To calculate them, we need to take into account all the proton transfer equilibria.

The prospect of doing so may seem daunting. However, we have already remarked that another scientific skill to learn—not just for chemistry, but for all scientific disciplines—is finding ways to simplify the calculations. That is, we start by using our insight to judge the relative concentration of each species in solution and identify terms that can be neglected. In this case, we use the general rule that concentrations of species present in the greater amount are not significantly affected by concentrations of species present in the smaller amount, especially if the differences are large. In practice, this rule means that we proceed as follows:

1 Calculate the molarities of the most abundant species.

2 Substitute these values into the equilibrium expressions for the less abundant species.

3 Solve the resulting expressions for the molarities of the less abundant species.

The concentrations of all species in a solution of a polyprotic acid can be calculated by assuming that species present in smaller amounts do not affect the concentrations of species present in larger amounts.

TOOLBOX 10.4 How to calculate the concentrations of all species in a polyprotic acid solution

Conceptual Basis

We assume that the polyprotic acid is the solute species present in largest amount. We also assume that only the first deprotonation contributes significantly to $[H_3O^+]$ and that the autoprotolysis of water does not contribute significantly to $[H_3O^+]$ or $[OH^-]$.

Procedure

Step 1 From the deprotonation equilibrium of H_3A (for a triprotic acid) or H_2A' (for a diprotic acid), determine the concentrations of the conjugate base, H_2A^- or HA'^-, respectively, and H_3O^+, as illustrated in Example 10.11.

Step 2 Find the concentration of HA^{2-} (or A'^{2-}) from the deprotonation equilibrium of H_2A^- (or HA'^-) by substituting the concentrations of H_3O^+ and H_2A^- (or HA'^-) from step 1 into the expression for K_{a2}.

Step 3 For a triprotic acid: find the concentration of A^{3-} from the deprotonation equilibrium of HA^{2-} by substituting the concentrations of H_3O^+ and HA^{2-} from step 2 into the expression for K_{a3}. The concentration of H_3O^+ is the same in all three calculations because only the first deprotonation makes a significant contribution to its value.

Step 4 Find the concentration of OH^- by dividing K_w by the concentration of H_3O^+.

EXAMPLE 10.13 Calculating the concentrations of all species in a polyprotic acid solution

Calculate the concentrations of all solute species in 0.10 M H_3PO_4(aq).

STRATEGY Assume that only H_3PO_4 significantly affects the pH and use the procedure in Toolbox 10.4.

SOLUTION ***Step 1*** The primary proton transfer equilibrium is

$$H_3PO_4(aq) + H_2O(l) \rightleftharpoons H_3O^+(aq) + H_2PO_4^-(aq)$$

and the first acidity constant, from Table 10.9, is 7.6×10^{-3}. The equilibrium table, with the concentrations in moles per liter, is

	Species		
	H_3PO_4	$H_2PO_4^-$	H_3O^+
Initial molarity	0.10	0	0
Change in molarity	$-x$	$+x$	$+x$
Equilibrium molarity	$0.10 - x$	x	x

Now substitute the equilibrium molarities into the acidity constant expression:

$$K_{a1} = 7.6 \times 10^{-3} = \frac{x^2}{0.10 - x}$$

If we were to assume that the hydronium ion concentration (x) is much smaller than 0.010, we could approximate the expression to

$$7.6 \times 10^{-3} \approx \frac{x^2}{0.10}$$

and find that $x = 0.028$. However, this value is not much smaller than 0.10, so we must use the quadratic equation instead:

$$(7.6 \times 10^{-3}) \times (0.10 - x) = x^2$$

This equation rearranges to

$$x^2 + (7.6 \times 10^{-3})x - 7.6 \times 10^{-4} = 0$$

From the quadratic formula, $x = 2.4 \times 10^{-2}$ or -3.2×10^{-2}. We reject the negative root, so $[H_3O^+] = [H_2PO_4^-] = 2.4 \times 10^{-2}\ mol{\cdot}L^{-1}$. It follows from the equilibrium table that

$$[H_3PO_4] \approx 0.10 - 0.024\ mol{\cdot}L^{-1} = 0.08\ mol{\cdot}L^{-1}$$

Step 2 Now we examine the second deprotonation of H_3PO_4 to find the concentration of HPO_4^{2-}. The second acidity constant of H_3PO_4, K_{a2}, is 6.2×10^{-8}. Because $K_{a2} \ll K_{a1}$, we can safely assume that the H_3O^+ concentration calculated in step 1 is unchanged in the second deprotonation. The proton transfer equilibrium is

$$H_2PO_4^-(aq) + H_2O(l) \rightleftharpoons H_3O^+(aq) + HPO_4^{2-}(aq)$$

We set up the equilibrium table with concentrations in moles per liter, using the results from step 1 for the concentrations of H_3O^+ and $H_2PO_4^-$.

	Species		
	$H_2PO_4^-$	HPO_4^{2-}	H_3O^+
Initial molarity (from step 1)	2.4×10^{-2}	0	2.4×10^{-2}
Change in molarity	$-x$	$+x$	$+x$
Equilibrium molarity	$2.4 \times 10^{-2} - x$	x	$2.4 \times 10^{-2} + x$

Substitute the equilibrium molarities into the acidity constant expression:

$$K_{a2} = 6.2 \times 10^{-8} = \frac{(2.4 \times 10^{-2} + x) \times x}{2.4 \times 10^{-2} - x}$$

Because K_{a2} is very small, we assume that $x \ll 2.4 \times 10^{-2}$ and write

$$6.2 \times 10^{-8} \approx \frac{(2.4 \times 10^{-2}) \times x}{2.4 \times 10^{-2}} = x$$

The assumption that $x \ll 2.4 \times 10^{-2}$ is justified; so we can conclude that $[HPO_4^{2-}] \approx K_{a2} = 6.2 \times 10^{-8}\ mol{\cdot}L^{-1}$. The second deprotonation proceeds only to a very small extent.

Step 3 The loss of the final proton from HPO_4^{2-} produces the phosphate ion, PO_4^{3-}. To find its concentration, we examine the proton transfer equilibrium for that loss:

$$HPO_4^{2-}(aq) + H_2O(l) \rightleftharpoons H_3O^+(aq) + PO_4^{3-}(aq)$$

The equilibrium constant is $K_{a3} = 2.1 \times 10^{-13}$, a very small value, so the concentration of PO_4^{3-} will be very small and the concentration of H_3O^+ will not be affected. We assume that the concentration of H_3O^+ calculated in step 1 and used in step 2 and the concentration of HPO_4^{2-} calculated in step 2 are unaffected by the additional deprotonation. We then set up an equilibrium table with concentrations in moles per liter.

	Species		
	HPO_4^{2-}	PO_4^{3-}	H_3O^+
Initial molarity (from step 2)	6.2×10^{-8}	0	2.4×10^{-2}
Change in molarity	$-x$	$+x$	$+x$
Equilibrium molarity	$6.2 \times 10^{-8} - x$	x	$2.4 \times 10^{-2} + x$

Now substitute the equilibrium molarities into the acidity constant expression:

$$K_{a3} = 2.1 \times 10^{-13} = \frac{(2.4 \times 10^{-2} + x) \times x}{6.2 \times 10^{-8} - x}$$

Because K_{a3} is so small, we can assume that $x \ll 6.2 \times 10^{-8}$ and simplify the equation to

$$2.1 \times 10^{-13} \approx \frac{(2.4 \times 10^{-2}) \times x}{6.2 \times 10^{-8}}$$

The solution of this equation is

$$x \approx \frac{(2.1 \times 10^{-13})(6.2 \times 10^{-8})}{2.4 \times 10^{-2}} = 5.4 \times 10^{-19}$$

Thus, $[PO_4^{3-}] \approx 5.4 \times 10^{-19}\ \text{mol}\cdot\text{L}^{-1}$. The HPO_4^{2-} ion is deprotonated only to a tiny extent, and our assumption that H_3O^+ is unaffected by the third deprotonation is justified.

Step 4 We find the concentration of OH^- from K_w and the concentration of H_3O^+:

$$[OH^-] = \frac{K_w}{[H_3O^+]} = \frac{1.0 \times 10^{-14}}{2.4 \times 10^{-2}}\ \text{mol}\cdot\text{L}^{-1} = 4.2 \times 10^{-13}\ \text{mol}\cdot\text{L}^{-1}$$

We can now summarize the concentrations of all the solute species in 0.10 M H_3PO_4(aq) in order of decreasing concentration:

Species	Concentration, $\text{mol}\cdot\text{L}^{-1}$
H_3PO_4	0.08
H_3O^+	2.4×10^{-2}
$H_2PO_4^-$	2.4×10^{-2}
HPO_4^{2-}	6.2×10^{-8}
OH^-	4.2×10^{-13}
PO_4^{3-}	5.4×10^{-19}

Self-Test 10.21A Calculate the concentrations of all solute species in the 0.20 M $H_2S(aq)$ solution explored in Example 10.11.

[*Answer:* With concentrations in $mol{\cdot}L^{-1}$: H_2S, 0.20; HS^-, 1.6×10^{-4}; S^{2-}, 7.1×10^{-15}; H_3O^+, 1.6×10^{-4}; OH^-, 6.2×10^{-11}]

Self-Test 10.21B Protonated glycine ($^+NH_3CH_2COOH$) is a diprotic acid with $K_{a1} = 4.5 \times 10^{-3}$ and $K_{a2} = 1.7 \times 10^{-10}$. Calculate the concentrations of all solute species in 0.50 M $NH_3CH_2COOHCl(aq)$.

10.20 Composition and pH

Sometimes we need to know how the concentrations of the ions present in a solution of a polyprotic acid vary with pH. This information is particularly important in the study of natural waters, such as rivers and lakes (Box 10.1). For example, if we were examining carbonic acid in rainwater, then at low pH, we would expect the fully protonated species (H_2CO_3) to be dominant; at high pH, we expect the fully deprotonated species (CO_3^{2-}) to be dominant; at intermediate pH, we expect the intermediate species (HCO_3^-, in this case) to be dominant (Fig. 10.21). We can verify these expectations quantitatively.

BOX 10.1 WHAT HAS THIS TO DO WITH . . . THE ENVIRONMENT?

Acid Rain and the Gene Pool

An example of human impact on the environment is the effect of acid rain on *biodiversity*, the diversity of living things. In the prairies that extend across the heartlands of North America and Asia, native plants have evolved that can survive even nitrogen-poor soil and drought. By studying prairie plants, scientists hope to breed food plants that will be hardy sources of food in times of drought. The research means studying many diverse plants, but that pool of plants is getting smaller and smaller. Some plants are becoming extinct as a result of the effects of cultivation and acid rain, which originates in distant cities.

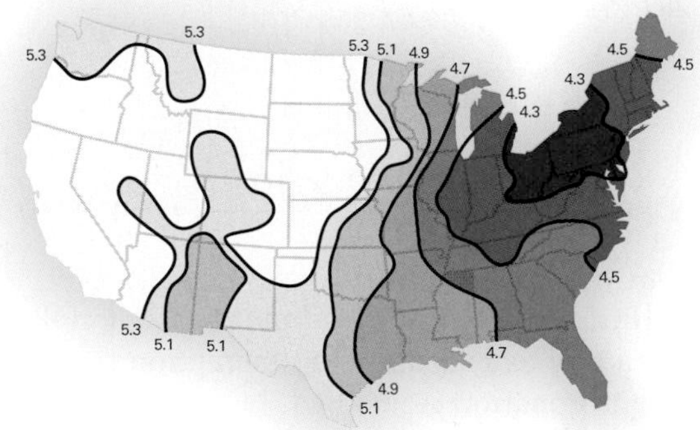

The lines on the map are pH isopleths, which identify regions in which the precipitation has the same pH, as indicated by the numbers. Notice that precipitation gradually becomes more acidic going from west to east, especially in industrialized areas of the Northeast. This acid rain may be a result of the release of nitrogen and sulfur oxides into the atmosphere.

Eastern gama grass is a variety of a prairie plant that produces seeds rich in protein. It is the subject of sustainable agriculture research because it produces an abundance of seeds and yet is a perennial plant that is resistant to drought.

Acid rain is a regional phenomenon. The lines on the map are *isopleths*, lines that indicate constant values of the pH of rain. Notice that the pH of rain decreases downwind (generally, east) of heavily populated areas. The low pH in heavily industrialized and

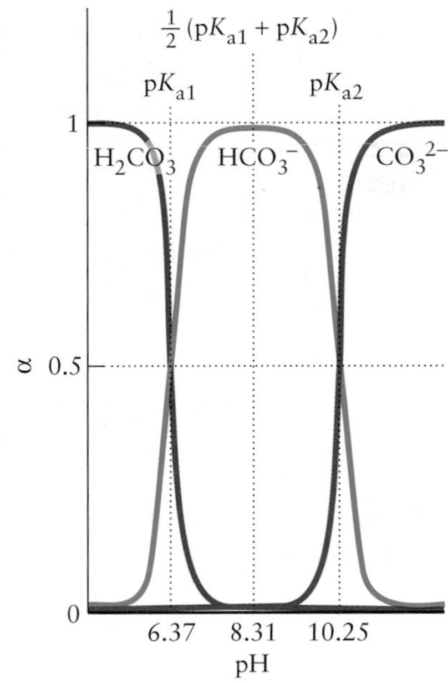

FIGURE 10.21 The fractional composition of the species in carbonic acid as a function of pH. Note that the more fully protonated species are dominant at lower pH.

populated areas is caused by the acidic oxides sulfur dioxide, SO_2, and the nitrogen oxides, NO and NO_2.

Rain unaffected by human activity contains mostly weak acids and has a pH of 5.7. The primary acid present is carbonic acid, H_2CO_3, which results when atmospheric carbon dioxide, an acidic oxide, dissolves in water. The major pollutants in acid rain are strong acids that arise from human activities. Atmospheric nitrogen and oxygen can react to form NO, but the endothermic reaction is spontaneous only at the high temperatures of automobile internal combustion engines and electrical power stations:

$$N_2(g) + O_2(g) \rightleftharpoons 2\ NO(g)$$

Nitric oxide, NO, is not very soluble in water, but it can be oxidized further in air to form nitrogen dioxide:

$$2\ NO(g) + O_2(g) \longrightarrow 2\ NO_2(g)$$

The NO_2 reacts with water, forming nitric acid and nitric oxide:

$$3\ NO_2(g) + 3\ H_2O(l) \longrightarrow 2\ H_3O^+(aq) + 2\ NO_3^-(aq) + NO(g)$$

The catalytic converters now used in automobiles can reduce NO to harmless N_2. They are required in the United States for all new cars and trucks (see Section 13.14).

Sulfur dioxide is produced as a by-product of the burning of fossil fuels. It may combine with water directly to form sulfurous acid, a weak acid:

$$SO_2(g) + H_2O(l) \longrightarrow H_2SO_3(aq)$$

Alternatively, in the presence of particulate matter and aerosols, sulfur dioxide may react with atmospheric oxygen to form sulfur trioxide, which forms sulfuric acid in water:

$$2\ SO_2(g) + O_2(g) \longrightarrow 2\ SO_3(g)$$

$$SO_3(g) + 2\ H_2O(l) \longrightarrow H_3O^+(aq) + HSO_4^-(aq)$$

Acid rain affects plants by changing the conditions in the soil. For example, nitric acid deposits nitrates, which fertilize the land. The nitrates allow fast-growing weeds such as quack grass to replace valuable prairie species. If these species were to become extinct, their genetic material would no longer be available for agricultural research.

Research on air pollution is complex. Forests and prairies cover vast areas, and the interplay of regional air pollutants is so subtle that it may take years to sort out all the environmental stresses. However, one message is already clear: if we can improve controls on acidic oxide emissions, we may be able to maintain our quality of life without losing our precious heritage of native plants. We can start today, by using automobiles less and bicycles more, or by taking public transportation when we can.

For Further Reading: J. P. Grime, "Biodiversity and ecosystem function: The debate deepens," *Science* **277**,1260–1261 (1997). R. A. Kerr, "Acid rain control: Success on the cheap," *Science* **282**,1024–1027 (1998). J. Raloff, "Pollution helps weeds take over prairies," *Science News* **150**, 356 (1996).

How do we do that?

To see how to calculate how the concentrations of species present in a solution vary with pH, we take the carbonic acid system as an example. Consider the following two proton transfer equilibria:

$$H_2CO_3(aq) + H_2O(l) \rightleftharpoons H_3O^+(aq) + HCO_3^-(aq)$$

$$K_{a1} = \frac{[H_3O^+][HCO_3^-]}{[H_2CO_3]}$$

$$HCO_3^-(aq) + H_2O(l) \rightleftharpoons H_3O^+(aq) + CO_3^{2-}(aq)$$

$$K_{a2} = \frac{[H_3O^+][CO_3^{2-}]}{[HCO_3^-]}$$

We can express the composition of the solution in terms of the fraction, $\alpha(X)$, of carbonate present as each species X, where

$$\alpha(X) = \frac{[X]}{[H_2CO_3] + [HCO_3^-] + [CO_3^{2-}]} \qquad (25)$$

It will turn out to be helpful to express $\alpha(X)$ in terms of the ratio of each species to the intermediate species, HCO_3^-. So we divide the numerator and denominator by $[HCO_3^-]$ and get

$$\alpha(X) = \frac{[X]/[HCO_3^-]}{([H_2CO_3]/[HCO_3^-]) + 1 + ([CO_3^{2-}]/[HCO_3^-])} \qquad (26)$$

All three concentration ratios can be written in terms of the hydronium ion concentration. We simply rearrange the expressions for the acidity constants:

$$\frac{[H_2CO_3]}{[HCO_3^-]} = \frac{[H_3O^+]}{K_{a1}} \qquad \frac{[CO_3^{2-}]}{[HCO_3^-]} = \frac{K_{a2}}{[H_3O^+]}$$

Therefore, the three fractions are

$$\alpha(H_2CO_3) = \frac{[H_3O^+]^2}{f}$$

$$\alpha(HCO_3^-) = \frac{[H_3O^+]K_{a1}}{f} \qquad (27a)$$

$$\alpha(CO_3^{2-}) = \frac{K_{a1}K_{a2}}{f}$$

where

$$f = [H_3O^+]^2 + [H_3O^+]K_{a1} + K_{a1}K_{a2} \qquad (27b)$$

These equations allow us to predict the composition at any pH.

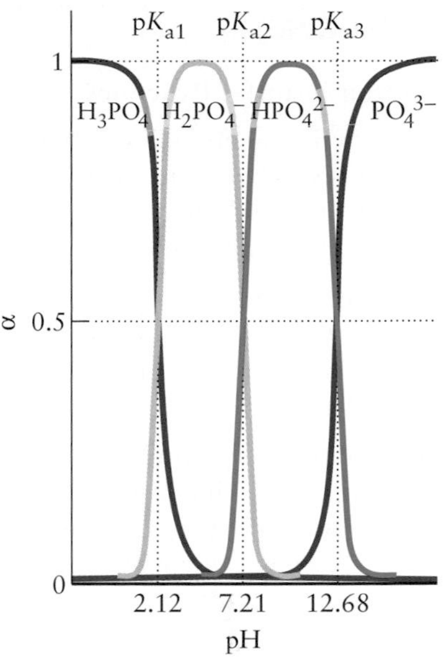

FIGURE 10.22 The fractional composition of the species in phosphoric acid as a function of pH. As in Fig. 10.21, the more fully protonated the species, the lower the pH at which it is dominant.

Let's see if the three expressions in Eq. 27a confirm our expectations. At high pH, the concentration of hydronium ions is very low, and therefore the numerators in $\alpha(H_2CO_3)$ and $\alpha(HCO_3^-)$ are very small, so these species are in very low abundance, as we expected. At low pH, the concentration of hydronium ions is high, and therefore the numerator in $\alpha(H_2CO_3)$ is large, and this species dominates.

If we substitute numerical values, we can see how the αs vary with pH. The shapes of the curves in Fig. 10.21 show us that $\alpha(HCO_3^-) \approx 1$ at intermediate pH. The maximum value of $\alpha(HCO_3^-)$ occurs at

$$\text{pH} = \tfrac{1}{2}(\text{p}K_{a1} + \text{p}K_{a2}) \qquad (28)$$

Note that the fully protonated form becomes dominant when $\text{pH} < \text{p}K_{a1}$ and the fully deprotonated form becomes dominant when $\text{pH} > \text{p}K_{a2}$. Similar calculations can be carried out for triprotic acid salts in water. Just as we should expect, we find that successively more deprotonated forms become dominant as the pH is raised (Fig. 10.22).

The fraction of deprotonated species increases as the pH is increased, as summarized in Figs. 10.21 and 10.22.

SKILLS YOU SHOULD HAVE MASTERED

1 Write the formulas for conjugate acids and bases, Self-Test 10.1.

2 Identify Brønsted and Lewis acids and bases in a chemical reaction, Self-Test 10.2.

3 Calculate the molarities of ions in a solution of a strong acid or base, Example 10.1.

4 Calculate the pH and pOH of a solution of a strong acid or base, Example 10.2 and Section 10.6.

5 Calculate the H_3O^+ molarity in a solution from its pH, Self-Test 10.5.

6 Show how the K_a and $\text{p}K_a$ of an acid are related to the K_b and $\text{p}K_b$ of its conjugate base, Section 10.8.

7 Use K_a values to predict the relative strengths of two acids or two bases, Self-Test 10.7.

8 Predict the relative strengths of acids from molecular structures, Sections 10.10 and 10.11.

9 Calculate the pH and percentage deprotonation of a weak acid, Toolbox 10.1 and Example 10.3.

10 Calculate the value of K_a for a weak acid, Example 10.4.

11 Calculate the pH and percentage protonation of a weak base, Toolbox 10.2 and Example 10.5.

12 Calculate the pH of an electrolyte solution, Toolbox 10.3 and Examples 10.6 and 10.7.

13 Find the pH of an aqueous solution of strong acid or base that is so dilute the autoprotolysis of water significantly affects the pH, Example 10.8.

14 Find the pH of an aqueous solution of a weak acid for which the autoprotolysis of water significantly affects the pH, Example 10.9.

15 Calculate the pH of a solution of sulfuric acid, Example 10.10.

16 Calculate the pH of a solution of a polyprotic acid, Example 10.11.

17 Estimate the pH of a solution of an amphiprotic salt, Example 10.12.

18 Calculate the concentrations of all species in a polyprotic acid solution, Toolbox 10.4 and Example 10.13.

EXERCISES

Proton Transfer Reactions

10.1 Write the formulas for the conjugate acids of (a) CH_3NH_2, methylamine; (b) hydrazine, NH_2NH_2; (c) HCO_3^-; and the conjugate bases of (d) HCO_3^-; (e) C_6H_5OH, phenol; (f) CH_3COOH.

10.2 Write the formulas for the conjugate acids of (a) H_2O; (b) OH^-; (c) $C_6H_5NH_2$, aniline; and the conjugate bases of (d) H_2S; (e) HPO_4^{2-}, hydrogen phosphate ion; (f) $HClO_4$, perchloric acid.

10.3 Write the proton transfer equilibria for the following acids in aqueous solution and identify the conjugate acid-base pairs in each one: (a) H_2SO_4; (b) $C_6H_5NH_3^+$, anilinium ion; (c) $H_2PO_4^-$, dihydrogen phosphate ion; (d) HCOOH, formic acid; (e) $NH_2NH_3^+$, hydrazinium ion.

10.4 Write the proton transfer equilibria for the following bases in aqueous solution and identify the conjugate acid-base pairs in each one: (a) CN^-; (b) NH_2NH_2, hydrazine; (c) CO_3^{2-}; (d) HPO_4^{2-}; (e) $CO(NH_2)_2$, urea.

10.5 Identify (a) the Brønsted acid and base in the following reaction, and (b) the conjugate base and acid formed: $HNO_3(aq) + HPO_4^{2-}(aq) \rightleftharpoons NO_3^-(aq) + H_2PO_4^-(aq)$.

10.6 Identify (a) the Brønsted acid and base in the following reaction, and (b) the conjugate base and acid formed: $HSO_3^-(aq) + NH_4^+(aq) \rightleftharpoons NH_3(aq) + H_2SO_3(aq)$.

10.7 Write the two proton transfer equilibria that demonstrate the amphiprotic character of (a) HCO_3^-; (b) HPO_4^{2-}, and identify the conjugate acid-base pairs in each equilibrium.

10.8 Write the two proton transfer equilibria that show the amphiprotic character of (a) $H_2PO_4^-$; (b) $HC_2O_4^-$, hydrogen oxalate ion; and identify the conjugate acid-base pairs in each equilibrium.

10.9 State whether the following oxides are acidic, basic, or amphoteric: (a) BaO; (b) SO_3; (c) As_2O_3; (d) Bi_2O_3.

10.10 State whether the following oxides are acidic, basic, or amphoteric: (a) SO_2; (b) CaO; (c) P_4O_{10}; (d) TeO_2.

10.11 Calculate the molarity of OH^- in solutions with the following H_3O^+ concentrations: (a) 0.20 $mol{\cdot}L^{-1}$ H_3O^+; (b) 1.0×10^{-4} $mol{\cdot}L^{-1}$ H_3O^+; (c) 3.1×10^{-2} $mol{\cdot}L^{-1}$ H_3O^+.

10.12 Estimate the molarity of H_3O^+ in solutions with the following OH^- concentrations: (a) 0.12 $mol{\cdot}L^{-1}$ OH^-; (b) 8.5×10^{-5} $mol{\cdot}L^{-1}$ OH^-; (c) 5.6×10^{-3} $mol{\cdot}L^{-1}$ OH^-.

10.13 The value of K_w for water at body temperature (37°C) is 2.1×10^{-14}. (a) What is the molarity of H_3O^+ ions and the pH of neutral water at 37°C? (b) What is the molarity of OH^- in neutral water at 37°C?

10.14 The concentration of H_3O^+ ions at the freezing point of water is 3.9×10^{-8} $mol{\cdot}L^{-1}$. (a) Calculate K_w and pK_w at 0°C. (b) What is the pH of neutral water at 0°C?

10.15 Calculate the initial molarity of $Ba(OH)_2$ and the molarities of Ba^{2+}, OH^-, and H_3O^+ in an aqueous solution that contains 0.50 g of $Ba(OH)_2$ in 0.100 L of solution.

10.16 Calculate the initial molarity of KNH_2 and the molarities of K^+, NH_2^-, OH^-, and H_3O^+ in an aqueous solution that contains 1.0 g of KNH_2 in 0.250 L of solution.

10.17 The pH of several solutions was measured in the research laboratories of a food company; convert each of the following pH values to the molarity of H_3O^+ ions: (a) 3.3 (the pH of sour orange juice); (b) 6.7 (the pH of a saliva sample); (c) 4.4 (the pH of beer); (d) 5.3 (the pH of a coffee sample).

10.18 The pH of several solutions was measured in a hospital laboratory; convert each of the following pH values to the molarity of H_3O^+ ions: (a) 5.0 (the pH of a urine sample); (b) 2.3 (the pH of a sample of lemon juice); (c) 7.4 (the pH of blood); (d) 10.5 (the pH of milk of magnesia).

10.19 Calculate the pH and pOH of each of the following aqueous solutions of strong acid or base: (a) 1.46 M HNO_3; (b) 0.11 M HCl; (c) 0.0092 M $Ba(OH)_2$; (d) 5.50 mL of 0.175 M KOH after dilution to 5.00×10^2 mL; (e) 27.9 mg of NaOH dissolved in 0.350 L of solution; (f) 75.0 mL of 3.5×10^{-4} M HBr after dilution to 0.250 L.

10.20 Calculate the pH and pOH of each of the following aqueous solutions of strong acid or base: (a) 0.0356 M HI; (b) 0.725 M HCl; (c) 3.46×10^{-3} M $Ba(OH)_2$; (d) 10.9 mg of KOH dissolved in 10.0 mL of solution; (e) 10.0 mL of 5.00 M NaOH after dilution to 0.250 L; (f) 5.0 mL of 3.5×10^{-4} M $HClO_4$ after dilution to 25.0 mL.

Weak Acids and Bases

10.21 Give the K_a values of the following acids: (a) phosphoric acid, H_3PO_4, $pK_{a1} = 2.12$; (b) phosphorous acid H_3PO_3, $pK_{a1} = 2.00$; (c) selenous acid, H_2SeO_3, $pK_{a1} = 2.46$; (d) hydrogen selenate ion, $HSeO_4^-$, $pK_{a2} = 1.92$. (e) List the acids in order of increasing strength.

10.22 Give the pK_b values of the following bases: (a) ammonia, NH_3, $K_b = 1.8 \times 10^{-5}$; (b) deuterated ammonia, ND_3, $K_b = 1.1 \times 10^{-5}$; (c) hydrazine, NH_2NH_2, $K_b = 1.7 \times 10^{-6}$; (d) hydroxylamine, NH_2OH, $K_b = 1.1 \times 10^{-8}$. (e) List the bases in order of increasing strength.

10.23 Write the names and formulas for (a) the strongest and (b) the weakest conjugate bases of the acids listed in Exercise 10.21. (c) What are the K_b values for these two bases? (d) Both bases are dissolved in water at the same concentration. Which solution will have the higher pH?

10.24 (a) Write the names and formulas for the strongest and weakest conjugate acids of the bases listed in Exercise 10.22. (b) What are the K_a values for the two acids? (c) Both acids are dissolved in water at the same concentration. Which solution will have the higher pH?

10.25 For each of the following weak acids, write the proton transfer equilibrium equation and give the mathematical expression for the equilibrium constant K_a. Identify the conjugate base and write the appropriate proton transfer equation. Write the mathematical expression corresponding to the basicity constant K_b. (a) $HClO_2$; (b) HCN; (c) C_6H_5OH.

10.26 For each of the following weak bases, write the proton transfer equilibrium equation and give the mathematical expression for the equilibrium constant, K_b. Identify the conjugate acid and write the appropriate proton transfer equation. Write the mathematical expression corresponding to the acidity constant, K_a. (a) $(CH_3)_2NH$, dimethylamine; (b) $C_{14}H_{10}N_2$, nicotine; (c) $C_6H_5NH_2$, aniline.

10.27 Using data available in Tables 10.1 and 10.2, place the following acids in order of increasing strength: HNO_2, $HClO_2$, $^+NH_3OH$, $(CH_3)_2NH_2^+$.

10.28 Using data available in Tables 10.1 and 10.2, place the following acids in order of increasing strength: HCOOH, $(CH_3)_3NH^+$, $N_2H_5^+$, HF.

10.29 Using data available in Tables 10.1 and 10.2, place the following bases in order of increasing strength: F^-, NH_3, $CH_3CO_2^-$, C_5H_5N (pyridine).

10.30 Using data available in Tables 10.1 and 10.2, place the following bases in order of increasing strength: CN^-, $(C_2H_5)_3N$, N_2H_4 (hydrazine), BrO^-.

10.31 Label each of the following species as a strong or weak acid. Refer to Tables 10.1, 10.2, and 10.3. (a) $HClO_3$; (b) H_2S; (c) HSO_4^-; (d) $CH_3NH_3^+$; (e) HCO_3^-; (f) HNO_3; (g) CH_4.

10.32 Label each of the following species as a strong or weak base. Refer to Tables 10.1, 10.2, and 10.3. (a) O^{2-}; (b) Br^-; (c) HSO_4^-; (d) HCO_3^-; (e) CH_3NH_2; (f) H^-; (g) CH_3^-.

10.33 The pK_a of HIO, hypoiodous acid, is 10.64 and that of HIO_3, iodic acid, is 0.77. Account for the difference in acid strength.

10.34 The pK_a of HClO, hypochlorous acid, is 7.53 and that of HBrO, hypobromous acid, is 8.69. Account for the difference in acid strength.

10.35 Determine which acid is stronger and explain why: (a) HF or HCl; (b) HClO or $HClO_2$; (c) $HBrO_2$ or $HClO_2$; (d) $HClO_4$ or H_3PO_4; (e) HNO_3 or HNO_2; (f) H_2CO_3 or H_2GeO_3.

10.36 Determine which acid is stronger and explain why: (a) H_3AsO_4 or H_3PO_4; (b) $HBrO_3$ or HBrO; (c) H_3PO_4 or H_3PO_3; (d) H_2Te or H_2Se; (e) H_2S or HCl; (f) HClO or HIO.

10.37 Suggest an explanation for the different strengths of (a) acetic acid and trichloroacetic acid; (b) acetic acid and formic acid.

10.38 Suggest an explanation for the different strengths of (a) ammonia and methylamine; (b) hydrazine and hydroxylamine.

10.39 The values of K_a for phenol and 2,4,6-trichlorophenol (see following structures) are 1.3×10^{-10} and 1.0×10^{-6}, respectively. Which is the stronger acid? Account for the difference in acid strength.

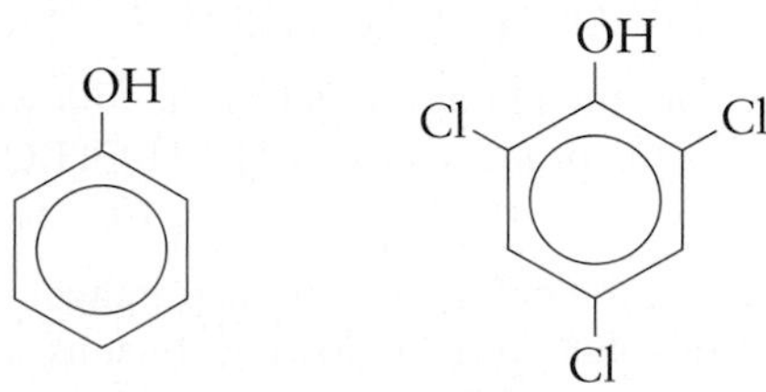

10.40 The value of pK_b for aniline is 9.37 and that for 4-chloroaniline is 9.85 (see the following structures).

Which is the stronger base? Account for the difference in base strength.

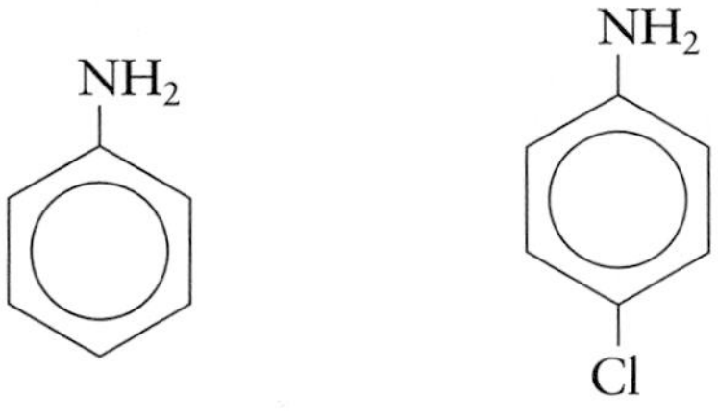

10.41 Arrange the following bases in order of increasing strength on the basis of the pK_a values of their conjugate acids, which are given in parentheses: (a) ammonia (9.26); (b) methylamine (10.56); (c) ethylamine (10.81); (d) aniline (4.63) (see Exercise 10.40). Is there a simple pattern of strengths?

10.42 Arrange the following bases in order of increasing strength on the basis of the pK_a values of their conjugate acids, which are given in parentheses: (a) aniline (4.63) (see Exercise 10.40); (b) 2-hydroxyaniline (4.72); (c) 3-hydroxyaniline (4.17); (d) 4-hydroxyaniline (5.47). Is there a simple pattern of strengths?

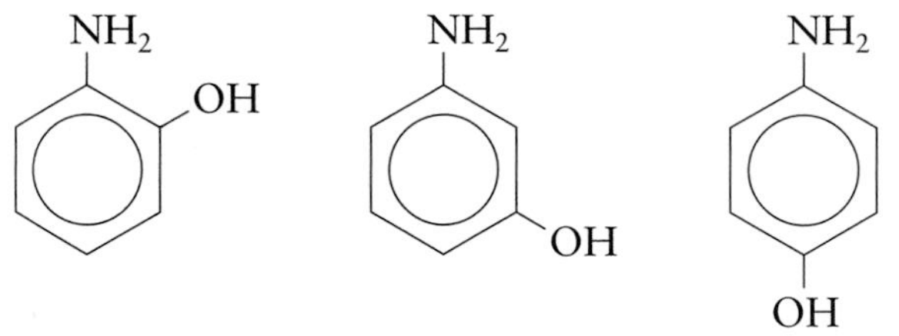

The pH of Solutions of Weak Acids and Bases

Refer to Tables 10.1 and 10.2 for the appropriate K_a and K_b values for the following exercises.

10.43 Calculate the pH and pOH of the following aqueous solutions: (a) 0.29 M $CH_3COOH(aq)$; (b) 0.29 M $CCl_3COOH(aq)$; (c) 0.29 M $HCOOH(aq)$.

10.44 Calculate the pH and pOH of the following aqueous solutions of lactic acid, $CH_3CH(OH)COOH$: (a) 0.12 M; (b) 1.2×10^{-3} M; (c) 1.2×10^{-5} M.

10.45 Calculate the pH, pOH, and percentage protonation of solute in the following aqueous solutions: (a) 0.057 M $NH_3(aq)$; (b) 0.162 M $NH_2OH(aq)$; (c) 0.35 M $(CH_3)_3N(aq)$; (d) 0.0073 M codeine, given that the pK_a of its conjugate acid is 8.21. Codeine, a cough suppressant, is extracted from opium.

10.46 Calculate the pOH, pH, and percentage protonation of solute in the following aqueous solutions: (a) 0.075 M $C_5H_5N(aq)$, pyridine; (b) 0.103 M $C_{10}H_{14}N_2(aq)$, nicotine; (c) 0.045 M quinine, given that the pK_a of its conjugate acid is 8.52; (d) 0.059 M strychnine, given that the K_a of its conjugate acid is 5.49×10^{-9}.

10.47 (a) When the pH of 0.10 M $HClO_2(aq)$ was measured, it was found to be 1.2. What are the values of K_a and pK_a of chlorous acid? (b) The pH of a 0.10 M propylamine, $C_3H_7NH_2$, aqueous solution was measured as 11.86. What are the values of K_b and pK_b of propylamine?

10.48 (a) The pH of a 0.015 M $HNO_2(aq)$ was measured as 2.63. What are the values of K_a and pK_a of nitrous acid? (b) The pH of 0.10 M $C_4H_9NH_2(aq)$, butylamine, was measured as 12.04. What are the values of K_b and pK_b of butylamine?

10.49 Find the initial concentration of the weak acid or base in each of the following aqueous solutions: (a) a solution of HClO with pH = 4.60; (b) a solution of hydrazine, NH_2NH_2, with pH = 10.20.

10.50 Find the initial concentration of the weak acid or base in each of the following aqueous solutions: (a) a solution of HCN with pH = 5.3; (b) a solution of pyridine, C_5H_5N, with pH = 8.8.

10.51 The percentage deprotonation of benzoic acid in a 0.110 M solution is 2.4%. What is the pH of the solution and the K_a of benzoic acid?

10.52 The percentage deprotonation of veronal (diethylbarbituric acid) in a 0.020 M aqueous solution is 0.14%. What is the pH of the solution and the K_a of veronal?

10.53 The percentage protonation of octylamine (an organic base) in a 0.100 M aqueous solution is 6.7%. What is the pH of the solution and the K_b of octylamine?

10.54 Cacodylic acid, $(CH_3)_2As(O)OH$, is used as a cotton defoliant. A 0.0110 M cacodylic acid solution is 0.77% deprotonated in water. What is the pH of the solution and the K_a of cacodylic acid?

10.55 Determine whether aqueous solutions of the following salts have a pH equal to, greater than, or less than 7. If pH > 7 or pH < 7, write a chemical equation to justify your answer. (a) NH_4Br; (b) Na_2CO_3; (c) KF; (d) KBr; (e) $AlCl_3$; (f) $Cu(NO_3)_2$.

10.56 Determine whether aqueous solutions of the following salts have a pH equal to, greater than, or less than 7. If pH > 7 or pH < 7, write a chemical equation to justify your answer. (a) $K_2C_2O_4$, potassium oxalate; (b) $Ca(NO_3)_2$; (c) CH_3NH_3Cl, methylamine hydrochloride; (d) K_3PO_4; (e) $FeCl_3$; (f) C_5H_5NHCl, pyridinium chloride.

10.57 Calculate the pH of the following solutions: (a) 0.63 M $NaCH_3CO_2$(aq); (b) 0.19 M NH_4Cl(aq); (c) 0.055 M $AlCl_3$(aq); (d) 0.65 M KCN(aq).

10.58 Calculate the pH of the following solutions: (a) 0.25 M CH_3NH_3Cl(aq); (b) 0.13 M Na_2SO_3(aq); (c) 0.071 M $FeCl_3$(aq).

10.59 (a) A 5.34-g sample of potassium acetate, KCH_3CO_2, is dissolved in 0.250 L of solution. What is the pH of the solution? (b) What is the pH of a solution resulting from the dissolution of 5.75 g of NH_4Br in 0.100 L of solution?

10.60 (a) A 0.755-g sample of sodium hydrogen sulfite, $NaHSO_3$, is dissolved in 50.0 mL of solution. What is the pH of the solution? (b) What is the pH of a solution resulting from the dissolution of 100 mg of $AgNO_3$ in 10.0 mL of solution?

10.61 (a) A 150-mL sample of 0.175 M $NaCH_3CO_2$(aq) is diluted to 0.500 L. What is the concentration of acetic acid at equilibrium? (b) What is the pH of a solution resulting from the dissolution of 4.32 g of NH_4Br in 0.400 L of solution?

10.62 (a) A 25.0-mL sample of 0.250 M KCN(aq) is diluted to 50.0 mL. What is the concentration of hydrocyanic acid present at equilibrium? (b) A 1.59-g sample of $NaHCO_3$ is dissolved in 200.0 mL of solution. What is the pH of the solution?

Autoprotolysis and pH

10.63 Calculate the pH of 6.55×10^{-7} M $HClO_4$(aq).

10.64 Calculate the pH of 7.49×10^{-8} M HI(aq).

10.65 Calculate the pH of 9.78×10^{-8} M KOH(aq).

10.66 Calculate the pH of 8.23×10^{-7} M $NaNH_2$(aq).

For Exercises 10.67 through 10.70, you may use a graphing calculator to solve the equation or suitable computer software to solve the set of simultaneous equations.

10.67 (a) Calculate the pH of 1.00×10^{-4} M and 1.00×10^{-6} M HBrO(aq) solution, neglecting the effect of the autoprotolysis of water. (b) Repeat the calculations, taking into account the autoprotolysis of water.

10.68 (a) Calculate the pH of 2.50×10^{-4} M and 2.50×10^{-6} M phenol(aq), neglecting the effect of the autoprotolysis of water. (b) Repeat the calculations, taking into account the autoprotolysis of water.

10.69 (a) Calculate the pH of 8.50×10^{-5} M and 7.37×10^{-6} M HCN(aq), neglecting the effect of the autoprotolysis of water. (b) Repeat the calculations, taking into account the autoprotolysis of water.

10.70 (a) Calculate the pH of 1.89×10^{-5} M and 9.64×10^{-7} M HClO(aq), neglecting the effect of the autoprotolysis of water. (b) Repeat the calculations, taking into account the autoprotolysis of water.

Polyprotic Acids and Bases

Refer to Table 10.9 for the K_a values needed for the following exercises.

10.71 Write the stepwise proton transfer equilibria for the deprotonation of (a) sulfuric acid, H_2SO_4; (b) arsenic acid, H_3AsO_4; (c) phthalic acid, $C_6H_4(COOH)_2$.

10.72 Write the stepwise proton transfer equilibria for the deprotonation of (a) phosphoric acid, H_3PO_4; (b) adipic acid; $(CH_2)_4(COOH)_2$; (c) succinic acid, $(CH_2)_2(COOH)_2$.

10.73 Calculate the pH of 0.15 M H_2SO_4(aq) at 25°C.

10.74 Calculate the pH of 0.010 M H_2SeO_4(aq), given that K_{a1} is very large and $K_{a2} = 1.2 \times 10^{-2}$.

10.75 Calculate the pH of the following diprotic acid solutions at 25°C, ignoring second deprotonations only when the approximation is justified: (a) 0.010 M H_2CO_3(aq); (b) 0.10 M $(COOH)_2$(aq); (c) 0.20 M H_2S(aq).

10.76 Calculate the pH of the following diprotic acid solutions at 25°C, ignoring second deprotonations only when that approximation is justified: (a) 0.10 M H_2S(aq); (b) 0.15 M $H_2C_4H_4O_6$(aq), tartaric acid; (c) 1.1×10^{-3} M H_2TeO_4(aq), telluric acid, for which $K_{a1} = 2.1 \times 10^{-8}$ and $K_{a2} = 6.5 \times 10^{-12}$.

10.77 Estimate the pH of (a) 0.15 M $NaHSO_3$(aq); (b) 0.050 M $NaHSO_3$(aq).

10.78 Estimate the pH of (a) 0.125 M $NaHCO_3$(aq); (b) 0.250 M $KHCO_3$(aq).

10.79 Citric acid, which is extracted from citrus fruits and pineapples, undergoes three successive deprotonations with pK_a values of 3.14, 5.95, and 6.39. Estimate the pH of (a) a 0.15 M aqueous solution of the monosodium salt; (b) a 0.075 M aqueous solution of the disodium salt.

10.80 Selenous acid, H_2SeO_3, undergoes two

deprotonations with pK_a values of 2.46 and 7.31. Estimate the pH of 0.100 M $NaHSeO_3(aq)$.

10.81 Calculate the concentrations of H_2CO_3, HCO_3^-, CO_3^{2-}, H_3O^+, and OH^- present in 0.0456 M $H_2CO_3(aq)$.

10.82 Calculate the concentrations of H_2SO_3, HSO_3^-, SO_3^{2-}, H_3O^+, and OH^- present in 0.125 M $H_2SO_3(aq)$.

10.83 Calculate the concentrations of H_2CO_3, HCO_3^-, CO_3^{2-}, H_3O^+, and OH^- present in 0.0456 M $Na_2CO_3(aq)$.

10.84 Calculate the concentrations of H_2SO_3, HSO_3^-, SO_3^{2-}, H_3O^+, and OH^- present in 0.125 M $Na_2SO_3(aq)$.

10.85 For each polyprotic acid, state which species (H_2A, HA^-, or A^{2-}) you expect to be the form present in highest concentration in aqueous solution at pH = 6.50: (a) phosphorous acid; (b) oxalic acid; (c) hydrosulfuric acid.

10.86 For each polyprotic acid, state which species (H_2A, HA^-, or A^{2-} or H_3A, H_2A^-, HA^{2-}, or A^{3-} in the case of triprotic acids) you expect to be the form present in highest concentration in aqueous solution at pH = 5.0: (a) tartaric acid; (b) hydrosulfuric acid; (c) phosphoric acid.

10.87 A large volume of 0.150 M $H_2SO_3(aq)$ is treated with a solid, strong base to adjust the pH to 5.50. Assume that the addition of the base does not significantly affect the volume of the solution. Estimate the concentrations of H_2SO_3, HSO_3^-, and SO_3^{2-} present in the final solution.

10.88 A large volume of 0.250 M $H_2S(aq)$ is treated with a solid, strong base to adjust the pH to 9.35. Assume that the addition of the base does not significantly affect the volume of the solution. Estimate the concentrations of H_2S, HS^-, and S^{2-} present in the final solution.

10.89 Calculate the pH of the following acid solutions at 25°C; ignore second deprotonations only when that approximation is justified. (a) 1.0×10^{-4} M $H_3BO_3(aq)$ (boric acid acts as a monoprotic acid); (b) 0.015 M $H_3PO_4(aq)$; (c) 0.10 M $H_2SO_3(aq)$.

10.90 Calculate the molarities of $(COOH)_2$, $HOOCCO_2^{-2}$, $(CO_2)_2^{2-}$, H_3O^+, and OH^- in 0.10 M $(COOH)_2(aq)$.

INTEGRATED EXERCISES

Media Link

10.91 (a) The value of K_w at 40°C is 3.8×10^{-14}. What is the pH of pure water at 40°C? (b) Using data from (a) and Exercise 10.13 and the values for the autoprotolysis of water at 25°C to determine graphically the enthalpy and entropy of the autoprotolysis of water. (c) Write an equation that describes the pH of pure water as a function of temperature.

10.92 Ammonium acetate is prepared by the reaction of equal amounts of ammonium hydroxide and acetic acid. Determine the concentration of all solute species present in 0.100 M $NH_4CH_3CO_2(aq)$.

10.93 The pH of a saturated solution of benzoic acid is 2.8 at 25°C and the K_a of benzoic acid is 6.5×10^{-5}. Use this information to calculate the molar concentration of a saturated solution of benzoic acid.

10.94 Decide on the basis of the information in Table 10.1 whether carbonic acid is a strong or weak acid in liquid ammonia solvent. Explain your answer.

10.95 Using the thermodynamic data available in Appendix 2A, calculate the acidity constant of HF(aq).

10.96 Calculate the pH of (a) 0.095 M $NaH_2AsO_4(aq)$ and (b) 0.148 M $Na_2HAsO_4(aq)$. For H_3AsO_4, $pK_{a1} = 2.25$, $pK_{a2} = 6.77$, and $pK_{a3} = 11.60$.

10.97 The autoprotolysis constant, K_{D_2O}, for heavy water at 25°C is 1.35×10^{-15}. (a) Write the chemical equation for the autoprotolysis of D_2O. (b) Evaluate pK_{D_2O} for D_2O at 25°C. (c) Calculate the molarities of D_3O^+ and OD^- in pure heavy water at 25°C. (d) Evaluate the pD and pOD of heavy water at 25°C. (e) Find the relation between pD, pOD, and pK_{D_2O}.

10.98 Although many chemical reactions take place in water, it is often necessary to use other solvents; liquid ammonia (b.p. −33°C) has been used extensively. Many of the reactions that occur in water have analogous reactions in liquid ammonia. (a) Write the chemical equation for the autoprotolysis of NH_3. (b) What are the formulas of the acid and base species that result from the autoprotolysis of liquid ammonia? (c) The autoprotolysis constant, K_{am}, of liquid ammonia has the value 1×10^{-33} at −35°C. What is the value of pK_{am} at that temperature? (d) What is the molarity of NH_4^+ ions in liquid ammonia? (e) Evaluate pNH_4 and pNH_2, which are the analogues of pH and pOH, in liquid ammonia at −35°C. (f) Determine the relation between pNH_4, pNH_2, and pK_{am}.

10.99 The percentage protonation of thiazole (an organic base) in a 0.0010 M solution is $5.2 \times 10^{-3}\%$. What is the pH of the solution and the K_b of thiazole?

10.100 Calculate the pH of a 0.020 M succinic acid solution at 25°C, taking into account both deprotonations ($K_{a1} = 6.9 \times 10^{-5}$, $K_{a2} = 2.5 \times 10^{-6}$).

10.101 Estimate the enthalpy of deprotonation of formic acid at 25°C, given that $K_a = 1.765 \times 10^{-4}$ at 20°C and 1.768×10^{-4} at 30°C.

10.102 Convert the van 't Hoff equation (see Section 9.11) for the temperature dependence of an equilibrium constant to an expression for the temperature dependence of pK_w. Estimate the value of pK_w at the normal boiling point of water from the enthalpy of the water autoprotolysis reaction, which is $+57\ kJ \cdot mol^{-1}$ for the reaction $2\ H_2O(l) \rightleftharpoons H_3O^+(aq) + OH^-(aq)$. What is the pH of water at that temperature?

10.103 Determine graphically the enthalpy of the deprotonation of heavy water, D_2O, given the following data on its autoprotolysis:

Temperature, °C	pK_{D_2O}
10	15.44
20	15.05
25	14.87
30	14.70
40	14.39
50	14.10

What is the pD (pD = $-\log[D_3O^+]$) of heavy water at 35°C? What would be the values of pOD, pK_{D_2O}, and K_{D_2O} at 35°C?

10.104 Twenty amino acids are commonly found in proteins. (a) Draw the chemical structures of alanine, glycine, phenylalanine, and cysteine (see Table 19.4). (b) From literature sources, find the pK_a values associated with the carboxylic acid function and the pK_b of the amine function of alanine. (c) Determine the equilibrium constant for the reaction of the carboxylic acid function of alanine with its amine function to give $R{-}CO_2^-$ and $R{-}NH_3^+$. (d) Look up the definition of the term *zwitterion*. Given the information determined in (c), how do you expect the concept of the zwitterion to apply to the amino acids?

10.105 Ammonium dihydrogen phosphate is prepared by the reaction of equal amounts of ammonium hydroxide and phosphoric acid. Determine the concentration of all solute species present in 0.150 M $NH_4H_2PO_4(aq)$.

10.106 The amino acid L-histidine has three acid-base equilibria, with pK_a values of 1.77, 6.10, and 10.53. (a) Using the Living Graph *Fractional Composition of Triprotic Acids,* plot the fraction of solute species present in solution over the pH range 0 to 14. Using the graph, determine (b) the major species present in solution at pH = 7.5; (c) the pH value at which equal amounts of the singly and doubly deprotonated forms are present; (d) the pH necessary to achieve greater than 99% fully deprotonated histidine; and (e) the pH range for which three different deprotonated forms are present in significant quantities.

N
HN
CH₂
CH—NH₂
HO—C
O

Histidine

10.107 (a) Using the Living Graph *Fractional Composition of Triprotic Acids,* plot the fraction of solute species for H_3PO_4. (b) On the graph, label the places where the system functions as a buffer solution. (c) For each region in (b), give the major phosphorus-containing species present in solution. (d) Repeat (a) for H_3AsO_4, using data from Exercise 10.96. (e) For pH = 7.00, determine the major species present in solution and compare the relative fractions of these species for H_3PO_4 and H_3AsO_4.

CHAPTER 11

These Iceland poppies contain organic dye molecules called anthocyanins, a fragment of one of which is shown in the insert. Because many such dyes respond to the presence of acids and bases by changing color, they can act as indicators for detecting acidic and basic solutions.

Why Do We Need to Know This Material?

The techniques described in this chapter provide the tools we need to analyze and control the concentrations of ions in solution. The ionic compounds released into waterways by individuals, industry, and agriculture can impair the quality of our water supplies, and hazardous ions must be identified and removed. Aqueous equilibria govern the stabilization of the pH in blood, seawater, and other solutions encountered in biology, medicine, and the environment.

What Do We Need to Know Already?

This chapter develops the ideas in Chapters 9 and 10 and applies them to equilibria involving ions in aqueous solution. To prepare for the sections on titrations, review Section L. For the discussion of solubility equilibria, review Sections I and 8.15. The discussion of Lewis acids and bases is based on Section 2.13.

AQUEOUS EQUILIBRIA

In Chapter 10, we learned about the properties of acids and bases. In this chapter, we apply the same ideas to solutions containing mixtures of acids and bases. We then see how to use solutions of weak acids or bases and their salts to maintain a desired pH. This kind of information is crucially important for understanding the ability of organisms to survive, because even minor drifts from the optimum value of the pH can result in disease. The information in this chapter is used in industry to control the pH of reaction mixtures and in agriculture to maintain the soil at an optimal pH. In this chapter, we also see how to identify the stoichiometric point of a titration, one of the most common techniques in chemical laboratories. In particular, we need to know the pH at the stoichiometric point of a titration involving a weak acid or a weak base and to predict the pH at any point during a titration. Finally, we need to understand solubility equilibria in order to appreciate another widely used chemical technique, that of **qualitative analysis,** the identification of the substances and ions present in a sample.

As we shall see, all these applications are examples of the equilibrium that occurs between species in solution, especially between species that can exchange protons, and all of them can be treated in a similar, uniform way. First, we identify the species present in solution. Then we identify the equilibrium relations between the species (often by setting up an equilibrium table) and use those relations to calculate the concentrations of the species present in the solution.

MIXED SOLUTIONS AND BUFFERS

We have seen how to estimate the pH of a solution of a weak acid or base (Chapter 10). Now we suppose that a salt of the acid or base is also present and investigate how that salt affects the pH. We shall see that "mixed solutions"—solutions that contain a weak acid or a weak base and one of its salts—provide a means of stabilizing the pH of aqueous solutions such as blood plasma, seawater, detergents, sap, and reaction mixtures.

11.1 Mixed Solutions

Suppose we want to predict the pH of a solution that contains appreciable concentrations of both an acid and its conjugate base. For instance, the solution might be hydrochloric acid to which sodium chloride has been added (the Cl^- ion supplied as NaCl is the conjugate base of HCl). In this case, the acid is strong, so its conjugate base is extremely weak and has no measurable effect on pH. The pH of a 0.10 M HCl(aq) solution will be 1.0, even after 0.10 mol NaCl is added to a liter of the solution. Now suppose the solution contains acetic acid to which sodium acetate has been added (the acetate ion, $CH_3CO_2^-$, is the conjugate base of CH_3COOH). The conjugate base of a weak acid is a weak base that does affect the pH. Therefore, we can predict *qualitatively* that adding acetate ions (as sodium acetate) to a solution of acetic acid will increase the pH of the solution.

We are disregarding the effect of added ions on the activity coefficient of the hydronium ions present in solution.

Similarly, suppose we have a solution of ammonia and add ammonium chloride to it. The ammonium ion is an acid; therefore, we can predict *qualitatively* that adding ammonium ions (as ammonium chloride, for instance) to a solution of ammonia will lower the pH of the solution.

To predict the pH of mixtures of weak acids or bases and their salts *quantitatively*, we set up an equilibrium table, as described in Toolboxes 10.1, 10.2, and 10.3, and use the acidity or basicity constant to calculate the concentration of H_3O^+ ions present in the solution. As before, we assume that the initial concentration of hydronium ions is 0. However, now *the conjugate acid and base are both present initially*, and we must take into account those initial concentrations, as illustrated in Example 11.1.

The pH of a solution of a weak acid increases when a salt containing its conjugate base is added. The pH of a solution of a weak base decreases when a salt containing its conjugate acid is added.

EXAMPLE 11.1 Calculating the pH of a solution of a weak acid and its salt

Calculate the pH of a solution that is 0.500 M $HNO_2(aq)$ and 0.100 M $KNO_2(aq)$. From Table 10.1, $K_a = 4.3 \times 10^{-4}$ for HNO_2.

STRATEGY The relatively high concentration of NO_2^-, a base, suggests that the pH will be higher than that of nitrous acid alone. The K^+ ion has no measurable effect on the pH of the solution (recall Section 10.14). Use the proton transfer equilibrium of HNO_2 to find the pH by means of an equilibrium table. Consider the initial molarity of HNO_2 to be 0.500 $mol \cdot L^{-1}$. Set the initial molarity of NO_2^- ions equal to the molarity of added salt (each KNO_2 formula unit supplies one NO_2^- anion). Then proceed as described in Toolbox 10.1. Because the concentrations of the added ions are much higher than 10^{-7} $mol \cdot L^{-1}$, we assume that we can ignore the contribution to the pH from the autoprotolysis of water. For a solution of a weak base and its salt, proceed as above and as described in Toolbox 10.2.

SOLUTION The proton transfer equilibrium to consider is

$$HNO_2(aq) + H_2O(l) \rightleftharpoons H_3O^+(aq) + NO_2^-(aq)$$

$$K_a = \frac{[H_3O^+][NO_2^-]}{[HNO_2]}$$

The equilibrium table, with all concentrations in moles per liter, is

	Species		
	HNO_2	H_3O^+	NO_2^-
Step 1 Initial molarity	0.500	0	0.100
Step 2 Change in molarity	$-x$	$+x$	$+x$
Step 3 Equilibrium molarity	$0.500 - x$	x	$0.100 + x$

Step 4 Substitution of the information from step 3 and $K_a = 4.3 \times 10^{-4}$ into the expression for K_a gives

$$4.3 \times 10^{-4} = \frac{x \times (0.100 + x)}{0.500 - x}$$

If we assume that x is less than 5% of 0.100 (and therefore also less than 5% of 0.500), then we can write

$$4.3 \times 10^{-4} \approx \frac{x \times 0.100}{0.500}$$

The solution of this equation is $x \approx 2.2 \times 10^{-3}$, which is only 2.2% of 0.100 and only 0.44% of 0.500. The approximations are valid, and the equilibrium molarity of H_3O^+ ions is 2.2×10^{-3} mol·L^{-1}. This concentration is much larger than that contributed by the autoprotolysis of water, so neglect of the autoprotolysis is also valid. It follows that the pH of the solution is

$$\text{pH} = -\log (2.2 \times 10^{-3}) = 2.66$$

or about 2.7. The pH of 0.500 M HNO_2(aq) alone is 1.8; so, as anticipated, the pH of the mixed solution is higher.

Self-Test 11.1A Calculate the pH of a solution that is 0.010 M HClO(aq) and 2.0×10^{-4} M NaClO(aq).

[*Answer:* 5.82]

Self-Test 11.1B Calculate the pH of a solution that is 0.300 M CH_3NH_2(aq) and 0.146 M CH_3NH_3Cl(aq). From Table 10.2, the K_b of CH_3NH_2 is 3.6×10^{-4}.

11.2 Buffer Action

The most important type of mixed solution is a **buffer,** a solution in which the pH resists change when strong acids or bases are added. Buffers, which consist of mixed weak acid and salt or weak base and salt solutions (just like those discussed in the preceding section), are used to calibrate pH meters, to culture bacteria, and to control the pH of solutions in which chemical reactions are taking place. They are also administered intravenously to critically ill patients. Our blood plasma is buffered to pH = 7.4; the ocean is buffered to about pH = 8.4 by a complex buffering process that depends on the presence of hydrogen carbonates and silicates.

An **acid buffer** is an aqueous solution of a weak acid and its conjugate base supplied as a salt. It buffers solutions on the acid side of neutrality (at pH < 7). An example of an acid buffer is a solution of acetic acid and sodium acetate. A **base buffer** is an aqueous solution of a weak base and its conjugate acid supplied as a salt. It buffers solutions on the basic side of neutrality (at pH > 7). An example of a base buffer is a solution of ammonia and ammonium chloride.

We can understand the action of acid buffers by considering the dynamic equilibrium between a weak acid and its conjugate base in water:

$$CH_3COOH(aq) + H_2O(l) \rightleftharpoons H_3O^+(aq) + CH_3CO_2^-(aq)$$

Suppose a strong acid is added to a solution that contains $CH_3CO_2^-$ ions and CH_3COOH molecules in about equal concentrations. The newly arrived H_3O^+ ions transfer protons to the $CH_3CO_2^-$ ions to form

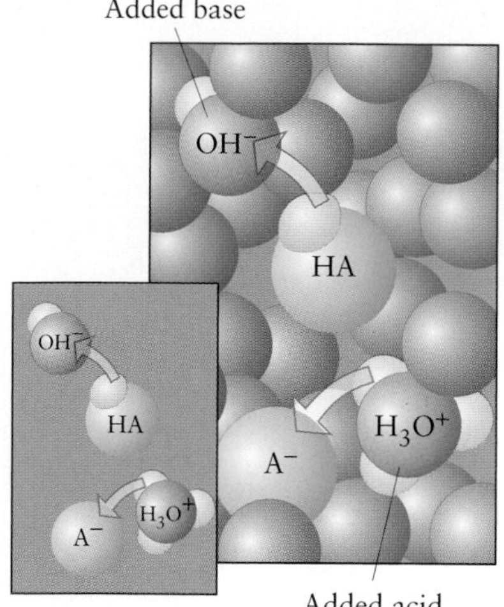

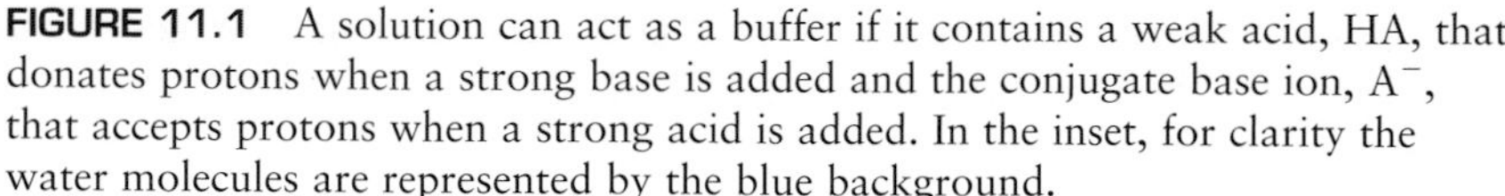
FIGURE 11.1 A solution can act as a buffer if it contains a weak acid, HA, that donates protons when a strong base is added and the conjugate base ion, A^-, that accepts protons when a strong acid is added. In the inset, for clarity the water molecules are represented by the blue background.

CH_3COOH and H_2O molecules (Fig. 11.1). Because the added hydronium ions are removed by the acetate ions, the pH remains almost unchanged. In effect, the acetate ions act as a "sink" for protons. If a small amount of strong base is added instead, the incoming OH^- ions remove protons from the CH_3COOH molecules to produce $CH_3CO_2^-$ ions and H_2O molecules. In this case, the acetic acid molecules act as a source of protons. As a result, the concentration of OH^- ions remains nearly unchanged. Consequently, the H_3O^+ concentration (and the pH) is also left nearly unchanged.

We can understand the action of base buffers along the same lines. Consider the ammonia equilibrium,

$$NH_3(aq) + H_2O(l) \rightleftharpoons NH_4^+(aq) + OH^-(aq)$$

in a solution that contains similar concentrations of $NH_3(aq)$ and $NH_4^+(aq)$ and has been prepared by adding ammonium chloride to aqueous ammonia. The NH_3 molecules act as a sink for protons, and the NH_4^+ ions act as a source of protons. When a strong base is added, the incoming OH^- ions remove protons from NH_4^+ ions to make NH_3 and H_2O molecules. When a strong acid is added, the incoming protons attach to NH_3 molecules to make NH_4^+ ions and hence are removed from the solution. In each case, the pH is left almost unchanged.

A buffer is a mixture of a weak conjugate acid and base that stabilizes the pH of a solution by providing both a source and a sink for protons (Fig. 11.2).

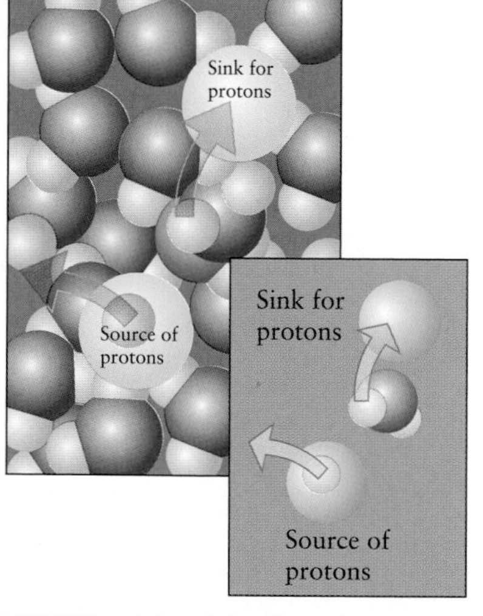

FIGURE 11.2 A buffer solution contains a sink for the protons supplied by a strong acid and a source of protons to supply to a strong base being added. The joint action of the source and sink keeps the pH constant when strong acid or base is added. The inset highlights the buffer action by representing the water molecules as a blue background.

11.3 Designing a Buffer

Suppose we need to make up a buffer with a particular pH. For instance, we might be culturing bacteria and need to maintain a precise pH to sustain their metabolism. To choose the most appropriate buffer system, we need to know the value of the pH at which a given buffer stabilizes a solution. Because buffer solutions are mixed solutions, we can calculate their pH by the same method we used in Example 11.1.

It will prove helpful to rewrite the expression for K_a to show how the pH of a solution relates to the pK_a of the acid present. Let's consider a solution containing a weak acid HA (such as CH_3COOH or NH_4^+) and the conjugate base anion A^- (such as $CH_3CO_2^-$ or NH_3, respectively). The proton transfer equilibrium is

$$HA(aq) + H_2O(l) \rightleftharpoons H_3O^+(aq) + A^-(aq)$$

If we rearrange the expression for K_a into

$$[H_3O^+] = K_a \times \frac{[HA]}{[A^-]}$$

and take the negative logarithm of both sides, we obtain

$$-\log [H_3O^+] = -\log K_a - \log \frac{[HA]}{[A^-]}$$

That is,

$$pH = pK_a - \log \frac{[HA]}{[A^-]} = pK_a + \log \frac{[A^-]}{[HA]}$$

The values of [HA] and $[A^-]$ in this expression are the equilibrium molarities of acid and base in the solution. However, a weak acid HA typically loses only a tiny fraction of its protons, so [HA] is negligibly different from the molarity of the acid used to prepare the buffer. Likewise, only a tiny fraction of the weakly basic anions A^- accept protons, so $[A^-]$ is negligibly different from the molarity of the base used to prepare the buffer. With these approximations, we obtain

$$pH \approx pK_a + \log \frac{[\text{base}]_{\text{initial}}}{[\text{acid}]_{\text{initial}}} \tag{1}$$

This expression is called the **Henderson-Hasselbalch equation.** The same expression is used for a base buffer, with pK_a that of the conjugate acid of the base.

Because so many chemical reactions in our bodies occur in buffered environments, biochemists commonly use the Henderson-Hasselbalch equation for quick estimates of pH. In practice, the equation is used to make rapid estimates of the pH of a mixed solution intended to be used as a buffer, and then the pH is adjusted to the precise value required by adding more acid or base and monitoring the solution with a pH meter.

EXAMPLE 11.2 Calculating the pH of a buffer solution

Suppose we are culturing bacteria that require an acid environment and want to prepare a buffer close to pH = 4. We first prepare a buffer solution that is 0.040 M $NaCH_3CO_2(aq)$ and 0.080 M $CH_3COOH(aq)$ at 25°C, to which we will later add more acid or base to adjust the pH to 4. What is the pH of the original buffer solution?

STRATEGY We could just substitute the data into the Henderson-Hasselbalch equation. However, we learn more, and can be more confident about our work, if we work out the pH by treating the mixed solution as an example of a dynamic equilibrium. So, we begin by identifying the weak acid and its conjugate base. Then we write the proton transfer equilibrium between them, rearrange the expression for K_a to give $[H_3O^+]$, and find the pH, as in the derivation of the Henderson-Hasselbalch equation described in the text.

SOLUTION The acid is CH_3COOH and its conjugate base is $CH_3CO_2^-$. The equilibrium to consider is

$$CH_3COOH(aq) + H_2O(l) \rightleftharpoons H_3O^+(aq) + CH_3CO_2^-(aq)$$

$$K_a = \frac{[H_3O^+][CH_3CO_2^-]}{[CH_3COOH]}$$

From Table 10.1, $pK_a = 4.75$. Rearranging the expression for K_a, we obtain first

$$[H_3O^+] = K_a \times \frac{[CH_3COOH]}{[CH_3CO_2^-]}$$

At this point, either we can calculate the value of $[H_3O^+]$ and then the pH, or we can take the negative logarithm of each side:

$$pH = pK_a + \log \frac{[CH_3CO_2^-]}{[CH_3COOH]}$$

As explained in the text, we now approximate the molarities of the acid and the base by their initial values:

$$pH \approx 4.75 + \log \frac{0.040}{0.080} = 4.45$$

That is, the solution acts as a buffer close to pH = 4.4.

Self-Test 11.2A Calculate the pH of a buffer solution that is 0.15 M $HNO_2(aq)$ and 0.20 M $NaNO_2(aq)$.

[*Answer:* 3.5]

Self-Test 11.2B Calculate the pH of a buffer solution that is 0.040 M $NH_4Cl(aq)$ and 0.030 M $NH_3(aq)$.

To lower the pH of a buffer solution, we add some acid; for a more basic buffer, we can add either more salt (which supplies the conjugate base) or some strong base such as NaOH to neutralize some of the acid and produce more conjugate base. Example 11.3 shows how to calculate the effect of added acid or base on the pH of a buffer.

EXAMPLE 11.3 Calculating the pH change when acid or base is added to a buffer solution

Suppose that we dissolve 1.2 g of sodium hydroxide (0.030 mol NaOH) in 500 mL of the buffer solution described in Example 11.2. Calculate the pH of the resulting solution and the change in pH. Assume the volume of the solution to be constant.

STRATEGY Once we know the molar concentrations of the acid and its conjugate base, we can rearrange the expression for K_a to obtain the pH of the solution. The 0.030 mol OH^- that is added to the buffer solution reacts with the acid of the buffer system, CH_3COOH, thereby decreasing its amount by 0.030 mol and increasing the amount of its conjugate base, $CH_3CO_2^-$, by 0.030 mol. There is no significant change in the volume of solution.

SOLUTION The amount of CH_3COOH in the solution initially (from the data in Example 11.2) is

$$\text{Amount of } CH_3COOH = (0.500\ L) \times (0.080\ mol{\cdot}L^{-1}) = 0.040\ mol$$

The added sodium hydroxide reacts with the acetic acid:

$$CH_3COOH(aq) + OH^-(aq) \longrightarrow CH_3CO_2^-(aq) + H_2O(aq)$$

The amount of CH_3COOH that reacts is

$$0.030 \text{ mol OH}^- \times \frac{1 \text{ mol CH}_3\text{COOH}}{1 \text{ mol OH}^-} = 0.030 \text{ mol CH}_3\text{COOH}$$

Therefore, the amount of CH_3COOH remaining is 0.040 − 0.030 mol = 0.010 mol. The resulting molarity of CH_3COOH is

$$\text{Molarity of CH}_3\text{COOH} = \frac{0.010 \text{ mol}}{0.500 \text{ L}} = 0.020 \text{ mol·L}^{-1}$$

Similarly, the initial amount of $CH_3CO_2^-$ in the 500 mL of solution (using data from Example 11.2) is

$$\text{Amount of CH}_3\text{CO}_2^- = (0.500 \text{ L}) \times (0.040 \text{ mol· L}^{-1}) = 0.020 \text{ mol}$$

The 0.030 mol OH^- that reacts with 0.030 mol CH_3COOH increases the amount of $CH_3CO_2^-$ to (0.020 + 0.030) mol = 0.050 mol. Hence, the molarity of the base $CH_3CO_2^-$ is increased to

$$\text{Molarity of CH}_3\text{CO}_2^- = \frac{0.050 \text{ mol}}{0.500 \text{ L}} = 0.10 \text{ mol·L}^{-1}$$

The proton transfer equilibrium is

$$CH_3COOH(aq) + H_2O(l) \rightleftharpoons H_3O^+(aq) + CH_3CO_2^-(aq)$$

$$K_a = \frac{[H_3O^+][CH_3CO_2^-]}{[CH_3COOH]}$$

As in the derivation of the Henderson-Hasselbalch equation, we can write

$$\text{pH} = \text{p}K_a + \log\frac{[CH_3CO_2^-]}{[CH_3COOH]} \approx 4.75 + \log\frac{0.10}{0.020} = 5.45$$

or about 5.4. That is, the pH of the solution changes from about 4.4 to about 5.4. The small change is the result of the buffering action. If the solution had originally contained only HCl at a pH of 4.4, the addition of the NaOH would have raised the pH to 12.8.

Self-Test 11.3A Suppose that 0.0200 mol NaOH(s) is dissolved in 300 mL of the buffer solution of Example 11.2. Calculate the pH of the resulting solution and the change in pH.

[*Answer:* 5.65, an increase of 1.25]

Self-Test 11.3B Suppose that 0.0100 mol HCl(g) is dissolved in 500 mL of the buffer solution of Example 11.2. Calculate the pH of the resulting solution and the change in pH.

Buffers are often prepared with equal concentrations of the conjugate acid and base. In these solutions, the term in the logarithm in Eq. 1 is equal to 1; and because $\log 1 = 0$, it follows that

$$\text{When [acid]} = \text{[base]}, \text{pH} = \text{p}K_a \quad (2)$$

This very simple result makes it easy to make an initial choice of a buffer: we just select an acid that has a $\text{p}K_a$ close to the pH we require and prepare an equimolar solution with its conjugate base. Two species are **equimolar** if they have the same molar concentrations. The same procedure used for acid buffers can also be used for base buffers, but we have to remember that the acid in this case is supplied by the salt, that the conjugate base is the base itself, and that the $\text{p}K_a$ is that of the conjugate acid (and hence related to the $\text{p}K_b$ of the base by $\text{p}K_a + \text{p}K_b = \text{p}K_w$). Mixtures in which the salt and acid (or base) have unequal molarities—like those considered in the examples—are also buffers, but they may be less effective than those in which the molarities are nearly equal. Specifically,

- To prepare and *acid buffer*, we select a weak acid (HA) with a $\text{p}K_a$ close to the desired pH and add its conjugate base (A^-) in the form of a salt.
- To prepare a *base buffer*, we select a weak base (B) with a $\text{p}K_a$ close to the desired pH and add its conjugate acid (BH^+) in the form of a salt.

An alternative method for preparing an acid buffer is to prepare a solution of the weak acid and neutralize half of it with strong base. A base buffer can be made in a similar way, by using a strong acid to neutralize half of the weak base in a solution. In any case, we establish the desired pH by adjusting the relative molarities of conjugate acid and base. Some typical buffer systems are listed in Table 11.1.

Self-Test 11.4A Which of the buffer systems listed in Table 11.1 would be a good choice to prepare a buffer with pH = 5 ± 1?

[*Answer:* $CH_3COOH/CH_3CO_2^-$]

Self-Test 11.4B Which of the buffer systems listed in Table 11.1 would be a good choice to prepare a buffer with pH = 10 ± 1?

TABLE 11.1 Typical Buffer Systems

Composition	$\text{p}K_a$
Acid buffers	
$CH_3COOH/CH_3CO_2^-$	4.75
HNO_2/NO_2^-	3.37
$HClO_2/ClO_2^-$	2.00
Base buffers	
NH_4^+/NH_3	9.25
$(CH_3)_3NH^+/(CH_3)_3N$	9.81
$H_2PO_4^-/HPO_4^{2-}$	7.21

Commercially available buffer solutions can be purchased for virtually any desired pH. A buffer solution commonly used to calibrate pH meters contains 0.025 M Na_2HPO_4(aq) and 0.025 M KH_2PO_4(aq) and has pH = 6.87 at 25°C. The method demonstrated in Example 11.2 would give pH = 7.2 for this solution. However, because these calculations use molarities, not activities, they ignore ion-ion interactions, so the values calculated are only approximate.

The pH of a buffer solution is close to the $\text{p}K_a$ of the weak acid component when the acid and base have similar concentrations.

EXAMPLE 11.4 Selecting the composition for a buffer solution with a given pH

Calculate the ratio of the molarities of CO_3^{2-} and HCO_3^- ions required to achieve buffering at pH = 9.50. The pK_{a2} of H_2CO_3 is 10.25.

STRATEGY Sometimes we need to make an initial estimate of buffer pH. So here we shall calculate the pH of a buffer from the Henderson-Hasselbalch equation, Eq. 1. First, we identify the acid and base, and then rearrange the equation to give the desired concentration ratio.

SOLUTION The acid is HCO_3^- and its conjugate base is CO_3^{2-}. Hence,

$$\log \frac{[CO_3^{2-}]}{[HCO_3^-]} = \text{pH} - \text{p}K_a = 9.50 - 10.25 = -0.75$$

The antilogarithm of −0.75 is 0.18, so the ratio of molarities is

$$\frac{[CO_3^{2-}]}{[HCO_3^-]} = 0.18$$

Therefore, the solution acts as a buffer with a pH close to 9.50 if it is prepared by mixing the solutes in the ratio 0.18 mol CO_3^{2-} to 1.0 mol HCO_3^-.

Self-Test 11.5A Calculate the ratio of the molarities of acetate ions and acetic acid needed to buffer a solution at pH = 5.25. The pK_a of CH_3COOH is 4.75.

[*Answer:* 3.2:1]

Self-Test 11.5B Calculate the ratio of the molarities of benzoate ions and benzoic acid (C_6H_5COOH) needed to buffer a solution at pH = 3.50. The pK_a of C_6H_5COOH is 4.19.

11.4 Buffer Capacity

Just as a sponge can hold only so much water, so a buffer can also run out of mopping-up power. Its proton sources and sinks become exhausted if too much strong acid or base is added to the solution. **Buffer capacity** is the amount of acid or base that can be added before the buffer loses its ability to resist the change in pH. A buffer with a high capacity can maintain its buffering action longer than can one with only a small capacity. The buffer is exhausted when most of the weak base has been converted to acid or when most of the weak acid has been converted to base. A concentrated buffer has a greater capacity than the same volume of a more dilute buffer.

Buffer capacity also depends on the relative concentrations of weak acid and base. Broadly speaking, it is found experimentally that a buffer has a high capacity when the amount of base present is about 10% or more of the amount of acid. Otherwise, the base is used up quickly as strong acid is added. Similarly, a buffer has a high capacity when the amount of acid present is about 10% or more of the amount of base, for otherwise the acid is used up quickly as strong base is added.

We can use these numbers to express the range of buffer action in terms of the pH of the solution. The Henderson-Hasselbalch equation shows us that when the acid is 10 times more abundant than the base ($[HA] = 10[A^-]$), the pH of the solution is

$$pH = pK_a + \log\frac{[A^-]}{10[A^-]} = pK_a + \log\frac{1}{10} = pK_a - 1 \quad (3a)$$

Likewise, when the base is 10 times more abundant than the acid ($[A^-] = 10[HA]$), the pH is

$$pH = pK_a + \log\frac{10[HA]}{[HA]} = pK_a + \log\frac{10}{1} = pK_a + 1 \quad (3b)$$

We see that the experimentally determined concentration range converts into a pH range of ± 1. That is, the buffer acts effectively within a range of ± 1 units of pK_a (Fig. 11.3). For instance, because the pK_a of $H_2PO_4^-$ is 7.21, a KH_2PO_4/K_2HPO_4 buffer can be expected to be effective between about 6.2 and 8.2.

The composition of blood plasma, in which the concentration of HCO_3^- ions is about 20 times that of H_2CO_3, seems to be outside the

BOX 11.1 WHAT HAS THIS TO DO WITH . . . STAYING ALIVE?*

Physiological Buffers

Buffer systems are so vital to the existence of living organisms that the most immediate threat to the survival of a person with severe injury or burns is a change in blood pH. One of a paramedic's first steps is to administer intravenous fluids.

Metabolic processes normally maintain the pH of human blood within a narrow range (7.35–7.45). To control blood pH, the body uses primarily the carbonic acid/hydrogen carbonate (bicarbonate) ion system. The normal ratio of HCO_3^- to H_2CO_3 in the blood is 20:1, with most of the carbonic acid in the form of dissolved CO_2. When the concentration of HCO_3^- increases further relative to that of H_2CO_3, blood pH rises. If the pH rises above the normal range, the condition is called *alkalosis*. Conversely, blood pH decreases when the ratio decreases; and when blood pH falls below the normal range, the condition is called *acidosis*. These conditions are life-threatening and death can result within minutes.

Blood carbonic acid and hydrogen carbonate concentrations are controlled by independent mechanisms. Carbonic acid concentration is controlled by respiration: as we exhale, we deplete our system of CO_2, and hence deplete it of H_2CO_3, too. This decrease in acid concentration raises the blood pH. Breathing faster and more deeply increases the amount of CO_2 exhaled and hence decreases the carbonic acid concentration in the blood, which in turn raises the blood pH. Hydrogen carbonate ion concentration is controlled by its rate of excretion in urine.

Respiratory acidosis results when decreased respiration raises the concentration of CO_2 in the blood. Asthma, pneumonia, emphysema, or inhaling smoke can all cause respiratory acidosis. So can any condition that reduces a person's ability to breathe. Respiratory acidosis is usually treated with a mechanical ventilator, to assist the victim's breathing. The improved exhalation increases the excretion of CO_2 and raises blood pH. In many cases of asthma, chemicals can also facilitate respiration by opening constricted bronchial passages.

Metabolic acidosis is caused by the release into the bloodstream of excessive amounts of lactic acid and other acidic by-products of metabolism. These acids enter the bloodstream, react with hydrogen carbonate ion to produce H_2CO_3, and shift the ratio of HCO_3^- to H_2CO_3 to a lower value. Heavy exercise, diabetes, and fasting can all produce metabolic acidosis. The normal response of the body is to increase the rate of breathing to eliminate some of the CO_2. Thus, we pant heavily when running uphill.

Metabolic acidosis can also result when a person is severely burned. Blood plasma leaks from the circulatory system into the injured area, producing edema (swelling) and reducing the blood volume. If the burned area is large, this loss of blood volume may be sufficient

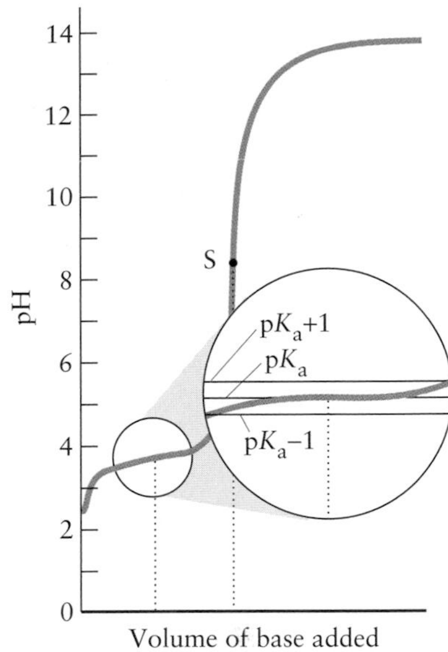

FIGURE 11.3 This plot shows how the pH of a weak acid changes as a strong base is added. When the conjugate acid and base are present at similar concentrations, the pH changes very little as more strong base (or strong acid) is added. This condition corresponds to the nearly horizontal region of the curve. As the inset shows, the pH lies between $pK_a \pm 1$ for a wide range of concentrations. S denotes the stoichiometric point (Section L).

range for optimum buffering. However, the principal waste products of living cells are carboxylic acids, such as lactic acid. Plasma, with its relatively high concentration of HCO_3^- ions, can absorb a significant surge of hydrogen ions from these carboxylic acids. The high proportion of HCO_3^- also helps us to withstand disturbances that lead to excess acid, such as disease and shock due to burns (Box 11.1).

The capacity of a buffer is determined by its concentration and pH. A more concentrated buffer can react with more added acid or base than a less concentrated one. A buffer solution is generally most effective in the range $pK_a \pm 1$.

to reduce blood flow and oxygen supply to all the body's tissues. Lack of oxygen, in turn, causes the tissues to produce an excessive amount of lactic acid and leads to metabolic acidosis. To minimize the decrease in pH, the injured individual breathes harder to eliminate the excess CO_2. However, if blood volume drops below levels for which the body can compensate, a vicious circle ensues in which blood flow decreases still further, blood pressure falls, CO_2 excretion diminishes, and acidosis becomes more severe. Individuals in this state are said to be in *shock* and will die if not treated promptly.

The dangers of shock are avoided or treated by intravenous infusion of large volumes of a salt-containing solution, usually one known as *lactated Ringer's solution*. The added liquid increases blood volume and blood flow, thereby improving oxygen delivery. The $[HCO_3^-]/[H_2CO_3]$ ratio then increases toward normal, and allows the severely injured person to survive.

Respiratory alkalosis is the rise in pH associated with excessive respiration. Hyperventilation, which can result from anxiety or high fever, is a common cause. The body may control blood pH in a hyperventilating individual by fainting, which results in slower respiration. An intervention that may avoid fainting is to have a hyperventilating person breathe into a paper bag, which allows much of the respired CO_2 to be taken up again.

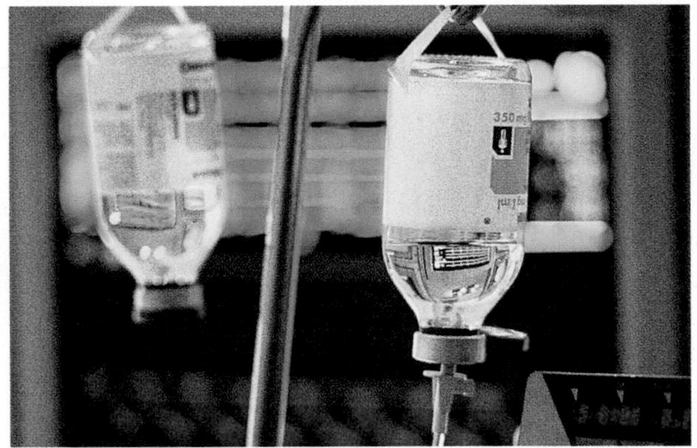

Patients who have suffered traumatic injuries must receive an immediate intravenous solution to combat the symptoms of shock and help maintain the pH of blood.

Metabolic alkalosis is the increase in pH resulting from illness or chemical ingestion. Repeated vomiting or the overuse of diuretics can cause metabolic alkalosis. Once again the body compensates, this time by decreasing the rate of respiration.

* This box is an adaptation of a contribution by B. A. Pruitt, M.D., and A. D. Mason, M.D., U.S. Army Institute of Surgical Research.

TITRATIONS

As we saw in Section L, titration involves adding a solution called the *titrant* from a buret to a flask containing the sample, called the *analyte*. For example, suppose an environmental chemist studying a lake in which fish are dying needs to know the concentration of acid in the water. A sample of the lake water would be the analyte and a solution of base of known concentration would be the titrant. At the stoichiometric point, the amount of OH^- (or H_3O^+) added as titrant is equal to the amount of H_3O^+ (or OH^-) initially present in the analyte. The success of the technique depends on our ability to detect this point. The techniques used in this chapter to determine the roles of different species in determining the pH will help us understand how the pH signals the stoichiometric point. They will also help us select the appropriate indicator for a titration.

We noted in Section L that the stoichiometric point is also called the equivalence point.

11.5 Strong Acid-Strong Base Titrations

A plot of the pH of the analyte solution as a function of the volume of titrant added during a titration is called a **pH curve.** The shape of the pH curve in Fig. 11.4 is typical of titrations in which a strong acid is added to a strong base. Initially, the pH falls slowly. Then, at the stoichiometric

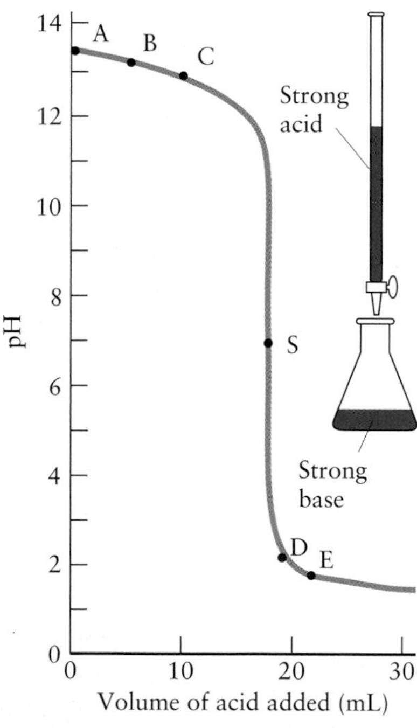

FIGURE 11.4 The variation of pH during the titration of a strong base, 25.00 mL of 0.250 M NaOH(aq), with a strong acid, 0.340 M HCl(aq). The stoichiometric point (S) occurs at pH = 7. The other points on the pH curve are explained in the text.

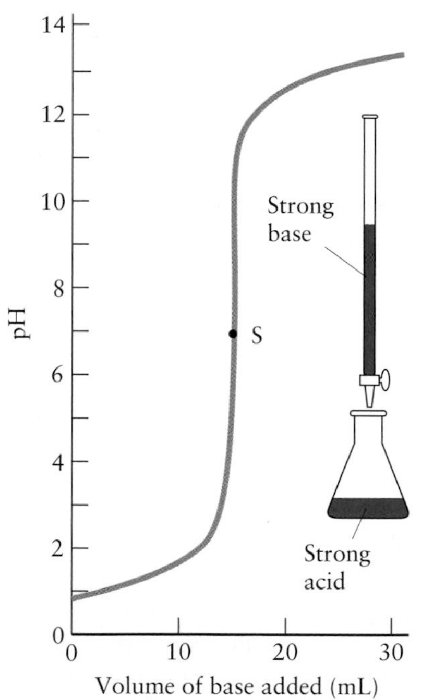

FIGURE 11.5 The variation of pH during a typical titration of a strong acid (the analyte) with a strong base (the titrant). The stoichiometric point (S) occurs at pH = 7.

point, there is a sudden decrease in pH through 7. At this point, an indicator changes color or an automatic titrator responds electronically to the sudden change in pH. Titrations typically end at this point. However, it is instructive to see what happens to the pH as more titrant is added. We find that the pH falls slowly toward the value of the acid itself as dilution by the original analyte solution becomes less and less important.

Figure 11.5 shows a pH curve for a titration in which the analyte is a strong acid and the titrant is a strong base. This curve is the mirror image of the curve for the titration of a strong base with a strong acid.

In the titration of a strong acid with a strong base, or a strong base with a strong acid, the pH changes slowly initially, changes rapidly through pH = 7 at the stoichiometric point, and then changes slowly again.

TOOLBOX 11.1 How to calculate the pH during a strong acid-strong base titration

Conceptual Basis

The pH during the titration of a strong acid with a strong base is governed by the major species in solution. Because the conjugate base of the strong acid has little effect on the pH, the pH is determined by whichever is in excess, the strong acid or strong base. We calculate the amount of acid or base in excess, then determine the concentration of the excess reagent from the reaction stoichiometry and the total volume of solution. The pH is then calculated from that concentration.

Procedure

First, use the reaction stoichiometry to find the amount of excess acid or base

Step 1 Calculate the amount of H_3O^+ ions (if the analyte is a strong acid) or OH^- ions (if the analyte is a strong base) in the original analyte solution from its molarity and its volume.

Step 2 Calculate the amount of OH^- ions (if the titrant is a strong base) or H_3O^+ ions (if the titrant is a strong acid) in the volume of titrant added.

Step 3 Write the chemical equation for the neutralization reaction and use the reaction stoichiometry to find the amount of H_3O^+ ions (or OH^- ions if the analyte is a strong base) that remains in the analyte solution. Each mole of H_3O^+ ions reacts with 1 mol OH^- ions; therefore, subtract the number of moles of H_3O^+ or OH^- ions that have reacted from the initial number of moles of H_3O^+ or OH^- ions.

Next, determine the concentration

Step 4 Divide the remaining amount of H_3O^+ (or OH^-) ions by the total volume of the combined solutions (analyte plus added titrant), which gives the molarity of the H_3O^+ (or OH^-) ions in the solution.

Finally, calculate the pH

Step 5 Take the negative logarithm of the relevant molarity to find the pH (or the pOH). Convert pOH to pH by using the relation pH + pOH = 14.00 derived from the autoprotolysis equilibrium.

EXAMPLE 11.5 Calculating points on the pH curve for a strong acid-strong base titration

Suppose we are carrying out a titration in which the analyte initially consists of 25.00 mL of 0.250 M NaOH(aq) and the titrant is 0.340 M HCl(aq). What is the pH of the analyte solution after the addition of 5.00 mL of titrant?

STRATEGY After the addition of 5.00 mL of the acid titrant, we can expect the pH to decrease from its initial value. The procedure is outlined as a series of steps in Toolbox 11.1.

SOLUTION ***Step 1*** Initially, the pOH of the analyte is pOH = −log 0.250 = 0.602, so the pH of the solution is pH = 14.00 − 0.602 = 13.40. This is point A in Fig. 11.4. The amount of OH^- ions initially present is

$$\text{Amount of OH}^- \text{ (from the base)} = (25.00 \times 10^{-3}\text{ L}) \times (0.250\text{ mol}\cdot\text{L}^{-1}) = 6.25 \times 10^{-3}\text{ mol}$$

or 6.25 mmol (because 1 mmol = 10^{-3} mol).

Step 2 The amount of H_3O^+ ions supplied by the titrant is

$$\text{Amount of H}_3\text{O}^+ \text{ (from the acid)} = (5.00 \times 10^{-3}\text{ L}) \times (0.340\text{ mol}\cdot\text{L}^{-1}) = 1.70 \times 10^{-3}\text{ mol} = 1.70\text{ mmol}$$

Step 3 Because 1 mol NaOH ≏ 1 mol HCl, after reaction of all the H_3O^+ ions added, the amount of OH^- remaining is 6.25 − 1.70 mmol = 4.55 mmol.

Step 4 Because the total volume of the solution is now 30.00 mL, or 0.030 00 L, the molarity of OH^- is

$$\text{Molarity of OH}^- = \frac{4.55 \times 10^{-3}\text{ mol}}{0.030\ 00\text{ L}} = 0.152\text{ mol}\cdot\text{L}^{-1}$$

Step 5 Hence, pOH = −log 0.152 = 0.82, so pH = 13.18, point B in Fig. 11.4. Note that the pH has fallen, as expected, but only by a very small amount. This small change is consistent with the shallow slope of the pH curve at the start of the titration.

Self-Test 11.6A What is the pH of the solution that results from the addition of a further 5.00 mL of the HCl titrant to the analyte?

[*Answer:* 12.91, point C]

Self-Test 11.6B What is the pH of the solution that results from the addition of another 2.00 mL of titrant to the analyte?

EXAMPLE 11.6 Calculating the pH after the stoichiometric point of a strong acid–strong base titration

Experimentally, we know that the pH changes abruptly close to the stoichiometric point. How much does the pH change if we reach the stoichiometric point in the titration described in Example 11.5 and then add an additional 1.00 mL of HCl(aq)? Evaluate the pH of the solution in Example 11.5 at that point.

STRATEGY First, we find the volume of titrant needed to reach the stoichiometric point. At this point, pH = 7. After the stoichiometric point, the amount of acid added exceeds the amount of base in the analyte, and we expect a pH of less than 7. We use the reaction stoichiometry to determine how much of the acid added remains after neutralization. Then we use the total volume of solution to find the molar concentration of H_3O^+ and convert it to pH. To implement the calculation, work through the steps in Toolbox 11.1.

SOLUTION Initially, 6.25×10^{-3} mol OH^- is present, so the volume of 0.340 M HCl(aq) required to reach the stoichiometric point is

$$\text{Volume of HCl} = \frac{6.25 \times 10^{-3}\text{ mol NaOH}}{0.340\text{ mol HCl}\cdot\text{L}^{-1}} \times \frac{1\text{ mol HCl}}{1\text{ mol NaOH}} = 0.0184\text{ L}$$

or 18.4 mL. If another 1.00 mL HCl(aq) is added, the total volume of acid added is 19.4 mL.

Step 1 The amount of OH^- ions initially present (from Example 11.5) is 6.25 mmol.

Step 2 The amount of H_3O^+ ions supplied by the titrant solution is

$$\text{Amount of H}_3\text{O}^+\text{ supplied} = (19.4 \times 10^{-3}\text{ L}) \times (0.340\text{ mol}\cdot\text{L}^{-1}) = 6.60 \times 10^{-3}\text{ mol} = 6.60\text{ mmol}$$

More moles of H_3O^+ have been supplied than there were moles of OH^- originally, so we know that we are past the stoichiometric point.

Step 3 After all the OH^- ions have reacted,

$$\text{Amount of H}_3\text{O}^+\text{ remaining} = (6.60 - 6.25)\text{ mmol} = 0.35\text{ mmol}$$

Step 4 Because the total volume of the solution is 25.0 mL + 19.4 mL, or 44.4 mL, which is 0.0444 L, the molarity of H_3O^+ is

$$\text{Molarity of H}_3\text{O}^+ = \frac{3.5 \times 10^{-4}\text{ mol}}{0.0444\text{ L}} = 7.9 \times 10^{-3}\text{ mol}\cdot\text{L}^{-1}$$

Step 5 Hence,

$$\text{pH} = -\log(7.9 \times 10^{-3}) = 2.1$$

A pH of 2.1 (point D in Fig. 11.4) is well below the pH (of 7) at the stoichiometric point, although only one more milliliter of acid has been added.

Self-Test 11.7A Calculate the pH of the solution after the addition of 20.4 mL of titrant.

[*Answer:* 1.82, point E]

Self-Test 11.7B Calculate the pH of the solution after the addition of 25.0 mL of titrant.

11.6 Strong Acid-Weak Base and Weak Acid-Strong Base Titrations

In many titrations, one solution—either the analyte or the titrant—contains a weak acid or base and the other solution contains a strong base or acid. For example, we might be interested in the concentration of formic

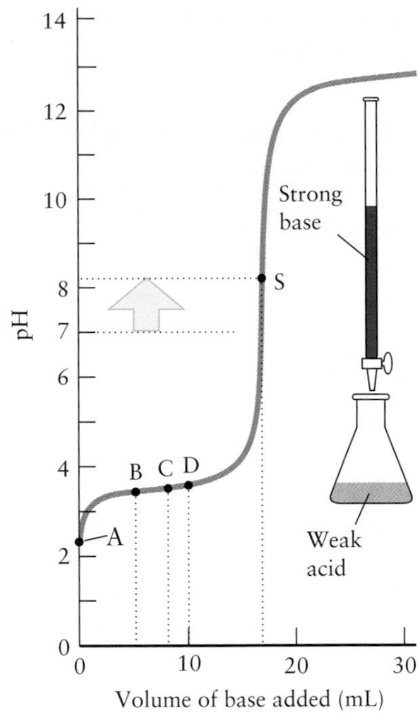

FIGURE 11.6 The pH curve for the titration of a weak acid with a strong base: 25.00 mL of 0.1 M $CH_3COOH(aq)$ with 0.150 M $NaOH(aq)$. The stoichiometric point (S) occurs on the basic side of pH = 7 because the anion $CH_3CO_2^-$ is a base.

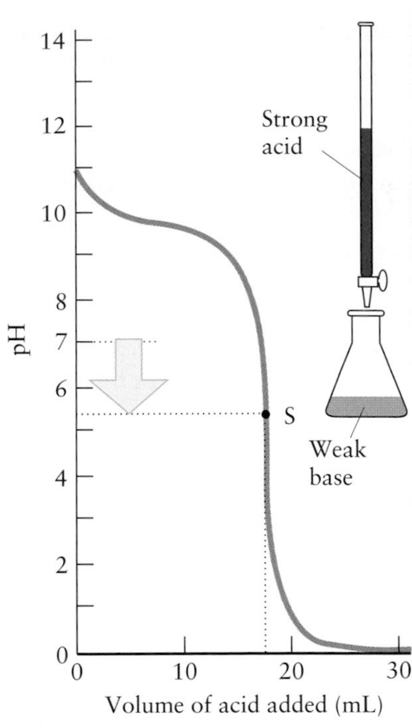

FIGURE 11.7 A typical pH curve for the titration of a weak base with a strong acid. The stoichiometric point (S) occurs on the acidic side of pH = 7 because the salt formed by the neutralization reaction has an acidic cation.

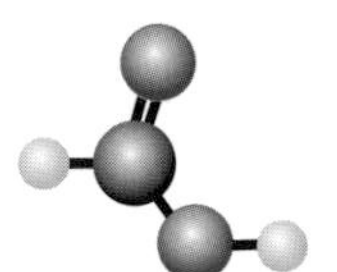

1 Formic acid, HCOOH

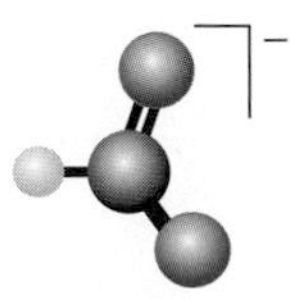

2 Formate ion, HCO_2^-

A rule of thumb is that *a strong base dominates a weak acid*, so a solution of a salt of a strong base and a weak acid is basic.

acid, a weak acid found in ant venom, and titrate it with sodium hydroxide. Alternatively, we might need to know the concentration of ammonia, a weak base, in a soil sample, and titrate it with hydrochloric acid. Figures 11.6 and 11.7 show the different pH curves that are found experimentally for these two types of solutions. The main feature to note is that the stoichiometric point does not occur at pH = 7. Moreover, although the pH changes reasonably sharply near the stoichiometric point, it does not change as abruptly as it does in a strong acid-strong base titration. Weak acids are normally not titrated with weak bases, because the stoichiometric point is difficult to locate.

The pH of a solution at the stoichiometric point of an acid-base titration depends on the properties of the salt formed in the neutralization reaction. At the stoichiometric point of the titration of formic acid, HCOOH (**1**), with sodium hydroxide, the solution consists of sodium formate, $NaHCO_2$, and water. Formate ion, HCO_2^- (**2**), is a base, and the Na^+ ions have virtually no effect on pH; so overall the solution is basic. Because the salt has a basic anion, we can expect the pH to be greater than 7 at the stoichiometric point. At the stoichiometric point of the titration of aqueous ammonia with hydrochloric acid, the solute is ammonium chloride. Because NH_4^+ is an acid, we expect pH < 7. The same is true for the stoichiometric point of the titration of any weak base and strong acid.

The rule of thumb is that *a strong acid dominates a weak base*, so the solution of a salt of a strong acid and a weak base is acidic.

EXAMPLE 11.7 Estimating the pH at the stoichiometric point of the titration of a weak acid with a strong base

Estimate the pH at the stoichiometric point of the titration of 25.00 mL of 0.100 M HCOOH(aq) with 0.150 M NaOH(aq).

STRATEGY The salt present at the stoichiometric point is sodium formate, which has a basic anion. The amount of HCO_2^- ions in the solution at the stoichiometric point is equal to the initial amount of HCOOH in the analyte, which we can calculate from the initial volume of analyte solution and its molarity. The molarity of HCO_2^- at the stoichiometric point is this amount of ions divided by the total volume of the solution (the volume of titrant solution needed to reach the stoichiometric point plus the initial volume of analyte). We then calculate the pH of the salt solution as described in Toolbox 10.3. The K_b of HCO_2^- is related to the K_a of its conjugate acid HCOOH by $K_a \times K_b = K_w$; K_a is listed in Table 10.1. Assume that the autoprotolysis of water has no significant effect on the pH, then check that assumption.

SOLUTION From Table 10.1, $K_a = 1.8 \times 10^{-4}$ for formic acid; therefore, $K_b = K_w/K_a = 5.6 \times 10^{-11}$. The amount of HCOOH in the initial analyte solution of volume 25.00 mL (2.500×10^{-2} L) is

$$\begin{aligned}\text{Amount of HCOOH} &= (2.500 \times 10^{-2}\ \text{L}) \times (0.100\ \text{mol}\cdot\text{L}^{-1}) \\ &= 2.50 \times 10^{-3}\ \text{mol} = 2.50\ \text{mmol}\end{aligned}$$

The amount of HCO_2^- at the stoichiometric point is therefore also 2.50 mmol. The stoichiometry of the reaction requires 1 mol OH^- for 1 mol HCOOH. So, we need the volume of 0.150 M NaOH(aq) containing that amount of OH^- ions:

$$\text{Volume} = \frac{2.50 \times 10^{-3}\ \text{mol}}{0.150\ \text{mol}\cdot\ \text{L}^{-1}} = 16.7 \times 10^{-3}\ \text{L}$$

or 16.7 mL. Therefore, the total volume of the combined solutions at the stoichiometric point is (25.00 + 16.7) mL = 41.7 mL. It follows that, at the stoichiometric point,

$$\text{Molarity of } HCO_2^- = \frac{2.50 \times 10^{-3}\ \text{mol}}{41.7 \times 10^{-3}\ \text{L}} = 0.0600\ \text{mol}\cdot\ \text{L}^{-1}$$

Now we can calculate the pH of the solution as described in Toolbox 10.3. The equilibrium to consider is

$$HCO_2^-(aq) + H_2O(l) \rightleftharpoons HCOOH(aq) + OH^-(aq)$$

$$K_b = \frac{[HCOOH][OH^-]}{[HCO_2^-]}$$

The equilibrium table, with all concentrations in moles per liter, is

	Species		
	HCO_2^-	HCOOH	OH^-
Step 1 Initial molarity	0.0600	0	0
Step 2 Change in molarity	$-x$	$+x$	$+x$
Step 3 Equilibrium molarity	$0.0600 - x$	x	x

Step 4 Substitution of the values from step 3 into the expression for K_b gives

$$5.6 \times 10^{-11} = \frac{[HCOOH][OH^-]}{[HCO_2^-]} = \frac{x \times x}{0.0600 - x}$$

If we suppose that x is less than 5% of 0.0600, we can write

$$5.6 \times 10^{-11} \approx \frac{x^2}{0.0600}$$

and obtain $x \approx 1.8 \times 10^{-6}$. It follows from step 3 that the molarity of OH^- is 1.8×10^{-6} mol·L^{-1}, which is about 18 times greater than the molarity of OH^- ions from the autoprotolysis of water (1.0×10^{-7} mol·L^{-1}), so neglect of the latter is reasonable. Now we can write

$$\text{pOH} = -\log(1.8 \times 10^{-6}) = 5.74$$

and therefore

$$\text{pH} = 14.00 - 5.74 = 8.26$$

or about 8.3. Notice that the pH at the stoichiometric point is greater than 7. The salt is basic, because it contains a basic anion (HCO_2^-) and a neutral cation. Conversely, in the titration of a weak base with a strong acid, we would expect pH < 7 at the stoichiometric point, because the salt formed will have an acidic cation (the conjugate acid of the weak base) and a neutral anion.

Self-Test 11.8A Calculate the pH at the stoichiometric point of the titration of 25.00 mL of 0.010 M HClO(aq) with 0.020 M KOH(aq). See Table 10.1 for K_a.

[*Answer:* 9.15]

Self-Test 11.8B Calculate the pH at the stoichiometric point of the titration of 25.00 mL of 0.020 M NH_3(aq) with 0.015 M HCl(aq). (For NH_4^+, $K_a = 5.6 \times 10^{-10}$.)

Notice that one way to prepare an acid buffer is to neutralize half the amount of weak acid present with strong base.

Now consider the overall shape of the pH curve. The slow change in pH about halfway to the stoichiometric point indicates that the solution acts as a buffer in that region (see Section 11.3). At the halfway point of the titration, [HA] = [A^-] and pH = pK_a. The flatness of the curve near pH = pK_a illustrates very clearly the ability of a buffer solution to stabilize the pH of the solution. Moreover, we can now see how to determine pK_a: all we need to do is plot the pH curve during a titration and then identify the pH halfway to the stoichiometric point (Fig. 11.8). To obtain the pK_b of a weak base, we find pK_a the same way but go on to use pK_a + pK_b = pK_w. The values recorded in Tables 10.1 and 10.2 were obtained in this way.

We have already seen how to estimate the pH of the initial analyte when only weak acid or weak base is present (point A in Fig. 11.6, for instance), and the pH at the stoichiometric point (point S). Other points between these two points correspond to a mixed solution, of some weak acid (or base) and some salt. We can therefore use the techniques described in Toolbox 11.2 and Examples 11.1 or 11.2 to account for the shape of the curve (see Example 11.8).

Well after the stoichiometric point in the titration of a weak acid with a strong base, the pH depends only on the concentration of excess strong base, as in Example 11.6. For example, suppose we went on to add several liters of strong base from a giant buret. The presence of salt—the product of the neutralization reaction—would be negligible relative to the concentration of excess base. The pH would be that of the nearly pure titrant (the original base solution).

Figure 11.8 summarizes the changes in pH of a solution during a titration of a weak acid by a strong base. Halfway to the stoichiometric point, the pH is equal to the pK_a of the acid. The pH is greater than 7 at the stoichiometric point of the titration of a weak acid and strong base. The pH is less than 7 at the stoichiometric point of the titration of a weak base and strong acid.

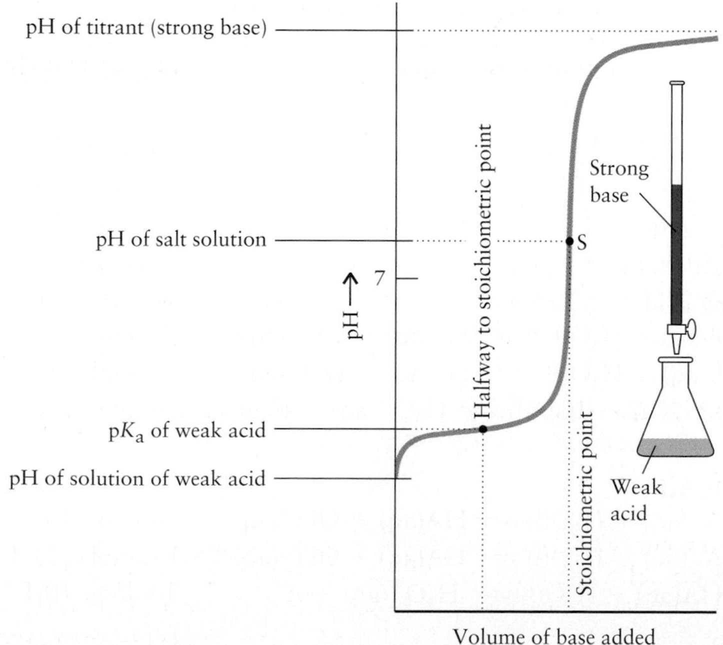

FIGURE 11.8 The pK_a of an acid can be determined by carrying out a titration of the weak acid with a strong base and locating the pH of the solution after the addition of half the volume of acid needed to reach the stoichiometric point. The pH at that point is equal to pK_a.

TOOLBOX 11.2 How to calculate the pH during a titration of a weak acid or a weak base

Conceptual Basis

The pH is governed by the major solute species present in solution. As strong base is added to a solution of a weak acid, a salt of the conjugate base of the weak acid is formed. This salt affects the pH and needs to be taken into account. Table 11.2 outlines the regions encountered during a titration and the primary equilibrium to consider in each region.

Procedure

The procedure is like that in Toolbox 11.1, except that an additional step is required to calculate the pH from the proton transfer equilibrium.

Use the reaction stoichiometry to find the number of moles of excess acid or base

Step 1 Calculate the amount of weak acid or base in the original analyte solution from its molarity and its volume.

Step 2 Calculate the amount of OH^- ions (or H_3O^+ ions if the analyte is a weak base) in the volume of titrant added.

Step 3 Write the chemical equation for the reaction between acid and base and use the reaction stoichiometry to calculate the following:

- *Weak acid with a strong base*: the amount of conjugate base formed by the reaction of the acid with added base, and the amount of weak acid remaining.
- *Weak base with a strong acid*: the amount of conjugate acid formed by the reaction of the base with added acid, and the amount of weak base remaining.

Determine the concentration

Step 4 Find the molarities of the conjugate acid and base in solution by dividing the amount of each by the total volume of the solution.

Set up an equilibrium table

Step 5 Use an equilibrium table to find the H_3O^+ concentration in a weak acid or the OH^- concentration in a weak base. Alternatively, if the concentrations of conjugate acid and base are both large relative to the concentration of hydronium ions, use the Henderson-Hasselbalch equation to determine the pH. In each case, assume that the contribution of the autoprotolysis of water to the pH is insignificant if the pH is less than 6 or greater than 8. If necessary, convert between K_a and K_b by using $K_w = K_a \times K_b$.

TABLE 11.2 Summary of Weak Acid and Weak Base Titration Equilibria

Point in titration	Primary species	Proton transfer equilibrium	Related text elements
1 Weak acid HA titrated with strong base			
Initial	HA	$HA(aq) + H_2O(l) \rightleftharpoons H_3O^+(aq) + A^-(aq)$	Toolbox 10.1
Buffer region	HA, A^-	$HA(aq) + H_2O(l) \rightleftharpoons H_3O^+(aq) + A^-(aq)$	Examples 11.1, 11.2
Stoichiometric point	A^-	$A^-(aq) + H_2O(l) \rightleftharpoons HA(aq) + OH^-(aq)$	Toolbox 10.3
(Note: This is a solution of a salt with a basic anion.)			
2a Neutral weak base B titrated with strong acid			
Initial	B	$B(aq) + H_2O(l) \rightleftharpoons BH^+(aq) + OH^-(aq)$	Toolbox 10.2
Buffer region	B, BH^+	$B(aq) + H_2O(l) \rightleftharpoons BH^+(aq) + OH^-(aq)$	Examples 11.1, 11.2
Stoichiometric point	BH^+	$BH^+(aq) + H_2O(l) \rightleftharpoons H_3O^+(aq) + B(aq)$	Toolbox 10.3
(Note: This is a solution of a salt with an acidic anion.)			
2b Weak base anion A^- titrated with strong acid			
Initial	A^-	$A^-(aq) + H_2O(l) \rightleftharpoons HA(aq) + OH^-(aq)$	Toolbox 10.3
Buffer region	A^-, HA	$A^-(aq) + H_2O(l) \rightleftharpoons HA(aq) + OH^-(aq)$	Examples 11.1, 11.2
Stoichiometric point	HA	$HA(aq) + H_2O(l) \rightleftharpoons H_3O^+(aq) + A^-$	Toolbox 10.1
(Note: This is a solution of a weak acid.)			

EXAMPLE 11.8 Calculating the pH before the stoichiometric point in a weak acid-strong base titration

Calculate the pH of the solution resulting when 5.00 mL of 0.150 M NaOH(aq) is added to 25.00 mL of 0.100 M HCOOH(aq). Use $K_a = 1.8 \times 10^{-4}$ for HCOOH.

STRATEGY As described in Toolbox 11.2, we view the procedure as a series of steps. The difference is that we must use the K_a to calculate the H_3O^+ ion concentration.

SOLUTION ***Step 1*** The initial amount of HCOOH in the analyte is

$$(25.00 \times 10^{-3}\ \text{L}) \times (0.100\ \text{mol}\cdot\text{L}^{-1}) = 2.50 \times 10^{-3}\ \text{mol}$$

or 2.50 mmol HCOOH.

Step 2 The amount of OH^- in 5.00 mL of the titrant is

$$(5.00 \times 10^{-3}\ \text{L}) \times (0.150\ \text{mol}\cdot\text{L}^{-1}) = 7.50 \times 10^{-4}\ \text{mol}$$

or 0.750 mmol OH^-.

Step 3 The chemical equation

$$\text{HCOOH(aq)} + \text{OH}^-\text{(aq)} \longrightarrow \text{HCO}_2{}^-\text{(aq)} + \text{H}_2\text{O(l)}$$

shows that 1 mol $OH^- \triangleq$ 1 mol HCOOH and 1 mol $OH^- \triangleq$ 1 mol $HCO_2{}^-$. So 0.750 mmol OH^- produces 0.750 mmol $HCO_2{}^-$ and leaves 2.50 − 0.750 mmol = 1.75 mmol HCOOH.

Step 4 The total volume of the solution at this stage of the titration is (25.00 + 5.00) mL = 30.00 mL, so the molarities of acid and conjugate base are

$$\text{Molarity of HCOOH} = \frac{1.75 \times 10^{-3}\ \text{mol}}{30.00 \times 10^{-3}\ \text{L}} = 0.0583\ \text{mol}\cdot\text{L}^{-1}$$

$$\text{Molarity of HCO}_2{}^- = \frac{7.50 \times 10^{-4}\ \text{mol}}{30.00 \times 10^{-3}\ \text{L}} = 0.0250\ \text{mol}\cdot\text{L}^{-1}$$

Step 5 The proton transfer equilibrium for HCOOH in water is

$$\text{HCOOH(aq)} + \text{H}_2\text{O(l)} \rightleftharpoons \text{H}_3\text{O}^+\text{(aq)} + \text{HCO}_2{}^-\text{(aq)}$$

$$K_a = \frac{[\text{H}_3\text{O}^+][\text{HCO}_2{}^-]}{[\text{HCOOH}]}$$

From Table 10.1, the pK_a for formic acid is 3.75. Assuming little change occurs in the concentrations of acid and base due to the deprotonation of HCOOH, we write

$$\text{pH} = \text{p}K_a + \log\frac{[\text{HCO}_2{}^-]}{[\text{HCOOH}]} \approx 3.75 + \log\frac{0.0250}{0.0583} = 3.38$$

This pH corresponds to $[H_3O^+] \approx 4.2 \times 10^{-4}$, point B in Fig. 11.6 and, as anticipated, the contribution of autoprotolysis is negligible. As expected, the pH of the mixed solution is higher than that of the original acid (2.4).

Self-Test 11.9A Calculate the pH of the solution after the addition of another 5.00 mL of 0.150 M NaOH(aq).

[*Answer:* 3.93, point D]

Self-Test 11.9B Calculate the pH of the solution after the addition of yet another 5.00 mL of 0.150 M NaOH(aq).

11.7 Acid-Base Indicators

The operation of this electrode is explained in Section 12.11.

A simple, reliable, and fast method of determining the pH of a solution and of monitoring a titration is with a pH meter, which uses a special electrode to measure H_3O^+ concentration. An automatic titrator monitors the pH of the analyte solution continuously: it uses a customized pH meter to detect the stoichiometric point by responding to the characteristic rapid change in pH (Fig. 11.9). Another common technique is to use an indicator to detect the stoichiometric point. An **acid-base indicator** is a water-soluble organic dye with a color that depends on the pH. The sudden change in pH that occurs at the stoichiometric point of a titration is signaled by a sharp change in color of the dye as it responds to the pH.

To understand the action of an acid-base indicator, we need to know that the dye is a weak acid that has one color in its acid form (HIn, where In stands for indicator) and another color in its conjugate base form (In^-). The color change results from the effect of the proton in HIn: it

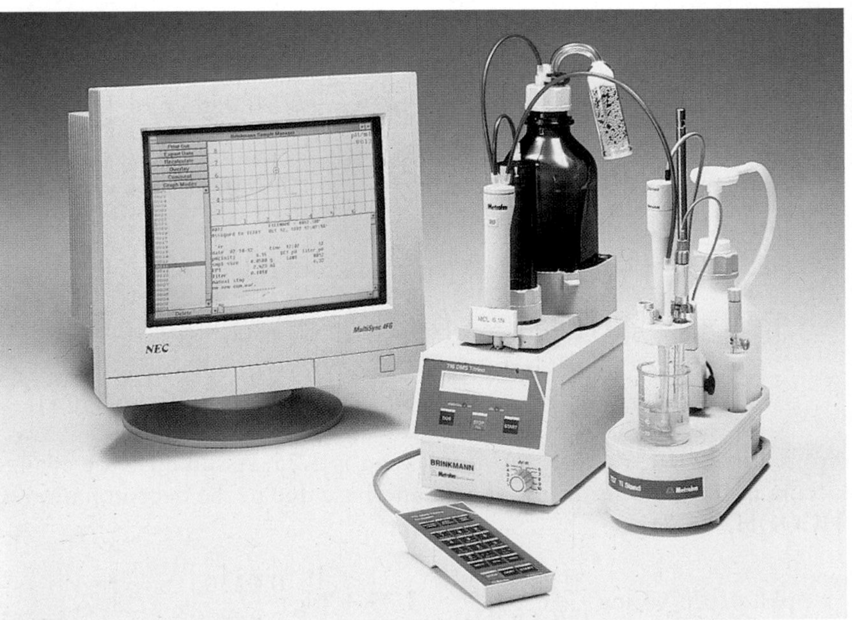

FIGURE 11.9 A commercially available automatic titrator. The stoichiometric point of the titration is detected by the sudden change in pH that occurs in its vicinity; the pH is monitored electronically. The pH curve can be plotted as the reaction proceeds, as shown on the monitor screen.

changes the structure of the molecule in such a way that the light absorption characteristics of HIn are different from those of In^-. When the concentration of HIn is much greater than that of In^-, the solution has the color of the acid form of the indicator. When the concentration of In^- is much greater than that of HIn, the solution has the color of the base form of the indicator.

Because it is a weak acid, an indicator takes part in a proton transfer equilibrium:

$$HIn(aq) + H_2O(l) \rightleftharpoons H_3O^+(aq) + In^-(aq)$$

$$K_{In} = \frac{[H_3O^+][In^-]}{[HIn]} \qquad (4)$$

The **end point** of a titration is the point at which the concentrations of the acid and base forms of the indicator are equal: $[HIn] = [In^-]$. When we substitute this equality into the expression for K_{In}, we see that at the end point $[H_3O^+] = K_{In}$. That is, the color change occurs when

$$pH = pK_{In} \qquad (5)$$

The color starts to change perceptibly about 1 pH unit before pK_{In} and is effectively complete about 1 pH unit after pK_{In}. Table 11.3 gives the values of pK_{In} for a number of common indicators.

One common indicator is phenolphthalein (Fig. 11.10). The acid form of this large molecule (**3**) is colorless; its conjugate base form (**4**) is pink. The structure of the base form of phenolphthalein allows electrons to be delocalized across all three of the benzenelike rings of carbon atoms, and the increase in delocalization is part of the reason for the change in color. The pK_{In} of phenolphthalein is 9.4, so the end point occurs in slightly

TABLE 11.3 Indicator Color Changes

Indicator	Color of acid form	pH range of color change	pK_{In}	Color of base form
thymol blue	red	1.2 to 2.8	1.7	yellow
methyl orange	red	3.2 to 4.4	3.4	yellow
bromophenol blue	yellow	3.0 to 4.6	3.9	blue
bromocresol green	yellow	3.8 to 5.4	4.7	blue
methyl red	red	4.8 to 6.0	5.0	yellow
litmus	red	5.0 to 8.0	6.5	blue
bromothymol blue	yellow	6.0 to 7.6	7.1	blue
phenol red	yellow	6.6 to 8.0	7.9	red
thymol blue	yellow	8.0 to 9.6	8.9	blue
phenolphthalein	colorless	8.2 to 10.0	9.4	pink
alizarin yellow R	yellow	10.1 to 12.0	11.2	red
alizarin	red	11.0 to 12.4	11.7	purple

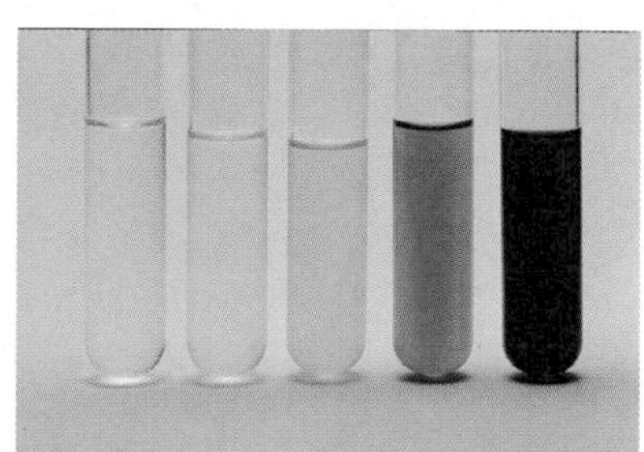

FIGURE 11.10 The stoichiometric point of an acid-base titration may be detected by the color change of an indicator. Here we see the colors of solutions containing a few drops of phenolphthalein at (from left to right) pH of 7.0, 8.5, 9.4 (its end point), 9.8, and 12.0. At the end point, the concentrations of the conjugate acid and base forms of the indicator are equal.

basic solution. The change from colorless to pink is noticeable at pH = 8.2 and is complete by pH = 10. Litmus, another well-known indicator, has pK_{In} = 6.5; it is red for pH < 5 and blue for pH > 8.

3 Phenolphthalein
(Acid form, colorless)

4 Phenolphthalein
(Base form, pink)

There are many naturally occurring indicators. For instance, a single compound is responsible for the colors of red poppies and blue cornflowers: the pH of the sap is different in the two plants. The color of hydrangeas also depends on the acidity of their sap and can be controlled by modifying the acidity of the soil (Fig. 11.11).

It is important to select an indicator with an end point close to the stoichiometric point of a titration (Fig. 11.12). In practice, the pK_{In} of the indicator should be within about 1 pH unit of the stoichiometric point of the titration:

$$pK_{In} \approx \text{pH(stoichiometric point)} \pm 1 \qquad (6)$$

Media Link **Lab video:** Fig. 11.11, Rose petal pigment: a natural indicator

FIGURE 11.11 The color of these hydrangeas depends on the acidity of the soil in which they are growing: acid soil results in blue flowers, alkaline soil results in pink flowers.

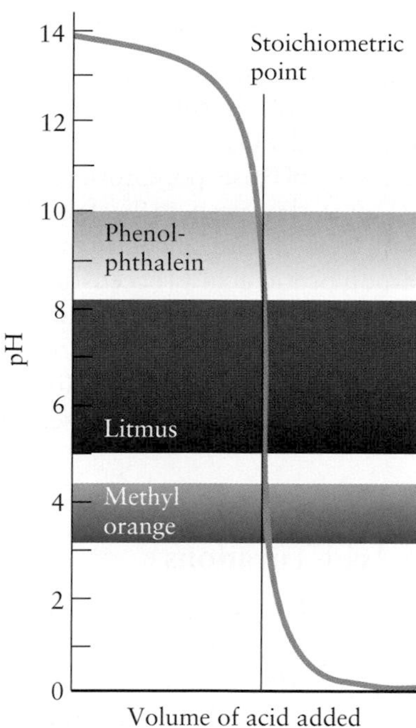

FIGURE 11.12 Ideally, an indicator should have a sharp color change close to the stoichiometric point of the titration, which is at pH = 7 for a strong acid-strong base titration. However, the change in pH is so abrupt that phenolphthalein can also be used. The color change of methyl orange, however, would give a less accurate result.

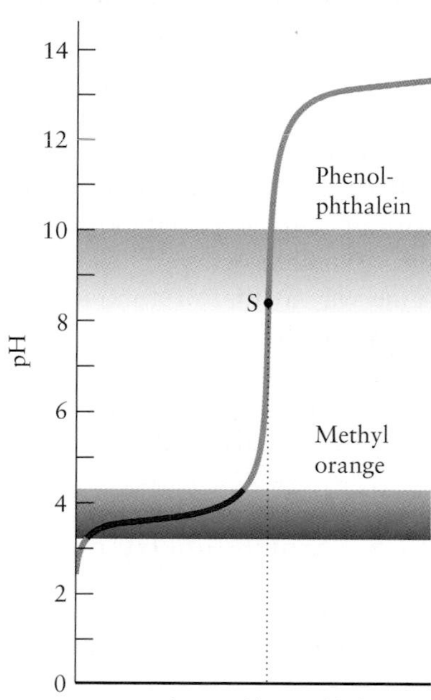

FIGURE 11.13 Phenolphthalein can be used to detect the stoichiometric point of a weak acid-strong base titration, but methyl orange would give a very inaccurate indication of the stoichiometric point. The pH curves are superimposed on approximations to the color of the indicators in the neighborhoods of their end points.

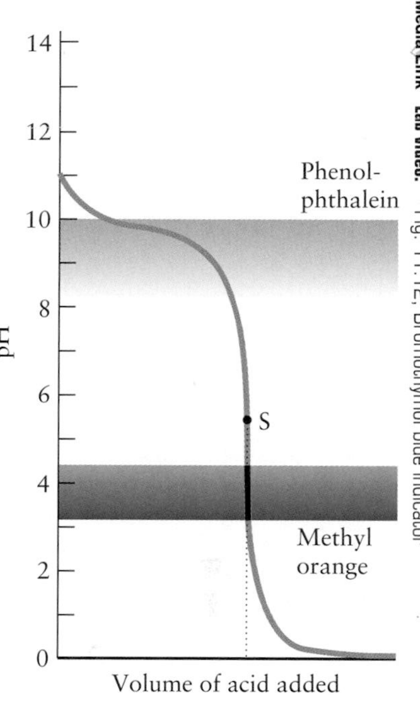

FIGURE 11.14 Methyl orange can be used for a weak base-strong acid titration. Phenolphthalein would be wholly unreliable because its color change occurs well away from the stoichiometric point. The pH curves are superimposed on approximations to the color of the indicators in the neighborhoods of their end points.

Media Link **Lab video:** Fig. 11.12, Bromothymol blue indicator

To select an indicator, first determine the pH of the salt solution formed at the stoichiometric point. Phenolphthalein can be used for titrations with a stoichiometric point near pH = 9, such as a titration of a weak acid with a strong base (Fig. 11.13). Methyl orange changes color between pH = 3.2 and pH = 4.4 and can be used in the titration of a weak base with a strong acid (Fig. 11.14). Ideally, indicators for strong acid-strong base titrations should have end points close to pH = 7; however, in strong acid-strong base titrations, the pH changes rapidly over several pH units, and even phenolphthalein can be used (see Fig. 11.12). The pH ranges over which several indicators can be used are included in Table 11.3.

Acid-base indicators are weak acids that change color close to pH = pK_{In}; an indicator should be chosen so that its end point is close to the stoichiometric point of the titration.

POLYPROTIC ACID TITRATIONS

As we saw in Section 10.19, solutions of polyprotic acids contain a variety of species that can affect the pH. A solution of phosphoric acid, for instance, contains H_3PO_4, $H_2PO_4^-$, HPO_4^{2-}, and PO_4^{3-} ions as well as hydronium ions. Because many biological systems use polyprotic acids and their anions to control pH, we may some day need to be familiar with pH curves for polyprotic titrations and to be able to calculate the pH during such a titration. The titration of a polyprotic acid proceeds in the same way as that of a monoprotic acid, with one exception: there are as many stoichiometric points in the titration as there are acidic hydrogen atoms. We therefore have to keep track of the major species in solution at each stage, as described in Sections 10.19 and 10.20 and summarized in Figs. 10.21 and 10.22.

11.8 Stoichiometry of Polyprotic Acid Titrations

Suppose we are titrating the triprotic acid H_3PO_4 with a solution of NaOH. The experimentally determined pH curve is shown in Fig. 11.15. Notice that there are three stoichiometric points (B, D, and F) and three buffer regions (A, C, and E). In pH calculations for these systems, we assume that as we add the hydroxide solution, initially NaOH reacts completely with the acid to form the diprotic conjugate base $H_2PO_4^-$. Once all the acid H_3PO_4 molecules have lost their first acidic protons, the system is at B and the primary species in solution are the diprotic conjugate base and sodium ion—we have a solution of $NaH_2PO_4(aq)$. Point B is the first stoichiometric point, and to reach it we need to supply 1 mol NaOH for each mole of H_3PO_4.

As we continue to add base, it reacts with the dihydrogen phosphate ion to form that acid's conjugate base, HPO_4^{2-}. Enough base takes us to the second stoichiometric point, D. The primary species in solution are now the monoprotic anions HPO_4^{2-} and sodium ions, which form a solution of $Na_2HPO_4(aq)$. To reach the second stoichiometric point, an additional mole of NaOH is required for each mole of H_3PO_4 originally present. A total of 2 mol NaOH for each mole of H_3PO_4 has now been added.

Additional base reacts with the monohydrogen phosphate ion to produce phosphate ion, PO_4^{3-}. When this reaction is complete, the primary species in solution are phosphate ions and sodium ions, which form a solution of $Na_3PO_4(aq)$. To reach this stoichiometric point, we have to add another mole of OH^- for each mole of H_3PO_4 initially present (F in the plot). At this point, a total of 3 mol OH^- has been added for each mole of H_3PO_4. Notice that the third stoichiometric point is not as distinct as the first two, largely because K_{a3} is comparable to K_w. As a result, it is not used for titrations.

Figure 11.16 shows the pH curve of a diprotic acid. Notice that there are two stoichiometric points (B and D) and that there are two buffer regions (A and C). The major species present in solution at each point are indicated. Note that it takes twice as much base to reach the second stoichiometric point as it does to reach the first.

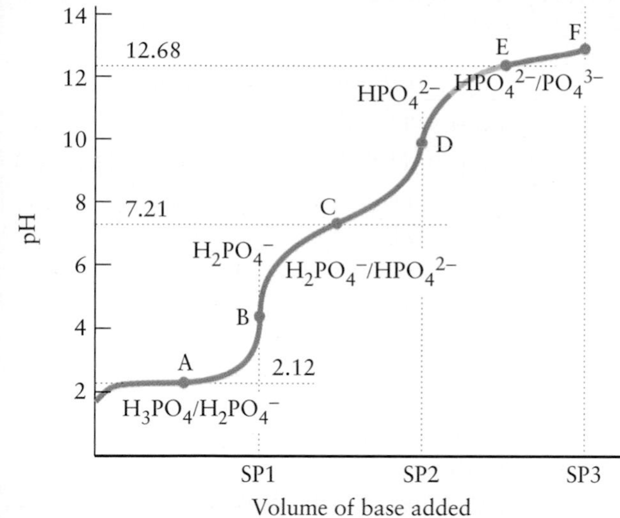

FIGURE 11.15 The variation of the pH of the analyte solution during the titration of a triprotic acid (phosphoric acid) and the major species present in solution at the first two stoichiometric points and at points when half the titrant required to reach a stoichiometric point has been added (A, C, and E). Compare this diagram with Fig. 10.22. The labels SP1, SP2, and SP3 denote the volumes of base required to reach the three stoichiometric points. Points A through F are explained in the text.

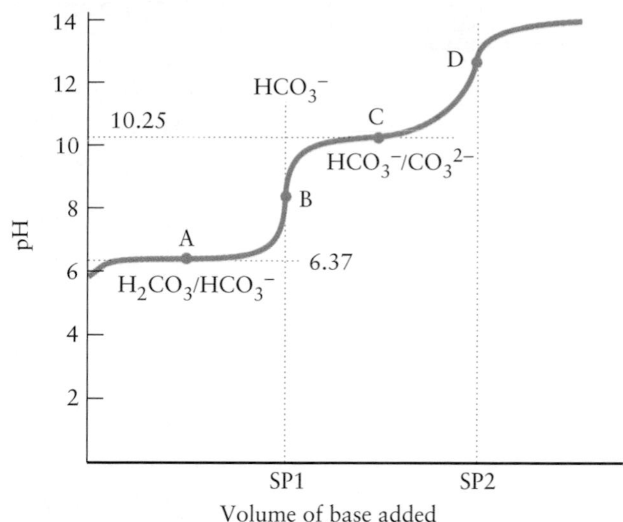

FIGURE 11.16 The variation of the pH of the analyte solution during the titration of a diprotic acid (carbonic acid) and the major species present in solution at the first stoichiometric point (B) and at points when half the titrant required to reach a stoichiometric point has been added (A and C). Compare this diagram with Fig. 10.21. The labels SP1 and SP2 denote the volumes of base required to reach the two stoichiometric points. Point D is the second stoichiometric point.

The titration of a polyprotic acid has a stoichiometric point corresponding to the removal of each acidic hydrogen atom. The product at each stoichiometric point is the salt of a conjugate base of the acid or one of its hydrogen-bearing anions.

Self-Test 11.10A What volume of 0.010 M NaOH(aq) is required to reach (a) the first stoichiometric point and (b) the second stoichiometric point in a titration of 25.00 mL of 0.010 M H_2SO_3(aq)?

[*Answer:* (a) 25 mL; (b) 50 mL]

Self-Test 11.10B What volume of 0.020 M NaOH(aq) is required to reach (a) the first stoichiometric point and (b) the second stoichiometric point in a titration of 30.00 mL of 0.010 M H_3PO_4(aq)?

11.9 pH Changes During Titrations of Polyprotic Acids

We can predict the pH at any point in the titration of a polyprotic acid with a strong base (see Toolbox 11.3). First, we have to consider the reaction stoichiometry to recognize what stage we have reached in the titration. Next we have to identify the principal solute species at that point and the proton transfer equilibrium that determines the pH. We then carry out

TABLE 11.4 Summary of Diprotic Acid Titration Equilibria

Point	Primary species	Proton transfer equilibrium
initial	H_2A	$H_2A(aq) + H_2O(l) \rightleftharpoons H_3O^+(aq) + HA^-(aq)$
buffer region 1	H_2A, HA^-	$H_2A(aq) + H_2O(l) \rightleftharpoons H_3O^+(aq) + HA^-(aq)$
stoichiometric point 1	HA^-	$HA^-(aq) + H_2O(l) \rightleftharpoons H_3O^+(aq) + A^{2-}(aq)$ $HA^-(aq) + H_2O(l) \rightleftharpoons H_2A(aq) + OH^-(aq)$
buffer region 2	HA^-, A^{2-}	$HA^-(aq) + H_2O(l) \rightleftharpoons H_3O^+(aq) + A^{2-}(aq)$
stoichiometric point 2	A^{2-}	$A^{2-}(aq) + H_2O(l) \rightleftharpoons OH^-(aq) + HA^-(aq)$

the calculation appropriate for the solution, referring to the previous worked examples if necessary. In this section, we see how to describe the solution at various stages of a titration; our conclusions are summarized in Tables 11.4 and 11.5.

The pH of a solution of a polyprotic acid undergoing a titration is estimated by considering the primary species in solution and the proton transfer equilibrium that determines the pH.

TOOLBOX 11.3 How to predict the pH during the titration of a polyprotic acid

Conceptual Basis

To predict the pH of the solution of a polyprotic acid at any stage of a titration, we consider the species present at that stage. As base reacts with a polyprotic acid, it successively deprotonates the acid and the pH changes to that characteristic of solutions of its succession of deprotonated forms.

Procedure

Initial acid solution The principal species in solution is the polyprotic acid, H_3A for a triprotic acid (for example, H_3PO_4) and H_2A for a diprotic acid (for example, oxalic acid, $H_2C_2O_4$). The proton transfer equilibrium is that between the acid and water, and the pH is determined as explained in Toolbox 10.1 and illustrated in Example 10.11.

First buffer region As base is added to the acid, the conjugate base of the acid is formed and accumulates in relatively large amounts. Midway between the solution of pure acid and the first stoichiometric point is a buffer region in which the acid and its conjugate base (for example, H_3PO_4 and $H_2PO_4^-$, or $H_2C_2O_4$ and $HC_2O_4^-$) are both present in appreciable amounts. They form a buffer system for which we predict the pH by using the procedures in Example 11.1 and the acidity constant K_{a1}. We must check assumptions before proceeding.

First stoichiometric point The first stoichiometric point is reached when the amount of OH^- added equals the amount of acid originally present. At this point, the main species in solution are cations from the added base and anions that have been formed as the conjugate base of the polyprotic acid (for example, NaH_2PO_4 or $NaHC_2O_4$). The pH is found as described in Example 10.12.

Second buffer region Additional base reacts with the amphiprotic anion (for example, $H_2PO_4^-$ or $HC_2O_4^-$) to produce the conjugate base of that anion (for example, HPO_4^{2-} or $C_2O_4^{2-}$) in sufficient concentration to form a second buffer region (consisting, for example, of $H_2PO_4^-$ and HPO_4^{2-} or $HC_2O_4^-$ and $C_2O_4^{2-}$, respectively). This system is also treated by the procedures in Example 11.1, using the acidity constant K_{a2}.

Second stoichiometric point The second stoichiometric point has been reached when the amount of OH^- added equals twice the amount of acid originally present. At this point, the main species in solution are cations from the added base and the conjugate base of the amphiprotic anion (for example, the solute is Na_2HPO_4 or $Na_2C_2O_4$). For a triprotic acid, the anion formed (for example, HPO_4^{2-}) is also amphiprotic and the pH is found as described in Example 10.12. For a diprotic acid, this anion (for example, $C_2O_4^{2-}$) is not amphiprotic, and the pH is determined by the procedures in Toolbox 10.3 and Example 10.7, using K_{a2} in the calculations.

Third buffer region If the acid is triprotic, there is a third buffer region between the anion with a single H atom and that with none (for example, between HPO_4^{2-} and PO_4^{3-}). This system is treated by the procedures in Example 11.1, using the acidity constant K_{a3}.

Third stoichiometric point The third stoichiometric point is reached when the amount of OH^- added is equal to three times the amount of the triprotic acid originally present. At this point, the main species in solution are cations from the added base and the anion with no H atoms (for example, the solute is Na_3PO_4). The pH of this solution is determined by the procedures in Toolbox 10.3 and Example 10.7, making sure to use K_{a3} in the calculations.

Note that, because ion-ion interactions are important in electrolyte solutions, the activities of ions may differ significantly from concentrations, even in a dilute solution. Therefore, the calculated values of the pH should be considered only as estimates of the true values.

EXAMPLE 11.9 Estimating the pH during the titration of a triprotic acid

A 10.00-mL sample of 0.100 M H_3PO_4(aq) was titrated with 0.100 M NaOH(aq). Identify the primary species in solution and the principal proton transfer equilibrium in solution after the addition of the following volumes of NaOH solution: (a) 10.00 mL; (b) 12.00 mL; (c) 26.00 mL.

STRATEGY Consider what point in the titration has been reached and identify the principal species in solution, as in Toolbox 11.3. Then write the chemical equation for the reaction between the acid and the base that represents the principal proton transfer equilibrium in the solution, as given in Table 11.4 or Table 11.5.

SOLUTION

Up to the first stoichiometric point Each OH^- ion from the base reacts with one molecule of H_3PO_4 until all the latter have been removed:

$$H_3PO_4(aq) + OH^-(aq) \longrightarrow H_2PO_4^-(aq) + H_2O(l) \quad \text{(A)}$$

TABLE 11.5 Summary of Triprotic Acid Titration Equilibria

Point	Primary species	Proton transfer equilibrium
initial	H_3A	$H_3A(aq) + H_2O(l) \rightleftharpoons H_3O^+(aq) + H_2A^-(aq)$
buffer region 1	H_3A, H_2A^-	$H_3A(aq) + H_2O(l) \rightleftharpoons H_3O^+(aq) + H_2A^-(aq)$
stoichiometric point 1	H_2A^-	$H_2A^-(aq) + H_2O(l) \rightleftharpoons H_3O^+(aq) + HA^{2-}(aq)$
		$H_2A^-(aq) + H_2O(l) \rightleftharpoons H_3A(aq) + OH^-(aq)$
buffer region 2	H_2A^-, HA^{2-}	$H_2A^-(aq) + H_2O(l) \rightleftharpoons H_3O^+(aq) + HA^{2-}(aq)$
stoichiometric point 2	HA^{2-}	$HA^{2-}(aq) + H_2O(l) \rightleftharpoons H_3O^+(aq) + A^{3-}(aq)$
		$HA^{2-}(aq) + H_2O(l) \rightleftharpoons H_2A^-(aq) + OH^-(aq)$
buffer region 3	HA^{2-}, A^{3-}	$HA^{2-}(aq) + H_2O(l) \rightleftharpoons H_3O^+(aq) + A^{3-}(aq)$
stoichiometric point 3	A^{3-}	$A^{3-}(aq) + H_2O(l) \rightleftharpoons OH^-(aq) + HA^{2-}(aq)$

We know that the amount of H_3PO_4 in the analyte is $(10.00 \times 10^{-3}\ L) \times (0.100\ mol{\cdot}L^{-1}) = 1.00\ mmol$. To reach the first stoichiometric point, we need an equal amount of NaOH, which is contained in the following volume of solution:

$$\frac{1.00 \times 10^{-3}\ \text{mol}}{0.100\ \text{mol}{\cdot}\text{L}^{-1}} = 10.00 \times 10^{-3}\ \text{L} = 10.00\ \text{mL}$$

At this point, all the acid has reacted, leaving 1.00 mmol $H_2PO_4^-$ in solution.

Between the first and second stoichiometric points The $H_2PO_4^-$ anion reacts with additional base until the second stoichiometric point is reached:

$$H_2PO_4^-(aq) + OH^-(aq) \longrightarrow HPO_4^{2-}(aq) + H_2O(l) \qquad \text{(B)}$$

Because 1 mol OH^- $\simeq$ 1 mol $H_2PO_4^-$, an additional 1.00 mmol NaOH(aq) is required. The volume of base needed to react with all the $H_2PO_4^-$ and reach the second stoichiometric point is 10.00 mL, so a total of 20.0 mL NaOH(aq) is needed to reach that point.

Between the second and third stoichiometric points At this stage, 1.00 mmol HPO_4^{2-} has formed, which reacts further with the base:

$$HPO_4^{2-}(aq) + OH^-(aq) \longrightarrow PO_4^{3-}(aq) + H_2O(l) \qquad \text{(C)}$$

An additional 1.00 mmol NaOH(aq) is required to reach the third stoichiometric point, so another 10.00 mL base must be added, for a total of 30.00 mL. After this stoichiometric point, the primary solute species are Na^+ and OH^- ions because the phosphate ions are increasingly diluted by the added base.

We can tell from the volume of NaOH(aq) added what stage of the titration has been reached. (a) At this point, 10.00 mL of 0.100 M NaOH(aq) has been added. The reaction of H_3PO_4 and NaOH has gone to completion, and we have reached the first stoichiometric point. The primary solute species is the amphiprotic salt, NaH_2PO_4 and the principal proton transfer equilibria are

$$H_2PO_4^-(aq) + H_2O(l) \rightleftharpoons H_3O^+(aq) + HPO_4^{2-}(aq)$$

$$H_2PO_4^-(aq) + H_2O(l) \rightleftharpoons H_3PO_4(aq) + OH^-(aq)$$

(b) At this point, 12.00 mL of 0.100 M NaOH(aq) has been added, which puts the solution between the first and second stoichiometric points. We assume that all the H_3PO_4 in the analyte has now been converted to $H_2PO_4^-$. The additional NaOH reacts with the $H_2PO_4^-$ to form HPO_4^{2-}, but some $H_2PO_4^-$ remains unreacted. Because the titration is between stoichiometric points, with appreciable concentrations of both $H_2PO_4^-$ and its conjugate base, the primary solute species are Na^+ from the base, $H_2PO_4^-$, and HPO_4^{2-}. The proton transfer equilibrium to study is

$$H_2PO_4^-(aq) + H_2O(l) \rightleftharpoons H_3O^+(aq) + HPO_4^{2-}(aq)$$

(c) After the addition of 26.00 mL NaOH, the titration is between the second and third stoichiometric points. The primary acid species in solution is now HPO_4^{2-} and the primary base species is PO_4^{3-}. The third solute species present is Na^+ from the added base. The principal proton transfer equilibrium is

$$HPO_4^{2-}(aq) + H_2O(l) \rightleftharpoons H_3O^+(aq) + PO_4^{3-}(aq)$$

Self-Test 11.11A A 25.00-mL sample of 0.200 M H_3PO_4(aq) was titrated with 0.100 M NaOH(aq). Identify the primary species in solution and the principal proton transfer equilibrium in solution after the addition of the following volumes of NaOH solution: (a) 70.00 mL; (b) 100.00 mL.

[*Answer:* (a) Between first and second stoichiometric points, Na^+ from the base, $H_2PO_4^-$, and HPO_4^{2-}, $H_2PO_4^-(aq) + H_2O(l) \rightleftharpoons H_3O^+(aq) + HPO_4^-(aq)$; (b) at the second stoichiometric point, Na_2HPO_4; $HPO_4^{2-}(aq) + H_2O(l) \rightleftharpoons H_3O^+(aq) + PO_4^{3-}(aq)$ and $HPO_4^{2-}(aq) + H_2O(l) \rightleftharpoons H_2PO_4^-(aq) + OH^-(aq)$.]

Self-Test 11.11B A 20.00-mL sample of 0.100 M H_2S(aq) was titrated with 0.300 M NaOH(aq). Identify the primary species in solution and the principal proton transfer equilibrium after the addition of the following volumes of NaOH solution: (a) 5.00 mL; (b) 13.34 mL.

SOLUBILITY EQUILIBRIA

Up to this point, we have focused only on aqueous equilibria involving proton transfer. Now we shall see how the same principles can be applied to the equilibrium that exists between a solid salt and its dissolved ions in a saturated solution. We can use the equilibrium constant for the dissolution of a substance to predict the solubility of a salt and to control precipitate formation. These methods are used in the laboratory to separate and analyze mixtures of salts. They also have important practical applications in municipal wastewater treatment and in the extraction of minerals from seawater.

11.10 The Solubility Product

The equilibrium constant for a solubility equilibrium between a solid and its dissolved form is called the **solubility product**, K_{sp}, of the solute. For example, the solubility product for bismuth sulfide, Bi_2S_3, is defined as

K_{sp} is also called the *solubility product constant* and, most simply, the *solubility constant*.

$$Bi_2S_3(s) \rightleftharpoons 2\ Bi^{3+}(aq) + 3\ S^{2-}(aq) \qquad K_{sp} = a_{Bi^{3+}}{}^2\, a_{S^{2-}}{}^3$$

Because the concentrations of ions in a solution of a sparingly soluble salt are small, we can approximate K_{sp} by

$$K_{sp} = [Bi^{3+}]^2[S^{2-}]^3$$

Solid Bi_2S_3 does not appear in the expression for K_{sp} because it is a pure solid and its activity is 1 (Section 9.2). A solubility product is used in the same way as any other equilibrium constant. However, because ion-ion interactions in even dilute electrolyte solutions can complicate its interpretation, a solubility product is generally meaningful only for sparingly soluble salts. Another complication that arises when dealing with almost insoluble compounds is that dissociation of the ions is rarely complete, and a saturated solution of PbI_2, for instance, contains substantial amounts of PbI^+ and PbI_2 ion clusters. At best, the calculations we are about to describe are only estimates.

One of the simplest ways to determine K_{sp} is to measure the **molar solubility** of the compound (the molarity of the compound in a saturated solution), but more advanced and accurate methods are also available. Table 11.6 gives some experimental values. In the following calculations, we use s to denote the numerical value of the molar solubility expressed in moles per liter; for example, if the molar solubility of a compound is 6.5×10^{-5} mol·L^{-1}, then we write $s = 6.5 \times 10^{-5}$.

One of these methods is electrochemical, as described in Section 12.9.

The solubility product is the equilibrium constant for the equilibrium between an undissolved salt and its ions in a saturated solution.

TABLE 11.6 Solubility Products at 25°C

Compound	Formula	K_{sp}
aluminum hydroxide	$Al(OH)_3$	1.0×10^{-33}
antimony sulfide	Sb_2S_3	1.7×10^{-93}
barium carbonate	$BaCO_3$	8.1×10^{-9}
fluoride	BaF_2	1.7×10^{-6}
sulfate	$BaSO_4$	1.1×10^{-10}
bismuth sulfide	Bi_2S_3	1.0×10^{-97}
calcium carbonate	$CaCO_3$	8.7×10^{-9}
fluoride	CaF_2	4.0×10^{-11}
hydroxide	$Ca(OH)_2$	5.5×10^{-6}
sulfate	$CaSO_4$	2.4×10^{-5}
chromium(III) iodate	$Cr(IO_3)_3$	5.0×10^{-6}
copper(I) bromide	CuBr	4.2×10^{-8}
chloride	CuCl	1.0×10^{-6}
iodide	CuI	5.1×10^{-12}
sulfide	Cu_2S	2.0×10^{-47}
copper(II) iodate	$Cu(IO_3)_2$	1.4×10^{-7}
oxalate	$Cu(C_2O_4)$	2.9×10^{-8}
sulfide	CuS	1.3×10^{-36}
iron(II) hydroxide	$Fe(OH)_2$	1.6×10^{-14}
sulfide	FeS	6.3×10^{-18}
iron(III) hydroxide	$Fe(OH)_3$	2.0×10^{-39}
lead(II) bromide	$PbBr_2$	7.9×10^{-5}
chloride	$PbCl_2$	1.6×10^{-5}
fluoride	PbF_2	3.7×10^{-8}
iodate	$Pb(IO_3)_2$	2.6×10^{-13}
iodide	PbI_2	1.4×10^{-8}
sulfate	$PbSO_4$	1.6×10^{-8}
sulfide	PbS	8.8×10^{-29}
magnesium		
ammonium phosphate	$MgNH_4PO_4$	2.5×10^{-13}
carbonate	$MgCO_3$	1.0×10^{-5}
fluoride	MgF_2	6.4×10^{-9}
hydroxide	$Mg(OH)_2$	1.1×10^{-11}
mercury(I) chloride	Hg_2Cl_2	2.6×10^{-18}
iodide	Hg_2I_2	1.2×10^{-28}
mercury(II) sulfide, black	HgS	1.6×10^{-52}
sulfide, red	HgS	1.4×10^{-53}
nickel(II) hydroxide	$Ni(OH)_2$	6.5×10^{-18}
silver bromide	AgBr	7.7×10^{-13}
carbonate	Ag_2CO_3	6.2×10^{-12}
chloride	AgCl	1.6×10^{-10}
hydroxide	AgOH	1.5×10^{-8}
iodide	AgI	8×10^{-17}
sulfide	Ag_2S	6.3×10^{-51}
zinc hydroxide	$Zn(OH)_2$	2.0×10^{-17}
sulfide	ZnS	1.6×10^{-24}

EXAMPLE 11.10 Determining the solubility product

The molar solubility of silver chromate, Ag_2CrO_4, is 6.5×10^{-5} mol·L^{-1}. Determine the value of K_{sp}.

STRATEGY First, we write the chemical equation for the equilibrium and the expression for the solubility product. To evaluate K_{sp}, we need to know the molarity of each type of ion formed by the salt. We can determine the molarities from the molar solubility, the chemical equation for the equilibrium, and the stoichiometric relations between the species. We assume complete dissociation.

SOLUTION The chemical equation and the expression for the solubility product are

$$Ag_2CrO_4(s) \rightleftharpoons 2\ Ag^+(aq) + CrO_4^{2-}(aq) \qquad K_{sp} = [Ag^+]^2[CrO_4^{2-}]$$

Because 2 mol $Ag^+ \simeq$ 1 mol Ag_2CrO_4 and 1 mol $CrO_4^{2-} \simeq$ 1 mol Ag_2CrO_4,

$$[Ag^+] = 2s = 2 \times (6.5 \times 10^{-5}) \qquad [CrO_4^{2-}] = s = 6.5 \times 10^{-5}$$

Therefore,

$$K_{sp} = [Ag^+]^2[CrO_4^{2-}] = (2s)^2(s) = 4s^3 = 4 \times (6.5 \times 10^{-5})^3 = 1.1 \times 10^{-12}$$

Self-Test 11.12A The molar solubility of lead(II) iodate, $Pb(IO_3)_2$, at 25°C is 4.0×10^{-5} mol·L^{-1}. What is the value of K_{sp} for lead(II) iodate?

[*Answer:* 2.6×10^{-13}]

Self-Test 11.12B The molar solubility of silver bromide, AgBr, at 25°C is 8.8×10^{-7} mol·L^{-1}. What is the value of K_{sp} for silver bromide?

EXAMPLE 11.11 Estimating the molar solubility from the solubility product

According to Table 11.6, $K_{sp} = 5.0 \times 10^{-6}$ for chromium(III) iodate in water. Estimate the molar solubility of the compound.

STRATEGY As usual, we begin by writing the chemical equation for the solubility equilibrium and the expression for K_{sp}. The molar solubility is the molarity of formula units in the saturated solution. Because each formula unit produces a known number of cations and anions in solution, we can express the molarities of the cations and anions in terms of s. Then we express K_{sp} in terms of s and solve for s. Assume complete dissociation.

SOLUTION The solubility equilibrium is

$$Cr(IO_3)_3(s) \rightleftharpoons Cr^{3+}(aq) + 3\ IO_3^-(aq) \qquad K_{sp} = [Cr^{3+}][IO_3^-]^3$$

Because 1 mol $Cr(IO_3)_3 \simeq$ 1 mol Cr^{3+} and 1 mol $Cr(IO_3)_3 \simeq$ 3 mol IO_3^-, we can write $[Cr^{3+}] = s$ and $[IO_3^-] = 3s$. It follows that

$$K_{sp} = [Cr^{3+}][IO_3^-]^3 = s \times (3s)^3 = 27s^4$$

This expression is easily solved for s:

$$s = \left(\frac{1}{27} K_{sp}\right)^{1/4}$$

Now we substitute the numerical value of K_{sp}:

$$s = \left\{\frac{1}{27} \times (5.0 \times 10^{-6})\right\}^{1/4} = 0.021$$

The molar solubility of $Cr(IO_3)_3$ is therefore 0.021 $mol{\cdot}L^{-1}$.

Self-Test 11.13A The solubility product of silver sulfate, Ag_2SO_4, is 1.4×10^{-5}. Estimate the molar solubility of the salt.

[*Answer:* 1.5×10^{-2} $mol{\cdot}L^{-1}$]

Self-Test 11.13B The solubility product of lead(II) fluoride, PbF_2, is 3.7×10^{-8}. Estimate the molar solubility of the salt.

11.11 The Common-Ion Effect

Sometimes we have to precipitate one ion of a sparingly soluble salt. For example, heavy metal ions such as lead and mercury can be removed from municipal wastewater by precipitating them as the hydroxides. However, because the ions are in dynamic equilibrium with the solid salt, some heavy metal ions remain in solution. How can we remove the remaining ions?

We can use Le Chatelier's principle as a guide. This principle tells us that if we add a second salt or an acid that supplies one of the same ions—a common ion—to a saturated solution, then the equilibrium will tend to adjust by decreasing the concentration of the added ions (Fig. 11.17). That is, the solubility of the original salt is decreased, and it precipitates. We can conclude that the addition of excess hydroxide ion to the water supply should precipitate more of the heavy metal ions. The decrease in solubility caused by the presence of a common ion is called the **common-ion effect.**

The excess hydroxide ion is neutralized after filtration of the precipitates.

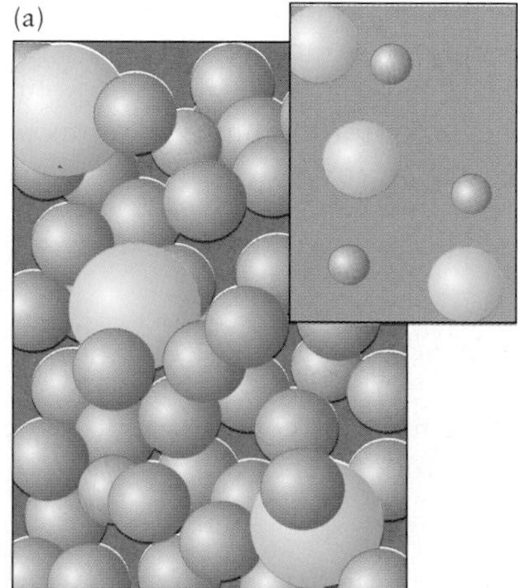

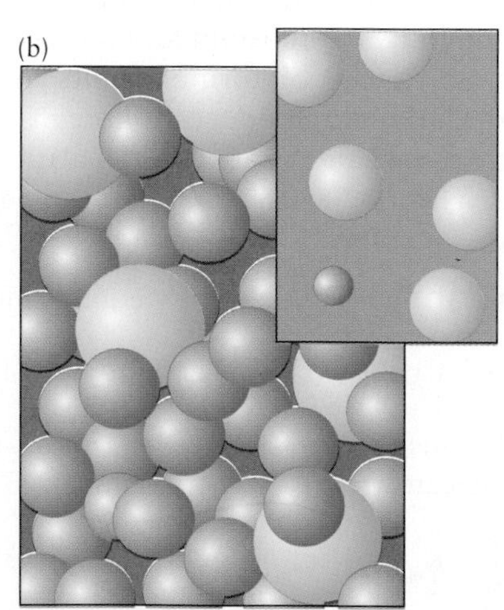

FIGURE 11.17 If the concentration of one of the ions in solution is increased, the concentration of the other decreases to maintain a constant value of K_{sp}. (a) The cations (pink) and anions (green) in solution. (b) When more anions are added (together with their accompanying spectator ions, which are not shown), the concentration of cations decreases. In other words, the solubility of the original compound is reduced by the presence of a common ion. In the insets, the blue background represents the solvent (water).

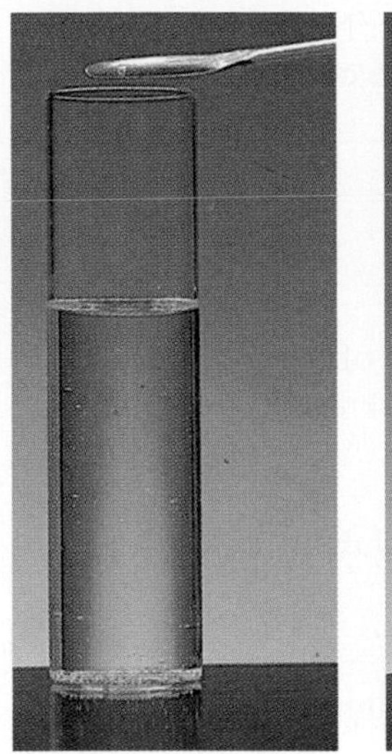
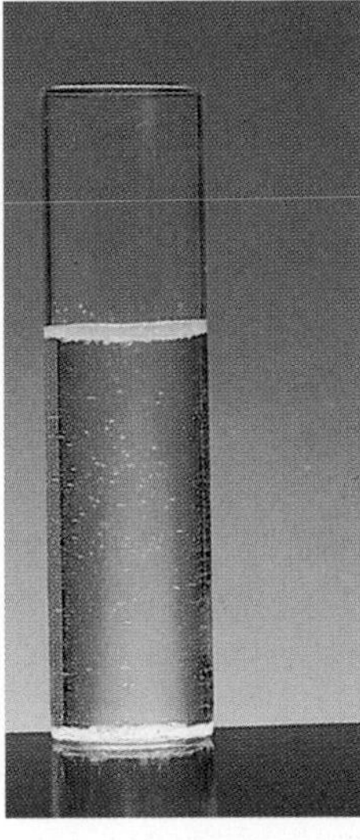

FIGURE 11.18 (a) A saturated solution of zinc acetate in water. (b) When acetate ions are added (as a single crystal of solid sodium acetate in the spatula shown in (a)), the solubility of the zinc acetate is significantly reduced, and more zinc acetate precipitates.

We stressed in Chapter 9 that Le Chatelier's principle is only a rule of thumb. We can work toward a deeper understanding and a quantitative understanding of the common-ion effect by considering the equilibrium constants involved. Suppose we have a saturated solution of silver chloride in water:

$$AgCl(s) \rightleftharpoons Ag^+(aq) + Cl^-(aq) \qquad K_{sp} = [Ag^+][Cl^-]$$

Experimentally, $K_{sp} = 1.6 \times 10^{-10}$ at 25°C, and the molar solubility of AgCl in water is 1.3×10^{-5} mol·L^{-1}. If we add sodium chloride to the solution, the concentration of Cl^- ions increases. For the equilibrium constant to remain constant, the concentration of Ag^+ ions must decrease. Because there is now less Ag^+ in solution, the solubility of AgCl is lower in a solution of NaCl than it is in pure water. A similar effect occurs whenever two salts have an ion in common (Fig. 11.18).

Predicting the magnitude of the common-ion effect quantitatively is difficult. Because ions interact with one another strongly, simple equilibrium calculations are rarely valid: the activities of ions differ markedly from their molarities. However, we can still get an idea of the size of the common-ion effect by rearranging the expression for K_{sp} into an equation for the concentration of the ion other than the common ion.

Removal of an ion has the opposite effect: it serves to dissolve precipitates, as described in Sections 11.14 and 11.15.

The common-ion effect is the reduction in the solubility of a sparingly soluble salt by the addition of a soluble salt that has an ion in common with it.

How can we use this information . . . to estimate the effect of a common ion on solubility?

For a given concentration of Cl^- ions, the concentration of Ag^+ ions must satisfy

$$[Ag^+] = \frac{K_{sp}}{[Cl^-]}$$

Suppose we try to dissolve silver chloride in 0.10 M NaCl(aq), in which $[Cl^-] = 0.10$. The silver chloride dissolves until the concentration of Ag^+ ions is given by

$$[Ag^+] = \frac{1.6 \times 10^{-10}}{0.10} = 1.6 \times 10^{-9}$$

The molarity of Ag^+ ions, and hence the solubility of AgCl formula units, rises only to 1.6×10^{-9} mol·L^{-1}. That is a ten-thousandfold decrease from the solubility of AgCl in pure water.

Self-Test 11.14A What is the approximate molar solubility of calcium carbonate in 0.20 M $CaCl_2$(aq)?

[*Answer:* 4.4×10^{-8} mol·L^{-1}]

Self-Test 11.14B What is the approximate molar solubility of silver bromide in 0.10 M $CaBr_2$(aq)?

11.12 Predicting Precipitation

Sometimes it is important to know when a precipitate will form. For example, if we are analyzing a mixture of ions, we may want to precipitate only one type of ion to separate it from the mixture. In Section 9.7, we saw how to predict the direction in which a reaction would occur by examining the relative values of Q, the reaction quotient, and K, the equilibrium constant. Exactly the same techniques can be used to decide whether a precipitate is likely to form from the ions in two electrolyte solutions that are mixed. In this case, the equilibrium constant is the solubility product, K_{sp}, and the reaction quotient is denoted Q_{sp}. Precipitation occurs when Q_{sp} is equal to or greater than K_{sp} (Fig. 11.19).

For example, we might want to know whether a precipitate of PbI_2 will form when equal volumes of 0.2 M solutions of lead(II) nitrate and potassium iodide are mixed. The overall equation for the reaction is

$$Pb(NO_3)_2\,(aq) + 2\,KI(aq) \longrightarrow 2\,KNO_3\,(aq) + PbI_2\,(s)$$

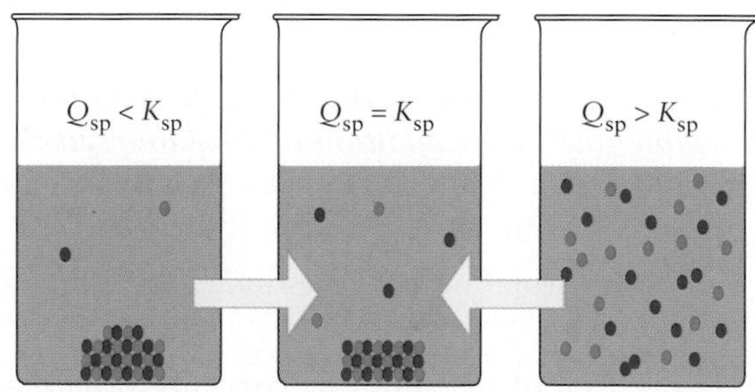

FIGURE 11.19 The relative magnitudes of the solubility quotient, Q_{sp}, and the solubility product K_{sp}, are used to decide whether a salt will precipitate (right) or dissolve (left). Q_{sp} is small when the concentrations of the ions are low (left) and large when the ion concentrations are small (right).

and the net ionic equation for the reaction is

$$Pb^{2+}(aq) + 2\,I^-(aq) \longrightarrow PbI_2(s)$$

The reverse of this equation is the dissolution of PbI_2; and at equilibrium,

$$PbI_2(s) \rightleftharpoons Pb^{2+}(aq) + 2\,I^-(aq) \qquad K_{sp} = [Pb^{2+}][I^-]^2$$

If $Q_{sp} > K_{sp}$ for the dissolution of PbI_2, the reaction will tend to produce the "reactant" PbI_2, so we need to assess the relative values of Q_{sp} and K_{sp}.

A salt precipitates if Q_{sp} is greater than K_{sp}.

Remember the point made earlier, that PbI_2 is not fully dissociated in solution, so these calculations are only estimates.

How can we use this information . . . to predict whether a precipitate will form when two solutions are mixed?

We know from Table 11.6 that $K_{sp} = 1.4 \times 10^{-8}$ for PbI_2 at 25°C. Suppose we mix the two 0.2 M solutions of $Pb(NO_3)_2$ and KI described above. The concentrations of the Pb^{2+} and I^- ions are high, and we can suspect that precipitation will be spontaneous. To verify this prediction, we note that, because equal volumes are mixed together, the new volume is twice each initial volume, so the new molarities are half their original values: 0.1 mol·L^{-1} for $Pb^{2+}(aq)$ and 0.1 mol·L^{-1} for $I^-(aq)$. Therefore, making the approximation that the salt is fully dissociated,

$$Q_{sp} = [Pb^{2+}][I^-]^2 = 0.1 \times (0.1)^2 = 1 \times 10^{-3}$$

This value is considerably higher than K_{sp}, so a precipitate forms (Fig. 11.20).

Sometimes the solid forms as particles so tiny that they form a colloidal dispersion (Section 15.13). The dispersion can often be precipitated by adding a soluble salt to increase the concentration of dissolved ions.

Self-Test 11.15A Does a precipitate of silver chloride form when 200 mL of 1.0×10^{-4} M $AgNO_3(aq)$ and 900 mL of 1.0×10^{-6} M KCl(aq) are mixed? Assume complete dissociation.

[*Answer:* No ($Q_{sp} = 1.5 \times 10^{-11} < K_{sp}$)]

Self-Test 11.15B Does a precipitate of barium fluoride form when 100 mL of 1.0×10^{-3} M $Ba(NO_3)_2(aq)$ is mixed with 200 mL of 1.0×10^{-3} M KF(aq)? Ignore possible protonation of F^-.

Media Link **Lab video:** Fig. 11.20, Formation of lead iodide

FIGURE 11.20 When a few drops of lead(II) nitrate solution are added to a solution of potassium iodide, yellow lead(II) iodide immediately precipitates.

11.13 Selective Precipitation

It is sometimes possible to separate different cations in a solution by adding a soluble salt containing an anion with which they form insoluble salts. For a good separation, the salts formed must differ greatly in their solubility. For example, seawater is a mixture of many different ions. It is possible to precipitate magnesium ion from seawater by adding hydroxide ion. However, other cations are also present in seawater. Their individual concentrations and the relative solubilities of their hydroxides determine which will precipitate first if a certain amount of hydroxide is added. Example 11.12 illustrates a strategy for predicting the order of precipitation.

A mixture of ions in solution can be separated by adding an anion with which they form salts having very different solubilities.

EXAMPLE 11.12 Predicting the order of precipitation

A sample of seawater contains, among other solutes, the following concentrations of soluble cations: 0.050 mol·L^{-1} Mg^{2+}(aq) and 0.010 mol·L^{-1} Ca^{2+}(aq). Use the information in Table 11.6 to determine the order in which each ion precipitates as solid NaOH is added, and give the concentration of OH^- when precipitation of each begins. Assume no volume change on addition of the NaOH.

STRATEGY A salt begins to precipitate when the concentrations of its ions are such that $Q_{sp} > K_{sp}$. Calculate the value of $[OH^-]$ required for each salt to precipitate by writing the expression for K_{sp} for each salt and then use the data provided to determine $[OH^-]$. The values of K_{sp} are found in Table 11.6.

SOLUTION For $Ca(OH)_2$, $K_{sp} = 5.5 \times 10^{-6} = [Ca^{2+}][OH^-]^2$. Therefore,

$$[OH^-] = \left(\frac{K_{sp}}{[Ca^{2+}]}\right)^{1/2} = \left(\frac{5.5 \times 10^{-6}}{0.010}\right)^{1/2} = 0.023$$

In other words, a hydroxide ion concentration of 0.023 mol·L^{-1} is required for calcium hydroxide to precipitate. For $Mg(OH)_2$, $K_{sp} = 1.1 \times 10^{-11} = [Mg^{2+}][OH^-]^2$. Therefore,

$$[OH^-] = \left(\frac{K_{sp}}{[Mg^{2+}]}\right)^{1/2} = \left(\frac{1.1 \times 10^{-11}}{0.050}\right)^{1/2} = 1.5 \times 10^{-5}$$

Magnesium hydroxide will begin to precipitate at a hydroxide ion concentration of 1.5×10^{-5} mol·L^{-1}. We conclude that the hydroxides precipitate in the order $Mg(OH)_2$ at 1.5×10^{-5} mol·L^{-1} OH^-(aq) and $Ca(OH)_2$ at 0.023 mol·L^{-1} OH^-(aq).

Self-Test 11.16A Potassium carbonate is added to a solution containing the following concentrations of soluble cations: 0.030 mol·L^{-1} Mg^{2+}(aq) and 0.0010 mol·L^{-1} Ca^{2+}(aq). Use the information in Table 11.6 to determine the order in which each ion precipitates as the concentration of K_2CO_3 is increased and give the concentration of CO_3^{2-} when precipitation of each begins.

[*Answer:* $CaCO_3$ precipitates first, at 8.7×10^{-6} mol·L^{-1} CO_3^{2-}, then $MgCO_3$ at 3.3×10^{-4} mol·L^{-1} CO_3^{2-}]

Self-Test 11.16B Chloride ion is added to a solution containing the following concentrations of soluble salts: 0.020 mol·L^{-1} $Pb(NO_3)_2$(aq) and 0.0010 mol·L^{-1} $AgNO_3$(aq). Use the information in Table 11.6 to determine the order in which each ion precipitates as the concentration of chloride ion is increased and give the concentration of Cl^- when precipitation of each begins.

11.14 Dissolving Precipitates

When a precipitate has been formed in the qualitative analysis of a solution, it may be necessary to dissolve it again to identify the cation or anion. One strategy is to remove one of the ions from the solubility equilibrium so that the precipitate will continue to dissolve in a fruitless chase for equilibrium. Suppose, for example, that a solid hydroxide such as iron(III) hydroxide is in equilibrium with its ions in solution:

$$Fe(OH)_3(s) \rightleftharpoons Fe^{3+}(aq) + 3\ OH^-(aq)$$

To dissolve more of the solid, we can add acid. The H_3O^+ ions from the acid remove the OH^- ions by converting them into water, and the $Fe(OH)_3$ dissolves

Many carbonate, sulfite, and sulfide precipitates can be dissolved by the addition of acid, because the anions react with the acid to form a gas that bubbles out of solution. For example, in a saturated solution of zinc carbonate, solid $ZnCO_3$ is in equilibrium with its ions:

$$ZnCO_3(s) \rightleftharpoons Zn^{2+}(aq) + CO_3^{2-}(aq)$$

The CO_3^{2-} ions react with acid to form CO_2:

$$CO_3^{2-}(aq) + 2\ HNO_3(aq) \longrightarrow CO_2(g) + H_2O(l) + 2\ NO_3^-(aq)$$

The dissolution of carbonates by acid is an undesired result of acid rain, which has damaged the appearance of many historic marble and limestone monuments (Fig. 11.21; marble and limestone are forms of calcium carbonate).

An alternative procedure for removing an ion from solution is to change its identity by changing its oxidation state. The metal ions in very insoluble heavy metal sulfide precipitates can be dissolved by oxidizing the sulfide ion to elemental sulfur. For example, copper(II) sulfide, CuS, takes part in the equilibrium

$$CuS(s) \rightleftharpoons Cu^{2+}(aq) + S^{2-}(aq)$$

However, when nitric acid is added, the sulfide ions are oxidized to elemental sulfur:

$$3\ S^{2-}(aq) + 8\ HNO_3(aq) \longrightarrow 3\ S(s) + 2\ NO(g) + 4\ H_2O(l) + 6\ NO_3^-(aq)$$

(a)

(b)

FIGURE 11.21 The state of the carving on Cleopatra's Needle has deteriorated as a result of the action of acid rain: (a) after 3500 y in Egypt; (b) after a further 90 y in Central Park, New York City.

This complicated oxidation removes the sulfide ions from the equilibrium, and Cu^{2+} ions dissolve as $Cu(NO_3)_2$.

The solubility of a solid can be increased by removing an ion from solution; acid can be used to dissolve a hydroxide, sulfide, sulfite, or carbonate precipitate, and nitric acid can be used to oxidize metal sulfides to sulfur and a soluble salt.

11.15 Complex Ion Formation

The formation of a complex can also remove an ion and disturb the solubility equilibrium until more solid dissolves. We first met complexes in Section 2.13, where we saw that they are formed by the reaction of a Lewis acid and a Lewis base. In this section, we consider complexes in which the Lewis acid is a metal cation, such as Ag^+. An example is the formation of $Ag(NH_3)_2^+$ when an aqueous solution of the Lewis base ammonia is added to a solution of silver ions:

$$Ag^+(aq) + 2\ NH_3(aq) \longrightarrow Ag(NH_3)_2^+(aq)$$

If a precipitate of silver halide is present and enough ammonia is added, all the precipitate dissolves. A similar procedure is used to remove the silver halide emulsion from an exposed photographic film after it has been developed. In this case, the reagent used to form the complex ion is the thiosulfate ion, $S_2O_3^{2-}$:

$$Ag^+(aq) + 2\ S_2O_3^{2-}(aq) \longrightarrow Ag(S_2O_3)_2^{3-}(aq)$$

Complex formation removes some of the Ag^+ ions from solution. As a result, to preserve the value of K_{sp}, more silver chloride dissolves. We can therefore conclude that the qualitative effect of complex formation is to *increase* the solubility of a sparingly soluble compound.

To treat complex formation quantitatively, we note that complex formation and the dissolving of the salt are both at equilibrium; so we can write

$$AgCl(s) \rightleftharpoons Ag^+(aq) + Cl^-(aq) \qquad K_{sp} = [Ag^+][Cl^-] \qquad \text{(D)}$$

$$Ag^+(aq) + 2\ NH_3(aq) \rightleftharpoons Ag(NH_3)_2^+(aq)$$
$$K_f = \frac{[Ag(NH_3)_2^+]}{[Ag^+][NH_3]^2} \qquad \text{(E)}$$

The equilibrium constant for the formation of a complex ion is called the **formation constant,** K_f. At 25°C, $K_f = 1.6 \times 10^7$ for reaction E. Values for other complexes are given in Table 11.7, and Example 11.13 shows how to use them.

A salt is more soluble if it can form a complex ion.

TABLE 11.7 Formation Constants in Water at 25°C

Equilibrium	K_f
$Ag^+(aq) + 2\ CN^-(aq) \rightleftharpoons Ag(CN)_2^-(aq)$	5.6×10^8
$Ag^+(aq) + 2\ NH_3(aq) \rightleftharpoons Ag(NH_3)_2^+(aq)$	1.6×10^7
$Au^+(aq) + 2\ CN^-(aq) \rightleftharpoons Au(CN)_2^-(aq)$	2.0×10^{38}
$Cu^{2+}(aq) + 4\ NH_3(aq) \rightleftharpoons Cu(NH_3)_4^{2+}(aq)$	1.2×10^{13}
$Hg^{2+}(aq) + 4\ Cl^-(aq) \rightleftharpoons HgCl_4^{2-}(aq)$	1.2×10^5
$Fe^{2+}(aq) + 6\ CN^-(aq) \rightleftharpoons Fe(CN)_6^{4-}(aq)$	7.7×10^{36}
$Ni^{2+}(aq) + 6\ NH_3(aq) \rightleftharpoons Ni(NH_3)_6^{2+}(aq)$	5.6×10^8

EXAMPLE 11.13 Calculating molar solubility in the presence of complex formation

Calculate the molar solubility of silver chloride in 0.10 M $NH_3(aq)$, given that $K_{sp} = 1.6 \times 10^{-10}$ for silver chloride and $K_f = 1.6 \times 10^7$ for the ammonia complex of Ag^+ ions, $Ag(NH_3)_2^+$.

STRATEGY First, we write the chemical equation for the equilibrium between the solid solute and the complex in solution. We find the value of K for this reaction by expressing the equation as the sum of the equations for the solubility and complex formation equilibria. The equilibrium constant for the overall equilibrium is therefore the product of the equilibrium constants for the two contributing processes. Then we set up an equilibrium table and solve for the equilibrium concentrations of ions in solution.

SOLUTION The overall equilibrium is the sum of reactions D and E:

$$AgCl(s) + 2\ NH_3(aq) \rightleftharpoons Ag(NH_3)_2^+(aq) + Cl^-(aq)$$

The equilibrium constant of the overall reaction is the product of the equilibrium constants for the two reactions:

$$K = K_{sp} \times K_f$$

Because 1 mol AgCl $\simeq$ 1 mol Cl^-, the molar solubility of AgCl is given by the equation $s = [Cl^-]$. From the balanced overall equation, we see that $[Cl^-] = [Ag(NH_3)_2^+]$ in the saturated solution. The equilibrium table, with all concentrations in moles per liter, is

	Species		
	NH_3	$Ag(NH_3)_2^+$	Cl^-
Step 1 Initial molarity	0.10	0	0
Step 2 Change in molarity	$-2x$	$+x$	$+x$
Step 3 Equilibrium molarity	$0.10 - 2x$	x	x

Step 4 We find K and substitute the information from step 3 for the equilibrium concentrations:

$$K = K_{sp} \times K_f = (1.6 \times 10^{-10}) \times (1.6 \times 10^{7}) = 2.6 \times 10^{-3}$$

It follows that

$$K = \frac{[Ag(NH_3)_2{}^+][Cl^-]}{[NH_3]^2} = \frac{x \times x}{(0.10 - 2x)^2} = 2.6 \times 10^{-3}$$

Now take the square root of each side:

$$\frac{x}{0.10 - 2x} = 5.1 \times 10^{-2}$$

Rearranging and solving for x gives $x = 4.6 \times 10^{-3}$. Therefore, from step 3, $[Ag(NH_3)_2{}^+] = 4.6 \times 10^{-3}$. Hence, the molar solubility of silver chloride is 4.6×10^{-3} mol·L^{-1}. The molar solubility of silver chloride in pure water is 1.3×10^{-5} mol·L^{-1}, lower by a factor of over 100 than the molar solubility of silver chloride in 0.10 M $NH_3(aq)$.

Self-Test 11.17A Use data from Tables 11.6 and 11.7 to calculate the molar solubility of silver bromide in 1.0 M $NH_3(aq)$.

[*Answer:* 3.5×10^{-3} mol·L^{-1}]

Self-Test 11.17B Use data from Tables 11.6 and 11.7 to calculate the molar solubility of copper(II) sulfide in 1.2 M $NH_3(aq)$.

11.16 Qualitative Analysis

Complex formation, selective precipitation, and control of the pH of a solution all play important roles in qualitative analysis. There are many different schemes of analysis, but they have common features and follow the same general principles. Let's think through a procedure for the identification of several cations following the steps that might be used in the laboratory. We shall see how each step makes sense in terms of solubility equilibria.

Suppose we have a solution that contains lead(II), mercury(I), silver, copper(II), and zinc ions. The method is outlined in Fig. 11.22, which includes additional cations, and is illustrated in Fig. 11.23. Most chlorides are soluble; so when hydrochloric acid is added to a mixture of salts, only certain chlorides precipitate (see Table 11.6). Silver and mercury(I) chlorides have such small values of K_{sp} that, even with low concentrations of Cl^- ions, the chlorides precipitate. Lead(II) chloride, which is slightly soluble, will also precipitate if the chloride ion concentration is high enough. The hydronium ions provided by the acid play no role in this step; they simply accompany the chloride ions. At this point, the precipitate can be separated from the solution by using a centrifuge to compact the solid, then decanting (pouring off) the solution. The solution now contains copper(II) and zinc ions, whereas the solid consists of $PbCl_2$, Hg_2Cl_2, and AgCl.

A large excess of hydrochloric acid must be avoided to prevent the formation of the soluble complex ions $AgCl_2{}^-$ and $PbCl_4{}^{2-}$.

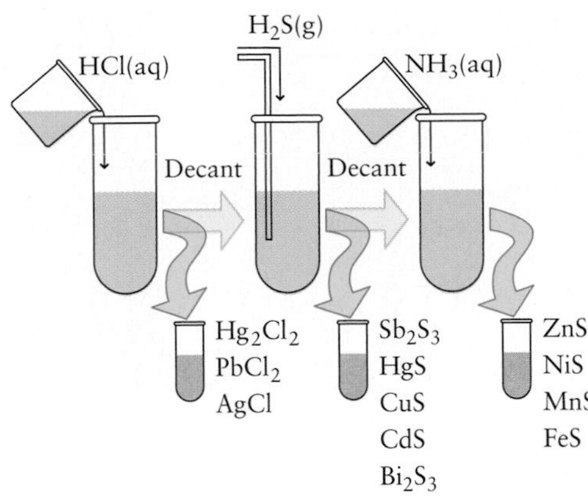

FIGURE 11.22 Part of a simple qualitative analysis scheme used to separate certain cations. In the first step, three cations are separated as insoluble chlorides. In the second step, cations that form highly insoluble sulfides are removed by precipitation; and in the third step, the remaining cations are precipitated as the sulfides.

Because $PbCl_2$ is slightly soluble, if the precipitate is rinsed with hot water, the lead(II) chloride dissolves and the solution is separated from the precipitate. If sodium chromate is then added to that solution, any lead(II) present will precipitate as lead(II) chromate:

$$Pb^{2+}(aq) + CrO_4^{\ 2-}(aq) \longrightarrow PbCrO_4(s)$$

At this point, the silver(I) and mercury(I) chlorides remain as precipitates. When ammonia is added to the solid mixture, the silver precipitate dissolves as a result of the formation of the complex ion $Ag(NH_3)_2^{\ +}$:

$$Ag^{+}(aq) + 2\ NH_3(aq) \longrightarrow Ag(NH_3)_2^{\ +}(aq)$$

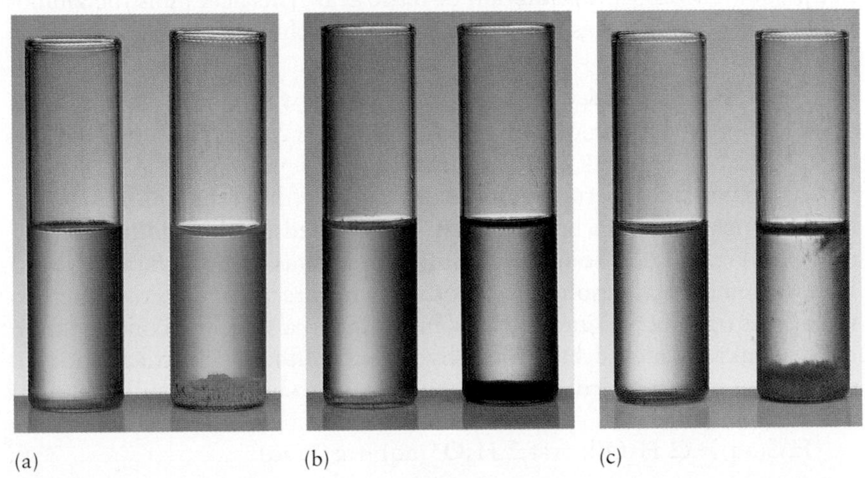

FIGURE 11.23 A sequence for the analysis of cations by selective precipitation. The original solution contains Pb^{2+}, $Hg_2^{\ 2+}$, Ag^{+}, Cu^{2+}, and Zn^{2+} ions. (a) Addition of HCl precipitates AgCl, Hg_2Cl_2, and $PbCl_2$, which can be removed by decanting or filtration. (b) Addition of H_2S to the remaining solution precipitates CuS, which can also be removed. (c) Making the solution basic by adding ammonia precipitates ZnS.

Media Link **Lab Video:** Fig. 11.23, Metal sulfides

FIGURE 11.24 When ammonia is added to a silver chloride precipitate, the precipitate dissolves. However, when ammonia is added to a precipitate of mercury(I) chloride, mercury metal and mercury(II) ions are formed in a redox reaction and the mass turns gray. Left to right: silver chloride in water, silver chloride in aqueous ammonia, mercury(I) chloride in water, and mercury(I) chloride in aqueous ammonia.

and mercury(I) reacts with ammonia to form a gray solid consisting of a mixture of mercury(II) ions precipitated as white $HgNH_2Cl(s)$ and black metallic mercury (Fig. 11.24):

$$Hg_2Cl_2(s) + 2\,NH_3(aq) \longrightarrow Hg(l) + HgNH_2Cl(s) + NH_4^+(aq) + Cl^-(aq)$$

At this point, any Hg_2^{2+} has precipitated and any Ag^+ present is in solution. The solution can be separated from the solid and the presence of silver ion in the solution verified by addition of nitric acid. This acid pulls the ammonia out of the complex as NH_4^+, allowing silver chloride to precipitate:

$$Ag(NH_3)_2^+(aq) + Cl^-(aq) + 2\,H_3O^+(aq) \longrightarrow AgCl(s) + 2\,NH_4^+(aq) + 2\,H_2O(l)$$

At a later stage, certain sulfides with widely different solubilities and solubility products can be selectively precipitated by the addition of S^{2-} ions (see Fig. 11.22). Some metal sulfides (such as CuS, HgS, and Sb_2S_3) have extremely small solubility products and are precipitated if there is the merest trace of S^{2-} ions in the solution. Such a very low concentration of S^{2-} ions is achieved by adding hydrogen sulfide, H_2S, to an acidified solution. A higher hydronium ion concentration shifts the equilibrium

$$H_2S(aq) + 2\,H_2O(l) \rightleftharpoons 2\,H_3O^+(aq) + S^{2-}(aq)$$

to the left, which ensures that almost all the H_2S is in its fully protonated form, and hence that very little S^{2-} is present. Nevertheless, that little will result in the precipitation of highly insoluble solids if the appropriate cations are present.

To verify the presence of Zn^{2+} ions in the solution remaining after the first two steps, we add H_2S followed by ammonia. The base removes the hydronium ion from the H_2S equilibrium, which shifts the equilibrium in favor of S^{2-} ions. The higher concentration of S^{2-} ions increases the Q_{sp} values of any remaining metal sulfides above their K_{sp} values, and they precipitate. This step detects the presence of metal cation sulfides that have larger solubility constants than those in the preceding step, including MnS and ZnS.

Qualitative analysis involves the separation and identification of ions by selective precipitation, complex formation, and the control of pH.

SKILLS YOU SHOULD HAVE MASTERED

1 Calculate the pH of a solution of a weak acid and its salt or a weak base and its salt, Example 11.1.

2 Calculate the pH of a buffer solution, Example 11.2.

3 Calculate the pH change when acid or base is added to a buffer solution, Example 11.3.

4 Specify the composition of a buffer solution with a given pH, Example 11.4.

5 Interpret the features of the pH curve for the titration of a strong or weak acid with a strong base, and a strong or weak base with a strong acid, Sections 11.5 and 11.6.

6 Calculate the pH at any point in a strong base-strong acid titration, Toolbox 11.1 and Examples 11.5 and 11.6.

7 Calculate the pH at any point in a strong base-weak acid and weak base-strong acid titration, Toolbox 11.2 and Examples 11.7 and 11.8.

8 Select an appropriate indicator for a given titration, Section 11.7.

9 Determine the primary species present at any point in the titration of a polyprotic acid with a strong base and write the chemical equation for the principal proton transfer equilibrium, Toolbox 11.3 and Example 11.9.

10 Estimate a solubility product from molar solubility and vice versa, Examples 11.10 and 11.11.

11 Describe the common-ion effect and assess its magnitude, Section 11.11.

12 Predict whether a salt will precipitate, given the concentrations of its ions in water, Section 11.12.

13 Predict the order of precipitation of a series of salts, Example 11.12.

14 Calculate molar solubility in the presence of complex ion formation, Example 11.13.

15 Use a simple qualitative analysis scheme and justify the steps in terms of solubility equilibria.

EXERCISES

K_a and K_b values for weak acids and bases may be found in Tables 10.1 and 10.2.

Mixed Solutions and Buffers

11.1 Explain what happens to (a) the concentration of H_3O^+ ions in an acetic acid solution when solid sodium acetate is added; (b) the percentage deprotonation of benzoic acid in a benzoic acid solution when hydrochloric acid is added; (c) the pH of the solution when solid ammonium chloride is added to an ammonia solution.

11.2 Explain what happens to (a) the pH of a phosphoric acid solution after the addition of solid sodium dihydrogen phosphate; (b) the percentage deprotonation of HCN in a hydrocyanic acid solution after the addition of hydrobromic acid; (c) the concentration of H_3O^+ ions when pyridinium chloride is added to a pyridine solution.

11.3 A solution of equal concentrations of lactic acid and sodium lactate was found to have pH = 3.08. (a) What are the values of pK_a and K_a of lactic acid? (b) What would the pH be if the acid had twice the concentration of the salt?

11.4 What is the concentration of hydronium ions in (a) a solution that is 0.075 M HCN(aq) and 0.045 M NaCN(aq); (b) a solution that is 0.15 M NH_2NH_2(aq) and 0.60 M NaCl(aq); (c) a solution that is 0.025 M HCN(aq) and 0.040 M NaCN(aq); (d) a solution that is 0.125 M NH_2NH_2(aq) and 0.125 M NH_2NH_3Br(aq)?

11.5 Determine the pH and pOH of (a) a solution that is 0.80 M $NaHSO_4$(aq) and 0.40 M Na_2SO_4(aq); (b) a solution that is 0.80 M $NaHSO_4$(aq) and 0.20 M Na_2SO_4(aq); (c) a solution that is 0.80 M $NaHSO_4$(aq) and 0.80 M Na_2SO_4(aq).

11.6 Calculate the pH and pOH of (a) a solution that is 0.17 M Na_2HPO_4(aq) and 0.25 M Na_3PO_4(aq); (b) a solution that is 0.66 M Na_2HPO_4(aq) and 0.42 M Na_3PO_4(aq); (c) a solution that is 0.12 M Na_2HPO_4(aq) and 0.12 M Na_3PO_4(aq).

11.7 The pH of 0.40 M HF(aq) is 1.93. Calculate the change in pH when 0.356 g of sodium fluoride is added to 50.0 mL of the solution. Ignore any volume change.

11.8 The pH of 0.50 M HBrO(aq) is 4.50. Calculate the change in pH when 7.50 g of sodium hypobromite is added to 100 mL of the solution. Ignore any volume change.

11.9 Calculate the pH of the solution that results from mixing (a) 20.0 mL of 0.050 M HCN(aq) with 80.0 mL of 0.030 M NaCN(aq); (b) 80.0 mL of 0.030 M HCN(aq) with 20.0 mL of 0.050 M NaCN(aq); (c) 25.0 mL of 0.105 M HCN(aq) with 25.0 mL of 0.105 M NaCN(aq).

11.10 Calculate the pH of the solution that results from mixing (a) 0.100 L of 0.020 M $(CH_3)_2NH$(aq) with 0.300 L of 0.030 M $(CH_3)_2NH_2Cl$(aq); (b) 65.0 mL of 0.010 M $(CH_3)_2NH$(aq) with 10.0 mL of 0.150 M $(CH_3)_2NH_2Cl$(aq); (c) 50.0 mL of 0.015 M $(CH_3)_2NH$(aq) with 125 mL of 0.015 M $(CH_3)_2NH_2Cl$(aq).

11.11 Sodium hypochlorite, NaClO, is the active ingredient in many bleaches. Calculate the ratio of the concentrations of ClO^- and HClO in a bleach solution having a pH adjusted to 6.50 by using strong acid or strong base.

11.12 Aspirin (shown below as acetylsalicylic acid) is a product of the reaction of salicylic acid with acetic anhydride, $CH_3{-}\overset{O}{\overset{\|}{C}}{-}O{-}\overset{O}{\overset{\|}{C}}{-}CH_3$. It has $K_a = 3.2 \times 10^{-4}$. Calculate the ratio of the concentrations of the acetylsalicylate ion to acetylsalicylic acid in a solution that has a pH adjusted to 4.67 by using strong acid or strong base.

COOH, OH — Salicylic acid

COOH, O—C(=O)—CH_3 — Acetylsalicylic acid

11.13 Predict the pH region in which each of the following buffers will be effective, assuming equal molarities of the acid and its conjugate base: (a) sodium lactate and lactic acid; (b) sodium benzoate and benzoic acid; (c) potassium hydrogen phosphate and potassium phosphate; (d) potassium hydrogen phosphate and potassium dihydrogen phosphate; (e) hydroxylamine and hydroxylammonium chloride.

11.14 Predict the pH region in which each of the following buffers will be effective, assuming equal molarities of the acid and its conjugate base: (a) sodium nitrite and nitrous acid; (b) sodium formate and formic acid; (c) sodium carbonate and sodium hydrogen carbonate; (d) ammonia and ammonium chloride; (e) pyridine and pyridinium chloride.

11.15 Use Tables 10.1, 10.2, and 10.9 to suggest a conjugate acid-base system that would be an effective buffer at a pH close to (a) 2; (b) 7; (c) 3; (d) 12.

11.16 Use Tables 10.1, 10.2, and 10.9 to suggest a conjugate acid-base system that would be an effective buffer at a pH close to (a) 4; (b) 9; (c) 5; (b) 11.

11.17 (a) What must be the ratio of the concentrations of CO_3^{2-} and HCO_3^- ions in a buffer solution having a pH of 11.0? (b) What mass of K_2CO_3 must be added to 1.00 L of 0.100 M $KHCO_3$(aq) to prepare a buffer solution with a pH of 11.0? (c) What mass of $KHCO_3$ must be added to 1.00 L of 0.100 M K_2CO_3(aq) to prepare a buffer solution with a pH of 11.0? (d) What volume of 0.200 M $K_2CO_3(aq)$ must be added to 100 mL of 0.100 M $KHCO_3(aq)$ to prepare a buffer solution with a pH of 11.0?

11.18 (a) What must be the ratio of the molarities of PO_4^{3-} and HPO_4^{2-} ions in a buffer solution having a pH of 12.0? (b) What mass of K_3PO_4 must be added to 1.00 L of 0.100 M $K_2HPO_4(aq)$ to prepare a buffer solution with a pH of 12.0? (c) What mass of K_2HPO_4 must be added to 1.00 L of 0.100 M $K_3PO_4(aq)$ to prepare a buffer solution with a pH of 12.0? (d) What volume of 0.150 M $K_3PO_4(aq)$ must be added to 50.0 mL of 0.100 M $K_2HPO_4(aq)$ to prepare a buffer solution with a pH of 12.0?

11.19 A 100-mL buffer solution is 0.100 M $CH_3COOH(aq)$ and 0.100 M $NaCH_3CO_2(aq)$. (a) What are the pH and the pH change resulting from the addition of 10.0 mL of 0.950 M NaOH(aq) to the buffer solution? (b) What are the pH and the pH change resulting from the addition of 20.0 mL of 0.100 M $HNO_3(aq)$ to the initial buffer solution?

11.20 A 100-mL buffer solution is 0.150 M $Na_2HPO_4(aq)$ and 0.100 M $KH_2PO_4(aq)$. (a) What are the pH and the pH change resulting from the addition of 80.0 mL of 0.0100 M NaOH(aq) to the buffer solution? (b) What are the pH and the pH change resulting from the addition of 10.0 mL of 1.0 M $HNO_3(aq)$ to the initial buffer solution?

Titrations

11.21 Sketch reasonably accurately the pH curve for the titration of 20.0 mL of 0.10 M HCl(aq) with 0.20 M KOH(aq). Mark on the curve (a) the initial pH; (b) the pH at the stoichiometric point.

11.22 Sketch reasonably accurately the pH curve for the titration of 20.0 mL of 0.10 M $Ba(OH)_2(aq)$ with 0.20 M HCl(aq). Mark on the curve (a) the initial pH; (b) the pH at the stoichiometric point.

11.23 Calculate the volume of 0.150 M HCl(aq) required to neutralize (a) one-half and (b) all the hydroxide ions in 25.0 mL of 0.110 M NaOH(aq). (c) What is the molarity of Na^+ ions at the stoichiometric point? (d) Calculate the pH of the solution after the addition of 20.0 mL of 0.150 M HCl(aq) to 25.0 mL of 0.110 M NaOH(aq).

11.24 Calculate the volume of 0.116 M HCl(aq) required to neutralize (a) one-half and (b) all the hydroxide ions in 25.0 mL of 0.215 M KOH(aq). (c) What is the molarity of Cl^- ions at the stoichiometric point? (d) Calculate the pH of the solution after the addition of 40.0 mL of 0.116 M HCl(aq) to 25.0 mL of 0.215 M KOH(aq).

11.25 (a) Suppose that we dissolved 1.33 g of NaOH in 50.0 mL of solution and titrated it with HCl(aq). What volume (in milliliters) of 0.135 M HCl(aq) is required to reach the stoichiometric point? (b) What is the molarity of Cl^- ions at the stoichiometric point?

11.26 (a) Suppose that we dissolved 2.88 g of KOH in 25.0 mL of solution and titrated it with $HNO_3(aq)$. What volume of 0.200 M $HNO_3(aq)$ is required to reach the stoichiometric point? (b) What is the molarity of NO_3^- ions at the stoichiometric point?

11.27 A 1.592-g sample of impure sodium hydroxide was dissolved in 300 mL of aqueous solution. A 25.0-mL portion of this solution was titrated to the stoichiometric point with 40.6 mL of 0.0695 M HCl(aq). What is the percentage purity of the original sample?

11.28 A 1.331-g sample of impure barium hydroxide was dissolved in enough water to produce 250 mL of solution. A 35.0-mL portion of this solution was titrated to the stoichiometric point with 17.6 mL of 0.0935 M HCl(aq). What is the percentage purity of the original sample?

11.29 Calculate the pH at each stage in the titration for the addition of 0.150 M HCl(aq) to 25.0 mL of 0.110 M NaOH(aq) (a) initially; (b) after the addition of 5.0 mL of acid; (c) after the addition of a further 5.0 mL; (d) at the stoichiometric point; (e) after the addition of 5.0 mL of acid beyond the stoichiometric point; (f) after the addition of 10 mL of acid beyond the stoichiometric point.

11.30 Calculate the pH at each stage in the titration in which 0.116 M HCl(aq) is added to 25.0 mL of 0.215 M KOH(aq) (a) initially; (b) after the addition of 5.0 mL of acid; (c) after the addition of a further 5.0 mL; (d) at the stoichiometric point; (e) after the addition of 5.0 mL of acid beyond the stoichiometric point; (f) after the addition of 10 mL of acid beyond the stoichiometric point.

11.31 A 25.0-mL sample of 0.10 M $CH_3COOH(aq)$ is titrated with 0.10 M NaOH(aq). (a) What is the initial pH of the 0.10 M $CH_3COOH(aq)$ solution? (b) What is the pH after the addition of 10.0 mL of 0.10 M NaOH(aq)? (c) What volume of

0.10 M NaOH(aq) is required to reach halfway to the stoichiometric point? (d) Calculate the pH at that halfway point. (e) What volume of 0.10 M NaOH(aq) is required to reach the stoichiometric point? (f) Calculate the pH at the stoichiometric point.

11.32 A 30.0-mL sample of 0.20 M $C_6H_5COOH(aq)$ solution is titrated with 0.30 M KOH(aq). (a) What is the initial pH of the 0.20 M $C_6H_5COOH(aq)$ solution? (b) What is the pH after the addition of 15.0 mL of 0.30 M KOH(aq)? (c) What volume of 0.30 M KOH(aq) is required to reach halfway to the stoichiometric point? (d) Calculate the pH at the halfway point. (e) What volume of 0.30 M KOH(aq) is required to reach the stoichiometric point? (f) Calculate the pH at the stoichiometric point.

11.33 A 15.0-mL sample of 0.15 M $NH_3(aq)$ is titrated with 0.10 M HCl(aq). (a) What is the initial pH of the 0.15 M $NH_3(aq)$? (b) What is the pH after the addition of 15.0 mL of 0.10 M HCl(aq)? (c) What volume of 0.10 M HCl(aq) is required to reach halfway to the stoichiometric point? (d) Calculate the pH at the halfway point. (e) What volume of 0.10 M HCl(aq) is required to reach the stoichiometric point? (f) Calculate the pH at the stoichiometric point.

11.34 A 50.0-mL sample of 0.25 M $CH_3NH_2(aq)$ is titrated with 0.35 M HCl(aq). (a) What is the initial pH of the 0.25 M $CH_3NH_2(aq)$? (b) What is the pH after the addition of 15.0 mL of 0.35 M HCl(aq)? (c) What volume of 0.35 M HCl(aq) is required to reach halfway to the stoichiometric point? (d) Calculate the pH at the halfway point. (e) What volume of 0.35 M HCl(aq) is required to reach the stoichiometric point? (f) Calculate the pH at the stoichiometric point.

For the pH ranges over which common indicators change color, see Table 11.3.

11.35 Which indicators could you use for a titration of 0.20 M $CH_3COOH(aq)$ with 0.20 M NaOH(aq): (a) methyl orange; (b) litmus; (c) thymol blue; (d) phenolphthalein? Explain your selections.

11.36 Which indicators could you use for a titration of 0.20 M $NH_3(aq)$ with 0.20 M HCl(aq): (a) bromocresol green; (b) methyl red; (c) phenol red; (d) thymol blue? Explain your selections.

11.37 Suggest suitable indicators for the titrations described in Exercises 11.31 and 11.33.

11.38 Suggest suitable indicators for the titrations described in Exercises 11.32 and 11.34.

Polyprotic Acid Titrations

11.39 What volume of 0.275 M KOH(aq) must be added to 75.0 mL of 0.137 M $H_3AsO_4(aq)$ to reach (a) the first stoichiometric point; (b) the second stoichiometric point; (c) the third stoichiometric point?

11.40 What volume of 0.123 M NaOH(aq) must be added to 125 mL of 0.197 M $H_2SO_3(aq)$ to reach (a) the first stoichiometric point; (b) the second stoichiometric point?

11.41 What volume of 0.255 M $HNO_3(aq)$ must be added to 35.5 mL of 0.158 M $Na_2HPO_3(aq)$ to reach (a) the first stoichiometric point; (b) the second stoichiometric point?

11.42 What volume of 0.0848 M HCl(aq) must be added to 88.8 mL of 0.233 M $Na_3PO_4(aq)$ to reach (a) the first stoichiometric point; (b) the second stoichiometric point; (c) the third stoichiometric point?

11.43 A 0.164-g sample of phosphorous acid, H_3PO_3, is dissolved in water, and the total volume of the solution is 50.0 mL. (a) Estimate the pH of this solution. (b) Estimate the pH of the solution that results when 6.50 mL of 0.175 M NaOH(aq) is added to the phosphorous acid solution. (c) Estimate the pH of the solution if an additional 4.93 mL of 0.175 M NaOH(aq) is added to the solution in part (b).

11.44 A 0.135-g sample of oxalic acid, $(COOH)_2$, is dissolved in 25.0 mL of water. (a) Estimate the pH of this solution. (b) Estimate the pH of the solution that results when 10.0 mL of 0.150 M NaOH(aq) is added to the oxalic acid solution. (c) Estimate the pH of the solution that results if an additional 5.00 mL of NaOH(aq) solution is added to the solution from part (b).

11.45 Estimate the pH of the solutions that result when the following solutions are added to 50.0 mL of 0.275 M $Na_2HPO_4(aq)$: (a) 50.0 mL of 0.275 M HCl(aq); (b) 75.0 mL of 0.275 M HCl(aq); (c) 25.0 mL of 0.275 M HCl(aq).

11.46 Estimate the pH of the solutions that result when 75.0 mL of 0.0995 M $Na_2CO_3(aq)$ is mixed with (a) 25.0 mL of 0.130 M $HNO_3(aq)$; (b) 65.0 mL of 0.130 M $HNO_3(aq)$.

Solubility Equilibria

The values for the solubility products of some sparingly soluble salts are listed in Table 11.6.

11.47 Determine K_{sp} for the following sparingly soluble substances, given their molar solubilities: (a) AgBr, 8.8×10^{-7} mol·L^{-1}; (b) $PbCrO_4$, 1.3×10^{-7} mol·L^{-1}; (c) $Ba(OH)_2$, 0.11 mol·L^{-1}; (d) MgF_2, 1.2×10^{-3} mol·L^{-1}.

11.48 Determine K_{sp} for the following sparingly soluble compounds, given their molar solubilities: (a) AgI, 9.1×10^{-9} mol·L^{-1}; (b) $Ca(OH)_2$, 0.011 mol·L^{-1}; (c) Ag_3PO_4, 2.7×10^{-6} mol·L^{-1}; (d) Hg_2Cl_2, 5.2×10^{-7} mol·L^{-1}.

11.49 Calculate the molar solubility in water of (a) Ag_2S; (b) CuS; (c) $CaCO_3$.

11.50 Determine the molar solubility in water of (a) $PbSO_4$; (b) Ag_2CO_3; (c) $Fe(OH)_2$.

11.51 The molarity of CrO_4^{2-} in a saturated Tl_2CrO_4 solution is 6.3×10^{-5} mol·L^{-1}. What is the K_{sp} of Tl_2CrO_4?

11.52 The molar solubility of cerium(III) hydroxide, $Ce(OH)_3$, is 5.2×10^{-6} mol·L^{-1}. What is the K_{sp} of cerium(III) hydroxide?

11.53 Use the data in Table 11.6 to calculate the molar solubility of each sparingly soluble substance in its respective solution: (a) silver chloride in 0.15 M NaCl(aq); (b) mercury(I) chloride in 0.225 M NaCl(aq); (c) lead(II) chloride in 0.050 M $CaCl_2$(aq); (d) iron(II) hydroxide in 2.5×10^{-3} M $FeCl_2$(aq).

11.54 Use the data in Table 11.6 to calculate the solubility of each sparingly soluble substance in its respective solution: (a) silver bromide in 0.050 M NaBr(aq); (b) magnesium carbonate in 1.0×10^{-3} M Na_2CO_3(aq); (c) lead(II) sulfate in 0.25 M Na_2SO_4(aq); (d) nickel hydroxide in 0.125 M $NiSO_4$(aq).

11.55 (a) What molarity of Ag^+ ions is required for the formation of a precipitate when added to 1.0×10^{-5} M NaCl(aq)? (b) What mass (in micrograms) of $AgNO_3$ needs to be added for the onset of precipitation in 100 mL of the solution in (a)?

11.56 It is necessary to add iodide ions to precipitate lead(II) ion from 0.0020 M $Pb(NO_3)_2$ (aq). (a) What minimum iodide ion concentration is required for the onset of PbI_2 precipitation? (b) What mass (in grams) of KI must be added for PbI_2 to form?

11.57 Determine the pH required for the onset of precipitation of $Ni(OH)_2$ from (a) 0.060 M $NiSO_4$(aq); (b) 0.030 M $NiSO_4$(aq).

11.58 Limestone is composed primarily of calcium carbonate. A 1.0-mm^3 chip of limestone was accidentally dropped into a water-filled swimming pool, measuring 10 m × 7 m × 2 m. Assuming that the carbonate ion does not function as a Brønsted base and that the pH of the water is 7, will the pebble dissolve entirely? The density of calcium carbonate is 2.71 g·cm^{-3}.

11.59 Decide whether a precipitate will form when the following solutions are mixed: (a) 27.0 mL of 0.0010 M NaCl(aq) and 73.0 mL of 0.0040 M $AgNO_3$(aq); (b) 1.0 mL of 1.0 M K_2SO_4(aq), 10.0 mL of 0.0030 M $CaCl_2$(aq), and 100 mL of water. Assume the volumes are additive when the solutions are mixed.

11.60 Decide whether a precipitate will form when the following solutions are mixed: (a) 5.0 mL of 0.10 M K_2CO_3(aq) and 1.00 L of 0.010 M $AgNO_3$(aq); (b) 3.3 mL of 1.0 M HCl(aq), 4.9 mL of 0.0030 M $AgNO_3$(aq), and enough water to dilute the solution to 50.0 mL.

11.61 Suppose that there are typically 20 average-sized drops in 1 mL of an aqueous solution. Will a precipitate form when 1 drop of 0.010 M NaCl(aq) is added to 10.0 mL of (a) 0.0040 M $AgNO_3$(aq)? (b) 0.0040 M $Pb(NO_3)_2$(aq)?

11.62 Assuming 20 drops per milliliter, will a precipitate form if (a) 7 drops of 0.0029 M K_2CO_3(aq) are added to 25.0 mL of 0.0018 M $CaCl_2$(aq); (b) 10 drops of 0.010 M Na_2CO_3(aq) are added to 10.0 mL of 0.0040 M $AgNO_3$(aq)?

11.63 The concentrations of magnesium, calcium, and nickel(II) ions in an aqueous solution are 0.0010 mol·L^{-1}. (a) In what order do they precipitate when solid KOH is added? (b) Determine the pH at which each salt precipitates.

11.64 Suppose that two hydroxides, MOH and $M'(OH)_2$, both have $K_{sp} = 1.0 \times 10^{-12}$ and that initially both cations are present in a solution at concentrations of 0.0010 mol·L^{-1}. Which hydroxide precipitates first, and at what pH, when solid NaOH is added?

11.65 We wish to separate magnesium ions and barium ions by selective precipitation. Which anion, fluoride or carbonate, would be the better choice for achieving this precipitation? Why?

11.66 We wish to separate barium ions from calcium ions by selective precipitation. Which anion, fluoride or carbonate, would be the better choice for achieving this precipitation? Why?

11.67 In the process of separating Pb^{2+} ions from Cu^{2+} ions as sparingly soluble iodates, what is the Pb^{2+} concentration when Cu^{2+} just begins to precipitate as sodium iodate is added to a solution that is initially 0.0010 M $Pb(NO_3)_2(aq)$ and 0.0010 M $Cu(NO_3)_2(aq)$?

11.68 A chemist attempts to separate barium ions from lead ions by using the sulfate ion as a precipitating agent. (a) What sulfate ion concentrations are required for the precipitation of $BaSO_4$ and $PbSO_4$ from a solution containing 0.010 M $Ba^{2+}(aq)$ and 0.010 M $Pb^{2+}(aq)$? (b) What is the concentration of barium ions when the lead sulfate begins to precipitate?

11.69 Use the data in Table 11.6 to calculate the solubility of each sparingly soluble substance in its respective solution: aluminum hydroxide at (a) pH = 7.0; (b) pH = 4.5; zinc hydroxide at (c) pH = 7.0; (d) pH = 6.0.

11.70 Use the data in Table 11.6 to calculate the solubility of each sparingly soluble compound in its respective solution: iron(III) hydroxide at (a) pH = 11.0; (b) pH = 3.0; iron(II) hydroxide at (c) pH = 8.0; (d) pH = 6.0.

11.71 Consider the two equilibria

$$CaF_2\,(s) \rightleftharpoons Ca^{2+}(aq) + 2\,F^-(aq) \qquad K_{sp} = 4.0 \times 10^{-11}$$

$$F^-(aq) + H_2O(l) \rightleftharpoons HF(aq) + OH^-(aq) \qquad K_b(F^-) = 2.9 \times 10^{-11}$$

(a) Write the chemical equation for the overall equilibrium and determine the corresponding equilibrium constant. Determine the solubility of CaF_2 at (b) pH = 7.0; (c) pH = 3.0.

11.72 Consider the two equilibria

$$BaF_2(s) \rightleftharpoons Ba^{2+}(aq) + 2\,F^-(aq) \qquad K_{sp} = 1.7 \times 10^{-6}$$

$$F^-(aq) + H_2O(l) \rightleftharpoons HF(aq) + OH^-(aq) \qquad K_b(F^-) = 2.9 \times 10^{-11}$$

(a) Write the chemical equation for the overall equilibrium and determine the corresponding equilibrium constant. Determine the solubility of BaF_2 at (b) pH = 7.0; (c) pH = 4.0.

11.73 Calculate the solubility of silver bromide in 0.10 M KCN(aq). Refer to Tables 11.6 and 11.7.

11.74 Precipitated silver chloride dissolves in ammonia solutions as a result of the formation of $Ag(NH_3)_2^+$. What is the solubility of silver chloride in 1.0 M $NH_3(aq)$?

11.75 A student finds on the shelf a bottle of a pure silver halide that could be AgCl or AgI. Develop a simple chemical test that would allow the student to distinguish which compound was in the bottle.

11.76 Which of the following compounds, if either, will dissolve in 1.00 M $HNO_3(aq)$: (a) Bi_2S_3; (b) FeS(s)? Substantiate your answer by giving an appropriate calculation.

11.77 A metal alloy sample is believed to contain silver, bismuth, and nickel. Explain how it could be determined qualitatively that all three of these metals are present.

11.78 Zinc(II) readily forms the complex ion $Zn(OH)_4^{2-}$. Explain how this fact can be used to distinguish a solution of $ZnCl_2$ from $MgCl_2$.

INTEGRATED EXERCISES

11.79 Determine the percentage deprotonation of the following aqueous hydroxylammonium chloride (more commonly called hydroxylamine hydrochloride) solutions: (a) 0.0240 M; (b) 0.0480 M; (c) 0.096 M. (d) How does the percentage deprotonation vary with concentration? (e) What is the pH of an optimal buffer solution prepared from hydroxylammonium chloride and its conjugate base?

11.80 Commercial solutions of concentrated aqueous ammonia are normally 29.9% NH_3 by mass and have a density of 0.89 $g \cdot cm^{-3}$. (a) What is the molarity of concentrated ammonia? (b) Concentrated ammonia solutions are sometimes referred to as aqueous ammonium hydroxide. What are the relative molar concentrations of NH_3 and NH_4OH in a concentrated solution of ammonium hydroxide? (c) When hydroxide ions are added to a solution of silver nitrate, a brownish black precipitate of silver oxide, Ag_2O, is formed. If one then adds concentrated ammonia, the precipitate will dissolve as a result of the formation of the complex ion $Ag(NH_3)_2{}^+$. If 1.00 L of solution contains 1.00 g $AgNO_3$, what mass of solid sodium hydroxide must be added just to precipitate all the silver ions (> 99%) as AgOH? Assume that the volume change is negligible. (d) What minimum amount of concentrated ammonium hydroxide must be added to get all the silver to dissolve and to have at least 99% of the silver ions present as the complex ion? Assume that the volume change is negligible and then verify that this is so at the end of the calculation.

11.81 A buffer solution is prepared from a mixture of equal amounts of acetic acid and sodium acetate. What concentration of the buffer must be prepared to prevent a change of the pH by more than 0.20 pH units following the addition of 1.00 mL of 6.00 M HCl(aq) to 100.0 mL of the buffer solution?

11.82 What is the pH of (a) 0.037 M $NaCH_3CO_2(aq)$; (b) 0.020 M $CH_3COOH(aq)$? (c) When 200 mL of 0.020 M $CH_3COOH(aq)$ is added to 150 mL of the solution in part (a), what is the pH of the mixed solution?

Media Link

11.83 (a) Using the Living Graph *Titration of a Strong Acid with a Strong Base*, prepare a plot for the titration of 15.0 mL of 0.0567 M $HNO_3(aq)$ by 0.0296 M KOH(aq). From the graph, determine (b) the volume of titrant needed to reach the stoichiometric point; (c) the starting pH of the acid solution. (d) What is the pH at the stoichiometric point?

Media Link

11.84 (a) Using the Living Graph *Titration of a Weak Acid with a Strong Base*, prepare a plot for the titration of 20.0 mL of 0.0329 M pyridinium chloride, $C_5H_5NH^+Cl^-(aq)$, by 0.0257 M KOH(aq). From the graph, determine (b) the volume of titrant needed to reach the stoichiometric point (note that the horizontal axis is given in terms of volume of titrant divided by the volume of the initial acid solution); (c) the starting pH of the acid solution. (d) What is the pH at the stoichiometric point? The K_a for pyridinium chloride is 5.6×10^{-6}.

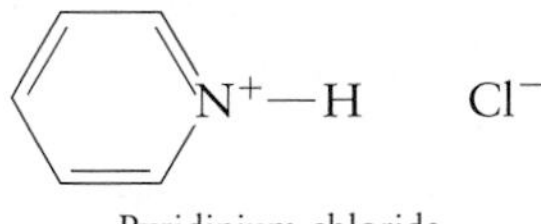

Pyridinium chloride

11.85 A 60.0-mL sample of 0.10 M $NaHCO_2(aq)$ is mixed with 4.0 mL of 0.070 M HCl(aq). Calculate the pH and the molarity of HCOOH in the mixed solution.

11.86 An old bottle labeled "Standardized 6.0 M NaOH" was found on the back of a shelf in the stockroom. Over time, some of the NaOH had reacted with the glass and the solution was no longer 6.0 M. To determine its purity, 5.0 mL of the solution was diluted to 100 mL and titrated to the stoichiometric point with 11.8 mL of 2.05 M HCl(aq). What is the molarity of the sodium hydroxide solution?

11.87 Novocaine, which is used by dentists as a local anesthetic, is a weak base with $pK_b = 5.05$. Blood has a pH of 7.4. What is the ratio of concentrations of novocaine to its conjugate acid in the bloodstream?

11.88 A buffer solution is prepared by mixing 50.0 mL of 0.022 M $C_6H_5COOH(aq)$ and 20.0 mL of 0.032 M $NaC_6H_5CO_2(aq)$. (a) What is the pH of the buffer solution? (b) What are the pH and the change in pH after the addition of 0.054 mmol HCl to the buffer solution? (c) What would be the pH change if the 0.054 mmol HCl had been added to pure water instead of the buffer solution? (d) What are the pH and the change in pH after the addition of 10.0 mL of 0.054 M HCl to the original buffer solution?

11.89 For 100 mL of a buffer solution that is 0.150 M $CH_3COOH(aq)$ and 0.50 M $NaCH_3CO_2(aq)$, what is the pH before and after adding (a) 10.0 mL of 1.2 M HCl(aq)? (b) 50.0 mL of 0.094 M NaOH(aq)?

11.90 To simulate blood conditions, a phosphate buffer system with a pH = 7.40 is desired. What mass of Na_2HPO_4 must be added to 0.500 L of 0.10 M $NaH_2PO_4(aq)$ to prepare such a buffer?

11.91 Fluoridation of city water supplies produces a fluoride ion concentration close to 5×10^{-5} $mol \cdot L^{-1}$.

If the city supply of water is hard (the water contains Ca^{2+} ions), CaF_2 may precipitate during the fluoridation process. What is the maximum concentration of Ca^{2+} ions that can be present in solution if no precipitation is to occur during fluoridation?

11.92 The fluoride ions in drinking water convert the hydroxyapatite, $Ca_5(PO_4)_3OH$, of teeth into fluorapatite, $Ca_5(PO_4)_3F$. The K_{sp} of the two compounds are 1.0×10^{-36} and 1.0×10^{-60}, respectively. (a) What are the molar solubilities of each substance? The solubility equilibria to consider are

$$Ca_5(PO_4)_3OH(s) \rightleftharpoons 5\ Ca^{2+}(aq) + 3\ PO_4^{\ 3-}(aq) + OH^-(aq)$$

$$Ca_5(PO_4)_3F(s) \rightleftharpoons 5\ Ca^{2+}+aq) + 3\ PO_4^{\ 3-}(aq) + F^-(aq)$$

(b) What is the reaction free energy for the conversion of hydroxyapatite into fluorapatite?

11.93 Will Ag_2CO_3 precipitate from a solution formed from a mixture of 100 mL of 1.0×10^{-4} M $AgNO_3(aq)$ and 100 mL of 1.0×10^{-4} M $Na_2CO_3(aq)$?

11.94 (a) To 500 mL of an aqueous solution adjusted to a pH of 8.00, a student adds 1.36 mg of $ZnCl_2$. Does a precipitate of $Zn(OH)_2$ form? (b) Would it precipitate if the pH were 12.00? (c) What minimum pH would be required to ensure that essentially all the zinc (> 99%) is present as the complex ion? The Zn^{2+} ion forms a soluble complex with hydroxide ions, $Zn(OH)_4^{\ 2-}$, which has a formation constant of 5×10^{14}.

11.95 Which sulfide precipitates first when sulfide ions are added to a solution containing equal amounts of Co^{2+}, Cu^{2+}, and Cd^{2+}? For CoS, $K_{sp} = 5 \times 10^{-22}$. Explain your conclusions.

11.96 We decide to use sulfide precipitation to separate the copper(II) ions from the manganese(II) ions in a solution that is 0.20 M $Cu^{2+}(aq)$ and 0.20 M $Mn^{2+}(aq)$. Determine the minimum sulfide ion concentration that will result in the precipitation of one cation (identify the cation), but not the other.

11.97 (a) Estimate the pH of the solution that results when 25.0 mL of 0.150 M NaOH(aq) is added to 25.0 mL of 0.125 M $H_2SO_3(aq)$. (b) If an additional 20.0 mL of the NaOH(aq) solution is added, what would you predict the pH of the resulting solution to be?

11.98 (a) Estimate the pH of the solution that results when 12.50 mL of 0.0250 M HCl(aq) is added to 25.0 mL of 0.0125 M $Na_2S(aq)$. (b) If an additional 12.50 mL of the HCl(aq) solution is added, what would you predict the pH of the resulting solution to be?

11.99 What is the pH at each stoichiometric point in the titration of 0.20 M $H_2SO_4(aq)$ with 0.20 M NaOH(aq)?

11.100 What volume (in liters) of a saturated mercury(II) sulfide, HgS, solution contains an average of one mercury(II) ion, Hg^{2+}?

11.101 It is often useful to know whether two ions can be separated by selective precipitation from a solution. Generally, a 99% separation is considered "separated." A solution is 0.010 M $Pb^{2+}(aq)$ and 0.010 M $Ag^+(aq)$. Chloride ions are added from a sodium chloride solution. (a) Determine the chloride ion concentration required for the precipitation of each cation. (b) Which cation precipitates first? (c) What is the molarity of the first cation that precipitates when the second cation begins to precipitate? (d) Determine the percentage of the first cation that remains in solution when the second cation begins to precipitate.

11.102 Consider the equilibria

$$ZnS(s) \rightleftharpoons Zn^{2+}(aq) + S^{2-}(aq)$$

$$S^{2-}(aq) + H_2O(l) \rightleftharpoons HS^-(aq) + OH^-(aq)$$

$$HS^-(aq) + H_2O(l) \rightleftharpoons H_2S(aq) + OH^-(aq)$$

(a) Write the chemical equation for the overall equilibrium and determine the corresponding equilibrium constant. (b) Determine the solubility of ZnS in a saturated H_2S solution (0.1 M $H_2S(aq)$) adjusted to pH = 7.0. (c) Determine the solubility of ZnS in a saturated H_2S solution (0.1 M $H_2S(aq)$) adjusted to pH = 10.0.

11.103 Using data available in the tables and appendices, calculate the free energy of formation of $PbF_2(s)$.

11.104 (a) To 1.00 L of 1.00 M $Pb(NO_3)_2(aq)$ is added 1.00 mol $Cu(IO_3)_2$. Do you expect the Pb^{2+} ions to replace the Cu^{2+} ions quantitatively (> 99.99%) to form lead(II) iodate? Why or why not? Support your answer with a calculation. (b) What will be the $IO_3^{\ -}$ concentration in this solution once it reaches equilibrium? (c) What is the standard free energy of the reaction of $Cu(IO_3)_2(s)$ with 1.00 M $Pb(NO_3)_2(aq)$?

11.105 (a) Which will have the greater molar solubility, Cu_2S or CuS? (b) What will be the ratio of Cu^+ ions to Cu^{2+} ions in a solution prepared by placing both solids in pure water?

11.106 (a) Write the equation for the dissolution of FeS in pure water. (b) What sulfur-containing species may possibly exist in solution? (c) Calculate the concentrations of all sulfur-containing species present in a solution when pure FeS dissolves in pure water. (d) What is the pH of a saturated aqueous solution of FeS? (e) Given the values calculated in part (c), is the equation as written in part (a) an accurate reflection of the process that is actually occurring in solution when FeS(s) dissolves? If not, what would be a better representation of the dissolution equilibrium?

Chapter 12

The interior of one compartment of a lead-acid battery while charging is in progress. The battery stores energy when it is charged and releases it when it is discharged. The bubbles of hydrogen gas form on the surface of the lead plates by reduction of hydrogen ions in solution when the charging rate is too high.

Why Do We Need to Know This Material?

The topics described in this chapter may one day unlock a virtually inexhaustible supply of clean energy supplied daily by the Sun. The key is electrochemistry, the study of the interaction of electricity and chemical reactions. An understanding of how electrons are transferred from one species to another helps us to design practical ways to use chemical reactions to generate electricity and to use electricity to bring about chemical reactions. Electrochemical measurements are also used to set up a scale of oxidizing and reducing power and to determine the values of thermodynamic quantities.

What Do We Need to Know Already?

This chapter extends the thermodynamic discussion presented in Chapter 7. In particular, it builds on the concept of free energy (Section 7.12), its relation to maximum nonexpansion work (Section 7.14), and the dependence of the reaction free energy on the reaction quotient (Section 9.3). For a review of redox reactions, see Section K. To prepare for the quantitative treatment of electrolysis, review stoichiometry in Section L.

ELECTROCHEMISTRY

The transfer of electrons from one species to another is one of the fundamental processes underlying life, photosynthesis, the launching of rockets, and the refining of metals. Finding a way to harness the energy released by the transfer of electrons may allow us to meet the energy needs of the future. We shall see in this chapter how chemical reactions—particularly redox reactions—can provide this energy.

Electrochemistry is the branch of chemistry that deals with the use of spontaneous chemical reactions to produce electricity and the use of electricity to drive nonspontaneous reactions forward. **Electrochemical techniques**—procedures based on electrochemistry—allow us to use electrical measurements to monitor chemical reactions and to determine the concentrations of ions in solution. We can use these techniques to monitor the pH of solutions and to determine the pK_a of acids. Electrochemistry even allows us to monitor the activity of our brain and heart (perhaps while we are trying to master chemistry), the pH of our blood, and the presence of pollutants in our water supply.

One of the key ideas in this chapter is the transfer of electrons, so we begin with a review of redox reactions. Redox reactions in aqueous solution will be particularly important in our discussion of the chemical production of electricity, which occupies the major part of this chapter. They are also important in the final part of the chapter, where we discuss the use of electricity to bring about chemical reactions in solution.

REPRESENTING REDOX REACTIONS

We first encountered redox reactions in Section K. Their chemical equations can be tricky to balance, so special techniques have been developed to simplify the procedure. These techniques, which we describe in Sections 12.1 and 12.2, are fundamental to the discussion of power generation and electrolysis.

12.1 Half-Reactions

The key to writing and balancing equations for redox reactions is to think of the reduction and oxidation processes individually. Let's consider oxidation (electron loss) first.

To show the removal of electrons from a species that is being oxidized in a redox reaction, we write the chemical equation for an oxidation half-reaction. A **half-reaction** is the oxidation or reduction considered alone. For example, to show the oxidation of magnesium, we write:

$$Mg(s) \longrightarrow Mg^{2+}(s) + 2\,e^-$$

An oxidation half-reaction is a *conceptual* way of reporting an oxidation: the electrons are never actually free. For example, when oxygen is used to oxidize magnesium—as occurs when magnesium burns in air—electrons move directly from magnesium atoms to adjacent oxygen atoms. In an equation for an oxidation half-reaction, the electrons released always appear on the right of the arrow. Their state is not given because they are in transit and do not have a definite physical state.

The reduced and oxidized species in a half-reaction jointly form a **redox couple.** In this example, the redox couple consists of Mg^{2+} and Mg, and is denoted Mg^{2+}/Mg. A redox couple has the form Ox/Red, where Ox is the oxidized form of the species and Red is the reduced form.

We can also write half-reactions for reduction (electron gain). For example, to show electron gain in the reduction of Fe^{3+} ions to Fe^{2+} ions, we write

$$Fe^{3+}(aq) + e^- \longrightarrow Fe^{2+}(aq)$$

This half-reaction too is conceptual: the electrons are not actually free. In the equation for a reduction half-reaction, the electrons gained always appear on the left of the arrow. In this example, the redox couple is Fe^{3+}/Fe^{2+}.

Half-reactions express the two contributions (oxidation and reduction) to an overall redox reaction.

12.2 Balancing Redox Equations

Balancing the chemical equation for a redox reaction can be a real challenge, especially for those taking place in aqueous solution, when water may be involved in the reaction and we must include H_2O, H^+, or OH^-. In such cases, balancing the equations for the reduction and oxidation half-reactions individually is easier, because the half-reactions are less complex than the equation for the overall reaction. Then we can add the two balanced equations together. When adding the equations for half-reactions, we match the number of electrons released by oxidation with the number used in reduction. The number of electrons in the two half-reactions must match, because electrons are neither created nor destroyed in chemical reactions. The procedure is outlined in Toolbox 12.1 and illustrated in Examples 12.1 and 12.2.

The chemical equation of a reduction half-reaction is added to the equation for an oxidation half-reaction to form the balanced chemical equation for the overall redox reaction.

TOOLBOX 12.1 How to balance complicated redox equations

Conceptual Basis

When balancing redox equations, especially those taking place in aqueous solution, we consider reduction (electron gain) and oxidation (electron loss) as distinct processes. Each of these processes is expressed as a half-reaction, which is balanced individually. Then the equations for the two half-reactions are combined by matching the number of electrons released in the oxidation with the number used in the reduction.

Procedure

The general procedure for balancing the chemical equation for a redox reaction is to work from the half-reactions.

Step 1 Identify the species being oxidized and the species being reduced from the changes in their oxidation numbers.

Step 2 Write the two skeletal (unbalanced) equations for the oxidation and reduction half-reactions.

Step 3 Balance all elements in the half-reactions except O and H.

Step 4 In acidic solution, balance O by using H_2O, then balance H by using H^+. In basic solution, balance O by using H_2O; then balance H by adding H_2O to the side of each half-reaction that needs H and adding OH^- to the other side.

When we add . . . OH^- . . . → . . . H_2O . . . , we are effectively adding one H atom to the right. When we add . . . H_2O . . . → . . . OH^- . . . , we are effectively adding one H atom to the left. Note that one H_2O molecule is added for each H atom needed.

Step 5 Balance the electric charges by adding electrons to the left for reductions and to the right for oxidations.

Step 6 Multiply all species in either one or both half-reactions by factors that result in equal numbers of electrons in the two half-reactions, and then add the two equations.

Finally, simplify the appearance of the equation by canceling species that appear on both sides of the arrow and check to make sure that charges as well as numbers of atoms balance.

EXAMPLE 12.1 Balancing a redox equation in acidic solution

Permanganate ions, MnO_4^-, react with oxalic acid, $H_2C_2O_4$, in acidic aqueous solution, producing manganese(II) ions and carbon dioxide. The skeletal equation is

$$MnO_4^-(aq) + H_2C_2O_4(aq) \longrightarrow Mn^{2+}(aq) + CO_2(g)$$

Balance this equation.

STRATEGY Work through the procedure set out in Toolbox 12.1. Use the strategy for an acidic solution in step 4.

SOLUTION

Step 1 Because the oxidation number of manganese decreases from +7 to +2, the MnO_4^- ion is reduced in the reaction. The oxidation number of carbon increases from +3 to +4, so oxalic acid is oxidized.

Step 2 We write skeletal equations for the half-reactions from the information in step 1:

$$\text{Reduction: } MnO_4^- \longrightarrow Mn^{2+}$$

$$\text{Oxidation: } H_2C_2O_4 \longrightarrow CO_2$$

Step 3 Balance all elements except H and O:

$$MnO_4^- \longrightarrow Mn^{2+}$$

$$H_2C_2O_4 \longrightarrow 2\ CO_2$$

Step 4 Balance the O atoms by using H_2O:

$$MnO_4^- \longrightarrow Mn^{2+} + 4\ H_2O$$

$$H_2C_2O_4 \longrightarrow 2\ CO_2$$

and balance the H atoms on either side of the half-reactions with H^+:

$$MnO_4^- + 8\,H^+ \longrightarrow Mn^{2+} + 4\,H_2O$$

$$H_2C_2O_4 \longrightarrow 2\,CO_2 + 2\,H^+$$

Step 5 Now balance the electric charge:

$$MnO_4^- + 8\,H^+ + 5\,e^- \longrightarrow Mn^{2+} + 4\,H_2O$$

$$H_2C_2O_4 \longrightarrow 2\,CO_2 + 2\,H^+ + 2\,e^-$$

To verify your equations at this point, check that the number of electrons lost or gained in each half-reaction is the same as the change in oxidation number of the element oxidized or reduced.

Step 6 Equalize the numbers of electrons transferred in the two half-reactions by multiplying the permanganate half-reaction by 2 and the oxalic acid half-reaction by 5, so that the same number (10) of electrons is transferred in each:

$$2\,MnO_4^- + 16\,H^+ + 10\,e^- \longrightarrow 2\,Mn^{2+} + 8\,H_2O$$

$$5\,H_2C_2O_4 \longrightarrow 10\,CO_2 + 10\,H^+ + 10\,e^-$$

Now add the two equations and cancel the electrons:

$$2\,MnO_4^- + 5\,H_2C_2O_4 + 16\,H^+ \longrightarrow 2\,Mn^{2+} + 8\,H_2O + 10\,CO_2 + 10\,H^+$$

Ten of the H^+ ions on the left are canceled by the 10 H^+ on the right; so after we have added the physical states of the species, we are left with

$$2\,MnO_4^-(aq) + 5\,H_2C_2O_4(aq) + 6\,H^+(aq) \longrightarrow 2\,Mn^{2+}(aq) + 8\,H_2O(l) + 10\,CO_2(g)$$

Mass and charge are both balanced; so this is the fully balanced net ionic equation.

Self-Test 12.1A Copper reacts with dilute nitric acid to form copper(II) nitrate and the gas nitric oxide, NO. Write the net ionic equation for the reaction.

[*Answer:* $3\,Cu(s) + 2\,NO_3^-(aq) + 8\,H^+(aq) \rightarrow 3\,Cu^{2+}(aq) + 2\,NO(g) + 4\,H_2O(l)$]

Self-Test 12.1B Acidified potassium permanganate solution reacts with sulfurous acid, $H_2SO_3(aq)$, to form sulfuric acid and manganese(II) ions. Write the net ionic equation for the reaction. In acidic aqueous solution, H_2SO_3 is present as electrically neutral molecules and sulfuric acid is present as HSO_4^- ions.

EXAMPLE 12.2 Balancing a redox equation in basic solution

The products of the reaction between bromide ions and permanganate ions, MnO_4^-, in basic aqueous solution are solid manganese(IV) oxide, MnO_2, and bromate ions. Balance the net ionic equation for the reaction.

STRATEGY Work through the procedure set out in Toolbox 12.1, using the rule for balancing in basic solution in step 4.

SOLUTION

Step 1 The oxidation number of manganese changes from +7 in MnO_4^- to +4 in MnO_2, so MnO_4^- is reduced. The oxidation number of bromine increases from −1 in Br^- to +5 in BrO_3^-, so Br^- is oxidized.

Step 2 The skeletal equations for the oxidation and reduction half-reactions are

Reduction: $MnO_4^- \longrightarrow MnO_2$

Oxidation: $Br^- \longrightarrow BrO_3^-$

Step 3 Both equations are already balanced with respect to Mn and Br.

Step 4 First, balance the O atoms by using H_2O:

$$MnO_4^- \longrightarrow MnO_2 + 2\,H_2O$$

$$Br^- + 3\,H_2O \longrightarrow BrO_3^-$$

Next, balance *each* H atom by adding one H_2O molecule to the side of each equation that needs hydrogen and one OH^- ion to the opposite side:

$$MnO_4^- + 4\,H_2O \longrightarrow MnO_2 + 2\,H_2O + 4\,OH^-$$

$$Br^- + 3\,H_2O + 6\,OH^- \longrightarrow BrO_3^- + 6\,H_2O$$

In the final part of this step, simplify each half-reaction by canceling like species on opposite sides of the arrow (in this case, H_2O):

$$MnO_4^- + 2\,H_2O \longrightarrow MnO_2 + 4\,OH^-$$

$$Br^- + 6\,OH^- \longrightarrow BrO_3^- + 3\,H_2O$$

Step 5 Now balance the electric charges by using electrons:

$$MnO_4^- + 2\,H_2O + 3\,e^- \longrightarrow MnO_2 + 4\,OH^-$$

$$Br^- + 6\,OH^- \longrightarrow BrO_3^- + 3\,H_2O + 6\,e^-$$

Step 6 To ensure that the same numbers of electrons (6) appear in both half-reactions, multiply the equation for the manganese reduction half-reaction by 2:

$$2\,MnO_4^- + 4\,H_2O + 6\,e^- \longrightarrow 2\,MnO_2 + 8\,OH^-$$

$$Br^- + 6\,OH^- \longrightarrow BrO_3^- + 3\,H_2O + 6\,e^-$$

Six electrons are transferred in the equation for this redox reaction. Now add the two equations and simplify the result by canceling species that appear on both sides of the arrow (in this case, electrons, H_2O, and OH^-) and attach state symbols. We obtain the following fully balanced equation:

$$2\,MnO_4^-(aq) + Br^-(aq) + H_2O(l) \longrightarrow 2\,MnO_2(s) + BrO_3^-(aq) + 2\,OH^-(aq)$$

Self-Test 12.2A An alkaline (basic) solution of hypochlorite ions reacts with solid chromium(III) hydroxide to produce aqueous chromate ions and chloride ions. Write the net ionic equation for the reaction.

[*Answer:* $2\,Cr(OH)_3(s) + 4\,OH^-(aq) + 3\,ClO^-(aq) \rightarrow 2\,CrO_4^{2-}(aq) + 5\,H_2O(l) + 3\,Cl^-(aq)$]

Self-Test 12.2B When iodide ions react with iodate ions in basic aqueous solution, triiodide ions, I_3^-, are formed. Write the net ionic equation for the reaction. (Note that the same product is obtained in each half-reaction.)

GALVANIC CELLS

One of the most familiar contributions of electrochemistry to everyday life is the battery used in a portable cassette player, compact disc player, or laptop computer. Such a battery is an example of an electrochemical cell. In general, an **electrochemical cell** is a device in which an electric current—a flow of electrons through a circuit—is either produced by a spontaneous chemical reaction or used to bring about a nonspontaneous reaction. A **galvanic cell** is an electrochemical cell in which a spontaneous chemical reaction is used to generate an electric current. Technically, a **battery** is a collection of galvanic cells joined in series, so the voltage it produces—its ability to push an electric current through a circuit—is the sum of the voltages of each cell.

Galvanic cells are also known as *voltaic cells.*

The formal term for "voltage" is potential difference, measured in volts: 1 V = 1 $J{\cdot}C^{-1}$ (see Section 12.4).

The structure of this part of the chapter is as follows. First, we explore the structure of galvanic cells and see how to express that structure in a compact manner. Then we see what information can be obtained from measurements of the voltage they produce. This discussion leads to one of the central concepts of electrochemistry, the voltage that can be ascribed to an individual electrode. We explore how that concept is defined and see its significance in chemistry for judging the spontaneity of reactions and calculating equilibrium constants. The voltage produced by a cell depends on the concentrations of **electroactive species,** the species participating in electron transfer, and we derive the quantitative relation between concentration and voltage. Finally, we describe an important application of this material, the determination of the concentration of electroactive species, particularly the concentration of H_3O^+, as expressed by pH.

12.3 The Structure of Galvanic Cells

Although there are some notable exceptions, most of the galvanic cells studied in chemistry involve chemical reactions in aqueous solution, and we concentrate on them. We shall consider how a spontaneous redox reaction can be used to generate an electric current. One example is the reaction of zinc metal and copper(II) ions

$$Zn(s) + Cu^{2+}(aq) \longrightarrow Zn^{2+}(aq) + Cu(s) \quad \text{(A)}$$

If we were to place a piece of zinc metal into an aqueous copper(II) sulfate solution, we would see a layer of metallic copper begin to deposit on the surface of the zinc (see Fig. K.5). If we could watch the reaction at the atomic level, we would see that, as the reaction takes place, electrons are transferred from the Zn atoms to Cu^{2+} ions nearby in the solution. These electrons reduce the Cu^{2+} ions to Cu atoms, which stick to the surface of the zinc or form a finely divided solid deposit in the beaker. The piece of zinc slowly disappears as its atoms give up electrons and form colorless Zn^{2+} ions that drift off into the solution. Electrons are transferred and the free energy of the system decreases as the reaction approaches equilibrium. However, although energy is released as heat, no electrical work is done.

Now suppose we separate the reactants but provide a means for the electrons to travel from the zinc to the copper(II) ions. This is what happens when the reaction takes place in a galvanic cell. A galvanic cell consists of two **electrodes,** or metallic conductors, that make electrical contact with the contents of the cell, and an **electrolyte,** an ionically conducting medium, inside the cell. The electrolyte is typically an aqueous solution of an ionic compound, although advanced cells make use of all kinds of exotic materials (see Box 12.1). Oxidation takes place at one electrode as the species being oxidized releases electrons into the electrode. Reduction takes place at the other electrode, where the species that is undergoing reduction collects electrons from the electrode (Fig. 12.1). We can think of the overall chemical reaction as pushing electrons on to one electrode and pulling them off the other electrode. This push-pull process sets up a flow of electrons in the external circuit joining the two electrodes, and that current can be used to do electrical work, such as powering an electric motor.

In a metallic conductor, charge is transported by electrons. Metallic conductors include metals and the nonmetal graphite.

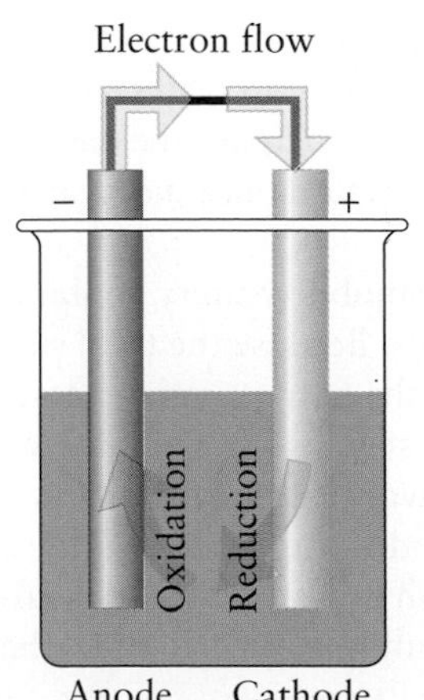

FIGURE 12.1 In an electrochemical cell, a reaction takes place in two separate regions. Oxidation occurs at one electrode, and the electrons released travel through the external circuit to the other electrode, where they cause reduction. The site of oxidation is called the anode, and the site of reduction is called the cathode. The circuit is completed by ions, which carry the electric charge through the solution.

Media Link **Lab video:** Fig. 12.1, A galvanic cell

BOX 12.1 FRONTIERS OF CHEMISTRY: PORTABLE ENERGY

The commercial galvanic cells we know as batteries are currently the subject of intense research by scientists who see their potential to solve environmental, health, communication, and transportation problems.

An ideal battery should be inexpensive, portable, safe to use, and environmentally benign. It should also maintain a potential difference that is stable over time (see Table 12.1). Both the mass and volume of a battery are critical parameters. The electrolyte in a battery uses as little water as possible, both to reduce leakage of the electrolyte and to keep the mass low. Much of the research on batteries is devoted to raising the *specific energy*, the reaction free energy per kilogram (typically expressed in kilowatt-hours per kilogram, $kW \cdot h \cdot kg^{-1}$), or the *energy density*, the reaction free energy per liter (in kilowatt-hours per liter, $kW \cdot h \cdot L^{-1}$).* The *power density* (in watts per liter, $W \cdot L^{-1}$) is also important in applications that require a surge of energy—starting an automobile engine, for example.

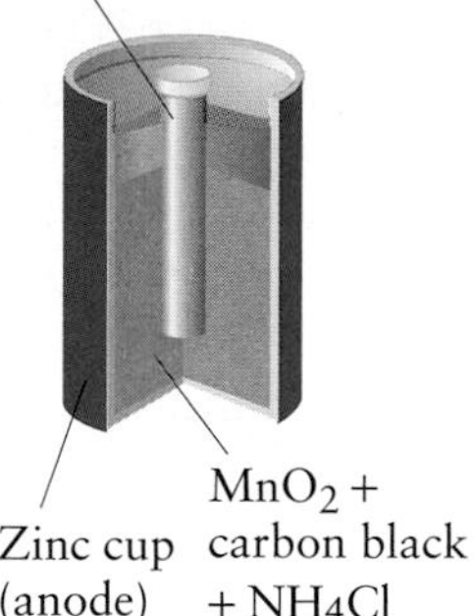

A commercial dry cell. The dry cell is also called the Leclanché cell for Georges Leclanché, the French engineer who invented it in about 1866. The electrolyte is not completely dry. Rather, it is a moist paste.

Primary cells Primary cells are galvanic cells with the reactants sealed inside at manufacture. They cannot be recharged; when they run down, they are discarded.

A *dry cell* is the workhorse of primary cells. Its familiar cylindrical zinc container serves as the anode; and in the center is the cathode, a carbon rod. The interior of the container is lined with paper that serves as the porous barrier. The electrolyte is a moist paste of ammonium chloride, manganese(IV) oxide, finely granulated carbon, and an inert filler, usually starch. The ammonia provided by the ammonium ions forms the complex ion $Zn(NH_3)_4^{2+}$ with the Zn^{2+} ions and prevents their buildup and a consequent reduction of the potential. Dry cells have many uses in devices such as portable electronic equipment, flashlights, and toys.

Potential: 1.5 V initially, falling with use to about 0.8 V as reaction products accumulate
Strengths: low cost, made of nontoxic materials
Weaknesses: cannot be recycled, prone to leaking, short shelf life, unstable potential and current

An *alkaline cell* is similar to a dry cell but uses an alkaline electrolyte, with which the zinc electrode does not readily react when the battery is not in use. As a result, alkaline cells have longer lives than dry cells. The uses are the same as those for the dry cell, plus applications that require a long-lasting charge, such as smoke detectors and backup power supplies.

A *silver cell* has a cathode made of Ag_2O. The relatively high potential of a silver cell, with its solid reactants and products, is maintained with great reliability over long periods of time, a feature making it desirable for medical implants such as pacemakers, for hearing aids, and for cameras.

Potential: 1.6 V
Strengths: can be made in very small sizes, applies a reliable steady potential over a long period of time, relatively low toxicity
Weaknesses: cannot be recycled, highly expensive

Secondary cells Secondary cells are galvanic cells that must be charged before they can be used; this type of cell is normally rechargeable. The batteries used in portable computers and automobiles are secondary cells. In the charging process, an external source of electricity reverses the spontaneous cell reaction and creates a nonequilibrium mixture of reactants. After charging, the cell can again produce electricity as the reaction once more sinks back toward equilibrium.

The *lead-acid cell* of an automobile battery contains several grids that act as electrodes. Because the total surface area of these grids is large, the battery can generate large currents for short periods, such as the time needed for starting an engine. In other words, this battery has a high power density. The electrodes are initially a hard lead-antimony alloy covered with a paste of lead(II) sulfate. The electrolyte is dilute sulfuric acid. During the

* $1\ kW \cdot h = (10^3\ J \cdot s^{-1}) \times (3600\ s) = 3.6 \times 10^6\ J = 3.6\ MJ$ exactly.

first charging, some of the lead(II) sulfate is reduced to lead on one of the electrodes; the same electrode will act as the anode during discharge. Simultaneously, during charging, lead(II) sulfate is oxidized to lead(IV) oxide on the electrode that will later act as the cathode during discharge. The lead-acid cell is used to provide power for starting the engine and running accessories in most vehicles and as the main power source in electric vehicles.

Potential: 2 V; in a series of six cells, 12 V
Strengths: inexpensive, high power density, long shelf life, recharges well
Weaknesses: releases lead to the environment when discarded; sometimes generates hydrogen gas at the cathode when charging and hence can generate an explosive mixture of gases; has a low specific energy

The *lithium-ion cell* is used in laptop computers, because it has a very long cycle life. This type of battery makes use of a special electrolyte in which polypropylene oxide or polyethylene oxide is dissolved in low melting point mixtures of lithium salts and then allowed to cool. The resulting materials are rubbery in texture, but serve as good conductors of Li^+ ions at normal temperatures. The low mass density of lithium gives it the highest available energy density, and lithium's very negative electrode potential provides a high voltage.

Potential: Up to 4 V
Strengths: holds a charge up to 10 years, charges rapidly, and can be recharged up to 1000 times
Weaknesses: expensive and sensitive to overcharging; lithium is highly reactive

A *sodium-sulfur cell* is one of the more startling batteries. It has liquid reactants (sodium and sulfur) and a solid electrolyte (a porous aluminum oxide ceramic); it must operate at a temperature of about 320°C; and it is highly dangerous in case of breakage. In its most common application—to power electric vehicles—2000 sodium-sulfur cells are placed inside an insulated container. Once the vehicle is operating, the heat generated by the battery is sufficient to maintain the temperature. This battery is the power source for the Ford Ecostar electric van.

Potential: 2.2 V
Strengths: very high specific energy due to the low mass density of sodium
Weaknesses: expensive, takes 6 h to recharge, short life; sodium is highly reactive

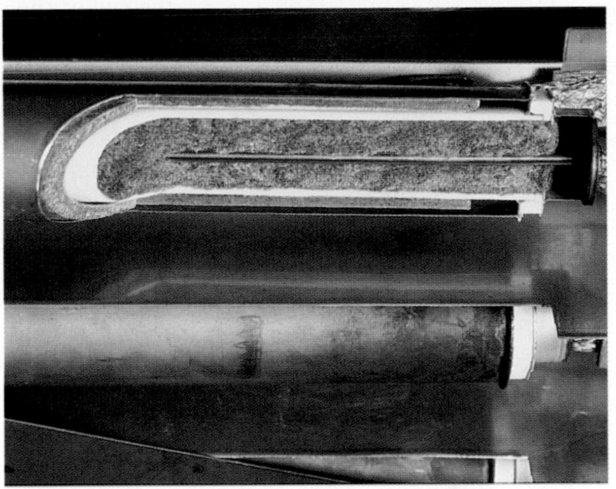
A sodium-sulfur battery used in an electric vehicle.

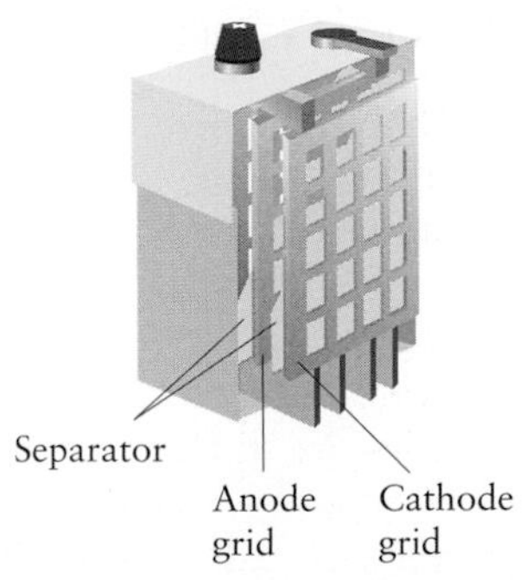

A lead-acid battery.

HOW MIGHT YOU CONTRIBUTE?

The search for rechargeable batteries for use in portable electronic equipment and electric vehicles is intense, but difficult. The need for miniature batteries is increasing, both for hand-held electronic equipment and for medical purposes. The reliability and rechargeability of these batteries must be increased and their size decreased.

For Further Reading: C. A. Vincent and B. Scrosati, *Modern Batteries: An Introduction to Electrochemical Power Sources*, Butterworth-Heinemann: 1998. S. L. Wilkinson, Electric vehicles gear up, *Chemical and Engineering News*, October 13, 1997, pp. 18–24.

The electrode at which oxidation occurs is called the **anode.** The electrode at which reduction occurs is called the **cathode.** Electrons entering the cell at the cathode are used in the reduction half-reaction. A commercial galvanic cell has its cathode marked with a + sign. The anode is marked with a − sign. In a commercial cell, the − electrode (the anode) is the source of the electrons released by the cell reaction: the reaction "pushes" the electrons out of the cell at the anode (Table 12.1). These electrons travel through the external circuit and are "pulled" back into the cell again at the + electrode (the cathode).

The *Daniell cell* is an early example of a galvanic cell that makes use of the oxidation of copper by zinc ions, as in reaction A. The British chemist John Daniell invented it in 1836, when the growth of telegraphy created an urgent need for a reliable, steady source of electric current. Daniell set up the arrangement shown in Fig. 12.2, in which the two reactants are separated. For the electrons to travel from Zn atoms to Cu^{2+} ions and bring about the spontaneous reaction, they must pass through the external circuit. The Cu^{2+} ions are converted into Cu atoms at the cathode by the reduction half-reaction

$$Cu^{2+}(aq) + 2\,e^- \longrightarrow Cu(s)$$

TABLE 12.1 Reactions in Commercial Batteries*

Primary cells	
dry	$Zn(s)|ZnCl_2(aq), NH_4Cl(aq)|MnO(OH)(s)|MnO_2(s)|graphite$, 1.5 V Anode: $Zn(s) \longrightarrow Zn^{2+}(aq) + 2\,e^-$ followed by $Zn^{2+}(aq) + 4\,NH_3(g) \longrightarrow Zn(NH_3)_4{}^{2+}(aq)$ Cathode: $MnO_2(s) + H_2O(l) + e^- \longrightarrow MnO(OH)(s) + OH^-(aq)$ followed by $NH_4{}^+(aq) + OH^-(aq) \longrightarrow H_2O(l) + NH_3(g)$
alkaline	$Zn(s)|ZnO(s)|OH^-(aq)|Mn(OH)_2(s)|MnO_2(s)|graphite$, 1.5 V Anode: $Zn(s) + 2\,OH^-(aq) \longrightarrow ZnO(s) + H_2O(l) + 2\,e^-$ Cathode: $MnO_2(s) + 2\,H_2O(l) + 2\,e^- \longrightarrow Mn(OH)_2(s) + 2\,OH^-(aq)$
silver	$Zn(s)|ZnO(s)|KOH(aq)|Ag_2O(s)|Ag(s)|steel$, 1.6 V Anode: $Zn(s) + 2\,OH^-(aq) \longrightarrow ZnO(s) + H_2O(l) + 2\,e^-$ Cathode: $Ag_2O(s) + H_2O(l) + 2\,e^- \longrightarrow 2\,Ag(s) + 2\,OH^-(aq)$
Secondary cells	
lead-acid	$Pb(s)|PbSO_4(s)|H^+(aq),HSO_4{}^-(aq)|PbO_2(s)|PbSO_4(s)|Pb(s)$, 2 V Anode: $Pb(s) + HSO_4{}^-(aq) \longrightarrow PbSO_4(s) + H^+(aq) + 2\,e^-$ Cathode: $PbO_2(s) + 3\,H^+(aq) + HSO_4{}^-(aq) + 2\,e^- \longrightarrow PbSO_4(s) + 2\,H_2O(l)$
nicad	$Cd(s)|Cd(OH)_2(s)|KOH(aq)|Ni(OH)_3(s)|Ni(OH)_2(s)|Ni(s)$, 1.25 V Anode: $Cd(s) + 2\,OH^-(aq) \longrightarrow Cd(OH)_2(s) + 2\,e^-$ Cathode: $2\,Ni(OH)_3(s) + 2\,e^- \longrightarrow 2\,Ni(OH)_2(s) + 2\,OH^-(aq)$
sodium-sulfur	$Na(l)|Na^+(solution)||S^{2-}(solution)|S_8(l)$, 2.2 V Anode: $2\,Na(l) \longrightarrow 2\,Na^+$ (solution) $+ 2\,e^-$ Cathode: $S_8(l) + 16\,e^- \longrightarrow 8\,S^{2-}$ (solution)

*Cell notation is described in Section 12.5.

Zinc atoms are converted to Zn^{2+} ions at the anode by the oxidation half-reaction

$$Zn(s) \longrightarrow Zn^{2+}(aq) + 2\ e^-$$

If the two solutions were completely separated, electrical charge would quickly build up at each electrode and prevent further reaction. To avoid charge buildup, Daniell allowed the solutions to come into contact through a porous wall. Ions provided by the electrolyte could then move between the two compartments and complete the electrical circuit. Anions travel toward the anode to balance the charges of the Zn^{2+} ions formed by the oxidation of the zinc electrode. Cations travel toward the cathode, to replace the charges of the Cu^{2+} ions that have been deposited as copper metal.

The electrodes in the Daniell cell are made of the metals involved in the reaction. However, not all electrode reactions involve a conducting solid directly. For example, to use the reduction $2\ H^+(aq) + 2\ e^- \rightarrow H_2(g)$ at an electrode, it is necessary to have a chemically inert metallic conductor, such as an unreactive metal or graphite, to carry the electrons into or out of the electrode compartment. Platinum is customarily used for the electrode: hydrogen gas is bubbled over the metal immersed in a solution that contains hydrogen ions. This arrangement is called a *hydrogen electrode.*

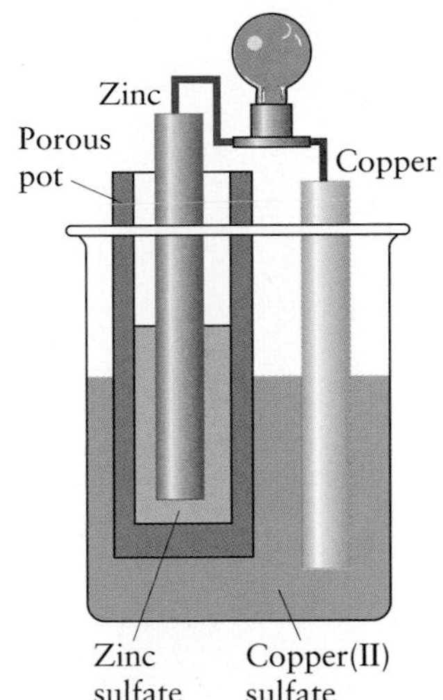

FIGURE 12.2 The Daniell cell consists of copper and zinc electrodes dipping into solutions of copper(II) sulfate and zinc sulfate, respectively. The two solutions make contact through the porous barrier, which allows ions to pass through and complete the electric circuit.

In a galvanic cell, a spontaneous chemical reaction draws electrons into the cell through the cathode, the site of reduction, and releases them at the anode, the site of oxidation.

12.4 Cell Potential and Reaction Free Energy

Galvanic cells differ in their abilities to generate an electric current. The **cell potential,** E, is a measure of the ability of a cell reaction to force electrons through a circuit. A reaction with a lot of pushing-and-pulling power generates a high cell potential (colloquially, a high voltage). A reaction with little pushing-and-pulling power, like a reaction close to equilibrium, generates only a small potential (a low voltage). An exhausted battery is a cell in which the reaction is at equilibrium, so it has lost its power to move electrons and has a potential of 0.

The SI unit of potential is the **volt** (V). A volt is defined so that a charge of one coulomb (1 C) falling through a potential difference of one volt (1 V) releases one joule (1 J) of energy:

$$1\ \text{V}\cdot\text{C} = 1\ \text{J}$$

One coulomb is the charge delivered by a current of one ampere flowing for one second: 1 C = 1 A·s.

Cell potential is analogous to gravitational potential. The maximum work a falling weight can do is equal to its mass times the difference in gravitational potential. Similarly, the maximum work an electron can do is equal to its charge times the difference in electrical potential through which it falls. To interpret this idea in terms of thermodynamics, recall from Section 7.14 that the value of ΔG gives the maximum nonexpansion work that can be obtained from a process at constant temperature and pressure. As explained there, *nonexpansion work* is any work other than that resulting from a change in volume. For example, electrical work and mechanical work, such as muscle contraction, are types of nonexpansion work. Our immediate task, therefore, is to find a relation between the thermodynamic property (the free energy of a reaction) and the electrochemical property (the cell potential produced by that reaction).

How do we do that?

To find the connection between cell potential and reaction free energy, we begin by recalling that in Section 7.14 (Eq. 22) we saw that the reaction free energy is the maximum nonexpansion work that a reaction can do at constant pressure and temperature:

$$\Delta G_r = w_e$$

The work done when an amount n of electrons (in moles) travels through a potential difference E is their total charge times the potential difference. The charge of one electron is $-e$; the charge per mole of electrons is $-eN_A$; so the total charge is $-neN_A$. Therefore, the work done is

$$w_e = -neN_AE$$

The **Faraday constant,** F, is the magnitude of the charge per mole of electrons (the product of the elementary charge e and the Avogadro constant N_A):

$$\begin{aligned} F &= eN_A \\ &= (1.602\,177 \times 10^{-19}\ \text{C}) \times (6.0223 \times 10^{23}\ (\text{mol e}^-)^{-1}) \\ &= 9.6485 \times 10^4\ \text{C}\cdot(\text{mol e}^-)^{-1} \end{aligned} \quad (1)$$

The Faraday constant is normally abbreviated to $9.6485 \times 10^4\ \text{C}\cdot\text{mol}^{-1}$. Therefore, we can write the previous expression for work as

$$w_e = -nFE$$

When this equation is combined with the thermodynamic equation, we obtain

$$\Delta G_r = -nFE \quad (2)$$

The n in this expression is the stoichiometric coefficient of the electrons in the oxidation and reduction half-reactions that are combined to make up the balanced equation for the cell reaction.

We see from Eq. 2 that, if the cell potential is positive, then the reaction free energy is negative, and the cell reaction has a spontaneous tendency to form products. If the cell potential is negative, then the *reverse* of the cell reaction is spontaneous, and the cell reaction has a spontaneous tendency to form reactants. Equation 2 is enormously important, because it relates the thermodynamic information that we have been compiling since Chapter 6 and the electrochemical information that we are starting to develop in this chapter.

There is a hidden assumption in the derivation of Eq. 2. *Maximum* nonexpansion work is obtained when a cell is operating *reversibly*. Therefore, Eq. 2 applies only when the pushing power of the cell is balanced against an external matching source of potential. To obtain the value of E to use in Eq. 2, we have to measure it in an arrangement in which an applied potential *just* balances the potential generated by the cell. The cell potential under these conditions is called the **electromotive force,** emf, of the cell. From now on, E will always be taken to represent this emf. A *working cell*, one actually producing current, like the cell in a cassette recorder, will produce a potential that is smaller than that predicted by Eq. 2. It cannot do as much work: the energy released may be the same, but some is released—wasted—as heat instead of work. However, despite this waste, galvanic cells are more efficient utilizers of fuel than combustion reactions, and research on their development is intense (see Box 12.2).

Remember the definition of reversibility in Section 6.7, which requires the pushing force of the system to be balanced against an equal and opposite force.

How can we use this information . . . to calculate the reaction free energy?

We can use Eq. 2 to determine a reaction free energy—a thermodynamic quantity—from a cell emf, an electrical quantity. From the chemical equation for the reaction in the Daniell cell (reaction A), we see that $n = 2$. The emf of the cell for certain concentrations of copper and zinc ions is 1.1 V, so under those conditions

$$\begin{aligned}\Delta G_r &= -nFE = -2 \times (9.6485 \times 10^4\ \text{C·mol}^{-1}) \times (1.1\ \text{V}) \\ &= -2.1 \times 10^5\ \text{C·V·mol}^{-1}\end{aligned}$$

Because 1 C·V = 1 J, we can conclude that the free energy of reaction A under these conditions is -2.1×10^2 kJ·mol^{-1}.

Self-Test 12.3A The reaction taking place in a nicad cell is Cd(s) + 2 $Ni(OH)_3$(s) $\longrightarrow$ $Cd(OH)_2$(s) + 2 $Ni(OH)_2$(s), and the emf of the cell when fully charged is 1.25 V. What is the reaction free energy?

[*Answer:* -241 kJ·mol^{-1}]

Self-Test 12.3B The reaction taking place in the silver cell used in some cameras and wristwatches is Ag_2O(s) + Zn(s) $\rightarrow$ 2 Ag(s) + ZnO(s), and the emf of the cell when new is 1.6 V. What is the reaction free energy?

We shall usually employ Eq. 2 for the *standard* reaction free energy, $\Delta G_r°$, when it becomes

$$\Delta G_r° = -nFE° \tag{3}$$

In this expression, $E°$ is the **standard emf** of the cell, the emf measured when all the species taking part are in their standard states. In practice, that means all gases are at 1 bar, all ions are at 1 mol·L^{-1}, and all liquids and solids are pure. For example, to measure the standard emf of the Daniell cell, we use 1 M $CuSO_4$(aq) and a pure copper electrode in one electrode compartment and 1 M $ZnSO_4$(aq) and a pure zinc electrode in the other.

More precisely, all the solutes should be at unit activity, not unit molarity. Activities differ appreciably from molarities in electrolyte solution because ions interact over long distances. However, we ignore this complication here.

BOX 12.2 FRONTIERS OF CHEMISTRY: FUEL CELLS

In the early years of the space program, life support in space was a critical dilemma. The cabin needed a source of electrical power and the astronauts needed enough water for drinking and washing. Because the mass of a spacecraft must be kept as low as possible, batteries—which usually provide energy from the oxidation of a metal—would be too heavy. Electricity can be obtained from combustion reactions by burning a fuel to create heat, which runs a generator. However, a generator is heavy, and producing electricity from burning fuels is very inefficient because energy is wasted as heat.

The problem was solved by Francis Bacon, a British scientist and engineer, who developed an idea proposed by Sir William Grove in 1839. A *fuel cell* generates electricity directly from a chemical reaction, as in a battery, but uses reactants that are supplied continuously, as in an engine. Fuel cells underwent intensive development by the aerospace industry, and eventually a fuel cell that runs on hydrogen and oxygen was produced for the space shuttle (see Fig. L.1). An advantage of this fuel cell is that the only product of the cell reaction, water, can be drunk by the crew.

The development of the fuel cell has led to an entirely new kind of technology for making use of redox reactions electrochemically and hence more efficiently. Fuel cells make efficient use of resources, because little waste heat is produced.

The development of fuel cell technology received a boost when California passed a law requiring that by 1998 2% of new motor vehicles sold in the state must be zero-emission vehicles, and by 2003, 10% of new vehicles must be zero-emission. "Zero-emission" means that they produce no pollution. Vehicles powered by rechargeable batteries or hydrogen fuel cells are considered to be the most promising solutions. Buses in Canada and in Chicago are being used to test the vehicular uses of the hydrogen fuel cell.

High-pressure hydrogen tanks run across the top of this bus provided by Ballard Power Systems for testing in Chicago. The bus uses an alkali fuel cell.

In a simple version of a fuel cell, a fuel such as hydrogen gas is passed over a platinum electrode, oxygen is passed over the other, similar electrode, and the electrolyte is aqueous potassium hydroxide. A porous membrane separates the two electrode compartments. Many varieties of fuel cells are possible, and in some the electrolyte is a solid polymer membrane or a ceramic (see Section 14.24). Three of the most promising fuel cells are

As we saw in Section 7.13, the value of $\Delta G_r°$ depends on the stoichiometric coefficients in the chemical equation, so multiplying all the coefficients by 2 doubles the value of $\Delta G_r°$. However, multiplying all the coefficients by 2 also doubles the value of n, so $E° = -\Delta G_r°/nF$ remains the same. That is, although the reaction free energy changes when the chemical equation is multiplied by a factor, $E°$ does not change:

$$Zn(s) + Cu^{2+}(aq) \longrightarrow Zn^{2+}(aq) + Cu(s) \qquad \Delta G_r° = -212\ kJ{\cdot}mol^{-1}\ E° = +1.10\ eV$$

$$2\ Zn(s) + 2\ Cu^{2+}(aq) \longrightarrow 2\ Zn^{2+}(aq) + 2\ Cu(s) \qquad \Delta G_r° = -414\ kJ{\cdot}mol^{-1}\ E° = +1.10\ eV$$

A practical implication of this conclusion is that the emf produced by a cell is independent of the size of the cell. To get a higher potential than the alkali fuel cell, the phosphoric acid fuel cell, and the molten carbonate fuel cell.

The hydrogen-oxygen cell used in the space shuttle is called an *alkali fuel cell*, because it has an alkaline electrolyte. Both electrodes are nickel, but in some versions a platinum catalyst is used.

Anode: $2\ H_2(g) + 4\ OH^-(aq) \longrightarrow 4\ H_2O(l) + 4\ e^-$
Electrolyte: $KOH(aq)$
Cathode: $O_2(g) + 4\ e^- + 2\ H_2O(l) \longrightarrow 4\ OH^-(aq)$

Although its cost prohibits its use in many applications, the alkali fuel cell is the primary fuel cell used in the aerospace industry, because it has a very high power density (see Box 12.1). One problem is the reaction of the alkali electrolyte with the carbon dioxide in air to form potassium carbonate, which clogs the openings in the electrode gauze.

If an acid electrolyte is used, water is produced at the cathode. An example is the *phosphoric acid fuel cell*.

Anode: $2\ H_2(g) \longrightarrow 4\ H^+(aq) + 4\ e^-$
Electrolyte: $H_3PO_4(aq)$
Cathode: $O_2(g) + 4\ H^+(aq) + 4\ e^- \longrightarrow 2\ H_2O(l)$

This fuel cell has shown promise for *combined heat and power systems* (CHP systems). In such systems, the waste heat is used to heat buildings or to do work. Efficiency in a CHP plant can reach 80%. These plants could replace heating plants and power sources in colleges and universities, hotels, and apartment buildings. Because fuel cells have few or no moving parts, little maintenance is required. However, hydrogen is still very expensive.

In the *molten carbonate fuel cell*, methane is used as the fuel. This cell runs at high temperatures and uses a molten mixture of lithium and potassium carbonates as electrolyte. In most such cells, the methane is reformed into hydrogen and carbon monoxide before reacting in the cell:

$$CH_4(g) + H_2O(g) \longrightarrow CO(g) + 3\ H_2(g)$$

However, in the *direct fuel cell*, methane is used directly.

Anode: $CH_4(g) + 4\ CO_3^{2-}(l) \longrightarrow 2\ H_2O(l) + 5\ CO_2(g) + 8\ e^-$
Electrolyte: $K_2CO_3(l)/Li_2CO_3(l)$
Cathode: $O_2(g) + 2\ CO_2(g) + 4\ e^- \longrightarrow 2\ CO_3^{2-}(l)$

Two difficulties with this cell are that it must run at a high temperature and that the electrolyte is highly corrosive.

HOW MIGHT YOU CONTRIBUTE?

There are many obstacles that must be overcome before fuel cells reach their potential to provide us with pollution-free energy. The hydrogen fuel cells are the most attractive, because of their use of a renewable fuel. Hydrogen can be obtained from the water in the oceans. The challenge is to extract it from seawater by a process using less energy than that given off in the fuel cell and to find safe means of transportation and storage. Many practical problems with fuel cells also need to be solved, such as controlling corrosion by electrolytes and reducing the operating temperature.

For Further Reading: T. Koppel, *Powering the Future: The Ballard Fuel Cell and the Race to Change the World*, Wiley: 1999. D. Sperling, The case for electric vehicles, *Scientific American*, November, 1996, pp. 54–59. P. Weiss, "Device ups hydrogen energy from sunlight," *Science News*, Vol. 158, September 16, 2000, p. 182.

predicted by Eq. 2, we have to construct a battery by joining cells in series: the potential is then the sum of the potentials of the individual cells.

Cell emf and reaction free energy are related by Eq. 2 ($\Delta G_r = -nFE$) and their standard values by Eq. 3 ($\Delta G_r° = -nFE°$). The magnitude of the emf is independent of how the chemical equation is written.

12.5 The Notation for Cells

Chemists use a special notation to specify the structure of electrode compartments in a galvanic cell. The two electrodes in the Daniell cell, for instance, are denoted $Zn(s)|Zn^{2+}(aq)$ and $Cu^{2+}(aq)|Cu(s)$. Each vertical line represents an interface between phases, in this case, between solid metal and ions in solution.

We report the structure of a cell in a symbolic **cell diagram,** using the conventions specified by IUPAC and used by chemists around the world. The diagram for the Daniell cell, for instance, is

$$Zn(s)|Zn^{2+}(aq)|Cu^{2+}(aq)|Cu(s)$$

In the Daniell cell, zinc sulfate and copper(II) sulfate solutions meet inside the porous barrier to complete the circuit. However, when different ions mingle together, they affect the voltage in ways that are difficult to predict. To keep solutions from mixing, chemists use a **salt bridge** to join the two electrode compartments and complete the electrical circuit. A salt bridge typically consists of a gel containing a concentrated aqueous salt solution in a bridge-shaped tube (Fig. 12.3). The bridge allows a flow of ions, so it completes the electrical circuit, but the ions are chosen so that they do not affect the cell reaction (often KCl is used). In a cell diagram, a salt bridge is shown by a double vertical line (||), so the arrangement in Fig. 12.3 is denoted

$$Zn(s)|Zn^{2+}(aq)||Cu^{2+}(aq)|Cu(s)$$

Any inert metallic component of an electrode is written as the outermost component of that electrode in the cell diagram. For example, a hydrogen electrode constructed with platinum is denoted $H^+(aq)|H_2(g)|Pt(s)$ when it is the right-hand electrode in a cell diagram and $Pt(s)|H_2(g)|H^+(aq)$ when it is the left-hand electrode. An electrode consisting of a platinum wire dipping into a solution of iron(II) and iron(III) ions is denoted either $Fe^{3+}(aq),Fe^{2+}(aq)|Pt(s)$ or $Pt(s)|Fe^{3+}(aq),Fe^{2+}(aq)$. In this case, the oxidized and reduced species are both in the same phase, so a comma rather than a line is used to separate them.

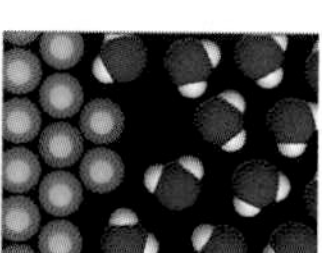

View an animation of ions and electrons carrying an electric current through a cell.

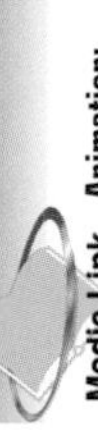

Media Link **Animation:** Fig. 12.3, Flow of current through an electrochemical cell

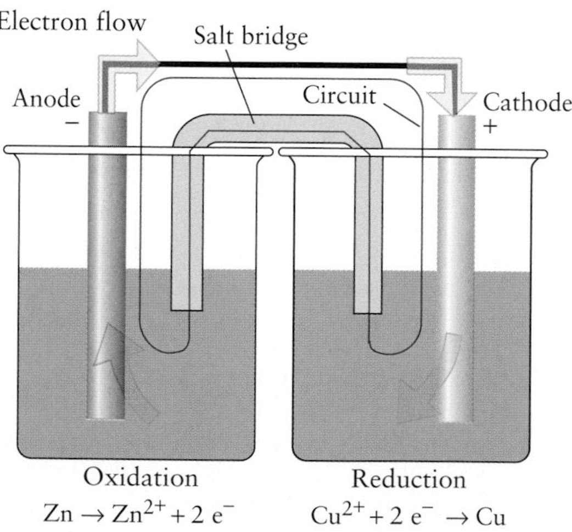

FIGURE 12.3 Electrons leave a galvanic cell at the anode (−), travel through the external circuit, and reenter the cell at the cathode (+). The source of the electrons is oxidation at the anode; the electrons pass through the external circuit and cause reduction at the cathode. The circuit is completed inside the cell by migration of ions through the salt bridge. A salt bridge is unnecessary when the two electrodes share a common electrolyte.

Self-Test 12.4A Write the diagram for a cell with a hydrogen electrode on the left and an iron(III)-iron(II) electrode on the right. The two electrodes are connected by a salt bridge.

[*Answer:* $Pt(s)|H_2(g)|H^+(aq)||Fe^{3+}(aq),Fe^{2+}(aq)|Pt(s)$]

Pairs of ions in solution are written in the order Ox,Red.

Self-Test 12.4B Write the diagram for a cell that has an electrode consisting of a manganese wire dipping into a solution of manganese(II) ions on the left, a salt bridge, and a copper(II)-copper(I) electrode on the right with a platinum wire.

The "left" and "right" of the diagram bear no relation to the way the cell is arranged in the laboratory. So the Daniell cell can be described as either

$$Zn(s)|Zn^{2+}(aq)||Cu^{2+}(aq)|Cu(s) \quad \text{or} \quad Cu(s)|Cu^{2+}(aq)||Zn^{2+}(aq)|Zn(s)$$

However, although both diagrams are equally valid depictions of the same cell, the IUPAC conventions described in the following paragraphs determine how we use the cell diagram to interpret measurements.

The emf of a cell is measured with an electronic voltmeter (Fig. 12.4). We identify the spontaneous direction of the reaction by connecting the voltmeter to the cell and seeing which terminal is positive. That terminal is functioning as the cathode. If the cathode is the electrode on the right *of the cell diagram*, then by convention, the emf of the cell described by the cell diagram is reported as positive, as in

Formerly, they were measured with a device called a *potentiometer*, but electronic devices are more reliable and easier to interpret.

$$\underset{\text{Anode }(-)}{Zn(s)|Zn^{2+}(aq)}||\underset{\text{Cathode }(+)}{Cu^{2+}(aq)|Cu(s)} \qquad E = +1.10\text{ V}$$

In this case, the electrons can be thought of as traveling through the external circuit from the left of the cell as written (the anode) to the right (the cathode). However, if the electrode on the *left* of the cell diagram turns

FIGURE 12.4 The cell potential is measured with an electronic voltmeter, a device designed to draw negligible current so that the composition of the cell does not change during the measurement. The display shows a positive value when the + terminal of the meter is connected to the cathode of the galvanic cell. The salt bridge completes the electric circuit within the cell.

Media Link **Lab video:** Fig. 12.4, The Daniell cell

out to be the cathode, then by convention, the emf of the cell described by the cell diagram is reported as negative, as in

$$\underset{\text{Cathode }(+)}{Cu(s)|Cu^{2+}(aq)}||\underset{\text{Anode }(-)}{Zn^{2+}(aq)|Zn(s)} \qquad E = -1.10\text{ V}$$

The flow of electrons in the external circuit is still from the anode to the cathode. We can think of the + sign on the cathode as indicating where electrons *enter* the cell from the external circuit and the − sign on the anode as showing where electrons *leave* the cell. Note that the sign of the emf is the same as that of the right-hand electrode *in the cell diagram.*

The chemical equation for the reaction corresponding to a given cell diagram is determined by the order of the diagram (see Toolbox 12.2). Regardless of which direction the reaction actually proceeds, we suppose that electrons travel from the left-hand electrode in the diagram and re-enter the cell at the right-hand electrode. In other words, *solely for the purpose of writing the cell reaction*, we suppose that the right-hand electrode in the cell diagram is the site of reduction.

Note that the right-hand electrode serves as a working cathode only if the cell reaction is spontaneous.

An electrode is designated by representing the interfaces between phases by a vertical line. A cell diagram depicts the physical arrangement of species and interfaces, with any salt bridge denoted by a double vertical line. The sign of the emf is the same as that of the right-hand electrode in the cell diagram.

TOOLBOX 12.2 How to write a cell reaction for a cell diagram

Conceptual Basis

A cell diagram implies a specific cell reaction, regardless of whether the reaction or its reverse is spontaneous. We suppose that the left-hand electrode in the cell diagram is the site of oxidation and that the right-hand electrode in the cell diagram is the site of reduction.

Procedure

Step 1 Write the equation for the half-reaction of the electrode on the right of the cell diagram as a reduction.

Step 2 Write the equation for the half-reaction of the electrode on the left of the cell diagram as an oxidation.

Step 3 Multiply one or both equations by a factor if necessary to equalize the number of electrons, and then add the two equations.

If the emf of the cell is positive when the electrode on the right of the cell diagram is attached to the positive terminal of a voltmeter, then the reaction is spontaneous as written. If the emf is negative, then the reverse reaction is spontaneous.

Self-Test 12.5A (a) Write the chemical equation for the reaction corresponding to the cell $Pt(s)|H_2(g)|H^+(aq)||Co^{3+}(aq),Co^{2+}(aq)|Pt(s)$. (b) Given that the cell has a positive emf, is the cell reaction spontaneous as written?

[*Answer:* (a) $H_2(g) + 2\ Co^{3+}(aq) \longrightarrow 2\ H^+(aq) + 2\ Co^{2+}(aq)$; (b) yes]

Self-Test 12.5B (a) Write the chemical equation for the reaction corresponding to the cell $Hg(l)|Hg_2Cl_2(s)|HCl(aq)||Hg_2(NO_3)_2(aq)|Hg(l)$. (b) Given that the cell has a positive emf, is the cell reaction spontaneous as written?

12.6 Standard Potentials

There are thousands of possible galvanic cells and therefore thousands of possible standard emfs. Instead of working with all these different cells, it is much simpler to think of each electrode as making a characteristic contribution called its **standard potential,** $E°$. Each standard potential is a measure of the electron-pulling power of a single electrode. In a galvanic cell, the electrodes pull in opposite directions, so the overall pulling power of the cell, the cell's standard emf, is the *difference* of the standard potentials of the two electrodes (Fig. 12.5). That difference is always written

$$E° = E°(\text{electrode on right of cell diagram}) - E°(\text{electrode on left of cell diagram}) \quad (4)$$

and may be positive or negative. If $E°$ is positive, then the right-hand electrode in the cell diagram is actually the cathode, the standard free energy of the cell reaction as written is negative, and the reaction is spontaneous under standard conditions. For example, the standard emf of the cell

$$Fe(s)|Fe^{2+}(aq)||Ag^{+}(aq)|Ag(s) \qquad E° = +1.24 \text{ V at } 25°C$$

is the difference in the potentials of the silver electrode, denoted $E°(Ag^{+}/Ag)$, and the iron electrode, denoted $E°(Fe^{2+}/Fe)$, under standard conditions (all solutes present at 1 mol·L^{-1} and all gases at 1 bar):

$$E° = E°(Ag^{+}/Ag) - E°(Fe^{2+}/Fe)$$

If $E°$ is negative, the right-hand electrode is actually the anode, the standard free energy of the reaction as written is positive, and the *reverse* of the reaction as written is spontaneous under standard conditions. For example, the cell diagram below represents a reaction that is not spontaneous:

$$Ag(s)|Ag^{+}(aq)||Fe^{2+}(aq)|Fe(s) \qquad E° = -1.24 \text{ V at } 25°C$$

A problem with compiling a list of standard potentials is that we know only the *overall* emf of the cell, not the contribution of a single electrode. A voltmeter placed between the two electrodes of a galvanic cell measures the *difference* of their potentials, not the individual values. To get around this difficulty, we arbitrarily set the standard potential of one particular electrode, the hydrogen electrode, equal to 0 at all temperatures:

$$2\ H^{+}(g) + 2\ e^{-} \longrightarrow H_2(g) \qquad E° = 0$$

In redox couple notation, $E°(H^{+}/H_2) = 0$ at all temperatures. The hydrogen electrode is then used to define the standard potentials of all other electrodes:

> The standard potential of a couple is the standard emf of a cell (including the sign) in which the couple forms the right-hand electrode in the cell diagram and a hydrogen electrode forms the left-hand electrode in the cell diagram.

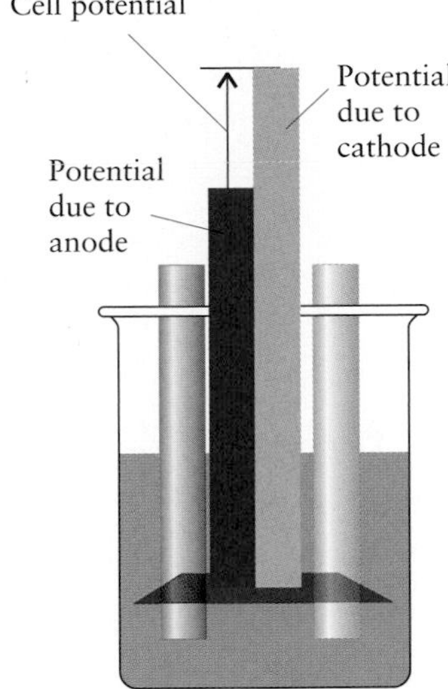

FIGURE 12.5 The cell potential can be thought of as being the difference of the two potentials produced by the two electrodes.

Standard potentials are also called *standard electrode potentials.* Because they are always written for reduction half-reactions, they are also sometimes called *standard reduction potentials.*

For example, at 25°C, we find that the standard emf of the cell

$$Pt(s)|H_2(g)|H^+(aq)||Zn^{2+}(aq)|Zn(s)$$

is −0.76 V. Because, according to convention, the hydrogen electrode contributes 0 to the standard emf of the cell, an emf of −0.76 V is attributed entirely to the zinc electrode, and we write $E°(Zn^{2+}/Zn) = -0.76$ V. Similarly, we find that the standard emf of the cell

$$Pt(s)|H_2(g)|H^+(aq)||Cu^{2+}(aq)|Cu(s)$$

is +0.34 V, so we conclude that $E°(Cu^{2+}/Cu) = +0.34$ V.

Because $E°(Cu^{2+}/Cu) > 0$, we know that the Cu^{2+}/Cu electrode is the cathode in the cell with H^+/H_2 as the other electrode, when all the species are in their standard states. Therefore, we know that the reduction of $Cu^{2+}(aq)$ by hydrogen,

$$Cu^{2+}(aq) + H_2(g) \longrightarrow Cu(s) + 2\,H^+(aq)$$

is spontaneous under standard conditions. We can also conclude that a positive standard potential means that the half-reaction $Cu^{2+}(aq) + 2\,e^- \rightarrow Cu(s)$ has a greater tendency to proceed under standard conditions than $2\,H^+(aq) + 2\,e^- \rightarrow H_2(g)$. In other words, Cu^{2+} ions are more strongly oxidizing than H^+ ions. In general, *the more positive the potential, the greater the electron-pulling power of the reduction half-reaction and, therefore, the more strongly oxidizing the species* (Fig. 12.6). A negative standard potential, like that of zinc, means that the Zn^{2+}/Zn electrode is the anode in a cell with H^+/H_2 as the other electrode and, therefore, that the reaction

$$Zn(s) + 2\,H^+(aq) \longrightarrow Zn^{2+}(aq) + H_2(g)$$

is spontaneous under standard conditions (that is, for this reaction, $K > 1$). We conclude that a negative standard potential means that the half-reaction $Zn(s) \rightarrow Zn^{2+}(aq) + 2\,e^-$ has a stronger tendency to proceed

Media Link Animation: Fig. 12.6, Reduction at a cathode

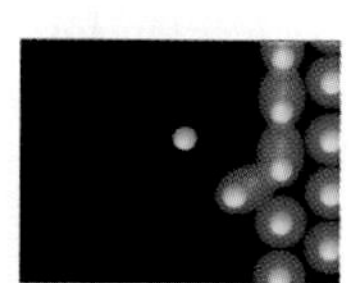

View an animation of the reduction of Zn^{2+} ions.

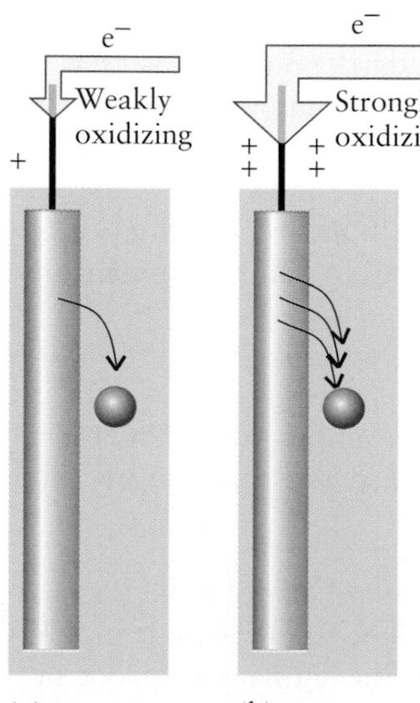

FIGURE 12.6 (a) A couple that has a small positive potential has a weak electron-pulling power (is a weak acceptor of electrons) relative to hydrogen ions and hence is a weak oxidizing agent. (b) A couple with a high positive potential has strong pulling power (is a strong acceptor of electrons) and is a strong oxidizing agent. The spheres represent cations formed as atoms are oxidized.

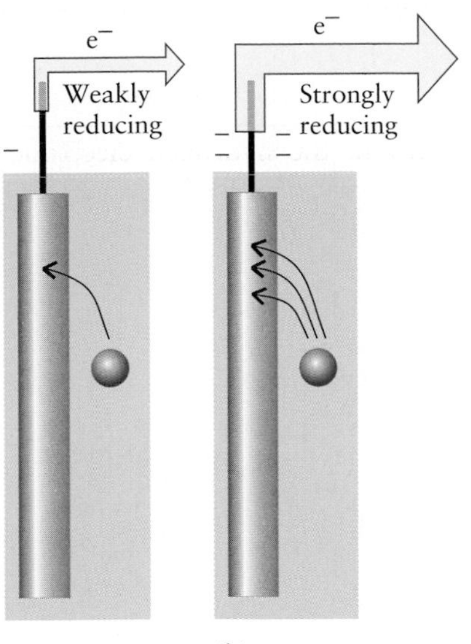

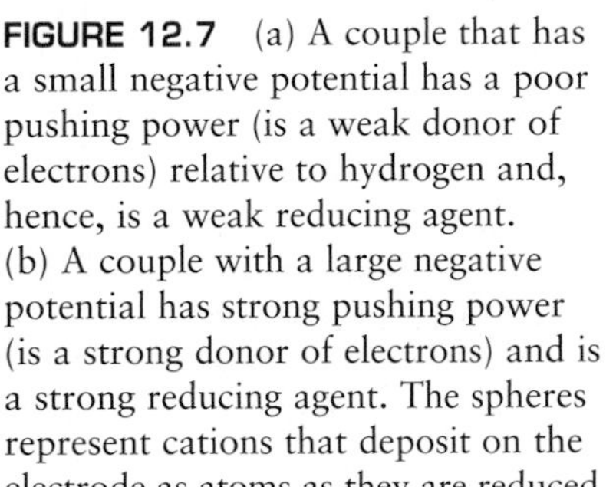

FIGURE 12.7 (a) A couple that has a small negative potential has a poor pushing power (is a weak donor of electrons) relative to hydrogen and, hence, is a weak reducing agent. (b) A couple with a large negative potential has strong pushing power (is a strong donor of electrons) and is a strong reducing agent. The spheres represent cations that deposit on the electrode as atoms as they are reduced.

View an animation of the oxidation of magnesium.

Media Link Animation: Fig. 12.7, Oxidation at an anode

than $H_2(g) \rightarrow 2\ H^+(aq) + 2\ e^-$ under standard conditions. In other words, zinc metal is a stronger reducing agent than hydrogen under standard conditions. In general, *the more negative the potential, the greater the electron-donating power of the oxidation half-reaction, and therefore the more strongly reducing the species* (Fig. 12.7).

Table 12.2 lists a number of standard electrode potentials measured at 25°C (the only temperature we consider); a longer list can be found in Appendix 2B. The standard potentials of elements vary in a complicated way through the periodic table (Fig. 12.8). However, the most negative—the most reducing—are usually found toward the left of the periodic table, and the most positive—the most oxidizing—are found toward the upper right.

Notice that in Appendix 2B, standard potentials are listed both by numerical value and alphabetically, to make it easy to find the one you want.

			H 0				He
Li −3.05	Be −1.85	B	C	N	O +1.23	F +2.87	Ne
Na −2.71	Mg −2.36	Al −1.66	Si	P	S −0.48	Cl +1.36	Ar
K −2.93	Ca −2.87	Ga −0.49	Ge	As	Se −0.67	Br +1.09	Kr
Rb −2.93	Sr −2.89	In −0.34	Sn −0.14	Sb	Te −0.84	I +0.54	Xe
Cs −2.92	Ba −2.91	Tl −0.34	Pb −0.13	Bi +0.20	Po	At	Rn
Fr	Ra −2.92						

FIGURE 12.8 The variation of standard potentials through the main groups of the periodic table. Note that the most negative values occur in the *s* block and that the most positive values occur close to fluorine.

TABLE 12.2 Standard Potentials at 25°C*

Species†	Reduction half-reaction	$E°$, V
Oxidized form is strongly oxidizing		
F_2/F^-	$F_2(g) + 2\,e^- \longrightarrow 2\,F^-(aq)$	+2.87
Au^+/Au	$Au^+(aq) + e^- \longrightarrow Au(s)$	+1.69
Ce^{4+}/Ce^{3+}	$Ce^{4+}(aq) + e^- \longrightarrow Ce^{3+}(aq)$	+1.61
$MnO_4^-, H^+/Mn^{2+}, H_2O$	$MnO_4^-(aq) + 8\,H^+(aq) + 5\,e^- \longrightarrow Mn^{2+}(aq) + 4\,H_2O(l)$	+1.51
Cl_2/Cl^-	$Cl_2(g) + 2\,e^- \longrightarrow 2\,Cl^-(aq)$	+1.36
$Cr_2O_7^{2-}, H^+/Cr^{3+}, H_2O$	$Cr_2O_7^{2-} + 14\,H^+(aq) + 6\,e^- \longrightarrow 2\,Cr^{3+}(aq) + 7\,H_2O(l)$	+1.33
$O_2, H^+/H_2O$	$O_2(g) + 4\,H^+(aq) + 4\,e^- \longrightarrow 2\,H_2O(l)$	+1.23; +0.82 at pH = 7
Br_2/Br^-	$Br_2(l) + 2\,e^- \longrightarrow 2\,Br^-(aq)$	+1.09
$NO_3^-, H^+/NO, H_2O$	$NO_3^-(aq) + 4\,H^+(aq) + 3\,e^- \longrightarrow NO(g) + 2\,H_2O(l)$	+0.96
Ag^+/Ag	$Ag^+(aq) + e^- \longrightarrow Ag(s)$	+0.80
Fe^{3+}/Fe^{2+}	$Fe^{3+}(aq) + e^- \longrightarrow Fe^{2+}(aq)$	+0.77
I_2/I^-	$I_2(s) + 2\,e^- \longrightarrow 2\,I^-(aq)$	+0.54
$O_2, H_2O/OH^-$	$O_2(g) + 2\,H_2O(l) + 4\,e^- \longrightarrow 4\,OH^-(aq)$	+0.40; +0.82 at pH = 7
Cu^{2+}/Cu	$Cu^{2+}(aq) + 2\,e^- \longrightarrow Cu(s)$	+0.34
$AgCl/Ag, Cl^-$	$AgCl(s) + e^- \longrightarrow Ag(s) + Cl^-(aq)$	+0.22
H^+/H_2	$2\,H^+(aq) + 2\,e^- \longrightarrow H_2(g)$	0, by definition
Fe^{3+}/Fe	$Fe^{3+}(aq) + 3\,e^- \longrightarrow Fe(s)$	−0.04
$O_2, H_2O/HO_2^-, OH^-$	$O_2(g) + H_2O(l) + 2\,e^- \longrightarrow HO_2^-(aq) + OH^-(aq)$	−0.08
Pb^{2+}/Pb	$Pb^{2+}(aq) + 2\,e^- \longrightarrow Pb(s)$	−0.13
Sn^{2+}/Sn	$Sn^{2+}(aq) + 2\,e^- \longrightarrow Sn(s)$	−0.14
Fe^{2+}/Fe	$Fe^{2+}(aq) + 2\,e^- \longrightarrow Fe(s)$	−0.44
Zn^{2+}/Zn	$Zn^{2+}(aq) + 2\,e^- \longrightarrow Zn(s)$	−0.76
$H_2O/H_2, OH^-$	$2\,H_2O(l) + 2\,e^- \longrightarrow H_2(g) + 2\,OH^-(aq)$	−0.83; −0.42 at pH = 7
Al^{3+}/Al	$Al^{3+}(aq) + 3\,e^- \longrightarrow Al(s)$	−1.66
Mg^{2+}/Mg	$Mg^{2+}(aq) + 2\,e^- \longrightarrow Mg(s)$	−2.36
Na^+/Na	$Na^+(aq) + e^- \longrightarrow Na(s)$	−2.71
K^+/K	$K^+(aq) + e^- \longrightarrow K(s)$	−2.93
Li^+/Li	$Li^+(aq) + e^- \longrightarrow Li(s)$	−3.05
Reduced form is strongly reducing		

* For a more extensive table, see Appendix 2B.

† In the notation X/Y, X is the oxidized species (the reactant, the oxidizing agent) and Y is the reduced species (the product, the reducing agent) in the half-reaction.

How can we use this information . . . to determine the standard potential of an electrode?

We can determine the standard potential of an electrode by setting up a standard cell in which the other electrode has a known standard potential. The standard potential of a zinc electrode is −0.76 V, and the standard emf of the cell

$$Zn(s)|Zn^{2+}(aq)||Sn^{4+}(aq),Sn^{2+}(aq)|Pt(s)$$

is +0.91 V. We can calculate the standard potential of the Sn^{4+}/Sn^{2+} electrode from this information, because (as stated in Eq. 4) the difference of the standard potentials, $E°(\text{right}) - E°(\text{left})$, is equal to the standard emf of the cell:

$$E° = E°(Sn^{4+}/Sn^{2+}) - E°(Zn^{2+}/Zn) = +0.91\text{ V}$$

Therefore,

$$\begin{aligned} E°(Sn^{4+}/Sn^{2+}) &= 0.91\text{ V} + E°(Zn^{2+}/Zn) \\ &= 0.91\text{ V} - 0.76\text{ V} = +0.15\text{ V} \end{aligned}$$

Self-Test 12.6A The standard potential of the Ag^{+}/Ag electrode is +0.80 V, and the standard emf of the cell $Pt(s)|I_2(s)|I^-(aq)||Ag^+(aq)|Ag(s)$ is +0.26 V at the same temperature. What is the standard potential of the I_2/I^- electrode?

[*Answer:* +0.54 V]

Self-Test 12.6B The standard potential of the Fe^{2+}/Fe electrode is −0.44 V and the standard emf of the cell $Fe(s)|Fe^{2+}(aq)||Pb^{2+}(aq)|Pb(s)$ is +0.31 V. What is the standard potential of the Pb^{2+}/Pb electrode?

In some cases, we find that the tables of data available do not contain precisely the standard potential we need but do contain closely related values for the same element; for instance, we might require the standard potential of the Ce^{4+}/Ce couple, whereas we know only the values for the Ce^{3+}/Ce and Ce^{4+}/Ce^{3+} couples. In such cases, we calculate the standard potential of the couple we need by combining the values for related couples. However, the potential of a couple cannot in general be determined by adding or subtracting the standard potentials directly. Instead, we calculate the values of $\Delta G_r°$ for each half-reaction and combine them into the $\Delta G_r°$ for the desired half-reaction. We then convert that value of $\Delta G_r°$ into the corresponding standard potential by using Eq. 3.

The standard hydrogen electrode is assigned zero potential. The standard potential of an electrode is the standard emf of a cell in which the electrode on the left in the cell diagram is a hydrogen electrode. The standard potential of a redox couple may be calculated from the free energies of redox couples for different oxidation states.

EXAMPLE 12.3 Calculating the standard potential of a couple from two related couples

Use the information in Appendix 2B to determine the standard potential for the redox couple Ce^{4+}/Ce, for which the reduction half-reaction is

$$Ce^{4+}(aq) + 4\,e^- \longrightarrow Ce(s) \qquad \text{(B)}$$

STRATEGY Use the alphabetical listing in Appendix 2B to find half-reactions that can be combined to give the desired half-reaction. Combine these half-reactions and their free energies of reaction. Convert free energies to standard potentials by using Eq. 3 and then simplify the resulting expressions.

SOLUTION From the data in Appendix 2B, we can write

$$Ce^{3+}(aq) + 3\,e^- \longrightarrow Ce(s) \qquad E° = -2.48\text{ V} \tag{C}$$

$$Ce^{4+}(aq) + e^- \longrightarrow Ce^{3+}(aq) \qquad E° = +1.61\text{ V} \tag{D}$$

We add the two half-reactions to generate the overall half-reaction (reaction B), so we need to do the same to the reaction free energies. We use Eq. 3 to convert from $E°$ to $\Delta G_r°$:

$$-4FE°(\text{B}) = -3FE°(\text{C}) + (-FE°(\text{D}))$$

Now we rearrange this expression to derive $E°$ for the overall half-reaction:

$$E°(\text{B}) = \frac{3E°(\text{C}) + E°(\text{D})}{4} = \frac{3(-2.48\text{ V}) + 1.61\text{ V}}{4} = -1.46\text{ V}$$

Notice that this value is *not* the same as the sum of the potentials for half-reactions C and D.

Self-Test 12.7A Use the data in Appendix 2B to calculate the standard potential of the couple $Au^{3+}(aq)/Au^{+}(aq)$.

[*Answer:* +1.26 V]

Self-Test 12.7B Use the data in Appendix 2B to calculate the standard potential of the couple $Mn^{3+}(aq)/Mn(s)$.

12.7 The Significance of Standard Potentials

We can now predict that magnesium, iron, indium, tin, lead, and other metals with negative standard potentials have a thermodynamic tendency to reduce hydrogen ions when the H_3O^+ concentration is 1 mol·L^{-1} (corresponding to 1 M strong acid). In other words, hydrogen ions can oxidize these metals. On the other hand, because metals with positive standard potentials (those above hydrogen in the table) cannot reduce hydrogen ions under standard conditions (H_2 has a greater reducing power), we know that they cannot produce hydrogen gas when acted on by 1 M strong acid. For example, copper and the noble metals silver, platinum, and gold are not oxidized by hydrogen ions. In fact, their cations can be *reduced* by hydrogen gas. We can summarize these conclusions for a metal M as follows.

We have written the equation for a metal that forms M^{2+} ions; for other metals, the chemical equation will be different.

- $M^{2+}(aq) + H_2(g) \rightarrow M(s) + 2\,H^+(aq)$ is spontaneous under standard conditions if $E°(M^{2+}/M) > 0$; that is, M^{2+} is more strongly oxidizing than H^+.
- $M(s) + 2\,H^+(aq) \rightarrow M^{2+}(aq) + H_2(g)$ is spontaneous under standard conditions if $E°(M^{2+}/M) < 0$; that is, M is more strongly reducing than H_2.

A thermodynamic tendency is not always realized in practice, often because the reaction is very slow or because a protective oxide is formed. For example, the standard potential of aluminum (−1.66 V) suggests that, like magnesium, it should produce hydrogen when added to 1 M strong

FIGURE 12.9 Aluminum has a negative standard potential, signifying that it can be oxidized by hydrogen ions. Oxidation does indeed occur when aluminum is placed into a solution of hydrochloric acid (left). However, nitric acid (right) stops reacting with it as soon as an impenetrable layer of aluminum oxide has formed on its surface. This resistance to further reaction is termed passivation of the metal.

acid. In accord with this prediction, aluminum can indeed be oxidized by 1 M HCl(aq). However, aluminum does not react with 1 M HNO_3(aq) because any Al^{3+} ions that are produced in nitric acid immediately form a layer of insoluble oxide on the surface of the metal. This layer prevents further reaction, and we say that the metal has been **passivated**, or protected from further reaction, by a surface film (Fig. 12.9). The passivation of aluminum is of great commercial importance, because it enables the otherwise reactive metal to be used, among other things, for airplanes and corrosion-resistant window frames in buildings.

In general, the oxide of any metal that takes up more space than the metal it replaces acts as a **protective oxide**, an oxide that protects the metal from further oxidation. Zinc and chromium metal are plated on iron to protect it from oxidation, although both zinc and chromium have more negative standard potentials than iron. Zinc and chromium both form low-density protective oxides that inhibit oxidation.

A metal with a negative standard potential has a thermodynamic tendency to reduce hydrogen ions in solution; the ions of a metal with a positive standard potential have a tendency to be reduced by hydrogen gas.

12.8 The Electrochemical Series

Standard potentials are used to arrange oxidizing and reducing agents in order of their strengths and to predict the thermodynamic tendency for a reaction to occur. As we have seen, *the more negative the standard*

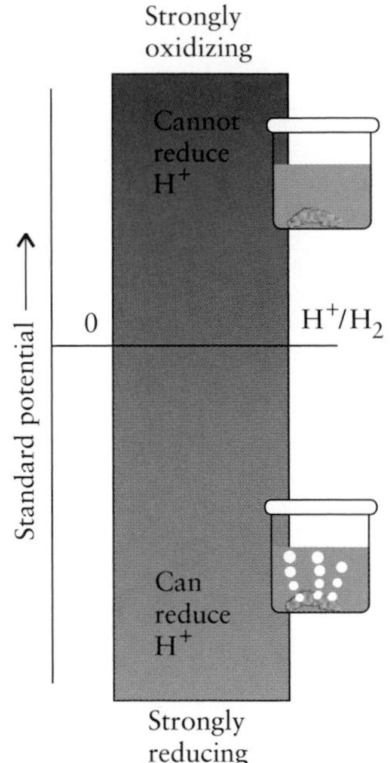

FIGURE 12.10 The significance of standard potentials. Only couples with negative standard potentials (and hence lying below hydrogen in the electrochemical series) can reduce hydrogen ions. The reducing power increases as the standard potential becomes more negative. Only the reduced form of species with negative standard potentials (and hence lying below hydrogen) can be oxidized by hydrogen ions under standard conditions.

potential of a couple, the greater its reducing strength. Only a couple with a negative potential can reduce hydrogen ions under standard conditions. A couple with a positive potential, such as Au^{3+}/Au, cannot reduce hydrogen ions under standard conditions (Fig. 12.10).

When Table 12.2 is viewed as a table of relative strengths of oxidizing and reducing agents, it is called the **electrochemical series**. The species on the left of each equation in Table 12.2 are potentially oxidizing agents. They can themselves be reduced. Species on the right of the equations are potentially reducing agents. An oxidized species in the list (on the left of the equation) has a tendency to oxidize a reduced species that lies below it. For example, Cu^{2+} ions oxidize zinc metal. A reduced species (on the right of the equation) has a tendency to reduce an oxidized species that lies above it. For example, zinc metal reduces H^+ ions.

The higher the position of a species on the left side of an equation in Table 12.2, the greater its oxidizing strength. For example, F_2 is a strong oxidizing agent, whereas Li^+ is a very, very poor oxidizing agent. It also follows that *the lower the standard potential, the greater the reducing strength of the reduced species on the right side of an equation in Table 12.2*. For example, lithium metal is the strongest reducing agent in the table.

Self-Test 12.8A Can mercury produce zinc metal from aqueous zinc sulfate under standard conditions?

[*Answer:* No, because mercury lies below zinc in the table.]

Self-Test 12.8B Can chlorine gas oxidize water to oxygen gas under standard conditions in basic solution?

Most commonly, when we write a cell diagram, we want the corresponding cell reaction to be spontaneous. A spontaneous cell reaction under standard conditions has a negative standard free energy and therefore a positive standard emf. We get a positive standard emf when the right-hand electrode in the cell diagram is the cathode and the left-hand electrode is the anode. Therefore, to write a cell diagram for a spontaneous reaction involving a given pair of redox couples, we identify the couple with the higher standard potential: that couple forms the cathode, so we put it on the right of the cell diagram.

The oxidizing and reducing power of a redox couple determine its position in the electrochemical series. The strongest oxidizing agents are at the top of the table; the strongest reducing agents are at the bottom of the table.

EXAMPLE 12.4 Predicting relative oxidizing strengths and standard cell emfs

Design an electrochemical cell that could be used to decide whether an acidified aqueous permanganate solution is a more powerful oxidizing agent than an acidified aqueous dichromate solution under standard conditions. Determine the standard emf of the cell. Specify the cell by using a cell diagram and write the net ionic equation for the spontaneous reaction between the two reagents.

STRATEGY Inspect the electrochemical series in Table 12.2 (or the longer version in Appendix 2B) and identify the couple with the higher potential. That couple will form the cathode. It will be the right-hand electrode in the cell diagram and undergo reduction in the overall cell reaction.

SOLUTION We find the following two half-reactions in Appendix 2B:

$$MnO_4^-(aq) + 8\,H^+(aq) + 5\,e^- \longrightarrow Mn^{2+}(aq) + 4\,H_2O(l) \qquad E° = +1.51\text{ V}$$

$$Cr_2O_7^{2-}(aq) + 14\,H^+(aq) + 6\,e^- \longrightarrow 2\,Cr^{3+}(aq) + 7\,H_2O(l) \qquad E° = +1.33\text{ V}$$

Because $E°(MnO_4^-,H^+/Mn^{2+},H_2O)$ lies above $E°(Cr_2O_7^{2-},H^+/Cr^{3+},H_2O)$, MnO_4^- is a stronger oxidizing agent than $Cr_2O_7^{2-}$ in acidic solution. Therefore, the manganese electrode is the cathode (an oxidizing agent is reduced). The standard emf of the cell

$$\underbrace{Pt(s)|Cr_2O_7^{2-}(aq), Cr^{3+}(aq), H^+(aq)}_{\text{Anode}}||\underbrace{H^+(aq), MnO_4^-(aq), Mn^{2+}(aq)|Pt(s)}_{\text{Cathode}}$$

is the difference of the potentials for the two electrodes (Eq. 4):

$$E° = 1.51\text{ V} - 1.33\text{ V} = +0.18\text{ V}$$

To construct the spontaneous cell reaction, we combine the two half-reactions. Because the manganese couple has the higher standard potential, it will form the cathode; we leave it unchanged at this stage. The chromium couple will form the anode. We therefore reverse its half-reaction

$$2\,Cr^{3+}(aq) + 7\,H_2O(l) \longrightarrow Cr_2O_7^{2-}(aq) + 14\,H^+(aq) + 6\,e^-$$

To match numbers of electrons, we multiply the manganese half-reaction by 6 and the chromium half-reaction by 5. Their net sum gives the spontaneous cell reaction:

$$6\,MnO_4^-(aq) + 11\,H_2O(l) + 10\,Cr^{3+}(aq) \longrightarrow 6\,Mn^{2+}(aq) + 22\,H^+(aq) + 5\,Cr_2O_7^{2-}(aq)$$

We can conclude that an acidified aqueous permanganate solution is a more powerful oxidizing agent than an acidified aqueous dichromate solution under standard conditions.

Self-Test 12.9A Which metal is the stronger reducing agent, zinc or nickel? Evaluate the standard emf of the appropriate cell, write the net ionic equation for the spontaneous reaction, and specify the cell with a cell diagram.

[*Answer:* Zinc; +0.53 V; $Zn(s) + Ni^{2+}(aq) \rightarrow Zn^{2+}(aq) + Ni(s)$, $Zn(s)|Zn^{2+}(aq)||Ni^{2+}(aq)|Ni(s)$]

Self-Test 12.9B Which is the stronger oxidizing agent, Cu^{2+} or Ag^+? Evaluate the standard emf of the appropriate cell, write the net ionic equation for the corresponding cell reaction, and specify the cell with a cell diagram.

12.9 Standard Potentials and Equilibrium Constants

One of the most useful applications of standard potentials is the calculation of equilibrium constants from electrochemical data. The techniques we develop here can be applied to neutralization, precipitation, and redox reactions, provided they can be expressed as the difference of two reduction half-reactions. It may seem puzzling at first that electrochemical data can be used to calculate the equilibrium constants for reactions that are not redox reactions, but we shall see that this is the case.

We saw in Section 9.3 that the standard reaction free energy, $\Delta G_r°$, is related to the equilibrium constant of the reaction by

$$\Delta G_r° = -RT \ln K$$

In this chapter, we have seen that the standard reaction free energy is related to the standard emf of a galvanic cell by

$$\Delta G_r° = -nFE°$$

When we combine the two equations, we get

$$-nFE° = -RT \ln K \tag{5}$$

This expression can be rearranged to allow us to calculate the equilibrium constant from the cell emf:

$$\ln K = \frac{nFE°}{RT} \tag{6}$$

Because we can calculate $E°$ from standard potentials, we can now also calculate equilibrium constants for any reaction that can be expressed in terms of two half-reactions. Toolbox 12.3 summarizes the steps involved and Example 12.5 shows the steps in action. Equation 6 also shows that the magnitude of $E°$ for a cell reaction is an indication of the equilibrium composition. It follows from the equation that a reaction with a large positive $E°$ has a very large K. A reaction with a large negative $E°$ has a very small K.

The equilibrium constant of a reaction can be calculated from standard potentials by combining the equations for the half-reactions to give the cell reaction of interest and determining the standard potential of the corresponding cell.

TOOLBOX 12.3 How to calculate equilibrium constants from electrochemical data

Conceptual Basis

The standard emf of a cell is proportional to the standard free energy of the cell reaction, which in turn is proportional to the logarithm of the equilibrium constant. A cell reaction with a positive emf has a negative standard free energy, which means that it has a strong drive to occur, producing a high proportion of products at equilibrium. Therefore, we expect $K > 1$. The opposite is true for a cell reaction with a negative emf.

Procedure

The procedure for calculating an equilibrium constant from electrochemical data is as follows.

Step 1 Write the balanced equation for the reaction of interest. Then find two reduction half-reactions to combine to give the equation of interest. Reverse one and add them together.

Step 2 Identify the value of n from the change in oxidation numbers, or by examining the half-reactions (after multiplication by appropriate factors) for the number of electrons transferred in the balanced equation.

Step 3 To obtain $E°$, subtract the standard potential of the half-reaction that was reversed from the standard potential of the half-reaction that was left as a reduction.

Step 4 Use the relation $\ln K = nFE°/RT$ to calculate the value of K.

At 25.00°C (298.15 K), $RT/F = 0.025\,693$ V; so, at that temperature,

$$\ln K = \frac{nE°}{0.025\,693\ \text{V}}$$

EXAMPLE 12.5 Calculating the equilibrium constant for a reaction

Calculate the equilibrium constant at 25°C for the reaction

$$AgCl(s) \longrightarrow Ag^{+}(aq) + Cl^{-}(aq)$$

The equilibrium constant for this reaction is actually the solubility product, $K_{sp} = [Ag^{+}][Cl^{-}]$, for silver chloride (Section 11.10).

STRATEGY It does not matter that overall the reaction is not a redox reaction, provided it can be expressed as the difference of two reduction half-reactions. Because silver chloride is almost insoluble, we expect K to be very small (and $E°$ to be negative). Follow the procedure in Toolbox 12.3.

SOLUTION *Step 1* The cell reaction is written above. The two reduction half-reactions required are

$$AgCl(s) + e^{-} \longrightarrow Ag(s) + Cl^{-}(aq) \qquad E° = +0.22\ \text{V}$$

$$Ag^{+}(aq) + e^{-} \longrightarrow Ag(s) \qquad E° = +0.80\ \text{V}$$

To obtain the reaction of interest, we need to reverse the second half-reaction:

$$Ag(s) \longrightarrow Ag^{+}(aq) + e^{-}$$

When we add this equation to the AgCl reduction, Ag(s) cancels and the result is the reaction we require:

$$AgCl(s) \longrightarrow Ag^{+}(aq) + Cl^{-}(aq)$$

Step 2 From the half-reactions that are added to obtain the balanced equation, we see that $n = 1$. Alternatively, note that the oxidation number of silver changes by 1.

Step 3 Subtract the standard potential of the half-reaction that we reversed from that of the reduction half-reaction. The standard emf is the difference

$$E° = 0.22\ \text{V} - 0.80\ \text{V} = -0.58\ \text{V}$$

The emf is negative, as expected. Now we find $\ln K_{sp}$ by using Eq. 6:

$$\ln K_{sp} = \frac{nFE^\circ}{RT} = \frac{nE^\circ}{(RT/F)} = \frac{(1) \times (-0.58\text{ V})}{0.025\ 693\text{ V}} = -\frac{0.58}{0.025\ 693}$$

Taking the antilogarithm ($e^{-0.58/0.025\ 693}$) gives $K_{sp} = 1.6 \times 10^{-10}$, the value listed in Table 11.6. Many of the solubility products listed in tables, such as those in Table 11.6, were determined from emf measurements.

Self-Test 12.10A Use Appendix 2B to calculate the solubility product of mercury(I) chloride, Hg_2Cl_2.

[*Answer:* 2.6×10^{-18}]

Self-Test 12.10B Use the tables in Appendix 2B to calculate the solubility product of cadmium hydroxide, $Cd(OH)_2$.

12.10 The Nernst Equation

As a reaction proceeds toward equilibrium, the concentrations of its reactants and products change and ΔG_r approaches 0. Therefore, as reactants are consumed in an electrochemical cell, the cell potential decreases until finally it reaches 0. A "dead" battery is one in which the cell reaction has reached equilibrium. At equilibrium, a cell generates zero potential difference across its electrodes and the reaction can no longer do work. To understand this behavior quantitatively, we need to find how the cell emf varies with the concentrations of species in the cell.

How do we do that?

To establish how the emf of a cell depends on concentration, we first note that, as the reactants in a chemical reaction are used up and the concentrations of products increase, ΔG_r changes until it reaches 0 at equilibrium. Because the emf is proportional to the reaction free energy (Eq. 2), it follows that E also changes as the reaction proceeds. We already know how ΔG_r varies with composition:

$$\Delta G_r = \Delta G_r^\circ + RT \ln Q$$

where Q is the reaction quotient for the cell reaction (this is Eq. 7 of Section 9.3). Because $\Delta G_r = -nFE$ and $\Delta G_r^\circ = -nFE^\circ$, it follows at once that

$$-nFE = -nFE^\circ + RT \ln Q$$

When we divide through by $-nF$, we get the **Nernst equation:**

$$E = E^\circ - \frac{RT}{nF} \ln Q \qquad (7)$$

At 298.15 K, $RT/F = 0.025\ 693$ V, so the Nernst equation takes the form

$$E = E^\circ - \frac{0.025\ 693\text{ V}}{n} \ln Q$$

It is sometimes convenient to use this equation with common logarithms, in which case, we make use of the relation log x = 2.303 ln x:

$$E = E° - \frac{2.303\ RT}{nF} \log Q = E° - \frac{0.05916\ \text{V}}{n} \log Q$$

Equation 7 was first derived by the German chemist Walther Nernst and is widely used to estimate the emf of cells. It is also used in biology to estimate the potential difference across biological cell membranes, such as those of neurons. Another application of the Nernst equation is the measurement of concentration. In a **concentration cell,** two electrodes with different concentrations of the same solute, one known and one unknown, are used to generate a potential. For such a cell, $E° = 0$ and the measured potential is used to determine the unknown concentration of the solute. For example, a concentration cell involving two Ag^+/Ag electrodes would have $n = 1$ and $Q = [Ag^+]_{anode}/[Ag^+]_{cathode}$. If the concentration of Ag^+ at the cathode is set to 1 mol·L^{-1}, the Nernst equation will have the form $E = -(0.025\ 693\ \text{V}) \ln[Ag^+]_{anode}$; so, by measuring E, we can infer the concentration of Ag^+ in the anode compartment.

The dependence of emf on composition is expressed by the Nernst equation, Eq. 7.

EXAMPLE 12.6 Using the Nernst equation to predict an emf

Calculate the emf at 25°C of a Daniell cell in which the concentration of Zn^{2+} ions is 0.10 mol·L^{-1} and that of the Cu^{2+} ions is 0.0010 mol·L^{-1}.

STRATEGY First, write the balanced equation and the corresponding expression for Q and note the value of n. Then determine $E°$ from the standard potentials in Table 12.2 (or Appendix 2B). Determine the value of Q for the stated conditions. Calculate the emf by substituting these values into the Nernst equation, Eq. 7. At 25.00°C, RT/F = 0.025 693 V.

SOLUTION The reaction in the Daniell cell is

$$Cu^{2+}(aq) + Zn(s) \longrightarrow Zn^{2+}(aq) + Cu(s) \qquad n = 2,\ E° = +1.10\ \text{V}$$

The reaction quotient is

$$Q = \frac{a_{Zn^{2+}}}{a_{Cu^{2+}}} = \frac{[Zn^{2+}]}{[Cu^{2+}]}$$

We are assuming that the solution is so dilute that activities can be replaced by molar concentrations. The Nernst equation gives

$$\begin{aligned} E &= 1.10\ \text{V} - \frac{0.025\ 693\ \text{V}}{2} \ln \frac{0.10}{0.0010} \\ &= 1.10\ \text{V} - 0.059\ \text{V} = +1.04\ \text{V} \end{aligned}$$

Self-Test 12.11A Calculate the emf of the cell $Zn(s)|Zn^{2+}(aq, 1.50\ mol\cdot L^{-1})||Fe^{2+}(aq, 0.10\ mol\cdot L^{-1})|Fe(s)$.

[*Answer:* +0.29 V]

Self-Test 12.11B Calculate the emf of the cell $Ag(s)|Ag^{+}(aq, 0.0010\ mol\cdot L^{-1})||Ag^{+}(aq, 0.010\ mol\cdot L^{-1})|Ag(s)$. *Hint:* $E° = 0$.

12.11 Ion-Selective Electrodes

One important application of the Nernst equation is the measurement of pH (and, through pH, acidity constants). The pH of a solution can be measured electrochemically with a device called a *pH meter*. The technique makes use of a cell in which one electrode is sensitive to the H_3O^+ concentration and the second electrode serves as a reference. An electrode sensitive to the concentration of a particular ion is called an **ion-selective electrode.**

One combination of electrodes that could be used to determine pH is a hydrogen electrode connected through a salt bridge to a *calomel electrode*. The reduction half-reaction for the calomel electrode is

Calomel is the common name for mercury(I) chloride, Hg_2Cl_2.

$$Hg_2Cl_2(s) + 2\,e^- \longrightarrow 2\,Hg(l) + 2\,Cl^-(aq) \qquad E° = +0.27\ V$$

The overall cell reaction is

$$Hg_2Cl_2(s) + H_2(g) \longrightarrow 2\,H^+(aq) + 2\,Hg(l) + 2\,Cl^-(aq) \qquad Q = \frac{[H^+]^2[Cl^-]^2}{P_{H_2}}$$

Provided the pressure of hydrogen is 1 bar, we can write the reaction quotient as $Q = [H^+]^2[Cl^-]^2$. To find the concentration of hydrogen ions, we write the Nernst equation:

$$\begin{aligned} E &= E° - \tfrac{1}{2}(0.0257\ V) \times \ln[H^+]^2[Cl^-]^2 \\ &= E° - \tfrac{1}{2}(0.0257\ V) \times \ln[Cl^-]^2 - \tfrac{1}{2}(0.0257\ V) \times \ln[H^+]^2 \\ &= E° - (0.0257\ V) \times \ln[Cl^-] - (0.0257\ V) \times \ln[H^+] \end{aligned}$$

The Cl^- concentration of a calomel electrode is fixed at the time of manufacture by saturating the solution with KCl, so $[Cl^-]$ is a constant. We can therefore combine the first two terms on the right into a single constant, $E' = E° - (0.0257\ V) \times \ln[Cl^-]$. Then, because $\ln x = 2.303 \log x$,

$$\begin{aligned} E &= E' - 2.303 \times (0.0257\ V) \times \log[H^+] \\ &= E' + (0.0592\ V) \times pH \end{aligned}$$

FIGURE 12.11 A glass electrode in a protective plastic sleeve (left) is used to measure pH. It is used in conjunction with a calomel electrode (right) in pH meters such as this one.

Therefore, by measuring the cell emf, E, we can determine the pH. The value of E' is established by calibrating the cell, which provides the value of E for a solution of known pH.

A **glass electrode,** a thin-walled glass bulb containing an electrolyte, is much easier to use than a hydrogen electrode and has a potential that is proportional to the pH outside the glass bulb (Fig. 12.11). Often there is a calomel electrode built into the probe that makes contact with the test solution through a miniature salt bridge. A pH meter therefore usually has only one probe, which contains a complete electrochemical cell. The meter is calibrated with a buffer of known pH, and the measured cell emf is then automatically converted into the pH of the solution, which is displayed.

Commercially available electrodes used in *pX meters* are sensitive to other ions, such as Na^+, Ca^{2+}, NH_4^+, CN^-, and S^{2-}. They are used to monitor industrial processes and in pollution control.

The pH or concentrations of ions can be measured by using an electrode that responds selectively to only one species of ion.

ELECTROLYSIS

Redox reactions that have a positive reaction free energy are not spontaneous, but electricity can be used to *make* them occur. For example, there is no common spontaneous chemical reaction in which fluorine is a product, so the element cannot be isolated by any common chemical reaction. It was not until 1886 that the French chemist Henri Moissan found a way to force the formation of fluorine by passing an electric current through an anhydrous molten mixture of potassium fluoride and hydrogen fluoride. Fluorine is still prepared commercially by the same process.

In this part of Chapter 12, we study **electrolysis,** the process of driving a reaction in a nonspontaneous direction by using an electric current. First, we see how electrochemical cells are constructed for electrolysis and how to predict the potential needed to bring electrolysis about. Then we examine the products of electrolysis and see how to predict the amount of products to expect for a given flow of electric current.

12.12 Electrolytic Cells

An **electrolytic cell** is an electrochemical cell in which electrolysis takes place. The arrangement of components in electrolytic cells is different from that in galvanic cells. Specifically, the two electrodes usually share the same compartment, there is usually only one electrolyte, and concentrations and pressures are usually far from standard.

Figure 12.12 shows the layout of an electrolytic cell used for the commercial production of magnesium metal from molten magnesium chloride (the *Dow process*). As in a galvanic cell, oxidation occurs at the anode and reduction occurs at the cathode, electrons travel through the external wire from anode to cathode, cations move through the electrolyte toward the cathode, and anions move toward the anode. But unlike the spontaneous current in a galvanic cell, a current must be supplied by an external electrical power source. That power source can be a galvanic cell, which provides the current that drives electrons through the wire in a predetermined direction (Fig. 12.13). The result is to force oxidation at one electrode and reduction at the other. For example, the following half-reactions are made to occur in Dow process:

The anode of an electrolytic cell is labeled + and the cathode −, the opposite of a galvanic cell.

Anode reaction: $2\ Cl^-(\text{melt}) \longrightarrow Cl_2(g) + 2\ e^-$

Cathode reaction: $Mg^{2+}(\text{melt}) + 2\ e^- \longrightarrow Mg(l)$

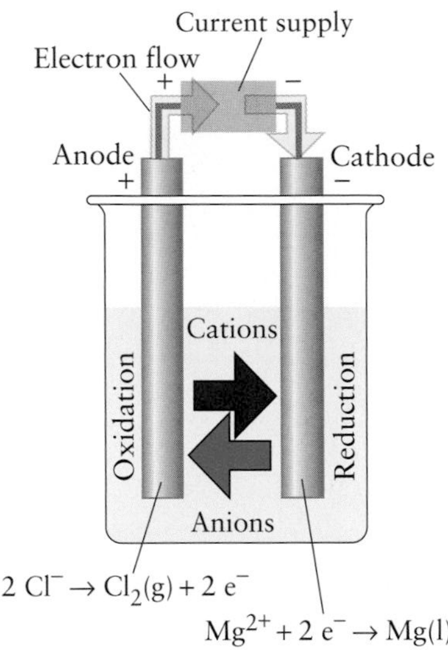

FIGURE 12.12 A schematic diagram of the electrolytic cell used in the Dow process for magnesium. The electrolyte is molten magnesium chloride. As the current generated by the external source passes through the cell, magnesium ions are reduced to magnesium metal at the cathode and chloride ions are oxidized to chlorine gas at the anode.

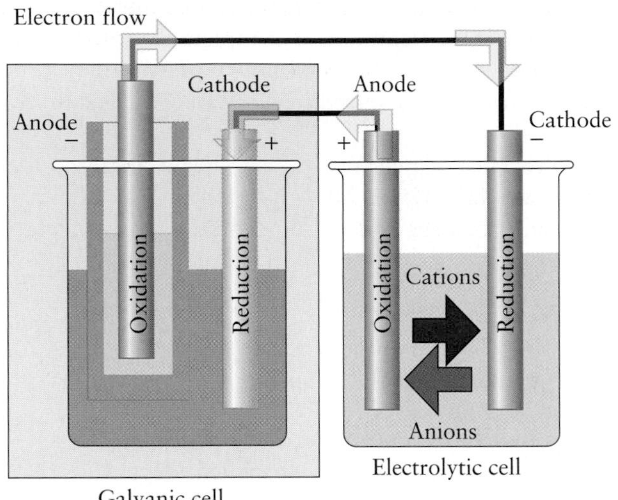

FIGURE 12.13 In this schematic diagram of an electrolysis experiment, a galvanic cell is used to power an electrolytic cell. Electrons emerge from the galvanic cell at its anode (−) and enter the electrolytic cell at its cathode (−), where they bring about reduction. Electrons are drawn out of the electrolytic cell through its anode (+) and into the galvanic cell at its cathode (+). If the cell reaction in the galvanic cell is more strongly spontaneous than the reaction in the electrolytic cell is nonspontaneous, then the overall process is spontaneous. This experiment is an example of one reaction driving another, to which it is coupled.

where "melt" signifies the molten salt. A rechargeable battery functions as a galvanic cell when it is doing work and as an electrolytic cell when it is being recharged.

> *In an electrolytic cell, current supplied by an external source is used to drive a nonspontaneous reaction.*

12.13 The Potential Needed for Electrolysis

To drive a reaction in a nonspontaneous direction, the external supply must generate a potential difference greater than the potential difference that would be produced by the reverse reaction. For example,

$$2\,H_2(g) + O_2(g) \longrightarrow 2\,H_2O(l) \qquad E = +1.23\text{ V at pH} = 7,\ \text{spontaneous}$$

To achieve the nonspontaneous reaction

$$2\,H_2O(l) \longrightarrow 2\,H_2(g) + O_2(g) \qquad E = -1.23\text{ V at pH} = 7,\ \text{nonspontaneous}$$

we must apply at least 1.23 V from the external source to overcome the reaction's natural pushing power in the opposite direction. In practice, the applied potential difference must usually be significantly greater than the cell emf in order to reverse a spontaneous cell reaction and achieve a significant rate of formation of product. The additional potential difference, which varies with the type of electrode, is called the **overpotential.** For platinum electrodes, the overpotential for the production of water from hydrogen and oxygen is about 0.6 V; so about 1.8 V (0.6 + 1.23 V) is actually required to electrolyze water when platinum electrodes are used. Much contemporary research on electrochemical cells involves attempts to reduce the overpotential and hence to increase the efficiency of electrolytic processes.

When carrying out an electrolysis in solution, we must consider the possibility that other species present might be oxidized or reduced by the electric current. For example, suppose that we want to electrolyze water to produce hydrogen and oxygen. Because pure water does not carry a current, we must add an ionic solute with ions that are less easily oxidized or reduced than water. The potential for the reduction of oxygen in water at pH = 7 is +0.82 V:

$$O_2(g) + 4\,H^+(aq) + 4\,e^- \longrightarrow 2\,H_2O(l) \qquad E = +0.82\text{ V at pH} = 7$$

To reverse this half-reaction and bring about the oxidation of water, we need an applied potential difference of at least 0.82 V. Suppose the added salt is sodium chloride. When Cl^- ions are present at 1 $mol{\cdot}L^{-1}$ in water, is it possible that they, and not the water, will be oxidized? From Table 12.2, the standard potential for the reduction of chlorine is +1.36 V:

$$Cl_2(g) + 2\,e^- \longrightarrow 2\,Cl^-(aq) \qquad E° = +1.36\text{ V}$$

To reverse this reaction and oxidize chloride ions, we have to supply at least 1.36 V. Because only 0.82 V is needed to force the oxidation of water, but 1.36 V is needed to force the oxidation of Cl^-, it appears that oxygen should be the product at the anode. However, the overpotential for oxygen production can be very high, and in practice chlorine might also be produced. At the cathode, we want the half-reaction $2\,H^+(aq) + 2\,e^- \rightarrow H_2(g)$ to occur. At pH = 7, the potential required for the reduction of hydrogen ions is 0.41 V. Hydrogen, rather than sodium metal will be produced at the cathode, because the potential required to reduce sodium ions is significantly higher (2.71 V).

The potential supplied to an electrolytic cell must be at least as great as that of the cell reaction to be reversed. If there is more than one reducible species in solution, the species with the greater potential for reduction is preferentially reduced. The same principle applies to oxidation.

How can we use this information . . . to decide which species will be produced at an electrode?

Let's suppose that an aqueous solution containing I^- ions at 1 $mol{\cdot}L^{-1}$ is being electrolyzed. Will O_2 or I_2 be produced at the anode? From Table 12.2,

$$I_2(s) + 2\,e^- \longrightarrow 2\,I^-(aq) \qquad E° = +0.54\text{ V}$$

so we know that at least 0.54 V is needed to oxidize I^-. We have seen that at least 0.82 V is needed to oxidize water:

$$O_2(g) + 4\,H^+(aq) + 4\,e^- \longrightarrow 2\,H_2O(l) \qquad E = +0.82 \text{ V at pH} = 7$$

Therefore, provided the overpotentials are similar, we expect I^- ions to be oxidized in preference to water.

Self-Test 12.12A Predict the products resulting from the electrolysis of 1 M $AgNO_3(aq)$.

[*Answer:* Cathode, Ag; anode, O_2]

Self-Test 12.12B Predict the products resulting from the electrolysis of 1 M NaBr(aq).

12.14 The Products of Electrolysis

Now we shall see how to calculate the amount of product formed by a given amount of electricity (**1**).

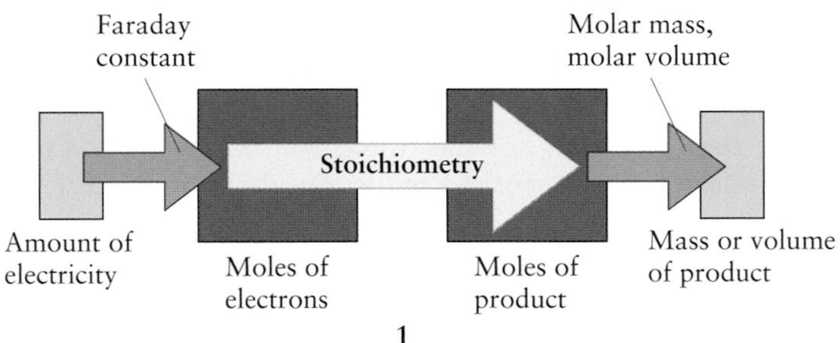

1

The calculation is based on observations made by Michael Faraday (Fig.12.14) and summarized—in more modern language than he used—as follows:

- **Faraday's law of electrolysis:** The amount of product formed or reactant consumed by an electric current is stoichiometrically equivalent to the amount of electrons supplied.

Once we know the amount (in moles) of product formed, we can calculate the masses of the products or, if they are gases, their volumes.

FIGURE 12.14 Michael Faraday (1791–1867).

How can we use this information . . . to calculate how much copper can be produced by electrolysis?

Copper is refined electrolytically by using an impure form of copper metal called blister copper as the anode in an electrolytic cell (Fig. 12.15). The current supply drives the oxidation of the blister copper to copper(II) ions,

$$Cu(s, \text{blister}) \longrightarrow Cu^{2+}(aq) + 2\,e^-$$

These ions are then reduced to pure copper metal at the cathode:

$$Cu^{2+}(aq) + 2\,e^- \longrightarrow Cu(s)$$

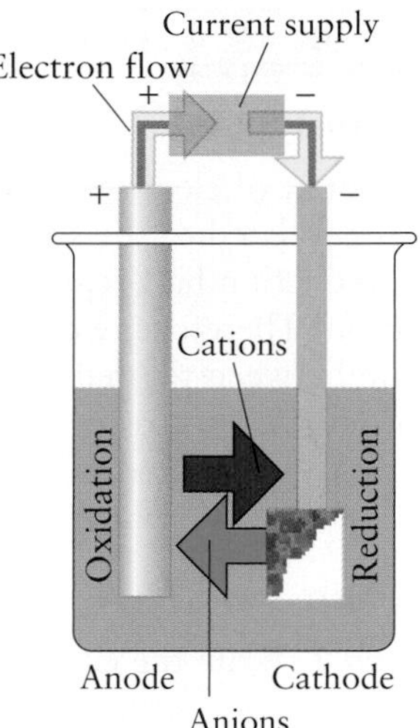

FIGURE 12.15 A schematic diagram showing the electrolytic process for refining copper. The anode is impure copper. The Cu^{2+} ions produced by oxidation of the anode migrate to the cathode, where they are reduced to pure copper metal. A similar arrangement is used for electroplating objects.

Media Link **Lab video:** Fig. 12.15, Electroplating copper

From the stoichiometry of these half-reactions, we know that 2 mol $e^- \mathrel{\widehat{=}}$ 1 mol Cu. Therefore, if 4.0 mol e^- is supplied, the amount of copper produced is

$$\text{Amount of Cu (mol)} = (4.0\ \text{mol e}^-) \times \frac{1\ \text{mol Cu}}{2\ \text{mol e}^-} = 2.0\ \text{mol Cu}$$

Self-Test 12.13A What amount in moles of Al(s) can be produced from Al_2O_3 if 5.0 mol e^- is supplied?

[*Answer:* 1.7 mol Al]

Self-Test 12.13B What amount in moles of Cr(s) can be produced from CrO_3 if 12.0 mol e^- is supplied?

The quantity of electricity passed through the electrolysis cell is determined by measuring the current, I, and the time, t, for which the current flows and is calculated from

$$\text{Charge supplied (C)} = \text{current (A)} \times \text{time (s)} \quad \text{or}$$
$$\text{Charge supplied} = It \qquad (8)$$

So if 2.00 A is passed for 125 s, the number of coulombs supplied to the cell is

$$\text{Charge supplied} = (2.00\ \text{A}) \times (125\ \text{s}) = 250\ \text{A·s} = 250\ \text{C}$$

We have used 1 A·s = 1 C. Next, we use the fact that the Faraday constant, F, is the magnitude of the charge per mole of electrons (Section 12.4). Because the charge supplied is nF, where n is the number of moles of electrons, it follows that

$$\text{Amount of e}^-\ \text{(mol)} = \frac{\text{charge supplied (C)}}{F} = \frac{\text{current (A)} \times \text{time (s)}}{F}$$

$$\text{or} \quad n_{e^-} = \frac{It}{F} \qquad (9)$$

So, by measuring the current and the time for which it flows, we can determine the amount of electrons supplied. By combining the amount of electrons supplied with the mole ratio from the stoichiometry of the electrode reaction, we can deduce the amount of product obtained (see **1**).

The amount of product in an electrolysis reaction is calculated from the stoichiometry of the half-reaction and the current and time for which the current flows.

TOOLBOX 12.4 How to predict the result of electrolysis

Conceptual Basis

The amount of electrons required to reduce a species is related to the stoichiometric coefficients in the reduction half-reaction. The same is true of oxidation. Therefore, we can set up a stoichiometric relation between the reduced or oxidized species and the amount of electrons supplied. The amount of electrons is calculated from the current and the length of time for which the current flows.

Procedure

Step 1 Write the equation for the half-reaction for the reduction or oxidation that results in the desired product (or consumes a specified reactant), and identify the stoichiometric relation between electrons and the species of interest.

Step 2 Calculate the amount (in moles) of electrons supplied from the current and the time for which it flows by using Eq. 9 ($n_{e^-} = It/F$).

Step 3 Use the stoichiometric relation from step 1 to convert the amount of electrons to the amount of product formed (or reactant consumed). If required, use the molar mass to convert to mass (or molar volume to convert to volume).

EXAMPLE 12.7 Calculating the amount of product formed by electrolysis

Aluminum is produced by electrolysis of its oxide dissolved in molten cryolite (Na_3AlF_6). Find the mass of aluminum that can be produced in 1.00 d in an electrolytic cell operating continuously at 1.00×10^5 A. The cryolite does not react.

STRATEGY Expect a large mass of aluminum for this industrial-scale process. Use the procedure in Toolbox 12.4. The Faraday constant is given inside the back cover. Convert moles of product to mass by using the molar mass.

SOLUTION *Step 1* The aluminum metal is produced from the reduction of Al^{3+} present in the molten Al_2O_3 according to the half-reaction $Al^{3+}(melt) + 3\ e^- \rightarrow Al(l)$. From this equation, it follows that 3 mol $e^- \simeq$ 1 mol Al.

Steps 2 and 3 Because the molar mass of aluminum is 26.98 $g \cdot mol^{-1}$, 3600 s = 1 h, and 24 h = 1 d, the conversions we need are

$$\begin{aligned}\text{Mass of aluminum (g)} &= n_{e^-} \times \frac{1 \text{ mol Al}}{3 \text{ mol e}^-} \times \frac{26.98 \text{ g Al}}{1 \text{ mol Al}} \\ &= \frac{(1.00 \times 10^5\ \text{C}\cdot\text{s}^{-1}) \times (24.0 \times 3600\ \text{s})}{9.65 \times 10^4\ \text{C}\cdot(\text{mol e}^-)^{-1}} \\ &\quad \times \frac{1 \text{ mol Al}}{3 \text{ mol e}^-} \times \frac{26.98 \text{ g Al}}{1 \text{ mol Al}} \\ &= 8.05 \times 10^5 \text{ g Al}\end{aligned}$$

The mass produced corresponds to 805 kg. The fact that the production of 1 mol Al requires 3 mol e^- accounts for the very high consumption of electricity characteristic of aluminum production plants.

Self-Test 12.14A Determine the time, in hours, required to obtain 7.00 g of magnesium metal from molten magnesium chloride, using a current of 7.30 A. What volume of chlorine gas at 25°C and 1.00 atm will be produced at the anode?

[*Answer:* 2.11 h; 7.04 L]

Self-Test 12.14B How many hours are required to obtain 12.00 g of chromium metal from a 1 M solution of CrO_3 in dilute sulfuric acid, using a current of 6.20 A?

THE IMPACT ON MATERIALS

Electrochemical cells play important roles in both the purification and the preservation of metallic materials. Redox reactions are used throughout the chemical industry to extract metals from their ores. However, redox reactions also corrode the artifacts that industry produces. What redox reactions achieve, redox reactions can destroy.

12.15 Applications of Electrolysis

We have already described the electrolytic extraction of aluminum, magnesium, and fluorine and the refining of copper. Another important industrial application of electrolysis is the production of sodium metal by the *Downs process*, the electrolysis of molten rock salt (Fig. 12.16):

Cathode reaction: $2\ Na^+(melt) + 2\ e^- \longrightarrow 2\ Na(l)$

Anode reaction: $2\ Cl^-(melt) \longrightarrow Cl_2(g) + 2\ e^-$

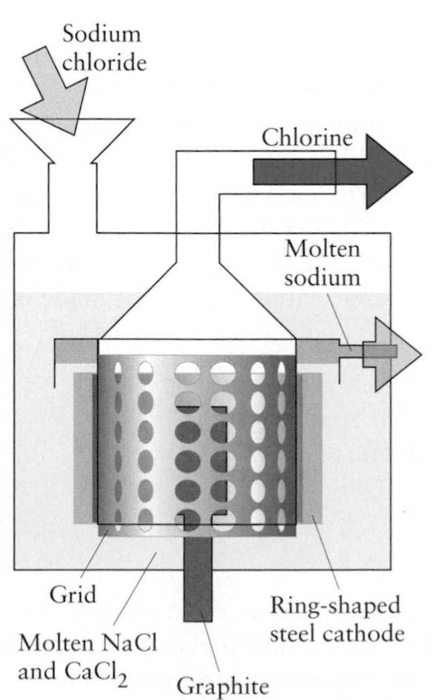

FIGURE 12.16 In the Downs process, molten sodium chloride is electrolyzed with a graphite anode (at which the Cl^- ions are oxidized to chlorine) and a steel cathode (at which the Na^+ ions are reduced to sodium). The sodium and chlorine are kept apart by the hoods surrounding the electrodes. Calcium chloride is present to lower the melting point of sodium chloride to a more economical temperature.

FIGURE 12.17 Chromium plating lends decorative flair as well as electrochemical protection to the steel of this motorcycle. Large quantities of electricity are needed for chromium plating because six electrons are required to produce each atom of chromium.

Sodium chloride is plentiful as rock salt, but the solid does not conduct electricity, because the ions are locked into place. Sodium chloride must be molten for electrolysis to occur. The electrodes in the cell are made of inert materials such as carbon, and the cell is designed to keep the sodium and chlorine produced by the electrolysis out of contact with each other and away from air. In a modification of the Downs process, the electrolyte is an aqueous solution of sodium chloride. The products of this *chloralkali process* are chlorine and aqueous sodium hydroxide.

Electroplating is the electrolytic deposition of a thin film of metal on an object. The object to be electroplated (either metal or plastic coated with graphite) constitutes the cathode, and the electrolyte is an aqueous solution of a salt of the plating metal. Metal is deposited on the cathode by reduction of ions in the electrolyte solution. These cations are supplied either by the added salt or from oxidation of the anode, which is made of the plating metal (Fig. 12.17).

Electrolysis is used industrially to produce aluminum and magnesium; to extract metals from their salts; to prepare chlorine, fluorine, and sodium hydroxide; to refine copper; and in electroplating.

12.16 Corrosion

Corrosion is the unwanted oxidation of a metal. It cuts short the lifetimes of steel products such as bridges and automobiles, and replacing corroded metal parts costs billions of dollars a year. Corrosion is an electrochemical process, and the electrochemical series gives us insight into why corrosion occurs and how to prevent it.

The main culprit in corrosion is water. One half-reaction we have to take into account is

$$2\,H_2O(l) + 2\,e^- \longrightarrow H_2(g) + 2\,OH^-(aq) \qquad E° = -0.83\ V$$

This standard potential is for an OH^- concentration of 1 mol·L^{-1}, which corresponds to pH = 14, a strongly basic solution. However, from the Nernst equation, we can calculate that at pH = 7 this couple has $E = -0.42$ V. Any metal with a standard potential more negative than -0.42 V can therefore reduce water at pH = 7: that is, at this pH, any such metal can be oxidized by water. Because $E° = -0.44$ V for $Fe^{2+}(aq) + 2\,e^- \rightarrow Fe(s)$, iron has only a very slight tendency to be oxidized by water at pH = 7. For this reason, iron can be used for pipes in water supply systems and can be stored in oxygen-free water without rusting (Fig. 12.18).

FIGURE 12.18 Iron nails stored in oxygen-free water (left) do not rust because the oxidizing power of water itself is weak. When oxygen is present (as a result of air dissolving in the water, right), oxidation is thermodynamically spontaneous and rust soon forms.

When iron is exposed to damp air, with both oxygen and water present, the half-reaction

$$O_2(g) + 4\,H^+(aq) + 4\,e^- \longrightarrow 2\,H_2O(l) \qquad E° = +1.23\ V$$

must be taken into account. The potential of this half-reaction at pH = 7 is +0.82 V, which lies well above the value for iron. Hence, iron can reduce oxygen in aqueous solution at pH = 7. In other words, oxygen and water can jointly oxidize iron metal to iron(II) ions. They can also subsequently work together to oxidize the iron(II) ions to iron(III), because $E° = +0.77$ V for $Fe^{3+}(aq) + e^- \rightarrow Fe^{2+}(aq)$.

Let's look in more detail at the processes involved. A drop of water on the surface of iron can act as the electrolyte for corrosion in a tiny electrochemical cell (Fig. 12.19). At the edge of the drop, dissolved oxygen oxidizes the iron. However, the electrons withdrawn from the metal by this oxidation can be restored from another part of the conducting metal—in particular, from iron lying beneath the oxygen-poor region in

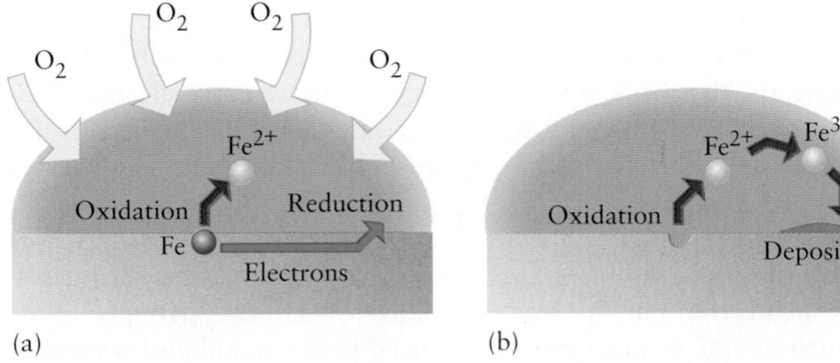

FIGURE 12.19 The mechanism of rust formation. (a) Oxidation of the iron occurs at a point out of contact with the oxygen of the air, and the surface of the metal acts as an anode in a tiny galvanic cell. (b) Further oxidation of Fe^{2+} to Fe^{3+} results in the deposition of rust on the surface.

FIGURE 12.20 Steel girders are galvanized by immersion in a bath of molten zinc.

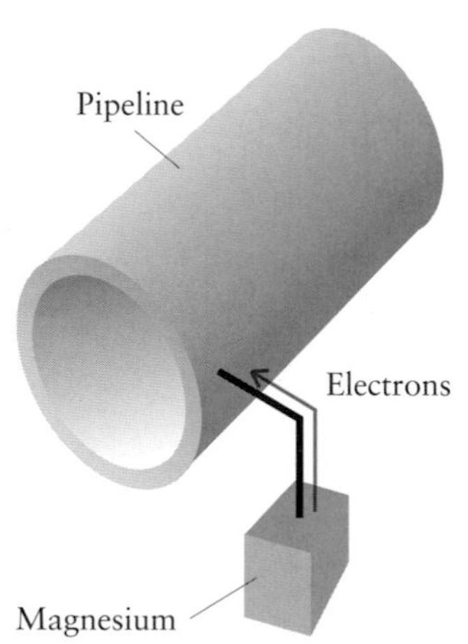

FIGURE 12.21 In the cathodic protection of a buried pipeline, or other large metal construction, the artifact is connected to a number of buried blocks of metal, such as magnesium or zinc. The sacrificial anodes (the magnesium block in this illustration) supply electrons to the pipeline (the cathode of the cell), thereby preserving it from oxidation.

the center of the drop. The iron atoms there give up their electrons, form Fe^{2+} ions, and drift away into the surrounding water. This process results in the formation of tiny pits in the surface. The Fe^{2+} ions are then oxidized further to Fe^{3+} by the dissolved oxygen. These ions then precipitate as a hydrated iron(III) oxide, $Fe_2O_3{\cdot}H_2O$, the brown, insoluble substance we call *rust*. Water is more highly conducting when it has dissolved ions, and the formation of rust is then accelerated. That is one reason why the salt air of coastal cities and salt used for deicing highways is so damaging to exposed metal.

Because corrosion is electrochemical, we can use our knowledge of redox reactions to combat it. The simplest way to prevent corrosion is to protect the surface of the metal from exposure to air and water by painting. A method that achieves greater protection is to **galvanize** the metal, which involves coating it with an unbroken film of zinc (Fig. 12.20). Zinc lies below iron in the electrochemical series, so if a scratch exposes the metal beneath, the more strongly reducing zinc releases electrons to the iron. As a result, the zinc, not the iron, is oxidized. The zinc itself survives exposure on the unbroken surface because, like aluminum, it is passivated by a protective oxide.

It is not possible to galvanize large structures—for example, ships, underground pipelines or gasoline storage tanks, and bridges—but **cathodic protection** can be used. A block of a more strongly reducing metal, such as zinc or magnesium, is buried in moist soil and connected to the underground pipeline (Fig. 12.21). The block of magnesium is oxidized preferentially and supplies electrons to the iron for the reduction of oxygen. The block, which is called a **sacrificial anode**, protects the iron pipeline and is cheaper to replace than the pipeline itself. For similar reasons, automobiles generally have negative ground systems as part of their electrical circuitry, which means that the body of the car is connected to the anode of the battery. The decay of the anode in the battery is the sacrifice that helps preserve the vehicle itself.

The corrosion of iron is accelerated by the presence of oxygen, moisture, and salt. Corrosion can be inhibited by coating the surface with paint or zinc, or by using cathodic protection.

Self-Test 12.15A Which of the following procedures helps to avoid the corrosion of an iron rod in water: (a) decreasing the concentration of oxygen in the water; (b) painting the rod?

[*Answer:* Both]

Self-Test 12.15B Which of the following can act as a sacrificial anode for iron: (a) copper; (b) zinc; (c) tin?

SKILLS YOU SHOULD HAVE MASTERED

1 Balance chemical equations for redox reactions by the half-reaction method, Toolbox 12.1 and Examples 12.1 and 12.2.

2 Estimate the reaction free energy from a cell emf, Section 12.4.

3 Write the cell diagram for a redox reaction, Self-Test 12.4.

4 Write the chemical equation for a cell reaction, given the cell diagram, Toolbox 12.2 and Self-Test 12.5.

5 Determine the standard potential of an electrode from a cell emf, Section 12.6.

6 Calculate the standard potential of a redox couple from two others relating to different oxidation states, Example 12.3.

7 Predict the spontaneous direction of a redox reaction by using the electrochemical series, Example 12.4.

8 Calculate the equilibrium constant for a reaction from the standard cell emf, Toolbox 12.3 and Example 12.5.

9 Use the Nernst equation to predict a cell emf, Example 12.6.

10 Predict the likely products of electrolysis of an aqueous solution from standard potentials, Section 12.13.

11 Calculate the amount of product produced from electrolysis, Toolbox 12.4 and Example 12.7.

12 Describe corrosion and means for protecting iron from corrosion, Section 12.16.

EXERCISES

Assume a temperature of 25°C (298 K) for the following exercises unless instructed otherwise.

Redox Equations

12.1 Balance the following skeletal equations, using oxidation and reduction half-reactions. All the reactions occur in acidic solution. Identify the oxidizing agent and reducing agent in each reaction.
(a) reaction of thiosulfate ion with chlorine gas:
$Cl_2(g) + S_2O_3^{2-}(aq) \rightarrow Cl^-(aq) + SO_4^{2-}(g)$
(b) action of the permanganate ion on sulfurous acid:
$MnO_4^-(aq) + H_2SO_3(aq) \rightarrow Mn^{2+}(aq) + HSO_4^-(aq)$
(c) reaction of hydrosulfuric acid with chlorine:
$H_2S(aq) + Cl_2(g) \rightarrow S(s) + Cl^-(aq)$
(d) reaction of chlorine in water:
$Cl_2(g) \rightarrow HClO(aq) + Cl^-(aq)$

12.2 Balance the following skeletal equations, using oxidation and reduction half-reactions. All the reactions occur in acidic solution. Identify the oxidizing agent and reducing agent in each reaction.

(a) conversion of iron(II) to iron(III) by dichromate ion:
$Fe^{2+}(aq) + Cr_2O_7^{2-}(aq) \rightarrow Fe^{3+}(aq) + Cr^{3+}(aq)$
(b) formation of acetic acid from ethanol by the action of permanganate ion:
$C_2H_5OH(aq) + MnO_4^-(aq) \rightarrow Mn^{2+}(aq) + CH_3COOH(aq)$
(c) reaction of iodide with nitric acid:
$I^-(aq) + NO_3^-(aq) \rightarrow I_2(aq) + NO(g)$
(d) reaction of arsenic(III) sulfide with nitric acid:
$As_2S_3(s) + NO_3^-(aq) \rightarrow H_3AsO_4(aq) + S(s) + NO(g)$

12.3 Balance the following skeletal equations, using oxidation and reduction half-reactions. All the reactions occur in basic solution. Identify the oxidizing agent and reducing agent in each reaction.
(a) action of ozone on bromide ions:
$O_3(aq) + Br^-(aq) \rightarrow O_2(g) + BrO_3^-(aq)$
(b) reaction of bromine in water:
$Br_2(l) \rightarrow BrO_3^-(aq) + Br^-(aq)$
(c) formation of chromate ions from chromium(III) ions:
$Cr^{3+}(aq) + MnO_2(s) \rightarrow Mn^{2+}(aq) + CrO_4^{2-}(aq)$
(d) reaction of elemental phosphorus to form phosphine, PH_3, a poisonous gas with the odor of decaying fish:
$P_4(s) \rightarrow H_2PO_2^-(aq) + PH_3(g)$

12.4 Balance the following skeletal equations, using oxidation and reduction half-reactions. All the reactions occur in basic solution. Identify the oxidizing agent and reducing agent in each reaction.
(a) production of chlorite ions from chlorine heptoxide:
$Cl_2O_7(g) + H_2O_2(aq) \rightarrow ClO_2^-(aq) + O_2(g)$
(b) action of permanganate ions on sulfide ions:
$MnO_4^-(aq) + S^{2-}(aq) \rightarrow S(s) + MnO_2(s)$
(c) reaction of hydrazine with chlorate ions:
$N_2H_4(g) + ClO_3^-(aq) \rightarrow NO(g) + Cl^-(aq)$
(d) reaction of plumbate ions and hypochlorite ions:
$Pb(OH)_4^{2-}(aq) + ClO^-(aq) \rightarrow PbO_2(s) + Cl^-(aq)$

12.5 Potassium permanganate is an excellent oxidizing agent for laboratory use and in sewage treatment. It readily reacts with the organic compounds in sewage to produce carbon dioxide and water. Write the equation for each half-reaction and the overall equation for the oxidation of glucose, $C_6H_{12}O_6$, the skeletal form of which is $MnO_4^{2-}(aq) + C_6H_{12}O_6(aq) \rightarrow Mn^{2+}(aq) + CO_2(g) + H_2O(l)$.

12.6 Nitric oxide gas can be produced from the reaction of nitrite ions with each other in a dilute sulfuric acid solution. The skeletal equation for the reaction is $NO_2^-(aq) \rightarrow NO(g) + NO_3^-(aq)$. Write the equation for each half-reaction and the overall equation for the reaction.

12.7 The molecule P_4S_3 is a reactive molecule that is oxidized by nitrate ions in acid solution to give phosphoric acid and sulfate ions. The nitrogen-containing product is nitric oxide, NO. Write the balanced equation for each half-reaction and the overall equation for the reaction.

12.8 Iron(II) hydrogen phosphite, $FeHPO_3$, is oxidized by hypochlorite ions in basic solution. The products are chloride ion, phosphate ion, and iron(III) hydroxide. Write the balanced equation for each half-reaction and the overall equation for the reaction.

Galvanic Cells

12.9 Write cathode and anode half-reactions and the balanced equation for the cell reaction for each of the following galvanic cells:
(a) $Ni(s)|Ni^{2+}(aq)||Ag^+(aq)|Ag(s)$
(b) $C(gr)|H_2(g)|H^+(aq)||Cl^-(aq)|Cl_2(g)|Pt(s)$
(c) $Cu(s)|Cu^{2+}(aq)||Ce^{4+}(aq),Ce^{3+}(aq)|Pt(s)$
(d) $Pt(s)|O_2(g)|H^+(aq)||OH^-(aq)|O_2(g)|Pt(s)$
(e) $Pt(s)|Sn^{4+}(aq),Sn^{2+}(aq)||Cl^-(aq)|Hg_2Cl_2(s)|Hg(l)$

12.10 Write cathode and anode half-reactions and the balanced equation for the cell reaction for each of the following galvanic cells:
(a) $Cu(s)|Cu^{2+}(aq)||Cu^+(aq)|Cu(s)$
(b) $Cr(s)|Cr^{2+}(aq)||Au^{3+}(aq)|Au(s)$
(c) $Ag(s)|AgI(s)|I^-(aq)||Cl^-(aq)|AgCl(s)|Ag(s)$
(d) $Hg(l)|Hg_2Cl_2(s)|Cl^-(aq)||Cl^-(aq)|AgCl(s)|Ag(s)$
(e) $Hg(l)|Hg_2^{2+}(aq)||MnO_4^-(aq),Mn^{2+}(aq),H^+(aq)|Pt(s)$

12.11 Write cathode and anode half-reactions, the balanced equation for the cell reaction, and the cell diagram for the following reactions:
(a) $Ni^{2+}(aq) + Zn(s) \rightarrow Ni(s) + Zn^{2+}(aq)$
(b) $Ce^{4+}(aq) + I^-(aq) \rightarrow I_2(s) + Ce^{3+}(aq)$
(c) $Cl_2(g) + H_2(g) \rightarrow HCl(aq)$
(d) $Au^+(aq) \rightarrow Au(s) + Au^{3+}(aq)$

12.12 Write cathode and anode half-reactions, the balanced equation for the cell reaction, and the cell diagram for the following reactions:
(a) $Mn(s) + Ti^{2+}(aq) \rightarrow Mn^{2+}(aq) + Ti(s)$
(b) $Fe^{3+}(aq) + H_2(g) \rightarrow Fe^{2+}(aq) + H^+(aq)$
(c) $Cu^+(aq) \rightarrow Cu(s) + Cu^{2+}(aq)$
(d) $MnO_4^-(aq) + H^+(aq) + Cl^-(aq) \rightarrow Cl_2(g) + Mn^{2+}(aq) + H_2O(l)$

12.13 Write cathode and anode half-reactions and devise a galvanic cell (write a cell diagram) to study each of the following reactions:
(a) $AgBr(s) \rightleftharpoons Ag^+(aq) + Br^-(aq)$, a solubility equilibrium

(b) $H_3O^+(aq) + OH^-(aq) \rightarrow 2\ H_2O(l)$, the Brønsted neutralization reaction
(c) $Cd(s) + 2\ Ni(OH)_3(s) \rightarrow Cd(OH)_2(s) + 2\ Ni(OH)_2(s)$, the reaction in the nickel-cadmium cell

12.14 Write balanced cathode and anode half-reactions and devise a galvanic cell (write a cell diagram) to study each of the following reactions:
(a) $AgNO_3(aq) + KI(aq) \rightarrow AgI(s) + KNO_3(aq)$, a precipitation reaction
(b) $H_3O^+(aq, concentrated) \rightarrow H_3O^+(aq, dilute)$
(c) $Zn(s) + Ag_2O(s) \rightarrow ZnO(s) + 2\ Ag(s)$, the reaction in a silver cell

12.15 (a) Write balanced cathode and anode half-reactions for the redox reaction of an acidified solution of potassium permanganate and iron(II) chloride. (b) Write the balanced equation for the cell reaction and devise a galvanic cell to study the reaction (write its cell diagram).

12.16 (a) Write balanced cathode and anode half-reactions for the redox reaction between sodium dichromate and mercury(I) nitrate in an acidic solution. (b) Write the balanced equation for the cell reaction and devise a galvanic cell to study the reaction (write its cell diagram).

12.17 Predict the standard potential of each of the following galvanic cells:
(a) $Pt(s)|Cr^{3+}(aq),Cr^{2+}(aq)||Cu^{2+}(aq)|Cu(s)$
(b) $Ag(s)|AgI(s)|I^-(aq)||Cl^-(aq)|AgCl(s)|Ag(s)$
(c) $Hg(l)|Hg_2Cl_2(s)|Cl^-(aq)||Hg_2^{2+}(aq)|Hg(l)$
(d) $C(gr)|Sn^{4+}(aq),Sn^{2+}(aq)||Pb^{4+}(aq),Pb^{2+}(aq)|Pt(s)$

12.18 Predict the standard potential of each of the following galvanic cells:
(a) $Pt(s)|Fe^{3+}(aq),Fe^{2+}(aq)||Ag^+(aq)|Ag(s)$
(b) $U(s)|U^{3+}(aq)||V^{2+}(aq)|V(s)$
(c) $Sn(s)|Sn^{2+}(aq)||Sn^{4+}(aq),Sn^{2+}(aq)|Pt(s)$
(d) $Cu(s)|Cu^{2+}(aq)||Au^+(aq)|Au(s)$

12.19 Predict the standard cell potential and calculate the standard free energy for the following galvanic cells (the standard potentials of these cells were obtained in Exercise 12.17):
(a) $Pt(s)|Cr^{3+}(aq),Cr^{2+}(aq)||Cu^{2+}(aq)|Cu(s)$
(b) $Ag(s)|AgI(s)|I^-(aq)||Cl^-(aq)|AgCl(s)|Ag(s)$
(c) $Hg(l)|Hg_2Cl_2(s)|Cl^-(aq)||Hg_2^{2+}(aq)|Hg(l)$
(d) $C(gr)|Sn^{4+}(aq),Sn^{2+}(aq)||Pb^{4+}(aq),Pb^{2+}(aq)|Pt(s)$

12.20 Predict the standard cell potential and calculate the standard free energy for galvanic cells having the following cell reactions:
(a) $3\ Zn(s) + 2\ Bi^{3+}(aq) \rightarrow 3\ Zn^{2+}(aq) + 2\ Bi(s)$
(b) $2\ H_2(g) + O_2(g) \rightarrow 2\ H_2O(l)$ in an acidic solution
(c) $2\ H_2(g) + O_2(g) \rightarrow 2\ H_2O(l)$ in a basic solution
(d) $3\ Au^+(aq) \rightarrow 2\ Au(s) + Au^{3+}(aq)$

12.21 A student was given a standard $Cu(s)|Cu^{2+}(aq)$ half-cell and another half-cell containing an unknown metal M immersed in a 1.00 M solution of its nitrate salt, $M(NO_3)_2$. When the copper was connected as the anode at 25°C, the cell emf was found to be −1.25 V. What is the reduction potential for the unknown M^{2+}/M couple?

12.22 A student was given a standard $Fe(s)|Fe^{2+}(aq)$ half-cell and another half-cell containing an unknown metal M immersed in a 1.00 M solution of the metal +1 nitrate salt, MNO_3. When these two cells were connected at 25°C, the complete cell functioned as a galvanic cell with $E = +1.24$ V. The reaction was allowed to continue overnight and the two electrodes were weighed. The iron electrode was found to be lighter and the unknown metal electrode was heavier. What is the standard potential of the unknown M^+/M couple?

12.23 Arrange the following metals in order of increasing strength as reducing agents: (a) Cu, Zn, Cr, Fe; (b) Li, Na, K, Mg; (c) U, V, Ti, Al; (d) Ni, Sn, Au, Ag.

12.24 Arrange the following species in order of increasing strength as oxidizing agents: (a) Co^{2+}, Cl_2, Ce^{4+}, In^{3+}; (b) NO_3^-, ClO_4^-, HBrO, $Cr_2O_7^{2-}$, all in acidic solution; (c) O_2, O_3, HClO, and HBrO, all in acidic solution; (d) O_2, O_3, ClO^-, BrO^-, all in basic solution.

12.25 Suppose that the following redox couples are joined to form a galvanic cell that generates a current under standard conditions. Identify the oxidizing agent and the reducing agent, write a cell diagram, and calculate the standard cell emf. (a) Co^{2+}/Co and Ti^{3+}/Ti^{2+}; (b) La^{3+}/La and U^{3+}/U; (c) H^+/H_2 and Fe^{3+}/Fe^{2+}; (d) $O_3/O_2,OH^-$ and Ag^+/Ag.

12.26 Suppose that the following redox couples are joined to form a galvanic cell that generates a current under standard conditions. Identify the oxidizing agent and the reducing agent, write a cell diagram, and calculate the standard cell emf. (a) Pt^{2+}/Pt and $AgF/Ag,F^-$, (b) Cr^{3+}/Cr^{2+} and I_3^-/I^-; (c) H^+/H_2 and Ni^{2+}/Ni; (d) $O_3/O_2,OH^-$ and $O_3,H^+/O_2$.

12.27 Answer the following questions and, for each "yes" response, write a balanced cell reaction and calculate the standard cell emf. (a) Can H_2 reduce $Ni^{2+}(aq)$ ions to nickel metal? (b) Can chromium metal reduce $Pb^{2+}(aq)$ ions to lead metal? (c) Can permanganate ions oxidize copper metal to $Cu^{2+}(aq)$ ions in an acidic solution? (d) Can $Fe^{3+}(aq)$ ions oxidize mercury metal to mercury(I)?

12.28 Answer the following questions and, for each "yes" response, write a balanced cell reaction and calculate the standard cell potential. (a) Will oxygen oxidize chromium metal to $Cr^{2+}(aq)$ ions in acidic solution? (b) Will oxygen oxidize chromium metal to $Cr^{3+}(aq)$ ions in acidic solution? (c) Will gallium metal reduce $Cr^{3+}(aq)$ ions to $Cr^{2+}(aq)$ ions? (d) Will gallium metal reduce $Cr^{2+}(aq)$ ions to chromium metal?

12.29 Identify the reactions with $K > 1$ in the following list and, for each such reaction, identify the oxidizing agent and calculate the standard cell emf.
(a) $Cl_2(g) + 2\ Br^-(aq) \rightarrow 2\ Cl^-(aq) + Br_2(l)$
(b) $MnO_4^-(aq) + 8\ H^+(aq) + 5\ Ce^{3+}(aq) \rightarrow 5\ Ce^{4+}(aq) + Mn^{2+}(aq) + 4\ H_2O(l)$
(c) $2\ Pb^{2+}(aq) \rightarrow Pb(s) + Pb^{4+}(aq)$
(d) $2\ NO_3^-(aq) + 4\ H^+(aq) + Zn(s) \rightarrow Zn^{2+}(aq) + 2\ NO_2(g) + 2\ H_2O(l)$

12.30 Identify the reactions with $K > 1$ among the following reactions and, for each such reaction, write balanced reduction and oxidation half-reactions. For those reactions, show that $K > 1$ by calculating the standard free energy of the reaction.
(a) $Mg^{2+} + Cu \rightarrow$?; (b) $Al + Pb^{2+} \rightarrow$?; (c) $Hg_2^{2+} + Ce^{3+} \rightarrow$?; (d) $Zn + Sn^{2+} \rightarrow$?; (e) $O_2 + H^+ + Hg \rightarrow$?.

12.31 Chlorine is used to displace bromine from brine that contains sodium bromide. Could oxygen in an acidified solution be used instead? If so, why is it not used?

12.32 A chemist is interested in the compounds formed by the *d*-block element manganese and wants to find a way to prepare Mn^{3+} from Mn^{2+}. Would an acidified solution of sodium dichromate be suitable?

12.33 (a) Using data available in Appendix 2B, write the disproportionation reaction for $Au^+(aq)$. (b) From the appropriate standard potentials, determine whether Au^+ will disproportionate spontaneously in aqueous solution.

12.34 (a) Using data available in Appendix 2B, write the disproportionation reaction for Sn^{2+}. (b) From the appropriate standard potentials, determine whether Sn^{2+} will spontaneously disproportionate in aqueous solution.

12.35 Using data available in Appendix 2B, calculate the standard potential for the half-reaction $U^{4+}(aq) + 4\ e^- \rightarrow U(s)$.

12.36 Using data available in Appendix 2B, calculate the standard potential for the half-reaction $Ti^{3+}(aq) + 3\ e^- \rightarrow Ti(s)$.

12.37 Determine the equilibrium constants for the following reactions and cells:
(a) $Pt(s)|Cr^{3+}(aq),Cr^{2+}(aq)||Cu^{2+}(aq)|Cu(s)$
(b) $Mn(s) + Ti^{2+}(aq) \rightarrow Mn^{2+}(aq) + Ti(s)$
(c) A Pb^{2+}/Pb redox couple in combination with a Hg_2^{2+}/Hg redox couple
(d) $In^{3+}(aq) + U^{3+}(aq) \rightarrow In^{2+}(aq) + U^{4+}(aq)$

12.38 Determine the equilibrium constants for the following reactions and cells:
(a) An $AgI/Ag,I^-$ redox couple in combination with an I_2/I^- redox couple
(b) $Pt(s)|Sn^{4+}(aq),Sn^{2+}(aq)||Cl^-(aq)|Hg_2Cl_2(s)|Hg(s)$
(c) $2\ Fe^{3+}(aq) + H_2(g) \rightarrow 2\ Fe^{2+}(aq) + 2\ H^+(aq)$
(d) $Cr(s) + Zn^{2+}(aq) \rightarrow Cr^{2+}(aq) + Zn(s)$

12.39 A chemist wants to make a range of silver(II) compounds. Could aqueous sodium persulfate be used to oxidize silver(I) compound to silver(II)? If so, what is the equilibrium constant for the reaction?

12.40 A chemist suspects that manganese(III) might be involved in an unusual biochemical reaction and wants to prepare some of its compounds. Could aqueous potassium permanganate be used to oxidize manganese(II) to manganese(III)? If so, what is the equilibrium constant for the reaction?

12.41 Calculate the reaction quotient, Q_c, for the cell reaction, given the measured values of the cell emf:
(a) $Pt(s)|Sn^{4+}(aq),Sn^{2+}(aq)||Pb^{4+}(aq), Pb^{2+}(aq)|C(gr)$, $E = +1.33$ V
(b) $Pt(s)|O_2(g)|H^+(aq)||Cr_2O_7^{2-}(aq), H^+(aq),Cr^{3+}(aq)|Pt(s)$, $E = +0.10$ V

12.42 Calculate the reaction quotient, Q_c, for the cell reaction, given the measured values of the cell emf:
(a) $Ag(s)|Ag^+(aq)||ClO_4^-(aq),H^+(aq),ClO_3^-(aq)|Pt(s)$, $E = +0.40$ V
(b) $C(gr)|Cl_2(g)|Cl^-(aq)||Au^{3+}(aq)|Au(s)$, $E = 0.00$ V

12.43 A concentration cell consists of the same redox couples at the anode and the cathode, with different concentrations of the ions in the respective compartments. Calculate E(cell) for the following concentration cells:
(a) $Cu(s)|Cu^{2+}(aq,\ 0.0010\ mol\cdot L^{-1})||Cu^{2+}(aq,\ 0.010\ mol\cdot L^{-1})|Cu(s)$
(b) $Pt(s)|H_2(g,\ 1\ bar)|H^+(aq,\ pH = 4.0)||H^+(aq,\ pH = 3.0)|H_2(g,\ 1\ bar)|Pt(s)$

12.44 A concentration cell consists of the same redox couples at the anode and the cathode, with different

concentrations of the ions in the respective compartments. Determine the unknown concentration of the ion in each of the following cells:
(a) $Pb(s)|Pb^{2+}(aq, ?)||Pb^{2+}(aq, 0.10\ mol\cdot L^{-1})|Pb(s)$, $E = 0.050$ V
(b) $Pt(s)|Fe^{3+}(aq, 0.10\ mol\cdot L^{-1}), Fe^{2+}(aq, 1.0\ mol\cdot L^{-1})||Fe^{3+}(aq, ?), Fe^{2+}(aq, 0.0010\ mol\cdot L^{-1})|Pt(s)$, $E = 0.10$ V

12.45 Determine the emfs of the following cells:
(a) $Pt(s)|H_2(g, 1.0\ bar)|HCl(aq, 0.075\ M)||HCl(aq, 1.0\ mol\cdot L^{-1})|H_2(g, 1.0\ bar)|Pt(s)$
(b) $Zn(s)|Zn^{2+}(aq, 0.37\ mol\cdot L^{-1})||Ni^{2+}(aq, 0.059\ mol\cdot L^{-1})|Ni(s)$
(c) $Pt(s)|Cl_2(g, 250\ Torr)|HCl(aq, 1.0\ M)||HCl(aq, 0.85\ M)|H_2(g, 125\ Torr)|Pt(s)$
(d) $Sn(s)|Sn^{2+}(aq, 0.277\ mol\cdot L^{-1})||Sn^{4+}(aq, 0.867\ mol\cdot L^{-1}), Sn^{2+}(aq, 0.55\ mol\cdot L^{-1})|Pt(s)$

12.46 Determine the emfs of the following cells:
(a) $Cr(s)|Cr^{3+}(aq, 0.37\ mol\cdot L^{-1})||Pb^{2+}(aq, 9.5 \times 10^{-3}\ mol\cdot L^{-1})|Pb(s)$
(b) $Pt(s)|H_2(g, 2.0\ bar)|H^+(pH = 3.5)||Cl^-(aq, 0.75\ mol\cdot L^{-1})|Hg_2Cl_2(s)|Hg(s)$
(c) $C(gr)|Sn^{4+}(aq, 0.059\ mol\cdot L^{-1}), Sn^{2+}(aq, 0.059\ mol\cdot L^{-1})||Fe^{3+}(aq, 0.15\ mol\cdot L^{-1}), Fe^{2+}(aq, 0.15\ mol\cdot L^{-1})|Pt(s)$
(d) $Ag(s)|AgI(s)|I^-(aq, 0.025\ mol\cdot L^{-1})||Cl^-(aq, 0.67\ mol\cdot L^{-1})|AgCl(s)|Ag(s)$

12.47 Determine the unknown quantity in the following cells:
(a) $Pt(s)|H_2(g, 1.0\ bar)|H^+(pH = ?)||Cl^-(aq, 1.0\ mol\cdot L^{-1})|Hg_2Cl_2(s)|Hg(l)$, $E = +0.33$ V
(b) $C(gr)|Cl_2(g, 1.0\ bar)|Cl^-; (aq, ?)||MnO_4^-(aq, 0.010\ mol\cdot L^{-1}), H^+(pH = 4.0), Mn^{2+}(aq, 0.10\ mol\cdot L^{-1})|Pt(s)$, $E = -0.30$ V

12.48 Determine the unknown quantity in the following cells:
(a) $Pt(s)|H_2(g, 1.0\ bar)|H^+(pH = ?)||Cl^-(aq, 1.0\ mol\cdot L^{-1})|AgCl(s)|Ag(s)$, $E = +0.30$ V
(b) $Pb(s)|Pb^{2+}(aq, ?)||Ni^{2+}(aq, 0.10\ mol\cdot L^{-1})|Ni(s)$, $E = +0.040$ V

12.49 (a) Using data from Appendix 2B, calculate the solubility product of Hg_2Cl_2. (b) Compare this number to the value listed in Table 11.6 and comment on the difference.

12.50 (a) The standard potential of Ag_2CrO_4 to Ag(s) and chromate ions is +0.446 V. Write the balanced half-reaction for the reduction of silver chromate. (b) Using the data from part (a) and Appendix 2B, calculate the solubility product of $Ag_2CrO_4(s)$.

12.51 Suppose the reference electrode for Table 12.2 were the standard calomel electrode, $Hg_2Cl_2/Hg, Cl^-([Cl^-] = 1.00\ mol\cdot L^{-1})$ with its $E°$ set equal to 0. Under this system, what would be the potential for (a) the standard hydrogen electrode; (b) the standard Cu^{2+}/Cu redox couple?

12.52 Consider the question posed in Exercise 12.51 in which a saturated calomel electrode (the solution is saturated with KCl instead of having 1.00 M $[Cl^-]$) is used in place of the standard calomel electrode. How will this change the answers given in Exercise 12.51? The solubility of KCl is 35 g/(100 mL H_2O).

12.53 A chromium-plated steel bicycle handlebar is scratched. Will rusting of the iron in the steel be encouraged or retarded by the chromium?

12.54 A solution is prepared by dissolving 1 mol each of $Cu(NO_3)_2$, $Ni(NO_3)_2$, and $AgNO_3$ in 1.0 L of water. Using only data from Appendix 2B, answer the following questions. Identify the metals that (a) will leave the Ni^{2+} ions unaffected but will cause Cu and Ag to plate out of solution; (b) will leave the Ni^{2+} and Cu^{2+} ions in solution but will cause Ag to plate out of solution; (c) will leave all three metals ions in solution; (d) will leave Ni^{2+} and Ag^+ ions in solution but will cause Cu to plate out of solution.

12.55 (a) What is the approximate chemical formula of rust? (b) What is the oxidizing agent in the formation of rust? (c) How does the presence of salt accelerate the rusting process?

12.56 (a) What is the electrolyte solution in the formation of rust? (b) How are steel (iron) objects protected by galvanizing and by sacrificial anodes? (c) Suggest two metals that could be used in place of zinc for galvanizing iron.

12.57 (a) Suggest two metals that could be used for the cathodic protection of a titanium pipeline. (b) What factors other than relative positions in the electrochemical series need to be considered in practice? (c) Often copper piping is connected to iron pipes in household plumbing systems. What is a possible effect of the copper on the iron pipes?

12.58 (a) Can aluminum be used for the cathodic protection of a steel underground storage container? (b) Which of the metals zinc, silver, copper, and magnesium cannot be used as a sacrificial anode in the protection of a buried iron pipeline? Explain your answer. (c) What is the electrolyte solution for the cathodic protection of an underground pipeline by a sacrificial anode?

Electrolysis

For the exercises in this section, base your answers on the potentials listed in Table 12.2 or Appendix 2B, with the exception of the reduction and oxidation of water at pH = 7:

$$2\ H_2O(l) + 2\ e^- \longrightarrow H_2(g) + 2\ OH^-(aq) \quad E = -0.42\ \text{V at pH} = 7$$

$$O_2(g) + 4\ H^+(aq) + 4\ e^- \longrightarrow 2\ H_2O(l) \quad E = +0.82\ \text{V at pH} = 7$$

Ignore other factors such as passivation or overpotential.

12.59 A 1 M $NiSO_4$(aq) solution was electrolyzed, using inert electrodes. Write (a) the cathode reaction; (b) the anode reaction. (c) Assuming no overpotential or passivity at the electrodes, what is the minimum potential that must be supplied to the cell for the onset of electrolysis?

12.60 A 1 M KBr(aq) solution was electrolyzed, using inert electrodes. Write (a) the cathode reaction; (b) the anode reaction. (c) Assuming no overpotential or passivity at the electrodes, what is the minimum potential that must be supplied to the cell for the onset of electrolysis?

12.61 Aqueous solutions of (a) Mn^{2+}; (b) Al^{3+}; (c) Ni^{2+}; (d) Au^{3+} are electrolyzed. Determine whether the metal ion or water will be reduced at the cathode.

12.62 The anode of an electrolytic cell was constructed from (a) Cr; (b) Pt; (c) Cu; (d) Ni. Determine whether oxidation of the electrode or of water will occur at the anode.

12.63 A total charge of 4.5 kC is passed through an electrolytic cell. Determine the quantity of substance produced in each case: (a) the mass (in grams) of bismuth metal from a bismuth nitrate solution; (b) the volume (in liters at 273 K and 1.00 atm) of hydrogen gas from a sulfuric acid solution; (c) the mass of cobalt (in grams) from a cobalt(III) chloride solution.

12.64 A total charge of 96.5 kC is passed through an electrolytic cell. Determine the quantity of substance produced in each case: (a) the mass (in grams) of silver metal from a silver nitrate solution; (b) the volume (in liters at 273 K and 1.00 atm) of chlorine gas from a brine solution (concentrated aqueous sodium chloride solution); (c) the mass of copper (in grams) from a copper(II) chloride solution.

12.65 (a) How much time is required to electroplate 1.00 g of silver from a silver nitrate solution, using a current of 17.8 mA? (b) When the same current is used for the same length of time, what mass of copper could be electroplated from a copper(II) sulfate solution?

12.66 (a) When a current of 375 mA is used for 6.3 h, what volume (measured in liters at 298 K and 1.0 atm) of fluorine gas can be produced from a molten mixture of potassium and hydrogen fluorides? (b) With the same current and time period, how many liters of oxygen gas at 298 K and 1.0 atm could be produced from the electrolysis of water?

12.67 (a) What current is required to produce 4.0 g of chromium metal from chromium(VI) oxide in 24 h? (b) What current is required to produce 4.0 g of sodium metal from molten sodium chloride during the same period?

12.68 What current is required to electroplate 6.66 μg of gold in 30.0 min from a gold(III) chloride aqueous solution? (b) How much time is required to electroplate 6.66 μg of chromium from a potassium dichromate solution, using a current of 100 mA?

12.69 When a ruthenium chloride solution was electrolyzed for 500 s with a 120-mA current, 31.0 mg of ruthenium was deposited. What is the oxidation number of ruthenium in the ruthenium chloride?

12.70 A 4.9-g sample of manganese was produced from a manganese nitrate aqueous solution when a current of 350 mA was passed for 13.7 h. What is the oxidation number of manganese in the manganese nitrate?

12.71 A current of 15.0 A electroplated 50.0 g of hafnium metal from an aqueous solution in 2.00 h. What was the oxidation number of hafnium in the solution?

12.72 A mass loss of 12.57 g occurred in 6.00 h at a titanium anode when a current of 4.70 A was used in an electrolytic cell. What is the oxidation number of the titanium in solution?

12.73 A metal forms the salt MCl_3. Electrolysis of the molten salt with a current of 0.700 A for 6.63 h produced 3.00 g of the metal. What is the molar mass of the metal?

12.74 Thomas Edison was faced with the problem of measuring the electricity that each of his customers had used. His first solution was to use a zinc "coulometer," an electrolytic cell in which the quantity of electricity is determined by measuring the mass of zinc deposited.

Only some of the current used by the customer passed through the coulometer. (a) What mass of zinc would be deposited in one month (of 31 days) if 1.0 mA of current passed through the cell continuously? (b) An alternative solution to this problem is to collect the hydrogen produced by electrolysis and measure its volume. What volume would be collected at 298 K and 1.00 bar under the same conditions? (c) Which method would be more practical?

For Exercises 12.75–12.80, refer to Boxes 12.1 and 12.2.

12.75 What is (a) the electrolyte; (b) the oxidizing agent in a mercury cell (shown below)? (c) Write the overall cell reaction for a mercury cell.

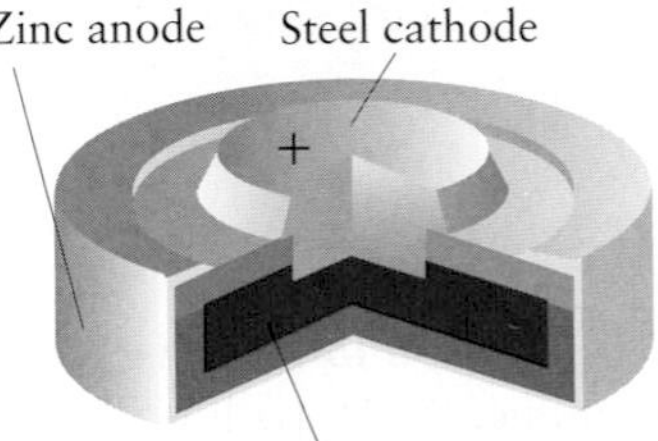

12.76 What is (a) the electrolyte; (b) the oxidizing agent during discharge in a lead-acid battery? (c) Write the reaction that occurs at the cathode during the charging of the lead-acid battery.

12.77 Explain how a dry cell generates electricity.

12.78 The density of the electrolyte in a lead-acid battery is measured to assess its state of charge. Explain how the density reflects the state of charge of the battery.

12.79 (a) What is the electrolyte in a nickel-cadmium cell? (b) Write the reaction that occurs at the anode when the cell is being charged.

12.80 (a) Why are lead-antimony grids used as electrodes in the lead-acid battery rather than smooth plates? (b) What is the reducing agent in the lead-acid battery? (c) The lead-acid cell potential is about 2 V. How, then, does a car battery produce 12 V for its electrical system?

INTEGRATED EXERCISES

12.81 A galvanic cell has the cell reaction, $M(s) + 2\,Zn^{2+}(aq) \rightarrow 2\,Zn(s) + M^{4+}(aq)$. The standard potential of the cell is 0.16 V. What is the standard potential of the M^{4+}/M redox couple?

12.82 Using data in Appendix 2B, calculate the standard potential for the half-reaction $Ti^{4+}(aq) + 4\,e^- \rightarrow Ti(s)$.

12.83 K_{sp} for $Cu(IO_3)_2$ is 1.4×10^{-7}. Using data available in Appendix 2B, calculate $E°$ for the half-reaction $Cu(IO_3)_2(s) + 2\,e^- \rightarrow Cu(s) + 2\,IO_3^-(aq)$.

12.84 K_{sp} for $Ni(OH)_2$ is 6.5×10^{-18}. Using this value and data from Appendix 2B, calculate $E°$ for the half-reaction $Ni(OH)_2(s) + 2\,e^- \rightarrow Ni(s) + 2\,OH^-(aq)$.

12.85 Answer the following questions and, for each "yes" response, write a balanced cell reaction and calculate the standard cell emf. (a) Can the oxygen present in air oxidize silver metal in acidic solution? (b) Can the oxygen in air oxidize silver metal in basic solution?

12.86 What is the standard potential for the reduction of oxygen to water in (a) an acidic solution? (b) a basic solution? (c) Is MnO_4^- more stable in an acidic or a basic aerated solution (a solution saturated with oxygen gas at 1 atm)? Explain your conclusion.

12.87 For each reaction that is spontaneous under standard conditions, write a cell diagram, determine the standard cell emf, and calculate $\Delta G°$ for the reaction:
(a) $2\,NO_3^-(aq) + 8\,H^+(aq) + 6\,Hg(l) \rightarrow 3\,Hg_2^{2+}(aq) + 2\,NO(g) + 4\,H_2O(l)$
(b) $2\,Hg^{2+}(aq) + 2\,Br^-(aq) \rightarrow Hg_2^{2+}(aq) + Br_2(l)$
(c) $Cr_2O_7^{2-}(aq) + 14\,H^+(aq) + 6\,Pu^{3+}(aq) \rightarrow 6\,Pu^{4+}(aq) + 2\,Cr^{3+}(aq) + 7\,H_2O(l)$

12.88 Dental amalgam, a solid solution of silver and tin in mercury, is used for filling tooth cavities. Two of the reduction half-reactions that the filling can undergo are

$$3\,Hg_2^{2+}(aq) + 4\,Ag(s) + 6\,e^- \longrightarrow 2\,Ag_2Hg_3(s), \quad E° = +0.85\text{ V}$$

$$Sn^{2+}(aq) + 3\,Ag(s) + 2\,e^- \longrightarrow Ag_3Sn(s), \quad E° = -0.05\text{ V}$$

Suggest a reason why, when you accidentally bite on a piece of aluminum foil with a tooth containing a silver filling, you may feel pain. Write a balanced chemical equation to support your suggestion.

12.89 Use the data in Appendix 2B and the fact that, for the half-reaction $F_2(g) + 2\ H^+(aq) + 2\ e^- \rightarrow 2\ HF(aq)$, $E° = +3.03$ V to calculate the value of K_a for HF.

12.90 The following items are obtained from the stockroom for the construction of a galvanic cell: two 250-mL beakers and a salt bridge, a voltmeter with attached wires and clips, 200 mL of 0.010 M $CrCl_3(aq)$, 200 mL of 0.16 M $CuSO_4(aq)$, a piece of copper wire, and a chrome-plated piece of metal. (a) Describe the construction of the galvanic cell. (b) Write the anode and cathode half-reactions. (c) Write the overall cell reaction. (d) Write the cell diagram for the galvanic cell. (e) What is the expected cell emf?

12.91 In a neuron (a nerve cell), the concentration of K^+ ions inside the cell is about 20 to 30 times greater than that outside. What potential difference between the inside and the outside of the cell would you expect to measure if the difference is due only to the imbalance of potassium ions?

12.92 (a) The emf of the cell $Zn(s)|Zn^{2+}(aq, ?)||Pb^{2+}(aq, 0.10\ mol{\cdot}L^{-1})|Pb(s)$ is 0.661 V. What is the molarity of Zn^{2+} ions? (b) Write an equation that gives the Zn^{2+} ion concentration as a function of emf, assuming all other aspects of the cell remain constant.

12.93 Calculate the standard emf and the actual emf of the galvanic cell $Pt(s)|Fe^{3+}(aq, 1.0\ mol{\cdot}L^{-1}),Fe^{2+}(aq, 0.0010\ mol{\cdot}L^{-1})||Ag^+(aq, 0.010\ mol{\cdot}L^{-1})|Ag(s)$. Comment on the difference in values.

12.94 Calculate the standard potential of the Cu^+/Cu couple from those of the Cu^{2+}/Cu^+ and Cu^{2+}/Cu couples.

12.95 Consider the galvanic cell $Pt(s)|Sn^{4+}(aq, 0.010\ mol{\cdot}L^{-1}),Sn^{2+}(aq, 0.10\ mol{\cdot}L^{-1})||H^+(aq, pH = 4.00)|O_2(g, 1\ bar)|C(gr)$. (a) What is the standard cell emf? (b) Write the cell reaction. (c) Calculate the actual cell emf. (d) Determine the equilibrium constant of the cell reaction. (e) Calculate the free energy of the cell reaction. (f) Suppose the cell emf is found to be +0.89 V. Assuming all other concentrations are precise, what is a more accurate pH of the solution?

12.96 Show how a silver-silver chloride electrode (silver in contact with solid AgCl and a solution of Cl^- ions) and a hydrogen electrode can be used to measure (a) pH; (b) pOH.

12.97 When a pH meter was standardized with a boric acid-borate buffer with a pH of 9.40, the cell emf was +0.060 V. When the buffer was replaced with a solution of unknown hydronium ion concentration, the cell emf was +0.22 V. What is the pH of the solution?

12.98 1,4-Benzoquinone, $C_6H_4O_2$, is reduced to 1,4-hydroquinone (1,4-dihydroxybenzene) with the standard potential $E° = +0.78$ mV. It is used in photography as an oxidizing agent.

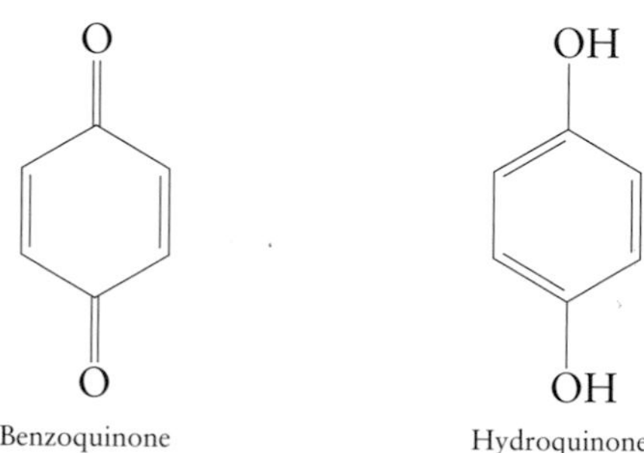

(a) Suggest a explanation for the ease of reduction of 1,4-benzoquinone. (b) Draw line structures for the following derivatives of benzoquinone and rationalize the trends in standard potentials:
methyl-1,4-benzoquinone, +23 mV;
2,3-dimethyl-1,4-benzoquinone, −74 mV;
2,5-dimethyl-1,4-benzoquinone, −67 mV;
2,6-dimethyl-1,4-benzoquinone, −80 mV;
2,3,5-trimethyl-1,4-benzoquinone, −165 mV;
2,3,5,6-tetramethyl-1,4-benzoquinone, −260 mV.

12.99 In the electrolytic refining of copper, blister copper is used as the anode and oxidized. The copper(II) ion that is produced from its oxidation is then reduced at the cathode to give a metal with a much higher purity. The impurities in the blister copper include iron, nickel, silver, gold, cobalt, and trace amounts of other metals. The material that is not oxidized at the anode falls to the bottom of the electrolytic cell and is called "anode mud." What are some of the components of anode mud? Explain your choices.

12.100 Using only data given in Appendix 2B, calculate the acidity constant for HBrO.

12.101 The reduction of ClO_4^- can be conducted in either basic or acidic solutions. The two half-reactions are

$$(1)\ ClO_4^-(aq) + 2\ H^+(aq) + 2\ e^- \longrightarrow ClO_3^-(aq) + H_2O(l) \qquad E° = +1.23\ V$$

(2) $ClO_4^-(aq) + H_2O(l) + 2\ e^- \longrightarrow$
$ClO_3^-(aq) + 2\ OH^-(l) \qquad E° = +0.36\ V$

(a) Show how these processes are related by deriving an expression that gives the pH dependence of the emf for each half-reaction. (b) What is the potential of each reaction in neutral solution?

12.102 (a) Derive an expression that gives the variation of the potential for the reduction of Ag^+ to Ag(s) as a function of $[Ag^+]$. (b) Show how this potential is related to the potentials for the reduction of the silver halides AgX (X = Cl, Br, I) and their K_{sp} values.

12.103 A mixture of 0.150 M $Mn(NO_3)_2(aq)$ and 0.100 M $Fe(NO_3)_2(aq)$ is placed in a beaker and the pH is adjusted to 5.00 by the addition of nitric acid. Two platinum electrodes are inserted, and current is passed through the solution to plate out the metals. (a) Write the half-reactions for the reduction of the metal ions and also for the likely oxidation process. (b) Using the answer to (a) and ignoring any overpotential, calculate the potential at which Mn and Fe will be plated out of these solutions. (c) Which metal will begin to deposit first? (d) Will the first metal be quantitatively precipitated by the time the second metal begins to be reduced? Determine this answer by calculating the residual concentration of the metal that plates out first when the second metal begins to be reduced.

12.104 Consider the electroplating of a metal +1 cation from a solution of unknown concentration according to the half-reaction $M^+(aq) + e^- \rightarrow M(s)$, with a standard potential $E°$. When the half-cell is connected to an appropriate oxidation half-cell and current is passed through it, the M^+ cation begins plating out at E_1. To what value (E_2) must the applied potential be adjusted, relative to E_1, if 99.99% of the metal is to be removed from the solution?

12.105 Consider the reduction of an organic molecule. (a) Describe what changes occur in the molecular orbital structure when an electron is added to the molecule. Where does the electron go? (b) Predict what will happen when an electron is added to CH_3X, where X = Cl, Br, or I. (c) For which halogen would the reaction described in (b) occur most readily?

CHAPTER

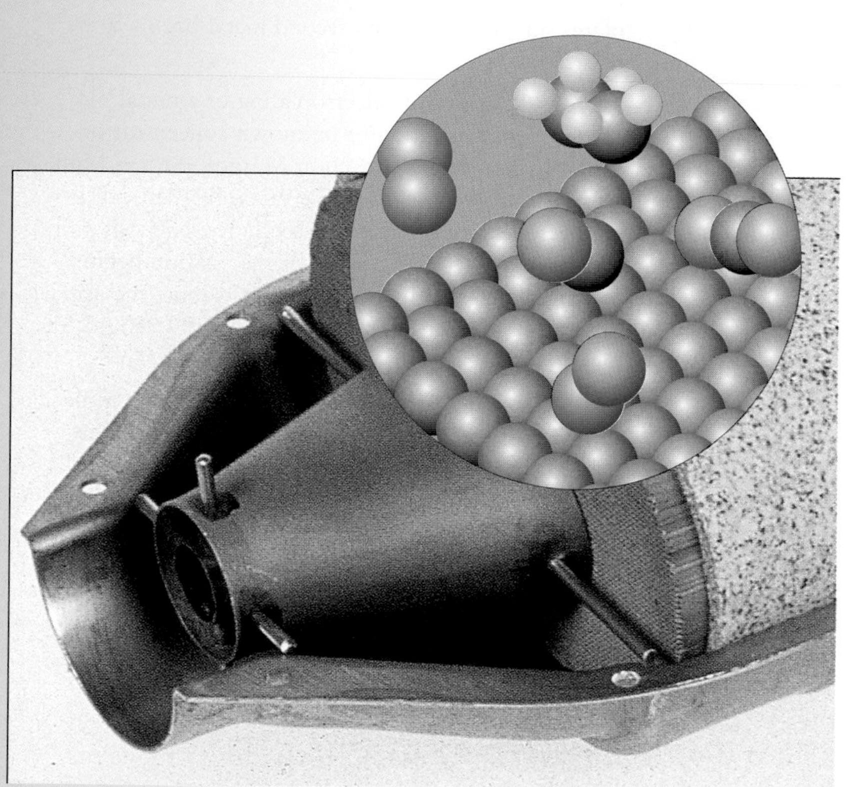

The processes taking place inside this ceramic catalytic converter include the reactions shown in the inset. Fragments of unburned hydrocarbons as well as carbon monoxide and nitric oxide molecules are converted to less noxious substances, such as nitrogen and carbon dioxide, by reactions at the surface of the catalyst.

Why Do We Need to Know This Material?

Chemical kinetics gives us insights into how chemical reactions take place at an atomic level, so it brings us to the heart of chemistry. It allows us, for instance, to model the atmosphere—the effects of damaging pollutants as well as the natural processes that influence our weather. Health represents a balance between large numbers of reactions going on inside organisms, and chemical kinetics contributes to our understanding of the network of reactions that keeps us alive. The development of catalysts, which are substances that speed up chemical reactions, is a branch of chemical kinetics that is crucial to the solution of major problems such as world hunger and the development of new fuels.

What Do We Need to Know Already?

Much of this chapter stands alone, but it would be helpful to review the kinetic model of gases (Section 4.13) and equilibrium constants (Section 9.2).

Chemical Kinetics

Although thermodynamics tells us the direction and extent of chemical change, it does not tell us how or how fast the reaction takes place. We have seen that some thermodynamically spontaneous reactions—such as the decomposition of benzene into carbon and hydrogen—do not seem to proceed at all, whereas other reactions—such as proton transfer reactions—reach equilibrium very rapidly. In this chapter, we examine the intimate details of how reactions proceed, what determines their rates, and how to control those rates. The study of the rates of chemical reactions is called **chemical kinetics.** When studying thermodynamics, we consider only the initial and final states of a chemical process and ignore what happens between them. In chemical kinetics, we are interested in the intermediate stages of the process—the details of the changes in atoms and molecules during reactions.

These changes help to account for the wide range of reaction rates and are used to devise ways of influencing rates. For instance, scientists are currently looking for economical ways of making hydrogen by the electrolysis of water: cheap hydrogen would mean cheap fuel and a reduction in carbon dioxide emissions. The electrolysis reaction is slow, but studies of the reaction rate have revealed that the slowness of the reaction arises from the great strength of the H—H bond. Chemists have been able to use this knowledge to devise special electrode materials that adsorb hydrogen in a way that weakens or even dissociates the H—H bond and facilitates the reaction.

This chapter begins by looking at how the rates of reactions are determined experimentally and summarized by succinct expressions known as *rate laws*. Then we use rate laws to predict the concentrations of reactants and products at any stage during a reaction and to gain insight about how reactions occur. Finally, we see how substances called catalysts accelerate reactions and control biological processes.

REACTION RATES

What do we mean by the "rate" of a chemical reaction? Informally, we know that a reaction is fast if the products are formed rapidly, as occurs in a precipitation reaction or an explosion (Fig. 13.1). A reaction is slow if the products are formed over a long period of time, as happens in corrosion or the decay of organic material (Fig. 13.2). Our first task is to set up a precise, quantitative definition of the rate of a chemical reaction. We use that definition to summarize experimentally determined rates by mathematical expressions and then discover that these expressions form a useful pattern.

13.1 Concentration and Reaction Rate

In everyday life, a rate is defined as the change in a property divided by the time it takes for that change to occur. For instance, the speed of an automobile, the rate of change of its position, is defined as distance

FIGURE 13.1 Reactions proceed at widely different rates. Some are very fast, like explosions of dynamite. Charges have been set off to demolish this old building. The chemical reaction in each explosion is over in fractions of a second; the gases produced expand more slowly.

FIGURE 13.2 Other reactions are very slow, as in the gradual buildup of corrosion on the prow of the *Titanic* on the cold floor of the Atlantic Ocean.

In chemical kinetics, the square brackets denote molar concentration, with the units $mol{\cdot}L^{-1}$ retained.

traveled divided by the time taken. We get the *average* speed if we divide the length of the journey by the total time for the journey; we get the *instantaneous* speed by reading the speedometer at some point on its journey. In chemistry, we want to know how quickly reactants are used up or products formed. We define the **reaction rate** as the change in concentration of one of the reactants or products divided by the time interval over which the change occurs. Because the rate may change as time passes, if we denote the change in molar concentration of a reactant R by $\Delta[R]$ and the time interval by Δt, the **average reaction rate** is

$$\text{Average rate of disappearance of R} = -\frac{\Delta[R]}{\Delta t} \qquad \text{(1a)}$$

where $\Delta[R] = [R]_{final} - [R]_{initial}$. Because reactants are used up in a reaction, the molar concentration of R decreases as time passes and $\Delta[R]$ is negative. The minus sign in Eq. 1a ensures that the rate is positive, which is the normal convention in chemical kinetics. If we follow the concentration of a product P, we would express the rate as

$$\text{Average rate of formation of P} = \frac{\Delta[P]}{\Delta t} \qquad \text{(1b)}$$

In this expression, $\Delta[P]$ is the change in molar concentration of P: it is a positive quantity because products are formed as time goes on.

How can we use this information . . . to calculate an average reaction rate?

Suppose we were studying the reaction 2 HI(g) $\longrightarrow$ H_2(g) + I_2(g) and found that, during an interval of 100 s, the concentration of HI decreased from 4.00 mmol·L^{-1} to 3.50 mmol·L^{-1}. We would report the average reaction rate as

$$\begin{aligned}\text{Average rate of disappearance of HI} &= -\frac{(3.50 - 4.00)\ (\text{mmol HI})\cdot\text{L}^{-1}}{100\ \text{s}} \\ &= 5.0 \times 10^{-3}\ (\text{mmol HI})\cdot\text{L}^{-1}\cdot\text{s}^{-1}\end{aligned}$$

Self-Test 13.1A When the reaction 2 HI(g) → H_2(g) + I_2(g) was carried out at a high temperature, the concentration of HI decreased from 6.00 mmol·L^{-1} to 4.20 mmol·L^{-1} in 200 s. What was the average reaction rate?

[*Answer:* 9.0×10^{-3} (mmol·L^{-1} HI)·L^{-1}·s^{-1}]

Self-Test 13.1B Hemoglobin (Hb) carries oxygen through our body by forming a complex with it: Hb(aq) + O_2(aq) → Hb·O_2(aq). In a solution of hemoglobin exposed to oxygen, the concentration of hemoglobin fell from 1.2×10^{-6} mmol·L^{-1} to 8.0×10^{-7} mmol·L^{-1} in 0.10 μs. What was the average rate at which hemoglobin reacted with oxygen in that solution, in millimoles per liter per microsecond (mmol·L^{-1}·μs^{-1})?

The calculation we have just done shows that the units used to report reaction rates are commonly moles per liter per second (mol·L^{-1}·s^{-1}), but other units for time (such as minutes or even hours) are commonly encountered for slower reactions. Note, too, that when we report a reaction rate, we have to be careful to specify the species to which the rate refers. For example, the rate of consumption of HI is twice the rate of formation of H_2 in the reaction above, because two HI molecules are used to make one H_2 molecule. The various ways of reporting the rate of a given reaction are related by the reaction stoichiometry. For example, the average rate of disappearance of H_2 is related to the rate of appearance of HI by

$$-\frac{\Delta[\text{H}_2]}{\Delta t} = \frac{1}{2}\frac{\Delta[\text{HI}]}{\Delta t}$$

It is possible to report a single unique average rate of a reaction without having to specify the species. The **unique average rate** of the reaction aA + bB → cC + dD is any of the following four equal quantities:

$$\text{Unique average reaction rate} = -\frac{1}{a}\frac{\Delta[\text{A}]}{\Delta t} = -\frac{1}{b}\frac{\Delta[\text{B}]}{\Delta t} = \frac{1}{c}\frac{\Delta[\text{C}]}{\Delta t} = \frac{1}{d}\frac{\Delta[\text{D}]}{\Delta t} \quad (2)$$

Division by the stoichiometric coefficients takes care of the stoichiometric relations between the reactants and products. There is no need to specify the species when reporting the unique average reaction rate.

BOX 13.1 HOW DO WE KNOW . . . WHAT HAPPENS TO ATOMS DURING A REACTION?

The fastest that anything happens in a chemical reaction is on a time scale of approximately 1 femtosecond (1 fs = 10^{-15} s). That is the time it takes for a bond to stretch or bend, and perhaps break. If we could follow atoms on that time scale, we could make a movie of the changes in molecules as they take part in a chemical reaction. The new field of *femtochemistry*, the study of very fast chemical processes, is bringing us closer to realizing that dream. Lasers can emit very short but intense pulses of electromagnetic radiation, so they can be used to study processes on very short time scales.

So far, the technique has been applied only to very simple reactions. For example, it is possible to watch the ion pair Na^+I^- decompose in the gas phase into separate Na and I atoms. At the start, the sodium ion and iodide ion are held together by the coulombic attraction of opposite charges. The pair is then hit by a femtosecond pulse of radiation from a laser. That pulse excites an electron from the I^- ion on to the Na^+ ion, thereby creating an NaI molecule in which the two atoms are held together by a covalent bond. The molecule has a lot of energy, and the bond length varies as the atoms swing in and out. At this point, a second femtosecond pulse is fired at the molecule. The radiation in the second pulse has a frequency that can be absorbed by the molecule only when the atoms have a particular separation. If the pulse is absorbed, we know that the atoms of the vibrating molecule have that special internuclear separation.

The illustration shows a typical result. The absorption reaches a maximum whenever the Na—I bond length returns to the length to which the second pulse is tuned. The peaks show that the sodium atom moves away from the iodine atom (corresponding to the dips in the curve), only to be recaptured (at the maxima) again. The separation of the maxima is about 1.3 ps, so the Na atom takes that long to swing out and be recaptured by the I atom. We can see that the peaks progressively decrease in intensity, showing that some Na atoms escape from their I atom partners on each swing. It takes about 10 swings outward before an Na atom can be sure of escaping. When sodium bromide is studied by these methods, the sodium atom escapes after about one swing, showing that an Na atom can escape more readily from a Br atom than from an I atom.

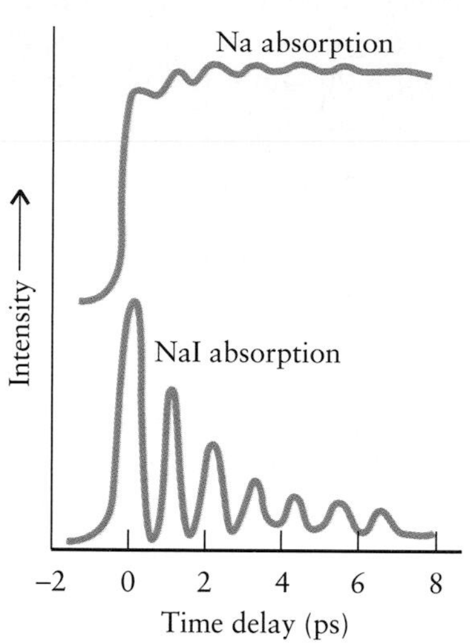

The femtosecond spectrum of the gas-phase NaI molecule as it dissociates into the separate atoms. A peak in the lower spectrum is observed whenever the bond distance in NaI reaches a certain value.

Self-Test 13.2A The average rate of the reaction $N_2(g) + 3\,H_2(g) \rightarrow 2\,NH_3(g)$ over a certain period is reported as 1.15 (mmol NH_3)$\cdot L^{-1}\cdot h^{-1}$. What is the average rate over the same period in terms of the disappearance of H_2?

[*Answer:* 1.73 (mmol H_2)$\cdot L^{-1}\cdot h^{-1}$]

Self-Test 13.2B What is (a) the average rate of the same reaction in terms of N_2; (b) the unique average rate, both over the same period?

The technique we employ to measure a rate of reaction depends on how rapidly the reaction takes place. Some biologically important reactions, such as the growth of hair, may take weeks to show significant changes, but other reactions, such as the explosion of nitroglycerin, are very rapid. Special techniques have to be used when the reaction is so fast that it is over in seconds. Using lasers, chemists can study reactions that are complete in a picosecond (1 ps = 10^{-12} s). The newest techniques can even monitor reactions that are complete after a few femtoseconds (1 fs = 10^{-15} s), as described in Box 13.1. On that time scale, atoms are hardly moving at all, and they are caught red-handed in the act of reaction.

All the techniques used to study fast reactions monitor concentration spectroscopically, as described in Major Technique 2, following Chapter 3. For instance, suppose we were studying the effect of a chlorofluorocarbon on the concentration of ozone, a blue gas. We could use a spectrometer to monitor the absorption responsible for the color of ozone and calculate the molar concentration of O_3 molecules from the intensity of absorption. In the **stopped-flow technique,** solutions of the reactants are forced into a mixing chamber very rapidly and the formation of products is observed spectroscopically (Fig. 13.3). This procedure is commonly used to study biologically important reactions.

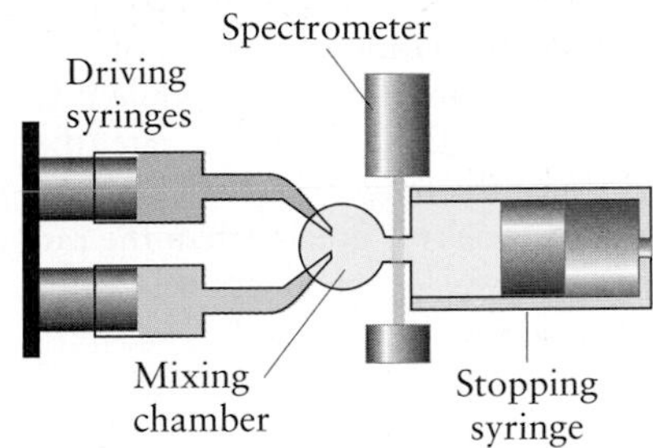

FIGURE 13.3 In a stopped-flow experiment, the driving syringes on the left push solutions of the reactants into the mixing chamber and the stopping syringe on the right stops the flow. The progress of the reaction is then monitored spectroscopically as a function of time.

The average rate of a reaction is the change in concentration of a species divided by the time it takes the change to occur. Spectroscopic techniques are widely used to monitor concentration, particularly for fast reactions.

13.2 The Instantaneous Rate of Reaction

Most reactions slow down as the reactants are used up. In other words, the average reaction rate changes from one interval to the next as the reaction proceeds. The reaction rate may also change over the time interval during which the change in concentration is being measured. To make sure that the reaction rate applies to a given instant in the course of the reaction, we should make our two concentration measurements as close together in time as possible. The best approximation to the rate at a single instant is obtained by drawing a tangent to the plot of concentration against time. The slope of this tangent is called the **instantaneous rate** of the reaction. From the graph in Fig. 13.4, the tangent is less steep at 10 weeks than it is at 5 weeks, so the instantaneous rate decreases with time. From now on, *whenever we speak of a reaction rate, we shall always mean an instantaneous rate.* The definitions in Eq. 1 can easily be adapted to refer to the instantaneous rate of a reaction.

How do we do that?

To set up expressions for the instantaneous rate of a reaction, we consider two times, t and $t + \Delta t$, to be a short time apart; we determine the concentration of a reactant or product at those times and find the average rate from Eq. 1. Then we decrease the interval and repeat the calculation. We can imagine continuing the process until the interval Δt has become

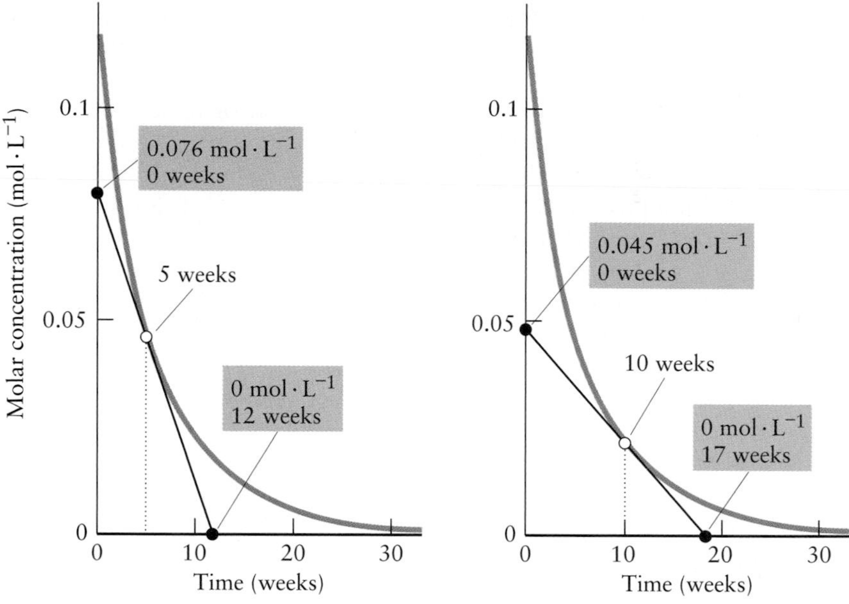

FIGURE 13.4 Two examples of the determination of the rate of deterioration of penicillin while it is being stored. Note that the rate (the slope of the tangent to the curve) after 5 weeks is greater than the rate after 10 weeks, when less penicillin is present.

infinitely small (denoted dt) and the change in molar concentration of a reactant R has become infinitesimal (denoted d[R]). Then we define the instantaneous rate as the ratio

$$\text{Rate of disappearance of R} = -\frac{d[R]}{dt} \tag{3a}$$

For a product P, we write

$$\text{Rate of formation of P} = \frac{d[P]}{dt} \tag{3b}$$

This chapter makes extensive use of expressions derived from calculus. For a discussion of derivatives and related functions, see Appendix 1F.

These expressions show that the rate is equal to the derivative of [P] with respect to time. The differential coefficients d[R]/dt and d[P]/dt are the mathematical expressions for the slope of the tangent drawn to a curve at the time of interest. Similarly, the unique *instantaneous* rate of a reaction is defined as in Eq. 2, but with differential coefficients in place of ratios of measurable differences:

$$\text{Reaction rate} = -\frac{1}{a}\frac{d[A]}{dt} = -\frac{1}{b}\frac{d[B]}{dt} = \frac{1}{c}\frac{d[C]}{dt} = \frac{1}{d}\frac{d[D]}{dt} \tag{3c}$$

Because it is difficult to draw a tangent accurately by eye, it is better to use a computer to analyze the curve. A superior method—which we meet in Section 13.4—is to report rates by using a procedure that avoids the use of tangents altogether.

The instantaneous reaction rate is the slope of a tangent drawn to the graph of concentration as a function of time; for most reactions, the rate decreases as the reaction proceeds.

13.3 Rate Laws and Reaction Order

Chemists often discover patterns in reaction rate data by examining the **initial rate** of reaction, the instantaneous rate of change in concentration of a species at the instant the reaction begins (Fig. 13.5). The advantage of examining the initial rate is that the products present later in the reaction may affect the rate; the interpretation of the rate is then quite complicated. There are no products present at the start of the reaction, so any pattern produced by the reactants is easier to find.

For example, suppose we were to measure different masses of solid dinitrogen pentoxide, N_2O_5, into different flasks of the same volume, immerse all the flasks in a water bath at 65°C to vaporize all the solid, and then, at regular time intervals, observe the concentrations of reactants and products in the reaction

$$2\ N_2O_5(g) \longrightarrow 4\ NO_2(g) + O_2(g)$$

Each flask has a different initial concentration of N_2O_5. We determine the initial rate of reaction in each flask by drawing the tangent to each curve at $t = 0$ (the black lines in Fig. 13.5). We find higher initial rates of decomposition of the vapor—steeper tangents—in the flasks with higher initial concentrations of N_2O_5. In fact, we would find that the initial rate is proportional to the initial concentration of N_2O_5 (Fig. 13.6), so we can write

$$\text{Initial rate} = k \times \text{initial concentration}$$

where k is a constant, called the **rate constant** for the reaction. The experimental value of k for this reaction, the slope of the straight line in Fig. 13.6, is $5.2 \times 10^{-3}\ s^{-1}$. The same rate constant applies to any initial concentration of N_2O_5, provided the temperature is constant.

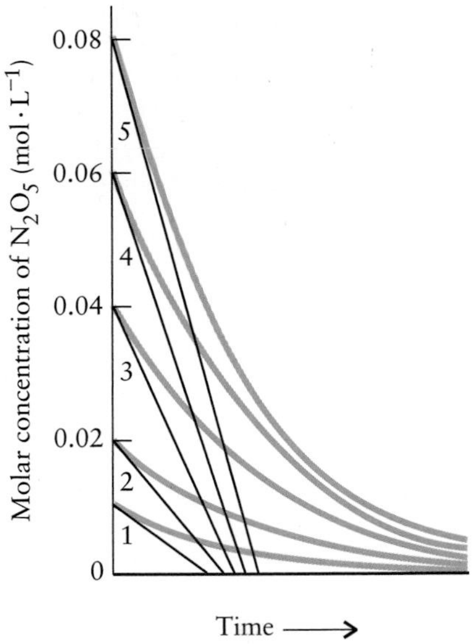

FIGURE 13.5 The definition of the initial rate of reaction. The initial rate of consumption of a reactant can be determined by drawing a tangent to the curve at the start of the reaction.

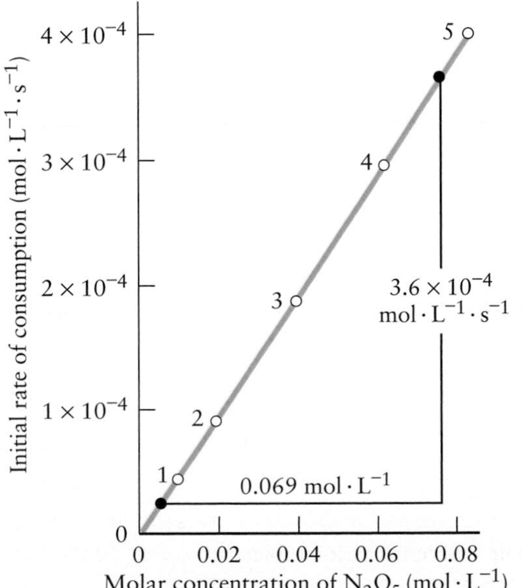

FIGURE 13.6 This graph was obtained by plotting the five initial rates from Fig. 13.5 against the initial concentration of N_2O_5. The initial rate is proportional to the initial concentration, as is shown by the straight line obtained when the rate is plotted against that concentration for the five samples in Fig. 13.5. This graph also illustrates how we obtain the values used to calculate the slope of the straight line.

The reaction rate is directly proportional to the concentration of N_2O_5 at all stages of the reaction, not just at the start. We therefore conclude that, at any stage of the reaction,

$$\text{Rate of disappearance of } N_2O_5 = k \times \text{concentration of } N_2O_5$$

This equation is an example of a **rate law,** an expression for the instantaneous reaction rate in terms of the concentration of a species at any instant.

Each reaction has its own characteristic rate law and rate constant k (Table 13.1). The rate constant is independent of the concentrations of the reactants but depends on the temperature. For the decomposition of N_2O_5 at 65°C, the rate law is

$$\text{Rate} = -\frac{d[N_2O_5]}{dt} = k[N_2O_5] \qquad k = 5.2 \times 10^{-3}\ \text{s}^{-1} \qquad (4)$$

where $[N_2O_5]$ denotes the molar concentration of N_2O_5.

TABLE 13.1 Rate Laws and Rate Constants

Reaction	Rate law*	Temperature, K	Rate constant
Gas phase			
$H_2 + I_2 \longrightarrow 2\,HI$	$k[H_2][I_2]$	500	4.3×10^{-7} L·mol^{-1}·s^{-1}
		600	4.4×10^{-4}
		700	6.3×10^{-2}
		800	2.6
$2\,HI \longrightarrow H_2 + I_2$	$k[HI]^2$	500	6.4×10^{-9} L·mol^{-1}·s^{-1}
		600	9.7×10^{-6}
		700	1.8×10^{-3}
		800	9.7×10^{-2}
$2\,N_2O_5 \longrightarrow 4\,NO_2 + O_2$	$k[N_2O_5]$	298	3.7×10^{-5} s^{-1}
		318	5.1×10^{-4}
		328	1.7×10^{-3}
		338	5.2×10^{-3}
$2\,N_2O \longrightarrow 2\,N_2 + O_2$	$k[N_2O]$	1000	0.76 s^{-1}
		1050	3.4
$2\,NO_2 \longrightarrow 2\,NO + O_2$	$k[NO_2]^2$	573	0.54 L·mol^{-1}·s^{-1}
$C_2H_6 \longrightarrow 2\,CH_3$	$k[C_2H_6]$	973	5.5×10^{-4} s^{-1}
cyclopropane $\longrightarrow$ propene	k[cyclopropane]	773	6.7×10^{-4} s^{-1}
Aqueous solution			
$H^+ + OH^- \longrightarrow H_2O$	$k[H^+][OH^-]$	298	1.5×10^{11} L·mol^{-1}·s^{-1}
$CH_3Br + OH^- \longrightarrow CH_3OH + Br^-$	$k[CH_3Br][OH^-]$	298	2.8×10^{-4} L·mol^{-1}·s^{-1}
$C_{12}H_{22}O_{11} + H_2O \longrightarrow 2\,C_6H_{12}O_6$	$k[C_{12}H_{22}O_{11}][H^+]$	298	1.8×10^{-4} L·mol^{-1}·s^{-1}

*For the rate of consumption or formation of the substance in bold type in the reaction column.

Not every rate law is quite so simple. If we made similar measurements on the reaction

$$2\ NO_2(g) \longrightarrow 2\ NO(g) + O_2(g)$$

we would not find a straight line when we plotted rate against concentration of NO_2 (Fig. 13.7a). However, when we plot the rate as a function of the *square* of the concentration of NO_2, we do get a straight line (Fig. 13.7b). This result shows that the rate is proportional to the square of the concentration and that we can write

$$\text{Rate of disappearance of } NO_2 = k \times (\text{concentration of } NO_2)^2$$

From the slope of the straight line in Fig. 13.7b, $k = 0.54\ \text{L}\cdot(\text{mol } NO_2)^{-1}\cdot\text{s}^{-1}$ at 300°C, so the rate law for the decomposition of NO_2 at that temperature is

$$\text{Rate} = -\frac{d[NO_2]}{dt} = k[NO_2]^2 \qquad k = 0.54\ \text{L}\cdot(\text{mol } NO_2)^{-1}\cdot\text{s}^{-1} \qquad (5)$$

Notice that the units of k are such that the units of rate are $\text{mol}\cdot\text{L}^{-1}\cdot\text{s}^{-1}$. For a first-order reaction, the units of k are s^{-1}; for a second-order reaction, $\text{L}\cdot\text{mol}^{-1}\cdot\text{s}^{-1}$.

The decomposition reactions of N_2O_5 and NO_2 have different rate laws, but both have the form

$$\text{Rate} = k \times (\text{concentration})^a \qquad (6)$$

with $a = 1$ for the N_2O_5 reaction and $a = 2$ for the NO_2 reaction. The N_2O_5 reaction is an example of a **first-order reaction**, because its rate is proportional to the *first* power of the concentration (that is, $a = 1$). The decomposition of NO_2 is an example of a **second-order reaction**, because its rate is proportional to the *second* power of the concentration (that is, $a = 2$). Doubling the concentration of a reactant in a first-order reaction doubles the reaction rate. Doubling the concentration of the reactant in any second-order reaction increases the reaction rate by a factor of $2^2 = 4$.

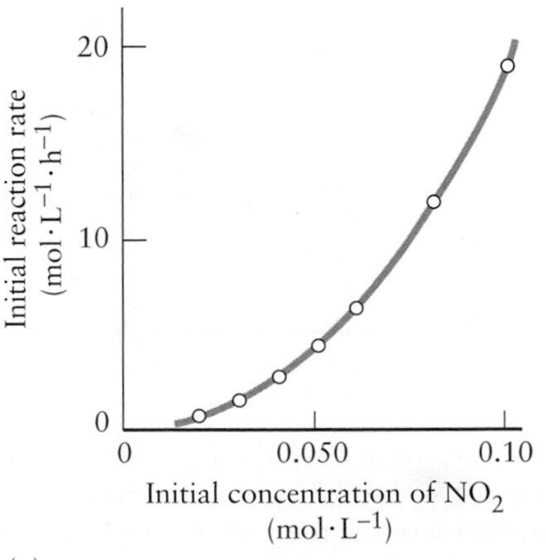

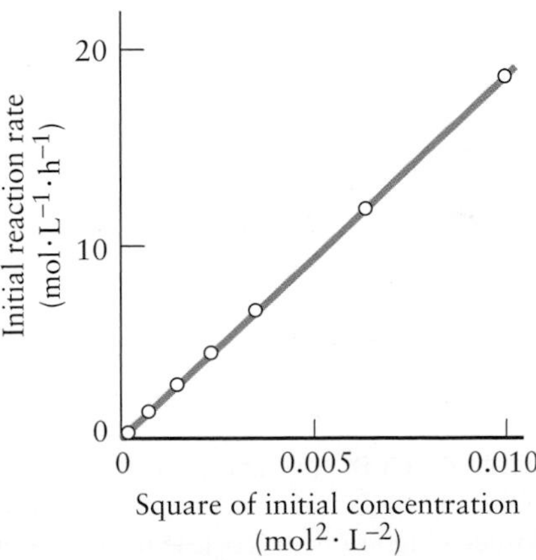

FIGURE 13.7 (a) When the rates of disappearance of NO_2 are plotted against its concentration, a straight line is not obtained. (b) However, a straight line is obtained when the rates are plotted against the square of the concentration.

Most of the reactions we shall consider are either first or second order in each reactant—but some reactions have other orders (different values of *a* in Eq. 6). For example, ammonia decomposes into nitrogen and hydrogen on a hot platinum wire:

$$2\ NH_3(g) \longrightarrow N_2(g) + 3\ H_2(g)$$

Experiments show that the decomposition occurs at a constant rate until all the ammonia has disappeared (Fig. 13.8). Its rate law is therefore

$$\text{Rate of disappearance of } NH_3 = k \qquad (7)$$

This decomposition is an example of a **zero-order reaction**, a reaction for which the rate is independent of concentration.

Zero-order reactions are so called because rate = $k \times$ (concentration)^0 = k.

The order of a reaction cannot in general be predicted from the chemical equation: a rate law is an *empirical* law. That is, *a rate law is an experimentally determined characteristic of the reaction and cannot in general be written down from the stoichiometry of the chemical equation for the reaction.* For instance, both the decomposition of N_2O_5 and that of NO_2 have a stoichiometric coefficient of 2 for the reactant, but one reaction is first order and the other is second order. The decomposition of ammonia also has a stoichiometric coefficient of 2 for the reactant, but its rate law is zero order.

Many reactions have rate laws that depend on the concentrations of more than one reactant. An example is the redox reaction between persulfate ions and iodide ions:

$$S_2O_8^{2-}(aq) + 3\ I^-(aq) \longrightarrow 2\ SO_4^{2-}(aq) + I_3^-(aq)$$

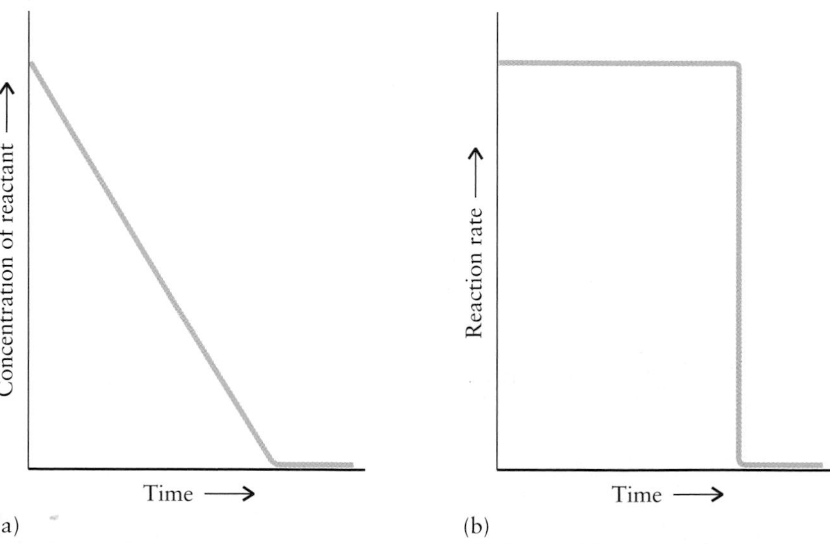

FIGURE 13.8 (a) The concentration of the reactant in a zero-order reaction falls at a constant rate until the reactant is exhausted. (b) The rate of a zero-order reaction is independent of the concentration of the reactant and remains constant until all the reactant has been consumed, when it falls abruptly to 0.

The rate of this reaction is found to be proportional to both the I^- and the $S_2O_8^{2-}$ concentrations. If we write the molar concentrations of these ions as $[I^-]$ and $[S_2O_8^{2-}]$, the rate law is

$$\text{Rate of disappearance of } S_2O_8^{2-} = k[S_2O_8^{2-}][I^-] \quad (8)$$

We say that the reaction is first order with respect to $S_2O_8^{2-}$ (or "in" $S_2O_8^{2-}$) and first order in I^-. Doubling either the $S_2O_8^{2-}$ ion concentration or the I^- ion concentration doubles the reaction rate. Doubling both concentrations would quadruple the reaction rate. We say that the *overall* order of the reaction is 2. In general, if

$$\text{Rate} = k[A]^a[B]^b \ldots \quad (9)$$

then the **overall order** is the sum of the powers $a + b + \ldots$.

Self-Test 13.3A When the NO concentration is doubled, the rate of the reaction $2\ NO(g) + O_2(g) \rightarrow 2\ NO_2(g)$ increases by a factor of 4. When both the O_2 and NO concentrations are doubled, the rate increases by a factor of 8. What are the reactant orders and overall order of the reaction?

[*Answer:* Second order in NO; first order in O_2; third order overall]

Self-Test 13.3B When the concentration of 2-bromo-2-methylpropane, C_4H_9Br, is doubled, the rate of the reaction $C_4H_9Br(aq) + OH^-(aq) \rightarrow C_4H_9OH(aq) + Br^-(aq)$ increases by a factor of 2. When both the C_4H_9Br and OH^- concentrations are doubled, the rate increase is the same, a factor of 2. What are the reactant orders and overall order of the reaction?

Orders can be negative numbers (as in $(\text{concentration})^{-1}$). A negative order implies that the concentration appears in the denominator of the rate law. Increasing the concentration of that species, usually a product, actually slows down the reaction because that species participates in a reverse reaction. An example is the decomposition of ozone, O_3, in the upper atmosphere:

$$2\ O_3(g) \longrightarrow 3\ O_2(g)$$

The experimentally determined rate law for this reaction is

Note that a rate law may depend on the concentration of products as well as those of reactants.

$$\text{Rate} = \frac{k[O_3]^2}{[O_2]} = k[O_3]^2[O_2]^{-1} \quad (10)$$

In Example 13.8 in Section 13.11, we will use this rate law to gain insight into how the reaction takes place.

Some reactions even have fractional orders (as in $(\text{concentration})^{1/2}$). For example, the oxidation of sulfur dioxide to sulfur trioxide in the presence of platinum in the reaction

$$2\ SO_2(g) + O_2(g) \longrightarrow 2\ SO_3(g)$$

is found to have the rate law

$$\text{Rate} = \frac{k[SO_2]}{[SO_3]^{1/2}} = k[SO_2][SO_3]^{-1/2} \tag{11}$$

and an overall order of $1 - \frac{1}{2} = \frac{1}{2}$. The presence of $[SO_3]$ in the denominator means that the reaction slows down as the concentration of product builds up.

EXAMPLE 13.1 Determining the reaction orders and rate law from experimental data

Four experiments were conducted to discover how the initial rate of consumption of BrO_3^- ions in the reaction $BrO_3^-(aq) + 5\ Br^-(aq) + 6\ H^+(aq) \rightarrow 3\ Br_2(aq) + 3\ H_2O(l)$ varies as the concentrations of the reactants are changed. (a) Use the experimental data in the following table to determine the order of the reaction with respect to each reactant and the overall order. (b) Write the rate law for the reaction and determine the value of *k*.

	Initial concentration, mol·L^{-1}			
Experiment	BrO_3^-	Br^-	H^+	Initial rate, (mol BrO_3^-)·L^{-1}·s^{-1}
1	0.10	0.10	0.10	1.2×10^{-3}
2	0.20	0.10	0.10	2.4×10^{-3}
3	0.10	0.30	0.10	3.5×10^{-3}
4	0.20	0.10	0.15	5.4×10^{-3}

STRATEGY Suppose that the concentration of a substance A is increased by a certain factor f and no other concentrations change. From the rate law, rate $= k[A]^a[B]^b \ldots$, we know that if the rate increases by f^a, then the reaction is of order a in A. To isolate the effect of each substance, compare experiments that differ in the concentration of only one substance at a time.

SOLUTION (a) Order in BrO_3^-: comparing experiment 1 with experiment 2, we see that, when the concentration of BrO_3^- is doubled, the rate also doubles. Therefore, the reaction is first order in BrO_3^- Order in Br^-: comparing experiment 1 with experiment 3, we see that when the concentration of Br^- is changed by a factor of 3.0, the rate changes by a factor of $3.5/1.2 = 2.9$. Allowing for experimental error, we can deduce that the reaction is also first order in Br^-. Order in H^+: when the concentration of hydrogen ions is increased from experiment 2 to experiment 4, by a factor of 1.5, the rate increases by a factor of $5.4/2.4 = 2.3$. Therefore, if the rate is proportional to $[H^+]^a$, the rate increases by a factor of 1.5^a when the concentration of H^+ ions increases by a factor of 1.5 and all other concentrations are held constant. We can find a from $1.5^a = 2.3$. To solve this relation, take logarithms of both sides, and use $\log 1.5^a = a \log 1.5$:

$$a \log 1.5 = \log 2.3$$

Therefore,

$$a = \frac{\log 2.3}{\log 1.5} = 2.0$$

The reaction is second order in H^+ and fourth order overall. (b) The rate law is therefore

$$\text{Rate of disappearance of } BrO_3^- = k[BrO_3^-][Br^-][H^+]^2$$

We find k by substituting the values from one of the experiments into the rate law and solving for k; for example, in experiment 4,

$$5.4 \times 10^{-3}\ \text{mol·L}^{-1}\text{·s}^{-1} = k \times (0.20\ \text{mol·L}^{-1}) \times (0.10\ \text{mol·L}^{-1}) \times (0.15\ \text{mol·L}^{-1})^2$$

so

Notice that the units of k depend on the order of the reaction.

$$k = \frac{5.4 \times 10^{-3}\ \text{mol·L}^{-1}\text{·s}^{-1}}{(0.20\ \text{mol·L}^{-1}) \times (0.10\ \text{mol·L}^{-1}) \times (0.15\ \text{mol·L}^{-1})^2} = 12\ \text{L}^3\text{·mol}^{-3}\text{·s}^{-1}$$

Self-Test 13.4A The reaction $S_2O_8^{2-}(aq) + 3\ I^-(aq) \rightarrow 2\ SO_4^{2-}(aq) + I_3^-(aq)$ is commonly used to produce triiodide ion in a "clock reaction," in which a sudden color change signifies that a reagent has been used up. Write the rate law for the consumption of persulfate ions with respect to each reactant and determine the value of k, given the following data:

	Initial concentration, mol·L^{-1}		
Experiment	**$S_2O_8^{2-}$**	**I^-**	**Initial rate, (mol $S_2O_8^{2-}$)·L^{-1}·s^{-1}**
1	0.15	0.21	1.14
2	0.22	0.21	1.70
3	0.22	0.12	0.98

[*Answer:* Rate = $k[S_2O_8^{2-}][I^-]$, using Experiment 1: k = 36 L·mol^{-1}·s^{-1}]

Self-Test 13.4B The highly toxic gas carbonyl chloride, $COCl_2$ (phosgene), is used to synthesize many organic compounds. Write the rate law and determine the value of k for the reaction used to produce carbonyl chloride: $CO(g) + Cl_2(g) \rightarrow COCl_2(g)$, given the following data collected at a certain temperature:

	Initial concentration, mol·L^{-1}		
Experiment	**CO**	**Cl_2**	**Initial rate, (mol $COCl_2$)·L^{-1}·s^{-1}**
1	0.12	0.20	0.121
2	0.24	0.20	0.241
3	0.24	0.40	0.682

Note: One of the orders is not an integer.

Chemists have found a way to simplify the problem of determining and analyzing rate laws that depend on the concentrations of more than one substance. To see what is involved, let's consider the rate law for the overall second-order oxidation of iodide ions, Eq. 8. Suppose we start with persulfate ions in such a high concentration that it barely changes in the course of the reaction. For instance, the concentration of persulfate ions might be 100 times greater than the concentration of iodide ions; so even when all the iodide ions have been oxidized, the persulfate concentration is almost the same as it was at the beginning of the reaction. Then, because $[S_2O_8^{2-}]$ is virtually constant, we can write Eq. 8 as

$$\text{Rate} = k[S_2O_8^{2-}][I^-] = k'[I^-] \qquad \textbf{(12)}$$

where $k' = k[S_2O_8^{2-}]$, another constant. We have turned the actual second-order reaction into a **pseudofirst-order reaction**, a reaction that is effectively first order. The rate law for a pseudofirst-order reaction is much easier to analyze than the true rate law because its rate depends on the concentration of only one substance.

Many reactions can be classified according to their order in a particular species, the power to which the concentration of a species occurs in the rate law, and by their overall order, the sum of the individual orders.

CONCENTRATION AND TIME

We often want to know how the concentration of a reactant or product varies with time. For example, how long will it take for a pollutant to decay? How much sulfur trioxide can be produced in an hour? How much penicillin will be left after six months? These questions can be answered by using formulas derived from the experimental rate law for the reactions. An **integrated rate law** gives the concentration of reactants or products at any time after the start of the reaction. Finding the integrated rate law from a rate law is very much like calculating the distance that a car will have traveled from a knowledge of its speed at each moment of the journey.

The integrated rate law for a zero-order reaction is easy to find. Because the rate is constant (at k), the difference in concentration of a reactant from its initial value, $[A]_0$, is proportional to the time for which the reaction is in progress, and we can write

$$[A]_0 - [A] = kt \quad \text{or} \quad [A] = [A]_0 - kt$$

As shown in Fig. 13.8a, a plot of concentration against time is linear, with slope $-k$.

In the next two sections, we concentrate on first- and second-order rate laws, but the techniques are applicable to other reaction orders, too. We also introduce the "half-life" of a substance, which is a useful measure of the lifetime of a reactant in a first-order reaction.

13.4 First-Order Integrated Rate Laws

One of the simplest integrated rate laws is for first-order reactions. Our aim is to find an expression for the concentration of a reactant A at a time t, given that the initial molar concentration of A is $[A]_0$.

How do we do that?

To find the concentration of reactant A in a first-order reaction at any time after it has begun, we begin by writing the rate law for the consumption of A as

$$-\frac{d[A]}{dt} = k[A]$$

Because an instantaneous rate is a derivative of concentration with respect to time, we can use the techniques of integral calculus to find the original function, the concentration [A], as a function of time. First, we divide both sides by [A] and multiply through by $-dt$:

$$\frac{d[A]}{[A]} = -k dt$$

Now we integrate both sides between the limits $t = 0$ (when $[A] = [A]_0$) and the time of interest, t (when $[A] = [A]_t$):

$$\int_{[A]_0}^{[A]_t} \frac{d[A]}{[A]} = -k \int_0^t dt$$

To evaluate the integral on the left, we use

$$\int \frac{dx}{x} = \ln x + \text{constant}$$

and obtain

$$\int_{[A]_0}^{[A]_t} \frac{d[A]}{[A]} = (\ln [A]_t + \text{constant}) - (\ln [A]_0 + \text{constant})$$

$$= \ln[A]_t - \ln[A]_0 = \ln \frac{[A]_t}{[A]_0}$$

The integral over time is simply equal to t, so we conclude that

$$\ln \frac{[A]_t}{[A]_0} = -kt \qquad (13a)$$

When we take antilogarithms of both sides, we obtain

$$[A]_t = [A]_0 e^{-kt} \qquad (13b)$$

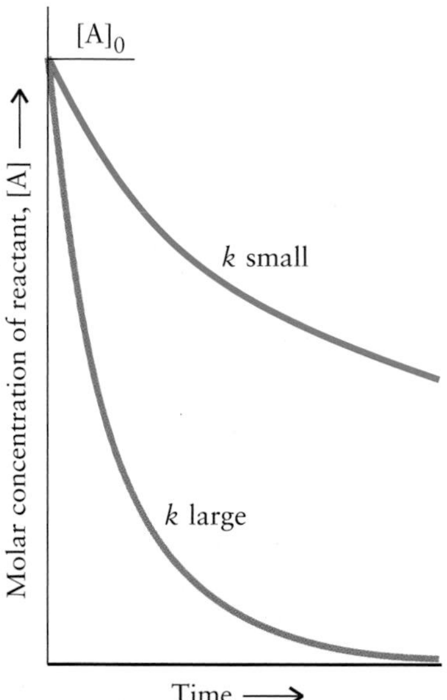

FIGURE 13.9 The graph of the concentration of a reactant in a first-order reaction is an exponential decay, as shown here. The larger the rate constant, the faster the decay from the same initial concentration.

Equations 13a and 13b are two forms of the integrated rate law for a first-order reaction. The variation of concentration with time predicted by Eq. 13b is shown in Fig. 13.9. This behavior is called an **exponential decay.** The change in concentration is initially rapid, but it changes more slowly as time goes on and the reactant is used up. We can use the integrated rate law to calculate the concentrations of reactants and products at different points in the reaction.

EXAMPLE 13.2 Calculating a concentration from a first-order integrated rate law

Calculate the concentration of N_2O_5 remaining 600 s (10 min) after the start of its decomposition at 65°C when its initial concentration was 0.040 $mol{\cdot}L^{-1}$. The reaction and its rate law are

$$2\,N_2O_5(g) \longrightarrow 4\,NO_2(g) + O_2(g)$$

$$\text{rate of disappearance of } N_2O_5 = k[N_2O_5]$$

with $k = 5.2 \times 10^{-3}\ s^{-1}$.

STRATEGY The concentration will be smaller than 0.040 $mol{\cdot}L^{-1}$ because some of the reactant decomposes as the reaction proceeds. Because the reaction is first order, use Eq. 13b to predict the concentration at any time after the reaction begins.

SOLUTION Substituting the data in Eq. 13b gives

$$\begin{aligned}[N_2O_5]_t &= [N_2O_5]_0\, e^{-kt} \\ &= (0.040\ mol{\cdot}L^{-1}) \times e^{-(5.2\times10^{-3}\ s^{-1})\times(600\ s)} \\ &= 0.0018\ mol{\cdot}L^{-1}\end{aligned}$$

That is, after 600 s, the concentration of N_2O_5 will have fallen from its initial value of 0.040 $mol{\cdot}L^{-1}$ to 0.0018 $mol{\cdot}L^{-1}$.

Self-Test 13.5A Calculate the concentration of N_2O remaining after the first-order decomposition

$$2\,N_2O(g) \longrightarrow 2\,N_2(g) + O_2(g) \qquad \text{rate} = k[N_2O]$$

has continued at 780°C for 100 ms, if the initial concentration of N_2O was 0.20 $mol{\cdot}L^{-1}$ and $k = 3.4\ s^{-1}$.

[*Answer:* 0.14 $mol{\cdot}L^{-1}$]

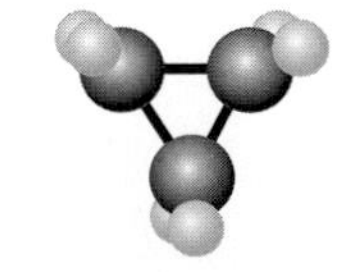

1 Cyclopropane, C_3H_6

Self-Test 13.5B Calculate the concentration of cyclopropane, C_3H_6 (**1**), remaining after the first-order isomerization to its isomer propene (**2**):

$$C_3H_6(g) \longrightarrow CH_3{-}CH{=}CH_2(g) \qquad \text{reaction rate} = k[C_3H_6]$$

has continued at 773 K for 200 s, if the initial concentration of C_3H_6 was 0.100 $mol{\cdot}L^{-1}$ and $k = 6.7 \times 10^{-4}\ s^{-1}$.

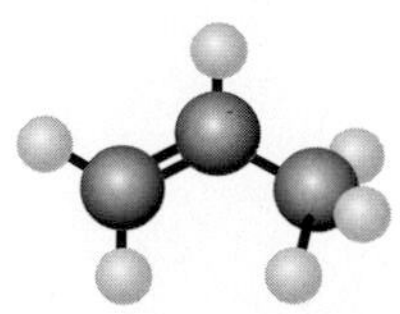

2 Propene, C_3H_6

We can use an integrated first-order rate law to confirm that a reaction is first order and to measure its rate constant. From Eq. 13a, we can write

$$\ln[A]_t = \ln[A]_0 - kt \qquad (14)$$

This equation has the form of an equation for a straight line (see Appendix 1E):

$$y = \text{intercept} + (\text{slope} \times x)$$

Therefore, if we plot $\ln[A]_t$ as a function of t, we should get a straight line with slope $-k$.

In a first-order reaction, the concentration of reactant decays exponentially with time. To verify that a reaction is first order, plot the natural logarithm of its concentration as a function of time and expect a straight line; the slope of the straight line is $-k$.

We mentioned in Section 13.2 that there is a better way of determining rate constants than trying to draw tangents to curves: this is it.

MEDIA EXAMPLE 13.3 Measuring a rate constant

When cyclopropane (C_3H_6, **1**) is heated to 500°C (773 K), it changes into an isomer, propene (**2**). The data in the table below show the concentration of cyclopropane at a series of times after the start of the reaction.

t, min	Concentration of C_3H_6, $mol{\cdot}L^{-1}$
0	1.5×10^{-3}
5	1.24×10^{-3}
10	1.00×10^{-3}
15	0.83×10^{-3}

Confirm that the reaction is first order in C_3H_6 and calculate the rate constant.

STRATEGY We need to plot $\ln[\text{cyclopropane}]_t$ as a function of t and see whether we obtain a straight line. If we do get a straight line, the reaction is first order and the slope of the graph is the value of $-k$. You can also use the Living Graph *Determination of Rate Constant (first-order rate law)* on the Web site for this book to make the plot.

SOLUTION For the graphical procedure, begin by setting up the following table:

t, min	$\ln[C_3H_6]_t$
0	−6.50
5	−6.69
10	−6.91
15	−7.09

The points are plotted in Fig. 13.10. The graph is a straight line, confirming that the reaction is first order in cyclopropane. The slope of the line is

$$\text{Slope} = \frac{(-7.02) - (-6.56)}{13.3\ \text{min} - 1.7\ \text{min}} = -0.040\ \text{min}^{-1}$$

Media Link Example 13.3: Measuring a rate constant

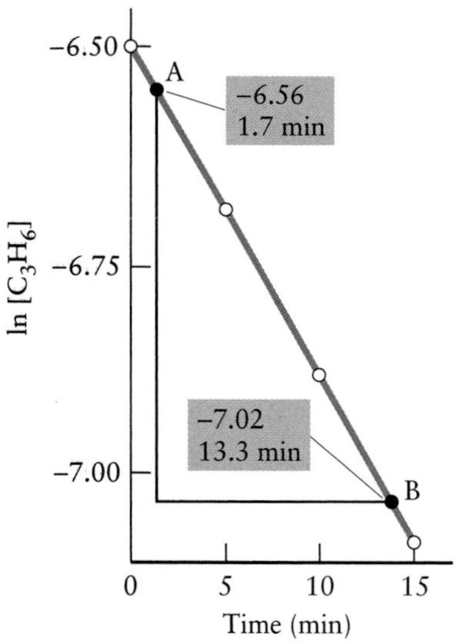

FIGURE 13.10 We can test for a first-order reaction by plotting the natural logarithm of the reactant concentration against time. The graph is linear if the reaction is first order. The slope of the line, which is calculated here by using the points A and B, is equal to the negative of the rate constant.

Media Link **Living graph:** Fig. 13.10, Determination of rate constant (first-order rate law)

Therefore, because $k = -\text{slope}$, $k = 0.040\ \text{min}^{-1}$. This value is equivalent to $k = 6.7 \times 10^{-4}\ \text{s}^{-1}$, the value in Table 13.1.

Self-Test 13.6A Data on the decomposition of N_2O_5 at 25°C are

t, min	Concentration of N_2O_5, mol·L^{-1}
0	15.0×10^{-3}
200	9.6×10^{-3}
400	6.2×10^{-3}
600	4.0×10^{-3}
800	2.5×10^{-3}
1000	1.6×10^{-3}

Confirm that the reaction is first order and find the value of k.

[*Answer:* $2.2 \times 10^{-3}\ \text{min}^{-1}$]

Self-Test 13.6B Azomethane, $CH_3N_2CH_3$, decomposes to ethane and nitrogen gas in the reaction $CH_3N_2CH_3(g) \rightarrow CH_3CH_3(g) + N_2(g)$. The reaction was followed at 460 K by measuring the partial pressure of azomethane over time:

t, s	Partial pressure of $CH_3N_2CH_3$, Torr
0	8.20×10^{-2}
1000	5.72×10^{-2}
2000	3.99×10^{-2}
3000	2.78×10^{-2}
4000	1.94×10^{-2}

Confirm that the reaction is first order and find the value of k.

EXAMPLE 13.4 Predicting how long it will take for a concentration to change

A sample of N_2O_5 is allowed to decompose. How long will it take for the concentration of N_2O_5 to decrease from 20 mol·L^{-1} to 2.0 mol·L^{-1} at 65°C? Use the data in Table 13.1.

STRATEGY Decide by referring to Table 13.1 whether the decomposition is first order. If it is, rearrange Eq. 13a into an equation for t in terms of the given concentrations:

$$t = \frac{1}{k} \ln \frac{[A]_0}{[A]_t}$$

Then substitute the numerical value of the rate constant and the concentration data.

SOLUTION The reaction is first order. Because $k = 5.2 \times 10^{-3}\ \text{s}^{-1}$ at 65°C (338 K), we can write

$$t = \frac{1}{5.2 \times 10^{-3}\ \text{s}^{-1}} \times \ln\left(\frac{20\ \text{mol·L}^{-1}}{2.0\ \text{mol·L}^{-1}}\right) = 4.4 \times 10^2\ \text{s}$$

The time required is approximately 7.3 min.

Self-Test 13.7A How long does it take for a concentration to decrease to 1.0% of its initial value in a first-order reaction with $k = 1.0\ s^{-1}$?

[*Answer:* 4.6 s]

Self-Test 13.7B Cyclopropane gas isomerizes to propene gas in a first-order process. How long is needed for the concentration of cyclopropane to decrease from 1.0 mol·L^{-1} to 0.0050 mol·L^{-1} at 500°C? Use the data in Table 13.1.

13.5 Half-Lives for First-Order Reactions

The **half-life,** $t_{1/2}$, of a substance is the time needed for its concentration to fall to one-half its initial value. Knowing the half-lives of pollutants such as chlorofluorocarbons allows us to assess their environmental impact. If their half-lives are short, they may not survive long enough to reach the stratosphere, where they can destroy ozone. Half-lives are also important in planning storage systems for radioactive materials, because the decay of radioactive nuclei is a first-order process.

We already know that the higher the value of k, the more rapid the disappearance of a reactant. Therefore, we should be able to deduce a relation for a first-order reaction that shows that the greater the rate constant, the shorter the half-life.

How do we do that?

To find the relation between the rate constant and the half-life, we begin by rearranging Eq. 13a into an expression for the time needed to reach a given concentration:

$$t = \frac{1}{k} \ln \frac{[A]_0}{[A]_t}$$

Now we set t equal to $t_{1/2}$ and $[A]_t = \frac{1}{2}[A]_0$, and obtain

$$t_{1/2} = \frac{1}{k} \ln \frac{[A]_0}{\frac{1}{2}[A]_0} = \frac{1}{k} \ln 2$$

Notice that the initial concentration has canceled.

We have shown that

$$t_{1/2} = \frac{\ln 2}{k} \qquad (15)$$

with $\ln 2 = 0.693\ldots$ As we anticipated, the greater the value of the rate constant k, the shorter the half-life for the reaction (Fig. 13.11). Note that the concentration does not appear in Eq. 15: for a first-order reaction, the half-life is independent of the initial concentration of the reactant.

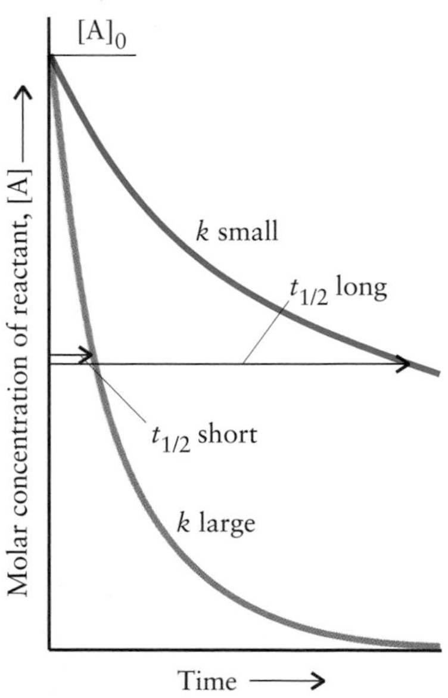

FIGURE 13.11 The change in concentration of the reactant in two first-order reactions plotted on the same graph. When the first-order rate constant is large, the half-life of the reactant is short, because the exponential decay of the concentration of the reactant is then fast.

Self-Test 13.8A Calculate the time needed for the concentration of N_2O to fall to (a) one-half; (b) one-eighth of its initial value in a first-order decomposition at 1000 K. Consult Table 13.1 for the rate constant.

[*Answer:* (a) 0.91 s; (b) 2.7 s]

Self-Test 13.8B Calculate the time needed for the concentration of C_2H_6 to fall to (a) one-half; (b) one-sixteenth of its initial value as it dissociates into CH_3 radicals at 973 K. Consult Table 13.1 for the rate constant.

The significance of the half-life of a first-order reaction becomes clear once we look at Fig. 13.12. We see that the concentration falls to half its initial value in the time $t_{1/2}$. The concentration falls to half its new value (one-fourth the initial value) in the same time, and then to half that new value (one-eighth the initial value) in another half-life. For each time interval equal to the half-life, the concentration falls to half its value at the start of the interval.

The half-life of a first-order reaction is characteristic of the reaction and independent of the initial concentration. A reaction with a large rate constant has a short half-life.

EXAMPLE 13.5 Using a half-life to calculate the amount of reactant remaining

Mercury(II) is eliminated from the body by a first-order process that has a half-life of 6 days (6 d). A farming family accidentally ingests mercury(II) by eating contaminated grain. What percentage of the mercury(II) would remain in their bodies after 30 d if therapeutic measures were not taken?

STRATEGY The fraction remaining after a given time can be calculated from the integrated first-order rate law, Eq. 13b. To use this equation, we need the rate constant, which can be calculated from the half-life by using Eq. 15. Therefore, start by calculating the rate constant and substitute the result into Eq. 13b. Express the answer as a percentage by multiplying the fraction by 100%.

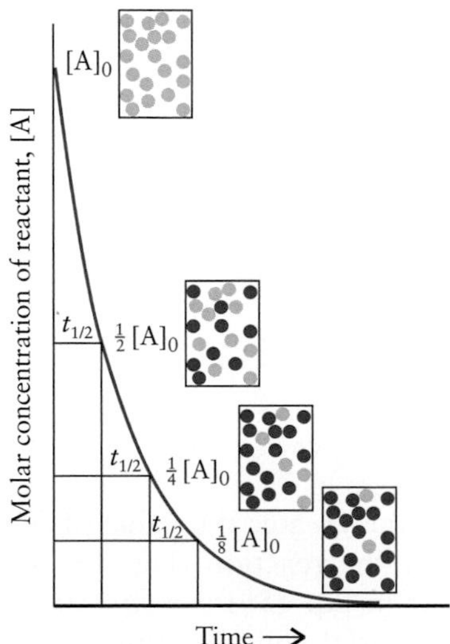

FIGURE 13.12 For first-order reactions, the half-life is the same whatever the concentration at the start of the chosen period. Therefore, it takes one half-life to fall to half the initial concentration, two half-lives to fall to one-fourth the initial concentration, three half-lives to fall to one-eighth, and so on. The molecular views portray the composition of the reaction mixture at the end of each half-life; the yellow circles represent the reactant A and the red circles represent the product.

SOLUTION From Eq. 15 and the data, we can write

$$k = \frac{\ln 2}{t_{1/2}} = \frac{\ln 2}{6\ \text{d}} = \frac{\ln 2}{6}\ \text{d}^{-1}$$

Then, from Eq. 13b, the fraction remaining is

$$\frac{[\text{A}]_t}{[\text{A}]_0} = \text{e}^{-kt} = \text{e}^{-((\ln 2)/6\ \text{d}^{-1})\times(30\ \text{d})} = \text{e}^{-((\ln 2)\times 30/6)} = 0.03$$

The percentage remaining is therefore $0.03 \times 100\% = 3\%$.

Self-Test 13.9A The half-life of methyl mercury in the body is 70 days. How many days are required for the amount of methyl mercury to drop to 10% of the original value after accidental ingestion?

[*Answer:* 230 days]

Self-Test 13.9B Soil near the Rocky Flats Nuclear Processing Facility in Colorado was found to be contaminated with radioactive plutonium-239, which has a half-life of 2.4×10^4 years. The soil was loaded into drums for storage. How many years must pass before the radioactivity drops to 20% of its initial value?

13.6 Second-Order Integrated Rate Laws

Now we drive the integrated rate law for second-order reactions with the rate law

$$\text{Rate} = k[\text{A}]^2 \qquad (16)$$

How do we do that?

To obtain the integrated rate law for a second-order reaction, we recognize that the rate law is a differential equation and write Eq. 16 as

$$\frac{\text{d}[\text{A}]}{\text{d}t} = -k[\text{A}]^2$$

After division by $[\text{A}]^2$ and multiplication by $\text{d}t$, this equation becomes

$$\frac{\text{d}[\text{A}]}{[\text{A}]^2} = -k\text{d}t$$

To solve this equation, we integrate it between the same limits used in the first-order case, but this time we need the integral

$$\int \frac{\text{d}x}{x^2} = -\frac{1}{x} + \text{constant}$$

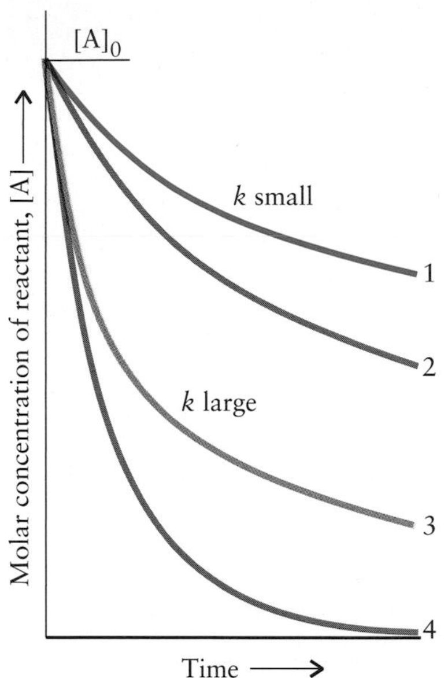

FIGURE 13.13 The characteristic shapes (lines 1 and 3) of the time dependence of the concentration of a reactant during a second-order reaction. The lower gray lines (2 and 4) are the curves for first-order reactions with the same initial rates as reactions 1 and 3, respectively. Note how the concentrations for second-order reactions fall away much less rapidly at longer times than those for first-order reactions do.

With this integral in mind, we write

$$\int_{[A]_0}^{[A]_t} \frac{d[A]}{[A]^2} = -k \int_0^t dt$$

and obtain

$$-\frac{1}{[A]_t} + \frac{1}{[A]_0} = -kt \tag{17a}$$

This expression can be rearranged into

$$[A]_t = \frac{[A]_0}{1 + [A]_0 kt} \tag{17b}$$

Equation 17b is plotted in Fig. 13.13. We see that the concentration of the reactant decreases rapidly at first, but then changes more slowly than a first-order reaction with the same initial rate. This slowing down of second-order reactions has important environmental consequences: because many pollutants disappear by second-order reactions, they remain at low concentration in the environment for long periods. Equation 17 can be written in the alternative form

$$\frac{1}{[A]_t} = kt + \frac{1}{[A]_0} \tag{17c}$$

This expression shows us that, to determine whether a reaction is second order in a reactant, we can plot the inverse of the concentration as a function of time to see whether a straight line results. The slope of the line is equal to k (Fig. 13.14).

A second-order reaction has a long tail of low concentration at long reaction times. The half-life of a second-order reaction is inversely proportional to the concentration of the reactant.

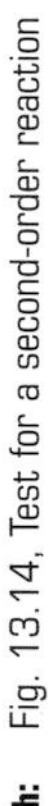

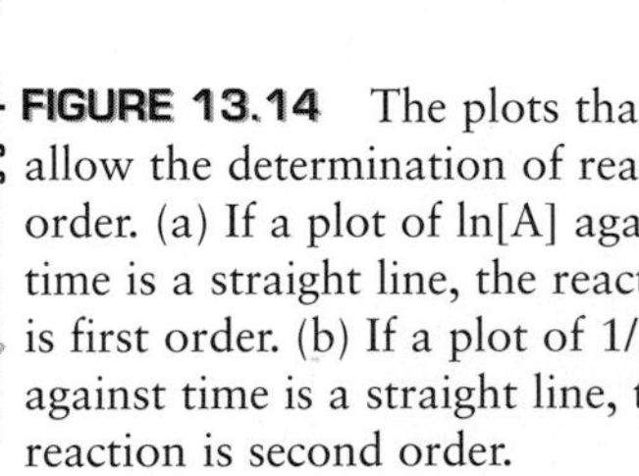

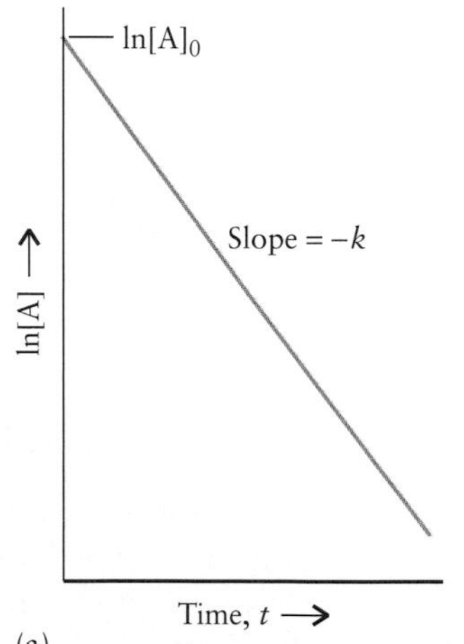

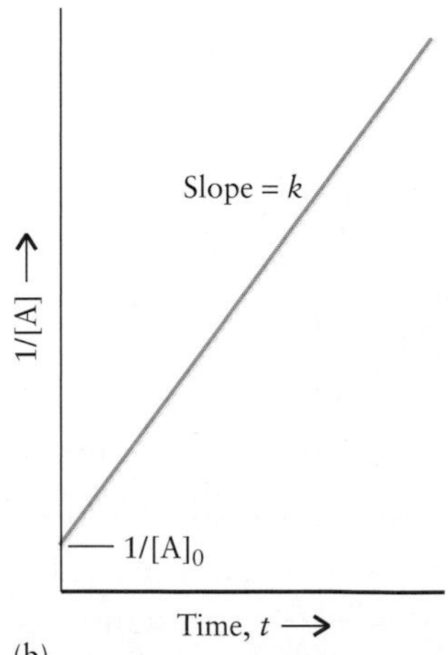

FIGURE 13.14 The plots that allow the determination of reaction order. (a) If a plot of ln[A] against time is a straight line, the reaction is first order. (b) If a plot of 1/[A] against time is a straight line, the reaction is second order.

MODELS OF REACTIONS

One of the principal reasons for studying reaction rates—apart from the obvious practical advantages of being able to predict the composition of a reaction mixture as it moves toward equilibrium—is to gain insight into the molecular processes that accompany reactions. Rate laws and rate constants are windows on chemical change. We shall explore the insights that chemical kinetics provides in two stages. In this part of the chapter, we look at the rate constants themselves and build molecular models to account for their values. In the remaining sections of the chapter, we see how to account for the different forms of the rate laws for reactions of increasing complexity. Finally, by discovering a relation between equilibrium constants and rate constants, we shall see that chemical kinetics deepens our insight into the topics of the preceding four chapters.

FIGURE 13.15 Reaction rates almost always increase with temperature. The beaker on the left contains magnesium in cold water and that on the right contains magnesium in hot water. An indicator has been added to show the formation of an alkaline solution as magnesium reacts.

13.7 The Effect of Temperature

A rate law summarizes the dependence of the rate on concentrations. However, rates also depend on temperature. The *qualitative* observation is that most reactions go faster as the temperature is raised (Fig. 13.15). An increase of 10°C from room temperature typically doubles the rate of reaction of organic species in solution. That is why we cook foods: heating accelerates reactions that lead to the breakdown of cell walls and the decomposition of proteins. We refrigerate foods to slow down the natural chemical reactions that lead to their decomposition.

The temperature dependence of a reaction rate is carried in the rate constant: if the rate constant increases with temperature, then for a given concentration of reactants, the reaction rate increases, too. As we shall now see, the temperature dependence of rate constants gives valuable insight into their origins.

The quantitative effect of temperature on a reaction rate was identified by the Swedish chemist Svante Arrhenius in the late nineteenth century. He found that he obtained a straight line if he plotted the logarithm of the rate constant against the inverse of the absolute temperature. In other words,

$$\ln k = \text{intercept} + \text{slope} \times \frac{1}{T}$$

The intercept here is denoted ln A; and for reasons that will become clear in Section 13.8, the slope is denoted $-E_a/R$, where R is the gas constant. The empirical **Arrhenius equation** is

$$\ln k = \ln A - \frac{E_a}{RT} \tag{18}$$

The use of the gas constant does not mean that the Arrhenius equation applies only to gases: it applies to a wide range of reactions in solution, too.

The two constants, A and E_a, are known as the **Arrhenius parameters** for the reaction and are found from experiment; A is called the **pre-exponential factor,** and E_a is the **activation energy.** Both A and E_a are nearly independent of temperature but do depend on the reaction being studied.

Media Link **Example 13.6:** Measuring an Activation Energy

MEDIA EXAMPLE 13.6 Measuring an activation energy

The rate constant for the second-order reaction between bromoethane and hydroxide ions in water, $C_2H_5Br(aq) + OH^-(aq) \rightarrow C_2H_5OH(aq) + Br^-(aq)$, was measured at several temperatures, with the following results:

Temperature, °C	k, $L \cdot mol^{-1} \cdot s^{-1}$
25	8.8×10^{-5}
30	1.6×10^{-4}
35	2.8×10^{-4}
40	5.0×10^{-4}
45	8.5×10^{-4}
50	1.40×10^{-3}

Find the activation energy of the reaction.

STRATEGY Activation energies are found by plotting ln k against $1/T$, with T in kelvins, and measuring the slope of the line. Multiply the slope of the graph by $-R$ to find the activation energy. Use $R = 8.3145\ J \cdot K^{-1} \cdot mol^{-1}$. A spreadsheet, curve-fitting program, or the Living Graph *Determination of Arrhenius Parameters* on the Web site for this book is very useful for this type of calculation.

SOLUTION The table to use for drawing the graph is

Temperature, °C	T, K	$1/T$, K^{-1}	k, $L \cdot mol^{-1} \cdot s^{-1}$	ln k
25	298	3.35×10^{-3}	8.8×10^{-5}	-9.34
30	303	3.30×10^{-3}	1.6×10^{-4}	-8.74
35	308	3.25×10^{-3}	2.8×10^{-4}	-8.18
40	313	3.19×10^{-3}	5.0×10^{-4}	-7.60
45	318	3.14×10^{-3}	8.5×10^{-4}	-7.07
50	323	3.10×10^{-3}	1.40×10^{-3}	-6.57

The points are plotted in Fig. 13.16. The slope can be calculated from two points, such as those marked A and B:

$$\text{Slope} = \frac{(-9.60) - (-6.38)}{(3.38 \times 10^{-3}) - (3.08 \times 10^{-3})} = -\frac{3.22}{0.30 \times 10^{-3}\ K^{-1}} = -1.07 \times 10^4\ K$$

Because the slope is equal to $-E_a/R$, we have

$$E_a = -(8.3145\ J \cdot K^{-1} \cdot mol^{-1}) \times (-1.07 \times 10^4\ K) = 89\ kJ \cdot mol^{-1}$$

Media Link **Living graph:** Fig. 13.16. Variation of rate constant with temperature

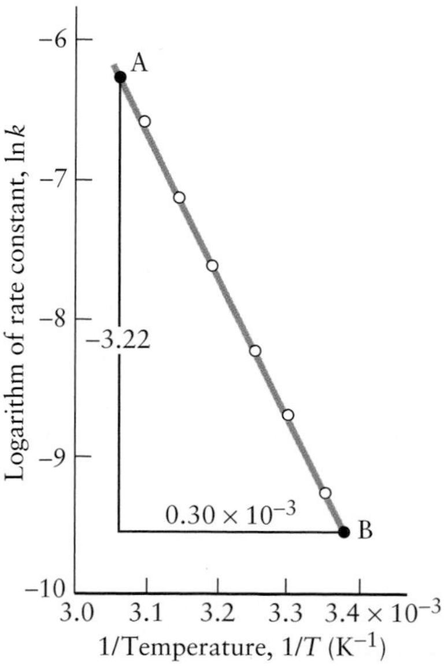

Logarithm of rate constant, ln k (−6 to −10); 1/Temperature, $1/T$ (K^{-1}) (3.0 to 3.4×10^{-3}); A; B; −3.22; 0.30×10^{-3}

FIGURE 13.16 An Arrhenius plot is a graph of ln k against $1/T$. If, as here, the line is straight, then the reaction is said to show Arrhenius behavior in the temperature range studied. This plot has been constructed from the data in Example 13.6. The slope of the line has been worked out by using the points A and B.

Self-Test 13.10A The rate constant for the second-order gas-phase reaction $HO(g) + H_2(g) \rightarrow H_2O(g) + H(g)$ varies with the temperature as follows:

Temperature, °C	k, $L \cdot mol^{-1} \cdot s^{-1}$
100	1.1×10^{-9}
200	1.8×10^{-8}
300	1.2×10^{-7}
400	4.4×10^{-7}

Determine the activation energy.

[*Answer:* 41 $kJ \cdot mol^{-1}$]

Self-Test 13.10B The rate of a reaction increased from 3.00 $mol \cdot L^{-1} \cdot s^{-1}$ to 4.35 $mol \cdot L^{-1} \cdot s^{-1}$ when the temperature was raised from 18°C to 30°C. What is the activation energy of the reaction?

Reactions that give a straight line when ln k is plotted against $1/T$ are said to show **Arrhenius behavior.** A wide range of reactions—not only those in the gas phase—show Arrhenius behavior. Even tropical fireflies flash more quickly on hot nights than on cold, and their rate of flashing is Arrhenius-like over a small range of temperatures. We can conclude that the biochemical reactions responsible for the flashing have rate constants that increase with temperature in accord with Eq. 18. Some Arrhenius parameters are listed in Table 13.2. Once the Arrhenius parameters have been measured for a reaction, it is easy to predict the rate constant of the reaction at any temperature that lies within the range of the original measurements.

We cannot be sure that the reaction is Arrhenius-like outside that range.

TABLE 13.2 Arrhenius Parameters

Reaction	A	E_a, $kJ \cdot mol^{-1}$
First order, gas phase		
cyclopropane $\longrightarrow$ propene	$1.6 \times 10^{15}\ s^{-1}$	272
$CH_3NC \longrightarrow CH_3CN$	$4.0 \times 10^{13}\ s^{-1}$	160
$C_2H_6 \longrightarrow 2\ CH_3$	$2.5 \times 10^{17}\ s^{-1}$	384
$N_2O \longrightarrow N_2 + O$	$8.0 \times 10^{11}\ s^{-1}$	250
$2\ N_2O_5 \longrightarrow 4\ NO_2 + O_2$	$4.0 \times 10^{13}\ s^{-1}$	103
Second order, gas phase		
$O + N_2 \longrightarrow NO + N$	$1 \times 10^{11}\ L \cdot mol^{-1} \cdot s^{-1}$	315
$OH + H_2 \longrightarrow H_2O + H$	$8 \times 10^{10}\ L \cdot mol^{-1} \cdot s^{-1}$	42
$2\ CH_3 \longrightarrow C_2H_6$	$2 \times 10^{10}\ L \cdot mol^{-1} \cdot s^{-1}$	0
Second order, in aqueous solution		
$C_2H_5Br + OH^- \longrightarrow C_2H_5OH + Br^-$	$4.3 \times 10^{11}\ L \cdot mol^{-1} \cdot s^{-1}$	90
$CO_2 + OH^- \longrightarrow HCO_3^-$	$1.5 \times 10^{10}\ L \cdot mol^{-1} \cdot s^{-1}$	38
$C_{12}H_{22}O_{11} + H_2O \longrightarrow 2\ C_6H_{12}O_6$	$1.5 \times 10^{15}\ L \cdot mol^{-1} \cdot s^{-1}$	108

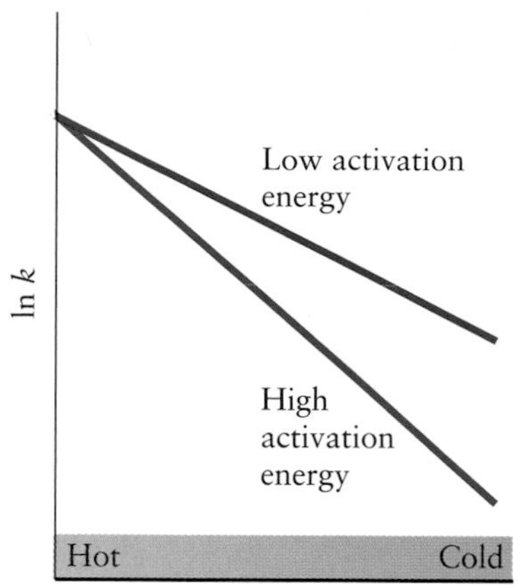

FIGURE 13.17 The dependence of the rate constant on temperature for two reactions with different activation energies. The higher the activation energy, the more strongly the rate constant depends on temperature.

An important point to remember is that *the higher the activation energy E_a, the stronger the temperature dependence of the rate constant.* Reactions with low activation energies (about 10 kJ·mol^{-1}, with not very steep Arrhenius plots) have rates that increase only slightly with temperature. Reactions with high activation energies (above approximately 60 kJ·mol^{-1}, with steep Arrhenius plots) have rates that depend strongly on the temperature (Fig. 13.17).

An Arrhenius plot of ln k against 1/T is used to determine the Arrhenius parameters of a reaction; a large activation energy signifies a high sensitivity of the rate constant to changes in temperature.

EXAMPLE 13.7 Using the activation energy to predict a rate constant

The hydrolysis of sucrose, in which a sucrose molecule is broken down into a glucose molecule and a fructose molecule, is a part of the digestive process. How strongly does the rate depend on our body temperature? Calculate the rate constant for the hydrolysis of sucrose at 35.0°C, given that $k = 1.0 \times 10^{-3}$ L·mol^{-1}·s^{-1} at 37.0°C (normal body temperature) and that the activation energy of the reaction is 108 kJ·mol^{-1}.

STRATEGY We expect a lower rate constant at the lower temperature. Write the Arrhenius equation, Eq. 18, for the two temperatures, which we call T' and T:

$$\text{At temperature } T': \quad \ln k' = \ln A - \frac{E_a}{RT'}$$

$$\text{At temperature } T: \quad \ln k = \ln A - \frac{E_a}{RT}$$

Eliminate ln A by subtracting the second equation from the first:

$$\ln k' - \ln k = -\frac{E_a}{RT'} + \frac{E_a}{RT}$$

That is,

$$\ln\frac{k'}{k} = \frac{E_a}{R}\left(\frac{1}{T} - \frac{1}{T'}\right)$$

At this point, substitute the data, remembering to express temperature in kelvins. Use $R = 8.3145 \times 10^{-3}$ kJ·K^{-1}·mol^{-1}.

SOLUTION We take $T = 310.0$ K and $T' = 308.0$ K. Then

$$\ln\frac{k'}{k} = \frac{108 \text{ kJ·mol}^{-1}}{8.3145 \times 10^{-3} \text{ kJ·K}^{-1}\text{·mol}^{-1}}\left(\frac{1}{310.0 \text{ K}} - \frac{1}{308.0 \text{ K}}\right)$$
$$= -0.27$$

Taking the natural antilogarithm of this expression, we obtain

$$\frac{k'}{k} = e^{-0.27}$$

Finally, because $k = 1.0 \times 10^{-3}$ L·mol^{-1}·s^{-1},

$$k' = (1.0 \times 10^{-3}\ \text{L·mol}^{-1}\text{·s}^{-1}) \times e^{-0.27} = 7.6 \times 10^{-4}\ \text{L·mol}^{-1}\text{·s}^{-1}$$

at 35°C. The high activation energy of the reaction means that its rate is very sensitive to temperature.

Self-Test 13.11A The rate constant for the second-order reaction between CH_3CH_2Br and OH^- in water is 2.8×10^{-4} L·mol^{-1}·s^{-1} at 35.0°C. What is the value of the rate constant at 50.0°C? See Table 13.2 for data.
[*Answer:* 1.4×10^{-3} L·mol^{-1}·s^{-1}]

Self-Test 13.11B The rate constant for the first-order isomerization of cyclopropane, C_3H_6, to propene, $CH_3CH{=}CH_2$, is 6.7×10^{-4} s^{-1} at 500°C. What is its value at 300°C? See Table 13.2 for data.

13.8 Collision Theory

We are now ready to build a model of how chemical reactions occur at the molecular level. Specifically, our model must account for the temperature dependence of rate constants, as expressed by the Arrhenius equation; it should also reveal the significance of the Arrhenius parameters A and E_a. Reactions in the gas phase are conceptually simpler than those in solution, so we begin with them.

First, we assume that a reaction can occur only if reactants meet. An encounter between two molecules in a gas is a collision, so the model we are about to build is called the **collision theory** of reactions. In this model, we suppose that molecules behave like defective billiard balls: they bounce apart if they collide at low speed, but they might smash into pieces when the impact is really energetic. If two molecules collide with less than a certain kinetic energy, they simply bounce apart. If they meet with more than that energy, bonds can break and new bonds can form (Fig. 13.18). We denote the minimum kinetic energy needed for reaction by E_{min}.

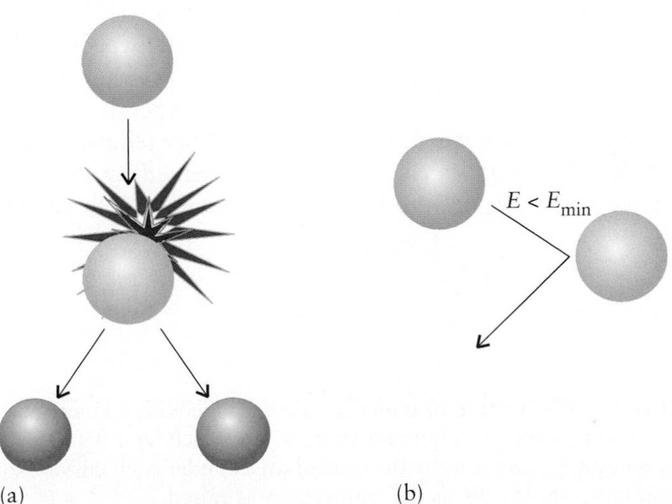

FIGURE 13.18 (a) In the collision theory of chemical reactions, reaction may occur only when two molecules collide with a kinetic energy at least equal to a minimum, value E_{min} (which later we identify with the activation energy). (b) Otherwise, they simply bounce apart.

To set up a quantitative theory based on this qualitative picture, we need to know the rate at which molecules collide and the fraction of those collisions that have at least the energy E_{min}. The rate of collisions (the number of collisions per second) between A and B molecules in a gas at a temperature T can be calculated from the kinetic model of a gas:

$$\text{Rate of collision} = \sigma \bar{c} N_A^2[\text{A}][\text{B}] \tag{19}$$

N_A is the Avogadro constant and $\bar{c}$ is the mean speed at which the molecules approach each other in a gas. The mean speed is calculated by multiplying each possible speed by the fraction of molecules that have that speed and then adding all the products together. When the temperature is T and the molar masses are M_A and M_B, this mean speed is

$$\bar{c} = \left(\frac{8RT}{\pi\mu}\right)^{1/2} \qquad \mu = \frac{M_A M_B}{M_A + M_B} \tag{20}$$

The rate of collision is greater the higher the relative speeds of the molecules, and therefore the higher the temperature. The quantity σ (sigma) is the **collision cross-section,** the area that a molecule presents as a target during a collision. The bigger the collision cross-section, the greater the collision rate, because big molecules are more likely than small molecules to hit one another as they hurtle through space.

Although the mean speed of the molecules increases with temperature, and the collision rate therefore increases, too, Eq. 20 shows that the mean speed increases only as the square root of the temperature. This is far too weak a dependence to account for observation. If we use Eq. 20 to predict the temperature dependence of reaction rates, we would conclude that an increase in temperature of 10°C at about room temperature (from 273 K to 283 K) increases the rate of collision by a factor of only 1.02, whereas experiments show that many reaction rates double over that range. Another factor must be affecting the rate.

To make progress, we need a measure of the fraction of molecules that collide with a kinetic energy equal to or greater than a certain minimum energy, E_{min}. Because kinetic energy is proportional to the square of the speed, this fraction can be obtained from the Maxwell distribution of speeds (Section 4.14). As indicated for a specific reaction by the shaded area under the blue curve in Fig. 13.19, at room temperature very few molecules have enough kinetic energy to react. At higher temperatures, a much larger fraction of molecules can react, as represented by the shaded area under the red curve. We need to bring this fraction into our equations.

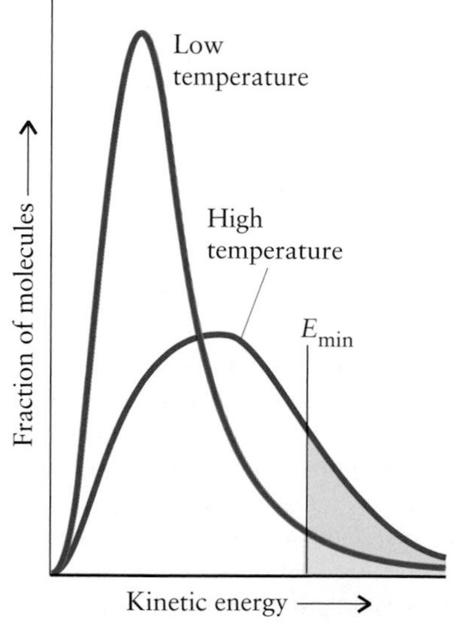

FIGURE 13.19 The fraction of molecules that collide with a kinetic energy that is at least equal to a certain minimum value, E_{min} (which later we show to be the activation energy, E_a), is given by the shaded areas under each curve. Note that the fraction increases rapidly as the temperature is raised.

How do we do that?

To take into account the fraction of molecules that collide with at least an energy E_{min}, we need to know that at a temperature T, the fraction of collisions with at least the energy E_{min} is proportional to $e^{-E_{min}/RT}$, where R is the gas constant. That result comes from an expression known as the *Boltzmann distribution*, which we do not derive here. The rate of reaction is the product of this factor and the rate of collision:

$$\begin{aligned}\text{Rate of reaction} &= (\text{rate of collision}) \times (\text{fraction with sufficient energy}) \\ &= \sigma\bar{c}N_A^{\,2}[A][B] \times e^{-E_{min}/RT}\end{aligned}$$

The rate law of a reaction that depends on collisions of A with B is rate = k[A][B], where k is the rate constant. We can therefore identify the expression for the rate constant as

$$k = \frac{\text{rate of reaction}}{[A][B]} = \sigma\bar{c}N_A^{\,2}\, e^{-E_{min}/RT} \qquad (21)$$

Equation 21 has an exponential temperature dependence, which is much stronger than the weak temperature dependence of the collision rate itself. Let's now confirm that our model is consistent with the Arrhenius equation. When we take logarithms of both sides, we obtain

$$\ln k = \ln(\sigma\bar{c}N_A^{\,2}\, e^{-E_{min}/RT}) = \ln(\sigma\bar{c}N_A^{\,2}) - \frac{E_{min}}{RT}$$

This expression has exactly the same form as the Arrhenius equation, so our model is consistent with observation. Moreover, we can now identify the term $\sigma\bar{c}N_A^{\,2}$ as the pre-exponential factor A and E_{min} as the activation energy, E_a. That is, *A is a measure of the rate at which molecules collide*, and the activation energy *E_a is the minimum kinetic energy required for a collision to result in reaction.*

Now that we have a model, we must check its consistency with various experiments. Sometimes such inconsistencies result in the complete rejection of a model. More often, they indicate that we need to refine the model. In the present case, careful experiments show that the collision model of reactions is not complete, because the experimental rate constant is normally smaller than predicted by collision theory. We can improve the model by realizing that the relative *direction* in which the molecules are moving when they collide might also matter. For example, experiments of the kind described in Box 13.2 have shown that, in the gas-phase reaction of chlorine atoms with HI molecules, $HI + Cl \longrightarrow HCl + I$, the Cl atom reacts with the HI molecule only if it approaches from a favorable direction (Fig. 13.20). A dependence on direction is called the **steric requirement** of the reaction. It is normally taken into account by introducing an empirical factor, P, called the **steric factor,** and changing Eq. 21 to

$$k = P\sigma\bar{c}N_A^{\,2}\, e^{-E_a/RT} \qquad (22)$$

This expression takes into account the frequency of collisions, their kinetic energy, and the steric requirement.

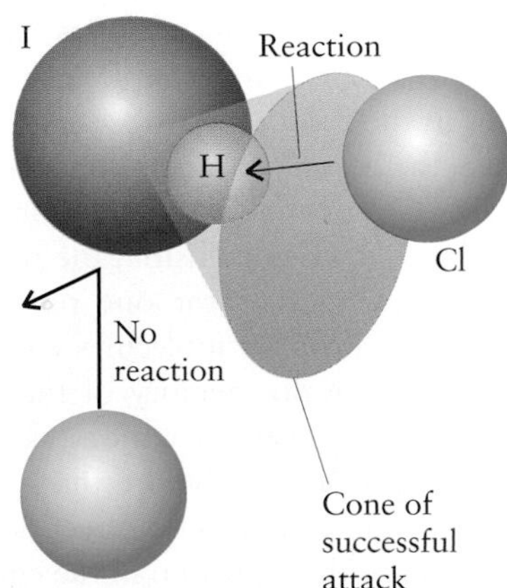

FIGURE 13.20 Whether or not a reaction occurs when two species collide in the gas phase also depends on their relative orientation. In the reaction between a Cl atom and an HI molecule, for example, only those collisions in which the Cl atom approaches the HI molecule along a line of approach that lies inside the cone indicated here lead to reaction, even though the energy of collisions in other orientations may exceed the activation energy.

BOX 13.2 HOW DO WE KNOW . . . WHAT HAPPENS DURING A MOLECULAR COLLISION?

When molecules collide, old bonds break and new bonds form. However, studying only the reactants and products gives us little insight into how the changes take place. At the climax of the reaction, old bonds are giving way to new ones and atoms in one molecule are becoming part of the second molecule. Is there any way of finding out *experimentally* what is happening during this highly stressful moment in the reaction?

Molecular beams allow chemists to determine what happens when molecules collide. We first met molecular beams in Box 4.2, where we saw how a velocity selector is constructed. A molecular beam consists of a stream of molecules moving in the same direction with the same speed. A beam may be directed at a gaseous sample or into the path of a second beam, consisting of molecules of a second reactant. The molecules may react when the beams collide, and the products of the collision are detected. Experimenters examine the direction at which the products emerge from the collision. They also use spectroscopic techniques to determine the vibrational and rotational excitation of the products.

By repeating the experiment with molecules having different speeds and different states of rotational or vibrational excitation, chemists can learn more about the collision itself. For example, experimenters have found that, in the reaction between a Cl atom and an HI molecule, the best direction of attack is within a cone of half-angle 30° surrounding the H atom.

In a "sticky" collision, the reactant molecules orbit around each other for one revolution or more. As a result, the product molecules emerge in random directions because no memory of the approach direction is retained. However, a rotation takes time—about 1 ps. If the reaction is over before that, the product molecules will emerge in a specific direction that depends on the direction of the collision. In the collision of K and I_2, for example, most of the products are thrown off in the forward direction. This observation is consistent with the "harpoon mechanism" that had been proposed for this reaction. In this mechanism, an electron flips across from the K atom to the I_2 molecule when they are quite far apart, and the resulting K^+ ion draws in the negatively charged I_2^- ion. We can think of the electron as the harpoon, the electrostatic attraction as the line attached to the harpoon, and I_2 as the whale. Because the crucial "harpooning" takes place at large distances, and

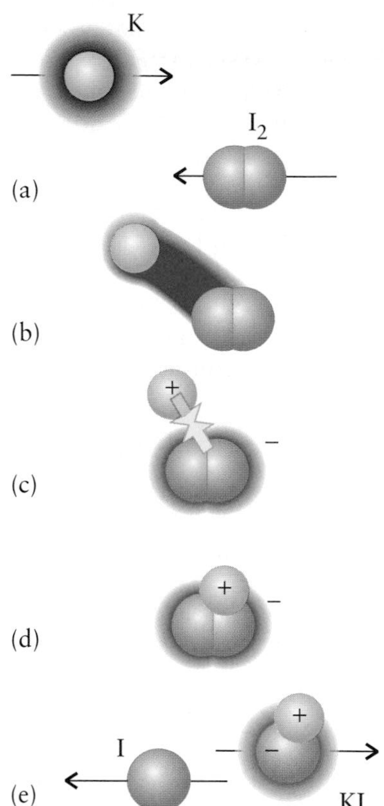

In the harpoon mechanism for the reaction between potassium and iodine to form potassium iodide, as a K atom approaches an I_2 molecule (a), an electron passes from the K atom to the I_2 molecule (b). The charge difference now tethers the two ions together (c) and (d) until an I^- ion separates and leaves with the K^+ ion (e).

there is no actual collision, the products are thrown off in approximately the same direction as the reactants were traveling.

The reaction between a K atom and a CH_3I molecule takes place by a different mechanism. A collision leads to reaction only if the two reactants approach each other very closely. In this mechanism, the K atom effectively bumps into a brick wall, and the KI product bounces out in the backward direction.

TABLE 13.3 The Steric Factor, *P*

Reaction	*P*
$NOCl + NOCl \longrightarrow NO + NO + Cl_2$	0.16
$NO_2 + NO_2 \longrightarrow NO + NO + O_2$	5.0×10^{-2}
$ClO + ClO \longrightarrow Cl_2 + O_2$	2.5×10^{-3}
$H_2 + C_2H_4 \longrightarrow C_2H_6$	1.7×10^{-6}

Table 13.3 lists some values of P. They are all less than 1 because the steric requirement reduces the probability of reaction. For collisions between complex species, the steric requirement may be severe, and P is very small. Then the reaction rate is considerably less than the rate at which high-energy collisions occur.

According to the collision theory of gas-phase reactions, a reaction occurs only if the reactant molecules collide with a kinetic energy of at least the activation energy, and they do so in the correct orientation.

13.9 Activated Complex Theory

Although collision theory applies to gas-phase reactions, we can extend some of its concepts to explain why the Arrhenius equation also applies to reactions in solution. There, molecules do not speed through space and collide, but jostle through the solvent and stay in one another's vicinity for relatively long periods. This more general theory is called **activated complex theory.** Activated complex theory applies to reactions in gases and solutions: it improves on collision theory by suggesting a way of calculating the rate constant even when steric requirements are significant.

This theory is also widely called *transition state theory.*

In activated complex theory, we imagine two molecules approaching each other and distorting as they meet. In the gas phase, that meeting and distortion are equivalent to the "collision" of collision theory. In solution, the approach is a jostling zigzag walk among solvent molecules, and the distortion might not take place until after the two reactant molecules have met and received a particularly vigorous kick from the solvent molecules around them (Fig. 13.21). In either case, the collision or the kick does not immediately tear the molecules apart. Instead, the encounter results in the formation of an **activated complex,** a combination of the two molecules that can either go on to form products or fall apart again into the unchanged reactants.

In the activated complex, the original bonds have lengthened and weakened, and the new bonds are only partially formed. For example, in the proton transfer reaction between the weak acid HCN and water, the activated complex can be pictured as consisting of an HCN molecule with its hydrogen atom involved in a hydrogen bond to the oxygen atom of a water molecule and is poised midway between the two molecules. At this point, the hydrogen atom could reform HCN or leave as the product, H_3O^+.

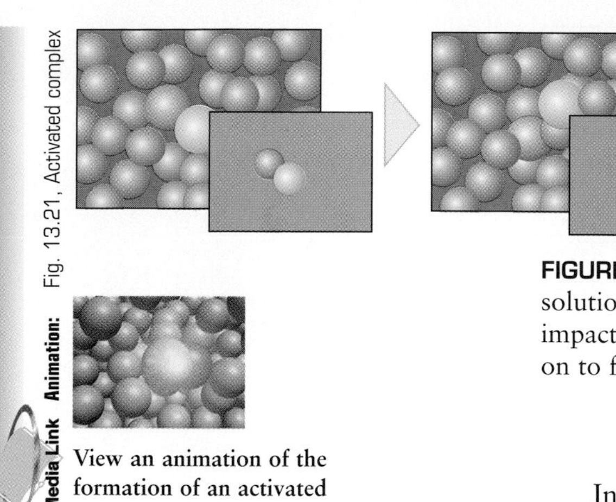

FIGURE 13.21 This sequence of images shows the reactant molecules in solution in the middle of their encounter. They may acquire enough energy by impacts from the solvent molecules to form an activated complex, which may go on to form products.

View an animation of the formation of an activated complex.

In activated complex theory, the activation energy is a measure of the energy of the activated complex relative to that of the reactants. A diagram called a **reaction profile** is used to represent the energy changes that occur during a reaction. The reaction profile in Fig. 13.22 shows how the energy changes as the reactants meet, form the activated complex, and then go on to form products. As the reactants approach each other, their energy climbs up the left side of the barrier. If the reactants have less energy than E_a, they roll back down on the left and separate again. If they have an energy of at least E_a, they can form the activated complex, pass over the top of the barrier, and roll down the other side, where they separate as products.

A **potential energy surface** can help us visualize the energy changes during a reaction as a function of the locations of the atoms. In this three-dimensional plot, the *z*-axis is a measure of potential energy and the *x*- and *y*-axes represent interatomic distances. For example, the plot in Fig. 13.23 shows the energy changes that occur during the attack of a bromine atom on a hydrogen molecule and the reverse process, the attack of a hydrogen atom on an HBr molecule:

$$H_2 + Br\cdot \longrightarrow HBr + H\cdot \qquad HBr + H\cdot \longrightarrow H_2 + Br\cdot$$

The low-energy locations corresponding to the reactants or products are separated by a barrier over which there is a path of minimum potential energy. The actual path of the encounter depends on the total energies of the particles, but we can gain insight into the reaction process by examining potential energy changes alone. For example, suppose the H—H bond remained the same length as the Br atom approached. That would take the system to point A, a state of very high potential energy. In fact, at normal temperatures, the colliding species might not have enough kinetic energy to reach this point. The path that involves lowest potential energy

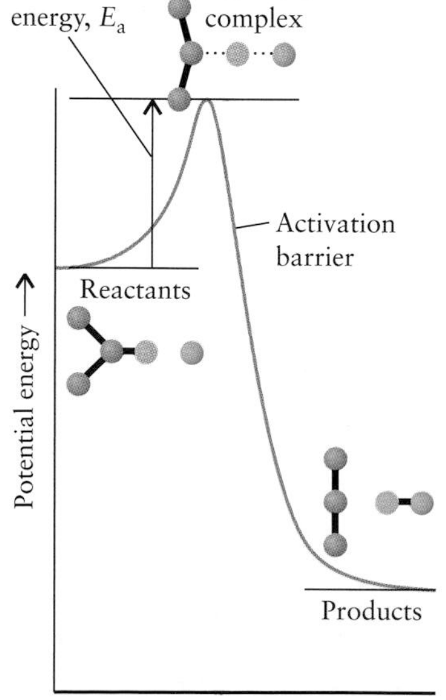

FIGURE 13.22 A reaction profile for an exothermic reaction. In the activated complex theory of reaction rates, it is supposed that the potential energy (the energy due to position) increases as the reactant molecules approach each other and reaches a maximum as they form an activated complex. It then decreases as the atoms rearrange into the bonding pattern characteristic of the products and these products separate. Only molecules with enough energy can cross the activation barrier and react to form products.

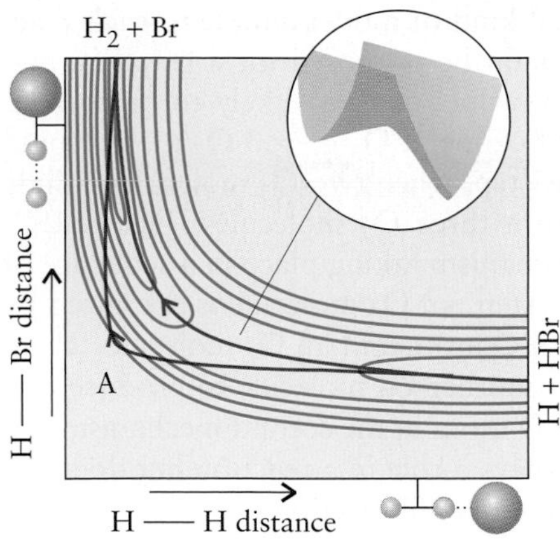

FIGURE 13.23 The contours of a potential energy surface for the reaction between a hydrogen molecule and a bromine atom. The atoms have been constrained to approach and depart in a straight line. The path of lowest potential energy (blue) is up one valley, across the pass—the saddle-shaped high point (see inset)—and down the floor of the other valley. The path shown in red would take the atoms to very high potential energies.

is the one up the foot of the valley, over the "saddle point" (a saddle-shaped region) at the top of the pass, and down the foot of the valley on the other side of the pass. To follow this path, the H—H bond must lengthen as the new H—Br bond begins to form.

As in collision theory, the rate of the reaction depends on the rate at which reactants can climb to the top of the barrier and form the activated complex. The resulting expression for the rate constant is very similar to the one given in Eq. 22, so this more general theory also accounts for the form of the Arrhenius equation and the observed dependence of the reaction rate on temperature.

In activated complex theory, a reaction occurs only if two molecules acquire enough energy, perhaps from the surrounding solvent, to form an activated complex and cross an energy barrier.

REACTION MECHANISMS

Now we turn to the origin of the rate laws themselves and ask why some reactions are first-order, some second-order, and others have more complicated rate laws. As we explore this question, we shall find that it answers other questions chemists are likely to ask. For example, what molecular events convert ozone into oxygen? What molecular events turn a mixture of fuel and air into carbon dioxide and water when it ignites in an engine? What is *really* going on in terms of atoms high in the atmosphere as chlorofluorocarbon molecules punch holes in the ozone layer? Chemical kinetics and particularly the rate law for a reaction provide the kind of information we need to answer questions like these.

13.10 Elementary Reactions

We stressed in Section 13.3 that we cannot in general write down a rate law from a chemical equation. The reason is that all but the simplest reactions are the outcome of several, and sometimes many, steps called **elementary reactions.** To understand how a reaction occurs, we have to propose a **reaction mechanism,** a sequence of elementary reactions describing the changes that we believe take place as reactants are transformed into products.

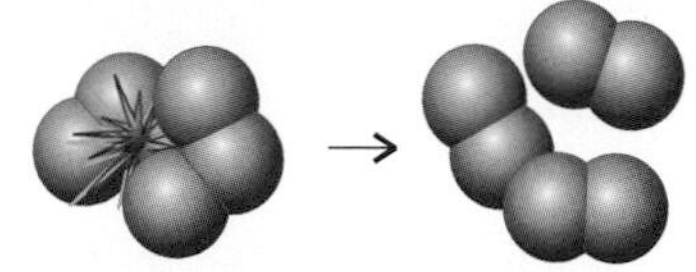

FIGURE 13.24 A representation of a proposed one-step mechanism for the decomposition of ozone in the atmosphere. This reaction occurs in a single bimolecular collision.

A reaction mechanism is a special kind of model unique to each reaction. Several different mechanisms might be proposed for a reaction, and rate measurements are used to decide which is most likely to be correct. For example, in the decomposition of ozone, $2\ O_3(g) \rightarrow 3\ O_2(g)$, we could imagine the reaction occurring in one step, when two O_3 molecules collide and rearrange their six atoms to form three O_2 molecules (Fig. 13.24). Alternatively, we could imagine a mechanism taking place in a sequence of two elementary reactions: in the first step, an O_3 molecule is energized by solar radiation and dissociates into an O atom and an O_2 molecule. Then, in a second step, the O atom attacks another O_3 molecule to produce two more O_2 molecules (Fig. 13.25). The O atom in the second mechanism is a **reaction intermediate,** a species that plays a role in a reaction but does not appear in the chemical equation for the overall reaction.

Elementary reactions are summarized by chemical equations written without the state symbols. These equations show how *individual* atoms and molecules take part in the reaction, as shown in Figs. 13.24 and 13.25. For example, one proposed step in the decomposition of ozone is

$$O_3 \longrightarrow O_2 + O$$

This equation tells us what happens to an individual O_3 molecule. It is an example of a **unimolecular reaction,** because only one reactant molecule participates. We can picture an ozone molecule acquiring energy from sunlight and vibrating so vigorously that it shakes itself apart.

If we want to propose that an O atom produced by the dissociation of O_3 goes on to attack another O_3 molecule, we write for that step

$$O + O_3 \longrightarrow O_2 + O_2$$

This elementary reaction is an example of a **bimolecular reaction,** because two reactant species come together to react. The **molecularity** of a reaction is the number of reactant molecules involved in a specified elementary reaction. The molecularity of a unimolecular reaction is 1 and that of a bimolecular reaction is 2.

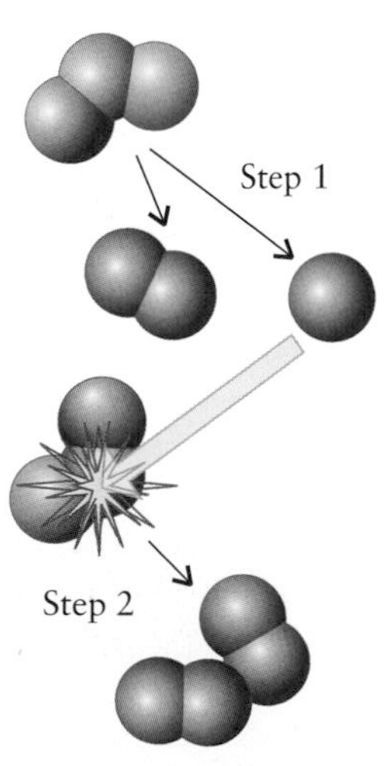

FIGURE 13.25 An alternative, two-step mechanism for the decomposition of ozone. In the first step, an energized ozone molecule shakes off an oxygen atom. In the second step, that oxygen atom attacks another ozone molecule.

Any forward reaction that can occur is also accompanied, in principle at least, by the corresponding reverse reaction. Therefore, the unimolecular decay of O_3 is accompanied by the formation reaction

$$O_2 + O \longrightarrow O_3$$

This step is bimolecular. Similarly, the bimolecular attack of O on O_3 is accompanied by its reverse reaction

$$O_2 + O_2 \longrightarrow O + O_3$$

This step is also bimolecular. These two equations add up to give the reverse of the overall equation, $3\ O_2(g) \rightarrow 2\ O_3(g)$. The forward and reverse reactions jointly provide a mechanism for reaching dynamic equilibrium between the reactants and products in the overall process.

The proposed one-step mechanism also yields a chemical equation that matches the overall reaction:

$$O_3 + O_3 \longrightarrow O_2 + O_2 + O_2$$

The reverse of this elementary reaction is a **termolecular reaction,** an elementary reaction requiring the simultaneous collision of three molecules:

$$O_2 + O_2 + O_2 \longrightarrow O_3 + O_3$$

Termolecular reactions are uncommon, because it is very unlikely that three molecules will collide simultaneously with one another under normal conditions.

Most reactions occur by a series of elementary reactions. The molecularity of an elementary reaction indicates how many reactant species are involved in the step.

Self-Test 13.12A What is the molecularity of the elementary reaction (a) $C_2N_2 \rightarrow CN + CN$; (b) $NO_2 + NO_2 \rightarrow NO + NO_3$?

[*Answer:* (a) Unimolecular; (b) bimolecular]

Self-Test 13.12B What is the molecularity of the elementary reaction (a) $C_2H_5Br + OH^- \rightarrow C_2H_5OH + Br^-$; (b) $Br_2 \rightarrow Br + Br$?

13.11 The Rate Laws of Elementary Reactions

We propose a mechanism for a reaction by taking into account the experimental data. Then, as we must do for any scientific model, we check the validity of the proposed mechanism. A valid reaction mechanism must be consistent with (1) the balanced equation for the reaction and (2) experimental data. To verify that a proposed reaction mechanism agrees with experimental data, we construct the overall rate law implied by the mechanism and check to see whether it is consistent with the experimentally determined rate law. However, it is important to realize that, although the constructed rate law and the experimental rate law may be the same, the proposed mechanism may still be incorrect because some other mechanism may also lead to the same rate law. Kinetic information can only support a proposed mechanism; it can never prove that a mechanism is correct. The acceptance of a suggested mechanism is more like the process of proof in an ideal court of law than a proof in mathematics, with evidence being assembled to give a convincing, consistent picture.

The first task in the construction of an overall rate law for a mechanism is to write the rate laws for the elementary reactions that have been proposed; the second task is to combine them into an overall rate law. The molecularity of an elementary reaction implies a specific rate law.

Writing rate laws from chemical equations is permissible only for one-step elementary reactions.

1 A unimolecular elementary reaction depends on a single energized particle shaking itself apart, so it has a first-order rate law: rate $= k[A]$

2 A bimolecular reaction depends on collisions between two particles, so its rate law is second order, as in collision theory: rate $= k[A][B]$ or rate $= k[A][A]$

3 A termolecular reaction requires a collision involving three particles, so it has a third-order rate law: rate $= k[A][B][C]$, rate $= k[A]^2[B]$, or rate $= k[A]^3$

To apply this procedure, let's consider the rate law for the gas-phase oxidation of NO to NO_2, which has been determined experimentally to be third-order overall:

$$2\ NO(g) + O_2(g) \longrightarrow 2\ NO_2(g) \qquad \text{rate of formation of } NO_2 = k[NO]^2[O_2]$$

The following mechanism has been proposed for the reaction.

A forward reaction and its reverse are often written with a double arrow, as in $NO + NO \rightleftharpoons N_2O_2$.

Step 1 A fast bimolecular dimerization and its reverse:

$$NO + NO \longrightarrow N_2O_2 \qquad \text{rate of formation of } N_2O_2 = k_1[NO]^2$$

$$N_2O_2 \longrightarrow NO + NO \qquad \text{rate of consumption of } N_2O_2 = k_1'[N_2O_2]$$

Step 2 A slow bimolecular reaction in which an O_2 molecule collides with the dimer:

$$O_2 + N_2O_2 \longrightarrow NO_2 + NO_2 \qquad \text{rate of consumption of } N_2O_2 = k_2[N_2O_2][O_2]$$

To evaluate the plausibility of this mechanism, we need to derive the corresponding overall rate law and compare it with the rate law found experimentally. First, we identify the elementary reactions—there may be more than one—that result in product and write down the equation for the net rate of product formation. In this case, NO_2 is formed only in step 2, so

$$\text{Rate of formation of } NO_2 = 2k_2[N_2O_2][O_2]$$

The factor of 2 appears in the rate law because two NO_2 molecules are formed in each collision; so the concentration of NO_2 increases at twice the rate that the concentration of N_2O_2 decays. This expression is not yet an acceptable rate law for the overall reaction because it includes the concentration of an intermediate, N_2O_2, and intermediates do not appear in the overall rate law for the reaction. We therefore write down the expression for the net rate of formation of this intermediate. According to the mechanism, N_2O_2 is formed in the forward reaction in step 1, removed in the reverse reaction, and removed in step 2. Therefore the net rate of its formation is

$$\text{Net rate of formation of } N_2O_2 = k_1[NO]^2 - k_1'[N_2O_2] - k_2[N_2O_2][O_2]$$

At this point, we make the **steady-state approximation** by setting the net rate of formation of any intermediate equal to 0. The justification for this approximation is that the reactions in step 1 are so fast that they keep the concentrations of intermediates constant. When we set this expression equal to 0, we can rearrange it and obtain an expression for the concentration of N_2O_2:

$$[N_2O_2] = \frac{k_1[NO]^2}{k_1' + k_2[O_2]}$$

This relation can now be substituted into the rate law:

$$\text{Rate of formation of NO}_2 = 2k_2[\text{N}_2\text{O}_2][\text{O}_2] = \frac{2k_1k_2[\text{NO}]^2[\text{O}_2]}{k_1' + k_2[\text{O}_2]} \quad (23)$$

The rate law we have derived is not the same as the experimental one. We have stressed that a model is plausible only if its predictions are in line with experimental results; so should we discard our proposed mechanism? Before doing so, it is always wise to explore whether *under certain conditions* the predictions do in fact agree with experimental data. In this case, if the rate of step 2 is very slow relative to the rapid equilibrium in step 1—so that $k_1'[\text{N}_2\text{O}_2] \gg k_2[\text{N}_2\text{O}_2][\text{O}_2]$, which implies that $k_1' \gg k_2[\text{O}_2]$—we can neglect the term $k_2[\text{O}_2]$ in the denominator. With this approximation, the rate law simplifies to

$$\text{Rate of formation of NO}_2 = \left(\frac{2k_1k_2}{k_1'}\right)[\text{NO}]^2[\text{O}_2] \quad (24)$$

which agrees with the experimentally determined rate law if we let

$$k = \frac{2k_1k_2}{k_1'} \quad (25)$$

The proposed mechanism is therefore consistent with experimental results.

The assumption that the consumption of the intermediate in the slow step is insignificant relative to its formation and decomposition in the first step is called a **pre-equilibrium condition.** A pre-equilibrium arises when an intermediate is formed in a rapid equilibrium reaction prior to a slow step in the mechanism. When we propose that the mechanism involves a pre-equilibrium, we can simplify the calculation of the rate law. For instance, if we propose that the first two steps in the NO oxidation give rise to a pre-equilibrium, then we would write

$$\text{NO} + \text{NO} \longrightarrow \text{N}_2\text{O}_2 \qquad K = \frac{[\text{N}_2\text{O}_2]}{[\text{NO}]^2}$$

This equilibrium implies that $[\text{N}_2\text{O}_2] = K[\text{NO}]^2$. The rate of formation of NO_2 is therefore

$$\text{Rate of formation of NO}_2 = 2k_2[\text{N}_2\text{O}_2][\text{O}_2] = 2k_2K[\text{NO}]^2[\text{O}_2]$$

As before, the mechanism gives rise to a third-order rate law, in agreement with experiment. Although this procedure is much simpler than the steady-state approach, it is less flexible: it is more difficult to extend to more complex mechanisms and it is not so easy to establish the conditions under which the approximation is valid.

The slowest elementary step in a sequence of reactions—in our example, the reaction between O_2 and N_2O_2—is called the **rate-determining step** of the reaction. The rate-determining step is so much slower than the rest that it governs the rate of the overall reaction (Fig. 13.26). A rate-determining step is like a slow ferry on the route between two cities. The rate at which

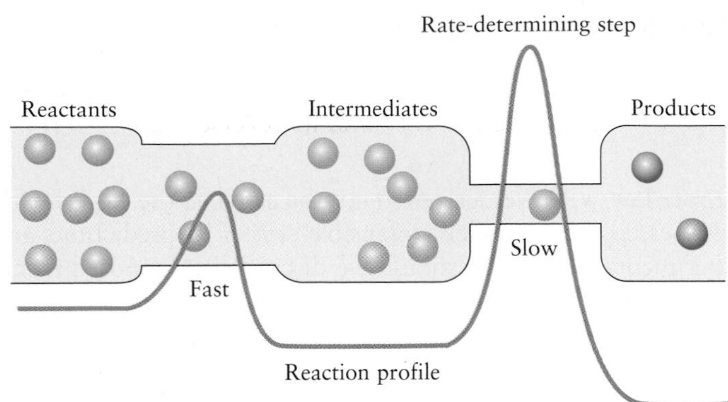

FIGURE 13.26 The rate-determining step in a reaction is an elementary reaction that governs the rate at which products are formed in a multistep series of reactions. The narrow neck representing the rate-determining step in the drawing is like a ferry that cannot handle the busy traffic on a highway. The superimposed reaction profile shows that the slow step has the highest activation energy.

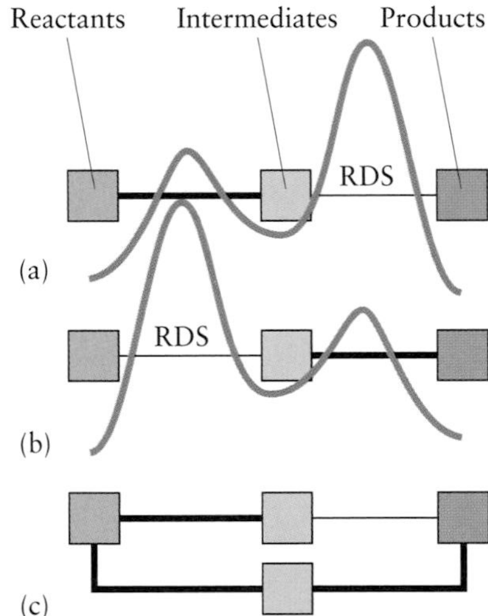

FIGURE 13.27 (a) If the rate-determining step (RDS) is the second step, the rate law for that step determines the rate law for the overall reaction. The orange curve shows the reaction profile for such a mechanism, with a high activation energy for the slow step. The rate law derived from such a mechanism takes into account steps that precede the RDS. (b) If the rate-determining step is the first step, the rate law for that step must match the rate law for the overall reaction. Later steps do not affect the rate or the rate law. (c) If two routes to the product are possible, the faster one (in this case, the lower one) determines the rate of the reaction; in the mechanism forming the upper route, the slow step (thinner line) is not an RDS.

the traffic arrives at its destination is governed by the rate at which it is ferried across the river, because this part of the journey is much slower than any of the others. Notice that a rate-determining step must be slow *and* govern the overall rate of formation of products. A very slow elementary reaction does not control the rate if it can be sidestepped by a faster one (Fig. 13.27).

In simple cases, we can identify the rate-determining step from the observed rate law. For example, the reaction

$$(CH_3)_3CBr(soln) + OH^-(soln) \longrightarrow (CH_3)_3COH(soln) + Br^-(soln)$$

takes place in an organic solvent in two steps:

$$(CH_3)_3CBr \longrightarrow (CH_3)_3C^+ + Br^-$$

$$(CH_3)_3C^+ + OH^- \longrightarrow (CH_3)_3COH$$

The observed rate law for the reaction is rate $= k[(CH_3)_3CBr]$, the same as the rate law for the first step. Because the reaction is zero order in hydroxide ion, the hydroxide ion concentration does not affect the rate and we can assume that the first step in the mechanism is the slow step. As soon as $(CH_3)_3C^+$ is formed, it is immediately converted into the alcohol.

Elementary unimolecular reactions have first-order rate laws; elementary bimolecular reactions have second-order rate laws. A rate law is often derived from a proposed mechanism by imposing the steady-state approximation. A proposed mechanism must be consistent with the experimental rate law.

EXAMPLE 13.8 Setting up an overall rate law from a proposed mechanism

The following rate law has been determined for the decomposition of ozone in the reaction $2\ O_3(g) \rightarrow 3\ O_2(g)$:

$$\text{Rate of decomposition of } O_3 = \frac{k[O_3]^2}{[O_2]}$$

The following mechanism has been proposed:

Step 1 $O_3 \longrightarrow O_2 + O$ and its reverse
Step 2 $O + O_3 \longrightarrow O_2 + O_2$ and its reverse

Measurements of the rates of the elementary forward reactions show that the slowest step is the second step, the attack of O on O_3. The reverse reaction $O_2 + O_2 \rightarrow O + O_3$ is so slow that it can be ignored. Derive the rate law implied by the mechanism, and confirm that it matches the observed rate law.

STRATEGY We construct the rate laws for the elementary reactions and combine them into the overall rate law for the decomposition of O_3. If necessary, use the steady-state approximation for any intermediates and simplify it by using arguments based on rapid pre-equilibria and the existence of a rate-determining step.

SOLUTION The rate laws for the elementary reactions are

Step 1	*(forward)*	$O_3 \longrightarrow O_2 + O$	rate $= k_1[O_3]$ (fast)
	(reverse)	$O_2 + O \longrightarrow O_3$	rate $= k_1'[O_2][O]$ (fast)
Step 2	*(forward)*	$O + O_3 \longrightarrow O_2 + O_2$	rate $= k_2[O][O_3]$ (slow)
	(reverse)	$O_2 + O_2 \longrightarrow O + O_3$	rate $= k'_2[O_2]^2$ (very slow, ignore)

The net rate of decomposition of ozone is

$$\text{Net rate of decomposition of } O_3 = k_1[O_3] - k'_1[O_2][O] + k_2[O][O_3]$$

To eliminate the O atom concentration from this expression, we consider the net rate of formation of O atoms and use the steady-state approximation to set that net rate equal to 0:

$$\text{Net rate of formation of O} = k_1[O_3] - k_1'[O_2][O] - k_2[O][O_3] = 0$$

This equation solves to

$$[O] = \frac{k_1[O_3]}{k_1'[O_2] + k_2[O_3]}$$

When we substitute this expression into the rate law for net rate of decomposition of ozone, we get

$$\begin{aligned}\text{Net rate of decomposition} &= k_1[O_3] - \frac{k_1k_1'[O_2][O_3]}{k_1'[O_2] + k_2[O_3]} + \frac{k_1k_2[O_3]^2}{k_1'[O_2] + k_2[O_3]} \\ &= \frac{2\,k_1k_2[O_3]^2}{k_1'[O_2] + k_2[O_3]}\end{aligned}$$

Because step 2 is slow relative to the fast pre-equilibrium, we can make the approximation $k_2[O][O_3] \ll k_1'[O_2][O]$, or equivalently, $k_2[O_3] \ll k_1'[O_2]$, and simplify this expression to

$$\text{Rate of disappearance of } O_3 = \left(\frac{2k_1k_2}{k_1'}\right)\frac{[O_3]^2}{[O_2]}$$

This equation has exactly the same form as the observed rate law. Moreover, we can recognize the experimental rate constant k as the combination $2k_1k_2/k_1'$ of the rate constants for the elementary steps. Because the mechanism is consistent with both the balanced equation and the experimental rate law, we can consider it plausible.

Self-Test 13.13A Consider the following mechanism for the formation of a DNA double helix from its strands A and B:

$$\text{A} + \text{B} \rightleftharpoons \text{unstable helix (fast, } k_1 \text{ and } k_1'\text{)}$$

$$\text{Unstable helix} \longrightarrow \text{stable double helix (slow, } k_2\text{)}$$

Derive the rate equation for the formation of the stable double helix and express the rate constant of the overall reaction in terms of the rate constants of the individual steps.

[*Answer:* Rate $= k[A][B]$, $k = k_1k_2/k_1'$]

Self-Test 13.13B The proposed mechanism for a reaction is $AH + B \rightarrow BH^+ + A^-$ and its reverse, both of which are fast, followed by $A^- + AH \rightarrow$ products, which is slow. Find the rate law with A^- treated as the intermediate.

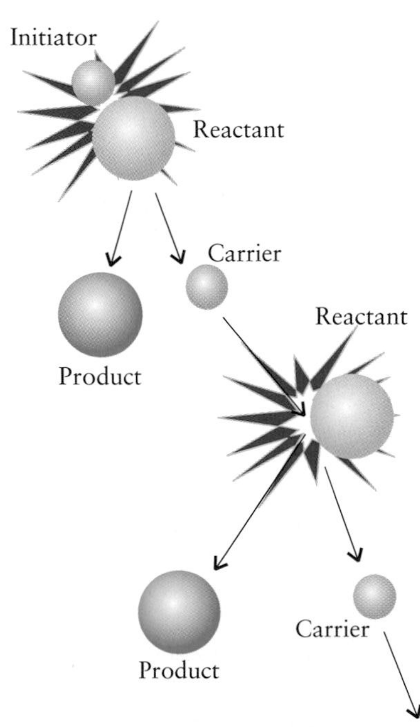

FIGURE 13.28 In a chain reaction, the product of one step in a reaction is a highly reactive reactant in a subsequent step, which in turn produces reactive species that can take part in subsequent reaction steps.

13.12 Chain Reactions

Reactions that occur explosively fast are often chain reactions. In a **chain reaction,** a highly reactive intermediate reacts to produce another highly reactive intermediate, which reacts to produce another, and so on (Fig. 13.28). In many cases, the reaction intermediate—which in this context is called a **chain carrier**—is a radical, and the reaction is called a **radical chain reaction.** In a radical chain reaction, one radical reacts with a molecule to produce another radical, that radical goes on to attack another molecule, and so on. The ideas presented in the preceding sections apply to chain reactions, too, but they often result in very complex rate laws, which we will not derive.

The formation of HBr in the reaction

$$H_2(g) + Br_2(g) \longrightarrow 2\ HBr(g)$$

takes place by a chain reaction. The chain carriers are hydrogen atoms (H·) and bromine atoms (Br·). The first step in any chain reaction is **initiation,** the formation of chain carriers from a reactant. Often heat (denoted Δ) or light (denoted $h\nu$, indicating the role of a photon) is used to generate the chain carriers:

$$Br_2 \xrightarrow{\Delta \text{ or } h\nu} Br\cdot + Br\cdot$$

Once chain carriers have been formed, the chain can propagate in a reaction in which one carrier reacts with a reactant molecule to produce another carrier. The elementary reactions for the **propagation** of the chain are

$$Br\cdot + H_2 \longrightarrow HBr + H\cdot$$

$$H\cdot + Br_2 \longrightarrow HBr + Br\cdot$$

The chain carriers—radicals here—produced in these reactions can go on to attack other reactant (H_2 and Br_2) molecules, thereby allowing the chain to continue. The elementary reaction that ends the chain, a process called **termination,** occurs when chain carriers combine to form products. One example of a termination reaction is

$$Br\cdot + Br\cdot \longrightarrow Br_2$$

In some cases, a chain propagates explosively. Explosions can be expected when **chain branching** occurs, that is, when more than one chain carrier is formed in a propagation step. The characteristic pop that occurs when a mixture of hydrogen and oxygen is ignited is a consequence of chain branching. The two gases combine in a radical chain reaction in which the initiation step may be the formation of hydrogen atoms:

$$\text{Initiation: } H_2 \xrightarrow{\Delta} H\cdot + H\cdot$$

We now see why a mixture of hydrogen and oxygen can survive for years: the activation energy for the production of radicals is very high, and no radicals are formed until a spark or flame is brought into contact with the mixture.

After the reaction has been initiated, two new radicals are formed when one hydrogen atom attacks an oxygen molecule:

$$\text{Branching: } H\cdot + O_2 \longrightarrow HO\cdot + \cdot O\cdot$$

The oxygen atom, with valence electron configuration $2s^2 2p_x^{\;2} 2p_y^{\;1} 2p_z^{\;1}$, has two electrons with unpaired spins (its Lewis symbol is $\cdot\ddot{\underset{\cdot\cdot}{O}}\cdot$, which we abbreviate to $\cdot O\cdot$). Two radicals are also produced when the oxygen atom attacks a hydrogen molecule:

$$\text{Branching: } \cdot O\cdot + H_2 \longrightarrow HO\cdot + H\cdot$$

As a result of these branching processes, the chain produces a large number of radicals that can take part in even more branching steps. The reaction rate increases rapidly, and an explosion typical of many combustion reactions may occur (Fig. 13.29).

Chain reactions begin with the initiation of a reactive intermediate that propagates the chain and concludes with termination when radicals combine. Branching chain reactions can be explosively fast.

FIGURE 13.29 This flame front was caught during the fast combustion—almost a miniature explosion—that occurs inside an internal combustion engine every time a spark plug ignites the mixture of fuel and air.

13.13 Rates and Equilibrium

At equilibrium, the rates of the forward and reverse reactions are equal. Because the rates are a function of rate constants and concentrations, we ought to be able to find a relation between rate constants for elementary reactions and equilibrium constants for the overall reaction.

How do we do that?

To deduce the relation between rate constants and equilibrium constants, we note that the equilibrium constant for a chemical reaction that has the form $A + B \rightleftharpoons C + D$ is

$$K = \frac{[C][D]}{[A][B]} \tag{26}$$

Suppose that experiments show that both the forward reaction and the reverse reaction are elementary second-order reactions, with the following rate laws:

$$A + B \longrightarrow C + D \qquad \text{rate} = k[A][B]$$

$$C + D \longrightarrow A + B \qquad \text{rate} = k'[C][D]$$

At equilibrium, these two rates are equal, so we can write

$$k[A][B] = k'[C][D]$$

It follows that, at equilibrium,

$$\frac{[C][D]}{[A][B]} = \frac{k}{k'}$$

Comparison of this expression with the expression for the equilibrium constant then shows that

$$K = \frac{k}{k'} \tag{27}$$

That is, *the equilibrium constant for a reaction is equal to the ratio of the rate constants for the forward and reverse elementary reactions that contribute to the overall reaction.*

We can now see in kinetic terms rather than thermodynamic (free energy) terms when to expect a large equilibrium constant. K will be much larger than 1 (and products favored) when k for the forward direction is much larger than k' for the reverse direction. In this case, the fast forward reaction builds up a high concentration of products

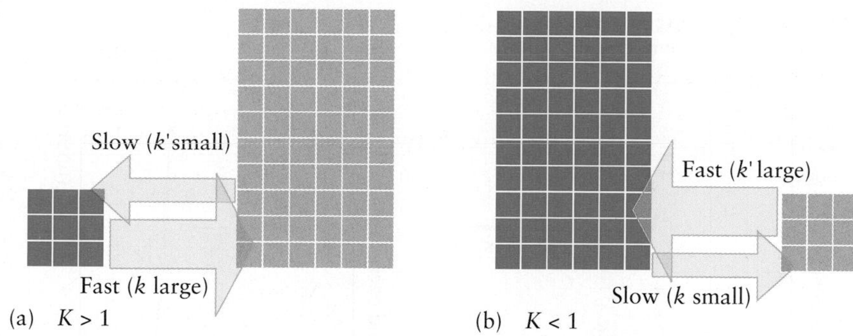

FIGURE 13.30 The equilibrium constant for a reaction is equal to the ratio of the rate constants for the forward and reverse reactions that continue in a state of dynamic equilibrium. (a) A relatively large forward rate constant (k) means that the forward rate can match the reverse rate even though the concentration of reactants is low. (b) Conversely, if the reverse rate constant (k') is relatively large, then the forward and reverse rates are equal when the concentration of products is low.

before reaching equilibrium (Fig. 13.30). In contrast, K is very small (and reactants are favored) when k is much smaller than k'. Now the reverse reaction dismantles the products rapidly, so their concentrations are very low.

We derived the relation between the equilibrium constant and the rate constant for a single-step reaction. However, suppose that a reaction has a complex mechanism in which the elementary reactions have rate constants $k_1, k_2, \ldots$ and the reverse elementary reactions have rate constants $k_1', k_2', \ldots$. Then, by an argument similar to that for the single-step reaction, the overall equilibrium constant is related to the rate constants as follows:

$$K = \frac{k_1}{k_1'} \times \frac{k_2}{k_2'} \times \cdots \tag{28}$$

Our expression for K in terms of rate constants helps to explain one application of Le Chatelier's principle (Section 9.11). According to Le Chatelier, an increase in temperature shifts the equilibrium composition in the endothermic direction. We can now see why. If the forward reaction is endothermic, the activation energy will be higher for the forward direction than for the reverse direction (Fig. 13.31). The higher activation energy means that the rate constant of the forward reaction depends more strongly on temperature than does the rate constant of the reverse reaction (recall Fig. 13.17). Therefore, when the temperature is raised, the rate constant for the forward reaction increases more than that of the reverse reaction. As a result, K will increase and the products will become more favored, just as Le Chatelier's principle predicts.

The equilibrium constant for an elementary reaction is equal to the ratio of the forward and reverse rate constants of the reaction.

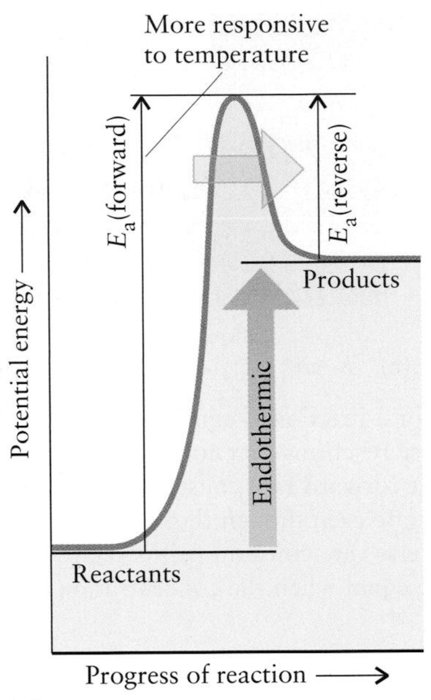

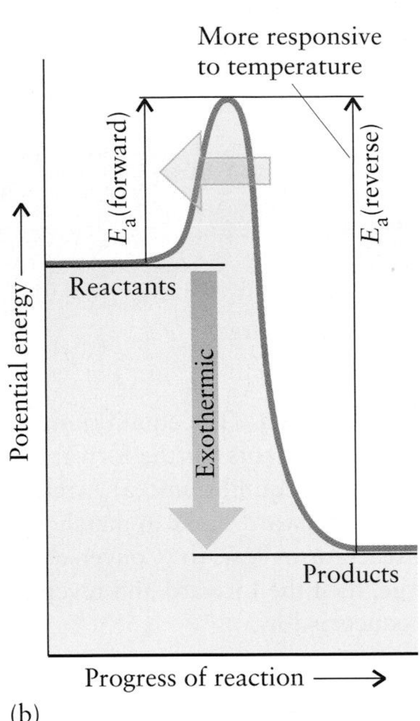

FIGURE 13.31 (a) The activation energy for an endothermic reaction is larger in the forward direction than in the reverse, so the rate of the forward reaction is more sensitive to temperature and the equilibrium shifts toward products as the temperature is raised. (b) The opposite is true for an exothermic reaction, in which case the reverse reaction is more sensitive to temperature and the equilibrium shifts toward reactants as the temperature is raised.

ACCELERATING REACTIONS

We have seen that the rates of many reactions increase if we increase the concentration of reactants. Similarly, the rate of a heterogeneous reaction can be increased by increasing the surface area of a reactant (Fig. 13.32). But suppose we want to increase the rate for a given concentration or surface area? One solution, as we have seen, is to raise the temperature because at higher temperatures more molecules have an energy in excess of the activation energy. In some cases, though, we may want to increase the rate of a reaction without raising the temperature. For instance, if the reaction is exothermic, then the equilibrium constant of the reaction is *decreased* by an increase in temperature (recall Section 9.11), and we would get a smaller yield of product if we raised the temperature. That was Haber's dilemma (Section 9.12): the synthesis of ammonia is exothermic, and although the reaction is faster at high temperatures, the equilibrium concentration of ammonia is much lower. Another consideration is that, in industry, high temperatures may also mean high energy costs, so accelerating reactions without resorting to high temperatures may be a more economical procedure.

13.14 Catalysis

One way to increase the rate of a reaction rate is to use a **catalyst,** a substance that increases the rate without being consumed in the reaction (Fig. 13.33). The name comes from the Greek words meaning "breaking down

Media Link **Lab Video** Fig. 13.32, The effect of surface area on combustion

FIGURE 13.32 Iron pots and pans can be heated in a flame without catching fire. However, a powder of finely divided iron filings oxidizes rapidly in air to form Fe_2O_3, because the powder presents a much greater surface area for reaction.

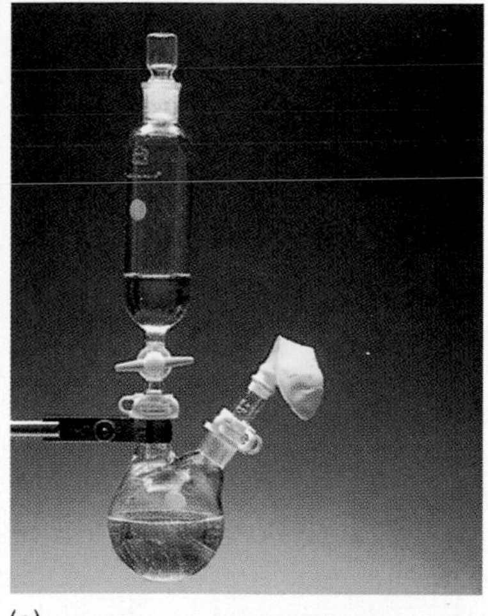

(a)

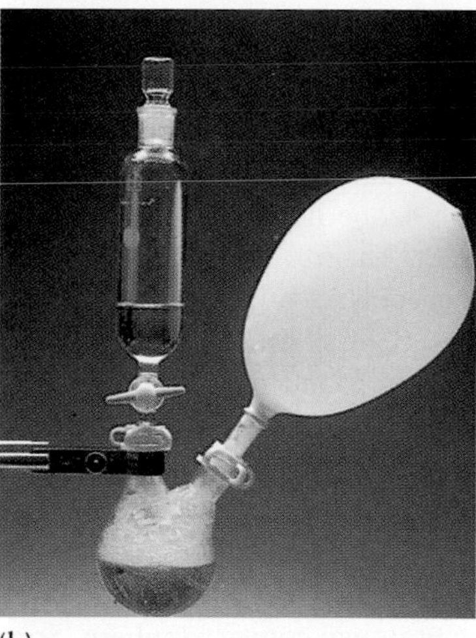

(b)

FIGURE 13.33 A small amount of catalyst—in this case, potassium iodide in aqueous solution—can accelerate the decomposition of hydrogen peroxide to water and oxygen. This effect is shown by (a) the slow inflation of the balloon when no catalyst is present and (b) its rapid inflation when a catalyst is present.

by coming together." In many cases, only a small amount of catalyst is necessary, because it acts over and over again. This is why small amounts of chlorofluorocarbons can have such a devastating effect on the ozone layer—in the stratosphere, they break down into radicals that catalyze the destruction of ozone (Box 13.3).

The Chinese characters for catalyst, which translate as "marriage broker," capture the sense quite well.

A catalyst speeds up a reaction by providing an alternative pathway—a different reaction mechanism—between reactants and products. This new pathway has a lower activation energy than the original pathway does (Fig. 13.34). At the same temperature, a greater fraction of reactant molecules can cross the lower barrier of the catalyzed path and turn into products than when no catalyst is present. Although the reaction occurs more quickly, a catalyst has no effect on the equilibrium composition. Both forward and reverse reactions are accelerated on the catalyzed path, leaving the equilibrium constant unchanged.

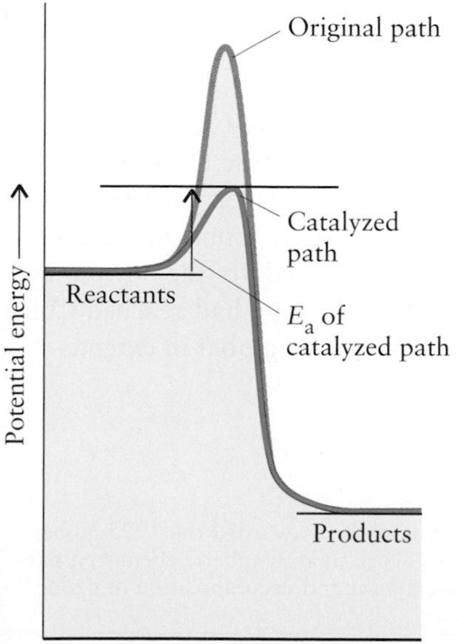

FIGURE 13.34 A catalyst provides a new reaction pathway with a lower activation energy, thereby allowing more reactant molecules to cross the barrier and form products.

BOX 13.3 WHAT HAS THIS TO DO WITH . . . THE ENVIRONMENT?

The Ozone Layer

Every year our planet is bombarded with enough energy from the Sun to destroy all life. Only the ozone in the stratosphere protects us from that onslaught. The ozone, though, is disappearing. Chemicals used as coolants and propellants, such as chlorofluorocarbons (CFCs) and the nitrogen oxides in jet exhausts, have been found to create holes in Earth's protective ozone layer. Because they act as catalysts, even small amounts of these chemicals can cause large changes in the vast reaches of the stratosphere.

Ozone forms in the stratosphere in two steps. First, oxygen molecules or another oxygen-containing compound are broken apart into atoms by sunlight, a process called *photodissociation*:

$$O_2 \xrightarrow{\text{sunlight, } \lambda < 340 \text{ nm}} O + O$$

Then the O atoms, which are reactive radicals with two unpaired electrons, react with the more abundant O_2 molecules to form ozone. The ozone molecules are created in such a high-energy state that their vibrational motions would quickly tear them apart unless another molecule, such as O_2 or N_2, collides with them first. The other molecule, indicated as M, carries off some of the energy:

$$O + O_2 \longrightarrow O_3{}^*$$

$$O_3{}^* + M \longrightarrow O_3 + M$$

where * denotes a high energy state. The net reaction, the sum of these two elementary reactions (after multiplying them by 2) and the reaction presented earlier, is $3\ O_2 \rightarrow 2\ O_3$. This exothermic reaction converts sunlight into heat, which raises the temperature of the stratosphere. Some of the ozone is decomposed by ultraviolet radiation:

$$O_3 \xrightarrow{\text{UV, } \lambda < 340 \text{ nm}} O + O_2$$

The oxygen atom produced in this step can react with oxygen molecules to produce more ozone, so normally the ozone concentration in the stratosphere remains constant, with seasonal variations. Because the decomposition of ozone absorbs ultraviolet radiation, it helps to shield the Earth from radiation damage.

Warnings of the possibility that *androgenic* (human-generated) chemicals can threaten the ozone in the stratosphere began to appear in 1970 and 1971, when Paul Crutzen determined experimentally that NO and NO_2 molecules catalyze the destruction of ozone. Nitrogen monoxide (nitric oxide), NO, is produced naturally in the atmosphere by lightning, but it is also a by-product of combustion in automobile and airplane engines. The NO is then oxidized by the oxygen in air to nitrogen dioxide, NO_2.

Mario Molina and Sherwood Rowland used Crutzen's work and other data in 1974 to build a model of the stratosphere that explained how chlorofluorocarbons could also threaten the ozone layer.[†] In 1985, ozone levels over Antarctica were indeed found to be continually decreasing and had dropped to the lowest ever observed; by the year 2000, the hole had reached Chile. These losses are now known to be global in extent.

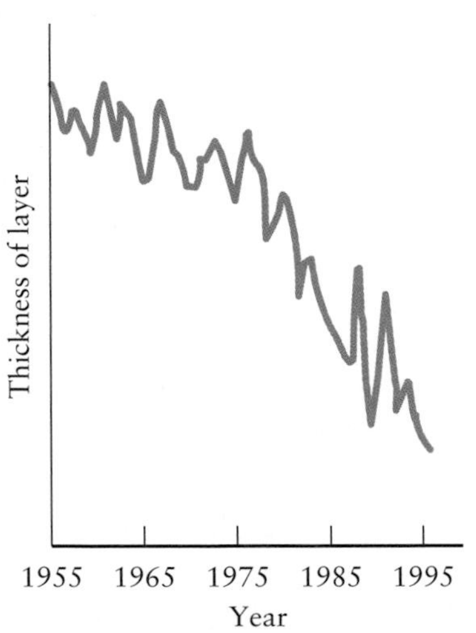

Thickness of the stratospheric ozone layer over Halley Bay, Antarctica, measured each October over a 40-year period. Notice the rapid decline that begins in the mid-1970s.

[†]Crutzen, Molina, and Rowland were awarded the 1995 Nobel prize in chemistry "for their work in atmospheric chemistry, particularly concerning the formation and decomposition of ozone."

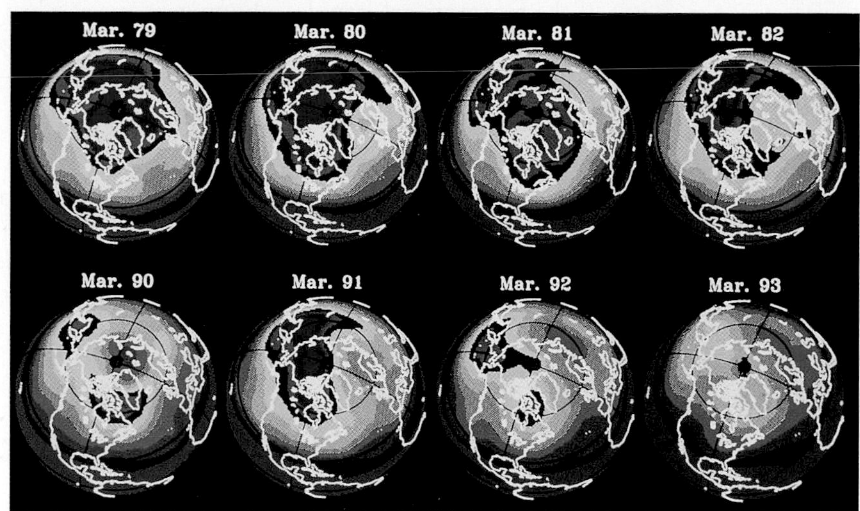

These maps of stratospheric ozone concentration over the North Pole show how the ozone was depleted from 1979 to 1993. Red areas represent ozone concentrations greater than 500 Dobson units (DU); the concentrations decrease through green, yellow, and blue, to purple at less than 270 DU. Normal ozone concentration at temperature latitudes is about 350 DU.

Susan Solomon and James Anderson then showed that CFCs produce chlorine atoms and chlorine oxide under the conditions of the ozone layer and identified the CFCs emanating from everyday objects, such as cans of hair spray, refrigerators, and air conditioners, as the primary culprits in the destruction of stratospheric ozone. The CFC molecules are not very polar, so they do not dissolve in rain or the oceans. Instead, they rise to the stratosphere, where they are exposed to ultraviolet radiation from the Sun. They readily dissociate in the presence of this radiation and form chlorine atoms, which destroy ozone by various mechanisms, one of which is

$$Cl\cdot + O_3 \xrightarrow{h\nu} ClO\cdot + O_2$$

$$ClO\cdot + \cdot O\cdot \longrightarrow Cl\cdot + O_2$$

The O atoms are produced when ozone is decomposed by ultraviolet light, as described previously. Notice that the net reaction, $O_3 + O \rightarrow 2\ O_2$, does not involve chlorine. Chlorine atoms act as continuously regenerated catalysts, so even a low abundance can do a lot of damage.

High-altitude planes, balloons, and satellites are now studying chemical reactions in stratospheric ozone. The most important information, the concentration of ozone, is measured by NASA's Total Ozone Mapping Spectrometer (TOMS). The ozone maps in the figure were obtained from TOMS data.

Large amounts of nitrogen oxides continue to be generated by jet engines, by automobiles, and by the degradation of nitrogen-based fertilizers. Most nations have signed the Montreal Protocol of 1987 and the amendments added in 1992, which required the more dangerous CFCs to be phased out by 1996. However, it takes years for CFC molecules to diffuse into the stratosphere. Therefore, much of the CFCs used during the last decade may still be on the way up. Currently, we are losing about 2% of the ozone in the stratosphere every 10 years.

For Further Reading: C. Baird, *Environmental Chemistry*, New York: W. H. Freeman and Company, 1999, pp. 17–84. P. J. Crutzen, "What Is Happening to Our Precious Air?" *Science Spectra*, **14**, 22–31, 1998. S. Perkins, "High-flying Science, with Strings Attached," *Science News*, 158: 186–188, September 16, 2000.

A **homogeneous catalyst** is a catalyst that is in the same phase as the reactants. For reactants that are gases, a homogeneous catalyst is also a gas. If the reactants are liquids, a homogeneous catalyst is a dissolved liquid or solid. Dissolved bromine is a homogeneous liquid-phase catalyst for the decomposition of aqueous hydrogen peroxide:

$$2\ H_2O_2(aq) \longrightarrow H_2O(l) + O_2(g)$$

In the absence of bromine or another catalyst, a solution of hydrogen peroxide can be stored for a long time at room temperature; however, bubbles of oxygen form as soon as a drop of bromine is added. Bromine's role in this reaction is believed to be its reduction to Br^- in one step, followed by oxidation back to Br_2 in a second step:

$$Br_2(aq) + H_2O_2(aq) \longrightarrow 2Br^-(aq) + 2\ H^+(aq) + O_2(g)$$

$$2\ Br^-(aq) + H_2O_2(aq) + 2\ H^+(aq) \longrightarrow Br_2(aq) + 2\ H_2O(l)$$

When we add the two steps, both the catalyst, Br_2 and the intermediate, Br^-, cancel, leaving the overall equation as the first of the three written above. Hence, although Br_2 molecules have participated in the reaction, they are not consumed and can be used over and over again.

Although a catalyst does not appear in the balanced equation for a reaction, the concentration of a homogeneous catalyst does appear in the rate law. For example, the reaction between the triiodide ion and the azide ion is very slow unless a catalyst such as carbon disulfide is present:

$$I_3^-(aq) + 2\ N_3^-(aq) \xrightarrow{CS_2} 3\ I^-(aq) + 3\ N_2(g)$$

The mechanism for the reaction has two stages:

Stage 1 Carbon disulfide takes part in the reaction by forming a reactive intermediate, a complex with azide ions:

$$CS_2 + N_3^- \longrightarrow S_2CN_3^-, \quad \text{slow}$$

Stage 2 The complex then reacts rapidly with triiodide ion in a series of fast steps that can be summarized as

$$2\ S_2CN_3^- + I_3^- \longrightarrow 2\ CS_2 + 3\ N_2 + 3\ I^-, \quad \text{fast}$$

The rate law derived from this mechanism is the same as the experimental rate law:

$$\text{Rate of disappearance of } I_3^- = k[CS_2][N_3^-]$$

Notice that the rate law is first order in the catalyst, carbon disulfide, but zero order in triiodide ion, which appears only in the fast step following the slow step.

A **heterogeneous catalyst** is a catalyst present in a phase different from that of the reactants. The most common heterogeneous catalysts are finely divided or porous solids used in gas-phase or liquid-phase reac-

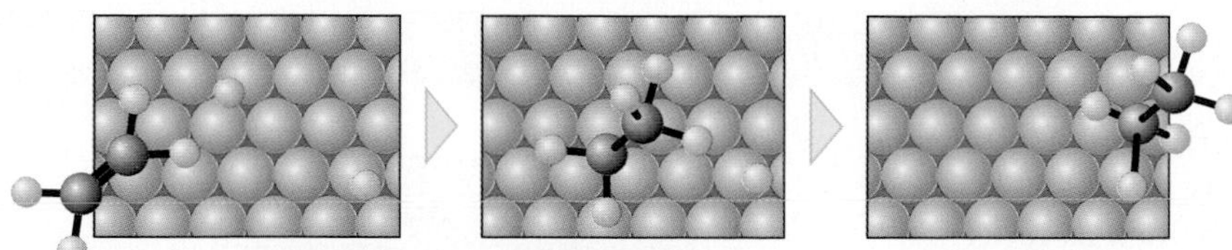

FIGURE 13.35 The reaction between ethene, $CH_2{=}CH_2$, and hydrogen on a catalytic metal surface. In this sequence of images, we see the ethene molecule approaching the metal surface to which hydrogen molecules have already adsorbed: when they adsorb, they dissociate and stick to the surface as hydrogen atoms. Next, after sticking to the surface, the ethene molecule meets a hydrogen atom and forms a bond. At this stage, the $\cdot CH_2CH_3$ radical is attached to the surface by one of its carbon atoms. Finally, the radical and another hydrogen atom meet; ethane is formed and escapes from the surface.

View an animation of a heterogeneous catalyst in action.

Media Link Animation: Fig. 13.35, Heterogeneous catalysis

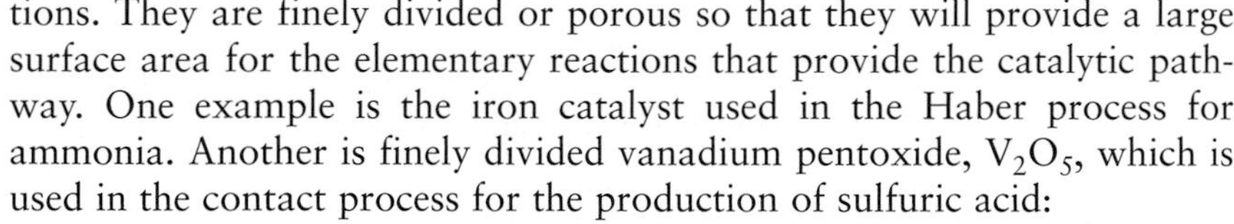

tions. They are finely divided or porous so that they will provide a large surface area for the elementary reactions that provide the catalytic pathway. One example is the iron catalyst used in the Haber process for ammonia. Another is finely divided vanadium pentoxide, V_2O_5, which is used in the contact process for the production of sulfuric acid:

$$2\ SO_2(g) + O_2(g) \xrightarrow{V_2O_5} 2\ SO_3(g)$$

The reactant **adsorbs**, or attaches, to the catalyst's surface. As a reactant molecule adsorbs, its bonds are weakened and the reaction can proceed more quickly because the bonds are more easily broken (Fig. 13.35). One important step in the reaction mechanism of the Haber process is the adsorption of N_2 molecules on the iron and the weakening of the strong $N{\equiv}N$ triple bond. As mentioned at the beginning of this chapter, adsorption onto a metal such as platinum is also used to dissociate the bond in H_2 to accelerate the electrolysis of water.

The catalytic converters of automobiles use catalysts to bring about the complete and rapid combustion of unburned fuel (Fig. 13.36). The mixture of gases leaving an engine includes carbon monoxide, unburned hydrocarbons, and the nitrogen oxides collectively referred to as NO_x. Air pollution is decreased if the carbon compounds are oxidized to carbon dioxide and the NO_x reduced, by another catalyst, to nitrogen. The challenge is to find a catalyst—or a mixture of catalysts—that will accelerate both the oxidation and the reduction reactions and be active when the car is first started and the engine is cool. *Zeolite catalysts* are heterogeneous catalysts used in catalytic converters and for many other specialized applications, because of their very high surface area and reaction specificity (Box 13.4).

Catalysts can be **poisoned**, or inactivated. A common cause of such poisoning is the adsorption of a molecule so tightly to the catalyst that it seals the surface of the catalyst against further reaction. Some heavy metals, especially lead, are very potent poisons for heterogeneous catalysts, which is why lead-free gasoline must be used in engines fitted with catalytic converters. The elimination of lead has the further benefit of decreasing the amount of poisonous lead in the environment.

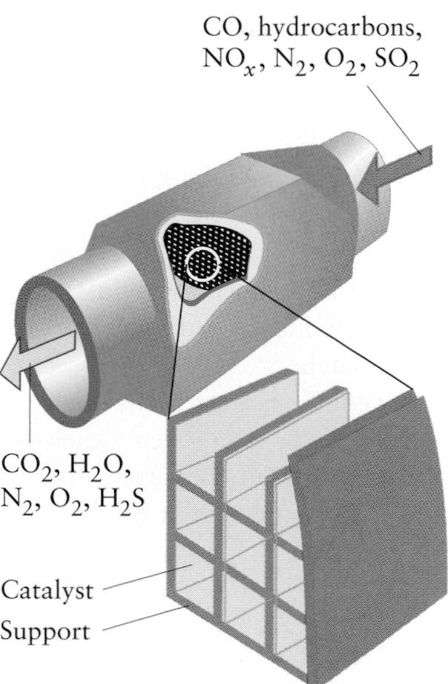

FIGURE 13.36 The catalytic converter of an automobile is made from a mixture of catalysts bonded to a honeycomb ceramic support.

BOX 13.4 FRONTIERS OF CHEMISTRY: ZEOLITE CATALYSTS

Many countries have no petroleum reserves, yet most have abundant organic resources, such as the alcohols ethanol and methanol. Alcohols can be burned as fuel, but because they are already partially oxidized forms of hydrocarbons, they do not have as high an energy density as gasoline (see Box 12.1). The compounds in gasoline are mainly hydrocarbons containing about eight carbon atoms. Finding a "magic wand" to convert alcohols into gasoline seems an impossible dream. It is easy to attach the carbon atoms of methanol molecules together into long chains, but it is difficult to get the process to stop at the right chain length. However, by emulating materials found in nature, scientists have developed heterogeneous catalysts that selectively produce the desired hydrocarbons. These powerful catalysts have transformed the chemical industry and affect many aspects of our lives.

Zeolites are aluminosilicates (see Section 14.21) with a highly porous, open structure.* These silicates have aluminum ions in place of some of the silicon atoms. Because an aluminum ion has a charge of +3, whereas silicon is present with oxidation number +4, additional cations must be present to maintain electrical neutrality. In zeolite catalysts, the cations are hydrogen ions. The three-dimensional structures of the solids are riddled with hexagonal channels connected by tunnels. The enclosed nature of the active sites in zeolites give them a special advantage over other heterogeneous catalysts, because an intermediate can be held in place inside the channels until the products form. Moreover, the channels allow products to grow only to a particular size.

One successful application of zeolites is the catalyst ZSM-5, which is used to convert methanol to gasoline. The pores of the zeolite are just large enough to allow product hydrocarbons of about eight carbons to escape, so the chains do not grow too long and equilibrium is never reached.

Zeolite catalysts that adsorb harmful pollutants preferentially have been developed for use in industry and in catalytic converters for vehicles (see Fig. 13.36 and the photograph at the beginning of this chapter). In catalytic converters, they trap nitrogen oxides so that they can be reduced to harmless nitrogen gas. Catalysts with a different formulation trap incompletely burned hydrocarbons and oxidize them to carbon dioxide. It is possible to imagine a world free of air pollution if all emissions could be passed through suitable zeolites.

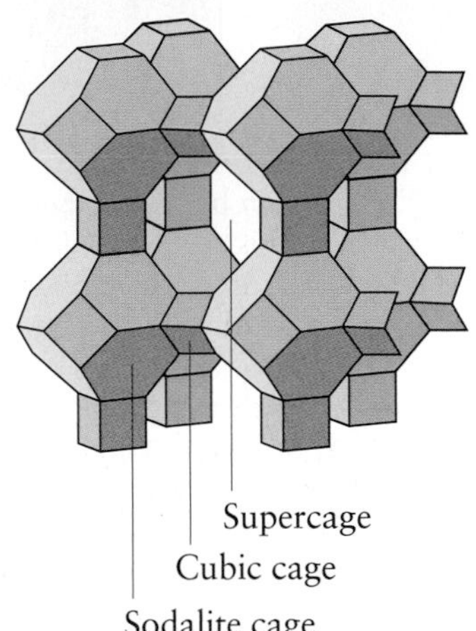

The structure of the ZSM-5 zeolite catalyst. Reactants diffuse through the channels, which are narrow enough to hold intermediates in positions favorable for reactions.

HOW MIGHT YOU CONTRIBUTE?

The mechanisms of reactions that take place inside zeolites are difficult to study and little is known about them. A knowledge of the mechanisms is needed to model the nature of the active sites within zeolites and predict useful new types of zeolites. New catalysts can then be designed to create custom polymers, control pollution, and synthesize medicines.

* The name *zeolite*, which means "boiling stone," was coined by the Scandinavian mineralogist Baron Axel Cronstedt, who observed that certain stones foamed and bubbled on heating.

For Further Reading: P. Ball, *Made to Measure: New Materials for the 21st Century*, Princeton, NJ: Princeton University Press, 1997, pp. 282–312. M. Thomas, "Solid acid catalysts," *Scientific American*, April 1992, pp. 112–118. J. Uppenbrink, "Materials Science: En Route to Cleaner Exhaust," *Science*, 289: 1257e–1259e, 2000.

Catalysts participate in reactions but are not themselves used up; they provide a reaction pathway with a lower activation energy. Catalysts are classified as either homogeneous or heterogeneous.

Self-Test 13.14A How does a catalyst affect (a) the rate of the reverse reaction; (b) ΔH for the reaction?

[*Answer:* (a) Increases it; (b) has no effect]

Self-Test 13.14B How does a homogeneous catalyst affect (a) the rate law; (b) the equilibrium constant?

13.15 Living Catalysts: Enzymes

Living cells contain thousands of different kinds of catalysts, each of which is necessary to life. Many of these catalysts are proteins called **enzymes** (Fig. 13.37). These proteins are large molecules with a slotlike active site, where reaction takes place. The **substrate**, the molecule on which the enzyme acts, fits into the slot as a key fits into a lock (Fig. 13.38). However, unlike an ordinary lock, a protein molecule distorts slightly as the substrate molecule approaches, and its ability to undergo the correct distortion also determines whether the "key" will fit. This refinement of the original lock-and-key model is known as the **induced-fit mechanism** of enzyme action.

Once in the active site, the substrate undergoes reaction. The product is then released for use in the next stage, which is controlled by another enzyme, and the original enzyme molecule is free to receive the next substrate molecule. One example of an enzyme is the amylase present in your mouth. The amylase in saliva helps to break down starches in food into the more easily digested glucose. If you chew a cracker long enough, you can notice the increased sweetness.

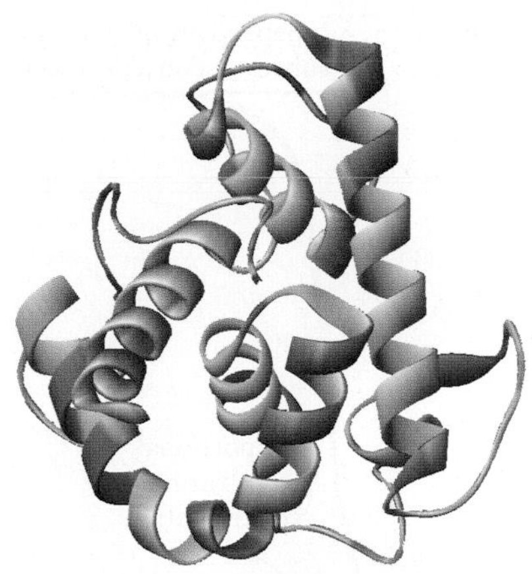

FIGURE 13.37 The lysozyme molecule is a typical enzyme molecule. Lysozyme occurs in a number of places, including tears and the mucus in the nose. One of its functions is to attack the cell walls of bacteria and destroy them. This "ribbon" representation shows only the general arrangement of the atoms to emphasize the overall shape of the molecule; the ribbon actually consists of amino acids linked together (Section 19.13).

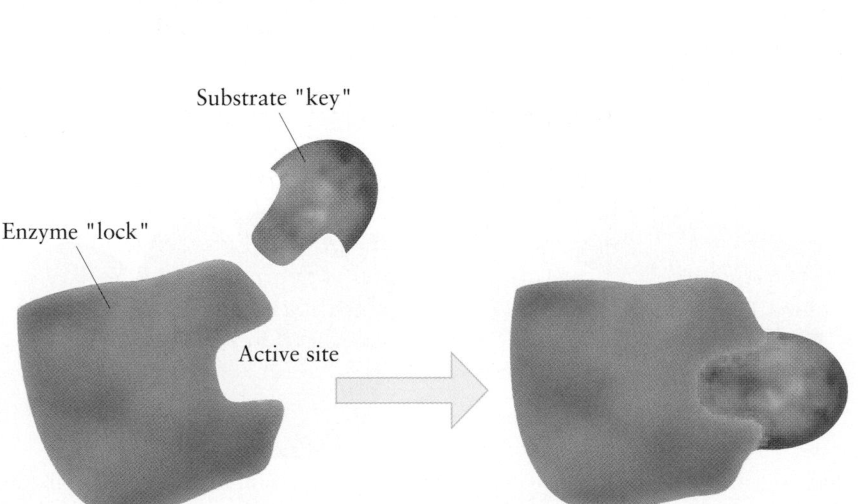

FIGURE 13.38 In the lock-and-key model of enzyme action, the correct substrate is recognized by its ability to fit into the active site like a key into a lock. In a refinement of this model, the enzyme changes its shape slightly as the key enters.

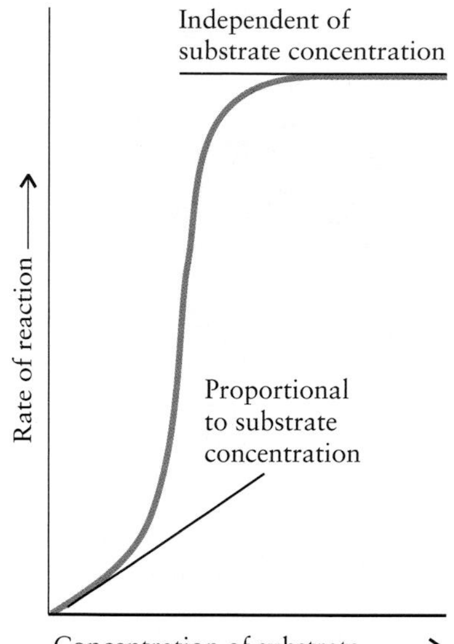

FIGURE 13.39 A plot of the rate of an enzyme-catalyzed reaction as a function of concentration of substrate. At low substrate concentrations, the rate of reaction is directly proportional to the substrate concentration. At high substrate concentrations, the rate is constant once the enzyme molecules are "saturated" with substrate.

The kinetics of enzyme reactions were first studied by the German chemists Leonor Michaelis and Maud Menten in the early part of the twentieth century. They found that, when the concentration of substrate is low, the rate of an enzyme-catalyzed reaction increases with the concentration of the substrate, as shown in the plot in Fig. 13.39. However, when the concentration of substrate is high, the reaction rate depends only on the concentration of the enzyme. In the **Michaelis-Menten mechanism** of enzyme reaction, the enzyme, E, and substrate, S, reach a rapid pre-equilibrium with the bound substrate-enzyme complex, ES:

$$\text{E} + \text{S} \rightleftharpoons \text{ES} \quad \text{forward: second order,} \quad \text{rate} = k_1[\text{E}][\text{S}]$$
$$\text{reverse: first order,} \quad \text{rate} = k_1'[\text{ES}]$$

The complex decays with first-order kinetics, releasing the enzyme to act again:

$$\text{ES} \longrightarrow \text{E} + \text{product} \qquad \text{forward: first order, rate} = k_2[\text{ES}]$$

When the overall rate law is worked out, it is found that

$$\text{Rate of formation of product} = \frac{k_2[\text{E}]_0[\text{S}]}{K_\text{M} + [\text{S}]} \tag{29a}$$

where $[\text{E}]_0$ is the total concentration of enzyme (bound plus unbound) and the **Michaelis constant,** K_M, is

$$K_\text{M} = \frac{k_1' + k_2}{k_1} \tag{29b}$$

When we plot the rate given by Eq. 29a against the concentration of substrate, we get a curve exactly like the one observed.

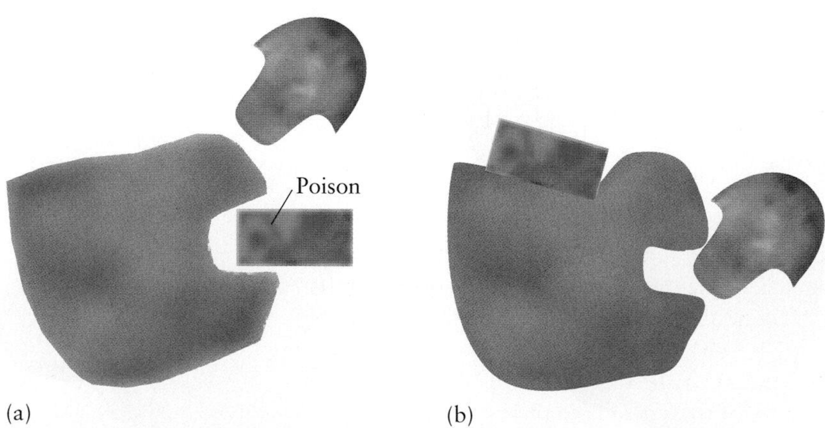

FIGURE 13.40 (a) An enzyme poison (represented by the spotted green rectangle) can act by attaching so strongly to the active site that it blocks the site, thereby taking the enzyme out of action. (b) Alternatively, the poison molecule may attach elsewhere, so distorting the enzyme molecule and its active site that the substrate no longer fits.

One form of biological poisoning mirrors the effect of lead on a catalytic converter. The activity of enzymes is destroyed if an alien substrate attaches too strongly to the reactive site, for then the site is blocked and made unavailable to the true substrate (Fig. 13.40). As a result, the chain of biochemical reactions in the cell stops, and the cell dies. The action of nerve gases is believed to stem from their ability to block the enzyme-controlled reactions that allow impulses to travel through nerves. Arsenic, that favorite of fictional poisoners, acts in a similar way. After ingestion as As(V) in the form of arsenate ions (AsO_4^{3-}), it is reduced to As(III), which binds to enzymes and inhibits their action.

Enzymes are biological catalysts that function by modifying substrate molecules to promote reaction.

SKILLS YOU SHOULD HAVE MASTERED

1 Determine the order of a reaction, its rate law, and its rate constant from experimental data, Example 13.1 and Self-Test 13.2.

2 Calculate a concentration, time, or rate constant by using an integrated rate law, Examples 13.2, 13.3, and 13.4.

3 For a first-order process, calculate the rate constant, elapsed time, and amount remaining from the half-life, Self-Test 13.8 and Example 13.5.

4 Use the Arrhenius equation and temperature dependence of a rate constant to determine an activation energy, Example 13.6.

5 Use the activation energy to find the rate constant at a given temperature, Example 13.7.

6 Show how collision theory and activated complex theory account for the temperature dependence of reactions, Sections 13.8 and 13.9.

7 Deduce a rate law from a mechanism, Example 13.8.

8 Show how the equilibrium constant is related to the forward and reverse rate constants of the elementary reactions contributing to an overall reaction, Section 13.13.

9 Describe the action of catalysts in terms of a reaction profile, Section 13.14.

EXERCISES

Reaction Rates

13.1 Complete the following statements relating to the production of ammonia by the Haber process, for which the overall reaction is $N_2(g) + 3\,H_2(g) \rightarrow 2\,NH_3(g)$. (a) The rate of disappearance of N_2 is ______ times the rate of disappearance of H_2. (b) The rate of formation of NH_3 is ______ times the rate of disappearance of H_2. (c) The rate of formation of NH_3 is ______ times the rate of disappearance of N_2.

13.2 Complete the following statements for the reaction $2\,N_2O_5(g) \rightarrow 4\,NO_2(g) + O_2(g)$. (a) The rate of disappearance of N_2O_5 is ______ times the rate of formation of O_2. (b) The rate of formation of NO_2 is ______ times the rate of disappearance of N_2O_5. (c) The rate of formation of NO_2 is ______ times the rate of formation of O_2.

13.3 (a) The rate of formation of dichromate ions is $0.14\ mol{\cdot}L^{-1}{\cdot}s^{-1}$ in the reaction $2\,CrO_4^{2-}(aq) + 2\,H^+(aq) \rightarrow Cr_2O_7^{2-}(aq) + H_2O(l)$. What is the rate of reaction of chromate ions in the reaction? (b) What is the unique rate of the reaction?

13.4 (a) In the reaction $3\,ClO^-(aq) \rightarrow 2\,Cl^-(aq) + ClO_3^-(aq)$, the rate of formation of Cl^- is $3.6\ mol{\cdot}L^{-1}{\cdot}min^{-1}$. What is the rate of reaction of ClO^-? (b) What is the unique rate of the reaction?

13.5 (a) Nitrogen dioxide, NO_2, decomposes at 6.5×10^{-3} mol·L^{-1}·s^{-1} by the reaction $2\ NO_2(g) \rightarrow 2\ NO(g) + O_2(g)$. Determine the rate of formation of O_2. (b) What is the unique rate of the reaction?

13.6 Manganate ions, MnO_4^{2-}, react at 2.0 mol·L^{-1}·min^{-1} in acidic solution to form permanganate ions and manganese(IV) oxide:

$$3\ MnO_4^{2-}(aq) + 4\ H^+(aq) \longrightarrow 2\ MnO_4^-(aq) + MnO_2(s) + 2\ H_2O(l)$$

(a) What is the rate of formation of permanganate ions? (b) What is the rate of reaction of $H^+(aq)$? (c) What is the unique rate of the reaction?

13.7 The decomposition of gaseous hydrogen iodide, $2\ HI(g) \rightarrow H_2(g) + I_2(g)$, gives the following data at 700 K:

Time, s	[HI], mmol·L^{-1}
0	10
1000	4.4
2000	2.8
3000	2.1
4000	1.6
5000	1.3

(a) Using standard graphing software, such as that on the Web site for this book, plot the concentration of HI as a function of time. (b) Estimate the rate of decomposition of HI at each time. (c) Plot the concentrations of H_2 and I_2 as a function of time on the same graph.

13.8 The decomposition of gaseous dinitrogen pentoxide in the reaction $2\ N_2O_5(g) \rightarrow 4\ NO_2(g) + O_2(g)$ gives the following data at 298 K:

Time, h	[N_2O_5], mmol·L^{-1}
0	2.15
1.11	1.88
2.22	1.64
3.33	1.43
4.44	1.25

(a) Using standard graphing software, such as that on the Web site for this book, plot the concentration of N_2O_5 as a function of time. (b) Estimate the rate of decomposition of N_2O_5 at each time. (c) Plot the concentrations of NO_2 and O_2 as a function of time on the same graph.

Rate Laws and Reaction Order

Rate constants may be found in Table 13.1.

13.9 Express the units for rate constants when the concentrations are in moles per liter and time is in seconds for (a) zero-order reactions; (b) first-order reactions; (c) second-order reactions.

13.10 Because partial pressures are proportional to concentrations, rate laws for gas-phase reactions can also be expressed in terms of partial pressures, for instance, as rate $= kP_X$ for a first-order reaction of a gas X. What are the units for the rate constants when partial pressures are expressed in Torr and time is expressed in seconds for (a) zero-order reactions; (b) first-order reactions; (c) second-order reactions?

13.11 Dinitrogen pentoxide, N_2O_5, decomposes by a first-order reaction. What is the initial rate for the decomposition of N_2O_5 when 3.45 g of N_2O_5 is confined in a 0.750-L container and heated to 65°C? For this reaction, $k = 5.2 \times 10^{-3}\ s^{-1}$.

13.12 Ethane, C_2H_6, dissociates into methyl radicals by a first-order reaction at elevated temperatures. If a 250-mg sample of ethane is confined to a 500-mL reaction vessel and heated to 700°C, what is the initial rate of ethane decomposition if $k = 5.5 \times 10^{-4}\ s^{-1}$ at that temperature?

13.13 A 0.45-g sample of H_2 and a 0.16-g sample of I_2 are confined to a 750-mL reaction vessel and heated to 700 K, when they react by a second-order process (first order in each reactant), with $k = 0.063$ L·mol^{-1}·s^{-1}. (a) What is the initial reaction rate? (b) By what factor does the reaction rate increase if the concentration of H_2 present in the mixture is doubled?

13.14 A 500-mg sample of NO_2, confined to a 150-mL reaction vessel, is heated to 300°C. At that temperature, it decomposes by a second-order process, with $k = 0.54$ L·mol^{-1}·s^{-1}. (a) What is the initial reaction rate? (b) How does the reaction rate change (and by what factor) if the mass of NO_2 present in the container is increased to 750 mg?

13.15 In the reaction $CH_3Br(aq) + OH^-(aq) \rightarrow CH_3OH(aq) + Br^-(aq)$, when the OH^- concentration alone was doubled, the rate doubled; when the CH_3Br concentration alone was increased by a factor of 1.2, the rate increased by a factor of 1.2. Write the rate law for the reaction.

13.16 In the reaction $2\ NO(g) + O_2(g) \rightarrow 2\ NO_2(g)$, when the NO concentration alone was doubled, the rate increased by a factor of 4; when both the NO and O_2 concentrations were increased by a factor of 2, the rate increased by a factor of 8. What is the rate law for the reaction?

13.17 The following kinetic data were obtained for the reaction $2\ ICl(g) + H_2(g) \rightarrow I_2(g) + 2\ HCl(g)$:

Experiment	Initial concentration, $mmol \cdot L^{-1}$		Initial rate, $mol \cdot L^{-1} \cdot s^{-1}$
	$[ICl]_0$	$[H_2]_0$	
1	1.5	1.5	3.7×10^{-7}
2	3.0	1.5	7.4×10^{-7}
3	3.0	4.5	22×10^{-7}
4	4.7	2.7	?

(a) Write the rate law for the reaction. (b) From the data, determine the value of the rate constant. (c) Use the data to predict the reaction rate for Experiment 4.

13.18 The following kinetic data were obtained for the reaction $NO_2(g) + O_3(g) \rightarrow NO_3(g) + O_2(g)$:

Experiment	Initial concentration, $mmol \cdot L^{-1}$		Initial rate, $mmol \cdot L^{-1} \cdot s^{-1}$
	$[NO_2]_0$	$[O_3]_0$	
1	0.21	0.70	6.3
2	0.21	1.39	12.5
3	0.38	0.70	11.4
4	0.66	0.18	?

(a) Write the rate law for the reaction. (b) What is the order of the reaction? (c) From the data, determine the value of the rate constant. (d) Use the data to predict the reaction rate for Experiment 4.

13.19 The following data were obtained for the reaction $A + B + C \rightarrow$ products:

Experiment	Initial concentration, $mmol \cdot L^{-1}$			Initial rate, $mmol \cdot L^{-1} \cdot s^{-1}$
	$[A]_0$	$[B]_0$	$[C]_0$	
1	1.25	1.25	1.25	8.7
2	2.50	1.25	1.25	17.4
3	1.25	3.02	1.25	50.8
4	1.25	3.02	3.75	457
5	3.01	1.00	1.15	?

(a) Write the rate law for the reaction. (b) What is the order of the reaction? (c) Determine the value of the rate constant. (d) Use the data to predict the reaction rate for Experiment 5.

13.20 For the reaction $A + 2\ B + C \rightarrow$ products, the following data were collected:

Experiment	Initial concentration, $mmol \cdot L^{-1}$			Initial rate, $mmol \cdot L^{-1} \cdot s^{-1}$
	$[A]_0$	$[B]_0$	$[C]_0$	
1	2.06	3.05	4.00	3.7
2	0.87	3.05	4.00	0.66
3	0.50	0.50	0.50	0.013
4	1.00	0.50	1.00	0.072

(a) Write the rate law for the reaction. (b) What is the order of the reaction? (c) Determine the value of the rate constant.

Integrated Rate Laws

13.21 Determine the rate constants for the following first-order reactions: (a) $A \rightarrow B$, given that the concentration of A decreases to one-half its initial value in 1000 s. (b) $A \rightarrow B$, given that the concentration of A decreases from $0.67\ mol \cdot L^{-1}$ to $0.53\ mol \cdot L^{-1}$ in 25 s. (c) $2\ A \rightarrow B + C$, given that $[A]_0 = 0.153\ mol \cdot L^{-1}$ and that after 115 s the concentration of B rises to $0.034\ mol \cdot L^{-1}$.

13.22 Determine the rate constants for the following first-order reactions: (a) $2\ A \rightarrow B + C$, given that the concentration of A decreases to one-fourth its initial value in 38 min. (b) $2\ A \rightarrow B + C$, given that $[A]_0 = 0.039\ mol \cdot L^{-1}$ and that after 75 s the concentration of B increases to $0.0095\ mol \cdot L^{-1}$. (c) $2\ A \rightarrow 3\ B + C$, given that $[A]_0 = 0.040\ mol \cdot L^{-1}$ and that after 8.8 min the concentration of B rises to $0.030\ mol \cdot L^{-1}$.

13.23 Dinitrogen pentoxide, N_2O_5, decomposes by first-order kinetics with a rate constant of $3.7 \times 10^{-5}\ s^{-1}$ at 298 K. (a) What is the half-life (in hours) for the decomposition of N_2O_5 at 298 K? (b) If $[N_2O_5]_0 = 0.0567\ mol \cdot L^{-1}$, what will be the concentration of N_2O_5 after 3.5 h? (c) How much time (in minutes) will elapse before the N_2O_5 concentration decreases from $0.0567\ mol \cdot L^{-1}$ to $0.0135\ mol \cdot L^{-1}$?

13.24 Dinitrogen pentoxide, N_2O_5, decomposes by first-order kinetics with a rate constant of $0.15\ s^{-1}$ at 353 K. (a) What is the half-life (in seconds) for the decomposition of N_2O_5 at 353 K? (b) If $[N_2O_5]_0 = 0.0567\ mol \cdot L^{-1}$, what will be the concentration of N_2O_5 after 2.0 s? (c) How much time (in minutes) will elapse before the N_2O_5 concentration decreases from $0.0567\ mol \cdot L^{-1}$ to $0.0135\ mol \cdot L^{-1}$?

13.25 The half-life for the first-order decomposition of A is 355 s. How much time must elapse for the concentration of A to decrease to (a) one-fourth; (b) 15% of its original value; (c) one-ninth of its initial concentration?

13.26 The first-order rate constant for the photodissociation of A is 5.74 h^{-1}. Calculate the time needed for the concentration of A to decrease to (a) $\frac{1}{8}[A]_0$; (b) 5% of its original value; (c) one-tenth of its initial concentration.

13.27 Sulfuryl chloride, SO_2Cl_2, decomposes by first-order kinetics, and $k = 2.81 \times 10^{-3}\ min^{-1}$ at a certain temperature. (a) Determine the half-life for the reaction. (b) Determine the time needed for the concentration of SO_2Cl_2 to decrease to 10% of its initial concentration. (c) If 14.0 g of SO_2Cl_2 is sealed in a 2500-L reaction vessel and heated to the specified temperature, what mass will remain after 1.5 h?

13.28 Ethane, C_2H_6, forms $\cdot CH_3$ radicals at 700°C in a first-order reaction, for which $k = 1.98\ h^{-1}$. (a) What is the half-life for the reaction? (b) Calculate the time needed for the amount of ethane to fall from 1.15×10^{-3} mol to 2.35×10^{-4} mol in a 500-mL reaction vessel at 700°C. (c) How much of a 6.88-mg sample of ethane in a 500-mL reaction vessel at 700°C will remain after 45 min?

13.29 For the first-order reaction $A \rightarrow 3\ B + C$, when $[A]_0 = 0.015\ mol{\cdot}L^{-1}$, the concentration of B increases to $0.018\ mol{\cdot}L^{-1}$ in 3.0 min. (a) What is the rate constant for the reaction? (b) How much more time would be needed for the concentration of B to increase to $0.030\ mol{\cdot}L^{-1}$?

13.30 In the first-order reaction $A \rightarrow 2\ B$, it was observed that, when the initial concentration of A was $0.0335\ mol{\cdot}L^{-1}$, the concentration of B rose to $0.0120\ mol{\cdot}L^{-1}$ in 95 s. How much more time would be needed for it to rise to $0.0190\ mol{\cdot}L^{-1}$?

13.31 The following data were collected for the reaction $2\ HI(g) \rightarrow H_2(g) + I_2(g)$ at 580 K:

Time, s	[HI], $mol{\cdot}L^{-1}$
0	1.00
1000	0.112
2000	0.061
3000	0.041
4000	0.031

(a) Using graphing software, such as that on the Web site for this book, plot the data in an appropriate fashion to determine the order of the reaction. (b) From the graph, determine the rate constant.

13.32 The following data were collected for the reaction $H_2(g) + I_2(g) \rightarrow 2\ HI(g)$ at 780 K:

Time, s	[HI], $10^{-3}\ mol{\cdot}L^{-1}$
0	1.00
1	0.43
2	0.27
3	0.20
4	0.16

(a) Using graphing software, such as that on the Web site for this book, plot the data in an appropriate fashion to determine the order of the reaction. (b) From the graph, determine the rate constant.

13.33 The half-life for the second-order reaction of a substance A is 50.5 s when $[A]_0 = 0.84\ mol{\cdot}L^{-1}$. Calculate the time needed for the concentration of A to decrease to (a) one-sixteenth; (b) one-fourth; (c) one-fifth of its original value.

13.34 Determine the rate constant for the following second-order reactions.
(a) $2\ A \rightarrow B + 2\ C$, given that the concentration of A decreases from $2.5 \times 10^{-3}\ mol{\cdot}L^{-1}$ to $1.25 \times 10^{-3}\ mol{\cdot}L^{-1}$ in 100 s. (b) $3\ A \rightarrow C + 2\ D$, given that $[A]_0 = 0.30\ mol{\cdot}L^{-1}$, and that the concentration of C increases to $0.010\ mol{\cdot}L^{-1}$ in 200 s.

13.35 Determine the time required for each of the following second-order reactions to occur. (a) $2\ A \rightarrow B + C$, for the concentration of A to decrease from $0.10\ mol{\cdot}L^{-1}$ to $0.080\ mol{\cdot}L^{-1}$, given that $k = 0.015\ L{\cdot}mol^{-1}{\cdot}min^{-1}$. (b) $A \rightarrow 2\ B + C$, when $[A]_0 = 0.15\ mol{\cdot}L^{-1}$, for the concentration of B to increase to $0.19\ mol{\cdot}L^{-1}$, given that $k = 0.0035\ L{\cdot}mol^{-1}{\cdot}min^{-1}$.

13.36 The second-order rate constant for the decomposition of NO_2 (to NO and O_2) at 573 K is $0.54\ L{\cdot}mol^{-1}{\cdot}s^{-1}$. Calculate the time for an initial NO_2 concentration of $0.20\ mol{\cdot}L^{-1}$ to decrease to (a) one-half; (b) one-sixteenth; (c) one-ninth of its initial concentration.

The Effect of Temperature

13.37 (a) Using standard graphing software, such as that on the Web site for this book, make an appropriate Arrhenius plot of the following data for the conversion of cyclopropane to propene and calculate the activation energy for the reaction.

T, K	k, s^{-1}
750	1.8×10^{-4}
800	2.7×10^{-3}
850	3.0×10^{-2}
900	0.26

(b) What is the value of the rate constant at 600°C?

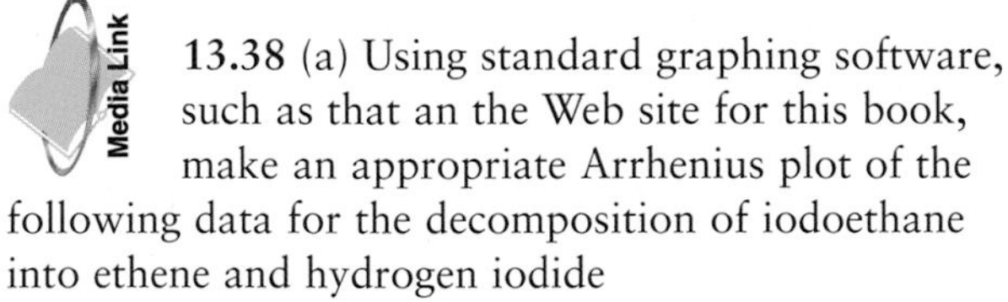

13.38 (a) Using standard graphing software, such as that an the Web site for this book, make an appropriate Arrhenius plot of the following data for the decomposition of iodoethane into ethene and hydrogen iodide

$$C_2H_5I(g) \longrightarrow C_2H_4(g) + HI(g)$$

and determine the activation energy for the reaction:

T, K	k, s^{-1}
660	7.2×10^{-4}
680	2.2×10^{-3}
720	1.7×10^{-2}
760	0.11

(b) What is the value of the rate constant at 400°C?

13.39 The rate constant of the first-order reaction $2\ N_2O(g) \rightarrow 2\ N_2(g) + O_2(g)$ is 0.76 s^{-1} at 1000 K and 0.87 s^{-1} at 1030 K. Calculate the activation energy of the reaction.

13.40 The rate constant of the second-order reaction $2\ HI(g) \rightarrow H_2(g) + I_2(g)$ is 2.4×10^{-6} $L{\cdot}mol^{-1}{\cdot}s^{-1}$ at 575 K and 6.0×10^{-5} $L{\cdot}mol^{-1}{\cdot}s^{-1}$ at 630 K. Calculate the activation energy of the reaction.

13.41 The rate constant of the reaction $O(g) + N_2(g) \rightarrow NO(g) + N(g)$, which occurs in the stratosphere, is 9.7×10^{10} $L{\cdot}mol^{-1}{\cdot}s^{-1}$ at 800°C. The activation energy of the reaction is 315 $kJ{\cdot}mol^{-1}$. Determine the rate constant at 700°C.

13.42 The rate constant of the reaction between CO_2 and OH^- in aqueous solution to give the HCO_3^- ion is 1.5×10^{10} $L{\cdot}mol^{-1}{\cdot}s^{-1}$ at 25°C. Determine the rate constant at blood temperature (37°C), given that the activation energy for the reaction is 38 $kJ{\cdot}mol^{-1}$.

13.43 The rate constant for the decomposition of N_2O_5 at 45°C is $k = 5.1 \times 10^{-4}$ s^{-1}. The activation energy for the reaction is 103 $kJ{\cdot}mol^{-1}$. Determine the value of the rate constant at 50°C.

13.44 Ethane, C_2H_6, dissociates into methyl radicals at 700°C with a rate constant, $k = 5.5 \times 10^{-4}$ s^{-1}. Determine the rate constant at 800°C, given that the activation energy of the reaction is 384 $kJ{\cdot}mol^{-1}$.

Reaction Mechanisms

13.45 Write the overall reaction for the mechanism proposed below and identify any reaction intermediates:

Step 1

$$H_2C{=}CH{-}C({=}O){-}O{-}H + HCl \rightarrow (H)(H)C{=}C(H){-}C({=}O^+{-}H){-}O{-}H + Cl^-$$

Step 2

$$(H)(H)C{=}C(H){-}C({=}O^+{-}H){-}O{-}H + Cl^- \rightarrow H{-}C(H)(Cl){-}C(H){=}C(O{-}H){-}O{-}H$$

Step 3

$$H{-}C(H)(Cl){-}C(H){=}C(O{-}H){-}O{-}H \rightarrow Cl{-}CH_2{-}CH_2{-}C({=}O){-}O{-}H + HCl$$

13.46 Lead(IV) acetate cleaves organic diols such as 2,3-dihydroxy-2,3-dimethylbutane in aqueous solution to give ketones. Write the overall reaction for the mechanism proposed below and identify any reaction intermediates:

Step 1

$$H_3C{-}C(OH)(CH_3){-}C(OH)(CH_3){-}CH_3 + Pb(O_2CCH_3)_4 \rightarrow (H_3C{-}C({=}O){-}O)(CH_3{-}C({=}O){-}O)Pb(O{-})(O{-}) \text{ bonded to } H_3C{-}C(CH_3){-}C(CH_3){-}CH_3 + 2\ CH_3CO_2H$$

Step 2

$$(H_3C{-}C({=}O){-}O)(CH_3{-}C({=}O){-}O)Pb(O{-})(O{-}) \text{ bonded to } H_3C{-}C(CH_3){-}C(CH_3){-}CH_3 \rightarrow 2\ H_3C{-}C({=}O){-}CH_3 + Pb(O_2CCH_3)_2$$

13.47 The following mechanism has been proposed for the reaction between nitric oxide and bromine:

Step 1 $NO + Br_2 \rightarrow NOBr_2$ (slow)
Step 2 $NOBr_2 + NO \rightarrow NOBr + NOBr$ (fast)

Write the rate law implied by this mechanism.

13.48 The following mechanism has been proposed for the reaction between chlorine and chloroform, $CHCl_3$:

Step 1 $Cl_2 \rightarrow Cl + Cl$, and its reverse, $Cl + Cl \rightarrow Cl_2$ (both fast, equilibrium)
Step 2 $CHCl_3 + Cl \rightarrow CCl_3 + HCl$ (slow)
Step 3 $CCl_3 + Cl \rightarrow CCl_4$ (fast)

Write the rate law implied by this mechanism.

13.49 The production of phosgene, $COCl_2$, from carbon monoxide and chlorine is believed to take place by the following mechanism:

Step 1 $Cl_2 \rightarrow Cl + Cl$, and its reverse $Cl + Cl \rightarrow Cl_2$ (both fast, equilibrium)
Step 2 $Cl + CO \rightarrow COCl$, and its reverse $COCl \rightarrow CO + Cl$ (both fast, equilibrium)
Step 3 $COCl + Cl_2 \rightarrow COCl_2 + Cl$ (slow)

Write the rate law implied by this mechanism.

13.50 The mechanism proposed for the oxidation of iodide ion by the hypochlorite ion in aqueous solution is as follows:

Step 1 $ClO^- + H_2O \rightarrow HClO + OH^-$, and its reverse $HClO + OH^- \rightarrow ClO^- + H_2O$ (both fast, equilibrium)
Step 2 $I^- + HClO \rightarrow HIO + Cl^-$ (slow)
Step 3 $HIO + OH^- \rightarrow IO^- + H_2O$ (fast)

Write the rate law implied by this mechanism.

13.51 Three mechanisms for the reaction $NO_2(g) + CO(g) \rightarrow CO_2(g) + NO(g)$ have been proposed:

(a) *Step 1* $NO_2 + CO \rightarrow CO_2 + NO$
(b) *Step 1* $NO_2 + NO_2 \rightarrow NO + NO_3$ (slow)
Step 2 $NO_3 + CO \rightarrow NO_2 + CO_2$ (fast)
(c) *Step 1* $NO_2 + NO_2 \rightarrow NO + NO_3$, and its reverse (both fast, equilibrium)
Step 2 $NO_3 + CO \rightarrow NO_2 + CO_2$ (slow)

Which mechanism agrees with the following rate law: rate $= k[NO_2]^2$? Explain your reasoning.

13.52 When the rate of the reaction $2\,NO(g) + O_2(g) \rightarrow 2\,NO_2(g)$ was studied, it was found that the rate doubled when the O_2 concentration alone was doubled, but quadrupled when the NO concentration alone was doubled. Which of the following mechanisms accounts for these observations? Explain your reasoning.

(a) *Step 1* $NO + O_2 \rightarrow NO_3$, and its reverse (both fast, equilibrium)
Step 2 $NO + NO_3 \rightarrow NO_2 + NO_2$ (slow)
(b) *Step 1* $NO + NO \rightarrow N_2O_2$ (slow)
Step 2 $O_2 + N_2O_2 \rightarrow N_2O_4$ (fast)
Step 3 $N_2O_4 \rightarrow NO_2 + NO_2$ (fast)

Rates and Equilibrium

13.53 Determine whether each of the following statements is true or false. If a statement is false, explain why. (a) For a reaction with a very large equilibrium constant, the rate constant of the forward reaction is much larger than the rate constant of the reverse reaction. (b) At equilibrium, the rate constants of the forward and reverse reactions are equal. (c) An increase in temperature does not affect the magnitude of the activation energy of the reaction. (d) Increasing the concentration of a reactant increases the rate of a reaction by increasing the rate constant in the forward direction.

13.54 Determine whether each of the following statements is true or false. If a statement is false, explain why. (a) The equilibrium constant for a reaction equals the rate constant for the forward reaction divided by the rate constant for the reverse reaction. (b) In a reaction that is a series of equilibrium steps, the overall equilibrium constant is equal to the product of all the forward rate constants divided by the product of all the reverse rate constants. (c) For an exothermic process, the rates of the forward and reverse reactions are affected in the same way by a rise in temperature. (d) For an endothermic reaction, increasing the temperature will increase the rate constant for the reverse reaction.

13.55 For the reversible, one-step reaction $2\,A \rightleftharpoons B + C$, the forward rate constant is 265 $L\cdot mol^{-1}\cdot min^{-1}$ and the rate constant for the reverse reaction is 392 $L\cdot mol^{-1}\cdot min^{-1}$. The activation energy for the forward reaction is 39.7 $kJ\cdot mol^{-1}$ and that of the reverse reaction is 25.4 $kJ\cdot mol^{-1}$. (a) What is the equilibrium constant for the reaction? (b) Is the reaction exothermic or endothermic? (c) What will be the effect of raising the temperature on the rate constants and the equilibrium constant?

13.56 For the reversible, one-step reaction $A + B \rightleftharpoons C + D$, the forward rate constant is 36.4 $L\cdot mol^{-1}\cdot h^{-1}$

and the rate constant for the reverse reaction is 24.3 $L{\cdot}mol^{-1}{\cdot}h^{-1}$. The activation energy for the forward reaction was found to be 33.8 $kJ{\cdot}mol^{-1}$ and for the reverse reaction 45.4 $kJ{\cdot}mol^{-1}$. (a) What is the equilibrium constant for the reaction? (b) Is the reaction exothermic or endothermic? (c) What will be the effect of raising the temperature on the rate constants and the equilibrium constant?

Catalysis

13.57 The presence of a catalyst provides a reaction pathway in which the activation energy of a certain reaction is reduced from 125 $kJ{\cdot}mol^{-1}$ to 75 $kJ{\cdot}mol^{-1}$. (a) By what factor does the rate of the reaction increase at 298 K, all other factors being equal? (b) By what factor would the rate change if the reaction were carried out at 350 K instead?

13.58 The presence of a catalyst provides a reaction pathway in which the activation energy of a certain reaction is reduced from 88 $kJ{\cdot}mol^{-1}$ to 62 $kJ{\cdot}mol^{-1}$. (a) By what factor does the rate of the reaction increase at 300 K, all other factors being equal? (b) By what factor would the rate change if the reaction were carried out at 500 K instead?

13.59 A reaction rate increases by a factor of 1000 in the presence of a catalyst at 25°C. The activation energy of the original pathway is 98 $kJ{\cdot}mol^{-1}$. What is the activation energy of the new pathway, all other factors being equal? In practice, the new pathway also has a different pre-exponential factor.

13.60 A reaction rate increases by a factor of 500 in the presence of a catalyst at 37°C. The activation energy of the original pathway is 106 $kJ{\cdot}mol^{-1}$. What is the activation energy of the new pathway, all other factors being equal? In practice, the new pathway also has a different pre-exponential factor.

13.61 Consider the following mechanism for the reaction of HBr with CH_3OH to produce CH_3Br. Write a complete balanced equation for the overall reaction and list any intermediates.

$$CH_3OH(aq) + H^+(aq) \longrightarrow CH_3OH_2{}^+(aq)$$

$$CH_3OH_2{}^+(aq) \longrightarrow CH_3{}^+(aq) + H_2O(l)$$

$$CH_3{}^+(aq) + Br^-(aq) \longrightarrow CH_3Br(aq)$$

13.62 Consider the following mechanism for the hydrolysis of ethyl acetate. Write a complete balanced equation for the overall reaction, list any intermediates and identify the catalyst in this reaction.

Step 1 $CH_3{-}C({=}O){-}O{-}CH_2CH_3 + OH^- \rightarrow CH_3{-}C({=}O){-}O{-}H + CH_3CH_2O^-$

Step 2 $CH_3CH_2O^- + H_2O \longrightarrow CH_3CH_2OH + OH^-$

13.63 Determine which of the following statements about catalysts are true. If the statement is false, explain why. (a) In an equilibrium process, a catalyst increases the rate of the forward reaction but leaves the rate of the reverse reaction unchanged. (b) A catalyst is not consumed during the course of a reaction. (c) The pathway for a reaction is the same in the presence of a catalyst as without the catalyst present, but the rate constants are decreased in both the forward and reverse directions. (d) A catalyst must be carefully chosen to shift the equilibrium toward the products.

13.64 Determine which of the following statements about catalysts are true. If the statement is false, explain why. (a) Heterogeneous catalysts work by binding one or more of the molecules undergoing reaction to the surface of the catalyst. (b) Enzymes are naturally occurring proteins that serve as catalysts in biological systems. (c) The equilibrium constant for a reaction is greater in the presence of a catalyst. (d) A catalyst changes the pathway of a reaction in such a way that the reaction becomes more exothermic.

INTEGRATED EXERCISES

13.65 When the concentration of 2-bromo-2-methylpropane, C_4H_9Br, is doubled, the rate of the reaction

$$C_4H_9Br(aq) + OH^-(aq) \longrightarrow C_4H_9OH(aq) + Br^-(aq)$$

increases by a factor of 2. When the C_4H_9Br and OH^- concentrations are both doubled, the rate increase is the same, a factor of 2. What are the reactant orders and overall order of the reaction?

13.66 The following data were collected for the decomposition of A:

$$2\,A \longrightarrow 4\,B + C$$

Time, s	[A], $mmol\cdot L^{-1}$
0	2.57
400	1.50
800	0.87
1200	0.51
1600	0.30

(a) Using standard graphing software, such as that on the Web site for this book, determine whether this reaction is most likely to be zero, first, or second order with respect to A. (b) Determine the rate constant k for this reaction.

13.67 (a) From the following data, determine the rate law and the rate constant for the reaction $H_2SeO_3(aq) + 6\,I^-(aq) + 4\,H^+(aq) \rightarrow Se(s) + 2\,I_3^-(aq) + 3\,H_2O(l)$

$[H_2SeO_3]$	$[I^-]$	$[H^+]$	Rate, $mol\cdot L^{-1}\cdot s^{-1}$
0.020	0.020	0.010	8.0×10^{-6}
0.020	0.010	0.020	4.0×10^{-6}
0.020	0.030	0.030	2.4×10^{-4}
0.010	0.020	0.020	1.6×10^{-5}

(b) Calculate the rate of the reaction when $[H_2SeO_3] = 0.035\ mol\cdot L^{-1}$, $[I^-] = 0.020\ mol\cdot L^{-1}$, and $[H^+] = 0.015\ mol\cdot L^{-1}$.

13.68 The decomposition of hydrogen peroxide, $2\,H_2O_2(aq) \rightarrow 2\,H_2O(l) + O_2(g)$, follows first-order kinetics with respect to H_2O_2 and has $k = 0.0410\ min^{-1}$. (a) If the initial concentration of H_2O_2 is $0.20\ mol\cdot L^{-1}$, what is its concentration after 10 min? (b) How much time will it take for the H_2O_2 concentration to decrease from $0.50\ mol\cdot L^{-1}$ to $0.10\ mol\cdot L^{-1}$? (c) How much time is needed for the H_2O_2 concentration to decrease by one-fourth? (d) Calculate the time needed for the H_2O_2 concentration to decrease by 75%.

13.69 (a) For the second-order reaction, $A \rightarrow B + C$, the concentration of the species A falls from $0.040\ mol\cdot L^{-1}$ to $0.0050\ mol\cdot L^{-1}$ in 12 h. What is the rate constant? (b) In the second-order reaction $EX_2 \rightarrow E + 2\,X$, the concentration of X increases to $0.070\ mol\cdot L^{-1}$ in 15 h when $[EX_2]_0 = 0.040\ mol\cdot L^{-1}$. What is the rate constant for the decomposition of EX_2?

13.70 The half-life for the (first-order) decomposition of azomethane, $CH_3N{=}NCH_3$, in the reaction $CH_3N{=}NCH_3(g) \rightarrow N_2(g) + C_2H_6(g)$ is 1.02 s at 300°C. A 45.0-mg sample of azomethane is placed in a 300-mL reaction vessel and heated to 300°C. (a) What mass (in milligrams) of azomethane remains after 10 s? (b) Determine the partial pressure of nitrogen in the reaction vessel after 3.0 s.

13.71 The mechanism of the reaction $A \rightarrow B$ consists of two steps, involving the formation of a reaction intermediate. Overall, the reaction is exothermic. (a) Sketch the reaction profile, labeling the activation energies for each step and the overall enthalpy of reaction. (b) Indicate on the same diagram the effect of a catalyst on the first step of the reaction.

13.72 An exothermic reaction for which $\Delta H_r = -200\ kJ\cdot mol^{-1}$ has an activation energy of $100\ kJ\cdot mol^{-1}$. Estimate the activation energy for the reverse reaction. For estimates, reaction enthalpies can be treated as energies.

13.73 The hydrolysis of an organic nitrile, a compound containing a —C≡N group in basic solution, is proposed to proceed by the mechanism shown below. Write the overall reaction, identify any intermediates, and determine the role of the hydroxide ion.

Step 1 $R{-}C{\equiv}N + OH^- \rightarrow R{-}C(OH){=}N^-$

Step 2 $R{-}C(OH){=}N^- + H_2O \rightarrow R{-}C(OH){=}N{-}H + OH^-$

Step 3 $R{-}C(OH){=}N{-}H \rightarrow R{-}C({=}O){-}NH_2$

13.74 Ketones can react with themselves in a process known as aldol condensation. The mechanism for this reaction in acidic solution is shown below. Write the overall reaction, identify any intermediates and determine the role of the hydrogen ion.

Step 1 $H_3C{-}C({=}O){-}CH_3 + H^+ \rightarrow H_3C{-}C({=}{}^+OH){-}CH_3$

Step 2 $H_3C{-}C({=}{}^+OH){-}CH_3 \rightleftharpoons H_3C{-}C(OH){=}CH_2 + H^+$

Step 3 $H_3C{-}C(OH){=}CH_2 + H_3C{-}C({=}{}^+OH){-}CH_3 \rightarrow H_3C{-}C({=}{}^+OH){-}CH_2{-}C(OH)(CH_3){-}CH_3$

Step 4 $H_3C{-}C({=}{}^+OH){-}CH_2{-}C(OH)(CH_3){-}CH_3 \rightleftharpoons H_3C{-}C({=}O){-}CH_2{-}C(OH)(CH_3){-}CH_3 + H^+$

13.75 All radioactive decay processes follow first-order kinetics. The half-life of the radioactive isotope, tritium (^3H or T), is 12.3 years. How much of a 10.0-mg sample of tritium would remain after 10.9 years?

13.76 Which of the following plots will be linear? (a) [A] as a function of time for a reaction first order in A. (b) [A] as a function of time for a reaction zero order in A. (c) ln[A] as a function of time for a reaction first order in A. (d) 1/[A] as a function of time for a reaction second order in A. (e) ln *k* against temperature. (f) Initial rate as a function of [A] for a reaction first order in A.

13.77 The rate law of the reaction $2\ NO(g) + 2\ H_2(g) \rightarrow N_2(g) + 2\ H_2O(g)$ is rate $= k[NO]^2[H_2]$, and the mechanism that has been proposed is

Step 1 $NO + NO \rightleftharpoons N_2O_2$
Step 2 $N_2O_2 + H_2 \rightarrow N_2O + H_2O$
Step 3 $N_2O + H_2 \rightarrow N_2 + H_2O$

(a) Which step in the mechanism is likely to be rate determining? Explain your answer. (b) Sketch a reaction profile for the overall reaction, which is known to be exothermic. Label the activation energies of each step and the overall reaction enthalpy.

13.78 (a) Using graphing software, such as that on the Web site for this book, calculate the activation energy for the acid hydrolysis of sucrose to give glucose and fructose from an Arrhenius plot of the following data:

T, °C	*k*, $L\cdot mol^{-1}\cdot s^{-1}$
24	4.8×10^{-3}
28	7.8×10^{-3}
32	13×10^{-3}
36	20×10^{-3}
40	32×10^{-3}

(b) Calculate the rate constant at 37°C (body temperature). (c) From data in Appendix 2A, calculate the enthalpy change for this reaction, assuming that the solvation enthalpies of the sugars are negligible. Draw an energy profile for the overall process.

13.79 Raw milk sours in about 4 h at 28°C but in about 48 h in a refrigerator at 5°C. What is the activation energy for the souring of milk?

13.80 (a) From the mechanism given below, derive Equation 29a that Michaelis and Menten proposed to represent the rate of formation of products in an enzyme-catalyzed reaction. (b) Show that the rate is independent of substrate concentration at high concentrations of substrate.

$$E + S \underset{k_1'}{\overset{k_1}{\rightleftharpoons}} ES$$

$$ES \xrightarrow{k_2} E + P$$

where E is the free enzyme, S is the substrate, ES is the enzyme-substrate complex, and P is the product. Note that the steady-state concentration of free enzyme will be equal to the initial concentration of the enzyme less the amount of enzyme that is present in the enzyme-substrate complex: $[E] = [E]_0 - [ES]$.

13.81 The following mechanism has been suggested to explain the contribution of chlorofluorocarbons to the destruction of the ozone layer:

Step 1 $O_3 + Cl \rightarrow ClO + O_2$
Step 2 $ClO + O \rightarrow Cl + O_2$

(a) What is the reaction intermediate and what is the catalyst? (b) Identify the radicals in the mechanism. (c) Identify the steps as initiating, propagating, or terminating. (d) Write a chain-terminating step for the reaction.

13.82 When a mixture containing equal amounts of hydrogen and chlorine is exposed to sunlight, the reaction $H_2(g) + Cl_2(g) \rightarrow 2\ HCl(g)$ occurs explosively. The chain-reaction mechanism is thought to be

Step 1 $Cl_2 \rightarrow Cl + Cl$, a chain-initiation step caused by light
Step 2 a chain-propagation step that results in the formation of a hydrogen atom
Step 3 a chain-propagation step in which the hydrogen atom reacts with Cl_2
Steps 4, 5, 6 three chain-termination steps that occur between the two radicals

Write the equations for the elementary reactions in steps 2 to 6.

13.83 At 328 K, the total pressure of the reaction mixture in which dinitrogen pentoxide decomposes to NO_2 and O_2 varied with time, as shown in the following table. (a) Use the data to find the rate in $mol\cdot L^{-1}\cdot min^{-1}$ at each time. (b) Determine the order and the rate constant *k* of the reaction.

Time, min	Pressure, kPa
0	27.3
5	43.7
10	53.6
15	59.4
20	63.0
30	66.3

13.84 The radioactive decay of carbon-14 is first order, and the half-life is 5800 y. While a plant or animal is living, it has a constant proportion of carbon-14 (relative to carbon-12) in its composition. When the organism dies, the proportion of carbon-14 decreases as a result of radioactive decay, and the age of the organism can be determined if the proportion of carbon-14 in its remains is measured. If the proportion of carbon-14 in an ancient piece of wood is found to be one-fourth that in living trees, how old is the sample?

13.85 The half-life of a substance taking part in a third-order reaction A $\rightarrow$ products is inversely proportional to the square of the initial concentration of A. How can this half-life be used to predict the time needed for the concentration to fall to (a) one-half; (b) one-fourth; (c) one-sixteenth of its initial value?

13.86 The reaction of many alkyl halides with hydroxide ion is second order, first order in alkyl halide, and first order in hydroxide. In aqueous solution, the rate of reaction of CH_3Br with OH^- to give CH_3OH and Br^- has a rate constant of $2.8 \times 10^{-4}\ L{\cdot}mol^{-1}{\cdot}s^{-1}$, whereas the reaction of C_2H_5Br with OH^- to give C_2H_5OH and Br^- has $k = 8.8 \times 10^{-5}\ L{\cdot}mol^{-1}{\cdot}s^{-1}$. The corresponding reaction between $(CH_3)_3CBr$ and OH^-, however, shows a first-order dependence on the concentration of $(CH_3)_3CBr$ but no dependence on the OH^- at low concentrations. Using your knowledge of molecular structure and properties, offer an explanation of these results.

13.87 The rate of a first-order reaction decreases as the reactant is consumed. However, for a first-order reaction that goes to equilibrium, as long as the system is undisturbed the concentration decreases to a certain value and then remains constant at that value. Explain why the concentration does not decrease to 0.

13.88 Suppose that a pollutant was entering the environment at a steady rate R and that, once there, its concentration decays by a first-order reaction. Derive an expression for (a) the concentration of the pollutant at equilibrium in terms of R and (b) the half-life of the pollutant species when $R = 0$.

13.89 The reaction of methane, CH_4, with chlorine gas occurs by a mechanism similar to that described in Exercise 13.82. Write the mechanism. (a) Identify the initiation, propagation, and termination steps. (b) What products are predicted from the mechanism?

13.90 Consider the general equilibrium reaction $A + B \rightleftharpoons C + D$. (a) If the rate of the forward reaction at T_1 is k_1 and the rate of the reverse reaction is k_1', what is the equilibrium constant for the reaction? (b) At a second temperature T_2, the corresponding rate constants are k_2 and k_2'. Write the Arrhenius expression relating k_2 to k_1 and k_2' to k_1'. (c) Starting with the relationships in part (b), show that the effect of temperature on reaction rate (the kinetics of the reaction) is consistent with the effect of temperature on the thermodynamics of the reaction (see the van 't Hoff equation in Chapter 9). (d) How are the activation energies of the forward and reverse reactions of an equilibrium process related to the thermodynamic quantities associated with the reaction?

13.91 From data given in Table 13.1, calculate the standard enthalpy and entropy changes of the reaction $H_2(g) + I_2(g) \rightarrow 2\ HI(g)$.

13.92 (a) From data in Table 13.1 and Appendix 2A, estimate the rate constant for the conversion of propene to cyclopropane at 773 K. (b) We expect the rate of reaction to be slower at lower temperatures. Which will be decreased more by lowering the reaction temperature to 500 K, the rate constant for the conversion of propene to cyclopropane or the rate constant for the conversion of cyclopropane to propene?

13.93 Reactions in which liquid- or gas-phase reactions are catalyzed by solid catalysts (heterogeneous catalysis) often exhibit kinetics that are zero order in the reactant species. Explain why this might be so.

13.94 The pre-equilibrium and the steady-state approximations are two different approaches to deriving a rate law from a proposed mechanism. (a) For the following mechanism, determine the rate law by the steady-state approximation. (c) Under what conditions do the two methods give the same answer? (d) What will the rate law become at high concentrations of Br^-?

$$CH_3OH + H^+ \rightleftharpoons CH_3OH_2^+ \quad \text{(fast in both directions)}$$

$$CH_3OH_2^+ + Br^- \longrightarrow CH_3Br + H_2O \quad \text{(slow)}$$

13.95 (a) What is the overall reaction for the following mechanism?

$$ClO^- + H_2O \rightleftharpoons HClO + OH^- \text{ (fast equilibrium)}$$

$$HClO + I^- \longrightarrow HIO + Cl^- \text{ (very slow)}$$

$$HIO + OH^- \rightleftharpoons IO^- + H_2O \text{ (fast equilibrium)}$$

(b) Write the rate law based on this mechanism. (c) Will the reaction rate be dependent on the pH of the solution? (d) How would the rate law differ if the reactions were carried out in an organic solvent?

13.96 The reaction between A and B has two possible pathways, which lead to different products:

$$A + B \longrightarrow C + D \qquad (1)$$

$$A + B \longrightarrow E + F \qquad (2)$$

Reaction 1 proceeds by a mechanism in which A first fragments into C and G. Mechanism for reaction 1:

$$A \longrightarrow C + G \text{ (slow)}$$

$$B + G \longrightarrow D \text{ (fast)}$$

Reaction 2 proceeds by a direct combination of A and B with a second-order, bimolecular rate law. (a) What can be done experimentally to favor reaction 2 over reaction 1? (b) Write an expression for the rate of disappearance of A. (c) The reaction is carried out under two sets of conditions and sampled at the end of 1.00 h. For the first reaction, the starting concentrations of A and B were both 0.100 $mol \cdot L^{-1}$; in the second experiment, the starting concentration of A was 0.100 $mol \cdot L^{-1}$ and that of B was 0.050 $mol \cdot L^{-1}$. In Experiment 1, analysis of the reaction solution at the end of 1.00 h indicated the presence of C with a concentration of 0.0045 $mol \cdot L^{-1}$ and E with a concentration of 0.0015 $mol \cdot L^{-1}$. For Experiment 2, the concentration of C at the end of the hour was 0.0045 $mol \cdot L^{-1}$ and that of E was 0.000 50 $mol \cdot L^{-1}$. What are the rate constants for reaction 1 and reaction 2?

CHAPTER

This photomicrograph of a limestone rock shows flakes of calcium carbonate that precipitated in an ancient ocean bed. Calcium forms a small, highly charged cation and consequently gives rise to rigid structures.

Why Do We Need to Know This Material?

All matter is made from elements and the compounds they form. In this chapter, we see how the principles that underlie the behavior of matter account for the properties and family relationships of the elements. This knowledge helps us comprehend how the world is made and allows us to design new materials and select new directions for research. The elements in the first four main groups include carbon, the central element of life; silicon and germanium, which are used to create artificial intelligence; calcium, which is responsible for the rigidity of our bones and teeth and the strength of concrete; the space-age metals aluminum and beryllium; the alkali metals; and hydrogen, the fuel of the stars—and perhaps our future energy source on Earth.

What Do We Need to Know Already?

The information in this chapter is organized around the principles of atomic structure and specifically the periodic table (Chapter 1). However, the chapter draws on all the preceding chapters, for it illustrates how those principles account for the properties of the elements.

THE ELEMENTS: THE FIRST FOUR MAIN GROUPS

Steel was the principal construction material of the nineteenth century. It opened up new technologies, especially for transport. But the twenty-first century demands novel materials. For example, to build a space ship, we need materials that are light enough to allow us to escape from Earth but strong enough to withstand the stress of acceleration. They might have to survive temperatures as high as 2000°C yet not become brittle in the bitter cold of outer space. To conceive and develop materials like these, scientists need a thorough knowledge of the elements and their compounds.

This and the following two chapters survey the properties of the elements and their compounds in relation to their locations in the periodic table. To prepare for this journey through the periodic table, we first review the trends in properties that we have discussed in earlier chapters. We then start the journey itself with the unique element hydrogen and move on to consider the elements of the main groups, working from left to right across the table. The same principles apply to the elements of the *d* and *f* blocks, but these elements have some unique characteristics (mainly their wide variety of oxidation states), so they are treated separately in Chapter 16.

Although we concentrate on periodic trends, we also need to know certain details about specific elements, such as where they are found in nature and how they are purified. In addition, some of their compounds have interesting properties or are of great commercial significance. So, as well as discussing the elements in the context of periodic trends, we shall also highlight some of the unique properties and applications that a well-informed chemist is expected to know. These topics are commonly called **descriptive chemistry**—the description of the preparation, properties, and applications of elements and their compounds. In part, the usefulness and availability of an element reflects its abundance on Earth and specifically in the Earth's crust, which may be quite different from its cosmic abundance. Figure 14.1 summarizes the abundances of the elements in three settings: the universe as a whole, the Earth's crust, and the human body.

PERIODIC TRENDS

Let's review what we already know about the periodic table. We saw in Chapter 1 that the members of each group have analogous valence-shell electron configurations, which give them characteristic properties. However, the element at the head of a group—the lightest element in the group—often has features that distinguish it from its **congeners,** the other members of the group. We also saw that there are diagonal relationships between elements—especially between Periods 2 and 3—and a trend toward metallic character to the left along a period and down a group.

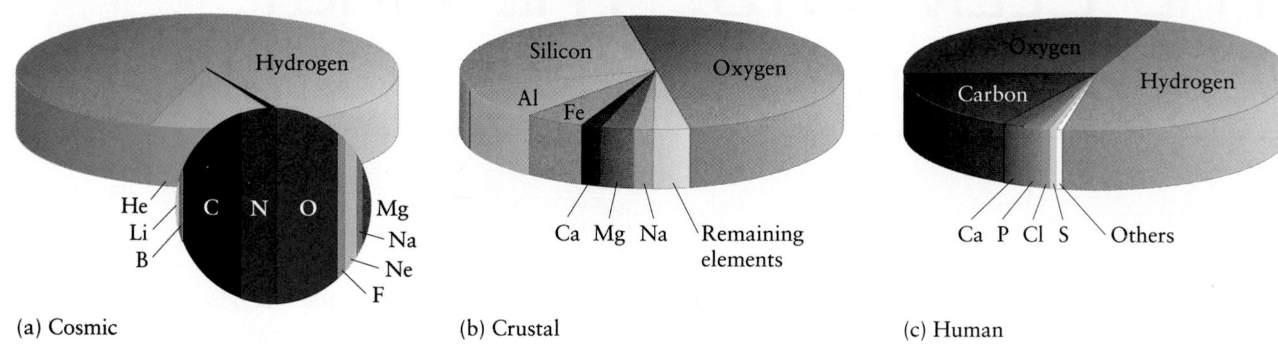

(a) Cosmic (b) Crustal (c) Human

FIGURE 14.1 These charts show the relative abundances of the principal elements in (a) the universe (the "cosmic abundances"); (b) the crust of the Earth, and (c) the human body.

In this introductory section, we review atomic properties that correlate with the position of the element in the periodic table. Then we investigate periodic trends in bonding and see how two types of compounds—hydrides and oxides—display the consequences of these trends.

14.1 Atomic Properties

The valence-shell electrons affect the chemical properties of an element most directly, and trends in the effective nuclear charge experienced by these electrons give rise to periodic trends. As we saw in Section 1.11 and Fig. 1.40, valence electrons experience increasing effective nuclear charge on going across a period from left to right. The attraction of a valence electron to the nucleus decreases on going down a group, because valence electrons belong to successive shells that are farther from the nucleus and surround electron-rich cores. These trends in effective nuclear charge help account for the diagonal trends in many atomic properties. The valence electrons of elements in the upper right of the periodic table, near fluorine, experience the greatest attraction to their nuclei and those of elements in the lower left, near cesium, experience the least.

All the elements in a main group share a characteristic valence electron configuration. The electron configuration controls the valence of the element (the number of bonds it can form) and affects its chemical and physical properties. Five atomic properties are principally responsible for the characteristic properties of each element: atomic radius, ionization energy, electron affinity, electronegativity, and polarizability. These five properties are all related to trends in the effective nuclear charge experienced by the valence electrons.

Atomic radii typically decrease from left to right across a period and increase down a group (Fig. 14.2; see also Fig. 1.41). As the nuclear charge experienced by the valence electrons increases, the electrons are pulled closer to the nucleus, so decreasing the atomic radius. Ionic radii follow similar periodic trends (see Fig. 1.43). We have also seen that cations are always smaller than their parent atoms and anions are always larger (see Fig. 1.44).

First ionization energies typically increase from left to right across a period and decrease down a group (Fig. 14.3; see also Figs. 1.45 and 1.46). The trend across a period is due to the increasing effective nuclear

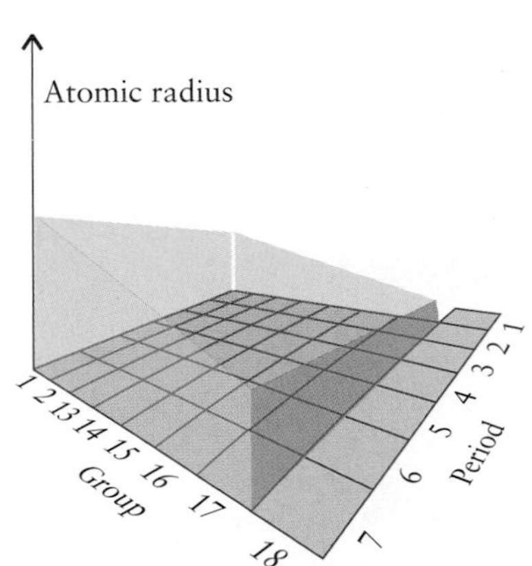

FIGURE 14.2 Atomic radius tends to decrease from left to right across a period and increase down a group. This diagram is a highly schematic representation of those trends. For clarity, only the Arabic group numbers are used in Figs. 14.2–14.6.

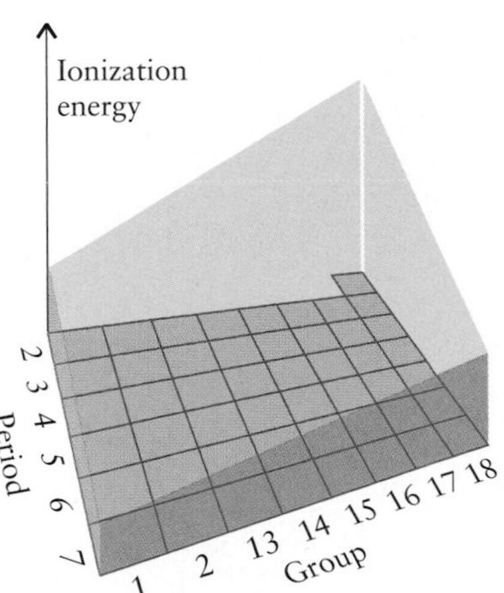

FIGURE 14.3 Ionization energy tends to increase across a period and decrease down a group. This diagram is a highly schematic representation of those trends.

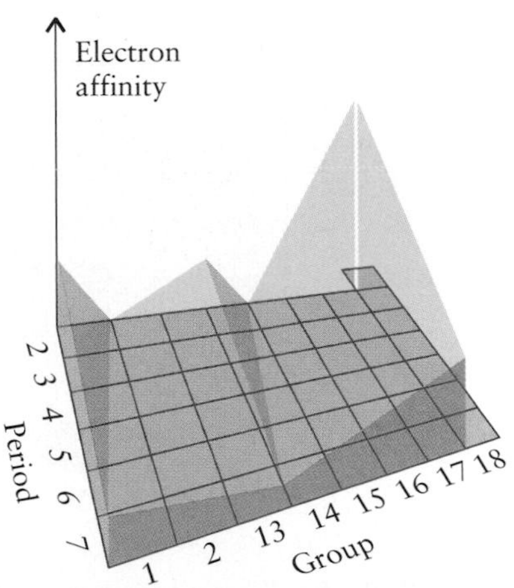

FIGURE 14.4 High values of electron affinity tend to be found close to fluorine. However, as this highly schematic representation shows, there is no simple trend.

charge, which grips the electrons more tightly. The decrease in ionization energy down a group shows that it is easier to remove valence electrons from shells that are farther from the nucleus and hence shielded by a greater number of core electrons.

The highest electron affinities are found at the top right of the periodic table (Fig. 14.4; see also Fig. 1.49). The electron affinity of an element is a measure of the energy released when an anion is formed. Except for the noble gases, elements near fluorine have the highest electron affinities, so we can expect them to be present as anions in compounds with metallic elements. For the same reason, these elements commonly have negative oxidation states in the covalent compounds they form with other nonmetallic elements. Once again, we see the effect of increasing effective nuclear charge: electrons are attracted more strongly to atoms in which they will occupy orbitals where they will experience greater effective nuclear charge.

Electronegativities typically increase from left to right across a period and decrease down a group (Fig. 14.5; see also Fig. 2.12). The greater the effective nuclear charge and the closer the electrons to the nucleus, the stronger the pull on the electrons in a bond. The electronegativity of an element—a measure of the tendency of an atom to attract electrons to itself when part of a compound—is a useful guide to the type of bond the element is likely to form. The relative electronegativities of two elements also indicates the likely polarity of a covalent bond: bonds between atoms of similar electronegativity are typically covalent, with the more electronegative element having the negative partial charge.

Polarizabilities typically decrease from left to right along a period and increase down a group (Fig. 14.6). Polarizability measures the ease with which an electron cloud can be distorted and is greatest for the electron-rich heavier atoms of a group. Anions are more polarizable than their

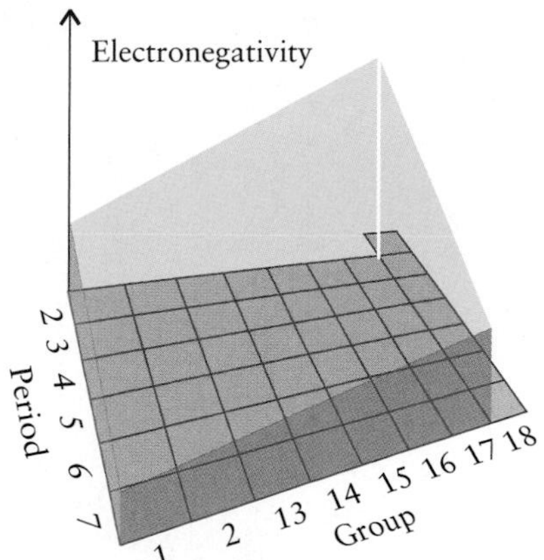

FIGURE 14.5 Electronegativity generally increases from left to right across a period and decreases down a group. This diagram is a highly schematic representation of those trends.

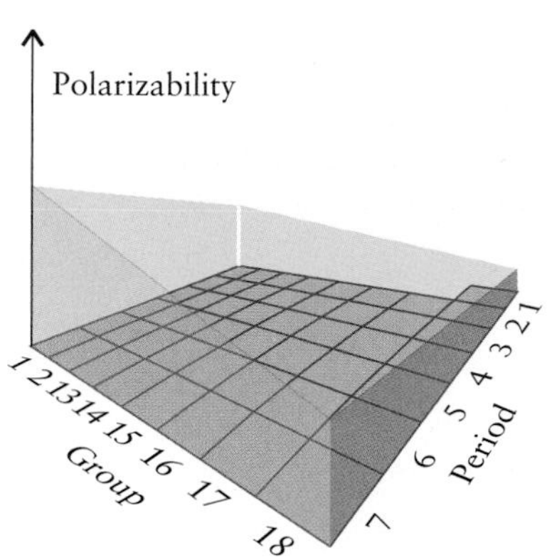

FIGURE 14.6 The general tendency of polarizability is to decrease from left to right across a period and increase down a group. This diagram is a highly schematic representation of those trends.

parent atoms. On the other hand, high polarizing power—the ability to distort the electron cloud of a neighboring atom or ion—is commonly associated with small size and high charge. We saw in Section 2.15 that a highly polarizable atom (such as iodine) in combination with an atom or ion of high polarizing power (such as magnesium) is likely to have considerable covalent character in its bonds. Periodic trends in polarizability mirror the diagonal relationships in the periodic table. Diagonal neighbors have similar polarizabilities or polarizing power and hence a similar degree of covalent character in the bonds they form.

Atomic radii decrease from left to right across a period and increase down a group; ionization energies increase across a period and decrease down a group; electron affinities and electronegativities are highest near fluorine; polarizabilities decrease from left to right across a period and increase down a group.

14.2 Bonding Trends

The oxidation states typical of a main-group element are the outcome of the trends in atomic properties and of the energy changes that accompany bond formation. We can usually predict the valence of elements in Period 2 from the number of electrons in the valence shell and the octet rule. For example, carbon, with four valence electrons, commonly forms four bonds; and oxygen, with six valence electrons and therefore needing two more to complete its octet, typically forms two bonds. Elements in Periods 3 and higher can reach higher oxidation states and have higher valences. These elements have access to empty *d*-orbitals and can use them to expand their valence shells past the usual octet of electrons, or they simply have atoms large enough to bond to more neighbors. Elements at the foot and toward the left of the *p* block also have variable valence. They can display the inert-pair effect and have an oxidation number 2 less than their group number suggests (Section 1.18).

The radius of an atom helps to determine how many other atoms can bond to it. The small radii of Period 2 atoms, for instance, are largely responsible for the differences between their properties and those of their congeners. One reason small atoms typically have low valences is that so few other atoms can pack around them. Nitrogen, for instance, never forms pentahalides, but phosphorus does. With few exceptions, only Period 2 elements form multiple bonds with themselves or other elements in the same period, for only they are small enough for their *p*-orbitals to have substantial π overlap (Fig. 14.7).

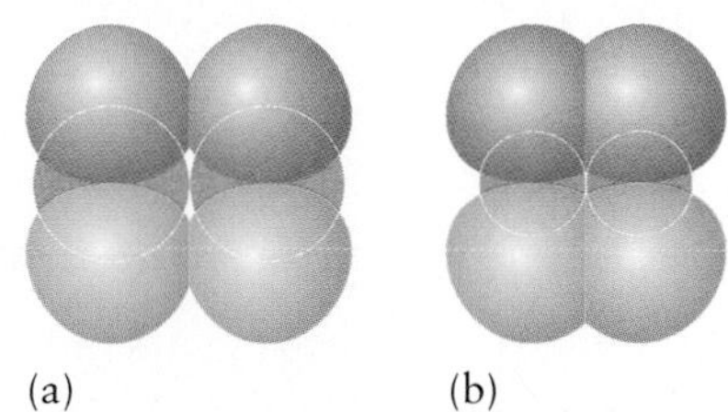

FIGURE 14.7 (a) The *p*-orbitals of Period 3 and heavier elements are held apart by the cores of the atoms (shown in gray) and have very little overlap with each other. (b) In contrast, the atoms of elements in Period 2 are small and can overlap effectively with each other and with elements in later periods.

With the exception of the noble gases, atoms also form bonds to one another in the uncombined elemental state. Elements with low ionization energies tend to have metallic bonds, whereas elements with high ionization energies are typically molecular and have covalent bonds. For example, the halogens, hydrogen, nitrogen, and oxygen form diatomic molecules, sulfur forms eight-membered rings, and phosphorus forms tetrahedral P_4 molecules or long chains of P atoms. Elements in or near the diagonal band of metalloids, which have intermediate ionization energies and three, four, or five valence electrons, have network structures in which each atom is bonded to three or more other atoms. For example, silicon forms diamondlike network structures.

Periodic trends in bonding among main-group elements become apparent when we compare the binary compounds they form with specific elements. All the main-group elements, with the exception of the noble gases and, possibly, indium and thallium, form binary compounds with hydrogen, so we can examine hydrides to look for periodic trends (Section 14.3). All the main-group elements except the noble gases also react with oxygen. Like hydrides, the resulting oxides also reveal periodic trends in the elements (Section 14.4). For instance, we can use their properties to classify elements as metals or nonmetals: oxides of main-group metals are basic, and oxides of nonmetals are acidic (see Section 10.3).

Many of these binary compounds have hydrogen in its +1 oxidation state, so the name "hydride" is not really appropriate. However, it is the conventional term.

Xenon forms an oxide, but by an indirect route.

Valence and oxidation state are directly related to the valence-shell electron configuration of a group.

14.3 Chemical Properties: Hydrides

The formulas of hydrides are related directly to the group number and reveal the typical valences of the elements (Fig. 14.8). For instance, carbon (Group 14/IV) forms CH_4, nitrogen (Group 15/V) forms NH_3, oxygen (Group 16/VI) forms H_2O, and fluorine (Group 17/VII) forms HF.

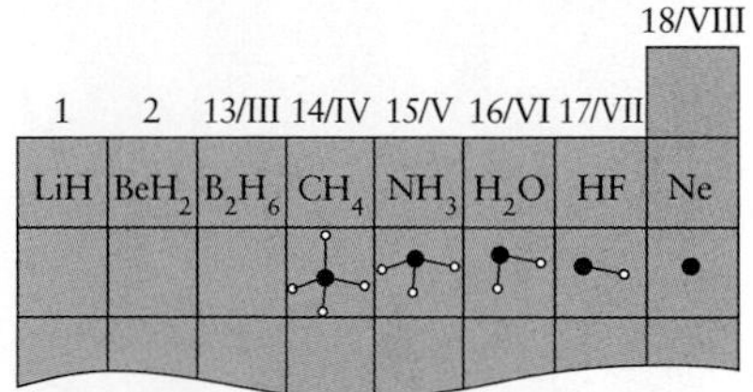

FIGURE 14.8 The chemical formulas of the elements in the main groups display the valence characteristic of each group. The formula of the simplest boron hydride, borane, shows it to be a dimer with empirical formula BH_3.

The nature of the binary hydride is related to the characteristics of the parent element (Fig. 14.9). Strongly electropositive metallic elements form ionic compounds with hydrogen in which the latter is present as a hydride ion, H^-, and has oxidation number −1. These ionic compounds are called *saline hydrides* (or saltlike hydrides). They are formed by all members of the *s* block, with the exception of beryllium, and are made by heating the metal in hydrogen:

$$2\,K(s) + H_2(g) \xrightarrow{\Delta} 2\,KH(s)$$

The saline hydrides are white, high-melting-point solids with crystal structures that resemble those of the corresponding halides. The alkali metal hydrides, for instance, have the rock-salt structure (Fig. 5.37).

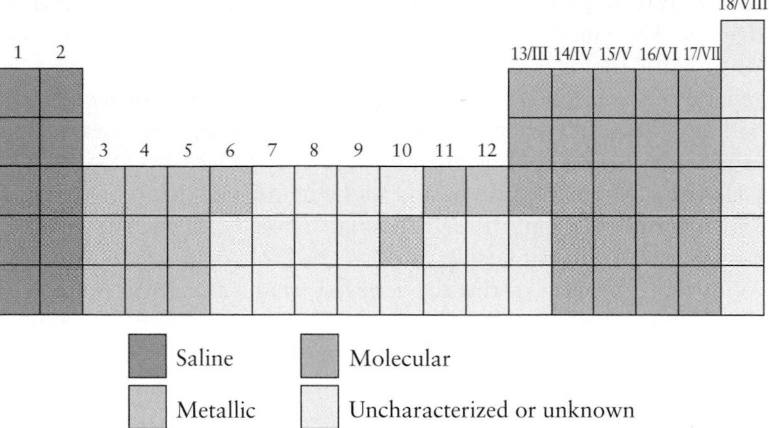

FIGURE 14.9 The different classes of binary hydrogen compounds and their distribution throughout the periodic table.

The *metallic hydrides* are black, powdery, electrically conducting solids formed by heating certain of the *d*-block metals in hydrogen (Fig. 14.10):

$$2\ Cu(s) + H_2(g) \xrightarrow{\Delta} 2\ CuH(s)$$

Because the metallic hydrides release their hydrogen (as H_2) when heated or treated with acid, they are being investigated for storing and transporting hydrogen. Both saline and metallic hydrides have the high enthalpy densities desirable in a portable fuel.

Nonmetals form covalent *molecular hydrides*, which consist of discrete molecules. These compounds are volatile and many are Brønsted acids. Some are gases; for example, ammonia, the hydrogen halides (HF, HCl, HBr, HI), and the lighter hydrocarbons such as methane, ethane, ethene, and ethyne. Liquid molecular hydrides include water and hydrocarbons such as octane and benzene.

Binary hydrides are classified as saline, metallic, or molecular. Some binary molecular compounds of hydrogen are acidic.

14.4 Chemical Properties: Oxides

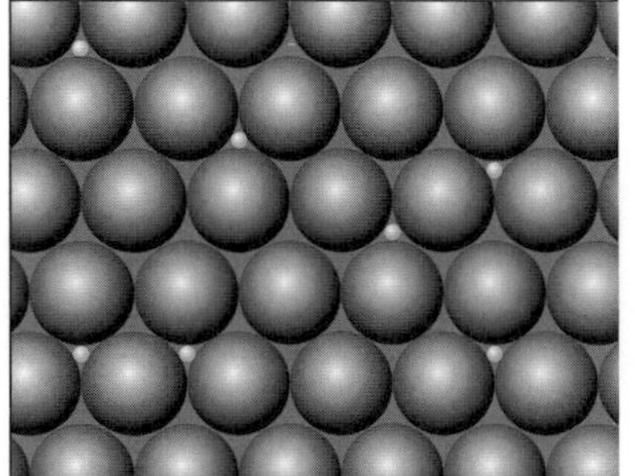

FIGURE 14.10 In a metallic hydride, the small hydrogen atoms (the small spheres) occupy gaps—called interstices—between the larger metal atoms (the large spheres).

The trend in bonding type is from soluble ionic oxides on the left of the periodic table, through insoluble, high-melting-point oxides on the left of the *p* block, to low-melting-point and often gaseous molecular oxides on the right.

Metallic elements with low ionization energies commonly form ionic oxides. As remarked in Section 10.1, the oxide ion is a strong base, so the oxides of most of these metals form basic solutions in water. Magnesium is an exception because its oxide, MgO, is insoluble in water. However, even this oxide reacts with acids, so it is regarded as basic. Elements with intermediate ionization energies, such as beryllium, boron, aluminum, and the metalloids, form amphoteric oxides. These oxides do not react with water, but they do dissolve in both acidic and basic solutions. The acidities of solutions of *d*-metal oxides vary with the oxidation state of the metal (see Section 16.2).

Many oxides of nonmetals are gaseous molecular compounds, such as CO_2, NO, and SO_3. Most can act as Lewis acids, because the electronegative oxygen atoms withdraw electrons from the central atom, enabling it to act as an electron-pair acceptor. For instance, carbon dioxide can react with the oxides of metals because the oxide ion in the metallic oxide is a strong Lewis base:

$$CO_2(g) + Na_2O(s) \longrightarrow Na_2CO_3(s)$$

Many oxides of nonmetals form acidic solutions in water and hence are called **acid anhydrides.** The familiar laboratory acids HNO_3 and H_2SO_4, for instance, are derived from the anhydrides N_2O_5 and SO_3, respectively. Even oxides that do not react with water can be regarded as the *formal* anhydrides of acids. A **formal anhydride** of an acid is the molecule obtained by striking out the elements of water (H, H, and O) from the molecular formula of the acid. Carbon monoxide, for instance, is the formal anhydride of formic acid, HCOOH, although CO does not react with cold water to form the acid.

The oxides of metalloids and some of the less electropositive elements are amphoteric (react with both acids and bases). Aluminum oxide, for instance, reacts with acids and with alkalis (aqueous solutions of strong bases). The oxides reveal a strong diagonal relationship between beryllium and aluminum, for beryllium oxide is also amphoteric.

The oxides of main-group elements show periodic trends in properties. Oxides of metals tend to be ionic and to form basic solutions in water. Oxides of nonmetals are molecular and the anhydrides of acids.

Self-Test 14.1A An element ("E") in Period 4 forms a molecular hydride with the formula H_2E. Identify the element.

[*Answer:* Selenium]

Self-Test 14.1B An element ("E") in Period 3 forms an amphoteric oxide with the formula E_2O_3. Identify the element.

HYDROGEN

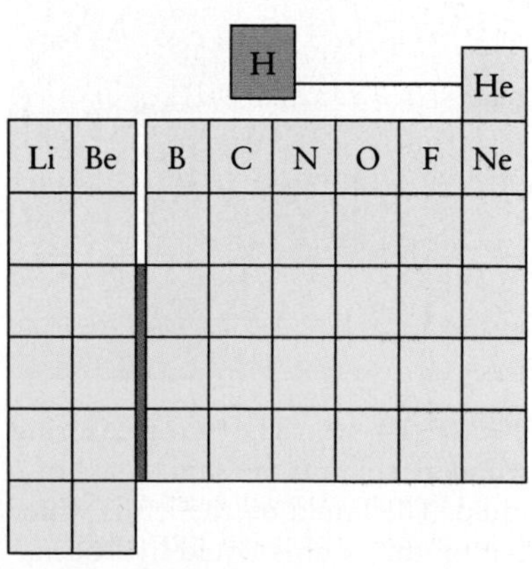

Hydrogen has the simplest atoms of all, for most of them are made up of just two subatomic particles, one proton and one electron. Although it has the same valence electron configuration as the Group 1 elements, ns^1, and forms +1 ions, hydrogen is a nonmetal. It has few other similarities to the alkali metals. Hydrogen resembles the halogens in that it needs only one electron to complete its valence electron configuration, it can form −1 ions, and it exists as diatomic molecules, H_2. However, the chemical properties of hydrogen are very different from those of the halogens. Because it has few resemblances to any group of elements, we do not assign it to any group.

14.5 The Element

Hydrogen is the most abundant element: it accounts for 89% of all atoms in the universe (see Fig. 14.1). These atoms were formed in the first few seconds after the Big Bang, the event that marked the beginning of the universe. However, there is little free hydrogen on Earth because H_2 molecules, being very light, move at such high average speeds that they tend to escape from the Earth's gravity. It takes heavier atoms to anchor hydrogen to the planet in compounds. Most of the Earth's hydrogen is present as water, either in the oceans or trapped inside minerals and clays. Hydrogen is also found in the hydrocarbons that make up the **fossil fuels:** coal, petroleum, and natural gas.

Because natural supplies of hydrogen gas are far too small to satisfy the needs of industry, it must be extracted from its compounds. Most commercial hydrogen is obtained as a by-product of petroleum refining in a sequence of two catalyzed reactions. The first is a *re-forming reaction*, in which a hydrocarbon and steam are converted to carbon monoxide and hydrogen over a nickel catalyst:

$$CH_4(g) + H_2O(g) \xrightarrow{Ni} CO(g) + 3\ H_2(g)$$

The mixture of products, called *synthesis gas*, is the starting point for the manufacture of many other compounds, including methanol. The re-forming reaction is followed by the *shift reaction*, in which the carbon monoxide in the synthesis gas reacts with more water:

$$CO(g) + H_2O(g) \xrightarrow{Fe/Cu} CO_2(g) + H_2(g)$$

Hydrogen is prepared in the laboratory by reducing hydrogen ions from a strong acid (such as hydrochloric acid) with a metal that has a negative standard potential, such as zinc:

$$Zn(s) + 2\ H^+(aq) \longrightarrow Zn^{2+}(aq) + H_2(g)$$

FIGURE 14.11 The two measuring cylinders contain the same mass of liquid. The liquid on the left is water, that on the right is liquid hydrogen, which is one-tenth as dense.

Hydrogen is a colorless, odorless, tasteless gas (Table 14.1). Because H_2 molecules are nonpolar, they can attract each other only by London forces. Each molecule has only two electrons, and fluctuations in their locations result in only a very small instantaneous electric dipole, so these forces are very weak. A consequence is that hydrogen does not condense to a liquid until it is cooled to a very low temperature (20 K, in fact). Another consequence is that hydrogen has only low solubility in many liquids, particularly polar liquids.

One striking physical property of liquid hydrogen is its very low density ($0.070\ g{\cdot}cm^{-3}$), which is less than one-tenth that of water (Fig. 14.11). This low density makes hydrogen a very lightweight fuel. Hydrogen has the highest specific enthalpy of any known fuel (the highest enthalpy of combustion per gram), so liquid hydrogen is used with liquid oxygen to power the space shuttle's main rocket engines.

Hydrogen may become the primary fuel of the future. It is abundant in compounds on Earth and burns without polluting the air or contributing significantly to the greenhouse effect (see Box 14.2), because water is the only combustion product. Petroleum, coal, and natural gas are becoming increasingly rare, but there is enough water in the oceans to

TABLE 14.1 Physical Properties of Hydrogen

Valence configuration: $1s^1$
*Normal form**: colorless, odorless gas

Z	Name	Symbol	Molar mass, $g \cdot mol^{-1}$	Abundance, %	Melting point, °C	Boiling point, °C	Density, $g \cdot L^{-1}$
1	hydrogen	H	1.008	99.98	−259 (14 K)	−253 (20 K)	0.089
1	deuterium	^2H or D	2.014	0.02	−254 (19 K)	−249 (24 K)	0.18
1	tritium	^3H or T	3.016	radioactive	−252 (21 K)	−248 (25 K)	0.27

**Normal form* means the state and appearance of the element at 25°C and 1 atm. The density refers to the same conditions.

generate all the hydrogen fuel we shall ever need. However, a drawback is that hydrogen must be extracted from its compounds, and that takes energy. Hydrogen is obtained from water by electrolysis, but that process is economical only where electricity is cheap. Chemists are currently seeking ways of using sunlight to drive the *water-splitting reaction*, the photochemical decomposition of water into its elements:

$$2\,H_2O(l) \xrightarrow{\text{light}} 2\,H_2(g) + O_2(g) \qquad \Delta G_r^\circ = +474\ kJ \cdot mol^{-1}$$

Our supply of hydrogen would be cheap and virtually inexhaustible if we could use solar energy to produce hydrogen from water. One promising route toward realizing this dream is a special kind of electrochemical cell powered by the Sun (Fig. 14.12). *Photoelectrochemical cells* are electrolytic cells that use the energy of light to drive a reaction in a nonspontaneous direction. Their electrodes are made of semiconductors, because light easily excites the valence electrons: these electrons leave their parent atoms and move through the solid, generating a current and producing oxygen gas from water. Hydrogen gas is generated at the cathode, which is usually an inert metal such as platinum.

Each year, about half the 3×10^8 kg of hydrogen used in industry is converted into ammonia by the Haber process (Section 9.12). Through the reactions of ammonia, hydrogen finds its way into numerous other important nitrogen compounds such as hydrazine and sodium amide. That story is taken up in Section 15.2.

About a third of the hydrogen manufactured is used in mining for the **hydrometallurgical extraction** of copper and other metals, the extraction of the metals from their ores by reduction in aqueous solution:

$$Cu^{2+}(aq) + H_2(g) \longrightarrow Cu(s) + 2\,H^+(aq) \qquad \Delta G_r^\circ = -65\ kJ \cdot mol^{-1}$$

In this process, ores containing copper(II) oxide and copper(II) sulfide are dissolved in sulfuric acid, and then hydrogen is bubbled through the solution. The reduction is thermodynamically favored, because the standard potential of the couple Cu^{2+}/Cu is positive ($E^\circ = +0.34$ V). Metals with negative standard potentials, such as zinc ($E^\circ = -0.76$ V) and nickel ($E^\circ = -0.23$ V), cannot be extracted hydrometallurgically.

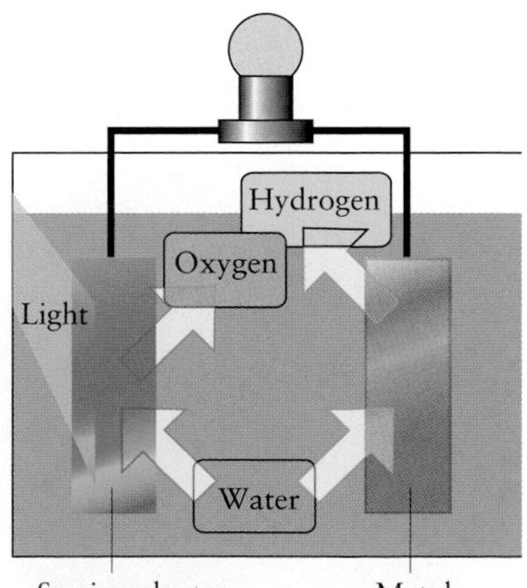

FIGURE 14.12 A photoelectrochemical cell uses solar energy to produce chemical fuels. This cell produces hydrogen and oxygen from water. The two gases are collected separately.

FIGURE 14.13 When the runny oil (top) is hydrogenated, it is converted into a solid—a fat (bottom). The hydrogen converts carbon-carbon double bonds into single bonds. The resulting more flexible molecules can pack together more closely and form a solid.

Hydrogen is used in the food industry to convert vegetable oils into shortening (Fig. 14.13). There it is added to carbon-carbon double bonds in a hydrogenation reaction:

$$H_2(g) + \cdots C{=}C \cdots \xrightarrow{\text{catalyst}} \cdots CH{-}CH \cdots$$

This reaction converts a C=C double bond into a C—C single bond. Oil and solid fat molecules both have long hydrocarbon chains, but oils have more double bonds. Because double bonds resist twisting, oil molecules do not pack together well, so the result is a liquid. When the double bonds are replaced by single bonds, the chains become much more flexible, so the molecules pack together better and form a solid.

Hydrogen is produced as a by-product of the refining of fossil fuels and by electrolysis of water. It has a low density and weak intermolecular forces. Hydrogen is a good reducing agent for species with positive standard potentials.

14.6 Compounds of Hydrogen

Hydrogen is unusual because it can form both a cation (H^+) and an anion (H^-). Moreover, its intermediate electronegativity (2.2) means that it can also form covalent bonds with all elements except metals. The hydrides were introduced in Section 14.3. Because hydrogen forms compounds with so many elements (Table 14.2), we shall meet more of its compounds when we study the other elements.

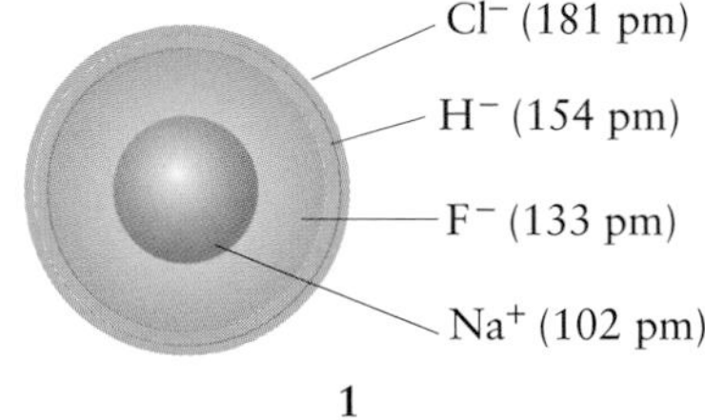

1

The hydride ion is large, with a radius of 154 pm (**1**), falling between the fluoride and chloride ions. The large radius of the ion and its low charge make it highly polarizable and contribute to covalent character in its bonds to cations. However, the single positive charge of the hydrogen atomic nucleus can barely manage to keep control over the two electrons in the H^- ion, so they are easily lost. The low energy required for a hydride ion to lose an electron causes the ionic hydrides to be very powerful reducing agents, and $E°(H_2/H^-) = -2.25$ V. This value is similar to the standard potential for Na^+/Na ($E° = -2.71$ V) and, like sodium metal, hydride ions react with water as soon as they come into contact with it:

$$NaH(s) + H_2O(l) \longrightarrow NaOH(aq) + H_2(g)$$

Because this reaction produces hydrogen, saline hydrides are potentially useful as transportable sources of hydrogen.

TABLE 14.2 Chemical Properties of Hydrogen

Reactant	Reaction with hydrogen
Group 1 metals (M)	$2\,M(s) + H_2(g) \longrightarrow 2\,MH(s)$
Group 2 metals (M, not Be or Mg)	$M(s) + H_2(g) \longrightarrow MH_2(s)$
some *d*-block metals (M)	$2\,M(s) + x\,H_2(g) \longrightarrow 2\,MH_x(s)$
oxygen	$O_2(g) + 2\,H_2(g) \longrightarrow 2\,H_2O(l)$
nitrogen	$N_2(g) + 3\,H_2(g) \longrightarrow 2\,NH_3(g)$
halogen (X_2)	$X_2(g,l,s) + H_2(g) \longrightarrow 2\,HX(g)$

The hydrides of nitrogen, oxygen, and fluorine participate in the very strong intermolecular force known as *hydrogen bonding* (Section 5.5). A hydrogen bond is about 5% as strong as covalent bond between the same types of atoms. For example, the H—O bond enthalpy is 463 kJ·mol^{-1}, whereas the O···H—O hydrogen bond enthalpy is about 20 kJ·mol^{-1}. We can visualize a hydrogen bond most simply as a coulombic interaction between the partial positive charge on a hydrogen atom and the partial negative charge of another atom, as in $O^{\delta-}\cdots{}^{\delta+}H—O$. A more sophisticated view can be expressed in terms of molecular orbitals. If we think of each of the H and O atoms of an O···H—O bond as contributing one atomic orbital, then from these three atomic orbitals we can build three molecular orbitals: one will be largely bonding, one largely nonbonding, and the third largely antibonding (Fig. 14.14). There are four electrons to accommodate: two from the original H—O bond and two from the lone pair of the other O atom. These four electrons occupy the bonding and nonbonding orbitals but leave the antibonding orbital empty. As a result, there is a net bonding character in the combination.

Energy
O H O
Antibonding
Nonbonding
Bonding

FIGURE 14.14 The molecular orbital energy-level scheme for a hydrogen bond between two oxygen atoms. The oxygen atoms each provide an atomic orbital; and together with the H1*s* orbital, they form three molecular orbitals. Only the bonding and nonbonding molecular orbitals are occupied by the four electrons (two from the O—H bond and two from the O lone pair).

The hydride ion has a large radius and is highly polarizable. Hydrides are powerful reducing agents. Hydrogen can form hydrogen bonds to lone pairs of electrons on highly electronegative elements.

Self-Test 14.2A What common binary compounds of hydrogen are powerful reducing agents?

[*Answer:* The saline (ionic) hydrides]

Self-Test 14.2B What are some of the properties of hydrogen that argue against its classification as a Group 1 element?

GROUP 1: THE ALKALI METALS

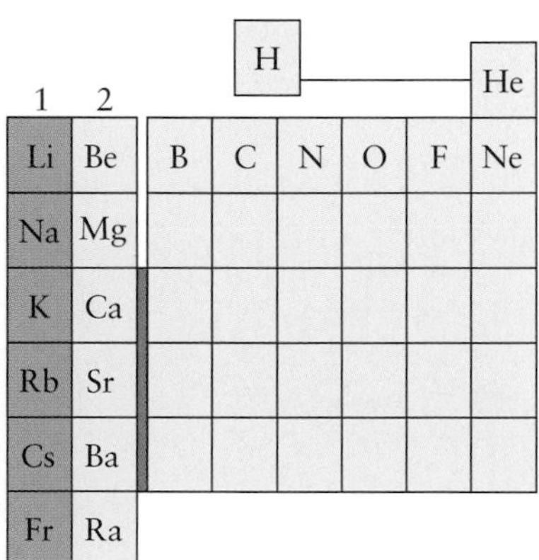

The members of Group 1 are called the *alkali metals*. One of the striking features of the group is the close similarity of the chemical properties of these elements. Nevertheless, there are differences, and the subtlety of some of these differences is the basis of the most subtle property of matter: consciousness. Our thinking, which relies on the transmission of signals along neurons, is achieved by the concerted action of sodium and potassium ions and their carefully regulated migration across membranes. So, even to learn about sodium and potassium, we have to make use of them in our brains.

As in the discussion of hydrogen, in this section we examine the properties of the alkali metals in the context of the periodic table and focus on particular aspects of the elements and selected compounds. The valence electron configuration of the alkali metals is ns^1, where n is the period number. Their physical and chemical properties are dominated by the ease with which the single valence electron can be removed (Tables 14.3 and 14.4).

14.7 The Group 1 Elements

The alkali metals are the most violently reactive of all the metals. They are too easily oxidized to be found in the free state in nature and cannot be extracted from their compounds by ordinary chemical reducing agents. The pure metals are obtained by electrolysis of their molten salts, as in the

TABLE 14.3 Group 1 Elements: The Alkali Metals

Valence configuration: ns^1
*Normal form**: soft, silver-gray metals

Z	Name	Symbol	Molar mass, $g \cdot mol^{-1}$	Melting point, °C	Boiling point, °C	Density, $g \cdot cm^{-3}$
3	lithium	Li	6.94	181	1347	0.53
11	sodium	Na	22.99	98	883	0.97
19	potassium	K	39.10	64	774	0.86
37	rubidium	Rb	85.47	39	688	1.53
55	cesium	Cs	132.91	28	678	1.87
87	francium	Fr	(223)	27	677	—

**Normal form* means the state and appearance of the element at 25°C and 1 atm.

electrolytic Downs process (Section 12.15) or, in the case of potassium, by exposing molten potassium chloride to sodium vapor:

$$KCl(l) + Na(g) \xrightarrow{750°C} NaCl(s) + K(g)$$

Although the equilibrium constant for this reaction is not particularly favorable, the reaction runs to the right because potassium is more volatile than sodium: the potassium vapor is driven off by the heat and condensed in a cooled collecting vessel.

All the Group 1 elements are soft, silver-gray metals (Fig. 14.15). Because the valence shell consists of a single electron, bonding in the metals is weak, leading to low melting points, boiling points, and densities. Lithium is the hardest alkali metal, but even so it is softer than lead. Francium is intensely radioactive, and little is known about the properties of this very rare element. The melting points of the alkali metals decrease down the group (Fig. 14.16). Cesium, which melts at 28°C, is barely a solid at room temperature.

TABLE 14.4 Chemical Properties of the Alkali Metals

Reactant	Reaction with alkali metal (M)
hydrogen	$2\,M(s) + H_2(g) \longrightarrow 2\,MH(s)$
oxygen	$4\,Li(s) + O_2(g) \longrightarrow 2\,Li_2O(s)$
	$2\,Na(s) + O_2(g) \longrightarrow Na_2O_2(s)$
	$M(s) + O_2(g) \longrightarrow MO_2(g)$, M = K, Rb, Cs
nitrogen	$6\,Li(s) + N_2(g) \longrightarrow 2\,Li_3N(s)$
halogen (X_2)	$2\,M(s) + X_2(g,l,s) \longrightarrow 2\,MX(s)$
water	$2\,M(s) + 2\,H_2O(l) \longrightarrow 2\,MOH(aq) + H_2(g)$

(a)

(b)

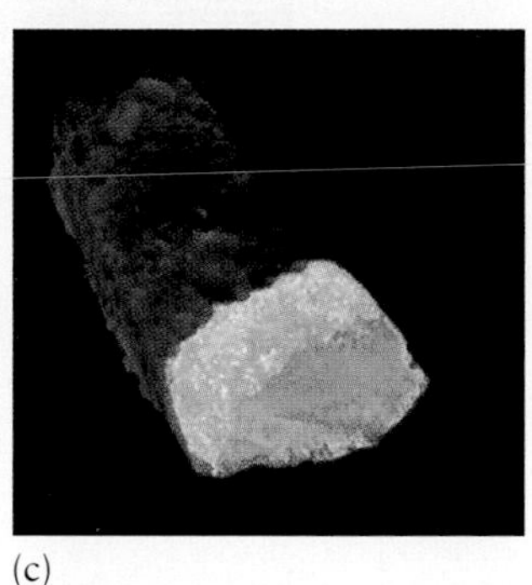
(c)

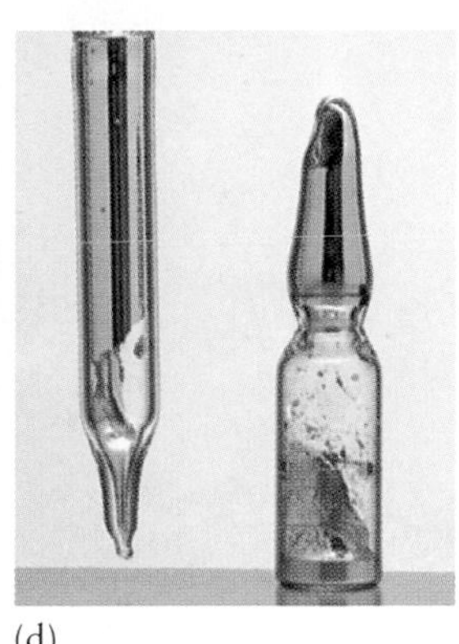
(d)

FIGURE 14.15 The alkali metals of Group 1: (a) lithium; (b) sodium; (c) potassium; (d) rubidium and cesium. Francium has never been isolated in visible quantities. The first three elements corrode rapidly in moist air; rubidium and cesium are even more reactive and have to be stored (and photographed) in sealed, airless containers.

Lithium metal had few uses until after World War II, when thermonuclear weapons were developed (see Section 17.10). This application has had an effect on the molar mass of lithium. Because only lithium-6 could be used in these weapons, the proportion of lithium-7 and, as a result, the molar mass of commercially available lithium has increased. A growing application of lithium is in the rechargeable lithium-ion battery. Because lithium has the most negative standard potential of all the elements, it can produce a high potential when used in a galvanic cell. Furthermore, because lithium has such a low density, lithium-ion batteries are light. They hold a stable charge over a long period of time and can be recharged many times. These properties make lithium-ion batteries a promising energy storage device for electric vehicles (see Box 12.1).

The alkali metals melt at low temperatures. They are the most reactive metals.

14.8 Chemical Properties of the Alkali Metals

The first ionization energies of the alkali metals are low, so in their compounds they exist as singly charged cations, such as Na^+. Most of their compounds are ionic (Table 14.5). The low ionization energies are the reason why the Group 1 metals are excellent reducing agents. This vigorous reducing power is used commercially in the extraction of some metals that otherwise would be difficult to obtain. Molten sodium metal, for instance, is used to produce zirconium and titanium from their chlorides:

$$TiCl_4(g) + 4\,Na(l) \longrightarrow 4\,NaCl(s) + Ti(s)$$

Because their standard potentials are so strongly negative, the alkali metals are strong reducing agents and react with water:

$$2\,Na(s) + 2\,H_2O(l) \longrightarrow 2\,NaOH(aq) + H_2(g)$$

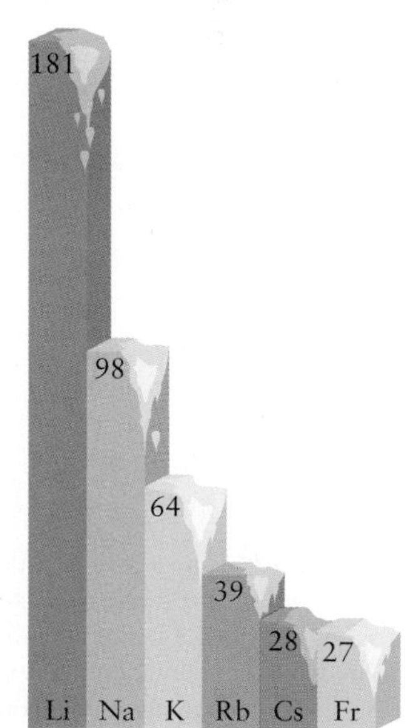

FIGURE 14.16 The melting points of the alkali metals decrease down the group. The numerical values shown here are degrees Celsius.

TABLE 14.5 Properties of Group 1 Compounds

Compound	Formula*	Comment
oxides	M_2O	formed by decomposition of carbonates; strong bases; react with water to form hydroxides
hydroxides	MOH	formed by reduction of water with the metal or from the oxide; strong bases
carbonates	M_2CO_3	soluble in water; most decompose into oxides when heated
hydrogen carbonates	$MHCO_3$	weak bases in water; can be obtained as solids
nitrates	MNO_3	decompose to nitrite and evolve oxygen when strongly heated

*M stands for any Group 1 metal.

The vigor of this reaction increases uniformly down the group (Fig. 14.17). The reaction of water with potassium is vigorous enough to ignite the hydrogen, and with rubidium and cesium is dangerously explosive. Rubidium and cesium are denser than water, so they sink and react beneath the surface; the rapidly evolved hydrogen gas then forms a shock wave that can shatter the vessel. Lithium is the least active of the Group 1 metals in the reaction with water, but molten lithium is one of the most active metals known; this difference shows that the main reason for the slow reaction of solid lithium is the strength with which the atoms are bound together.

The alkali metals also release their valence electrons when they dissolve in liquid ammonia, but the outcome is different. The electrons occupy cavities formed by groups of NH_3 molecules and give ink-blue *metal-ammonia solutions* (Fig. 14.18). These solutions of solvated electrons (and cations of the metal) are often used to reduce organic compounds. As the metal concentration is increased, the blue gives way to a metallic bronze, and the solutions begin to conduct electricity like liquid metals.

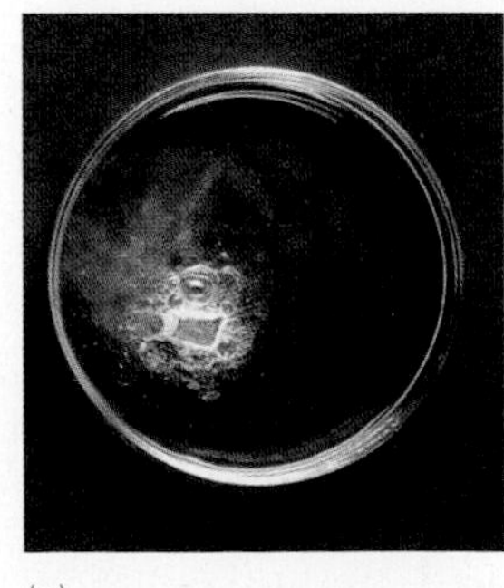
(a)

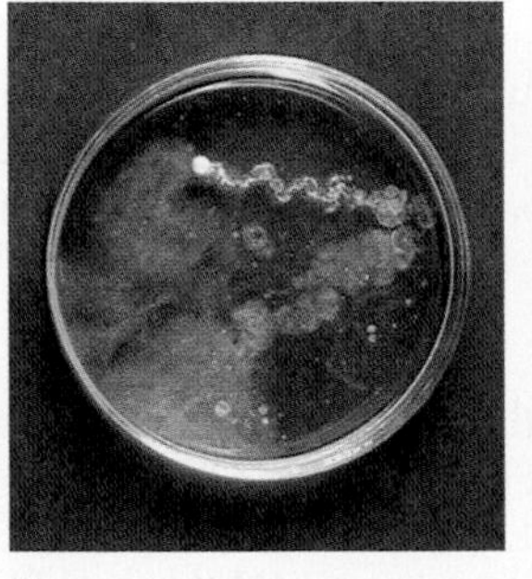
(b)

(c)

FIGURE 14.17 The alkali metals react with water, producing gaseous hydrogen and a solution of the alkali metal hydroxide. (a) Lithium reacts quietly. (b) Sodium reacts so vigorously that the heat released melts the unreacted metal. (c) Potassium reacts even more vigorously, producing so much heat that the hydrogen produced in the reaction ignites.

All alkali metals react directly with almost all nonmetals except the noble gases. However, only lithium reacts with nitrogen, which it reduces to the nitride ion:

$$6\ Li(s) + N_2(g) \longrightarrow 2\ Li_3N(s)$$

The principal product of the reaction of the alkali metals with oxygen varies systematically down the group (Fig. 14.19). It is commonly found that ionic compounds formed from cations and anions of similar radius are more stable than those formed from ions with markedly different radii. Such is the case here. Lithium forms mainly the oxide, Li_2O. Sodium, which has a larger cation, forms predominantly the very pale yellow sodium peroxide, Na_2O_2. Potassium, with an even bigger cation, forms mainly the superoxide, KO_2, which contains the superoxide ion, O_2^-. Potassium superoxide, a yellow-orange solid, is used in closed-system breathing apparatus (gas masks, submarines, and space vehicles). The KO_2 removes the exhaled water vapor and generates oxygen gas:

$$4\ KO_2(s) + 2\ H_2O(g) \longrightarrow 4\ KOH(s) + 3\ O_2(g)$$

The KOH produced in this reaction removes the exhaled CO_2:

$$KOH(s) + CO_2(g) \longrightarrow KHCO_3(s)$$

Much of the potassium currently produced is used to manufacture potassium superoxide for this purpose.

The alkali metals are usually found as singly charged cations. They react with water with increasing vigor down the group. Cations tend to form their most stable compounds with anions of similar size.

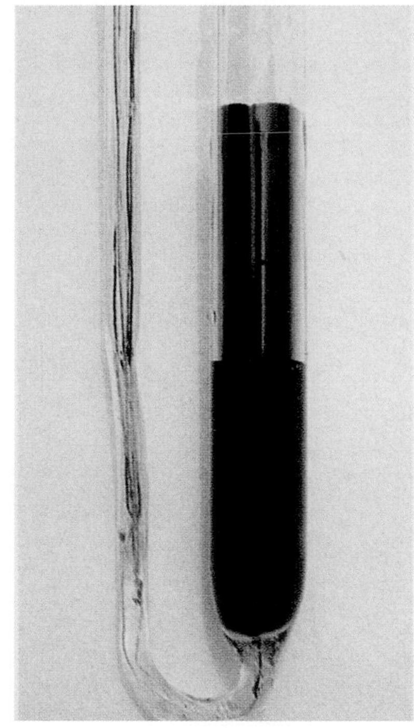

FIGURE 14.18 Sodium dissolves in liquid ammonia to form the deep blue solution, which appears nearly black, in the lower half of the tube. At higher concentrations, the metal ammonia solution becomes bronze in color, as in the top half of the tube.

FIGURE 14.19 Although the alkali metals form a mixture of products when they react with oxygen, lithium gives mainly the oxide (left), sodium the very pale yellow peroxide (center), and potassium the yellow superoxide (right).

14.9 Compounds of Lithium, Sodium, and Potassium

Lithium is typical of an element at the head of its group, in that it differs significantly from the other elements in the group. The differences stem in part from the small size of the Li^+ cation, which gives it a strong polarizing power and hence a tendency to form covalent bonds. A further consequence of the small size of a Li^+ ion is strong ion-dipole interactions, with the result that many lithium salts are hydrated in water. Lithium has a diagonal relationship with magnesium and is found in magnesium minerals, where it replaces the magnesium ion.

Lithium is not very abundant in the universe, because it is not favored in the nuclear reactions that produce the elements in the stars. Lithium compounds are used in ceramics, lubricants, and medicine. Small daily doses of lithium carbonate are an effective treatment for bipolar (manic-depressive) disorder; but scientists still do not fully understand why. Lithium soaps—the lithium salts of long-chain carboxylic acids—are used as thickeners in lubricating greases for high-temperature applications because they have higher melting points than more conventional sodium and potassium soaps.

Sodium compounds are important largely because they are inexpensive and soluble in water. Sodium chloride is readily mined as rock salt, which is a deposit of sodium chloride left as ancient oceans evaporated; it is also obtained from the evaporation of brine (salt water) from present-day seas and salt lakes (Fig. 14.20). Sodium chloride is used in large quantities in the electrolytic production of chlorine and sodium hydroxide from brine.

Sodium hydroxide, NaOH, is a soft, waxy, white, corrosive solid. It is an important industrial chemical because it is an inexpensive starting

FIGURE 14.20 An evaporation pond. The blue color is due to a dye added to the brine to increase heat absorption and hence speed up evaporation.

material for the production of other sodium salts. The amount of electricity used to electrolyze brine to produce NaOH in the chloralkali process (Section 12.15) is second only to the amount used to extract aluminum from its ores. The process produces chlorine and hydrogen as well as sodium hydroxide. The net ionic reaction is

$$2\ Cl^-(aq) + 2\ H_2O(l) \xrightarrow{\text{electrolysis}} Cl_2(g) + 2\ OH^-(aq) + H_2(g)$$

Most modern production uses a diaphragm cell, in which compartments containing steel and titanium electrodes are separated by porous diaphragms to isolate the products (Fig. 14.21).

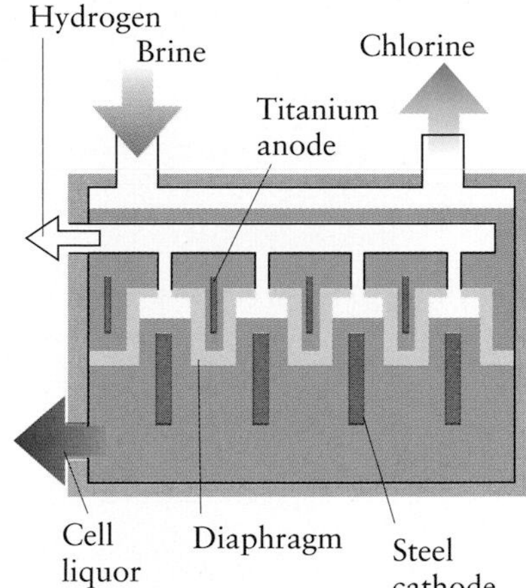

FIGURE 14.21 A diaphragm cell for the electrolytic production of sodium hydroxide from brine (aqueous sodium chloride solution), represented by the blue color. The diaphragm prevents the chlorine produced at the titanium anodes from mixing with the hydrogen and the sodium hydroxide formed at the steel cathodes. The liquid is drawn off and the water is partially evaporated. The unconverted sodium chloride crystallizes, leaving the sodium hydroxide in solution.

Sodium hydrogen carbonate, $NaHCO_3$ (sodium bicarbonate), is commonly called *bicarbonate of soda* or *baking soda*. The rising action of baking soda in bread and doughs depends on the reaction of a weak acid, HA, with the hydrogen carbonate ions:

$$HCO_3^-(aq) + HA(aq) \longrightarrow A^-(aq) + H_2O(l) + CO_2(g)$$

The release of gas causes the dough to rise. The weak acids are provided by the recipe, generally in the form of lactic acid from sour milk or buttermilk, citric acid from lemons, or the acetic acid in vinegar. *Baking powder* contains a solid weak acid as well as the hydrogen carbonate, and carbon dioxide is released when water is added (Fig. 14.22).

Sodium carbonate decahydrate, $Na_2CO_3 \cdot 10H_2O$, was once widely used as *washing soda*. It is still sometimes added to water to precipitate Mg^{2+} and Ca^{2+} ions as carbonates, as in

$$Ca^{2+}(aq) + CO_3^{2-}(aq) \longrightarrow CaCO_3(s)$$

and to provide an alkaline environment that helps to remove grease from fabrics. Anhydrous sodium carbonate, or *soda ash*, is used in large amounts in the glass industry as a source of sodium oxide, into which it decomposes when heated.

The principal mineral sources of potassium are *sylvite*, KCl, and *carnallite*, $KCl \cdot MgCl_2 \cdot 6H_2O$. Potassium chloride is incorporated directly into fertilizers as a source of essential potassium, but potassium nitrate is used instead for crops such as potatoes and tobacco, which cannot tolerate high chloride ion concentrations.

Potassium compounds are generally more expensive than the corresponding sodium compounds, but in some applications their advantages outweigh their expense. Potassium nitrate, KNO_3, releases oxygen when heated, in the reaction

$$2\ KNO_3(s) \xrightarrow{\Delta} 2\ KNO_2(s) + O_2(g)$$

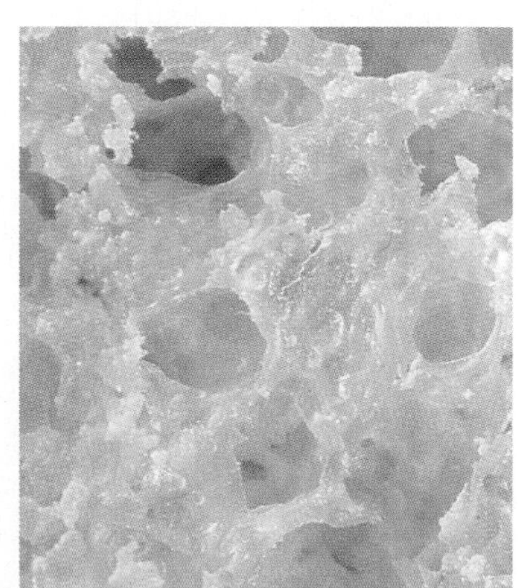

FIGURE 14.22 When moistened, double-acting baking powder generates some carbon dioxide gas that forms small cavities in the dough. The cavities are later inflated by a second release of carbon dioxide during baking.

and is used to facilitate the ignition of matches. It is less hygroscopic (water absorbing) than the corresponding sodium compounds, because the potassium cation is larger and is less strongly hydrated by H_2O molecules.

Potassium nitrate is also the oxidizing agent in black gunpowder, a mixture of 75% potassium nitrate, 15% charcoal, and 10% sulfur. It reacts as follows:

$$2\,KNO_3(s) + 4\,C(s) \longrightarrow K_2CO_3(s) + 3\,CO(g) + N_2(g)$$

$$2\,KNO_3(s) + 2\,S(s) \longrightarrow K_2SO_4(s) + SO_2(g) + N_2(g)$$

The two reactions result in the rapid formation of 6 mol of gas molecules from 10 mol of solid reactants, causing a sudden explosive expansion.

Lithium resembles magnesium, and its compounds show covalent character. Sodium compounds are soluble in water, plentiful, and inexpensive. Potassium compounds are generally less hygroscopic than sodium compounds.

Self-Test 14.3A Use Table 12.2 to determine the minimum potential that must be supplied to carry out the chloralkali process, in which NaOH and Cl_2 are obtained from aqueous NaCl (under standard conditions).

[*Answer:* 2.19 V]

Self-Test 14.3B Cesium superoxide, CsO_2, can also be used to remove exhaled CO_2 and generate oxygen from water. Explain why KO_2 is preferred for this purpose on spacecraft.

GROUP 2: THE ALKALINE EARTH METALS

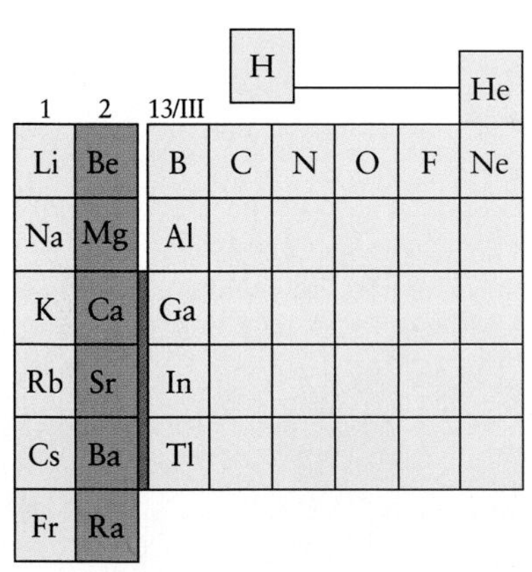

Calcium, strontium, and barium are called the *alkaline earth metals*, because their earths—the old name for oxides—are basic (alkaline). The name alkaline earth metals is often extended to all the members of Group 2 (Table 14.6).

Magnesium and calcium are by far the most important members of the group. Magnesium is, in effect, the doorway to life: it occurs in every chlorophyll molecule and hence enables photosynthesis to take place. Calcium is the element of rigidity and construction: it is the cation in the bones of our skeletons, the shells of shellfish, and the concrete, mortar, and limestone of our buildings.

14.10 The Group 2 Elements

The valence electron configuration of the atoms of the Group 2 elements is ns^2, where n is the period number. The second ionization energy is low enough to be recovered from the increased lattice enthalpy (Fig. 14.23). Hence, the Group 2 elements occur with an oxidation number of +2, as the cation M^{2+}, in all their compounds. Apart from a tendency toward nonmetallic character in beryllium, the elements have all the chemical characteristics of metals, such as forming basic oxides and hydroxides.

TABLE 14.6 Group 2 Elements

Valence configuration: ns^2
*Normal form**: soft, silver-gray metals

Z	Name	Symbol	Molar mass, $g{\cdot}mol^{-1}$	Melting point, °C	Boiling point, °C	Density, $g{\cdot}cm^{-3}$
4	beryllium	Be	9.01	1285	2470	1.85
12	magnesium	Mg	24.31	650	1100	1.74
20	calcium	Ca	40.08	840	1490	1.53
38	strontium	Sr	87.62	770	1380	2.58
56	barium	Ba	137.33	710	1640	3.59
88	radium	Ra	(226)	700	1500	5.00

**Normal form* means the state and appearance of the element at 25°C and 1 atm.

All the Group 2 elements are too reactive to occur in the uncombined state in nature (Fig. 14.24). Instead, they are generally found in compounds as doubly charged cations. The element beryllium occurs mainly as *beryl*, $3BeO{\cdot}Al_2O_3{\cdot}6SiO_2$, sometimes in crystals so big that they weigh several tons. The gemstone *emerald* is a form of beryl; its green color is caused by the Cr^{3+} ions present as impurities (Fig. 14.25). Magnesium occurs in seawater and as the mineral *dolomite*, $CaCO_3{\cdot}MgCO_3$. Calcium also occurs as $CaCO_3$ in compressed deposits of the shells of ancient marine organisms and exoskeletons of tiny one-celled organisms; these deposits include *limestone, calcite*, and *chalk* (a softer variety of calcium carbonate).

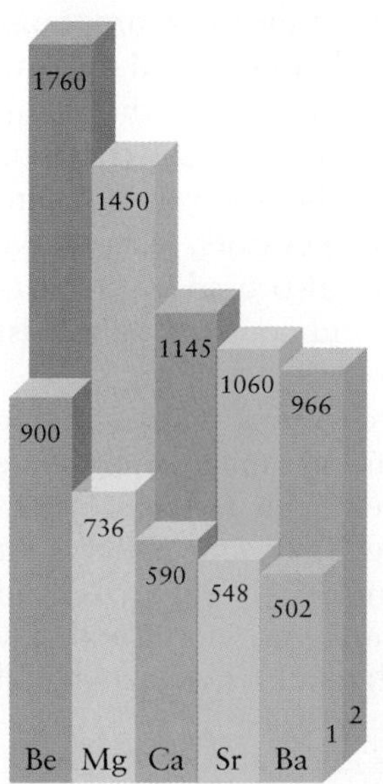

FIGURE 14.23 The first (front) and second ionization energies (in kilojoules per mole) of the Group 2 elements. Although the second ionization energies are larger than the first, they are not enormous, and both valence electrons are lost from each atom in all the compounds of these elements.

(a)

(b)

(c)

(d)

(e)

FIGURE 14.24 The elements of Group 2: (a) beryllium; (b) magnesium; (c) calcium; (d) strontium; and (e) barium. The four central elements of the group (magnesium through barium) were discovered by Humphry Davy in a single year (1808). The two outer elements were discovered later: beryllium in 1828 (by Friedrich Wöhler) and radium (which is not shown here) in 1898 (by Pierre and Marie Curie).

FIGURE 14.25 An emerald is a crystal of beryl with some Cr^{3+} ions, which are responsible for the color.

Beryllium is obtained by electrolytic reduction of molten beryllium chloride. The element's low density makes it useful for the construction of missiles and satellites. Beryllium is also used as windows for x-ray tubes: because Be atoms have so few electrons, thin sheets of the metal are transparent to x-rays and allow the rays to escape. Beryllium is added in small amounts to copper: the small Be atoms pin the Cu atoms together in an interstitial alloy that is more rigid than pure copper but still conducts electricity well. These hard, electrically conducting alloys are formed into nonsparking tools for use in oil refineries and grain elevators, where there is a risk of explosion. Beryllium-copper alloys are also used in the electronics industry to form tiny nonmagnetic parts and contacts that resist deformation and corrosion.

Metallic magnesium is produced by either chemical or electrolytic reduction of its compounds. In chemical reduction, first magnesium oxide is obtained from the decomposition of dolomite. Then ferrosilicon, an alloy of iron and silicon, is used to reduce the MgO at about 1200°C. At this temperature, the magnesium produced is immediately vaporized and carried away. The electrolytic method uses seawater as its principal raw material. The first stage is the precipitation of magnesium hydroxide with slaked lime (calcium hydroxide):

$$Mg^{2+}(aq) + Ca(OH)_2(aq) \longrightarrow Mg(OH)_2(s) + Ca^{2+}(aq)$$

The lime is produced by the thermal decomposition of calcium carbonate in shells dredged up from the ocean floor. The precipitated magnesium hydroxide is filtered off and treated with hydrochloric acid:

$$Mg(OH)_2(s) + 2\ H_3O^+(aq) \longrightarrow Mg^{2+}(aq) + 4\ H_2O(l)$$

Finally, the magnesium chloride is dried and the molten salt is added to an electrolytic cell. Magnesium is produced at the cathode and chlorine is produced at the anode:

$$MgCl_2(l) \xrightarrow{\text{electrolysis}} Mg(s) + Cl_2(g)$$

Magnesium is a silver-white metal that is protected from extensive oxidation in air by a film of white oxide, which gives the metal a dull gray coat. Its density is only two-thirds that of aluminum, and it is very soft. However, its alloys have great strength, and they are used in applications requiring lightness and toughness—in airplanes, for instance. The use of magnesium alloys in automobiles increased by a factor of 5 between 1989 and 1995 as manufacturers sought to reduce the weight of vehicles. However, don't expect to be driving a magnesium car soon: magnesium is more expensive than steel and more difficult to work. It melts at a low temperature and is deformed by the heat produced when it is machined. Water cannot be used to cool the tools because magnesium reacts with hot water to produce hydrogen.

The true alkaline earth metals—calcium, strontium, and barium—are obtained either by electrolysis or by reduction with aluminum in a version of the thermite process (see Fig. 6.12):

$$3\ BaO(s) + 2\ Al(s) \xrightarrow{\Delta} Al_2O_3(s) + 3\ Ba(s)$$

Table 14.7 summarizes some important reactions of the Group 2 elements. Magnesium burns vigorously in air with a brilliant white flame, partly because it reacts with the nitrogen and carbon dioxide in air as well as with oxygen. Because the reaction is accelerated when burning magnesium is sprayed with water or exposed to carbon dioxide, neither water nor CO_2 fire extinguishers should ever be used on a magnesium fire. They will only make the fire worse!

As in Group 1, reactions of the Group 2 metals with oxygen and water increase in vigor going down the group. Beryllium, magnesium, calcium,

TABLE 14.7 Chemical Properties of the Group 2 Metals

Reactant	Reaction with Group 2 metal (M)
hydrogen	$M(s) + H_2(g) \longrightarrow MH_2(s)$, not Be or Mg
oxygen	$2\ M(s) + O_2(g) \longrightarrow 2\ MO(s)$
nitrogen	$3\ M(s) + N_2(g) \longrightarrow M_3N_2(s)$
halogen (X_2)	$M(s) + X_2(g,l,s) \longrightarrow MX_2(s)$
water	$M(s) + 2\ H_2O(l) \longrightarrow M(OH)_2(aq) + H_2(g)$, not Be

FIGURE 14.26 Calcium reacts gently with cold water to produce hydrogen and calcium hydroxide, $Ca(OH)_2$.

and strontium are partially passivated in air by a protective surface layer of oxide. Barium, however, does not form a protective oxide and may ignite in moist air.

All the Group 2 elements with the exception of beryllium react with water; for example,

$$Ca(s) + 2\ H_2O(l) \longrightarrow Ca^{2+}(aq) + 2\ OH^-(aq) + H_2(g)$$

Beryllium does not react with water, even when red hot: its protective oxide film survives even at high temperatures. Magnesium reacts with hot water (see Fig. 13.15), and calcium reacts with cold water (Fig. 14.26). The metals reduce hydrogen ions to hydrogen gas, but neither beryllium nor magnesium dissolves in nitric acid because both become passivated by a film of oxide.

The alkaline earth metals can be detected in burning compounds by the colors they give to flames. Calcium burns orange-red, strontium crimson, and barium yellow-green. Fireworks are often made from their salts (typically nitrates and chlorates, because the anions then provide an additional supply of oxygen) together with magnesium powder. Sparklers are made of aluminum powder in place of magnesium, so that they burn more slowly.

Beryllium shows a hint of nonmetallic character, but the other Group 2 elements are all typical metals. The vigor of reaction with water and oxygen increases down the group.

14.11 Compounds of Beryllium and Magnesium

The properties of a selection of Group 2 compounds are summarized in Table 14.8. Beryllium, at the head of Group 2, resembles its diagonal neighbor aluminum in its chemical properties. It is the least metallic element of the group, and many of its compounds have properties commonly attributed to covalent bonding. Beryllium is amphoteric and reacts with both acids and alkalis. Like aluminum, beryllium reacts with water in the presence of sodium hydroxide; the products are the *beryllate ion*, $Be(OH)_4^{2-}$, and hydrogen:

$$Be(s) + 2\ OH^-(aq) + 2\ H_2O(l) \longrightarrow Be(OH)_4^{2-}(aq) + H_2(g)$$

TABLE 14.8 Properties of Group 2 Compounds

Compound	Formula*	Comment
oxides	MO	formed by decomposition of carbonates; strong bases (BeO is amphoteric), react with water to form hydroxides; withstand high temperatures
hydroxides	$M(OH)_2$	formed by action of water on oxides or by precipitation from salt solutions; sparingly soluble in water (except Ba); strong bases (Be is amphoteric)
carbonates	MCO_3	very slightly soluble in water; most decompose into oxides when heated
hydrogen carbonates	$M(HCO_3)_2$	unstable as solids; more soluble than carbonates
nitrates	$M(NO_3)_2$	decompose when heated; soluble in water

*M stands for any Group 2 metal.

Beryllium compounds are very toxic and must be handled with great caution. Their properties are dominated by the highly polarizing character of the Be^{2+} ion and its small size. The strong polarizing power results in moderately covalent compounds, and its small size limits to four the number of groups that can attach to the ion. These two features together are responsible for the prominence of the tetrahedral BeX_4 unit (**2**), like that in the beryllate ion. A tetrahedral unit is also found in the solid chloride (**3**) and hydride (**4**). The chloride is made by the action of chlorine on the oxide in the presence of carbon:

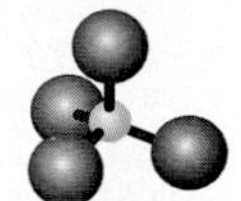

2 BeX_4 unit

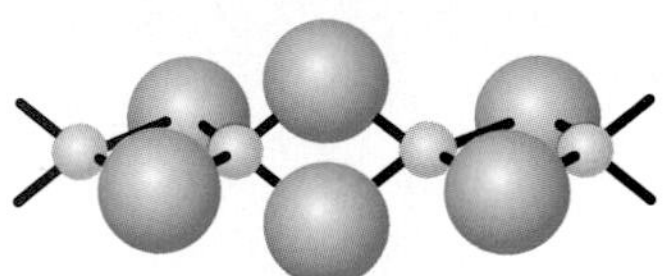

3 Beryllium chloride, $BeCl_2$

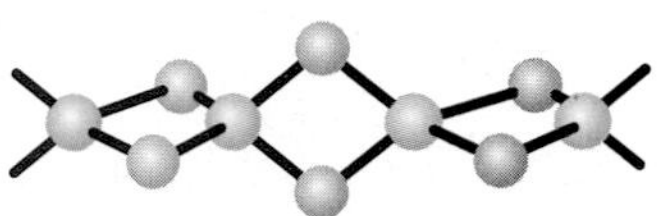

4 Beryllium hydride, BeH_2

$$BeO(s) + C(s) + Cl_2(g) \xrightarrow{600\text{–}800°C} BeCl_2(g) + CO(g)$$

The Be atoms in $BeCl_2$ act as Lewis acids and accept electron pairs from the Cl atoms of the neighboring linear $BeCl_2$ groups, forming a chain of tetrahedral $BeCl_4$ units in the solid.

Magnesium has more pronounced metallic properties than does beryllium; its compounds are primarily ionic, with some covalent character. Magnesium oxide, MgO, is formed when magnesium burns in air, but the product is contaminated by magnesium nitride. To prepare the pure oxide, the hydroxide or the carbonate is heated. Magnesium oxide dissolves only very slightly in water. One of its most striking properties is that is *refractory* (able to withstand high temperatures), for it melts only at 2800°C. This high stability can be traced to the small ionic radii of the Mg^{2+} and O^{2-} ions, and hence to their very strong electrostatic interaction with each other. The oxide has two other useful characteristics: it conducts heat very well, and it conducts electricity poorly. All three properties lead to its use as an insulator in electric heaters.

Magnesium hydroxide, $Mg(OH)_2$, is a base. It is not very soluble in water but forms instead a white colloidal suspension, a mist of small particles dispersed through a liquid (see Section 15.13), which is known as *milk of magnesia* and used as a stomach antacid. Because this base is relatively insoluble, it is not absorbed from the stomach but remains to act on whatever acids are present. Hydroxides have the advantage over hydrogen carbonates (which are also used as antacids) in that their neutralization does not lead to the formation of carbon dioxide and its consequence, belching. Milk of magnesia must be used sparingly as an antacid, because the salt produced by the neutralization of milk of magnesia in the stomach is magnesium chloride, which can act as a purgative. Magnesium sulfate, as *Epsom salts*, $MgSO_4 \cdot 7H_2O$, is another common purgative. The magnesium ions inhibit the absorption of water from the intestine. The resulting increased flow of water through the intestine triggers the mechanism that results in defecation.

Arguably the most important compound of magnesium is chlorophyll. This green organic compound consists of large molecules that capture light from the Sun and channel its energy into photosynthesis. One function of the Mg^{2+} ion, which lies just above the plane of the ring (**5**), appears to be to keep the ring rigid. This rigidity helps to ensure that the energy captured from the incoming photon is not lost as heat before it has been used to bring about a chemical reaction. Magnesium also plays an important role in energy generation in living cells. For example, it is involved in the contraction of muscles.

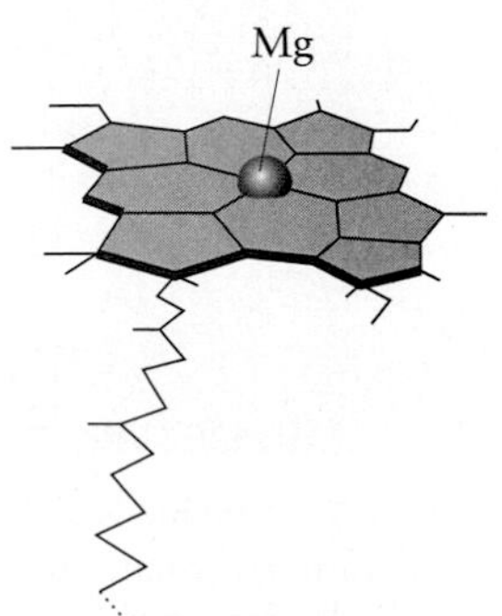

5 Chlorophyll

Beryllium compounds have a pronounced covalent character, and the structural unit is commonly tetrahedral. The small size of the magnesium cation results in a thermally stable oxide with low solubility in water.

14.12 Compounds of Calcium

The most common compound of calcium is calcium carbonate, $CaCO_3$, which occurs naturally as chalk and limestone. Marble is a dense form of calcium carbonate that can be given a high polish; it is often colored by impurities, most commonly iron cations (Fig. 14.27). The two most common forms of pure calcium carbonate are calcite and aragonite. All these carbonates are the fossilized remains of marine life. Calcium carbonate decomposes to calcium oxide, CaO, or *quicklime*, when heated:

$$CaCO_3(s) \xrightarrow{\Delta} CaO(s) + CO_2(g)$$

Calcium oxide is called quicklime because it reacts so exothermically and rapidly with water:

$$CaO(s) + H_2O(l) \longrightarrow Ca^{2+}(aq) + 2\ OH^-(aq)$$

The product, calcium hydroxide, is commonly known as *slaked lime* because, as calcium hydroxide, the thirst of lime for water has been quenched (slaked). Slaked lime is the form in which lime is normally sold because quicklime can set fire to moist wood and paper. In fact, the wooden boats that were once used to transport quicklime sometimes caught fire in the heat of reaction when water seeped into their holds. An aqueous solution of calcium hydroxide, which is sparingly soluble in water, is called *lime water*. It is used as a test for carbon dioxide, with which it reacts to form a suspension of the even less soluble calcium carbonate:

$$Ca(OH)_2(aq) + CO_2(g) \longrightarrow CaCO_3(s) + H_2O(l)$$

Quicklime is produced in enormous quantities throughout the world. About 40% of this output is used in metallurgy. In ironmaking (Section 16.14), it is used as a Lewis base; its O^{2-} ion reacts with silica, SiO_2, impurities in the ore to form a liquid *slag*:

$$CaO(s) + SiO_2(s) \xrightarrow{\Delta} CaSiO_3(l)$$

About 50 kg of lime is needed to produce 1 tonne (10^3 kg) of iron.

Slaked lime is used as an inexpensive base in industry, as well as to adjust the pH of soils in agriculture. Perhaps surprisingly, it is also used to *remove* Ca^{2+} ions from hard water containing $Ca(HCO_3)_2$. Its role here is to convert HCO_3^- to CO_3^{2-} by providing OH^- ions:

$$HCO_3^-(aq) + OH^-(aq) \longrightarrow CO_3^{2-}(aq) + H_2O(l)$$

FIGURE 14.27 Marble is a dense form of calcium carbonate. It is often colored by impurities, such as iron ions.

As the concentration of CO_3^{2-} ions rises, Ca^{2+} ions precipitate in the reaction

$$Ca^{2+}(aq) + CO_3^{2-}(aq) \longrightarrow CaCO_3(aq)$$

This reaction removes the Ca^{2+} ions that were present initially and those that were added as lime; overall, the Ca^{2+} ion concentration is reduced.

A reaction that acts as a bridge between inorganic and organic compounds is the conversion of calcium oxide into calcium carbide:

$$CaO(s) + 3\ C(s) \longrightarrow CaC_2(s) + CO(g)$$

The carbide ion ($:C{\equiv}C:^{2-}$) is a Brønsted base and is readily protonated by water to form the organic compound ethyne (acetylene):

Carbides are treated at greater length in Section 14.22.

$$CaC_2(s) + 2\ H_2O(l) \longrightarrow C_2H_2(g) + Ca(OH)_2(aq)$$

This reaction was once used to provide a portable source of fuel for lamps. Water was allowed to drip onto chunks of calcium carbide, and the acetylene produced was burned.

As we have seen, calcium compounds are often used as structural materials in organisms, buildings, and civil engineering. The rigidity of calcium compounds stems from the strength with which the small, highly charged Ca^{2+} cation interacts with its neighbors. *Concrete* is a strong building material that consists of a binder and a filler. The filler is usually gravel, but sometimes polymer or vermiculite pellets are added to lower the density. The binder is *cement*, usually *Portland cement*, which is made by heating a mixture of crushed limestone, clay or shale, sand, and oxides such as iron ore in a kiln. The hard pellets that result are called "clinkers." The clinkers contain many compounds but can be thought of as a mixture of calcium oxide, calcium silicates, and calcium aluminum silicates. The pellets are ground together with *gypsum*, $CaSO_4{\cdot}2H_2O$, into a powder that sets into a hard mass when mixed with water. The water reacts with the mixture to produce hydrates and hydroxides that bind the salts together into a three-dimensional network. *Mortar* consists of about one part cement and three parts sand (largely silica, SiO_2). It sets to a hard mass as the lime reacts with the carbon dioxide of the air to form the carbonate (Fig. 14.28).

FIGURE 14.28 An electron micrograph of the surface of mortar, showing the growth of tiny interlocking crystals as carbon dioxide reacts with calcium oxide and silica.

Calcium is also found in the rigid structural components of living things, either as the calcium carbonate of the shells of shellfish or the calcium phosphate of bone. About a kilogram of calcium is present in an adult human body, mostly in the form of insoluble calcium phosphate, but also as Ca^{2+} ions in other fluids inside our cells. The calcium in newly formed bone is in dynamic equilibrium with the calcium ions in the body fluids, so calcium must be part of our daily diet to maintain bone strength.

Tooth enamel is a *hydroxyapatite*, $Ca_5(PO_4)_3OH$. Tooth decay begins when acids attack the enamel:

$$Ca_5(PO_4)_3OH(s) + 4\ H_3O^+(aq) \longrightarrow 5\ Ca^{2+}(aq) + 3\ HPO_4^{2-}(aq) + 5\ H_2O(l)$$

The principal agents of tooth decay are the carboxylic acids produced when bacteria act on the remains of food. A more resistant coating forms when the OH^- ions in the apatite are replaced by F^- ions. The resulting mineral is called *fluorapatite:*

$$Ca_5(PO_4)_3OH(s) + F^-(aq) \longrightarrow Ca_5(PO_4)_3F(s) + OH^-(aq)$$

The addition of fluoride ions to domestic water supplies (in the form of NaF) is now widespread and has resulted in a dramatic decrease in dental cavities. Fluoridated toothpastes, containing either tin(II) fluoride or sodium monofluorophosphate (MFP, Na_2FPO_3), are also recommended to strengthen tooth enamel.

Calcium compounds are common in structural materials because the small, highly charged Ca^{2+} ion results in rigid structures.

Self-Test 14.4A Explain why beryllium compounds have covalent characteristics.

[*Answer:* The small size and high charge on the beryllium ion make it highly polarizing.]

Self-Test 14.4B Explain why calcium compounds are such good construction materials.

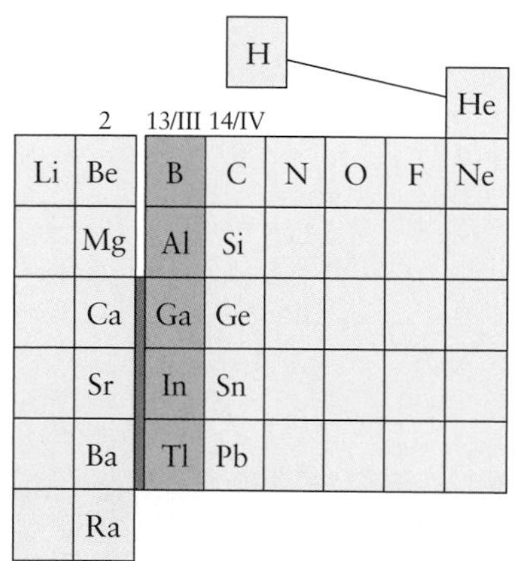

GROUP 13/III: THE BORON FAMILY

Now we move into the *p* block of the periodic table and encounter the complex but fascinating world of the nonmetals. Here, close to the center of the periodic table, we meet strange properties, for the elements are neither so electropositive that they easily lose electrons nor so electronegative that they easily gain them.

Group 13/III is the first group of the *p* block. Its members have an ns^2np^1 electron configuration (Table 14.9), so we expect a maximum oxidation number of +3. The oxidation numbers of B and A1 are +3 in almost all their compounds (Table 14.10). However, the heavier elements

TABLE 14.9 Group 13/III Elements

Valence configuration: ns^2np^1

Z	Name	Symbol	Molar mass, $g \cdot mol^{-1}$	Melting point, °C	Boiling point, °C	Density, $g \cdot cm^{-3}$	Normal form*
5	boron	B	10.81	2300	3931	2.47	powdery brown metalloid
13	aluminum	Al	26.98	660	2467	2.70	silver-white metal
31	gallium	Ga	69.72	30	2403	5.91	silver metal
49	indium	In	114.82	156	2080	7.29	silver-white metal
81	thallium	Tl	204.38	304	1457	11.87	soft metal

**Normal form* means the state and appearance of the element at 25°C and 1 atm.

TABLE 14.10 Chemical Properties of the Group 13/III Elements

Reactant	Reaction with Group 13/III element (E)
oxygen	$4\ E(s) + 3\ O_2(g) \longrightarrow 2\ E_2O_3(s)$
nitrogen	$2\ E(s) + N_2(g) \longrightarrow 2\ EN(s)$, E = B, Al
halogen (X_2)	$2\ B(s) + 3\ X_2(g,l,s) \longrightarrow 2\ BX_3(g)$
	$2\ E(s) + 3\ X_2(g,l,s) \longrightarrow E_2X_6(g)$, E = Al, Ga, In
	$2\ Tl(s) + X_2(g,l,s) \longrightarrow 2\ TlX(s)$
water	$2\ Tl(s) + 2\ H_2O(l) \longrightarrow 2\ TlOH(aq) + H_2(g)$
acid	$2\ E(s) + 6\ H_3O^+(aq) \longrightarrow 2\ E^{3+}(aq) + 6\ H_2O(l) + 3\ H_2(g)$, E = Al, Ga, Tl
base	$2\ E(s) + 6\ H_2O(l) + 2\ OH^-(aq) \longrightarrow 2\ E(OH)_4^-(aq) + 3\ H_2(g)$, E = Al, Ga

in the group are more likely to keep their *s*-electrons (the inert-pair effect, Section 1.18), so the oxidation number +1 becomes increasingly important down the group, and thallium(I) compounds are as common as thallium(III) compounds. We shall concentrate on the two most important members of the group, boron and aluminum.

14.13 The Group 13/III Elements

Boron forms perhaps the most extraordinary structures of all the elements. It has a relatively high ionization energy and is a metalloid that forms covalent bonds, like its diagonal neighbor silicon. However, because it has only three electrons in its valence shell and has a small atomic radius, it forms unusual compounds. Some of its compounds have incomplete octets (Section 2.12) and others are electron deficient (Section 3.9). These unusual bonding characteristics lead to the remarkable properties that have made boron an essential element of modern technology.

Boron is mined as *borax* and *kernite*, $Na_2B_4O_7{\cdot}xH_2O$, with $x = 10$ and 4, respectively. Large deposits from ancient hot springs are found in volcanic regions, such as the Mojave Desert region of California. In the extraction process, the ore is converted into boron oxide with acid and then reduced with magnesium to an impure brown, amorphous form of boron:

$$B_2O_3(s) + 3\ Mg(s) \xrightarrow{\Delta} 2\ B(s) + 3\ MgO(s)$$

A purer product is obtained by reducing a volatile boron compound, such as BCl_3 or BBr_3, with hydrogen on a heated filament (tantalum is used for the filament because it has a very high melting point):

$$2\ BBr_3(g) + 3\ H_2(g) \longrightarrow 2\ B(s) + 6\ HBr(g)$$

Boron production remains quite low despite the element's desirable properties of hardness and low density.

Here, perhaps, is an opportunity for another young chemist like Hall (see below) to transform the production of boron as Hall did for aluminum.

Elemental boron exists in a variety of allotropic forms in which each atom strives to find ways to share eight electrons despite being so small and having only three electrons to contribute. It is most commonly found as either a gray-black nonmetallic, high-melting-point solid or a dark

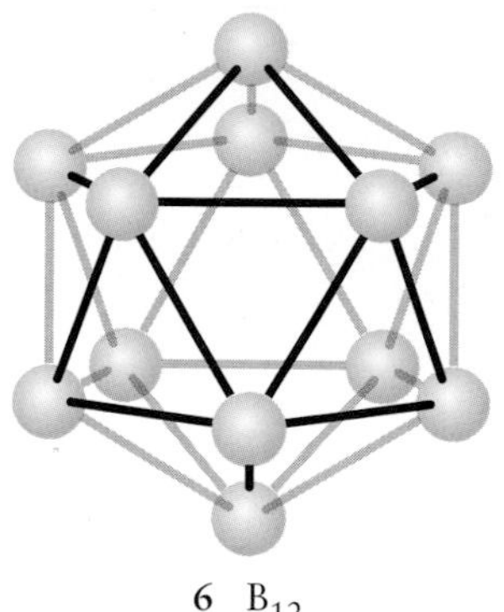

6 B_{12}

brown powder with an icosahedral (twenty-faced) structure based on clusters of 12 atoms (6). Because of the three-dimensional network formed by these bonds, boron is very hard. When boron fibers are incorporated in plastics, the result is a very tough material that is stiffer than steel yet lighter than aluminum; this material is used in aircraft, missiles, and body armor. The element is attacked by only the strongest oxidizing agents.

Aluminum is the most abundant metallic element in the Earth's crust and, following oxygen and silicon, the third most abundant element (see Fig. 14.1). However, the aluminum content in most minerals is low, and the commercial source of aluminum, *bauxite*, is a hydrated, impure oxide, $Al_2O_3{\cdot}xH_2O$, where x can range up to 3. The bauxite ore, which is red from the iron oxides it contains (Fig. 14.29), is processed to obtain alumina, Al_2O_3, in the *Bayer process*. In this process, the ore is first treated with aqueous sodium hydroxide, which dissolves the amphoteric alumina as the aluminate ion, $Al(OH)_4^-(aq)$:

$$Al_2O_3{\cdot}xH_2O(s) + 2\ OH^-(aq) + 3\ H_2O(l) \xrightarrow{170°C} 2\ Al(OH)_4^-(aq) + x\ H_2O(l)$$

The insoluble iron and titanium oxides react with water and precipitate in a "red mud," which is filtered off:

$$Fe_2O_3{\cdot}xH_2O(s) + 3\ H_2O(l) \xrightarrow{170°C} 2\ Fe(OH)_3(s) + x\ H_2O(l)$$

Carbon dioxide is then bubbled through the solution of aluminate ions. The CO_2 reacts with the OH^- ions, which lowers the pH of the solution, causing a partial decomposition of the aluminate ions and the precipitation of aluminum hydroxide:

$$Al(OH)_4^-(aq) + CO_2(g) \longrightarrow Al(OH)_3(s) + HCO_3^-(aq)$$

The aluminum hydroxide is removed and dehydrated to the oxide by heating to 1200°C:

$$2\ Al(OH)_3(s) \xrightarrow{1200°C} Al_2O_3(s) + 3\ H_2O(g)$$

FIGURE 14.29 The iron oxides in bauxite ore give it a red tint.

Obtaining aluminum metal from the oxide was a challenge to early scientists and engineers. When it was first isolated, aluminum was a rare and expensive metal. During the nineteenth century, it so symbolized modern technology that the Washington Monument was given an expensive aluminum tip. That rarity and expense was transformed by electrochemistry. Aluminum metal is now obtained on a huge scale by the *Hall process*. In 1886, Charles Hall discovered that by mixing the mineral *cryolite*, Na_3AlF_6, with alumina he got a mixture that melted at a much more economical temperature, 950°C, instead of the 2050°C of pure alumina. The melt is electrolyzed in a cell that uses graphite (or carbonized petroleum) anodes and a carbonized steel-lined vat that serves as the cathode (Fig. 14.30). The electrolysis half-reactions are

Napoleon reserved his aluminum plates for his special guests: the rest had to make do with gold.

$$\text{Cathode reaction:}\quad Al^{3+}(\text{melt}) + 3\,e^- \longrightarrow Al(l)$$

$$\text{Anode reaction:}\quad 2\,O^{2-}(\text{melt}) + C(s, gr) \longrightarrow CO_2(g) + 4\,e^-$$

The overall reaction is

$$4\,Al^{3+}(\text{melt}) + 6\,O^{2-}(\text{melt}) + 3\,C(s, gr) \longrightarrow 4\,Al(l) + 3\,CO_2(g)$$

Note that the carbon electrode takes part in the reaction. From the reaction stoichiometry, we can calculate that a current of 1 A must flow for 80 h to produce 1 mol Al (27 g of aluminum, about enough for two soft-drink cans). The very high energy consumption can be greatly reduced by recycling, which requires less than 5% of the electricity needed to extract aluminum from bauxite. Although a 12-ounce beverage can contains less than 14 g of aluminum, the energy that we are throwing away when we discard an aluminum can is equivalent to burning the amount of gasoline that would fill half the can. Note also that the production of 1 tonne of aluminum from alumina is accompanied by the release of more than 1 tonne of carbon dioxide into the atmosphere.

Aluminum has a low density; it is a strong metal and an excellent electrical conductor. Although it is strongly reducing and therefore easily oxidized, aluminum is resistant to corrosion because its surface is passivated in air by a stable oxide film. The thickness of the oxide layer can be increased by making aluminum the anode of an electrolytic cell; the result is called *anodized aluminum*. Dyes may be added to the dilute sulfuric acid electrolyte used in the anodizing process to produce surface layers with different colors.

Aluminum's low density, wide availability, and corrosion resistance make it ideal for construction and for the aerospace industry. Aluminum is a soft metal, so it is usually alloyed with copper and silicon for greater strength. Its lightness and good electrical conductivity have also led to its use for overhead power lines, and its negative electrode potential has led to its use in fuel cells. Perhaps one day your automobile will not only be made of aluminum but fueled by it too.

Boron, at the head of Group 13/III, is best regarded as a nonmetal in most of its chemical properties. It has acidic oxides and forms an interesting and extensive range of binary molecular hydrides. Metallic character increases down the group, and even boron's immediate neighbor aluminum

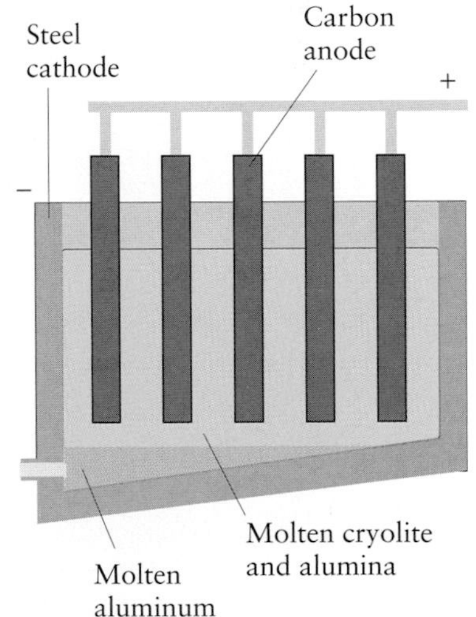

FIGURE 14.30 In the Hall process, aluminum oxide is dissolved in molten cryolite and the mixture is electrolyzed in a cell with carbon anodes and a steel cathode. The molten aluminum flows out of the bottom of the cell.

is a metal. Nonetheless, aluminum is sufficiently far to the right in the periodic table to show a hint of nonmetallic character. Aluminum is amphoteric, reacting both with nonoxidizing acids (such as hydrochloric acid) to form aluminum ions (see Fig. 12.9):

An oxidizing acid is an oxoacid in which the anion is an oxidizing agent; examples are HNO_3 and $HClO_3$.

$$2\ Al(s) + 6\ H^+(aq) \longrightarrow 2\ Al^{3+}(aq) + 3\ H_2(g)$$

and with hot aqueous alkali to form aluminate ions:

$$2\ Al(s) + 2\ OH^-(aq) + 6\ H_2O(l) \longrightarrow 2\ Al(OH)_4^-(aq) + 3\ H_2(g)$$

Boron is a hard metalloid with pronounced nonmetallic properties. Aluminum is a light, strong, amphoteric, reactive metallic element with a surface that becomes passivated when exposed to air.

14.14 Group 13/III Oxides

Boron, a metalloid with largely nonmetallic properties, has acidic oxides. Aluminum, its metallic neighbor, has amphoteric oxides (like its diagonal neighbor in Group 2, beryllium). The oxides of both elements are important in their own right, as sources of the elements and as the starting point for the manufacture of other compounds.

Boric acid, $B(OH)_3$, is a white solid that melts at 171°C. It is toxic to bacteria and many insects as well as humans and has long been used as a mild antiseptic and pesticide. Because the boron atom in $B(OH)_3$ has an incomplete octet, it can act as a Lewis acid and form a bond by accepting a lone pair of electrons from an H_2O molecule acting as a Lewis base:

$$(OH)_3B + :OH_2 \longrightarrow (OH)_3B{-}OH_2$$

The compound so formed is a weak *mono*protic acid:

$$B(OH)_3OH_2(aq) + H_2O(l) \rightleftharpoons H_3O^+(aq) + B(OH)_4^-(aq) \qquad pK_a = 9.14$$

The major use of boric acid is as the starting material for its anhydride, boron oxide, B_2O_3. Because it melts (at 450°C) to a liquid that dissolves many metal oxides, boron oxide (often as the acid) is used as a *flux*, a substance that cleans metals as they are soldered or welded. Boron oxide is also used to make fiberglass and borosilicate glass, a glass with a very low thermal expansion, such as Pyrex.

Aluminum oxide, Al_2O_3, is known almost universally as *alumina*. It exists with a variety of crystal structures, many of which form important ceramic materials (see Section 14.24). As α-alumina, it is the very hard, stable, crystalline substance *corundum*; impure microcrystalline corundum is the purple-black abrasive known as *emery*. Some impure forms of alumina are beautiful, rare, and highly prized (Fig. 14.31). A less dense and more reactive form of the oxide is γ-alumina. This form absorbs water and is used as the stationary phase in chromatography.

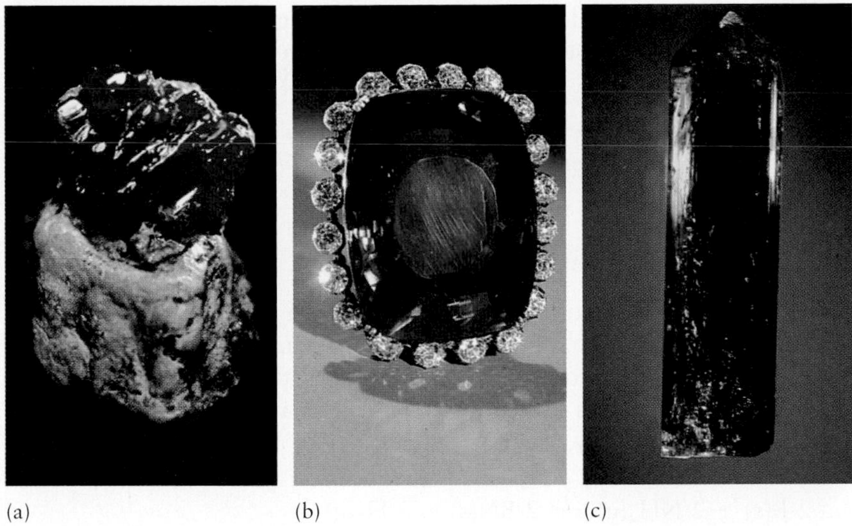

FIGURE 14.31 Some of the impure forms of α-alumina are prized as gems. (a) Ruby is alumina with Cr^{3+} in place of some Al^{3+} ions. (b) Sapphire is alumina with Fe^{3+} and Ti^{4+} impurities. (c) Topaz is alumina with Fe^{3+} impurities.

γ-Alumina is produced by heating aluminum hydroxide. It is moderately reactive and is amphoteric, dissolving readily in bases to produce the aluminate ion and in acids to produce the hydrated Al^{3+} ion:

$$Al_2O_3(s) + 2\ OH^-(aq) + 3\ H_2O(l) \longrightarrow 2\ Al(OH)_4^-(aq)$$

$$Al_2O_3(s) + 6\ H_3O^+(aq) + 3\ H_2O(l) \longrightarrow 2\ Al(H_2O)_6^{3+}(aq)$$

The strong polarizing effect of the small, highly charged Al^{3+} ion on the water molecules around it gives the $Al(H_2O)_6^{3+}$ ion acidic properties:

$$Al(H_2O)_6^{3+}(aq) + H_2O(l) \rightleftharpoons H_3O^+(aq) + Al(OH)(H_2O)_5^{2+}(aq)$$

One of the most important aluminum salts prepared by the action of an acid on alumina is aluminum sulfate, $Al_2(SO_4)_3$:

$$Al_2O_3(s) + 3\ H_2SO_4(aq) \longrightarrow Al_2(SO_4)_3(aq) + 3\ H_2O(l)$$

Aluminum sulfate is called *papermaker's alum* and is used in the paper industry to coagulate cellulose fibers into a hard, nonabsorbent surface. True alums (from which aluminum takes its name) are mixed sulfates of formula $M^+M'^{3+}(SO_4)_2\cdot 12H_2O$. They include potassium alum, $KAl(SO_4)_2\cdot 12H_2O$ (which is used in water and sewage treatment), and ammonium alum, $NH_4Al(SO_4)_2\cdot 12H_2O$ (which is used for pickling cucumbers and as the Brønsted acid in baking powders). Other alums are used for waterproofing fabrics and as *mordants*, which are compounds that help attach dyes to textiles.

Sodium aluminate, $NaAl(OH)_4$, is used along with aluminum sulfate in water purification. When mixed with aluminate ions, the acidic hydrated Al^{3+} cation from the aluminum sulfate produces aluminum hydroxide:

$$Al^{3+}(aq) + 3\ Al(OH)_4^-(aq) \longrightarrow 4\ Al(OH)_3(s)$$

FIGURE 14.32 Aluminum hydroxide, $Al(OH)_3$, forms as a white, fluffy precipitate. The fluffy form of the solid captures impurities and is used in the purification of water.

The aluminum hydroxide is formed as a fluffy, gelatinous network that entraps impurities as it settles, and this precipitate can be removed by filtration (Fig. 14.32).

Boron oxide is an acid anhydride. Aluminum shows some nonmetallic character in that its oxide is amphoteric.

14.15 Carbides, Nitrides, and Halides

When boron is heated to high temperatures with carbon, it forms boron carbide, $B_{12}C_3$, a solid that has a high melting point and is almost as hard as diamond. The solid consists of B_{12} groups that are pinned together by C atoms. When boron is heated to white heat in ammonia, boron nitride, BN, is formed as a fluffy, slippery powder:

$$2\ B(s) + 2\ NH_3(g) \xrightarrow{\Delta} 2\ BN(s) + 3\ H_2(g)$$

Its structure resembles that of graphite, but the latter's flat planes of carbon hexagons are replaced in boron nitride by planes of hexagons of alternating B and N atoms (Fig. 14.33). Unlike graphite, boron nitride is white and does not conduct electricity. Under high pressure, boron nitride is converted to a very hard, diamondlike crystalline form called Borazon. In recent years, boron nitride nanotubes similar to those formed by carbon have been synthesized (Section 14.18), and they have been found to be semiconducting (see Box 14.1).

The boron halides are made either by direct reaction of the elements at a high temperature or from boron oxide. The most important is boron trifluoride, BF_3, an industrial catalyst produced by the reaction between boron oxide, calcium fluoride, and sulfuric acid:

$$B_2O_3(s) + 3\ CaF_2(s) + 3\ H_2SO_4(l) \xrightarrow{\Delta} 2\ BF_3(g) + 3\ CaSO_4(s) + 3\ H_2O(l)$$

Boron trichloride, BCl_3, which is also widely used as a catalyst, is produced commercially by the action of chlorine gas on the oxide in the presence of carbon:

$$B_2O_3(s) + 3\ C(s) + 3\ Cl_2(g) \xrightarrow{500°C} 2\ BCl_3(g) + 3\ CO(g)$$

The B atom has an incomplete octet in all its trihalides. The compounds consist of trigonal planar molecules with an empty 2*p*-orbital perpendicular to the molecular plane. The empty orbital allows the molecules to act as Lewis acids, which accounts for the catalytic action of BF_3 and BCl_3.

Aluminum chloride, $AlCl_3$, another major industrial catalyst, is made by the action of chlorine on aluminum or on alumina in the presence of carbon:

$$2\ Al(s) + 3\ Cl_2(g) \longrightarrow 2\ AlCl_3(s)$$

$$Al_2O_3(s) + 3\ C(s) + 3\ Cl_2(g) \longrightarrow 2\ AlCl_3(s) + 3\ CO(g)$$

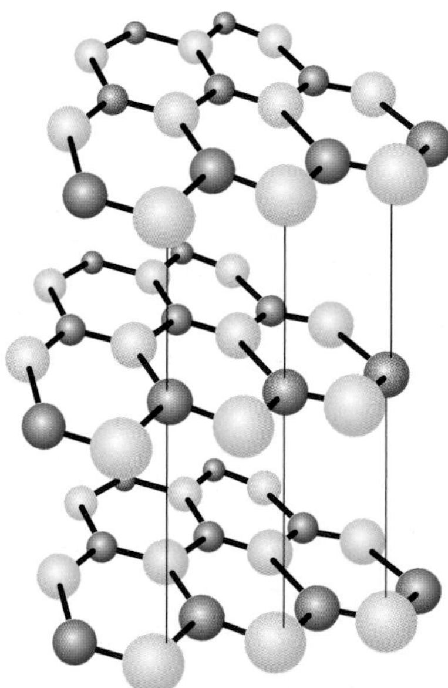

FIGURE 14.33 The structure of boron nitride, BN, resembles that of graphite, consisting of flat planes of hexagons. In boron nitride, however, the hexagons consist of alternating B and N atoms (in place of C atoms) and are stacked differently. (Compare with Fig. 14.36.)

Aluminum chloride is an ionic solid in which each Al^{3+} ion is surrounded by six Cl^- ions. However, it sublimes at 192°C to a vapor of Al_2Cl_6 molecules (7). In this molecule, the Al atom of one $AlCl_3$ fragment acts as a Lewis acid and accepts an electron pair from a Cl atom of the other $AlCl_3$ fragment, which is acting as a Lewis base.

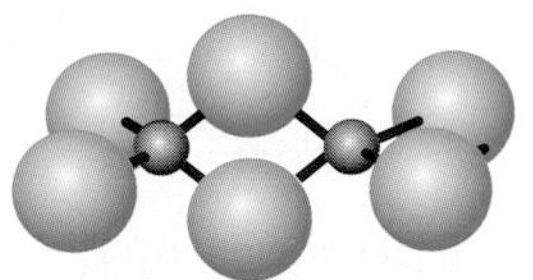

7 Aluminum chloride dimer, Al_2Cl_6

Aluminum halides react with water with a considerable evolution of heat. When the anhydrous chloride is exposed to moist air, it produces fumes of hydrochloric acid (Fig. 14.34). Aluminum chloride hexahydrate, $AlCl_3{\cdot}6H_2O$, the chloride salt of $Al(H_2O)_6{}^{3+}$, is prepared by dissolving alumina in concentrated hydrochloric acid. It is used as a deodorant and antiperspirant, one of its roles being to kill the bacteria that feed on perspiration and convert it into unpleasant-smelling compounds. Unlike most other hydrates, the hexahydrate does not form the anhydrous chloride when it is heated. The Al—O bonds are too strong to be replaced by Al—Cl bonds, so instead, heating produces the oxide:

$$2\ AlCl_3{\cdot}6\ H_2O(s) \longrightarrow Al_2O_3(s) + 6\ HCl(g) + 9\ H_2O(g)$$

Boron and aluminum halides have incomplete octets and act as Lewis acids.

14.16 Boranes, Borohydrides, and Borides

Boron forms a remarkable series of binary compounds with hydrogen—the *boranes*. These compounds include diborane, B_2H_6, and more complex compounds such as decaborane, $B_{10}H_{14}$. Anionic versions of these compounds, the *borohydrides*, are also known; the most important is $BH_4{}^-$ as sodium borohydride, $NaBH_4$.

Sodium borohydride is a white crystalline solid produced from the reaction between sodium hydride and boron trichloride dissolved in a nonaqueous solvent:

$$4\ NaH + BCl_3 \longrightarrow NaBH_4 + 3\ NaCl$$

Sodium borohydride is a very useful reducing agent. At pH = 14 (strongly alkaline conditions), the potential of the half-reaction

$$H_2BO_3{}^-(aq) + 5\ H_2O(l) + 8\ e^- \longrightarrow BH_4{}^-(aq) + 8\ OH^-(aq)$$

is −1.24 V. Because this potential is well below that of the Ni^{2+}/Ni couple (−0.23 V), borohydride ions can reduce Ni^{2+} ions to metallic nickel. This reduction is the basis of the "chemical plating" of nickel. The advantage of this chemical plating over electroplating is that the item being plated does not have to be an electrical conductor.

FIGURE 14.34 When anhydrous aluminum chloride is left exposed to moist air, it reacts to form hydrochloric acid. Here a white "smoke" of ammonium chloride particles forms as the hydrochloric acid fumes react with ammonia released in the vicinity.

The boranes are an extensive series of binary compounds of boron and hydrogen, somewhat analogous to the hydrocarbons. The starting point for borane production is the reaction (in an organic solvent) of sodium borohydride with boron trifluoride:

$$4\ BF_3 + 3\ BH_4{}^- \longrightarrow 3\ BF_4{}^- + 2\ B_2H_6$$

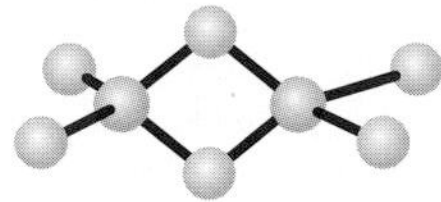

8 Diborane, B_2H_6

The product B_2H_6 is diborane (8), a colorless gas that bursts into flame in air. On contact with water, it immediately reduces the hydrogen in the water:

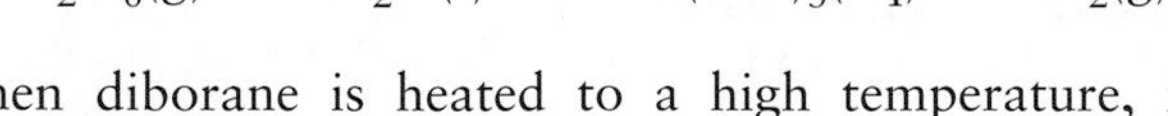

$$B_2H_6(g) + 6\ H_2O(l) \longrightarrow 2\ B(OH)_3(aq) + 6\ H_2(g)$$

When diborane is heated to a high temperature, it decomposes into hydrogen and pure boron:

$$B_2H_6(g) \xrightarrow{\Delta} 2\ B(s) + 3\ H_2(g)$$

This sequence of reactions is a useful route to the pure element, but more complex boranes form when the heating is less severe. When diborane is heated to 100°C, for instance, it forms decaborane, $B_{10}H_{14}$, a solid that melts at 100°C. Decaborane is stable in air, is oxidized by water only slowly, and is an example of the general rule that heavier boranes are less flammable than boranes of low molar mass.

The boranes are electron-deficient compounds (Section 3.9): we cannot write valid Lewis structures for them because too few electrons are available. For instance, there are 8 atoms in diborane, so we need at least 7 bonds; however, there are only 12 valence electrons, so we can form at most 6 electron-pair bonds. In molecular orbital theory, these electron pairs are regarded as delocalized over the entire molecule, and their bonding power is shared by several atoms. In diborane, for instance, a single electron pair is delocalized over a B—H—B unit. It binds all three atoms together with bond orders of $\frac{1}{2}$ for each of the B—H bridging bonds. The molecule has two such bridging **three-center bonds** (9).

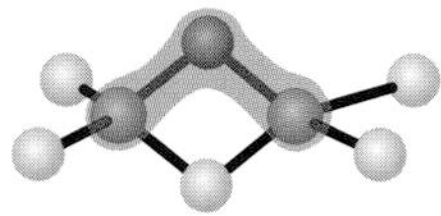

9 Three-center bond

Numerous borides of the metals and nonmetals are known. Their formulas are typically unrelated to their location in the periodic table and include AlB_2, CaB_6, $B_{13}C_2$, $B_{12}S_2$, Ti_3B_4, TiB, and TiB_2. In some metal borides, the boron atoms are found at the centers of clusters of metal atoms. More commonly, the boron atoms form extended structures such as zigzag chains, branched chains, or networks of hexagonal rings of boron atoms. In titanium diboride, for instance, the boron atoms form an extended chickenwire network, and the titanium atoms lie over the openings in the network.

The boranes are an extensive series of highly reactive electron-deficient binary compounds of boron and hydrogen. Some borides contain chains or rings of boron atoms.

Self-Test 14.5A Compare the reactions for the formation of BF_3 and BCl_3. Which reaction is a redox reaction?

[*Answer:* The formation of BCl_3]

Self-Test 14.5B Boric acid does not itself lose any protons in water. Why, then, is it called an acid?

GROUP 14/IV: THE CARBON FAMILY

Carbon is central to life and natural intelligence. Silicon and germanium are central to electronic technology and artificial intelligence (Fig. 14.35, Table 14.11). The unique properties of Group 14/IV elements make both types of intelligence possible. The half-filled valence shell of these elements

TABLE 14.11 Group 14/IV Elements

Valence configuration: ns^2np^2

Z	Name	Symbol	Molar mass, $g{\cdot}mol^{-1}$	Melting point, °C	Boiling point, °C	Density, $g{\cdot}cm^{-3}$	Normal form*
6	carbon	C	12.01	3370s†	—	1.9 to 2.3	black nonmetal (graphite)
						3.2 to 3.5	transparent nonmetal (diamond)
							orange nonmetal (fullerite)
14	silicon	Si	28.09	1410	2620	2.33	gray metalloid
32	germanium	Ge	72.61	937	2830	5.32	gray-white metalloid
50	tin	Sn	118.71	232	2720	7.29	white lustrous metal
82	lead	Pb	207.2	328	1760	11.34	blue-white lustrous metal

**Normal form* means the state and appearance of the element at 25°C and 1 atm.
†The symbol s denotes that the element sublimes.

gives them special properties that straddle the line between metals and nonmetals. Carbon, at the head of the group, forms so many compounds that it has its own branch of chemistry, organic chemistry (Chapters 18 and 19).

14.17 The Group 14/IV Elements

The elements show increasing metallic character down the group (Table 14.12). Carbon has definite nonmetallic properties: it forms covalent compounds with nonmetals and ionic compounds with metals. The oxides of carbon and silicon are acidic. Germanium is a typical metalloid in that it exhibits metallic or nonmetallic properties according to the other element present in the compound. Tin and, even more so, lead have definite metallic properties. However, even though tin is classified as a metal, it is not far from the metalloids in the periodic table, and it does

FIGURE 14.35 The elements of Group 14/IV. Back row, from left to right: silicon, tin. Front row: carbon (graphite), germanium, lead.

TABLE 14.12 Chemical Properties of the Group 14/IV Elements

Reactant	Reaction with Group 14/IV element (E)
hydrogen	$C(s) + 2\,H_2(g) \longrightarrow CH_4(g)$ and other hydrocarbons
oxygen	$E(s) + O_2(g) \longrightarrow EO_2(s)$, E = C, Si, Ge, Sn
	$2\,Pb(s) + O_2(g) \longrightarrow 2\,PbO(s)$
halogen (X_2)	$E(s) + 2\,X_2(g,l,s) \longrightarrow EX_4(s,l,g)$, E = C, Si, Ge, Sn
	$Pb(s) + X_2(g,l,s) \longrightarrow PbX_2(s)$
water	$C(s) + H_2O(g) \xrightarrow{\Delta} CO(g) + H_2(g)$
	$Si(s) + 2\,H_2O(l) \xrightarrow{\Delta} SiO_2(s) + 2\,H_2(g)$
acid	$E(s) + 2\,H_3O^+(aq) \longrightarrow E^{2+}(aq) + 2\,H_2O(l) + H_2(g)$, E = Sn, Pb
base	$E(s) + 2\,H_2O(l) + 2\,OH^-(aq) \longrightarrow E(OH)_4{}^{2-}(aq) + H_2(g)$, E = Sn, Pb

have some amphoteric properties. For example, tin reacts with both hot concentrated hydrochloric acid and hot alkali:

$$Sn(s) + 2\,H_3O^+(aq) \longrightarrow Sn^{2+}(aq) + H_2(g) + 2\,H_2O(l)$$

$$Sn(s) + 2\,OH^-(aq) + 2\,H_2O(l) \longrightarrow Sn(OH)_4{}^{2-}(aq) + H_2(g)$$

Because carbon stands at the head of its group, we expect it to differ from the other members of the group. In fact, the differences are more pronounced in Group 14/IV than anywhere else in the periodic table. Some of the differences between carbon and silicon stem from the smaller atomic radius of carbon, which explains the wide occurrence of C=C and C=O double bonds relative to the rarity of Si=Si and Si=O double bonds. Silicon atoms are too large for the side-by-side overlap of *p*-orbitals necessary for π-bonds to form between them. Carbon dioxide, which consists of discrete O=C=O molecules, is a gas that we exhale. Silicon dioxide (silica), which consists of networks of —O—Si—O— groups, is a mineral we stand on.

Silicon compounds can also act as Lewis acids, whereas carbon compounds typically cannot. Because a silicon atom is bigger than a carbon atom and can expand its valence shell by using its *d*-orbitals, it can accommodate the lone pair of an attacking Lewis base. A carbon atom is smaller and has no available *d*-orbitals, so in general it cannot act as a Lewis acid. An exception to this behavior occurs when the carbon atom has multiple bonds, for then a π-bond can give way to the formation of a σ-bond, as in the reaction between CO_2 and O^{2-} to form $CO_3{}^{2-}$:

$$:\ddot{O}:^{2-} \quad \ddot{O}{=}C{=}\ddot{O} \longrightarrow \left[\,:\ddot{O}{-}C(:\ddot{O}:){=}\ddot{O}:\,\right]^{2-}$$

The valence electron configuration is ns^2np^2 for all members of the group. All four electrons are approximately equally available for bonding in the lighter elements, and carbon and silicon are characterized by their ability to form four covalent bonds. On descending the group, however, the energy separation between the *s*- and *p*-orbitals increases and the *s*-electrons become progressively less available for bonding; in fact, the most common oxidation number for lead is +2.

Carbon is the only member of Group 14/IV that commonly forms multiple bonds with itself; singly bonded silicon atoms can act as Lewis acids because a silicon atom can expand its valence shell.

14.18 The Different Forms of Carbon

Solid carbon exists as graphite, diamond, and other phases such as the fullerenes, which have structures related to that of graphite. Graphite is the thermodynamically most stable of these allotropes under ordinary conditions. In this section, we see how the properties of the different allotropes of carbon are related to differences in bonding.

Soot and *carbon black* contain very small crystals of graphite. Carbon black, which is produced by heating gaseous hydrocarbons to nearly 1000°C in the absence of air, is used for reinforcing rubber, for pigments, and for printing inks, such as the ink on this page. *Activated carbon*, which is also called *activated charcoal*, consists of granules of microcrystalline carbon. It is produced by heating waste organic matter in the absence of air and then processing it to increase the porosity. The very high specific surface area (up to about 2000 $m^2 \cdot g^{-1}$) of the porous carbon enables it to remove organic impurities from liquids and gases by adsorption. It is used in air purifiers, gas masks, and aquarium water filters. On a larger scale, activated carbon is used in water purification plants to remove organic compounds from drinking water.

We can understand the differences in properties between the carbon allotropes by comparing their structures. Graphite consists of planar sheets of sp^2 hybridized carbon atoms in a hexagonal network (Fig. 14.36). Electrons are free to move from one carbon atom to another through a delocalized π-network formed by the overlap of unhybridized *p*-orbitals on each carbon atom. This network spreads across the entire plane. Because of the electron delocalization, graphite is a black, lustrous,

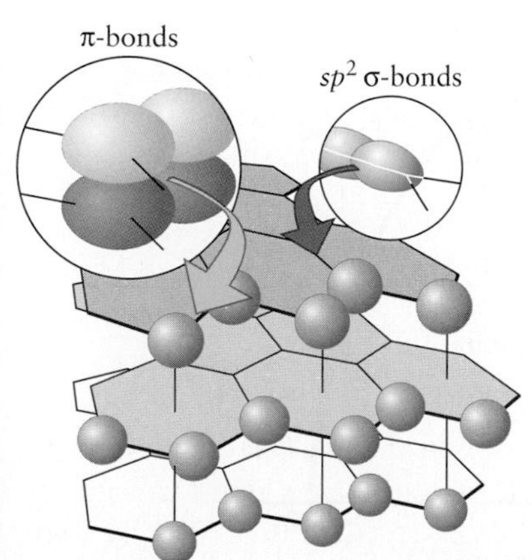

FIGURE 14.36 Graphite consists of flat planes of hexagons lying above one another. When impurities are present, the planes can slide over one another quite easily. Graphite conducts electricity well within the planes but less well perpendicular to the planes.

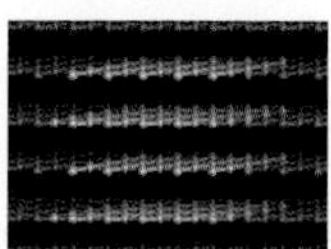

View the internal structure of the planes of graphite.

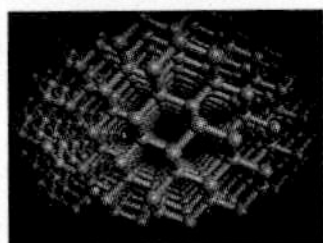

View the internal structure of a diamond.

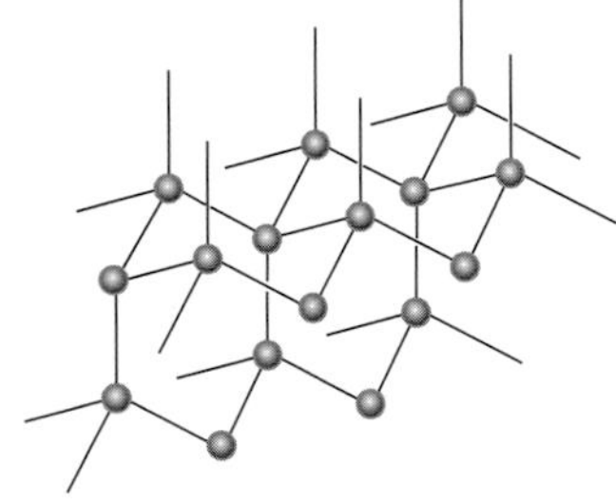

Media Link Animation: Fig. 14.37, Diamond

FIGURE 14.37 Structure of diamond. Each carbon atom is sp^3 hybridized and forms tetrahedral σ-bonds to four neighbors. This pattern is repeated throughout the crystal and accounts for diamond's great hardness.

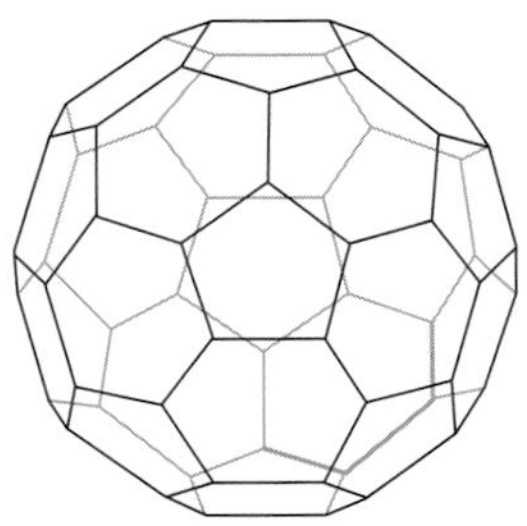

10 Buckminsterfullerene, C_{60}

electrically conducting solid; indeed, graphite is used as an electrical conductor in industry and as electrodes in electrochemical cells and batteries. Its slipperiness, which results from the ease with which the flat planes move past one another when impurities are present, leads to its use as a lubricant and, when mixed with clay, as the "lead" in lead pencils. The graphite found in nature is the result of changes in ancient organic remains. Pure graphite is produced in industry by passing a heavy electric current through rods of coke for several days.

In diamond, each carbon atom is sp^3 hybridized and linked tetrahedrally to its four neighbors, with all electrons in C—C σ-bonds (Fig. 14.37). Diamond is a rigid, transparent, electrically insulating solid. It is the hardest substance known and the best conductor of heat, being about five times better than copper. These last two properties make it an ideal abrasive, for it can scratch all other substances, yet the heat generated by friction is quickly conducted away.

In nature, diamond is found embedded in a soft rock called *kimberlite*. This rock rises in columns from deep in the Earth, where the diamonds are formed under intense pressure. One method used to make synthetic industrial diamonds recreates the geological conditions that produce natural diamonds by compressing graphite at pressures over 80 kbar and temperatures above 1500°C (Fig. 14.38). Small amounts of metals such as chromium and iron are added to the graphite. It is thought that the molten metals dissolve the graphite and then, as they cool, deposit crystals of diamond, which are less soluble than graphite in the molten metal. Most synthetic diamonds are produced by thermal decomposition of methane. In this technique, the carbon atoms settle on a cool surface as both graphite and diamond. However, because hydrogen atoms produced in the decomposition react more quickly with graphite, forming volatile hydrocarbons, more diamond than graphite survives.

Chemists were greatly surprised when soccer-ball-shaped carbon molecules were first identified in 1985, particularly because they might be even more abundant than graphite and diamond! The C_{60} molecule (**10**) is named buckminsterfullerene after the American architect R. Buckmin-

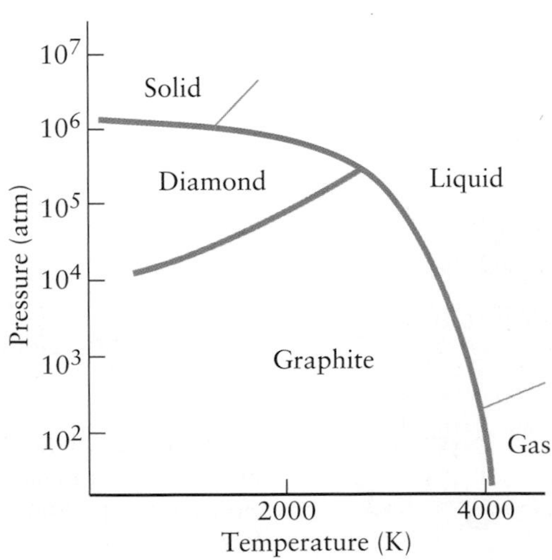

FIGURE 14.38 The phase diagram of carbon, showing the regions of phase stability.

ster Fuller, whose geodesic domes they resemble. Within two years, scientists had succeeded in making crystals of them: the solid samples are called *fullerite* (Fig. 14.39). Like the discovery of benzene, the discovery of these molecules opened up the prospect of a whole new field of chemistry. For instance, the interior of a C_{60} molecule is big enough to hold an atom of another element, and chemists are now busily preparing a whole new periodic table of these shrink-wrapped atoms. The *fullerenes* are members of the family of molecules resembling (and including) buckminsterfullerene. They are formed in smoky flames and by red giants (stars with low surface temperatures and large diameters), so the universe might contain huge numbers of them.

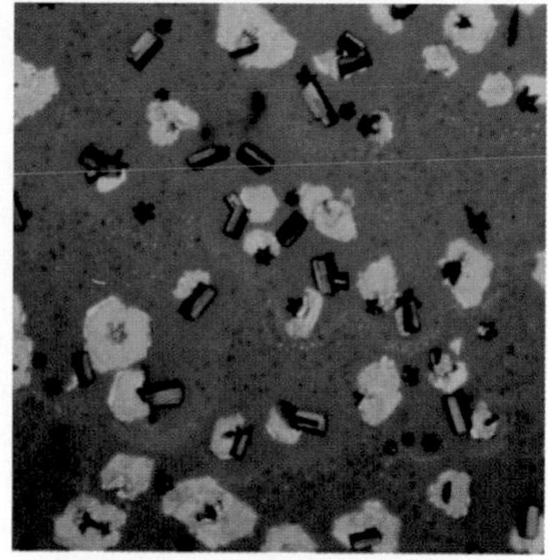

FIGURE 14.39 These small crystals are fullerite, in which buckminsterfullerene molecules are packed together in a close-packed lattice.

Graphite and diamond are network solids that are insoluble in all liquid solvents except some liquid metals. However, the fullerenes, which are molecular, can be dissolved by suitable solvents (such as benzene); buckminsterfullerene itself forms a red-brown solution. Fullerite currently has few uses, but some of the compounds of the fullerenes have great promise. For example, K_3C_{60} is a superconductor below 18 K, and other compounds appear to be active against cancer and diseases such as AIDS.

A newly discovered form of fibrous carbon consists of concentric tubes with walls like sheets of graphite rolled into cylinders. These tiny structures, called *nanotubes*, hold out the promise of forming strong, conducting fibers (Box 14.1).

Carbon has an important series of allotropes: diamond, graphite, and the fullerenes.

14.19 Silicon, Germanium, Tin, and Lead

Silicon is the second most abundant element in the Earth's crust. It occurs widely in rocks as *silicates*, compounds containing the silicate ion, SiO_3^{2-}, and as the silica, SiO_2, of sand (Fig. 14.40). Pure silicon is obtained from *quartzite*, a granular form of *quartz*, by reduction with high-purity carbon in an electric arc furnace:

$$SiO_2(s) + 2\,C(s) \xrightarrow{\Delta} Si(s) + 2\,CO(g)$$

(a)

(b)

(c)

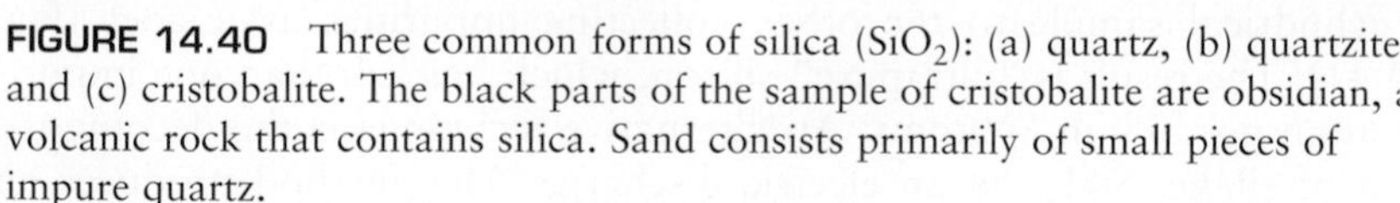

FIGURE 14.40 Three common forms of silica (SiO_2): (a) quartz, (b) quartzite, and (c) cristobalite. The black parts of the sample of cristobalite are obsidian, a volcanic rock that contains silica. Sand consists primarily of small pieces of impure quartz.

BOX 14.1 FRONTIERS OF CHEMISTRY: NANOTUBES, NATURE'S SMALLEST PIPES

Imagine a tiny pipet only a few nanometers in diameter that can be used to inject a drug into a single cell, or tiny wires and semiconducting devices in a microprocessor too small to be visible to the naked eye. Then consider huge arrays of these same lilliputian fibers wound into ropes so strong and flexible that they can support bridges and earthquake-proof buildings. These are only a few of the astonishing applications envisioned for *nanotubes*, tiny cylinders made of boron nitride or carbon.

In 1991, a group of scientists identified the first carbon nanotubes. They sealed two graphite rods inside a container of helium gas and sent an electric discharge from one rod to the other. Much of one rod evaporated, but out of the inferno some amazing structures emerged (see illustrations). As well as the tiny 60-atom carbon spheres known as buckminsterfullerene—which had been known since 1985—long, hollow, perfectly straight carbon nanotubes were also detected.

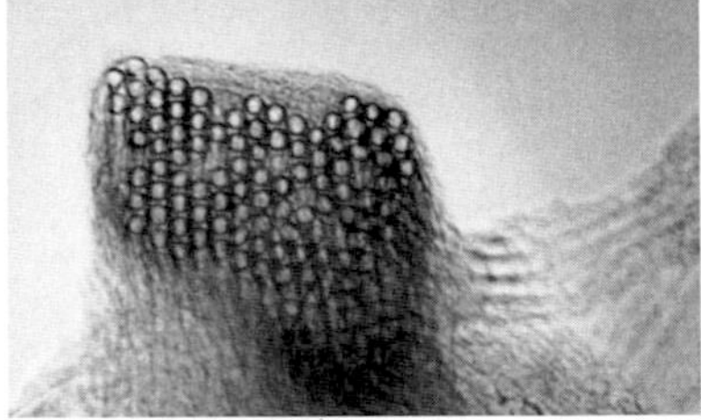

A cluster of carbon nanotubes that has formed a "rope." The surrounding material below the rope in the photograph consists of fullerenes and other carbon structures.

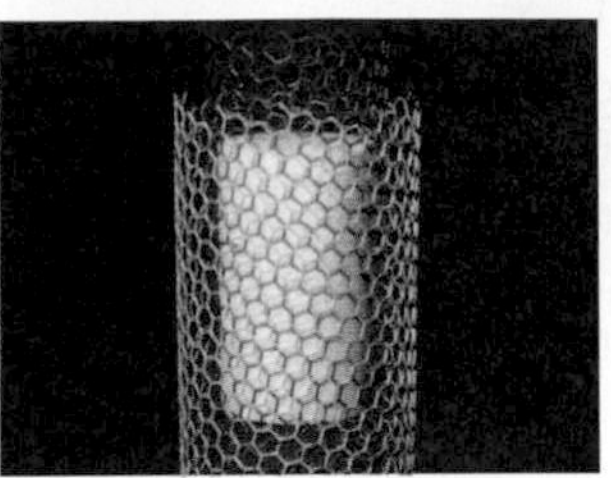

This model of a carbon nanotube shows that it consists of several concentric tubes. Such layered structures are very strong.

The tubes were found to have a bonding structure similar to that of graphite. They can be thought of as narrow sheets of a million or more carbon atoms linked together in six-membered benzenelike rings resembling chicken wire and rolled into a very long cylinder only 1–3 nm in diameter.

Carbon nanotubes conduct electricity because of the extended network of delocalized π-bonds that runs from one end of the tube to the other. Along the long axis of the tube, their conductivity is high enough to be considered metallic. The tubes are very strong, and their tensile strength parallel to the axis of the tube is the greatest of any material that has been measured. Because they have a very low density, their strength-to-mass ratio is 40 times that of steel! This may seem surprising at first, because we do not associate graphite with great strength. However, graphite forms flat sheets, not tubes. Think of

The crude product is exposed to chlorine to form silicon tetrachloride, which is then distilled and reduced with hydrogen to a purer form of the element:

$$SiCl_4(l) + 2\,H_2(g) \longrightarrow Si(s) + 4\,HCl(g)$$

Further purification is necessary before silicon can be used in semiconductors. In one process, a large single crystal is grown by pulling a solid rod of the element slowly from the melt. The silicon is then purified by **zone refining,** in which a hot, molten zone is dragged from one end of a cylindrical sample to the other, collecting impurities as it goes (Fig. 14.41). The result is "ultrapure" silicon, which has less than one impurity atom per billion Si atoms. An alternative technique is the decomposition of silane, SiH_4, by an electric discharge. This method produces an how flexible a sheet of paper is. You would have difficulty standing it on edge, for instance. But if you roll it into a tight cylinder, it forms a strong structure that stands on end and can even support a relatively large weight.

Boron nitride forms nanotubes with properties different from those of carbon nanotubes. The electrical conductivity of carbon nanotubes varies from metallic to semiconducting, depending on factors such as the diameter of the tube. However, boron nitride nanotubes have stable electrical characteristics that do not depend on the diameter of the tube. They have a fairly large band gap, so they are considered to be only weakly semiconducting. Carbon nanotubes grow until they are "capped" by bonds to another element or by five-membered rings. The five-membered rings pucker the structure, forming bends along the length of tubes and domes at their ends. In the case of BN nanotubes, however, five-membered rings cannot form, because they would involve high-energy boron-boron or nitrogen-nitrogen bonds. Instead, BN nanotubes are capped by the atom of a metal such as tungsten.

Their great strength and conductivity have led to the use of nanotubes in submicroscopic electronic components such as transistors. The rigidity of both types of nanotubes may also allow them to be used as molds for other elements. For example, they can be filled with molten lead to create lead wires one atom in diameter. Nanotubes that are filled with biomolecules such as cytochrome *c* hold the promise of acting as nanosensors for medical applications.

The very large surface area provided by the highly porous nanotubes means that atoms of gases are readily adsorbed on the inner surfaces of the tubes. Nanotubes carrying hydrogen molecules could therefore become a high-enthalpy-density storage medium for hydrogen-powered vehicles (see Box 12.2). Such a material would solve a major obstacle to the use of hydrogen fuel cells—the problem of a safe, compact storage medium for hydrogen. Hydrogen adsorbed into nanotubes can be stored in a volume much smaller than that required to store the gas.

HOW MIGHT YOU CONTRIBUTE?

The discovery of nanotubes and other nanostructures has opened up an exciting new field of research. But just what other shapes are possible and what other materials will form nanotubes? To find out, we will need to predict the effect of different configurations. There are also many experimental problems to be solved. For example, how would you form an electrical connection to a nanotube? Methods for synthesizing the large amounts of nanotubes needed in large-scale applications of nanotube assemblies also need to be developed.

For Further Reading: R. F. Service, "Atom-Scale Research Gets Real," *Science*, November 24, 2000, 290: 1524–1531. B. I. Yakobson and R. E. Smalley, "Fullerene nanotubes: $C_{1,000,000}$ and beyond," *American Scientist*, July–August, 1997, 324–337.

amorphous form of silicon with a significant hydrogen content. *Amorphous silicon* is used in photovoltaic devices, which produce electricity from sunlight.

Germanium was not known until 1886 and has the distinction of having been predicted by Mendeleev before it was discovered (see Box 1.2). It is recovered from the flue dust of industrial plants processing zinc ores (in which it occurs as an impurity). It is used mainly—and increasingly—in the semiconductor industry.

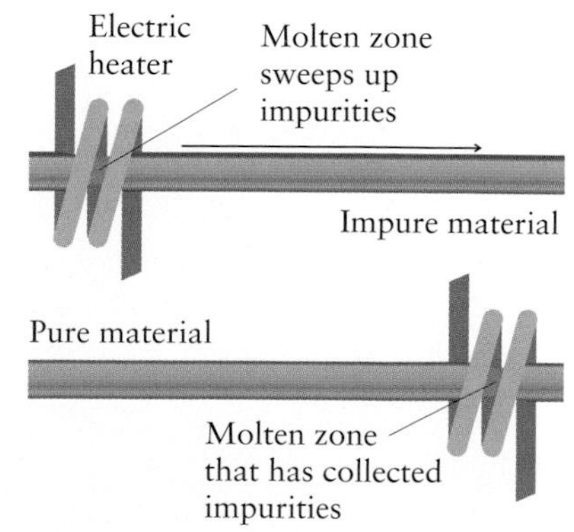

FIGURE 14.41 In the technique of zone refining, a molten zone is passed repeatedly from one end of the solid sample to the other. The impurities collect in the zone.

Tin and lead are obtained very easily from their ores and have been known since antiquity. Tin occurs chiefly as the mineral *cassiterite*, SnO_2, and is obtained from it by reduction with carbon at 1200°C:

$$SnO_2(s) + C(s) \xrightarrow{1200°C} Sn(l) + CO_2(g)$$

The principal lead ore is *galena*, PbS. It is roasted in air, which converts it to PbO, and this oxide is then reduced with coke:

$$2\ PbS(s) + 3\ O_2(g) \xrightarrow{\Delta} 2\ PbO(s) + 2\ SO_2(g)$$

$$PbO(s) + C(s) \longrightarrow Pb(s) + CO(g)$$

Tin is expensive and not very strong, but it is resistant to corrosion. Its main use is in tinplating, which accounts for about 40% of its consumption. Tin is also used in alloys such as bronze (with copper) and pewter (with antimony and copper).

Lead's durability (its chemical inertness) and malleability make it useful in the construction industry. The inertness of lead under normal conditions can be traced to the passivation of its surface by oxides, chlorides, and sulfates (see Section 12.7). Passivated lead containers can be used for transporting hot concentrated sulfuric acid but not nitric acid, because lead nitrate is soluble. Another important property of lead is its high density, which makes it useful as a radiation shield because its numerous electrons absorb high-energy radiation. The main use of lead today is for the electrodes of rechargeable storage batteries (see Box 12.1).

Metallic character increases significantly down Group 14/IV.

14.20 Oxides of Carbon

We have already met carbon dioxide, CO_2, many times throughout this book. It is formed when organic matter burns in a plentiful supply of air and during animal respiration. It is normally present in the atmosphere; but there is widespread and well-founded concern that an increase in atmospheric carbon dioxide due to the combustion of fossil fuels is contributing to global warming (see Box 14.2).

Carbon dioxide is the acid anhydride of carbonic acid, H_2CO_3, which forms when the gas dissolves in water. However, not all the dissolved molecules react to form the acid, and a solution of carbon dioxide in water is an equilibrium mixture of CO_2, H_2CO_3, HCO_3^-, and a very small amount of CO_3^{2-}:

$$CO_2(g) + H_2O(l) \rightleftharpoons H_2CO_3(aq)$$

Carbonated beverages are made by using high partial pressures of CO_2 to produce high concentrations of carbon dioxide in water. When the partial pressure of the CO_2 is reduced by removing a bottle cap or the seal, the equilibrium shifts from H_2CO_3 toward CO_2 and the liquid effervesces.

Carbon dioxide is formed by the action of acids on carbonates and as a by-product of *fermentation*, the conversion of carbohydrates to ethanol by yeast. Because its triple point is at 5.1 atm, carbon dioxide cannot exist as a liquid at ordinary atmospheric pressure. Solid carbon dioxide, which is sold as "dry ice," sublimes directly to the gas—a property that makes it convenient as a portable refrigerant.

Carbon monoxide, CO, is produced when carbon or organic compounds burn in a limited supply of air, as happens in cigarettes and badly tuned automobile engines. It is produced commercially as synthesis gas by the re-forming reaction (Section 14.5). Carbon monoxide is the formal anhydride of formic acid, HCOOH, and the gas can be produced in the laboratory by the dehydration of formic acid with hot, concentrated sulfuric acid:

$$HCOOH(l) \xrightarrow{150°C,\ H_2SO_4} CO(g) + H_2O(l)$$

Although the reverse of this reaction cannot be carried out directly, carbon monoxide does react with hydroxide ions in hot alkali to produce formate ions:

$$CO(g) + OH^-(aq) \longrightarrow HCO_2^-(aq)$$

Carbon monoxide is a colorless, odorless, flammable, almost insoluble, very toxic gas that condenses to a colorless liquid at −90°C. It is not very reactive, largely because its bond enthalpy (1074 $kJ \cdot mol^{-1}$) is higher than that of any other molecule. However, it is a Lewis base, and the lone pair on the carbon atom forms covalent bonds with *d*-block atoms and ions. Carbon monoxide is also a Lewis acid, because its empty antibonding π-orbitals can accept electron density from a metal (Fig. 14.42). This dual character makes carbon monoxide very useful for forming complexes, and numerous metal carbonyls are known. An example of this behavior is its reaction with nickel to give nickel carbonyl, a toxic, volatile liquid:

$$Ni(s) + 4\ CO(g) \xrightarrow{50°C,\ 1\ atm} Ni(CO)_4(l)$$

Although nickel carbonyl is intensely poisonous, it is used in the Mond process for the refinement of nickel (see Section 16.3). Complex formation is also responsible for carbon monoxide's toxicity: it attaches more strongly than oxygen to the iron in hemoglobin and prevents it from accepting oxygen from the air in the lungs. As a result, the victim suffocates.

Carbon monoxide is a reducing agent used in the production of a number of metals, most notably iron in blast furnaces (Section 16.14):

$$Fe_2O_3(s) + 3\ CO(g) \xrightarrow{\Delta} 2\ Fe(l) + 3\ CO_2(g)$$

Carbon has two important oxides, carbon dioxide and carbon monoxide. The former is the acid anhydride of carbonic acid, the parent acid of the hydrogen carbonates and the carbonates.

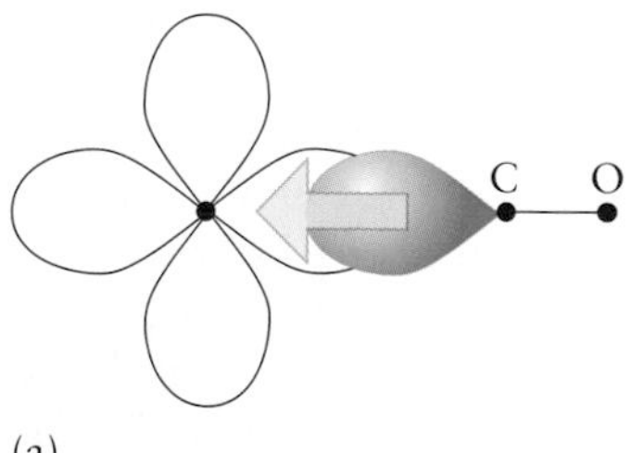

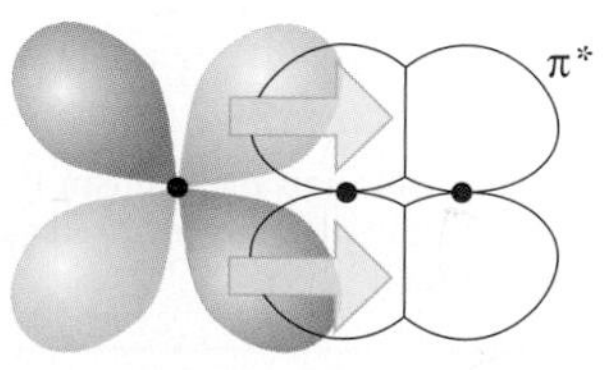

FIGURE 14.42 Carbon monoxide can bind to a *d*-metal atom (a) by using the lone pair on the C atom to form a σ-bond and (b) by using one of its empty antibonding π-orbitals to accept electrons donated from a *d*-orbital on the metal atom.

BOX 14.2 WHAT HAS THIS TO DO WITH . . . THE ENVIRONMENT?

The Greenhouse Effect

A visitor to our solar system is likely to be struck with the abundance of life on Earth, the mild surface temperatures, and the presence of liquid water. Every year our planet receives from the Sun more than enough radiant energy to supply all our energy needs. About 55% of solar radiation is reflected away or used in natural processes. The remaining 45% is converted to thermal motion (heat), most of which escapes as infrared radiation with wavelengths between 4 and 50 μm.

The *greenhouse effect* is the trapping of this heat by certain gases in the atmosphere. This effect warms the Earth, as if the entire planet were enclosed in a huge greenhouse.* Oxygen and nitrogen, which make up roughly 99% of the atmosphere, do not trap heat. However, water vapor and CO_2 do. Even though these two gases make up only about 1% of the atmosphere, they trap enough heat to raise the temperature of the Earth by 33°C. Without this naturally occurring greenhouse effect, the average surface temperature of the Earth would be well below the freezing point of water.

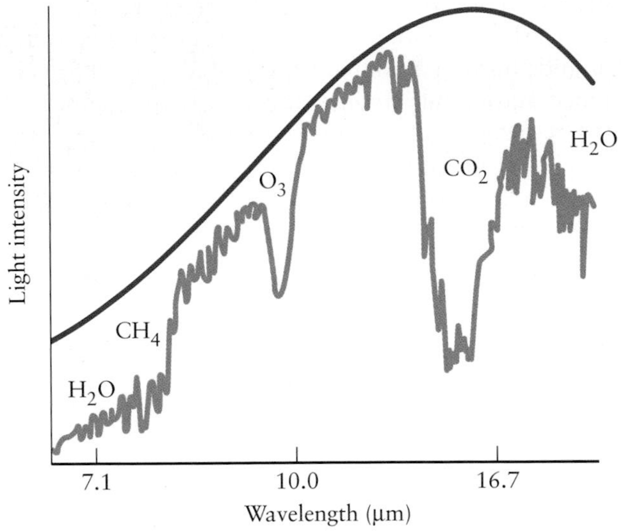

The intensity of infrared radiation at various wavelengths that would be lost from Earth in the absence of greenhouse gases is shown by the smooth line. The jagged line is the intensity of the radiation actually emitted. The maximum wavelength of radiation absorbed by each greenhouse gas is indicated.

The major greenhouse gases include water vapor, carbon dioxide, methane, dinitrogen oxide, ozone, and certain chlorofluorocarbons. Water is the most important greenhouse gas. It absorbs strongly near 6.3 μm and also at wavelengths longer than 12 μm, as shown on the plot. The carbon dioxide in the atmosphere absorbs about half the infrared radiation with wavelengths of 14–16 μm (see graph at right).

The concentration of water vapor in the atmosphere is thought to have remained steady over time, but concentrations of some other greenhouse gases are rising. From the year 1000 or earlier until about 1750, the CO_2 concentration in the atmosphere remained fairly stable at about 280 ± 10 parts per million by volume (ppmv). Since then, the CO_2 concentration has increased by 28% to 360 ppmv (see graph on next page, top). The concentration of methane, CH_4, has more than doubled during this time and is now at its highest level in 160 000 years. Studies of air pockets in ice cores taken from Antarctica show that changes in the concentrations of both atmospheric carbon dioxide and methane over the past 160 000 years correlate well with changes in the global surface temperature. The rising concentrations of carbon dioxide and methane are therefore of some concern.

Where is the additional CO_2 coming from? Human activities are responsible. Some is generated when limestone, $CaCO_3$, is heated and decomposed in cement making (see Section 14.12). Large amounts of CO_2 are also released to the atmosphere by deforestation, which involves burning large areas of brush and trees. However, most of it comes from the burning of fossil fuels, which began on a large scale after 1850. The additional methane is coming mainly from the petroleum industry and from agriculture.

The temperature of the surface of the Earth has increased about 0.5°C since the late nineteenth century (see graph on next page, bottom). If current trends in population growth and energy use continue, then by the middle of the twenty-first century, the concentration of CO_2 in the atmosphere will be about twice its value prior

* The actual insulation mechanism is different in a greenhouse. The glass not only absorbs infrared radiation but also keeps the warm air inside.

to the Industrial Revolution. What are the likely consequences of this doubling of the CO_2 concentration?

The Intergovernmental Panel on Climate Change (IPCC) is an international body of scientists who are studying these changes to the atmosphere and the climate changes that are likely to result. The IPCC's best estimate as of 1995 is that, by the year 2100, the Earth will undergo an increase in temperature of 3°C, with a rise in sea level of about 0.5 m. A rise of 3°C may not sound like much. However, the temperature during the last ice age was only 6°C colder than at present. Furthermore, the *rate* of temperature change is likely to be faster than at any time in the last 10 000 years. Rapid climate changes may have detrimental effects on many of the Earth's ecosystems.

Computer projections of atmospheric CO_2 concentration for the next 200 years predict escalating increases in CO_2 concentration. Only about half the CO_2 released by humans is absorbed by Earth's natural systems. The other half causes the CO_2 concentration in the atmosphere to increase by about 1.5 ppmv per year. Two conclusions can be drawn from the these facts. First, even if CO_2 emissions were reduced to the amount emitted in 1990 and held constant at that level, the concentration of CO_2 in the atmosphere would continue to increase at about 1.5 ppmv per year for the next century. Second, to maintain CO_2 at its current concentration of 360 ppmv, we would have to immediately reduce fossil fuel consumption by about 50%.

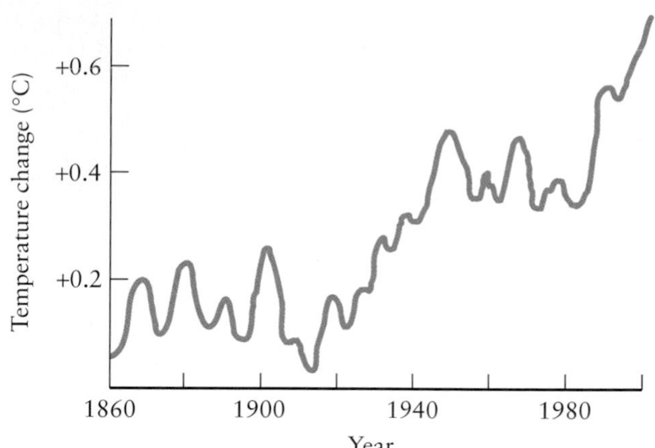

The average change in surface temperature of the Earth from 1860 to 1995.

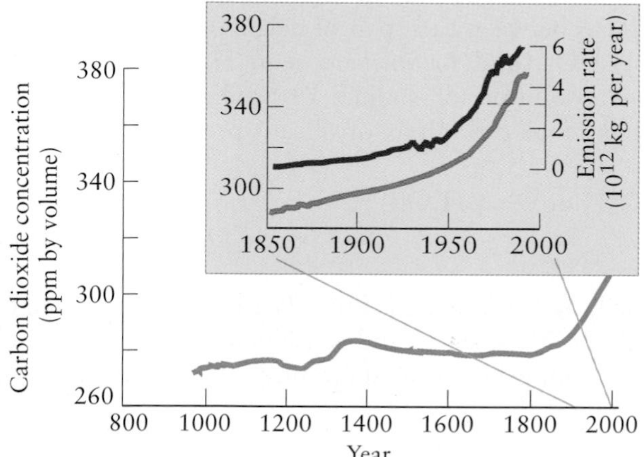

Concentration of carbon dioxide in the atmosphere, over the past 1000 years as determined from ice-core measurements. The blue line shows the emission of CO_2.

Alternatives to fossil fuels, such as hydrogen, are explored in Box 6.2 and Section 14.5. Coal, which is mostly carbon, can also be converted into fuels with a lower proportion of carbon. Its conversion to methane, CH_4, for instance, would reduce CO_2 emissions per unit of energy. We can also work with nature by accelerating the uptake of carbon by the natural processes of the carbon cycle. For example, one proposed solution is to pump CO_2 exhaust deep into the ocean, where it would dissolve to form carbonic acid and bicarbonate ions. Carbon dioxide can also be removed from power plant exhaust gases by passing the exhaust through an aqueous slurry of calcium silicate to produce harmless solid products:

$$2\,CO_2(g) + H_2O(l) + CaSiO_3(s) \longrightarrow SiO_2(s) + Ca(HCO_3)_2(s)$$

For Further Reading: J. Houghton, *Global Warming: The Complete Briefing* (New York: Cambridge University Press, 1997); A brief summary by the co-chair of the IPCC Scientific Assessment Working Group. F. W. Zwiers and A. J. Weaver, "The Causes of 20th Century Warming," *Science*, December 15, 2000, 290: 2081–2083.

FIGURE 14.43 Impure forms of silica: amethyst (left), in which the color is due to Fe^{3+} impurities; agate (center); and onyx (right).

14.21 Oxides of Silicon: The Silicates

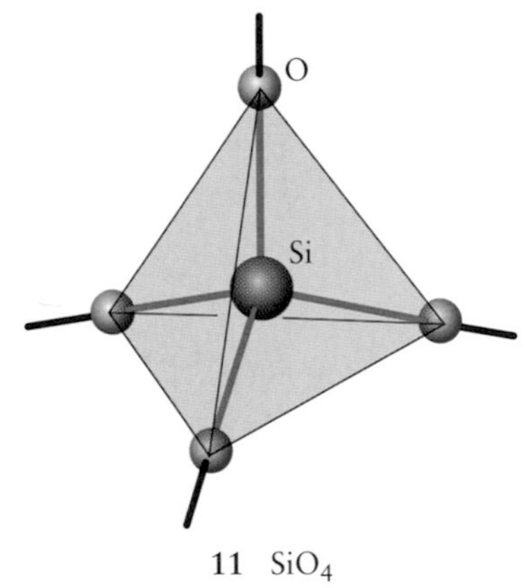

11 SiO_4

Silica, SiO_2, is a hard, rigid network solid that is insoluble in water. It occurs naturally as quartz and as sand, which consists of small fragments of quartz, usually colored golden brown by iron oxide impurities. Some precious and semiprecious stones are impure silica (Fig. 14.43). *Flint* is silica colored black by carbon impurities.

Silica gets its strength from its covalently bonded network structure. In silica itself, each silicon atom is at the center of a tetrahedron of oxygen atoms, and each corner O atom is shared by two Si atoms (**11**). Hence, each tetrahedron contributes one Si atom and $4 \times \frac{1}{2} = 2$ O atoms to the solid, which has the empirical formula SiO_2. Quartz has a complicated structure; it is built from helical chains of SiO_4 units wound around one another. When it is heated to about 1500°C, it changes to another arrangement, that of the mineral *cristobalite* (Fig. 14.44). This structure is easier to describe: its Si atoms are arranged like the C atoms in diamond; but in cristobalite, an O atom lies between each pair of neighboring Si atoms.

Metasilicic acid, H_2SiO_3, and *orthosilicic acid*, H_4SiO_4, are weak acids. However, when a solution of sodium orthosilicate is acidified, instead of H_4SiO_4, a gelatinous precipitate of silica is produced:

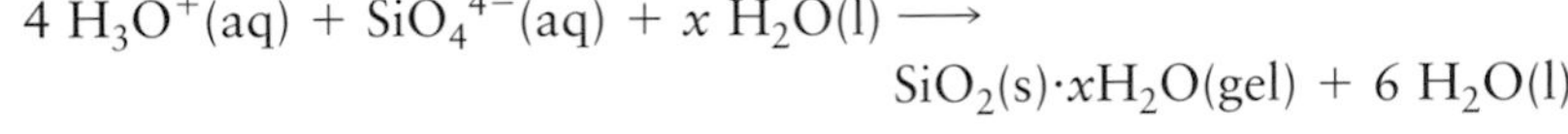

$$4\ H_3O^+(aq) + SiO_4{}^{4-}(aq) + x\ H_2O(l) \longrightarrow SiO_2(s)\cdot xH_2O(gel) + 6\ H_2O(l)$$

After it is washed, dried, and granulated, this *silica gel* has a very high specific surface area (about 700 $m^2\cdot g^{-1}$) and is useful as a drying agent, a support for catalysts, a packing for chromatography columns, and a thermal insulator.

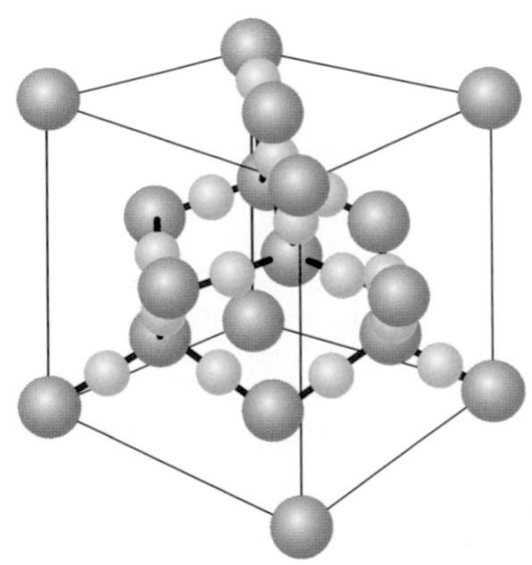

FIGURE 14.44 The structure of cristobalite is like that of diamond except that an O atom (light red) lies between each Si atom (purple). The arrangement about each Si atom is shown in structure **11**.

Silicates can be viewed as various arrangements of tetrahedral oxoanions of silicon in which each Si—O bond has considerable covalent character. The differences in properties between the various silicates are related to the number of negative charges on each tetrahedron, the number of corner O atoms shared with other tetrahedra, and the manner in which chains and sheets of the linked tetrahedra lie together. Differences in the internal structures of these highly regular network solids lead to a wide array of materials, ranging from gemstones to fibers.

The simplest silicates, the *orthosilicates*, are built from SiO_4^{4-} ions. They are not very common but include the mineral *zircon*, $ZrSiO_4$, which is used as a substitute for diamond in costume jewelry. The *pyroxenes* consist of chains of SiO_4 units in which two corner O atoms are shared by neighboring units (Fig. 14.45); the repeating unit is the metasilicate ion, SiO_3^{2-}. Electrical neutrality is provided by cations regularly spaced along the chain. The pyroxenes include *jade*, $NaAl(SiO_3)_2$.

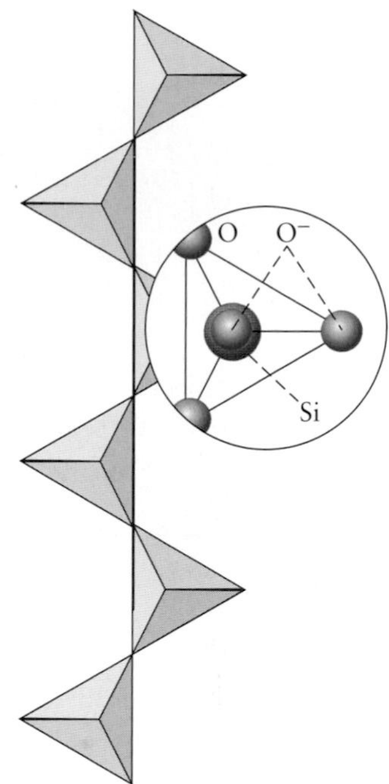

FIGURE 14.45 The basic structural unit of the minerals called pyroxenes. Each tetrahedron is an SiO_4 unit (like that in structure **11**), and a shared corner represents a shared O atom (red in the inset where one O atom lies directly over an Si atom). The two unshared O atoms each carry a negative charge.

Chains of silicate units can link together to form ladderlike structures that include *tremolite*, $Ca_2Mg_5(Si_4O_{11})_2(OH)_2$. Tremolite is one of the fibrous minerals called *asbestos*, which can withstand extreme heat (Fig. 14.46). Their fibrous quality reflects the way the ladders of SiO_4 units lie together but can easily be torn apart. Because of their great resistance to fire, asbestos fibers were once widely used for heat insulation in buildings. However, these fibers can lodge in lung tissue, where fibrous scar tissue forms around them, giving rise to asbestosis and a susceptibility to lung cancer. In some minerals, the SiO_4 tetrahedra link together to form sheets. An example is *talc*, a hydrated magnesium silicate, $Mg_3(Si_2O_5)_2(OH)_2$. Talc is soft and slippery because the silicate sheets slide over one another.

More complex (and more common) structures result when some of the silicon(IV) in silicates is replaced by aluminum(III) to form the *aluminosilicates*. The missing positive charge is made up by extra cations. These cations account for the difference in properties between the silicate talc and the aluminosilicate *mica*. One form of mica is $KMg_3(Si_3AlO_{10})(OH)_2$. In this mineral, the sheets of tetrahedra are held together by extra K^+ ions. Although it cleaves neatly into transparent layers when the sheets are torn apart, mica is not slippery like talc (Fig. 14.47). Sheets of mica are used for windows in furnaces.

FIGURE 14.46 The minerals commonly called asbestos (from the Greek words meaning "not burning") are fibrous because they consist of long chains based on SiO_4 tetrahedra linked through shared oxygen atoms.

FIGURE 14.47 The aluminosilicate mica cleaves into thin transparent sheets with high melting points. These properties allow it to be used for windows in furnaces.

FIGURE 14.48 The mineral granite is a compressed mixture of mica, quartz, and feldspar.

The *feldspars* are aluminosilicates in which up to half the silicon(IV) has been replaced by aluminum(III). They are the most abundant silicate materials on Earth and are a major component of *granite*, a compressed mixture of mica, quartz, and feldspar (Fig. 14.48). When some of the ions between the crystal layers are washed away as these rocks weather, the structure crumbles to clay, one of the main inorganic components of soil. A typical feldspar has the formula $KAlSi_3O_8$. Its weathering by carbon dioxide and water can be described by the equation

$$2\,KAlSi_3O_8(s) + 2\,H_2O(l) + CO_2(g) \longrightarrow K_2CO_3(aq) + Al_2Si_2O_5(OH)_4(s) + 4\,SiO_2(s)$$

The potassium carbonate is soluble and washes away, but the aluminosilicate remains as the clay. *Cements* are obtained when aluminosilicates are roasted with limestone and other minerals and then allowed to solidify.

Silicones are synthetic materials that consist of long —O—Si—O—Si— chains with the remaining silicon bonding positions occupied by organic groups, such as the methyl group, $-CH_3$ (Fig. 14.49). Silicones are used to waterproof fabrics because their oxygen atoms attach to the fabric, leaving the hydrophobic (water-repelling) methyl groups like tiny, inside-out umbrellas sticking up out of the fabric's surface.

Silicate structures are based on SiO_4 tetrahedral units with different negative charges and different numbers of shared O atoms.

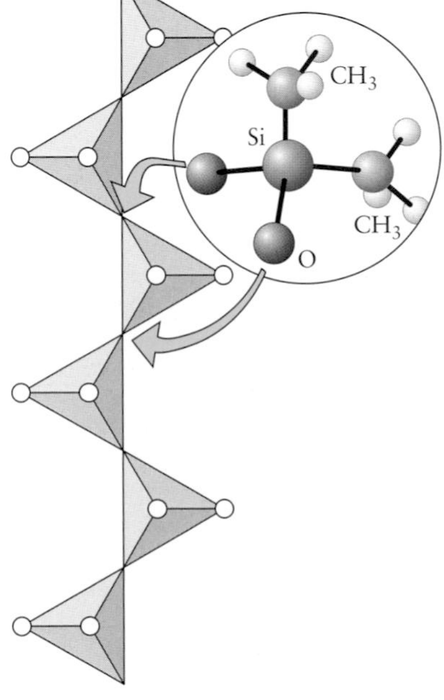

FIGURE 14.49 A typical silicone structure. The hydrocarbon groups give the substance a water-repelling quality. Note the similarity of this structure to that of the purely inorganic pyroxenes in Fig. 14.45.

14.22 Other Important Group 14/IV Compounds

Carbon is the only Group 14/IV element that forms anions. There are three classes of carbides: saline carbides (or saltlike carbides), covalent carbides, and interstitial carbides.

The *saline carbides* are formed most commonly from the metals of Groups 1 and 2, aluminum, and a few other metals. The *s*-block metals form saline carbides when their oxides are heated with carbon. The anions present in saline carbides are either C_2^{2-} or C^{4-}. All the C^{4-} carbides, which are called *methides*, produce methane and the corresponding hydroxide in water:

$$Al_4C_3(s) + 12\ H_2O(l) \longrightarrow 4\ Al(OH)_3(s) + 3\ CH_4(g)$$

This reaction shows that the methide ion is a very strong Brønsted base. The species C_2^{2-} is the *acetylide ion*, and the carbides that contain it are called *acetylides*. The acetylide ion is also a strong Brønsted base, and acetylides react with water to produce ethyne (acetylene, the conjugate acid of the acetylide ion) and the corresponding hydroxide (Section 14.12). Calcium carbide, CaC_2, is the most common saline carbide.

The *covalent carbides* include silicon carbide, SiC, which is sold as *carborundum*:

$$SiO_2(s) + 3\ C(s) \xrightarrow{2000°C} SiC(s) + 2\ CO(g)$$

Pure silicon carbide is colorless, but iron impurities normally impart an almost black color to the crystals. Carborundum is an excellent abrasive because it is very hard, with a diamondlike structure that fractures into pieces with sharp edges (Fig. 14.50).

FIGURE 14.50 Carborundum crystals, showing the sharp fractured edges that give the substance its abrasive power.

The *interstitial carbides* are compounds formed by the direct reaction of a *d*-block metal and carbon at temperatures above 2000°C. In these compounds, the C atoms occupy the gaps between the metal atoms, as do the H atoms in metallic hydrides (Fig. 14.51). Here, however, the C atoms pin the metal atoms together into a rigid structure, resulting in very hard substances with melting points often well above 3000°C. Tungsten carbide, WC, is used for the cutting surfaces of drills, and iron carbide, Fe_3C, is an important component of steel.

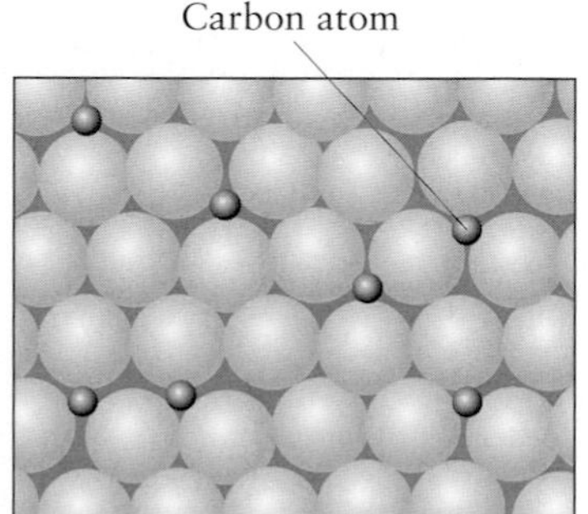

FIGURE 14.51 The structure of an interstitial carbide, in which the carbon atoms (represented by the small black spheres) lie between metal atoms (the gray spheres), thereby producing a rigid structure.

All Group 14/IV elements form liquid molecular tetrachlorides. The least stable is $PbCl_4$, which decomposes to solid $PbCl_2$ when it is warmed to about 50°C. Carbon tetrachloride, CCl_4 (tetrachloromethane), was widely used as an industrial solvent; however, now that it is known to be carcinogenic, it is used primarily as the starting point for the manufacture of chlorofluorocarbons. Carbon tetrachloride is formed by the action of chlorine on methane:

$$CH_4(g) + 4\ Cl_2(g) \xrightarrow{650°C} CCl_4(g, l \text{ when cooled}) + 4\ HCl(g)$$

Silicon reacts directly with chlorine to form silicon tetrachloride, $SiCl_4$ (this reaction was introduced in Section 14.19, as one step in the purification of silicon). This compound differs strikingly from CCl_4 in that it reacts readily with water as a Lewis acid, accepting a lone pair of electrons from H_2O:

$$SiCl_4(l) + 2\ H_2O(l) \longrightarrow SiO_2(s) + 4\ HCl(aq)$$

The cyanide ion, CN^-, is the conjugate base of hydrogen cyanide, HCN. This acid is made by heating ammonia, methane, and air in the presence of a platinum catalyst:

$$2\ CH_4(g) + 2\ NH_3(g) + 3\ O_2(g) \xrightarrow{1100°C,\ Pt} 2\ HCN(g) + 6\ H_2O(g)$$

When HCN reacts with ethyne (acetylene), HC≡CH, using a copper(I) chloride catalyst, *acrylonitrile*, $H_2C{=}CHCN$, is produced:

$$HCN(g) + HC{\equiv}CH(g) \xrightarrow{CuCl} H_2C{=}CHCN(l)$$

This compound is the starting point for the production of the synthetic fibers Acrilan and Orlon.

Cyanides are strong Lewis bases that form a range of complexes with *d*-block metal ions. They are also famous as poisons. When they are ingested, they combine with protein molecules—the cytochromes—involved in the transfer of electrons and the supply of energy in cells, and the victim dies.

Because carbon bonds so readily with itself, there are many hydrocarbons (see Section C). Silicon forms a much smaller number of compounds with hydrogen, called the *silanes*. The simplest silane is silane itself, SiH_4, the analogue of methane. Silane is formed by the action of lithium aluminum hydride on silicon halides in ether:

$$SiCl_4 + LiAlH_4 \longrightarrow SiH_4 + LiCl + AlCl_3$$

Silane is much more reactive than methane and bursts into flame on contact with air. Although it survives in pure water, it forms SiO_2 when a trace of alkali is present:

$$SiH_4(g) + 2\ H_2O(l) \xrightarrow{OH^-} SiO_2(s) + 4\ H_2(g)$$

The more complicated silanes, such as $SiH_3{-}SiH_2{-}SiH_3$, the analogue of propane, decompose on standing.

The sulfur analogue of carbon dioxide, carbon disulfide, CS_2, is a volatile, very poisonous liquid. Its reputation for having a vile smell is undeserved, because the odor is in fact due not to CS_2 but to impurities. It is prepared by the reaction between methane and sulfur:

$$CH_4(g) + 4\ S(g) \xrightarrow{600°C,\ Al_2O_3} CS_2(g) + 2\ H_2S(g)$$

In the laboratory, it is sometimes used as a nonpolar solvent.

Carbon forms ionic carbides with the metals of Groups 1 and 2, covalent carbides with nonmetals, and interstitial carbides with d-block metals. Silicon compounds are more reactive than carbon compounds. They can act as Lewis acids.

Self-Test 14.6A Explain why graphite can conduct electricity but diamond cannot.

[*Answer:* Graphite has a network of π-bonds, through which electrons can be delocalized. In diamond, electrons are localized in σ-bonds.]

Self-Test 14.6B Explain the fact that SiH_4 reacts with water containing OH^- ions but CH_4 does not.

THE IMPACT ON MATERIALS

Within your lifetime, extraordinary changes will sweep through modern technology. These changes are made possible by advanced materials, especially glasses and ceramics. Optical fibers—which are already playing a major role in communication—will control our computers, and our automobiles will be much lighter and more economical. Glasses and ceramics are diverse materials, but silicon compounds can be used to make both.

14.23 Glasses

The growing use of optical fibers to provide wideband telecommunications networks has led to considerable advances in the technology of glassmaking. A **glass** is an ionic solid with an amorphous structure resembling that of a liquid. Glass has a network structure based on a nonmetal oxide, usually silica, SiO_2, that has been melted together with metal oxides acting as "network modifiers," which alter the arrangement of bonds in the solid.

In a glass factory, silica in the form of sand is heated to about 1600°C. Metal oxides, such as MO (where M is a metal cation) are added to the silica. As the mixture melts, many of the Si—O bonds break and the orderly structure of the individual crystals is lost. When the melt cools, the Si—O bonds reform but are prevented from forming a crystalline lattice because some of the silicon atoms bond with the O^{2-} ions of the metal oxide to give $—Si—O^{-}—M^{+}$ groups in place of some of the —Si—O—Si— links present in pure silica. Silicate glasses are generally transparent and durable, and can be formed in flat sheets, blown into bottles, or molded into desired shapes. Optical fibers are made by drawing a thin fiber from an optically pure glass rod heated until it softens. The fiber is then coated with plastic (Fig. 14.52).

Almost 90% of all manufactured glass combines sodium and calcium oxides with silica to form *soda-lime glass*. This glass, which is used for windows and bottles, contains about 12% Na_2O prepared by the action of heat on sodium carbonate (the soda) and 12% CaO (the lime). When the proportions of soda and lime are reduced and 16% B_2O_3 is added, a *borosilicate glass*, such as Pyrex, is produced. Because borosilicate glasses do not expand much when heated, they survive rapid heating and cooling and are used for ovenware and laboratory beakers.

Glass is resistant to attack by most chemicals. However, the silica in glass reacts with the strong Lewis base F^- from hydrofluoric acid to form fluorosilicate ions:

$$SiO_2(s) + 6\,HF(aq) \longrightarrow SiF_6^{2-}(aq) + 2\,H_3O^+(aq)$$

It is also attacked by the Lewis base OH^- in hot, molten sodium hydroxide and by O^{2-} from the carbonate anion of hot molten sodium carbonate:

$$SiO_2(g) + Na_2CO_3(l) \xrightarrow{1400°C} Na_2SiO_3(s) + CO_2(g)$$

The process by which silica is removed from glass by the ions F^- (from HF), OH^-, and CO_3^{2-} is called *etching*.

Silicate glasses have amorphous structures produced by addition of salts that disrupt the crystalline structure. They can be attacked by strong base and hydrofluoric acid.

FIGURE 14.52 Glass fibers, such as this one, about the diameter of a human hair, are used in communication networks when large amounts of information must be transmitted in a short time.

FIGURE 14.53 The layers of clay particles can be seen in this micrograph. Because the surfaces of these layers have like charges, they repel one another and easily slide past one another, making clay soft and malleable.

14.24 Ceramics

Many of the materials used in the most advanced technologies are made from common clay. Most clays used commercially are oxides of silicon, aluminum, and magnesium. *China clay* contains primarily *kaolinite*, a form of aluminum aluminosilicate that can be obtained reasonably free of the iron impurities that make many clays look reddish brown, so it is white. However, other clays contain the iron oxides that cause the orange color of terra cotta tiles and flower pots. The appearance of a flake of clay reflects its internal structure, which is something like an untidy stack of papers (Fig. 14.53). Sheets of tetrahedral silicate units or octahedral units of aluminum or magnesium oxides are separated by layers of water molecules that serve to bind the layers of the flake together. Each flake of clay is surrounded by a double layer of ions that separates the flakes by repelling the like charges on the other flakes. This repulsion allows the flakes to slide past one another and gives the clay some flexibility in response to stress. As a result, clays can easily be molded. When clay is baked in a kiln, it forms the hard, tough "ceramic" materials used in firebricks, tiles, and pots as the water is driven out and strong chemical bonds form between the flakes.

A **ceramic** is an inorganic material that has been hardened by heating to a high temperature. Typically, a ceramic material is very hard, insoluble in water, and stable to corrosion and high temperatures. These characteristics are the key to their value. Although many ceramics tend to be brittle, they can be used at high temperatures without failing, and they resist deformation. Ceramics are often the oxides of elements that are on the border between metals and nonmetals, but *d*-metal oxides and some compounds of boron and silicon with carbon and nitrogen are also ceramic materials. Most ceramics are electrical insulators, but some are semiconductors and some are superconductors (see Box 5.2).

Many aluminosilicate ceramics are created by heating aluminosilicate clays to drive the water out from between the sheets of tetrahedra. This procedure leaves a rigid heterogeneous mass of tiny interlocking crystals bound together by glassy silica. Large amounts of china clay, which is used to make ceramics such as porcelain and china, are applied in the coating of paper (such as this page) to give a smooth, nonabsorbent surface.

Aluminum oxide in the form of corundum makes up about 80% of the advanced ceramics used in high technology applications. Its hardness, rigidity, thermal conductivity, stability at high temperatures, and electrical insulating ability make it suitable for a wide range of applications, including acting as the substrate for microchips. Corundum is prepared from powdered Al_2O_3 dispersed in liquid. Granules that form in the dispersion are compressed in a mold and *sintered* (heated at a high temperature until hard). Single-crystal forms of aluminum oxide are known and large single-crystal sapphires, which derive their color from iron and titanium impurities, are grown for specialized applications such as acoustic microscopes and as heat-resistant windows on heat-seeking missiles.

The challenge presented by corundum and most other ceramics is to find a way to overcome their brittleness. One route used with some silicon dioxide ceramics is the *sol-gel process*. In this process, an organic silicon compound is dissolved in water and polymerized into a network structure. Many cross-links are formed as the gel forms. These cross-links form a rigid, strong matrix that has few of the tiny cracks that can initiate breakage in brittle ceramics. If the solvent is removed at high temperatures and low pressures, an *aerogel*, a synthetic solid foam that has a density about the same as that of air but is a good insulator, is formed (Fig. 14.54).

Another means of strengthening ceramics is to create a composite material. Flakes of very strong, tough materials such as titanium diboride, TiB_2, are distributed throughout the ceramic material. Even if a crack did form, it would be unable to propagate past the first flake it encountered.

The stability of ceramic materials at high temperatures has made them useful as furnace liners and has led to interest in ceramic automobile engines, which could endure overheating. Currently, a typical automobile contains about 35 kg of ceramic materials such as spark plugs, pressure and vibration sensors, brake linings, catalytic converters, and thermal and electrical insulation. Some fuel cells make use of a porous solid electrolyte such as zirconia, ZrO_2, that contains a small amount of calcium oxide. It is an electronic insulator, so electrons do not flow through it, but oxide ions do.

Ceramics are inorganic materials that have been hardened by heating.

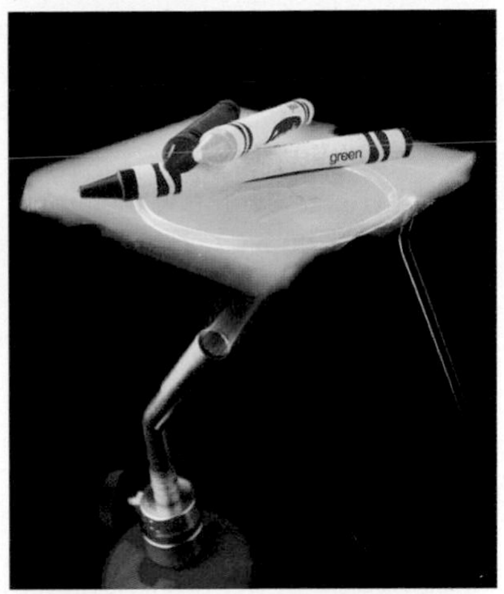

FIGURE 14.54 An aerogel is a ceramic foam. Its low density and low thermal conductivity, combined with great strength, make it an ideal insulating material. Here a thin piece protects three wax crayons from the heat of a flame. Aerogels were used to insulate the Mars Rover.

SKILLS YOU SHOULD HAVE MASTERED

1 Predict and explain trends in the properties and formulas of the main-group elements.

2 Describe the names, properties, and reactions of the principal compounds of hydrogen and the Period 1–3 elements in Groups 1, 2, 13/III, and 14/IV.

3 Describe and write balanced equations for the principal reactions used to produce hydrogen and the Period 1–3 elements in Groups 1, 2, 13/III, and 14/IV.

4 Describe the major uses of hydrogen, sodium, potassium, beryllium, magnesium, boron, aluminum, carbon, and silicon.

5 Describe the reactions of the alkali metals with water and with nonmetals.

6 Distinguish the allotropes of carbon by their structures and show how their structures affect their properties.

7 Explain the differences in reactivity between compounds of carbon and those of silicon.

8 Distinguish the principal silicate structures and describe their properties.

9 Describe the nature and properties of glasses and ceramics.

EXERCISES

Periodic Trends

These exercises are a review of principles covered in Chapter 1. See also Exercises 1.77–1.88.

14.1 State which atom of each of the following pairs is larger: (a) carbon, oxygen; (b) lithium, beryllium; (c) indium, tellurium; (d) bromine, iodine.

14.2 State which atom of each of the following pairs is larger: (a) thallium, lead; (b) arsenic, bismuth; (c) germanium, tin; (d) aluminum, boron.

14.3 State which atom of each of the following pairs is the more electronegative: (a) sulfur, phosphorus; (b) selenium, tellurium; (c) sodium, cesium; (d) silicon, oxygen.

14.4 State which atom of each of the following pairs is the more electropositive: (a) potassium, rubidium; (b) germanium, bromine; (c) oxygen, selenium; (d) aluminum, silicon.

14.5 Arrange the following atoms in order of increasing first ionization energy: chlorine, bromine, iodine.

14.6 Arrange the following atoms in order of increasing first ionization energy: nitrogen, arsenic, bismuth.

14.7 Which atom has the greatest electron affinity: chlorine, bromine, iodine?

14.8 Which atom has the greatest electron affinity: arsenic, fluorine, sulfur?

14.9 Which of the following atoms will have the greatest polarizability: nitrogen, oxygen, antimony?

14.10 Which of the following atoms will have the greatest polarizability: germanium, bromine, or tin?

14.11 Which bond distance will be longer: (a) the Na—Cl distance in solid sodium chloride or the K—Cl distance in solid potassium chloride; (b) the Na—O distance in solid sodium oxide or the Mg—O distance in solid magnesium oxide; (c) the Tl-Cl distance in solid thallium(I) chloride or in thallium(III) chloride?

14.12 Which of the following bonds do you expect to be longer: (a) the Sn—Cl bond in $SnCl_2$ or in $SnCl_4$; (b) the Cl—O bond in ClO^- or in ClO_2^-; (c) the N—H bond in NH_3 or the P—H bond in PH_3?

Hydrides

14.13 Write the equation for the reaction between potassium and hydrogen.

14.14 Write the balanced equation for the reaction between calcium and hydrogen.

14.15 Classify each of the following compounds as a saline, molecular, or metallic hydride: (a) LiH; (b) NH_3; (c) HBr; (d) UH_3.

14.16 Classify each of the following compounds as a saline, molecular, or metallic hydride: (a) B_2H_6; (b) SiH_4; (c) CaH_2; (d) PdH_x, $x < 1$.

Oxides

14.17 For the following oxides, state whether the compound is acidic, basic, or amphoteric: (a) NO_2; (b) Al_2O_3; (c) $B(OH)_3$; (d) MgO.

14.18 For the following oxides, state whether the compound is acidic, basic, or amphoteric: (a) Li_2O; (b) Tl_2O_3; (c) Sb_2O_3; (d) SO_3.

14.19 Give the formula for the formal anhydride of each acid: (a) H_2CO_3; (b) $B(OH)_3$.

14.20 Give the formula for the acid corresponding to each of the following formal anhydrides: (a) N_2O_5; (b) P_4O_{10}; (c) SeO_3.

Hydrogen

14.21 Write balanced chemical equations for (a) the hydrogenation of ethyne (acetylene, C_2H_2) to ethane (C_2H_4) by hydrogen (give the oxidation number of the carbon atoms in the reactant and product); (b) the shift reaction (sometimes called the water gas shift reaction, WGSR); (c) the reaction of barium hydride with water.

14.22 Write balanced chemical equations for (a) the reaction between sodium hydride and water; (b) the formation of synthesis gas; (c) the hydrogenation of ethene, $H_2C{=}CH_2$, and give the oxidation number of the carbon atoms in the reactant and product; (d) the reaction of magnesium with hydrochloric acid.

14.23 Use Appendix 2A to determine (a) the standard reaction enthalpy; (b) the entropy; (c) the free energy at 25°C for the re-forming reaction of methane.

14.24 Use Appendix 2A to determine (a) the standard reaction enthalpy; (b) the entropy; (c) the free energy at 25°C for the shift reaction.

14.25 Calcium hydride is used as a portable source of hydrogen because of its reaction with water:

$$CaH_2(s) + 2\ H_2O(l) \longrightarrow Ca(OH)_2(s) + 2\ H_2(g)$$

(a) What volume of H_2 gas (at 273 K and 1.00 atm) can be produced from 10.0 g of CaH_2? (b) How many milliliters of water should be supplied for the reaction? Assume the density of water to be 1.0 $g{\cdot}mL^{-1}$.

14.26 What volume of hydrogen gas (at 273 K and 1.00 atm) is required for the reduction of 35.9 g of WO_3 in the reaction $WO_3(s) + 3\ H_2(g) \rightarrow W(s) + 3\ H_2O(l)$?

14.27 Identify the products and write a balanced equation for the reaction of hydrogen with (a) chlorine; (b) sodium; (c) phosphorus; (d) copper metal.

14.28 Identify the products and write a balanced equation for the reaction of hydrogen with (a) nitrogen; (b) fluorine; (c) cesium; (d) copper(II) ions.

14.29 (a) Calculate the maximum potential that could be produced by a fuel cell that uses the reaction between hydrogen and oxygen in aqueous solution under standard conditions. (b) What technical problems would need to be overcome in the design of such a cell?

14.30 (a) What is the maximum potential that could be produced from a cell using the reaction of lithium with Cu^{2+} ions under standard conditions? (b) What technical problems would need to be overcome in the design of such a cell?

Group 1: The Alkali Metals

14.31 Explain why lithium differs from the other Group 1 elements in its chemical and physical properties. Give two examples to support your explanation.

14.32 (a) Write the valence electron configuration for the alkali metal atoms. (b) Explain why the alkali metals are strong reducing agents in terms of electron configurations and ionization energies.

14.33 Write the chemical equations for the reactions between (a) lithium and oxygen; (b) lithium and nitrogen; (c) sodium and water; (d) potassium superoxide and water.

14.34 Write the chemical equations for the reactions between (a) potassium and oxygen; (b) sodium oxide and water; (c) lithium and hydrochloric acid; (d) cesium and iodine.

14.35 Sodium carbonate is often supplied as the decahydrate, $Na_2CO_3 \cdot 10H_2O$. What mass of this solid should be used to prepare 500 mL of 0.135 M $Na_2CO_3(aq)$?

14.36 Sodium metal is produced from the electrolysis of molten sodium chloride in the Downs process (Section 12.15). Determine (a) the standard free energy of the reaction $2\ NaCl(s) \rightarrow 2\ Na(s) + Cl_2(g)$ and (b) the current needed to produce 500 g of sodium in 4.50 h.

Group 2: The Alkaline Earth Metals

14.37 Write the equation for the reaction between magnesium and hot water.

14.38 Complete the equation: $Ba(s) + H_2(g) \rightarrow$

14.39 CaO and BaO are sometimes used as drying agents for organic solvents such as pyridine, C_5H_5N. The drying agent and the product of the drying reaction are both insoluble in the organic solvent. (a) Write the balanced chemical equation corresponding to the reaction that occurs to dry the solvent. (b) For CaO, determine the standard free energy of the drying reaction.

14.40 Metal carbonates are often used as reagents in solid-state reactions because heating them creates finely divided metal oxides due to the release of $CO_2(g)$ from the crystal lattice. These finely divided particles, which are made freshly in situ, are often more reactive in heterogeneous solid-state reactions than the metal oxides found on the laboratory shelf because their freshness gives them a higher surface area that is more reactive. Determine the minimum temperature necessary to achieve spontaneous thermal decomposition of $BaCO_3$.

14.41 Predict and explain the trend in strengths of the Group 2 metals as reducing agents.

14.42 Explain the trend of decreasing lattice enthalpies of the chlorides of the Group 2 metals down the group.

14.43 Aluminum and beryllium have a diagonal relationship. Compare the chemical equations for the reaction of aluminum with aqueous sodium hydroxide to that of beryllium with aqueous sodium hydroxide.

14.44 Write the chemical equations for (a) the industrial preparation of magnesium metal from the magnesium chloride in seawater; (b) the action of water on calcium metal.

14.45 Predict the products of the following reactions and then balance the equations:
(a) $Mg(OH)_2 + HCl(aq) \rightarrow$
(b) $Ca(s) + H_2O(l) \rightarrow$
(c) $BaCO_3(s) \xrightarrow{\Delta}$

14.46 Predict the products of the following reactions and then balance the equations:
(a) $Mg(s) + Br_2(l) \rightarrow$
(b) $BaO(s) + Al(s) \rightarrow$
(c) $CaO(s) + SiO_2(s) \rightarrow$

14.47 (a) Write the Lewis structures for $BeCl_2$ and $MgCl_2$. How does the Lewis structure of $MgCl_2$ differ from that of $BeCl_2$? (b) Predict the Cl—Be—Cl bond angle. (c) What hybrid orbitals are used in the bonding in $BeCl_2$?

14.48 (a) Calculate the heat released when 50 kg of lime, the amount needed to produce 1000 kg of iron, reacts with water. (b) Describe a chemical process that could be used to reconvert the slaked lime into quicklime.

14.49 (a) Calculate the standard reaction enthalpy and entropy for the thermal decomposition of calcium carbonate as calcite (see Appendix 2A). (b) Determine the temperature at which the equilibrium constant for the thermal decomposition becomes greater than 1.

14.50 (a) Calculate the standard reaction enthalpy and entropy for the thermal decomposition of magnesium carbonate to MgO and CO_2 (see Appendix 2A). (b) Determine the temperature at which the equilibrium pressure of CO_2 above a sample of magnesium carbonate is 0.500 bar.

Group 13/III: The Boron Family

14.51 Write a balanced equation for the industrial preparation of aluminum from its oxide.

14.52 Write a balanced equation for the industrial preparation of impure boron.

14.53 Complete and balance the following skeletal equations:
(a) $B_2O_3(s) + Mg(l) \rightarrow$
(b) $Al(s) + Cl_2(g) \rightarrow$
(c) $Al(s) + O_2(g) \rightarrow$

14.54 Complete and balance the following skeletal equations:
(a) $Al_2O_3(s) + OH^-(aq) \rightarrow$
(b) $Al_2O_3(s) + H_3O^+(aq) + H_2O(l) \rightarrow$
(c) $B(s) + NH_3(g) \rightarrow$

14.55 Identify a use for (a) $AlCl_3$; (b) α-alumina; (c) $B(OH)_3$.

14.56 Identify a use for (a) BF_3; (b) $NaBH_4$; (c) $Al_2(SO_4)_3$.

14.57 What mass of aluminum can be produced by the Hall process in a period of 24.0 h, using a current of 3.5 MA?

14.58 The annual production of aluminum in the United States in 2000 was 3.6 Mt (1 Mt $= 1.0 \times 10^9$ kg). What mass of carbon, lost from the anode of the electrolysis cell, was required to produce this amount of aluminum by the Hall process?

14.59 (a) The standard free energy of formation of $Tl^{3+}(aq)$ is $+215\ kJ\cdot mol^{-1}$ at 25°C. Calculate the standard potential of the Tl^{3+}/Tl couple. (b) Will Tl^+ disproportionate in aqueous solution?

14.60 The standard potential of the Al^{3+}/Al couple is -1.66 V. Calculate the standard free energy of formation for $Al^{3+}(aq)$. Account for any differences between the standard free energy of formation of $Tl^{3+}(aq)$ (see Exercise 14.59) and that of $Al^{3+}(aq)$.

Group 14/IV: The Carbon Family

14.61 Describe the sources of silicon and write balanced equations for the three steps in the industrial preparation of silicon.

14.62 Describe the sources of carbon and how carbon is converted to graphite.

14.63 Compare the hybridization and structure of carbon in diamond and graphite. How do these features explain the physical properties of the two allotropes?

14.64 Explain why the size of the silicon atom does not permit a silicon analogue of the graphite structure.

14.65 Complete and balance the equations for the following reactions:
(a) $SiCl_4(l) + H_2(g) \rightarrow$
(b) $SiO_2(s) + C(s) \rightarrow$
(c) $Ge(s) + F_2(g) \rightarrow$
(d) $CaC_2(s) + H_2O(l) \rightarrow$

14.66 Complete and balance the equations for the following reactions:
(a) $Sn(s) + H_2O(l) + \text{excess } OH^-(aq) \rightarrow$
(b) $C(s) + H_2O(g) \rightarrow$
(c) $CH_4(g) + H_2O(g) \rightarrow$
(d) $Al_4C_3(s) + H_2O(l) \rightarrow$

14.67 Write a Lewis structure for the orthosilicate anion, SiO_4^{4-}, and deduce the formal charges and oxidation numbers of the atoms. Use the VSEPR model to predict the shape of the ion.

14.68 Use the VSEPR model to estimate the Si-O-Si bond angle in silica.

14.69 Determine the values of $\Delta H_r°$, $\Delta S_r°$, and $\Delta G_r°$ for the production of high-purity silicon by the reaction $SiO_2(s) + 2\ C(s, \text{graphite}) \rightarrow Si(s) + 2\ CO(g)$ at 25°C and estimate the temperature at which the equilibrium constant becomes greater than 1.

14.70 Determine the values of $\Delta H_r°$, $\Delta S_r°$, and $\Delta G_r°$ for the reaction $2\ CO(g) + O_2(g) \rightarrow 2\ CO_2(g)$ at 25°C and estimate the temperature at which the equilibrium constant becomes less than 1.

14.71 Determine the mass of HF as hydrofluoric acid that is required to etch 2.00 mg of SiO_2 from a glass plate in the reaction $SiO_2(s) + 6\ HF(aq) \rightarrow 2\ H_3O^+(aq) + SiF_6^{2-}(aq)$.

14.72 Explain why silicon tetrachloride reacts with water to produce SiO_2 but carbon tetrachloride does not react with water.

14.73 Describe the structures of a silicate in which the silicate tetrahedra share (a) one O atom; (b) two O atoms.

14.74 What is the empirical formula of a potassium silicate in which the silicate tetrahedra share (a) two O atoms and form a chain or (b) three O atoms and form a sheet? In each case, there are single negative charges on the unshared O atoms.

Glasses

14.75 What complex ion is formed when glass is etched by HF?

14.76 A typical procedure for etching glass involves covering the glass surface with a mask (a protective coating). The mask is then removed from the areas that are to be etched and a paste made of fluorite and sulfuric acid is placed on the surface. From standard reference sources, determine the chemical formula of fluorite and describe the chemical reactions that take place to achieve the etching.

14.77 Why is it unwise to store metal fluorides in glass bottles?

14.78 Solutions of strong bases stored in glass bottles react slowly with their containers. Write the balanced chemical equation for four possible reactions between OH^- and SiO_2.

Ceramics

14.79 Why is it desirable to remove iron from ceramic clays?

14.80 Draw a simple chemical picture to show how the removal of water helps to make aluminosilicates into rigid ceramics.

INTEGRATED EXERCISES

14.81 Describe evidence for the statement that hydrogen can act as both a reducing agent and an oxidizing agent. Give chemical equations to support your evidence.

14.82 (a) Reactions of lithium with various oxidizing agents have been examined for use in batteries. A particularly well-studied case is that of the lithium-sulfur battery. Determine the potential that is possible for a battery that operates on the reaction of Li(s) with S(s).

Media Link

14.83 (a) Using graphing software, such as that on the Web site for this book, plot standard potential as a function of atomic number for the elements of Groups 1 and 2 (refer to Appendix 2B for data). (b) What generalizations can be deduced from the graph?

Media Link

14.84 Using graphing software, such as that on the Web site for this book, and data from Fig. 1.45 and Appendix 2B, plot ionization energy as a function of standard potential for the elements of Groups 1 and 2. What generalizations can be drawn from the graph?

14.85 (a) Write equations for the reactions of hydrogen gas with the halogens, from fluorine to iodine. (b) Predict the relative vigor of the reactions. (c) Name the products when they are dissolved in aqueous solutions.

14.86 In aqueous solution, the beryllium(II) ion exists as $Be(H_2O)_4{}^{2+}$ ions. Write a chemical equation that illustrates the acidic character of this ion.

14.87 (a) Name the type of reaction that occurs between calcium oxide and silica in a blast furnace. (b) Write the chemical equation for a related reaction, that between calcium oxide and carbon dioxide.

14.88 Very pure boron is obtained by reducing boron trichloride with hydrogen gas; a by-product is hydrogen chloride gas. What volume of hydrogen gas (at 273 K and 1.00 atm) is needed for the production of 50.0 g of boron from the reduction of boron trichloride?

14.89 (a) What is the formula of the conjugate base of boric acid in water? (b) Write an equation showing that $B(OH)_3$ is an acid in water.

14.90 Arrange the elements aluminum, gallium, indium, thallium, tin, and germanium in order of increasing electronegativity (see Fig. 2.12) and increasing reducing strength.

14.91 (a) State the trends in first ionization energies and atomic radii down Groups 13/III and 14/IV. (b) Account for the trends. (c) How do the trends correlate with the properties of the elements?

14.92 (a) Suggest a reason for the observations that methane survives in an aqueous alkaline solution, but silane, SiH_4, reacts rapidly in the same solution. (b) Write a balanced chemical equation for the reaction of silane with water in an aqueous alkaline solution.

14.93 Hydrogen can be produced from the electrolysis of water. (a) Write the equation for the half-reaction for the production of hydrogen (refer to Appendix 2B, if necessary). (b) Is the hydrogen produced at the anode or the cathode? (c) Calculate the volume of hydrogen (at 273 K and 1.00 atm) that is produced if a current of 10.0 A is passed through an electrolytic cell for 30 min.

14.94 The reduction of tin(IV) oxide to (white) tin metal by graphite proceeds at moderately low temperatures. More commonly, tin(IV) oxide is reduced to the metal by an excess of carbon monoxide at temperatures above 980 K. (a) Write the two chemical equations for the reduction of SnO_2. (b) Determine the value of $\Delta G_r°$ for each reaction at 25°C. (c) How is the equilibrium constant affected by temperature in each case?

14.95 Is there any chemical support for the view that hydrogen should be classified as a member of Group 1? Would it be better to consider hydrogen a member of Group 17/VII? Give evidence that supports each view.

14.96 What justification is there for regarding the ammonium ion as an analogue of a Group 1 metal cation? Consider properties such as solubility, charge, and radius. The radius of NH_4^+ is 137 pm.

14.97 It has been proposed that the ability of a cation to polarize anions is proportional to its charge divided by its radius (see Section 2.15). (a) Use this criterion to arrange the *s*-block elements in order of increasing polarizing power. (b) Do the resulting values support the diagonal relationships within the block?

14.98 The first ionization energies of the Group 2 elements decrease smoothly down the group, but, in Group 13/III, the values for gallium and thallium are both higher than for aluminum. Suggest a reason.

14.99 Hydrogen burns in an atmosphere of bromine to give hydrogen bromide. If 120 mL of H_2 gas at 273 K and 1.00 atm combines with a stoichiometric amount of bromine and the resulting hydrogen bromide dissolves to form 150 mL of an aqueous solution, what is the molar concentration of the resulting hydrobromic acid solution?

14.100 The standard enthalpies of formation of BH_3(g) and diborane, B_2H_6 (8), are $+100\ kJ\cdot mol^{-1}$ and $+36\ kJ\cdot mol^{-1}$, respectively, and the enthalpies of formation of B(g) and H(g) are $+563\ kJ\cdot mol^{-1}$ and $+218\ kJ\cdot mol^{-1}$, respectively. (a) Use these values to calculate the mean bond enthalpies of the B—H bonds in each case (b) Assume that terminal B—H bonds have the same strengths in each compound, and estimate the bond enthalpy of the three-center B—H—B bonds in diborane. (c) Which bonds would you expect to be longer, the terminal B—H bonds or the three-center bonds? Explain your answer.

14.101 Suppose that the stability of carbonates when heated depends on the ability of the metal cation to polarize the carbonate ion and remove an oxide ion from it, so releasing carbon dioxide. Predict the order of thermal stability of the Groups 1 and 2 metal carbonates. Comment on the likely stability of aluminum carbonate.

14.102 When the mineral dolomite, $CaCO_3\cdot MgCO_3$, is heated, it gives off carbon dioxide and forms a mixture of a metal oxide and a metal carbonate. Which oxide is formed, CaO or MgO? Which carbonate remains, $CaCO_3$ or $MgCO_3$? Justify your answer.

14.103 (a) Write a balanced chemical equation to predict the products you would expect to be formed when sodium hydride reacts with methanol. (b) Draw the Lewis structure of the organic product. (c) Write a balanced equation for the reaction of that organic product with water.

14.104 (a) What is the hybridization of carbon atoms in carbon nanotubes? (b) Simple nanotubes are made up of sheets resembling a layer of graphite (similar to chickenwire), each of which has been curled up and bonded to itself. How many hexagons must be strung together (the hexagons denoted by the highlighted bonds in the diagram below) around the circumference of a nanotube in order to form a tube that is approximately 1.3 nm in diameter? The diagram shows the orientation of hexagons with respect to the curvature of the nanotube. The C—C bond distance in carbon nanotubes is 142 pm.

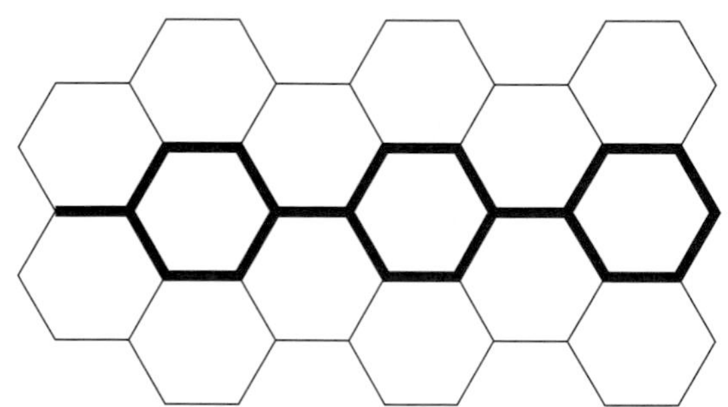

14.105 Boron nitride has two crystalline forms, hexagonal and cubic. These forms are similar to the two common forms of carbon, graphite and diamond. (a) The nitrogen atoms of BN form a face-centered cubic unit cell in which half of the tetrahedral interstitial sites are occupied by B atoms (if all B and N atoms were replaced by C atoms, the diamond structure would result). Calculate the density of cubic BN if the edge length of the unit cell is 361.5 pm. (b) Which form of BN is favored at high pressures, hexagonal or cubic? The density of hexagonal BN is $2.29\ g\cdot cm^{-3}$.

14.106 Both calcium and strontium crystallize in face-centered cubic unit cells. Determine which metal is more dense, calcium or strontium, given that their radii are 197 pm and 215 pm, respectively.

14.107 Draw simple molecular orbital energy-level diagrams to indicate how the bonding in the saline hydrides, such as NaH or KH, differs from that for bonds from hydrogen to a light *p*-block element such as carbon or nitrogen.

14.108 Buckminsterfullerene, C_{60}, can be hydrogenated, but, to date, a compound with the formula $C_{60}H_{60}$ has not been prepared. The most hydrogenated form of C_{60} known is $C_{60}H_{36}$. Suggest an explanation for why the hydrogenation stops at this point.

14.109 (a) Examine the structures of diborane, B_2H_6, and $Al_2Cl_6(g)$, which may be found on the Web site for this book. Compare the bonding in these two compounds. How are they similar? (b) What are the differences, if any, in the types of bonds formed? (c) What is the hybridization of the Group 13/III element? (d) Are the molecules planar? If not, describe their shapes.

14.110 The standard heat of formation of $SiCl_4(g)$ is $-662.75\ kJ\cdot mol^{-1}$ and its standard entropy is $+330.86\ J\cdot mol^{-1}\cdot K^{-1}$. Calculate the temperature range over which the reduction of $SiCl_4(g)$ to Si(s) with hydrogen gas is spontaneous.

14.111 Use data from Appendix 2A to calculate the standard potential for a galvanic cell based on the reaction of tin(II) oxide with oxygen gas to form tin(IV) oxide.

14.112 The crystal structure of Li_3N is a structure based on two types of alternating layers. One layer is made up of nitrogen atoms surrounded by six lithium atoms. The Li atoms are each connected to three nitrogen atoms, as shown at the top of the next column. The Li—N distances within the layers are all equal at 213 pm. The Li—N—Li bond angles are 60° and the N—Li—N angles are 120°. The second type of layer has only lithium atoms in it. Each of these lithium atoms bonds to two nitrogen atoms ($d_{Li\text{-}N} = 1.94$) in the layers above and below. These N—Li—N angles are 180°. (a) From this information, draw a reasonable unit cell for the crystal structure of lithium nitride. (b) To what crystal system does this structure belong? (c) How many formula units of Li_3N are present in each unit cell? (d) Determine the density of Li_3N from this data.

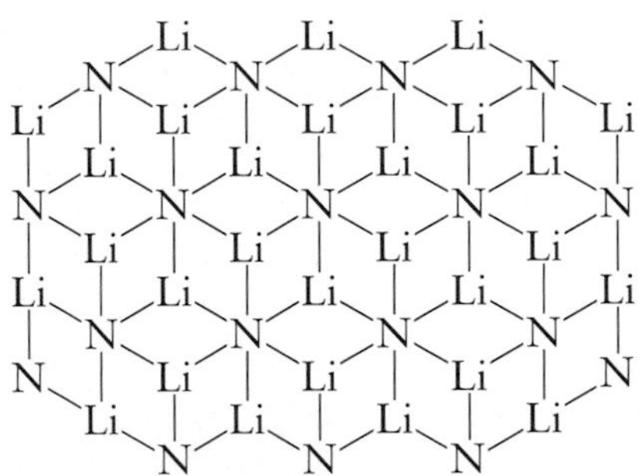

14.113 A unit cell for the calcite structure may be found on the Web site for this book. From this structure, determine (a) the crystal system and (b) the number of formula units present in the unit cell.

14.114 Using the Web site for this book, examine the unit cells of calcite and dolomite. (a) How are these two structures the same? (b) What makes them different? (c) Where are the magnesium and calcium ions located in dolomite?

14.115 How many calcium ions, sulfate ions, and water molecules are found in the unit cell of gypsum (the unit cell may be viewed on the Web site for this book)? What is the empirical formula of the compound?

14.116 (a) Consider the reaction of methane with ammonia in the presence of oxygen to produce hydrogen cyanide and water.

$$2\ CH_4(g) + 2\ NH_3(g) + 3\ O_2(g) \longrightarrow 2\ HCN(g) + 6\ H_2O(g)$$

What are the standard enthalpy, entropy, and free energy changes associated with this reaction? (b) The reaction is carried out at 1100°C, using platinum metal as a catalyst. Is the high temperature necessary for thermodynamic or kinetic reasons? Explain.

CHAPTER 15

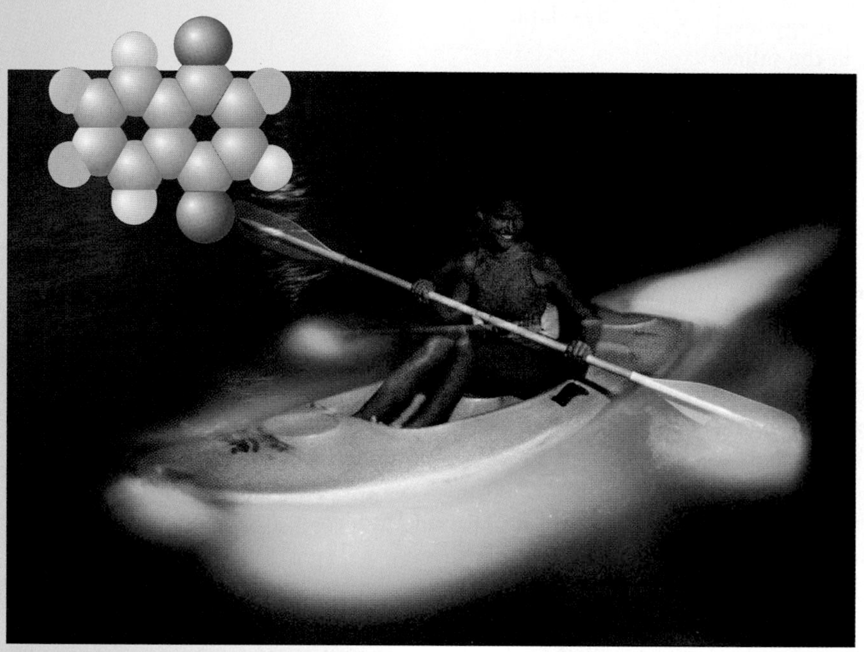

A bay on the island of Vieques, off the coast of Puerto Rico, is famous for the bioluminescent bacteria it contains. These bacteria luminesce when they are disturbed by motion. The bacteria contain one of the varieties of compounds called luciferin. A model compound for the generation of chemiluminescence is luminol, shown superimposed here, which luminesces in the presence of hydrogen peroxide.

Why Do We Need to Know This Material?

The elements in the last four groups of the periodic table illustrate the rich variety of the properties of the nonmetals and, hence, demonstrate many of the principles of chemistry. These elements include some that are vital to life, such as the nitrogen of proteins, the oxygen of the air, and the phosphorus of our bones, so a familiarity with their properties helps us to understand the world around us. Many of these elements are also central to the materials that provide the backbone of emerging technology.

What Do We Need to Know Already?

The prerequisites for this chapter are essentially the same as for Chapter 14. It would be a good idea to review the section on periodic trends at the start of that chapter. Because the nonmetals form molecular compounds, it would also be helpful to review covalent bonding in Chapters 2 and 3. The bulk properties of nonmetallic materials are affected by intermolecular forces (Sections 5.1–5.5).

THE ELEMENTS: THE LAST FOUR MAIN GROUPS

In the next breath you take, 99.99% of the atoms you inhale will be of elements in the final four groups of the periodic table. Except for the gases containing carbon and hydrogen, air is made up solely of elements from this part of the *p* block, some as elements and some as compounds. The *p*-block elements are present in many of the compounds responsible for life. They are also used to create fascinating and useful modern materials, such as superconductors, electroluminescent materials that glow when excited by an electrical current, and high-performance ceramics.

We can understand the wide range of properties of these fascinating elements by considering their atomic structures. Here, on the far right of the periodic table, the electron configurations are only a few electrons short of a closed-shell, noble-gas configuration, and the effective nuclear charge is relatively high. As a result, the chemistry of the elements in Groups 15/V–18/VIII is dominated by their high ionization energies and—apart from the noble gases—their high electronegativities. When the elements in this part of the periodic table form compounds with other nonmetals, they do so by sharing electrons to form covalent bonds. Metallic character increases down each group, but the only element in this region considered to be metallic is bismuth, at the bottom of Group 15/V. Arsenic, antimony, tellurium, and polonium are metalloids.

The general structure of this chapter is the same as that of Chapter 14. We work systematically across the remainder of the main-group elements, highlighting particularly important elements and compounds. As in Chapter 14, we highlight periodic trends, the production of the elements, and the properties and applications of the elements and their important compounds.

GROUP 15/V: THE NITROGEN FAMILY

Atoms of the Group 15/V elements have the valence electron configuration ns^2np^3 (Table 15.1). The properties of the elements vary sharply in this group, from the almost unreactive gas nitrogen, through the soft nonmetal phosphorus, which is so reactive it ignites in air, to the important semiconducting materials arsenic and antimony, and the largely metallic bismuth (Fig. 15.1). This range of behavior is reflected in the chemical properties of the elements and their compounds (Table 15.2). For example, all the oxides of nitrogen and phosphorus are acidic, but bismuth's oxide is basic.

15.1 The Group 15/V Elements

Nitrogen is rare in Earth's crust, but elemental nitrogen is the principal component of our atmosphere (76% by mass). This abundance of elemental nitrogen is unusual in our solar system. Because the atmosphere of

TABLE 15.1 The Group 15/V Elements

Valence configuration: ns^2np^3

Z	Name	Symbol	Molar mass, $g \cdot mol^{-1}$	Melting point, °C	Boiling point, °C	Density, $g \cdot cm^{-3}$ at 25°C	Normal form*
7	nitrogen	N	14.01	−210	−196	1.04‡	colorless gas
15	phosphorus	P	30.97	44	280	1.82	white or red nonmetal
33	arsenic	As	74.92	613s†	—	5.78	gray metalloid
51	antimony	Sb	121.76	631	1750	6.69	blue-white lustrous metalloid
83	bismuth	Bi	208.98	271	1650	8.90	white-pink metal

**Normal form* means the appearance and state of the element at 25°C and 1 atm.
†The symbol s denotes that the element sublimes.
‡For the liquid at its boiling point.

Mars is only 2.6% nitrogen, colonization of Mars would depend on the availability of nitrogen-bearing rocks in the planet's crust. The atmosphere of Venus is 96% carbon dioxide, the remainder mostly nitrogen. Jupiter has an atmosphere rich in ammonia, but the outer planets are so cold that the ammonia has condensed. Even nitrogen would condense on Uranus, Neptune, and Pluto, where the average surface temperatures are below −196°C.

Lavoisier named the element *azote*, which means "lifeless." Ironically, we now know that there would be no life as we know it without nitrogen.

Plants need nitrogen to grow. However, they cannot use N_2 directly because the strength of the N≡N bond (944 $kJ \cdot mol^{-1}$) makes nitrogen gas almost as inert as the noble gases. To be available for organisms, nitrogen must first be **fixed,** or combined with other elements into more useful compounds. Once fixed, nitrogen can be converted to other compounds for use as medicines, fertilizers, explosives, and plastics. Lightning converts some nitrogen to its oxides, which rain then washes into the soil. Some bacteria also fix nitrogen in nodules on the roots of clover, beans, peas, alfalfa, and other legumes (Fig. 15.2). An intensely active field of research is the search for catalysts that can mimic these bacteria and fix

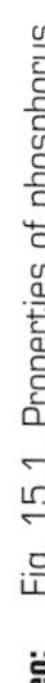

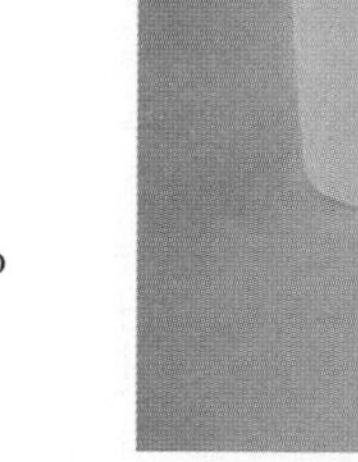

FIGURE 15.1 The elements of Group 15/V. Back row, from left to right: liquid nitrogen, red phosphorus, arsenic. Front row: antimony and bismuth.

TABLE 15.2 Chemical Properties of the Group 15/V Elements

Reactant	Reaction with Group 15/V Element (E)
hydrogen	$N_2(g) + 3\ H_2(g) \longrightarrow 2\ NH_3(g)$ $P_4(s) + 6\ H_2(g) \longrightarrow 4\ PH_3(g)$
oxygen	$N_2(g) + x\ O_2(g) \longrightarrow 2\ NO_x(g)$ $P_4(s) + 3$ or $5\ O_2(g) \longrightarrow P_4O_6(s)$ or $P_4O_{10}(s)$ $4\ As(s) + 3\ O_2(g) \longrightarrow As_4O_6(s)$ $4\ E(s) + 3\ O_2(g) \longrightarrow 2\ E_2O_3(s)$, E = Sb, Bi
water	no reaction
halogen	$2\ E(s) + 3\ X_2(s, l, g) \longrightarrow 2\ EX_3(s, l)$, E = P, As, Sb, Bi $2\ E(s) + 5\ X_2(s, l, g) \longrightarrow 2\ EX_5(s, l)$, E = P, As, Sb

FIGURE 15.2 The bacteria that inhabit these nodules on the roots of a pea plant are responsible for fixing atmospheric nitrogen and making it available to the plant.

nitrogen at ordinary temperatures. At present, the Haber synthesis of ammonia is the main industrial route for fixing nitrogen, but it requires costly high pressures (see Section 9.12).

Pure nitrogen gas is obtained by the fractional distillation of liquid air. Air is cooled to below −196°C by repeated expansion and compression in a refrigerator like that described in Section 4.16. The liquid mixture is then warmed, and the nitrogen (b.p. −196°C) boils off while most of the argon (b.p. −186°C) and oxygen (b.p. −183°C) remain liquid. Nitrogen gas produced industrially is used primarily as a raw material for the synthesis of ammonia in the Haber process. The gas is also used to provide a nonoxidizing atmosphere in the electronics industry. Nitrogen glove boxes—glass-sided enclosures fitted with rubber gloves for manipulating the contents—are used when oxygen-sensitive materials are being studied. Liquid nitrogen is used as a cooling agent when low temperatures (down to about 77 K) are required.

Like other elements at the head of their group, nitrogen differs in many ways from its congeners. For example, nitrogen is highly electronegative ($\chi = 3.0$, about the same as for chlorine). Because of its high electronegativity, nitrogen is the only element in Group 15/V that forms hydrides capable of hydrogen bonding. Because its atoms are small, nitrogen can form multiple bonds with other Period 2 atoms by using its *p*-orbitals. As we shall see, its small size and the unavailability of *d*-orbitals account for many of the differences between the chemical and physical properties of nitrogen and its congeners. Another important chemical property of nitrogen is its wide range of oxidation numbers: compounds are known for each whole-number oxidation number from −3 (in NH_3) to +5 (in nitric acid and the nitrates). It also occurs with fractional oxidation numbers, such as $-\frac{1}{3}$ in the azide ion, N_3^-.

The properties of phosphorus, in Period 3, differ significantly from those of nitrogen, in large part because atoms of phosphorus are bigger and have 3*d*-orbitals available, and the electronegativity of phosphorus is lower. The atomic radius of phosphorus is nearly 50% larger than that of nitrogen, so two phosphorus atoms are too big to approach each other closely enough for their 3*p*-orbitals to overlap and form multiple bonds.

Thus, whereas nitrogen can form multiply bonded structures such as N_2O_3 (**1**), phosphorus forms additional single bonds, as in P_4O_6 (**2**). The size of its atoms and the availability of 3*d*-orbitals means that phosphorus may form as many as six bonds, whereas nitrogen can form only four (see Section 2.11).

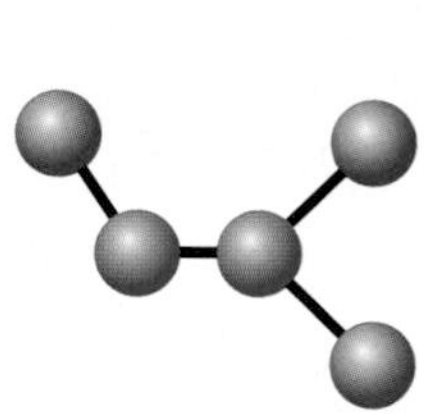

1 Dinitrogen trioxide, N_2O_3

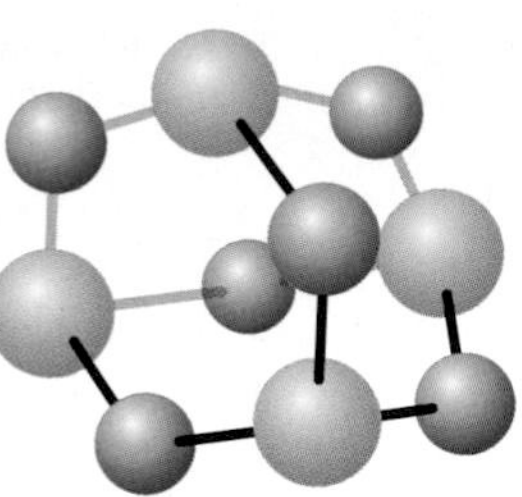

2 Phosphorus(III) oxide, P_4O_6

Phosphorus is obtained from the *apatites*, which are mineral forms of calcium phosphate, $Ca_3(PO_4)_2$. The rocks are heated in an electric furnace with carbon and sand:

$$2\,Ca_3(PO_4)_2(s) + 6\,SiO_2(s) + 10\,C(s) \xrightarrow{\Delta} P_4(g) + 6\,CaSiO_3(l) + 10\,CO(g)$$

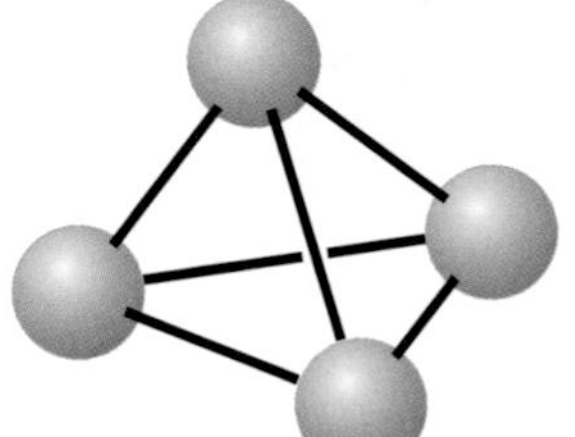

3 Phosphorus, P_4

The phosphorus vapor condenses as *white phosphorus*, a soft, white, poisonous molecular solid consisting of tetrahedral P_4 molecules (**3**). This allotrope is highly reactive, in part because of the strain associated with the 60° angles between the bonds. It is very dangerous to handle because it bursts into flame on contact with air and can cause severe burns. White phosphorus is normally stored under water. It changes into *red phosphorus* when heated in the absence of air. Red phosphorus is less reactive than the white allotrope; however, it can be ignited by friction and is used in the striking surfaces of matchbooks. The friction created by rubbing a match across the surface ignites the phosphorus, which, in turn, lights the highly flammable material in the match head. Red phosphorus is thought to consist of chains of linked P_4 tetrahedra.

The name *phosphorus* means "light bringer."

Vapor rising from white phosphorus in moist air glows with a yellow-green light. The oxides produced by the reaction of phosphorus with the oxygen in air are formed in electronically excited states, and light is given off as electrons fall back into the ground state.

Arsenic and antimony are metalloids. They have been known in the pure state since ancient times because they are easily obtained from their ores (Fig. 15.3). In the elemental state, they are used primarily in the semiconductor industry and in the lead alloys employed as electrodes in storage batteries and in the semiconductor industry. Gallium arsenide is used in lasers, including the lasers used in CD players. Metallic bismuth, with its large, weakly bonded atoms, has a low melting point and is used in alloys that serve as fire detectors in sprinkler systems: the alloy melts when a fire breaks out nearby, and the sprinkler system is activated. Bismuth is also used to make low-temperature castings. Like ice, solid bismuth is less dense than the liquid. As a result, molten bismuth does not shrink when it solidifies in molds.

FIGURE 15.3 The minerals (from left to right) orpiment, As_2S_3, stibnite, Sb_2S_3, and realgar, As_4S_4, are all ores that have been used as sources of Group 15/V elements.

Nitrogen is highly unreactive as an element, largely because of its strong triple bond. White phosphorus is highly reactive. The differences in properties between the nonmetals nitrogen and phosphorus can be traced to the larger atomic radius of phosphorus and the availability of d-orbitals in its valence shell. Metallic character increases down the group.

15.2 Compounds with Hydrogen and the Halogens

By far the most important hydrogen compound of a Group 15/V element is ammonia, NH_3, which is prepared in huge amounts by the Haber process. Small quantities of ammonia are present naturally in the atmosphere as a result of the bacterial decomposition of organic matter in the absence of air. This decomposition typically occurs in lake and river beds, in swamps, and in cattle feedlots.

Ammonia is a pungent, toxic gas that condenses to a colorless liquid at −33°C. The liquid resembles water in its physical properties, including its ability to act as a solvent for a wide range of substances. Because the dipole moment of the NH_3 molecule (1.47 D) is lower than that of the H_2O molecule (1.85 D), salts with strong ionic character, such as KCl, cannot dissolve in ammonia. Salts with polarizable anions tend to be more soluble in ammonia than salts with stronger ionic character. For example, iodides are more soluble than chlorides in ammonia. Autoprotolysis occurs to a much smaller extent in ammonia than in water:

$$2\,NH_3(am) \rightleftharpoons NH_4^+(am) + NH_2^-(am) \qquad K_{am} = 1 \times 10^{-33} \text{ at } -35°C$$

Consequently, very strong bases that would be protonated in water, such as the cyclopentadienide anion, $C_5H_5^-$, survive in ammonia.

Ammonia is very soluble in water because the NH_3 molecules can form hydrogen bonds to H_2O molecules. Ammonia is a weak Brønsted base in water; it is also a reasonably strong Lewis base, particularly toward *d*-block elements. For example, it reacts with Cu^{2+}(aq) ions to give a deep blue complex (Fig. 15.4):

$$Cu^{2+}(aq) + 4\ NH_3(aq) \longrightarrow Cu(NH_3)_4{}^{2+}(aq)$$

Ammonium salts decompose when heated:

$$(NH_4)_2CO_3(s) \xrightarrow{\Delta} 2\ NH_3(g) + CO_2(g) + H_2O(g)$$

The pungent smell of decomposing ammonium carbonate made it an effective "smelling salt," a stimulant used to revive people who had fainted.

The pungency of heated ammonium chloride was known in antiquity to the Ammonians, the worshippers of the Egyptian god Ammon.

The ammonium cation of an ammonium salt may be oxidized by an anion that has oxidizing character, such as a nitrate. The products depend on the temperature of the reaction:

$$NH_4NO_3(s) \xrightarrow{250^\circ C} N_2O(g) + 2\ H_2O(g)$$

$$2\ NH_4NO_3(s) \xrightarrow{> 300^\circ C} 2\ N_2(g) + O_2(g) + 4\ H_2O(g)$$

The explosive violence of the second reaction is the reason ammonium nitrate is used as a component of dynamite. Ammonium nitrate has a high nitrogen content (33.5% by mass) and is highly soluble in water. These characteristics make it attractive as a fertilizer, which is its principal use.

Media Link Animation: Fig. 15.4, Formation of a copper-ammonia complex

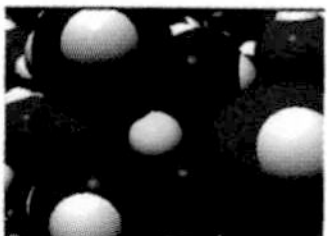

View an animation of the formation of a copper-ammonia complex.

FIGURE 15.4 When aqueous ammonia is added to a copper(II) sulfate solution, first a light-blue precipitate of $Cu(OH)_2$ forms (the cloudy region at the top, which appears dark because it is backlit). The precipitate disappears when more ammonia is added to form the dark blue complex $Cu(NH_3)_4{}^{2+}$ by a Lewis acid-base reaction.

Hydrazine, NH_2NH_2, is an oily, colorless liquid. It is prepared by the gentle oxidation of ammonia with alkaline hypochlorite solution:

$$2\ NH_3(aq) + ClO^-(aq) \xrightarrow{\text{aqueous alkali}} N_2H_4(aq) + Cl^-(aq) + H_2O(l)$$

Its physical properties are very similar to those of water; for instance, its melting point is 1.5°C and its boiling point is 113°C. However, its chemical properties are very different. It is dangerously explosive and is normally stored and used in aqueous solution. Hydrazine is used as a rocket fuel and to eliminate dissolved, corrosive oxygen from the water used in high-pressure, high-temperature steam furnaces:

$$N_2H_4(aq) + O_2(g) \longrightarrow N_2(g) + 2\ H_2O(l)$$

Nitrogen has oxidation number +3 in the nitrogen halides. Nitrogen trifluoride, NF_3, is the most stable halide; it does not react with water. However, NCl_3 does react with water, forming ammonia and hypochlorous acid. Nitrogen triiodide, NI_3, which is known only in combination with ammonia as an "ammoniate" (the analogue of a hydrate) is so unstable that it decomposes explosively when lightly touched.

Nitrides are solids that contain the nitride ion, N^{3-}. Nitrides are stable only for small cations such as lithium or magnesium. Magnesium nitride, Mg_3N_2, is formed together with the oxide when magnesium is burned in air (Fig. 15.5):

$$3\ Mg(s) + N_2(g) \xrightarrow{\Delta} Mg_3N_2(s)$$

Magnesium nitride, like all nitrides, dissolves in water to produce ammonia and the corresponding hydroxide:

$$Mg_3N_2(s) + 6\ H_2O(l) \longrightarrow 3\ Mg(OH)_2(s) + 2\ NH_3(g)$$

In this reaction, the nitride ion acts as a strong base, accepting protons from water to form ammonia.

The azide ion is a highly reactive polyatomic anion of nitrogen, N_3^-. Its most common salt, sodium azide, NaN_3, is prepared from dinitrogen oxide and molten sodium amide:

$$N_2O(g) + 2\ NaNH_2(l) \xrightarrow{175°C} NaN_3(l) + NaOH(l) + NH_3(g)$$

Sodium azide, like most azide salts, is shock sensitive. It is used in automobile air bags, where it decomposes to elemental sodium and nitrogen when detonated (see Section 4.10):

$$2\ NaN_3(s) \longrightarrow 2\ Na(s) + 3\ N_2(g)$$

The azide ion is a weak base and accepts a proton to form its conjugate acid, hydrazoic acid, HN_3. Hydrazoic acid is a weak acid similar in strength to acetic acid.

The hydrogen compounds of other members of Group 15/V are much less stable than ammonia and decrease in stability down the group. Phosphine, PH_3, is a poisonous gas that smells faintly of garlic and bursts into

FIGURE 15.5 This sample of magnesium nitride was formed by burning magnesium in an atmosphere of nitrogen. When magnesium burns in air, the product is the oxide as well as the nitride.

Media Link Lab video: Fig. 15.5, Reaction of magnesium and nitrogen flame in air if it is slightly impure. It is much less soluble than ammonia in water because PH_3 cannot form hydrogen bonds to water. Aqueous solutions of phosphine are neutral, for the electronegativity of phosphorus is so low that the lone pair in PH_3 is spread over the hydrogen atoms as well as the phosphorus atom; consequently, the molecule has only a very weak tendency to accept a proton ($pK_b = 27.4$).

Because PH_3 is the very weak parent acid of the strong Brønsted base P^{3-}, we can form phosphine by protonating phosphide ions with a Brønsted acid. Even water is a sufficiently strong proton donor:

$$2\ P^{3-}(s) + 6\ H_2O(l) \longrightarrow 2\ PH_3(g) + 6\ OH^-(aq)$$

Phosphorus trichloride, PCl_3, and phosphorus pentachloride, PCl_5, are the two most important halides of phosphorus. The former is prepared by direct chlorination of phosphorus. Phosphorus trichloride, a liquid, is a major intermediate for the production of pesticides, oil additives, and flame retardants. Phosphorus pentachloride, a solid, is made by allowing the trichloride to react with more chlorine (recall Fig. 2.10).

A typical reaction of the nonmetal halides is their reaction with water to give oxoacids, without a change in oxidation number:

$$PCl_3(l) + 3\ H_2O(l) \longrightarrow H_3PO_3(s) + 3\ HCl(g)$$

This reaction is an example of a **hydrolysis reaction,** a reaction with water in which new element-oxygen bonds are formed. Another example is the reaction of PCl_5 (phosphorus oxidation state +5) with water to produce phosphoric acid, H_3PO_4 (also phosphorus oxidation state +5):

$$PCl_5(s) + 4\ H_2O(l) \longrightarrow H_3PO_4(l) + 5\ HCl(g)$$

This reaction is violent and dangerous.

An interesting feature of phosphorus pentachloride is that it is an ionic solid of tetrahedral PCl_4^+ cations and octahedral PCl_6^- anions, but it vaporizes to a gas of trigonal bipyramidal PCl_5 molecules (see Section 2.11). Phosphorus pentabromide is also molecular in the vapor and ionic as the solid; but in the solid, the anions are simply Br^- ions, presumably because six bulky Br atoms simply do not fit around a central P atom.

The important compounds of nitrogen with hydrogen are ammonia, hydrazine, and hydrazoic acid, the parent of the shock-sensitive azides. Phosphine forms neutral solutions in water; reaction of nonmetal halides with water—hydrolysis—produces oxoacids without change in oxidation number.

15.3 Nitrogen Oxides and Oxoacids

There are several nitrogen oxides, with properties that are related to the oxidation numbers of nitrogen, which range from +1 to +5. All nitrogen oxides are acidic and some are the acid anhydrides of the nitrogen oxoacids (Table 15.3). In atmospheric chemistry, where the oxides play an important two-edged role in both maintaining and polluting the atmosphere, they are referred to collectively as NO_x (read "nox").

TABLE 15.3 The Oxides and Oxoacids of Nitrogen

Oxidation number	Oxide		Oxoacid	
	Formula	Name	Formula	Name
5	N_2O_5	dinitrogen pentoxide	HNO_3	nitric acid
4	NO_2*	nitrogen dioxide	—	
	N_2O_4	dinitrogen tetroxide	—	
3	N_2O_3	dinitrogen trioxide	HNO_2	nitrous acid
2	NO	nitrogen monoxide nitric oxide	—	
1	N_2O	dinitrogen monoxide nitrous oxide	$H_2N_2O_2$	hyponitrous acid

*$2\ NO_2 \rightleftharpoons N_2O_4$.

Dinitrogen oxide, N_2O (oxidation number +1), is commonly called nitrous oxide. It is formed by gently heating ammonium nitrate:

Like many of the reactions in this book, this reaction is extremely hazardous. Don't try it.

$$NH_4NO_3(s) \xrightarrow{250°C} N_2O(g) + 2\ H_2O(g)$$

Because it is tasteless, is nontoxic in small amounts, and dissolves readily in fats, N_2O is sometimes used as a foaming agent and propellant for whipped cream. Dinitrogen oxide is unreactive and does not react at ordinary temperatures with the halogens, the alkali metals, or ozone. However, organic matter burns in it after being ignited because the heat released in the reaction decomposes N_2O into its elements and hence provides a supply of oxygen that sustains the combustion.

Nitrogen oxide (or nitrogen monoxide), NO (oxidation number +2), is commonly called nitric oxide. It is prepared industrially by the catalytic oxidation of ammonia:

$$4\ NH_3(g) + 5\ O_2(g) \xrightarrow{1000°C,\ Pt} 4\ NO(g) + 6\ H_2O(g)$$

The conversion of atmospheric nitrogen into NO in hot airplane and automobile engines contributes to the problem of acid rain and the formation of smog. Small amounts of nitrogen oxide occur naturally in our bodies, where it acts as a neurotransmitter, helps to dilate blood vessels, and participates in other physiological changes.

Nitrogen oxide contributes to the destruction of the ozone layer:

$$NO + O_3 \longrightarrow NO_2 + O_2$$

$$NO_2 + O \longrightarrow NO + O_2$$

The net result of these two reactions ($O_3 + O \rightarrow 2\ O_2$) is the destruction of an ozone molecule. Because the NO is regenerated, a small amount of nitrogen oxide can eliminate a large amount of ozone.

In the laboratory, nitrogen oxide can be prepared by reducing a nitrite with a mild reducing agent such as I^-:

$$2\,NO_2^-(aq) + 2\,I^-(aq) + 4\,H^+(aq) \longrightarrow 2\,NO(g) + I_2(aq) + 2\,H_2O(l)$$

Nitrogen oxide is a colorless gas that is rapidly oxidized to nitrogen dioxide on exposure to air:

$$2\,NO(g) + O_2(g) \longrightarrow 2\,NO_2(g)$$

Nitrogen dioxide, NO_2 (oxidation number +4), is a choking, poisonous, brown gas that contributes to the color and odor of smog. The molecule has an odd number of electrons, and in the gas phase it exists in equilibrium with its colorless dimer N_2O_4. Only the dimer exists in the solid, so the brown gas condenses to a colorless solid. When it dissolves in water, NO_2 disproportionates into nitric acid (oxidation number +5) and nitrogen oxide (oxidation number +2):

$$3\,NO_2(g) + H_2O(l) \longrightarrow 2\,HNO_3(aq) + NO(g)$$

Nitrogen dioxide in the atmosphere undergoes the same reaction and contributes to the formation of acid rain. It also initiates a complex sequence of smog-forming photochemical reactions.

Both NO and NO_2 are odd-electron, paramagnetic molecules. When a mixture of the odd-electron molecules NO and NO_2 is cooled to −20°C, the single electrons pair and the two radicals combine to form the blue liquid dinitrogen trioxide, N_2O_3 (Fig. 15.6, also see **1**). This compound, in which the oxidation number of nitrogen is +3, is unstable as a gas. It is the anhydride of nitrous acid, HNO_2, and forms that acid when it dissolves in water:

$$N_2O_3(g) + H_2O(l) \longrightarrow 2\,HNO_2(aq)$$

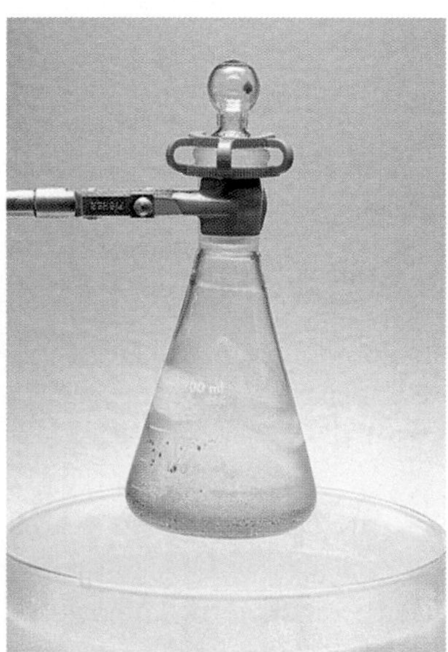

FIGURE 15.6 Dinitrogen trioxide, N_2O_3, condenses to a deep blue liquid that freezes at −100°C to a pale blue solid, as shown here. On standing, it turns green as a result of partial decomposition into nitrogen dioxide, a yellow-brown gas.

Nitrous acid has not been isolated in pure form but is widely used in aqueous solution.

Nitrites are produced by the reduction of nitrates with hot metal:

$$KNO_3(s) + Pb(s) \xrightarrow{350°C} KNO_2(s) + PbO(s)$$

Most nitrites are soluble in water and mildly toxic. Despite their toxicity, nitrites are used in the processing of meat products because they form a pink complex with hemoglobin and inhibit the oxidation of blood (a reaction that turns the meat brown). Nitrites are responsible for the pink color of ham, sausages, and other cured meat, in which nitrites are used to retard bacterial growth.

Nitric acid, HNO_3 (oxidation number +5) is used extensively in the production of fertilizers and explosives. It is made by the three-stage *Ostwald process*:

Step 1 Oxidation of ammonia, nitrogen's oxidation number increases from −3 to +2:

$$4\ NH_3(g) + 5\ O_2(g) \xrightarrow{850°C,\ 5\ atm,\ Pt/Rh} 4\ NO(g) + 6\ H_2O(g)$$

Step 2 Oxidation of nitrogen oxide, nitrogen's oxidation number increases from +2 to +4:

$$2\ NO(g) + O_2(g) \longrightarrow 2\ NO_2(g)$$

Step 3 Disproportionation in water, nitrogen's oxidation number changes from +4 to +5 and +2:

$$3\ NO_2(g) + H_2O(l) \longrightarrow 2\ HNO_3(aq) + NO(g)$$

Nitric acid, a colorless liquid that boils at 83°C, is normally used in aqueous solution. Concentrated nitric acid is often pale yellow as a result of partial decomposition of the acid to NO_2. Because nitrogen has its highest oxidation number (+5) in HNO_3, nitric acid is an oxidizing agent as well as an acid. It is used to make the explosives nitroglycerin and trinitrotoluene (TNT). Much of its strength as an oxidizing agent—the strongly negative value of the reaction free energy when it takes part in an oxidizing reaction—is due to the strong bond in the N_2 molecule, which is a product of the reactions.

Nitrogen forms oxides in each of its integer oxidation states from +1 to +5; the properties of the oxides and oxoacids can be explained in terms of the oxidation number of nitrogen.

15.4 Phosphorus Oxides and Oxoacids

The oxoacids and oxoanions of phosphorus are among the most heavily manufactured chemicals. Phosphate fertilizer production consumes two-thirds of all the sulfuric acid produced in the United States.

The structures of the phosphorus oxides are based on the tetrahedral PO_4 unit, which is similar to the unit that occurs in the oxides of its neighbor silicon. White phosphorus burns in a limited supply of air to form phosphorus(III) oxide, P_4O_6 (**2**):

$$P_4(s, \text{white}) + 3\ O_2(g) \longrightarrow P_4O_6(s)$$

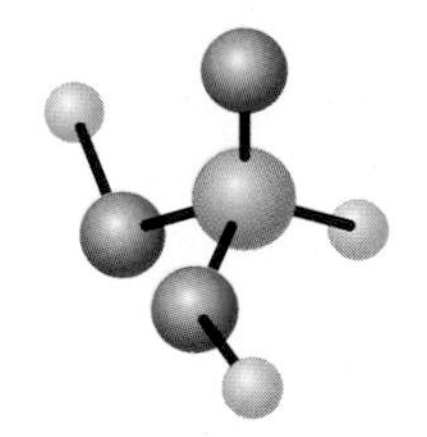

4 Phosphorous acid, H_3PO_3

The molecules are tetrahedral, like P_4, but an O atom lies between each pair of P atoms. Phosphorus(III) oxide is the anhydride of phosphorous acid, H_3PO_3 (**4**), and is converted to the acid by cold water:

$$P_4O_6(s) + 6\ H_2O(l) \longrightarrow 4\ H_3PO_3(aq)$$

Although its formula suggests that it should be a triprotic acid, H_3PO_3 is in fact diprotic because one of the H atoms is attached directly to the P atom and the P—H bond is nonpolar (Section 10.11).

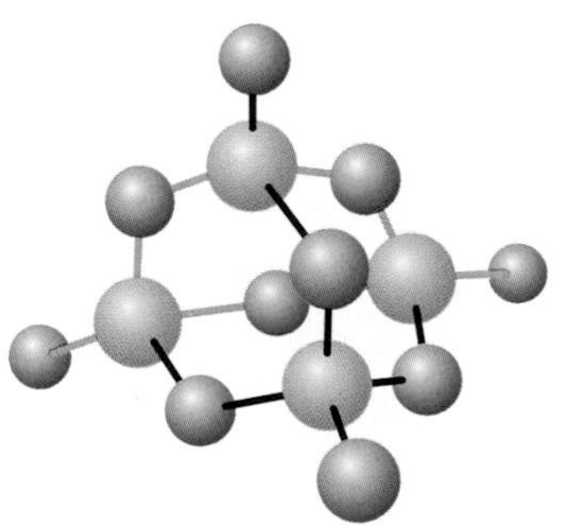

5 Phosphorus(V) oxide, P_4O_{10}

When phosphorus burns in an ample supply of air, it forms phosphorus(V) oxide, P_4O_{10} (**5**). This white solid reacts so vigorously with water that it is widely used in the laboratory as a drying agent. Phosphorus(V) oxide is the anhydride of phosphoric acid, H_3PO_4 (**6**):

$$P_4O_{10}(s) + 6\ H_2O(l) \longrightarrow 4\ H_3PO_4(aq)$$

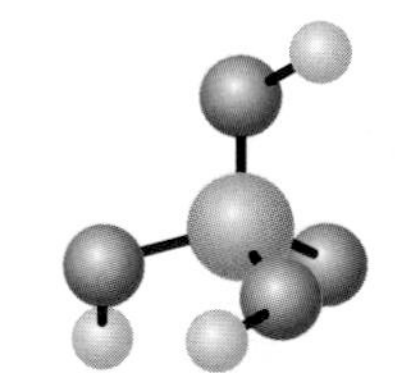

6 Phosphoric acid, H_3PO_4

Phosphoric acid is used primarily for the production of fertilizers, as a food additive, and in detergents. Many soft drinks owe their tart taste to the presence of small amounts of phosphoric acid. Pure phosphoric acid, H_3PO_4, is a colorless solid with a melting point of 42°C; but in the laboratory, it is normally a syrupy liquid on account of the water it has absorbed. Phosphoric acid is usually purchased as 85% H_3PO_4. Its high viscosity can be traced to extensive hydrogen bonding. Although phosphorus has a high oxidation number (+5), the acid shows appreciable oxidizing power only at temperatures above about 350°C; it may be used where nitric acid and sulfuric acid would be too oxidizing.

Phosphoric acid is the parent of the phosphates, which contain the tetrahedral PO_4^{3-} anion and are of great commercial importance. Phosphate rock is mined in huge quantities in Florida and Morocco. After being crushed, it is treated with sulfuric acid to give a mixture of sulfates and phosphates called *superphosphate*, a major fertilizer:

$$Ca_3(PO_4)_2(s) + 2\ H_2SO_4(l) \longrightarrow 2\ CaSO_4(s) + Ca(H_2PO_4)_2(s)$$

When phosphoric acid is heated, it undergoes a **condensation reaction,** a reaction in which two molecules combine together with the elimination of a small molecule, normally that of water:

$$\text{HO}-\underset{\text{OH}}{\overset{\text{O}}{\overset{\|}{\underset{|}{\text{P}}}}}-\text{O}-\text{H} + \text{HO}-\underset{\text{OH}}{\overset{\text{O}}{\overset{\|}{\underset{|}{\text{P}}}}}-\text{OH} \longrightarrow \text{HO}-\underset{\text{OH}}{\overset{\text{O}}{\overset{\|}{\underset{|}{\text{P}}}}}-\text{O}-\underset{\text{OH}}{\overset{\text{O}}{\overset{\|}{\underset{|}{\text{P}}}}}-\text{OH} + H_2O$$

The product, $H_4P_2O_7$, is pyrophosphoric acid. Further heating gives even more complicated products that have chains and rings of PO_4 groups. The products are called polyphosphoric acids. Polyphosphoric acids are far from being of only academic interest: they empower our actions and thoughts. The most important polyphosphate is adenosine triphosphate, ATP (7), which is found in every living cell.

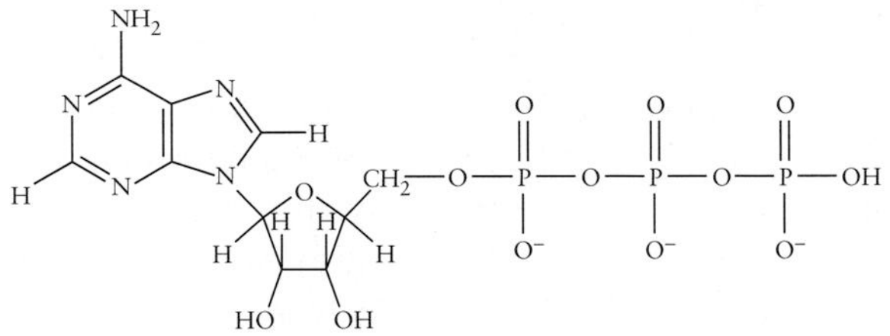

7 Adenosine triphosphate (ATP)

The triphosphate part of this molecule is a chain of three phosphate groups. Its conversion to adenosine diphosphate, ADP, in the reaction

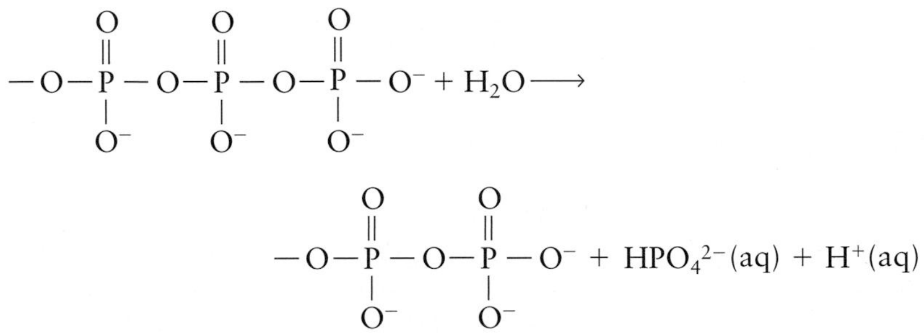

$$-O-\overset{\overset{\displaystyle O}{\|}}{\underset{\underset{\displaystyle O^-}{|}}{P}}-O-\overset{\overset{\displaystyle O}{\|}}{\underset{\underset{\displaystyle O^-}{|}}{P}}-O-\overset{\overset{\displaystyle O}{\|}}{\underset{\underset{\displaystyle O^-}{|}}{P}}-O^- + H_2O \longrightarrow$$

$$-O-\overset{\overset{\displaystyle O}{\|}}{\underset{\underset{\displaystyle O^-}{|}}{P}}-O-\overset{\overset{\displaystyle O}{\|}}{\underset{\underset{\displaystyle O^-}{|}}{P}}-O^- + HPO_4^{2-}(aq) + H^+(aq)$$

releases a large amount of energy ($\Delta G_r^\circ = -30\ \text{kJ}\cdot\text{mol}^{-1}$ at pH = 7), which is used to power energy-demanding processes in cells.

The oxides of phosphorus have structures based on the tetrahedral PO_4 unit; P_4O_6 and P_4O_{10} are the anhydrides of phosphorous and phosphoric acid, respectively. Polyphosphates are extended structures used (as ATP) by living cells to store and transfer energy.

Self-Test 15.1A How does (a) the acidity of a nitrogen oxoacid and (b) its strength as an oxidizing agent change as the oxidation number of N increases from +1 to +5?

[*Answer:* (a) Increases; (b) increases]

Self-Test 15.1B What is the molarity of phosphoric acid in 85% H_3PO_4(aq), which has a density of 1.7 $g\cdot mL^{-1}$?

GROUP 16/VI: THE OXYGEN FAMILY

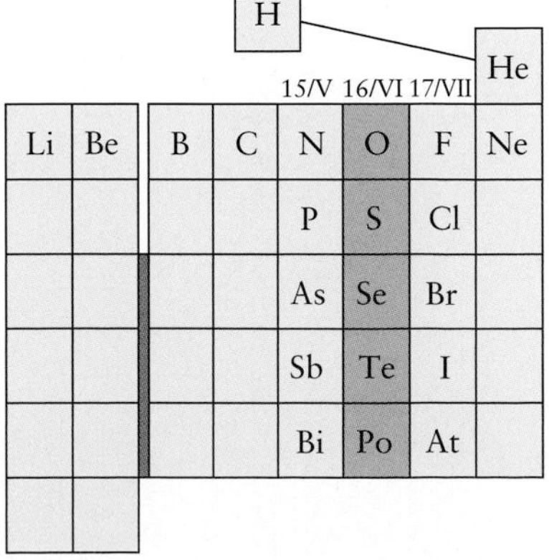

Elements become increasingly nonmetallic toward the right of the periodic table; and by Group 16/VI, even polonium, at the foot of the group, is best regarded as a metalloid (Table 15.4). The atoms have the valence electron configuration ns^2np^4, so they need only two more electrons to complete their valence shells. The members of the group are collectively called the **chalcogens**.

TABLE 15.4 The Group 16/VI Elements

Valence configuration: ns^2np^4

Z	Name	Symbol	Molar mass, $g \cdot mol^{-1}$	Melting point, °C	Boiling point, °C	Density, $g \cdot cm^{-3}$ at 25°C	Normal form*
8	oxygen	O	16.00	−218	−183	1.14†	colorless paramagnetic gas (O_2)
				−192	−112	1.35†	blue gas (ozone, O_3)
16	sulfur	S	32.07	115	445	2.09	yellow nonmetallic solid (S_8)
34	selenium	Se	78.96	220	685	4.79	gray nonmetallic solid
52	tellurium	Te	127.60	450	990	6.25	silver-white metalloid
84	polonium‡	Po	(209)	254	960	9.40	gray metalloid

**Normal form* means the appearance and state of the element at 25°C and 1 atm.
†For the liquid at its boiling point.
‡Radioactive.

15.5 The Group 16/VI Elements

Oxygen is the most abundant element in the Earth's crust and accounts for 23% of the mass of the atmosphere. In fact, Earth is the only planet in the solar system with an oxidizing atmosphere. On Mars, elemental oxygen provides only 0.15% of the atmospheric mass; and in the atmospheres of the outer planets, oxygen is essentially nonexistent. In the hot atmosphere of Venus, the oxygen has reacted and is present mainly as carbon dioxide.

At one time, the concentration of oxygen in the atmosphere was very low (Fig. 15.7). A small amount of atmospheric oxygen was produced by the photochemical action of sunlight on water. The oxygen level of the atmosphere did not begin to rise significantly until plants began to produce oxygen by photosynthesis:

$$6\ CO_2(g) + 6\ H_2O(l) \longrightarrow C_6H_{12}O_6(s) + 6\ O_2(g)$$

Like nitrogen, oxygen is present as the free element in air, but it is much more reactive than nitrogen. The combustion of all living things in oxygen is thermodynamically spontaneous; however, we do not burst into flame at normal temperatures because combustion has a high activation energy.

More than 2×10^{10} kg of liquid oxygen is produced each year in the United States (about 80 kg per inhabitant) by fractional distillation of liquid air. The biggest consumer of oxygen is the steel industry, which needs about 1 t of oxygen (1 t = 10^3 kg) to produce 1 t of steel. In steelmaking, oxygen is blown into molten iron to oxidize any impurities, particularly carbon (see Section 16.14). Elemental oxygen is also used for welding (to create a very hot flame in oxyacetylene torches) and in medicine. Physicians administer oxygen to relieve strain on the heart and lungs, and as a stimulant.

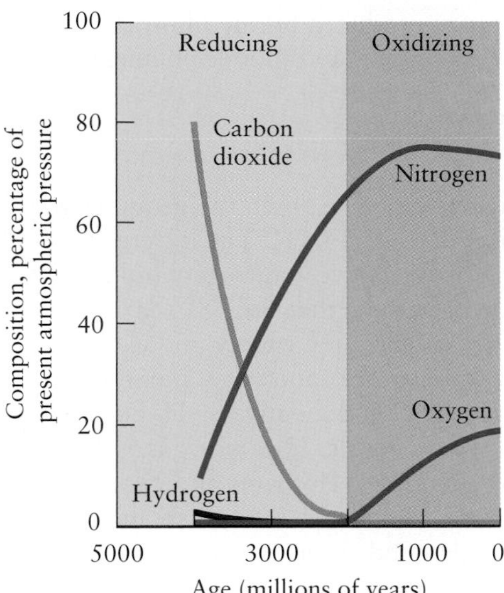

FIGURE 15.7 This speculative history of the atmosphere shows a change from a predominantly reducing to a predominantly oxidizing atmosphere about 2 billion years ago when photosynthetic plants began to appear. Hydrogen has since been lost to compound formation and to outer space. Earth is the only planet in the solar system with an oxidizing atmosphere.

Oxygen is a colorless, tasteless, odorless gas of O_2 molecules; the gas condenses to a pale blue liquid at −183°C (Fig. 15.8). Although O_2 has an even number of electrons, two of them are unpaired, making the molecule paramagnetic; in other words, O_2 behaves like a tiny magnet and is attracted into a magnetic field.

Recall Section 3.11, where we saw that in O_2 two electrons occupy a pair of antibonding π-orbitals.

An allotrope of oxygen, ozone, O_3 (8), forms in the stratosphere by the effect of solar radiation on O_2 molecules. Its total abundance in the atmosphere is equivalent to a layer that, at normal temperature and pressure, would cover the Earth to a thickness of only 3 mm. Ozone can be made in the laboratory by passing an electric discharge through oxygen. It is a blue gas that condenses at −112°C to an explosive blue liquid that looks like ink (Fig. 15.9). Its pungent smell can often be detected near electrical equipment and after lightning. Ozone is also present in smog, where it is produced by the reaction of oxygen molecules with oxygen atoms:

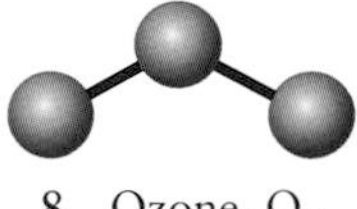

8 Ozone, O_3

$$O + O_2 \longrightarrow O_3$$

FIGURE 15.8 Liquid oxygen is pale blue. (The gas itself is colorless.) The inset shows that the gas consists of diatomic molecules known formally as dioxygen.

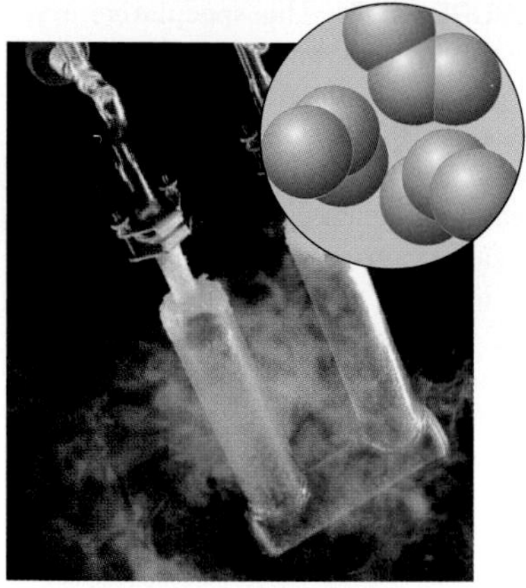

FIGURE 15.9 Ozone is a blue gas that condenses to a dark blue, highly unstable liquid. The inset shows that the gas consists of triatomic molecules.

The oxygen atoms are produced by the photochemical decomposition of NO_2, an emission product of automobile engines:

$$NO_2 \xrightarrow{\text{UV radiation}} NO + O$$

Electronegativities decrease down the group (Fig. 15.10), and atomic and ionic radii increase (Fig. 15.11). The differences between oxygen and sulfur are similar to those between nitrogen and phosphorus, and for similar reasons: sulfur has atoms that are 58% larger than those of oxygen, and it has a lower first ionization energy and electronegativity. The bonds sulfur forms to hydrogen are much less polar than the bonds between oxygen and hydrogen. Consequently, S—H bonds do not take part in hydrogen bonding; as a result, H_2S is a gas, whereas H_2O is a liquid, despite the smaller number of electrons and hence weaker London forces. Sulfur also has weaker tendencies to form multiple bonds, tending to form additional single bonds to as many as six other atoms and using its *d*-orbitals to do so.

Sulfur has a striking ability to **catenate,** or form chains of atoms. Oxygen's ability to form chains is very limited, with H_2O_2, O_3, and the anions O_2^-, O_2^{2-} and O_3^- the only examples. Sulfur's ability is much more pronounced. It appears, for instance, in the existence of S_8 rings, their fragments, and the long strands of "plastic sulfur" that form when sulfur is heated to about 200°C and suddenly cooled. The —S—S— links that connect different parts of the chains of amino acids in proteins are another example of catenation. These "disulfide links" contribute to the shapes of proteins, including the keratin of our hair; thus, sulfur helps to keep us alive and, perhaps, curly haired.

Sulfur is widely distributed as sulfide ores, which include *galena*, PbS; *cinnabar*, HgS; *iron pyrite*, FeS_2; and *sphalerite*, ZnS (Fig. 15.12). The mineral *molybdenite*, MoS_2, is a soft rock with a metallic sheen and properties similar to those of graphite. Because these ores are so common, sulfur is a by-product of the extraction of a number of metals, especially copper. Sulfur is also found as deposits of the native element (called *brimstone*), which are formed by bacterial action on H_2S. It is also commonly found in petroleum, and extracting it chemically has been made inexpensive and safe by

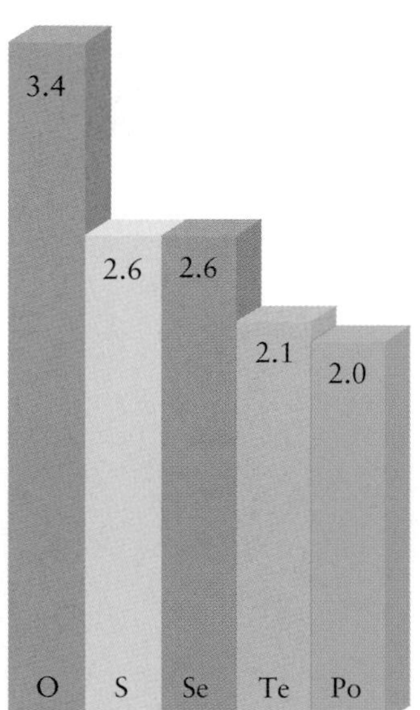

FIGURE 15.10 The electronegativities of the Group 16/VI elements decrease steadily down the group.

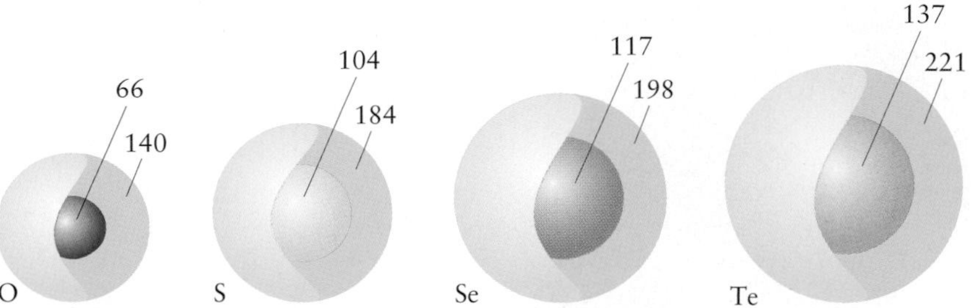

FIGURE 15.11 The atomic and ionic radii of the Group 16/VI elements increase down the group. The values shown are in picometers, and the anion (shown green in each case) is substantially larger than the neutral parent atom.

FIGURE 15.12 A collection of sulfide ores. From left to right galena, PbS; cinnabar, HgS; pyrite, FeS_2; sphalerite, ZnS. Pyrite has a lustrous golden color and has frequently been mistaken for gold; hence, it is also known as fool's gold. Gold and fool's gold are readily distinguished by their densities.

the use of heterogeneous catalysts. One method is the **Claus process,** in which some of the H_2S that occurs in oil and natural gas wells is first oxidized to sulfur dioxide:

$$2\ H_2S(g) + 3\ O_2(g) \longrightarrow 2\ SO_2(g) + 2\ H_2O(l)$$

The SO_2 is then used to oxidize the remainder of the hydrogen sulfide:

$$2\ H_2S(g) + SO_2(g) \xrightarrow{300°C,\ Al_2O_3} 3\ S(s) + 2\ H_2O(l)$$

Sulfur is of major industrial importance. Most of the sulfur that is produced is used to make sulfuric acid, but an appreciable amount is used to vulcanize rubber (Section 19.12).

Elemental sulfur is a yellow, tasteless, almost odorless, insoluble, nonmetallic molecular solid of crownlike S_8 rings (**9**). The two common crystal forms of sulfur are monoclinic sulfur and rhombic sulfur. The more stable form under normal conditions is rhombic sulfur, which forms beautiful yellow crystals (Fig. 15.13). At low temperatures, sulfur vapor consists mainly of S_8 molecules. At temperatures above 720°C, the vapor has a blue tint from the S_2 molecules that form. The latter are paramagnetic, like O_2.

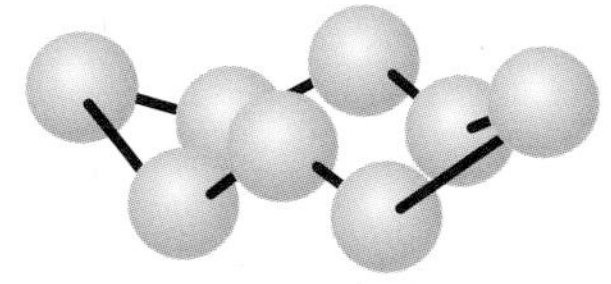

9 Sulfur, S_8

The names are taken from the shapes of the crystals they form.

Media Link **Lab video:** Fig. 15.13, Properties of sulfur

(a)

(b)

FIGURE 15.13 One of the two most common forms of sulfur is the blocklike rhombic form (a). It differs from the needlelike monoclinic sulfur (b) in the manner in which the S_8 rings are stacked together.

FIGURE 15.14 Two of the Group 16/VI elements: selenium on the left and tellurium on the right.

Selenium and tellurium occur in sulfide ores; they are also recovered from the anode sludge formed during the electrolytic refining of copper. Both elements have several allotropes, the most stable consisting of long zigzag chains of atoms. Although these allotropes look like silver-white metals, they are poor electrical conductors (Fig. 15.14). The conductivity of selenium is increased by exposure to light, so it is used in solar cells, photoelectric devices, and photocopying machines. Selenium also occurs as a deep red solid consisting of Se_8 molecules.

Polonium is a radioactive, low-melting metalloid. It is a useful source of α particles and is used in antistatic devices in textile mills: the α particles reduce static by counteracting the negative charges that tend to build up on the fast-moving fabric.

Metallic character increases down Group 16/VI as electronegativity decreases. Oxygen and sulfur occur naturally in the elemental state; sulfur forms chains and rings with itself, but oxygen does not.

15.6 Compounds with Hydrogen

By far the most important compound of oxygen and hydrogen is water, H_2O. Water is essential to life as we know it and possibly to life anywhere in the universe. A lot of water arrived on Earth from outer space in the form of comets, which are like huge dirty snowballs. When the Earth was young, comets collided with it frequently, and each collision brought great amounts of water to the planet. Another source of water was the rocks from which the young Earth initially formed. The water locked up inside these solids was released when they melted in the depths of the Earth. The emission from volcanoes—even today—contains vast quantities of water.

Municipal water supplies normally undergo several stages of purification (Fig. 15.15). Raw water is aerated by bubbling air through it to remove foul-smelling dissolved gases such as H_2S, to oxidize some organic compounds to CO_2, and to add oxygen and nitrogen. Adding slaked lime, $Ca(OH)_2$, reduces acidity, precipitates Mg^{2+}, Fe^{3+}, Cu^{2+}, and other metal ions as hydroxides, and softens "hard" water (see Section 8.14).

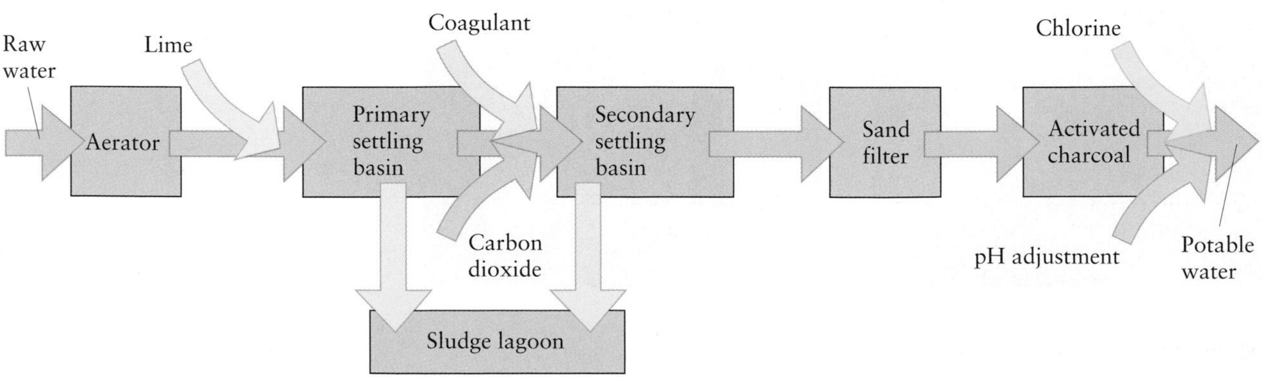

FIGURE 15.15 Steps in the purification of drinking water.

After the lime is added, the water is pumped into a primary settling basin. Because the precipitates tend to form as a very fine powder that remains suspended in the water, either $Fe_2(SO_4)_3$ or alum (specifically $Al_2(SO_4)_3 \cdot 18H_2O$) is added to coagulate and flocculate the precipitate so that it can be filtered. **Coagulation** is the aggregation of small particles into larger particles; **flocculation** is the loose aggregation of particles to form a fluffy gel. Aluminum is amphoteric and forms soluble $Al(OH)_4^-$ ions in basic solution (Section 14.14). Carbon dioxide is often added to raise the acidity of the water, which promotes precipitation of the aluminum as $Al(OH)_3$ so that it can be removed by filtration.

Other types of compounds called alums are described in Section 14.14.

As the precipitate settles slowly in a secondary basin, it adsorbs any remaining suspended $CaCO_3$, bacteria, and other particles, such as dirt and algae. Precipitates from the primary and secondary basins are combined in a sludge lagoon for disposal. The clear water is then passed through a sand filter to remove any remaining suspended particles.

The pH of the water is checked again and made slightly basic to reduce acid corrosion of the pipes. At this point, a disinfectant, usually chlorine, is added. In the United States, the chlorine level is required to be greater than 1 g of Cl_2 per 1000 kg of water at the point of consumption. In water, chlorine forms hypochlorous acid, which is highly toxic to bacteria:

$$Cl_2(g) + 2\,H_2O(l) \longrightarrow H_3O^+(aq) + Cl^-(aq) + HClO(aq)$$

Depending on the source and condition of the water, additional water purification steps, such as ion exchange and reverse osmosis may be required (Section 8.19).

High-purity water for special applications is obtained by distillation or by **ion exchange,** the exchange of one type of ion in a solution for another. In ion exchange, water passes through a column packed with a material such as a zeolite, an aluminosilicate with a very open structure that can capture ions such as Mg^{2+} and Ca^{2+} and exchange them for H^+ ions (see Box 13.4).

We might think of water as a solvent and hence merely as a passive medium in which chemical reactions occur. However, water is a reactive compound, and an alien raised in a nonaqueous environment might consider it aggressively corrosive and be surprised at our survival. For instance, water is an oxidizing agent:

$$2\,H_2O(l) + 2\,e^- \longrightarrow 2\,OH^-(aq) + H_2(g) \qquad E = -0.42 \text{ V at pH} = 7$$

One example is its reaction with the alkali metals, as in

$$2\,Na(s) + 2\,H_2O(l) \longrightarrow 2\,NaOH(aq) + H_2(g)$$

However, unless the other reactant is a strong reducing agent, water acts as an oxidizing agent only at high temperatures, as in the *re-forming reaction*:

$$CH_4(g) + H_2O(g) \xrightarrow{\Delta} CO(g) + 3\,H_2(g)$$

Water is a very mild reducing agent:

$$4\,H^+(aq) + O_2(g) + 4\,e^- \longrightarrow 2\,H_2O(l) \qquad E = +0.82 \text{ V at pH} = 7$$

However, few substances besides fluorine are strong enough oxidizing agents to accept the electrons released in the reverse of this half-reaction.

Water is also a Lewis base, for an H_2O molecule can donate one of its lone pairs to a Lewis acid and form complexes such as $Fe(H_2O)_6^{3+}$. Water's ability to act as a Lewis base is also the origin of its ability to hydrolyze substances. The reaction between water and phosphorus pentachloride mentioned in Section 15.2 is an example.

The hydrolysis of PCl_5 leaves the oxidation number of phosphorus unchanged. However, some hydrolyses do occur with change of oxidation number. Such is the case in the reaction between chlorine and water, which can be pictured as a hydrolysis in which the —OH group trades partners with one of the Cl atoms:

$$\text{Cl—Cl(g)} + H_2O(l) \longrightarrow \text{Cl—OH(aq)} + \text{HCl(aq)}$$

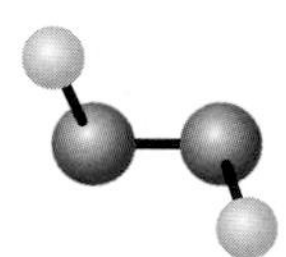

10 Hydrogen peroxide, H_2O_2

This reaction is a disproportionation in which the oxidation number of Cl changes from 0 to +1 (in HClO) and −1 (in HCl).

Hydrogen peroxide, H_2O_2 (**10**), is a very pale blue liquid that is appreciably denser than water (1.44 $g \cdot mL^{-1}$ at 25°C) but similar in other physical properties: its melting point is −0.4°C, and its boiling point is 152°C. Chemically, though, hydrogen peroxide and water differ greatly. The presence of the second oxygen atom makes H_2O_2 a very weak acid ($pK_{a1} = 11.75$). Hydrogen peroxide is a stronger oxidizing agent than water. For example, H_2O_2 oxidizes Fe^{2+} and Mn^{2+} in acidic and basic solutions. It can also act as a reducing agent in the presence of more powerful oxidizing agents such as permanganate ions and chlorine (usually in basic solution).

Hydrogen peroxide is normally sold for industrial use as a 30% by mass aqueous solution. When used as a hair bleach (as a 6% solution), it acts by oxidizing the pigments in the hair. A 3% H_2O_2 aqueous solution is used as a mild antiseptic in the home. Contact with blood catalyzes its disproportionation into water and oxygen gas, which cleanses the wound. Because it oxidizes unpleasant effluents without producing any harmful by-products, H_2O_2 is also increasingly used as an oxidizing agent to control pollution.

The conjugate base of H_2O_2 is the hydrogen peroxide ion, HO_2^-, and that ion's conjugate base is the peroxide ion, O_2^{2-}. The latter ion occurs in sodium peroxide, which is used to bleach textiles and wood pulp for paper. In an aqueous solution, all peroxide salts from hydrogen peroxide:

$$O_2^{2-}(aq) + 2\,H_2O(l) \longrightarrow 2\,OH^-(aq) + H_2O_2(aq)$$

Self-Test 15.2A From the Lewis structures of the following molecules, determine which are paramagnetic and explain how you decided: (a) N_2O_3; (b) NO; (c) N_2O.

[*Answer:* (b) is paramagnetic, because it has an unpaired electron.]

Self-Test 15.2B Write the half-reactions and the overall reaction for the oxidation of water by F_2. Determine the standard potential and $\Delta G_r°$ for the reaction.

Except for water, all the Group 16/VI binary compounds with hydrogen (the compounds H_2E, where E is a Group 16/VI element) are toxic gases with offensive odors. They are insidious poisons because they paralyze the olfactory nerve, and soon after exposure, the victim cannot smell them. Rotten eggs smell of hydrogen sulfide, H_2S, because egg proteins contain sulfur and give off the gas when they decompose. Another sign of the formation of sulfides in eggs is the pale green discoloration sometimes seen in hard-boiled eggs where the white meets the yolk: this discoloration is a deposit of iron(II) sulfide. Hydrogen sulfide, H_2S, is prepared either by protonation of the sulfide ion, a Brønsted base, in a reaction such as

$$FeS(s) + 2\,HCl(aq) \longrightarrow FeCl_2(aq) + H_2S(g)$$

or by direct reaction of the elements at 600°C.

Hydrogen sulfide dissolves in water to give a solution that, as a result of its oxidation by dissolved air, slowly becomes cloudy as S_8 molecules form and then coagulate. Hydrogen sulfide is a weak diprotic acid and the parent acid of the hydrogen sulfides (which contain the HS^- ion) and the sulfides (which contain the S^{2-} ion). The sulfides of the *s*-block elements are moderately soluble, whereas the sulfides of the heavy *p*- and *d*-block metals are generally very insoluble.

The sulfur analogue of hydrogen peroxide also exists and is an example of a *polysulfane*, a catenated molecular compound of composition $HS{-}S_n{-}SH$, where n can take on values from 0 through 6. The polysulfide ions obtained from the polysulfanes include two ions found in lapis lazuli (Fig. 15.16).

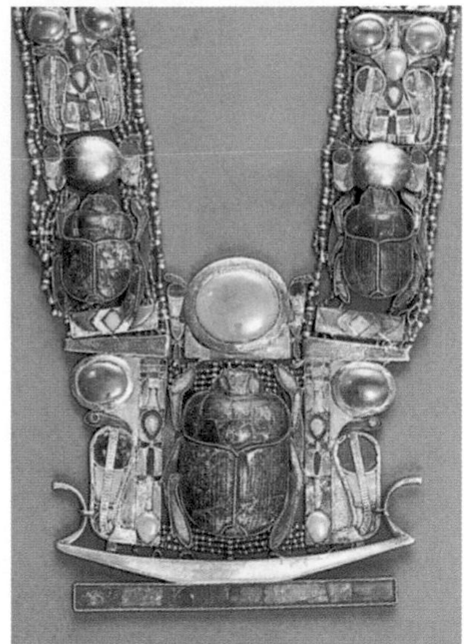

FIGURE 15.16 The blue stones in this ancient Egyptian ornament are lapis lazuli. This semiprecious stone is an aluminosilicate colored by S_2^- and S_3^- impurities. The blue color is due to S_3^- and its hint of green to S_2^-.

Water can act as a Lewis base, an oxidizing agent, and a weak reducing agent; hydrogen peroxide is a stronger oxidizing agent than water. Hydrogen sulfide is a weak acid; the polysulfanes exhibit the ability of sulfur to catenate.

15.7 Sulfur Oxides and Oxoacids

Sulfur forms several oxides that in atmospheric chemistry are referred to collectively as SO_x (read "sox"). The most important oxides and oxoacids of sulfur are the dioxide and trioxide and the corresponding sulfurous and sulfuric acids. Sulfur burns in air to form sulfur dioxide, SO_2 (**11**), a colorless, choking, poisonous gas (recall Fig. C.1). About 7×10^{10} kg of sulfur dioxide are produced annually from the decomposition of vegetation and from volcanic emissions. In addition, approximately 1×10^{11} kg of naturally occurring hydrogen sulfide is oxidized to the dioxide by atmospheric oxygen:

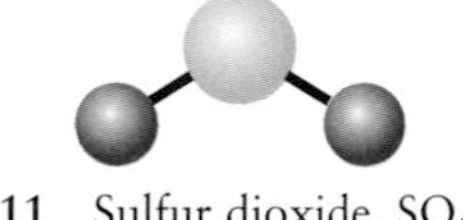

11 Sulfur dioxide, SO_2

$$2\,H_2S(g) + 3\,O_2(g) \longrightarrow 2\,SO_2(g) + 2\,H_2O(g)$$

Industry and transport contribute another 1.5×10^{11} kg of the dioxide, of which about 70% comes from oil and coal combustion—mainly in electricity-generating plants. The average concentration of SO_x in the atmosphere in rural areas in the northern hemisphere is found to be about 1 μmol·L^{-1}. However, the concentration is much higher in industrialized areas (see Box 10.1).

Sulfur dioxide is the anhydride of sulfurous acid, H_2SO_3, which is the parent acid of the hydrogen sulfites (or bisulfites) and the sulfites:

$$SO_2(g) + H_2O(l) \longrightarrow H_2SO_3(aq)$$

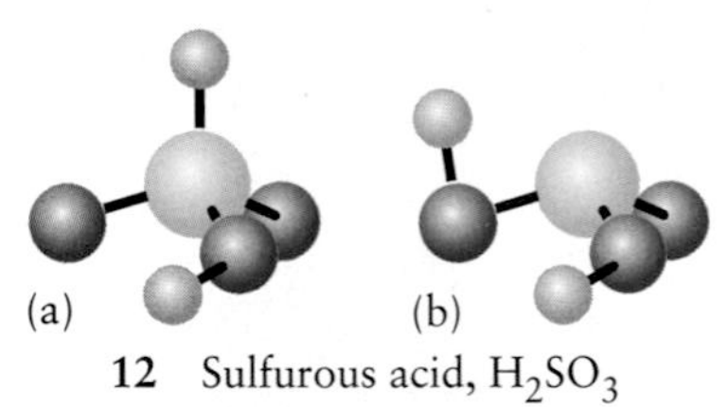

12 Sulfurous acid, H_2SO_3

Sulfurous acid is an equilibrium mixture of two molecules (**12a** and **12b**); in the former, it resembles phosphorous acid, with one of the H atoms attached directly to the S atom. These molecules are also in equilibrium with molecules of SO_2, each of which is surrounded by a cage of water molecules. The evidence for this equilibrium is that crystals of composition $SO_2{\cdot}xH_2O$, with x about 7, are obtained when the solution is cooled. Substances like this, in which a molecule occupies a cage formed by other molecules, are called **clathrates.** Methane, carbon dioxide, and the noble gases also form clathrates with water.

The name *clathrate* comes from the Greek word for "cage."

Sulfur dioxide is easily liquified under pressure and can therefore be used as a refrigerant. It is also a preservative for dried fruit and a bleach for textiles and flour, but its most important use is in the production of sulfuric acid.

The oxidation number of sulfur in sulfur dioxide and the sulfites is +4, an intermediate value in sulfur's range from −2 to +6. Hence, these compounds can act as either oxidizing agents or reducing agents. By far the most important reaction of sulfur dioxide is its slow oxidation to sulfur trioxide, SO_3 (**13**), in which sulfur has the oxidation number +6:

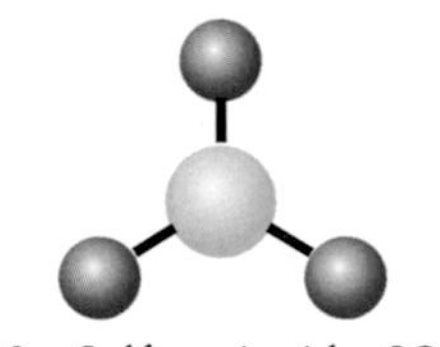
13 Sulfur trioxide, SO_3

$$2\,SO_2(g) + O_2(g) \longrightarrow 2\,SO_3(g)$$

At normal temperatures, sulfur trioxide is a volatile liquid (its boiling point is 45°C), composed of trigonal planar SO_3 molecules. In the solid, and to some extent in the liquid, these molecules form trimers (unions of three molecules) of composition S_3O_9 (**14**) as well as larger clusters.

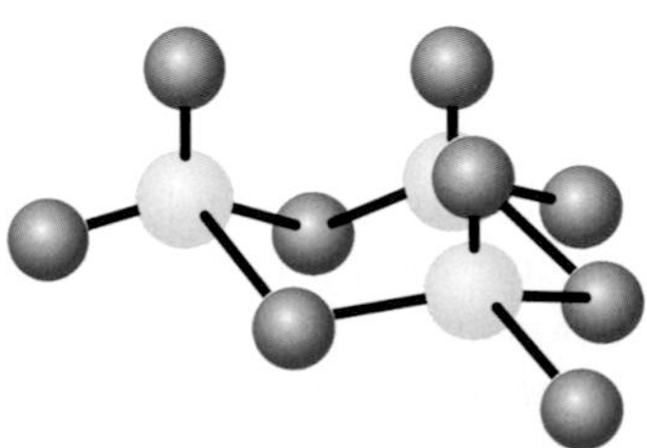
14 Sulfur trioxide trimer, S_3O_9

Sulfuric acid, H_2SO_4, is produced commercially in the *contact process*, in which sulfur is first burned in oxygen and the SO_2 produced is oxidized to SO_3 over a V_2O_5 catalyst:

$$S(g) + O_2(g) \xrightarrow{1000°C} SO_2(g)$$

$$2\,SO_2(g) + O_2(g) \xrightarrow{500°C,\ V_2O_5} 2\,SO_3(g)$$

Because sulfur trioxide forms a corrosive acid mist with water vapor, it is absorbed instead in 98% concentrated sulfuric acid to give the dense, oily liquid called *oleum*:

$$SO_3(g) + H_2SO_4(l) \longrightarrow H_2S_2O_7(l)$$

Oleum is then converted to the acid by reaction with water:

$$H_2S_2O_7(l) + H_2O(l) \longrightarrow 2\ H_2SO_4(l)$$

Sulfuric acid is the most heavily produced inorganic chemical worldwide, the annual production being over 4×10^{10} kg in the United States alone. The low cost of sulfuric acid leads to its widespread use in industry, particularly for the production of fertilizers, petrochemicals, dyestuffs, and detergents. About two-thirds is used in the manufacture of phosphate and ammonium sulfate fertilizers (see Section 15.4).

Sulfuric acid is a colorless, corrosive, oily liquid that boils (and decomposes) at about 300°C. It has three chemically important properties: it is a strong Brønsted acid, a dehydrating agent, and an oxidizing agent (Fig. 15.17). Sulfuric acid is a strong acid in the sense that its first deprotonation is almost complete at normal concentrations in water. However, its conjugate base HSO_4^- is only a weak acid, with $pK_a = 1.92$ (Section 10.17).

FIGURE 15.17 Sulfuric acid is an oxidizing agent. When some concentrated acid is poured on solid sodium bromide, NaBr, the bromide ions are oxidized to bromine, which colors the mixture red-brown.

The powerful dehydrating ability of sulfuric acid is seen when a little concentrated acid is poured on sucrose, $C_{12}H_{22}O_{11}$. A black, frothy mass of carbon forms as a result of the extraction of H_2O (Fig. 15.18):

$$C_{12}H_{22}O_{11}(s) \longrightarrow 12\ C(s) + 11\ H_2O(l)$$

The froth is caused by CO and CO_2 generated in side reactions. Sulfuric acid forms strong hydrogen bonds with water. The energy released when these bonds form can cause the mixture to boil violently. Sulfuric acid should always be diluted slowly and with great care, making sure to add the acid to the water, instead of the reverse, to avoid splashing concentrated acid.

FIGURE 15.18 Sulfuric acid is a dehydrating agent. (a) When concentrated sulfuric acid is poured on to sucrose, (b) the sucrose, a carbohydrate, is dehydrated, (c) leaving a frothy black mass of carbon.

Media Link **Lab video:** Fig. 15.18, Dehydration of sucrose by sulfuric acid

Sulfur dioxide is the anhydride of sulfurous acid, and sulfur trioxide the anhydride of sulfuric acid. Sulfuric acid is a strong acid, a dehydrating agent, and an oxidizing agent.

15.8 Sulfur Halides

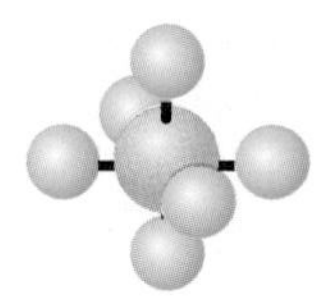

15 Sulfur hexafluoride, SF_6

Sulfur reacts directly with all the halogens except iodine. It ignites spontaneously in fluorine and burns brightly to give sulfur hexafluoride, SF_6 (**15**), a dense, colorless, odorless, nontoxic, thermally stable, insoluble gas. Despite the high oxidation number (+6) of the sulfur atom, SF_6 is not a good oxidizing agent. Its low reactivity is in large part due to the F atoms that surround the central S atom like armor and protect it from attack. The ionization energy of SF_6 is very high because any electron that is removed must come from the highly electronegative F atoms. Even quite strong electric fields cannot strip off these electrons, so SF_6 is a good gas-phase electrical insulator. It is, in fact, a much better insulator than air and is used in switches on high-voltage power lines.

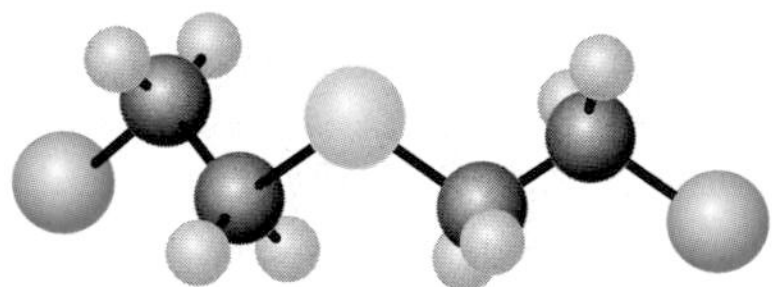

16 Mustard gas, $S(CH_2CH_2Cl)_2$

One of the products of the reaction of sulfur with chlorine is disulfur dichloride, S_2Cl_2, a yellow liquid with a nauseating smell; it is used for the vulcanization of rubber. When disulfur dichloride reacts with more chlorine in the presence of iron(III) chloride as a catalyst, the evil-smelling red liquid sulfur dichloride, SCl_2, is produced. Sulfur dichloride reacts with ethene to give mustard gas (**16**), which has been used in chemical warfare. Mustard gas causes blisters, discharges from the nose, and vomiting; it also destroys the cornea of the eye. All in all, it is easy to see why ancient civilizations associated sulfur with the underworld.

Sulfur hexafluoride is a thermally stable nontoxic gas that is a good insulator; sulfur chlorides are toxic liquids.

Self-Test 15.3A What is the oxidation number of sulfur in (a) the persulfate ion, $S_2O_8^{2-}$; (b) the thiosulfate ion, $S_2O_3^{2-}$?

[*Answer:* (a) +7; (b) +2]

Self-Test 15.3B When concentrated sulfuric acid and water are mixed, the total volume is less than the sum of the individual volumes before mixing. Explain this volume contraction in terms of intermolecular interactions.

GROUP 17/VII: THE HALOGENS

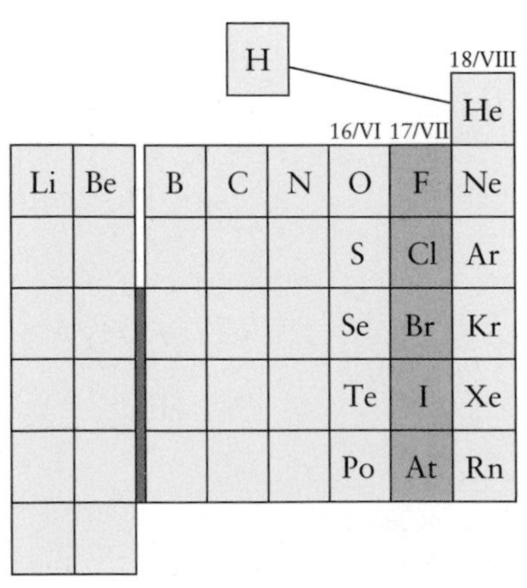

The unique properties of the halogens (Table 15.5), the members of Group 17/VII, can be traced to their valence configuration, ns^2np^5, which needs only one more electron to reach a closed-shell configuration. To complete the octet of valence electrons, in the elemental state all halogen atoms combine to form diatomic molecules, such as F_2 and I_2. The elements form a family showing the smooth trends in physical properties that we would expect when London forces between molecules are the dominant intermolecular force. Because electronegativity decreases down the group (Fig. 15.19) and atomic and ionic radii (Fig. 15.20) increase smoothly down the group, chemical properties show smooth trends as well, with the exception of some properties of fluorine.

TABLE 15.5 The Group 17/VII Elements

Valence configuration: ns^2np^5

Z	Name	Symbol	Molar mass, $g \cdot mol^{-1}$	Melting point, °C	Boiling point, °C	Density, $g \cdot cm^{-3}$ at 25°C	Normal form*
9	fluorine	F	19.00	−220	−188	1.51†	almost colorless gas
17	chlorine	Cl	35.45	−101	−34	1.66†	yellow-green gas
35	bromine	Br	79.90	−7	59	3.12	red-brown liquid
53	iodine	I	126.90	114	184	4.95	purple-black nonmetallic solid
85	astatine‡	At	(210)	300	350	—	nonmetallic solid

**Normal form* means the appearance and state of the element at 25°C and 1 atm.
†For the liquid at its boiling point.
‡Radioactive.

15.9 The Group 17/VII Elements

Fluorine, the first element of the group, is the halogen of greatest abundance in the Earth's crust. It occurs widely in many minerals, including *fluorspar*, CaF_2; *cryolite*, Na_3AlF_6; and the *fluorapatites*, $Ca_5F(PO_4)_3$. Because fluorine is the most strongly oxidizing element ($E° = +2.87$ V), it cannot be obtained from its compounds by oxidation with another element. Fluorine is produced by electrolyzing an anhydrous molten mixture of potassium fluoride and hydrogen fluoride at about 75°C with a carbon anode.

Fluorine is a reactive, almost colorless gas of F_2 molecules. Most of the fluorine produced by industry is used to make the volatile solid UF_6 used for processing nuclear fuel (Section 17.12). Much of the rest is used in the production of SF_6 for electrical equipment.

Fluorine has a number of peculiarities that stem from its high electronegativity, small size, and lack of available *d*-orbitals. It is the most electronegative element of all and has an oxidation number of −1 in all its compounds. Its high electronegativity and small size allow it to oxidize other elements to their highest oxidation numbers. The small size helps in this, for it allows several F atoms to pack around a central atom, as in IF_7.

Because the fluoride ion is so small, the lattice enthalpies of its ionic compounds tend to be high (see Table 6.6). One result is the lower solubility of fluorides relative to those of other halides. This difference in solubility is one of the reasons why the oceans are salty with chlorides rather than fluorides, even though fluorine is more abundant than chlorine in the Earth's crust. Chlorides are more readily dissolved and washed out to sea. There are some exceptions to this trend in solubilities, including AgF, which is soluble; the other silver halides are insoluble. The exception arises because the covalent character of the silver halides increases from AgCl to AgI as the anion becomes larger and more polarizable. Silver fluoride, which contains the small and almost unpolarizable fluoride ion, is freely soluble in water because it is predominantly ionic.

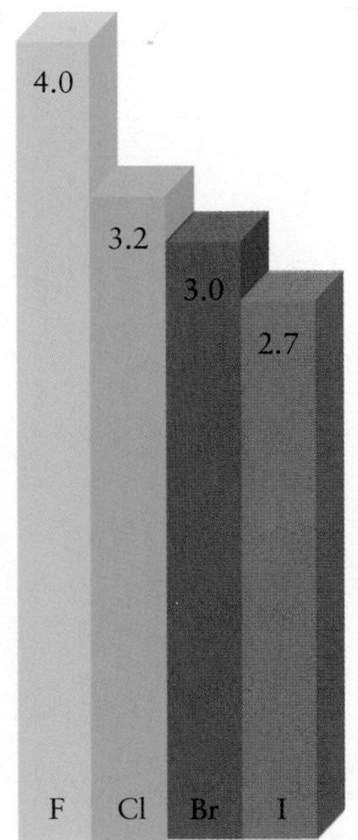

FIGURE 15.19 The electronegativities of the halogens decrease steadily down the group.

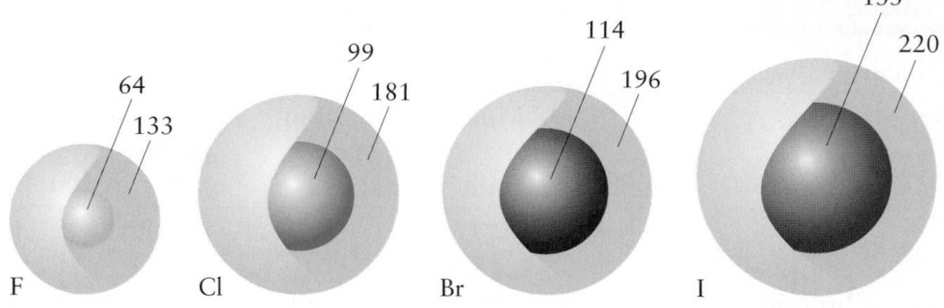

FIGURE 15.20 The atomic and ionic radii of the halogens increase steadily down the group as electrons occupy outer shells of the atoms. The values shown are in picometers. In all cases, ionic radii (represented by the green spheres) are larger than atomic radii.

Covalent fluorides are often more volatile than the corresponding covalent chlorides because the former have fewer electrons, so the London forces between them are weaker. However, a striking exception is the low volatility of HF relative to that of HCl; the difference is due to fluorine's ability to take part in hydrogen bonding, which allows the HF molecules to stick together more strongly in the liquid.

Chlorine is one of the most heavily manufactured chemicals. It is obtained from sodium chloride by electrolysis of molten rock salt or brine (recall Sections 12.15 and 14.9). It is a pale yellow-green gas of Cl_2 molecules that condenses at $-34°C$. It reacts directly with nearly all the elements (the exceptions being carbon, nitrogen, oxygen, and the noble gases). It is a strong oxidizing agent and oxidizes metals to high oxidation states; for example, anhydrous iron(III) chloride, not iron(II) chloride, is formed when chlorine reacts with iron (Fig. 15.21):

$$2\,Fe(s) + 3\,Cl_2(g) \longrightarrow 2\,FeCl_3(s)$$

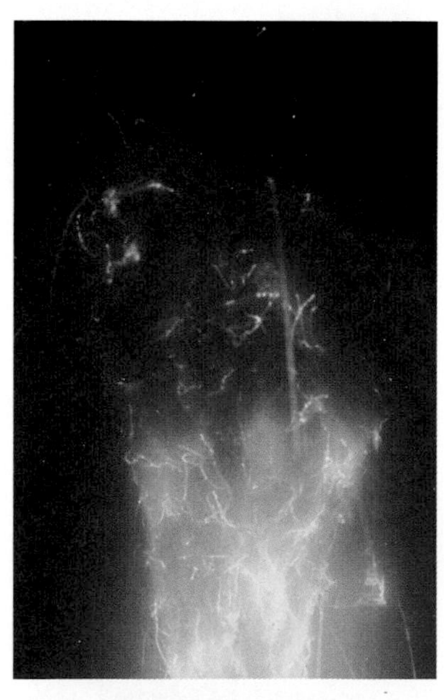

FIGURE 15.21 Iron reacts vigorously and exothermically with chlorine to form anhydrous iron(III) chloride.

Chlorine is used in a number of industrial processes, including the manufacture of plastics, solvents, and pesticides. It is also used as a bleach in the paper and textile industries and as a disinfectant in water treatment. The use of chlorine to provide potable water has made life in large cities and our modern lifestyles feasible.

Chlorine is also used to produce bromine from brine wells through the oxidation of Br^- ions (see Fig. K.2):

$$2\,Br^-(aq) + Cl_2(g) \longrightarrow Br_2(l) + 2\,Cl^-(aq)$$

Air is bubbled through the solution to vaporize the bromine and drive it out.

Bromine is a corrosive, red-brown fuming liquid of Br_2 molecules. It has a penetrating odor. Bromine is used widely in synthetic organic chemistry because of the ease with which it can be added to and removed from organic chemicals that are being used to carry out complicated syntheses. Organic bromides are incorporated into textiles as fire retardants and are used as pesticides; inorganic bromides, particularly silver bromide, are used in photographic emulsions. Saturated aqueous zinc bromide has a very high density and is used in the oil industry to control the escape of oil from deep wells. The tall column of liquid that results when this solution is poured down a well exerts a very high pressure at its base.

FIGURE 15.22 Solutions of iodine in a variety of solvents. From left to right, the solvents are tetrachloromethane (carbon tetrachloride), water, and potassium iodide solution. In the solution of the far right, some starch has been added to a solution of iodine in potassium iodide solution. The intense blue color that results has led to the use of starch as an indicator for the presence of iodine.

Iodine occurs as iodide ions in brines and as an impurity in Chile saltpeter. It was once obtained from seaweed, which contains high concentrations accumulated from seawater: 2000 kg of seaweed produce about 1 kg of iodine. The best modern source is the brine from oil wells, for the oil itself was produced by the decay of marine organisms that had accumulated the iodine while they were alive. Elemental iodine is produced by oxidation with chlorine:

$$Cl_2(g) + 2\ I^-(aq) \longrightarrow I_2(aq) + 2\ Cl^-(aq) \qquad \Delta G° = -54.54\ kJ$$

The blue-black lustrous solid sublimes easily and forms a purple vapor.

When iodine dissolves in organic solvents, it produces solutions having a variety of colors. These colors arise from the different interactions between the I_2 molecules and the solvent (Fig. 15.22). The element is only slightly soluble in water, unless I^- ions are present, in which case the soluble, brown triiodide ion, I_3^-, is formed. Iodine itself has few direct uses; but dissolved in alcohol, it is familiar as a mild oxidizing antiseptic. It is an essential trace element for living systems; a deficiency in humans leads to a swelling of the thyroid gland in the neck. Iodides are added to table salt (to produce "iodized salt") to prevent this deficiency.

Astatine is a radioactive element that occurs in nature in uranium and thorium ores, but only to a minute extent. Samples are made by bombarding bismuth with α particles in a cyclotron, which accelerates the particles to a very high speed. Astatine isotopes do not exist long enough for its properties to be studied, but it is thought from spectroscopic measurements to have properties similar to those of iodine.

The halogens show smooth trends in chemical properties down the group; fluorine has some anomalous properties, such as its strength as an oxidizing agent and the lower solubilities of most fluorides.

15.10 Compounds of the Halogens

The halogens form compounds among themselves. These *interhalogens* have the formulas XX′, XX'_3, XX'_5, and XX'_7, where X is the heavier (and larger) of the two halogens. Only some of the possible combinations

TABLE 15.6 Known Interhalogens

Interhalogen	Normal form*
XF_n	
ClF	colorless gas
ClF_3	colorless gas
ClF_5	colorless gas
BrF	pale brown gas
BrF_3	pale yellow liquid
BrF_5	colorless liquid
IF	unstable
IF_3	yellow solid
IF_5	colorless liquid
IF_7	colorless gas
XCl_n	
BrCl	red-brown gas
ICl	red solid
I_2Cl_6	yellow solid
XBr_n	
IBr	black solid

**Normal form* means the appearance and state of the compound at 25°C and 1 atm.

have been prepared (Table 15.6). They are all prepared by direct reaction of the two halogens, the product formed being determined by the proportions of reactants used. For example,

$$Cl_2(g) + 3\,F_2(g) \longrightarrow 2\,ClF_3(g)$$

$$Cl_2(g) + 5\,F_2(g) \longrightarrow 2\,ClF_5(g)$$

The physical properties of the interhalogens are intermediate between those of their parent halogens. Trends in the chemistry of the interhalogen fluorides can be related to the decrease in bond dissociation energy as the other halogen becomes heavier. The fluorides of the heavier halogens are all very reactive: bromine trifluoride gas is so reactive that even asbestos burns in it.

The hydrogen halides, HX, can be prepared by the direct reaction of the elements:

$$H_2(g) + X_2(g) \longrightarrow 2\,HX(g)$$

Fluorine reacts explosively by a radical chain reaction as soon as the gases are mixed. A mixture of hydrogen and chlorine explodes when exposed to light. Bromine and iodine react with hydrogen much more slowly. A less hazardous laboratory source of the hydrogen halides is the action of a nonvolatile acid on a metal halide, as in

$$CaF_2(s) + 2\,H_2SO_4(aq, conc) \longrightarrow Ca(HSO_4)_2(aq) + 2\,HF(g)$$

Because Br^- and I^- are oxidized by sulfuric acid, phosphoric acid is used in the preparation of HBr and HI:

$$KI(s) + H_3PO_4(aq) \xrightarrow{\Delta} KH_2PO_4(aq) + HI(g)$$

All the hydrogen halides are colorless, pungent gases except hydrogen fluoride, which is a liquid at temperatures below 20°C (Table 15.7). Its low volatility is a sign of extensive hydrogen bonding, and short zigzag chains of hydrogen-bonded molecules, up to about $(HF)_5$, survive to some extent in the vapor. All the hydrogen halides dissolve in water to give acidic solutions. Hydrofluoric acid has the distinctive property of attacking glass and silica. The interiors of lamp bulbs are frosted by the vapors from a solution of hydrofluoric acid and ammonium fluoride (Fig. 15.23).

TABLE 15.7 The Hydrogen Halides

Compound	Molar mass, $g\cdot mol^{-1}$	Melting point, °C	Boiling point, °C	pK_a in water	Bond enthalpy, $kJ\cdot mol^{-1}$	Bond length, pm
HF	20.01	−83	20	3.45	565	92
HCl	36.46	−115	−85	strong	421	127
HBr	80.91	−89	−67	strong	366	141
HI	127.91	−51	−35	strong	299	161

FIGURE 15.23 When a mixture of hydrofluoric acid and ammonium fluoride is swirled inside a flask (a), the reaction with the silica in the cover glass frosts its surface as well as the walls of the flask (b).

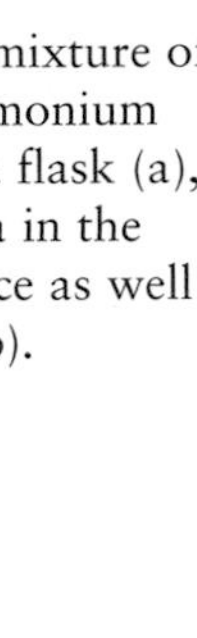

Media Link Lab video: Fig. 15.23, Etching of glass by hydrogen fluoride

Hydrogen fluoride is used to make fluorinated carbon compounds, such as Teflon (polytetrafluoroethylene). Most fluoro-substituted hydrocarbons are relatively inert chemically: they are inert to oxidation by air, hot nitric acid, concentrated sulfuric acid, and other strong oxidizing agents.

Anhydrous metal halides may be formed by direct reaction:

$$2\ Fe(s) + 3\ Cl_2(g) \longrightarrow 2\ FeCl_3(s)$$

They are also formed by reaction of the halogen with an oxide of the metal in the presence of a reducing agent, as in

$$Cr_2O_3(s) + 3\ C(s) + 3\ Cl_2(g) \xrightarrow{\Delta} 2\ CrCl_3(s) + 3\ CO(g)$$

Halides of metals tend to be ionic unless the metal has an oxidation number greater than +2. For example, sodium chloride and copper(II) chloride are ionic compounds and have high melting points, but $TiCl_4$ and $FeCl_3$ sublime as molecules.

The acid strengths and oxidizing abilities of the halogen oxoacids (Table 15.8) increase with the oxidation number of the halogen. The hypohalous acids, HXO (halogen oxidation number +1), are prepared by direct reaction of the halogen with water. For example, chlorine gas disproportionates in water to produce hypochlorous acid and hydrochloric acid:

Hypofluorous acid is so unstable that it survives only below the freezing point of water.

$$Cl_2(g) + H_2O(aq) \longrightarrow HClO(aq) + HCl(aq)$$

Hypohalite ions, XO^-, are formed when a halogen is added to the aqueous solution of a base. Sodium hypochlorite, NaClO, is produced from the electrolysis of brine when the electrolyte is rapidly stirred, and the chlorine gas produced at the anode reacts with the hydroxide ion generated at the cathode. The chlorine gas disproportionates to produce hypochlorite and chloride ions:

$$Cl_2(g) + 2\ OH^-(aq) \longrightarrow ClO^-(aq) + Cl^-(aq) + H_2O(l)$$

TABLE 15.8 Halogen Oxoacids

Oxidation number	General formula	General acid name	Known examples	pK_a in water
7	HXO_4	perhalic acid	$HClO_4$	strong
			$HBrO_4$	strong
			HIO_4	1.64
5	HXO_3	halic acid	$HClO_3$	strong
			$HBrO_3$	strong
			HIO_3	0.77
3	HXO_2	halous acid	$HClO_2$	2.00
			$HBrO_2$	unstable
1	HXO	hypohalous acid	HFO	unstable
			HClO	7.53
			HBrO	8.69
			HIO	10.64

Calcium hypochlorite is the main component of bleach powders and is used for purifying the water in home swimming pools. It is produced by passing chlorine gas over dry calcium oxide (quicklime). Calcium hypochlorite is used for swimming pools in preference to sodium hypochlorite because the Ca^{2+} ions form insoluble calcium carbonate, which is removed by filtration; sodium would remain in solution and make the water too salty.

Because hypochlorites oxidize organic material, they are used in liquid household bleaches and as disinfectants. Their action as oxidizing agents stems partly from the decomposition of hypochlorous acid in solution:

$$2\ HClO(aq) \longrightarrow 2\ H^+(aq) + 2\ Cl^-(aq) + O_2(g)$$

The oxygen either bubbles out of the solution or attacks oxidizable material. Similar reactions occur when chlorine is used to purify water. The reverse of the disproportionation of chlorine

$$ClO^-(aq) + Cl^-(aq) + H_2O(l) \longrightarrow Cl_2(g) + 2\ OH^-(aq)$$

occurs if the pH of the solution is decreased; chlorine gas is evolved. The odor of chlorine may therefore be detected if swimming pool water becomes acidic, although much of the "chlorine" smell is due to the presence of chlorinated amines.

Chlorate ions, ClO_3^- (oxidation state +5), form when chlorine reacts with hot concentrated aqueous alkali:

$$3\ Cl_2(g) + 6\ OH^-(aq) \xrightarrow{\Delta} ClO_3^-(aq) + 5\ Cl^-(aq) + 3\ H_2O(l)$$

They decompose when heated, to an extent that depends on whether or not a catalyst is present:

$$4\ KClO_3(s) \xrightarrow{\Delta} 3\ KClO_4(s) + KCl(s)$$

$$2\ KClO_3(s) \xrightarrow{\Delta,\ MnO_2} 2\ KCl(s) + 3\ O_2(g)$$

The latter reaction is a convenient laboratory source of oxygen.

Chlorates are useful oxidizing agents. Potassium chlorate is used as an oxygen supply in fireworks and in matches. The heads of safety matches consist of a paste of potassium chlorate, antimony sulfide, sulfur, and powdered glass to create friction when the match is struck; the striking strip contains red phosphorus, which ignites the match head.

The principal use of sodium chlorate is as a source of chlorine dioxide, ClO_2. The chlorine in ClO_2 has oxidation number +4, so the chlorate must be reduced to form it. Sulfur dioxide is a convenient reducing agent for this reaction:

$$2\ NaClO_3(aq) + SO_2(g) + H_2SO_4(aq,\ dilute) \longrightarrow 2\ NaHSO_4(aq) + 2\ ClO_2(g)$$

Chlorine dioxide has an odd number of electrons and is a paramagnetic yellow gas. It is used to bleach paper pulp, because it can oxidize the various pigments in the pulp without degrading the wood fibers.

The perchlorates, ClO_4^- (oxidation state +7), are prepared by electrolytic oxidation of aqueous chlorates:

$$ClO_3^-(aq) + H_2O(l) \longrightarrow ClO_4^-(aq) + 2\ H^+(aq) + 2\ e^-$$

Perchloric acid, $HClO_4$, is prepared by the action of concentrated hydrochloric acid on sodium perchlorate, followed by distillation. It is a colorless liquid and the strongest of all common acids. Because chlorine has its highest oxidation number, +7, in these compounds, they are powerful oxidizing agents; contact between perchloric acid and even a small amount of organic material can result in a dangerous explosion.

One spectacular example of the oxidizing ability of perchlorates is their use in the booster rockets of the space shuttle. The solid fuel consists of aluminum powder (the fuel), ammonium perchlorate (the oxidizing agent as well as a fuel), and iron(III) oxide (the catalyst). These reactants are mixed into a liquid polymer, which sets to a solid inside the rocket shell. A variety of products can be produced when the mixture is ignited. One of the reactions that occurs is

$$3\ NH_4ClO_4(s) + 3\ Al(s) \xrightarrow{Fe_2O_3} Al_2O_3(s) + AlCl_3(s) + 6\ H_2O(g) + 3\ NO(g)$$

The solid products form the thick clouds of white powder emitted by the solid rocket boosters during liftoff (Fig. 15.24).

FIGURE 15.24 The white smoke emitted by the shuttle booster rockets consists of powdered aluminum oxide and aluminum chloride.

The interhalogens have properties intermediate between those of the constituent halogens. Nonmetals form covalent halides; metals tend to form ionic halides. The oxoacids of chlorine are all oxidizing agents; acidity and oxidizing strength of oxoacids both increase as the oxidation number of the halogen increases.

Self-Test 15.4A Predict the trend in oxidizing strength of the halogens in aqueous solution.

[*Answer:* Decreases down the group: F > Cl > Br > I]

Self-Test 15.4B Which oxoacid of bromine is the strongest acid?

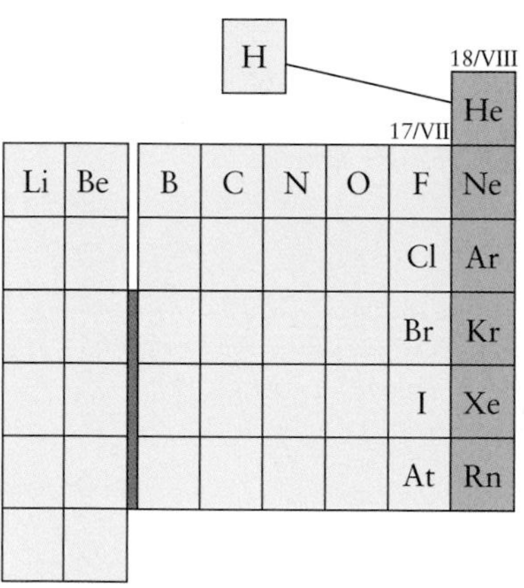

GROUP 18/VIII: THE NOBLE GASES

The elements in Group 18/VIII, the noble gases, get their group name from their very low reactivity (Table 15.9). Experiments on the gases and, later, recognition of their closed-shell electron configurations (ns^2np^6), prompted the belief that these elements were chemically inert. Indeed, no compounds of the noble gases were known until 1962. That year, the British chemist Neil Bartlett synthesized the first noble-gas compound, xenon hexafluoroplatinate, $XePtF_6$ by reacting xenon with platinum hexafluoride vapor. Soon after, chemists at Argonne National Laboratory made xenon tetrafluoride, XeF_4, from a high-temperature mixture of xenon and fluorine.

15.11 The Group 18/VIII Elements

All the Group 18/VIII elements occur in the atmosphere as monatomic gases; together they make up about 1% of its mass. Argon is the third most abundant gas in the atmosphere after nitrogen and oxygen (discounting the variable amount of water vapor). All the noble gases except helium and radon are obtained by the fractional distillation of liquid air.

TABLE 15.9 The Group 18/VIII Elements (the Noble Gases)

Valence configuration: ns^2np^6
Normal form: colorless monatomic gas

Z	Name	Symbol	Molar mass, g·mol^{-1}	Melting point, °C	Boiling point, °C
2	helium	He	4.00	—	−269 (4.2 K)
10	neon	Ne	20.18	−249	−246
18	argon	Ar	39.95	−189	−186
36	krypton	Kr	83.80	−157	−153
54	xenon	Xe	131.29	−112	−108
86	radon*	Rn	(222)	−71	−62

*Radioactive.

Helium, the second most abundant element in the universe after hydrogen, is rare on Earth because its atoms are so light that a large proportion of them reach high speeds and escape from the atmosphere; unlike hydrogen, they cannot be anchored to compounds. However, helium is found as a component of natural gases trapped under rock formations (notably in Texas), where it has collected as a result of the emission of α particles by radioactive elements. An α particle is a helium-4 nucleus ($^4He^{2+}$), and an atom of the element forms when the particle picks up two electrons from its surroundings.

Helium gas is twice as dense as hydrogen under the same conditions. Nevertheless, because its density is still very low and it is nonflammable, it is used to provide buoyancy in airships such as blimps. Helium is also used to dilute oxygen for use in hospitals and in deep-sea diving, to pressurize rocket fuels, as a coolant, and in helium-neon lasers. The element has the lowest boiling point of any substance (4.2 K), and it does not freeze to a solid at any temperature unless pressure is applied to hold the light, mobile atoms together. These properties and its chemical inertness make helium useful for **cryogenics,** the study of matter at very low temperatures. Helium is the only substance known to have more than one liquid phase. Its phase diagram indicates the temperature and pressure ranges over which each phase is stable (Fig. 15.25). Below 2 K, liquid helium-II shows the remarkable property of **superfluidity,** the ability to flow without viscosity.

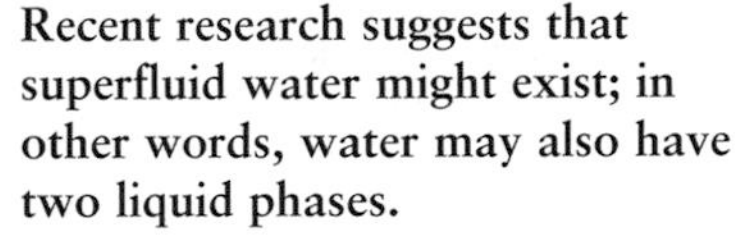
Recent research suggests that superfluid water might exist; in other words, water may also have two liquid phases.

Neon, which emits an orange-red glow when an electric current flows through it, is widely used in advertising signs and displays (Fig. 15.26). Argon is used to provide an inert atmosphere for welding (to prevent oxidation) and to fill some types of light bulbs, where it conducts heat away from the filament. Krypton gives an intense white light when an electric discharge is passed through it, so it is used in airport runway lighting. Because krypton is produced by nuclear fission, its atmospheric abundance is one measure of worldwide nuclear activity. Xenon is used in halogen lamps for automobile headlights and in high-speed photographic flash tubes; it is also being investigated as an anesthetic.

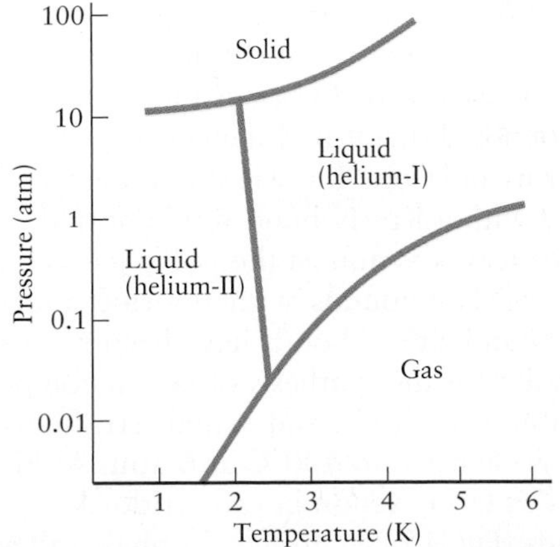

FIGURE 15.25 The phase diagram for helium-4 shows the two liquid phases of helium. Helium-II, the low-temperature liquid phase, is a superfluid.

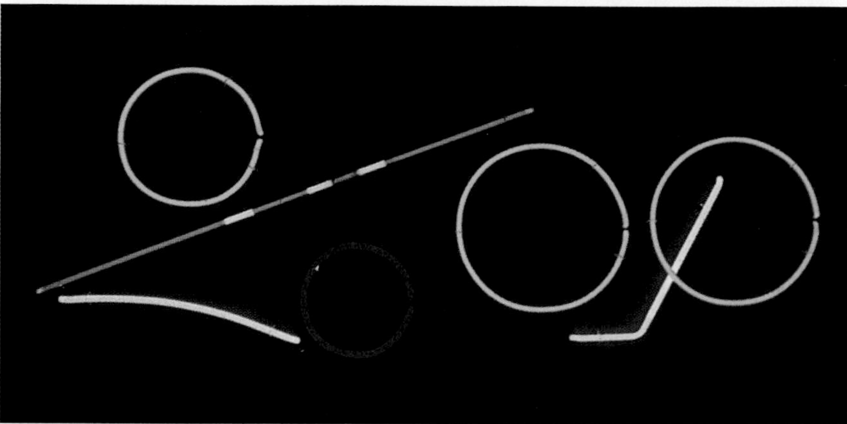

FIGURE 15.26 The colors of the fluorescent lighting art by Tom Anthony are due to emission from noble gas atoms. Neon is responsible for the red light; when it is mixed with a little argon, the color becomes blue-green. The yellow color is achieved by coating the inside of the glass with phosphors that give off yellow light when excited.

The radioactive gas radon seeps out of the ground as a product of radioactive processes deep in the Earth. There is now some concern that its accumulation in buildings, and its nuclear decay products, can lead to dangerously high levels of radiation.

> *The noble gases are all found naturally as unreactive monatomic gases. Helium has two liquid phases; the lower-temperature liquid phase exhibits superfluidity.*

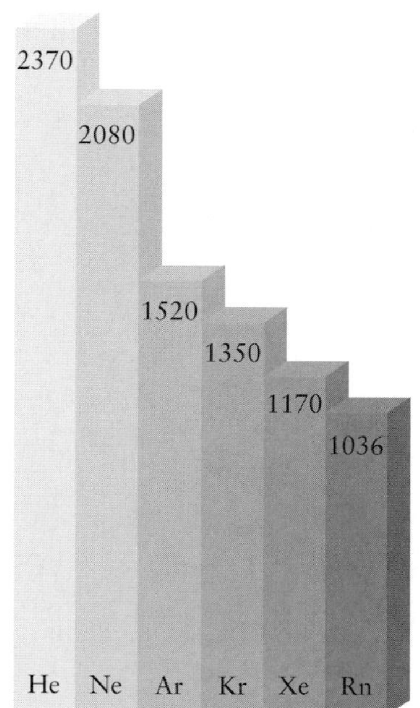

FIGURE 15.27 The ionization energies of the noble gases decrease steadily down the group. The values shown are in kilojoules per mole.

15.12 Compounds of the Noble Gases

The ionization energies of the noble gases are very high but decrease down the group (Fig. 15.27). Xenon's ionization energy is low enough for electrons to be lost to very electronegative elements. No compounds of helium, neon, and argon exist, except under very special conditions. Radon is known to react with fluorine; but its radioactivity makes the study of its compounds difficult and dangerous, so little is known about them. Krypton forms only one known stable neutral molecule, KrF_2. In 1988, a compound with a Kr—N bond was reported, but it is stable only below −50°C. This leaves xenon as the noble gas with the richest chemistry. It forms several compounds with fluorine and oxygen, and compounds with Xe—N and Xe—C bonds have been reported.

The starting point for the synthesis of xenon compounds is the preparation of xenon difluoride, XeF_2, and xenon tetrafluoride, XeF_4, by heating a mixture of the elements to 400°C at 6 atm. At higher pressures, fluorination proceeds as far as xenon hexafluoride, XeF_6. All three fluorides are crystalline solids (Fig. 15.28). In the gas phase, all are molecular compounds. Solid xenon hexafluoride, however, is ionic, with a complex structure consisting of XeF_5^+ cations bridged by F^- anions.

FIGURE 15.28 Crystals of xenon tetrafluoride, XeF_4. This compound was first prepared in 1962 by the reaction of xenon and fluorine at 6 atm and 400°C.

The xenon fluorides are used as powerful fluorinating agents (reagents for attaching fluorine atoms to other substances). The tetrafluoride will even fluorinate platinum metal:

$$Pt(s) + XeF_4(s) \longrightarrow Xe(g) + PtF_4(s)$$

The xenon fluorides are used to prepare the xenon oxides and oxoacids and, in a series of disproportionations, to bring the oxidation number of xenon up to +8. First, xenon tetrafluoride is hydrolyzed to xenon trioxide, XeO_3 in a disproportionation reaction:

$$6\ XeF_4(s) + 12\ H_2O(l) \longrightarrow 2\ XeO_3(aq) + 4\ Xe(g) + 3\ O_2(g) + 24\ HF(aq)$$

Xenon trioxide, XeO_3 (**17**), is the anhydride of xenic acid, H_2XeO_4 (**18**). It reacts with aqueous alkali to form a hydrogen xenate ion, $HXeO_4^-$. This ion slowly disproportionates into xenon and the octahedral perxenate ion, XeO_6^{4-} (**19**), in which the oxidation number of xenon is +8. Aqueous perxenate solutions are yellow and are very powerful oxidizing agents as a result of the high oxidation number of xenon. When barium perxenate is treated with sulfuric acid, it is dehydrated to the anhydride of perxenic acid, xenon tetroxide, XeO_4 (**20**), an explosively unstable gas.

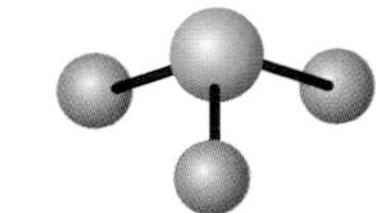

17 Xenon trioxide, XeO_3

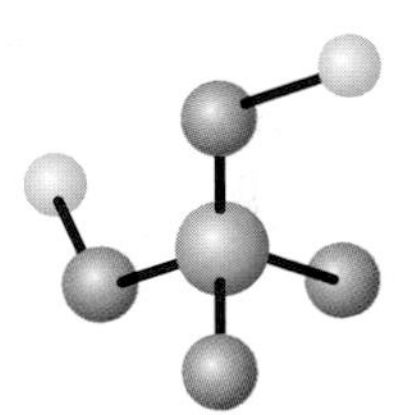

18 Xenic acid, H_2XeO_4

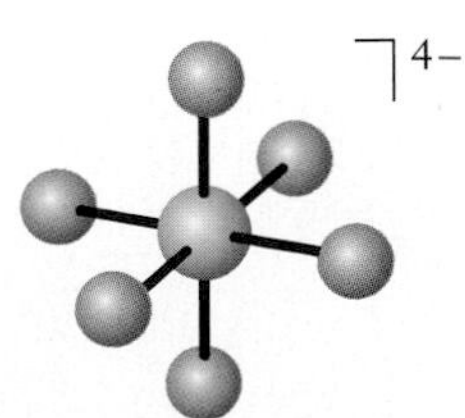

19 Perxenate ion, XeO_6^{4-}

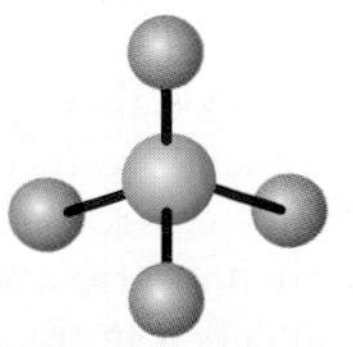

20 Xenon tetroxide, XeO_4

Xenon is the only noble gas known to form an extensive series of compounds with fluorine and oxygen; xenon fluorides are powerful fluorinating agents and xenon oxides are powerful oxidizing agents.

Self-Test 15.5A Explain why the ease of compound formation increases down Group 18/VIII.

[*Answer:* Ionization energy decreases down the group, so it is easier for the heavier elements to share their electrons.]

Self-Test 15.5B Draw the Lewis structure for $XeOF_4$ and predict its molecular shape.

THE IMPACT ON MATERIALS

Artificial joints and muscles are becoming more realistic as new lightweight "soft technologies" replace the steels of the industrial age and even the plastics of the twentieth century. Some new materials flex when an electrical impulse is passed through, and others expand more than 100 times when the temperature is raised by 1°C. The nonmetals and metalloids play an important role in these new materials, especially in gels, composite materials, ceramics, polymers, biomaterials, and luminescent materials.

15.13 Soft Materials: Colloids, Gels, and Biomaterials

A **colloid** is a *dispersion* of large particles (1 nm to 1 μm in diameter) in a solvent. Colloids differ from solutions in that the suspended particles are large enough to scatter light. Many foods are colloids, as are clay particles, smoke, foams, and the fluids in living cells. Many new lightweight materials are also colloidal. "Smart gels" are colloids that change their properties dramatically in response to their environment (Box 15.1). Aerogels, insulating ceramic foams that have densities nearly as low as that of air, are another type of colloidal material (Section 14.24). Colloidal particles are much larger than most molecules but are too small to be seen with an optical microscope. The small particles give the colloid a homogeneous appearance but are large enough to scatter light. The light scattering explains why milk is white, not transparent, and why searchlight and laser beams are more visible in fog, smoke, and clouds than in clear, dry air (Fig. 15.29).

Colloids are classified according to the phases of their components (Table 15.10). A colloid that is a suspension of solids in a liquid is called a **sol,** and a suspension of one liquid in another is called an **emulsion.** For example, muddy water is a sol in which tiny flakes of clay are dispersed in water; mayonnaise is an emulsion in which small droplets of water are suspended in vegetable oil. **Foams** are suspensions of a gas in a liquid or solid. Foam rubber, Styrofoam, soapsuds, and aerogels are foams. Zeolites (Box 13.4) are a type of solid foam in which the openings in the solid are comparable in size to molecules.

FIGURE 15.29 Laser beams are invisible. However, they can be traced when they pass through smoky or misty environments because the light scatters from the particles suspended in the air.

A **solid emulsion** is a suspension of a liquid or solid phase in a solid. For example, opals are solid emulsions in which partly hydrated silica fills the interstices between close-packed microspheres of silica aggregates. Gelatin desserts are a type of solid emulsion called a **gel,** which is soft but holds its shape. Photographic emulsions are gels that also contain solid colloidal particles of light-sensitive materials such as silver bromide. Many liquid crystalline arrays can be considered colloids. Cell membranes form a two-dimensional colloidal structure (see the illustration in Box 8.1).

Aqueous colloids can be classified as hydrophilic or hydrophobic. Suspensions of fat in water (such as milk) and water in fat (such as mayonnaise and hand lotions) are hydrophobic colloids, because fat molecules have little attraction for water molecules. Gels and puddings are examples of hydrophilic colloids. The macromolecules of the proteins in

TABLE 15.10 The Classification of Colloids*

Dispersed phase	Dispersion medium	Technical name	Examples
solid	gas	aerosol	smoke
liquid	gas	aerosol	hairspray, mist, fog
solid	liquid	sol or gel	printing ink, paint
liquid	liquid	emulsion	milk, mayonnaise
gas	liquid	foam	fire-extinguisher foam
solid	solid	solid dispersion	ruby glass (Au in glass); some alloys
liquid	solid	solid emulsion	bituminous road paving; ice cream
gas	solid	solid foam	insulating foam

*Based on R. J. Hunter, *Foundations of Colloid Science*, Vol. 1 (Oxford: Oxford University Press, 1987).

gelatin and the starch in pudding have many hydrophilic groups that attract water. The giant protein molecules in gelatin uncoil in hot water, and their numerous polar groups form hydrogen bonds with the water. When the mixture cools, the protein chains link together again, but now they have entwined to form a three-dimensional web that encloses many water molecules within themselves, as well as molecules of sugar, dye, and flavoring agents. The result is a gel: an open network of protein chains that holds the water in a flexible solid structure.

Clusters of metal atoms can also form colloidal suspensions. Colloidal clusters of copper, silver, and gold in glass are responsible for some of the vivid colors of stained glass in medieval cathedrals. Even aqueous suspensions of metal clusters are known (Fig. 15.30). The color of the metal colloid depends on the size of the particle. The smaller the particle, the lower the frequency of light absorbed.

Many precipitates, such as $Fe(OH)_3$, form initially as colloidal suspensions. The tiny particles are kept from settling out by **Brownian motion,** the motion of small particles resulting from constant bombardment by solvent molecules. The sol is further stabilized by the adsorption of ions on the surfaces of the particles. The ions attract a layer of water molecules that prevents the particles from adhering to one another.

Colloids are suspensions of particles too small to be seen with a microscope but large enough to scatter light.

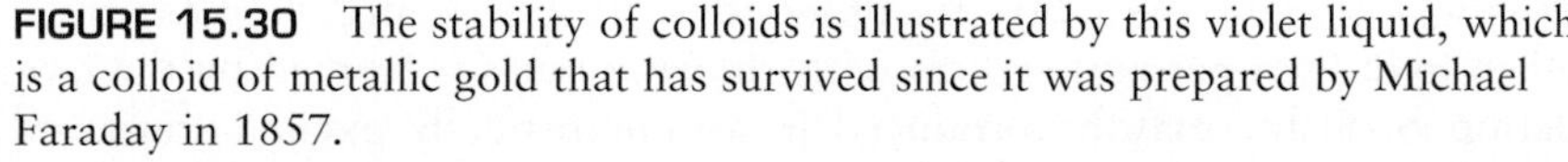

FIGURE 15.30 The stability of colloids is illustrated by this violet liquid, which is a colloid of metallic gold that has survived since it was prepared by Michael Faraday in 1857.

BOX 15.1 FRONTIERS OF CHEMISTRY: SMART GELS

When we think of materials, we usually think of rigid solids like steel, glass, and ceramics. Conventionally, the term "materials" refers to substances used in industry and research laboratories on account of their desirable physical properties; and in most cases, materials are indeed solids. However, there is a whole new class of *soft materials*. For example, materials like gels (semisolid emulsions) and flexible polymers that can be stretched and twisted are finding their way into daily use. Some can be made to crawl on their own like tiny nanometer worms, others pulsate to an internally generated rhythm, and still others respond in seemingly intelligent ways to stimuli.

You might actually be using one of these new materials the next time you use in-line skates. One brand of these skates contains a smart gel that is a liquid at normal temperatures but sets to a firm, rubbery gel when exposed to body temperature. The gel fills the space between the lining of the skate boot and the sides, and once the skate has been secured, it sets to conform to the exact shape of the foot. The gel has gone through a phase transition.

A gel like this generally consists of an aqueous suspension of organic molecules with long chains containing different regions, some of which have strong intermolecular forces and some of which have weak intermolecular forces. The temperature determines how the chains are coiled. At low temperatures, the regions with strong intermolecular forces are turned inward, so the molecules remain in liquid solution. However, at higher temperatures, those regions are turned outward, where they can attract other chains, and a flexible, but firm, network is created.

Another type of gel expands and contracts according to electrical signals and is being investigated for use in artificial limbs that would respond and feel like real ones. One type of gel being studied for use in artificial muscle contains a mixture of polymers, silicone oil (a polymer with a $-(O-Si-O-Si-)_n-$ backbone and hydrocarbon side chains), and salts. When exposed to an electric field, the soft gel contracts and stiffens. If struck, the stiffened gel can break but, on softening, the gel is reformed. The transition between gel and solid state is therefore reversible, a characteristic suggesting other applications for this type of gel, such as shock absorbers and circuit breakers.

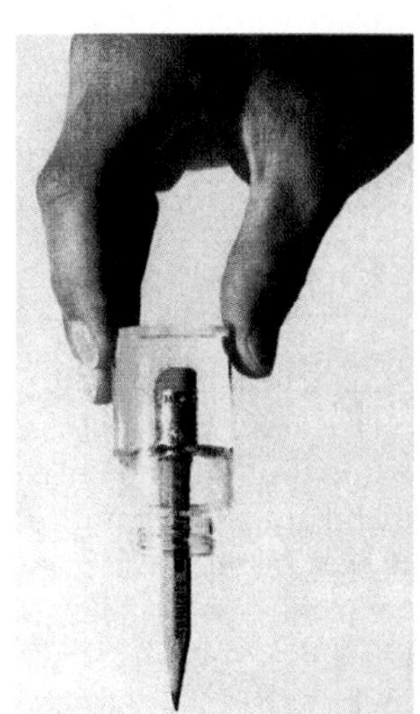

When heated, this smart gel is transformed from a liquid to a soft but firm solid that expands in response to an increase in temperature.

15.14 Phosphors and Other Luminescent Materials

Incandescence is light emitted by a hot body, such as the filament in a lamp or the particles of hot soot in a candle flame. **Luminescence** is the emission of light by a process other than incandescence. For example, when hydrogen peroxide reacts with chlorine, the O_2 formed by the oxidation of H_2O_2 may be produced in an energetically excited state and emit light as it discards its excess energy. This process is an example of

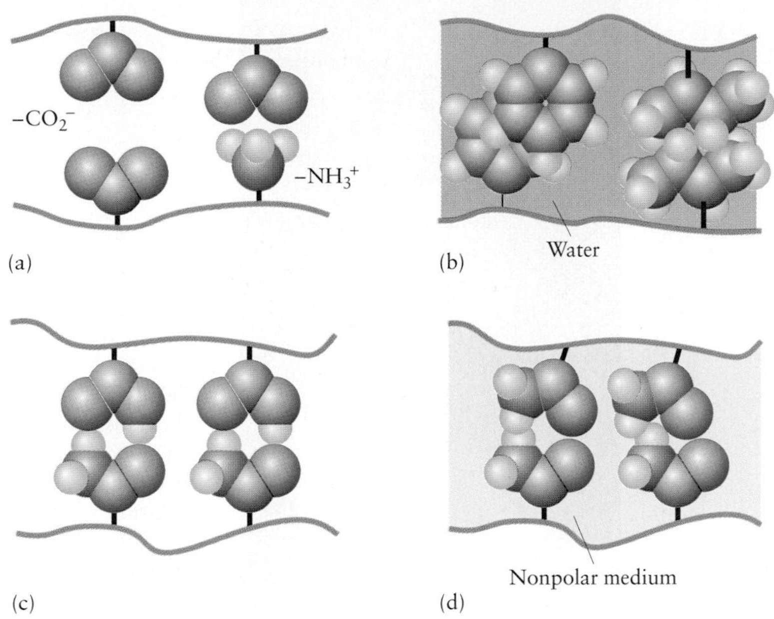

These four types of forces are responsible for the adaptive behavior of smart gels. The different forces come into play when the network of polymer chains composing the gel is disturbed. (a) Charged ionic regions can attract or repel each other. (b) Nonpolar hydrophobic regions exclude water. (c) Hydrogen bonds may form from one chain to another. (d) Dipole-dipole interactions can attract or repel chains.

HOW MIGHT YOU CONTRIBUTE?

The properties of gel-based materials need to be studied and categorized. In addition, new types of gels with possibly even more startling properties remain to be formulated. The performance of these materials is another area of intense research. For example, the response time of artificial muscle must be very short for the material to function in a natural manner. Natural muscle contracts within 100 ms of receiving the instruction from the brain, which is faster than most responsive gels can perform.

For Further Reading: P. Ball, "Clever stuff: Smart materials," Chapter 3, *Made to Measure* (Princeton: Princeton University Press, 1997), pp. 103–142. R. Dagani, "Intelligent gels," *Chemical and Engineering News* (June 9, 1997) pp. 26–27. R. S. Langer and J. P. Vacanti, "Tissue engineering: The challenges ahead," *Scientific American* (April, 1999), 86–89.

chemiluminescence, the emission of light by products formed in energetically excited states when reagents are mixed (Fig. 15.31). The light sticks used for emergency lighting glow with light from a chemiluminescent process. *Bioluminescence* is a form of chemiluminescence generated by a living organism. For example, the enzyme *luciferase* produces the luminescent compound *luciferin* in fireflies and some bacteria (see the photograph at the beginning of this chapter).

Media Link **Lab video:** Fig. 15.31, Light-emitting organic compound

FIGURE 15.31 Chemiluminescence, the emission of light as the result of a chemical reaction, occurs when hydrogen peroxide is added to a solution of the organic compound perylene. Although hydrogen peroxide itself can fluoresce, in this case the light is emitted by the perylene.

Fluorescence is the emission of light from molecules excited by radiation of higher frequency. For example, fluorescence in the visible region of the spectrum is often observed when a substance is illuminated with ultraviolet radiation. The fluorescence lasts only as long as the illumination lasts. **Phosphorescence** is the emission of light from a molecule that remains excited after the stimulus has ceased. Glow-in-the-dark toys are phosphorescent. **Triboluminescence** is luminescence that results from mechanical shock to a crystal. It is observed in many crystalline substances and can readily be detected by striking or grinding sugar crystals in a dark room. The luminescence is given off by trapped nitrogen gas escaping from the crystal while in an excited state.

The development of flat-panel computer and television displays depends on the design of bright and stable **phosphors,** phosphorescent materials that glow when activated by the impact of fast electrons. The electron impact excites the phosphor into a state of higher energy, which it then discards as photons of electromagnetic radiation. The phosphors used in cathode-ray tubes, such as those in television sets and computer displays, must stop glowing as soon as the electron beam sweeps away, so that the picture can change rapidly.

Clusters of three phosphors are used for each dot in a color television or computer display screen. Zinc sulfide doped by a *d*-block metal is commonly used in these phosphors. Silver-activated ZnS is often used for the blue phosphor and copper-activated ZnS for the green phosphor. Solid zinc sulfide is an insulator, with a large energy difference between the conductance and valence bands, and the activating metal has an energy level between those two bands. When the zinc sulfide is struck by electrons of

Yttrium orthovanadate activated by europium is often used for the red phosphor.

sufficient energy, one of its electrons is excited to the conduction band, leaving a positively charged hole in the lattice. The hole "travels" to an atom of the activating metal. The excited electron then drops back from the conduction band into that hole, emitting light of a wavelength corresponding to the energy separation between the activating metal and the excited state.

Neon lamps glow when the neon atoms are excited by an electrical discharge: the entire volume of gas glows. Fluorescent lamps make use of fluorescent materials, and the light is generated at the surface of the tube. The inside of the tube is coated with a thin layer of a phosphor such as a calcium halophosphate, $Ca_5(PO_4)_3F_{1-x}Cl_x$, doped with manganese(II) and antimony(III) ions. The tube is evacuated and a small amount of mercury vapor introduced before it is sealed. When the lamp is turned on, a high-voltage discharge excites the atoms of mercury, which emit radiation, mostly in the ultraviolet region at 254 nm and 185 nm. The doped phosphors absorb the radiation and emit visible light of lower wavelengths. The antimony(III) emits blue light and the manganese(II) emits yellow light. The result is a light that looks approximately white.

Fluorescent materials are very important in medical research. Dyes such as fluorescein (**21**) can be attached to protein molecules, and the protein can be traced in a biological system by exciting the fluorescein and looking for its emissions. By using a fluorescent material, it is possible to detect much smaller concentrations than would otherwise be possible. Because fluorescent materials can be activated by radioactivity, they are also used in scintillation counters to measure radiation (see Box 17.3).

Luminescent materials release energy as light as they return from excited states.

HO O O COOH

21 Fluorescein

SKILLS YOU SHOULD HAVE MASTERED

1 Predict and explain periodic trends in the properties of the elements in Groups 15/V, 16/VI, 17/VII, and 18/VIII.

2 Rationalize the properties of the oxides and oxoacids of the nonmetals in terms of the oxidation number of the nonmetal and the identification of acid anhydrides.

3 Describe how nitrogen is obtained from air and converted into useful compounds.

4 Describe the structures and properties of the nitrogen, phosphorus, and sulfur oxides.

5 Describe the acidic and basic character of water and hydrogen peroxide and compare their functions as oxidizing and reducing agents.

6 Describe the names, properties, and reactions of the principal compounds of the members of Groups 15/V, 16/VI, 17/VII, and 18/VIII.

7 Describe and write balanced equations for the principal reactions used to obtain the elements in Groups 15/V, 16/VI, and 17/VII.

8 Explain the low reactivity of the noble gases and the fact that the ease of compound formation increases down the group.

9 Explain the properties of colloidal and luminescent materials.

EXERCISES

Group 15/V

15.1 Nitrogen can be found in compounds with oxidation numbers ranging from -3 to $+5$. Give one example of a nitrogen compound or ion in which the nitrogen atom is found with each of the possible integral oxidation numbers.

15.2 Give the oxidation number of phosphorus in (a) white phosphorus; (b) red phosphorus; (c) phosphine, PH_3; (d) diphosphine, P_2H_4; (e) PCl_3; (f) PF_5; (g) phosphate ion, PO_4^{3-}; (h) hydrogen phosphite ion, HPO_3^{2-}.

15.3 Urea, $CO(NH_2)_2$, reacts with water to form ammonium carbonate. Write the chemical equation and calculate the mass of ammonium carbonate that can be obtained from 5.0 kg of urea.

15.4 (a) Nitrous acid reacts with hydrazine in acidic solution to form hydrazoic acid, HN_3. Write the chemical equation and determine the mass of hydrazoic acid that can be produced from 20.0 g of hydrazine. (b) Suggest a method for preparing sodium azide, NaN_3. (c) Is the production of hydrazoic acid an oxidation or a reduction of hydrazine?

15.5 Lead azide, $Pb(N_3)_2$, is used as a detonator. (a) What volume of nitrogen at STP does 1.0 g of lead azide produce when it decomposes into lead metal and nitrogen gas? (b) Would 1.0 g of mercury(II) azide, $Hg(N_3)_2$, which is also used as a detonator, produce a larger or smaller volume, given that its decomposition products are elemental mercury and nitrogen gas? (c) Metal azides in general are potent explosives. Why?

15.6 Sodium azide is used to inflate protective air bags in automobiles. What mass of solid sodium azide is needed to provide 100 L of N_2 at 1.5 atm and 20°C?

15.7 The common acid anhydrides of nitrogen are N_2O, N_2O_3, and N_2O_5. Write the formulas of their corresponding acids and chemical equations for the formation of the acids by the reaction of the anhydrides with water.

15.8 The common acid anhydrides of phosphorus are P_4O_6 and P_4O_{10}. Write the formulas of their corresponding acids and chemical equations for the formation of the acids by the reaction of the anhydrides with water.

15.9 (a) Solid phosphorus pentachloride exists as $PCl_4^+PCl_6^-$. Write the Lewis structures for the two ions and predict their shapes from VSEPR theory. (b) In contrast, PBr_5 is found as $PBr_4^+Br^-$ in the solid state. This difference is attributed to the relatively small size of the phosphorus atom, which gives rise to Br—Br repulsions. (b) Given that average P—Cl and P—Br distances are 204 pm and 220 pm, respectively, calculate what the separation of the bromine atoms would be in a hypothetical PBr_6^- ion and compare this value to the atomic and van der Waals radii for a Br atom. (c) Repeat the calculation and comparison for a PCl_6^- ion. The van der Waals radius for Br is 195 pm and that of Cl is 181 pm.

15.10 Examine the structure of phosphorus(III) oxide found on the Web site of this book. (a) Draw the Lewis structure of this molecule. (b) Does it obey the octet rule? (c) What hydrocarbon fragments are isolobal (see Exercise 2.95) to the phosphorus and oxygen atoms in this structure? (d) Predict the formula of a stable hydrocarbon with the same structure as phosphorus(III) oxide.

15.11 (a) Calculate the standard enthalpy, entropy, and free energy changes associated with the oxidation of ammonia to produce nitric oxide and gaseous water at 1000°C. The reaction is conducted in the presence of a platinum catalyst. (b) Over what temperature range is the decomposition spontaneous?

15.12 (a) Calculate the standard reaction enthalpy, entropy, and free energy associated with the thermal decomposition of ammonium nitrate to produce nitrous oxide gas and water vapor at 298 K. (b) Over what temperature range is the decomposition spontaneous under standard conditions?

15.13 (a) Calculate the standard enthalpy for the reaction of nitrogen dioxide with water to produce aqueous nitric acid and nitric oxide. (b) Calculate the standard enthalpy of solution of nitric acid.

15.14 Is the reaction of PCl_5(g) with water to give an aqueous mixture of hydrochloric and phosphoric acids spontaneous?

Group 16/VI

15.15 Write equations for (a) the burning of lithium in oxygen; (b) the reaction of sodium metal with water; (c) the reaction of fluorine gas with water; (d) the oxidation of water at the anode of an electrolytic cell.

15.16 Write equations for the reaction of (a) sodium oxide and water; (b) sodium peroxide and water;

(c) sulfur dioxide and water; (d) sulfur dioxide and oxygen, using a vanadium pentoxide catalyst.

15.17 Complete and balance the following skeletal equations:
(a) $H_2S(aq) + O_2(g) \rightarrow$
(b) $CaO(s) + H_2O(l) \rightarrow$
(c) $H_2S(g) + SO_2(g) \rightarrow$

15.18 Complete and balance the following skeletal equation:
(a) $FeS(s) + HCl(aq) \rightarrow$
(b) $H_2(g) + S(s) \rightarrow$
(c) $S_2Cl_2(l) + Cl_2(g) \rightarrow$

15.19 (a) Write the Lewis structure of H_2O_2 and predict the approximate H-O-O bond angle. Which of the following ions will be oxidized by hydrogen peroxide in acidic solution? (b) Cu^+; (c) Mn^{2+}; (d) Ag^+; (e) F^-.

15.20 (a) Draw the Lewis structure for oleum, $H_2S_2O_7$, prepared by treating sulfuric acid with SO_3. (b) Determine the formal charges on the sulfur and oxygen atoms. (c) What is the oxidation number for sulfur in this compound?

15.21 Hydrogen peroxide is unstable with respect to decomposition into water and oxygen when exposed to light, heat, or a catalyst. If all the hydrogen peroxide in 500 mL of a 3% H_2O_2 aqueous solution (like that sold in drugstores) decomposes, what volume of oxygen at 273 K and 1.00 atm will be produced? Assume the density of the solution to be 1.0 $g \cdot mL^{-1}$.

15.22 (a) Calculate the volume of pure (concentrated) sulfuric acid (of density 1.84 $g \cdot mL^{-1}$) that can be produced from 100 g of sulfur. (b) Can 1.0 M $H_2SO_4(aq)$ oxidize copper metal?

15.23 If 2.00 g of sodium peroxide is dissolved to form 200 mL of an aqueous solution, what would be the pH of the solution? For H_2O_2, $K_{a1} = 1.8 \times 10^{-12}$ and K_{a2} is negligible.

15.24 What is the pH of 0.050 M NaHS(aq)?

15.25 Describe the trend in acidity of the binary hydrogen compounds of the Group 16/VI elements and account for the trend in terms of bond strength.

15.26 Write chemical equations that represent the acid or base character of (a) two metal oxides and (b) two nonmetal oxides.

15.27 (a) Calculate the standard enthalpy and entropy of reaction for the formation of ozone from oxygen. (b) Is the reaction more favorable at high or low temperatures? (c) Does the reaction entropy favor the spontaneous formation of ozone? Explain your conclusions.

15.28 When lead(II) sulfide is treated with hydrogen peroxide, the possible products are either lead(II) sulfate or lead(IV) oxide and sulfur dioxide. (a) Write balanced equations for the two reactions. (b) Using data available in Appendix 2A, determine which possibility is more likely.

15.29 (a) Calculate the standard enthalpy of solution of $H_2SO_4(l)$ in water. (b) Estimate how hot the solution would become if 10.00 g of $H_2SO_4(l)$ were dissolved in 500.0 mL of water at 25°C. Assume all the heat goes into raising the temperature of the water and that the volume change of the solution is negligible.

15.30 (a) Calculate the standard potential for the reaction of hydrogen peroxide with Fe^{2+} in acidic solution. Will this oxidation be spontaneous? See Appendix 2B. (b) What will the potential for this reaction be in neutral solution? (c) What problem would be encountered in attempting to carry out the oxidation in basic solution?

Group 17/VII

15.31 Describe the preparation of elemental fluorine, chlorine, bromine, and iodine.

15.32 (a) In what form is chlorine found in swimming pools? (b) To what chemical property does this substance owe its disinfecting capabilities?

15.33 Identify the oxidation number of the halogen in (a) hypoiodous acid; (b) ClO_2; (c) dichlorine heptoxide; (d) $NaIO_3$.

15.34 Identify the oxidation number of the halogen atoms in (a) iodine heptafluoride; (b) sodium periodate; (c) hypobromous acid; (d) sodium chlorite.

15.35 Write the balanced chemical equations for (a) the thermal decomposition of potassium chlorate without a catalyst; (b) the reaction of bromine with water; (c) the reaction between sodium chloride and concentrated sulfuric acid. (d) Identify each reaction as a Brønsted acid-base, Lewis acid-base, or redox reaction.

15.36 Write the balanced chemical equations for the reaction of chlorine in (a) a neutral aqueous solution; (b) a dilute basic solution; (c) a concentrated basic solution. (d) Verify that each reaction is a disproportionation reaction.

15.37 (a) Arrange the chlorine oxoacids in order of increasing oxidizing strength. (b) Suggest an interpretation of that order in terms of oxidation numbers.

15.38 (a) Arrange the hypohalous acids in order of increasing acid strength. (b) Suggest an interpretation of that order in terms of electronegativities.

15.39 Write the Lewis structure for Cl_2O. Predict the shape of the Cl_2O molecule and estimate the Cl—O—Cl bond angle.

15.40 Write the Lewis structure for BrF_3. What is the hybridization of the bromine atom in the molecule?

15.41 Plot a graph of the standard free energy of formation of the hydrogen halides against the period number of the halogens. What conclusions can be drawn from the graph?

15.42 Use data from Table 15.7 to plot the normal boiling points of the hydrogen halides against the period number of the halogens. Account for the trend revealed by the graph.

15.43 Use data from Appendix 2B to determine whether chlorine gas will oxidize Mn^{2+} to form the permanganate ion in an acidic solution.

15.44 (a) Use the data in Appendix 2B to determine which is the stronger oxidizing agent, ozone or fluorine. (b) Does your answer depend on whether the reaction is carried out in acidic or basic solution?

15.45 The standard free energy of formation of HI is $+1.70\ kJ\cdot mol^{-1}$ at 25°C. Determine the equilibrium constant for the formation of HI from its elements according to the equation

$$\tfrac{1}{2}\,H_2(g) + \tfrac{1}{2}\,I_2(s) \rightleftharpoons HI(g)$$

15.46 Determine the equilibrium constant for the formation of HCl(g) according to the equation $\tfrac{1}{2}\,H_2(g) + \tfrac{1}{2}\,Cl_2(g) \rightleftharpoons HCl(g)$. Use data in Appendix 2A.

15.47 The concentration of F^- ions can be measured by adding an excess of lead(II) chloride solution and weighing the lead(II) chlorofluoride (PbClF) precipitate. Calculate the molarity of F^- ions in 25.00 mL of a solution that gave a lead chlorofluoride precipitate of mass 0.765 g.

15.48 Suppose 25.00 mL of an aqueous solution of iodine was titrated with 0.025 M $Na_2S_2O_3(aq)$, using starch as the indicator. The blue color of the starch-iodine complex disappeared when 28.45 mL of the thiosulfate solution had been added. What was the molar concentration of I_2 in the original solution? The titration reaction is $I_2(aq) + 2\,S_2O_3^{2-}(aq) \rightarrow 2\,I^-(aq) + S_4O_6^{2-}(aq)$.

15.49 Calculate the heat released when ammonium perchlorate (1.00 kg) reacts with aluminum powder (1.00 kg) as rocket fuel: $3\,NH_4ClO_4(s) + 3\,Al(s) \rightarrow Al_2O_3(s) + AlCl_3(s) + 6\,H_2O(g) + 3\,NO(g)$.

15.50 (a) Using data available in the Appendices, calculate the standard free energy of the reaction $Cl_2(s) + H_2O(l) \rightarrow HClO(aq) + HCl(aq)$. (b) What is the equilibrium constant for this process?

Group 18/VIII

15.51 What are the sources for the production of helium and argon?

15.52 What are the sources for the production of krypton and xenon?

15.53 Determine the oxidation number of the noble gases in (a) KrF_2; (b) XeF_6; (c) KrF_4; (d) XeO_4^{2-}.

15.54 Determine the oxidation number of the noble gases in (a) XeO_3; (b) XeO_6^{4-}; (c) XeF_2; (d) $HXeO_4^-$.

15.55 Xenon tetrafluoride is a powerful oxidizing agent. In an acidic solution, it is reduced to xenon. Write the corresponding half-reaction.

15.56 Complete and balance the following:
(a) $XeF_6(s) + H_2O(l) \rightarrow XeO_3(aq) + HF(aq)$
(b) $Pt(s) + XeF_4(s) \rightarrow$
(c) $Kr(g) + F_2(g) \xrightarrow{\text{electric discharge}}$

15.57 Predict the relative acid strengths of H_2XeO_4 and H_4XeO_6. Explain your conclusions.

15.58 Predict the relative oxidizing strengths of H_2XeO_4 and H_4XeO_6. Explain your conclusions.

Colloids, Clays, and Gels

15.59 Distinguish between a foam and a sol. Give at least one example of each.

15.60 Distinguish between an emulsion and a gel. Give at least one example of each.

15.61 Classify these foods by type of colloid: (a) yogurt; (b) marshmallow; (c) orange juice.

15.62 What types of intermolecular forces are primarily responsible for the formation of the colloid in (a) gelatin dessert; (b) a starch-based pudding; (c) meringue?

Phosphors and Other Luminescent Materials

15.63 Distinguish between fluorescence and phosphorescence.

15.64 In fluorescence, how does the energy of the emitted radiation compare with the energy of the exciting radiation?

15.65 In a mercury vapor lamp, what is the role of the mercury vapor in producing the light seen?

15.66 What is the primary difference between the ways in which a neon light and a fluorescent light work?

INTEGRATED EXERCISES

15.67 (a) Write the Lewis structure for the azide ion, N_3^-. (b) Assign formal charges to each atom. (c) The text indicates that the oxidation number for nitrogen in the azide ion is $-\frac{1}{3}$. Discuss the appropriateness of this assignment in light of the Lewis structure. (d) In general, what circumstances give rise to nonintegral oxidation numbers?

15.68 A chemist has a sealed glass bulb containing a mixture of $NO_2(g)$ and $N_2O_4(g)$ at 1.00 bar at 298 K. The gases are related by the simple equilibrium $2\ NO_2(g) \rightleftharpoons N_2O_4(g)$. At what temperature must the container be maintained to produce equal amounts of NO_2 and N_2O_4 in the flask at equilibrium?

15.69 Refer to Appendix 2B and arrange the halogens in order of increasing oxidizing strength in water.

15.70 Refer to Appendix 2B and arrange S, O_2 (to H_2O), Cl_2, and H_2O_2 (to H_2O) in order of increasing oxidizing strength in water.

15.71 Arsenic(III) sulfide is oxidized by acidic hydrogen peroxide solution to the arsenate ion, AsO_4^{3-}. Write the chemical equation and the reduction and oxidation half-reactions for the reaction.

15.72 Determine the values of $\Delta H_r°$ and $\Delta S_r°$ for each stage of the Ostwald process for the production of nitric acid. Predict the conditions of pressure and temperature that favor the formation of products in each case.

Stage 1: $4\ NH_3(g) + 5\ O_2(g) \rightarrow 4\ NO(g) + 6\ H_2O(g)$
Stage 2: $2\ NO(g) + O_2(g) \rightarrow 2\ NO_2(g)$
Stage 3: $3\ NO_2(g) + H_2O(l) \rightarrow 2\ HNO_3(aq) + NO(g)$

15.73 When the enthalpy of vaporization of water is divided by its boiling point (on the Kelvin scale), the result is $110\ J{\cdot}K^{-1}{\cdot}mol^{-1}$. For hydrogen sulfide, the same calculation gives $88\ J{\cdot}K^{-1}{\cdot}mol^{-1}$. Explain why the value for water is greater than that for hydrogen sulfide.

15.74 The annual production of sulfuric acid in the United States is approximately 4×10^{10} kg. (a) If the acid were all produced from elemental sulfur, what mass of sulfur would be used for the production of sulfuric acid? (b) What volume of sulfur trioxide (at 25°C and 5.0 atm) is required for the annual production of sulfuric acid?

15.75 Determine the volume (in liters) of concentrated sulfuric acid (density, $1.84\ g{\cdot}mL^{-1}$) that is required for the production of 1000 kg of phosphoric acid from phosphate rock by the reaction $Ca_3(PO_4)_2(s) + 3\ H_2SO_4(l) \longrightarrow 2\ H_3PO_4(l) + 3\ CaSO_4(s)$.

15.76 Balance each skeletal equation and classify the reaction as acid-base, Lewis acid-base, or redox:
(a) $KClO_3(s) \rightarrow KCl(s) + O_2(g)$
(b) $CaF_2(s) + H_2SO_4(conc) \rightarrow Ca(HSO_4)_2(aq) + 2\ HF(g)$
(c) $OF_2(g) + OH^-(aq) \rightarrow O_2(g) + F_2(g) + H_2O(l)$
(d) $H_2S(g) + O_2(g) \rightarrow SO_2(g) + H_2O(l)$

15.77 Using the molecule database available on the Web site for this text, compare the structures of phosphorus(III) oxide and phosphorus(V) oxide. Draw Lewis structures for the two molecules. Include in your discussion an analysis of the formal charges, the phosphorus-oxygen bond orders, and the phosphorus-oxygen bond distances. Find the P—O bond distances from a standard library reference source. Do the bond distances agree with your predictions from the Lewis structures?

15.78 Mustard gas consists of sulfur-containing organic molecules with a structure that may be viewed on the Web site for this text. (a) Write a balanced chemical equation for the production of mustard gas. (b) What is the hybridization of the sulfur atom? (c) Suggest a mechanism by which the reaction could proceed.

15.79 Balance each skeletal equation and classify the reaction as Brønsted acid-base, Lewis acid-base, or redox:
(a) $SO_2(g) + H_2O(l) \rightarrow H_2SO_3(aq)$
(b) $F_2(g) + NaOH(aq) \rightarrow OF_2(g) + NaF(aq) + H_2O(l)$
(c) $S_2O_3^{2-}(aq) + Cl_2(g) + H_2O(l) \rightarrow HSO_4^-(aq) + H_3O^+(aq) + Cl^-(aq)$
(d) $XeF_6(s) + OH^-(aq) \rightarrow XeO_6^{4-}(aq) + Xe(g) + O_2(g) + F^-(aq) + H_2O(l)$

15.80 Predict the formulas and Lewis structures of the anhydrides that correspond to the following acids: (a) HClO; (b) HIO_3; (c) $HClO_4$.

15.81 Complete the following skeletal equations:
(a) $I_2(s) + F_2(g) \rightarrow$ (any one of a number of products)
(b) $I_2(aq) + I^-(aq) \rightarrow$
(c) $Cl_2(g) + H_2O(l) \rightarrow$
(d) $F_2(g) + H_2O(l) \rightarrow$

15.82 The concentration of Cl^- ions can be measured gravimetrically by precipitating silver chloride, with silver nitrate as the precipitating reagent in the presence of dilute nitric acid. The white precipitate is filtered off and its mass is determined. (a) Calculate the Cl^- ion concentration in 50.00 mL of a solution that gave a silver chloride precipitate of mass 2.798 g. (b) Why is the method inappropriate for measuring the concentration of fluoride ions?

15.83 Discuss the chemistry of a safety match: identify the chemicals that are used and the function of each. (See Section 14.9, too.)

15.84 (a) When Xe and Pt are treated with fluorine gas under pressure at 200°C, the ionic solid $XeF_5^+PtF_6^-$ is formed. Write the Lewis structure for XeF_5^+ and, from VSEPR theory, predict its shape. (b) When two molecules of XeF_2 react with one molecule of AsF_5, the resulting compound has the formula $Xe_2F_3^+\ AsF_6^-$. Predict the Lewis structure of the $Xe_2F_3^+$ ion.

15.85 From standard reference sources, determine the chemical composition and color of the minerals orpiment and realgar. Mention a use of these substances.

15.86 Chapters 14 and 15 have been organized by group. Discuss whether it would be helpful to organize the elements according to period and give examples of trends that could be displayed helpfully in that system.

15.87 The azide ion has an ionic radius of 148 pm and forms many ionic and covalent compounds that are similar to those of the halides. (a) Write the Lewis formula for the azide ion and predict the N—N—N bond angle. (b) On the basis of its ionic radius, where in Group 17/VII would you place the azide ion? (c) Compare the acidity of hydrazoic acid with those of the hydrohalic acids and explain any differences (for HN_3, $K_a = 1.7 \times 10^{-5}$). (d) Write the formulas of three ionic or covalent azides.

15.88 The concentration of nitrate ion in a basic solution can be determined by the following sequence of steps: (1) zinc metal reduces nitrate ions to ammonia in a basic aqueous solution; (2) the ammonia is passed into a solution containing a known, but excess, amount of HCl(aq); (3) the unreacted HCl(aq) is titrated with a standard NaOH(aq) solution. (a) Write balanced chemical equations for the three reactions. (b) A 25.00-mL sample of water from a rural well contaminated with $NO_3^-(aq)$ was treated with base and an excess of zinc metal. The evolved ammonia gas was passed into 50.00 mL of 0.250 M HCl(aq). The unreacted HCl(aq) was titrated to the stoichiometric point with 28.22 mL of 0.150 M NaOH(aq). What is the molar concentration of nitrate ion in the well water?

15.89 The concentration of hypochlorite ions in a solution can be determined by adding a sample of known volume to a solution containing excess I^- ions, which are oxidized to iodine:

$$ClO_2(aq) + 2\,I^-(aq) + H_2O(l) \longrightarrow I_2(aq) + Cl^-(aq) + 2\,OH^-(aq)$$

The iodine concentration is then measured by titration with sodium thiosulfate:

$$I_2(aq) + 2\,S_2O_3^{2-}(aq) \longrightarrow 2\,I^-(aq) + S_4O_6^{2-}(aq)$$

In one experiment, 10.00 mL of ClO^- solution was added to a KI solution, which in turn required 28.34 mL of 0.110 M $Na_2S_2O_3(aq)$ to reach the stoichiometric point. Calculate the molar concentration of ClO^- ions in the original solution.

15.90 Account for the observation that melting and boiling points generally decrease from fluoride to iodide for ionic halides, but increase from fluoride to iodide for molecular halides.

15.91 Account for the observation that solubility in water generally increases from chloride to iodide for ionic halides with low covalent character (such as the potassium halides), but decreases from chloride to iodide for ionic halides in which the bonds are significantly covalent (such as the silver halides).

15.92 The sodium iodate found as an impurity in Chile saltpeter was once the major source of iodine.

The element was obtained by the reduction of an acidic solution of the iodate ion with sodium hydrogen sulfite. (a) Write the chemical equation for the reaction, assuming the oxidized product to be HSO_4^-. (b) Calculate the mass of sodium hydrogen sulfite needed to produce 50.0 g iodine.

15.93 A common procedure for drying slightly damp compounds is to store them in the presence of a suitable drying agent inside a closed container, sometimes under vacuum. Both calcium oxide and P_4O_{10} are used for this purpose. Using data available in this book, determine which of these two substances would be the better drying agent. The pertinent reactions are

$$CaO(s) + H_2O(g) \longrightarrow Ca(OH)_2(s)$$

$$P_4O_{10}(s) + 6\,H_2O(g) \longrightarrow 4\,H_3PO_4(s)$$

The free energy of formation of solid H_3PO_4 is $-1119.2\ kJ{\cdot}mol^{-1}$.

15.94 Write the half-reactions that show how hydrogen peroxide can function as an oxidizing and as a reducing agent in both acidic and basic solution. Give the appropriate potentials for these half-reactions.

15.95 (a) Using data available in Appendix 2B, determine the electrochemical potential for the redox reaction $2\,Ag^+ + 2\,I^- \longrightarrow 2\,Ag + I_2$. Is this a spontaneous process? (b) When equal volumes of 2.0 M $AgNO_3(aq)$ and 2.0 M NaI(aq) are mixed, the redox reaction described in (a) does not occur. Instead a precipitate of AgI is formed. Why doesn't the redox reaction take place? Support your answer with an appropriate calculation.

15.96 (a) Draw the molecular orbital energy-level diagram for O_2. Using the diagram determine the bond order and magnetic properties of O_2. (b) What molecular property of oxygen is explained by its molecular orbital diagram but not by its Lewis structure? (c) Describe the nature of the highest occupied molecular orbital—is it bonding, antibonding, or nonbonding? (d) Predict the bond order and magnetic properties of the peroxide ion and the superoxide ion based upon molecular orbital theory.

15.97 (a) The nitrosyl ion, NO^+, is isoelectronic with N_2 and has two fewer electrons than O_2. The ion is stable and can be purchased as its hexafluorophosphate (PF_6^-) or tetrafluoroborate salts (BF_4^-). Draw its molecular orbital diagram and compare it with those of N_2 and O_2. (b) NO^+ is diamagnetic. Can we tell which is higher in energy, the σ_{2p} or the π_{2p} orbital, from this information?

15.98 (a) Draw the Lewis structure of N_2O_3. (b) What is the N—N bond order? (c) Explain why N_2O_3 adopts this structure whereas the analogous compound for phosphorus is P_4O_6.

Media Link

15.99 Pyrite is a sulfur-based mineral with the formula FeS_2. View the unit cell of pyrite on the Web site for this text. (a) What is the type of unit cell adopted by pyrite? (b) Describe the locations of the iron and sulfur atoms in this unit cell (unit cell faces, corners, edges, etc.). (c) What is the coordination number of iron? (d) To what is each sulfur atom bonded? (e) On the basis of these observations, how would you best describe the oxidation numbers of iron and sulfur in pyrite?

15.100 The structure of apatite $Ca_5(PO_4)_3(X)$ (where X is a monovalent anion such as OH^-, F^-, or Cl^-) can be viewed on the Web site for this text. From this model, determine (a) the type of unit cell adopted by apatite; (b) the number of Ca^{2+} ions and PO_4^{3-} ions in the unit cell; (c) the locations of the Ca^{2+} and PO_4^{3-} ions (unit cell corners, faces, edges, etc.). (d) What is missing from the Web model for the structure of apatite?

CHAPTER 16

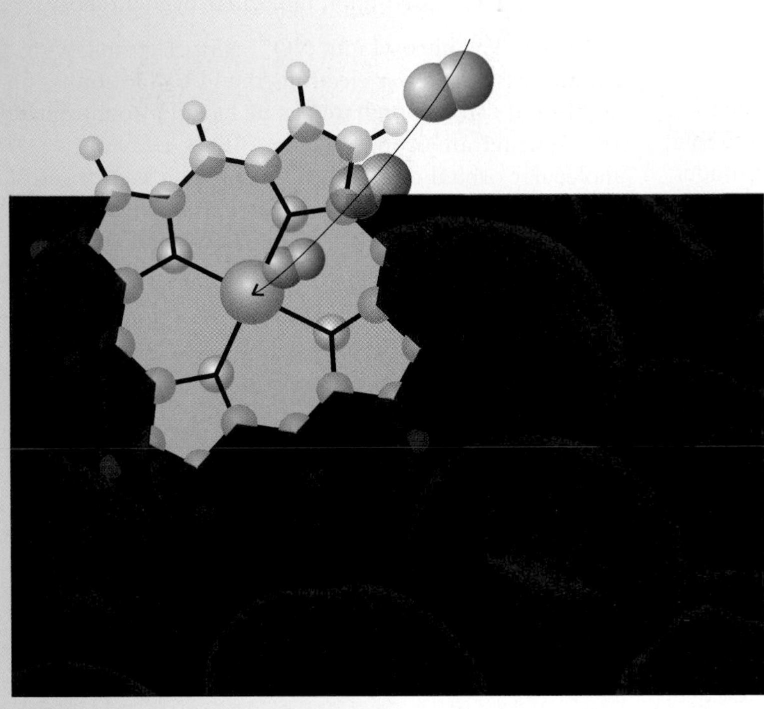

The molecule responsible for the color of these red blood cells is hemoglobin. Hemoglobin is a protein molecule consisting of four long chains of amino acids. In the center of each one is a "heme" group like the one shown here. This group is a complex of iron: the iron atom is the point of attachment of the incoming oxygen molecule.

Why Do We Need to Know This Material?

The *d*-block metals are the workhorse elements of the periodic table. Iron and copper helped civilization rise from the Stone Age and are still our most important industrial metals. Other members of the block include the metals of new technologies, such as titanium for the aerospace industry and vanadium for catalysts in the petrochemical industry. The precious metals—silver, platinum, and gold—are prized as much for their appearance, rarity, and durability as for their usefulness. Compounds of *d*-block metals give color to paint, turn sunlight into electricity, serve as powerful oxidizing agents, and form the basis of some cancer treatments.

What Do We Need to Know Already?

This chapter draws on many of the principles introduced in the preceding chapters. In particular, it makes use of the electron configurations of atoms and ions (Sections 1.12 and 2.3) and the classification of species as Lewis acids and bases (Section 2.13). Molecular orbital theory (Sections 3.9–3.12) plays an important role in Section 16.13.

THE *d* BLOCK: METALS IN TRANSITION

The Industrial Revolution was built on steel, which was used for construction and for transportation. Iron, the principal component of steel, transformed our past, but other *d*-block metals are transforming our present. Titanium, for instance, is used in jet engines. Vanadium and platinum are used in the development of catalysts to reduce pollution and in the continuing effort to make hydrogen the fuel of our future.

The compounds of the *d*-block elements also show a wide range of interesting properties. Some are vital to life. Iron is an essential component of mammalian blood. Compounds of cobalt, molybdenum, and zinc are found in vitamins and essential enzymes. Other compounds simply make life more interesting. The beautiful color of cobalt blue glass, the brilliant greens and blues of kiln-baked pottery, and many pigments used by artists make use of *d*-block compounds.

We begin this chapter by summarizing the major periodic trends exhibited by the *d*-block elements and their compounds. Then we describe some of the properties and key reactions of selected elements. One characteristic class of compounds formed by these elements consists of their numerous complexes, and in the second half of the chapter, we describe the structures of these complexes and the two principal theories of their bonding. We end the chapter by examining the contribution of *d*-block elements to modern materials.

THE *d*-BLOCK ELEMENTS AND THEIR COMPOUNDS

The elements in Groups 3 through 11 represent a transition from the highly reactive metals of the *s* block to the much less reactive metals of Group 12 and the *p* block; they are called the *transition metals* (Fig. 16.1). Note that the transition metals do not extend all the way across the *d* block because the Group 12 elements (zinc, cadmium, and mercury) are not normally considered to be transition elements. Because their *d*-orbitals are full, the Group 12 elements have properties that are more like those of main-group metals than those of transition metals. Just after the third row of the *d* block has been started, at lanthanum, electrons begin to occupy the seven 4*f*-orbitals and the lanthanides (the "rare earths") delay the completion of Period 6. These elements, together with the actinides, the analogous series in Period 7, are sometimes referred to as the *inner transition metals*.

This distinction is not always made: take the advice of your instructor.

The incompletely filled *d*-subshell is responsible for the wide range of colors shown by compounds of the *d*-block elements. Furthermore, many *d*-metal compounds are paramagnetic (Box 3.2). One of the challenges we face in this chapter is to build a model of bonding that accounts for color and magnetism in a unified way. First, though, we consider the physical and chemical properties of the elements themselves.

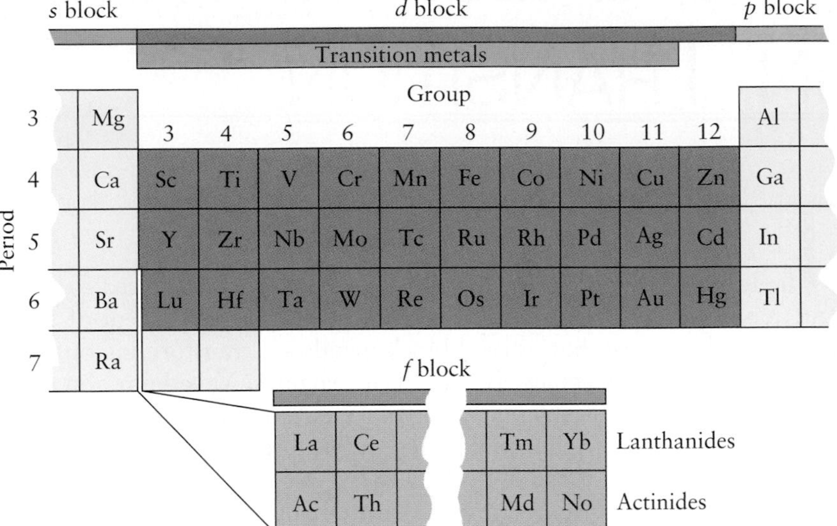

FIGURE 16.1 The elements in the *d* block of the periodic table. Note that the *f* block, consisting of the inner transition metals, intervenes in Periods 6 and 7 as indicated by the purple bar. The horizontal gray bar indicates the groups containing the elements regarded as transition metals.

16.1 Trends in Physical Properties

In the previous two chapters, we saw that elements in the main groups have physical properties that depend greatly on their valence electron configurations, resulting in large differences between groups. However, ground-state electron configurations of the atoms of *d*-block elements differ primarily in the occupation of the $(n - 1)d$-orbitals that lie slightly lower in energy than the outer *ns*-orbitals. According to the building-up principle, the *d*-orbitals of these elements are the last orbitals to be occupied. There are five *d*-orbitals in a given shell, and each one can accommodate up to two electrons, so there are 10 elements in each row of the *d* block. Because the differences in electron configurations of the *d*-block metals are primarily in these inner *d*-orbitals, from which electrons can be lost in chemical reactions, the physical properties of the *d*-block metals tend to be similar.

All the *d*-block elements are metals. Most of these "*d*-metals" are good electrical conductors. In fact, silver is the best electrical conductor of all elements at room temperature. Most of the *d*-block metals are malleable, ductile, lustrous, and silver-white in color. Generally, their melting and boiling points are higher than those of the main-group elements. There are a few notable exceptions: copper is red-brown, gold is yellow, and mercury has such a low melting point that it is a liquid at room temperature.

We can trace some of the properties of the *d*-block elements to the shapes of their *d*-orbitals (see Fig. 1.31). There are two points to keep in mind. First, the lobes of two *d*-orbitals on the same atom occupy markedly different regions of space. Because they are relatively far apart, electrons in different *d*-orbitals repel one another only weakly. Second, the electron density in *d*-orbitals is low near the nucleus, so they are not very effective at shielding other electrons from the positive nuclear charge.

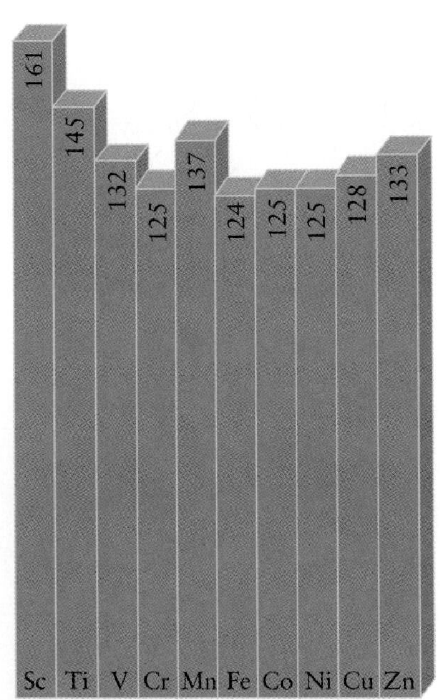

FIGURE 16.2 The atomic radii (in picometers) of the elements of the first row of the *d* block.

One consequence of these two characteristics is the trend in atomic radii of the *d*-block metals (Fig. 16.2). Nuclear charge and number of *d*-electrons both increase from left to right across each row (from scandium to zinc, for instance). Because the repulsion between *d*-electrons is weak, the increasing nuclear charge can draw them inward, so the atoms become smaller. Further across the block, the radii begin to increase slightly as the atomic number increases: there are now so many *d*-electrons that electron-electron repulsion increases more rapidly than the nuclear charge. However, because these attractions and repulsions are finely balanced, the range of *d*-metal atomic radii is not very great. In fact, some of the atoms of one *d*-metal can easily replace atoms of another *d*-metal in a crystal lattice (Fig. 16.3). The *d*-metals can therefore form a wide range of alloys, including the many varieties of steel (see Sections 16.14 and 16.15).

The atomic radii of the second row of *d*-metals (Period 5) are typically greater than those in the first row (Period 4). The atomic radii in the third series (Period 6), however, are about the same as those in the second row and smaller than expected. This effect is due to the **lanthanide contraction**, the decrease in radius along the first row of the *f* block (Fig. 16.4). This decrease is due to the increasing nuclear charge along the period coupled with the poor shielding ability of *f*-electrons. When the *d* block resumes (at lutetium), the atomic radius has fallen from 224 pm for barium to 172 pm for lutetium.

One technologically important effect of the lanthanide contraction is the high density of the Period 6 elements (Fig. 16.5). The atomic radii of these elements are comparable to those of the Period 5 elements, but their atomic masses are about twice as large; so more mass is packed into the same volume. A block of iridium, for example, contains about as many atoms as a block of rhodium of the same volume. However, each iridium atom is nearly twice as heavy as a rhodium atom, so the density of the sample is nearly twice as great. In fact, iridium is one of the two densest elements; its neighbor osmium is the other. Another effect of the contraction is the low reactivity—the "nobility"—of gold and platinum. Because their valence electrons are relatively close to the nucleus, they are tightly bound and not readily available for chemical reactions.

The presence of unpaired *d*-electrons in the ground states of the *d*-block elements explains why some of these metals—notably, iron, cobalt, and nickel—make good permanent magnets. To understand this property, we need to recall some facts about magnetism (see Box 3.2). Paramagnetism is the tendency of a substance to move into a magnetic field. It arises

FIGURE 16.3 Because the atomic radii of the *d*-block elements are so similar, the atoms of one element can replace the atoms of another element with minor modification of the atomic locations; consequently, *d*-block metals form a wide range of alloys.

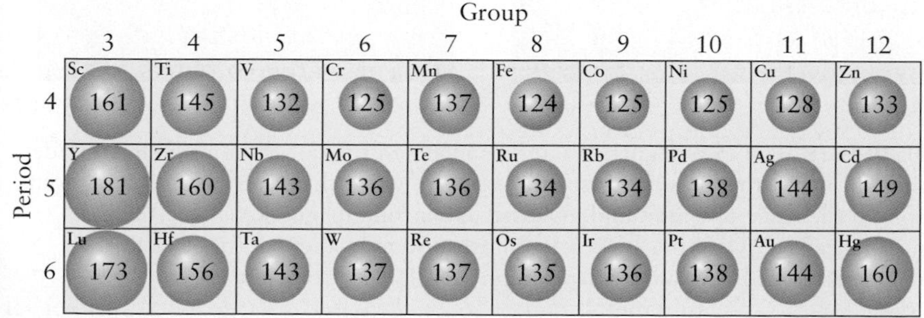

FIGURE 16.4 The atomic radii of the *d*-block elements (in picometers). Notice the similarity of all the values and, in particular, the close similarity between the second and third rows as a result of the lanthanide contraction.

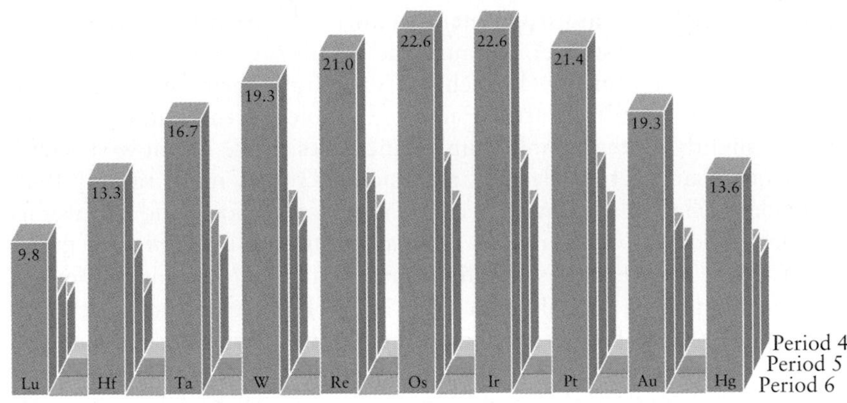

FIGURE 16.5 The densities (in grams per centimeter cubed, $g\cdot cm^{-3}$) of the *d*-metals at 25°C. The lanthanide contraction has a pronounced effect on the densities of the elements in Period 6 (front row in this illustration), which are among the densest of all the elements.

when an atom or molecule has at least one unpaired electron. However, because the spins on neighboring atoms or molecules are aligned almost randomly, paramagnetism is very weak. In some *d*-metals, however, the unpaired electrons of many neighboring atoms can align with one another, producing the much stronger effect of **ferromagnetism.** The aligned spins form **domains,** or regions of aligned spins (Fig. 16.6), which survive even after the applied field is turned off. Ferromagnetism also occurs in some compounds of the *d*-metals; oxides of iron and chromium, for instance, are used to record signals on tapes. The electromagnetic recording heads align large numbers of spins as the tape passes underneath, and the spin alignment in the domains remains for years.

Magnetic materials are discussed further in Section 16.16.

The atomic radii of the d-block metals are all very similar but typically decrease across a series. The lanthanide contraction accounts for the smaller than expected radii and higher densities of the d-block atoms in Period 6.

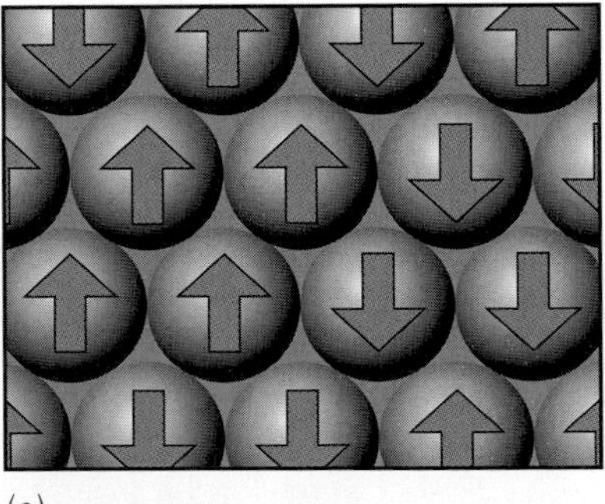
(a)

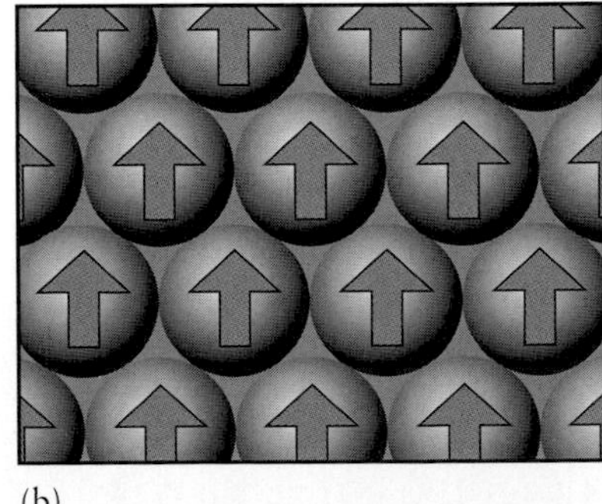
(b)

FIGURE 16.6 Ferromagnetic materials include iron, cobalt, and the iron oxide mineral magnetite. They consist of crystals in which the electrons of many atoms spin in the same direction and give rise to a strong magnetic field. (a) Before magnetization, when the spins are almost randomly aligned; (b) after magnetization. The orange arrows represent the electron spins.

16.2 Trends in Chemical Properties

The *d*-block elements tend to lose their valence *s*-electrons when they form compounds. Most of them can also lose a variable number of *d*-electrons and show variable valence. The only elements of the block that do not use their *d*-electrons in compound formation are the members of Group 12 (zinc, cadmium, and mercury). The ability to exist in different oxidation states is responsible for many of the special chemical properties of these elements and plays a role in the action of many biomolecules.

Although several of the less reactive *d*-block elements toward the right of the block are found "native" (in the uncombined elemental state), only gold is mined primarily as the metal. The elements on the left of the *d* block are commonly found as oxides, halides, or oxoanion minerals, including carbonates and silicates. For example, manganese is found as *pyrolusite*, MnO_2, and lanthanum as *bastnaesite*, $LaCO_3F$. Those on the right of the block commonly occur as sulfides. For example, zinc is found as *zinc blende*, ZnS, and copper as *chalcopyrite*, $CuFeS_2$.

Most *d*-block elements have more than one common oxidation state (Fig. 16.7). The distribution of oxidation states looks daunting at first sight, but there is a pattern. Except for mercury, the elements at the ends of each row of the *d* block occur in only one oxidation state other than 0. Scandium, for example, is found only in oxidation state +3, and zinc only as +2. All the other elements of each row have at least two oxidation states. Copper, for example, is found in two oxidation states, +1 (as in CuCl) and +2 (as in $CuCl_2$). Elements close to the center of each row have the widest range of oxidation states. Manganese, at the center of its row, has seven oxidation states. Elements in the second and third rows of the block are more likely to reach higher oxidation states than those in the first row.

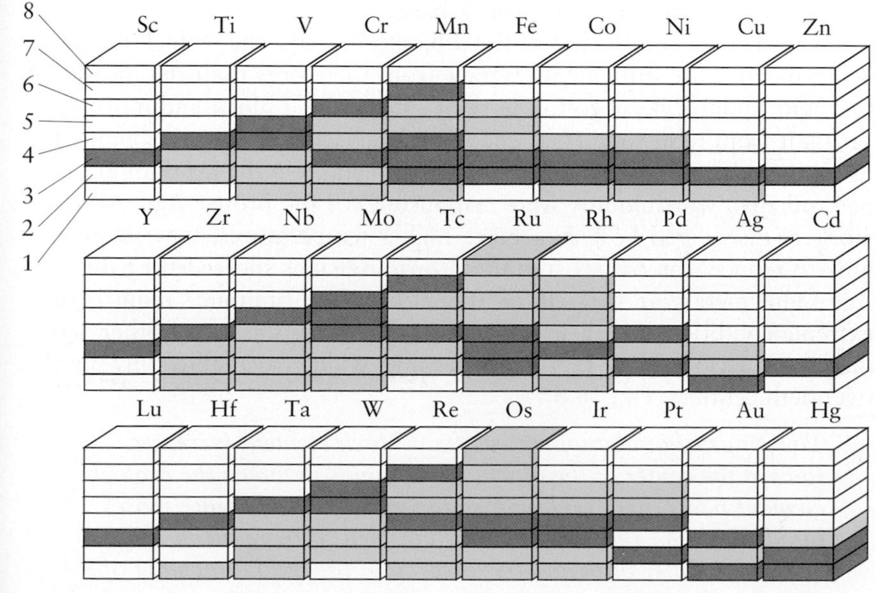

FIGURE 16.7 The oxidation numbers of the *d*-block elements. The orange blocks mark the common oxidation numbers for each element; the green blocks mark the other known states.

The pattern of oxidation states underlies trends in the chemical properties of the *d*-block elements. When an element has a high oxidation state, the compound is likely to be a good oxidizing agent. For example, manganese has oxidation number +7 in the permanganate ion, MnO_4^-, and this ion is a good oxidizing agent in acidic solution:

$$MnO_4^-(aq) + 8\,H^+(aq) + 5\,e^- \longrightarrow Mn^{2+}(aq) + 4\,H_2O(l) \qquad E° = +1.51\text{ V}$$

Compounds that contain the element in a low oxidation state, such as Cr^{2+}, are often good reducing agents:

$$Cr^{3+}(aq) + e^- \longrightarrow Cr^{2+}(s) \qquad E° = -0.41\text{ V}$$

The pattern of oxidation states correlates with the pattern of acid-base behavior of *d*-metal oxides. Although most *d*-metal oxides are basic, the oxides of a given element show a shift toward acidic character with increasing oxidation number. The family of chromium oxides is a good example:

CrO	+2	basic
Cr_2O_3	+3	amphoteric
CrO_3	+6	acidic

Chromium(VI) oxide, CrO_3, is the anhydride of chromic acid, H_2CrO_4, the parent acid of the chromates. In this highly oxidized state, chromium is electron poor and the oxygen atoms attached to it are less likely to share electrons with a proton. The conjugate base of chromic acid, $HCrO_4^-$, does not attract protons strongly, despite its overall negative charge, and is therefore a very weak base.

The elements on the left of the *d* block resemble the *s*-block metals in being much more difficult to extract from their ores than the metals on the right. Indeed, if we go to the right side of the *d* block and move back across it from right to left, we encounter the elements in very roughly the order in which they were put to use as the metal. On the far right are copper and zinc, which jointly were responsible for the Bronze Age. As metal workers discovered how to achieve higher temperatures, they found out how to reduce iron oxides, and the Bronze Age was succeeded by the Iron Age. The metals at the left of the block, like titanium, require such extreme conditions for their extraction—including the use of other active metals or electrolysis—that they became widely available only in the twentieth century (Fig. 16.8).

The range of oxidation states of a d-block element increases toward the center of the block. Compounds in which the d-block element has a high oxidation state tend to be oxidizing; those in which it has a low oxidation state tend to be reducing. The acidic character of oxides increases with the oxidation state of the element.

(a) (b) (c)

FIGURE 16.8 These three artifacts represent the progress that has been made in the extraction of *d*-metals. (a) An ancient bronze chariot axle cap from China made from an alloy of metals that are easy to extract. (b) An early iron steam engine made from a metal that was moderately easy to extract once high temperatures could be achieved. (c) A twenty-first-century airplane engine with titanium components that had to await advanced, high-temperature technology before the element became widely available.

Self-Test 16.1A Predict trends in ionization energies of the *d*-block metals.

[*Answer:* Ionization energy increases from left to right across a row and decreases down a group.]

Self-Test 16.1B Predict trends in the densities of the *d*-block metals.

SELECTED ELEMENTS: A SURVEY

The physical properties of the *d*-block elements are similar, but the chemical properties of these elements are so diverse that it is impossible to summarize them fully. We can, however, observe some of the major trends in properties within the *d* block by considering the properties of certain representative elements, particularly those in the first row of the *d* block.

16.3 Scandium Through Nickel

Table 16.1 summarizes the physical properties of the elements from scandium to nickel. Notice the similarities in their melting and boiling points, but the gradual increase in density.

Scandium, Sc, which was first isolated in 1937, is a reactive metal: it reacts with water about as vigorously as calcium does. It has few uses and is not thought to be essential to life. The small, highly charged Sc^{3+} ion is strongly hydrated in water (like Al^{3+}), and the resulting $[Sc(H_2O)_6]^{3+}$ complex is about as strong a Brønsted acid as acetic acid.

Square brackets are commonly used to indicate the presence of a *d*-metal complex.

TABLE 16.1 Properties of the *d*-Block Elements Scandium Through Nickel

Z	Name	Symbol	Valence electron configuration	Melting point, °C	Boiling point, °C	Density, $g \cdot cm^{-3}$
21	scandium	Sc	$3d^1 4s^2$	1540	2800	2.99
22	titanium	Ti	$3d^2 4s^2$	1660	3300	4.55
23	vanadium	V	$3d^3 4s^2$	1920	3400	6.11
24	chromium	Cr	$3d^5 4s^1$	1860	2600	7.19
25	manganese	Mn	$3d^5 4s^2$	1250	2120	7.47
26	iron	Fe	$3d^6 4s^2$	1540	2760	7.87
27	cobalt	Co	$3d^7 4s^2$	1494	2900	8.80
28	nickel	Ni	$3d^8 4s^2$	1455	2150	8.91

Titanium, Ti, a light, strong metal, is used where these properties are vital—in widely diverse applications such as jet engines and dental appliances (Fig. 16.9). Although titanium is relatively reactive, unlike scandium it is resistant to corrosion because it is passivated by a protective skin of oxide on its surface. The principal sources of the metal are the ores *ilmenite*, $FeTiO_3$, and *rutile*, TiO_2.

Titanium requires strong reducing agents for extraction from its ores. It was not widely exploited commercially until demand from the aerospace industry increased in the latter part of the twentieth century. The metal is obtained first by treating the ores with chlorine in the presence of coke to form titanium(IV) chloride; the volatile chloride is then reduced by passing it through liquid magnesium:

$$TiCl_4(g) + 2\ Mg(l) \xrightarrow{700°C} Ti(s) + 2\ MgCl_2(s)$$

FIGURE 16.9 This piece of titanium has been formed and then machined for use in a high-performance military jet plane. Titanium has excellent corrosion resistance, low density and high strength, making it ideal for aerospace applications.

The most common oxidation state of titanium is +4, with both its 4*s*-electrons and its two 3*d*-electrons lost. Its most important compound is titanium(IV) oxide, TiO_2, which is almost universally known as titanium dioxide. This oxide is a brilliantly white, nontoxic, stable solid used as the white pigment in paints and paper. It acts as a semiconductor in the presence of light, so it can be used as an electrode in photoelectrochemical cells.

Titanium forms a series of oxides called *titanates*, which are prepared by heating TiO_2 with a stoichiometric amount of the oxide or carbonate of a second metal. Barium titanate, $BaTiO_3$, is **piezoelectric,** which means that it becomes electrically charged when it is mechanically distorted. This property leads to its use for underwater sound detection, in which a mechanical vibration is converted into an electrical signal.

Vanadium, V, a soft silver-gray metal, is produced by reducing its oxide or chloride. For example, vanadium(V) oxide is reduced by calcium:

$$V_2O_5(s) + 5\ Ca(l) \xrightarrow{\Delta} 2\ V(s) + 5\ CaO(s)$$

Vanadium is used to make tough steels for automobile and truck springs. Because it is not economical to add the pure metal to iron, a **ferroalloy** of the metal, an alloy with iron and carbon, that is less expensive to produce, is used instead.

Vanadium(V) oxide, V_2O_5, commonly known as vanadium pentoxide, is the most important compound of vanadium. This orange-yellow solid is used as an oxidizing agent and as an oxidizing catalyst in the contact process for the manufacture of sulfuric acid (Section 15.7). The wide range of colors of vanadium compounds, including the blue of the vanadyl ion, VO^{2+} (Fig. 16.10), has led to their use as glazes in the ceramics industry. Although its role is not fully known, vanadium is thought to be an essential element for life and is concentrated in some sea animals. Some chemists have speculated that these high concentrations arise because the organism unwittingly mistakes vanadates, VO_4^{3-}, for phosphates, PO_4^{3-}.

Chromium, Cr, is a bright, lustrous, corrosion-resistant metal. Its name, which comes from the Greek word for "color," was inspired by its colorful compounds. The metal is obtained from the mineral *chromite*, $FeCr_2O_4$, by reduction with carbon in an electric arc furnace:

$$FeCr_2O_4(s) + 4\,C(s) \xrightarrow{\Delta} Fe(l) + 2\,Cr(l) + 4\,CO(g)$$

Chromium metal is also produced by the thermite process (see Fig. 6.12):

$$Cr_2O_3(s) + 2\,Al(s) \xrightarrow{\Delta} Al_2O_3(s) + 2\,Cr(l)$$

Chromium metal is important in metallurgy, because it is used to make stainless steel and for chromium plating (Section 12.15). Chromium(IV) oxide, CrO_2, is a ferromagnetic material that is used for coating "chrome" recording tapes; they respond better to rapidly oscillating magnetic fields than do conventional "ferric", Fe_2O_3, tapes.

Sodium chromate, Na_2CrO_4, a yellow solid, is the source of most other chromium compounds. The chromate ion changes into the orange dichromate ion, $Cr_2O_7^{2-}$, in the presence of acid (Fig. 16.11):

$$2\,CrO_4^{2-}(aq) + 2\,H^+(aq) \longrightarrow Cr_2O_7^{2-}(aq) + H_2O(l)$$

FIGURE 16.10 Many vanadium compounds form vividly colored solutions in water. They are also used in pottery glazes. The blue colors here are due to the vanadyl ion, VO^{2+}.

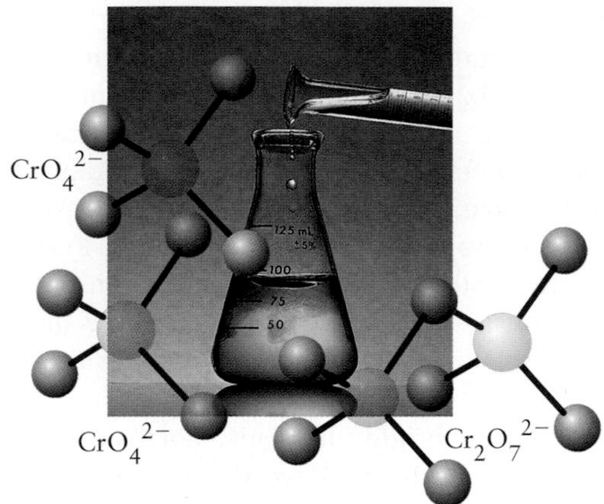

FIGURE 16.11 The chromate ion, CrO_4^{2-}, is yellow. When acid is added to a chromate solution, the ions form orange dichromate ions, $Cr_2O_7^{2-}$.

In the laboratory, acidified solutions of dichromates, in which the oxidation state of chromium is +6, are useful oxidizing agents:

$$Cr_2O_7^{2-}(aq) + 14\ H^+(aq) + 6\ e^- \longrightarrow 2\ Cr^{3+}(aq) + 7\ H_2O(l) \qquad E° = +1.33\ V$$

Sodium chromate and sodium dichromate are both starting points for the production of pigments, corrosion inhibitors, fungicides, and ceramic glazes. Chromium(III) is essential to human health and plays a role in the regulation of glucose metabolism.

Manganese, Mn, is a gray metal that resembles iron. It is much less resistant to corrosion than chromium is and becomes coated with a thin brown layer of oxide when exposed to air. The metal is rarely used alone, but it is an important component of alloys. When it is added to iron as ferromanganese to make steel, it removes sulfur by forming a sulfide. It also increases hardness, toughness, and resistance to abrasion (Table 16.2). Another useful alloy is *manganese bronze* (39% by mass Zn, 1% Mn, a small amount of iron and aluminum, and the rest copper), which is very resistant to corrosion and is used for the propellers of ships. Manganese is also alloyed with aluminum to increase the stiffness of beverage cans, allowing them to be manufactured with thinner walls.

TABLE 16.2 Composition of Different Steels

Element blended into iron	Typical amount, %	Effect
manganese	0.5 to 1.0	increases strength and hardness but lowers ductility
	13	increases wear resistance
nickel	< 5	increases strength and shock resistance
	> 5	increases corrosion resistance (stainless) and hardness
chromium	variable	increases hardness and wear resistance
	> 12	increases corrosion resistance (stainless)
vanadium	variable	increases hardness
tungsten	< 20	increases hardness, especially at high temperatures

A rich supply of manganese lies in nodules of ore that litter the ocean floors (Fig. 16.12). These nodules range in diameter from millimeters to meters and are lumps of the oxides of iron, manganese, and other elements. However, because this source is technically difficult to exploit, manganese is currently obtained by the thermite process from *pyrolusite*, a mineral form of manganese dioxide:

$$3\ MnO_2(s) + 4\ Al(s) \xrightarrow{\Delta} 3\ Mn(l) + 2\ Al_2O_3(s)$$

FIGURE 16.12 Manganese nodules litter the ocean floor and are potentially a valuable source of the element.

Manganese lies near the center of its row (in Group 7) and occurs in a wide variety of oxidation states. The most stable state is +2, but +4, +7, and, to some extent, +3 are also common in manganese compounds. Its most important compound is manganese(IV) oxide, MnO_2, commonly called manganese dioxide. This compound is a brown-black solid used in dry cells, as a decolorizer to conceal the green tint of glass, and as the starting point for the production of other manganese compounds.

Potassium permanganate is a strong oxidizing agent in acidic solution and is used to oxidize organic compounds and as a mild disinfectant. Its usefulness stems not only from its thermodynamic tendency to oxidize other species but also from its ability to act by a variety of mechanisms; hence, it is likely to be able to find a path with low activation energy and act rapidly.

Iron, Fe, the most widely used of all the *d*-metals, is the most abundant element on Earth and the second most abundant metal in the Earth's crust (after aluminum). Its principal ores are the oxides *hematite*, Fe_2O_3, and *magnetite*, Fe_3O_4. The sulfide mineral *pyrite*, FeS_2 (see Fig. 15.12), is also widely available, but it is not used in steelmaking because the sulfur is difficult to remove.

Steelmaking is described in Section 16.14.

Iron is quite reactive and corrodes in moist air. It reacts with acids that have nonoxidizing anions, evolving hydrogen and forming iron(II) salts. The colors of these salts vary from pale yellow to dark green-brown. Iron(II) salts are readily oxidized to iron(III) salts. The oxidation is slow in acidic solution but rapid in basic solution, where insoluble iron(III) hydroxide, $Fe(OH)_3$, is precipitated. Although $[Fe(H_2O)_6]^{3+}$ ions are pale purple and Fe^{3+} ions give amethyst its purple color, the colors of aqueous solutions of iron(III) salts are dominated by the conjugate base of $[Fe(H_2O)_6]^{3+}$, the yellow $[FeOH(H_2O)_5]^{2+}$ ion:

$$[Fe(H_2O)_6]^{3+}(aq) + H_2O(l) \rightleftharpoons H_3O^+(aq) + [FeOH(H_2O)_5]^{2+}(aq)$$

When iron is heated in carbon monoxide, it reacts to form *iron pentacarbonyl*, $Fe(CO)_5$, a yellow molecular liquid that boils at 103°C. A CO group is not very electronegative, and the number of electrons on the Fe atom is barely changed when a carbonyl is formed. Hence, the iron is ascribed an oxidation number of 0 in this and similar carbonyl compounds.

Although electronegativities are defined for individual atoms, it is sometimes convenient to think of a group of atoms as having a collective electronegativity.

A healthy adult human body contains around 3 g of iron, mostly as hemoglobin. Around 1 mg is lost daily (in sweat, feces, and hair), and women lose about 20 mg during menstruation, so iron must be ingested daily to maintain the balance. Iron deficiency, or anemia, results in reduced transport of oxygen to the brain and muscles, and an early symptom is chronic tiredness.

Cobalt ores are often found in association with copper(II) sulfide. Cobalt is a silver-gray metal and is used mainly for alloying with iron. *Alnico steel*, an alloy of iron, nickel, cobalt, and aluminum, is used to make permanent magnets such as those used in loudspeakers. Cobalt steels are hard enough to be used as surgical steels, drill bits, and lathe tools. The blue color of cobalt glass is due to cobalt(II) oxide, which produces a blue pigment when heated with silica and alumina. We need cobalt in our diet, for it is a component of vitamin B_{12}.

Nickel, Ni, is also used to alloy with iron. About 70% of the western world's supply comes from iron and nickel sulfide ores that were brought close to the surface by the impact of a meteor at Sudbury, Ontario. The ore is first *roasted* (heated in air) to form nickel(II) oxide, which is reduced to the metal either electrolytically or by reaction with hydrogen gas in the first step of the *Mond process:*

$$NiO(s) + H_2(g) \xrightarrow{\Delta} Ni(s) + H_2O(g)$$

The impure nickel is then refined by first exposing it to carbon monoxide, with which it forms nickel tetracarbonyl, $Ni(CO)_4$:

$$Ni(s) + 4\ CO(g) \longrightarrow Ni(CO)_4(g)$$

Nickel tetracarbonyl is a volatile, poisonous liquid that boils at 43°C and can therefore be removed from impurities. Nickel metal is then obtained by heating the pure nickel tetracarbonyl to about 200°C, when it decomposes.

Nickel is a hard, silver-white metal used mainly for the production of stainless steel and for alloying with copper to produce *cupronickels*, the alloys used for nickel coins (which consist of about 25% Ni and 75% Cu). Cupronickels are slightly yellow but are whitened by the addition of small amounts of cobalt. Nickel is also used in nicad batteries and as a catalyst, especially for the addition of hydrogen to organic compounds, for example, in the manufacture of edible solid fats from vegetable oils (Section 14.5). Nickel's most common oxidation state is +2, and the green color of aqueous solutions of nickel salts is due to the presence of $[Ni(H_2O)_6]^{2+}$ ions.

The Period 4 d-block elements titanium through nickel are obtained chemically from their ores, with the ease of reduction increasing to the right in the periodic table. They have many industrial uses, particularly as alloys.

16.4 Groups 11 and 12

The elements close to the right-hand edge of the *d* block have full *d*-orbitals. Group 11 contains the **coinage metals**—copper, silver, and gold—which have $(n-1)d^{10}ns^1$ valence electron configurations (Table 16.3). Group 12 contains zinc, cadmium, and mercury, with valence configurations $(n-1)d^{10}ns^2$. The low reactivity of the coinage metals is partly due to the poor shielding abilities of the *d*-electrons and hence the tight grip that the nucleus can exert on the outermost electrons. This effect is

TABLE 16.3 Properties of Elements in Groups 11 and 12

Z	Name	Symbol	Valence electron configuration	Melting point, °C	Boiling point, °C	Density, $g·cm^{-3}$
29	copper	Cu	$3d^{10}4s^1$	1083	2567	8.93
47	silver	Ag	$4d^{10}5s^1$	962	2212	10.50
79	gold	Au	$5d^{10}6s^1$	1064	2807	19.28
30	zinc	Zn	$3d^{10}4s^2$	420	907	7.14
48	cadmium	Cd	$4d^{10}5s^2$	321	765	8.65
80	mercury	Hg	$5d^{10}6s^2$	−39	357	13.55

enhanced in Period 6 by the lanthanide contraction, which helps to account for the inertness of gold.

Copper, Cu, is unreactive enough for some to be found native, but most is produced from its sulfides, particularly the ore *chalcopyrite*, $CuFeS_2$ (Fig. 16.13). The crushed and ground ore is separated from excess rock by the process called *froth flotation*, which depends on the ability of sulfide ores to be wetted by oils but not by water. In this process, the powdered ore is combined with oil, water, and detergents (Fig. 16.14). Then air is blown through the mixture; the oil-coated sulfide mineral floats to the surface with the froth, and the unwanted copper-poor residue, which is called *gangue*, sinks to the bottom.

FIGURE 16.13 Three important copper ores (from left to right): chalcopyrite, $CuFeS_2$; malachite, $CuCO_3·Cu(OH)_2$; and chalcocite, Cu_2S.

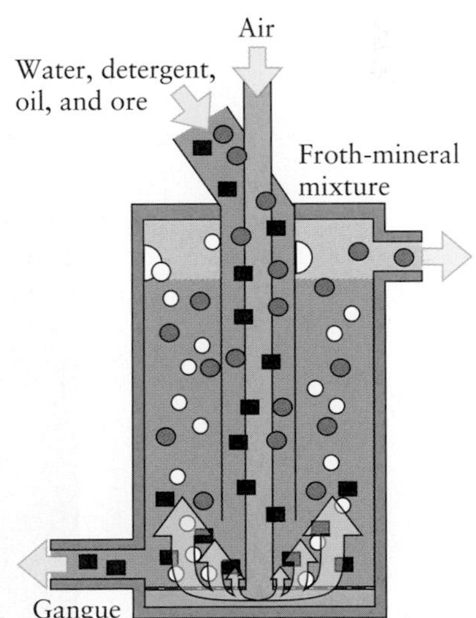

FIGURE 16.14 In the froth flotation process, a stream of bubbles is passed through a mixture of ore (orange circles), rock (brown rectangles), and detergent. The ore is buoyed up by the froth of bubbles and is removed from the top of the chamber. The unwanted gangue is washed away through the bottom of the container.

Processes for extracting metals from their ores are generally classified as **pyrometallurgical**, when high temperatures are used (Fig. 16.15), or **hydrometallurgical**, when aqueous solutions are used. Copper is extracted by both methods. In the pyrometallurgical process for the extraction of copper, the enriched ore is roasted:

$$2\ CuFeS_2(s) + 3\ O_2(g) \xrightarrow{\Delta} 2\ CuS(s) + 2\ FeO(s) + 2\ SO_2(g)$$

The CuS is then *smelted*, a process in which metal ions are reduced by melting with another compound. At the same time, the sulfur is oxidized to SO_2. This oxidation is accomplished by blowing compressed air through the mixture of ore, limestone, and sand:

$$CuS(s) + O_2(g) \xrightarrow{\Delta} Cu(l) + SO_2(g)$$

These processes could contribute a damaging amount of SO_2 to the atmosphere if precautions were not taken to remove it. The limestone and sand form a molten slag that removes many of the impurities as well as the SO_2. For example, calcium oxide (a basic oxide) from the limestone reacts with the SO_2 (an acidic oxide) to produce calcium sulfite in a Lewis acid-base reaction:

$$CaO(s) + SO_2(g) \longrightarrow CaSO_3(s)$$

The copper product is known as *blister copper* because of the appearance of air bubbles in the solidified metal.

In the hydrometallurgical processes for extracting copper, soluble Cu^{2+} ions are formed by the action of sulfuric acid on the ores. Then the metal is obtained by reducing these ions in aqueous solution. The reduction is carried out either electrolytically or chemically, using an inexpen-

FIGURE 16.15 In this industrial-scale copper refinery, the molten impure copper produced by smelting is poured into molds. Next, the copper will be purified by electrolysis.

sive reducing agent that has a more negative standard potential than copper, such as hydrogen or iron (see Section 14.5).

The impure copper from either process is refined electrolytically: it is made into anodes and plated onto cathodes of pure copper. The rare metals—most notably, platinum, silver, and gold—obtained from the anode sludge are sold to recover much of the cost of the electricity used in the electrolysis.

Copper alloys such as brass and bronze, which are harder and more resistant to corrosion than copper, are important construction materials. Copper corrodes in moist air in the presence of oxygen and carbon dioxide:

$$2\ Cu(s) + H_2O(l) + O_2(g) + CO_2(g) \longrightarrow Cu_2(OH)_2CO_3(s)$$

The pale green product is called *basic copper carbonate* and is responsible for the green patina of copper and bronze objects (Fig. 16.16). The patina adheres to the surface, protects the metal, and has a pleasing appearance.

FIGURE 16.16 Copper corrodes in air to form an attractive pale green layer of basic copper carbonate. This patina, or incrustation, passivates the surface.

Like all the coinage metals, copper forms compounds with oxidation number +1. However, in water, copper(I) salts disproportionate into metallic copper and copper(II) ions. The latter exist as pale blue $[Cu(H_2O)_6]^{2+}$ ions in water.

Copper is essential in animal metabolism. In some animals, such as the octopus and certain arthropods, it transports oxygen through the blood, a role performed by iron in mammals. As a result, the blood of these animals is green rather than red. In mammals, copper-bearing enzymes are necessary for healthy nerves and connective tissue.

Silver, Ag, is rarely found native. Most is obtained as a by-product of the refining of copper and lead, and a considerable amount is recycled through the photographic industry. Silver has a positive standard potential, so it does not reduce $H^+(aq)$ to hydrogen. Silver reacts readily with sulfur and sulfur compounds, producing the familiar black tarnish on silver dishes and cutlery.

Silver(I) does not disproportionate in aqueous solution; and in almost all its compounds, silver has oxidation number +1. Apart from silver nitrate and silver fluoride, silver salts are generally only sparingly soluble in water. Silver nitrate, $AgNO_3$, is the most important compound of silver and the starting point for the manufacture of silver halides for use in photography.

Gold, Au, is so inert that most of it is found in nature as the metal. Pure gold is classified as 24-carat gold. Its alloys with silver and copper, which differ in hardness and hue, are classified according to the proportion of gold they contain (Fig. 16.17). For example, 10- and 14-carat golds contain, respectively, $\frac{10}{24}$ and $\frac{14}{24}$ parts by mass of gold. Gold is a highly malleable metal, and 1 g of gold can be worked into a leaf covering an area of about 1 m^2 or pulled out into a wire over 2 km long. Gold leaf provides a valuable decorative protection, like that on dishes and books, on electrical contacts, and in cathedrals and temples (Fig. 16.18).

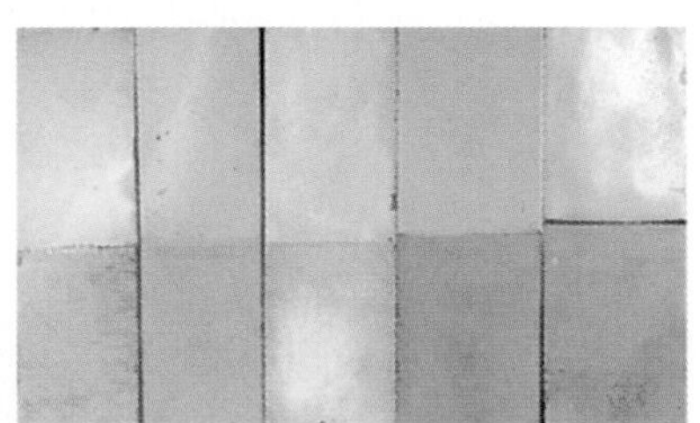

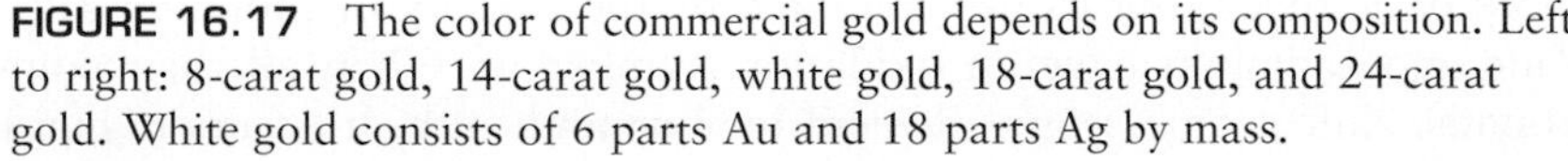

FIGURE 16.17 The color of commercial gold depends on its composition. Left to right: 8-carat gold, 14-carat gold, white gold, 18-carat gold, and 24-carat gold. White gold consists of 6 parts Au and 18 parts Ag by mass.

FIGURE 16.18 The gold leaf coating these cathedral domes not only adds long-lasting beauty but also adds protection from corrosion.

Gold is too noble to react even with strong oxidizing agents such as nitric acid. Both the gold couples

$$Au^{+}(aq) + e^{-} \longrightarrow Au(s) \qquad E^{\circ} = +1.69\ V$$

$$Au^{3+}(aq) + 3\ e^{-} \longrightarrow Au(s) \qquad E^{\circ} = +1.40\ V$$

lie above H^{+}/H_2 and $NO_3^{-},H^{+}/NO,H_2O$:

$$NO_3^{-}(aq) + 4\ H^{+}(aq) + 3\ e^{-} \longrightarrow NO(g) + 2\ H_2O(l) \qquad E^{\circ} = +0.96\ V$$

However, gold does react with *aqua regia*, a mixture of concentrated nitric and hydrochloric acids, because the complex ion $[AuCl_4]^{-}$ forms:

$$Au(s) + 6\ H^{+}(aq) + 3\ NO_3^{-}(aq) + 4\ Cl^{-}(aq) \longrightarrow [AuCl_4]^{-}(aq) + 3\ NO_2(g) + 3\ H_2O(l)$$

Even though the equilibrium constant for the formation of Au^{3+} from gold is very unfavorable, the reaction proceeds because any Au^{3+} ions formed are immediately captured by chloride ions and removed from the equilibrium.

Metallic gold has no known role in human health (except as artificial teeth), but its compounds are used to treat arthritis. It is a rare element in the crust of the Earth, but large deposits are mined in North America, Siberia, and South Africa.

Zinc, Zn, is found mainly as its sulfide, ZnS, in *sphalerite*, often in association with lead ores (see Fig. 15.12). The ore is concentrated by froth flotation, and the metal is extracted by roasting and then smelting with coke:

$$2\ ZnS(s) + 3\ O_2(g) \xrightarrow{\Delta} 2\ ZnO(s) + 2\ SO_2(g)$$

$$ZnO(s) + C(s) \xrightarrow{\Delta} Zn(l) + CO(g)$$

Cadmium, Cd, is obtained in a similar manner. Zinc and cadmium are silvery, reactive metals. Zinc is used mainly for galvanizing iron; like copper, it is protected by a hard film of basic carbonate, $Zn_2(OH)_2CO_3$, which forms on contact with air.

Zinc and cadmium are similar to each other but differ sharply from mercury. Zinc is amphoteric (like its main-group neighbor aluminum). It reacts with acids to form Zn^{2+} ions and with alkalis to form the zincate ion, $[Zn(OH)_4]^{2-}$:

$$Zn(s) + 2\ OH^{-}(aq) + 2\ H_2O(l) \longrightarrow [Zn(OH)_4]^{2-}(aq) + H_2(g)$$

Galvanized containers should therefore not be used for transporting alkalis. Cadmium, which is lower down the group and is more metallic, has a more basic oxide, but its use is limited by the toxicity of cadmium salts. Zinc and cadmium have an oxidation number of +2 in all their compounds. Zinc is an essential element for human health. It occurs in many enzymes and plays a role in the expression of DNA and in growth. Zinc is toxic only in very high amounts. However, cadmium is a deadly poison that disrupts metabolism by substituting for other essential metals in the body such as zinc and calcium, leading to soft bones and to kidney and lung disorders.

Mercury, Hg, occurs mainly as HgS in the mineral *cinnabar* (see Fig. 15.12), from which it is separated by froth flotation and then roasting in air:

$$HgS(s) + O_2(g) \xrightarrow{\Delta} Hg(g) + SO_2(g)$$

The volatile metal is separated by distillation and condensed. Mercury is the only metallic element that is liquid at room temperature (gallium and cesium are liquids on warm days). It has a long liquid range, from its melting point of −39°C to its boiling point of 357°C, so it is well suited for its use in thermometers, silent electrical switches, and high-vacuum pumps.

Because mercury lies above hydrogen in the electrochemical series, it is not oxidized by hydrogen ions. However, it does react with nitric acid:

$$3\ Hg(l) + 8\ H^+(aq) + 2\ NO_3^-(aq) \longrightarrow 3\ Hg^{2+}(aq) + 2\ NO(g) + 4\ H_2O(l)$$

In compounds, mercury has the oxidation number +1 or +2. Its compounds with oxidation number +1 are unusual in that the mercury(I) cation is the covalently bonded diatomic ion $(Hg–Hg)^{2+}$, written Hg_2^{2+}.

Mercury compounds, particularly its organic compounds, are acutely poisonous. Mercury vapor is also an insidious poison because its effect is cumulative. Frequent exposure to low levels of mercury vapor can allow mercury to accumulate in the body. The effects include impaired neurological function, hearing loss, and other ailments.

Metals in Groups 11 and 12 are easily reduced from their compounds and have low reactivity as a result of poor shielding of the nuclear charge by the d-electrons. Copper is extracted from its ores either by pyrometallurgical or hydrometallurgical processes.

Self-Test 16.2A Use standard free energies of formation to calculate $\Delta G_r°$ (per mole of Cu) at 298 K for the reaction in which solid copper is obtained by smelting CuS ore with oxygen. ($\Delta G_f°(CuS) = -49.0\ kJ·mol^{-1}$.)

[*Answer:* $\Delta G_r° = -251.2\ kJ·mol^{-1}$]

Self-Test 16.2B Calculate $E°$ for a cell powered by the reaction of mercury and nitric acid to form mercury(I) and NO.

COORDINATION COMPOUNDS

Many of the *d*-block elements form characteristically colored solutions in water. For example, although solid copper(II) chloride is brown and copper(II) bromide is black, their aqueous solutions are both light blue. The blue color is caused by the hydrated copper(II) ions of composition $[Cu(H_2O)_6]^{2+}$ that form when the solids dissolve. As the formula

BOX 16.1 WHAT HAS THIS MATERIAL TO DO WITH . . . STAYING ALIVE?

Why We Need to Eat *d*-Metals

We think of ourselves as organic creatures, but without the *d*-metals we would be dead within minutes. Some of the critical enzymes in our cells are *metalloproteins*, large organic molecules made up of folded polymerized chains of amino acids that also include at least one metal atom. These metalloproteins are intensely studied by biochemists, for they control life and protect against disease. They have even been used to trace evolutionary paths. The *d*-block metals catalyze redox reactions, form components of membrane, muscle, skin, and bone, catalyze acid-base reactions, control the flow of energy and oxygen, and carry out nitrogen fixation.

Hemoglobin, in which an iron atom lies as iron(II) at the center of a heme group, is the most familiar metalloprotein. Four nitrogen atoms from amine groups in the heme serve as the ligands in a square planar arrangement. The oxygen molecule serves as a fifth ligand, attaching itself directly to the iron atom and producing a modified square pyramidal shape about the iron atom. Carbon monoxide forms a much stronger bond to the iron atom than oxygen does, hence its high toxicity: the attached CO ligand prevents an O_2 molecule from attaching to the iron, and the victim suffocates.

Cobalt is at the center of a coenzyme required for the activity of vitamin B_{12} (sometimes called cobalamin) and gives the vitamin its red color. The cobalt atom is found in an octahedral complex in which five of the ligands are nitrogen atoms from organic amine groups and one ligand attaches through a —CH_2— group. Cobalamin is the only biomolecule known to have a metal-carbon bond. The ease with which this bond is broken and the ability of the cobalt ion to change from one oxidation state to another is responsible for the importance of cobalamin as a biological catalyst. Vitamin B_{12} deficiency is associated with mental illness and with pernicious anemia.

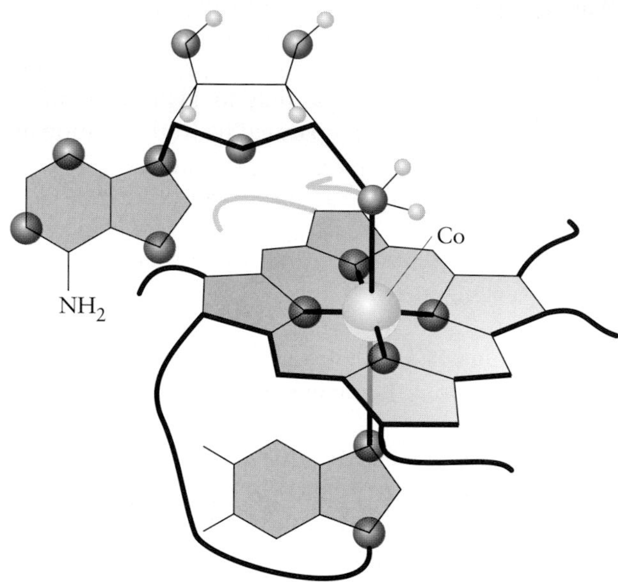

In cobalamin, vitamin B_{12}, one of the six ligands forming an octahedral structure around a cobalt atom is an organic molecule attached through a carbon-cobalt bond. The bond is weak and easily broken.

Zinc enzymes play several important roles in metabolism, including the expression of our genes, the digestion of food, the storage of insulin, and the building of collagen. In fact, zinc has so many functions in our systems that it has been called a "master hormone." Its concentration in our bodies is about the same as that of iron, but the intracellular concentration of zinc declines with aging. It has even been suggested that aging is actually a result of the decline in intracellular zinc concentration. However, this decline cannot be slowed by eating more zinc; intracellular zinc concentrations are controlled by enzymes, not by the lack of zinc in the diet.

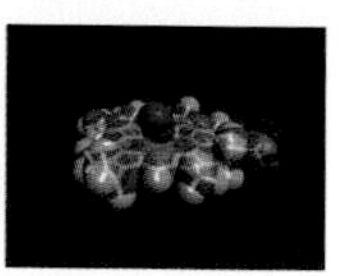

View an animation of oxygen transport by hemoglobin.

Media Link Animation: Box 16.1, Oxygenation of hemoglobin

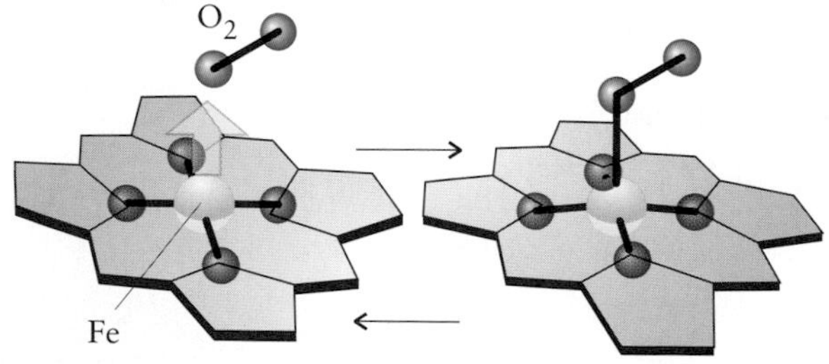

Molecular oxygen is transported throughout the body by attaching to the iron(II) atom in the heme groups of hemoglobin. The iron(II) atom lies at the center of a square planar complex formed by nitrogen atoms. When the O_2 molecule attaches to the iron, the plane of the heme group becomes distorted.

For Further Reading: J. J. R. Fraústo da Silva and R. J. P. Williams, *The Biological Chemistry of the Elements: The Inorganic Chemistry of Life* (Oxford: Oxford University Press, 1991).

suggests, these hydrated ions have a specific composition; they also have definite shapes and properties. They can be regarded as the outcome of a reaction in which the water molecules act as Lewis bases (electron pair donors, Section 2.13) and the Cu^{2+} ion acts as a Lewis acid (an electron pair acceptor). This type of Lewis acid-base reaction is characteristic of many cations of both main-group and *d*-block elements.

The hydrated ion $[Cu(H_2O)_6]^{2+}$ is an example of a **complex,** a species consisting of a central metal atom or ion to which is attached a number of molecules or ions by coordinate-covalent bonds. A **coordination compound** is an electrically neutral compound in which at least one of the ions present is a complex. The terms *coordination compound* (the overall neutral compound) and *complex* (one or more of the ions or neutral species present in the compound), however, are often used interchangeably. Coordination compounds include complexes in which the central metal atom is electrically neutral, such as $Ni(CO)_4$, and ionic compounds such as $K_4[Fe(CN)_6]$.

Much chemical research focuses on the structures, properties, and uses of the complexes formed between *d*-metal ions acting as Lewis acids and a variety of Lewis bases, partly because they participate in many biological reactions. Hemoglobin and vitamin B_{12}, for example, are both complexes—the former of iron and the latter of cobalt (Box 16.1). Complexes of the *d*-metals are often brightly colored and magnetic, and are used in chemistry for analysis, to dissolve ions (Section 11.15), in the electroplating of metals, and for catalysis. They are also the target of current research in solar-energy conversion, in atmospheric nitrogen fixation, and in pharmaceuticals.

Complexes are also formed by main-group metal ions. Chlorophyll, for instance, which gathers energy from the Sun in photosynthesis, is a complex of magnesium. However, the *d*-metals form the richest variety of complexes, and we concentrate on them.

16.5 Coordination Complexes

The molecules or ions (typically, four or six) that act as Lewis bases attached to a central metal atom or ion in a *d*-metal complex are known as **ligands.** One example is the hexacyanoferrate(II) ion, $[Fe(CN)_6]^{4-}$, in which the Lewis acid Fe^{2+} forms bonds by sharing electron pairs provided by the CN^- ions. In the neutral complex $Ni(CO)_4$, the Ni atom acts as the Lewis acid and the ligands, the CO molecules, act as Lewis bases.

Each ligand in a complex has at least one lone pair of electrons with which it bonds to the central atom or ion by forming a coordinate covalent bond. We say that the ligands **coordinate** to the metal when they form the complex in this way. In the chemical formula for a coordination complex, we represent the ligands directly attached to the central ion by enclosing them and the central ion within brackets. These ligands make up the **coordination sphere** of the central ion. The number of points at which ligands are attached to the central metal atom is called the **coordination number** of the complex (Fig. 16.19). The coordination number is 4 in $Ni(CO)_4$ and 6 in $[Fe(CN)_6]^{4-}$.

The formation of coordinate covalent bonds is described in Section 2.13.

Because water is a Lewis base, it forms complexes with most *d*-block ions when they dissolve in it. Aqueous solutions of *d*-metal ions are usually solutions of their H_2O complexes: $Fe^{2+}(aq)$, for instance, is more

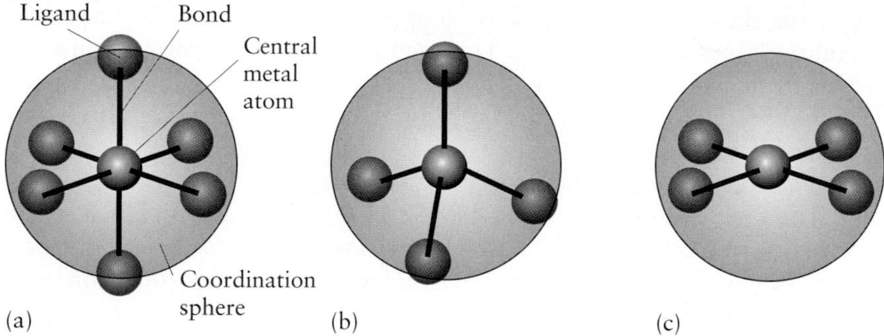

FIGURE 16.19 The various components of a complex may be either neutral or charged. (a) Almost all six-coordinate complexes are octahedral. Four-coordinate complexes are either (b) tetrahedral or (c) square planar.

accurately $[Fe(H_2O)_6]^{2+}$. Many complexes are prepared simply by mixing aqueous solutions of a *d*-metal ion with the appropriate Lewis base (Fig. 16.20); for example,

$$[Fe(H_2O)_6]^{2+}(aq) + 6\,CN^-(aq) \longrightarrow [Fe(CN)_6]^{4-}(aq) + 6\,H_2O(l)$$

This is an example of a **substitution reaction,** a reaction in which one Lewis base takes the place of another. Here the CN^- ions drive out H_2O molecules from the coordination sphere of the $[Fe(H_2O)_6]^{2+}$ complex and take their place. A less complete replacement occurs when certain other ions, such as Cl^-, are added to an iron(II) solution:

$$[Fe(H_2O)_6]^{2+}(aq) + Cl^-(aq) \longrightarrow [FeCl(H_2O)_5]^{+}(aq) + H_2O(l)$$

Because the color of a *d*-metal complex depends on the identity of the ligands as well as that of the metal, impressive changes of color often accompany substitution reactions (Fig. 16.21).

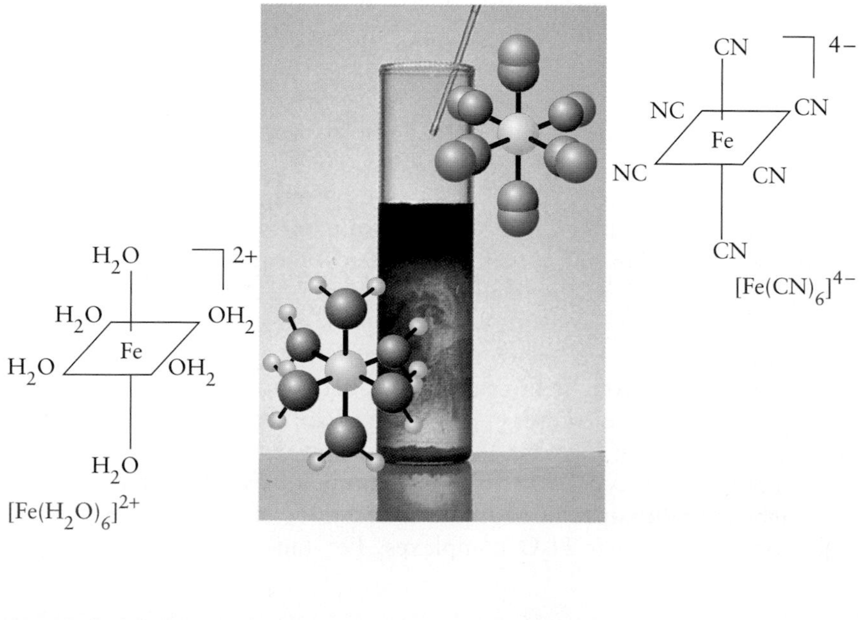

FIGURE 16.20 When cyanide ions (in the form of potassium cyanide) are added to a solution of iron(II) sulfate, they replace the H_2O ligands of the $[Fe(H_2O)_6]^{2+}$ complex (left) and produce a new complex, the more strongly colored blue hexacyanoferrate(II) ion, $[Fe(CN)_6]^{4-}$ (right).

TABLE 16.4 Common Ligands

Formula	Name
Neutral ligands	
H_2O	aqua
NH_3	ammine
NO	nitrosyl
CO	carbonyl
$NH_2CH_2CH_2NH_2$	ethylenediamine (en)*
$NH_2CH_2CH_2NHCH_2CH_2NH_2$	diethylenetriamine (dien)†
Anionic ligands	
F^-	fluoro
Cl^-	chloro
Br^-	bromo
I^-	iodo
OH^-	hydroxo
O^{2-}	oxo
CN^-	cyano (as M—CN)
NC^-	isocyano (as M—NC)
SCN^-	thiocyanato (as M—SCN)
NCS^-	isothiocyanato (as M—NCS)
NO_2^-	nitrito (as M—ONO)
NO_2^-	nitro (as $M—NO_2$)
CO_3^{2-}	carbonato
$C_2O_4^{2-}$	oxalato (ox)*
$(^-O(O{=})C{-}CH_2)_2N{-}CH_2{-}CH_2{-}N(CH_2{-}C({=}O)O^-)_2$	ethylenediaminetetraacetato (edta)‡
SO_4^{2-}	sulfato

*Bidentate (attaches to two sites).
†Tridentate (attaches to three sites).
‡Hexadentate (attaches to six sites).

FIGURE 16.21 Some of the highly colored compounds that result when complexes are formed. Left to right: aqueous solutions of $[Fe(SCN)(H_2O)_5]^{2+}$, $[Co(SCN)_4(H_2O)_2]^{2-}$, $[Cu(NH_3)_4(H_2O)_2]^{2+}$, and $[CuBr_4]^{2-}$.

The names of coordination compounds can become alarmingly long because each of the ligands must be included in the name. In most cases, chemists avoid the issue by using the chemical formula rather than the name itself. For instance, it is much easier to refer to $[FeCl(H_2O)_5]^+$ than to pentaaquachloroiron(II) ion, its formal name. However, names are sometimes needed, and they can be constructed and interpreted, in simple cases at least, by using the rules set out in Toolbox 16.1. Table 16.4 gives the names of common ligands and their abbreviations.

A complex is formed between a Lewis acid (the metal atom or ion) and a number of Lewis bases (the ligands).

TOOLBOX 16.1 How to name *d*-metal complexes and coordination compounds

Conceptual Basis

The aim of nomenclature is to be succinct but unambiguous. *d*-Metal complexes are identified by giving the names and numbers of the individual ligands. Because some names can be quite long, interpreting the names of coordination compounds is rather like eating a large bun: nibble it, don't try to swallow it in one gulp.

Procedure

The following rules are adequate for most common complexes; more elaborate rules are needed if the complex contains more than one metal atom.

1 When naming a *d*-metal complex, name the ligands first, then the metal atom or ion.

2 Neutral ligands have the same name as the molecule, such as $H_2NCH_2CH_2NH_2$ (ethylenediamine), except for H_2O (aqua), NH_3 (ammine), CO (carbonyl), and NO (nitrosyl).

3 Anionic ligands end in -o; for anions that end in -ide (such as chloride), -ate (such as sulfate), and -ite (such as nitrite), change the endings as follows:

ide ⟶ -o -ate ⟶ -ato -ite ⟶ -ito

Examples: chloro, cyano, sulfato, carbonato, sulfito, and nitrito.

4 Greek prefixes are used to denote the number of each type of ligand in the complex ion:

2	3	4	5	6	. . .
di-	tri-	tetra-	penta-	hexa-	. . .

If the ligand already contains a Greek prefix (as in ethylenediamine) or if it is polydentate (able to attach at more than one binding site), then the following prefixes are used instead:

2	3	4	. . .
bis-	tris-	tetrakis-	. . .

5 Ligands are named in alphabetical order, ignoring the Greek prefix that indicates the number of each one present. (Notice that Cl_2 in the coordination sphere of the second complex in rule 7 below represents two chloride ligands, named as a dichloro, and not a Cl_2 molecular ligand.)

6 The chemical symbols of anionic ligands precede those of neutral ligands in the chemical formula of the complex (but not necessarily in its name). Thus, Cl^- precedes H_2O and NH_3.

7 A Roman numeral denotes the oxidation number of the central metal ion:

$[FeCl(H_2O)_5]^+$
pentaaquachloroiron(II) ion

$[CrCl_2(NH_3)_4]^+$
tetraamminedichlorochromium(III) ion

$[Co(en)_3]^{3+}$
tris(ethylenediamine)cobalt(III) ion

8 If the complex has an overall negative charge (an anionic complex), the suffix -ate is added to the stem of the metal's name. If the symbol of the metal originates from a Latin name (as listed in Appendix 2D), then the Latin stem is used. For example, the symbol for iron is Fe, from the Latin *ferrum*. Therefore, any anionic complex of iron ends with -ferrate followed by the oxidation number of the metal in Roman numerals:

$[Fe(CN)_6]^{4-}$ hexacyanoferrate(II) ion

$[Ni(CN)_4]^{2-}$ tetracyanonickelate(II) ion

9 The name of a coordination compound (as distinct from a complex cation or anion) is built in the same way as that of a simple compound, with the (possibly complex) cation named before the (possibly complex) anion:

$NH_4[PtCl_3(NH_3)]$
ammonium amminetrichloroplatinate(II)

$[Cr(OH)_2(NH_3)_4]Br$
tetraamminedihydroxochromium(III) bromide

EXAMPLE 16.1 **Naming complexes and coordination compounds**

(a) Name the coordination compound $[Co(NH_3)_3(H_2O)_3]_2(SO_4)_3$. (b) Write the formula of sodium dichlorobis(oxalato)platinate(IV).

STRATEGY (a) Determine the oxidation number of the cobalt ion, then follow the procedure in Toolbox 16.1: the cation is a complex and the anion is a sulfate ion. (b) The complex is an anion. Determine its charge and follow rule 6 in Toolbox 16.1.

SOLUTION (a) First, we note that there are three SO_4^{2-} ions for every two complex ions, so the charge on the complex cation must be +3 to ensure charge neutrality of the compound. Therefore, the complex cation is $[Co(NH_3)_3(H_2O)_3]^{3+}$. Because all the ligands are neutral, the cobalt must be present as cobalt(III). It follows that the name of the cation is triamminetriaquacobalt(III), so the compound is triamminetriaquacobalt(III) sulfate. (b) Two chloride ligands, each with a charge of −1, and two oxalate ions, each with charge −2, are attached to a platinum(IV) ion, so the charge on the complex is −2. From Table 16.4, the symbol for oxalate is ox, so the formula for the complex anion is $[PtCl_2(ox)]^{2-}$ and that of the compound it forms is $Na_2[PtCl_2(ox)]$.

Self-Test 16.3A (a) Name the compound $[FeOH(H_2O)_5]Cl_2$. (b) Write the formula of potassium diaquabis(oxalato)chromate(II).

[*Answer:* (a) Pentaaquahydroxoiron(III) chloride; (b) $K_2[Cr(ox)_2(H_2O)_2]$]

Self-Test 16.3B (a) Name the compound $[CoBr(NH_3)_5]SO_4$. (b) Write the formula of tetraamminediaquachromium(III) bromide.

16.6 The Shapes of Complexes

The richness of coordination chemistry is enhanced by the variety of shapes that complexes can adopt. The most common complexes have coordination number 6. Almost all these species have their ligands at the vertices of a regular octahedron, with the metal ion at the center, and are called **octahedral complexes** (**1**). An example of an octahedral complex is the hexacyanoferrate(II) ion, $[Fe(CN)_6]^{4-}$. Chemists commonly represent the structures of these octahedral complexes by a simplified diagram that emphasizes the arrangement of the ligands (**2**).

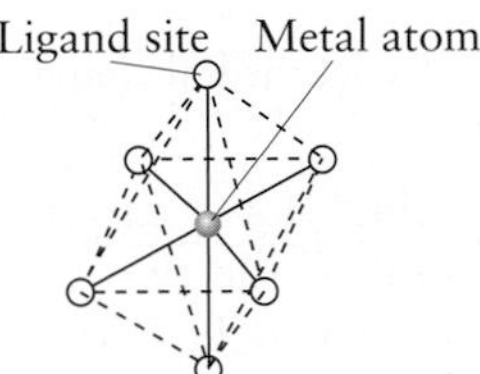

1 An octahedral complex

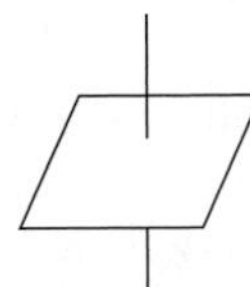
2 Simplified representation

The next most common complexes have coordination number 4. Two shapes are typically found for this coordination number. In a **tetrahedral complex,** the four ligands are found at the vertices of a tetrahedron, as in the tetrachlorocobaltate(II) ion, $[CoCl_4]^{2-}$ (**3**). An alternative arrangement, most notably for atoms and ions with d^8 electron configurations such as Pt^{2+} and Au^{3+}, is for the ligands to lie at the corners of a square, giving a **square planar complex** (**4**).

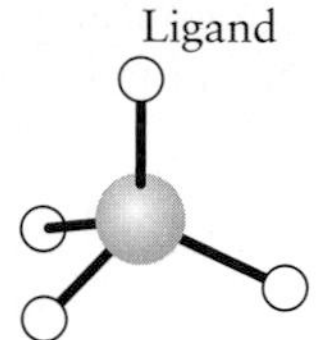

3 Tetrahedral complex

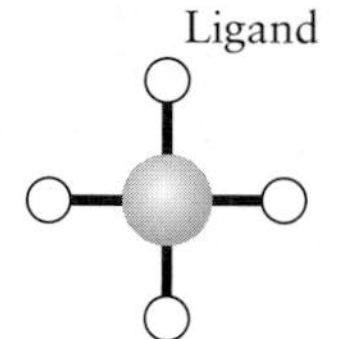

4 Square planar complex

5 Dimethylmercury(0), $Hg(CH_3)_2$

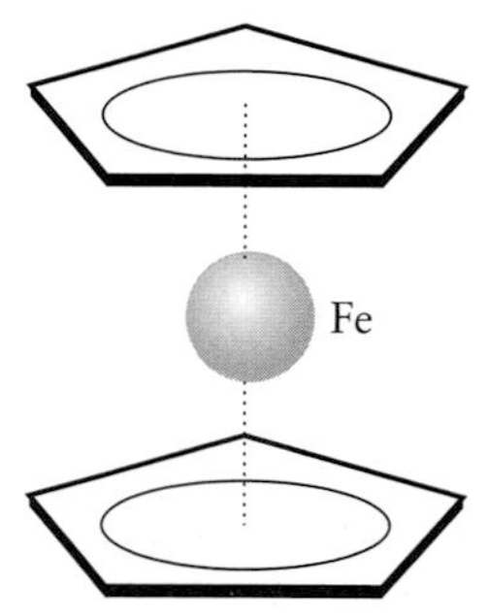

6 Ferrocene, $Fe(C_5H_5)_2$

Many other shapes are possible for complexes. The simplest are linear, with coordination number 2. An example is dimethylmercury(0), $[Hg(CH_3)_2]$ (**5**), which is a toxic compound formed by bacterial action on aqueous solutions of Hg^{2+} ions. Coordination numbers of up to 12 are found for members of the *f* block, but they are rare in the *d* block. One interesting type of *d*-metal compound in which the coordination number is formally only 2 but there are 10 links between the ligands and the central metal ion is ferrocene, dicyclopentadienyliron(0), $[Fe(C_5H_5)_2]$ (**6**). Ferrocene is an aptly named "sandwich compound," with the two cyclopentadienyl ligands the "bread" and the metal atom the "filling." The formal name for a sandwich compound is a **metallocene.**

Complexes of molybdenum and tungsten with eight ligands are known. These complexes have antiprismatic (**7**) or dodecahedral shapes (**8**). However, complexes with more than six ligands are rare.

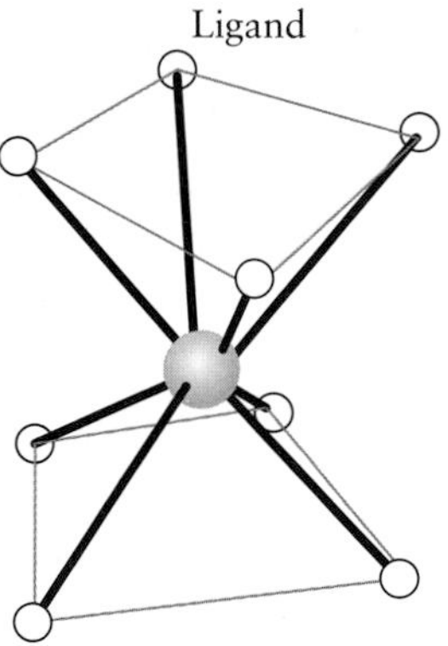

7 Square antiprism

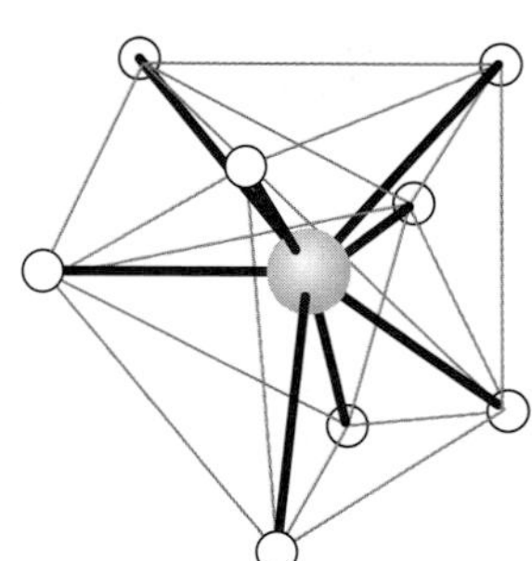

8 Dodecahedral complex

Some ligands are **polydentate** ("many toothed") and can occupy more than one binding site simultaneously. Each end of the two-toothed (that is, bidentate) ethylenediamine molecule, $NH_2CH_2CH_2NH_2$ (**9**), has a nitrogen atom with a lone pair of electrons. This ligand is widely used in coordination chemistry and is abbreviated to en, as in tris(ethylenediamine)cobalt(III), $[Co(en)_3]^{3+}$ (**10**). The metal ion in $[Co(en)_3]^{3+}$ lies at the center of the three ligands as though pinched by three molecular claws. It is an example of a **chelate,** a complex containing one or more ligands that form a ring of atoms that includes the central metal atom.

There are few hexadentate ligands, but a common example is the ethylenediaminetetraacetate ion, edta (**11**). This ligand forms complexes with many metal ions, including Pb^{2+}; hence, it is used as an antidote to lead poisoning.

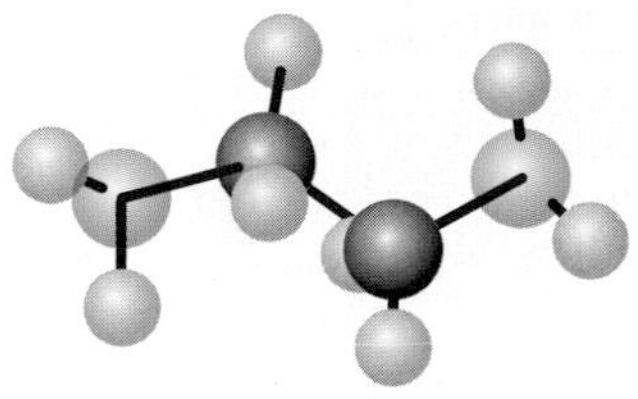

9 Ethylenediamine, $NH_2CH_2CH_2NH_2$

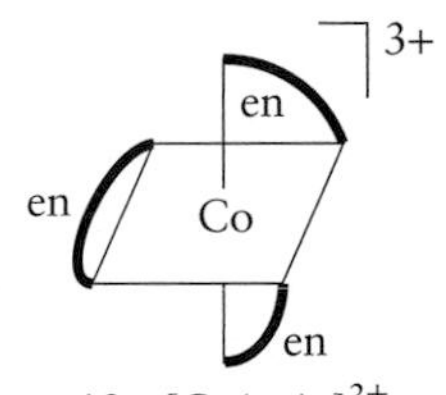

10 $[Co(en)_3]^{3+}$

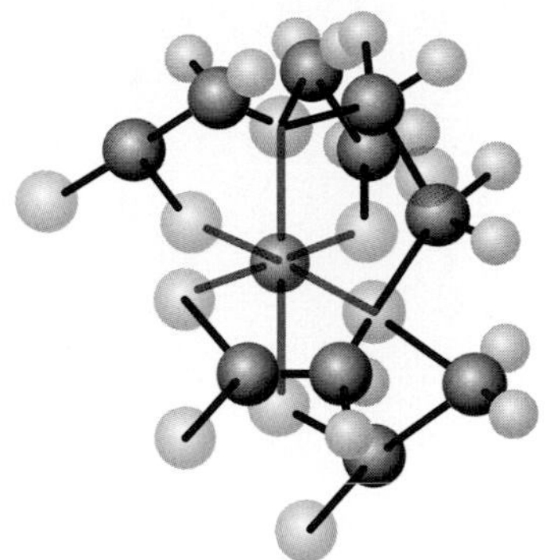

11 An edta complex

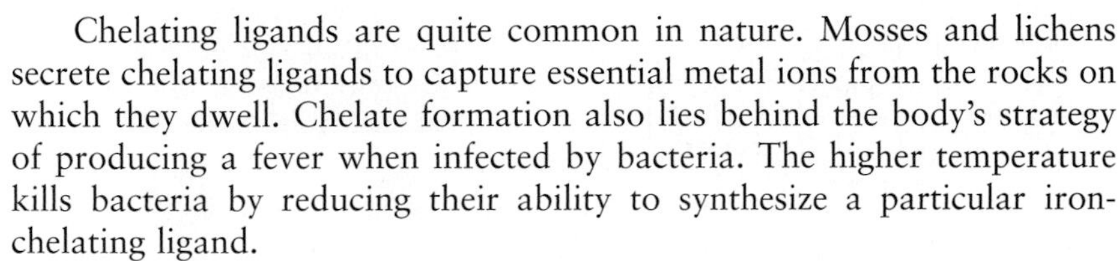

Chelating ligands are quite common in nature. Mosses and lichens secrete chelating ligands to capture essential metal ions from the rocks on which they dwell. Chelate formation also lies behind the body's strategy of producing a fever when infected by bacteria. The higher temperature kills bacteria by reducing their ability to synthesize a particular iron-chelating ligand.

Complexes with coordination number 6 are octahedral; those with coordination number 4 are either tetrahedral or square planar. Polydentate ligands can form chelates.

16.7 Isomers

Many complexes and coordination compounds exist as isomers. **Isomers** are compounds that contain the same numbers of the same atoms but in different arrangements. For example, the ions shown in (**12a**) and (**12b**) differ only in the positions of the Cl^- ligands, but they are distinct species, for they have different physical and chemical properties. Isomerism is of more than academic interest: some anticancer drugs based on complexes of platinum are active only when they are the correct isomer. The complex needs to have a particular shape to interact with DNA molecules.

Isomerism is very important in organic chemistry, and the topic is developed again in Section 18.3.

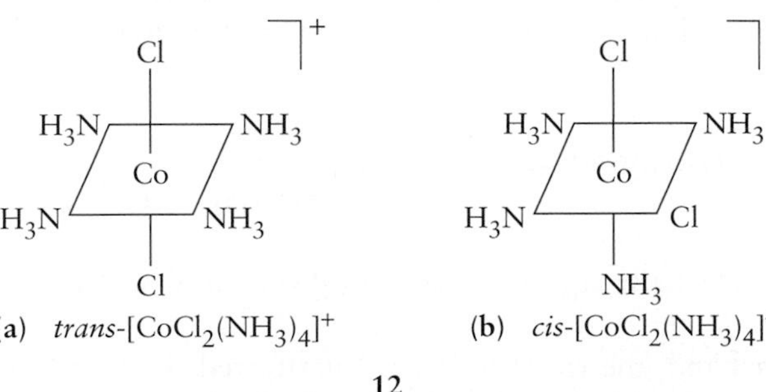

(a) *trans*-$[CoCl_2(NH_3)_4]^+$ **(b)** *cis*-$[CoCl_2(NH_3)_4]^+$

12

Figure 16.22 summarizes the types of isomerism. The two major classes of isomers are **structural isomers,** in which the atoms have different partners, and **stereoisomers,** in which the atoms have the same partners but are arranged differently in space. Structural isomers of coordination compounds are subdivided into ionization, hydrate, linkage, and coordination isomers.

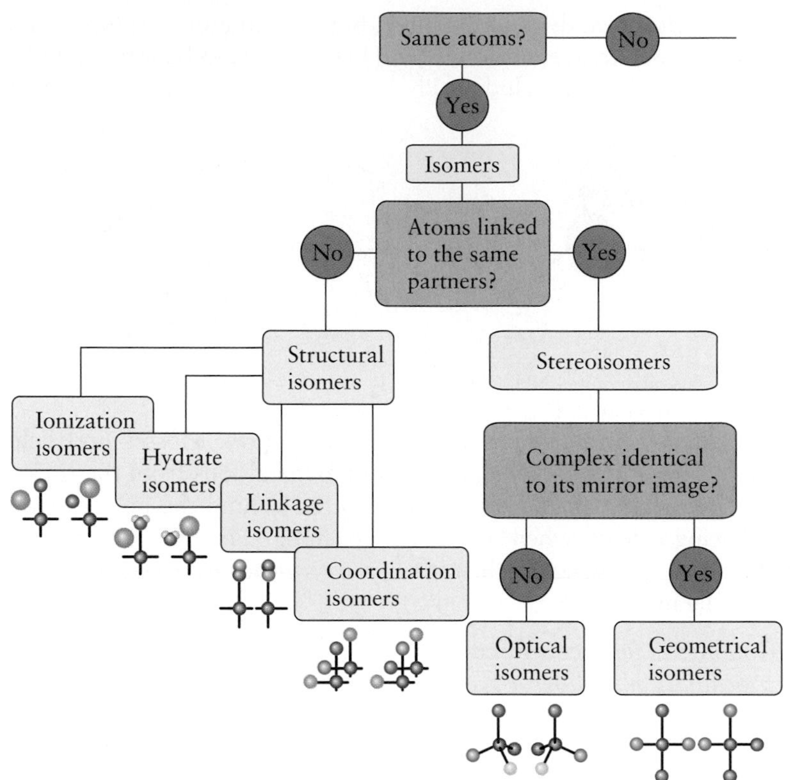

FIGURE 16.22 The various types of isomerism in coordination compounds.

Ionization isomers differ by the exchange of a ligand with an anion or a neutral molecule outside the coordination sphere. For instance, $[CoBr(NH_3)_5]SO_4$ and $[CoSO_4(NH_3)_5]Br$ are ionization isomers because the Br^- ion is a ligand of the cobalt in the former but an accompanying anion in the latter. The two compounds have different physical properties: the bromo complex, $[CoBr(NH_3)_5]^{2+}$, is violet and the sulfato complex, $[CoSO_4(NH_3)_5]^+$, is red. The chemical properties of the two ionization isomers are also different: the red isomer forms an off-white precipitate of AgBr when Ag^+ ions are added, but no precipitate forms after the addition of Ba^{2+}:

$$[CoSO_4(NH_3)_5]Br(aq, red) + Ag^+(aq) \longrightarrow AgBr(s) + [CoSO_4(NH_3)_5]^+(aq)$$

$$[CoSO_4(NH_3)_5]Br(aq, red) + Ba^{2+}(aq) \longrightarrow \text{no reaction}$$

On the other hand, the violet isomer is unaffected by the addition of low concentrations of Ag^+ ions, but it forms a white precipitate of $BaSO_4$ after the addition of Ba^{2+} ions:

$$[CoBr(NH_3)_5]SO_4(aq, violet) + Ag^+(aq) \longrightarrow \text{no reaction}$$

$$[CoBr(NH_3)_5]SO_4(aq, violet) + Ba^{2+}(aq) \longrightarrow BaSO_4(s) + [CoBr(NH_3)_5]^{2+}(aq)$$

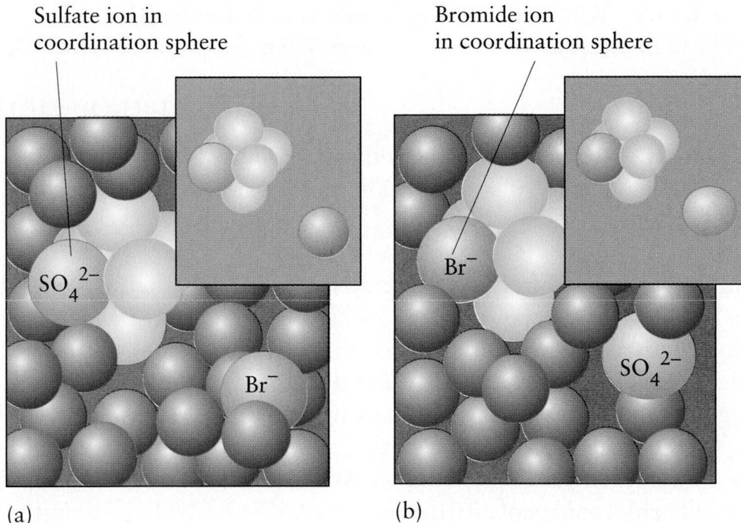

FIGURE 16.23 (a) When $[CoSO_4(NH_3)_5]Br$ dissolves, the bromide ion (the orange sphere) is not part of the coordination sphere; it is free to move and is available for reaction. (b) In contrast, when $[CoBr(NH_3)_5]SO_4$ dissolves, the bromide ion remains attached to the cobalt ion in the complex and is not available for reaction. In this case, the sulfate ion (purple sphere) is free to react.

These observations show that ions bound in a coordination sphere are not free to react. When the red sulfato complex dissolves in water, bromide ions become free to move through the solution (Fig. 16.23a), and when they encounter a silver ion, they form a precipitate. Conversely, when the violet bromo complex dissolves in water, the bromide ion remains attached to the cobalt ion and is not free to react with silver ions (Fig. 16.23b). However, the sulfate ion is not a part of the coordination sphere and is free to react with added barium ions.

Hydrate isomers are a special type of ionization isomer. They differ by the exchange of an H_2O molecule with another ligand in the coordination sphere (Fig. 16.24). For example, the solid hexahydrate of chromium(III) chloride, $CrCl_3 \cdot 6H_2O$, may be any of the three compounds

hexaaquachromium(III) chloride
$[Cr(H_2O)_6]Cl_3$ violet

pentaaquachlorochromium(III) chloride monohydrate
$[CrCl(H_2O)_5]Cl_2 \cdot H_2O$ blue-green

tetraaquadichlorochromium(III) chloride dihydrate
$[CrCl_2(H_2O)_4]Cl \cdot 2H_2O$ green

The isomers can be distinguished by adding $AgNO_3$ to solutions of the compounds. Different amounts of AgCl precipitate from each of them, because only the chloride ions outside the coordination sphere react with the silver ions.

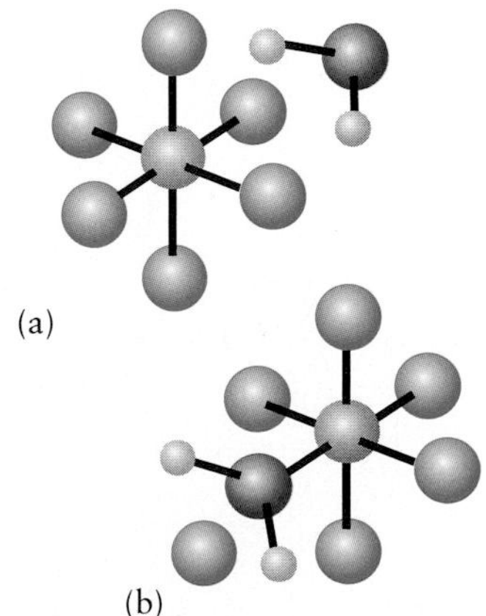

FIGURE 16.24 Hydrate isomers. In (a), the water molecule is simply part of the surrounding solvent; in (b), the water molecule is present in the coordination sphere and a ligand (green sphere), is now present in the solution.

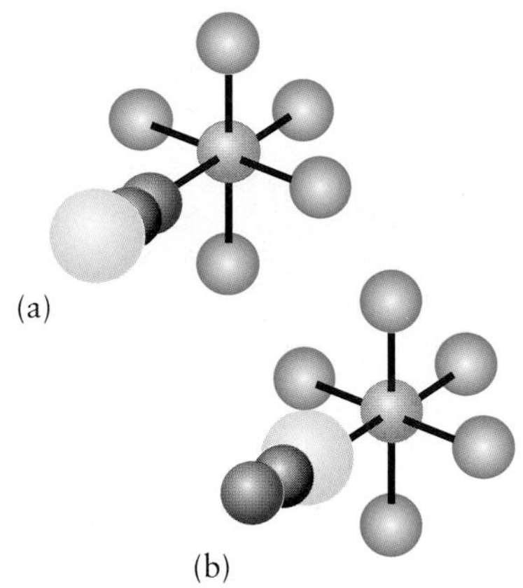

FIGURE 16.25 Linkage isomers. In (a), the ligand (here NCS^-) is attached through its N atom; but in (b), it is attached through its S atom.

Self-Test 16.4A When excess silver nitrate is added to 0.0010 mol $CrCl_3 \cdot 6H_2O$ in aqueous solution, 0.0010 mol AgCl is formed. Which isomer is present?

[*Answer:* $[CrCl_2(H_2O)_4]Cl \cdot 2H_2O$]

Self-Test 16.4B When excess silver nitrate is added to 0.0010 mol $CrCl_3 \cdot 6H_2O$ in aqueous solution, 0.0030 mol AgCl is formed. Which isomer is present?

Linkage isomers differ in the identity of the atom that a given ligand uses to attach to the metal ion (Fig. 16.25). A ligand may have more than one atom with a lone pair that can bond to the metal ion, but because of its size or shape, only one atom can bond to the metal ion at a time. Common ligands that show linkage isomerism are SCN^- versus NCS^-, NO_2^- versus ONO^-, and CN^- versus NC^-, where the coordinating atom is written first in each pair. For example, NO_2^- can form the yellow complex tetraamminechloronitrocobalt(III) ion, $[CoCl(NO_2)(NH_3)_4]^+$, or the red complex tetraamminechloronitritocobalt(III) ion, $[CoCl(ONO)(NH_3)_4]^+$. Note that the ligand has a different name in each case: nitro signifies that it is linked through the N atom; nitrito, that it is linked through an O atom. Table 16.4 shows the names to use for these so-called **ambidentate ligands** that can attach through atoms of different elements.

Coordination isomers differ by the exchange of one or more ligands between a cationic complex and an anionic complex (Fig. 16.26). An example of a pair of coordination isomers is

hexaamminechromium(III) hexacyanoferrate(III)
$[Cr(NH_3)_6][Fe(CN)_6]$

hexaammineiron(III) hexacyanochromate(III)
$[Fe(NH_3)_6][Cr(CN)_6]$

Self-Test 16.5A Identify the type of isomers represented by the following pairs: (a) $[Cu(NH_3)_4][PtCl_4]$ and $[Pt(NH_3)_4][CuCl_4]$; (b) $[Cr(OH)_2(NH_3)_4]Br$ and $[CrBr(OH)(NH_3)_4]OH$.

[*Answer:* (a) Coordination; (b) ionization]

Self-Test 16.5B Identify the type of isomers represented by the following pairs: (a) $[Co(NCS)(NH_3)_5]Cl_2$ and $[Co(SCN)(NH_3)_5]Cl_2$; (b) $[CrCl(H_2O)_5]Cl_2 \cdot H_2O$ and $[CrCl_2(H_2O)_4]Cl \cdot 2H_2O$.

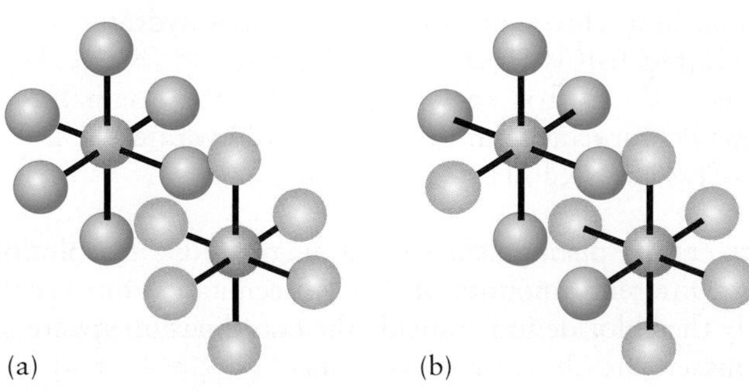

FIGURE 16.26 The compounds in (a) and (b) are coordination isomers. In these compounds, a ligand has been exchanged between the cationic and anionic complexes.

Although they are built from the same numbers and kinds of atoms, *structural isomers* have different chemical formulas, because the formulas show how the atoms are grouped in or outside the coordination sphere. *Stereoisomers*, on the other hand, have the same formulas because their atoms have the same partners in the coordination spheres; only the arrangements of the ligands in space differ from one isomer to the other. There are two types of stereoisomers, geometrical and optical. In **geometrical isomers,** atoms are bonded to the same neighbors but have different orientations. The two complexes of formula $[CoCl_2(NH_3)_4]^{2+}$ shown in (**12a**) and (**12b**) are geometrical isomers because the coordination spheres differ only in the geometry of the ligands. The isomer with the Cl^- ligands on opposite sides of the central atom is called the *trans isomer*, and the isomer with the ligands on the same side is called the *cis isomer*. Geometrical isomers can occur for square planar and octahedral complexes. They cannot occur for tetrahedral complexes because any pair of vertices is equivalent to any other pair.

The prefix *trans* comes from the Latin word for "across" and the prefix *cis* from the Latin word for "on this side."

The chemical and physiological properties of geometrical isomers can differ greatly. For example, *cis*-$[PtCl_2(NH_3)_2]$ (**13a**) is pale orange-yellow, has a solubility of 0.252 g per 100 g of water, and is used for chemotherapy treatment of cancer patients. Conversely, *trans*-$[PtCl_2(NH_3)_2]$ (**13b**) is dark yellow, has a solubility of 0.037 g per 100 g of water, and exhibits no chemotherapeutic effect. Antitumor activity is also found in other cis isomers of platinum(II), such as *cis*-$[PtCl_2(en)]$, *cis*-$[Pt(ox)(NH_3)_2]$, and *cis*-$[Pt(NO_2)_2(C_6H_4(NH_2)_2)]$.

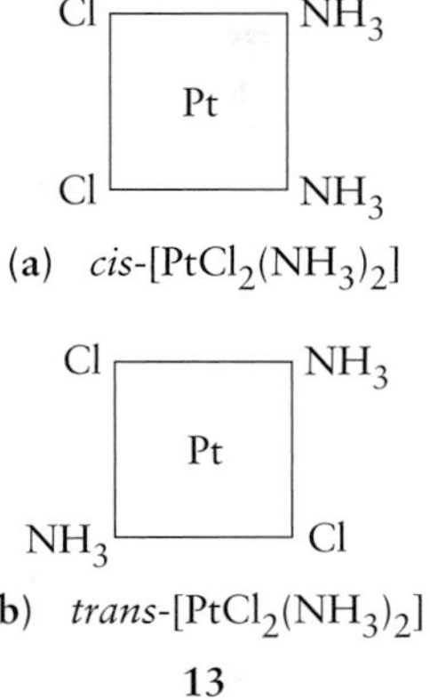

(a) *cis*-$[PtCl_2(NH_3)_2]$

(b) *trans*-$[PtCl_2(NH_3)_2]$

13

Optical isomers are special kinds of stereoisomers: they are nonsuperimposable mirror images of each other (Fig. 16.27). Both geometrical and optical isomerism can occur in an octahedral complex, as in $[CoCl_2(en)_2]^+$: the trans isomer is green (**14a**) and the two alternative cis isomers (**14b**) and (**14c**), which are optical isomers of one another, are violet.

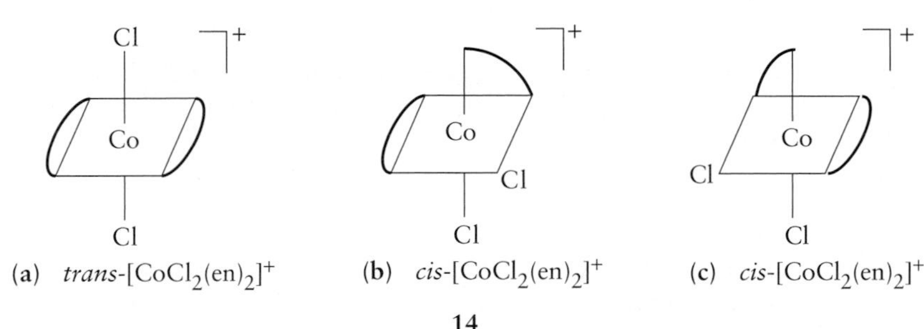

(a) *trans*-$[CoCl_2(en)_2]^+$ (b) *cis*-$[CoCl_2(en)_2]^+$ (c) *cis*-$[CoCl_2(en)_2]^+$

14

To appreciate the difference between the two cis isomers, try to visualize the complex and its mirror image in three dimensions (or better, build models). No matter how we twist and turn the molecules, we cannot superimpose the mirror image molecule on the original molecule so that the location of each type of atom is the same. It is like trying to superimpose your right hand on your left hand. The two molecules are, in fact, two distinct compounds. Optical isomers can also occur whenever four different groups form a tetrahedral complex, but not if they form a square planar complex.

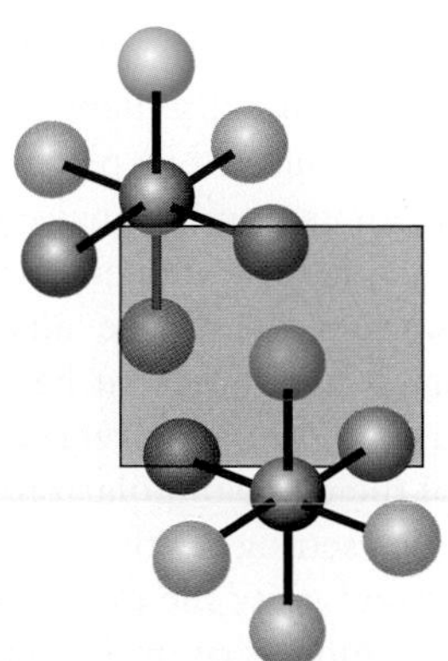

FIGURE 16.27 Optical isomers. The two complexes are each other's mirror image; no matter how we rotate them, one complex cannot be superimposed on the other.

A **chiral complex** is one that is not identical to its mirror image. Thus, all optical isomers are chiral. The cis isomers of $[CoCl_2(en)_2]^+$ are chiral,

The name *enantiomer* comes from the Greek words for "both parts."

and a chiral complex and its mirror image form a pair of **enantiomers.** The trans isomer is superimposable on its mirror image; complexes with this property are called **achiral.**

Enantiomers differ in one physical property: chiral molecules display **optical activity,** the ability to rotate the plane of polarization of light (Box 16.2). In ordinary light, the wave motion lies at random directions around the direction of travel. In plane-polarized light, the wave lies in a single plane (Fig. 16.28). Plane-polarized light can be prepared by passing ordinary light through a special filter, like the material used to make polarized sunglasses. One enantiomer of a chiral complex rotates the plane of polarization clockwise; its mirror image partner rotates it by the same amount the other way. Achiral complexes are not optically active: they do not rotate the plane of polarization of polarized light.

Some complexes are synthesized in the laboratory as **racemic mixtures,** or mixtures of enantiomers in equal proportions. Because enantiomers rotate the plane of polarization of light in opposite directions, a racemic sample is not optically active.

BOX 16.2 HOW DO WE KNOW . . . THAT A COMPLEX IS OPTICALLY ACTIVE?

In plane-polarized light, the electric field of the radiation oscillates in a single plane as it passes through space and any transparent material it encounters. We can prepare plane-polarized light by passing ordinary, unpolarized light through a polarizer, which consists of a material that allows the light to pass only if the electric field is aligned in a certain direction. The same kind of material is used in polarized sunglasses.

An optically active substance rotates the plane of polarization of a beam of light. The light is passed through a sample cell about 10 cm long. The cell contains a solution of the chiral complex at a known concentration. When the light emerges from the far end of the cell, the angle of its plane of polarization may be different from its original angle. To determine the angle, the light is passed through an analyzer that contains another polarizing filter. The analyzer is rotated until the intensity of the light that has passed through the polarizer, sample, and analyzer reaches its maximum. The angle of the plane of polarization is determined from this maximum setting. If the sample is not optically active, the analyzer gives the maximum intensity at an angle of 0°. The sample is optically active if the angle of rotation is different from 0°. The actual value depends on the identity of the complex, its concentration, and the length of the sample cell.

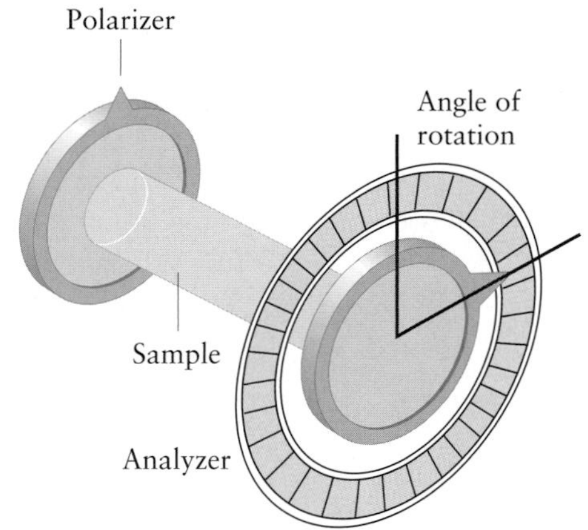

A polarimeter determines the optical activity of a substance by measuring the angle by which plane-polarized light is rotated by a sample.

The determination of the angle of rotation is called *polarimetry*. In some cases, it can help a chemist follow a reaction. For example, if a reaction destroys the chirality of a complex, then the angle of optical rotation decreases with time as the concentration of the complex falls.

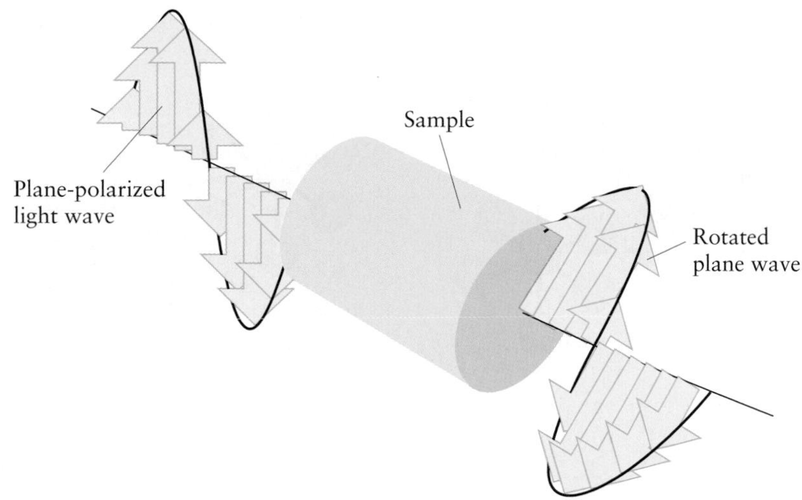

FIGURE 16.28 Plane-polarized light consists of radiation in which all the wave motion lies in one plane (as represented by the arrows on the left). When such light passes through a solution of an optically active substance, the plane of the polarization is rotated through a characteristic angle that depends on the identity and concentration of the solute and the length of the path through the sample.

The varieties of isomerism are summarized in Fig. 16.22. Optical isomers rotate the plane of polarization of light in opposite directions.

EXAMPLE 16.2 **Identifying optical isomerism**

Which of the complexes in Fig. 16.29 are chiral, and which form enantiomeric pairs?

STRATEGY Draw the mirror image of each complex and mentally rotate it; judge whether any rotation will allow the mirror image to be superimposed on the original molecule. If not, then the complex is chiral. Determine which complexes form enantiomeric pairs by finding pairs that are the nonsuperimposable mirror images of each other. If imagining the three-dimensional structure is difficult, build simple paper models of the complexes.

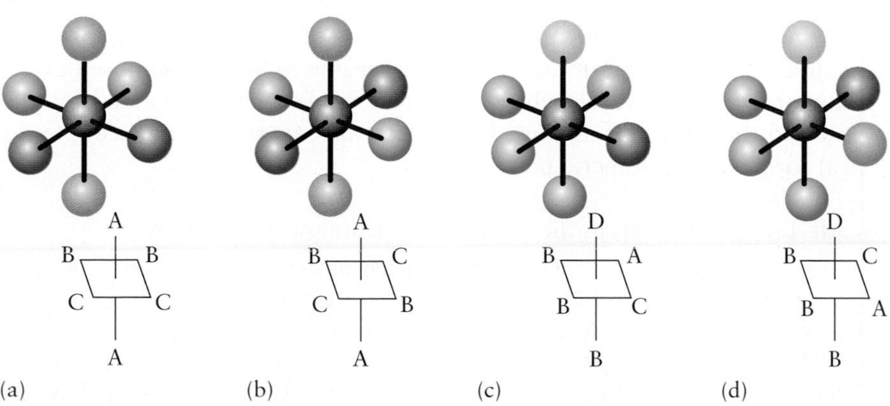

FIGURE 16.29 The complexes referred to in Example 16.2. Note that we also display the abbreviated forms of their structures.

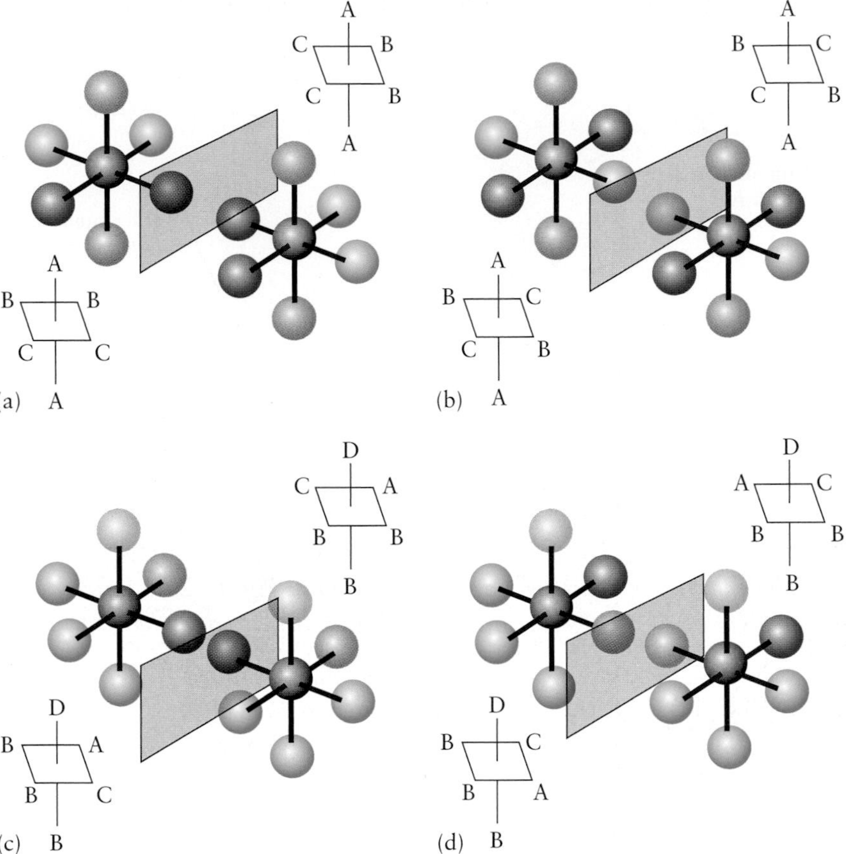

FIGURE 16.30 The complexes and their mirror images referred to in the solution of Example 16.2.

SOLUTION In Fig. 16.30, the original complexes are on the left of each pair, and the mirror image of each complex is on the right. (a, b) Neither is chiral. (a) If we rotate the mirror image about A–A, we obtain a structure identical to the original and so superimposable on it. (b) The mirror image is superimposable on the original. (c, d) The original complexes are chiral because no rotation will allow either one to be superimposed on its mirror image. However, when the mirror image of (c) is rotated by 90° around the vertical B–D axis, it becomes the complex (d); hence (c) and (d) form an enantiomeric pair.

Self-Test 16.6A Repeat the exercise for the following complexes:

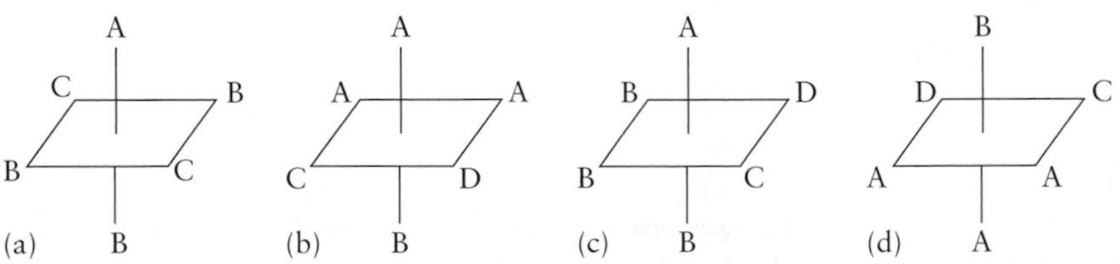

[*Answer:* (a) Not chiral; (c) chiral; (b, d) chiral and enantiomeric]

Self-Test 16.6B Repeat the exercise for the following complexes:

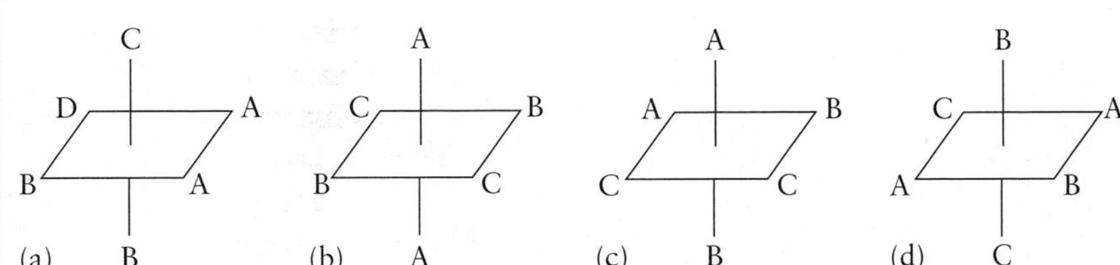

THE ELECTRONIC STRUCTURES OF COMPLEXES

The striking feature of many coordination compounds is that they are colored, paramagnetic, or both. How do these properties arise? To find out, we need to understand the electronic structures of complexes, the details of the bonding, and the distribution of their electrons.

There are two major theories of bonding in *d*-metal complexes. *Crystal field theory* was devised to explain the colors of solids, particularly ruby, which owes its color to Cr^{3+} ions. Crystal field theory is simple to apply and enables us to make useful predictions with very little labor. However, it does not account for all the properties of complexes. A more sophisticated approach, *ligand field theory* (Section 16.13), is based on molecular orbital theory.

16.8 Crystal Field Theory

Crystal field theory takes a very simple view of the environment of the central metal atom (or ion): it supposes that the atomic orbitals on the central atom are the unhybridized atomic orbitals and that each ligand can be represented by a point negative charge. These negative charges represent the ligand lone pairs directed toward the central metal atom (Fig. 16.31). Because the metal atom at the center of a complex is usually positively charged, the negative charges representing the ligands are attracted to it. This attraction results in the formation of the complex. The $[Ti(H_2O)_6]^{3+}$ complex, for instance, exists because there is a strong attraction between the Ti^{3+} ion and the six negative point charges representing a lone pair on each of the six H_2O ligands.

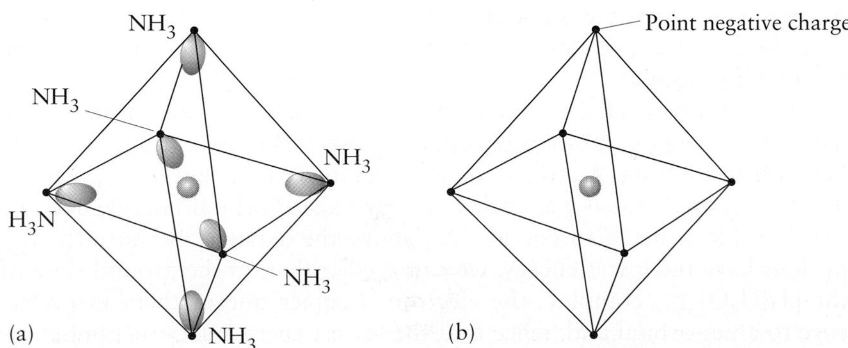

FIGURE 16.31 In the crystal field theory of complexes, the lone pairs of electrons (the Lewis base sites) on the ligands (a) are treated as equivalent to point negative charges (b).

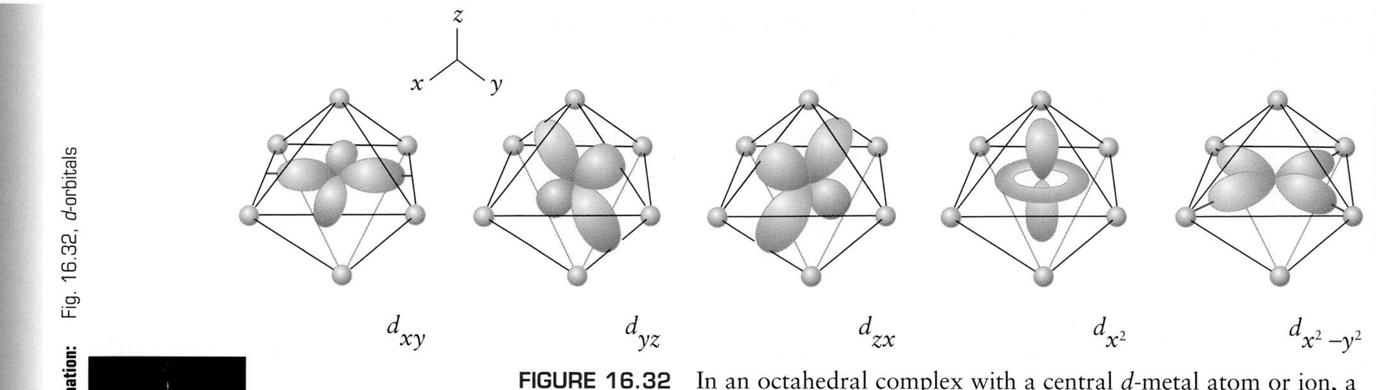

FIGURE 16.32 In an octahedral complex with a central *d*-metal atom or ion, a d_{xy}-orbital is directed between the ligand sites, and an electron that occupies it has a relatively low energy. The same lowering of energy occurs for d_{yz}- and d_{zx}-orbitals. A d_{z^2}-orbital points directly toward two ligands, and an electron that occupies it has a relatively high energy. The same rise in energy occurs for a $d_{x^2-y^2}$ electron. (b)

Media Link Animation: Fig. 16.32, *d*-orbitals

View animations of the five *d*-orbitals.

Let's consider the interaction between these point charges and the outer electrons of the metal ion in the $[Ti(H_2O)_6]^{3+}$ complex. Note that the electron configuration of the Ti^{3+} ion is [Ar]$3d^1$, so there is only one *d*-electron to consider. That single 3*d*-electron can interact in different ways with the point charges, depending on which of the five 3*d*-orbitals it occupies. In an octahedral complex such as $[Ti(H_2O)_6]^{3+}$, the six ligands (represented by point charges) lie on opposite sides of the central metal ion along the *x*-, *y*-, and *z*-axes. From Fig. 16.32, we can see that three of the orbitals (d_{xy}, d_{yz}, and d_{zx}) have their lobes directed between the point charges. These three *d*-orbitals are called **t_{2g}-orbitals** in crystal field theory. The other two *d*-orbitals (d_{z^2} and $d_{x^2-y^2}$), which point directly toward the point charges, are called **e_g-orbitals.**

The labels t_{2g} and e_g are derived from group theory, the mathematical theory of symmetry. The g indicates that the orbital does not change sign when we start from any point, pass through the nucleus, and end at the corresponding point on the other side of the nucleus.

Because the point charges representing the ligands repel the electrons in the *d*-orbitals, the energy of these orbitals is increased, but not by the same amount. The t_{2g}-orbitals are further from the ligands than the e_g-orbitals, so electrons in t_{2g}-orbitals are repelled less by the negative point charges of the ligands than are electrons in e_g-orbitals. As a result, the energies of the t_{2g}-orbitals are lower than those of the e_g-orbitals. The difference in the energies of the t_{2g}-orbitals and the e_g-orbitals is typically about 10% of the total interaction energy between the central ion and its ligands. This difference is therefore only a minor contribution to the overall stability of the complex, but we shall see that it is a major factor in the complex's properties.

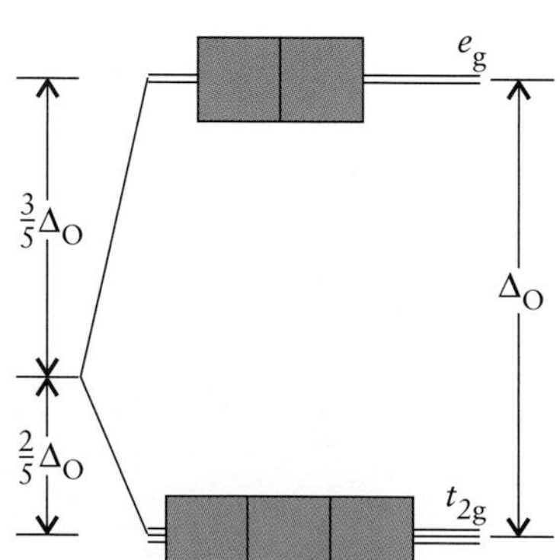

FIGURE 16.33 The energy levels of the *d*-orbitals in an octahedral complex with the ligand field splitting Δ_O. Each orbital (represented by a box) can hold two electrons.

The energy-level diagram in Fig. 16.33 helps to summarize these ideas. The energy separation between the two sets of orbitals is called the **ligand field splitting,** Δ_O (the O denotes octahedral). The three t_{2g}-orbitals lie at an energy that is $\frac{2}{5}\Delta_O$ below the average *d*-orbital energy, and the two e_g-orbitals lie at an energy $\frac{3}{5}\Delta_O$ above the average. Because the t_{2g}-orbitals have the lower energy, we can predict that, in the ground state of the $[Ti(H_2O)_6]^{3+}$ complex, the electron occupies one of them in preference to an e_g-orbital and hence that the lowest energy electron configuration of the complex is $t_{2g}^{\ 1}$. This ground-state configuration is represented by the box diagram in (15).

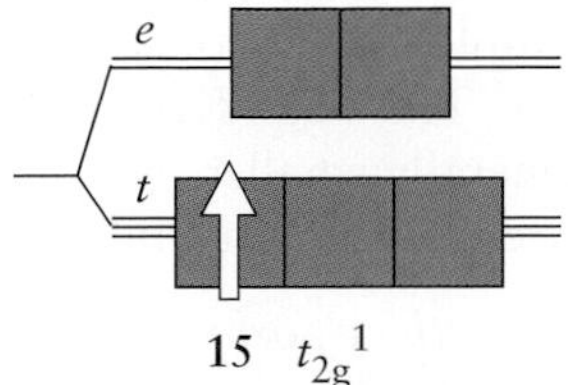

15 $t_{2g}^{\ 1}$

The single d-electron of an octahedral $[Ti(H_2O)_6]^{3+}$ complex can be excited from the t_{2g}-orbital into one of the e_g-orbitals if it absorbs a photon of energy Δ_O (Fig. 16.34). The greater the splitting, the shorter the wavelength of the light that is absorbed by the complex. Therefore, we can use the wavelength of the electromagnetic radiation absorbed by a complex to determine the ligand field splitting.

EXAMPLE 16.3 Determining the ligand field splitting

The complex $[Ti(H_2O)_6]^{3+}$ absorbs light of wavelength 510 nm. What is the ligand field splitting in the complex?

STRATEGY Because a photon carries an energy $h\nu$, where h is the Planck constant and ν (nu) is the frequency of the radiation, it can be absorbed if $h\nu = \Delta_O$. The wavelength, λ (lambda), of light is related to the frequency by $\lambda = c/\nu$, where c is the speed of light (Section 1.1). Therefore, the wavelength of light absorbed and the ligand field splitting are related by

$$\Delta_O = \frac{hc}{\lambda}$$

Ligand field splittings are normally reported as a molar energy, so we need to multiply this expression by the Avogadro constant:

$$\Delta_O = \frac{N_A hc}{\lambda}$$

We can then substitute the data.

SOLUTION Because the wavelength absorbed is 510 nm (corresponding to 5.10×10^{-7} m), it follows that the ligand field splitting is

$$\Delta_O = \frac{(6.022 \times 10^{23}\ \text{mol}^{-1}) \times (6.626 \times 10^{-34}\ \text{J·s}) \times (2.998 \times 10^{8}\ \text{m·s}^{-1})}{5.10 \times 10^{-7}\ \text{m}}$$
$$= 2.35 \times 10^{5}\ \text{J·mol}^{-1} = 235\ \text{kJ·mol}^{-1}$$

Self-Test 16.7A The complex $[Fe(H_2O)_6]^{3+}$ absorbs light of wavelength 700 nm (3 sf). What is the value (in kilojoules per mole) of the ligand field splitting?

[*Answer:* 171 kJ·mol^{-1}]

Self-Test 16.7B The complex $[Fe(CN)_6]^{4-}$ absorbs light of wavelength 305 nm. What is the value (in kilojoules per mole) of the ligand field splitting?

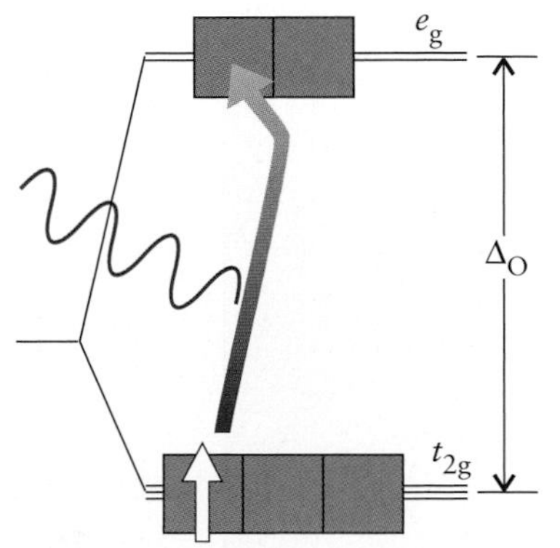

FIGURE 16.34 When a complex is exposed to light with the correct frequency, an electron can be excited to a higher energy orbital and a photon of light is absorbed.

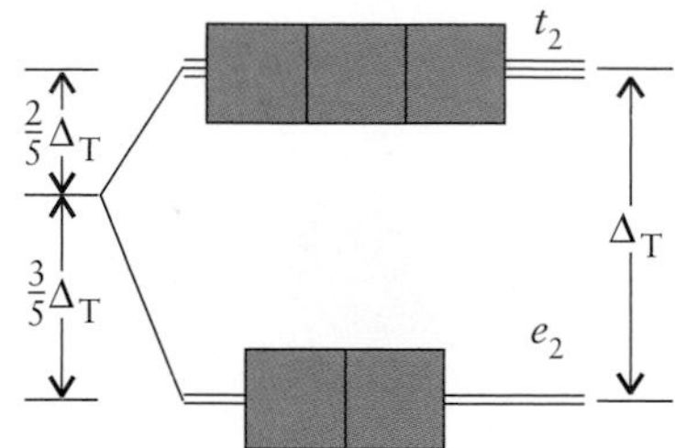

FIGURE 16.35 The energy levels of the *d*-orbitals in a tetrahedral complex with the ligand field splitting Δ_T. Each box (that is, orbital) can hold two electrons. The subscript g is not used to label the orbitals in a tetrahedral complex.

The relative energies of the *d*-orbitals are different in complexes with different shapes. For example, in a tetrahedral complex, the three t_2-orbitals point more directly at the ligands than the two *e*-orbitals do. As a result, in a tetrahedral complex, the t_2-orbitals have a higher energy than the *e*-orbitals (Fig. 16.35). The ligand field splitting, Δ_T (where the T denotes tetrahedral), is generally smaller than in octahedral complexes, in part because there are fewer repelling ligands.

In octahedral complexes, the e_g-orbitals (d_{z^2} and $d_{x^2-y^2}$) lie higher in energy than the t_{2g}-orbitals (d_{xy}, d_{yz}, and d_{zx}). The opposite is true in a tetrahedral complex, for which the ligand field splitting is smaller.

16.9 The Spectrochemical Series

Different ligands affect the *d*-orbitals of a given metal atom or ion to different degrees and thus produce different values of the ligand field splitting. For example, the ligand field splitting is much greater in $[Fe(CN)_6]^{4-}$ than it is in $[Fe(H_2O)_6]^{2+}$. The relative strengths of the splitting produced by a given ligand are much the same regardless of the identity of the *d*-metal in the complex. Thus, ligands can be arranged in a **spectrochemical series** according to the relative magnitudes of the ligand field splittings they produce (Fig. 16.36). Those ligands appearing below the horizontal line in the figure produce only a small ligand field splitting, so they are called **weak-field ligands;** those above it produce a larger splitting and are called **strong-field ligands.** A CN^- ion is therefore a strong-field ligand, whereas an H_2O molecule is a weak-field ligand.

Knowledge of relative ligand strengths allows us to control the magnetism and color of a complex ion. Substituting weak-field ligands for strong-field ligands (or vice versa) acts like a chemical switch for turning paramagnetism on and off. Replacing one ligand by another also gives us chemical control over color.

The spectrochemical series summarizes the relative strengths of ligand field splittings.

16.10 High-Spin and Low-Spin Complexes

In an isolated metal atom, all five *d*-orbitals have the same energy. Electrons therefore occupy each orbital separately until each has one electron. Only after five electrons have been accommodated do we get more than one electron in a *d*-orbital. When the atom is part of a complex, however, we have to take into account the different energies of the *t*- and *e*-orbitals, and the order in which the orbitals are occupied is more subtle.

First, consider the metal atom or ion at the center of an octahedral complex. The energies of its *d*-orbitals are split by the ligands as shown in Fig. 16.33. All three t_{2g}-orbitals have the same energy and lie below the

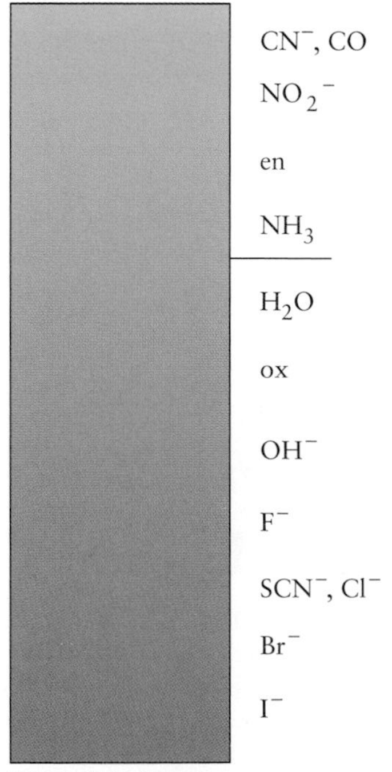

FIGURE 16.36 The spectrochemical series. Strong-field ligands give rise to a large splitting between the *t*- and *e*-orbitals, whereas weak-field ligands give rise to only a small splitting. The horizontal line marks the approximate frontier between the two kinds of ligands. The changing color represents the increasing energy of light absorbed as the field strength increases.

two e_g-orbitals. The single d-electron of a d^1 complex occupies one of the t_{2g}-orbitals, so the ground-state configuration is $t_{2g}^{\,1}$ (see **15**). The two d-electrons of a d^2 complex occupy separate t_{2g}-orbitals and give rise to the configuration $t_{2g}^{\,2}$ (**16**). Similarly, a d^3 complex will have the ground-state configuration $t_{2g}^{\,3}$ (**17**). According to Hund's rule (Section 1.12), all these electrons have parallel spins.

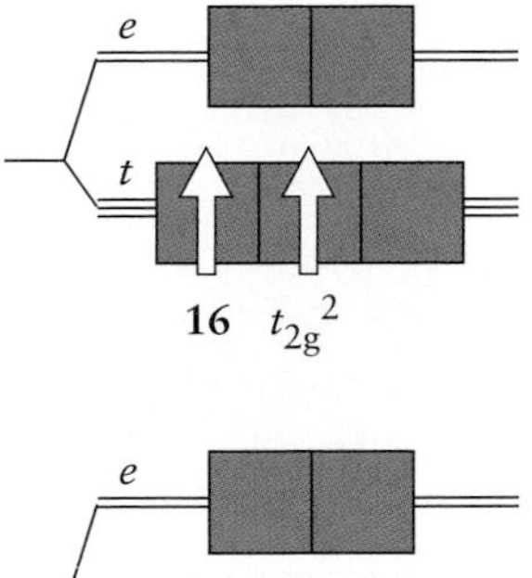

16 $t_{2g}^{\,2}$

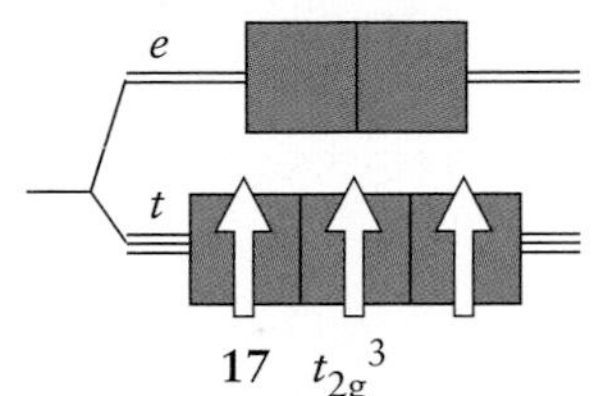

17 $t_{2g}^{\,3}$

A d^4 complex presents a problem. The fourth electron could enter a t_{2g}-orbital, resulting in a $t_{2g}^{\,4}$ configuration. However, to do so, it would have to enter an orbital that is already half full; hence, it would experience a strong repulsion from the electron already there (**18**). To avoid this repulsion, it could occupy an empty e_g-orbital to give a $t_{2g}^{\,3}e_g^{\,1}$ configuration (**19**), but now it experiences a strong repulsion from the ligands. Which configuration has the lower energy depends on the ligands present. If Δ_O is large (as it is for strong-field ligands), the energy difference between the t_{2g}- and e_g-orbitals will be large and the configuration $t_{2g}^{\,4}$ will have a lower energy than $t_{2g}^{\,3}e_g^{\,1}$. If Δ_O is small (as it is for weak-field ligands), $t_{2g}^{\,3}e_g^{\,1}$ will be the lower energy configuration and hence the one adopted by the complex.

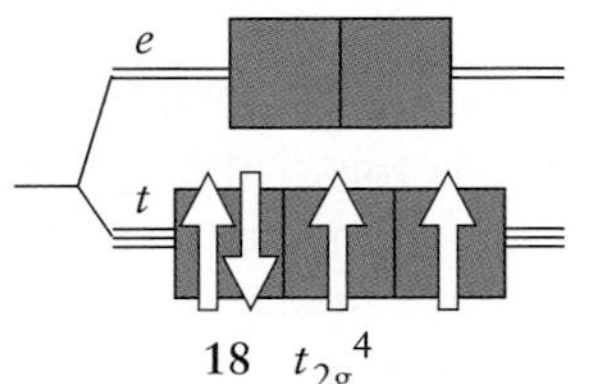

18 $t_{2g}^{\,4}$

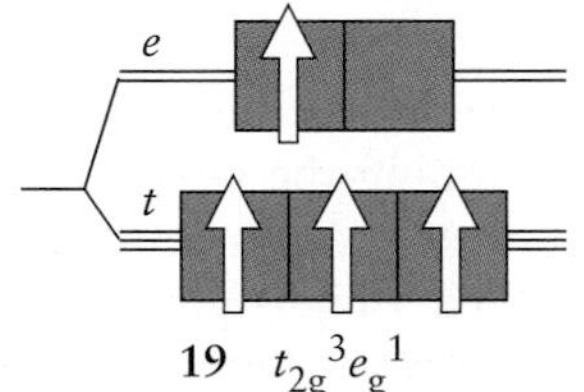

19 $t_{2g}^{\,3}e_g^{\,1}$

EXAMPLE 16.4 Predicting the electron configuration of a complex

Predict the electron configuration of an octahedral d^5 complex with (a) strong-field ligands and (b) weak-field ligands, and state the number of unpaired electrons in each case.

STRATEGY Electrons occupy the orbitals that result in the lowest energy configuration. For weak-field ligands, Δ_O is small. In this case, electrons are likely to occupy all the vacant orbitals, even those at the higher energy, before pairing. For strong-field ligands, Δ_O is large and electrons are likely to pair in the lower energy orbitals and fill them completely before occupying any of the higher energy orbitals.

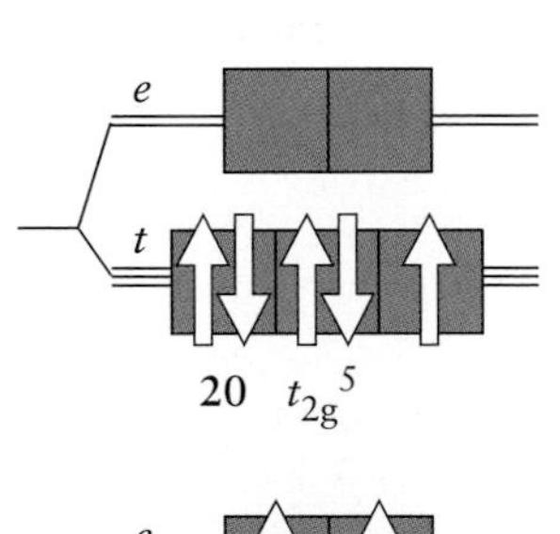

20 $t_{2g}^{\,5}$

SOLUTION (a) In the strong-field case, all five electrons enter the t_{2g}-orbitals; to do so, some of them must pair (**20**). There is one unpaired electron in this configuration. (b) In the weak-field case, the five electrons occupy all five orbitals without pairing (**21**). There are now five unpaired electrons.

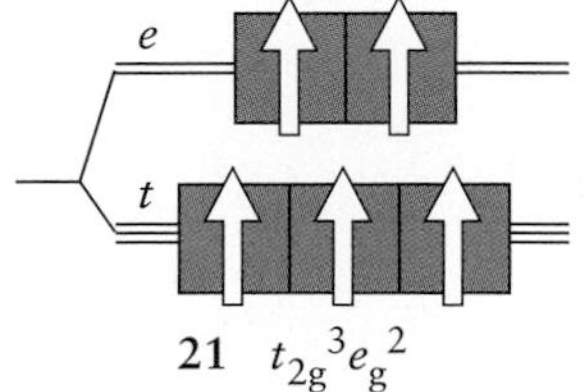

21 $t_{2g}^{\,3}e_g^{\,2}$

Self-Test 16.8A Predict the electron configurations and the number of unpaired electrons of an octahedral d^6 complex with (a) strong-field ligands and (b) weak-field ligands.

[*Answer:* (a) $t_{2g}^{\,6}$ (0); (b) $t_{2g}^{\,4}e_g^{\,2}$ (4)]

Self-Test 16.8B Predict the electron configurations and the number of unpaired electrons of an octahedral d^7 complex with (a) strong-field ligands and (b) weak-field ligands.

TABLE 16.5 Electron Configurations of d^n Complexes

Number of *d*-electrons	Configuration: Octahedral complexes			Configuration: Tetrahedral complexes
	Low spin		High spin	
d^1		$t_{2g}{}^1$		e^1
d^2		$t_{2g}{}^2$		e^2
d^3		$t_{2g}{}^3$		$e^2t_2{}^1$
d^4	$t_{2g}{}^4$		$t_{2g}{}^3e_g{}^1$	$e^2t_2{}^2$
d^5	$t_{2g}{}^5$		$t_{2g}{}^3e_g{}^2$	$e^2t_2{}^3$
d^6	$t_{2g}{}^6$		$t_{2g}{}^4e_g{}^2$	$e^3t_2{}^3$
d^7	$t_{2g}{}^6e_g{}^1$		$t_{2g}{}^5e_g{}^2$	$e^4t_2{}^3$
d^8		$t_{2g}{}^6e_g{}^2$		$e^4t_2{}^4$
d^9		$t_{2g}{}^6e_g{}^3$		$e^4t_2{}^5$
d^{10}		$t_{2g}{}^6e_g{}^4$		$e^4t_2{}^6$

Table 16.5 lists the configurations for d^1 through d^{10} octahedral complexes, including the alternative configurations for d^4 through d^7 octahedral complexes. A d^n complex with the maximum number of unpaired spins is called a **high-spin complex.** High-spin complexes are expected for weak-field ligands because the electrons can easily occupy both the t_{2g}- and e_g-orbitals, and the greatest number of electrons then have parallel spins. A d_n complex with the minimum number of unpaired spins is called a **low-spin complex.** We should expect a low-spin complex when the ligands are strong-field, because electrons enter the t_{2g}-orbitals until they are completely full, even though they have to pair their spins. We can predict whether an octahedral complex is likely to be a high-spin or low-spin complex by noting where the ligands lie in the spectrochemical series. If they are strong-field ligands, we expect a low-spin complex; if they are weak-field ligands, we expect a high-spin complex.

Tetrahedral complexes are almost always high spin. We saw that ligand field splittings are smaller for tetrahedral than for octahedral complexes. Therefore, even if the ligands are classified as strong-field ligands for octahedral complexes, the splitting is so small in the corresponding tetrahedral complex that the t_2-orbitals are energetically accessible.

The electron configurations of d-block metal atoms and ions in complexes are obtained by applying the building-up principle to the d-orbitals, taking into account the strength of the ligand field splitting.

16.11 The Colors of Complexes

White light is a mixture of all wavelengths of electromagnetic radiation from about 400 nm (violet) to about 800 nm (red). When some of these wavelengths are removed from a beam of white light by passing the light through a sample, the emerging light is no longer white to the eye. For example, if red light is absorbed from white light, then the light that remains appears green. Conversely, if green is removed, then the light appears red. We say that red and green are each other's **complementary color**—each is the color that white light appears when the other is removed.

Complementary colors are shown on the color wheel in Major Technique 2, following Chapter 3. We can see from the color wheel that, if a substance looks blue (as does a copper(II) sulfate solution, for instance), then it is absorbing orange (580 to 620 nm) light. Conversely, if we know the wavelength (and therefore the color) of the light that a substance absorbs, then we can predict the color of the substance by noting the complementary color on the color wheel. Because $[Ti(H_2O)_6]^{3+}$ absorbs 510-nm light, which is yellow-green light, the complex looks violet (Fig. 16.37).

The prediction of color from the formula of a complex is, in fact, quite difficult. One problem is that compounds absorb light over a range of wavelengths and may in fact absorb in several regions of the spectrum. Chlorophyll, for example, absorbs both red and blue light, leaving only the wavelengths near green to be reflected from vegetation (see Major Technique 2).

Because weak-field ligands generate small splittings, the complexes they form absorb low-energy, long-wavelength radiation. The long wavelengths correspond to red light, so these complexes exhibit colors near green. Because strong-field ligands give large splittings, the complexes they form should absorb high-energy, short-wavelength radiation, corresponding to the violet end of the visible spectrum. Such complexes can therefore be expected to have colors near orange and yellow (Fig. 16.38).

The colors we have described arise from ***d–d* transitions,** in which an electron is excited from one *d*-orbital into another. In octahedral complexes, the excitation is from a t_{2g}-orbital to an e_g-orbital. In a tetrahedral complex, the excitation is from an *e*-orbital to a t_2-orbital, because the order of orbitals is reversed. However, there is an alternative type of transition called a **charge-transfer transition,** in which an electron is excited from the ligands on to the metal atom or vice versa. Charge-transfer transitions are often very intense. An example of a ligand-to-metal charge-transfer transition is the one responsible for the deep purple of permanganate ions, MnO_4^- (Fig. 16.39).

Transitions between d-orbitals or between the ligands and the metal atom in complexes give rise to color; the wavelength can be correlated with the magnitude of the ligand field splitting.

FIGURE 16.37 Because $[Ti(H_2O)_6]^{3+}$ absorbs yellow-green light (of wavelengths close to 510 nm), it looks violet in white light.

FIGURE 16.38 The effect of changing the ligands in octahedral cobalt(III) complexes in aqueous solution. The ligand field strengths increase from left to right.

FIGURE 16.39 In a ligand-to-metal charge-transfer transition, an energetically excited electron migrates from the ligands to the central metal ion. This type of transition is responsible for the intense purple of the permanganate ion, MnO_4^-.

Self-Test 16.9A Predict which of the following complexes absorbs light of the shorter wavelength and explain your reasoning: (a) $[Co(H_2O)_6]^{3+}$ or (b) $[Co(en)_3]^{3+}$.

[*Answer:* (b); en is a stronger-field ligand than H_2O.]

Self-Test 16.9B Predict which of the following complexes absorbs light of the shorter wavelength and explain your reasoning: (a) $[Fe(CN)_6]^{4-}$ or (b) $[Fe(NH_3)_6]^{2+}$.

16.12 Magnetic Properties of Complexes

As we saw in Box 3.2, a substance with unpaired electrons is paramagnetic and is pulled into a magnetic field. A substance without unpaired electrons is diamagnetic and is pushed out of a magnetic field. The apparatus shown in Fig. 16.40 allows us to distinguish paramagnetism and diamagnetism experimentally. A sample is hung from a balance so that it lies between the poles of an electromagnet. When the magnet is turned on, a paramagnetic substance is pulled into the field and appears to weigh more than when the magnet is off. A diamagnetic substance is pushed out of the field and appears to weigh less.

Many *d*-metal complexes have unpaired *d*-electrons and are therefore paramagnetic. We have just seen that a high-spin d^n complex has more unpaired electrons than a low-spin d^n complex. The high-spin complex is therefore more strongly paramagnetic and drawn more strongly into a magnetic field. Moreover, whether a complex is high spin or low spin depends on the ligands present. Strong-field ligands create a large energy difference between the t_{2g}- and e_g-orbitals of an octahedral complex (Fig. 16.41). Their d^4–d^7 complexes therefore tend to be low spin and diamagnetic or only weakly paramagnetic. Weak-field ligands create a small energy gap, so electrons fill the higher energy orbitals rather than pairing in the lower energy orbitals. Their d^4–d^7 complexes therefore tend to be high spin and strongly paramagnetic.

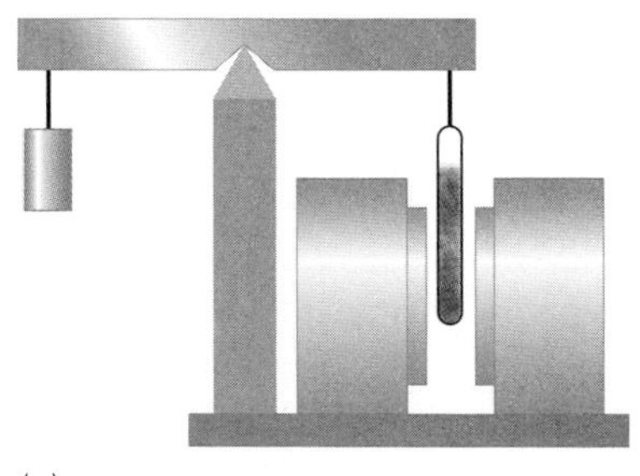

(a)

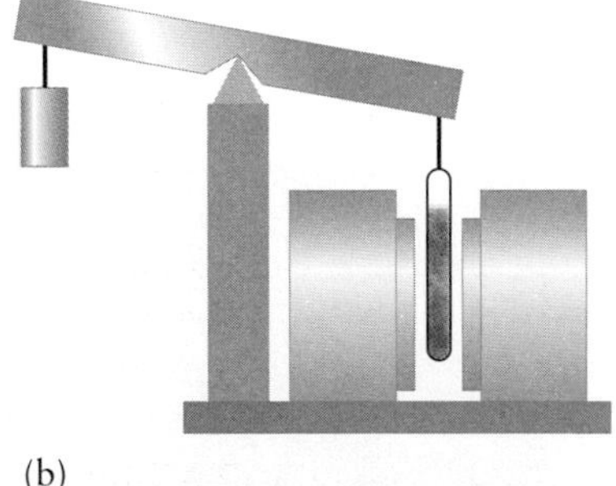

(b)

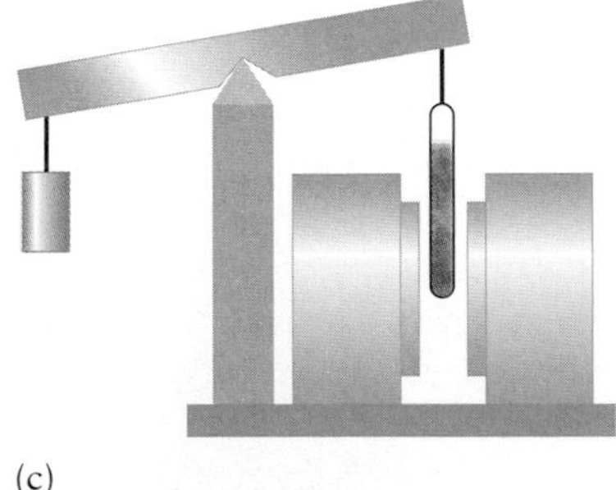

(c)

FIGURE 16.40 The magnetic character of a complex can be studied with a Gouy balance. (a) A sample is hung from a balance so that it lies partly between the poles of an electromagnet. (b) When the magnetic field is turned on, a paramagnetic sample is drawn into it, so the sample seems to weigh more. (c) In contrast, a diamagnetic sample is pushed out of the field when the field is turned on, and seems to weigh less.

The magnetic properties of a complex depend on the magnitude of the ligand field splitting. Strong-field ligands tend to form low-spin, weakly paramagnetic complexes; weak-field ligands tend to form high-spin, strongly paramagnetic complexes.

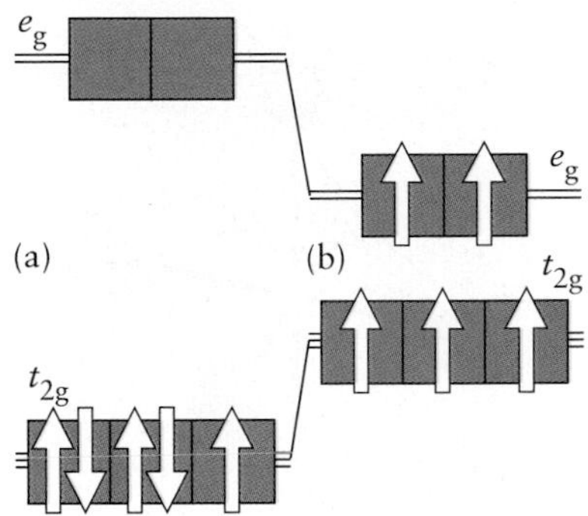

FIGURE 16.41 (a) A strong-field ligand is likely to lead to a low-spin complex (in this case, the configuration is that of Fe^{3+}). (b) A weak-field ligand is likely to result in a high-spin complex.

EXAMPLE 16.5 Predicting the magnetic properties of a complex

Compare the magnetic properties of $[Fe(H_2O)_6]^{2+}$ and $[Fe(CN)_6]^{4-}$.

STRATEGY Decide from their positions in the spectrochemical series whether the ligands in the complex are weak-field or strong-field ligands. Then judge whether the complex is high-spin or low-spin.

SOLUTION The Fe^{2+} ion is a d^6 ion. Because H_2O is a weak-field ligand, we predict a high-spin configuration with four unpaired electrons for $[Fe(H_2O)_6]^{2+}$. The ion is therefore predicted (and found) to be paramagnetic. When cyanide ions are added to an aqueous solution of Fe^{2+} ions, they form the $[Fe(CN)_6]^{4-}$ ion. Now the ligands are strong-field ligands, and the resulting complex is a low-spin $t_{2g}^{\ 6}$ complex; it has no unpaired electrons, so it is not paramagnetic.

Self-Test 16.10A What change in magnetic properties can be expected when NO_2^- ligands in an octahedral complex are replaced by Cl^- ligands in (a) a d^6 complex; (b) a d^3 complex?

[*Answer:* (a) The complex becomes paramagnetic; (b) there is no change in magnetic properties.]

Self-Test 16.10B Compare the magnetic properties of $[Ni(en)_3]^{2+}$ with those of $[Ni(H_2O)_6]^{2+}$.

16.13 Ligand Field Theory

Crystal field theory is based on a very primitive model of bonding. For instance, ligands are not point charges: they are actual molecules or ions. The theory also leaves several questions unanswered. Why, for instance, is the electrically neutral molecule CO a strong-field ligand but the negatively charged ion Cl^- a weak-field ligand?

To improve the model of bonding in complexes, chemists have turned to the general theory of chemical bonding, molecular orbital theory (Section 3.10). **Ligand field theory** describes bonding in complexes in terms of molecular orbitals built from the metal atom *d*-orbitals and ligand orbitals. In contrast to crystal field theory, which assumes an ionic bond, ligand field theory assumes that ligands are attached to the central metal atom or ion by a covalent bond. As we shall see, much of the work we have done in connection with crystal field theory can be transferred into ligand field theory: the only major difference is the origin of the ligand field splitting.

To describe the electronic structure of a complex, first we set up molecular orbitals from the available atomic orbitals in the complex, just as we would for a molecule. Consider an octahedral complex of a *d*-metal in Period 4, such as iron, cobalt, or copper. We need to consider the nine 4*s*-, 4*p*-, and 3*d*-orbitals of the central metal ion, because all these orbitals have similar energies. To simplify the discussion, we use only one atomic

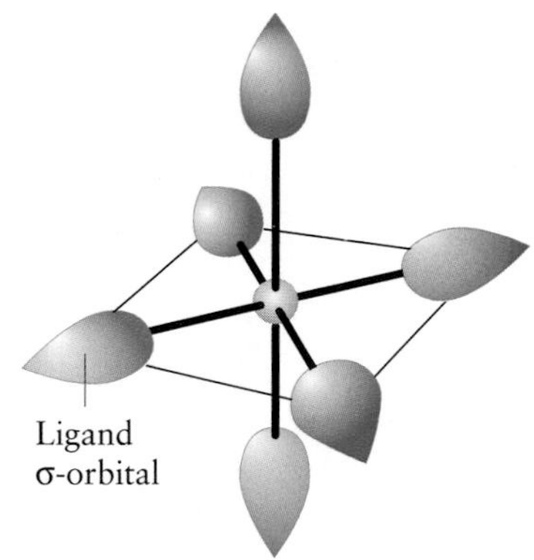

FIGURE 16.42 The tear-shaped objects are representations of the ligand atomic orbitals that are used to build molecular orbitals in ligand field theory. They might represent *s*- or *p*-orbitals on the ligands, or hybrids of the two.

orbital on each of the ligands. For instance, for a Cl^- ligand, we use the Cl3*p*-orbital directed toward the metal atom; for an NH_3 ligand, we use the sp^3 lone-pair orbital of the nitrogen atom. These six ligand orbitals are represented by tear-shaped objects in Fig. 16.42. The orbitals have cylindrical symmetry around the metal-ligand axis, so they are ready to form σ-orbitals.

There are 9 orbitals on the metal atom and 6 on the ligands, giving 15 in all. We can therefore expect to find 15 molecular orbitals: six are bonding, six are antibonding, and three are nonbonding. The energies of all 15 are displayed in Fig. 16.43, together with the labels that they are commonly given. Notice in Fig. 16.43 that the t_{2g}-orbitals on the metal atom have no partners among the ligands. There are simply no ligand orbitals to match them. Therefore, these three orbitals are nonbonding orbitals in the complex.

Now we proceed exactly as we did in the discussion of diatomic molecules (Section 3.11): we use the building-up principle to work out the ground-state electron configuration. First, we need to count the available electrons. In a d^n complex, there are *n* electrons supplied by the metal; each ligand orbital held two electrons, so there are 12 electrons supplied by the ligands, giving $12 + n$ in all. The first 12 electrons fill the 6 bonding orbitals. That leaves *n* electrons to be accommodated. At this point, notice that the next available orbitals (those in the blue box in Fig. 16.43), lie in exactly the same pattern as we encountered in crystal field theory. The only difference is that in ligand field theory we recognize the t_{2g}-orbitals as nonbonding orbitals and the e_g-orbitals as antibonding between the metal and the ligands. That is, the ligand field splitting can be recognized as the energy separation between nonbonding and antibonding orbitals.

From this point on, the analysis is the same as in crystal field theory. The order of filling these two sets of orbitals is exactly the same as in crystal field theory, and so is the discussion of optical and magnetic properties. If the ligand field splitting is large, the t_{2g}-orbitals are occupied first, and we can expect a low-spin complex. If the ligand field splitting is small, the e_g-orbitals are occupied before spin-pairing begins in the t_{2g}-orbitals, and we expect a high-spin complex.

Although ligand field theory puts the discussion of bonding on a firmer basis, it has not yet explained all the puzzling features of crystal field theory. In particular, why is CO a strong-field ligand? Why is Cl^- a weak-field ligand despite its negative charge?

We need to develop the model by considering the effects of other ligand orbitals. When we constructed the molecular orbitals, we considered only ligand orbitals that pointed directly at the central metal atom. Ligands also have orbitals perpendicular to the metal-ligand axis, which we

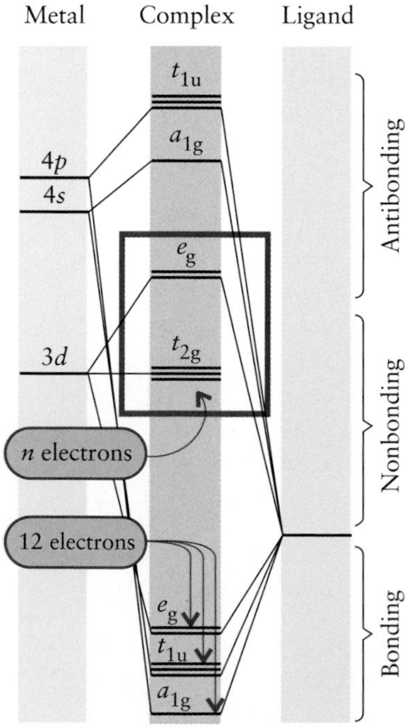

FIGURE 16.43 The molecular orbital energy-level diagram for an octahedral complex. The 12 electrons provided by the six ligands fill the lowest six orbitals, which are all bonding orbitals. The *n* *d*-electrons provided by the metal atom or ion are accommodated in the orbitals inside the blue box. The ligand field splitting is the energy separation of the nonbonding and antibonding orbitals in the box.

can expect to form bonding and antibonding π-orbitals. As Fig. 16.44 shows, a p-orbital on the ligand perpendicular to the axis of the metal-ligand bond can overlap with one of the t_{2g}-orbitals. The resulting bonding combination lies below the energy of the original t_{2g}-orbitals; the antibonding combination lies above them.

Now we count the number of electrons to accommodate. If the ligand is Cl^-, the Cl3p-orbital that we have used to build the metal-ligand π-orbital is full. It provides two electrons, which occupy the bonding metal-ligand combination, as in Fig. 16.44a, for example. The n d-electrons provided by the metal must therefore occupy the *antibonding* metal-ligand combination. Because this molecular orbital is higher in energy than the original e_g-orbitals, the ligand field splitting is decreased by π-bonding. We see how Cl^- can be a weak-field ligand despite its negative charge.

Now suppose the ligand is CO. The orbital that overlaps with the metal t_{2g}-orbitals in this case is either the full bonding π-orbital or the empty antibonding π-orbital of the molecule. It turns out that the latter orbital is closer in energy to the metal orbitals, so it plays the dominant role in bond formation to the metal, as in Fig. 16.44b, for example. There are no electrons from the ligand to accommodate because its π^*-orbital is empty. The n d-electrons therefore enter the *bonding* metal-ligand combination. Because this molecular orbital is lower in energy than the original t_{2g}-orbitals, the ligand field splitting is increased by π-bonding and CO is a strong-field ligand despite being electrically neutral.

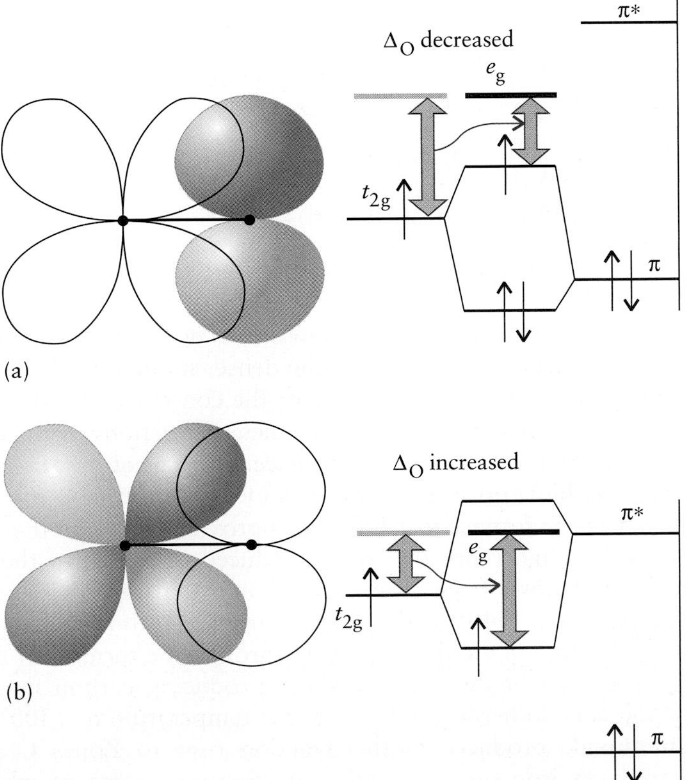

FIGURE 16.44 The effect of π-bonding on the ligand field splitting. (a) In this case, the antibonding π^*-orbital of the ligand is too high in energy to take part in bonding, or is absent (as is the case with a monatomic ligand, such as Cl^-). (b) In this case, the antibonding π^*-orbital of a ligand such as CO matches the metal orbital in energy, and bonding and antibonding metal-ligand combinations are formed.

According to ligand field theory, the ligand field splitting is the energy separation between nonbonding and antibonding molecular orbitals built principally from d-orbitals. When π-bonding is possible, the ligand field splitting is decreased if the ligand supplies p-electrons and is increased if the ligand does not supply p-electrons.

THE IMPACT ON MATERIALS

We have already described important roles that the *d*-block metals play in virtually every aspect of our lives. Steel is a construction material of huge importance in buildings and transport, and will remain so for the foreseeable future. Modern steels are alloys based on iron, but there are also numerous "nonferrous" alloys based on other metals, most notably, copper. The careful blending of different metals—many drawn from the *d* block—can result in materials with essential properties, such as corrosion resistance and tensile strength. The *d*-block metals are also used in the fabrication of magnetic materials.

16.14 Steel

The production of steel begins when iron ore is fed into a blast furnace (Fig. 16.45). The furnace, which is approximately 40 m high, is continuously replenished from the top with a mixture of ore, coke (impure carbon obtained by heating coal in the absence of air), and limestone. Each kilogram of iron produced requires about 1.75 kg of ore, 0.75 kg of coke, and 0.25 kg of limestone. The limestone, which is primarily calcium carbonate, undergoes thermal decomposition to calcium oxide (lime) and carbon dioxide. The calcium oxide, which contains the Lewis base O^{2-}, helps to remove the acidic (nonmetal oxide) and amphoteric impurities from the ore:

$$CaO(s) + SiO_2(s) \xrightarrow{\Delta} CaSiO_3(l)$$

$$CaO(s) + Al_2O_3(s) \xrightarrow{\Delta} Ca(AlO_2)_2(l)$$

$$6\ CaO(s) + P_4O_{10}(s) \xrightarrow{\Delta} 2\ Ca_3(PO_4)_2(l)$$

The mixture of products, which is known as *slag*, is molten at the temperatures in the furnace and floats on the denser molten iron. It is drawn off and used to make rocklike material for the construction industry.

Molten iron is produced through a series of reactions in three of the four main temperature zones of the furnace. At the bottom, in Zone A, preheated air is blown into the furnace under pressure, and the coke is oxidized to heat the furnace to 1900°C and provide carbon in the form of carbon dioxide. Higher up, the iron is reduced in stages to the metal, which melts and flows from Zone C to Zone A. Although the melting point of pure iron is 1540°C, iron mixed with 4% carbon melts at about 1015°C. As the carbon dioxide moves up through the furnace to Zone B, it reacts with some of the added carbon, producing carbon monoxide. This reaction is endothermic and lowers the temperature to 1300°C. The carbon monoxide produced in this reaction rises to Zones C and D, where it reduces iron ore in a series of reactions, some of which are shown in Fig. 16.45. Molten iron (of density 7.9 $g\cdot cm^{-3}$), runs off as *pig*

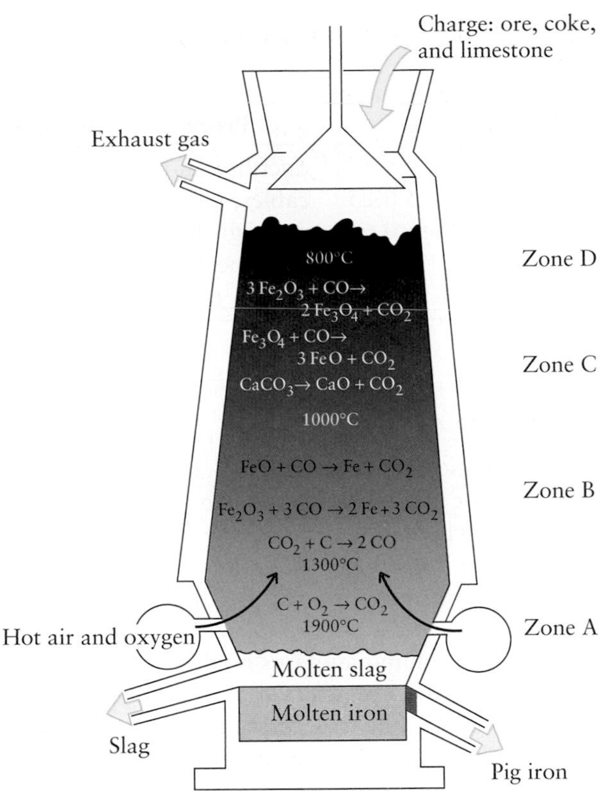

FIGURE 16.45 The reduction of iron ore takes place in a blast furnace containing a mixture of the ore with coke and limestone. Different reactions occur in different zones when the blast of air and oxygen is admitted. The ore, an oxide, is reduced to the metal by reduction with carbon monoxide produced in the furnace.

iron, consisting of 90–95% iron, 3–5% carbon, 2% silicon, and trace amounts of other elements found in the original ore. Cast iron is similar to pig iron, with a carbon content that is usually greater than 2% and fewer impurities.

Pure iron is relatively flexible and malleable, but the carbon atoms make cast iron very hard and brittle. It is used for objects that experience little mechanical and thermal shock, such as ornamental railings, engine blocks, brake drums, and transmission housings. Steel is a homogeneous alloy, a solid solution of 2% or less carbon in iron.

Steels have various hardnesses, tensile strengths, and ductilities, and the higher the carbon content, the harder and more brittle the steel (Table 16.6). Heat treatment can greatly increase the strength and toughness of steel by controlling the size of the crystals that form. The corrosion resistance of iron is significantly improved by alloying with other elements to form a variety of steels (recall Table 16.2). Stainless steels are highly corrosion resistant; they typically contain about 15% chromium by mass.

Pig iron is processed further to produce steel. The first stage is to lower the carbon content of the iron and to remove the remaining impurities, which include silicon, phosphorus, and sulfur. In the *basic oxygen process*, oxygen and powdered limestone are forced through the molten metal (Fig. 16.46). In the second stage, steels are produced by adding the appropriate metals, often as ferroalloys, to the molten iron.

Iron is produced in a blast furnace by reduction of ores, and its properties are modified by blending with other metals.

FIGURE 16.46 In the basic oxygen process, a blast of oxygen and powdered limestone is used to purify molten iron by, respectively, oxidizing and combining with the impurities present in it.

TABLE 16.6 The Effect of Carbon Content on Steel

Type of steel	Carbon content, %	Properties and applications
low-carbon steel	< 0.15	ductility and low hardness, iron wire
mild-carbon steel	0.15 to 0.25	cables, nails, chains, and horseshoes
medium-carbon steel	0.20 to 0.60	nails, girders, rails, and structural purposes
high-carbon steel	0.61 to 1.5	knives, razors, cutting tools, drill bits

16.15 Nonferrous Alloys

Alloys are solid metallic mixtures designed to meet specific needs. For example, the frames of racing bicycles can be made of a steel that contains manganese, molybdenum, and carbon to give them the stiffness needed to resist mechanical shock. Titanium frames are also used, but not the pure metal. Titanium metal stretches easily, so much so that it may not keep its shape under stress. However, when alloyed with metals such as tin and aluminum, titanium maintains its flexibility but keeps its shape.

The properties of alloys are affected by their composition and structure. Not only is the crystalline structure important, but the size and texture of the individual grains also contribute to the properties of an alloy. Some metal alloys are one-phase homogeneous solutions. Examples are brass, bronze, and the gold coinage alloys. Other alloys are heterogeneous mixtures of different crystalline phases, such as tin-lead solder and the mercury amalgams used to fill teeth.

Homogeneous alloys of metals with atoms of similar radius are substitutional alloys. For example, in brass, zinc atoms readily replace copper atoms in the crystalline lattice, because they are nearly the same size (Fig. 16.47). However, the presence of the solute atoms changes the lattice parameters and distorts the local electronic structure. This distortion lowers the electrical and thermal conductivity of the solvent, but it also increases hardness and strength. Coinage alloys are usually substitutional alloys. They are selected for durability—a coin must last for at least three years—and electrical resistance, so that genuine coins can be identified by vending machines.

Alloys are mixtures of metals and other elements formulated to achieve desired properties. Generally, an alloy is harder and stronger than the pure metal but has lower electrical conductivity.

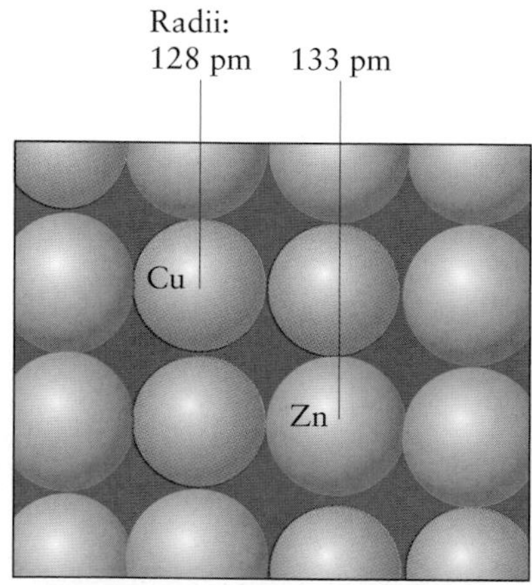

FIGURE 16.47 The atomic radii of copper and zinc are similar, and atoms of one element fit reasonably comfortably into the lattice of the other element to give the range of alloys we know as brass.

16.16 Magnetic Materials

Magnetism was observed by ancient people, who used lodestones, a form of magnetite, to steer ships by Earth's magnetic field. The property arises from the presence of unpaired electrons in elements and compounds (Section 16.1). Ferromagnetism is much stronger than paramagnetism, and ferromagnetic materials are used to make permanent magnets. They are also used in the coatings of cassette tapes and computer disks. In an **antiferromagnetic** material, neighboring spins are locked into an *antiparallel* arrangement; so the magnetic moments cancel. Manganese is antiferromagnetic.

(a)

(b)

FIGURE 16.48 (a) When a magnet is lowered over this viscous ferrofluid, the particles of iron(III) oxide begin to align themselves with the magnetic field. (b) Because strong attractions exist between the particles and the detergent molecules in the oil, the liquid is pulled into the field along with the particles. Some ferrofluid clings to the magnet from an earlier immersion.

Ceramic magnets, which are used in refrigerator magnets, are made of barium ferrite ($BaO \cdot nFe_2O_3$) or strontium ferrite ($SrO \cdot nFe_2O_3$). They are made by compressing the powdered ferrite in a magnetic field and heating it until it hardens. Because these magnets are ceramic, they are hard and brittle, with low densities. However, they are the most widely used magnets because of their low cost.

Ferrofluids are ferromagnetic liquids. They are suspensions of finely powdered magnetite, Fe_3O_4, in a viscous, oily liquid (such as mineral oil) that contains a detergent (such as oleic acid, a long-chain carboxylic acid). The iron oxide particles do not settle out because they are attracted to the polar ends of the detergent molecules, which form micelles that are dispersed in the oil. When a magnet is brought near a ferrofluid, the particles in the liquid try to align themselves with the magnetic field, but they are kept in place by the oil (Fig. 16.48). As a result, it is possible to control the flow and position of the ferrofluid by means of an applied magnetic field. One application of ferrofluids is in the braking systems of exercise machines. The stronger the magnetic field, the greater the resistance to motion.

Magnetic materials can be paramagnetic, ferromagnetic, or antiferromagnetic. In ferromagnetic materials, large domains of electrons have the same magnetic moment.

SKILLS YOU SHOULD HAVE MASTERED

1 Explain trends in chemical and physical properties among the *d*-block elements, Sections 16.1 and 16.2.

2 Identify common *d*-metals that are ferromagnetic, Sections 16.1 and 16.16.

3 Describe and write balanced equations for the principal reactions used to produce the elements in the first row (Period 4) of the *d* block and in Groups 11 and 12, Sections 16.3 and 16.4.

4 Describe the names, properties, and reactions of some of the principal compounds of the elements in the first row of the *d* block, Sections 16.3 and 16.4.

5 Name and write formulas for *d*-metal complexes, Toolbox 16.1 and Example 16.1.

6 Identify pairs of ionization, linkage, hydrate, coordination, geometrical, and optical isomers, Example 16.2 and Self-Tests 16.4 and 16.5.

7 Determine the ligand field splitting from the wavelength of light absorbed by a complex, Example 16.3.

8 Use the spectrochemical series to predict the effect of a ligand on the color, electron configuration, and magnetic properties of a *d*-metal complex, Examples 16.4 and 16.5.

9 Describe bonding in *d*-metal complexes in terms of ligand field theory, Section 16.13.

10 Describe the operation of a blast furnace and how steel is made by the basic oxygen process, Section 16.14.

11 Distinguish the types of magnetism, Section 16.6.

EXERCISES

Trends in Properties

16.1 Which members of the *d* block, those at the left or at the right of block, tend to have strongly negative standard potentials?

16.2 Name five elements in the *d* block that have positive standard potentials.

16.3 Identify the element with the larger atomic radius in each of the following pairs: (a) scandium and titanium; (b) copper and gold; (c) vanadium and niobium; (d) ruthenium and osmium.

16.4 Identify the element with the larger atomic radius in each of the following pairs: (a) cobalt and manganese; (b) copper and zinc; (c) silver and gold; (d) chromium and molybdenum.

16.5 Identify the element with the higher first ionization energy in each of the following pairs: (a) scandium and titanium; (b) nickel and copper; (c) iron and zinc; (d) iron and ruthenium; (e) ruthenium and osmium.

16.6 Identify the element with the higher first ionization energy in each of the following pairs: (a) iron and cobalt; (b) manganese and iron; (c) vanadium and chromium; (d) copper and silver; (e) silver and gold.

16.7 Explain why the density of mercury ($13.55\ g{\cdot}cm^{-3}$) is significantly higher than that of cadmium ($8.65\ g{\cdot}cm^{-3}$), whereas the density of cadmium is only slightly greater than that of zinc ($7.14\ g{\cdot}cm^{-3}$).

16.8 Explain why the density of vanadium ($6.11\ g{\cdot}cm^{-3}$) is significantly less than that of chromium ($7.19\ g{\cdot}cm^{-3}$). Both vanadium and chromium crystallize in a body-centered cubic lattice.

16.9 (a) Describe the trend in the stability of oxidation states moving down a group in the *d* block (for example, from chromium to molybdenum to tungsten). (b) How does this trend compare with the trend in the stabilities of oxidation states observed for the *p*-block elements upon moving down a group?

16.10 Which oxoanion, MnO_4^- or ReO_4^-, is expected to be the stronger oxidizing agent? Explain your choice.

16.11 Which of the elements vanadium, chromium, and manganese is most likely to form an oxide with the formula MO_3? Explain your answer.

16.12 Which of the elements zirconium, chromium, and iron is most likely to form a chloride with the formula MCl_4? Explain your answer.

Selected Elements: A Survey

16.13 Predict the major products of each of the following reactions and then balance the skeletal equations:
(a) $TiCl_4(s) + Mg(s) \xrightarrow{\Delta}$
(b) $CoCO_3(s) + HNO_3(aq) \rightarrow$
(c) $V_2O_5(s) + Ca(l) \xrightarrow{\Delta}$

16.14 Predict the major products of each of the following reactions and then balance the skeletal equations:
(a) $FeCr_2O_4(s) + C(s) \xrightarrow{\Delta}$
(b) $CrO_4^{2-}(s) + H_3O^+(aq) \rightarrow$
(c) $MnO_2(s) + Al(s) \xrightarrow{\Delta}$

16.15 Give the systematic names and chemical formulas of the principal component of (a) rutile; (b) hematite; (c) pyrolusite.

16.16 Give the systematic names and chemical formulas of the principal component of (a) magnetite; (b) pyrite; (c) ilmenite; (d) chromite.

16.17 Use Appendix 2B to predict the products of the reaction of 1 M HCl(aq) with the following metals: (a) vanadium; (b) mercury; (c) cobalt.

16.18 Use Appendix 2B to predict the products of the following reactions: (a) nickel with 1 M HCl(aq); (b) titanium with 1 M HCl(aq); (c) platinum with 1 M $KMnO_4$(aq) in 1 M HCl(aq).

16.19 Write chemical equations that describe the following processes: (a) the reduction of hematite in a blast furnace; (b) the reduction of titanium tetrachloride to titanium metal; (c) the removal of silica, SiO_2, from iron ore.

16.20 Write chemical equations that describe the following processes: (a) the production of chromium by the thermite reaction; (b) the corrosion of copper metal by carbon dioxide in moist air; (c) the purification of nickel, using carbon monoxide.

16.21 (a) Explain why the dissolution of a chromium(III) salt produces an acidic solution. (b) Explain why the slow addition of hydroxide ions to a solution containing chromium(III) ions first produces a gelatinous precipitate that subsequently dissolves with further addition of hydroxide ions. Write chemical equations showing these aspects of the behavior of chromium(III) ions.

16.22 (a) What is generally the first step is obtaining a metal from its sulfide ore? (b) What metal may be used to remove impurities of sulfur from metallic iron?

16.23 Outline a process, using chemical equations where possible, by which (a) zinc and (b) mercury can be produced.

16.24 Outline a process, using chemical equations where possible, by which copper is extracted and purified from chalcopyrite by the pyrometallurgical process.

16.25 Use the information in Appendix 2B to determine the equilibrium constant for the disproportionation of copper(I) ions in aqueous solution at 25°C to copper metal and copper(II) ions.

16.26 Use information in Appendix 2B to determine the equilibrium constant for the disproportionation of mercury(I) ions in aqueous solution at 25°C to mercury metal and mercury(II) ions.

Coordination Compounds

16.27 Name the following complex ions and determine the oxidation number of the metal: (a) $[Fe(CN)_6]^{4-}$; (b) $[Co(NH_3)_6]^{3+}$; (c) $[Co(CN)_5(H_2O)]^{2-}$; (d) $[Co(SO_4)(NH_3)_5]^+$.

16.28 Name the following complex ions and determine the oxidation number of the metal: (a) $[Fe(CN)_6]^{3-}$; (b) $[Fe(OH)(H_2O)_5]^{2+}$; (c) $[CoCl(NH_3)_4(H_2O)]^{2+}$; (d) $[Ir(en)_3)]^{3+}$.

16.29 Use the information in Table 16.4 to write the formula for each of the following coordination compounds: (a) potassium hexacyanochromate(III); (b) pentaamminesulfatocobalt(III) chloride; (c) tetraamminediaquacobalt(III) bromide; (d) sodium diaquabis(oxalato)ferrate(III).

16.30 Use the information in Table 16.4 to write the formula for each of the following coordination compounds: (a) triammineaquadihydroxochromium(III) chloride; (b) potassium tetrachloroplatinate(II); (c) tetraaquadichloronickel(IV) sulfate; (d) potassium tris(oxalato)rhodium(III); (e) chlorohydroxobis(oxalato)rhodate(III) octahydrate.

16.31 Which of the following ligands may be polydentate? If the ligand can be polydentate, give the maximum number of places on the ligand that can bind simultaneously to a single metal center. (a) $HN(CH_2CH_2NH_2)_2$; (b) CO_3^{2-}; (c) H_2O; (d) oxalate.

16.32 Which of the following ligands may be polydentate? If the ligand can be polydentate, give the maximum number of places on the ligand that can bind simultaneously to a single metal center. (a) Chloride ion; (b) cyanide ion; (c) ethylenediaminetetraacetate; (d) $N(CH_2CH_2NH_2)_3$.

16.33 Which of the following isomers of diaminobenzene can form chelating complexes?

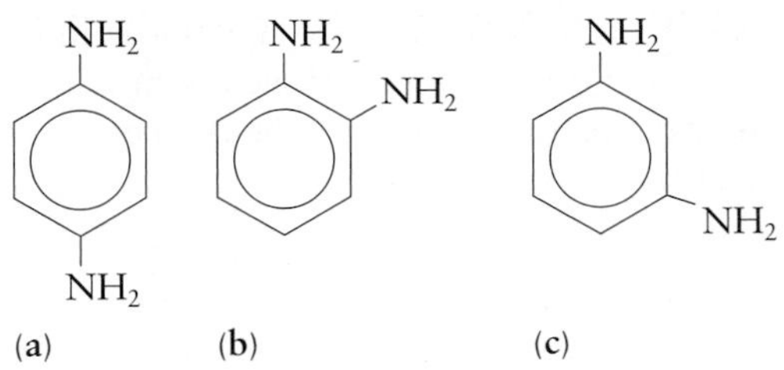

16.34 Which of the following ligands do you expect to form chelating complexes?

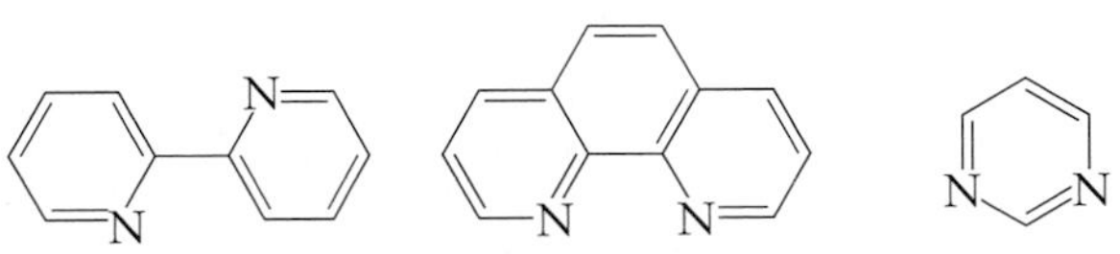

(a) Bipyridine (b) *ortho*-Phenanthroline (c) Pyrimidine

16.35 With the help of Table 16.4, determine the coordination number of the metal ion in the following complexes: (a) $[NiCl_4]^{2-}$; (b) $[Ag(NH_3)_2]^+$; (c) $[PtCl_2(en)_2]^{2+}$; (d) $[Cr(EDTA)]^-$.

16.36 With the help of Table 16.4, determine the coordination number of the metal ion in the following complexes: (a) $[Ir(en)_3)]^{3+}$; (b) $[Fe(ox)_3)]^{3-}$ (the oxalato ligand is bidentate); (c) $[PtCl_2(NH_3)_2]$; (d) $[Fe(CO)_5]$.

16.37 Identify the type of structural isomerism that exists in the following pairs of compounds:
(a) $[Co(NO_2)(NH_3)_5]Br_2$ and $[Co(ONO)(NH_3)_5]Br_2$
(b) $[Pt(SO_4)(NH_3)_4](OH)_2$ and $[Pt(OH)_2(NH_3)_4]SO_4$
(c) $[CoCl(SCN)(NH_3)_4]Cl$ and $[CoCl(NCS)(NH_3)_4]Cl$
(d) $[CrCl(NH_3)_5]Br$ and $[CrBr(NH_3)_5]Cl$

16.38 Identify the type of structural isomerism that exists in the following pairs of compounds:
(a) $[Cr(en)_3][Co(ox)_3]$ and $[Co(en)_3][Cr(ox)_3]$
(b) $[CoCl_2(NH_3)_4]Cl\cdot H_2O$ and $[CoCl(NH_3)_4(H_2O)]Cl_2$
(c) $[Co(CN)_5(NCS)]^{3-}$ and $[Co(CN)_5(SCN)]^{3-}$
(d) $[Pt(NH_3)_4][PtCl_6]$ and $[PtCl_2(NH_3)_4][PtCl_4]$

16.39 Write the formulas for the hydrate isomers of a compound having the formula $CoCl_3\cdot 6H_2O$ and a coordination number of six.

16.40 Write the formula of (a) a coordination isomer of $[Co(NH_3)_6][Cr(NO_2)_6]$; (b) a linkage isomer of $[CoCl(NO_2)(en)_2]Cl$; (c) an ionization isomer of $[CoCl(NO_2)(en)_2]Cl$.

16.41 Which of the following coordination compounds can have cis and trans isomers? If such isomerism exists, draw the two structures and name the compound. (a) $[CoCl_2(NH_3)_4]Cl\cdot H_2O$; (b) $[CoCl(NH_3)_5]Br$; (c) $[PtCl_2(NH_3)_2]$, a square planar complex.

16.42 Which of the following coordination compounds can have cis and trans isomers? If such isomerism exists, draw the two structures and label the compounds. (a) $[Fe(OH)(H_2O)_5]^{2+}$; (b) $[RuBr_2(NH_3)_4]^+$; (c) $[Co(NH_3)_4(H_2O)_2]^{3+}$.

16.43 (a) Draw the structure of the *cis*-diammine-*cis*-diaqua-*cis*-dichlorochromium(III) ion and comment on its isomerism. (b) Is this compound optically active?

16.44 (a) Draw the geometrical isomers of $[Cr(ox)_2(H_2O)_2]^-$. (b) Are any of these isomers optically active? If so, which ones?

16.45 Is either of the following complexes chiral? If both complexes are chiral, do they form an enantiomeric pair?

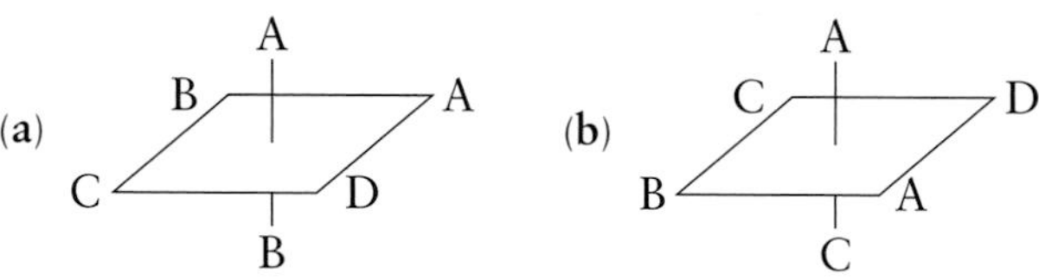

16.46 Is either of the following complexes chiral? If both complexes are chiral, do they form an enantiomeric pair?

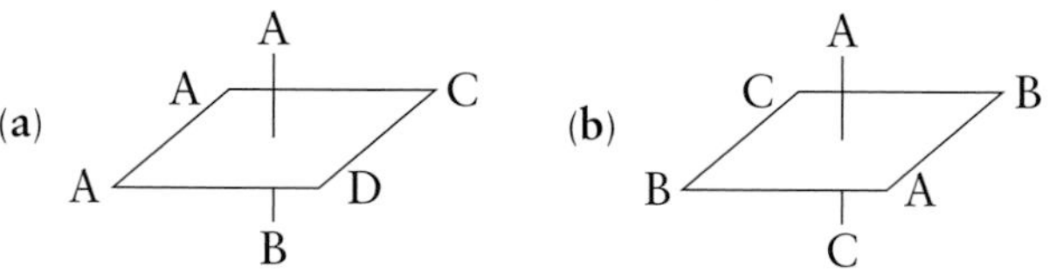

The Electronic Structures of Complexes

16.47 Determine the number of valence electrons present in the following metal ions: (a) Ti^{3+}; (b) Fe^{2+}; (c) Mn^{2+}; (d) Cr^{3+}; (e) Os^{2+}; (f) Pd^{4+}.

16.48 Determine the number of valence electrons present in the following metal ions: (a) V^{2+}; (b) Ru^{2+}; (c) Rh^{3+}; (d) Cu^{2+}; (e) Os^{4+}; (f) Co^{3+}.

16.49 Determine the number of valence electrons present in the following metal ions: (a) Ti^{2+}; (b) Tc^{2+}; (c) Ir^+; (d) Ag^+; (e) Y^{3+}; (f) Zn^{2+}.

16.50 Determine the number of valence electrons present in the following metal ions: (a) Nb^{3+}; (b) Mo^{3+}; (c) Rh^+; (d) Ni^{2+}; (e) Re^{4+}; (f) Au^{3+}.

16.51 Draw an orbital energy-level diagram (like those in Figs. 16.33 and 16.35), showing the configuration of *d*-electrons on the metal ion in the following complexes: (a) $[Co(NH_3)_6]^{3+}$; (b) $[NiCl_4]^{2-}$ (tetrahedral); (c) $[Fe(H_2O)_6]^{3+}$; (d) $[Fe(CN)_6]^{3-}$. Predict the number of unpaired electrons for each complex.

16.52 Draw an orbital energy-level diagram (like those in Figs. 16.33 and 16.35), showing the configuration of *d*-electrons on the metal ion in the following complexes: (a) $[Zn(H_2O)_6]^{2+}$; (b) $[CoCl_4]^{2-}$ (tetrahedral); (c) $[Co(CN)_6]^{3-}$; (d) $[CoF_6]^{3-}$. Predict the number of unpaired electrons for each complex.

16.53 The complexes (a) $[Co(en)_3]^{3+}$ and (b) $[Mn(CN)_6]^{3-}$ have low-spin electron configurations. How many unpaired electrons, if any, are there in each complex?

16.54 The complexes (a) $[FeF_6]^{3-}$ and (b) $[Co(ox)_3]^{4-}$ have high-spin electron configurations. How many unpaired electrons are there in each complex?

16.55 Explain the difference between a weak-field ligand and a strong-field ligand. What measurements can be used to classify them as such?

16.56 Describe the changes that may occur in a compound's properties when weak-field ligands are replaced by strong-field ligands.

16.57 Of the two complexes (a) $[CoF_6]^{3-}$ and (b) $[Co(en)_3]^{3+}$, one appears yellow and the other appears blue. Match the complex to the color and explain your choice.

16.58 In aqueous solution, water competes effectively with bromide ions for coordination to Cu^{2+} ions. The hexaaquocopper(II) ion is the predominant species in solution. However, in the presence of a large concentration of bromide ions, the solution becomes deep violet. This violet color is due to the presence of the tetrabromocuprate(II) ions, which are tetrahedral. This process is reversible, so the solution becomes light blue again upon dilution with water. (a) Write the formulas of the two complex ions of copper(II) that form. (b) Is the change in color from violet to blue upon dilution expected? Explain your reasoning.

16.59 Suggest a reason why Zn^{2+}(aq) ions are colorless. Would you expect zinc compounds to be paramagnetic? Explain your answer.

16.60 Suggest a reason why copper(II) compounds are often colored but copper(I) compounds are colorless. Which oxidation number results in paramagnetic compounds?

16.61 Estimate the ligand field splitting for (a) $[CrCl_6]^{3-}$ (λ_{max} = 740 nm); (b) $[Cr(NH_3)_6]^{3+}$ (λ_{max} = 460 nm); (c) $[Cr(H_2O)_6]^{3+}$ (λ_{max} = 575 nm), where λ_{max} is the wavelength of the most intensely absorbed light. (d) Arrange the ligands in order of increasing ligand field strength.

16.62 Estimate the ligand field splitting for (a) $[Co(CN)_6]^{3-}$ (λ_{max} = 295 nm); (b) $[Co(NH_3)_6]^{3+}$

(λ_{max} = 435 nm); (c) $[Co(H_2O)_6]^{3+}$ (λ_{max} = 540 nm), where λ_{max} is the wavelength of the most intensely absorbed light. (d) Arrange the ligands in order of increasing ligand field strength.

16.63 Which *d*-orbitals are used in forming σ-bonds between octahedral metal ions and ligands?

16.64 Which *d*-orbitals are used in forming π-bonds between octahedral metal ions and ligands?

16.65 Common terms used to describe ligands that can interact with a metal center by forming a π-type bond are *π-acid* and *π-base*. The definitions of π-acid and π-base are similar to those used for Lewis acidity. A π-acid is a ligand that can accept electrons by using π-type orbitals and a π-base donates electrons by using π-type orbitals. On the basis of these definitions, describe the following ligands as either π-acids, π-bases, or neither: (a) CN^-; (b) Cl^-; (c) H_2O; (d) en. (e) Place them in order of increasing ligand field splitting.

16.66 Describe the following ligands as either π-acids, π-bases, or neither (see Exercise 16.65): (a) NH_3; (b) ox; (c) F^-; (d) CO. (e) Place them in order of increasing ligand field splitting.

16.67 Is the best description of the t_{2g}-orbitals in $[Co(H_2O)_6]^{3+}$ (a) bonding; (b) antibonding; or (c) nonbonding? Explain how you reached that conclusion.

16.68 Is the best description of the t_{2g}-orbitals in $[Fe(CN)_6]^{3-}$ (a) bonding; (b) antibonding; or (c) nonbonding? Explain how you reached that conclusion.

16.69 Is the best description of the e_g-orbitals in $[Fe(H_2O)_6]^{3+}$ (a) bonding; (b) antibonding; or (c) nonbonding? Explain how you reached that conclusion.

16.70 Is the best description of the t_{2g}-orbitals in $[CoF_6]^{3-}$ (a) bonding; (b) antibonding; or (c) nonbonding? Explain how you reached that conclusion.

16.71 Using concepts of ligand field theory, explain why water is a weaker field ligand than ammonia.

16.72 Using concepts of ligand field theory, explain why ethylenediamine is a weaker field ligand than CO.

The Impact on Materials

16.73 (a) What reducing agent is used in the production of iron from its ore? (b) Write chemical equations for the production of iron in a blast furnace. (c) What is the major impurity in the product of the blast furnace?

16.74 (a) What is the purpose of adding limestone to a blast furnace? (b) Write chemical equations that show the reactions of lime in a blast furnace.

16.75 How is the impurity present in pig iron removed to create high-quality steel?

16.76 What metal is generally added to steel to make it corrosion resistant?

16.77 What metals are the primary components of bronze? (You might want to refer to Chapter 5.)

16.78 What metals are the primary components of brass? (You might want to refer to Chapter 5.)

16.79 How do alloys differ from the pure metals from which they are made?

16.80 What is the oxidation state of iron in the alkaline earth ferrites?

16.81 (a) Define ferrofluid. (b) Assign the oxidation number for the iron in Fe_3O_4, which is used to make ferrofluids.

16.82 Describe the differences between diamagnetism, paramagnetism, ferromagnetism, and antiferromagnetism.

INTEGRATED EXERCISES

16.83 *Magnetic susceptibility*, χ, is a measure of the number of unpaired electrons in a compound and varies with temperature, *T*. A graph of magnetic susceptibility as a function of temperature for a typical ferromagnetic material is shown on the right. The temperature labeled T_C is known as the Curie temperature. On one side of the Curie temperature, the compound is a simple paramagnetic material, and on the other side, the substance is ferromagnetic. From the graph, determine whether this compound is ferromagnetic above or below the Curie temperature. Explain your reasoning.

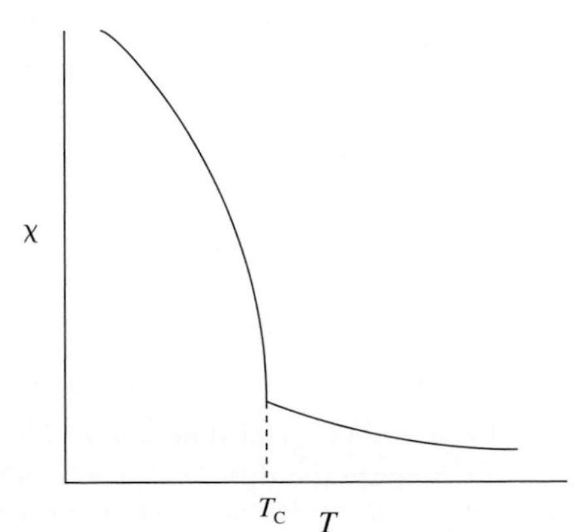

16.84 By considering electron configurations, suggest a reason why iron(III) compounds are readily prepared from iron(II), but the conversion of nickel(II) and cobalt(II) to nickel(III) and cobalt(III) is much more difficult.

16.85 (a) Draw all the possible isomers of the square-planar complex $[PtBrCl(NH_3)_2]$ and name each isomer. (b) How can the existence of these isomers be used to show that the complex is square planar rather than tetrahedral?

16.86 Draw the structures of the isomeric forms of $[CrClBrI(NH_3)_3]$. Which isomers are chiral?

16.87 Suggest a chemical test for distinguishing between (a) $[Ni(SO_4)(en)_2]Cl_2$ and $[NiCl_2(en)_2]SO_4$; (b) $[NiI_2(en)_2]Cl_2$ and $[NiCl_2(en)_2]I_2$.

16.88 Trigonal bipyramidal complexes in which a metal ion is surrounded by five ligands are more rare than octahedral or tetrahedral complexes, but many are known. Will trigonal bipyramidal compounds of the formula MX_3Y_2 exhibit isomerism? If so what types of isomerism are possible?

16.89 (a) Sketch the orbital energy-level diagrams for $[MnCl_6]^{4-}$ and $[Mn(CN)_6]^{4-}$. (b) How many unpaired electrons are present in $[Mn(CN)_6]^{4-}$? (c) Which complex transmits (that is, does not absorb) the longer wavelengths of incident electromagnetic radiation? Explain your reasoning.

16.90 The complex $[Co(CN)_6]^{3-}$ is pale yellow. (a) Is short- or long-wavelength visible radiation absorbed? (b) Is the ligand field splitting strong or weak? (c) How many unpaired electrons are present in the complex? (d) If ammonia molecules are substituted for cyanide ions as ligands, will the shift in absorbance of radiation be toward the blue or the red regions of the spectrum?

16.91 It makes sense to use the terms *low spin* and *high spin* only with metal ions having certain numbers of *d*-electrons. For other number of *d*-electrons only one electron configuration is possible. For what number of *d*-electrons are both high spin and low spin octahedral complexes possible?

16.92 (a) Which types of ligands are π-acid ligands in general (See Exercise 16.65), strong field or weak field? (b) Which types of ligands are π-base ligands in general, strong field or weak field?

16.93 Explain why high spin Mn^{2+} ions form complexes that are only faintly colored.

16.94 The hydride ion, H^-, can function as a ligand to *d*-block metals. Compare the differences in bonding to metal ions that you would expect between H^- and the halide ions.

16.95 Polydentate ligands do not necessarily need to bind a metal at all possible locations. In some cases, only one site is bound to the metal and the other one or more are left dangling. Such complexes are rare, and it has been necessary to develop special synthetic methods to prepare them. Explain in terms of thermodynamic arguments why polydentate ligands almost always form chelating complexes rather than leave dangling arms.

16.96 Suggest the form that the orbital energy-level diagram would take for a square-planar complex, and discuss how the building-up principle applies. *Hint*: The d_{z^2} orbital has more electron density in the *xy* plane than the d_{xz}- or d_{yz}-orbitals, but less than the d_{xy}-orbital.

16.97 Before the structures of octahedral complexes had been determined, various means were used to explain the fact that transition metal ions could bind to a greater number of ligands than expected on the basis of their charges. For example, Co^{3+} can bind to six ligands, not just three. An early theory attempted to explain this behavior by postulating that once three ligands had attached to a metal ion of charge +3, the others bonded to the attached ligands, forming chains of ligands. Thus, the compound $[Co(NH_3)_6]Cl_3$, which we now know to have the octahedral structure, would have been described as $Co(NH_3–NH_3–Cl)_3$. Show how the chain theory is not consistent with at least two properties of coordination compounds.

16.98 The relative thermodynamic stability of two complexes can be predicted from a comparison of their standard potentials. Determine which complex of the following pair is the more stable and state your conclusions about the relationship between the stability of a complex and (a) the oxidation number of the central atom; (b) the field strength of the attached ligands. Give your reasoning for each conclusion and suggest an explanation for the relationship.

$$[Co(NH_3)_6]^{3+}(aq) + e^- \longrightarrow [Co(NH_3)_6]^{2+}(aq) \qquad E^\circ = +0.11\text{ V}$$

$$[Co(H_2O)_6]^{3+}(aq) + e^- \longrightarrow [Co(H_2O)_6]^{2+}(aq) \qquad E^\circ = +1.81\text{ V}$$

16.99 Is there a correlation between the ligand field strength of the halide ions F^-, Cl^-, Br^-, and I^- and the electronegativity of the halogen? If so, can this correlation be explained by ligand field theory?

16.100 The measurement of optical rotation is a way of determining the purity of a solution of a given enantiomer. Such measurements are important in the pharmaceutical industrial, where it is often imperative that the appropriate enantiomer of a compound be prepared free of its mirror image because the mirror image molecule is at best ineffective or, in the worst case, highly toxic. A complex **A** has an optical rotation of 48.0 degrees$\cdot$(mol$\cdot$L^{-1})$^{-1}\cdot$cm^{-1} when pure. The units demonstrate that the rotation depends on the concentration of the sample and the pathlength of the cell in which the measurement is made. A student prepares **A** and separates it from its mirror image by using a selective crystallization technique. Upon dissolving **A** in solution, the sample is found to have a rotation of 46.5 degrees$\cdot$(mol$\cdot$L^{-1})$^{-1}\cdot$cm^{-1}. If the sample contains only **A** and its mirror image **A***, calculate the percentage of **A** in the sample.

16.101 Ligand field theory predicts that different types of metal ions will form more stable complexes with certain types of ligands. From your understanding of ligand field theory, predict what types of ligands (weak-field or strong-field) would form the most stable complexes with the early transition metals in their highest oxidation states. Likewise, predict the types of ligands that would form the most stable complexes of the late transition metals (those to the right of the *d* block) in their lowest oxidation states. Explain the reasoning behind your choices.

16.102 Explain as fully as possible, with diagrams of the structures of the hydrated ions, the changes that occur when (a) anhydrous copper(II) sulfate is moistened; (b) it is dissolved in water; (c) an excess of aqueous ammonia is added to the resulting solution.

16.103 View the unit cell contents for the 123 high-temperature superconductor $YBa_2Cu_3O_7$ available on the Web site. (a) Show that the empirical formula for this compound is correct by describing the locations of the various atoms. (b) Measure the dimensions of the unit cell. This may be done by downloading the .pdb file and examining it, using appropriate molecular modeling software. (c) What type of unit cell is adopted by this material?

16.104 View the unit cell for hematite available on the Web site. (a) What are the coordination numbers of the iron and oxygen atoms in hematite? (b) What are the coordination geometries of the iron and oxygen atoms? (c) How many iron atoms and oxygen atoms are present in the unit cell? (d) What type of unit cell does magnetite adopt? (e) Calculate the density of magnetite. The unit cell dimensions may be obtained by viewing the .pdb file with appropriate molecular modeling software.

16.105 *cis*-Platin is an anticancer drug with a structure that may be viewed on the Web site. (a) What is the formula and systematic name for the compound *cis*-Platin. (b) Draw any isomers that are possible for this compound. Label any isomers that are optically active. (c) What is the coordination geometry of the platinum atom?

16.106 The Web site contains a molecular model of an iridium complex with formula $IrCl(CO)_2\{P(C_6H_5)_3\}_2$. (a) View this structure and determine the coordination geometry of the iridium atom. (b) What is the oxidation number of the iridium atom? (c) Are isomers of this compound possible? If so, draw them. Label any that are optical isomers.

16.107 Zirconia is a very hard mineral that has been used as a synthetic gemstone. (a) From the unit cell, which may be viewed on the Web site, determine the empirical formula of zirconia. (b) What type of unit cell is this? (c) What is the coordination number of the zirconium atom? (d) What is the coordination number and coordination geometry of the oxygen atoms? (e) The structure of zirconia can be compared with that of diamond. In what ways are these two structures similar and in what ways do they differ?

16.108 In terms of Δ_O, how stable are the following transition metal complexes: (a) $[CoF_6]^{3-}$; (b) $[Co(CN)_6]^{4-}$; (c) $[Co(CN)_6]^{3-}$? Place the compounds in order of decreasing stability.

CHAPTER 17

Uranium hexafluoride, UF_6, shown here, is a relatively volatile molecular compound that is used in the separation of the naturally occurring isotopes uranium-235 and uranium-236. There is only one naturally occurring isotope of fluorine, so any difference in the molar mass of the molecule is due to the uranium.

Why Do We Need to Know This Material?

Nuclear chemistry is central to the development of nuclear energy. Chemistry provides techniques for purifying and recycling of nuclear fuels and for the disposal of hazardous radioactive waste. Nuclear chemistry is used in medicine to treat cancer and to produce images of organs inside living bodies. It is used in chemistry research to investigate reaction mechanisms and in archaeology to date ancient objects; nuclear chemistry also underlies the military strategies of many nations.

What Do We Need to Know Already?

Nuclear processes can be understood in terms of atomic structure (Section B and Chapter 1) and energy changes (Chapter 6). The section on rates of radioactive decay builds on chemical kinetics (particularly Sections 13.4 and 13.5).

NUCLEAR CHEMISTRY

Nuclear Decay

Nuclear Radiation

Nuclear Energy

So far, we have regarded the atomic nucleus as an unchanging passenger in chemical reactions and have concentrated on the changes that electrons undergo. However, nuclei can change, too; and **nuclear chemistry** explores the chemical consequences of those changes. One kind of change used as a source of energy is **nuclear fission,** the fragmentation of large nuclei into smaller nuclei, but the resulting radioactive waste presents awesome hazards to life. Another kind of change is **nuclear fusion,** the merging of small nuclei into larger ones, but the technique is technologically challenging and therefore expensive to develop. These changes in nuclear structure are critically important for the future of humanity, for they may answer one of the great challenges for the present and indefinitely into the future—the development of adequate energy supplies. The issues involved and chemistry's contributions to their solution are the subjects of this chapter.

This chapter has three main themes. First, we consider the pattern of nuclear stability and the manner in which unstable nuclei undergo change. Nuclear change is the source of all the elements except hydrogen, so we also discuss the origin of the elements. Next, we consider the consequences of nuclear change, and in particular the uses to which nuclear radiation is put in chemistry. Finally, we consider the origins of nuclear energy and how chemistry is used to resolve the problems associated with its use.

NUCLEAR DECAY

Atomic nuclei are extraordinary particles. They contain all the protons in the atom crammed together in a tiny volume, despite their positive charges (Fig. 17.1). Yet most nuclei survive indefinitely despite the immense repulsive forces that exist between the protons they contain. In some nuclei, though, the repulsions that protons exert on one another overcome the force that holds the nucleons together. Fragments of the nucleus are then ejected, and the nucleus is said to decay.

In this part of the chapter, we concentrate on the events accompanying the decay of nuclei, and specifically the types of radiation emitted by an unstable nucleus. Nuclear decay exhibits a variety of patterns, which we can use to predict the type of decay a nucleus is likely to undergo and the elements it forms as a result of its decay.

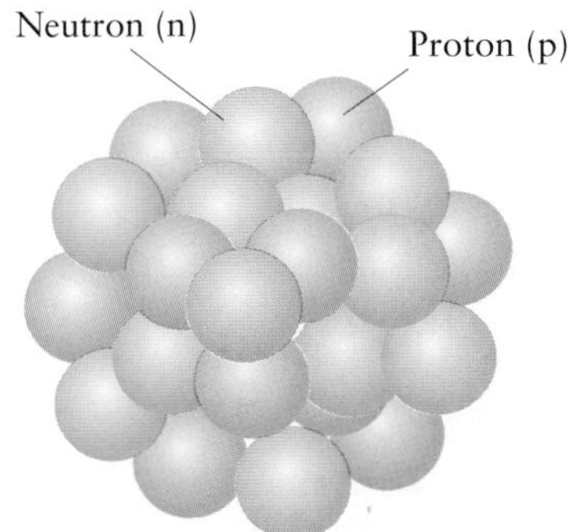

FIGURE 17.1 A nucleus can be pictured as a collection of tightly bonded protons (pink) and neutrons (gray). The diameter of a nucleus is about 10 fm (1 fm = 10^{-15} m).

17.1 The Evidence for Spontaneous Nuclear Decay

The first indications of nuclear instability came from the radiation nuclei emit as they decay. In 1896, the French scientist Henri Becquerel happened to store a sample of uranium oxide in a drawer that contained some photographic plates (Fig. 17.2). He was astonished to find that the uranium compound darkened the plates even though they were stored in an opaque covering. Becquerel realized that some kind of radiation must

FIGURE 17.2 Henri Becquerel discovered radioactivity when he noticed that an unexposed photographic plate left near some uranium oxide became fogged. This photograph shows one of his original plates annotated with his record of the event.

FIGURE 17.3 Marie Sklodowska Curie (1867–1934).

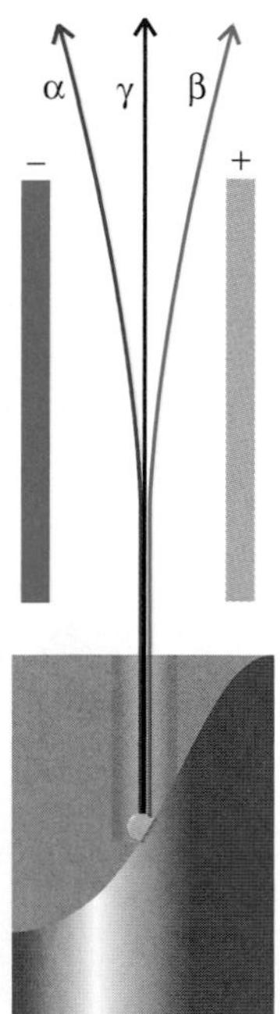

FIGURE 17.4 The effects of an electric field on nuclear radiation. The deflection identifies α rays as positively charged, β rays as negatively charged, and γ rays as uncharged.

be given off by the uranium compound. Marie Sklodowska Curie (Fig. 17.3), a young Polish doctoral student, showed that the radiation, which she called **radioactivity,** was consistently emitted by uranium regardless of the compound formed. She concluded that the source must be the uranium atoms themselves. Together with her husband, Pierre, she went on to show that thorium, radium, and polonium are also radioactive.

The origin of the rays was initially a mystery, for the existence of the atomic nucleus was unknown at the time. However, in 1898, Ernest Rutherford took the first step to discover their origin when he identified three different types of radioactivity by observing the effect of electric fields on radioactive emissions (Fig. 17.4). Rutherford called the three types α (alpha), β (beta), and γ (gamma) radiation.

When Rutherford allowed the radiation to pass between two electrically charged electrodes, he found that one type was attracted to the negatively charged electrode. He proposed that it consists of positively charged particles, which he called **α particles.** From the charge and mass of the particles, he was able to identify them as helium atoms that had lost two electrons. Once Rutherford had identified the atomic nucleus (in 1911, Section B), he realized that an α particle must be a helium nucleus, ${}^{4}_{2}He^{2+}$. An α particle is denoted ${}^{4}_{2}\alpha$, or simply α. We can think of it as a tightly bound cluster of two protons and two neutrons (Fig. 17.5).

Rutherford found that a second type of radiation was attracted to the positively charged electrode, and he proposed that it consists of a stream of negatively charged particles. By measuring the charge and mass of these particles, he showed that they are electrons. The rapidly moving electrons emitted by nuclei are called **β particles** and denoted β^-. Because a β particle has no protons or neutrons, its mass number is 0.

The third common type of radiation that Rutherford identified, **γ radiation,** is not affected by an electrical field. It is electromagnetic radiation like light, but of much higher frequency—greater than about 10^{20} Hz, and corresponding to wavelengths less than about 1 pm. Gamma radiation can be regarded as a stream of very high energy photons; each photon is emitted by a single nucleus as that nucleus discards energy. The frequencies, ν, of γ rays are given by the relation $\nu = \Delta E/h$ (see Section 1.2), where ΔE is the energy discarded by the nucleus. The frequency is very high because the energy difference between the excited and ground nuclear states is very large.

We shall see that heavy elements are more likely to give off α radiation, whereas β radiation is more typical of lighter elements. Both α and β radiation are often accompanied by γ radiation, because the new nucleus may be formed with its nucleons in a high-energy arrangement (Fig. 17.6). The γ ray photon is emitted when the nucleons settle down into a state of lower energy.

Since Rutherford's work, scientists have identified other types of nuclear radiation. Some consist of rapidly moving particles, such as neutrons or protons. Others consist of rapidly moving **antiparticles,** particles with a mass equal to that of one of the subatomic particles, but with an opposite charge. For example, the positron has the same mass as an electron but a positive charge; it is denoted β^+. When an antiparticle encounters the corresponding particle, both particles are annihilated—completely converted to energy. Table 17.1 summarizes the properties of particles commonly found in nuclear radiation.

The most common types of radiation emitted by, radioactive nuclei are α particles (the nuclei of helium atoms), β particles (fast electrons ejected from the nucleus), and γ rays (high-energy electromagnetic radiation).

FIGURE 17.5 An α particle has two positive charges and a mass number of 4. It consists of two protons and two neutrons, and is the same as the nucleus of a helium-4 atom.

The annihilation of a positron by an electron is utilized in some medical applications (see Box 17.2).

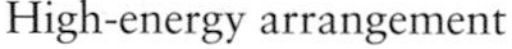

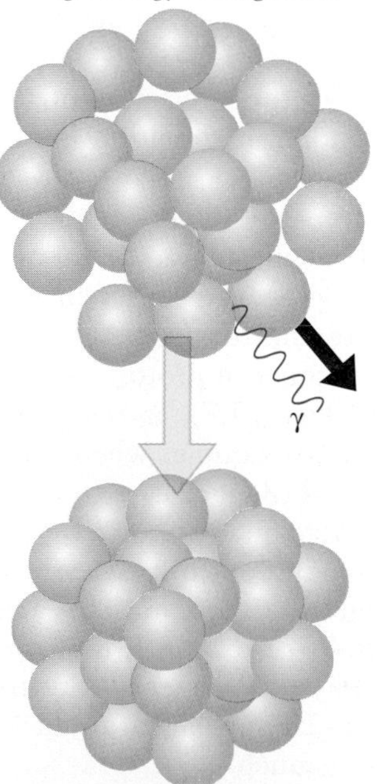

FIGURE 17.6 After a nucleus decays, the nucleons remaining in the nucleus may be left in a high-energy state, as shown by the loose arrangement in the upper part of the illustration. As the nucleons adjust to a lower energy arrangement (bottom), the excess energy is released as a γ ray photon.

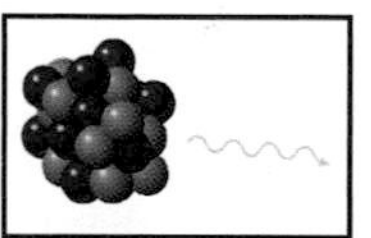

View an animation of gamma decay.

Media Link **Animation:** Fig. 17.6, Gamma decay

TABLE 17.1 Nuclear Radiation

Type	Degree of penetration	Speed*	Particle†	Example
α	not penetrating but damaging	10% of c	helium-4 nucleus $^{4}_{2}He^{2+}$, $^{4}_{2}\alpha$, α	$^{226}_{88}Ra \rightarrow ^{222}_{86}Rn + ^{4}_{2}\alpha$
β	moderately penetrating	less than 90% of c	electron $^{0}_{-1}e$, β^{-}, β	$^{3}_{1}H \rightarrow ^{3}_{2}He + ^{0}_{-1}e$
γ	very penetrating; often accompanies other radiation	c	photon	$^{60}_{27}Co^{\ddagger} \rightarrow ^{60}_{27}Co + \gamma$
β^{+}	moderately penetrating	less than 90% of c	positron $^{0}_{+1}e$, β^{+}	$^{22}_{11}Na \rightarrow ^{22}_{10}Ne + ^{0}_{+1}e$
p	moderate/low penetration	10% of c	proton $^{1}_{1}H^{+}$, $^{1}_{1}p$, p	$^{53}_{27}Co \rightarrow ^{52}_{26}Fe + ^{1}_{1}p$
n	very penetrating	less than 10% of c	neutron $^{1}_{0}n$, n	$^{137}_{53}I \rightarrow ^{136}_{53}I + ^{1}_{0}n$

*c is the speed of light.
†Alternative symbols are given for the particles; often it is sufficient to use the simplest (the one on the right).
‡An energetically excited state of a nucleus is usually denoted by an asterisk.

17.2 Nuclear Reactions

Recall from Section B that the isotopes of an element have nuclei with the same number of protons but different numbers of neutrons.

Becquerel's, Curie's, and Rutherford's discoveries showed that radioactivity is produced by **nuclear decay,** the partial breakup of a nucleus. The change in the composition of a nucleus is called a **nuclear reaction.** Recall from Section B that nuclei are composed of protons and neutrons that are collectively called *nucleons*; a specific nucleus with a given atomic number and mass number is called a **nuclide.** Thus, ^{1}H, ^{2}H, and ^{16}O are three different nuclides; the first two being isotopes of the same element. Nuclei that change their structure spontaneously and emit radiation are called **radioactive.** Often the result is a different nuclide.

Nuclear reactions differ in some important ways from chemical reactions. First, different isotopes of the same element undergo essentially the same chemical reactions, but their nuclei undergo very different nuclear reactions. Second, when α or β particles are emitted from the nucleus, they leave behind a nucleus with a different number of protons. The product, which is called the **daughter nucleus** (Fig. 17.7), is therefore the nucleus of an atom of a different element. For example, when a radon-222 nucleus emits an α particle, a polonium-218 nucleus is formed. In this case, a **nuclear transmutation,** the conversion of one element into another, has taken place. Another important difference between nuclear and chemical reactions is that energy changes are very much greater for nuclear reactions than for chemical reactions. For example, the combustion of 1.0 g of methane produces about 52 kJ of energy as heat. In contrast, a nuclear reaction of 1.0 g of uranium-235 produces about 8.2×10^{7} kJ of energy, more than a million times as much.

To predict the identity of a daughter nucleus, we note how the atomic number and mass number change when the parent nucleus ejects a particle. For example, when a radium-226 nucleus, with $Z = 88$, undergoes **α decay,** it emits an α particle. The fragment remaining is a nucleus of atomic number 86 (radon) and mass number 222; so the daughter nucleus is radon-222:

$$^{226}_{88}\mathrm{Ra} \longrightarrow {}^{222}_{86}\mathrm{Rn} + {}^{4}_{2}\alpha$$

The expression for these changes is called a **nuclear equation.** Toolbox 17.1 describes how to use nuclear equations to identify daughter nuclei.

Nuclear reactions may result in the formation of different elements. The transmutation of a nucleus can be predicted by noting the atomic numbers and the mass numbers in the nuclear equation for the process.

TOOLBOX 17.1 How to identify the products of a nuclear reaction

Conceptual Basis

Once the numbers and types of particles emitted from a nucleus are known, the product nucleus can be identified by balancing mass numbers and atomic numbers in the nuclear equation.

Procedure

Step 1 Identify the type of decay and write the nuclear reaction, with the mass number and the atomic number of the daughter nucleus written as A and Z, respectively.

Step 2 Find the values of A and Z for the daughter nucleus from the requirement that the sums of the mass numbers and the atomic numbers, remain unchanged in the decay.

Step 3 Once A and Z are known, identify the daughter nucleus by referring to the periodic table.

The masses and charges of the particles commonly ejected or captured by nuclei are as follows:

Process	Mass number of particle	Charge	Example	Illustration
α decay	4	+2	$^{211}_{84}\mathrm{Po} \rightarrow {}^{207}_{82}\mathrm{Pb} + {}^{4}_{2}\alpha$	Fig. 17.7
β decay	0	−1	$^{24}_{11}\mathrm{Na} \rightarrow {}^{24}_{12}\mathrm{Mg} + {}^{0}_{-1}\mathrm{e}$	Fig. 17.8
electron capture	0	−1	$^{44}_{22}\mathrm{Ti} + {}^{0}_{-1}\mathrm{e} \rightarrow {}^{44}_{21}\mathrm{Sc}$	Fig. 17.9
positron emission	0	+1	$^{43}_{22}\mathrm{Ti} \rightarrow {}^{43}_{21}\mathrm{Sc} + {}^{0}_{+1}\mathrm{e}$	Fig. 17.10
proton emission	1	+1	$^{57}_{30}\mathrm{Zn} \rightarrow {}^{56}_{29}\mathrm{Cu} + {}^{1}_{1}\mathrm{p}$	—
neutron emission	1	0	$^{91}_{34}\mathrm{Se} \rightarrow {}^{90}_{34}\mathrm{Se} + {}^{1}_{0}\mathrm{n}$	—

The following examples show how to balance nuclear equations involving each of these types of decay. More information on these particles is given in Table 17.1.

How can we use this information . . . to predict the outcome of α decay?

To identify the nuclide produced by α decay of polonium-211, we write the nuclear equation for the reaction:

$$^{211}_{84}\mathrm{Po} \longrightarrow {}^{A}_{Z}\mathrm{E} + {}^{4}_{2}\alpha$$

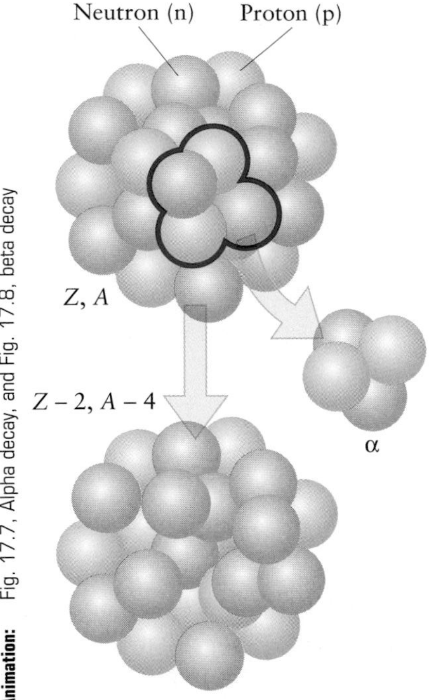

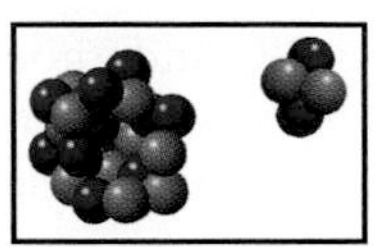

View an animation of alpha decay.

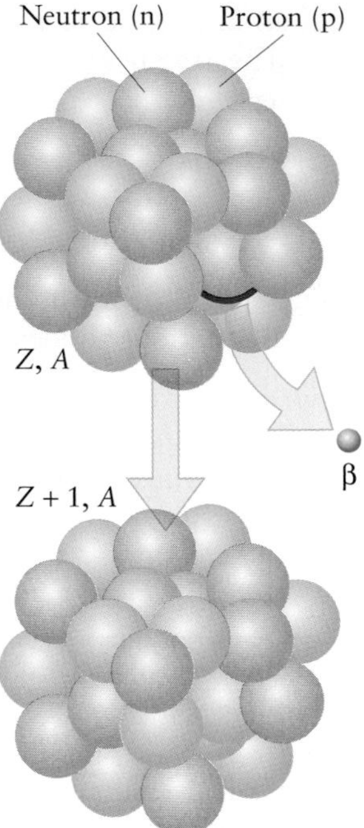

View an animation of beta decay.

Media Link Animation: Fig. 17.7, Alpha decay, and Fig. 17.8, beta decay

FIGURE 17.7 When a nucleus ejects an α particle, the atomic number of the atom decreases by 2 and the mass number decreases by 4. The nucleons ejected from the upper nucleus have been indicated by the blue boundary.

FIGURE 17.8 When a nucleus ejects a β particle, the atomic number of the atom increases by 1 and the mass number remains unchanged. The neutron that we can regard as the source of the electron is indicated by the blue boundary in the upper nucleus.

Because 211 = $A + 4$, we can write $A = 207$; because 84 = $Z + 2$, $Z = 82$. The element with $Z = 82$ is lead, Pb, so the nuclide produced is lead-207; the nuclear equation for the α decay is therefore

$$^{211}_{84}\text{Po} \longrightarrow {}^{207}_{82}\text{Pb} + {}^{4}_{2}\alpha$$

Self-Test 17.1A Identify the nuclide produced by α decay of uranium-235.

[***Answer:*** Thorium-231]

Self-Test 17.1B Identify the nuclide produced by α decay of thorium-232.

How can we use this information . . . to predict the outcome of β decay?

The loss of one negative charge when an electron is ejected from the nucleus (see Fig. 17.8) can be interpreted as the conversion of a neutron into a proton within the nucleus.

β decay is accompanied by the emission of an antineutrino (a particle with zero charge and very small mass). We are ignoring this complication.

$$^{1}_{0}\text{n} \longrightarrow {}^{1}_{1}\text{p} + {}^{0}_{-1}\text{e}, \quad \text{or more simply, n} \longrightarrow \text{p} + \beta$$

Because the daughter nucleus is left with an additional proton, its atomic number is greater by 1 than that of the parent nucleus. The mass

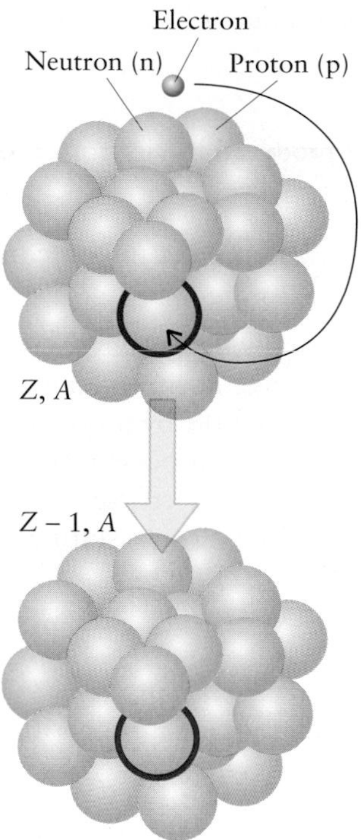

FIGURE 17.9 In electron capture, a nucleus captures one of the surrounding electrons. The effect is to convert a proton (outlined in red) into a neutron (outlined in blue). As a result, the atomic number decreases by 1 but the mass number remains the same.

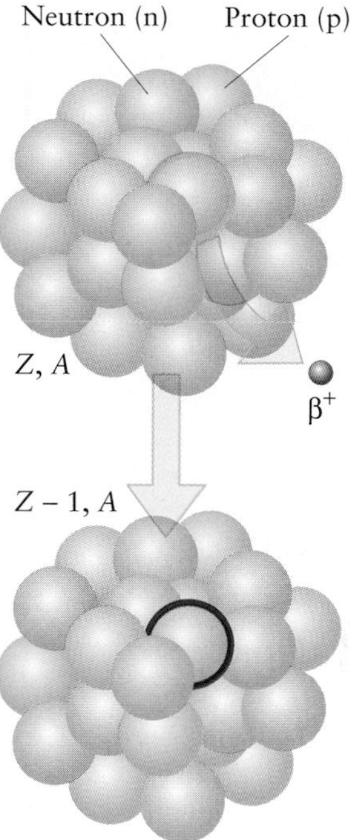

FIGURE 17.10 In positron (β^+) emission, the nucleus ejects a positron. The effect is to convert a proton into a neutron (outlined in blue). As a result, the atomic number decreases by 1 but the mass number remains the same.

number is unchanged, because the total number of nucleons in the nucleus is unchanged. For example, when sodium-24 undergoes β decay, the daughter nucleus is an atom of the element with atomic number 12 (magnesium), but with the same mass number as the parent nucleus:

$$^{24}_{11}\text{Na} \longrightarrow {}^{24}_{12}\text{Mg} + {}^{0}_{-1}\text{e}$$

Self-Test 17.2A Identify the nuclide produced by β decay of lithium-9.

[*Answer:* $^{9}_{4}\text{Be}$]

Self-Test 17.2B Identify the nuclide produced by β decay of radium-228.

How can we use this information . . . to predict the outcome of electron capture?

In electron capture, a nucleus captures one of the surrounding electrons; a proton is turned into a neutron and, although there is no change in mass number, the atomic number is reduced by 1 (see Fig. 17.9). An example is the nuclear reaction

$$^{41}_{20}\text{Ca} + {}^{0}_{-1}\text{e} \longrightarrow {}^{41}_{19}\text{K}$$

Self-Test 17.3A Identify the nuclide produced by electron capture by beryllium-7.

[*Answer:* ${}^{7}_{3}Li$]

Self-Test 17.3B Identify the nuclide produced by electron capture by iron-55.

How can we use this information . . . to predict the outcome of positron emission?

A positron is denoted ${}^{0}_{+1}e$ or, more simply, β^+; it has the same tiny mass as an electron but a single positive charge. Positron emission can be thought of as the positive charge shrugged off by a proton as it is converted into a neutron (the reverse of the process responsible for β^- generation). As a result, the atomic number decreases by 1 but there is no change in mass number (see Fig. 17.10). The resulting change in the nucleus is the same as that for electron capture:

$${}^{15}_{8}O \longrightarrow {}^{15}_{7}N + {}^{0}_{+1}e$$

Self-Test 17.4A Identify the nuclide that results from positron emission by sodium-22.

[*Answer:* ${}^{22}_{10}Ne$]

Self-Test 17.4B Identify the nuclide that results from positron emission by carbon-11.

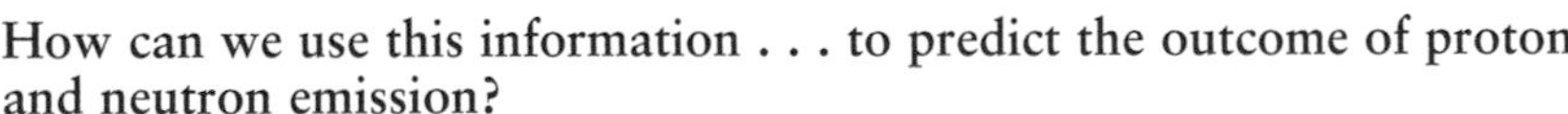

How can we use this information . . . to predict the outcome of proton and neutron emission?

Proton and neutron emission are less common and tend to occur only in special cases. Loss of a proton decreases both mass number and atomic number by 1. Loss of a neutron decreases only the mass number by 1:

$${}^{53}_{27}Co \longrightarrow {}^{52}_{26}Fe + {}^{1}_{1}p$$

$${}^{137}_{53}I \longrightarrow {}^{136}_{53}I + {}^{1}_{0}n$$

Self-Test 17.5A Identify the nuclide that results from proton emission by scandium-39.

[*Answer:* Calcium-38]

Self-Test 17.5B Identify the nuclide that results from proton emission by iodine-109.

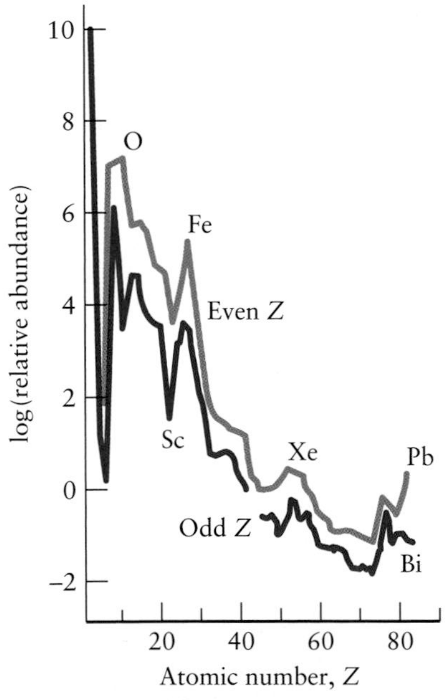

FIGURE 17.11 The variation in cosmic nuclear abundance with atomic number. Note that elements with even atomic numbers (brown curve) are consistently more abundant than neighboring elements with odd atomic numbers (blue curve).

17.3 The Pattern of Nuclear Stability

Is there a pattern to the stabilities and instabilities of nuclei? The existence of a pattern would allow us to make predictions about the modes of nuclear decay. One clue is that elements with even atomic numbers are consistently more abundant than neighboring elements with odd atomic numbers. We can see this difference in Fig. 17.11, which is a plot of the cosmic abundance of the elements against atomic number. The same pattern occurs on Earth. Of the eight elements present as 1% or more of the mass of the Earth, only one, aluminum, has an odd atomic number.

Nuclei with even numbers of both protons and neutrons are more stable than those with any other combination. Conversely, nuclei with odd numbers of both protons and neutrons are the least stable (Fig. 17.12). Nuclei are more likely to be stable if they are built from certain numbers of either kind of nucleons. These numbers, namely, 2, 8, 20, 50, 82, 114, 126, and 184, are called **magic numbers**. For example, there are 10 stable isotopes of tin ($Z = 50$), the most of any element, but only two stable isotopes of its neighbor antimony ($Z = 51$). The α particle itself is a "doubly magic" nucleus, with two protons and two neutrons. We shall see later that a number of actinides decay through a series of steps until they reach $^{208}_{82}Pb$, another doubly magic nuclide, with 126 neutrons and 82 protons. This pattern of nuclear stability is similar to the pattern of electron stability in atoms: the noble gas atoms have 2, 10, 18, 36, 54, and 86 electrons.

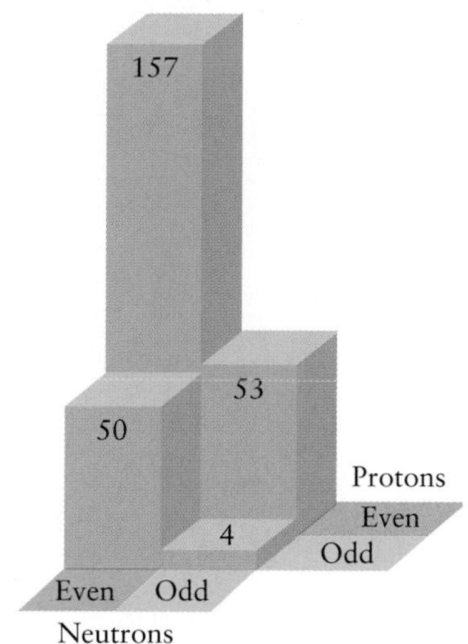

FIGURE 17.12 The numbers of stable nuclides for even and odd numbers of neutrons and protons. With the exception of hydrogen, by far the greatest number of stable nuclides have even numbers of protons and of neutrons.

Figure 17.13 is a plot of mass number against atomic number for known nuclides. Stable nuclei are found in a **band of stability** surrounded by a **sea of instability**, the region of unstable nuclides that decay with the emission of radiation. For atomic numbers up to about 20, the stable nuclides have approximately equal numbers of neutrons and protons, so A is close to $2Z$. For higher atomic numbers, all known nuclides—both stable and unstable—have more neutrons than protons, so $A > 2Z$.

The increase in the ratio of neutrons to protons with increasing atomic number can be explained by considering the role of the neutrons and the repulsions between protons. The **strong force** that holds the protons and neutrons together in a nucleus is powerful enough to overcome the repulsion between the protons, but it can act only over a very short distance—about the diameter of a nucleus (Fig. 17.14). Because neutrons are

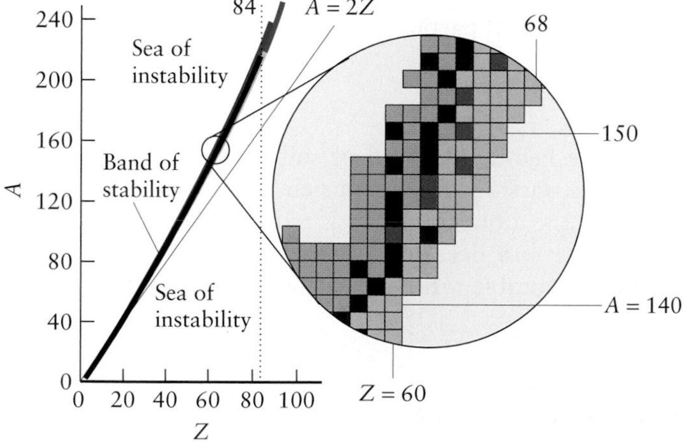

FIGURE 17.13 The manner in which nuclear stability depends on the atomic number and the mass number. Nuclides along the narrow black band (the band of stability) are generally stable. Nuclides in the blue region are likely to emit a β particle, and those in the red region are likely to emit an α particle (provided $Z \geq 84$). Nuclei in the pink region are likely either to emit positrons or to undergo electron capture. (b) The inset shows a magnified view of the diagram near $Z = 60$.

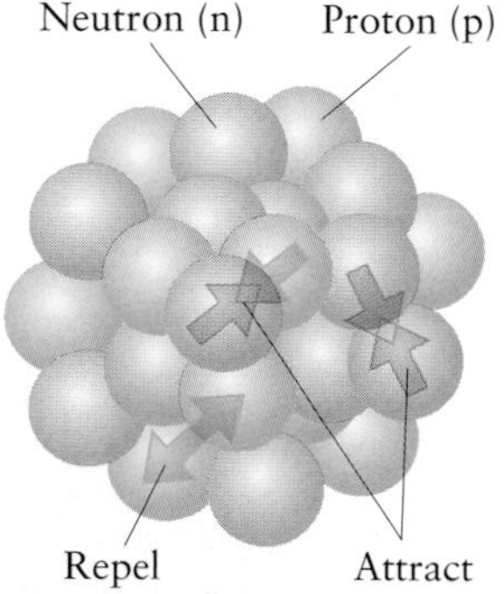

FIGURE 17.14 The protons in a nucleus repel each other electrically, but the strong force, which acts between all nucleons, holds the nucleus together.

uncharged, they can contribute to the strong force without adding to the electrostatic repulsion. A lot of neutrons are needed to overcome the mutual repulsion of the protons in a nucleus of high atomic number, which explains why the band of stability curves upward.

Models of nuclei have grown in sophistication as knowledge has accumulated. One of the simplest models was suggested by Niels Bohr, the Danish scientist who contributed a great deal to our understanding of atomic structure. Bohr compared the nucleus to a drop of liquid. His **liquid drop model** proposes that nucleons are packed together like the molecules in a liquid. Nucleons at the surface of the nucleus are less strongly bound and more easily lost. A slightly more sophisticated model explains the existence of magic numbers by treating nucleons like electrons in an atom. In this **independent particle model**, the nucleons are described by quantum numbers and occupy shells within the nucleus. Like electrons in atomic orbitals, most of the nucleons occupy inner shells and those in the outermost shell are most easily lost through radioactive decay. The **collective model** combines aspects of the two previous theories. In this model, the nucleons occupy quantized energy levels but interact with one another by the strong force and the electrostatic force.

Nuclei with certain even numbers of protons and of neutrons are the most stable. One model of the nucleus is the collective model, which pictures nucleons as occupying quantized energy levels and interacting with one another strongly.

17.4 Predicting the Type of Nuclear Decay

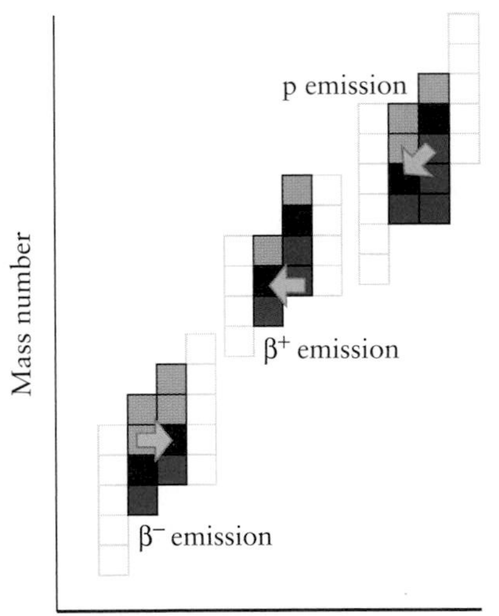

FIGURE 17.15 Three different ways of reaching the band of stability (black). Nuclei that are neutron rich (blue) tend to convert neutrons into protons by β emission; nuclei that are proton rich (red) tend to reach stability (black) by emitting a positron, capturing an electron, or emitting a proton.

We can use Fig. 17.13 to predict the type of disintegration a radioactive nuclide is likely to undergo. Nuclei that lie above the band of stability are **neutron rich**: they have a high proportion of neutrons. These nuclei tend to decay in such a way that the final n/p ratio is closer to that found in the band of stability. For example, a ${}^{14}_{6}C$ nucleus can reach safety (a more stable state) by ejecting a β particle, which reduces the n/p ratio (Fig. 17.15):

$${}^{14}_{6}C \longrightarrow {}^{14}_{7}N + {}^{0}_{-1}e$$

Nuclides that lie below the band of stability have a low proportion of neutrons and are classified as **proton rich**. These isotopes tend to decay in such a way that the atomic number is reduced. For example, proton-rich ${}^{29}_{15}P$ is radioactive and decays by emitting a positron, which gives the nucleus a more favorable ratio of neutrons to protons:

$${}^{29}_{15}P \longrightarrow {}^{29}_{14}Si + {}^{0}_{+1}e$$

Electron capture also decreases the proton count of proton-rich nuclides:

$${}^{7}_{4}Be + {}^{0}_{-1}e \longrightarrow {}^{7}_{3}Li$$

Proton emission is a third way of reducing the proton count:

$${}^{43}_{21}Sc \longrightarrow {}^{42}_{20}Ca + {}^{1}_{1}p + \gamma$$

Very few nuclides with $Z < 60$ emit α particles. All nuclei with $Z > 83$ are unstable and decay mainly by α particle emission. Nuclides of elements with $Z > 83$ must discard protons to reduce their atomic number, and they generally need to lose neutrons, too. These nuclei decay in a stepwise manner and give rise to a **radioactive series**, a characteristic sequence of nuclides (Fig. 17.16). First, one α particle is ejected, then another α particle or a β^- particle is ejected, and so on, until a stable nucleus is formed—usually the final nuclide is an isotope of lead (the element with the magic atomic number 82). For example, the uranium-238 series ends at lead-206, the uranium-235 series ends at lead-207, and the thorium-232 series ends at lead-208.

The pattern of nuclear stability can be used to predict the likely mode of radioactive decay: neutron-rich nuclei tend to reduce their neutron count; proton-rich nuclei tend to reduce their proton count. In general, only heavy nuclides emit α particles.

Self-Test 17.6A Which of the processes, (a) electron capture, (b) proton emission, (c) β^- emission, (d) β^+ emission, might a $^{145}_{64}Gd$ nucleus undergo to begin to reach stability? Refer to Figs. 17.13 and 17.15.

[*Answer:* a, b, d]

Self-Test 17.6B Which of the same set of processes listed in Self-Test 17.6A might a $^{148}_{58}Ce$ nucleus undergo to begin to reach stability?

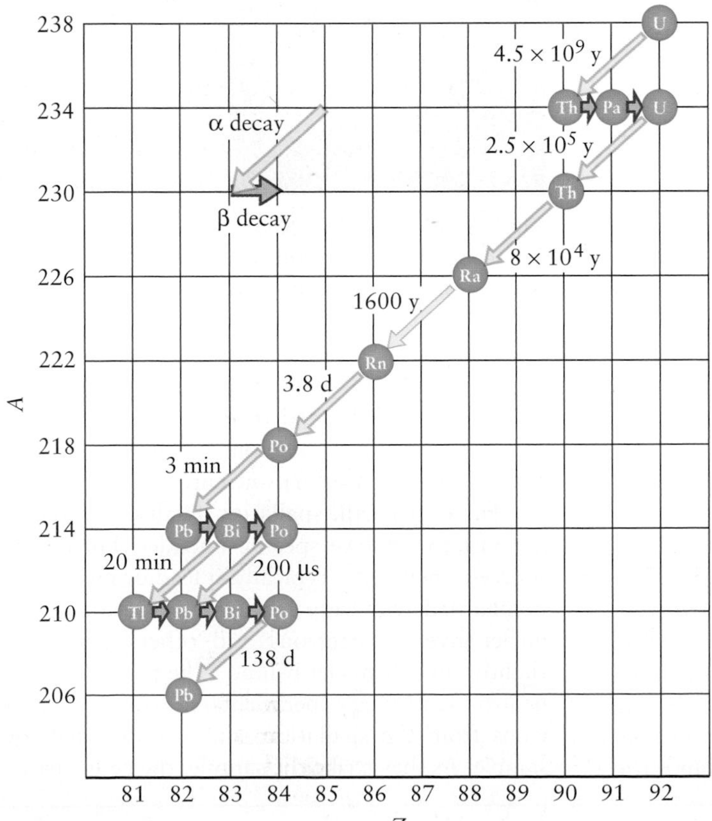

FIGURE 17.16 The uranium-238 decay series. The times are the half-lives of the nuclides (see Section 17.7).

17.5 Nucleosynthesis

Nucleosynthesis is the formation of elements. Chemical synthesis involves the rearrangement of atoms; nucleosynthesis involves the rearrangement of nucleons. Hydrogen and helium were produced in the Big Bang; all other elements are descended from these two, either as a result of nuclear reactions in stars or as a result of nuclear reactions taking place in space (Box 17.1). Only traces of some elements—among them technetium and promethium—are found on Earth. Although these elements were made in stars, their short lifetimes did not allow them to survive long enough to contribute to the formation of our planet. However, nuclides that are too unstable to be found on Earth can be made by artificial techniques, and scientists have added about 2200 different nuclides to the 300 or so that occur naturally.

BOX 17.1 FRONTIERS OF CHEMISTRY: SPALLATION

A process used by astronomers to determine the age of the universe is also making possible superconducting train tracks, weather satellites, compact discs, adjustable automobile seats, artificial blood vessels, and better ice cream. The process is *spallation*, in which very high energy particles slam into nuclei, splitting off smaller nuclei and subatomic particles. The term *spallation* comes from the little-used English word *spall*, meaning "a chip or splinter." In astronomy, it is the smaller nuclei produced by spallation that are of interest; in terrestrial research, it is often the neutrons ejected by spallation that are of interest.

The discovery of spallation has led to clues to the age of the universe. The story begins with the puzzling chemical composition of the universe. Why is it that some of the lightest elements, such as lithium, beryllium, and boron exist at all? Their very fragile nuclei cannot survive the high temperatures inside stars, where heavier elements, such as carbon, nitrogen, and oxygen, are formed. Scientists believe that the nuclei of these three light elements, and possibly deuterium too, are chips that cosmic rays have struck from bigger nuclei in the process of spallation.

The observed abundance of light elements can be used to deduce some of the properties of cosmic rays, which are fast-moving particles such as electrons and protons. The abundances of elements such as lithium, beryllium, and boron suggest that each proton has to travel 3 million light years and along the way pass through a quantity of matter equivalent to an ordinary brick wall before it reaches the Earth. The distance traveled suggests that the path of the proton is highly convoluted as it twists and turns in response to magnetic and electric fields throughout our galaxy, and that cosmic ray protons are very ancient.

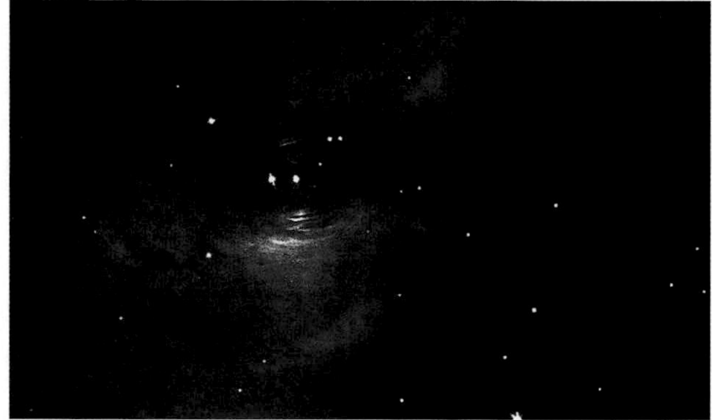

This image of the Crab Nebula taken by the Hubble Space Telescope shows enormous interstellar gas clouds, in which spallation reactions may be taking place.

Here on Earth, spallation facilities are being built, not just to produce specific nuclides, but to provide a source of neutrons. Typically, a lead or mercury target is bombarded with very high energy protons. The target nuclei give off neutrons and other particles, leaving slightly smaller nuclei behind. The neutrons are used in neutron scattering experiments, in which a beam of neutrons from the spallation source is directed toward a sample. As they strike the sample, the neutrons scatter in

To overcome the energy barriers to nuclear synthesis, particles must collide vigorously with one another (Fig. 17.17), as they do in stars. Therefore, to make elements artificially, we need to simulate the conditions found inside a star. If it is traveling fast enough, a proton, an α particle, or another positively charged nucleus has enough kinetic energy to overcome the electrostatic repulsion from the nucleus. The incoming particle penetrates into the nucleus, where it is captured by the strong force. The high speeds required can be achieved in a particle accelerator.

The **transmutation** of elements, the conversion of one element into another, particularly of lead into gold, was the dream of the alchemists and one of the roots of modern chemistry. However, the alchemists had access only to chemical techniques, which proved ineffective because they involved energy changes far too weak to force nucleons into nuclei.

The reflector shield in the nuclear spallation facility at Los Alamos National Laboratory is used to optimize and direct the flow of neutrons.

all directions. The characteristics of the scattering depend on the sample, allowing its structure to be determined in much the same way as in x-ray diffraction. Because neutrons are sensitive to light atoms such as hydrogen and oxygen atoms, they are particularly useful in determining the structures of biological materials and high-temperature ceramic superconductors. They can also be used to study polymers such as those used in the manufacture of compact discs, the complex structures of emulsions (hence, their contribution to the improvement of ice cream), and cell membranes.

HOW MIGHT YOU CONTRIBUTE?

The structures of many materials and biomolecules can be studied with neutron scattering. For example, the shape of the active site on an enzyme could be determined and used to manufacture new drugs. Spallation facilities are used to study the mineralization of bone and the development of soft materials such as artificial blood vessels and viscous liquids such as epoxies, glues, and paints. In addition, a study of intergalactic spallation events could also help us understand the nature of the universe.

For Further Reading: E. Parizot, "Superbubbles and the galactic evolution of Li, Be, and B," *Astron. Astrophys.* (2000), 362:786–798. Information about the Spallation Neutron Source at Oak Ridge National Laboratory and the booklet, *Advancing the Frontiers of Science and Technology*, U. S. Department of Energy: 2000, are available at www.ornl.gov/sns/.

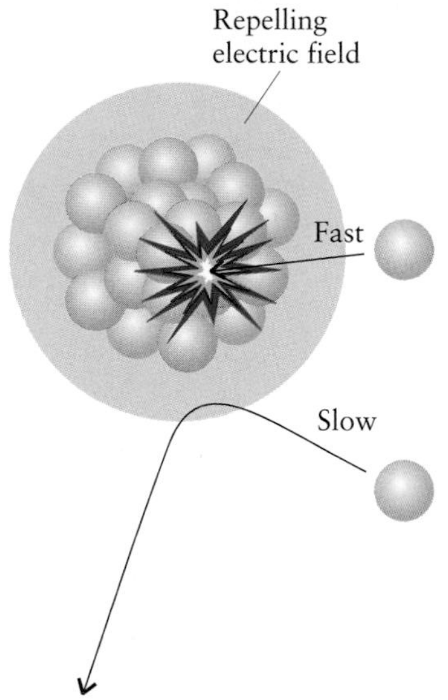

FIGURE 17.17 When a positively charged particle approaches a nucleus, it is repelled strongly. However, a very fast particle can reach the nucleus before the repulsion turns it aside, and a nuclear reaction may result.

Transmutation has now been recognized in nature and achieved in the laboratory, but by using methods undreamed of by the early alchemists. Rutherford achieved the first artificial nuclear transmutation in 1919. He bombarded nitrogen-14 nuclei with high-speed α particles. The products of the transmutation were oxygen-17 and a proton:

$$^{14}_{7}\text{N} + {}^{4}_{2}\alpha \longrightarrow {}^{17}_{8}\text{O} + {}^{1}_{1}\text{P}$$

Processes like this are commonly written simply as

$$^{14}_{7}\text{N}(\alpha,\text{p})^{17}_{8}\text{O}$$

or, in general,

Target(incoming particle, ejected particle)product

A similar process occurs in stars and results in the formation of oxygen-16 from carbon-12:

$$^{12}_{6}\text{C} + {}^{4}_{2}\alpha \longrightarrow {}^{16}_{8}\text{O} + \gamma, \quad \text{or} \quad {}^{12}_{6}\text{C}(\alpha,\gamma)^{16}_{8}\text{O}$$

A large number of nuclides have been synthesized on Earth. For instance, technetium was prepared (as technetium-97) for the first time on Earth in 1937 by the reaction between molybdenum and deuterium nuclei:

$$^{97}_{42}\text{Mo} + {}^{2}_{1}\text{H} \longrightarrow {}^{97}_{43}\text{Tc} + 2\,{}^{1}_{0}\text{n}$$

Technetium is now moderately abundant because it accumulates in the decay products of nuclear power plants. Another isotope, technetium-99, has pharmaceutical applications, particularly for bone scans (Box 17.2).

A neutron can get close to a target nucleus more easily than a proton can. Because a neutron has no charge and hence is not repelled by the nuclear charge, it need not be accelerated to such high speeds. An example of **neutron-induced transmutation** is the formation of cobalt-60, which is used in the radiation treatment of cancer. The three-step process starts from iron-58. First, iron-59 is produced:

$$^{58}_{26}\text{Fe} + {}^{1}_{0}\text{n} \longrightarrow {}^{59}_{26}\text{Fe}$$

The second step is the β decay of iron-59 to cobalt-59:

$$^{59}_{26}\text{Fe} \longrightarrow {}^{59}_{27}\text{Co} + {}^{0}_{-1}\text{e}$$

In the final step, cobalt-59 absorbs another neutron from the incident neutron beam and is converted into cobalt-60:

$$^{59}_{27}\text{Co} + {}^{1}_{0}\text{n} \longrightarrow {}^{60}_{27}\text{Co}$$

The overall reaction is

$$^{58}_{26}\text{Fe} + 2\,{}^{1}_{0}\text{n} \longrightarrow {}^{60}_{27}\text{Co} + {}^{0}_{-1}\text{e} \quad \text{or} \quad {}^{58}_{26}\text{Fe}(2\text{n},\beta^{-})^{60}_{27}\text{Co}$$

BOX 17.2 WHAT HAS THIS TO DO WITH . . . STAYING ALIVE?

NUCLEAR MEDICINE

Medical x-rays provided one of the first applications of radioisotopes. In 1914, when the wounded from World War I were pouring into Paris, Marie Curie converted a Renault car into the first mobile radiological unit and drove it from hospital to hospital. Radioisotopes are now widely used in medicine to diagnose, study, and treat illness. A physician can determine, for example, how and at what rate the thyroid gland takes up iodine by using iodine-131 as a radioactive tracer, and cobalt-60 is used to kill rapidly growing cancer cells.

Nuclear chemistry has transformed medical diagnosis. Radioactive tracers are used to measure organ function, sodium-24 to monitor blood flow, and strontium-87 to study bone growth. However, the most dramatic impact of radioisotopes on diagnosis has been in the field of imaging. Technetium-99 is the most widely used radioactive nuclide in medicine, especially for bone scans. This isotope is highly active and gives off γ rays that pass quickly through the body. Gamma rays cause much less damage than α particles would, and the isotope is so short-lived that the risk to the patient is minimal.

Positron emission tomography (PET) makes use of a positron emitter such as fluorine-18 to image human tissue in a degree of detail not possible with x-rays. They have been used extensively to study brain function (see illustration) and in medical diagnosis. For example, when the hormone estrogen is labeled with fluorine-18 and injected into a cancer patient, the fluorine-bearing compound is preferentially absorbed by the tumor. The positrons given off by the fluorine atoms are quickly annihilated when they meet electrons. The resulting γ rays are detected by a scanner that moves slowly over the part of the body containing the tumor. The growth of the tumor can be quickly and accurately estimated with this technique. A PET imaging facility must be located near a cyclotron, so that the positron emitters can be quickly incorporated into the desired compounds as soon as they are created.

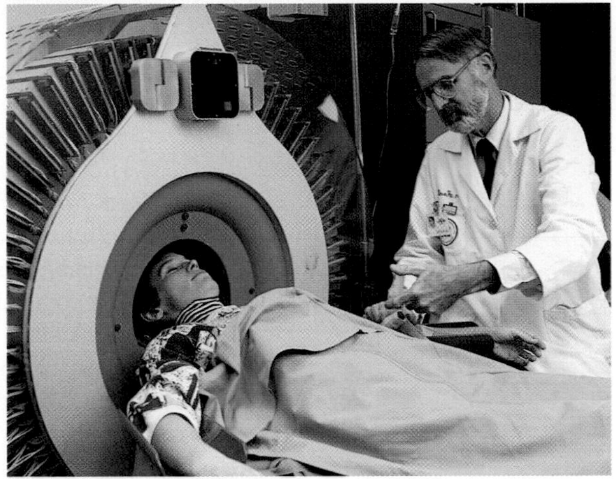

This patient is about to undergo a PET scan of brain function.

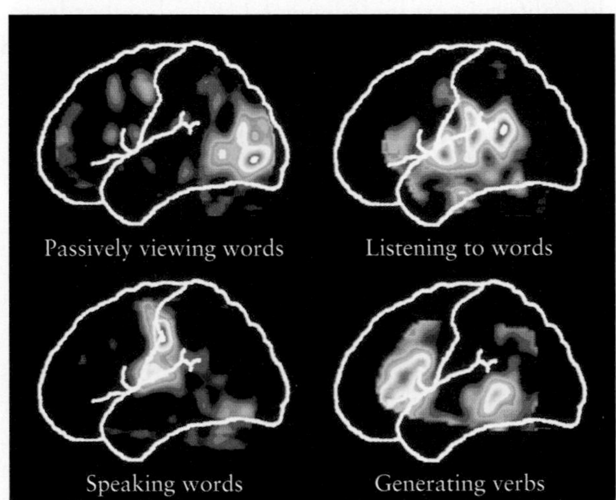

These four PET scans show how blood flow to different parts of the brain is affected by various activities. In this case, an oxygen isotope that is taken up by the hemoglobin in blood is used as a source of positrons.

Several kinds of cancer therapy use radiation to destroy malignant cells. *Boron neutron capture therapy* is unusual in that boron-10, the isotope injected, is not radioactive. However, when boron-10 is bombarded with neutrons, it gives off highly destructive α particles. In boron neutron capture therapy, the boron-10 is incorporated into a compound that is preferentially absorbed by tumors. The patient is then exposed to brief periods of neutron bombardment. As soon as the bombardment ceases, the boron-10 stops generating α particles.

For Further Reading: R. F. Barth, A. H. Soloway, and R. G. Fairchild, "Boron neutron capture therapy for cancer," *Scientific American* (October, 1990) 100. J. Raloff, "Wanted: medical isotopes," *Science News* (October 23, 1999), 156: 264.

TABLE 17.2 Notation for the Systematic Nomenclature of Elements*

Digit	Prefix	Abbreviation
0	nil	n
1	un	u
2	bi	b
3	tri	t
4	quad	q
5	pent	p
6	hex	h
7	sep	s
8	oct	o
9	enn	e

*For instance, element 123 would be named unbitriium, Ubt.

Self-Test 17.7A Complete the following nuclear reactions: (a) ? + $^{4}_{2}\alpha \rightarrow$ $^{243}_{96}Cm + ^{1}_{0}n$; (b) $^{242}_{96}Cm + ^{4}_{2}\alpha \rightarrow ^{245}_{98}Cf$ + ?.

[*Answer:* (a) $^{240}_{94}Pu$; (b) $^{1}_{0}n$]

Self-Test 17.7B Complete the following nuclear reactions: (a) $^{250}_{98}Cf$ + ? $\rightarrow ^{257}_{103}Lr + 4\,^{1}_{0}n$; (b) ? + $^{12}_{6}C \rightarrow ^{254}_{102}No + 4\,^{1}_{0}n$.

The **transuranium elements** are the elements following uranium in the periodic table. The elements from rutherfordium (Rf, $Z = 104$) through meitnerium (Mt, $Z = 109$) were formally named in 1997. The **transmeitnerium elements,** the elements beyond meitnerium (including hypothetical nuclides that have not yet been made) are named systematically, at least until they have been identified and there is international agreement on a permanent name. Their systematic names use the prefixes in Table 17.2, which identify their atomic numbers, with the ending *-ium*. For example, the element with $Z = 111$, one atom of which was first made in 1994, will be called unununium, Uuu, until it is finally named.

All the elements beyond $Z = 94$ are synthetic and are produced by the bombardment of target nuclei with a smaller projectile. For example, in 1996, German scientists made two atoms of element 112, ununbiium (Uub), by bombarding lead-208 atoms with zinc-70 atoms in a heavy-ion accelerator:

$$^{208}_{82}Pb + ^{70}_{30}Zn \longrightarrow ^{277}_{112}Uub + ^{1}_{0}n$$

The zinc atoms were accelerated to a relatively low speed so that the huge lead nucleus would not be torn apart by the impact. As we might expect from lighter nuclides, elements with even atomic numbers are more easily produced than those with odd atomic numbers. Element 114 was first reported by a team of Russian and American scientists in 1999. Later that year, nuclear chemists in California created element 118 by bombardment of lead-208 with krypton-86 atoms. The unstable atoms of element 118 then decayed by α emission to element 116, then element 114. In one process, two new elements were created.

New elements and isotopes of known elements are made by nucleosynthesis; the repulsive electrical forces of like-charged particles are overcome when very fast particles collide.

NUCLEAR RADIATION

Nuclear radiation is sometimes called **ionizing radiation**, because it is energetic enough to eject electrons from atoms. Hospitals use nuclear radiation to kill unwanted tissue, such as cancerous cells (see Box 17.2). Yet the same powerful effects that help to diagnose and cure illness can also damage healthy tissue. The damage depends on the strength of the source, the type of radiation, and the length of exposure.

In the following three sections, we examine the effects of nuclear radiation on living tissue. That application raises the questions of how we measure the exposure of organisms to radiation and how the intensity of radiation varies with time. To answer the latter question, we need to use some of the information about rate laws that we encountered in the discussion of chemical kinetics. Finally, we explore some of the applications of radioactive isotopes and see the kind of information they are used to obtain.

17.6 The Biological Effects of Radiation

The three principal types of nuclear radiation penetrate matter to different extents. Table 17.3 compares the penetration power of each type.

Relatively massive, highly charged α particles interact so strongly with matter that they slow down, capture electrons from surrounding matter, and change into bulky helium atoms before traveling very far. They penetrate only the first layer of skin and can be stopped even by a sheet of paper. Most α radiation is absorbed by the surface layer of dead skin, where it can do little harm. However, α particles can be extremely dangerous if inhaled or ingested. The energy of their impact can knock atoms out of molecules, which can lead to serious illness and death. For example, plutonium, considered to be one of the most toxic radioactive materials, is an α emitter and can be handled safely with minimal shielding. However, it is easily oxidized to Pu^{4+}, which has chemical properties similar to those of Fe^{3+}. Plutonium can take the place of iron in the body and become absorbed into bone, where it destroys the body's ability to produce red blood cells. The result is radiation sickness, cancer, or even death.

Second in penetrating power is β radiation. These fast electrons can penetrate about 1 cm into flesh before their electrostatic interactions with the electrons and nuclei of molecules bring them to a standstill.

Most penetrating of all is γ radiation. The high-energy γ ray photons can pass right through buildings and bodies, and cause damage by ionizing the molecules in their path. Protein molecules and DNA that have been damaged in this way can no longer function, and the result can be radiation sickness and cancer. Intense sources of γ rays must be surrounded by shields built from lead bricks or thick concrete to absorb this penetrating radiation.

The **absorbed dose** of radiation is the energy deposited in a sample (in particular, the human body) when it is exposed to radiation. The SI unit of absorbed dose is the **gray**, Gy, which corresponds to an energy deposit of $1\ J{\cdot}kg^{-1}$. The original unit used for reporting dose was the **rad**, the amount of radiation that deposits 10^{-2} J of energy per kilogram of tissue, so 1 rad $= 10^{-2}$ Gy. A dose of 1 rad corresponds to a 65-kg person absorbing a total of 0.65 J. This is not very much energy: it is enough to

The name *rad* stands for "radiation absorbed dose."

TABLE 17.3 Shielding Requirements of α, β, and γ Radiation

Radiation	Relative penetrating power	Shielding required
α	1	paper, skin
β	100	3 mm aluminum
γ	10 000	concrete, lead

boil only 0.2 mg of water. However, the energy of a particle of nuclear radiation is highly localized, like the impact of a subatomic bullet. As a result, the incoming particles can break individual bonds as they collide with molecules in their path.

The extent of radiation damage to living tissue depends on the type of radiation and the type of tissue. We must therefore include the **relative biological effectiveness**, *Q*, when assessing the damage that a given dose of each type of radiation may cause. For β and γ radiation, *Q* is set arbitrarily at about 1; but for α radiation, *Q* is close to 20. A dose of 1 Gy of γ radiation causes about the same amount of damage as 1 Gy of β radiation, but 1 Gy of α particles is about 20 times more damaging (even though these particles are the least penetrating). The precise figures depend on the total dose, the rate at which the dose accumulates, and the type of tissue, but these values are typical.

A further factor of 5 (that is, 5*Q*) is used for bone under certain circumstances.

The **dose equivalent** is the actual dose modified to take into account the different destructive powers of the various types of radiation in combination with various types of tissue. It is obtained by multiplying the actual dose (in gray) by the value of *Q* for the radiation type. The result is expressed in the SI unit called a **sievert** (Sv):

$$\text{Dose equivalent (Sv)} = Q \times \text{absorbed dose (Gy)} \qquad (1)$$

The former (non-SI) unit of dose equivalent was the **roentgen equivalent man** (rem), which was defined in the same way as the sievert but with the absorbed dose in rads; thus, 1 rem = 10^{-2} Sv.

A dose of 0.3 Gy (30 rad) of γ radiation corresponds to a dose equivalent of 0.3 Sv (30 rem), enough to cause a reduction in the number of white blood cells (the cells that fight infection), but 0.3 Gy of α radiation corresponds to 6 Sv (600 rem), which is enough to cause death. A typical average annual dose equivalent that we each receive from natural sources, called **background radiation**, is about 2 $\text{mSv}\cdot\text{y}^{-1}$, but this figure varies, depending on our lifestyle and where we live. About 20% of background radiation comes from our own bodies. About 30% comes from cosmic rays (a mix of γ rays and high-energy subatomic particles from outer space) that continuously bombard the Earth, and 40% comes from radon seeping out of the ground. The remaining 10% is a result largely of medical diagnoses (a typical chest x-ray gives a dose equivalent of about 0.07 mSv). Emissions from nuclear power plants and other nuclear facilities contribute about 0.1% in countries where they are widely used.

The principal source of radioactivity in the human body is potassium-40. About 35 000 potassium-40 nuclei disintegrated in your body while you were reading this sentence.

Human health in the presence of radiation is monitored by reporting the absorbed dose and the dose equivalent; the latter takes into account the effects of different types of radiation on tissues.

17.7 Measuring the Rate of Nuclear Decay

Geiger counters and scintillation counters are used to measure the rate at which a radioactive nucleus decays (Box 17.3). Each click of a Geiger counter or flash of a phosphor in a scintillation counter indicates that one nuclear disintegration has been detected. The **activity** of a sample is the number of nuclear disintegrations that occur per second. The SI unit of

BOX 17.3 HOW DO WE KNOW . . . HOW RADIOACTIVE A MATERIAL IS?

The ability of nuclear radiation to eject electrons from atoms and ions can be used to measure its intensity. Becquerel first gauged the intensity of radiation by determining the degree to which it blackened a photographic film. The blackening results from the same redox processes as those occurring in ordinary photography, such as

$$Ag^+ + Br^- \xrightarrow{\text{light}} Ag + Br$$

except that the initial oxidation of the bromide ions is caused by nuclear radiation instead of light. Becquerel's technique is still used in film badges like the one pictured, which monitor the exposure of workers to radiation.

A *Geiger counter* monitors radiation by detecting the ionization of a low-pressure gas, as shown in the illustration. The radiation ionizes atoms of the gas inside a cylinder and allows a brief flow of current between the electrodes. The resulting electrical signal can be recorded directly or converted into an audible click. The frequency of the clicks indicates the intensity of the radiation. A limitation of Geiger counters is that they do not respond well to γ rays. Only about 1% of the γ ray photons are detected, whereas all the β particles incident on the counter are detected. Because the efficiency of a Geiger counter depends on the size of the tube, a counter used to monitor a wide range of activities usually contains two tubes of different sizes.

A *scintillation counter* makes use of the fact that phosphors, phosphorescent substances such as sodium iodide and zinc sulfide (see Section 15.14), give a flash of light—a scintillation—when exposed to radiation. The counter also contains a *photomultiplier tube*, which converts light into an electrical signal. The intensity of the radiation is determined from the strength of the electronic signal.

A *dosimeter* is used to collect cumulative evidence of exposure to radiation and is worn as a badge. Dosimeters contain a *thermoluminescent* material such as lithium fluoride. Any incident radiation ionizes the fluoride ions by knocking electrons out of their orbitals. The electrons migrate away from the fluorine atoms but are trapped in the crystal. When the crystal is heated, they fall back into the fluorine atoms and give off the difference of energy as light. The dose of radiation received is determined by the intensity of the light. Dosimeters can be used over a period of time ranging from one day to several weeks, because the excited electrons accumulate with continued exposure, allowing a long-term dosage to be evaluated.

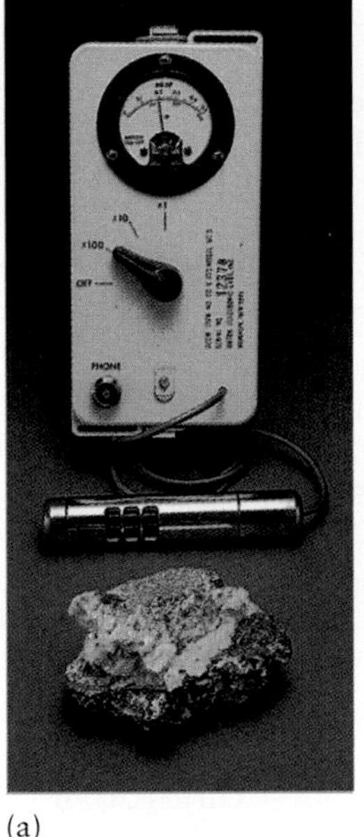
(a)

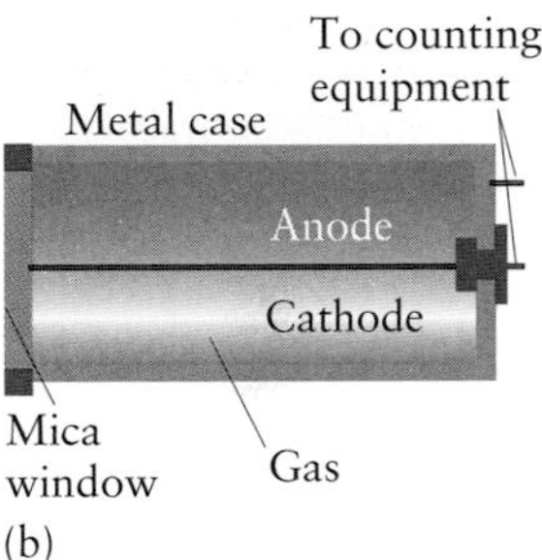

(b)

(a) A Geiger counter with a piece of uranium ore. (b) The detector in a Geiger counter contains a gas (often argon and a little ethanol vapor, or neon with some bromine vapor) in a cylinder with a high potential difference (500–1200 V) between the central wire and the walls. When radiation ionizes the gas, the ions allow current to flow briefly, resulting in the characteristic click.

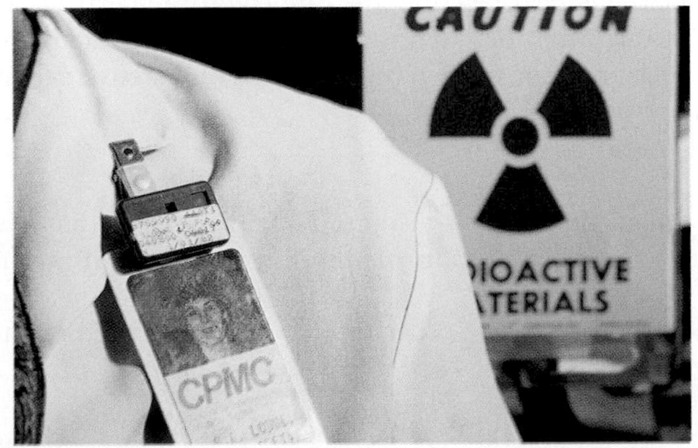

A film badge worn by workers in areas where radiation exposure may occur indicates the extent of the exposure, should any have occurred.

TABLE 17.4 Radiation Units*

Property	Unit name	Symbol	Definition
activity	curie	Ci	3.7×10^{10} disintegrations per second
	becquerel	Bq	1 disintegration per second
absorbed dose	radiation absorbed dose	rad	10^{-2} J·kg^{-1}
	gray	Gy	1 J·kg^{-1}
dose equivalent	roentgen equivalent man	rem	$Q \times$ absorbed dose†
	sievert	Sv	100 rem

*The SI units are becquerel, gray, and sievert.
†Q is the relative biological effectiveness of the radiation. Normally, $Q \approx 1$ for γ, β, and most other radiation, but $Q \approx 20$ for α radiation and fast neutrons. A further factor of 5 (that is, $5Q$) is used for bone under certain circumstances.

activity is the **becquerel** (Bq): 1 Bq is equal to one nuclear disintegration per second. Another commonly used (non-SI) unit of radioactivity is the **curie** (Ci). It is equal to 3.7×10^{10} nuclear disintegrations per second, the radioactive output of 1 g of radium-226. Because the curie is a very large unit, most activities are expressed in millicuries or microcuries. Table 17.4 summarizes these units.

The equation for the decay of a nucleus is

$$\text{Parent nucleus} \longrightarrow \text{daughter nucleus} + \text{radiation}$$

This expression has exactly the same form as a unimolecular elementary reaction (Section 13.10), with an unstable nucleus taking the place of an excited molecule. This type of decay is expected for a process that does not depend on any external factors, but only on the instability of the nucleus. The rate of nuclear decay depends only on the identity of the isotope, not on its chemical form or temperature.

As in a unimolecular chemical reaction, the rate law for nuclear decay is first order. That is, the relation between the rate of decay and the number N of radioactive nuclei present is given by the **law of radioactive decay:**

$$\text{Activity} = \text{rate of decay} = k \times N \qquad (2)$$

In this context, k is called the **decay constant.** The law tells us that the activity of a radioactive sample is proportional to the number of atoms in the sample.

As we saw in Section 13.4, a first-order rate law implies an exponential decay. It follows that the number N of nuclei remaining after a time t is given by

$$N = N_0 e^{-kt} \qquad (3)$$

where N_0 is the number of radioactive nuclei present initially (at $t = 0$); this expression is plotted in Fig. 17.18.

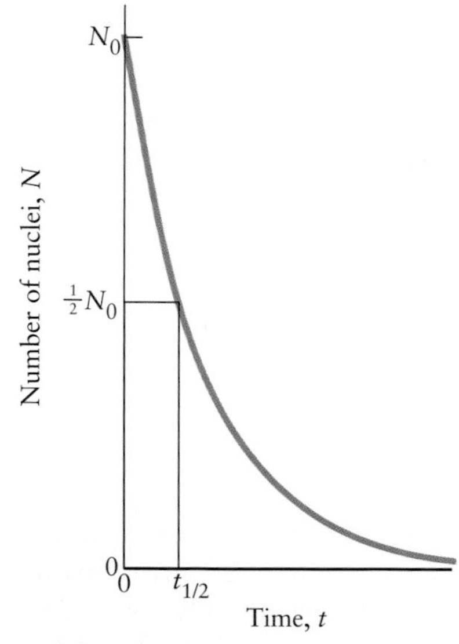

FIGURE 17.18 The exponential decay of the number of radioactive nuclei in a sample implies that the activity of the sample also decays exponentially with time. The curve is characterized by the half-life, $t_{1/2}$.

EXAMPLE 17.1 Using the law of radioactive decay

One of the reasons why thermonuclear weapons have to be serviced regularly is the nuclear decay of the tritium they contain. Suppose a sample of tritium of mass 1.0 g is stored. What mass of that isotope will remain after 5.0 y? The decay constant of tritium is 0.0564 y^{-1}.

STRATEGY The number of tritium nuclei in the sample is proportional to the total mass m of tritium remaining after 5.0 y, so Eq. 3 can be expressed as

$$m = m_0 e^{-kt}$$

where m_0 is the initial mass of tritium.

SOLUTION Substitution of the data gives

$$m = (1.0 \text{ g}) \times e^{-(0.0564 \text{ y}^{-1}) \times (5.0 \text{ y})} = 0.75 \text{ g}$$

Self-Test 17.8A The decay constant for fermium-254 is 210 s^{-1}. What mass of the isotope will be present if a sample of mass 1.00 μg is kept for 10 ms?
[*Answer:* 0.12 μg]

Self-Test 17.8B The decay constant for the nuclide neptunium-237 is 3.3×10^{-7} y^{-1}. What mass of the isotope will be present if a sample of mass 5.0 μg survives for 1.0×10^6 y?

Radioactive decay is normally discussed in terms of the half-life, $t_{1/2}$, the time needed for half the initial number of nuclei to disintegrate. Just as we did in Section 13.5, we find the relation between $t_{1/2}$ and k by setting $N = \frac{1}{2}N_0$ and $t = t_{1/2}$ in Eq. 3:

$$t_{1/2} = \frac{\ln 2}{k} \qquad (4)$$

This equation shows that the larger the value of k, the shorter the half-life of the nuclide. Nuclides with short half-lives are less stable than nuclides with long half-lives. They are more likely to decay in a given period of time and are "hotter" (more intensely radioactive) than nuclides with long half-lives.

Half-lives span a very wide range (Table 17.5). Consider strontium-90, for which the half-life is 28 y. This nuclide occurs in nuclear **fallout,** the fine dust that settles from clouds of airborne particles after a nuclear bomb test, and may also occur in the accidental release of radioactive materials into the air. Because it is chemically very similar to calcium, strontium may accompany that element through the environment and become incorporated into bones; once there, it continues to emit radiation for many years. Even after three half-lives (84 y), one-eighth of the original strontium-90 still survives, provided it has not been removed from the body by normal chemical processes. About 10 half-lives (for strontium-90, 280 y) must pass before the activity of a sample has fallen to 1/1000 of its initial value. Iodine-131, which was released in the accidental fire at the Chernobyl nuclear power plant, has a half-life of only 8.05 d, but it accumulates in the thyroid gland. Several cases of

TABLE 17.5 Half-Lives of Radioactive Isotopes*

Nuclide	Half-life, $t_{1/2}$
tritium	12.3 y
carbon-14	5.73×10^3 y
carbon-15	2.4 s
potassium-40	1.26×10^9 y
cobalt-60	5.26 y
strontium-90	28.1 y
iodine-131	8.05 d
cesium-137	30.17 y
radium-226	1.60×10^3 y
uranium-235	7.1×10^8 y
uranium-238	4.5×10^9 y
fermium-244	3.3 ms

*d = day, y = year.

thyroid cancer have been linked to iodine-131 exposure from the accident. Plutonium-239 has a half-life of 2.4×10^4 y. Consequently, very long term storage facilities are required for plutonium waste, and land contaminated with plutonium cannot be inhabited again for thousands of years without expensive remediation efforts.

The constant half-life of a nuclide is put to practical use in determining the ages of archaeological artifacts. In **isotopic dating,** we measure the activity of the radioactive isotopes they contain. Isotopes used for dating objects include uranium-238, potassium-40, and tritium. However, the most important example is **radiocarbon dating,** which uses the β decay of carbon-14, for which the half-life is 5730 y.

Carbon-12 is the principal isotope of carbon, but a small proportion of carbon-14 is present in all living things. Its nuclei are produced when nitrogen nuclei in the atmosphere are bombarded by neutrons formed in the collisions of cosmic rays with other nuclei:

$$^{14}_{7}N + {}^{1}_{0}n \longrightarrow {}^{14}_{6}C + {}^{1}_{1}p$$

Nuclear testing has increased the amount of carbon-14 in the air, and sensitive radiocarbon dating techniques take this increase into account.

The carbon-14 atoms are produced in the atmosphere at an almost constant rate and, as a result, the proportion of carbon-14 to carbon-12 atoms in the atmosphere tends to be constant over time. The carbon-14 atoms enter living organisms as $^{14}CO_2$ through photosynthesis and digestion. They leave the organisms by the normal processes of excretion and respiration, and also because the nuclei decay at a steady rate. As a result, all living things have a fixed ratio (of about 1 to 10^{12}) of carbon-14 atoms to carbon-12 atoms, and 1.0 g of naturally occurring carbon has an activity of 15 disintegrations per minute.

When an organism dies, it no longer exchanges carbon with its surroundings. However, carbon-14 nuclei already inside the organism continue to decay with a constant half-life, so the ratio of carbon-14 to carbon-12 decreases. The ratio observed in a sample of dead tissue can therefore be used to estimate the time since death.

In the technique developed by Willard Libby in Chicago in the late 1940s, the proportion of carbon-14 is determined by monitoring the β radiation from CO_2 obtained by burning the sample. This procedure is illustrated in the following example. In the modern version of the technique, which requires only a few milligrams of sample, the carbon atoms are converted into C^- ions by bombardment of the sample with cesium atoms. The C^- ions are then accelerated with electric fields, and the carbon isotopes are separated and counted with a mass spectrometer (Fig. 17.19).

The law of radioactive decay implies that the number of radioactive nuclei decreases exponentially with time with a characteristic half-life. Radioactive isotopes are used to determine the ages of objects.

EXAMPLE 17.2 Interpreting carbon-14 dating

A sample of carbon of mass 1.00 g from wood found in an archaeological site in Arizona underwent 7900 carbon-14 disintegrations in a period of 20 h. In the same period, 1.00 g of carbon from a modern source underwent 18 400 disintegrations. Calculate the age of the sample.

FIGURE 17.19 In the modern version of the carbon-14 dating technique, a mass spectrometer is used to determine the proportion of carbon-14 nuclei in the sample relative to the number of carbon-12 nuclei.

STRATEGY First, we rearrange Eq. 3 to give an expression for the time

$$t = -\frac{1}{k}\ln\left(\frac{N}{N_0}\right)$$

Then we express k in terms of $t_{1/2}$ by using Eq. 4:

$$t = -\frac{t_{1/2}}{\ln 2}\ln\left(\frac{N}{N_0}\right)$$

The number of disintegrations reported in a given interval is proportional to the number of carbon-14 nuclei present for each sample. If we suppose that the proportion of carbon-14 in the atmosphere is the same now as it was when the tissue composing the ancient sample was alive, then the carbon-14 activity of the modern sample can be assumed to be the same as the original activity in the ancient sample. We can therefore set N/N_0 equal to the ratio of the number of disintegrations in the ancient and modern samples.

SOLUTION From the expression above, we obtain

$$t = -\left(\frac{5.73 \times 10^3\ \text{y}}{\ln 2}\right) \times \ln\left(\frac{7900}{18\ 400}\right) = 6.99 \times 10^3\ \text{y}$$

We conclude that approximately 7000 years have elapsed since the piece of wood was part of a living tree.

Self-Test 17.9A A sample of carbon of mass 250 mg from wood found in a tomb in Israel underwent 2480 carbon-14 disintegrations in 20 h. Estimate the time since death, assuming the same activity for a modern sample as in Example 17.2.

[*Answer:* 5.1×10^3 y]

Self-Test 17.9B A sample of carbon of mass 1.00 g from scrolls found near the Dead Sea underwent 1.4×10^4 carbon-14 disintegrations in 20 h. Estimate the approximate time since the sheepskins composing the scrolls were removed from the sheep, assuming the same activity for a modern sample as in Example 17.2.

17.8 Uses of Radioisotopes

Radioisotopes are radioactive isotopes. They are used not only to cure disease (as we saw in Box 17.2), but also to preserve food, to trace mechanisms of reactions, to power spacecraft, and to locate the sources of our water supplies.

Radioactive **tracers** are isotopes that are used to track changes and locations. For example, a sample of sugar can be *labeled* with carbon-14; that is, some carbon-12 atoms in sugar molecules are replaced by carbon-14 atoms, which can be detected by radiation counters. In this way, the progress of the sugar molecules through the body can be monitored. Fertilizers labeled with radioactive nitrogen, phosphorus, and potassium are used to follow the mechanism of plant growth and the passage of these elements through the environment. Chemists and biochemists use tracers to help determine the mechanisms of reactions (see Box 19.1).

Radioisotopes have important commercial applications. For example, americium-243 is used in smoke detectors. Its role is to ionize any smoke particles, which then allow a current to flow and set off the alarm. Carbon-14 is incorporated into steel samples to test for wear and into plastic samples to check for uniformity. Exposure to radiation is also used to sterilize food and inhibit the sprouting of potatoes. The radiation kills the bacteria that spoil food but does not produce harmful substances. Radioisotopes that give off a lot of energy as heat are also used to provide power in remote locations, where refueling of generators is not possible. Unmanned spacecraft, such as Voyager 2, are powered by plutonium produced in nuclear reactors.

Isotopes are also used to determine properties of the environment. Just as carbon-14 is used to date organic materials, geologists can determine the age of very old substances such as rocks by using radioisotopes with longer half-lives. Uranium-238 ($t_{1/2} = 4.5 \times 10^9$ y) and potassium-40 ($t_{1/2} = 1.26 \times 10^9$ y) are used to date very old rocks. Uranium-238 decays through a series of α and β emissions to lead-206 (recall Fig. 17.16); to determine the age of a rock that contains uranium, geologists measure the ratio of ^{238}U to ^{206}Pb present in it. Potassium-40 decays by electron capture to form argon-40. The rock is placed under vacuum and crushed, and a mass spectrometer is used to measure the amount of argon gas that escapes. This technique was used to determine the age of rocks collected on the surface of the Moon; they were found to be 3.5–4 billion years old, about the same age as the Earth.

Tritium ($t_{1/2}$ = 12.3 y) is used to date relatively recent samples in a technique similar to carbon-14 dating. For example, tritium is used to determine the age of groundwater. The ^{3}H abundance in rain and snow is stable over time. However, once the water seeps into the ground, it cannot mix with the water in the atmosphere, and the tritium nuclei that decay are not replaced. To determine the age of groundwater, we can measure the concentrations of both tritium and its daughter nuclide, helium-3. The age of the groundwater can help determine properties such as the depth of its source.

Radioisotopes are used as long-lasting heat sources, to ionize smoke particles, to study the environment, and to track movement. They are used in biology to trace metabolic pathways, in chemistry to trace reaction mechanisms, and in geology to determine the age and source of rocks and groundwater.

NUCLEAR ENERGY

Nuclear reactions can release huge amounts of energy, and nuclear reactors are widely used to provide power. The benefits of nuclear power include the large quantity of energy that can be obtained from a small mass of fuel and the absence of chemical pollution of the kind associated with fossil fuels. However, like other powerful resources, nuclear energy presents us with great technical challenges. The following sections outline some of the principles involved.

17.9 Mass-Energy Conversion

The energy released when the nucleons in a nucleus adopt a more stable arrangement can be detected by comparing the masses of the nuclear reactants and the nuclear products. Einstein's theory of relativity tells us that the mass of an object is a measure of its energy content: the greater the mass of an object, the greater its energy. Specifically, the total energy, E, and the mass, m, are related by Einstein's famous equation

$$E = mc^2 \tag{5}$$

where c is the speed of light ($3.00 \times 10^8\ \text{m·s}^{-1}$). It follows from this relation that loss of energy is always accompanied by loss of mass.

The mass loss that always accompanies the loss of energy is normally far too small to detect. Even in a strongly exothermic chemical reaction, such as one that releases 10^3 kJ of energy, the mass of the products and the reactants differ by only 10^{-8} g. In a nuclear reaction, the energy changes are very large, the mass loss is measurable, and we can calculate the energy released from the observed change in mass.

The **nuclear binding energy,** E_{bind}, is the energy *released* when protons and neutrons come together to form a nucleus. A positive value of the binding energy means that a nucleus has a lower energy than its constituent nucleons; the greater the binding energy, the lower the energy of the nuclide. We can use Einstein's equation to calculate the nuclear binding energy from the difference in mass, Δm, between the nucleus and the separated nucleons. For example, iron-56 has 26 protons, each of mass m_p, and 30 neutrons, each of mass m_n. The difference in mass between the nucleus and the separate nucleons is

$$\begin{aligned}\Delta m &= \sum m(\text{products}) - \sum m(\text{reactants}) \\ &= m(^{56}_{26}\text{Fe nucleus}) - (26m_p + 30m_n)\end{aligned}$$

Then we can calculate the binding energy from

$$E_{\text{bind}} = |\Delta m| \times c^2 \qquad (6)$$

we use the absolute value of the mass difference because all binding energies are positive.

Before we see how to use Eq. 6 (in Example 17.3), we need to note three points. One is that the masses of nuclides are normally reported in **atomic mass units,** u (formerly, amu). One atomic mass unit is defined as exactly $\frac{1}{12}$ the mass of one atom of carbon-12:

$$1\ \text{u} = 1.6605 \times 10^{-27}\ \text{kg}$$

Therefore, the mass of an atom in atomic mass units is the same as the molar mass in grams per mole. The second point is that it is more convenient to use tables of isotope masses for calculations of binding energies than to use the masses of the bare nuclei. However, an isotope mass is the mass of the neutral atom, not the nucleus alone; consequently, it includes the mass of the electrons and the mass equivalent of their energy in the atom. We can take into account the mass of the electrons by substituting the mass of a hydrogen atom for the mass of each proton. With this approach, the masses of electrons will be the same on each side of the equation. The electron-nucleus binding energy is equivalent to only about 0.000 001 u per proton, so it can be ignored in elementary calculations. The third point is that binding energies are commonly reported in **electronvolts** (eV) or, more specifically, millions of electronvolts (1 MeV = 10^6 eV), where 1 eV is the change in potential energy of an electron when it is moved through a potential difference of 1 V:

$$1\ \text{eV} = 1.602\ 18 \times 10^{-19}\ \text{J}$$

EXAMPLE 17.3 Calculating the nuclear binding energy

Calculate the nuclear binding energy in electronvolts for a helium-4 nucleus, given the following masses: ^{4}He, 4.0026 u; ^{1}H, 1.0078 u; n, 1.0087 u.

STRATEGY The nuclear binding energy is the energy released in the formation of the nucleus from its nucleons. Use H atoms instead of protons to account for the masses of the electrons in the He atom produced:

$$2\ ^1\text{H} + 2\ ^1\text{n} \longrightarrow {}^4\text{He}$$

Begin by calculating the difference in masses between the products and the reactants; then convert the result from atomic mass units to kilograms. Finally, use the Einstein relation to calculate the energy corresponding to this loss of mass.

SOLUTION The change in mass is

$$\Delta m = 4.0026 \text{ u} - \{(2 \times 1.0078) + (2 \times 1.0087)\} \text{ u} = -0.0304 \text{ u}$$

In kilograms, this mass difference is

$$\begin{aligned}\Delta m &= (-0.0304 \text{ u}) \times \left(\frac{1.6605 \times 10^{-27} \text{ kg}}{1 \text{ u}}\right) \\ &= -0.0304 \times 1.6605 \times 10^{-27} \text{ kg}\end{aligned}$$

Then, from Eq. 6,

$$\begin{aligned}E_{\text{bind}} &= |-0.0304 \times 1.6605 \times 10^{-27} \text{ kg}| \times (3.00 \times 10^{8} \text{ m·s}^{-1})^2 \\ &= 4.54 \times 10^{-12} \text{ kg·m}^2\text{·s}^{-2} = 4.54 \times 10^{-12} \text{ J}\end{aligned}$$

(In the last line, we have used 1 $\text{kg·m}^2\text{·s}^{-2}$ = 1 J.) The value of the binding energy shows that 4.54×10^{-12} J is released when one He nucleus forms from its nucleons. Multiplying by the Avogadro constant, we find that the molar binding energy of helium is 2.7×10^9 kJ·mol^{-1}, which is 10 million times larger than the energy of a typical chemical bond. In electronvolts, the binding energy for one nucleus is

$$E_{\text{bind}} = (4.54 \times 10^{-12} \text{ J}) \times \frac{1 \text{ eV}}{1.602\,18 \times 10^{-19} \text{ J}} = 2.83 \times 10^{7} \text{ eV}$$

or 28.3 MeV.

Self-Test 17.10A Calculate the binding energy of a carbon-12 nucleus in electronvolts.

[*Answer:* 92.3 MeV]

Self-Test 17.10B Calculate the binding energy of a uranium-235 nucleus in electronvolts. The mass of one uranium atom is 235.0439 u.

Figure 17.20 shows the binding energy per nucleon, E_{bind}/A, for the elements. The graph shows that nucleons are bonded together most strongly in the elements near iron and nickel. This high binding energy is one of the reasons why iron and nickel are so abundant in meteorites and on a rocky planet like Earth. The binding energy per nucleon is less for all other nuclides. We can infer that nuclei of light atoms become more stable when they "fuse" together and that heavy nuclei become more stable when they undergo "fission" and split into lighter nuclei.

Nuclear binding energies are determined from the mass difference between the nucleus and its components and the use of Einstein's formula. Iron and nickel have the highest binding energy per nucleon.

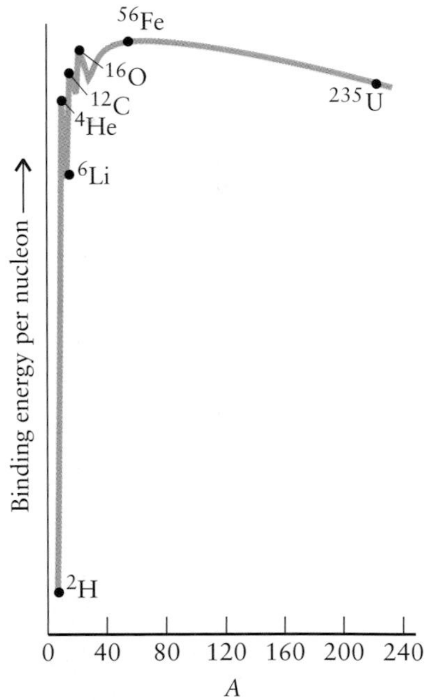

FIGURE 17.20 The variation of the nuclear binding energy per nucleon. The maximum binding energy per nucleon occurs near iron and nickel. Their nuclei have the lowest energy of all because their nucleons are most tightly bound. (The vertical axis is E_{bind}/A.)

17.10 Nuclear Fission

When nuclear fission takes place, the original nucleus breaks up into two or more smaller nuclei and a large amount of energy is released. The energy generated by fission reactions is so huge that they are used to generate electricity in nuclear power plants. We can calculate the energy released by subtracting the total mass of the starting materials from the mass of the fission products and converting the difference in mass into a difference in energy by using Einstein's equation.

Notice that we do not cancel neutrons, although they appear on both sides of the equation. Like equations for elementary chemical reactions, nuclear equations show the mechanism by which the process occurs.

EXAMPLE 17.4 Calculating the energy released during fission

When uranium-235 nuclei are bombarded with neutrons, they can split apart in a variety of ways, like glass balls that shatter into pieces of different sizes. In one process, uranium-235 forms barium-142 and krypton-92.

$$^{235}_{92}\text{U} + {}^{1}_{0}\text{n} \longrightarrow {}^{142}_{56}\text{Ba} + {}^{92}_{36}\text{Kr} + 2\,{}^{1}_{0}\text{n}$$

Calculate the energy (in joules) released when 1.0 g of uranium-235 undergoes this fission reaction. The masses of the particles are $^{235}_{92}\text{U}$, 235.04 u; $^{142}_{56}\text{Ba}$, 141.92 u; $^{92}_{36}\text{Kr}$, 91.92 u; n, 1.0087 u.

STRATEGY If we know the mass loss, we can find the energy released by using Einstein's equation. Therefore, we must calculate the total mass of the particles on each side of the equation and then substitute the difference into Eq. 6.

SOLUTION

$$\begin{aligned}\text{Mass of products} &= m(\text{Ba}) + m(\text{Kr}) + 2m_{\text{n}}\\ &= 141.92 + 91.92 + (2 \times 1.0087)\text{ u} = 235.86\text{ u}\end{aligned}$$

$$\begin{aligned}\text{Mass of reactants} &= m(\text{U}) + m_{\text{n}}\\ &= 235.04 + 1.0087\text{ u} = 236.05\text{ u}\end{aligned}$$

The change in mass is

$$\begin{aligned}\Delta m &= 235.86 - 236.05\text{ u} = -0.19\text{ u}\\ &= (-0.19\text{ u}) \times \frac{1.6605 \times 10^{-27}\text{ kg}}{1\text{ u}}\\ &= -(0.19 \times 1.6605) \times 10^{-27}\text{ kg}\end{aligned}$$

The energy change accompanying the fission of one ^{235}U nucleus to these products is therefore

$$\begin{aligned}\Delta E &= -(0.19 \times 1.6605 \times 10^{-27}\text{ kg}) \times (3.00 \times 10^{8}\text{ m}\cdot\text{s}^{-1})^2\\ &= -0.19 \times 1.6605 \times 10^{-27} \times (3.00 \times 10^{8})^2\text{ J}\end{aligned}$$

This energy release accompanies the fission of one nucleus. The number of atoms in 1.0 g (1.0×10^{-3} kg) of uranium-235 is

$$\begin{aligned}\text{Number of atoms} &= \frac{\text{mass of sample}}{\text{mass of one atom}}\\ &= \frac{1.0 \times 10^{-3}\text{ kg}}{(235.04\text{ u}) \times \left(\dfrac{1.6605 \times 10^{-27}\text{ kg}}{1\text{ u}}\right)}\\ &= \frac{1.0 \times 10^{-3}}{235.04 \times 1.6605 \times 10^{-27}}\end{aligned}$$

The total energy change is therefore

$$\Delta E = \left(\frac{1.0 \times 10^{-3}}{235.04 \times 1.6605 \times 10^{-27}}\right) \times \{-0.19 \times 1.6605 \times 10^{-27} \times (3.00 \times 10^8)^2 \text{ J}\}$$
$$= -7.3 \times 10^{10} \text{ J}$$

That is, 73 GJ of energy is released.

Self-Test 17.11A Another mode in which uranium-235 can undergo fission is

$${}^{235}_{92}\text{U} + {}^{1}_{0}\text{n} \longrightarrow {}^{135}_{52}\text{Te} + {}^{100}_{40}\text{Zr} + {}^{1}_{0}\text{n}$$

Calculate the energy change when 1.0 g of uranium-235 undergoes fission in this way. The masses needed are ${}^{235}_{92}\text{U}$, 235.04 u; n, 1.0087 u; ${}^{135}_{52}\text{Te}$, 134.92 u; ${}^{100}_{40}\text{Zr}$, 99.92 u.

[*Answer:* -7.7×10^{10} J]

Self-Test 17.11B A nuclear reaction that can cause great destruction is one of many that take place in the ${}^{235}\text{U}$ atomic bomb:

$${}^{235}_{92}\text{U} + {}^{1}_{0}\text{n} \longrightarrow {}^{138}_{56}\text{Ba} + {}^{86}_{36}\text{Kr} + 12\,{}^{1}_{0}\text{n}$$

How much energy is released when 1.0 g of uranium-235 undergoes fission in this manner? The additional masses needed are ${}^{138}_{56}\text{Ba}$, 137.91 u; ${}^{86}_{36}\text{Kr}$, 85.91 u.

Spontaneous nuclear fission takes place when the natural oscillations of a heavy nucleus cause it to break into two nuclei of similar mass (Fig. 17.21). In terms of the liquid drop model, we can think of the nucleus as distorting into a dumbbell shape and then breaking into two smaller droplets. An example is the disintegration of americium-244 into iodine and molybdenum:

$${}^{244}_{95}\text{Am} \longrightarrow {}^{134}_{53}\text{I} + {}^{107}_{42}\text{Mo} + 3\,{}^{1}_{0}\text{n}$$

Fission does not occur in precisely the same way in every instance. For example, the fission products of americium-244 may include other isotopes of iodine and molybdenum and also isotopes of other elements. More than 200 isotopes of 35 different elements have been identified

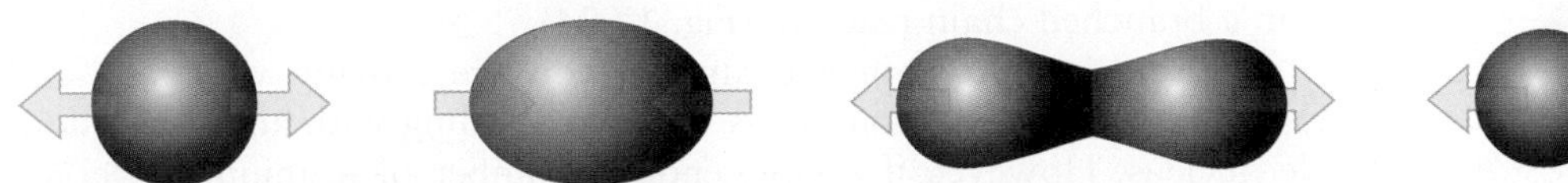

FIGURE 17.21 In spontaneous nuclear fission, the oscillations of the heavy nucleus in effect tear the nucleus apart, thereby forming two or more smaller nuclei of similar mass. Here, two neutrons are also released. This picture is based on the liquid drop model of the nucleus.

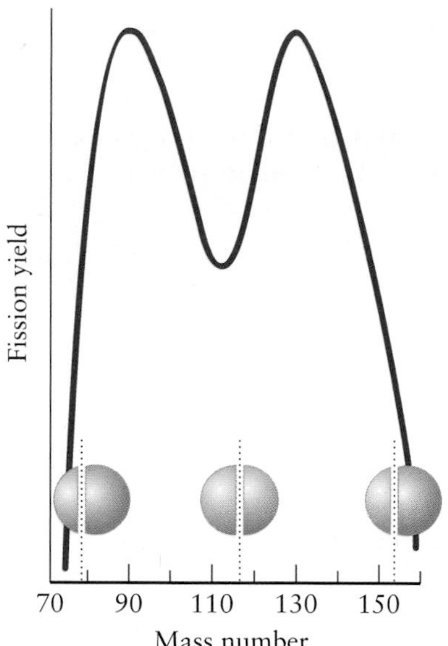

FIGURE 17.22 The fission yield of uranium-235. Note that the majority of fission products lie in the regions close to $A = 90$ and 130, and that relatively few nuclides corresponding to symmetrical fission (A close to 117) are formed.

among the fission products of uranium-235, with most products having mass numbers close to 90 or 130 (Fig. 17.22).

Induced nuclear fission is fission caused by bombarding a heavy nucleus with neutrons (Fig. 17.23). In terms of the liquid drop model, the nucleus breaks into two droplets when struck by a projectile. Nuclei that can undergo induced fission are called **fissionable.** For most nuclei, fission takes place only if the impinging neutrons travel so rapidly that they can smash into the nucleus and drive it apart with the shock of impact; uranium-238 undergoes fission in this way. **Fissile nuclei,** however, are nuclei that can be nudged into breaking apart even by slow neutrons. They include uranium-235, uranium-233, and plutonium-239, the fuels of nuclear power plants.

Once nuclear fission has been induced, it can continue, even if the supply of neutrons from outside is discontinued, provided the fission produces more neutrons. Such self-sustaining fission takes place in uranium-235, which undergoes numerous fission processes, including

$$^{235}_{92}\text{U} + ^{1}_{0}\text{n} \longrightarrow ^{141}_{56}\text{Ba} + ^{92}_{36}\text{Kr} + 3\,^{1}_{0}\text{n}$$

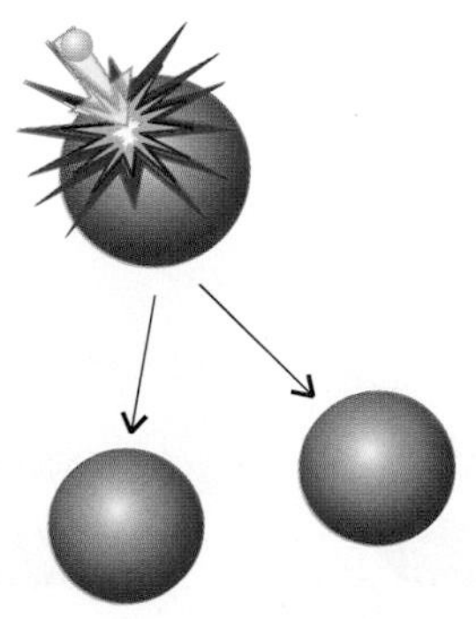

FIGURE 17.23 In induced nuclear fission, the impact of an incoming neutron causes the nucleus to break apart.

If the three product neutrons strike three other fissile nuclei, then after the next round of fission there will be nine neutrons, which can induce fission in nine more nuclei. In the language of Section 13.12, neutrons are *carriers* in a branched chain reaction (Fig. 17.24).

Neutrons produced in a chain reaction are moving very fast, and most escape into the surroundings without colliding with another fissionable nucleus. However, if a large enough number of uranium nuclei are present in the sample, enough neutrons can be captured to sustain the chain reaction. In that case, there is a **critical mass,** a mass of fissionable material above which so few neutrons escape from the sample that the fission chain reaction is sustained. If a sample is **supercritical,** with a mass in excess of the critical value, then the reaction is self-sustaining and may result in an explosion. The critical mass for a solid sphere of pure plutonium of normal density is about 15 kg, a sphere about the size of a grapefruit. The critical mass is smaller if the metal is compressed by detonating a conventional explosive that surrounds it. Then the nuclei are pressed closer together and become more effective at blocking the escape of neutrons. The critical mass can be as low as 5 kg for highly compressed plutonium. A sample that has less than the critical mass for its density is said to be **subcritical.**

In a nuclear weapon, the fissile material is initially subcritical. The challenge is to produce a supercritical mass so rapidly that the chain reaction occurs uniformly throughout the metal. Supercriticality can be achieved by shooting two subcritical blocks toward each other (as was done in the bomb that fell on Hiroshima) or by implosion of a single subcritical mass (the technique used in the bomb that destroyed Nagasaki). A strong neutron emitter, typically polonium, helps to initiate the chain reaction.

Explosive fission cannot occur in a nuclear reactor because the fuel is not dense enough. Instead, reactors sustain a much slower, controlled chain reaction by making efficient use of a limited supply of neutrons and slowing the neutrons down. The fuel is shaped into long rods and inserted into a **moderator,** a material that slows down the neutrons as they pass between fuel rods; the slower neutrons have a greater probability of collision with a nucleus (Fig. 17.25). Slow neutrons have three significant roles: they do not induce the fission of fissionable (as distinct from fissile) material; they are most effectively absorbed by the fissile uranium-235; and they are more easily controlled. The first moderator used was graphite. Light-water reactors (LWRs) use ordinary water as a moderator. They are the most common type of nuclear reactor in the United States (Fig. 17.26).

Media Link Animation: Fig. 17.24, Nuclear chain reaction

View an animation of a fission chain reaction.

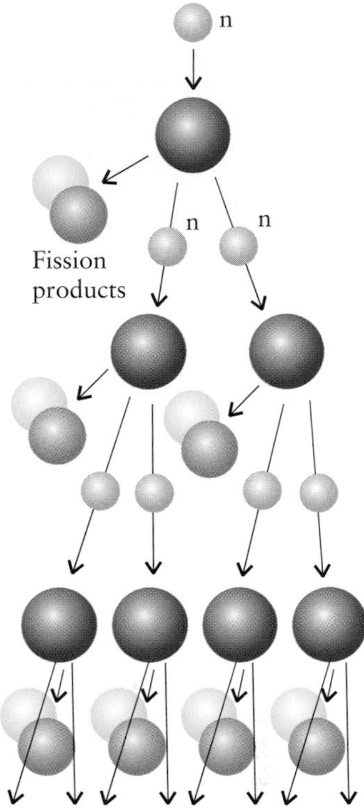

FIGURE 17.24 A self-sustaining chain reaction, in which neutrons are the chain carriers, occurs when induced fission produces more than one neutron per fission event. These newly produced neutrons can go on to stimulate fission in other nuclei.

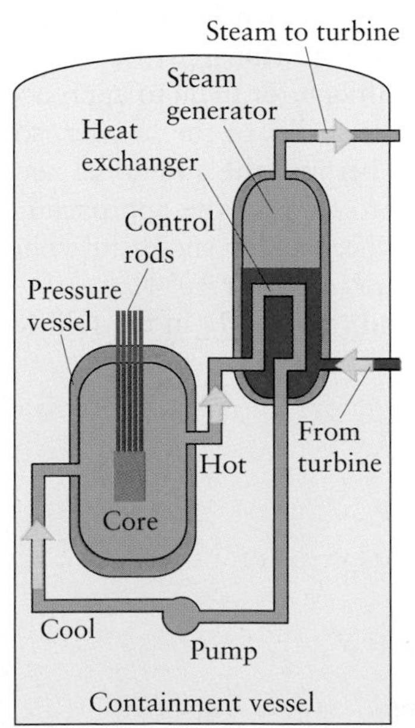

FIGURE 17.25 A schematic diagram of one type of nuclear reactor in which water acts as a moderator for the nuclear reaction. In this pressurized water reactor (PWR), the coolant is water under pressure. The fission reactions produce heat, which boils water in the steam generator; the resulting steam turns the turbines that generate electricity.

FIGURE 17.26 The reactor core of a typical light-water reactor (LWR) nuclear power plant is immersed in water.

If the rate of the chain reaction exceeds a certain level, the reactor will become too hot and begin to melt. Control rods—made from neutron-absorbing elements, such as boron or cadmium, and inserted between the fuel rods—help to control the number of available neutrons and the rate of nuclear reaction. This control is effective: all explosions in nuclear power plants have been *chemical* explosions, generally from overheating.

One of the many problems of nuclear power is the availability of fuel: uranium-235 reserves are only about 0.7% those of the nonfissile uranium-238, and the separation of the isotopes is costly (Section 17.12). One solution is to synthesize fissile nuclides from other elements. In a **breeder reactor,** a reactor that is used to create nuclear fuel, the neutrons are not moderated. Their high speeds result in the formation of not only uranium-235 but also some fissile plutonium-239, which can be used as fuel (or for warheads). However, breeder reactors are more hazardous to operate than nuclear power plants. They run very hot, and the fast reactions require more careful control than a reactor used for nuclear power generation. Because of safety concerns, their use is controversial.

Nuclear energy can be extracted by arranging for a nuclear chain reaction to occur in a critical mass of fissionable material, with neutrons as the chain carriers. A moderator is used to reduce the speeds of the neutrons in a reactor that uses fissile material.

17.11 Nuclear Fusion

Although fission reactors do not generate chemical pollution, they do produce highly hazardous radioactive waste. However, another kind of nuclear reaction being investigated for power generation is essentially free of long-lived waste products, and its abundant fuel is readily extracted from seawater. The reaction is the fusion of hydrogen nuclei to form helium nuclei.

From the plot in Fig. 17.20, we can see that there is a large increase in nuclear binding energy on going from light elements such as deuterium to heavier elements. Consequently, energy is released when hydrogen nuclei fuse together to give nuclei of those elements. Unfortunately, the strong electrical repulsion between protons makes it difficult for them to approach each other closely enough to fuse together. The nuclei of the heavier isotopes of hydrogen fuse together more readily, because the additional neutrons contribute to the strong force and help to capture the approaching protons. To achieve the high kinetic energies needed for successful collisions, fusion reactors need to operate at temperatures over 10^8 K.

One fusion scheme uses deuterium (D) and tritium (T) in the following sequence of nuclear reactions

$$\mathrm{D + D \longrightarrow {}^3He + n}$$

$$\mathrm{D + D \longrightarrow T + p}$$

$$\mathrm{D + T \longrightarrow {}^4He + n}$$

$$\mathrm{D + {}^3He \longrightarrow {}^4He + p}$$

Overall reaction: $$\mathrm{6\,D \longrightarrow 2\,{}^4He + 2\,p + 2\,n}$$

The overall reaction releases 3×10^8 kJ for each gram of deuterium consumed. That energy corresponds to the energy generated when the Hoover Dam operates at full capacity for about an hour. Additional tritium is supplied to facilitate the process. Because tritium has a very low natural abundance and is radioactive, it is generated by bombarding lithium-6 with neutrons in the immediate surroundings of the reaction zone:

$$^{6}\text{Li} + \text{n} \longrightarrow \text{T} + {}^{4}\text{He}$$

Nuclear fusion is very difficult to achieve because the charged nuclei must be hurled at each other with enormous kinetic energies. One way of accelerating nuclei to sufficiently high speeds is to heat them with a fission explosion: this method is used to produce a **thermonuclear explosion**, an explosion due to nuclear fusion. In the "hydrogen bombs" used in warfare, a fission bomb (using uranium or plutonium) ignites a lithium-6 fusion bomb. Fusion bombs can be of varying destructive capacities, because there is no critical mass to maintain. Typical fusion bombs have a destructive capacity greater than 200 times that of the fission bombs that were dropped on Hiroshima and Nagasaki in World War II.

A more constructive approach to nuclear fusion—one that achieves a controlled release of nuclear energy—involves heating a **plasma**, or ionized gas, by passing an electric current through it. The very fast ions in the plasma are kept away from the walls of the container with magnetic fields. This method of achieving fusion is the subject of intense research and is beginning to show signs of success (Fig. 17.27). Research is also focused on fusing nuclei with lasers. In this approach, frozen pellets of mixed deuterium and tritium are blasted from all directions by as many as 192 lasers. The sudden evaporation of the outer layer of the pellet simultaneously compresses the material inside the pellet and heats it, thereby triggering fusion.

Nuclear fusion makes use of the energy released when light nuclei fuse together to form heavier nuclei.

FIGURE 17.27 Research into controlled nuclear fusion is being carried out in several countries. Here we see the Tokomak fusion test reactor at the Princeton Plasma Physics Laboratory.

FIGURE 17.28 Pitchblende is a common uranium ore. It is a variety of uranite, UO_2.

17.12 The Chemistry of Nuclear Power

Chemistry is the key to the safe use of nuclear power. It is used in the preparation of the fuel itself, the recovery of important fission products, and the safe disposal or utilization of nuclear waste.

Uranium is the fuel of nuclear reactors. The most important of its minerals is *pitchblende*, UO_2 (Fig. 17.28), much of which is obtained from strip mines in New Mexico and Wyoming. Uranium is refined to reduce the ore to the metal and to **enrich** it, that is, to increase the abundance of a specific isotope—in this case, uranium-235. The natural abundance of uranium-235 is about 0.7%; for use in a nuclear reactor, this fraction must be increased to about 3%.

The enrichment procedure uses the small mass difference between the hexafluorides of uranium-235 and uranium-238 to separate them. The first procedure to be developed converts the uranium into uranium hexafluoride, UF_6, which can be vaporized readily. The different effusion rates of the two isotopic fluorides are then used to separate them. From Graham's law of effusion (rate of effusion $\propto 1/(\text{molar mass})^{1/2}$; Section 4.12), the rates of effusion of $^{235}UF_6$ (molar mass, 349.0 g·mol^{-1}) and $^{238}UF_6$ (molar mass, 352.1 g·mol^{-1}) should be in the ratio

$$\frac{\text{Rate of effusion of } ^{235}\text{UF}_6}{\text{Rate of effusion of } ^{238}\text{UF}_6} = \sqrt{\frac{352.1}{349.0}} = 1.004$$

Because the ratio is so close to 1, the vapor must be allowed to effuse repeatedly through porous barriers consisting of screens with large numbers of minute holes. In practice, it is allowed to do so thousands of times.

Because the effusion process is technically demanding and uses a lot of energy, scientists and engineers continue to look for alternative enrichment procedures. One of these approaches uses a centrifuge that rotates samples of uranium hexafluoride vapor at very high speed. This rotation causes the heavier $^{238}UF_6$ molecules to be thrown outward and collected as a solid on the outer parts of the rotor, leaving a high proportion of $^{235}UF_6$ closer to the axis of the rotor, from where it can be removed.

Spent nuclear fuel remains radioactive and consists of a mixture of uranium and fission products. Nuclear reactor waste can be processed and some of it reused, but the percentage processed depends on the price of uranium. When the price is low, as it was in the late 1990s, most nuclear waste is simply stored without further processing.

The processing of nuclear waste is complex. Any remaining uranium-235 must be recovered, any plutonium produced must be extracted, and the largely useless but radioactive fission products must be stored safely (Fig. 17.29). The highly radioactive fission (HRF) products from used nuclear fuel rods must be stored until their level of radioactivity is no longer dangerous (about 10 half-lives). Generally they are buried underground, but even burial of radioactive wastes is not without problems. Metal storage drums can corrode and allow liquid radioactive waste to seep into aquifers that supply drinking water (Fig. 17.30). Leakage can be minimized by incorporating the HRF products into a glass—a solid, complex network of silicon and oxygen atoms. Most of the fission products

FIGURE 17.29 Containers of high-level waste products, including cesium-137 and strontium-90, glow under a protective layer of water. If the canisters were unshielded, the radiation they emit would be great enough to cause death within about 4 s.

FIGURE 17.30 This 35-year-old drum of radioactive waste has corroded and leaked radioactive materials into the soil. The drum was located in one of the nuclear waste disposal sites at the U.S. Department of Energy's Hanford, Washington, nuclear manufacturing and research facility. Several storage sites at this facility have become seriously contaminated.

are oxides of the type that form one of the components of glass—they are network formers (see Section 5.13); that is, they promote the formation of a relatively disorderly Si–O network rather than inducing crystallization into an orderly array of atoms. Crystallization is dangerous because crystalline regions easily fracture, exposing the incorporated radioactive materials to moisture. Water could dissolve them and carry them away from the storage area. An alternative is to incorporate radioactive waste into hard ceramic materials. An example is Synroc, a titanate-based ceramic that can accept radioactive waste products into its lattice.

The partial elimination of nuclear weapons has created an additional disposal problem. The thousands of weapons being dismantled contain thousands of kilograms of plutonium. Some of it is suitable for use in nuclear power plants. However, the less useful low-grade plutonium must be disposed of in a way that would not allow its use in weapons. One strategy being investigated is first to convert the plutonium into a hydride, which would allow its separation from other elements in the warhead or bomb. The hydride would then be converted to an oxide, which could be incorporated into a ceramic material containing neutron absorbers to stabilize the material further.

Uranium is extracted by a series of reactions that lead to uranium hexafluoride; the isotopes are then separated by a variety of procedures. Some radioactive waste is currently converted to glass or ceramic materials for storage underground.

SKILLS YOU SHOULD HAVE MASTERED

1 Write, complete, and balance nuclear equations, Toolbox 17.1 and Self-Tests 17.1 to 17.5.

2 Use the band of stability to predict the types of decay a certain radioactive nucleus is likely to undergo, Self-Test 17.6.

3 Distinguish α, β, and γ radiation by their response to an electric field, penetrating power, and relative biological effectiveness, Sections 17.1 and 17.6.

4 Predict the amount of a radioactive sample that will remain after a given time period, given the decay constant or half-life of the sample, Example 17.1.

5 Use the half-life of an isotope to determine the age of an object, Example 17.2.

6 Calculate the nuclear binding energy of a given nuclide, Example 17.3.

7 Calculate the energy released during a nuclear reaction, Example 17.4.

8 Distinguish nuclear fission from nuclear fusion and predict which nuclides undergo each type of process, Sections 17.10 and 17.11.

EXERCISES

Nuclear Stability

17.1 Calculate the wavelength and the energy per mole of photons of γ radiation of frequency (a) 6.7×10^{20} Hz; (b) 3.2×10^{22} Hz; (c) 4.1×10^{21} Hz; (d) 8.6×10^{19} Hz.

17.2 Calculate the frequency and wavelength of the γ radiation emitted as a result of a rearrangement of nucleons in a single daughter nucleus through an energy (a) 4.8×10^{-13} J; (b) 1.1×10^{-14} J; (c) 9.3×10^{-16} kJ; (d) 7.5×10^{-16} kJ.

17.3 When the nucleons rearrange in the following daughter nuclei, the energy changes by the amount shown, and a γ ray is emitted. Determine the frequency and wavelength of the γ ray in each case: (a) cobalt-60, 1.33 MeV; (b) arsenic-80, 1.64 MeV; (c) iron-59, 1.10 MeV. (1 MeV = 1.602×10^{-13} J.)

17.4 Determine the frequency and wavelength of the γ ray emitted in the decay of the following nuclides: (a) carbon-15, 5.30 MeV; (b) scandium-50, 0.26 MeV; (c) bromine-87, 5.4 MeV. (1 MeV = 1.602×10^{-13} J.)

Radioactive Decay

17.5 Write the balanced nuclear equation for each of the following decays: (a) β^- decay of tritium; (b) β^+ decay of yttrium-83; (c) β^- decay of krypton-87; (d) α decay of protactinium-225.

17.6 Write the balanced nuclear equation for each of the following decays: (a) β^- decay of actinium-228; (b) α decay of radon-212; (c) α decay of francium-221; (d) electron capture by protactinium-230.

17.7 Write the balanced nuclear equation for the following radioactive decays: (a) β^+ decay of boron-8; (b) β^- decay of nickel-63; (c) α decay of gold-185; (d) electron capture by beryllium-7.

17.8 Write the balanced nuclear equation for the following radioactive decays: (a) β^- decay of uranium-233; (b) proton emission of cobalt-56; (c) β^+ decay of holmium-158; (d) α decay of polonium-212.

17.9 Determine the particle emitted and write the balanced nuclear equation for the following nuclear transformations: (a) sodium-24 to magnesium-24; (b) ^{128}Sn to ^{128}Sb; (c) lanthanum-140 to barium-140; (d) ^{228}Th to ^{224}Ra.

17.10 Determine the particle emitted and write the balanced nuclear equation for the following nuclear transformations: (a) carbon-14 to nitrogen-14; (b) neon-19 to fluorine-19; (c) gold-188 to platinum-188; (d) uranium-229 to thorium-225.

17.11 Complete the following equations for nuclear reactions:
(a) $^{11}\text{B} + ? \rightarrow 2\ \text{n} + {}^{13}\text{N}$
(b) $? + \text{D} \rightarrow \text{n} + {}^{36}\text{Ar}$
(c) $^{96}\text{Mo} + \text{D} \rightarrow ? + {}^{97}\text{Tc}$
(d) $^{45}\text{Sc} + \text{n} \rightarrow \alpha + ?$

17.12 Write balanced nuclear equations for the radioactive decay of the following nuclides: (a) ^{74}Kr, β^+ emission; (b) ^{174}Hf, α emission; (c) ^{98}Tc, β^- emission; (d) ^{41}Ca, electron capture.

The Pattern of Nuclear Stability

17.13 The following nuclides lie outside the band of stability. Predict whether they are most likely to undergo β^- decay, β^+ decay, or α decay and identify the daughter nucleus; (a) copper-68; (b) cadmium-103; (c) berkelium-243; (d) dubnium-260.

17.14 The following nuclides lie outside the band of stability. Predict whether they are most likely to undergo β^- decay, β^+ decay, or α decay and identify the daughter nucleus: (a) copper-60; (b) xenon-140; (c) americium-246; (d) neptunium-240.

17.15 Identify the daughter nuclides in each step of the radioactive decay of uranium-235, if the string of particle emissions is α, β, α, β, α, α, α, β, α, β, α. Write a balanced nuclear equation for each step.

17.16 Neptunium-237 undergoes an α, β, α, α, β, α, α, α, β, α, β sequence of radioactive decays. Write a balanced nuclear equation for each step.

17.17 Identify the daughter nuclides in each step of the radioactive decay of uranium-238, if the string of particle emissions is α, β, β, α, α, α, α, α, β, β, α, β, β, α.

17.18 Thorium-232 undergoes an α, β, β, α, α, α, α, β, β, α sequence of radioactive decays. Write a balanced nuclear equation for each step.

Nucleosynthesis

17.19 Complete the following nuclear equations:
(a) ${}^{14}_{7}\text{N} + ? \rightarrow {}^{17}_{8}\text{O} + {}^{1}_{1}\text{p}$
(b) $? + {}^{1}_{0}\text{n} \rightarrow {}^{249}_{97}\text{Bk} + {}^{0}_{-1}\text{e}$
(c) ${}^{243}_{95}\text{Am} + {}^{1}_{0}\text{n} \rightarrow {}^{244}_{96}\text{Cm} + ? + \gamma$
(d) ${}^{13}_{6}\text{C} + {}^{1}_{0}\text{n} \rightarrow ? + \gamma$

17.20 Complete the following nuclear equations:
(a) $? + {}^{1}_{1}\text{p} \rightarrow {}^{21}_{11}\text{Na} + \gamma$
(b) ${}^{1}_{1}\text{H} + {}^{1}_{1}\text{p} \rightarrow {}^{2}_{1}\text{H} + ?$
(c) ${}^{15}_{7}\text{N} + {}^{1}_{1}\text{p} \rightarrow {}^{12}_{6}\text{C} + ?$
(d) ${}^{20}_{10}\text{Ne} + ? \rightarrow {}^{24}_{12}\text{Mg} + \gamma$

17.21 Complete the following equations for nuclear transmutation:
(a) ${}^{20}_{10}\text{Ne} + {}^{4}_{2}\alpha \rightarrow ? + {}^{16}_{8}\text{O}$
(b) ${}^{20}_{10}\text{Ne} + {}^{20}_{10}\text{Ne} \rightarrow {}^{16}_{8}\text{O} + ?$
(c) ${}^{44}_{20}\text{Ca} + ? \rightarrow \gamma + {}^{48}_{22}\text{Ti}$
(d) ${}^{27}_{13}\text{Al} + {}^{2}_{1}\text{H} \rightarrow ? + {}^{28}_{13}\text{Al}$

17.22 Complete the following equations for nuclear transmutation:
(a) $? + \gamma \rightarrow {}^{0}_{-1}\text{e} + {}^{20}_{10}\text{Ne}$
(b) ${}^{44}_{22}\text{Ti} + {}^{0}_{-1}\text{e} \rightarrow {}^{0}_{+1}\text{e} + ?$
(c) ${}^{241}_{95}\text{Am} + ? \rightarrow 4\,{}^{1}_{0}\text{n} + {}^{248}_{100}\text{Fm}$
(d) $? + {}^{1}_{0}\text{n} \rightarrow {}^{0}_{-1}\text{e} + {}^{244}_{96}\text{Cm}$

17.23 Write nuclear equations for the following processes: (a) oxygen-17 produced by α particle bombardment of nitrogen-14; (b) americium-240 produced by neutron bombardment of plutonium-239.

17.24 Write nuclear equations for the following transformations: (a) ${}^{257}\text{Rf}$ produced by the bombardment of californium-245 with carbon-12 nuclei; (b) the first synthesis of ${}^{266}\text{Mt}$ by the bombardment of bismuth-209 with iron-58 nuclei. Given that the first decay of meitnerium is by α emission, what is the daughter nucleus?

17.25 What will be the systematic name and atomic symbol given to (a) element 134; (b) element 145; (c) element 201?

17.26 What will be the systematic name and atomic symbol given to (a) element 117; (b) element 129; (c) element 222?

Nuclear Radiation

17.27 The activity of a certain radioactive source is 5.8×10^5 Bq. Express this activity in curies.

17.28 The activity of a sample containing carbon-14 is 10.7 Bq. Express this activity in microcuries.

17.29 Determine the number of disintegrations per second for radioactive sources of the following activities: (a) 2.5 μCi; (b) 117 Ci; (c) 7.2 mCi.

17.30 A certain Geiger counter is known to respond to only 1 of every 1000 radiations emitted from a sample. Calculate the activity of each radioactive source in curies, given the following data: (a) 590 clicks in 100 s; (b) 2.7×10^4 clicks in 1.5 h; (c) 159 clicks in 1.0 min.

17.31 A 1.0-kg sample absorbs an energy of 1.0 J as a result of its exposure to β radiation. Calculate the dose in rads and the dose equivalent in rems and in sieverts.

17.32 A 5.0-g sample of muscle tissue absorbs 2.0 J of energy as a result of its exposure to α radiation. Calculate the dose in rads and the dose equivalent in rems and in sieverts.

17.33 Someone is exposed to a source of β radiation that results in a dose rate of 1.0 rad$\cdot$d^{-1}. Given that nausea begins after a dose equivalent of about 100 rem, after what period will that symptom of radiation sickness be apparent?

17.34 Someone is exposed to a source of α radiation that results in a dose rate of 2.0 mrad·d^{-1}. If nausea begins after a dose equivalent of about 100 rem, after what period will nausea become apparent?

Measuring the Rate of Nuclear Decay

17.35 Determine the decay constant for (a) tritium, $t_{1/2}$ = 12.3 y; (b) lithium-8, $t_{1/2}$ = 0.84 s; (c) nitrogen-13, $t_{1/2}$ = 10.0 min.

17.36 Determine the half-life of (a) potassium-40, $k = 5.3 \times 10^{-10}$ y^{-1}; (b) cobalt-60, k = 0.132 y^{-1}; (c) nobelium-255, $k = 3.85 \times 10^{-3}$ s^{-1}.

17.37 Use Table 17.5 to calculate the time needed for the activity of each source to change as indicated: (a) a 1.0-Ci radium-226 source to decay to 0.10 Ci; (b) a 1.0-μCi potassium-40 source to decay to 10 nCi; (c) a 10-Ci cobalt-60 source to decay to 8 Ci.

17.38 Use Table 17.5 to calculate the time needed for the activity of each source to change as indicated: (a) a 0.010-mCi strontium-90 source to decay to 0.0010 mCi; (b) a 1.0-Ci iodine-131 source to decay to 1.0 mCi; (c) a 1.0-mCi thorium-234 source (half-life 24.1 d) to decay to 0.10 mCi.

17.39 Estimate the activity of a 4.4-Ci cobalt-60 source, $t_{1/2}$ = 5.26 y, after 50 y has passed.

17.40 The activity of a strontium-90 source is 3.0×10^4 Bq. What is its activity after 50 y has passed?

17.41 (a) What percentage of a carbon-14 sample remains after 2000 y? (b) Determine the percentage of a tritium sample that remains after 11.0 y.

17.42 (a) What percentage of a strontium-90 sample remains after 7.5 y? (b) Determine the percentage of an iodine-131 sample that remains after 7.0 d.

17.43 (a) What fraction of the original activity of ^{238}U remains after 4.5 Gy (1 Gy = 10^9 y)? (b) Potassium-40, which is presumed to exist at the formation of the Earth, is used for dating minerals. If three-fifths of the original potassium-40 exists in a rock, how old is the rock?

17.44 (a) A sample of krypton-85 ($t_{1/2}$ = 10.8 y) is released into the atmosphere. What fraction of the krypton-85 remains after 50.0 y? (b) A piece of wood, found in an archaeological dig, has a carbon-14 activity that is 75% of the current carbon-14 activity. How old is the piece of wood?

17.45 A 250-mg sample of carbon from a piece of cloth excavated from an ancient tomb in Nubia undergoes 1500 disintegrations in 10.0 h. If a current 1.00-g sample of carbon shows 920 disintegrations per hour, how old is the piece of cloth?

17.46 A current 1.00-g sample of carbon shows 920 disintegrations per hour. If 1.00 g of charcoal from an archaeological dig in a limestone cave in Slovenia shows 5500 disintegrations in 24.0 h, what is the age of the charcoal sample?

17.47 Use the law of radioactive decay to determine the activity of (a) a 1.0-mg sample of radium-226 ($t_{1/2} = 1.60 \times 10^3$ y); (b) a 2.0-mg sample of strontium-90 ($t_{1/2}$ = 28.1 y); (c) a 0.43-mg sample of promethium-147 ($t_{1/2}$ = 2.6 y). The mass of each nuclide in atomic mass units (u) is equal to its mass number, within two significant figures.

17.48 Use the law of radioactive decay to determine the activity of (a) a 1.0-g sample of $^{235}UO_2$ ($t_{1/2} = 7.1 \times 10^8$ y); (b) a 1.0-g sample of cobalt containing 1.0% ^{60}Co ($t_{1/2}$ = 5.26 y); (c) a 5.0-mg sample of thallium-200 ($t_{1/2}$ = 26.1 h). The mass of each nuclide in atomic mass units (u) is equal to its mass number, within two significant figures.

17.49 The activity of an iodine-131 source ($t_{1/2}$ = 8.05 d), which is used to monitor the functioning of the thyroid gland, is 500 Bq. How long will it be before the activity is 10 disintegrations per second?

17.50 A sample of fermium-244 ($t_{1/2}$ = 3.3 ms) having an activity of 0.10 mCi was produced in a nuclear reactor. What will the activity of the ^{244}Fm be after 1.0 s?

17.51 (a) A sample of phosphorus-32 has an initial activity of 58 counts per second. After 12.3 days, the activity was 32 counts per second. What is the half-life of phosphorus-32? (b) If phosphorus-32 is used in an experiment to monitor the consumption of phosphorus by plants, what fraction of the nuclide will remain after 30 d?

17.52 A cobalt-60 source purchased for the radiation therapy of cancer patients has an activity of 1.20 Ci. What will the activity of the source be after 5.0 y?

17.53 A sample containing ^{35}S ($t_{1/2}$ = 88 d) that is being used to study the reactions by which sulfur is utilized by bacteria has an activity of 10.0 Ci. What mass of sulfur-35 is present? The mass of a sulfur-35 atom is 35 u.

17.54 What is the activity (in curies) of a 22-mg sample of ^{210}Po ($t_{1/2}$ = 138.4 d)? The mass of a polonium-210 atom is about 210 u.

Uses of Radioisotopes

17.55 A chemist would like to study the mechanism of the hydrolysis reaction of the organic ester methyl acetate, as shown below:

$$H_3C{-}C({=}O){-}O{-}CH_3 + H_2O \longrightarrow H_3C{-}C({=}O){-}O{-}H + CH_3OH$$

The question arises as to whether the oxygen atom present in the product methanol derives from the starting methyl acetate or from the added water. Propose an experiment using isotopes that would allow one to determine the origin of the oxygen atom.

17.56 It is not possible to measure the blood volume of an animal directly without damaging the animal. Suggest a nondestructive method of determining blood volume by using radioisotopes.

17.57 What would you expect to happen to the vibrational frequency of the C—H bond of methane if the protons ^1H are replaced by ^2H? See Major Technique 1, Infrared Spectroscopy.

17.58 Normal water, H_2O, is necessary for life, but heavy water, D_2O, is toxic. Suggest a reason for this difference. The pK_a of deuterium oxide is 14.955 at 25°C.

Nuclear Energy

17.59 Calculate the mass loss or gain for each of the following processes: (a) a 250-g sample of copper (specific heat capacity, $0.39\ J\cdot(°C)^{-1}\cdot g^{-1}$) is heated from 35°C to 250°C; (b) a 50.0-g sample of water freezes at 0°C ($\Delta H_{melt}° = 6.01\ kJ\cdot mol^{-1}$); (c) 2.0 mol $PCl_5(g)$ is formed from its elements under standard conditions at 25°C.

17.60 Calculate the mass loss or gain for each of the following processes: (a) a 50.0-g block of iron (specific heat capacity, $0.45\ J\cdot(°C)^{-1}\cdot g^{-1}$) cools from 600°C to 25°C; (b) a 100-g sample of ethanol vaporizes at its normal boiling point ($\Delta H_{vap}° = 43.5\ kJ\cdot mol^{-1}$); (c) 10.0 g of $SO_2(g)$ is formed from its elements under standard conditions at 25°C.

17.61 Calculate the energy in joules that is equivalent to (a) 1.0 g of matter; (b) one electron; (c) 1.0 pg of matter; (d) one proton.

17.62 Calculate the energy in joules that is equivalent to (a) 1.00 kg of matter; (b) 1.00 lb of matter (1 lb = 454 g); (c) one neutron; (d) one hydrogen atom.

17.63 The Sun emits radiant energy at the rate of $3.9 \times 10^{26}\ J\cdot s^{-1}$. What is the rate of mass loss (in kilograms per second) of the Sun?

17.64 (a) For the fusion reaction $6\ D \rightarrow 2\ ^4He + 2\ ^1H + 2\ n$, 3×10^8 kJ of energy is released by a certain sample of deuterium. What is the mass loss (in grams) for the reaction? (b) What was the mass of deuterium converted? The molar mass of deuterium is $2.014\ g\cdot mol^{-1}$.

17.65 Calculate the binding energy per nucleon ($J\cdot nucleon^{-1}$) for (a) ^{62}Ni, 61.928 346 u; (b) ^{239}Pu, 239.0522 u; (c) ^2H, 2.0141 u; (d) ^3H, 3.016 05 u. (e) Which nuclide is the most stable? ($1\ u = 1.6605 \times 10^{-27}$ kg.)

17.66 Calculate the binding energy per nucleon ($J\cdot nucleon^{-1}$) for (a) ^{98}Mo, 97.9055 u; (b) ^{151}Eu, 150.9196 u; (c) ^{56}Fe, 55.9349 u; (d) ^{232}Th, 232.0382 u. (e) Which nuclide is the most stable? ($1\ u = 1.6605 \times 10^{-27}$ kg.)

17.67 Calculate the energy released per gram of starting material in the fusion reaction represented by each of the following equations. ($1\ u = 1.6605 \times 10^{-27}$ kg.)
(a) $D + D \rightarrow {^3He} + n$ (D, 2.0141 u; ^3He, 3.0160 u)
(b) $^3He + D \rightarrow {^4He} + {^1H}$ (^1H, 1.0078 u; ^4He, 4.0026 u)
(c) $^7Li + {^1H} \rightarrow 2\ ^4He$ (^7Li, 7.0160 u)
(d) $D + T \rightarrow {^4He} + {^1H}$ (T, 3.0160 u)

17.68 Calculate the energy released per gram of starting material in the nuclear reaction represented by each of the following equations. ($1\ u = 1.6605 \times 10^{-27}$ kg.)
(a) $^7Li + {^1H} \longrightarrow n + {^7Be}$ (^1H, 1.0078 u; ^7Li, 7.0160 u; ^7Be, 7.0169 u)
(b) $^{59}Co + D \longrightarrow {^1H} + {^{60}Co}$ (^{59}Co, 58.9332 u; ^{60}Co, 59.9529 u; D, 2.0141 u)
(c) $^{40}K + \beta \longrightarrow {^{40}Ar}$ (^{40}K, 39.9640 u; ^{40}Ar, 39.9624 u; β, 0.0005 u)
(d) $^{10}B + n \longrightarrow {^4He} + {^7Li}$ (^{10}B, 10.0129 u; ^4He, 4.0026 u; ^7Li, 7.0160 u)

17.69 Sodium-24 (23.990 96 u) decays to magnesium-24 (23.985 04 u). Determine (a) the change in energy that accompanies the decay; (b) the change in the binding energy per nucleon.

17.70 (a) How much energy is emitted in each α decay of plutonium-234? (^{234}Pu, 234.0433 u; ^{230}U, 230.0339 u). (b) The half-life of plutonium-234 is 8.8 h. How much heat is released by the α decay of a 1.00-μg sample of plutonium-234 in a 24-h period?

Nuclear Fission

17.71 Each of the following equations represents a fission reaction. Complete and balance the equations:
(a) ${}^{244}_{95}\text{Am} \rightarrow {}^{134}_{53}\text{I} + {}^{107}_{42}\text{Mo} + 3?$
(b) ${}^{235}_{92}\text{U} + {}^{1}_{0}\text{n} \rightarrow ? + {}^{138}_{52}\text{Te} + 2\ {}^{1}_{0}\text{n}$
(c) ${}^{235}_{92}\text{U} + {}^{1}_{0}\text{n} \rightarrow {}^{101}_{42}\text{Mo} + {}^{132}_{50}\text{Sn} + ?$

17.72 Complete each of the following nuclear equations for fission reactions:
(a) ${}^{239}_{94}\text{Pu} + {}^{1}_{0}\text{n} \longrightarrow {}^{98}_{42}\text{Mo} + {}^{138}_{52}\text{Te} + ?$
(b) ${}^{239}_{94}\text{Pu} + {}^{1}_{0}\text{n} \longrightarrow {}^{100}_{43}\text{Tc} + ? + 4\ {}^{1}_{0}\text{n}$
(c) ${}^{239}_{94}\text{Pu} + {}^{1}_{0}\text{n} \longrightarrow ? + {}^{139}_{49}\text{In} + 3\ {}^{1}_{0}\text{n}$

INTEGRATED EXERCISES

17.73 (a) How many radon-222 nuclei ($t_{1/2} = 3.82$ d) decay per minute to produce an activity of 4 pCi? (b) A bathroom in the basement of a home measures $2.0 \times 3.0 \times 2.5$ m. If the activity of radon-222 in the room is $4.0\ \text{pCi}\cdot\text{L}^{-1}$, how many nuclei decay during a shower lasting 5.0 min?

17.74 (a) How long will it take for the loss of 99.0% of a sample of radon-222 gas ($t_{1/2} = 3.82$ d)? (b) Comment on radon's lifetime and chances of reaching the surface if formed deep in the Earth's crust. (c) How might a homeowner reduce the rate at which radon enters a home?

17.75 It is found that 1.0×10^{-5} mol of radon-222 atoms ($t_{1/2} = 3.82$ d) has seeped into a closed basement with a volume of 2000 m^3. (a) What is the initial activity of the radon in picocuries per liter ($\text{pCi}\cdot\text{L}^{-1}$)? (b) How many atoms of ${}^{222}\text{Rn}$ will remain after one day (24 h)? (c) How long will it take for the radon to decay to below the EPA recommended level of $4\ \text{pCi}\cdot\text{L}^{-1}$?

17.76 Tritium undergoes β decay, and the emitted β particle has an energy of 0.0186 MeV (1 MeV = 1.602×10^{-13} J). If a 1.0-g sample of tissue absorbs 10% of the decay products of 1.0 mg of tritium, what dose equivalent does the tissue absorb?

17.77 In the mid-1940s, it was proposed that atomic bombs could be used to simplify excavation work, such as that required to dig large canals, and the issue has been raised again. (a) Which type of bomb would be more suitable for such work, fusion or fission? Explain your reasoning. (b) Present arguments for and against this proposal.

17.78 Explain why containers of radioactive waste are more susceptible to corrosion than containers of nonradioactive waste with the same chemical reactivity.

17.79 Uranium-238 decays through a series of α and β emissions to lead-206, with an overall half-life for the entire process of 4.5×10^9 y. If a uranium-bearing ore is found to have a ${}^{238}\text{U}/{}^{206}\text{Pb}$ ratio of (a) 1.00; (b) 1.25, how old is the ore sample?

17.80 The age of a bottle of wine was determined by monitoring the tritium level in the wine. The activity of the tritium is determined to be 8.3% that of a sample of fresh grape juice from the same region from which the wine was bottled. How old is the wine?

17.81 Sodium-24 is used for monitoring blood circulation. (a) If a 2.0-mg sample of sodium-24 has an activity of 17.3 Ci, what is its decay constant and its half-life? (b) What mass of a 2.0-mg sample remains after 2.0 days? The mass of a sodium-24 atom is 24 u.

17.82 Breeder reactors run so hot that a high-boiling-point mixture of molten sodium and potassium is used as the coolant, instead of water. Unfortunately, the steel cooling pipes are susceptible to embrittlement and fracture. It is suspected that the steel becomes brittle because the liquid metals leach carbon from it. Suppose that you had at your disposal the following materials: a fully equipped research laboratory, a small research foundry for making steel, a gas chromatograph, a scintillation counter, and supplies of sodium, potassium, iron, molybdenum, and carbon-14. Design an experiment to test the hypothesis that liquid sodium and potassium mixtures can dissolve carbon from steel.

17.83 Radiopharmaceuticals have one of two general functions: (1) They may be used to *detect* or *image* biological problems such as tumors. (2) They may be used to *treat* an illness. Which type of radiation (α, β, or γ) would be the best suited for (a) detection and (b) therapy. Justify your selections. (c) From standard literature sources, find at least two radionuclides that have been used for imaging body tissues. (d) What are the half-lives of these radionuclides?

17.84 Radioactive metal ions that have short half-lives are being intensely studied as pharmaceuticals. The strategy is to attach a well-designed ligand to the metal ion so that the complex very selectively aggregates in one particular type of body tissue. What properties of the ligand are important to the design of effective radiopharmaceutical therapeutic agents?

Media Link

17.85 (a) Using a standard graphing program such as that found on the Web site for this text, plot the fraction of ^{14}C remaining in a 40 000-year-old archaeological sample as a function of time. For convenience, use 1000-y intervals. (b) Plot the natural logarithm of the fraction of ^{14}C remaining as a function of time. (c) After what period of time, will there be less than 1.00% of the original ^{14}C in the sample remaining?

17.86 One difficulty in removing plutonium from the human body by chelation therapy is that the normal agents also remove iron. (a) What are the common oxidation states of plutonium? (b) From standard literature sources, find the ionic radii of the most important plutonium and iron ions. (c) Suggest an explanation, given these data, as to why chelating ligands appropriate for removal of plutonium will also remove iron.

Chapter 18

Lubricating oil, like this automobile engine oil, consists primarily of hydrocarbons of high molar mass. Their long chains coil about one another, giving the oil a high viscosity.

Why Do We Need to Know This Material?

All life on Earth is based on carbon; so is the fuel we burn, our food, and the clothes we wear. Therefore, to understand a major part of the everyday world, we need to be familiar with the chemistry of this element. Compounds of carbon and hydrogen provide electricity, heat our homes, and serve as the foundation of the petrochemical industry. They are also used to make the flexible, strong polymeric and composite materials that make modern communication and transportation possible.

What Do We Need to Know Already?

This chapter draws on the introduction to organic formulas and nomenclature in Sections C and D, the structure of benzene (Sections 2.8 and 3.13), the formation of σ- and π-bonds (Section 3.4), hybridization of carbon atoms (Sections 3.5–3.7), intermolecular forces (Sections 5.3–5.5), reaction enthalpy (Section 6.14), and reaction mechanism (Sections 13.10–13.12).

ORGANIC CHEMISTRY I: THE HYDROCARBONS

Organic chemistry is the study of the enormously varied range of compounds of carbon. Materials based on carbon have properties ranging from the softness of artificial skin and cartilage to the tough composite materials used in experimental aircraft. What makes carbon atoms so adaptable is their ability to string together to form chains and rings of endless variety. This versatility allows carbon to form the thousands of complicated molecules that are needed to carry out the intricate processes we associate with being alive. The inclusion of elements other than carbon and hydrogen into organic compounds increases their diversity enormously, as we shall see in Chapter 19.

This chapter and the next introduce you to the chemistry of carbon and its compounds. This chapter deals primarily with **hydrocarbons,** compounds built from carbon and hydrogen. Hydrocarbons fall into two distinct categories, those that contain a benzene ring and those that do not. These two classes, which are called *aromatic* and *aliphatic,* respectively, have very different properties, and we deal with them separately in this chapter, aliphatic first, then aromatic. In the following chapter, we see that both types of hydrocarbons can be regarded as the basic framework for all organic compounds. This chapter also introduces you to the principles of organic nomenclature, and we see that the vast number of hydrocarbons can easily be distinguished by naming them systematically.

ALIPHATIC HYDROCARBONS

Petroleum, the raw material of the petrochemical industry, consists mostly of members of the simplest family of organic compounds, the hydrocarbons. The different types of hydrocarbons are characterized by different reactions. We can understand the chemical properties of hydrocarbons in terms of the mechanisms of their reactions, the details of the steps that turn one compound into another. Once the mechanism of a reaction is known, we can use the knowledge to create new substances in a rational way, rather than just by trial and error.

18.1 Types of Hydrocarbons

An **aromatic hydrocarbon** has a benzene ring as a part of its molecular structure; an **aliphatic hydrocarbon** does not. The compound shown as (**1**) is aromatic; compound (**2**) is aliphatic. We can often speak of a molecule as having an "aromatic region" consisting of benzene rings and an "aliphatic region" consisting of chains of carbon atoms.

Many organic molecules are very complicated, so we need a simple way to represent their structures. It is often enough to give a **condensed structural formula,** which shows how the atoms are grouped together (see

CH_2CH_3

1 Ethylbenzene, $C_6H_5CH_2CH_3$

CH_3 CH_2 CH_2 CH_2 CH_3

2 Pentane, C_5H_{12}

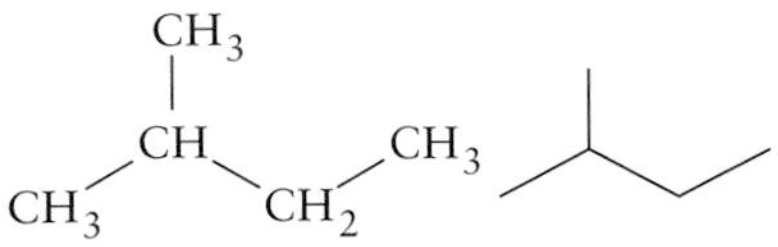

Section C). For instance, we write $CH_3CH_2CH_2CH_3$ for butane and $CH_3CH(CH_3)CH_3$ for methylpropane. The parentheses around the CH_3 group show that it is attached to the carbon atom on its left (or if the formula begins with the parenthetical group, the carbon atom on its right). When several groups of atoms are repeated, we collect them together; so we can write butane as $CH_3(CH_2)_2CH_3$ and methylpropane as $(CH_3)_3CH$.

A line structure (introduced in Section C) represents a chain of carbon atoms as a zigzag line. The end of each short line in the zigzag represents a carbon atom. Because carbon nearly always has a valence of 4 in organic compounds, we do not need to show the C—H bonds. We just fill in the correct number of hydrogen atoms mentally, as we see for methylbutane (**3**), isoprene (**4**), and propyne (**5**). As explained in Section 2.8, a benzene ring is represented by a circle inside a hexagon, and we need to remember that one hydrogen atom is attached to each carbon atom.

3 Methylbutane, $(CH_3)_2CHCH_2CH_3$

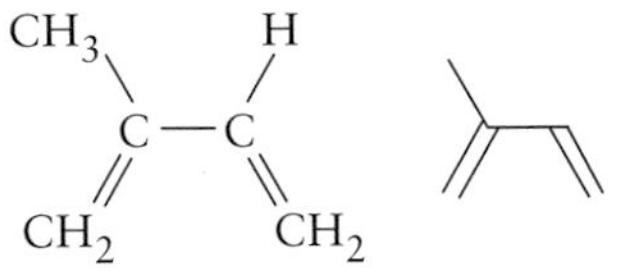

4 Isoprene, $CH_2C(CH_3)CHCH_2$

$CH_3—C{\equiv}CH$

5 Propyne, CH_3CCH

Self-Test 18.1A Draw (a) the line structure of aspirin (**6a**) and (b) the structural formula for $CH_3(CH_2)_2C(CH_3)_2CH_2C(CH_3)_3$.

[*Answer:* (a) (**6b**); (b) (**7**)]

Self-Test 18.1B Write (a) the structural formula of carvone (**8**) and (b) the condensed structural formula for

$$CH_3-\underset{\displaystyle CH_3}{\underset{|}{CH}}-CH_2-CH_2-CH_2-\underset{\displaystyle CH_3}{\underset{|}{CH}}-CH_2-CH_3$$

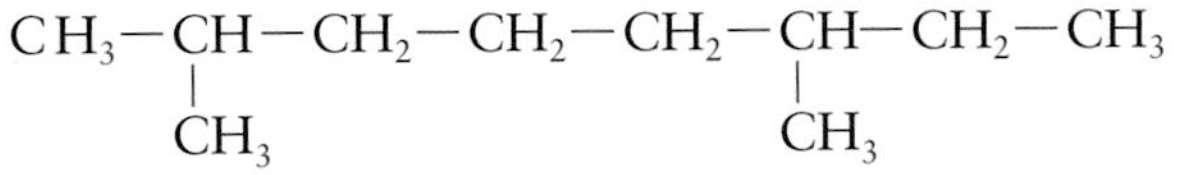

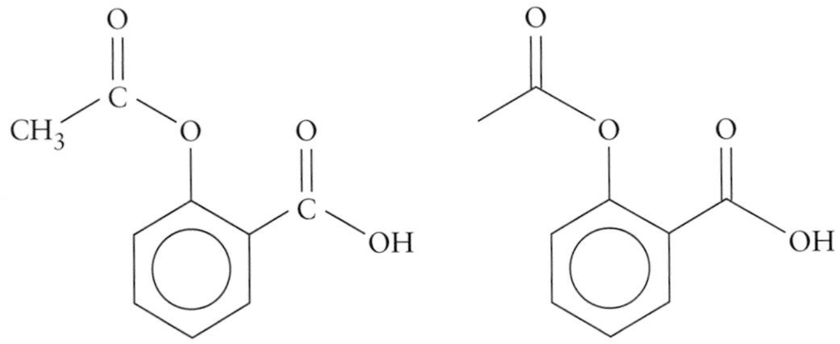

(a) (b)

6 Acetylsalicylic acid (aspirin)

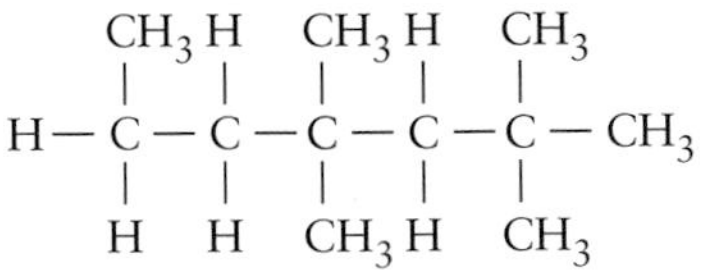

7

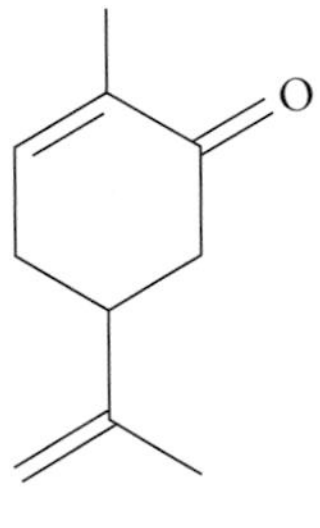

8 Carvone

A **saturated hydrocarbon** is an aliphatic hydrocarbon with no multiple carbon-carbon bonds; an **unsaturated hydrocarbon** has one or more double or triple carbon-carbon bonds. More hydrogen can be added to compounds in which there are multiple bonds, but compounds with only single bonds are "saturated" with hydrogen. Compound (**9**) is saturated; compounds (**10**) and (**11**) are unsaturated.

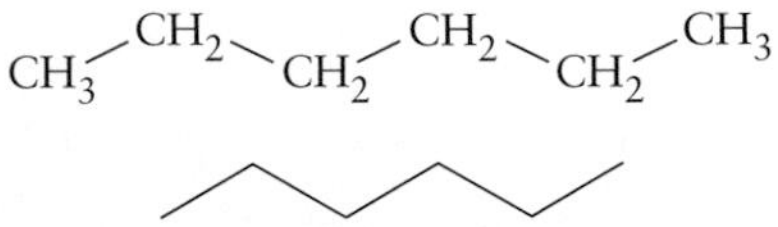

9 Hexane, C_6H_{14}

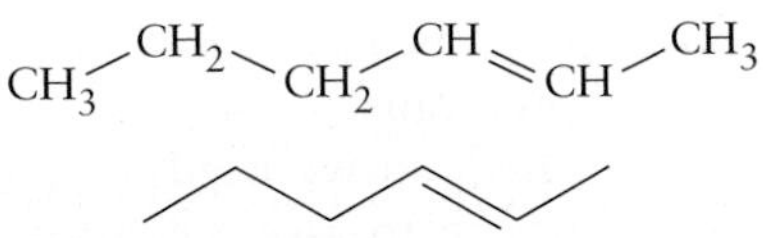

10 2-Hexene, C_6H_{12}

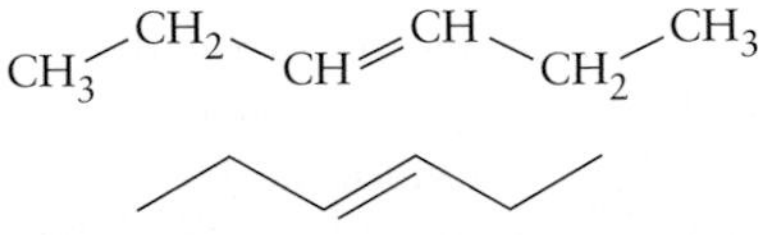

11 3-Hexene, C_6H_{12}

Aromatic hydrocarbons contain benzene rings; aliphatic hydrocarbons do not. Saturated hydrocarbons have only single bonds; unsaturated hydrocarbons have one or more multiple bonds.

18.2 Nomenclature of Aliphatic Hydrocarbons

Saturated hydrocarbons are called **alkanes.** The simplest alkane is methane, CH_4 (**12**). The formulas of other alkanes are derived from CH_4 by inserting CH_2 groups between pairs of atoms. To name an alkane in which the carbon atoms form a single chain, we combine a prefix denoting the number of carbon atoms with the suffix- *ane* (Table 18.1). For example, $CH_3{-}CH_2{-}H$ (more simply, CH_3CH_3) is ethane and $CH_3CH_2{-}CH_2{-}H$ (that is, $CH_3CH_2CH_3$) is propane. Cyclopropane, C_3H_6 (**13**), and cyclohexane, C_6H_{12} (**14**), are **cycloalkanes,** alkanes that contain rings of carbon atoms.

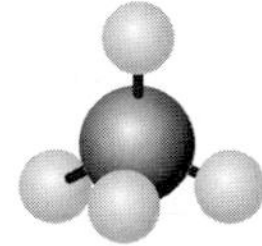

12 Methane, CH_4

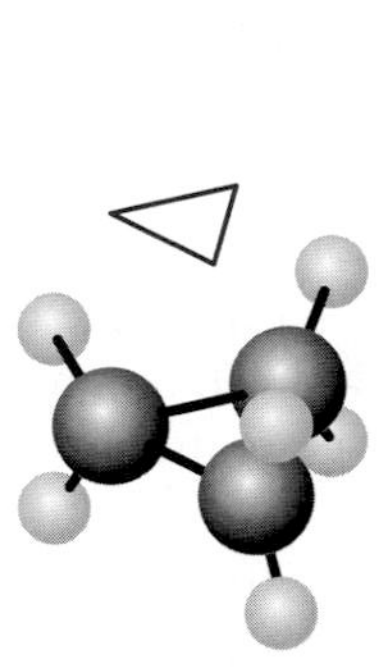

13 Cyclopropane, C_3H_6

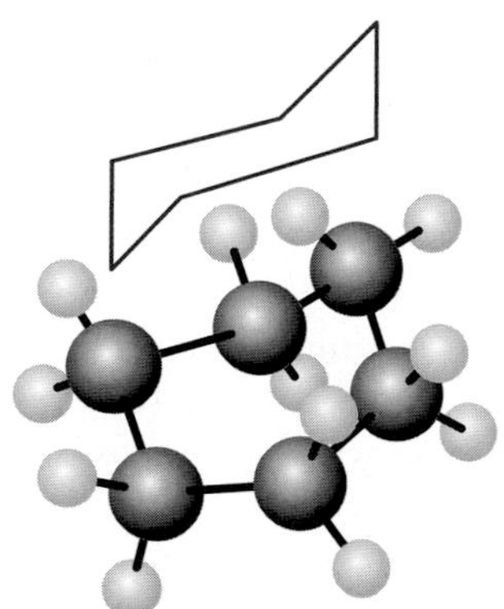

14 Cyclohexane, C_6H_{12}

As we saw in Section 3.5, the bonds to each carbon atom in an alkane lie in a tetrahedral arrangement, with sp^3 hybridization. Although we write alkanes as linear structures, they really consist of three-dimensional

TABLE 18.1 Alkane Nomenclature

Number of carbon atoms	Formula	Name of alkane	Name of alkyl group	Formula
1	CH_4	methane	methyl	$CH_3{-}$
2	CH_3CH_3	ethane	ethyl	$CH_3CH_2{-}$
3	$CH_3CH_2CH_3$	propane	propyl	$CH_3CH_2CH_2{-}$
4	$CH_3(CH_2)_2CH_3$	butane	butyl	$CH_3(CH_2)_2CH_2{-}$
5	$CH_3(CH_2)_3CH_3$	pentane	pentyl	$CH_3(CH_2)_3CH_2{-}$
6	$CH_3(CH_2)_4CH_3$	hexane	hexyl	$CH_3(CH_2)_4CH_2{-}$
7	$CH_3(CH_2)_5CH_3$	heptane	heptyl	$CH_3(CH_2)_5CH_2{-}$
8	$CH_3(CH_2)_6CH_3$	octane	octyl	$CH_3(CH_2)_6CH_2{-}$
9	$CH_3(CH_2)_7CH_3$	nonane	nonyl	$CH_3(CH_2)_7CH_2{-}$
10	$CH_3(CH_2)_8CH_3$	decane	decyl	$CH_3(CH_2)_8CH_2{-}$
11	$CH_3(CH_2)_9CH_3$	undecane	undecyl	$CH_3(CH_2)_9CH_2{-}$
12	$CH_3(CH_2)_{10}CH_3$	dodecane	dodecyl	$CH_3(CH_2)_{10}CH_2{-}$

tetrahedral arrangements of bonds at each carbon atom. Moreover, because all the C—C bonds are single, the different parts of an alkane molecule can rotate relative to each other. In liquids and gases, the chains of alkanes are in constant motion, often rolling up into a ball (**15**) and sometimes stretching out into a zigzag (**16**).

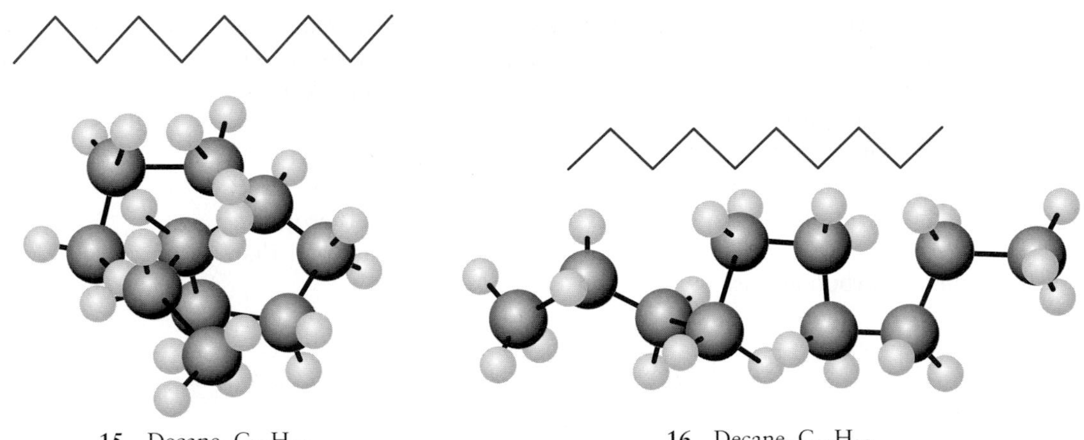

15 Decane, $C_{10}H_{22}$

16 Decane, $C_{10}H_{22}$

The electronegativities of carbon and hydrogen (2.6 and 2.2, respectively) are so similar and rotation about bonds so free that hydrocarbon molecules are best regarded as nonpolar. The dominant interaction between alkane molecules is therefore the London force (Section 5.4). Because the strength of this interaction increases with the number of electrons in the molecule, we can expect the alkanes to become less volatile with increasing molar mass and therefore to be separable by fractional distillation (Fig. 18.1). The lightest members, methane through butane, are gases at room temperature. Pentane is a volatile liquid, and hexane through undecane ($C_{11}H_{24}$) are moderately volatile liquids that are present in gasoline (Table 18.2). *Kerosene*, a fuel used in jet and diesel engines, contains a number of alkanes in the range C_{10} to C_{16}. Lubricating oils are mixtures in the range C_{17} to C_{22}. The heavier members of the series include the *paraffin waxes* and *asphalt*. All alkanes are insoluble in water. Because their densities are lower than that of water, they float on the surface of water. Even a small oil spill at sea can form an organic layer on the surface of the ocean and spread over a huge area.

With C_4H_{10}, we find another reason for the variety of compounds that carbon can form: the same atoms can bond together in different

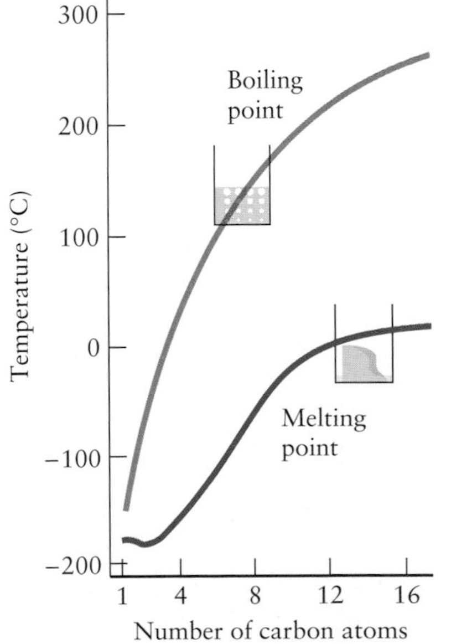

FIGURE 18.1 The melting and boiling points of the unbranched alkanes from CH_4 to $C_{16}H_{34}$.

TABLE 18.2 Hydrocarbon Constituents of Petroleum

Hydrocarbons	Boiling range, °C	Fraction
C_1 to C_4	−160 to 0	natural gas and propane
C_5 to C_{11}	30 to 200	gasoline
C_{10} to C_{16}	180 to 400	kerosene, fuel oil
C_{17} to C_{22}	350 and above	lubricants
C_{23} to C_{34}	low-melting-point solids	paraffin wax
C_{35} upward	soft solids	asphalt

arrangements. Four carbon atoms can link together in a chain to form butane (**17**) or in a Y-shape to form methylpropane (**18**). Different compounds with the same molecular formula are called **isomers.** Thus, butane and methylpropane are isomers with the same molecular formula, C_4H_{10}.

The name *isomer* comes from the Greek words for "equal parts," suggesting that isomers are built from the same kit of parts. Isomers of organic compounds are discussed in more detail in Section 18.3.

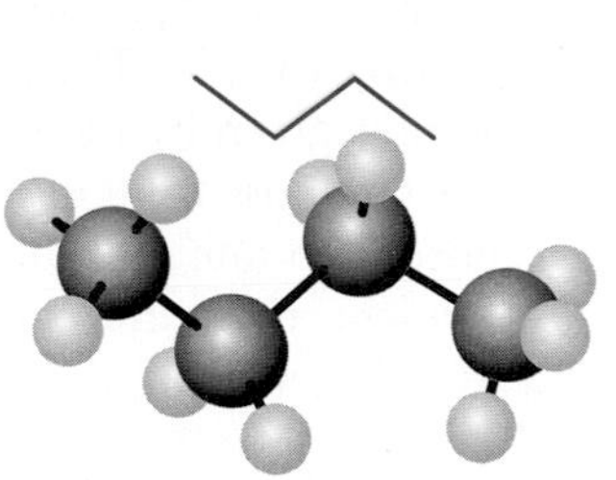

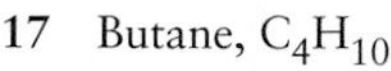

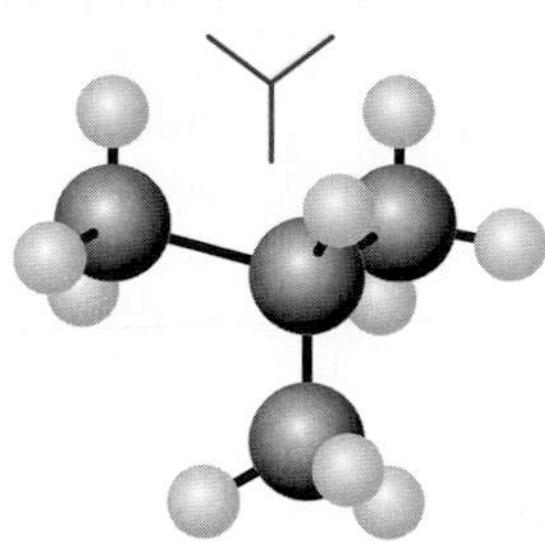

17 Butane, C_4H_{10} **18** Methylpropane, C_4H_{10}

The simplest unsaturated hydrocarbon is ethene, C_2H_4 or $H_2C{=}CH_2$, commonly called ethylene (**19**). It is the parent of the **alkenes,** a series of compounds with formulas derived from $H_2C{=}CH_2$ by inserting CH_2 groups. For example, the next member of the family is propene, $H{-}CH_2{-}CH{=}CH_2$ (more simply, $CH_3CH{=}CH_2$). The name of an alkene is the same as the name of the corresponding alkane, except that it ends in *-ene*. The location of the double bond is specified by numbering the carbon atoms in the chain and writing the lower of the two numbers of the carbon atoms joined by the double bond. Thus, $CH_3CH_2CH{=}CH_2$ is 1-butene and $CH_3CH{=}CHCH_3$ is 2-butene (see Toolbox 18.1).

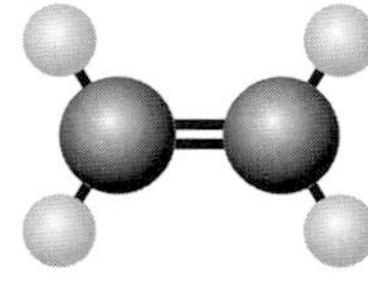

19 Ethene, C_2H_4

The **alkynes** are hydrocarbons that have at least one carbon-carbon triple bond. The simplest is ethyne, $HC{\equiv}CH$, which is commonly called acetylene (**20**). Alkynes are named like the alkenes but with the suffix *-yne.*

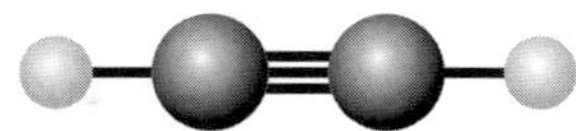

20 Ethyne, C_2H_2

Alkanes are saturated hydrocarbons. Alkenes and alkynes are unsaturated hydrocarbons: the former have a carbon-carbon double bond and the latter have a triple bond.

TOOLBOX 18.1 How to name aliphatic hydrocarbons

Conceptual Basis

The aim here, as throughout chemical nomenclature, is to be succinct but unambiguous.

Procedure

The following rules have been adopted by the International Union for Pure and Applied Chemistry (IUPAC).

Alkanes

The names of straight-chain alkanes are given in Table 18.1: they all end in *-ane*. To name a branched alkane, we treat the side chains as substituents. Name a substituent formed by the removal of one hydrogen atom from an alkane by changing the ending *-ane* to *-yl* (as in the last two columns in Table 18.1)

Example: $CH_3CH_2{-}$ is the ethyl group

The names of branched-chain hydrocarbons and hydrocarbon derivatives are based on the name of the *longest* continuous carbon chain in the molecule (which may not be shown in a horizontal line).

Example:

$$\begin{array}{l} CH_3{-}\underset{\displaystyle |}{C}H{-}CH_3 \\ \quad\;\; CH_2{-}CH_3 \end{array}$$

is 2-methylbutane

A cyclic (ring) hydrocarbon is designated by the prefix *cyclo-*.

Example:

$$\begin{array}{c} CH_2 \\ / \;\backslash \\ H_2C{-}CH_2 \end{array}$$

is cyclopropane

To indicate the position of a branch or substituent, the carbons in the longest chain are numbered consecutively from one end to the other, starting at the end that will give the *lower* number(s) to the substituent(s).

Examples:

```
5    4    3    2  1
CH3CH2CH2CHCH3
          |
          CH3
```

2-Methylbutane

```
1    2   3          4    5    6
CH3CH2C(CH3)2CH2CH2CH3
```

3,3-Dimethylhexane

The prefixes di-, tri-, tetra-, penta-, hexa-, etc., indicate how many of each substituent are in the molecule. Numbers set off by hyphens specify to which carbon atoms the groups are attached.

Examples:

```
     CH3 CH3
      |   |
H3C — C — C — CH3
      |   |
     CH3  H
```

2,2,3-Trimethylbutane

```
     H   CH2CH3
      \ /
       C      H
      / \    /
  H2C     C
     \   / \
   H2C—CH2  CH3
```

1-Ethyl-2-methylcyclopentane

The substituents are listed in alphabetical order (disregarding the Greek prefixes) and attached to the root name. The names of substituents other than hydrocarbon groups are discussed more fully in Chapter 19 (see Toolbox 19.1).

Alkenes and alkynes

Double bonds in hydrocarbons are indicated by changing the suffix *-ane* to *-ene*, and triple bonds are indicated by changing the suffix to *-yne*. The position of the multiple bond is given by the number of the first (lower numbered) carbon atom involved in the multiple bond. If more than one multiple bond of the same type is present, the number of these bonds is indicated by a Greek prefix.

Examples:

$CH_3{-}CH_2{-}CH{=}CH{-}CH_3$

2-Pentene

$CH_3{-}CH_2{-}CH_2{-}C{\equiv}CH$

1-Pentyne

$CH_2{=}CH{-}CH{=}CH_2$

1,3-Butadiene

When numbering atoms in the chain, the lowest numbers are given preferentially to (a) functional groups named by suffixes (see Toolbox 19.1), (b) double bonds, (c) triple bonds, and (d) groups named by prefixes. Example 18.2 illustrates this rule.

```
      CH3           CH3
      |            /
CH3 — C — CH2 — CH
      |            \
      CH3           CH3
```

21

EXAMPLE 18.1 Naming alkanes and cycloalkanes

(a) Name the compound shown as (**21**) and (b) write the structural formula of 2-ethyl-1,1-dimethylcyclohexane.

STRATEGY Use the rules in Toolbox 18.1 for naming alkanes. To write a structural formula, begin with the longest chain or ring and add any multiple bonds or groups. Satisfy carbon's valence of four by adding hydrogen atoms where they are needed.

SOLUTION (a) First, identify the longest unbranched chain of carbon atoms and give it the name of the corresponding alkane. It follows that (**21**) is a substituted pentane, because its longest carbon chain (in red type) has five carbon atoms. Next, we identify the substituent groups: there are three methyl groups (CH_3—). The molecule is therefore a trimethylpentane. Finally, we denote the locations of the substituents by numbering the backbone carbon atoms from whichever end of the molecule results in the lowest numbered locations for the substituents:

```
        CH3         CH3
   1   2|    3     4|   5
H3C — C — CH2 — C — CH3
        |           |
        CH3         H
```

The molecule is 2,2,4-trimethylpentane. (b) Draw the longest chain of carbon atoms first. The last part of the name, cyclohexane, tells us that the molecule has a ring of six carbon atoms:

```
    C
  /   \
 C     C
 |     |
 C     C
  \   /
    C
```

Add two methyl groups to one carbon atom, which becomes carbon atom 1. Then add an ethyl group to carbon atom 2. Add hydrogen atoms as needed to give each carbon atom a valence of four:

```
     H3C  CH3
        \ /
         C         CH2CH3
       /   \      /
   H2C      CH
    |        |
   H2C      CH2
      \    /
       CH2
```

Self-Test 18.2A (a) Name the compound shown as (**22**) and (b) write the structural formula of 5-ethyl-2,2-dimethyloctane.

[*Answer:* (a) 4-Ethyl-3-methyloctane; (b) (**23**)]

Self-Test 18.2B Name the compound $(CH_3)_2CHCH_2CH(CH_2CH_3)_2$ and write the structural formula of 3,3,5-triethylheptane.

```
CH3—CH2—CH2
         |
        CH2—CH—CH2—CH3
             |
             CH—CH3
            /
     CH3—CH2

           22
```

```
     CH3
      |
CH3—C—CH2—CH2—CH—CH2—CH2—CH3
      |            |
     CH3          CH2CH3

           23
```

EXAMPLE 18.2 Naming alkenes

(a) Name the alkene $CH_3CH_2CH{=}CH_2$ and (b) write the condensed structural formula for 5-methyl-1,3-hexadiene.

STRATEGY Use the rules in Toolbox 18.1 for naming alkenes. In all cases except ethene (ethylene) and propene (propylene) and the corresponding alkynes, it is necessary to specify the location of the multiple bond. The numbering of the multiple bonds starts at the end of the backbone that results in the lower numbers and has priority over the numbering of substituents.

SOLUTION (a) $CH_3CH_2CH{=}CH_2$ is 1-butene. (b) 5-Methyl-1,3-hexadiene has two double bonds, at the first and third carbon atoms, with a methyl group on the fifth carbon atom: $CH_2{=}CHCH{=}CHCH(CH_3)CH_3$.

Self-Test 18.3A (a) Name the alkene $(CH_3)_2CHCH{=}CH_2$ and (b) write the condensed structural formula for 2-methylpropene.

[*Answer:* (a) 3-Methyl-1-butene; (b) $CH_2{=}C(CH_3)_2$]

Self-Test 18.3B (a) Name the alkene $(CH_3CH_2)_2CHCH{=}CHCH_3$ and (b) write the structural formula for cyclopropene.

18.3 Isomers

Isomers were first introduced in Section 16.7.

Figure 18.2 summarizes the types of isomerism that are found in organic compounds. Molecules that are **structural isomers** are built from the same atoms but the atoms are connected differently; that is, the molecules have a different **connectivity.** For example, as we saw in Section 18.2, we can insert a $-CH_2-$ group into the C_3H_8 molecule in two different ways to give two different compounds with the formula C_4H_{10}:

$CH_3CH_2CH_2CH_3$
Butane

$CH(CH_3)_3$
Methylpropane

Although the $-CH_2-$ group could be inserted in other places, the resulting molecules can be twisted into one or the other of these two isomers. Both compounds are gases, but butane (**24**) condenses at -1°C, whereas methylpropane (**25**) condenses at -12°C. Two molecules that differ only by rotation about one or more bonds may look different, but they are not isomers of each other; they are different **conformations** of the same molecule.

24 Butane, C_4H_{10}

25 Methylpropane, C_4H_{10}

Structural isomers have identical molecular formulas, but their atoms are linked to different neighbors. Different conformations of the same molecule differ only by rotation about one or more bonds.

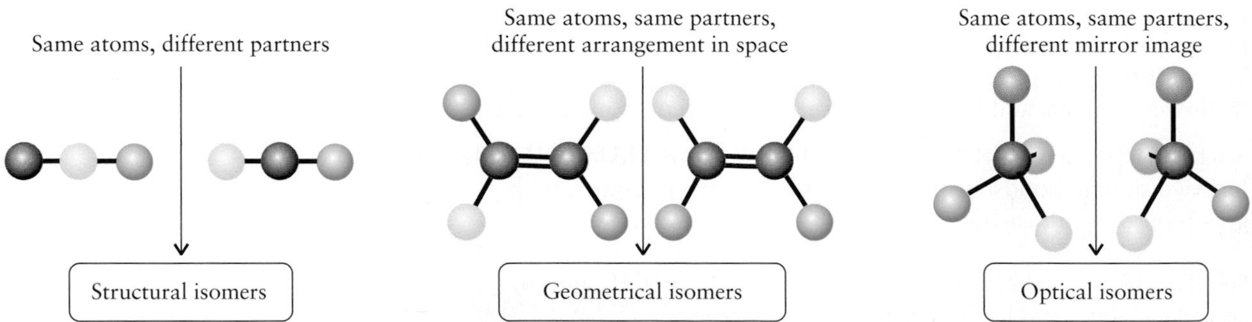

FIGURE 18.2 A summary of the various types of isomerism that are encountered in molecular compounds.

EXAMPLE 18.3 Writing the formulas of structural isomers

Draw two-dimensional structural formulas for all the isomeric alkanes of formula C_5H_{12}.

STRATEGY Decide which isomers can be constructed by inserting one $—CH_2—$ group into the formulas of the two C_4H_8 molecules given earlier. Isomers cannot be changed into one another simply by rotating either the entire formula or parts of the formula on the page. One approach is to insert $—CH_2—$ groups into different parts of the molecule, then discard formulas that repeat those already obtained. It is often easier to identify different conformations of the same isomer by building molecular models that allow rotation about single bonds.

SOLUTION From butane, we can form

```
                                          H
                                          | H
                                          |/
    H   H   H   H   H            H   H    C   H
    |   |   |   |   |            |   |  H |   |
H - C - C - C - C - C - H    H - C - C -- C - C - H
    |   |   |   |   |            |   |    |   |
    H   H   H   H   H            H   H    H   H
```

(a) (b)

From methylpropane, we can form

```
          H                       H                            H   H
          | H                     | H                          | H | H
          |/                      |/                           |/  |/
  H   H   C   H               H   C   H                    H   C   C
  |   | H |   |               | H |   |                    | H | H |
H-C - C - C - C - H       H - C - C - C - H            H - C - C - C - H
  |   |   |   |               |   | H |                    |   |   |
  H   H   H   H               H   C   H                    H   H   H
                                H |
                                  H
```

(c) (d) (e)

Molecules (b) and (c) are the same. The atoms of molecule (e) are joined together in the same arrangement as (b) and (c), but the structure looks different because it has been rotated and twisted; so (b), (c), and (e) are the same. There are therefore only three distinct isomers with formula C_5H_{12}: (a), (b), and (d).

Self-Test 18.4A Write the condensed structural formulas for the five isomeric alkanes of molecular formula C_6H_{14}.

[*Answer:* $CH_3(CH_2)_4CH_3$; $CH_3(CH_2)_2CH(CH_3)CH_3$; $CH_3CH_2CH(CH_3)CH_2CH_3$; $CH_3CH_2C(CH_3)_2CH_3$; $CH_3CH(CH_3)CH(CH_3)CH_3$]

Self-Test 18.4B Write the condensed structural formulas for the four isomers with the molecular formula C_4H_9Br.

In **stereoisomers,** the molecules have the same connectivity but the atoms are arranged differently in space. One class of stereoisomers are **geometrical isomers,** in which atoms have different arrangements on either side of a double bond or above and below the ring of a cycloalkane

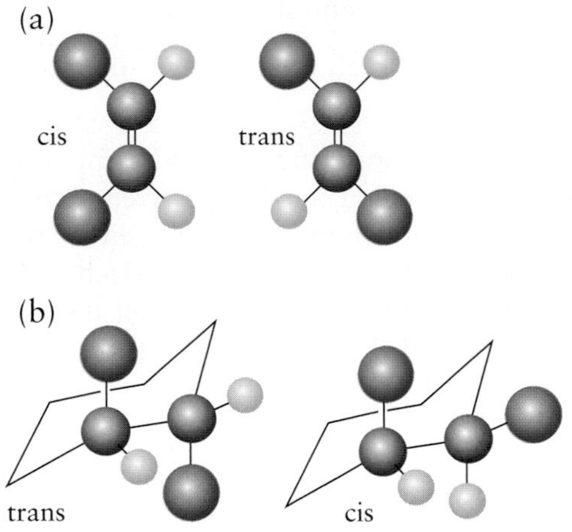

FIGURE 18.3 Two pairs of geometrical isomers. In (a), the isomerism is related to the position of two groups across a double bond; and in (b), to the position of two groups above and below a ring. Note that the neighbors of each atom are the same in both cases, but the arrangement of the atoms in space is different. When the purple spheres denote CH_3, the compound in (a) is butene.

(Fig. 18.3). For example, as we see in the upper part of the illustration, there are two different 2-butenes: they have the same molecular formula and structural formula but different properties. Geometrical isomers of organic molecules are distinguished by the prefixes cis and trans.

Self-Test 18.5A Identify (**26a**) and (**26b**) as cis or trans:

[*Answer:* (**26a**) is *trans*-2-pentene; (**26b**) is *cis*-2-pentene]

Self-Test 18.5B Identify (**27a**) and (**27b**) as cis or trans.

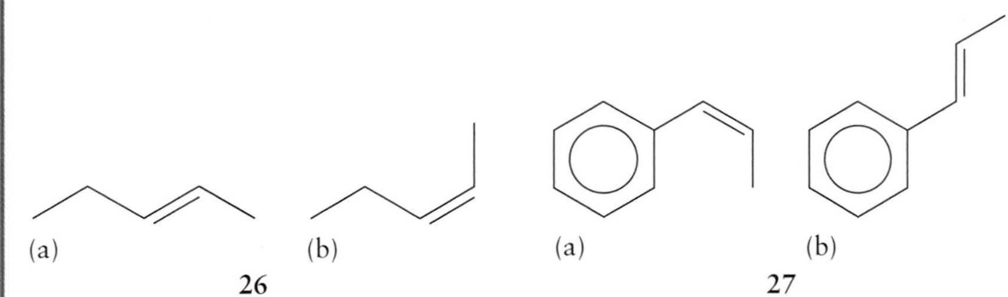

Another type of stereoisomerism is optical isomerism. **Optical isomers** are each other's nonsuperimposable mirror image. To understand this definition, consider 3-methylhexane, $CH_3CH_2CH(CH_3)CH_2CH_2CH_3$, and its mirror image (Fig. 18.4). No matter how we twist and turn the molecules, we cannot superimpose the mirror image molecule on the original molecule. It is like trying to superimpose your right hand on your left hand. A **chiral molecule,** such as 3-methylhexane, is a molecule that is not identical to its mirror image. A chiral molecule and its mirror image form a pair of **enantiomers,** or mirror image isomers. Although they have the same composition, the two enantiomers are, in fact, two distinct compounds. In organic compounds, optical isomers occur whenever four different groups are attached to a carbon atom. The alkane 3-methylpentane (**28**), does not have four different groups attached to the central carbon atom. Its mirror image can be rotated and superimposed on the original molecule. 3-Methylpentane is an example of an **achiral molecule,** a molecule that is superimposable on its mirror image.

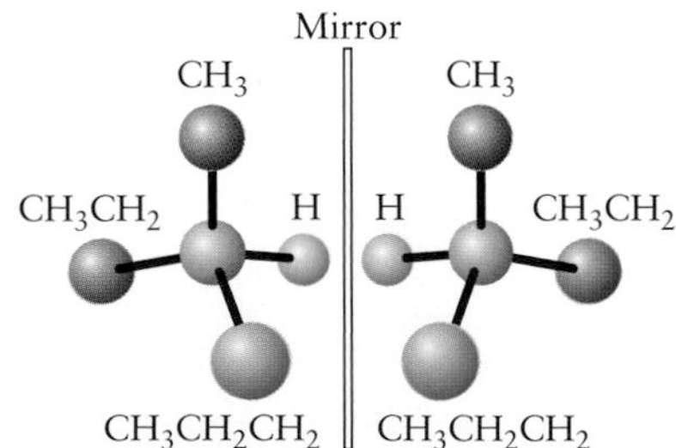

FIGURE 18.4 The molecule of 3-methylhexane on the right is the mirror image of the one on the left. The molecule on the left cannot be superimposed on the one on the right, so these two molecules are distinct optical isomers.

Enantiomers have identical chemical properties, except when they react with other chiral compounds. Because many biochemical substances are chiral, one consequence of this difference in reactivity is that enantiomers may have different odors and pharmacological activities. The molecule has to fit into a cavity, or slot, of a certain shape, either in an odor receptor in the nose or in an enzyme. Only one member of the enantiomeric pair may be able to fit.

Enantiomers differ in one physical property: chiral molecules display optical activity, the ability to rotate the plane of polarization of light (Section 16.7 and Box 16.2). If a chiral molecule rotates the plane of polarization clockwise, then its mirror image partner rotates it through the same angle in the opposite direction.

Often organic compounds synthesized in the laboratory are **racemic mixtures,** or mixtures of enantiomers in equal proportions. Because enantiomers rotate the plane of polarization of light in opposite directions, their effects cancel and a racemic sample is not optically active. In contrast, reactions in living cells commonly lead to only one enantiomer. It is a remarkable feature of nature that almost all naturally occurring amino acids in animals have the same handedness.

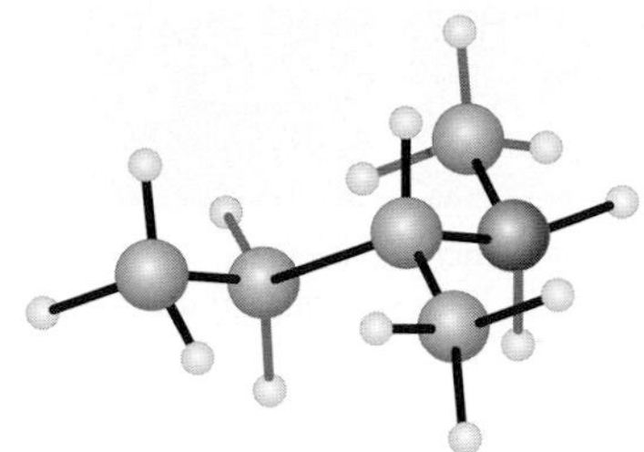

28 3-Methylpentane

Geometrical isomers have the same molecular and structural formulas but different arrangements in space. Molecules with four different groups attached to a single carbon atom are chiral and optically active.

EXAMPLE 18.4 Predicting whether a compound is chiral

Decide whether or not the bromoalkanes (a) $CH_3CH_2CHBrCH_3$ and (b) $CH_3CHBrCH_3$ are chiral.

STRATEGY To decide whether a compound is chiral, we look for a carbon atom with four different groups attached.

SOLUTION The two compounds are (**29**) and (**30**). We see that (**29**) contains a carbon atom that has four different groups attached; hence it is chiral. However, (**30**) does not have such a carbon atom; this compound is identical to its mirror image and therefore achiral.

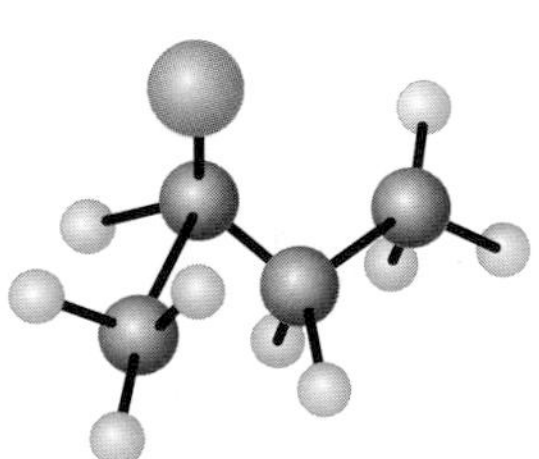

29 2-Bromobutane

Self-Test 18.6A A chlorofluorocarbon molecule contains Cl and F atoms as well as C and H. Which of the following chlorofluorocarbons is chiral: (a) CH_3CF_2Cl; (b) CH_3CHFCl; (c) CH_2FCl?

[*Answer:* (b)]

Self-Test 18.6B In an alcohol, an —OH group is attached to a C atom. Which of the following alcohols is chiral: (a) CH_3CH_2OH; (b) $CH_3CH(OH)CH_3$; (c) $CH_3CH(OH)CH_2CH_3$?

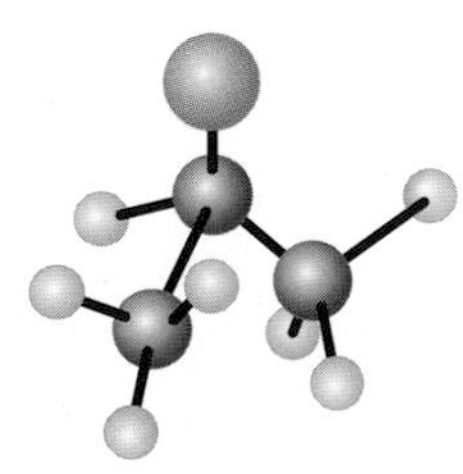

30 2-Bromopropane

18.4 Properties of Alkanes

Alkanes with long, unbranched chains tend to have higher melting points, boiling points, and enthalpies of vaporization than their branched isomers. The difference arises because the atoms of neighboring branched

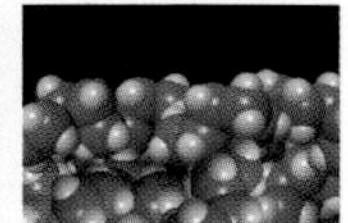

See how branching affects intermolecular forces in pentane isomers.

Media Link Animation: Fig. 18.5, Pentanes

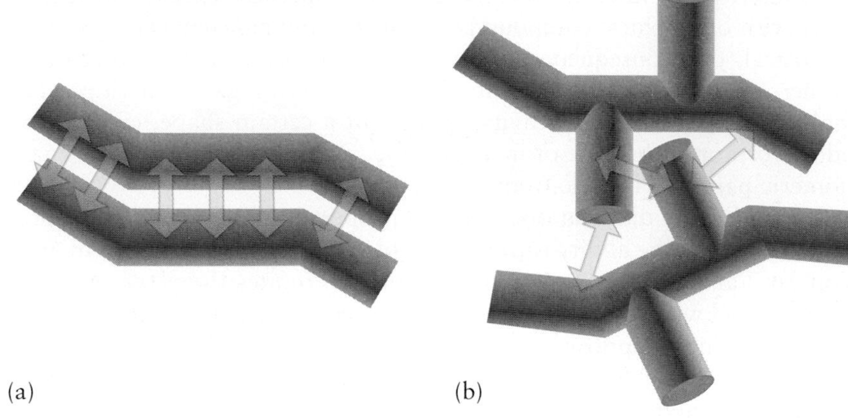

FIGURE 18.5 (a) The atoms in neighboring straight-chain alkanes, represented by the tubelike structures, can lie close together. (b) Fewer of the atoms of neighboring branched alkane molecules can get so close together overall, so the London forces (represented by double-headed arrows) are weaker and branched alkanes are more volatile.

molecules cannot get as close together as atoms of neighboring unbranched molecules can (Fig. 18.5). As a result, molecules with branched chains have weaker intermolecular forces than their isomers with unbranched chains.

Self-Test 18.7A Which compound has the higher boiling point: (a) $CH_3CH_2CH_2CH_2CH_3$ or (b) $(CH_3)_2CHCH_2CH_3$? Why?

[*Answer:* (a), because it is not branched and so has stronger London forces.]

Self-Test 18.7B Which compound has the higher boiling point: (a) $CH_3CH_2CH_2CH_2CH_3$ or (b) $CH_3CH_2CH_2CH_3$? Why?

The alkanes were once called the *paraffins*, from the Latin words for "little affinity." As this name suggests, they are not very reactive. They are unaffected by concentrated sulfuric acid, by boiling nitric acid, by strong oxidizing agents like potassium permanganate, and by boiling aqueous sodium hydroxide. One reason for their resistance to chemical onslaught is thermodynamic: the C—C and C—H bonds are strong (their mean bond enthalpies are 348 $kJ \cdot mol^{-1}$ and 412 $kJ \cdot mol^{-1}$, respectively), so there is little energy advantage in replacing them with most other bonds. The most notable exceptions are C=O (743 $kJ \cdot mol^{-1}$), C—OH (360 $kJ \cdot mol^{-1}$), and C—F (484 $kJ \cdot mol^{-1}$) bonds.

Alkanes undergo two common reactions: oxidation and substitution (Section 18.5). The alkanes are commonly used as fuels, because their combustion to carbon dioxide and water gives off a lot of heat (Section 6.18):

$$CH_4(g) + 2\ O_2(g) \longrightarrow CO_2(g) + 2\ H_2O(g) \qquad \Delta H^\circ = -890\ kJ$$

In this reaction, the strong carbon-hydrogen bonds are replaced by the even stronger O—H bonds (463 $kJ \cdot mol^{-1}$), and the oxygen-oxygen bond (496 $kJ \cdot mol^{-1}$) is replaced by two very strong C=O bonds. The excess energy is given off as heat.

The strength of the London forces between alkane molecules increases as the molar mass of the molecules increases. Alkanes are not very reactive, but they do undergo oxidation (combustion) and substitution.

18.5 Alkane Substitution Reactions

Alkanes are abundant in petroleum (Box 18.1) and are used as the raw materials for the synthesis of many more reactive compounds. Starting with alkanes, organic chemists introduce reactive groups of atoms into the molecules in the process called **functionalization.** Functionalization of alkanes can be achieved by a **substitution reaction,** a reaction in which an atom or group of atoms replaces an atom (hydrogen, in this case) in the original molecule (Fig. 18.6). An example of a substitution reaction is that between methane and chlorine. A mixture of these two gases survives indefinitely in the dark, but when exposed to ultraviolet radiation or heated to more than 300°C, the gases react explosively:

$$CH_4(g) + Cl_2(g) \xrightarrow{\text{light or heat}} CH_3Cl(g) + HCl(g)$$

Chloromethane, CH_3Cl, is only one of the products; dichloromethane, CH_2Cl_2; trichloromethane, $CHCl_3$; and tetrachloromethane, CCl_4, also form, especially at high concentrations of chlorine. Trichloromethane, better known as *chloroform*, was an early anesthetic. Tetrachloromethane, which is commonly called carbon tetrachloride, is nonflammable; it was used in fire extinguishers and as a nonpolar solvent until its carcinogenic character became known.

Organic chemists use chemical kinetic studies to build up models of the mechanisms of reactions, basing their suggested mechanisms on the response of the rate to changes in concentration, catalyst, solvent, and temperature. Kinetic studies of this kind suggest that alkane substitution reactions typically proceed by a radical chain mechanism (Section 13.12).

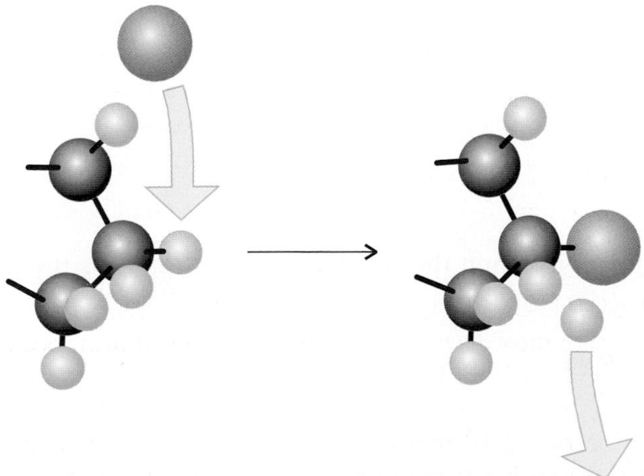

FIGURE 18.6 In an alkane substitution reaction, an incoming atom or group of atoms (represented by the orange sphere) replaces a hydrogen atom in the alkane molecule.

BOX 18.1 WHAT HAS THIS TO DO WITH . . . THE ENVIRONMENT?

Fossil Fuels

The thought of an automobile that runs on coal is strange, but the use of gasoline substitutes derived from coal is a real possibility. As reserves of petroleum are declining worldwide, interest in using coal is growing. Unfortunately, the increased usage of coal raises environmental concerns. Crude petroleum is a mixture of aliphatic and aromatic hydrocarbons, together with some organic compounds containing sulfur and nitrogen. Coal is also a fossil fuel, but coal contains much less hydrogen than petroleum does and is harder to purify and to work. Although both coal and petroleum contribute to the greenhouse effect, the use of coal is potentially more damaging.

The hydrocarbons in petroleum are separated by fractional distillation (Section 8.21). However, the primary use of petroleum is for gasoline production, and the gasoline fraction (C_5 to C_{11} hydrocarbons) is too small. In addition, the straight-chain alkanes make poor gasoline. Therefore, petroleum is refined to increase both the quantity and quality of gasoline.

The quantity of gasoline that can be obtained from petroleum is increased by *cracking*, or breaking down long hydrocarbon chains, and by *alkylation*, or combining small molecules to make larger ones. In cracking, the less volatile fractions are exposed to higher temperatures in the presence of a catalyst, often a modified zeolite (see Box 13.4). For example, fuel oil can be converted to a mixture of octene and octane isomers:

$$C_{16}H_{34} \xrightarrow{\Delta,\text{ catalyst}} C_8H_{16} + C_8H_{18}$$

Alkylation also requires a catalyst to achieve the desired chain length. Octane, for instance, can be synthesized from a mixture of butane and butene:

$$C_4H_{10} + C_4H_8 \xrightarrow{\text{catalyst}} C_8H_{18}$$

The quality of gasoline, which determines how smoothly it burns, is measured by the *octane rating*. For example, the straight-chain molecule octane, $CH_3(CH_2)_6CH_3$, burns so poorly it has an octane rating of −19, but its isomer 2,2,4-trimethylpentane, which is commonly called isooctane

$$\begin{array}{c} \quad\;\; CH_3 \qquad\quad CH_3 \\ \quad\;\; | \qquad\qquad\;\; | \\ H_3C-C-CH_2-CH-CH_3 \\ \quad\;\; | \qquad\qquad\qquad \\ \quad\;\; CH_3 \qquad\qquad\quad \end{array}$$

Isooctane, 2,2,4-trimethylpentane

has an octane rating of 100. The octane rating is improved by increasing the branching of the molecules and by introducing unsaturation and rings. In *isomerization*, straight-chain hydrocarbons are converted into their branched-chain isomers:

$$CH_3(CH_2)_6CH_3 \xrightarrow{AlCl_3} CH_3C(CH_3)_2CH_2CH(CH_3)CH_3$$

Aromatization is the conversion of an alkane into an arene:

$$CH_3(CH_2)_5CH_3 \xrightarrow{Al_2O_3,\ Cr_2O_3} CH_3C_6H_5 + 4\ H_2$$

The product of this reaction, toluene, has an octane rating of 120.

The quality of gasoline is also improved by the addition of ethanol, which has an octane rating of 120. The use of ethanol helps to reduce the demand for petroleum. Unlike petroleum, ethanol is a renewable fuel that can be regenerated every year (see Box 6.2).

The initiation step in the chlorination of methane is the dissociation of chlorine:

$$Cl_2 \xrightarrow{\text{light or heat}} Cl\cdot + \cdot Cl$$

Recall from Section 2.10 that radicals tend to be very reactive. Chlorine atoms (each of which has one unpaired electron) are highly reactive and go on to attack methane molecules and extract a hydrogen atom:

$$Cl\cdot + CH_4 \longrightarrow HCl + \cdot CH_3$$

Coal is more widely distributed than petroleum. Because it is a solid and often very impure, it is no longer used as a fuel for transportation (it was once used for ships and trains). Coal does not burn cleanly, so it releases a large amount of pollution in the form of particulate matter (primarily ash) and sulfur and nitrogen oxides. Because its carbon content is high, coal burning also contributes significantly to the greenhouse effect. Much of the research on coal is aimed at converting it into more useful fuels.

Because fossil fuel reserves are limited, they must be extracted from wherever they are found. This platform is used to extract petroleum from beneath the ocean. The natural gas accompanying it cannot easily be transported, so it is burned off.

Coal, a carbonaceous sedimentary rock, is the end product of the decay of swamp vegetation in anaerobic conditions (with a very low supply of oxygen). Oxygen and hydrogen are gradually lost through the *coalification* process. The further coalification has proceeded, the higher the carbon content and the greater the value of the coal. *Peat* has a low carbon content and a high moisture content. It is soft and has a low enthalpy density. *Lignite,* which is found in the western part of the United States, is very abundant, but gives off only half as much heat per kilogram as anthracite coals. *Bituminous* coal has been further coalified, but it is contaminated with sulfur and other pollutants. *Anthracite* coal, which is the most highly coalified, is a very hard, clean-burning fuel with a high enthalpy density. Unfortunately, it is the least common.

Coal contains many aromatic rings (see Fig. 18.12). As coalification proceeds, hydrogen is released and the aromatic structures increase. All types of coal are used as raw materials for other chemicals. When coal is heated, the volatile compounds produced as the extended aromatic structures are broken up, are collected and used by industry. *Coal gas*, which is given off first, contains carbon monoxide, hydrogen, methane, and small amounts of other gases. A large number of pharmaceuticals, dyes, and fertilizers come from coal tar. Benzene is the raw material for nylon, detergents, and pesticides. Naphthalene is used to make synthetic indigo (the dye in blue jeans), ammonia is used as fertilizer, and pitch, which contains the heaviest tars, is used for waterproofing and rustproofing.

For Further Reading: C. Baird "Energy Reserves." *Environmental Chemistry* (New York: W. H. Freeman and Co., 1999), pp. 228–231.

Because one of the products is another radical, this reaction is a *propagation* step (a step in which one reactive radical intermediate produces another). In a second propagation step, the methyl radical may react with a chlorine molecule:

$$Cl_2 + \cdot CH_3 \longrightarrow CH_3Cl + Cl\cdot$$

The chlorine atom so produced can take part in the other propagation step. However, we can see that other reactions are possible, such as

$$Cl\cdot + CH_3Cl \longrightarrow CH_2Cl_2 + H\cdot$$

A termination step occurs when two radicals combine to form a nonradical product, as in

$$Cl\cdot + \cdot Cl \longrightarrow Cl_2$$

The product of substitution reactions of alkanes with the halogens is typically a complex mixture of *haloalkanes* (halogenated alkanes). One way to limit the production of the more highly substituted alkanes is to use a large excess of the alkane; then most reactions take place with the original hydrocarbon rather than with any haloalkanes produced in the reaction.

Alkane substitution takes place by a radical chain mechanism.

18.6 Properties of Alkenes

The properties of alkenes reflect the presence of the carbon-carbon double bond. As we saw in Section 3.4, a double bond consists of a σ-bond and a π-bond. Each carbon atom is sp^2 hybridized and uses one of the hybrid orbitals to form a σ-bond. The unhybridized *p*-orbitals on each atom overlap with each other and form a π-bond.

Section 3.8 describes the basis for the resistance to rotation of the C=C bond.

All four atoms attached to the C=C group lie in the same plane and are locked into that arrangement by the resistance to twisting of the π-bond (Fig. 18.7). Because alkene molecules cannot roll up into a ball as compactly as alkanes, they cannot pack together as closely as alkanes; so alkenes have lower melting points. As we saw in Section 3.7, the carbon-carbon π-bond is relatively weak because the overlap responsible for the formation of the π-bond is less extensive than that responsible for the formation of the σ-bond. A consequence of this weakness is the reaction most characteristic of alkenes, the replacement of the π-bond by two new σ-bonds.

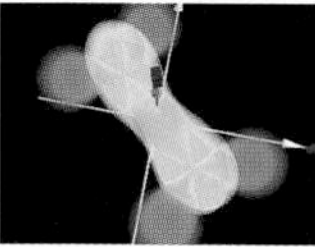

View an animation of the bonding in ethene.

One of the first steps in the refining of petroleum is to convert some of the abundant alkanes into the more reactive alkenes:

$$CH_3CH_3(g) \xrightarrow{Cr_2O_3} CH_2{=}CH_2(g) + H_2(g)$$

This is an example of an **elimination reaction,** a reaction in which two groups or atoms on neighboring carbon atoms are removed from a molecule, leaving a multiple bond (Fig. 18.8). The mechanism of this particular elimination reaction is very complicated and involves reaction between the alkane and the oxide ions on the surface of the catalyst.

Another route commonly used in the laboratory to produce alkenes is the **dehydrohalogenation** of haloalkanes, the removal of a hydrogen atom and a halogen atom from neighboring carbon atoms:

$$CH_3CH_2CHBrCH_3 + CH_3CH_2O^- \xrightarrow{\text{ethanol at 70°C}} CH_3CH{=}CHCH_3 + CH_3CH_2OH + Br^-$$

The reaction is another example of an elimination reaction and is carried out in hot ethanol, with sodium ethoxide, $NaCH_3CH_2O$, as the reagent. Some $CH_3CH_2CH{=}CH_2$ is also formed in this reaction.

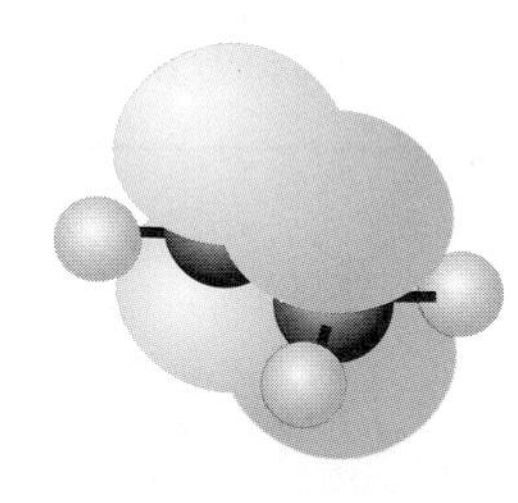

Media Link Animation: Fig. 18.7, Ethene bonding

FIGURE 18.7 The π-bond (represented by the yellow electron clouds) in an alkene molecule makes the molecule resistant to twisting around a double bond, so all six atoms (the two C atoms that form the bond and the four atoms attached to them) lie in the same plane.

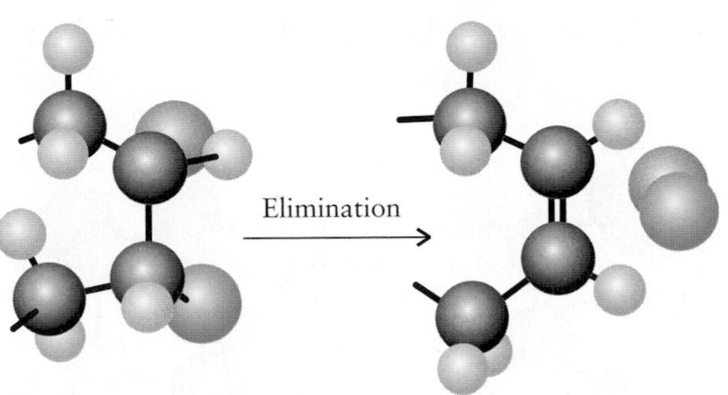

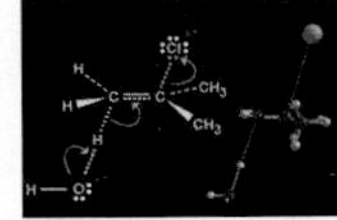

Media Link Animation: Fig. 18.8, Elimination mechanism

View an animation of an elimination reaction.

FIGURE 18.8 In an elimination reaction, two atoms or (groups the orange and purple spheres) attached to neighboring carbon atoms are removed from the molecule, leaving a double bond in their place.

Kinetic studies suggest that dehydrohalogenation occurs by the attack of the ethoxide ion on a hydrogen atom attached to the carbon atom next to the carbon carrying the Br atom. That hydrogen atom is pulled out of the molecule as a proton as $CH_3CH_2O{-}H$ is formed, and the bromine atom is effectively squeezed out of the molecule as a Br^- ion when the carbon atom forms a second bond to its neighbor:

$CH_3CH_2O^-$ CH_3CH_2OH

H Br $\longrightarrow$ $+\ Br^-$

Curved arrows like the ones here are often used when illustrating organic reaction mechanisms to show the direction in which electron pairs move as they form new bonds. The dashed and solid wedge-shaped bonds are commonly used when displaying organic structures to convey a sense of the three-dimensional shapes. The dashed wedge-shaped bonds go into the paper and the solid wedge-shaped bonds come toward us.

The double bonds in alkenes can be generated by elimination reactions.

18.7 Electrophilic Addition to Alkenes

The most characteristic chemical reaction of an alkene is an **addition reaction,** in which atoms supplied by the reactant form σ-bonds to the two atoms joined by the double bond (Fig. 18.9). In the process, the π-bond is lost. An example is **halogenation,** the addition of two halogen atoms at a double bond, as in the formation of 1,2-dichloroethane:

$$CH_2{=}CH_2 + Cl_2 \longrightarrow CH_2Cl{-}CH_2Cl$$

The addition of hydrogen chloride to give chloroethane is an example of a **hydrohalogenation** reaction:

$$CH_2{=}CH_2 + HCl \longrightarrow CH_3{-}CH_2Cl$$

The model that organic chemists have built up of the mechanism of alkene addition reactions is quite different from that of alkane substitution

Media Link Animation: Fig. 18.9, Addition mechansim

View an animation of an addition reaction.

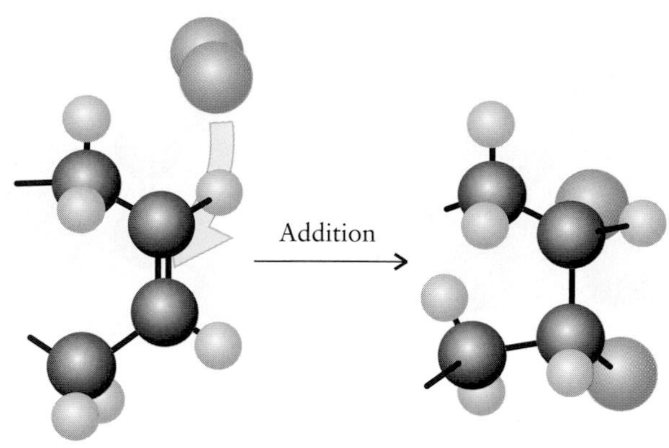

FIGURE 18.9 In an addition reaction, the atoms provided by an incoming molecule are attached to the carbon atoms originally joined by a multiple bond. Addition is the reverse of elimination.

reactions. The crucial difference between the two types of molecules is the high density of electrons in the region of the double bond (Fig. 18.10). Because electrons are negatively charged, this region represents an accumulation of negative charge that can attract a positively charged reactant. A reactant that is attracted to a region of high electron density is called an **electrophile.** The mechanism of alkene addition is by electrophilic attack on the carbon atoms that make up the double bond. An electrophile may be a positively charged species or it may be a species that has or can acquire a partial positive charge in the course of the reaction.

An example is the bromination of ethene. When ethene (or any other alkene) is bubbled through a solution of bromine, the solution is decolorized as the bromine reacts to form dibromoethane (this reaction was illustrated in Fig. 2.9). Bromine molecules are polarizable; and as a Br_2 molecule approaches the region of high electron density of an alkene double bond, a partial positive charge is formed on the Br atom closer to the double bond (Fig. 18.11). This separation of charges means that a Br_2 molecule can act as an electrophile. As it moves in for the attack, the partially positively charged Br atom becomes more and more like Br^+, and its partner becomes more like Br^-. The bond between the two atoms snaps, and the Br^+ ion forms a bridge between the two carbon atoms of the alkene, giving a cyclic "bromonium ion":

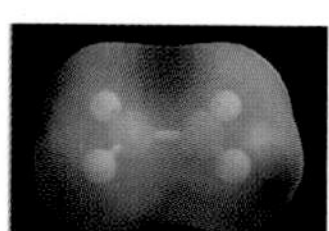

Media Link Animation: Fig. 18.10, Ethene structure

View multiple representations of the ethene molecule.

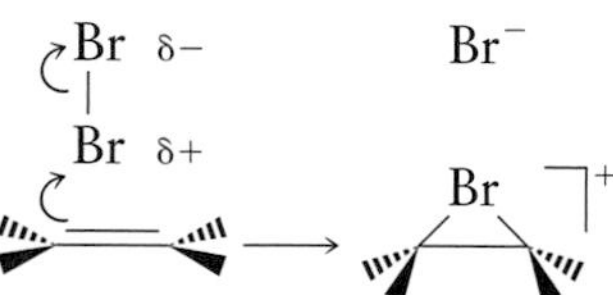

Almost immediately, a Br^- ion swoops in for the attack, attracted by the positive charge of the bromonium ion. It forms a bond to one carbon atom, and the bromine atom already present forms another bond, giving 1,2-dibromoethane:

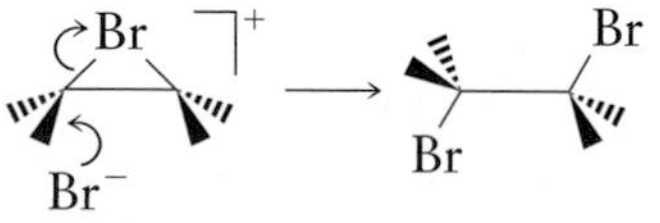

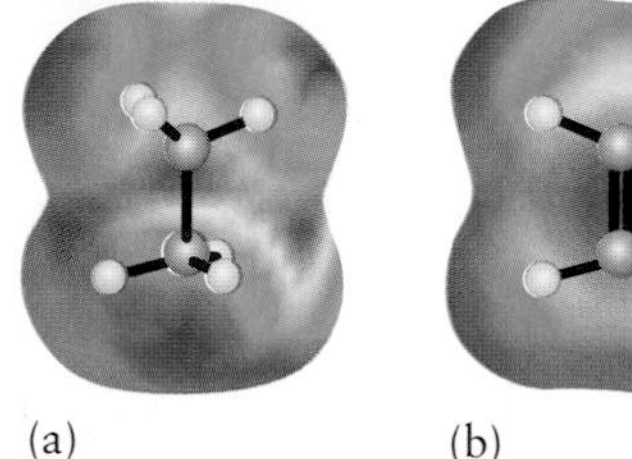

(a) (b)

FIGURE 18.10 (a) The electrostatic potential diagram for an ethane molecule: the blue regions show where the positive charge of the nuclei outweighs the negative charge of the electrons and the red regions show where the opposite is true. (b) The electron distribution in an ethene molecule shows the negative region of charge associated with the accumulation of electrons in the region of the double bond.

$$\text{[Br}^+\text{ bridged ion]} + Br^- \longrightarrow \text{BrCH}_2\text{–CH}_2\text{Br}$$

It is possible to verify this mechanism by experiment.

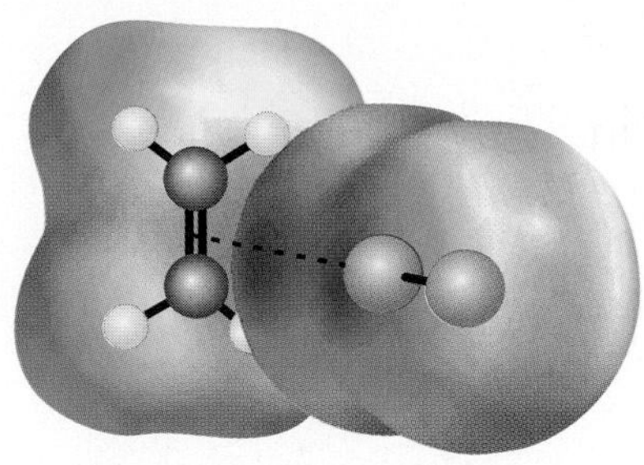

FIGURE 18.11 As a bromine molecule approaches a double bond in an alkene, the atom closer to the ethene molecule acquires a partial positive charge (the blue region). The computation that produced this image was carried out for the point at which the bromine molecule is so close to the double bond that a carbon-bromine bond is starting to form.

How do we do that?

We can verify the mechanism for the addition of bromine to an alkene if we repeat the experiment with *cis*-2-butene (**31**) instead of ethene. If the bromine molecule simply added across the same side of the double bond, the product would be the 2,3-dibromobutane shown as (**32**). However, the product actually obtained is (**33**), a pair of enantiomers. We can account for the formation of the observed product if the bromine attaches in two stages and if the first bromine atom protects one side of the molecule from attack by the second atom. The incoming Br^- ion must come in from the opposite face of the bromonium ion.

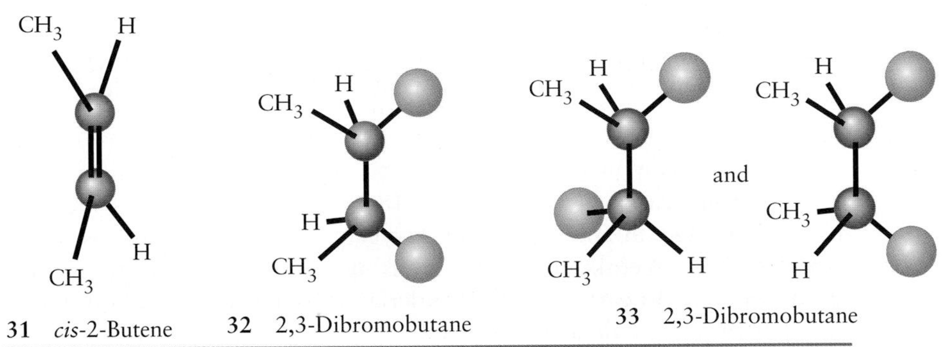

31 *cis*-2-Butene **32** 2,3-Dibromobutane **33** 2,3-Dibromobutane

The mechanism of addition to alkenes is electrophilic attack.

Self-Test 18.8A Write the condensed structural formula of the product of the addition of hydrogen to butene: $CH_3CH{=}CHCH_3 + H_2 \rightarrow$ product.

[*Answer:* $CH_3CH_2CH_2CH_3$]

Self-Test 18.8B Write the condensed structural formula of the compound formed by the addition of hydrogen bromide to ethene.

AROMATIC COMPOUNDS

Aromatic hydrocarbons are called **arenes.** An abundant source of arenes is coal, which is a very complex mixture of compounds; some of these compounds consist of extensive networks containing regions in which

FIGURE 18.12 A highly schematic representation of a portion of the structure of coal. When coal is heated in the absence of oxygen, the structure breaks up and a complex mixture of products—many of them aromatic—is obtained.

aromatic rings can be identified (Fig. 18.12). When coal is destructively distilled—heated in the absence of oxygen so that it decomposes and vaporizes—its sheetlike molecules break up, and the fragments include the aromatic hydrocarbons and their derivatives. The complex liquid mixture that results is called *coal tar.*

18.8 Nomenclature of Arenes

The parent compound of aromatic compounds is benzene, C_6H_6 (**34**); and when the benzene ring is named as a substituent, it is called the *phenyl* group, as in 2-phenylbutane, $CH_3CH(C_6H_5)CH_2CH_3$. In general, aromatic hydrocarbon groups are called *aryl* groups. Aromatic compounds also include the fused-ring analogues of benzene, such as naphthalene, $C_{10}H_8$ (**35**), and anthracene, $C_{14}H_{10}$ (**36**), obtained by the distillation of coal.

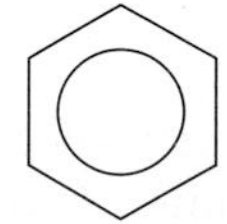

34 Benzene, C_6H_6

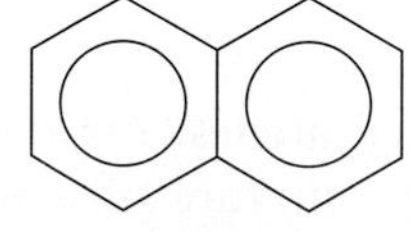

35 Naphthalene, $C_{10}H_8$

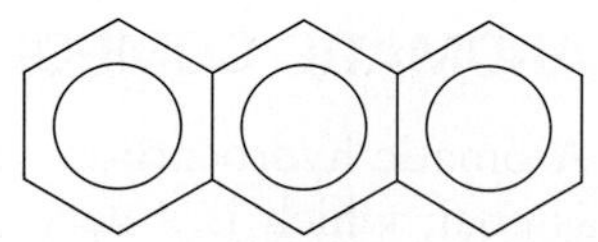

36 Anthracene, $C_{14}H_{10}$

In an older but still widely used system of nomenclature, when there is a substituent in location 1, locations 2, 3, and 4 are denoted *ortho-* (abbreviated *o-*), *meta-* (*m-*), and *para-* (*p-*), respectively. Thus, (37) is *ortho*-dinitrobenzene. In systematic nomenclature, the locations of substituents on the aromatic ring are identified by numbering the carbon atoms from 1 to 6 around the ring, selecting the direction that corresponds to the lower substituent numbers. In this system, (37) is 1,2-dinitrobenzene ($-NO_2$ is the nitro group) and (38) is 2,4,6-trinitrophenol (the compound C_6H_5OH is phenol).

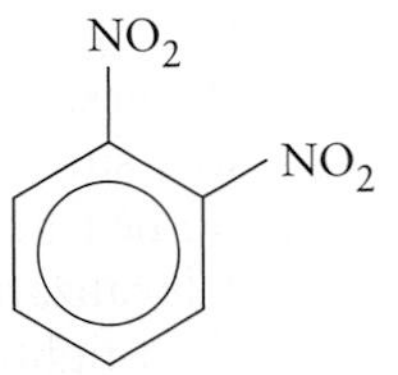

NO_2 NO_2

37 1,2-Dinitrobenzene

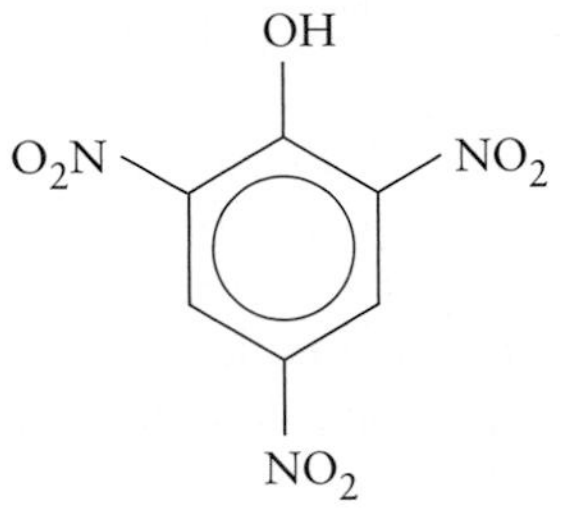

OH O_2N NO_2 NO_2

38 2,4,6-Trinitrophenol

How can we use this information . . . to name an aromatic compound?

(a) The compound (39) is named by counting around the ring in the direction that gives the smallest numbers to the substituents:

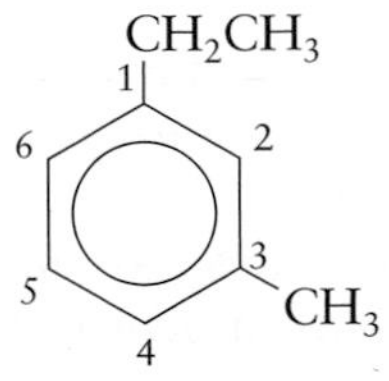

CH_2CH_3 1 2 3 CH_3 4 5 6

The compound is therefore 1-ethyl-3-methylbenzene. (b) The three dimethylbenzenes, which are also known as xylenes, would therefore be named

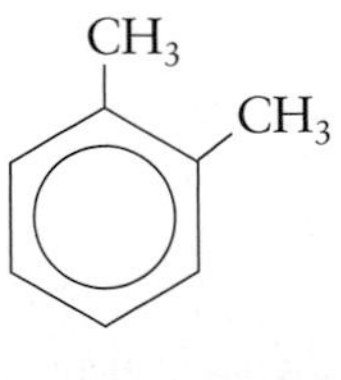

CH_3 CH_3

1,2-Dimethylbenzene
o-Xylene

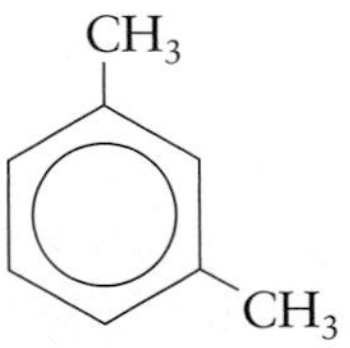

CH_3 CH_3

1,3-Dimethylbenzene
m-Xylene

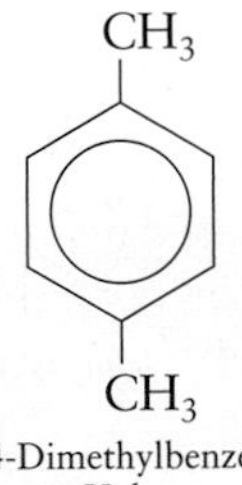

CH_3 CH_3

1,4-Dimethylbenzene
p-Xylene

Self-Test 18.9A Name the compound (40).

[*Answer:* 1-Ethyl-3-propylbenzene]

Self-Test 18.9B Name the compound (41).

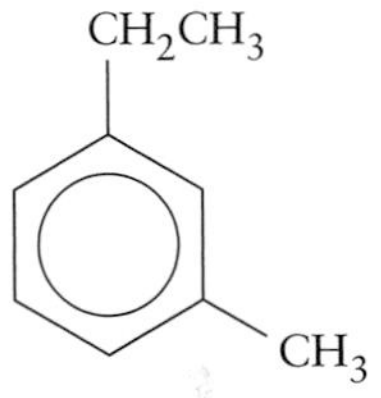

CH_2CH_3 CH_3

39

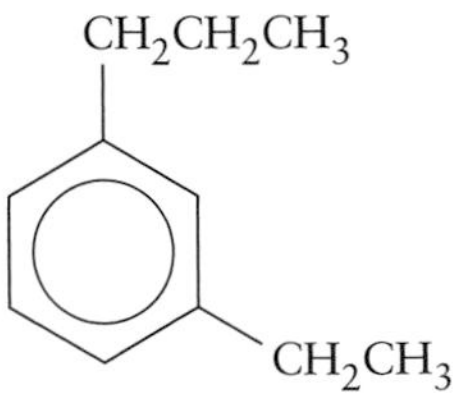

$CH_2CH_2CH_3$ CH_2CH_3

40

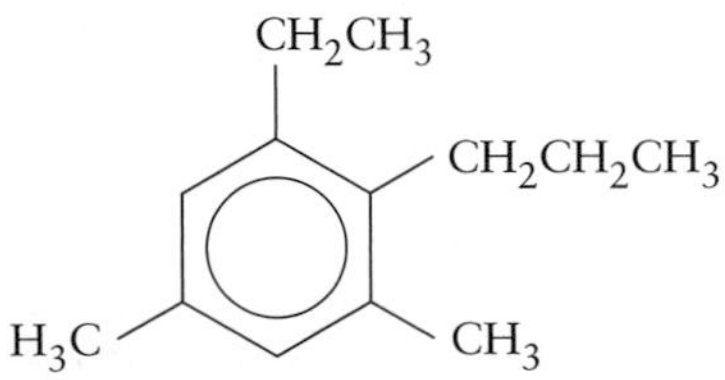

CH_2CH_3 $CH_2CH_2CH_3$ H_3C CH_3

41

18.9 Reactions of Arenes: Electrophilic Substitution

Arenes are unsaturated, but they are not very reactive. Unlike alkenes, arenes undergo predominantly *substitution* reactions, with the π-bonds of the ring left intact. As an example, whereas bromine immediately adds to a double bond of an alkene, it reacts with benzene only in the presence of a catalyst—typically, iron(III) bromide—and it does not affect the bonding in the ring. Instead, one of the bromine atoms substitutes for a hydrogen atom to give bromobenzene, C_6H_5Br:

$$C_6H_6 + Br_2 \xrightarrow{FeBr_3} C_6H_5Br + HBr$$

The mechanism of the substitution of a benzene ring is electrophilic attack, just as in the addition to double bonds in alkenes, so the reaction is called **electrophilic substitution**. Why, though, does substitution occur, not addition to the double bonds? The difference can be traced to the stability of the delocalized π-electrons in the ring. Delocalization gives the electrons such low energy—that is, they are bound so tightly—that they are unavailable for forming new σ-bonds (see Sections 2.8 and 3.13).

The bromination of benzene illustrates the difference between addition to alkenes and substitution of arenes. First, to achieve bromination, we have to use a catalyst (a requirement that underscores again the stability of the benzene ring). Iron(III) bromide is used. It acts as a Lewis acid, binding to the bromine molecule (a Lewis base) and ensuring that the outer bromine atom has a pronounced partial positive charge:

$$:Br{-}Br: + FeBr_3 \longrightarrow {}^{\delta+}:Br{-}Br{-}^{\delta-}FeBr_3$$

(We show only the lone pair involved in the reaction.) The outer bromine atom of the complex is now primed to act as a strong electrophile (Fig. 18.13).

Electrophilic substitution begins like electrophilic addition, with an attack on a region of high electron density to form a positively charged intermediate:

$$C_6H_5{-}H \; {}^{\delta+}Br{-}Br{-}FeBr_3{}^{\delta-} \longrightarrow [C_6H_5(H)Br]^{+} + FeBr_4^{-}$$

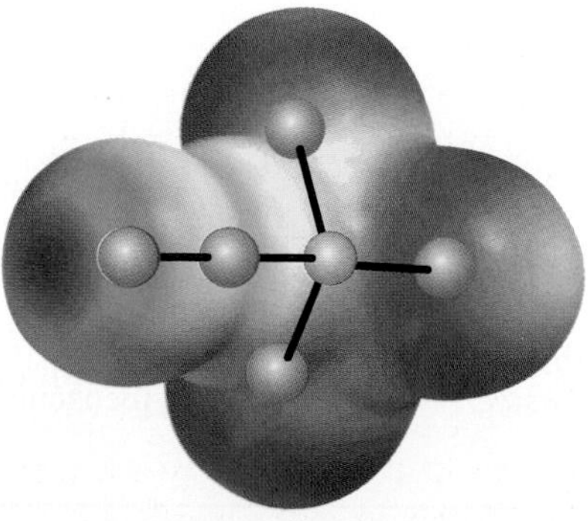

FIGURE 18.13 The catalyst $FeBr_3$ acts by forming a complex with the bromine molecule. As a result, the bromine atom not directly attached to the iron atom acquires a partial positive charge (the blue region). This partial charge enhances the ability of the bromine molecule to act as an electrophile.

However, because the delocalized π-electrons constitute such a stable arrangement, the hydrogen atom shown can be pulled reasonably easily from the ring by a bromine atom in the $FeBr_4^-$ complex, for in that way delocalization is regained:

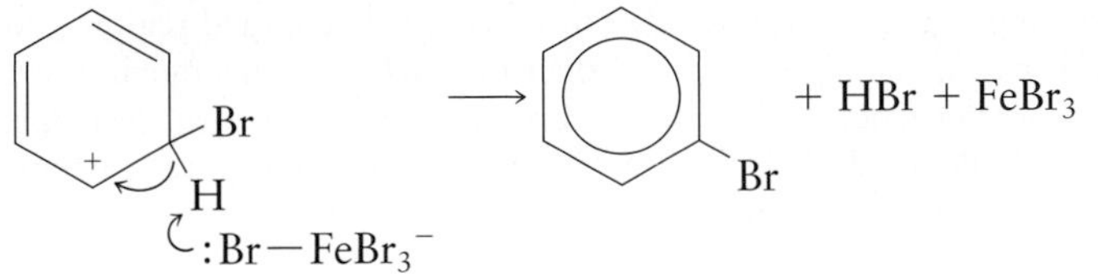

$$\text{C}_6\text{H}_5\text{BrH}^+ + :\text{Br}-\text{FeBr}_3^- \longrightarrow \text{C}_6\text{H}_5\text{Br} + \text{HBr} + \text{FeBr}_3$$

The iron(III) bromide is released in this step and is free to activate another bromine molecule.

One of the most extensively studied examples of electrophilic substitution is the nitration of benzene. This reaction sheds further light on the mechanism and helps us to build a more detailed model of the steps involved. A mixture of nitric acid and concentrated sulfuric acid converts benzene slowly into nitrobenzene. The actual nitrating agent is the electrophile NO_2^+ (the nitronium ion), a linear triatomic ion:

$$HNO_3 + H_2SO_4 \longrightarrow NO_2^+ + HSO_4^- + H_2O$$

The accepted mechanism of the reaction is

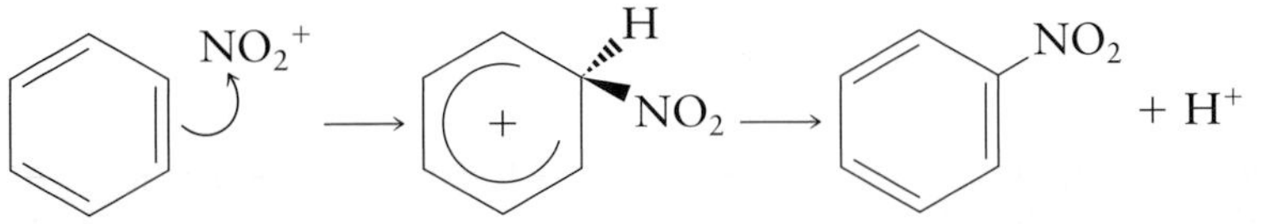

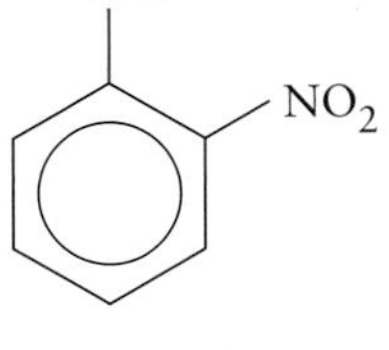

$$\text{C}_6\text{H}_6 + NO_2^+ \longrightarrow [\text{C}_6\text{H}_6NO_2]^+ \longrightarrow \text{C}_6\text{H}_5NO_2 + H^+$$

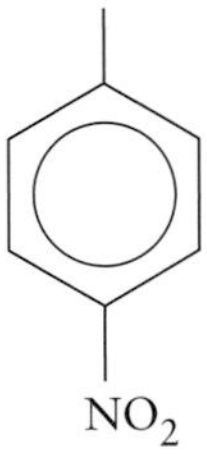

In the second step, the hydrogen ion is pulled out of the ring by the HSO_4^- ion acting as a Brønsted base. As in the bromination reaction, its removal is facilitated by the restoration of the delocalization of the π-electrons.

Certain groups attached to an aromatic ring can donate electrons into its delocalized molecular orbitals. Examples of these electron-donating substituents include $-NH_2$ and $-OH$. Electrophilic substitution of benzene is much faster when an electron-donating substituent is present. The nitration of phenol, C_6H_5OH, proceeds so quickly that it requires no catalyst. Moreover, when the products are analyzed, it is found that the only products are 2-nitrophenol (*ortho*-nitrophenol, 42) and 4-nitrophenol (*para*-nitrophenol, 43).

OH, NO_2

42 *o*-Nitrophenol

OH, NO_2

43 *p*-Nitrophenol

Why is the meta position such an unattractive location for substitution, and why does phenol react much faster than benzene? An electrophile is attracted to regions of high electron density. Therefore, to account for the fast reaction of phenol, the electron density must be greater in the ring when the electron-donating $-OH$ substituent is present. To account for the dominance of the ortho and para products, electron density must be relatively high at the ortho and para positions. A molecular orbital calculation shows that there is indeed a higher concentration of electrons in the ring in phenol, especially at the ortho and para positions, than in benzene, largely because the O atom has lone pairs of electrons that can participate in π-bonding with the carbon atoms (Fig. 18.14).

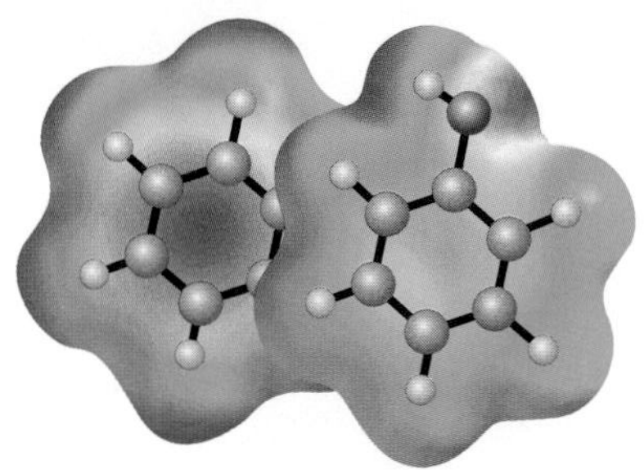

FIGURE 18.14 The presence of an —OH group in phenol (foreground) alters the distribution of electrons in the benzene ring (background). The blue regions denote relatively positively charged parts of the molecule and the green, yellow, and red regions denote progressively more negatively charged regions. Note how the electron-rich green regions spread over the carbon atoms much more in phenol than in benzene, so these atoms are more susceptible to electrophilic attack. A more detailed analysis shows that the ortho- and para- carbon atoms have the greatest accumulation of negative charge.

Long before their theories were supported by computations, organic chemists found a way to use resonance structures to explain the product distribution in electrophilic substitution. To focus on the essentials, only the lone pairs of electrons involved in resonance are shown. Thus, the Lewis structure for phenol, is regarded as a resonance hybrid of the following structures:

:Ö—H ⟷ $^+$Ö—H ⟷ $^+$Ö—H ⟷ $^+$Ö—H

The curved arrows show how one resonance structure relates to another. Notice that the formal negative charge is located on the ortho and para positions, exactly where reaction occurs most quickly. Other ortho- and para-directing groups include NH_2, Cl, and Br. All have an atom with a lone pair of electrons next to the ring, and all accelerate reaction.

We have seen how an —OH group can accelerate a reaction. Are there substituents that can slow down the electrophilic substitution of benzene? One way to reduce the electron density in the benzene ring and to make it less attractive to electrophiles is to substitute the benzene ring with a highly electronegative substituent, which can withdraw electrons partially. Alternatively, we could use a substituent that removes electrons by resonance. A number of substituents, such as the carboxyl group (—COOH), have both effects. For instance, it is found that the nitration of benzoic acid, C_6H_5COOH, is much slower than the nitration of benzene; moreover, most of the product is the meta nitro compound:

COOH ⟶ COOH, NO_2 (ortho) + COOH, NO_2 (meta) + COOH, NO_2 (para)

18.5% 80.0% 1.5%

The electronegative O atoms of the carboxylic acid group withdraw electrons from the whole ring, thereby reducing its overall electron density. Moreover, resonance preferentially removes electrons from the ortho and para positions:

As a result of both effects, the reaction rate is lowered, especially at the ortho and para positions, and the meta position is left as the most likely place for attack. Other meta-directing electron-withdrawing substituents include $-NO_2$, $-CF_3$, and $-C{\equiv}N$. Notice that none of these substituents has a lone pair of electrons on the atom next to the ring.

Aromatic rings are much less reactive than their double bond character would suggest; they commonly undergo substitution rather than addition. Electrophilic substitution of benzene with electron-donating substituents is accelerated and occurs at the ortho and para positions preferentially. Electrophilic substitution of benzene with electron-withdrawing substituents occurs at a reduced rate and primarily at the meta positions.

SKILLS YOU SHOULD HAVE MASTERED

1 Distinguish alkanes, alkenes, alkynes, and arenes by differences in bonding, structure, and reactivity.

2 Name simple hydrocarbons, Toolbox 18.1 and Examples 18.1, 18.2, and Self-Test 18.9.

3 Identify two molecules, given their structural formulas, as structural, geometrical, or optical isomers, Section 18.3.

4 Write the formulas of isomeric molecules, Example 18.3.

5 Judge whether a compound is chiral, Example 18.4.

6 Describe general trends in the physical properties of alkanes, Sections 18.2 and 18.4.

7 Predict the products of given elimination, addition, and substitution reactions, Sections 18.5, 18.7, and 18.9.

8 Explain why ortho- and para-directing groups accelerate electrophilic substitution on the benzene ring and identify such groups, Section 18.9.

EXERCISES

Types of Hydrocarbons

18.1 Draw Lewis structures of the following molecules and identify them as an alkane, alkene, or alkyne: (a) $CH_3CH_2CCCH_2CH_3$; (b) $CH_3CH_2CH_2CH_2CH_2CH_3$; (c) $CH_2CHCH_2CH_3$; (d) $CH_3CHCHCH_2CCCH_3$; (e) $CH_3CH_2CH_2CHCH_2$.

18.2 Give the molecular formula and identify the following molecules as being an alkane, alkene, or alkyne:

(a) (b) (c) H_2CCCH_2 (d)

18.3 Give the molecular formulas corresponding to the following stick figures and label each as an alkane, an alkene, an alkyne, or an aromatic hydrocarbon:

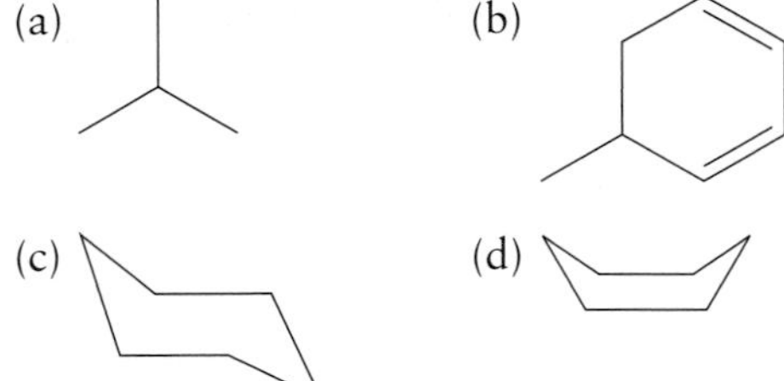

18.4 Give the molecular formulas corresponding to the following stick figures and label each as an alkane, an alkene, an alkyne, or an aromatic hydrocarbon:

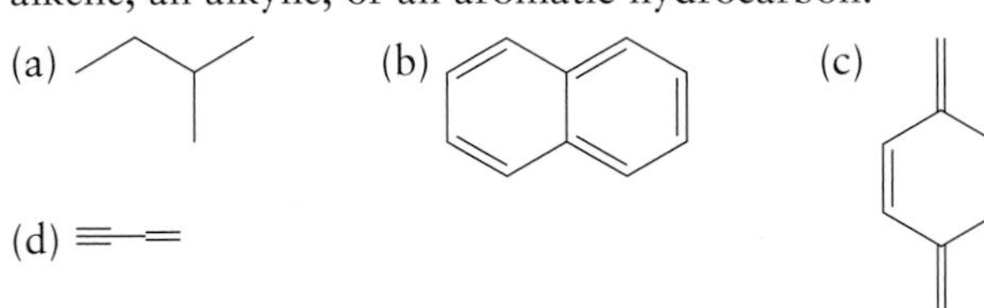

18.5 Give the molecular formulas corresponding to the following stick figures and label each as an alkane, an alkene, an alkyne, or an aromatic hydrocarbon:

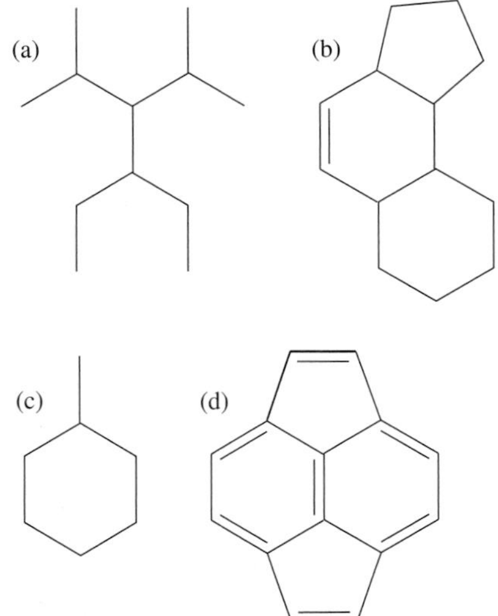

18.6 Give the molecular formulas corresponding to the following stick figures and label each as an alkane, an alkene, an alkyne, or an aromatic hydrocarbon:

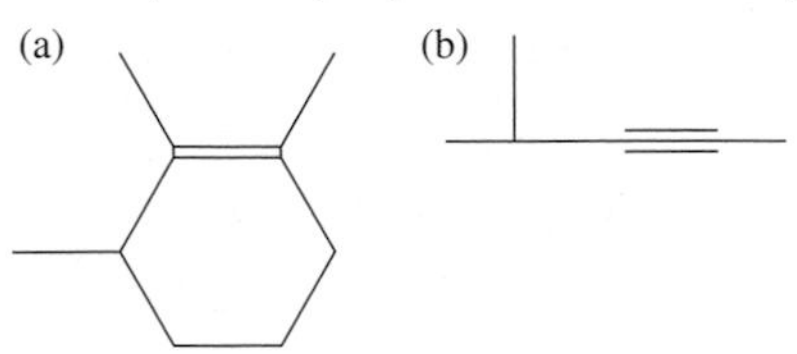

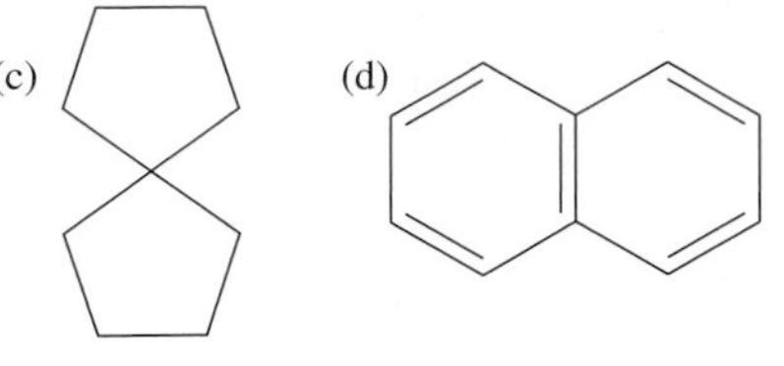

Nomenclature of Hydrocarbons

18.7 Name each of the following as an unbranched alkane: (a) C_3H_8; (b) C_4H_{10}; (c) C_7H_{16}; (d) $C_{10}H_{22}$.

18.8 Name each of the following as an unbranched alkane: (a) C_9H_{20}; (b) C_5H_{12}; (c) C_6H_{14}; (d) C_8H_{18}.

18.9 Name the following substituents: (a) CH_3—; (b) $CH_3(CH_2)_3CH_2$—; (c) $CH_3CH_2CH_2$—; (d) $CH_3CH_2CH_2CH_2CH_2CH_2$—.

18.10 Name the following substituents: (a) Cl—; (b) CH_3CH_2—; (c) $CH_3CH_2CH_2CH_2$—; (d) $CH_3CH_2CH_2CH_2CH_2CH_2CH_2CH_2$—.

18.11 Give the systematic name of (a) $CH_3CH_2CH_3$; (b) CH_3CH_3; (c) $CH_3(CH_2)_3CH_3$; (d) $(CH_3)_2CHCH(CH_3)_2$.

18.12 Give the systematic name of (a) $CH_3CH_2CH_2CH_2CH_2CH_3$; (b) $CH_3CH(CH_2CH_3)C(CH_3)_3$; (c) $(CH_3)_3CCH_2CH(CH_3)_2$; (d) $CH_3C(CH_3)_2CH_2CH_3$.

18.13 Name the compounds (a) $CH_3CH{=}CHCH(CH_3)_2$; (b) $CH_3CH_2CH(CH_3)C(C_6H_5)(CH_3)_2$.

18.14 Name the compounds (a) $CH_2{=}CHCH(C_6H_5)(CH_2)_4CH_3$; (b) $(CH_3)_2CHCH(CH_3)CHClC{\equiv}CH$.

18.15 Write the shortened (condensed) structural formula of (a) 3 methyl-1-pentene; (b) 4-ethyl-3,3-dimethylheptane; (c) 5,5-dimethyl-1-hexyne; (d) 3-ethyl-2,4-dimethylpentane.

18.16 Write the shortened (condensed) structural formula of (a) 4-ethyl-2-methylhexane; (b) 3,3-dimethyl-1-pentene; (c) *cis*-4-methyl-2-hexene; (d) *trans*-2-butene.

18.17 Write the structural formula of (a) 4,4-dimethylnonane; (b) 4-propyl-5,5-diethyl-1-decyne; (c) 2,2,4-trimethylpentane; (d) *trans*-3-hexene.

18.18 Write the structural formula of (a) 3,4,9-trimethyldecane; (b) 2,3,5-trimethyl-4-propylheptane; (c) 1,3-dimethylcyclohexane; (d) 5-ethyl-3-heptene.

18.19 Draw stick structures to represent each of the following molecules: (a) nonane, $CH_3(CH_2)_7CH_3$; (b) cyclopropane, C_3H_6; (c) cyclohexene, C_6H_{10}.

18.20 Draw stick structures to represent each of the following species: (a) 2,2,3,3-tetramethylhexane, $CH_3C(CH_3)_2C(CH_3)_2CH_2CH_2CH_3$; (b) the trityl cation, $(C_6H_5)_3C^+$; (c) butadiene, $CH_2CHCHCH_2$.

Isomers

18.21 Write the structural formulas and name (a) at least 10 *alkene* isomers having the formula C_6H_{12}; (b) at least 10 *cycloalkane* isomers have the formula C_6H_{12}.

18.22 Write the structural formulas for and name all the isomers (including geometrical isomers) of the alkenes (a) C_4H_8; (b) C_5H_{10}.

18.23 Identify each of the following pairs as structural isomers, geometrical isomers, or not isomers: (a) butane and cyclobutane; (b) cyclopentane and pentene;

(c) 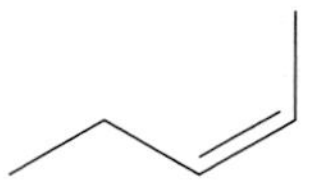 and 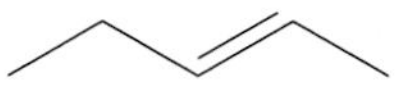 ;

(d) 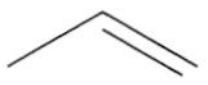 and  .

18.24 Identify each of the following pairs as structural isomers, geometrical isomers, or not isomers: (a) 1-chlorohexane and chlorocyclohexane;

(b)  and  ;

(c) 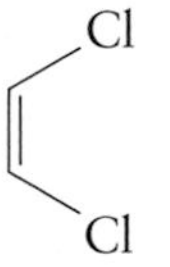 Cl Cl and Cl Cl .

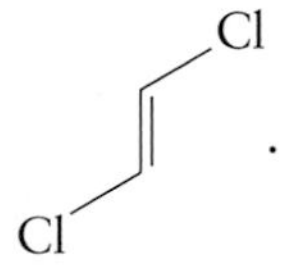

18.25 A branched hydrocarbon C_4H_{10} reacts with chlorine in the presence of light to give two branched structural isomers with the formula C_4H_9Cl. Write the structural formulas of (a) the hydrocarbon; (b) the isomeric products.

18.26 A branched hydrocarbon C_6H_{14} reacts with chlorine in the presence of light to give only two structural isomers with the formula $C_6H_{13}Cl$. Write the structural formulas of (a) the hydrocarbon; (b) the two isomeric products.

18.27 Indicate which molecules display optical activity and identify the chiral carbon atoms in those that do: (a) $CH_3CHBrCH_2CH_3$; (b) $CH_3CH_2CHCl_2$; (c) 1-bromo-2-chloropropane; (d) 1,2-dichloropentane.

18.28 Indicate which molecules display optical activity and identify the chiral carbon atoms in those that do: (a) $CH_3CHBrCl$; (b) $CH_3CH_2CHClCH_3$; (c) 2-bromo-2-chloropropane; (d) 1,2-dibromobutane.

Alkanes

18.29 Why do branched-chain alkanes have lower melting points and boiling points than unbranched alkanes with the same number of carbon atoms?

18.30 Arrange the following compounds in order of increasing boiling point: $CH_3CH_2CH_2CH_2CH_3$, $CH_3CH_2CH(CH_3)_2$, $C(CH_3)_4$. Explain your reasoning.

18.31 Using data available in Appendix 2A, write balanced equations and calculate the heat released when (a) 1.00 mol and (b) 1.00 g of each of the following compounds is burned in excess oxygen: propane, butane, and pentane. Is there a trend in the amount of heat released per mole of compound or per gram of compound? If so, what is it?

18.32 Write the balanced chemical equation for the complete fluorination of methane to tetrafluoromethane. Using bond enthalpies, estimate the enthalpy of this reaction. The corresponding reaction using chlorine is much less exothermic. To what can this difference be attributed?

18.33 How many different products containing two carbon atoms are possible in the reaction of chlorine with ethane? Do any of the products exist as optical isomers?

18.34 How many different products that retain the three-membered ring are possible in the reaction of chlorine with cyclopropane? Do any of the products exist as stereoisomers?

Alkenes

18.35 Two structural isomers can result when hydrogen bromide reacts with 2-pentene. (a) Write their structural formulas. (b) What name is given to this type of reaction?

18.36 (a) Write a balanced equation for the production of 2,3-dibromobutane from 2-butyne. (b) What name is given to this type of reaction?

18.37 (a) Write a balanced equation for the reaction of bromocyclohexane with sodium ethoxide in ethanol. (b) Draw structural formulas of the cyclic reactant and product. (c) What name is given to this type of reaction?

18.38 Draw stick structures of the possible products for the reaction of sodium ethoxide with (a) 1-bromobutane; (b) 2-bromobutane.

18.39 Compare the reaction enthalpies for the halogenation of ethene using chlorine, bromine, and iodine. What trend, if any, exists in these numbers? Use bond enthalpies to estimate the enthalpies of reaction.

18.40 Compare the reaction enthalpies for the hydrohalogenation of ethane using HX, where X = Cl, Br, I. What trend, if any, exists in these numbers? Use bond enthalpies to estimate the enthalpies of reaction.

Aromatic Compounds

18.41 Name the following compounds:

(a) CH_2CH_3 CH_3

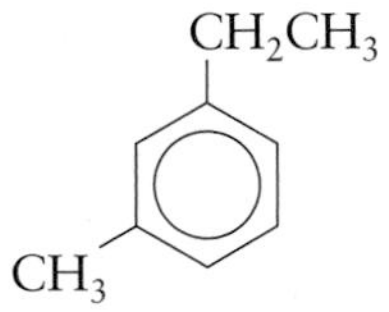

(b) CH_3 CH_3 CH_3 CH_3 CH_3

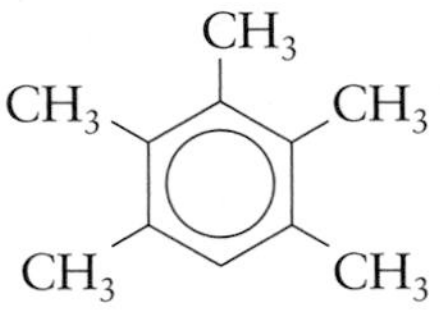

18.42 Name the following compounds:

(a) CH_3 CH_2CH_3 $CH_2CH_2CH_3$

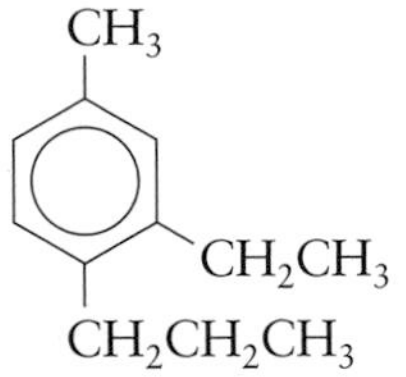

(b) CH_2CH_3 $CH_2CH_2CH_3$

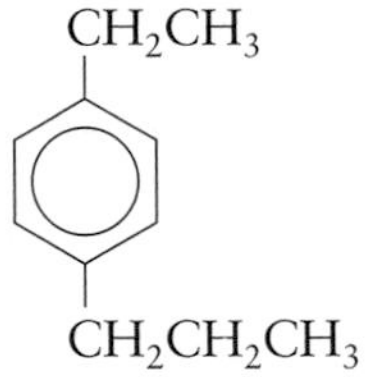

18.43 Write the structural formulas of (a) methylbenzene, more commonly known as toluene; (b) *p*-chlorotoluene; (c) 1,3-dimethylbenzene; (d) 4-chloromethylbenzene.

18.44 Write the structural formulas of (a) *p*-xylene; (b) 1,2-dibromobenzene; (c) 3-phenylpropene; (d) 2-ethyl-1,4-dimethylbenzene.

18.45 (a) Draw all the isomeric dichloromethylbenzenes. (b) Name each one and indicate which are polar and which are nonpolar.

18.46 (a) Draw all the isomeric diaminodichlorobenzenes. (b) Name each one and indicate which are polar and which are nonpolar.

18.47 Draw resonance structures of cyanobenzene that show how it functions as a meta-directing substituent.

18.48 One can prepare bromonitrobenzenes either by nitrating bromobenzene or by brominating nitrobenzene. Will these two reactions give the same product (or product distribution)? If not, how will the reactions differ?

18.49 How many different compounds can be produced when naphthalene, $C_{10}H_8$, undergoes electrophilic aromatic substitution by a single electrophile (denoted E)? Draw stick figures to represent them.

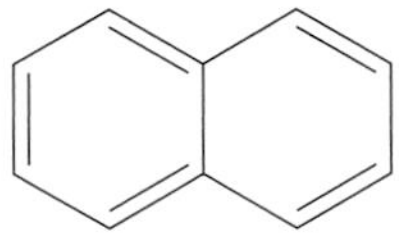

18.50 The structure of benzaldehyde is

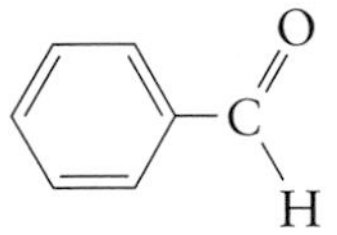

O C H

Do you expect the aldehyde functional group (CHO) to act as a meta-directing or as an ortho,para-directing group? Explain.

INTEGRATED EXERCISES

18.51 Identify the type and number of bonds on carbon atom 2 in (a) pentane; (b) 2-pentene; (c) 2-pentyne.

18.52 Predict the geometry and hybridization of the orbitals used in bonding on carbon atom 2 in (a) pentane; (b) 2-pentene; (c) 2-pentyne.

18.53 Classify each of the following reactions as addition or substitution and write its chemical equation: (a) chlorine reacts with methane when exposed to light; (b) bromine reacts with ethene in the absence of light.

18.54 Classify each of the following reactions as addition or substitution and write its chemical equation: (a) hydrogen reacts with 2-pentene in the presence of a nickel catalyst; (b) hydrogen chloride reacts with propene.

18.55 Classify as addition or substitution and write the chemical equations for the reactions that occur (if any) when chlorine gas is mixed with (a) ethane and exposed to sunlight; (b) ethene; (c) ethyne.

18.56 Classify as addition or substitution and write the chemical equations for the reactions that occur (if any) when an excess of hydrogen is mixed in the presence of a platinum catalyst with (a) 2-methylpropene; (b) propyne; (c) propane.

18.57 Why do alkenes have lower melting points and boiling points than alkanes with the same number of carbon atoms?

18.58 The *torsion angle* is a measure of the twist about a given bond. As seen in the illustration below, the torsion angle is defined between four atoms. Figure i looks down on the molecule from the top, whereas Figure ii views the molecule along the B—C bond.

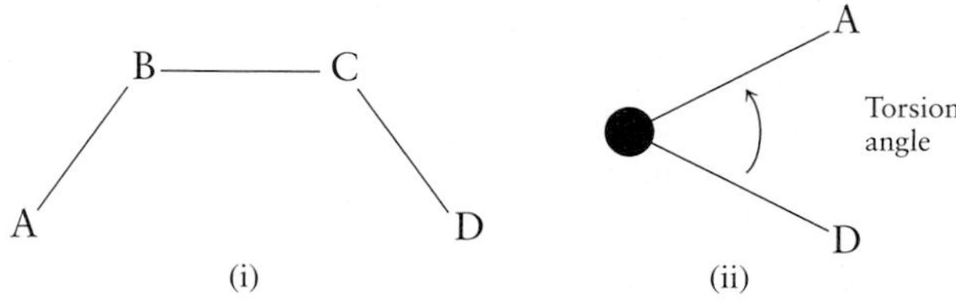

The measurement of torsion angles is valuable because it provides a measure of the amount of distortion that may be present in a molecule. (a) On the basis of this definition, predict the C—C—C—C torsion angle for carbon atoms 2–5 in *trans*-3-hexene. (b) On the Web site, examine the structures of *trans*-3-hexene, cyclohexene, and cyclohexane. Download these structures and use a suitable software package to view them, measure the C—C—C—C torsion angles about the double bonds in *trans*-3-hexene and cyclohexene and any of the C—C—C—C torsion angles in cyclohexane. How do these numbers compare? (c) Compare the C—C—C bond angles for the sp^3-hybridized carbon atoms in cyclohexene and cyclohexane. What feature is most notable?

18.59 Give the systematic name of the following compounds. If geometrical isomers are possible, write the names of each one: (a) $CH_2{=}C(CH_3)_2$; (b) $CH_3CH{=}C(CH_3)CH_2CH_3$; (c) $HC{\equiv}CCH_2CH_2CH_2CH_3$; (d) $CH_3CH_2C{\equiv}CCH_2CH_3$; (e) $CH_3CH_2CH_2C{\equiv}CCH_3$.

18.60 Give the systematic name of the following compounds. If geometrical isomers are possible, write the names of each one: (a) $CH_3CH_2CH_2CH{=}CH_2$; (b) $CH_3CH_2C(CH_3){=}CHCH_3$ (with CH_3 on the third carbon); (c) $(CH_3)_2C{=}CHCH_2(CH_3)_2$; (d) $CH_3C{\equiv}CCH_2CH(CH_2CH_3)CH_3$; (e) $CH_3CH_2C{\equiv}CCH(CH_2CH_3)CH_3$.

18.61 Draw stick structures to represent the different straight-chain heptenes. These molecules have the formula C_7H_{14}. Include geometrical isomers.

18.62 Spiropentane has the structural formula

$$\begin{array}{ccc} CH_2 & - & CH_2 \\ & \diagdown\ \diagup & \\ & C & \\ & \diagup\ \diagdown & \\ CH_2 & - & CH_2 \end{array}$$

(a) Draw the three-dimensional structure of the molecule. (b) Identify the type of hybridization at each carbon atom. (c) Estimate the bond angles.

18.63 The structure of decalin may be found on the Web site. (a) From an examination of this structure, determine the chemical formula of decalin. (b) From what aromatic hydrocarbon may decalin be produced by complete hydrogenation? (c) Are isomers possible for decalin? If so, draw the appropriate Lewis structures.

18.64 Nonsystematic names for organic compounds may still be found in the chemical literature and chemical supply catalogs, so it is important to be somewhat familiar with these as well as with the IUPAC rules. For example, isobutane and isopentane are commonly used. Structures for these compounds may be found on the Web site. Give the systematic name for (a) isobutane and (b) isopentane. (c) Formulate a rule for the usage of the prefix *iso*- and predict the structure of isohexane.

18.65 In a combustion experiment, 2.14 g of a hydrocarbon formed 3.32 g of water and 6.48 g of carbon dioxide. Deduce its empirical formula and state whether it is likely to be an alkane, an alkene, or an alkyne. Explain your reasoning.

18.66 In a combustion experiment, 3.69 g of a hydrocarbon formed 4.50 g of water and 11.7 g of carbon dioxide. Deduce its empirical formula and state whether it is likely to be an alkane, an alkene, or an alkyne. Explain your reasoning.

18.67 The following names of organic molecules are incorrect. Draw line structures and determine the correct systematic name for (a) 4-methyl-3-propylheptane; (b) 4,6-dimethyloctane; (c) 2,2-dimethyl-4-propylhexane; (d) 2,2-dimethyl-3-ethylhexane.

18.68 Alkanes undergo substitution of hydrogen atoms when treated with halogens. Bromination of which of the following compounds could give rise to chiral monosubstituted products: (a) ethane; (b) propane; (c) butane; (d) pentane? Include in your answer the names of the chiral compounds that would be formed.

18.69 Why are the hydrocarbons with carbon numbers between 1 and 4 not suitable for gasoline (see Box 18.1)?

18.70 What are the major problems associated with using coal as a fuel (see Box 18.1)?

18.71 In the production of gasoline, both *cracking* and *alkylation* are employed, although they are chemical opposites of each other. Define cracking and alkylation as they apply to the production of gasoline (see Box 18.1).

18.72 What reasons are there for using ethanol as a part of gasoline mixtures (see Box 18.1)?

18.73 Concentrated sulfuric acid is a strong dehydrating agent, in the sense that it can remove the elements of water, HOH, from a compound. What principal product results from the reaction of $CH_3(CH_2)_2CH(OH)CH_3$ with concentrated sulfuric acid at 120°C?

18.74 A hydrocarbon contains 90% carbon by mass and 10% hydrogen by mass and has molar mass 40 $g{\cdot}mol^{-1}$. It decolorizes bromine water, and 0.73 g of the hydrocarbon reacts with 800 mL of hydrogen (measured at STP) in the presence of a nickel catalyst. Write the molecular formula of the hydrocarbon and the structural formulas of two possible isomers.

18.75 The compound 1-bromo-4-nitrobenzene (*p*-bromonitrobenzene) may be brominated. What do you expect to be the major product and why? Draw appropriate Lewis structures to support your answer.

18.76 Consider the nitration of

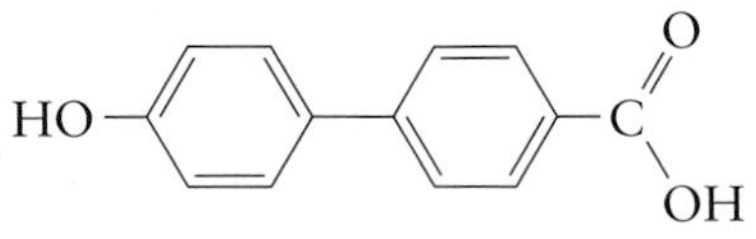

If the reaction can be controlled so that one NO_2 group replaces one H atom of the molecule, where do you expect the nitro group to end up in the product?

18.77 A mixture of ethane (0.500 bar) and hydrogen (0.300 bar) was placed over a metal catalyst and equilibrium was established at 500°C. The dominant reaction is the hydrogenation of ethane to methane. What will be the ratio of methane to ethane at equilibrium? Assume that other side reactions are negligible relative to the formation of methane.

18.78 Consider the line structure of the following molecule:

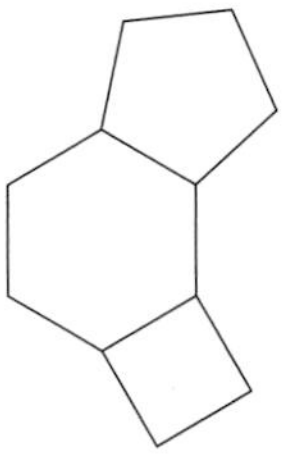

(a) How many different chiral carbon atoms are present in the molecule? (b) How many compounds can be generated if you draw all the possible structures that differ in the orientation at each carbon atom but leave the carbon skeleton intact? In other words, consider how many stereoisomers you would have if you produced every possibility. (c) Do any of these isomers form enantiomeric pairs?

18.79 It is possible for molecules to have carbon atoms that have four different substituents but still do not exist as a set of optically active enantiomers. How is this possible? Give an example to illustrate your answer.

18.80 Trinitrotoluene, TNT, is a well-known explosive. The structure of TNT may be found on the Web site of this textbook. (a) Using the structure available on the Web site, determine the systematic name for TNT. (b) TNT is made by nitrating toluene (methylbenzene) with concentrated nitric acid dissolved in sulfuric acid. Explain why this particular trisubstituted isomer is the one that is formed during the nitration.

18.81 A chemist obtains the mass spectrum of 1,2-dichloro-4-ethylbenzene. Give at least four possible fragments and the masses at which you would expect them to occur. See Major Technique 5, Mass Spectrometry, which follows this set of exercises. Chlorine has two naturally occurring isotopes: ^{35}Cl, 34.969 u, 75.53% and ^{37}Cl, 36.966 u, 24.47%. The mass of ^{1}H is 1.0078 u.

18.82 How will the mass spectrum of D_2O differ from that described in the text for H_2O? See Major Technique 5, Mass Spectrometry, which follows this set of exercises.

18.83 The occurrence of isotopes of some elements allows the presence of these elements to be identified readily in a mass spectrum of a compound. For example, Br has two naturally occurring isotopes with masses 79 u and 81 u, in the relative abundances 50.5% and 49.5%. How will this make the determination of the number of bromine atoms in an organic molecule relatively easy? For example, in the bromination of an alkane, it is possible that a number of products with more than one bromine atom might be produced. See Major Technique 5, Mass Spectrometry, which follows this set of exercises.

18.84 A chemist treats toluene (methylbenzene) with chlorine in the presence of a metal catalyst. A sample of the reaction mixture is injected into a GC-MS unit (gas chromatograph-mass spectrometer). One of the compounds that is separated by using gas chromatography gives a mass spectrum with peaks at 128 u, 126 u, 113 u, 111 u, and 91 u, among others. The peak at 128 u is approximately one-third the size of the peak at 126 u, and the peak at 113 u is approximately one-third the size of the peak at 111 u. What is the likely composition of this compound, and what fragments may be assigned to these peaks? See Major Technique 5, Mass Spectrometry, which follows this set of exercises.

MASS SPECTROMETRY

In spectroscopy, we use a diffraction grating to separate photons of light by their wavelengths. In *mass spectrometry*, we use a magnetic field to separate molecular ions by their masses. This technique is one of our most powerful analytical methods because it provides both quantitative and qualitative information about the substance being analyzed, it can be conducted on tiny samples, and it can be conducted in the field, using lightweight, portable units. Mass spectrometry is used to determine isotopic abundance (Section B), to study the composition of bone and other body tissues, to analyze the blood of newborns for congenital diseases, to detect vanishingly small concentrations of drugs in urine, and to determine the structure of the human genome.

1 The Technique

We met a classic type of mass spectrometer in Section B. Another common type of mass spectrometer is the *large magnet mass spectrometer* (Fig. 1). In each case, the spectrometer uses the degree of deflection of charged particles by a magnetic field to determine the relative masses of the particles. In a mass spectrometer, the sample is first vaporized and then ionized. The resulting ions are accelerated by an electric field into a narrow beam that, as it passes between a set of magnets, is deflected toward a detector. The heavier the particle, the less it is deflected, and the degree of deflection allows the relative masses of the particles to be determined. The results are displayed as a series of peaks, with the height of each peak being proportional to the relative number of particles with a given mass.

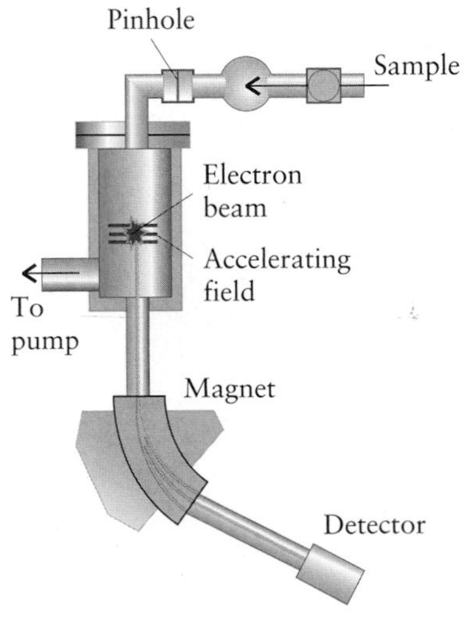

FIGURE 1 A large magnet mass spectrometer.

The entire spectrometer must first be evacuated (have all the air pumped out) to ensure that no gas molecules can collide with the ions produced from the sample and deflect them in unpredictable ways. Then the sample is introduced in the form of a vapor into the sample inlet chamber and allowed into the ionization chamber. In this chamber, rapidly moving electrons collide violently with the molecules of the vapor. When one of the accelerated electrons collides with a molecule, the colliding electron knocks another electron out of the molecule, leaving a molecular cation. The molecular ion may split apart into two or more charged fragments. The positive ions—the original molecular ions and their fragments—are accelerated out of the chamber by a strong electric field applied across a series of metal grids. The speeds reached by the ions depend on their masses: light ions reach higher speeds than heavier ones.

The rapidly moving ions pass between the poles of an electromagnet. The magnetic field bends the paths of the ions to an extent that depends on their speed and on the strength of the field. In the large magnet mass spectrometer, the magnetic field is kept constant but the strength of the electric field is varied, so the molecular ions are accelerated to different speeds. At a given electric field strength, ions of only one particular mass reach the detector. Any ions with different masses collide with the walls of the chamber. The strength of the electric field is varied and the result is a mass spectrum, a plot of the detector signal against the strength of the electric field. The positions of the peaks in the mass spectrum give the mass-to-charge ratios of the ions. If all the ions have charges of +1, then the peaks give the mass ratios and their relative heights give the proportions of ions of various masses.

2 Analyzing a Mass Spectrum

If the sample consists of atoms of one element, the mass spectrum gives the isotopic distribution of the sample. The relative molar masses of the isotopes can be determined by comparison with atoms of carbon-12. If the sample is a compound, the formula and structure of the compound can be determined by studying the fragments. For example, the ions that CH_4 could produce are CH_4^+, CH_3^+, CH_2^+, CH^+, C^+, and H^+. Some of the particles that strike the detector are those that result when the molecule simply loses an electron (for example, to produce CH_4^+ from methane):

$$\text{Molecule} + e^- \longrightarrow \text{molecule}^+ + 2\,e^-$$

The mass of this molecular ion, the *parent ion*, is called the *parent mass*. The parent mass has essentially the same molar mass as the compound itself. However, if the electron beam is moving at a very high speed, only fragments of the molecule may survive. Some fragments are almost always produced, and other information must be used to help with the interpretation. For example, a sample of a known compound can be analyzed and the two spectra compared to look for similar fragments. The sample can also be passed through a gas chromatograph first (see Major Technique 4). The chromatogram allows comparison with a large number of possible compounds, and the mass spectrum verifies the identification.

The identification of a molecular structure from a mass spectrum requires good chemical detective work. Let's see how that is done by trying to identify a simple compound, the one that produced the mass spectrum summarized in the table. The parent mass is 18, 19, or 20. If the parent mass is 20, the compound could be HF. However, fluorine has only one stable isotope, fluorine-19, so we could not explain the large peaks at 17 and 18. Because of the very small abundances of the peaks at 19 and 20, we can also guess that they represent the less abundant isotopes of one or more of the elements in the compound. We can conclude that the parent mass is 18 and the compound is probably water. The peaks at 19 and 20 would be due to the mass of parent molecules containing naturally occurring isotopes such as oxygen-18 and deuterium. The peak at 17 represents the OH^+ fragment, the peak at 16 would be O^+ ions, and the peak at 1 the few H^+ ions that form in the ionization process.

Mass Spectrum of an Unknown Compound

*m/e**	Peak intensity as a percentage of largest peak
1	<0.1
16	1.0
17	21.0
18	100.0
19	0.08
20	0.22

**m/e* is the mass-to-charge ratio of the molecular ions.

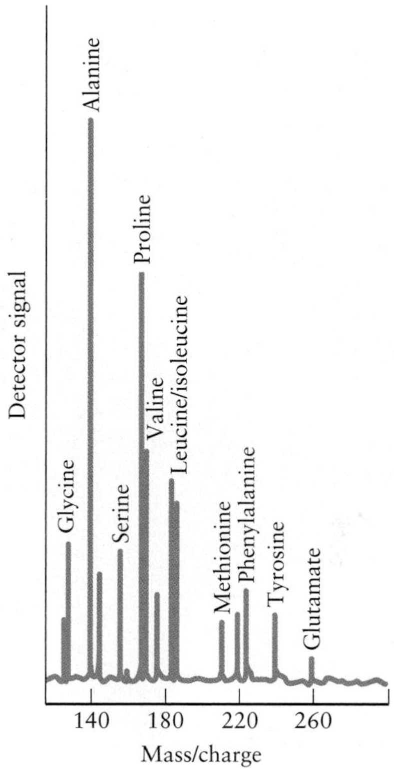

FIGURE 2 The two phenylalanine peaks include the radioactive tracer (the peak on the right) and the phenylalanine in the blood (the smaller peak 5 mass units to the left). The presence of the latter peak shows that the baby from which the blood was drawn has phenylketonuria.

More complex detective work is required to analyze large biomolecules and drugs. However, fragmentation generally follows predictable patterns, and one compound can be identified by comparing its mass spectrum with those of other known compounds with similar structures. In the spectrum shown in Fig. 2, we see a sample of blood from a newborn infant. The small spot of blood is being analyzed to determine whether the child has phenylketonuria. The presence of the compound phenylalanine is a positive indication of the condition. Some phenylalanine labeled with a radioactive isotope has been injected along with the sample to identify the parent mass. The radioactive compound, which is 5 atomic mass units (5 u) heavier than the phenylalanine, appears in the spectrum 5 u above it, thereby allowing a positive identification of the presence of the chemical.

Related Exercises: 18.81–18.84.

CHAPTER

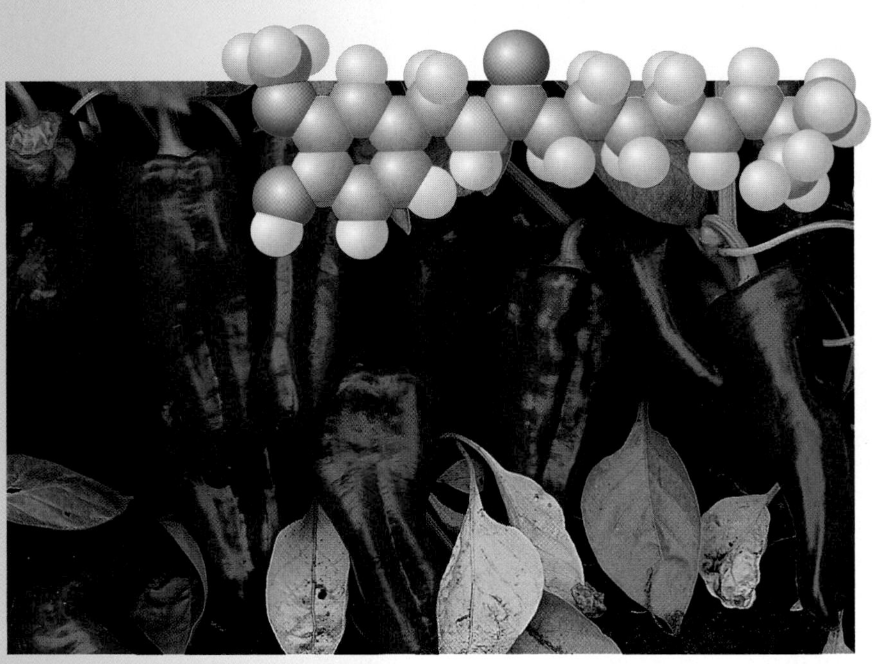

In common with chili peppers, these capsicums contain the compound capsaicin, which is responsible for the hot taste. Capsaicin has a number of functional groups: the model shows the presence of a hydroxyl group, an amide group, and an ether. The aromatic ring can also be regarded as a functional group attached to the alkane chain.

Why Do We Need to Know This Material?

Organic chemistry is the foundation of the pharmaceutical industry and, through that, of medicine. Organic chemistry is also the basis of biochemistry and molecular biology; through those disciplines, it provides an understanding of life. The great chemical diversity of the complex organic molecules of living systems is made possible by the presence of groups of atoms with characteristic functions. Once we understand these groups, we can predict the properties of other organic molecules, even those of the giant biomolecules of proteins, carbohydrates, and genetic material, and the polymeric materials that have revolutionized our lives.

What Do We Need to Know Already?

This chapter builds on Chapter 18 and requires the same preparation as that chapter as well as the concepts of acid-base and redox reactions (Sections J and K). It also makes use of the concepts of molecular polarity (Section 3.3) and solubility (Section 8.10).

ORGANIC CHEMISTRY II: FUNCTIONAL GROUPS

Despite their immense variety, organic compounds can be understood in terms of **functional groups,** which are small groups of atoms with characteristic properties. Whereas the hydrocarbons are built only from carbon and hydrogen, functional groups may include atoms of other elements and accordingly bring strikingly different properties to organic compounds. Moreover, because functional groups have characteristic chemical properties, they help to make organic chemistry a highly systematic subject: we do not have to learn all the properties of each different compound—once we identify the functional groups that are present, we have gone a long way toward understanding and predicting the likely chemical properties of the compound. In this chapter, we shall look at the structure and properties of some of the common functional groups.

This chapter has three parts. In the first part, we look at the structure and properties of some of the common functional groups and see something of the mechanisms by which the groups are thought to react. Then we examine the impact that the presence of functional groups has on modern polymers. As in so many spheres, Nature has preceded chemists' explorations and has made full and subtle use of functional groups in the development of the processes of life. In the final part of the chapter, we see three examples of how functional groups sustain us, feed us, and replicate our genetic material.

COMMON FUNCTIONAL GROUPS

Functional groups are either attached to the carbon backbone of a molecule or form part of that chain. Examples are the chlorine atom in chloroethane, CH_3CH_2Cl; the $\rangle C{=}O$ group in acetone (propanone), CH_3COCH_3; and the carbon-carbon double bond in 2-butene. Table 19.1 lists the most common functional groups. Toolbox 19.1 at the end of Section 19.7 summarizes the rules for naming compounds with functional groups.

TABLE 19.1 Common Functional Groups

Group	Class of compound	Group	Class of compound
—X	halide (X = F, Cl, Br, or I)	—COOH	carboxylic acid
		—COOR	ester
—OH	alcohol, phenol	$-N\langle$	amine
—O—	ether		
—CHO	aldehyde	$-CO-N\langle$	amide
—CO—	ketone		

In the following eight sections, we examine some of the more important functional groups, look at their characteristic chemical properties, and in some cases see how those properties come about in terms of reaction mechanisms. We also see how compounds containing these functional groups are named, and Toolbox 19.1 at the end of Section 19.8 summarizes the nomenclature rules. Many organic compounds have common names that are widely used; we mention these names in the following sections.

19.1 Haloalkanes

The **haloalkanes** (also called alkyl halides) are alkanes in which at least one hydrogen atom has been replaced by a halogen atom. Although they have important uses, many haloalkanes are highly toxic and a threat to the environment. The haloalkane 1,2-dichloro-1-fluoroethane, $CHClFCH_2Cl$, is an example of a chlorofluorocarbon (CFC), one of the compounds held responsible for the depletion of the ozone layer (see Box 13.3). Many pesticides are aromatic compounds with several halogen atoms. Because pesticides have low overall polarity, they tend not to be washed away by rain. They also tend to be fat soluble; so when we eat food containing them, they migrate into fatty tissues. Compounds stored in fat can remain for a long time and accumulate as more is ingested.

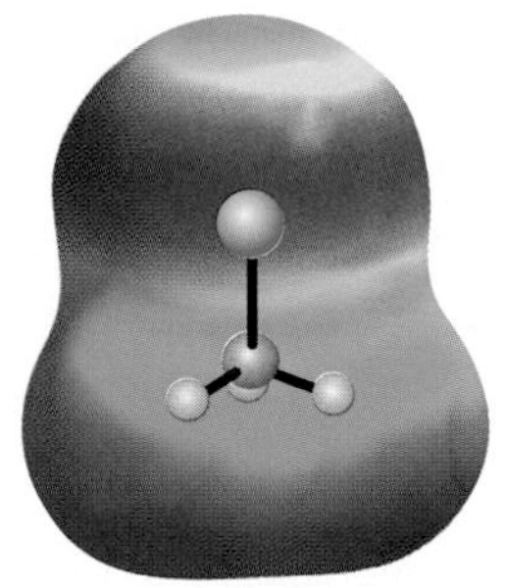

1 Chloromethane, CH_3Cl

A nucleophile is the opposite of an electrophile, which seeks out centers of negative charge.

Carbon-halogen bonds are polar (**1**). Although haloalkane molecules are not sufficiently polar to dissolve in water, the polarity of the C—X bond governs their chemical properties. The carbon atom to which the halogen atom is attached is partially positive, making it susceptible to **nucleophilic substitution** reactions, in which a nucleophile replaces the halogen atom. A **nucleophile** is a reactant that seeks out centers of positive charge in a molecule. A typical example is the hydroxide ion, OH^-: its negative charge is attracted to regions of a molecule that have a partial positive charge. Another example of a nucleophile is H_2O: a lone pair on the oxygen atom is attracted to regions of partial positive charge. Water acts as a nucleophile in a **hydrolysis reaction,** a reaction with water in which a carbon-element bond is replaced by a carbon-oxygen bond. For example, bromomethane undergoes hydrolysis in water to methanol and hydrobromic acid:

$$CH_3Br + H_2O \longrightarrow CH_3OH + HBr$$

Chemists have discovered that there are two basic mechanisms for nucleophilic substitution. These mechanisms are denoted S_N1 and S_N2. The S indicates substitution, the N stands for nucleophilic, and the 1 and 2 give the molecularity of the rate-determining step.

Molecularity (Section 13.10) is the number of reactant molecules involved in a specified elementary reaction.

In an **S_N1 reaction,** the rate-determining step is unimolecular. We can envision this stage as the **heterolytic dissociation** (dissociation into ions) of the haloalkane into a halide anion and a positively charged molecular ion, a **carbocation.** For example, the first step in the hydrolysis of $(CH_3)_3CI$ is

$$(H_3C)_3C{-}I \longrightarrow \left[C(CH_3)_3 \right]^+ + I^-$$

This carbocation is immediately attacked by the nucleophile:

$$(CH_3)_3C^+ + :OH_2 \longrightarrow [(H_3C)_3C{-}OH_2]^+$$

The product of this reaction is the conjugate acid of an alcohol, which immediately donates a proton to a nearby H_2O molecule:

$$[(H_3C)_3C{-}OH_2]^+ + :OH_2 \longrightarrow (H_3C)_3C{-}OH + H_3O^+$$

In an S_N2 **reaction**, the rate-determining stage is bimolecular. As the **leaving group**—the group displaced in the reaction—departs, the **entering group**—the group replacing the displaced group—simultaneously forms a bond:

$$HO^- + CH_3{-}I \longrightarrow [HO\cdots CH_3\cdots I]^- \longrightarrow HO{-}CH_3 + I^-$$

Bimolecular complex

The reaction takes place as a single, concerted step.

When should we expect an S_N1 reaction and when an S_N2 reaction? An S_N1 reaction is favored by haloalkanes that have bulky groups around the reaction center. An S_N2 reaction is favored by haloalkanes that have an unobstructed reaction center. In a molecule such as $(CH_3)_3CI$ (**2**), the three methyl groups protect the back side of the central carbon atom from attack by the entering group. Because a bimolecular complex cannot form, the reaction proceeds by an S_N1 mechanism. In contrast, the carbon atom attached to the iodine atom in $CH_3CH_2CH_2CH_2I$ (**3**) is largely unobstructed, and nucleophilic substitution proceeds by an S_N2 mechanism.

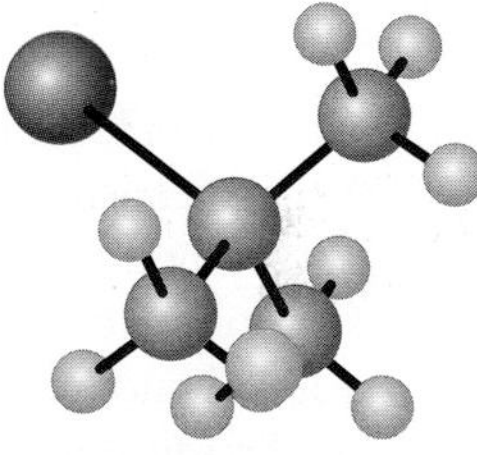

2 $(CH_3)_3CI$

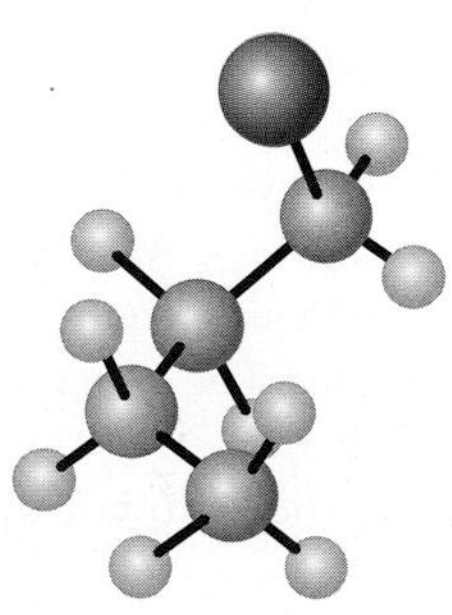

3 $CH_3CH_2CH_2CH_2I$

If the carbon atom bonded to the halogen atom is chiral, we can use stereochemical techniques to determine whether the mechanism is S_N1 or S_N2. For example, suppose we hydrolyze $(CH_3)(CH_3CH_2)CHBr$, in which the bromine atom is attached to a chiral carbon atom. If the mechanism is S_N1, the planar carbocation survives for an appreciable time and the hydroxide ion is free to attack it from either face. As a result, both enantiomeric forms of the alcohol can be expected, and the racemic mixture formed is optically inactive (Fig. 19.1). On the other hand, if the mechanism is S_N2, the hydroxide ion can attack only from the face opposite the leaving group:

$$HO^- + (H_3C)(H)(CH_2CH_3)C{-}Br \longrightarrow [HO\cdots C(CH_3)(H)(CH_2CH_3)\cdots Br]^- \longrightarrow HO{-}C(CH_3)(H)(CH_2CH_3) + Br^-$$

Media Link Animation: Fig. 19.1 S_N1 mechanism

View an animation for an S_N1 mechanism.

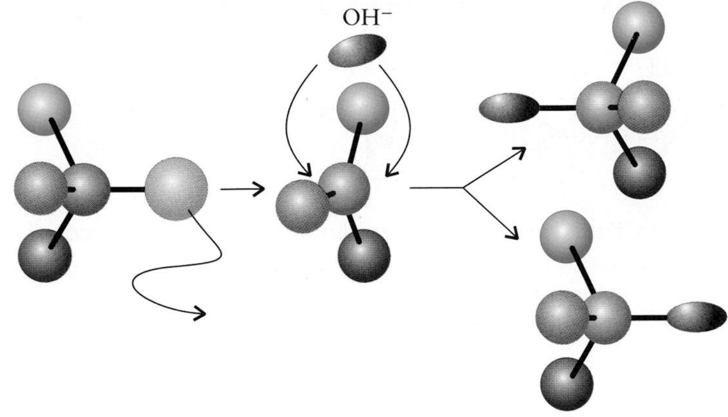

FIGURE 19.1 The events taking place when a chiral haloalkane undergoes nucleophilic substitution by an S_N1 mechanism. Different groups are represented by spheres of different colors. When the OH^- ion attacks, it may do so on either side of the planar intermediate, and the two enantiomers are formed in equal abundance.

This process inverts the molecular structure, much like turning an umbrella inside out. Only one of the two possible enantiomers is formed, and the product is optically active. We say that an S_N2 reaction proceeds with **inversion** of configuration, meaning that an optically active reactant gives rise to an optically active product with the reverse chirality. Box 19.1 describes other techniques used to determine organic reaction mechanisms.

BOX 19.1 HOW DO WE KNOW . . . THE PATHWAY OF A REACTION?

Chemists draw on a wide range of techniques and evidence to unravel the mechanism by which a chemical reaction takes place. A common technique is *isotopic substitution*, in which one atom in a reactant is replaced by an isotope. If the isotope is radioactive, it can act as a tracer, and we can monitor its presence in intermediates and products.

Radioactive tracers were used to determine the mechanism of esterification. Oxygen-18, which is a positron emitter, was used to track the fate of the singly bonded oxygen atom in a carboxylic acid group undergoing esterification. The hydroxyl oxygen atoms were replaced by oxygen-18 atoms (shown in red in the following equations). After esterification with ethanol, all the oxygen-18 was found in the water—none was present in the ester. The experiment showed that, in an esterification reaction, the oxygen of the eliminated water comes from the acid, not the alcohol:

$$CH_3COOH(l) + CH_3CH_2OH(l) \longrightarrow CH_3COOCH_2CH_3(l) + H_2O \quad (1)$$

A similar experiment was used to show that, in photosynthesis, the oxygen released comes from water:

$$6\ CO_2(g) + 6\ H_2O(l) \longrightarrow C_6H_{12}O_6(aq) + 3\ O_2(g)$$

Chemists can identify the reaction pathway of nucleophilic substitutions by combining chemical kinetic data with considerations of stereochemistry. For instance, if the substitution reaction is first order in the organic compound undergoing substitution, then there is a high chance that the reaction is S_N1. On the other hand, if the reaction is overall second order (first order in the organic compound and first order in the entering group), then the rate-determining stage is likely to be bimolecular, hence an S_N2 reaction.

Two kinds of stereochemical experiments can be used to study nucleophilic substitutions. One, discussed in Section 19.1, studies substitution at chiral carbon atoms. Another studies the role of the groups attached to the carbon atoms undergoing substitution: an S_N2 reaction is very sensitive to the size of the neighboring groups because it is hard for the transition state to accommodate five bulky groups. On the other hand, an S_N1 reaction is much less sensitive to this kind of congestion. In a typical experiment, the groups about the atom undergoing substitution are replaced by bulky groups and substitution is carried out again. If the rate of reaction is unaffected, then the mechanism of the reaction is probably S_N1. If the rate is reduced by the presence of the bulky groups, then the reaction mechanism of the original compound is probably S_N2.

Haloalkanes are alkanes in which at least one hydrogen atom has been replaced by a halogen atom; they undergo nucleophilic substitution reactions. A haloalkane may react with nucleophiles by either an S_N1 or an S_N2 mechanism; the latter proceeds with inversion of configuration.

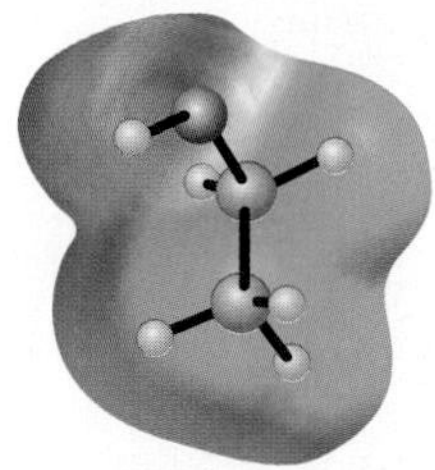

FIGURE 19.2 The charge distribution in an ethanol molecule. As in structure (**1**), red denotes a region of net negative charge and blue, regions of net positive charge. The polarity of the molecule and its ability to form hydrogen bonds result in high solubility in water.

19.2 Alcohols

The **hydroxyl group,** —OH, is an —O—H group covalently bonded to a carbon atom. An **alcohol** is an organic compound that contains a hydroxyl group not connected directly to a benzene ring or to a >C=O group. One of the best-known organic compounds is ethanol, CH_3CH_2OH, which is also called *ethyl alcohol* and *grain alcohol*. Alcohol molecules can give up their hydroxyl protons in certain solvents (Fig. 19.2), but their conjugate bases are so strong that they are not acids in water.

Alcohols are named by adding the suffix *-ol* to the stem of the parent hydrocarbon, as in methanol and ethanol. When the location of the —OH group needs to be specified, the number of the carbon atom to which it is attached is given, as in 1-propanol for $CH_3CH_2CH_2OH$ and 2-propanol for $CH_3CH(OH)CH_3$. Sometimes the hydrocarbon chain of an alcohol is named as a group, as in methyl alcohol, CH_3OH, and ethyl alcohol,

2-Propanol is often known by its common name, isopropyl alcohol.

The second type of experiment looks at the "stereochemical course" of the reaction. An S_N2 reaction is like the inversion of an umbrella, so a chiral reactant gives rise to a chiral product. On the other hand, as illustrated in Fig. 19.1 and discussed in the text, the transition state of an S_N1 reaction, provided it lives long enough, can be attacked from either side of its plane, and a chiral reactant will result in a racemic mixture of enantiomers. So, monitoring the optical activity of a reaction system can reveal the reaction path. If a substitution reaction involving a chiral carbon atom results in a chiral product, then the mechanism is probably S_N2. If the product is a racemic mixture, then the mechanism is probably S_N1.

As always in chemistry, we must interpret results with caution. Not all S_N1 reactions give a racemic mixture of products. First, the transition state might be so short lived that the leaving group continues to protect one face, requiring the entering group to attack from the opposite face and giving a chiral product. Alternatively, another group in the molecule might swing across into the path of an incoming group and protect one face of the transition state. This *neighboring group participation* can result in the retention of the stereochemistry of the reactant, as shown in the illustration, even though the rate law indicates that the reaction is first order. Chemistry is the art of the prudent interpretation of sometimes seemingly conflicting evidence.

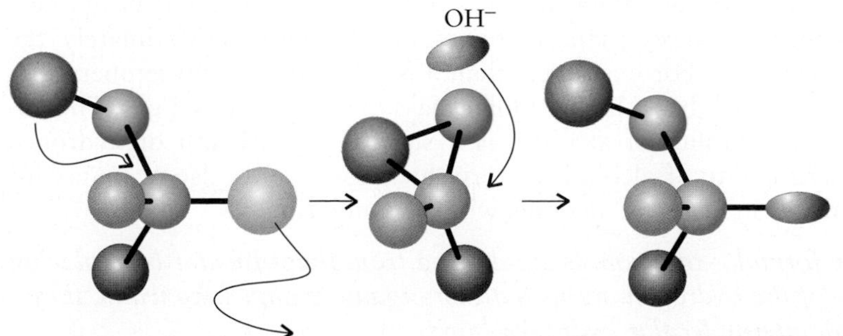

Neighboring group participation can result in the retention of configuration. As the gold-colored group leaves, the red group forms a bond with the central carbon atom and protects it from nucleophilic attack from that side. Nucleophilic attack occurs only on the side vacated by the leaving group.

CH_3CH_2OH. The —OH group may also be named as a substituent, in which case the name *hydroxy* is used, as in 2-hydroxybutane for $CH_3CH(OH)CH_2CH_3$.

Alcohols are divided into three classes according to the number of organic groups attached to the carbon atom connected to the —OH group. A **primary alcohol** has the form RCH_2—OH, where R can be any group. A **secondary alcohol** has the form R_2CH—OH, and a **tertiary alcohol** has the form R_3C—OH. The R groups need not all be the same. Ethanol is a primary alcohol and 2-butanol is a secondary alcohol. An example of a tertiary alcohol is 2-methyl-2-propanol, $(CH_3)_3COH$.

The common names of 2-methyl-2-propanol are *tert*-butanol and *tert*-butyl alcohol.

Methanol is usually prepared industrially from synthesis gas, which is obtained from coal (Section 14.5):

$$CO(g) + 2\ H_2(g) \xrightarrow{\text{catalyst, 250°C, 50–100 atm}} CH_3OH(g)$$

The catalyst is a mixture of copper, zinc oxide, and chromium(III) oxide. Ethanol is produced in large quantities all over the world by the fermentation of carbohydrates. It is also prepared by the hydration of ethene in an addition reaction:

$$CH_2{=}CH_2(g) + H_2O(g) \xrightarrow{\text{catalyst, 300°C}} CH_3CH_2OH(g)$$

The catalyst is phosphoric acid. The laboratory synthesis of alcohols is by nucleophilic substitution of haloalkanes.

Ethylene glycol, or 1,2-ethanediol, $HOCH_2CH_2OH$ (**4**), is an example of a **diol,** which is a compound with two hydroxyl groups. Ethylene glycol is used as a component of antifreeze mixtures and in the manufacture of some synthetic fibers.

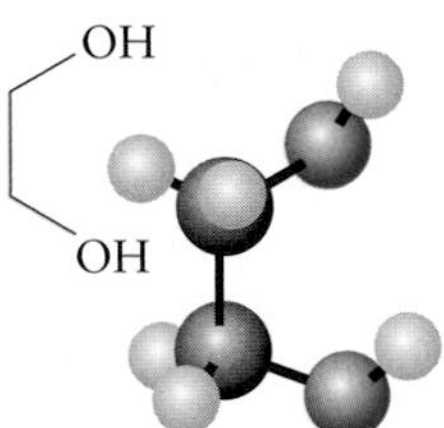

4 1,2-Ethanediol, $HOCH_2CH_2OH$

Alcohols with low molar masses are liquids, and alcohols have much lower vapor pressures than do hydrocarbons with approximately the same molar mass. For example, ethanol is a liquid at room temperature, but butane, which has a higher molar mass than ethanol, is a gas. The relatively low volatility of alcohols is a sign of the strength of hydrogen bonds. The ability of alcohols to form hydrogen bonds also accounts for the solubility in water of alcohols with low molar mass.

The formulas of alcohols are derived from that of water by replacing one of the hydrogen atoms with an organic group. Like water, they form intermolecular hydrogen bonds.

19.3 Ethers

An **ether** is an organic compound of the form R—O—R, where R is an alkyl group (the two R groups need not be the same). We can think of an ether as an HOH molecule in which both H atoms have been replaced by alkyl groups:

H—O—H	CH_3CH_2—O—H	CH_3CH_2—O—CH_2CH_3
Water	Ethanol (an alcohol)	Diethyl ether (an ether)

Ethers are more volatile than alcohols of the same molar mass because their molecules do not form hydrogen bonds to one another (Fig. 19.3). Because ethers are not very reactive and have low molecular polarity, they are useful solvents for other organic compounds. However, ethers are flammable; diethyl ether is easily ignited and must be used with great care.

Cyclic ethers (ethers in the form of a ring) with alternating —CH_2CH_2—O— units are called **crown ethers.** The name reflects the crownlike shape of the molecules (**5**).

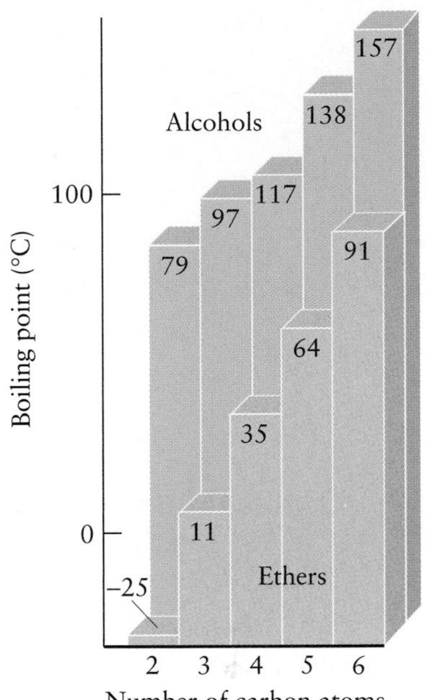

FIGURE 19.3 The boiling points of ethers are lower than those of isomeric alcohols, because hydrogen bonding occurs in alcohols but not in ethers. All the molecules represented here are unbranched.

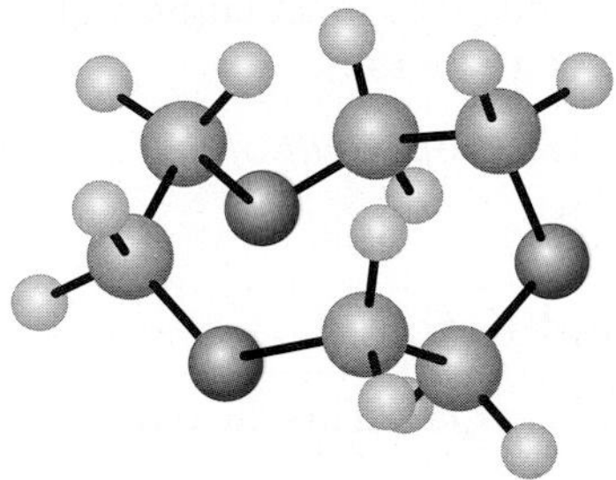

5 A crown ether

Crown ethers bind very strongly to metal cations such as K^+, effectively encasing them in a shell with the O atoms pointing inward toward the K^+ ion and the hydrocarbon chain wrapped around the outside. The crown ether acts as a sort of "Trojan horse" for the K^+ ions, which in this form are soluble in nonpolar solvents. Potassium permanganate, for example, dissolves readily in benzene when a crown ether with six O atoms is added to it. The resulting solution can be used to carry out oxidation reactions in organic solvents.

Ethers are not very reactive. They are more volatile and much less reactive than alcohols with similar molar masses because their molecules cannot form hydrogen bonds with each other. Crown ethers adopt shapes that can enclose ions and carry them into nonpolar solvents.

19.4 Phenols

In a **phenol,** a hydroxyl group is attached directly to an aromatic ring. The parent compound, phenol itself, C_6H_5OH (**6**), is a white, crystalline, molecular solid. It was once obtained from the distillation of coal tar, but now it is mainly synthesized from benzene. Many substituted phenols occur naturally, some being responsible for the fragrances of plants. They are often components of *essential oils*, the oils that can be distilled from

OH

6 Phenol, C_6H_5OH

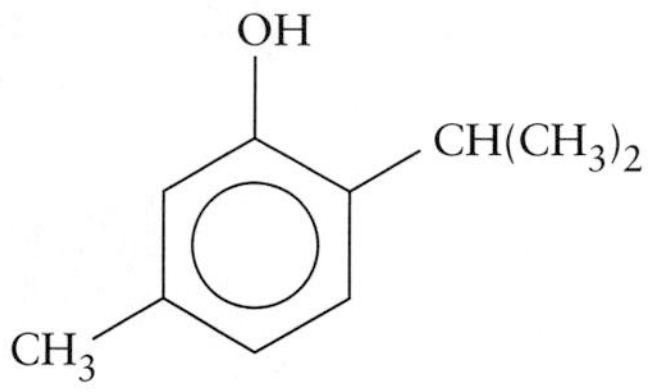

OH

CH(CH$_3$)$_2$

CH$_3$

7 Thymol

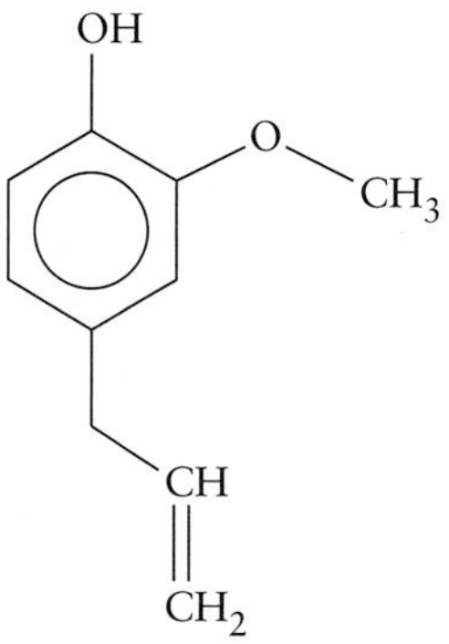

OH

O

CH$_3$

CH

CH$_2$

8 Eugenol

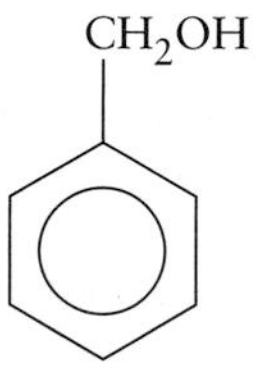

CH$_2$OH

9 Benzyl alcohol, $C_6H_5CH_2OH$

flowers and leaves. Thymol (7), for instance, is the active ingredient of oil of thyme, and eugenol (8) provides most of the scent and flavor of oil of cloves.

Phenols differ from alcohols in that they are weak acids. The resonance

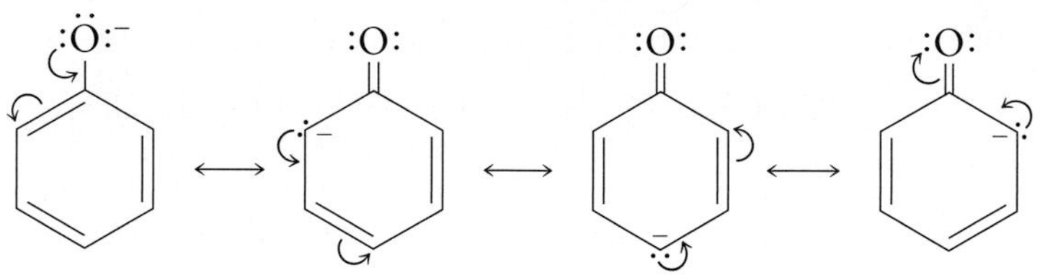

delocalizes the negative charge of the conjugate base of phenol and hence stabilizes the anion. As a result, $C_6H_5O^-$ is a weaker conjugate base than the corresponding conjugate base of an alcohol, such as $C_2H_5O^-$ from ethanol. Consequently, phenol, C_6H_5OH, is a stronger acid than ethanol, C_2H_5OH, and phenols that are otherwise insoluble in water are soluble in basic solutions. However, even a $-CH_2-$ group can insulate the O atom from the benzene ring; and phenylmethanol, $C_6H_5CH_2OH$ (benzyl alcohol, 9), is an alcohol, not a phenol.

Phenols are weak acids as a result of delocalization and stabilization of the conjugate base.

19.5 Aldehydes and Ketones

The **carbonyl group,** $>C{=}O$, occurs in two closely related families of compounds:

Aldehydes are compounds of the form R—C(=O)—H

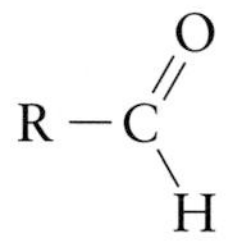

Ketones are compounds of the form R—C(=O)—R

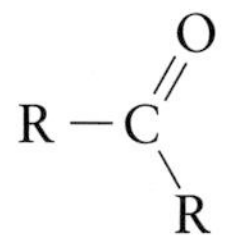

In an aldehyde, the carbonyl group is found at the end of the carbon chain, whereas in a ketone, it is at an intermediate position. The R groups may be either aliphatic or aromatic. The group characteristic of the ketones is written —CO—, as in CH_3COCH_3, propanone (acetone), a common laboratory solvent. The group characteristic of the aldehydes is normally written —CHO, as in HCHO (formaldehyde), the simplest aldehyde. *Formalin,* the liquid used to preserve biological specimens, is an aqueous solution of formaldehyde. Wood smoke contains formaldehyde, and formaldehyde's destructive effect on bacteria is one reason why smoking food helps to preserve it.

The systematic names of aldehydes are obtained by replacing the ending *-e* by *-al,* as in methanal for HCHO (formaldehyde) and ethanal for CH_3CHO (acetaldehyde). Note that the carbon atom of the carbonyl group is included in the count of carbon atoms when determining the alkane from which the aldehyde is derived. Ketones are given the suffix -*one*, as in propanone, CH_3COCH_3 (acetone). To avoid ambiguity, a number is used to denote the carbon atom that has become the carbonyl group; thus, $CH_3CH_2CH_2COCH_3$ is 2-pentanone and $CH_3CH_2COCH_2CH_3$ is 3-pentanone.

Aldehydes occur naturally in essential oils and contribute to the flavors of fruits and the odors of plants. Benzaldehyde, C_6H_5CHO (**10**), contributes to the characteristic aroma of cherries and almonds. Cinnamaldehyde (**11**) is found in cinnamon, and vanilla extract contains vanillin (**12**), which occurs in oil of vanilla. Ketones can also be fragrant. For example, carvone (Section 18.1) is the essential oil of spearmint.

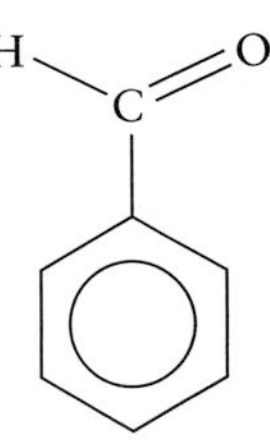

10 Benzaldehyde, C_6H_5CHO

CHO

11 Cinnamaldehyde

OH
OCH_3
CHO

12 Vanillin

Formaldehyde is prepared industrially (for the manufacture of phenol-formaldehyde resins) by the catalytic oxidation of methanol:

$$2\ CH_3OH(g) + O_2(g) \xrightarrow{600°C,\ Ag} 2\ HCHO(g) + 2\ H_2O(g)$$

Further oxidation of the product to a carboxylic acid (Section 19.6) is avoided by using a mild oxidizing agent. There is less risk of further oxidation for ketones than for aldehydes, because a C—C bond would have to be broken for a carboxylic acid to form.

Dichromate oxidation of secondary alcohols produces ketones in good yield, with little additional oxidation. For example, $CH_3CH_2CH(OH)CH_3$ can be oxidized to $CH_3CH_2COCH_3$. The difference between the ease of oxidation of aldehydes and that of ketones is used to distinguish them. Aldehydes can reduce silver ions to form a silver mirror—a coating of silver on test tube walls—with *Tollens reagent*, a solution of Ag^+ ions in aqueous ammonia (Fig. 19.4):

Aldehydes: $CH_3CH_2CHO + Ag^+$ (in Tollens reagent) gives CH_3CH_2COOH and Ag(s)

Ketones: $CH_3COCH_3 + Ag^+$ (in Tollens reagent) gives no reaction

Aldehydes and ketones can be prepared by the oxidation of alcohols. Aldehydes are reducing agents; ketones are not.

Media Link **Lab video:** Fig. 19.4, Formation of a silver mirror

FIGURE 19.4 An aldehyde (left) produces a silver mirror with Tollens reagent, but a ketone (right) does not.

19.6 Carboxylic Acids

The **carboxyl group,** $-\text{C}(=\text{O})-\text{O}-\text{H}$, normally abbreviated —COOH, is the functional group found in **carboxylic acids,** which are weak acids of the form R—COOH. The simplest carboxylic acid is formic acid, HCOOH, the acid in ant venom. Another common carboxylic acid is acetic acid, CH_3COOH, the acid of vinegar. It forms when the ethanol in wine is oxidized by air:

$$CH_3CH_2OH \xrightarrow{O_2} H_3C-\overset{\overset{\displaystyle O}{\|}}{C}-OH + H_2O$$

Carboxylic acids are named systematically by replacing the *-e* of the parent hydrocarbon by the suffix *-oic acid*; the carbon atom of the carboxyl group is included in the count of atoms to determine the parent hydrocarbon molecule. Thus, formic acid is formally methanoic acid and acetic acid is ethanoic acid.

Carboxylic acids can be prepared by oxidizing primary alcohols and aldehydes with acidified aqueous potassium permanganate solution. In some cases, an alkyl group can be oxidized directly to a carboxyl group. This process is very important industrially and, among other applications, is carried out with a cobalt(III) catalyst to oxidize the methyl groups on *p*-xylene:

$$CH_3C_6H_4CH_3 + 3\,O_2 \xrightarrow{Co(III)} HOOCC_6H_4COOH + 2\,H_2O$$

The product, terephthalic acid, is used for the production of artificial fibers (Section 19.10).

Carboxylic acids have an —OH group attached to a carbonyl group.

$CH_3—C(=O)—O—CH_2—CH_3$

13 Ethyl acetate, $CH_3COOCH_2CH_3$

19.7 Esters

The product of the reaction between a carboxylic acid and an alcohol is called an **ester** (**13**). Acetic acid and ethanol, for example, react when heated to about 100°C in the presence of a strong acid. The products of this esterification are the fragrant liquid ethyl acetate and water:

$$H_3C{-}C(=O){-}OH + H{-}O{-}CH_2CH_3 \longrightarrow H_3C{-}C(=O){-}OCH_2CH_3 + H_2O$$

Many esters have fragrant odors and contribute to the flavors of fruits. Other naturally occurring esters include fats and oils. For example, the animal fat tristearin (**14**), which is a component of beef fat, is an ester formed from glycerol and stearic acid.

14 Tristearin, $C_{57}H_{110}O_6$

FIGURE 19.5 In a condensation reaction, two molecules are linked as a result of removing two atoms or groups (the orange and purple spheres) as a small molecule. When the groups are H and OH, water is eliminated. If they are H and Cl, HCl is eliminated.

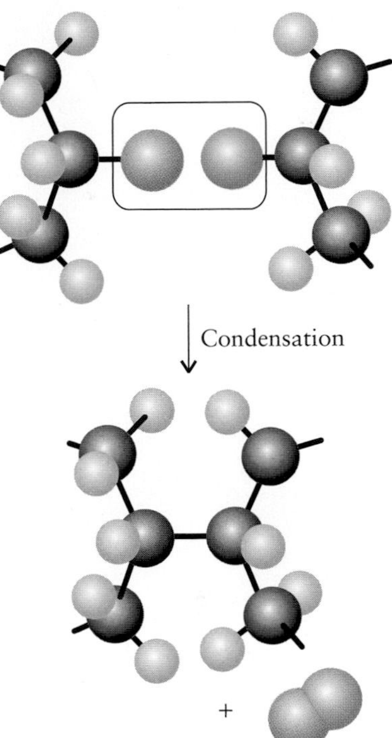

Ester formation is an example of a **condensation reaction** in which two molecules combine to form a larger one and a small molecule is eliminated (Fig. 19.5). The reaction is catalyzed by a small amount of strong acid, such as sulfuric acid. In an esterification involving a carboxylic acid and an alcohol, the eliminated molecule is H_2O.

Alcohols condense with carboxylic acids to form esters.

Self-Test 19.1A (a) Write the condensed structural formula and the name of the ester formed from the reaction between propanoic acid, CH_3CH_2COOH, and methanol, CH_3OH. (b) Write the condensed structural formulas of the acid and alcohol that react to form pentyl ethanoate, $CH_3COOC_5H_{11}$, a contributor to the flavor of bananas. See Toolbox 19.1.

[*Answer:* (a) $CH_3CH_2COOCH_3$, methyl propanoate; (b) CH_3COOH and $CH_3(CH_2)_3CH_2OH$]

Self-Test 19.1B (a) Write the condensed structural formula and the name of the ester formed from the reaction between formic acid, HCOOH, and ethanol, CH_3CH_2OH. (b) Write the condensed structural formulas of the acid and alcohol that react to form methyl butanoate, $CH_3(CH_2)_2COOCH_3$, a contributor to the flavor of apples. See Toolbox 19.1.

19.8 Amines, Amino Acids, and Amides

An **amine** is a compound with a formula derived from NH_3 by replacing various numbers of H atoms with organic groups. Amines are classified as primary, secondary, or tertiary according to the number of R groups on the nitrogen atom:

H | H—N—H	H | H—N—CH_3	CH_3 | H—N—CH_3	CH_3 | H_3C—N—CH_3
Ammonia	Methylamine (primary)	Dimethylamine (secondary)	Trimethylamine (tertiary)

In each case, the N atom is sp^3 hybridized, with one lone pair of electrons and three σ-bonds. A *quaternary ammonium ion* is a tetrahedral ion of the form R_4N^+, such as the tetramethylammonium ion, $(CH_3)_4N^+$. The **amino group,** the parent functional group of amines, is $-NH_2$.

Amines are widespread naturally. Many of them have a pungent, often unpleasant odor. Because proteins are organic polymers containing nitrogen, amines are present in the decomposing remains of living matter and, together with sulfur compounds, are responsible for the stench of decaying flesh. The common names of two diamines—putrescine, $NH_2(CH_2)_4NH_2$, and cadaverine, $NH_2(CH_2)_5NH_2$—speak for themselves. Like ammonia, amines are weak bases (Section 10.7), but quaternary ammonium ions are generally acidic.

An **amino acid** is a carboxylic acid that contains an amino group as well as a carboxyl group. The simplest example is glycine, NH_2CH_2COOH (**15**). Notice that an amino acid has both a basic group ($-NH_2$) and an acidic group ($-COOH$) in the same molecule.

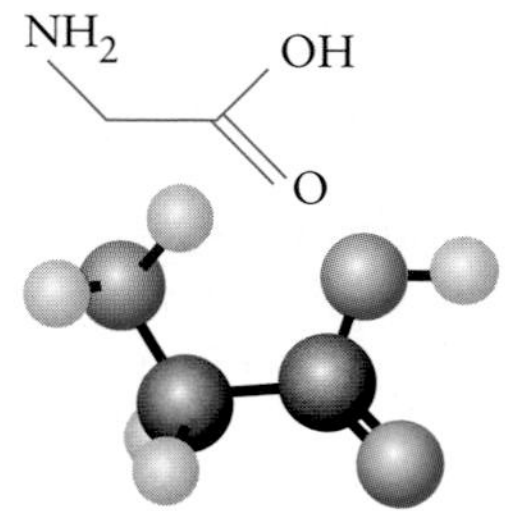

15 Glycine, NH_2CH_2COOH

Like alcohols, amines condense with carboxylic acids:

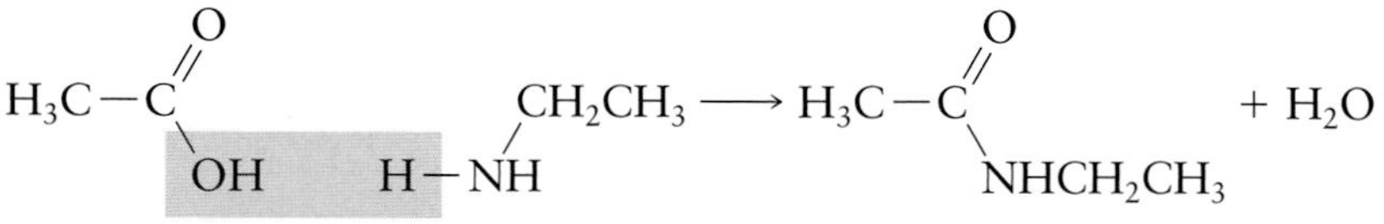

$$H_3C-C(=O)-OH + H-NH-CH_2CH_3 \longrightarrow H_3C-C(=O)-NHCH_2CH_3 + H_2O$$

When the reactant is a primary amine, the product is an **amide,** a molecule of the form R—(CO)—NH—R. At first sight, we might expect an amine to act as a base and simply accept a proton from the carboxylic acid. Indeed, that does happen, and a quaternary ammonium salt is formed when the reagents are mixed in the absence of a solvent:

$$CH_3COOH + CH_3NH_2 \longrightarrow CH_3CO_2^- + CH_3NH_3^+$$

However, on heating to about 200°C, a thermodynamically more favorable reaction takes place. The proton transfer is reversed, and the amine acts as a nucleophile and attacks the carbon atom of the carboxylic acid. An example of this process is the condensation of ammonia with a carboxylic acid:

The arrow on the second structure shows the migration of a proton.

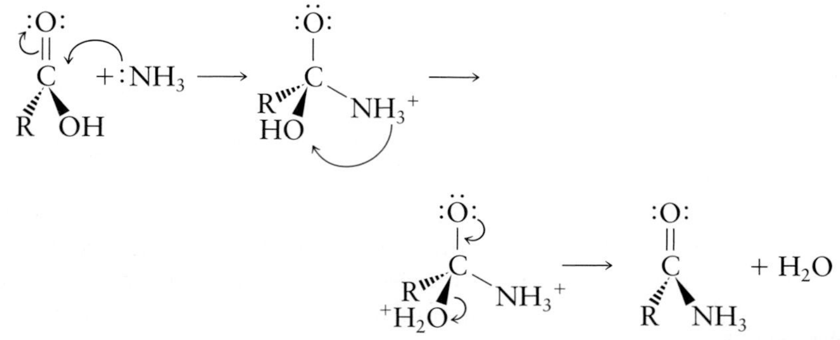

$$RC(=O)OH + :NH_3 \longrightarrow R(HO)C(-O^-)NH_3^+ \longrightarrow R(^+H_2O)C(-O^-)NH_3^+ \longrightarrow RC(=O)NH_3 + H_2O$$

In practice, amides are best prepared by the reaction between an amine and an acid chloride, a compound that contains the $-COCl$ group (which is like a carboxyl group, $-COOH$, but with $-Cl$ in place of

—OH). For example, the reaction between C_6H_5COCl and diethylamine, $(CH_3CH_2)_2NH$, yields diethylbenzamide:

$$C_6H_5C(=O)Cl + HN(CH_2CH_3)_2 \longrightarrow C_6H_5C(=O)N(CH_2CH_3)_2 + HCl$$

The HCl produced in the reaction is neutralized by sodium hydroxide added to the reaction mixture. The mechanism of the reaction is similar to the reaction of an amine with a carboxylic acid and consists of nucleophilic attack by the nitrogen atom of the amine on the carbon atom of the carbonyl group, followed by elimination of HCl.

One example of an amide is the pain-relieving drug sold as Tylenol (**16**); we shall see another important example when we consider the polymer known as nylon in Section 19.10. Many amides have N—H bonds that can take part in hydrogen bonding, so the intermolecular forces between their molecules are relatively strong.

16 Tylenol®

Amines are derived from ammonia by replacing hydrogen atoms with organic groups. Amides result from the condensation of amines with carboxylic acids. Amines and amides take part in hydrogen bonding.

Self-Test 19.2A Predict whether an ester or a primary amine of the same molar mass is likely to have the higher boiling point and explain why.

[*Answer:* The amine; it can form hydrogen bonds with the $—NH_2$ group, but the ester group cannot form hydrogen bonds in the pure state.]

Self-Test 19.2B What are the hybridization schemes of the C and N atoms in formamide, $HCONH_2$?

TOOLBOX 19.1 How to name simple compounds with functional groups

Conceptual Basis

In general, the names of compounds containing functional groups follow the same conventions and numbering system as do the names of hydrocarbons (Toolbox 18.1). However, usually the ending of the name is changed to indicate the functional group. The aim here, as there, is to be succinct but unambiguous. Here we include a limited set of the most common functional groups.

Procedure

Alcohols To form the systematic name, we identify the parent hydrocarbon and replace the ending *-e* by *-ol*. The location of the hydroxyl group is denoted by numbering the C atoms of the backbone, starting at the end of the chain that results in the lowest number for the —OH group location. When —OH is named as a substituent, it is called *hydroxy*. For example, $CH_3CH(OH)CH_2CH_2CH_3$ is 2-pentanol or 2-hydroxypentane.

Aldehydes and ketones For aldehydes, identify the parent hydrocarbon: include the C of —CHO in the count of carbon atoms. Then change the final *-e* of the hydrocarbon name to *-al*. The —CHO group can occur only at the end of a carbon chain and has priority in numbering over other groups. For ketones, change the *-e* of the parent hydrocarbon to *-one*. The location of the group is identified by selecting a numbering order that gives it the lowest number. Thus, $CH_3COCH_2CH_2CH_3$ is 2-pentanone.

Carboxylic acids Change the *-e* of the parent hydrocarbon to *-oic acid*. To identify the parent acid, include the C atom of the —COOH group when counting carbon atoms. Thus, $CH_3CH_2CH_2COOH$ is butanoic acid.

Esters Change the *-anol* of the alcohol to *-yl* and the *-oic acid* of the parent acid to *-oate*. Thus, CH_3COOCH_3 is methyl ethanoate.

Amines Amines are named systematically by specifying the groups attached to the nitrogen atom in alphabetical order (disregarding prefixes), followed by the suffix *-amine*. Amines with two amino groups are called *diamines*. The $—NH_2$ group is called *amino-* when it is a substituent. Thus, $(CH_3CH_2)_3N$ is triethylamine and $CH_3CHNH_2CH_3$ is 2-aminopropane.

Halides Name the halogen atom as a substituent by changing the *-ine* part of its name to *-o*, as in 2-chlorobutane, $CH_3CHClCH_2CH_3$

EXAMPLE 19.1 Naming compounds with functional groups

Name the compounds (a) $CH_3CH(CH_3)CH(OH)CH_3$; (b) $CH_3CHClCH_2COCH_3$; and (c) $(CH_3CH_2)_2NCH_2CH_2CH_3$.

STRATEGY Follow the procedure set out in Toolbox 19.1. A numerical prefix indicates the carbon atom to which the group is attached, numbering from the end of the chain that gives the lower number.

SOLUTION (a) We determine the number of carbon atoms in the longest chain and replace the ending of the name of that chain with *-ol*. The functional group has priority in numbering over the methyl group. The alcohol is 3-methyl-2-butanol. (b) The compound is both a ketone and a haloalkane. Identify the hydrocarbon chain and number the chain to give the ketone group the lower number. The chain has five carbon atoms; the ketone group is on the second carbon from one end and the chlorine atom on the fourth, so the name of the compound is 4-chloro-2-pentanone. (c) The compound $(CH_3CH_2)_2NCH_2CH_2CH_3$ has two ethyl groups and one propyl group attached to a nitrogen atom. Give the ethyl groups the prefix *di-* to indicate their number and list the groups in alphabetical order. The compound is diethylpropylamine.

Self-Test 19.3A Name (a) $CH_3CH(CH_2CH_2OH)CH_3$; (b) $CH_3CH(CHO)CH_2CH_3$; (c) $(C_6H_5)_3N$.

[*Answer:* (a) 3-Methyl-1-butanol; (b) 2-methylbutanal; (c) triphenylamine]

Self-Test 19.3B Name (a) $CH_3CH_2CH(OH)CH_2CH_3$; (b) $CH_3CH_2COCH_2CH_3$; (c) $CH_3CH_2NHCH_3$.

THE IMPACT ON MATERIALS

Many modern synthetic materials are **polymers,** compounds in which chains or networks of small repeating units form giant molecules, such as polypropylene and polytetrafluoroethylene (sold as Teflon). Plastics are simply polymers that can be molded into shape. Despite their enormous size, polymers are built by using reactions like the ones we have discussed earlier in the chapter.

Polymers are made by two main types of reactions: by *addition reactions* and by *condensation reactions*. Which type of reaction is used depends on the functional groups present in the starting materials. We begin by describing the two types of reactions used to prepare polymers and then examine the characteristic physical properties of the two types of polymers formed.

19.9 Addition Polymerization

Alkenes can react with themselves to form long chains in a process called **addition polymerization.** For example, an ethene molecule may form a bond to another ethene molecule; another ethene molecule may add to that, and so on, forming a long hydrocarbon chain. The original alkene, such as ethene, is called the **monomer.** Each monomer becomes one **repeating unit,** the structure that repeats over and over to produce the polymer chain. The product, the chain of covalently linked repeating units, is the polymer. The simplest addition polymer is polyethylene, $-(CH_2CH_2)_n-$, which consists of long chains of repeating $-CH_2CH_2-$ units. Many addition polymer molecules also have a number of branches, which are generated as new chains sprout from intermediate points along the "backbone" chain.

The plastics industry has developed polymers from a number of monomers with the formula $CHX{=}CH_2$, where X is a single atom (such as the Cl in vinyl chloride, $CHCl{=}CH_2$) or a group of atoms (such as the CH_3 in propene). These substituted alkenes give polymers of formula $-(CHXCH_2)_n-$, including polyvinyl chloride (PVC), $-(CHClCH_2)_n-$, and polypropylene, $-(CH(CH_3)CH_2)_n-$ (Table 19.2). They differ in appearance, rigidity, transparency, and resistance to weathering. Many plastic materials made of addition polymers can be recycled by melting and reprocessing (Table 19.3).

TABLE 19.2 Common Addition Polymers

Monomer name	Formula	Polymer formula	Common name
ethene*	$CH_2{=}CH_2$	$-(CH_2-CH_2)_n-$	polyethylene
vinyl chloride	$CHCl{=}CH_2$	$-(CHCl-CH_2)_n-$	polyvinyl chloride
styrene	$CH(C_6H_5{=}CH_2$	$-(CH(C_6H_5)-CH_2)_n-$	polystyrene
acrylonitrile	$CH(CN){=}CH_2$	$-(CH(CN)-CH_2)_n-$	Orlon, Acrilan
propene*	$CH(CH_3){=}CH_2$	$-(CH(CH_3)-CH_2)_n-$	polypropylene
methyl methacrylate	$CH_3OC(=O)C(CH_3){=}CH_2$	$-(C(CH_3)(COOCH_3)-CH_2)_n-$	Plexiglas, Lucite
tetrafluoroethene*	$CF_2{=}CF_2$	$-(CF_2-CF_2)_n-$	Teflon, PTFE†

*The suffix *-ene* is replaced by *-ylene* in the common names of these compounds, hence the names of the corresponding polymers.
†PTFE, polytetrafluoroethylene.

TABLE 19.3 Recycling Codes

Recycling code	Polymer
1 PET(E)	polyethylene terephthalate
2 HDPE	high-density polyethylene
3 PVC or VC	polyvinyl chloride
4 LDPE	low-density polyethylene
5 PP	polypropylene
6 PS	polystyrene

A widely used synthetic procedure is **radical polymerization,** polymerization by a radical chain reaction (Section 13.12). In a typical procedure, a monomer (such as ethene) is compressed to about 1000 atm and heated to 100°C in the presence of a small amount of an organic peroxide (a compound of formula R—O—O—R, where R is an organic group). The reaction is initiated by dissociation of the O—O bond, giving two radicals:

$$\text{R—O—O—R} \longrightarrow \text{R—O}\cdot + \cdot\text{O—R}$$

Once it has been initiated, the chain reaction can propagate as the radicals attack the monomer molecules $CHX{=}CH_2$ (with X = H for ethene itself) and form a new, highly reactive radical:

$$\text{R—O}\cdot + \text{H}_2\text{C}{=}\text{C(H)(X)} \longrightarrow \text{R—O—CH}_2\text{—C}\cdot\text{(H)(X)}$$

This radical attacks another monomer molecule, and the chain grows longer:

$$\text{R—O—CH}_2\text{—C}\cdot\text{(H)(X)} + \text{H}_2\text{C}{=}\text{C(H)(X)} \longrightarrow \text{R—O—CH}_2\text{—C(H)(X)—CH}_2\text{—C}\cdot\text{(H)(X)}$$

The reaction continues until all the monomer has been used up, or until it terminates when pairs of chains have linked together into single nonradical species. The product, polyethylene for instance, consists of long chains of formula $—(CH_2CH_2)_n—$ in which n can reach many thousands.

Strong, tough polymers have chains that pack together well. One problem with early attempts to make polypropylene was that the orientations of the H and CH_3 groups on each C atom were random, thereby preventing the chains from packing well. The resulting material was amorphous, sticky, and nearly useless. Now, however, the stereochemistry of the chains can be controlled by using a **Ziegler-Natta catalyst,** a catalyst consisting of the compounds titanium tetrachloride, $TiCl_4$, and triethylaluminum, $(CH_3CH_2)_3Al$. A polymer in which each unit or pair of repeating units has the same relative orientation is described as being

stereoregular; the stereoregularity arises from the manner in which the chains grow on the catalyst. The chains of stereoregular polymers produced by Ziegler-Natta catalysts pack together well to form highly crystalline, dense materials (Fig. 19.6).

Stereoregular polymers may be either isotactic or syndiotactic. In an **isotactic polymer,** all the carbon atoms forming the chain have the same stereochemistry, so all the methyl groups in isotactic polypropylene are on the same side of the chain. In a **syndiotactic polymer,** the carbon atoms in the chain alternate in handedness, so the methyl groups in syndiotactic polypropylene alternate in direction. In an **atactic polymer,** the side groups are attached randomly on either side of the chain (Fig. 19.7).

Rubber is a polymer of isoprene (**17**). Natural rubber is obtained from the bark of the rubber tree as a milky white liquid, which is called *latex* (Fig. 19.8) and consists of a suspension of rubber particles in water. The rubber itself is a soft white solid that becomes even softer when warm. It is used for pencil erasers and has been used as crepe rubber for the soles of shoes.

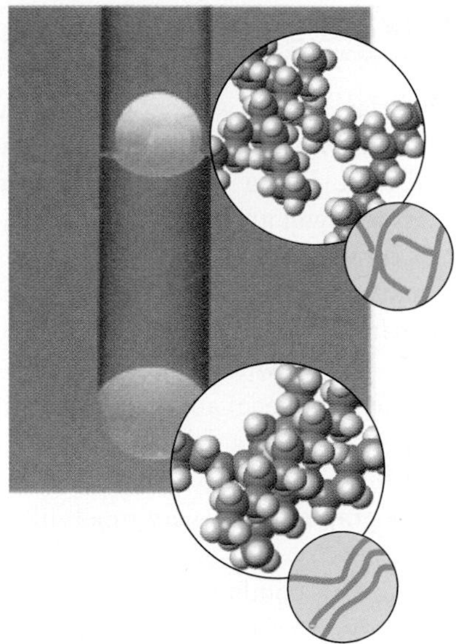

FIGURE 19.6 The two samples of polyethylene in the test tube were produced by different processes. The floating (low-density) one was produced by high-pressure polymerization. The one at the bottom (high-density) was produced with a Ziegler-Natta catalyst. As the insets show, the latter has a higher density because the polymer chains pack together better.

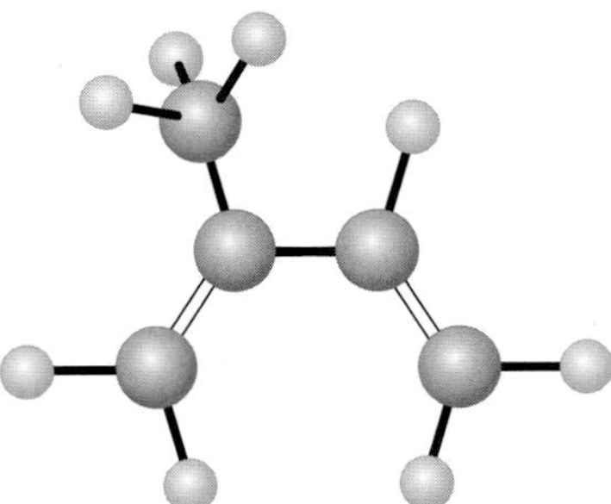

17 Isoprene

Chemists were unable to synthesize rubber for a long time, even though they knew it was a polymer of isoprene. The enzymes in the rubber tree produce a stereoregular polymer in which all the links between monomers are in a cis arrangement (Fig. 19.9); straightforward radical polymerization, however, produces a random mixture of cis and trans links and a sticky, useless product. The stereoregular polymer was achieved with a Ziegler-Natta catalyst, and almost pure, rubbery *cis*-polyisoprene can now be produced. *trans*-Polyisoprene, in which all the links are trans, is the hard, naturally occurring material *gutta-percha*, once used inside golf balls and still used to fill root canals in teeth.

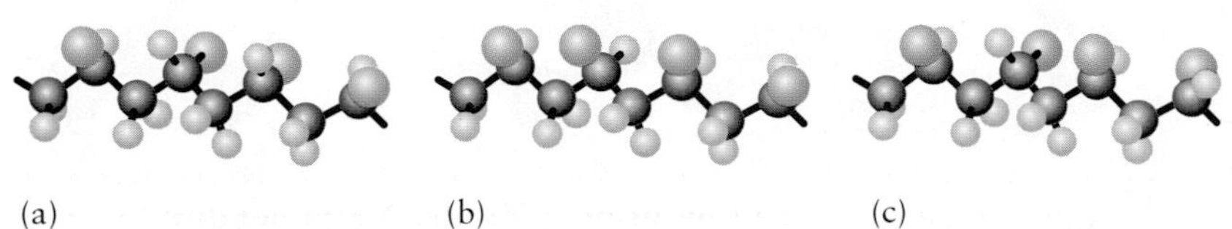

FIGURE 19.7 (a) In an atactic polymer, the substituents lie on random sides of the chain. The stereoregular polymers produced by using Ziegler-Natta catalysts may be (b) isotactic (all on one side) or (c) syndiotactic (alternating).

FIGURE 19.8 Collecting latex from a rubber tree in Malaysia, a principal producer of latex.

BOX 19.2 FRONTIERS OF CHEMISTRY: CONDUCTING POLYMERS*

One day, if you want to check your e-mail, you might unroll a plastic sheet with a tiny microprocessor embedded in it. When you activate the microprocessor, your messages will appear and you might reply by writing on the screen with a special pen or speaking to it. The remarkable material of this flat computer already exists: an early form of it was discovered by accident in the early 1970s when a chemist who was polymerizing ethyne (acetylene) added a thousand times too much catalyst. Instead of a synthetic rubber, he made a thin, flexible film. It looked like metal foil tinted pink (see the photograph), and—very much like a metal—it conducted electricity.

Metals conduct electricity because their valence electrons move easily from atom to atom. Most covalently bonded solids do not conduct electricity because their valence electrons are locked into individual bonds and are not free to move. The exceptions, like graphite and carbon nanotubes, have delocalized π-bonds in connected aromatic rings through which electrons can move freely because there are empty molecular orbitals close in energy to occupied orbitals (Section 14.18). However, a disadvantage is that commercial graphite is fragile and brittle.

This flexible polyacetylene sheet was peeled from the walls of the reaction flask in which it was made.

Conducting polymers provide a new and exciting alternative. They do not rust and they have low densities. They can be molded or drawn into shells, fibers, or thin plastic sheets, but they can still act like metallic conductors. They can be made to glow with almost any color and to change conductivity with conditions. Imagine cases of food labeled with polymer tags that change in conductivity when the cases are left unrefrigerated for too long.

All conducting polymers have a common feature: a long chain of sp^2 hybridized carbon atoms, often with nitrogen or sulfur atoms included in the chains. Polyacetylene, the first conducting polymer, is also the simplest, consisting of thousands of —CH=CH— units:

H H H H H

H H H H H

The alternation of double and single bonds means that each C atom has an unhybridized p-orbital that can overlap with the p-orbital on either side. This arrangement allows electrons to be delocalized along the entire chain like a one-dimensional version of graphite.

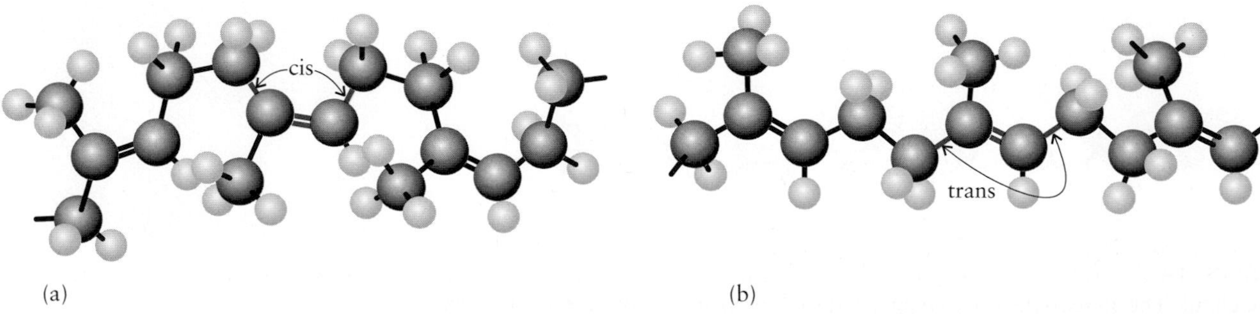

FIGURE 19.9 (a) In natural rubber, the isoprene units are polymerized to be all cis. (b) The harder material, gutta-percha, is the all-trans polymer.

One conducting polymer, polypyrrole,

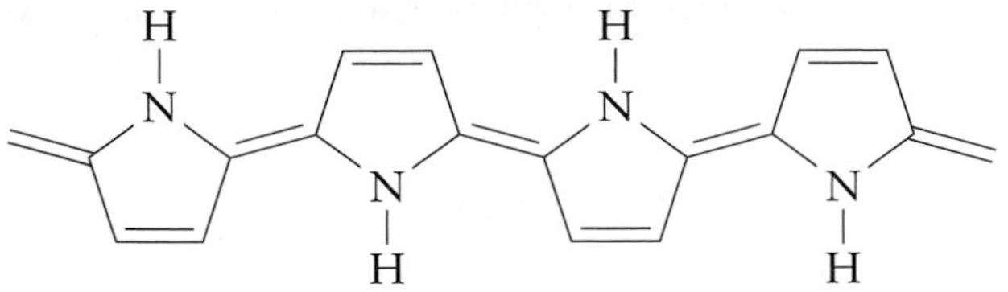

has been used in smart windows, which darken from a transparent yellow-green to a nearly opaque blue-black in bright sunlight. Fibers of polypyrrole are also woven into radar camouflage cloth that absorbs microwaves. Because it does not reflect microwaves back to their source, the cloth appears on radar as a patch of empty space.

Polyaniline, which has the structure

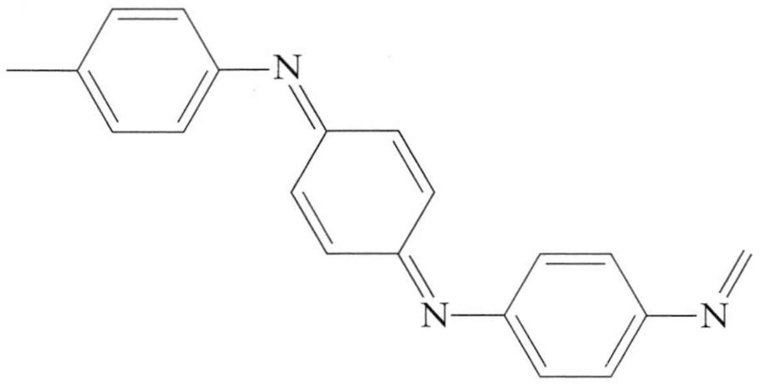

is being used in flexible coaxial cable; rechargeable, flat, buttonlike batteries; and laminated, rolled films that could be used as flexible computer or television displays. Thin films of poly-*p*-phenylenevinylene, PPV,

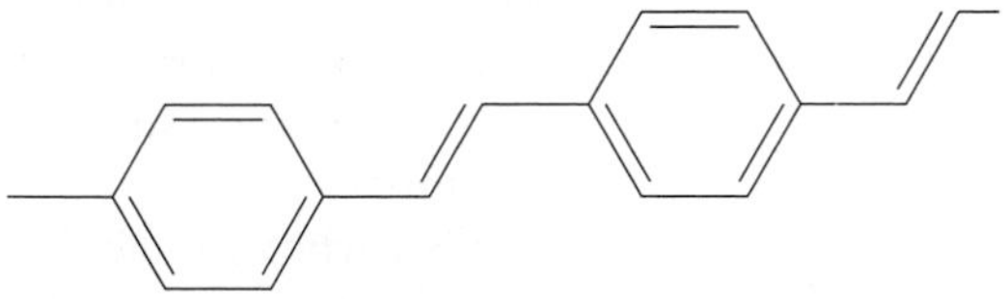

give off light when exposed to an electric field, a process called *electroluminescence*. By varying the composition of the polymer, scientists have coaxed it to glow in a wide range of colors. Such multicolor light-emitting diode (LED) displays could be as bright as fluorescent ones.

HOW MIGHT YOU CONTRIBUTE?

Current PPV displays last only about 10% as long as fluorescent displays. If their longevity grows, however, they could begin to replace the computer and television screens we use today. Polyaniline could one day serve both as a nonmetallic solder for the printed circuit boards inside a computer and as electrical shielding for the case. All-plastic transistors and other electronic components can be miniaturized to an amazing degree, raising the possibility of microscopic microprocessors and computers that could survive highly corrosive conditions, such as those at marine research sites. Because conducting polymers can also be designed to change shape with the level of electrical current, they could serve as artificial muscles or give flexibility of movement to robots. For these goals to be achieved, the characteristics of these polymers must be known and their responses to various conditions have to be studied.

For Further Reading: R. E. Gleason, "How far will circuits shrink?" *Science Spectra* (2000), 20: 32–40. M. G. Kanatzidis, "Conductive polymers," *Chemical and Engineering News* (Dec. 3, 1990), pp. 36–54. P. Yam, "Plastics get wired," *Scientific American* (July 1995), pp. 83–89. Information (advanced) on the Nobel Prize 2000, at www.nobel.se/announcement/2000.

* The 2000 Nobel prize in chemistry was awarded to A. J. Heger, A. G. MacDiarmid, and H. Shirakawa for their discovery of conducting polymers.

Polymers with Lewis structures that have alternating double bonds along the chain can be used to conduct electricity (Box 19.2). These conducting polymers tend to have long chains with few branches.

Alkenes undergo addition polymerization. When a Ziegler-Natta catalyst is used, the polymer is stereoregular and has a relatively high density.

19.10 Condensation Polymerization

Polymers formed by linking together monomers that have carboxylic acid groups with those that have alcohol groups are called **polyesters.** Polymers of this type are widely used to make artificial fibers. They are examples of

condensation polymers, which are polymers formed by a series of condensation reactions. A typical polyester is Dacron, or Terylene, a polymer produced from the esterification of terephthalic acid with ethylene glycol (see **4,** 1,2-ethanediol): its technical name is polyethylene terephthalate. The first condensation is

$$\text{HOOC}-\text{C}_6\text{H}_4-\text{COOH} + \text{HO}-\text{CH}_2\text{CH}_2\text{OH} \longrightarrow \text{HOOC}-\text{C}_6\text{H}_4-\text{C}(=\text{O})-\text{O}-\text{CH}_2\text{CH}_2\text{OH} + \text{H}_2\text{O}$$

A new ethylene glycol molecule can condense with the carboxyl group on the left of the product, and another terephthalic acid molecule can condense with the hydroxyl group on the right. As a result, the polymer grows at both ends and becomes

$$-\text{C}_6\text{H}_4-\text{C}(=\text{O})\left(-\text{O}-\text{CH}_2\text{CH}_2-\text{O}-\text{C}(=\text{O})-\text{C}_6\text{H}_4-\text{C}(=\text{O})-\right)_n\text{O}-\text{CH}_2\text{CH}_2\text{OH}$$

In the radical addition polymerization of alkenes, side chains can grow out from the main chain. However, in condensation polymerization, growth can occur only at the functional groups, so chain branching is much less likely. As a result, polyester molecules make good fibers, because they can be made to lie side by side by stretching the heated product and forcing it through a small hole (Fig. 19.10). The fibers produced can then be spun into yarn (Fig. 19.11). On the other hand, because many monomers can initiate the reaction simultaneously, the chains of condensation polymers are typically shorter than those of addition polymers. Polyesters also can be molded and used in surgical implants, such as artificial hearts, or made into thin films for cassette tapes.

Condensation polymerization of amines with carboxylic acids leads to the **polyamides,** substances more commonly known as *nylons*. A common polyamide is nylon-66, which is a polymer of 1,6-diaminohexane, $H_2N(CH_2)_6NH_2$, and adipic acid, $HOOC(CH_2)_4COOH$. The 66 in the name indicates the numbers of C atoms in the two monomers.

For condensation polymerization, it is necessary to have two functional groups on each monomer and to mix stoichiometric amounts of the reactants. In polyamide production, the starting materials first form "nylon salt" by proton transfer:

$$\text{HOOC(CH}_2)_4\text{COOH} + \text{H}_2\text{N(CH}_2)_6\text{NH}_2 \longrightarrow {}^{-}\text{O}_2\text{C(CH}_2)_4\text{CO}_2{}^{-} + {}^{+}\text{H}_3\text{N(CH}_2)_6\text{NH}_3{}^{+}$$

At this point, the excess acid or amine can be removed. Then, when the nylon salt—an ammonium salt of a carboxylic acid—is heated, the

FIGURE 19.10 Synthetic fibers are made by extruding liquid polymer from small holes in an industrial version of a spider's spinneret.

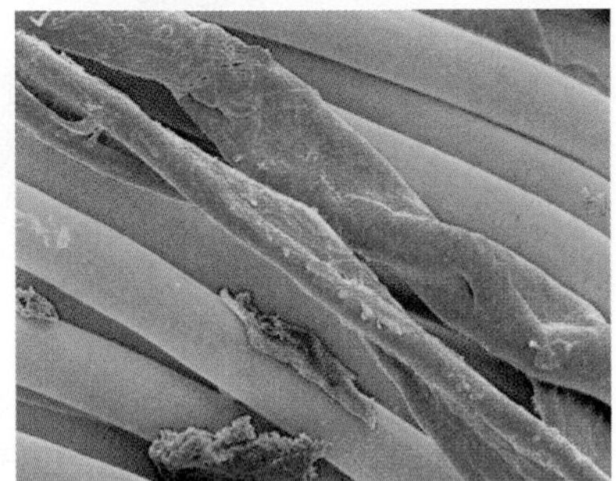

FIGURE 19.11 A scanning electron micrograph of Dacron polyester and cotton fibers in a blended shirt fabric. The cotton fibers have been colored green. Compare the smooth cylinders of polyester with the irregular surface of the cotton. The smooth polyester fibers resist wrinkles, and the irregular cotton fibers produce a more comfortable and absorbent texture.

condensation begins, just as in the preparation of simple amides. The first step is

$$^{-}O-\overset{\displaystyle O}{\overset{\|}{C}}-(CH_2)_4-\overset{\displaystyle O}{\overset{\|}{C}}-O^{-} + {}^{+}NH_3(CH_2)_6NH_3^{+} \longrightarrow$$

$$^{-}O-\overset{\displaystyle O}{\overset{\|}{C}}-(CH_2)_4-\overset{\displaystyle O}{\overset{\|}{C}}-NH(CH_2)_6NH_3^{+} + H_2O$$

The amide grows at both ends by further condensations (Fig. 19.12), and the final product is

$$^{-}O-\overset{\displaystyle O}{\overset{\|}{C}}-(CH_2)_4-\overset{\displaystyle O}{\overset{\|}{C}}-HN\left[(CH_2)_6-NH-\overset{\displaystyle O}{\overset{\|}{C}}-(CH_2)_4-\overset{\displaystyle O}{\overset{\|}{C}}-HN\right]_n(CH_2)_6NH_3^{+}$$

The long polyamide (nylon) chains can be spun into fibers (like polyesters) or molded. The N—H···O═C hydrogen bonding that occurs between neighboring chains accounts for much of the strength of nylon fibers (Fig. 19.13). However, this ability to form hydrogen bonds also accounts for the tendency of nylon to absorb moisture, because H_2O molecules can hydrogen bond to the polymer and worm their way in among the chains.

Media Link **Lab video:** Fig. 19.12, Formation of a polymer, nylon-6,6

FIGURE 19.12 A rather crude nylon fiber can be made by dissolving the salt of an amine in water and dissolving the acid in a layer of hexane, which floats on the water. The polymer forms at the interface of the two layers, and a long string can be slowly pulled out.

Media Link Animation: Fig. 19.13, Nylon

View an animation of hydrogen bonding between chains of nylon.

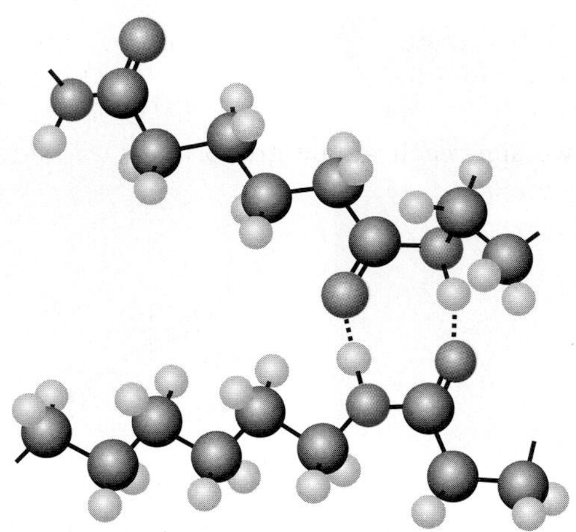

FIGURE 19.13 The strength of nylon fibers is yet another sign of the strength of hydrogen bonds, this time present between neighboring polyamide chains.

Condensation polymers are formed by condensing a carboxylic acid with an alcohol to form a polyester or with an amine to form a polyamide. They tend to have shorter chains than addition polymers and properties that depend on the structures of their monomers.

EXAMPLE 19.2 Determining the formulas of polymers and monomers

Write the formulas of (a) the monomers of Kevlar, a strong fiber used to make bulletproof vests:

$$\left(\overset{O}{\overset{\|}{C}}-C_6H_4-\overset{O}{\overset{\|}{C}}-HN-C_6H_4-NH\right)_n$$

(b) two repeating units of the polymer that is formed when peroxides are added to $CH_3CH_2CH{=}CH_2$.

STRATEGY (a) Look at the backbone of the polymer, the long chain to which the other groups are attached. If the atoms are all carbon atoms, then the compound is an addition polymer. If ester groups are present in the backbone, then the polymer is a polyester and the monomers will be an acid and an alcohol. If the backbone contains amide groups, then the polymer is a polyamide and the monomers will be an acid and an amine. If all the polyester or polyamide groups face in the same direction, then both groups are on the same molecule and there is only one monomer; otherwise there are two different monomers. (b) If the monomer is an alkene or alkyne, then the monomers will add to one another; a π-bond will be replaced by new σ-bonds between the monomers. If the monomers are an acid and an alcohol or amine, then a condensation polymer forms. Draw the ester or amide groups that would result from the loss of a molecule of water.

SOLUTION (a) Amide groups are present in the backbone, so the polymer is a polyamide. The amide groups face in opposite directions, so there are two different monomers, one with two acid groups and one with two amine groups. We split each amide group apart and add a molecule of water to each amide link:

$$\text{---}C(=O)\text{-}C_6H_4\text{-}C(=O)\text{-}HN\text{-}C_6H_4\text{-}NH\text{---} \xrightarrow{H_2O} \text{------} + HOOC\text{-}C_6H_4\text{-}COOH + NH_2C_6H_4NH_2 + \text{------}$$

(b) The monomer is an alkene, so it forms an addition polymer. Replace the π-bond by two additional σ-bonds, one to each adjacent monomer:

$$\text{---}CH_3CH_2CH{=}CH_2 \;\; CH_3CH_2CH{=}CH_2\text{---} \longrightarrow \text{---}CH(CH_2CH_3)\text{-}CH_2\text{-}CH(CH_2CH_3)\text{-}CH_2\text{---}$$

Self-Test 19.4A (a) Write the formula for the monomer of the polymer sold as Teflon, $-(CF_2CF_2)_n-$. (b) The polymer of lactic acid (**18**) is used in surgical sutures that dissolve in the body. Write the formula for a repeating unit of this polymer.

[*Answer:* (a) $CF_2{=}CF_2$; (b) $(OCH(CH_3)CO)_n$]

Self-Test 19.4B Write the formula for (a) the monomer of poly(methyl methacrylate), used in contact lenses (**19**); (b) two repeating units of polyalanine, the polymer of the amino acid alanine, $CH_3CH(NH_2)COOH$.

$$CH_3\text{-}CH(OH)\text{-}C(=O)OH$$

18 Lactic acid, $CH_3CH(OH)COOH$

$$\left(CH_2-C(CH_3)(COOCH_3)\right)_n$$

19 Poly(methyl methacrylate)

19.11 Copolymers and Composites

Copolymers are polymers made up of more than one type of repeating unit (Fig. 19.14). One example is nylon-66, in which the repeating units are formed from 1,6-diaminohexane, $H_2N(CH_2)_6NH_2$, and adipic acid, $HOOC(CH_2)_4COOH$. They form an **alternating copolymer,** in which acid and amine monomers alternate.

In a **block copolymer,** a long segment in which the repeating unit is one monomer is followed by a segment formed from the other monomer. One example is the block copolymer formed from styrene and butadiene. Pure polystyrene is a transparent, brittle material that is easily broken; polybutadiene is a synthetic rubber that is very resilient, but soft. A block copolymer of the two monomers produces *high-impact polystyrene,* a material that is a durable, strong, yet transparent plastic. A different

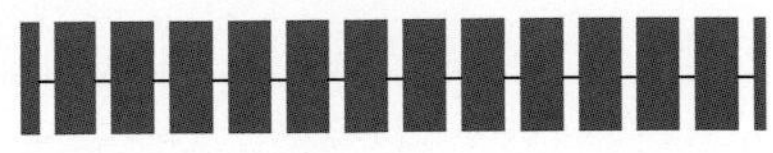
(a) Simple polymer

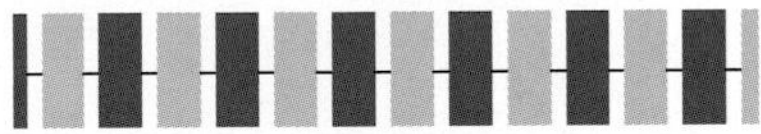
(b) Alternating copolymer

(c) Block copolymer

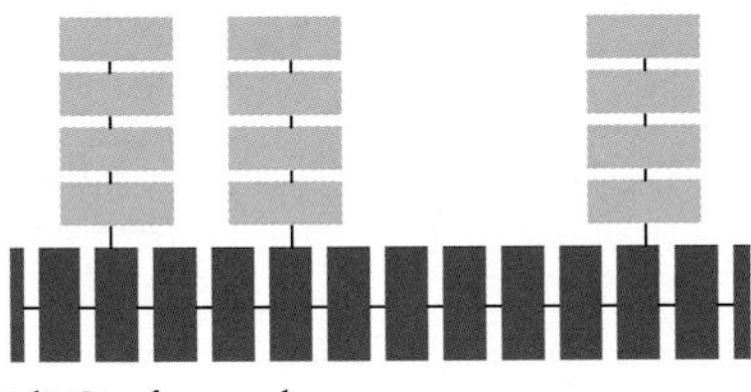
(d) Graft copolymer

FIGURE 19.14 The classification of copolymers. (a) A simple polymer made from a single monomer represented by the red rectangles. (b) An alternating copolymer of two monomers, represented by the red and yellow rectangles. (c) A block copolymer. (d) A graft copolymer.

formulation of the two polymers produces *styrene-butadiene rubber* (SBR), which is used mainly for automobile tires and running shoes, but also in chewing gum.

In a **random copolymer,** different monomers are linked in no particular order. A **graft copolymer** consists of long chains of one monomer with chains of the other monomer attached as side groups. For example, the polymer used to make hard contact lenses is a nonpolar hydrocarbon that repels water. The polymer used to make soft contact lenses is a graft copolymer that has a backbone of nonpolar monomers but side groups of a different water-absorbing monomer. So much water is absorbed by the side chains that 50% of the volume of the contact lens is water, which makes the lens pliable, soft, and more comfortable than a hard contact lens.

Self-Test 19.5A Use Fig. 19.14 to identify the type of copolymer formed by monomers A and B: —AAAABBBBB—.

[*Answer:* Block copolymer]

Self-Test 19.5B Identify the type of copolymer formed by monomers A and B: —ABABABAB—.

A **composite material** consists of two or more materials that have been solidified together. Composite materials are found in nature. For instance, seashells owe their strength to a tough organic matrix and their hardness to the calcium carbonate crystals imbedded in the matrix (Fig. 19.15). Bones have low density, but they are strong because of their composite nature. In bone tissue, crystals of phosphate salts are embedded in fibers of a natural polymer called *collagen* that hold them in place but allow a small amount of flexibility. Artificial bone and cartilage are now

FIGURE 19.15 A photomicrograph of the cross-section of the mother-of-pearl lining a mollusk shell. The flat crystals of calcium carbonate are embedded in a tough, flexible organic matrix that resists cracking.

becoming realities through the preparation of synthetic composite materials. Synthetic composite materials like fiberglass mimic bone in that they have inorganic solids mixed into a crack-resistant polymer matrix. The result is a material that has great strength yet remains flexible. Some lightweight composites, like the graphite composite used for tennis rackets and the body of the space shuttle and high-performance aircraft, can have three times the strength-to-density ratio of steel (Fig. 19.16).

FIGURE 19.16 This high-performance car is made of composite materials that are stronger than steel.

Composite materials and copolymers combine the advantages of more than one component material. Copolymers contain more than one type of monomer.

19.12 Physical Properties of Polymers

Because synthetic polymers consist of molecules with different lengths, they do not have definite molar masses. We can speak only of the *average* molar mass and the *average* chain length of a polymer. Nor do polymers have definite melting points; rather, they soften gradually as the temperature is raised. The viscosity of a polymer, its ability to flow when molten (Section 5.7), depends on its chain length. The longer the chains, the more tangled together they may be, and hence the slower their flow.

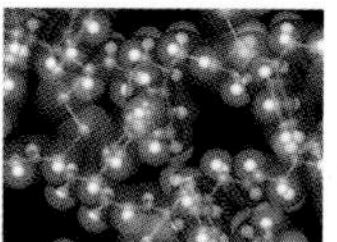

View an animation of the disulfide links in rubber.

Media Link Animation: Fig. 19.17, Rubber

The polarity of the functional groups in a polymer also affects the strength of intermolecular forces, which in turn affects mechanical strength. For example, nylon is a polyamide, and its —NH and —CO groups can take part in hydrogen bonding. Polyethylene has only London forces holding its molecules together and so is not as strong as nylon. For chains of the same average length, stronger intermolecular forces result in greater mechanical strength.

Chain-packing arrangements that maximize intermolecular contact also maximize intermolecular forces and result in greater strength as well as greater density. Long, unbranched chains can line up next to one another like raw spaghetti and form crystalline regions that allow maximum interaction between chains and strong, dense materials. Branched polymer chains cannot fit together as closely and form weaker, less dense materials (recall Fig. 19.6).

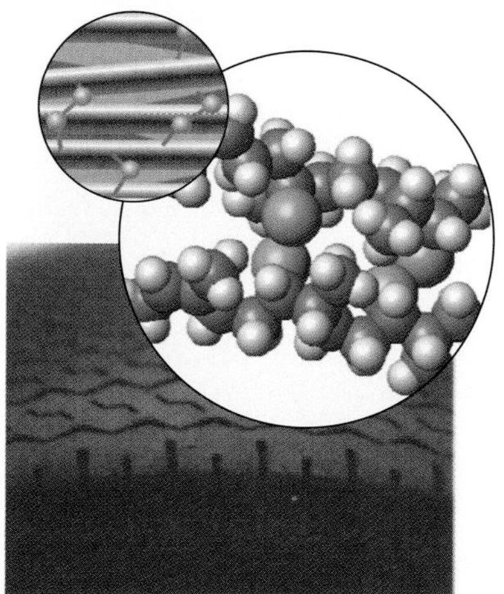

FIGURE 19.17 The gray cylinders in the small inset represent polyisoprene molecules, and the beaded yellow strings represent disulfide (—S—S—) links that are introduced when the rubber is vulcanized, or heated with sulfur. These cross-links increase the resilience of the rubber and make it more useful than natural rubber. Automobile tires are made of vulcanized rubber and a number of additives, including carbon.

The **elasticity** of a polymer is its ability to return to its original shape after being stretched. Natural rubber has low elasticity and is easily softened by heating. However, "vulcanizing" rubber produces the desired properties. In vulcanization, rubber is heated with sulfur. The sulfur atoms form cross-links between the polyisoprene chains and produce a three-dimensional network of atoms (Fig. 19.17). Because the chains are linked together, vulcanized rubber does not soften as much as natural rubber does as the temperature is raised. Vulcanized rubber is also much more resistant to deformation when stretched, because the cross-links pull it back. Materials that return to their original shapes after stretching are called **elastomers**. Extensive cross-linking can produce a rigid network that resists stretching. For example, high concentrations of sulfur result in very extensive cross-linking and the hard material called *ebonite*.

The physical properties of more than one material can be combined in copolymers and composite materials. For example, butyl rubber is a copolymer of a small proportion of isoprene with isobutylene,

$(CH_3)_2C{=}CH_2$. The advantage of butyl rubber over natural rubber stems from the presence in isobutylene of only one double bond; because this polymer has fewer double bonds in the chain than natural rubber has, it is less likely to be oxidized and degraded by the oxygen in air.

Polymers melt over a range of temperatures, and polymers consisting of long chains tend to have high viscosities. Polymer strength increases with increasing chain length, polarity of functional groups in the polymer, and the extent of crystallization. Elastomers contain cross-links that increase their strength and elasticity.

THE IMPACT ON BIOLOGY

Organic chemistry is the basis of the functioning of all living organisms; at one level, life can be regarded as one hugely complex chemical reaction taking place in oddly shaped containers. Many of the organic compounds found in organisms are polymers, including the cellulose of wood, natural fibers such as cotton and silk, the proteins and carbohydrates in our food, and the nucleic acids of our genes. The molecules of some naturally occurring polymers—the proteins, carbohydrates, and nucleic acids—have thousands of functional groups. Nevertheless, the properties we have already met apply to these complicated molecules, too.

19.13 Proteins

The molecules of proteins are condensation copolymers of up to 20 different naturally occurring amino acids, which are distinguished only by their side chains (Table 19.4). There are literally billions of possible combinations and thousands of different kinds of proteins in our bodies, each kind with a specific function to perform. Our bodies can synthesize 11 amino acids in sufficient amounts for our needs. But we cannot produce the proteins necessary for life unless we ingest the other nine, which are known as the **essential amino acids**.

A molecule formed from two or more amino acids is called a **peptide**. An example is the combination of glycine and alanine, denoted Gly-Ala:

```
            O
           //
    H2C — C        CH3
    /       \      /
 H2N        HN — CH
                    \
                    COOH
```

The —CO—NH— link shown in the gray box is called a **peptide bond,** and each amino acid in a peptide is called a **residue.** A typical protein is a polypeptide chain of more than a hundred residues joined through peptide bonds and arranged in a strict order. When only a few amino acid residues are present, we call the molecule an **oligopeptide.** The artificial sweetening agent aspartame is a type of oligopeptide called a **dipeptide** because it has two residues.

TABLE 19.4 The Naturally Occurring Amino Acids, $X{-}CH(NH_2)COOH$

X	Name	Abbreviation	−X	Name	Abbreviation
$-H$	glycine	Gly	$-CH_2(CH_2)_3NH_2$	lysine*	Lys
$-CH_3$	alanine	Ala	$-CH_2(CH_2)_2NH-C(=NH)-NH_2$	arginine	Arg
$-CH_2-$[benzene ring]	phenylalanine*	Phe	$-CH_2-$[imidazole ring: N, NH]	histidine*	His
$-CH(CH_3)_2$	valine*	Val	$-CH_2-$[indole ring: NH]	tryptophan*	Trp
$-CH_2CH(CH_3)_2$	leucine*	Leu	$-CH_2CONH_2$	asparagine	Asn
$-CH(CH_3)CH_2CH_3$	isoleucine*	Ile	$-CH_2CH_2CONH_2$	glutamine	Gln
$-CH_2OH$	serine	Ser	[pyrrolidine ring: NH, COOH]	proline†	Pro
$-CH(OH)CH_3$	threonine*	Thr			
$-CH_2-$[benzene ring]$-OH$	tyrosine	Tyr			
$-CH_2COOH$	aspartic acid	Asp			
$-CH_2CH_2COOH$	glutamic acid	Glu			
$-CH_2SH$	cysteine	Cys			
$-CH_2CH_2SCH_3$	methionine*	Met			

*Essential amino acids for humans.
†The entire amino acid is shown.

A protein's **primary structure** is the sequence of residues in the peptide chain. Aspartame consists of phenylalanine (Phe) and aspartic acid (Asp), so its primary structure is Phe-Asp. Three fragments of the primary structure of human hemoglobin are

Leu-Ser-Pro-Ala-Asp-Lys-Thr-Asn-Val-Lys- . . .
. . . -Val-Lys-Gly-Trp-Ala-Ala- . . .
-Ser-Thr-Val-Leu-Thr-Ser-Lys-Ser-Lys-Tyr-Arg

The determination of the primary structures of proteins is a very demanding analytical task, but, thanks to automated procedures, many of these structures are now known. Any modification of the primary structure of a protein—the replacement of one amino acid residue by another—may lead to the malfunction we call congenital disease. Even one wrong amino acid in the chain can disrupt the normal function of the molecule (Fig. 19.18).

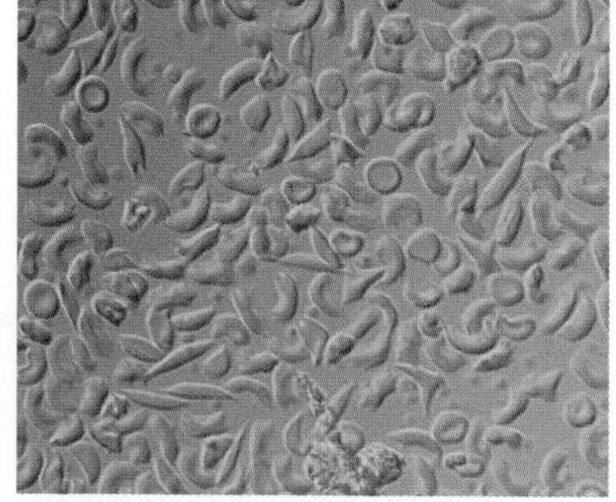

FIGURE 19.18 Sickle-shaped red blood cells form when only one amino acid (glutamic acid) in a polypeptide chain is replaced by another amino acid (valine). These cells are less able to take up oxygen than normal cells.

The **secondary structure** of a protein molecule is the shape adopted by the polypeptide chain—in particular, how it coils or forms sheets. The order of the amino acids in the chain controls the secondary structure, because their intermolecular forces hold the chains together. The most common secondary structure in animal proteins is the **α helix,** a helical conformation of a polypeptide chain held in place by hydrogen bonds between residues (Fig. 19.19). An alternative secondary structure is the **β sheet,** which is characteristic of the protein we know as silk. In silk, protein molecules lie side by side to form nearly flat sheets. The molecules of many other proteins consist of alternating α-helical and β sheet regions (Fig. 19.20).

Protein molecules also adopt a specific **tertiary structure,** the shape into which the α helix, β sheet, and other regions are folded as a result of interactions between residues lying in different parts of the primary structure. The globular form of each chain in hemoglobin is an example. One important type of link responsible for tertiary structure is the **disulfide link,** —S—S—, between amino acids containing sulfur. The chains of proteins are long and can twist in many directions. The study of exactly how proteins fold into the most favorable arrangements is an important research area; this information is essential to the synthesis of new enzymes and the study of disease (Box 19.3).

Proteins may also have a **quaternary structure,** in which neighboring polypeptide units stack together in a specific arrangement. The hemoglobin molecule, for example, has a quaternary structure of four polypeptide units, one of which is shown in Fig. 19.20.

The loss of structure by a protein is called **denaturation.** This structural change may be a loss of quaternary, tertiary, or secondary structure; it may also be degradation of the primary structure by cleavage of the peptide bonds. Even mild heating can cause irreversible denaturation. When we cook an egg, the protein called albumen denatures into a white mass. The "permanent waving" of hair, which consists primarily of long α helices of the protein keratin, is a result of partial denaturation.

Natural protein-based materials include fibers such as silk and wool, and certain adhesives. Some animals, such as spiders, also use proteins as

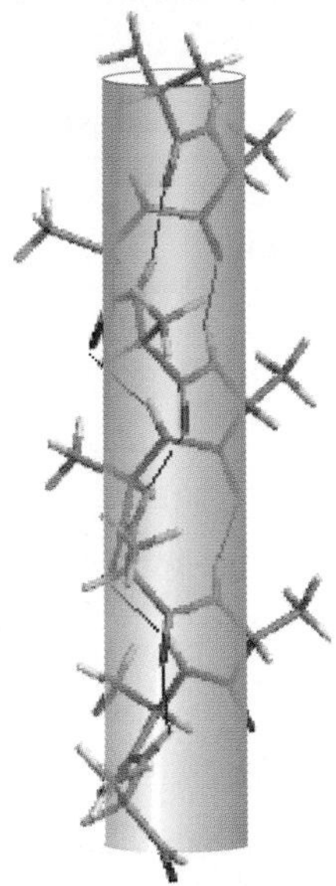

FIGURE 19.19 A representation of part of an α helix, one of the secondary structures adopted by polypeptide chains. The tubes represent the atoms and their bonds. The cylinder encloses the "backbone" of the polypeptide chain, and the side groups project through it. The thin lines represent the hydrogen bonds that hold the helix in shape.

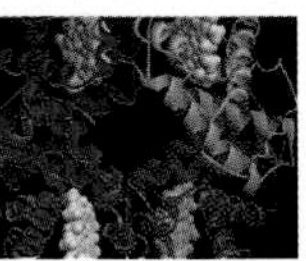

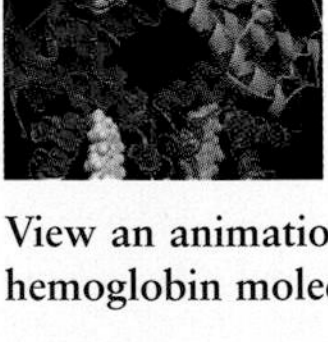

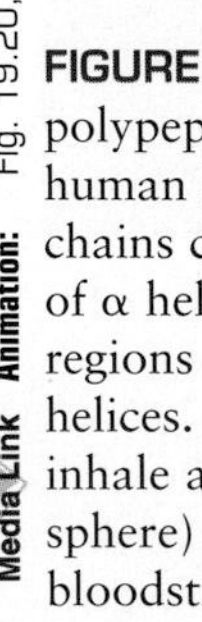

Media Link Animation: Fig. 19.20, Hemoglobin

View an animation of the hemoglobin molecule.

FIGURE 19.20 One of the four polypeptide chains that make up the human hemoglobin molecule. The chains consist of alternating regions of α helix and β sheet. The α helix regions are represented by red helices. The oxygen molecules we inhale attach to the iron atom (blue sphere) and are carried through the bloodstream.

BOX 19.3 FRONTIERS OF CHEMISTRY: PROTEIN FOLDING

Suppose we try to make a protein. We could assemble all the amino acids, one after the other, as is done in our cells by the RNA (Section 19.15). But the action of many proteins, particularly the enzymes that catalyze the reactions in our cells, depends on their three-dimensional shapes. How can we expect that the long chain of hundreds of amino acids will coil and fold up into the exact shape needed?

The amazing fact is that, in most cases, a given protein will fold into a precise conformation. That shape is determined by the locations of the hydrophobic and hydrophilic groups on the chain. The chain coils into α helices or folds into β sheets, then the coils and sheets fold into one another to create the overall three-dimensional shape of the protein. Sometimes, however, the proteins do not fold properly. When this happens in the body, it can lead to illnesses such as Alzheimer's disease. In Alzheimer's disease, brain function is constricted by deposits of proteins that have misfolded and can no longer perform their functions. If scientists can solve the puzzle of why proteins misfold and discover how to correct that, then diseases now considered untreatable could be cured.

Relating the three-dimensional structure of a protein to its sequence of amino acids is one of the most important unsolved problems in biochemistry. In recent research, scientists have found that the shape of a certain protein can be reproduced by using only 5 amino acids, not the 20 found in our cells. The synthetic protein folded into the same shape as a protein made from 18 amino acids. The scientists concluded that the presence of hydrophilic or hydrophobic groups in the side chains was the determining factor in the folding of the chain. However, the other amino acids are necessary when the identity of the amino acids on the surface of the protein or within the active site of an enzyme are critical to its function.

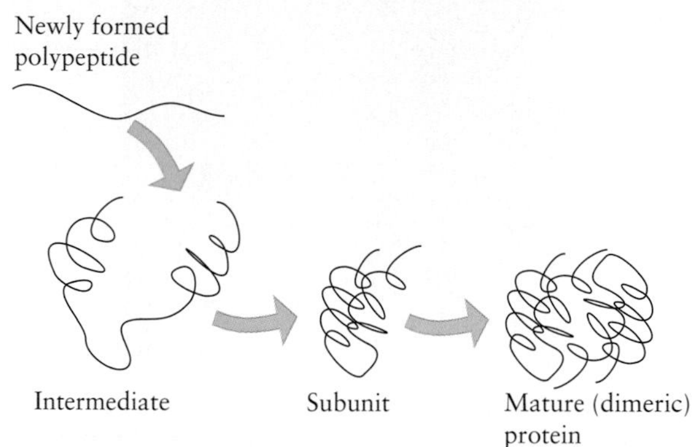

These structures show how a protein first forms α helices and β sheets, and then how the coils and sheets fold together to form the shape of a protein. Finally, the protein subunits stack together in pairs.

HOW MIGHT YOU CONTRIBUTE?

Computer simulations of protein folding have revealed important new information about how a protein folds and how misfolding can occur. These simulations need to be refined and improved. In addition, the effect on shape of individual amino acids needs to be determined and the mechanism by which misfolding occurs in the body needs to be identified.

For Further Reading: I. Peterson, "Simulations nab protein-folding mistakes," *Science News* (March 6, 1999), 155: 150. G. Taubes, "Misfolding the way to disease," *Science* (1996), 271: 1493–1495. C. Wu, "An alphabet for a letter-perfect protein," *Science News* (October 4, 1997), 152: 214.

a structural material of remarkable strength (Fig. 19.21). Chemists are duplicating nature by making artificial spider silk (Fig. 19.22), which is one of the strongest fibers known.

Proteins are polymers made of amino acid units. The primary structure of a polypeptide is the sequence of amino acid residues; the secondary structure is the formation of helices and sheets; the tertiary structure is the folding into a compact unit; the quaternary structure is the packing of individual protein units together.

FIGURE 19.21 The protein produced by spiders to make a web is a form of silk that can be exceptionally strong.

FIGURE 19.22 Artificial spider silk can now be made in bulk. It can be spun into thin, tough thread, like that on the spools shown here, or wound into cables strong enough to support suspension bridges.

19.14 Carbohydrates

The **carbohydrates** are so called because they often have the empirical formula CH_2O, which suggests a hydrate of carbon. They include starches, cellulose, and sugars like glucose, $C_6H_{12}O_6$ (**20**), which contains an aldehyde group, and fructose (fruit sugar), a structural isomer of glucose that is a ketone (**21**). Carbohydrates have many —OH groups, so they may also be regarded as alcohols. The presence of these —OH groups allows them to form numerous hydrogen bonds with one another and with water.

```
 H      O                          CH2OH
  \\   //                            |
    C                                C=O
    |                                |
H—C—OH                         HO—C—H
    |                                |
HO—C—H                          H—C—OH
    |                                |
H—C—OH                          H—C—OH
    |                                |
H—C—OH                            CH2OH
    |
  CH2OH
```

20 Glucose, $C_6H_{12}O_6$ **21** Fructose, $C_6H_{12}O_6$

The **polysaccharides** are polymers of glucose. They include starch, which we can digest, and cellulose, which we cannot. Starch is made up of two

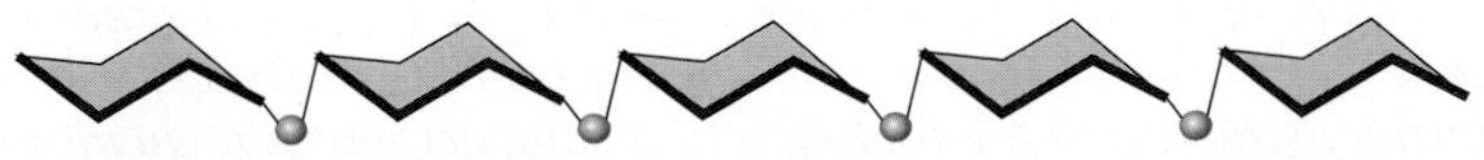

FIGURE 19.23 The amylose molecule, one component of starch, is a polysaccharide, a polymer of glucose. It consists of glucose units linked together to give a structure like this but with a moderate degree of branching.

Media Link Animation: Fig. 19.23, Starch

View an animation of a starch molecule.

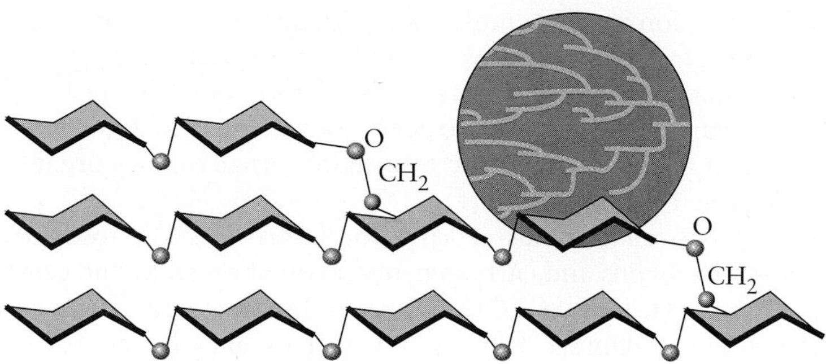

FIGURE 19.24 The amylopectin molecule is another component of starch. It has a more highly branched structure than amylose.

components, amylose and amylopectin. Amylose, which makes up about 20% to 25% of most starches, consists of chains made up of several thousand glucose units linked together (Fig. 19.23). Amylopectin is also made up of glucose chains (Fig. 19.24), but its chains are linked into a branched structure and its molecules are much larger. Each molecule consists of about a million glucose units.

Cellulose is the structural material of plants. It is a polymer with the same monomer (glucose) that starch has, but the units in cellulose are linked differently, forming flat, ribbonlike strands (Fig. 19.25). Hydrogen bonds between these strands lock them together into a rigid structure that for us (but not for termites) is indigestible. Cellulose is the most abundant organic chemical in the world, and billions of tons of it are produced annually by photosynthesis. In research on alternative fuels, enzymes are used to break down the cellulose in waste biomass into glucose, which is then fermented to make ethanol for use as a fuel (Box 6.2).

Glucose is an alcohol and an aldehyde that polymerizes to form starch and cellulose.

19.15 Nucleic Acids

People have used materials to store information at least since the time of the first cave paintings. Today, we can speculate about the possibility of

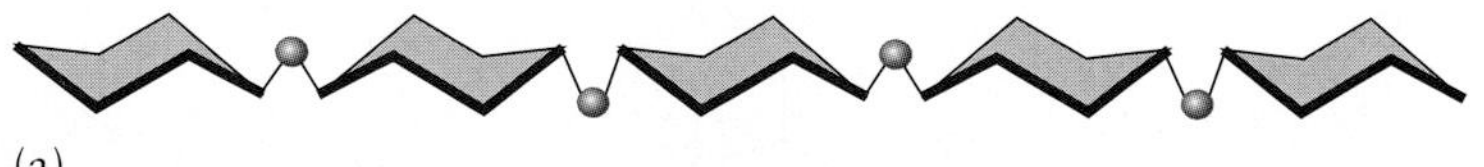

(a)

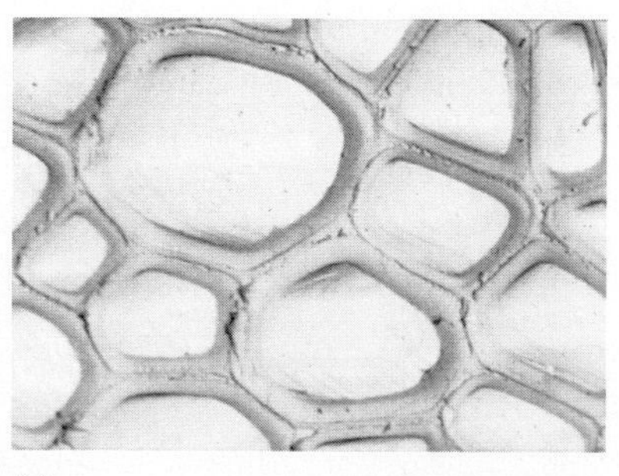

(b)

FIGURE 19.25 (a) Cellulose is yet another polysaccharide constructed from glucose units. However, the linking between the units in cellulose results in long, flat ribbons that can produce a fibrous material through hydrogen bonding. (b) These long tubes of cellulose (shown in cross section) form the structural material of an aspen tree.

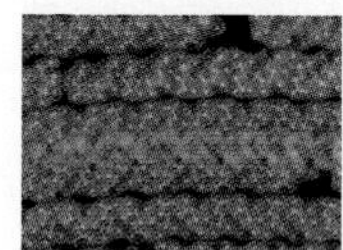

View an animation of cellulose chains.

Media Link Animation: Fig. 19.25, Cellulose

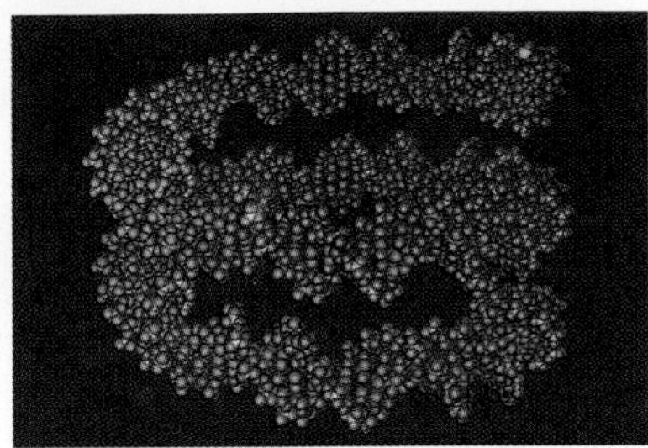

FIGURE 19.26 A computer graphics image of a short section of a DNA molecule, which consists of two entwined helices. In this illustration, the double helix is also coiled around itself in a shape called a superhelix.

storing information in single molecules. A dream of computer designers is that one day arrays of such molecules will serve as data storage devices of enormous capacity. Nature, however, has already used this technique for millions of years. It uses a "nucleic acid" molecule called deoxyribonucleic acid (DNA) to store the genetic information that enables organisms to reproduce (Fig. 19.26).

Every living cell contains at least one DNA molecule to control the production of proteins and carry genetic information from one generation of cells to the next. Human DNA molecules are immense: if one could be extracted without damage from a cell nucleus and drawn out straight from its highly coiled natural shape, it would be about 2 m long (Fig. 19.27). The ribonucleic acid (RNA) molecule is closely related to DNA. One of its functions is to carry information stored by DNA to a region of the cell where that information can be used in protein synthesis.

DNA is a polymer built of repeating units derived from the sugar ribose (**22**). For DNA, the ribose molecule has been modified by removing the oxygen atom at carbon atom 2, the second carbon atom clockwise from the ether oxygen atom in the five-membered ring. Therefore, the repeating unit—the monomer—is called deoxyribose (**23**).

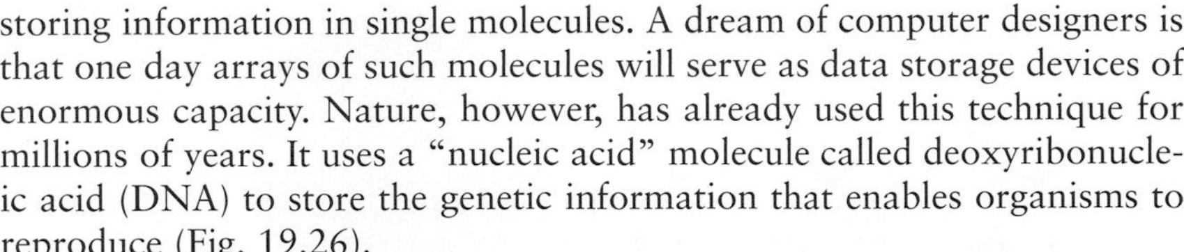

22 Ribose, $C_5H_{10}O_5$ **23** Deoxyribose, $C_5H_{10}O_4$

Attached by a covalent bond to carbon atom 1 of the deoxyribose ring is an amine (and therefore a base), which may be adenine, A (**24**); guanine, G (**25**); cytosine, C (**26**); or thymine, T (**27**). In RNA, uracil, U (**28**), occurs in place of thymine. The base bonds to carbon atom 1 of deoxyribose through the nitrogen of the —NH group (printed in red); and the compound so formed is called a **nucleoside.** All nucleosides have a similar structure, which we can summarize as the shape shown in (**29**); the lens-shaped object represents the attached amine.

24 Adenine **25** Guanine **26** Cytosine

27 Thymine **28** Uracil **29** A nucleoside

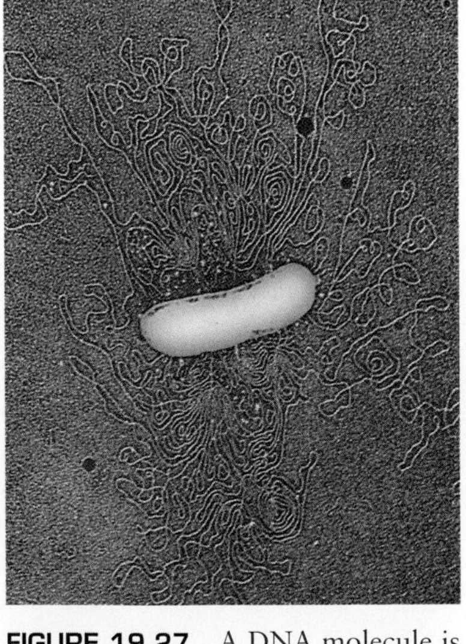

FIGURE 19.27 A DNA molecule is very large, even in bacteria. In this micrograph, a DNA molecule has spilled out through the damaged cell wall of a bacterium.

The DNA monomers are each completed by a phosphate group, $—O—PO_3^{2-}$, covalently bonded to carbon atom 5 of the ribose unit to give a compound called a **nucleotide** (30). Because there are four possible nucleoside monomers (one for each base), there are four possible nucleotides in each type of nucleic acid.

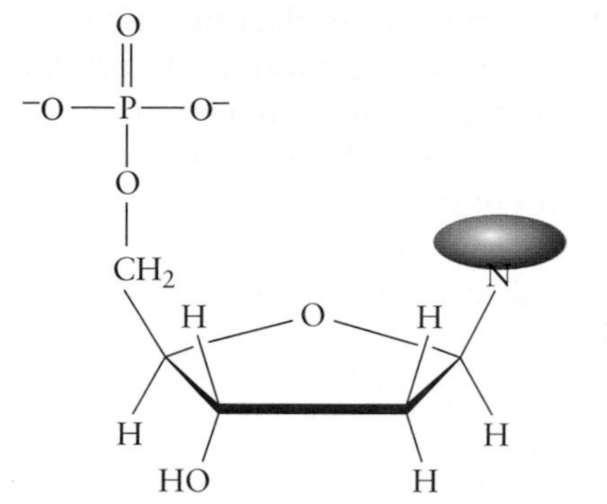

30 A nucleotide

Molecules of DNA and RNA are **polynucleotides,** polymeric species built from nucleotide units. Polymerization occurs when the phosphate group of one nucleotide (which is the conjugate base of an organic phosphoric acid) condenses with the —OH group on carbon atom 3 of the ribose unit of another nucleotide, thereby forming an ester link and releasing a molecule of water. As this condensation continues, it results in a structure like that shown in Fig. 19.28, a compound known as a **nucleic acid.** The DNA molecule itself is a double helix in which two long nucleic acid strands are wound around each other.

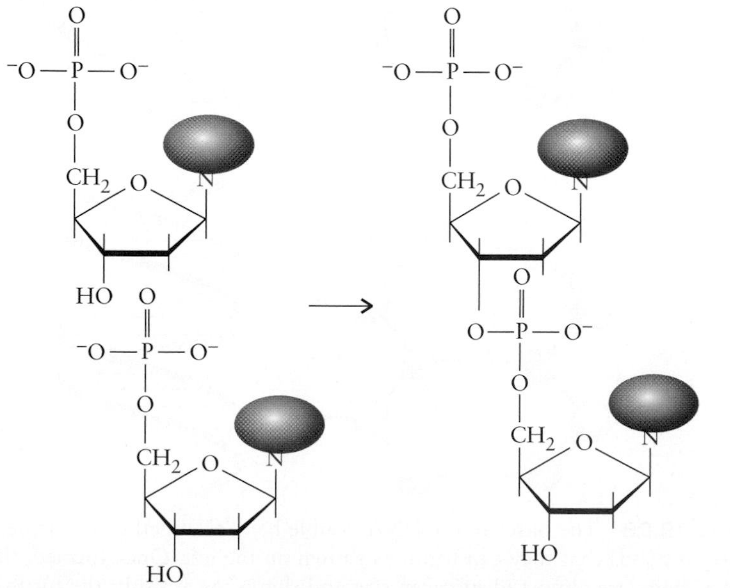

FIGURE 19.28 The condensation of nucleotides that leads to the formation of a nucleic acid, a polynucleotide.

The ability of DNA to replicate lies in its double helical structure. There is a precise correspondence between the bases in the two strands. Adenine in one strand always forms two hydrogen bonds to thymine in the other, and guanine always forms three hydrogen bonds to cytosine; so, across the helix, the base pairs are always AT and GC (Fig. 19.29). Any other combination would not be held together as well.

As well as replication—the production of copies of itself for reproduction and cell division—DNA governs the production of proteins by serving as the template during the synthesis of molecules of RNA. These new molecules, with U in place of T, carry information about segments of the genetic message out of the animal or plant cell nucleus to the locations where protein synthesis takes place. In this way, the chemical reactions of functional groups, and in a broader sense, the principles of chemistry, bring matter to life.

Nucleic acids are copolymers of four nucleotides joined by phosphate ester links. The nucleotide sequence stores all genetic information.

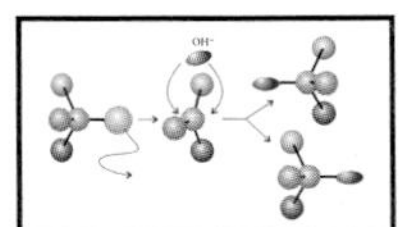

View an animation of DNA strands interacting.

Media Link **Animation:** Fig. 19.29, DNA

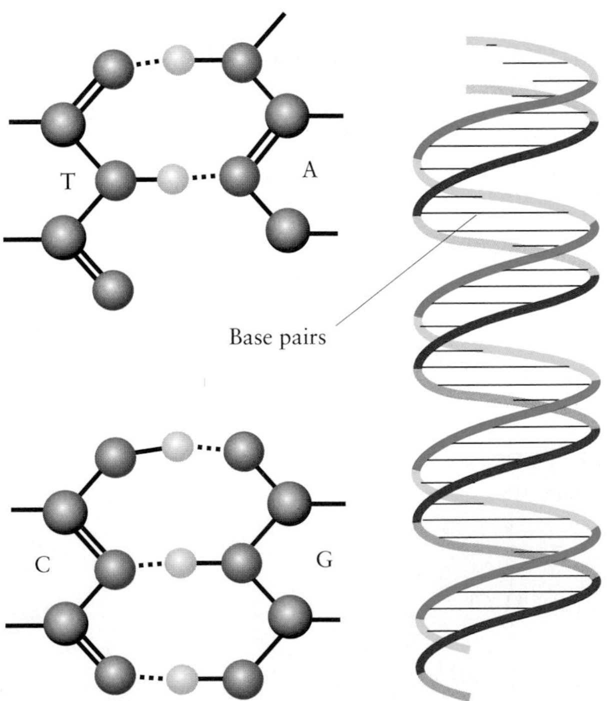

FIGURE 19.29 The bases in the DNA double helix fit together by virtue of the hydrogen bonds that they can form, as shown on the left. Once formed, the AT and GC pairs are almost identical in size and shape. As a result, the turns of the helix shown on the right are regular and consistent.

SKILLS YOU SHOULD HAVE MASTERED

1 Recognize a simple haloalkane, alcohol, ether, phenol, aldehyde, ketone, carboxylic acid, amine, amide, or ester, given a molecular structure.

2 Write the structural formula of an ester or an amide formed from the condensation reaction of a given carboxylic acid with a given alcohol or amine, Self-Test 19.1.

3 Predict the influence of hydrogen bonding on the physical properties of organic compounds, Self-Test 19.2.

4 Name simple functional groups, Toolbox 19.1 and Example 19.1.

5 Predict the oxidation products of aldehydes and ketones, Section 19.5.

6 Predict the type of polymer a given monomer can form and identify monomers, given the repeating unit of a polymer, Example 19.2.

7 Distinguish the various types of copolymers, Self-Test 19.5.

8 Explain the role of chain length, crystallinity, network formation, cross-linking, and intermolecular forces in determining the physical properties of polymers, Section 19.11.

9 Describe the composition of proteins and distinguish their primary, secondary, tertiary, and quaternary structures, Section 19.13.

10 Describe the composition of carbohydrates, Section 19.14.

11 Describe the structures and functions of nucleic acids, Section 19.15.

EXERCISES

Identifying Functional Groups and Naming Their Compounds

19.1 Write the general formula of each of the following types of compounds, using R to denote an organic group: (a) amine; (b) alcohol; (c) carboxylic acid; (d) aldehyde.

19.2 Write the general formula of each of the following types of compounds, using R to denote an organic group: (a) ether; (b) ketone; (c) ester; (d) amide.

19.3 Identify each type of compound: (a) R—O—R; (b) R—CO—R; (c) $R—NH_2$; (d) R—COOR.

19.4 Identify each type of compound: (a) R—CHO; (b) R—COOH; (c) R—CONHR; (d) R—OH.

19.5 Name the following compounds:
(a) $CH_3CHClCH_3$;
(b) $CH_3CH_2C(CH_3)ClCH_2CHClCH_3$;
(c) CH_3CI_3; (d) CH_2Cl_2.

19.6 Name the following compounds: (a) CH_2CHCH_2Br; (b) $CF_2CHCHCH_2$; (c) $CH_3CH_2CCCH_2Br$; (d) $Cl_3CCOCH_2CH_3$.

19.7 Write the formulas of the following compounds and state whether each one is a primary, secondary, or tertiary alcohol or a phenol: (a) 1-chloro-2-hydroxybenzene; (b) 2-methyl-3-pentanol; (c) 2,4-dimethyl-1-hexanol; (d) 2-methyl-2-butanol.

19.8 Write the formulas of the following compounds and state whether each one is a primary, secondary, or tertiary alcohol or a phenol: (a) 2-methyl-1-propanol; (b) 2-propanol; (c) *p*-hydroxytoluene; (d) 1-bromo-2-(hydroxymethyl)benzene.

19.9 Write the formula of (a) ethyl methyl ether, (b) ethyl propyl ether; (c) dimethyl ether.

19.10 Write the formula of (a) diethyl ether; (b) dibutyl ether; (c) methyl propyl ether.

19.11 Name the compound
(a) $CH_3CH_2CH_2OCH_2CH_2CH_2CH_3$;
(b) $C_6H_5OCH_3$;
(c) $CH_3CH_2CH_2OCH_2CH_2CH_2CH_2CH_3$.

19.12 Name the compound
(a) $CH_3CH_2CH_2OCH_2CH_3$;
(b) $CH_3OCH_2CH_2CH_2CH_3$;
(c) $CH_3CH_2CH_2OCH_2CH_2CH_3$.

19.13 Identify each compound as an aldehyde or a ketone and give its systematic name: (a) CH_3CHO; (b) CH_3COCH_3; (c) $(CH_3CH_2)_2CO$.

19.14 Identify each compound as an aldehyde or a ketone and give its systematic name:
(a) CH_3CH_2CHO;

(b) H_3C—C_6H_4—CHO (para-substituted benzene ring);

(c) $(CH_3CH_2)_2CHCH_2COCH_3$.

19.15 Write the structural formulas of (a) methanal; (b) propanone; (c) 2-heptanone.

19.16 Write the structural formulas of (a) 2-ethyl-2-methylpentanal; (b) 3,5-dihydroxy-4-octanone; (c) octanal.

19.17 Give the systematic names of (a) CH_3COOH; (b) $CH_3CH_2CH_2COOH$; (c) $CH_2(NH_2)COOH$.

19.18 Give the systematic names of (a) CH_3CH_2COOH; (b) $CHCl_2COOH$; (c) $CH_3(CH_2)_7COOH$.

19.19 Draw the structures of (a) benzoic acid, C_6H_5COOH; (b) 2-chloro-3-methylpentanoic acid; (c) hexanoic acid; (d) propenoic acid.

19.20 Draw the structures of (a) 2-methylpropanoic acid; (b) 2,2-dichlorobutanoic acid; (c) 2,2,2-trifluoroethanoic acid; (d) 4,4-dimethylpentanoic acid.

19.21 Give the systematic names of the following amines: (a) CH_3NH_2; (b) $(CH_3CH_2)_2NH$; (c) *o*-$CH_3C_6H_4NH_2$.

19.22 Give the systematic names of the following amines: (a) $CH_3CH_2CH_2NH_2$; (b) $(CH_3CH_2)_4N^+$; (c) *p*-$ClC_6H_4NH_2$.

19.23 Write the structural formulas of the following amines: (a) *o*-methylphenylamine; (b) triethylamine; (c) tetramethylammonium ion.

19.24 Write the structural formulas of the following amines: (a) methylpropylamine; (b) dimethylamine; (c) *m*-methylphenylamine.

Reactivity Patterns and Functional Groups

19.25 A pure, optically active sample of one isomer of $C_6H_5CHClCH_3$ is hydrolyzed by water, and the product is optically inactive. (a) Write the formula of the product. (b) Is the mechanism S_N1 or S_N2? Explain your conclusion.

19.26 Consider the haloalkane compounds that undergo nucleophilic substitution by an S_N1 mechanism. For a given organic group R, which would you expect to proceed more rapidly, substitution of RF, RCl, RBr, or RI? Explain your conclusion.

19.27 Which of the following may function as nucleophiles in a nucleophilic substitution reaction: (a) NH_3; (b) CO_2; (c) Br^-; (d) SiH_4?

19.28 Which of the following may function as nucleophiles in a nucleophilic substitution reaction: (a) CH_3O^-; (b) CN^-; (c) NH_4^+; (d) PH_3?

19.29 Suggest an alcohol that could be used for the preparation of each of the following compounds and indicate how the reaction would be carried out; (a) ethanal; (b) 2-octanone; (c) 5-methyloctanal.

19.30 Suggest an alcohol that could be used for the preparation of each of the following compounds and indicate how the reaction would be carried out: (a) methanal; (b) propanone; (c) 6-methyl-5-decanone.

19.31 Draw the structures of the principal products formed from the following condensation reactions: (a) butanoic acid with 2-propanol; (b) ethanoic acid with 1-pentanol; (c) hexanoic acid with methylethyl amine; (d) ethanoic acid with propylamine.

19.32 Draw the structures of the principal products formed from the following condensation reactions: (a) octanoic acid with methanol; (b) propanoic acid with ethanol; (c) propanoic acid with methylamine; (d) methanoic acid with diethylamine.

19.33 You are given samples of propanal, 2-propanone, and ethanoic acid. Describe how you would use chemical tests, such as acid-base indicators and oxidizing agents, to distinguish between the three compounds.

19.34 You are given samples of 1-propanol, pentane, and ethanoic acid. Describe how you would use chemical tests, such as aqueous solubility and acid-base indicators, to distinguish among the three compounds.

Polymers

19.35 Sketch three repeating units of the polymer formed from (a) $CH_2{=}C(CH_3)_2$; (b) $CH_2{=}CHCN$; (c) isoprene,

$H_2C{=}C(H){-}C(CH_3){=}CH_2$

19.36 Sketch three repeating units of the polymers formed from (a) tetrafluoroethene; (b) phenylethene; (c) $CH_3CH{=}CHCH_3$.

19.37 Write the structural formulas of the monomers of the following polymers, for which one repeating unit is shown: (a) polyvinylchloride (PVC), $-(CHClCH_2)_n-$; (b) Kel-F, $-(CFClCF_2)_n-$

19.38 Write the structural formulas of the monomers of the following polymers, for which one repeating unit is shown: (a) a polymer used to make carpets, $-(OC(CH_3)_2-CO-)_n-$; (b) $-(CH(CH_3)CH_2)_n-$; (c) a polypeptide, $-(NHCH_2CO)_n-$.

19.39 Write the structural formula of two units of the polymer formed from (a) the reaction of oxalic acid (ethanedioic acid), HOOCCOOH, with 1,4-diaminobutane, $H_2NCH_2CH_2CH_2CH_2NH_2$; (b) the polymerization of the amino acid alanine (2-aminopropanoic acid).

19.40 Write the structural formula of two units of the polymer formed from (a) the reaction of terephthalic acid (see below) with 1,2-diaminoethane, $H_2NCH_2CH_2NH_2$; (b) the polymerization of 4-hydroxybenzoic acid (see below).

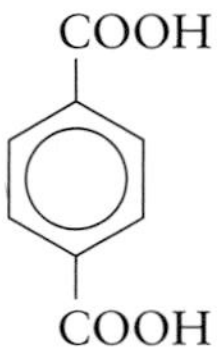

COOH	COOH
(benzene ring)	(benzene ring)
COOH	OH
Terephthalic acid	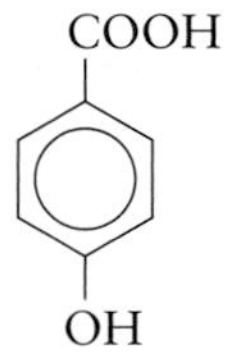 4-Hydroxybenzoic acid

19.41 Distinguish between an isotactic polymer, a syndiotactic polymer, and an atactic polymer.

19.42 Explain why a syndiotactic polymer is stronger than an atactic polymer.

19.43 Identify the type of copolymer formed by monomers A and B: —BBBBAA—.

19.44 Identify the type of copolymer formed by monomers A and B: —AABABBAA—.

19.45 How does average molar mass affect the following polymer characteristics: (a) softening point; (b) viscosity; (c) strength?

19.46 How does the polarity of the functional groups in a polymer affect the following characteristics of the polymer: (a) softening point; (b) viscosity; (c) strength?

19.47 Describe how the linearity of the polymer chain affects polymer strength.

19.48 Describe how cross-linking affects the elasticity and rigidity of a polymer.

19.49 Which of the following molecules is an organic peroxide that can be used to initiate radical reactions: (a) $CH_3CH_2CH_2COOH$; (b) $CH_3CH_2CH_2COOCH_3$; (c) $CH_3CH_2CH_2CH_2OOH$; (d) $CH_3CH_2CH_2COCH_3$?

19.50 Which of the following phrases best describes Ziegler-Natta catalysis: (a) a radical chain polymerization of alkenes leading to stereoregular polymers; (b) a polymerization of alkenes catalyzed by a *d*-metal complex activated by a Lewis acid; (c) a self-condensation polymerization reaction in which one monomer contains both functional groups necessary for the condensation reaction; (d) a condensation reaction in which the equilibrium is shifted toward polymer formation by removal of the product water as it is formed?

Biological Compounds

19.51 (a) Draw the structure of the peptide bond that links the amino acids in proteins. (b) Identify the functional group formed. (c) Identify the type of polymer formed (addition or condensation).

19.52 (a) Draw the structure of the link between glucose units that creates amylose. (b) Identify the functional group formed. (c) Identify the type of polymer formed (addition or condensation).

19.53 Name the amino acids in Table 19.4 that contain side groups capable of forming hydrogen bonds. This interaction contributes to the tertiary structure of a protein.

19.54 Name the amino acids in Table 19.4 that contain nonpolar side groups. These groups contribute to the tertiary structure of a protein, by avoiding contact with water.

19.55 Draw the structure of the peptide formed from the reaction of the acid group of tyrosine with the amine group of glycine.

19.56 Draw the structure of the peptide formed from the reaction of the acid group of glycine with the amine group of tyrosine.

19.57 Identify (a) the functional groups and (b) the chiral carbon atoms in the mannose molecule shown below.

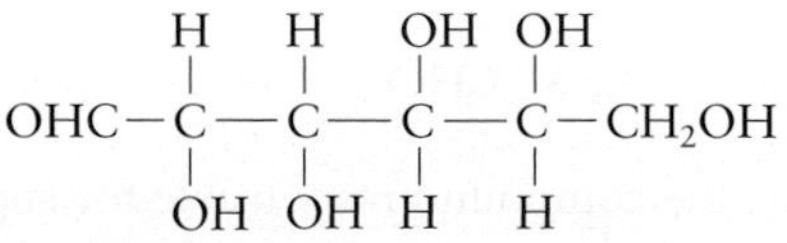

```
      H   H   OH  OH
      |   |   |   |
OHC—C—C—C—C—CH2OH
      |   |   |   |
      OH  OH  H   H
```

19.58 Identify (a) the functional groups and (b) the chiral carbon atoms in the histidine molecule shown below.

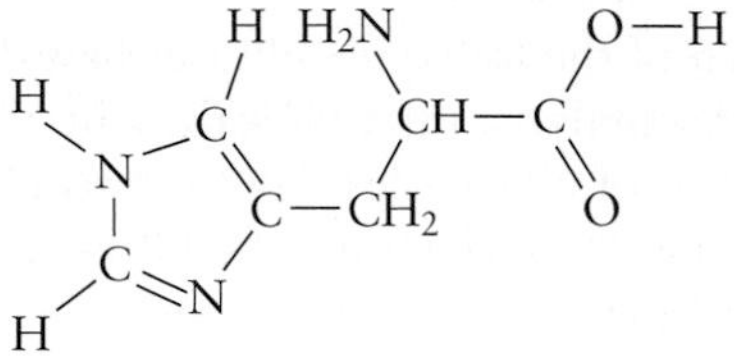

19.59 Write the complementary nucleic acid sequence that would pair with the following DNA sequences: (a) CATGAGTTA; (b) TGAATTGCA.

19.60 Write the complementary nucleic acid sequence that would pair with the following DNA sequences: (a) GGATCTCAG; (b) CTAGCCTGT.

INTEGRATED EXERCISES

19.61 Write the chemical formula of the compound represented by each of the following line structures

(a) Guanine, a base found in DNA

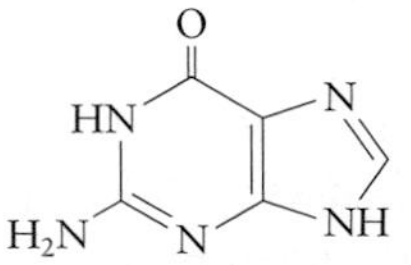

(b) D-Glucose, a monosaccharide

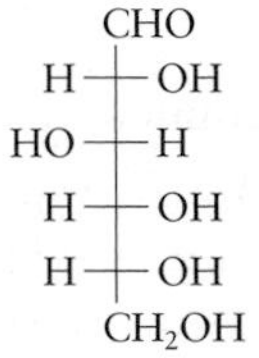

(c) Alanine, an amino acid

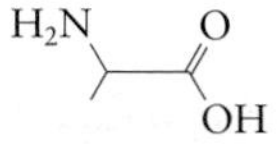

19.62 Write the chemical formula of the compound represented by each of the following line structures:

(a) Cysteine, an amino acid

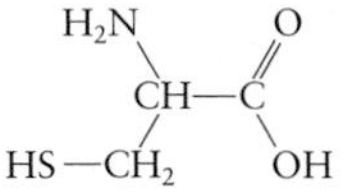

(b) Adenine, a base found in DNA

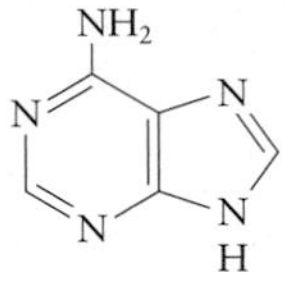

(c) Eugenol

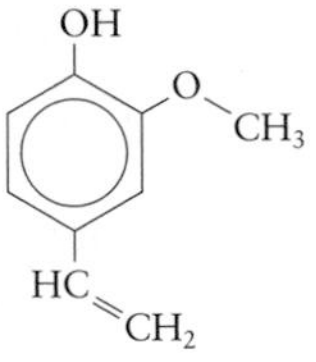

19.63 Identify all the functional groups in the following compounds: (a) vanillin, the compound responsible for vanilla flavor

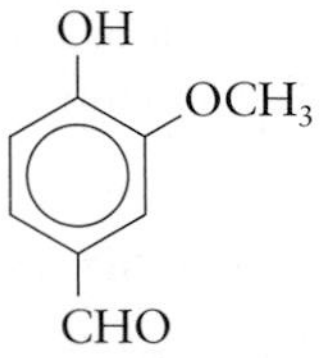

(b) carvone, the compound responsible for spearmint flavor

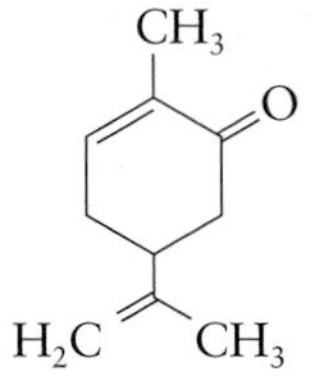

(c) caffeine, the stimulant in coffee, tea, and cola drinks

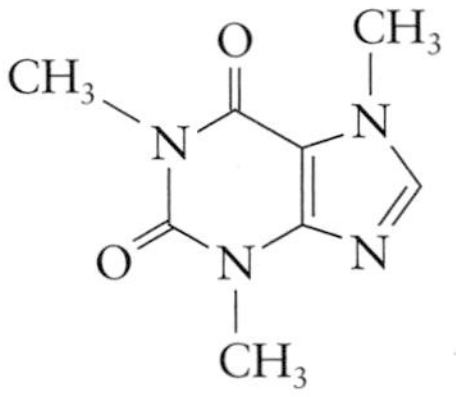

19.64 Identify all the functional groups in the following compounds: (a) zingerone, the pungent, hot component of ginger

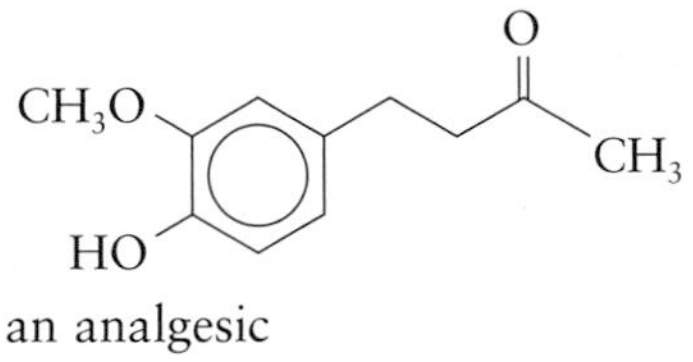

(b) Tylenol, an analgesic

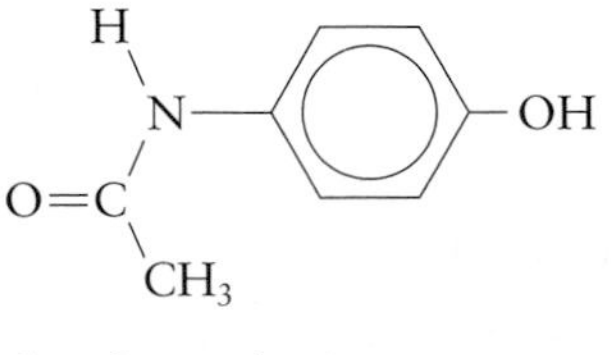

(c) procaine, a local anesthetic,

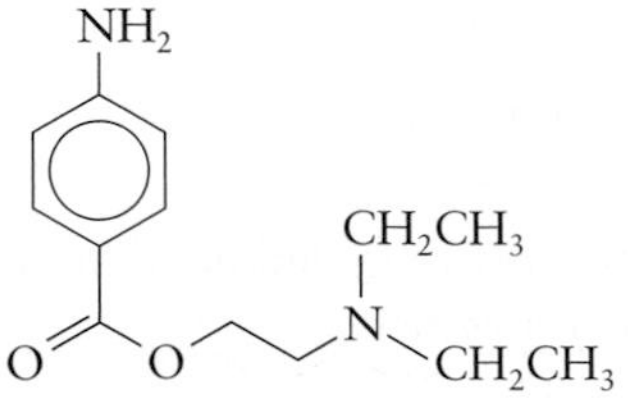

19.65 Identify all the functional groups in the following compounds (their structures may be found on the Web site for this text): (a) aspirin; (b) capsaicin, the biting hot component of chili peppers; (c) nicotine, an addictive agent found in tobacco and used as an insecticide; (d) cortisone, a hormone used as an antiinflammatory agent.

19.66 Identify all the functional groups in the following compounds (their structures can be found on the Web site for this text): (a) vitamin A, a vitamin found in carrots and other vegetables and necessary for good night vision; (b) vitamin E, a vitamin that is found in wheat germ, sunflower seed, soybean oils, and lettuce, and acts as an antioxidant, prevents muscle degeneration, and promotes fertility; (c) strychnine, a poison.

19.67 Identify each chiral carbon in the following compounds: (a) camphor, used in cooling salves

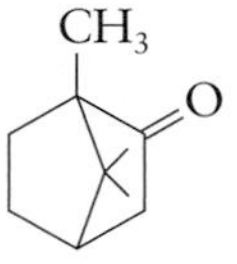

(b) testosterone, a male sex hormone

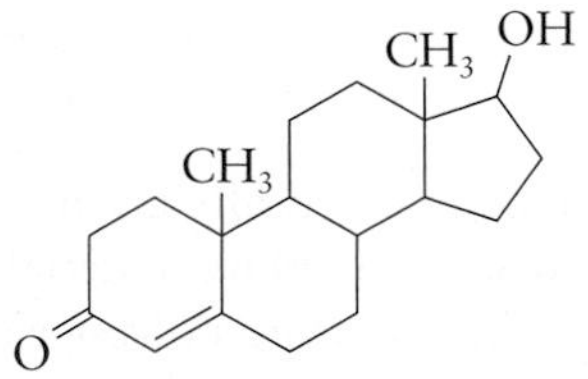

19.68 Identify each chiral carbon atom in the following compounds: (a) menthol, the flavor of peppermint

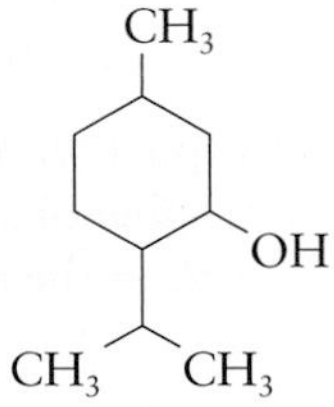

(b) estradiol, a female sex hormone

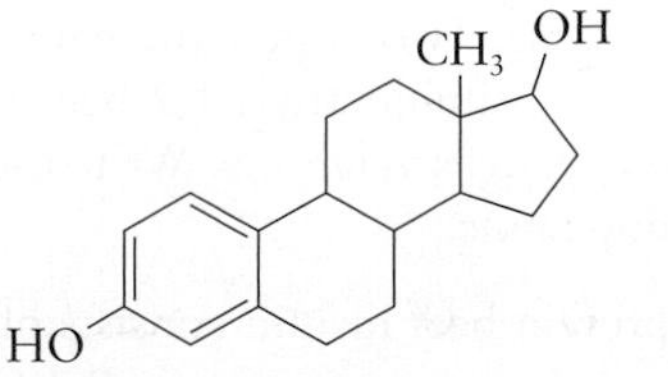

19.69 The structures of the following molecules can be found on the Web site for this text. Draw the structure of each and identify the chiral carbon atoms: (a) cocaine, a narcotic and local anesthetic; (b) aflatoxin B_2, a toxin and carcinogen that has been reported to occur naturally in peanuts as a by-product of the growth of fungi in genus *Aspergillus*.

19.70 The structures of the following molecules, may be found on the Web site for this text. Draw the line structure of each and identify the chiral carbon atoms: (a) cephalosporin C, toxic to penicillin-resistant staphylococci; (b) thromboxane A_2, a substance that promotes clotting of blood.

19.71 (a) Write the structural formulas of diethylether and 1-butanol (note that they are isomers). (b) Whereas the solubility of both compounds in water is about 8 g per 100 mL, the boiling point of 1-butanol is 117°C, but that of diethyl ether is 35°C. Account for these observations.

19.72 Consider the following organic molecules that have roughly the same molar masses but contain different functional groups: $CH_3CH_2CH_2CH_2CH_2CH_3$, $CH_3CH_2CH_2CH_2CHCH_2$, $CH_3CH_2CH_2CH_2CH_2OH$, $CH_3CH_2CH_2CH_2CHO$, $CH_3CH_2CH_2COOH$, $CH_3CH_2COOCH_3$, $CH_3CH_2COCH_2CH_3$, $CH_3CH_2CH_2CH_2OCH_3$. (a) Draw a Lewis structure for each molecule, name it, and classify it by functional group. (b) Which molecules are isomers of each other? Are any chiral? If so, which ones? (c) For each molecule, list the types of intermolecular forces that will be present. (d) From the answers to (a) and (b), predict the relative boiling points, from lowest to highest.

19.73 Write the structural formulas for the products of (a) the reaction of glycerol (1,2,3-trihydroxypropane with stearic acid, $CH_3(CH_2)_{16}COOH$, to produce a saturated fat; (b) the oxidation of 4-hydroxybenzyl alcohol by sodium dichromate in an acidic organic solvent.

19.74 Pheromones are commonly called sex attractants, although they have more complex signaling functions. The structure of a pheromone in the queen bee is *trans*-$CH_3CO(CH_2)_5CH{=}CHCOOH$. (a) Write the structural formula of the pheromone. (b) Identify and name the functional groups in the molecule.

19.75 Classify the following reactions as (1) an addition reaction, (2) a nucleophilic substitution reaction, (3) an electrophilic substitution reaction, or (4) a condensation reaction: (a) the reaction of butene with chlorine; (b) the polymerization of the amino acid glycine; (c) the hydrogenation of butyne; (d) the polymerization of styrene, $CH_2CHC_6H_5$, by tertiary butyl hydroperoxide, $(CH_3)_3COOH$; (e) the reaction of methylamine with butanoic acid.

19.76 Classify the following reactions as (1) an addition reaction, (2) a nucleophilic substitution reaction, (3) an electrophilic substitution reaction, or (4) a condensation reaction: (a) the reaction of terephthalic acid with 1,2-ethanediol; (b) the reaction of 3-chlorohexane with concentrated sodium hydroxide; (c) the reaction of water with 1-iodo-1,1-dimethylethane; (d) the reaction of propanoic acid with ethanol; (e) the reaction of toluene with bromine in the presence of $FeBr_3$.

19.77 From the viewpoint of the organic molecule involved in each of the following transformations, state whether the reaction is an oxidation reaction, a reduction reaction, or neither: (a) the conversion of propanol into propanal; (b) the conversion of bromomethane into methanol; (c) the conversion of ethyne into ethane; (d) the reaction of Ag^+ ions with propanal; (e) the conversion of methane into bromomethane.

19.78 From the viewpoint of the organic molecule involved in each of the following transformations, state whether the reaction is an oxidation reaction, a reduction reaction, or neither: (a) the conversion of butanal into butanoic acid; (b) the reaction of butanoic acid with ethanol; (c) the reaction of potassium dichromate with propanol; (d) the conversion of wine into vinegar (see Section 19.6); (e) the formation of 2-propanol from acetone.

19.79 Write the condensed structural formulas of the principal products of the reaction that occurs when (a) ethylene glycol, 1,2-ethanediol, is heated with stearic acid, $CH_3(CH_2)_{16}COOH$; (b) ethanol is heated with oxalic acid, HOOCCOOH; (c) 1-butanol is heated with propanoic acid.

19.80 Why do polymers not have definite molar masses? How does the fact that polymers have average molar masses affect their melting points?

19.81 Rank these polymers according to increasing value as fibers: polyesters, polyamides, polyalkenes. Explain your reasoning.

19.82 A certain polyester has the repeating unit $-(OCH_2-C_6H_4-CH_2OOC-C_6H_4-CO)_n-$. Identify the monomers of the polyester.

19.83 (a) Draw Lewis structures that show how the monomer known as styrene, $C_6H_5CH{=}CH_2$, may be polymerized to form polystyrene. (b) Repeat this exercise for propene, C_3H_6 (also known as propylene), to form polypropylene. (c) Compare the properties of polyethylene (plastic wrap), polystyrene (used to make transparent plastic boxes and as packing material), and polypropylene (used as the material to hold aluminum drink cans together in six-packs) and suggest an explanation for why the polymers differ in their physical properties.

19.84 (a) Describe the structure of one of the four monomers of the DNA molecule. (b) What functional group links the monomers of a single DNA strand together? (c) In the monomer structure you have drawn, identify the functional groups that are involved in holding the two strands of the double helix together.

19.85 (a) Explain the differences among the primary, secondary, tertiary, and quaternary structures of a protein. (b) Identify the forces holding each structure together as covalent bonds or primarily intermolecular forces.

19.86 Haloalkanes may react with hydroxide ions, undergoing nucleophilic displacement of the halide ion to form an alcohol. A complication of such reactions is competition from elimination reactions rather than substitution (see Section 18.6). (a) Predict the possible products from the reaction of 2-bromopentane with sodium hydroxide. (b) What can be done to favor the substitution reaction over the elimination pathway, or vice versa?

19.87 Radon, which is naturally given off by concrete, cinder block, and stone building materials, is absorbed by polyester. A visitor wearing polyester clothing set off a sensitive radiation-monitoring device while touring a nuclear power plant. The fabric had absorbed radon naturally emitted from the cinder blocks of the building. Why is polyester more likely to absorb radon than fabrics such as wool, silk, and nylon?

19.88 The pK_a values for phenol, *o*-nitrophenol, *m*-nitrophenol, and *p*-nitrophenol are 9.89, 7.17, 8.28, and 7.15, respectively. Explain the origin of these differences in pK_a.

19.89 Dopamine is a neurotransmitter in the central nervous system, and dopamine deficiency is related to Parkinson's disease. Two systematic names for dopamine are 4-(2-aminoethyl)-1,2-benzenediol and 3,4-dihydroxyphenylethylamine. Write the structural formula of dopamine.

19.90 The protein beef insulin consists of two chains in which the chain segments -(Leu)$_2$(Ile)(Cys)$_4$Arg- and -Pro(Phe)$_3$(Cys)$_2$Arg- occur. (a) What could be the nature of the interaction between the chains? (b) The chains can be separated by adding a reducing agent. Suggest a reason.

19.91 Explain the process of condensation polymerization. How might the polymer obtained from benzene-1,2-dicarboxylic acid and ethylene glycol differ from Dacron?

For the following exercises, see Major Technique 6, Nuclear Magnetic Resonance, which follows this chapter.

19.92 Predict the features of the ^{1}H NMR spectrum of propane.

19.93 Predict the features of the ^{1}H NMR spectrum of ethanal, CH_3CHO.

19.94 Predict the features of the ^{1}H NMR spectra of benzene, and the three isomeric dichlorobenzenes. Is it possible to distinguish the isomeric dichlorobenzenes on the basis of their NMR spectra alone?

19.95 The ^{1}H NMR spectrum of ethyl 3,3-dimethylbutanoate is shown below. Explain the features of this spectrum.

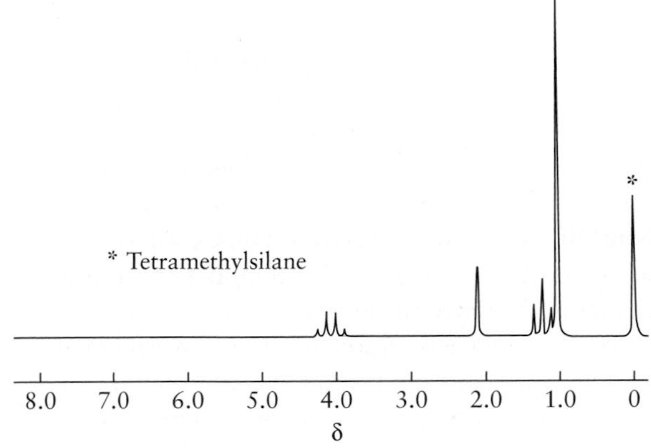

19.96 The ^{1}H NMR spectra of (a) 2-butanone and (b) ethyl acetate are shown below. The spectra are very similar yet have some important differences. Explain the observed similarities and differences.

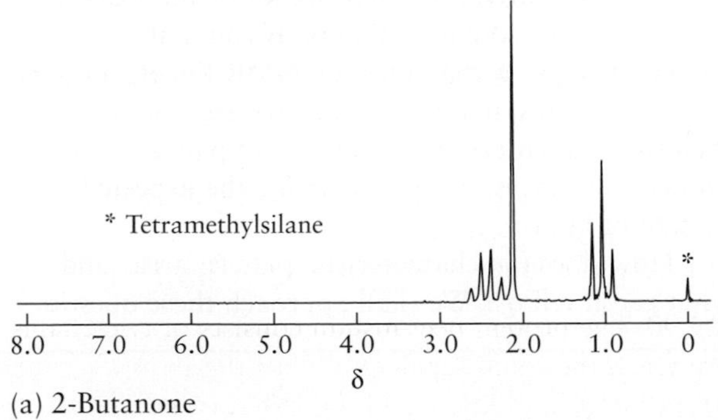

(a) 2-Butanone

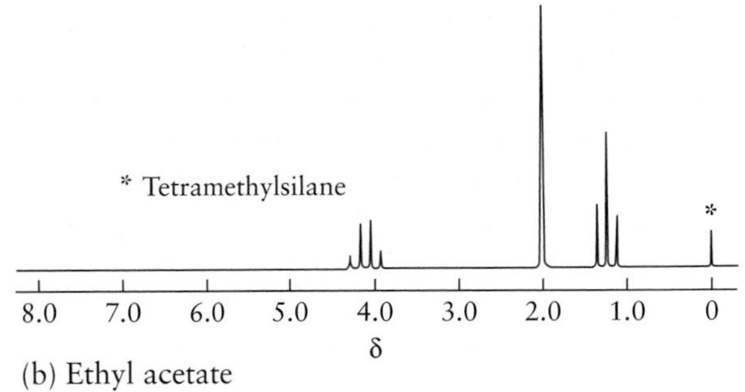

(b) Ethyl acetate

19.97 When propane is treated with chlorine gas, a mixture of compounds results. The mixture is separated and analysis of one of the components gives the ^{1}H NMR spectrum shown below. What is this product?

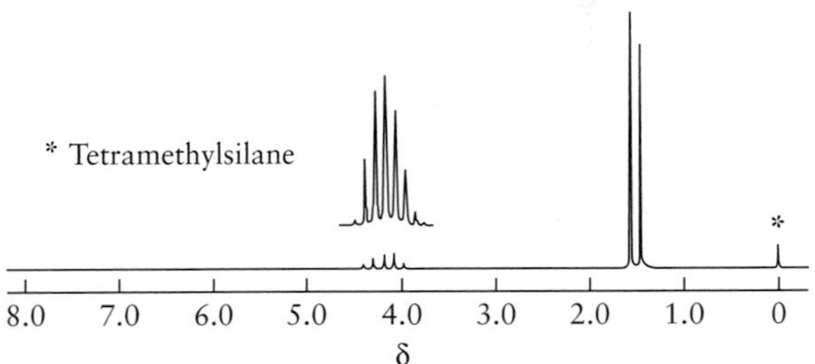

19.98 Other nuclei besides the proton have a nuclear spin of $\frac{1}{2}$ and, in principle at least, may be suitable for nuclear magnetic resonance spectroscopy. From standard literature sources, find at least five other naturally occurring, nonradioactive elements that have spin of $\frac{1}{2}$ and that can be examined by NMR spectroscopy.

19.99 One naturally occurring isotope of carbon is suitable for NMR spectroscopy. (a) From literature sources, determine which isotope of carbon has a spin of $\frac{1}{2}$. (b) What is the natural abundance of this isotope of carbon? (c) Given that only nuclei having a nuclear spin may interact to produce fine structure in an NMR spectrum, do you expect to see the splitting due to interactions between adjacent carbon atoms in an organic molecule such as propane in its carbon NMR spectrum? (d) Do you expect to see splitting of the proton signal in the ^{1}H NMR spectrum of propane due to the interactions between the carbon and proton nuclei? (e) Do you expect to see splitting of the carbon signal in the carbon NMR spectrum of propane due to the interactions between the carbon and proton nuclei?

Nuclear magnetic resonance (NMR) is the principal technique for the identification of organic compounds and is among the leading techniques for the determination of their structures. The technique has also been developed, as *magnetic resonance imaging* (MRI), as a diagnostic procedure in medicine.

Many atomic nuclei behave like small bar magnets, with energies that depend on their orientation in a magnetic field. An NMR spectrometer detects transitions between these energy levels. The nucleus most widely used for NMR is the proton, and we shall concentrate on it. Two other very common nuclei, those of carbon-12 and oxygen-16, are nonmagnetic, so they are invisible in NMR.

1 The Technique

Like an electron, a proton possesses the property of spin, which for elementary purposes can be thought of as an actual spinning motion. Because a proton has an electric charge and because a moving charge generates a magnetic field, a proton acts as a tiny bar magnet that can have either of two orientations, which are denoted ↑ (or α) and ↓ (or β). If an external magnetic field is applied, these two orientations correspond to different energies, and the separation between them is proportional to the strength of the applied field (Fig. 1).

If the sample is exposed to electromagnetic radiation, the nuclei are flipped from one orientation to another when the energy of the incoming photons matches the energy separation of the two spin orientations. (Recall from Section 1.1 that the photon's energy is equal to $h\nu$, where ν (nu) is the frequency of the radiation.) The matching of the radio frequency to the energy separation of the two spin orientations is called the *resonance condition*. Superconducting magnets are used to generate very high magnetic fields, and resonance requires radio-frequency radiation at about 400 MHz. When the resonance condition is met, the radiation is strongly absorbed and a sharp peak appears in the detector output.

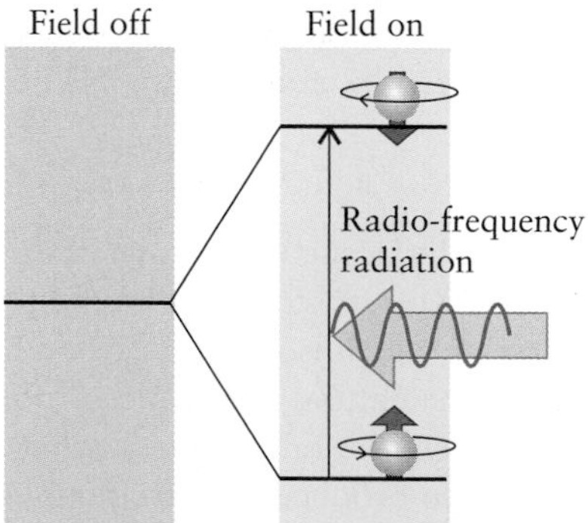

FIGURE 1 The two orientations of a nuclear spin have the same energy in the absence of a magnetic field. When a field is applied, the energy of the α spin falls and that of the β rises. When the separation between the two energy levels is equal to the energy of a radio-frequency photon, there is a strong absorption of radiation, giving a peak in the NMR spectrum.

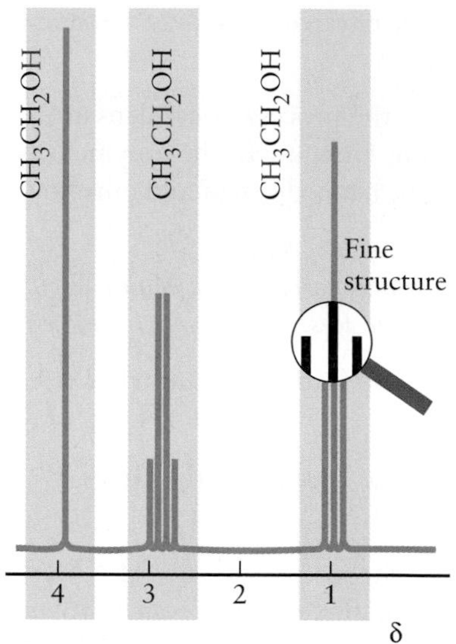

FIGURE 2 The NMR spectrum of ethanol. The red letters denote the protons that give rise to the associated peaks.

Nuclear magnetic resonance is a powerful analytical technique because each proton in a compound comes into resonance at a frequency related to its environment in the molecule. Figure 2, for instance, shows the NMR spectrum of ethanol; and we see that there are three groups of peaks and a characteristic pattern of splitting within the groups. The heights of the peaks tell us how many protons resonate at each frequency. If we see this pattern in an NMR spectrum, we know immediately that the sample contains ethanol. Because all compounds give a characteristic NMR fingerprint, or pattern of peaks, many can be recognized by comparing the observed pattern with a library of patterns from known substances, or by calculating the expected pattern of lines.

How does the characteristic pattern arise, and what can it tell us? We shall approach these questions in two steps. First, we look for the origin of the separation of the spectrum into groups of lines. Then we account for the splitting of the groups into individual narrow peaks.

2 The Chemical Shift

The separation of the absorption into groups of lines is due to the different environments within the molecule in which the protons are found. Thus, in ethanol, CH_3CH_2OH, three protons are present in the methyl group (CH_3), two are present in the methylene group (CH_2), and one is present in the hydroxyl group (OH). The applied magnetic field acts on the electrons in these three groups and causes them to circulate throughout the nuclear framework. These circulating charged particles give rise to an additional magnetic field, so the protons in each of the three groups experience a local magnetic field that is not the same as the applied magnetic field. Moreover, because the electronic structure in each of the groups is different, the protons in each group experience slightly different local magnetic fields. As a result, slightly different radio-frequency fields are needed to bring them into resonance, and three groups of peaks are observed in the NMR spectrum.

We say that each group of protons has a characteristic *chemical shift* from a standard sample. The measurement of the chemical shift helps to identify the type of group responsible for the absorption and indicates what groups are present in the molecule. The chemical shift of a group of lines is expressed in terms of the δ *scale* (delta scale), where δ is defined as

$$\delta = \frac{\nu - \nu^\circ}{\nu^\circ} \times 10^6$$

Here ν is the resonance frequency at the center of the group of lines and ν° is the resonance frequency of a standard. The latter is typically tetramethylsilane, $Si(CH_3)_4$, which has a lot of protons and dissolves in many samples without reaction. Each group has a characteristic chemical shift, although the precise value depends on the other groups attached to the group of interest. For instance, if we observe a resonance at $\delta = 1$, we can be reasonably confident that it arises from a methyl group in an alcohol.*

Another aid to identification is the relative intensities of the peaks, which are proportional to the numbers of protons they contain. The three peaks in the ethanol spectrum, for example, have overall intensities in the ratio 3:2:1, which is what we would expect for the three methyl, two methylene, and one hydroxyl protons.

3 The Fine Structure

Now we look at an individual group of lines and see what accounts for its *fine structure*, the splitting of the resonance into sharp peaks. Note that the methyl resonance at $\delta = 1$ consists of three peaks with intensities in the ratio 1:2:1.

The fine structure arises from the presence of other magnetic nuclei close to the protons undergoing resonance. The fine structure of the methyl group in ethanol, for instance, arises from the presence of the protons in the neighboring methylene group.

Let's imagine that we are a proton in the methyl group in ethanol and can sense the spins of the two neighboring methylene ($-CH_2-$) group protons. There are four possible orientations of these two spins: αα, αβ, βα, and ββ. Suppose our neighbors have αα spins. This arrangement gives rise to a magnetic field that we experience in addition to the applied magnetic field; so we (the methyl protons) come into resonance at the corresponding frequency. If we are in a molecule in which our neighbors are ββ, we experience a different local magnetic field, and we come into resonance at a different frequency. If our neighbors happen to be either αβ or βα, the magnetic field arising from the α spin cancels the field arising from the β spin, so the local field we experience is the same as it would be in the absence of any neighbors, and we resonate at the characteristic frequency. Because there are two arrangements that give this resonance (αβ and βα), the central line of the resonance will be twice as intense as the two outer lines (which arise when our neighbors are αα and ββ, respectively). Thus, we expect a 1:2:1 fine structure, just as observed.

We can predict the pattern of fine structure arising from *N* equivalent protons in a nearby group by referring to the appropriate line in *Pascal's triangle* (Fig. 3). If there are three equivalent protons in a nearby group (as there are for ethanol's methylene group resonance, because it has a methyl group as a

* You will often see chemical shifts expressed in parts per million, or ppm. So, the methyl resonance might be reported as lying at $\delta = 1$ ppm. However, the use of ppm is unnecessary.

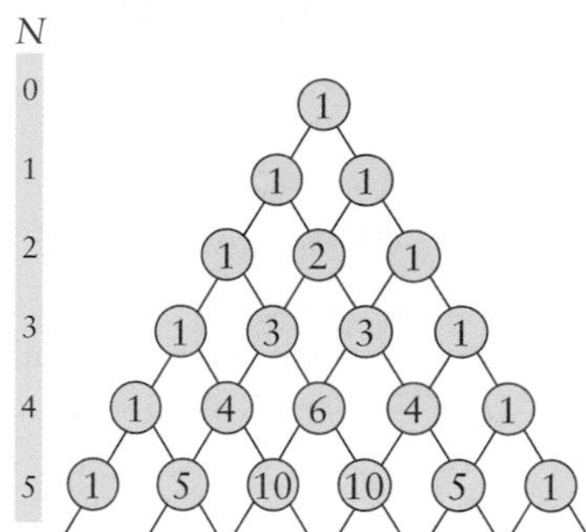

FIGURE 3 Part of Pascal's triangle. The rows give the intensity distributions for the coupling of a proton spin to N other equivalent protons, with $N = 0$ for the apex of the triangle.

neighbor), we expect four lines in the intensity ratio 1:3:3:1, just as observed.

So what about the hydroxyl resonance: why is that resonance a single line and not split by the other protons in the molecule? The answer is that this proton is very mobile: it can jump from one ethanol molecule to another ethanol molecule or to any water molecules present. As a result, the proton does not stay on one molecule long enough to show any characteristic splitting, or to give rise to splitting in other groups.

4 Magnetic Resonance Imaging

Magnetic resonance imaging (MRI) is a noninvasive structural technique for complex systems of molecules, such as people. In its simplest form, MRI portrays the concentration of protons in a sample. If the sample—which may be a living human body—is exposed to a uniform magnetic field in an NMR spectrometer and if we work at a resolution that does not show any chemical shifts or fine structure, then the protons give rise to a single resonance line. However, if the magnetic field varies linearly across the sample, the protons will resonate at different frequencies according to their location in the field. Moreover, the intensity of the resonance at a given field will be proportional to the number of protons at the spatial location corresponding to that particular field value. If the field gradient is rotated into different orientations, another portrayal of the concentration of protons through the sample is obtained. After several such measurements of the absorption intensity, the data can be analyzed on a computer, which constructs a two-dimensional image of any section through the sample. Figure 4 shows the distribution of protons—in large measure, the distribution of water in a brain—and the different regions can be identified.

A major advantage of MRI over x-rays is that the patient is exposed only to radio-frequency radiation, so the damage caused by x-rays is avoided. In addition, tumors invisible to x-rays can be detected. Another advantage is that, with MRI, a slice of the body can be viewed without interference from structures in the front or back of it. If a series of slices are obtained, they can even be assembled into a three-dimensional image that yields a much more accurate image than is possible with x-rays.

Related Exercises: 19.92–19.99.

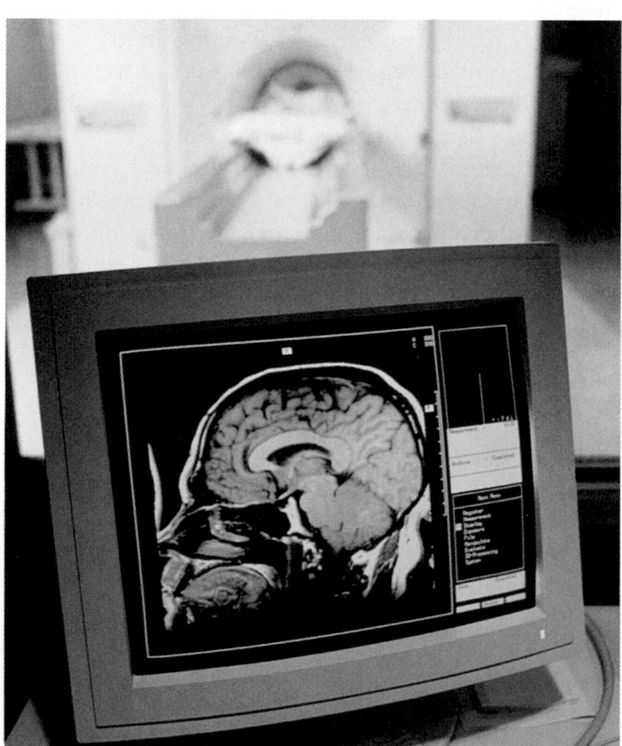

FIGURE 4 An MRI image of a human brain. The patient must lie within the strong magnetic field (background). The detectors can be rotated around the patient's head, thereby allowing many different views to be recorded.

APPENDIX 1 Symbols, Units, and Mathematical Techniques

1A Symbols

Each physical quantity is given a symbol that is either italic or Greek. Table 1 lists most of the symbols used in this text together with their units (see also Appendix 1B). The symbols may be modified by attaching subscripts, as set out in Table 2. Fundamental constants are not included in the lists, but can be found inside the back cover of the text.

TABLE 1 Common Symbols and Units

Symbol	Physical quantity	SI unit
α (alpha)	polarizability	$C^2 \cdot m^2 \cdot J^{-1}$
γ (gamma)	surface tension	$N \cdot m^{-1}$
δ (delta)	chemical shift	—
θ (theta)	colatitude	degree, rad
λ (lambda)	wavelength	m
μ (mu)	dipole moment	C·m
ν (nu)	frequency	Hz
π (pi)	osmotic pressure	Pa
σ (sigma)	cross section	m^2
ϕ (phi)	azimuth	degree, rad
χ (chi)	electronegativity	—
ψ (psi)	wavefunction	$m^{-3/2}$
a	activity	—
	van der Waals parameter	$L^2 \cdot atm \cdot mol^{-2}$
	unit cell parameter	m
A	area	m^2
	mass number	—
	Madelung constant	—
b	van der Waals parameter	$L \cdot mol^{-1}$
B	second virial coefficient	$L \cdot mol^{-1}$
C	heat capacity	$J \cdot K^{-1}$
	third virial coefficient	$L^2 \cdot mol^{-2}$
c	molar concentration, molarity	$mol \cdot L^{-1}$, M
c_2	second radiation constant	K·m
d	density	$kg \cdot m^{-3}$ ($g \cdot cm^{-3}$)
	length of unit cell diagonal	m
E	energy	J
	electrode potential	V
	electromotive force	V
E_a	activation energy	$J \cdot mol^{-1}$ ($kJ \cdot mol^{-1}$)
E_{bind}	nuclear binding energy	J
E_{ea}	electron affinity	$J \cdot mol^{-1}$ ($kJ \cdot mol^{-1}$)
E_K	kinetic energy	J
E_P	potential energy	J
e	elementary charge	C
F	force	N
G	Gibbs energy, free energy	J
H	enthalpy	J

(continued)

TABLE 1 Common Symbols and Units *(continued)*

Symbol	Physical quantity	SI unit*
h	height	m
I	ionization energy	$J \cdot mol^{-1}$ ($kJ \cdot mol^{-1}$)
	electric current	A ($C \cdot s^{-1}$)
i	i factor	—
[J]	molarity	$mol \cdot L^{-1}$, M
k	rate constant	(depends on order)
k	decay constant	s^{-1}
k_b	boiling-point constant	$K \cdot kg \cdot mol^{-1}$
k_f	freezing-point constant	$K \cdot kg \cdot mol^{-1}$
k_H	Henry's law constant	$mol \cdot L^{-1} \cdot atm^{-1}$
K	equilibrium constant	—
K_a	acidity constant	—
K_b	basicity constant	—
K_c	equilibrium constant	—
K_f	formation constant	—
K_m	Michaelis constant	—
K_p	equilibrium constant	—
K_{sp}	solubility product	—
K_w	water autoprotolysis constant	—
l, L	length	m
m	mass	kg
—	molality	$mol \cdot kg^{-1}$, m
M	molar mass	$kg \cdot mol^{-1}$ ($g \cdot mol^{-1}$)
N	number of entities	—
n	amount	mol
p	linear momentum	$kg \cdot m \cdot s^{-1}$
P	pressure	Pa
P_X	partial pressure	Pa
q	heat	J
	electric charge	C
Q	reaction quotient	—
	relative biological effectiveness	—
r	radius	m
R	radial wavefunction	$m^{-3/2}$
S	entropy	$J \cdot K^{-1}$
	molar solubility	$mol \cdot L^{-1}$
s	dimensionless molar solubility	—
t	time	s
$t_{1/2}$	half-life	s
T	absolute temperature	K
U	internal energy	J
v	velocity	$m \cdot s^{-1}$
V	volume	m^3, L
w	work	J
x	mole fraction	—
Y	angular wavefunction	—
Z	compression factor	—
	atomic number	—

*Units in parentheses are the commonly encountered forms.

TABLE 2 Subscripts for Symbols

Subscript	Meaning	Example (units)
a	acid	acidity constant, K_a
b	base	basicity constant, K_b
B	bond	bond enthalpy, ΔH_B ($kJ\cdot mol^{-1}$)
c	concentration	equilibrium constant, K_c
	combustion	enthalpy of combustion, ΔH_c ($kJ\cdot mol^{-1}$)
	critical	critical temperature, T_c (K)
e	nonexpansion (extra) work	electrical work, w_e (J)
f	formation	enthalpy of formation, ΔH_f ($kJ\cdot mol^{-1}$)
		formation constant, K_f
fus	fusion	enthalpy of fusion, ΔH_{fus} ($kJ\cdot mol^{-1}$)
In	indicator	indicator constant, K_{In}
J	substance	partial pressure, P_J (Pa)
K	kinetic	kinetic energy, E_K (J)
L	lattice	lattice enthalpy, ΔH_L ($kJ\cdot mol^{-1}$)
m	molar	molar volume, $V_m = V/n$ ($L\cdot mol^{-1}$)
mix	mixing	enthalpy of mixing, ΔH_{mix} ($kJ\cdot mol^{-1}$)
P	potential	potential energy, E_P (J)
P	constant pressure	heat capacity at constant pressure, C_P ($J\cdot K^{-1}$)
r	reaction	reaction enthalpy, ΔH_r ($kJ\cdot mol^{-1}$)
s	specific	specific heat capacity, $C_s = C/m$ ($J\cdot K^{-1}\cdot g^{-1}$)
sol	solution	enthalpy of solution, ΔH_{sol} ($kJ\cdot mol^{-1}$)
sp	solubility product	solubility product, K_{sp}
sub	sublimation	enthalpy of sublimation, ΔH_{sub} ($kJ\cdot mol^{-1}$)
surr	surroundings	entropy of surroundings, S_{surr} ($J\cdot K^{-1}$)
tot	total	total entropy, S_{tot} ($J\cdot K^{-1}$)
V	constant volume	heat capacity at constant volume, C_V ($J\cdot K^{-1}$)
vap	vaporization	enthalpy of vaporization, ΔH_{vap} ($kJ\cdot mol^{-1}$)
w	water	water autoprotolysis constant, K_w

1B Units and Unit Conversions

Each physical quantity is reported as a multiple of a defined unit:

$$\text{Physical quantity} = \text{numerical value} \times \text{unit}$$

For instance, a length may be expressed as a multiple of the unit of length, the meter, m; so we write $l = 2.0$ m if the length is the product of 2.0 (the multiple) times the unit. All units are denoted by Roman letters, such as m for meter and s for second, that distinguish them from the physical quantity to which they refer.

The Système International (SI), which is a development of the metric system, is an internationally accepted set of units and rules for their manipulation. It defines seven **base units** in terms of which all physical quantities can be expressed:

meter, m The meter, the unit of length, is the length of the path traveled by light during a time interval of 1/299 792 458 of a second.

kilogram, kg The kilogram, the unit of mass, is the mass of a standard cylinder maintained at a laboratory in France.

second, s The second, the unit of time, is 9 192 631 770 periods of a certain spectroscopic transition in a cesium-133 atom.

ampere, A The ampere, the unit of electric current, is defined in terms of the force exerted between two parallel wires carrying the current.

kelvin, K The kelvin, the unit of temperature, is 1/273.16 of the absolute temperature of the triple point of water.

mole, mol The mole, the unit of chemical amount, is the amount of substance that contains as many specified entities as there are atoms in exactly 12 g of carbon-12.

candela, cd The candela, the unit of luminous intensity, is defined in terms of a carefully specified source. We do not use the candela in this book

Any unit may be modified by one of the prefixes given in Table 3, which denote multiplication or division by a power of 10 of the unit. Thus, 1 mm = 10^{-3} m and 1 MK = 10^{6} K.

TABLE 3 Typical SI Prefixes

Prefix:	deka	kilo	mega	giga	tera				
Symbol:	da	k	M	G	T				
Factor:	10	10^{3}	10^{6}	10^{9}	10^{12}				
Prefix:	deci	centi	milli	micro	nano	pico	femto	atto	zepto
Symbol:	d	c	m	μ(mu)	n	p	f	a	z
Factor:	10^{-1}	10^{-2}	10^{-3}	10^{-6}	10^{-9}	10^{-12}	10^{-15}	10^{-18}	10^{-21}

Derived units are multiples of the base units and are used to express physical quantities that are built up from the basic physical observables. For example, the volume, the product of three lengths, is expressed in terms of a product of the base unit meter for each occurrence of the length; therefore, the derived unit of volume is $(\text{meter})^3$, denoted m^3. Similarly, density, the mass of a sample divided by its volume, is expressed in terms of the base unit for mass divided by the derived unit for volume, namely, $\text{kilogram}/(\text{meter})^3$, denoted kg/m^3 or, equivalently, $kg{\cdot}m^{-3}$. The SI convention is that a power, such as the 3 in cm^3, refers to the unit and its multiple. That is, cm^3 should be interpreted as $(cm)^3$. Many of the more common derived units have symbols of their own, such as joule, J, for the combination $kg{\cdot}m^2{\cdot}s^{-2}$. Table 4 lists some of these derived units. Note that the names of units derived from the names of people all begin with a lowercase letter but their symbols are uppercase.

TABLE 4 Derived Units with Special Names

Physical quantity	Name of unit	Symbol	Definition
absorbed dose	gray	Gy	$J{\cdot}kg^{-1}$
dose equivalent	sievert	Sv	$J{\cdot}kg^{-1}$
electric charge	coulomb	C	$A{\cdot}s$
electric potential	volt	V	$J{\cdot}C^{-1}$
energy	joule	J	$N{\cdot}m$, $kg{\cdot}m^2{\cdot}s^{-2}$
force	newton	N	$kg{\cdot}m{\cdot}s^{-2}$
frequency	hertz	Hz	s^{-1}
power	watt	W	$J{\cdot}s^{-1}$
pressure	pascal	Pa	$N{\cdot}m^{-2}$, $kg{\cdot}m^{-1}{\cdot}s^{-2}$
volume	liter	L	dm^3

A common requirement for solving a problem is to convert one set of units (for instance, common units such as calories for energy and inches for length) into SI units. Table 5 lists some common conversions; the values in bold are exact.

TABLE 5 Relations Between Units

Physical quantity	Common unit	Symbol	SI equivalent
mass	pound	lb	0.453 592 37 kg
	tonne	t	10^3 kg (1 Mg)
	ton (short, US)	ton	907.184 74 kg
	ton (long, UK)	ton	1016.046 kg
length	inch	in.	2.54 cm
	foot	ft	30.48 cm
volume	US quart	qt	0.946 352 5 L
	US gallon	gal	3.785 41 L
	Imperial quart	qt	1.136 522 5 L
	Imperial gallon	gal	4.546 09 L
time	minute	min	60 s
	hour	h	3600 s
energy	calorie (thermochemical)	cal	4.184 J
	electronvolt	eV	$1.602\ 177 \times 10^{-19}$ J
	kilowatt hour	kWh	3.6×10^6 J
	liter-atmosphere	L·atm	101.325 J
pressure	torr	Torr	133.322 Pa
	atmosphere	atm	101 325 Pa (= 760 Torr)
	bar	bar	10^5 Pa
	pounds/square inch	psi	6894.76 Pa
power	horsepower	hp	745.7 W
dipole moment	debye	D	$3.335\ 64 \times 10^{-30}$ C·m

To convert between units, we use a **conversion factor** of the form

$$\text{Conversion factor} = \frac{\text{units required}}{\text{units given}}$$

constructed from the relation

$$\text{Units given} = \text{units required}$$

taken from information like that in Table 5 and applied as follows:

$$\text{Information required} = \text{information given} \times \text{conversion factor}$$

When using a conversion factor, the units are treated just like algebraic quantities: they are multiplied or canceled in the normal way. Thus, the units in the denominator of the conversion factor cancel the units in the original data, leaving the units in the numerator of the conversion factor. The same procedure can be used to convert decimal multiples or fractions of units.

EXAMPLE 1 Converting units

Suppose we want to convert a volume of 1.7 qt into liters. First we identify the relation between the two units from Table 5:

$$1 \text{ qt} = 0.946\ 352\ 5 \text{ L}$$

The conversion factor from the units given (qt) to the units required (L) is

$$\text{Conversion factor} = \frac{0.946\ 352\ 5 \text{ L}}{1 \text{ qt}}$$

Therefore,

$$\text{Volume (L)} = (1.7 \text{ qt}) \times \frac{0.946\ 352\ 5 \text{ L}}{1 \text{ qt}} = 1.6 \text{ L}$$

We have rounded the answer to two significant figures, as explained in Appendix 1C.

Self-Test 1A Express the height of a person 6.00 ft tall in centimeters.

[*Answer:* 183 cm]

Self-Test 1B Express the mass in ounces $\left(1 \text{ oz} = \frac{1}{16} \text{ lb}\right)$ of a 250-g package of breakfast cereal.

It is often necessary to convert a unit that is raised to a power (including negative powers). In such cases, the conversion factor is raised to the same power. For example, to convert a density of 11 700 $\text{kg}\cdot\text{m}^{-3}$ (3 sf) into grams per centimeter cubed ($\text{g}\cdot\text{cm}^{-3}$), we use the two relations

$$1 \text{ kg} = 10^3 \text{ g} \qquad 1 \text{ cm} = 10^{-2} \text{ m}$$

as follows:

$$\begin{aligned}\text{Density } (\text{g}\cdot\text{cm}^{-3}) &= (11700 \text{ kg}\cdot\text{m}^{-3}) \times \frac{10^3 \text{ g}}{1 \text{ kg}} \times \left(\frac{1 \text{ cm}}{10^{-2} \text{ m}}\right)^{-3} \\ &= \left(11700 \frac{\text{kg}}{\text{m}^3}\right) \times \frac{10^3 \text{ g}}{1 \text{ kg}} \times \frac{10^{-6} \text{ m}^3}{1 \text{ cm}^3} \\ &= 11.7 \frac{\text{g}}{\text{cm}^3} = 11.7 \text{ g}\cdot\text{cm}^{-3}\end{aligned}$$

Self-Test A2A Express a density of 6.5 $\text{g}\cdot\text{mm}^{-3}$ in micrograms per nanometer cubed ($\mu\text{g}\cdot\text{nm}^{-3}$).

[*Answer:* 6.5×10^{-12} $\mu\text{g}\cdot\text{nm}^{-3}$]

Self-Test A2B Express an acceleration of 9.81 $\text{m}\cdot\text{s}^{-2}$ in kilometers per hour squared.

The conversion of temperatures is carried out slightly differently. Because the Fahrenheit degree (°F) is smaller by a factor of 5/9 than a Celsius degree (because there are 180°F between the freezing point and boiling point of water but only 100°C between the same two points), and because 0°C coincides with 32°F, we use

$$\text{Temperature (°F)} = \left\{\tfrac{9}{5} \times \text{Temperature (°C)}\right\} + 32$$

(The 32 is exact.) For example, to convert a temperature of 37°C (blood temperature) to degrees Fahrenheit, we write

$$\text{Temperature (°F)} = \left(\tfrac{9}{5} \times 37\right) + 32 = 99$$

and the temperature is reported as 99°F. A more sophisticated way of expressing the same relation is to write

$$\text{Temperature/°F} = \left(\tfrac{9}{5} \times \text{Temperature/°C}\right) + 32$$

In this expression, we treat temperature units like numbers and cancel them when it is appropriate. The same conversion then becomes

$$\begin{aligned}\text{Temperature/°F} &= \left\{\tfrac{9}{5} \times (37\text{°C})/\text{°C}\right\} + 32 \\ &= \left(\tfrac{9}{5} \times 37\right) + 32 = 99\end{aligned}$$

and multiplication through by °F gives

$$\text{Temperature} = 99\text{°F}$$

The corresponding expression for conversion between the Celsius and Kelvin scales is

$$\text{Temperature/°C} = \text{Temperature/K} + 273.15$$

(The 273.15 is exact.) Note that the size of the degree Celsius is the same as that of the kelvin.

1C Scientific Notation

In **scientific notation**, a number is written as $A \times 10^a$. Here A is a decimal number with one nonzero digit in front of the decimal point and a is a whole number. For example, 333 is written 3.33×10^2 in scientific notation, because $10^2 = 10 \times 10 = 100$:

$$333 = 3.33 \times 100 = 3.33 \times 10^2$$

We use

$$\begin{aligned}10^1 &= 10 \\ 10^2 &= 10 \times 10 = 100 \\ 10^3 &= 10 \times 10 \times 10 = 1000 \\ 10^4 &= 10 \times 10 \times 10 \times 10 = 10\,000\end{aligned}$$

and so on. Note that the number of zeros following 1 is equal to the power of 10.

Numbers between 0 and 1 are expressed in the same way, but with a negative power of 10; they have the form $A \times 10^{-a}$, with $10^{-1} = \frac{1}{10} = 0.1$, and so on. Thus, 0.0333 in decimal notation is 3.33×10^{-2} because

$$\begin{aligned}10^{-2} &= \tfrac{1}{10} \times \tfrac{1}{10} = \tfrac{1}{100} \\ 0.0333 &= 3.33 \times \tfrac{1}{100} = 3.33 \times 10^{-2}\end{aligned}$$

We use

$$\begin{aligned}10^{-2} &= 10^{-1} \times 10^{-1} = 0.01 \\ 10^{-3} &= 10^{-1} \times 10^{-1} \times 10^{-1} = 0.001 \\ 10^{-4} &= 10^{-1} \times 10^{-1} \times 10^{-1} \times 10^{-1} = 0.0001\end{aligned}$$

When a negative power of 10 is written out as a decimal number, the number of zeros following the decimal point is one less than the number (disregarding the sign) to which 10 is raised. Thus, 10^{-6} is written as a decimal point followed by $6 - 1 = 5$ zeros and then a 1:

$$10^{-6} = 10^{-1} \times 10^{-1} \times 10^{-1} \times 10^{-1} \times 10^{-1} \times 10^{-1}$$
$$= 0.000\ 001$$

The digits in a reported measurement are called the **significant figures.** There are two significant figures (written 2 sf) in 1.2 cm^3 and 3 sf in 1.78 g. To find the number of significant figures in a measurement, we first express the data in scientific notation, leaving one nonzero digit to the left of the decimal point. Then we count the total number of digits. For instance, 0.0025 kg is written as 2.5×10^{-3} kg, a value with 2 sf.

Some zeros are legitimately measured digits, but other zeros serve only to mark the place of the decimal point. Trailing zeros (the last ones after a decimal point), as in 22.0 mL, are significant, because they were measured. Thus, 22.0 mL has 3 sf. The "captive" zero in 80.1 kg is also a measured digit, so 80.1 kg has 3 sf. However, the leading digits in 0.0025 g are not significant; they are only placeholders used to indicate powers of 10, not measured numbers. We can see that they are only placeholders by reporting the mass as 2.5×10^{-3} g, which has 2 sf.

We distinguish between the results of measurements, which are always uncertain, and the results of *counting*, which are exact. For example, the report "12 eggs" means that there are *exactly* 12 eggs present, not a number somewhere between 11.5 and 12.5.

Some ambiguity arises with whole numbers ending in zero. Does a length reported as 400 m have 3 sf (4.00×10^2), 2 sf (4.0×10^2), or only 1 sf (4×10^2)? In such cases, the use of scientific notation removes any ambiguity. Throughout this text, we shall use the convention that all trailing zeros are significant (so 400 g has three significant figures), unless indicated otherwise by the context.

Different rounding off rules are needed for addition (and its reverse, subtraction) and multiplication (and its reverse, division). In both procedures, we round off the answers to the correct number of significant figures.

Rounding off In calculations, round *up* if the last digit is above 5 and round *down* if it is below 5. For numbers ending in 5, always round to the nearest even number. For example, 2.35 rounds to 2.4 and 2.65 rounds to 2.6. The correct procedure is to round off only at the final stage of the calculation and to carry all digits in the memory of the calculator until that stage.

Addition and subtraction When adding or subtracting, the number of decimal places in the result should be the same as the *smallest number of decimal places* in the data.

Multiplication and division When multiplying or dividing, the number of significant figures in the result should be the same as *the smallest number of significant figures* in the data.

Integers and exact numbers When multiplying or dividing by an integer or an exact number, the uncertainty of the result is determined by the measured value. Some unit conversion factors are defined exactly, even though they are not whole numbers. For example, 1 in. is defined as *exactly* 2.54 cm and the 273.15 in the conversion between Celsius and Kelvin temperatures is also exact, so 100.000°C converts to 373.150 K.

Logarithms and exponentials The mantissa of a common logarithm (see Appendix 1D) has the same number of significant figures as the original number. Thus, $\log 2.45 = 0.389$. A common antilogarithm of a number has the same number of significant figures as the mantissa of the original number (the digits following the decimal point). Thus, $10^{0.389} = 2.45$ and $10^{12.389} = 2.45 \times 10^{12}$. There is no simple rule for assessing the correct number of significant figures when natural logarithms are used: one way is to convert natural logarithms to common logarithms and then to use the rules specified above.

1D Exponents and Logarithms

To multiply numbers in scientific notation, we multiply the decimal parts of the numbers and add the powers of 10:

$$(A \times 10^a) \times (B \times 10^b) = (A \times B) \times 10^{a+b}$$

An example is

$$(1.23 \times 10^2) \times (4.56 \times 10^3) = 1.23 \times 4.56 \times 10^{2+3} = 5.61 \times 10^5$$

This rule holds even if the powers of 10 are negative:

$$(1.23 \times 10^{-2}) \times (4.56 \times 10^{-3}) = 1.23 \times 4.56 \times 10^{-2-3} = 5.61 \times 10^{-5}$$

The results of such calculations are then adjusted so that there is one digit in front of the decimal point:

$$(4.56 \times 10^{-3}) \times (7.65 \times 10^{6}) = 34.88 \times 10^{3} = 3.488 \times 10^{4}$$

When dividing two numbers in scientific notation, we divide the decimal parts of the numbers and subtract the powers of 10:

$$\frac{A \times 10^{a}}{B \times 10^{b}} = \frac{A}{B} \times 10^{a-b}$$

An example is

$$\frac{4.31 \times 10^{5}}{9.87 \times 10^{-8}} = \frac{4.31}{9.87} \times 10^{5-(-8)} = 0.437 \times 10^{13} = 4.37 \times 10^{12}$$

Before adding and subtracting numbers in scientific notation, we rewrite the numbers as decimal numbers multiplied by the same power of 10:

$$1.00 \times 10^{3} + 2.00 \times 10^{2} = 1.00 \times 10^{3} + 0.200 \times 10^{3} = 1.20 \times 10^{3}$$

When raising a number in scientific notation to a particular power, we raise the decimal part of the number to the power and multiply the power of 10 by the power:

$$(A \times 10^{a})^{b} = A^{b} \times 10^{a \times b}$$

For example, 2.88×10^{4} raised to the third power is

$$(2.88 \times 10^{4})^{3} = 2.88^{3} \times (10^{4})^{3} = 2.88^{3} \times 10^{3 \times 4} = 23.9 \times 10^{12} = 2.39 \times 10^{13}$$

The rule follows from the fact that

$$(10^{4})^{3} = 10^{4} \times 10^{4} \times 10^{4} = 10^{4+4+4} = 10^{3 \times 4}$$

The **common logarithm** of a number x, denoted $\log x$, is the power to which 10 must be raised to equal x. Thus, the logarithm of 100 is 2, written $\log 100 = 2$, because $10^{2} = 100$. The logarithm of 1.5×10^{2} is 2.18 because

$$10^{2.18} = 10^{0.18+2} = 10^{0.18} \times 10^{2} = 1.5 \times 10^{2}$$

The number in front of the decimal point in the logarithm (the 2 in $\log(1.5 \times 10^{2}) = 2.18$) is called the **characteristic** of the logarithm: it is the power of 10 in the original number (the power 2 in 1.5×10^{2}). The decimal fraction (the numbers following the decimal point; such as 0.18 in our example) is called the **mantissa.** It is the logarithm of the decimal number written with one nonzero digit in front of the decimal point (the 1.5 in the example).

The distinction between the characteristic and the mantissa is important when we have to decide how many significant figures to retain in a calculation that involves logarithms (as in the calculation of pH). Just as the power of 10 in a decimal number indicates only the location of the decimal point and plays no role in the determination of significant figures, so the characteristic of a logarithm is not included in the count of significant figures in a logarithm (see Appendix 1C). The number of significant figures in the mantissa is equal to the number of significant figures in the decimal number.

The **common antilogarithm** of a number x is the number that has x as its common logarithm. In practice, the common antilogarithm of x is simply another name for 10^{x}, so the common antilogarithm of 2 is $10^{2} = 100$ and that of 2.18 is

$$10^{2.18} = 10^{0.18+2} = 10^{0.18} \times 10^{2} = 1.5 \times 10^{2}$$

The logarithm of a number greater than 1 is positive, and the logarithm of a number smaller than 1 (but greater than 0) is negative:

If $x > 1$, $\log x > 0$

If $x = 1$, $\log x = 0$

If $x < 1$, $\log x < 0$

Logarithms are not defined either for 0 or for negative numbers.

The **natural logarithm** of a number x, denoted $\ln x$, is the power to which the number $e = 2.718\ldots$ must be raised to equal x. Thus, $\ln 10.0 = 2.303$, signifying that $e^{2.303} = 10.0$. The value of e may seem a peculiar choice, but it occurs naturally in a number of mathematical expressions, and its use simplifies many formulas. Common and natural logarithms are related by the expression

$$\ln x = \ln 10 \times \log x = 2.303\ldots \times \log x$$

The **natural antilogarithm** of x is normally called the **exponential** of e; it is the value of e raised to the power of x. Thus, the natural antilogarithm of 2.303 is $e^{2.303} = 10.0$.

The following relations between logarithms are useful. They are written here mainly for common logarithms, but they apply to natural logarithms as well.

Relation	Example
$\log 10^x = x$	$\log 10^{-7} = -7$
$\ln e^x = x$	$\ln e^{-kt} = -kt$
$\log x + \log y = \log xy$	$\log[Ag^+] + \log[Cl^-] = \log[Ag^+][Cl^-]$
$\log x - \log y = \log(x/y)$	$\log A_0 - \log A = \log(A_0/A)$
$x \log y = \log y^x$	$2 \log[H^+] = \log([H^+]^2)$
$\log(1/x) = -\log x$	$\log(1/[H^+]) = -\log[H^+]$

Logarithms are useful for solving expressions of the form

$$a^x = b$$

for the unknown x. (This type of calculation can arise in the study of chemical kinetics when the order of a reaction is being determined.) We take logarithms of both sides

$$\log a^x = \log b$$

and from a relation given in the table, we write the expression as

$$x \log a = \log b$$

Therefore,

$$x = \frac{\log b}{\log a}$$

1E Equations and Graphs

A **quadratic equation** is an equation of the form

$$ax^2 + bx + c = 0$$

The two **roots** of the equation (the solutions) are given by the expression

$$x = \frac{-b \pm \sqrt{b^2 - 4ac}}{2a}$$

The roots can also be determined graphically (using a graphing calculator, for instance) by noting where the graph of $y(x) = ax^2 + bx + c$ against x passes through $y = 0$ (Fig. 1). When a quadratic equation arises in connection with a chemical calculation, we accept only the root that leads to a physically plausible result. For example, if x is a concentration, then it must be a positive number, and a negative root can be ignored. However, if x is a *change* in concentration, then it may be either positive or negative. In such a case, we would have to determine which root led to an acceptable (positive) final concentration.

On occasion, an equilibrium table (or some other type of calculation) results in a cubic equation:

$$ax^3 + bx^2 + cx + d = 0$$

Cubic equations are often very tedious to solve exactly, so it is better to use mathematical software, a graphing calculator, or a plotter, such as the one on the Web site for this book, and to identify the locations where the graph of $y(x)$ against x passes through $y = 0$ (Fig. 2).

Experimental data can often be analyzed most effectively from a graph. In many cases, the best procedure is to find a way of plotting the data as a straight line. It is easier to tell whether or not the data in fact fall on the line, whereas small deviations from a curve are harder to detect. Moreover, it is also easy to calculate the slope of a straight line, to **extrapolate** (or extend) a straight line beyond the range of the data, and to **interpolate** between the data points (that is, find a value between two measured values).

The formula of a straight-line graph of y (the vertical axis) plotted against x (the horizontal axis) is

$$y = mx + b$$

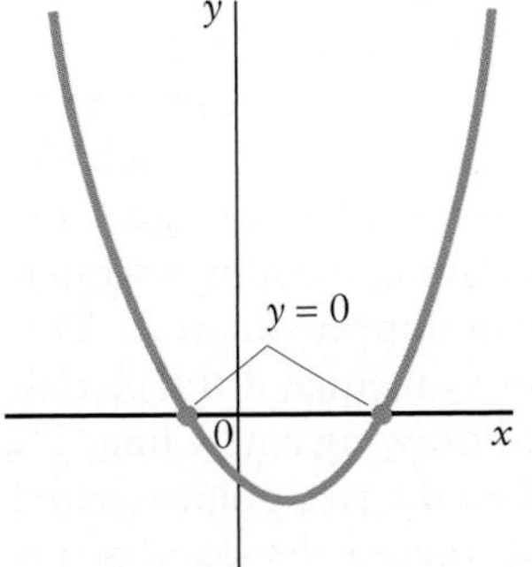

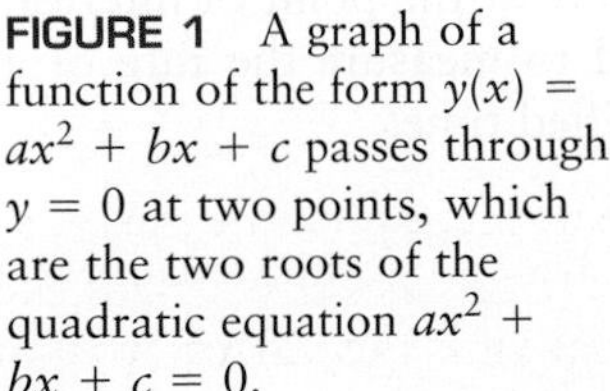

FIGURE 1 A graph of a function of the form $y(x) = ax^2 + bx + c$ passes through $y = 0$ at two points, which are the two roots of the quadratic equation $ax^2 + bx + c = 0$.

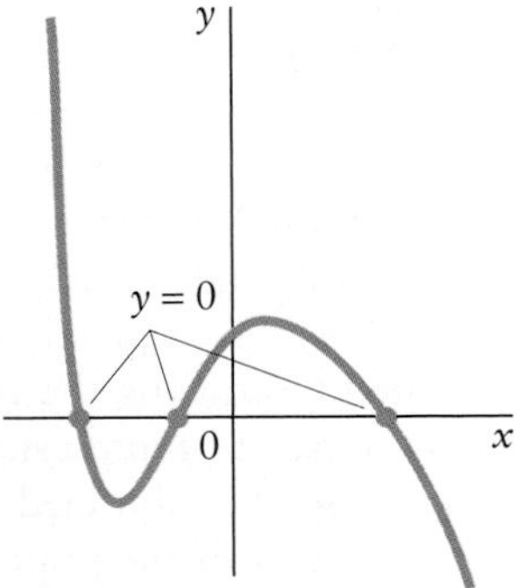

FIGURE 2 A graph of a function of the form $y(x) = ax^3 + bx^2 + cx + d$ passes through $y = 0$ at three points, which are the three roots of the cubic equation $ax^3 + bx^2 + cx + d = 0$.

Here b is the **intercept** of the graph with the y-axis (Fig. 3), the value of y where the graph cuts through the vertical axis at $x = 0$. The **slope** of the graph, its gradient, is m. The slope can be calculated by choosing two points, x_1 and x_2, and their corresponding values on the y-axis, y_1 and y_2, and substituting the values into the formula

$$m = \frac{y_2 - y_1}{x_2 - x_1}$$

Because b is the intercept and m is the slope, the equation of the straight line is equivalent to

$$y = (\text{slope} \times x) + \text{intercept}$$

Many of the equations we meet in the text can be rearranged to give a straight line graph when plotted as shown in the following table.

Application	y	= slope × x	+ intercept
temperature scale conversions	temp./°C	$= 1 \times T/\text{K}$	-273.15
	temp./°F	$= \frac{9}{5} \times$ temp./°C	$+32$
ideal gas law	P	$= nRT \times (1/V)$	
first-order integrated rate law	$\ln[\text{A}]$	$= -k \times t$	$+ \ln[\text{A}]_0$
second-order integrated rate law	$1/[\text{A}]$	$= k \times t$	$+ 1/[\text{A}]_0$
Arrhenius law	$\ln k$	$= (-E_a/R) \times (1/T)$	$+ A$

The slope of a straight line is the same at all points. On a curve, however, the slope changes from point to point. Its slope at a specified point is given by the slope of the line tangent to the curve at that point. The tangent can be found by a series of approximations, as shown in Fig. 4. We might start (approximation 1) by drawing a point on the curve on each side of the point of interest (corresponding to equal distances along the x-axis) and joining them by a straight line. A better approximation (approximation 2) is obtained by moving the points an equal distance closer to the point of interest and drawing a new line. The exact tangent is obtained when the two points virtually coincide with the point of interest. Its slope is then equal to the slope of the curve at the point of interest. This technique can be used to measure the rate of a chemical reaction at a specified time.

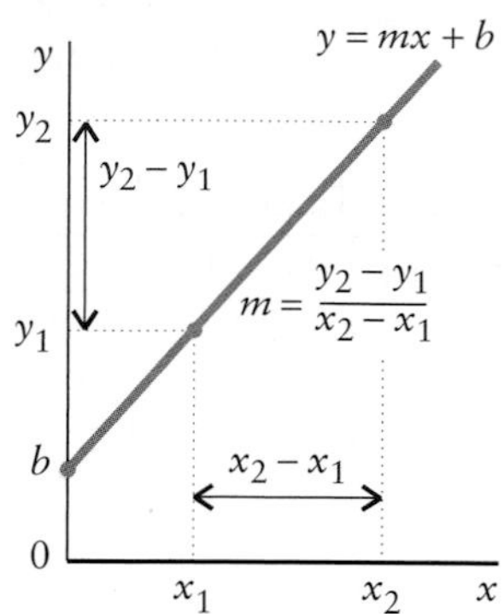

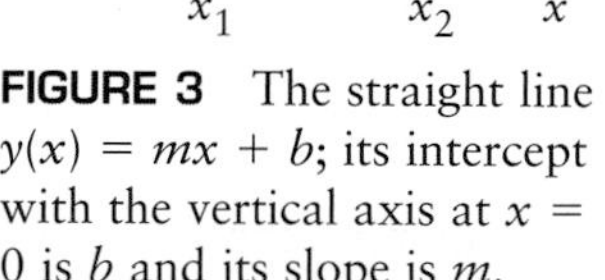
FIGURE 3 The straight line $y(x) = mx + b$; its intercept with the vertical axis at $x = 0$ is b and its slope is m.

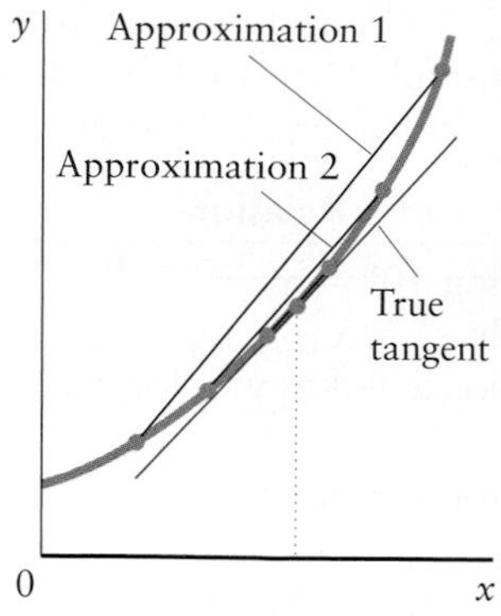

FIGURE 4 Successive approximations to the true tangent are obtained as the two points defining the straight line come close together and finally coincide.

1F Calculus

Differential calculus is the part of mathematics that deals with the slopes of curves and with infinitesimal quantities. Suppose we are studying a function $y(x)$. As explained in Appendix 1E, the slope of its graph at a point can be calculated by considering the straight line joining in two points x and $x + \delta x$, where δx is small. The slope of this line is

$$\text{Slope} = \frac{y(x + \delta x) - y(x)}{\delta x}$$

In differential calculus, the slope of the curve is found by letting the separation of the points become infinitesimally small. The **first derivative** of the function y with respect to x is then defined as

$$\frac{\mathrm{d}y}{\mathrm{d}x} = \lim_{\delta x \to 0} \frac{y(x + \delta x) - y(x)}{\delta x}$$

where "lim" means the limit of whatever follows. For example, if $y(x) = x^2$,

$$\begin{aligned}\frac{\mathrm{d}y}{\mathrm{d}x} &= \lim_{\delta x \to 0} \frac{(x + \delta x)^2 - x^2}{\delta x} \\ &= \lim_{\delta x \to 0} \frac{x^2 + 2x\delta x + (\delta x)^2 - x^2}{\delta x} \\ &= \lim_{\delta x \to 0} \frac{2x\delta x + (\delta x)^2}{\delta x} \\ &= \lim_{\delta x \to 0} (2x + \delta x) = 2x\end{aligned}$$

Therefore, the slope of the graph at any point x is $2x$. The same procedure can be applied to other functions. In practice, we do not keep returning to the basic definition in terms of limits but simply use tables of first derivatives. A selection of common functions and their first derivatives are given below.

Function, $y(x)$	Derivative, dy/dx
x^n	nx^{n-1}
$\ln x$	$1/x$
e^{ax}	ae^{ax}
$\sin ax$	$a \cos ax$
$\cos ax$	$-a \sin ax$

The **second derivative** of a function, denoted d^2y/dx^2, is defined like the first derivative, but applied to the function obtained by taking the first derivative. For example, the second derivative of the function x^2 is the derivative of the function $2x$, which is the constant 2. Likewise, the second derivative of $\sin ax$ is $-a^2 \sin ax$, as shown in the table of derivatives. The second derivative is an indication of the curvature of the function. Where d^2y/dx^2 is positive, the graph has a ∪ shape; where it is negative, the graph has a ∩ shape. The greater the magnitude of d^2y/dx^2, the sharper the curvature of the graph.

Integral calculus provides a way to determine the original function, given its first derivative. Thus, if we know that the first derivative is $2x$, then the integral calculus allows us to deduce that the function itself is $y = x^2$. Formally, we write

$$\int (2x)\, dx = x^2$$

It follows that the functions in the left-hand column of the table above are the integrals of the functions in the right-hand column. More formally, they are the **indefinite integrals** of the function, in contrast to the "definite" integrals described below. Table of indefinite integrals may be consulted for more complex examples, or mathematical software can be used to evaluate them.

An integral has a further important interpretation: the integral of a function evaluated between two points is the *area* beneath the graph of the function between the two points (Fig. 5). For example, the area beneath the curve $y(x) = \sin x$ between $x = 0$ and $x = \pi$ is

$$\text{Area} = \int_0^{\pi} \sin x\, dx = (-\cos x)\Big|_0^{\pi}$$
$$= (-\cos \pi) - (-\cos 0) = 1 + 1 = 2$$

An integral with limits attached, as in this example, is called a **definite integral**.

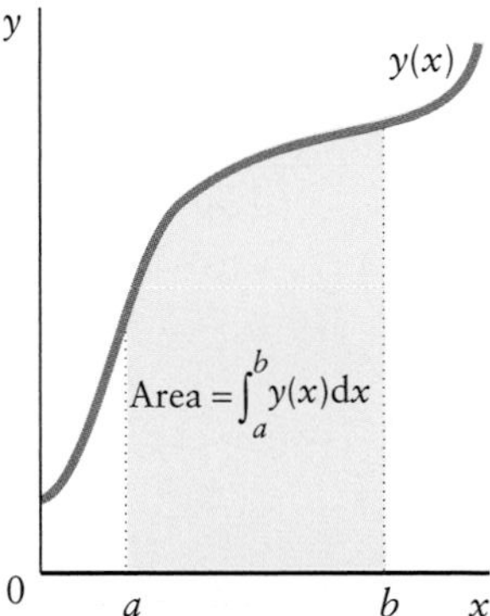

FIGURE 5 The definite integral of the function $y(x)$ between $x = a$ and $x = b$ is equal to the area bounded by the curve, the x-axis, and the two vertical lines through a and b.

APPENDIX 2 Experimental Data

2A Thermodynamic Data at 25°C

Inorganic Substances

Substance	Molar mass, M, g·mol^{-1}	Enthalpy of formation, $\Delta H_f°$, kJ·mol^{-1}	Free energy of formation, $\Delta G_f°$, kJ·mol^{-1}	Molar heat capacity, $C_{P,m}$, J·K^{-1}·mol^{-1}	Molar entropy,* $S_m°$, J·K^{-1}·mol^{-1}
Aluminum					
Al(s)	26.98	0	0	24.35	28.33
Al^{3+}(aq)	26.98	−524.7	−481.2	—	−321.7
Al_2O_3(s)	101.96	−1675.7	−1582.3	79.04	50.92
$Al(OH)_3$(s)	78.00	−1276	—	—	—
$AlCl_3$(s)	133.33	−704.2	−628.8	91.84	110.67
Antimony					
Sb(s)	121.76	0	0	25.23	45.69
SbH_3(g)	124.78	+145.11	+147.75	41.05	232.78
$SbCl_3$(g)	228.11	−313.8	−301.2	76.69	337.80
$SbCl_5$(g)	299.01	−394.34	−334.29	121.13	401.94
Arsenic					
As(s), gray	74.92	0	0	24.64	35.1
As_2S_3(s)	246.05	−169.0	−168.6	116.3	163.6
AsO_4^{3-}(aq)	138.92	−888.14	−648.41	—	−162.8
Barium					
Ba(s)	137.33	0	0	28.07	62.8
Ba^{2+}(aq)	137.33	−537.64	−560.77	—	+9.6
BaO(s)	153.33	−553.5	−525.1	47.78	70.42
$BaCO_3$(s)	197.34	−1216.3	−1137.6	85.35	112.1
$BaCO_3$(aq)	197.34	−1214.78	−1088.59	—	−47.3
Boron					
B(s)	10.81	0	0	11.09	5.86
B_2O_3(s)	69.62	−1272.8	−1193.7	62.93	53.97
BF_3(g)	67.81	−1137.0	−1120.3	50.46	254.12
Bromine					
Br_2(l)	159.80	0	0	75.69	152.23
Br_2(g)	159.80	+30.91	+3.11	36.02	245.46
Br(g)	79.90	+111.88	+82.40	20.79	175.02
Br^-(aq)	79.90	−121.55	−103.96	—	+82.4
HBr(g)	80.91	−36.40	−53.45	29.14	198.70
Calcium					
Ca(s)	40.08	0	0	25.31	41.42
Ca(g)	40.08	+178.2	+144.3	20.79	154.88
Ca^{2+}(aq)	40.08	−542.83	−553.58	—	−53.1
CaO(s)	56.08	−635.09	−604.03	42.80	39.75
$Ca(OH)_2$(s)	74.10	−986.09	−898.49	87.49	83.39
$Ca(OH)_2$(aq)	74.10	−1002.82	−868.07	—	−74.5

(continued)

Inorganic Substances *(continued)*

Substance	Molar mass, M, g·mol^{-1}	Enthalpy of formation, $\Delta H_f°$, kJ·mol^{-1}	Free energy of formation, $\Delta G_f°$, kJ·mol^{-1}	Molar heat capacity, $C_{P,m}$, J·K^{-1}·mol^{-1}	Molar entropy,* $S_m°$, J·K^{-1}·mol^{-1}
$CaCO_3$(s), calcite	100.09	−1206.9	−1128.8	81.88	92.9
$CaCO_3$(s), aragonite	100.09	−1207.1	−1127.8	81.25	88.7
$CaCO_3$(aq)	100.09	−1219.97	−1081.39	—	−110.0
CaF_2(s)	78.08	−1219.6	−1167.3	67.03	68.87
CaF_2(aq)	78.08	−1208.09	−1111.15	—	−80.8
$CaCl_2$(s)	110.98	−795.8	−748.1	72.59	104.6
$CaCl_2$(aq)	110.98	−877.1	−816.0	—	59.8
$CaBr_2$(s)	199.88	−682.8	−663.6	72.59	130
CaC_2(s)	64.10	−59.8	−64.9	62.72	69.96
$CaSO_4$(s)	136.15	−1434.11	−1321.79	99.66	106.7
$CaSO_4$(aq)	136.15	−1452.10	−1298.10	—	−33.1
Carbon (for organic compounds, see the next table)					
C(s), graphite	12.01	0	0	8.53	5.740
C(s), diamond	12.01	+1.895	+2.900	6.11	2.377
C(g)	12.01	+716.68	+671.26	20.84	158.10
CO(g)	28.01	−110.53	−137.17	29.14	197.67
CO_2(g)	44.01	−393.51	−394.36	37.11	213.74
CO_3^{2-}(aq)	60.01	−677.14	−527.81	—	−56.9
CCl_4(l)	153.81	−135.44	−65.21	131.75	216.40
CS_2(l)	76.15	+89.70	+65.27	75.7	151.34
HCN(g)	27.03	+135.1	+124.7	35.86	201.78
HCN(l)	27.03	+108.87	+124.97	70.63	112.84
HCN(aq)	27.03	+107.1	+119.7	—	124.7
Cerium					
Ce(s)	140.12	0	0	26.94	72.0
Ce^{3+}(aq)	140.12	−696.2	−672.0	—	−205
Ce^{4+}(aq)	140.12	−537.2	−503.8	—	−301
Chlorine					
Cl_2(g)	70.90	0	0	33.91	223.07
Cl(g)	35.45	+121.68	+105.68	21.84	165.20
Cl^-(aq)	35.45	−167.16	−131.23	—	+56.5
HCl(g)	36.46	−92.31	−95.30	29.12	186.91
HCl(aq)	36.46	−167.16	−131.23	—	56.5
Copper					
Cu(s)	63.55	0	0	24.44	33.15
Cu^+(aq)	63.55	+71.67	+49.98	—	+40.6
Cu^{2+}(aq)	63.55	+64.77	+65.49	—	−99.6
Cu_2O(s)	143.09	−168.6	−146.0	63.64	93.14
CuO(s)	79.55	−157.3	−129.7	42.30	42.63
$CuSO_4$(s)	159.61	−771.36	−661.8	100.0	109
$CuSO_4 \cdot 5H_2O$(s)	249.69	−2279.7	−1879.7	280	300.4
Deuterium					
D_2(g)	4.028	0	0	29.20	144.96
D_2O(g)	20.028	−249.20	−234.54	34.27	198.34
D_2O(l)	20.028	−294.60	−243.44	34.27	75.94

Substance	Molar mass, M, g·mol^{-1}	Enthalpy of formation, $\Delta H_f°$, kJ·mol^{-1}	Free energy of formation, $\Delta G_f°$, kJ·mol^{-1}	Molar heat capacity, $C_{P,m}$, J·K^{-1}·mol^{-1}	Molar entropy,* $S_m°$, J·K^{-1}·mol^{-1}
Fluorine					
$F_2(g)$	38.00	0	0	31.30	202.78
$F^-(aq)$	19.00	−332.63	−278.79	—	−13.8
HF(g)	20.01	−271.1	−273.2	29.13	173.78
HF(aq)	20.01	−330.08	−296.82	—	88.7
Hydrogen (see also Deuterium)					
$H_2(g)$	2.0158	0	0	28.82	130.68
H(g)	1.0079	+217.97	+203.25	20.78	114.71
$H^+(aq)$	1.0079	0	0	0	0
$H_2O(l)$	18.02	−285.83	−237.13	75.29	69.91
$H_2O(g)$	18.02	−241.82	−228.57	33.58	188.83
$H_2O_2(l)$	34.02	−187.78	−120.35	89.1	109.6
$H_2O_2(aq)$	34.02	−191.17	−134.03	—	143.9
$H_3O^+(aq)$	19.02	−285.83	−237.13	75.29	+69.91
Iodine					
$I_2(s)$	253.80	0	0	54.44	116.14
$I_2(g)$	253.80	+62.44	+19.33	36.90	260.69
$I^-(aq)$	126.90	−55.19	−51.57	—	+111.3
HI(g)	127.91	+26.48	+1.70	29.16	206.59
Iron					
Fe(s)	55.85	0	0	25.10	27.28
$Fe^{2+}(aq)$	55.85	−89.1	−78.90	—	−137.7
$Fe^{3+}(aq)$	55.85	−48.5	−4.7	—	−315.9
$Fe_3O_4(s)$, magnetite	231.55	−1118.4	−1015.4	143.43	146.4
$Fe_2O_3(s)$, hematite	159.70	−824.2	−742.2	103.85	87.40
FeS(s, α)	87.92	−100.0	−100.4	50.54	60.29
FeS(aq)	87.92	—	+6.9	—	—
$FeS_2(s)$	119.99	−178.2	−166.9	62.17	52.93
Lead					
Pb(s)	207.2	0	0	26.44	64.81
$Pb^{2+}(aq)$	207.2	−1.7	−24.43	—	+10.5
$PbO_2(s)$	239.2	−277.4	−217.33	64.64	68.6
$PbSO_4(s)$	303.3	−919.94	−813.14	103.21	148.57
$PbBr_2(s)$	367.0	−278.7	−261.92	80.12	161.5
$PbBr_2(aq)$	367.0	−244.8	−232.34	—	175.3
Magnesium					
Mg(s)	24.31	0	0	24.89	32.68
Mg(g)	24.31	+147.70	+113.10	20.79	148.65
$Mg^{2+}(aq)$	24.31	−466.85	−454.8	—	−138.1
MgO(s)	40.31	−601.70	−569.43	37.15	26.94
$MgCO_3(s)$	84.32	−1095.8	−1012.1	75.52	65.7
$MgBr_2(s)$	184.11	−524.3	−503.8	—	117.2

(continued)

Inorganic Substances *(continued)*

Substance	Molar mass, M, g·mol^{-1}	Enthalpy of formation, $\Delta H_f°$, kJ·mol^{-1}	Free energy of formation, $\Delta G_f°$, kJ·mol^{-1}	Molar heat capacity, $C_{P,m}$, J·K^{-1}·mol^{-1}	Molar entropy,* $S_m°$, J·K^{-1}·mol^{-1}
Mercury					
Hg(l)	200.59	0	0	27.98	76.02
Hg(g)	200.59	+61.32	+31.82	20.79	174.96
HgO(s)	216.59	−90.83	−58.54	44.06	70.29
Hg_2Cl_2(s)	472.08	−265.22	−210.75	102	192.5
Nitrogen					
N_2(g)	28.02	0	0	29.12	191.61
NO(g)	30.01	+90.25	+86.55	29.84	210.76
N_2O(g)	44.02	+82.05	+104.20	38.45	219.85
NO_2(g)	46.01	+33.18	+51.31	37.20	240.06
N_2O_4(g)	92.02	+9.16	+97.89	77.28	304.29
HNO_3(l)	63.02	−174.10	−80.71	109.87	155.60
HNO_3(aq)	63.02	−207.36	−111.25	—	146.4
NO_3^-(aq)	62.02	−205.0	−108.74	—	+146.4
NH_3(g)	17.03	−46.11	−16.45	35.06	192.45
NH_3(aq)	17.03	−80.29	−26.50	—	111.3
NH_4^+(aq)	18.04	−132.51	−79.31	—	+113.4
NH_2OH(s)	33.03	−114.2	—	—	—
HN_3(g)	43.04	+294.1	+328.1	98.87	238.97
N_2H_4(l)	32.05	+50.63	+149.34	139.3	121.21
NH_4NO_3(s)	80.05	−365.56	−183.87	84.1	151.08
NH_4Cl(s)	53.49	−314.43	−202.87	—	94.6
NH_4ClO_4(s)	117.49	−295.31	−88.75	—	186.2
Oxygen					
O_2(g)	32.00	0	0	29.36	205.14
O_3(g)	48.00	+142.7	+163.2	39.29	238.93
OH^-(aq)	17.01	−229.99	−157.24	—	−10.75
Phosphorus					
P(s), white	30.97	0	0	23.84	41.09
P_4(g)	123.88	+58.91	+24.44	67.15	279.98
PH_3(g)	33.99	+5.4	+13.4	37.11	210.23
P_4O_{10}(s)	283.88	−2984.0	−2697.0	—	228.86
H_3PO_3(aq)	81.99	−964.8	—	—	—
H_3PO_4(l)	97.99	−1266.9	—	—	—
H_3PO_4(aq)	97.99	−1288.34	−1142.54	—	158.2
PCl_3(l)	137.32	−319.7	−272.3	—	217.18
PCl_3(g)	137.32	−287.0	−267.8	71.84	311.78
PCl_5(g)	208.22	−374.9	−305.0	112.8	364.6
PCl_5(s)	208.22	−443.5	—	—	—
Potassium					
K(s)	39.10	0	0	29.58	64.18
K(g)	39.10	+89.24	+60.59	20.79	160.34
K^+(aq)	39.10	−252.38	−283.27	—	+102.5
KOH(s)	56.11	−424.76	−379.08	64.9	78.9

Substance	Molar mass, M, g·mol^{-1}	Enthalpy of formation, ΔH_f°, kJ·mol^{-1}	Free energy of formation, ΔG_f°, kJ·mol^{-1}	Molar heat capacity, $C_{P,m}$, J·K^{-1}·mol^{-1}	Molar entropy,* S_m°, J·K^{-1}·mol^{-1}
KOH(aq)	56.11	−482.37	−440.50	—	91.6
KF(s)	58.10	−567.27	−537.75	49.04	66.57
KCl(s)	74.55	−436.75	−409.14	51.30	82.59
KBr(s)	119.00	−393.80	−380.66	52.30	95.90
KI(s)	166.00	−327.90	−324.89	52.93	106.32
$KClO_3$(s)	122.55	−397.73	−296.25	100.25	143.1
$KClO_4$(s)	138.55	−432.75	−303.09	112.38	151.0
K_2S(s)	110.27	−380.7	−364.0	—	105
K_2S(aq)	110.27	−471.5	−480.7	—	190.4
Silicon					
Si(s)	28.09	0	0	20.00	18.83
SiO_2(s, α)	60.09	−910.94	−856.64	44.43	41.84
Silver					
Ag(s)	107.87	0	0	25.35	42.55
Ag^+(aq)	107.87	+105.58	+77.11	—	+72.68
Ag_2O(s)	231.74	−31.05	−11.20	65.86	121.3
AgBr(s)	187.77	−100.37	−96.90	52.38	107.1
AgBr(aq)	187.77	−15.98	−26.86	—	155.2
AgCl(s)	143.32	−127.07	−109.79	50.79	96.2
AgCl(aq)	143.32	−61.58	−54.12	—	129.3
AgI(s)	234.77	−61.84	−66.19	56.82	115.5
AgI(aq)	234.77	+50.38	+25.52	—	184.1
$AgNO_3$(s)	169.88	−124.39	−33.41	93.05	140.92
Sodium					
Na(s)	22.99	0	0	28.24	51.21
Na(g)	22.99	+107.32	+76.76	20.79	153.71
Na^+(aq)	22.99	−240.12	−261.91	—	+59.0
NaOH(s)	40.00	−425.61	−379.49	59.54	64.46
NaOH(aq)	40.00	−470.11	−419.15	—	48.1
NaCl(s)	58.44	−411.15	−384.14	50.50	72.13
NaBr(s)	102.89	−361.06	−348.98	51.38	86.82
NaI(s)	149.89	−287.78	−286.06	52.09	98.53
Sulfur					
S(s), rhombic	32.07	0	0	22.64	31.80
S(s), monoclinic	32.07	+0.33	+0.1	23.6	32.6
S^{2-}(aq)	32.07	+33.1	+85.8	—	−14.6
SO_2(g)	64.07	−296.83	−300.19	39.87	248.22
SO_3(g)	80.07	−395.72	−371.06	50.67	256.76
H_2SO_4(l)	98.09	−813.99	−690.00	138.9	156.90
SO_4^{2-}(aq)	96.07	−909.27	−744.53	—	+20.1
HSO_4^-(aq)	97.08	−887.34	−755.91	—	+131.8
H_2S(g)	34.09	−20.63	−33.56	34.23	205.79
H_2S(aq)	34.09	−39.7	−27.83	—	121
SF_6(g)	146.07	−1209	−1105.3	97.28	291.82

(continued)

Inorganic Substances *(continued)*

Substance	Molar mass, M, g·mol^{-1}	Enthalpy of formation, $\Delta H_f°$, kJ·mol^{-1}	Free energy of formation, $\Delta G_f°$, kJ·mol^{-1}	Molar heat capacity, $C_{P,m}$, J·K^{-1}·mol^{-1}	Molar entropy,* $S_m°$, J·K^{-1}·mol^{-1}
Tin					
Sn(s), white	118.71	0	0	26.99	51.55
Sn(s), gray	118.71	−2.09	+0.13	25.77	44.14
SnO(s)	134.71	−285.8	−256.9	44.31	56.5
SnO_2(s)	150.71	−580.7	−519.6	52.59	52.3
Zinc					
Zn(s)	65.39	0	0	25.40	41.63
Zn^{2+}(aq)	65.39	−153.89	−147.06	—	−112.1
ZnO(s)	81.39	−348.28	−318.30	40.25	43.64

*The entropies of individual ions in solution are determined by setting the entropy of H^+ in water equal to 0 and then defining the entropies of all other ions relative to this value; hence a negative entropy is one that is lower than the entropy of H^+ in water. All *absolute* entropies are positive, and no sign need be given; all entropies of ions are relative to that of H^+, and are listed here with a sign (either + or −).

Organic Compounds

Substance	Molar mass, M, g·mol^{-1}	Enthalpy of combustion, $\Delta H_c°$, kJ·mol^{-1}	Enthalpy of formation, $\Delta H_f°$, kJ·mol^{-1}	Free energy of formation, $\Delta G_f°$, kJ·mol^{-1}	Molar heat capacity, $C_{P,m}$, J·K^{-1}·mol^{-1}	Molar entropy, $S°$, J·K^{-1}·mol^{-1}
Hydrocarbons						
CH_4(g), methane	16.04	−890	−74.81	−50.72	35.31	186.26
C_2H_2(g), ethyne (acetylene)	26.04	−1300	+226.73	+209.20	43.93	200.94
C_2H_4(g), ethene (ethylene)	28.05	−1411	+52.26	+68.15	43.56	219.56
C_2H_6(g), ethane	30.07	−1560	−84.68	−32.82	52.63	229.60
C_3H_6(g), propene (propylene)	42.08	−2058	+20.42	+62.78	63.89	266.6
C_3H_6(g), cyclopropane	42.08	−2091	+53.30	+104.45	55.94	237.4
C_3H_8(g), propane	44.09	−2220	−103.85	−23.49	73.5	270.2
C_4H_{10}(g), butane	58.12	−2878	−126.15	−17.03	97.45	310.1
C_5H_{12}(g), pentane	72.14	−3537	−146.44	−8.20	120.2	349
C_6H_6(l), benzene	78.11	−3268	+49.0	+124.3	136.1	173.3
C_6H_6(g)	78.11	−3302	+82.9	+129.72	81.67	269.31

Organic Compounds *(continued)*

Substance	Molar mass, M, $g \cdot mol^{-1}$	Enthalpy of combustion, $\Delta H_c°$, $kJ \cdot mol^{-1}$	Enthalpy of formation, $\Delta H_f°$, $kJ \cdot mol^{-1}$	Free energy of formation, $\Delta G_f°$, $kJ \cdot mol^{-1}$	Molar heat capacity, $C_{P,m}$, $J \cdot K^{-1} \cdot mol^{-1}$	Molar entropy, $S°$, $J \cdot K^{-1} \cdot mol^{-1}$
$C_7H_8(l)$, toluene	92.13	−3910	+12.0	+113.8	—	221.0
$C_7H_8(g)$	92.13	−3953	+50.0	+122.0	103.6	320.7
$C_6H_{12}(l)$, cyclohexane	84.15	−3920	−156.4	+26.7	156.5	204.4
$C_6H_{12}(g)$	84.15	−3953	—	—	—	—
$C_8H_{18}(l)$, octane	114.22	−5471	−249.9	+6.4	—	358
Alcohols and phenols						
$CH_3OH(l)$, methanol	32.04	−726	−238.86	−166.27	81.6	126.8
$CH_3OH(g)$	32.04	−764	−200.66	−161.96	43.89	239.81
$C_2H_5OH(l)$, ethanol	46.07	−1368	−277.69	−174.78	111.46	160.7
$C_2H_5OH(g)$	46.07	−1409	−235.10	−168.49	65.44	282.70
$C_6H_5OH(s)$, phenol	94.11	−3054	−164.6	−50.42	—	144.0
Carboxylic acids						
$HCOOH(l)$, formic acid	46.02	−255	−424.72	−361.35	99.04	128.95
$CH_3COOH(l)$, acetic acid	60.05	−875	−484.5	−389.9	124.3	159.8
$CH_3COOH(aq)$	60.05	—	−485.76	−396.46	—	86.6
$(COOH)_2(s)$, oxalic acid	90.04	−254	−827.2	−697.9	117	120
$C_6H_5COOH(s)$, benzoic acid	122.12	−3227	−385.1	−245.3	146.8	167.6
Aldehydes and ketones						
$HCHO(g)$, methanal (formaldehyde)	30.03	−571	−108.57	−102.53	35.40	218.77
$CH_3CHO(l)$, ethanal (acetaldehyde)	44.05	−1166	−192.30	−128.12	—	160.2
$CH_3CHO(g)$	44.05	−1192	−166.19	−128.86	57.3	250.3
$CH_3COCH_3(l)$, propanone (acetone)	58.08	−1790	−248.1	−155.4	124.7	200
Sugars						
$C_6H_{12}O_6(s)$, glucose	180.15	−2808	−1268	−910	—	212
$C_6H_{12}O_6(aq)$	180.15	—	—	−917	—	—
$C_6H_{12}O_6(s)$, fructose	180.15	−2810	−1266	—	—	—
$C_{12}H_{22}O_{11}(s)$, sucrose	342.29	−5645	−2222	−1545	—	360
Nitrogen compounds						
$CO(NH_2)_2(s)$, urea	60.06	−632	−333.51	−197.33	93.14	104.60
$C_6H_5NH_2(l)$, aniline	93.13	−3393	+31.6	+149.1	—	191.3
$NH_2CH_2COOH(s)$, glycine	75.07	−969	−532.9	−373.4	99.2	103.51
$CH_3NH_2(g)$, methylamine	31.06	−1085	−22.97	+32.16	53.1	243.41

2B Standard Potentials at 25°C

Potentials in Electrochemical Order

Reduction half-reaction	$E°$, V
Strongly oxidizing	
$H_4XeO_6 + 2\,H^+ + 2\,e^- \longrightarrow XeO_3 + 3\,H_2O$	+3.0
$F_2 + 2\,e^- \longrightarrow 2\,F^-$	+2.87
$O_3 + 2\,H^+ + 2\,e^- \longrightarrow O_2 + H_2O$	+2.07
$S_2O_8^{2-} + 2\,e^- \longrightarrow 2\,SO_4^{2-}$	+2.05
$Ag^{2+} + e^- \longrightarrow Ag^+$	+1.98
$Co^{3+} + e^- \longrightarrow Co^{2+}$	+1.81
$H_2O_2 + 2\,H^+ + 2\,e^- \longrightarrow 2\,H_2O$	+1.78
$Au^+ + e^- \longrightarrow Au$	+1.69
$Pb^{4+} + 2\,e^- \longrightarrow Pb^{2+}$	+1.67
$2\,HClO + 2\,H^+ + 2\,e^- \longrightarrow Cl_2 + 2\,H_2O$	+1.63
$Ce^{4+} + e^- \longrightarrow Ce^{3+}$	+1.61
$2\,HBrO + 2\,H^+ + 2\,e^- \longrightarrow Br_2 + 2\,H_2O$	+1.60
$MnO_4^- + 8\,H^+ + 5\,e^- \longrightarrow Mn^{2+} + 4\,H_2O$	+1.51
$Mn^{3+} + e^- \longrightarrow Mn^{2+}$	+1.51
$Au^{3+} + 3\,e^- \longrightarrow Au$	+1.40
$Cl_2 + 2\,e^- \longrightarrow 2\,Cl^-$	+1.36
$Cr_2O_7^{2-} + 14\,H^+ + 6\,e^- \longrightarrow 2\,Cr^{3+} + 7\,H_2O$	+1.33
$O_3 + H_2O + 2\,e^- \longrightarrow O_2 + 2\,OH^-$	+1.24
$O_2 + 4\,H^+ + 4\,e^- \longrightarrow 2\,H_2O$	+1.23
$MnO_2 + 4\,H^+ + 2\,e^- \longrightarrow Mn^{2+} + 2\,H_2O$	+1.23
$ClO_4^- + 2\,H^+ + 2\,e^- \longrightarrow ClO_3^- + H_2O$	+1.23
$Pt^{2+} + 2\,e^- \longrightarrow Pt$	+1.20
$Br_2 + 2\,e^- \longrightarrow 2\,Br^-$	+1.09
$Pu^{4+} + e^- \longrightarrow Pu^{3+}$	+0.97
$NO_3^- + 4\,H^+ + 3\,e^- \longrightarrow NO + 2\,H_2O$	+0.96
$2\,Hg^{2+} + 2\,e^- \longrightarrow Hg_2^{2+}$	+0.92
$ClO^- + H_2O + 2\,e^- \longrightarrow Cl^- + 2\,OH^-$	+0.89
$Hg^{2+} + 2\,e^- \longrightarrow Hg$	+0.85
$NO_3^- + 2\,H^+ + e^- \longrightarrow NO_2 + H_2O$	+0.80
$Ag^+ + e^- \longrightarrow Ag$	+0.80
$Hg_2^{2+} + 2\,e^- \longrightarrow 2\,Hg$	+0.79
$AgF + e^- \longrightarrow Ag + F^-$	+0.78
$Fe^{3+} + e^- \longrightarrow Fe^{2+}$	+0.77
$BrO^- + H_2O + 2\,e^- \longrightarrow Br^- + 2\,OH^-$	+0.76
$MnO_4^{2-} + 2\,H_2O + 2\,e^- \longrightarrow MnO_2 + 4\,OH^-$	+0.60
$MnO_4^- + e^- \longrightarrow MnO_4^{2-}$	+0.56
$I_2 + 2\,e^- \longrightarrow 2\,I^-$	+0.54
$I_3^- + 2\,e^- \longrightarrow 3\,I^-$	+0.53
$Cu^+ + e^- \longrightarrow Cu$	+0.52
$Ni(OH)_3 + e^- \longrightarrow Ni(OH)_2 + OH^-$	+0.49
$O_2 + 2\,H_2O + 4\,e^- \longrightarrow 4\,OH^-$	+0.40
$ClO_4^- + H_2O + 2\,e^- \longrightarrow ClO_3^- + 2\,OH^-$	+0.36
$Cu^{2+} + 2\,e^- \longrightarrow Cu$	+0.34
$Hg_2Cl_2 + 2\,e^- \longrightarrow 2\,Hg + 2\,Cl^-$	+0.27
$AgCl + e^- \longrightarrow Ag + Cl^-$	+0.22
$Bi^{3+} + 3\,e^- \longrightarrow Bi$	+0.20
$SO_4^{2-} + 4\,H^+ + 2\,e^- \longrightarrow H_2SO_3 + H_2O$	+0.17
$Cu^{2+} + e^- \longrightarrow Cu^+$	+0.15
$Sn^{4+} + 2e^- \longrightarrow Sn^{2+}$	+0.15
$AgBr + e^- \longrightarrow Ag + Br^-$	+0.07
$NO_3^- + H_2O + 2\,e^- \longrightarrow NO_2^- + 2\,OH^-$	+0.01
$Ti^{4+} + e^- \longrightarrow Ti^{3+}$	0.00
$2\,H^+ + 2\,e^- \longrightarrow H_2$	0, by definition
$Fe^{3+} + 3\,e^- \longrightarrow Fe$	−0.04
$O_2 + H_2O + 2\,e^- \longrightarrow HO_2^- + OH^-$	−0.08
$Pb^{2+} + 2\,e^- \longrightarrow Pb$	−0.13
$In^+ + e^- \longrightarrow In$	−0.14
$Sn^{2+} + 2\,e^- \longrightarrow Sn$	−0.14
$AgI + e^- \longrightarrow Ag + I^-$	−0.15
$Ni^{2+} + 2\,e^- \longrightarrow Ni$	−0.23
$V^{3+} + e^- \longrightarrow V^{2+}$	−0.26
$Co^{2+} + 2\,e^- \longrightarrow Co$	−0.28
$In^{3+} + 3\,e^- \longrightarrow In$	−0.34
$Tl^+ + e^- \longrightarrow Tl$	−0.34
$PbSO_4 + 2\,e^- \longrightarrow Pb + SO_4^{2-}$	−0.36
$Ti^{3+} + e^- \longrightarrow Ti^{2+}$	−0.37
$In^{2+} + e^- \longrightarrow In^+$	−0.40
$Cd^{2+} + 2\,e^- \longrightarrow Cd$	−0.40
$Cr^{3+} + e^- \longrightarrow Cr^{2+}$	−0.41
$Fe^{2+} + 2\,e^- \longrightarrow Fe$	−0.44
$In^{3+} + 2\,e^- \longrightarrow In^+$	−0.44
$S + 2\,e^- \longrightarrow S^{2-}$	−0.48
$In^{3+} + e^- \longrightarrow In^{2+}$	−0.49
$Ga^+ + e^- \longrightarrow Ga$	−0.53
$O_2 + e^- \longrightarrow O_2^-$	−0.56
$U^{4+} + e^- \longrightarrow U^{3+}$	−0.61
$Se + 2\,e^- \longrightarrow Se^{2-}$	−0.67
$Cr^{3+} + 3\,e^- \longrightarrow Cr$	−0.74
$Zn^{2+} + 2\,e^- \longrightarrow Zn$	−0.76
$Cd(OH)_2 + 2\,e^- \longrightarrow Cd + 2\,OH^-$	−0.81
$2\,H_2O + 2\,e^- \longrightarrow H_2 + 2\,OH^-$	−0.83
$Te + 2\,e^- \longrightarrow Te^{2-}$	−0.84
$Cr^{2+} + 2\,e^- \longrightarrow Cr$	−0.91
$Mn^{2+} + 2\,e^- \longrightarrow Mn$	−1.18
$V^{2+} + 2\,e^- \longrightarrow V$	−1.19
$Ti^{2+} + 2\,e^- \longrightarrow Ti$	−1.63
$Al^{3+} + 3\,e^- \longrightarrow Al$	−1.66
$U^{3+} + 3\,e^- \longrightarrow U$	−1.79
$Be^{2+} + 2\,e^- \longrightarrow Be$	−1.85
$Mg^{2+} + 2\,e^- \longrightarrow Mg$	−2.36
$Ce^{3+} + 3\,e^- \longrightarrow Ce$	−2.48
$La^{3+} + 3\,e^- \longrightarrow La$	−2.52
$Na^+ + e^- \longrightarrow Na$	−2.71
$Ca^{2+} + 2\,e^- \longrightarrow Ca$	−2.87
$Sr^{2+} + 2\,e^- \longrightarrow Sr$	−2.89
$Ba^{2+} + 2\,e^- \longrightarrow Ba$	−2.91
$Ra^{2+} + 2\,e^- \longrightarrow Ra$	−2.92
$Cs^+ + e^- \longrightarrow Cs$	−2.92
$Rb^+ + e^- \longrightarrow Rb$	−2.93
$K^+ + e^- \longrightarrow K$	−2.93
$Li^+ + e^- \longrightarrow Li$	−3.05
Strongly reducing	

Potentials in Alphabetical Order

Reduction half-reaction	$E°$, V
$Ag^+ + e^- \longrightarrow Ag$	+0.80
$Ag^{2+} + e^- \longrightarrow Ag^+$	+1.98
$AgBr + e^- \longrightarrow Ag + Br^-$	+0.07
$AgCl + e^- \longrightarrow Ag + Cl^-$	+0.22
$AgF + e^- \longrightarrow Ag + F^-$	+0.78
$AgI + e^- \longrightarrow Ag + I^-$	−0.15
$Al^{3+} + 3\ e^- \longrightarrow Al$	−1.66
$Au^+ + e^- \longrightarrow Au$	+1.69
$Au^{3+} + 3\ e^- \longrightarrow Au$	+1.40
$Ba^{2+} + 2\ e^- \longrightarrow Ba$	−2.91
$Be^{2+} + 2\ e^- \longrightarrow Be$	−1.85
$Bi^{3+} + 3\ e^- \longrightarrow Bi$	+0.20
$Br_2 + 2\ e^- \longrightarrow 2\ Br^-$	+1.09
$BrO^- + H_2O + 2\ e^- \longrightarrow Br^- + 2\ OH^-$	+0.76
$Ca^{2+} + 2\ e^- \longrightarrow Ca$	−2.87
$Cd^{2+} + 2\ e^- \longrightarrow Cd$	−0.40
$Cd(OH)_2 + 2\ e^- \longrightarrow Cd + 2\ OH^-$	−0.81
$Ce^{3+} + 3\ e^- \longrightarrow Ce$	−2.48
$Ce^{4+} + e^- \longrightarrow Ce^{3+}$	+1.61
$Cl_2 + 2\ e^- \longrightarrow 2\ Cl^-$	+1.36
$ClO^- + H_2O + 2\ e^- \longrightarrow Cl^- + 2\ OH^-$	+0.89
$ClO_4^- + 2\ H^+ + 2\ e^- \longrightarrow ClO_3^- + H_2O$	+1.23
$ClO_4^- + H_2O + 2\ e^- \longrightarrow ClO_3^- + 2\ OH^-$	+0.36
$Co^{2+} + 2\ e^- \longrightarrow Co$	−0.28
$Co^{3+} + e^- \longrightarrow Co^{2+}$	+1.81
$Cr^{2+} + 2\ e^- \longrightarrow Cr$	−0.91
$Cr_2O_7^{2-} + 14\ H^+ + 6\ e^- \longrightarrow 2\ Cr^{3+} + 7\ H_2O$	+1.33
$Cr^{3+} + 3\ e^- \longrightarrow Cr$	−0.74
$Cr^{3+} + e^- \longrightarrow Cr^{2+}$	−0.41
$Cs^+ + e^- \longrightarrow Cs$	−2.92
$Cu^+ + e^- \longrightarrow Cu$	+0.52
$Cu^{2+} + 2\ e^- \longrightarrow Cu$	+0.34
$Cu^{2+} + e^- \longrightarrow Cu^+$	+0.15
$F_2 + 2\ e^- \longrightarrow 2\ F^-$	+2.87
$Fe^{2+} + 2\ e^- \longrightarrow Fe$	−0.44
$Fe^{3+} + 3\ e^- \longrightarrow Fe$	−0.04
$Fe^{3+} + e^- \longrightarrow Fe^{2+}$	+0.77
$Ga^+ + e^- \longrightarrow Ga$	−0.53
$2\ H^+ + 2\ e^- \longrightarrow H_2$	0, by definition
$2\ HBrO + 2\ H^+ + 2\ e^- \longrightarrow Br_2 + 2\ H_2O$	+1.60
$2\ HClO + 2\ H^+ + 2\ e^- \longrightarrow Cl_2 + 2\ H_2O$	+1.63
$2\ H_2O + 2\ e^- \longrightarrow H_2 + 2\ OH^-$	−0.83
$H_2O_2 + 2\ H^+ + 2\ e^- \longrightarrow 2\ H_2O$	+1.78
$H_4XeO_6 + 2\ H^+ + 2\ e^- \longrightarrow XeO_3 + 3\ H_2O$	+3.0
$Hg_2^{2+} + 2\ e^- \longrightarrow 2\ Hg$	+0.79
$Hg^{2+} + 2\ e^- \longrightarrow Hg$	+0.85
$2\ Hg^{2+} + 2\ e^- \longrightarrow Hg_2^{2+}$	+0.92
$Hg_2Cl_2 + 2\ e^- \longrightarrow 2\ Hg + 2\ Cl^-$	+0.27
$I_2 + 2\ e^- \longrightarrow 2\ I^-$	+0.54
$I_3^- + 2\ e^- \longrightarrow 3\ I^-$	+0.53
$In^+ + e^- \longrightarrow In$	−0.14
$In^{2+} + e^- \longrightarrow In^+$	−0.40
$In^{3+} + e^- \longrightarrow In^{2+}$	−0.49
$In^{3+} + 2\ e^- \longrightarrow In^+$	−0.44
$In^{3+} + 3\ e^- \longrightarrow In$	−0.34
$K^+ + e^- \longrightarrow K$	−2.93
$La^{3+} + 3\ e^- \longrightarrow La$	−2.52
$Li^+ + e^- \longrightarrow Li$	−3.05
$Mg^{2+} + 2\ e^- \longrightarrow Mg$	−2.36
$Mn^{2+} + 2\ e^- \longrightarrow Mn$	−1.18
$Mn^{3+} + e^- \longrightarrow Mn^{2+}$	+1.51
$MnO_2 + 4\ H^+ + 2\ e^- \longrightarrow Mn^{2+} + 2\ H_2O$	+1.23
$MnO_4^- + e^- \longrightarrow MnO_4^{2-}$	+0.56
$MnO_4^- + 8\ H^+ + 5\ e^- \longrightarrow Mn^{2+} + 4\ H_2O$	+1.51
$MnO_4^{2-} + 2\ H_2O + 2\ e^- \longrightarrow MnO_2 + 4\ OH^-$	+0.60
$NO_3^- + 2\ H^+ + e^- \longrightarrow NO_2 + H_2O$	+0.80
$NO_3^- + 4\ H^+ + 3\ e^- \longrightarrow NO + 2\ H_2O$	+0.96
$NO_3^- + H_2O + 2\ e^- \longrightarrow NO_2^- + 2\ OH^-$	+0.01
$Na^+ + e^- \longrightarrow Na$	−2.71
$Ni^{2+} + 2\ e^- \longrightarrow Ni$	−0.23
$Ni(OH)_3 + e^- \longrightarrow Ni(OH)_2 + OH^-$	+0.49
$O_2 + e^- \longrightarrow O_2^-$	−0.56
$O_2 + 4\ H^+ + 4\ e^- \longrightarrow 2\ H_2O$	+1.23
$O_2 + H_2O + 2\ e^- \longrightarrow HO_2 + OH^-$	−0.08
$O_2 + 2\ H_2O + 4\ e^- \longrightarrow 4\ OH^-$	+0.40
$O_3 + 2\ H^+ + 2\ e^- \longrightarrow O_2 + H_2O$	+2.07
$O_3 + H_2O + 2\ e^- \longrightarrow O_2 + 2\ OH^-$	+1.24
$Pb^{4+} + 2\ e^- \longrightarrow Pb$	−0.13
$Pb^{4+} + 2\ e^- \longrightarrow Pb^{2+}$	+1.67
$PbSO_4 + 2\ e^- \longrightarrow Pb + SO_4^{2-}$	−0.36
$Pt^{2+} + 2\ e^- \longrightarrow Pt$	+1.20
$Pu^{4+} + e^- \longrightarrow Pu^{3+}$	+0.97
$Ra^{2+} + 2\ e^- \longrightarrow Ra$	−2.92
$Rb^+ + e^- \longrightarrow Rb$	−2.93
$S + 2\ e^- \longrightarrow S^{2-}$	−0.48
$SO_4^{2-} + 4\ H^+ + 2\ e^- \longrightarrow H_2SO_3 + H_2O$	+0.17
$S_2O_8^{2-} + 2\ e^- \longrightarrow 2\ SO_4^{2-}$	+2.05
$Se + 2\ e^- \longrightarrow Se^{2-}$	−0.67
$Sn^{2+} + 2\ e^- \longrightarrow Sn$	−0.14
$Sn^{4+} + 2\ e^- \longrightarrow Sn^{2+}$	+0.15
$Sr^{2+} + 2\ e^- \longrightarrow Sr$	−2.89
$Te + 2\ e^- \longrightarrow Te^{2-}$	−0.84
$Ti^{2+} + 2\ e^- \longrightarrow Ti$	−1.63
$Ti^{3+} + e^- \longrightarrow Ti^{2+}$	−0.37
$Ti^{4+} + e^- \longrightarrow Ti^{3+}$	0.00
$Tl^+ + e^- \longrightarrow Tl$	−0.34
$U^{3+} + 3\ e^- \longrightarrow U$	−1.79
$U^{4+} + e^- \longrightarrow U^{3+}$	−0.61
$V^{2+} + 2\ e^- \longrightarrow V$	−1.19
$V^{3+} + e^- \longrightarrow V^{2+}$	−0.26
$Zn^{2+} + 2\ e^- \longrightarrow Zn$	−0.76

2C Ground-State Electron Configurations*

Z	Symbol	Configuration	Z	Symbol	Configuration
1	H	$1s^1$	29	Cu	$[Ar]3d^{10}4s^1$
2	He	$1s^2$	30	Zn	$[Ar]3d^{10}4s^2$
3	Li	$[He]2s^1$	31	Ga	$[Ar]3d^{10}4s^24p^1$
4	Be	$[He]2s^2$	32	Ge	$[Ar]3d^{10}4s^24p^2$
5	B	$[He]2s^22p^1$	33	As	$[Ar]3d^{10}4s^24p^3$
6	C	$[He]2s^22p^2$	34	Se	$[Ar]3d^{10}4s^24p^4$
7	N	$[He]2s^22p^3$	35	Br	$[Ar]3d^{10}4s^24p^5$
8	O	$[He]2s^22p^4$	36	Kr	$[Ar]3d^{10}4s^24p^6$
9	F	$[He]2s^22p^5$	37	Rb	$[Kr]5s^1$
10	Ne	$[He]2s^22p^6$	38	Sr	$[Kr]5s^2$
11	Na	$[Ne]3s^1$	39	Y	$[Kr]4d^15s^2$
12	Mg	$[Ne]3s^2$	40	Zr	$[Kr]4d^25s^2$
13	Al	$[Ne]3s^23p^1$	41	Nb	$[Kr]4d^45s^1$
14	Si	$[Ne]3s^23p^2$	42	Mo	$[Kr]4d^55s^1$
15	P	$[Ne]3s^23p^3$	43	Tc	$[Kr]4d^55s^2$
16	S	$[Ne]3s^23p^4$	44	Ru	$[Kr]4d^75s^1$
17	Cl	$[Ne]3s^23p^5$	45	Rh	$[Kr]4d^85s^1$
18	Ar	$[Ne]3s^23p^6$	46	Pd	$[Kr]4d^{10}$
19	K	$[Ar]4s^1$	47	Ag	$[Kr]4d^{10}5s^1$
20	Ca	$[Ar]4s^2$	48	Cd	$[Kr]4d^{10}5s^2$
21	Sc	$[Ar]3d^14s^2$	49	In	$[Kr]4d^{10}5s^25p^1$
22	Ti	$[Ar]3d^24s^2$	50	Sn	$[Kr]4d^{10}5s^25p^2$
23	V	$[Ar]3d^34s^2$	51	Sb	$[Kr]4d^{10}5s^25p^3$
24	Cr	$[Ar]3d^54s^1$	52	Te	$[Kr]4d^{10}5s^25p^4$
25	Mn	$[Ar]3d^54s^2$	53	I	$[Kr]4d^{10}5s^25p^5$
26	Fe	$[Ar]3d^64s^2$	54	Xe	$[Kr]4d^{10}5s^25p^6$
27	Co	$[Ar]3d^74s^2$	55	Cs	$[Xe]6s^1$
28	Ni	$[Ar]3d^84s^2$	56	Ba	$[Xe]6s^2$

Z	Symbol	Configuration	Z	Symbol	Configuration
57	La	$[Xe]5d^1 6s^2$	85	At	$[Xe]4f^{14} 5d^{10} 6s^2 6p^5$
58	Ce	$[Xe]4f^1 5d^1 6s^2$	86	Rn	$[Xe]4f^{14} 5d^{10} 6s^2 6p^6$
59	Pr	$[Xe]4f^3 6s^2$	87	Fr	$[Rn]7s^1$
60	Nd	$[Xe]4f^4 6s^2$	88	Ra	$[Rn]7s^2$
61	Pm	$[Xe]4f^5 6s^2$	89	Ac	$[Rn]6d^1 7s^2$
62	Sm	$[Xe]4f^6 6s^2$	90	Th	$[Rn]6d^2 7s^2$
63	Eu	$[Xe]4f^7 6s^2$	91	Pa	$[Rn]5f^2 6d^1 7s^2$
64	Gd	$[Xe]4f^7 5d^1 6s^2$	92	U	$[Rn]5f^3 6d^1 7s^2$
65	Tb	$[Xe]4f^9 6s^2$	93	Np	$[Rn]5f^4 6d^1 7s^2$
66	Dy	$[Xe]4f^{10} 6s^2$	94	Pu	$[Rn]5f^6 7s^2$
67	Ho	$[Xe]4f^{11} 6s^2$	95	Am	$[Rn]5f^7 7s^2$
68	Er	$[Xe]4f^{12} 6s^2$	96	Cm	$[Rn]5f^7 6d^1 7s^2$
69	Tm	$[Xe]4f^{13} 6s^2$	97	Bk	$[Rn]5f^9 7s^2$
70	Yb	$[Xe]4f^{14} 6s^2$	98	Cf	$[Rn]5f^{10} 7s^2$
71	Lu	$[Xe]4f^{14} 5d^1 6s^2$	99	Es	$[Rn]5f^{11} 7s^2$
72	Hf	$[Xe]4f^{14} 5d^2 6s^2$	100	Fm	$[Rn]5f^{12} 7s^2$
73	Ta	$[Xe]4f^{14} 5d^3 6s^2$	101	Md	$[Rn]5f^{13} 7s^2$
74	W	$[Xe]4f^{14} 5d^4 6s^2$	102	No	$[Rn]5f^{14} 7s^2$
75	Re	$[Xe]4f^{14} 5d^5 6s^2$	103	Lr	$[Rn]5f^{14} 6d^1 7s^2$
76	Os	$[Xe]4f^{14} 5d^6 6s^2$	104	Rf	$[Rn]5f^{14} 6d^2 7s^2$(?)
77	Ir	$[Xe]4f^{14} 5d^7 6s^2$	105	Db	$[Rn]5f^{14} 6d^3 7s^2$(?)
78	Pt	$[Xe]4f^{14} 5d^9 6s^1$	106	Sg	$[Rn]5f^{14} 6d^4 7s^2$(?)
79	Au	$[Xe]4f^{14} 5d^{10} 6s^1$	107	Bh	$[Rn]5f^{14} 6d^5 7s^2$(?)
80	Hg	$[Xe]4f^{14} 5d^{10} 6s^2$	108	Hs	$[Rn]5f^{14} 6d^6 7s^2$(?)
81	Tl	$[Xe]4f^{14} 5d^{10} 6s^2 6p^1$	109	Mt	$[Rn]5f^{14} 6d^7 7s^2$(?)
82	Pb	$[Xe]4f^{14} 5d^{10} 6s^2 6p^2$			
83	Bi	$[Xe]4f^{14} 5d^{10} 6s^2 6p^3$			
84	Po	$[Xe]4f^{14} 5d^{10} 6s^2 6p^4$			

*The electron configurations followed by a question mark are speculative.

2D The Elements

Element	Symbol	Atomic number	Molar mass,* $g \cdot mol^{-1}$	Normal state†	Density, $g \cdot cm^{-3}$	Melting point, °C
actinium (Greek *aktis*, ray)	Ac	89	(227)	s, m	10.07	1230
aluminum (from alum, salts of the form $KAl(SO_4)_2 \cdot 12H_2O$)	Al	13	26.98	s, m	2.70	660
americium (the Americas)	Am	95	(243)	s, m	13.67	990
antimony (probably a corruption of an old Arabic word; Latin *stibium*)	Sb	51	121.76	s, md	6.69	631
argon (Greek *argos*, inactive)	Ar	18	39.95	g, nm	1.66‡	−189
arsenic (Greek *arsenikos*, male)	As	33	74.92	s, md	5.78	613§
astatine (Greek *astatos*, unstable)	At	85	(210)	s, nm	—	300
barium (Greek *barys*, heavy)	Ba	56	137.33	s, m	3.59	710
berkelium (Berkeley, California)	Bk	97	(247)	s, m	14.79	986
beryllium (from the mineral beryl, $Be_3Al_2SiO_{18}$)	Be	4	9.01	s, m	1.85	1285
bismuth (German *weisse Masse*, white mass)	Bi	83	208.98	s, m	8.90	271
bohrium (Niels Bohr)	Bh	107	(262)	—	—	—
boron (Arabic *buraq*, borax, $Na_2B_4O_7 \cdot 10H_2O$; *bor*(ax) + (carb)*on*	B	5	10.81	s, md	2.47	2300
bromine (Greek *bromos*, stench)	Br	35	79.90	l, nm	3.12	−7
cadmium (Greek *Cadmus*, founder of Thebes)	Cd	48	112.41	s, m	8.65	321
calcium (Latin *calx*, lime)	Ca	20	40.08	s, m	1.53	840
californium (California)	Cf	98	(251)	s, m	—	—
carbon (Latin *carbo*, coal or charcoal)	C	6	12.01	s, nm	2.27	3700§
cerium (the asteroid Ceres, discovered 2 days earlier)	Ce	58	140.12	s, m	6.71	800

Boiling point, °C	Ionization energies, $kJ \cdot mol^{-1}$	Electron affinity, $kJ \cdot mol^{-1}$	Electronegativity	Principal oxidation states	Atomic radius, pm	Ionic radius,‖ pm
3200	499, 1170, 1900	—	1.1	+3	188	118(3+)
2467	577, 1817, 2744	+43	1.6	+3	143	57(3+)
2600	578	—	1.3	+3	173	107(3+)
1750	834, 1794, 2443	+103	2.1	−3, +3, +5	182	89(3+)
−186	1520	<0	—	0	174	—
—	947, 1798	+78	2.2	−3, +3, +5	125	222(3−)
350	1037, 1600	+270	2.0	−1	—	227(1−)
1640	502, 965	+14	0.89	+2	217	143(2+)
—	601	—	1.3	+3	—	87(4+)
2470	900, 1757	<0	1.6	+2	113	34(2+)
1650	703, 1610, 2466	+91	2.0	+3, +5	155	96(3+)
—	660	—	—	+5	128#	83(5+)#
3931	799, 2427, 3660	+27	2.0	+3	83	23(3+)
59	1140, 2104	+325	3.0	−1, +1, +3, +4, +5, +7	114	196(1−)
765	868, 1631	<0	1.7	+2	149	103(2+)
1490	590, 1145, 4910	+2	1.3	+2	197	106(2+)
—	608	—	1.3	+3	169	117(2+)
—	1090, 2352, 4620	+122	2.6	−4, −1, +2, +4	77	260(4−)
3000	527, 1047, 1949	<50	1.1	+3, +4	183	107(3+)

(continued)

Element	Symbol	Atomic number	Molar mass,* $g \cdot mol^{-1}$	Normal state†	Density, $g \cdot cm^{-3}$	Melting point, °C
cesium (Latin *caesius*, sky blue)	Cs	55	132.91	s, m	1.87	28
chlorine (Greek *chloros*, yellowish green)	Cl	17	35.45	g, nm	1.66‡	−101
chromium (Greek *chroma*, color)	Cr	24	52.00	s, m	7.19	1860
cobalt (German *Kobold*, evil spirit; Greek *kobalos*, goblin)	Co	27	58.93	s, m	8.80	1494
copper (Latin *cuprum*, from Cyprus)	Cu	29	63.55	s, m	8.93	1083
curium (Marie Curie)	Cm	96	(247)	s, m	13.30	1340
dubnium (Dubna)	Db	105	(262)	s, m	29	—
dysprosium (Greek *dysprositos*, hard to get at)	Dy	66	162.50	s, m	8.53	1410
einsteinium (Albert Einstein)	Es	99	(252)	s, m	—	—
erbium (Ytterby, a town in Sweden)	Er	68	167.26	s, m	9.04	1520
europium (Europe)	Eu	63	151.96	s, m	5.25	820
fermium (Enrico Fermi, an Italian physicist)	Fm	100	(257)	s, m	—	—
fluorine (Latin *fluere*, to flow)	F	9	19.00	g, nm	1.51‡	−220
francium (France)	Fr	87	(223)	s, m	—	27
gadolinium (Johann Gadolin, a Finnish chemist)	Gd	64	157.25	s, m	7.87	1310
gallium (Latin *Gallia*, France; also a pun on the discoverer's forename, Le Coq)	Ga	31	69.72	s, m	5.91	30
germanium (Latin *Germania*, Germany)	Ge	32	72.61	s, md	5.32	937
gold (Anglo-Saxon *gold*; Latin *aurum*, gold)	Au	79	196.97	s, m	19.28	1064
hafnium (Latin *Hafnia*, Copenhagen)	Hf	72	178.49	s, m	13.28	2230
hassium (Hesse, the German state)	Hs	108	(265)	—	—	—
helium (Greek *helios*, the sun)	He	2	4.00	g, nm	0.12‡	—

Boiling point, °C	Ionization energies, kJ·mol^{-1}	Electron affinity, kJ·mol^{-1}	Electronegativity	Principal oxidation states	Atomic radius, pm	Ionic radius,$^{\|}$ pm
678	376, 2420	+46	0.79	+1	265	165(1+)
−34	1255, 2297	+349	3.2	−1, +1, +3, +4, +5, +6, +7	99	181(1−)
2600	653, 1592, 2987	+64	1.7	+2, +3	125	84(2+)
2900	760, 1646, 3232	+64	1.9	+3, +6	125	64(3+)
2567	785, 1958, 3554	+118	1.9	+1, +2	128	72(2+)
—	581	—	1.3	+3	174	99(3+)
—	640	—	—	+5	139#	68(5+)#
2600	572, 1126, 2200	—	1.2	+3	177	91(3+)
—	619	<50	1.3	+3	203	98(3+)
2600	589, 1151, 2194	<50	1.2	+3	176	89(3+)
1450	547, 1085, 2404	<50	—	+3	204	98(3+)
—	627	—	1.3	+3	—	91(3+)
−188	1680, 3374	+328	4.0	−1	71	133(1−)
677	400	+44	0.7	+1	270	180(1+)
3000	592, 1167, 1990	<50	1.2	+2, +3	180	97(3+)
2403	577, 1979, 2963	+29	1.6	+1, +3	122	62(3+)
2830	784, 1557, 3302	+116	2.0	+2, +4	123	90(2+)
2807	890, 1980	+223	2.5	+1, +3	144	91(3+)
5300	642, 1440, 2250	0	1.3	+4	156	84(3+)
—	750	—	—	+3	126#	80(4+)#
−269	2370, 5250	<0	—	0	128	—

(continued)

Element	Symbol	Atomic number	Molar mass,* g·mol^{-1}	Normal state†	Density, g·cm^{-3}	Melting point, °C
holmium (Latin *Holmia*, Stockholm)	Ho	67	164.93	s, m	8.80	1470
hydrogen (Greek *hydro* + *genes*, water-forming)	H	1	1.0079	g, nm	0.089‡	−259
indium (from the bright indigo line in its spectrum)	In	49	114.82	s, m	7.29	156
iodine (Greek *ioeidēs*, violet)	I	53	126.90	s, nm	4.95	114
iridium (Greek and Latin *iris*, rainbow)	Ir	77	192.22	s, m	22.56	2447
iron (Anglo-Saxon *iron*; Latin *ferrum*)	Fe	26	55.85	s, m	7.87	1540
krypton (Greek *kryptos*, hidden)	Kr	36	83.80	g, nm	3.00‡	−157
lanthanum (Greek *lanthanein*, to lie hidden)	La	57	138.91	s, m	6.17	920
lawrencium (Ernest Lawrence, an American physicist)	Lr	103	(262)	s, m	—	—
lead (Anglo-Saxon *lead*; Latin *plumbum*)	Pb	82	207.2	s, m	11.34	328
lithium (Greek *lithos*, stone)	Li	3	6.94	s, m	0.53	181
lutetium (*Lutetia*, ancient name of Paris)	Lu	71	174.97	s, m	9.84	1700
magnesium (Magnesia, a district in Thessaly, Greece)	Mg	12	24.31	s, m	1.74	650
manganese (Greek and Latin *magnes*, magnet)	Mn	25	54.94	s, m	7.47	1250
meitnenium (Lise Meitner)	Mt	109	(266)	—	—	—
mendelevium (D. Mendeleev)	Md	101	(258)	—	—	—
mercury (the planet Mercury; Latin *hydrargyrum*, liquid silver)	Hg	80	200.59	l, m	13.55	−39
molybdenum (Greek *molybdos*, lead)	Mo	42	95.94	s, m	10.22	2620
neodymium (Greek *neos* + *didymos*, new twin)	Nd	60	144.24	s, m	7.00	1024
neon (Greek *neos*, new)	Ne	10	20.18	g, nm	1.44‡	−249

Boiling point, °C	Ionization energies, kJ·mol^{-1}	Electron affinity, kJ·mol^{-1}	Electronegativity	Principal oxidation states	Atomic radius, pm	Ionic radius,‖ pm
2300	581, 1139	<50	1.2	+3	177	89(3+)
−253	1310	+73	2.2	−1, +1	78	154(1−)
2080	556, 1821	+29	1.8	+1, +3	163	92(3+)
184	1008, 1846	+295	2.7	−1, +1, +3, +5, +7	133	196(1−)
4550	880	+151	2.2	+3, +4	136	75(3+)
2760	759, 1561, 2957	+16	1.8	+2, +3	124	82(2+)
−153	1350, 2350	<0	—	+2	189	169(1+)
3450	538, 1067, 1850	+50	1.1	+3	188	122(3+)
—	—	—	1.3	+3	—	88(3+)
1760	716, 1450	+35	2.3	+2, +4	175	132(2+)
1347	519, 7298	+60	1.0	+1	152	78(1+)
3400	524, 1340, 2022	<50	1.3	+3	173	85(3+)
1100	736, 1451	<0	1.3	+2	160	79(2+)
2120	717, 1509	<0	1.6	+2, +3, +4, +7	137	91(2+)
—	840	—	—	+2	—	83(2+)
—	635	—	1.3	+3	—	90(3+)
357	1007, 1810	−18	2.0	+1, +2	160	112(2+)
4830	685, 1558, 2621	+72	2.2	+4, +5, +6	136	92(2+)
3100	530, 1035	<0	1.1	+3	182	104(3+)
−246	2080, 3952	0	—	0	—	—

(continued)

Element	Symbol	Atomic number	Molar mass,* $g \cdot mol^{-1}$	Normal state†	Density, $g \cdot cm^{-3}$	Melting point, °C
neptunium (the planet Neptune)	Np	93	(237)	s, m	20.45	640
nickel (German *Nickel*, Old Nick, Satan)	Ni	28	58.69	s, m	8.91	1455
niobium (Niobe, daughter of Tantalus; see tantalum)	Nb	41	92.91	s, m	8.57	2425
nitrogen (Greek *nitron* + *genes*, soda-forming)	N	7	14.01	g, nm	1.04‡	−210
nobelium (Alfred Nobel, the founder of the Nobel prizes)	No	102	(259)	s, m	—	—
osmium (Greek *osme*, a smell)	Os	76	190.23	s, m	22.58	3030
oxygen (Greek *oxys* + *genes*, acid-forming)	O	8	16.00	g, nm	1.14	−218
palladium (the asteroid Pallas, discovered at about the same time)	Pd	46	106.42	s, m	12.00	1554
phosphorus (Greek *phosphoros*, light-bearing)	P	15	30.97	s, nm	1.82	44
platinum (Spanish *plata*, silver)	Pt	78	195.08	s, m	21.45	1772
plutonium (the planet Pluto)	Pu	94	(244)	s, m	19.81	640
polonium (Poland)	Po	84	(209)	s, md	9.40	254
potassium (from potash; Latin *kalium* and Arabic *qali*, alkali)	K	19	39.10	s, m	0.86	64
praseodymium (Greek *prasios* + *didymos*, green twin)	Pr	59	140.91	s, m	6.78	935
promethium (Prometheus, the Greek god)	Pm	61	(145)	s, m	7.22	1168
protactinium (Greek *protos* + *aktis*, first ray)	Pa	91	231.04	s, m	15.37	1200
radium (Latin *radius*, ray)	Ra	88	(226)	s, m	5.00	700
radon (from radium)	Rn	86	(222)	g, nm	4.40‡	−71
rhenium (Latin *Rhenus*, Rhine)	Re	75	186.21	s, m	21.02	3180
rhodium (Greek *rhodon*, rose; its aqueous solutions are often rose-colored)	Rh	45	102.91	s, m	12.42	1963

Boiling point, °C	Ionization energies, kJ·mol^{-1}	Electron affinity, kJ·mol^{-1}	Electronegativity	Principal oxidation states	Atomic radius, pm	Ionic radius,[‖] pm
—	597	—	1.4	+5	150	88(5+)
2150	737, 1753	+156	1.9	+2, +3	125	78(2+)
5000	664, 1382	+86	1.6	+5	143	69(5+)
−196	1400, 2856	−7	3.0	−3, +3, +5	71	171(3−)
—	642	—	1.3	+2	—	113(2+)
5000	840	+106	2.2	+3, +4	135	81(3+)
−183	1310, 3388	+141, −844	3.4	−2	66	132(2−)
3000	805, 1875	+54	2.2	+2, +4	138	86(2+)
280	1011, 1903, 2912	+72	2.2	−3, +3, +5	115	212(3−)
3720	870, 1791	+205	2.3	+2, +4	138	85(2+)
3200	585	—	1.3	+3, +4	151	108(3+)
960	812	+174	2.0	+2, +4	167	65(4+)
774	418, 3051	+48	0.82	+1	227	133(1+)
3000	523, 1018	<50	1.1	+3	183	106(3+)
3300	536, 1052	<50	—	+3	181	106(3+)
4000	568	—	1.5	+5	161	89(5+)
1500	509, 979	—	0.9	+2	223	152(2+)
−62	1036, 1930	<0	—	+2	—	—
5600	760, 1260	+14	1.9	+4, +7	137	72(4+)
3700	720, 1744	+110	2.3	+3	134	75(3+)

(continued)

Element	Symbol	Atomic number	Molar mass,* g·mol^{-1}	Normal state†	Density, g·cm^{-3}	Melting point, °C
rubidium (Latin *rubidus*, deep red, "flushed")	Rb	37	85.47	s, m	1.53	39
ruthenium (Latin *Ruthenia*, Russia)	Ru	44	101.07	s, m	12.36	2310
rutherfordium (Ernest Rutherford)	Rf	104	(261)	—	—	—
samarium (from samarskite, a mineral)	Sm	62	150.36	s, m	7.54	1060
scandium (Latin *Scandia*, Scandinavia)	Sc	21	44.96	s, m	2.99	1540
seaborgium (Glenn Seaborg)	Sg	106	(263)	—	—	—
selenium (Greek *selēnē*, the moon)	Se	34	78.96	s, nm	4.79	220
silicon (Latin *silex*, flint)	Si	14	28.09	s, md	2.33	1410
silver (Anglo-Saxon *seolfor*; Latin *argentum*)	Ag	47	107.87	s, m	10.50	962
sodium (English soda; Latin *natrium*)	Na	11	22.99	s, m	0.97	98
strontium (Strontian, Scotland)	Sr	38	87.62	s, m	2.58	770
sulfur (Sanskrit *sulvere*)	S	16	32.07	s, nm	2.09	115
tantalum (Tantalos, Greek mythological figure)	Ta	73	180.95	s, m	16.65	3000
technetium (Greek *technētos*, artificial)	Tc	43	(98)	s, m	11.50	2200
tellurium (Latin *tellus*, earth)	Te	52	127.60	s, md	6.25	450
terbium (Ytterby, a town in Sweden)	Tb	65	158.93	s, m	8.27	1360
thallium (Greek *thallos*, a green shoot)	Tl	81	204.38	s, m	11.87	304
thorium (Thor, Norse god of thunder, weather, and crops)	Th	90	232.04	s, m	11.73	1700
thulium (*Thule*, early name for Scandinavia)	Tm	69	168.93	s, m	9.33	1550
tin (Anglo-Saxon *tin*; Latin *stannum*)	Sn	50	118.71	s, m	7.29	232
titanium (Titans, Greek mythological figures, sons of the Earth)	Ti	22	47.87	s, m	4.55	1660

Boiling point, °C	Ionization energies, kJ·mol^{-1}	Electron affinity, kJ·mol^{-1}	Electronegativity	Principal oxidation states	Atomic radius, pm	Ionic radius,[‖] pm
688	402, 2632	+47	0.82	+1	250	149(1+)
4100	711, 1617	+101	2.2	+2, +3, +4	134	77(3+)
—	490	—	—	+4	150[#]	67(4+)[#]
1600	543, 1068	<50	1.2	+3	180	100(3+)
2800	631, 1235	+18	1.4	+3	161	83(3+)
—	730	—	—	+6	132[#]	86(5+)[#]
685	941, 2044	+195	2.6	−2, +4, +6	215	191(2−)
2620	786, 1577	+134	1.9	+4	117	26(4+)
2212	731, 2073	+126	1.9	+1	144	113(1+)
883	494, 4562	+53	0.93	+1	154	98(1+)
1380	548, 1064	+5	0.95	+2	215	127(2+)
445	1000, 2251	+200, −532	2.6	−2, +4, +6	104	184(2−)
5400	761	+14	1.5	+5	143	72(3+)
4600	702, 1472	+96	1.9	+4, +7	136	72(4+)
990	870, 1775	+190	2.1	−2, +4	143	211(2−)
2500	565, 1112	<50	—	+3	178	97(3+)
1457	590, 1971	+19	2.0	+1, +3	170	105(3+)
4500	587, 1110	—	1.3	+4	180	99(4+)
2000	597, 1163	<50	1.2	+3	175	94(3+)
2720	707, 1412	+116	2.0	+2, +4	141	93(2+)
3300	658, 1310	+7.6	1.5	+4	145	69(4+)

(continued)

Element	Symbol	Atomic number	Molar mass,* $g \cdot mol^{-1}$	Normal state†	Density, $g \cdot cm^{-3}$	Melting point, °C
tungsten (Swedish *tung* + *sten*, heavy stone; from wolframite)	W	74	183.84	s, m	19.30	3387
uranium (the planet Uranus)	U	92	238.03	s, m	18.95	1135
vanadium (Vanadis, Scandinavian mythological figure)	V	23	50.94	s, m	6.11	1920
xenon (Greek *xenos*, stranger)	Xe	54	131.29	g, nm	3.56‡	−112
ytterbium (Ytterby, a town in Sweden)	Yb	70	173.04	s, m	6.97	824
yttrium (Ytterby, a town in Sweden)	Y	39	88.91	s, m	4.48	1510
zinc (Anglo-Saxon *zinc*)	Zn	30	65.39	s, m	7.14	420
zirconium (Arabic *zargun*, gold color)	Zr	40	91.22	s, m	6.51	1850

Boiling point, °C	Ionization energies, kJ·mol^{-1}	Electron affinity, kJ·mol^{-1}	Electronegativity	Principal oxidation states	Atomic radius, pm	Ionic radius,[‖] pm
5420	770	+79	2.4	+5, +6	137	62(6+)
4000	584, 1420	—	1.4	+6	154	80(6+)
3400	650, 1414	+51	1.6	+4, +5	132	61(4+)
−108	1170, 2046	<0	2.6	+2, +4, +6	218	190(1+)
1500	603, 1176	<50	—	+3	194	86(3+)
3300	616, 1181	+30	1.2	+3	181	106(3+)
907	906, 1733	+9	1.6	+2	133	83(2+)
4400	660, 1267	+41	1.3	+4	160	87(4+)

*Parentheses around molar mass indicate the most stable isotope of a radioactive element.
†The normal state is the state of the element at normal temperature and pressure (20°C and 1 atm). s denotes solid; l, liquid; and g, gas; m denotes metal; nm, nonmetal; and md, metalloid.
‡The density quoted is for the liquid.
§The solid sublimes.
‖Charge in parentheses.
#Atomic and ionic radii are estimated.

2E The Top 25 Chemicals by Industrial Production in the United States in 1999

Production data are compiled annually by the American Chemical Society and published in *Chemical and Engineering News*. This table is based on the information about production in 1999 that was published in the June 26, 2000 issue. Water, sodium chloride, and steel traditionally are not included and would outrank the rest if they were. Hydrogen is heavily used but almost always "on site" as soon as it has been prepared.

Rank	Name	Annual production, 10^9 kg	Comment on source
1	sulfuric acid	45.1	contact process
2	ethene (ethylene)	25.3	thermal cracking of petroleum
3	calcium hydroxide	22.6	decomposition of limestone
4	ammonia	19.0	Haber process
5	diammonium hydrogen phosphate	15.6	processing of phosphate rocks
6	phosphoric acid	13.8	from phosphate rocks
7	polyethylene	13.4	polymerization of ethene
8	chlorine	13.3	electrolysis
9	propene (propylene)	13.2	thermal cracking of petroleum
10	sulfur	12.2	Claus process
11	sodium carbonate	11.1	mining
12	dichloroethane	10.4	chlorination of ethene
13	sodium hydroxide	10.4	electrolysis of brine
14	nitric acid	9.0	Ostwald process
15	urea	8.5	ammonia + carbon dioxide
16	ammonium nitrate	8.2	ammonia + nitric acid
17	polypropylene	7.0	polymerization of propylene
18	polyvinyl chloride and copolymers	6.8	polymerization of vinyl chloride
19	ethylbenzene	5.9	Friedel-Crafts alkylation of benzene
20	styrene	5.4	dehydration of ethylbenzene
21	hydrogen chloride	4.5	by-product of hydrocarbon chlorination
22	ethylene oxide	4.0	addition of O_2 to ethene
23	*p*-xylene	4.0	catalytic reforming of petroleum
24	monoammonium dihydrogen phosphate	4.0	processing of phosphate rocks
25	cumene (isopropylbenzene)	3.2	alkylation of benzene

APPENDIX 3 Nomenclature

3A The Nomenclature of Polytomic Ions

Charge number	Chemical formula	Name	Oxidation number of central element
+2	Hg_2^{2+}	mercury(I)	+1
	UO_2^{2+}	uranyl	+6
	VO^{2+}	vanadyl	+4
+1	NH_4^+	ammonium	−3
	PH_4^+	phosphonium	−3
−1	$CH_3CO_2^-$	acetate (ethanoate)	0(C)
	HCO_2^-	formate (methanoate)	+2(C)
	CN^-	cyanide	+2(C)
	ClO_4^-	perchlorate*	+7
	ClO_3^-	chlorate*	+5
	ClO_2^-	chlorite*	+3
	ClO^-	hypochlorite*	+1
	MnO_4^-	permanganate	+7
	NO_3^-	nitrate	+5
	NO_2^-	nitrite	+3
	N_3^-	azide	$-\frac{1}{3}$
	O_3^-	ozonide	$-\frac{1}{3}$
	OH^-	hydroxide	
	SCN^-	thiocyanate	
−2	C_2^{2-}	carbide (acetylide)	−1
	CO_3^{2-}	carbonate	+4
	$C_2O_4^{2-}$	oxalate	+3
	CrO_4^{2-}	chromate	+6
	$Cr_2O_7^{2-}$	dichromate	+6
	O_2^{2-}	peroxide	−1
	S_2^{2-}	disulfide	−1
	SiO_3^{2-}	metasilicate	+4
	SO_4^{2-}	sulfate	+6
	SO_3^{2-}	sulfite	+4
	$S_2O_3^{2-}$	thiosulfate	+2
−3	AsO_4^{3-}	arsenate	+5
	BO_3^{3-}	borate	+3
	PO_4^{3-}	phosphate	+5

* These names are representative of the halogen oxoanions.

When a hydrogen ion bonds to a −2 or −3 anion, add "hydrogen" before the name of the anion. For example, HSO_3^- is hydrogen sulfite (or hydrogensulfite). If two hydrogen ions bond to a −3 anion, add "dihydrogen" before the name of the anion. For example, $H_2PO_4^-$ is dihydrogen phosphate.

Oxoacids and Oxoanions

The names of oxoanions and their parent acids can be determined by noting the oxidation number of the central atom and then referring to the table on the right. For example, the nitrogen in $N_2O_2^{2-}$ has an oxidation number of +1; because nitrogen belongs to Group 15/V, the ion is a hyponitrite ion.

Group number					
14/IV	15/V	16/VI	17/VII	Oxoanion	Oxoacid
—	—	—	+7	per . . . ate	per . . . ic acid
+4	+5	+6	+5	. . . ate	. . . ic acid
—	+3	+4	+3	. . . ite	. . . ous acid
—	+1	+2	+1	hypo . . . ite	hypo . . . ous acid

3B Common Names of Chemicals

Many chemicals have acquired common names, sometimes as a result of their use over hundreds of years and sometimes because they appear on the labels of consumer products, such as detergents, beverages, and antacids. The following substances are just a few that have found their way into the language of everyday life.

Common name	Formula	Chemical name
baking soda	$NaHCO_3$	sodium hydrogen carbonate (sodium bicarbonate)
bleach, laundry	$NaClO$	sodium hypochlorite
borax	$Na_2B_4O_7 \cdot 10H_2O$	sodium tetraborate decahydrate
brimstone	S_8	sulfur
calamine	$ZnCO_3$	zinc carbonate
chalk	$CaCO_3$	calcium carbonate
Epsom salts	$MgSO_4 \cdot 7H_2O$	magnesium sulfate heptahydrate
fool's gold	FeS_2	iron(II) disulfide
gypsum	$CaSO_4 \cdot 2H_2O$	calcium sulfate dihydrate
lime (quicklime)	CaO	calcium oxide
lime (slaked lime)	$CaOH$	calcium hydroxide
limestone	$CaCO_3$	calcium carbonate
lye, caustic soda	$NaOH$	sodium hydroxide
marble	$CaCO_3$	calcium carbonate
milk of magnesia	$Mg(OH)_2$	magnesium hydroxide
plaster of Paris	$CaSO_4 \cdot \frac{1}{2}H_2O$	calcium sulfate hemihydrate
potash*	K_2CO_3	potassium carbonate
quartz	SiO_2	silcon dioxide
table salt	$NaCl$	sodium chloride
vinegar	CH_3COOH	acetic acid
washing soda	$Na_2CO_3 \cdot 10H_2O$	sodium carbonate decahydrate

*Potash also refers collectively to K_2CO_3, KOH, K_2SO_4, KCl, and KNO_3.

3C Names of Some Common Cations with Variable Charge Numbers

Modern nomenclature includes the oxidation number of elements with variable oxidation states in the names of their compounds, as in cobalt(II) chloride. However, the traditional nomenclature, in which the suffixes -ous and -ic are used, is still encountered. The table on the right translates from one system to the other for some common elements.

Element	Cation	Old style name	Modern name
cobalt	Co^{2+}	cobaltous	cobalt(II)
	Co^{3+}	cobaltic	cobalt(III)
copper	Cu^{+}	cuprous	copper(I)
	Cu^{2+}	cupric	copper(II)
iron	Fe^{2+}	ferrous	iron(II)
	Fe^{3+}	ferric	iron(III)
lead	Pb^{2+}	plumbous	lead(II)
	Pb^{4+}	plumbic	lead(IV)
manganese	Mn^{2+}	manganous	manganese(II)
	Mn^{3+}	manganic	manganese(III)
mercury	Hg_2^{2+}	mercurous	mercury(I)
	Hg^{2+}	mercuric	mercury(II)
tin	Sn^{2+}	stannous	tin(II)
	Sn^{4+}	stannic	tin(IV)

GLOSSARY

***ab initio* method** The calculation of molecular structure by solving the *Schrödinger equation* numerically. Compare with *semiempirical method.*

absolute zero ($T = 0$; that is, 0 on the *Kelvin scale*) The lowest possible temperature (−273.15°C).

absorb To accept one substance into and throughout the bulk of another substance. Compare with *adsorb.*

absorbance A A measure of the extent of absorption of radiation by a sample: $A = \log(I_0/I)$.

absorbed dose (of radiation) The energy deposited in a given mass of sample when it is exposed to radiation (particularly but not exclusively nuclear radiation). Absorbed dose is measured in *rad* or *gray.*

absorption spectrum The wavelength dependence of the absorption of a sample, determined by measuring the extent to which the sample absorbs electromagnetic radiation as the wavelength is varied over a range.

abundance (of an isotope) The percentage (in terms of the numbers of atoms) of the isotope present in a sample of the element. See also *natural abundance.*

acceleration The rate of change of velocity (either its direction or magnitude).

acceleration of free fall g The acceleration experienced by a body due to the gravitational field at the surface of the Earth.

accuracy Freedom from systematic error. Compare with *precision.*

accurate measurements Measurements that have small systematic error and give a result close to the accepted value of the property.

achiral, achiral molecule Not chiral: identical to its mirror image. See also *chiral.*

acid See *Arrhenius acid*; *Brønsted acid*; *Lewis acid.* Used alone, "acid" normally means a Brønsted acid.

acid anhydride A compound that forms an oxoacid when it reacts with water. A *formal anhydride* is a compound that has the formula of an acid minus the elements of water but does not react with water to produce the acid. *Examples:* SO_3, the anhydride of sulfuric acid; CO, the formal anhydride of formic acid, HCOOH.

acid-base indicator See *indicator.*

acid-base titration See *titration.*

acid buffer See *buffer.*

acidic hydrogen atom A hydrogen atom (more exactly, the proton of that hydrogen atom) that can be donated to a base.

acidic ion An ion that acts as a Brønsted acid. *Examples:* NH_4^+; $[Al(H_2O)_6]^{3+}$.

acidic oxide An oxide that reacts with water to give an acid; the oxides of nonmetallic elements generally are acidic oxides. *Examples:* CO_2; SO_3.

acidic solution A solution with pH < 7.

acid ionization (dissociation) constant K_a See *acidity constant.*

acidity The strength of the tendency to donate a proton.

acidity constant K_a The equilibrium constant for proton transfer to water; for an acid HA, $K_a = [H_3O^+][A^-]/[HA]$ at equilibrium.

actinide A member of the second row of the f block (actinium through nobelium).

activated complex An unstable combination of reactant molecules that can either go on to form products or fall apart into the unchanged reactants.

activated complex theory A theory of reaction rates in which it is supposed that the reactants form an activated complex.

activation barrier The potential energy barrier between reactants and products; the height of the barrier is the activation energy of the reaction.

activation energy E_a (1) The minimum energy needed for reaction. (2) The height of the activation barrier. (3) An empirical parameter that describes the temperature dependence of the rate constant of a reaction.

active site (1) The region of an enzyme molecule where the substrate reacts. (2) The effective catalytic site on the surface of a heterogeneous catalyst.

activity (1) In thermodynamics, a_J, the effective concentration or pressure of a species J expressed as the partial pressure or concentration of the species relative to its standard value; (2) in radioactivity, the number of nuclear disintegrations per second.

activity coefficient γ_J The proportionality factor between the activity of a species and its molar concentration or partial pressure. *Example:* for a solute J, $a_J = \gamma_J[J]/[J]°$.

addition polymerization The polymerization, usually of alkenes, by an addition reaction propagated by radical or ionic intermediates.

addition reaction A chemical reaction in which atoms or groups bond to two atoms joined by a multiple bond. The product of the reaction is a single molecule that contains all the reactant atoms. *Example:* $CH_3CH{=}CH_2 + HBr \rightarrow CH_3CH_2CH_2Br$.

adhesion Binding to a surface.

adhesive forces Forces that bind a substance to a surface.

adiabatic Without the transfer of energy as heat.

adsorb To bind a substance to a surface; the surface *adsorbs* the substance. Distinguish from *absorb*.

aerosol A fine mist of solid particles or droplets of liquid suspended in a gas.

alcohol An organic molecule containing an —OH group attached to a carbon atom that is not part of a carbonyl group or an aromatic ring. Alcohols are classified as *primary*, *secondary*, and *tertiary* according to the number of carbon atoms attached to the C—OH carbon atom. *Examples:* CH_3CH_2OH (primary); $(CH_3)_2CHOH$ (secondary); $(CH_3)_3COH$ (tertiary).

aldehyde An organic compound containing the —CHO group. *Examples:* CH_3CHO, ethanal (acetaldehyde); C_6H_5CHO, benzaldehyde.

aliphatic hydrocarbon A hydrocarbon that does not have benzene rings in its structure.

alkali An aqueous solution of a strong base. *Example:* aqueous NaOH.

alkali metal A member of Group 1 of the periodic table (the lithium family).

alkaline cell A dry cell in which the electrolyte contains sodium or potassium hydroxide.

alkaline earth metal Calcium, strontium, and barium; more informally, a member of Group 2 of the periodic table (the beryllium family).

alkaline solution An aqueous solution with pH >7.

alkane (1) A hydrocarbon with no carbon-carbon multiple bonds. (2) A saturated hydrocarbon. (3) A member of a series of hydrocarbons derived from methane by the repetitive insertion of $—CH_2—$ groups; alkanes have molecular formula C_nH_{2n+2}. *Examples:* CH_4, CH_3CH_3, $CH_3(CH_2)_6CH_3$.

alkene (1) A hydrocarbon with at least one carbon-carbon double bond. (2) A member of a series of hydrocarbons derived from ethene by the repetitive insertion of $—CH_2—$ groups; alkenes with one double bond have molecular formula C_nH_{2n}. *Examples:* $CH_2{=}CH_2$; $CH_3CH{=}CH_2$; $CH_3CH{=}CHCH_2CH_3$.

alkyl group R A group of atoms that can be regarded as derived from an alkane by loss of a hydrogen atom. *Examples:* $—CH_3$, methyl; $—CH_2CH_3$, ethyl.

alkyne (1) A hydrocarbon with at least one carbon-carbon triple bond. (2) A member of a series of hydrocarbons derived from ethyne by the repetitive insertion of $—CH_2—$ groups; alkynes have molecular formula C_nH_{2n-2}. *Examples:* $CH{\equiv}CH$; $CH_3C{\equiv}CCH_3$.

allotropes Alternative forms of an element that differ in the way the atoms are linked. *Examples:* O_2 and O_3; white and gray tin.

alloy A mixture of two or more metals formed by melting, mixing, and then cooling. A *substitutional alloy* is an alloy in which atoms of one metal are substituted for atoms of another metal. An *interstitial alloy* is an alloy in which atoms of one metal lie in the gaps in the lattice formed by atoms of another metal. A *homogeneous alloy* is an alloy in which the atoms of the elements are distributed uniformly. A *heterogeneous alloy* is an alloy that consists of (micro)crystalline phases with different compositions.

alpha (α) decay Nuclear decay that involves the emission of an α particle.

alpha (α) helix One type of secondary structure adopted by a polypeptide chain, in the form of a right-handed helix.

alpha (α) particle Positively charged, subatomic particle emitted from some radioactive nuclei; nucleus of a helium atom ($^4_2He^{2+}$).

alternating copolymer See *copolymer*.

ambidentate ligand A ligand that can coordinate to a metal atom by using atoms of different elements. *Example:* SCN^-, which may coordinate via S or N.

amide An organic compound formed by the reaction of an amine and a carboxylic acid in which the acidic —OH group has been replaced by an amino group or a substituted amino group. An amide contains the group $—CONR_2$. *Example:* CH_3CONH_2, acetamide.

amine A compound derived from ammonia by replacing various numbers of H atoms with organic groups; the number of hydrogen atoms replaced determines the classification as *primary*, *secondary*, or *tertiary*. *Examples*: CH_3NH_2 (primary); $(CH_3)_2NH$ (secondary); $(CH_3)_3N$ (tertiary); $(CH_3)_4N^+$, quaternary ammonium ion.

amino acid A carboxylic acid that also contains an amino group. The *essential amino acids* are amino acids that must be ingested as a part of the diet. *Example:* NH_2CH_2COOH, glycine (nonessential).

amino group The functional group $—NH_2$ characteristic of *amines*.

amorphous solid A solid in which the atoms, ions, or molecules lie in a random jumble with no long-range order. *Examples:* glass, butter. Compare with *crystalline solid*.

amount of substance n The number of entities in a sample divided by the Avogadro constant. Also referred to as *chemical amount*. See *mole*.

ampere A The SI unit of electric current. See also Appendix 1B.

amphiprotic Having the ability both to donate and accept protons. Amphiprotic species are often called *amphoteric*. Examples: H_2O; HCO_3^-.

amphoteric Having the ability to react with both acids and bases. *Examples:* Al; Al_2O_3.

amplitude The height of a mathematical function above 0. On a graph depicting a wave, the height of the wave above the center line.

analysis See *chemical analysis*.

analyte The solution of unknown concentration in a titration. Normally, the analyte is in the flask, not the buret.

analytical chemistry The study and application of the techniques for identifying substances and measuring their amounts.

angular wavefunction $Y(\theta, \phi)$ The angular part of a wavefunction, particularly the angular component of the wavefunctions of the hydrogen atom; the probability amplitude of an electron as a function of orientation around the nucleus.

anhydride See *acid anhydride.*

anhydrous Lacking water. *Example:* $CuSO_4$, the anhydrous form of copper(II) sulfate. Compare with *hydrate.*

anion A negatively charged ion. *Examples:* F^-; SO_4^{2-}.

anisotropic Depending on orientation.

anode The electrode at which oxidation occurs.

antibonding orbital A molecular orbital that, when occupied, contributes to an overall raising of the energy of a molecule.

antiferromagnetic material A substance in which electron spins on neighboring atoms are locked into an antiparallel arrangement over large regions. *Example:* manganese.

antilogarithm If the logarithm to the base B is x, then the antilogarithm of x is B^x. The *common antilogarithm* of x is 10^x. The *natural antilogarithm* of x is the exponential e^x.

antioxidant A substance that reacts with radicals and so prevents the oxidation of another substance.

antiparticle A particle with the same mass as a subatomic particle but with opposite charge. *Example:* positron, the antiparticle of an electron.

applied research Investigations directed toward the solution of real-world problems. See also *basic research.*

aqueous solution A solution in which the solvent is water.

arene An aromatic hydrocarbon.

aromatic hydrocarbon An organic hydrocarbon that includes a benzene ring as part of its structure. *Examples:* C_6H_6 (benzene); $C_6H_5CH_3$ (toluene); $C_{10}H_8$ (naphthalene).

Arrhenius acid A compound that contains hydrogen and releases hydrogen ions (H^+) in water. *Examples:* HCl; CH_3COOH; but not CH_4.

Arrhenius base A compound that produces hydroxide ions (OH^-) in water. *Examples:* NaOH; NH_3; but not Na.

Arrhenius behavior A reaction shows Arrhenius behavior if a plot of $\ln k$ against $1/T$ is a straight line. See *Arrhenius equation.*

Arrhenius equation The equation $\ln k = \ln A - E_a/RT$ for the commonly observed temperature dependence of a rate constant k. An *Arrhenius plot* is a graph of $\ln k$ against $1/T$; if the plot is a straight line, then the reaction is said to show *Arrhenius behavior.*

Arrhenius parameters The *pre-exponential factor A* (also called the *frequency factor*) and the *activation energy* E_a. See also *Arrhenius equation.*

aryl group An aromatic group. *Example:* $-C_6H_5$, phenyl.

atactic polymer See *polymer.*

atmosphere (1) The layer of gases surrounding a planet (specifically, the air for the planet Earth). (2) A unit of pressure (1 atm = $1.013\ 25 \times 10^5$ Pa).

atom (1) The smallest particle of an element that has the chemical properties of that element. (2) An electrically neutral species consisting of a nucleus and its surrounding electrons.

atomic hypothesis The proposal advanced by John Dalton that matter is composed of atoms.

atomic mass unit u (formerly amu) One-twelfth the mass of one atom of carbon-12.

atomic nucleus See *nucleus.*

atomic number Z The number of protons in the nucleus of an atom; this number determines the identity of the element and the number of electrons in the neutral atom.

atomic orbital A region of space in which there is a high probability of finding an electron in an atom. An *s-orbital* is a spherical region; a *p-orbital* has two lobes, on opposite sides of the nucleus; a *d-orbital* typically has four lobes, with the nucleus at the center; an *f-orbital* has a more complicated arrangement of lobes.

atomic radius Half the distance between the centers of neighboring atoms in a solid or a homonuclear molecule.

atomic structure The arrangement of electrons around the nucleus of an atom.

atomic weight See *molar mass.*

***Aufbau* principle** See *building-up principle.*

autoionization See *autoprotolysis.*

autoprotolysis A reaction in which a proton is transferred between two molecules of the same substance. The products are the conjugate acid and conjugate base of the substance. *Example:* $2\ H_2O(l) \rightleftharpoons H_3O^+(aq) + OH^-(aq)$

autoprotolysis constant The equilibrium constant for an autoprotolysis reaction. *Example:* For water, K_w, with $K_w = [H_3O^+][OH^-]$.

average bond enthalpy $\Delta H_B(A{-}B)$ See *mean bond enthalpy.*

average reaction rate The reaction rate calculated by measuring the change in concentration of a reactant or product over a finite time interval (and hence an average of the changing rate within that interval). See also *unique average rate.*

Avogadro constant The number of objects per mole of objects ($N_A = 6.022\ 14 \times 10^{23}\ mol^{-1}$). *Avogadro's number* is the number of objects *in* one mole of objects (that is, the dimensionless number $6.022\ 14 \times 10^{23}$).

Avogadro's principle The volume of a sample of gas at a given temperature and pressure is proportional to the amount of gas molecules in the sample: $V \propto n$.

axial bond A bond that is perpendicular to the molecular plane in a bipyramidal molecule.

axial lone pair A lone pair lying on the axis of a bipyramidal molecule.

azeotrope A mixture of liquids that boils without change of composition. A *minimum-boiling azeotrope* has a boiling point lower than that of either component; a *maximum-boiling azeotrope* has a boiling point higher than that of either component.

azimuthal quantum number l See *orbital angular momentum quantum number*.

background radiation The average nuclear radiation to which the Earth's inhabitants are exposed daily.

balanced equation See *chemical equation*.

ball-and-stick model A depiction of a molecule in which atoms are represented by balls and bonds are represented by sticks.

Balmer series A family of spectral lines (some of which lie in the visible region) in the spectrum of atomic hydrogen.

band gap A range of energies for which there are no orbitals in a solid. Unless stated otherwise, the band gap refers to the gap between the valence band and the conduction band.

band of stability A region of a plot of mass number against atomic number corresponding to the existence of stable nuclei.

bar A unit of pressure: 1 bar = 10^5 Pa.

barometer An instrument for measuring the atmospheric pressure.

base See *Arrhenius base*; *Brønsted base*; *Lewis base*. Used alone, "base" normally means a Brønsted base.

base buffer See *buffer*.

base ionization constant See *basicity constant*.

base pair Two specific nucleotides that link one complementary strand of a DNA molecule to the other by means of hydrogen bonding: adenine pairs with thymine and guanine pairs with cytosine.

base units The units of measurement in the International System (SI) in terms of which all other units are defined. *Examples: kilogram* for mass; *meter* for length; *second* for time; *kelvin* for temperature; *ampere* for electric current.

basic ion An ion that acts as a Brønsted base. *Example:* $CH_3CO_2^-$.

basicity constant K_b The equilibrium constant for proton transfer from water to a base: for a base B, $K_b = [BH^+][OH^-]/[B]$.

basic oxide An oxide that is a Brønsted base. The oxides of metallic elements are generally basic. *Examples:* Na_2O; MgO.

basic oxygen process The production of iron by forcing oxygen and powdered limestone through the molten metal.

basic research Investigations aimed at discovering the reasons for phenomena and the fundamental principles of chemistry, the synthesis of new materials, and the investigation of their properties. See also *applied research*.

basic solution A solution with pH > 7.

battery A collection of galvanic cells joined in series, so the voltage it produces is the sum of the voltages of each cell.

becquerel Bq The SI unit of radioactivity (1 disintegration per second).

Beer's law The absorbance of a sample is proportional to the molar concentration of the absorbing species and the length of the sample through which the radiation passes.

beta (β) decay Nuclear decay that involves the emission of a β particle.

beta (β) particle A fast electron emitted from a nucleus in a radioactive decay.

beta (β) sheet One type of planar secondary structure adopted by a polypeptide, in the form of a pleated sheet.

bimolecular reaction An elementary reaction in which two molecules, atoms, or ions come together and form a product. *Example:* $O + O_3 \rightarrow O_2 + O_2$.

binary Consisting of two components, as in *binary mixture* and *binary (ionic or molecular) compound*. *Examples:* acetone and water (a binary mixture); HCl, $CaCl_2$, C_6H_6 (binary compounds; $CaCl_2$ is ionic, HCl and C_6H_6 are molecular).

biochemistry The study of biologically important substances, processes, and reactions.

bioenergetics The deployment and utilization of energy in living cells.

biomass The organic material of the planet produced annually by photosynthesis.

biradical A species with two unpaired electrons. *Example:* $\cdot CH_2CH_2CH_2\cdot$.

black body An object that absorbs and emits all frequencies of radiation without favor.

black-body radiation The radiation emitted by a *black body*.

block (*s* block, *p* block, *d* block, *f* block) The region of the periodic table containing elements for which, according to the building-up principle, the corresponding subshell is currently being filled.

block copolymer See *copolymer*.

body-centered cubic structure bcc A crystal structure with a unit cell in which a central atom lies at the center of a cube formed by eight others.

Bohr frequency condition The relation between the change in energy of an atom or molecule and the frequency of radiation emitted or absorbed: $\Delta E = h\nu$.

Bohr radius a_0 In an early model of the hydrogen atom, the radius of the lowest energy orbit; now a specific combination of fundamental constants ($a_0 = 4\pi\varepsilon_0\hbar^2/m_e e^2 \approx 52.9$ pm) that occurs in the description of the wavefunctions of hydrogen.

boiling Rapid vaporization taking place throughout a liquid. See *boiling temperature*.

boiling point b.p. See *boiling temperature*; *normal boiling point*.

boiling-point constant k_b The constant of proportionality between the boiling-point elevation and the molality of a solute.

boiling-point elevation The increase in normal boiling point of a solvent caused by the presence of a solute (a *colligative property*).

boiling temperature (1) The temperature at which a liquid boils. (2) The temperature at which a liquid is in equilibrium with its vapor at the pressure of the surroundings; vaporization then occurs throughout the liquid, not only at the liquid's surface.

Boltzmann constant k The value of R/N_A, where R is the gas constant and N_A is the Avogadro constant; $k = 1.380\,66 \times 10^{-23}\ \text{J·K}^{-1}$.

Boltzmann formula (for the entropy) The formula $S = k \ln W$, where k is the Boltzmann constant and W is the number of atomic arrangements that correspond to the same energy.

bond A link between atoms. See also *covalent bond*; *double bond*; *ionic bond*; *triple bond*.

bond angle In an A—B—C molecule or part of a molecule, the angle between the B—A and B—C bonds.

bond enthalpy ΔH_B(X—Y) The enthalpy change accompanying the dissociation of a bond. *Example:* $H_2(g) \rightarrow 2\ H(g)$, $\Delta H_B(\text{H—H}) = +436\ \text{kJ·mol}^{-1}$.

bonding orbital A molecular orbital that, when occupied, results in an overall lowering of the energy of a molecule.

bond length The distance between the nuclei of two atoms joined by a bond.

bond order The number of electron-pair bonds that link a specific pair of atoms.

bond parameters The characteristic features of a bond. *Examples:* bond length; bond enthalpy.

Born-Haber cycle A closed series of reactions used to express the enthalpy of formation of an ionic solid in terms of contributions that include the lattice enthalpy.

Born interpretation The interpretation of the square of the wavefunction, ψ, of a particle as the probability density for finding the particle in a region of space.

boundary conditions Constraints on the values of the wavefunction of a particle.

boundary surface The surface showing the region of space within which there is about 90% probability of finding an electron when it occupies a specific orbital.

Boyle's law At constant temperature, and for a given sample of gas, the volume is inversely proportional to the pressure: $P \propto 1/V$.

Bragg equation An equation relating the angle of diffraction of x-rays to the spacing of layers of atoms in a crystal ($\lambda = 2d \sin\theta$).

branched alkane An alkane with hydrocarbon side chains.

branching Description of a step in a chain reaction in which more than one chain carrier is formed in a propagation step. *Example:* $\cdot\text{O}\cdot + H_2 \rightarrow \cdot\text{OH} + \cdot\text{H}$.

Bravais lattices The 14 basic patterns of unit cells from which a crystal may be built.

breeder reactor A reactor used to generate nuclear fuel by making use of neutrons that are not moderated.

Brønsted acid A proton donor (a source of hydrogen ions, H^+). *Examples:* HCl; CH_3COOH; HCO_3^-; NH_4^+.

Brønsted base A proton acceptor (a species to which hydrogen ions, H^+, can bond). *Examples:* OH^-; Cl^-; $CH_3CO_2^-$; HCO_3^-; NH_3.

Brønsted equilibrium The proton transfer equilibrium $\text{acid}_1 + \text{base}_2 \rightleftharpoons \text{acid}_2 + \text{base}_1$. *Example:* $CH_3COOH(aq) + H_2O(l) \rightleftharpoons H_3O^+(aq) + CH_3CO_2^-(aq)$.

Brønsted-Lowry definition A definition of acids and bases in terms of the ability of molecules and ions to participate in proton transfer.

Brownian motion The ceaseless jittering motion of colloidal particles caused by the impact of solvent molecules.

buffer A solution that resists any change in pH when small amounts of acid or base are added. An *acid buffer* stabilizes solutions at pH < 7 and a *base buffer* stabilizes solutions at pH > 7. *Examples:* a solution containing CH_3COOH and $CH_3CO_2^-$ (acid buffer); a solution containing NH_3 and NH_4^+ (base buffer).

buffer capacity An indication of the amount of acid or base that can be added before a buffer loses its ability to resist the change in pH.

building-up principle The procedure for arriving at the ground-state electron configurations of atoms and molecules.

bulk matter Matter composed of large numbers of atoms. See *bulk properties*.

bulk properties Properties that depend on the collective behavior of large numbers of atoms. *Examples:* melting point; vapor pressure; internal energy.

buret A narrow, graduated tube fitted with a stopcock, used to measure the volume of liquid delivered into another vessel.

calibration Interpretation of an observation by comparison with known information.

calorie cal A unit of energy. The unit is now defined in terms of the joule by 1 cal = 4.184 J exactly. The dietary Calorie is 1 kcal.

calorimeter An apparatus used to determine the heat released or absorbed in a process by measuring the temperature change.

calorimetry The use of a calorimeter to measure the thermochemical properties of reactions.

candela cd The SI unit of luminous intensity. See also Appendix 1B.

capillary action The rise of liquids up narrow tubes.

carbide A binary compound of a metal or metalloid and carbon: carbides may be *saline* (saltlike), *covalent*, and *interstitial*. *Examples:* CaC_2 (saline); SiC (covalent); WC (interstitial).

carbocation A positively charged organic molecular ion. *Example:* $C(CH_3)_3^+$.

carbohydrate A compound of general formula $C_m(H_2O)_n$, although small deviations from this general formula are often encountered. Carbohydrates include cellulose, starches, and sugars. *Examples:* $C_6H_{12}O_6$, glucose; $C_{12}H_{22}O_{11}$, sucrose.

carbon cycle The sequence of physical processes and chemical reactions by means of which carbon atoms circulate through the environment.

carbonyl group A $>$CO group in an inorganic or organic compound.

carboxyl group The functional group —COOH. See *carboxylic acid.*

carboxylic acid An organic compound containing the carboxyl group, —COOH. *Examples:* CH_3COOH, acetic acid; C_6H_5COOH, benzoic acid.

catalyst A substance that increases the rate of a reaction without being consumed in the reaction. A catalyst is *homogeneous* if it is present in the same phase as the reactants and *heterogeneous* if it is in a different phase from the reactants. *Examples:* homogeneous, $Br^-(aq)$ for the decomposition of $H_2O_2(aq)$; heterogeneous, Pt in the Ostwald process.

catenate To form chains or rings of atoms. *Examples:* O_3; S_8.

cathode The electrode at which reduction occurs.

cathodic protection Protection of a metal object by connecting it to a more strongly reducing metal, which acts as a sacrificial anode.

cation A positively charged ion. *Examples:* Na^+; NH_4^+; Al^{3+}.

cell diagram A statement of the arrangement of electrodes in an electrochemical cell. *Example:* $Zn(s)|Zn^{2+}(aq)||Cu^{2+}(aq)|Cu(s)$.

cell potential See *electromotive force.*

Celsius scale A temperature scale on which the freezing point of water is at 0 degrees and its normal boiling point is at 100 degrees. Units on this scale are degrees Celsius, °C.

cement A solid obtained by the action of heat on silicates and aluminosilicates, with gypsum and lime added.

ceramic (1) A solid obtained by the action of heat on clay. (2) A noncrystalline inorganic solid usually containing oxides, borides, and carbides.

cesium-chloride structure A crystal structure the same as that of solid cesium chloride.

chain branching A propagation step in a chain reaction when more than one chain carrier is formed.

chain carrier An intermediate in a chain reaction.

chain reaction A reaction in which an intermediate reacts to produce another intermediate in a series of elementary reactions. *Example:* $Br\cdot + H_2 \rightarrow HBr + H\cdot$; $H\cdot + Br_2 \rightarrow HBr + Br\cdot$.

chalcogens Oxygen, sulfur, selenium, and tellurium in Group 16/VI of the periodic table.

change of state The change of a substance from one of its physical states to another of its physical states. *Example:* melting, solid → liquid.

characteristic (of a logarithm) The number in front of the decimal point.

charge A measure of the strength with which a particle can interact electrostatically with another particle.

charge balance The requirement that, because a solution is neutral overall, the concentration of positive charge due to cations must equal the concentration of negative charge due to anions.

charge-transfer transition A transition in which an electron is excited from the ligands of a complex to the metal atom, or vice versa.

Charles's law The volume of a given sample of gas at constant pressure is directly proportional to its absolute temperature: $V \propto T$.

chelate A complex containing at least one polydentate ligand that forms a ring of atoms including the central metal atom. *Example:* $[Co(en)_3]^{3+}$.

chemical analysis The determination of the chemical composition of a sample. See also *qualitative*, *quantitative*.

chemical bond See *bond.*

chemical change The conversion of one or more substances into different substances.

chemical element See *element.*

chemical engineering The study of industrial chemical processes.

chemical equation A statement in terms of chemical formulas summarizing the qualitative information about the chemical changes taking place in a reaction and the quantitative information that atoms are neither created nor destroyed in a chemical reaction. In a *balanced chemical equation* (commonly called a "chemical equation"), the same number of atoms of each element appear on both sides of the equation.

chemical equilibrium A dynamic equilibrium between reactants and products in a chemical reaction.

chemical formula A collection of chemical symbols and subscripts that shows the composition of a substance.

See also *condensed structural formula*; *empirical formula*; *molecular formula*; *structural formula*.

chemical kinetics The study of the rates of reactions and the steps by which they occur.

chemical nomenclature The systematic naming of compounds.

chemical plating The deposition of a metal surface on an object by making use of a chemical reduction reaction.

chemical property The ability of a substance to participate in a chemical reaction.

chemical reaction A chemical change in which one substance responds to the presence of another, to a change of temperature, or to some other influence.

chemical symbol One- or two-letter abbreviation of an element's name.

chemiluminescence The emission of light by products formed in energetically excited states during a chemical reaction.

chemistry The branch of science concerned with the study of matter and the changes that matter can undergo.

chiral (molecule or complex) Not able to be superimposed on its own mirror image. *Examples:* $CH_3CH(NH_2)COOH$; CHBrClF; $[Co(en)_3]^{3+}$.

chloralkali process The production of chlorine and sodium hydroxide by the electrolysis of aqueous sodium chloride.

cholesteric phase A liquid crystal phase in which layers of parallel molecules are twisted relative to one another in such a way that the orientations of the molecules form a spiral structure.

chromatogram A record of the signal from the detector, or the paper record, obtained in a chromatographic analysis of a mixture.

chromatography A separation technique that relies on the ability of different phases to adsorb substances to different extents.

cis-trans isomerization The conversion of a cis isomer into a trans isomer, and vice versa. *Example:* *cis*-butene → *trans*-butene.

classical mechanics The laws of motion proposed by Isaac Newton in which particles travel in definite paths in response to forces.

clathrate A structure in which a molecule of one substance sits in a cage made up of molecules of another substance, typically water. *Example:* SO_2 in water.

Clausius-Clapeyron equation An equation that gives the quantitative dependence of the vapor pressure of a substance on the temperature.

Claus process A process for obtaining sulfur from the H_2S in oil wells by the oxidation of H_2S with SO_2; the latter is formed by the oxidation of H_2S with oxygen.

closed shell (or subshell) A shell (or subshell) containing the maximum number of electrons allowed by the exclusion principle. *Example:* the neonlike core $1s^22s^22p^6$.

closed system See *system*.

close-packed structure A crystal structure in which atoms occupy the smallest total volume with the least empty space. *Examples:* hexagonal close packing and cubic close packing of identical spheres.

coagulation The formation of aggregates from colloidal particles.

cohesion The act or state in which the particles of a substance stick to one another.

cohesive forces The forces that bind the molecules of a substance together to form a bulk material and that are responsible for condensation.

coinage metals The elements copper, silver, and gold.

collective model A model of the nucleus in which the nucleons occupy quantized energy levels but interact with one another.

colligative property A property that depends only on the relative number of solute and solvent particles present in a solution, and not on the chemical identity of the solute. *Examples:* elevation of boiling point; depression of freezing point; osmosis.

collision cross-section The area that a molecule presents as a target during a collision.

collision theory The theory of elementary gas-phase bimolecular reactions in which it is assumed that molecules react only if they collide with a characteristic minimum kinetic energy.

colloid (or *colloidal suspension*) A dispersion of tiny particles with diameters between 1 nm and 1 μm in a gas, liquid, or solid. *Example:* milk.

combined gas law A combination of Boyle's law and Charles's law that allows the pressure, volume, or temperature of a sample of an ideal gas to be predicted after a change in state. $P_1V_1/T_1 = P_2V_2/T_2$.

combustion A reaction in which an element or compound burns in oxygen. *Example:* $CH_4(g) + 2\,O_2(g) \rightarrow CO_2(g) + 2\,H_2O(g)$.

combustion analysis The determination of the composition of a sample by measuring the masses of the products of its combustion.

common antilogarithm See *antilogarithm*.

common-ion effect Reduction of the solubility of one salt by the presence of another salt with one ion in common. *Example:* the lower solubility of AgCl in NaCl(aq) than in pure water.

common logarithm See *logarithm*.

common name An informal name for a compound that may give little or no clue to the compound's composition. *Examples:* water; aspirin; acetic acid.

competing reaction A reaction occurring at the same time as the reaction of interest and using some of the same reactants.

complementarity The impossibility of knowing the position of a particle with arbitrarily great precision if its linear momentum is known precisely.

complementary color The color that white light becomes when one of the colors present in it is removed.

complete ionic equation A balanced chemical equation expressed in terms of the cations and anions present in solution. *Example:* $Ag^+(aq) + NO_3^-(aq) + Na^+(aq) + Cl^-(aq) \rightarrow AgCl(s) + NO_3^-(aq) + Na^+(aq)$.

complex (1) The combination of a Lewis acid and a Lewis base linked by a coordinate covalent bond. (2) A species consisting of several ligands (the Lewis bases) that have an independent existence bonded to a single central metal atom or ion (the Lewis acid). *Examples:* (1) $H_3N{-}BF_3$; (2) $[Fe(H_2O)_6]^{3+}$; $[PtCl_4]^-$.

composite material A synthetic material composed of a polymer and one or more other substances that have been solidified together.

compound (1) A specific combination of elements that can be separated into its elements by chemical techniques but not physical techniques. (2) A substance consisting of atoms of two or more elements in a definite, unchangeable ratio.

compress To reduce the volume of a sample.

compressibility The ability to be confined into a smaller volume.

compression factor Z The ratio of the actual molar volume of a gas to the molar volume of an ideal gas under the same conditions.

concentration The quantity of a substance in a given volume. See also *molar concentration*.

concentration cell A galvanic cell in which the electrodes have the same composition but are at different concentrations.

condensation The formation of a liquid or solid phase from the gas phase of the substance.

condensation polymer A polymer formed by a chain of condensation reactions. *Examples:* polyesters; polyamides (nylon).

condensation reaction A reaction in which two molecules combine to form a larger one and a small molecule is eliminated. *Example:* $CH_3COOH + C_2H_5OH \rightarrow CH_3COOC_2H_5 + H_2O$.

condensed phase A solid or liquid phase; not a gas.

condensed structural formula A compact version of the structural formula, showing how the atoms are grouped together. *Example:* $CH_3CH(CH_3)CH_3$ for methylpropane.

conduction band An incompletely occupied band of energy levels in a solid.

configuration See *electron configuration*.

conformations Molecular shapes that can be interchanged by rotation about bonds, without bond breakage and reformation.

congeners Elements in the same group of the periodic table.

conjugate acid The Brønsted acid formed when a Brønsted base has accepted a proton. *Example:* NH_4^+ is the conjugate acid of NH_3.

conjugate acid-base pair A Brønsted acid and its conjugate base. *Examples:* HCl and Cl^-; NH_4^+ and NH_3.

conjugate base The Brønsted base formed when a Brønsted acid has donated a proton. *Example:* NH_3 is the conjugate base of NH_4^+.

connectivity (of atoms in a molecule) The pattern in which the atoms in a molecule are bonded to one another.

constructive interference Interference that results in an increased amplitude of a wave. Compare with *destructive interference*.

contact process The production of sulfuric acid by the combustion of sulfur and the catalyzed oxidation of sulfur dioxide to sulfur trioxide.

convection The bulk motion of regions of a fluid, often as a result of differences in density brought about by differences in temperature.

conversion factor A factor that is used to convert a measurement from one unit to another.

cooling curve A graph of the variation of the temperature of a sample over time as it loses energy at a constant rate.

coordinate Use of a lone pair to form a coordinate covalent bond. *Examples:* $F_3B + {:}NH_3 \rightarrow F_3B{-}NH_3$; $Ni + 4\,CO \rightarrow Ni(CO)_4$.

coordinate covalent bond A bond formed between a Lewis base and a Lewis acid by sharing an electron pair originally belonging to the Lewis base.

(6,6)-coordination A crystal structure in which each cation is surrounded by six anions and each anion is surrounded by six cations.

coordination compound A neutral complex or an ionic compound in which at least one of the ions is a complex. *Examples:* $Ni(CO)_4$; $K_3[Fe(CN)_6]$.

coordination isomers Isomers that differ by the exchange of one or more ligands between a cationic complex and an anionic complex.

coordination number (1) The number of nearest neighbors of an atom in a solid. (2) For ionic solids, the coordination number of an ion is the number of nearest neighbors of opposite charge. (3) For complexes, the number of ligands attached to the central metal ion.

coordination sphere The ligands directly attached to the central ion in a complex.

copolymer A polymer formed from a mixture of different monomers. In *random copolymers*, the sequence of monomers has no particular order; in *alternating copolymers*, two monomers alternate; in *block copolymers*, regions of one monomer alternate with regions of another; in *graft copolymers*, chains of one monomer are attached to a backbone chain of a second monomer.

core The inner closed shells of an atom.

core electrons The electrons that belong to an atom's core.

corrosion The unwanted reaction of a material that results in the dissolution or consumption of the material. *Example*: the unwanted oxidation of a metal.
corrosive (1) A reagent that can cause corrosion. (2) Having a high reactivity, such as the reactivity of a strong oxidizing agent or a concentrated acid or base.
coulombic attraction The attraction between opposite electric charges.
Coulomb potential energy The potential energy of an electric charge in the vicinity of another electric charge; the potential energy is inversely proportional to the separation of the charges.
Coulomb's law The potential energy of a pair of electric charges is inversely proportional to the distance between them and proportional to the product of the charges.
couple See *redox couple.*
covalent bond A pair of electrons shared between two atoms.
covalent carbide See *carbide.*
covalent radius The contribution of an atom to the length of a covalent bond.
cracking The process of converting petroleum fractions into smaller molecules with more double bonds. *Example:* $CH_3(CH_2)_6CH_3 \rightarrow CH_3(CH_2)_3CH_3 + CH_3CH{=}CH_2$.
critical mass The mass of fissionable material above which so few neutrons escape from a sample of nuclear fuel that the fission chain reaction is sustained; a greater mass is *supercritical* and a smaller mass is *subcritical.*
critical pressure P_c The vapor pressure of a liquid at its critical temperature.
critical temperature T_c The temperature at and above which a substance cannot exist as a liquid.
crown ether A cyclic ether with alternating $-CH_2CH_2O-$ units.
cryogenics The study of matter at very low temperatures.
cryoscopy The measurement of molar mass by using the depression of freezing point.
crystal face A flat plane forming an edge of a crystal.
crystal field The electrostatic influence of the ligands (modeled as point negative charges) on the central ion of a complex. *Crystal field theory* is a rationalization of the optical, magnetic, and thermodynamic properties of complexes in terms of the crystal field of their ligands.
crystalline solid A solid in which the atoms, ions, or molecules lie in an orderly array. *Examples:* NaCl; diamond; graphite. Compare with *amorphous solid.*
crystallization The process in which a solute comes out of solution as crystals.
cubic close-packed structure ccp A close-packed structure with an ABCABC . . . pattern of layers.
curie Ci A unit of activity (for radioactivity).
current I The rate of supply of charge; current is measured in *amperes* (A), with $1\ A = 1\ C{\cdot}s^{-1}$.
cycle (1) In thermodynamics, a sequence of changes that begins and ends at the same state. (2) In spectroscopy, one complete reversal of the direction of the electromagnetic field and its return to the original direction.
cycloalkane A saturated aliphatic hydrocarbon in which the carbon atoms form a ring. *Example:* C_6H_{12}, cyclohexane.

Dalton's law of partial pressures The total pressure of a mixture of gases is the sum of the partial pressures of its components.
data The information provided or obtained from experiments.
daughter nucleus A nucleus that is the product of a nuclear decay.
***d–d* transition** A transition in which an electron is excited from one *d*-orbital to another.
de Broglie relation The proposal that every particle has wavelike properties and that its wavelength, λ, is related to its mass by $\lambda = h/(\text{mass} \times \text{velocity})$.
debye D The unit used to report electric dipole moments: $1\ D = 3.336 \times 10^{-30}\ C{\cdot}m$.
decant To pour off a liquid from on top of another, denser liquid or from a solid.
decay constant k The rate constant for radioactive decay.
decomposition A reaction in which a substance is broken down into simpler substances; *thermal decomposition* is decomposition brought about by heat. *Example:* $CaCO_3(s) \xrightarrow{\Delta} CaO(s) + CO_2(g)$.
decomposition vapor pressure The pressure of the gaseous decomposition product of a solid at equilibrium.
definite integral An *integral* with limits attached. See also Appendix 1F.
degenerate Having the same energy. *Example*: atomic orbitals in the same subshell.
degrees of freedom (1) In phase equilibria, the number of intensive variables that can be changed independently. (2) In mechanics, the number of independent modes of motion.
dehydrating agent A reagent that removes water or the elements of water from a compound. *Example:* H_2SO_4.
dehydrogenation The removal of a hydrogen atom from each of two neighboring carbon atoms, resulting in the formation of a carbon-carbon multiple bond.
delocalized Spread over a region. In particular, *delocalized electrons* are electrons that spread over several atoms in a molecule.
delta Δ (in a chemical equation) A symbol that signifies that the reaction occurs at elevated temperatures.

delta *X* ΔX The difference between the final and initial values of a property, $\Delta X = X_{final} - X_{initial}$. *Examples:* ΔT; ΔE.

denaturation The loss of structure of a large molecule, such as a protein.

density *d* The mass of a sample of a substance divided by its volume: $d = m/V$.

density isosurface A drawing that represents a molecular structure as a surface showing the distribution of electrons in a molecule; the surface corresponds to locations with the same electron density.

deposition The condensation of a vapor directly to a solid. Deposition is the reverse of *sublimation*.

deprotonation Loss of a proton from a Brønsted acid. *Example:* $NH_4^+(aq) + H_2O(l) \rightarrow H_3O^+(aq) + NH_3(aq)$.

derived unit A combination of base units. *Examples:* centimeters cubed (cm^3); joules ($kg{\cdot}m^2{\cdot}s^{-2}$).

descriptive chemistry The description of the preparation, properties, and applications of the elements and their compounds.

destructive interference Interference that results in a reduced amplitude of a wave. Compare with *constructive interference*.

deuteron The nucleus of a deuterium atom, $^2H^+$, consisting of a proton and a neutron.

diagonal relationship A similarity in properties between diagonal neighbors in the periodic table, especially for main-group elements in Periods 2 and 3 at the left of the table. *Examples:* Li and Mg; Be and Al.

diamagnetic (substance) A substance that tends to be pushed out of a magnetic field; consisting of atoms, ions, or molecules with no unpaired electrons. *Examples:* most common substances.

diamine An organic compound that contains two $-NH_2$ groups.

diathermic Able to permit the passage of energy as heat.

diatomic ion An ion that consists of two atoms with a net charge.

diatomic molecule A molecule that consists of two atoms. *Examples:* H_2; CO.

differential calculus The part of mathematics that deals with the slopes of curves and with infinitesimal quantities. See also Appendix 1F.

diffraction The deflection of waves and the resulting interference caused by an object in their path. See also *x-ray diffraction*.

diffraction pattern The pattern of bright spots against a dark background resulting from diffraction.

diffusion The spreading of one substance through another substance.

dilute To reduce the concentration of a solute by adding more solvent.

dimer The union of two identical molecules. *Example:* Al_2Cl_6 formed from two $AlCl_3$ molecules.

diol An organic compound with two $-OH$ groups.

dipeptide An *oligopeptide* formed by the condensation of two amino acids.

dipole See *electric dipole*; *instantaneous dipole moment*.

dipole-dipole interaction The interaction between two electric dipoles: like partial charges repel and opposite partial charges attract.

dipole-induced dipole interaction The interaction between an electric dipole and the instantaneous dipole it induces in a nonpolar molecule.

dipole moment See *electric dipole moment*.

diprotic An acid with two acidic hydrogen atoms. See also *polyprotic acid or base*.

disaccharide A carbohydrate molecule that is composed of two saccharide units. *Example:* $C_{12}H_{22}O_{11}$, sucrose.

dispersion See *suspension*.

dispersion force See *London force*.

disproportionation A redox reaction in which a single element is simultaneously oxidized and reduced. *Example:* $2\ Cu^+(aq) \rightarrow Cu(s) + Cu^{2+}(aq)$.

dissociation (1) The breaking of a bond. (2) The separation of ions that occurs when an ionic solid dissolves.

dissociation constant See *acidity constant*.

dissociation energy *D* The energy required to separate bonded atoms.

distillate A liquid obtained by distillation.

distillation The separation of the components of a mixture by making use of their different volatilities.

distribution (of molecular speeds) The fraction of gas molecules moving at each speed at any instant.

disulfide link An —S—S— link that contributes to the secondary and tertiary structures of polypeptides.

domain A region of a metal in which the electron spins of the atoms are aligned, so resulting in *ferromagnetism*.

doping The addition of a known, small amount of a second substance to an otherwise pure solid substance.

***d*-orbital** See *atomic orbital*.

dose equivalent The actual dose of radiation experienced by a sample modified to take into account the *relative biological effectiveness* of the radiation. The dose equivalent is measured in *rem*. See also *roentgen equivalent man*; *sievert*.

double bond (1) Two electron pairs shared by neighboring atoms. (2) One σ-bond and one π-bond between neighboring atoms.

Downs process The production of sodium and chlorine by the electrolysis of molten sodium chloride.

Dow process The electrolytic production of magnesium from molten magnesium chloride.

dry cell A galvanic cell in which the electrolyte is a moist paste. Commercial dry cells contain a graphite cathode and a zinc anode.

drying agent A substance that absorbs water and thus maintains a dry atmosphere. *Example:* phosphorus(V) oxide.

ductility Ability to be drawn out into a wire (as for a metal).

duplet The $1s^2$ electron pair of the heliumlike electron configuration.

dynamic equilibrium The condition in which a forward process and its reverse are occurring simultaneously at equal rates. *Examples:* vaporizing and condensing; chemical reactions at equilibrium.

effective nuclear charge Z_{eff} The net nuclear charge after taking into account the shielding caused by other electrons in the atom.

effervesce To bubble out of solution as a gas.

effusion The escape of a substance (particularly a gas) through a small hole.

elasticity The ability to return to the original shape after distortion.

elastomer An elastic polymer. *Example:* rubber (polyisoprene).

electrical conduction The conduction of electric charge through matter. See also *electronic conductor*; *ionic conduction.*

electric current See *current.*

electric dipole A positive charge next to an equal but opposite negative charge.

electric dipole moment μ The magnitude of the electric dipole (in debye).

electric field A region of influence that affects charged particles.

electric potential surface A molecular structure in which the net charge is calculated at each point of the density isosurface and depicted by different colors; an "elpot" surface.

electroactive species The species in a redox reaction participating in electron transfer.

electrochemical cell A system consisting of two electrodes in contact with an electrolyte. A *galvanic cell (voltaic cell)* is an electrochemical cell used to produce electricity, and an *electrolytic cell* is an electrochemical cell in which an electric current is used to cause chemical change.

electrochemical series Redox couples arranged in order of oxidizing and reducing strengths; usually arranged with strong oxidizing agents at the top of the list and strong reducing agents at the bottom.

electrochemical techniques Chemical procedures based on electrochemistry. *Examples:* electrolysis, analysis.

electrochemistry The branch of chemistry that deals with the use of chemical reactions to produce electricity, the relative strengths of oxidizing and reducing agents, and the use of electricity to produce chemical change.

electrode A metallic conductor that makes contact with an electrolyte in an electrochemical cell.

electrolysis (1) A process in which a chemical change is produced by passing an electric current through a liquid. (2) The process of driving a reaction in a nonspontaneous direction by passing an electric current through a solution.

electrolyte A substance that dissolves to give an electrically conducting solution. A *strong electrolyte* is a substance that is fully ionized in solution. A *weak electrolyte* is a molecular substance that is only partially ionized in solution. A *nonelectrolyte* does not ionize in solution. *Examples:* NaCl is a strong electrolyte; CH_3COOH is a weak electrolyte; $C_6H_{12}O_6$ is a nonelectrolyte.

electrolyte solution An ionically conducting (usually aqueous) solution.

electrolytic cell See *electrochemical cell.*

electromagnetic field The region of influence generated by accelerated charged particles.

electromagnetic radiation A wave of oscillating electric and magnetic fields; includes light, x-rays, and γ rays.

electromotive force emf, E (1) The pushing and pulling power of a reaction in an electrochemical cell. (2) The potential difference between the electrodes of an electrochemical cell when it is producing no current.

electron e^- A negatively charged subatomic particle found outside the nucleus of an atom.

electron affinity E_{ea} The energy released when an electron is added to a gas-phase atom or monatomic ion.

electron arrangement (VSEPR model) The three-dimensional geometry of the arrangement of bonds and lone pairs about a central atom in a molecule or ion.

electron capture The capture by a nucleus of one of its own atom's *s*-electrons.

electron configuration The occupancy of orbitals in an atom or molecule. *Example:* N, $1s^22s^22p^3$.

electron-deficient compound A compound with too few valence electrons for it to be assigned a valid Lewis structure. *Example:* B_2H_6.

electronegative element An element with a high electronegativity. *Examples:* O; F.

electronegativity χ (chi) The ability of an atom to attract electrons to itself when it is part of a compound.

electronic conductor A substance that conducts electricity by the motion of electrons.

electronic structure The details of the distribution of the electrons that surround the nuclei in atoms and molecules.

electronvolt eV A unit of energy; the change in potential energy of an electron when it moves through a potential difference of 1 V; 1 eV = $1.602\ 18 \times 10^{-19}$ J.

electrophile A reactant that is attracted to a region of high electron density. *Examples:* Br_2, NO_2^+.

electrophilic substitution Substitution that takes place as a result of an attack by an *electrophile. Example:* nitration of benzene.

electroplating The deposition of a thin film of metal on an object by electrolysis.

electropositive element An element that has a low electronegativity and is likely to give up electrons to another element on compound formation. *Examples:* Cs; Mg.

element (1) A substance that cannot be separated into simpler components by using chemical techniques. (2) A substance consisting of atoms of the same atomic number. *Examples:* hydrogen; gold; uranium.

elementary reaction An individual reaction step in a mechanism.

elimination reaction A reaction in which two groups or atoms on neighboring carbon atoms are removed from a molecule, thereby leaving a multiple bond between the carbon atoms. *Example:* $CH_3CHBrCH_3 + OH^- \rightarrow CH_3CH{=}CH_2 + H_2O + Br^-$.

empirical Determined by experiment.

empirical formula A chemical formula that shows the relative numbers of atoms of each element in a compound by using the simplest whole-number subscripts. *Examples:* P_2O_5; CH for benzene.

emulsion A suspension of droplets of one liquid dispersed throughout another liquid.

enantiomers A pair of optical isomers that are mirror images of, but not superimposable on, each other.

endothermic process A process that absorbs heat ($\Delta H > 0$). *Examples:* vaporization; $N_2O_4(g) \rightarrow 2\ NO_2(g)$.

end point The stage in a titration at which enough titrant has been added to bring the indicator to a color halfway between its initial and final colors.

energy *E* The capacity of a system to do work or supply heat. *Kinetic energy* is the energy of motion, and *potential energy* is the energy arising from position. The *total energy* is the sum of the kinetic and potential energies.

energy level A permitted value of the energy in a quantized system such as an atom or a molecule.

enrich In nuclear chemistry, to increase the abundance of a specific isotope.

ensemble A collection of hypothetical replications of a system.

entering group The group displacing the *leaving group* in a substitution reaction.

enthalpy *H* A state property; $H = U + PV$. A change in enthalpy is equal to the heat transferred at constant pressure.

enthalpy density (of a fuel) The enthalpy of combustion per liter (without the negative sign).

enthalpy of freezing The negative of the *enthalpy of fusion.*

enthalpy of fusion ΔH_{fus} The enthalpy change per mole accompanying fusion (melting).

enthalpy of hydration ΔH_{hyd} The enthalpy change accompanying the hydration of gas-phase ions.

enthalpy of ionization The change in enthalpy for the process $E(g) \rightarrow E^+(g) + e^-(g)$.

enthalpy of melting ΔH_{melt} See *enthalpy of fusion.*

enthalpy of mixing ΔH_{mix} The change in enthalpy that occurs when two fluids (liquids or gases) mix together.

enthalpy of solution ΔH_{sol} The change in enthalpy that occurs when a substance dissolves. The *limiting enthalpy of solution* is the enthalpy of solution for the formation of an infinitely dilute solution.

enthalpy of sublimation ΔH_{sub} The enthalpy change per mole accompanying sublimation (the direct conversion of a solid to a vapor).

enthalpy of vaporization ΔH_{vap} The enthalpy change per mole accompanying vaporization (the conversion of a substance from the liquid state to the vapor state).

entropy *S* (1) A measure of the disorder of a system. (2) A change in entropy is equal to the heat supplied reversibly to a system divided by the temperature at which the transfer occurs.

entropy of vaporization ΔS_{vap} The entropy change per mole accompanying vaporization (the conversion of a substance from the liquid state to the vapor state).

enzyme A biological catalyst.

***e*-orbital** One of the orbitals d_{z^2} or $d_{x^2-y^2}$ in an octahedral or tetrahedral complex. In an octahedral complex, the orbitals are designated e_g.

equation of state A mathematical expression relating the pressure, volume, temperature, and amount of substance present in a sample. *Example:* ideal gas law, $PV = nRT$.

equatorial bond A bond perpendicular to the axis of a molecule (particularly trigonal bipyramidal and octahedral molecules).

equatorial lone pair A lone pair in the plane perpendicular to the molecular axis.

equilibrium See *chemical equilibrium*; *dynamic equilibrium.*

equilibrium constant *K* An expression characteristic of the equilibrium composition of the reaction mixture, with a form given by the law of mass action. *Example:* $N_2(g) + 3\ H_2(g) \rightleftharpoons 2\ NH_3(g)$, $K = P_{NH_3}{}^2/P_{N_2}P_{H_2}{}^3$.

equilibrium table A table used to calculate the composition of a reaction mixture at equilibrium, given the initial composition. The columns are headed by the species and the rows are, successively, the initial composition, the change to reach equilibrium, and the equilibrium composition.

equimolar Having the same molar concentration or the same number of moles.

equipartition theorem The average energy of each degree of freedom of a molecule in a sample at a temperature *T* is equal to $\frac{1}{2}kT$ (where *k* is the Boltzmann constant).

equivalence point See *stoichiometric point.*

essential amino acid An amino acid that is an essential component of the diet because it cannot be synthesized in the body.

essential oil An oil that can be distilled from flowers and leaves (and that conveys the "essence" of a plant).

ester The product (other than water) of the reaction between a carboxylic acid and an alcohol and having the formula RCOOR′. *Example:* $CH_3COOC_2H_5$, ethyl acetate.

esterification The formation of an ester.

ether An organic compound of the form R—O—R′. *Examples:* $CH_3OC_2H_5$, ethylmethyl ether; $C_2H_5OC_2H_5$, diethyl ether.

evaporate To vaporize completely.

excited state A state other than the state of lowest energy.

exclusion principle No more than two electrons can occupy any given orbital; and when two electrons do occupy one orbital, their spins must be paired.

exothermic process A process that releases heat ($\Delta H < 0$). *Examples:* freezing; $N_2(g) + 3\ H_2(g) \rightarrow 2\ NH_3(g)$.

expanded valence shell A valence shell containing more than eight electrons. Also called an *expanded octet.* *Examples:* the valence shells of P and S in PCl_5 and SF_6.

expansion work See *work*.

experiment A test carried out under carefully controlled conditions.

exponential The exponential of x is the *natural antilogarithm* of x, namely, e^x.

exponential decay A variation with time of the form e^{-kt}. *Example*: $[A] = [A]_0e^{-kt}$.

extensive property A physical property of a substance that depends on the mass of the sample. *Examples:* volume, internal energy; enthalpy; entropy.

extrapolate To extend a graph outside the region covered by the data.

face See *crystal face*.

face-centered cubic structure fcc A crystal structure built from a cubic unit cell in which there is an atom at the center of each face and one at each corner.

Fahrenheit scale A temperature scale on which the freezing point of water is at 32 degrees and the normal boiling point is at 212 degrees. Units on this scale are degrees Fahrenheit, °F.

fallout The fine dust that settles from clouds of airborne particles after a nuclear bomb test.

Faraday constant F The magnitude of the charge per mole of electrons; $F = N_Ae = 96.485\ \text{kC·mol}^{-1}$.

Faraday's law of electrolysis The amount of product formed by an electric current is chemically equivalent to the amount of electrons supplied.

fat An ester of glycerol and carboxylic acids with long hydrocarbon chains; fats act as long-term energy storage.

fatty acid A carboxylic acid with a long hydrocarbon chain. *Example:* $CH_3(CH_2)_{16}COOH$, stearic acid.

ferroalloy An alloy of a metal with iron and, often, carbon. *Example:* ferrovanadium.

ferrofluid A magnetic liquid that is a suspension of a finely powdered magnetic material such as magnetite, Fe_3O_4, in a viscous, oily liquid (such as mineral oil) that contains a detergent.

ferromagnetism The ability of some substances to be permanently magnetized. *Examples:* iron; magnetite, Fe_3O_4.

field An influence spreading over a region of space. *Examples:* an *electric field* from a charge; a *magnetic field* from a magnet or a moving charge.

filtration The separation of a heterogeneous mixture of a solid and liquid by passing the mixture through a fine mesh.

first derivative dy/dx A measure of the slope of a curve. See also Appendix 1F.

first ionization energy I_1 See *ionization energy*.

first law of thermodynamics The internal energy of an isolated system is constant.

first-order reaction A reaction in which the rate is proportional to the first power of the concentration of a substance.

fissile nucleus A nucleus having the ability to undergo fission induced by slow neutrons. *Example:* ^{235}U is fissile.

fission (nuclear) The breakup of a nucleus into two smaller nuclei of similar mass; fission may be *spontaneous* or *induced* (particularly by the impact of neutrons).

fissionable Having the ability to undergo induced nuclear fission.

fixation of nitrogen Conversion of elemental nitrogen to its compounds, particularly ammonia.

flammability The ability of a substance to burn in air.

flocculation The reversible aggregation of colloidal particles into larger particles that can be filtered.

foam (1) A frothy collection of bubbles formed by a liquid. (2) A type of *colloid* formed by a gas of tiny bubbles in a liquid or solid.

***f*-orbital** See *atomic orbital.*

force F An influence that changes the state of motion of an object. *Examples:* an electrostatic force from an electric charge; a mechanical force from an impact.

formal anhydride See *acid anhydride*.

formal charge (1) The electric charge of an atom in a molecule assigned on the assumption that the bonding is nonpolar covalent. (2) Formal charge (FC) = number of valence electrons in the free atom − (number of lone-pair electrons + $\frac{1}{2}$ × number of shared electrons).

formation constant K_f The equilibrium constant for complex formation. The *overall formation constant* is the product of stepwise formation constants. The inverse of the formation constant ($1/K_f$) is called the *stability constant*.

formula unit The group of ions that matches the formula of the smallest unit of an ionic compound. *Example:* NaCl, one Na^+ ion and one Cl^- ion.

formula weight See *molar mass.*

fossil fuels The partially decomposed remains of vegetable and marine life (mainly coal, oil, and natural gas).

fraction Samples of distillate obtained in different ranges of boiling temperatures.

fractional distillation Separation of the components of a liquid mixture by repeated distillation, making use of their differing volatilities.

Frasch process A process for mining sulfur that uses superheated water to melt the sulfur and compressed air to force it to the surface.

free energy $G = H - TS$ The energy of a system that is free to do work at constant temperature and pressure. The direction of spontaneous change at constant pressure and temperature is the direction of decreasing free energy.

free expansion Expansion against zero opposing pressure.

freezing-point constant k_f The constant of proportionality between the freezing-point depression and the solute molality.

freezing-point depression The lowering of the freezing point of a solvent caused by the presence of a solute (a colligative property).

freezing temperature The temperature at which a liquid freezes. The *normal freezing point* is the freezing temperature under a pressure of 1 atm.

frequency (of radiation) ν (nu) The number of cycles (repeats of the waveform) per second (unit: *hertz*, Hz).

froth flotation A process for separating the mineral from unwanted rock in an ore by blowing air through a mixture that contains oil, water, and detergents to generate a froth.

fuel cell A primary electrochemical cell in which the reactants are supplied continuously from outside while the cell is in use.

functional group A group of atoms that brings a characteristic set of chemical properties to an organic molecule. *Examples:* —OH ; —Br ; —COOH.

functionalization The introduction of functional groups into an alkane molecule.

fusion (1) Melting. (2) The merging of nuclei to form the nucleus of a heavier element.

galvanic cell See *electrochemical cell.*

galvanize Coat a metal with an unbroken film of zinc.

gamma (γ) radiation Very high frequency, short wavelength electromagnetic radiation emitted by nuclei.

gas A fluid form of matter that fills the container it occupies and can easily be compressed into a much smaller volume. (A gas differs from a vapor in that a gas is a substance at a higher temperature than its critical temperature; a *vapor* is a gaseous form of matter at a temperature below its critical temperature.)

gas constant R The constant that appears in the ideal gas law. See inside the back cover of the text for values.

gas-liquid chromatography A version of chromatography in which a gas carries the sample over a stationary liquid phase.

gauge pressure The pressure inside a container less that outside the container.

Geiger counter A device that is used to detect and measure radioactivity by relying on ionization caused by incident radiation.

gel A soft, solid colloid.

geometrical isomers Stereoisomers that differ in the spatial arrangement of the atoms. Geometrical isomers exhibit cis-trans isomerism.

Gibbs energy See *free energy.*

glass An ionic solid with an amorphous structure resembling that of a liquid.

glass electrode A thin-walled glass bulb containing an electrolyte solution and a metallic contact; used for measuring pH.

graft copolymer See *copolymer.*

Graham's law of effusion The rate of effusion of a gas is inversely proportional to the square root of its molar mass.

gravimetric analysis An analytical method using measurements of mass.

gray Gy The SI unit of *absorbed dose*; 1 Gy corresponds to an energy deposit of 1 $J \cdot kg^{-1}$. See also *rad.*

greenhouse effect The blocking by some atmospheric gases (notably carbon dioxide) of the radiation of heat from the surface of the Earth back into space, leading to the possibility of a worldwide rise in temperature.

greenhouse gas A gas that contributes to the greenhouse effect.

ground state The state of lowest energy.

group A vertical column in the periodic table.

Haber process (Haber-Bosch process) The catalyzed synthesis of ammonia at high pressure.

half-life $t_{1/2}$ (1) In chemical kinetics, the time needed for the concentration of a substance to fall to half its initial value. (2) In radioactivity, the time needed for half the initial number of radioactive nuclei to disintegrate.

half-reaction A hypothetical oxidation or reduction reaction showing either electron loss or electron gain. *Examples:* $Na(s) \rightarrow Na^+(aq) + e^-$; $Cl_2(g) + 2\,e^- \rightarrow 2\,Cl^-(aq)$.

halide ion An anion formed from a halogen atom. *Examples:* F^-; I^-.

Hall process The production of aluminum by the electrolysis of aluminum oxide dissolved in molten cryolite.

haloalkane An alkane with a halogen substituent. *Example:* CH_3Cl, chloromethane.

halogenation The incorporation of a halogen into a compound (particularly, into an organic compound).

halogens The elements in Group 17/VII.
hamiltonian The operator H in the Schrödinger equation; $H\psi = E\psi$.
hard water Water that contains dissolved calcium and magnesium salts.
heat q The energy that is transferred as the result of a temperature difference between a system and its surroundings.
heat capacity C The ratio of heat supplied to the temperature rise produced. The *heat capacity at constant pressure*, C_P, and the *heat capacity at constant volume*, C_V, are normally distinguished. See also *molar heat capacity*; *specific heat capacity*.
heating The act of transferring energy as heat.
heating curve A graph of the variation of the temperature of a sample as it is heated at a constant rate.
Heisenberg uncertainty principle If the location of a particle is known to within an uncertainty Δx, then the linear momentum parallel to the x-axis can be known only to within an uncertainty Δp, where $\Delta p \Delta x \geq \frac{1}{2}\hbar$.
Henderson-Hasselbalch equation An approximate equation for estimating the pH of a solution containing a conjugate acid and base. See also Section 11.3.
Henry's constant The constant k_H that appears in Henry's law.
Henry's law The solubility of a gas in a liquid is proportional to its partial pressure above the liquid: solubility $= k_H \times$ partial pressure.
hertz Hz The SI unit of frequency: 1 Hz is one complete cycle per second; $1\ \text{Hz} = 1\ \text{s}^{-1}$.
Hess's law A reaction enthalpy is the sum of the enthalpies of any sequence of reactions (at the same temperature and pressure) into which the overall reaction can be divided.
heterogenous alloy See *alloy*.
heterogeneous catalyst See *catalyst*.
heterogeneous equilibrium An equilibrium in which at least one substance is in a different phase from that of the rest. *Example:* $AgCl(s) \rightleftharpoons Ag^+(aq) + Cl^-(aq)$.
heterogeneous mixture A mixture in which the individual components, although mixed together, lie in distinct regions that can be distinguished on a microscopic scale. *Example:* a mixture of sand and sugar.
heterolytic dissociation Dissociation into ions. *Example:* $CH_3I \rightarrow CH_3^+ + I^-$. Compare with *homolytic dissociation*.
heteronuclear diatomic molecule A molecule consisting of two atoms of different elements. *Examples:* HCl; CO.
hexagonal close-packed structure hcp A close-packed structure with an ABABAB . . . pattern of layers.
high-boiling azeotrope See *azeotrope*.
highest occupied molecular orbital HOMO The highest energy molecular orbital in the ground state of a molecule occupied by at least one electron.
high-spin complex A d^n complex with the maximum number of unpaired electron spins.
high-temperature superconductor A material that becomes superconducting at temperatures well above the transition temperature for the first generation of superconductors, typically 100 K and above.
homogeneous alloy See *alloy*.
homogeneous catalyst See *catalyst*.
homogeneous equilibrium A chemical equilibrium in which all the substances taking part are in the same phase. *Example:* $H_2(g) + I_2(g) \rightleftharpoons 2\ HI(g)$.
homogeneous mixture A mixture in which the individual components are uniformly mixed, even on a molecular scale. *Examples:* air; solutions.
homolytic dissociation Dissociation into radicals. *Example:* $CH_3I \rightarrow CH_3 + I$. Compare with *heterolytic dissociation*.
homonuclear diatomic molecule A molecule consisting of two atoms of the same element. *Examples:* H_2; N_2.
Hund's rule If more than one orbital in a subshell is available, add electrons with parallel spins to different orbitals of that subshell.
hybridization The formation of hybrid orbitals.
hybrid orbital A mixed orbital formed by blending together atomic orbitals on the same atom. *Example:* an sp^3 hybrid orbital.
hydrate A solid compound containing H_2O molecules. *Example:* $CuSO_4{\cdot}5H_2O$.
hydrated Having water molecules attached.
hydrate isomers Isomers that differ by an exchange of an H_2O molecule and a ligand in the coordination sphere.
hydration (1) (of ions) The attachment of water molecules to a central ion. (2) (of organic compounds) The addition of water across a multiple bond (H to one carbon atom, OH to the other). *Example:* $CH_2{=}CH_2 + H_2O \rightarrow CH_3CH_2OH$.
hydride A binary compound of a metal or metalloid with hydrogen; the term is often extended to include all binary compounds of hydrogen. A *saline* or *saltlike hydride* is a compound of hydrogen and a strongly electropositive metal; a *molecular hydride* is a compound of hydrogen and a nonmetal; a *metallic hydride* is a compound of certain d-block metals and hydrogen.
hydrocarbon A binary compound of carbon and hydrogen. *Examples:* CH_4; C_6H_6.
hydrogenation The addition of hydrogen to multiple bonds. *Example:* $CH_3CH{=}CH_2 + H_2 \rightarrow CH_3CH_2CH_3$.
hydrogen bond A link formed by a hydrogen atom lying between two strongly electronegative atoms (O, N, or F). The electronegative atoms may be located on different molecules or in different regions of the same molecule.
hydrogen economy The widespread use of hydrogen as a fuel.

hydrohalogenation The addition of a hydrogen halide to an alkene to form a haloalkane. *Example:* $CH_3CH{=}CH_2 + HCl \rightarrow CH_3CHClCH_3$.

hydrolysis reaction The reaction of water with a substance, resulting in the formation of a new element-oxygen bond. *Example:* $PCl_5(s) + 4\ H_2O(l) \rightarrow H_3PO_4(aq) + 5\ HCl(aq)$.

hydrolyze To undergo hydrolysis. See also *hydrolysis reaction.*

hydrometallurgical extraction The extraction of metals by reduction of their ions in aqueous solution. *Example:* $Cu^{2+}(aq) + Fe(s) \rightarrow Cu(s) + Fe^{2+}(aq)$.

hydronium ion The ion H_3O^+.

hydrophilic Water-attracting. *Example:* hydroxyl groups are hydrophilic.

hydrophobic Water-repelling. *Example:* hydrocarbon chains are hydrophobic.

hydrostatic pressure The pressure exerted by a column of fluid.

hydroxyl group An —OH group in an organic compound.

hygroscopic Water-absorbing.

hypothesis A suggestion put forward to account for a series of observations. *Example:* Dalton's atomic hypothesis.

ideal gas A gas that satisfies the ideal gas law and is described by the *kinetic model.*

ideal gas law $PV = nRT$ All gases obey the law more and more closely as the pressure is reduced to very low values.

ideal solution A solution that obeys Raoult's law at any concentration; all solutions behave ideally as the concentration approaches 0. *Example:* benzene and toluene form an almost ideal system.

***i* factor** A factor that takes into account the existence of ions in an electrolyte solution, particularly for the interpretation of colligative properties. It indicates the number of particles formed from one formula unit of the solute. *Example:* $i \approx 2$ for very dilute NaCl(aq).

incandescence Light emitted by a hot body.

incomplete octet A valence shell of an atom that has fewer than eight electrons. *Example:* the valence shell of B in BF_3.

indefinite integral An *integral* without limits attached. See also Appendix 1F.

independent particle model A model of the nucleus in which the nucleons are described by quantum numbers and occupy shells within the nucleus.

indicator A substance that changes color when it goes from its acidic to its basic form (an *acid-base indicator*) or from its oxidized to its reduced form (a *redox indicator*).

induced dipole moment An electric dipole moment produced in a polarizable molecule by a neighboring charge or partial charge.

induced-fit mechanism A model of the action of an enzyme in which the enzyme molecule adjusts its shape to accommodate the incoming substrate molecule. A modification of the *lock-and-key mechanism* of enzyme action.

induced nuclear fission See *fission.*

inert (1) Unreactive. (2) Thermodynamically unstable but survives for long periods (*nonlabile*).

inert-pair effect The observation that an element displays a valence lower than expected from its group number. An *inert pair* is a pair of valence shell *s*-electrons that are tightly bound to the atom and might not participate in bond formation.

infrared radiation Electromagnetic radiation with a lower frequency (longer wavelength) than that of red light but a higher frequency (shorter wavelength) than microwave radiation.

initial concentration (of a weak acid or base) The concentration as prepared, as if no deprotonation or protonation had occurred.

initial rate The rate at the start of the reaction when products are present in concentrations too low to affect the rate.

initiation The formation of reactive intermediates that serve as chain carriers from a reactant at the start of a chain reaction. *Example:* $Br_2 \rightarrow Br\cdot + Br\cdot$.

inner transition metal A member of the *f* block of the periodic table (the *lanthanides* and *actinides*).

inorganic chemistry The study of the elements other than carbon and their compounds.

inorganic compound A compound that is not organic. See also *organic compound.*

insoluble substance A substance that does not dissolve in a specified solvent. When the solvent is not specified, water is generally meant.

instantaneous dipole moment A dipole moment that arises from a transient redistribution of charge and is responsible for the London force.

instantaneous rate The slope of the tangent of a graph of concentration against time.

insulator (electrical) A substance that does not conduct electricity. *Examples:* nonmetallic elements; molecular solids.

integral (1) The sum of infinitesimal quantities. (2) The inverse of a derivative in the sense that the integral of the first derivative of a function is the original function. See *definite integral, indefinite integral.*

integral calculus The part of mathematics that deals with the combination of infinitesimal quantities and the areas under curves. See also Appendix 1F.

integrated rate law An expression for the concentration of a reactant or product in terms of the time, obtained from the rate law of the reaction. *Example:* $[A] = [A]_0e^{-kt}$.

intensity The brightness of electromagnetic radiation. The intensity of a wave of electromagnetic radiation is proportional to the square of its *amplitude.*

intensive property A physical property of a substance that is independent of the mass of the sample. *Examples:* density; molar volume; temperature.

intercept (of a graph) The value at which the graph cuts the vertical axis. See also Appendix 1E.

interference Interaction between waves, leading to a greater amplitude (*constructive interference*) or to a smaller one (*destructive interference*).

interhalogen A binary compound of two halogens. *Example:* IF_3.

intermediate See *reaction intermediate.*

intermolecular Between molecules.

intermolecular forces The forces of attraction and repulsion between molecules; *van der Waals forces. Examples:* hydrogen bonding; dipole-dipole force; London force.

internal energy *U* The total energy of a system.

International System See *SI.*

internuclear axis The straight line between the nuclei of two bonded atoms.

interpolate To find a value between two measured values.

interstice A hole or gap in a crystal lattice.

interstitial alloy See *alloy.*

interstitial carbide See *carbide.*

interstitial compound A compound in which one type of atom occupies the gaps in a lattice of other atoms. *Example:* an interstitial carbide.

intramolecular Within a molecule.

inversion (a) In symmetry, projecting each point of an object through the center and out a corresponding distance on the other side of the center. (b) In a chemical reaction, an umbrellalike modification of the bonds about an atom in a molecule that changes the stereochemistry of a reaction center but results in an optically active product from an optically active reactant.

ion An electrically charged atom or group of atoms. *Examples:* Al^{3+}; SO_4^{2-}. See also *anion*; *cation.*

ion-dipole interaction The attraction between an ion and the opposite partial charge of the electric dipole of a polar molecule.

ion exchange The exchange of one type of ion in solution for another.

ionic bond The attraction between the opposite charges of cations and anions.

ionic character The extent to which ionic structures contribute to the resonance of a molecule or ion.

ionic compound A compound that consists of ions. *Examples:* NaCl; KNO_3.

ionic conduction Electrical conduction in which the charge is carried by ions. See also *electrical conduction; electronic conductor.*

ionic equation See *complete ionic equation.*

ionic model The description of bonding in terms of ions.

ionic radius The contribution of an ion to the distance between neighboring ions in a solid ionic compound. In practice, the radius of an ion is defined as its contribution to the distance between the centers of neighboring ions, with the radius of the O^{2-} ion set equal to 140 pm.

ionic solid A solid built from cations and anions. *Examples:* NaCl; KNO_3.

ionization (1) (of atoms and molecules) Conversion to cations by the removal of electrons. *Example:* $K(g) \rightarrow K^+(g) + e^-(g)$. (2) (of acids and bases) See *protonation* and *deprotonation.*

ionization constant See *acidity constant.*

ionization energy *I* The minimum energy required to remove an electron from the ground state of a gaseous atom, molecule, or ion. The *first ionization energy* is the ionization energy for the removal of the first electron from an atom. The *second ionization energy* is the ionization energy for removal of a second electron, and so on.

ionization isomers Isomers that differ by the exchange of a ligand with an anion or neutral molecule outside the coordination sphere.

ionizing radiation High-energy radiation (typically but not necessarily nuclear radiation) that can cause ionization.

ion pair A cation and anion in close proximity.

ion-selective electrode An electrode sensitive to the concentration of a particular ion.

irreversible process A process that is not reversed by an infinitesimal change in a variable.

isoelectronic species Species with the same number of atoms and the same number of valence electrons. *Examples:* F^- and Ne; SO_2 and O_3; CN^- and CO.

isolated system See *system.*

isomer One of two or more compounds that contain the same number of the same atoms in different arrangements. In *structural isomers*, the atoms have different partners or lie in a different order; in *stereoisomers*, the atoms have the same partners but are in different arrangements in space. *Optical isomers* are related like an object and its mirror image; they are types of stereoisomers. *Examples:* CH_3–O–CH_3 and CH_3CH_2–OH (structural isomers); *cis*- and *trans*-2-butene (stereoisomers).

isomerization A reaction in which a compound is converted into one of its isomers. *Example: cis*-butene → *trans*-butene.

isotactic polymer See *polymer.*

isotherm A line of constant temperature on a graph.

isothermal process A change that occurs at constant temperature.

isotope One of two or more atoms that have the same atomic number but different atomic masses. *Example:* ^1H, ^2H, and ^3H are all isotopes of hydrogen.

isotopic abundance See *abundance*.
isotopic dating The determination of the age of objects by measuring the activity of the radioactive isotopes they contain, particularly ^{14}C.
isotopic label See *tracer*.
isotopomers Molecules that differ in their isotopic composition. *Examples:* $^{12}CH_4$ and $^{13}CH_4$.
isotropic Not depending on orientation.

joule J The SI unit of energy ($1\ J = 1\ kg \cdot m^2 \cdot s^{-2}$).
Joule-Thomson effect The cooling of a gas as it expands.

Kekulé structures Two Lewis structures of benzene, consisting of alternating single and double bonds.
kelvin K The SI unit of temperature. See also Appendix 1B.
Kelvin scale A fundamental scale of temperature on which the triple point of water lies at 273.16 K and the lowest attainable temperature is at 0. The unit on the Kelvin scale is the kelvin, K.
ketone An organic compound containing a carbonyl group between two carbon atoms, having the form R—CO—R′. *Example:* CH_3—CO—CH_2CH_3, butanone.
kilogram kg The SI unit of mass. See also Appendix 1B.
kinetic energy The energy of a particle due to its motion. Kinetic energy may be *translational* (arising from motion through space), *rotational* (arising from rotation about a center of mass), or *vibrational* (arising from the oscillating motion of the atoms in a molecule). *Example:* the translational kinetic energy of a particle of mass m and speed v is $\frac{1}{2}mv^2$.
kinetic model A model of the properties of an ideal gas in which pointlike molecules are in continuous random motion in straight lines until collisions occur between them.
kinetic molecular theory The theory that discusses the *kinetic model* of gases.
Kirchhoff's law The relation between the standard reaction enthalpies at two temperatures in terms of the temperature difference and the difference in heat capacities (at constant pressure) of the products and reactants.

labile A species that survives only for short periods.
lachrymator A substance that stimulates the production of tears. *Example:* PAN.
lanthanide A member of the first row of the *f* block (lanthanum through ytterbium).
lanthanide contraction The reduction of atomic radius of the elements following the lanthanides below the values that would be expected by extrapolation of the trend down a group (and arising from the poor shielding ability of *f*-electrons).
lattice An orderly array of atoms, molecules, or ions in a crystal.
lattice energy The difference in potential energy of the ions in a crystal and that of the same widely separated ions in a gas.
lattice enthalpy The standard enthalpy change for the conversion of an ionic solid to a gas of ions.
law A summary of an extensive series of observations.
law of conservation of energy Energy can be neither created nor destroyed.
law of conservation of mass Matter (and specifically atoms) is neither created nor destroyed in a chemical reaction.
law of constant composition A compound has the same composition whatever its source.
law of mass action For an equilibrium of the form $aA + bB \rightleftharpoons cC + dD$, the ratio $[C]^c[D]^d/[A]^a[B]^b$ evaluated at equilibrium is equal to a constant K_c, which has a specific value for a given reaction and temperature.
law of partial pressures See *Dalton's law of partial pressures*.
law of radioactive decay The rate of decay is proportional to the number of radioactive nuclides in the sample.
LCAO See *linear combination of atomic orbitals*.
lead-acid cell A secondary cell in which the electrodes are lead and the electrolyte is dilute sulfuric acid.
leaving group The group displaced in a substitution reaction.
Le Chatelier's principle When a stress is applied to a system in dynamic equilibrium, the equilibrium adjusts to minimize the effect of the stress. *Example:* a reaction at equilibrium tends to proceed in the endothermic reaction when the temperature is raised.
leveling The observation that strong acids all have the same strength in water, and all behave as though they were solutions of H_3O^+ ions.
Lewis acid An electron pair acceptor. *Examples:* H^+; Fe^{3+}; BF_3.
Lewis base An electron pair donor. *Examples:* OH^-; H_2O; NH_3.
Lewis formula (for an ionic compound) A representation of the structure of an ionic compound showing the formula unit of ions in terms of their Lewis diagrams.
Lewis structure A diagram showing how electron pairs are shared between atoms in a molecule. *Examples:* H—C̈l:; Ö=C=Ö.
Lewis symbol (for atoms and ions) The chemical symbol of an element, with a dot for each valence electron.
ligand A group attached to the central metal ion in a complex; a *polydentate ligand* occupies more than one binding site. See also *ambidentate ligand*.
ligand field splitting Δ The energy separation of the *e*- and *t*-orbitals in a complex induced by the presence of ligands.

ligand field theory The theory of bonding in *d*-metal complexes, a more complete version of *crystal field theory*. See also *crystal field*.
light See *visible radiation*.
limiting enthalpy of solution See *enthalpy of solution*.
limiting law A law that is accurately obeyed only at the limit of a property, such as when a property (the pressure of a gas, for example) is made very small.
limiting reactant The reactant that governs the theoretical yield of product in a given reaction.
linear combination of atomic orbitals LCAO A molecular orbital formed by superimposing atomic orbitals.
linear momentum p The product of mass and velocity.
line structure A representation of the structure of an organic molecule in terms of lines representing the bonds; carbon atoms and the hydrogen atoms attached to them are not usually shown explicitly.
linkage isomers Isomers that differ in the identity of the atom that a ligand uses to attach to the metal ion.
lipid A naturally occurring organic compound that dissolves in hydrocarbons but not in water. *Examples:* fats; steroids; terpenes; the molecules that form cell membranes.
liquid A fluid form of matter that has a well-defined surface and takes the shape of the part of the container it occupies.
liquid crystal A substance that flows like a liquid but has molecules that lie in a moderately orderly array. Liquid crystals may be *nematic*, *smectic*, or *cholesteric*, depending on the arrangement of the molecules.
liquid-drop model A model of the nucleus based on the view that nucleons are packed together like the molecules in a liquid.
lock-and-key mechanism A model of enzyme action in which the enzyme is thought of as a lock and its substrate as a matching key.
logarithm If a number x is written as B^y, then y is the logarithm of x to the base B. For *common logarithms* (denoted $\log x$), $B = 10$. For *natural logarithms* (denoted $\ln x$), $B = e$. See also Appendix 1D.
London force The force of attraction that arises from the interaction between instantaneous electric dipoles on neighboring molecules.
lone pair A pair of valence electrons that is not involved in bonding.
long period A period of the periodic table with more than eight members.
long-range order An orderly arrangement of atoms or molecules that is repeated over long distances.
lowest unoccupied molecular orbital LUMO The lowest energy molecular orbital that is unoccupied in the ground state.
low-spin complex A d^n complex with the minimum number of unpaired electron spins.
luminescence The emission of light from a process, other than incandescence, that results in the formation of an excited state.
Lyman series A series of lines in the spectrum of atomic hydrogen in which the transitions are to orbitals with $n = 2$.
lyotropic liquid crystal A liquid crystal that results from the action of a solvent on a solute.

macroscopic level The level at which visible objects can be observed directly.
Madelung constant A A number that appears in the expression for the lattice energy and which depends on the type of crystal lattice. *Example:* $A = 1.748$ for the rock-salt structure.
magic numbers The numbers of protons or neutrons that correlate with enhanced nuclear stability. *Examples:* 2, 8, 20, 50, 82, and 126.
magnetic field A region of influence that affects the motion of moving charged particles.
magnetic quantum number m_l The quantum number that identifies the individual orbitals of a subshell of an atom and determines their orientation in space.
main group Any one of the groups forming the *s* and *p* blocks of the periodic table (Groups 1, 2, and 13/III through 18/VIII).
malleability The ability to be deformed by striking with a hammer (as a metal).
manometer An instrument used for measuring the pressure of a gas confined inside a container.
mantissa (of a logarithm) The numbers following the decimal point.
many-electron atom An atom with more than one electron.
mass m The quantity of matter in a sample.
mass concentration The mass of solute per liter of solution.
mass number A The total number of nucleons (protons plus neutrons) in the nucleus of an atom. *Example:* $^{14}_{6}C$, with mass number 14, has 14 nucleons (6 protons and 8 neutrons).
mass percentage composition The mass of a substance present in a sample, expressed as a percentage of the total mass of the sample.
mass spectrometer An instrument used in *mass spectrometry*.
mass spectrometry Technique for measuring the masses and abundances of atoms and molecules by passing a beam of ions through a magnetic field.
mass spectrum The display of the relative number of particles with each specified mass; the output generated by the detector of a mass spectrometer. See also *mass spectrometry*.
material balance The requirement that the sum of the concentrations of all forms of a solute in a solution be

equal to the initial concentration of the solute. *Example:* The sum of the concentrations of HCN and CN^- in an aqueous solution of HCN is equal to the initial concentration of HCN.

materials science The study of the chemical structures, compositions, and properties of materials.

matter Anything that has mass and takes up space.

maximum-boiling azeotrope See *azeotrope*.

Maxwell distribution of molecular speeds The formula for calculating the percentage of molecules that move at any given speed in a gas at a specified temperature.

mean bond enthalpy ΔH_B The mean of A—B bond enthalpies for a number of different molecules containing the A—B bond. See also *bond enthalpy*.

mean free path The average distance that a molecule travels between collisions.

mechanical equilibrium The state in which the pressure of a system is equal to that of its surroundings.

mechanism See *reaction mechanism*.

melting temperature The temperature at which a substance melts. The *normal melting point* is the melting point under a pressure of 1 atm.

meniscus The curved surface that a liquid forms in a narrow tube.

mesophase A state of matter showing some of the properties of both a liquid and a solid (a liquid crystal).

metal (1) A substance that conducts electricity, has a metallic luster, is malleable and ductile, forms cations, and has basic oxides. (2) A metal consists of cations held together by a sea of electrons. *Examples:* iron; copper; uranium.

metallic conductor An electronic conductor with a resistance that increases as the temperature is raised.

metallic hydride See *hydride*.

metallic radius (of a metallic element) Half the distance between the centers of neighboring atoms in a solid sample.

metallic solid See *metal*.

metallocene A compound in which a metal atom lies between two cyclic ligands, thus resembling a sandwich. *Example:* ferrocene (dicyclopentadienyliron(0), $[Fe(C_5H_5)_2]$).

metalloid An element that has the physical appearance and properties of a metal but behaves chemically like a nonmetal. *Examples:* arsenic; polonium.

meter m The SI unit of length. See also Appendix 1B.

micelle A compact, often nearly spherical, cluster of oriented detergent (surfactant) molecules.

Michaelis constant K_M A constant that occurs in the *Michaelis-Menten mechanism*.

Michaelis-Menten mechanism A model of enzyme catalysis in which the enzyme and its substrate reach a rapid *pre-equilibrium* with the bound substrate-enzyme complex.

microscopic level A level of description that refers to the very small, such as atoms.

microstate A permissible arrangement of molecules in a sample (in the context of statistical thermodynamics and the statistical definition of entropy).

microwaves Electromagnetic radiation with wavelengths close to 1 cm.

minerals Substances that are mined; more generally, inorganic substances.

minimum-boiling azeotrope See *azeotrope*.

mixture A type of matter that consists of more than one substance and can be separated into its components by making use of the different physical properties of the substances present.

model A simplified description of nature.

moderator A substance that slows neutrons. *Examples:* graphite; heavy water.

molality The amount of solute (in moles) per kilogram of solvent.

molar Referring to the quantity per mole. *Examples: molar mass*, the mass per mole; *molar volume*, the volume per mole. (*Molar concentration* and some related quantities are exceptions.)

molar absorption coefficient The constant of proportionality between the absorbance of a sample and the product of its molar concentration and path length. See also *Beer's law*.

molar concentration [J] The amount (in moles) of solute per liter of solution.

molarity [J] Molar concentration.

molar heat capacity C_m The heat capacity per mole of substance.

molar mass (1) The mass per mole of atoms of an element (formerly, *atomic weight*). (2) The mass per mole of molecules of a compound (formerly, *molecular weight*). (3) The mass per mole of formula units of an ionic compound (formerly, *formula weight*).

molar solubility S The molar concentration of a saturated solution of a substance. The *dimensionless* molar solubility is denoted s ($s = S/\text{mol}\cdot\text{L}^{-1}$).

molar volume The volume of a sample divided by the amount (in moles) of atoms or molecules it contains.

mole mol The SI unit of chemical amount. See also Appendix 1B.

molecular biology The study of the functions of living organisms in terms of their molecular composition.

molecular compound A compound that consists of molecules. *Examples:* water; sulfur hexafluoride; benzoic acid.

molecular formula A combination of chemical symbols and subscripts showing the actual numbers of atoms of each element present in a molecule. *Examples:* H_2O; SF_6; C_6H_5COOH.

molecular hydride See *hydride*.

molecularity The number of reactant molecules (or free atoms) taking part in an *elementary reaction*. See also *bimolecular reaction*; *termolecular reaction*; *unimolecular reaction*.

molecular orbital A one-electron wavefunction that spreads throughout the molecule and gives the probability (through its square) that an electron will be found at each location.
molecular orbital energy-level diagram A portrayal of the relative energies of the molecular orbitals in a molecule.
molecular orbital theory The description of molecular structure in which electrons occupy orbitals that spread throughout the molecule.
molecular solid A solid consisting of a collection of individual molecules held together by intermolecular forces. *Examples:* glucose; aspirin; sulfur.
molecular weight See *molar mass.*
molecule (1) The smallest particle of a compound that possesses the chemical properties of the compound. (2) A definite and distinct, electrically neutral group of bonded atoms. *Examples:* H_2; NH_3; CH_3COOH.
mole fraction x The amount of particles (molecules, atoms, or ions) of a substance in a mixture expressed as a fraction of the total amount of particles in the mixture.
mole ratio The stoichiometric relation between two species in a chemical reaction written as a conversion factor. *Example:* (2 mol H_2)/(1 mol O_2) in the reaction $2\ H_2(g) + O_2(g) \rightarrow 2\ H_2O(l)$.
momentum p (linear) The product of mass and velocity.
monatomic gas A gas composed of single atoms. *Examples:* helium; radon.
monatomic ion An ion formed from a single atom. *Examples:* Na^+; Cl^-.
Mond process The purification of nickel by the formation and decomposition of nickel carbonyl.
monomer A small molecule from which a polymer is formed. *Examples:* $CH_2{=}CH_2$ for polyethylene; $NH_2(CH_2)_6NH_2$ for nylon.
monoprotic acid A Brønsted acid with one acidic hydrogen atom. *Example:* CH_3COOH.
monosaccharide An individual unit from which carbohydrates are considered to be composed. *Example:* $C_6H_{12}O_6$, glucose.
multiple bond A double or triple bond between two atoms.

native Occurring in an uncombined state as the element itself.
natural abundance (of an isotope) The abundance of an isotope in a sample of a naturally occurring material.
natural antilogarithm See *exponential.*
natural logarithm See *logarithm.*
naturally occurring Found in nature without needing to be synthesized.
natural product An organic substance that occurs naturally in the environment.
negative deviation (from Raoult's law) The tendency of a nonideal solution to have a vapor pressure lower than predicted by *Raoult's law.*
nematic phase A liquid-crystal phase in which rod-shaped molecules have axes arranged parallel to but staggered with respect to one another in other directions.
Nernst equation The equation expressing the emf of an electrochemical cell in terms of the concentrations of the reagents taking part in the cell reaction; $E = E° - (RT/nF) \ln Q$.
net ionic equation The equation showing the net change in a chemical reaction, obtained by canceling the spectator ions in a complete ionic equation. *Example:* $Ag^+(aq) + Cl^-(aq) \rightarrow AgCl(s)$.
network solid A solid consisting of atoms linked together covalently throughout its extent. *Examples:* diamond; silica.
neutralization reaction The reaction of an acid with a base to form a salt and water or another molecular compound. *Example:* $HCl(aq) + NaOH(aq) \rightarrow NaCl(aq) + H_2O(l)$.
neutron n An electrically neutral subatomic particle found in the nucleus of an atom; it has approximately the same mass as a proton.
neutron emission A nuclear decay process in which a neutron is emitted. In neutron emission, the mass number decreases by 1, but the charge number remains the same.
neutron-induced transmutation The conversion of one nucleus into another by the impact of a neutron. *Example:* ${}^{58}_{26}Fe + 2\ {}^{1}_{0}n \rightarrow {}^{60}_{27}Co + {}^{0}_{-1}e$.
neutron-rich nuclei Nuclei with such a high proportion of neutrons that they lie above the *band of stability.*
nickel-cadmium cell nicad cell A rechargeable cell in which the reaction is the reduction of nickel(III) and the oxidation of cadmium.
NO_x An oxide, or mixture of oxides, of nitrogen, typically in atmospheric chemistry.
noble gas A member of Group 18/VIII of the periodic table (the helium family).
nodal plane A plane on which an electron will not be found.
node A point or surface at which an electron occupying an orbital will not be found.
nomenclature See *chemical nomenclature.*
nonaqueous solution A solution in which the solvent is not water. *Example:* sulfur in carbon disulfide.
nonbonding orbital A valence-shell atomic orbital that has not been used to form a bond to another atom.
nonelectrolyte A substance that dissolves to give a solution that does not conduct electricity. *Example:* sucrose.
nonexpansion work w_e See *work.*
nonideal solution A solution that does not obey Raoult's law. Compare with *ideal solution.*
nonlabile Thermodynamically unstable but survives for a long period.
nonmetal A substance that does not conduct electricity and is neither malleable nor ductile. *Examples:* all gases; phosphorus; sodium chloride.

nonpolar bond (1) A covalent bond between two atoms that have zero partial charges. (2) A covalent bond between two atoms with the same or nearly the same electronegativity.

nonpolar molecule A molecule with zero electric dipole moment.

normal boiling point T_b (1) The boiling temperature when the pressure is 1 atm. (2) The temperature at which the vapor pressure of a liquid is 1 atm.

normal form The form of a substance under typical everyday conditions (for instance, close to 1 atm, 25°C).

normal freezing point T_f The temperature at which a liquid freezes at 1 atm.

normal melting point T_f The melting point of a substance at a pressure of 1 atm.

***ns*-orbital** An atomic orbital with principal quantum number n and $l = 0$.

n-type semiconductor See *semiconductor*.

nuclear atom The structure of the atom proposed by Rutherford: a central, small, very dense, positively charged nucleus surrounded by electrons.

nuclear binding energy E_{bind} The energy released when Z protons and $A - Z$ neutrons come together to form a nucleus. The greater the binding energy per nucleon, the lower the energy of the nucleus.

nuclear chemistry The study of the chemical consequences of nuclear reactions.

nuclear decay The spontaneous partial breakup of a nucleus (including its fission). Nuclear decay is also referred to as *nuclear disintegration. Example:* ${}^{226}_{88}Ra \rightarrow {}^{222}_{86}Rn + {}^{4}_{2}\alpha$.

nuclear equation A summary of the changes in a nuclear reaction, written in a form resembling a chemical equation.

nuclear fission See *fission*.

nuclear fusion See *fusion*.

nuclear model A model of the atom in which the electrons surround a minute central nucleus.

nuclear reaction A change that a nucleus undergoes (such as a nuclear transmutation).

nuclear reactor A device for achieving controlled self-sustaining nuclear fission.

nuclear transmutation The conversion of one element into another. *Example:* ${}^{12}_{6}C + {}^{4}_{2}\alpha \rightarrow {}^{16}_{8}O + \gamma$.

nucleic acid (1) The product of a condensation of nucleotides. (2) A molecule containing an organism's genetic information.

nucleon A proton or a neutron; thus, either of the two principal components of a nucleus.

nucleophile A reactant that seeks out centers of positive charge in a molecule. *Examples:* H_2O, OH^-.

nucleophilic substitition Substitution that takes place by the attack of a nucleophile. See also *S_N1 and S_N2 reactions. Examples:* Hydrolysis of haloalkanes, $CH_3Br + H_2O \rightarrow CH_3OH + HBr$.

nucleoside A combination of an organic base and a ribose or deoxyribose molecule.

nucleosynthesis The formation of elements.

nucleotide A nucleoside with a phosphate group attached to the carbohydrate ring; one of the units from which nucleic acids are made.

nucleus The small, positively charged particle at the center of an atom that is responsible for most of its mass.

nuclide A specific nucleus. *Examples:* ${}^{2}_{1}H$; ${}^{16}_{8}O$.

occupy Have the characteristics of the wavefunction of a specified state; to be in a specified state.

octahedral complex A complex in which six ligands are arranged at the corners of a regular octahedron, with the metal atom at the center. *Example:* $[Fe(CN)_6]^{4-}$.

octahedral hole A cavity in a (usually close packed) crystal lattice formed from six spheres at the corners of a regular octahedron.

octet An s^2p^6 valence electron configuration.

octet rule When atoms form bonds, they proceed as far as possible toward completing their octets by sharing electron pairs.

oligopeptide A short chain of amino acids connected by amide (peptide) bonds.

one-component phase diagram See *phase diagram*.

open system See *system*.

optical activity The ability of a substance to rotate the plane of polarized light passing through it.

optical isomers Isomers that are related like an object and its mirror image. *Optical isomerism* is the existence of optical isomers. Optical isomerism is a type of *stereoisomerism*.

orbital See *atomic orbital*.

orbital angular momentum quantum number l The quantum number that specifies the subshell of a given shell in an atom and determines the shapes of the orbitals in the subshell.; $l = 0,1,2, \ldots, n - 1$. *Examples:* $l = 0$ for the *s*-subshell; $l = 1$ for the *p*-subshell. (The quantum number l also specifies the magnitude of the angular momentum of the electron around the nucleus.)

order of reaction The power to which the concentration of a single substance is raised in a rate law. *Example:* if rate $= k[SO_2][SO_3]^{-1/2}$, then the reaction is first order in SO_2 and of order $-\frac{1}{2}$ in SO_3.

ore The natural mineral source of a metal. *Example:* Fe_2O_3, hematite, an iron ore.

organic chemistry The branch of chemistry that deals with organic compounds.

organic compound A compound containing the element carbon and usually hydrogen. (The carbonates are normally excluded.)

organometallic compound A compound that contains a metal-carbon bond. *Example:* $Ni(CO)_4$ (complexes of CN^- are normally excluded).

oscillating Varying in a periodic manner with time.
osmometry The measurement of the molar mass of a solute from observations of osmotic pressure.
osmosis The tendency of a solvent to flow through a semipermeable membrane into a more concentrated solution (a colligative property).
osmotic pressure Π (pi) The pressure needed to stop the flow of solvent through a semipermeable membrane. See also *osmosis*.
Ostwald process The production of nitric acid by the catalytic oxidation of ammonia.
overall formation constant See *formation constant*.
overall order The sum of the powers to which individual concentrations are raised in the rate law of a reaction. *Example:* If the rate $= k[SO_2][SO_3]^{-1/2}$, then the overall order is $\frac{1}{2}$.
overall reaction The net outcome of a sequence of reactions.
overlap The merging of orbitals belonging to different atoms of a molecule.
overpotential The additional potential difference that must be applied beyond the cell potential to cause appreciable electrolysis.
oxidation (1) Combination with oxygen. (2) A reaction in which an atom, ion, or molecule loses an electron. (3) A reaction in which the oxidation number of an element is increased. *Examples:* (1, 2, 3) $Mg(s) + O_2(g) \rightarrow MgO(s)$; (2, 3) $Mg(s) \rightarrow Mg^{2+}(s) + 2\ e^-$.
oxidation number The effective charge on an atom in a compound, calculated according to a set of rules (see Toolbox K.1). An increase in oxidation number corresponds to oxidation and a decrease, to reduction.
oxidation-reduction reaction See *redox reaction*.
oxidation state The actual condition of a species with a specified oxidation number.
oxidizing agent A species that removes electrons from a species being oxidized (and is itself reduced) in a redox reaction. *Examples:* O_2; O_3; MnO_4^-; Fe^{3+}.
oxoacid An acid that contains oxygen. *Examples:* H_2CO_3; HNO_3; HNO_2; $HClO$.
oxoanion An anion of an oxoacid. *Examples:* HCO_3^-; CO_3^{2-}.

paired electrons Two electrons with opposed spins (↓↑).
parallel spins Electrons with spin in the same direction (↑↑).
paramagnetic Having the tendency to be pulled into a magnetic field; a paramagnetic substance is composed of atoms or molecules with unpaired electrons. *Examples:* O_2; $[Fe(CN)_6]^{3-}$.
parent nucleus In a nuclear reaction, the nucleus that undergoes disintegration or transmutation.
partial charge A charge arising from small shifts in the distributions of electrons. A partial charge can be either *positive* ($\delta+$) or *negative* ($\delta-$).
partial pressure P_X The pressure a gas (X) in a mixture would exert if it alone occupied the container.
parts per million ppm (1) The ratio of the mass of a solute to the mass of the solvent, multiplied by 10^6. (2) The *mass percentage composition* multiplied by 10^4. (*Parts per billion*, ppb, the mass ratio multiplied by 10^9, may also be used.)
pascal Pa The SI unit of pressure: $1\ Pa = 1\ kg{\cdot}m^{-1}{\cdot}s^{-2}$. See also Appendix 1B.
passivation Protection from further reaction by a surface film. *Example:* aluminum in air.
Pauli exclusion principle See *exclusion principle*.
penetration The possibility that an *s*-electron may be found inside the inner shells of an atom and hence close to the nucleus.
peptide A molecule formed by a condensation reaction between amino acids; often described in terms of the number of units, for example, *dipeptide, oligopeptide, polypeptide*.
peptide bond The —CONH— group.
percentage composition See *mass percentage composition*; *volume percentage composition*.
percentage deprotonation The fraction of a weak acid, expressed as a percentage, that is present as its conjugate base in a solution.
percentage ionization The fraction of molecules of a substance, expressed as a percentage, that have ionized.
percentage protonated The fraction of a base, expressed as a percentage, that is present as its conjugate acid in a solution.
percentage yield The percentage of the theoretical yield of a product achieved in practice.
period A horizontal row in the periodic table; the number of the period is equal to the principal quantum number of the valence shell of the atoms.
periodic table A chart in which the elements are arranged in order of atomic number and divided into groups and periods in a manner that shows the relationships between the properties of the elements.
pH The negative logarithm of the hydronium ion molarity in a solution: $pH = -\log[H_3O^+]$. $pH < 7$ indicates an acidic solution; $pH = 7$, a neutral solution; and $pH > 7$, a basic solution.
pH curve A graph of the pH of a reaction mixture against volume of titrant added in an acid-base titration.
phase A specific physical state of matter. A substance may exist in solid, liquid, and gaseous phases and, in certain cases, in more than one solid or liquid phase. *Examples:* white and gray tin are two solid phases of tin; ice, liquid, and vapor are three phases of water.
phase boundary A line separating two areas in a phase diagram; the points on a phase boundary represent the conditions at which the two adjoining phases are in dynamic equilibrium.
phase diagram A summary in graphical form of the conditions of temperature and pressure at which the various solid, liquid, and gaseous phases of a substance

exist. A *one-component phase diagram* is a phase diagram for a single substance.

phase rule The relation between the degrees of freedom, f, the number of phases, p, and the number of components, c, for a system at equilibrium: $f = c - p + 2$.

phase transition The conversion of one phase of a substance to another phase. *Examples:* vaporization; white tin → gray tin.

phenol An organic compound in which a hydroxyl group is attached directly to an aromatic ring (Ar—OH). *Example:* C_6H_5OH, phenol.

phenolic resin A polymer resulting from the condensation reaction between phenol and formaldehyde.

phosphor A phosphorescent material that emits light after excitation to higher energy states.

phosphorescence Long-lasting luminescence.

photochemical reaction A reaction caused by light. *Example:* $H_2(g) + Cl_2(g) \xrightarrow{\text{light}} 2\,HCl(g)$.

photoelectric effect The emission of electrons from the surface of a metal when electromagnetic radiation strikes it.

photon A particlelike packet of electromagnetic radiation. The energy of a photon of frequency ν is $E = h\nu$.

physical chemistry The study of the principles of chemistry.

physical equilibrium A state in which two or more phases of a substance coexist without a tendency to change. *Example:* ice and water at 0°C and 1 atm.

physical property A characteristic that we observe or measure without changing the identity of the substance.

physical state The condition of being a solid, a liquid, or a gas at a particular temperature.

pi (π) bond A bond formed by the side-to-side overlap of two *p*-orbitals.

piezoelectric Having the property of becoming electrically charged when mechanically distorted. *Example:* $BaTiO_3$.

pi (π) orbital A molecular orbital that has one nodal plane cutting through the internuclear axis.

pipet A narrow tube, sometimes with a central bulb, calibrated to deliver a specified volume.

pK_a and pK_b The negative logarithms of the acid and base ionization constants: $pK = -\log K$. The larger the value of pK_a or pK_b, the weaker the acid or base, respectively.

Planck constant *h* A fundamental constant of nature with the value $6.626\,08 \times 10^{-34}$ J·s.

plasma (1) An ionized gas. (2) In biology, the colorless component of blood in which the red and white blood cells are dispersed.

p-n junction An interface between a p-type and an n-type semiconductor.

pOH The negative logarithm of the hydroxide ion molarity in a solution; $pOH = -\log[OH^-]$.

poison To inactivate a catalyst.

polar covalent bond A covalent bond between atoms that have partial electric charges. *Examples:* H—Cl; O—S.

polarizability α The ease with which the electron cloud of a molecule can be distorted.

polarizable Easily polarized species.

polarize To distort the electron cloud of an atom or ion.

polarized light Plane-polarized light is light in which the wave motion occurs in a single plane.

polarizing power The ability of an ion to polarize a neighboring atom or ion.

polar molecule A molecule with a nonzero electric dipole moment. *Examples:* HCl; NH_3.

polyamide A polymer in which the monomers are linked by amide bonds formed by condensation polymerization. Often the monomers are a dicarboxylic acid and a diamine. *Example:* nylon.

polyatomic ion An ion in which more than two atoms are linked by covalent bonds. *Examples:* NH_4^+; NO_3^-; SiF_6^{2-}.

polyatomic molecule A molecule that consists of more than two atoms. *Examples:* O_3; $C_{12}H_{22}O_{11}$.

polycyclic compound An aromatic compound containing two or more benzene rings that share two neighboring carbon atoms. *Example:* naphthalene.

polydentate ligand A ligand that can attach at several binding sites.

polyester A polymer in which the monomers are linked by ester groups formed by condensation polymerization.

polymer A substance with large molecules consisting of chains of covalently linked repeating units formed from small molecules known as *monomers*. *Examples:* polyethylene; nylon. An *atactic polymer* is a polymer in which substituents are attached to each side of the chain at random. An *isotactic polymer* is a polymer in which the substituents are all on the same side of the chain. A *syndiotactic polymer* is a polymer in which the substituents alternate on either side of the chain. See also *copolymer*.

polynucleotide A polymer built from nucleotide units. *Examples:* DNA; RNA.

polypeptide A polymer formed by the condensation of amino acids.

polyprotic acid or base A Brønsted acid or base that can donate or accept more than one proton. (A polyprotic acid is sometimes called a polybasic acid.) *Examples:* H_3PO_4, triprotic acid; N_2H_4, diprotic base.

polysaccharide A chain of many saccharide units, such as glucose, linked together. *Examples:* cellulose; amylose.

***p*-orbital** See *atomic orbital*.

positional disorder Disorderly locations of molecules; a contribution to the entropy.

positive deviation (from Raoult's law) Behavior of a nonideal solution that has a vapor pressure higher than predicted by *Raoult's law*.

positive ion An ion formed by loss of one or more electrons from an atom or molecule; a *cation*.

positron A fundamental particle with the same mass as an electron but with opposite charge.

positron emission A mode of radioactive decay in which a nucleus emits a positron.

potential difference An electric potential difference between two points is a measure of the work that must be done to move an electric charge from one point to the other. Potential difference is measured in volts, V, and is commonly called the *voltage*.

potential energy E_P The energy arising from position. *Example:* the Coulomb potential energy of a charge is inversely proportional to its distance from another charge.

potential energy surface A surface showing the variation of the potential energy with the relative location of the atoms in a polyatomic cluster of atoms (as in a collision between a diatomic molecule and an atom).

power The rate of supply of energy. The SI unit of power is the watt, W ($1\ W = 1\ J{\cdot}s^{-1}$). See also Appendix 1B.

precipitate The solid formed in a precipitation reaction.

precipitation The process in which a solute comes out of solution rapidly as a finely divided powder, called a *precipitate*.

precipitation reaction A reaction in which a solid product is formed when two solutions are mixed. *Example:* $KBr(aq) + AgNO_3(aq) \rightarrow KNO_3(aq) + AgBr(s)$.

precise measurements (1) Measurements with a large number of significant figures. (2) A series of measurements with small random error and hence in close mutual agreement.

precision Freedom from random error. Compare with *accuracy*.

pre-equilibrium condition A pre-equilibrium arises (or is assumed) when an intermediate is formed in a rapid equilibrium reaction prior to a slow step in a reaction mechanism.

pre-exponential factor A The constant obtained from the y intercept in an Arrhenius plot.

pressure P Force divided by the area to which it is applied.

primary alcohol See *alcohol*.

primary cell A *galvanic cell* that produces electricity from chemicals sealed within it at the time of manufacture. It cannot be recharged.

primary pollutant A pollutant directly introduced into the environment. *Example:* SO_2.

primary structure The sequence of amino acids in the polypeptide chain of a protein.

primitive cubic structure A structure in which the unit cell consists of spheres (representing atoms or ions) at the corners of a cube.

principal quantum number n The quantum number that specifies the energy of an electron in a hydrogen atom and labels the shells of the atom.

probability density (of a particle) A function that, when multiplied by the volume of the region, gives the probability that the particle will be found in that region of space. See also *Born interpretation*.

product A species formed in a chemical reaction.

promotion (of an electron) The conceptual excitation of an electron to an orbital of higher energy in the description of bond formation.

propagation A series of steps in a chain reaction in which one chain carrier reacts with a reactant molecule to produce another carrier. *Examples:* $Br\cdot + H_2 \rightarrow HBr + H\cdot$; $H\cdot + Br_2 \rightarrow HBr + Br\cdot$.

properties The characteristics of matter. *Examples:* vapor pressure; color; density; temperature. See also *chemical property; physical property*.

protective oxide An oxide that protects a metal from oxidation. *Example:* aluminum oxide.

proton p A positively charged subatomic particle found in the nucleus of an atom.

protonation Proton transfer to a Brønsted base. *Example:* $2\ H_3O^+(aq) + S^{2-}(s) \rightarrow H_2S(g) + 2\ H_2O(l)$.

proton emission A nuclear decay process in which a proton is emitted. In proton emission, the mass and charge numbers of the nucleus both decrease by 1.

proton-rich nuclei Nuclei that have a low proportion of neutrons and lie below the *band of stability*.

proton transfer equilibrium The equilibrium involving the transfer of a hydrogen ion between an acid and a base.

proton transfer reaction See *proton transfer equilibrium*.

pseudo–first-order rate law A rate law that is effectively first order because all but one substance have virtually constant concentrations.

p-type semiconductor See *semiconductor*.

pX The quantity $-\log X$. *Example:* $pOH = -\log[OH^-]$.

pyrometallurgical process The extraction of metals by using reactions at high temperatures. *Example:* $Fe_2O_3(s) + 3\ CO(g) \xrightarrow{\Delta} 2\ Fe(l) + 3\ CO_2(g)$.

quadratic equation An equation of the form $ax^2 + bx + c = 0$. See also Appendix 1E.

qualitative A nonnumerical description of the properties of a substance, system, or process. *Example: qualitative analysis*, the identification of the substances present in a sample.

quanta The plural of *quantum*.

quantitative A numerical description of the properties of a substance, system, or process. *Example: quantitative analysis*, the determination of the amounts or concentrations of substances present in a sample.

quantization The restriction of a property to certain values. *Examples:* the quantization of energy and angular momentum.

quantum A packet of energy.
quantum mechanics The description of matter that takes into account the wave-particle duality of matter and the fact that the energy of an object may be changed only in discrete steps.
quantum number An integer (sometimes, a half-integer) that labels a wavefunction and specifies the value of a property. *Example:* principal quantum number, n.
quaternary ammonium ion An ion of the form NR_4^+, where R denotes hydrogen or an alkyl group (the four groups may be different).
quaternary structure The manner in which neighboring polypeptide units stack together to form a protein molecule.

racemic mixture A mixture containing equal amounts of two enantiomers.
rad A (non-SI) unit of *absorbed dose* of radiation; 1 rad corresponds to an energy deposit of 0.01 $J{\cdot}kg^{-1}$. See also *gray*.
radial wavefunction $R(r)$ The radial part of a wavefunction, particularly the radial component of the wavefunctions of the hydrogen atom; the probability amplitude of an electron as a function of distance from the nucleus.
radical An atom, molecule, or ion with at least one unpaired electron. *Examples:* ·NO; ·O·; $\cdot CH_3$.
radical chain reaction A chain reaction propagated by radicals.
radical polymerization A polymerization procedure that utilizes a radical chain reaction.
radical scavenger Impurities that react with radicals and inhibit a chain reaction.
radioactive A nucleus is radioactive if it can change its structure spontaneously and emit radiation.
radioactive series A stepwise nuclear decay path in which α and β particles are successively ejected and which terminates at a stable nuclide (often a nuclide of lead).
radioactivity The spontaneous emission of radiation by nuclei. Such nuclei are radioactive.
radiocarbon dating *Isotopic dating* based specifically on the use of carbon-14.
radioisotope Radioactive isotope.
radius ratio The ratio of the radius of the smaller ion in an ionic solid to the radius of the larger ion. The radius ratio controls which crystal structure is adopted by a simple ionic solid.
random copolymer See *copolymer*.
random error An error that varies randomly from measurement to measurement, sometimes giving a high value and sometimes a low one.
Raoult's law The vapor pressure of a liquid solution of a nonvolatile solute is directly proportional to the mole fraction of the solvent in the solution: $P = x_{solvent}P_{pure}$, where P_{pure} is the vapor pressure of the pure solvent.
rate The change in a property divided by the time interval.
rate constant k The constant of proportionality in a rate law.
rate-determining step The elementary reaction that governs the rate of the overall reaction. *Example:* the step $O + O_3 \rightarrow O_2 + O_2$ in the decomposition of ozone.
rate law An equation expressing the instantaneous reaction rate in terms of the concentrations, at that instant, of the substances taking part in the reaction. *Example:* rate $= k[NO_2]^2$.
rate of reaction See *instantaneous rate; reaction rate.*
reactant A species acting as a starting material in a chemical reaction; a reagent taking part in a specified reaction.
reaction enthalpy ΔH_r The change of enthalpy for the reaction exactly as the chemical equation is written. *Example:* $CH_4(g) + 2\,O_2(g) \rightarrow CO_2(g) + 2\,H_2O(l)$, $\Delta H_r = -890\ kJ{\cdot}mol^{-1}$.
reaction free energy ΔG_r The change in free energy associated with a chemical reaction. The reaction free energy is equal to the difference in molar free energies of the products and reactants, each value multiplied by the stoichiometric coefficient in the chemical equation.
reaction Gibbs energy See *reaction free energy*.
reaction intermediate A species that is produced and consumed during a reaction but does not occur in the overall chemical equation.
reaction mechanism The pathway that is proposed for an overall reaction and accounts for the experimental rate law.
reaction order See *order of reaction.*
reaction profile The variation in potential energy that occurs as two reactants meet, form an activated complex, and separate as products.
reaction quotient Q The ratio of the molar concentrations or partial pressures of the products to those of the reactants, each raised to a power equal to the stoichiometric coefficient (as in the definition of the equilibrium constant, but at an arbitrary stage of a reaction). *Example:* for $N_2(g) + 3\,H_2(g) \rightarrow 2\,NH_3(g)$, $Q = P_{NH_3}^{\,2}/P_{N_2}P_{H_2}^{\,3}$.
reaction rate The unique rate of a chemical reaction calculated by dividing the change in concentration of a substance by the interval during which the change occurs, and taking into account the stoichiometric coefficient of the substance. See also *unique average reaction rate*.
reaction sequence A series of reactions in which the products of one reaction take part as reactants in the next. *Example:* $2\,C(s) + O_2(g) \rightarrow 2\,CO(g)$, followed by $2\,CO(g) + O_2(g) \rightarrow 2\,CO_2(g)$.
reaction stoichiometry The quantitative relation between the amounts of reactants consumed and products formed in chemical reactions as expressed by the balanced chemical equation for the reaction.

reagent A substance or a solution that reacts with other substances.
real gas An actual gas; a gas that differs from an ideal gas in its behavior.
recrystallization Purification by repeated dissolving and crystallization.
redox couple The oxidized and reduced forms of a substance taking part in a reduction or oxidation half-reaction. The notation is oxidized species/reduced species. *Example:* H^+/H_2.
redox indicator See *indicator*.
redox reaction A reaction in which oxidation and reduction occur. *Example:* $S(s) + 3\ F_2(g) \rightarrow SF_6(g)$.
redox titration See *titration*.
reducing agent The species that supplies electrons to a substance being reduced (and is itself oxidized) in a redox reaction. *Examples:* H_2; H_2S; SO_3^{2-}.
reduction (1) The removal of oxygen from, or the addition of hydrogen to, a compound. (2) A reaction in which an atom, ion, or molecule gains an electron. (3) A reaction in which the oxidation number of an element is decreased. *Example:* $Cl_2(g) + 2\ e^- \rightarrow 2\ Cl^-(aq)$.
re-forming reaction A reaction in which a hydrocarbon is converted to carbon monoxide and hydrogen over a nickel catalyst.
refractory Able to withstand high temperatures.
relative biological effectiveness Q A factor used when assessing the damage caused by a given dose of radiation.
rem See *roentgen equivalent man*.
repeating unit The combination of atoms in a polymer that repeats over and over again in the polymer chain.
residual entropy The nonzero entropy at $T = 0$ in certain systems, which is due to a surviving disorder in the orientation of molecules.
residue An amino acid in a polypeptide chain.
resistance (electrical) A measure of the ability of matter to conduct electricity: the lower the resistance, the better it conducts.
resonance A blending of Lewis structures into a single composite, hybrid structure. *Example:* $:\ddot{O}-\ddot{S}=\ddot{O} \leftrightarrow \ddot{O}=\ddot{S}-\ddot{O}:$.
resonance hybrid The composite structure that results from resonance.
reverse osmosis The passage of solvent out of a solution when a pressure greater than the osmotic pressure is applied on the solution side of a semipermeable membrane.
reversible, isothermal expansion Expansion at constant temperature done against an external pressure matched to the pressure of the system.
reversible process A process that can be reversed by an infinitesimal change in a variable.
roast To heat a metal ore in air. *Example:* $2\ CuFeS_2(s) + 3\ O_2(g) \rightarrow 2\ CuS(s) + 2\ FeO(s) + 2\ SO_2(g)$ in the extraction of copper.
rock-salt structure A crystal structure the same as that of a mineral form of sodium chloride.
roentgen equivalent man rem The (non-SI) unit for reporting *dose equivalent*. See also *sievert*.
root mean square speed c The square root of the average value of the squares of the speeds of the molecules in a sample.
roots (of an equation) The solutions of the equation $f(x) = 0$. See also Appendix 1E.
rotational kinetic energy See *kinetic energy*.
Rydberg constant R The constant that occurs in the formula for the frequencies of the lines in the spectrum of atomic hydrogen; $R = 3.289\ 84 \times 10^{15}$ Hz.

sacrificial anode See *cathodic protection*.
saline carbide See *carbide*.
saline hydride See *hydride*.
salt (1) An ionic compound. (2) The product (other than water) of the reaction between an acid and a base. *Examples:* NaCl; K_2SO_4.
salt bridge A bridge-shaped tube containing a concentrated salt (potassium chloride or potassium nitrate) in a gel that acts as an electrolyte and provides a conducting path between two compartments of an electrochemical cell.
sample A representative part of a whole.
saturated Unable to take up further material.
saturated hydrocarbon A hydrocarbon with no carbon-carbon multiple bonds. *Example:* CH_3CH_3.
saturated solution A solution in which the dissolved and undissolved solute are in dynamic equilibrium.
Schrödinger equation An equation for calculating the wavefunction of a particle, especially for an electron in an atom or molecule. See also *wavefunction*.
science The systematically collected and organized body of knowledge based on experiment, observation, and careful reasoning.
scientific method A set of procedures employed to develop a scientific understanding of nature.
scientific notation The expression of numbers in the form $A \times 10^a$.
scintillation counter A device for detecting and measuring radioactivity that makes use of the fact that certain substances give a flash of light when they are exposed to radiation.
sea of instability A region in a graph of mass number against atomic number corresponding to unstable nuclei that decay with the emission of radiation. See also *band of stability*.
second s The SI unit of time. See also Appendix 1B.
secondary alcohol See *alcohol*.
secondary cell A *galvanic cell* that must be charged (or recharged) by using an externally supplied current before it can be used.

secondary pollutant A pollutant formed by the chemical reaction of another species in the environment. *Example:* SO_3 from the oxidation of SO_2.

secondary structure The manner in which a polypeptide chain is coiled. *Examples:* α helix; β sheet.

second derivative d^2y/dx^2 A measure of the curvature of a function. See also Appendix 1F.

second ionization energy I_2 The energy needed to remove an electron from a singly charged gas-phase cation. *Example:* $Cu^+(g) \rightarrow Cu^{2+}(g) + e^-(g)$, $I_2 = 1955\ kJ \cdot mol^{-1}$.

second law of thermodynamics A spontaneous change is accompanied by an increase in the total entropy of the system and its surroundings.

second-order reaction (1) A reaction with a rate that is proportional to the square of the molar concentration of a reactant. (2) A reaction with an overall order of 2.

second radiation constant c_2 The proportionality constant in *Wien's law*.

second virial coefficient *B*. See *virial equation*.

selective precipitation The precipitation of one compound in the presence of other, more soluble compounds.

self-sustaining fission Induced nuclear fission that, once it is initiated, can continue even if the supply of neutrons from outside is discontinued.

semiconductor An *electronic conductor* with a resistance that decreases as the temperature is raised. In an *n-type semiconductor*, the current is carried by electrons in a largely empty band; in a *p-type semiconductor*, the conduction is a result of electrons missing from otherwise filled bands.

semiempirical method The calculation of molecular structure that draws on experimental information to simplify the procedure. Compare with *ab initio method*.

semipermeable membrane A membrane that allows only certain types of molecules or ions to pass.

sequestration The formation of a complex between a cation and a bulky molecule or ion that enwraps the central ion. *Example:* Ca^{2+} and $O_3POPO_2OPO_5{}^{3-}$.

series (in spectroscopy) A family of spectral lines arising from transitions that have one state in common. *Example:* the Balmer series in the spectrum of atomic hydrogen.

shell All the orbitals of a given principal quantum number. *Example:* the single 2*s*- and three 2*p*-orbitals of the shell with $n = 2$.

shielding The repulsion that is experienced by an electron in an atom; it arises from the other electrons present and opposes the attraction exerted by the nucleus.

shift reaction A reaction between carbon monoxide and water: $CO(g) + H_2O(g) \rightarrow CO_2(g) + H_2(g)$; the reaction is used in the manufacture of hydrogen.

short-range order Atoms or molecules lying in a regular arrangement that does not extend very far past their nearest neighbors.

SI (Système International) The International System of units; a collection of definitions of units and symbols and their deployment. It is an extension and rationalization of the metric system. See also Appendix 1B.

side chain A hydrocarbon substituent on a hydrocarbon chain.

sievert Sv The SI unit of *dose equivalent*: 1 Sv = 100 rem.

sigma (σ) bond Two electrons in a cylindrically symmetric cloud between two atoms.

sigma (σ) orbital A molecular orbital that has no nodal plane on the internuclear axis.

significant figures sf (in a measurement) The digits in the measurement, up to and including the first uncertain digit in scientific notation. *Example:* 0.0260 mL (i.e., 2.60×10^{-2} mL), a measurement with three significant figures (3 sf).

single bond A shared electron pair.

skeletal equation An unbalanced equation that summarizes the qualitative information about the reaction. *Example:* $H_2 + O_2 \rightarrow H_2O$.

slip plane A layer of atoms in a solid that may slide relatively easily across a neighboring layer.

slope (of graph) The gradient of a graph. See also Appendix 1E.

smectic phase A liquid-crystal phase in which molecules lie parallel to one another and form layers.

smelt To melt a metal ore with a reducing agent. *Example:* $CuS(l) + O_2(g) \xrightarrow{\Delta} Cu(l) + SO_2(g)$.

S_N1 and S_N2 reactions Two varieties of *nucleophilic substitution*, with unimolecular and bimolecular rate-determining stages, respectively.

sol A colloidal dispersion of solid particles in a liquid.

solid A rigid form of matter that maintains the same shape whatever the shape of its container.

solid emulsion A colloidal dispersion of a liquid in a solid. *Example:* butter, an emulsion of water in butterfat.

solid solution A solid homogeneous mixture of two or more substances.

solubility The concentration of a saturated solution of a substance.

solubility constant See *solubility product*.

solubility product K_{sp} The product of ionic molar concentrations in a saturated solution; the dissolution equilibrium constant. *Example:* $Hg_2Cl_2(s) \rightleftharpoons Hg_2{}^{2+}(aq) + 2\ Cl^-(aq)$, $K_{sp} = [Hg_2{}^{2+}][Cl^-]^2$.

solubility rules A summary of the solubility pattern of a range of common compounds in water. See also Table I.1.

soluble substance A substance that dissolves to a significant extent in a specified solvent. When the solvent is not specified, water is generally meant.
solute A dissolved substance.
solution A homogeneous mixture. See also *solute*; *solvent*.
solvated Surrounded by and linked to solvent molecules. (Hydration is a special case in which the solvent is water.)
solvent The most abundant component of a solution.
solvent extraction A process for separating a mixture of substances that makes use of their differing solubilities in various solvents.
***s*-orbital** See *atomic orbital*.
space-filling model A depiction of a molecule in which atoms are represented by spheres that indicate the space each atom occupies.
sp^3d^n hybrid orbital A hybrid orbital formed from one *s*-orbital, three *p*-orbitals, and *n* *d*-orbitals.
species An atom, ion, or molecule.
specific enthalpy (of a fuel) The enthalpy of combustion per gram (without the negative sign).
specific heat capacity The heat capacity of a sample divided by its mass.
spectator ion An ion that is present but remains unchanged during a reaction. *Examples:* Na^+ and NO_3^- in $NaCl(aq) + AgNO_3(aq) \rightarrow NaNO_3(aq) + AgCl(s)$.
spectral line Radiation of a single wavelength emitted or absorbed by an atom or molecule.
spectrochemical series Ligands ordered according to the strength of the ligand field splitting they produce.
spectrometer An instrument for recording the spectrum of a sample.
spectrophotometer An instrument for measuring and recording electronically the intensity of radiation passing through a sample and hence recording its spectrum.
spectroscopy The analysis of the electromagnetic radiation emitted or absorbed by substances.
spectrum The frequencies or wavelengths of the electromagnetic radiation emitted or absorbed by substances.
speed The magnitude of the velocity; the rate of change of position.
sphalerite structure See *zinc-blende structure*.
spherically symmetrical Independent of orientation around a central point.
spherical polar coordinates The coordinates of a point expressed in terms of the radius r, the colatitude θ, and the azimuth ϕ.
sp^n hybrid A hybrid orbital constructed from an *s*-orbital and *n* *p*-orbitals. There are two *sp hybrids*, three *sp^2 hybrids*, and four *sp^3 hybrids*.
spin The intrinsic angular momentum of an electron; the spin cannot be eliminated and can occur in only two senses, denoted ↑ and ↓.
spin magnetic quantum number m_s The quantum number that distinguishes the two spin states of an electron: $m_s = +\frac{1}{2}$ (↑) and $m_s = -\frac{1}{2}$ (↓).
spontaneous change A natural change, one that has a tendency to occur without needing to be driven by an external influence. *Examples:* a gas expanding into a vacuum; a hot object cooling; methane burning.
spontaneous neutron emission The decay of *neutron-rich nuclei* by the emission of neutrons without an external stimulus to do so.
spontaneous nuclear fission See *fission*.
square planar complex A complex in which four ligands lie at the corners of a square with the metal atom at the center.
stability constant See *formation constant*.
stable See *thermodynamically stable compound*.
standard cell potential See *standard emf*.
standard emf $E°$ The emf when the concentration of each solute taking part in the cell reaction is 1 $mol{\cdot}L^{-1}$ (strictly, unit activity) and all the gases are at 1 bar. The standard emf of a galvanic cell is the difference between its two standard potentials: $E° = E°(\text{cathode}) - E°(\text{anode})$.
standard enthalpy of combustion $\Delta H_c°$ The change of enthalpy per mole of substance when it burns (reacts with oxygen) completely under standard conditions.
standard enthalpy of formation $\Delta H_f°$ The standard reaction enthalpy per mole of compound for the compound's synthesis from its elements in their most stable form at 1 bar and the specified temperature.
standard entropy of fusion $\Delta S_{fus}°$ The standard entropy change per mole accompanying fusion (the conversion of a substance from the solid state to the liquid state).
standard entropy of vaporization $\Delta S_{vap}°$ The standard entropy change per mole accompanying vaporization (the conversion of a substance from the liquid state to the vapor state).
standard Gibbs free energy of formation $\Delta G_f°$ The standard reaction free energy per mole for the formation of a compound from its elements in their most stable form.
standard Gibbs free energy of reaction See *standard reaction Gibbs energy*.
standard free energy of vaporization See *standard Gibbs energy of vaporization*.
standard Gibbs energy of formation See *standard Gibbs free energy of formation*.
standard Gibbs energy of fusion $\Delta G_{fus}°$ The standard Gibbs energy change per mole accompanying fusion (the conversion of a substance from the solid state to the liquid state).
standard Gibbs energy of vaporization $\Delta G_{vap}°$ The standard Gibbs energy change per mole accompanying vaporization (the conversion of a substance from the liquid state to the vapor state).

standard hydrogen electrode SHE A hydrogen electrode that is in its standard state [hydrogen ions at concentration 1 $mol \cdot L^{-1}$ (strictly, unit activity) and hydrogen pressure 1 bar] and is defined as having $E° = 0$.

standard molar entropy $S°$ The entropy per mole of a pure substance at 1 bar.

standard potential $E°$ (1) The contribution of an electrode to the standard emf of a cell. (2) The standard emf of a cell when the left-hand electrode is a standard hydrogen electrode and the right-hand electrode is the electrode of interest.

standard reaction enthalpy $\Delta H_r°$ The difference between the molar enthalpies of the products of a reaction in their standard states and the molar enthalpies of the reactants in their standard states (in kilojoules per mole); in practice, $\Delta H_r° = \Sigma n\Delta H_f°(\text{products}) - \Sigma n\Delta H_f°(\text{reactants})$, where n represents the stoichiometric coefficients.

standard reaction entropy $\Delta S_r°$ The difference between the molar entropies of the products of a reaction in their standard states and the molar entropies of the reactants in their standard states (in joules per kelvin per mole). $\Delta S_r° = \Sigma nS_m°(\text{products}) - \Sigma nS_m°(\text{reactants})$, where n represents the stoichiometric coefficients.

standard reaction free energy *See standard reaction Gibbs energy.*

standard reaction Gibbs energy $\Delta G_r°$ The difference between the molar Gibbs energy of the products of a reaction in their standard states and that of the reactants in their standard states (in kilojoules per mole); in practice, $\Delta G_r° = \Sigma n\Delta G_f°(\text{products}) - \Sigma n\Delta G_f°(\text{reactants})$, where n denotes the stoichiometric coefficients.

standard state The pure form of a substance at 1 bar; for a solute, the concentration 1 $mol \cdot L^{-1}$.

standard temperature and pressure STP 0°C (273.15 K) and 1 atm (101.325 kPa).

standing wave A wave stable over time; a wave with peaks and troughs that do not migrate with time.

state function A property of a substance that is independent of how the sample was prepared. *Examples:* pressure; enthalpy; entropy; color.

state of matter The physical condition of a sample. The most common states of a pure substance are solid, liquid, or gas (vapor).

state property See *state function.*

state symbol A symbol denoting the state of a species. *Examples:* s (solid); l (liquid); g (gas); aq (aqueous solution).

statistical entropy The entropy calculated from statistical thermodynamics; $S = k \ln W$.

statistical thermodynamics The interpretation of the laws of thermodynamics in terms of the behavior of large numbers of atoms and molecules.

steady-state approximation The assumption that the net rate of formation of reaction intermediates is 0.

Stefan-Boltzmann law The total intensity of radiation emitted by a heated black body is proportional to the fourth power of the absolute temperature.

stereoisomers Isomers in which atoms have the same partners arranged differently in space.

stereoregular polymer A polymer in which each unit or pair of repeating units has the same relative orientation.

steric factor P An empirical factor that takes into account the steric requirement of a reaction.

steric requirement A constraint on an elementary reaction in which the successful collision of two molecules depends on their relative orientation.

Stern-Gerlach experiment The demonstration of the quantization of electron spin by passing a beam of atoms through a magnetic field.

stick structure See *line structure.*

stock solution A solution stored in concentrated form.

stoichiometric coefficients The numbers multiplying chemical formulas in a chemical equation. *Examples:* 1, 1, and 2 in $H_2 + Br_2 \rightarrow 2\ HBr$.

stoichiometric point The stage in a titration when exactly the right volume of solution needed to complete the reaction has been added.

stoichiometric proportions Reactants in the same proportions as their coefficients in the chemical equation. *Example:* equal amounts of H_2 and Br_2 in the formation of HBr.

stoichiometric relation An expression that equates the relative amounts of reactants and products that participate in a reaction. *Example:* 1 mol $H_2 \simeq$ 2 mol HBr.

stoichiometry See *reaction stoichiometry.*

stopped-flow technique A procedure for observing fast reactions, involving the spectrometric analysis of a reaction mixture immediately following the rapid injection of reactants into a mixing chamber.

STP See *standard temperature and pressure.*

strong acids and bases Acids and bases that are fully deprotonated or protonated, respectively, in solution. *Examples:* HCl, $HClO_4$ (strong acids); NaOH, $Ca(OH)_2$ (strong bases).

strong electrolyte See *electrolyte.*

strong-field ligand A ligand that produces a large *ligand field splitting* and lies above H_2O in the spectrochemical series.

strong force A short-range but very strong force that acts between nucleons and binds them together to form a nucleus.

structural formula A chemical formula that shows how atoms in a compound are attached to one another. See also *condensed structural formula.*

structural isomers Isomers in which the atoms have different partners.

subatomic particle A particle smaller than an atom. *Examples:* electron; proton; neutron.
subcritical mass See *critical mass*.
sublimation The direct conversion of a solid to a vapor without first forming a liquid.
sublimation vapor pressure The vapor pressure of a solid.
subshell All the atomic orbitals of a given shell of an atom that have the same value of the quantum number *l*. *Example:* the five 3*d*-orbitals of an atom.
substance A single, pure type of matter; either a compound or an element.
substituent Atoms or groups that have replaced hydrogen atoms in an organic molecule.
substitutional alloy See *alloy*.
substitution reaction (1) A reaction in which an atom (or a group of atoms) replaces an atom in the original molecule. (2) In complexes, a reaction in which one Lewis base expels another and takes its place. *Examples:* (1) $C_6H_5OH + Br_2 \rightarrow BrC_6H_4OH + HBr$; (2) $[Fe(H_2O)_6]^{3+}(aq) + 6\ CN^-(aq) \rightarrow [Fe(CN)_6]^{3-}(aq) + 6\ H_2O(l)$.
substrate The chemical species on which an enzyme acts.
superconductor An *electronic conductor* that conducts electricity with zero resistance. See also *high-temperature superconductor*.
supercooled A liquid cooled to below its freezing point but not yet frozen.
supercritical fluid A substance above its *critical temperature* and *critical pressure*.
supercritical mass See *critical mass*.
superfluidity The ability to flow without viscosity.
surface-active agent See *surfactant*.
surface tension γ The tendency of molecules at the surface of a liquid to be pulled inward, so resulting in a smooth surface.
surfactant A substance that accumulates at the surface of a solution and affects the surface tension of the solvent; a component of detergents. *Example:* the stearate ion of soaps.
surroundings The region outside a system.
suspension A mist of small particles.
symbolic language The expression of chemical phenomena in terms of chemical symbols and mathematical equations.
syndiotactic polymer See *polymer*.
synthesis A reaction in which a substance is formed from simpler starting materials. *Example:* $N_2(g) + 3\ H_2(g) \rightarrow 2\ NH_3(g)$.
synthesis gas A mixture of carbon monoxide and hydrogen produced by the catalyzed reaction of a hydrocarbon and water.
system The object of study, usually a reaction vessel and its contents. An *open system* can exchange both matter and energy with the surroundings. A *closed system* has a fixed amount of matter but can exchange energy with the surroundings. An *isolated system* has no contact with its surroundings.
systematic error An error that persists in a series of measurements and does not average out. See also *accuracy*.
systematic name The name of a compound that reveals which elements are present (and, in its most complete form, how the atoms are arranged). *Example:* methylbenzene is the systematic name for toluene.
Système International d'Unités See *SI*.

temperature *T* (1) How hot or cold a sample is. (2) The intensive property that determines the direction in which heat will flow between two objects in contact.
temperature-composition diagram A *phase diagram* showing how the normal boiling point of a liquid mixture varies with composition.
termination A step in a *chain reaction* in which chain carriers combine to form products. *Example:* $Br\cdot + Br\cdot \rightarrow Br_2$.
termolecular reaction An *elementary reaction* involving the simultaneous collision of three species.
tertiary alcohol See *alcohol*.
tertiary structure The shape into which the α helix and β sheet sections of a polypeptide are twisted as a result of interactions between peptide groups lying in different parts of the primary structure.
tetrahedral complex A complex in which four ligands lie at the corners of a regular tetrahedron with the metal atom at the center. *Example:* $[Cu(NH_3)_4]^{2+}$.
tetrahedral hole A cavity in a (usually close-packed) crystal structure formed by one sphere lying in the dip formed by three others.
theoretical chemistry The study of molecular structure and properties in terms of mathematical models.
theoretical yield The maximum quantity of product that can be obtained, according to the reaction stoichiometry, from a given quantity of a specified reactant.
theory A collection of ideas and concepts used to account for a scientific law.
thermal decomposition See *decomposition*.
thermal equilibrium The state in which a system is at the same temperature as its surroundings.
thermal motion The random, chaotic motion of atoms.
thermal pollution The damage caused to the environment by the waste heat of an industrial process.
thermite reaction (thermite process) The reduction of a metal oxide by aluminum. *Example:* $2\ Al(s) + Fe_2O_3(s) \rightarrow Al_2O_3(s) + 2\ Fe(l)$.
thermochemical equation An expression consisting of the balanced chemical equation and the reaction enthalpy for the chemical reaction exactly as written.

thermochemistry The study of the heat released or absorbed by chemical reactions; a branch of thermodynamics.
thermodynamically stable compound (1) A compound with no thermodynamic tendency to decompose into its elements. (2) A compound with a negative free energy of formation.
thermodynamically unstable compound (1) A compound with a thermodynamic tendency to decompose into its elements. (2) A compound with a positive free energy of formation.
thermodynamics The study of the transformations of energy from one form to another. See also *first law of thermodynamics*; *second law of thermodynamics, third law of thermodynamics.*
thermonuclear explosion An explosion resulting from uncontrolled nuclear fusion.
thermotropic liquid crystal A liquid crystal prepared by melting the solid phase.
third law of thermodynamics The entropies of all perfect crystals are the same at the absolute zero of temperature.
third virial coefficient See *virial equation.*
three-center bond A chemical bond in which a hydrogen atom lies between two other atoms (typically boron atoms) and one electron pair binds all three atoms together.
tie line (in a temperature-composition phase diagram) A line joining the point indicating the boiling point of a mixture of given composition with the corresponding composition of the vapor at that temperature.
titrant The solution of known concentration added from a buret in a titration.
titration The analysis of composition by measuring the volume of one solution needed to react with a given volume of another solution. In an *acid-base titration,* an acid is titrated with a base; in a *redox titration,* an oxidizing agent is titrated with a reducing agent.
***t*-orbital** One of the orbitals d_{xy}, d_{yz}, and d_{zx} in an octahedral or tetrahedral complex. In an octahedral complex, these orbitals are designated t_{2g}.
total energy See *energy.*
tracer (in nuclear chemistry) An isotope that can be tracked from compound to compound in the course of a sequence of reactions.
trajectory The path of a particle on which location and linear momentum are specified at each instant.
transition A change of state. (1) In thermodynamics, a change of physical state. (2) In spectroscopy, a change of quantum state.
transition metal An element that belongs to Groups 3 through 11. *Examples:* vanadium; iron; gold.
translational kinetic energy See *kinetic energy.*
transmeitnerium elements The elements beyond meitnerium; those with $Z > 109$.
transmutation See *nuclear transmutation.*
transuranium elements The elements beyond uranium; those with $Z > 92$.
triple bond (1) Three electron pairs shared by two neighboring atoms. (2) One σ-bond and two π-bonds between neighboring atoms.
triple point The point where three phase boundaries meet in a phase diagram. Under the conditions represented by the triple point, all three adjoining phases coexist in dynamic equilibrium.
triprotic See *polyprotic acid or base.*
Trouton's rule The empirical observation that the entropy of vaporization at the boiling point (the enthalpy of vaporization divided by the boiling temperature) is approximately 85 $J{\cdot}K^{-1}{\cdot}mol^{-1}$ for many liquids.
tube structure A representation of molecular structure that uses tubes to indicate the lengths and distances of bonds. The ends of each tube are colored to indicate the identities of the elements forming the bond.

ultraviolet catastrophe The classical prediction that any black body at any temperature should emit intense ultraviolet radiation.
ultraviolet radiation Electromagnetic radiation with a higher frequency (shorter wavelength) than that of violet light.
unbranched alkane An alkane with no side chains, in which all the carbon atoms lie in a linear chain.
uncertainty principle See *Heisenberg uncertainty principle.*
unimolecular reaction An elementary reaction in which a single reactant molecule changes into products. *Example:* $O_3 \rightarrow O_2 + O$.
unique average rate The rate of change in the concentration of a reactant or product divided by its stoichiometric coefficient in the balanced equation. All unique average rates are reported as positive quantities. See also *average reaction rate.*
unit See *base units.*
unit cell The smallest unit that, when stacked together repeatedly without any gaps, can reproduce an entire crystal.
unsaturated hydrocarbon A hydrocarbon with at least one carbon-carbon multiple bond. *Examples:* $CH_2{=}CH_2$; C_6H_6.

valence The number of bonds that an atom can form.
valence band In the theory of solids, a band of energy levels fully occupied by electrons.
valence-bond theory The description of bond formation in terms of the pairing of spins in the atomic orbitals of neighboring atoms.
valence electrons The electrons that belong to the valence shell.
valence shell The outermost shell of an atom. *Example:* The $n = 2$ shell of Period 2 atoms.

valence-shell electron-pair repulsion model VSEPR model A model for predicting the shapes of molecules, using the fact that electron pairs repel one another.
van der Waals equation An approximate equation of state for a real gas in which two parameters represent the effects of intermolecular forces.
van der Waals forces See *intermolecular forces.*
van der Waals parameters The experimentally determined coefficients that appear in the van der Waals equation and are unique for each real gas. The parameter a is an indication of the strength of attractive intermolecular forces, and the parameter b is an indication of the strength of repulsive intermolecular forces. See also *van der Waals equation.*
van der Waals radius Half the distance between the centers of non-bonded, touching atoms in a solid.
van 't Hoff equation (1) The equation for the osmotic pressure in terms of the molarity, $\Pi = i[\mathrm{J}]RT$. (2) An equation that shows how the equilibrium constant varies with temperature.
van 't Hoff *i* factor See *i factor.*
vapor The gaseous phase of a substance (specifically, of a substance that is a liquid or solid at the temperature in question). See also *gas.*
vaporization The formation of a gas or a vapor from a liquid.
vapor pressure The pressure exerted by the vapor of a liquid (or a solid) when the vapor and the liquid (or solid) are in dynamic equilibrium.
variable covalence The ability of an element to form different numbers of covalent bonds. *Example:* S in SO_2 and SO_3.
variable valence The ability of an element to form ions with different charges. *Example:* In^+ and In^{3+}.
velocity The rate of change of position.
vibrational kinetic energy See *kinetic energy.*
virial equation An equation of state expressed in powers of $1/V_m$: specifically, $PV = nRT(1 + B/V_m + C/V_m^2 + \ldots)$, where B is the *second virial coefficient* and C is the *third virial coefficient.*
viscosity The resistance of a fluid (a gas or a liquid) to flow: the higher the viscosity, the slower the flow.
visible light See *visible radiation.*
visible radiation Electromagnetic radiation that can be detected by the human eye, with wavelengths in the range 700 nm to 400 nm. Visible radiation is also called *visible light* or simply *light.*
volatile Having a high vapor pressure at ordinary temperatures. A substance is typically regarded as volatile if its boiling point is below 100°C.
volatility The readiness with which a substance vaporizes. See also *volatile.*
volt V The SI unit of electric potential. See also Appendix 1B.
voltage The *potential difference* measured in volts.
voltaic cell See *electrochemical cell.*
volume *V* The amount of space a sample occupies.
volume percentage composition The volume of a substance present in a mixture, expressed as a percentage of the total volume.
volumetric analysis An analytical method using measurement of volume.
volumetric flask A flask calibrated to contain a specified volume.

water autoprotolysis constant K_w The equilibrium constant for the autoprotolysis (autoionization) of water, $2\,H_2O(l) \rightleftharpoons H_3O^+(aq) + OH^-(aq)$, $K_w = [H_3O^+][OH^-]$.
water of hydration See *hydration.*
water-splitting reaction The photochemical decomposition of water into hydrogen and oxygen.
wavefunction ψ A solution of the Schrödinger equation; the probability amplitude.
wavelength λ The peak-to-peak distance of a wave.
wave-particle duality The combined wavelike and particlelike character of both radiation and matter.
weak acids and bases Acids and bases that are only incompletely deprotonated or protonated, respectively, in aqueous solutions at normal concentrations. *Examples:* HF, CH_3COOH (weak acids); NH_3, CH_3NH_2 (weak bases).
weak electrolyte See *electrolyte.*
weak-field ligand A ligand that produces a small *ligand field splitting* and that lies below NH_3 in the *spectrochemical series.*
weight The gravitational force on a sample.
Wien's law The wavelength corresponding to the maximum in the radiation emitted by a heated *black body* is inversely proportional to the absolute temperature.
work w The energy expended during the act of moving an object against an opposing force. In *expansion work* the system expands against an opposing pressure; *nonexpansion work* is work that does not involve a change in volume.
work function Φ The energy required to remove an electron from a metal.

x-ray Electromagnetic radiation with wavelengths from about 10 pm up to about 1000 pm.
x-ray diffraction The analysis of crystal structures by studying the interference pattern in a beam of x-rays.

yield See *percentage yield*; *theoretical yield.*

zero-order reaction A reaction with a rate that is independent of the concentration of the reactant. *Example:* the catalyzed decomposition of ammonia.

zero-point energy The lowest possible energy of a system. *Example:* $E = h^2/8mL^2$ for a particle of mass m in a box of length L.

Ziegler-Natta catalyst A stereospecific catalyst for polymerization reactions, consisting of titanium tetrachloride and triethylaluminum.

zinc-blende structure A crystal structure in which the cations occupy half the tetrahedral holes in a nearly close-packed cubic lattice of anions; also known as *sphalerite structure.*

zone refining A method for purifying a solid by repeatedly passing a molten zone along the length of a sample.

ANSWERS

Odd-Numbered Exercises

Fundamentals

A.1 (a) chemical; (b) physical; (c) physical
A.3 The temperature, humidity, and evaporation of water are physical properties. The ripening of oranges is a chemical change.
A.5 (a) intensive; (b) intensive; (c) extensive; (d) extensive
A.7 $19.30\ g\cdot cm^{-3}$
A.9 $0.0427\ cm^3$
A.11 $35.9\ g\cdot cm^{-3}$; the atoms do not occupy all the space
A.13 32 J
A.15 7.8×10^{13} J
A.17 15 J
A.19 238 J
A.21 1.1×10^{-24} N
B.1 1.4×10^{22} atoms
B.3 (a) Radiation may pass through a metal foil. (b) All light travels at the same speed; the slower speed supports the particle model. (c) This observation supports the radiation model. (d) This observation supports the particle model; electromagnetic radiation has no mass and no charge.
B.5 (a) 1 proton, 1 neutron, and 1 electron. (b) 53 protons, 74 neutrons, and 53 electrons. (c) 7 protons, 8 neutrons, and 7 electrons. (d) 83 protons, 126 neutrons, and 83 electrons.
B.7 (a) ^{175}Lu; (b) ^{118}Sn; (c) ^{6}Li
B.9 (a) ^{12}C, ^{13}C, and ^{14}C all have 6 protons and 6 electrons. (b) They have different numbers of neutrons: an atom of ^{12}C has 6 neutrons, an atom of ^{13}C has 7 neutrons, and an atom of ^{14}C has 8 neutrons.
B.11 (a) Atoms of ^{55}Mn, ^{56}Fe, and ^{58}Ni all have 30 neutrons. (b) They have different masses, numbers of protons, and numbers of electrons.
B.13 (a) 10 protons, 7 neutrons, and 10 electrons total (b) mass of protons $= 1.673 \times 10^{-23}$ g; mass of neutrons $= 1.172 \times 10^{-23}$ g; mass of electrons $= 9.109 \times 10^{-27}$ g; (c) 44.4%
B.15 (a) Silicon, Group 14/IV metalloid; (b) strontium, Group 2 metal; (c) tin, Group 14/IV metal; (d) antimony, Group 15/V metalloid
B.17 (a) Ag, metal; (b) Ge, metalloid; (c) Kr, nonmetal
B.19 Li, 3; Na, 11; K, 19; Rb, 37; Cs, 55. All the alkali metals react with water to produce hydrogen. Lithium reacts gently, but the reaction becomes more vigorous down the group. The melting points decrease down the group.
B.21 (a) Metals are malleable, conduct electricity, and have luster. (b) Nonmetals are not malleable or ductile and do not conduct electricity.
C.1 (a) An ionic compound is made up of ions. (b) A molecular compound is made up of molecules. Ionic compounds typically have higher melting points than molecular compounds.
C.3

```
          CH3       CH3
           |         |
(a)  H3C—C—CH2—CH—CH3
           |
          CH3

          CH3
           |
(b)  H3C—CH—CH2—OH
```

C.5 (a) Cesium is a metal in Group 1; it will form Cs^+ ions. (b) Iodine is a nonmetal in Group 17/VII and will form I^- ions. (c) Selenium is a Group 16/VI nonmetal and will form Se^{2-} ions. (d) Calcium is a Group 2 metal and will form Ca^{2+} ions.
C.7 (a) $^{4}He^{2+}$ has 2 protons, 2 neutrons, and no electrons. (b) $^{15}N^{3-}$ has 7 protons, 8 neutrons, and 10 electrons. (c) $^{127}I^-$ has 53 protons, 74 neutrons, and 54 electrons. (d) $^{80}Se^{2-}$ has 34 protons, 46 neutrons, and 36 electrons.
C.9 (a) $^{19}F^-$; (b) $^{24}Mg^{2+}$; (c) $^{128}Te^{2-}$; (d) $^{86}Rb^+$
C.11 (a) Al^{3+} ions, Te^{2-} ions, Al_2Te_3 (b) Mg^{2+} ions, O^{2-} ions, MgO (c) Na^+ ions, S^{2-} ions, Na_2S (d) Rb^+ ions, I^- ions, RbI
D.1 (a) Al_2O_3; (b) $Sr_3(PO_4)_2$; (c) $Al_2(CO_3)_3$; (d) Li_3N
D.3 (a) calcium phosphate; (b) tin(II) fluoride, stannous fluoride; (c) vanadium(V) oxide; (d) copper(I) oxide, cuprous oxide
D.5 (a) TiO_2; (b) $SiCl_4$; (c) CS_2; (d) SF_4; (e) Li_2S; (f) SbF_5; (g) N_2O_5; (h) IF_7
D.7 (a) sulfur hexafluoride; (b) dinitrogen pentoxide; (c) nitrogen triiodide; (d) xenon tetrafluoride; (e) arsenic tribromide; (f) chlorine dioxide
D.9 (a) hydrochloric acid; (b) sulfuric acid; (c) nitric acid; (d) acetic acid; (e) sulfurous acid; (f) phosphoric acid

D.11 (a) Na_2O; (b) K_2SO_4; (c) AgF; (d) $Zn(NO_3)_2$; (e) Al_2S_3
D.13 (a) sodium sulfite; (b) iron(III) oxide, ferric oxide; (c) iron(II) oxide, ferrous oxide; (d) magnesium hydroxide; (e) nickel(II) sulfate hexahydrate; (f) phosphorus pentachloride or phosphorus(V) chloride; (g) chromium(III) dihydrogen phosphate; (h) diarsenic trioxide; (i) ruthenium(II) chloride
D.15 (a) heptane; (b) propane; (c) butanol; (d) chloropropane
D.17 (a) telluric acid; (b) sodium arsenate; (c) calcium selenite; (d) barium antimonate; (e) arsenic acid; (f) cobalt(III) tellurate
E.1 (a) 1.0×10^{-14} mol; (b) 3.2×10^6 years
E.3 (a) 6.94 $g{\cdot}mol^{-1}$; (b) 6.96 $g{\cdot}mol^{-1}$
E.5 26.2% ^{10}B; 73.8% ^{11}B
E.7 (a) 80 g of tellurium contains a greater amount of atoms than 75 g of indium. (b) 15.0 g of P has slightly more atoms than 15.0 g of S. (c) Because the two samples have the same number of atoms, they will have the same number of moles.
E.9 (a) 4.1×10^2 g Rh; (b) 63 g Rh
E.11 (a) 0.0981 mol Al_2O_3, 5.91×10^{22} molecules; (b) 1.30 mmol HF, 7.83×10^{20} molecules; (c) 45.6 μmol H_2O_2, 2.75×10^{19} molecules; (d) 6.94 mol $C_6H_{12}O_6$, 4.18×10^{24} molecules; (e) 0.312 mol N, 0.156 mol N_2, 1.88×10^{23} atoms, 9.39×10^{22} molecules
E.13 (a) 0.0140 mol Ag^+; (b) 2.10 mol UO_3; (c) 7.75×10^{-5} mol Cl^-; (d) 5.89×10^{-3} mol H_2O
E.15 (a) 4.52×10^{23} formula units; (b) 124 mg; (c) 3.036×10^{25} formula units
E.17 (a) 2.992×10^{-23} $g{\cdot}molecule^{-1}$; (b) 3.34×10^{25} molecules
E.19 (a) 2.49×10^{-2} mol; (b) 4.98×10^{-2} mol Br^-; (c) 6.00×10^{22} molecules H_2O; (d) 0.2166
E.21 (a) 4.11×10^{23} S atoms; (b) 5.14×10^{22} molecules; (c) 21.9 g S
E.23 (a) 1.38×10^{23} O atoms; (b) 1.26×10^{22} formula units; (c) 0.146 mol H_2O
F.1 79.12% C, 9.78% H, 11.10% O
F.3 (a) Na_3AlF_6; (b) $KClO_3$; (c) NH_6PO_4 or $[NH_4][H_2PO_4]$
F.5 PCl_5
F.7 $C_9H_{11}O_4N$
F.9 (a) $Os(CO)_4$; (b) $Os_3(CO)_{12}$
F.11 $C_8H_{10}O_2N_4$
F.13 The elemental compositions are close, but the C and O values differ sufficiently to make them distinguishable.
G.1 (a) solubility; (b) the abilities of the components to adsorb; (c) boiling points
G.3 (a) homogeneous, distillation; (b) heterogeneous, dissolving followed by filtration; (c) homogeneous, distillation
G.5 15.2 g
G.7 (a) 13.5 mL; (b) 62.5 mL; (c) 5.92 mL
G.9 (a) Weigh 1.6 g $KMnO_4$ into a 1.0-L volumetric flask and add water to give a total volume of 1.0 L. (b) Add 40 mL of water to 10 mL of 0.050 $mol{\cdot}L^{-1}$ $KMnO_4$.
G.11 (a) 4.51 mL; (b) To prepare 60.0 mL of solution, we will need 4 parts water to 1 part NaOH solution. Dividing 60.0 by 5 gives 12.0 mL of 2.5 $mol{\cdot}L^{-1}$ NaOH solution needed for dilution by enough water to give 60.0 mL solution.
G.13 (a) 8.0 g; (b) 12 g
G.15 (a) 0.004 34 M $Cl^-(aq)$; (b) 0.274 M $Ni^{2+}(aq)$
G.17 (a) 21.8 g; (b) 0.24 g; (c) 60 g
H.1 (a) $BCl_3(g) + 3\,H_2O(l) \rightarrow B(OH)_3(aq) + 3\,HCl(aq)$
(b) $2\,NaNO_3(s) \rightarrow 2\,NaNO_2(s) + O_2(g)$
(c) $2\,Ca_3(PO_4)_2(s) + 6\,SiO_2(s) + 10\,C(s) \rightarrow 6\,CaSiO_3(s) + 10\,CO(g) + P_4(s)$
(d) $4\,Fe_2P(s) + 18\,S(s) \rightarrow P_4S_{10}(s) + 8\,FeS(s)$
H.3 (a) $2\,Na(s) + 2\,H_2O(l) \rightarrow H_2(g) + 2\,NaOH(aq)$
(b) $Na_2O(s) + H_2O(l) \rightarrow 2\,NaOH(aq)$
(c) $6\,Li(s) + N_2(g) \rightarrow 2\,Li_3N(s)$
(d) $Ca(s) + 2\,H_2O(l) \rightarrow H_2(g) + Ca(OH)_2(aq)$
H.5 (I) $3\,Fe_2O_3(s) + CO(g) \rightarrow 2\,Fe_3O_4(s) + CO_2(g)$
(II) $Fe_3O_4(s) + 4\,CO(g) \rightarrow 3\,Fe(s) + 4\,CO_2(g)$
H.7 (I) $N_2(g) + O_2(g) \rightarrow 2\,NO(g)$
(II) $2\,NO(g) + O_2(g) \rightarrow 2\,NO_2(g)$
H.9 $4\,HF(aq) + SiO_2(s) \rightarrow SiF_4(aq) + 2\,H_2O(l)$
H.11 $2\,C_8H_{18}(l) + 25\,O_2(g) \rightarrow 16\,CO_2(g) + 18\,H_2O(g)$
H.13 $4\,C_{10}H_{15}N + 55\,O_2(g) \rightarrow 40\,CO_2(g) + 30\,H_2O(l) + 2\,N_2(g)$
H.15 (I) $H_2S(g) + 2\,NaOH(aq) \rightarrow Na_2S(aq) + 2\,H_2O$
(II) $4\,H_2S(g) + Na_2S(alc) \xrightarrow{\text{alcohol}} Na_2S_5(alc) + 4\,H_2(g)$
(III) $2\,Na_2S_5(alc) + 9\,O_2(g) + 10\,H_2O(l) \rightarrow 2\,Na_2S_2O_3{\cdot}5H_2O(s) + 6\,SO_2(g)$
I.1 (a) CH_3OH, nonelectrolyte; (b) $CaBr_2$, strong electrolyte; (c) KI, strong electrolyte
I.3 (a) soluble; (b) slightly soluble; (c) insoluble; (d) insoluble
I.5 (a) $Na^+(aq)$ and $I^-(aq)$; (b) solid Ag_2CO_3; (c) $NH_4^+(aq)$, $PO_4^{3-}(aq)$; (d) $Fe^{2+}(aq)$ and $SO_4^{2-}(aq)$
I.7 (a) $Fe(OH)_3$, precipitate; (b) Ag_2CO_3, precipitate forms; (c) No precipitate will form because both possible products are soluble in water.
$Pb(NO_3)_2(aq) + 2\,CH_3CO_2Na(aq) \rightarrow 2\,NaNO_3(aq) + Pb(CH_3CO_2)_2(aq)$
I.9 (a) net ionic equation:
$Fe^{2+}(aq) + S^{2-}(aq) \rightarrow FeS(s)$; spectator ions: Na^+, Cl^-;
(b) net ionic equation:
$Pb^{2+}(aq) + 2\,I^-(aq) \rightarrow PbI_2(s)$; spectator ions: K^+, NO_3^-;
(c) net ionic equation:
$Ca^{2+}(aq) + SO_4^{2-}(aq) \rightarrow CaSO_4(s)$; spectator ions: $NO_3^-(aq)$, $K^+(aq)$;

(d) net ionic equation:
$Pb^{2+}(aq) + CrO_4^{2-}(aq) \rightarrow PbCrO_4(s)$; spectator ions: Na^+, NO_3^-;
(e) net ionic equation:
$Hg_2^{2+}(aq) + SO_4^{2-}(aq) \rightarrow Hg_2SO_4(s)$; spectator ions: $K^+(aq)$, $NO_3^-(aq)$
I.11 (a) $(NH_4)_2CrO_4(aq) + BaCl_2(aq) \rightarrow BaCrO_4(s) + 2\ NH_4Cl(aq)$
complete ionic equation:
$2\ NH_4^+(aq) + CrO_4^{2-}(aq) + Ba^{2+}(aq) + 2\ Cl^-(aq) \rightarrow BaCrO_4(s) + 2\ NH_4^+(aq) + 2\ Cl^-(aq)$
net ionic equation: $Ba^{2+}(aq) + CrO_4^{2-}(aq) \rightarrow BaCrO_4(s)$; spectator ions: NH_4^+, Cl^-
(b) $CuSO_4(aq) + Na_2S(aq) \rightarrow CuS(s) + Na_2SO_4(aq)$
complete ionic equation:
$Cu^{2+}(aq) + SO_4^{2-}(aq) + 2\ Na^+(aq) + S^{2-}(aq) \rightarrow CuS(s) + 2\ Na^+(aq) + SO_4^{2-}(aq)$
net ionic equation: $Cu^{2+}(aq) + S^{2-}(aq) \rightarrow CuS(s)$;
spectator ions: $Na^+(aq)$, $SO_4^{2-}(aq)$
(c) $3\ FeCl_2(aq) + 2\ (NH_4)_3PO_4(aq) \rightarrow Fe_3(PO_4)_2(s) + 6\ NH_4^+(aq) + 6\ Cl^-(aq)$
complete ionic equation:
$3\ Fe^{2+}(aq) + 6\ Cl^-(aq) + 6\ NH_4^+(aq) + 2\ PO_4^{3-} \rightarrow Fe_3(PO_4)_2(s) + 6\ NH_4^+(aq) + 6\ Cl^-(aq)$
net ionic equation:
$3\ Fe^{2+}(aq) + 2\ PO_4^{3-}(aq) \rightarrow Fe_3(PO_4)_2(s)$;
spectator ions: $Cl^-(aq)$, $NH_4^+(aq)$
(d) $K_2C_2O_4(aq) + Ca(NO_3)_2(aq) \rightarrow CaC_2O_4(s) + 2KNO_3(aq)$
complete ionic equation:
$2\ K^+(aq) + C_2O_4^{2-}(aq) + Ca^{2+}(aq) + 2\ NO_3^-(aq) \rightarrow CaC_2O_4(s) + 2\ K^+(aq) + 2\ NO_3^-(aq)$
net ionic equation: $Ca^{2+}(aq) + C_2O_4^{2-}(aq) \rightarrow CaC_2O_4(s)$
spectator ions: K^+, NO_3^-
(e) $NiSO_4(aq) + Ba(NO_3)_2(aq) \rightarrow Ni(NO_3)_2(aq) + BaSO_4(s)$
complete ionic equation:
$Ni^{2+}(aq) + SO_4^{2-}(aq) + Ba^{2+}(aq) + 2\ NO_3^-(aq) \rightarrow Ni^{2+}(aq) + 2\ NO_3^-(aq) + BaSO_4(s)$
net ionic equation: $Ba^{2+}(aq) + SO_4^{2-}(aq) \rightarrow BaSO_4(s)$
spectator ions: Ni^{2+}, NO_3^-
I.13 (a) $Ba(CH_3CO_2)_2(aq) + Li_2CO_3(aq) \rightarrow BaCO_3(s) + 2\ LiCH_3CO_2(aq)$
complete ionic equation:
$Ba^{2+}(aq) + 2\ CH_3CO_2^-(aq) + 2\ Li^+(aq) + CO_3^{2-}(aq) \rightarrow BaCO_3(s) + 2\ Li^+(aq) + 2\ CH_3CO_2^-(aq)$
net ionic equation: $Ba^{2+}(aq) + CO_3^{2-}(aq) \rightarrow BaCO_3(s)$
(b) $2\ NH_4Cl(aq) + Hg_2(NO_3)_2(aq) \rightarrow 2\ NH_4NO_3(aq) + Hg_2Cl_2(s)$
complete ionic equation:
$2\ NH_4^+(aq) + 2\ Cl^-(aq) + Hg_2^{2+}(aq) + 2\ NO_3^-(aq) \rightarrow Hg_2Cl_2(s) + 2\ NH_4^+(aq) + 2\ NO_3^-(aq)$
net ionic equation: $Hg_2^{2+}(aq) + 2\ Cl^-(aq) \rightarrow Hg_2Cl_2(s)$
(c) $Cu(NO_3)_2(aq) + Ba(OH)_2(aq) \rightarrow Cu(OH)_2(s) + Ba(NO_3)_2(aq)$
complete ionic equation:
$Cu^{2+}(aq) + 2\ NO_3^-(aq) + Ba^{2+}(aq) + 2\ OH^-(aq) \rightarrow Cu(OH)_2(s) + Ba^{2+}(aq) + 2\ NO_3^-(aq)$
net ionic equation: $Cu^{2+}(aq) + 2\ OH^-(aq) \rightarrow Cu(OH)_2(s)$
I.15 (a) $AgNO_3$ and Na_2CrO_4
(b) $CaCl_2$ and Na_2CO_3
(c) $Cd(ClO_4)_2$ and $(NH_4)_2S$
I.17 (a) $2\ Ag^+(aq) + SO_4^{2-}(aq) \rightarrow Ag_2SO_4(s)$
(b) $Hg^{2+}(aq) + S^{2-}(aq) \rightarrow HgS(s)$
(c) $3\ Ca^{2+}(aq) + 2\ PO_4^{3-}(aq) \rightarrow Ca_3(PO_4)_2(s)$
(d) $AgNO_3$ and Na_2SO_4; Na^+, NO_3^-
$Hg(CH_3CO_2)_2$ and Li_2S; Li^+, $CH_3CO_2^-$
$CaCl_2$ and K_3PO_4; K^+, Cl^-
J.1 (a) base; (b) acid; (c) base; (d) acid; (e) base
J.3 (a) overall equation: $HF(aq) + NaOH(aq) \rightarrow NaF(aq) + H_2O(l)$
total ionic equation:
$HF(aq) + Na^+(aq) + OH^-(aq) \rightarrow Na^+(aq) + F^-(aq) + H_2O(l)$
net ionic equation:
$HF(aq) + OH^-(aq) \rightarrow F^-(aq) + H_2O(l)$
(b) overall equation: $(CH_3)_3N(aq) + HNO_3(aq) \rightarrow (CH_3)_3NHNO_3(aq)$
total ionic equation:
$(CH_3)_3N(aq) + H^+(aq) + NO_3^-(aq) \rightarrow (CH_3)_3NH^+(aq) + NO_3^-(aq)$
net ionic equation: $(CH_3)_3N(aq) + H^+(aq) \rightarrow (CH_3)_3NH^+(aq)$
(c) overall equation: $LiOH(aq) + HI(aq) \rightarrow LiI(aq) + H_2O(l)$
total ionic equation:
$Li^+(aq) + OH^-(aq) + H_3O^+(aq) + I^-(aq) \rightarrow Li^+(aq) + I^-(aq) + 2\ H_2O(l)$
J.5 (a) $HBr(aq) + KOH(aq) \rightarrow KBr(aq) + H_2O(l)$
(b) $Zn(OH)_2(aq) + 2\ HNO_2(aq) \rightarrow Zn(NO_2)_2(aq) + 2\ H_2O(l)$
(c) $Ca(OH)_2(aq) + 2\ HCN(aq) \rightarrow Ca(CN)_2(aq) + 2\ H_2O(l)$
(d) $3\ KOH(aq) + H_3PO_4(aq) \rightarrow K_3PO_4(aq) + 3\ H_2O(l)$
J.7 (a) acid: $H_3O^+(aq)$; base: $CH_3NH_2(aq)$
(b) acid: $HCl(aq)$; base: $C_2H_5NH_2(aq)$
(c) acid: $HI(aq)$; base: $CaO(s)$
K.1 (a) $2\ NO_2(g) + O_3(g) \rightarrow N_2O_5(g) + O_2(g)$
(b) $S_8(s) + 16\ Na(s) \rightarrow 8\ Na_2S(s)$
(c) $2\ Cr^{2+}(aq) + Sn^{4+}(aq) \rightarrow 2\ Cr^{3+}(aq) + Sn^{2+}(aq)$
(d) $2\ As(s) + 3\ Cl_2(g) \rightarrow 2\ AsCl_3(l)$
K.3 (a) $Mg(s) + Cu^{2+}(aq) \rightarrow Mg^{2+}(aq) + Cu(s)$
(b) $Fe^{2+}(aq) + Ce^{4+}(aq) \rightarrow Fe^{3+}(aq) + Ce^{3+}(aq)$
(c) $H_2(g) + Cl_2(g) \rightarrow 2\ HCl(g)$
(d) $4\ Fe(s) + 3\ O_2(g) \rightarrow 2\ Fe_2O_3(s)$
K.5 (a) +4; (b) +4; (c) +2; (d) +5; (e) +1; (f) 0
K.7 (a) +2; (b) +2; (c) +6; (d) +4; (e) +1
K.9 (a) Methanol is oxidized, $O_2(g)$ is reduced.
(b) Mo is reduced, and *some* sulfur is oxidized.
(c) Tl^+ is both oxidized and reduced.

K.11 (a) Cl_2 will more easily be reduced and is therefore a stronger oxidizing agent than Cl^-. (b) N_2O_5 will be a stronger oxidizing agent because it will readily be reduced. N^{5+} will accept electrons more readily than N^+. The higher the oxidation number of an element, the more easily it gains electrons. The lower its oxidation number, the more easily it loses them.
K.13 (a) oxidizing agent: HCl(aq); reducing agent: Zn(s); (b) oxidizing agent: SO_2(g); reducing agent: H_2S(g); (c) oxidizing agent: B_2O_3(s); reducing agent: Mg(s)
K.15 (a) $ClO_3^- \rightarrow ClO_2$; Cl goes from $+5 \rightarrow +4$, reducing agent
(b) $SO_4^{2-} \rightarrow S^{2-}$; S goes from $+6 \rightarrow -2$, reducing agent
(c) $Mn^{2+} \rightarrow MnO_2$; Mn goes from $+2 \rightarrow +4$, oxidizing agent
(d) HCHO $\rightarrow$ HCOOH; C goes from 0 to +2, oxidizing agent
K.17 (a) oxidizing agent: WO_3(s); reducing agent: H_2(g); (b) oxidizing agent: HCl(aq); reducing agent: Zn(s); (c) oxidizing agent: SnO_2(s); reducing agent: C(s); (d) oxidizing agent: N_2O_4(g); reducing agent: N_2H_4(g)
L.1 (a) 1.1×10^{-5} mol; (b) 6.2 g AgBr
L.3 (a) 507 g Al; (b) 6.613×10^3 kg Al_2O_3
L.5 (a) 505 g H_2O; (b) 1.33 kg O_2
L.7 4.246 kg H_2O
L.9 (a) 0.271 M; (b) 0.163 g
L.11 (a) 0.209 $mol \cdot L^{-1}$; (b) 0.329 g
L.13 63.0 $g \cdot mol^{-1}$
L.15 (a) $Na_2CO_3(aq) + 2\,HCl(aq) \rightarrow 2\,NaCl(aq) + H_2CO_3(aq)$
(b) 0.126 M HCl(aq)
M.1 93.1%
M.3 (a) 14.6 g O_2; (b) 30%
M.5 (a) O_2; (b) 5.77 g; (c) 5.7 g P_4O_6
M.7 (a) 0.3424 g CO_2; 0.097 92 g H_2O
M.9 empirical formula: $C_4H_5N_2O$; molecular formula: $C_8H_{10}N_4O_2$; $2\,C_8H_{10}N_4O_2(s) + 21\,O_2(g) \rightarrow 16\,CO_2(g) + 10\,H_2O(g) + 4\,N_2(g)$
M.11 (a) $Ca_3(PO_4)_2$; (b) 130 g
M.13 93.0%

Chapter 1

1.1 ultraviolet radiation < visible light < infrared radiation < radio waves
1.3 microwaves < visible light < ultraviolet radiation < x-rays ≈ gamma rays
1.5 (a) $3.24 \times 10^{14}\ s^{-1}$; (b) $7.22 \times 10^{10}\ s^{-1}$
1.7 1.59×10^3 nm
1.9 (a) 3.37×10^{-19} J; (b) 44.1 J; (c) 203 kJ
1.11 8.8236 pm
1.13 (a) false; (b) true; (c) false
1.15 (a) 3.3×10^{-10} m; (b) 6.63×10^{-19} J; (c) 68 nm; (d) far ultraviolet, almost to the x-ray region
1.17 1.10×10^{-34} m
1.19 $3.96 \times 10^3\ m \cdot s^{-1}$
1.21 3.37×10^{-19} J
1.23 (a) 410 nm; (b) Balmer series; (c) violet
1.25 (a) 1.48×10^{-8} m; (b) 8.25×10^{-9} m
1.27 n for lower energy level
1.29 (a) See Figs. 1.27, 1.30, and 1.31. (b) where the wavefunction passes through 0; (c) s, 0 nodes; p, 1; d, 2; (d) f, 3
1.31 The p_x-orbital has its lobes oriented along the x-axis, the p_y-orbital has its lobes oriented along the y-axis, and the p_z-orbital has its lobes oriented along the z-axis.
1.33 0.50
1.35 (a) 2; (b) 3; (c) 0, 1, 2
1.37 (a) 1; (b) 5; (c) 3; (d) 7
1.39 (a) 7; (b) 5; (c) 3; (d) 4
1.41 (a) $n = 7, l = 0$; (b) $n = 5, l = 3$; (c) $n = 3, l = 2$; (d) $n = 2, l = 1$
1.43 (a) 0; (b) −3, −2, −1, 0, +1, +2, +3; (c) −2, −1, 0, +1, +2; (d) −1, 0, +1
1.45 (a) 6 electrons; (b) 10 electrons; (c) 2 electrons; (d) 14 electrons
1.47 (a) 5d, five; (b) 1s; one; (c) 6f, seven; (d) 2p, three
1.49 (a) $n = 2, l = 1$; (b) $n = 5, l = 3$; (c) $n = 3, l = 0$; (d) $n = 4, l = 2$
1.51 (a) 6; (b) 2; (c) 8; (d) 2
1.53 (a) and (c) cannot exist
1.55 (a) $V(r) =$

$$\left(\frac{-3e^2}{4\pi\epsilon_0}\right)\left(\frac{1}{r_1} + \frac{1}{r_2} + \frac{1}{r_3}\right) + \frac{e^2}{4\pi\epsilon_0}\left(\frac{1}{r_{12}} + \frac{1}{r_{13}} + \frac{1}{r_{23}}\right)$$

(b) The first term represents the coulombic attractions between the nucleus and each electron, and the second term represents the coulombic repulsions between each pair of electrons.
1.57 (a) false; (b) true; (c) false; (d) true
1.59 Only (d) is the configuration expected for a ground-state atom; the others all represent excited-state configurations.
1.61 (b) not possible because $l = 0$, so m_l must be 0; (c) not possible because $n = 4$, so $l = 0, 1, 2, 3$ only.
1.63 Ga: $[Ar]3d^{10}4s^24p^1$, one unpaired electron; Ge: $[Ar]3d^{10}4s^24p^2$, two unpaired electrons; As: $[Ar]3d^{10}4s^24p^3$, three unpaired electrons; Se: $[Ar]3d^{10}4s^24p^4$, two unpaired electrons; Br: $[Ar]3d^{10}4s^24p^5$, one unpaired electron
1.65 (a) $[Kr]4d^{10}5s^1$; (b) $[He]2s^2$; (c) $[Kr]4d^{10}5s^25p^3$; (d) $[Ar]3d^{10}4s^24p^1$; (e) $[Xe]4f^{14}5d^46s^2$; (f) $[Kr]4d^{10}5s^25p^5$
1.67 (a) Fe; (b) P; (c) At; (d) Pu
1.69 (a) 4p; (b) 4s; (c) 6s; (d) 6s
1.71 (a) 5; (b) 3; (c) 6; (d) 8
1.73 (a) 3; (b) 2; (c) 3; (d) 2
1.75 (a) ns^1; (b) ns^2np^1; (c) $(n-1)d^5ns^2$; (d) $(n-1)d^{10}ns^1$

1.77 (a) O > Se > Te; (b) Au > Os > Ta; (c) Pb > Ba > Cs
1.79 Sc, 164; Ti, 147; V, 135; Cr, 129; Mn, 137; Fe, 124; Co, 125; Ni, 125; Cu 128; Zn, 137; decreasing radius with increasing nuclear charge; Cu and Zn are exceptions because they show the effects of electron-electron repulsion
1.81 $Cl^- < S^{2-} < P^{3-}$
1.83 (a) F; (b) C; (c) Cl; (d) Li
1.85 A diagonal relationship is a similarity in chemical properties between an element in the periodic table and one lying one period lower and one group to the right. It is caused by the similarity in polarizing power of the ions. The lower right element in the pair would generally be larger because it lies in a higher period, but it also will have a higher charge, which will cause the ion to have a similar polarizing power. For example, Al^{3+} and some Ge^{4+} compounds will show the diagonal relationship, as will Li and Mg.
1.87 (a) N and S; (b) Li and Mg
1.89 lower ionization energies
1.91 (a) metal; (b) nonmetal; (c) metal; (d) metalloid; (e) metalloid; (f) metal
1.93 (a) 6.424×10^{13} Hz; (b) 4.257×10^{-20} J; (c) 25.64 $kJ \cdot mol^{-1}$
1.95 Half-filled and completely filled shells give lower energy configurations.
1.97 inert-pair effect
1.99 No. Ultraviolet radiation is needed.
1.101 (a) The energy of the photon entering, E_{total}, must be equal to the energy required to eject the electron, $E_{ejection}$, plus the energy that ends up as kinetic energy, E_K, in the movement of the electron: $E_{total} = E_{ejection} + E_K$, but E_{total} for the photon $= h\nu$ and $E_K = \frac{1}{2}mv^2$, where m is the mass of the object and v is its speed. $E_{ejection}$ corresponds to the ionization energy, I, so we arrive at the final relation desired. (b) 6.7×10^{-19}
1.103 By the time we get to the lanthanides and actinides—the two series of *f*-orbital filling elements—the energy levels become so close together that minor changes in the environment can cause the energy order of the different types of orbitals to reverse. For the elements mentioned, the electron configurations are as follows:

La	$[Xe]6s^25d^1$	Lu	$[Xe]6s^24f^{14}5d^1$
Ac	$[Rn]7s^26d^1$	Lr	$[Rn]7s^25f^{14}6d^1$

As can be seen, all these elements have one electron in a *d*-orbital, so placement in the third column of the periodic table could be considered appropriate for any depending on what aspects of the chemistry of these elements we are comparing. The choice is controversial and is discussed by W. B. Jensen, *J. Chem. Ed.* **59**, 634 (1982).

1.105 (a) Points fit on the line described by the equation

$$f(x) = \frac{(2 - x)^2 \, e^{-x}}{4}$$

where x is the multiplication factor of a_0 ($r = xa_0$). (c) The node occurs at $r = 2a_0$.
1.107 The probability of finding the particle somewhere in the box is given by the integral

$$\int_0^L \left(\frac{2}{L}\right) \sin^2\left(\frac{n\pi x}{L}\right) dx$$

which can be shown to be equal to 1.

Chapter 2

2.1 (b)
2.3 Li^+ has a smaller radius
2.5 (a) 5; (b) 4; (c) 7; (d) 3
2.7 (a) [Ar]; (b) $[Ar]3d^{10}4s^2$; (c) $[Kr]4d^5$; (d) $[Ar]3d^{10}4s^2$
2.9 (a) [Kr]; (b) $[Xe]4f^{14}5d^{10}6s^2$; (c) $[Ar]3d^{10}$; (d) $[Xe]3d^2$
2.11 (a) $[Xe]4f^{14}5d^8$; (b) $[Ar]3d^3$; (c) $[Ar]3d^9$; (d) $[Ar]3d^2$
2.13 (a) Co^{2+}; (b) Fe^{2+}; (c) Mo^{2+}; (d) Nb^{2+}
2.15 (a) Co^{3+}; (b) Fe^{3+}; (c) Ru^{3+}; (d) Mo^{3+}
2.17 (a) 4*s*; (b) 3*p*; (c) 3*p*; (d) 4*s*
2.19 (a) −1; (b) −2; (c) +1; (d) +3, although +1 is sometimes observed; (e) +1, because of the inert-pair effect; +3 is also observed
2.21 (a) 3; (b) 6; (c) 6; (d) 2
2.23 (a) $[Kr]4d^{10}5s^2$, no unpaired electrons; (b) $[Kr]4d^{10}$, no unpaired electrons; (c) $[Xe]4f^{14}5d^4$, four unpaired electrons; (d) [Kr], no unpaired electrons; (e) $[Ar]3d^8$, two unpaired electrons
2.25 (a) 3*d*; (b) 5*s*; (c) 5*p*; (d) 4*d*
2.27 (a) +7; (b) −1; (c) [Kr] for +7, [Xe] for −1; (d) electrons are lost or added to give noble-gas configuration
2.29 (a) Mg_3As_2; (b) In_2S_3; (c) AlH_3; (d) H_2Te; (e) BiF_3 (BiF_5 is also known, but BiF_3 is the more stable compound because of the inert-pair effect.)
2.31 (a) GaAs; (b) MgO; (c) Al_2Te_3; (d) RuO_2; (e) V_2O_5
2.33 (a) Bi_2O_3 ; (b) PbO_2; (c) Tl_2O_3
2.35 (a) :C̈l–C(–C̈l:)(–C̈l:)–C̈l: (b) :C̈l–C(=Ö:)–C̈l: (c) Ö=N̈–F̈: (d) :F̈–N̈(–F̈:)–F̈:

2.37 (a) 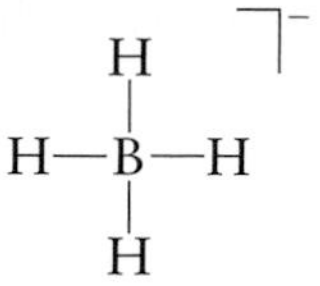 (b) 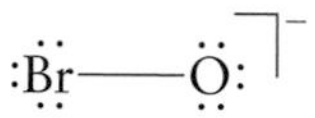

(c) 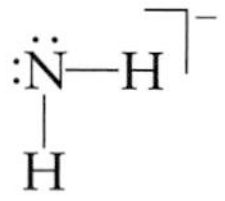

2.39 (a) 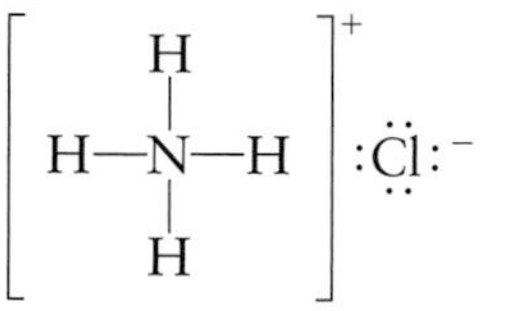 (b) 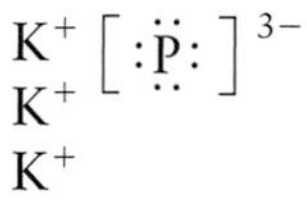

(c) 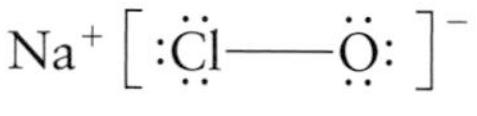

2.41 (a) 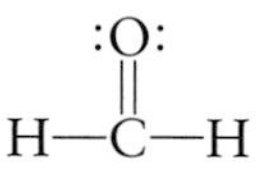 (b) 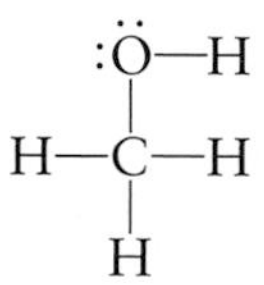

(c) 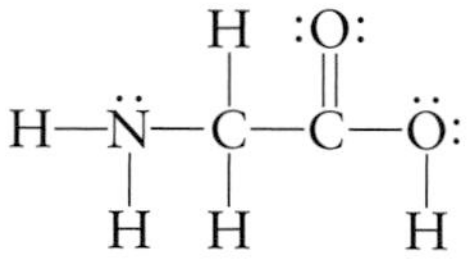

2.43

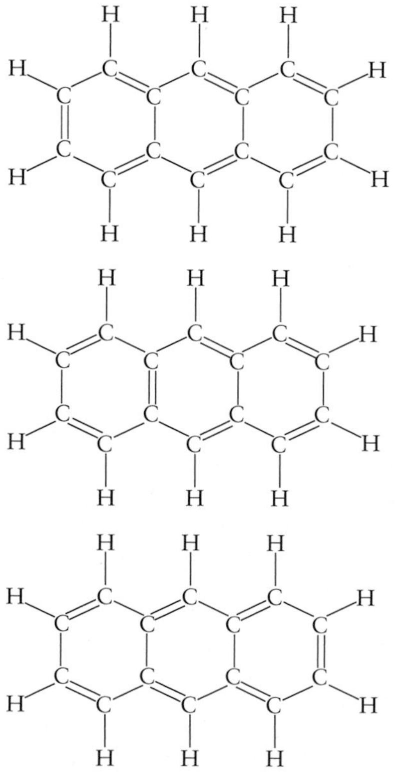

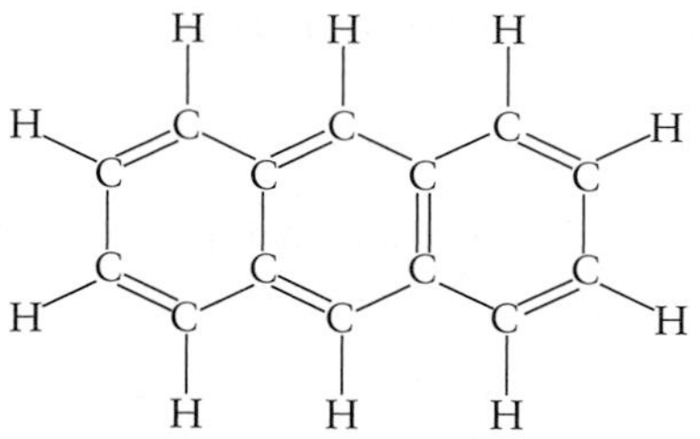

2.45 

2.47 (a) 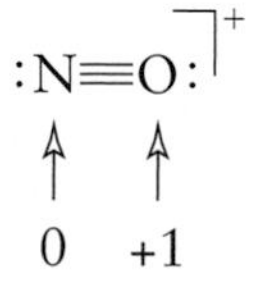 (b) 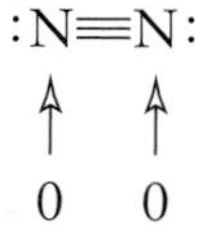 (c) 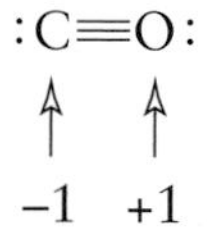

(d) 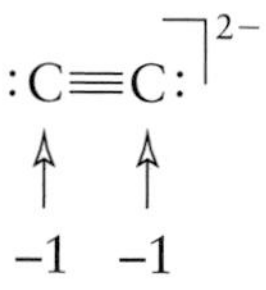 (e) 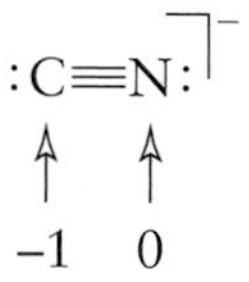

2.49 (a) The first structure has a formal charge of 0 on S and −1, 0, 0 on the three O atoms; the second structure has +1 on S, and −1, −1, 0 on the O atoms. The first structure has lower charges and therefore lower energy. (b) The first structure has +1 on S and −1, −1, 0, 0 on the four O atoms; the second has 0 on S and −1, 0, 0, 0 on the O atoms. The second is expected to have lower energy.

2.51 (a) 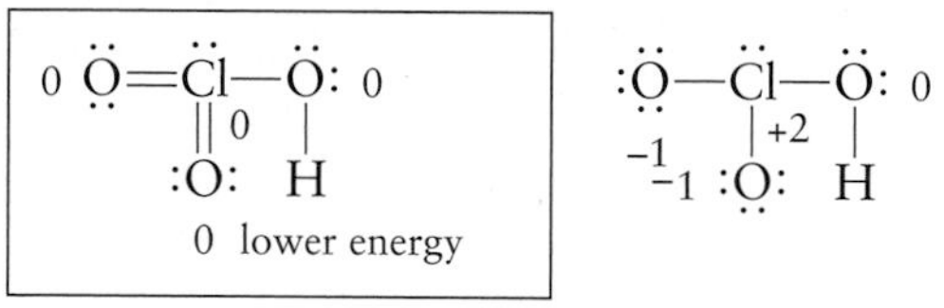

(b) 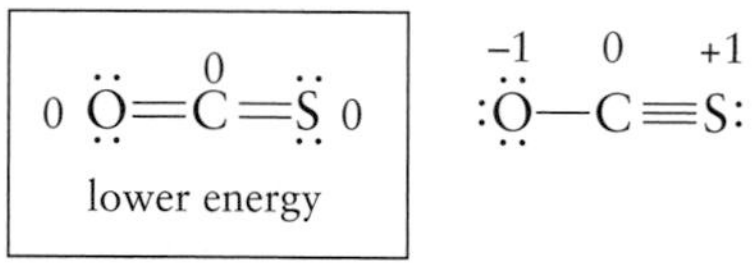

(c) 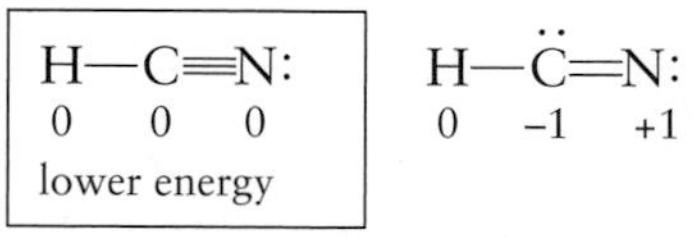

2.53 (a) The formal charges at Xe and F are 0 in the first structure; whereas in the second structure, Xe is −1, one F is 0 and one F is +1. The first structure is favored on the basis of the formal charges. (b) In the first structure, all the atoms have formal charges of 0; whereas in the second, one oxygen has a formal charge of +1 and one has a formal charge of −1. The first structure is therefore preferred.

2.55 (a) The sulfite ion has one Lewis structure that obeys the octet rule:

$[\mathrm{O{-}S({-}O){-}O}]^{2-}$

and three with an expanded octet:

$[\mathrm{O{=}S({-}O){-}O}]^{2-}$ $[\mathrm{O{-}S({-}O){=}O}]^{2-}$ $[\mathrm{O{-}S({=}O){-}O}]^{2-}$

The structures with expanded octets have lower formal charges.
(b) There is one Lewis structure that obeys the octet rule:

$[\mathrm{O{-}S(OH){-}O}]^{-}$

The formal charge at sulfur can be reduced to 0 by including one double bond contribution. This change gives rise to two expanded octet structures.

$[\mathrm{O{-}S(OH){=}O}]^{-}$ $[\mathrm{O{=}S(OH){-}O}]^{-}$

Notice that, unlike the sulfite ion, which has three resonance forms, the presence of the hydrogen ion restricts the electrons to the oxygen atom to which it is attached. Because H is electropositive, its placement near an oxygen atom makes it less likely for that oxygen to donate a lone pair to an adjacent atom.
(c) For the perchlorate ion, there is one Lewis structure that obeys the octet rule:

$[\mathrm{O{-}Cl(O)(O){-}O}]^{-}$ (all single bonds)

The formal charge at Cl can be reduced to 0 by including three double bond contributions, thereby giving rise to four resonance forms.

$[\mathrm{O{=}Cl({=}O)({-}O){=}O}]^{-}$ $[\mathrm{O{-}Cl({=}O)({=}O){=}O}]^{-}$

$[\mathrm{O{=}Cl({=}O)({=}O){-}O}]^{-}$ $[\mathrm{O{=}Cl({-}O)({=}O){=}O}]^{-}$

(d) For the nitrite ion, there are two resonance forms, both of which obey the octet rule:

$[\mathrm{O{-}N{=}O}]^{-}$ $[\mathrm{O{=}N{-}O}]^{-}$

2.57 (b) and (c)

2.59 (a) :Cl—O· radical (b) :Cl—O—O—Cl: not a radical

(c) O=N—O—O—Cl (O—Cl bonded to N via O) not a radical (d) :Cl—O—O· radical

2.61 (a) SF_6 12 electrons (b) XeF_2 10 electrons

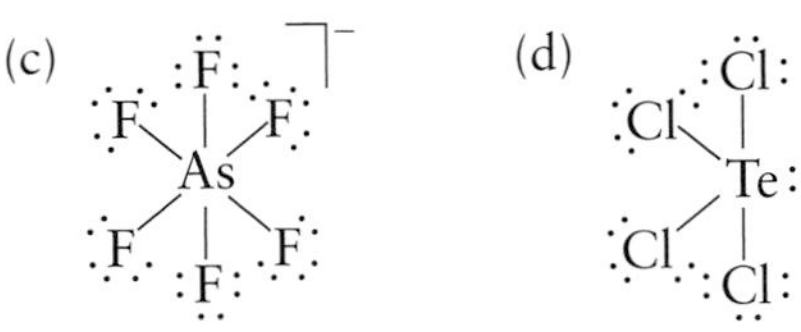

12 electrons 10 electrons

2.63 (a) $F{-}Xe({=}O){-}F$ 2 lone pairs (b) XeF_4 2 lone pairs (c) $XeOF_4$ 1 lone pair

2.65 (a) Lewis base; (b) Lewis acid; (c) Lewis base; (d) Lewis acid
2.67 (a) PF_5 is the Lewis acid, F^- is the Lewis base, the product is PF_6^-; (b) SO_2 is the Lewis acid, Cl^- is the Lewis base, the product is SO_2Cl^-; (c) SO_3 is the Lewis acid, OH^- is the Lewis base, the product is HSO_4^-
2.69 (a) CH_3^- is more basic than CH_4 because it has a lone pair of electrons, whereas CH_4 does not. (b) H_2O is more basic than H_2S because O is more electronegative than S and has a higher partial negative charge. (c) NH_2^- is more basic than NH_3 because of its higher negative charge.
2.71 I (2.7) $<$ Br (3.0) $<$ Cl (3.2) $<$ F (4.0). Electronegativity generally increases as one goes higher in the periodic table.
2.73 In (1.8) $<$ Sn (2.0) $<$ Sb (2.1) $<$ Se (2.6). Generally, electronegativity increases from left to right across the periodic table and from heavier to lighter elements within a group.
2.75 (a) HCl; (b) CF_4; (c) CO_2

2.77 $Rb^+ < Sr^{2+} < Be^{2+}$; smaller, more highly charged cations have greater polarizing power
2.79 $O^{2-} < N^{3-} < Cl^- < Br^-$; polarizability increases with radius of the ions
2.81 (a) $CO_3^{2-} > CO_2 > CO$; (b) $SO_3^{2-} > SO_2 \approx SO_3$; (c) $CH_3NH_2 > CH_2NH > HCN$
2.83 (a) 150 pm; (b) 127 pm; (c) CO, 127 pm; CN 152 pm; (d) 120 pm
2.85 (a) 149 pm; (b) 183 pm; (c) 213 pm. Bond distance reflects size of the Group 14/IV element.

2.87 (a) Four resonance structures of $C_2O_4^{2-}$, each with two C—C-bonded carbons, each carbon bearing one C=O and one C—O⁻; formal charges: 0 on C and double-bonded O, −1 on single-bonded O.

(b) $\overset{+1}{Br}{=}\overset{0}{O}$ $]^+$ (c) $:\overset{-1}{C}{\equiv}\overset{-1}{C}:$ $]^{2-}$

2.89 (a) $CH_3—C(=O)—O:^-$ ⟷ $CH_3—C(—O^-)=O$

(b) $H—C(H)=C(CH_3)—O:^-$ ⟷ $H—C^-(H)—C(CH_3)=O$

(c) $[H—CH=CH—CH_2]^+$ ⟷ $[H—CH_2—CH=CH]^+$ (H—C=C—C—H ⟷ H—C—C=C—H)

(d) $CH_3—C(=O)—N^-H$ ⟷ $CH_3—C(—O^-)=NH$

2.91 The chemical formula for phosgene is $COCl_2$, as determined by the method described in Section F. The Lewis structure is

$:Cl—C(=O:)—Cl:$

2.93 P and S are larger atoms that are less able to form multiple bonds to themselves, unlike the small N and O atoms. All bonds in P_4 and S_8 are single bonds, whereas N_2 has a triple bond and O_2 a double bond.

2.95 (a) H—C≡C—H H—C≡Si—H
H—Si≡Si—H H—C≡N: :N≡N:

(b) Six-membered rings: benzene (C_6H_6), and rings in which CH units are successively replaced by N: pyridine, the diazines, the triazines, the tetrazines, pentazine, and the N_6 ring.

+ suitable resonance forms

Note: The ability to draw the structures suggests that the compounds are feasible, but it does not indicate stability. For example, the N_6 ring gives a reasonable Lewis structure but would be very unstable.

2.97 The Lewis structures for NO and NO_2 are

$\dot{N}=O$ $O=\dot{N}—O:$ ⟷ $:O—\dot{N}=O$

Both compounds are radicals.
(a) NO has a double bond, but NO_2 has N—O bonds that are the average of a single bond and a double bond. Thus, NO would be expected to have a shorter, stronger bond, and this fact is indicated by the bond energies.
(b) The fact that the two N—O bonds in NO_2 are equal is a result of the two available resonance forms.

2.99 (a) Alkali metal iodide lattice energies versus d_{M-I}:

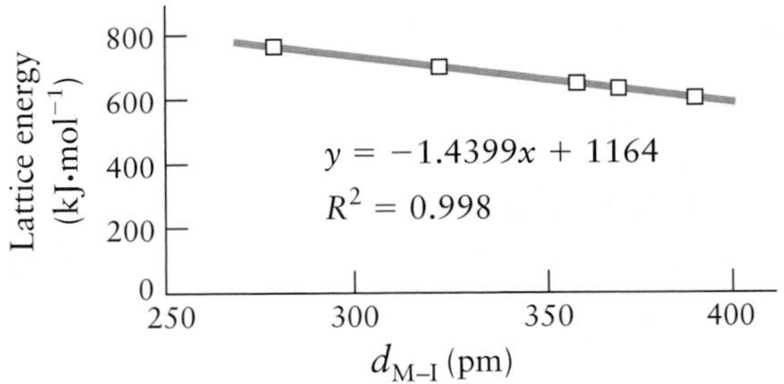

Agreement is excellent. (b) 656 kJ·mol^{-1}; (c) The agreement is not good. AgI is more covalent than the alkali metal iodides.

2.101 (a)

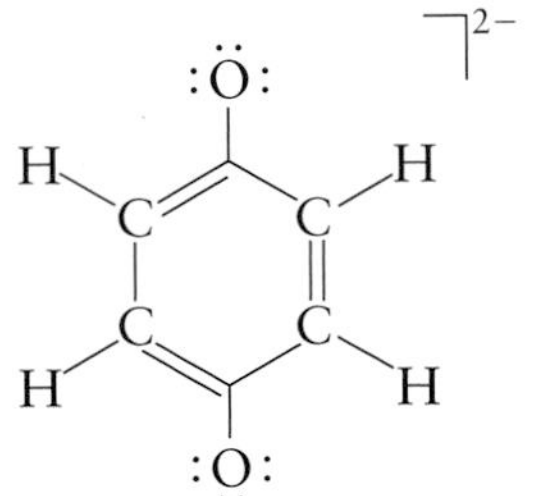

(b) All the atoms have formal charge 0 except the two oxygen atoms, which are −1. The negative charge is most likely to be concentrated at the oxygen atoms. (c) Because the oxygen atoms are the most negative sites and the lone pairs on the oxygen atoms are good Lewis bases, the protons will bond to the oxygen atoms. This compound is known as hydroquinone.

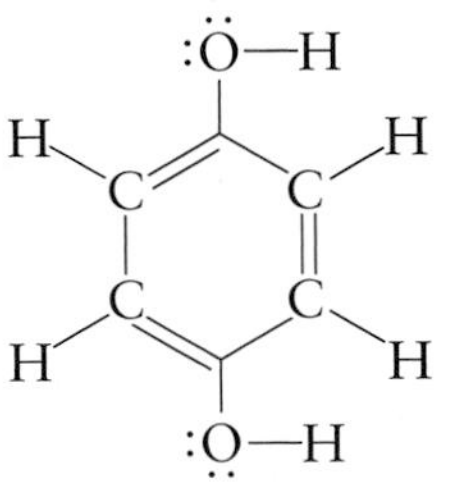

Hydroquinone

2.103

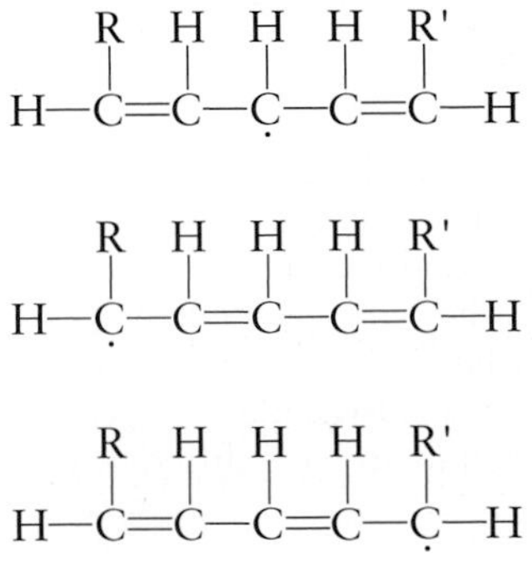

2.105 (a) 1; (b) 1; (c) 3; (d) 1; (e) 1; rotation about the C—C bond means that all the H atoms in $-CH_3$ are equivalent.

2.107 The alkyne group will have the stiffer C—H bond because a large force constant k results in a higher frequency absorption.

Chapter 3

3.1 (a) trigonal pyramidal; (b) one; (c) slightly less than 109.5°

3.3 (a) bent or angular; (b) slightly less than 120°

3.5 (a) trigonal planar; (b) one; (c) 120°

3.7 (a) angular or bent; (b) slightly less than 120°

3.9 (a) seesaw, AX_4E (b) T-shaped, AX_3E_2

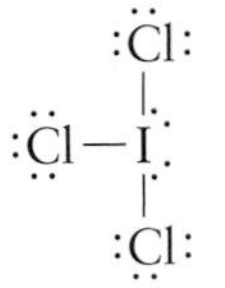

(c) square planar, AX_4E_2 (d) trigonal pyramidal, AX_3E

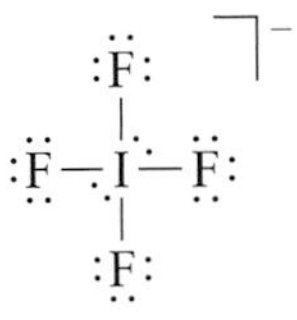

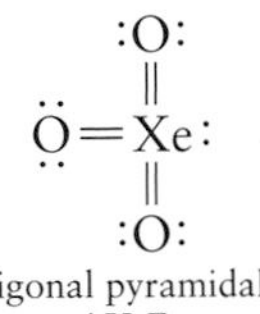

3.11 (a) linear, 180°, AX_2E_3; (b) T-shaped, 90° and 180°, AX_3E_2; (c) tetrahedral, 109.5°, AX_4; (d) octahedral, 90° and 180°, AX_6

3.13 (a) (b)

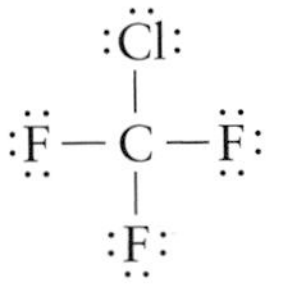

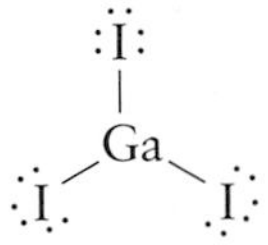

(c) 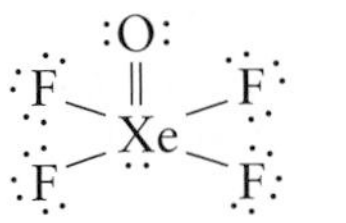 (d)

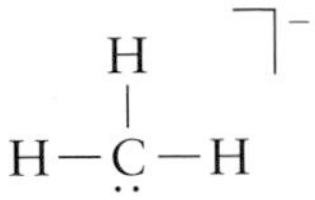

(a) tetrahedral, all bond angles 109.5°, AX_4; (b) trigonal planar; bond angles of approximately 120°, AX_3; (c) square pyramidal, O—Xe—F angles of 90° and F—Xe—F angles of 90° and 180°, AX_5E; (d) trigonal pyramidal, angles of slightly less than 109.5°, AX_3E

3.15 (a) a and b ≈ 120°, c ≈ 109.5° in 2,4-pentanedione, all angles approximately 120° in acetylacetonate ion; (b) The major differences arise in the hybidization of the originally sp^3 hybridized CH_2 group, which becomes sp^2 hybridized upon deprotonation.

3.17 (a) slightly less than 120°; (b) 180°; (c) 180°; (d) slightly less than 109.5°

3.19 (a) 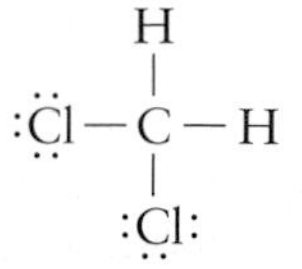 H—C(—Cl:)(—:Cl:)—H [Lewis structure: central C bonded to H, H, :Cl:, :Cl:] (b) [Lewis structure: central C bonded to four :Cl:]

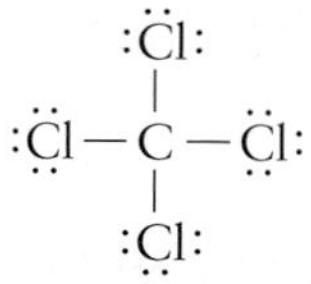

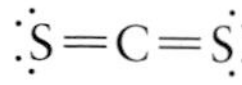

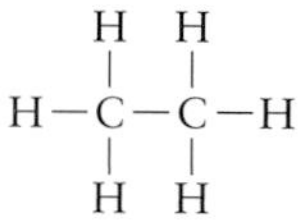

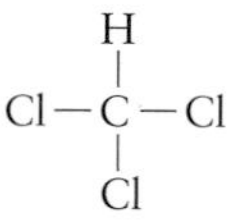

(c) :S=C=S: (d) 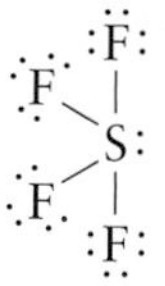 [Lewis structure: S bonded to four :F: atoms]

Molecules (a) and (d) will be polar and (b) and (c) will be nonpolar.

3.21

(a) 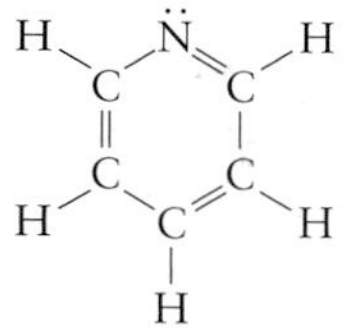 [ring structure: C₅H₅N ring with H on each C] polar (b) H—C(H)(H)—C(H)(H)—H nonpolar (c) Cl—C(H)(Cl)—Cl polar

3.23 (a) **1** and **2** are polar, **3** is nonpolar; (b) **1**

3.25 [structure: $H_2C=CH-C\equiv N$; sp^2 carbons with 120° angles, sp carbon with 180° angle]

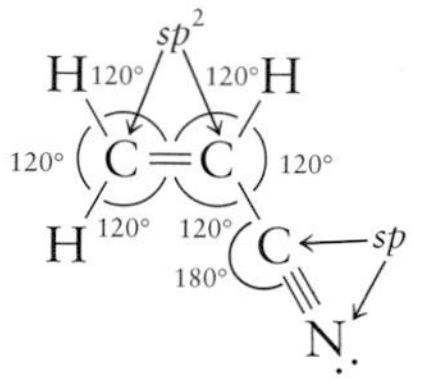

3.27 (a) tetrahedral, 109.5°; (b) linear, 180°; (c) trigonal bipyramidal, 90° and 180°; (d) trigonal planar, 120°
3.29 (a) sp^3d; (b) sp^2; (c) sp^3; (d) sp
3.31 (a) sp^2; (b) sp^3; (c) sp^3d; (d) sp^3
3.33 (a) sp^3; (b) sp^3d^2; (c) sp^3d; (d) sp^3
3.35 (a) Thiosulfate is tetrahedral, with bond angles of 109.5°. (b) Dimethyl beryllium is linear, with a C—Be—C bond angle of 180°. The H—C—H bond angles are approximately 109.5°. (c) BH_2^- is angular, with an H—B—H bond angle of slightly less than 120°. (d) $SnCl_2$ is angular, with a Cl—Sn—Cl bond angle of slightly less than 120°.

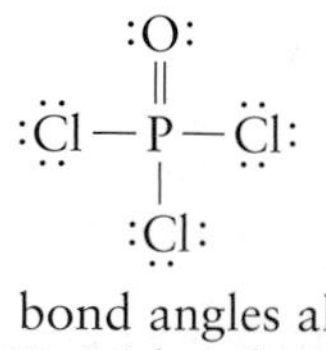

3.37 (a) 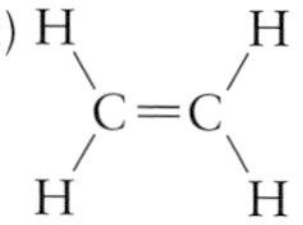 [structure: $H_2C=CH_2$] (b) 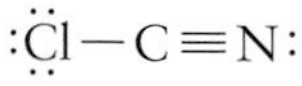 :Cl—C≡N:

(c) [Lewis structure: P with =O: and three :Cl:] (d) 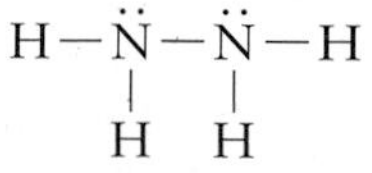 H—N(H)—N(H)—H [each N with one lone pair]

(a) bond angles all approximately 120°; (b) bond angle 180°; (c) bond angles 109.5°; (d) bond angles about both nitrogen atoms are slightly less than 109.5°

3.39 [Lewis structures: :Cl—Sb(=O:)(—:Cl:)—Cl: ; :Cl—S(=O:)(—:Cl:)=O ; $[IO_2F_2]^-$ with I bonded to two O and two F]

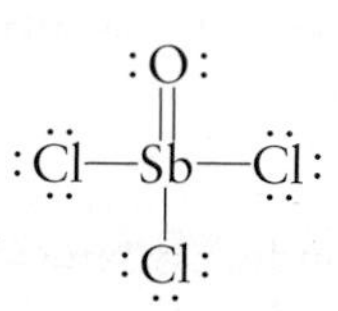

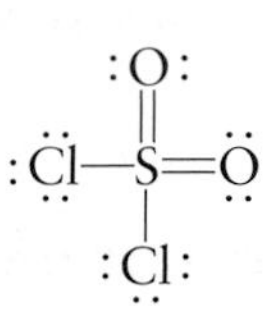

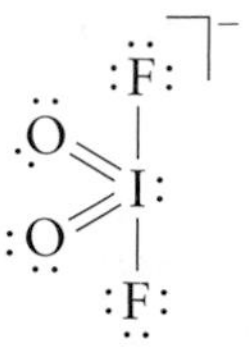

(a) tetrahedral, bond angles all approximately 109.5°; (b) tetrahedral, bond angles all approximately 109.5°; (c) seesaw-shaped, angles are 90°, 180°, and 120°
3.41 (a) Li_2, BO $= \frac{1}{2}(2 + 2 - 2) = 1$

diamagnetic, no unpaired electrons

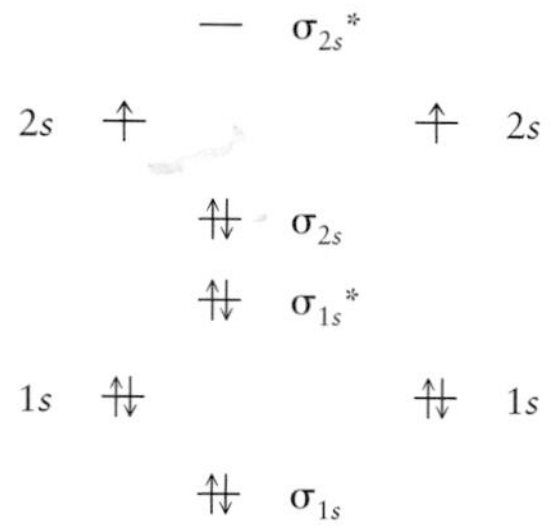

Li	MO	Li
	— σ_{2s}^*	
2s ↑		↑ 2s
	⇅ σ_{2s}	
	⇅ σ_{1s}^*	
1s ⇅		⇅ 1s
	⇅ σ_{1s}	

(b) Li_2^+, BO $= \frac{1}{2}(2 - 2 + 1) = \frac{1}{2}$

paramagnetic, one unpaired electron

Li	MO	Li⁺
	— σ_{2s}^*	
2s ↑		— 2s
	↑ σ_{2s}	
	⇅ σ_{1s}^*	
1s ⇅		⇅ 1s
	⇅ σ_{1s}	

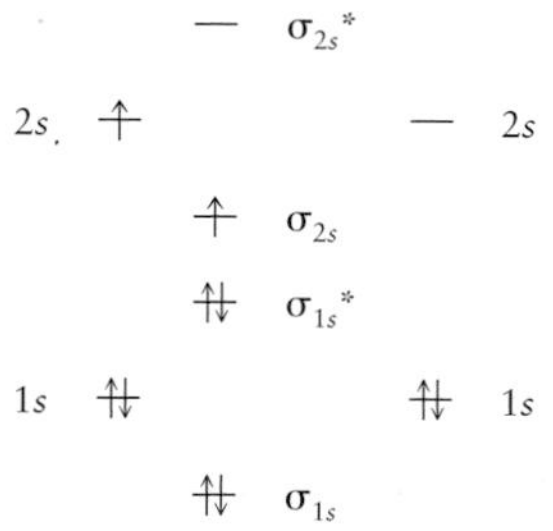

(c) Li_2^-, BO $= \frac{1}{2}(2 + 2 - 2 - 1) = \frac{1}{2}$

paramagnetic, one unpaired electron

Li	MO	Li⁻
	↑ σ_{2s}^*	
2s ↑		⇅ 2s
	⇅ σ_{2s}	
	⇅ σ_{1s}^*	
1s ⇅		⇅ 1s
	⇅ σ_{1s}	

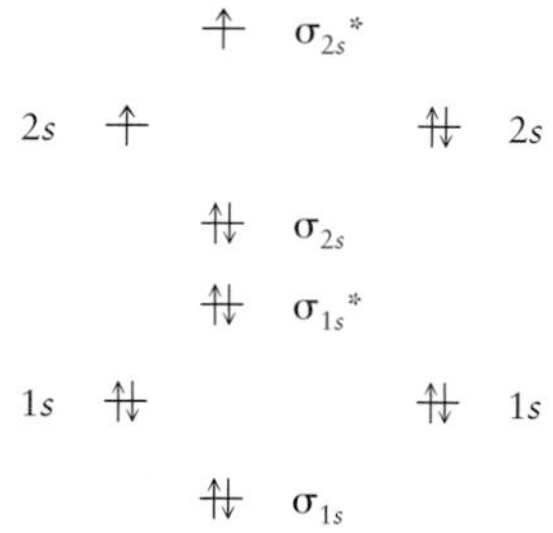

3.43 (a) (1) $\sigma_{2s}^2\sigma_{2s}^{*2}\sigma_{2p}^2\pi_{2p}^4\pi_{2p}^{*2}\pi_{2p}^{*1}$
(2) $\sigma_{2s}^2\sigma_{2s}^{*2}\sigma_{2p}^2\pi_{2p}^4\pi_{2p}^{*1}$
(3) $\sigma_{2s}^2\sigma_{2s}^{*2}\sigma_{2p}^2\pi_{2p}^4\pi_{2p}^{*4}$
(b) (1) 1.5, (2) 2.5, (3) 1; (c) (1) and (2) are paramagnetic, with one unpaired electron each; (d) π in all three cases
3.45 An N=N double bond is composed of a σ-bond and a π-bond whereas the N—N single bond is a σ-bond. Because the types of orbitals that compose the σ- and π-bonds are different, it is reasonable to expect the resulting molecular orbitals to have different energies.

3.47 (a) See (**51**) for the energy-level diagram for N_2. (b) The nitrogen atom is more electronegative; hence, its orbitals are lower in energy than those of C. The revised energy-level diagram is shown below. Because N is more electronegative, all the bonding orbitals will be closer to N than to C in energy and all the antibonding orbitals will be closer to C than to N in energy.

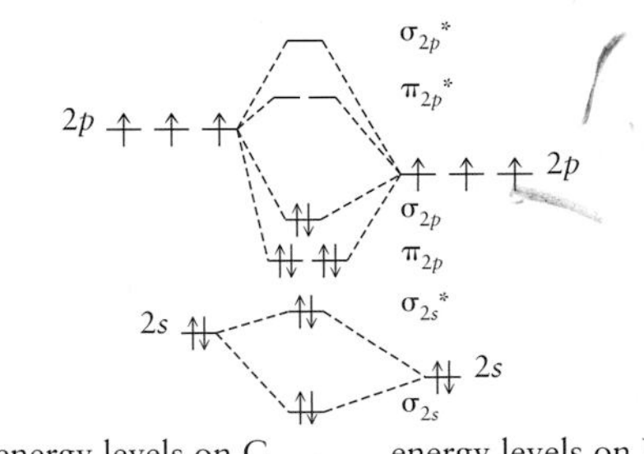

(c) The electrons in the bonding orbitals will have a higher probability of being at N because it is the more electronegative atom and its orbitals are lower in energy.

3.49 (a) C_2 (8 valence electrons): $\sigma_{2s}^{2}\sigma_{2s}^{*2}\pi_{2p}^{4}$; bond order = 2; (b) Be_2 (4 valence electrons): $\sigma_{2s}^{2}\sigma_{2s}^{*2}$, bond order = 0; (c) Ne_2 (16 valence electrons): $\sigma_{2s}^{2}\sigma_{2s}^{2}\sigma_{2p}^{2}\pi_{2p}^{4}\pi_{2p}^{*4}\sigma_{2p}^{*2}$; bond order = 0

3.51 All these species are paramagnetic. O_2^- and O_2^+ have an odd number of electrons and must, therefore, have at least one unpaired electron. (a) O_2 has an even number of electrons, but in its molecular orbital energy-level diagram, the HOMO is a degenerate set of orbitals and each is singly occupied. Thus, this molecule has two unpaired electrons. (b) For O_2^-, one more electron will be placed in this degenerate set of orbitals, thereby causing one of the original unpaired electrons to be paired. O_2^- will therefore have one unpaired electron. (c) Likewise O_2^+ will have one less electron than O_2; thus, one of the originally unpaired electrons will be removed, thereby leaving one unpaired electron in this species.

3.53 (a) F_2 bond order = 1; F_2^- bond order = 0.5; F_2 will have the stronger bond. (b) B_2 bond order = 1, B_2^+ bond order = 0.5. B_2, therefore, will have the stronger bond.

3.55 The conductivity of a semiconductor increases with temperature as increasing numbers of electrons are promoted into the conduction band, whereas the conductivity of a metal decreases as the increasing motion of the atoms slows down the migration of electrons.

3.57 (a) In and Ga; (b) P and Sb

3.59 (a) $[GaCl_4]^-$ (Cl—Ga—Cl Lewis structure): tetrahedral, sp^3, all Cl—Ga—Cl bond angles = 109.5°, nonpolar

(b) TeF_4 (Lewis structure): seesaw, sp^3d, F—Te—F bond angles = 90°, 120°, and 180°, polar

(c) $[SbCl_4]^-$ (Lewis structure): seesaw, sp^3d, Cl—Sb—Cl bond angles = 90°, 120°, and 180°, polar

(d) $SiCl_4$ (Lewis structure): tetrahedral, sp^3, all Cl—Si—Cl bond angles = 109.5°, nonpolar

3.61 (a) $[InCl_4]^-$ (Lewis structure): tetrahedral, sp^3, all Cl—In—Cl bond angles = 109.5°, nonpolar

(b) $[PCl_4]^+$ (Lewis structure): tetrahedral, sp^3, all Cl—P—Cl bond angles = 109.5°, nonpolar

(c) $[ICl_4]^+$ (Lewis structure): seesaw, sp^3d, Cl—I—Cl bond angles = 90°, 120°, and 180°, polar

(d) $SeOF_2$ (O=Se, Se—F, Se—F Lewis structure): trigonal pyramidal, bond angles slightly less than 109.5°, polar

3.63 (a) SiF_4: SiF_4 is nonpolar (tetrahedral, AX_4) whereas PF_3 is polar (trigonal pyramidal, AX_3E); (b) SF_6: SF_6 is nonpolar (octahedral, AX_6) whereas SF_4 is polar (seesaw, AX_4E); (c) AsF_5: IF_5 is polar (square pyramidal, AX_5E) whereas AsF_5 is nonpolar (AX_5, trigonal bipyramidal).

3.65 The elemental composition gives an empirical formula of CH_4O, which agrees with the molar mass. There is only one reasonable Lewis structure, which corresponds to the compound methanol. Both carbon and oxygen are sp^3 hybridized. All the bond angles at carbon should be 109.5°. The bond angles about oxygen should be close to 109.5° but will be somewhat less as a result of the repulsions by the lone pairs.

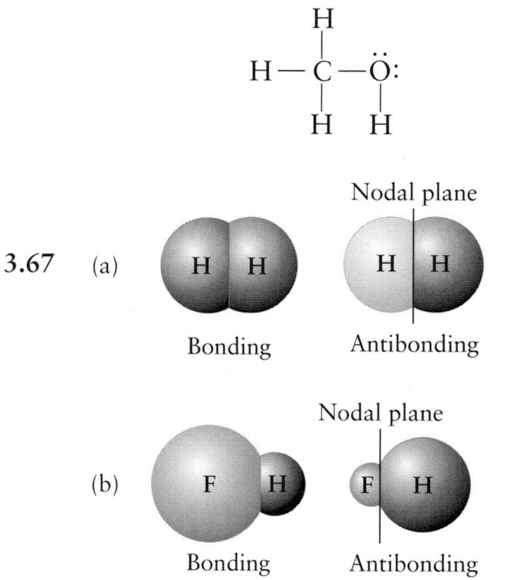

The sizes denote the relative contributions of the orbitals.

3.69 (a) The expected molecular orbital diagram for CF is

σ_{2p}^*
π_{2p}^*
C2p
F2p
π_{2p}
σ_{2p}
C2s
σ_{2s}^*
F2s
σ_{2s}

The bond order for the neutral species is 2.5 because one electron occupies a π_{2p}^* orbital. Adding an electron to form CF^- reduces the bond order to 2, whereas removing an electron to form CF^+ increases the bond order to 3. The bond lengths increase as the bond order decreases: $CF^+ < CF < CF^-$. (b) The CF^+ ion is diamagnetic, but both CF and CF^- have unpaired electrons (one in the case of CF and two in the case of CF^-).

3.71 The Lewis structure of borazine is nearly identical to that of benzene. It is obtained by replacing alternating C atoms in the benzene structure with B and N as shown. The orbitals at each atom are sp^2 hybridized.

[Borazine structure: six-membered ring of alternating B and N atoms, each bearing an H]

3.73 (a)

$[CH_3]^+$ CH_4 $[CH_3]^-$

CH_2 $[CH_2]^{2+}$ $[CH_2]^{2-}$

(b) None of these species is a radical.
(c) $CH_2^{2-} < CH_3^- < CH_4 < CH_2 < CH_3^+ < CH_2^{2+}$

3.75

H—C≡C—H ⟶ —CH=CH—CH=CH—CH=CH—CH=CH—

Polyacetylene retains multiple bonds along the chain. Electrons can be conducted through the series of π-orbitals. A resonance form of the Lewis structure can be drawn to show that the electrons may be delocalized along the polyacetylene chain. No such resonance form is possible for polyethylene, which has the structure $-(CH_2-CH_2)_n-$.

3.77 (a)

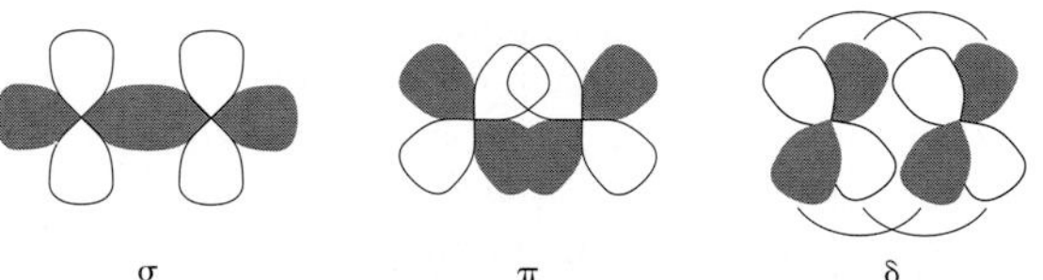

σ π δ

(b) Bond strengths decrease in the order δ < π < σ

3.79 (a) Two electrons will be removed from the π-orbitals, thereby causing the number of π-bonds to fall from 3 to 2. (b) The number of π-bonds would also fall from 3 to 2, but for a different reason: two electrons are added to antibonding orbitals; a net total of one π-bond will be lost. On the basis of this simple model, we predict that the ions formed will be paramagnetic.

3.81 (a) The Lewis structure of benzyne is

(b) Benzyne would be highly reactive because the two carbons that are *sp* hybridized are constrained to have a very unfavorable structure relative to their linear hybridization. Instead of 180° angles at these carbon atoms, the angles, by necessity of being part of the six-membered ring, are constrained to be close to 120°. One possible structure that allows the carbons to adopt more reasonable angles is a biradical:

Radicals are also highly reactive in general because orbitals in which the unpaired electrons reside can easily accommodate another electron, and this process can result in bond formation, which is energetically favored.

3.83 (a) $[ClO_2]^+[AsF_6]^-$

(b)

The formal charges are oxygen, 0; chlorine, +1; arsenic, −1; fluorine, 0. The ClO_2^+ ion is angular (based on a trigonal planar arrangement of electrons), with sp^2 hybridization and an expected O—Cl—O bond angle of slightly less than 120°. The AsF_6^- ion will be a regular octahedron with sp^3d^2 hybridization and F—As—F angles of 90° and 180°.

3.85 (a) sp^3; (b) 109.5°; (c) 60°; (d) A σ-bond will have the electron density localized in the region between the two nuclei. (e) The orbitals may not point directly between the atoms.

(f) The orbitals can still overlap, even if they do not point directly between the atoms. Such bonds are known as "bent" bonds or "banana" bonds.

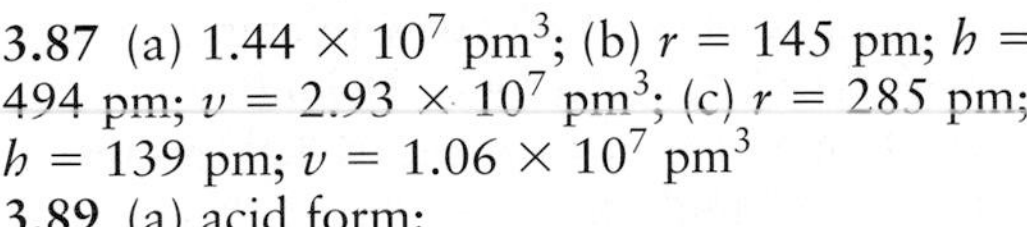

3.87 (a) 1.44×10^7 pm^3; (b) $r = 145$ pm; $h = 494$ pm; $v = 2.93 \times 10^7$ pm^3; (c) $r = 285$ pm; $h = 139$ pm; $v = 1.06 \times 10^7$ pm^3

3.89 (a) acid form:

Phenolphthalein (acid form)

base form:

(b) sp^3 in acid form, sp^2 in base form;
(c) more extensive conjugation

3.91 (a) and (d) have the possibility of *n*-to-π* transitions because these molecules possess both an atom with a lone pair of electrons (on O in HCOOH and on N in HCN) and a π-bond to that atom. The other molecules have either a lone pair or a π-bond but not both.

Chapter 4

4.1 (a) 8×10^9 Pa; (b) 80 kbar; (c) 6×10^7 Torr; (d) 1×10^6 lb·inch^{-2}

4.3 (a) 158 mm; (b) side open to the atmosphere; (c) 596 Torr

4.5 902 cm

4.7 2.9×10^3 lb

4.9 (a) The plots are straight lines going through the origin, with slopes (*V*, in L, and slope, in atm·K^{-1}): 0.01, 8.21; 0.02, 4.1; 0.03, 2.74; 0.04, 2.05; 0.05, 1.64. (b) slope = nR/V; (c) 0

4.11 (a) 3.6×10^3 kPa; (b) 3.8×10^3 Torr

4.13 3.31 atm

4.15 1.6 atm

4.17 increase *V* by 10%

4.19 (a) 11.8 mL; (b) 0.969 atm; (c) 199 K
4.21 (a) 1.06×10^3 kPa; (b) 4.06×10^2 mL; (c) 0.20 g; (d) 3.24×10^5 mol; (e) 9.2×10^{14} atoms
4.23 (a) 1.83×10^{15} molecules; (b) 4.80×10^{22} molecules; (c) 1.21×10^{18} molecules
4.25 (a) 63.4 L; (b) 6.32 L
4.27 18.1 L
4.29 4.18 atm
4.31 2.52×10^{-3} mol
4.33 (a) 3.6 m^3; (b) 1.8×10^2 m^3
4.35 2.63×10^3 Torr
4.37 342°C
4.39 $CO < H_2S < CO_2$
4.41 (a) 1.28 $g{\cdot}L^{-1}$; (b) 3.90 $g{\cdot}L^{-1}$
4.43 (a) 70 $g{\cdot}mol^{-1}$; (b) CHF_3; (c) 2.9 $g{\cdot}L^{-1}$
4.45 $C_2H_2Cl_2$
4.47 132 $g{\cdot}mol^{-1}$
4.49 (a) 3.00×10^5 L; (b) 1.00×10^4 L
4.51 (b)
4.53 2.4×10^4 L
4.55 (a) 0.047 g; (b) 58 mL
4.57 (a) 4.25×10^{-3} g; (b) 0.0757 atm HCl
4.59 (a) H_2(g), 2.0 atm; N_2(g), 1.0 atm; (b) 3.0 atm
4.61 (a) 0.787 atm; (b) 1.03 g
4.63 (a) 739.2 Torr; (b) $H_2O(l) \rightarrow H_2(g) + \frac{1}{2} O_2(g)$; (c) 0.142 g
4.65 $C_{10}H_{10}$
4.67 (a) 154 s; (b) 123 s; (c) 33.0 s; (d) 186 s
4.69 110 $g{\cdot}mol^{-1}$; C_8H_{12}
4.71 (a) 4103.2 $J{\cdot}mol^{-1}$; (b) 4090.7 $J{\cdot}mol^{-1}$; (c) 12.5 $J{\cdot}mol^{-1}$
4.73 (a) 627 $m{\cdot}s^{-1}$; (b) 458 $m{\cdot}s^{-1}$; (c) 378 $m{\cdot}s^{-1}$
4.75 (a) The most probable speed is the one that corresponds to the maximum on the distribution curve. (b) The percentage of molecules having the most probable speed decreases as the temperature is raised (the distribution spreads out).
4.77 Hydrogen bonding is important in HF. At low temperatures, this hydrogen bonding causes the molecules of HF to be attracted to each other more strongly, thus lowering the pressure. As the temperature is increased, the hydrogen bonds are broken and the pressure goes up more quickly than that for an ideal gas.
4.79 ideal gas: (a) 1.63 atm, (b) 48.9 atm, (c) 489 atm, van der Waals gas: (a) 1.62 atm, (b) 38.9 atm, (c) 1.88×10^3 atm. The deviation is largest when the pressure of the gas is high; under that condition, the assumptions about ideal gas conditions break down.
4.81 (a)

n	Percentage deviation
0.100	0.8
0.200	2.1
0.300	3.2
0.400	4.3
0.500	5.5

(b) The term due to intermolecular attractions dominates. (c) The gas starts to deviate from ideality by more than 5% at pressures above about 10 atm.
4.83 All three curves have a similar shape, but the plots for ammonia and oxygen have considerably higher values for P at low V. The ammonia curve is higher than the oxygen curve owing to its stronger intermolecular attractions. The deviation is expected because at higher P and low V we do not expect ideal behavior. The van der Waals parameters are larger for ammonia than for oxygen.
4.85 (a) 9.32 atm; (b) 0.025; (c) 0.2095; (d) $x_{N_2} = 0.72$; $x_{octane} = 0.008$; $x_{CO_2} = 0.13$; $x_{H_2O} = 0.14$.
4.87 (a) $N_2O_4(g) \rightarrow 2\ NO_2(g)$; (b) 2.33 atm; (c) 4.65 atm; (d) $x_{NO_2} = 0.426$; $x_{N_2O_4} = 0.574$
4.89 (a) N_2H_4 (b) H—N̈—N̈—H (each N also bonded to H below)
(c) 4.25×10^{-4} mol
4.91 (a)

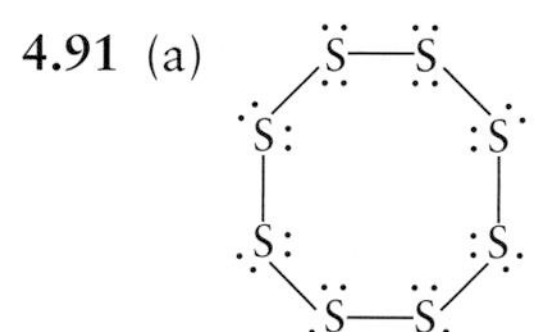

(b) bent or angular; (c) yes; (d) The overly high pressure obtained when S_8 is vaporized is an indication that S_8 units have decomposed to a greater number of units, S_6, S_4, S_2, etc. In other words, the number of moles of gas molecules increased upon vaporization.
4.93 (a) $x_{COCl_2} = 0.090$; $x_{CO} = 0.527$; $x_{Cl_2} = 0.383$; (b) 7.20 $g{\cdot}L^{-1}$
4.95 254 $g{\cdot}mol^{-1}$; OsO_4
4.97 12.7 g
4.99 (a) 3.94×10^7 pm^3; $r = 211$ pm; (b) 8.78×10^6 pm^3; (c) One would not expect the values to be the same; the van der Waals parameter b is an empirical constant derived from a fit to the observed gas behavior and is probably not simply the molecular volume. Also the van der Waals radius is different from the atomic radius because the van der Waals forces represent weaker, longer range interactions.

Chapter 5

5.1 (a) dipole-dipole, London forces; (b) dipole-dipole, London forces; (c) London forces, dipole-dipole, hydrogen bonds; (d) London forces
5.3 (b), (c), and (d)
5.5 (b) ≈ (c) < (e) < (a) < (d)
5.7 only (c)
5.9 (a) NaCl (ionic compound); (b) butanol (hydrogen bonding); (c) HF (hydrogen bonding); (d) H_2O (can form more hydrogen bonds than CH_3OH)

5.11 (a) both are trigonal pyramidal, PCl_3; (b) SO_2 is bent and CO_2 is linear, SO_2; (c) both are trigonal planar, BCl_3; (d) AsF_3 is trigonal pyramidal and AsF_5 is trigonal bipyramidal, AsF_3
5.13 0.39; Be^{2+}
5.15 Al^{3+}:Ga^{3+}, 1.4:1; water will be more strongly attracted to Al^{3+}
5.17 (a) xenon has stronger London forces; (b) hydrogen bonding in water; (c) dimethylpropane is more compact
5.19 (a) *cis*-dichlorethene, because it is more polar; (b) benzene at 20°C, because at higher temperatures the intermolecular forces are disrupted
5.21 $C_6H_6 < C_6H_5SH < C_6H_5OH$
5.23 water
5.25 (a) 2; (b) 8; (c) 543 pm
5.27 (a) 21.4 g·cm^{-3}; (b) 1.78 g·cm^{-3}
5.29 (a) 139 pm; (b) 143 pm
5.31 (a) fcc; (b) 12
5.33 (a) 5.61×10^{18} unit cells; (b) 3.72×10^{-5} mol
5.35 (a) 1 Cs^+, 1 Cl^-, 1 formula unit; (b) 2 Ti^{4+} ions, 4 O^{2-} ions, 2 formula units per cell; (c) the coordination number of Ti^{4+} is 6 and that of O^{2-} is 3.
5.37 $YBa_2Cu_3O_7$
5.39 (a) 8; (b) 6; (c) 6
5.41 (a) 3.36 g·cm^{-3}; (b) 4.67 g·cm^{-3}
5.43 (a) Glucose will be held in the solid by London forces, dipole-dipole interactions, and hydrogen bonds. Benzophenone will be held in the solid by dipole-dipole interactions and London forces. Methane will be held together by London forces only. London forces are strongest in benzophenone, but glucose can experience hydrogen bonding, which is a strong interaction and dominates intermolecular forces. Methane has few electrons, so it experiences only weak London forces. (b) We would expect the melting points to increase in the order CH_4 (m.p. = −182°C) < benzophenone (m.p. = 48°C) < glucose (m.p. = 148°–155°C).
5.45 one form of boron nitride, silicon dioxide (and many others)
5.47 (a) interstitial; atomic radius of nitrogen is much smaller than that of iron; (b) become harder and stronger, with a lower electrical conductivity
5.49 metallic conductor parallel to planes; semiconductor perpendicular to planes
5.51 (a) 2.8 Cu per Ni; (b) 15 Sn: 1.2 Sb: 1 Cu
5.53 Because there are so many ways that these molecules can rotate and twist, they do not remain rodlike. The molecular backbone of the stretched-out molecule is rodlike, but the molecules tend to curl up on themselves, thereby destroying any possibility of long-range order with neighboring molecules. This flexibility is partly due to the fact that the molecules have only single bonds, which allow rotation about the bonds and the adoption of many conformations. If multiple bonds are present, the structure is more rigid.
5.55 The use of a nonpolar solvent, such as hexane or benzene, in place of water, should give rise to the formation of inverse micelles.
5.57 (a) approximately 8.11×10^5 pm^2 for 2,2-dimethylpropane and 9.91×10^5 pm^2 for pentane; (b) pentane
5.59 (a) anthracene, $C_{14}H_{10}$, London forces

(b) phosgene, $COCl_2$, dipole-dipole forces, London forces

(c) glutamic acid, $C_5H_9NO_4$, hydrogen bonding, dipole-dipole forces, London forces

5.61 neon, 170 pm; argon, 203 pm; krypton, 225 pm; xenon, 239 pm; radon, 246 pm
5.63 21.0 g·cm^{-3}
5.65 (a) (b) fraction of Ti^{3+} = 0.36, fraction of Ti^{2+} = 0.64
5.67 (a) true; (b) false; (c) true; (d) false
5.69 4.25 g·cm^{-3}
5.71 There are several ways to draw unit cells which will repeat to generate the entire lattice. Some examples are shown below. The choice of unit cell is determined by conventions that are beyond the scope of this text (the smallest unit cell that indicates all of the symmetry present in the lattice is typically the one of choice).

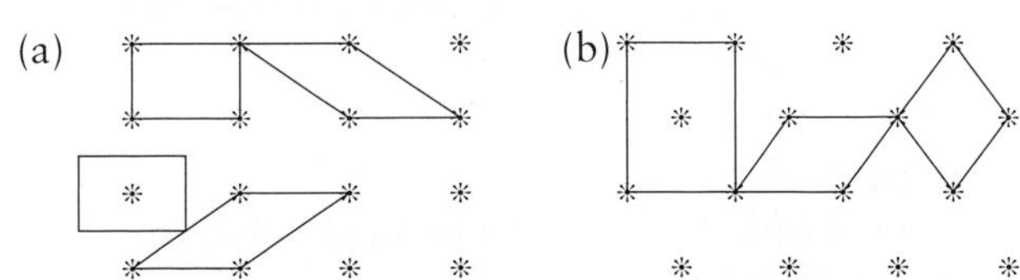

5.73 In a salt such as sodium chloride or sodium bromide, all the interactions will be ionic. In sodium acetate or sodium methoxide, however, the organic anions $CH_3CO_2^-$ and CH_3O^- will have the negative charges localized largely on the oxygen atoms, which are the most electronegative atoms. The carbon part of the molecule will interact with other parts of the crystal lattice via London forces rather than ionic interactions. In general, solid organic salts will not be as hard as purely ionic substances and will also have lower melting points.
5.75 Fused silica, also known as fused quartz, is predominantly SiO_2 with very few impurities. This glass can be used at high temperatures. Quartz vessels are routinely used for reactions that must be carried out at temperatures up to 1000°C. Vycor, which is 96% SiO_2, can be used up to about 900°C and normal borosilicate glasses up to approximately 550°C. The glasses melt at lower temperatures as the amount of materials other than SiO_2 increases. Borosilicate glasses (Pyrex or Kimax) are commonly used because the low softening points allow them to be molded easily and shaped into different types of glass objects, such as reaction flasks, beakers, and other types of laboratory and technical glassware.
5.77 MA_2
5.79 35%
5.81 (a) face-centered cubic; its density is 6.214 $g{\cdot}cm^{-3}$ whereas the density of the bcc cell is only 5.97 $g{\cdot}cm^{-3}$. The denser phase is more stable under high pressure. (b) No. Because both the face-centered cubic and hexagonal faces are close packed, they should have the same density.
5.83 (a) 546 pm; (b) 236 pm; (c) good agreement, sum of radii = 233 pm
5.85 5.30×10^5 layers
5.87 The solution lies in showing that the extra distance traveled by x-ray 2 must be equal to an integral number of wavelengths in order for x-rays 1 and 2 to still be in phase as they leave the crystal.
5.89 5.074°
5.91 (a) Because the bisulfate anions occupy spaces between the planes of the graphite layers, they will separate the layers more than they are separated in pure graphite. The conduction between layers will be reduced. The conduction within the layers, however, is enhanced because the partial oxidation produces some "holes" in the occupied delocalized molecular orbitals that extend over the graphite sheet. The presence of these holes makes it easier for the electrons to migrate through the solid along the planar carbon networks. (b) The x-ray diffracted by the planes of the graphite will occur at smaller values of θ.

Chapter 6

6.1 (a) isolated; (b) closed; (c) isolated; (d) open; (e) closed; (f) open
6.3 +732 kJ
6.5 (a) work was done on the system; (b) +490 J
6.7 +5500 kJ
6.9 −1626 kJ
6.11 +7 kJ
6.13 (a) −226 J; (b) −326 J
6.15 (a) −12.1 kJ; (b) +102 kJ
6.17 (a) 24 $J{\cdot}(°C)^{-1}{\cdot}mol^{-1}$; (b) 5.9 kg
6.19 (a) 155 kJ; (b) 91.1%.
6.21 25°C
6.23 14.8 $kJ{\cdot}(°C)^{-1}$
6.25 (a) 305 J; (b) 507 J; (c) At constant pressure, more heat is needed because some of the heat escapes into the surroundings in the form of expansion work.
6.27 NO_2. The heat capacity increases with molecular complexity. The more atoms there are in the molecule, the greater the number of ways the molecule can vibrate and absorb energy.
6.29 (a) 19.4 J; (b) 11.7 J
6.31 (a) $\frac{5}{2}R$; (b) $3R$; (c) $\frac{3}{2}R$; (d) $\frac{5}{2}R$
6.33 (a) 8.21 $kJ{\cdot}mol^{-1}$; (b) 43.5 $kJ{\cdot}mol^{-1}$
6.35 (a) +226 kJ; (b) +203 kJ
6.37 21.9 kJ
6.39 31°C
6.41 (a) 448 kJ; (b) 1.47×10^2 kJ; (c) 352 g
6.43 (a) 23.0 g; (b) -1.3×10^5 kJ
6.45 (a) 53 g; (b) 8.6 h
6.47 (a) −286 kJ; (b) −257 kJ; (c) +227 kJ; (d) −1 kJ
6.49 +1.90 kJ
6.51 −197.78 kJ
6.53 −1.775 MJ
6.55 −312 kJ
6.57 −2987 $kJ{\cdot}mol^{-1}$
6.59 −184.6 kJ
6.61 (a) $2\ C(s) + 2\ H_2(g) + \frac{1}{2}\ O_2(g) \rightarrow CH_3CHO(l)$, $\Delta H° = -192.30\ kJ{\cdot}mol^{-1}$
(b) $Cu(s) + S(s) + \frac{9}{2}\ O_2(g) + 5\ H_2(g) \rightarrow CuSO_4{\cdot}5H_2O(s)$, $\Delta H° = -2279.7\ kJ{\cdot}mol^{-1}$
(c) $\frac{1}{2}\ N_2(g) + 2\ H_2(g) + 2\ O_2(g) + \frac{1}{2}\ Cl_2(g) \rightarrow NH_4ClO_4(s)$, $\Delta H° = -295.31\ kJ{\cdot}mol^{-1}$
6.63 (a) $\Delta H_f°\ (N_2O_5, g) = +11.3\ kJ{\cdot}mol^{-1}$
6.65 −444 $kJ{\cdot}mol^{-1}$
6.67 +3011 kJ
6.69 2564 $kJ{\cdot}mol^{-1}$
6.71 (a) −412 $kJ{\cdot}mol^{-1}$; (b) 673 $kJ{\cdot}mol^{-1}$; (c) +63 $kJ{\cdot}mol^{-1}$
6.73 (a) −1272 $kJ{\cdot}mol^{-1}$; (b) −460 $kJ{\cdot}mol^{-1}$; (c) approximately 0
6.75 (a) +248 kJ; (b) −45 kJ; (c) −115 kJ
6.77 approximately 228 kJ; lower energy
6.79 (a) 33.93 $kJ{\cdot}mol^{-1}$; (b) 30.94 $kJ{\cdot}mol^{-1}$; (c) values agree closely; heat capacities are not strictly constant with temperature
6.81 88.7 kJ
6.83 +12.71 $kJ{\cdot}mol^{-1}$

6.85 (a) $n =$ 1.0 $T =$ 300 $V_f/V_i =$ 1 to 10

5.74 × 10³
2.87 × 10³
w (J)
0
0 5 10
V_f/V_i

(b) The work done is greater at the higher temperature because work is directly proportional to the temperature at which the expansion takes place. (c) The second expansion requires about one-third the work that the first expansion requires.
6.87 3 g
6.89 (a) $C_6H_5NH_2(l) + \frac{31}{2} O_2(g) \rightarrow 6\ CO_2(g) + \frac{7}{2} H_2O(l) + \frac{1}{2} N_2(g)$; (b) 0.4873 g $CO_2(g)$, 0.1188 g $H_2O(l)$, 0.026 39 g $N_2(g)$; (c) 0.999 atm
6.91 (a) +131.29 kJ, endothermic; (b) +623 kJ
6.93 (a) The standard enthalpy of combustion of $H_2(g)$ is the same as the standard enthalpy of formation of liquid water: $\Delta H_f°(H_2O, l) = -285.84\ kJ \cdot mol^{-1}$, whereas the standard enthalpy of combustion of octane is $-5471\ kJ \cdot mol^{-1}$. For hydrogen, that is $-142\ kJ \cdot g^{-1}$, whereas for octane it is $47.89\ kJ \cdot g^{-1}$. Hydrogen produces more heat per gram than the organic liquid and is significantly cleaner to burn. (b) The use of hydrogen as a fuel is limited by its storage problems. Normally hydrogen is a gas, and to use it as a fuel requires carrying a large quantity of hydrogen under pressure. Much research is aimed at creating safe but also economical hydrogen storage tanks (certain metals such as palladium absorb hydrogen). Safety is a major issue because $H_2(g)$ can react explosively with air, a serious concern for automobiles involved in accidents.
6.95 (a) −6.19 kJ; (b) −3135.5 kJ; (c) −3147.7 kJ
6.97 (a) −1.33 kJ; (b) −53.2 kJ
6.99 (a)

(b) −372 kJ; (c) $-205.4\ kJ \cdot mol^{-1}$; (d) The hydrogenation of benzene is much less exothermic than predicted from estimates of bond enthalpies. Part of this can be due to the inherent inaccuracy of using average values, but the difference is so large that this cannot be the complete explanation. The resonance energy of benzene makes it more stable than would be expected by treating it as a set of three isolated double and three isolated single bonds. The difference in these two values, −205 kJ − (−372 kJ) = 167 kJ, is a measure of how much more stable benzene is than the Kekulé structure implies.
6.101 $689\ kJ \cdot mol^{-1}$
6.103 (a) $+2326\ kJ \cdot mol^{-1}$; (b) = +3547 kJ; (c) *more* stable by 1221 kJ than predicted by the isolated bond model; (d) (1221 kJ)/60 = 20 kJ per carbon atom; (e) (150 kJ)/6 = 25 kJ per carbon atom; (f) Although the comparison of the stabilization of benzene with that of C_{60} should be treated with caution, it does appear that there is slightly less stabilization per carbon atom in C_{60} than in benzene. This calculation fits expectations, because the C_{60} molecule is forced by its geometry to be curved, a conformation causing the overlap of the *p*-orbitals, which gives rise to the delocalization that results in resonance, to be less favorable in C_{60} than it is in the planar benzene molecule. Another perspective on this result is obtained by noting that the C atoms in C_{60} are forced to be partially sp^3 hybridized because they cannot be rigorously planar, as required by sp^2 hybridization.
6.105 The oxidation of 1,3,5-triaminobenzene should be more exothermic by about −1715.0 kJ.

Chapter 7

7.1 (a) $0.341\ J \cdot K^{-1} \cdot s^{-1}$; (b) $29.5\ kJ \cdot K^{-1} \cdot day^{-1}$; (c) less, because, in the equation $\Delta S = -\Delta H/T$, if T is larger, ΔS is smaller
7.3 (a) $0.22\ J \cdot K^{-1}$; (b) $0.17\ J \cdot K^{-1}$; (c)) The entropy change is smaller at higher temperatures because the matter is already more disordered.
7.5 (a) $6.80\ J \cdot K^{-1}$; (b) $4.08\ J \cdot K^{-1}$
7.7 $88.1\ J \cdot K^{-1}$
7.9 (a) $-22.0\ J \cdot K^{-1}$; (b) $+134\ J \cdot K^{-1}$
7.11 (a) 290 K; (b) 20.8°C, or 293.9 K; (c) These values are in very good agreement. (d) Differences arise partly because the enthalpy and entropy of vaporization are slightly different from the values calculated at 298 K which is not the boiling point of ethanal.
7.13 (a) 253 K; (b) 248 K; agreement is reasonably good
7.15 (a) $30\ kJ \cdot mol^{-1}$; (b) $-11\ J \cdot K^{-1}$
7.17 COF_2. Both molecules are trigonal planar, but the atoms attached to C in COF_2 are different and allow positional (rotational) disorder in the solid.
7.19 $14.9\ J \cdot K^{-1} \cdot mol^{-1}$
7.21 (a) HBr(g); (b) $NH_3(g)$; (c) $I_2(l)$; (d) 1.0 mol Ar(g) at 1.00 atm
7.23 $C(s) < H_2O(s) < H_2O(l) < H_2O(g)$. The last two inequalities arise from the general fact that the solid phase has a lower entropy than the liquid phase, and the liquid phase has a lower entropy than the gas

phase. Solid carbon (diamond) has a very rigid lattice and hence a very low entropy.
7.25 (a) bromine, because it is a liquid; (b) pentene, because its molecules are more flexible; (c) ethene, because it is a gas and polyethylene is a solid of linked $-CH_2-CH_2-$ molecules
7.27 The Cl_2 molecule not only has translational energy but also rotational and vibrational energy. The disordering due to this energy contributes to the overall disorder. It also has twice as many atoms. However, matter is more organized in the molecule, so the molar entropy of the molecule is not twice that of the atom.
7.29 (a) decreases; (b) increases; (c) decreases
7.31 (a)Entropy should decrease, because the number of moles of gas molecules is less on the product side of the reaction. (b)Entropy should increase, because the dissolution of the solid copper phosphate will increase the disorder of the copper and phosphate ions.
(c) Entropy should decrease, because the number of moles of molecules decreases.
7.33 (a) $-163.34\ J{\cdot}K^{-1}{\cdot}mol^{-1}$. The entropy change is negative because the number of moles of gas molecules has decreased by a factor of 1.5. Note that the absolute entropies of the elements are not 0 and that the entropy of the formation reaction of a compound is also not 0.
(b) $-86.5\ J{\cdot}K^{-1}{\cdot}mol^{-1}$. The entropy change is negative because the number of moles of gas molecules has decreased by a factor of one-half.
(c) $+160.6\ J{\cdot}K^{-1}{\cdot}mol^{-1}$. The entropy change is positive because the number of moles of gas has increased.
(d) $-36.8\ J{\cdot}K^{-1}{\cdot}mol^{-1}$. It is not immediately obvious, but the 4 mol of solid products is more ordered than the 4 mol of solid reactants.
7.35 (a) $5 \times 10^3\ J{\cdot}K^{-1}$; (b) $3 \times 10^{-3}\ J{\cdot}K^{-1}$; (c) $5.7\ J{\cdot}K^{-1}$
7.37 (a) $\Delta S_{surr} = -73\ J{\cdot}K^{-1}$, $\Delta S_{sys} = +73\ J{\cdot}K^{-1}$; (b) $\Delta S_{surr} = -29.0\ J{\cdot}K^{-1}$, $\Delta S_{sys} = +29.0\ J{\cdot}K^{-1}$; (c) $\Delta S_{surr} = +29.0\ J{\cdot}K^{-1}$, $\Delta S_{sys} = -29.0\ J{\cdot}K^{-1}$
7.39 (a) For a reversible isothermal process, $\Delta S_{tot} = 0$; therefore, $\Delta S_{surr} = -\ \Delta S = -3.84\ J{\cdot}K^{-1}$. (b) $\Delta S = +3.84\ J{\cdot}K^{-1}$; $\Delta S_{surr} = 0$; $\Delta S_{tot} = +3.84\ J{\cdot}K^{-1}$
7.41 $-6.58\ J{\cdot}K^{-1}$. The total change in entropy will be negative, so we predict that the process is not spontaneous.
7.43 Exothermic reactions tend to be spontaneous because they result in an increase in the entropy of the surroundings. The relation $\Delta G_r = \Delta H_r - T\Delta S_r$ implies that, if ΔH_r is large compared to $T\Delta S_r$, then the reaction is likely to be spontaneous.
7.45 (a) $\Delta H_r° = -235.8\ kJ{\cdot}mol^{-1}$, $\Delta S_r° = -133.17\ J{\cdot}K^{-1}{\cdot}mol^{-1}$, $\Delta G_r° = -196.1\ kJ{\cdot}mol^{-1}$; (b) $\Delta H_r° = +11.5\ kJ{\cdot}mol^{-1}$, $\Delta S_r° = -149.7\ J{\cdot}K^{-1}{\cdot}mol^{-1}$, $0\Delta G_r° = +56.1\ kJ{\cdot}mol^{-1}$; (c) $\Delta H_r° = -57.20\ kJ{\cdot}mol^{-1}$, $\Delta S_r° = -175.83\ J{\cdot}K^{-1}{\cdot}mol^{-1}$, $\Delta G_r° = -4.80\ kJ{\cdot}mol^{-1}$
7.47 (a) $\frac{1}{2}\ N_2(g) + \frac{3}{2}\ H_2(g) \rightarrow NH_3(g)$, $\Delta G_r° = -16.49\ kJ{\cdot}mol^{-1}$; $\Delta S_r° = -99.38\ J{\cdot}K^{-1}{\cdot}mol^{-1}$; $S_m°(NH_3) = 192.45\ J{\cdot}K^{-1}{\cdot}mol^{-1}$ (b) $H_2\ (g) + \frac{1}{2}\ O_2(g) \rightarrow H_2O(g)$, $\Delta G_r° = -228.58\ kJ{\cdot}mol^{-1}$; $\Delta S_r° = -44.42\ J{\cdot}K^{-1}{\cdot}mol^{-1}$; $S_m°\ (H_2O_2\ g) = 188.83\ J{\cdot}K^{-1}{\cdot}mol^{-1}$ (c) C(s), graphite $+ \frac{1}{2}\ O_2\ (g) \rightarrow CO\ (g)$; $\Delta G_r° = -137.2\ kJ{\cdot}mol^{-1}$; $\Delta S_r° = +89.36\ J{\cdot}K^{-1}{\cdot}mol^{-1}$; $S_m°\ (CO) = 197.67\ J{\cdot}K^{-1}{\cdot}mol^{-1}$, (d) $\frac{1}{2}\ N_2(g) + O_2(g) \rightarrow NO_2(g)$; $\Delta G_r° = +51.33\ kJ{\cdot}mol^{-1}$; $\Delta S_r° = -60.89\ J{\cdot}K^{-1}{\cdot}mol^{-1}$, $S_m°\ (NO_2) = 240.06\ J{\cdot}K^{-1}{\cdot}mol^{-1}$. For reactions a, b, and d, $\Delta S_f° < 0$ because the formation reactions involve a number of gaseous molecules combining to form one molecule. The $S_m°$ values, however, are positive, as expected, because the entropies of all substances are positive.
7.49 (a) $-141.74\ kJ{\cdot}mol^{-1}$, spontaneous; (b) $+130.41\ kJ{\cdot}mol^{-1}$, spontaneous; (c) $-10\ 590.9\ kJ{\cdot}mol^{-1}$, spontaneous
7.51 (a) and (d) are thermodynamically stable with respect to the elements, whereas (b) and (c) are unstable.
7.53 $\Delta G_r° = -133.41\ kJ{\cdot}mol^{-1}$, $\Delta S_r° = -36.8\ J{\cdot}K^{-1}{\cdot}mol^{-1}$. The reaction has a tendency to occur at 25°C, but the tendency will be less as the temperature is raised because $\Delta S°$ is negative.
7.55 (a) less stable; (b) more stable; (c) more stable; (d) more stable
7.57 (a) $\Delta S_r° = +125.8\ J{\cdot}K^{-1}{\cdot}mol^{-1}$; $\Delta H_r° = -196.10\ kJ{\cdot}mol^{-1}$; $\Delta G_r° = -233.56\ kJ{\cdot}mol^{-1}$; (b) $\Delta S_r° = +14.6\ J{\cdot}K^{-1}{\cdot}mol^{-1}$; $\Delta H_r° = -748.66\ kJ{\cdot}mol^{-1}$; $\Delta G_r° = -713.02\ kJ{\cdot}mol^{-1}$
7.59 (a) $-98.42\ kJ{\cdot}mol^{-1}$, spontaneous below 612.9 K; (b) $-283.7\ kJ{\cdot}mol^{-1}$, spontaneous at all temperatures; (c) $+3.082\ J{\cdot}K^{-1}{\cdot}mol^{-1}$, nonspontaneous at all temperatures
7.61 (a) In the liquid phase, one might expect 1-propanol to have a lower entropy than 2-propanone because the hydrogen bonding present in 1-propanol would make its solution more ordered. This is consistent with the observed data (193 $J{\cdot}K^{-1}{\cdot}mol^{-1}$ versus 200 $J{\cdot}K^{-1}{\cdot}mol^{-1}$, respectively). (b) In the gas phase, hydrogen bonding should be less important.
7.63 The cis form should have a higher residual entropy.
7.65 Pb, 8.5 $J{\cdot}K^{-1}{\cdot}mol^{-1}$; Hg, 9.8 $J{\cdot}K^{-1}{\cdot}mol^{-1}$; Na, 7.1 $J{\cdot}K^{-1}{\cdot}mol^{-1}$. These values are reasonably close but clearly much smaller than the value associated with Trouton's rule.
7.67 Fe_2O_3, because the formation of Fe_2O_3 from Fe_3O_4 in air has a large, negative $\Delta G°$.
7.69 (a) $+91.40\ kJ{\cdot}mol^{-1}$; (b) 980 K
7.71 (a) no, non-spontaneous; (b) no, nonspontaneous; (c) no effect
7.73 (a) Because the enthalpy change for dissolution is positive, the entropy change of the surroundings must be negative. Because spontaneous processes are

accompanied by an increase in entropy, the change in enthalpy does not favor dissolution. (b) We know the process is spontaneous because it occurs readily. For the process to be spontaneous, the reaction entropy must be positive because the reaction enthalpy is positive. (c) Positional disorder is dominant. (d) Because the surroundings participate in the dissolution process only as a source of heat, the entropy change of the surroundings is primarily a result of the dispersal of thermal motion. (e) The driving force for the dissolution is the dispersal of matter, resulting in an overall positive ΔS.
7.75 $\Delta H° = +1.895$ kJ·mol^{-1}; $\Delta S° = -3.363$ J·K^{-1}·mol^{-1}; $\Delta G° = +2.900$ kJ·mol^{-1}. The entropy of graphite is higher than that of diamond in part because the former has a less rigid lattice. The mobility of the electrons also contributes.
7.77 The entries all correspond to aqueous ions. The fact that they are negative is due to the reference point that has been established. Because ions cannot actually be separated and measured independently, a reference point that defines $S_m°(H^+, aq) = 0$ has been established. This definition is then used to calculate the standard entropies for the other ions. The fact that they are negative arises in part because the solvated ion $M(H_2O)_x^{n+}$ is more ordered than the isolated ion and solvent molecules ($M^{n+} + x\ H_2O$).
7.79 In this reaction, we expect the change in enthalpy to be close to 0 (it should not be exactly 0) because the number and types of bonds that are broken are equal to the number and types of bonds that are formed. The free energy change should therefore be dominated by the entropy change of the reaction. We would expect the entropy change of the reaction to be positive because the mixed complex $InCl_2Br$(s) would be expected to have a larger disorder than a molecule composed of all the same types of halide ions, InX_3. Consequently, we predict a positive value for $\Delta S_r°$ and hence a negative value of $\Delta G_r°$.
7.81 (a) at 298 K, $\Delta G_r° = +8.57$ kJ·mol^{-1}; at 373 K, $\Delta G_r° = -0.35$ kJ·mol^{-1}; at 423 K, $\Delta G_r° = -6.29$ kJ·mol^{-1}. (b) The value at 100°C should be exactly 0, because this is the normal boiling point of water. At this point, we expect the vaporization to go from being nonspontaneous to being spontaneous. (c) The discrepancy arises because the enthalpy and entropy values calculated from the tables are not rigorously constant with temperature. Better values would be obtained by using the actual enthalpy and entropy of vaporization measured at the boiling point.
7.83 $\Delta G_r°$ for the dehydrogenation of cyclohexane to benzene is +97.6 kJ·mol^{-1}. If the reaction is coupled to the hydrogenation of ethene: $C_6H_{12}(l) + 3\ C_2H_2(g) \rightarrow C_6H_6(l) + 3\ C_2H_6(g)$; $\Delta G_r°$ becomes favorable, with a value of −205.13 kJ·mol^{-1}.
7.85 (a)

$CH_3(H)C{=}C(H)CH_3$ (cis); $CH_3(H)C{=}C(CH_3)H$ (trans); $(CH_3)_2C{=}CH_2$

cis-2-Butene; *trans*-2-Butene; 2-Methylpropene

(b) *cis*-2-butene ⇌ *trans*-2-butene; $\Delta G_r° = -2.89$ kJ·mol^{-1}, $\Delta H_r° = -4.18$ kJ·mol^{-1}, $\Delta S_r° = -4.33$ J·K^{-1}·mol^{-1}; *cis*-2-butene ⇌ 2-methylpropene: $\Delta G_r° = -7.79$ kJ·mol^{-1}, $\Delta H_r° = -9.91$ kJ·mol^{-1}, $\Delta S_r° = -7.11$ J·K^{-1}·mol^{-1}; *trans*-2-butene ⇌ 2-methylpropene: $\Delta G_r° = -4.90$ kJ·mol^{-1}, $\Delta H_r° = -5.73$ kJ·mol^{-1}, $\Delta S_r° = -2.78$ J·K^{-1}·mol^{-1}; (b) most stable of the three compounds, 2-methylpropene; (c) $S_m°$ (2-methylpropene) < $S_m°$ (*trans*-2-butene) < $S_m°$ (*cis*-2-butene)
7.87 (a) $HCOOH(l) \rightarrow H_2(g) + CO_2(g)$; $\Delta H_r° = +31.21$ kJ·mol^{-1}, $\Delta S_r° = +215.47$ J·K^{-1}·mol^{-1}, $T = 145$ K; (b) $CH_3COOH(l) \rightarrow CH_4(g) + CO_2(g)$; $\Delta H_r° = +16.18$ kJ·mol^{-1}, $\Delta S_r° = +240.2$ J·K^{-1}·mol^{-1}, $T = 67$ K; (c) $C_6H_5COOH(s) \rightarrow C_6H_6(l) + CO_2(g)$; $\Delta H_r° = +40.6$ kJ·mol^{-1}; $\Delta S_r° = +219.4$ J·K^{-1}·mol^{-1}; $T = 185$ K; (d) The reactions are all driven by the entropy increase in the production of the gas because the enthalpy of the reaction in all cases is positive (endothermic).

Chapter 8

8.1 0.017 g of H_2O
8.3 1.8×10^3 g
8.5 (a) 99.2°C; (b) 99.7°C
8.7 (a) 28.3 kJ·mol^{-1}; (b) 91.3 J·K^{-1}·mol^{-1}; (c) +1.1 kJ·mol^{-1}; (d) 36°C
8.9 (a) 25.6 kJ·mol^{-1}; (b) 87.5 J·K^{-1}·mol^{-1}; (c) 7.6×10^2 Torr
8.11 (a) 17.7 kJ·mol^{-1}; (b) 84.5 J·K^{-1}·mol^{-1}; (c) −7.3 kJ·mol^{-1}; (d) 211 K
8.13 (a) 28.6 kJ·mol^{-1}; (b) 90.6 J·K^{-1}; (c) 4.0×10^2 Torr
8.15 1.5×10^2 Torr
8.17 7.6 Torr
8.19 (a) vapor; (b) liquid; (c) vapor; (d) vapor
8.21 (a) 2.4 K; (b) ~10 atm; (c) 5.5 K; (d) no
8.23 (a) At the lower pressure triple point, liquid helium-I and helium-II are in equilibrium with helium gas; at the higher pressure triple point, liquid helium-I and helium-II are in equilibrium with solid helium. (b) helium-I
8.25 The pressure increase would bring CO_2 into the solid region.
8.27 (a) water; (b) benzene; (c) water
8.29 (a) hydrophilic, because $-NH_2$ is polar and has a lone pair and H atoms that can participate in hydrogen bonding to water molecules; (b) hydrophobic, because $-CH_3$ is not very polar; (c) hydrophobic, because

the C—Br bond is not very polar; (d) hydrophilic, because the carboxylic acid group has lone pairs on oxygen and an acidic proton that can participate in hydrogen bonding to water molecules

8.31 (a) 6.4×10^{-4} mol·L^{-1}; (b) 1.5×10^{-2} mol·L^{-1}; (c) 2.3×10^{-3} mol·L^{-1}

8.33 0.59 L

8.35 (a) By Henry's law, the concentration of CO_2 in solution will double. (b) No change in the equilibrium will occur; the partial pressure of CO_2 is unchanged and the concentration is unchanged.

8.37 (a) 4 ppm; (b) 0.1 atm; (c) 0.5 atm

8.39 Ion hydration enthalpies for ions of the same charge within a group of elements become less negative down a group. This trend parallels increasing ion size. Hydration energy is an ion-dipole interaction, so it should be higher for small, highly charged ions.

8.41 (a) negative; (b) $Li_2SO_4(s) \rightleftharpoons 2\ Li^+(aq) + SO_4^{2-}(aq)$ + heat; (c) enthalpy of hydration

8.43 (a) $+6.7 \times 10^2$ J; (b) −0.06 kJ; (c) −24.7 kJ; (d) +3.21 kJ

8.45 −51 kJ·mol^{-1}

8.47 All the enthalpies of solution are positive. Those of the alkali metal chlorides increase as the cation becomes larger and less strongly hydrated by water. All the alkali metal chlorides are soluble in water, but AgCl is not. AgCl has a relatively large positive enthalpy of solution. When dissolving is highly endothermic, the small increase in disorder due to solution formation may not be enough to compensate for the decrease in disorder of the surroundings and a solution does not form. This is the case for AgCl. AgCl is also thought to have covalent character in its bonds, which reduces the disorder on solution formation.

8.49 (a) 0.684 *m*; (b) 3.0 g; (c) 0.0498 *m*

8.51 (a) 0.248 *m*; (b) 0.246 M

8.53 (a) 1.98 *m*; (b) 0.548 *m*; (c) 28.43 *m*

8.55 (a) 19.5 g; (b) 45 g

8.57 (a) 684 Torr; (b) 758 Torr

8.59 (a) 4.55 Torr; (b) 354.3 Torr; (c) 0.17 Torr

8.61 (a) 0.052; (b) 115 g·mol^{-1}

8.63 (a) 0.051°C, 100.051°C; (b) 0.22°C, 100.22°C; (c) 0.090°C, 100.090°C

8.65 (a) 100.34°C; (b) 81.0°C

8.67 1.7×10^2 g·mol^{-1}

8.69 (a) 0.19°C, −0.19°C; (b) 0.82°C, −0.82°C; (c) 0.172°C, −0.172°C

8.71 1.6×10^2 g·mol^{-1}

8.73 (a) 1.7°C; (b) 1.63 mol·kg^{-1}; (c) 0.209 mol

8.75 (a) 1.84; (b) 0.318 mol·kg^{-1}; (c) 83.8%

8.77 −0.20°C

8.79 (a) 0.24 atm; (b) 48 atm; (c) 0.72 atm

8.81 2.0×10^3 g·mol^{-1}

8.83 3.4×10^3 g·mol^{-1}

8.85 8.3×10^{-5} atm

8.87 2.5×10^5 g·mol^{-1}

8.89 (a) P = 78.2 Torr, $x_{benzene} = 0.91$, $x_{toluene} = 0.09$; (b) P = 43.0 Torr, $x_{benzene} = 0.469$; $x_{toluene} = 0.531$

8.91 63 g

8.93 Raoult's law applies to the vapor pressure of the mixture, so a positive deviation means that the vapor pressure is higher than expected for an ideal solution. Negative deviation means that the vapor pressure is lower than expected for an ideal solution. Negative deviation will occur when the interactions between the different molecules are somewhat stronger than the interactions between molecules of the same kind. (a) For methanol and ethanol, we expect the types of intermolecular attractions in the mixture to be similar to those in the component liquids, so an ideal solution is predicted. (b) For HF and H_2O, the possibility of intermolecular hydrogen bonding between water and HF would suggest that negative deviation would be observed, which is the case. HF and H_2O form an azeotrope that boils at 111°C, a temperature higher than the boiling point of either HF (19.4°C) or water. (c) Because hexane is nonpolar and water is polar with hydrogen bonding, we would expect a mixture of these two to exhibit positive deviation (the interactions between the different molecules would be weaker than the intermolecular forces between like molecules). Hexane and water do form an azeotrope that boils at 61.6°C, a temperature below the boiling point of either hexane or water.

8.95 (a) stronger; (b) low; (c) high; (d) weaker; (e) weak, low; (f) low; (g) strong, high

8.97 (a, b) Viscosity and surface tension decrease with increasing temperature; at high temperatures, the molecules readily move away from their neighbors because of increased kinetic energy. (c, d) Evaporation rate and vapor pressure increase with increasing temperature because the kinetic energy of the molecules increases with temperature and the molecules are more likely to escape into the gas phase.

8.99 If the external pressure is lowered, water boils at a lower temperature. The boiling temperature is the temperature at which the vapor pressure equals the external pressure. Vapor pressure increases with temperature. If the external pressure is reduced, the vapor pressure of the water will reach that value at a lower temperature; thus, lowering the external pressure lowers the boiling temperature.

8.101 (a) no; (b) yes; (c) rhombic sulfur; (d) monoclinic sulfur; (e) no

8.103 (a) 78.5%. (b) Some of the water vapor in the air would condense as dew or fog.

8.105 (a) 0.497 mol·L^{-1}; (b) 17.4 Torr; (c) 100.28°C

8.107 at 30°C, 31.26 Torr; at 100°C, 747 Torr; at 0°C, 4.50 Torr

8.109 (a) $1.5 \times 10^2\ g \cdot mol^{-1}$; (b) $C_6H_4Cl_2$; (c) actual molar mass, $146.99\ g \cdot mol^{-1}$
8.111 (a) $C_9H_{13}O_3N$; (b) $1.8 \times 10^2\ g \cdot mol^{-1}$; (c) $C_9H_{13}O_3N$
8.113 Hypotonic solutions will cause a net flow of solvent into the cells to equalize the osmotic pressure. The cells will burst and die (hemolysis). Hypertonic solutions will cause a net flow of solvent out of the cells to equalize the osmotic pressure. The cells will shrink and die.
8.115 (a) 3.3×10^2 Torr; $x_{\text{diethyl ether}} = 0.34$, $x_{\text{ethyl methyl ether}} = 0.66$; (b) 419 Torr, $x_{\text{diethyl ether}} = 0.162$; $x_{\text{ethyl methyl ether}} = 0.838$
8.117 (a) partial pressure of ethanol = 58.9 Torr because some of the ethanol will condense to reestablish the equilbrium condition; (b) 1.44×10^3 Torr
8.119 (a) 14.0 atm; (b) 21 atm; (c) 61 atm
8.121 (a) urea solution; (b) urea solution; (c) urea solution; (d) 0.73 atm
8.123 (a) $\ln P = -\dfrac{3.70 \times 10^3\ K}{T} + 10.5$;
(b) plot ln *P* versus 1/*T*:

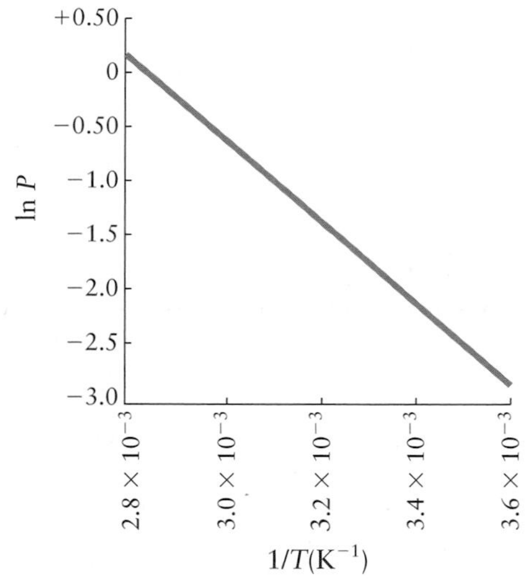

(c) 66°C; (d) $260.6\ J \cdot K^{-1} \cdot mol^{-1}$
8.125 6.26 Torr
8.127 The critical pressures and temperatures of the compounds are as follows:

Compound	T_c, °C	P_c, atm
methane, CH_4	−82.1	45.8
methylamine, CH_3NH_2	156.9	40.2
ammonia, NH_3	132.5	112.5
tetrafluoromethane, CF_4	−45.7	41.4

To achieve the supercritical fluid state, the temperature and pressure must exceed their critical values. Given the pressure rating of the autoclave, ammonia would not be suitable because the supercritical state could not be achieved. Methylamine would not be suitable for a room temperature extraction because its T_c is too high. Either methane or tetrafluoromethane would be suitable for this application.
8.129 (a) If there is enough diethyl ether present, then the pressure will be due to the vapor pressure of diethyl ether at that temperature. If there is insufficient diethyl ether present, then it will all convert to gas and the pressure will be determined from the ideal gas law. We can calculate the amount of diethyl ether needed to achieve the vapor pressure from the ideal gas equation, using the vapor pressure as the pressure of the gas. There is sufficient diethyl ether in this case, so the pressure equals the vapor pressure (57 Torr). (b) 376 Torr; (c) 57 Torr; (d) If flask B is cooled by liquid nitrogen, its temperature will approach −196°C. The vapor pressure of ether at that temperature is negligible and the ether will solidify in flask B. As the ether condenses in flask B, the liquid ether in flask A will continue to vaporize and will do so until all the ether has condensed in flask B. At the end of this process, there will be no ether left in flask A and the pressure in the apparatus will be 0.
8.131 $x_{\text{pentane, liquid}} = 0.228$; $x_{\text{hexane, liquid}} = 0.772$
8.133 (a) The vapor pressures over the two solutions are different. The volatile component, ethanol, will transfer from the more concentrated solution to the less concentrated solution until the vapor pressures (and therefore concentrations) are equal. The vapor pressure is lower over the more concentrated solution, so ethanol will condense on that side of the apparatus. As this takes place, there will be a net transfer of ethanol from the less concentrated side to the more concentrated side of the apparatus. (b) The mole fraction of sucrose in the solutions on both sides of the apparatus will be 0.0045, corresponding to a 0.098 *m* solution. The vapor pressure will be essentially 60 Torr.
8.135 compound (a), because it is more polar and hence will be more strongly adsorbed.

Chapter 9

9.1 (a) False. Equilibrium is dynamic. At equilibrium, the concentrations of reactants and products will not change, but the reaction will continue to proceed in both directions. (b) False. Equilibrium reactions are affected by the presence of both products and reactants. (c) False. The value of the equilibrium constant is not affected by the amounts of reactants or products added as long as the temperature is constant. (d) True

9.3

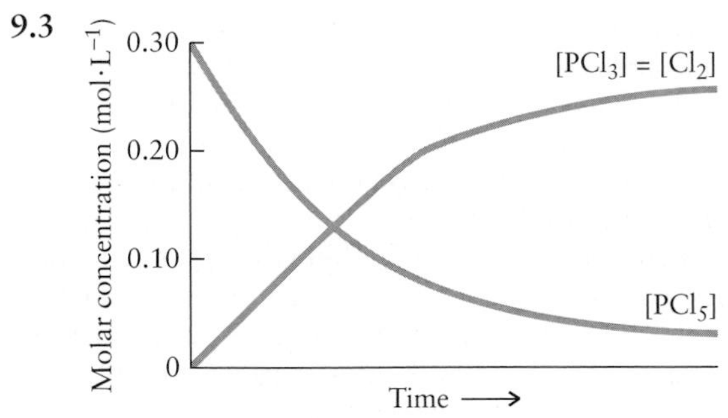

9.5 (a) $+9.69\ kJ \cdot mol^{-1}$; (b) $-1.72\ kJ \cdot mol^{-1}$; (c) $-1.72\ kJ \cdot mol^{-1}$; (d) $-1.72\ kJ \cdot mol^{-1}$
9.7 (a) $-0.623\ kJ \cdot mol^{-1}$; (b) $-0.623\ kJ \cdot mol^{-1}$; (c) $+1.72\ kJ \cdot mol^{-1}$; (d) $+1.72\ kJ \cdot mol^{-1}$
9.9 (a) 1×10^{80}; (b) 1×10^{90}; (c) 1×10^{-23}
9.11 (a) -19 kJ; (b) $+68$ kJ
9.13 form reactants
9.15 (a) $-32\ kJ \cdot mol^{-1}$; (b) The reaction will proceed to produce I.
9.17 (a) $-6.5\ kJ \cdot mol^{-1}$; (b) forms products
9.19 (a) $K_c = \dfrac{[COCl][Cl]}{[CO][Cl_2]}$; (b) $K_c = \dfrac{[HBr]^2}{[H_2][Br_2]}$; (c) $K_c = \dfrac{[SO_2]^2[H_2O]^2}{[H_2S]^2[O_2]^3}$
9.21 (a) 4.3×10^{-4}; (b) 1.87
9.23 (a) Because the volume is the same, the number of moles of O_2 is larger in the second experiment. (b) Because K_c is a constant and the denominator is larger in the second case, the numerator must also be larger. So the concentration of O_2 is larger in the second case. (c) Although $[O_2]^3/[O_3]^2$ is the same, $[O_2]/[O_3]$ will be different, a result seen by solving for K_c in each case. (d) Because K_c is a constant, $[O_2]^3/[O_3]^2$ is the same. (e) Because $[O_2]^3/[O_3]^2$ is the same, its reciprocal must be the same.
9.25 (a) 0.024; (b) 6.4; (c) 1.7×10^3
9.27 $K_c = 48.8$ for case 1 and 48.9 for cases 2 and 3. These numbers are equivalent within experimental error.
9.29 (a) $1/P_{O_2}{}^3$; (b) $P_{H_2O}{}^7$; (c) $(P_{NO}P_{NO_2})/P_{N_2O_3}$
9.31 2.1×10^{-5}
9.33 15 bar
9.35 (a) 0.50; (b) no; (c) products
9.37 (a) 6.9; (b) yes
9.39 6.6×10^{-3}
9.41 1.58×10^{-8}
9.43 $[H_2O] = [CH_3COOC_2H_5]$; $K_c = 4$, 4.01, and 3.9 for the three cases, respectively. These values are identical within the experimental error.
9.45 (a) $[Cl] = 1.1 \times 10^{-5}\ mol \cdot L^{-1}$; $[Cl_2] = 0.0010\ mol \cdot L^{-1}$, percentage dissociation = 0.55%; (b) $[F] = 3.2 \times 10^{-4}\ mol \cdot L^{-1}$; $[F_2] = 8 \times 10^{-4}\ mol \cdot L^{-1}$; percentage dissociation = 16%; (c) chlorine
9.47 $[Br_2] = [H_2] = 1.1 \times 10^{-8}\ mol \cdot L^{-1}$, $[HBr] = 1.2 \times 10^{-3}\ mol \cdot L^{-1}$; % decomposition = 1.8×10^{-3}%
9.49 $[PCl_3] = [Cl_2] = 0.010\ mol \cdot L^{-1}$; $[PCl_5] = 0.009\ mol \cdot L^{-1}$; % decomposition = 53%
9.51 $[NH_3] = 0.200\ mol \cdot L^{-1}$, $[H_2S] = 8 \times 10^{-4}\ mol \cdot L^{-1}$
9.53 $[PCl_5] = 3 \times 10^{-4}\ mol \cdot L^{-1}$; $[PCl_3] = 0.200\ mol \cdot L^{-1}$; $[Cl_2] = 0.0497\ mol \cdot L^{-1}$
9.55 $[NO] = 3.6 \times 10^{-4}\ mol \cdot L^{-1}$; the concentrations of N_2 and O_2 remain essentially unchanged at $0.114\ mol \cdot L^{-1}$
9.57 1.1
9.59 $[CO_2] = 8.6 \times 10^{-5}\ mol \cdot L^{-1}$; $[CO] = 4.9 \times 10^{-3}\ mol \cdot L^{-1}$; $[O_2] = 4.6 \times 10^{-4}\ mol \cdot L^{-1}$
9.61 $0.32\ mol \cdot L^{-1}$
9.63 $9.4\ mol \cdot L^{-1}$
9.65 3.88
9.67 $[SO_2] = 0.0011\ mol \cdot L^{-1}$; $[NO_2] = 0.0211\ mol \cdot L^{-1}$; $[NO] = 0.0389\ mol \cdot L^{-1}$; $[SO_3] = 0.0489\ mol \cdot L^{-1}$
9.69 (a) $[N_2O] = 0.0100\ mol \cdot L^{-1}$; $[O_2] = 0.0410\ mol \cdot L^{-1}$; $[NO_2] = 0.0200\ mol \cdot L^{-1}$; (b) $K_c = 23.2$
9.71 (a) no; (b) form products; (c) $[PCl_5] = 3.07\ mol \cdot L^{-1}$; $[PCl_3] = 5.93\ mol \cdot L^{-1}$; $[Cl_2] = 0.93\ mol \cdot L^{-1}$
9.73 $P_{HCl} = 0.22$ bar; $P_{H_2} = P_{Cl_2} = 3.9 \times 10^{-18}$ bar
9.75 (a) decrease, because $Q \gg K_c$; hence, there is too much CH_3OH relative to reactants and reactants will be favored; (b) $[CO] = 0.156\ mol \cdot L^{-1}$; $[H_2] = 0.155\ mol \cdot L^{-1}$; $[CH_3OH] = 2.6 \times 10^{-4}\ mol \cdot L^{-1}$
9.77 (a) decreases; (b) decreases; (c) increases; (d) no change
9.79 (a) decreases; (b) increases; (c) increases; (d) increases; (e) no change; (f) decreases; (g) decreases
9.81 (a) reactants; (b) reactants; (c) reactants; (d) no change (there are the same number of moles of gas on both sides of the equation); (e) reactants
9.83 (a) increases; (b) no, it increases
9.85 (a) products; (b) products; (c) reactants; (d) reactants
9.87 No, there is less ammonia at 700 K.
9.89 No, the reaction proceeds to form more products.
9.91 (a) 1×10^{-16} at 25°C, 1×10^{-7} at 150°C; (b) 7.8×10^{-2} at 25°C, 0.22 at 150°C
9.93 (a) $P_{NOCl} = 92.9$ bar; $P_{NO} = 2.69$ bar; $P_{Cl_2} = 1.34$ bar; (b) 2.8%
9.95 (a) reactants; (b) no effect; (c) products; (d) products; (e) products; (f) no effect; (g) reactants
9.97 (a) 2.9; (b) $\Delta G_r° = -10\ kJ \cdot mol^{-1}$; $K_3 = 2$
9.99 $\alpha = \sqrt{\dfrac{K}{P + K}}$; (a) 0.953; (b) 0.912
9.101 (a) $\Delta G° = 0$; (b) 978 K (or 705°C); (c) All pressures are equal to 7.50 bar. (d) $P_{CO} = 7.04$ bar; $P_{H_2O} = 5.04$ bar; $P_{CO_2} = 8.96$ bar; $P_{H_2} = 3.96$ bar
9.103 (a) F(g), $19.2\ kJ \cdot mol^{-1}$; Cl(g), $47.8\ kJ \cdot mol^{-1}$; Br(g), $42.8\ kJ \cdot mol^{-1}$; I(g), $5.6\ kJ \cdot mol^{-1}$. (b) The bond dissociation energy generally increases as the free energy of formation of the atomic species increases, but

the relation is not linear. For the heavier three halogens, there is a trend to decreasing free energy of formation of the atoms as the element becomes heavier, but fluorine is anomalous. The F—F bond energy is lower than expected owing to repulsions of the lone pairs of electrons on the adjacent F atoms because the F—F bond distance is so short.
9.105 (a) $K = 2.6 \times 10^{-29}$; (b) $K = 4.5 \times 10^{-29}$. For equilibrium constant calculations, this is reasonably good agreement with the value obtained from part (a), especially if one considers that $\Delta H°$ will not be perfectly constant over so large a temperature range. (c) 0.285 bar or 0.289 atm; (d) $P_{Br(g)} = 3.6 \times 10^{-15}$ bar or 3.6×10^{-15} atm; (e) 846 mL
9.107 $P(N_2) = 4.33$ bar or 4.39 atm; $P(H_2) = 1.11$ bar or 1.12 atm; $P(NH_3) = 24.07$ bar or 24.39 atm
9.109 The greater change in equilibrium constant as temperature changes is seen for reactions that have the larger values of $\Delta G°$.

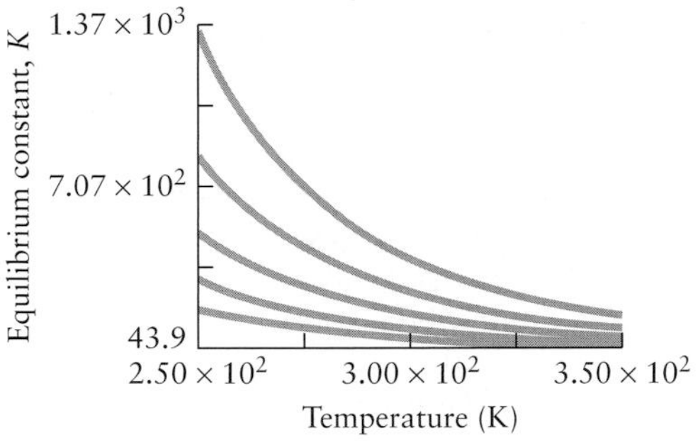

9.111 (a) 11.2. (b) If NO_2 is added, the equilibrium will shift to produce more N_2O_4. The amount of NO_2 will be greater than initially present but less than the 3.13 mol·L^{-1} present immediately upon making the addition. K_c will not be affected. (c) $[N_2O_4] =$ 0.64 mol·L^{-1}; $[NO_2] = 2.67$ mol·L^{-1}
9.113 For H_2O, 24 Torr; for D_2O, 21 Torr. Because D is heavier than H, it has a lower zero-point vibrational energy, so the O—D···O bond is stronger than the O—H···O bond, and hence the liquid is less volatile.
9.115 (a) +1.70 kJ·mol; (b)−137.15 kJ·mol^{-1}; (c) +124.7 kJ·mol^{-1}; (d) −22.2 kJ·mol^{-1}
9.117 (a) $CH_3Cl(g)$, −63.01 kJ·mol^{-1}; $CH_3Br(g)$, −28.16 kJ·mol^{-1}; (b) 584; (c) BCl_3 and BCl_2Br; (d) $P_{BCl_3} = 0.191$ bar; $P_{BBr_3} = 0.257$ bar; $P_{BCl_2Br} =$ 0.291 bar; $P_{BClBr_2} = 0.262$ bar; $P_{CH_3Cl} = 0.002$ bar; $P_{CH_3Br} = 0.998$ bar

Chapter 10

10.1 (a) $CH_3NH_3^+$; (b) $NH_2NH_3^+$; (c) H_2CO_3; (d) CO_3^{2-}; (e) $C_6H_5O^-$; (f) $CH_3CO_2^-$
10.3 For all parts (a)–(e), H_2O and H_3O^+ form a conjugate acid-base pair in which H_2O is the base and H_3O^+ is the acid.
(a) $H_2SO_4(aq) + H_2O(l) \rightleftharpoons H_3O^+(aq) + HSO_4^-(aq)$
H_2SO_4 and HSO_4^- form a conjugate acid-base pair in which H_2SO_4 is the acid and HSO_4^- is the base.
(b) $C_6H_5NH_3^+(aq) + H_2O(l) \rightleftharpoons C_6H_5NH_2(aq) + H_3O^+(aq)$
$C_6H_5NH_3^+$ and $C_6H_5NH_2$ form a conjugate acid-base pair in which $C_6H_5NH_3^+$ is the acid and $C_6H_5NH_2$ is the base.
(c) $H_2PO_4^-(aq) + H_2O(l) \rightleftharpoons H_3O^+(aq) + HPO_4^{2-}(aq)$
$H_2PO_4^-$ and HPO_4^{2-} form a conjugate acid-base pair in which $H_2PO_4^-$ is the acid and HPO_4^{2-} is the base.
(d) $HCOOH(aq) + H_2O(l) \rightleftharpoons HCO_2^-(aq) + H_3O^+(aq)$
HCOOH and HCO_2^- form a conjugate acid-base pair in which HCOOH is the acid and HCO_2^- is the base.
(e) $NH_2NH_3^+(aq) + H_2O(l) \rightleftharpoons NH_2NH_2(aq) + H_3O^+(aq)$
$NH_2NH_3^+$ and NH_2NH_2 form a conjugate acid-base pair in which $NH_2NH_3^+$ is the acid and NH_2NH_2 is the base.
10.5 (a) Brønsted acid, HNO_3; Brønsted base, HPO_4^{2-}; (b) conjugate base to HNO_3, NO_3^-; conjugate acid to HPO_4^{2-}, $H_2PO_4^-$
10.7 (a) HCO_3^- as an acid: $HCO_3^-(aq) + H_2O(l) \rightleftharpoons CO_3^{2-}(aq) + H_3O^+(aq)$. HCO_3^- and CO_3^{2-} form a conjugate acid-base pair in which HCO_3^- is the acid and CO_3^{2-} is the base. H_2O and H_3O^+ form a conjugate acid-base pair in which H_2O is the base and H_3O^+ is the acid.
HCO_3^- as a base: $HCO_3^-(aq) + H_2O(l) \rightleftharpoons H_2CO_3(aq) + OH^-(aq)$. HCO_3^- and H_2CO_3 form a conjugate acid-base pair in which HCO_3^- is the base and H_2CO_3 is the acid. H_2O and OH^- form a conjugate acid-base pair in which H_2O is the acid and OH^- is the base.
(b) HPO_4^{2-} as an acid: $HPO_4^{2-}(aq) + H_2O(l) \rightleftharpoons PO_4^{3-}(aq) + H_3O^+(aq)$. HPO_4^{2-} and PO_4^{3-} form a conjugate acid-base pair in which HPO_4^{2-} is the acid and PO_4^{3-} is the base. H_2O and H_3O^+ form a conjugate acid-base pair in which H_2O is the base and H_3O^+ is the acid.
HPO_4^{2-} as a base: $HPO_4^{2-}(aq) + H_2O(l) \rightleftharpoons H_2PO_4^-(aq) + OH^-(aq)$. HPO_4^{2-} and $H_2PO_4^-$ form a conjugate acid-base pair in which HPO_4^{2-} is the base and $H_2PO_4^-$ is the acid. H_2O and OH^- form a conjugate acid-base pair in which H_2O is the acid and OH^- is the base.
10.9 (a) basic; (b) acidic; (c) amphoteric; (d) basic
10.11 (a) 5.4×10^{-14} mol·L^{-1}; (b) 1.0×10^{-10} mol·L^{-1}; (c) 3.2×10^{-13} mol·L^{-1}
10.13 (a) $[H_3O^+] = 1.4 \times 10^{-7}$ mol·L^{-1}; pH = 6.80; (b) 1.4×10^{-7} mol·L^{-1}
10.15 The initial concentration of $Ba(OH)_2$ is 2.9×10^{-2} mol·L^{-1}; Because $Ba(OH)_2$ is a strong base, this is the concentration of Ba^{2+} in the solution; the concentration of OH^- ions is 5.8×10^{-2} mol·L^{-1}, and the concentration of H_3O^+ ions is 1.7×10^{-13} mol·L^{-1}.

10.17 (a) 5.0×10^{-4} mol·L^{-1}; (b) 2×10^{-7} mol·L^{-1}; (c) 4×10^{-5} mol·L^{-1}; (d) 5×10^{-6} mol·L^{-1}
10.19 (a) pH = −0.164, pOH = 14.16; (b) pH = 0.96, pOH = 13.04; (c) pOH = 1.74, pH = 12.26; (d) pOH = 2.72, pH = 11.28; (e) pOH = 2.70, pH = 11.30; (f) pH = 4.00, pOH = 10.00
10.21 (a) 7.6×10^{-3}; (b) 0.010; (c) 3.5×10^{-3}; (d) 0.012; (e) $H_2SeO_3 < H_3PO_4 < H_3PO_3 < HSeO_4^-$
10.23 (a) $HSeO_3^-$; (b) SeO_4^{2-}; (c) K_b ($HSeO_3^-$) = 2.9×10^{-12}. K_b (SeO_4^{2-}) = 8.3×10^{-13} (d) $HSeO_3^-$
10.25
(a) $HClO_2(aq) + H_2O(l) \rightleftharpoons H_3O^+(aq) + ClO_2^-(aq)$

$$K_a = \frac{[H_3O^+][ClO_2^-]}{[HClO_2]}$$

$ClO_2^-(aq) + H_2O(l) \rightleftharpoons HClO_2(aq) + OH^-(aq)$

$$K_b = \frac{[HClO_2][OH^-]}{[ClO_2^-]}$$

(b) $HCN(aq) + H_2O(l) \rightleftharpoons H_3O^+(aq) + CN^-(aq)$

$$K_a = \frac{[H_3O^+][CN^-]}{[HCN]}$$

$CN^-(aq) + H_2O(l) \rightleftharpoons HCN(aq) + OH^-(aq)$

$$K_b = \frac{[HCN][OH^-]}{[CN^-]}$$

(c) $C_6H_5OH\ (aq) + H_2O(l) \rightleftharpoons H_3O^+(aq) + C_6H_5O^-(aq)$

$$K_a = \frac{[H_3O^+][C_6H_5O^-]}{[C_6H_5OH]}$$

$C_6H_5O^-\ (aq) + H_2O(l) \rightleftharpoons C_6H_5OH\ (aq) + OH^-(aq)$

$$K_b = \frac{[C_6H_5OH][OH^-]}{[C_6H_5O^-]}$$

10.27 $(CH_3)_2NH_2^+$ (pK_a = 14.00 − 3.27 = 10.73) < $^+NH_3OH$ (pK_a = 14.00 − 7.97 = 6.03) < HNO_2 (pK_a = 3.37) < $HClO_2$ (pK_a = 2.00)
10.29 F^- (pK_b = 14.00 − 3.45 = 10.55) < $CH_3CO_2^-$ (pK_b = 14.00 − 4.75 = 9.25) < C_5H_5N (pK_b = 8.75) ≪ NH_3 (pK_b = 4.75)
10.31 (a) strong; (b) weak; (c) weak; (d) weak; (e) weak; (f) strong; (g) weak
10.33 For oxoacids, the greater the number of highly electronegative O atoms attached to the central atom, the stronger the acid. This effect is related to the increased oxidation number of the central atom as the number of O atoms increases. Therefore, HIO_3 is the stronger acid, with the lower pK_a.
10.35 (a) HCl is stronger because its bond is much weaker than the bond in HF. (b) $HClO_2$ is stronger because there is one more oxygen attached to the Cl atom, which helps to pull electrons out of the H—O bond. (c) $HClO_2$ is stronger because Cl has a greater electronegativity than Br. (d) $HClO_4$ is stronger because Cl has a greater electronegativity than P. (e) HNO_3 is stronger for the same reason as given in (b). (f) H_2CO_3 is stronger because C has a greater electronegativity than Ge.
10.37 (a) the $—CCl_3$ group is more electron-withdrawing than the $—CH_3$ group, thus making trichloroacetic acid the stronger acid. (b) The $—CH_3$ group has electron-donating properties, which means that it is less electron withdrawing than the —H attached to the carboxyl group in formic acid. Thus, formic acid is slightly stronger than acetic acid.
10.39 The larger the K_a, the stronger the corresponding acid. 2,4,6-Trichlorophenol is the stronger acid because the chlorine atoms have a greater electron-withdrawing power than the hydrogen atoms present in the unsubstituted phenol.
10.41 (d) < (a) < (b) < (c). It appears that amines in which the nitrogen atom of the amine group is attached to an aromatic ring such as benzene are weaker than ammonia, which is weaker than alkylamines.
10.43 (a) pH = 2.64, pOH = 11.36; (b) pH = 0.74, pOH = 13.26; (c) pH = 2.14, pOH = 11.86
10.45 (a) pOH = 3.00, pH = 11.00, 1.8% protonated; (b) pOH = 4.38, pH = 9.62, 0.026% protonated; (c) pOH = 2.32, pH = 11.68, 1.4% protonated; (d) pOH = 3.96, pH = 10.04, 2.5% protonated
10.47 (a) K_a = 0.09, pK_a = 1.0; (b) $K_b = 5.6 \times 10^{-4}$, pK_b = 3.25
10.49 (a) 0.021 mol·L^{-1}; (b) 0.015 mol·L^{-1}
10.51 pH = 2.58; $K_a = 6.3 \times 10^{-5}$
10.53 pH = 11.83, $K_b = 4.8 \times 10^{-4}$
10.55 (a) less than 7, $NH_4^+(aq) + H_2O(l) \rightleftharpoons H_3O^+(aq) + NH_3(aq)$
(b) greater than 7, $H_2O(l) + CO_3^{2-}(aq) \rightleftharpoons HCO_3^-(aq) + OH^-(aq)$
(c) greater than 7, $H_2O(l) + F^-(aq) \rightleftharpoons HF(aq) + OH^-(aq)$
(d) neutral
(e) less than 7, $Al(H_2O)_6^{3+}(aq) + H_2O(l) \rightleftharpoons H_3O^+(aq) + Al(H_2O)_5(OH)^{2+}(aq)$
(f) less than 7, $Cu(H_2O)_6^{2+}(aq) + H_2O(l) \rightleftharpoons H_3O^+(aq) + Cu(H_2O)_5(OH)^+(aq)$
10.57 (a) 9.28; (b) 5.00; (c) 3.06; (d) 11.04
10.59 (a) 9.04; (b) 4.74
10.61 (a) 5.4×10^{-6} mol·L^{-1}; (b) 5.11
10.63 6.174
10.65 7.205
10.67 (a) for 1.00×10^{-4} mol·L^{-1}, 6.35; for 1.00×10^{-6} mol·L^{-1}, 7.35; (b) for 1.00×10^{-4} mol·L^{-1}, 6.34; for 1.00×10^{-6} mol·L^{-1}, 6.96
10.69 (a) for 8.5×10^{-5} mol·L^{-1}, 6.69; for 7.37×10^{-6} mol·L^{-1}, 7.22; (b) for 8.5×10^{-5} mol·L^{-1}, 6.64; for 7.37×10^{-6} mol·L^{-1}, 6.92
10.71 (a) $H_2SO_4(aq) + H_2O(l) \rightleftharpoons H_3O^+(aq) + HSO_4^-(aq)$
$HSO_4^-(aq) + H_2O(l) \rightleftharpoons H_3O^+(aq) + SO_4^{2-}(aq)$
(b) $H_3AsO_4(aq) + H_2O(l) \rightleftharpoons H_3O^+(aq) + H_2AsO_4^-(aq)$
$H_2AsO_4^-(aq) + H_2O(l) \rightleftharpoons H_3O^+(aq) + HAsO_4^{2-}(aq)$
$HAsO_4^{2-}(aq) + H_2O(l) \rightleftharpoons H_3O^+(aq) + AsO_4^{3-}(aq)$

(c) $C_6H_4(COOH)_2(aq) + H_2O(l) \rightleftharpoons H_3O^+(aq) + C_6H_4(COOH)CO_2^-(aq)$
$C_6H_4(COOH)CO_2^-(aq) + H_2O(l) \rightleftharpoons H_3O^+(aq) + C_6H_4(CO_2)_2^{2-}(aq)$

10.73 0.80

10.75 (a) 4.18; (b) 1.28; (c) 3.80

10.77 (a) 4.37; (b) 4.37

10.79 (a) 4.55; (b) 6.17

10.81 $[H_2CO_3] = 0.0455\ mol{\cdot}L^{-1}$; $[H_3O^+] = [HCO_3^-] = 1.4 \times 10^{-4}\ mol{\cdot}L^{-1}$; $[CO_3^{2-}] = 5.6 \times 10^{-11}\ mol{\cdot}L^{-1}$; $[OH^-] = 7.1 \times 10^{-11}\ mol{\cdot}L^{-1}$

10.83 $[H_2CO_3] = 2.3 \times 10^{-8}\ mol{\cdot}L^{-1}$; $[OH^-] = [HCO_3^-] = 0.0028\ mol{\cdot}L^{-1}$; $[CO_3^{2-}] = 0.0428\ mol{\cdot}L^{-1}$; $[H_3O^+] = 3.6 \times 10^{-12}\ mol{\cdot}L^{-1}$

10.85 (a) HA^-; (b) A^{2-}; (c) H_2A

10.87 $[HSO_3^-] = 0.14\ mol{\cdot}L^{-1}$; $[H_2SO_3] = 3.2 \times 10^{-5}\ mol{\cdot}L^{-1}$; $[SO_3^{2-}] = 0.0054\ mol{\cdot}L^{-1}$

10.89 (a) 6.54; (b) 2.12; (c) 1.49

10.91 (a) 6.72; (b) a plot of $\ln K_w$ versus T^{-1} gives a straight line with a slope of -6502 K and an intercept of -10.43; (c) $pH = \dfrac{1411}{T} + 2.264$

10.93 0.041 M

10.95 $K_a = 6.9 \times 10^{-4}$

10.97 (a) $2\ D_2O(l) \rightleftharpoons D_3O^+(aq) + OD^-(aq)$; (b) 14.870; (c) $[D_3O^+] = [OD^-] = 3.67 \times 10^{-8}\ mol{\cdot}L^{-1}$; (d) pD = pOD = 7.435; (e) $pD + pOD = pK_{D_2O} = 14.870$

10.99 pH = 7.11; $pK_b = 6.8 \times 10^{-12}$

10.101 $1.3 \times 10^2\ J{\cdot}mol^{-1}$

10.103 A plot of $\ln K$ versus T^{-1} gives a straight line with a slope of -7030 K and an intercept of -10.67. From the slope $\Delta H°$ can be calculated (slope $= -\Delta H°/R$). $\Delta H° = 58.4\ kJ{\cdot}mol^{-1}$, pD = 7.28. pOD = 7.28, $pK_{D_2O} = 14.55$, $K_{D_2O} = 2.8 \times 10^{-15}$

10.105 All concentrations are moles per liter. $[H_3PO_4] = 8.5 \times 10^{-4}$; $[H_2PO_4^-] = 0.148$; $[HPO_4^{2-}] = 8.9 \times 10^{-4}$; $[PO_4^{3-}] = 4.3 \times 10^{-12}$; $[NH_4^+] = 0.150$; $[NH_3] = 1.9 \times 10^{-6}$; $[H_3O^+] = 4.3 \times 10^{-5}$; $[OH^-] = 2.3 \times 10^{-10}$

10.107 (a) and (b) Buffer regions are marked A, B, and C.

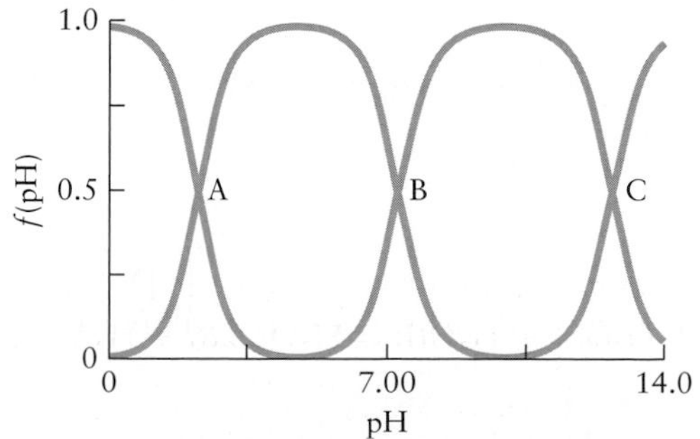

(c) Region A: H_3PO_4 and $H_2PO_4^-$
Region B: $H_2PO_4^-$ and HPO_4^{2-}
Region C: HPO_4^{2-} and PO_4^{3-}

(d)

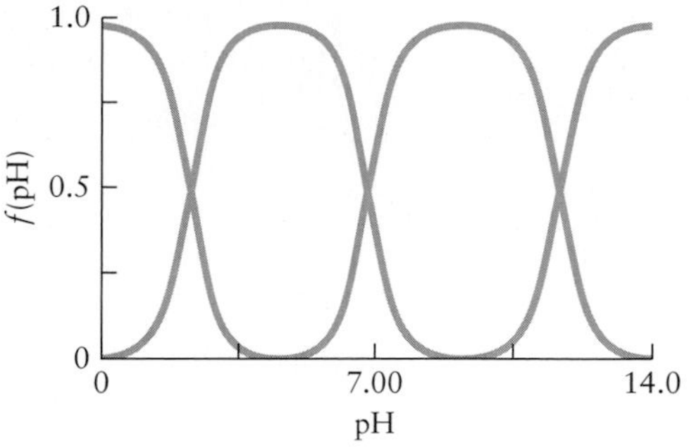

(e) The major species present are similar for both H_3PO_4 and H_3AsO_4: $H_2EO_4^-$ and HEO_4^{2-} where E = P or As. For As, there is more $HAsO_4^{2-}$ than $H_2AsO_4^-$, with a ratio of approximately 0.63 to 0.37, or 1.7:1. For P, the situation is reversed, with more $H_2PO_4^-$ than HPO_4^{2-}, in a ratio of about 0.61 to 0.39, or 2.2:1.

Chapter 11

11.1 (a) When solid sodium acetate is added to an acetic acid solution, the concentration of H_3O^+ decreases because the equilibrium $CH_3COOH(aq) + H_2O(l) \rightleftharpoons H_3O^+(aq) + CH_3CO_2^-(aq)$ shifts to the left to relieve the stress imposed by the increase of $[CH_3CO_2^-]$. (b) When HCl is added to a benzoic acid solution, the percentage of benzoic acid that is deprotonated decreases because the equilibrium $C_6H_5COOH(aq) + H_2O(l) \rightleftharpoons H_3O^+(aq) + C_6H_5CO_2^-(aq)$ shifts to the left to relieve the stress imposed by the increased $[H_3O^+]$. (c) When solid NH_4Cl is added to an ammonia solution, the pH decreases as the concentration of OH^- decreases, because the equilibrium $NH_3(aq) + H_2O(l) \rightleftharpoons NH_4^+(aq) + OH^-(aq)$ shifts to the left to relieve the stress imposed by the increased $[NH_4^+]$.

11.3 (a) p $K_a = 3.08$, $K_a = 8.3 \times 10^{-4}$; (b) 2.77

11.5 (a) pH = 1.62, pOH = 12.38; (b) pH = 1.32, pOH = 12.68; (c) pH = 1.92, pOH = 12.08

11.7 The pH increases by 1.16 units.

11.9 (a) 9.70; (b) 8.92; (c) 9.31

11.11 $[ClO^-]/[HClO] = 9.3 \times 10^{-2}$

11.13 (a) 2–4; (b) 3–5; (c) 11.5–13.5; (d) 6–8; (e) 5–7

11.15 (a) $HClO_2$ and $NaClO_2$; (b) NaH_2PO_4 and Na_2HPO_4; (c) $CH_2ClCOOH$ and $NaCH_2ClCO_2$; (d) Na_2HPO_4 and Na_3PO_4 (there are other possibilities)

11.17 (a) $[CO_3^{2-}]/[HCO_3^-] = 5.6$; (b) 77 g K_2CO_3; (c) 1.8 g $KHCO_3$; (d) 2.8×10^2 mL

11.19 (a) pH = 6.3, change of 1.5; (b) pH = 4.58, change of -0.17

11.21

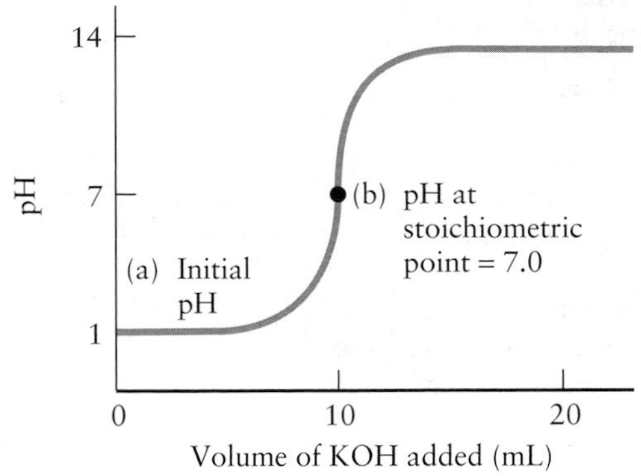

14
7
1
pH
(a) Initial pH
(b) pH at stoichiometric point = 7.0
0 10 20
Volume of KOH added (mL)

11.23 (a) 9.17×10^{-3} L HCl(aq); (b) 0.0183 L; (c) 0.0635 mol·L^{-1}; (d) 2.25
11.25 (a) 246 mL; (b) 0.112 mol·L^{-1}
11.27 84.8%
11.29 (a) 13.04; (b) 12.83; (c) 12.56; (d) 7.00; (e) 1.80 (f) 1.55
11.31 (a) 2.89; (b) 4.56; (c) 12.5 mL; (d) 4.75; (e) 25.0 mL; (f) 8.72
11.33 (a) 11.20; (b) 8.99; (c) 11.25 mL; (d) 9.25; (e) 22.5 mL; (f) 5.24
11.35 Use (d) phenolphthalein, because the pH at the stoichiometric point of the titration is 8.88.
11.37 Exercise 11.31, thymol blue or phenolphthalein; Exercise 11.33, methyl red or bromcresol green
11.39 (a) 37.4 mL; (b) 74.8 mL; (c) 112 mL
11.41 (a) 22.0 mL; (b) 44.0 mL
11.43 (a) 1.80; (b) 2.36; (c) 4.30
11.45 (a) 4.66; (b) 2.80; (c) 7.21
11.47 (a) 7.7×10^{-13}; (b) 1.7×10^{-14}; (c) 5.3×10^{-3}; (d) 6.9×10^{-9}
11.49 (a) 1.2×10^{-17} mol·L^{-1}; (b) 1.1×10^{-18} mol·L^{-1}; (c) 9.3×10^{-5} mol·L^{-1}
11.51 1.0×10^{-12}
11.53 (a) 1.1×10^{-9} mol·L^{-1}; (b) 5.8×10^{-18} mol·L^{-1}; (c) 1.6×10^{-3} mol·L^{-1}; (d) 1.3×10^{-6} mol·L^{-1}
11.55 (a) 1.6×10^{-5} mol·L^{-1}; (b) 2.7×10^{2} μg $AgNO_3$
11.57 (a) 6.00; (b) 6.18
11.59 (a) will precipitate; (b) will not precipitate
11.61 (a) will precipitate; (b) will not precipitate
11.63 (a) first to last: $Ni(OH)_2$, $Mg(OH)_2$, $Ca(OH)_2$; (b) $Ni(OH)_2$ precipitates around pH = 7, $Mg(OH)_2$ around pH = 10, $Ca(OH)_2$ around pH = 13
11.65 carbonate, because the solubility difference is larger
11.67 1.8×10^{-9} mol·L^{-1}
11.69 (a) 1.0×10^{-12} mol·L^{-1}; (b) 3.1×10^{-5} mol·L^{-1}; (c) 2.0×10^{-3} mol·L^{-1}; (d) 2.0×10^{-1} mol·L^{-1}
11.71 (a) $CaF_2(s) + 2\,H_2O(l) \rightleftharpoons Ca^{2+}(aq) + 2\,HF(aq) + 2\,OH^-(aq)$, $K = 3.4 \times 10^{-32}$; (b) 2.2×10^{-4} mol·L^{-1}; (c) 4×10^{-4} mol·L^{-1}
11.73 2.0×10^{-3} mol·L^{-1}
11.75 The two salts can be distinguished by their solubility in 1.00 M NH_3 (or some other suitable concentration). The AgCl will dissolve, but most of the AgI will not.
11.77 To use qualitative analysis, the sample must first be dissolved. This can be accomplished by heating the sample with concentrated HNO_3 and then diluting the resulting solution. HCl or H_2SO_4 could not be used because some of the metal compounds formed would be insoluble, whereas all the nitrates would dissolve. Once the sample is dissolved and diluted, an aqueous solution containing chloride ions can be introduced. The solution should precipitate the Ag^+ as AgCl but leave the bismuth and nickel in solution as long as the solution is acidic. The remaining solution can then be treated with H_2S. In acidic solution, Bi_2S_3 will precipitate but NiS will not. Once the Bi_2S_3 has been precipitated, the pH of the solution can be raised by addition of base. Once this is done, NiS will precipitate.
11.79 (a) 0.62%; (b) 0.44%; (c) 0.31%; (d) The percentage deprotonation decreases with increasing concentration. (e) 6.0
11.81 0.026 M in acetic acid and in sodium acetate
11.83 (a)

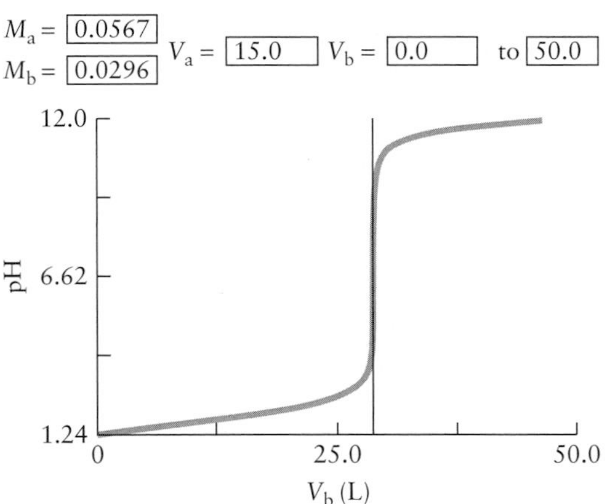

(b) 28.6 mL; (c) 1.24. (d) Because this is a titration of a strong acid with a strong base, the pH at the stoichiometric point will be 7.00.
11.85 pH = 5.05, [HCOOH] = 4.4×10^{-3} mol·L^{-1}
11.87 2.8×10^{-2}
11.89 (a) initial pH = 5.27, after adding HCl, pH = 4.90; (b) 5.47
11.91 The maximum concentration of Ca^{2+} allowed is 0.016 mol·L^{-1}.
11.93 no
11.95 The K_{sp} values in Table 11.6 are Cu^{2+}, 1.3×10^{-36}; Co^{2+} 5×10^{-22}; Cd^{2+}, 4×10^{-29}. All the salts have the same expression for K_{sp}, $K_{sp} = [M^{2+}][S^{2-}]$, so the compound with the smallest K_{sp} will precipitate first, in this case, CuS.
11.97 (a) 6.34; (b) 11.86
11.99 The pH at the first stoichiometric point is 1.54; at the second stoichiometric point, the pH is 7.46.

11.101 (a) A Cl^- concentration of 4.0×10^{-2} mol·L^{-1} will precipitate lead(II) ions. A Cl^- concentration of 1.6×10^{-8} mol·L^{-1} will precipitate Ag^+ ions. (b) Ag^+ ions will precipitate first, as AgCl. (c) When $PbCl_2$ begins to precipitate, the concentration of Ag^+ is 4.0×10^{-9} mol·L^{-1}. (d) 4.0×10^{-5}%
11.103 −345.65 kJ·mol^{-1}
11.105 (a) Cu_2S; (b) 4.5×10^4

Chapter 12

12.1 (a) $4\,Cl_2(g) + S_2O_3^{2-}(aq) + 5\,H_2O(l) \rightarrow 8\,Cl^-(aq) + 2\,SO_4^{2-}(aq) + 10\,H^+(aq)$; Cl_2 is the oxidizing agent and $S_2O_3^{2-}$ is the reducing agent. (b) $2\,MnO_4^-(aq) + H^+(aq) + 5\,H_2SO_3(aq) \rightarrow 2\,Mn^{2+}(aq) + 3\,H_2O(l) + 5\,HSO_4^-$; MnO_4^- is the oxidizing agent and H_2SO_3 is the reducing agent. (c) $Cl_2(g) + H_2S(aq) \rightarrow 2\,Cl^-(aq) + S(s) + 2\,H^+(aq)$; Cl_2 is the oxidizing agent and H_2S is the reducing agent. (d) $H_2O(l) + Cl_2(g) \rightarrow HOCl(aq) + H^+(aq) + Cl^-(aq)$; Cl_2 is both oxidizing agent and reducing agent.
12.3 (a) $3\,O_3(g) + Br^-(aq) \rightarrow 3\,O_2(g) + BrO_3^-(aq)$; O_3 is the oxidizing agent and Br^- is the reducing agent. (b) $6\,Br_2(l) + 12\,OH^-(aq) \rightarrow 10\,Br^-(aq) + 2\,BrO_3^-(aq) + 6\,H_2O(l)$; Br_2 is both oxidizing agent and reducing agent. (c) $2\,Cr^{3+}(aq) + 4\,OH^-(aq) + 3\,MnO_2(s) \rightarrow 2\,CrO_4^{2-}(aq) + 2\,H_2O(l) + 3\,Mn^{2+}(aq)$. Cr^{3+} is the reducing agent and MnO_2 is the oxidizing agent. (d) $P_4(s) + 3\,H_2O(l) + 3\,OH^-(aq) \rightarrow 3\,H_2PO_2^-(aq) + PH_3(g)$. P_4 is both oxidizing agent and reducing agent.
12.5 $MnO_4^-(aq) + 8\,H^+(aq) + 5\,e^- \rightarrow Mn^{2+}(aq) + 4\,H_2O(l)$ and $C_6H_{12}O_6(aq) + 6\,H_2O(l) \rightarrow 6\,CO_2(g) + 24\,H^+(aq) + 24\,e^-$; overall, $24\,MnO_4^-(aq) + 72\,H^+(aq) + 5\,C_6H_{12}O_6(aq) \rightarrow 24\,Mn^{2+}(aq) + 66\,H_2O(l) + 30\,CO_2(g)$
12.7 half reactions:
$P_4S_3(aq) + 28\,H_2O(l) \rightarrow 4\,H_3PO_4(aq) + 3\,SO_4^{2-}(aq) + 44\,H^+(aq) + 38\,e^-$
$NO_3^-(aq) + 4\,H^+(aq) + 3\,e^- \rightarrow NO(aq) + 2\,H_2O(l)$
overall reaction:
$3\,P_4S_3(aq) + 38\,NO_3^-(aq) + 20\,H^+(aq) + 8\,H_2O(l) \rightarrow 12\,H_3PO_4(aq) + 9\,SO_4^{2-}(aq) + 38\,NO(g)$
12.9 (a) anode, $Ni(s) \rightarrow Ni^{2+}(aq) + 2\,e^-$; cathode, $Ag^+(aq) + e^- \rightarrow Ag(s)$; overall, $2\,Ag^+(aq) + Ni(s) \rightarrow Ag(s) + 2\,Ni^{2+}(aq)$ (b) anode, $H_2(g) \rightarrow 2\,H^+(aq) + 2\,e^-$; cathode, $Cl_2(g) + 2\,e^- \rightarrow 2\,Cl^-(aq)$; overall, $Cl_2(g) + H_2(g) \rightarrow 2\,H^+(aq) + 2\,Cl^-(aq)$ (c) anode, $Cu(s) \rightarrow Cu^{2+}(aq) + 2\,e^-$; cathode, $Ce^{4+}(aq) + e^- \rightarrow Ce^{3+}(aq)$; overall, $2\,Ce^{4+}(aq) + Cu(s) \rightarrow 2\,Ce^{3+}(aq) + Cu^{2+}(aq)$ (d) anode, $2\,H_2O(l) \rightarrow O_2(g) + 4\,H^+(aq) + 4\,e^-$; cathode, $O_2(g) + 2\,H_2O(l) + 4\,e^- \rightarrow 4\,OH^-(aq)$; overall, $H_2O(l) \rightarrow H^+(aq) + OH^-(aq)$ (e) anode, $Sn^{2+}(aq) \rightarrow Sn^{4+}(aq) + 2\,e^-$; cathode, $Hg_2Cl_2(s) + 2\,e^- \rightarrow 2\,Hg(l) + 2\,Cl^-(aq)$; overall, $Hg_2Cl_2(s) + Sn^{2+}(aq) \rightarrow 2\,Hg(l) + 2\,Cl^-(aq) + Sn^{4+}(aq)$
12.11 (a) anode, $Zn(s) \rightarrow Zn^{2+}(aq) + 2\,e^-$; cathode, $Ni^{2+}(aq) + 2\,e^- \rightarrow Ni(s)$; overall, $Ni^{2+}(aq) + Zn(s) \rightarrow Ni(s) + Zn^{2+}(aq)$; $Zn(s)|Zn^{2+}(aq)||Ni^{2+}(aq)|Ni(s)$ (b) anode, $2\,I^-(aq) \rightarrow I_2(s) + 2\,e^-$; cathode, $Ce^{4+}(aq) + e^- \rightarrow Ce^{3+}(aq)$; overall, $2\,Ce^{4+}(aq) + 2\,I^-(aq) \rightarrow I_2(s) + 2\,Ce^{3+}(aq)$; $Pt(s)|I^-(aq)|I_2(s)||Ce^{4+}(aq),Ce^{3+}(aq)|Pt(s)$ (c) anode, $H_2(g) \rightarrow 2\,H^+(aq) + 2\,e^-$; cathode, $Cl_2(g) + 2\,e^- \rightarrow 2\,Cl^-(aq)$; overall, $H_2(g) + Cl_2(g) \rightarrow 2\,H^+(aq) + 2\,Cl^-(aq)$; $Pt(s)|H_2(g)|H^+(aq)||Cl^-(aq)|Cl_2(g)|Pt(s)$ (d) anode, $Au(s) \rightarrow Au^{3+}(aq) + 3\,e^-$; cathode, $Au^+(aq) + e^- \rightarrow Au(s)$; overall, $3\,Au^+(aq) \rightarrow 2\,Au(s) + Au^{3+}(aq)$; $Au(s)|Au^{3+}(aq)||Au^+(aq)|Au(s)$
12.13 (a) anode, $Ag(s) + Br^-(aq) \rightarrow AgBr(s) + e^-$; cathode, $Ag^+(aq) + e^- \rightarrow Ag(s)$; $Ag(s)|AgBr(s)|Br^-(aq)||Ag^+(aq)|Ag(s)$ (b) anode, $4\,OH^-(aq) \rightarrow O_2(g) + 2\,H_2O(l) + 4\,e^-$; cathode, $O_2(g) + 4\,H^+(aq) + 4\,e^- \rightarrow 2\,H_2O(l)$; $Pt(s)|O_2(g)|OH^-(aq)||H^+(aq)|O_2(g)|Pt(s)$ (c) anode, $Cd(s) + 2\,OH^-(aq) \rightarrow Cd(OH)_2(s) + 2\,e^-$; cathode, $Ni(OH)_3(s) + e^- \rightarrow Ni(OH)_2(s) + OH^-(aq)$; $Cd(s)|Cd(OH)_2(s)|KOH(aq)||Ni(OH)_3(s)|Ni(OH)_2(s)|Ni(s)$
12.15 (a) anode, $Fe^{2+}(aq) \rightarrow Fe^{3+}(aq) + e^-$; cathode, $MnO_4^-(aq) + 8\,H^+(aq) + 5\,e^- \rightarrow Mn^{2+}(aq) + 4\,H_2O(l)$ (b) $MnO_4^-(aq) + 5\,Fe^{2+}(aq) + 8\,H^+(aq) \rightarrow Mn^{2+}(aq) + 5\,Fe^{3+}(aq) + 4\,H_2O(l)$; $Pt(s)|Fe^{3+}(aq),Fe^{2+}(aq)||H^+(aq),MnO_4^-(aq),Mn^{2+}(aq)|Pt(s)$
12.17 (a) +0.75V; (b) +0.37V; (c) +0.52V; (d) +1.52 V
12.19 (a) +0.75 V, −145 kJ·mol^{-1}; (b) +0.37 V, −36 kJ·mol^{-1}; (c) +0.52 V, −100 kJ·mol^{-1}; (d) +1.52 V, −293 kJ·mol^{-1}
12.21 −0.91 V
12.23 (a) Cu < Fe < Zn < Cr; (b) Mg < Na < K < Li; (c) V < Ti < Al < U; (d) Au < Ag < Sn < Ni
12.25 (a) Co^{2+} is the oxidizing agent and Ti^{2+} is the reducing agent; +0.09 V; $Pt(s)|Ti^{3+}(aq),Ti^{2+}(aq)||Co^{2+}(aq)|Co(s)$; (b) U^{3+} is the oxidizing agent and La is the reducing agent; +0.73 V; $La(s)|La^{3+}(aq)||U^{3+}(aq)|U(s)$; (c) Fe^{3+} is the oxidizing agent and H_2 is the reducing agent; +0.77 V; $Pt(s)|H_2(g)|H^+(aq)||Fe^{3+}(aq),Fe^{2+}(aq)|Pt(s)$; (d) O_3 is the oxidizing agent and Ag is the reducing agent; +0.44 V; $Ag(s)|Ag^+(aq)||OH^-(aq)|O_3(g),O_2(g)|Pt(s)$
12.27 (a) no; (b) yes, +0.61 V; $3\,Pb^{2+}(aq) + 2\,Cr(s) \rightarrow 3\,Pb(s) + 2\,Cr^{3+}(aq)$; (c) yes, +1.17 V; $5\,Cu(s) + 2\,MnO_4^-(aq) + 16\,H^+(aq) \rightarrow 5\,Cu^{2+}(aq) + 2\,Mn^{2+}(aq) + 8\,H_2O(l)$; (d) no
12.29 (a) $Cl_2(g)$ is the oxidizing agent, $E° = +0.27$ V; (b) and (c) are not spontaneous as written; (d) NO_3^- is the oxidizing agent, $E° = +1.56$ V.
12.31 Yes, because $E°(O_2, H^+, H_2O/H_2) > E°(Br_2, Br^-)$. It is not used because that reaction is so much slower than the one with Cl_2.

12.33 (a) $3\ Au^+ \rightarrow 2\ Au + Au^{3+}$; (b) yes
12.35 -1.50 V
12.37 (a) 10^{25}; (b) 10^{-15}; (c) 10^{31}; (d) 10^2
12.39 yes; $K = 10^2$
12.41 (a) 10^6; (b) 1.0
12.43 (a) $+0.030$ V; (b) $+0.06$ V
12.45 (a) $+0.067$ V; (b) $+0.51$ V; (c) -1.33 V; (d) $+0.31$ V
12.47 (a) pH = 1; (b) $[Cl^-] = 0.1\ mol{\cdot}L^{-1}$
12.49 (a) 1.2×10^{-17}; (b) The value is a factor of 10 greater than the tabulated value.
12.51 (a) -0.27 V; (b) $+0.07$ V
12.53 Comparison of the standard potentials shows that Cr is more easily oxidized than Fe, so the presence of Cr retards the rusting of Fe.
12.55 (a) $Fe_2O_3{\cdot}H_2O$; (b) H_2O and O_2 jointly oxidize iron. (c) Water is more highly conducting if it contains dissolved ions, so the rate of rusting is increased.
12.57 (a) aluminum or magnesium; (b) cost, availability, and toxicity of products in the environment; (c) Fe can act as the anode of an electrochemical cell if Cu^{2+} or Cu^+ are present; hence, it would be oxidized at the point of contact. Water with dissolved ions acts as the electrolyte.
12.59 (a) $Ni^{2+}(aq) + 2\ e^- \rightarrow Ni(s)$; (b) $2\ H_2O(l) \rightarrow O_2(g) + 4\ H^+(aq) + 4\ e^-$; (c) $E_{supplied} = +1.04$ V
12.61 (a) water; (b) water; (c) Ni^{2+}; (d) Au^{3+}
12.63 (a) 3.3 g; (b) 0.59 L; (c) 0.94 g
12.65 (a) 14 h; (b) 0.29 g Cu
12.67 (a) 0.52 A; (b) 0.19 A
12.69 +2
12.71 +4
12.73 $52.0\ g{\cdot}mol^{-1}$
12.75 (a) a moist paste of KOH(aq) and HgO(s); (b) HgO(s); (c) $HgO(s) + Zn(s) \rightarrow Hg(l) + ZnO(s)$
12.77 The reaction at the anode, $Zn(s) \rightarrow Zn^{2+}(aq) + 2\ e^-$, supplies the electrons to the external circuit. The cathode reaction is $MnO_2(s) + H_2O(l) + e^- \rightarrow MnO(OH)_2(s) + OH^-(aq)$. The $OH^-(aq)$ produced reacts with $NH_4^+(aq)$ from the $NH_4Cl(aq)$: $NH_4^+(aq) + OH^-(aq) \rightarrow H_2O(l) + NH_3(g)$. The $NH_3(g)$ produced complexes with the $Zn^{2+}(aq)$ produced in the anode reaction: $Zn^{2+}(aq) + 4\ NH_3(g) \rightarrow Zn(NH_3)_4^{2+}(aq)$.
12.79 (a) KOH(aq); (b) $Ni(OH)_2(s) + OH^-(aq) \rightarrow Ni(OH)_3(s) + e^-$
12.81 -0.92 V
12.83 $+0.14$ V
12.85 (a) yes, $O_2(g) + 4H^+(aq) + 4Ag(s) \rightarrow 4Ag^+(aq) + 2H_2O(l)$, $+0.42$ V; (b) yes, $O_2(g) + 2H_2O(l) + 4Ag(s) \rightarrow 4Ag^+(aq) + 4\ OH^-(aq)$, $+ 0.05$ V
12.87 (a) $Hg(l)|Hg_2^{2+}(aq)||NO_3^-(aq), H^+(aq)|NO(g)|Pt(s)$; $+0.17$ V, $-98\ kJ{\cdot}mol^{-1}$; (b) not spontaneous; (c) $Pt(s)|Pu^{4+}(aq), Pu^{3+}(aq)||Cr_2O_7^{2-}(aq), Cr^{3+}(aq), H^+(aq)|Pt(s)$; $+0.36$ V, $-208\ kJ{\cdot}mol^{-1}$
12.89 10^{-3}
12.91 $+0.08$ V to $+0.09$ V
12.93 $E° = +0.03$ V; $E = -0.27$ V. The cell changes from spontaneous to nonspontaneous when the concentrations are changed from standard values
12.95 (a) $+1.08$ V; (b) $O_2(g) + 4\ H^+(aq) + 2\ Sn^{2+}(aq) \rightarrow 2\ Sn^{4+}(aq) + 2\ H_2O(l)$; (c) $+ 0.87$ V; (d) 6×10^{60}; (e) $-3.4 \times 10^2\ kJ{\cdot}mol^{-1}$; (f) 3.72
12.97 12
12.99 Copper cannot reduce Ni^{2+}, Co^{2+}, and Fe^{2+}, thus leaving Au and Ag as "anode mud."
12.101 (a) $E'(1) = +1.23\ V - (0.059\ 16)(pH)$, $E'(2) = +0.36\ V + (0.059\ 16)(pOH) = +1.19\ V - (0.059\ 16)(pH)$; (b) $E'(1) = 0.82$ V; $E'(2) = 0.77$ V; potentials should be identical, but differ because of the limited number of significant figures used to derive and evaluate the equations.
12.103 (a) reduction half-reactions: $Fe^{2+} + 2\ e^- \rightarrow Fe$, $Mn^{2+} + 2\ e^- \rightarrow Mn$; half-reaction for oxidation of water: $O_2 + 4\ H^+ + 4\ e^- \rightarrow 2\ H_2O$; (b) for Fe: $+1.39$ V; for Mn: $+2.13$ V; (c) Fe; (d) yes
12.105 (a) The first electron added to any molecule should enter the molecule's lowest unoccupied molecular orbital (LUMO). (b) For CH_3X, the LUMO is the antibonding orbital between C and one of the atoms attached to it. Because the C—H bond strength ($412\ kJ{\cdot}mol^{-1}$) is greater than all of the C—X bond strengths given (C—Cl, $338\ kJ{\cdot}mol^{-1}$; C—Br, $276\ kJ{\cdot}mol^{-1}$; C—I, $238\ kJ{\cdot}mol^{-1}$), we expect the LUMO to be the antibonding orbital for the C—X bond. Adding an electron to this orbital should then result in a weakening of the C—X bond. The result is elimination of X^- and formation of a CH_3 radical:

$$CH_3X + e^- \rightarrow {\cdot}CH_3 + X^-$$

(c) We expect the ease of this reduction process to follow the C—X bond strengths, so the formation of X^- and the generation of $CH_3{\cdot}$ radicals would be easiest for X = I, followed by Br, then Cl.

Chapter 13

13.1 (a) one-third; (b) two-thirds; (c) two
13.3 (a) $0.28\ (mol\ CrO_4^{2-}){\cdot}L^{-1}{\cdot}s^{-1}$; (b) $0.14\ mol{\cdot}L^{-1}{\cdot}s^{-1}$
13.5 (a) $3.3 \times 10^{-3}\ (mol\ O_2){\cdot}L^{-1}{\cdot}s^{-1}$; (b) $3.3 \times 10^{-3}\ mol{\cdot}L^{-1}{\cdot}s^{-1}$
13.7 (a) and (c)

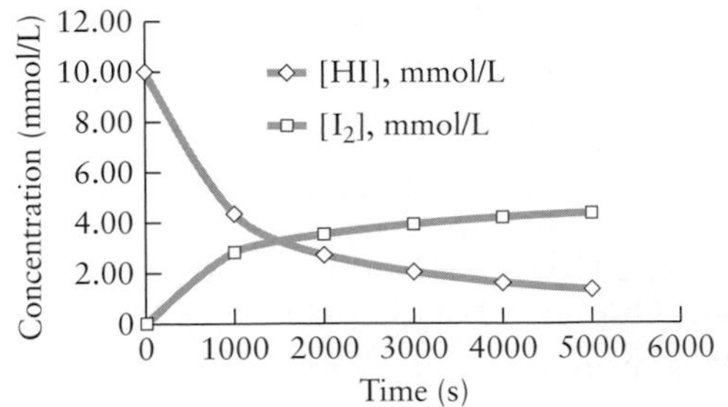

Note that the curves for $[I_2]$ and $[H_2]$ are identical and only the $[I_2]$ curve is shown.

(b)

Time, s	Rate, $mmol \cdot L^{-1} \cdot s^{-1}$
0	−0.0060
1000	−0.003
2000	−0.000 98
3000	−0.000 61
4000	−0.000 40
5000	−0.000 31

13.9 (a) $mol \cdot L^{-1} \cdot s^{-1}$; (b) s^{-1}; (c) $L \cdot mol^{-1} \cdot s^{-1}$
13.11 2.2×10^{-4} $(mol\ N_2O_5) \cdot L^{-1} \cdot s^{-1}$
13.13 (a) 1.6×10^{-5} $mol \cdot L^{-1} \cdot s^{-1}$; (b) factor of 2
13.15 rate $= k[CH_3Br][OH^-]$
13.17 (a) rate $= k[ICl][H_2]$; (b) 0.16 $L \cdot mol^{-1} \cdot s^{-1}$; (c) 2.0×10^{-6} $mol \cdot L^{-1} \cdot s^{-1}$
13.19 (a) rate $= k[A][B]^2[C]^2$; (b) overall order is 5; (c) 2.85×10^{12} $L^4 \cdot mol^{-4} \cdot s^{-1}$; (d) 1.13×10^{-2} $mol \cdot L^{-1} \cdot s^{-1}$
13.21 (a) 6.93×10^{-4} s^{-1}; (b) 9.4×10^{-3} s^{-1}; (c) 5.1×10^{-3} s^{-1}
13.23 (a) 5.2 h; (b) 3.5×10^{-2} $mol \cdot L^{-1}$; (c) 6.5×10^2 min
13.25 (a) 710 s; (b) 9.7×10^2 s; (c) 1.1×10^3 s
13.27 (a) 247 min; (b) 819 min; (c) 10.9 g
13.29 (a) $k = 0.17$ min^{-1}; (b) an additional 3.5 min
13.31 (a)

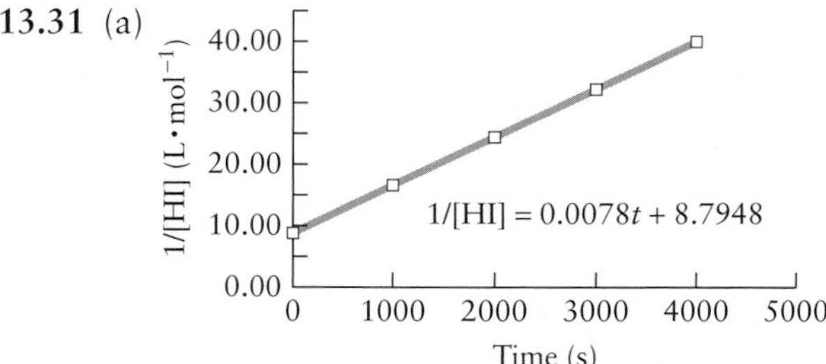

(b) $k = 7.8 \times 10^{-3}$ $L \cdot mol^{-1} \cdot s^{-1}$
13.33 Using the equations modified from Eq. 17b yields $k = 1/(t_{1/2}[A]_0)$ and $t = \{(1/[A]) - (1/[A]_0)\}/k$. Then (a) 7.4×10^2 s; (b) 1.5×10^2 s; (c) 2.0×10^2 s
13.35 Using the equations from Exercise 13.33 yields (a) 1.7×10^2 min; (b) 3.3×10^3 min
13.37 (a)

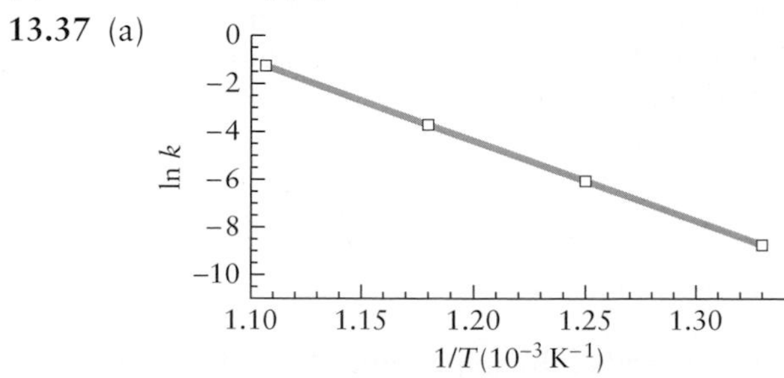

2.7×10^2 $kJ \cdot mol^{-1}$; (b) $k = 8.8 \times 10^{-2}$ s^{-1}
13.39 40 $kJ \cdot mol^{-1}$
13.41 2.5×10^9 $L \cdot mol^{-1} \cdot s^{-1}$
13.43 9.2×10^{-4} s^{-1}
13.45 The overall reaction is $CH_2{=}CHCOOH + HCl \rightarrow ClCH_2CH_2COOH$. The intermediates include Cl^-, $CH_2{=}CHC(OH)_2^+$, and $ClCH_2CHC(OH)_2$.
13.47 The first elementary reaction is the rate-determining step because it is the slow step. The second elementary reaction is fast and does not affect the overall reaction order. Rate $= k[NO][Br_2]$.
13.49 rate $= k_3(k_2/k_2')(k_1/k_1')^{1/2}[CO][Cl_2]^{3/2} = k[CO][Cl_2]^{3/2}$
13.51 If mechanism (a) were correct, the rate law would be rate $= k[NO_2][CO]$. This expression does not agree with the experimental result and can be eliminated as a possibility. Mechanism (b) has rate $= k[NO_2]^2$ from the slow step. Step 2 does not influence the overall rate, but it is necessary to achieve the correct overall reaction; thus this mechanism agrees with the experimental data. Mechanism (c) is not correct, as seen from the rate expression for the slow step, rate $= k[NO_3][CO]$; the experimental expression does not contain [CO].
13.53 (a) true; (b) false; at equilibrium, the *rates* of the forward and reverse reactions are equal, not the rate *constants*; (c) true; (d) false; increasing the concentration of a reactant causes the rate to increase by providing more reacting molecules. It does not affect the rate constant of the reaction.
13.55 (a) 0.676; (b) endothermic; (c) Raising the temperature will increase the rate constant of the reaction with the higher activation barrier relatively more than that of the reaction with the lower energy barrier. In this case, we expect the rate of the forward reaction to go up substantially more than that for the reverse reaction. k will increase more than k', and, consequently, the equilibrium constant K will increase.
13.57 (a) 6×10^8; (b) 3×10^7
13.59 $E_{a,cat} = 81$ $kJ \cdot mol^{-1}$
13.61 balanced equation: $CH_3OH(aq) + HBr(aq) \rightarrow CH_3Br(aq) + H_2O(aq)$; intermediates: $CH_3OH_2^+$, CH_3^+
13.63 (a) false; a catalyst increases the rate of both the forward and reverse reactions by providing a completely different pathway; (b) true, although a catalyst may be poisoned and lose activity; (c) false; there is a completely different pathway provided for the reaction in the presence of a catalyst; (d) false; the position of the equilibrium is unaffected by the presence of a catalyst.
13.65 first order in C_4H_9Br, zero order in OH^-, first order overall
13.67 (a) rate law: rate $= k[H_2SeO_3][I^-]^3[H^+]^2$; $k = 5 \times 10^5$ $L^5 \cdot mol^{-5} \cdot s^{-1}$; (b) rate $= 3.2 \times 10^{-5}$ $mol \cdot L^{-1} \cdot s^{-1}$
13.69 (a) 15 $L \cdot mol^{-1} \cdot h^{-1}$; (b) 10 $L \cdot mol^{-1} \cdot h^{-1}$ (1 sf)

13.71

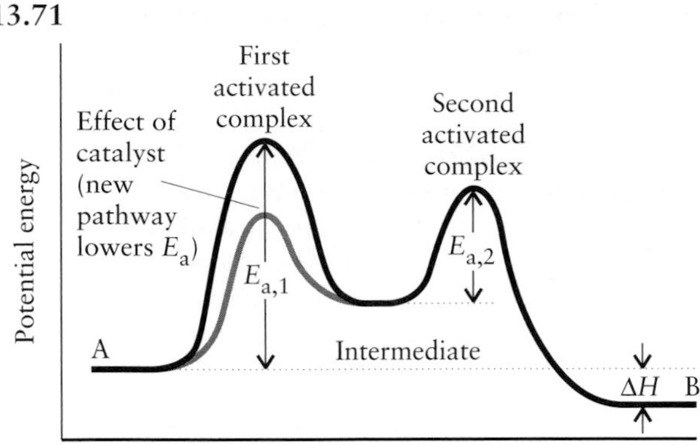

13.73 The overall reaction is $RCN + H_2O \rightarrow RC(=O)NH_2$. The intermediates include

$$R-\underset{\displaystyle OH}{\underset{|}{C}}=N^- \quad \text{and} \quad R-\underset{\displaystyle OH}{\underset{|}{C}}=N-H$$

The hydroxide ion serves as a catalyst for the reaction.
13.75 5.4 mg
13.77 (a) If step 2 is the slow step, if step 1 is a rapid equilibrium, and if step 3 is fast also, then our proposed rate law will be rate $= k_2[N_2O_2][H_2]$. Consider the equilibrium of step 1:

$$k_1[NO]^2 = k_1'[N_2O_2]$$

$$[N_2O_2] = \frac{k}{k_1'}[NO]^2$$

Substituting in our proposed rate law, we have

$$\text{Rate} = k^2\left(\frac{k}{k_1'}\right)[NO]^2[H_2] = k[NO]^2[H_2]$$

where $k = k_2(k_1/k_1')$. The assumptions made above reproduce the observed rate law; therefore, step 2 is the slow step and rate determining.
(b)

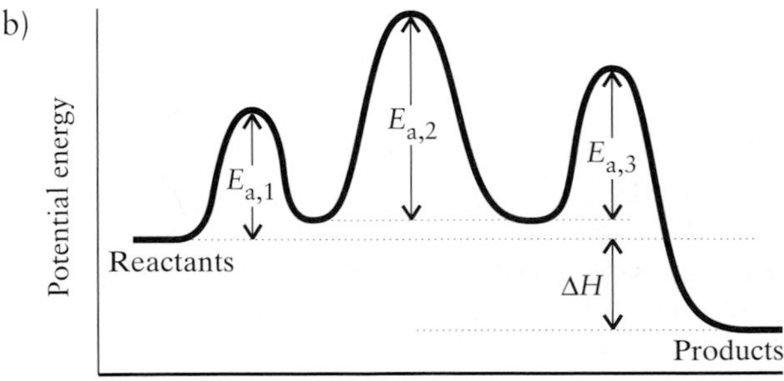

13.79 75 $kJ\cdot mol^{-1}$
13.81 (a) ClO is the reaction intermediate and Cl is the catalyst; (b) Cl, ClO, O, O_2; (c) step 1 is propagating; step 2 is also propagating; (d) $Cl + Cl \rightarrow Cl_2$
13.83 (a) rate(0 min) $= 1.01 \times 10^{-3}\ mol\cdot L^{-1}\cdot min^{-1}$; rate(5 min) $= 6.07 \times 10^{-4}\ mol\cdot L^{-1}\cdot min^{-1}$; rate(10 min) $= 3.65 \times 10^{-4}\ mol\cdot L^{-1}\cdot min^{-1}$; rate(15 min) $= 2.2 \times 10^{-4}\ mol\cdot L^{-1}\cdot min^{-1}$; rate(20 min) $= 1.3 \times 10^{-4}\ mol\cdot L^{-1}\cdot min^{-1}$; rate(30 min) $= 4.8 \times 10^{-5}\ mol\cdot L^{-1}\cdot min^{-1}$; (b) first order, $k = 0.101\ min^{-1}$
13.85 For a third-order reaction, $t_{1/2} = \text{constant}/[A_0]^2$. (a) For the first half-life, $t_{1/2} = \text{constant}/[A_0]^2 = t_1$. (b) The second half-life, $t_{1/4} = t_2 = 4t_1$, so the total time is $t_1 + t_2 = t_1 + 4t_1 = 5t_1$. (c) $t_{1/16}$ requires $t_3 = 16t_1$ and $t_4 = 64t_1$, so $t_{1/16} = 85t_1$ by the same process as in (b).
13.87 Consider, for example, $A \rightarrow B$. Assume that the reaction is first order in both directions: rate(forward) $= k[A]$ and rate(reverse) $= k'[B]$. At equilibrium, rate(forward) = rate(reverse), so $k[A] = k'[B]$, or $k[A] = k'([A]_0 - [A])$ or $(k + k')[A] = k'[A]_0$ or $[A] = (k'/(k + k'))[A]_0$, which is not 0. The reactant concentration drops to its equilibrium value and remains at that value.
13.89 (a) initiation: $Cl_2 \rightarrow Cl + Cl$; propagation: $Cl + CH_4 \rightarrow CH_3Cl + H$, $H + Cl_2 \rightarrow HCl + Cl$; termination $Cl + Cl \rightarrow Cl_2$, $H + H \rightarrow H_2$, $H + Cl \rightarrow HCl$. (b) CH_3Cl and HCl
13.91 $\Delta H_r^\circ = +9.8\ kJ\cdot mol^{-1}$; $\Delta S_r^\circ = -15\ J\cdot K^{-1}\cdot mol^{-1}$
13.93 The rate is determined by the surface area of the catalyst and not by the concentrations or pressures of reactants.
13.95 (a) The overall reaction is $ClO^- + I^- \rightarrow IO^- + Cl^-$; (b) rate $= \dfrac{k_2k_1}{k_1'}\dfrac{[ClO^-][I^-]}{[OH^-]}$; (c) yes; (d) If the reaction is carried out in an organic solvent, then H_2O is no longer the solvent and its concentration must be included in calculating the equilibrium concentration of HClO: rate $= \dfrac{k_2k_1}{k_1'}\dfrac{[ClO^-][I^-][H_2O]}{[OH^-]}$
The rate of reaction will then show a dependence on the concentration of water, which is obscured when the reaction is carried out with water as the solvent.

Chapter 14

14.1 (a) carbon; (b) lithium; (c) indium; (d) iodine
14.3 (a) sulfur; (b) selenium; (c) sodium; (d) oxygen
14.5 iodine < bromine < chlorine
14.7 chlorine
14.9 antimony
14.11 (a) KCl; (b) Na_2O; (c) thallium(I) chloride
14.13 $2\ K(s) + H_2(g) \rightarrow 2\ KH(s)$
14.15 (a) saline; (b) molecular; (c) molecular; (d) metallic
14.17 (a) acidic; (b) amphoteric; (c) acidic; (d) basic
14.19 (a) CO_2; (b) B_2O_3
14.21 (a) $C_2H_2(g) + H_2(g) \rightarrow H_2C{=}CH_2(g)$. Oxidation

number of C in $C_2H_2 = -1$; of C in $H_2C{=}CH_2 = -2$, carbon has been reduced (b) $CO(g) + H_2O(g) \rightarrow CO_2(g) + H_2(g)$ (c) $BaH_2(s) + 2\ H_2O(l) \rightarrow Ba(OH)_2 + 2\ H_2(g)$

14.23 (a) +206.10 $kJ{\cdot}mol^{-1}$; (b) +214.62 $J{\cdot}K^{-1}{\cdot}mol^{-1}$; (c) +142.12 $kJ{\cdot}mol^{-1}$

14.25 (a) 10.6 L H_2; (b) 8.56 mL H_2O

14.27 (a) $H_2(g) + Cl_2(g) \xrightarrow{\text{light}} 2\ HCl(g)$

(b) $H_2(g) + 2\ Na(l) \xrightarrow{\Delta} 2\ NaH(s)$

(c) $P_4(s) + 6\ H_2(g) \rightarrow 4\ PH_3(g)$

(d) $Cu(s) + H_2(g) \rightarrow$ no reaction

14.29 (a) 1.23 V; (b) The difficulty is isolating the two half-cells while maintaining electrical contact. Ions need to flow through the system to maintain charge balance in the reaction. In this case, a material that allows hydrogen ions but not hydrogen gas or oxygen gas to pass through would be necessary.

14.31 Lithium exhibits the diagonal relationship that is common to many first members of a group. Li is similar in many of its compounds to the compounds of Mg. This is related to the small ionic radius of Li^+, 58 pm, which is close to the ionic radius of Mg^{2+}, 72 pm, but substantially less than that of Na^+, 102 pm. Lithium is the only Group 1 element that reacts directly with nitrogen. It forms lithium nitride:

$$6\ Li(s) + N_2(g) \rightarrow 2\ Li_3N(s)$$

Lithium reacts with oxygen to form mainly the oxide:

$$4\ Li(s) + O_2(g) \rightarrow 2\ Li_2O(s)$$

The other members of the group form mainly the peroxide or superoxide.

14.33 (a) $4\ Li(s) + O_2(g) \rightarrow 2\ Li_2O(s)$

(b) $6\ Li(s) + N_2(g) \rightarrow 2\ Li_3N(s)$

(c) $2\ Na(s) + 2\ H_2O(l) \rightarrow 2\ NaOH(aq) + H_2(g)$

(d) $4\ KO_2(s) + 2\ H_2O(g) \rightarrow 4\ KOH(s) + 3\ O_2(g)$

14.35 19.3 g $Na_2CO_3{\cdot}10H_2O$

14.37 $Mg(s) + 2\ H_2O(l) \rightarrow Mg(OH)_2(s) + H_2(g)$

14.39 (a) $CaO(s) + H_2O(l) \rightarrow Ca(OH)_2(s)$; (b) −57.33 $kJ{\cdot}mol^{-1}$

14.41 Be is the weakest reducing agent; Mg is stronger, but weaker than the remaining members of the group, all of which have approximately the same reducing strength. This trend is related to the very small radius of the Be^{2+} ion, 27 pm; its strong polarizing power introduces much covalent character into its compounds. Thus, Be attracts electrons more strongly and does not release them as readily as the other members of the group. Mg^{2+} is also a small ion, 72 pm, so the same reasoning applies to it also, but to a lesser extent. The remaining ions of the group are considerably larger, release electrons more readily, and are better reducing agents.

14.43 $2\ Al(s) + 2\ OH^-(aq) + 6\ H_2O(l) \rightarrow 2\ Al(OH)_4^-(aq) + 3\ H_2(g)$

$Be(s) + 2\ OH^-(aq) + 2\ H_2O(l) \rightarrow Be(OH)_4^{2-}(aq) + H_2(g)$

The similarity of these two reactions demonstrates the diagonal relationship between Be and Al.

14.45 (a) $Mg(OH)_2(s) + 2\ HCl(aq) \rightarrow MgCl_2(aq) + 2\ H_2O(l)$

(b) $Ca(s) + 2\ H_2O(l) \rightarrow Ca(OH)_2(aq) + H_2(g)$

(c) $BaCO_3(s) \xrightarrow{\Delta} BaO(s) + CO_2(g)$

14.47 (a) $:\ddot{Cl}{-}Be{-}\ddot{Cl}:$ $Mg^{2+}\ 2\left[:\ddot{Cl}:\right]^-$; $MgCl_2$ is ionic and $BeCl_2$ is a molecular compound. (b) 180°; (c) *sp*

14.49 (a) $\Delta H_r^\circ = +178.3\ kJ{\cdot}mol^{-1}$; $\Delta S_r^\circ = +160.6\ J{\cdot}K^{-1}{\cdot}mol^{-1}$; (b) 1110 K

14.51 $4\ Al^{3+}\text{(melt)} + 6\ O_2^-\text{(melt)} + 3\ C(s, gr) \rightarrow 4\ Al(s) + 3\ CO_2(g)$

14.53 (a) $B_2O_3(s) + 3\ Mg(l) \rightarrow 2\ B(s) + 3\ MgO(s)$

(b) $2\ Al(s) + 3\ Cl_2(g) \rightarrow 2\ AlCl_3(s)$

(c) $4\ Al(s) + 3\ O_2(g) \rightarrow 2\ Al_2O_3(s)$

14.55 (a) The hydrate of $AlCl_3$, $AlCl_3{\cdot}6H_2O$, functions as a deodorant and antiperspirant. (b) Corundum, α-alumina, is used as an abrasive in sandpaper. (c) $B(OH)_3$ is an antiseptic and insecticide.

14.57 2.8×10^7 g of Al

14.59 (a) +0.743 V; (b) no

14.61 Silicon occurs widely in the Earth's crust in the form of silicates in rocks and as silicon dioxide in sand. It is obtained from quartzite, a form of quartz, by the following processes: (1) reduction in an electric arc furnace: $SiO_2(s) + 2\ C(s) \rightarrow Si(s, crude) + 2\ CO(g)$; (2) purification of the crude product in two steps: $Si(s, crude) + 2\ Cl_2(g) \rightarrow SiCl_4(l)$, followed by reduction with hydrogen to the pure element, $SiCl_4(l) + 2\ H_2(g) \rightarrow Si(s, pure) + 4\ HCl(g)$.

14.63 In diamond, carbon is sp^3 hybridized and forms a tetrahedral, three-dimensional network structure, which is very rigid. Graphite carbon is sp^2 hybridized and planar, and its application as a lubricant results from the fact that the two-dimensional sheets can "slide" across one another, thereby reducing friction. In graphite, electrons in the unhybridized *p*-orbitals are free to move from one carbon atom to another, a property resulting in its high electrical conductivity. In diamond, all electrons are localized in sp^3 hybridized C—C σ-bonds, so diamond does not conduct electricity well.

14.65 (a) $SiCl_4(l) + 2\ H_2(g) \rightarrow Si(s) + 4\ HCl(g)$

(b) $SiO_2(s) + 3\ C(s) \xrightarrow{2000^\circ C} SiC(s) + 2\ CO(g)$

(c) $Ge(s) + 2\ F_2(g) \rightarrow GeF_4(s)$

(d) $CaC_2(s) + 2\ H_2O(l) \rightarrow Ca(OH)_2(s) + C_2H_2(g)$

14.67 $\left[SiO_4 \right]^{4-}$: Si bonded to four :Ö: atoms (Lewis structure)

formal charges: Si = 0, O = −1; oxidation numbers: Si = +4, O = −2; This is an AX_4 VSEPR structure; therefore the shape is tetrahedral.

14.69 $\Delta H_r° = +689.88\ kJ\cdot mol^{-1}$; $\Delta S_r° = +360.85\ J\cdot K^{-1}\cdot mol^{-1}$; $\Delta G_r° = +582.29\ kJ\cdot mol^{-1}$. The equilibrium constant becomes greater than 1 at 1912 K.
14.71 4.00 mg of HF
14.73 (a) The $Si_2O_7^{6-}$ ion is built from two SiO_4^{4-} tetrahedral ions in which the silicate tetrahedra share one O atom. See **11.** (b) Pyroxenes (e.g., jade) consist of chains of SiO_4 units in which two O atoms are shared by neighboring units. The repeating unit has the formula SiO_3^{2-}. See Fig. 14.45.
14.75 SiF_6^{2-}
14.77 Ionic fluorides react with water to liberate HF, which then reacts with the glass. Glass bottles used to store metal fluorides become brittle and may disintegrate upon standing on the shelf.
14.79 The iron ions impart a deep red color to the clay that is not desirable in the manufacture of fine china, for which a white base is aesthetically more pleasing.
14.81 In the majority of its reactions, hydrogen acts as a reducing agent. Examples are $2\ H_2(g) + O_2(g) \rightarrow 2\ H_2O(l)$ and various reduction processes, such as $NiO(s) + H_2(g) \rightarrow Ni(s) + H_2O(g)$. With highly electropositive elements, such as the alkali metals, $H_2(g)$ acts as an oxidizing agent and forms metal hydrides, for example, $2\ K(s) + H_2(g) \rightarrow 2\ KH(s)$.
14.83 (a)

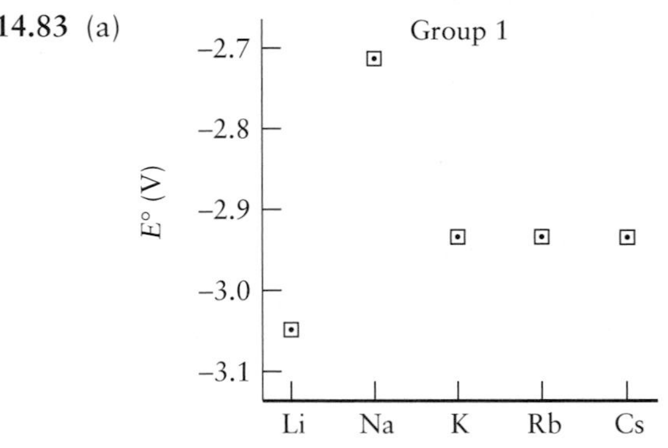

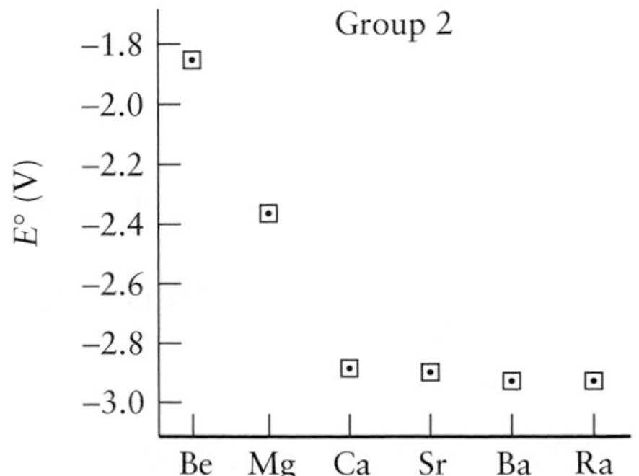

(b) For both groups, the trend in standard potentials with increasing atomic number is overall a downward trend (they become more negative), but lithium is anomalous. This overall downward trend makes sense, because, theoretically, it is easier to remove electrons that are farther away from the nuclei. However, because there are several factors that influence ease of removal, the trend is not smooth. The potentials are a net composite of the free energies of sublimation of solids, dissociation of gaseous molecules, ionization enthalpies, and enthalpies of hydration of gaseous ions. The origin of the anomalously strong reducing power of Li is the strongly exothermic energy of hydration of the very small Li^+ ion, which favors the ionization of the element in aqueous solution.
14.85 (a, b) $H_2(g) + F_2(g) \rightarrow 2\ HF(g)$, explosive; $H_2(g) + Cl_2(g) \rightarrow 2\ HCl(g)$, explosive; $H_2(g) + Br_2(g) \rightarrow 2\ HBr(g)$, vigorous; $H_2(g) + I_2(g) \rightarrow 2\ HI(g)$, less vigorous
(c) hydrofluoric acid, hydrochloric acid, hydrobromic acid, hydroiodic acid
14.87 (a) Lewis acid-base reaction; (b) $CaO(s) + CO_2(g) \rightarrow CaCO_3(s)$
14.89 (a) $B(OH)_4^-$; (b) $B(OH)_3(aq) + 2\ H_2O(l) \rightarrow H_3O^+(aq) + B(OH)_4^-(aq)$
14.91 (a) Ionization energies decrease (become less positive) down Groups 13/III (B → Tl) and 14/IV (C → Pb). Atomic radii increase down Groups 13/III (B → Tl) and 14/IV (C → Pb). (b) The ionization energies generally decrease down a group. As the atomic number of an element increases, atomic shells and subshells that are farther from the nucleus are filled. The outermost valence electrons are consequently easier to remove. The radii increase down a group for the same reason. The radii are primarily determined by the outer shell electrons, which are farther from the nucleus in the heavy elements. (c) The trends correlate well with elemental properties; for example, the greater ease of outermost electron removal correlates with increased metallic character, that is, ability to form positive ions by losing one or more electrons.
14.93 (a) $2\ H_2O(l) + 2\ e^- \rightarrow H_2(g) + 2\ OH^-(aq)$; (b) cathode; (c) 2.1 L H_2
14.95 In the majority of its reactions, hydrogen acts as a reducing agent, like the alkali metals. However, as described in the text and in the answer to Exercise 14.81, it may also act as an oxidizing agent, like the halogens. Consequently, H_2 will oxidize elements with standard reduction potentials more negative than −2.25 V, such as the alkali and alkaline earth metals (except Be). The compounds formed are hydrides and contain the H^- ion; the singly charged negative ion is reminiscent of the halide ions. Hydrogen also forms diatomic molecules and covalent bonds like the halogens. Consequently, hydrogen could be placed in either Group 1 or Group 17/VII. But it is probably best to think of hydrogen as a unique element that has properties in common with both metals and nonmetals; therefore, it should probably be centered in the periodic table as indicated in the text.

14.97 (a) (polarizing ability × 1000 follows the ion) $Cs^+(5.9) < Rb^+(6.7) < K^+(7.2) < Na^+(10) < Li^+(17)$ and $Ba^{2+}(15) < Sr^{2+}(17) < Ca^{2+}(20) < Mg^{2+}(28) < Be^{2+}(74)$; (b) These data weakly support the diagonal relationship. Li^+ is more like Mg^{2+} than Be^{2+} and Na^+ more like Ca^{2+} than Mg^{2+}; but further down the group, the correlation fails.
14.99 0.0713 $mol{\cdot}L^{-1}$
14.101 The smaller the cation, the greater the ability of the cation to polarize and weaken the carbonate ion. On that basis, we would predict that within a group the carbonates of the first members of the group are less stable than those of the later members. Thus $Li_2CO_3 < Na_2CO_3 < K_2CO_3 < Rb_2CO_3 < Cs_2CO_3$ and $BeCO_3 < MgCO_3 < CaCO_3 < SrCO_3 < BaCO_3$. Between groups, we would expect the stability of the carbonates in one period to decrease from Group 1 to Group 13/III because of the smaller size of Group 13/III ions. Thus $Al_2(CO_3)_3 < MgCO_3 < Na_2CO_3$. Carbonates of Group 13/III (+3 ions) are, in fact, so unstable that they do not exist.
14.103 $NaH(s) + CH_3OH(l) \rightarrow NaOCH_3(alc) + H_2(g)$; sodium methoxide (also known as sodium methylate), $NaOCH_3$, is a strong base; (b)

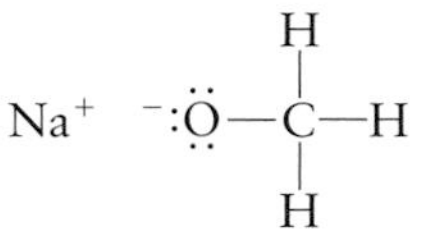

Na^+ $^-$:Ö—C(H)(H)—H

(c) $NaOCH_3(alc) + H_2O(l) \rightarrow NaOH(alc) + HOCH_3(l)$
14.105 (a) 3.491 $g{\cdot}cm^{-3}$; (b) cubic form is favored at high pressures
14.107 The metal hydride compounds have a molecular orbital structure that is very asymmetric. Because the hydrogen atom is much more electronegative than the metal atom, its orbital lies much lower in energy. Consequently, when a bond is formed, it is a strongly ionic bond with the electrons heavily localized on the H atom.

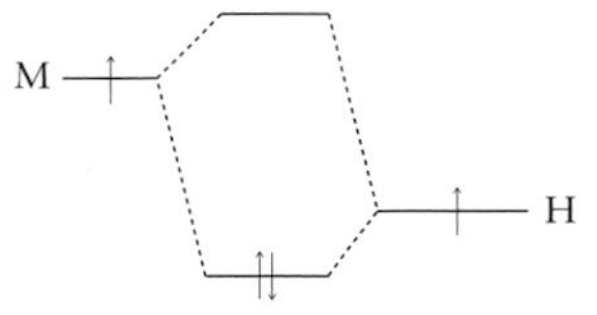

M H

However, in the case of the lighter *p*-block elements, the electronegativity difference is not so great and the bonding is much more covalent. If anything, the *p*-block element is more electronegative than hydrogen and the bond polarity would lie in the other direction, but much less pronounced than in the saline hydrides.

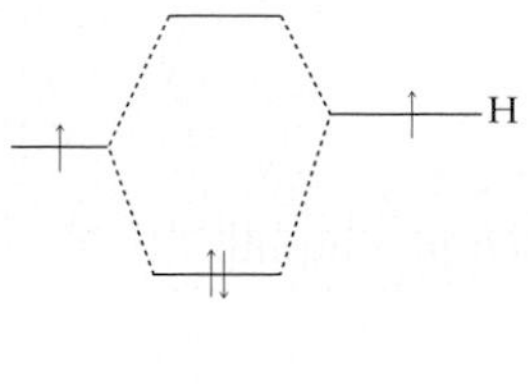

H

Light *p*-block element

14.109 (a) Diborane, B_2H_6, and $Al_2Cl_6(g)$ have the same basic structure in the way in which the atoms are arranged in space. (b) The bonding between the boron atoms and the bridging hydrogen atoms is electron deficient. There are three atoms and only two electrons to hold them together in a 3-center-2-electron bond. The bonding is Al_2Cl_6 is conventional in that all the bonds involve two atoms and two electrons. Here the lone pair of a Cl atom is donated to an adjacent Al. (c) sp^3 at the B and Al atoms; (d) The molecules are not planar. The Group 13/III element and the terminal atoms to which it is bound lie in a plane that is perpendicular to the plane that contains the main-group element and the bridging atoms.
14.111 +1.36 V
14.113 (a) hexagonal; (b) The total number of carbonate ions in the unit cell is 6. The total number of calcium ions in the cell is 6, agreeing with the overall stoichiometry of calcite, $CaCO_3$.
14.115 There is a total of 4 calcium ions and 4 sulfate ions in the unit cell. There are 8 water molecules completely inside the unit cell. The overall formula is $Ca_4(SO_4)_4(H_2O)_8$ or $CaSO_4{\cdot}2H_2O$.

Chapter 15

15.1 −3: NH_3, Li_3N, $LiNH_2$, NH_2^-; −2: H_2NNH_2; −1: N_2H_2, NH_2OH; 0: N_2; +1: N_2O, N_2F_2; +2: NO; +3: NF_3, NO_2^-, NO^+; +4: NO_2, N_2O_4; +5: HNO_3, NO_3^-, NO_2F
15.3 (a) $CO(NH_2)_2(aq) + 2\ H_2O(l) \rightarrow (NH_4)_2CO_3(aq)$; 8.0 kg of $(NH_4)_2CO_3$
15.5 (a) 0.23 L N_2; (b) $Hg(N_3)_2$ would produce a larger volume, because its molar mass is less. (c) Metal azides are good explosives because the azide ion is thermodynamically unstable with respect to the production of $N_2(g)$. The azide ion is unstable because the N≡N triple bond is so strong and also because the production of a gas is favored by the increase in entropy.
15.7 N_2O: $H_2N_2O_2$; $N_2O(g) + H_2O(l) \rightarrow H_2N_2O_2(aq)$; N_2O_3: HNO_2; $N_2O_3(g) + H_2O(l) \rightarrow 2\ HNO_2(aq)$; N_2O_5: HNO_3; $N_2O_5(g) + H_2O(l) \rightarrow 2\ HNO_3(aq)$

15.9 (a) $[PCl_4]^+$ $[PCl_6]^-$

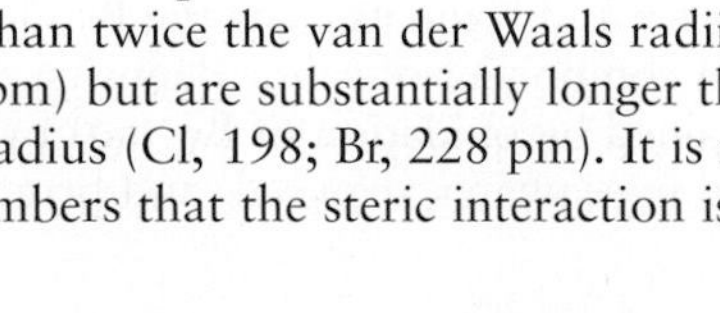

AX_4, tetrahedral AX_6, octahedral

(b and c) The separation of the Br atoms is 311 pm. For Cl, it is 288 pm. These distances are both much shorter than twice the van der Waals radii (Cl, 362 pm; Br, 390 pm) but are substantially longer than twice the atomic radius (Cl, 198; Br, 228 pm). It is not clear from these numbers that the steric interaction is the main consideration preventing the formation of the PBr_6^- ion. Weaker P—Br bond energies may also play a role in the nonobservation of the PBr_6^- ion.

15.11 $\Delta H_r° = -905.48\ kJ\cdot mol^{-1}$; $\Delta S_r° = +180.52\ J\cdot K^{-1}\cdot mol^{-1}$; $\Delta G_r° = -959.42\ kJ\cdot mol^{-1}$ at 298 K. At 1000°C (1273 K), $\Delta G_r° = -1135.38\ kJ\cdot mol^{-1}$. (b) The reaction is spontaneous at all temperatures.

15.13 (a) $3\ NO_2(g) + H_2O(l) \rightarrow 2\ HNO_3(aq) + NO(g)$; $\Delta H_r° = -138.18\ kJ\cdot mol^{-1}$
(b) $HNO_3(l) \rightarrow HNO_3(aq)$; $\Delta H_r° = -33.26\ kJ\cdot mol^{-1}$

15.15 (a) $4\ Li(s) + O_2(g) \xrightarrow{\Delta} 2\ Li_2O(s)$
(b) $2\ Na(s) + 2\ H_2O(l) \rightarrow 2\ NaOH(aq) + H_2(g)$
(c) $2\ F_2(g) + 2\ H_2O(l) \rightarrow 4\ HF(aq) + O_2(g)$
(d) $2\ H_2O(l) \rightarrow O_2(g) + 4\ H^+(aq) + 4\ e^-$

15.17 (a) $2\ H_2S(g) + 3\ O_2(g) \rightarrow 2\ SO_2(g) + 2\ H_2O(g)$
(b) $CaO(s) + H_2O(l) \rightarrow Ca(OH)_2(aq)$
(c) $2\ H_2S(g) + SO_2(g) \xrightarrow{300°C,\ Al_2O_3} 3\ S(s) + 2\ H_2O(l)$

15.19 (a)

```
  H
  |   ..
 :O—O:
  ..  |
      H
```

Each O in H_2O_2 is an AX_2E_2 structure; hence, the bond angle is predicted to be less than 109.5°. The experimental value is 97°.

(b)–(e), The standard potential of H_2O_2 is +1.78 V in acidic solution. It should, therefore, be able to oxidize any ion that has a standard potential less than +1.78 V. For the ions listed, Cu^+ and Mn^{2+} will be oxidized. It would require an input of 1.98 V to oxidize Ag^+ to Ag^{2+} and 2.87 V to oxidize F^-.

15.21 5 L of $O_2(g)$

15.23 13.11 (13.12 if K_{a2} is not ignored)

15.25 See Fig. 2.18 for bond strengths. The weaker the H—X bond, the stronger the acid. H_2Te has the weakest bonds, H_2O the strongest. Therefore, acid strengths are $H_2Te > H_2Se > H_2S > H_2O$.

15.27 (a) $\Delta H_f° = +142.7\ kJ\cdot mol^{-1}$; $\Delta S_f° = -68.78\ J\cdot mol^{-1}\cdot K^{-1}$ (b) Because $\Delta G_f°$ (O_3, g) is positive at all temperatures, the reaction is not spontaneous at any temperature. It is less favored at high temperatures. (c) Because the reaction entropy is negative, the $-T\Delta S_f°$ term is always positive, so the entropy contribution to $\Delta G_f°$ is always positive. The entropy change does not favor the spontaneous formation of ozone.

15.29 (a) $-73.35\ kJ\cdot mol^{-1}$; (b) 28.6°C

15.31 Fluorine comes from the minerals fluorspar, CaF_2, cryolite, Na_3AlF_6, and the fluorapatites, $Ca_5(PO_4)_3F$. The free element is prepared from HF and KF by electrolysis, but the HF and the KF needed are prepared in the laboratory. Chlorine comes primarily from the mineral rock salt, NaCl. The pure element is obtained by electrolysis of liquid NaCl. Bromine is found in seawater and brine wells as the Br^- ion; it is also found as a component of saline deposits; the pure element is obtained by oxidation of $Br^-(aq)$ by $Cl_2(g)$. Iodine is found in seawater, seaweed, and brine wells as the I^- ion; the pure element is obtained by oxidation of $I^-(aq)$ by $Cl_2(g)$.

15.33 (a) +1; (b) +4; (c) +7; (d) +5

15.35 (a) $4\ KClO_3(l) \xrightarrow{\Delta} 3\ KClO_4(s) + KCl(s)$
(b) $Br_2(l) + H_2O(l) \rightarrow HBrO(aq) + HBr(aq)$
(c) $NaCl(s) + H_2SO_4(aq) \rightarrow NaHSO_4(aq) + HCl(g)$
(d) Reactions (a) and (b) are redox reactions, (c) is a Brønsted acid-base reaction.

15.37 (a) $HClO < HClO_2 < HClO_3 < HClO_4$;
(b) The oxidation number of Cl increases from HClO to $HClO_4$. In $HClO_4$, chlorine has its highest oxidation number of +7, so $HClO_4$ will be the strongest oxidizing agent.

15.39 AX_2E_2, angular, slightly less than 109°. The actual bond angle is 110.9°.

15.41

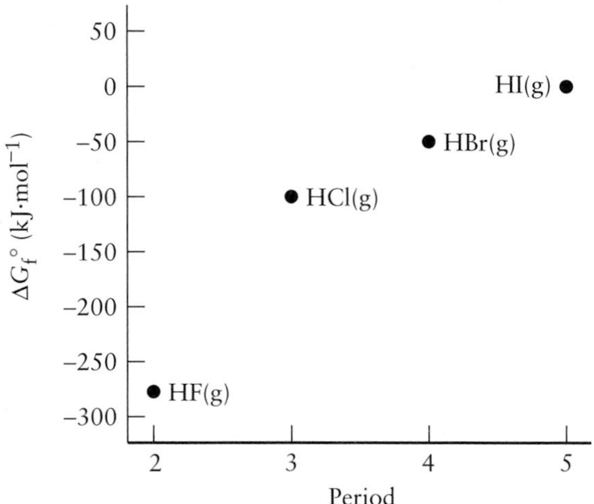

The $\Delta G_f°$ values of HCl, HBr, and HI fit nicely on a straight line, but HF is anomalous. In other properties, HF is also the anomalous member of the group, in particular with respect to its acidity. Thermodynamic stability of the hydrogen halides decreases down the group.

15.43 Because $E_{cell}°$ is negative, Cl_2 will not oxidize Mn^{2+} to form permanganate ion in an acidic solution.

15.45 $K = 0.504$

15.47 $0.117\ mol\cdot L^{-1}$

15.49 7.59×10^3 kJ of heat released

15.51 Helium occurs as a component of natural gases found under rock formations in certain locations, especially in Texas. Argon is obtained by distillation of liquid air.

15.53 (a) +2; (b) +6; (c) +4; (d) +6

15.55 $XeF_4 + 4\ H^+ + 4\ e^- \rightarrow Xe + 4\ HF$

15.57 Because H_4XeO_6 has more oxygen atoms, which are highly electronegative, bonded to Xe, we predict that H_4XeO_6 would be more acidic than H_2XeO_4.

15.59 A sol is a colloid comprising solid particles suspended in a liquid. Muddy water is a type of sol. A foam is a suspension of a gas in a solid or liquid. Styrofoam, foam rubber, soapsuds, and aerogels are all types of foams.

15.61 (a) both a sol and an emulsion; (b) a foam; (c) a sol

15.63 In fluorescence, the light that is emitted is of lower energy than the light that is absorbed and the fluorescence stops as soon as the exciting radiation is stopped. In phosphorescence, the phosphorescing molecules remain excited for a period of time after the stimulus has stopped.

15.65 A mercury atom is an energy transfer agent. It absorbs the energy from a high-voltage discharge and emits ultraviolet radiation in the regions of 254 nm and 185 nm. This emitted radiation then excites a fluorescent material that emits radiation in the visible region of the spectrum, which is the light observed when the lamp is turned on.

15.67 (a) and (b). The Lewis structure of N_3^- is (formal charges are given under each atom)

$$[\ddot{\underset{..}{N}}=N=\ddot{\underset{..}{N}}]^-$$
$$-1 \quad +1 \quad -1$$

(c) The value of $-\frac{1}{3}$ is an average oxidation number based solely on the number of nitrogen atoms and the overall charge. From the Lewis structure, we can see that the nitrogen atoms are not equivalent, so they may have different oxidation numbers. It would be extreme to state that the terminal nitrogens have oxidation numbers of -1 and the central nitrogen atom has an oxidation number of $+1$, but assigning the same oxidation number of $-\frac{1}{3}$ to all three atoms is also not strictly accurate. (d) This situation most often arises when an element bonds to other atoms of the same type.

15.69 $I_2 < Br_2 < Cl_2 < F_2$

15.71 oxidation: $As_2S_3(s) + 8\ H_2O(l) \rightarrow 2\ H_3AsO_4\ (aq) + 3\ H_2S\ (aq) + 4\ H^+(aq) + 4\ e^-$; reduction: $H_2O_2(aq) + 2\ e^- + 2\ H^+(aq) \rightarrow 2\ H_2O(l)$; overall: $As_2S_3(s) + 2\ H_2O_2(aq) + 4\ H_2O(l) \rightarrow 2\ H_3AsO_4(aq) + 3\ H_2S\ (aq)$

15.73 This ratio, $\Delta H_{vap}/T_b$, is the entropy of vaporization (see Section 7.4). Hydrogen bonding is much stronger in $H_2O(l)$ than in $H_2S(l)$. Thus, $H_2O(l)$ has a more ordered arrangement than $H_2S(l)$ does. Consequently, the change in entropy upon transformation to the gaseous state is greater for H_2O than for H_2S.

15.75 8.16×10^2 L of concentrated H_2SO_4

15.77 See the Web site for the structures of the phosphorus(III) and phosphorus(V) oxides. The basic structure of the two molecules is the same. The phosphorus atoms lie in a tetrahedral arrangement in which there are bridging oxygen atoms to the other phosphorus atoms. In phosphorus(V) oxide, there is an additional terminal oxygen atom bonded to each phosphorus atom. In phosphorus(III) oxide each oxygen atom has a formal charge of 0, as does each phosphorus atom. In phosphorus(V) oxide, this condition is also true. According to the Lewis structures, all P—O bonds in phosphorus(III) oxide have a bond order of 1, whereas in phosphorus(V) oxide, the terminal oxygen atoms have a bond order of 2 between them and the phosphorus atoms to which they are attached. When we examine the molecular parameters, we see that all the $P—O_{bridging}$ distances for phosphorus(III) oxide are slightly longer than those in phosphorus(V) oxide (163.8 pm versus 160.4 pm). This difference is expected because the radius of phosphorus(V) should be smaller than that of phosphorus(III). The terminal P═O distances in phosphorus(V) oxide are considerably shorter (142.9 pm), a finding that agrees with the higher bond order between phosphorus and these atoms.

15.79 (a) $SO_2(g) + H_2O(l) \rightarrow H_2SO_3(l)$, Lewis acid-base reaction; (b) $2\ F_2(g) + 2\ NaOH(aq) \rightarrow OF_2(g) + 2\ NaF(aq) + H_2O(l)$, redox reaction in basic solution; (c) $S_2O_3^{2-}(aq) + 4\ Cl_2(g) + 13\ H_2O(l) \rightarrow 2\ HSO_4^-(aq) + 8\ H_3O^+(aq) + 8\ Cl^-(aq)$, redox reaction; (d) $2\ XeF_6(s) + 16\ OH^-(aq) \rightarrow XeO_6^{4-}(aq) + Xe(g) + 12\ F^-(aq) + 8\ H_2O(g) + O_2(g)$, redox reaction

15.81 (a) $I_2(s) + 3\ F_2(g) \rightarrow 2\ IF_3(s)$; (b) $I_2(aq) + I^-(aq) \rightarrow I_3^-(aq)$; (c) $Cl_2(g) + H_2O(l) \rightarrow HCl(aq) + HOCl(aq)$, but there are competing reactions, such as $Cl_2(g) + H_2O(l) \rightarrow 2\ HCl(aq) + O_2(g)$. The predominant reaction is determined by the temperature and pH. (d) $2\ F_2(g) + 2\ H_2O(l) \rightarrow 4\ HF(aq) + O_2(g)$

15.83 The heads of matches consist of a paste of potassium chlorate, $KClO_3$, antimony sulfide, Sb_2S_3, sulfur, and powdered glass. The striking strip contains red phosphorus. When the match is struck against the red phosphorus surface, a reaction of the red phosphorus and potassium chlorate causes the match to ignite. The Sb_2S_3 and sulfur are the fuels that are consumed by combustion after the ignition. The powdered glass helps to produce the friction required for ignition.

15.85 Orpiment is As_2S_3 and realgar is As_4S_4. Orpiment is yellow and realgar is orange-red. They are both used as pigments.

15.87 (a) $[\ddot{\underset{..}{N}}=N=\ddot{\underset{..}{N}}]^-$, AX_2, linear, 180°; (b) between fluoride and chloride; (c) HCl, HBr, and HI are all strong acids. For HF, $K_a = 3.5 \times 10^{-4}$, so HF is slightly more acidic than HN_3. The small size of the azide ion suggests that the H—N bond in HN_3 is similar in strength to that of the H—F bond, so it is expected to be a weak acid. (d) ionic: NaN_3, $Pb(N_3)_2$, AgN_3; covalent: HN_3, $B(N_3)_3$, FN_3

15.89 $0.156\ mol \cdot L^{-1}$

15.91 The solubilities of the ionic halides are determined by a variety of factors, especially the lattice enthalpy and enthalpy of hydration. There is a delicate balance between the two factors, with the lattice enthalpy usually being the determining one. Lattice enthalpies decrease from chloride to iodide, so water molecules can more readily separate the ions in the latter. Less ionic halides, such as silver halides, generally have a much lower solubility, and the trend in solubility is the reverse of the more ionic halides. For the less ionic halides, the covalent character of the bond allows ion pairs to persist in water. The ions are not easily hydrated, making them less soluble. The polarizability of the halide ions and the covalency of their bonding increases down the group.

15.93 P_4O_{10} is the better drying agent.

15.95 (a) +0.26 V. The process should be spontaneous. (b) The formation of AgI precipitate means that the concentration of Ag^+ ions is never high enough to achieve the conditions necessary for the redox reaction to take place. The solubility product, K_{sp}, limits the concentrations in solution, so the actual redox potential is not the value calculated, which represents the values when $[Ag^+] = 1\ mol \cdot L^{-1}$ and $[I^-] = 1\ mol \cdot L^-$. If we use the concentrations established by the solubility equilibrium and the Nernst equation, we can calculate the actual redox potential to be −0.68 V.

15.97 (a) The molecular orbital diagram for NO^+ should have the oxygen orbitals slightly lower in energy than the nitrogen orbitals because oxygen is more electronegative. This will cause the bonding to be more ionic than in either N_2 or O_2. There is an ambiguity, however, in that the MO diagram could be similar to either that of N_2 or that of O_2. Refer to Figs. 3.34 and 3.35, where you will see that the σ_{2p} and the π_{2p} levels have different relative energies. There are consequently two possibilities for the orbital energy diagram. One is shown here:

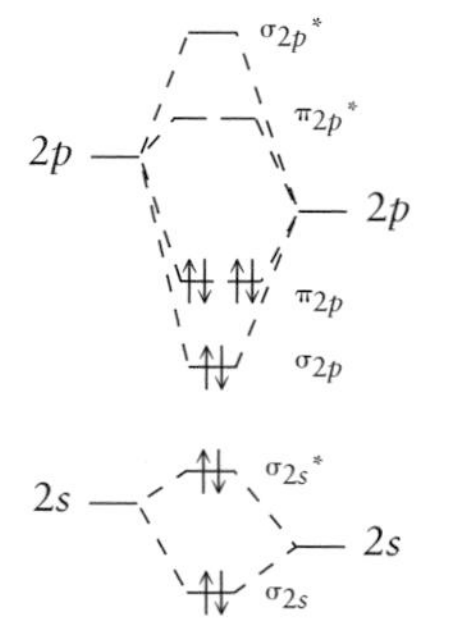

Nitrogen Oxygen

(b) The two orbital diagrams predict the same bond order (3) and same magnetic properties (diamagnetic), so these properties cannot be used to determine which diagram is the correct one. That must be determined by more complex spectroscopic measurements.

15.99 (a) Pyrite adopts a face-centered cubic unit cell. (b) The iron atoms lie at the corners and at the centers of the faces of the unit cell. Eight sulfur atoms lie completely within the unit cell. (c) The coordination number of iron is 6 (octahedral). (d) Each sulfur atom is bonded to one other sulfur atom and three iron atoms. (e) The locations of the sulfur atoms can be considered in one of two ways. An examination of the structure shows that these are best thought of as S_2^{2-} ions. The locations of the midpoints of the S—S bonds are at the centers of each edge of the unit cell $(6\ S_2^{2-} \times \frac{1}{2})$ plus 1 S_2^{2-} ion in the center of the unit cell (only half the ions have sulfur atoms within a given cell). This gives a total of 4 S_2^{2-} ions in the unit cell. Alternatively, identify eight S atoms within the unit cell.

Chapter 16

16.1 Those on the left have strongly negative potentials.

16.3 (a) Sc; (b) Au; (c) Nb; (d) Ru and Os are about the same size

16.5 (a) Ti; (b) Cu; (c) Zn; (d) Fe; (e) Os

16.7 Hg is much more dense than Cd, because the shrinkage in atomic radius that occurs between $Z = 58$ and $Z = 71$ (lanthanide contraction) causes the atoms following the rare earths to be smaller than might have been expected for their atomic masses and atomic numbers. Zn and Cd have densities that are not too dissimilar, because the radius of Cd is subject to only a smaller *d*-block contraction.

16.9 (a) Proceeding down a group in the *d* block (say, from Cr to Mo to W), there is an increasing probability of finding the elements in higher oxidation states. Thus, higher oxidation numbers become more stable going down a group. (b) The trend for the *p*-block elements is reversed. Because of the inert-pair effect, the higher oxidation states tend to be less stable as one descends a group.

16.11 In MO_3, M has an oxidation number of +6. Of these three elements, the +6 state is most stable for Cr. See Fig. 16.7.

16.13 (a) Ti(s), $MgCl_2$(s); $TiCl_4(g) + 2\ Mg(l) \xrightarrow{\Delta} Ti(s) + 2\ MgCl_2(s)$; (b) $Co^{2+}(aq)$, $HCO_3^-(aq)$, $NO_3^-(aq)$; $CoCO_3(s) + HNO_3(aq) \rightarrow Co^{2+}(aq) + HCO_3^-(aq) + NO_3^-(aq)$; (c) V(s), CaO(s), V_2O_5(s); $V_2O_5(s) + 5\ Ca(l) \xrightarrow{\Delta} 2\ V(s) + 5\ CaO(s)$

16.15 (a) titanium (IV) oxide, TiO_2; (b) iron(III) oxide, Fe_2O_3; (c) manganese(IV) oxide, MnO_2

16.17 (a) V^{3+}, H_2, and Cl^-; (b) no reaction; (c) Co^{2+}, H_2, and Cl^-

16.19 (a) $Fe_2O_3(s) + 3\ CO(g) \rightarrow 2\ Fe(s) + 3\ CO_2(g)$ (Zone C); $3\ Fe_2O_3(s) + CO(g) \rightarrow 2\ Fe_3O_4(s) + CO_2(g)$ and $Fe_3O_4(s) + CO(g) \rightarrow 3\ FeO(s) + CO_2(g)$; these combine to $Fe_2O_3(s) + CO(g) \rightarrow 2\ FeO(s) + CO_2(g)$ (Zone D); $FeO(s) + CO(g) \rightarrow Fe(s) + CO_2(g)$ (Zone C) (b) $TiCl_4(g) + 2\ Mg(l) \xrightarrow{\Delta} Ti(s) + 2\ MgCl_2(s)$ (c) $CaO(s) + SiO_2(s) \xrightarrow{\Delta} CaSiO_3(s)$

16.21 (a) Cr^{3+} ions in water form the complex $[Cr(H_2O)_6]^{3+}(aq)$, which behaves as a Brønsted acid:

$[Cr(H_2O)_6]^{3+}(aq) + H_2O(l) \rightarrow [Cr(H_2O)_5OH]^{2+}(aq) + H_3O^+(aq)$; (b) The gelatinous precipitate is $Cr(OH)_3$. The precipitate dissolves as the $[Cr(OH)_4]^-$ complex is formed. $Cr^{3+}(aq) + 3\ OH^-(aq) \rightarrow Cr(OH)_3(s)$ and $Cr(OH)_3(s) + OH^- \rightarrow [Cr(OH)_4]^-(aq)$

16.23 (a) $2\ ZnS(s) + 3\ O_2(g) \xrightarrow{\Delta} 2\ ZnO(s) + 2\ SO_2(g)$, followed by $ZnO(s) + C(s) \xrightarrow{\Delta} Zn(l) + CO(g)$; (b) $HgS(s) + O_2(g) \xrightarrow{\Delta} Hg(l) + SO_2(g)$

16.25 10^6

16.27 (a) hexacyanoferrate(II) ion, +2; (b) hexaamminecobalt(III) ion, +3; (c) aquapentacyanocobaltate(III) ion, +3; (d) pentaamminesulfatocobalt(III) ion, +3

16.29 (a) $K_3[Cr(CN)_6]$; (b) $[Co(NH_3)_5(SO_4)]Cl$; (c) $[Co(NH_3)_4(H_2O)_2]Br_3$; (d) $Na[Fe(H_2O)_2(C_2O_4)_2]$

16.31 (a), (b), and (d) can be polydentate; (a) 3; (b) 1 or 2; (d) 2

16.33 only (b)

16.35 (a) 4; (b) 2; (c) 6; (d) 6

16.37 (a) linkage isomers; (b) ionization isomers; (c) linkage isomers; (d) ionization isomers

16.39 $[Co(H_2O)_6]Cl_3$, $[CoCl(H_2O)_5]Cl_2 \cdot H_2O$, $[CoCl_2(H_2O)_4]Cl \cdot 2H_2O$; $[CoCl_3(H_2O)_3] \cdot 3H_2O$

16.41 (a) yes,

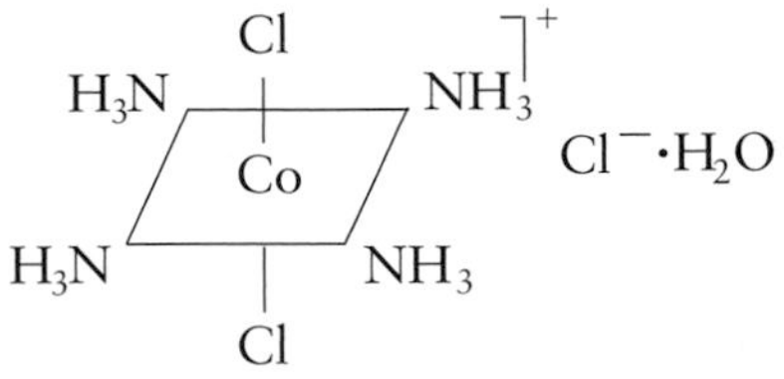

trans-Tetraamminedichlorocobalt(III) chloride monohydrate

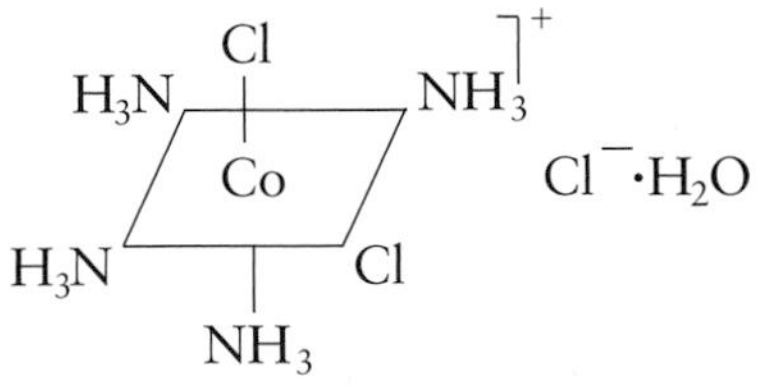

cis-Tetraamminedichlorocobalt(III) chloride monohydrate

(b) no; (c) yes,

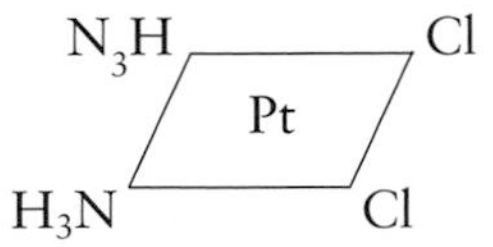

cis-Diamminedichloroplatinum(II)

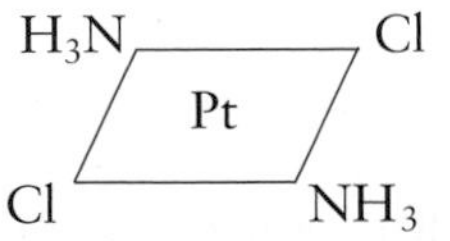

trans-Diamminedichloroplatinum(II)

16.43 (a)

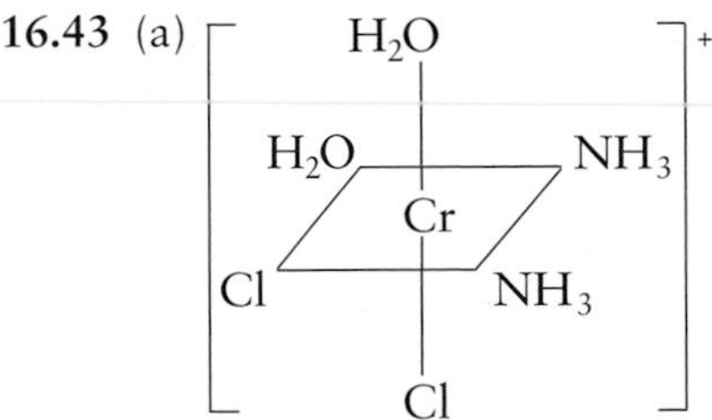

(b) yes

16.45 (a) chiral; (b) not chiral

16.47 (a) 1; (b) 6; (c) 5; (d) 3; (e) 6; (f) 6

16.49 (a) 2; (b) 5; (c) 8; (d) 10; (e) 0; (f) 0

16.51

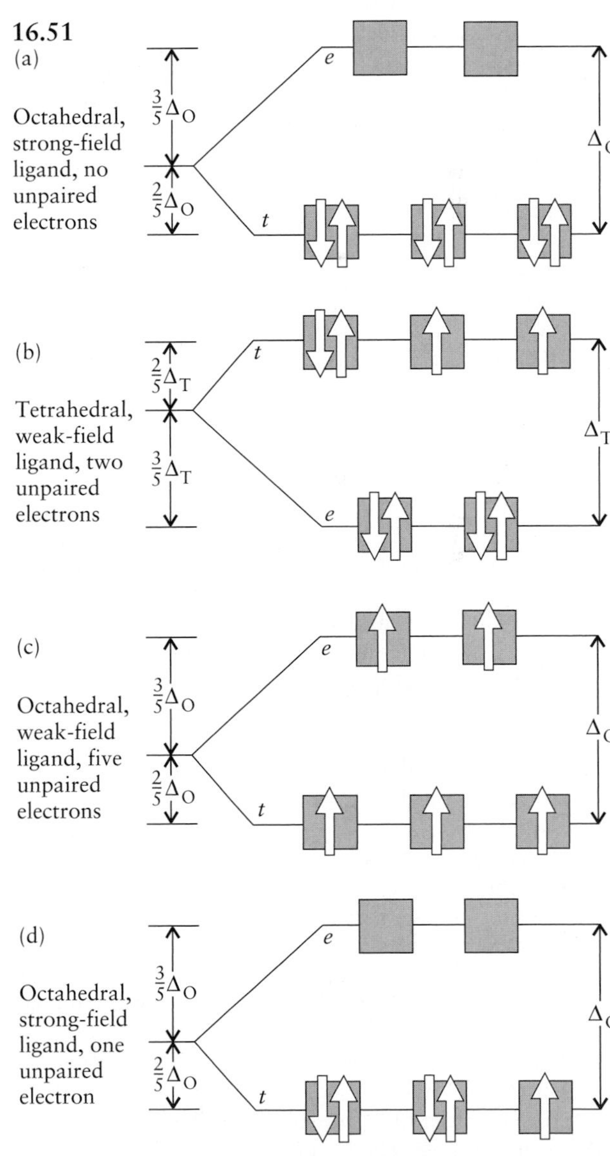

16.53 (a) no unpaired electrons; (b) two unpaired electrons

16.55 Weak-field ligands do not interact strongly with the *d*-electrons in the metal ion, so they produce only a small crystal field splitting of the *d*-electron energy states. The opposite is true of strong-field ligands. With weak-field ligands, unpaired electrons remain unpaired if there are unfilled orbitals; hence, a weak-field ligand is likely to lead to a high-spin complex. Strong-field ligands cause electrons in excess of three to pair up with electrons in lower energy orbitals. A strong-field ligand is likely to lead to a low-spin complex. Arranging ligands in the spectrochemical series helps to distinguish strong-field and weak-field ligands. Measurement of magnetic susceptibility (paramagnetism) can be used to determine the number of unpaired electrons, which in turn establishes whether the associated ligand is weak-field or strong-field in nature.
16.57 (a) blue, because F^- is a weak-field ligand; (b) yellow, because the en ligand is a strong-field ligand. The splitting between levels is less in (a) than in (b), therefore (a) will absorb light of a longer wavelength than (b) and consequently display a shorter wavelength color.
16.59 In Zn^{2+}, the *d*-orbitals are filled (d^{10}). There can be no electronic transitions between the *t* and *e* levels; hence no visible light is absorbed and the aqueous ion is colorless. The d^{10} configuration has no unpaired electrons, so Zn compounds would not be paramagnetic.
16.61 (a) 162 $kJ\cdot mol^{-1}$; (b) 260 $kJ\cdot mol^{-1}$; (c) 208 $kJ\cdot mol^{-1}$; (d) $Cl < H_2O < NH_3$
16.63 The e_g set, which comprises the $d_{x^2-y^2}$ and d_{z^2} orbitals.
16.65 (a) The CN^- ion is a π-acid ligand accepting electrons into the empty π^*-orbital created by the C—N multiple bond. (b) The Cl^- ion has extra lone pairs in addition to the one that is used to form the σ-bond to the metal, so it can act as a π-base, donating electrons in a *p*-orbital to an empty *d*-orbital on the metal. (c) H_2O, like Cl^-, also has an "extra" lone pair of electrons that can be donated to a metal center, thus making it a weak π-base; (d) en is neither a π-acid nor a π-base because it does not have any empty π-type antibonding orbitals nor does it have any extra lone pairs of electrons to donate. (e) $Cl^- < H_2O <$ en $< CN^-$. Note that the spectrochemical series orders the ligands generally as π-bases < ligands that form only σ-bonds < π-acceptors.
16.67 Nonbonding or slightly antibonding. If a complex forms only σ-bonds, the t_{2g} set of orbitals is nonbonding. If the ligands can function as weak π-donors (those close to the middle of the spectrochemical series, such as H_2O), the t_{2g} set becomes slightly antibonding by interacting with the filled *p*-orbitals on the ligands.
16.69 Antibonding. The e_g set of orbitals on an octahedral metal ion are always antibonding because of interactions with ligand orbitals that form the σ-bonds. This is true regardless of whether the ligands are π-acceptors, π-donors, or neither.
16.71 Water has two lone pairs of electrons. Once one of these is used to form the σ-bond to the metal ion, the second may be used to form a π-bond. This arrangement causes the t_{2g} set of orbitals to move up in energy, thereby making Δ_O smaller; hence water is a weak-field ligand. Ammonia does not have this extra lone pair of electrons and consequently cannot function as a π-donor ligand.
16.73 (a) CO; (b) $Fe_3O_4(s) + CO(g) \rightarrow 3\ FeO(s) + CO_2(g)$ (Zone D); followed by $FeO(s) + CO(g) \rightarrow Fe(s) + CO_2(g)$ (Zone C); $Fe_2O_3(s) + 3\ CO(g) \rightarrow 2\ Fe(s) + 3\ CO_2(g)$ (Zone C); (c) carbon
16.75 The major impurity is carbon and it is removed by oxidation of the carbon to CO_2, followed by capture of the CO_2 by base to form a slag.
16.77 copper and zinc
16.79 Alloys are usually (1) harder and more brittle and (2) poorer conductors of electricity than the metals from which they are made.
16.81 (a) A ferrofluid is a mixture containing a viscous, nonpolar liquid (oil), a detergent that can form micelles and a finely powdered magnetic material such as Fe_3O_4. When placed in a magnetic field, the magnetic particles in the fluid tend to align with the field; this behavior means that the position and motion of the liquid can be controlled by the applied magnetic field. (b) At first glance, the oxidation state in Fe_3O_4 appears to be the nonintegral value of $+\frac{8}{3}$. Another way to view it is that the compound is composed of both Fe^{2+} and Fe^{3+} in the ratio 1:2. The compound could also be written as $FeO\cdot Fe_2O_3$.
16.83 The compound is ferromagnetic below T_C because the magnetization is higher. Above the Curie temperature, the compound is a simple paramagnet with randomly oriented spins, but below that temperature, the spins align and the magnetization increases.

16.85 (a)

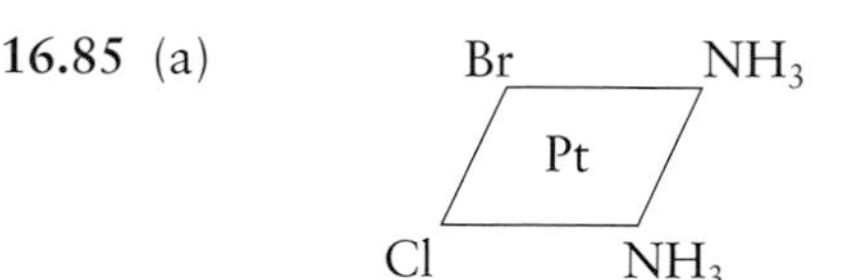

cis-Diamminebromochloroplatinum(II)

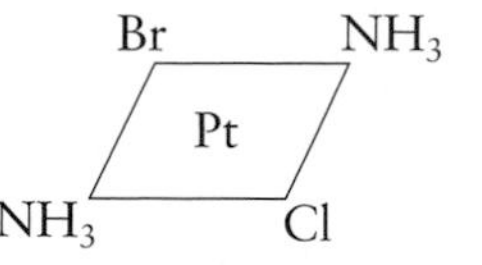

trans-Diamminebromochloroplatinum(II)

(b) If the compound were tetrahedral, there would be only one compound, not two.
16.87 The first, $[Ni(SO_4)(en)_2]Cl_2$, will give a precipitate of AgCl when $AgNO_3$ is added; the second

will not. (b) The second, $[NiCl_2(en)_2]I_2$, will show free I_2 when mildly oxidized with, for example, Br_2, but the first will not.

16.89 (a) $[MnCl_6]^{4-}$ ↑ ↑ / ↑ ↑ ↑ $[Mn(CN)_6]^{4-}$ ⇅ ⇅ ↑ (b) one; (c) Strong-field ligands absorb shorter wavelength light, therefore $[Mn(CN)_6]^{4-}$ transmits longer wavelengths.

16.91 4, 5, 6, and 7

16.93 High-spin Mn^{2+} ions have a d^5 configuration with 5 unpaired electrons. For light to be absorbed in the visible region of the spectrum, an electron from the t_{2g} set has to be moved into the e_g set of orbitals. Because these orbitals already each have an electron in them, it would be necessary to pair the electrons, a process that requires additional energy. This extra energy requirement makes the absorption of the visible light less probable, so the complexes usually are only faintly colored.

16.95 If one considers the formation reaction of a metal complex from the metal ion and its ligands, $M^{x+} + n\,L \rightarrow ML_n{}^{x+}$, it can be seen that there is a considerable loss in entropy on creating the complex. Polydentate ligands, which have two or more places on a ligand that can bind to a metal, do not have nearly so unfavorable an entropy change—the coordination of the second site on the ligand occurs intramolecularly once the first part of the ligand is attached. Thus, in general, metal complexes with chelating ligands have much higher formation constants than complexes with all monodentate ligands. This observation is known as the chelation effect, and it is so strong that, once one end of a polydentate ligand binds to a metal center, the other end usually binds extremely rapidly.

16.97 The correct structure for $[Co(NH_3)_6]Cl_3$ consists of four ions, $[Co(NH_3)_6]^{3+}$ and 3 Cl^- in aqueous solution. The chloride ions can be easily precipitated as AgCl. This reaction would not be possible if they were bonded to the other (NH_3) ligands. If the structure were $Co(NH_3{-}NH_3{-}Cl)_3$, VSEPR theory would predict that the Co^{3+} ion would have a trigonal planar ligand arrangement. The splitting of the *d*-orbital energies would not be the same as the octahedral arrangement and would lead to different spectroscopic and magnetic properties inconsistent with the experimental evidence. In addition, neither optical isomers nor geometrical isomers would be observed.

16.99 The ligand field strengths of the halide ions lie in the order $I^- < Br^- < Cl^- < F^-$, which shows that the less electronegative the ligand is, the smaller the value of Δ_O for the complex. The value of Δ_O correlates with the ability of the ligand's extra lone pairs of electrons to interact with the t_{2g} set of the octahedral metal ion. This relation means that the less electronegative the ligand is, the easier it is for that ligand to donate electrons to the metal ion, in this case, through π-donation.

16.101 To determine these relationships, we need to consider the types of interactions that the ligands on the ends of the spectrochemical series will have with the metal ions. Those that are weak field (form high-spin complexes, π-bases) are those that have extra lone pairs of electrons that can be donated to a metal ion possessing empty orbitals with π symmetry. The strong-field ligands (form low-spin complexes, π-acids) accept electrons from the metals. The complexes that will be more stable will be produced in general by the match between ligand and metal. Thus, the early transition metals in high oxidation states will have few or no electrons in the *d*-orbitals. These metal ions will become stabilized by ligands that can donate more electrons to the metal—the *d*-orbitals that are empty can readily accept electrons. The more stable complexes will be formed with the weak-field ligands. The opposite is true for metals with lots of electrons, which are the elements at the right of the periodic table in low oxidation states. These metals generally have most of the *d*-orbitals filled, so they would, in fact, be destabilized by π-donation. They form instead more stable complexes with the π-acceptor ligands that can remove some of the electron density from the metal ions.

16.103 (a) Cu atoms: $\frac{1}{8} \times 8$ atoms at the corners of the unit cell $= 1$ atom; $\frac{1}{4} \times 8$ atoms on the edges of the unit cell $= 2$ atoms, total $= 3$ atoms; Y atoms: 1 atom in the cell center; Ba atoms: 2 atoms completely within the unit cell; O atoms: $\frac{1}{4} \times 12$ atoms on the edges of the unit cell $= 3$ atoms, $\frac{1}{2} \times 8$ atoms on the faces of the unit cell $= 4$ atoms, total $= 7$ atoms; (b) 382 pm $\times$ 388 pm $\times$ 1351 pm; (c) orthorhombic ($a \neq b \neq c$; $\alpha = \beta = \delta = 90°$). The cell is close to tetragonal; but, for a tetragonal cell, two of the edge lengths must be exactly equal, and here they are slightly different.

16.105 (a) $PtCl_2(NH_3)_2$, *cis*-diamminedichloroplatinum(II); (b) The only isomer is the trans form. Neither the cis nor the trans form is optically active.

```
      NH3
       |
  Cl — Pt — Cl
       |
      NH3
```

(c) square planar

16.107 (a) ZrO_2: there are 4 zirconium atoms in the unit cell ($\frac{1}{8} \times 8$ corner atoms $+ \frac{1}{2} \times 6$ face-centered atoms) and 8 oxygen atoms (all O atoms lie completely inside the unit cell); (b) face-centered cubic; (c) 8; (d) coordination number $= 4$, tetrahedral; (e) An obvious difference is that zirconia is composed of two different types of atoms, whereas diamond is made up solely of carbon; however, both structures are based on a face-centered cubic unit cell. In zirconia, all the tetrahedral holes created in the cubic close-packed arrangement of zirconium atoms are occupied by oxide ligands. In the case of diamond, only half of the tetrahedral holes are filled. The consequence is that the zirconium atoms are connected to twice as many oxygen atoms as carbon is connected to other carbon atoms.

Chapter 17

17.1 (a) 4.5×10^{-13} m, 2.7×10^{11} J·mol^{-1}; (b) 9.4×10^{-15} m, 1.3×10^{13} J·mol^{-1}; (c) 7.3×10^{-14} m, 1.6×10^{12} J·mol^{-1}; (d) 3.5×10^{-12} m, 3.4×10^{10} J·mol^{-1}

17.3 (a) 3.21×10^{20} Hz, 9.35×10^{-13} m; (b) 3.97×10^{20} Hz, 7.59×10^{-13} m; (c) 2.65×10^{20} Hz, 1.13×10^{-12} m

17.5 (a) ${}^{3}_{1}T \rightarrow {}^{0}_{-1}e + {}^{3}_{2}He$; (b) ${}^{83}_{39}Y \rightarrow {}^{0}_{1}e + {}^{83}_{38}Sr$; (c) ${}^{87}_{36}Kr \rightarrow {}^{0}_{-1}e + {}^{87}_{37}Rb$; (d) ${}^{225}_{91}Pa \rightarrow {}^{4}_{2}\alpha + {}^{221}_{89}Ac$

17.7 (a) ${}^{8}_{5}B \rightarrow {}^{0}_{1}e + {}^{8}_{4}Be$; (b) ${}^{63}_{28}Ni \rightarrow {}^{0}_{-1}e + {}^{63}_{29}Cu$; (c) ${}^{185}_{79}Au \rightarrow {}^{4}_{2}\alpha + {}^{181}_{77}Ir$; (d) ${}^{7}_{4}Be + {}^{0}_{-1}e \rightarrow {}^{7}_{3}Li$

17.9 (a) β, ${}^{24}_{11}Na \rightarrow {}^{0}_{-1}e + {}^{24}_{12}Mg$; (b) β, ${}^{128}_{50}Sn \rightarrow {}^{0}_{-1}e + {}^{128}_{51}Sb$; (c) β^+, ${}^{140}_{57}La \rightarrow {}^{0}_{1}e + {}^{140}_{56}Ba$; (d) α, ${}^{228}_{90}Th \rightarrow {}^{4}_{2}\alpha + {}^{224}_{88}Ra$

17.11 (a) ${}^{11}_{5}B + {}^{4}_{2}\alpha \rightarrow {}^{13}_{7}N + 2\,{}^{1}_{0}n$; (b) ${}^{35}_{17}Cl + {}^{2}_{1}D \rightarrow {}^{36}_{18}Ar + {}^{1}_{0}n$; (c) ${}^{96}_{42}Mo + {}^{2}_{1}D \rightarrow {}^{97}_{43}Tc + {}^{1}_{0}n$; (d) ${}^{45}_{21}Sc + {}^{1}_{0}n \rightarrow {}^{42}_{19}K + {}^{4}_{2}\alpha$

17.13 (a) β decay, ${}^{68}_{30}Zn$; (b) β^+ decay, ${}^{103}_{47}Ag$; (c) α decay, ${}^{239}_{95}Am$; (d) α decay, ${}^{256}_{103}Lr$

17.15 ${}^{235}_{92}U \rightarrow {}^{4}_{2}\alpha + {}^{231}_{90}Th$, ${}^{231}_{90}Th \rightarrow {}^{0}_{-1}e + {}^{231}_{91}Pa$, ${}^{231}_{91}Pa \rightarrow {}^{4}_{2}\alpha + {}^{227}_{89}Ac$, ${}^{227}_{89}Ac \rightarrow {}^{0}_{-1}e + {}^{227}_{90}Th$, ${}^{227}_{90}Th \rightarrow {}^{4}_{2}\alpha + {}^{223}_{88}Ra$, ${}^{223}_{88}Ra \rightarrow {}^{4}_{2}\alpha + {}^{219}_{86}Rn$, ${}^{219}_{86}Rn \rightarrow {}^{4}_{2}\alpha + {}^{215}_{84}Po$, ${}^{215}_{84}Po \rightarrow {}^{0}_{-1}e + {}^{215}_{85}At$, ${}^{215}_{85}At \rightarrow {}^{4}_{2}\alpha + {}^{211}_{83}Bi$, ${}^{211}_{83}Bi \rightarrow {}^{0}_{-1}e + {}^{211}_{84}Po$, ${}^{211}_{84}Po \rightarrow {}^{4}_{2}\alpha + {}^{207}_{82}Pb$

17.17 ${}^{238}_{92}U \rightarrow {}^{234}_{90}Th + {}^{4}_{2}\alpha$, ${}^{234}_{90}Th \rightarrow {}^{234}_{91}Pa + {}^{0}_{-1}e$, ${}^{234}_{91}Pa \rightarrow {}^{234}_{92}U + {}^{0}_{-1}e$, ${}^{234}_{92}U \rightarrow {}^{230}_{90}Th + {}^{4}_{2}\alpha$, ${}^{230}_{90}Th \rightarrow {}^{226}_{88}Ra + {}^{4}_{2}\alpha$, ${}^{226}_{88}Ra \rightarrow {}^{222}_{86}Rn + {}^{4}_{2}\alpha$, ${}^{222}_{86}Rn \rightarrow {}^{218}_{84}Po + {}^{4}_{2}\alpha$, ${}^{218}_{84}Po \rightarrow {}^{214}_{82}Pb + {}^{4}_{2}\alpha$, ${}^{214}_{82}Pb \rightarrow {}^{214}_{83}Bi + {}^{0}_{-1}e$, ${}^{214}_{83}Bi \rightarrow {}^{214}_{84}Po + {}^{0}_{-1}e$, ${}^{214}_{84}Po \rightarrow {}^{210}_{82}Pb + {}^{4}_{2}\alpha$, ${}^{210}_{82}Pb \rightarrow {}^{210}_{83}Bi + {}^{0}_{-1}e$, ${}^{210}_{83}Bi \rightarrow {}^{210}_{84}Po + {}^{0}_{-1}e$, ${}^{210}_{84}Po \rightarrow {}^{206}_{82}Pb + {}^{4}_{2}\alpha$

17.19 (a) ${}^{14}_{7}N + {}^{4}_{2}\alpha \rightarrow {}^{17}_{8}O + {}^{1}_{1}p$; (b) ${}^{248}_{96}Cm + {}^{1}_{0}n \rightarrow {}^{249}_{97}Bk + {}^{0}_{-1}e$; (c) ${}^{243}_{95}Am + {}^{1}_{0}n \rightarrow {}^{244}_{96}Cm + {}^{0}_{-1}e + \gamma$; (d) ${}^{13}_{6}C + {}^{1}_{0}n \rightarrow {}^{14}_{6}C + \gamma$

17.21 (a) ${}^{20}_{10}Ne + {}^{4}_{2}\alpha \rightarrow {}^{8}_{4}Be + {}^{16}_{8}O$; (b) ${}^{20}_{10}Ne + {}^{20}_{10}Ne \rightarrow {}^{24}_{12}Mg + {}^{16}_{8}O$; (c) ${}^{44}_{20}Ca + {}^{4}_{2}\alpha \rightarrow {}^{48}_{22}Ti + \gamma$; (d) ${}^{27}_{13}Al + {}^{2}_{1}H \rightarrow {}^{28}_{13}Al + {}^{1}_{1}p$

17.23 (a) ${}^{14}_{7}N + {}^{4}_{2}\alpha \rightarrow {}^{17}_{8}O + {}^{1}_{1}p$; (b) ${}^{239}_{94}Pu + {}^{1}_{0}n \rightarrow {}^{240}_{95}Am + {}^{0}_{-1}e$

17.25 (a) untriquadium, Utq; (b) unquadpentium, Uqp; (c) binilunium, Bnu

17.27 1.6×10^{-5} Ci

17.29 (a) 9.2×10^{4} Bq (1 Bq = 1 disintegration per second); (b) 4.3×10^{12} Bq; (c) 2.7×10^{8} Bq

17.31 1.0×10^{2} rad, 1.0×10^{2} rem, 1.0 Sv

17.33 100 d

17.35 (a) 5.63×10^{-2} y^{-1}; (b) 0.83 s^{-1}; (c) 0.0693 min^{-1}

17.37 (a) 5.3×10^{3} y; (b) 8.4×10^{9} y; (c) 2 y

17.39 6.0×10^{-3} Ci

17.41 (a) 78.5%; (b) 53.8%

17.43 (a) 0.50; (b) 9.3×10^{8} y

17.45 3.53×10^{3} y

17.47 (a) 1.0×10^{-3} Ci; (b) 2.7×10^{-4} Ci; (c) 0.41 Ci

17.49 45 d

17.51 (a) 14 d; (b) 0.25

17.53 2.4×10^{-4} g

17.55 If isotopically enriched water, such as $H_2{}^{18}O$, is used in the reaction, the label can be followed. Once the products are separated, a suitable technique such as vibrational spectroscopy or mass spectrometry can be used to determine whether the product has incorporated the ^{18}O or not. For example, if the methanol ends up with the O atom from the water molecules, then its molar mass would be 34 g·mol^{-1} rather than 32 g·mol^{-1} found for methanol with elements present at their natural isotopic abundance.

17.57 The vibrational frequency for the C—D bond will be approximately 0.73 times (that is, lower in energy) the value for the C—H bond.

17.59 (a) 2.3×10^{-10} g gained; (b) 1.86×10^{-10} g lost; (c) 8.33×10^{-9} g lost

17.61 (a) 9.0×10^{13} J; (b) 8.20×10^{-14} J; (c) 90 J; (d) 1.51×10^{-10} J

17.63 -4.3×10^{9} kg·s^{-1}

17.65 (a) 1.1421×10^{-12} J·nucleon^{-1}; (b) 1.212×10^{-12} J·nucleon^{-1}; (c) 1.8×10^{-13} J·nucleon^{-1}; (d) 4.57×10^{-13} J·nucleon^{-1}. (e) ^{62}Ni is the most stable, because it has the largest binding energy per nucleon.

17.67 (a) -7.8×10^{10} J·g^{-1}; (b) -3.52×10^{11} J·g^{-1}; (c) -2.09×10^{11} J·g^{-1}; (d) -3.52×10^{11} J·g^{-1}

17.69 (a) -8.85×10^{-13} J per decay; (b) -3.69×10^{-14} J·nucleon^{-1}

17.71 (a) ${}^{244}_{95}Am \rightarrow {}^{134}_{53}I + {}^{107}_{42}Mo + 3\,{}^{1}_{0}n$; (b) ${}^{235}_{92}U + {}^{1}_{0}n \rightarrow {}^{96}_{40}Zr + {}^{138}_{52}Te + 2\,{}^{1}_{0}n$; (c) ${}^{235}_{92}U + {}^{1}_{0}n \rightarrow {}^{101}_{42}Mo + {}^{132}_{50}Sn + 3\,{}^{1}_{0}n$

17.73 (a) 9 dpm; (b) 7×10^{5} decays

17.75 (a) 1.7×10^{8} pCi·L^{-1}; (b) 5.0×10^{18} atoms; (c) 1×10^{2} d

17.77 (a) At first thought, it might seem that a fusion bomb would be more suitable for excavation work, because the fusion process itself does not generate harmful radioactive waste products. However, in practice, fusion cannot be initiated in a bomb in the

absence of the high temperatures that can be generated only by a fission bomb. So there is no environmental advantage to the use of a fusion bomb. The fission bomb has the advantage that its destructive power can be more carefully controlled. It is possible to make small fission bombs such that the destructive effect could be contained within a small area. (b) The principal argument for the use of bombs in excavation is speed, and hence cost effectiveness, of the process. The principal argument against their use is environmental damage.
17.79 (a) 4.5×10^9 y; (b) 3.9×10^9 y
17.81 (a) $k = 1.1\ \text{d}^{-1}$, $t_{1/2} = 0.63$ d; (b) 0.22 mg
17.83 (a) Radioactive substances that emit γ radiation are most effective for diagnosis because they are the least destructive of the types of radiation listed. Additionally, γ rays pass easily through body tissues and can be counted, whereas α and β particles are stopped by the body tissues. (b) α particles tend to be best for this application because they cause the most destruction. (c) and (d) ^{131}I, 8 d (used to image the thyroid); ^{67}Ga, 78 h (used most often as the citrate complex); ^{99m}Tc, 6 h (used for various body tissues by varying the ligands attached to the Tc atom).

17.85 (a)

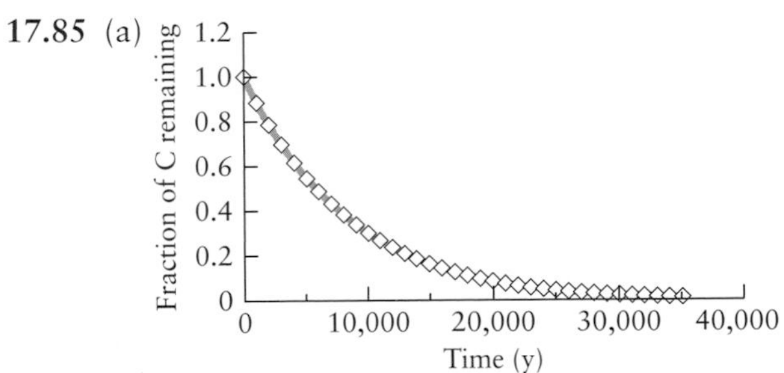

(b) The equation to plot is $-\ln(N/N_0) = 1.21 \times 10^{-4}\, t$.

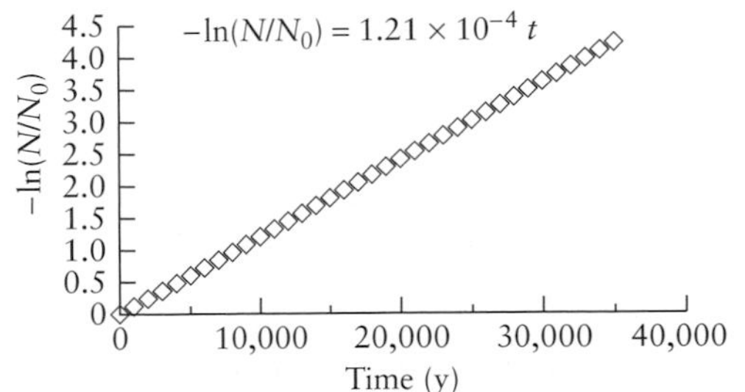

(c) The information can be obtained from the graphs or from the equation

$$-\ln \frac{N}{N_0} = kt$$

If we want to have less than 1% of the original amount of ^{14}C present, then we will want the value for which N/N_0 is 0.01 or less. $-\ln 0.01 = (1.21 \times 10^{-4}\text{y}^{-1})(t)$, so $t = 3.8 \times 10^4$ y.

Chapter 18

18.1 (a)

```
    H   H           H   H
    |   |           |   |
H — C — C — C ≡ C — C — C — H
    |   |           |   |
    H   H           H   H
```

Alkyne

(b)

```
    H   H   H   H   H   H
    |   |   |   |   |   |
H — C — C — C — C — C — C — H
    |   |   |   |   |   |
    H   H   H   H   H   H
```

Alkane

(c)

```
            H   H
            |   |
H — C = C — C — C — H
    |   |   |   |
    H   H   H   H
```

Alkene

(d)

```
    H   H   H   H               H
    |   |   |   |               |
H — C — C = C — C — C ≡ C — C — H
    |           |               |
    H           H               H
```

Alkene and alkyne

(e)

```
    H   H   H   H   H
    |   |   |   |   |
H — C — C — C — C = C — H
    |   |   |
    H   H   H
```

Alkene

18.3 (a) $(CH_3)_3CH$, or C_4H_{10}, alkane; (b) $C_6H_7CH_3$ or C_7H_{10}, alkene; (c) C_6H_{12}, alkane; (d) C_6H_{12}, alkane
18.5 (a) $C_{12}H_{26}$, alkane; (b) $C_{13}H_{20}$, alkene; (c) C_7H_{14}, alkane; (d) $C_{14}H_8$, aromatic hydrocarbon
18.7 (a) propane; (b) butane; (c) heptane; (d) decane
18.9 (a) methyl; (b) pentyl; (c) propyl; (d) hexyl
18.11 (a) propane; (b) ethane; (c) pentane; (d) 2,3-dimethylbutane
18.13 (a) 4-methyl-2-pentene; (b) 2,3-dimethyl-2-phenylpentane
18.15 (a) $CH_2{=}CHCH(CH_3)CH_2CH_3$; (b) $CH_3CH_2C(CH_3)_2CH(CH_2CH_3)(CH_2)_2CH_3$; (c) $HC{\equiv}C(CH_2)_2C(CH_3)_3$; (d) $CH_3CH(CH_3)CH(CH_2CH_3)CH(CH_3)_2$
18.17 (a)

```
    H   H   H   CH3 H   H   H   H   H
    |   |   |   |   |   |   |   |   |
H — C — C — C — C — C — C — C — C — C — H
    |   |   |   |   |   |   |   |   |
    H   H   H   CH3 H   H   H   H   H
```

(b)

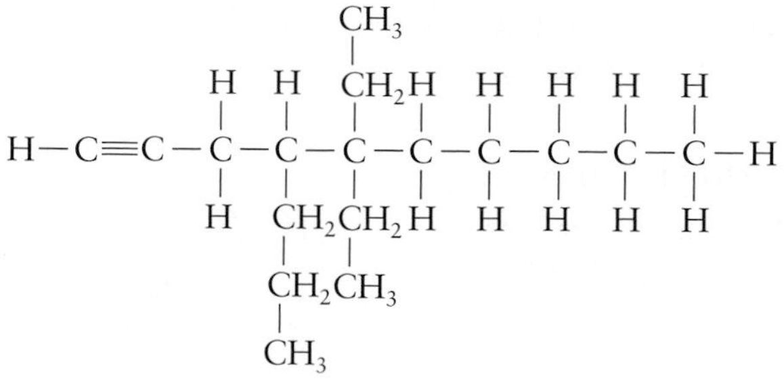

CH_3

H H CH_2H H H H H

H—C≡C—C—C—C—C—C—C—C—C—H

H CH_2CH_2H H H H H

CH_2CH_3

CH_3

(c) H CH_3H CH_3H

H—C—C—C—C—C—H

H CH_3H H H

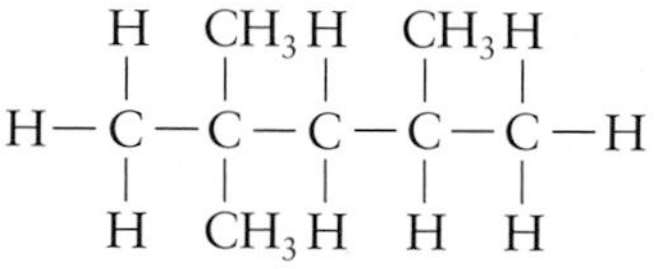

(d) $H_3C—CH_2$ H

C=C

H $CH_2—CH_3$

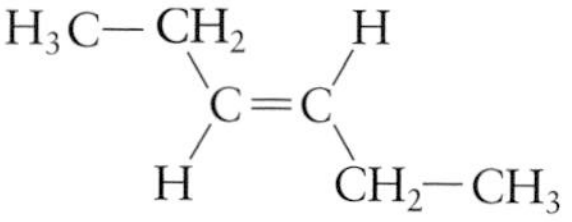

18.19 (a)

(b) (c)

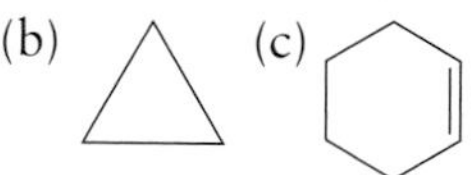

18.21 (a) hexenes:

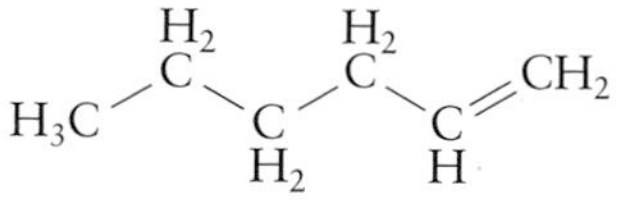

H_3C—CH_2—CH_2—CH_2—CH=CH_2

1-Hexene

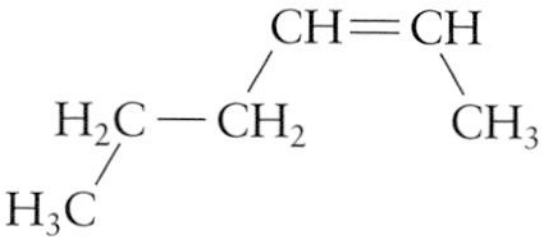

CH=CH, $H_2C—CH_2$, CH_3, H_3C

cis-2-Hexene

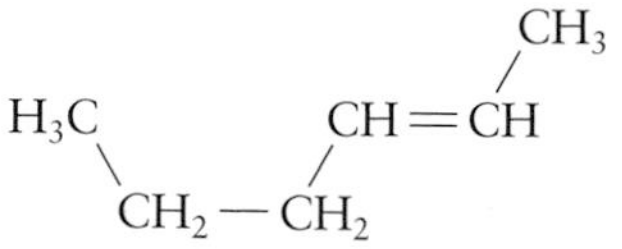

H_3C, CH=CH, CH_3, $CH_2—CH_2$

trans-2-Hexene

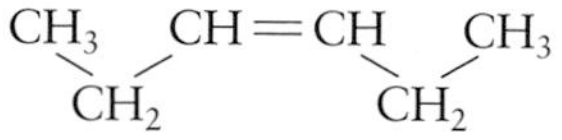

CH_3 CH=CH CH_3, CH_2 CH_2

cis-3-Hexene

CH_3—CH_2—CH=CH—CH_2—CH_3

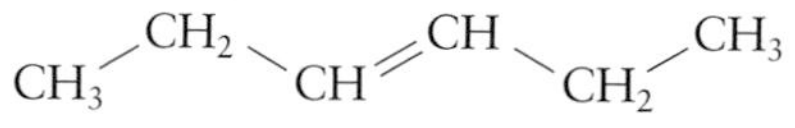

trans-3-Hexene

pentenes:

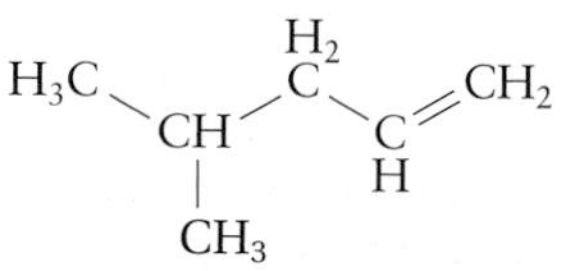

H_3C—CH(CH_3)—CH_2—CH=CH_2

4-Methyl-1-pentene

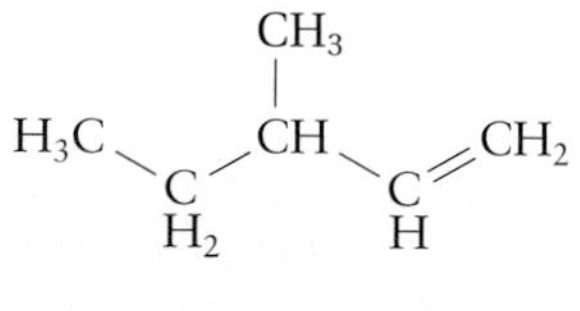

H_3C—CH_2—CH(CH_3)—CH=CH_2

3-Methyl-1-pentene

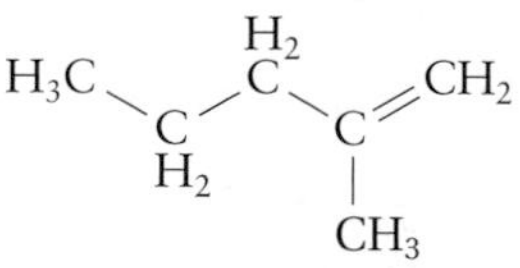

H_3C—CH_2—CH_2—C(CH_3)=CH_2

2-Methyl-1-pentene

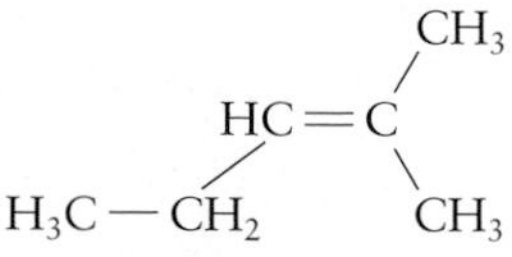

$H_3C—CH_2$—HC=C(CH_3)CH_3

2-Methyl-2-pentene

H_3C, C=CH, $H_3C—CH_2$, CH_3

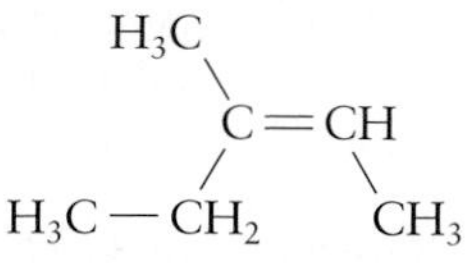

cis-3-Methyl-2-pentene

(+ trans isomer)

HC=CH, $H_3C—CH$, CH_3, CH_3

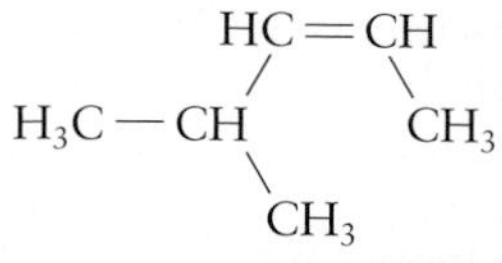

cis-4-Methyl-2-pentene

(+ trans isomer)

butenes:

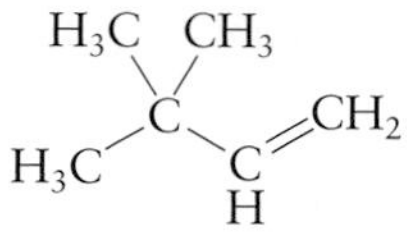

H_3C CH_3, C, H_3C, C(H)=CH_2

3,3-Dimethyl-1-butene

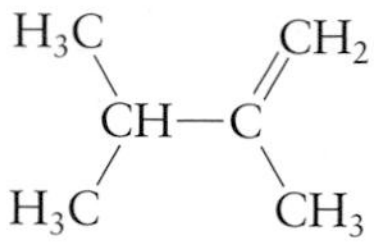

H_3C, CH_2, CH—C, H_3C, CH_3

2,3-Dimethyl-1-butene

H_3C CH_3, C=C, H_3C CH_3

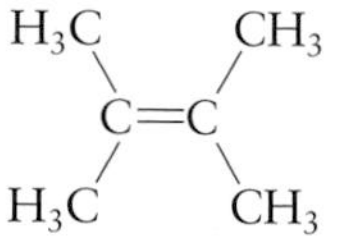

2,3-Dimethyl-2-butene

(b) cyclic molecules:

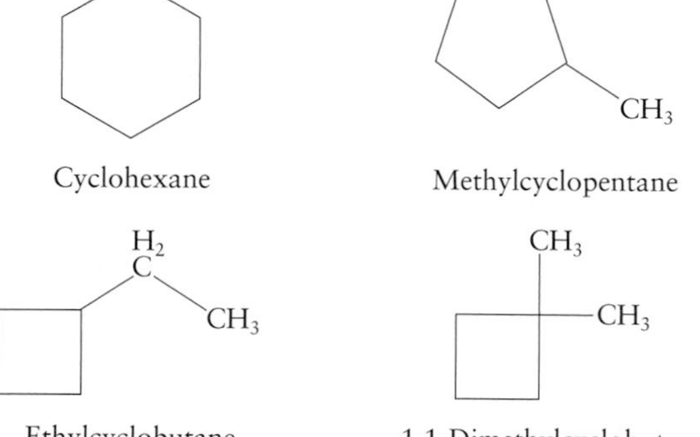

Cyclohexane

Methylcyclopentane

Ethylcyclobutane

1,1-Dimethylcyclobutane

The following structures are drawn to emphasize the stereochemistry

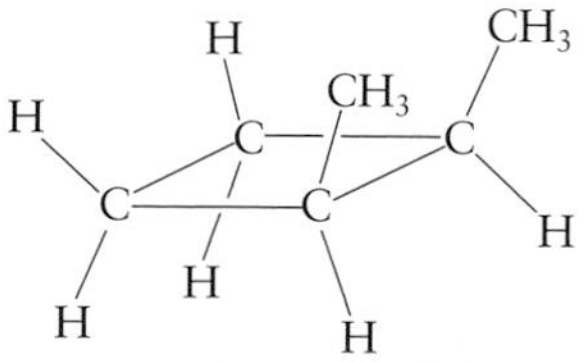

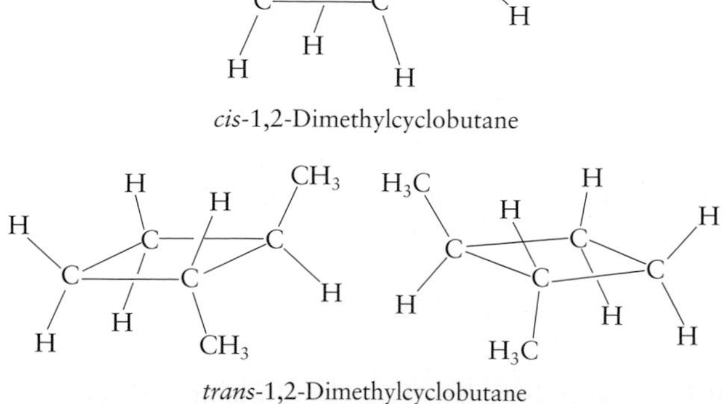

cis-1,2-Dimethylcyclobutane

trans-1,2-Dimethylcyclobutane

(nonsuperimposable mirror images)

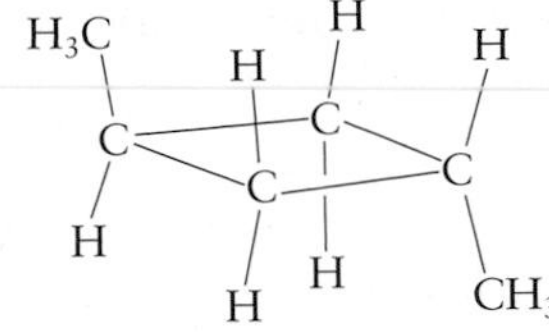

trans-1,3-Dimethylcyclobutane

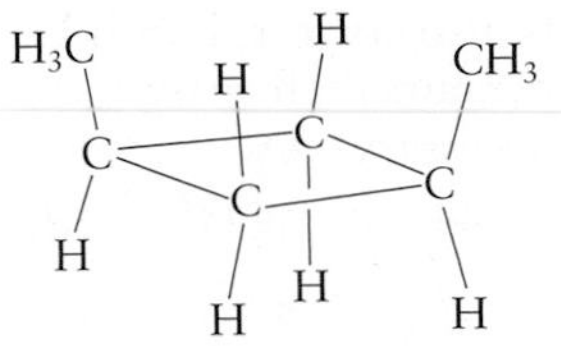

cis-1,3-Dimethylcyclobutane

1,2,3-Trimethylcyclopropane (all cis isomer)

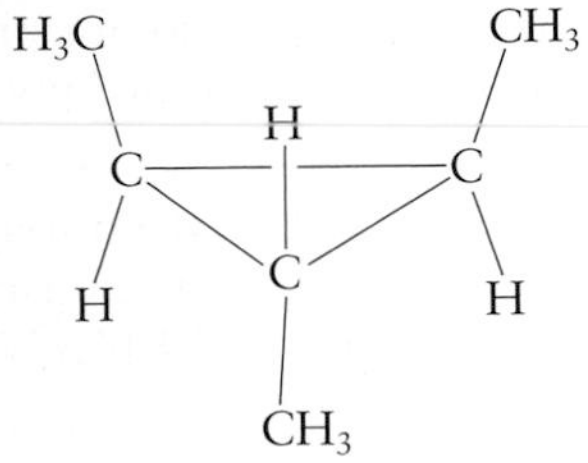

1,2,3-Trimethylcyclopropane (cis, trans isomer)

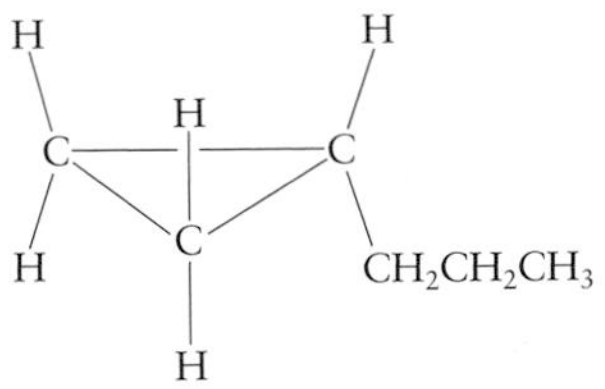

Propylcyclopropane

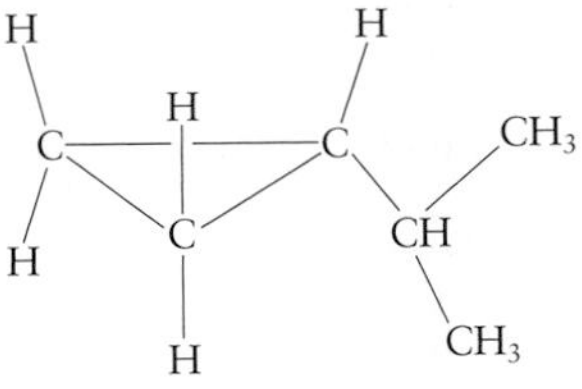

Isopropylcyclopropane or 2-cyclopropylpropane

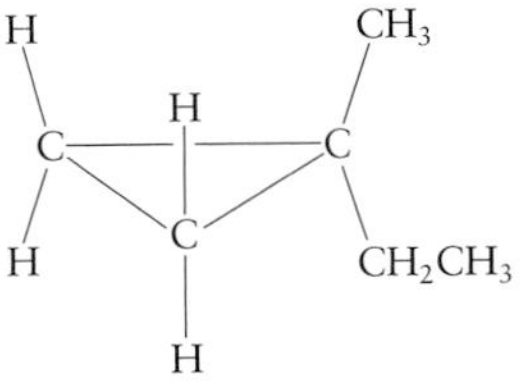

1-Ethyl-1-methylcyclopropane

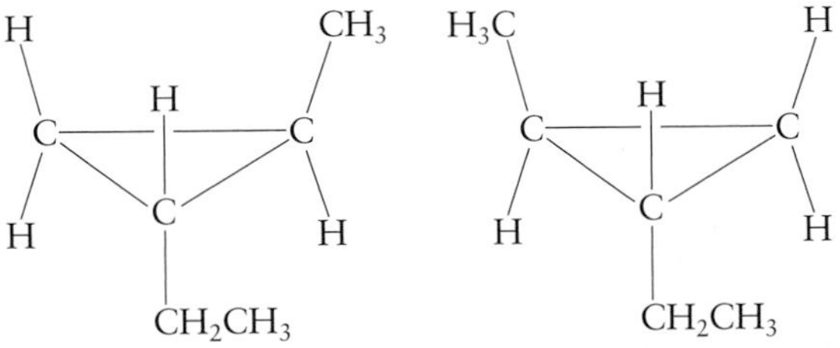

trans-1-Ethyl-2-methylcyclopropane (nonsuperimposable mirror images)

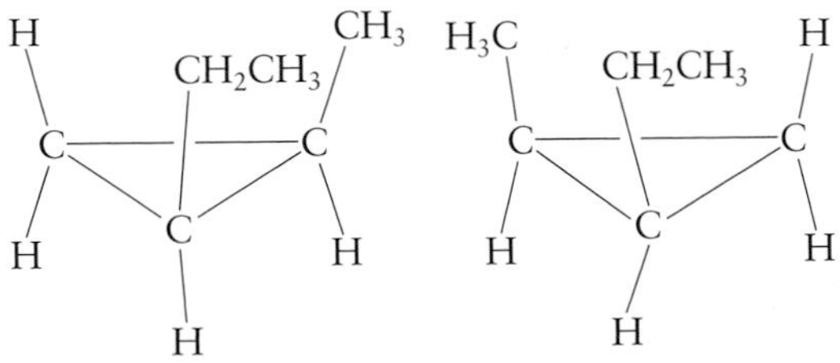

cis-1-Ethyl-2-methylcyclopropane (nonsuperimposable mirror images)

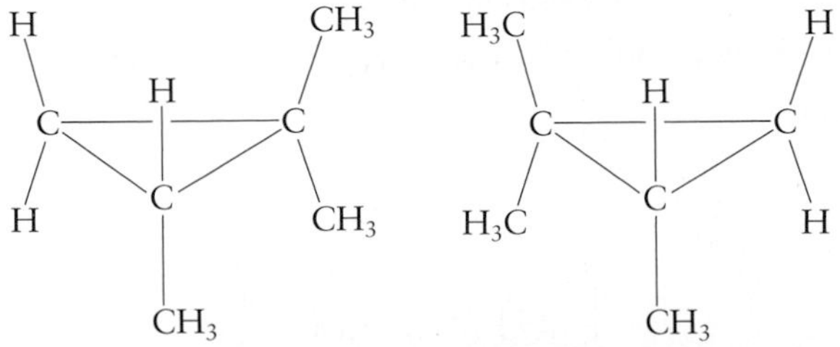

1,1,2-Trimethylcyclopropane (nonsuperimposable mirror images)

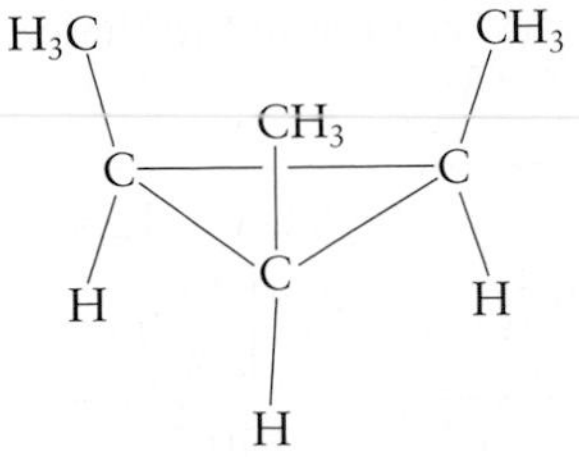

18.23 (a) not isomers; (b) structural isomers; (c) geometrical isomers; (d) not isomers, just different views of the same molecule

18.25 (a) $H_3C-C(H)(CH_3)-CH_3$

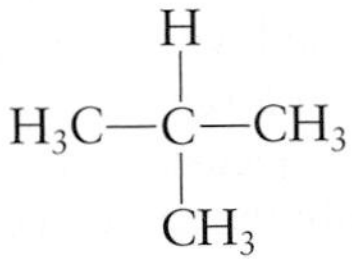

(b) $H_3C-C(Cl)(CH_3)-CH_3$ $\quad$ $H_2C-C(Cl)(H)-CH_3$ with CH_3 below

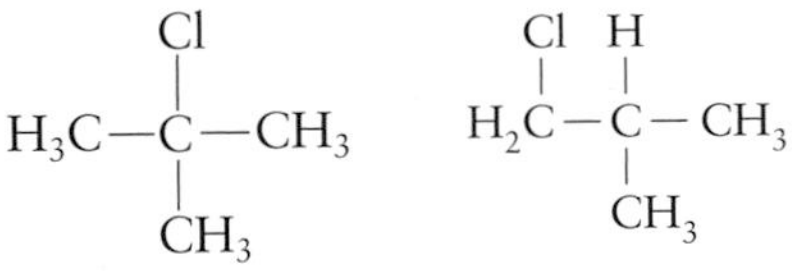

18.27 An* designates a chiral carbon.

(a) optically active, $H_3C-C^{*}(H)(Br)-CH_2CH_3$

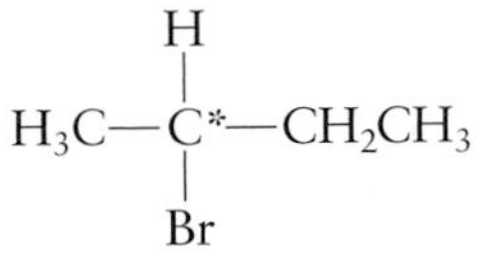

(b) not optically active, $Cl-C(H)(Cl)-C(H)(H)-CH_3$

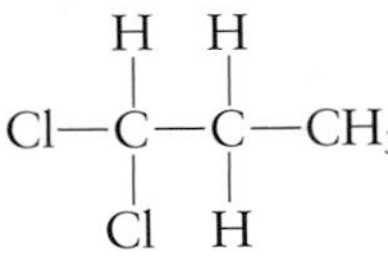

(c) optically active, $H-C(Br)(H)-C^{*}(Cl)(H)-CH_3$

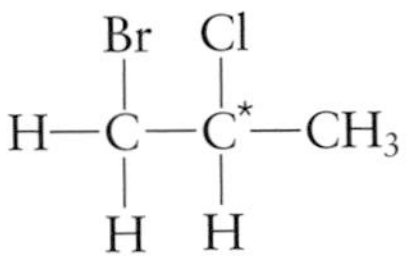

(d) optically active, $H-C(Cl)(H)-C^{*}(Cl)(H)-CH_2CH_2CH_3$

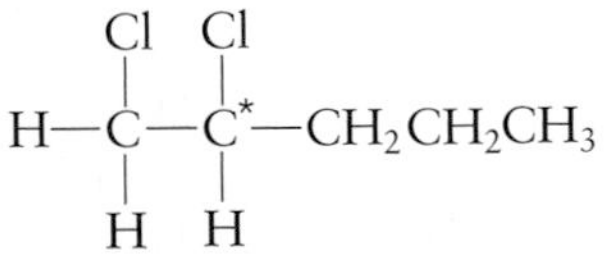

18.29 The difference can be traced to the weaker London forces that exist in branched molecules. Atoms in neighboring branched molecules cannot lie as close together as they can in unbranched isomers.

18.31 The balanced equations are

$$C_3H_8(g) + 5\ O_2(g) \rightarrow 3\ CO_2(g) + 4\ H_2O(l)$$

$$C_4H_{10}(g) + \tfrac{13}{2}\ O_2(g) \rightarrow 4\ CO_2(g) + 5\ H_2O(l)$$

$$C_5H_{12}(g) + 8\ O_2(g) \rightarrow 5\ CO_2(g) + 6\ H_2O(1)$$

The enthalpies of combustion that correspond to these reactions are listed in Appendix 2A:

Compound	(a) Enthalpy of combustion, $kJ \cdot mol^{-1}$	(b) Heat released per g, $kJ \cdot g^{-1}$
propane	−2220	50.3
butane	−2878	49.5
pentane	−3537	49.0

The molar enthalpy of combustion increases with molar mass, as might be expected, because the number of moles of CO_2 and H_2O formed will increase as the number of carbon and hydrogen atoms in the compounds increases. The heat released per gram of these hydrocarbons is essentially the same because the ratio of H atoms to C atoms is similar in the three hydrocarbons.

18.33 nine, none of which are optical isomers

18.35 (a)

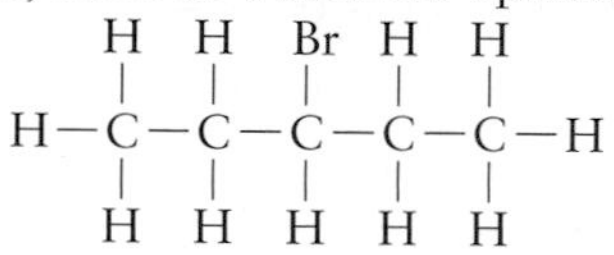

```
    H  H  Br H  H
    |  |  |  |  |
 H—C—C—C—C—C—H
    |  |  |  |  |
    H  H  H  H  H
```

3-Bromopentane

```
    H  Br H  H  H
    |  |  |  |  |
 H—C—C—C—C—C—H
    |  |  |  |  |
    H  H  H  H  H
```

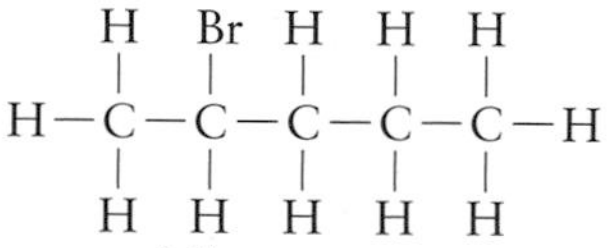

2-Bromopentane

(b) addition reaction

18.37 (a) $C_6H_{11}Br + NaOCH_2CH_3 \rightarrow C_6H_{10} + NaBr + HOCH_2CH_3$

(b) (bromocyclohexane) + $NaOCH_2CH_3 \longrightarrow$ (cyclohexene) + $NaBr + HOCH_2CH_3$

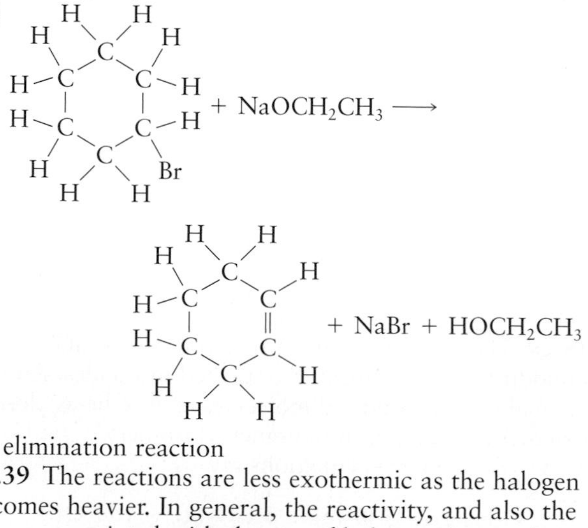

(c) elimination reaction

18.39 The reactions are less exothermic as the halogen becomes heavier. In general, the reactivity, and also the danger associated with the use of halogens in reactions, decreases as one descends the periodic table.

18.41 (a) 1-ethyl-3-methylbenzene;
(b) pentamethylbenzene (1,2,3,4,5-pentamethylbenzene is also correct, but, because there is only one possible pentamethylbenzene, the use of the numbers is not necessary)

18.43 (a) CH_3 (b) CH_3 … Cl (c) CH_3 … CH_3 (d) CH_3 … Cl

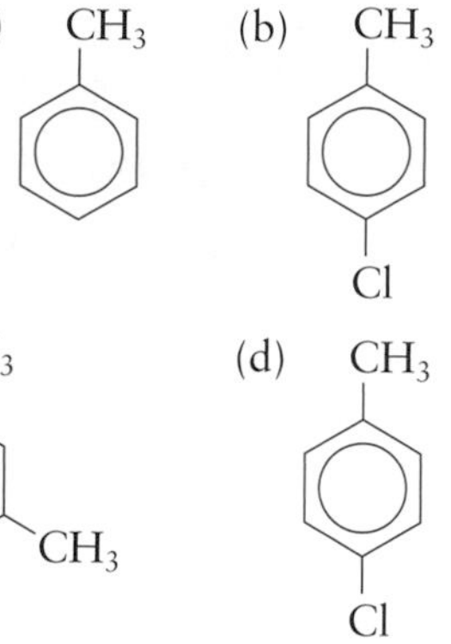

18.45

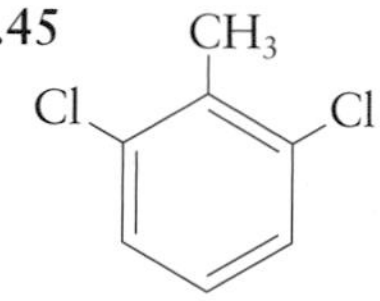

1,3-Dichloro-2-methylbenzene

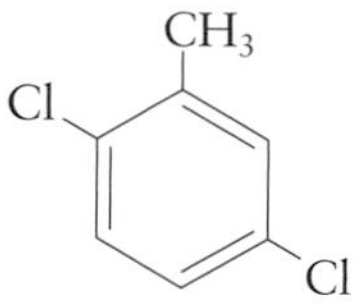

1,4-Dichloro-2-methylbenzene

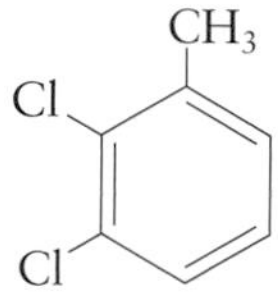

1,2-Dichloro-3-methylbenzene

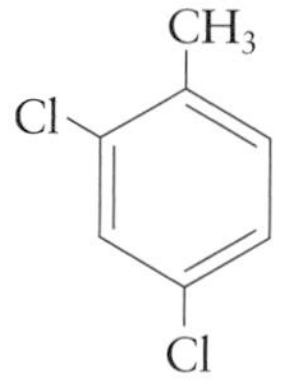

1,5-Dichloro-2-methylbenzene

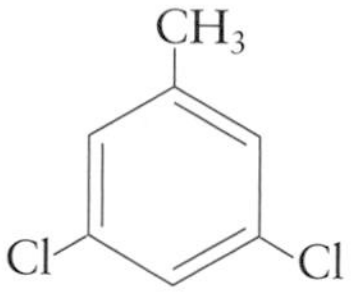

1,3-Dichloro-5-methylbenzene

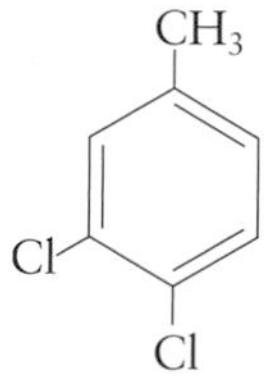

1,2-Dichloro-4-methylbenzene

(b) All these molecules are at least slightly polar.

18.47 (resonance structures of benzonitrile: N≡C– attached to the benzene ring, and three N=C structures with a negative charge on N and a positive charge on the ring)

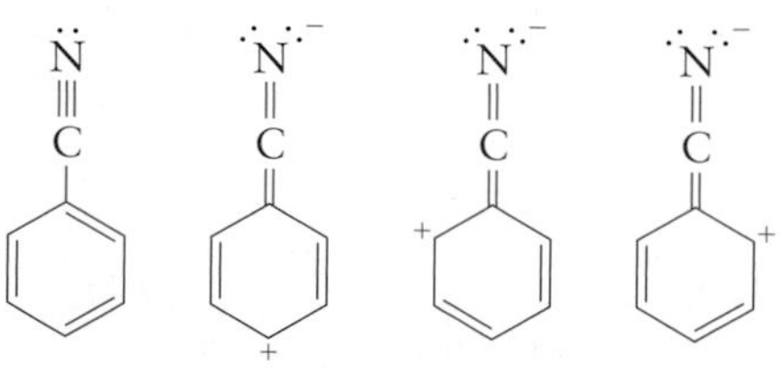

18.49 Two compounds can be produced. Resonance makes positions 1, 4, 6, and 9 equivalent. It also makes positions 2, 3, 7, and 8 equivalent. Positions 5 and 10 are equivalent but have no H atom.

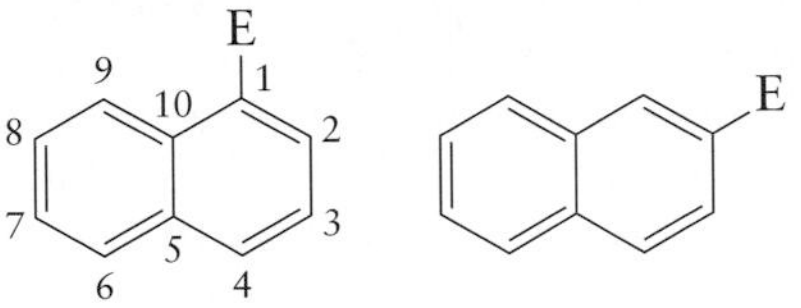

E = electrophile

18.51 (a) 4 σ-type single bonds; (b) 2 σ-type single bonds and 1 double bond with a σ-and a π-bond; (c) 1 σ-type single bond and 1 triple bond with a σ-bond and 2 π-bonds
18.53 (a) substitution, $CH_4 + Cl_2 \rightarrow CH_3Cl + HCl$; (b) addition, $CH_2{=}CH_2 + Br_2 \rightarrow CH_2Br{-}CH_2Br$
18.55 (a) substitution, $CH_3CH_3 + 2\ Cl_2 \rightarrow CH_2ClCH_2Cl + 2\ HCl$ (CH_2ClCH_3 and other chloroalkanes are also produced); (b) addition, $CH_2{=}CH_2 + Cl_2 \rightarrow CH_2ClCH_2Cl$; (c) addition, $HC{\equiv}CH + 2\ Cl_2 \rightarrow CHCl_2CHCl_2$
18.57 The double bond in alkenes makes them more rigid than alkanes. Some of the atoms of alkene molecules tend to be locked into a planar arrangement by the π-bond; hence, they cannot roll up into a ball as compactly as alkanes can. Because they do not pack together as compactly as alkanes do, they have lower boiling and melting points.
18.59 (a) 2-methyl-1-propene, no geometrical isomers; (b) *cis*-3-methyl-2-pentene, *trans*-3-methyl-2-pentene; (c) 1-hexyne, no geometrical isomers; (d) 3-hexyne, no geometrical isomers; (e) 2-hexyne, no geometrical isomers
18.61

1-Heptene

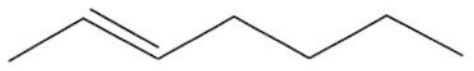

trans-2-Heptene

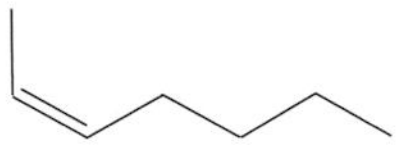

cis-2-Heptene

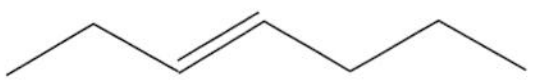

trans-3-Heptene

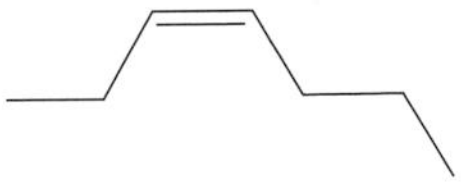

cis-3-Heptene

18.63 (a) $C_{10}H_{18}$; (b) naphthalene, 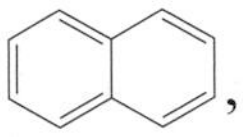 , $C_{10}H_8$; (c) Yes. Cis and trans forms (relative to the C—C bond common to the two six-membered rings) are possible.

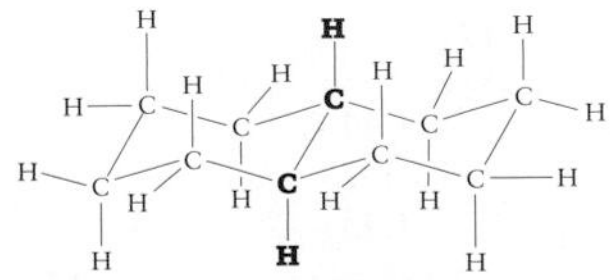

trans-Decalin

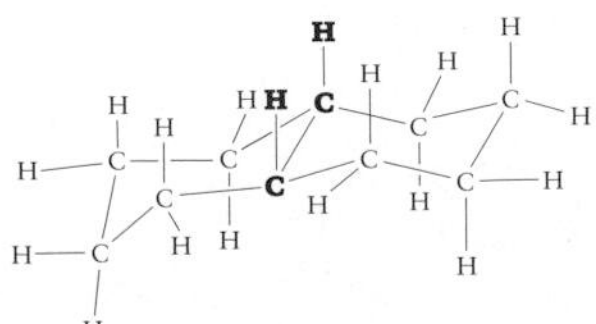

cis-Decalin

18.65 The empirical formula is C_2H_5. The compound is an alkane; the molecular formula might be C_4H_{10}, which matches the formula for alkanes (C_nH_{2n+2}). It is not an alkene or alkyne, because they both have (mole H):(mole C) ratios less than 2.5.
18.67 (a)

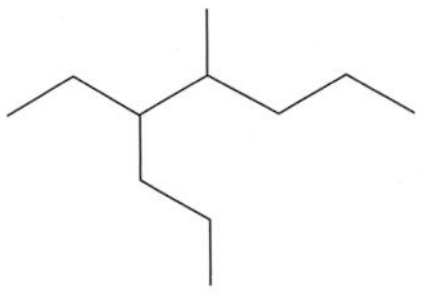

4-ethyl-5-methyloctane;

(b)

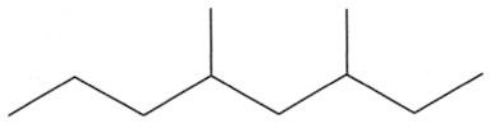

3,5-dimethyloctane;

(c)

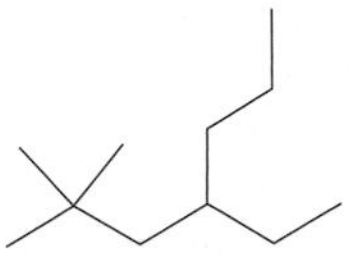

2,2-dimethyl-4-ethylheptane;

(d)

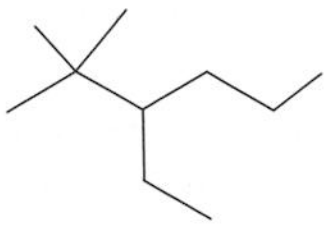

3-ethyl-2,2-dimethylhexane

18.69 These hydrocarbons are too volatile (they are all gases at room temperature) and would not remain liquid.
18.71 Cracking is the process of breaking down hydrocarbons with many carbon atoms into smaller units, whereas alkylation is the process of combining smaller hydrocarbons into larger units. Both processes are carried out catalytically, and both are used to convert hydrocarbons into units having from 5 to 11 carbon atoms suitable for use in gasoline.
18.73 $CH_3(CH_2)_2CH{=}CH_2$ (minor product: $CH_3CH_2CH{=}CHCH_3$)
18.75 The NO_2 group is a meta-directing group and the Br atom is an ortho, para-directing group. Because the position para to Br is already substituted with the NO_2 group, further bromination will not occur there. The resonance forms show that the bromine atom will activate the position ortho to it as expected. The NO_2 group will *deactivate* the group ortho to itself, thus in essence enhancing the reactivity of the position meta to the NO_2 group. This position is ortho to the Br atom, so the effects of the Br and NO_2 groups reinforce each other. Bromination is thus expected to occur as shown:

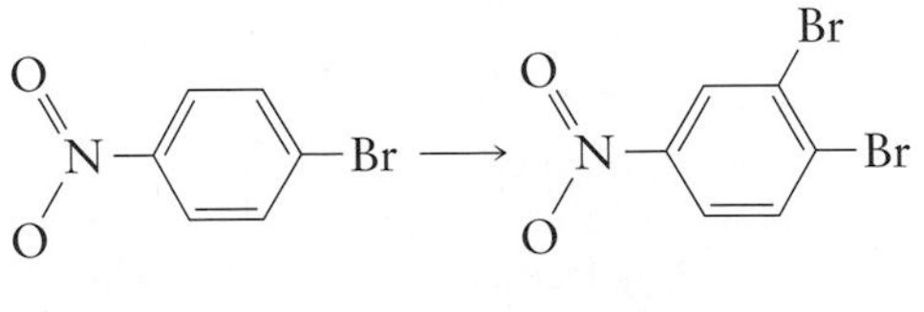

18.77 3:2
18.79 If the molecule contains two carbon centers that have four different substituents attached but are arranged such that they are mirror images of one another *within* the molecule, the molecule will not be optically active. Such an example is shown below in general for a 1,2-X-1,2-Y-1,2,-Z substituted ethane. Many other examples are possible, the only criterion

being that the carbon atoms that have four substituents must have a mirror image carbon center within the molecule.

Z Z
Y C—C Y
X X

A mirror plane exists in the molecule

18.81 For a molecule such as 1,2-dichloro-4-diethylbenzene, $C_6H_3Cl_2(CH_2CH_3)$, 175.04 u, it is relatively easy to lose heavy atoms such as chlorine and groups of atoms such as methyl and ethyl fragments. Molecules can also lose hydrogen atoms. In mass spectroscopy, *P* is used to represent the *parent ion,* which is the ion formed from the molecule without fragmentation. Fragments are then represented as $P - x$, where x is the particular fragment lost from the parent ion to give the observed mass. Because the mass spectrum will measure the masses of individual molecules, the mass of carbon used will be 12.00 u (by definition) because the large majority of the molecules will have all ^{12}C. The mass of H is 1.0078 u. Some representative peaks that may be present are listed below.

Fragment formula	Relation to parent ion	Mass, u
$C_6H_3{}^{35}Cl_2(CH_2CH_3)$	P	174.00
$C_6H_3{}^{35}Cl^{37}Cl(CH_2CH_3)$	P	176.00
$C_6H_3{}^{37}Cl_2(CH_2CH_3)$	P	177.99
$C_6H_3{}^{35}Cl(CH_2CH_3)$	$P - Cl$	139.03
$C_6H_3{}^{37}Cl(CH_2CH_3)$	$P - Cl$	141.03
$C_6H_3{}^{35}Cl_2(CH_2)$	$P - CH_3$	158.98
$C_6H_3{}^{35}Cl^{37}Cl(CH_2)$	$P - CH_3$	160.97
$C_6H_3{}^{37}Cl_2(CH_2)$	$P - CH_3$	162.97
$C_6H_3{}^{35}Cl_2$	$P - CH_2CH_3$	144.96
$C_6H_3{}^{35}Cl^{37}Cl$	$P - CH_2CH_3$	146.96
$C_6H_3{}^{37}Cl_2$	$P - CH_2CH_3$	148.96
$C_6H_3{}^{35}Cl$	$P - CH_2CH_3 - Cl$	109.99
$C_6H_3{}^{37}Cl$	$P - CH_2CH_3 - Cl$	111.99
etc.		

18.83 The presence of one bromine atom will produce in the ions that contain Br companion peaks that are separated by 2 u. Any fragment that contains Br will show this "doublet," in which the peaks are nearly but not exactly equal in intensity. Thus, a mass spectrum of a compound that is known to have Br or was involved in a reaction in which Br could have been added or substituted showing such doublets is a good indication that Br is present in the compound. It is also fairly easy to detect Br atoms in the mass spectrum at 79 and 81 u, thus confirming their presence. If more than one Br atom is present, then a more complicated pattern is observed for the presence of the two isotopes. The possible combinations for a molecule of unknown formula with two Br atoms is $^{79}Br^{79}Br$, $^{79}Br^{81}Br$, $^{81}Br^{79}Br$, and $^{81}Br^{81}Br$. Thus a set of three peaks (the two possibilities $^{79}Br^{81}Br$ and $^{81}Br^{79}Br$ have identical masses) will be generated and will differ in mass by two units. The center peak, which is produced by the $^{79}Br^{81}Br$ and $^{81}Br^{79}Br$ combinations will have twice the intensity of the outer two peaks because statistically there are twice as many combinations that produce this mass. All modern mass spectrometers have spectral simulation programs that can readily calculate and print out the relative isotopic distribution pattern expected for any compound formulation, so it is possible to match the experimental pattern for a particular ion with the experimental result quite easily.

Chapter 19

19.1 (a) RNH_2, R_2NH, R_3N; (b) ROH; (c) RCOOH; (d) RCHO
19.3 (a) ether; (b) ketone; (c) amine; (d) ester
19.5 (a) 2-iodo-2-butene; (b) 2,4-dichloro-4-methylhexane; (c) 1,1,1-triiodoethane; (d) dichloromethane
19.7 (a) [benzene ring with OH]—$ClC_6H_4Cl(OH)$, phenol;
OH
(b) $CH_3CH(CH_3)CH(OH)CH_2CH_3$, secondary alcohol; (c) $CH_3CH_2CH(CH_3)CH_2CH(CH_3)CH_2OH$, primary alcohol; (d) $CH_3C(CH_3)(OH)CH_2CH_3$, tertiary alcohol
19.9 (a) $CH_3CH_2OCH_3$; (b) $CH_3CH_2OCH_2CH_2CH_3$; (c) CH_3OCH_3
19.11 (a) butyl propyl ether; (b) methyl phenyl ether; (c) pentyl propyl ether
19.13 (a) aldehyde, ethanal; (b) ketone, propanone; (c) ketone, 3-pentanone
19.15 (a) H, C=O, H (b) CH_3, C=O, CH_3

(c) CH_3, C=O, $CH_3(CH_2)_3CH_2$

19.17 (a) ethanoic acid; (b) butanoic acid; (c) 2-aminoethanoic acid
19.19 (a) [benzene ring]—C(OH)=O (b) H_3C—C(H_2)—CH(CH_3)—CH(Cl)—C(=O)—O—H

(c) 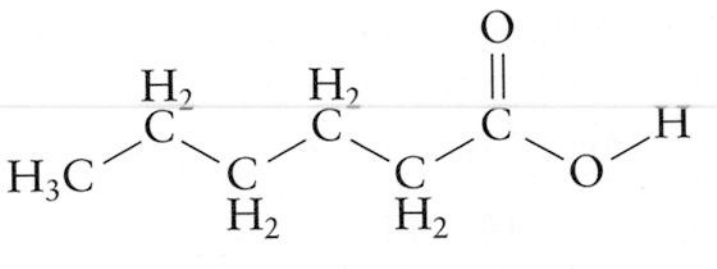

(d) 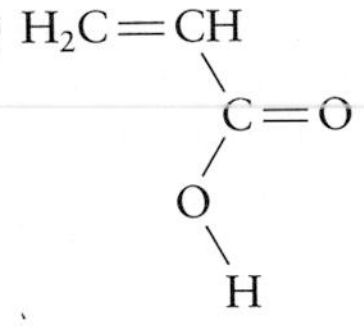

19.21 (a) methylamine; (b) diethylamine; (c) 2-methylaniline, *o*-methylaniline, *o*-methylphenylamine, 1-amino-2-methylbenzene

19.23 (a) 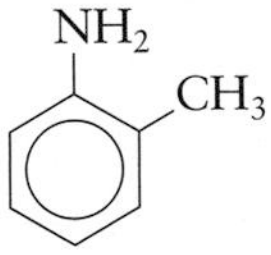

(b) 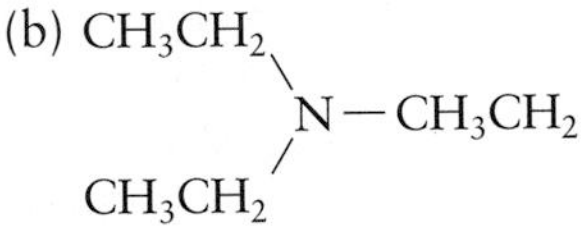

(c) 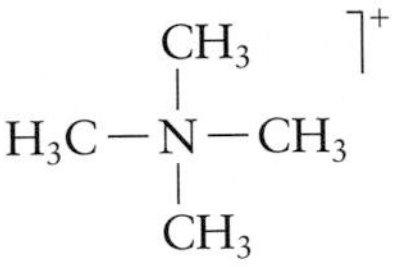

19.25 (a) The product is $C_6H_5CH(OH)CH_3$ (b) The mechanism must be S_N1 because an S_N2 mechanism would produce inversion of configuration and an optically active product. The dissociation of the halide ion prior to addition of water would allow the intermediate ion to form a racemic product.

19.27 (a) and (c)

19.29 (a) ethanol; (b) 2-octanol; (c) 5-methyl-1-octanol. These reactions can be accomplished with an oxidizing agent such as acidified sodium dichromate, $Na_2Cr_2O_7$.

19.31 (a) 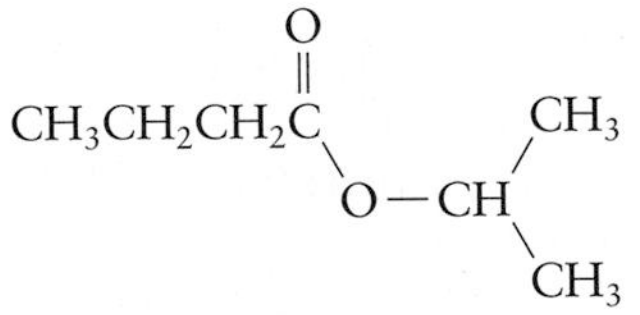

(b) 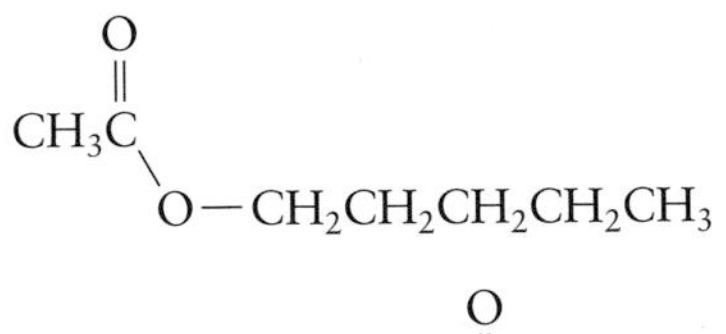

(c) 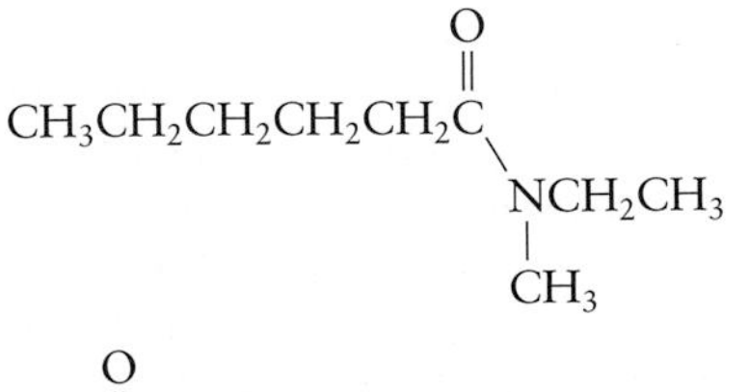

(d) 

19.33 The following procedures can be used:
(1) Dissolve the compounds in water and use an acid-base indicator to look for a color change.
(2) $CH_3CH_2CHO \xrightarrow{\text{Tollens reagent}} CH_3CH_2COOH + Ag(s)$
(3) $CH_3COCH_3 \xrightarrow{\text{Tollens reagent}}$ no reaction
Procedure (1) will distinguish ethanoic acid; (2) and (3) will distinguish propanal from 2-propanone.

19.35 (a) 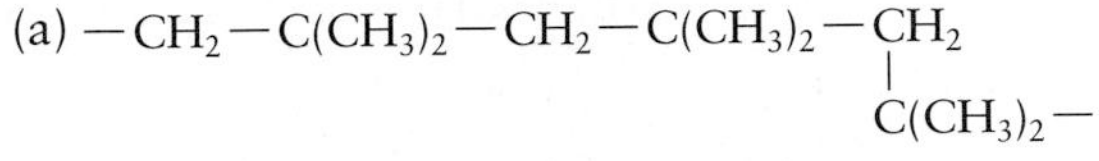

(b) 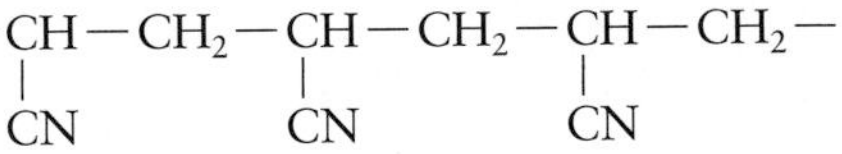

(c) 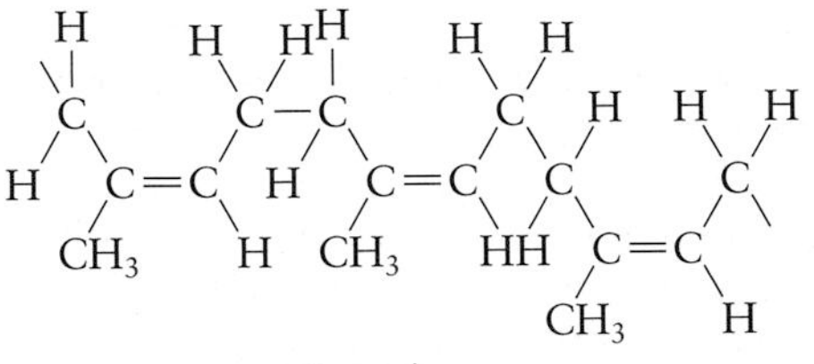

cis version

(d) 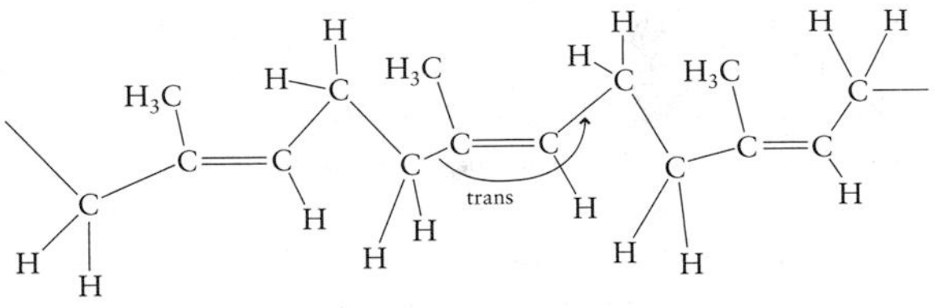

trans version

19.37 (a) $CHCl{=}CH_2$; (b) $CFCl{=}CCF_2$

19.39 (a) $-OCCONH(CH_2)_4NHCOCONH(CH_2)_4NH-$
(b) $-OC-CH(CH_3)-NH-OC-CH(CH_3)-NH-$

19.41 An isotactic polymer is a polymer in which the substituents are all on the same side of the chain. A syndiotactic polymer is a polymer in which the substituent groups alternate, one side of the chain to the other. An atactic polymer is a polymer in which the groups are randomly attached, to one side or the other, along the chain.

19.43 block copolymer

19.45 Characteristics (a), (b), and (c) all increase with increasing molar mass.

19.47 Highly linear, unbranched chains allow maximum interaction between chains. More interactions lead to stronger forces between chains and thus a stronger material.

19.49 (c)

19.51 (a) 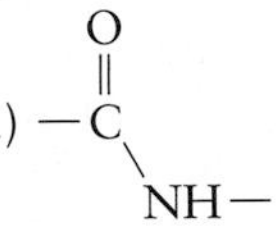 (b) amide; (c) condensation

19.53 Serine, threonine, tyrosine, aspartic acid, glutamic acid, lysine, arginine, histidine, asparagine, and glutamine contribute to tertiary structure through hydrogen bonding. Proline and tryptophan generally do not contribute through hydrogen bonding because they are typically found in hydrophobic regions of proteins.

19.55

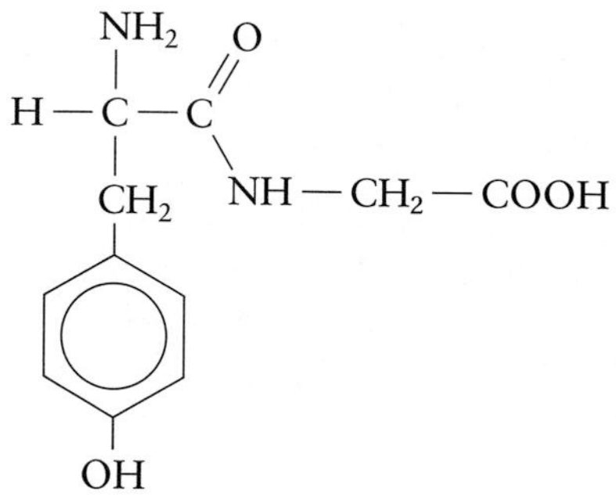

19.57 (a) alcohol, aldehyde; (b) The chiral C atoms are marked with asterisks (*):

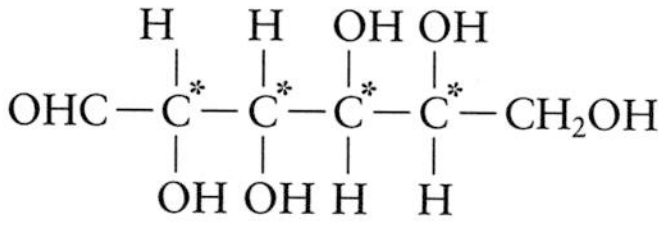

$OHC-C^{*}H(OH)-C^{*}H(OH)-C^{*}(OH)H-C^{*}(OH)H-CH_2OH$

19.59 (a) GTACTCAAT; (b) ACTTAACGT
19.61 (a) $C_5H_5N_5O$; (b) $C_6H_{12}O_6$; (c) $C_3H_7NO_2$
19.63 (a) alcohol, ether, aldehyde; (b) ketone; (c) amine, amide
19.65 (a) carboxylic acid, ester; (b) ether, ketone, phenol, alkene; (c) aromatic amine, tertiary amine; (d) ketone, alcohol, alkene
19.67 An asterisk (*) indicates a chiral carbon atom.

(a) (b)

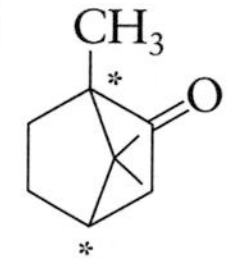

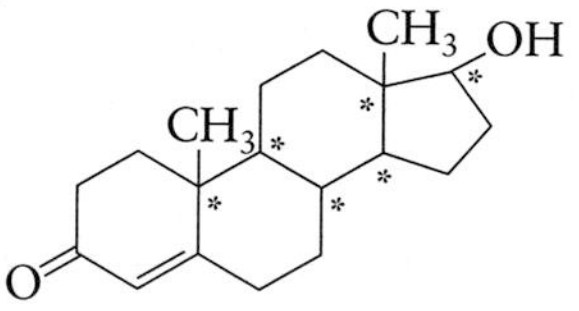

19.69 An asterisk (*) denotes a chiral carbon atom.

(a)

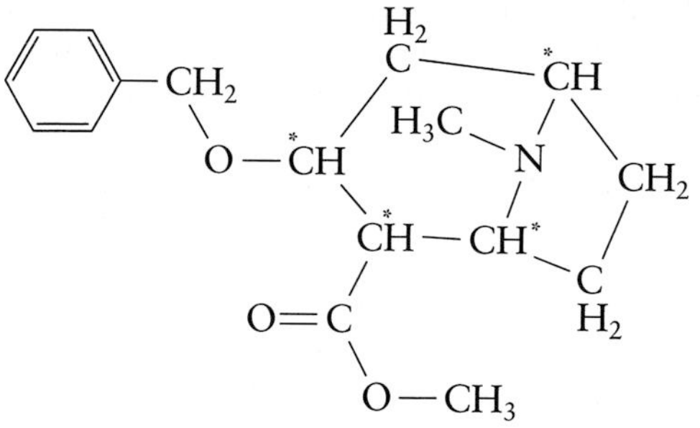

(b)

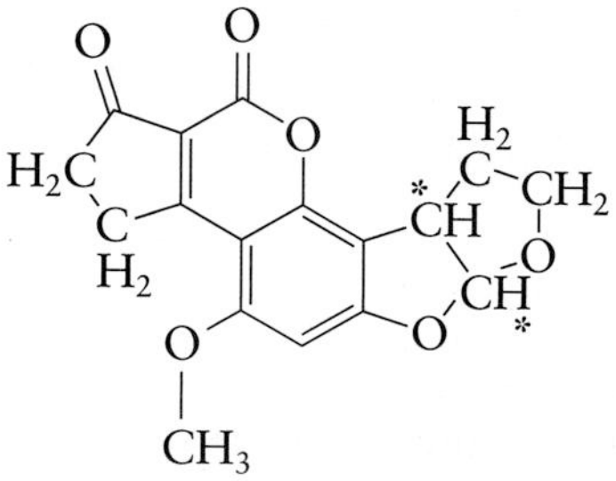

19.71a (a) $CH_3-CH_2-O-CH_2-CH_3$

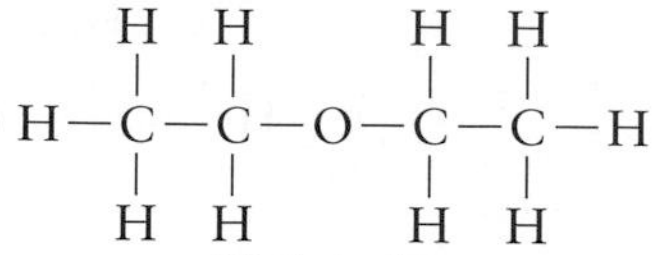

Diethyl ether

$CH_3-CH_2-CH_2-CH_2-OH$

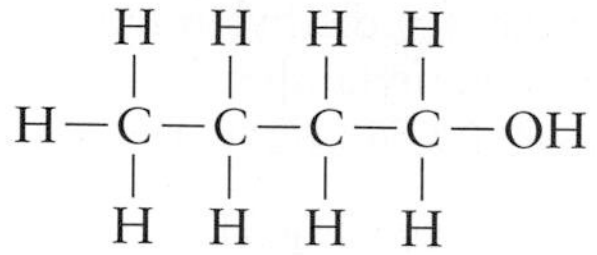

1-Butanol

(b) 1-Butanol can hydrogen bond with itself but diethyl ether cannot, so 1-butanol molecules are held together more strongly in the liquid and therefore 1-butanol has the higher boiling point. Both compounds can form hydrogen bonds with water and therefore have similar solubilities.

19.73 (a)

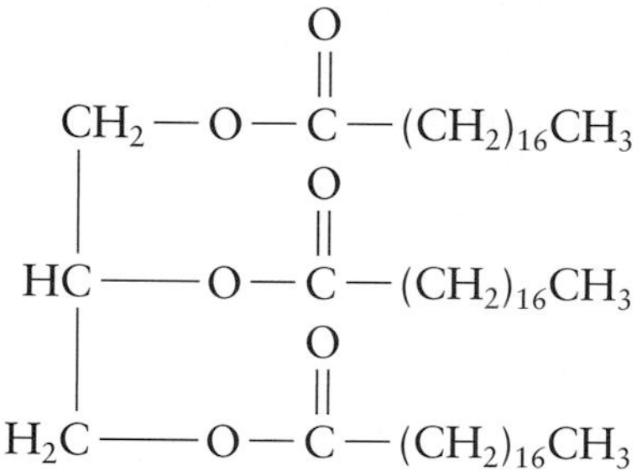

(b) HO–C_6H_4–CHO

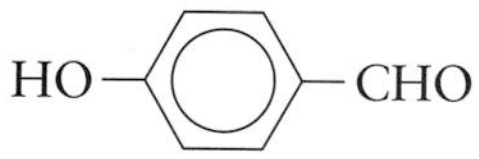

19.75 (a) addition; (b) condensation; (c) addition; (d) addition; (e) condensation
19.77 (a) oxidation; (b) neither; (c) reduction; (d) oxidation; (e) oxidation
19.79 (a) $HOCH_2CH_2OH + 2\ CH_3(CH_2)_{16}COOH \rightarrow CH_3(CH_2)_{16}COOCH_2CH_2OOC(CH_2)_{16}CH_3 + 2\ H_2O$; (b) $2\ CH_3CH_2OH + HOOCCOOH \rightarrow CH_3CH_2OOCCOOCH_2CH_3 + 2\ H_2O$; (c) $CH_3CH_2CH_2CH_2OH + CH_3CH_2COOH \xrightarrow{\Delta} CH_3CH_2COOCH_2CH_2CH_2CH_3 + H_2O$
19.81 increasing value as fibers: polyalkenes < polyesters < polyamides, as a result of the increasing strength of intermolecular forces between the chains. The three types of polymers have similar London forces if their chains are about the same length. However, polyesters also have dipole forces contributing to the strength of intermolecular forces, and polyamides form very strong hydrogen bonds between their chains.

19.83

(a)

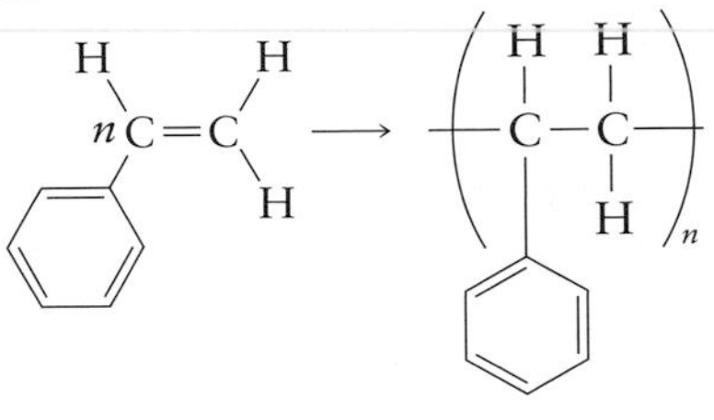

$$n\,\mathrm{C_6H_5CH{=}CH_2} \longrightarrow \left[\begin{array}{c} \mathrm{H\;\;H} \\ -\mathrm{C}-\mathrm{C}- \\ \mathrm{C_6H_5\;\;H} \end{array}\right]_n$$

(b)

$$n\,\mathrm{H_3CCH{=}CH_2} \longrightarrow \left[\begin{array}{c} \mathrm{H\;\;\;H} \\ -\mathrm{C}-\mathrm{C}- \\ \mathrm{CH_3\,H} \end{array}\right]_n$$

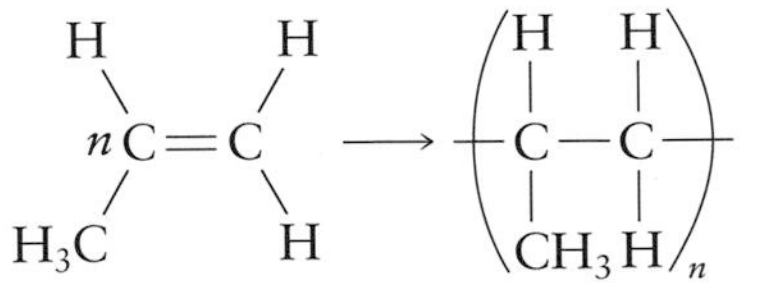

(c) The backbones of all three polymers are the same. They differ simply in what is attached to the long carbon chain. Both styrene and propylene form polymers that have organic groups attached to the polymer backbone, in the first case a phenyl ring and in the other a methyl group. These groups are known as *side groups* or *side chains*. For ethylene, there is only one way in which the pure monomer can be polymerized, but for propylene and styrene, the asymmetry in the monomer means that the final polymer will be more complex. This complexity is illustrated by the more extended structures shown below for a general side group R. Notice that the polymer may be very regular or somewhat irregular in the way the R groups become attached with respect to each other. The presence of regular or irregular structures affects the properties of the polymers as does the nature of the side chains–not only because of their size but also because of their shape, polarity, etc.

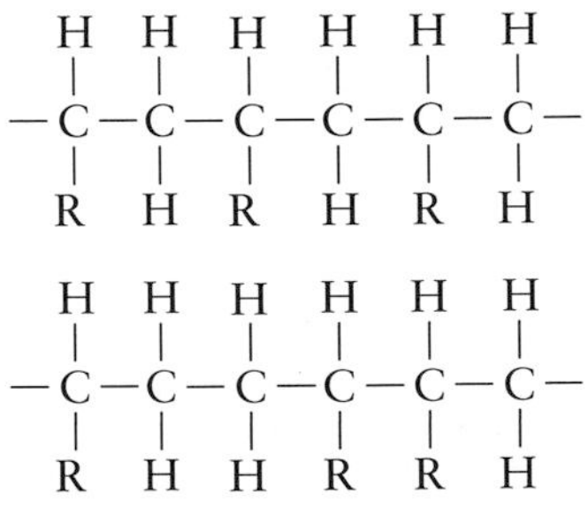

$$\begin{array}{c} \mathrm{H\;\;H\;\;H\;\;H\;\;H\;\;H} \\ -\mathrm{C-C-C-C-C-C}- \\ \mathrm{R\;\;H\;\;R\;\;H\;\;R\;\;H} \end{array}$$

$$\begin{array}{c} \mathrm{H\;\;H\;\;H\;\;H\;\;H\;\;H} \\ -\mathrm{C-C-C-C-C-C}- \\ \mathrm{R\;\;H\;\;H\;\;R\;\;R\;\;H} \end{array}$$

19.85 (a) Primary structure is the sequence of amino acids along a protein chain. Secondary structure is the conformation of the protein or the manner in which the chain is coiled or layered as a result of interactions between amide groups. Tertiary structure is the shape into which sections of the proteins twist and intertwine as a result of interactions between side groups of the amino acids in the protein. If the protein consists of several polypeptide units, then the manner in which the units stick together is the quaternary structure. (b) The primary structure is held together by covalent bonds. Secondary structure is stabilized by "intermolecular" forces (between chains, in this instance). Tertiary and quaternary structures are maintained by disulfide bonds and the intermolecular forces between the chains.

19.87 The basic difference between polyester fabrics and fabrics such as wool, silk, and nylon is that the latter group are all polyamides. The presence of the amide functional group allows for extensive cross-linking between polymer chains by hydrogen bonding with the amine group on one chain to the carbonyl group on another. Intermolecular attractions are thus greater between polyamide molecules than between polyester molecules. The less polar polyester molecules can more readily accept radon atoms among their chains.

19.89 (HO)$_2$C$_6$H$_3$—$CH_2-CH_2-NH_2$ (HO on the ring para to the side chain and a second HO adjacent to it)

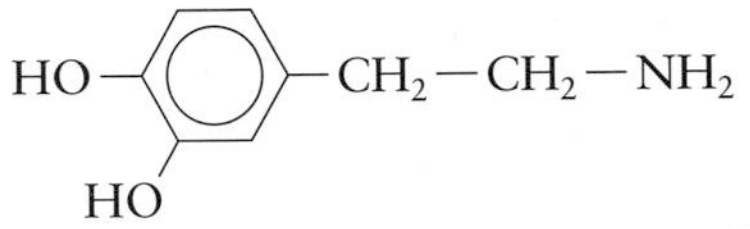

19.91 Condensation polymerization involves the loss of a small molecule, often water or HCl, when monomers are combined. Dacron is more linear than the polymer obtained from benzene-1,2-dicarboxylic acid and ethylene glycol, so Dacron can be more readily spun into yarn.

19.93 Two peaks with relative overall intensities 3:1. The larger peak is due to the three methyl protons and is split into two lines with equal intensities. The smaller peak is due to the proton on the carbonyl carbon atom and is split into three lines with relative intensities 1:3:3:1.

19.95 The peaks in the spectrum can be assigned on the basis of the intensities and the coupling to other peaks. The hydrogen atoms of the CH_3 unit of the ethyl group will have an intensity of 3 and will be split into a triplet by the two protons on the CH_2 unit. This peak is found at about $\delta = 1.2$. The CH_2 group will have an intensity of 2 and will be split into a quartet by the three protons on the methyl group. This peak is found at $\delta = 4.1$. The CH_2 group that is part of the butyl function will have an intensity of two but will appear as a singlet because there are no protons on adjacent carbon atoms. This is the signal found at $\delta = 2.1$. The remaining CH_3 groups are equivalent and also will not show coupling. They can be attributed to the signal at $\delta = 1.0$. Notice that the peak that is furthest downfield is the one for the CH_2 group attached directly to the electronegative oxygen atom, and that the second furthest downfield peak is the one attached to the carbonyl group.

19.97 2-chloropropane

19.99 (a) ^{13}C; (b) 1.11%; (c) No. The probability of finding two ^{13}C nuclei next to each other is very low. A ^{12}C nucleus next to a ^{13}C nucleus will not interact with the ^{13}C nucleus because the ^{12}C nucleus has no spin.

Because the natural abundance of ^{13}C is 1.11%, the probability of finding two ^{13}C atoms next to each other in an organic molecule will be 0.0111×0.0111 or 1.23×10^{-4}. Although such coupling is possible, it is generally not observed because the signal is so much weaker than the signal due to the molecules with a single ^{13}C nucleus. (d) Maybe. Because most of the carbon atoms to which the protons are attached are ^{12}C, the bulk of the signal will not be split. The protons that are attached to the ^{13}C atoms will be split, but this amounts to only 1.11% of the sample, so the peaks are very small. Peaks that result from coupling to a small percentage of a magnetically active isotope are referred to as satellites and may be observed if one has a very good spectrum. (e) Yes. Even though the splitting of protons by ^{13}C may not be observed because the amount of ^{13}C present is low, the opposite situation is not true. If a ^{13}C is attached to H atoms, the large majority of those H atoms will all have a spin, so the ^{13}C will show fine structure due to splitting by the H atoms.

Self-Test B

Fundamentals

A.1B $V = d/m = (0.176\ 85\ g{\cdot}L^{-1})/(10.0\ g) = 17.7\ cm^3$
A.2B $E_K = mv^2/2 = 1/2 \times (1.5\ kg) \times (3.0\ m{\cdot}s^{-1})^2 = 6.8\ J$
A.3B $E_K = mgh = (0.350\ kg) \times (9.81\ m{\cdot}s^{-2}) \times (443\ m) \times (10^{-3}\ kJ/J) = 1.52\ kJ$
B.1B number of Au atoms = mass (sample)/mass (one atom) = $(0.0123\ kg)/(3.27 \times 10^{-25}\ kg) = 3.76 \times 10^{22}$ Au atoms
B.2B (a) 8, 8, 8; (b) 92, 144, 92

C.1B (a)
```
              H
              |
           H—C—H
      H   H   |   H
      |   |   |   |
  H—C—C—C—C—S—H
      |   |   |   |
      H   H   H   H
```
2-Methyl-1-butanethiol

(b)
```
      H   H       H   H
      |   |       |   |
  H—C—C—O—C—C—H
      |   |       |   |
      H   H       H   H
```
Diethyl ether

C.2B (a) sulfur is a nonmetal in Group 16/VI: anion, $16 - 18 = -2$, so S^{2-}; (b) potassium is a Group 1 metal: cation, +1, so K^+
C.3B (a) Li_3N; (b) $SrBr_2$
D.1B (a) H_3PO_4 is derived from PO_4^{3-}, phosphate ion, so it is called phosphoric acid. (b) Chloric acid is derived from chlorate ion, ClO_3^-, so its formula is $HClO_3$.
D.2B (a) gold (III) chloride; (b) calcium sulfide; (c) manganese (III) oxide
D.3B (a) phosphorus trichloride; (b) sulfur trioxide; (c) hydrobromic acid
D.4B (a) $Cs_2S{\cdot}4H_2O$; (b) Mn_2O_7; (c) HCN; (d) S_2Cl_2
D.5B (a) pentane; (b) a carboxylic acid
E.1B atoms H = $(3.14\ mol\ H_2O) \times (2\ mol\ H/1\ mol\ H_2O) \times (6.022 \times 10^{23}\ atoms\ H/mol\ H) = 3.78 \times 10^{24}$ H atoms
E.2B (a) $n = m/M = (5.4 \times 10^3\ g)/(26.98\ g{\cdot}mol^{-1}) = 2.0 \times 10^2$ mol; (b) $N = N_A \times n = (6.022 \times 10^{23}\ Al\ atom/mol) \times (2.0 \times 10^2\ mol) = 1.2 \times 10^{26}$ Al atoms
E.3B copper-63: $(62.94\ g{\cdot}mol^{-1}) \times 0.69.17 = 43.536\ g{\cdot}mol^{-1}$; copper-65: $(64.93\ g{\cdot}\ mol^{-1}) \times 0.3083 = 20.018\ g{\cdot}mol^{-1}$. $43.536\ g \cdot mol^{-1} + 20.018\ g{\cdot}mol^{-1} = 63.55\ g{\cdot}mol^{-1}$
E.4B (a) phenol: 6 C, 6 H, 1 O; $6(12.01\ g{\cdot}mol^{-1}) + 6(1.008\ g{\cdot}mol^{-1}) + (16.00\ g{\cdot}mol^{-1}) = 94.11\ g{\cdot}mol^{-1}$; (b) $Na_2CO_3{\cdot}10H_2O$: 2 Na, 1 C, 13 O, 20 H; $2(22.99\ g{\cdot}mol^{-1}) + (12.01\ g{\cdot}mol^{-1}) + 13(16.00\ g{\cdot}mol^{-1}) + 20(1.008\ g{\cdot}mol^{-1}) = 286.15\ g{\cdot}mol^{-1}$
E.5B $Ca(OH)_2$: 1 Ca, 2 O, 2 H; $(40.08\ g{\cdot}mol^{-1}) + 2(16.00\ g{\cdot}mol^{-1}) + 2(1.008\ g{\cdot}mol^{-1}) = 74.10\ g{\cdot}mol^{-1}$; $(1.0 \times 10^3\ g\ lime)/(74.10\ g{\cdot}mol^{-1}) = 13.5$ mol lime
E.6B CH_3COOH: 2 C, 4 H, 2 O; $2(12.01\ g{\cdot}mol^{-1}) + 4(1.008\ g{\cdot}mol^{-1}) + 2(16.00\ g{\cdot}mol^{-1}) = 60.05\ g{\cdot}mol^{-1}$; $(60.05\ g{\cdot}mol^{-1})(1.5\ mol) = 90\ g$
F.1B $\%C = (6.61\ g/7.50\ g) \times 100\% = 88.1\%$; $\%H = (0.89\ g/7.50\ g) \times 100\% = 11.9\%$
F.2B $AgNO_3$: $(107.87\ g{\cdot}mol^{-1}) + (14.01\ g{\cdot}mol^{-1}) + 3(16.00\ g{\cdot}mol^{-1}) = 169.88\ g{\cdot}mol^{-1}$; $\%Ag = (107.87\ g{\cdot}mol^{-1})/(169.88\ g{\cdot}mol^{-1}) \times 100\% = 63.50\%$
F.3B carbon: $0.778 \times (100\ g\ eucalyptol) = 77.8\ g\ C$; $(77.8\ g\ C)/(12.01\ g{\cdot}mol^{-1}) = 6.48\ mol\ C$; hydrogen: $(11.8\ g\ H)/(1.0079\ g{\cdot}mol^{-1}) = 11.7\ mol\ H$; oxygen: $(10.4\ g\ O)/(16.00\ g{\cdot}mol^{-1}) = 0.650\ mol\ O$
F.4B mol O = $(18.59\ g)/(16.00\ g{\cdot}mol^{-1}) = 1.162$ mol O; mol S = $(37.25\ g)/(32.07\ g{\cdot}mol^{-1}) = 1.162$ mol; mol F = $(44.16\ g)/(19.00\ g{\cdot}mol^{-1}) = 2.324$ mol; 1:1:2 ratio, formula is SOF_2
F.5B molar mass of CHO_2 is $45.012\ g{\cdot}mol^{-1}$; $(90.0\ g{\cdot}mol^{-1})/(45.012\ g{\cdot}mol^{-1}) = 2.00$; $2 \times (CHO_2) = C_2H_2O_4$
G.1B molar mass of Na_2SO_4 is $142.05\ g{\cdot}mol^{-1}$; $(15.5\ g)/(142.05\ g{\cdot}mol^{-1}) = 0.109$ mol; $(0.109\ mol)/(0.350\ L) = 0.312$ M $Na_2SO_4(aq)$
G.2B $(1.45 \times 10^{-2}\ mol{\cdot}L^{-1}) \times (0.100\ L) = 1.45 \times 10^{-3}$ mol urea
G.3B $(0.125\ mol{\cdot}L^{-1}) \times (0.050\ L) = 0.006\ 25$ mol oxalic acid; molar mass of oxalic acid is $90.036\ g{\cdot}mol^{-1}$; $(0.006\ 25\ mol) \times (90.036\ g{\cdot}mol^{-1}) = 0.563$ g oxalic acid
G.4B $(2.55 \times 10^{-3}\ mol\ HCl)/(0.358\ mol{\cdot}L^{-1}\ HCl) = 7.12$ mL HCl
G.5B $V_{init} = (c_{final} \times V_{final})/c_{init} = (1.59 \times 10^{-5}\ mol{\cdot}L^{-1}) \times (0.025\ L)/(0.0152\ mol{\cdot}L^{-1}) = 2.6 \times 10^{-3}$ mL
H.1B $C_3H_8(g) + 5\ O_2(g) \rightarrow 3\ CO_2(g) + 4\ H_2O(l)$
I.1B (a) nonelectrolyte, does not conduct electricity; (b) strong electrolyte, conducts electricity
I.2B $3\ Hg_2^{2+}(aq) + 2\ PO_4^{3-}(aq) \rightarrow (Hg_2)_3(PO_4)_2(s)$
I.3B $SrCl_2$ and Na_2SO_4; $Sr^{2+}(aq) + SO_4^{2-}(aq) \rightarrow SrSO_4(s)$
J.1B (a) neither; (b) acid; (c) acid; (d) base
J.2B $3\ Ca(OH)_2(aq) + 2\ H_3PO_4(aq) \rightarrow Ca_3(PO_4)_2(s) + 6\ H_2O(l)$
K.1B Cu^+ is oxidized, I_2 is reduced
K.2B (a) $x + 3(-2) = -2$; $x = +4$ for S; (b) $x + 2(-2) = -1$; $x = +3$ for N; (c) $x + 1 + 3(-2) = 0$; $x = +5$ for Cl
K.3B H_2SO_4 is the oxidizing agent (S is reduced from +6 to +4); NaI is the reducing agent (I is oxidized from −1 to +5).
K.4B $2\ Ce^{4+}(aq) + 2\ I^-(aq) \rightarrow 2\ Ce^{3+}(aq) + I_2(s)$
L.1B (a) 2 mol N_2/3 mol O_2; (b) 2 mol N_2/1 mol O_2
L.2B $(2\ mol\ Fe)/(1\ mol\ Fe_2O_3) \times 25\ mol\ Fe_2O_3 = 50$ mol Fe

L.3B 2 mol CO_2/1 mol $CaSiO_3$; amount CO_2 = (44.01 g·mol^{-1}) × (300 g) = 1.32 × 10^4 mol; (1 mol $CaSiO_3$/2 mol CO_2) × (1.32 × 10^4 mol CO_2) = 6.60 × 10^3 mol $CaSiO_3$; (6.60 × 10^3 mol $CaSiO_3$) × (116.17 g·mol^{-1} $CaSiO_3$) = 770 kg $CaSiO_3$
L.4B 2 KOH + $H_2SO_4 \rightarrow K_2SO_4$ + 2 H_2O; 2 mol KOH/1 mol H_2SO_4; (0.255 mol KOH/L) × (0.025 L) = 6.375 × 10^{-3} mol KOH; (6.375 × 10^{-3} mol KOH) × (1 mol/2 mol KOH) = 3.19 × 10^{-3} mol H_2SO_4; (3.19 × 10^{-3} mol H_2SO_4)/(0.01645 L) = 0.194 M H_2SO_4
L.5B (0.100 × 0.02815) mol $KMnO_4$ × (5 mol As_4O_6)/(8 mol $KMnO_4$) × 395.28 g·mol^{-1} = 6.962 × 10^{-2} g As_4O_6
M.1B 2 $Fe_2O_3 \rightarrow$ 4 Fe^0 + 3 O_2; (15 kg Fe_2O_3)/159.69 g·mol^{-1}) × (2 mol Fe)/(1 mol Fe_2O_3) × (55.85 g·mol^{-1}) = 10.5 kg Fe; 8.8 kg/10.5 kg × 100% = 84% yield
M.2B 2 NH_3 + $CO_2 \rightarrow OC(NH_2)_2$ + H_2O; amount NH_3 = (14.5 × 10^3 g)/(17.034 g · mol^{-1}) = 851 mol NH_3; amount CO_2 = (22.1 × 10^3 g)/(44.01 g·mol^{-1}) = 502 mol NH_3; 2 mol NH_3/1 mol CO_2 (a) NH_3 is the limiting reagent. [851 mol NH_3/2] < 502 mol CO_2); (b) 2 mol NH_3/1 mol urea. 426 mol urea can be produced = 25.6 kg urea; (c) (502 − 426) mol = 76 mol CO_2 excess = 3.3 kg CO_2
M.3B The sample contains 0.0118 mol C (0.142 g C) and 0.0105 mol H (0.0106 g H). Mass of O = 0.236 − (0.142 + 0.0105)g = 0.0834 g O (0.00521 mol O). The C:H:O mole ratios are 0.0118:0.0105:0.005 21, or 2.26:2.02:1. Multiplying these numbers by 4 gives 9:8:4 and the empirical formula $C_9H_8O_4$

Chapter 1

1.1B $\lambda = c/\nu$ = (3.00 × 10^8 m·s^{-1})/(98.4 × 10^6 Hz) = 3.05 m
1.2B $T = c_2/5\lambda_{max}$ = (1.44 × 10^{-2} K·m)/(5 × 700 × 10^{-9} m) = 4.11 × 10^3 K
1.3B $E = h\nu$ = (6.626 × 10^{-34} J·s) × (4.8 × 10^{14} Hz) = 3.2 × 10^{-19} J
1.4B $\lambda = (hc)$ × (number of photons)/E_{tot} = (6.626 × 10^{-34} J·s) × (3.00 × 10^8 m·s^{-1}) × (5.5 × 10^{19} photons)/25 J = 440 nm
1.5B $\lambda = h/mv$ = (6.626 × 10^{-34} J·s)/(0.0050 kg × 331 m·s^{-1}) = 4.0 × 10^{-34} m
1.6B $\Delta v = \hbar/2m\Delta x$ = (1.054 × 10^{-34} J)/(2 × 2.0 × 10^3 kg × 1 m) = 2.6 × 10^{-38} m·s^{-1}
1.7B $E_3 - E_2 = 5h^2/8m_eL^2 = h\nu$; $\nu = 5h/8m_eL^2$; $\lambda = c/\nu = 8m_ecL^2/5h$ = [8 × (9.109 × 10^{-31} kg) × (3.00 × 10^8 m·s^{-1}) × (1.50 × 10^{-10} m)^2]/(5 × 6.626 × 10^{-34} J·s) = 14.8 × 10^{-9} m, or 14.8 nm
1.8B 4 nodes
1.9B $\nu = \mathcal{R}(1/2^2 - 1/5^2) = 21\mathcal{R}/100$
$\lambda = c/\nu = 100c/21\mathcal{R}$ = (100 × 3 × 10^8 m·s^{-1})/(21 × 3.29 × 10^{15} s^{-1}) = 434 nm; violet line
1.10B $\psi(\frac{1}{2}a_0, \theta, \phi) = (2e^{-a_0/2a_0}/a_0^{3/2})/(2\pi^{1/2}) = e^{-1/2}/a_0^{3/2}\pi^{1/2}$
$\psi^2(\frac{1}{2}a_0, \theta, \phi) = e^{-1}/a_0^3\pi$
$\psi^2(0, \theta, \phi) = 1/a_0^3\pi$
$\psi^2(\frac{1}{2}a_0, \theta, \phi)/\psi^2(0, \theta, \phi) = (e^{-1}/a_0^3\pi)/(1/a_0^3\pi) = e^{-1} = 0.37$
1.11B number of orbitals, n^2
1.12B occupied orbital, 3p
1.13B $1s^22s^22p^63s^23p^1$ or [Ne]$3s^23p^1$
1.14B [Ar]$3d^{10}4s^24p^3$
1.15B (a) $r(Ca^{2+}) < r(K^+)$; (b) $r(Cl^-) < r(S^{2-})$
1.16B Formation of Be^{3+} requires removing an electron from an ion with noble-gas configuration (a filled shell), whereas an electron is removed from the outer shell to form B^{3+} (which itself has only filled shells).
1.17B In fluorine, a member of Group 17/VII, an additional electron fills the single vacancy in the valence shell; the shell now has the noble-gas configuration of neon and is complete. In neon, an additional electron would have to enter a new shell, where it would be farther from the attraction of the nucleus.

Chapter 2

2.1B KCl, because Cl^- has a smaller radius than Br^-
2.2B (a) [Ar]$3d^5$; (b) [Xe]$4f^{14}5d^{10}$
2.3B I^-, [Kr]$5s^25p^64d^{10}$
2.4B H—B̈r: H has no lone pairs, Br has 3 lone pairs.
2.5B H—N̈—H (with H below N)
2.6B H—N̈—N̈—H (with H below each N)
2.7B [:Ö—N=Ö]$^-$ ⟷ [Ö=N—Ö:]$^-$
2.8B :F̈—Ö—F̈: (formal charges 0 0 0)
2.9B Ö=N—Ö: ⟷ :Ö—N=Ö
2.10B [:Ï—Ï—Ï:]$^-$
2.11B :Ö—Ö=Ö (formal charges −1 +1 0)
2.12B See structure (42). One chlorine atom on an $AlCl_3$ molecule acts as a Lewis base when it donates an electron pair to an Al atom (the Lewis acid) on a second $AlCl_3$ molecule, and one Cl atom on the second molecule donates an electron pair to the Al on the first.
2.13B CO_2
2.14B CaS

Chapter 3

3.1B trigonal planar
3.2B (a) trigonal planar; (b) angular
3.3B (a) AX_2E_2; (b) tetrahedral; (c) angular
3.4B square planar
3.5B (a) nonpolar; (b) polar
3.6B (a) 3 σ, no π; (b) 2 σ, 2 π
3.7B two σ-bonds formed from C2*sp* hybrids and O2p_z-orbitals in a linear arrangement; two π-bonds, one between C2p_x-orbital and O2p_x-orbital on one oxygen, the other between C2p_y-orbital and O2p_y-orbital on the other oxygen
3.8B (a) octahedral; (b) square planar; (c) sp^3d^2
3.9B Carbon of CH_3 group is sp^3 hybridized and forms four σ-bonds with bond angles of about 109.5°. The other two carbons are both sp^2 hybridized and form three σ-bonds and one π-bond; bond angles are about 120°.
3.10B NO: $\sigma_{2s}{}^2\sigma_{2s}{}^{*2}\pi_{2p}{}^4\sigma_{2p}{}^2\pi_{2p}{}^{*1}$; BO = (8 − 3)/2 = 2.5
3.11B p-type

Chapter 4

4.1B $h = P/dg = (1.01 \times 10^5\ \text{kg·m}^{-1}\text{·s}^{-2})/[(998\ \text{kg·m}^{-3}) \times (9.806\ 65\ \text{m·s}^{-2})] = 10.3$ m
4.2B P = (10 cm Hg) × (10 mm)/(1 cm) = 100 mmHg
4.3B $(1.013\ 25 \times 10^5\ \text{Pa/atm}) \times (1\ \text{atm}/760\ \text{Torr}) \times (630\ \text{Torr}) = 8.4 \times 10^4$ Pa or 84 kPa
4.4B $V_2 = P_1V_1/P_2 = (1.00\ \text{atm}) \times (80\ \text{cm}^3)/(3.20\ \text{atm}) = 25\ \text{cm}^3$
4.5B $P_2 = P_1T_2/T_1 = (760\ \text{mmHg}) \times (573\ \text{K})/(293\ \text{K}) = 1.49 \times 10^3$ mmHg
4.6B $P_2 = P_1n_2/n_1 = (1.20\ \text{atm}) \times (300\ \text{mol})/(200\ \text{mol}) = 1.80$ atm
4.7B $V/\text{min} = (n/\text{min}) \times (RT)/P = (1.00\ \text{mol·min}^{-1}) \times (8.206 \times 10^{-2}\ \text{L·atm·K}^{-1}\text{·mol}^{-1}) \times (300\ \text{K})/(1.00\ \text{atm}) = 24.6\ \text{L·min}^{-1}$
4.8B $P_2 = P_1V_1T_2/V_2T_1 = [(1.00\ \text{atm}) \times (250\ \text{L}) \times (243\ \text{K})]/[(800\ \text{L}) \times (293\ \text{K})] = 0.26$ atm
4.9B n = (1 mol He/4.003 g He) × (2.0 g He) = 0.500 mol He
$V = nRT/P = (0.500\ \text{mol}) \times (8.206 \times 10^{-2}\ \text{L·atm·K}^{-1}\text{·mol}^{-1}) \times (298.15\ \text{K})/(1.0\ \text{atm}) = 12$ L
4.10B $M = dRT/P = (1.04\ \text{g·L}^{-1}) \times (62.364\ \text{L·Torr·K}^{-1}\text{·mol}^{-1}) \times (450\ \text{K})/(200\ \text{Torr}) = 146\ \text{g·mol}^{-1}$
4.11B $2\ H_2(g) + O_2(g) \rightarrow 2\ H_2O(l)$, so 2 mol $H_2O \simeq$ 1 mol O_2
$n_{O_2} = PV/RT = [(1.00\ \text{atm}) \times (100.0\ \text{L})]/[(8.206 \times 10^{-2}\ \text{L·atm·K}^{-1}\text{·mol}^{-1}) \times (298\ \text{K})] = 4.09$ mol O_2; amount H_2O = 2 (4.09 mol O_2) = 8.18 mol; g H_2O = (8.18 mol) × $(18.02\ \text{g·mol}^{-1})$ = 147 g H_2O
4.12B $2\ H_2O(l) \rightarrow 2\ H_2(g) + O_2(g)$
(2 mol H_2/3 mol gas) × (720 Torr) = 480 Torr H_2
(1 mol O_2/3 mol gas) × (720 Torr) = 240 Torr O_2
4.13B amount O_2 = (141.2 g O_2)/$(32.00\ \text{g·mol}^{-1})$ = 4.412 mol O_2
amount Ne = (335.0 g Ne)/$(20.179\ \text{g·mol}^{-1})$ = 16.60 mol Ne
P_{O_2} = (4.412 mol O_2/21.01 mol total) × (50 atm) = 10.5 atm
4.14B $t_{CH_4} = t_{He} \times (M_{CH_4}/M_{He})^{1/2} = (10\ \text{s}) \times [(16.04\ \text{g·mol}^{-1})/(4.00\ \text{g·mol}^{-1})]^{1/2} = 20$ s
4.15B $c = (3RT/M)^{1/2} = [3 \times (8.3145\ \text{J·K}^{-1}\text{·mol}^{-1}) \times (298\ \text{K})/(16.04 \times 10^{-3}\ \text{kg})]^{1/2} = 681\ \text{m·s}^{-1}$
4.16B $P = [nRT/(V - nb)] - an^2V^2 = [20 \times (8.206 \times 10^{-2}\ \text{L·atm·K}^{-1}\text{·mol}^{-1}) \times (293\ \text{K})]/[100 - 20 \times (4.267 \times 10^{-2}\ \text{L·mol}^{-1})] - (3.640\ \text{L}^2\text{·atm·mol}^{-2}) \times 20^2/100^2 = 4.7$ atm

Chapter 5

5.1B 1,1-dichloroethane, because it has a dipole moment
5.2B Unlike CF_4, CHF_3 has a net dipole moment. The resulting dipole-dipole interactions account for the higher boiling point of CHF_3, even though one might expect the CF_4 molecule (with more electrons) to exhibit stronger London interactions.
5.3B (a) CH_3OH and (c) HClO
5.4B coordination number, 6
5.5B $8(\frac{1}{8}) + 2(\frac{1}{2}) + 2(1) = 4$ atoms
5.6B $d_{ccp} = (4M)/(8^{3/2}N_Ar^3) = (4 \times 55.85\ \text{g·mol}^{-1})/[8^{3/2} \times 6.022 \times 10^{23}\ \text{mol}^{-1} \times (1.24 \times 10^{-8}\ \text{cm})^3 = 8.60\ \text{g·cm}^{-3}]$; $d_{bcc} = (3^{3/2}\ M)/(32N_Ar^3) = (3^{3/2} \times 55.85\ \text{g·mol}^{-1})/[32 \times 6.022 \times 10^{23}\ \text{mol}^{-1} \times (1.24 \times 10^{-8}\ \text{cm})^3] = 7.90\ \text{g·cm}^{-3}$; The observed density is closest to that predicted for a body-centered cubic structure.
5.7B 8 corner ions = $8 \times \frac{1}{8}$ = 1 chloride ion per unit cell of CsCl
5.8B $d = M/[N_A \times (b/3^{1/2})^3] = 259.81\ \text{g·mol}^{-1}/[6.022 \times 10^{23}\ \text{mol}^{-1} \times (7.80 \times 10^{-8}\ \text{cm}/3^{1/2})^3] = 4.72\ \text{g·cm}^{-3}$

Major Technique 3: X-Ray Diffraction

MT3.1B d = (152 pm)/(2 sin 12.1°) = 392 pm

Chapter 6

6.1B $w = \Delta U - q = -150\ \text{J} - (+300\ \text{J}) = -450$ J; $w < 0$ (i.e., the system does the work)
6.2B $w = -P\Delta V = -(9.60\ \text{atm})(2.2\ \text{L} - 0.22\ \text{L}) \times (101.325\ \text{J}/1\ \text{L·atm}) = -1926\ \text{J} = -1.9$ kJ
6.3B q = +1.00 kJ (heat is absorbed);
$w = -(2.00\ \text{atm})(3.00\ \text{L} - 1.00\ \text{L}) = -4.00\ \text{L·atm} \times (101.325\ \text{J}/1\ \text{L·atm}) = -405$ J;
$\Delta U = q + w = 1.00\ \text{kJ} + (-0.405\ \text{kJ}) = +0.60$ kJ

6.4B $q_{cal} = C_{cal}\Delta T = [3.0\ kJ\cdot(°C)^{-1}] \times (5.1°C) = +15\ kJ$;
$q_{rxn} = -q_{cal} = -15\ kJ$

6.5B $m_{alloy}C_{s\ (alloy)}\Delta T_{alloy} = m_{H_2O}C_{s(H_2O)}\Delta T_{H_2O}$;
$(25.0\ g)(C_s)(21.3°C - 88.6°C) = -(61.2\ g)(4.184\ J\cdot g^{-1}\cdot(°C)^{-1})(21.3°C - 19.6°C)$;
$C_{s\ (alloy)} = 0.26\ J\cdot g^{-1}\cdot(°C)^{-1}$

6.6B $q_{cal} = -q_{rxn} = +4.16\ kJ$;
$C_{cal} = 4.16\ kJ/\Delta T = 4.16\ kJ/3.24°C = 1.28\ kJ\cdot(°C)^{-1}$

6.7B (a) $\Delta H = +30\ kJ$;
(b) $\Delta U = q + w = 30\ kJ + 40\ kJ = +70\ kJ$

6.8B (a) $w = 0$; $\Delta U = q = +1.20\ kJ$;
$\Delta T = q/nC_V = \{(1.20\ kJ)(10^3\ J/kJ)\}/\{(1\ mol) \times (5/2) \times (8.3145\ J\cdot K^{-1}\cdot mol^{-1})\} = 57.7\ K$;
$T_f = 298 + 57.7\ K = 356\ K$; (b) (step 1) constant volume, find final temperature: $q = nC_P\Delta T = 1.20\ kJ$;

$$\Delta T = \frac{\{1.20\ kJ \times 10^3 J/kJ\}}{(1.0\ mol) \times (7/2) \times (8.3145\ J\cdot K^{-1}\cdot mol^{-1})} = 41.2\ K;$$

$T_f = 339\ K$; $w = 0$, so $\Delta U = q = nC_V\Delta T = (1\ mol)(2.5)(8.3145\ J\cdot mol^{-1}\cdot K^{-1})(41.2\ K) = 856\ J$
(step 2) isothermal, $\Delta T = 0 \Rightarrow \Delta U = 0$. Therefore, after both steps, $\Delta U = +856\ J$, $T_f = 339\ K$

6.9B $\Delta H_{vap} = \dfrac{22\ kJ}{23\ g} \times 46.07\ g\cdot mol^{-1} = 44\ kJ\cdot mol^{-1}$

6.10B $\Delta H_{sub} = \Delta H_{vap} + \Delta H_{fus} = 38\ kJ\cdot mol^{-1} + 3\ kJ\cdot mol^{-1} = 41\ kJ\cdot mol^{-1}$

6.11B $q_r = -q_{cal} = -(216\ J\cdot(°C)^{-1})(76.7°C) = -1.66 \times 10^3\ J$;
$\Delta H_r = (-1.66 \times 10^4\ J/0.338\ g) \times (72.15\ g\cdot mol^{-1}) \times (1\ kJ/10^3\ J) = -3.54 \times 10^3\ kJ$;
$C_5H_{12}(l) + 8\ O_2(g) \rightarrow 5\ CO_2(g) + 6\ H_2O(l)$,
$\Delta H = -3.54 \times 10^3\ kJ$

6.12B $\Delta U = \Delta H - \Delta n_{(gas)}RT = -3378\ kJ - (-3\ mol)(8.3145\ J\cdot mol^{-1}\cdot K^{-1})(1273\ K)(1\ kJ/10^3\ J) = -3.35 \times 10^3\ kJ$

6.13B

$CH_4(g) + H_2O(g) \rightarrow CO\ (g) + 3\ H_2(g)$	$\Delta H = +206.10\ kJ$
$2\ H_2(g) + CO(g) \rightarrow CH_3OH(l)$	$\Delta H = -128.33\ kJ$
$H_2(g) + \frac{1}{2}\ O_2(g) \rightarrow H_2O(g)$	$\Delta H = \frac{1}{2}(-483.64)\ kJ$
$CH_4(g) + \frac{1}{2}\ O_2(g) \rightarrow CH_3OH(l)$	$\Delta H = -164.05\ kJ$

6.14B mass(C_2H_5OH) = 350 kJ × (1 mol C_2H_5OH/1368 kJ) × (46.07 $g\cdot mol^{-1}$) = 11.8 g C_2H_5OH

6.15B C(diamond) + $O_2(g) \rightarrow CO_2(g)$, $\Delta H_r° = \Delta H_f°(CO_2) - \Delta H_f°(C, diamond) - \Delta H_f°(O_2, g) = -393.51\ kJ\cdot mol^{-1} - (+1.895\ kJ\cdot mol^{-1}) - 0\ kJ\cdot mol^{-1} = -395.41\ kJ\cdot mol^{-1}$

6.16B $CO(NH_2)_2 + \frac{5}{2}\ O_2 \rightarrow CO_2(g) + 2\ H_2O(l) + N_2(g)$; $\Delta H_r° = \Delta H_f°(CO_2) + 2\ \Delta H_f°(H_2O) + \Delta H_f°(N_2) - \Delta H_f°(CO(NH_2)_2)$; $-632\ kJ = -393.51\ kJ + 2(-285.83\ kJ) + 0\ kJ - \Delta H_f°(CO(NH_2)_2)$;
$\Delta H_f°(CO(NH_2)_2) = -333\ kJ\cdot mol^{-1}$

6.17B

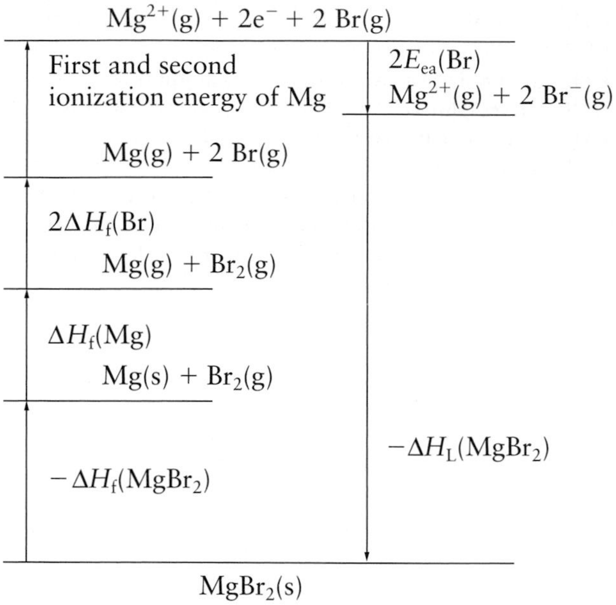

$Mg^{2+}(g) + 2e^- + 2\ Br(g)$

First and second ionization energy of Mg

$2E_{ea}(Br)$

$Mg^{2+}(g) + 2\ Br^-(g)$

$Mg(g) + 2\ Br(g)$

$2\Delta H_f(Br)$

$Mg(g) + Br_2(g)$

$\Delta H_f(Mg)$

$Mg(s) + Br_2(g)$

$-\Delta H_L(MgBr_2)$

$-\Delta H_f(MgBr_2)$

$MgBr_2(s)$

$[524.3 + 147.70 + 2(118.88) + 736 + 1451 - 2(325)]\ kJ\cdot mol^{-1} - \Delta H_L = 0$; $\Delta H_L = +2433\ kJ\cdot mol^{-1}$

6.18B $CH_4(g) + 2\ F_2(g) \rightarrow CH_2F_2(g) + 2\ HF(g)$;
bonds broken [2(C—H), 2(F—F)]: $2(412\ kJ\cdot mol^{-1}) + 2(158\ kJ\cdot mol^{-1}) = 1140\ kJ\cdot mol^{-1}$;
bonds formed [2(C—F), 2(H—F)]: $2(484\ kJ\cdot mol^{-1}) + 2(565\ kJ\cdot mol^{-1}) = 2098\ kJ\cdot mol^{-1}$;
$\Delta H = 1140\ kJ\cdot mol^{-1} - 2098\ kJ\cdot mol^{-1} = -958\ kJ\cdot mol^{-1}$

6.19B $\Delta H_{523K} = \Delta H_{298K} + \Delta C_P\Delta T$
$= -365.56\ kJ\cdot mol^{-1} + [84.1 - 29.12 - 2(28.82) - (3/2)(29.36)]\ J\cdot K^{-1}\cdot mol^{-1} \times (523\ K - 298\ K) \times (1\ kJ/10^3\ J) = -365.56 + (-46.7)(225)(10^{-3}) = -376.1\ kJ\cdot mol^{-1}$

Chapter 7

7.1B $\Delta S = -50\ J/1373\ K = -0.036\ J\cdot K^{-1}$

7.2B $\Delta S = (5.5\ g)(0.51\ J\cdot K^{-1}\cdot g^{-1})\ \ln(373/293) = +0.68\ J\cdot K^{-1}$

7.3B $\Delta S = nR\ \ln(V_2/V_1) = (8.314\ J\cdot K^{-1}\cdot mol^{-1})\ \ln(10) = +19\ J\cdot K^{-1}\cdot mol^{-1}$

7.4B $\Delta S = nR\ \ln(P_1/P_2) = \{70.9\ g \times (1\ mol/70.9\ g)\} \times (8.3145\ J\cdot mol^{-1}\cdot K^{-1})\ \ln(3.00\ kPa/24.00\ kPa) = -17.3\ J\cdot K^{-1}$

7.5B $\Delta S_{vap} = \Delta H_{vap}/T_{bp} = (40.7\ kJ\cdot mol^{-1}/373.2\ K) \times (10^3\ J/1\ kJ) = 109\ J\cdot K^{-1}\cdot mol^{-1}$

7.6B $\Delta H_{vap} = (85\ J\cdot K^{-1}\cdot mol^{-1})(307.5\ K) \times (1\ kJ/10^3\ J) = 26.1\ kJ\cdot mol^{-1}$

7.7B $\Delta S_{fus} = \Delta H_{fus}/T_{mp} = (10.59 \times 10^3\ J\cdot mol^{-1})/278.6\ K = 38.01\ J\cdot K^{-1}\cdot mol^{-1}$

7.8B $\Delta S = (1.3807 \times 10^{-23}\ J\cdot K^{-1})\ \ln(6)^{6.022 \times 10^{23}} = (1.3807 \times 10^{-23}\ J\cdot K^{-1})(6.022 \times 10^{23})\ \ln(6) = 15\ J\cdot K^{-1}$
7.9B Nonzero entropy at $T = 0$ reflects disorder. This disorder results when a molecule can orient itself more than one way in the crystal. In ice, each O atom is surrounded by four H atoms. There are two types of H atoms. Two of the H atoms are covalently bonded to the O atom and the other two H atoms, which belong to neighboring water molecules, are interacting with the central O atom through hydrogen bonds. Thus, more than one orientation is possible in the crystal, and entropy will not be zero at $T = 0$.
7.10B (a) $\Delta S = S_{gray} - S_{white} = (44.14 - 51.55)\ J\cdot K^{-1}\cdot mol^{-1} = -7.41\ J\cdot K^{-1}\cdot mol^{-1}$; the gray form; (b) $\Delta S = S_{graphite} - S_{diamond} = (5.7 - 2.4)\ J\cdot K^{-1}\cdot mol^{-1} = +3.3\ J\cdot K^{-1}\cdot mol^{-1}$; diamond has less thermal disorder than graphite
7.11B $\Delta S_r^\circ = S_{C_2H_6(g)} - S_{C_2H_4(g)} - S_{H_2(g)} = 229.60\ J\cdot K^{-1}\cdot mol^{-1} - 219.56\ J\cdot K^{-1}\cdot mol^{-1} - 130.68\ J\cdot K^{-1}\cdot mol^{-1} = -120.64\ J\cdot K^{-1}\cdot mol^{-1}$
7.12B $\Delta S_{surr} = -\Delta H/T = \{-(1.00\ mol)(-23.4\ kJ\cdot mol^{-1})/243\ K\} \times (10^3\ J/1\ kJ) = +96.3\ J\cdot K^{-1}$
7.13B $\Delta S_{surr} = -\Delta H/T = -(49.0\ kJ/298\ K) \times (10^3\ J/1\ kJ) = -164\ J\cdot K^{-1}$; $\Delta S_{tot} = -164\ J\cdot K^{-1} + (-253.18\ J\cdot K^{-1}) = -417\ J\cdot K^{-1}$; no
7.14B $\Delta S = nR\ln(V_2/V_1) = (2.00\ mol)(8.314\ J\cdot K^{-1}\cdot mol^{-1})\ \ln(0.200\ L/4.00\ L) = -49.8\ J\cdot K^{-1}$; $\Delta S_{surr} = +49.8\ J\cdot K^{-1}\cdot mol^{-1}$; $\Delta S_{tot} = 0$
7.15B $\Delta S_{surr} = -\Delta H_{vap}/T = (-30.8 \times 10^3\ kJ\cdot mol^{-1})/353.3\ K = -87.2\ J\cdot K^{-1}$; $\Delta S = +87.2\ J\cdot K^{-1}$; $\Delta S_{tot} = \Delta S + \Delta S_{surr} = 87.2\ J\cdot K^{-1} + (-87.2\ J\cdot K^{-1}) = 0$
7.16B Yes. $\Delta G = \Delta H - T\Delta S$, so process becomes spontaneous: $\Delta G < 0$, $\Delta S > 0$; therefore, $T\Delta S > 0$. Hence, as temperature increases, $-T\Delta S$ becomes more negative.
7.17B (a) $\Delta G = \Delta H - T\Delta S = 59.3\ kJ\cdot mol^{-1} - (623\ K)(0.0942\ kJ\cdot K^{-1}\cdot mol^{-1}) = +0.6\ kJ\cdot mol^{-1}$, vaporization is nonspontaneous; (b) $\Delta G = \Delta H - T\Delta S = 59.3\ kJ\cdot mol^{-1} - (643\ K)(0.0942\ kJ\cdot K^{-1}\cdot mol^{-1}) = -1.3\ kJ\cdot mol^{-1}$, vaporization is spontaneous
7.18B $3\ H_2(g) + 3\ C(s, graphite) \rightarrow C_3H_6(g)$; $\Delta S_r = 237.4\ J\cdot K^{-1}\cdot mol^{-1} - [3(130.68\ J\cdot K^{-1}\cdot mol^{-1}) + 3(5.740\ J\cdot K^{-1}\cdot mol^{-1})] = -171.86\ J\cdot K^{-1}\cdot mol^{-1}$; $\Delta G = \Delta H - T\Delta S = +53.30\ kJ\cdot mol^{-1} - (298\ K)(-171.86\ J\cdot K^{-1}\cdot mol^{-1}) \times (1\ kJ/10^3\ J) = +104.5\ kJ\cdot mol^{-1}$
7.19B From Appendix 2A, $\Delta G_f^\circ(CH_3NH_2, g) = +32.1\ kJ\cdot mol^{-1}$ at 298 K. Because $\Delta G_f^\circ > 0$, CH_3NH_2 is less stable than its elements at the stated conditions.
7.20B $6\ CO_2(g) + 6\ H_2O(l) \rightarrow C_6H_{12}O_6(s) + 6\ O_2(g)$; $\Delta G = [-910 + 6\ (0)\ ] - [6(-394.36) + 6(-237.13)] = +2879\ kJ\cdot mol^{-1}$
7.21B $H_2(g) + \frac{1}{2}\ O_2(g) \rightarrow H_2O(l)$; $\Delta G_r = \Delta G_f(H_2O, l) = -237.13\ kJ\cdot mol^{-1}$; $w_e = \Delta G = (1.00\ g)/(2.0158\ g\cdot mol^{-1}) \times (-237.13\ kJ\cdot mol^{-1}) = -117.6\ kJ$
7.22B $MgCO_3(s) \rightarrow MgO(s) + CO_2(g)$ $\Delta H_r^\circ = -601.70 + (-393.51) - (-1095.8) = +100.6\ kJ\cdot mol^{-1}$; $\Delta S = 26.94 + 213.74 - 65.7 = 175.0\ J\cdot K^{-1}\cdot mol^{-1}$; $T = \Delta H/\Delta S = (100.6\ kJ\cdot mol^{-1}/175.0\ J\cdot mol^{-1}\cdot K^{-1}) \times (10^3\ J/1\ kJ) = 574.9\ K$

Chapter 8

8.1B

```
   H  H              H  H  H
   |  |              |  |  |
H—C—C=O          H—C—C—C—H
   |                 |  |  |
   H                 H  H  H

    polar             nonpolar
```

$CH_3CH_2CH_3$; molecules of the two substances have the same molar mass, thus the same number of electrons and comparable London forces. However, CH_3CHO is polar and also experiences dipole-dipole forces.

8.2B $\ln\left(\frac{P_2}{94.6\ Torr}\right) = \frac{30.8 \times 10^3\ J\cdot mol^{-1}}{8.3145\ J\cdot mol^{-1}\cdot K^{-1}}\left(\frac{1}{298\ K} - \frac{1}{308\ K}\right)$

$\ln\left(\frac{P_2}{94.6\ Torr}\right) = 0.404$; $\ln(P_2) = 4.95$; $P_2 = 142\ Torr$

8.3B $\ln\left(\frac{760\ Torr}{400\ Torr}\right) = \frac{35.3 \times 10^3\ J\cdot mol^{-1}}{8.3145\ J\cdot mol^{-1}\cdot K^{-1}}\left(\frac{1}{323\ K} - \frac{1}{T_2}\right)$

$\left(\frac{1}{323\ K} - \frac{1}{T_2}\right) = 1.51 \times 10^{-4}$;

$T_2 = 340\ K$ (actual: 337.9 K)
8.4B Liquid sulfur is more dense than monoclinic sulfur over the temperature range in which monoclinic sulfur is stable, because the liquid is more stable at high pressure.
8.5B Carbon dioxide is liquid at 60 atm and 25°C. When it is released into a room at 1 atm and 25°C, as the pressure falls, the system reaches the liquid-vapor boundary, at which point the liquid is changed to vapor. The vaporization absorbs sufficient heat to cool the CO_2 to a temperature below its sublimation temperature at 1 atm. As a result, fine particles of solid CO_2 "snow" are produced.
8.6B Critical temperature increases with the strength of intermolecular forces. For example, CH_4 cannot form hydrogen bonds, so it has a lower critical temperature than either NH_3 or H_2O, which can form hydrogen bonds.
8.7B $s = k_H P = (2.3 \times 10^{-2}\ mol\cdot L^{-1}\cdot atm^{-1})(1.00\ atm) = 2.3 \times 10^{-2}\ mol\cdot L^{-1}$; $n(CO_2) = 0.900\ L \times 2.3 \times 10^{-2}\ mol\cdot L^{-1} = 0.0207\ mol$

8.8B $\Delta H_{hyd}(ZnI_2) = \Delta H_{hyd}(Zn^{2+}) + 2\Delta H_{hyd}(I^-)$; $\Delta H_{hyd}(Zn^{2+}) = -2638\ kJ \cdot mol^{-1} - (2)(-296\ kJ \cdot mol^{-1}) = -2046\ kJ \cdot mol^{-1}$
8.9B Sr^{2+} has a more negative hydration enthalpy than Rb^+ because Sr^{2+} has a greater charge and is slightly smaller than Rb^+.
8.10B molality = (7.36 g/0.200 kg) × (1 mol/122.55 g) = $0.300\ mol \cdot kg^{-1}$
8.11B amount of $H_2O = 1 - n_{CH_3OH} = 1 - 0.250$ mol = 0.750 mol;
mass of H_2O = 0.750 mol $H_2O \times 18.02\ g \cdot mol^{-1} \times 1\ kg/10^3$ g = 0.0135 kg H_2O
$$\text{molality} = \frac{0.250\ \text{mol}}{0.0135\ \text{kg}} = 18.5\ \text{mol}\cdot\text{kg}^{-1}$$
8.12B Assume a 1 L solution. mass (NaCl) = 1 L × (1.83 mol NaCl/1 L) × $58.44\ g \cdot mol^{-1}$ = 106.9 g NaCl; mass (solution) = 1 L × (10^3 mL/1 L) × (1.07 g/1 mL) = 1070 g; mass (H_2O) = 1070 g − 107 g = 963 g; 1.83 mol/0.963 kg = $1.90\ mol \cdot kg^{-1}$
8.13B $n_{C_2H_5OH} = 50.0\ g/46.07\ g \cdot mol^{-1} = 1.09$ mol; $n_{C_9H_8O}$ = 2.00 g × (1 mol/132.16 g) = 0.0151 mol; $x = 1.09/(1.09 + 0.0151) = 0.986$; P = (0.986)(5.3 k Pa) = 5.2 kPa
8.14B $\Delta T_f = k_f m = (39.7\ K \cdot kg \cdot mol^{-1})(0.050\ mol \cdot kg^{-1}) = 1.99\ K = 1.99°C$; T_f = 179.8°C − 1.99°C = 177.8°C
8.15B $CoCl_3 \rightarrow Co^{3+} + 3\ Cl^-$; total amount of ions =
$$0.25\ \text{mol CoCl}_3 \times \frac{4\ \text{mol ions}}{1\ \text{mol CoCl}_3} = 1.0\ \text{mol ions};$$
$i = 4$ (4 ions are formed)
8.16B $i = 1$, because sucrose is a nonelectrolyte and does not dissociate. $\Pi = (1)(0.082\ 06\ L \cdot atm \cdot mol^{-1} \cdot K^{-1})(298\ K)(0.120\ mol \cdot L^{-1})$ = 2.93 atm
8.17B molality = $0.51\ K/39.7\ kg \cdot K \cdot mol^{-1} = 0.0128\ mol \cdot kg^{-1}$;
amount of linalool = 0.100 kg × (0.0128 mol/1 kg) = 1.28×10^{-3} mol;
molar mass = $0.200\ g/1.28 \times 10^{-3}\ mol = 156\ g \cdot mol^{-1}$
8.18B $n = cV$ =
$\{2.11\ kPa/(1)(8.3145\ L \cdot kPa \cdot K^{-1} \cdot mol^{-1})(293\ K)\} \times 0.175\ L = 1.52 \times 10^{-4}$ mol;
molar mass = $1.50\ g/1.52 \times 10^{-4}$ mol = $9.9 \times 10^3\ g \cdot mol^{-1}$ or $9.9\ kg \cdot mol^{-1}$
8.19B equal masses; therefore, assume 50 g of each; amount of C_6H_5 = 50 g/(78.11 $g \cdot mol^{-1}$) = 0.640 mol; mol $CH_3C_6H_5$ = 50 g/(92.13 $g \cdot mol^{-1}$) = 0.543 mol;
$x_{C_6H_6} = 0.640/(0.640 + 0.543) = 0.541$;
$x_{CH_3C_6H_5} = 0.543/(0.640 + 0.543) = 0.459$;
P_{total} = (0.541)(94.6 Torr) + (0.459)(29.1 Torr) = 64.5 Torr
8.20B (a) P = (0.500)(94.6 Torr) + (0.500)(29.1 Torr) = 61.8 Torr;
(b) $x_{C_6H_6,vap}$ = (0.500)(94.6 Torr)/61.8 Torr = 0.765;
$x_{CH_3C_6H_6,vap} = 1 - 0.765 = 0.235$

Major Technique 4: Chromatography

MT4.1B The relative values of the capacity factor k will be in the same order as the solubilities: $CuCl_2 > CoCl_2 > NiCl_2$.

Chapter 9

9.1B $K = \dfrac{P_{SO_2}{}^2 P_{H_2O}{}^2}{P_{H_2S}{}^2 P_{O_2}{}^3}$

9.2B $K = \dfrac{1}{P_{O_2}{}^5}$

9.3B $\Delta G_m = RT[\ln(0.250) - \ln(0.0500)]$
$= (8.314\ J \cdot mol^{-1} \cdot K^{-1})(303\ K)\ln(5)$
$= +4054\ J \cdot mol^{-1} = +4.05\ kJ \cdot mol^{-1}$
9.4B $\Delta G_r = +4.73\ kJ \cdot mol^{-1} + (8.314\ J \cdot mol^{-1} \cdot K^{-1})(298\ K)$
$\ln[(2.10)^2/0.80] \times 1\ kJ/10^3\ J$
$= +8.96\ kJ \cdot mol^{-1}$; $\Delta G_r > 0$;
therefore reaction proceeds toward reactants.
9.5B $\Delta G_r° = 2\Delta G_f°(NO_2; g) - \Delta G_f°(O_2, g) - 2\Delta G_f°(NO, g)$
$= 2(51.31) - [0 + 2(86.55)]\ kJ \cdot mol^{-1}$
$= -70.48\ kJ \cdot mol^{-1}$
$\ln K = -(-70.48 \times 10^3\ J \cdot mol^{-1})/(8.314\ kJ \cdot K^{-1} \cdot mol^{-1} \times 298\ K) = 28.45$; $K = 2.3 \times 10^{12}$
9.6B $K = P_{O_2}{}^3/P_{O_3}{}^2$; $K_c = [O_2]^3/[O_3]^2$
9.7B $K_c = K(RT)^{-\Delta n}$, $\Delta n = 2 - 1 = +1$
$K_c = (47.9)[(8.3145 \times 10^{-2}\ L \cdot bar \cdot K^{-1} \cdot mol^{-1})(400\ K)]^{-1}$ = 1.44
9.8B $K_c = (3.51)(0.156)/(0.131)^2 = 31.9$;
$\Delta n = 2 - 2 = 0$; therefore, $K = K_c = 31.9$
9.9B $K_c = (7.3 \times 10^{-13})^{1/2} = 8.5 \times 10^{-7}$
9.10B $\Delta n = 2 - 2 = 0$; $K_c = K(RT)^0 = K = 3.4 \times 10^{-21} = [NO]^2/(0.100)^2$; $[NO] = 5.8 \times 10^{-12}\ mol \cdot L^{-1}$
9.11B $Q = (P_{NO_2})_0{}^2/(P_{N_2O_4})_0 = (1.2)^2/2.4 = 0.60$; $K = 0.98$, so $Q < K = 0.98$ and P_{NO_2} will increase.
9.12B $K_c = (2x)^2(x)/(0.012 - 2x)^2 = 3.5 \times 10^{-32}$; $4x^3/(0.012)^2 = 3.5 \times 10^{-32}$; $x = 1.08 \times 10^{-12}$. At equilibrium: [HCl] = $0.012\ mol \cdot L^{-1}$; [HI] = $2.2 \times 10^{-12}\ mol \cdot L^{-1}$; $[Cl_2] = 1.1 \times 10^{-12}\ mol \cdot L^{-1}$; some solid I_2 remains as well.
9.13B $K_c = [ClF]^2/[Cl_2][F_2] = 20$; $Q = (0.100)^2/(0.100)(0.200) = 0.5 < K$
$K_c = 20 = (0.100 + 2x)^2/(0.200 - x)(0.100 - x)$;
$x = 0.075$
$[ClF]_{equil} = 0.100 + 2x = 0.25\ mol \cdot L^{-1}$
9.14B The equilibrium tends to shift toward (a) products; (b) products; (c) reactants.
9.15B $Q = (1.30)^2/(0.080)(0.050)^3 = 1.7 \times 10^5$; $Q < K$.
Initially, $P_{NH_3} = \frac{1}{2}(2.60\ atm) = 1.30$ atm

	N_2	H_2	NH_3
Initial	0.080	0.050	1.30
Change	$-x$	$-3x$	$+2x$
Equil.	$0.080 - x$	$0.050 - 3x$	$1.30 + 2x$

$$K = \frac{(1.30 + 2x)^2}{(0.080 - x)(0.050 - 3x)^3}; x = 5.86 \times 10^{-3}$$

$P_{N_2} = 0.074$ atm; $P_{NH_3} = 1.31$ atm; $P_{H_2} = 0.032$ atm

9.16B Compression affects gaseous species only. In the forward reaction, 1 mol $CO_2(g)$ combines to form an aqueous species. Therefore, compression favors the formation of $H_2CO_3(aq)$.

9.17B $\Delta H_r° = 2\Delta H_f°(CO_2, g) - 2\Delta H_f°(CO, g) - \Delta H_f°(O_2, g) = 2(-393.51\ kJ\cdot mol^{-1}) - 2(-110.53\ kJ\cdot mol^{-1}) - 0 = -565.96$ kJ. The reaction is exothermic; therefore, lowering the temperature shifts the reaction toward products. CO_2 is favored.

9.18B $\Delta H_r° = \Delta H_f°(Cl_2, g) + \Delta H_f°(PCl_3, g) - \Delta H_f°(PCl_5, g) = 0 + (-287.0\ kJ\cdot mol^{-1}) - (-374.9\ kJ\cdot mol^{-1}) = +87.9$ kJ

$$\ln\left(\frac{K_2}{K_1}\right) = \frac{87.9 \times 10^3\ J\cdot mol^{-1}}{8.3145\ J\cdot mol^{-1}\cdot K^{-1}}\left(\frac{1}{523\ K} - \frac{1}{800\ K}\right) = 7.0$$

$K_2 = K_1 e^{7.0} = (78.3)e^{7.0} = 8.59 \times 10^4$

Chapter 10

10.1B (a) H_3O^+ has one more H^+ than H_2O; therefore, the conjugate acid of H_2O is H_3O^+. (b) NH_2^- has one less H^+ than NH_3; therefore, the conjugate base of NH_3 is NH_2^-.

10.2B (a) Brønsted acids: $NH_4^+(aq)$, $H_2CO_3(aq)$; Brønsted bases: $HCO_3^-(aq)$, $NH_3(aq)$; (b) Lewis acid: $H^+(aq)$; Lewis bases: $NH_3(aq)$, $HCO_3^-(aq)$

10.3B (a) $[H_3O^+] = (1.0 \times 10^{-14})/(2.2 \times 10^{-3}) = 4.5 \times 10^{-12}\ mol\cdot L^{-1}$; (b) $[OH^-] = [NaOH] = 2.2 \times 10^{-3}\ mol\cdot L^{-1}$

10.4B $[OH^-] = [NaOH] = 0.077\ mol\cdot L^{-1}$; $[H_3O^+] = (1.0 \times 10^{-14})/(0.077) = 1.30 \times 10^{-13}\ mol\cdot L^{-1}$; pH $= -\log(1.3 \times 10^{-13}) = 12.89$

10.5B $[H_3O^+] = 10^{-8.2} = 6.3 \times 10^{-9}\ mol\cdot L^{-1}$

10.6B The conjugate base of HIO_3 is IO_3^-; $pK_b = 14.00 - pK_a = 14.00 - 10.64 = 3.36$

10.7B (a) $1.8 \times 10^{-9} = K_b(C_5H_5N) < K_b(NH_2NH_2) = 1.7 \times 10^{-6}$. Therefore, NH_2NH_2 is the stronger base. (b) From (a), $K_a(C_5H_5NH^+) > K_a(NH_2NH_3^+)$. Therefore, $C_5H_5NH^+$ is the stronger acid. (c) $1.7 \times 10^{-1} = K_a(HIO_3) < K_a(HClO_2) = 1.0 \times 10^{-2}$. Therefore, HIO_3 is the stronger acid. (d) $K_b(HSO_3^-) = 1.0 \times 10^{-14}/1.5 \times 10^{-2} = 6.7 \times 10^{-13}$; $6.7 \times 10^{-13} = K_b(HSO_3^-) < K_b(ClO_2^-) = 1.0 \times 10^{-14}/1.0 \times 10^{-2} = 1.0 \times 10^{-12}$. Therefore, ClO_2^- is the stronger base.

10.8B (a) weak base; (b) strong base; (c) weak base

10.9B $CH_3COOH < CH_2ClCOOH < CHCl_2COOH$ (The electronegativity of Cl is greater than that of H. Therefore, the acidity increases with increasing number of chlorine atoms attached.)

10.10B $K_a = 1.4 \times 10^{-3} = x^2/(0.22 - x)$; we must use a quadratic equation: $x^2 + (1.4 \times 10^{-3}x) - (3.1 \times 10^{-4})$; $x = 1.7 \times 10^{-2} = [H_3O^+]$; pH $= 1.77$

10.11B $[H_3O^+] = 10^{-2.35} = 4.5 \times 10^{-3}$; $K_a = (4.5 \times 10^{-3})^2/[0.50 - (4.5 \times 10^{-3})] = 4.1 \times 10^{-5}$

10.12B $1.0 \times 10^{-6} = x^2/(0.012 - x) \approx x^2/0.012$; $x = 1.1 \times 10^{-4} = [OH^-]$; pOH $= 3.96$; pH $= 14.00 - 3.96 = 10.04$; percentage protonated $= (1.1 \times 10^{-4}/0.012) \times 100 = 0.92\%$

10.13B (a) CO_3^{2-} is the conjugate base of HCO_3^-, which is a weak acid; therefore, the aqueous solution is basic. (b) $K_a(Al(H_2O)_6^{3+}) = 1.4 \times 10^{-5}$. Therefore, the solution is acidic. (c) K^+ is a "neutral cation" and NO_3^- is a conjugate base of a strong acid; therefore, the aqueous solution is neutral.

10.14B $NH_4^+(aq) + H_2O(l) \rightleftharpoons NH_3(aq) + H_3O^+(aq)$; $K_b(NH_3) = 1.8 \times 10^{-5}$; $K_a(NH_4^+) = (1.0 \times 10^{-14})/(1.8 \times 10^{-5}) = 5.6 \times 10^{-10}$; $K_a = 5.6 \times 10^{-10} = x^2/(0.10 - x) \approx x^2/0.10$; $x = 7.5 \times 10^{-6}$; pH $= -\log(7.5 \times 10^{-6}) = 5.12$

10.15B $F^-(aq) + H_2O(l) \rightleftharpoons HF(aq) + OH^-(aq)$; $K_a(HF) = 3.5 \times 10^{-4}$; $K_b(F^-) = (1.0 \times 10^{-14})/(3.5 \times 10^{-4}) = 2.9 \times 10^{-11}$; $K_b = 2.9 \times 10^{-11} = x^2/(0.20 - x) \approx x^2/0.20$; $x = 2.39 \times 10^{-6}$; pOH $= 5.62$; pH $= 14.00 - 5.62 = 8.38$

10.16B $[H_3O^+]^2 + [NaOH]_{initial}[H_3O^+] - K_w = 0$; $x^2 + (2.0 \times 10^{-7})x - (1.0 \times 10^{-14}) = 0$; $[H_3O^+] = x = 4.1 \times 10^{-8}\ mol\cdot L^{-1}$; pH $= -\log(4.1 \times 10^{-8}) = 7.38$

10.17B $K_a(HIO) = 2.3 \times 10^{-11}$; $[H_3O^+] = [(1.0 \times 10^{-14}) + (2.3 \times 10^{-11}) \times (1.0 \times 10^{-2})]^{1/2}$; $[H_3O^+] = 4.8 \times 10^{-7}$; pH $= 6.32$

10.18B $K_{a2} = [H_3O^+][SO_4^{2-}]/[HSO_4^-]$; $1.2 \times 10^{-2} = (0.10 + x)(x)/(0.10 - x)$; $x^2 + 0.112x - (1.2 \times 10^{-3}) = 0$; $x = 9.8 \times 10^{-3}\ mol\cdot L^{-1}$; $[H_3O^+]_{total} = 0.10 + (9.8 \times 10^{-3})\ mol\cdot L^{-1} = 0.11\ mol\cdot L^{-1}$; pH $= 0.96$

10.19B H_3PO_3 is a weak polyprotic acid; therefore, only the first deprotonation is significant. $K_a = [H_3O^+][H_2PO_4^-]/[H_3PO_4]$; $7.6 \times 10^{-3} = x^2/(0.50 - x) \approx x^2/0.50$; $[H_3O^+] = x = 6.2 \times 10^{-2}\ mol\cdot L^{-1}$; pH $= 1.21$

10.20B $pK_{a1}(H_3PO_4) = 2.12$; $pK_{a2}(H_3PO_4) = 7.21$; pH $= \frac{1}{2}(2.12 + 7.21) = 4.66$

10.21B $[Cl^-] = 0.50\ mol\cdot L^{-1}$; $^+NH_3CH_2COOH(aq) + H_2O(l) \rightleftharpoons NH_2CH_2COOH(aq) + H_3O^+(aq)$; $K_{a1} = 4.5 \times 10^{-3}$; $NH_2CH_2COOH(aq) + H_2O(l) \rightleftharpoons NH_2CH_2CO_2^-(aq) + H_3O^+(aq)$; $K_{a2} = 1.7 \times 10^{-10}$; $K_a = 4.5 \times 10^{-3} = x^2/(0.50 - x)$; we must solve the quadratic equation $x^2 + (4.5 \times 10^{-3})x - 0.002\,25 = 0$. $x = 0.045\ mol\cdot L^{-1} = [H_3O^+] = [NH_2CH_2COOH]$; $[^+NH_3CH_2COOH] = 0.50 - 0.045\ mol\cdot L^{-1} = 0.45\ mol\cdot L^{-1}$. $K_{a2} = 1.7 \times 10^{-10} = (0.045)(x)/(0.045)$; $x = 1.7 \times 10^{-10} = [NH_2CH_2CO_2^-]$; $[OH^-] = K_w/[H_3O^+] = (1.0 \times 10^{-14})/(0.045) = 2.2 \times 10^{-13}\ mol\cdot L^{-1}$. Concentrations of all species: $[Cl^-] = 0.50\ mol\cdot L^{-1}$; $[^+NH_3CH_2COOH] = 0.45\ mol\cdot L^{-1}$; $[H_3O^+] = 0.045\ mol\cdot L^{-1}$; $[NH_2CH_2COOH] = 0.045\ mol\cdot L^{-1}$; $[NH_2CH_2CO_2^-] = 1.7 \times 10^{-10}\ mol\cdot L^{-1}$; $[OH^-] = 2.2 \times 10^{-13}\ mol\cdot L^{-1}$.

Chapter 11

11.1B $K_b = [OH^-][CH_3NH_3^+]/[CH_3NH_2] = (x)(0.146 + x)/(0.300 - x) = 3.6 \times 10^{-4}$; $[OH^-] = x = 7.40 \times 10^{-4}$; pOH = 3.13; pH = 14.00 − 3.13 = 10.87

11.2B $K_a = (1.0 \times 10^{-14})/(1.8 \times 10^{-5}) = 5.6 \times 10^{-10}$; pH = $-\log(5.6 \times 10^{-10}) + \log(0.030/0.040) = 9.13$

11.3B $[CH_3CO_2^-] = 0.040 - 0.020 = 0.020\ mol\cdot L^{-1}$; $[CH_3COOH] = 0.080 + 0.020 = 0.100\ mol\cdot L^{-1}$; pH = 4.75 + log(0.020/0.100) = 4.05; $pH_{final} - pH_{initial} = 4.05 - 4.45 = -0.4$ (a decrease of 0.4)

11.4B $(CH_3)_3NH^+/(CH_3)_3N$, because the $pK_a = 9.81$, which is close to 10.

11.5B $pH - pK_a = 3.50 - 4.19 = -0.69$; $[C_6H_5CO_2^-]/[C_6H_5COOH] = 10^{-0.69} = 0.20$

11.6B amount of H_3O^+ added = 0.012 L × (0.340 mol/1 L) = 0.004 08 mol; amount of OH^- remaining = 0.006 25 − 0.004 08 mol = 0.002 17 mol. $[OH^-] = 0.00217\ mol/(0.0120 + 0.0250)\ L = 0.0586\ mol\cdot L^{-1}$; pOH = 1.23; pH = 14.00 − 1.23 = 12.77

11.7B amount of H_3O^+ added = 0.025 L × (0.340 mol/1 L) = 0.008 50 mol HCl; amount of OH^- remaining = 0.008 50 − 0.006 25 mol = 0.002 25 mol. $[H_3O^+] = 0.00225\ mol/(0.250 + 0.250)\ L = 0.045\ mol\cdot L^{-1}$; pH = 1.35

11.8B initial amount of NH_3 = 0.025 L × $(0.020\ mol\cdot L^{-1}) = 5.0 \times 10^{-4}$ mol; volume of HCl added = $(5.0 \times 10^{-4}\ mol)/(0.015\ mol\cdot L^{-1}) = 0.0333$ L. $[NH_4^+] = 5.0 \times 10^{-4}\ mol/(0.0333 + 0.025\ 00)\ L = 0.0086\ mol\cdot L^{-1}$; $K_a = [H_3O^+][NH_3]/[NH_4^+] = 5.6 \times 10^{-10} = (x)^2/(0.0086 - x) \approx x^2/0.0086$; $[H_3O^+] = x = 2.2 \times 10^{-6}\ mol\cdot L^{-1}$; pH = 5.66

11.9B final volume of solution = 25.00 + 15.00 mL = 40.00 mL; amount of HCO_2^- formed = amount of OH^- added = $(0.150\ mol\cdot L^{-1}) \times (15.00\ mL)$ = 2.25 mmol; amount of HCOOH remaining = 2.50 − 2.25 mmol = 0.25 mmol. [HCOOH] = $(0.25 \times 10^{-3}\ mol)/(0.040\ 00\ L) = 0.062\ mol\cdot L^{-1}$; $[HCO_2^-] = (2.25 \times 10^{-3}\ mol)/(0.040\ 00\ L) = 0.0562\ mol\cdot L^{-1}$; pH = 3.75 + log(0.0562/0.0062) = 4.71

11.10B (a) amount of H_3PO_4 = (0.030 L) × $(0.010\ mol\cdot L^{-1}) = 3.0 \times 10^{-4}$ mol; at first stoichiometric point, initial amount of H_3PO_4 = amount of NaOH added; volume of NaOH = $3.0 \times 10^{-4}\ mol/0.020\ mol\cdot L^{-1} = 0.015$ L, or 15 mL; (b) 2(15 mL) = 30 mL

11.11B $(0.020\ L)(0.100\ mol\cdot L^{-1}) = 0.0020$ mol H_2S; $0.0020\ mol/(0.300\ mol\cdot L^{-1}) = 0.0067$ L, or 6.7 mL NaOH for first stoichiometric point and 2 × 6.7 mL = 13.4 mL for the second stoichiometric point. (a) Before the first stoichiometric point, $H_2S(aq) + H_2O(l) \rightleftharpoons HS^-(aq) + H_3O^+(aq)$; the primary species present are Na^+, H_2S, and HS^-. (b) At the second stoichiometric point, the primary equilibrium is $S^{2-}(aq) + H_2O(l) \rightleftharpoons HS^-(aq) + OH^-(aq)$. The species present are Na^+ and S^{2-} (also HS^- and OH^- because S^{2-} has a relatively large K_b).

11.12B $K_{sp} = [Ag^+][Br^-] = (8.8 \times 10^{-7})^2 = 7.7 \times 10^{-13}$

11.13B $K_{sp} = [Pb^{2+}][F^-]^2 = (s)(2s)^2 = 4s^3$; $3.7 \times 10^{-8} = 4s^3$; $s = 2.1 \times 10^{-3}\ mol\cdot L^{-1}$

11.14B In 0.10 M $CaBr_2(aq)$, $[Br^-] = 2(0.10\ mol\cdot L^{-1}) = 0.20\ mol\cdot L^{-1}$; $s = [Ag^+] = K_{sp}/[Br^-] = (7.7 \times 10^{-13})/(0.20) = 3.8 \times 10^{-12}\ mol\cdot L^{-1}$

11.15B $[Ba^{2+}] = (1.0 \times 10^{-3}\ mol\cdot L^{-1})(100\ mL/300\ mL) = 3.3 \times 10^{-4}\ mol\cdot L^{-1}$; $[F^-] = (1.0 \times 10^{-3}\ mol\cdot L^{-1})(200\ mL/300\ mL) = 6.7 \times 10^{-4}\ mol\cdot L^{-1}$; $Q = (3.3 \times 10^{-4})(6.7 \times 10^{-4})^2 = 1.5 \times 10^{-10} < K_{sp} = 1.7 \times 10^{-6}$; therefore, no precipitate of BaF_2 forms.

11.16B For $PbCl_2$ to precipitate, $[Cl^-] = (K_{sp}/[Pb^{2+}])^{1/2} = (1.6 \times 10^{-5}/0.020)^{1/2} = 2.8 \times 10^{-2}\ mol\cdot L^{-1}$. For AgCl to precipitate, $[Cl^-] = (K_{sp}/[Ag^+]) = (1.6 \times 10^{-10}/0.0010) = 1.6 \times 10^{-7}\ mol\cdot L^{-1}$. Therefore, AgCl precipitates first at $[Cl^-] = 1.6 \times 10^{-7}\ mol\cdot L^{-1}$; then $PbCl_2$ precipitates at $[Cl^-] = 2.8 \times 10^{-2}\ mol\cdot L^{-1}$.

11.17B $CuS(s) + 4\ NH_3(aq) \rightleftharpoons Cu(NH_3)_4^{2+}(aq) + S^{2-}(aq)$; $K = (1.3 \times 10^{-36}) \times (1.2 \times 10^{13}) = 1.6 \times 10^{-23}$; $K = [Cu(NH_3)_4^{2+}][S^{2-}]/[NH_3]^4 = 1.6 \times 10^{-23} = x^2/(1.2 - 4x)^4$; $x/(1.2 - 4x)^2 \approx x/1.2 = 4.0 \times 10^{-12}$; $s = [S^{2-}] = x = 5.8 \times 10^{-12}\ mol\cdot L^{-1}$.

Chapter 12

12.1B reduction: $(8\ H^+ + MnO_4^- + 5\ e^- \rightarrow Mn^{2+} + 4\ H_2O) \times 2$; oxidation: $(H_2O + H_2SO_3 \rightarrow HSO_4^- + 2\ e^- + 3\ H^+) \times 5$; net reaction: $H^+(aq) + 2\ MnO_4^-(aq) + 5\ H_2SO_3(aq) \rightarrow 2\ Mn^{2+}(aq) + 5\ HSO_4^-(aq) + 3\ H_2O(aq)(l)$

12.2B oxidation: $(3\ I^- \rightarrow I_3^- + 2\ e^-) \times 8$; reduction: $9\ H_2O + 16\ e^- + 3\ IO_3^- \rightarrow I_3^- + 18\ OH^-$; net reaction: $3\ H_2O(l) + IO_3^-(aq) + 8\ I^-(aq) \rightarrow 3\ I_3^-(aq) + 6\ OH^-(aq)$

12.3B Half-reactions for the silver battery are in Table 12.1. From these half-reactions, $n = 2$. $\Delta G_r = -(2)(9.6485 \times 10^4\ C\cdot mol^{-1})(1.6\ V) = -3.09 \times 10^5\ C\cdot V\cdot mol^{-1} = -310\ kJ\cdot mol^{-1}$

12.4B $Mn(s)|Mn^{2+}(aq)||Cu^{2+}(aq), Cu^+(aq)|Pt(s)$

12.5B (a) left: $2\ Hg(l) + 2\ HCl(aq) \rightarrow Hg_2Cl_2(s) + 2\ e^-$; right: $2\ e^- + Hg_2(NO_3)_2(aq) \rightarrow 2\ Hg(l) + 2\ NO_3^-(aq)$; reaction: $2\ HCl(aq) + Hg_2(NO_3)_2(aq) \rightarrow Hg_2Cl_2(s) + 2\ HNO_3(aq)$. The cell reaction is spontaneous as written.

12.6B $E_{Pb^{2+}/Pb} = E_{Fe^{2+}/Fe} - E_{cell} = -0.44\ V - 0.31\ V = -0.75\ V$

12.7B $Mn^{3+}(aq) + e^- \rightarrow Mn^{2+}(aq)$, $E_1° = +1.51$ V; $Mn^{2+}(aq) + 2\ e^- \rightarrow Mn(s)$, $E_2° = -1.18$ V; net half reaction: $Mn^{3+}(aq) + 3\ e^- \rightarrow Mn(s)$, $E_3° = ?$

$$E_3° = \frac{n_1E_1° + n_2E_2°}{n_3} = \frac{(1\ mol)(1.51\ V) + (2\ mol)(-1.18\ V)}{3\ mol} = -0.28\ V$$

12.8B $O_2(g) + 2\ H_2O(l) + 4\ e^- \rightarrow 4\ OH^-(aq)$, $E° = +0.40$ V; $Cl_2(g) + 2\ e^- \rightarrow 2\ Cl^-(aq)$, $E° = +1.36$ V. Yes, $Cl_2(g)$ can oxidize H_2O to $O_2(g)$ in basic solution, because the Cl_2 reduction half-reaction has a more positive standard potential than the O_2 reduction half-reaction.
12.9B oxidation half reaction: $Cu^{2+}(aq) + 2\ e^- \rightarrow Cu(s)$, $E° = +0.34$ V; reduction half reaction: $Ag^+(aq) + e^- \rightarrow Ag(s)$, $E° = +0.80$ V; net reaction: $Cu(s) + 2\ Ag^{2+}(aq) \rightarrow Cu^{2+}(aq) + 2\ Ag(s)$, $E° = +0.46$ V. Therefore, Ag^+ is the stronger oxidizing agent. Cell diagram: $Cu(s)|Cu^{2+}(aq)||Ag^+(aq)|Ag(s)$
12.10B $Cd(OH)_2(s) + 2\ e^- \rightarrow Cd(s) + 2\ OH^-(aq)$, $E° = -0.81$ V; $Cd^{2+}(aq) + 2\ e^- \rightarrow Cd(s)$, $E° = -0.40$ V; net: $Cd(OH)_2(s) \rightarrow Cd^{2+}(aq) + 2\ OH^-(aq)$, $E° = -0.41$ V; $\ln K_{sp} = (2)(-0.41\text{ V})/0.025\ 693\text{ V} = -31.92$; $K_{sp} = 1.4 \times 10^{-14}$
12.11B cell reaction: $Ag^+(aq, 0.010\ mol{\cdot}L^{-1}) \rightarrow Ag^+(aq, 0.0010\ mol{\cdot}L^{-1})$, $E° = 0$, $n = 1$. $E = 0 - 0.025\ 693\text{ V} \ln (0.0010/0.010) = +0.059$ V.
12.12B The reduction half reactions are $Br_2(l) + 2\ e^- \rightarrow 2\ Br^-(aq)$, $E° = +1.09$ V; $O_2(g) + 4\ H^+(aq) + 4\ e^- \rightarrow 2\ H_2O(l)$, $E° = +1.23$ V; $2\ H^+(aq) + 2\ e^- \rightarrow 2\ H_2(g)$, $E° = 0.00$ V. product at the cathode: H_2; product at the anode: O_2 and Br_2. (On the basis of standard potentials alone, we would predict that H_2O will be oxidized preferentially to Br^-. However, because of the high overpotential for oxygen production, bromine may also be produced.)
12.13B 12.0 mol $e^- \times$ (1 mol Cr)/(6 mol e^-) = 2 mol Cr
12.14B $12.00\text{ g Cr} \times \dfrac{1\text{ mol Cr}}{52.00\text{ g Cr}} \times \dfrac{6\text{ mol e}^-}{1\text{ mol Cr}} = 1.385\text{ mol e}^-$

$$t = \frac{(1.385\text{ mol e}^-)(9.6485 \times 10^4\text{ C}{\cdot}\text{mol}^{-1})}{6.20\text{ C}{\cdot}\text{s}^{-1}} \times \frac{1\text{ h}}{3600\text{ s}} = 5.99\text{ h}$$

12.15B (b) zinc; it is the only one of the metals with a standard potential (+0.76 V) greater than that of iron (+0.44 V), so it is the only one more easily oxidized than iron.

Chapter 13

13.1B average rate of disappearance of Hb = $\{-(8.0 \times 10^{-7} - 1.2 \times 10^{-6})\text{ mmol}{\cdot}\text{L}^{-1}\}/0.10\ \mu\text{s} = 4 \times 10^{-6}\text{ mmol}{\cdot}\text{L}^{-1}{\cdot}\mu\text{s}^{-1}$
13.2B (a) $\frac{1}{2}(1.15\text{ mmol NH}_3){\cdot}\text{L}^{-1}{\cdot}\text{h}^{-1} = 0.575\ (\text{mmol N}_2){\cdot}\text{L}^{-1}{\cdot}\text{h}^{-1}$; (b) unique average rate = $-\Delta[N_2]/dt = -(1/3)(\Delta[H_2]/dt) = +(1/2)(\Delta[NH_3]/dt) = 0.575\text{ mmol}{\cdot}\text{L}^{-1}{\cdot}\text{h}^{-1}$
13.3B rate = $k\ [C_4H_9Br]^1\ [OH^-]^0$. The overall order of the reaction is first-order.
13.4B rate = $k\ [CO]^m[Cl_2]^n$
rate (2)/rate (1) = 0.241/0.121 = $(0.24/0.12)^m(0.20/0.20)^n$; $2 = (2)^m$; $m = 1$.
rate 3/rate 2 = 0.682/0.241 = $(0.24/0.24)^m(0.40/0.20)^n$; $2.8 = (2)^n$; $n = \log 2.8/\log 2 = 1.5$.
Therefore, rate = $k\ [CO][Cl_2]^{3/2}$. From experiment 1,

$$k = \frac{0.121\text{ mol}{\cdot}\text{L}^{-1}{\cdot}\text{s}^{-1}}{(0.12\text{ mol}{\cdot}\text{L}^{-1})(0.20\text{ mol}{\cdot}\text{L}^{-1})^{3/2}} = 11\text{ L}^{3/2}{\cdot}\text{mol}^{-3/2}{\cdot}\text{s}^{-1}$$

13.5B $[C_3H_6]_t = [C_3H_6]_0\ e^{-kt} = (0.100\text{ mol}{\cdot}\text{L}^{-1})e^{-(6.7\times10^{-4}{\cdot}\text{s}^{-1})(200\text{ s})} = 0.087\text{ mol}{\cdot}\text{L}^{-1}$
13.6B A plot of $\ln[CH_3N_2CH_3]$ against time is linear, showing that the reaction must be first order; $k = 3.60 \times 10^{-4}\text{ s}^{-1}$.
13.7B reaction is first order, with $k = 6.7 \times 10^{-4}\text{ s}^{-1}$ at 500°C

$$t = \frac{1}{k}\ln\frac{[C_3H_6]_0}{[C_3H_6]_t} = \frac{1}{6.7 \times 10^{-4}\text{s}^{-1}}\ln\left(\frac{1.0\text{ mol}{\cdot}\text{L}^{-1}}{0.005\text{ mol}{\cdot}\text{L}^{-1}}\right) = 7.9 \times 10^3\text{ s} = 2.2\text{ h}$$

13.8B reaction is first order, with $k = 5.5 \times 10^{-4}\text{ s}^{-1}$ at 973 K
(a) $t_{1/2} = (\ln 2)/k = (\ln 2)/(5.5 \times 10^{-4}\text{ s}^{-1}) = 1.3 \times 10^3\text{ s} = 21$ min; (b) It requires four half-lives for the concentration to fall to one-sixteenth of its initial value, so $t = 4t_{1/2} = 4 \times 21\text{ min} = 84$ min.
13.9B $k = \ln 2/2.4 \times 10^4$ y;
$t = (2.4 \times 10^4\text{ y}/\ln 2) \ln(1.0/0.20) = 5.6 \times 10^4$ y
13.10B $\ln(4.35/3.00) = E_a/(8.3145\text{ J}{\cdot}\text{mol}^{-1}{\cdot}\text{K}^{-1})[(1/291\text{ K}) - (1/303\text{ K})]$;
$E_a = 2.2 \times 10^5\text{ J}{\cdot}\text{mol}^{-1} = 22\text{ kJ}{\cdot}\text{mol}^{-1}$
13.11B $\ln(k'/k) = (272\text{ kJ}{\cdot}\text{mol}^{-1}/8.3145 \times 10^{-3}\text{ kJ}{\cdot}\text{mol}^{-1}{\cdot}\text{K}^{-1})[(1/773\text{ K}) - (1/573\text{ K})] = -14.77$;
$k' = (6.7 \times 10^{-4}\text{ s}^{-1})e^{-14.77} = 2.6 \times 10^{-10}\text{ s}^{-1}$
13.12B (a) bimolecular (two reactants); (b) unimolecular (one reactant)
13.13B net rate of disappearance of AH = $k_1[AH][B] - k_1'[BH^+][A^-] + k_2[A^-][AH]$
rate of appearance of $[A^-] = k_1[AH][B] - k_1'[A^-][BH^+] - k_2[A^-][AH] = 0$

$$[A^-] = \frac{k_1[AH][B]}{k_1 \times [BH^+] + k_2[AH]}$$

Substitute:
net rate of disappearance of AH

$$= k_1\ [AH][B] - \frac{k_1'k_1[AH][B][BH^+]}{k_1'[BH^+] + k_2[AH]} + \frac{k_2k_1[AH]^2[B]}{k_1'[BH^+] + k_2[AH]} = \frac{2k_1k_2[AH]^2[B]}{k_1'[BH^+] + k_2[AH]}$$

Because $k_2[AH] \ll k_1'[BH^+]$,

$$\text{net rate of disappearance of AH} = \frac{2\ k_1k_2}{k_1'}\frac{[AH]^2[B]}{[BH^+]} = k[AH]^2[B][BH^+]^{-1},$$

where $k = 2\ k_1k_2/k_1'$.

13.14B (a) The homogeneous catalyst changes the reaction pathway and therefore changes the rate law. (b) Because a catalyst does not change the thermodynamics of the reaction, a homogeneous catalyst does not change the equilibrium constant.

Chapter 14

14.1B Aluminum forms an amphoteric oxide in which it has the oxidation state +3; therefore, aluminum is the element.
14.2B Hydrogen is a nonmetal, and it is a diatomic gas at room temperature. It has an intermediate electronegativity ($\chi = 2.2$), so it forms covalent bonds with all elements except metals and forms an anion. However, Group 1 elements are solid metals that have low electronegativity and form cations.
14.3B The molar mass of potassium is smaller than that of cesium. Thus, the same amount of O_2 can be obtained from a smaller mass of KO_2 than can be obtained from a larger mass of CsO_2.
14.4B Calcium compounds are good construction materials because they are rigid. The Ca^{2+} cation is small and highly charged. These characteristics result in strong interactions with its neighbors in the solid state.
14.5B Boric acid acts as a Lewis acid. The boron atom in $B(OH)_3$ has an incomplete octet and forms a bond by accepting a lone pair of electrons from a water molecule, which is acting as a Lewis base. The complex formed is a weak Brønsted acid in which an acidic proton can be lost from one of the H_2O molecules in the complex.
14.6B SiH_4 is a Lewis acid that can react with the Lewis base OH^-. CH_4 is not a Lewis acid, because the C atom is much smaller than the Si atom and has no accessible *d*-orbitals to accommodate added electron pairs.

Chapter 15

15.1B Assume 1 L solution.

$$\text{Molarity} = 1\text{ L soln} \times \frac{10^3\text{ mL}}{1\text{ L}} \times \frac{1.7\text{ g soln}}{1\text{ mL soln}} \times \frac{85\text{ g H}_3\text{PO}_4}{100\text{ g soln}} \times \frac{1\text{ mol H}_3\text{PO}_4}{97.99\text{ g H}_3\text{PO}_4} = 15\text{ M H}_3\text{PO}_4(\text{aq})$$

15.2B

$$2\,(2\,e^- + F_2 \rightarrow 2\,F^-)$$
$$2\,H_2O(l) \rightarrow O_2(g) + 4\,H^+ + 4\,e^-$$
$$2\,H_2O(l) + 2\,F_2(g) \rightarrow O_2(g) + 4\,H^+(aq) + 4\,F^-(aq)$$

$E° = 2.87\text{ V} - 1.23\text{ V} = +1.64\text{ V}$
$\Delta G° = -nFE° = -(4\text{ mol})(96.485\text{ kC·mol}^{-1})(1.64\text{ V}) = -633\text{ kJ}$
15.3B The volume of the solution is less than the sum of the individual volumes because the intermolecular interactions of H_2SO_4 with H_2O are stronger than the intermolecular interaction of H_2O with itself and H_2SO_4 with itself.
15.4B Acid strength of halogen oxoacids increases as the oxidation number of the halogen increases. Therefore, perbromic acid, $HBrO_4$, in which Br has an oxidation number of 7, is the strongest acid.
15.5B

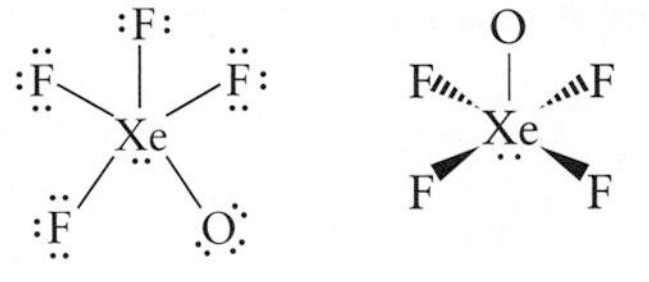

Square pyramidal

Chapter 16

16.1B Density increases from left to right across a row until the *d*-shell is half-filled and then starts to decrease. Density increases down a group. See Fig. 16.5.
16.2B oxidation: $2\,Hg(l) \rightarrow Hg_2^{2+}(aq) + 2\,e^-$
reduction: $4\,H^+(aq) + NO_3^-(aq) + 3\,e^- \rightarrow NO(g) + 2\,H_2O(l)$

net: $6\,Hg(l) + 8\,H^+(aq) + 2\,NO_3^-(aq) \rightarrow 3\,Hg_2^{2+}(aq) + 2\,NO(g) + 4\,H_2O(l)$
$E° = 0.96\text{ V} - 0.79\text{ V} = +1.75\text{ V}$
16.3B (a) pentaamminebromocobalt(III) sulfate; (b) $[Cr(NH_3)_4(H_2O)_2]Br_3$
16.4B There is a 1:3 ratio of $CrCl_3·6H_2O$ to Cl; therefore, the isomer $[Cr(H_2O)_6]Cl_3$ is present.
16.5B (a) NCS^- is replaced by SCN^-; therefore, the pair are linkage isomers. (b) H_2O is exchanged for Cl^-; therefore, the pair are hydrate isomers.
16.6B (a) chiral; (b) not chiral; (c) chiral; (d) not chiral; no entantiomeric pairs
16.7B $\Delta_O = hc/\lambda = \{(6.022 \times 10^{23}\text{ mol}^{-1}) \times (6.626 \times 10^{-34}\text{ J·s}) \times (2.998 \times 10^8\text{ m·s}^{-1})/(305 \times 10^{-9}\text{ nm})\} \times (1\text{ kJ}/10^3\text{ J}) = 392\text{ kJ·mol}^{-1}$
16.8B (a) Strong-field ligands have large Δ_O. The configuration is $t_{2g}^{\,6}e_g^{\,1}$ and there is one unpaired electron. (b) Weak-field ligands have small Δ_O. The configuration is $t_{2g}^{\,5}e_g^{\,2}$ and there are three unpaired electrons.
16.9B The bigger the splitting, the greater the energy of the photon of light absorbed, and the shorter the wavelength of light absorbed. The splitting increases as the field strength of the ligand increases. CN^- is a stronger-field ligand than NH_3; therefore, $[Fe(CN)_6]^{4-}$ absorbs at the shorter wavelength.
16.10B (a) en is a strong-field ligand, and H_2O is a weak-field ligand. However, because Ni^{2+} has eight *d*-electrons it is a d^8 complex, and there is no empty e_g-orbital. Therefore, both $[Ni(en)_3]^{2+}$ and $[Ni(H_2O)_6]^{2+}$ have a $t_{2g}^{\,6}e_g^{\,2}$ configuration and are paramagnetic.

Chapter 17

17.1B ${}^{232}_{90}Th \rightarrow {}^{A}_{Z}E + {}^{4}_{2}\alpha$; $A = 232 - 4 = 228$; $Z = 90 - 2 = 88$; therefore, ${}^{228}_{88}Ra$ is produced.
17.2B ${}^{228}_{88}Ra \rightarrow {}^{A}_{Z}E + {}^{0}_{-1}e$; $A = 228 - 0 = 228$; $Z = 88 - (-1) = 89$; therefore, ${}^{228}_{89}Ac$ is produced.

17.3B ${}^{55}_{26}\text{Fe} + {}^{0}_{-1}\text{e} \rightarrow {}^{A}_{Z}\text{E}$; $A = 55 + 0 = 55$; $Z = 26 + (-1) = 25$; therefore, ${}^{55}_{25}\text{Mn}$ is produced.
17.4B ${}^{11}_{6}\text{C} \rightarrow {}^{A}_{Z}\text{E} + {}^{0}_{+1}\text{e}$; $A = 11 - 0 = 11$; $Z = 6 - 1 = 5$; therefore, ${}^{11}_{5}\text{B}$ is produced.
17.5B ${}^{109}_{53}\text{I} \rightarrow {}^{A}_{Z}\text{E} + {}^{1}_{1}\text{p}$; $A = 109 - 1 = 108$; $Z = 53 - 1 = 52$; Therefore, ${}^{108}_{52}\text{Te}$ is produced.
17.6B (c) ${}^{148}_{58}\text{Ce}$ is neutron rich (in the blue band) and therefore might undergo β emission, which converts a neutron to a proton, to reach stability.
17.7B (a) Balance mass number: 257 + 4(1) − 250 = 11; balance atomic number: 103 + 4(0) − 98 = 5; the missing nuclide is ${}^{11}_{5}\text{B}$. (b) Balance mass number: 254 + 4(1) − 12 = 246; balance atomic number: 102 + 4(0) − 6 = 96; the missing nuclide is ${}^{246}_{96}\text{Cm}$.
17.8B $m = (5.0\ \mu\text{g})\, e^{-(3.3\times10^{-7}\,\text{y}^{-1})\times(1.0\times10^{6}\,\text{y})} = 3.6\ \mu\text{g}$
17.9B $t = -\left(\dfrac{5.73 \times 10^{3}\ \text{y}}{\ln 2}\right) \times \ln\left(\dfrac{1.4 \times 10^{4}}{18\ 400}\right) = 2.2 \times 10^{3}\ \text{y}$
17.10B $\Delta m = \{(235.0439\ \text{u}) - 92(1.0078\ \text{u}) - 143(1.0087\ \text{u})\} \times \{(1.6605 \times 10^{-27}\ \text{kg})/1\ \text{u}\} = -3.1845 \times 10^{-27}\ \text{kg}$
$E_{\text{bind}} = |3.1845 \times 10^{-27}\ \text{kg}| \times (3.00 \times 10^{8}\ \text{m·s}^{-1})^2 \times \{1\ \text{eV}/(1.602\ 18 \times 10^{-19}\ \text{J})\} = 1.79 \times 10^{9}\ \text{eV}$
17.11B $\Delta m = \{137.91\ \text{u} + 85.91\ \text{u} + 12(1.0087\ \text{u}) - (235.04\ \text{u} + 1.0087\ \text{u})\} \times \{(1.6605 \times 10^{-27}\ \text{kg})/1\ \text{u}\} = -2.06 \times 10^{-28}$ kg per nucleus
$\Delta E = \left(\dfrac{1.0\ \text{g}}{235.0\ \text{u}} \times \dfrac{1\ \text{kg}}{10^{3}\ \text{g}} \times \dfrac{1\ \text{u}}{1.6605 \times 10^{-27}\ \text{kg}}\right) \times (-2.06 \times 10^{-28}\ \text{kg}) \times (3.00 \times 10^{8}\ \text{m·s}^{-1})^2 = -4.75 \times 10^{10}\ \text{J}$

Chapter 18

18.1B (a)

CH_3 on C(=C) of ring: HC=C(CH₃)–C(=O)–CH₂–CH(–C(=CH₂)CH₃)–CH₂–

(ring atoms: C(CH₃), C=O, CH_2, CH, H_2C, HC; CH bears C with H_2C= and CH_3)

(b) $(CH_3)_2CH(CH_2)_3CH(CH_3)CH_2CH_3$
18.2B (a) 4-ethyl-2-methylhexane;
(b)

CH_3–CH_2–C(CH_2CH_3)(CH_2CH_3)–CH_2–CH(CH_2CH_3)–CH_2–CH_3

18.3B (a) 4-ethyl-2-hexene; (b) CH_2 bridging HC=CH (cyclopropene)

18.4B $CH_3(CH_2)_2CH_2Br$; $CH_3CH_2CHBrCH_3$; $(CH_3)_2CHCH_2Br$; $(CH_3)_3CBr$
18.5B 27a is cis; 27b is trans
18.6B (c) is chiral, because the second C atom is attached to four different groups.
18.7B (a) has the higher boiling point because it has a longer chain than (b), and therefore has stronger London forces.
18.8B CH_3CH_2Br
18.9B 1-ethyl-3,5-dimethyl-2-propylbenzene

Chapter 19

19.1B (a) HCO<u>OH</u> + CH_3CH_2O<u>H</u> → $HCOOCH_2CH_3$; ethyl methanoate; (b) $CH_3(CH_2)_2COOCH_3$; CH_3O<u>H</u> + $CH_3(CH_2)_2CO$<u>OH</u> → $CH_3(CH_2)_2$ $COOCH_3$
19.2B H–C(=O)–N(H)–H (N with lone pair) C: sp^2 hybridization; N: sp^3 hybridization
19.3B (a) CH_3–CH_2–C(H)(OH)–CH_2–CH_3

3-Pentanol

(b) CH_3–CH_2–C(=O)–CH_2–CH_3

3-Pentanone

(c) CH_3–CH_2–N(H)–CH_3 (N with lone pair)

Ethylmethylamine

19.4B (a) CH_2=C(CH_3)$COOCH_3$;
(b) –N(H)–C(H)(CH_3)–C(=O)–N(H)–C(H)(CH_3)–C(=O)–

19.5B The A and B monomers alternate; therefore, —ABABAB— copolymer is an alternating copolymer.

Appendix 1

A-1B 250 g × (1 lb/453.59 g) × (16 oz/1 lb) = 8.82 oz
A-2B $(9.81\ \text{m/s}^2) \times (1\ \text{km}/10^{3}\ \text{m}) \times (3600\ \text{s}/1\ \text{h})^2 = 1.27 \times 10^{5}\ \text{km·h}^{-2}$

ILLUSTRATION CREDITS

Contents and Preface **vii:** Charlie Westerman/Gamma Liaison. **viii:** Shaun Egan/Stone. **ix:** Courtesy Ford Motor Company. **x:** Sinclair Stammers/Science Photo Library/Photo Researchers. **xi:** Gary Yeowell/Stone. **xii:** Kim Reacox/Peter Arnold. **xiii:** Patrick J. LaCroix/The Image Bank. **xiv:** Darrell Gulin/Corbis. **xv:** Peter Arnold. **xvi:** BIOS (W. Fautre)/Peter Arnold. **xvii:** Frank L. Losa/frankly.com, McLean, Virginia. **xviii:** David M. Phillips, Science Source/Photo Researchers. **xix:** Zefa-Boddenberg/ The Stock Market. **xx:** David Cavagnaro/Peter Arnold. **xxiii:** W. H. Freeman photo by Ken Karp. **xxiv:** Gary Retherford/Science Source/Photo Researchers. **xxvi:** Chip Clark. **Fundamentals** **F1:** Fig. 1, Erich Lessing/Art Resource. **F2:** Fig. 2, David Cavagnaro/Peter Arnold; Fig. 3, Richard Megna/Fundamental Photographs. **F5:** Fig. 7, Warren Gretz/NREL (National Renewable Energy Laboratory). **F7:** Fig. A.1, W. H. Freeman photo by Ken Karp. **F9:** Fig A.2, W. H. Freeman photo by Ken Karp. **F15:** Fig. B.1, W. H. Freeman photo by Ken Karp; Fig. B.2, Burndy Library. **F16:** Fig. B.3, Randall M. Feenstra, IBM Thomas J. Watson Research Center, Yorktown Heights, New York. **F17:** Fig. B.4, Cavendish Laboratory; Fig. B.5, Donald Clegg, Champaign, Illinois. **F18:** Fig. B.7, Corbis-Bettmann. **F19:** Fig. B.10, Walter Schmid/Stone. **F24:** Fig. B.15, Alexander Boden. **F25:** Fig. B.16 & Fig. B.17, Chip Clark. **F28:** Fig. C.1, Chip Clark. **F41:** Fig. D.1, W. H. Freeman photo by Ken Karp. **F48:** Fig. E.2, Chip Clark; Fig. E.3, Burndy Library. **F53:** Fig. E.5, Chip Clark. **F 56:** Fig. F.1, Kenneth L. Rinehart, University of Illinois at Champaign-Urbana. **F62:** Fig. G.1, Peter Kresan. **F63:** Fig. G.4, Chip Clark. **F65:** Fig. G.6, W. H. Freeman photo by Ken Karp. **F67:** Fig. G.8, Chip Clark. **F69:** Fig. G.9, Richard Megna/Fundamental Photographs. **F74:** Fig. H.1, Chip Clark. **F76:** Fig. H.3b, Pacific Gas and Electric. **F81:** Fig. I.3, Fundamental Photographs. **F82:** Fig I.4b, Chip Clark. **F84:** Fig. I.5, Richard Megna/Fundamental Photographs. **F88:** Fig. J.1, W. H. Freeman photo by Ken Karp. **F95:** Fig. K.1, W. H. Freeman photo by Ken Karp. **F96:** Fig. K.2, Chip Clark. **F99:** Fig. K.5a, W. H. Freeman photo by Ken Karp. **F101:** Fig. K.6 a & b, Peticolas/Megna/Fundamental Photographs. **F104:** Fig. L.1, NASA. **F108:** L.2 & L.3, W. H. Freeman photos by Ken Karp. **Chapter 1** **0:** Opener, W. H. Freeman photo by Ken Karp. **11:** Fig. 1.12, Science Museum, London. **14:** Fig. 1.15, UPI/Corbis-Bettmann. **23:** Fig 1.23, D. C. Harris, *Quantitative Chemical Analysis*, 5th Ed. (New York: W. H. Freeman & Co., 1999). **42:** BOX 1.2, Edgar Fahs Smith Collection, ACS Center for History of Chemistry, University of Pennsylvania. **52:** Fig. 1.50, W. H. Freeman photo by Ken Karp. **53:** Fig. 1.53 & Fig. 1.54, Chip Clark. **54:** Fig. 1.55 & Fig. 1.56, Chip Clark. **55:** Fig. 1.57, Chip Clark. **56:** Fig. 1.58, W. H. Freeman photo by Ken Karp. **Chapter 2** **66:** Opener, Charlie Westerman/Gamma Liaison. **70:** Fig. 2.4 d & e, W. H. Freeman photo by Ken Karp. **71:** Fig. 2.5a, P. Motta, Department of Anatomy, University "La Sapienza"/SPL/Photo Researchers. **81:** Fig. 2.9, W. H. Freeman photo by Ken Karp. **86:** BOX 2.1, Breck P. Kent. **88:** Fig. 2.10, W. H. Freeman photo by Ken Karp. **100:** Fig. 2.19, Ken Eward/BioGrafx-Science Source/Photo Researchers. **Chapter 3** **112:** Opener, Shaun Egan/Stone. **118:** BOX 3.1, Gary Retherford/Science Source/Photo Researchers. **142:** Fig. 3.27, Donald Clegg, Champaign, Illinois. **157:** Fig. 3.42, Railway Technical Research Institute, Tokyo. **Chapter 4** **170:** Opener, Courtesy Ford Motor Company. **171:** Fig. 4.1, NASA. **176:** Fig. 4.6, Weather Services International. www.lutellicast.com. **186:** Fig. 4.15, W. H. Freeman photo by Ken Karp. **190:** Fig. 4.16, W. H. Freeman photo by Ken Karp. **191:** Fig. 4.17, U.S. Department of the Interior, Bureau of Mines. **192:** Fig. 4.18, TRW Inc. **208:** Fig. 4.31, W. H. Freeman photo by Ken Karp. **Chapter 5** **220:** Opener, Sinclair Stammers/Science Photo Library/Photo Researchers. **229:** Fig. 5.7, W. H. Freeman photo by Ken Karp. **233:** Fig. 5.10, Brenda Tharp/Photo Researchers. **235:** Fig. 5.15, W. H. Freeman photo by Ken Karp. **236:** Fig. 5.16, Peter Pfletschinger/Peter Arnold; Fig. 5.17, NASA. **236:** Fig. 5.18, Chip Clark. **237:** Fig. 5.19, Chip Clark; Fig. 5.20a, Steven Smale; b, W. H. Freeman photo by Ken Karp. **238:** BOX 5.1 left & right, Courtesy Don Eigler, IBM Research Division, Almaden Research Center, San Jose, California. **240:** Fig. 5.21, Michael Heron/Woodfin Camp & Associates. **249:** Fig. 5.38, Andrew Syred/Science Photo Library/Photo Researchers. **252:** Fig. 5.44, Michael Fairchild/Peter Arnold; Fig. 5.45, John Shaw/Bruce Coleman. **253:** Fig. 5.47, Chip Clark. **254:** Fig. 5.48, Rick Gayle/Stock Market. **255:** BOX 5.2, Courtesy American Superconductor, Westerborough, Massachusetts. **257:** Fig. 5.50b, Sandia National Laboratories/NREL, National Renewable Energy Laboratory. **258:** Fig 5.51b, Chip Clark. **262:** Fig. 5.57, Jessica Wecher/Photo Researchers. **Chapter 6** **274:** Opener, John Kaprielian/Photo Researchers. **276:** Fig. 6.1 & Fig. 6.2, Corbis-Bettmann. **291:** Fig. 6.12 & Fig. 6.13, W. H. Freeman photos by Ken Karp. **306:** Fig. 6.23, USGS, Flagstaff, Arizona. **313:** Fig. 6.26, Thiokol Propulsion. **319:** BOX 6.2 bottom, Warren Gretz/NREL (National Renewable Energy Laboratory); top, Telegraph Colour Library/FPG International. **Chapter 7** **340:** Opener, Gary Yeowell/Stone. **342:** Fig. 7.1 top & bottom, W. H. Freeman photos by Ken Karp. **343:** Fig. 7.2 top & bottom, W. H. Freeman photos by Ken Karp. **352:** Fig. 7.4, Dieter Flamm. **360:** Fig. 7.13, International Tin Research Institute. **371:** Fig. 7.22, Yale University. **Chapter 8** **390:** Opener, Kim Reacox/Peter Arnold. **409:** Fig. 8.14, W. H. Freeman photo by Ken Karp. **411:** Fig. 8.17, Chip Clark. **415:** BOX 8.1a, Sam Ogden/Patrick Aebischer; BOX 8.1b, Sam Ogden/Theracyte; BOX 8.1c, Sam Ogden/Cytotherapeutics; BOX 8.1d, Sam Ogden/Circe Biomedical; BOX 8.1e, Patrick Aebischer. **416:** Fig. 8.21 a & b, E. B. Smith, Physical Chemistry Laboratory, University of Oxford. **418:** Fig. 8.23, W. H. Freeman photo by Ken Karp. **425:** Fig. 8.31, W. H. Freeman photo by Ken Karp. **427:** Fig. 8.32, W. H. Freeman photo by Ken Karp. **433:** Fig. 8.37, W. H. Freeman photo by Ken Karp. **434:** Fig. 8.38 a, b, c, from "Biological Membranes as Bilayer Couples," by M. Sheetz, R. Painter, and S. Singer, 1976, *Journal of Cell Biology*, 70:193. **Chapter 9** **460:** Opener, Richard Megna/Fundamental Photographs. **462:** Fig. 9.1a, Pacific Gas & Electric; Fig 9.1b, W. H. Freeman photo by Ken Karp; Fig 9.1c,

Richard Megna/Fundamental Photographs. **491:** Fig. 9.10, Smithsonian Institution. **495:** Fig. 9.14, M. W. Kellogg Company. **497:** Fig. 9.15, JPL/NASA. **501:** Fig. 9.16, American Institute for Physics, Niels Bohr Library/Stein Collection; Fig. 9.17, Corbis-Bettmann. **Chapter 10** **512:** Opener, Patrick J. LaCroix/The Image Bank. **515:** Fig. 10.3, Jay Laurie/Bruce Coleman. **525:** Fig. 10.11 a & b, W. H. Freeman photos by Ken Karp. **528:** Fig. 10.14, W. H. Freeman photo by Ken Karp. **548:** Fig. 10.19, W. H. Freeman photo by Ken Karp. **568:** BOX 10.1, D. Cavagnaro/Visuals Unlimited. **Chapter 11** **578:** Opener, Darrell Gulin/Corbis. **589:** BOX 11.1, Ben Edwards/Stone. **600:** Fig. 11.9, Brinkmann Instruments. **601:** Fig. 11.10, Chip Clark. **602:** Fig. 11.11, Darrell Gulin/Corbis. **613:** Fig. 11.18 a & b, W. H. Freeman photos by Ken Karp. **615:** Fig. 11.20, Chip Clark. **617:** Fig. 11.21 a & b, Mitchell Bleier/Peter Arnold. **621:** Fig. 11.23 a, b, c, W. H. Freeman photos by Ken Karp. **622:** Fig. 11.24, W. H. Freeman photo by Ken Karp. **Chapter 12** **632:** Opener, Paul Brierley. **649:** Fig. 12.4, W. H. Freeman photo by Ken Karp. **641:** BOX 12.1, Takeshi Takahara/Photo Researchers. **646:** BOX 12.2, Courtesy Ballard Power Systems, Inc., Burnaby, B. C., Canada. **657:** Fig. 12.9, W. H. Freeman photo by Ken Karp. **664:** Fig. 12.11, W. H. Freeman photo by Ken Karp. **668:** Fig. 12.14, The National Museum of Science & Industry. **672:** Fig. 12.17, Comstock. **673:** Fig. 12.18, W. H. Freeman photo by Ken Karp. **674:** Fig. 12.20, St. Joe Zinc Company/American Hot Dip Galvanizers Association. **Chapter 13** **684:** Opener, Peter Arnold. **686:** Fig. 13.1, Betsy R. Strasser/Visuals Unlimited; Fig. 13.2, Woods Hole Oceanographic Institute. **707:** Fig. 13.15, W. H. Freeman photo by Ken Karp. **725:** Fig. 13.29, W. H. Freeman photo by Ken Karp. **728:** Fig. 13.32, W. H. Freeman photo by Ken Karp. **729:** Fig. 13.33 a & b, W. H. Freeman photos by Ken Karp. **731:** BOX 13.3, NASA/Science Photo Library/Photo Researchers. **Chapter 14** **748:** Opener, BIOS (W. Fautre)/Peter Arnold. **756:** Fig. 14.11, W. H. Freeman photo by Ken Karp. **758:** Fig. 14.13, W. H. Freeman photo by Ken Karp. **761:** Fig. 14.15 a-d, W. H. Freeman photos by Ken Karp. **762:** Fig. 14.17 a-c, W. H. Freeman photos by Ken Karp. **763:** Fig. 14.18, P. P. Edwards, University of Birmingham; Fig. 14.19, W. H. Freeman photo by Ken Karp. **764:** Fig. 14.20, Texasgulf. **765:** Fig. 14.22, Photo Researchers. **768:** Fig. 14.24 a-e, W. H. Freeman photos by Ken Karp; Fig. 14.25, Houston Museum of Natural Science. **770:** Fig. 14.26, Alexander Boden. **772:** Fig. 14.27, Field Museum of Natural History, Chicago. **773:** Fig. 14.28, Aalborg Portland Betonforsknings Laboratorium, Karlslunde. **776:** Fig. 14.29, L. Linkhart/Visuals Unlimited. **779:** Fig. 14.31 a & c, Houston Museum of Natural Science; Fig. 14.31 b, Lee Boltin Picture Library. **780:** Fig. 14.32, Chip Clark. **781:** Fig. 14.34, W. H. Freeman photo by Ken Karp. **783:** Fig. 14.35, Chip Clark. **787:** Fig. 14.39, Max Planck/Institute fur Kenphysik; Fig. 14.40 a-c, Field Museum of Natural History, Chicago. **788:** BOX 14.1 bottom left & top right, Richard E. Smalley, Rice University, Center for Nanoscale Science and Technology, from *Nature*, January 28, l993. **794:** Fig.14.43, W. H. Freeman photo by Ken Karp. **795:** Fig.14.46, Chip Clark. **796:** Fig. 14.47, Chip Clark; Fig. 14.48, Phillip Hayson/Photo Researchers. **797:** Fig.14.50, Chip Clark. **799:** Fig. 14.52, Corning Glass. **800:** Fig. 14.53, R. L. Kugler & J. C. Pashim. **801:** Fig. 14.54, Courtesy of NASA/JPL/Caltech. **Chapter 15** **808:** Opener, Frank L. Losa/frankly.com, McLean, Virginia. **810:** Fig. 15.1, Chip Clark. **811:** Fig. 15.2, Photo Researchers. **813:** Fig. 15.3, Chip Clark. **814:** Fig. 15.4, W. H. Freeman photo by Ken Karp. **815:** Fig. 15.5, Chip Clark. **818:** Fig. 15.6, W. H. Freeman photo by Ken Karp. **823:** Fig. 15.8, Chip Clark. **824:** Fig. 15.9, Ross Chapple. **825:** Fig. 15.12, Chip Clark; Fig. 15.13 a, Chip Clark; Fig. 15.13 b, National Park Services. **826:** Fig. 15.14, Chip Clark. **829:** Fig. 15.16, Lee Boltin Picture Library. **831:** Fig. 15.17, W. H. Freeman photo by Ken Karp; Fig. 15.18, all, Chip Clark. **834:** Fig. 15.21, Chip Clark. **835:** Fig. 15.22, Chip Clark. **837:** Fig. 15.23 a & b, W. H. Freeman photos by Ken Karp. **839:** Fig. 15.24, Johnson Space Center/NASA. **842:** Fig. 15.26, Greater Pittsburgh Neon/Tom Anthony. **843:** Fig. 15.28, Argonne National Laboratory. **844:** Fig. 15.29, Brian W. Thomas/Visuals Unlimited. **845:** Fig. 15.30, Courtesy of the Royal Institute of Great Britain. **Chapter 16** **856:** Opener, David M. Phillips, Science Source/Photo Researchers. **863:** Fig. 16.8 a, Field Museum of Natural History, Chicago; b, Hagley Museum and Library; c, Oremet Titanium. **864:** Fig. 16.9, Tom Carroll/FPG International. **865:** Fig. 16.10, W. H. Freeman photo by Ken Karp. **866:** Fig. 16.11, W. H. Freeman photo by Ken Karp. **867:** Fig. 16.12, Institute of Oceanographic Sciences/NERC/SPL/Photo Researchers. **869:** Fig. 16.13, Chip Clark. **870:** Fig. 16.15, Paul Chesley/Stone. **871:** Fig. 16.16, Lee Boltin Picture Library; Fig. 16.17, Field Museum of Natural History, Chicago. **872:** Fig. 16.18, Thierry Borredon/Stone. **876:** Fig. 16.20, W. H. Freeman photo by Ken Karp. **877:** Fig. 16.21, W. H. Freeman photo by Ken Karp. **895:** Fig. 16.37 & Fig. 16.38, Carol Moralejo, Concordia University. **896:** Fig. 16.39, Richard Megna/Fundamental Photographs. **901:** Fig. 16.46, U.S. Steel. **903:** Fig 16.48 a & b, S. Odenbach, ZARM, University of Bremen, Germany. **Chapter 17** **910:** Opener, Eddy Cervo/Cogema. **912:** Fig. 17.2 & Fig. 17.3, The Granger Collection. **922:** BOX 17.1, NASA/Science Source/Photo Researchers. **923:** BOX 17.1, Los Alamos National Laboratory. **925:** BOX 17.2, top right, Marcus E. Raichle and Washington University School of Medicine; bottom left, adapted from "Positron Emission Tomographic Studies of the Cortical Anatomy of Single Word Processing," by S. E. Peterson, P. T. Fox, M. I. Posner, M. A. Mintum, M. E. Raichle, 1988, *Nature*, 331:585–589. **929:** BOX 17.3, bottom left, Yoav Levy/Phototake NYC; top right, Chip Clark. **933:** Fig. 17.19, University of Arizona. **942:** Fig. 17.26, Tom Carroll/Phototake NYC. **943:** Fig. 17.27, Princeton Plasma Physics Laboratory. **944:** Fig. 17.28, Chip Clark. **945:** Fig. 17.29, Phil Schofield; Fig. 17.30, U.S. Department of Energy. **Chapter 18** **952:** Opener, Zefa-Boddenberg/The Stock Market. **967:** BOX 18.1, Chevron Corporation. **Chapter 19** **986:** Opener, David Cavagnaro/Peter Arnold. **995:** Fig. 19.4, W. H. Freeman photo by Ken Karp. **1003:** Fig. 19.6, W. H. Freeman photo by Ken Karp; Fig. 19.8, W. H. Freeman photo by Ken Karp. **1004:** BOX 19.2, James Kilkelly. **1007:** Fig. 19.10, Fibers Division, Monsanto Chemical Company; Fig. 19.11, Andrew Syred/Science Photo Library/Photo Researchers; Fig. 19.12, Chip Clark. **1010:** Fig. 19.15, Courtesy of Professor Mehmet Sarikaya, University of Washington, Seattle, WA. **1011:** Fig. 19.16, Gamma Liaison; Fig. 19.17, Steven Gottlieb/FPG International. **1013:** Fig. 19.18, A. Marmont and E. Damasio, Division of Hematology, S. Martin's Hospital, Genoa, Italy. **1016:** Fig. 19.21, Tom Bean/Stone; Fig. 19.22, John P. O'Brien, DuPont. **1017:** Fig. 19.25, Institute of Paper Chemistry. **1018:** Fig. 19.26 & Fig. 19.27, Photo Researchers. **Major Technique 6** **1031:** Fig. 4, Owen Franken/Stone.

INDEX

THE ELEMENTS

Element	Symbol	Atomic number	Molar mass,* $g \cdot mol^{-1}$	Element	Symbol	Atomic number	Molar mass,* $g \cdot mol^{-1}$
Actinium	Ac	89	(227)	Mendelevium	Md	101	(258)
Aluminium	Al	13	26.98	Mercury	Hg	80	200.59
Americium	Am	95	(243)	Molybdenum	Mo	42	95.94
Antimony	Sb	51	121.76	Neodymium	Nd	60	144.24
Argon	Ar	18	39.95	Neon	Ne	10	20.18
Arsenic	As	33	74.92	Neptunium	Np	93	(237)
Astatine	At	85	(210)	Nickel	Ni	28	58.69
Barium	Ba	56	137.33	Niobium	Nb	41	92.91
Berkelium	Bk	97	(247)	Nitrogen	N	7	14.01
Beryllium	Be	4	9.01	Nobelium	No	102	(259)
Bismuth	Bi	83	208.98	Osmium	Os	76	190.23
Bohrium	Bh	107	(262)	Oxygen	O	8	16.00
Boron	B	5	10.81	Palladium	Pd	46	106.42
Bromine	Br	35	79.90	Phosphorus	P	15	30.97
Cadmium	Cd	48	112.41	Platinum	Pt	78	195.08
Calcium	Ca	20	40.08	Plutonium	Pu	94	(244)
Californium	Cf	98	(251)	Polonium	Po	84	(209)
Carbon	C	6	12.01	Potassium	K	19	39.10
Cerium	Ce	58	140.12	Praseodymium	Pr	59	140.91
Cesium	Cs	55	132.91	Promethium	Pm	61	(145)
Chlorine	Cl	17	35.45	Protactinium	Pa	91	231.04
Chromium	Cr	24	52.00	Radium	Ra	88	(226)
Cobalt	Co	27	58.93	Radon	Rn	86	(222)
Copper	Cu	29	63.55	Rhenium	Re	75	186.21
Curium	Cm	96	(247)	Rhodium	Rh	45	102.91
Dubnium	Db	105	(262)	Rubidium	Rb	37	85.47
Dysprosium	Dy	66	162.50	Ruthenium	Ru	44	101.07
Einsteinium	Es	99	(252)	Rutherfordium	Rf	104	(261)
Erbium	Er	68	167.26	Samarium	Sm	62	150.36
Europium	Eu	63	151.96	Scandium	Sc	21	44.96
Fermium	Fm	100	(257)	Seaborgium	Sg	106	(263)
Fluorine	F	9	19.00	Selenium	Se	34	78.96
Francium	Fr	87	(223)	Silicon	Si	14	28.09
Gadolinium	Gd	64	157.25	Silver	Ag	47	107.87
Gallium	Ga	31	69.72	Sodium	Na	11	22.99
Germanium	Ge	32	72.61	Strontium	Sr	38	87.62
Gold	Au	79	196.97	Sulfur	S	16	32.07
Hafnium	Hf	72	178.49	Tantalum	Ta	73	180.95
Hassium	Hs	108	(265)	Technetium	Tc	43	(98)
Helium	He	2	4.00	Tellurium	Te	52	127.60
Holmium	Ho	67	164.93	Terbium	Tb	65	158.93
Hydrogen	H	1	1.0079	Thallium	Tl	81	204.38
Indium	In	49	114.82	Thorium	Th	90	232.04
Iodine	I	53	126.90	Thulium	Tm	69	168.93
Iridium	Ir	77	192.22	Tin	Sn	50	118.71
Iron	Fe	26	55.85	Titanium	Ti	22	47.87
Krypton	Kr	36	83.80	Tungsten	W	74	183.84
Lanthanum	La	57	138.91	Uranium	U	92	238.03
Lawrencium	Lr	103	(262)	Vanadium	V	23	50.94
Lead	Pb	82	207.2	Xenon	Xe	54	131.29
Lithium	Li	3	6.94	Ytterbium	Yb	70	173.04
Lutetium	Lu	71	174.97	Yttrium	Y	39	88.91
Magnesium	Mg	12	24.31	Zinc	Zn	30	65.39
Manganese	Mn	25	54.94	Zirconium	Zr	40	91.22
Meitnerium	Mt	109	(266)				

*Parentheses around molar mass indicate the most stable isotope of a radioactive element.